the biology of
CANCER

SECOND EDITION

the biology of

CANCER

SECOND EDITION

Robert A. Weinberg

GS Garland Science
Taylor & Francis Group

NEW YORK AND LONDON

Garland Science
Vice President: Denise Schanck
Assistant Editor: Allie Bochicchio
Production Editor and Layout: EJ Publishing Services
Text Editor: Elizabeth Zayatz
Copy Editor: Richard K. Mickey
Proofreader: Sally Huish
Illustrator: Nigel Orme
Designer: Matthew McClements, Blink Studio, Ltd.
Permissions Coordinator: Becky Hainz-Baxter
Indexer: Bill Johncocks
Director of Digital Publishing: Michael Morales
Editorial Assistant: Lamia Harik

About the Author
Robert A. Weinberg is a founding member of the Whitehead Institute for Biomedical Research. He is the Daniel K. Ludwig Professor for Cancer Research and the American Cancer Society Research Professor at the Massachusetts Institute of Technology (MIT). Dr. Weinberg is an internationally recognized authority on the genetic basis of human cancer and was awarded the U.S. National Medal of Science in 1997.

Front Cover
A micrograph section of a human *in situ* ductal carcinoma with α-smooth muscle actin stained in *pink*, cytokeratins 5 and 6 in *red-orange*, and cytokeratins 8 and 18 in *green*. (Courtesy of Werner Böcker and Igor B. Buchwalow of the Institute for Hematopathology, Hamburg, Germany.)

ISBNs: 978-0-8153-4219-9 (hardcover); 978-0-8153-4220-5 (softcover).

Library of Congress Cataloging-in-Publication Data

Weinberg, Robert A. (Robert Allan), 1942-
  The biology of cancer. -- Second edition.
    pages cm
  Includes bibliographical references.
  ISBN 978-0-8153-4219-9 (hardback) -- ISBN 978-0-8153-4220-5 (pbk.) 1.
Cancer--Molecular aspects. 2. Cancer--Genetic aspects. 3. Cancer cells.
I. Title.
  RC268.4.W45 2014
  616.99'4--dc23
                        2013012335

Published by Garland Science, Taylor & Francis Group, LLC,
an informa business, 711 Third Avenue, New York, NY 10017,
USA, and 3 Park Square, Milton Park, Abingdon, OX14 4RN, UK.

Printed in the United States of America

15 14 13 12 11 10 9 8 7 6 5 4

Garland Science
Taylor & Francis Group

Visit our website at http://www.garlandscience.com

SUSTAINABLE FORESTRY INITIATIVE

Certified Chain of Custody
Promoting Sustainable Forestry

www.sfiprogram.org
SFI-01681

This SFI Logo applies to the text stock only.

Dedication

I dedicate this second edition, as the first one, to my dear wife, Amy Shulman Weinberg, who endured long hours of inattention, hearing from me repeatedly the claim that the writing of this edition was almost complete, when in fact years of work lay ahead. She deserved much better! With much love.

Preface

Compared with other areas of biological research, the science of molecular oncology is a recent arrival; its beginning can be traced with some precision to a milestone discovery in 1975. In that year, the laboratory of Harold Varmus and J. Michael Bishop in San Francisco, California demonstrated that normal cell genomes carry a gene—they called it a proto-oncogene—that has the potential, following alteration, to incite cancer. Before that time, we knew essentially nothing about the molecular mechanisms underlying cancer formation; since that time an abundance of information has accumulated that now reveals in outline and fine detail how normal cells become transformed into tumor cells, and how these neoplastic cells collaborate to form life-threatening tumors.

The scientific literature on cancer pathogenesis has grown explosively and today encompasses millions of research publications. So much information would seem to be a pure blessing. After all, knowing more is always better than knowing less. In truth, it represents an embarrassment of riches. By now, we seem to know too much, making it difficult to conceptualize cancer research as a single coherent body of science rather than a patchwork quilt of discoveries that bear only a vague relationship with one another.

This book is written in a far more positive frame of mind, which holds that this patchwork quilt is indeed a manifestation of a body of science that has some simple, underlying principles that unify these diverse discoveries. Cancer research is indeed a field with conceptual integrity, much like other areas of biomedical research and even sciences like physics and chemistry, and the bewildering diversity of the cancer research literature can indeed be understood through these underlying principles.

Prior to the pioneering findings of 1975, we knew almost nothing about the molecular and cellular mechanisms that create tumors. There were some intriguing clues lying around: We knew that carcinogenic agents often, but not always, operate as mutagens; this suggested that mutant genes are involved in some fashion in programming the abnormal proliferation of cancer cells. We knew that the development of cancer is often a long, protracted process. And we knew that individual cancer cells extracted from tumors behave very differently than their counterparts in normal tissues.

Now, almost four decades later, we understand how mutant genes govern the diverse traits of cancer cells and how the traits of these individual cells determine the behavior of tumors. Many of these advances can be traced to the stunning improvements in experimental tools. The techniques of genetic analysis, which were quite primitive at the beginning of this period, have advanced to the stage where we can sequence entire tumor cell genomes in several days. (This is in sharp contrast to the state of affairs in 1975, when the sequencing of oligonucleotides represented a formidable task!) Given the critical role of genotype in determining phenotype, we now understand, as least in outline, why cancer cells behave the way that they do. On the one hand, the molecular differences among individual cancers suggest hundreds of distinct types of human cancer. On the other, molecular and biochemical analyses reveal that this bewildering diversity really manifests a small number of underlying common biochemical traits and molecular processes.

Amusingly, much of this unification was preordained by decisions made 600 million years ago. Once the laws and mechanisms of organismic development were established, they governed all that followed, including the behavior of both normal and neoplastic cells. Modern cancer researchers continue to benefit from this rigid adherence to the fundamental, evolutionarily conserved rules of life. As is evident repeatedly throughout this book, much of what we understand about cancer cells, and thus about the disease of cancer, has been learned by studying the cells of worms and fruit flies and frogs. These laws and principles are invoked repeatedly to explain the complex behaviors of human tumors. By providing context and perspective, they can be used to help us understand all types of human cancer.

While these basic principles are now in clear view, critical details continue to elude us. This explains why modern cancer research is still in active ferment, and why new, fascinating discoveries are being reported every month. While they create new perspectives, they do not threaten the solidity of the enduring truths, which this book attempts to lay out. These principles were already apparent seven years ago when the first edition of this book appeared and, reassuringly, their credibility has not been undermined by all that has followed.

In part, this book has been written as a recruiting pamphlet, as new generations of researchers are needed to move cancer research forward. They are so important because the lessons about cancer's origins, laid out extensively in this book, have not yet been successfully applied to make major inroads into the prevention and cure of this disease. This represents the major frustration of contemporary cancer research: the lessons of disease causation have rarely been followed, as day follows night, by the development of definitive cures.

And yes, there are still major questions that remain murky and poorly resolved. We still do not understand how cancer cells create the metastases that are responsible for 90% of cancer-associated mortality. We understand rather little of the role of the immune system in preventing cancer development. And while we know much about the individual signaling molecules operating inside individual human cells, we lack a clear understanding of how the complex signaling circuitry formed by these molecules makes the life-and-death decisions that determine the fate of individual cells within our body. Those decisions ultimately determine whether or not one of our cells begins the journey down the long road leading to cancerous proliferation and, finally, to a life-threatening tumor.

Contemporary cancer research has enriched numerous other areas of modern biomedical research. Consequently, much of what you will learn from this book will be useful in understanding many aspects of immunology, neurobiology, developmental biology, and a dozen other biomedical research fields. Enjoy the ride!

Robert A. Weinberg
Cambridge, Massachusetts
March 2013

A Note to the Reader

The second edition of this book is organized, like the first, into 16 chapters of quite different lengths. The conceptual structure that was established in the first edition still seemed to be highly appropriate for the second, and so it was retained. What has changed are the contents of these chapters: some have changed substantially since their first appearance seven years ago, while others—largely early chapters—have changed little. The unchanging nature of the latter is actually reassuring, since these chapters deal with early conceptual foundations of current molecular oncology; it would be most unsettling if these foundational chapters had undergone radical revision, which would indicate that the earlier edition was a castle built on sand, with little that could be embraced as well-established, unchanging certainties.

The chapters are meant to be read in the order that they appear, in that each builds on the ideas that have been presented in the chapters before it. The first chapter is a condensed refresher course for undergraduate biology majors and pre-doctoral students; it lays out many of the background concepts that are assumed in the subsequent chapters.

The driving force of these two editions has been a belief that modern cancer research represents a conceptually coherent field of science that can be presented as a clear, logical progression. Embedded in these discussions is an anticipation that much of this information will one day prove useful in devising novel diagnostic and therapeutic strategies that can be deployed in oncology clinics. Some experiments are described in detail to indicate the logic supporting many of these concepts. You will find numerous schematic drawings, often coupled with micrographs, that will help you to appreciate how experimental results have been assembled, piece-by-piece, generating the syntheses that underlie molecular oncology.

Scattered about the text are "Sidebars," which consist of commentaries that represent detours from the main thrust of the discussion. Often these Sidebars contain anecdotes or elaborate on ideas presented in the main text. Read them if you are interested, or skip over them if you find them too distracting. They are presented to provide additional interest—a bit of extra seasoning in the rich stew of ideas that constitutes contemporary research in this area. The same can be said about the "Supplementary Sidebars," which have been relegated to the DVD-ROM that accompanies this book. These also elaborate upon topics that are laid out in the main text and are cross-referenced throughout the book. Space constraints dictated that the Supplementary Sidebars could not be included in the hardcopy version of the textbook.

Throughout the main text you will find extensive cross-references whenever topics under discussion have been introduced or described elsewhere. Many of these have been inserted in the event that you read the chapters in an order different from their presentation here. These cross-references should not provoke you to continually leaf through other chapters in order to track down cited sections or figures. If you feel that you will benefit from earlier introductions to a topic, use these cross-references; otherwise, ignore them.

Each chapter ends with a forward-looking summary entitled "Synopsis and Prospects." This section synthesizes the main concepts of the chapter and often addresses ideas that remain matters of contention. It also considers where research might go in the future. This overview is extended by a list of key concepts and a set of questions. Some of the questions are deliberately challenging and we hope they will provoke you to think more deeply about many of the issues and concepts developed. Finally, most chapters have an extensive list of articles from research journals. These will be useful if you wish to explore a particular topic in detail. Almost all of the cited references are review articles, and many contain detailed discussions of various subfields of research as well as recent findings. In addition, there are occasional references to older publications that will clarify how certain lines of research developed.

Perhaps the most important goal of this book is to enable you to move beyond the textbook and jump directly into the primary research literature. This explains why some of the text is directed toward teaching the elaborate, specialized vocabulary of the cancer research literature, and many of its terms are defined in the glossary. Boldface type has been used throughout to highlight key terms that you should understand. Cancer research, like most areas of contemporary biomedical research, is plagued by numerous abbreviations and acronyms that pepper the text of many published reports. The book provides a key to deciphering this alphabet soup by defining these acronyms. You will find a list of such abbreviations in the back.

Also contained in the book is a newly compiled List of Key Techniques. This list will assist you in locating techniques and experimental strategies used in contemporary cancer research.

The DVD-ROM that accompanies the book also contains a PowerPoint® presentation for each chapter, as well as a companion folder that contains individual JPEG files of the book images including figures, tables, and micrographs. In addition, you will find on this disc a variety of media for students and instructors: movies and audio recordings. There is a selection of movies that will aid in understanding some of the processes discussed; these movies are referenced on the first page of the corresponding chapter in a blue box. The movies are available in QuickTime and WMV formats, and can be used on a computer or transferred to a mobile device. The author has also recorded mini-lectures on the following topics for students and instructors: Mutations and the Origin of Cancer, Growth Factors, p53 and Apoptosis, Metastasis, Immunology and Cancer, and Cancer Therapies. These are available in MP3 format and, like the movies, are easy to transfer to other devices. These media items, as well as future media updates, are available to students and instructors at: http://www.garlandscience.com. On the website, qualified instructors will be able to access a newly created Question Bank. The questions are written to test various levels of understanding within each chapter. The instructor's website also offers access to instructional resources from all of the Garland Science textbooks. For access to instructor's resources please contact your Garland Science sales representative or e-mail science@garland.com.

The poster entitled "The Pathways of Human Cancer" summarizes many of the intracellular signaling pathways implicated in tumor development. This poster has been produced and updated for the Second Edition by Cell Signaling Technology.

Because this book describes an area of research in which new and exciting findings are being announced all the time, some of the details and interpretations presented here may become outdated (or, equally likely, proven to be wrong) once this book is in print. Still, the primary concepts presented here will remain, as they rest on solid foundations of experimental results.

The author and the publisher would greatly appreciate your feedback. Every effort has been made to minimize errors. Nonetheless, you may find them here and there, and it would be of great benefit if you took the trouble to communicate them. Even more importantly, much of the science described herein will require reinterpretation in coming years as new discoveries are made. Please email us at science@garland.com with your suggestions, which will be considered for incorporation into future editions.

PowerPoint is a registered trademark of the Microsoft Corporation in the United States and/or other countries.

Acknowledgments

The science described in this book is the opus of a large, highly interactive research community stretching across the globe. Its members have moved forward our understanding of cancer immeasurably over the past generation. The colleagues listed below have helped the author in countless ways, large and small, by providing sound advice, referring me to critical scientific literature, analyzing complex and occasionally contentious scientific issues, and reviewing individual chapters and providing much-appreciated critiques. Their scientific expertise and their insights into pedagogical clarity have proven to be invaluable. Their help extends and complements the help of an equally large roster of colleagues who helped with the preparation of the first edition. These individuals are representatives of a community, whose members are, virtually without exception, ready and pleased to provide a helping hand to those who request it. I am most grateful to them. Not listed below are the many colleagues who generously provided high quality versions of their published images; they are acknowledged through the literature citations in the figure legends. I would like to thank the following for their suggestions in preparing this edition, as well as those who helped with the first edition. (Those who helped on this second edition are listed immediately, while those who helped with the first edition follow.)

Second edition Eric Abbate, Janis Abkowitz, Julian Adams, Peter Adams, Gemma Alderton, Lourdes Aleman, Kari Alitalo, C. David Allis, Claudia Andl, Annika Antonsson, Paula Apsell, Steven Artandi, Carlos Arteaga, Avi Ashkenazi, Duncan Baird, Amy Baldwin, Frances Balkwill, Allan Balmain, David Bartel, Josep Baselga, Stephen Baylin, Philip Beachy, Robert Beckman, Jürgen Behrens, Roderick Beijersbergen, George Bell, Robert Benezra, Thomas Benjamin, Michael Berger, Arnold Berk, René Bernards, Rameen Beroukhim, Donald Berry, Timothy Bestor, Mariann Bienz, Brian Bierie, Leon Bignold, Walter Birchmeier, Oliver Bischof, John Bixby, Jenny Black, Elizabeth Blackburn, Maria Blasco, Matthew Blatnik, Günter Blobel, Julian Blow, Bruce Boman, Gareth Bond, Katherine Borden, Lubor Borsig, Piet Borst, Blaise Bossy, Michael Botchan, Nancy Boudreau, Henry Bourne, Marina Bousquet, Thomas Brabletz, Barbara Brandhuber, Ulrich Brandt, James Brenton, Marta Briarava, Cathrin Brisken, Jacqueline Bromberg, Myles Brown, Patrick Brown, Thijn Brummelkamp, Ferdinando Bruno, Richard Bucala, Janet Butel, Eliezer Calo, Eleanor Cameron, Ian Campbell, Judith Campbell, Judith Campisi, Lewis Cantley, Yihai Cao, Mario Capecchi, Robert Carlson, Peter Carmeliet, Kermit Carraway, Oriol Casanovas, Tom Cech, Howard Cedar, Ann Chambers, Eric Chang, Mark Chao, Iain Cheeseman, Herbert Chen, Jen-Tsan Chi, Lewis Chodosh, Gerhard Christofori, Inhee Chung, Karen Cichowski, Daniela Cimini, Tim Clackson, Lena Claesson-Welsh, Michele Clamp, Trevor Clancy, Rachael Clark, Bayard Clarkson, James Cleaver, Don Cleveland, David Cobrinik, John Coffin, Philip Cohen, Robert Cohen, Michael Cole, Hilary Coller, Kathleen Collins, Duane Compton, John Condeelis, Simon Cook, Christopher Counter, Sara Courtneidge, Lisa Coussens, Charles Craik, James Darnell, Mark Davis, George Daley, Titia de Lange, Pierre De Meyts, Hugues de Thé, Rik Derynck, Mark Dewhirst, James DeCaprio, Mark Depristo, Channing Der, Tom DiCesare, John Dick, Daniel DiMaio, Charles Dimitroff, Nadya Dimitrova, Charles Dinarello, Joseph DiPaolo, Peter Dirks, Vishwa Dixit, Lawrence Donehower, Philip Donoghue, Martin Dorf, David Dornan, Gian Paolo Dotto, Steven Dowdy, James Downing, Harry Drabkin, Brian Druker, Crislyn D'Souza-Schorey, Eric Duell, Patricia Duffner, Michel DuPage, Robert Duronio, Michael Dyer, Nick Dyson, Connie Eaves, Michael Eck, Mikala Egeblad, Charles Eigenbrot, Steve Elledge, Robert Eisenman, Susan Erster, Manel Esteller, Mark Ewen, Patrick Eyers, Doriano Fabbro, Reinhard Fässler, Mark Featherstone, David Felser, Karen Ferrante, Soldano Ferrone, Isaiah Fidler, Barbara Fingleton, Zvi Fishelson, Ignacio Flores, Antonio Foji, David Foster, A. Raymond Frackelton jr., Hervé Wolf Fridman, Peter Friedl, Kenji Fukasawa, Priscilla A. Furth, Vladimir Gabai, Brenda Gallie, Jerome Galon, Sanjiv Sam Gambhir, Levi Garraway, Yan Geng, Bruce Gelb, Richard Gelber, Frank Gertler, Gad Getz, Edward Giovannucci, Michael Gnant, Sumita Gokhale, Leslie Gold, Alfred Goldberg, Richard Goldsby, Jesus Gomez-Navarro, David Gordon, Eyal Gottlieb, Stephen Grant, Alexander Greenhough, Christoph Kahlert, Florian Greten, Jay Grisolano, Athur Grollman, Bernd Groner, Wenjun Guo, Piyush Gupta, Daniel Haber, William Hahn, Kevin Haigis, Marcia Haigis, William Hait, Thanos Halazonetis, John Haley, Stephen Hall, Douglas Hanahan, Steven Hanks, J. Marie Hardwick, Iswar Hariharan, Ed Harlow, Masanori Hatakeyama, Georgia Hatzivassiliou, Lin He, Matthias Hebrok, Stephen Hecht,

Kristian Helin, Samuel Hellman, Michael Hemann, Linda Hendershot, Meenhard Herlyn, Julian Heuberger, Philip Hinds, Susan Hilsenbeck, Michelle Hirsch, Andreas Hochwagen, H. Robert Horvitz, Susan Horwitz, Peter Howley, Ralph Hruban, Peggy Hsu, David Huang, Paul Huang, Robert Huber, Honor Hugo, Tony Hunter, Richard Hynes, Tan Ince, Yoko Irie, Mark Israel, Jean-Pierre Issa, Yoshiaki Ito, Michael Ittmann, Shalev Itzkovitz, Tyler Jacks, Stephen Jackson, Rudolf Jaenisch, Rakesh Jain, Katherine Janeway, Ahmedin Jemal, Harry Jenq, Kim Jensen, Josef Jiricny, Claudio Joazeiro, Bruce Johnson, Candace Johnson, David Jones, Peter Jones, Nik Joshi, Johanna Joyce, William Kaelin, Kong Jie Kah, Nada Kalaany, Raghu Kalluri, Lawrence Kane, Antoine Karnoub, John Katzenellenbogen, Khandan Keyomarsi, Katherine Janeway, William Kaelin jr., Andrius Kazlauskas, Joseph Kelleher, Elliott Kieff, Nicole King, Christian Klein, Pamela Klein, Frederick Koerner, Richard Kolesnick, Anthony Komaroff, Konstantinos Konstantopoulos, Jordan Krall, Igor Kramnik, Wilhelm Krek, Guido Kroemer, Eve Kruger, Genevieve Kruger, Madhu Kumar, Charlotte Kuperwasser, Thomas Kupper, Bruno Kyewski, Sunil Lakhani, Eric Lander, Lewis Lanier, Peter Lansdorp, David Largaespada, Michael Lawrence, Emma Lees, Jacqueline Lees, Robert Lefkowitz, Mark Lemmon, Stanley Lemon, Arnold Levine, Beth Levine, Ronald Levy, Ephrat Levy-Lahad, Kate Liddell, Stuart Linn, Marta Lipinski, Joe Lipsick, Edison Liu, David Livingston, Harvey Lodish, Lawrence Loeb, Jay Loeffler, David Louis, Julie-Aurore Losman, Scott Lowe, Haihui Lu, Kunxin Luo, Mathieu Lupien, Li Ma, Elisabeth Mack, Alexander MacKerell jr., Ben Major, Tak Mak, Shiva Malek, Scott Manalis, Sridhar Mani, Matthias Mann, Alberto Mantovani, Richard Marais, Jean-Christophe Marine, Sanford Markowitz, Ronen Marmorstein, Lawrence Marnett, Chris Marshall, G. Steven Martin, Joan Massagué, Lynn Matrisian, Massimilano Mazzone, Sandra McAllister, Grant McArthur, David McClay, Donald McDonald, David Glenn McFadden, Wallace McKeehan, Margaret McLaughlin-Drubin, Anthony Means, René Medema, Cornelis Melief, Craig Mermel, Marek Michalak, Brian Miller, Nicholas Mitsiades, Sibylle Mittnacht, Holger Moch, Ute Moll, Deborah Morrsion, Aristides Moustakis, Gregory Mundy, Cornelius Murre, Ruth Muschel, Senthil Muthuswamy, Jeffrey Myers, Harikrishna Nakshatri, Inke Näthke, Geoffrey Neale, Ben Neel, Joel Neilson, M. Angela Nieto, Irene Ng, Ingo Nindl, Larry Norton, Roel Nusse, Shuji Ogino, Kenneth Olive, Andre Oliveira, Gilbert Omenn, Tamer Onder, Moshe Oren, Barbara Osborne, Liliana Ossowski, David Page, Klaus Pantel, David Panzarella, William Pao, Jongsun Park, Paul Parren, Ramon Parsons, Dhavalkumar Patel, Mathias Pawlak, Tony Pawson, Daniel Peeper, Mark Peifer, David Pellman, Tim Perera, Charles Perou, Mary Ellen Perry, Manuel Perucho, Richard Pestell, Julian Peto, Richard Peto, Stefano Piccolo, Jackie Pierce, Eli Pikarsky, Hidde Ploegh, Nikolaus Pfanner, Kristy Pluchino, Heike Pohla, Paul Polakis, Michael Pollak, John Potter, Carol Prives, Lajos Pusztai, Xuebin Qin, Priyamvada Rai, Terence Rabbitts, Anjana Rao, Julia Rastelli, David Raulet, John Rebers, Roger Reddel, Peter Reddien, Danny Reinberg, Michael Retsky, Jeremy Rich, Andrea Richardson, Tim Richmond, Gail Risbridger, Paul Robbins, James Roberts, Leonardo Rodriguez, Veronica Rodriguez, Mark Rolfe, Michael Rosenblatt, David Rosenthal, Theodora Ross, Yolanda Roth, David Rowitch, Brigitte Royer-Pokora, Anil Rustgi, David Sabatini, Erik Sahai, Jesse Salk, Leona Samson, Yardena Samuels, Bengt Samuelsson, Christopher Sansam, Richard Santen, Van Savage, Andrew Sharrocks, Brian Schaffhausen, Pepper Schedin, Christina Scheel, Rachel Schiff, Joseph Schlessinger, Ulrich Schopfer, Hubert Schorle, Deborah Schrag, Brenda Schulman, Wolfgang Schulz, Bert Schutte, Hans Schreiber, Robert Schreiber, Martin Schwartz, Ralph Scully, John Sedivy, Helmut Seitz, Manuel Serrano, Jeffrey Settleman, Kevin Shannon, Phillip Sharp, Norman Sharpless, Jerry Shay, Stephen Sherwin, Yigong Shi, Tsukasa Shibuye, Ben-Zion Shilo, Piotr Sicinski, Daniel Silver, Arun Singh, Michail Sitkovsky, George Sledge, Jr., Mark Sliwkowski, David I. Smith, Eric Snyder, Pierre Sonveaux, Jean-Charles Soria, Ben Stanger, Sheila Stewart, Charles Stiles, Jayne Stommel, Shannon Stott, Jenny Stow, Michael Stratton, Ravid Straussman, Jonathan Strosberg, Charles Streuli, Herman Suit, Peter Sun, Thomas Sutter, Kathy Svoboda, Alejandro Sweet-Cordero, Mario Sznol, Clifford Tabin, Wai Leong Tam, Hsin-Hsiung Tai, Makoto Taketo, Wai Leong Tam, Filemon Tan, Michael Tangrea, Masae Tatematsu, Steven Teitelbaum, Sabine Tejpar, Adam Telerman, Jennifer Temel, David Tenenbaum, Mine Tezal, Jean Paul Thiery, Craig Thompson, Michael Thun, Thea Tlsty, Rune Toftgård, Nicholas Tonks, James Trager, Donald L. Trump, Scott Valastyan, Linda van Aelst, Benoit van den Eynde, Matthew Vander Heiden, Maarten van Lohuizen, Eugene van Scott, Peter Vaupel, Laura van't Veer, George Vassiliou, Inder Verma, Gabriel Victora, Christoph Viebahn, Danijela Vignjevic, Bert Vogelstein, Robert Vonderheide, Daniel von Hoff, Dorien Voskuil, Karen Vousden, Geoffrey Wahl, Lynne Waldman, Herbert Waldmann, Graham Walker, Rongfu Wang, Patricia Watson, Bill Weis, Stephen Weiss, Irv Weissman, Danny Welch, H. Gilbert Welch, Zena Werb, Marius Wernig, Bengt Westermark, John Westwick, Eileen White, Forest White, Max Wicha, Walter Willett, Catherine Wilson, Owen Witte, Alfred Wittinghofer, Norman Wolmark, Sopit Wongkham, Richard Wood, Nicholas Wright, Xu Wu, David Wynford-Thomas, Michael Yaffe, Jing Yang, James Yao, Yosef Yarden, Robert Yauch, Xin Ye, Sam Yoon, Richard Youle, Richard Young, Patrick Zarrinkar, Ann Zauber, Jiri Zavadil, Lin Zhang, Alicia Zhou, Ulrike Ziebold, Kai Zinn, Johannes Zuber, James Zwiebel.

Special thanks to **Makoto Mark Taketo** of Kyoto University and **Richard A. Goldsby** of Amherst College.

First edition Joan Abbott, Eike-Gert Achilles, Jerry Adams, Kari Alitalo, James Allison, David Alpers, Fred Alt, Carl Anderson, Andrew Aprikyan, Jon Aster, Laura Attardi, Frank Austen, Joseph Avruch, Sunil Badve, William Baird, Frances Balkwill, Allan Balmain, Alan Barge, J. Carl Barrett, David Bartel, Renato Baserga, Richard Bates, Philip Beachy, Camille Bedrosian, Anna Belkina, Robert Benezra, Thomas Benjamin, Yinon Ben-Neriah, Ittai Ben-Porath, Bradford Berk, René Bernards, Anton Berns, Kenneth Berns, Monica Bessler, Neil Bhowmick, Marianne Bienz, Line Bjørge, Harald von Boehmer, Gareth Bond, Thierry Boon, Dorin-Bogdan Borza, Chris Boshoff, Noël Bouck, Thomas Brabletz, Douglas Brash, Cathrin Brisken, Garrett Brodeur, Patrick Brown, Richard Bucala, Patricia Buffler, Tony Burgess, Suzanne Bursaux, Randall Burt, Stephen Bustin, Janet Butel, Lisa Butterfield, Blake Cady, John Cairns, Judith Campisi, Harvey Cantor, Robert Cardiff, Peter Carroll, Arlindo Castelanho, Bruce Chabner, Ann Chambers, Howard Chang, Andrew Chess, Ann Cheung, Lynda Chin, Francis Chisari, Yunje Cho, Margaret Chou, Karen Cichowski, Michael Clarke, Hans Clevers, Brent Cochran, Robert Coffey, John Coffin, Samuel Cohen, Graham Colditz, Kathleen Collins, Dave Comb, John Condeelis, Suzanne Cory, Christopher Counter, Sara Courtneidge, Sandra Cowan-Jacob, John Crispino, John Crissman, Carlo Croce, Tim Crook, Christopher Crum, Marcia Cruz-Correa, Gerald Cunha, George Daley, Riccardo Dalla-Favera, Alan D'Andrea, Chi Dang, Douglas Daniels, James Darnell, Jr., Robert Darnell, Galina Deichman, Titia de Lange, Hugues de Thé, Chuxia Deng, Edward Dennis, Lucas Dennis, Ronald DePinho, Theodora Devereaux, Tom DiCesare, Jules Dienstag, John DiGiovanni, Peter Dirks, Ethan Dmitrovsky, Daniel Donoghue, John Doorbar, G. Paolo Dotto, William Dove, Julian Downward, Glenn Dranoff, Thaddeus Dryja, Raymond DuBois, Nick Duesbery, Michel DuPage, Harold Dvorak, Nicholas Dyson, Michael Eck, Walter Eckhart, Argiris Efstratiadis, Robert Eisenman, Klaus Elenius, Steven Elledge, Elissa Epel, John Eppig, Raymond Erikson, James Eshleman, John Essigmann, Gerard Evan, Mark Ewen, Guowei Fang, Juli Feigon, Andrew Feinberg, Stephan Feller, Bruce Fenton, Stephen Fesik, Isaiah Fidler, Gerald Fink, Alain Fischer, Zvi Fishelson, David Fisher, Richard Fisher, Richard Flavell, Riccardo Fodde, M. Judah Folkman, David Foster, Uta Francke, Emil Frei, Errol Friedberg, Peter Friedl, Stephen Friend, Jonas Frisen, Elaine Fuchs, Margaret Fuller, Yuen Kai (Teddy) Fung, Kyle Furge, Amar Gajjar, Joseph Gall, Donald Ganem, Judy Garber, Frank Gertler, Charlene Gilbert, Richard Gilbertson, Robert Gillies, Doron Ginsberg, Edward Giovannucci, Inna Gitelman, Steve Goff, Lois Gold, Alfred Goldberg, Mitchell Goldfarb, Richard Goldsby, Joseph Goldstein, Susanne Gollin, Mehra Golshan, Todd Golub, Jeffrey Gordon, Michael Gordon, Siamon Gordon, Martin Gorovsky, Arko Gorter, Joe Gray, Douglas Green, Yoram Groner, John Groopman, Steven Grossman, Wei Gu, David Guertin, Piyush Gupta, Barry Gusterson, Daniel Haber, James Haber, William Hahn, Kevin Haigis, Senitiroh Hakomori, Alan Hall, Dina Gould Halme, Douglas Hanahan, Philip Hanawalt, Adrian Harris, Curtis Harris, Lyndsay Harris, Stephen Harrison, Kimberly Hartwell, Leland Hartwell, Harald zur Hausen, Carol Heckman, Ruth Heimann, Samuel Hellman, Brian Hemmings, Lothar Hennighausen, Meenhard Herlyn, Glenn Herrick, Avram Hershko, Douglas Heuman, Richard Hodes, Jan Hoeijmakers, Robert Hoffman, Robert Hoover, David Hopwood, Gabriel Hortobagyi, H. Robert Horvitz, Marshall Horwitz, Alan Houghton, Peter Howley, Robert Huber, Tim Hunt, Tony Hunter, Stephen Hursting, Nancy Hynes, Richard Hynes, Antonio Iavarone, J. Dirk Iglehart, Tan Ince, Max Ingman, Mark Israel, Kurt Isselbacher, Tyler Jacks, Rudolf Jaenisch, Rakesh Jain, Bruce Johnson, David Jones, Richard Jones, William Kaelin, Jr., Raghu Kalluri, Alexander Kamb, Barton Kamen, Manolis Kamvysselis, Yibin Kang, Philip Kantoff, Paul Kantrowitz, Jan Karlsreder, Michael Kastan, Michael Kauffman, William Kaufmann, Robert Kerbel, Scott Kern, Khandan Keyomarsi, Marc Kirschner, Christoph Klein, George Klein, Yoel Kloog, Alfred Knudson, Frederick Koerner, Anthony Komaroff, Kenneth Korach, Alan Korman, Eva Kramarova, Jackie Kraveka, Wilhelm Krek, Charlotte Kuperwasser, James Kyranos, Carole LaBonne, Peter Laird, Sergio Lamprecht, Eric Lander, Laura Landweber, Lewis Lanier, Andrew Lassar, Robert Latek, Lester Lau, Derek Le Roith, Chung Lee, Keng Boon Lee, Richard Lee, Jacqueline Lees, Rudolf Leibel, Mark Lemmon, Christoph Lengauer, Jack Lenz, Gabriel Leung, Arnold Levine, Beth Levine, Jay Levy, Ronald Levy, Fran Lewitter, Frederick Li, Siming Li, Frank Lieberman, Elaine Lin, Joachim Lingner, Martin Lipkin, Joe Lipsick, David Livingston, Harvey Lodish, Lawrence Loeb, Edward Loechler, Michael Lotze, Lawrence Lum, Vicky Lundblad, David MacPherson, Sendurai Mani, Alberto Mantovani, Sandy Markowitz, Larry Marnett, G. Steven Martin, Seamus Martin, Joan Massagué, Patrice Mathevet, Paul Matsudaira, Andrea McClatchey, Frank McCormick, Patricia McManus, Mark McMenamin, U. Thomas Meier, Matthew Meyerson, George Miller, Nathan Miselis, Randall Moon, David Morgan, Rebecca Morris, Simon Conway Morris, Robert Moschel, Bernard Moss, Paul Mueller, Anja Mueller-Homey, William A. Muller, Gregory Mundy, Karl Münger, Lance Munn, Ruth Muschel, Lee Nadler, David G. Nathan, Jeremy Nathans, Sergei Nedospasov, Benjamin Neel, David Neuhaus, Donald Newmeyer, Leonard Norkin, Lloyd Old, Kenneth Olive, Tamer Onder, Moshe Oren, Terry Orr-Weaver, Barbara Osborne, Michele Pagano, David Page, Asit Parikh, Chris Parker, William Paul, Amanda Paulovich, Tony Pawson, Mark Peifer, David Pellman, David Phillips, Jacqueline Pierce, Malcolm Pike, John Pintar, Maricarmen Planas-Silva, Roland Pochet, Daniel Podolsky, Beatriz Pogo, Roberto Polakiewicz, Jeffrey Pollard, Nicolae Popescu, Christoph Poremba, Richmond Prehn, Carol Prives, Vito Quaranta, Peter Rabinovitch, Al Rabson, Priyamvada Rai, Klaus Rajewsky, Sridhar Ramaswamy, Anapoorni Rangarajan, Jeffrey Ravetch, Ilaria Rebay, John Reed, Steven Reed, Alan Rein, Ee Chee Ren, Elizabeth Repasky, Jeremy Rich, Andrea Richardson, Dave Richardson, Darrell Rigel, James Roberts, Diane Rodi, Clifford Rosen, Jeffrey Rosen, Neal Rosen, Naomi Rosenberg, Michael Rosenblatt, Theodora Ross, Martine Roussel, Steve Rozen, Jeffrey Ruben, José Russo, David Sabatini, Julien Sage, Ronit Sarid, Edward Sausville, Charles Sawyers, David Scadden, David Schatz, Christina Scheel, Joseph Schlessinger, Anja Schmidt, Stuart Schnitt, Robert Schoen, Robert Schreiber, Edward Scolnick, Ralph Scully, Harold

Seifried, William Sessa, Jeffrey Settleman, Fergus Shanahan, Jerry Shay, James Sherley, Charles Sherr, Ethan Shevach, Chiaho Shih, Frank Sicheri, Peter Sicinski, Sandy Simon, Dinah Singer, Arthur Skarin, Jonathan Skipper, Judy Small, Gilbert Smith, Lauren Sompayrac, Holger Sondermann, Gail Sonenshein, Deborah Spector, Michael Sporn, Eric Stanbridge, E. Richard Stanley, Louis Staudt, Philipp Steiner, Ralph Steinman, Gunther Stent, Sheila Stewart, Charles Stiles, Jonathan Stoye, Michael Stratton, Bill Sugden, Takashi Sugimura, John Sullivan, Nevin Summers, Calum Sutherland, Clifford Tabin, John Tainer, Jussi Taipale, Shinichiro Takahashi, Martin Tallman, Steven Tannenbaum, Susan Taylor, Margaret Tempero, Masaaki Terada, Satvir Tevethia, Jean Paul Thiery, William Thilly, David Thorley-Lawson, Jay Tischfield, Robertus Tollenaar, Stephen Tomlinson, Dimitrios Trichopoulos, Elaine Trujillo, James Umen, Alex van der Eb, Wim van Egmond, Diana van Heemst, Laura van't Veer, Harold Varmus, Alexander Varshavsky, Anna Velcich, Ashok Venkitaraman, Björn Vennström, Inder Verma, Shelia Violette, Bert Vogelstein, Peter Vogt, Olga Volpert, Evan Vosburgh, Geoffrey Wahl, Graham Walker, Gernot Walter, Jack Wands, Elizabeth Ward, Jonathan Warner, Randolph Watnick, I. Bernard Weinstein, Robin Weiss, Irving Weissman, Danny Welch, H. Gilbert Welch, Zena Werb, Forest White, Michael White, Raymond White, Max Wicha, Walter Willet, Owen Witte, Richard Wood, Andrew Wyllie, John Wysolmerski, Michael Yaffe, Yukiko Yamashita, George Yancopoulos, Jing Yang, Moshe Yaniv, Chun-Nan Yeh, Richard Youle, Richard Young, Stuart Yuspa, Claudio Zanon, David Zaridze, Patrick Zarrinkar, Bruce Zetter, Drazen Zimonjic, Leonard Zon, Weiping Zou

Readers: Through their careful reading of the text, these graduate students provided extraordinarily useful feedback in improving many sections of this book and in clarifying sections that were, in their original versions, poorly written and confusing.

Jamie Weyandt (Duke University), Matthew Crowe (Duke University), Venice Calinisan Chiueh (University of California, Berkeley), Yvette Soignier (University of California, Berkeley)

Question Bank: Jamie Weyandt also produced the accompanying question bank available to qualified adopters on the instructor resource site.

Whitehead Institute/MIT: Christine Hickey was responsible over several years' time in helping to organize the extensive files that constituted each chapter. Her help was truly extraordinary.

Dave Richardson of the Whitehead Institute library helped on countless occasions to retrieve papers from obscure corners of the vast scientific literature, doing so with lightning speed!

Garland: While this book has a single recognized author, it really is the work of many hands. The prose was edited by Elizabeth Zayatz and Richard K. Mickey, two editors who are nothing less than superb. To the extent that this book is clear and readable, much of this is a reflection of their dedication to clarity, precision of language, graceful syntax, and the use of images that truly serve to enlighten rather than confound. I have been most fortunate to have two such extraordinary people looking over my shoulder at every step of the writing process. And, to be sure, I have learned much from them. I cannot praise them enough!

Many of the figures are the work of Nigel Orme, an illustrator of great talent, whose sense of design and dedication to precision and detail are, once again, nothing less than extraordinary.

Garland Science determined the structure and design and provided unfaltering support and encouragement through every step of the process that was required to bring this project to fruition. Denise Schanck gave guidance and cheered me on every step of the way. Unfailingly gracious, she is, in every sense, a superb publisher, whose instincts for design and standards of quality publishing are a model. All textbook authors should be as fortunate as I have been to have someone of her qualities at the helm!

The editorial and logistical support required to organize and assemble a book of this complexity was provided first by Janete Scobie and then over a longer period by Allie Bochicchio, both of whom are multitalented and exemplars of ever-cheerful competence, thoroughness, and helpfulness. Without the organizational skills of these two in the Garland office, this text would have emerged as an incoherent jumble.

The truly Herculean task of procuring permissions for publication of the myriad figures fell on the shoulders of Becky Hainz-Baxter. This remains a daunting task, even in this age of Internet and email. Without her help, it would have been impossible to share with the reader many of the images that have created the field of modern cancer research.

The layout is a tribute to the talents of Emma Jeffcock, once again an exemplar of competence, who has an unerring instinct for how to make images and the pages that hold them accessible and welcoming to the reader; she also provided much-valued editorial help that resulted in many improvements of the prose.

The electronic media associated with this book are the work of Michael Morales, whose ability to organize clear and effective visual presentations are indicated by the electronic files that are carried in the accompanying DVD-ROM. He and his editorial assistant, Lamia Harik, are recognized and thanked for their dedication to detail, thoroughness, and their great talent in providing accessible images that inform the reader and complement the written text.

Additional, highly valuable input into the organization and design were provided by Adam Sendroff, Alain Mentha, and Lucy Brodie.

Together, the Garland team, as cited above, represents a unique collection of gifted people whose respective talents are truly peerless and, to say so a second time, individuals who are unfailingly gracious and helpful. Other textbook authors should be as fortunate as I have been in receiving the support that I have enjoyed in the preparation of this second edition!

Contents

List of Key Techniques

● Can be found on the DVD-ROM accompanying the book.

Detailed Contents

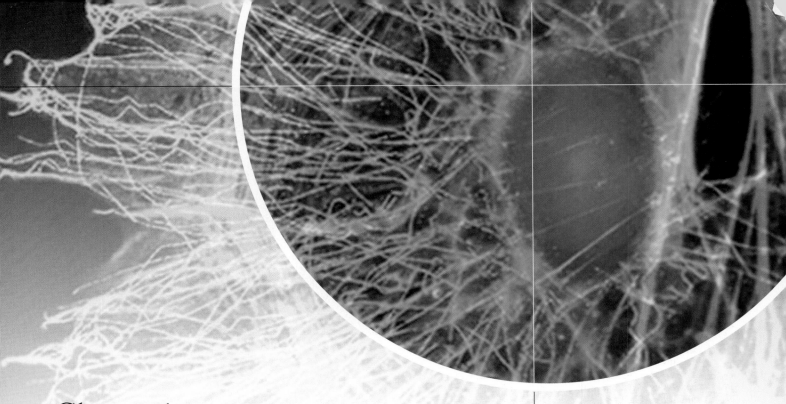

Chapter 1

The Biology and Genetics of Cells and Organisms

> Protoplasm, simple or nucleated, is the formal basis of all life... Thus it becomes clear that all living powers are cognate, and that all living forms are fundamentally of one character. The researches of the chemist have revealed a no less striking uniformity of material composition in living matter.
>
> Thomas Henry Huxley, evolutionary biologist, 1868

> Anything found to be true of *E. coli* must also be true of elephants.
>
> Jacques Monod, pioneer molecular biologist, 1954

The biological revolution of the twentieth century totally reshaped all fields of biomedical study, cancer research being only one of them. The fruits of this revolution were revelations of both the outlines and the minute details of genetics and heredity, of how cells grow and divide, how they assemble to form tissues, and how the tissues develop under the control of specific genes. Everything that follows in this text draws directly or indirectly on this new knowledge.

This revolution, which began in mid-century and was triggered by Watson and Crick's discovery of the DNA double helix, continues to this day. Indeed, we are still too close to this breakthrough to properly understand its true importance and its long-term ramifications. The discipline of molecular biology, which grew from this discovery, delivered solutions to the most profound problem of twentieth-century biology—how does the genetic constitution of a cell or organism determine its appearance and function?

Without this molecular foundation, modern cancer research, like many other biological disciplines, would have remained a descriptive science that cataloged diverse biological phenomena without being able to explain the mechanics of how they occur.

Figure 1.1 Darwin and Mendel
(A) Charles Darwin's 1859 publication of *On the Origin of Species by Means of Natural Selection* exerted a profound effect on thinking about the origin of life, the evolution of organismic complexity, and the relatedness of species. (B) Darwin's theory of evolution lacked a genetic rationale until the work of Gregor Mendel. The synthesis of Darwinian evolution and Mendelian genetics is the foundation for much of modern biological thinking. (A, from the Grace K. Babson Collection, the Henry E. Huntington Library, San Marino, California. Reproduced by permission of The Huntington Library, San Marino, California. B, courtesy of the Mendelianum Museum Moraviae, Brno, Czech Republic.)

(A)

(B)

Today, our understanding of how cancers arise is being continually enriched by discoveries in diverse fields of biological research, most of which draw on the sciences of molecular biology and genetics. Perhaps unexpectedly, many of our insights into the origins of malignant disease are not coming from the laboratory benches of cancer researchers. Instead, the study of diverse organisms, ranging from yeast to worms to flies, provides us with much of the intellectual capital that fuels the forward thrust of the rapidly moving field of cancer research.

Those who fired up this biological revolution stood on the shoulders of nineteenth-century giants, specifically, Darwin and Mendel (**Figure 1.1**). Without the concepts established by these two, which influence all aspects of modern biological thinking, molecular biology and contemporary cancer research would be inconceivable. So, throughout this chapter, we frequently make reference to evolutionary processes as proposed by Charles Darwin and genetic systems as conceived by Gregor Mendel.

1.1   Mendel establishes the basic rules of genetics

Many of the basic rules of genetics that govern how genes are passed from one complex organism to the next were discovered in the 1860s by Gregor Mendel and have come to us basically unchanged. Mendel's work, which tracked the breeding of pea plants, was soon forgotten, only to be rediscovered independently by three researchers in 1900. During the decade that followed, it became clear that these rules—we now call them Mendelian genetics—apply to virtually all sexual organisms, including **metazoa** (multicellular animals), as well as **metaphyta** (multicellular plants).

Mendel's most fundamental insight came from his realization that genetic information is passed in particulate form from an organism to its offspring. This implied that the entire repertoire of an organism's genetic information—its genome, in today's terminology—is organized as a collection of discrete, separable information packets, now called genes. Only in recent years have we begun to know with any precision how many distinct genes are present in the genomes of mammals; many current analyses of the human genome—the best studied of these—place the number in the range of 21,000, somewhat more than the 14,500 genes identified in the genome of the fruit fly, *Drosophila melanogaster*.

Mendel's work also implied that the constitution of an organism, including its physical and chemical makeup, could be divided into a series of discrete, separable entities. Mendel went further by showing that distinct anatomical parts are controlled by distinct genes. He found that the heritable material controlling the smoothness of peas behaved independently of the material governing plant height or flower color. In

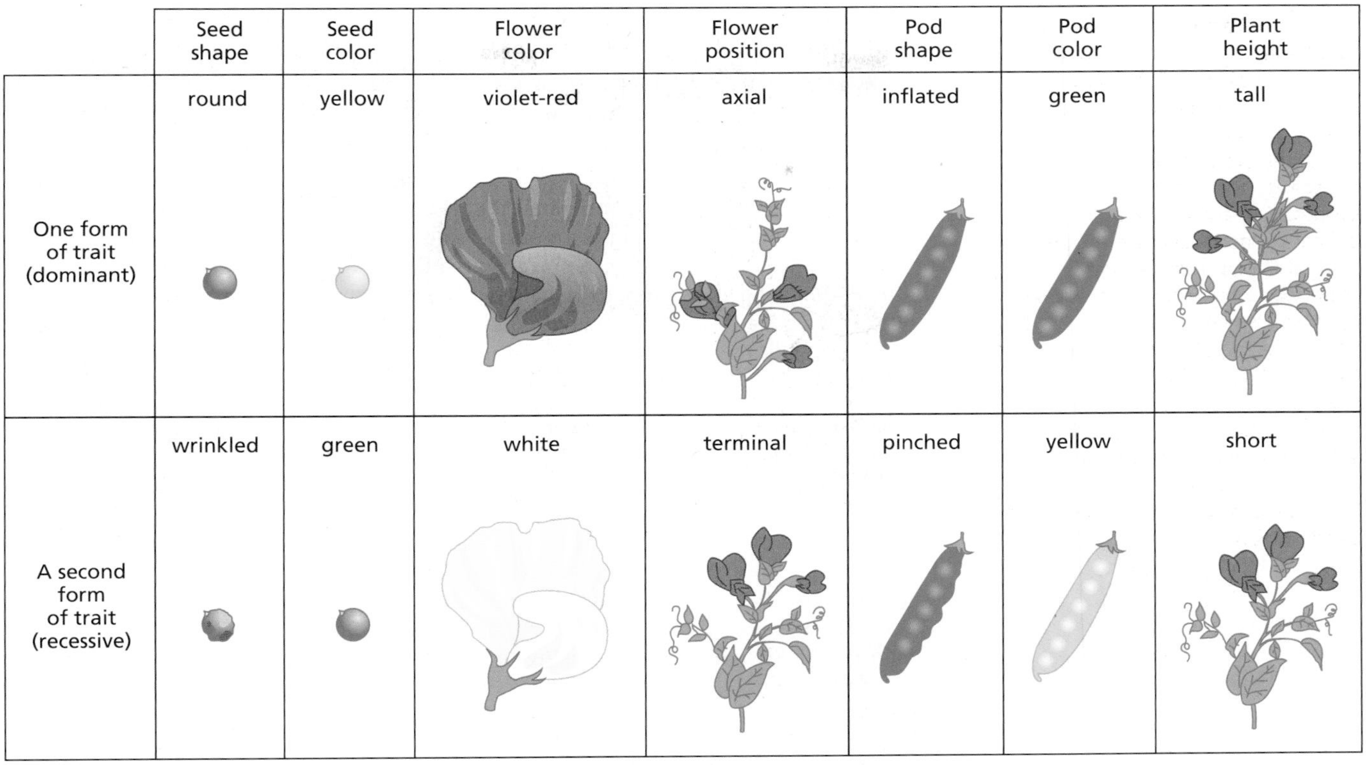

	Seed shape	Seed color	Flower color	Flower position	Pod shape	Pod color	Plant height
One form of trait (dominant)	round	yellow	violet-red	axial	inflated	green	tall
A second form of trait (recessive)	wrinkled	green	white	terminal	pinched	yellow	short

Figure 1.2 A particulate theory of inheritance One of Gregor Mendel's principal insights was that the genetic content of an organism consists of discrete parcels of information, each responsible for a distinct observable trait. Shown are the seven pea-plant traits that Mendel studied through breeding experiments. Each trait had two observable (phenotypic) manifestations, which we now know to be specified by the alternative versions of genes that we call alleles. When the two alternative alleles coexisted within a single plant, the "dominant" trait (*above*) was always observed while the "recessive" trait (*below*) was never observed. (Courtesy of J. Postlethwait and J. Hopson.)

effect, each observable trait of an individual might be traceable to a separate gene that served as its blueprint. Thus, Mendel's research implied that the genetic constitution of an organism (its **genotype**) could be divided into hundreds, perhaps thousands of discrete information packets; in parallel, its observable, outward appearance (its **phenotype**) could be subdivided into a large number of discrete physical or chemical traits (**Figure 1.2**).

Mendel's thinking launched a century-long research project among geneticists, who applied his principles to studying thousands of traits in a variety of experimental animals, including flies (*Drosophila melanogaster*), worms (*Caenorhabditis elegans*), and mice (*Mus musculus*). In the mid-twentieth century, geneticists also began to apply Mendelian principles to study the genetic behavior of single-celled organisms, such as the bacterium *Escherichia coli* and baker's yeast, *Saccharomyces cerevisiae*. The principle of genotype governing phenotype was directly transferable to these simpler organisms and their genetic systems.

While Mendelian genetics represents the foundation of contemporary genetics, it has been adapted and extended in myriad ways since its embodiments of 1865 and 1900. For example, the fact that single-celled organisms often reproduce asexually, that is, without mating, created the need for adaptations of Mendel's original rules. Moreover, the notion that each attribute of an organism could be traced to instructions carried in a single gene was realized to be simplistic. The great majority of observable traits of an organism are traceable to the cooperative interactions of a number of genes. Conversely, almost all the genes carried in the genome of a complex organism play roles in the development and maintenance of multiple organs, tissues, and physiologic processes.

Mendelian genetics revealed for the first time that genetic information is carried redundantly in the genomes of complex plants and animals. Mendel deduced that there were two copies of a gene for flower color and two for pea shape. Today we know that this twofold redundancy applies to the entire genome with the exception of the genes carried in the sex chromosomes. Hence, the genomes of higher organisms are termed **diploid**.

Mendel's observations also indicated that the two copies of a gene could convey different, possibly conflicting information. Thus, one gene copy might specify rough-surfaced and the other smooth-surfaced peas. In the twentieth century, these different versions of a gene came to be called **alleles**. An organism may carry two identical alleles of a gene, in which case, with respect to this gene, it is said to be **homozygous**. Conversely, the presence of two different alleles of a gene in an organism's genome renders this organism **heterozygous** with respect to this gene.

Because the two alleles of a gene may carry conflicting instructions, our views of how genotype determines phenotype become more complicated. Mendel found that in many instances, the voice of one allele may dominate over that of the other in deciding the ultimate appearance of a trait. For example, a pea genome may be heterozygous for the gene that determines the shape of peas, carrying one round and one wrinkled allele. However, the pea plant carrying this pair of alleles will invariably produce round peas. This indicates that the round allele is **dominant**, and that it will invariably overrule its **recessive** counterpart allele (wrinkled) in determining phenotype (see Figure 1.2). (Strictly speaking, using proper genetic parlance, we would say that the phenotype encoded by one allele of a gene is dominant with respect to the phenotype encoded by another allele, the latter phenotype being recessive.)

In fact, classifying alleles as being either dominant or recessive oversimplifies biological realities. The alleles of some genes may be **co-dominant**, in that an expressed phenotype may represent a blend of the actions of the two alleles. Equally common are examples of **incomplete penetrance**, in which case a dominant allele may be present but its phenotype is not manifested because of the actions of other genes within the organism's genome. Therefore, the dominance of an allele is gauged by its interactions with other allelic versions of its gene, rather than its ability to dictate phenotype.

With such distinctions in mind, we note that the development of tumors also provides us with examples of dominance and recessiveness. For instance, one class of alleles that predispose cells to develop cancer encode defective versions of enzymes involved in DNA repair and thus in the maintenance of genomic integrity (discussed again in Chapter 12). These defective alleles are relatively rare in the general population and function recessively. Consequently, their presence in the genomes of many **heterozygotes** (of a wild-type/mutant genotype) is not apparent. However, two heterozygotes carrying recessive defective alleles of the same DNA repair gene may mate. One-fourth of the offspring of such mating pairs, on average, will inherit two defective alleles, exhibit a specific DNA repair defect in their cells, and develop certain types of cancer at greatly increased rates (**Figure 1.3**).

1.2 Mendelian genetics helps to explain Darwinian evolution

In the early twentieth century, it was not apparent how the distinct allelic versions of a gene arise. At first, this variability in information content seemed to have been present in the collective gene pool of a species from its earliest evolutionary beginnings. This perception changed only later, beginning in the 1920s and 1930s, when it became apparent that genetic information is corruptible; the information content in genetic texts, like that in all texts, can be altered. **Mutations** were found to be responsible for changing the information content of a gene, thereby converting one allele into another or creating a new allele from one previously widespread within a species. An allele that is present in the great majority of individuals within a species is usually termed **wild type**, the term implying that such an allele, being naturally present in large numbers of apparently healthy organisms, is compatible with normal structure and function.

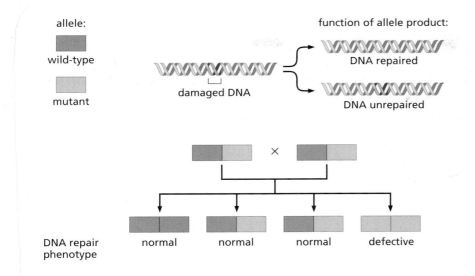

allele:

wild-type

mutant

function of allele product:

damaged DNA

DNA repaired

DNA unrepaired

DNA repair phenotype

normal   normal   normal   defective

Figure 1.3 Discrepancy between genotype and phenotype The phenotype of an individual often does not indicate genotype. For example, individuals who are phenotypically normal for a trait may nevertheless, at the level of genotype, carry one wild-type (normal) and one mutant (defective) allele of the gene that specifies this trait; this mutant allele will be recessive to the wild-type allele, the latter being dominant. Such individuals are heterozygotes with respect to this gene. In the example shown here, two individuals mate, both of whom are phenotypically normal but heterozygous for a gene specifying a DNA repair function. On average, of their four children, three will be phenotypically normal and their cells will exhibit normal DNA repair function: one of these children will receive two wild-type alleles (be a homozygote) and two will be heterozygotes like their parents. A fourth child, however, will receive two mutant alleles (i.e., be a homozygote) and will be phenotypically mutant, in that this child's cells will lack the DNA repair function specified by this gene. Individuals whose cells lack proper DNA repair function are often cancer-prone, as described in Chapter 12.

Mutations alter genomes continually throughout the evolutionary life span of a species, which usually extends over millions of years. They strike the genome and its constituent genes randomly. Mutations provide a species with a method for continually tinkering with its genome, for trying out new versions of genes that offer the prospect of novel, possibly improved phenotypes. The result of the continuing mutations on the genome is a progressive increase during the evolutionary history of a species in the genetic diversity of its members. Thus, the collection of alleles present in the genomes of all members of a species—the **gene pool** of this species—becomes progressively more heterogeneous as the species grows older.

This means that older species carry more distinct alleles in their genomes than younger ones. Humans, belonging to a relatively young species (<150,000 years old), have one-third as many alleles and genetic diversity as chimpanzees, allowing us to infer that they have been around as a species three times longer than we have.

The continuing diversification of alleles in a species' genome, occurring over millions of years, is countered to some extent by the forces of natural selection that Charles Darwin first described. Some alleles of a gene may confer more advantageous phenotypes than others, so individuals carrying these alleles have a greater probability of leaving numerous descendants than do those members of the same species that lack them. Consequently, natural selection results in a continual discarding of many of the alleles that have been generated by random mutations. In the long run, all things being equal, disadvantageous alleles are lost from the pool of alleles carried by the members of a species, advantageous alleles increase in number, and the overall fitness of the species improves incrementally.

Now, more than a century after Mendel was rediscovered and Mendelian genetics revived, we have come to realize that the great bulk of the genetic information in our own genome—indeed, in the genomes of all mammals—does not seem to specify phenotype and is often not associated with specific genes. Reflecting the discovery in 1944 that genetic information is encoded in DNA molecules, these "noncoding" stretches in the genome are often called **junk DNA** (**Figure 1.4**). Only about 1.5% of a mammal's genomic DNA carries sequence information that encodes the structures of proteins. Recent sequence comparisons of human, mouse, and dog genomes suggest that another ~2% encodes important information regulating gene expression and mediating other, still-poorly understood functions.

Because mutations act randomly on a genome, altering true genes and junk DNA indiscriminately, the great majority of mutations alter genetic information—nucleotide sequences in the DNA—that have no effect on cellular or organismic phenotype. These mutations remain silent phenotypically and are said, from the point of view of natural selection, to be **neutral mutations**, being neither advantageous nor

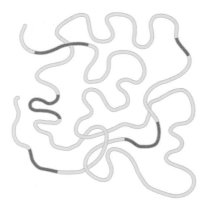

Figure 1.4 Biologically important sequences in the human genome The human genome can be characterized as a collection of relatively small islands of biologically important sequences (~3.5% of the total genome; *red*) floating amid a sea of "junk" DNA (*yellow*). The proportion of sequences carrying biological information has been greatly exaggerated for the sake of illustration. (With the passage of time, genes that appear to play important roles in cell and organismic physiology and specify certain noncoding RNA species have been localized to these intergenic regions; hence the blanket classification of all genomic sequences localized between a human cell's ~21,000 protein-coding genes as useless junk is simplistic.)

disadvantageous (**Figure 1.5**). Since the alleles created by these mutations are silent, their existence could not be discerned by early geneticists whose work depended on gauging phenotypes. However, with the advent of DNA sequencing techniques, it became apparent that hundreds of thousands, even a million functionally silent

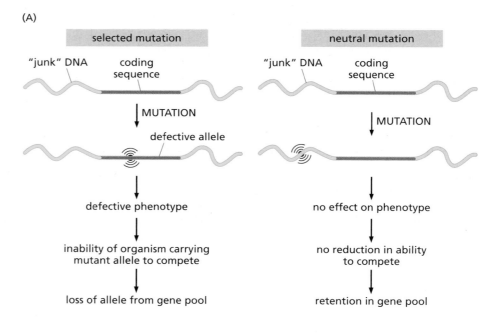

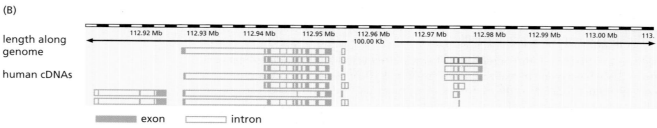

Figure 1.5 Neutral mutations and evolution (A) The coding sequences (*red*) of most genes were optimized in the distant evolutionary past. Hence, many mutations affecting amino acid sequence and thus protein structure (*left*) create alleles that compromise the organism's ability to survive. For this reason, these mutant alleles are likely to be eliminated from the species' gene pool. In contrast, mutations striking "junk" DNA (*yellow*) have no effect on phenotype and are therefore often preserved in the species' gene pool (*right*). This explains why, over extended periods of evolutionary time, coding DNA sequences change slowly, while noncoding DNA sequences change far more rapidly. (B) Depicted is a physical map of a randomly chosen 0.1-megabase segment of human Chromosome 1 (from base pair 112,912,286 to base pair 113,012,285) containing four genes. Each consists of a few islands (*solid rectangles*) that are known or likely to specify segments of mRNA molecules (i.e., exons) and large stretches of intervening sequences (i.e., introns) that do not appear to specify biological information (see Figure 1.16). The large stretches of DNA sequence between genes have not been associated with any biological function. (B, courtesy of The Wellcome Trust Sanger Institute.)

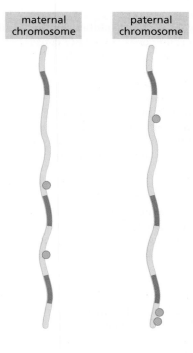

Figure 1.6 Polymorphic diversity in the human gene pool Because the great majority of human genomic DNA does not encode biologically important information (*yellow*), it has evolved relatively rapidly and has accumulated many subtle differences in sequences—polymorphisms—that are phenotypically silent (see Figure 1.5). Such polymorphisms are transmitted like Mendelian alleles, but their presence in a genome can be ascertained only by molecular techniques such as DNA sequencing. The dots (*green*) indicate where the sequence on this chromosome differs from the sequence that is most common in the human gene pool. For example, the prevalent sequence in one stretch may be TAACTGG, while the variant sequence TTACTGG may be carried by a minority of humans and constitute a polymorphism. The presence of a polymorphism in one chromosome but not the other represents a region of heterozygosity, even though a nearby gene (*red*) may be present in the identical allelic version on both chromosomes and therefore be in a homozygous configuration.

mutations can be found scattered throughout the genomes of organisms such as humans. The genome of each human carries its own unique array of these functionally silent genetic alterations. The term *polymorphism* was originally used to describe variations in shape and form that distinguish normal individuals within a species from each other. These days, geneticists use the term **genetic polymorphisms** to describe the inter-individual, functionally silent differences in DNA sequence that make each human genome unique (**Figure 1.6**).

During the course of evolution, the approximately 3.5% of the genome that does encode biological function behaves much differently from the junk DNA. Junk DNA sequences suffer mutations that have no effect on the viability of an organism. Consequently, countless mutations in the noncoding sequences of a species' genome survive in its gene pool and accumulate progressively during its evolutionary history. In contrast, mutations affecting the coding sequences usually lead to loss of function and, as a consequence, loss of organismic viability; hence, these mutations are weeded out of the gene pool by the hand of natural selection, explaining why genetic sequences that do specify biological phenotypes generally change very slowly over long evolutionary time periods (**Sidebar 1.1**).

1.3 Mendelian genetics governs how both genes and chromosomes behave

In the first decade of the twentieth century, Mendel's rules of genetics were found to have a striking parallel in the behavior of the chromosomes that were then being visualized under the light microscope. Both Mendel's genes and the chromosomes were found to be present in pairs. Soon it became clear that an identical set of chromosomes is present in almost all the cells of a complex organism. This chromosomal array, often termed the **karyotype**, was found to be duplicated each time a cell went through a cycle of growth and division.

The parallels between the behaviors of genes and chromosomes led to the speculation, soon validated in hundreds of different ways, that the mysterious information packets called genes were carried by the chromosomes. Each chromosome was realized to carry its own unique set of genes in a linear array. Today, we know that as many as several thousand genes may be arrayed along a mammalian chromosome. (Human Chromosome 1—the largest of the set—holds at least 3148 distinct genes.) Indeed, the length of a chromosome, as viewed under the microscope, is roughly proportional to the number of genes that it carries.

Each gene was found to be localized to a specific site along the length of a specific chromosome. This site is often termed a genetic **locus**. Much effort was expended by geneticists throughout the twentieth century to map the sites of genes—genetic loci—along the chromosomes of a species (**Figure 1.8**).

Sidebar 1.1 Evolutionary forces dictate that certain genes are highly conserved Many genes encode cellular traits that are essential for the continued viability of the cell. These genes, like all others in the genome, are susceptible to the ever-tinkering hand of mutation, which is continually creating new gene sequences by altering existing ones. Natural selection tests these novel sequences and determines whether they specify phenotypes that are more advantageous than the preexisting ones.

Almost invariably, the sequences in genes required for cell and therefore organismic viability were already optimized hundreds of millions of years ago. Consequently, almost all subsequently occurring changes in the sequence information of these genes would have been deleterious and would have compromised the viability of the cell and, in turn, the organism. These mutant alleles were soon lost, because the mutant organisms carrying them failed to leave descendants. This dynamic explains why the sequences of many genes have been highly conserved over vast evolutionary time periods. Stated more accurately, the structures of their encoded proteins have been highly conserved.

In fact, the great majority of the proteins that are present in our own cells and are required for cell viability were first developed during the evolution of single-cell **eukaryotes**. This is indicated by numerous observations showing that many of our proteins have clearly recognizable counterparts in single-cell eukaryotes, such as baker's yeast. Another large repertoire of highly conserved genes and proteins is traceable to the appearance of the first multicellular animals (metazoa); these genes enabled the development of distinct organs and of organismic physiology. Hence, another large group of our own genes and proteins is present in counterpart form in worms and flies (**Figure 1.7**).

By the time the ancestor of all mammals first appeared more than 150 million years ago, virtually all the biochemical and molecular features present in contemporary mammals had already been developed. The fact that they have changed little in the intervening time points to their optimization long before the appearance of the various mammalian orders. This explains why the embryogenesis, physiology, and biochemistry of all mammals is very similar, indeed, so similar that lessons learned through the study of laboratory mice are almost always transferable to an understanding of human biology.

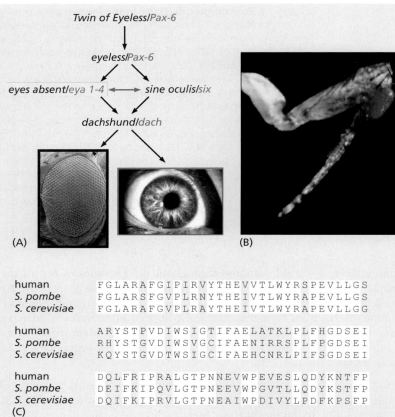

human	FGLARAFGIPIRVYTHEVVTLWYRSPEVLLGS
S. pombe	FGLARSFGVPLRNYTHEIVTLWYRAPEVLLGS
S. cerevisiae	FGLARAFGVPLRAYTHEIVTLWYRAPEVLLGG
human	ARYSTPVDIWSIGTIFAELATKLPLFHGDSEI
S. pombe	RHYSTGVDIWSVGCIFAENIRRSPLFPGDSEI
S. cerevisiae	KQYSTGVDTWSIGCIFAEHCNRLPIFSGDSEI
human	DQLFRIPRALGTPNNEVWPEVESLQDYKNTFP
S. pombe	DEIFKIPQVLGTPNEEVWPGVTLLQDYKSTFP
S. cerevisiae	DQIFKIPRVLGTPNEAIWPDIVYLPDFKPSFP

(C)

Figure 1.7 Extraordinary conservation of gene function
The last common ancestor of flies and mammals lived more than 600 million years ago. Moreover, fly (i.e., arthropod) eyes and mammalian eyes show totally different architectures. Nevertheless, the genes that orchestrate their development (*eyeless* in the fly, *Pax-6/small eye* in the mouse) are interchangeable—the gene from one organism can replace the corresponding mutant gene from the other and restore wild-type function. (A) Thus, the genes encoding components of the signal transduction cascades that operate downstream of these master regulators to trigger eye development (*black* for flies, *pink* for mice) are also highly conserved and interchangeable. (B) The expression of the mouse *Pax-6/small eye* gene, like the *Drosophila eyeless* gene, in an inappropriate (*ectopic*) location in a fly embryo results in the fly developing a fly eye on its leg, demonstrating the interchangeability of the two genes. (C) The conservation of genetic function over vast evolutionary distances is often manifested in the amino acid sequences of homologous proteins. Here, the amino acid sequence of a human protein is given together with the sequences of the corresponding proteins from two yeast species, *S. pombe* and *S. cerevisiae*. (A, courtesy of I. Rebay. B, courtesy of Walter Gehring. C, adapted from B. Alberts et al., Essential Cell Biology, 3rd edition New York: Garland Science, 2010.)

The diploid genetic state that reigns in most cells throughout the body was found to be violated in the *germ cells*, sperm and egg. These cells carry only a single copy of each chromosome and gene and thus are said to be **haploid**. During the formation of germ cells in the testes and ovaries, each pair of chromosomes is separated and one of the pair (and thus associated genes) is chosen at random for incorporation into the sperm or egg. When sperm and egg combine subsequently during fertilization,

genetic
distance — 00 00+ 00++ 01 06 08 15   30   55   69 75
gene —— y ac sc svc br pn w   fa   ec   bi rb
name

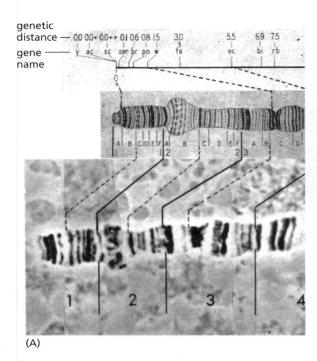

(A)

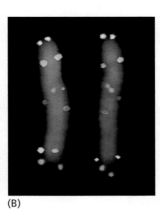

(B)

Figure 1.8 Localization of genes along chromosomes (A) The physical structure of *Drosophila* chromosomes was mapped by using the fly's salivary gland chromosomes, which exhibit banding patterns resulting from alternating light (sparse) and dark (condensed) chromosomal regions (*bottom*). Independently, genetic crosses yielded linear maps (*top*) of various genetic loci arrayed along the chromosomes. These loci were then aligned with physical banding maps, like the one shown here for the beginning of the left arm of *Drosophila* chromosome 1. (B) The availability of DNA probes that hybridize specifically to various genes now makes it possible to localize genes along a chromosome by tagging each probe with a specific fluorescent dye or combination of dyes. Shown are six genes that were localized to various sites along human Chromosome 5 by using fluorescence *in situ* hybridization (FISH) during metaphase. (There are two dots for each gene because chromosomes are present in duplicate form during metaphase of mitosis.) (A, from M. Singer and P. Berg, Genes and Genomes. Mill Valley, CA: University Science Books, 1991, as taken from C.B. Bridges, *J. Hered.* 26:60, 1935. B, courtesy of David C. Ward.)

the two haploid genomes fuse to yield the new diploid genome of the fertilized egg. All cells in the organism descend directly from this diploid cell and, if all goes well, inherit precise replicas of its diploid genome. In a large multicellular organism like the human, this means that a complete copy of the genome is present in almost all of the approximately 3×10^{13} cells throughout the body!

With the realization that genes reside in chromosomes, and that a complete set of chromosomes is present in almost all cell types in the body, came yet another conclusion that was rarely noted: genes create the phenotypes of an organism through their ability to act locally by influencing the behavior of its individual cells. The alternative— that a single set of genes residing at some unique anatomical site in the organism controls the entire organism's development and physiology—was now discredited.

The rule of paired, similarly appearing chromosomes was found to be violated by some of the sex chromosomes. In the cells of female placental mammals, there are two similarly appearing X chromosomes, and these behave like the **autosomes** (the nonsex chromosomes). But in males, an X chromosome is paired with a Y chromosome, which is smaller and carries a much smaller repertoire of genes. In humans, the X chromosome is thought to carry about 900 genes, compared with the 78 distinct genes on the Y chromosome, which, because of redundancy, specify only 27 distinct proteins (**Figure 1.9**).

This asymmetry in the configuration of the sex chromosomes puts males at a biological disadvantage. Many of the 900 or so genes on the X chromosome are vital to normal organismic development and function. The twofold redundancy created by the paired X chromosomes guarantees more robust biology. If a gene copy on one of the X chromosomes is defective (that is, a nonfunctional mutant allele), chances are that the second copy of the gene on the other X chromosome can continue to carry out the task of the gene, ensuring normal biological function. Males lack this genetic fail-safe system in their sex chromosomes. One of the more benign consequences of this is color blindness, which strikes males frequently and females infrequently, due to the localization on the X chromosome of the genes encoding the color-sensing proteins of the retina.

This disparity between the genders is mitigated somewhat by the mechanism of X-inactivation. Early in embryogenesis, one of the two X chromosomes is randomly

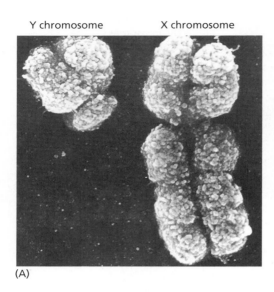

(A)

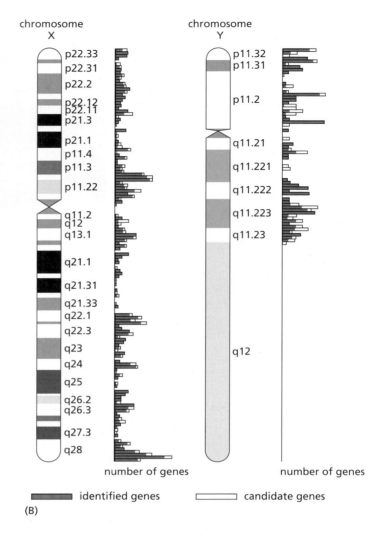

number of genes   number of genes

identified genes   candidate genes

(B)

Figure 1.9 Physical maps of human sex chromosomes (A) Shown is a scanning electron micrograph of human X and Y chromosomes. Like the 22 autosomes (nonsex chromosomes), they have been sequenced. (B) This has allowed the *cytologic* maps of these chromosomes (determined by microscopically examining stained chromosomes at the metaphase of mitosis) to be matched with their DNA sequence. Note that the short arm of a human chromosome is the "p" arm, while the long arm is the "q" arm. Each chromosome has been divided into regions on the basis of the observed banding pattern, and distinct genes have been assigned on the basis of the sequence analyses (*histograms to right of each chromosome*). Identified genes are filled bars (*red*), while sequences that appear to encode still-to-be-identified genes are in open bars; in most chromosomal regions the latter represent a small minority. The human Y chromosome is ~57 megabases (Mb) long, compared with the X chromosome's ~155 Mb. (A, courtesy of Indigo® Instruments. B, courtesy of The Wellcome Trust Sanger Institute. Ensembl genome browser http://www.ensembl.org.)

inactivated in each of the cells of a female embryo. This inactivation silences almost all of the genes on this chromosome and causes it to shrink into a small particle termed the **Barr body**. Subsequently, all descendants of that cell will inherit this pattern of chromosomal inactivation and will therefore continue to carry the same inactivated X chromosome. Accordingly, the female advantage of carrying redundant copies of X chromosome–associated genes is only a partial one (Supplementary Sidebar 1.1).

Color blindness reveals the virtues of having two redundant gene copies around to ensure that biological function is maintained. If one copy is lost through mutational inactivation, the surviving gene copy is often capable of specifying a wild-type phenotype. Such functional redundancy operates for the great majority of genes carried by the autosomes. As we will see later, this dynamic plays an important role in cancer development, since virtually all of the genes that operate to prevent runaway proliferation of cells are present in two redundant copies, both of which must be inactivated in a cell before their growth-suppressing functions are lost and malignant cell proliferation can occur.

1.4   Chromosomes are altered in most types of cancer cells

Individual genes are far too small to be seen with a light microscope, and subtle mutations within a gene are smaller still. Consequently, the great majority of the mutations that play a part in cancer cannot be visualized through microscopy. However, the examination of chromosomes through the light microscope can give evidence of large-scale alterations of the cell genome. Indeed, such alterations were noted as early as 1892, specifically in cancer cells.

Today, we know that cancer cells often exhibit aberrantly structured chromosomes of various sorts, the loss of entire chromosomes, the presence of extra copies of others, and the fusion of the arm of one chromosome with part of another. These changes in overall chromosomal configuration expand our conception of how mutations can affect the genome: since alterations of overall chromosomal structure and number also constitute types of genetic change, these changes must be considered to be the consequences of mutations (**Sidebar 1.2**). And importantly, the abnormal chromosomes seen initially in cancer cells provided the first clue that these cells might be genetically aberrant, that is, that they were mutants (see Figure 1.11).

The normal configuration of chromosomes is often termed the **euploid** karyotypic state. Euploidy implies that each of the autosomes is present in normally structured pairs and that the X and Y chromosomes are present in the numbers appropriate for the sex of the individual carrying them. Deviation from the euploid karyotype—the state termed **aneuploidy**—is seen, as mentioned above, in many cancer cells. Often this aneuploidy is merely a consequence of the general chaos that reigns within a cancer cell. However, this connection between aneuploidy and malignant cell proliferation also hints at a theme that we will return to repeatedly in this book: the acquisition of extra copies of one chromosome or the loss of another can create a genetic configuration that somehow benefits the cancer cell and its agenda of runaway proliferation.

1.5 Mutations causing cancer occur in both the germ line and the soma

Mutations alter the information content of genes, and the resulting mutant alleles of a gene can be passed from parent to offspring. This transmission from one generation to the next, made possible by the germ cells (sperm and egg), is said to occur via the **germ line** (**Figure 1.10**). Importantly, the germ-line transmission of a recently created mutant allele from one organism to its offspring can occur only if a precondition has been met: the responsible mutation must strike a gene carried in the genome of sperm or egg or in the genome of one of the cell types that are immediate precursors of the sperm or egg within the gonads. Mutations affecting the genomes of cells everywhere else in the body—which constitute the **soma**—may well affect the particular cells in which such mutations strike but will have no prospect of being transmitted to the offspring of an organism. Such **somatic mutations** cannot become incorporated into the vehicles of generation-to-generation genetic transmission—the chromosomes of sperm or eggs.

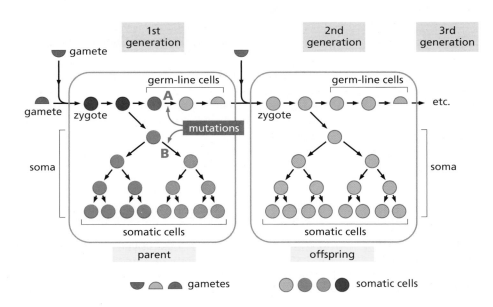

Figure 1.10 Germ-line versus somatic mutations Mutation A, which occurs in the genome of a germ-line cell in the gonads, can be passed from parent (*above left*) to offspring via gametes—sperm or egg (*half circles*). Once incorporated into the fertilized egg (zygote), the mutant alleles can then be transmitted to all of the cells in the body of the offspring (*middle*) outside of the gonads, i.e., its soma, as well as being transmitted via germ-line cells and gametes to a third generation (*not shown*). However, mutation B (*left*), which strikes the genome of a somatic cell in the parent, can be passed only to the lineal descendants of that mutant cell within the body of the parent and cannot be transmitted to offspring. (Adapted from B. Alberts et al., Essential Cell Biology, 3rd ed. New York: Garland Science, 2010.)

Sidebar 1.2 Cancer cells are often aneuploid The presence of abnormally structured chromosomes and changes in chromosome number provided the first clue, early in the twentieth century, that changes in cell genotype often accompany and perhaps cause the uncontrolled proliferation of malignant cells. These deviations from the normal euploid karyotype can be placed into a number of categories. Chromosomes that seem to be structurally normal may accumulate in extra copies, leading to three, four, or even more copies of these chromosomes per cancer cell nucleus (**Figure 1.11**); such deviations from normal chromosome number are manifestations of *aneuploidy*.

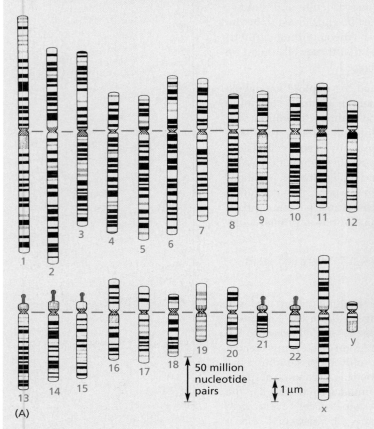

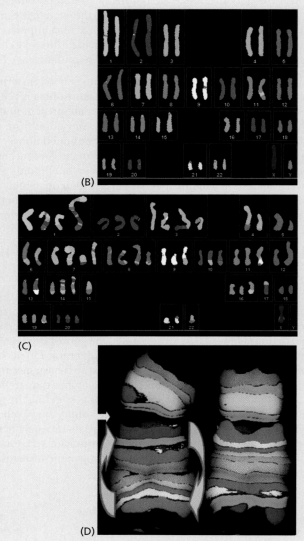

Figure 1.11 Normal and abnormal chromosomal complements (A) Staining of metaphase chromosomes reveals a characteristic light and dark banding pattern for each. The full array of human chromosomes is depicted; their centromeres are aligned (*pink line*). (B) The techniques of spectral karyotype (SKY) analysis and multicolor fluorescence *in situ* hybridization (mFISH) allow an experimenter to "paint" each metaphase chromosome with a distinct color (by hybridizing chromosome-specific DNA probes labeled with various fluorescing dyes to the chromosomes). The actual colors in images such as these are generated by computer. The diploid karyotype of a normal human male cell is presented. (The small regions in certain chromosomes that differ from the bulk of these chromosomes represent hybridization artifacts.) (C) The aneuploid karyotype of a human pancreatic cancer cell, in which some chromosomes are present in inappropriate numbers and in which numerous translocations (exchanges of segments between chromosomes) are apparent. (D) Here, mFISH was used to label intrachromosomal subregions with specific fluorescent dyes, revealing that a large portion of an arm of normal human Chromosome 5 (*right*) has been inverted (*left*) in cells of a worker who had been exposed to plutonium in the nuclear weapons industry of the former Soviet Union. (A, adapted from U. Francke, *Cytogenet. Cell Genet.* 31:24–32, 1981. B and C, courtesy of M. Grigorova, J.M. Staines and P.A.W. Edwards. D, from M.P. Hande et al., *Am. J. Hum. Genet.* 72:1162–1170, 2003.)

Alternatively, chromosomes may undergo changes in their structure. A segment may be broken off one chromosomal arm and become fused to the arm of another chromosome, resulting in a chromosomal **translocation** (Figure 1.11C). Moreover, chromosomal segments may be exchanged between chromosomes from different chromosome pairs, resulting in **reciprocal translocations**. A chromosomal segment may also become inverted, which may affect the regulation of genes that are located near the breakage-and-fusion points (Figure 1.11D).

A segment of a chromosome may be copied many times over, and the resulting extra copies may be fused head-to-tail in long arrays within a chromosomal segment that is termed an HSR (**homogeneously staining region; Figure 1.12A**). A segment may also be cleaved out of a chromosome, replicate

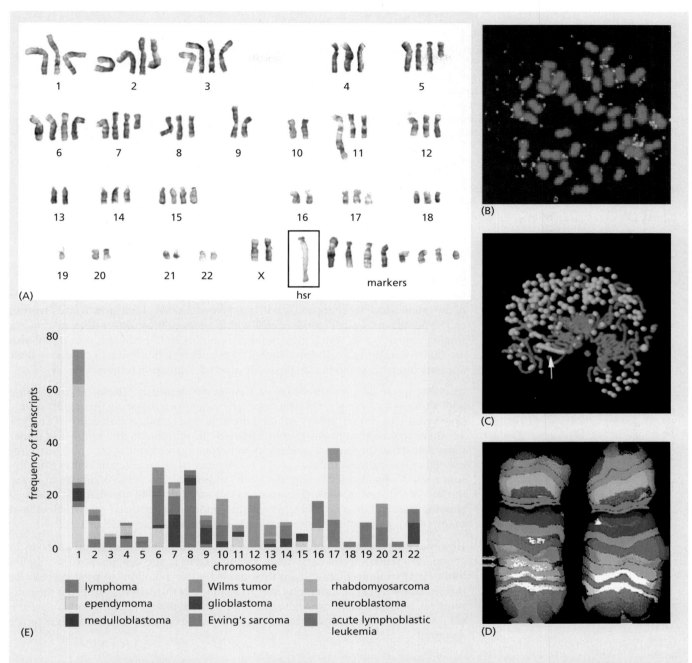

Figure 1.12 Increases and decreases in copy number of chromosomal segments (A) The amplification in the copy number of the *myc* oncogene (see Section 8.9) in a human neuroendocrinal tumor has caused an entire stretch of chromosome to stain *white* (rectangle), creating a homogeneously staining region (HSR). (B) Double-minute chromosomes (DMs) derive from chromosomal segments that have broken loose from their original sites and have been replicated repeatedly as extrachromosomal genetic elements; like normal chromatids, these structures are doubled during metaphase of mitosis. FISH reveals the presence of amplified copies of the *HER2/neu* oncogene borne on DMs (*yellow dots*) in a mouse breast cancer cell. (C) Occasionally, an amplified gene may be found both in an HSR (nested within a chromosome) and in DMs. Here, analysis of COLO320 cells reveals multiple copies of the *myc* oncogene (*yellow*), amid the chromosomes (*red*). One HSR is indicated by the arrow, while many dozens of DMs are apparent. (D) The use of multicolor FISH (mFISH) revealed that a segment within normal human Chromosome 5 (*paired arrows, left*) has been deleted (an interstitial deletion, *right*) following extensive exposure to radiation from plutonium. (E) A survey of nine different types of pediatric cancer indicates that each cancer type has characteristic gene amplification and deletion patterns with corresponding changes in the expression of the altered genes. For example, neuroblastomas (*pink*) often have changes in the copy numbers of genes on chromosomes 1 and 17 and corresponding changes in the levels of the transcripts expressed by these genes. (A, from J.-M. Wen et al., *Cancer Genet. Cytogenet.* 135:91–95, 2002. B, from C. Montagna et al., *Oncogene* 21:890–898, 2002. C, from N. Shimizu et al., *J. Cell Biol.* 140:1307–1320, 1998. D, from M.P. Hande et al., *Am. J. Hum. Genet.* 72:1162–1170, 2003. E, from G. Neale et al., *Clin. Cancer Res.* 14:4572–4583, 2008.)

as an autonomous, extrachromosomal entity, and increase to many copies per nucleus, resulting in the appearance of subchromosomal fragments termed DMs (**double minutes**; Figure 1.12B). These latter two changes cause increases in the copy number of genes carried in such segments, resulting in **gene amplification**. Sometimes, both types of amplification coexist in the same cell (Figure 1.12C). Gene amplification can favor the growth of cancer cells by increasing the copy number of growth-promoting genes.

On some occasions, certain growth-inhibiting genes may be discarded by cancer cells during their development. For example, when a segment in the middle of a chromosomal arm is discarded and the flanking chromosomal regions are joined, this results in an **interstitial deletion** (Figure 1.12D).

These descriptions of copy-number changes in genes, involving both amplifications and deletions, might suggest widespread chaos in the genomes of cancer cells, with gene amplifications and deletions occurring randomly. However, as the karyotypes and genomes of human tumors have been examined more intensively, it has become clear that certain regions of the genome tend to be lost characteristically in certain tumor types but not in others (Figure 1.12E). This suggests a theme that we will pursue in great detail throughout this book—that the gains and losses of particular genes favor the proliferation of specific types of tumors. This indicates that different tumor types undergo different genetic changes as they develop progressively from the precursor cells in normal tissues.

Somatic mutations are of central importance to the process of cancer formation. As described repeatedly throughout this book, a somatic mutation can affect the behavior of the cell in which it occurs and, through repeated rounds of cell growth and division, can be passed on to all descendant cells within a tissue. These direct descendants of a single progenitor cell, which may ultimately number in the millions or even billions, are said to constitute a cell **clone**, in that all members of this group of cells trace their ancestry directly back to the single cell in which the mutation originally occurred.

An elaborate repair apparatus within each cell continuously monitors the cell's genome and, with great efficiency, eradicates mutant sequences, replacing them with appropriate wild-type sequences. We will examine this repair apparatus in depth in Chapter 12. This apparatus maintains genomic integrity by minimizing the number of mutations that strike the genome and are then perpetuated by transmission to descendant cells. One stunning indication of the efficiency of genome repair comes from the successes of organismic cloning: the ability to generate an entire organism from the nucleus of a differentiated cell (prepared from an adult) indicates that this adult cell genome is essentially a faithful replica of the genome of a fertilized egg, which existed many years and many cell generations earlier (Supplementary Sidebar 1.2).

However, no system of damage detection and repair is infallible. Some mistakes in genetic sequence survive its scrutiny, become fixed in the cell genome, are copied into new DNA molecules, and are then passed on as mutations to progeny cells. In this sense, many of the mutations that accumulate in the genome represent the consequences of occasional oversights made by the repair apparatus. Yet others are the results of catastrophic damage to the genome that exceeds the capacities of the repair apparatus.

1.6 Genotype embodied in DNA sequences creates phenotype through proteins

The genes studied in Mendelian genetics are essentially mathematical abstractions. Mendelian genetics explains their transmission, but it sheds no light on how genes create cellular and organismic phenotypes. Phenotypic attributes can range from complex, genetically templated behavioral traits to the **morphology** (shape, form) of cells and subcellular organelles to the biochemistry of cell metabolism. This mystery of how genotype creates phenotype represented the major problem of twentieth-century biology. Indeed, attempts at forging a connection between these two became the obsession of many molecular biologists during the second half of the twentieth century and continue as such into the twenty-first, if only because we still possess an incomplete understanding of how genotype influences phenotype.

Molecular biology has provided the basic conceptual scaffold for understanding the connection between genotype and phenotype. In 1944, DNA was proven to be the

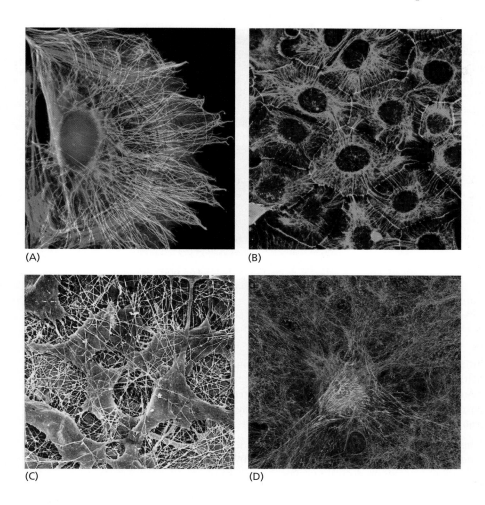

(A)　(B)　(C)　(D)

Figure 1.13 Intracellular and extracellular scaffolding The cytoskeleton is assembled from complex networks of intermediate filaments, actin microfilaments, and microtubules. Together, they generate the shape of a cell and enable its motion. (A) In this cultured cell, microfilaments composed of actin (*orange*) form bundles that lie parallel to the cell surface while microtubules composed of tubulin (*green*) radiate outward from the nucleus (*blue*). Both types of fibers are involved in the formation of protrusions from the cell surface. (B) Here, an important intermediate filament of epithelial cells—keratin—is detected using an anti-keratin-specific antibody (*green*). The boundaries of cells are labeled with a second antibody that reacts with a plasma membrane protein (*blue*). (C) Cells secrete a diverse array of proteins that are assembled into the extracellular matrix (ECM). A scanning electron micrograph reveals the complex meshwork of collagen fibers, glycoproteins, hyaluronan, and proteoglycans, in which fibroblasts (connective tissue cells) are embedded. (D) A cell of the NIH 3T3 cell line, which is used extensively in cancer cell biology, is shown amid an ECM network of fibronectin fibers (*green*). The points of cellular attachment to the fibronectin are mediated by integrin receptors on the cell surface (*orange, yellow*). (A, courtesy of Albert Tousson, High-Resolution Imaging Facility, University of Alabama at Birmingham. B, courtesy of Kathleen Green and Evangeline Amargo. C, courtesy of T. Nishida. D, from E. Cukierman et al., *Curr. Opin. Cell Biol.* 14:633–639, 2002.)

chemical entity in which the genetic information of cells is carried. Nine years later, Watson and Crick elucidated the double-helical structure of DNA. A dozen years after that, in 1964, it became clear that the sequences in the bases of the DNA double helix determine precisely the sequence of amino acids in proteins. The unique structure and function of each type of protein in the cell is determined by its sequence of amino acids. Therefore, the specification of amino acid sequence, which is accomplished by base sequences in the DNA, provides almost all the information that is required to construct a protein.

Once synthesized within cells, proteins proceed to create phenotype, doing so in a variety of ways. Proteins can assemble within the cell to create the components of the **cytoarchitecture**, or more specifically, the **cytoskeleton** (**Figure 1.13A and B**). When secreted into the space between cells, such proteins form the **extracellular matrix** (ECM); it ties cells together, enabling them to form complex tissues (Figure 1.13C and D). As we will see later, the structure of the ECM is often disturbed by malignant cancer cells, enabling them to migrate to sites within a tissue and organism that are usually forbidden to them.

Many proteins function as enzymes that catalyze the thousands of biochemical reactions that together are termed **intermediary metabolism**; without the active intervention of enzymes, few of these reactions would occur spontaneously. Proteins can also contract and create cellular movement (**motility; Figure 1.14**) as well as muscle contraction. Cellular motility plays a role in cancer development by allowing cancer cells to spread through tissues and migrate to distant organs.

And most important for the process of cancer formation, proteins can convey signals between cells, thereby enabling complex tissues to maintain the appropriate numbers of constituent cell types. Within individual cells, certain proteins receive signals

Figure 1.14 Cell motility (A) The movement of individual cells in a culture dish can be plotted at intervals and scored electronically. This image traces the movement of a human vascular endothelial cell (the cell type that forms the lining of blood vessels) toward two attractants located at the bottom—vascular endothelial growth factor (VEGF) and sphingosine-1-phosphate (S1P). Such locomotion is presumed to be critical to the formation of new blood vessels within a tumor. Each point represents a position plotted at 10-minute intervals. This motility is made possible by complex networks of proteins that form the cells' cytoskeletons. (B) The advancing cell is a fish keratocyte; its leading edge (*green*) is pushed forward by an actin filament network, such as the one shown in C. (C) Seen here is the network of actin filaments that is assembled at the leading edge of a motile cell. (A, courtesy of C. Furman and F. Gertler. B and C, from T. Svitkina and G. Borisy, *J. Cell Biol.* 145:1009–1026, 1999. © The Rockefeller University Press.)

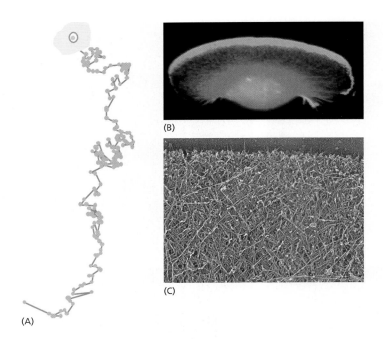

(A)

(B)

(C)

from an extracellular source, process these signals, and pass them on to other proteins within the cell; such signal-processing functions, often termed intracellular signal **transduction**, are also central to the creation of cancers, since many of the abnormal-growth phenotypes of cancer cells are the result of aberrantly functioning intracellular signal-transducing molecules.

The functional versatility of proteins makes it apparent that almost all aspects of cell and organismic phenotype can be created by their actions. Once we realize this, we can depict genotype and phenotype in the simplest of molecular terms: genotype resides in the sequences of bases in DNA, while phenotype derives from the actions of proteins. (In fact, this depiction is simplistic, because it ignores the important role of RNA molecules as intermediaries between DNA sequences and protein structure and the recently discovered abilities of some RNA molecules to function as enzymes and others to act as regulators of the expression of certain genes.)

In the complex **eukaryotic** (nucleated) cells of animals, as in the simpler **prokaryotic** cells of bacteria, DNA sequences are copied into RNA molecules in the process termed **transcription**; a gene that is being transcribed is said to be actively **expressed**, while a gene that is not being transcribed is often considered to be **repressed**. In its simplest version, the transcription of a gene yields an RNA molecule of length comparable to the gene itself. Once synthesized, the base sequences in the RNA molecule are **translated** by the protein-synthesizing factories in the cell, its **ribosomes**, into a sequence of amino acids. The resulting macromolecule, which may be hundreds, even thousands of amino acids long, folds up into a unique three-dimensional configuration and becomes a functional protein (**Figure 1.15**).

Post-translational modification of the initially synthesized protein may result in the covalent attachment of certain chemical groups to specific amino acid residues in the protein chain; included among these modifications are, notably, phosphates, complex sugar chains, and methyl, acetyl, and lipid groups (**Sidebar 1.3**). Thus, the extracellular domains of most cell-surface proteins and almost all secreted proteins are glycosylated, having one or more covalently attached sugar side chains; proteins of the Ras family, which are located in the cytoplasm and play important roles in cancer development, contain lipid groups attached to their carboxy termini. An equally important post-translational modification involves the cleavage of one protein by a second protein termed a **protease**, which has the ability to cut amino acid chains at certain sites. Accordingly, the final, mature form of a protein chain may include far fewer amino acid residues than were present in the initially synthesized protein. Following their synthesis, many proteins are dispatched to specific sites within the cell or are exported

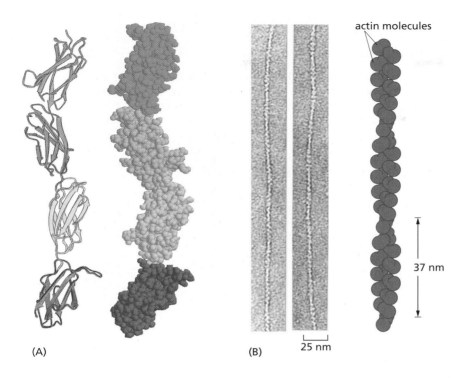

actin molecules

37 nm

25 nm

(A)   (B)

Figure 1.15 Structures of proteins and multiprotein assemblies (A) The three-dimensional structure of part of fibronectin, an important extracellular matrix protein (see Figure 1.13D), is depicted as a ribbon diagram (*left*), which illustrates the path taken by its amino acid chain; alternatively, the space-filling model (*right*) shows the positions of the individual atoms. One portion of fibronectin is composed of four distinct, similarly structured domains, which are shown here with different colors. (B) The actin fibers (*left*), which constitute an important component of the cytoskeleton (see Figures 1.13 and 1.14), are composed of assemblies of individual protein molecules, each of which is illustrated here as a distinct two-lobed body (*right*). (A, adapted from D.J. Leahy, I. Aukhil and H.P. Erickson, *Cell* 84:155–164, 1996. B, left, courtesy of Roger Craig; right, from B. Alberts et al., Molecular Biology of the Cell, 5th ed. New York: Garland Science, 2008.)

from the cell through the process of secretion; these alternative destinations are specified in the newly synthesized proteins by short amino acid (oligopeptide) sequences that function, much like postal addresses, to ensure the diversion of these proteins to specific intracellular sites.

In eukaryotic cells—the main subject of this book—the synthesis of RNA is itself a complex process. An RNA molecule transcribed from its parent gene may initially be almost as long as that gene. However, while it is being elongated, segments of the RNA

Sidebar 1.3 How many distinct proteins can be found in the human body? While some have ventured to provide estimates of the total number of human genes (a bit more than 21,000), it is difficult to extrapolate from this number to the total number of distinct proteins encoded in the human genome. The simplest estimate comes from the assumption that each gene encodes the structure of a single protein. But this assumption is naive, because it ignores the fact that the pre-mRNA transcript deriving from a single gene may be subjected to several *alternative splicing* patterns, yielding multiple, distinctly structured mRNAs, many of which may in turn encode distinct proteins (see Figure 1.16). Thus, in some cells, splicing may include certain exons in the final mRNA molecule made from a gene, while in other cells, these exons may be absent. Such alternative splicing patterns can generate mRNAs having greatly differing structures and protein-encoding sequences. In one, admittedly extreme case, a single *Drosophila* gene has been found to be capable of generating 38,016 distinct mRNAs and thus proteins through various alternative splices of its pre-mRNA; genes having similarly complex alternative splicing patterns are likely to reside in our own genome.

An additional dimension of complexity derives from the post-translational modifications of proteins. The proteins that are exported to the cell surface or released in soluble form into the extracellular space are usually modified by the attachment

of complex trees of sugar molecules during the process of **glycosylation**. Intracellular proteins often undergo other types of chemical modifications. Proteins involved in transducing the signals that govern cell proliferation often undergo **phosphorylation** through the covalent attachment of phosphate groups to serine, threonine, or tyrosine amino acid residues. Many of these phosphorylations affect some aspect of the functioning of these proteins. Similarly, the histone proteins that wrap around DNA and control its access by the RNA polymerases that synthesize hnRNA are subject to methylation, acetylation, and phosphorylation, as well as more complex post-translational modifications.

The polypeptide chains that form proteins may also undergo cleavage at specific sites following their initial assembly, often yielding small proteins showing functions that were not apparent in the uncleaved precursor proteins. Later, we will describe how certain signals may be transmitted through the cell via a cascade of the protein-cleaving enzymes termed proteases. In these cases, protein A may cleave protein B, activating its previously latent protease activity; thus activated, protein B may cleave protein C, and so forth. Taken together, alternative splicing and post-translational modifications of proteins generate vastly more distinct protein molecules than are apparent from counting the number of genes in the human genome.

molecule, some very small and others enormous, will be cleaved out of the growing RNA molecule. These segments, termed **introns**, are soon discarded and consequently have no impact on the subsequent coding ability of the RNA molecule (**Figure 1.16**).

Flanking each intron are two retained sequences, the **exons**, which are fused together during this process of **splicing**. The initially synthesized RNA molecule and its derivatives found at various stages of splicing, together with nuclear RNA transcripts being processed from other genes, collectively constitute the **hnRNA** (**heterogeneous**

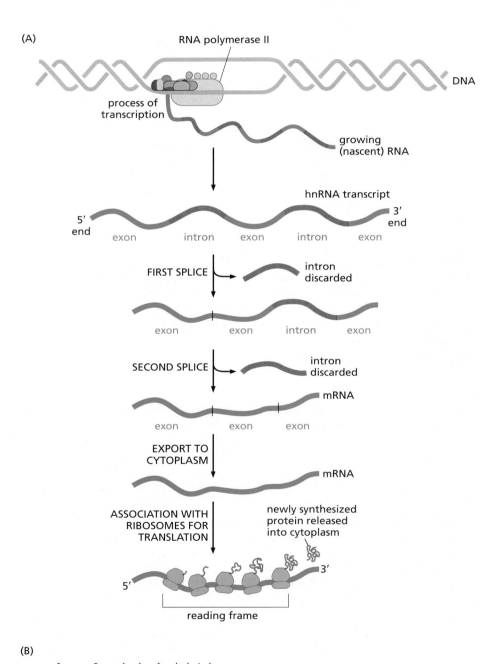

Figure 1.16 Processing of pre-mRNA
(A) By synthesizing a complementary RNA copy of one of the two DNA strands of a gene, RNA polymerase II creates a molecule of heterogeneous nuclear RNA (hnRNA) (*red and blue*). Those hnRNA molecules that are processed into mRNAs are termed pre-mRNA. The progressive removal of the introns (*red*) leads to a processed mRNA containing only exons (*blue*). (B) A given pre-mRNA molecule may be spliced in a number of alternative ways, yielding distinct mRNAs that may encode distinct protein molecules. Illustrated here are the tissue-specific alternative splicing patterns of the α-tropomyosin pre-mRNA molecule, whose mRNA products specify important components of cell (and thus muscle) contractility. In this case, the introns are indicated as *black carets* while the exons are indicated as *blue rectangles*.
(B, adapted from B. Alberts et al., Molecular Biology of the Cell, 5th ed. New York: Garland Science, 2008.)

nuclear RNA). The end product of these post-transcriptional modifications may be an RNA molecule that is only a small fraction of the length of its initially synthesized, hnRNA precursor. This final, mature RNA molecule is likely to be exported into the cytoplasm, where, as an **mRNA** (**messenger RNA**) molecule, it serves as the template on which ribosomes assemble the amino acids that form the proteins. (The term **pre-mRNAs** is often used to designate those hnRNAs that are known precursors of cytoplasmic mRNAs.) Some mature mRNAs may be less than 1% of the length of their pre-mRNA precursor. The complexity of post-transcriptional modification of RNA and post-translational modification of proteins yields an enormous array of distinct protein species within the cell (see Sidebar 1.3).

Of note, an initially transcribed pre-mRNA may be processed through **alternative splicing** into a series of distinct mRNA molecules that retain different combinations of exons (see Figure 1.16B). Indeed, the pre-mRNAs arising from more than 95% of the genes in our genome are subject to alternative splicing. The resulting alternatively spliced mRNAs may carry altered reading frames, explaining, for example, the distinct isoforms of certain proteins that are found in cancer cells but not in their normal counterparts. Alternatively, these splicing events may affect untranslated regions of mRNAs, such as those targeted by microRNAs (miRNAs; Section 1.10); these interactions with miRNAs can alter the function of an mRNA, by regulating either its translation or its stability. Interestingly, a protein that specifies an alternative splicing pattern of pre-mRNAs has been reported, when expressed in excessively high levels in cells, to favor their **transformation** (conversion) from a normal to a cancerous growth state. Such an effect is surprising, since one might imagine that proteins that regulate splicing would mediate the processing of many or all pre-mRNAs within the cell rather than affecting only a subset of genes involved in a specific cell-biological function, such as cell transformation. Moreover, a 2008 survey of alternatively spliced mRNAs found 41 that showed a distinct pattern of alternative splicing in human breast cancer cells compared with normal mammary cells; indeed, these alternatively spliced mRNAs could be used as diagnostic markers of the cancerous state of these cells. Even more dramatic, in 2010 as many as 1000 pre-mRNAs were found to undergo alternative splicing as cells passed through an epithelial–mesenchymal transition (EMT), an important transdifferentiation step that carcinoma cells utilize to acquire traits of high-grade malignancy, as will be discussed in Chapter 14.

1.7   Gene expression patterns also control phenotype

The 21,000 or so genes in the mammalian genome, acting combinatorially within individual cells, are able to create the extraordinarily complex organismic phenotypes of the mammalian body. A central goal of twenty-first-century biology is to relate the functioning of this large repertoire of genes to organismic physiology, developmental biology, and disease development. The complexity of this problem is illustrated by the fact that there are at least several hundred distinct cell types within the mammalian body, each with its own behavior, its own distinct metabolism, and its own physiology.

This complexity is acquired during the process of organismic development, and its study is the purview of developmental biologists. They wrestle with a problem that is inherent in the organization of all multicellular organisms. All of the cells in the body of an animal are the lineal descendants of a fertilized egg. Moreover, almost all of these cells carry genomes that are reasonably accurate copies of the genome that was initially present in this fertilized egg (see Supplementary Sidebar 1.2). The fact that cells throughout the body are phenotypically quite distinct from one another (e.g., a skin cell versus a brain cell) while being genetically identical creates this central problem of developmental biology: how do these various cell types acquire distinct phenotypes if they all carry identical genetic templates? The answer, documented in thousands of ways over the past three decades, lies in the selective reading of the genome by different cell types (**Figure 1.17**).

As cells in the early embryo pass through repeated cycles of growth and division, the cells located in different parts of the embryo begin to assume distinct phenotypes, this being the process of **differentiation**. Differentiating cells become committed to form

Figure 1.17 Global surveys of gene expression arrays Gene expression microarrays make it possible to survey the expression levels of thousands of genes within a given type of cell. In this image, higher-than-average levels of expression are indicated as *red* pixels, while lower-than-average levels are indicated by *green* pixels. Average-level expression is indicated by *black* pixels. The mRNAs from 142 different human tumors (*arrayed left to right*) were analyzed. In each case, the expression levels of 1800 human genes were measured (*top to bottom*). Each class of tumors has its characteristic spectrum of expressed genes. In this case, a tumor of unknown type (*yellow label*) was judged to be a lung cancer because its pattern of gene expression was similar to those of a series of already-identified lung cancers. (Courtesy of P.O. Brown, D. Botstein and The Stanford Expression Collaboration.)

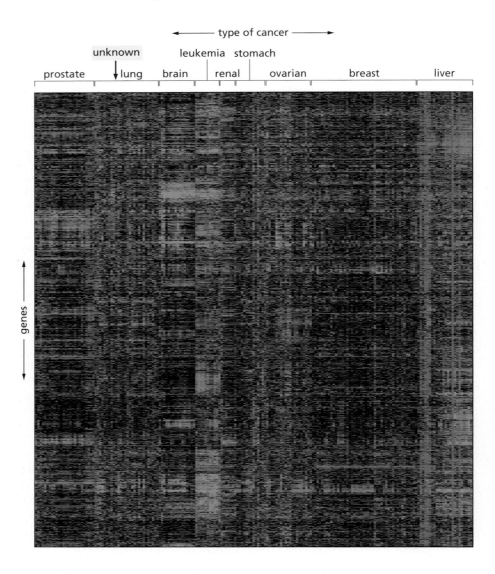

one type of tissue rather than another, for example, gut as opposed to nervous system. All the while, they retain the same set of genes. This discrepancy leads to a simplifying conclusion: sooner or later, differentiation must be understood in terms of the sets of genes that are expressed (that is, transcribed) in some cells but not in others.

By being expressed in a particular cell type, a suite of genes dictates the synthesis of a cohort of proteins and RNA molecules that collaborate to create a specific cell phenotype. Accordingly, the phenotype of each kind of differentiated cell in the body should, in principle, be understandable in terms of the specific subset of genes that is expressed in that cell type.

The genes within mammalian cells can be grouped into two broad functional classes—the **housekeeping** and the **tissue-specific** genes. Many genes encode proteins that are required universally to maintain viability of all cell types throughout the body or to carry out certain biological functions common to all cell types. These commonly expressed genes are classified as housekeeping genes. Within a given differentiated cell type, housekeeping genes represent the great majority of expressed genes.

A minority of genes within a differentiated cell—the tissue-specific genes—are dedicated to the production of proteins and thus phenotypes that are associated specifically with this cell type. It may be, for example, that 3000–5000 housekeeping genes are expressed by the cell while far fewer than 1000 tissue-specific genes are responsible for the distinguishing, differentiated characteristics of the cell. By implication, in each type of differentiated cell, a significant proportion of the 21,000 or so genes in the genome are unexpressed, since they are not required either for the cell's specific differentiation program or for general housekeeping purposes.

1.8   Histone modification and transcription factors control gene expression

The foregoing description of differentiation makes it clear that large groups of genes must be coordinately expressed while other genes must be repressed in order for cells to display complex, tissue-specific phenotypes. Such coordination of expression is the job of **transcription factors** (TFs; **Figure 1.18**). Many of these proteins bind to specific DNA sequences in the control region of each gene and determine whether or not the gene will be transcribed. The specific stretch of nucleotide sequence to which the TFs bind, often called a **sequence motif**, is usually quite short, typically 5–10 nucleotides long. In ways that are still incompletely understood at the molecular level, some TFs provide the RNA polymerase enzyme (RNA polymerase II in the case of pre-mRNAs) with access to a gene. Yet other TFs may block such access and thereby ensure that a gene is transcriptionally repressed.

Transcription factors can exercise great power, since a single type of TF can simultaneously affect the expression of a large cohort of downstream responder genes, each of which carries the recognition sequence that allows this TF to bind its promoter (see Figure 1.18). This ability of a single TF (or a single gene that specifies this TF) to elicit multiple changes within a cell or organism is often termed **pleiotropy**. In the case of cancer cells, a single malfunctioning, pleiotropically acting TF may simultaneously orchestrate the expression of a large cohort of responder genes that together proceed to create major components of the cancer cell phenotype. One enumeration of the genes in the human genome that are likely to encode TFs listed 1445 distinct genes (about 7% of the genes carried in the human genome). Not included in this list were variant versions of these proteins arising through alternative splicing of pre-mRNAs.

The transcription of most genes is dependent upon the actions of several distinct TFs that must sit down together, each at its appropriate sequence site (that is, **enhancer**) in or near the gene promoter, and collaborate to activate gene expression. This means that the expression of a gene is most often the result of the combinatorial actions of several TFs. Therefore, the coordinated expression of multiple genes within a cell, often called its gene **expression program**, is dependent on the actions of multiple TFs acting in combination on large numbers of gene promoters.

Figure 1.18 implies that modulation of gene expression is achieved by controlling initiation of transcription by RNA polymerase II (pol II) and that transcription proceeds in one direction. In fact, for many genes, possibly the majority, pol II molecules sit

Figure 1.18 Regulation of gene expression The control region of a gene includes specific segments of DNA to which gene regulatory proteins known as transcription factors (TFs) bind, often as multiprotein complexes; in this case TFs, functioning as activators (*light brown*), bind to *enhancer* sequences (*orange*) located some distance upstream of the promoter. In addition, the *promoter* of the gene (*dark, light green*) contains sequences to which RNA polymerase II (pol II) can bind, together with associated general transcription factors. The bound TFs, interacting with the transcription initiation complex via *mediator* proteins, influence the structure of chromatin (notably the histone proteins that package DNA; see Figures 1.19 and 1.20), creating a localized chromatin environment that enables pol II to produce an RNA transcript (*orange-red arrow*). (The general TFs are involved in initiating the transcription of many genes throughout the genome, while the specialized ones regulate the expression of subsets of genes.) Although in general a gene can be separated into two functionally significant regions—the nontranscribed control sequences and the transcribed sequences represented in pre-mRNA and mRNA molecules—some of the regulatory sequences (enhancers) may be located within the transcribed region of a gene, often in introns. (From B. Alberts et al., Molecular Biology of the Cell, 5th ed. New York: Garland Science, 2008.)

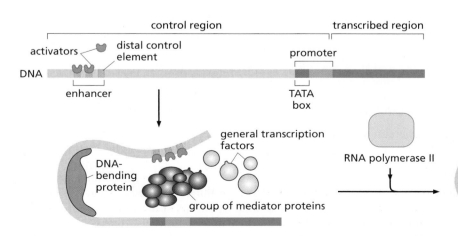

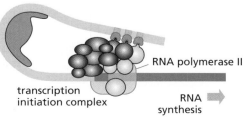

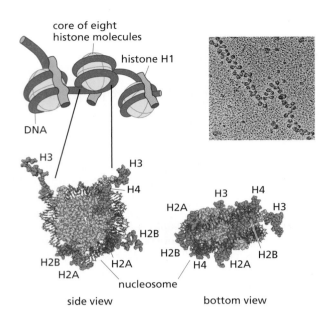

Figure 1.19 Organization of chromatin structure Examination of chromatin under the electron microscope (*above right*) reveals that DNA is associated with small globes of proteins termed nucleosomes, giving the appearance of beads on a string. The DNA double-helix (*above left, red*) is wrapped ~1.7 times around each nucleosome, which consists of a core (*yellow*) formed as an octamer of four different histone molecules (each present in two copies); often an additional histone, H1 (*green*), is located on the outside. X-ray crystallography has revealed (*below*) that the core of the nucleosome (*yellow*) is disc-shaped and that the N-terminal tail (*green*) of each of the four histones extends beyond this core. (Upper schematic, from W.K. Purves et al., Life: The Science of Biology, 5th ed. Sunderland, MA: Sinauer, 1998. Lower schematic from B. Alberts et al., Molecular Biology of the Cell, 5th ed. New York: Garland Science, 2008. Micrograph from F. Thoma, T. Koller and A. Klug, *J. Cell Biol.* 83:403–427, 1979.)

down on the promoter of a gene and proceed to transcribe the DNA in both directions. After extending nascent RNA transcripts for 60–80 nucleotides, pol II halts—the process termed **transcriptional pausing.** A subset of the stalled polymerase complexes that have initiated in the appropriate transcriptional direction are then induced by physiologic signals to resume elongation, resulting in full-length pre-mRNA transcripts, while other pol II complexes remain stalled and never resume transcription. The factors that permit stalled pol II to proceed with elongation of transcripts are incompletely understood but would seem to be as important as the conventionally defined TFs in regulating gene expression. One important cancer-causing protein, termed Myc, has been found to act as an anti-pausing protein whose actions permit thousands of cellular genes to be fully transcribed.

Figure 1.18 also implies that both TFs and RNA polymerase interact only with DNA. In fact, in eukaryotic cells, DNA is packaged in a complex mixture of proteins that, together with the DNA, form the **chromatin (Figure 1.19)**. These chromatin proteins are responsible for controlling the interactions of TFs and RNA polymerases with DNA and therefore play critical roles in governing gene expression.

The core of chromatin is formed by DNA bound to nucleosomes, the latter being octamers consisting of two copies of each of four distinct histone species (H2A, H2B, H3, and H4) with a fifth histone species—H1—bound to some but not all nucleosome octamers. This basic organization of chromatin structure, which resembles beads on a string, is found throughout the chromosomes.

The globular core of the nucleosome represents the basic scaffold of chromatin that is modified in two ways. First, some of the standard histones, such as histones H2A and H3, may be replaced in a minority of nucleosomes by variant forms, for example, histones H2AZ and H3.3 (specified by genes distinct from those encoding the standard histones). Indeed, a number of such variant histones can be found scattered here and there throughout the chromatin; their precise contributions to the regulation of chromatin structure and transcription remain poorly understood.

Second, chromatin structure and transcription is strongly affected by post-translation modifications of the standard four histones. These modifications do not directly alter the globular core of the nucleosome. Instead, they affect the N-terminal tails of the core histones (**Figure 1.20A**), which extend outward from the globular core and undergo a variety of covalent modifications, prominent among these being methylation, acetylation, phosphorylation, and ubiquitylation. For example, one type of histone phosphorylation is associated with the condensation of chromatin that occurs during mitosis and the related global shutdown of gene expression. At other times in the cell cycle, acetylation of core histones is generally associated with active gene

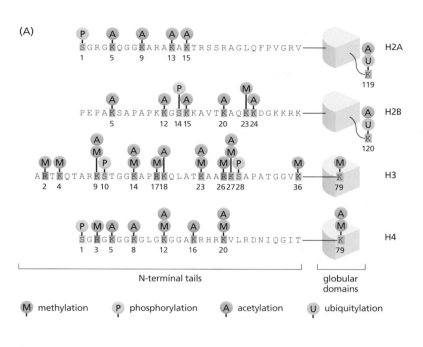

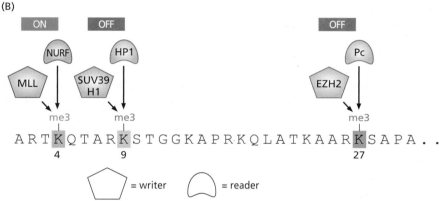

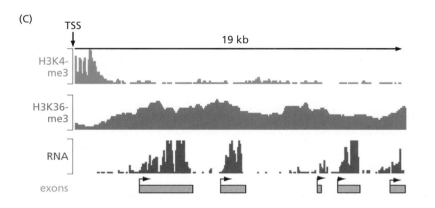

Figure 1.20 Post-translational modification of histone tails (A) Each of these N-terminal histone tails can be modified by the covalent attachment of a variety of chemical groups, most commonly methyl, acetyl, phosphate, and ubiquitin groups. These modifications are attached by histone "writers," which thereby alter the structure and the function of the chromatin, and are removed by histone "erasers." (B) One example of histone modification is provided by three of the lysine (K) residues in the amino-terminal domain of histone H3. (Amino terminus is at *left*; numbers below each K indicate residue number.) Each of these can be trimethylated (indicated by "me3") through the actions of histone methyltransferase writers (HMTs). Trimethylation at the K4 residue is carried out by the MLL1 HMT; the resulting methyl mark is recognized by a NURF (nucleosome remodeling factor) "reader" complex, which contributes to gene activation (*green*). Conversely, trimethylation of the K9 and K27 residues by the SUV39H1 and EZH2 HMT writers, respectively, results in gene repression (*red*). The methylation marks made by the latter two HMTs are recognized by the HP1 and Pc readers, respectively. Once bound, the HP1 reader can trigger the formation of heterochromatin and thereby block transcription. Not shown are other methyltransferase writers that make mono- and dimethyl marks, and histone demethylase erasers that remove the marks made by HMTs on these residues. (C) The locations of various modified histones can be mapped across a gene by using an antibody that specifically immunoprecipitates a modified histone species followed by DNA sequencing of the precipitate. In this fashion, the locations of the nucleosomes containing trimethylated lysine 4 of histone H3 (H3K4me3, *green*) and H3K36me3 (*blue*) have been mapped relative to the transcription start site (TSS) of this gene. Correlations like these indicate that nucleosomes containing H3K4me3 are associated with TSSs, while those containing H3K36me3 are found along the lengths of actively transcribed genes. When the RNA molecules are analyzed (*red*), those that map to known exons of the gene are found in greater abundance, consistent with their long lifetime relative to the short lifetimes of rapidly degraded intron sequences. The function of the gene studied here is not known. (A, from H. Santos-Rosa and C. Caldas, *Eur. J. Cancer* 41:2381–2402, 2005. B, from S.B. Hake, A. Xiao and C.D. Allis, *Brit. J. Cancer* 90:761–769, 2007. C, from M. Guttman et al., *Nature* 458:223–227, 2009.)

expression, while methylation is generally correlated with gene repression. However, as is seen in Figure 1.20B, which presents only one example of a bewildering variety of histone modifications, methylation of histone H3 is correlated with both gene repression and expression, depending on the position of the affected lysine residue.

Rapidly growing evidence indicates that these various histone modifications are functionally important in permitting or preventing transcription by RNA polymerases of specific regions of chromosomal DNA (see Figure 1.20C). Moreover, the modification state of chromatin can be passed from mother to daughter cells through mechanisms that are still unresolved. This area of research is in great flux: as many as 60 distinct histone-modifying enzymes have been discovered, whose roles in transcriptional regulation and cell biology are largely obscure, and there are likely an even larger number of proteins that form complexes with these enzymes and direct them toward distinct substrates within the chromatin. As more effective sequencing techniques are applied to cancer cell genomes, mutant alleles of the genes encoding these enzymes are being uncovered with ever-increasing frequency.

1.9 Heritable gene expression is controlled through additional mechanisms

The descriptions above of the mechanisms controlling gene expression provide only a partial explanation of how gene expression programs that are established in one human cell are transmitted to its lineal descendants. For example, the specific gene expression program of a fibroblast grown in culture will continue to be expressed by its lineal descendants 10 and 20 cell generations later. Since decisions to express or repress a gene within a fibroblast are not imprinted in the gene's DNA sequence, this implies alternative means of maintaining such decisions in a stable fashion and transmitting them faithfully from one cell generation to the next via biochemical mechanisms that mediate **epigenetic** inheritance.

In addition to the transmission of histone modifications described above, the other key mechanism that enables epigenetic inheritance of gene expression depends on covalent modification of DNA, specifically by DNA methyltransferases—enzymes that attach methyl groups directly to cytosine bases of CpG dinucleotides in the DNA double helix. (The designation CpG indicates that the sequence is a cytidine positioned 5' immediately before a guanosine.) The affected CpG dinucleotides are often located near transcriptional promoters, and the resulting methylation generally causes repression of nearby genes. The biochemical mechanism of maintenance methylation is well understood: maintenance DNA methyltransferase enzymes recognize **hemi-methylated** segments of recently replicated DNA and proceed to methylate any unmethylated CpG dinucleotides that are complementary to already methylated CpGs in the other DNA strand (**Figure 1.21**).

The mechanism(s) that lead to *de novo* methylation of previously unmethylated CpGs are still elusive. However, recent research reveals how the reverse process occurs: The Tet (ten eleven translocation) enzymes oxidize the methyl group of 5-methyl-cytidine to hydoxymethyl, formyl, and carboxy groups. The altered nucleotides may then be excised by DNA repair enzymes (Chapter 12) and replaced by cytidine; alternatively, when DNA bearing an oxidized cytidine is replicated, the maintenance methylase may fail to methylate the complementary strand. This research has not yet identified how the Tet enzymes are controlled.

The methyl CpG groups do not, on their own, directly block transcription. Instead, they appear to affect the structure of the chromatin proteins that are responsible for packaging chromosomal DNA and presenting it to RNA polymerases for transcription,

Figure 1.21 Maintenance of DNA methylation following replication
When a DNA double helix that is methylated (*green* Me groups, *left*) at complementary CpG sites undergoes replication, the newly synthesized daughter helices will initially lack methyl groups attached to CpGs in the recently synthesized daughter strands (*purple, red*) and will therefore be *hemi-methylated*. Shortly after their synthesis, however, a maintenance DNA methyltransferase will detect the hemi-methylated DNA and attach methyl groups (*green*) to these CpGs, thereby regenerating the same configuration of methyl groups that existed in the parental helix prior to replication. CpG sites that are unmethylated in the parental helix (*not shown*) will be ignored by the maintenance methyltransferase and will therefore remain so in the newly synthesized strands.

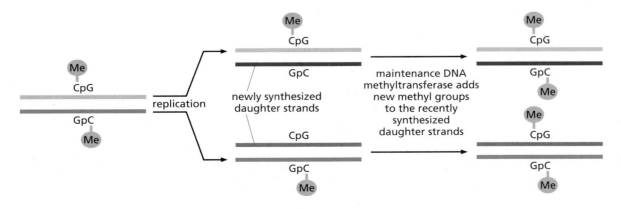

as described above. In particular, methyl-CpG–binding proteins associate specifically with methylated dinucleotides and influence the structure of the nearby chromatin in still-poorly understood ways. There is also evidence that the modification of certain histones can operate in the opposite direction to influence the state of DNA methylation.

1.10 Unconventional RNA molecules also affect the expression of genes

The Central Dogma of molecular biology, developed in the decade after the 1953 discovery of the DNA double helix, proposed that information flows in cells from DNA via mRNA to proteins. In addition, non-informational RNA molecules—ribosomal and transfer RNAs—were implicated as components of the translational machinery, and small nuclear RNAs were found to play key roles in the splicing and maturation of pre-mRNAs. In the 1980s, the view of RNA's functions was expanded through the discovery that certain RNA species can act as enzymes, thereby taking their place alongside proteins as catalysts of certain biochemical reactions.

The 1990s revealed an entirely new type of RNA molecule that functions to control either the levels of certain mRNAs in the cytoplasm, the efficiency of translating these mRNAs, or both. These **microRNAs** (miRNAs) are only 21 to 25 nucleotides long and are generated as cleavage products of far larger nuclear RNA precursors. As outlined in **Figure 1.22**, the post-transcriptional processing of a primary miRNA transcript results in the formation in the cytoplasm of a miRNA that is part of a RISC (RNA-induced silencing complex) nucleoprotein. This complex associates with a spectrum of mRNA

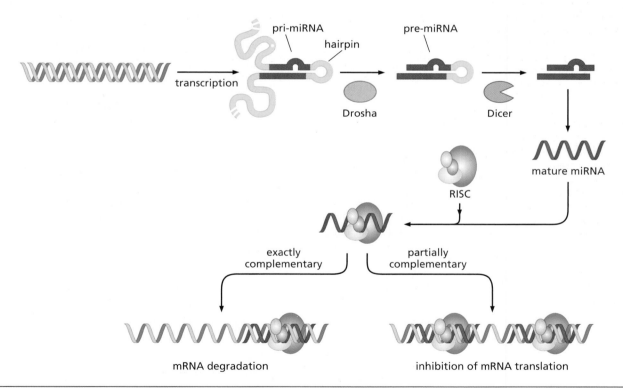

Figure 1.22 MicroRNAs and gene regulation A primary microRNA (pri-miRNA) is transcribed from a gene, and an enzyme complex involving the Drosha protein excises a small segment of the pri-miRNA that has formed a double-stranded RNA hairpin because of the self-complementarity of nucleotide sequences. The resulting pre-miRNA is exported to the cytoplasm, where it is further processed by the Dicer enzyme to generate a mature miRNA of 21 to 25 nucleotides. This miRNA binds to a nucleoprotein complex termed RISC (RNA-induced silencing complex) and associates with mRNAs in the cytoplasm with which it has precise or partial sequence complementarity, resulting in either degradation of the mRNA or inhibition of its translation. Several dozen miRNAs have been found to regulate various steps of tumor formation, either favoring or blocking critical steps of this process. Loss of the Dicer enzyme has been associated with cancer progression, and analyses of miRNA expression patterns, much like expression array analyses of mRNAs (see Figure 1.17), have proved useful in classifying various types of cancer. (Courtesy of P.A. Sharp.)

targets that contain, usually in their untranslated region, a sequence that is partially or completely complementary to the miRNA in the complex. Such association can result in either the inhibition of translation of the mRNA or its degradation, or both.

More than 650 distinct miRNA species have been found in human cells, and this roster continues to grow. Although it is unclear how many of these miRNAs are actually involved in regulating the translation and stability of mRNAs, those that do affect mRNA function are thought to regulate expression of at least one-third of all genes in the human genome. Moreover, a single miRNA species can target and thus regulate the expression of dozens of distinct mRNA species, enabling it to act pleiotropically on a variety of cellular processes.

The potential importance of miRNAs in regulating gene expression is suggested by one survey of mRNAs and corresponding proteins in a group of 76 lung cancers. Only about 20% of the genes studied showed a close correlation between mRNA expression and protein expression levels. Hence, in the remaining 80%, the rate of protein synthesis (which can be strongly influenced by miRNAs) and the post-translational lifetime of proteins (see Supplementary Sidebar 7.4) strongly influenced actual protein levels. Since proteins, rather than mRNA, are responsible for creating cell phenotypes, this also reveals the limitations of studying mRNA levels as indicators of gene activity.

Let-7, an miRNA expressed by the *C. elegans* worm, was one of two initially characterized miRNAs. It was found to suppress expression of the *ras* gene in worms and later in mammals. As we will read later (Chapters 4 through 6), the Ras proteins play critical roles in the development of many types of common human cancers. Since this pioneering work, the overexpression or loss of more than a dozen miRNA species has been associated with the formation of a variety of human cancers and the acquisition by tumors of malignant traits. The list of these miRNAs, which have garnered the term "oncoMiRs," continues to lengthen (see Supplementary Sidebar 1.3). In addition, loss of the Dicer processing enzyme (see Figure 1.22), which is involved in creating mature miRNA, has been found to facilitate the formation of tumors in mice, doing so through still-unknown mechanisms. Interestingly, inheritance of a variant of the *K-ras* gene, which causes a single nucleotide change in the 3′ untranslated region (3′ UTR) of its mRNA, prevents recognition by *Let-7* and is associated with higher levels of the growth-promoting K-Ras protein and as much as a twofold increased risk of certain forms of lung and ovarian cancers.

A decade after the discovery of microRNAs, yet another unusual class of RNAs appeared on the scene: a diverse assay of lncRNA molecules (long non-coding RNAs) were found in the nucleus and cytoplasm to be involved in still-poorly understood ways in regulating gene expression. The discovery of these came from the realization that 4 to 9% of the human genome is transcribed into relatively long (>200 nucleotide) RNA molecules that have no identifiable protein-coding sequences and thus no readily ascertainable functions. Some lncRNAs are polyadenylated while others are not. The few lncRNAs that have been characterized seem to function by associating with proteins that are involved in one fashion or another in regulating transcription, often by serving as scaffolds to hold certain chromatin-modifying proteins together. There may be several thousand distinct lncRNA species encoded by the human genome and they are increasingly viewed as key molecular components of the cell's regulatory machinery.

The role of lncRNAs in cancer development is only beginning to be uncovered. For example, elevated expression of the *HOTAIR* lncRNA has been found to be correlated with metastatic behavior of human breast and colorectal carcinomas. More importantly, forced expression of *HOTAIR* in carcinoma cells causes localization of a transcription-repressing protein complex, termed PRC2, to certain chromosomal sites, altered methylation of histone H3 lysine 27 (see Figure 1.20), and increased cancer invasiveness and metastasis.

The actions of miRNAs and lncRNAs provide a glimpse of the complexity of gene expression and its regulation in mammalian cells. Thus, after the transcription of a gene is permitted, a number of mechanisms may then intervene to control the accumulation of its ultimate product—a protein that does the actual work of the gene. Among these mechanisms are (1) post-transcriptional processing of pre-mRNA transcripts, including alternative splicing patterns; (2) stabilization or degradation of the mRNA product; (3) regulation of mRNA translation; and (4) post-translational modification, stabilization, or degradation of the protein product. These mechanisms reinforce the notion, cited above, that the rate of transcription of a gene often provides little insight into the levels of its protein product within a cell. Hence, as we will see, distinct patterns of mRNA expression may help us to distinguish various neoplastic cells from one another but, on their own, tell us rather little about how these cells are likely to behave.

1.11 Metazoa are formed from components conserved over vast evolutionary time periods

These descriptions of cell biology, genetics, and evolution are informed in part by our knowledge of the history of life on Earth. Metazoa probably arose only once during the evolution of life on this planet, perhaps 700 million years ago. Once the principal mechanisms governing their genetics, biochemistry, and embryonic development were developed, these mechanisms remained largely unchanged in the descendant organisms up to the present (**Figure 1.23**; see also Figure 1.7). This sharing of conserved traits among various animal phyla has profound consequences for cancer research, since many lessons learned from the study of more primitive but genetically tractable organisms, such as flies and worms, have proven to be directly transferable to our understanding of how mammalian tissues, including those of humans, develop and function.

Upon surveying the diverse organisms grouped within the mammalian class, one finds that the differences in biochemistry and cell biology are minimal. For this reason, throughout this book we will move effortlessly back and forth between mouse biology and human biology, treating them as if they are essentially identical. On occasion, where species-specific differences are important, these will be pointed out.

The complex signaling circuits operating within cells seem to be organized in virtually identical fashion in all types of mammals. Even more stunning is the interchangeability of the component parts. It is rare that a human protein cannot function in place of its counterpart *orthologous* protein (**Sidebar 1.4**) in mouse cells. In the case of many types of proteins, this conservation of both function and structure is so profound that proteins can be swapped between organisms that are separated by far greater evolutionary distances. A striking example of this, noted earlier (see Figure 1.7), is provided by the gene and thus protein that specifies eye formation in mammals and in flies. Extending even further back in our evolutionary history are the histones and the mechanisms of chromatin remodeling discussed earlier. In fact, the counterparts of many molecules and biochemical mechanisms that operate in mammalian cells are already apparent in protozoa.

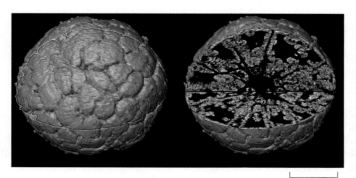

100 μm

Figure 1.23 Visual evidence of the conservation of metazoan biological traits A stunning visual demonstration that contemporary metazoa develop through pathways that have changed little since the Cambrian era has come from the use of synchrotron-generated X-rays to visualize microscopic fossils at sub-micron resolution, yielding this image of an early Cambrian (~530 million years ago) blastula related either to the modern cnidarian or arthropod phylum. Its resemblance to the blastulas of contemporary metazoa indicates that, in addition to conserved molecular and biochemical mechanisms, certain features of embryonic development have changed relatively little since the emergence of modern metazoan phyla during the Cambrian era. Both the surface (*left*) and the interior cleavage pattern (*right*) are shown. (From P.C.J. Donoghue et al., *Nature* 442:680–683, 2006.)

Sidebar 1.4 Orthologs and homologs All higher vertebrates (birds and mammals) seem to have comparable numbers of genes—in the range of 21,000. Moreover, almost every gene present in the bird genome seems to have a closely related counterpart in the human genome. The correspondence between mouse and human genes is even stronger, given the closer evolutionary relatedness of these two mammalian species.

Within the genome of any single species, there are genes that are clearly related to one another in their information content and in the related structures of the proteins they specify. Such genes form a **gene family**. For example, the group of genes in the human genome encoding globins constitutes such a group. It is clear that these related genes arose at some point in the evolutionary past through repeated cycles of the process in which an existing gene is duplicated followed by the divergence of the two duplicated nucleotide sequences from one another (**Figure 1.24**). More directly related to cancer development are the more than 500 protein kinases encoded by the human genome. Kinases attach phosphate groups to their protein substrates, and almost all of these enzymes are specified by members of a single gene family that underwent hundreds of cycles of gene duplication and divergence during the course of evolution (see Supplementary Figure 16.5).

Figure 1.24 Evolutionary development of gene families The evolution of organismic complexity has been enabled, in part, by the development of increasingly specialized proteins. New proteins are "invented" largely through a process of gene duplication followed by diverging evolution of the two resulting genes. Repeated cycles of such gene duplications followed by divergence have led to the development of large numbers of multi-gene families. During vertebrate evolution, an ancestral globin gene, shown here, which encoded the protein component of hemoglobin, was duplicated repeatedly, leading to the large number of distinct globin genes in the modern mammalian genome that are present on two human chromosomes. Because these globins have distinct amino acid sequences, each can serve a specific physiologic function. (From B. Alberts et al., Molecular Biology of the Cell, 5th ed. New York: Garland Science, 2008.)

Genes that are related to one another within a single species' genome or genes that are related to one another in the genomes of two distinct species are said to be **homologous** to one another. Often the precise counterpart of a gene in a human can be found in the genome of another species. These two closely related genes are said to be **orthologs** of one another. Thus, the precise counterpart—the ortholog—of the c-*myc* gene in humans is the c-*myc* gene in chickens. To the extent that there are other *myc*-like genes harbored by the human genome (that is, N-*myc* and L-*myc*), the latter are members of the same gene family as c-*myc* but are not orthologs of one another or of the c-*myc* gene in chickens.

Throughout this book we will often refer to genes without making reference to the species from which they were isolated. This is done consciously, since in the great majority of cases, the functioning of a mouse gene (and encoded protein) is indistinguishable from that of its human or chicken ortholog.

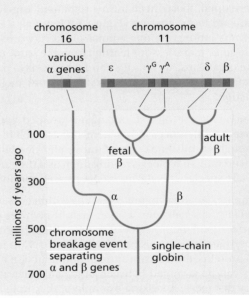

1.12 Gene cloning techniques revolutionized the study of normal and malignant cells

Until the mid-1970s, the molecular analysis of mammalian genes was confined largely to the genomes of DNA tumor viruses, indeed the viruses described later in Chapter 3. These viruses have relatively simple genomes that accumulate to a high copy number (that is, number of molecules) per cell. This made it possible for biologists to readily purify and study the detailed structure and functioning of viral genes that operate much like the genes of the host cells in which these viruses multiplied. In contrast, molecular analysis of cellular genes was essentially impossible, since there are so many of them (tens of thousands per haploid genome) and they are embedded in a genome of daunting complexity (~3.2 billion base pairs of DNA per haploid cellular genome).

All this changed with the advent of gene cloning. Thereafter, cellular genomes could be fragmented and used to create the collections of DNA fragments known as genomic **libraries**. Various DNA hybridization techniques could then be used to identify the genomic fragments within these libraries that were of special interest to the experimenter, in particular the DNA fragment that carried part or all of a gene under study. The retrieval of such a fragment from the library and the amplification of this retrieved fragment into millions of identical copies yielded a purified, **cloned** fragment of DNA and thus a cloned gene (see Supplementary Sidebar 1.4). Yet other techniques were used to generate DNA copies of the mRNAs that are synthesized in the nucleus and exported to the cytoplasm, where they serve as the templates for protein synthesis. Discovery of the enzyme **reverse transcriptase** (RT; see Figure 3.18) was of central importance here. Use of this enzyme made it possible to synthesize *in vitro* (that is, in the test tube) **complementary DNA** copies of mRNA molecules. These DNA molecules, termed cDNAs, carry the sequence information that is present in an mRNA molecule after the process of splicing has removed all introns. While we will refer frequently throughout this book to DNA clones of the genomic (that is, chromosomal) versions of genes and to cDNAs generated from the mRNA transcripts of such genes, space limitations preclude any detailed descriptions of the cloning procedures *per se.*

For cancer researchers, gene cloning arrived just at the right time. As we will see in the next chapters, research in the 1970s diminished the candidacy of tumor viruses as the cause of most human cancers. As these viruses moved off center stage, cellular genes took their place as the most important agents responsible for the formation of human tumors. Study of these genes would have been impossible without the newly developed gene cloning technology, which became widely available in the late 1970s, just when it was needed by the community of scientists intent on finding the root causes of cancer.

Additional reading

Amaral PP, Diner ME, Mercer TR & Mattick JS (2008) The eukaryotic genome as an RNA machine. *Science* 319, 1787–1789.

Bartel DP (2004) MicroRNAs: genomics, biogenesis, mechanism and function. *Cell* 116, 281–297.

Bartel DP (2009) MicroRNAs: target recognition and regulatory functions. *Cell* 136, 215–233.

Bhaumik SR, Smith E & Shilatifard A (2007) Covalent modifications of histones during development and disease pathogenesis. *Nat. Struct. Mol. Biol.* 14, 1008–1016.

Cedar H & Bergman Y (2009) Linking DNA methylation and histone modification: patterns and paradigms. *Nat. Rev. Genet.* 10, 295–304.

Chi P, Wang GG & Allis CD (2010) Covalent histone modifications: miswritten, misinterpreted and mis-erased in human cancers. *Nat. Rev. Cancer* 10, 457–469.

Deng S, Calin GA, Croce CM et al. (2008) Mechanisms of microRNA deregulation in human cancer. *Cell Cycle* 7, 2643–2646.

Esquela-Kerscher A & Slack FJ (2006) Oncomirs: microRNAs with a role in cancer. *Nat. Rev. Cancer* 6, 259–269.

Esteller M (2008) Epigenetics in cancer. *N. Engl. J. Med.* 358, 1148–1159.

Fernandez AF, Assenov Y & Martin-Subero J (2011) A DNA methylation fingerprint of 1,628 human samples. *Genome Res.* 22, 407–419.

Frese KK & Tuveson DA (2007) Maximizing mouse cancer models. *Nat. Rev. Cancer* 7, 654–658.

Füllgrabe J, Kavanagh E & Joseph B (2011) Histone onco-modifications. *Oncogene* 30, 3391–3403.

Gancz D & Fishelson Z (2009) Cancer resistance to complement-dependent cytotoxicity (CDC): Problem-oriented research and development. *Mol. Immunol.* 46, 2794–2800.

Kim E, Goren A & Ast G (2008) Insights into the connection between cancer and alternative splicing. *Trends Genet.* 24, 7–10.

Lander ES (2011) Initial impact of sequencing the human genome. *Nature* 470, 187–197.

Melo SA & Esteller M (2010) Dysregulation of microRNAs in cancer: playing with fire. *FEBS Lett.* 13, 2087–2099.

Nabel CS, Manning SA & Kohli RM (2012) The curious chemical biology of cytosine: deamination, methylation, and oxidation of genomic potential. *ACS Chem. Biol.* 7, 20–23.

Ponting CP, Oliver PL & Reik W (2009) Evolution and functions of long noncoding RNAs. *Cell* 136, 629–641.

Rodriguez-Paredes M & Esteller M (2011) Cancer epigenetics reaches mainstream oncology. *Nature Med.* 17, 330–339.

Sparman A & van Lohuizen M (2006) Polycomb silencers control cell fate, development, and cancer. *Nat. Rev. Cancer* 6, 846–856.

Ting AH, McGarvey KM & Baylin SB (2006) The cancer epigenome—components and functional correlates. *Genes Dev.* 20, 3215–3231.

Ventura A & Jacks T (2009). MicroRNAs and cancer: short RNAs go a long way. *Cell* 136, 586–591.

Wang GG, Allis CD & Chi P (2007) Chromatin remodeling and cancer, part I: covalent histone modifications. *Trends Mol. Med.* 13, 363–372.

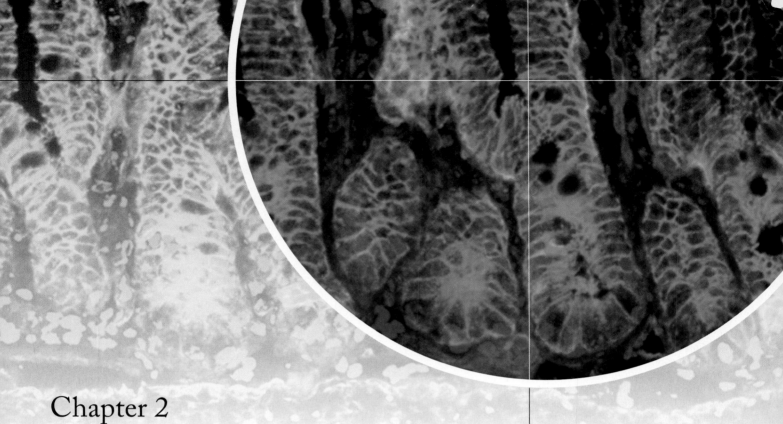

Chapter 2

The Nature of Cancer

When I published the results of my experiments on the development of double-fertilized sea-urchin eggs in 1902, I added the suggestion that malignant tumors might be the result of a certain abnormal condition of the chromosomes, which may arise from multipolar mitosis. ... So I have carried on for a long time the kind of experiments I suggested, which are so far without success, but my conviction remains unshaken.

Theodor Boveri, pathologist, 1914

Tumors destroy man in a unique and appalling way, as flesh of his own flesh which has somehow been rendered proliferative, rampant, predatory and ungovernable. They are the most concrete and formidable of human maladies, yet despite more than 70 years of experimental study they remain the least understood.

Francis Peyton Rous, tumor virologist, Nobel lecture, 1966

The cellular organization of metazoan tissues has made possible the evolution of an extraordinary diversity of anatomical designs. Much of this plasticity in design can be traced to the fact that the building blocks of tissue and organ construction—individual cells—are endowed with great autonomy and versatility. Most types of cells in the metazoan body carry a complete organismic genome—far more information than any one of these cells will ever require. And many cells retain the ability to grow and divide long after organismic development has been completed. This retained ability to proliferate and to participate in tissue **morphogenesis** (the creation of shape) makes possible the maintenance of adult tissues throughout the life span of an organism. Such maintenance may involve the repair of wounds and the replacement of cells that have suffered attrition after extended periods of service.

At the same time, this versatility and autonomy pose a grave danger, in that individual cells within the organism may gain access to information in their genomes that is normally denied to them and assume roles that are inappropriate for normal tissue

Figure 2.1 Normal versus neoplastic tissue (A) This histological section of the lining of the ileum in the small intestine, viewed at low magnification, reveals the continuity between normal and cancerous tissue. At the far left is the normal epithelial lining, the mucosa. In the middle is mucosal tissue that has become highly abnormal, or "dysplastic." To the right is an obvious tumor—an adenocarcinoma—which has begun to invade underlying tissues. (B) This pair of sections, viewed at high magnification, shows how normal tissue architecture becomes deranged in tumors. In the normal human mammary gland (*upper panel*), a milk duct is lined by epithelial cells (*dark purple nuclei*). These ducts are surrounded by mesenchymal tissue (see Figure 2.7) termed "stroma," which consists of connective tissue cells, such as fibroblasts and adipocytes, and collagen matrix (*pink*). In an invasive ductal breast carcinoma (*lower panel*), the cancer cells, which arise from the epithelial cells lining the normal ducts, exhibit abnormally large nuclei (*purple*), no longer form well-structured ducts, and have invaded the stroma (*pink*). (A, from A.T. Skarin, Atlas of Diagnostic Oncology, 3rd ed. Philadelphia: Elsevier Science Ltd., 2003; B, courtesy of A. Orimo.)

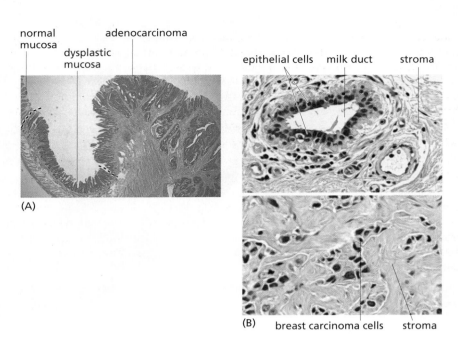

maintenance and function. Moreover, their genomic sequences are subject to corruption by various mechanisms that alter the structure and hence information content of the genome. The resulting mutated genes may divert cells into acquiring novel, often highly abnormal phenotypes. Such changes may be incompatible with the normally assigned roles of these cells in organismic structure and physiology. Among these inappropriate changes may be alterations in cellular proliferation programs, and these in turn can lead to the appearance of large populations of cells that no longer obey the rules governing normal tissue construction and maintenance.

When portrayed in this way, the renegade cells that form a tumor are the result of normal development gone awry. In spite of extraordinary safeguards taken by the organism to prevent their appearance, cancer cells somehow learn to thrive. Normal cells are carefully programmed to collaborate with one another in constructing the diverse tissues that make possible organismic survival. Cancer cells have a quite different and more focused agenda. They appear to be motivated by only one consideration: making more copies of themselves.

2.1 Tumors arise from normal tissues

A confluence of discoveries in the mid- and late nineteenth century led to our current understanding of how tissues and complex organisms arise from fertilized eggs. The most fundamental of these was the discovery that all tissues are composed of cells and cell products, and that all cells arise through the division of preexisting cells. Taken together, these two revelations led to the deduction, so obvious to us now, that all the cells in the body of a complex organism are members of cell lineages that can be traced back to the fertilized egg. Conversely, the fertilized egg is able to spawn all the cells in the body, doing so through repeated cycles of cell growth and division.

These realizations had a profound impact on how tumors were perceived. Previously, many had portrayed tumors as foreign bodies that had somehow taken root in an afflicted person. Now, tumors, like normal tissues, could be examined under the microscope by researchers in the then-new science of **histology**. These examinations of tissue **sections** (thin slices) revealed that tumors, like normal tissues, were composed of masses of cells (**Figure 2.1**). Contemporary cancer research makes frequent use of a variety of histological techniques; the most frequently used of these are illustrated in Supplementary Sidebar 2.1.

Evidence accumulated that tumors of various types, rather than invading the body from the outside world, often derive directly from the normal tissues in which they are first discovered. However, tumors did seem to be capable of moving within the confines of the human body: in many patients, multiple tumors were discovered at anatomical sites quite distant from where their disease first began, a consequence of the tendency of cancers to spread throughout the body and to establish new colonies of cancer cells (**Figure 2.2**). These new settlements, termed **metastases**, were often traceable directly back to the site where the disease of cancer had begun—the founding or **primary tumor**.

Invariably, detailed examination of the organization of cells within tumor masses gave evidence of a tissue architecture that was less organized than the architecture of nearby normal tissues (Figure 2.1). These **histopathological** comparisons provided the first seeds of an idea that would take the greater part of the twentieth century to prove: tumors are created by cells that have lost the ability to assemble and create tissues of normal form and function. Stated more simply, cancer came to be viewed as a disease of malfunctioning cells.

While the microarchitecture of tumors differed from that of normal tissue, tumors nevertheless bore certain histological features that resembled those of normal tissue.

(A)

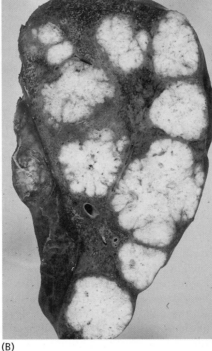

(B)

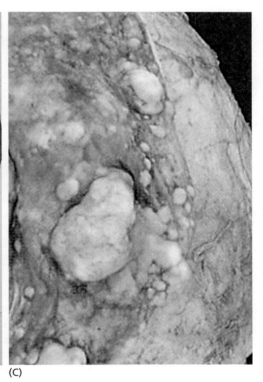

(C)

Figure 2.2 Metastasis of cancer cells to distant sites Many types of tumors eventually release cancer cells that migrate to distant sites in the body, where they form the secondary tumors known as metastases. (A) Melanoma metastases can be quickly identified in mice because of their distinctive dark pigmentation. Seen here are the lungs of two mice, in one of which the formation of metastases was almost entirely blocked (*left*) and one in which hundreds of metastases (*black spots*) were allowed to form (*right*), as observed two weeks after B16 mouse melanoma cells were injected into the tail veins of these mice. This injection route causes many of the cells to become mechanically trapped in the lungs, where they seed numerous colonies. (B) Metastases (*white*) in the liver often arise in patients with advanced colon carcinomas. The portal vein, which drains blood from the colon into the liver (see Figure 14.45), provides a route for metastasizing colon cancer cells to migrate directly into the liver. (C) Breast cancer often metastasizes to the brain. Here, large metastases are revealed post mortem in the right side of a brain where the dura (membrane covering; shown intact at *right*) of the brain has been removed. (A, from F. Nimmerjahn et al., *Immunity* 23:41–51, 2005. B, courtesy of Peter Isaacson. C, from H. Okazaki and B.W. Scheithauer, Atlas of Neuropathology. Gower Medical Publishing, 1988.)

This suggested that all tumors should, in principle, be traceable back to the specific tissue or organ site in which they first arose, using the histopathological analyses of tumor sections to provide critical clues. This simple idea led to a new way of classifying these growths, which depended on their presumed tissues of origin. The resulting classifications often united under one roof cancers that arise in tissues and organs that have radically different functions in the body but share common types of tissue organization.

The science of histopathology also made it possible to understand the relationship between the clinical behavior of a tumor (that is, the effects that the tumor had on the patient) and its microscopic features. Most important here were the criteria that segregated tumors into two broad categories depending on their degree of aggressive growth. Those that grew locally without invading adjacent tissues were classified as **benign**. Others that invaded nearby tissues and spawned metastases were termed **malignant**.

In fact, the great majority of primary tumors arising in humans are benign and are harmless to their hosts, except in the rare cases where the expansion of these localized masses causes them to press on vital organs or tissues. Some benign tumors, however, may cause clinical problems because they release dangerously high levels of hormones that create physiologic imbalances in the body. For example, thyroid **adenomas** (premalignant epithelial growths) may cause excessive release of thyroid hormone into the circulation, leading to hyperthyroidism; pituitary adenomas may release growth hormone into the circulation, causing excessive growth of certain tissues—a condition known as **acromegaly**. Nonetheless, deaths caused by benign tumors are relatively uncommon. The vast majority of cancer-related mortality derives from malignant tumors. More specifically, it is the metastases spawned by these tumors that are responsible for some 90% of deaths from cancer.

2.2   Tumors arise from many specialized cell types throughout the body

The majority of human tumors arise from epithelial tissues. **Epithelia** are sheets of cells that line the walls of cavities and channels or, in the case of skin, serve as the outside covering of the body. By the first decades of the twentieth century, detailed histological analyses had revealed that normal tissues containing epithelia are all structured similarly. Thus, beneath the epithelial cell layers in each of these tissues lies a **basement membrane** (sometimes called a **basal lamina**); it separates the epithelial cells from the underlying layer of supporting connective tissue cells, termed the **stroma** (**Figure 2.3**).

The basement membrane is a specialized type of extracellular matrix (ECM) and is assembled from proteins secreted largely by the epithelial cells. Another type of basement membrane separates **endothelial** cells, which form the inner linings of capillaries and larger vessels, from an outer layer of specialized smooth muscle cells. In all cases, these basement membranes serve as a structural scaffolding of the tissue. In addition, as we will learn later, cells attach a variety of biologically active signaling molecules to basement membranes.

Epithelia are of special interest here, because they spawn the most common human cancers—the **carcinomas**. These tumors are responsible for more than 80% of the cancer-related deaths in the Western world. Included among the carcinomas are tumors arising from the epithelial cell layers of the gastrointestinal tract—which includes mouth, esophagus, stomach, and small and large intestines—as well as the skin, mammary gland, pancreas, lung, liver, ovary, uterus, prostate, gallbladder, and urinary bladder. Examples of normal epithelial tissues are presented in **Figure 2.4**.

This group of tissues encompasses cell types that arise from all three of the primitive cell layers in the early vertebrate embryo. Thus, the epithelia of the lungs, liver, gallbladder, pancreas, esophagus, stomach, and intestines all derive from the inner cell layer, the **endoderm**. Skin arises from the outer embryonic cell layer, termed the

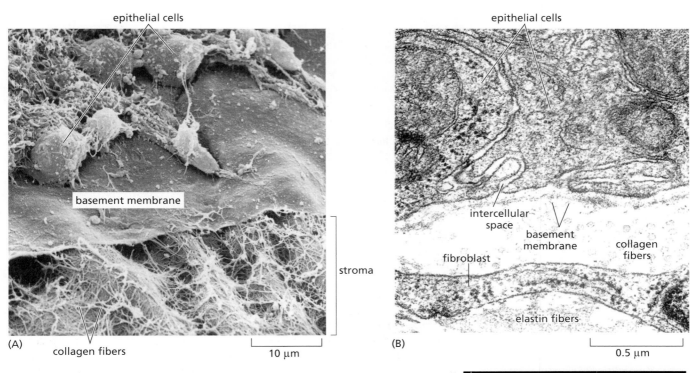

epithelial cells

epithelial cells

basement membrane

intercellular space

fibroblast

basement membrane

collagen fibers

stroma

elastin fibers

(A)   collagen fibers   10 µm

(B)   0.5 µm

Figure 2.3 Basement membranes (A) This scanning electron micrograph of a chick corneal epithelium illustrates the basic plan of epithelial tissues, in which epithelial cells are tethered to one side of the basement membrane, sometimes termed "basal lamina." Seen here as a continuous sheet, it is formed as meshwork of extracellular matrix proteins. A network of collagen fibers anchors the underside of the basement membrane to the extracellular matrix (ECM) of the stroma. (B) The epithelium of the mouse trachea is viewed here at far higher magnification through a transmission electron microscope. Several epithelial cells are seen above the basement membrane, while below are collagen fibrils, a fibroblast, and elastin fibers. Note that the basement membrane is not interrupted at the intercellular space between the epithelial cells. (C) While basement membranes cannot be detected using conventional staining techniques, use of immunofluorescence with an antibody against a basement membrane protein—in this case laminin 5 (*red*)—allows its visualization. The epithelial cells coating the villi of the mouse small intestine have been stained with an antibody against E-cadherin (*green*), while all cell nuclei are stained *blue*. Here the convoluted basement membrane separates the outer villus layer of epithelial cells, termed *enterocytes*, from the mesenchymal cells forming the core of each villus (not stained). [A, courtesy of Robert Trelstad. B, from B. Young et al., Wheater's Functional Histology, 4th ed. Edinburgh: Churchill Livingstone, 2003. C, from Z.X. Mahoney et al., *J. Cell Sci.* 121:2493–2502, 2008 (cover image).]

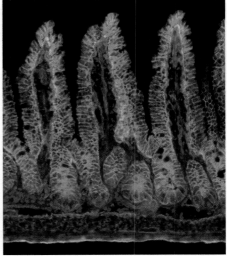

(C)

ectoderm, while the ovaries originate embryologically from the middle layer, the **mesoderm** (**Figure 2.5**). Therefore, in the case of carcinomas, histopathological classification is not informed by the developmental history of the tissue of origin.

The epithelial and stromal cells of these various tissues collaborate in forming and maintaining the epithelial sheets. When viewed from the perspective of evolution, it now seems that the embryologic mechanisms for organizing and structuring epithelial tissues were invented early in metazoan evolution, likely more than 600 million years ago, and that these mechanistic principles have been exploited time and again during metazoan evolution to construct tissues and organs having a wide array of physiologic functions.

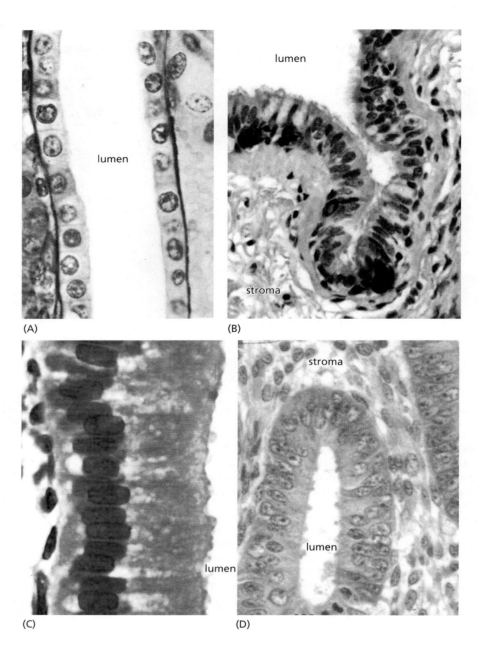

Figure 2.4 Architecture of epithelial tissues A common organizational plan describes most of the epithelial tissues in the body: The mature, differentiated epithelial cells are at the exposed surface of an epithelium. In many tissues, underlying these epithelia are less differentiated epithelial cells, not seen in this figure. Beneath the epithelial cell layer lies a basement membrane (see Figure 2.3), which is usually difficult to visualize in the light microscope. Shown here are epithelia of (A) a collecting tubule of the kidney, (B) the bronchiole of the lung, (C) the columnar epithelium of the gallbladder, and (D) the endometrium of the uterus. In each case, the epithelial cells protect the underlying tissue from the contents of the *lumen* (cavity) that they are lining. Panel C illustrates another property that is characteristic of the epithelial cells forming an epithelium: the state of *apico-basal* polarity, in which individual epithelial cells are organized to present their **apical** surface toward the lumen *(right)* and their **basal** surface toward the underlying basement membrane. This polarization involves the asymmetric localization of the nuclei, which are more basally located, along with hundreds of cell-surface (and associated cytoskeletal) proteins *(not shown)* that are specifically localized either to the apical or basal surfaces of these cells. In addition, the lateral surfaces of the epithelial cells establish several distinct types of junctions with their adjacent epithelial neighbors. (From B. Young et al., Wheater's Functional Histology, 4th ed. Edinburgh: Churchill Livingstone, 2003.)

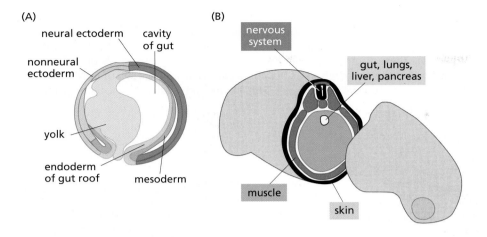

Figure 2.5 Embryonic cell layers (A) The tissues of more complex metazoa develop from three embryonic cell compartments—ectoderm (*blue*), mesoderm (*green*), and endoderm (*yellow*). Each of the three embryonic cell layers is precursor to distinct types of differentiated cells. (B) In an early-stage tadpole, the skin and nervous system develop from the ectoderm (*black*), while the connective tissue, including bone, muscle, and blood-forming cells, develops from the mesoderm (*red*). The gut and derived outpouchings, including lung, pancreas, and liver, develop from the endoderm (*green*). The development of all chordates follows this plan. (Adapted from T. Mohun et al., *Cell* 22:9–15, 1980.)

Most carcinomas fall into two major categories that reflect the two major biological functions associated with epithelia (**Table 2.1**). Some epithelial sheets serve largely to seal the cavity or channel that they line and to protect the underlying cell populations (**Figure 2.6**). Tumors that arise from epithelial cells forming these protective cell layers are termed **squamous cell carcinomas**. For example, the epithelial cells lining the skin (**keratinocytes**) and most of the oral cavity spawn tumors of this type.

Many epithelia also contain specialized cells that secrete substances into the ducts or cavities that they line. This class of epithelial cells generates **adenocarcinomas**. Often these secreted products are used to protect the epithelial cell layers from the contents of the cavities (**lumina**) that they surround (see Figure 2.6). Thus, some epithelial cells lining the lung and stomach secrete mucus layers that protect them, respectively, from the air (and airborne particles) and from the corrosive effects of high concentrations of acid. The epithelia in some organs such as the lung, uterus, and cervix have the capacity to give rise to pure adenocarcinomas or pure squamous cell carcinomas; quite frequently, however, tumors in these organs are found in which both types of carcinoma cells coexist.

Table 2.1 Carcinomas

Tissue sites of more common types of adenocarcinoma	Tissue sites of more common types of squamous cell carcinoma	Other types of carcinoma
lung	skin	small-cell lung carcinoma
colon	nasal cavity	large-cell lung carcinoma
breast	oropharynx	hepatocellular carcinoma
pancreas	larynx	renal cell carcinoma
stomach	lung	transitional-cell carcinoma
esophagus	esophagus	(of urinary bladder)
prostate	cervix	
endometrium		
ovary		

The remainder of malignant tumors arise from nonepithelial tissues throughout the body. The first major class of nonepithelial cancers derive from the various connective tissues, all of which share a common origin in the mesoderm of the embryo (**Table 2.2**). These tumors, the **sarcomas**, constitute only about 1% of the tumors encountered in the oncology clinic. Sarcomas derive from a variety of **mesenchymal** cell types. Included among these are **fibroblasts** and related connective tissue cell

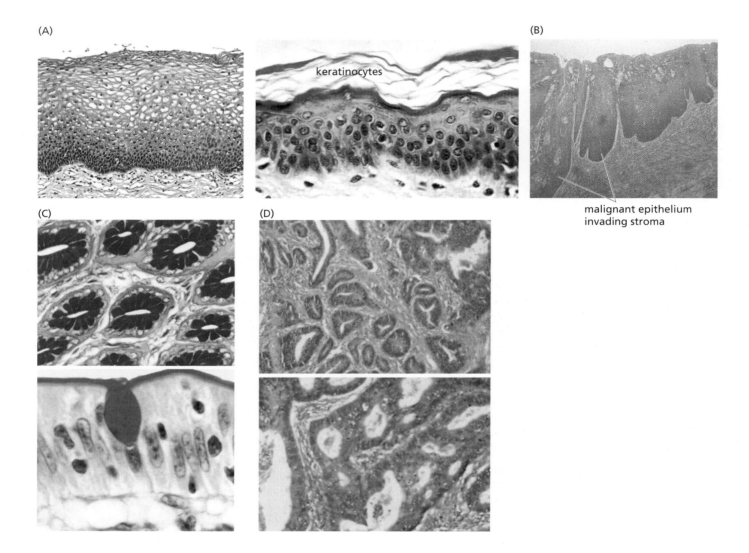

(A)

keratinocytes

(B)

malignant epithelium invading stroma

(C)

(D)

Figure 2.6 Epithelia and derived carcinomas Epithelia can be classified into subtypes depending on the shape and function of the normal epithelial cells and the carcinomas arising from them. The origins of squamous cell carcinomas and adenocarcinomas are seen here. (A) Normal squamous cells are often flattened and function to protect the epithelium and underlying tissue from the contents of the lumen or, in the case of skin, from the outside world. The squamous epithelia of the cervix of the uterus (*left*) and the skin (*right*) are organized quite similarly, with mature flattened cells at the surface being continually shed (for example, the dead keratinocytes of the skin) and replaced by less differentiated cells that move upward and proceed to differentiate. (B) In this carcinoma of the esophagus, large tongues of malignant squamous epithelial cells are invading the underlying stromal/mesenchymal tissue. (C) In some tissues, the glandular cells within epithelia secrete mucopolysaccharides to protect the epithelium; in other tissues, they secrete proteins that function within the *lumina* (cavities) of ducts or are distributed to distant sites in the body. Pits in the stomach wall are lined by mucus-secreting cells (*dark red, upper panel*). In the epithelium of the small intestine (*lower panel*) a single mucus-secreting goblet cell (*purple*) is surrounded by epithelial cells of a third type—columnar cells, which are involved in the absorption of water. (D) These adenocarcinomas of the stomach (*upper panel*) and colon (*lower panel*) show multiple ductal elements, which are clear indications of their derivation from secretory epithelia such as those in panel C. (A and C, from B. Young et al., Wheater's Functional Histology, 4th ed. Edinburgh: Churchill Livingstone, 2003. B and D, from A.T. Skarin, Atlas of Diagnostic Oncology, 3rd ed. Philadelphia: Elsevier Science Ltd., 2003.)

Table 2.2 Various types of more common sarcomas

Type of tumor	Presumed cell lineage of founding cell
Osteosarcoma	osteoblast (bone-forming cell)
Liposarcoma	adipocyte (fat cell)
Leiomyosarcoma	smooth muscle cell (e.g., in gut)
Rhabdomyosarcoma	striated/skeletal muscle cell
Malignant fibrous histiocytoma	adipocyte/muscle cell
Fibrosarcoma	fibroblast (connective tissue cell)
Angiosarcoma	endothelial cells (lining of blood vessels)
Chondrosarcoma	chondrocyte (cartilage-forming cell)

types that secrete collagen, the major structural component of the extracellular matrix of tendons and skin; **adipocytes**, which store fat in their cytoplasm; **osteoblasts**, which assemble calcium phosphate crystals within matrices of collagen to form bone; and **myocytes**, which assemble to form muscle (**Figure 2.7**). Hemangiomas, which are relatively common in children, arise from precursors of the endothelial cells. The stromal layers of epithelial tissues include some of these mesenchymal cell types.

The second group of nonepithelial cancers arise from the various cell types that constitute the blood-forming (**hematopoietic**) tissues, including the cells of the immune system (**Table 2.3** and **Figure 2.8**); these cells also derive from the embryonic mesoderm. Among them are cells destined to form **erythrocytes** (red blood cells), antibody-secreting (**plasma**) cells, as well as T and B **lymphocytes**. The term **leukemia** (literally "white blood") refers to malignant derivatives of several of these hematopoietic cell lineages that move freely through the circulation and, unlike the red blood cells, are nonpigmented. **Lymphomas** include tumors of the **lymphoid** lineages (B and T lymphocytes) that aggregate to form solid tumor masses, most frequently found in lymph nodes, rather than the dispersed, single-cell populations of tumor cells seen in leukemias. This class of tumors is responsible for ~7% of cancer-associated mortality in the United States.

The third and last major grouping of nonepithelial tumors arises from cells that form various components of the central and peripheral nervous systems (**Table 2.4**). These are often termed **neuroectodermal** tumors to reflect their origins in the outer cell

Table 2.3 Various types of more common hematopoietic malignancies

Acute lymphocytic leukemia (ALL)
Acute myelogenous leukemia (AML)
Chronic myelogenous leukemia (CML)
Chronic lymphocytic leukemia (CLL)
Multiple myeloma (MM)
Non-Hodgkin's lymphoma[a] (NHL)
Hodgkin's lymphoma (HL)

[a]The non-Hodgkin's lymphoma types, also known as lymphocytic lymphomas, can be placed in as many as 15–20 distinct subcategories, depending upon classification system.

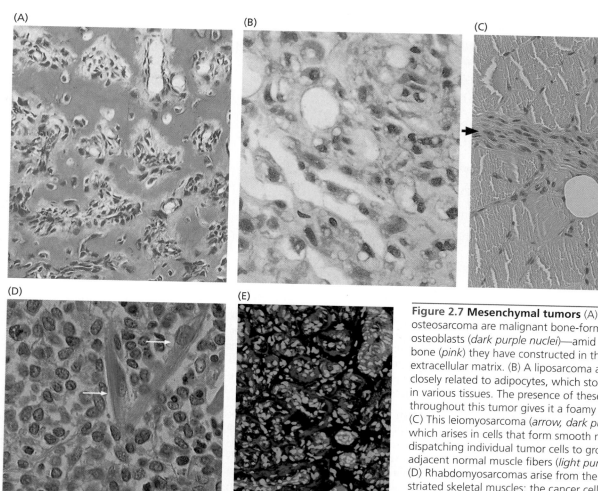

Figure 2.7 Mesenchymal tumors (A) Seen in this osteosarcoma are malignant bone-forming cells—osteoblasts (*dark purple nuclei*)—amid the mineralized bone (*pink*) they have constructed in the surrounding extracellular matrix. (B) A liposarcoma arises from cells closely related to adipocytes, which store lipid globules in various tissues. The presence of these globules throughout this tumor gives it a foamy appearance. (C) This leiomyosarcoma (*arrow, dark purple nuclei*), which arises in cells that form smooth muscle, is dispatching individual tumor cells to grow among adjacent normal muscle fibers (*light purple*). (D) Rhabdomyosarcomas arise from the cells forming striated skeletal muscles; the cancer cells (*dark red nuclei*) are seen here amid several normal muscle cells (*arrows*). (E) Hemangiomas—common tumors in infants—derive from the endothelial cells that form the lining of the lumina of small and large blood vessels. The densely packed capillaries in this particular tumor are formed from endothelial cells with cell nuclei stained *green* and cytoplasms stained *red*. Like epithelial cells, endothelial cells form basement membranes to which they attach, seen here in *blue*. (A–C, from A.T. Skarin, Atlas of Diagnostic Oncology, 3rd ed. Philadelphia: Elsevier Science Ltd., 2003. D, from H. Okazaki and B.W. Scheithauer, Atlas of Neuropathology. Gower Medical Publishing, 1988. E, from M.R Ritter et al., *Proc. Natl. Acad. Sci. USA* 99:7455–7460, 2002.)

layer of the early embryo. Included here are **gliomas**, **glioblastomas**, **neuroblastomas**, **schwannomas**, and **medulloblastomas** (**Figure 2.9**). While comprising only 1.3% of all diagnosed cancers, these are responsible for about 2.5% of cancer-related deaths.

2.3 Some types of tumors do not fit into the major classifications

Not all tumors fall neatly into one of these four major groups. For example, **melanomas** derive from melanocytes, the pigmented cells of the skin and the retina. The melanocytes, in turn, arise from a primitive embryonic structure termed the **neural crest**. While having an embryonic origin close to that of the neuroectodermal cells,

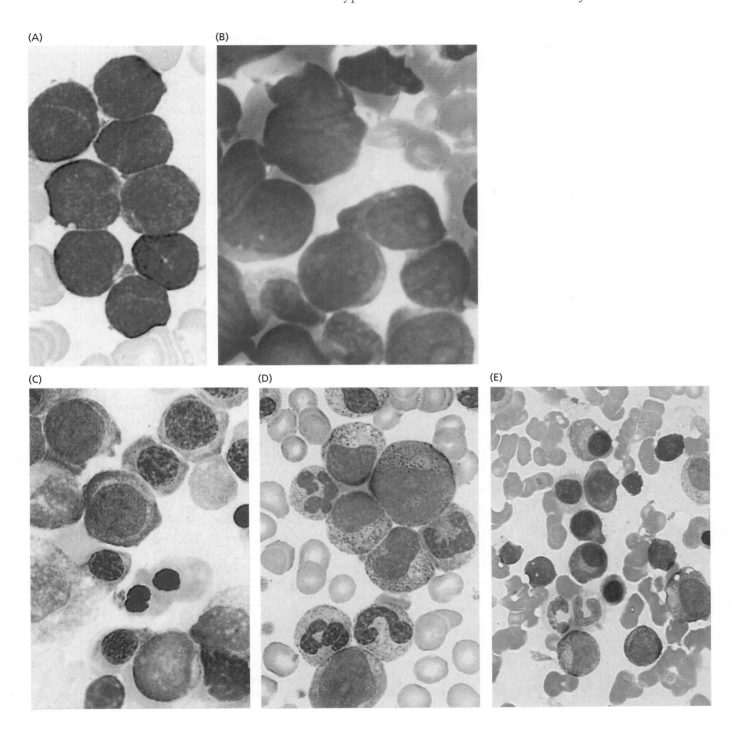

Figure 2.8 Hematopoietic malignancies (A) Acute lymphocytic leukemias (ALLs) arise from both the B-cell (80%) and T-cell (20%) lineages of lymphocytes (see Section 15.1). The cells forming this particular tumor (*red-purple*) exhibited the antigenic markers indicating origin from pre-B cells. (B) As in many hematopoietic malignancies, these acute myelogenous leukemia (AML) cells (*blue*) have only a small rim of cytoplasm around their large nuclei. They derive from precursor cells of the lineage that forms various types of granulocytes as well as monocytes, the latter developing, in turn, into macrophages, dendritic cells, osteoclasts, and other tissue-specific phagocytic cells. (C) The large erythroblasts in this erythroleukemia (*red-purple*) closely resemble the precursors of differentiated red blood cells—erythrocytes. (D) In chronic myelogenous leukemia (CML), a variety of leukemic cells of the myeloid (marrow) lineage are apparent (*red nuclei*), suggesting the differentiation of myeloid stem cells into several distinct cell types. (E) Multiple myeloma (MM) is a malignancy of the plasma cells of the B-cell lineage, which secrete antibody molecules, explaining their relatively large cytoplasms in which proteins destined for secretion are processed and matured. Seen here are plasma cells of MM at various stages of differentiation (*purple nuclei*). In some of these micrographs, numerous lightly staining erythrocytes are seen in the background. (From A.T. Skarin, Atlas of Diagnostic Oncology, 4th ed. Philadelphia: Elsevier Science Ltd., 2010.)

Table 2.4 Various types of neuroectodermal malignancies

Name of tumor	Lineage of founding cell
Glioblastoma multiforme	highly progressed astrocytoma
Astrocytoma	astrocyte (type of glial cell)[a]
Meningioma	arachnoidal cells of meninges[b]
Schwannoma	Schwann cell around axons[c]
Retinoblastoma	cone cell in retina[d]
Neuroblastoma[e]	cells of peripheral nervous system
Ependymoma	cells lining ventricles of brain[f]
Oligodendroglioma	oligodendrocyte covering axons[g]
Medulloblastoma	granular cells of cerebellum[h]

[a]Nonneuronal cell of central nervous system that supports neurons.
[b]Membranous covering of brain.
[c]Constructs insulating myelin sheath around axons in peripheral nervous system.
[d]Photosensor for color vision during daylight.
[e]These tumors arise from cells of the sympathetic nervous system.
[f]Fluid-filled cavities in brain.
[g]Similar to Schwann cells but in brain.
[h]Cells of the lower level of cerebellar cortex (for example, see Figure 2.9B).

the melanocytes end up during development as wanderers that settle in the skin and the eye, provide pigment to these tissues, but acquire no direct connections with the nervous system (**Figure 2.10**).

Small-cell lung carcinomas (SCLCs) contain cells having many attributes of **neurosecretory** cells, such as those of neural crest origin in the **adrenal** glands that sit above the kidneys. Such cells, often in response to neuronal signaling, secrete biologically active peptides. It remains unclear whether the SCLCs, frequently seen in tobacco users, arise from neuroectodermal cells that have insinuated themselves during normal development into the developing lung. According to a more likely alternative, these tumors originate in endodermal cell populations of the lung that have shed some of their epithelial characteristics and taken on those of a neuroectodermal lineage.

This switching of tissue lineage and resulting acquisition of an entirely new set of differentiated characteristics is often termed **transdifferentiation**. The term implies that the commitments cells have made during embryogenesis to enter into one or another tissue and cell lineage are not irreversible, and that under certain conditions, cells can move from one differentiation lineage to another. Such a change in phenotype may affect both normal and cancer cells. For example, at the borders of many carcinomas, epithelial cancer cells often change shape and gene expression programs and take on attributes of the nearby stromal cells of mesenchymal origin. This dramatic shift in cell phenotype, termed the **epithelial–mesenchymal transition**, or simply EMT, implies great plasticity on the part of cells that normally seem to be fully committed to behaving like epithelial cells. As described later (Chapters 13 and 14), this transition may often accompany and enable the invasion by carcinoma cells into adjacent normal tissues.

Of the atypical tumor types, **teratomas** are arguably the most bizarre of all, in part because they defy all attempts at classification. While only ~10,000 cases are diagnosed worldwide annually, teratomas deserve mention because they are unique and shed light on the biology of embryonic stem (ES) cells, which have become so important to biologists; ES cells enable genetic manipulation of the mouse germ line and are central to certain types of stem cell therapies currently under development. Teratomas

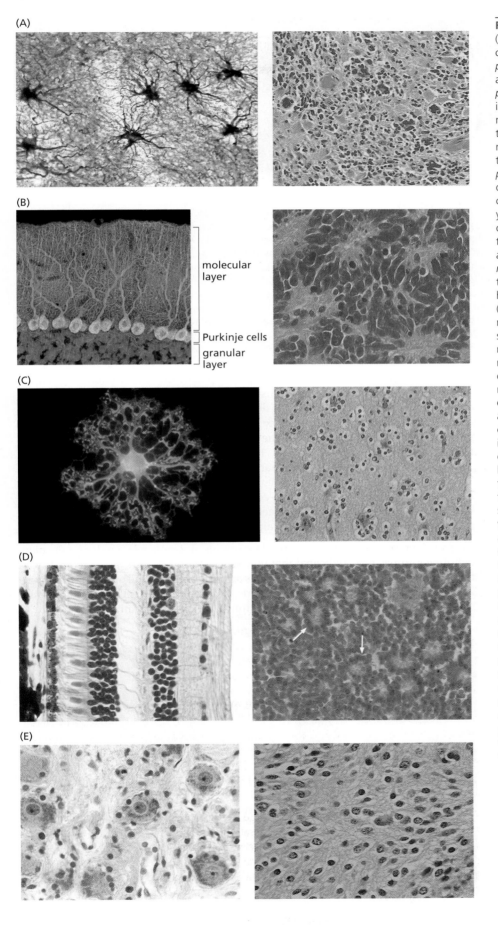

Figure 2.9 Neuroectodermal tumors
(A) Astrocytes—nonneuronal, supporting cells of the brain (*dark purple, left panel*)—are the presumed precursors of astrocytomas and glioblastomas (*right panel*). Glioblastoma multiforme takes its name from the multiple distinct neuroectodermal cell types that constitute the tumor. The tumor cells are seen to have nuclei of various sizes (*purple*). (B) Cells of the granular layer of the cerebellum (*left panel*) reside below Purkinje cells and cells of the molecular layer in the cortex of the cerebellum. The precursors of granular cells yield medulloblastomas (*right panel*), the cells of which are notable for their ability to differentiate into neurons, glial cells, and pigmented neuroepithelial cells (*purple nuclei, pink cytoplasms*). About one-third of these tumors show the rosettes of cells seen here. (C) Shown is an oligodendroglioma (*right*), which derives from oligodendrocytes, nonneuronal cells of ectodermal origin that support and insulate axons in the central nervous system. Each of the neoplastic cell nuclei here has a halo around it, which is characteristic of this tumor. The cultured normal oligodendrocyte shown here (*left*) exhibits a number of branching (dendritic) arms—each of which associates with one or several axons and proceeds to form an insulating myelin sheath around a segment of each of these axons. The cell body has been immunostained (*yellow/orange*) for the O4 oligodendrocyte marker, while the tips of the dendritic arms (*green*) have been stained for CNPase, an enzyme associated with myelination of axons. (D) Rods, cones, and other neuronal cell types (*left panel*) constitute important components of the normal retina. Retinoblastomas (*right panel*) arise from cells with attributes of the cone precursors present in the normal developing retina. Retinoblastomas often show the characteristic rosettes, indicated here with *arrows*. (E) Cells of the sympathetic ganglia of the peripheral nervous system (*larger cells, left panel*) give rise to neuroblastomas (*right panel*), which are usually seen in children. The individual tumor cells here are surrounded by dense fibrillary webs, which are derived from neurites—cytoplasmic processes used by neurons to communicate with one another. (A, D, and E, left panels, from B. Young et al., Wheater's Functional Histology, 4th ed. Edinburgh: Churchill Livingstone, 2003. A–C, right panels, from H. Okazaki, B.W. Scheithauer, Atlas of Neuropathology. Gower Medical Publishing, 1988. B, left panel, Thomas Deerinck, NCMIR/Science Source. C, left panel, courtesy of R. Hardy and R. Reynolds. D, E, right panels, from A.T. Skarin, Atlas of Diagnostic Oncology, 3rd ed. Philadelphia: Elsevier Science Ltd., 2003.)

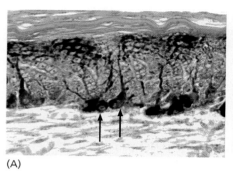

(A)

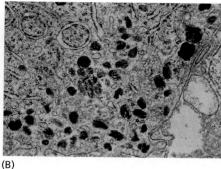

(B)

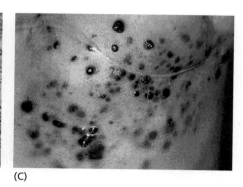

(C)

Figure 2.10 Melanocytes and melanomas (A) Melanocytes (*arrows*), which form melanin pigment granules, are normally scattered among the basal keratinocytes of the skin. They extend long, thin cytoplasmic processes through which they deposit these granules in the cytoplasm of keratinocytes, which form the bulk of the epithelium (see Figure 2.6A). Layers of dead keratinocytes at the surface of the skin (*above*) and stroma cells (*below*) are also apparent. (B) The pigment granules, visualized here by transmission electron microscopy, have made melanomas favored objects of research because the metastases that they form are easily visualized. (for example, see Figure 2.2A). Once melanomas have begun to invade vertically from the superficial layers of the skin into the underlying stroma, they have a high tendency to metastasize to distant tissues. (C) This case of cutaneous melanoma dramatizes the metastatic nature of the disease and the readily observed, pigmented metastases. (A, from W.J. Bacha Jr. et al., Color Atlas of Veterinary Histology, 3rd ed. Ames, IA: Wiley–Blackwell, 2012. B and C, from A.T. Skarin, Atlas of Diagnostic Oncology, 3rd ed. Philadelphia: Elsevier Science Ltd., 2003.)

seem to arise from **germ cell** (egg and sperm) precursors (see Section 1.3) that fail to migrate to their proper destinations during embryonic development and persist at **ectopic** (inappropriate) sites in the developing fetus. They retain the **pluripotency** of early embryonic cells—the ability to generate most and possibly all of the tissues present in the fully developed fetus. The cells in different sectors of common "mature" teratomas—which are largely benign, localized growths—differentiate to create tissues that are very similar to those found in a variety of adult tissues (**Figure 2.11**). Typically, representatives of the three cell layers of the embryo—endoderm, mesoderm, and ectoderm (see Figure 2.5)—coexist within a single tumor and often develop into recognizable structures, such as teeth, hair, and bones. Occasionally these tumors progress to become highly malignant and thus life-threatening.

Of special interest is the fact that careful karyotypic and molecular analyses of benign, mature teratomas have indicated that the associated tumor cells are genetically wild type. This suggests that such teratoma cells are unique, being the only type of tumorigenic cell whose genomes are truly wild type, in contrast to the cells of all other tumor types described in this book, which carry multiple genetic aberrations.

Figure 2.11 Teratomas This teratoma was created by implanting human embryonic stem (ES) cells into a mouse, yielding a tumor that is a **phenocopy** of the spontaneous teratomas found in children; such "mature" teratomas contain fully differentiated cells and are localized, noninvasive tumors. The two sections of this teratoma (A, B) indicate the typical behavior of these tumors, in that different sectors of this tumor have formed differentiated tissues deriving from all three cell layers of the early embryo depicted in Figure 2.5. (Courtesy of Sumita Gokhale.)

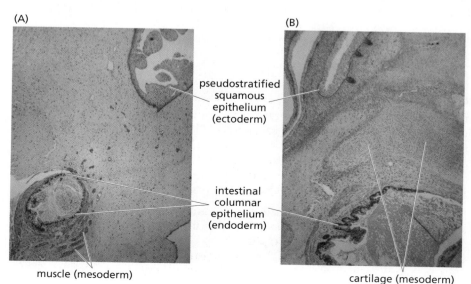

(A)

(B)

pseudostratified squamous epithelium (ectoderm)

intestinal columnar epithelium (endoderm)

muscle (mesoderm)

cartilage (mesoderm)

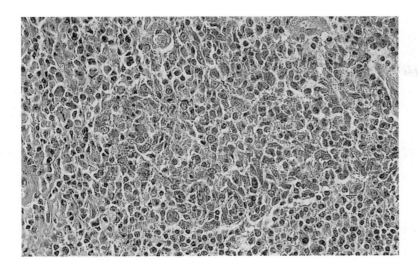

Figure 2.12 Anaplastic tumors of obscure origin The histological appearance of an anaplastic tumor, such as that shown here, gives little indication of its tissue of origin. Attempts to determine the origin of these cells with an antibody stain that specifically recognizes one or another tissue-specific protein marker may also prove uninformative. (From A.T. Skarin, Atlas of Diagnostic Oncology, 3rd ed. Philadelphia: Elsevier Science Ltd., 2003.)

The occasional rule-breaking exceptions, such as those represented by teratomas and the products of the EMT, do not detract from one major biological principle that seems to govern the vast majority of cancers: while cancer cells deviate substantially in behavior from their normal cellular precursors, they almost always retain some of the distinctive attributes of the normal cell types from which they have arisen. These attributes provide critical clues about the origins of most tumors; they enable pathologists to examine tumor biopsies under the microscope and assign a tissue of origin and tumor classification, even without prior knowledge of the anatomical sites from which these biopsies were prepared.

In a small minority of cases (2–4%), the tumors given to pathologists for analysis have shed virtually all of the tissue-specific, differentiated traits of their normal precursor tissues. The cells in such tumors are said to have **dedifferentiated**, and the tumors as a whole are **anaplastic**, in that it is no longer possible to use histopathological criteria to identify the tissues from which they have arisen (**Figure 2.12**). A tumor of this type is often classified as a **cancer of unknown primary** (CUP)**,** reflecting the difficulty of identifying the original site of tumor formation in the patient.

2.4 Cancers seem to develop progressively

Between the two extremes of fully normal and highly malignant tissue architectures lies a broad spectrum of tissues of intermediate appearance. The different gradations of abnormality may well reflect cell populations that are evolving progressively toward greater degrees of aggressive and invasive behavior. Thus, each type of abnormal growth within a tissue may represent a distinct step along this evolutionary pathway. If so, these architectures suggest, but hardly prove, that the development of tumors is a complex, multi-step process, a subject that is discussed in great detail in Chapter 11.

Some growths contain cells that deviate only minimally from those of normal tissues but may nevertheless be abnormal in that they contain excessive *numbers* of cells. Such growths are termed **hyperplastic** (**Figure 2.13**). In spite of their apparently deregulated proliferation, the cells forming hyperplastic growths have retained the ability to assemble into tissues that appear reasonably normal.

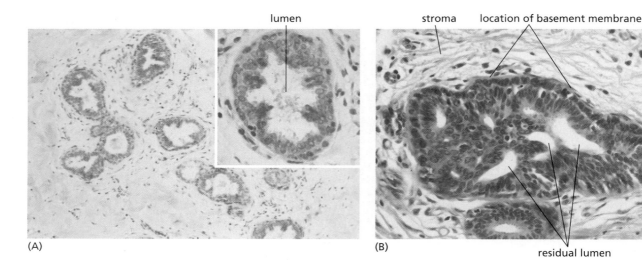

(A)   (B)

Figure 2.13 Normal versus hyperplastic epithelium The morphology of the normal ductal epithelium of the mammary gland (see Figure 2.1B) can be compared with different degrees of hyperplasia. (A) In these mildly hyperplastic milk ducts, shown at low magnification and high magnification (*inset*), mammary epithelial cells have begun to form piles that protrude into the lumina. (B) A more advanced hyperplastic mammary duct shows epithelial cells that are crowded together and almost completely fill the lumen. However, they have not penetrated the basement membrane (*not visible*) and invaded the surrounding stroma. (From A.T. Skarin, Atlas of Diagnostic Oncology, 3rd ed. Philadelphia: Elsevier Science Ltd., 2003.)

An equally minimal deviation from normal is seen in **metaplasia**, where one type of normal cell layer is displaced by cells of another type that are not normally encountered in this site within a tissue. These invaders, although present in the wrong location, often appear completely normal under the microscope. Metaplasia is most frequent in epithelial transition zones where one type of epithelium meets another. Transition zones like these are found at the junction of the cervix with the uterus and the junction of the esophagus and the stomach. In both locations, a squamous epithelium normally undergoes an abrupt transition into a mucus-secreting epithelium. For example, an early indication of premalignant change in the esophagus is a metaplastic condition termed **Barrett's esophagus**, in which the normally present squamous epithelium is replaced by secretory epithelial cells of a type usually found within the stomach (**Figure 2.14**). Even though these gastric cells have a quite normal appearance, this metaplasia is considered an early step in the development of esophageal adenocarcinomas. Indeed, patients suffering from Barrett's esophagus have a thirty-fold increased risk of developing these highly malignant tumors.

Figure 2.14 Metaplastic conversion of epithelia In certain precancerous conditions, the normally present epithelium is replaced by an epithelium from a nearby tissue—the process of metaplasia. For example, in Barrett's esophagus (sometimes termed Barrett's esophagitis), the squamous cells that normally line the wall of the esophagus (*residual squamous mucosa*) are replaced by secretory cells that migrate from the lining of the stomach (*metaplastic Barrett's epithelium*). This particular metaplasia, which is provoked by chronic acid reflux from the stomach, can become a precursor lesion to an esophageal carcinoma, which has developed here from cells of gastric origin (*ulcerated adenocarcinoma*). (Adapted from A.T. Skarin, Atlas of Diagnostic Oncology, 3rd ed. Philadelphia: Elsevier Science Ltd., 2003.)

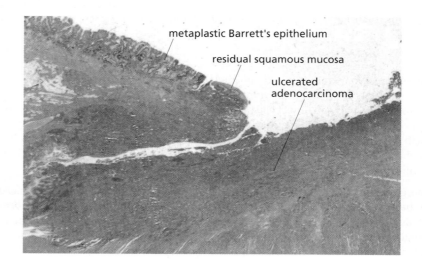

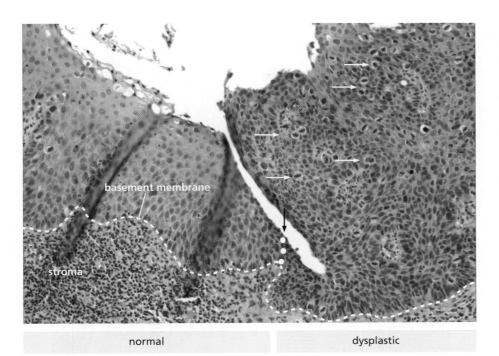

normal   dysplastic

Figure 2.15 Formation of dysplastic epithelium In this intraepithelial squamous neoplasia of the cervix (*to right of white dotted line, black arrow*), the epithelial cells have not broken through the basement membrane (*not visible, indicated by white dashed line*) and invaded the underlying stroma. The cells in this dysplasia continue to be densely packed all the way to the luminal surface (*above*), in contrast to the more diffuse distribution of cells in the normal epithelium (*left*), whose cytoplasms (*light pink*) increase in size as the cells differentiate. Numerous mitotic figures are also apparent in the dysplasia (*white arrows*), indicating extensive cell proliferation. (Courtesy of Tan A. Ince.)

A slightly more abnormal tissue is said to be **dysplastic**. Cells within a dysplasia are usually abnormal **cytologically**; that is, the appearance of individual cells is no longer normal. The cytological changes include variability in nuclear size and shape, increased nuclear staining by dyes, increased ratio of nuclear versus cytoplasmic size, increased mitotic activity, and lack of the cytoplasmic features associated with the normal differentiated cells of the tissue (**Figure 2.15**). In dysplastic growths, the relative numbers of the various cell types seen in the normal tissue are no longer observed. Together, these changes in individual cells and in cell numbers have major effects on the overall tissue architecture. Dysplasia is considered to be a transitional state between completely benign growths and those that are premalignant.

Even more abnormal are the growths that are seen in epithelial tissues and termed variously adenomas, **polyps**, adenomatous polyps, **papillomas**, and, in skin, warts (**Figure 2.16**). These are often large growths that can be readily detected with the naked eye. They contain all the cell types found in the normal epithelial tissue, but this assemblage of cells has launched a program of substantial expansion, creating a macroscopic mass. Under the microscope, the tissue within these adenomatous growths is seen to be dysplastic. These tumors usually grow to a certain size and then stop growing, and they respect the boundary created by the basement membrane, which continues to separate them from the underlying stroma. Since adenomatous growths do not penetrate the basement membrane and invade underlying tissues, they are considered to be benign.

A further degree of abnormality is represented by growths that do invade underlying tissues. In the case of carcinoma cells, this incursion is signaled the moment carcinoma cells break through a basement membrane and invade into the adjacent stroma (**Figure 2.17**). Here, for the first time, we encounter malignant cells that have a substantial potential of threatening the life of the individual who carries them. Clinical oncologists and surgeons often reserve the word **cancer** for these and even more abnormal growths. However, in this book, as in much of contemporary cancer research, the word cancer is used more loosely to include all types of abnormal growths. (In the case of epithelial tissues, the term "carcinoma" is usually applied to growths that have acquired this degree of invasiveness.) This disparate collection of growths—both benign and malignant—are called collectively **neoplasms**, that is, new types of tissue.

Figure 2.16 Pre-invasive adenomas and carcinomas Adenomatous growths, termed polyps in certain organs, have a morphology that sets them clearly apart from normal and dysplastic epithelium. (A) In the colon, pre-invasive growths appear as either flat thickenings of the colonic wall (sessile polyps, *not shown*) or as the stalk-like growths (pedunculated polyps) shown here in a photograph (*left*) and a micrograph (*right*). These growths, also termed "adenomas," have not penetrated the basement membrane and invaded the underlying stroma. (B) The lobules of the normal human breast (*purple islands, left half of figure*), each containing numerous small alveoli in which milk is produced, are surrounded by extensive fibrous stroma (*pink*). The cells of an intraductal carcinoma, often called a ductal carcinoma *in situ* (DCIS; *purple, to right of dashed line*), fill and distend ducts but have not invaded through the basement membrane surrounding the ducts into the stroma. In the middle of one of these ducts is an island of necrotic carcinoma cells (*dark red*) that have died, ostensibly because of inadequate access to the circulation. (A, left, courtesy of John Northover and Cancer Research, UK; right, courtesy of Anne Campbell. B, courtesy of Tan A. Ince.)

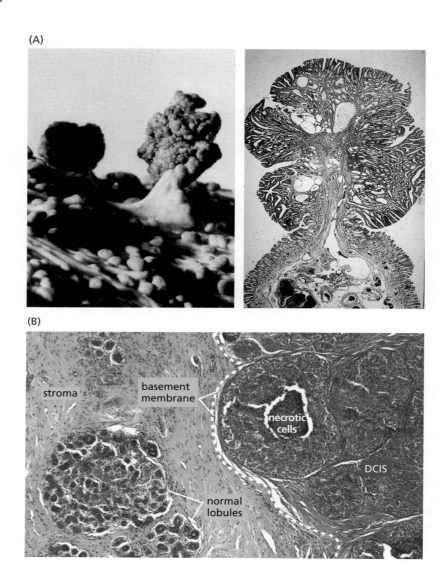

Figure 2.17 Invasive carcinomas Tumors are considered malignant only after they have breached the basement membrane and invaded the surrounding stroma. (A) These breast cancer cells (*dark red*), which previously constituted a ductal carcinoma *in situ* (DCIS; see Figure 2.16B), have now broken through on a broad front (*dashed line*) the layer of myoepithelial cells (*dark brown*) and underlying attached basement membrane (*not visible*) into the stroma; this indicates that they have acquired a new trait: invasiveness. (B) After breaching the basement membrane, invasive cancer cells can appear in various configurations amid the stroma.

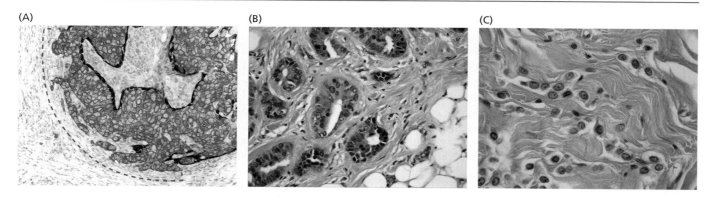

In this invasive ductal carcinoma of the breast, islands of epithelial cancer cells (*dark purple*) are interspersed amid the stroma (*dark pink*). The ductal nature of this carcinoma is revealed by the numerous rudimentary ducts formed by the breast cancer cells. (C) In this invasive lobular carcinoma of the breast, individual carcinoma cells (*dark purple nuclei*) have ventured into the stroma (*red-orange*), often doing so in single-file formation. (A, from F. Koerner, Diagnostic Problems in Breast Pathology. Philadelphia: Saunders/Elsevier, 2008. B and C, courtesy of Tan A. Ince.)

(Some reserve the term "neoplasm" for malignant tumors.) A summary of the overall pathological classification scheme of tumors is provided in **Figure 2.18**. A short discussion of the organizing principles underlying these classifications can be found in Supplementary Sidebar 2.2.

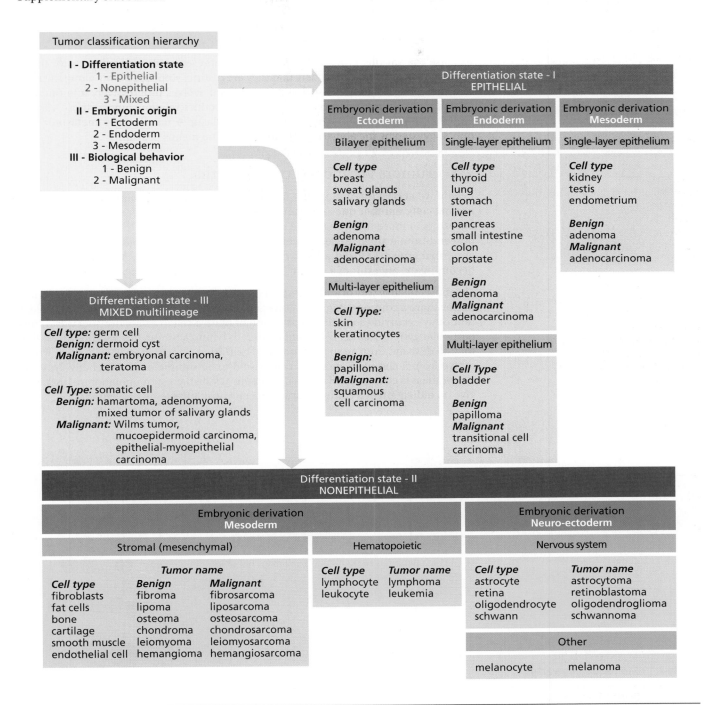

Figure 2.18 Classification scheme of tumors A clear understanding of the histopathological classification of tumors is essential for the study of cancer. However, the entire spectrum of tumors arising in various organs and tissues has been difficult to capture in a single classification scheme that is either purely morphologic or purely molecular. This has necessitated the use of histological features of tumor cells together with information about their respective tissues-of-origin, differentiation states, and biological behaviors; together these make it possible to develop a taxonomy of human tumors that has proven useful for the diagnosis and clinical management of most tumors. The scheme for classifying tumors presented here responds to three critical determinants of tumor biology: the embryonic tissue-of-origin and normal cell-of-origin of the tumor, the phenotype of the cell that has undergone transformation (for example, epithelial vs. mesenchymal), and the extent of progression to a highly malignant state. This scheme allows classification of the great majority of, but not all, human tumors. (Courtesy of Tan A. Ince.)

As mentioned above, cells in an initially formed primary tumor may seed new tumor colonies at distant sites in the body through the process of metastasis. This process is itself extraordinarily complex, and it depends upon the ability of cancer cells to invade adjacent tissues, to enter into blood and lymph vessels, to migrate through these vessels to distant anatomical sites, to leave the vessels and invade underlying tissue, and to found a new tumor cell colony at the distant site. These steps are the subject of detailed discussion in Chapter 14.

Because the various growths cataloged here represent increasing degrees of tissue abnormality, it would seem likely that they are distinct stopping points along the road of **tumor progression**, in which a normal tissue evolves progressively into one that is highly malignant. However, the precursor–product relationships of these various growths (that is, normal → hyperplastic → dysplastic → neoplastic → metastatic) are only suggested by the above descriptions but by no means proven.

2.5 Tumors are monoclonal growths

Even if we accept the notion that tumors arise through the progressive alteration of normal cells, another question remains unanswered: how many normal cells are the ancestors of those that congregate to form a tumor (**Figure 2.19**)? Do the tumor cells descend from a single ancestral cell that crossed over the boundary from normal to abnormal growth? Or did a large cohort of normal cells undergo this change, each becoming the ancestor of a distinct subpopulation of cells within a tumor mass?

The most effective way of addressing this issue is to determine whether all the cells in a tumor share a common, highly unique genetic or biochemical marker. For example, a randomly occurring somatic mutation might mark a cell in a very unusual way. If this particular genetic marker is present in all cells within a tumor, this would suggest that they all descend from an initially mutated cell. Such a population of cells, all of which derive from a common ancestral cell, is said to be **monoclonal**. Alternatively, if the tumor mass is composed of a series of genetically distinct subpopulations of cells that give no indication of a common origin, it can considered to be **polyclonal**.

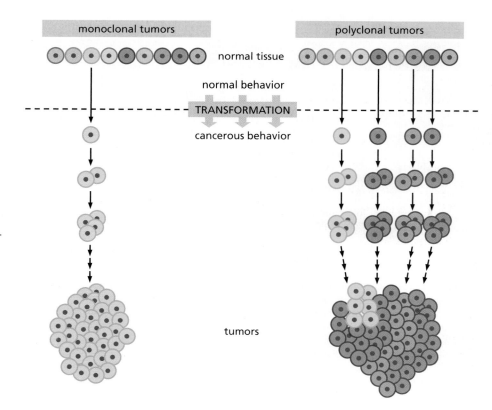

Figure 2.19 Monoclonality versus polyclonality of tumors In theory, tumors may be polyclonal or monoclonal in origin. In a polyclonal tumor (*right*), multiple cells cross over the border from normalcy to malignancy to become the ancestors of several, genetically distinct subpopulations of cells within a tumor mass. In a monoclonal tumor (*left*), only a single cell is transformed from normal to cancerous behavior to become the ancestor of the cells in a tumor mass.

The first experiments designed to measure the clonality of tumor cell populations actually relied on a naturally occurring, nongenetic (**epigenetic**) marking event. As described in Chapter 1, in the somatic cells of early embryos of female placental mammals, one of the two X chromosomes in each cell is selected randomly for silencing. This silencing causes almost all genes on one X chromosome in a cell to be repressed transcriptionally and is manifested karyotypically through the condensation of the silenced X chromosome into a small particle termed the Barr body (see Supplementary Sidebar 1.1). Once an X chromosome (of maternal or paternal origin) has been inactivated in a cell, all descendant cells in adult tissues appear to respect this decision and thus continue to inactivate the same X chromosome.

Thus, the lineage of a cell can be followed *in vivo* from its embryonic ancestor, a term called **lineage tracing**. The gene for glucose-6-phosphate dehydrogenase (G6PD) is located on the X chromosome, and more than 30% of African American women are heterozygous at this locus. Thus, they carry two alleles specifying forms of this enzyme that can be distinguished either by starch gel electrophoresis or by susceptibility to heat inactivation. Because of X-chromosome silencing, each of the cells in these heterozygous women will express only one or the other allele of the *G6PD* gene, which is manifested in turn in the variant of the G6PD protein that these cells synthesize (**Figure 2.20**). In most of their tissues, half of the cells make one variant enzyme, while the other half make the other variant. In 1965, observations were reported on a number of **leiomyomas** (benign tumors of the uterine wall) in African American heterozygotes. Each leiomyoma invariably expressed either one or the other variant form of the G6PD enzyme. This meant that, with great likelihood, its component cancer cells all descended from a single founding progenitor that expressed only that particular allele.

This initial demonstration of the monoclonality of human tumors was followed by many other confirmations of this concept. One proof came from observations of **myelomas**, which derive from the B-cell precursors of antibody-producing plasma

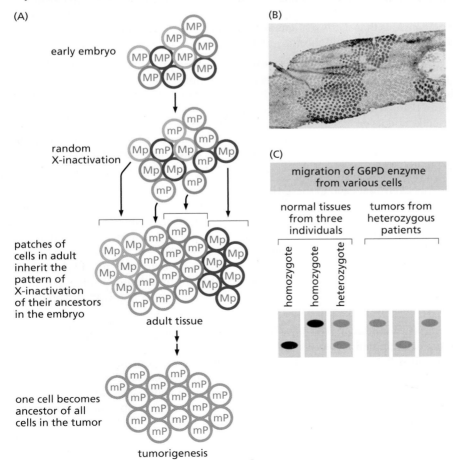

Figure 2.20 X-chromosome inactivation patterns and the monoclonality of tumors (A) While the female embryo begins with both X chromosomes in an equally active state, either the X chromosome inherited from the mother (M) or the one from the father (P) soon undergoes inactivation at random. Such inactivation silences expression of almost all genes on that chromosome. In the adult, all of the lineal descendants of a particular embryonic cell continue to inactivate the same X chromosome. Hence, the adult female body is made of patches (clones) of cells of the type Mp and patches of the type mP, where the *lowercase letter* denotes an inactivated state. (B) The two allelic forms of glucose-6-phosphate dehydrogenase (G6PD), which is encoded by a gene on the X chromosome, have differing sensitivities to heat inactivation. Hence, gentle heating of tissue from a heterozygote—in this case a section of intestine—reveals patches of cells that carry the heat-resistant, still-active enzyme variant (*dark blue spots*) among patches that do not. The cells in each patch are the descendants of an embryonic cell that had inactivated either its maternal or paternal X chromosome. (C) Use of starch gel electrophoresis to resolve the two forms of G6PD showed that all of the cancer cells in a tumor from a *G6PD* heterozygous patient express the same version of this enzyme. This indicated their likely descent from a common ancestral cell that already had this particular pattern of X-inactivation, suggesting that the cancer cells within a tumor mass constitute a monoclonal growth. (B, from M. Novelli et al., *Proc. Natl. Acad. Sci. USA* 100:3311–3314, 2003. C, adapted from P.J. Fialkow, *N. Engl. J. Med.* 291:26–35, 1974.)

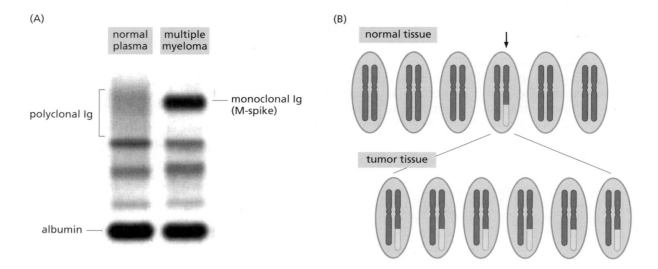

Figure 2.21 Additional proofs of tumor monoclonality (A) In normal plasma, the immunoglobulin (Ig) molecules (for example, antibodies) migrate as a heterogeneous collection of molecules upon gel electrophoresis (polyclonal Ig, *top of left channel*); this heterogeneity is indicative of the participation of a diverse spectrum (a polyclonal population) of plasma cells in antibody production. However, in multiple myeloma, this heterogeneity is replaced by a single antibody species (termed an M-spike) that is produced by a single clonal population of antibody-secreting tumor cells. (B) Illustrated is an unusual translocation (*arrow*) that involves exchange of segments between two different (nonhomologous) chromosomes—a *red* and a *yellow* chromosome. (Only one of the two chromosomal products of the translocation is shown here.) The translocation creates a characteristic "signature" that distinguishes the affected cell from the surrounding population of karyotypically normal cells (*top row*). Since all of the cancer cells within a subsequently arising tumor carry the identical, rare translocation (*bottom row*), this indicates their descent from a common progenitor in which this translocation initially occurred. (A, courtesy of S. Chen-Kiang and S. Ely.)

cells. Normally, the pool of these B-cell precursors consists of hundreds of thousands, likely millions of distinct subpopulations, each expressing its own specific antibody molecules as a consequence of a particular **immunoglobulin** (antibody) gene rearrangement. In contrast, all the myeloma cells in a patient produce the identical antibody molecule, indicating their descent from a single, common ancestor that was present years earlier in this complex, heterogeneous cell population (**Figure 2. 21A**).

Perhaps the most vivid demonstrations of tumor monoclonality have come from cancer cells sporting a variety of chromosomal aberrations that can be visualized microscopically when chromosomes condense during metaphase of mitosis. Often, a very peculiar chromosomal abnormality—the clear result of a rare genetic accident—is seen in all the cancer cells within a tumor mass (see Figure 2.21B). This observation makes it obvious that all the malignant cells within this tumor descend from the single ancestral cell in which this chromosomal restructuring originally occurred.

While such observations seem to provide compelling proof that tumor populations are monoclonal, tumorigenesis may actually be more complex. Let us imagine, as a counterexample, that 10 normal cells in a tissue simultaneously crossed over the border from being normal to being malignant (or at least premalignant) and that each of these cells, and its descendants in turn, proliferated uncontrollably (see Figure 2.19). Each of these founding cells would spawn a large monoclonal population, and the tumor mass, as a whole, consisting of a mixture of these 10 cell populations, would be polyclonal.

It is highly likely that each of these 10 clonal populations varies subtly from the other 9 in a number of characteristics, among them the time required for their cells to double. Simple mathematics indicates that a cell population that exhibits a slightly shorter doubling time will, sooner or later, outgrow all the others, and that the descendants of these cells will dominate in the tumor mass, creating what will appear to be a monoclonal tumor. In fact, many tumors seem to require decades to develop, which is plenty of time for one clonal subpopulation to dominate in the overall tumor cell population. Hence, the monoclonality of the cells in a large tumor mass hardly proves that this tumor was strictly monoclonal during its early stages of development.

A second confounding factor derives from the genotypic and phenotypic instability of tumor cell populations. As we will discuss in great detail in Chapter 11, the population of cells within a tumor may begin as a relatively homogeneous collection of cells (thus constituting a monoclonal growth) but soon may become quite heterogeneous because of the continual acquisition of new mutant alleles by some of its cells, a term called genetic instability. The resulting genetic heterogeneity may mask the true monoclonal origin of this cell population, since many of the genetic markers in these descendant cells will be present only in specific subpopulations of cells within the tumor mass.

These caveats complicate our assessment of the monoclonal origins of tumors. Nonetheless, it is a widespread consensus that the vast majority of advanced human tumors are monoclonal growths descended from single normal progenitor cells that took the first small steps to becoming cancerous. Such progenitors are often termed **cells-of-origin**, and it is increasingly appreciated that the differentiation programs of these cells continue to influence the behavior of derived tumor cell populations decades later. Indeed, in the great majority of human tumor types, one can identify the tissues in which these cells-of-origin resided, but the precise identities of these normal cells, including their state of differentiation, often remain obscure.

2.6 Cancer cells exhibit an altered energy metabolism

The monoclonality of tumor cell populations was first demonstrated in 1965. Another equally interesting peculiarity of tumors was already appreciated more than four decades earlier: the energy metabolism of most cancer cells differs markedly from that of normal cells, a trait first reported in 1924 by Otto Warburg, the Nobelist later honored for discovering the respiratory enzyme now known as cytochrome c oxidase. As was documented in the decades that followed, normal cells that experience aerobic conditions break down glucose into pyruvate in the cytosol through the process of glycolysis and then dispatch the pyruvate into mitochondria, where it is broken down further into carbon dioxide in the citric acid cycle (known also as the Krebs cycle; **Figure 2.22A**). Under anaerobic or **hypoxic** (low oxygen tension) conditions, however, normal cells are limited to using only glycolysis, generating pyruvate that is reduced to lactate, which is then secreted from cells. Warburg discovered that even when exposed to ample oxygen, many types of cancer cells rely largely on glycolysis, generating lactate as the breakdown product of glucose (see Figure 2.22B).

The use by cancer cells of "aerobic glycolysis," as Warburg called it, would seem to make little sense energetically, since the breakdown of one molecule of glucose yields only two molecules of ATP through glycolysis. In contrast, when under aerobic conditions glycolysis is followed by oxidation of pyruvate in the citric acid cycle, as many as 36 ATPs per glucose molecule are generated. In fact, most types of normal cells in the body have continuous access to O_2 conveyed by the blood and therefore metabolize glucose through this energetically far more efficient route. The tendency of cancer cells to limit themselves to glycolysis, even when provided with adequate oxygen, stands out as exceedingly unusual behavior.

The fact that cancer cells metabolize glucose so inefficiently requires them to compensate by importing enormous amounts of glucose. This behavior is seen in many types of cancer cells, including both carcinomas and hematopoietic tumors; they express greatly elevated levels of glucose transporters, particularly GLUT1, which span the plasma membrane and drive the high rates of glucose uptake by these cells. Radiologists take advantage of this elevated glucose uptake by injecting into the circulation radiolabeled glucose [2-deoxy-2-(^{18}F)fluoro-D-glucose, FDG] and observing its rapid concentration in tumors (see Figure 2.22C).

In the 1950s, Warburg proposed that this altered energy metabolism was the driving force in the formation of cancer cells, a notion that was discredited in the decades that followed. However, the process of aerobic glycolysis that he discovered was ultimately found to operate in a wide variety of human cancer cells and is now thought to represent one of the many consequences of cell transformation.

Aerobic glycolysis, sometimes called the **Warburg effect**, remains a subject of much contention, as its rationale in cancer cell biology has never been fully resolved: why do as many as 80% of cancer cells metabolize most of their glucose via glycolysis when completion of glucose degradation in mitochondria by the citric acid cycle would afford them vastly more ATP to fuel their own growth and proliferation? Is aerobic glycolysis required for maintenance of the cancer cell phenotype, or does it represent nothing more than a side effect of cell transformation that plays no causal role in cell transformation and tumor growth?

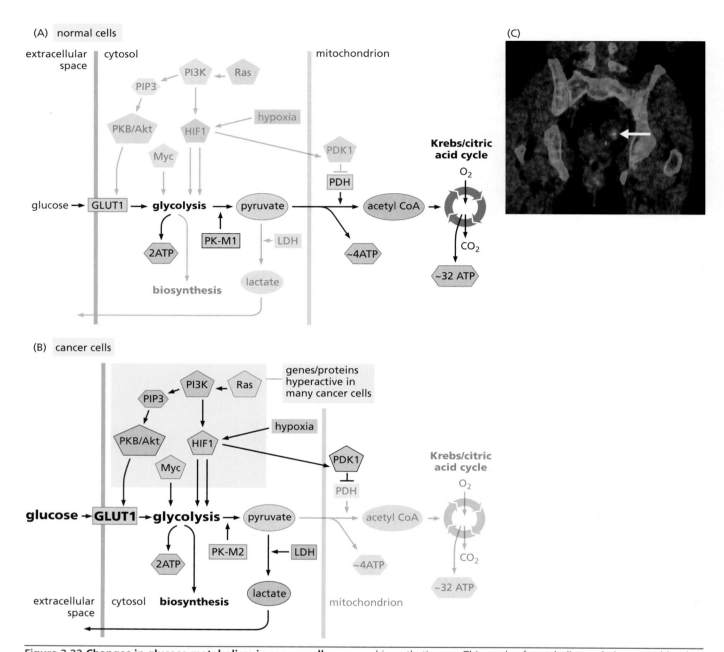

Figure 2.22 Changes in glucose metabolism in cancer cells
(A) In most normal nonproliferating cells having access to adequate oxygen, glucose is imported into the cells by glucose transporters (GLUTs) and then broken down by glycolysis and the citric acid cycle. During the last step of glycolysis, pyruvate kinase form M1 (PK-M1) ensures that its product, pyruvate, is imported into the mitochondria, where it is oxidized by pyruvate dehydrogenase (PDH) into acetyl CoA for processing in the citric acid cycle. Altogether, the mitochondria can generate as much as 36 ATP molecules per glucose molecule. (B) In cancer cells, including those with access to ample oxygen, the GLUT1 glucose transporter imports large amounts of glucose into the cytosol, where it is processed by glycolysis. However, as the last step of glycolysis, pyruvate kinase M2 (PK-M2) causes its pyruvate product to be diverted to lactate dehydrogenase (LDH-A), yielding the lactate that is secreted in abundance by cancer cells. Because relatively little of the initially imported glucose is metabolized by the mitochondria, as few as 2 ATPs are generated per glucose molecule. Moreover, many of the intermediates generated during glycolysis are diverted toward biosynthetic uses. This mode of metabolic regulation resembles the metabolic state of normal, rapidly dividing cells, which also divert a significant portion of their glycolytic intermediates to biosynthetic pathways. Enzymes are in *rectangles*, glucose metabolites are in *ovals*, low–molecular-weight compounds are in *hexagons*, regulatory proteins are in *pentagons*. (C) 2-Deoxy-2-(^{18}F)fluoro-D-glucose positron-emission tomography (FGD-PET) makes it possible to visualize tumors in the body that have concentrated large amounts of glucose because of the hyperactivity of the GLUT1 transporter in the associated cancer cells. In the case shown here, FDG-PET revealed a small tumor (*bright orange; arrow*) in the region near an ovary of a woman who was under treatment for breast cancer but was otherwise without symptoms. X-ray-computed tomography (CT) was used at the same time to image the outlines of the tissues of this patient. This highly sensitive technology provided the first indication of an incipient ovarian cancer in this patient. (C, from R.A. Milam, M.R. Milam and R.B. Iyer, *J. Clin. Oncol.* 25:5657–5658, 2007.)

One explanation of aerobic glycolysis comes from the observation that the cancer cells within a tumor often have inadequate access to oxygen, as we will discuss in detail in Chapter 13. The resulting hypoxic state limits cancer cells to glycolysis and thus to inefficient ATP production—just as normal cells would be limited under these conditions. Because of the Warburg effect, cancer cells would seem to be well adapted to this oxygen starvation, since glycolysis operates normally under hypoxic conditions. Still, this fails to explain why cancer cells, even when provided with abundant oxygen, do not take advantage of this oxygen to generate ATP in far larger quantities.

Another rationale for aerobic glycolysis derives from the fact that glycolysis actually serves a second role independent of ATP generation: the intermediates in the glycolytic pathway function as precursors of many molecules involved in cell growth, including the biosynthesis of nucleotides and lipids. By blocking the last step of glycolysis (see below), cancer cells ensure the accumulation of earlier intermediates via feedback reactions in this pathway. These glycolytic intermediates can then be diverted into critically important biosynthetic reactions. This behavior contrasts with that of normal cells, which are generally not actively proliferating, do not require large-scale biosynthetic reactions, and depend largely on ATP to sustain their metabolic activity. (By some estimates, normal cells use more than 30% of their imported glucose to make ATP, while cancer cells use only ~1% of their glucose for this purpose—a striking contrast in metabolic organization.)

A complete rationale for why cancer cells use aerobic glycolysis is still not in hand. However, independent of how this question is resolved, there is yet another: how do cancer cells actually manage to avoid mitochondrial processing of glucose metabolites? Pyruvate kinase (PK) catalyzes the last step of glycolysis—the conversion of phosphoenolpyruvate (PEP) to pyruvate. As noted earlier, this end product of glycolysis is normally destined for import into the mitochondria, where it is broken down in the citric acid cycle (see Figure 2.22). The M1 isoform of PK typically is expressed in most adult tissues, while the M2 isoform is expressed by early embryonic cells, rapidly growing normal cells, and cancer cells. For reasons that are still poorly understood, the commonly expressed M1 isoform of PK ensures that its product, pyruvate, is dispatched from the cytosol into the mitochondria, while the M2 isoform that is expressed instead in cancer cells causes its pyruvate product to be reduced to lactate in the cytosol. Relative to the M1 form of PK, the M2 enzyme has a very slow **turnover number**, which results in a backup of glycolytic intermediates and their diversion into biosynthetic pathways. Importantly, the relative inactivity of the citric acid cycle in cancer cells is not due to defects in the mitochondria: they are normal and fully capable of receiving pyruvate and processing it in the citric acid cycle.

Experimental evidence indicates that the growth of tumors actually depends on the expression of the M2 form of PK and on the elevated expression of the glucose importer GLUT1 and lactate dehydrogenase-A (LDH-A), the latter being involved in reducing pyruvate to lactate, which is then secreted (see Figure 2.22B). When any one of these is inhibited, tumor growth slows down, sometimes dramatically. Observations like these provide the first indications that the bizarre glucose metabolism of cancer cells creates a physiologic state on which cancer cell growth and proliferation depend.

2.7 Cancers occur with vastly different frequencies in different human populations

The nature of cancer suggests that it is a disease of chaos, a breakdown of existing biological order within the body. More specifically, the disorder seen in cancer appears to derive directly from malfunctioning of the controls that are normally responsible for determining when and where cells throughout the body will multiply. In fact, there is ample opportunity for the disorder of cancer to strike a human body. Most of the more than 10^{13} cells in the body continue to carry the genetic information that previously allowed them to come into existence and might, in the future, allow them to multiply once again. This explains why the risk of uncontrolled cell proliferation in countless sites throughout the body is substantial throughout the lives of mammals like ourselves.

To be more accurate, the risk of cancer is far greater than the $>10^{13}$ population size would suggest, since this number represents the average, steady-state population of cells in the body at any point in time during adulthood. The aggregate number of cells that are formed during an average human lifetime is about 10^{16}, a number that testifies to the enormous amount of cell turnover—involving cell death and replacement (almost 10^7 events per second)—that occurs continuously in many tissues in the body. As discussed in Chapters 9 and 12, each time a new cell is formed by the complex process of cell growth and division, there are many ways for things to go awry. Hence, the chance for disaster to strike, including the inadvertent formation of cancer cells, is great.

Since a normal biological process (incessant cell division) is likely to create a substantial risk of cancer, it would seem logical that human populations throughout the world would experience similar frequencies of cancer. However, when cancer **incidence** rates (that is, the rates with which the disease is diagnosed) are examined in various countries, we learn that the risks of many types of cancer vary dramatically (**Table 2.5**), while other cancers (not indicated in Table 2.5) do indeed show comparable incidence rates across the globe. So, our speculation that all cancers should strike different human populations at comparable rates is simply wrong. Some do and some don't. This realization forces us to reconsider our thinking about how cancers are formed.

Table 2.5 Geographic variation in cancer incidence and death rates

Countries showing highest and lowest incidence of specific types of cancer[a]			
Cancer site	**Country of highest risk**	**Country of lowest risk**	**Relative risk H/L[b]**
Skin (melanoma)	Australia (Queensland)	Japan	155
Lip	Canada (Newfoundland)	Japan	151
Nasopharynx	Hong Kong	United Kingdom	100
Prostate	U.S. (African American)	China	70
Liver	China (Shanghai)	Canada (Nova Scotia)	49
Penis	Brazil	Israel (Ashkenazic)	42
Cervix (uterus)	Brazil	Israel (non-Jews)	28
Stomach	Japan	Kuwait	22
Lung	U.S. (Louisiana, African American)	India (Madras)	19
Pancreas	U.S. (Los Angeles, Korean American)	India	11
Ovary	New Zealand (Polynesian)	Kuwait	8
Geographic areas showing highest and lowest death rates from specific types of cancer[c]			
Cancer site	**Area of highest risk**	**Area of lowest risk**	**Relative risk H/L[b]**
Lung, male	Eastern Europe	West Africa	33
Esophagus	Southern Africa	West Africa	16
Colon, male	Australia, New Zealand	Middle Africa	15
Breast, female	Northern Europe	China	6

[a]See C. Muir, J. Waterhouse, T. Mack et al., eds., Cancer Incidence in Five Continents, vol. 5. Lyon: International Agency for Research on Cancer, 1987. Excerpted by V.T. DeVita, S. Hellman and S.A. Rosenberg, Cancer: Principles and Practice of Oncology. Philadelphia: Lippincott, 1993.
[b]Relative risk: age-adjusted incidence or death rate in highest country or area (H) divided by age-adjusted incidence or death rate in lowest country or area (L). These numbers refer to age-adjusted rates, for example, the relative risk of a 60-year-old dying from a specific type of tumor in one country compared with a 60-year-old in another country.
[c]See P. Pisani, D.M. Parkin, F. Bray and J. Ferlay, Int. J. Cancer 83:18–29, 1999. This survey divided the human population into 23 geographic areas and surveyed the relative mortality rates of various cancer types in each area.

Some of the more than 100 types of human cancers do seem to have a high proportion of tumors that are caused by random, unavoidable accidents of nature and thus occur with comparable frequencies in various human populations. This seems to be true for certain pediatric tumors. In addition to this relatively constant "background rate" of some specific cancers, yet other factors appear to intervene in certain populations to increase dramatically the total number of cancer cases. The two obvious contributory factors here are heredity and environment.

Which of these two alternatives—heredity or environment—is the dominant determinant of the country-to-country variability of cancer incidence? While many types of disease-causing alleles are distributed unequally in the gene pools of different human populations, these alleles do not seem to explain the dramatically different incidence rates of various cancers throughout the world. This point is demonstrated most dramatically by measuring cancer rates in migrant populations. For example, Japanese experience rates of stomach cancer that are 6 to 8 times higher than those of Americans (**Figure 2.23**). However, when Japanese settle in the United States, within a generation their offspring exhibit a stomach cancer rate that is comparable to that of the surrounding population. For the great majority of cancers, disease risk therefore seems to be "environmental," where this term is understood to include both physical environment and lifestyle.

As indicated in Table 2.5, the incidence of some types of cancer may vary enormously from one population to the next. Thus, breast cancer in China is about one-sixth as common as in the United States or Northern Europe. Having excluded genetic contributions to this difference, we might then conclude that as many as 85% of the breast cancers in the United States might in theory be avoidable, if only American women were to experience an environment and lifestyle comparable to those of their Chinese counterparts. Even within the American population, there are vast differences in cancer mortality: the Seventh-Day Adventists, whose religion discourages smoking, heavy drinking, and the consumption of pork, die from cancer at a rate that is only about three-quarters that of the general population.

For those who wish to understand the **etiologic** (causative) mechanisms of cancer, these findings lead to an inescapable conclusion: the great majority of the commonly occurring cancers are caused by factors or agents that are external to the body, enter into the body, and somehow attack and corrupt its tissues. In a minority of cancers, substantial variations in cancer risk may be attributable to differences in reproductive behavior and the resulting dramatic effects on the hormonal environment within the human female body.

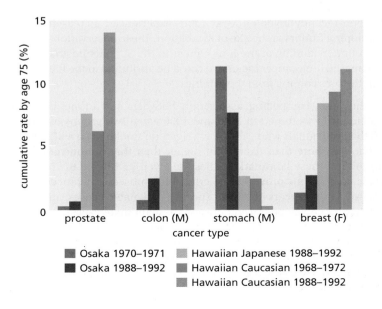

Figure 2.23 Country-to-country comparisons of cancer incidence Public health records reveal dramatic differences in the incidence of certain cancers in different countries. Here, the relative incidences of a group of cancers in Japan and in the American island of Hawaii are presented. Invariably, after Japanese have immigrated to Hawaii, within a generation their cancer rates approach those of the population that settled there before them. This indicates that the differing cancer rates are not due to genetic differences between the Japanese and the American populations. (From J. Peto, *Nature* 411:390–395, 2001.)

Let us imagine, for the sake of argument, that avoidance of certain obvious cancer-causing factors in diet and lifestyle resulted in a 50% reduction in the risk of dying from cancer in the West, leaving the disease of cancer as the cause of about 10% of overall mortality in this population. Under these conditions, given the approximately 10^{16} mitoses occurring in each human body during a normal life span, we calculate that only 1 in 10^{17} cell divisions—the total number of cell divisions occurring in the bodies of 10 individuals during their lifetimes—would lead directly or indirectly to a clinically detectable cancer. Now, we become persuaded that in spite of the enormous intrinsic risk of developing cancer, the body must be able to mount highly effective defenses that usually succeed in holding off the disease for the 70 or 80 years that most of us spend on this planet. These defenses are the subject of many discussions throughout this book.

2.8 The risks of cancers often seem to be increased by assignable influences including lifestyle

Evidence that certain kinds of cancers are associated with specific exposures or lifestyles is actually quite old, predating modern epidemiology by more than a century. The first known report comes from the observations of the English physician John Hill, who in 1761 noted the connection between the development of nasal cancer and the excessive use of tobacco snuff. Fourteen years later, Percivall Pott, a surgeon in London, reported that he had encountered a substantial number of skin cancers of the scrotum in adolescent men who, in their youth, had worked as chimney sweeps. Within three years, the Danish sweepers guild urged its members to take daily baths to remove the apparently cancer-causing material from their skin. This practice was likely the cause of the markedly lower rate of scrotal cancer in continental Europe when compared with Britain even a century later.

Beginning in the mid-sixteenth century, silver was extracted in large quantities from the mines in St. Joachimsthal in Bohemia, today Jáchymov in the Czech Republic. By the first half of the nineteenth century, lung cancer was documented at high rates in the miners, a disease that was otherwise almost unheard of at the time. Once again, an occupational exposure had been correlated with a specific type of cancer.

In 1839, an Italian physician reported that breast cancer was a scourge in the nunneries, being present at rates that were six times higher than among women in the general population who had given birth multiple times. By the end of the nineteenth century, it was clear that occupational exposure and lifestyle were closely connected to and apparently causes of a number of types of cancer.

The range of agents that might trigger cancer was expanded with the discovery in the first decade of the twentieth century that physicians and others who experimented with the then-recently invented X-ray tubes experienced increased rates of cancer, often developing tumors at the site of irradiation. These observations led, many years later, to an understanding of the lung cancer in the St. Joachimsthaler miners: their greatly increased lung cancer incidence could be attributed to the high levels of radioactivity in the ores coming from these mines.

Perhaps the most compelling association between environmental exposure and cancer incidence was forged in 1949 and 1950 when two groups of epidemiologists reported that individuals who were heavy cigarette smokers ran a lifetime risk of lung cancer that was more than twentyfold higher than that of nonsmokers. The initial results of one of these landmark studies are given in **Table 2.6**. These various epidemiologic correlations proved to be critical for subsequent cancer research, since they suggested that cancers often had specific, assignable causes, and that a chain of causality might one day be traced between these ultimate causes and the cancerous changes observed in certain human tissues. Indeed, in the half century that followed the 1949–1950 reports, epidemiologists identified a variety of environmental and lifestyle factors that were strongly correlated with the incidence of certain cancers (**Table 2.7**); in some of these cases, researchers have been able to discover the specific biological mechanisms through which these factors act.

Specific chemical agents can induce cancer   59

Table 2.6 Relative risk of lung cancer as a function of the number of cigarettes smoked per day[a]

	Lifelong nonsmoker	Smokers			
Most recent number of cigarettes smoked (by subjects) per day before onset of disease	—	≥1, <5	≥5, <15	≥15, <25	≥25
Relative risk	1	8	12	14	27

[a]The relative risk indicates the risk of contracting lung cancer compared with that of a nonsmoker, which is set at 1.
From R. Doll and A.B. Hill, *BMJ* 2:739–748, 1950.

2.9 Specific chemical agents can induce cancer

Coal tar condensates, much like those implicated in cancer causation by Percivall Pott's work, were used in Japan at the beginning of the twentieth century to induce skin cancers in rabbits. Repeated painting of localized areas of the skin of their ears resulted, after many months, in the outgrowth of carcinomas. This work, first reported by Katsusaburo Yamagiwa in 1915, was little noticed in the international scientific community of the time (**Figure 2.24**). In retrospect, it represented a stunning advance, because it directly implicated chemicals (those in coal tar) in cancer causation. Equally important, Yamagiwa's work, together with that of Peyton Rous (to be described in Chapter 3), demonstrated that cancer could be induced at will in laboratory animals. Before these breakthroughs, researchers had been forced to wait for tumors to appear spontaneously in wild or domesticated animals. Now, cancers could be produced according to a predictable schedule, often involving many months of experimental treatment of animals.

By 1940, British chemists had purified several of the components of coal tar that were particularly **carcinogenic** (that is, cancer-causing), as demonstrated by the ability of these compounds to induce cancers on the skin of laboratory mice. Compounds such as 3-methylcholanthrene, benzo[*a*]pyrene, and 1,2,4,5-dibenz[*a,h*]anthracene were common products of combustion, and some of these hydrocarbons, notably benzo[*a*] pyrene, were subsequently found in the condensates of cigarette smoke as well

(A)

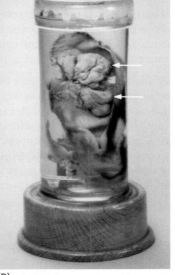

(B)

Figure 2.24 The first induction of tumors by chemical carcinogens (A) In 1915, Katsusaburo Yamagiwa reported the first successful induction of cancer by repeated treatment of rabbit ears with a chemical carcinogen, in this case coal tars. (B) The skin carcinomas (*arrows*) that he induced on the ears of these rabbits are preserved to this day in the medical museum of the University of Tokyo. This particular carcinoma was harvested and fixed following 660 days of painting with coal tar. (Courtesy of T. Taniguchi.)

Table 2.7 Known or suspected causes of human cancers

Environmental and lifestyle factors known or suspected to be etiologic for human cancers in the United States[a]	
Type	% of total cases[b]
Cancers due to occupational exposures	1–2
Lifestyle cancers	
Tobacco-related (sites: e.g., lung, bladder, kidney)	34
Diet (low in vegetables, high in nitrates, salt) (sites: e.g., stomach, esophagus)	5
Diet (high fat, low fiber, broiled/fried foods) (sites: e.g., bowel, pancreas, prostate, breast)	37
Tobacco plus alcohol (sites: mouth, throat)	2
Specific carcinogenic agents implicated in the causation of certain cancers[c]	
Cancer	Exposure
Scrotal carcinomas	chimney smoke condensates
Liver angiosarcoma	vinyl chloride
Acute leukemias	benzene
Nasal adenocarcinoma	hardwood dust
Osteosarcoma	radium
Skin carcinoma	arsenic
Mesothelioma	asbestos
Vaginal carcinoma	diethylstilbestrol
Oral carcinoma	snuff
ER+ breast cancer[d]	hormone replacement therapy (E + P)[e]

[a]Adapted from American Cancer Society. Cancer Facts & Figures 1990. Atlanta: American Cancer Society, Inc.
[b]A large number of cancers are thought to be provoked by a diet high in calories (see Sidebar 9.10) acting in combination with many of these lifestyle factors.
[c]Adapted from S. Wilson, L. Jones, C. Coussens and K. Hanna, eds., Cancer and the Environment: Gene–Environment Interaction. Washington, DC: National Academy Press, 2002.
[d]ER+, estrogen receptor–positive.
[e]E + P, therapy containing both estrogen and progesterone.

(**Figure 2.25**). These findings suggested that certain chemical species that entered into the human body could perturb tissues and cells and ultimately provoke the emergence of a tumor. The same could be said of X-rays, which were also able to produce cancers, ostensibly through a quite different mechanism of action.

While these discoveries were being reported, an independent line of research developed that portrayed cancer as an infectious disease. As described in detail in Chapter 3, researchers in the first decade of the twentieth century found that viruses could cause leukemias and sarcomas in infected chickens. By mid-century, a wide variety of viruses had been found able to induce cancer in rabbits, chickens, mice, and rats. As a consequence, those intent on uncovering the origins of human cancer were pulled in three different directions, since the evidence of cancer causation by chemical, viral, and radioactive agents had become compelling.

2.10 Both physical and chemical carcinogens act as mutagens

The confusion caused by the three competing theories of carcinogenesis was reduced significantly by discoveries made in the field of fruit fly genetics. In 1927, Hermann Muller discovered that he could induce mutations in the genome of *Drosophila*

dibenz[*a*,*h*]anthracene　　　　benzo[*a*]pyrene　　　　3-methylcholanthrene　　　7,12-dimethylbenz[*a*]-anthracene

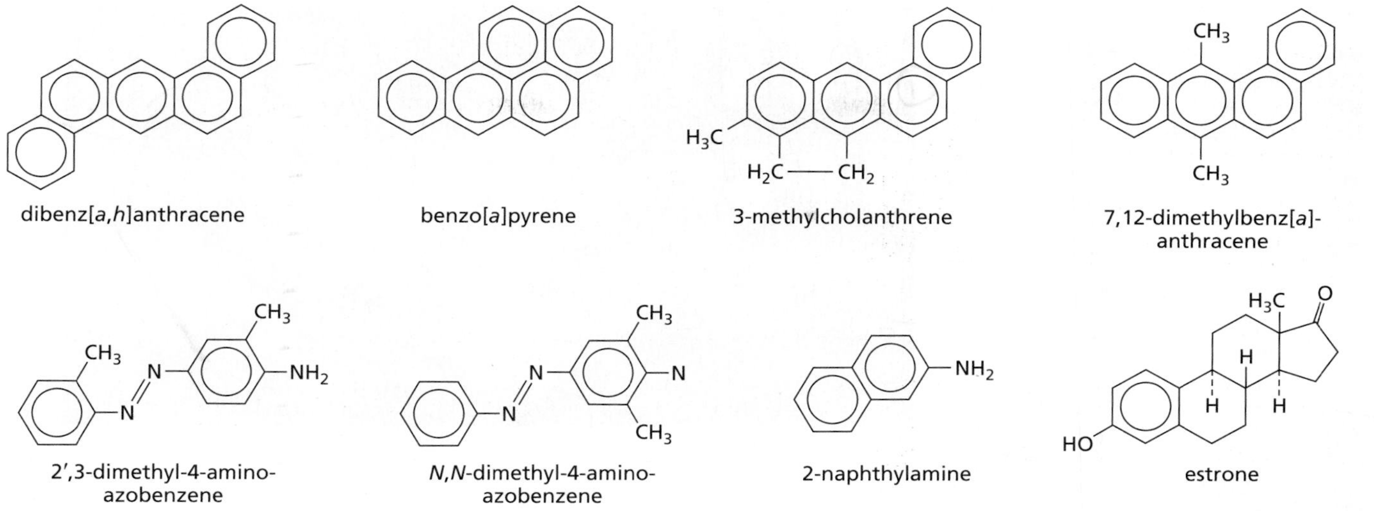

2′,3-dimethyl-4-amino-azobenzene　　　*N*,*N*-dimethyl-4-amino-azobenzene　　　2-naphthylamine　　　estrone

melanogaster by exposing these flies to X-rays. Most important, this discovery revealed that the genome of an animal was mutable, that is, that its information content could be changed through specific treatments, notably irradiation. At the same time, it suggested at least one mechanism by which X-rays could induce cancer: perhaps radiation was able to mutate the genes of normal cells, thereby creating mutant cells that grew in a malignant fashion.

By the late 1940s, a series of chemicals, many of them alkylating agents of the type that had been used in World War I mustard gas warfare, were also found to be **mutagenic** for fruit flies. Soon thereafter, some of these same compounds were shown to be carcinogenic in laboratory animals. These findings caused several geneticists to speculate that cancer was a disease of mutant genes, and that carcinogenic agents, such as X-rays and certain chemicals, succeeded in inducing cancer through their ability to mutate genes.

These speculations were hardly the first ones of this sort. As early as 1914, the German biologist Theodor Boveri, drawing on yet older observations of others, suggested that chromosomes, which by then had been implicated as carriers of genetic information, were aberrant within cancer cells, and that cancer cells might therefore be mutants. Boveri's notion, along with many other speculations on the origin of cancer, gained few adherents, however, until the discovery in 1960 of an abnormally configured chromosome in a large proportion of cases of chronic myelogenous leukemia (CML). This chromosome, soon called the Philadelphia chromosome after the place of its discovery, was clearly a distinctive characteristic of this type of cancer (**Figure 2.26**). Its reproducible association with this class of tumor cells suggested, but hardly proved, that it played a causal role in tumorigenesis.

In 1975 Bruce Ames, a bacterial geneticist working at the University of California in Berkeley, reported experimental results that lent great weight to the theory that carcinogens can function as mutagens. Decades of experiments with laboratory mice and rats had demonstrated that chemical carcinogens acted with vastly different potencies, differing by as much as 1 million-fold in their ability to induce cancers. Such experiments showed, for example, that one microgram of aflatoxin, a compound produced by molds growing on peanuts and wheat, was as potently carcinogenic as a 10,000 times greater weight of the synthetic compound benzidine. Ames posed the question whether these various compounds were also mutagenic, more specifically, whether compounds that were potent carcinogens also happened to be potent mutagens.

The difficulty was that there were no good ways of measuring the relative mutagenic potencies of various chemical species. So Ames devised his own method. It consisted of applying various carcinogenic chemicals to a population of *Salmonella* bacteria growing in Petri dishes and then scoring for the abilities of these carcinogens to mutate the bacteria. The readout here was the number of colonies of *Salmonella* that grew out following exposure to one or another chemical.

Figure 2.25 Structures of carcinogenic hydrocarbons These chemical species arise from the incomplete combustion of organic (for example, carbon-containing) compounds. Each of the chemical structures shown here, which were already determined before 1940, represents a chemical species that was found, following purification, to be potently carcinogenic. The four compounds shown in the top row are all polycyclic aromatic hydrocarbons (PAHs). (From E.C. Miller, *Cancer Res.* 38: 1479–1496, 1978.)

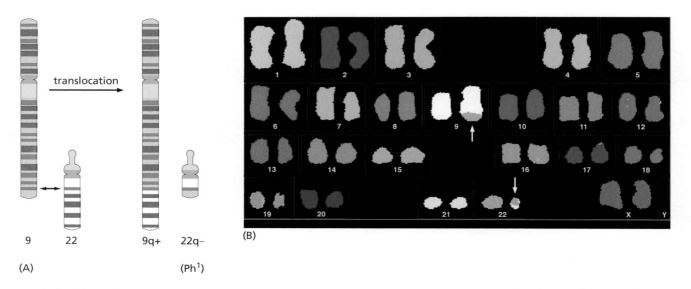

Figure 2.26 Structure of the Philadelphia chromosome
Analyses of the banding patterns of stained metaphase chromosomes of chronic myelogenous leukemia (CML) cells first revealed the characteristic tiny chromosome (called the "Philadelphia chromosome" or Ph[1]) that is present in the leukemia cells of most CML patients. (A) This banding pattern, determined through light-microscopic surveys, is illustrated here schematically. While the chromosomal translocation generating the two altered chromosomes (9q+,22q–) is *reciprocal* (for example, involving a loss and a gain by both), the sizes of the exchanged chromosomal arms are unequal, leading to the greatly truncated Chromosome 22 (for example, 22q–). The small arrow indicates the point of crossing over, known as the translocation *breakpoint*. (B) The relatively minor change to the tumor cell karyotype that is created by the CML translocation is apparent in this SKY analysis, in which chromosome-specific probes are used, together with fluorescent dyes and computer-generated coloring, to visualize the entire chromosomal complement of CML cells. As is apparent, one of the two Chromosomes 9 has acquired a *light purple* segment (a color assigned to Chromosome 22) at the end of its long arm. Reciprocally, one of the two Chromosomes 22 has acquired a white region (characteristic of Chromosome 9) at the end of its long arm (*arrows*). (A, from B. Alberts et al., Molecular Biology of the Cell, 5th ed. New York: Garland Science, 2008. B, courtesy of Thomas Ried and Nicole McNeil.)

In detail, Ames used a mutant strain of *Salmonella* that was unable to grow in medium lacking the amino acid histidine. The mutant allele that caused this phenotype was susceptible to back-mutation to a wild-type allele. Once the wild-type allele was formed in response to exposure to a mutagen, a bacterium carrying this allele became capable of growing in Ames's selective medium, multiplying until it formed a colony that could be scored by eye (**Figure 2.27**).

In principle, Ames needed only to introduce a test compound into a Petri dish containing his special *Salmonella* strain and count the bacterial colonies that later appeared. There remained, however, one substantial obstacle to the success of this mutagenesis assay. Detailed studies had shown that after carcinogenic molecules entered the tissues of laboratory animals, they were metabolized into yet other chemical species. In many cases, the resulting products of metabolism, rather than the initially introduced chemicals, seemed to be the agents that were directly responsible for the observed cancer induction. These metabolized compounds were found to be highly reactive chemically and able to form covalent bonds with the various macromolecules known to be present in cells—DNA, RNA, and protein.

The original, unmodified compounds that were introduced into laboratory animals came to be called **procarcinogens** to indicate their ability to become converted into actively carcinogenic compounds, which were labeled **ultimate carcinogens**. This chemical conversion complicated the design of Ames's mutagenesis assay. If many compounds required metabolic activation before their carcinogenicity was apparent, it seemed plausible that their mutagenic powers would also be evident only after such conversion. Given the radically different metabolisms of bacteria and mammalian cells, it was highly unlikely that Ames's *Salmonella* would be able to accomplish the metabolic activation of procarcinogens that occurred in the tissues of laboratory animals.

Figure 2.27 The Ames test for gauging mutagenicity The Ames test makes it possible to quantitatively assess the mutagenic potency of a test compound. To begin, the liver of a rat (or other species) is homogenized to release its enzymes. The liver homogenate (*brown dots*) is then mixed with the test compound (*orange*), which often results in the liver enzymes metabolically converting the test compound to a chemically activated, mutagenic state (*red*). This mixture (still containing the liver homogenate, *not shown*) is applied to a dish of mutant *Salmonella* bacteria (*small green dots*) that require the amino acid histidine in their culture medium in order to grow. Since histidine is left out of the medium, only those bacteria that are mutated to a histidine-independent genotype (and phenotype) will be able to grow, and each of these will yield a large colony (*green*) that can be counted with the naked eye, indicating how many mutant bacteria (and thus mutant alleles) were generated by the brief exposure to the activated compound.

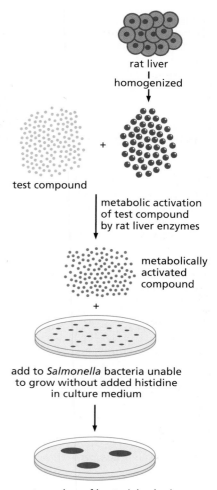

rat liver
|
homogenized

test compound   +

metabolic activation
of test compound
by rat liver enzymes

metabolically
activated
compound

+

add to *Salmonella* bacteria unable
to grow without added histidine
in culture medium

count number of bacterial colonies
that have undergone mutation
enabling them to grow without
added histidine

Earlier work of others had shown that a great many chemicals introduced into the body undergo metabolic conversion, specifically in the liver. Moreover, many of these conversions could be achieved in the test tube simply by mixing such chemicals with homogenized liver. So Ames mixed rat liver homogenates with his test compounds and then introduced this mixture into the Petri dishes carrying *Salmonella*. (We now know that the metabolic activation of procarcinogens in the liver is often mediated by enzymes that are normally involved, paradoxically, in the **detoxification** of compounds introduced into the body; see Section 12.6.)

With the addition of this extra step, Ames's assay revealed that a number of known carcinogens were also actively mutagenic. Even more important were the correlations that Ames found. Chemicals that were potently mutagenic were also powerful carcinogens. Those that were weakly mutagenic induced cancer poorly. These correlations, as plotted by others, extended over five orders of magnitude of potency (**Figure 2.28**).

As we have read, the notion that carcinogens are mutagens predated Ames's work by a quarter of a century. Nonetheless, his analyses galvanized researchers interested in the origins of cancer, since the results addressed the carcinogen–mutagen relationship so directly. Their reasoning went like this: Ames had demonstrated the mutagenic powers of certain chemical compounds in bacteria. Since the genomes of bacterial and animal cells are both made of the same chemical substance—double-stranded DNA—it was likely that the compounds that induced mutations in the *Salmonella* genome were similarly capable of inducing mutations in the genomes of animal cells. Hence, the "Ames test," as it came to be known, should be able to predict the mutagenicity of these compounds in mammals. And in light of the correlation between mutagenic and carcinogenic potency, the Ames test could be employed to screen various substances for their carcinogenic powers, and thus for their threat to human health. By 1976, Ames and his group reported on the mutagenic potencies of 300 distinct organic compounds. Yet other tests for mutagenic potency were developed in the years that followed (**Sidebar 2.1**).

Ames's results led to the next deduction, really more of a speculation: if, as Ames argued, carcinogens are mutagens, then it followed that the carcinogenic powers of various agents derived directly from their ability to induce mutations in the cells of target tissues. As a further deduction, it seemed inescapable that the cancer cells created by chemical carcinogens carry mutated genes. These mutated genes, whatever their identity, must in some way be responsible for the aberrant growth phenotypes of such cancer cells.

This logic was transferable to X-ray carcinogenesis as well. Since X-rays were mutagens and carcinogens, it followed that they also induced cancer through their ability to mutate genes. This convergence of cancer research with genetics had a profound effect on researchers intent on puzzling out the origins of cancer. Though still unproven, it appeared likely that the disease of cancer could be understood in terms of the mutant genes carried by cancer cells.

Figure 2.28 Mutagenic versus carcinogenic potency On this log–log plot, the relative carcinogenic potencies of a group of chemicals (*ordinate*) that have been used to treat laboratory animals (rats and mice) are plotted as a function of their mutagenic potencies (*abscissa*) as gauged by the Ames test (see Figure 2.27). Since both the ordinate and abscissa are plotted as the amount of compound required to elicit an observable effect (yielding tumors in 50% of treated animals or 100 colonies of mutant *Salmonella* bacteria, termed here "revertants"), the compounds that are the most potent mutagens and most potent carcinogens appear in the lower left of this graph. Note that both parameters vary by five orders of magnitude. moca—4,4′-methylenebis(2-chloroaniline), used in manufacture of polyurethane; mms—methyl methanesulfonate, an alkylating mutagen. (Adapted from M. Meselson et al., in H.H. Hiatt et al., eds., Origins of Human Cancer, Book C: Human Risk Assessment. Cold Spring Harbor, NY: Cold Spring Harbor Laboratory Press, 1977.)

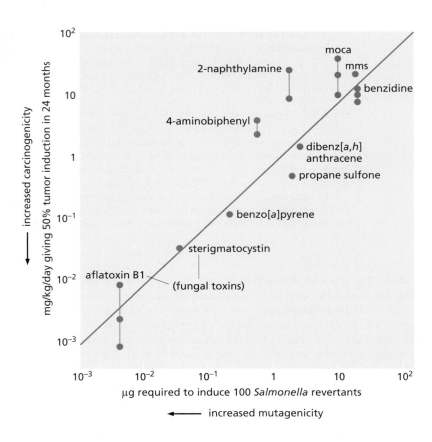

2.11 Mutagens may be responsible for some human cancers

The connection between carcinogenesis and mutagenesis seemed to shed light on how human tumors arise. Perhaps many of these neoplasms were the direct consequence of the mutagenic actions of chemical and physical carcinogens. The mutagenic chemicals, specifically, procarcinogens, need not derive exclusively from the combustion of carbon compounds and the resulting formation of coal tars. It seemed plausible that chemical species present naturally in foodstuffs or generated during cooking could also induce cancer. Even if many foods did not contain ultimate carcinogens, chemical conversions carried out by liver cells or by the abundant bacteria in the colon might well succeed in creating actively mutagenic and thus carcinogenic chemical species.

Sidebar 2.1 Other tests for mutagenicity help assess possible carcinogenicity The Ames test is only one of a number of biological assay systems that can be used to assess the mutagenic potency of suspected carcinogenic chemicals. Many of these other assays depend upon exposing mammalian cells directly to the chemical compounds being tested and the subsequent use of a diverse array of biological readouts. For example, a test for sister chromatid exchange (SCE) measures crossing over between the two paired chromatids that are formed by DNA replication during the S phase and persist in paired form during the late (that is, G_2) phase of a cell's growth-and-division cycle. Many mutagenic agents have been shown to provoke this SCE. Mutagenic agents may also register as being capable of inducing the formation of fragmented cell nuclei, that is, **micronuclei**. Use of genetics has made it possible to select mammalian cells that have lost by mutation their thymidine kinase or HGPRT (hypoxanthine guanine phosphoribosyl transferase) enzymes. The ability to examine under the light microscope the chromosomal array (that is, the karyotype; see Figure 1.11) of cells in metaphase of mitosis makes it possible to screen for chromosomal aberrations inflicted by test compounds. Yet another assay gauges the degree of DNA labeling in those cells that are in the G_1 or G_2 phase of the cell cycle (described in Chapter 8); since cellular DNA synthesis normally occurs in the S phase, such non-S-phase labeling, which is sometimes referred to as "unscheduled DNA synthesis," has also been shown to be a good indicator of the genomic damage that has been inflicted on a cultured cell, since this type of DNA synthesis represents one key step in the process used by cells to repair damaged DNA.

None of these tests has proven to be ideal as a predictor of the carcinogenicity of a test substance. The Ames test, as an example, has been found by some to have a sensitivity (% of established carcinogens identified as mutagens) of about 54% and a specificity (% of noncarcinogens identified as nonmutagens) of 70%.

As this research on the causes of human cancer proceeded, it became apparent that virtually all compounds that are mutagenic in human cells are likely to be carcinogenic as well. However, the converse does not seem to hold: chemical compounds that are carcinogenic are not necessarily mutagenic. Thus, by the 1990s, extensive use of the Ames test showed that as many as 40% of the compounds that were known to be carcinogenic in rodents showed no obvious mutagenicity in the *Salmonella* mutation assay. So, the conclusions drawn from the initial applications of Ames's test required major revision: some carcinogens act through their ability to mutate DNA, while others promote the appearance of tumors through nongenetic mechanisms. We will encounter these nonmutagenic carcinogens, often called tumor **promoters**, again in Chapter 11.

Ames and others eventually used his test to catalog the mutagenic powers of a diverse group of chemicals and natural foodstuffs, including many of the plants that are common and abundant in the Western diet. As Ames argued, the presence of such compounds in foodstuffs derived from plants was hardly surprising, since plants have evolved thousands, possibly millions of distinct toxic chemical compounds in order to defend themselves from predation by insects and larger animals. Some of these naturally toxic compounds, initially developed as anti-predator defenses, might also, as an unintended side effect, be mutagenic (Table 2.8).

A diverse set of discoveries led to the model, which remains unproven in many of its aspects to this day, that a significant proportion of human cancer is attributable directly to the consumption of foodstuffs that are mutagenic and hence carcinogenic.

Table 2.8 **A sampling of Bruce Ames's roster of carcinogens identified in the normal diet[a]**

Foodstuff	Compound	Concentration in foodstuff
Black pepper	piperine	100 mg/g
Common mushroom	agaritine	3 mg/g
Celery	furocoumarins, psoralens[b]	1 µg/g, 0.8 µg/g
Rhubarb	anthraquinones	varies
Cocoa powder	theobromine	20 mg/g
Mustard, horseradish	allyl isothiocyanate	varies
Alfalfa sprouts	canavanine[c]	15 mg/g
Burnt materials[d]	large number	varies
Coffee	caffeic acid	11.6 mg/g

[a]Ames has cited 37 naturally occurring compounds that have registered as carcinogens in laboratory animals; one or more have been found in each of the following foodstuffs: absinthe, allspice, anise, apple, apricot, banana, basil, beet, broccoli, Brussels sprouts, cabbage, cantaloupe, caraway, cardamom, carrot, cauliflower, celery, cherries, chili pepper, chocolate, cinnamon, cloves, coffee, collard greens, comfrey herb tea, coriander, corn, currants, dill, eggplant, endive, fennel, garlic, grapefruit, grapes, guava, honey, honeydew melon, horseradish, kale, lemon, lentils, lettuce, licorice, lime, mace, mango, marjoram, mint, mushrooms, mustard, nutmeg, onion, orange, paprika, parsley, parsnip, peach, pear, peas, pepper (black), pineapple, plum, potato, radish, raspberries, rhubarb, rosemary, rutabaga, sage, savory, sesame seeds, soybean, star anise, tarragon, tea, thyme, tomato, turmeric, and turnip.
[b]The levels of these can increase 100-fold in diseased plants.
[c]Canavanine is indirectly genotoxic because of oxygen radicals that are released, perhaps during the inflammatory reactions associated with elimination of canavanine-containing proteins.
[d]On average, several grams of burnt material are consumed daily in the form of bread crusts, burnt toast, and burnt surfaces of meats cooked at high temperature.

Adapted from B.N. Ames, *Science* 221:1256–1264, 1983; B.N. Ames and L.S. Gold, *Proc. Natl. Acad. Sci. USA* 87:7777–7781, 1990; and L.S. Gold, B.N. Ames and T.H. Slone, Misconceptions about the causes of cancer, in D. Paustenbach, ed., Human and Environmental Risk Assessment: Theory and Practice, New York: John Wiley & Sons, 2002, pp. 1415–1460.

Included among these foodstuffs is, for example, red meat, which upon cooking at high temperatures generates compounds such as heterocyclic amines, which are potently mutagenic (see Section 12.6).

The difficulties in proving this model derive from several sources. Each of the plant and animal foodstuffs in our diet is composed of thousands of diverse chemical species present in vastly differing concentrations. Almost all of these compounds undergo metabolic conversions once ingested, first in the gastrointestinal tract and often thereafter in the liver. Accordingly, the number of distinct chemical species that are introduced into our bodies is incalculable. Each of these introduced compounds may then be concentrated in some cells or quickly metabolized and excreted, creating a further dimension of complexity.

Moreover, the actual mutagenicity of various compounds in different cell types may vary enormously because of metabolic differences in these cells. For example, some cells, such as **hepatocytes** in the liver, express high levels of biochemical species designed to scavenge and inactivate mutagenic compounds, while others, such as fibroblasts, express far lower levels. In sum, the ability to relate the mutagenicity of foodstuffs to actual rates of mutagenesis and carcinogenesis in the human body is far beyond our reach at present—a problem of intractable complexity (**Sidebar 2.2**).

2.12 Synopsis and prospects

The descriptions of cancer and cancer cells developed during the second half of the nineteenth century and the first half of the twentieth indicated that tumors were nothing more than normal cell populations that had run amok. Moreover, many tumors seemed to be composed largely of the descendants of a single cell that had crossed over the border from normalcy to malignancy and proceeded to spawn the billions of descendant cells constituting these neoplastic masses. This model drew attention to the nature of the cells that founded tumors and to the mechanisms that led to their transformation into cancer cells. If one could understand why a cell multiplied uncontrollably, somehow other pieces of the cancer puzzle were likely to fall into place.

Still, existing observations and experimental techniques offered little prospect of revealing precisely why a cell altered its behavior, transforming itself from a normal into a malignant cell. The carcinogen = mutagen theory seemed to offer some clarification, since it implicated mutant cellular genes as the agents responsible for disease development and, therefore, for the aberrant behavior of cancer cells. Perhaps there were mutant genes operating inside cancer cells that programmed the runaway proliferation of these cells, but the prospects for discovering such genes and understanding their actions seemed remote. No one knew how many genes were present in the human genome and how to analyze them. If mutant genes really did play a major part in cancer causation, they were likely to be small in number and dwarfed by the apparently vast number of genes present in the genome as a whole. They seemed to be the proverbial needles in the haystack, in this case a vast haystack of unknown size.

This theorizing about cancer's origins was further complicated by two other important considerations. First, many apparent carcinogens failed the Ames test, suggesting that they were nonmutagenic. Second, certain viral infections seemed to be closely connected to the incidence of a small but significant subset of human cancer types. Somehow, their carcinogenic powers had to be reconciled with the actions of mutagenic carcinogens and mutant cellular genes.

By the mid-1970s, recombinant DNA technology, including gene cloning, began to influence a wide variety of biomedical research areas. While appreciating the powers of this new technology to isolate and characterize genes, cancer researchers were unable, at least initially, to exploit it to track down the elusive mutant genes that were responsible for cancer. One thing was clear, however. Sooner or later, the process of cancer **pathogenesis** (disease development) needed to be explained and understood in molecular terms. Somehow, the paradigm of DNA, RNA, and proteins, so powerful in elucidating a vast range of biological processes, would need to be brought to bear on the cancer problem.

In the end, the breakthrough came from study of the tumor viruses, which by most accounts were minor players in human cancer development. Tumor viruses were genetically simple, and yet they possessed potent carcinogenic powers. To understand these viruses and their import, we need to move back, once again, to the beginning of the twentieth century and confront another of the ancient roots of modern cancer research. This is the subject of Chapter 3.

A major challenge for the future is to understand how various biological and environmental factors, the latter including lifestyle, contribute to the incidence of cancers, many of them quite common ones. For example, as indicated in part in Table 2.5, the incidence of cancers, such as those of the colon, breast, and prostate, shows enormous geographic variation—dramatic differences that cannot be ascribed to differing genetic susceptibilities. In fact, epidemiologists have uncovered many correlations between the frequencies of these and other cancer types and various lifestyle factors (for example, those listed in Table 2.9). However, with rare exception, our understanding of the biological and biochemical mechanisms by which these factors increase (or reduce) disease incidence is either incomplete or nonexistent. Indeed, these correlations represent one of the major unsolved mysteries confronting contemporary cancer researchers.

Until we understand how various biological and lifestyle factors succeed in triggering or preventing tumor development, our ability to prevent new cancers (which is usually far more effective than trying to cure them after they have been diagnosed) will be limited. Many of the chapters that follow provide critical information that may ultimately help to unravel these mysteries of cancer etiology.

Table 2.9 **Examples of etiologic mysteries: epidemiologic correlations between environmental/lifestyle factors and cancer incidence that lack a clear explanation of causal mechanism[a]**

Lifestyle, dietary factor, or medical condition	Altered cancer risk
High birth weight	premenopausal breast cancer ↑ infant acute leukemia ↑
Processed red meat[b]	ER+ breast cancer ↑ squamous cell and adenocarcinoma of lung ↑
Childhood soy consumption	breast cancer ↓
Well-done red meat	prostate cancer ↑
Western diet—high in fat, high in red meat	colorectal, esophageal, liver, and lung cancer ↑
Exercise	hormone-responsive breast cancer ↓
Diet with cruciferous vegetables	prostate cancer ↓
High body-mass index (BMI)	multiple cancer types ↑
Higher ratio of number of daughters to number of sons born to a woman	ovarian carcinoma ↑
Parkinson's disease	melanoma ↑
Low circulating vitamin D	breast cancer incidence, CRC mortality ↑
Periodontal disease	esophageal carcinoma ↑
Coffee consumption	hepatocellular carcinoma ↓

[a]Relative risk (RR) is not given, because not all studies used the same criteria to gauge RR. ↑ = increased risk; ↓ = decreased risk.
[b]Processed red meat generally refers to meat that has been preserved by smoking, curing, salting or adding chemical preservatives.
Abbreviations: ER+ = estrogen receptor–positive; CRC = colorectal cancer.

Key concepts

- The nineteenth-century discovery that all cells of an organism descend from the fertilized egg led to the realization that tumors are not foreign bodies but growths derived from normal tissues. The comparatively disorganized tissue architecture of tumors pointed toward cancer as being a disease of malfunctioning cells.

- Tumors can be either benign (localized, noninvasive) or malignant (invasive, metastatic). The metastases spawned by malignant tumors are responsible for almost all deaths from cancer.

- With some exceptions, most tumors are classified into four major groups according to their origin (epithelial, mesenchymal, hematopoietic, and neuroectodermal).

- Virtually all cell types can give rise to cancer, but the most common human cancers are of epithelial origin—the carcinomas. Most carcinomas fall into two categories: squamous cell carcinomas arise from epithelia that form protective cell layers, while adenocarcinomas arise from secretory epithelia.

- Nonepithelial malignant tumors include (1) sarcomas, which originate from mesenchymal cells; (2) hematopoietic cancers, which arise from the precursors of blood cells; and (3) neuroectodermal tumors, which originate from components of the nervous system.

- If a tumor's cells have dedifferentiated (lost all tissue-specific traits), its origin cannot be readily identified; such tumors are said to be anaplastic.

- Cancers seem to develop progressively, with tumors demonstrating different gradations of abnormality along the way from benign to metastatic.

- Benign tumors may be hyperplastic or metaplastic. Hyperplastic tissues appear normal except for an excessive number of cells, whereas metaplastic tissues show displacement of normal cells by normal cell types not usually encountered at that site. Metaplasia is most frequent in epithelial transition zones.

- Dysplastic tumors contain cells that are cytologically abnormal. Dysplasia is a transitional state between completely benign and premalignant. Adenomatous growths (adenomas, polyps, papillomas, and warts) are dysplastic epithelial tumors that are considered to be benign because they respect the boundary created by the basement membrane.

- Tumors that breach the basement membrane and invade underlying tissue are malignant. An even further degree of abnormality is metastasis, the seeding of tumor colonies to other sites in the body. Metastasis requires not only invasiveness but also such newly acquired traits as motility and adaptation to foreign tissue environments.

- Biochemical and genetic markers seem to indicate that human tumors are monoclonal (descended from one ancestral cell) rather than polyclonal (descended from multiple ancestral cells, each of which independently spawned a population of cancer cells).

- Most normal cells start metabolizing glucose through glycolysis, and then transfer pyruvate (the product of glycolysis) into the mitochondria, where it is further processed to yield 36 ATPs and CO_2. Most cancer cells rely largely on glycolysis alone, which yields lactate and only 2 ATPs.

- The incidence of many (but not all) cancers varies dramatically by country, an indication that they cannot be due simply to a normal biologic process gone awry by chance. While differences in either heredity or environment could explain these variations, epidemiologic studies show that environment (including lifestyle factors) is the dominant determinant of the country-by-country variations in cancer incidence.

- Laboratory research supported the epidemiologic studies by directly implicating chemical and physical agents (tobacco, coal dust, X-rays) as causes of cancers. However, the possibility of cancer as an infectious disease arose when viruses were found to cause leukemias and sarcomas in chickens.

- A possible mechanism that supported carcinogenesis by physical and chemical agents surfaced when mutations were induced in fruit flies by exposing them to either X-rays or chemicals, indicating that they were mutagenic. Since these agents were also known to be carcinogenic in laboratory animals, this led to the speculation that cancer was a disease of mutant genes and that carcinogenic agents induced cancer through their ability to mutate genes.

- In 1975 the Ames test provided support for this idea by showing that many carcinogens can act as mutagens. Additional research showed that although almost all compounds that are mutagenic are likely to be carcinogens, the converse does not hold true. So, some carcinogens act through their ability to mutate DNA, while others promote tumorigenesis through nongenetic mechanisms. Such nonmutagenic carcinogens are called tumor promoters.

- The Ames test combined with other discoveries led to the model, still unproven, that a significant portion of human cancers are attributable to the consumption of foodstuffs that are directly or indirectly mutagenic and hence carcinogenic.

Thought questions

1. What types of observation allow a pathologist to identify the tissue of origin of a tumor? And why are certain tumors extremely difficult to assign to a specific tissue of origin?

2. Under certain circumstances, all tumors of a class can be traced to a specific embryonic cell layer, while in other classes of tumors, no such association can be made. What tumors would fit into each of these two groupings?

3. What evidence persuades us that a cancer arises from the native tissues of an individual rather than invading the body from outside and thus being of foreign origin?

4. How compelling are the arguments for the monoclonality of tumor cell populations, and what logic and observations undermine the conclusion of monoclonality?

5. How can we estimate what percentage of cancers in a population are avoidable (through virtuous lifestyles) and what percentage occur independently of lifestyle?

6. What limitations does the Ames test have in predicting the carcinogenicity of various agents?

7. In the absence of being able to directly detect mutant genes within cancer cells, what types of observation allow one to infer that cancer is a disease of mutant cells?

Additional reading

Ames BN (1983) Dietary carcinogens and anticarcinogens. *Science* 231, 1256–1264.

Bickers DR & Lowy DR (1989) Carcinogenesis: a fifty-year historical perspective. *J. Invest. Dermatol.* 92, 121S–131S.

Greenlee RT, Murray T, Bolden S & Wingo PA (2000) Cancer statistics 2000. *CA Cancer J. Clin.* 50, 7–33.

Greenwald P & Dunn BK (2009) Landmarks in the history of cancer epidemiology. *Cancer Res.* 69, 2151–2162.

Hoque A, Parnes HL, Stefanek ME et al. (2007) Meeting Report, Fifth Annual AACR Frontiers of Cancer Prevention Research. *Cancer Res.* 67, 8989–8993.

Jemal A, Center MM, DeSantis C & Ward EM (2010) Global patterns of cancer incidence and mortality rates and trends. *Cancer Epidemiol. Biomarkers Prev.* 19, 1893–1907.

Loeb LA & Harris CC (2008) Advances in chemical carcinogenesis: a historical review and prospective. *Cancer Res.* 68, 6863–6872.

Lu W, Pelicano H & Huang P (2010) Cancer metabolism: is glutamine sweeter than glucose? *Cancer Cell* 18, 199–200.

Mazurek S, Boschek CB, Hugo F & Eigenbrodt E (2005) Pyruvate kinase type M2 and its role in tumor growth and spreading. *Semin. Cancer Biol.* 15, 300–308.

Peto J (2001) Cancer epidemiology in the last century and the next decade. *Nature* 411, 390–395.

Pisani P, Parkin DM, Bray F & Ferlay J (1999) Estimates of worldwide mortality from 25 cancers in 1990. *Int. J. Cancer* 83, 18–29.

Preston-Martin S, Pike MC, Ross RK et al. (1990) Increased cell division as a cause of human cancer. *Cancer Res.* 50, 7415–7421.

Vander Heiden MG, Cantley LC & Thompson CB (2009) Understanding the Warburg effect: the metabolic requirements of cell proliferation. *Science* 324, 1029–1033.

Wilson S, Jones L, Coussens C & Hanna K (eds) (2002) Cancer and the Environment: Gene-Environment Interaction. Washington, DC: National Academy Press.

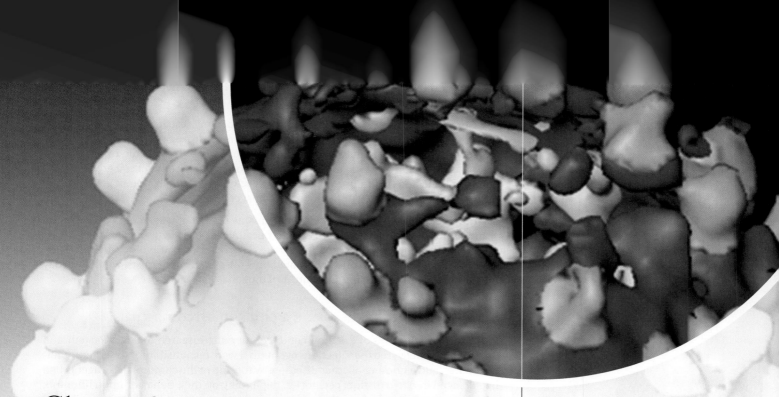

Chapter 3

Tumor Viruses

A tumor of the chicken ... has been propagated in this laboratory since October, 1909. The behavior of this new growth has been throughout that of a true neoplasm, for which reason the fact of its transmission by means of a cell-free filtrate assumes exceptional importance.

Francis Peyton Rous, cancer biologist, 1911

Viruses are capable of causing a wide variety of human diseases, ranging from rabies to smallpox to the common cold. The great majority of these infectious agents do harm through their ability to multiply inside infected host cells, to kill these cells, and to release progeny virus particles that proceed to infect other hosts nearby. The **cytopathic** (cell-killing) effects of viruses, together with their ability to spread rapidly throughout a tissue, enable these agents to leave a wide swath of destruction in their wake.

But the peculiarities of certain viral replication cycles may on occasion yield quite another outcome. Rather than killing infected cells, some viruses may, quite paradoxically, force their hosts to thrive, indeed, to proliferate uncontrollably. In so doing, such viruses—often called tumor viruses—can create cancer.

At one time, beginning in the early 1970s, tumor viruses were studied intensively because they were suspected to be the cause of many common human cancers. This notion was eventually rejected based on the evidence subsequently gathered during that decade, which indicated that virus-induced cancers represent only a minority of the cancer types afflicting humans. Nonetheless, this line of research proved to be invaluable for cancer biologists: study of various tumor viruses provided the key for opening many of the long-hidden secrets of human cancers, including the great majority of cancers that have no connection with tumor virus infections.

As we will see, tumor virus research had a highly variable history over the course of the last century. These infectious agents were discovered in the first decade of the twentieth century and then retreated from the center stage of science. Half a century

Movie in this chapter

3.1   Contact Inhibition

later, interest in these agents revived, culminating in the frenetic pace of tumor virus research during the 1970s.

The cancer-causing powers of tumor viruses drove many researchers to ask precisely how they succeed in creating disease. Most of these viruses possess relatively simple genomes containing only a few genes, yet some were found able to overwhelm an infected cell and its vastly more complex genome and to redirect cell growth. Such behavior indicated that tumor viruses have developed extremely potent genes to perturb the complex regulatory circuitry of the host cells that they infect.

By studying tumor viruses and their mechanisms of action, researchers changed the entire mindset of cancer research. Cancer became a disease of genes and thus a condition that was susceptible to analysis by the tools of molecular biology and genetics. When this story began, no one anticipated how obscure tumor viruses would one day revolutionize the study of human cancer pathogenesis.

3.1 Peyton Rous discovers a chicken sarcoma virus

In the last two decades of the nineteenth century, the research of Louis Pasteur and Robert Koch uncovered the infectious agents that were responsible for dysentery, cholera, rabies, and a number of other diseases. By the end of the century, these agents had been placed into two distinct categories, depending on their behavior upon filtration. Solutions of infectious agents that were trapped in the pores of filters were considered to contain bacteria. The other agents, which were small enough to pass through the filters, were classified as viruses. On the basis of this criterion, the agents for rabies, foot-and-mouth disease, and smallpox were categorized as viruses.

Cancer, too, was considered a candidate infectious disease. As early as 1876, a researcher in Russia reported the transmission of a tumor from one dog to another: chunks of tumor tissue from the first dog were implanted into the second, whereupon a tumor appeared several weeks later. This success was followed by many others using rat and mouse tumors.

The significance of these early experiments remained controversial. Some researchers interpreted these outcomes as proof that cancer was a transmissible disease. Yet others dismissed these transplantation experiments, since in their eyes, such work showed only that tumors, like normal tissues, could be excised from one animal and forced to grow as a graft in the body of a second animal.

In 1908, two researchers in Copenhagen reported extracting a filterable agent from chicken leukemia cells and transmitting this agent to other birds, which then contracted the disease. The two Danes did not follow up on their initial discovery, and it remained for Peyton Rous, working at the Rockefeller Institute in New York, to found the discipline of tumor virology (**Figure 3.1**).

In 1909, Rous began his study of a sarcoma that had appeared in the breast muscle of a hen. In initial experiments, Rous succeeded in transmitting the tumor by implanting small fragments of it into other birds of the same breed. Later, as a variation of this experiment, he ground up a sarcoma fragment in sand and filtered the resulting homogenate (**Figure 3.2**). When he injected the resulting filtrate into young birds, they too developed tumors, sometimes within several weeks. He subsequently found that these induced tumors could also be homogenized to yield, once again, an infectious agent that could be transmitted to yet other birds, which also developed sarcomas at the sites of injection.

These serial passages of the sarcoma-inducing agent from one animal to another yielded a number of conclusions that are obvious to us now but at the time were nothing less than revolutionary. The carcinogenic agent, whatever its nature, was clearly very small, since it could pass through a filter. Hence, it was a virus (**Sidebar 3.1**). This virus could cause the appearance of a sarcoma in an injected chicken, doing so on a predictable timetable. Such an infectious agent offered researchers the unique opportunity to induce cancers at will rather than relying on the spontaneous and unpredictable appearance of tumors in animals or humans. In addition to its ability to induce

(A)

(B)

Figure 3.1 Peyton Rous and the hen that launched modern cancer research (A) Francis Peyton Rous began his work in 1910 that led to the discovery of Rous sarcoma virus (RSV) *(left)*. More than 50 years later (1966), he received the Nobel Prize in Medicine and Physiology for this seminal work—a tribute to his persistence and longevity *(right)*. (B) His good fortune began when a Long Island, NY chicken farmer brought Rous, then working at the Rockefeller Institute in New York, a prized barred Plymouth Rock hen. The farmer wanted Rous to treat the large tumor growing in its chest muscle; Rous saw experimental opportunity and dispatched the hen, extracting the tumor. The arthritic hands are likely those of the chicken farmer. (A, courtesy of the Rockefeller University Archives. B, from P. Rous, *J. Exp. Med.* 12:696–705, 1910.)

cancer, this agent, which came to be called Rous sarcoma virus (RSV), was capable of multiplying within the tissues of the chicken; far more virus could be recovered from an infected tumor tissue than was originally injected.

In 1911, when Rous finally published his work, yet another report appeared on a transmissible virus of rabbit tumors, called myxomas. Soon thereafter, Rous and his collaborators found two other chicken viruses, and yet another chicken sarcoma virus was reported by others in Japan. Then, there was only silence for two decades until other novel tumor viruses were discovered. The molecular nature of viruses and the means by which they multiplied would remain mysteries for more than half a century after Rous's initial discovery.

Still, his finding of a sarcoma virus reinforced the convictions of those who believed that virtually all human diseases were provoked by infectious agents. In their eyes, cancer could be added to the lengthening list of diseases, such as cholera, tuberculosis, rabies, and sepsis, whose causes could be associated with a specific microbial agent. By 1913, the Dane Johannes Grib Fibiger reported that stomach cancers in rats could be traced to spiroptera worms that they harbored. His work, for which he received the

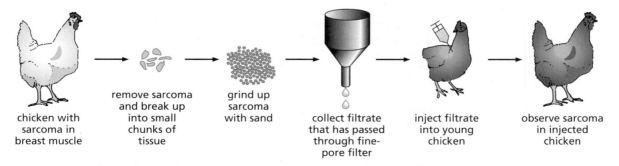

chicken with sarcoma in breast muscle → remove sarcoma and break up into small chunks of tissue → grind up sarcoma with sand → collect filtrate that has passed through fine-pore filter → inject filtrate into young chicken → observe sarcoma in injected chicken

Figure 3.2 Rous's protocol for inducing sarcomas in chickens
Rous removed a sarcoma from the breast muscle of a chicken, ground it with sand, and passed the resulting homogenate through a fine-pore filter. He then injected the filtrate (the liquid that passed through the filter) into the wing web of a young chicken and observed the development of a sarcoma many weeks later. He then ground up this new sarcoma and repeated the cycle of homogenization, filtration, and injection, once again observing a tumor in another young chicken. These cycles could be repeated indefinitely; after repeated serial passaging, the virus produced sarcomas far more rapidly than the original viral isolate.

Sidebar 3.1 Viruses have simple life cycles The term "virus" refers to a diverse array of particles that infect and multiply within a wide variety of cells, ranging from bacteria to the cells of plants and metazoa. Relative to the cells that they infect, individual virus particles, often termed **virions**, are tiny. Virions are generally simple in structure, with a nucleic acid (DNA or RNA) genome wrapped in a protein coat (a **capsid**) and, in some cases, a lipid membrane surrounding the capsid. In isolation, viruses are metabolically inert. They can multiply only by infecting and parasitizing a suitable host cell. The viral genome, once introduced into the cell, provides instructions for the synthesis of progeny virus particles. The host cell, for its part, provides the low–molecular-weight precursors needed for the synthesis of viral proteins and nucleic acids, the protein-synthetic machinery, and, in many cases, the polymerases required for replicating and transcribing the viral genome.

The endpoint of the resulting infectious cycle is the production of hundreds, even thousands of progeny virus particles that can then leave the infected cell and proceed to infect other susceptible cells. The interaction of the virus with the host cell can be either a **virulent** one, in which the host cell is destroyed during the infectious cycle, or a **temperate** one, in which the host cell survives for extended periods, all the while harboring the viral genome and releasing progeny virus particles.

Many viruses carrying double-stranded DNA (dsDNA) replicate in a fashion that closely parallels the macromolecular metabolism of the host cell (**Figure 3.3**). This allows them to use host-cell DNA polymerases to replicate their DNA, host-cell RNA polymerases to transcribe the viral mRNAs from double-stranded viral DNA templates, and host ribosomes to translate the viral mRNAs. Once synthesized, viral proteins are used to coat (**encapsidate**) the newly synthesized viral genomes, resulting in the assembly of complete progeny virions, which then are released from the infected cell.

Since cells do not express enzymes that can replicate RNA molecules, the genomes of many RNA-containing virus particles encode their own RNA-dependent RNA polymerases to replicate their genomes. Poliovirus, as an example, makes such an enzyme, as does rabies virus. RNA tumor viruses like Rous sarcoma virus, as we will learn later in this chapter, follow a much more circuitous route for replicating their viral RNA.

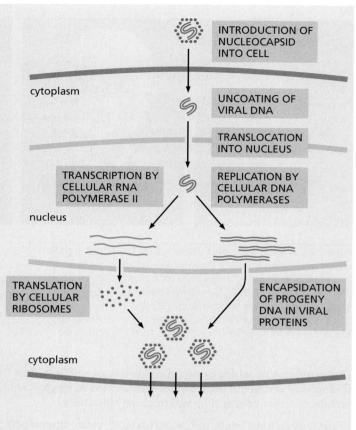

Figure 3.3 Life cycle of viruses with dsDNA genomes The life cycle of viruses with double-stranded DNA (dsDNA) genomes closely parallels that of the host cell. Almost all of the steps leading to the synthesis of viral DNA, RNA, and proteins can be achieved by using the synthetic machinery provided by the infected host cell. Not shown here is the release of progeny virus particles from the infected cell. (Adapted from B. Alberts et al., Molecular Biology of the Cell, 5th ed. New York: Garland Science, 2008.)

1926 Nobel Prize in Physiology or Medicine, represented direct and strong validation of the idea, first indicated by Rous's work, that cancer was yet another example of an infectious disease.

A year after Fibiger's 1926 Nobel award, he passed away and his scientific opus began to disintegrate. The stomach tumors that he had described were not tumors at all. Instead, they were found to be metaplastic stomach epithelia that were the result of the profound vitamin deficiencies suffered by these rats; they lived in sugar refineries and ate sugar cane almost exclusively. Fibiger's Nobel Prize became an embarrassment to the still-small community of cancer researchers. They threw the proverbial baby out with the bathwater, discrediting both his work and the notion that cancer could ever be caused by infectious agents.

Interest in the origins of cancer shifted almost totally to chemically induced cancers. Chemicals had been discovered in the early twentieth century that were clearly carcinogenic (see Section 2.9). Study of Rous sarcoma virus and the other tumor viruses languished and entered into a deep sleep for several decades.

3.2 Rous sarcoma virus is discovered to transform infected cells in culture

The rebirth of Rous sarcoma virus research began largely at the California Institute of Technology in Pasadena, in the laboratory of Renato Dulbecco. Dulbecco's post-doctoral fellow Harry Rubin found that when stocks of RSV were introduced into Petri dishes carrying cultures of chicken embryo fibroblasts, the RSV-infected cells survived, apparently indefinitely. It seemed that RSV parasitized these cells, forcing them to produce a steady stream of progeny virus particles for many days, weeks, even months (**Figure 3.4**). Most other viruses, in contrast, were known to enter into host cells, multiply, and quickly kill their hosts; the multitude of progeny virus particles

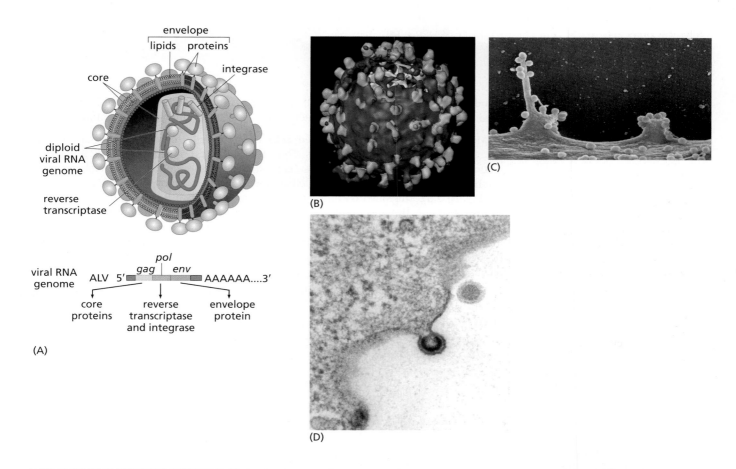

Figure 3.4 The virion of RSV and related viruses (A) An artist's reconstruction of the structure of a *retrovirus* virion, such as that of Rous sarcoma virus, which has four major types of viral proteins. The glycoprotein spikes (encoded by the viral *env* gene) that protrude from the lipid bilayer enable the virion to *adsorb* to the surface of a cell and to introduce its contents into the cytoplasm. Beneath this envelope lies a protein shell formed by the several core proteins encoded by the viral *gag* gene. Within this protein shell are two identical copies of the viral genomic RNA and a number of reverse transcriptase and integrase molecules specified by the viral *pol* gene. (B) Cryoelectron microscopy and complex image-processing algorithms have produced the first high-resolution reconstructed image of a murine leukemia virus (MLV) virion; MLV is related to RSV, and its virion has a structure similar to that of RSV. The glycoprotein spikes are seen here in *magenta*, while the lipid bilayer of the virion is shown in *purple*; parts of the underlying nucleocapsid, revealed at lower resolution, are shown in *yellow*. (C) Scanning electron micrograph of human immunodeficiency virus (HIV) nucleocapsid cores budding from the surface of baby hamster kidney (BHK) cells; HIV-infected lymphocytes generate similar images. HIV is a member of the same family of viruses as RSV and MLV. (D) Transmission electron micrograph showing murine leukemia virus (MLV) particles budding from the surface of an infected cell. As the nucleocapsid cores (containing the gag proteins, the virion RNA, and the reverse transcriptase and integrase enzymes) leave the cell, they wrap themselves with a patch of lipid bilayer taken from the plasma membrane of the infected cell. (A, adapted from H. Fan et al., The Biology of AIDS. Boston, MA: Jones and Bartlett Publishers, 1989. B, from F. Förster et al., *Proc. Natl. Acad. Sci. USA* 102:4729–4734, 2005. C, courtesy of P. Roingeard. D, courtesy of Laboratoire de Biologie Moleculaire.)

Figure 3.5 An RSV-induced focus This phase-contrast micrograph reveals a focus of Rous sarcoma virus–transformed chicken embryo fibroblasts surrounded by a monolayer of uninfected cells. The focus stands out because it is composed of rounded, refractile cells that are beginning to pile up on one another, in contrast to the flattened morphology of the surrounding normal cells, which stop proliferating after they have produced a layer of cells one cell thick. [Courtesy of P.K. Vogt; from J. Hauber and P.K. Vogt (eds.). Nuclear Export of Viral RNAs. Current Topics in Microbiology and Immunology, Vol. 259. Berlin: Springer-Verlag, 2001.]

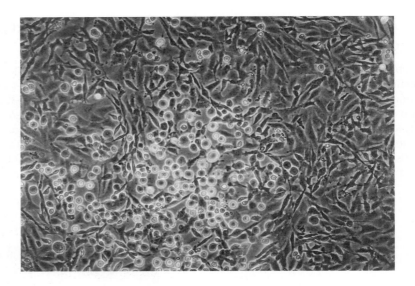

released from dying cells could then proceed to infect yet other susceptible cells in the vicinity, repeating the cycle of infection, multiplication, and cell destruction.

Most important, the RSV-infected cells in these cultures displayed many of the traits associated with cancer cells. Thus, **foci** (clusters) of cells appeared after infection. Under the microscope, these cells strongly resembled the cells isolated from chicken sarcomas, exhibiting the characteristic rounded morphology of cancer cells (**Figure 3.5**). They also had a metabolism reminiscent of that seen in cells isolated from tumors. This resemblance led Rubin, Howard Temin (a student in Dulbecco's laboratory; **Figure 3.6**), and others to conclude that the process of cell **transformation**—conversion of a normal cell into a tumor cell—could be accomplished within the confines of a Petri dish, not just in the complex and difficult-to-study environment of a living tissue.

These simple observations radically changed the course of twentieth-century cancer research, because they clearly demonstrated that cancer formation could be studied at the level of individual cells whose behavior could be tracked closely under the microscope. This insight suggested the further possibility that the entire complex biology of tumors could one day be understood by studying the transformed cells forming tumor masses. So, an increasing number of biologists began to view cancer as a disease of malfunctioning cells rather than abnormally developing tissues.

Temin and Rubin, soon followed by many others, used this experimental model to learn some basic principles about cell transformation. They were interested in the fate of a cell that was initially infected by RSV. How did such a cell proliferate when compared with uninfected neighboring cells? After exposure of cells to a solution of virus particles (often called a **virus stock**), the two researchers would place a layer of agar above the cell layer growing at the bottom of the Petri dish, thereby preventing free virus particles from spreading from initially infected cells to uninfected cells in other parts of the dish. Hence, any changes in cell behavior were the direct result of the initial infection by virus particles of chicken embryo fibroblasts.

Figure 3.6 Howard Temin Howard Temin, pictured here in 1964, began his work by demonstrating the ability of Rous sarcoma virus to transform cells *in vitro*, and showed the persistence of the virus in the infected, transformed cells. He postulated the existence of a DNA provirus and subsequently shared the 1975 Nobel Prize in Physiology or Medicine with David Baltimore for their simultaneous discoveries of the enzyme reverse transcriptase. (Courtesy of the University of Wisconsin Archives.)

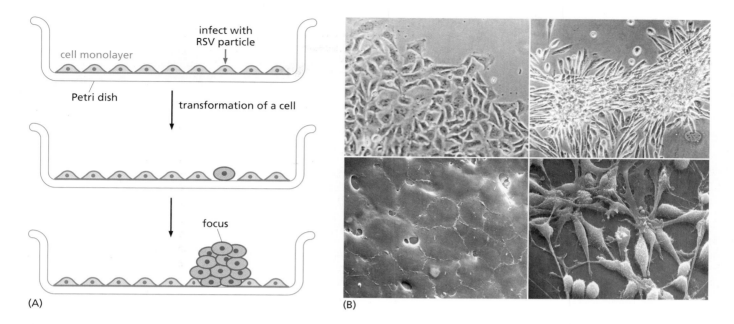

(A)

(B)

The foci that Temin and Rubin studied revealed dramatic differences in the behavior of normal versus transformed cells. When first introduced into a Petri dish, normal cells formed islands scattered across the bottom of the dish. They then proliferated and eventually filled up all the space in the bottom of the dish, thereby creating **confluent** cultures. Once they reached confluence, however, these normal cells stopped proliferating, resulting in a one-cell-thick (or slightly thicker) layer of cells, often called a cell **monolayer** (**Figure 3.7**).

The cessation of growth of these normal cells after forming confluent monolayers was the result of a process that came to be called **contact inhibition**, **density inhibition**, or **topoinhibition**. Somehow, high cell density or contact with neighbors caused these cells to stop dividing. This behavior of normal cells contrasted starkly with that of the **transformants** within the RSV-induced foci. The latter clearly had lost contact (density) inhibition and consequently continued to proliferate, piling up on top of one another and creating multilayered clumps of cells so thick that they could often be seen with the naked eye.

Under certain experimental conditions, it could be shown that all the cells within a given focus were the descendants of a single progenitor cell that had been infected and presumably transformed by an RSV particle. Today, we would term such a flock of descendant cells a cell **clone** and the focus as a whole, a clonal outgrowth.

The behavior of these foci gave support to one speculation about the possible similarities between the cell transformation triggered by RSV in the Petri dish and the processes that led to the appearance of tumors in living animals, including humans: maybe all the cells within a spontaneously arising human tumor mass also constitute a clonal outgrowth and therefore are the descendants of a single common progenitor cell that somehow underwent transformation and then launched a program of replication that led eventually to the millions, even billions, of descendant cells that together formed the mass. As discussed earlier (see Section 2.5), detailed genetic analyses of human tumor cells were required, in the end, to test this notion in a truly definitive way.

Figure 3.7 Transformed cells forming foci (A) Normal chicken embryo fibroblasts (*blue*) growing on the bottom of a Petri dish form a layer one cell thick—a monolayer. This is formed because these cells cease proliferating when they touch one another—the behavior termed "*contact inhibition.*" However, if one of these cells is infected by RSV prior to reaching confluence, this cell (*red*) and its descendants acquire a rounded morphology and lose contact inhibition. As a consequence, they continue to proliferate in spite of touching one another and eventually accumulate in a multilayered clump of cells (a *focus*) visible to the naked eye. (B) The effects of RSV infection and transformation can be observed through both phase-contrast microscopy (*upper panels*) and scanning electron microscopy (*lower panels*). Normal chicken embryo fibroblasts are spread out and form a continuous monolayer (*left panels*). However, upon transformation by RSV, they round up, become refractile (*white halos, upper right panel*), and pile up upon one another (*right panels*). (B, courtesy of L.B. Chen.)

3.3 The continued presence of RSV is needed to maintain transformation

The behavior of the cells within an RSV-induced focus indicated that the transformation phenotype was transmitted from an initially infected, transformed chicken cell to its direct descendants. This transmission provoked another set of questions: Did an

Figure 3.8 Temperature-sensitive mutant and the maintenance of transformation by RSV When chicken embryo fibroblasts were infected with a temperature-sensitive mutant of RSV and cultured at the *permissive* temperature (37°C), the viral transforming function could be expressed and the cells became transformed. However, when the cultures containing these infected cells were shifted to the *nonpermissive* temperature (41°C), the cells reverted to a normal, nontransformed morphology. Later, with a shift of the cultures back to the permissive temperature, the cells once again exhibited a transformed morphology. The loss of the transformed phenotype upon temperature shift-up demonstrated that the continuous action of some temperature-sensitive viral protein was required in order to maintain this phenotype. The reacquisition of the transformed phenotype after temperature shift-down many days later indicated that the viral genome continued to be present in these cells at the high temperature in spite of their normal appearance.

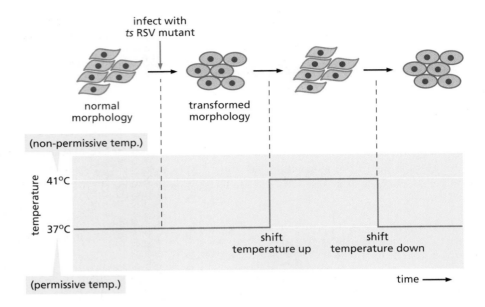

RSV particle infect and transform the progenitor cell of the focus and, later on, continue to influence the behavior of all of its direct descendants, ensuring that they also remained transformed? Or, as an alternative, did RSV act in a "hit-and-run" fashion by striking the initially infected progenitor cell, altering its behavior, and then leaving the scene of the crime? According to this second scenario, the progenitor cell could somehow transmit the phenotype of cancerous growth to its descendants without the continued presence of RSV.

Temin and Rubin's work made it clear that the descendants of an RSV-infected cell continued to harbor copies of the RSV genome, but that evidence, on its own, settled little. The real question was: Did the transformed state of the descendant cells actually *depend* on some continuing influence exerted by the RSV genomes that they carried?

An experiment performed in 1970 at the University of California, Berkeley, settled this issue unambiguously. A mutant of RSV was developed that was capable of transforming chicken cells when these cells were cultured at 37°C but not at 41°C (the latter being the normal temperature at which chicken cells grow). **Temperature-sensitive** (*ts*) mutants like this one were known to encode partially defective proteins, which retain their normal structure and function at one temperature and lose their function at another temperature, presumably through thermal **denaturation** of the structure of the mutant protein.

After the chicken embryo fibroblasts were infected with the *ts* mutant of RSV, these cells became transformed if they were subsequently cultured at the lower (**permissive**) temperature of 37°C, as anticipated. Indeed, these cells could be propagated for many cell generations at this lower temperature and continued to grow and divide just like cancer cells, showing their characteristic transformed morphology (see Figure 3.7B). But weeks later, if the temperature of these infected cultures was raised to 41°C (the *nonpermissive* temperature), these cells lost their transformed shape and quickly reverted to the shape and growth pattern of cells that had never experienced an RSV infection (**Figure 3.8**).

The Berkeley experiments led to simple and yet profoundly important conclusions. Since the cells that descended from a *ts* RSV–infected cell continued to show the temperature-sensitive growth trait, it was obvious that copies of the genome of the infecting virus persisted in these cells for weeks after the initial infection. These copies of the RSV genome in the descendant cells continued to make some temperature-sensitive protein (whose precise identity was not known). Most important, the *continuing actions* of this protein were required in order to maintain the transformed growth phenotype of the RSV-infected cells.

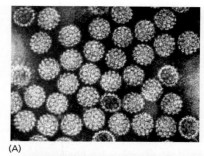

(A)

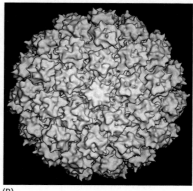

(B)

Figure 3.9 Shope papillomavirus (A) Electron micrograph of Shope papillomavirus. Virus particles of this class carry dsDNA genomes and lack a lipid envelope like that of retroviruses. (B) This cryoelectron micrograph, together with image enhancement, reveals the structure of a human papillomavirus particle. (A, courtesy of D. DiMaio. B, from B.L. Trus et al., *Nat. Struct. Biol.* 4:413–420, 1997.)

This work showed that cell transformation, at least that induced by RSV, was not a hit-and-run affair. In the language of the tumor virologists, the viral transforming gene was required to both *initiate* and *maintain* the transformed phenotype of virus-infected cells.

3.4 Viruses containing DNA molecules are also able to induce cancer

RSV was only one of a disparate group of viruses that were found able to induce tumors in infected animals. By 1960, four other classes of tumor viruses had become equally attractive agents for study by cancer biologists. One of them was a tumor virus that was discovered almost a quarter century after Rous's pioneering work. This virus, discovered by Richard Shope in rabbits, caused **papillomas** (warts) on the skin. These were really benign lesions, which on rare occasions progressed to true tumors—squamous cell carcinomas of the skin.

By the late 1950s, it became clear that Shope's virus was constructed very differently from RSV. The papillomavirus particles carried DNA genomes, whereas RSV particles were known to carry RNA molecules. Also, the Shope virus particles were sheathed in a protein coat, whereas RSV clearly had, in addition, a lipid membrane coating on the outside. In the decades that followed, more than 100 distinct human papillomavirus (HPV) types, all related to the Shope virus, would be discovered (**Figure 3.9**).

In the 1950s and 1960s, various other DNA tumor viruses were isolated (**Table 3.1**). Polyomavirus, named for its ability to induce a variety of distinct tumor types in mice, was discovered in 1953. Closely related to polyomavirus in its size and chemical makeup was SV40 virus (the 40th simian *v*irus in a series of isolates). This monkey virus had originally been discovered as a contaminant of the poliovirus vaccine

Table 3.1 Tumor virus genomes

	Virus family	Approximate size of genome (kb)
DNA viruses		
Hepatitis B virus (HBV)	hepadna	3
SV40/polyoma	papova	5
Human papilloma 16 (HPV)	papova	8
Human adenovirus 5	adenovirus	35
Human herpesvirus 8 (HSV-8; KSHV)	herpesviruses	165
Shope fibroma virus	poxviruses	160
RNA viruses		
Rous sarcoma virus (RSV)	retrovirus	9
Human T-cell leukemia virus (HTLV-I)	retrovirus	9

Adapted in part from G.M. Cooper, Oncogenes, 2nd ed. Boston: Jones and Bartlett Publishers, 1995.

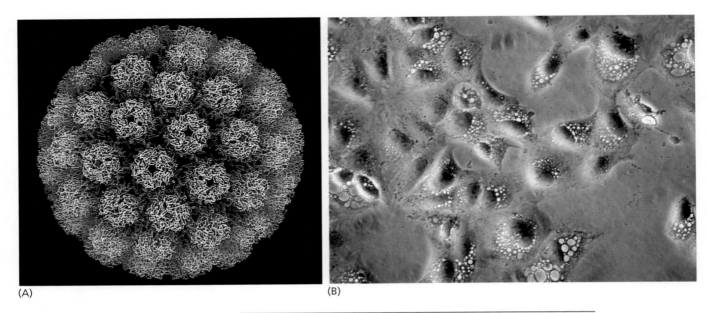

(A)   (B)

Figure 3.10 SV40 virus (A) As determined by X-ray diffraction, the protein capsid of the SV40 DNA tumor virus consists of three virus-encoded proteins that are assembled into pentamers and hexamers with icosahedral symmetry. The dsDNA genome of SV40 is carried within this capsid. (B) SV40 launches a lytic cycle in permissive host cells, such as the kidney cells of several monkey species. The resulting cytopathic effect, seen here, involves the formation of large cytoplasmic vacuoles prior to the death of the cell and the release of tens of thousands of progeny virus particles. (A, courtesy of Robert Grant, Stephan Crainic, and James M. Hogle. B, from A. Gordon-Shaag et al., *J. Virol.* 77:4273–4282, 2003.)

stocks prepared in the mid- and late 1950s (**Figure 3.10A**). Clever virological sleuthing revealed that SV40 particles often hid out in cultures of the rhesus and cynomolgus monkey kidney cells used to propagate poliovirus during the preparation of vaccine. In fact, the presence of SV40 was not initially apparent in these cell cultures. However, when poliovirus stocks that had been propagated in these monkey cells were later used to infect African green monkey kidney (AGMK) cells, SV40 revealed itself by inducing a very distinctive cytopathic effect—numerous large **vacuoles** (fluid-filled bubble-like structures) in the cytoplasm of infected cells (see Figure 3.10B). Within a day after the vacuoles formed in an SV40-infected cell, this cell would lyse, releasing tens of thousands of progeny virus particles. (Because of SV40 contamination, occurring during the course of poliovirus vaccine production, some poliovirus-infected cell cultures yielded far more SV40 virus particles than poliovirus particles!)

This **lytic cycle** of SV40 contrasted starkly with its behavior in cells prepared from mouse, rat, or hamster embryos. SV40 was unable to replicate in these rodent cells, which were therefore considered to be *nonpermissive* hosts. But on occasion, in one cell out of thousands in a nonpermissive infected cell population, a transformant grew out that shared many characteristics with RSV-transformed cells, that is, a cell that had undergone changes in morphology and loss of contact inhibition, and had acquired the ability to seed tumors *in vivo*. On the basis of this, SV40 was classified as a tumor virus.

By some estimates, between one-third and two-thirds of the polio vaccines—the oral, live Sabin vaccine and the inactivated, injected Salk vaccine—administered between 1955 and 1963 contained SV40 virus as a contaminant, and between 10 and 30 million people were exposed to this virus through vaccination. In 1960, the fear was first voiced that the SV40 contaminant might trigger cancer in many of those who were vaccinated. Reassuringly, epidemiologic analyses conducted over the succeeding four decades indicated little, if any, increased risk of cancer among those exposed to these two vaccines (Supplementary Sidebar 3.1).

Shope's papillomavirus, the mouse polyomavirus, and SV40 were grouped together as the **papovavirus** class of DNA tumor viruses, the term signifying *pa*pilloma, *po*lyoma, and the *va*cuoles induced by SV40 during its lytic infection. By the mid-1960s, it was apparent that the genomes of the papovaviruses were all formed from circular double-stranded DNA molecules (**Figure 3.11**). This represented a great convenience for experimenters studying the viral genomes, since there were several techniques in use at the time that made it possible to separate these relatively small, circular DNA molecules [about 5–8 kilobases (kb) in length] from the far larger, linear DNA molecules present in the chromosomes of infected host cells.

The group of DNA tumor viruses grew further with the discovery that human adenovirus, known to be responsible for upper respiratory infections in humans, was able to induce tumors in infected hamsters. Here was a striking parallel with the behavior of SV40. The two viruses could multiply freely in their natural host cells, which were therefore considered to be permissive. During the resulting lytic cycles of the virus, permissive host cells were rapidly killed in concert with the release of progeny virus particles. But when introduced into nonpermissive cells, both adenovirus and SV40 failed to replicate and instead left behind, albeit at very low frequency, clones of transformants.

Other entrants into the class of DNA tumor viruses were members of the herpesvirus group. While human herpesvirus types 1 and 2 were apparently not **tumorigenic** (capable of inducing tumors), a distantly related herpesvirus of *Saimiri* monkeys provoked rapid and fatal lymphomas when injected into monkeys from several other species. Another distantly related member of the herpesvirus family—Epstein–Barr virus (EBV)—was discovered to play a causal role in provoking Burkitt's lymphomas in young children in Equatorial Africa and New Guinea as well as nasopharyngeal carcinomas in Southeast Asia. Finally, at least two members of the poxvirus class, which includes smallpox virus, were found to be tumorigenic: Shope fibroma virus and Yaba monkey virus cause benign skin lesions in rabbits and rhesus monkeys, respectively. The tumorigenic powers of these viruses, which have very large genomes (135–160 kb), remain poorly understood to this day.

Researchers found that adenovirus and herpesvirus particles contain long, linear double-stranded DNA (dsDNA) molecules, which, like the genomes of RSV, carry the information required for both viral replication and virus-induced cell transformation. Compared with the papovaviruses, the herpesviruses had genomes of enormous size (see Table 3.1), suggesting that they carried a proportionately larger number of genes. In the end, it was the relatively small sizes of papovavirus genomes that made them attractive objects of study.

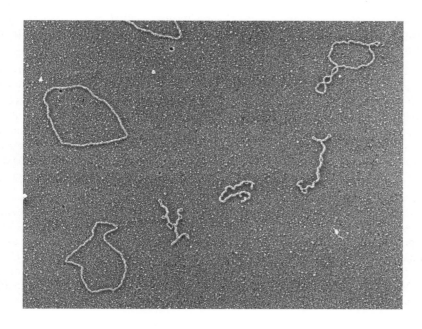

Figure 3.11 Papovavirus genomes
Electron-microscopic analyses revealed the structure of DNA genomes of papovaviruses to be double-stranded and circular. Shown are the closed-circular, supercoiled form of SV40 DNA (form I; kinked DNAs) and the relaxed, nicked circular form (form II; open, spread circles). Form III, which is a linear form of the viral DNA resulting from a double-strand DNA break, is not shown. (Courtesy of Jack D. Griffith, University of North Carolina.)

Table 3.2 Properties of transformed cells

Altered morphology (rounded shape, refractile in phase-contrast microscope)
Loss of contact inhibition (ability to grow over one another)
Ability to grow without attachment to solid substrate (anchorage independence)
Ability to proliferate indefinitely (immortalization)
Reduced requirement for mitogenic growth factors
High saturation density (ability to accumulate large numbers of cells in culture dish)
Inability to halt proliferation in response to deprivation of growth factors
Increased transport of glucose
Tumorigenicity

Adapted in part from S.J. Flint, L.W. Enquist, R.M. Krug et al., Principles of Virology. Washington, DC: ASM Press, 2000.

Most of the small group of genes in a papovavirus genome were apparently required to program viral replication; included among these were several genes specifying the proteins that form the capsid coat of the virus particle. This dictated that papovaviruses could devote only a small number of their genes to the process of cell transformation.

This realization offered the prospect of greatly simplifying the cancer problem by reducing the array of responsible genes and causal mechanisms down to a very small number. Without such simplification, cancer biologists were forced to study the can-cer-causing genes that were thought to be present in the genomes of cells transformed by nonviral mechanisms. Cellular genomes clearly harbored large arrays of genes, possibly more than a hundred thousand. At the time, the ability to analyze genomes of such vast complexity and to isolate individual genes from these genomes was a distant prospect.

3.5 Tumor viruses induce multiple changes in cell phenotype including acquisition of tumorigenicity

Like RSV-transformed cells, the cells transformed by SV40 showed profoundly altered shape and piled up on one another. This loss of contact inhibition was only one of a number of changes exhibited by virus-transformed cells (**Table 3.2**). As discussed later in this book, normal cells in culture will not proliferate unless they are provided with serum and serum-associated growth-stimulating factors. Cells transformed by a variety of tumor viruses were often found to have substantially reduced requirements for these factors in their culture medium.

Yet another hallmark of the transformed state is an ability to proliferate in culture for unusually long time periods. Normal cells have a limited proliferative potential in culture and ultimately stop multiplying after a certain, apparently predetermined number of cell divisions. Cancer cells seemed to be able to proliferate indefinitely in culture, and hence undergo **immortalization** (as discussed in Chapter 10).

When transformed cells were suspended in an agar gel, they were able to proliferate into spherical colonies containing dozens, even hundreds of cells (**Figure 3.12**). This ability to multiply without attachment to the solid substrate provided by the bottom of the Petri dish was termed the trait of **anchorage independence**. Normal cells, in con-trast, demonstrated an absolute requirement for tethering to a solid substrate before they would grow and were therefore considered to be **anchorage-dependent**. The ability of cells to grow in an anchorage-independent fashion *in vitro* usually served as a good predictor of their ability to form tumors *in vivo* following injection into appro-priate host animals.

Figure 3.12 Anchorage-independent growth A photomicrograph of colonies of cells growing in an anchorage-independent fashion. (Each of the larger colonies seen here may contain several hundred cells.) This test is usually performed by suspending cells in a semi-solid medium, such as agarose or methylcellulose, to prevent their attachment to a solid substrate, specifically, the bottom of the Petri dish. The ability of cells to proliferate while held in suspension—the phenotype of anchorage independence—is usually a good (but hardly infallible) predictor of their ability to form tumors *in vivo*. (Courtesy of A. Orimo.)

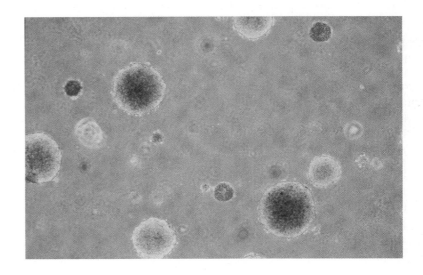

Sidebar 3.2 The powers of the immune system dictate whether transformed cells can form tumors Over the past century, repeated cycles of inbreeding were used to create distinct strains of mice and rats that are genetically identical within a strain (aside from the gender-specific differences created by the Y chromosome). Included among these mouse strains are, for example, the BALB/c, C3H, and C57/Bl6 strains. The shared genetic identity of all individuals within a strain means that tissue fragments (including tumors) from one animal of a strain can be transplanted to other animals of that strain and become established in the recipient animals.

This ability to transplant tissues from one individual to another does not exist in outbred populations, such as humans and natural populations of mice. Except for the rare instances of identical twins, tissues that are introduced from one animal (or human) to another are recognized as being foreign by the recipient's immune system. The immune system then attempts to eliminate the foreign tissue and usually succeeds in doing so rapidly. This means that true tests of the tumorigenicity of *in vitro* transformed mouse cells can be undertaken only when such cells derive from the same strain of mice as the hosts into which these cells are introduced. The lack of inbred strains of chickens severely limited the type of cancer research that could be carried out with these birds.

This barrier to transplantation can be circumvented in several ways. Often the immune system of a very young animal is more tolerant of genetically foreign tissue than the more robust, well-developed immune system of an adult. This allowed Rous to carry out some of his experiments. More common and practical, however, is the use of *immunocompromised* strains of mice, which lack fully functional immune systems and therefore tolerate engrafted tissues of foreign genetic origin, including cells and tissues originating in other species (termed **xenografts**). Frequently used strains of immunocompromised mice include the Nude, RAG, and NOD/SCID mice, each of which has a defect in one or more key components of its immune system.

This tumor-forming ability—the phenotype of tumorigenicity—represented the acid test of whether cells were fully transformed, that is, had acquired the full repertoire of neoplastic traits. One test for tumorigenicity could be performed by injecting mouse cells that had been transformed in an *in vitro* transformation experiment into host mice of the same strain (**Sidebar 3.2**). Since the host and injected cells came from the same genetic strain, the immune systems of such **syngeneic** host mice would not recognize the transformed cells as being foreign bodies and therefore would not attempt to eliminate them—the process of **tumor rejection** (a process to which we will return in Chapter 15). This allowed injected cells to survive in their animal hosts, enabling them to multiply into large tumors if, indeed, they had acquired the tumorigenic phenotype. (Populations of injected non-tumorigenic cells, in contrast, might survive in a syngeneic host as small clumps for weeks, even months, without increasing in size.)

Often, it was impossible to test the tumorigenicity of tumor virus–infected cells in a syngeneic host animal, simply because the cells being studied came from a species in which inbred syngeneic hosts were not available. This forced the use of **immunocompromised** hosts whose immune systems were tolerant of a wide variety of foreign cell types, including those from other species (see Sidebar 3.2). Mice of the Nude strain soon became the most commonly used hosts to test the tumorigenicity of a wide variety of cells, including those of human origin. Quite frequently, candidate tumor cells are injected **subcutaneously**, that is, directly under the skin of these animals. Since they also lack the ability to grow hair, Nude mice provide the additional advantage of allowing the experimenter to closely monitor the behavior of implanted tumor cells, specifically the growth of any tumors that they may spawn (**Figure 3.13**).

As mentioned, the small size of the genomes of RNA tumor viruses (for example, RSV) and papovaviruses dictated that each of these use only a small number of genes (perhaps as few as one) to elicit multiple changes in the cells that they infect and transform. Recall that the ability of a gene to concomitantly induce a number of distinct alterations in a cell is termed pleiotropy. Accordingly, though little direct evidence about gene number was yet in hand, it seemed highly likely that the genes used by tumor viruses to induce cell transformation were acting pleiotropically on a variety of molecular targets within cells.

3.6 Tumor virus genomes persist in virus-transformed cells by becoming part of host-cell DNA

The Berkeley experiments (see Section 3.3) provided strong evidence that the continued actions of the RSV genome were required to maintain the transformed state of cells, including those that were many cell generations removed from an initially

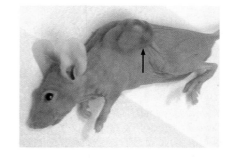

Figure 3.13 Nude mice Mice of the Nude strain offer two advantages in tests of tumorigenicity. Lacking a thymus, they are highly immunocompromised and therefore relatively receptive to engrafted cells from genetically unrelated sources, including those from foreign species. In addition, because these mice are hairless, it is easy to monitor closely the progress of tumor formation (*arrow*) after transformed cells have been injected under their skin. (From X.G. Wang et al., *PLoS ONE* 2(10): e1114, 2007.)

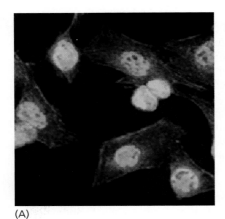

(A)

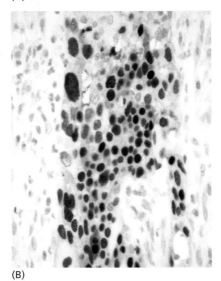

(B)

Figure 3.14 Expression of the SV40 T antigen When SV40-transformed cells are introduced into a syngeneic host, they provoke a strong immune response against the SV40-encoded T antigen (tumor antigen), which is usually termed "large T" (LT). Anti-LT serum, together with appropriate immunofluorescent tags, can be used to stain virus-transformed cells. (A) In the light micrograph shown here, an anti-LT monoclonal antibody (rather than anti-tumor serum) has been used to stain mouse cells that have been transformed by SV40, revealing the nuclear localization of LT (*green*); the cytoplasm of each cell has been stained with phalloidin (*red*), which reveals the actin fibers of the cytoskeleton. (B) While SV40 itself rarely if ever causes human tumors (see Section 3.4), immunosuppression can allow an otherwise latent close cousin of SV40, termed BK virus, to become active and occasionally oncogenic. Seen here is a urothelial (bladder) carcinoma arising in a kidney transplant patient. The tumor cells show strong staining for T antigen (*brown*), while the nearby normal urothelial cells are unstained. (The tumor cell nuclei show dramatic variations in size.) (A, courtesy of S.S. Tevethia. B, from I.S.D. Roberts et al., *Brit. J. Cancer* 99: 1383–1386, 2008.)

infected progenitor cell. This meant that some or all of the viral genetic information needed to be perpetuated in some form, being passed from a transformed mother cell to its two daughters and, further on, to descendant cells through many cycles of cell growth and division. Conversely, a failure to transmit these viral genes to descendant cells would result in their reversion to cells showing normal growth behavior.

Paralleling the behavior of RSV, cell transformation achieved by two intensively studied DNA tumor viruses—SV40 and polyomavirus—also seemed to depend on the continued presence of viral genomes in the descendants of an initially transformed cell. The evidence proving this came in a roundabout way, largely from the discovery of tumor-associated proteins (T antigens) that were found in cancers induced by these two viruses. For example, sera prepared from mice carrying an SV40-induced tumor showed strong reactivity with a nuclear protein that was present characteristically in tumors triggered by SV40 and absent in tumors induced by polyomavirus or by other carcinogenic agents. The implication was that the viral genome residing in tumor cells encoded a protein (in this case, the SV40 T antigen) that induced a strong immunological response in the tumor-bearing mouse or rat host (**Figure 3.14**).

The display of the virus-induced T antigen correlated directly with the transformed state of these cells. Therefore, cells that lost the T antigen would also lose the transformation phenotype induced by the virus. This correlation suggested, but hardly proved, that the viral gene sequences responsible for transformation were associated with or closely linked to viral sequences encoding the T antigen.

The cell-to-cell transmission of viral genomes over many cell generations represented a major conceptual problem. Cellular genes were clearly transmitted with almost total fidelity from mother cells to daughter cells through the carefully programmed processes of chromosomal DNA replication and mitosis that occur during each cellular growth-and-division cycle. How could viral genomes succeed in being replicated and transmitted efficiently through an unlimited number of cell generations? This was especially puzzling, since viral genomes seemed to lack the genetic elements, specifically those forming centromeres, that were thought to be required for proper allocation of chromosomes to daughter cells during mitosis.

Adding to this problem was the fact that the DNA metabolism of papovaviruses, such as SV40 and polyomavirus, was very different from that of the host cells that they preyed upon. When SV40 and polyomavirus infected permissive host cells, the viral DNAs were replicated as autonomous, extrachromosomal molecules. Both viruses could form many tens of thousands of circular, double-stranded DNA genomes of about 5 kb in size from a single viral DNA genome initially introduced by infection (see Figure 3.11). While the viral DNA replication exploited a number of host-cell DNA replication enzymes, it proceeded independently of the infected cells' chromosomal

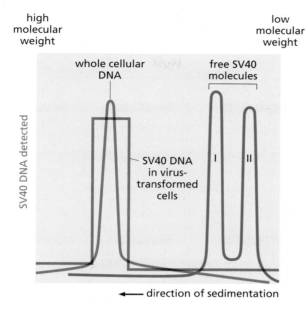

Figure 3.15 Integration of SV40 DNA DNA molecules from SV40-transformed cells were isolated and sedimented by centrifugation through an alkaline solution stabilized by a sucrose gradient (used to prevent mixing of different fluid strata within the centrifuge tube). Under these conditions, the high–molecular-weight cellular DNA (*blue*) sedimented a substantial distance down the sucrose gradient (*left side of graph*). In contrast, the SV40 DNA isolated from virus particles (*green*) sedimented more slowly, indicative of its lower molecular weight. Forms I and II viral DNA refer to the closed circular and nicked circular DNAs of SV40, respectively (see Figure 3.11). Use of nucleic acid hybridization revealed that the SV40 DNA sequences in SV40 virus–transformed cells co-sedimented with the high–molecular-weight chromosomal DNA of the virus-transformed cells, indicating covalent association of the viral genome with that of the host cell (*red*). (Adapted from J. Sambrook et al., *Proc. Natl. Acad. Sci. USA* 60:1288–1295, 1968.)

DNA replication. This nonchromosomal replication occurring during the lytic cycles of SV40 and polyomavirus shed no light on how these viral genomes were perpetuated in populations of virus-transformed cells. The latter were, after all, nonpermissive hosts and therefore prevented these viruses from replicating their DNA.

A solution to this puzzle came in 1968, when it was discovered that the viral DNA in SV40-transformed mouse cells was tightly associated with their chromosomal DNA. Using centrifugation techniques to gauge the molecular weights of DNA molecules, it became clear that the SV40 DNA in these cells no longer sedimented like a small (~5 kb) viral DNA genome. Instead, the SV40 DNA sequences in virus-transformed cells co-sedimented with the high–molecular-weight (>50 kb) chromosomal DNA of the host cells (**Figure 3.15**). In fact, the viral DNA in these transformed cells could not be separated from the cells' chromosomal DNA by even the most stringent methods of dissociation, including the harsh treatment of exposure to alkaline pH.

These results indicated that SV40 DNA in virus-transformed cells was inserted into the cells' chromosomes, becoming covalently linked to the chromosomal DNA. Such **integration** of the viral genome into a host-cell chromosome solved an important problem in viral transformation: transmission of viral DNA sequences from a mother cell to its offspring could be guaranteed, since the viral DNA would be co-replicated with the cell's chromosomal DNA during the S (DNA synthesis) phase of each cell cycle. In effect, by integrating into the chromosome, the viral DNA sequences became as much a part of a cell's genome as the cell's own native genes (**Figure 3.16**).

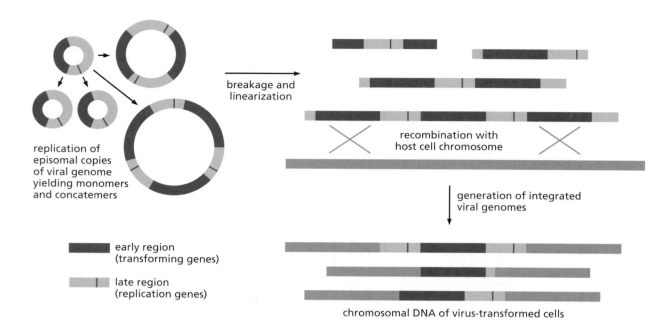

early region
(transforming genes)

late region
(replication genes)

chromosomal DNA of virus-transformed cells

Figure 3.16 Formation of integrated SV40 genomes When unintegrated (that is, *episomal*) SV40 genomes (*top leftmost circle*) replicate, they often generate large, tandemly repeated genomes termed **concatemers**. Monomeric and concatemerized SV40 DNA genomes can undergo linearization and then recombine at low efficiency with the host-cell chromosomal DNA (*gray*), doing so at random sites in the host genome; this leads to the covalent linkage of viral and cellular DNA sequences. Since the viral and host-cell genomes lack significant sequence identity, this type of recombination is termed *nonhomologous* or *illegitimate recombination*. Use of restriction enzyme cleavage-site mapping revealed the configurations of SV40 genomes integrated into the chromosomal DNA of virus-transformed cells. In some cells, only a portion of the viral genome was present; in others, full, multiple, head-to-head tandem arrays were present. Cell transformation by SV40 requires only the genes in the "early region" of the viral genome (*dark red*). Consequently, all virus-transformed cells contained at least one uninterrupted copy of this early region but often lacked segments of the viral late region (*pink*), which is unimportant for transformation. (Adapted from M. Botchan, W. Topp and J. Sambrook, *Cell* 9:269–287, 1976.)

Some years later, this ability of papovavirus genomes to integrate into host-cell genomes became highly relevant to the pathogenesis of one common form of human cancer—cervical carcinoma. Almost all (>99.7%) of these tumors have been found to carry fragments of human papillomavirus (HPV) genomes integrated into their chromosomal DNA. Provocatively, intact viral genomes are rarely discovered to be present in integrated form in cancer cell genomes. Instead, only the portion of the viral genome that contains **oncogenic** (cancer-causing) information has been found in the chromosomal DNA of these cancer cells, while the portion that enables these viruses to replicate and construct progeny virus particles is almost always absent or present in only fragmentary form (see also Figure 3.16). Interestingly, among the HPV genes discarded or disrupted during the chromosomal integration process is the viral *E2* gene, whose product serves to repress transcription of viral **oncogenes** (genes contributing to cell transformation); in the absence of *E2*, the viral *E6* and *E7* oncogenes (discussed in Chapters 8 and 9) are actively transcribed, thereby driving cervical epithelial cells toward a neoplastic phenotype.

More recently, herpesviruses have been found to use at least two different molecular strategies to maintain intact viral genomes in infected cells over extended periods of time. Certain **lymphotropic** herpesviruses (which infect lymphocytes), specifically human herpesvirus 6 (HHV-6) and Marek's disease virus (MDV) of chickens, can integrate their genomes into the telomeric repeats (see Chapter 10) of the chromosomal DNA of their cellular hosts. This strategy enables these viruses to establish long-term latent infections, in which the viral genomes are maintained indefinitely while the viruses shut down active replication; after great delay, the viral genomes can excise themselves and launch active replication.

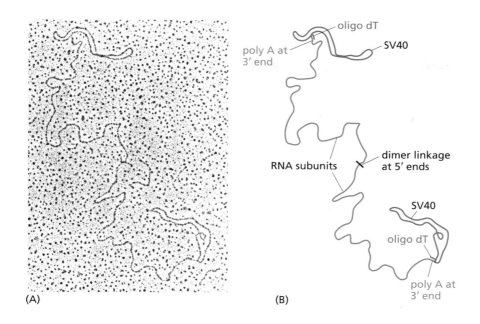

(A)   (B)

Figure 3.17 Genome structure of RNA tumor viruses Viruses like RSV carry single-stranded RNA (ssRNA) genomes. Uniquely, the genomes of this class of viruses were discovered to be diploid, that is, to carry two identical copies of the viral genetic sequence. (A) This electron micrograph shows the genome of baboon endogenous virus, a distant relative of RSV. (B) The identities of the molecules shown in panel A are revealed here. As indicated, circular SV40 dsDNA molecules (*red*), to which oligo dT (oligodeoxythymidine) tails were attached enzymatically, have been used to visually label the 3′ termini of the two single-stranded viral RNA molecules (*blue*). The viral RNA molecules (like cellular mRNA molecules) contain polyadenylate at their 3′ termini and therefore anneal to the oligo dT tails attached to the circular SV40 molecules. Note also that the two ssRNAs are associated at their 5′ ends. (Adapted from W. Bender and N. Davidson, *Cell* 7:595–607, 1976.)

More common, however, are molecular strategies that allow DNA tumor viruses to maintain their genomes as **episomes** (unintegrated genetic elements) over many cell generations. Like chromosomal integration, these strategies enable certain herpesviruses, such as Kaposi's sarcoma herpesvirus (KSHV/HHV-8), as well as HPV, to maintain chronic infections. Rather than covalently linking their genomes to cellular DNA, these viruses deploy viral proteins that serve as bridges between the episomal viral genomes and cellular chromatin. This allows the viral genomes to "hitchhike" with cellular chromosomes, ensuring nuclear localization during mitosis and proper allocation of viral genomes to daughter cells (Supplementary Sidebar 3.2).

3.7 Retroviral genomes become integrated into the chromosomes of infected cells

The ability of SV40 and polyomavirus to integrate copies of their genomes into host-cell chromosomal DNA solved one problem but created another that seemed much less solvable: how could RSV succeed in transmitting its genome through many generations within a cell lineage? The genome of RSV is made of single-stranded RNA (**Figure 3.17**), which clearly could not be integrated directly into the chromosomal DNA of an infected cell. Still, RSV succeeded in transmitting its genetic information through many successive cycles of cell growth and division (**Sidebar 3.3**).

This puzzle consumed Temin in the mid- and late-1960s and caused him to propose a solution so unorthodox that it was ridiculed by many, landing him in the scientific wilderness. Temin argued that after RSV particles (and those of related viruses) infected a cell, they made double-stranded DNA (dsDNA) copies of their RNA genomes. It was these dsDNA versions of the viral genome, he said, that became established in the chromosomal DNA of the host cell. Once established, the DNA version of the viral

Sidebar 3.3 Re-infection could not explain the stable transmission of RSV genomes The transmission of RSV genomes to the descendants of initially infected cells could be explained, in principle, by a mechanism in which the virus-infected cells continually release virus particles that subsequently infect daughter cells. Such repeated cycles of re-infection can ensure that an RSV genome can be perpetuated indefinitely in a population of cells without the need for chromosomal integration of viral genetic information—the strategy used by SV40 virus (Section 3.6). In fact, stable transformation of cell populations by RSV was also observed using strains of RSV that could infect and transform a cell but were unable to replicate in that cell. In addition, certain cell populations were found that could be transformed by an initial RSV infection but resisted subsequent cycles of re-infection. Together, such observations effectively ruled out re-infection as the means by which RSV perpetuated itself in lineages of cells over multiple cell generations.

genome—which he called a **provirus**—then assumed the molecular configuration of a cellular gene and would be replicated each time the cell replicated its chromosomal DNA. In addition, the proviral DNA could then serve as a template for transcription by cellular RNA polymerase, thereby yielding RNA molecules that could be incorporated into progeny virus particles or, alternatively, could function as messenger RNA (mRNA) that was used for the synthesis of viral proteins (**Figure 3.18**).

The process of **reverse transcription** that Temin proposed—making DNA copies of RNA—was without precedent in the molecular biology of the time, which recognized information flow only in a single direction, specifically, DNA→RNA→proteins. But the idea prevailed, receiving strong support from Temin's and David Baltimore's simultaneous discoveries in 1970 that RSV and related virus particles carry the enzyme reverse transcriptase. As both research groups discovered, this enzyme has the capacity to execute the key copying step that Temin had predicted—the step required in order for RSV to transmit its genome through many cycles of cell growth and division.

It soon became apparent that RSV was only one of a large group of similarly constructed viruses, which together came to be called **retroviruses** to reflect the fact that their cycle of replication depends on information flowing "backward" from RNA to DNA. Within a year of the discovery of reverse transcriptase, the presence of proviral DNA was detected in the chromosomal DNA of RSV-infected cells. Hence, like SV40

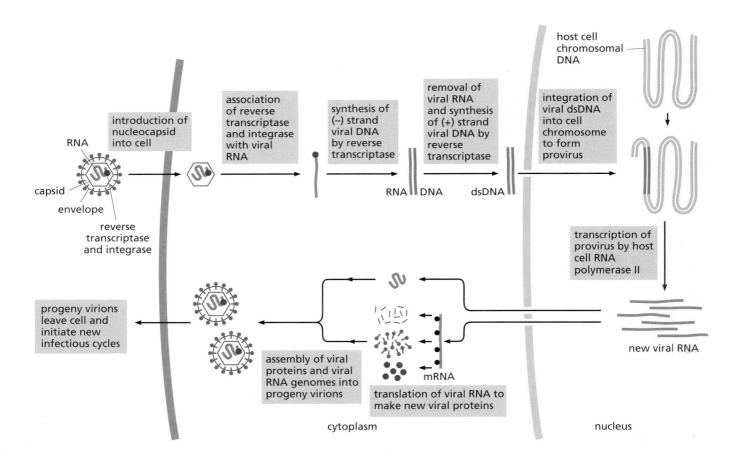

Figure 3.18 The life cycle of an RNA tumor virus like RSV An infecting virion introduces its single-stranded (ss) RNA genome (*blue*) into the cytoplasm of a cell together with the reverse transcriptase enzyme (RT, *purple*). The RT makes a single-stranded DNA using the viral RNA as template, then a double-stranded DNA copy (*red*) of the viral RNA. The reverse-transcribed DNA then moves into the nucleus, where it becomes integrated into the cellular chromosomal DNA (*orange*). The resulting integrated provirus is then transcribed by host-cell RNA polymerase II into progeny viral RNA molecules (*blue*). The progeny RNA molecules are exported to the cytoplasm, where they either serve as mRNAs to make viral proteins or are packaged into progeny virus particles that leave the cell and initiate a new round of infection. (Adapted from B. Alberts et al., Essential Cell Biology, 3rd ed. New York: Garland Science, 2010.)

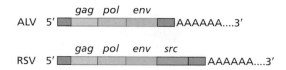

Figure 3.19 Structure of the Rous sarcoma virus genome The RNA genome of RSV is closely related to that of a relatively common infectious agent of chickens, avian leukosis virus (ALV). Both genomes include the *gag* gene, which encodes the proteins that, together with the viral RNA, form each virion's nucleoprotein core (see Figure 3.4); the *pol* gene, which encodes the reverse transcriptase and integrase; and the *env* gene, which specifies the glycoprotein spikes of the virion. The RSV genome carries, in addition, a gene (*src, brown*) that specifies the Src protein, which causes cell transformation. Not indicated here is the specialized cap structure at the 5′ end of these RNA molecules.

and polyomavirus, retroviruses rely on integration of their genomes into the chromosome to ensure the stable retention and transmission of their genomes.

There is, however, an important distinction between the integration mechanisms used by retroviruses like RSV (see Figure 3.18) and the DNA tumor viruses such as SV40 and polyomavirus (see Figure 3.16). Integration is a normal, essential part of the replication cycle of retroviruses. Indeed, retrovirus genomes carry the gene specifying an **integrase**. This enzyme is encoded by the viral *pol* gene (**Figure 3.19**) and is synthesized together with the reverse transcriptase as part of a larger polyprotein. Integrase is dedicated to mediating the process of chromosomal integration. Accordingly, during each cycle of infection and replication, the retroviral genome—its provirus—is integrated into the host chromosomal DNA through a precise, ordered process that ensures the presence of the entire viral genome and thus the retention of all viral genes. This contrasts with the behavior of DNA tumor viruses: chromosomal integration of their genomes is a very rare accident (<<1 per 1000 infections) that enables the perpetuation of viral genomes in the descendants of an initially infected cell; the rare SV40 genomes that do succeed in becoming established in chromosomal DNA are found integrated in a haphazard fashion that often includes only fragments of the wild-type genome

3.8 A version of the *src* gene carried by RSV is also present in uninfected cells

Because the genomes of retroviruses, like those of papovaviruses, were found to be quite small (<10 kb), it seemed likely that the coding capacity of the retroviral genomes was limited to a small number of genes, probably far fewer than ten. Using this small repertoire of genes, retroviruses nevertheless succeeded in specifying some viral proteins needed for viral genome replication, others required for the construction and assembly of progeny virus particles, and yet other proteins used to transform infected cells.

In the case of RSV, the use of mutant viruses revealed that the functions of viral replication (including reverse transcription and the construction of progeny virions) required one set of genes, while the function of viral transformation required another. Thus, some mutant versions of RSV could replicate perfectly well in infected cells, producing large numbers of progeny virus particles, yet such mutants lacked transforming function. Conversely, other mutant derivatives of RSV could transform cells but had lost the ability to replicate and make progeny virions in these transformed cells.

At least three retroviral genes were implicated in viral replication. Two of these encode structural proteins that are required for assembly of virus particles; a third specifies both the reverse transcriptase (RT) enzyme, which copies viral RNA into double-stranded DNA shortly after retrovirus particles enter into host cells, and the **integrase** enzyme, which is responsible for integrating the newly synthesized viral DNA into the host-cell genome (see Figure 3.18). A comparison of the RNA genome of RSV with the

Sidebar 3.4 The making of a *src*-specific DNA probe In order to make a *src*-specific DNA probe, a researcher in Bishop and Varmus's laboratory exploited two types of RSV strains: a wild-type RSV genome that carried all of the sequences needed for viral replication and transformation, and a mutant RSV genome that was able to replicate but had lost, by deletion, the *src* sequences required for transformation (**Figure 3.20**). Using reverse transcriptase, he made a DNA copy of the wild-type sequences, yielding single-stranded DNA molecules complementary to the viral RNA genome. He fragmented the complementary DNA (often termed cDNA) and hybridized it to the RNA genome of the RSV deletion mutant that lacked *src* sequences, creating DNA–RNA hybrid molecules. The bulk of the cDNA fragments annealed to the RNA genome of the deletion mutant, with the exception of cDNA fragments derived from the region encoding the *src* sequences. He then retrieved the ssDNA molecules that failed to form DNA–RNA hybrids (discarding the hybrids). The result was ssDNA fragments that specifically recognized sequences contained within the deleted portion of the mutant RSV genome, that is, those lying within the *src* gene.

Because the initial reverse transcription of the wild-type RSV RNA was carried out in the presence of radiolabeled deoxyribonucleoside triphosphates, the *src*-specific DNA fragments (which constituted the *src* "probe") were also labeled with radioisotope. This made it possible to discover whether DNAs of interest (such as the DNAs prepared from virus-infected or uninfected cells) also carried *src* sequences by determining whether or not the *src* probe (with its associated radioactivity) was able to hybridize to these cellular DNAs.

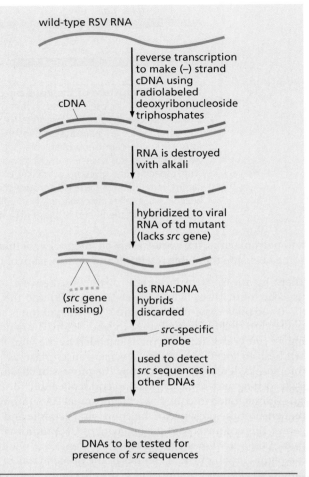

Figure 3.20 The construction of a *src*-specific DNA probe Wild-type (wt) RSV RNA (*blue*) was reverse-transcribed under conditions where only a single-stranded (ss) complementary DNA molecule (a cDNA; *red*) was synthesized. The wt single-stranded viral DNA was then annealed (hybridized) to viral RNA (*green*) of a transformation-defective (td) RSV mutant, which had lost its transforming function and apparently deleted its *src* gene. The resulting ds RNA:DNA hybrids were discarded, leaving behind the ssDNA fragment of the wtDNA that failed to hybridize to the RNA of the td mutant. This surviving DNA fragment, if radiolabeled, could then be used as a *src*-specific probe to detect *src*-related sequences in various cellular DNAs (*orange*).

genomes of related retroviruses lacking transforming ability suggested that there was rather little information in the RSV genome devoted to encoding the remaining known viral function—transformation. Consequently, geneticists speculated that all the viral transforming functions of RSV resided in a single gene, which they termed *src* (pronounced "sark"), to indicate its role in triggering the formation of sarcomas in infected chickens (see Figure 3.19).

In 1974, the laboratory run jointly by J. Michael Bishop and Harold Varmus at the University of California, San Francisco, undertook to make a DNA **probe** that specifically recognized the transformation-associated (that is, *src*) sequences of the RSV genome in order to understand its origins and functions (**Sidebar 3.4**). This *src*-specific probe was then used to follow the fate of the *src* gene after cells were infected with RSV. The notion here was that uninfected chicken cells would carry no *src*-related DNA sequences in their genomes. However, following RSV infection, *src* sequences would become readily detectable in cells, having been introduced by the infecting viral genome.

The actual outcome of this experiment was, however, totally different from expectation. In 1975, this research group, using their *src*-specific probe, found that *src* sequences were clearly present among the DNA sequences of uninfected chicken cells. These *src*

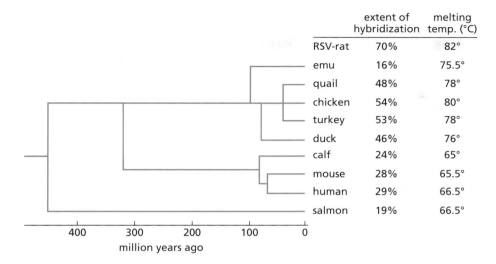

	extent of hybridization	melting temp. (°C)
RSV-rat	70%	82°
emu	16%	75.5°
quail	48%	78°
chicken	54%	80°
turkey	53%	78°
duck	46%	76°
calf	24%	65°
mouse	28%	65.5°
human	29%	66.5°
salmon	19%	66.5°

Figure 3.21 Evolutionary tree of the *src* gene The *src*-specific cDNA probe (see Figure 3.20) was annealed with the DNAs of a variety of vertebrate species—whose relatedness to the chicken is indicated on this evolutionary tree—as well as to a rat cell that had recently been infected and transformed by Rous sarcoma virus (RSV). The abscissa indicates the time when the last common ancestor shared by the chicken and each other vertebrate species lived. The best hybridization of the RSV *src* probe was with the DNA of an RSV-transformed rat cell (RSV-rat) carrying an RSV provirus (used as a positive control), as indicated by the percentage of the probe radioactivity that annealed to this cellular DNA and the higher melting temperature of the resulting hybrid. (Melting temperature refers to the temperature at which two nucleic acid strands separate from one another, that is, the temperature at which a double-helix is denatured. The higher the melting temperature, the greater the sequence complementarity and thus base-pairing between two annealing DNA strands; hence, genes borne by distantly related species, because of sequence divergence accumulated during evolution, should hybridize only imperfectly with the chicken-derived *src* probe and resulting hybrids should be disrupted at relatively low temperature.) The results for chicken DNA indicate that the sequence of the *src* gene of RSV deviates slightly from the related gene in the chicken genome. The *src* proto-oncogenes of species that were further removed evolutionarily from chickens showed lesser extents of hybridization. Subsequent work, not shown here, revealed the presence of *src* sequences in the DNA of organisms belonging to other metazoan phyla, such as arthropods (for example, *Drosophila melanogaster*) and even a sponge, whose ancestors diverged from those of vertebrates and other metazoa ~600 million years ago. The extensive evolutionary conservation of the *src* gene indicates that it played a vital role in the lives of all the organisms that carried it in their genomes. (Courtesy of H.E. Varmus and adapted from D.H. Spector, H.E. Varmus and J.M. Bishop, *Proc. Natl. Acad. Sci. USA* 75:4102–4106, 1978.)

sequences were present as single-copy cellular genes; that is, two copies of the *src*-related DNA sequences were present per diploid chicken cell genome—precisely the representation of the great majority of genes in the cellular genome.

The presence of *src* sequences in the chicken cell genome could not be dismissed as some artifact of the hybridization procedure used to detect them. Moreover, careful characterization of these *src* sequences made it unlikely that they had been inserted into the chicken genome by some retrovirus. For example, *src*-related DNA sequences were readily detectable in the genomes of several related bird species, and, more distantly on the evolutionary tree, in the DNAs of several mammals (**Figure 3.21**). The more distant the evolutionary relatedness of a species was to chickens, the weaker was the reactivity of the *src* probe with its DNA. This was precisely the behavior expected of a cellular gene that had been present in the genome of a common ancestral species and had acquired increasingly divergent DNA sequences as descendant species evolved progressively away from one another over the course of millions of years.

The evidence converged on the idea that the *src* sequences present in the genome of an uninfected chicken cell possessed all the properties of a normal cellular gene, being present in a single copy per haploid genome, evolving slowly over tens of millions of years, and being present in species that were ancestral to all modern vertebrates. This realization created a revolution in thinking about the origins of cancer.

3.9 RSV exploits a kidnapped cellular gene to transform cells

The presence of a highly conserved *src* gene in the genome of a normal organism implied that this cellular version of *src*, sometimes termed c-*src* (that is, cell *src*), played some role in the life of this organism (the chicken) and its cells. How could this role be reconciled with the presence of a transforming *src* gene carried in the genome of RSV? This viral transforming gene (v-*src*) was closely related to the c-*src* gene of the chicken, yet the two genes had drastically different effects and apparent functions. When ensconced in the cellular genome, the actions of c-*src* were apparently compatible with normal cellular behavior and normal organismic development. In contrast, the very similar v-*src* gene borne by the RSV genome acted as a potent **oncogene**—a gene capable of transforming a normal chicken cell into a tumor cell.

One solution to this puzzle came from considering the possibility that perhaps the *src* gene of RSV was not naturally present in the retrovirus ancestral to RSV. This hypothetical viral ancestor, while lacking *src* sequences, would be perfectly capable of replicating in chicken cells. In fact, such a *src*-negative retrovirus—avian leukosis virus (ALV)—was common in chickens and was capable of infectious spread from one chicken to another. This suggested that during the course of infecting a chicken cell, an ancestral virus, similar to this common chicken virus, somehow acquired sequences from the genome of an infected cell (**Figure 3.22**), doing so through some genetic trick. The acquired cellular sequences (e.g., the *src* sequences) were then incorporated into the viral genome, thereby adding a fourth gene to the existing three genes that this retrovirus used for its replication in infected cells (see Figure 3.19). Once present in the genome of RSV, the kidnapped *src* gene could then be altered and exploited by this virus to transform subsequently infected cells.

This scheme attributed great cleverness to retroviruses by implying that they had the ability to pick up and exploit preexisting cellular genes for their own purposes. Such behavior is most unusual for a virus, since virtually all other types of viruses carry genes that have little if any relatedness to DNA sequences native to the cells that they infect (**Sidebar 3.5**).

But there was an even more important lesson to be learned here, this one concerning the c-*src* gene. This cellular gene, one among tens of thousands in the chicken cell genome, could be converted into a potent viral oncogene following some slight remodeling by a retrovirus such as RSV. Because it was a precursor to an active oncogene, c-*src* was called a **proto-oncogene**. The very concept of a proto-oncogene was revolutionary: it implied that the genomes of normal vertebrate cells carry a gene that has the potential, under certain circumstances, to induce cell transformation and thus cancer.

The structures of the c-*src* proto-oncogene and the v-*src* oncogene were worked out rapidly in the years that followed these discoveries in 1975 and 1976. Just as the viral geneticists had speculated, all of the viral transforming sequences resided in a single viral oncogene. Within the RSV RNA genome, the v-*src* gene was found at the 3' end of the genome, added to the three preexisting retroviral genes that were involved in viral replication (see Figure 3.19).

This scenario of acquisition and activation of c-*src* by a retrovirus led to three further ideas. First, if retroviruses could activate this proto-oncogene into a potent oncogene, perhaps other types of mutational mechanisms might reshape the normal c-*src* gene and yield a similar outcome. Maybe such mechanisms could activate a cellular proto-oncogene without removing the normal gene from its regular roosting site on the cellular chromosome. Maybe the information for inducing cancer was already present in the normal cell genome, waiting to be unmasked.

Second, it became clear that all of the transforming powers of RSV derived from the presence of a single gene—v-*src*—in its genome. This was of great importance, because it implied that a single oncogene could, as long suspected, elicit a large number of

Figure 3.22 Capture of *src* by avian leukosis virus The precise mechanism by which an avian leukosis virus (ALV) captured a cellular *src* gene (c-*src*) is not known. According to the most accepted model, ALV proviral DNA (*red*) became integrated (by chance) next to a c-*src* proto-oncogene (*green*) in an infected chicken cell. The ALV provirus and adjacent c-*src* gene were then co-transcribed into a single hybrid RNA transcript (*blue and brown*). After splicing out c-*src* introns (not shown), this hybrid viral RNA was packaged into a virus particle that became the ancestor of Rous sarcoma virus (RSV). (Not shown is the attachment of ALV sequences at the 3' side of v-*src*, with the result that this viral oncogene became flanked on both sides by ALV sequences.)

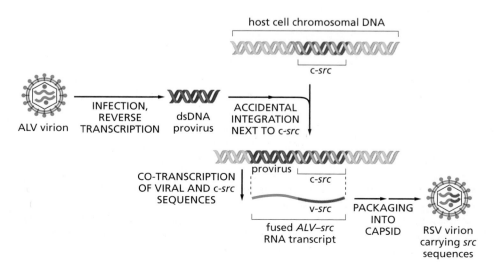

Sidebar 3.5 Where do viruses and their genes originate? The discovery by Varmus and Bishop's laboratory of the cellular origins of the RSV *src* gene provoked the question of where most viral genes come from and how viruses originate. Viral genome replication—in both DNA and RNA virus replication cycles—is executed with far less fidelity than is the replication of cellular DNA. The consequence is that viral genome sequences evolve far more rapidly than the genomic sequences of metazoan cells. Hence, any traces of precursor gene sequences that may have been present in other viruses or in cells were erased by this rapid evolution hundreds of millions of years ago.

We will probably never know where and how most viral genomes originated. Presumably, viruses have been around since the cells that they parasitize first appeared. Some types of viruses may have begun as renegade genes that broke away from cellular genomes and struck out on their own. RNA viruses may bear vestiges of early cellular life forms that used RNA molecules as genomes and, in the case of retroviruses, of a later stage of cellular evolution when a reverse transcriptase-like cellular enzyme enabled the transition from RNA to DNA genomes.

Whatever their origins, the great majority of viral genes have no obvious relatedness with the genes of their hosts. This highlights the uniqueness of the Varmus and Bishop finding that retroviruses can apparently acquire and **transduce** (carry in their genomes) cellular genes, seemingly promiscuously. This flexibility of their genomes suggested a novel application of retroviruses, which began to flourish in the mid-1980s: a variety of interesting genes were introduced into retroviral genomes using recombinant DNA techniques. The resulting viral genomes were then used as **vectors** to transduce these genes into cultured cells *in vitro* and into living tissues *in vivo* (the latter application often being referred to as "gene therapy"; Supplementary Sidebar 3.3).

changes in the shape, metabolism, and growth behavior of a cell. More generally, this suggested that other cancer-causing genes could also act pleiotropically. Accordingly, if a transformed, tumorigenic cell differed from a normal cell in 20 or 30 distinct traits, perhaps these multiple changes were not dependent on the alteration of 20 or 30 different genes; instead, maybe a small number of genes would suffice to transform a normal into a tumorigenic cell.

Third, RSV and its v-*src* oncogene might represent a model for the behavior of other types of retroviruses that were similarly capable of transforming infected cells *in vitro* and inducing tumors *in vivo*. Perhaps such retroviruses had acquired other cellular genes unrelated to *src*. While c-*src* was certainly the first cellular proto-oncogene to be discovered, maybe other cellular proto-oncogenes were hiding in the vertebrate cellular genome, waiting to be picked up and activated by some passing retrovirus.

3.10 The vertebrate genome carries a large group of proto-oncogenes

An accident of history—an encounter in 1909 between a Long Island chicken farmer and Peyton Rous—made RSV the first tumorigenic retrovirus to be isolated and characterized in detail. Consequently, RSV was favored initially with the most detailed molecular and genetic analysis. However, in the 1950s and 1960s, a group of other chicken and rodent tumor viruses were found that were subsequently realized to be members, like RSV, of the retrovirus class.

The diversity of these other transforming retroviruses and the diseases that they caused suggested that they, like RSV, might be carrying kidnapped cellular proto-oncogenes and might be using these acquired genes to transform infected cells. So, within a year of the discovery of v-*src* and c-*src*, the race began to find additional viruses that had traveled down a similar genetic path and picked up other, potentially interesting proto-oncogenes.

Another chicken retrovirus—the MC29 myelocytomatosis virus—which was known to be capable of inducing a bone marrow malignancy in chickens, was one of this class. MC29 was also found to carry an acquired cellular gene in its genome, termed the v-*myc* oncogene, which this virus exploited to induce rapidly growing tumors in infected chickens. As was the case with v-*src*, the origin of the v-*myc* gene could be traced to a corresponding proto-oncogene residing in the normal chicken genome. Like *src*, *myc* underwent some remodeling after being incorporated into the retroviral genome. This remodeling imparted potent oncogenic powers to a gene that previously had played a benign and apparently essential role in the life of normal chicken cells.

Of additional interest was the discovery that MC29, like RSV, descended from avian leukosis virus (ALV). This reinforced the notion that retroviruses like ALV were adept at acquiring random pieces of a cell's genome. The biological powers of the resulting hybrid viruses would presumably depend on which particular cellular genes had been picked up. In the case of the large majority of acquired cellular genes, hybrid viruses carrying such genes would show no obvious phenotypes such as tumor-inducing potential. Only on rare occasion, when a growth-promoting cellular gene—a proto-oncogene—was acquired, might the hybrid virus exhibit a cancer-inducing phenotype that could lead to its discovery and eventual isolation by a virologist.

Mammals were also found to harbor retroviruses that are distantly related to ALV and, like ALV, are capable of acquiring cellular proto-oncogenes and converting them into potent oncogenes. Among these is the feline leukemia virus, which acquired the *fes* oncogene in its genome, yielding feline sarcoma virus, and a hybrid rat–mouse leukemia virus, which on separate occasions acquired two distinct proto-oncogenes: the resulting transforming retroviruses, Harvey and Kirsten sarcoma viruses, carry the H-*ras* and K-*ras* oncogenes, respectively, in their genomes. Within a decade, the repertoire of retrovirus-associated oncogenes had increased to more than two dozen, many named after the viruses in which they were originally discovered (Table 3.3). By now, more than thirty distinct vertebrate proto-oncogenes have been discovered.

In each case, a proto-oncogene found in the DNA of a mammalian or avian species was readily detectable in the genomes of all other vertebrates. There were, for example, chicken, mouse, and human versions of c-*myc*, and these genes seemed to function identically in their respective hosts. The same could be said of all the other proto-oncogenes that were uncovered. Soon it became clear that this large repertoire of proto-oncogenes must have been present in the genome of the vertebrate that was the common ancestor of all mammals and birds, and that this group of genes, like most others in the vertebrate genome, was inherited by all of the modern descendant species.

Now, a third of a century later, we realize that these transforming retroviruses provided cancer researchers with a convenient window through which to view the cellular genome and its cohort of proto-oncogenes. Without these retroviruses, the discovery of proto-oncogenes would have been exceedingly difficult. By fishing these genes out of the cellular genome and revealing their latent powers, these viruses catapulted cancer research forward by decades.

3.11 Slowly transforming retroviruses activate proto-oncogenes by inserting their genomes adjacent to these cellular genes

As described above, each of the various tumorigenic retroviruses arose when a nontransforming retrovirus, such as avian leukosis virus (ALV) or murine leukemia virus (MLV), acquired a proto-oncogene from the genome of an infected host cell. In fact, the "nontransforming" precursor viruses could also induce cancers, but they were able to do so only on a much more extended timetable; often months passed before these viruses succeeded in producing cancers. The oncogene-bearing retroviruses, in contrast, often induced tumors within days or weeks after they were injected into host animals.

When the rapidly transforming retroviruses were used to infect cells in culture, the cells usually responded by undergoing the changes in morphology and growth behavior that typify the behavior of cancer cells (see Table 3.2). In contrast, when the slowly tumorigenic, nontransforming viruses, such as ALV and MLV, infected cells, these cells released progeny virus particles but did not show any apparent changes in shape or growth behavior. This lack of change in cell phenotypes was consistent with the fact that these viruses lacked oncogenes in their genomes.

These facts, when taken together, represented a major puzzle. How could viruses like MLV or ALV induce a malignancy if they carried no oncogenes?

Table 3.3 Acutely transforming retroviruses and the oncogenes that they have acquired[a]

Name of virus	Viral oncogene	Species	Major disease	Nature of oncoprotein
Rous sarcoma	src	chicken	sarcoma	non-receptor TK
Y73/Esh sarcoma	yes	chicken	sarcoma	non-receptor TK
Fujinami sarcoma	fps[b]	chicken	sarcoma	non-receptor TK
UR2	ros	chicken	sarcoma	RTK; unknown ligand
Myelocytomatosis 29	myc	chicken	myeloid leukemia[c]	transcription factor
Mill Hill virus 2	mil[d]	chicken	myeloid leukemia	ser/thr kinase
Avian myeloblastosis E26	myb	chicken	myeloid leukemia	transcription factor
Avian myeloblastosis E26	ets	chicken	myeloid leukemia	transcription factor
Avian erythroblastosis ES4	erbA	chicken	erythroleukemia	thyroid hormone receptor
Avian erythroblastosis ES4	erbB	chicken	erythroleukemia	EGF RTK
3611 murine sarcoma	raf[e]	mouse	sarcoma	ser/thr kinase
SKV770	ski	chicken	endothelioma (?)	transcription factor
Reticuloendotheliosis	rel	turkey	immature B-cell lymphoma	transcription factor
Abelson murine leukemia	abl	mouse	pre-B-cell lymphoma	non-receptor TK
Moloney murine sarcoma	mos	mouse	sarcoma, erythroleukemia	ser/thr kinase
Harvey murine sarcoma	H-ras	rat, mouse	sarcoma	small G protein
Kirsten murine sarcoma	K-ras	mouse	sarcoma	small G protein
FBJ murine sarcoma	fos	mouse	osteosarcoma	transcription factor
Snyder–Theilen feline sarcoma	fes[f]	cat	sarcoma	non-receptor TK
McDonough feline sarcoma	fms	cat	sarcoma	CSF-1 RTK
Gardner–Rasheed feline sarcoma	fgr	cat	sarcoma	non-receptor TK
Hardy–Zuckerman feline sarcoma	kit	cat	sarcoma	steel factor RTK
Simian sarcoma	sis	woolly monkey	sarcoma	PDGF
AKT8	akt	mouse	lymphoma	ser/thr kinase
Avian virus S13	sea	chicken	erythroblastic leukemia[g]	RTK; unknown ligand
Myeloproliferative leukemia	mpl	mouse	myeloproliferation	TPO receptor
Regional Poultry Lab v. 30	eyk	chicken	sarcoma	RTK; unknown ligand
Avian sarcoma virus CT10	crk	chicken	sarcoma	SH2/SH3 adaptor
Avian sarcoma virus 17	jun	chicken	sarcoma	transcription factor
Avian sarcoma virus 31	qin	chicken	sarcoma	transcription factor[h]
AS42 sarcoma virus	maf	chicken	sarcoma	transcription factor
Cas NS-1 virus	cbl	mouse	lymphoma	SH2-dependent ubiquitylation factor

[a]Not all viruses that have yielded these oncogenes are indicated here. "Species" denotes the animal species from which the virus was initially isolated.
[b]Ortholog of the mammalian fes oncogene.
[c]Also causes carcinomas and endotheliomas.
[d]Ortholog of the mammalian raf oncogene.
[e]Ortholog of the avian mil oncogene.
[f]Ortholog of the avian fps oncogene.
[g]Also causes granulocytic leukemias and sarcomas.
[h]Functions as a transcriptional repressor.

Abbreviations: CSF, colony-stimulating factor; EGF, epidermal growth factor; G, GTP-binding; PDGF, platelet-derived growth factor; RTK, receptor tyrosine kinase; ser/thr, serine/threonine; SH, src-homology segment; TK, tyrosine kinase; TPO, thrombopoietin.

Adapted in part from S.J. Flint, L.W. Enquist, R.M. Krug et al., Principles of Virology. Washington, DC: ASM Press, 2000. Also in part from G.M. Cooper, Oncogenes. Boston: Jones and Bartlett Publishers, 1995.

The solution came in 1981 from study of the leukemias that ALV induced in chickens, more specifically, from detailed analysis of the genomic DNAs of the leukemic cells. These cells invariably carried copies of the ALV provirus integrated into their genomes. By the time these experiments began, a decade after Temin and Baltimore's discovery, it had become clear that the integration of proviruses occurs at random sites throughout the chromosomal DNA of infected host cells. Given the size of the chicken genome, there might be many millions of distinct chromosomal sites used by ALV to integrate its provirus; moreover, each infected leukemia cell might carry one or several proviruses in its genome. Knowing all this, researchers nevertheless undertook to map the sites in the leukemia cell chromosomal DNA where ALV proviruses had integrated.

This molecular analysis of a series of ALV-induced leukemias, each arising independently in a separate chicken, revealed a picture very different from the expected collection of random integration sites. In a majority (>80%) of these leukemia cell genomes, the ALV provirus was found to be integrated into the chromosomal DNA immediately adjacent to the c-*myc* proto-oncogene (**Figure 3.23**)! This observation, on its own, was difficult to reconcile with the notion that provirus integration occurs randomly at millions of sites throughout the genomes of infected cells.

It soon became clear that the close physical association of an integrated viral genome and the cellular *myc* gene led to a functional link between these two genetic elements. The viral transcriptional promoter, nested within the ALV provirus, disrupted the

Figure 3.23 Insertional mutagenesis The oncogenic actions of viruses, such as avian leukosis virus (ALV), that lack acquired oncogenes could be explained by the integration of their proviral DNA adjacent to a cellular proto-oncogene. (A) Analysis of numerous B-cell lymphomas that were induced by ALV revealed that a large proportion of the ALV proviruses were integrated (*filled triangles*) into the DNA segment carrying the c-*myc* proto-oncogene, usually in the same transcriptional orientation as that of the c-*myc* gene; the majority were integrated between the first noncoding exon of c-*myc* and the second exon, in which the *myc* reading frame begins. (B) ALV's oncogenic behavior could be rationalized as follows. In the course of ALV infection of chicken lymphocytes, ALV proviruses (*green*) become integrated randomly at millions of different sites in the chromosomal DNA of these cells. (Chromosomal DNA maps of only four are illustrated schematically here.) On rare occasions, an ALV provirus becomes integrated (by chance) within the c-*myc* proto-oncogene (*red*). This may then cause transcription of the c-*myc* gene to be driven by the strong, constitutively acting ALV promoter. Because high levels of the Myc protein are potent in driving cell proliferation, the cell carrying this particular integrated provirus and activated c-*myc* gene will now multiply uncontrollably, eventually spawning a large host of descendants that will constitute a lymphoma. (Adapted from S.J. Flint, L.W. Enquist, R.M. Krug et al., Principles of Virology. Washington, DC: ASM Press, 2000.)

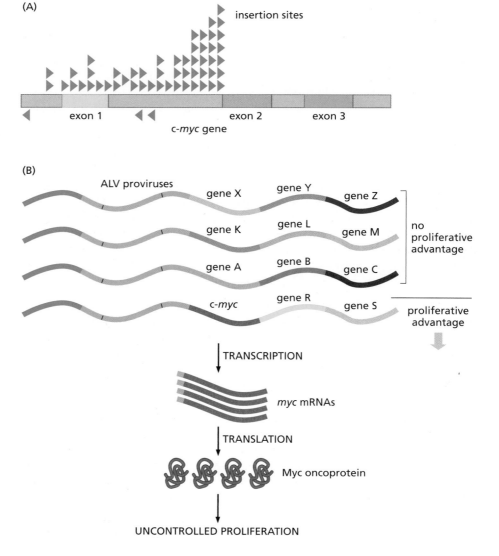

control mechanisms that normally govern expression of the c-*myc* gene (see Figure 3.23A). Now, instead of being regulated by its own native gene promoter, the cellular *myc* gene was placed directly under viral transcriptional control. As a consequence, rather than being regulated up and down by the finely tuned control circuitry of the cell, c-*myc* expression was taken over by a foreign usurper that drove its expression unceasingly and at a high rate. In essence, this hybrid viral–cellular gene arising in the chromosomes of leukemic cells now functioned much like the v-*myc* oncogene carried by avian myelocytomatosis virus.

Suddenly, all the clues needed to solve the puzzle of **leukemogenesis** (leukemia formation) by ALV fell into place. The solution went like this. During the course of infecting a chicken, ALV spread to thousands, then millions of cells in the hematopoietic system of this bird. Soon, the infection was so successful that the bird would become **viremic**, that is, its bloodstream carried high concentrations of virus particles. Each of these tens of millions of infections resulted in the insertion of an ALV provirus at some random location in the genome of an infected cell. In the vast majority of cases, this provirus integration had no effect on the infected host cell, aside from forcing the host to produce large numbers of progeny virus particles. But on rare occasions, perhaps in 1 out of 10 million infections, a provirus became integrated by chance next to the c-*myc* gene (Figure 3.23B). This jackpot event led to an explosive outcome—conversion of the c-*myc* gene into a potent oncogene whose unceasing expression was now driven by the adjacently integrated provirus and its transcriptional promoter. The rare cell carrying this deregulated *myc* gene then began uncontrolled proliferation, and within weeks, some of the progeny cells evolved further into more aggressive cancer cells that constituted a leukemia.

This scenario explains the slow kinetics with which these leukemias arise after initial viral infection of a bird. Since activation of the c-*myc* gene through provirus integration is a low-probability event, many weeks and many millions of infectious events are required before these malignancies are triggered. This particular mechanism of proto-oncogene activation came to be called **insertional mutagenesis**; it explains, as well, the leukemogenic powers of other slowly acting retroviruses, such as MLV. By now, study of avian and murine retrovirus-induced infections has demonstrated integration events next to more than 25 distinct cellular proto-oncogenes. Indeed, insertional mutagenesis can be used as a powerful strategy to find new proto-oncogenes (**Sidebar 3.6**).

Unvoiced by those who uncovered insertional mutagenesis was another provocative idea: maybe it was possible that nonviral carcinogens could achieve the same end result as ALV did. Perhaps these other carcinogens, including X-rays and mutagenic chemicals, could alter cellular proto-oncogenes while these genes resided in their normal sites in cellular chromosomes. The result might be a disruption of cellular growth control that was just as destabilizing as the events that led to ALV-induced leukemias.

3.12 Some retroviruses naturally carry oncogenes

The descriptions of retroviruses in this chapter indicate that they essentially fall into two classes. Some, such as ALV and MLV, carry no oncogenes but can induce tumors that erupt only after a long latent period (that is, many weeks) following the initial infection of a host animal. Other viruses, such as RSV, can induce cancer rapidly (that is, in days or several weeks), having acquired an oncogene from a cellular proto-oncogene precursor.

In reality, there is a third class of retroviruses that conforms to neither of these patterns. Human T-cell leukemia virus (HTLV-I) infects about 1% of the inhabitants of Kyushu, the south island of Japan. An endemic infection is also present, albeit at a lower rate, in some islands of the Caribbean. Lifelong HTLV-I infection carries a 3–4% risk of developing adult T-cell leukemia, and the virus seems to be maintained in the population via milk-borne, mother-to-infant transmission.

There are no indications, in spite of extensive molecular surveys, that HTLV-I provirus integration sites are clustered in certain chromosomal regions. Accordingly, it

Sidebar 3.6 Insertional mutagenesis uncovers novel proto-oncogenes As described earlier (see Section 3.10), the analysis of the genomes of rapidly transforming retroviruses enabled investigators to identify a large cohort of proto-oncogenes. The phenomenon of insertional mutagenesis, first discovered through the insertion of an ALV genome adjacent to the c-*myc* proto-oncogene, offered an alternative strategy for discovering these cellular genes. Thus, a researcher could study a series of independently arising tumors, all of which had been induced by a retrovirus, such as ALV or MLV, that was known to lack its own oncogene. More specifically, this researcher could analyze the host-cell sequences that lay immediately adjacent to the integrated proviruses in the chromosomal DNA of tumor cells. The hope was that the proviruses might be found to be integrated repeatedly next to a (possibly still-unknown) cellular gene whose activation was triggered by the transcriptional promoter of the provirus. The adjacent gene could be readily cloned, since it was effectively tagged by the closely linked proviral DNA.

The initial fruits of this strategy came from studying the breast cancers induced by mouse mammary tumor virus (MMTV),

another retrovirus that lacked an oncogene in its genome. Researchers mapped the integration sites of MMTV proviruses in the genomes of mouse breast cancers that had been induced by this virus. Most of the proviruses were found to be integrated in one of three chromosomal locations, clustering next to cellular genes that were then called *int-1*, *int-2*, and *int-3* (**Table 3.4**). Each of these genes was later discovered to encode a protein involved in stimulating cell proliferation in one way or another. The deregulated expression of each of these genes, resulting from nearby MMTV provirus integration, seemed to be responsible for triggering the cell proliferation that led to the appearance of mammary tumors.

The *int-1* gene, which was found to be homologous to the *wingless* gene of *Drosophila*, was renamed *Wnt-1*, and was the forerunner of a whole series of *Wnt* genes that have proven to be important vertebrate mitogens and **morphogens**, that is, factors controlling morphogenesis. More recently, this search strategy has uncovered a large group of other cellular genes, each of which, when activated by insertional mutagenesis mediated by MLV, triggers leukemias in mice (see Table 3.4).

Table 3.4 Examples of cellular genes found to be activated by insertional mutagenesis

Gene	Insertional mutagen	Tumor type	Species	Type of oncoprotein
myc	ALV	B-cell lymphoma	chicken	transcription factor
myc	ALV, FeLV	T-cell lymphoma	chicken, cat	transcription factor
nov	ALV	nephroblastoma	chicken	growth factor
erbB	ALV	erythroblastosis	chicken	receptor TK
mos	IAP	plasmacytoma	mouse	ser/thr kinase
int-1[a]	MMTV	mammary carcinoma	mouse	growth factor
int-2[b]	MMTV	mammary carcinoma	mouse	growth factor
int-3	MMTV	mammary carcinoma	mouse	receptor[c]
int-H/int-5	MMTV	mammary carcinoma	mouse	enzyme[d]
pim-1	Mo-MLV	T-cell lymphoma	mouse	ser/thr kinase
pim-2	Mo-MLV	B-cell lymphoma	mouse	ser/thr kinase
bmi-1	Mo-MLV	T-cell lymphoma	mouse	transcription repressor
tpl-2	Mo-MLV	T-cell lymphoma	mouse	ser/thr kinase
lck	Mo-MLV	T-cell lymphoma	mouse	non-receptor TK
p53	Mo-MLV	T-cell lymphoma	mouse	transcription factor
GM-CSF	IAP	myelomonocytic leukemia	mouse	growth factor
IL2	GaLV	T-cell lymphoma	gibbon ape	cytokine[e]
IL3	IAP	T-cell lymphoma	mouse	cytokine
K-ras	F-MLV	T-cell lymphoma	mouse	small G protein
CycD1	F-MLV	T-cell lymphoma	mouse	G1 cyclin
CycD2	Mo-MLV	T-cell lymphoma	mouse	G1 cyclin

Abbreviations: ALV, avian leukosis virus; FeLV, feline leukemia virus; F-MLV, Friend murine leukemia virus; GaLV, gibbon ape leukemia virus; GF, growth factor; IAP, intracisternal A particle (a retrovirus-like genome that is endogenous to cells); Mo-MLV, Moloney murine leukemia virus; MMTV, mouse mammary tumor virus; ser/thr, serine/threonine; TK, tyrosine kinase.

[a]Subsequently renamed *Wnt-1*.
[b]Subsequently identified as a gene encoding a fibroblast growth factor (FGF).
[c]Related to Notch receptors.
[d]Enzyme that converts androgens to estrogens.
[e]Cytokines are GFs that largely regulate various types of hematopoietic cells.

Adapted in part from J. Butel, *Carcinogenesis* 21:405–426, 2000; and from N. Rosenberg and P. Jolicoeur, in J.M. Coffin, S.H. Hughes and H.E. Varmus (eds.), Retroviruses. Cold Spring Harbor, NY: Cold Spring Harbor Laboratory Press, 1997. Also in part from G.M. Cooper, Oncogenes, 2nd ed. Boston: Jones and Bartlett Publishers, 1995.

appears highly unlikely that HTLV-I uses insertional mutagenesis to incite leukemias. Instead, its leukemogenic powers seem to be traceable to one or more viral proteins that are naturally encoded by the viral genome. The best understood of these is the viral *tax* gene, whose product is responsible for activating transcription of proviral DNA sequences, thereby enabling production of progeny RNA genomes. At the same time, the *tax* gene product appears to activate transcription of two cellular genes that specify important growth-stimulating proteins—IL-2 (interleukin-2) and GM-CSF (granulocyte macrophage colony-stimulating factor). These "growth factors," to which we will return in Chapter 5, are released by virus-infected cells and proceed to stimulate the proliferation of several types of hematopoietic cells. While such induced proliferation, on its own, does not directly create a leukemia, it seems that populations of these HTLV-I–stimulated cells may progress at a low but predictable frequency to spawn variants that are indeed neoplastic. In this instance, the expression of certain viral oncogenes, notably *tax*, appears to be an intrinsic and essential component of a retroviral replication cycle within host animals, rather than the consequence of rare genetic accidents that yield unusual hybrid genomes, such as the genome of RSV.

3.13 Synopsis and prospects

By studying tumors in laboratory and domesticated animals, cancer biologists discovered a wide array of cancer-causing viruses during the twentieth century. Many of these viruses, having either DNA or RNA genomes, were found able to infect cultured cells and transform them into tumorigenic cells. These transforming powers pointed to the presence of powerful oncogenes in the genomes of the viruses, indeed, oncogenes that were potent enough to induce many of the phenotypes associated with cancer cells (see Table 3.2). Moreover, the ability of these viruses to create transformed cells in the culture dish shed light on the mechanisms by which such viruses could induce cancers in the tissues of infected host animals.

A major conceptual revolution came from the detailed study of RNA tumor viruses, specifically, Rous sarcoma virus (RSV). Its oncogene, termed v-*src*, was found to have originated in a normal cellular gene, c-*src*. This discovery revealed the ability of nontransforming, slowly tumorigenic retroviruses, such as ALV (avian leukosis virus), to acquire normal cellular genes and convert these captured genes into potently transforming oncogenes. The hybrid viruses that arose following these genetic acquisitions were now able to rapidly induce tumors in infected hosts.

Even more important were the implications of finding the c-*src* gene. Its presence in uninfected cells demonstrated that the normal cellular genome carries a proto-oncogene that can be converted into an oncogene by altering its sequences. (The details of these alterations will be described in the next several chapters.) Soon a number of retroviruses of both avian and mammalian origin were discovered to carry other oncogenes that had been acquired in similar fashion from the genomes of infected cells. While each of these proto-oncogenes was found initially in the genome of one or another vertebrate species, we now know that all of these genes are represented in the genomes of all vertebrates. Consequently, the generic vertebrate genome carries dozens of such normal genes, each of which has the potential to become converted into an active oncogene.

Yet other proto-oncogenes were discovered by studying the integration sites of proviruses in the genomes of tumors that had been induced by nontransforming retroviruses, such as murine leukemia virus (MLV) and ALV. The random integration of these proviruses into chromosomal DNA occasionally yielded, through the process of insertional mutagenesis, the conversion of a proto-oncogene into an activated oncogene that could readily be isolated because of its close linkage to the provirus. On many occasions, insertional mutagenesis led to rediscovery of a proto-oncogene that was already known because of its presence in an acutely transforming retrovirus; *myc* and avian myelocytomatosis virus (AMV) exemplify this situation. On other occasions, truly novel proto-oncogenes were discovered through study of provirus integration sites; the *int-1* gene activated by MMTV provides a striking example of this route of discovery. In fact, the process of insertional mutagenesis remained little more than a laboratory curiosity, of interest to only a small cadre of cancer biologists, until it was reported to lead to tumors in several patients being treated by gene therapy (**Sidebar 3.7**).

The discoveries of proto-oncogenes and oncogenes, as profound as they were, provoked as many questions as they answered. It remained unclear how the retrovirus-encoded oncogene proteins (called **oncoproteins**) differed functionally from the closely related proteins encoded by corresponding proto-oncogenes. The biochemical mechanisms used by these oncoproteins to transform cells were also obscure.

The molecular mechanisms used by DNA tumor viruses to transform infected cells were even more elusive, since these viruses seemed to specify oncoproteins that were very different from the proteins made by their host cells. Such differences suggested that these viral oncoproteins could not insinuate themselves into the cellular growth-regulating machinery in any easy, obvious way. Only in the mid-1980s, ten years after this research began, did their transforming mechanisms become apparent, as we will see in Chapters 8 and 9.

For many cancer researchers, and for the public that supported this research, there was a single overriding issue that had motivated much of this work in the first place: did any of these viruses and the proto-oncogenes that they activated play key roles in causing human cancers? As we will learn, about one-fifth of the human cancer burden worldwide is associated with infectious agents. Hepatitis B and C viruses (HBV, HCV), as well as human papillomaviruses (HPVs), play key roles in triggering commonly occurring cancers. Indeed, even infrequently occurring human tumors that seem to be familial have been traced in recent years to viral infections (Supplementary Sidebar 3.4). So the recognized role of viruses in cancer pathogenesis is substantial and growing.

Still, even if RNA and DNA tumor viruses were not responsible for inciting a single case of human cancer, the research into their transforming mechanisms would have been justified. This research opened the curtain on the genes in our genome that play central roles in all types of human cancer. It accelerated by decades our understanding of cancer pathogenesis at the level of genes and molecules. It catapulted cancer research from a descriptive science into one where complex phenomena could finally be understood and explained in precise, mechanistic terms.

Sidebar 3.7 Gene therapy can occasionally have tragic consequences Gene therapy has been found to be most applicable to diseases of the hematopoietic system. Thus, children who are born with a severe immunodeficiency due to a germ-line–specified defect in one or another critical component of the immune system can, in principle, be cured if the missing gene is transduced into their bone marrow stem cells using retrovirus vectors (see Supplementary Sidebar 3.3). After infection *in vitro* by a gene-transducing retroviral vector, stem cells are introduced into the afflicted children, in whose bone marrow these cells become stably engrafted. The differentiated progeny of these engrafted, genetically altered stem cells are then able to supply missing immune functions, thereby reversing the congenital immunodeficiency.

Just such a therapeutic approach was launched in France. Bone marrow stem cells from ten children suffering from a congenital severe combined immunodeficiency (SCID), which involves the absence of T cells and NK cells (to be described in Chapter 15), were infected with a Moloney murine leukemia virus (MLV)–derived retroviral vector; the vector transduced a gene specifying the gene product that the children lacked—the γc protein, a component of the interleukin receptors that allow proper development and regulation of the immune system. Nine of these children responded by showing a dramatic reconstitution of their immunological function. However, in the years after the inception of this trial, four of the children developed T-cell acute lymphoblastic leukemia.

In three of the four cases, analyses of the DNA of leukemic cells revealed proviruses derived from the viral vector that had become integrated within several kilobase pairs of the first exon of the *LMO2* gene, a proto-oncogene known to be activated in human T-cell leukemias. DNA of cells from the fourth patient showed a provirus integration near another proto-oncogene, *Bmi-1*. Three of these four leukemic children were eventually cured, and the nine surviving children showed restitution of their immune functions for periods as long as a decade following original treatment.

Given the known role of the *LMO2* and *Bmi-1* oncogenes in leukemogenesis, the inserted proviruses were almost certainly responsible for triggering the four leukemias. Hence, insertional mutagenesis leading to oncogenesis, which had long been feared as a possible but remote risk incurred by such a gene therapy strategy, became a grim reality. In response to these findings, researchers have begun to develop new types of viral vectors that are less likely to activate proto-oncogenes via insertional mutagenesis, giving hope to those who are interested in correcting inborn defects through gene therapy.

Key concepts

- Decades after Peyton Rous's 1910 discovery that a virus could induce tumors in chickens, the transformation of cultured cells into tumor cells upon infection with Rous sarcoma virus (RSV) resurrected tumor virus research and led to the realization that cancer could be studied at the level of the cell.

- Transformed cells in culture have numerous unusual characteristics, including altered morphology, lack of contact inhibition, anchorage independence, proliferation in the absence of growth factors, immortalization, and tumorigenicity, the latter being the acid test for full cellular transformation.

- The transformation phenotype induced by RSV infection was found to be transmitted to progeny cells and to depend on the continued activity of an RSV gene product.

- In addition to RNA viruses like RSV, several classes of DNA viruses—including papovavirus, human adenovirus, herpesvirus, and poxvirus—were found to induce cancers in laboratory animals or humans.

- While the genomes of RNA tumor viruses consist of single-stranded RNA, the genomes of DNA tumor viruses consist of double-stranded DNA (dsDNA).

- Since replication of viral DNA genomes normally occurs independently of the host cells' DNA and since viral genomes lack the DNA segments (centromeres) to properly segregate during mitosis, the transmission of DNA tumor virus genomes from one cell generation to the next posed a conceptual problem, until it was discovered that DNA tumor virus genomes integrate into host-cell chromosomal DNA.

- Since the genomes of RNA tumor viruses consist of single-stranded RNA that cannot be incorporated into host DNA, and since re-infection does not explain the persistence of the transformed state in descendant cells, Howard Temin postulated that RNA viruses make double-stranded DNA copies of their genomes—the process of reverse transcription—and that these DNA copies are integrated into the host's chromosomal DNA as a part of the normal viral replication cycle. This is a major distinction from DNA tumor viruses, for which integration is a very rare, haphazard event and not an integral part of viral replication.

- Because their replication cycle depends on information flowing backward (that is, from RNA to DNA), RNA viruses came to be called retroviruses and the DNA version of their viral genomes was called a provirus.

- Working with RSV, researchers found that viral replication and cell transformation were specified by separate genes, with the transforming function residing in a single gene called *src*.

- Use of a DNA probe that specifically recognized the transformation-associated (that is, *src*) sequences of the RSV genome led to the unexpected discovery that *src*-related sequences were present in the DNA of uninfected chicken cells. Further research indicated that the *src* gene was a normal, highly conserved gene of all vertebrate species (as later proved true of many other such genes).

- The difference between the actions of the cellular version of *src* (c-*src*), which supports normal cell function, and the viral version (v-*src*), which acts as an oncogene, can be explained if v-*src* were an altered version of a c-*src* that had originally been kidnapped from a cellular genome by an ancestor of RSV.

- Because it can serve as a precursor to an oncogene, c-*src* was called a proto-oncogene, a term denoting that normal vertebrate cells contain genes that have the intrinsic potential to become converted into oncogenes that can induce cancer.

- The actions of v-*src* indicated that a single oncogene could act pleiotropically to evoke a multiplicity of changes in cellular traits. In addition, the discovery of c-*src* suggested the possibility that other mutational mechanisms might activate proto-oncogenes that continued to reside in their normal sites in cellular chromosomes.

In contrast to RSV, nontransforming, slowly tumorigenic retroviruses lack oncogenes and work by integrating their genomes adjacent to proto-oncogenes in cellular chromosomes, a process called insertional mutagenesis. This chance occurrence places the proto-oncogene under the control of the viral transcriptional promoter, which deregulates the gene's expression and leads to uncontrolled cell proliferation. Insertional mutagenesis can be exploited to find new proto-oncogenes.

• The formation of most oncogene-bearing, rapidly transforming retroviruses involves the replacement of viral replicative genes by acquired cellular oncogenic sequences; this results in the formation of a potently transforming virus that cannot, however, proliferate on its own because of the loss of viral genes that are essential for replication (Supplementary Sidebar 3.5).

• In addition to the nontransforming retroviruses (which work via insertional mutagenesis) and the acutely transforming ones (which work via acquired oncogenes), retroviruses exist whose carcinogenic powers are traceable to their own normal gene products. Needed for viral replication, these viral proteins have the side effect of also activating the expression of cellular genes involved in cell growth and proliferation.

Thought questions

1. What observations favor or argue against the notion that cancer is an infectious disease?

2. How can one prove that tumor virus genomes must be present in order to maintain the transformed state of a virus-induced tumor? What genetic mechanisms, do you imagine, might enable this process to become "hit-and-run," in which the continued presence of a tumor virus is not required to maintain the tumorigenic phenotype after a certain time?

3. Why are oncogene-bearing viruses like Rous sarcoma virus so rarely encountered in wild populations of chickens?

4. What evidence suggests that the phenotypes of cells transformed by tumor viruses *in vitro* reflect comparable phenotypes of tumor cells *in vivo*?

5. What logic suggests that the chromosomal integration of tumor virus genomes is an intrinsic, obligatory part of the replication cycle of RNA tumor viruses but an inadvertent side product of DNA tumor virus replication?

6. What evidence suggests that a proto-oncogene like c-*src* is actually a normal cellular gene rather than a gene that has been inserted into the germ line by an infecting retrovirus?

7. How do you imagine that DNA tumor viruses and retroviruses like avian leukosis virus arose in the distant evolutionary past?

8. Why do retroviruses like avian leukosis virus take so long to induce cancer?

Additional reading

Brower V (2004) Accidental passengers or perpetrators? Current virus-cancer research. *J. Natl. Cancer Inst.* 96, 257–258.

Brower V (2004) Connecting viruses to cancer: how research moves from association to causation. *J. Natl. Cancer Inst.* 96, 256–257.

Bruix J et al. (2004) Focus on hepatocellular carcinoma. *Cancer Cell* 5, 215–219.

Carbone M, Pass HI, Miele L & Bocchetta M (2003) New developments about the association of SV40 with human mesothelioma. *Oncogene* 22, 5173–5180.

Coffin JM, Hughes SH & Varmus HE (eds) (1997) Retroviruses. Cold Spring Harbor, NY: Cold Spring Harbor Laboratory Press.

DiMaio D & Liao JB (2006) Human papillomaviruses and cervical cancer. *Adv. Virus Res.* 66, 125–159.

Flint SJ, Enquist LW, Krug RM et al. (2000) Principles of Virology. Washington, DC: ASM Press.

Hsu JL & Glaser SL (2000) Epstein-Barr virus-associated malignancies: epidemiologic patterns and etiologic implications. *Crit. Rev. Oncol. Hematol.* 34, 27–53.

Kung HJ, Boerkoel C & Cater TH (1991) Retroviral mutagenesis of cellular oncogenes: a review with insights into the mechanisms of insertional activation. *Curr. Top. Microbiol. Immunol.* 171, 1–15.

Martin D & Gutkind JS (2008) Human tumor-associated viruses and new insights into the molecular mechanisms of cancer. *Oncogene* 27, S31–S42.

Moore PS & Chang Y (2010) Why do viruses cause cancer? Highlights of the first century of human tumor virology. *Nat. Rev. Cancer* 10, 878–889.

Newton R, Beral V & Weiss R (1999) Human immunodeficiency virus infection and cancer. *Cancer Surv.* 33, 237–262.

Parsonnet J (ed) (1999) Microbes and Malignancy: Infection as a Cause of Human Cancers. Oxford, UK: Oxford University Press.

Phillips AC & Vousden KH (1999) Human papillomavirus and cancer: the viral transforming genes. *Cancer Surv.* 33, 55–74.

Pisani P, Parkin DM, Muñoz N & Ferlay J (1997) Cancer and infection: estimates of the attributable fraction in 1990. *Cancer Epidemiol. Biomarkers Prevent.* 6, 387–400.

Stehelin D, Varmus HE, Bishop JM & Vogt PK (1976) DNA related to the transforming gene(s) of avian sarcoma viruses is present in normal avian DNA. *Nature* 260, 170–173.

Weiss R, Teich N, Varmus H & Coffin J (eds) (1985) Molecular Biology of Tumor Viruses: RNA Tumor Viruses, 2nd ed. Cold Spring Harbor, NY: Cold Spring Harbor Laboratory Press.

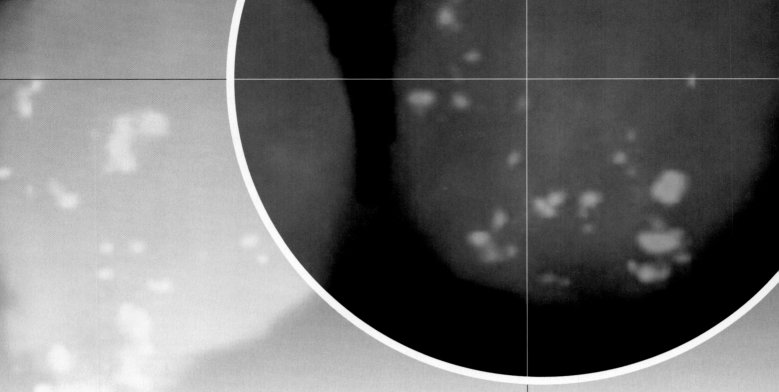

Chapter 4

Cellular Oncogenes

The viral origin of the majority of all malignant tumors ... has now been documented beyond any reasonable doubt. It ... would be rather difficult to assume a fundamentally different etiology for human tumors.

Ludwik Gross, tumor virologist, 1970

The DNA and RNA tumor viruses characterized in the 1970s provided cancer biologists with a simple and powerful theory of how human tumors could arise. Viruses that occurred commonly in the human population might, with some frequency, infect susceptible tissues and cause the transformation of infected cells. These cells, in turn, would begin to multiply and, sooner or later, form the large cell masses that were encountered frequently in the oncology clinic. Since tumor viruses succeeded in transforming normal rodent and chicken cells into tumor cells with only a small number of introduced genes, these viruses might have similar powers in transforming human cells as well.

With the passage of time, this scenario, attractive as it was, became increasingly difficult to reconcile with the biology and epidemiology of human cancer. Most types of human cancer clearly did not spread from one individual to another as an infectious disease. Significant clusters of cancer cases—mini-epidemics of disease—were hard to find. Even more important, attempts undertaken during the 1970s to isolate viruses from most types of human tumors were unsuccessful. Of the hundred and more tumor types encountered in the oncology clinic, only two commonly occurring tumor types in the Western world—cervical carcinomas and **hepatomas** (liver carcinomas)—could clearly be tied to specific viral causative agents.

These realizations evoked two responses. Those who hung tenaciously to tumor viruses as causative agents of all human cancers argued that chemical and physical carcinogens interacted with viruses that normally hid within the body's cells, activating their latent cancer-causing powers. Other researchers responded by jettisoning viruses entirely and began looking at another potential source of the genes responsible for human cancers—the cellular genome with its tens of thousands of genes. This second tack eventually triumphed, and by the late 1980s, the cell genome was recognized to be a rich source of the genes that drive human cancer cell proliferation.

So, tumor viruses, once viewed as the key agents triggering all human cancers, failed to live up to these high expectations. Ironically, however, tumor virus research proved to be critical in uncovering the cellular genes that are indeed responsible for the neoplastic cell phenotype. The large catalog of cellular cancer-causing genes assembled over the ensuing decades—oncogenes and tumor suppressor genes—derives directly from these early efforts to find infectious cancer-causing agents in human populations.

4.1 Can cancers be triggered by the activation of endogenous retroviruses?

Research begun in Japan by Katsusaburo Yamagiwa in the first decade of the twentieth century revealed that chemical agents could induce cancers in laboratory animals (see Section 2.9). As mentioned earlier, his work showed that repeated painting of coal tars on the ears of rabbits yielded skin carcinomas after several months' time. By the middle of the following decade, a PhD thesis in Paris documented more than a hundred cases of human cancer, largely of the skin, in individuals who had worked with X-ray tubes. In both cases, it was clear that the agents that directly provoked the tumors were nonbiological, being either organic chemicals or radiation (see Sections 2.9 and 2.10).

These discoveries were well known to all cancer researchers by the mid-twentieth century and were hard to reconcile with the theory that all cancers are triggered in one way or another by the actions of infectious agents, that is, tumor viruses. Responding to this, some adherents of the virus theory of cancer, especially those working with retroviruses, proposed a new mechanism in the early 1970s. Their model explained how tumor viruses could participate in the formation of the many cancers that had no outward signs of viral infection.

This new scheme derived from the peculiar biology of retroviruses. On occasion, retrovirus genomes become integrated into the germ-line chromosomes of various vertebrate species, and the resulting proviruses are then transmitted like Mendelian alleles from one generation to the next (Supplementary Sidebar 4.1). More often than not, these **endogenous proviruses** are transcriptionally silent, and their presence in all of the cells of an organism is not apparent. On rare occasions, however, it is possible to awaken the expression of such latent endogenous proviruses, which often retain the ability to encode infectious retrovirus particles.

Activation of an endogenous retroviral (ERV) genome in fibroblasts prepared from certain strains of mice can be accomplished by culturing these cells in the presence of the thymidine analog bromodeoxyuridine (BrdU). In response, these connective tissue cells, which were ostensibly free of retroviral infection, suddenly begin to release retrovirus particles, due to the transcriptional de-repression of their normally silent, endogenous proviruses. Similarly, latent endogenous proviruses may be activated spontaneously *in vivo* in a small number of cells in a mouse. Once infectious virus particles are released from these few cells, they can multiply by cell-to-cell infection, spread rapidly throughout the body, and induce leukemias in these animals.

Knowing this behavior of endogenous retroviruses, some suspected that human cancers might arise in a similar fashion. For example, mutagenic carcinogens, such as those present in tobacco tar, might provoke the activation of previously latent endogenous retroviruses. The resulting virus particles would then begin multiplying, spread throughout an individual's body, and, like the endogenous retroviruses in some mouse strains, cause cancers to form in one or another susceptible tissue. At the same time, while capable of spreading throughout a person's tissues, such endogenous viruses might be unable to spread horizontally to another individual, explaining the repeated observations that cancer does not behave like a communicable disease. Another related scheme postulated that retroviruses had inserted viral oncogenes into the germ lines of various species, and these latent viral oncogenes became activated by various types of carcinogens.

While attractive in concept, these models of human cancer causation soon collapsed because supportive evidence was not forthcoming. Reports of infectious retroviral particles in human tumors could not be verified. Even reverse transcriptase–containing virus particles were difficult to find in human tumors.

It became clear that most endogenous retroviral genomes present in the human genome are relics of ancient germ-line infections that occurred 5 million years ago and earlier in ancestral primates. Since that time, these proviruses mutated progressively into sequences that were no longer capable of specifying infectious retrovirus particles and thereby joined the ranks of the junk DNA sequences that form the bulk of our genome (see Figure 1.4). Even though as much as 8% of the human genome derives from endogenous retroviral genomes, only several of the approximately 40,000 retrovirus-derived segments have ever been shown to be genetically intact and capable, in principle, of specifying infectious virus particles. One subfamily of these viruses, termed HERV-K, has entered into the human germ line relatively recently, and several of its proviruses are seemingly intact, but to date, even these have not been found to produce infectious viruses or to become mobilized in cancer cells. (It remains unclear why the germ lines of other mammalian species have continued to acquire new, functional endogenous proviruses during recent evolutionary times while ours has not.) So cancer researchers began to look elsewhere for the genetic elements that might be triggering human cancer formation.

4.2   Transfection of DNA provides a strategy for detecting nonviral oncogenes

For those researchers intent on understanding nonviral carcinogenesis, the demise of the endogenous retrovirus theory left one viable theory on the table. According to this theory, carcinogens function as mutagens (see Section 2.10). Whether physical (for example, X-rays) or chemical (for example, tobacco tars), these agents induce cancer through their ability to mutate critical growth-controlling genes in the genomes of susceptible cells. Such growth-controlling genes might be, for example, normal cellular genes, such as the proto-oncogenes discovered by the retrovirologists. Once these genes were mutated, the resulting mutant alleles might function as active oncogenes, driving the cancerous growth of the cells that carried them.

Stated differently, this model—really a speculation—predicted that chemically transformed cells carried mutated genes and that these genes were responsible for programming the aberrant growth of these cells. In fact, the notion that cancer cells were mutant cells traced its origins back to the beginning of the twentieth century and the speculations of the pathologist David von Hansemann and the embryologist Theodor Boveri (Supplementary Sidebar 4.2). It was impossible to predict the number of such mutated genes present in the genomes of these cells. More important, experimental proofs of the existence of these cancer-causing genes represented a daunting challenge. If they were really present in the genomes of chemically transformed cells, including perhaps human tumor cells, how could they possibly be found? If these genes were mutant versions of normal cellular genes, then they were embedded in cancer cell genomes together with tens of thousands, perhaps even a hundred thousand other genes, each present in at least one copy per haploid genome. These cancer genes, if they existed, were clearly tiny needles buried in very large haystacks.

To determine whether nonviral oncogenes existed in chemically transformed cells, a novel experimental strategy was devised. It involved introducing DNA (and thus the genes) of cancer cells into normal cells, and then determining whether the recipient cells became transformed in response to the introduced tumor cell DNA. Such transformation could be scored by the appearance of foci in the cultures of recipient cells several weeks after their exposure to tumor cell DNA—essentially the assay that Howard Temin had used to score for the presence of infectious transforming Rous sarcoma virus particles in monolayers of chick embryo fibroblasts (Section 3.2). This

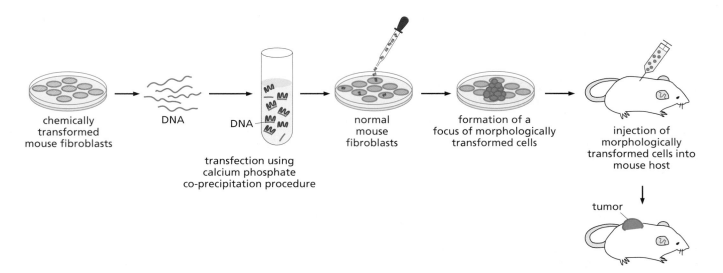

chemically transformed mouse fibroblasts → DNA → DNA

transfection using calcium phosphate co-precipitation procedure

normal mouse fibroblasts

formation of a focus of morphologically transformed cells

injection of morphologically transformed cells into mouse host

tumor

Figure 4.1 Transfection of DNA of transformed cells The procedure of transfection can be used to detect oncogenes in the DNA of cancer cells. DNA is extracted from cancer cells (*pink*) growing in a Petri dish. (For simplicity, the double-stranded DNA is depicted as single lines.) DNA is then introduced into a phosphate buffer. When calcium ions are added, a co-precipitate of DNA and calcium phosphate crystals is formed (*pink and purple*). These crystals are added to a monolayer culture of normal cells (*green*). The calcium phosphate crystals facilitate the uptake of DNA fragments by cells. If a transforming gene (oncogene) is present in the donor DNA, it may become incorporated into the genome of one of the recipient cells and transform the latter. This transformed cell will now proliferate, and its descendants will form a clump (focus, *blue*) of cells that is visible to the naked eye. Injection of these cells into a host mouse and resulting tumor formation can be used to confirm the transformed state of these cells.

strategy depended on several experimental advances, including the development of an effective gene transfer procedure. Additionally, appropriate cancer cells from which to extract DNA and suitable recipient cells needed to be identified.

In 1972, a highly effective gene transfer procedure, soon termed **transfection**, was developed. This procedure, which made possible the introduction of naked DNA molecules directly into mammalian cells (**Figure 4.1**), was based on co-precipitation of DNA with calcium phosphate crystals. For reasons that remain poorly understood to this day, these co-precipitates could overcome the natural resistance of cells to taking up foreign genetic material. Cells of the NIH 3T3 cell line, derived originally from mouse embryo fibroblasts, turned out to be especially adept at taking up and integrating into their genomes the foreign DNA introduced in this fashion.

The final issue to be settled before this experimental plan could proceed was the identity of the donor cancer cells from which DNA would be prepared. Here the researchers were working blind. It was not clear whether all types of cancer cells possessed transforming genes like the *src* oncogene borne by RSV. Also, it was not known whether a cellular transforming gene—a cellular oncogene—that had been responsible for transforming a normal epithelial cell into a carcinoma cell would also be able to function in the unfamiliar intracellular environment of connective tissue (fibroblastic) cells, like that of the NIH 3T3 cells. There were yet other possible problems. For example, an oncogene that was responsible for transforming normal human cells into cancer cells might fail to transform normal mouse cells because of some interspecies incompatibilities.

With these concerns in mind, researchers chose donor tumor cells derived from mouse fibroblasts. These particular cancer cells originated with mouse fibroblasts of the C3H10T1/2 mouse cell line that had been treated repeatedly with the potent carcinogen and mutagen 3-methylcholanthrene (3-MC), a known component of coal tars. Importantly, these cells bore no traces of either tumor virus infection or activated endogenous retroviral genomes. Hence, any transforming oncogenes detected in the genome of these cells would, with great likelihood, be of cellular origin, that is, mutant versions of normal cellular genes.

In 1978–1979, DNAs extracted from several such 3-MC–transformed mouse cell lines were transfected into cultures of NIH 3T3 recipient cells, yielding large numbers of foci after several weeks. The cells plucked from the resulting foci were later found to be both anchorage-independent and tumorigenic (see Table 3.2). This simple experiment proved that the donor tumor DNA carried one or several genetic elements that were able to convert a non-tumorigenic NIH 3T3 recipient cell into a cell that was strongly tumorigenic.

DNA extracted from normal, untransformed C3H10T1/2 cells was unable to induce foci in the NIH 3T3 cell monolayers. This difference made it highly likely that previous exposure of normal C3H10T1/2 cells to the 3-MC carcinogen had altered the genomes of these cells in some way, resulting in the creation of novel genetic sequences that possessed transforming powers. In other words, it seemed likely that the 3-MC carcinogen had converted a previously normal C3H10T1/2 gene (or genes) into a mutant allele that now could function as a transforming oncogene when introduced into NIH 3T3 cells.

At first, it seemed quite difficult to determine whether the donor tumor cells carried a single oncogene in their genomes or several distinct oncogenes that acted in concert to transform the recipient cells. Careful analysis of the transfection procedure soon resolved this issue. Researchers discovered that when cellular DNA was applied to a recipient cell, only about 0.1% of a cell genome's worth of donor DNA became established in the genome of each transfected recipient cell. The probability of two independent, genetically unlinked donor genes both being introduced into a single recipient cell was therefore $10^{-3} \times 10^{-3} = 10^{-6}$, that is, a highly unlikely event. From this calculation they could infer that only a single gene was responsible for the transformation of NIH 3T3 cells following transfection of donor tumor cell DNA. This led, in turn, to the conclusion that years earlier exposure of normal C3H10T1/2 mouse cells to the 3-MC carcinogen had caused the formation of a single mutant oncogenic allele; this allele was able, on its own, to transform both the C3H10T1/2 cells and, later on, the recipient NIH 3T3 cells into which this allele was introduced by gene transfer.

These transfection experiments were highly important, in that they provided strong indication that oncogenes can arise in the genomes of cells through mechanisms that have no apparent connection with viral infection. Perhaps human tumor cells, which likewise appeared to arise via nonviral mechanisms, also carried transfectable oncogenes. Would human oncogenes, if present in the genomes of these cells, also be able to alter the behavior of mouse cells?

Both of these questions were soon answered in the affirmative. DNAs extracted from cell lines derived from human bladder, lung, and colon carcinomas, as well as DNA from a human promyelocytic leukemia, were all found capable of transforming recipient NIH3T3 mouse fibroblasts (**Figure 4.2**). This meant that the oncogenes in these cell lines, whatever their nature, were capable of acting across species and tissue boundaries to induce cell transformation.

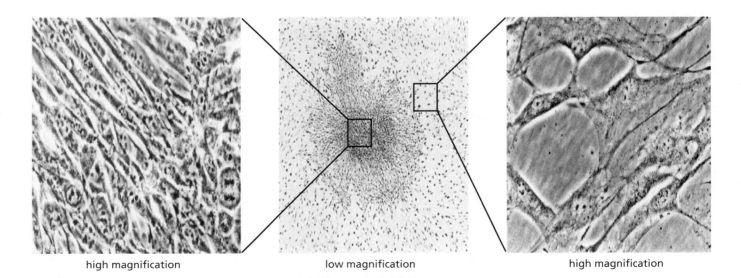

high magnification   low magnification   high magnification

Figure 4.2 Transformation of mouse cells by human tumor DNA The introduction via transfection of various human tumor DNAs into NIH 3T3 cells yielded foci of transformants. Seen at low magnification (*middle*) is a focus generated by transfection of DNA from the T24 human bladder carcinoma cell line. At high magnification the transformed cells within this focus (*left*), like many transformed fibroblasts, are spindle-shaped, refractile, and piled up densely on one another. At the same magnification, the cells in the surrounding monolayer of untransformed NIH 3T3 cells (*right*), like normal fibroblasts, have wide, extended cytoplasms and are not piled on one another. (From M. Perucho et al., *Cell* 27:467–476, 1981.)

4.3   Oncogenes discovered in human tumor cell lines are related to those carried by transforming retroviruses

The oncogenes detected by transfection in the genomes of various human tumor cells were ostensibly derived from preexisting normal cellular genes that lacked oncogenic function. This seemed to parallel the process that led to the appearance of transforming retroviruses (see Section 3.9). Recall that during the formation of these viruses, preexisting normal cellular genes—proto-oncogenes—became activated into potent oncogenes, albeit through an entirely different genetic mechanism.

These apparent parallels led to an obvious question: Could the same group of cellular proto-oncogenes become activated into oncogenes by marauding retroviruses in one context and by nonviral mutagens in another? Or did the retrovirus-associated oncogenes and those activated by nonviral mechanisms arise from two very distinct groups of cellular proto-oncogenes?

Use of DNA probes specific for the retrovirus-associated oncogenes provided the answers in short order. Using the Southern blot procedure (Supplementary Sidebar 4.3), a DNA probe derived from the H-*ras* oncogene present in Harvey rat sarcoma virus was able to recognize and form hybrids with the oncogene detected by transfection in the DNA of a human bladder carcinoma cell (**Figure 4.3**). A related oncogene, termed K-*ras* from its presence in the genome of Kirsten sarcoma virus, was able to anneal with the oncogene detected by transfection of DNA from a human colon carcinoma cell line.

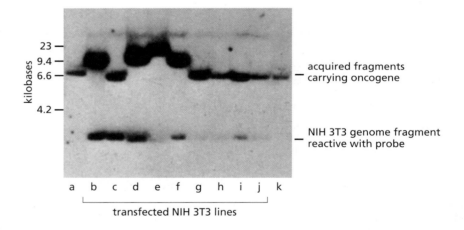

Figure 4.3 Homology between transfected oncogenes and retroviral oncogenes The Southern blot procedure (see Supplementary Sidebar 4.3) was used to determine whether there was any relatedness between retrovirus-associated oncogenes and those discovered by transfection of tumor cell DNA. Cloned retroviral oncogene DNAs were used to make radiolabeled probes, while the restriction enzyme–cleaved genomic DNAs from transfected cells were analyzed by the Southern blot procedure. Shown here is the annealing between a radiolabeled H-*ras* oncogene probe (cloned from the genome of Harvey murine sarcoma virus) and the genomic DNAs from a series of nine lines of NIH 3T3 cells (*lanes b–j*) that had been transformed by transfection of DNA extracted from the EJ human bladder carcinoma cell line; the DNA of untransfected NIH 3T3 cells was analyzed in channel a and the genomic DNA of T24 human bladder carcinoma cells was analyzed in *lane k*. In many of the lines of transfected NIH 3T3 cells (*lanes b–j*), in addition to the acquired DNA fragments carrying the transfected oncogene, a lower–molecular-weight fragment representing the endogenous H-*ras* proto-oncogene was also detected. In many of these transfected cell lines, the size of the DNA fragment carrying the transfected H-*ras* oncogene was affected by random DNA breakage events that had occurred in prior rounds of transfection and separated restriction enzyme cleavage sites from the oncogene. The human tumor DNAs in channels a and k indicate the size of the restriction fragment carrying the human H-*ras* proto-oncogene. (From L.F. Parada et al., *Nature* 297:474–478, 1982.)

Table 4.1 Examples of retrovirus-associated oncogenes that have been discovered in altered form in human cancers

Name of virus	Species	Oncogene	Type of oncoprotein	Homologous oncogene found in human tumors
Rous sarcoma	chicken	src	non-receptor TK	colon carcinoma[a]
Abelson leukemia	mouse	abl	non-receptor TK	CML
Avian erythroblastosis	mouse	erbB	receptor TK	gastric, lung, breast[b]
McDonough feline sarcoma	cat	fms	receptor TK	AML[c]
H-Z feline	cat	kit	receptor TK[d]	gastrointestinal stromal
Murine sarcoma 3611	mouse	raf	ser/thr kinase[e]	bladder carcinoma
Simian sarcoma	monkey	sis	platelet-derived growth factor (PDGF)	many types[f]
Harvey sarcoma	mouse/rat	H-ras[g]	small G protein	bladder carcinoma
Kirsten sarcoma	mouse/rat	K-ras[g]	small G protein	many types
Avian erythroblastosis	chicken	erbA	nuclear receptor[h]	liver, kidney, pituitary
Avian myeloblastosis E26	chicken	ets	transcription factor	leukemia[i]
Avian myelocytoma	chicken	myc[j]	transcription factor	many types
Reticuloendotheliosis	turkey	rel[k]	transcription factor	lymphoma

[a]Mutant forms found in a small number of these tumors.
[b]Receptor for EGF; the related erbB2/HER2/Neu protein is overexpressed in 30% of breast cancers.
[c]Fms, the receptor for colony-stimulating factor (CSF-1), is found in mutant form in a small number of AMLs; the related Flt3 (Fms-like tyrosine kinase-3) protein is frequently found in mutant form in these leukemias.
[d]Receptor for stem cell factor.
[e]The closely related B-Raf protein is mutant in the majority of melanomas.
[f]Protein is overexpressed in many types of tumors.
[g]The related N-ras gene is found in mutant form in a variety of human tumors.
[h]Receptor for thyroid hormone.
[i]27 distinct members of the Ets family of transcription factors are encoded in the human genome. Ets-1 is overexpressed in many types of tumors; others are involved in chromosomal translocations in AML, Ewing's sarcoma, and prostate carcinoma.
[j]The related N-myc gene is overexpressed in pediatric neuroblastomas and small-cell lung carcinomas.
[k]Rel is a member of a family of proteins that constitute the NF-κB transcription factor, which is constitutively activated in a wide range of human tumors.

Abbreviations: AML, acute myelogenous leukemia; CML, chronic myelogenous leukemia; TK, tyrosine kinase.

Adapted in part from J. Butel, *Carcinogenesis* 21:405–426, 2000; and G.M. Cooper, Oncogenes, 2nd ed. Boston and London: Jones and Bartlett, 1995.

The list of connections between the retrovirus-associated oncogenes and oncogenes present in non-virally induced human tumors soon grew by leaps and bounds (**Table 4.1**). In these cases, the connections were often forged following the discovery that the retrovirus-associated oncogenes were present in increased copy number in human tumor cell genomes. The *myc* oncogene, originally known from its presence in avian myelocytomatosis virus (AMV; see Section 3.10) was found to be present in multiple copies in the DNA of the HL-60 human promyelocytic leukemia cell line. These extra copies of the *myc* gene (about 10–20 per diploid genome) were the result of the process of **gene amplification** and the apparent cause of the proportionately increased levels of its protein product, the Myc oncoprotein; the excess Myc protein somehow favored the proliferation of the cancer cells. The *erbB* gene, first discovered through its presence in the genome of avian erythroblastosis virus (AEV; refer to Table 3.3), was discovered to be present in increased copy number in the DNAs of human stomach, breast, and brain tumor cells. (**Erythroblastosis** is a malignancy of red blood cell precursors.) Elevated expression of the homolog of the *erbB* gene is now thought to be present in the majority of human carcinomas.

Figure 4.4 Amplification of the *erbB2/HER2/neu* oncogene in breast cancers
(A) The Southern blotting procedure (see Supplementary Sidebar 4.3) was used to determine whether the DNA of five human breast carcinomas carried extra (that is, amplified) copies of the *erbB2/neu* oncogene (also termed *HER2*), a close relative of the *erbB* oncogene. As indicated by the dark bands representing restriction enzyme fragments, some human breast carcinomas carried extra copies of this gene. Subsequent work indicated that while the *erbB2/HER2* oncogene was amplified in some tumors ("DNA," *lanes 1, 2, and 3*), others overexpressed the mRNA without gene amplification ("RNA," *lane 5*). Levels of the encoded ErbB2/HER2 protein could also be gauged via immunoprecipitation and immunohistochemistry (see Supplementary Sidebar 2.1) by using an antibody that reacts with this protein; these levels generally correlated with levels of the encoding mRNA. (B) This gene amplification is correlated with a poor prognosis for the breast cancer patient, as indicated by this Kaplan–Meier plot. Those patients whose tumors carried more than five copies of the *erbB2/neu* gene were far more prone to experience a relapse in the first 18 months after diagnosis and treatment than were those patients whose tumors lacked this amplification. (All patients in this study had breast cancer cells that had spread to the lymph nodes draining the involved breast. *n*, number of patients.) (C) Use of fluorescence *in situ* hybridization (FISH) has shown that in some human breast cancers, there are multiple spots in each cell nucleus that react with a fluorescence-labeled DNA probe that recognizes the *erbB2/HER2* gene (*light green*) and yet other spots that react with a DNA probe specific for the *CCND1/cyclin D1* gene (*orange/pink*). Each of these probes generates only two spots when used to analyze nuclei of normal diploid cells. (As described in Chapter 8, cyclin D1 promotes progression of cells through their growth-and-division cycles.) (A, courtesy of D.J. Slamon. B, from D.J. Slamon et al., *Science* 235:177–182, 1987. C, from J. Hicks et al., *Genome Res.* 16:1465–1479, 2006.)

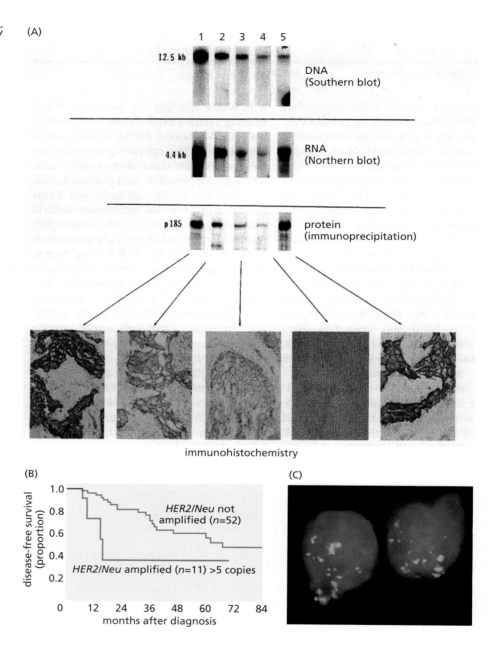

In 1987, amplification of the *erbB*-related gene known variously as *erbB2*, *neu*, or *HER2* was reported in many breast cancers (**Figure 4.4A**). Increases in gene copy number of more than five copies per cancer cell were found to correlate with a decrease in the number of patients who, following initial treatment, survived without recurrence of disease (see Figure 4.4B). (**Kaplan–Meier plots** like this one, will be used throughout this book. In each case, the status of patients—either disease-free survival, overall survival, or another clinical parameter—is plotted as a function of the time elapsed following initial diagnosis or treatment.) Significantly, the observed amplification of the *erbB2/HER2* gene was often but not always correlated with an increased expression of its encoded protein (see Figure 4.4A). Among a large group of breast cancer patients, those whose tumors expressed normal levels of this protein showed a median survival of 6 to 7 years after diagnosis, while those patients whose tumors expressed elevated levels had a median survival of only 3 years. This inverse correlation between *erbB2/HER2* expression levels and long-term patient survival provided a strong indication that this gene, in amplified form, was causally involved in driving the malignant growth of the breast cancer cells (but see **Sidebar 4.1**).

Sidebar 4.1 Gene amplifications may be difficult to interpret
The discovery that the *erbB2/neu/HER2* gene is amplified in about 30% of human breast cancers, and that this amplification is correlated with poor prognosis (see Figure 4.4), would seem to explain how many highly malignant breast cancers acquire their aggressive phenotypes. It is known that elevated signaling by this protein drives cells into endless rounds of growth and division and also protects them from programmed cell death—apoptosis. However, analyses of gene expression patterns (**Figure 4.5**) yield more complex interpretations. In the expression array analysis shown here, the expression levels of a cohort of 160 genes that flank this gene (labeled here *ErbB2/HER2*) on both sides along human Chromosome 17q, together with the expression of this gene itself, were monitored in a series of 360 human breast cancers. Elevated expression is indicated in *red* while normal expression is indicated in *green*. As is apparent, in about one-fourth of these breast cancers (*right quarter of array*), expression of *erbB2/neu/HER2* RNA was elevated, as

might be expected from the amplification that this gene had undergone in many of these tumors. At the same time, in many of these tumors, expression of closely linked genes mapping to both sides of this gene was also elevated. This reflects the fact that the unit of DNA amplification—the **amplicon**—almost always included a stretch of chromosomal DNA that was far longer than the *erbB2/neu/HER2* gene itself, leading to co-amplification of these neighboring genes. Among these genes are several that may also positively influence cell proliferation and survival, including *GRB7* and *PPARB*, whose protein products interact with ErbB2 (see Chapter 6) and with the apoptosis circuitry (see Chapter 9), respectively. Hence, in such cases, a number of co-amplified genes may be collaborating to orchestrate the malignant phenotype of human breast cancer cells, and it becomes difficult to ascribe specific cancer cell phenotypes to the elevated expression of only a single gene, such as the *erbB2/neu/HER2* discussed here.

Figure 4.5 Elevated expression of 17q genes together with overexpression of *erbB2/HER2*
The amplification of a gene, such as *erbB2/HER2*, occurs as a consequence of the amplification of an entire chromosomal segment—an amplicon—that usually extends beyond this gene on both sides for several megabases. Because an amplicon of this size encompasses many additional genes, these other genes will also be amplified and some of them may also affect tumor cell phenotype (in this case that of breast cancer cells). The map of some of the genes identified that flank *HER2* on both sides is provided (*red vertical bar, right*). In this case, RNA samples from 360 primary breast tumors were analyzed (*columns, left to right*), while probes for 160 distinct genes in this chromosomal region were arrayed in the order of their location along human Chromosome 17q (that is, the long arm of Chromosome 17) (*rows, top to bottom*). Those tumors with similar patterns of gene expression, including elevated *HER2* expression, were clustered together by a computer and are grouped on the *right*. As is apparent, genes flanking *HER2* were also overexpressed in a number of these tumors. In some tumors, genes flanking *HER2* on one side tended to be co-amplified with *HER2*, while in others, genes flanking on the other side were co-amplified. (Courtesy of Lance D. Miller, Genome Institute of Singapore.)

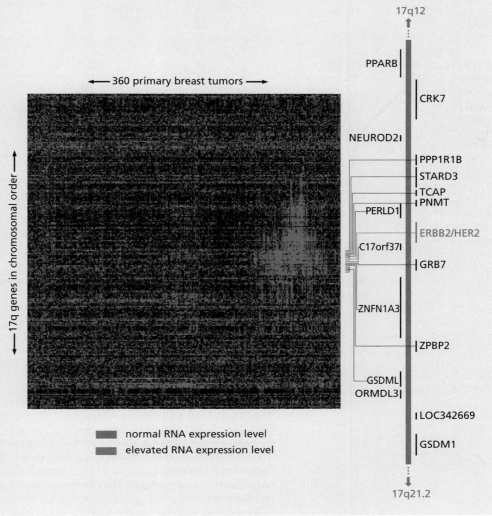

← 360 primary breast tumors →

← 17q genes in chromosomal order →

17q12

PPARB
CRK7
NEUROD2
PPP1R1B
STARD3
TCAP
PNMT
PERLD1
ERBB2/HER2
C17orf37
GRB7
ZNFN1A3
ZPBP2
GSDML
ORMDL3
LOC342669
GSDM1

17q21.2

▬ normal RNA expression level
▬ elevated RNA expression level

In the years that followed, yet other techniques, including fluorescence *in situ* hybridization (FISH), have been used to determine gene amplification in these and other tumor types (Figure 4.4C). By 2010, a variety of techniques had been brought to bear on validating the oncogenic roles of genes that are amplified in a variety of human cancers; included among these techniques were tests of the transforming powers of these genes. This work resulted in a roster of 77 genes that are, with great likelihood, contributing to tumor development when they are present in amplified form in human cancer cell genomes.

As genetic technology has improved, the ability to survey entire tumor cell genomes for chromosomal regions that have suffered changes in copy number has improved immeasurably, and genome-wide surveys of both gene amplifications and deletions have become almost routine. In the case of cancer cells, the amplifications presumably result in the overexpression of growth-promoting genes (that is, proto-oncogenes), while the deletions involve the loss of putative growth-retarding tumor suppressor genes, which are discussed in Chapter 7. As **Figure 4.6** makes clear, different cell types undergo distinct sets of genomic alterations during their transformation into full-fledged tumor cells.

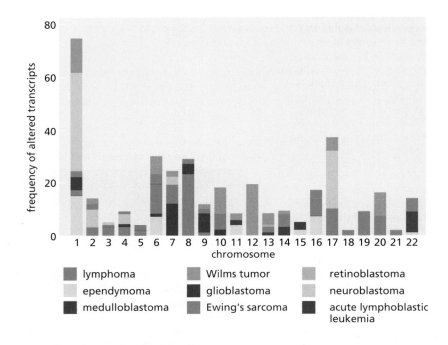

Figure 4.6 Nonrandom amplifications and deletions of chromosomal regions In a diverse array of human pediatric cancers, measurable changes in the copy number of 493 genes were found to be paralleled by changes in mRNA expression levels—that is, when DNA copy number of a chromosomal region was amplified or reduced, there was a parallel change in the level of the corresponding RNA, this being observed in multiple tumors of a given type. (Of these 493 chromosomal sites, which were scattered across the human genome, 440 could be associated with genes of known biological function.) This parallel behavior made it possible to assess changes in gene copy number by measuring levels of RNA transcripts in these tumors rather than by directly measuring DNA copy number of the corresponding genes. (The "altered transcripts" indicated in the ordinate denotes RNAs whose concentrations were significantly increased above or below normal in the tumors being examined.) As is apparent, Chromosome 1 carries regions that are most commonly found to be amplified or deleted in a variety of these tumors, which might be expected from its relatively large size. This does not explain, however, the relatively infrequent amplification or deletion of regions of Chromosomes 2, 3, 4, and 5, which are almost as large. Some tumors, such as lymphomas (*medium green*), show changes at diverse chromosomal sites, while others, such as neuroblastomas (*pink*), usually show changes mapping to one or two chromosomes. (From G. Neale et al., *Clin. Cancer Res.* 14:4572–4583, 2008.)

Ironically, mutant alleles of the *src* oncogene, the first cellular oncogene to be discovered, proved to be elusive in human tumor cell genomes. Finally, in 1999—almost a quarter of a century after the *src* gene was first cloned—mutant forms of the *src* gene were found in the genomes of human tumor cells, specifically, in the genomes of 12% of advanced human colon carcinomas.

The lesson taught by these numerous cross connections was simple and clear: many of the oncogenes originally discovered through their association with avian and mammalian retroviruses could be found in a mutated, activated state in human tumor cell genomes. This meant that a common set of cellular proto-oncogenes might be activated either by retroviruses (in animals) or, alternatively, by nonviral mutational mechanisms operating during the formation of human cancers.

4.4   Proto-oncogenes can be activated by genetic changes affecting either protein expression or structure

While a number of proto-oncogenes were found in activated, oncogenic form in human tumor genomes, the precise genetic alterations that led to many of these activations remained unclear. In the case of retrovirus-associated oncogenes, one mechanism became obvious once the organization of the transforming retrovirus genomes was known. In the normal cell, the expression of each proto-oncogene was regulated by its own transcriptional promoter—the DNA sequence that controls the level of its transcription. The promoter of each proto-oncogene enabled the gene to respond to a variety of physiologic signals. Often the needs of the cell, communicated through these signals, caused a proto-oncogene to be expressed at very low levels; however, on certain occasions, when required by the cell, expression of the gene might be strongly induced.

A quite different situation resulted after a proto-oncogene was acquired by a retrovirus. After insertion into the retrovirus genome, expression of this captured gene was controlled by a retroviral transcriptional promoter (see Figures 3.19 and 3.22), which invariably drove the gene's expression unceasingly and at high levels. Transcription of this virus-associated gene, now an oncogene, was therefore no longer responsive to the cellular signals that had previously regulated its expression. For example, in the case of c-*myc*, expression or **repression** (that is, shutdown) of this gene is normally tightly controlled by the changing levels of extracellular signals, such as those conveyed by mitogenic growth factors (to be discussed in Chapter 5). Once present in the genome of avian myelocytomatosis virus (AMV), expression of this gene (now called v-*myc*) is found to be at far higher levels than are seen normally in cells, and this expression occurs at a constant (sometimes termed **constitutive**) level.

But how did a normal human H-*ras* proto-oncogene become converted into the potent oncogene that was detected by transfection of human bladder carcinoma DNA (see Section 4.2)? Gene amplification could not be invoked to explain its activation, since this oncogene seemed to be present in bladder carcinoma DNA as a single-copy gene. The puzzle grew when this H-*ras* bladder carcinoma oncogene was isolated by molecular cloning (**Sidebar 4.2**, see Figure 4.7). It was localized to a genomic DNA fragment of 6.6 kilobases in length. Provocatively, an identically sized DNA fragment was found in normal human DNAs. The latter fragment clearly represented the human H-*ras* proto-oncogene—the normal gene that suffered some type of mutation that converted it into an oncogene during the formation of the bladder carcinoma.

While their overall DNA structures were very similar, these two versions of the H-*ras* gene performed in dramatically different ways. The oncogene that had been cloned from human bladder carcinoma cells caused transformation of NIH 3T3 cells, while its normal proto-oncogene counterpart lacked this ability. The mystery deepened when more detailed mapping of the physical structures of these two DNA segments—achieved by making maps of the cleavage sites of various restriction enzymes—revealed that the two versions of the gene had overall physical structures that were indistinguishable from one another.

Sidebar 4.2 Cloning of transfected oncogenes The oncogene of the T24/EJ human bladder carcinoma cell line was cloned by two research groups before its relationship with the H-*ras* oncogene of Harvey sarcoma virus was known. These groups faced the challenge of isolating a gene without knowing anything about its sequence or structure. One group of researchers transfected DNA of the human bladder carcinoma cells into NIH 3T3 mouse cells (**Figure 4.7**). They used Southern blotting (see Supplementary Sidebar 4.3) to detect the donor DNA by exploiting probes that were specific for the **Alu repeat** sequences, which are scattered randomly about the human genome in about 1 million sites but are absent from the mouse genome. (More precisely, distantly related mouse repeat sequences are not recognized by DNA probes that are specific for the human *Alu* repeats.) Thus, *Alu* sequences are present in the human genome, on average, about every 5 kb or so. Accordingly, it was likely that the human bladder carcinoma oncogene carried with it some linked human *Alu* sequences into recipient mouse cells during the transfection procedure.

Indeed, the researchers found a relatively small number of human *Alu* sequences (about 0.1% of the total *Alu* sequences in a human genome; see Section 4.2) in the genomes of transfected, transformed mouse cells. They then prepared whole genomic DNA from these transformed NIH 3T3 cells and transfected it once again into fresh NIH 3T3 cells and isolated transformed cells that arose following this second cycle of transfection. Once again, only about 0.1% of the donor DNA was transferred from donor to recipient (see Figure 4.7). In these secondarily transfected cells, the only human *Alu* sequences that had survived the two cycles of transfection were those that were closely linked to the oncogene responsible for the observed transformation. These researchers then used a human *Alu*-specific sequence probe to identify the presence of an *Alu*-containing DNA fragment in a collection (a genomic library) of DNA fragments prepared from the genomic DNA of the secondarily transfected cells. They then retrieved this fragment using standard gene cloning procedures. The cloned *Alu*-containing DNA segment was found to carry, in addition, the long-sought bladder carcinoma oncogene.

The other research group used an elegant procedure that caused the bladder carcinoma gene to become closely linked to a bacterial gene during the initial transfection event. They then followed the fate of this bacterial segment through another cycle of transfection and, by using a probe specific for it, were able to isolate both this bacterial segment and the linked bladder carcinoma cell from transfected cells by molecular cloning.

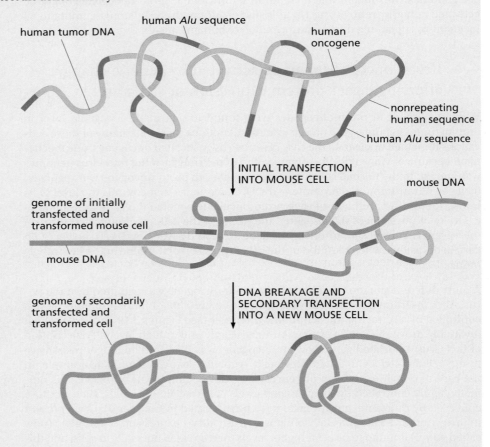

Figure 4.7 Cloning of transfected human oncogenes This strategy for cloning the oncogene present in a human bladder carcinoma depended on *Alu* sequences (*red segments*), which are present in almost a million copies scattered throughout the human genome (*orange segments*). As a consequence, virtually all human genes are closely linked to one or more of the *Alu* sequences. If the genomic DNA of a human tumor cell bearing an oncogene (*blue segment*) is transfected into a mouse cell [whose DNA (*light brown line*) lacks sequences closely related to human *Alu* sequences], the introduced human DNA can be detected by use of an *Alu*-specific probe in the Southern blotting procedure (see Supplementary Sidebar 4.3). Only about 0.1% of the donor DNA genome became established in the initially transfected, transformed mouse cells, indicating that these cells carried both the transforming oncogene and perhaps 10^3 co-transfected human *Alu* sequences. Because so many human *Alu* sequences were co-introduced with the human oncogene into these cells, the DNA of cells descended from them was extracted, fragmented, and used in a second cycle of transfection into mouse cells. Once again, only about 0.1% of the donor DNA (from the initially transfected cells) survived in the secondary transfectants, dictating that the only human DNA and associated *Alu* sequences that were present in the secondarily transformed cells were those that were closely linked to the human oncogene (whose presence was detected because of the transformed phenotype that it caused in these cells). A genomic library (see Supplementary Sidebar 1.5) could then be made from the DNA of these secondary transformants, and the DNA clone containing the bladder carcinoma oncogene could be identified (using an *Alu*-specific probe) and retrieved.

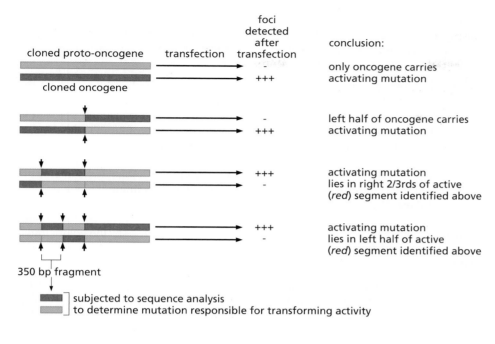

Figure 4.8 Localization of an oncogene-activating mutation The cloned DNAs of a human bladder carcinoma oncogene (*red segments*) and the closely related human H-*ras* proto-oncogene (*green segments*) were cleaved by restriction enzymes at the sites indicated (*vertical arrows*) and recombinant genes were made by ligating (linking) the resulting DNA fragments from the two sources and testing the hybrid DNA molecules for their transforming activity using the transfection-focus assay (see Figure 4.1). This made it possible to progressively localize the mutation responsible for oncogene activation to a small 350-base-pair segment, which could then be subjected to sequence analysis in order to determine the precise sequence change that distinguishes the two allelic versions of this gene. (From C.J. Tabin et al., *Nature* 300:143–149, 1982.)

Yet clearly, the two versions of the H-*ras* gene had some significant difference in their sequences, because they functioned so differently. The critical sequence difference was initially localized by recombining segments of the cloned proto-oncogene with other segments deriving from the oncogene (**Figure 4.8**). This made it possible to narrow down the critical difference to a segment only 350 base pairs long.

The puzzle was finally solved when the corresponding 350-bp segments from the proto-oncogene and oncogene were subjected to DNA sequence analysis. The critical difference was extraordinarily subtle—a single base substitution in which a G (guanosine) residue in the proto-oncogene was replaced by a T (thymidine) in the oncogene. This single base-pair replacement—a **point mutation**—appeared to be all that was required to convert the normal gene into a potent oncogene (**Figure 4.9**)! This important discovery was made simultaneously in three laboratories, eliminating all doubt about its correctness.

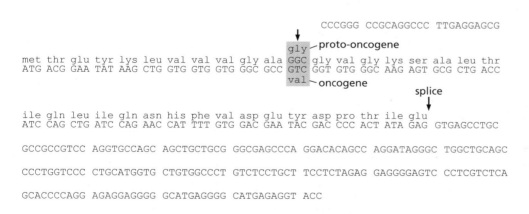

Figure 4.9 Mutation responsible for H-ras oncogene activation As indicated in Figure 4.8, the critical difference between the human bladder carcinoma oncogene and its proto-oncogene could be localized to a subgenic fragment of 350 base pairs. The sequences of the two 350-nucleotide-long DNA fragments from the oncogene and proto-oncogene were then determined. The two differed at a single nucleotide, which affected the 12th codon of the H-*ras* reading frame (*arrow*), converting the normally present glycine-encoding codon to one specifying valine. (From C.J. Tabin et al., *Nature* 300:143–149, 1982.)

The discovery of this point mutation represented a significant milestone in cancer research. It was the first time that a mutation was discovered in a gene that contributed causally to the neoplastic growth of a human cancer. Equally important, it seemed that this genetic change arose as a somatic mutation.

With this information in hand, researchers could devise a likely explanation for the origin of the bladder carcinoma and, by extension, other similar tumors. The particular bladder carcinoma from which the H-*ras* oncogene had been cloned was said to have arisen in a middle-aged man who had been smoking for four decades. During this time, carcinogens present in cigarette smoke were introduced in large amounts into his lungs and passed from there through the bloodstream to his kidneys, which excreted these chemical species with the urine. While in the bladder, some of the carcinogen molecules present in the urine had entered cells lining the bladder and attacked their DNA. On one occasion, a mutagenic carcinogen introduced a point mutation in the H-*ras* proto-oncogene of an epithelial cell. Thereafter, this mutant cell and its descendants proliferated uncontrollably, being driven by the potent transforming action of the H-*ras* oncogene that they carried. The result, years later, was the large tumor mass that was eventually diagnosed in this patient.

Importantly, this base-pair substitution occurred in the **reading frame** of the H-*ras* gene—the portion of the gene dedicated to encoding amino acid sequence (see Figure 4.9). In particular, this point mutation caused the substitution of a glycine residue present in the normal H-*ras*–encoded protein by a valine residue. The effects of this amino acid substitution on the function of the Ras oncoprotein will be discussed later in Chapters 5 and 6.

The discovery of this point mutation established a mechanism for oncogene activation that was quite different from that responsible for the creation of *myc* oncogenes. In the case of H-*ras*, a change in the structure of the encoded protein appeared to be critical. In the contrasting case of *myc*, deregulation of its expression seemed to be important for imparting oncogenic powers to this gene.

Within a decade, a large number of human tumors were found that carried point mutations in one of the three *ras* genes present in the mammalian genome: H-*ras*, K-*ras*, and N-*ras*. Significantly, in each of these tumors, the point mutation that was uncovered was present in one of three specific codons in the reading frame of a *ras* gene. Consequently, all Ras oncoproteins (whether made by the H-, K-, or N-*ras* gene) were found to carry amino acid substitutions in residue 12 or, less frequently, residues 13 and 61 (**Figure 4.10**). (This striking concentration of mutations affecting only a small number of residues in the Ras protein occurs in spite of the fact that all of the coding sequences in the *ras* proto-oncogenes presumably have similar vulnerability to mutation. The resolution of this puzzle comes from studying the structure and function of the Ras protein, topics that we will address in the next chapter.) A survey of 40,000 human tumor genomes conducted many years later revealed that 22%, 8.2%, and 3.7% of these tumors had activating mutations in the K-, N-, and H-*ras* genes, respectively.

Figure 4.10 Concentration of point mutations leading to activation of the K-*ras* oncogene The sequencing of almost 10,000 K-*ras* oncogenes in human tumors has revealed that, rather than occurring randomly throughout the reading frame of the gene, which encompasses 189 codons, mutations affect almost exclusively the nucleotides forming codon 12; mutations affect codon 13 only about 12% as frequently and codon 61 only on rare occasions. Since all the codons of the K-*ras* reading frame are likely to suffer mutations at comparable rates during tumor development, this suggests that an extraordinarily strong selective pressure favors the outgrowth of cells having mutations affecting codons 12, 13, and 61, while mutations altering other codons in the reading frame either have no effect on cell phenotype or are actively disadvantageous for the cells bearing them. (From Catalogue of Somatic Mutations in Cancer (COSMIC), Sanger Institute/Wellcome Trust, http://www.sanger.ac.uk, accessed March, 2009.)

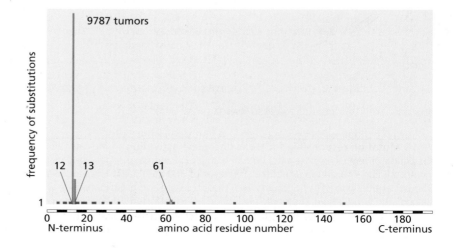

Table 4.2 A list of point-mutated *ras* oncogenes carried by a variety of human tumor cells

Tumor type	Proportion (%) of tumors carrying a point-mutated *ras* gene[a]
Pancreas	90 (K)
Thyroid (papillary)	60 (H, K, N)
Thyroid (follicular)	55 (H, K, N)
Colorectal	45 (K)
Seminoma	45 (K, N)
Myelodysplasia	40 (N, K)
Lung (non-small-cell)	35 (K)
Acute myelogenous leukemia	30 (N)
Liver	30 (N)
Melanoma	15 (N)
Bladder	10 (H, K)
Kidney	10 (H)

[a]H, K, and N refer to the human *H-RAS*, *K-RAS*, and *N-RAS* genes, respectively.

Adapted from J. Downward, *Nature Rev. Cancer* 3:11–22, 2003.

For unknown reasons, the frequency with which these mutant oncogenes are found varies dramatically from one tissue site to another (**Table 4.2**). One clue, however, has come from genetically engineered mice in which the reading frame of the K-*ras* proto-oncogene has been replaced ("**knocked-in**"; see Supplementary Sidebar 7.7) with the reading frame of an H-*ras* oncogene. Wild-type mice usually develop lung tumors carrying K-*ras* oncogenes in response to treatment by a chemical carcinogen; however, the vast majority of the knock-in mice develop lung tumors carrying H-*ras* oncogenes following this treatment. This indicates that the *regulatory sequences* controlling the expression of the K-*ras* proto-oncogene, rather than the *structure* of the encoded protein, determine which *ras* proto-oncogene becomes activated in response to a mutagenic carcinogen.

Both activation mechanisms—regulatory and structural—might collaborate to create an active oncogene. In the case of the *myc* oncogene carried by avian myelocytomatosis virus, for example, expression of this gene was found to be strongly deregulated by the viral transcription promoter. At the same time, some subtle alterations in the reading frame of the *myc* oncogene (and thus changes in the structure of its encoded oncoprotein, Myc) further enhanced the already-potent transforming powers of this oncogene. Similarly, the H-*ras* oncogene carried by Harvey sarcoma virus was found to carry a point mutation in its reading frame (like that discovered in the bladder carcinoma oncogene); at the same time, this gene was greatly overexpressed, being driven by the retroviral transcriptional promoter. And in many animal and human tumors carrying a *ras* oncogene, the initially arising oncogene carrying a point mutation is found to subsequently undergo the process of gene amplification, enabling the tumor cells to acquire increased signaling from the mutant oncoprotein.

4.5 Variations on a theme: the *myc* oncogene can arise via at least three additional distinct mechanisms

The observation that the v-*myc* oncogene of avian myelocytomatosis virus (AMV) arose largely through deregulation of its expression only hints at the diverse mechanisms

Sidebar 4.3 N-*myc* amplification and childhood neuroblastomas Amplification of the N-*myc* gene occurs in about 30% of advanced pediatric neuroblastomas, which are tumors of the peripheral nervous system. This amplification, which is associated with the formation of either double minutes (DMs) or homogeneously staining regions (HSRs), represents a bad prognosis for the patient (**Figure 4.11**). The HSRs, which contain multiple copies of the genomic region encompassing the N-*myc* gene, are often found to have broken away from the normal chromosomal mapping site of N-*myc* and, in one study, to have become associated with at least 18 other different chromosomal regions. While N-*myc* amplification was originally thought to be a peculiarity of neuroblastomas (and thus a specific diagnostic marker for this particular disease), it has now been found in a variety of neuroectodermal tumors, including astrocytomas and retinoblastomas; in addition, small-cell lung carcinomas, which have neuroendocrine traits, also often exhibit amplified N-*myc* genes.

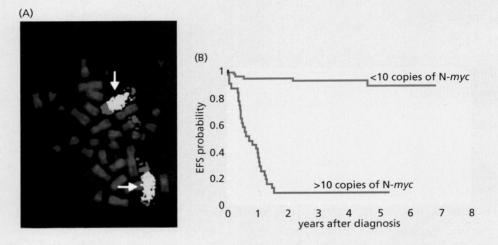

Figure 4.11 N-*myc* amplification and neuroblastoma prognosis (A) The N-*myc* gene is often amplified in human childhood neuroblastomas. Here, use of fluorescence *in situ* hybridization (FISH) has revealed multiple copies of this gene (*yellow*). The fact that these N-*myc* gene copies are present as tandem arrays within chromosomes means that they constitute homogeneous staining regions (HSRs) rather than extrachromosomal particles—double minutes (DMs)—which are also frequently seen in these tumors. (B) This Kaplan–Meier plot illustrates the event-free survival (EFS) of children suffering from neuroblastoma, that is, no clinically significant cancer-related symptoms in the indicated years following initial diagnosis and treatment. Those whose tumor cells have minimal or no N-*myc* amplification have a very good prognosis and minimal clinical events, while those whose tumors have extensive N-*myc* amplification have a dramatically poorer prognosis and therefore short survival times after diagnosis. (A, from C. Lengauer et al., *Nature* 396:643–649, 1998. B, from M.L. Schmidt et al., *J. Clin. Oncol.* 18:1260–1268, 2000.)

that are capable of creating this oncogene. As cited in Section 4.3, in some human tumors, expression of the *myc* gene continues to be driven by its own natural transcriptional promoter but the copy number of this gene is found to be elevated to levels many times higher than the two copies present in the normal human genome. In 30% of childhood neuroblastomas, a close relative of c-*myc*, termed N-*myc*, has also been found to be amplified, specifically in the more aggressive tumors of this type (**Sidebar 4.3**). In both instances, the increased gene copy numbers result in corresponding increases in the level of gene products—the Myc and N-Myc proteins. As we will discuss later in Chapter 8, proteins of the Myc family possess potent growth-promoting powers. Consequently, when present at excessive levels, these proteins seem to drive uncontrolled cell proliferation.

Note, by the way, the notations that are used here and throughout this text. Nonhuman oncogenes are usually written as uncapitalized three-letter words in italics (for example, *myc*), while their protein products are written in roman font with an initial capital (for example, Myc). The *myc* proto-oncogene itself is often termed c-*myc* to distinguish it from its two cousin genes, N-*myc* and L-*myc*. To make matters more confusing, human genes follow a different nomenclature, so that the human *myc* gene is denoted as *MYC* and its protein product is written as MYC. We will generally use the nonhuman acronym conventions in this book.

The gene amplification process, which is responsible for increases in *myc* copy number, occurs through the preferential replication of a limited region of chromosomal DNA, leaving the more distantly located chromosomal regions unaffected (see Figure 4.5). Since the region of chromosomal DNA that undergoes amplification—the **amplicon**—usually includes a stretch of DNA far longer than the c-*myc* or N-*myc* gene (for example, typically including 0.5 to 10 megabases of DNA), the amplified chromosomal regions are often large enough to be observed at the metaphase of mitosis through a light microscope. Often gene amplification yields large, repeating end-to-end linear arrays of the chromosomal region, which appear as homogeneously staining regions (HSRs) in the microscope (see Figure 1.12A). Alternatively, the chromosomal region carrying a *myc* or N-*myc* gene may break away from the chromosome and can be seen as small, independently replicating, extrachromosomal particles (double minutes; see Figure 1.12B,C). Indeed, we now know that a number of proto-oncogenes can be found in amplified gene copy number in various types of human tumors (**Table 4.3**).

Table 4.3 Some frequently amplified chromosomal regions and the genes they are known to carry

Name of oncogene[a]	Human chromosomal location	Human cancers	Nature of protein[b]
MDM4/MDMX	1q32	breast, colon, lung, pre-B leukemias	p53 inhibitor
PIK3CA	3q26.3	lung SCC, ovarian, breast	PI kinase
erbB1/EGFR	7q12–13	glioblastomas (50%); squamous cell carcinomas (10–20%)	RTK
cab1–erbB2–grb7	17q12	gastric, ovarian, breast carcinomas (10–25%)	RTK, adaptor protein
k-sam	7q26	gastric, breast carcinomas (10–20%)	RTK
FGF-R1	8p12	breast carcinomas (10%)	RTK
met	7q31	gastric carcinomas (20%)	RTK
K-*ras*	12p12.1	lung, ovarian, colorectal, bladder carcinomas (5–20%)	small G protein
N-*ras*	1p13	head and neck cancers (30%)	small G protein
H-*ras*	11p15	colorectal carcinomas (30%)	small G protein
c-*myc*	8q24	various leukemias, carcinomas (10–50%)	TF
L-*myc*	1p32	lung carcinomas (10%)	TF
N-*myc–DDX1*	2p24–25	neuroblastomas, lung carcinomas (30%)	TF
akt-1	14q32–33	gastric cancers (20%)	ser/thr kinase
akt-2	19q13	ovarian carcinomas	ser/thr kinase
cyclin D1–exp1–hst1–ems1	(11q13)	breast and squamous cell carcinomas (25–50%)	G1 cyclin
cdk4–mdm2–sas–gli	12q13	sarcomas (10–30%), HNSCC (40%), B-cell lymphomas (25%)	CDK, p53 antagonist
cyclin E	19q12	gastric cancers (15%)	cyclin
akt2	(19q13)	pancreatic, ovarian cancers (30%)	ser/thr kinase
AIB1, BTAK	(20q12–13)	breast cancers (15%)	receptor co-activator
cdk6	(19q21–22)	gliomas (5%)	CDK
myb	6q23–24	colon carcinoma (5–20%), leukemias	TF
ets-1	11q23	lymphoma	TF
gli	12q13	glioblastomas	TF

[a]The listing of several genes indicates the frequent co-amplification of a number of closely linked genes; only the products of the most frequently amplified genes are described in the right column.
[b]Abbreviations: TF, transcription factor; RTK, receptor tyrosine kinase; CDK, cyclin-dependent kinase; G protein, guanine nucleotide-binding protein; HNSCC, head-and-neck squamous cell carcinomas.

Courtesy of M. Terada, Tokyo, and adapted from G.M. Cooper, Oncogenes, 2nd ed. Boston and London: Jones and Bartlett, 1995.

In addition, regions of various human chromosomes have been found to undergo amplification in a variety of tumors. As Figure 4.6 indicates, some tumors, such as Wilms tumor, show multiple regions of chromosomal amplification, while others, such as Ewing's sarcoma, usually exhibit amplification of only a single chromosomal region. In most cases, the identities of the critical growth-promoting genes in these chromosomal regions remain elusive. Even within a single tumor, such as the breast cancers described above, multiple chromosomal regions that have undergone gene amplification are often found (see Table 4.3). As discussed later (Chapter 11), these multiple changes are thought to collaborate with one another to create the fully neoplastic phenotypes of cancer cells. The molecular mechanisms responsible for generating gene amplifications remain poorly understood; however, in the case of tandem duplications that yield HSRs, detailed structural analyses of amplified genes have provided good clues as to how these may occur (see Supplementary Sidebar 4.4).

An even more unusual way of deregulating *myc* expression levels has already been cited (see Section 3.11). Recall that the insertional mutagenesis mechanism causes the expression of the c-*myc* proto-oncogene to be placed under the transcriptional control of an ALV provirus that has integrated nearby in the chromosomal DNA. The resulting constitutive overexpression of c-*myc* RNA and thus Myc protein results, once again, in flooding of the cell with excessive growth-promoting signals.

Such activation by provirus integration hinted at a more general mode of activation of the c-*myc* proto-oncogene: even while continuing to reside at its normal chromosomal site, c-*myc* can become involved in cancer development if it happens to come under the control of foreign transcriptional promoters. In the disease of Burkitt's lymphoma (BL), this principle was validated in a most dramatic way. This tumor occurs with some frequency among young children in East and Central Africa (**Figure 4.12**). The etiologic agents of this disease include chronic infections both by Epstein–Barr virus (EBV, a distant relative of human herpesviruses; Section 3.4) and by malarial parasites.

Neither of these etiologic factors shed light on the nature of a critical genetic change inside Burkitt's lymphoma cells that is responsible for their runaway proliferation. Careful examination of metaphase chromosome spreads of these tumor cells did, however, uncover a striking clue: the tumor cells almost invariably carried chromosomal **translocations** (see also Figure 2.26). Such alterations fuse a region from one chromosome with a region from a second, unrelated chromosome (**Figure 4.13**). Translocations are often found to be *reciprocal*, in the sense that a region from chromosome A lands on chromosome B, while the displaced segment of chromosome B ends up being linked to chromosome A. In the case of Burkitt's lymphomas, three distinct, alternative chromosomal translocations were found, involving human Chromosomes 2, 14, or 22. The three translocations were united by the fact that in each case, a region from one of these three chromosomes was fused to a section of Chromosome 8.

Figure 4.12 Burkitt's lymphoma incidence in Africa (A) The geographic distribution in Africa of the *Anopheles gambiae* mosquito, a known vector in the transmission of malaria. (B) The geographic distribution of childhood Burkitt's lymphoma (BL), as originally documented by Dennis Burkitt. The rough congruence of these two distribution maps suggested that malarial infection was likely an etiologic factor in this disease. In addition, the invariable presence of the Epstein–Barr virus (EBV) genome in the BL tumor cells indicates a second etiologic factor. As the tumor cells evolve, a chromosomal translocation involving the c-*myc* gene develops, which contributes to the neoplastic proliferation of the EBV-infected B lymphocytes. (A and B, from A.J. Haddow, in D.P. Burkitt and D.H. Wright (eds.), Burkitt's Lymphoma. Edinburgh and London: E. & S. Livingstone, 1970.)

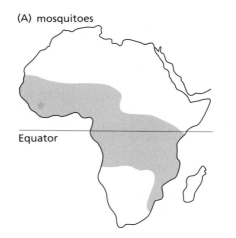

(A) mosquitoes

Equator

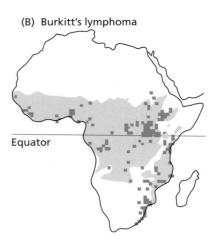

(B) Burkitt's lymphoma

Equator

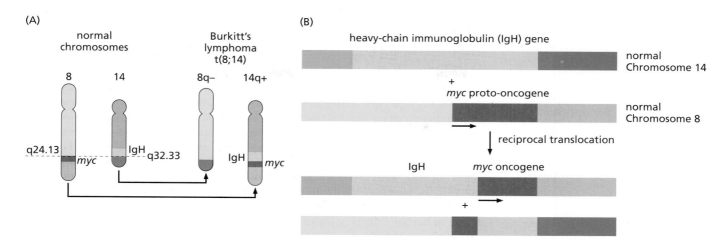

Figure 4.13 Chromosomal translocations in Burkitt's lymphomas (A) In the genomes of Burkitt's lymphoma (BL) cells, the expression of the c-*myc* gene is placed under control of the transcription-controlling enhancer sequences (see Figure 1.18) of an immunoglobulin gene as a direct consequence of reciprocal chromosomal translocations. These translocations juxtapose immunoglobulin genes on Chromosomes 2, 14, or 22 with the *myc* gene on Chromosome 8. (Translocations involving the κ, λ, or heavy immunoglobulin chain occur in 9%, 16%, and 75% of BLs, respectively.) The most common translocation, t(8;14), is shown here. (B) Depicted is a genetic map of the translocation event that places the c-*myc* gene (*red rectangle*) on Chromosome 8 under the control of the immunoglobulin heavy-chain sequences (IgH; *gray rectangle*) on human Chromosome 14. Because the immunoglobulin enhancer sequences (see Figure 1.18) direct high, constitutive expression, the normal modulation of *myc* expression in response to physiologic signals is abrogated. The resulting *myc* oncogene initially makes structurally normal Myc protein but in abnormally high amounts. (Subsequently occurring point mutations in the *myc* reading frame may further potentiate the function of the Myc oncoprotein.) (Data from P. Leder et al., *Science* 222:765–771, 1983.)

In 1983 researchers realized that the *myc* proto-oncogene could be found in the region on human Chromosome 8 that is involved in these three distinct translocations. On the other side of the fusion site (often termed the chromosomal **breakpoint**) were found the transcription-promoting sequences from any one of three distinct **immunoglobulin** (antibody) genes. Thus, the immunoglobulin heavy-chain gene cluster is found on Chromosome 14, the κ antibody light-chain gene is located on Chromosome 2, and the λ antibody light-chain gene is found on Chromosome 22. There is clear evidence that the enzymes responsible for rearranging the sequences of antibody genes during the development of the immune system (see Supplementary Sidebar 15.1) occasionally lose specificity and, instead of creating a rearranged antibody gene, inadvertently fuse part of an antibody gene with the *myc* proto-oncogene. Parenthetically, none of this explains the role of EBV in the pathogenesis of Burkitt's lymphoma (**Sidebar 4.4**).

Suddenly, the grand design underlying these complex chromosomal changes became clear, and it was a simple one: these translocations separate the *myc* gene from its normal transcriptional promoter and place it, instead, under the control of one of three highly active transcriptional regulators, each part of an immunoglobulin gene (see Figure 4.13). Once its expression is subjugated by the antibody gene promoters, *myc*

Sidebar 4.4 How does Epstein–Barr virus (EBV) cause cancer? The 1982 discovery of the Burkitt's lymphoma–associated chromosomal translocations shed little light on the precise mechanism by which the etiologic factors (chronic EBV and malarial parasite infections) favor the formation of these tumors. The contributions of each of these agents to lymphoma pathogenesis are still not totally clear. It appears that chronic malarial infection together with malnutrition can compromise the immune defenses of children and thereby render them susceptible to runaway EBV infections that further debilitate their already-weakened immune systems. These EBV infections can lead to the accumulation of large pools of EBV-immortalized B lymphocytes that are driven to proliferate continuously by the virus. In some of these cells, the enzymatic machinery dedicated to organizing normal immunoglobulin gene rearrangement (see Supplementary Sidebar 15.1) misfires and, in so doing, occasionally creates inappropriate juxtapositions of segments of antibody genes with the c-*myc* proto-oncogene. The resulting *myc* oncogenes, working together with several genes (possible oncogenes) expressed by EBV, then drive the cell proliferation that leads eventually to the eruption of lymphomas. An additional puzzle comes from the involvement of EBV in nasopharyngeal carcinomas (NPCs) in Southeast Asia, where this viral infection, together with certain lifestyle factors (possibly the consumption of Chinese-style salted fish early in life), has been implicated as an etiologic agent. Even more puzzling is the fact that in Western populations, EBV infection is common and causes mononucleosis in **immunocompetent** individuals (that is, those with a fully functional immune system), triggering malignancies only in rare instances.

becomes a potent oncogene and drives the relentless proliferation of lymphoid cells in which these transcriptional promoters are highly active. Hence, the proliferation of the rare cell that happens to acquire such a deregulated *myc* gene will be strongly favored.

Since these discoveries, several dozen distinct chromosomal translocations have been found to cause deregulated expression of known proto-oncogenes; most of these genes remain poorly characterized (**Table 4.4**). Altogether, more than 300 distinct recurring translocations (that is, those that have been encountered in multiple, independently arising human tumors) have been cataloged, and more than 100 of the novel hybrid genes created by these translocations have been isolated by molecular cloning. More often than not, each of these translocations can be found in a small set of hematopoietic tumors. However, solid tumors are increasingly found to harbor specific translocations, some quite common. For example, in 2005, ~70% of prostate carcinomas were found to harbor chromosomal translocations that lead to deregulated expression of one of two transcription factors (termed ERG and ETV1); given the high incidence of prostate cancer in Western populations, these may represent the most frequent translocations in human cancer. (Indeed, the dearth of translocations discovered to date in solid tumors may simply be a consequence of the difficulties of dissociating solid tumors and studying the karyotypes of their constituent cells.)

The discovery of microRNAs and the roles they play in gene regulation (see Section 1.10) has made it possible to understand certain chromosomal translocations whose molecular mechanisms-of-action were previously obscure. In one case, a small non-histone nuclear protein, termed HMGA2, is known to be overexpressed in a variety of benign and malignant tumors and to be capable of functioning as an oncoprotein. Its gene is frequently found to undergo translocations that sever the sequences specifying the 3′ untranslated region (3′UTR) from the reading frame of the *HMGA2* mRNA while joining the reading frame of the *HMGA2* gene with that of a second gene (**Figure 4.14**). It was long assumed that the resulting fusion protein enhanced the oncogenic functions of HMGA2 (similar to situations that we will encounter shortly in Section 4.6). However, deletion of the protein sequences that became fused to HMGA2 did not affect the transforming powers of the hybrid protein. This led to the realization

Table 4.4 Translocations in human tumors that deregulate proto-oncogene expression and thereby create oncogenes

Oncogene	Neoplasm
myc	Burkitt's lymphoma; other B- and T-cell malignancies
bcl-2	follicular B-cell lymphoma
bcl-3	chronic B-cell lymphoma
bcl-6	diffuse B-cell lymphoma
hox1	acute T-cell leukemia
lyl	acute T-cell leukemia
rhom-1	acute T-cell leukemia
rhom-2	acute T-cell leukemia
tal-1	acute T-cell leukemia
tal-2	acute T-cell leukemia
tan-1	acute T-cell leukemia
ETV-1, ETV-4	prostate carcinoma
ERG	prostate carcinoma

Adapted from G.M. Cooper, Oncogenes, 2nd ed. Sudbury, MA: Jones and Bartlett, 1995. www.jbpub.com. Reprinted with permission.

that the translocation served another purpose: it caused deletion of seven recognition sites for the *Let-7* microRNA (see Section 1.10) that are normally present in the 3'UTR of the *HMGA2* mRNA. This loss, in turn, enables the resulting mRNA to escape *Let-7*–mediated translational inhibition and eventual degradation. The end result is greatly increased levels of the *HMGA2* mRNA and protein, which proceeds, via still poorly understood mechanisms, to alter chromatin configuration (see Section 1.8) and thereby to facilitate cell transformation. In this instance, translocations serve to liberate a proto-oncogene from negative regulation rather than fusing it with a positive regulator.

In another case, chromosomal translocations, like those involving *myc*, have been found to drive enhanced expression of a microRNA by fusing the miRNA-encoding gene with a foreign transcriptional promoter. In tumors with such translocations—myelodysplasia and associated acute myelogenous leukemias—the resulting overexpressed miRNA blocks differentiation of cells, trapping them in a state where they can evolve into highly aggressive tumors. This discovery, along with others that followed, resolves the long-standing mystery of how it is possible for chromosomal translocations that do not involve protein-encoding genes to affect cell behavior.

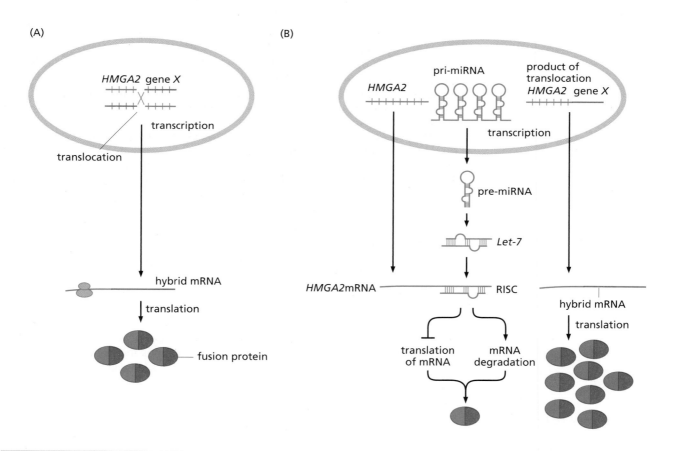

Figure 4.14 Translocations liberating an mRNA from miRNA inhibition The *HMGA2* gene encodes a chromatin-modifying protein whose function is still poorly understood; this protein (often present as a hybrid protein fused to the protein encoded by a second gene) is expressed at elevated levels in a variety of cancer cell types in the absence of detectable gene amplification. (A) A chromosomal translocation that fuses the *HMGA2* (*aquamarine*) gene with a second gene (gene X, *red*) was initially assumed to create a hybrid fusion protein whose two domains collaborated to form a functional oncoprotein. (B) The true functional consequence of such a translocation was later realized to derive from severing the reading frame of the *HMGA2* mRNA from its 3' untranslated region (3'UTR). In the 3'UTR of this mRNA there are normally seven recognition sites that are bound by the *Let-7* microRNA, which reduces translation of this mRNA and drives its degradation, resulting in suppression of HMGA2 protein synthesis. (Only one binding site is illustrated here.) These recognition sites are absent in the novel hybrid mRNA, allowing this mRNA to escape inhibition by *Let-7*; this results in synthesis of greatly increased amounts of the hybrid protein *(right)*. Importantly, the domain of this novel protein that is encoded by the fusion partner gene (gene X, *red*) does not contribute to the oncogenic functions of the resulting protein. (Adapted from A.R. Young and M. Narita, *Genes Dev.* 21:1005–1009, 2007.)

Returning to *myc*, we see that there are three alternative ways of activating the c-*myc* proto-oncogene—through provirus integration, gene amplification, or chromosomal translocation—all converging on a common mechanistic theme. Invariably, the gene is deprived of its normal physiologic regulation and is forced instead to be expressed at high, constitutive levels.

In general, the mechanisms leading to the overexpression of genes in cancer cells remain poorly understood. Some overexpression, as indicated here, is achieved through gene amplification and chromosomal translocation. But even more frequently, genes that are present in normal configuration and at normal copy number are transcribed at excessively high levels in cancer cells through the actions of deregulated transcription factors; the latter are largely uncharacterized. To complicate matters, gene amplification does not always lead to overexpression. Instead, only 40 to 60% of the genes that are found to be amplified in cancer cell genomes show corresponding increases in their RNA transcripts (and thus proteins). Such observations indicate that the expression levels of many genes are regulated by complex negative-feedback mechanisms that ensure physiologically appropriate levels of expression even in the presence of excessive copies of these genes.

4.6 A diverse array of structural changes in proteins can also lead to oncogene activation

The point mutation discovered in *ras* genes was the first of many mutations that were found to affect the structures of proto-oncogene–encoded proteins, converting them into active oncoproteins. As an important example, the formation of certain human tumors, such as gastric and mammary carcinomas and glioblastoma brain tumors involves the protein that serves as the cell surface **receptor** for epidermal growth factor (EGF). As we will discuss in detail in the next chapter, this receptor protein extends from the extracellular space through the plasma membrane of cells into their cytoplasm. Normally, the EGF receptor, like almost 60 similarly structured receptors, recognizes the presence of its cognate **ligand** (that is, EGF) in the extracellular space and, in response, informs the cell interior of this encounter. In about one-third of glioblastomas, however, the EGF receptor has been found to be decapitated, lacking most of its extracellular domain (**Figure 4.15**). We now know that such truncated receptors send growth-stimulatory signals into cells, even in the absence of any EGF. In so doing, they act as oncoproteins to drive cell proliferation. In Chapter 5, we will describe more precisely how structural alterations of growth factor receptors convert them into potent oncoproteins.

The chromosomal translocations cited in the previous section result, in one way or another, in *deregulated expression* of oncogenic proteins or microRNAs; even more common are translocations that lead to the formation of *novel hybrid proteins*. The best known of these are the chromosomal translocations found in more than 95% of the cases of chronic **myelogenous** leukemia (CML; see Figure 2.26). These CML translocations, which result in the synthesis of hybrid proteins, contrast with those

Figure 4.15 Deregulated firing of growth factor receptors Normally, growth factor receptors displayed on the plasma membrane of a cell release signals into the cell interior only when the extracellular domain of the receptor has bound the appropriate growth factor (GF; that is, the *ligand* of the receptor). However, if the extracellular domain of certain receptors is deleted because of a mutation in the receptor-encoding gene or alternative splicing of the receptor pre-mRNA, the resulting truncated receptor protein then emits signals into the cell without binding its growth factor ligand.

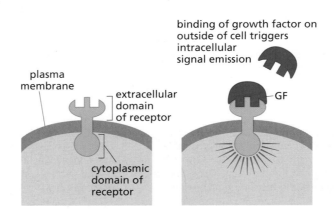

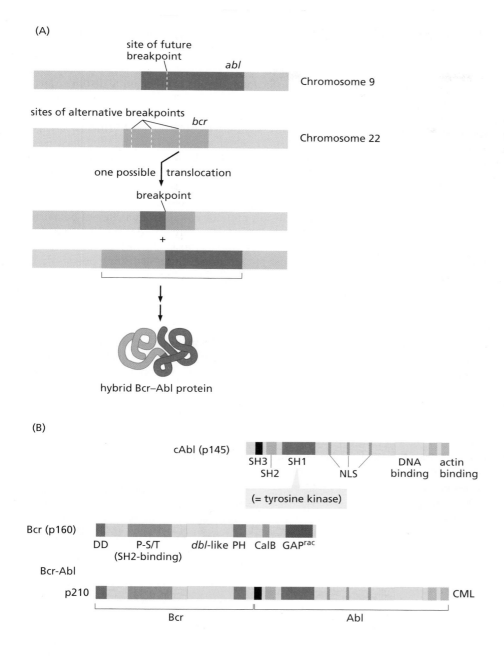

(A)

site of future breakpoint

abl

Chromosome 9

sites of alternative breakpoints

bcr

Chromosome 22

one possible | translocation

breakpoint

+

hybrid Bcr–Abl protein

(B)

cAbl (p145)

SH3   SH1   DNA   actin
SH2   NLS   binding   binding

(= tyrosine kinase)

Bcr (p160)

DD   P-S/T   *dbl*-like PH CalB GAP^rac
(SH2-binding)

Bcr-Abl

p210   CML

Bcr   Abl

Figure 4.16 Formation of the *bcr–abl* oncogene (A) Reciprocal chromosomal translocations between human Chromosomes 9 and 22, which carry the *abl* and *bcr* genes, respectively, result in the formation of fused, hybrid genes that encode hybrid Bcr–Abl proteins commonly found in chronic myelogenous leukemias (CML). Two other alternative breakpoint sites in the *bcr* gene are involved in *bcr–abl* translocations (*not shown here*) arising in certain other hematopoietic malignancies. (B) The domain structures of the proteins encoded by the normal *bcr* and *abl* genes are shown here (*above and middle*). Each is a multidomain, multifunctional protein, as indicated by the labels attached to the colored areas. The chromosomal translocation found in almost all cases of CML results in the fusion of most of the reading frame of the *Bcr* gene with most of the reading frame of the *Abl* gene (*below*). The fusion causes deregulated firing of the tyrosine kinase of the fusion protein, which is equivalent to the SH1 domain; this firing (to be discussed in Section 6.3) is responsible for most of the transforming effects of the newly formed oncoprotein. (B, adapted from A.M. Pendergast, in A.M. Carella, Chronic Myeloid Leukaemia: Biology and Treatment. London: Martin Dunitz, 2001.)

found in Burkitt's lymphomas (see Figure 4.13), in which the transcription-regulating sequences of one gene are juxtaposed with the reading frame of a second gene.

To the right of the CML translocation breakpoint illustrated in **Figure 4.16** are sequences that encode the protein made by the *abl* proto-oncogene. The *abl* gene was originally discovered by virtue of its presence as an acquired oncogene in **Abel**son murine leukemia virus, a rapidly tumorigenic retrovirus (refer to Table 3.3). The *abl* gene, which maps to Chromosome 9q34 (that is, the fourth band of the third region of the long arm of human Chromosome 9), was found, upon examination of various CMLs, to be fused with sequences that are clustered in a narrow region in Chromosome 22q11. [The standard notation used here is t(9;22)(q34;q11), where t signifies a translocation, q the long arm, and p the short arm of a chromosome.] This area on Chromosome 22 was called the **b**reakpoint **c**luster **r**egion, hence *bcr*. Subsequently, all of these breakpoints were found to lie within the *bcr* gene. The resulting fusion of Abl with Bcr amino acid sequences deregulates the normally well-controlled Abl protein, causing it to emit growth-promoting signals in a strong, deregulated fashion. (Two other alternative breakpoints in the *bcr* gene result in the formation of the slightly different Bcr–Abl proteins found in other hematopoietic malignancies; these alternative Bcr–Abl proteins are not illustrated in Figure 4.16.)

Since the discovery of the *bcr-abl* translocation, dozens of other, quite distinct translocations have been documented that also result in the formation of hybrid proteins, most but not all encountered in small subsets of hematopoietic tumors (for examples, see **Table 4.5**). Possibly the greatest diversity derives from the *MLL1* (mixed-lineage leukemia) gene, occasionally termed *ALL1*, which encodes a histone methylase (see Section 1.8); it participates in translocations with at least 50 distinct fusion-partner genes, resulting in a diverse array of hybrid proteins, all of which appear to affect chromatin structure and function. Because each type of translocation is found only rarely, the function of many of the resulting hybrid proteins remains obscure. In the case of solid tumors, an example of such fusion proteins comes from a class of pediatric tumors—alveolar rhabdomyosarcomas—80% of which are found to harbor a translocation that fuses the domains of two transcription factors (PAX3 or PAX7 and FKHR), thereby yielding a protein with far more potent transcription-activating powers than either of its two normal precursors possessed. (This novel transcription factor seems to expand the population of skeletal muscle progenitor cells, thereby setting the stage for acquisition of additional mutations that then lead to the eruption of highly aggressive tumors.)

Every translocation, even those occurring repeatedly in many tumors of a given type, appears to represent the product of an exceedingly rare genetic accident. Some of these accidents generate deregulated gene or protein expression while others encode novel hybrid proteins. However, once formed, some of these translocations favor the outgrowth of the cells that carry them, this being indicated by the presence of these chromosomal translocations in all of the neoplastic cells within resulting tumors.

Table 4.5 Translocations in human tumors that cause the formation of oncogenic fusion proteins of novel structure and function

Oncogene	Neoplasm
bcr/abl	chronic myelogenous leukemia; acute lymphocytic leukemia
dek/can	acute myeloid leukemia
E2A/pbx1	acute pre-B-cell leukemia
PML/RAR	acute promyelocytic leukemia
tls/erg	myeloid leukemia
irel/urg	B-cell lymphoma
CBFβ/MYH11	acute myeloid leukemia
aml1/mtg8	acute myeloid leukemia
ews/fli	Ewing's sarcoma
lyt-10/Cα1	B-cell lymphoma
hrx/enl	acute leukemias
hrx/af4	acute leukemias
NPM/ALK	large-cell lymphomas
PAX3/FKHR	alveolar rhabdomyosarcoma
EML4/ALK	non-small-cell lung cancer
MLL/various	acute leukemias

Adapted from G.M. Cooper, Oncogenes, 2nd ed. Sudbury, MA: Jones and Bartlett, 1995. www.jbpub.com. Reprinted with permission.

4.7 Synopsis and prospects

By the late 1970s, disparate lines of evidence concerning cancer genes coalesced into a relatively simple idea. The genomes of mammals and birds contain a cohort of proto-oncogenes, which function to regulate normal cell proliferation and differentiation. Alterations of these genes that affect either the control of their expression or the structure of their encoded proteins can lead to overly active growth-promoting genes, which appear in cancer cells as activated oncogenes. Once formed, such oncogenes proceed to drive cell multiplication and, in so doing, play a central role in the pathogenesis of cancer.

Many of these cellular genes were originally identified because of their presence in the genomes of rapidly transforming retroviruses, such as Rous sarcoma virus, avian erythroblastosis virus, and Harvey sarcoma virus. Subsequently, transfection experiments revealed the presence of potent transforming genes in the genomes of cells that had been transformed by exposure to chemical carcinogens and cells derived from spontaneously arising human tumors. These tumor cells had no associations with retrovirus infections. Nonetheless, the oncogenes that they carried were found to be related to those carried by transforming retroviruses. This meant that a common repertoire of proto-oncogenes could be activated by two alternative routes: retrovirus acquisition or somatic mutation.

The somatic mutations that caused proto-oncogene activation could be divided into two categories—those that caused changes in the structure of encoded proteins and those that led to elevated, deregulated expression of these proteins. Mutations affecting structure included the point mutations affecting *ras* proto-oncogenes and the chromosomal translocations that yielded hybrid genes such as *bcr–abl*. Elevated expression could be achieved in human tumors through gene amplification or chromosomal translocations, such as those that place the *myc* gene under the control of immunoglobulin enhancer sequences.

These revelations about the role of mutant cellular genes in cancer pathogenesis eclipsed for some years the observations that certain human cancers, some of them quite common, are associated with and likely caused by infectious agents, notably viruses and bacteria. With the passage of time, however, the importance of infections in human cancer pathogenesis became clear. We now know that about one-fifth of human deaths from cancer worldwide are associated with tumors triggered by infectious agents. Thus, the 9% of worldwide cancer mortality caused by stomach cancer is associated with long-term infections by *Helicobacter pylori* bacteria. Six percent of cancer mortality is caused by carcinomas of the liver (hepatomas), almost all of which are associated with chronic hepatitis B and C virus infections. And the 5% of cancer mortality from cervical carcinomas is almost entirely attributable to human papillomavirus (HPV) infections (**Table 4.6**).

Somehow, we must integrate the carcinogenic mechanisms activated by these various infectious agents into the larger scheme of how human cancers arise—the scheme that rests largely on the discoveries, described in this chapter, of cellular oncogenes. In the case of the cancer-causing viruses, it is clear that they introduce viral oncogenes into cells that contribute to the transformation phenotype. Included here are HPV, Epstein–Barr virus (EBV), human herpesvirus type 8 (HHV-8) and the two hepatitis viruses, HBV and HCV.

Each of these viruses carries at least one gene that alters the physiology of the cells that it infects. In each case, it seems that these virus-induced changes are driven by the need to create an intracellular environment that is more hospitable for viral replication. Of note, the viral genes that induce these changes, which can fairly be termed viral oncogenes, have resided in viral genomes for millions of years rather than being rare aberrations of the sort that have led to the formation of transforming retroviruses like Rous sarcoma virus (RSV). Hence, cancer, to the extent that it ensues from infections by these viruses, is an unintended side-effect of the changes favoring viral replication. In Chapters 8 and 9 we shall see precisely how several of these viral oncogenes function to deregulate cell proliferation and protect cells from apoptosis (programmed cell death), two critical changes that contribute to the formation of human cancers.

Table 4.6 Viruses implicated in human cancer causation

Virus[a]	Virus family	Cells infected	Human malignancy	Transmission route
EBV[b]	Herpesviridae	B cells	Burkitt's lymphoma	saliva
		oropharyngeal epithelial cells	nasopharyngeal carcinoma	saliva
		lymphoid	Hodgkin's disease[c]	
HTLV-I	Retroviridae	T cells	non-Hodgkin's lymphoma	parenteral, venereal[d]
HHV-8[e]	Herpesviridae	endothelial cells	Kaposi's sarcoma, body cavity lymphoma	venereal, vertical[d]
HBV	Hepadnaviridae	hepatocytes	hepatocellular carcinoma	parenteral, venereal
HCV	Flaviviridae	hepatocytes	hepatocellular carcinoma	parenteral
HPV	Papillomaviridae	cervical epithelial	cervical carcinoma	venereal
JCV[f]	Polyomaviridae	central nervous system	astrocytoma, glioblastoma	?

[a]Most of these viruses carry one or more growth-promoting genes/oncogenes in their genomes.
[b]EBV is also involved in transplant-related lymphoproliferative disorder, gastric cancer, and cutaneous leiomyosarcoma (in patients with AIDS).
[c]These lymphomas, which bear copies of EBV genomes, appear in immunosuppressed patients.
[d]Parenteral, blood-borne; venereal, via sexual intercourse; vertical, parent to child.
[e]Also known as KSHV, Kaposi's sarcoma herpesvirus.
[f]JCV (JC virus, a close relative of SV40) infects more than 75% of the human population by age 15, but the listed virus-containing tumors are not common. Much correlative evidence supports the role of JCV in the transformation of human central nervous system cells, but evidence of a causal role in tumor formation is lacking.

Adapted in part from J. Butel, *Carcinogenesis* 21:405–426, 2000.

In the next two chapters, however, we focus on the cellular genes that trigger cancer as a consequence of somatic mutations and the various biochemical mechanisms-of-action of the encoded oncoproteins. At first, the discoveries of cellular oncogenes seemed, on their own, to provide the definitive answers about the origins and growth of human tumors. But soon it became clear that cellular oncogenes and their actions could explain only a part of human cancer development. Yet other types of genetic elements were clearly involved in programming the runaway proliferation of cancer cells. Moreover, a definitive understanding of cancer development depended on understanding the complex machinery inside cells that enables them to respond to the growth-promoting signals released by oncoproteins. So, the discovery of cellular oncogenes was a beginning, a very good one, but still only a beginning. Two decades of further work were required before a more complete understanding of cancer pathogenesis could be assembled.

Key concepts

- Unable to find tumor viruses in the majority of human cancers, researchers in the mid-1970s were left with one main theory of how most human cancers arise: the mutation by carcinogens of normal growth-controlling genes, converting them into oncogenes.

- To verify this model's prediction that transformed cells carry mutated genes functioning as oncogenes, a novel experimental strategy was devised: DNA from chemically transformed cells was introduced into normal cells—the procedure of transfection—and the recipient cells were then monitored to determine whether they too had become transformed.

- Cultures of NIH 3T3 cells that had been transfected with DNA from chemically transformed mouse cells yielded numerous transformants, which proved to be both anchorage-independent and tumorigenic; this indicated that the chemically transformed cells carried genes that could function as oncogenes and that oncogenes could arise in the genomes of cells independently of viral infections.

- Further experiments using human tumor donor cells transfected into murine cells showed that the oncogenes could act across species and tissue boundaries to induce cell transformation.

- The oncogenes detected in human tumor cells by transfection experiments and the oncogenes of transforming retroviruses were both found to derive from the same group of preexisting normal cellular genes. This meant that these normal proto-oncogenes could be activated either by retroviral modifications or by somatic mutations.

- Retrovirus-associated oncogenes were often found to be present in increased copy number in human tumor cell genomes, which suggested that gene amplification resulted in increased levels of protein products that favored the proliferation of cancer cells, an example being the amplification of the *myc* gene in a variety of human cancer types.

- The activation of retrovirus-associated genes via deregulation by viral transcriptional promoters was well documented. However, the mechanism(s) by which normal human proto-oncogenes became converted into oncogenes in the absence of viruses was not evident until the discovery of a point mutation in the H-*ras* gene that yielded a structurally altered protein with aberrant behavior. This established a new mechanism for oncogene activation based on a change in the structure of an oncogene-encoded protein rather than the expression levels of such a protein.

- Both activation mechanisms—regulatory and structural—might collaborate to create an active oncogene.

- The *myc* oncogene was initially discovered in the genome of an avian retrovirus. This oncogene was found to be activated through several alternative mechanisms in cancer cells: provirus integration, gene amplification, and chromosomal translocation.

- Gene amplification occurs through preferential replication of a segment (an amplicon) of chromosomal DNA. The result may be repeating end-to-end linear arrays of the segment, which appear under the light microscope as homogeneously staining regions (HSRs) of a chromosome. Alternatively, the region carrying the amplified segment may break away from the chromosome and be seen as small, independently replicating, extrachromosomal particles (double minutes, DMs).

- Translocation involves the fusion of a region from one chromosome to a nonhomologous chromosome. Translocation can place a gene under the control of a foreign transcriptional promoter and lead to its overexpression, as is the case with the *myc* oncogenes in Burkitt's lymphomas. Translocation may also free an mRNA from inhibition by a microRNA by removing mRNA sequences normally recognized by the miRNA. Alternatively, a translocation can result in the fusion of two protein-coding sequences, leading to a hybrid protein that functions differently than the two normal proteins from which it arose, as is seen in the Bcr–Abl protein encountered in chronic myelogenous leukemias.

- Besides the amino acid substitutions that activate signaling by the Ras oncoprotein and the fusion of protein domains, as is seen in the Bcr–Abl protein, yet other changes in protein structure can lead to oncogene activation. For example, truncation of the EGF receptor leads this protein to emit growth-promoting signals in an unremitting fashion.

Thought questions

1. What evidence do we have that suggests that endogenous retrovirus genomes play little if any role in the development of human cancers?

2. Why might an assay like the transfection-focus assay fail to detect certain types of human tumor–associated oncogenes?

3. What molecular mechanisms might cause a certain region of chromosomal DNA to accidentally undergo amplification?

4. How many distinct molecular mechanisms might be responsible for converting a single proto-oncogene into a potent oncogene?

5. How many distinct molecular mechanisms might allow chromosomal translocations to activate proto-oncogenes into oncogenes?

6. What experimental search strategies would you propose if you wished to launch a systematic screening of a vertebrate genome in order to enumerate all of the proto-oncogenes that it harbors?

7. Since proto-oncogenes represent distinct liabilities for an organism, in that they can incite cancer, why have these genes not been eliminated from the genomes of chordates?

Additional reading

Barbacid M (1987) *ras* genes. *Annu. Rev. Biochem.* 56, 779–827.

Bos JL (1989) *Ras* oncogenes in human cancer: a review. *Cancer Res.* 49, 4682–4689.

Bos JL, Rehmann H & Wittinghofer A (2007) GEFs and GAPs: critical elements in the control of small G proteins. *Cell* 129, 865–877.

Brodeur G (2003) Neuroblastoma: biological insights into a clinical enigma. *Nat. Rev. Cancer* 3, 203–216.

Butel J (2000) Viral carcinogenesis: revelation of molecular mechanisms and etiology of human disease. *Carcinogenesis* 21, 405–426.

Edwards PAW (2010) Fusion genes and chromosome translocations in the common epithelial cancers. *J. Pathol.* 220, 244–254.

Lowy DR & Willumsen BM (1993) Function and regulation of ras. *Annu. Rev. Biochem.* 62, 851–891.

Martin GS (2001) The hunting of the Src. *Nat. Rev. Mol. Cell Biol.* 2, 467–475.

Mitelman F, Johansson B & Mertens F (eds) (2004) Mitelman Database of Chromosome Aberrations in Cancer. http://cgap.nci.nih.gov/Chromosomes/Mitelman

Morris DS, Tomlins SA, Montie JE & Chinnaiyan AM (2008) The discovery and application of gene fusions in prostate cancer. *BJU Int.* 102, 276–281.

Rabbitts TH (1991) Translocations, master genes, and differences between the origins of acute and chronic leukemias. *Cell* 67, 641–644.

Rowley JD (2001) Chromosome translocations: dangerous liaisons revisited. *Nat. Rev. Cancer* 1, 245–250.

Santarius T, Shipley J, Brewers D et al. (2010) A census of amplified and overexpressed human cancer genes. *Nat. Rev. Cancer* 10, 59–64.

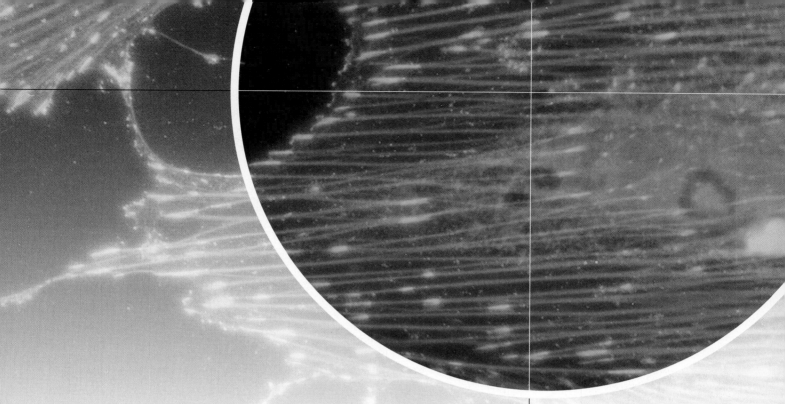

Chapter 5

Growth Factors, Receptors, and Cancer

> The ability of certain fetal serums to stimulate cell growth and the decreased requirement for such factors by transformed cells may be due to the fact that these serum factors are the same or similar to the transforming factors synthesized by some embryonic or neoplastic cells.
>
> David E. Comings, geneticist, 1973

The discovery of oncogenes and their precursors, the proto-oncogenes, stimulated a variety of questions. The most pointed of these centered on the issue of how the oncogenes, acting through their encoded protein products, succeed in perturbing cell behavior so profoundly. A variety of cell phenotypes were concomitantly altered by the actions of oncoproteins such as Src and Ras, the products of the *src* and *ras* oncogenes. How could a single protein species succeed in changing so many different cellular regulatory pathways at the same time?

The vital clues about oncoprotein functioning came from detailed studies of how normal cells regulate their growth and division. Normal cells receive growth-stimulatory signals from their surroundings. These signals are processed and integrated by complex circuits within the cell, which decide whether cell growth and division is appropriate or not (**Figure 5.1**).

This need to receive extracellular signals at the cell surface and to transfer them into the cytoplasm creates a challenging biochemical problem. The extra- and intracellular spaces are separated by the lipid bilayer that is the plasma membrane. This membrane is a barrier that effectively blocks the movement of virtually all but the smallest molecules through it, resulting in dramatically differing concentrations of many types of molecules (including ions) on its two sides. How have cells managed to solve the problem of passing (**transducing**) signals through a membrane that is almost impervious? And given this barrier, how can the inside of the cell possibly know what is going on in the surrounding extracellular space?

131

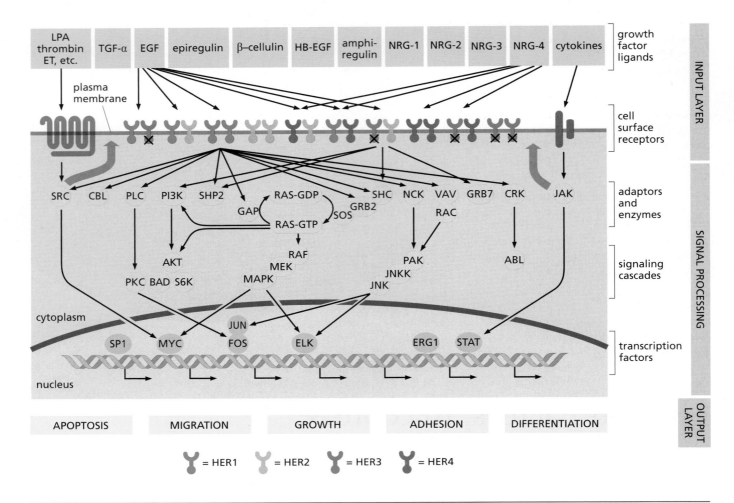

Figure 5.1 The human EGF receptor (HER) signaling network: how cells communicate with their surroundings

A variety of protein messengers (growth factor ligands, *light green rectangles, top*) interact with a complex array of cell surface receptors, which transduce signals across the plasma membrane (*gray*) into the cytoplasm. There, a complex network of signal-transducing proteins processes these signals, funnels signals into the nucleus (*bottom*), and ultimately evokes a variety of biological responses ("output layer," *yellow rectangles, bottom*). Many of the components of this circuitry, both at the cell surface and in the cell interior, are involved in cancer pathogenesis. This cartoon focuses on a small subset of the receptors—the EGF receptor and its cousins—that are displayed on the surfaces of mammalian cells. Receptors like these are the main topic of this chapter; the adaptors and signaling cascades will be covered in the next chapter. The X's associated with the cytoplasmic domain of HER3 indicate the absence of detectable tyrosine kinase activity in contrast to the readily detectable kinase activity of the other three members of this family of receptors. (From Y. Yarden and M.X. Sliwkowski, *Nature Rev. Mol. Cell Biol.* 2:127–137, 2001.)

These signaling processes are part of the larger problem of cell-to-cell communication. Indeed, the evolution of the first multicellular animals (metazoa) 600 to 700 million years ago depended on the development of biochemical mechanisms that allow cells to receive and process signals arising from their neighbors within tissues. Without effective intercellular communication, the behavior of individual cells could not be coordinated, and the formation of architecturally complex tissues and organisms was inconceivable. Obviously, such communication depended on the ability of some cells to emit signals and of others to receive them and respond in specific ways.

In very large part, the signals passed between cells are carried by proteins. Hence, signal emission requires an ability by some cells to release proteins into the extracellular space. Such release—the process of protein secretion—is also complicated by the imperviousness of the plasma membrane. After these signaling proteins are released into the extracellular space, the designated recipient cells must be able to sense the presence of these proteins in their surroundings. Much of this chapter is focused on this second problem—how normal cells receive signals from the environment that surrounds them. As we will see, the deregulation of this signaling is central to the formation of cancer cells.

5.1 Normal metazoan cells control each other's lives

As already implied, the normal versions of oncogene-encoded proteins often serve as components of the machinery that enables cells to receive and process biochemical signals regulating cell proliferation. Therefore, to truly appreciate the complexities of oncogene and oncoprotein function, we need to understand the details of how normal cell proliferation is governed.

Our entrée into this discussion comes from a basic and far-reaching principle. Proper tissue architecture depends absolutely on maintaining appropriate proportions of different constituent cell types within a tissue, on the replacement of missing cells, and on discarding extra, unneeded cells (Figure 5.2). Wounds must be repaired, and attacks by foreign infectious agents must be warded off through the concerted actions of many cells within the tissue.

All of these functions depend upon cooperation among large groups of cells. This explains why cells in a living tissue are constantly talking to one another. Much of this incessant chatter is conveyed by **growth factors (GFs)**. These are relatively small proteins that are released by some cells, make their way through intercellular space, and eventually impinge on yet other cells, carrying with them specific biological messages. Growth factors convey many of the signals that tie the cells within a tissue together into a single community, all members of which are in continuous communication with their neighbors.

Decisions about growth versus no-growth must be made for the welfare of the entire tissue and whole organism, not for the benefit of its individual component cells. For this reason, no single cell within the condominium of cells that is a living tissue can be granted the autonomy to decide on its own whether it should proliferate or remain in a nongrowing, quiescent state. This weighty decision can be undertaken only after

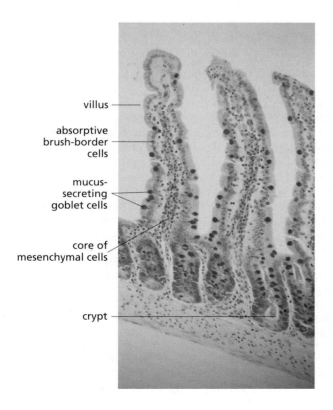

villus

absorptive brush-border cells

mucus-secreting goblet cells

core of mesenchymal cells

crypt

Figure 5.2 Maintenance of tissue architecture The epithelial lining of the small intestine shown here illustrates the fact that a number of distinct cell types coexist within a tissue. The relative numbers and positions of each cell type must be tightly controlled in order to ensure proper tissue structure and function. This control is largely achieved via the exchange of signals between neighboring cells within the tissue. In this particular epithelium, cell-to-cell signaling also ensures that new epithelial cells—termed *enterocytes* in the intestine—are continually being generated in the crypts (at the bases of fingerlike villi) in order to replace others that have migrated up the sides of the villi to the tips, where they are sloughed off. Included among the epithelial cells are mucus-secreting goblet cells (*dark red*) and absorptive cells, as well as enteroendocrine cells and Paneth cells (*not shown*). In addition, in the middle of each villus is a core of mesenchymal cell types, which together constitute the stroma and include fibroblasts, endothelial cells, pericytes, and macrophages. (From B. Alberts et al., Molecular Biology of the Cell, 5th ed. New York: Garland Science, 2008.)

Figure 5.3 Storage of PDGF by platelets Platelets carry numerous secretory (*exocytotic*) vesicles termed α-granules (*arrows*), which contain important signaling molecules, prominent among them being several types of platelet-derived growth factor (PDGF). When platelets become activated during clot formation, these vesicles fuse with the plasma membrane, releasing previously stored growth and survival factors into the extracellular space. (Courtesy of S. Israels.)

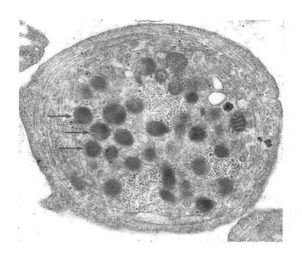

consultation with other cells within the tissue. These neighbors may provide a particular cell with growth factors that stimulate its proliferation or release growth-inhibitory factors that discourage it. In the end, all the decisions made by an individual cell about its proliferation must, by necessity, represent a consensus decision shared with the cells that reside in its neighborhood.

The dependence of individual cells on their surroundings is illustrated nicely by the behavior of normal cells when they are removed from living tissue and propagated in a Petri dish. Even though the liquid medium placed above the cells contains all the nutrients required to sustain their growth and division, including amino acids, vitamins, glucose, and salts, such medium, on its own, does not suffice to induce these cells to proliferate. Instead, this decision depends upon the addition to the medium of serum, usually prepared from the blood of calves or fetal calves. Serum contains the growth factors that persuade cells to multiply.

Serum is produced when blood is allowed to clot. The blood platelets adhere to one another and form a matrix that gradually contracts and traps most of the cellular components of the blood, including both the white and red cells. In the context of a wounded tissue, this clot formation is designed to stanch further bleeding. The clear fluid that remains after clot formation and retraction constitutes the serum.

While the platelets in a wound site are in the process of aggregating as part of clot formation, they also initiate the wound-healing process, doing so through the release of growth factors, notably platelet-derived growth factor (PDGF), into the medium around them (**Figure 5.3**). PDGF is a potent stimulator of fibroblasts, which form much of the connective tissue including the cell layers beneath epithelia (see Figure 2.3). Growth-stimulating factors such as PDGF are often termed **mitogens**, to indicate their ability to induce cells to proliferate. More specifically, PDGF attracts fibroblasts into the wound site and then stimulates their proliferation (for example, **Figure 5.4A**). [Other growth factors, such as epidermal growth factor (EGF), can cause dramatic changes in cell shape in addition to stimulating cell proliferation (see Figure 5.4B).] Without stimulation by serum-derived PDGF, cultured fibroblasts will remain viable and metabolically active for weeks in a Petri dish, but they will not grow and divide.

This dependence of fibroblast on growth-stimulating signals released by a second cell type, in this case blood platelets, mirrors hundreds of similar cell-to-cell communication routes that operate within living tissues to encourage or discourage cell proliferation. Moreover, as discussed in greater detail below, PDGF and EGF are only two of a large and disparate group of growth factors that help to convey important growth-controlling messages from one cell to another.

As developed in the rest of this chapter, the signaling molecules that enable cells to sense the presence of growth factors in their surroundings, to convey this information into the cell interior, and to process this information are usurped by oncogene-encoded

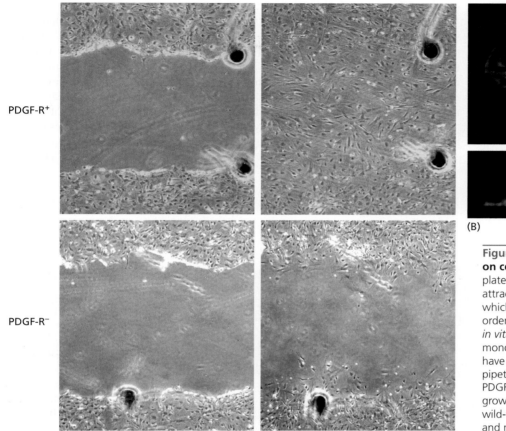

PDGF-R⁺

PDGF-R⁻

(A) control + PDGF

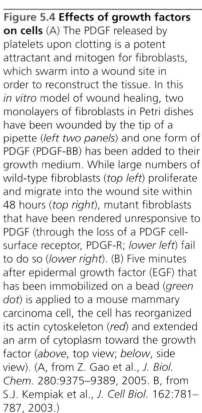

(B)

Figure 5.4 Effects of growth factors on cells (A) The PDGF released by platelets upon clotting is a potent attractant and mitogen for fibroblasts, which swarm into a wound site in order to reconstruct the tissue. In this *in vitro* model of wound healing, two monolayers of fibroblasts in Petri dishes have been wounded by the tip of a pipette (*left two panels*) and one form of PDGF (PDGF-BB) has been added to their growth medium. While large numbers of wild-type fibroblasts (*top left*) proliferate and migrate into the wound site within 48 hours (*top right*), mutant fibroblasts that have been rendered unresponsive to PDGF (through the loss of a PDGF cell-surface receptor, PDGF-R; *lower left*) fail to do so (*lower right*). (B) Five minutes after epidermal growth factor (EGF) that has been immobilized on a bead (*green dot*) is applied to a mouse mammary carcinoma cell, the cell has reorganized its actin cytoskeleton (*red*) and extended an arm of cytoplasm toward the growth factor (*above*, top view; *below*, side view). (A, from Z. Gao et al., *J. Biol. Chem.* 280:9375–9389, 2005. B, from S.J. Kempiak et al., *J. Cell Biol.* 162:781–787, 2003.)

proteins. By taking charge of the natural growth-stimulating machinery of the cell, oncoproteins are able to delude a cell into believing that it has encountered growth factors in its surroundings. Once this deception has taken place, the cell responds slavishly by beginning to proliferate, just as it would have done if abundant growth factors were indeed present in the medium around it.

5.2 The Src protein functions as a tyrosine kinase

The first clues to how cell-to-cell signaling via growth factors operates came from biochemical analysis of the v-*src* oncogene and the protein that it specifies. The trail of clues led, step-by-step, from this protein to the receptors used by cells to detect growth factors and in turn to the intracellular signaling pathways that control normal and malignant cell proliferation.

The initial characterization of the v-*src*–encoded oncoprotein attracted enormous interest. This was the first cellular oncoprotein to be studied and therefore had the prospect of giving cancer researchers their first view of the biochemical mechanisms of cell transformation. After the molecular cloning of the *src* oncogene, the amino acid sequence of its protein product was deduced directly from the nucleotide sequence of the cloned gene. The encoded protein was made as a polypeptide chain of 533 amino acid residues and had a mass of almost precisely 60 kilodaltons (kD).

The amino acid sequence of Src provided few clues as to how this oncoprotein functions to promote cell proliferation and transformation. Avian and mammalian cells transformed by the v-*src* oncogene were known to exhibit a radically altered shape, to pump in glucose from the surrounding medium more rapidly than normal cells, to grow in an anchorage-independent fashion, to lose contact inhibition, and to form tumors. Whatever the precise mechanisms of action, it was obvious that Src affected, directly or indirectly, a wide variety of cellular targets.

Clever biochemical sleuthing in 1977–1978 solved the puzzle of how Src operates. Antibodies were produced that specifically recognized and bound Src (Supplementary Sidebar 5.1). The antibody molecules were found to become phosphorylated when they were incubated in solution with both Src and adenosine triphosphate (ATP), the universal phosphate donor of the cell. Phosphorylation was known to involve the covalent attachment of phosphate groups to the side chains of specific amino acid residues. Hence, it was clear that Src operates as a protein **kinase**—an enzyme that removes a high-energy phosphate group from ATP and transfers it to a suitable protein substrate, in this instance, an antibody molecule (**Figure 5.5**). While Src does not normally phosphorylate antibody molecules, its ability to do so in these early experiments suggested that its usual mode of action also involved the phosphorylation of certain protein substrates within cells.

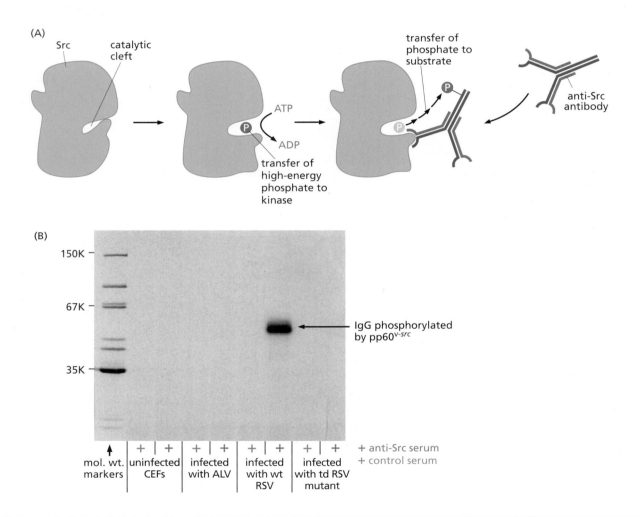

Figure 5.5 Phosphorylation of a precipitating antibody molecule by Src Protein kinases operate by removing the high-energy γ phosphate group from ATP and attaching it to the hydroxyl groups in the side chains of serine, threonine, or tyrosine residues of substrate proteins. (A) The antibody molecule that was used to immunoprecipitate Src molecules (see Supplementary Sidebar 5.1) also happened to serve as a substrate for phosphorylation by this kinase. In cells, Src phosphorylates a wide range of protein substrates. (B) This experiment revealed, for the first time, the biochemical activity associated with an oncoprotein. Normal rabbit serum (*blue + signs*) or serum from a rabbit bearing an RSV-induced, Src-expressing tumor (which contained antibodies against Src; *red + signs*) was used to immunoprecipitate cell lysates that were incubated with ^{32}P-radiolabeled ATP. Lysates were prepared from uninfected chicken embryo fibroblasts (CEFs); CEFs infected with avian leukosis virus (ALV), which lacks a *src* oncogene; CEFs infected with wild-type (wt) RSV; or CEFs infected with a transformation-defective (td) mutant of RSV, which lacks a functional *src* gene. Only the combination of tumor-bearing rabbit serum and a lysate from wt RSV–infected CEFs yielded a strongly ^{32}P-labeled protein that co-migrated with the precipitating antibody, indicating that the antibody molecule had become phosphorylated by the Src oncoprotein. (B, from M.S. Collett and R.L. Erikson, *Proc. Natl. Acad. Sci. USA* 75:2021–2024, 1978.)

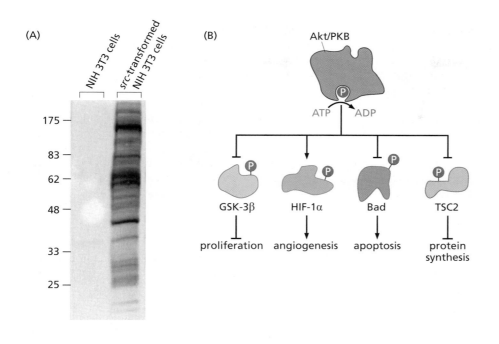

Figure 5.6 Actions of protein kinases
(A) This autoradiograph of proteins resolved by gel electrophoresis illustrates the fact that the phosphorylation state of a number of distinct protein species within a cell can be altered by the actions of the Src kinase expressed in cells by Rous sarcoma virus. This immunoblot analysis, using an anti-phosphotyrosine antibody, specifically detects proteins that carry phosphotyrosine residues, the known products of Src kinase action (see Figure 5.7). The left channel contains proteins from a lysate of untransformed mouse NIH 3T3 fibroblasts and is essentially blank, indicating the relative absence of phosphotyrosine-containing proteins in cells that lack Src expression. The right channel shows the phosphorylated proteins in cells expressing an active Src kinase protein. (B) The pleiotropic actions of a protein kinase usually derive from its ability to phosphorylate and thereby modify the functional state of a number of distinct substrate proteins. Illustrated here are the actions of the Akt/PKB kinase, a serine/threonine kinase that can influence a wide variety of biological processes through phosphorylation of the indicated major control proteins. (A large number of other known substrates of this kinase are not shown here.) Thus, Akt/PKB can inactivate the antiproliferative actions of GSK-3β, the pro-apoptotic powers of Bad, and the translation-inhibiting actions of TSC2, and it can activate the angiogenic (blood vessel–inducing) powers of HIF-1α. In this diagram and throughout this book, arrowheads denote stimulatory signals, while lines at right angles (crossbars) denote an inhibitory signal. (A, from S.M. Ulrich et al., *Biochemistry* 42:7915–7921, 2003.)

Independent of this kinase activity, Src was itself a **phosphoprotein**, that is, it carried phosphate groups attached covalently to one or more of its amino acid side chains. This indicated that Src also served as a substrate for phosphorylation by a protein kinase—either phosphorylating itself (**autophosphorylation**) or serving as the substrate of yet another kinase. Its molecular weight and phosphoprotein status caused it initially to be called pp60[src], although hereafter we will refer to it simply as Src.

The fact that Src functions as a kinase was a major revelation. In principle, a protein kinase can phosphorylate multiple, distinct substrate proteins within a cell. (In the case of Src, more than 50 distinct substrates have since been enumerated.) Once phosphorylated, each of these substrate proteins may be functionally altered and proceed, in turn, to alter the functions of its own set of downstream targets. This mode of action (**Figure 5.6**) seemed to explain how a protein like Src could act pleiotropically to perturb multiple cell phenotypes. At the time of this discovery, other protein kinases had already been found able to regulate complex circuits, notably those involved in carbohydrate metabolism.

Src was soon found to be quite different from all other previously discovered protein kinases. These other kinases were known to attach phosphate groups to the side chains of serine and threonine amino acid residues. Src, in contrast, phosphorylated certain tyrosine residues of its protein substrates (**Figure 5.7**). Careful quantification of the phosphorylated amino acid residues in cells revealed how unusual this enzymatic activity is. More than 99% of the phosphoamino acids in normal cells are phosphothreonine or phosphoserine; phosphotyrosine constitutes as little as 0.05 to 0.1% of these cells' total phosphoamino acids.

After transformation of cells by the v-*src* oncogene, the level of phosphotyrosine was found to rise dramatically, becoming as much as 1% of the total phosphoamino acids in these cells. (This helps to explain the dramatic differences between the two channels shown in Figure 5.6A). When the same cells were transformed by other oncogenes, such as H-*ras*, their phosphoamino acid content hardly changed at all. Therefore, the creation of phosphotyrosine residues was a specific attribute of Src and was not associated with all mechanisms of cell transformation. Subsequent studies yielded a second, related conclusion: signaling through tyrosine phosphorylation is a device that is used largely by mitogenic signaling pathways in mammalian cells, whereas the kinases that are involved in thousands of other signaling processes rely almost exclusively on serine and threonine phosphorylation to convey their messages.

Figure 5.7 Substrate specificity of the Src kinase Electrophoresis can be used to resolve the three types of commonly encountered phosphoamino acids—phosphotyrosine, phosphothreonine, and phosphoserine. In the images shown, electrophoresis was performed toward the cathode at pH 1.9 (*right to left*) and pH 3.5 (*bottom to top*). The right panel shows the expected locations of these three phosphoamino acids following electrophoresis. In normal chicken embryo fibroblasts (CEFs; *left panel*), phosphoserine and phosphothreonine are present in abundance, while almost no phosphotyrosine is evident (*arrow*). However, in *src*-transformed CEFs (*center panel*), the level of phosphotyrosine (*arrow*) rises dramatically. Unrelated phosphorylated compounds and inorganic phosphate are seen at the lower right of each electrophoresis. (Courtesy of T. Hunter.)

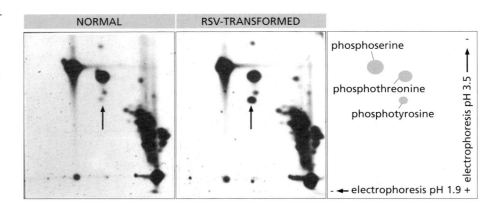

Mutant forms of Src that had lost the ability to phosphorylate substrate proteins also lost their transforming powers. So the accumulating evidence converged on the idea that Src succeeds in transforming cells through its ability to act as a tyrosine kinase (TK) and phosphorylate a still-uncharacterized group of substrate proteins within these cells. While an important initial clue, these advances only allowed a reformulation of the major question surrounding this research area. Now the problem could be restated as follows: How did the phosphorylation of tyrosine residues result in cell transformation?

5.3   The EGF receptor functions as a tyrosine kinase

After the cloning and sequencing of the v-*src* oncogene, the oncogenes of a number of other acutely transforming retroviruses were isolated by molecular cloning and subjected to DNA sequencing. Unfortunately, in the great majority of cases, the amino acid sequences of their protein products provided few clues about biochemical function. The next major insights came instead from another area of research. Cell biologists interested in how cells exchange signals with one another isolated a variety of proteins involved in cell-to-cell signaling and determined the amino acid sequences of these proteins. Quite unexpectedly, close connections were uncovered between these signaling proteins and the protein products of certain oncogenes.

This line of research began with epidermal growth factor (EGF), the first of the growth factors to be discovered. EGF was initially characterized because of its ability to provoke premature eye opening in newborn mice. Soon after, EGF was found to have mitogenic effects when applied to a variety of epithelial cell types.

EGF was able to bind to the surfaces of the cells whose growth it stimulated; cells to which EGF was unable to bind were unresponsive to its mitogenic effects. Taken together, such observations suggested the involvement of a cell surface protein, an EGF *receptor* (EGF-R), that was able to specifically recognize EGF in the extracellular space, bind to it, and inform the cell interior that an encounter with EGF had occurred. In the language of biochemistry, EGF served as the *ligand* for its cognate receptor—the still-hypothetical EGF receptor.

Isolation of the EGF-R protein proved challenging, because this receptor, like many others, is usually expressed at very low levels in cells. Researchers circumvented this problem by taking advantage of a human tumor cell line (an epidermoid carcinoma of the uterus) that expresses the EGF-R at elevated levels—as much as 100-fold higher than normal. These cells yielded an abundance of the receptor protein. It could then be purified biochemically and subjected to amino acid sequence analysis.

The sequence of the EGF-R protein provided important insights into its overall structural features and how this structure enables it to function. Its large N-terminal domain of 621 amino acid residues, which protrudes into the extracellular space (and is therefore termed its **ectodomain**), was clearly involved in recognizing and binding the EGF ligand. The EGF receptor possessed a second distinctive domain that is common

(A)

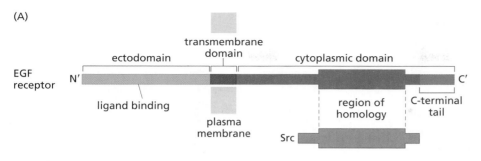

(B)

Src	RESLRLEAK**LGGGCFGEVVWMG**TWNDTTR-----**VAIKTLKPGTMSPE**--**AFLQEAQVMKK**
Abl	RTDITMKHKLGGGQYGEVVYEGVWKKYSLT----VAVKTLKEDTMEVE--EFLKEAAVMKE
Fes	HEDLVLGEQIGRGNFGEVFSGRLRADNTL----VAVKSCRETLPPDIKAKFLQEAKILKQ
ErbB	ETEFKKVKV**LGSGAFG**TIYKG**LW**IPEGEKVKI**VAIKEL**REA**T**SPKANKEIL**DEAYVM**AS
Raf	ASEVMLSTRIGSGSFGIVYKGKWHGD-------VAVKILKVVDPTPEQLQAFRNEVAVLR
Mos	WEQVCLMHRLGSGGFGSVYKATYHGVP------VAIKQVNKCTEDLRASQRSFWAELNIA
Src	LR-HEKLVQL**YAVV**SEEP-IYIVIEY**M**SK-----**GSLLD**FLKGEMGKY--------LRLP
Abl	IK-HPNLVQLLGVCTREPPFYIITEFMTY-----GNLLDYLRECNRQE--------VSAV
Fes	YS-HPNIVRLIGVCTQKQPIYIVMELVQG-----GDFLTFLRTEGAR--------LRMK
ErbB	VD-NPHVCR**LL**GICLTST-VQLITQLM**PY**-----**GCLLD**YIREHKDN---------IGSQ
Raf	KTRHVNI--LLFMGYMTKDNLAIVTQWCEGSSLYKHLHVQETK-----------FQMF
Mos	GLRHDNIVRVVAASTRTPEDSNSLGTIIMEFGGNVTLHQVIYDATRSPEPLSCRKQLSLG
Src	**QLV**DMAA**QIASGMA**YVERMNY**VHRDL**RAA**NILV**GENLVCK**VADFGLARL**IEDN**EY**TARQG
Abl	VLLYMATQISSAMEYLEKKNFIHRDLAARNCLVGENHLVKVADFGLSRLMTGDTYTAHAG
Fes	TLLQMVGDAAAGMEYLESKCCIHRDLAARNCLVTEKNVLKISDFGMSREEADGVYAASGG
ErbB	Y**LL**NWCV**QIAKGMN**Y**LE**ERRL**VHRDL**AARN**VLV**KTPQHV**KITDFGLAK**LLGADE**KE**YHAE
Raf	QLIDIARQTAQGMDYLHAKNIIHRDMKSNNLFLHEGLTVKIGDFGLATVKSRWSGSQQVE
Mos	KCLKYSLDVVNGLLFLHSQSILHLDLKPANILISEQDVCKISDFGCSQKLQDLRGRQASP
Src	AKF--**PIKWTAPEA**ALYGR---FTIK**SDVWS**FGILL**TELTT**KG**RV**PYPG**M**VNRE**VL**DQVE
Abl	AKF--PIKWTAPESLAYNK---FSIKSDVWAFGVLLWEIATYGMSPYPGIDLSQVYELLE
Fes	LRLV-PVKWTAPEALNYGR---YSSESDVWSFGILLWETFSLGASPYPNLSNQQTREFVE
ErbB	GGKV-**PIKWMALE**SILHRI---YTHQ**SDVWS**YGV**TVWELM**TFGSKPYDGIPASE**ISSVLE**.
Raf	QPTG-SVLWMAPEVIRMQDDNPFSFQSDVYSYGIVLYELMA-GELPYAHINNRDQIIFMV
Mos	PHIGGTYTHQAPEILKGEI---ATPKADIYSFGITLWQMTT-REVPYSGEPQYVQYAVVA
Src	R--**GYRMPCPPECP**ESLHD----**LMCQCW**RK**DP**EERP**TFKYL**GAQLLPA
Abl	K--DYRMERPEGCPEKVYE----LMRACWQWNPSDRPSFAEIHQAFETM
Fes	K--GGRLPCPELCPDAVYH----LMEQCWAYEPGQRPSFSAFYQELQSI
ErbB	K--**GERLPQPP**IC**T**IDVYM----I**MVKCWM**IDADS**RP**K**FREL**IAEFSKM
Raf	GR-GYASPDLSRLYKNCPKAIKRLVADCVKKVKEERPLFPQILSSIELL
Mos	YNLRPSLAGAVFTASLTGKALQNIIQSCWEARGLQRPSAELLGRDLKAF

Figure 5.8 Structure of the EGF receptor (A) The receptor for epidermal growth factor (EGF-R) is a complex protein with an extracellular domain (ectodomain, *green*), a transmembrane domain that threads its way through the plasma membrane (*brown*), and a cytoplasmic domain (*red, blue*). The ligand-binding domain is responsible for binding EGF. Similarity in amino acid sequences (see panel B) demonstrates that one region of the cytoplasmic domain (*wide red rectangle*) is related to a region in the Src oncoprotein (*gray*). (B) Comparison of the amino acid sequences (using the single-letter code) of the cytoplasmic domains of the EGF-R (labeled "ErbB" here) and Src revealed areas of sequence identity (*green*), suggesting that the EGF-R, like Src, emits signals by functioning as a tyrosine kinase. While the sequence identities seem to be quite scattered, the shared residues nonetheless indicate clear evolutionary relatedness (homology) between Src and ErbB. Yet other viral oncoproteins, such as Abl and Fes, some of whose sequences are also shown here, were found to share some sequence similarity with these two. The viral oncoproteins specified by the *raf* and *mos* oncogenes, which function as serine/threonine kinases, are seen to be more distantly related, sharing even fewer sequences with Src and the EGF-R. The dashes indicate amino acid residues that are missing in one protein but are present in one or more of its homologs; these have been introduced to maximize the sequence alignment between homologs. (Adapted from G.M. Cooper, Oncogenes, 2nd ed. Sudbury, MA: Jones and Bartlett, 1995. www.jbpub.com. Reprinted with permission.)

to many cell surface glycoproteins; this **transmembrane** domain of 23 amino acid residues threads its way from outside the cell through the lipid bilayer of the plasma membrane into the cytoplasm. The existence of this transmembrane domain could be deduced from the presence of a continuous stretch of **hydrophobic** amino acid residues located in the middle of the protein's sequence; this domain of EGF-R protein resides comfortably in the highly hydrophobic environment of the lipid bilayer that forms the plasma membrane. Finally, a third domain of 542 residues at the C-terminus (that is, the end of the chain synthesized last) of the EGF receptor was found to extend into the cytoplasm (**Figure 5.8A**).

The overall structure of the EGF receptor suggested in outline how it functions. After its ectodomain binds EGF, a signal is transmitted through the plasma membrane to activate the cytoplasmic domain of the receptor; once activated, the latter emits signals that induce a cell to grow and divide. Importantly, examination of this cytoplasmic domain revealed a clear sequence similarity with the already-known sequence of the Src protein (see Figure 5.8B).

Suddenly, it became clear how the EGF receptor emits signals in the cell interior: once its ectodomain binds EGF, somehow the Src-like kinase in its cytoplasmic domain becomes activated, proceeds to phosphorylate tyrosines on certain cytoplasmic proteins, and thereby causes a cell to proliferate. Subsequent sequencing efforts revealed overall sequence similarities among a variety of tyrosine kinases, many of which can function as oncoproteins (see Figure 5.8B).

Figure 5.9 Structure of tyrosine kinase receptors The EGF receptor (Figure 5.8A) is only one of many similarly structured receptors that are encoded by the human genome. These tyrosine kinase receptors (RTKs) can be placed into distinct families, depending on the details of their structure. Representatives of most of these families are shown here. All have in common quite similar cytoplasmic tyrosine kinase domains (*red*), although in some cases (e.g., the PDGF receptor) these domains are interrupted by small "insert" regions. The RTK ectodomains (which protrude into the extracellular space, *green, gray*) have highly variable structures, reflecting the fact that they recognize and bind a wide variety of extracellular ligands. (From B. Alberts et al., Molecular Biology of the Cell, 5th ed. New York: Garland Science, 2008.)

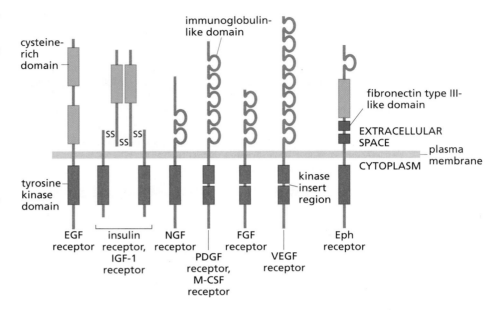

Since these initial analyses of the EGF receptor structure, a large number of other, similarly structured receptors have been characterized (**Figure 5.9**). As discussed later, each of these receptors has its own growth factor ligand or set of ligands (**Table 5.1**). Depending on the particular growth factor–receptor pair, the binding of a ligand to its

Table 5.1 Growth factors (GFs) and tyrosine kinase receptors that are often involved in tumor pathogenesis

Name of GF	Name of receptor	Cells responding to GF
PDGF[a]	PDGF-R	endothelial, VSMCs, fibroblasts, other mesenchymal cells, glial cells
EGF[b]	EGF-R[c]	many types of epithelial cells, some mesenchymal cells
NGF	Trk	neurons
FGF[d]	FGF-R[e]	endothelial, fibroblasts, other mesenchymal cells, VSMCs, neuroectodermal cells
HGF/SF	Met	various epithelial cells
VEGF[f]	VEGF-R[g]	endothelial cells in capillaries, lymph ducts
IGF[h]	IGF-R1	wide variety of cell types
GDNF	Ret	neuroectodermal cells
SCF	Kit	hematopoietic, mesenchymal cells

[a]PDGF is represented by four distinct polypeptides, PDGF-A, -B, -C, and -D. The PDGF-Rs consist of at least two distinct species, α and β, that can homodimerize or heterodimerize and associate with these ligands in different ways.

[b]The EGF family of ligands, all of which bind to the EGF-R (ErbB1) and/or heterodimers of erbB1 and one of its related receptors (footnote c), includes—in addition to EGF—TGF-α, HB-EGF, amphiregulin, betacellulin, and epiregulin. In addition, other related ligands bind to heterodimers of ErbB2 and ErbB3 or ErbB4; these include epigen and a variety of proteins generated by alternatively spliced neuregulin (NRG) mRNAs, including heregulin (HRG), glial growth factor (GGF), and less well-studied factors such as sensory and motor neuron–derived factor (SMDF).

[c]The EGF-R family of receptors consists of four distinct proteins, ErbB1 (EGF-R), ErbB2 (HER2, Neu), ErbB3 (HER3), and ErbB4 (HER4). They often bind ligands as heterodimeric receptors, for example, ErbB1 + ErbB3, ErbB1 + ErbB2, or ErbB2 + ErbB4; ErbB3 is devoid of kinase activity and is phosphorylated by ErbB2 when the two form heterodimers. ErbB2 has no ligand of its own but does have strong tyrosine kinase activity. ErbB3 and ErbB4 bind neuregulins, a family of more than 15 ligands that are generated by alternative splicing.

[d]FGFs constitute a large family of GFs. The prototypes are acidic FGF (aFGF) and basic FGF (bFGF); in addition there are other known members of this family.

[e]There are four well-characterized FGF-Rs.

[f]There are four known VEGFs. VEGF-A and -B are involved in angiogenesis, while VEGF-C and -D are involved predominantly in lymphangiogenesis.

[g]There are three known VEGF-Rs: VEGF-R1 (also known as Flt-1) and VEGF-R2 (also known as Flk-1/KDR), involved in angiogenesis; and VEGF-R3, involved in lymphangiogenesis.

[h]The two known IGFs, IGF-1 and IGF-2, both related in structure to insulin, stimulate cell growth (that is, increase in size) and survival; they also appear to be weakly mitogenic.

Abbreviation: VSMC, vascular smooth muscle cell.
Adapted in part from B. Alberts et al., Molecular Biology of the Cell, 5th ed. New York: Garland Science, 2008.

receptor can trigger multiple biological responses in the cell in addition to stimulating growth and division. Important among these are changes in cell shape, cell survival, and cell **motility**.

Interestingly, these receptors and their ligands represent relatively recent evolutionary inventions, and their antecedents cannot be found among the proteins made by most single-cell eukaryotes. Many were invented just before metazoan life first arose and have been retained in clearly recognizable form in most and probably all modern metazoa. This explains why research on the cell-to-cell communication systems of simpler metazoan organisms, such as worms and flies, has so greatly enriched our understanding of growth factors and how they signal. Unexpectedly, recent research has revealed the presence of a tyrosine kinase receptor gene in the genome of at least one single-cell eukaryote; this organism may be closely related to the common protozoan ancestor of all metazoa and its receptor may be the prototype of many dozens of tyrosine kinase receptors present in modern metazoa including ourselves (Supplementary Sidebar 5.2).

5.4 An altered growth factor receptor can function as an oncoprotein

A real bombshell fell in 1984 when the sequence of the EGF receptor was recognized to be closely related to the sequence of a known oncoprotein specified by the *erbB* oncogene. This oncogene had been discovered originally in the genome of avian erythroblastosis virus (AEV), a transforming retrovirus that rapidly induces a leukemia of the red blood cell precursors (an **erythroleukemia**). In one stroke, two areas of cell biology were united. A protein used by the cell to sense the presence of a growth factor in its surroundings had been appropriated (in its avian form) and converted into a potent retrovirus-encoded oncoprotein.

Once examined in detail, the oncoprotein specified by the *erbB* oncogene of avian erythroblastosis virus was found to lack sequences present in the N-terminal ectodomain of the EGF receptor (**Figure 5.10**). Without these N-terminal sequences, the ErbB oncoprotein clearly cannot recognize and bind EGF, and yet it functions as a potent stimulator of cell proliferation. This realization led to an interesting speculation that was soon validated: somehow, deletion of the ectodomain enables the resulting truncated EGF receptor protein to send growth-stimulating signals into cells in a constitutive fashion, fully independent of EGF. Years later, such truncated EGF receptors were discovered to be present in one-third of human glioblastomas and in smaller proportions of a variety of other human tumors. Still later, the presence in breast cancers of similarly truncated versions of ErbB/EGF-R's close cousin, termed variously ErbB2, HER2, or Neu, were found to be associated with particularly poor prognosis.

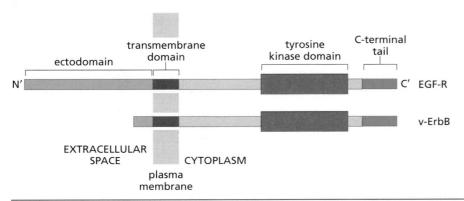

Figure 5.10 The EGF receptor and v-ErbB The EGF receptor and the v-ErbB oncoprotein of avian erythroblastosis virus are closely related. More specifically, the v-ErbB protein is specified by an altered version of the gene encoding the chicken EGF receptor, which encodes a truncated form of EGF-R that lacks most of the normally present ectodomain (*green*). Such a truncated receptor can emit mitogenic signals constitutively, i.e., without stimulation by EGF ligand.

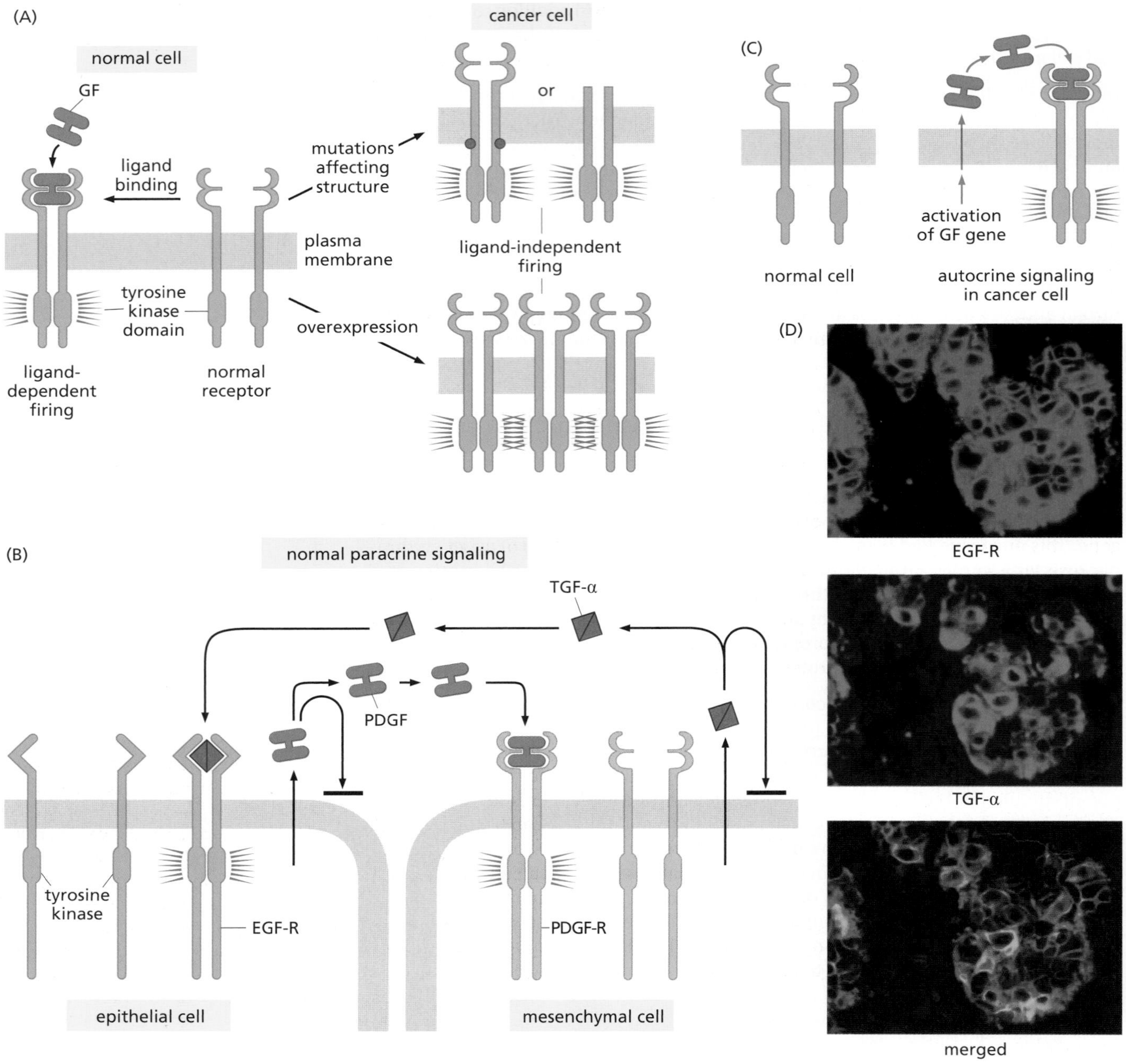

Figure 5.11 Deregulation of receptor firing (A) Normally functioning growth factor receptors emit cytoplasmic signals (*red spikes*) in response to binding ligand (*blue, left*). However, mutations in the genes encoding the receptor molecules (*upper right*) can cause subtle alterations in protein structure, such as amino acid substitutions (*red dots*), that cause ligand-independent firing. More drastic alterations in receptor structure, including truncation of the ectodomain (see, for example, Figure 5.10), may also yield such deregulated signaling. In many human tumors, receptor proteins are overexpressed (*lower right*). Excessive numbers of normally structured receptor molecules can also drive ligand-independent receptor firing by causing these molecules to frequently collide and thereby spontaneously dimerize and release signals (*red spikes*). (B) In general, normal cells do not synthesize and release a growth factor ligand whose cognate receptor they also display. Shown here is the tendency of epithelial cells to make PDGF, the ligand for the PDGF-R displayed by mesenchymal cells; conversely mesenchymal cells often make TGF-α, an EGF-related ligand which binds the EGF-R displayed by epithelial cells. However, the PDGF made by an epithelial cell fails to find its cognate receptor on the surface of that cell (*horizontal bar*) preventing inadvertent activation of autocrine signaling. (C) In contrast, in many types of cancer, tumor cells acquire the ability to make a ligand for a growth factor receptor that they also display (*right*). This creates an auto-stimulatory or autocrine signaling loop. (D) An example of autocrine signaling is seen in these successive sections of an invasive human breast carcinoma, in which islands of cancer cells are surrounded by non-staining stroma (*black*). The top section has been immunostained for expression of the EGF receptor (*red*), while the middle section has been stained for expression of TGF-α (*green*), an EGF-like ligand of the EGF receptor. When these two images are superimposed (*bottom panel*), cells that express both receptor and ligand are stained *yellow*. Nuclei are stained *blue*. (D, from J.S. de Jong et al., *J. Pathol.* 184:44–52, 1998.)

As indicated in **Figure 5.11A**, mutations in the genes encoding growth factor receptors, including those specifying truncated receptors, may trigger ligand-independent firing by these receptors. Indeed, a variety of mutations, including those creating amino acid substitutions in the transmembrane domain (shown), or in the ectodomain and cytoplasmic domains (not shown) of some receptors can provoke ligand-independent firing. (In fact, Figure 5.11 introduces yet another idea that we will explore in detail shortly: growth factors often activate receptors by triggering receptor dimerization.) In Chapter 16, we will see how structurally altered growth factor receptors influence the responsiveness of human tumors to anti-cancer therapeutic drugs.

These insights into receptor function provided one solution to a long-standing problem in cancer cell biology. As mentioned earlier, normal cells had long been known to require growth factors in their culture medium in order to grow. Cancer cells, in contrast, were known to have a greatly reduced dependence on growth factors for their growth and survival. The discovery of the ErbB–EGF-R connection yielded a simple, neat explanation of this particular trait of cancer cells: the ErbB oncoprotein releases signals very similar to those emitted by a ligand-activated EGF receptor. However, unlike the EGF receptor, the ErbB oncoprotein can send a constant, unrelenting stream of growth-stimulating signals into the cell, thereby persuading the cell that substantial amounts of EGF are present in its surroundings when none might be there at all.

As mentioned in Section 4.6, we now realize that truncated versions of the EGF receptor are found in a number of human tumor cell types. In many glioblastomas and lung cancers, for example, the EGF-R mRNA lacks the coding sequences carried by exons 2 through 7. This loss derives from the deletion of the corresponding DNA sequences in the gene that specifies this receptor. More generally, a variety of growth factor receptors that are configured much like the EGF receptor have been found in human tumors to be overexpressed (**Sidebar 5.1**; see also Figure 5.11A) or synthesized in a structurally altered form (**Table 5.2**).

Sidebar 5.1 Receptor overexpression results from a variety of molecular mechanisms Receptor overexpression may cause a cell to become hyper-responsive to low concentrations of growth factor in the extracellular space; alternatively, overexpression may cause ligand-independent receptor firing, simply because **mass-action** effects may drive spontaneous receptor dimerization (see Figure 5.11A).

In some cancer cells, receptors are overexpressed at the cell surface because the encoding gene is transcribed at an elevated rate, resulting in correspondingly increased levels of mRNA and protein. Alternatively, in a number of types of human cancers, these receptor genes are often amplified, leading once again to proportionately increased levels of mRNA transcript and receptor protein.

Yet other, more subtle mechanisms may also be responsible for increased numbers of receptor molecules being displayed at the cell surface. Some of these derive from the mechanisms that govern the *lifetime* of receptor molecules at the cell surface. Many receptor molecules are displayed at the surface for only a limited amount of time before they are internalized via **endocytosis**, in which a patch of plasma membrane together with associated proteins is pulled into the cytoplasm, where it forms a vesicle. Thereafter, the contents of this vesicle may be dispatched to **lysosomes**, in which they are degraded, or they may be recycled back to the cell surface.

One protein, termed huntingtin-interacting protein-1 (HIP1), is part of a complex of proteins that facilitate endocytosis. HIP1 has been found to be overexpressed in a variety of human carcinomas. For reasons that are poorly understood, overexpression of HIP1 prevents the normal endocytosis of a number of cell surface proteins, including the EGF receptor (EGF-R). As a consequence, cells overexpressing HIP1 accumulate excessive numbers of EGF-R molecules at the cell surface and, in the case of NIH 3T3 mouse fibroblasts, become hyper-responsive to EGF. This explains why such cells can grow vigorously in the presence of very low levels of growth factor–containing serum (0.1%) that would otherwise cause normal NIH 3T3 cells to retreat into a nongrowing, quiescent state. Moreover, these cells become transformed in the presence of 10% serum in the tissue culture medium—the serum concentration that is routinely used to propagate untransformed NIH 3T3 cells; both behaviors demonstrate the hyper-responsiveness to GFs resulting from receptor overexpression.

Another protein, called cyclin G–associated kinase (GAK), is a strong promoter of EGF-R endocytosis. When its expression is suppressed, levels of cell surface EGF-R increase by as much as fiftyfold. A third protein, called c-Cbl, is responsible for tagging ligand-activated EGF-R, doing so via a covalent modification termed mono-ubiquitylation (described in Supplementary Sidebar 7.5). This tagging causes the endocytosis of the EGF-R and its subsequent degradation in lysosomes. The v-Cbl viral oncoprotein, as well as several cellular proteins (Sts-1, Sts-2), can block this endocytosis, resulting, once again, in accumulation of elevated levels of EGF-R at the cell surface. In a more general sense, the spontaneous receptor dimerization or hyper-responsiveness to growth factors resulting from receptor overexpression is likely to drive the proliferation of a variety of human cancer cell types *in vivo*.

Table 5.2 Tyrosine kinase GF receptors altered in human tumors[a]

Name of receptor	Main ligand	Type of alteration	Types of tumor
EGF-R/ErbB1	EGF, TGF-α	overexpression	non-small cell lung cancer; breast, head and neck, stomach, colorectal, esophageal, prostate, bladder, renal, pancreatic, and ovarian carcinomas; glioblastoma
EGF-R/ErbB1		truncation of ectodomain	glioblastoma; lung and breast carcinomas
ErbB2/HER2/Neu	NRG, EGF	overexpression	30% of breast adenocarcinomas
ErbB3, 4	various	overexpression	oral squamous cell carcinoma
Flt-3	FL	tandem duplication	acute myelogenous leukemia
Kit	SCF	amino acid substitutions	gastrointestinal stromal tumor
Ret	GFL	fusion with other proteins; point mutations	papillary thyroid carcinomas; multiple endocrine neoplasias 2A and 2B
FGF-R2	FGF	amino acid substitutions	breast, gastric, endometrial carcinomas
FGF-R3	FGF	overexpression; amino acid substitutions; translocations	multiple myeloma; bladder and cervical carcinomas; acute myelogenous leukemia
PDGF-Rβ	PDGF	translocations	chronic myelomonocytic leukemia

[a]See also Figure 5.16.

5.5 A growth factor gene can become an oncogene: the case of *sis*

The notion that oncoproteins can activate mitogenic signaling pathways received another big boost when the platelet-derived growth factor (PDGF) protein was isolated and its amino acid sequence determined. In 1983, the B chain of PDGF was found to be closely related in sequence to the oncoprotein encoded by the v-*sis* oncogene of simian sarcoma virus. Once again, study of the oncogenes carried by rapidly transforming retroviruses paid off handsomely.

The PDGF protein was discovered to be unrelated in structure to EGF and to stimulate proliferation of a different set of cells. PDGF stimulates largely mesenchymal cells, such as fibroblasts, adipocytes, and smooth muscle cells; in the brain, it can also stimulate the growth and survival of certain types of glial cells. By contrast, the mitogenic activities of EGF are focused largely (but not entirely) on epithelial cells (for example, see Figure 5.11B). This specificity of action could be understood once the PDGF receptor was isolated: the PDGF-R was found to be expressed on the surfaces of mesenchymal cells and is not usually displayed by epithelial cells, while the EGF-R largely shows the opposite pattern of expression. (Like the EGF receptor, the PDGF-R uses a tyrosine kinase in its cytoplasmic domain to broadcast signals into the cell; see Figure 5.9.)

The connection between PDGF and the *sis*-encoded oncoprotein suggested another important mechanism by which oncoproteins could transform cells: when simian sarcoma virus infects a cell, its *sis* oncogene causes the infected cell to release copious amounts of a PDGF-like Sis protein into the surrounding extracellular space. There, the PDGF-like molecules attach to the PDGF-R displayed by the same cell that just synthesized and released them. The result is strong activation of this cell's PDGF receptors and, in turn, a flooding of the cell with an unrelenting flux of the growth-stimulating signals released by the ligand-activated PDGF-R.

These discoveries also resolved a long-standing puzzle. Most types of acutely transforming retroviruses are able to transform a variety of infected cell types. Simian sarcoma virus, however, was known to be able to transform fibroblasts, but it failed to transform epithelial cells. The cell type–specific display of the PDGF-R explained the differing susceptibilities to transformation by simian sarcoma virus.

Sidebar 5.2 **Autocrine signaling solves the mystery of Friend leukemia virus leukemogenesis** At one time, the biological powers of the Friend murine leukemia virus (FV) represented a major mystery. This retrovirus was found to induce erythroleukemias rapidly in mice, yet careful examination of the viral genome failed to reveal any apparent oncogene that had been acquired from the cellular genome, as was the case with all other rapidly oncogenic retroviruses. Thus, FV seemed to contain only the *gag*, *pol*, and *env* genes that are required for replication by all retroviruses (see Figure 3.19).

A clue came from a smaller retroviral genome that co-proliferated together with FV. This smaller genome, which apparently resulted from deletion of most of the genes of a wild-type retrovirus genome, carried only the gene for a viral glycoprotein, that is, a gene similar to and derived from the retroviral *env* gene. (Usually the *env*-encoded glycoproteins are used by retrovirus virions to attach to target cells and to fuse the virion membrane with the plasma membrane of these cells.) Detailed characterization of this glycoprotein, termed gp55, solved the puzzle of FV-induced erythroleukemias. gp55 was found to act as a mimic of the growth factor **erythropoietin** (EPO). Normally, when oxygen tension in the blood is less than normal, EPO is released from the kidneys and binds to the EPO receptors displayed by cells in the bone marrow that are the immediate precursors of erythroblasts (precursors of the mature red blood cells). This activation of the EPO-R causes these cells to increase in number and stimulates them to differentiate into **erythrocytes** (red blood cells). In mice infected by FV, the gp55 released from an infected erythroblast acts in an autocrine manner on the infected cell by binding to and stimulating its EPO receptors, thereby driving its proliferation. (The resulting erythrocyte precursor cells then accumulate in large numbers and suffer additional mutations that cause them to become neoplastic.) It is still unclear how Friend leukemia virus has succeeded in remodeling a retroviral glycoprotein into an effective mimic of EPO.

Once again, a close connection had been forged between a protein involved in mitogenic signaling and a viral oncoprotein. In this instance, a virus-infected cell was forced to make and release a growth factor to which it could also respond. Rather than a growth factor signal sent from one type of cell to another cell type located nearby (often termed **paracrine** signaling), or sent via the circulation from cells in one tissue in the body to others in a distant tissue (**endocrine** signaling), this represented an auto-stimulatory, or **autocrine**, signaling loop in which a cell manufactured its own mitogens (see Figure 5.11C and D). (An autocrine signaling mechanism also explains how Friend leukemia virus, another retrovirus of mice, provokes disease; see Sidebar 5.2.)

In fact, a variety of human tumor cells are known to produce and release substantial amounts of growth factors to which these cells can also respond (**Table 5.3**). Some

Table 5.3 Examples of human tumors making autocrine growth factors

Ligand	Receptor	Tumor type(s)
HGF	Met	miscellaneous endocrinal tumors, invasive breast and lung cancers, osteosarcoma
IGF-2	IGF-1R	colorectal
IL-6	IL-6R	myeloma, HNSCC
IL-8	IL-8R A	bladder cancer
NRG	ErbB2[a]/ErbB3	ovarian carcinoma
PDGF-BB	PDGF-Rα/β	osteosarcoma, glioma
PDGF-C	PDGF-Rα/β	Ewing's sarcoma
PRL	PRL-R	breast carcinoma
SCF	Kit	Ewing's sarcoma, SCLC
VEGF-A	VEGF-R (Flt-1)	neuroblastoma, prostate cancer, Kaposi's sarcoma
TGF-α	EGF-R	squamous cell lung, breast and prostate adenocarcinoma, pancreatic, mesothelioma
GRP	GRP-R	small-cell lung cancer

[a]Also known as HER2 or Neu receptor

human cancers, such as certain lung cancers, produce at least three distinct growth factors, specifically transforming growth factor-α (TGF-α), stem cell factor (SCF), and insulin-like growth factor-1 (IGF-1); at the same time, the tumors express the receptors for these three ligands, thereby establishing three autocrine signaling loops simultaneously. These signaling loops seem to be functionally important for the growth of tumors. For example, in one study of small-cell lung cancer (SCLC) patients, those whose tumors expressed Kit, the receptor for stem cell factor (SCF), survived for an average of only 71 days after diagnosis of their disease, while those whose tumors lacked Kit expression survived an average of 288 days.

Possibly the champion autocrine tumor is Kaposi's sarcoma, a tumor of cells closely related to the endothelial cells that form lymph ducts (see Supplementary Sidebar 3.4). To date, Kaposi's tumors have been documented to produce PDGF, TGF-β, IGF-1, Ang2, CCl8/14, CXCL11, and endothelin—all ligands of cellular origin—as well as the receptors for these ligands. At the same time, the causal agent of this disease, the human herpesvirus-8 (HHV-8) genome that is present in Kaposi's tumor cells, produces two additional ligands—vIL6 and vMIP—whose cognate receptors are also expressed by the endothelial cell precursors that generate these tumors.

In most of these cases, expression of the cellular genes encoding various mitogenic growth factors has somehow become deregulated, resulting in the production of mitogens in cells that do not normally express significant levels of these cellular proteins. This pattern of behavior is compounded by HHV-8, which forces infected cells to synthesize two novel growth factors that are not encoded in the cellular genome.

In a more general sense, autocrine signaling loops seem to represent potential perils for tissues and organisms. In normal tissues, the proliferation of individual cells almost always depends on signals received from other cells; such interdependence ensures the stability of cell populations and the constancy of tissue architecture. A cell that has gained the ability to control its own proliferation (by making its own mitogens) therefore creates an imminent danger, since self-reinforcing, positive feedback loops often lead to gross physiologic imbalances. (Nonetheless, autocrine signaling loops are clearly used in certain normal biological situations, such as those that operate to stably maintain the unique properties of stem cells, as we will see in Chapter 14. Accordingly, it is plausible that some autocrine loops operate normally to maintain specific states of differentiation, while others conveying mitogenic signals are confined largely to neoplastic cells.)

We should note that the activation of autocrine mitogenic signaling loops yields an outcome very similar to that occurring when a structurally altered receptor protein such as ErbB/EGF-R is expressed by a cell. In both cases, the cell generates its own mitogenic signals and its dependence upon exogenous mitogens is greatly reduced. Interestingly, these dynamics also have important implications for the development of anti-cancer drugs (**Sidebar 5.3**).

5.6 Transphosphorylation underlies the operations of receptor tyrosine kinases

The actions of oncogenes such as *sis* and *erbB* provide a satisfying biological explanation of how a cell can become transformed. By supplying cells with a continuous flux of growth-stimulatory signals, oncoproteins are able to drive the repeated rounds of cell growth and division that are needed in order for large populations of cancer cells to accumulate and for tumors to form. Still, this biological explanation sidesteps an important biochemical question lying at the heart of mitogenic signaling: How do growth factor receptors containing tyrosine kinases (often called receptor tyrosine kinases or RTKs) succeed in transducing signals from the extracellular space into the cytoplasm of cells?

Knowing the presence of the tyrosine kinase enzymatic activity borne by the cytoplasmic domains of these proteins allows us to rephrase this question: How do growth factor receptors use their tyrosine kinase domains to emit signals in response to ligand binding?

Sidebar 5.3 Autocrine signaling influences the development of anti-cancer therapies Testing of the efficacy of a novel anti-cancer drug usually begins with analyses of the drug's ability to affect the growth and/or survival of cancer cells grown in culture (*in vitro*). These tests are then followed up by the seeding of such cancer cells in host mice, where they are allowed to form tumors (*in vivo*). The responsiveness of these tumor grafts to drug treatment of the tumor-bearing mice is often used to predict whether the drug under study will have clinical efficacy, that is, an ability to shrink a tumor or stop its further growth in patients.

On some occasions, the *in vitro* drug responsiveness of a tumor cell population may predict its behavior *in vivo*; in other cases, it will not. The experience of some drug developers indicates that these contrasting behaviors correlate with the production of autocrine growth factors by tumor cell lines. Those cancer cells that produce abundant autocrine growth factors usually respond to drug treatment similarly *in vitro* and *in vivo*, while those that do not secrete growth factors behave very differently under these two conditions. It seems that the tumor cells that secrete autocrine factors create their own growth factor environment both *in vitro* and *in vivo*. Conversely, those that do not secrete these factors depend on serum-associated mitogens *in vitro* and on the growth factors supplied by nearby host cells *in vivo*; the latter may differ dramatically from the spectrum of mitogens present in serum, hence the difference in response.

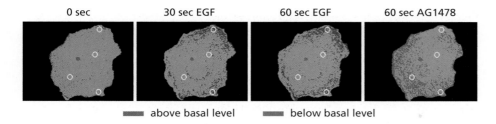

0 sec   30 sec EGF   60 sec EGF   60 sec AG1478

━━━ above basal level   ━━━ below basal level

The solution to this problem came from a detailed examination of the proteins that become phosphorylated within seconds after a growth factor such as EGF is applied to cells expressing its cognate receptor, in this case the EGF-R (**Figure 5.12**). It would seem reasonable that a variety of cytoplasmic proteins become phosphorylated on their tyrosine residues following ligand binding to a growth factor receptor. Indeed, there are a number of such proteins. But the most prominent among these phospho-tyrosine-bearing proteins is often the receptor molecule itself (**Figure 5.13**)! Hence, these receptors seem to be capable of autophosphorylation.

Another clue came from the structure of many growth factor ligands; they were often found to be **dimeric**, being composed either of two identical protein subunits (**homodimers**) or of very similar but nonidentical subunits (**heterodimers**). A third clue derived from observations that many transmembrane proteins constructed like the EGF and PDGF receptors have lateral mobility in the plane of the plasma membrane. That is, as long as the transmembrane domains of these receptor proteins remain embedded within the lipid bilayer of this membrane, they are relatively free to wander back and forth across the surface of the cell.

Taken together, these facts led to a simple model (**Figure 5.14**). In the absence of ligand, a growth factor receptor exists in a **monomeric** (single-subunit) form, embedded as always in the plasma membrane. When presented with its growth factor ligand, which is a homodimer, a receptor molecule will bind to one of the two subunits of this ligand. Thereafter, the ligand–receptor complex will wander around the plasma membrane until it encounters another receptor molecule, to which the second, still-unengaged subunit of the ligand will bind. The result is a cross-linking of the two receptor molecules achieved by the dimeric ligand that forms a bridge between them.

In fact, X-ray crystallographic studies of growth factors bound to the ectodomains of their cognate receptors have revealed a variety of mechanisms by which these ligands are able to induce receptor dimerization (**Figure 5.15**). For example, in the case of the EGF-R-related receptors, individual monomeric ligand molecules bind separately to

Figure 5.12 Formation of phosphotyrosine on the EGF-R following ligand addition The use of a fluorescent reagent that binds specifically to a phosphotyrosine residue on the EGF-R enables the visualization of receptor activation following ligand binding. Here, receptor activation is measured on a monkey kidney cell at a basal level (0 second), as well as 30 and 60 seconds after EGF addition. In addition, following a 2-minute stimulation by EGF, the effects of a 60-second treatment by a chemical inhibitor of the EGF-R kinase (AG1478) are shown (*right*), indicating that the EGF-induced activation of the EGF-R can be rapidly reversed. Receptor activity above the basal level is indicated in *blue*, while activity below the basal level is indicated in *red*. The response to AG1478 treatment indicates that a significant basal level of EGF-R activity was present (at 0 second) even before EGF addition. (From M. Offterdinger et al., *J. Biol. Chem.* 279:36972–36981, 2004.)

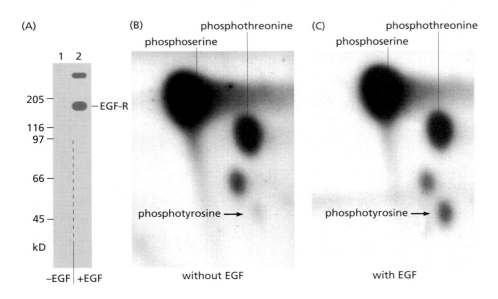

(A)

1   2

205 —   —EGF-R

116 —
97 —

66 —

45 —

kD

−EGF | +EGF

(B)   phosphothreonine
phosphoserine

phosphotyrosine ⟶

without EGF

(C)   phosphothreonine
phosphoserine

phosphotyrosine ⟶

with EGF

Figure 5.13 Apparent autophosphorylation of the EGF receptor (A) When human A431 epidermoid carcinoma cells, which greatly overexpress the EGF-R, are incubated in $^{32}PO_4$-containing medium and then exposed to EGF, a radiolabeled protein can be immunoprecipitated with an anti-phosphotyrosine antiserum (*lane 2*); this protein co-migrates upon gel electrophoresis with the EGF-R and is not detectable in the absence of prior EGF treatment (*lane 1*). (B) The $^{32}PO_4$-labeled phosphoamino acids borne by the proteins in such cells in the absence of EGF (as resolved by electrophoresis; see Figure 5.7) are seen here. (C) Following the addition of EGF to A431 cells, a spot in the lower right becomes darker; internal markers indicate that this is phosphotyrosine. (A, from A.B. Sorokin et al., *FEBS Lett.* 268:121–124, 1990. B and C, courtesy of A.R. Frackelton Jr.)

Figure 5.14 Receptor dimerization, ligand binding, transphosphorylation and glioblastoma pathogenesis (A) In the absence of ligand, many growth factor receptor molecules (left, green) are free to move laterally (diverging arrows) in the plane of the plasma membrane. The consequences of these lateral migrations have been studied in the greatest detail in the case of the EGF receptor depicted here. When two receptor molecules encounter one another, they form a homodimer that persists for a brief period; at the same time, the ectodomains of these receptors change conformation and acquire increased affinity for binding growth factor ligands (Step 1). Subsequent binding of EGF ligand molecules results in further stabilization of the dimer and activation of the linked tyrosine kinase (TK) domains (Step 2). Thus, the TK domain of each receptor subunit phosphorylates the C'-terminal cytoplasmic tail of the other subunit—the process of transphosphorylation (arrows). Other GF receptors, not yet studied in such detail, seem to behave in a similar fashion. (B) The dynamics of receptor function depicted in panel A shed light on how amplification of *EGFR* genes and resulting EGF-R protein overexpression confers proliferative advantage on glioblastoma cells. Sequence analyses of 206 human glioblastomas has revealed chromosomal regions that are repeatedly amplified (focal amplifications), resulting in copy-number alterations (CNAs) of the genes embedded in these chromosomal regions. As seen here, the chromosomal region harboring the *EGFR* gene is amplified in ~90 of these tumors out of a total cohort of 206 tumors that were analyzed; in contrast, chromosomal regions harboring other growth factor receptors (*PDGFRA, MET*) and oncogenes are found to be amplified far less frequently. (The majority of the amplified *EGFR* genes encode wild-type protein, indicating that any effects of receptor amplification must derive from the increased number of receptor molecules expressed at the cell surface rather than the deregulated firing of structurally altered receptor molecules.) The excessive numbers of EGF-R molecules that are displayed on the surfaces of many of these brain tumor cells must drive an increased frequency of random collisions between these molecules and thus dimerization (Step 1 of panel A). The numbers given below each receptor indicates the number of genes that are co-amplified with the receptor gene in the particular amplicon in which this gene resides. (B, from The Cancer Genome Atlas Research Network, *Nature* 455:1061–1068, 2008.)

the ectodomains of monomeric receptors. The latter respond by undergoing a steric shift that allows them to dimerize with one another via sites quite distant from the sites of ligand binding (see Figure 5.15C).

Once dimerization of the ectodomains of two receptor molecules has been achieved by ligand binding, the cytoplasmic portions are also pulled together. Now each kinase domain phosphorylates tyrosine residues present in the cytoplasmic domain of the other receptor. So the term "autophosphorylation" (see Figure 5.13) is actually a misnomer. This bidirectional, reciprocal phosphorylation is best described as a process of **transphosphorylation** (see Figure 5.14). (Indeed, it is unclear whether a single isolated kinase molecule is ever able to phosphorylate itself.)

The phosphorylation of these tyrosine residues can have at least two consequences. The catalytic cleft of a kinase—the region of the protein where its enzymatic function occurs—may normally be partially obstructed by a loop of the protein, preventing the kinase from interacting effectively with its substrates. Transphosphorylation of a critical tyrosine in the obstructing "activation loop" causes the loop to swing out of the way, thereby providing the catalytic cleft with direct access to substrate molecules. In addition, transphosphorylation results in the phosphorylation of an array of tyrosine residues present in cytoplasmic portions of the growth factor receptor outside the kinase domain, as indicated in Figure 5.14A. As we will learn in the next chapter, these phosphorylation events enable the receptor to activate a diverse array of downstream signaling pathways.

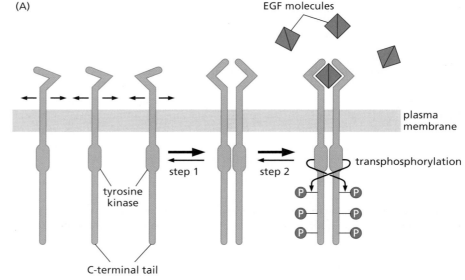

(A)

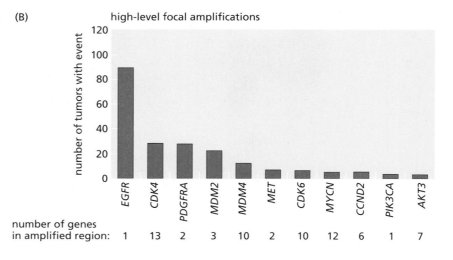

(B)

As suggested earlier, some receptors may heterodimerize with others. For example, the HER2/ErbB2/Neu receptor, which is overexpressed in human breast cancers (see Figure 4.4), often forms heterodimers with its cousins, either the HER1/EGF-R or HER3 receptor (see Figure 5.1). Since HER2 lacks its own identified ligand, it may be that a single ligand molecule binding either HER1 or HER3 suffices to trigger ligand-activated heterodimerization and signaling. HER3 has a defective kinase domain, but its C-terminal tail can serve as a substrate for phosphorylation by the HER2-borne kinase. The resulting phosphotyrosines then allow HER3 to release signals into the cell.

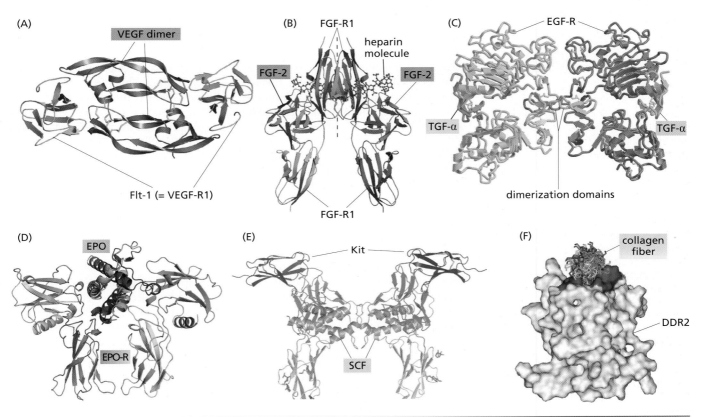

Figure 5.15 Alternative mechanisms of growth factor–induced receptor dimerization X-ray crystallography has revealed how a number of growth factor ligands cause receptor dimerization. (A) Vascular endothelial growth factor (VEGF) binds as a homodimer (*red, blue*) to two monomers of its receptor, called Flt-1 (*green*), thereby bringing them together. This complex is seen as a bottom view, i.e., from the plasma membrane looking outward, and exemplifies the dimerization mechanism illustrated in Figure 5.14. (B) In the case of the fibroblast growth factor receptor-1 (FGF-R1), the ligands are growth factor monomers (FGF-2) that do not contact each other directly but instead are linked by a heparin molecule to which both are bound (*stick figure*). (Heparin is a glycosaminoglycan component of the extracellular matrix.) Each FGF-2 monomer (*dark orange*) then attracts and binds the ligand-binding domain of a receptor monomer (*green, aquamarine*). Each ligand-binding domain of the receptor subunit, seen here in side view, is composed of two subdomains, one binding above, the other below the FGF-2 ligand. The interface between the two dimerized receptor molecules is indicated by the *dashed line*. (C) In the case of the EGF-R, molecules of tumor growth factor-α (TGF-α), one of the alternative EGF-related ligands of this receptor, bind as individual monomers (*green, left; purple, right*) to the "front sides" of two receptor subunits (*dark yellow, dark pink*). As seen in a *top view* facing the plasma membrane, receptor monomers respond to ligand binding

by exposing finger-like domains that mediate their "back-to-back" dimerization. Hence, there is no direct connection between the two ligand molecules, since each is actually far removed from the region of the receptor monomer that is directly involved in dimerization. (D) Some growth factors function as single polypeptide chains. Thus, a single molecule of erythropoietin (EPO) (*blue/aquamarine*) is able to simultaneously contact two receptor molecules (EPO-R) (*orange/red* and *green/yellow*), causing their dimerization. (E) Like VEGF, stem cell factor (SCF) is a homodimeric ligand that binds to two molecules of the Kit receptor; in the side view here, one molecule of SCF (*magenta*) binds in front of the *left* Kit molecule while the other passes behind the *right* Kit molecule. (F) Collagen, an extracellular matrix protein, is an unusual ligand for an RTK since it is not presented in a soluble form to its two tyrosine kinase receptors, discoidin domain receptors (DDR1 and -2). Seen here is a space-filling representation of the collagen-binding domain (*red*) of DDR2, which by binding the triple-helix fibrillar collagen protein (in cross section, *blue, green, yellow* stick figures) causes receptor clustering in a still-poorly understood manner. (A, from C. Wiesmann et al., *Cell* 91:695–704, 1997; B, from A. Plotnikov et al., *Cell* 98:641–650, 1999; C, from T.P.J. Garrett et al., *Cell* 110:763–773, 2002; D, courtesy of J.S. Finer-Moore, from R.S. Syed et al., *Nature* 395:511–516, 1998; E, from S. Yuzawa et al., *Cell* 130:323–334, 2007; F, from O. Ichikawa et al., *EMBO J.* 26:4168–4176, 2007.)

The receptor dimerization model explains how growth factor receptors can participate in the formation of cancers when the receptor molecules are overexpressed, that is, displayed on the cell surface at levels that greatly exceed those seen in normal cells (see Sidebar 5.1). Since these receptors are free to move laterally in the plane of the plasma membrane, their high numbers cause them to collide frequently, and these encounters, like the dimerization events triggered by ligand binding, can result in transphosphorylation, receptor activation, and signal emission. For instance, as Table 5.2 indicates, the EGF receptor is overexpressed in a wide variety of human tumors, mostly carcinomas. In such tumors, this overexpression may result in ligand-independent receptor dimerization and firing.

Actually, the mechanisms that govern receptor behavior are slightly more complex and more variable. In the case of the EGF receptor, as Figure 5.14 indicates, spontaneous receptor dimerization actually precedes ligand binding and confers greater ligand-binding affinity on the receptor ectodomains; tyrosine kinase firing, however, only occurs *after* ligand binding. This suggests that in the many human tumors that overexpress the EGF receptor, the numbers of spontaneously dimerized receptors is greatly increased because of more frequent collisions occurring as the EGF-R molecules travel laterally in the plane of the plasma membrane; cancer cell proliferation is then favored by the increased numbers of high-affinity receptors formed on the surfaces of these cells. Stated differently, excessive EGF receptor expression appears to make cancer cells hyper-responsive to the low levels of EGF and TGF-α ligands that are present in their surroundings.

However, this scenario cannot explain how all overexpressed receptors succeed in driving cancer cell proliferation. Experiments with the Met receptor demonstrate an alternative mechanism—ligand-independent dimerization and firing. In these experiments, the human Met receptor [whose ligand is hepatocyte growth factor, (HGF), also known as scatter factor (SF)] was overexpressed in liver cells of a genetically altered mouse strain; this led, in turn, to the development of hepatocellular carcinomas (HCCs). Since the human Met receptor cannot bind the mouse HGF/SF ligand (the only ligand available in these mice), this dictated that the observed receptor activation and cell transformation could only be ascribed to a ligand-independent process, specifically spontaneous dimerization of overexpressed Met receptor molecules.

A diverse array of structural alterations can also cause ligand-independent receptor dimerization, doing so in the absence of receptor overexpression. This lesson is driven home most dramatically by a mutant version of Neu, the rat ortholog of the human HER2/ErbB2 receptor. Exposure of pregnant rats to the mutagen ethylnitrosourea (ENU) results in the birth of pups that often succumb to neuroectodermal tumors 3 to 6 months later. The tumors almost invariably carry a point mutation in the sequence specifying the transmembrane domain of this receptor. The resulting substitution of a valine by a glutamic acid residue favors the constitutive dimerization of the receptors, even in the absence of ligand binding, thereby creating a potent oncoprotein.

Like HER2/ErbB2/Neu, the EGF and FGF (fibroblast growth factor) receptors may be affected by point mutations or small deletions that generate constitutively active receptors; the responsible somatic mutations alter the cytoplasmic domains of these receptors, perturbing the domains that regulate the receptor-associated tyrosine kinases. In a series of non-small-cell lung carcinomas, about 10% were found to have mutant EGF receptors, and these turned out to be particularly responsive to therapeutic drugs directed against the receptor molecule, as we will learn in Chapter 16. Structurally altered receptors have been found to be far more common in bladder carcinomas: at least 60% of these tumors have been found to harbor subtly altered versions of the FGF receptor-3. The precise mechanism by which truncated, mutant versions of the EGF receptor lacking ectodomains become constitutively activated (see Figure 5.10) is less well understood.

Quite different major structural alterations of RTKs can also generate oncoproteins. For example, Met, which functions as the receptor for hepatocyte growth factor (HGF), and TrkA, which serves as the receptor for nerve growth factor (NGF), have both been found to have become converted into oncoproteins in some tumors as a consequence of gene fusion events. In addition to truncated ectodomains, the consequences of gene fusions may be receptors whose other parts are fused to unrelated proteins that are normally prone to dimerize or oligomerize (**Figure 5.16**). This results in a constitutive, ligand-independent dimerization of these receptors, explaining the oncogenic powers of these fusion proteins. Yet other, subtle changes in receptor structure can also encourage ligand-independent firing in diverse cell types (**Sidebar 5.4**).

Although we have placed great emphasis here on receptor dimerization, it represents only one of the molecular changes that tyrosine kinase receptor molecules can undergo after they bind their ligands. There is increasing evidence that, in addition to this dimerization, the ligand-binding ectodomain undergoes some type of rotation or stereochemical shift, and that this shift is transmitted through the plasma membrane to the cytoplasmic kinase domains, which themselves are moved into new

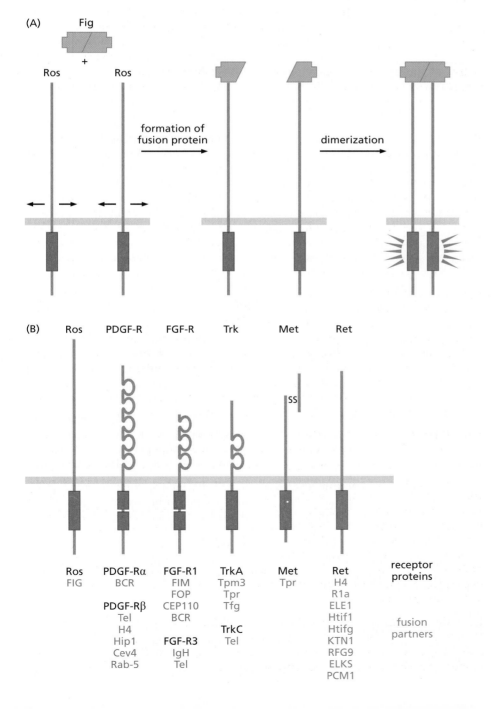

Figure 5.16 Gene fusion causing constitutively dimerized receptors (A) The Ros receptor (*lower left*), like other RTKs, normally is free to move laterally in the plane of the plasma membrane and requires ligand binding for dimerization. However, in a glioblastoma, the reading frame specifying the ectodomain of the *ROS* gene has become fused to the reading frame of a second, unrelated neighboring gene, termed *FIG*, which encodes a protein that normally dimerizes spontaneously (*green, upper left*). In the resulting fusion protein (*right*), the dimerized Fig domain pulls two Ros receptors together, resulting in constitutive, ligand-independent dimerization and signaling. (B) Similarly, in a diverse array of malignant tumors, receptor dimerization occurs when the genes encoding growth factor receptors (*black type*) become fused to unrelated genes that happen to specify proteins that normally dimerize or form higher-order oligomers (*red type*), dragging together pairs of receptor monomers. Listed here are a number of such mutant, fused receptors found in human tumors; in certain cases, a given receptor may be found fused to multiple alternative oligomerizing proteins. (B, courtesy of A. Charest.)

Sidebar 5.4 Mutant forms of a single tyrosine kinase receptor may play a causal role in very different types of cancer The Kit protein was originally discovered as an oncoprotein encoded by a feline sarcoma retrovirus. The *kit* proto-oncogene encodes a tyrosine kinase receptor whose ligand is stem cell factor (SCF), a growth factor that is important in stimulating the formation of various types of blood cells (the process of **hematopoiesis**), as well as the development of a variety of nonhematopoietic cell types, including melanocytes and the cells mediating gut motility. Mutant forms of the Kit receptor, which fire constitutively in a ligand-independent fashion, are found in a diverse array of malignancies. In many leukemias of dogs, mutant Kit receptors carrying amino acid substitutions in the cytoplasmic **juxtamembrane** residues—those close to the transmembrane domain—are common.

The pacemaker cells of the intestine, which control the peristaltic contractions of the smooth muscles of this organ, also rely on the Kit receptor for their proper development, and indeed, these cells are absent if the *kit* gene is inactivated in the germ line (and thus in all cells) of a mouse. In human gastrointestinal stromal tumors (GISTs), which seem to arise from these same cells, juxtamembrane amino acid substitutions similar to those seen in dog leukemias are commonly observed. Analyses of the structure of the Kit receptor and related tyrosine kinase receptors indicate that the juxtamembrane domain plays an *inhibitory* role in regulating the receptor, and that its ability to suppress receptor firing is compromised by the amino acid substitutions and deletions in this domain that are commonly seen in GISTs (**Figure 5.17**). In one group of GIST patients, those whose tumors carried subtle (missense) mutations affecting exon 11 of the *Kit* gene were all alive four years after initial diagnosis and treatment, while only half of those whose GISTs carried deletions of this entire exon survived that long.

Figure 5.17 Multiple structural alterations affect Kit receptor firing (A) As seen here, the Kit receptor may be found in a constitutively active, mutant form in a number of different types of human tumors (*red type*). Most of the alterations in the indicated exons involve missense point mutations in the reading frame of the Kit mRNA. GIST, gastrointestinal stromal tumor; AML, acute myelogenous leukemia. (B) The cytoplasmic domain of the Flt-3 receptor, which is very closely related to that of Kit, is shown here as a ribbon structure. Like almost all protein kinases, this one is composed of an N-terminal lobe (*red*) and a C-terminal lobe (*blue*), with a catalytic cleft in between. The catalytic cleft is normally obstructed by both an "activation loop" (*green*) and a juxtamembrane domain (*yellow*). In the activation loop, note the tyrosine residue, whose phosphorylation causes this loop to swing out of the way, allowing the catalytic cleft access to substrates. (C) The Kit receptor, which is found in mutant form in almost all gastrointestinal stromal tumors (GISTs), often carries alterations affecting the juxtamembrane (JM) domain, depicted here as a *yellow* cylinder that sits above the N-terminal lobe of the two-lobed Kit tyrosine kinase [*side view, plasma membrane (not shown) above*]. Ligand binding to the Kit receptor results initially in transphosphorylation of tyrosine residues in the JM domain, which causes it to dissociate from the side of the N-terminal lobe; the resulting phosphorylated structure (*yellow zigzag*) is depicted here as sitting above the kinase domain, out of the way. Subsequent transphosphorylation of the key tyrosine residue in the activation loop (*green*; see panel B), which normally obstructs the catalytic cleft, also contributes to activation of kinase function. Mutant Kit receptors with defective JM domains are therefore partially activated even in the absence of ligand binding. (A, courtesy of D. Emerson. B, from J. Griffith, *J. Mol. Cell* 13:169–178, 2004. C, from P.M. Chan et al., *Mol. Cell. Biol.* 23:3067–3078, 2003.)

configurations that favor transphosphorylation. In fact, in some tyrosine kinase receptors, constitutive dimerization may exist even without ligand binding, and the ligand works only to effect a concurrent rotation of both the ectodomains and cytoplasmic domains of the receptor molecules that facilitates signaling by their kinase domains.

Additional details are being uncovered about the behavior of homodimeric receptors: initially, in response to ligand binding, receptors seem to form asymmetric complexes in which the cytoplasmic domain of one subunit serves as a kinase to phosphorylate tyrosine residues of the other subunit; only later does the second subunit transphosphorylate the first one. Moreover, the multiple phosphate groups that are attached to the cytoplasmic domain of a receptor subunit appear to be phosphorylated in a specific sequence rather than stochastically. (The functional roles of these phosphates will be addressed later, in Section 6.3.)

The sequencing of the human genome has revealed the full repertoire of tyrosine kinase receptors that are likely to be expressed in human tissues. Of the approximately 20,000 genes in our genome, at least 59 encode proteins having the general structures of the EGF and PDGF receptors, that is, a ligand-binding ectodomain, a transmembrane domain, and a tyrosine kinase cytoplasmic domain. These receptors are further

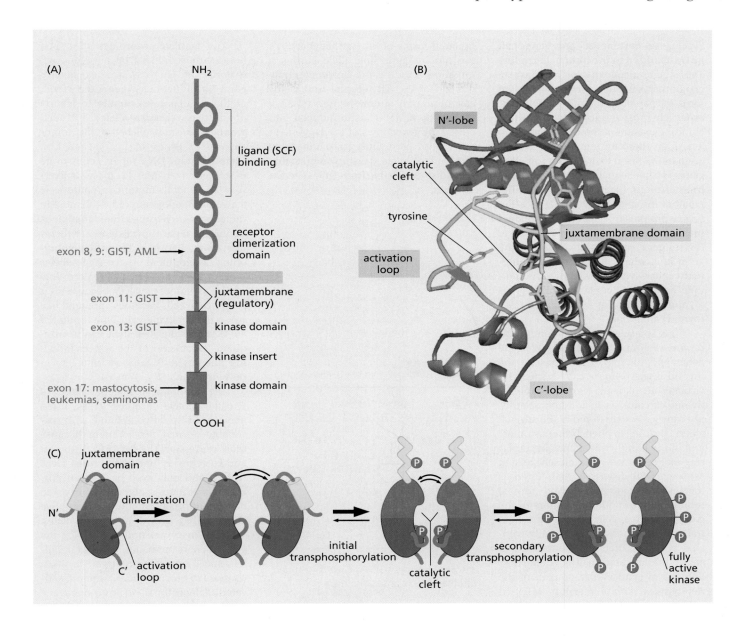

grouped into subclasses according to their structural similarities and differences (see Figure 5.9). To date, only a minority of these receptors have been implicated as agents contributing to human cancer formation. It is likely that the involvement of other receptors of this class in human cancer will be demonstrated in the future. Significantly, mutant alleles of certain tyrosine kinase receptor genes may also be transmitted in the human germ line. This explains the origins of a number of familial cancer syndromes, in which affected family members show greatly increased risks of contracting certain types of cancer (**Sidebar 5.5**).

5.7 Yet other types of receptors enable mammalian cells to communicate with their environment

Cancer is ultimately a failure of individual cells to communicate appropriately with the surrounding tissue microenvironment. Given the variety of biochemical signals that cells receive from their surroundings, this suggests that numerous receptor types are involved in cancer pathogenesis. Indeed, as early as the 1980s, the discovery of a number of other distinct types of metazoan receptors made it clear that the tyrosine

Sidebar 5.5 Mutant receptor genes can be transmitted in the human germ line The *ret* gene encodes a tyrosine kinase receptor that is frequently found in mutant form in papillary thyroid carcinomas, especially those induced by inadvertent clinical exposure of the thyroid gland to X-rays. In these tumors, the *ret* gene is often found fused to one of several other genes as a consequence of chromosomal translocations (see Figure 5.16B). The result is the constitutive firing of a Ret fusion protein in a ligand-independent manner. More subtly mutated alleles of the *ret* gene can be transmitted in the germ line, where they cause the inherited cancer syndromes known as multiple endocrine neoplasia (MEN) types 2A and 2B, as well as familial medullary thyroid carcinoma.

MEN is a relatively uncommon syndrome, seen in 1 in about 30,000 in the general population. In MEN type 2A, the thyroid is the primary cancer site, with virtually all patients carrying papillary thyroid carcinomas. In addition, secondary cancer sites are found in the adrenal glands (leading to pheochromocytomas) and in the parathyroid gland (hyperplasias or adenomas). The closely related MEN type 2B disease is manifested as increased risk of thyroid and adrenal tumors and, in addition, tumors of the ganglion nerve cells in the intestinal tract. MEN-2A cases are usually initiated by an inherited point mutation that causes the substitution of a cysteine residue in the extracellular domain of the Ret tyrosine kinase receptor. MEN-2B cases are caused by a substitution in the tyrosine kinase domain of this receptor.

Hereditary papillary renal cancer is due to inherited mutant alleles in the *met* gene, which encodes the receptor for hepatocyte growth factor (HGF). These mutant germ-line alleles of *met* usually carry point mutations that cause amino acid substitutions in the tyrosine kinase domain of Met that result in constitutive, ligand-independent firing by the receptor. (In the Rottweiler breed of dogs, a point mutation affecting the Met juxtamembrane region is transmitted through the germ line of ~70% of these pets and likely contributes to their high risk of developing cancer.)

Members of a Japanese family (**Figure 5.18**) have been reported to transmit a mutant allele of the Kit receptor in their germ line. Like the somatically

mutated forms of *kit* (see Sidebar 5.4), the mutant germ-line allele carries a mutation affecting the juxtamembrane domain of the Kit receptor and yields gastrointestinal stromal tumors (GISTs). When this mutation is introduced into one of the two copies of the *kit* gene in the mouse germ line, mice inheriting this mutant allele develop tumors that are indistinguishable from GISTs seen in humans.

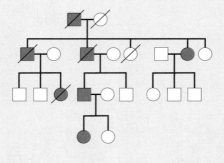

Figure 5.18 Familial gastrointestinal stromal tumors (GISTs) Gastrointestinal stromal tumors arise from the mesenchymal pacemaker cells of the lower gastrointestinal tract and contrast with the far more frequent colon carcinomas, which derive from the epithelial cell layer lining the lumen of the gut. In the case of the kindred depicted here, an allele encoding a mutant, constitutively active Kit tyrosine kinase receptor has been carried by at least four generations, afflicting them with GISTs and/or intestinal obstructions. The transmission of this mutant allele, which carries a deletion of a single amino acid residue in the juxtamembrane domain of the receptor protein (see Figure 5.17), through so many generations indicates that phenotypic expression of this mutant germ-line allele is often delayed until relatively late in life after childbearing years. Affected individuals, *red*; males, *squares*; females, *circles*; deceased, *strikethroughs*. (From T. Nishida et al., *Nat. Genet.* 19:323–324, 1998.)

The familial syndromes described here are most unusual, because the great majority of such syndromes involve germ-line mutations of genes that result, at the cellular level, in the formation of inactive, recessive alleles. Included among these are alleles of the tumor suppressor genes to be discussed in Chapter 7 and DNA repair genes to be described in Chapter 12. It seems likely that, in general, mutant, constitutively active oncogenes cannot be tolerated in the germ line, because they function as dominant alleles in cells and are therefore highly disruptive of normal embryonic development. Recessive alleles present in single copy in the germ line can be tolerated, however, because their presence in cells becomes apparent only when the surviving wild-type dominant allele is lost. Because such loss is infrequent, expression of the cancer-inducing phenotype is delayed until late in development.

The dominant, oncogenic alleles of tyrosine kinase receptor genes listed above seem to be compatible with reasonably normal development because their expression (in the case of the *ret* and *kit* genes) is limited to a small set of tissues, and may even be delayed during embryogenesis, allowing embryos to develop normally without the disruptive effects of these growth-promoting alleles.

This logic was undermined by the genetic characterization of a rare familial cancer syndrome reported in 2005: contrary to preconception, mutant, activated alleles of the H-*ras* proto-oncogene present in a zygote are not incompatible with reasonably normal human development. Individuals inheriting these alleles are afflicted with Costello syndrome, which is associated with mental retardation, high birth weight, **cardiomyopathy** (pathological defects in the heart muscle), and a tendency to develop tumors; together these pathologies prevent these individuals from surviving to adulthood. (This explains why the 12 cases that have been described all represented *de novo* mutations, that is, new mutations that arose during the formation of a sperm or egg rather than being inherited from a similarly afflicted parent.) Strikingly, most organ systems in these individuals develop relatively normally in spite of the presence of a highly penetrant allele of an activated H-*ras* oncogene (see Section 4.4) in all of their cells.

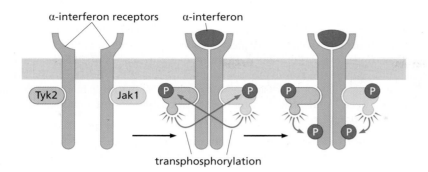

α-interferon receptors   α-interferon

Tyk2   Jak1

transphosphorylation

Figure 5.19 Structure of cytokine receptors The interferon receptor (IFN-R, *dark green*), like the receptors for erythropoietin and thrombopoietin, carries noncovalently attached tyrosine kinases (TKs) of the Jak family (in this case, Tyk2 and Jak1). Upon ligand binding (in this case α-interferon), the two TKs transphosphorylate and thereby activate one another. Subsequently, they phosphorylate the C-terminal tails of the receptor subunits, placing the receptor in an actively signaling configuration, much like the receptor tyrosine kinases described in Section 5.6.

kinase receptors (RTKs) described above represent only one class of receptors among a far larger group of structurally diverse receptors. As is the case with RTKs, each of these other receptor types is specialized to detect the presence of specific extracellular ligands or groups of related ligands and to respond to this encounter by transducing signals into the cell.

One class of receptors that is especially important in receiving signals controlling the development of various hematopoietic cell types also uses tyrosine kinases. In this case, however, the responsible kinases, termed Jaks (Janus kinases), are separate polypeptides that associate with the cytoplasmic domains of these receptors through noncovalent links. Included here are the receptor for erythropoietin (EPO), which regulates the development of erythrocytes (red blood cells), and the **thrombopoietin** (TPO) receptor, which controls the development of the precursors of blood platelets, called **megakaryocytes**. The receptors for the important immune regulator interferon are also members of this class, as is the large group of interleukin receptors that regulate diverse immune responses. When these various **cytokine** receptor molecules dimerize in response to ligand binding, the associated Jaks phosphorylate and activate each other (**Figure 5.19**); the activated Jaks then proceed to phosphorylate the C-terminal tails of the receptor molecules, thereby creating receptors that are activated to emit signals, much like the RTKs described in the previous section. We will return to the details of Jak signaling in the next chapter.

Transforming growth factor-β (TGF-β) and related ligands have receptors that are superficially similar to the tyrosine kinase receptors, in that they have an extracellular ligand-binding domain, a transmembrane domain, and a cytoplasmic kinase domain. However, these receptors invariably function as heterodimers, such as the complex of the type I and type II TGF-β receptors (**Figure 5.20**). Importantly, their kinase domains phosphorylate serine and threonine rather than tyrosine residues. TGF-βs play centrally important roles in cancer pathogenesis, since they suppress the proliferation of normal epithelial cells while promoting the acquisition of invasive properties by already-transformed cells. Upon ligand binding, the type II TGF-β receptor subunit, which has a constitutively active serine/threonine kinase, is brought into close

Figure 5.20 Structure of the TGF-β receptor The structure of the TGF-β receptor is superficially similar to that of tyrosine kinase receptors (RTKs), in that both types of receptors signal through cytoplasmic kinase domains. However, the kinase domains of TGF-β receptors specifically phosphorylate serine and threonine, rather than tyrosine residues. After the type II receptor (TGF-βRII) is brought into contact with the type I receptor (TGF-βRI) through the binding of TGF-β ligand, it phosphorylates and activates the kinase carried by the type I receptor. The activated type I receptor kinase then emits signals by phosphorylating certain cytoplasmic substrates.

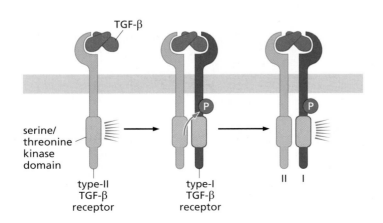

TGF-β

serine/
threonine
kinase
domain

type-II
TGF-β
receptor

type-I
TGF-β
receptor

II   I

proximity with the type I TGF-β receptor subunit, which it then phosphorylates. The kinase belonging to the type I TGF-β receptor subunit becomes activated as a consequence, and proceeds to phosphorylate cytosolic proteins that migrate to the nucleus, where they trigger expression of certain target genes. (A series of TGF-β–related factors, including activin and bone morphogenetic proteins, or BMPs, use similar receptors for signaling; they are not discussed further here, as their role in cancer development has not been extensively documented.) We postpone more detailed discussion of the TGF-β signaling system until Chapter 8.

A far more primitive form of a transmembrane signaling system is embodied in the Notch receptor (termed simply Notch) and its multiple alternative ligands (NotchL, Delta, Jagged; **Figure 5.21**). Importantly, these ligands are immobilized cell surface proteins that are displayed by the signal-emitting cell. Hence, ligand–receptor interactions require close physical association between the signal-emitting and signal-receiving cells, unlike other examples described here, where the ligands are secreted as soluble proteins into the extracellular space by cells that may be located at some distance from one another. This type of signal transmission has been termed **juxtacrine** signaling. (Indeed, certain GFs of the EGF family are also presented as cell surface proteins to the receiving cells, which in this case display the EGF-R.)

Interestingly, considerable evidence points to a mechanism in which the Notch ligand–presenting cell must attach its ligand to the Notch receptor and then use mechanical force to tear the Notch ectodomain away from the signal-receiving cell. (This mechanical force is generated by the endocytosis of the ligand, which pulls it into the cytoplasm of the ligand-presenting cell!). The mechanical tension makes the Notch ectodomain vulnerable to cleavage by a protease; this cleavage is followed by a second proteolytic cleavage that releases a fragment from the Notch cytoplasmic domain. The fragment then migrates to the nucleus, where it functions as part of a complex of transcription factors that activate expression of a cohort of responder genes. Mutant, constitutively active forms of Notch, which fire in a ligand-independent fashion, have been found in half of adult T-cell leukemias.

Yet another receptor, termed Patched, is constructed from multiple transmembrane domains that weave their way back and forth through the plasma membrane

Figure 5.21 Structure of the Notch receptor The Notch receptor (*green*) appears to embody a very primitive type of signaling. It is composed of two portions that are associated via noncovalent interactions. After it has bound a ligand (e.g., NotchL or, as shown here, Delta, *pink*) that is being displayed on the surface of an adjacent cell, the Delta ligand undergoes endocytosis into the ligand-presenting cell, which triggers two successive proteolytic cleavages of the Notch receptor. The resulting C-terminal cytoplasmic fragment of the Notch receptor is thereby freed, allowing it to migrate to the cell nucleus, where it alters the expression of certain genes. Mammals have five distinct Notch ligands and four distinct Notch receptors.

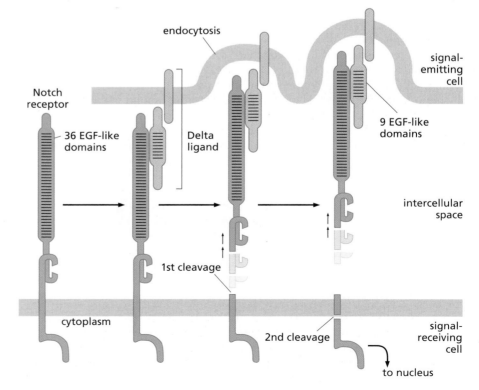

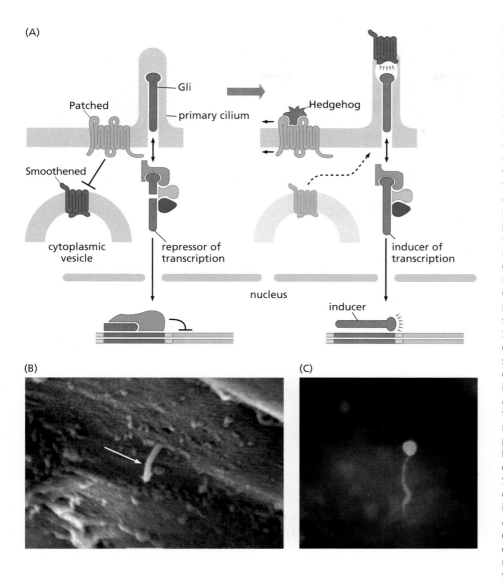

(A)

(B)

(C)

Figure 5.22 **The Patched–Smoothened signaling system** The Hedgehog pathway utilizes its own, unique signaling system. (A) Normally (*left*), Smoothened (Smo), a seven-membrane-spanning protein (*dark green*), is functionally inert, being confined by Patched (Ptc; *light green*), which contains 12 membrane-spanning domains (*top left*), in cytoplasmic vesicles; this sequestration seems to depend on the catalytic actions of Patched on Smoothened, which prevents Smoothened from moving into primary cilia. Under these conditions, Gli (*red*) is cleaved into a protein that moves into the nucleus, where it operates as a repressor of transcription. However, when a Hedgehog (Hh) ligand (*right*) binds to Patched, the latter releases Smoothened from its site of sequestration in the cytoplasm, and Smoothened moves onto the surface of a primary cilium and proceeds, in some unknown fashion, to prevent cleavage of Gli, enabling Gli to move into the nucleus, where it can act as an inducer of transcription. According to an alternative version, Smoothened routinely cycles in and out of the primary cilium. In the absence of Hh binding, Patched, which is associated closely with the primary cilium, prevents Smoothened accumulation in the cilium; in the presence of Hh, Patched is inactivated, thereby allowing the accumulation of Smoothened in the cilium. (B) While most discussions of cell surface receptors imply that such receptors are dispersed randomly across the surface of the plasma membrane, in fact many receptors are displayed only in localized, specialized domains of the cell surface. The Smoothened receptor (like the PDGF-α receptor) is displayed on the surface of primary cilia (*white arrow*), which protrude from the plasma membrane of many and perhaps all mammalian cells, as is seen here on the surface of an epithelial cell. (C) The cycling of Gli back and forth between the cytosol and the tip of the primary cilium can be blocked experimentally, trapping Gli (*green*) at the tip of the primary cilium (*magenta stalk*) and revealing dramatically one phase of its biochemical cycle. (A, courtesy of D. Thomas and adapted from S. Goetz and K.V. Anderson, *Nature Rev. Genet.* 11:331–344, 2010. B, from C.J. Haycraft, et al., *PLoS Genet.* 1(4):e53, 2005. C, from J. Kim, M. Kato and P.A. Beachy, *Proc. Natl. Acad. Sci. USA* 106:21666–21671, 2009.)

(**Figure 5.22A**); receptors configured in this way are often termed *serpentine* receptors. When the ligands of Patched—proteins of the Hedgehog (Hh) class—bind to Patched (Ptc), the latter moves away from a second serpentine G protein called Smoothened (Smo).

Curiously, the Patched–Smoothened signaling system depends on a structure that was once dismissed as a vestige of our early origins as ciliated or flagellated single-cell eukaryotes. A primary cilium protrudes from the surfaces of many if not all mammalian cell types (see Figure 5.22B). Normally, Smoothened (Smo) and the Gli transcription factor cycle back and forth out of the primary cilium (Figure 5.22A and C). However, when a Hedgehog (Hh) ligand binds its cognate receptor, Patched, in the plasma membrane, Smoothened and Gli accumulate in the primary cilium, where Gli is converted from a transcriptional repressor into an inducer of transcription and is dispatched to the nucleus.

The Wnt factors represent yet another, independent signaling system. As described earlier, their discovery is traced back to a *Drosophila* mutant termed **Wingless** and a related gene termed *int-1* in mice; the latter was discovered because of insertional mutagenesis by mouse mammary tumor virus (see Sidebar 3.6). These Wnt proteins—humans make at least 19 distinct Wnt species—are tethered tightly to the extracellular matrix (ECM) and, through a lipid tail, to cell membranes, and thus do not appear

Figure 5.23 Canonical Wnt signaling via Frizzled receptors Receptors of the Wnt proteins are members of the Frizzled (Frz) family of transmembrane proteins. In the canonical Wnt signaling pathway, and in the absence of Wnt ligand binding (*left*), a complex of axin (*brown*) and Apc (*light blue-green*) allows glycogen synthase kinase-3β (GSK-3β, *pink*) to phosphorylate β-catenin (*blue*). This marks β-catenin for rapid destruction by *proteolysis*. However, when a Wnt ligand binds to certain Frizzled receptors (*right, dark green*) and the LRP co-receptor (*purple*), the resulting activated Frizzled receptor, acting via the Dishevelled protein (*light green*) and axin (*dark brown*), causes inhibition of GSK-3β. This spares β-catenin, which may accumulate in the nucleus and promote cell proliferation.

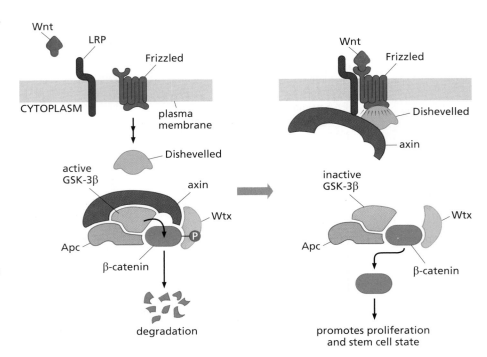

to be freely diffusible like several of the growth factors that we discussed earlier. (In fact, many growth factors probably become attached to the ECM soon after being secreted and are usually freed by proteases produced by cells that also carry receptors for them.) Growth factors of the Wnt class activate receptors of the Frizzled (Frz) family, which, like Patched, are complex receptors that weave back and forth through the plasma membrane multiple times (**Figure 5.23**).

In the absence of Wnt signaling, the cytoplasmic enzyme glycogen synthase kinase-3β (GSK-3β) phosphorylates several key proliferation-promoting proteins, including β-catenin, tagging them for destruction. However, the binding of Wnts to Frizzled receptors triggers a cascade of steps that shut down GSK-3β firing, allowing these proteins to escape degradation and promote cell proliferation (see Figure 5.23A). This pathway also plays a critical role in cancer pathogenesis, and we will discuss it in great detail later in this book.

The control of β-catenin levels by Wnts, often termed "canonical Wnt signaling," is only one of two ways by which Wnts and their Frizzled receptors affect cell behavior. Certain Wnts, acting together with a subset of the Frizzled receptors, signal through an entirely different biochemical pathway, termed "non-canonical Wnt signaling" (**Figure 5.24A**). This more recently discovered signaling channel is now realized to play a key role in the control by Wnts of various cancer cell phenotypes, including motility, invasiveness, and the maintenance of self-renewal potential. It depends on the ability of non-canonical Frizzled receptors to activate guanine nucleotide-binding proteins (**G-proteins**, for short), which function, much like Ras, as binary on–off switches that cycle between a GTP-bound active state and a GDP-bound inactive state (see Figure 5.30). Because these non-canonical Wnt proteins regulate G-proteins, they are grouped in the class of G-protein–coupled receptors (GPCRs).

The G-proteins are structurally complex, being formed as heterotrimers composed of α, β, and γ subunits. As indicated in Figure 5.24A, once a GPCR binds its ligand, a poorly understood shift in the configuration of the receptor causes it to trigger the release of GDP by the α subunit of the G-protein and the binding of GTP instead. The α subunit then dissociates from its β and γ partners, and both the α subunit and the complexed β + γ subunits may then proceed to interact with their own respective **effector** proteins, that is, proteins that transmit signals further into the cell interior. In the case of the non-canonical Frizzled receptors, the effector of the activated α subunit is a phosphodiesterase (PDE) enzyme, while the β + γ complex activates a distinct effector—phospholipase C-β (PLC-β). Eventually, the α subunit turns itself off

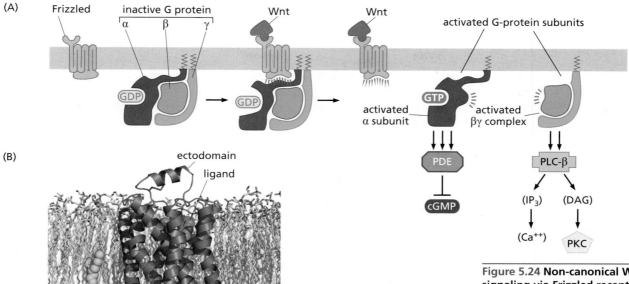

Figure 5.24 Non-canonical Wnt signaling via Frizzled receptors (A) In the non-canonical Wnt signaling pathway, Wnts bind to a second subset of Frizzled receptors. The receptor–ligand complex then binds a heterotrimeric G-protein, which is composed of α, β, and γ subunits. The ligand-activated receptor then stimulates the α subunit of the G protein to release its GDP (guanosine diphosphate) and bind GTP (guanosine triphosphate) instead (*light blue*). The α subunit then dissociates from the β + γ pair. Now, the α subunit (*brown*) and, independently, the β + γ complex (*blue, light green*), can each proceed to regulate key enzymes: the α subunit activates a phosphodiesterase (PDE), which reduces cyclic guanosine monophosphate (cGMP) levels, while the β + γ pair activates phospholipase C-β (PLC-β), which generates products that increase intracellular calcium and activate one form of protein kinase C (PKC). By hydrolyzing GTP to GDP, the α subunit eventually shuts down signaling, acting much like the Ras protein (see Figure 5.30). (B) The structures of G-protein–coupled receptors (GPCRs) have been elusive because of difficulties in crystallizing them. A structure of rhodopsin produced in 2000 was followed in 2007 by the elucidation of the structure of the β2-adrenergic receptor, which is composed of seven α-helices (*red*). A low–molecular-weight ligand (*green*) is shown binding in the receptor's core, while a molecule of cholesterol (*yellow*) is shown embedded in the lipid bilayer of the surrounding plasma membrane. Some GPCRs are specialized to bind low–molecular-weight ligands, while others, like Frizzled receptors, bind higher–molecular-weight polypeptide ligands. (B, from V. Cherezov et al., *Science* 318:1258–1265, 2007.)

by hydrolyzing its bound GTP (converting it to GDP) and reassociates with its β and γ partners. These heterotrimeric complexes then retreat into their inactive, nonsignaling state, awaiting signals once again from a ligand-activated GPCR.

A recent survey of the human genome sequence indicates that it encodes, in addition to the tyrosine kinase receptors, eleven TGF-β–like receptors, four Notch receptors, one or two Patched receptors, and ten Frizzled receptors. Far more numerous than these are a class of serpentine receptors that, like the Patched and Frizzled proteins, weave back and forth through the plasma membrane. Like the Frizzled proteins, these numerous receptor molecules traverse the plasma membrane precisely seven times (see Figure 5.24B). And like the non-canonical Frizzled receptors, these are GPCRs that signal through their ability to activate G-proteins.

There are more than a thousand GPCR-encoding genes in the mammalian genome; the majority of these (600) are inactive pseudogene versions of the 400 or so GPCRs that function as olfactory receptors. The functional versatility of these receptors is illustrated by their diverse functions beyond olfaction, as some underlie mechanisms of taste in the mouth, detection of photons in the retina, and signaling by dozens of distinct neurotransmitters in the brain. While vast in number, these receptors have to date been found to contribute directly to the formation of only a relatively small number of human cancers (described in Chapter 6). However, a subset of the GPCRs, termed **chemokine** receptors, play an important role in recruiting cells into the tumor-associated stroma, which provides essential physiologic support for a variety of cancer cells, as described in Chapter 13.

5.8 Nuclear receptors sense the presence of low–molecular–weight lipophilic ligands

The ability of cells to sense various extracellular signals is dictated by the chemical properties of signaling ligands and the impermeability of the plasma membrane to polar compounds, including most polypeptides. These conditions necessitated the development of two of the major classes of receptors that we have encountered:

Figure 5.25 Nuclear receptors
(A) While the estrogen receptor (ER) often functions as a homodimer, many other nuclear receptors function as heterodimers, with each monomer binding its own ligand and its own recognition site in the DNA. Seen here is the structure of the only nuclear receptor complex that has been elucidated in its entirety—the heterodimeric complex of PPAR-γ (peroxisome proliferator-activated receptor-γ, *light purple*) and the RXR (retinoid X receptor, *blue*). The LXLL domain (where L = leucine and | X = variable amino acid residue) of the co-activators of each receptor (*red*) enable the co-activators to bind directly to the receptor; these then recruit chromatin-modifying enzymes to the nearby chromatin. The ligands (*green*) of the two receptors are rosiglitazone (bound by PPAR) and 9-*cis*-retinoic acid (bound by RXR). Zinc ions are seen as *red dots*. (B) The DNA-binding domains of the two subunits of the receptor shown in (A) are seen here in greater detail, in this case binding to two copies of the sequence AGGTCA (separated by a single base pair spacer sequence). Other nuclear receptors have paired DNA recognition sites with spacers of various lengths. (C) Like the ligands of other nuclear receptors, 17β-estradiol, otherwise called estrogen or simply E2, causes a conformation shift of the estrogen receptor (ER) that enables the receptor to associate physically with a co-activator, resulting in activation of transcription. Binding of pseudoligands, such as the drug 4-hydroxytamoxifen (4-OHT), causes an alternative stereochemical shift that results in the loss of contact with co-activators and the acquisition of contact with co-repressors; this explains why tamoxifen is often used therapeutically to antagonize estrogen signaling. (D) Analysis of the effects of the E2 mimic diethylstilbestrol (DES; *white, left*) versus 4-OHT (*white, right*) reveals a shift in the location of helix 12 (*red*). Following DES binding, the ER (*dark purple/blue*) is able to associate with a domain of the GRIP-1 (glucocorticoid receptor-interacting protein-1) transcriptional co-activator (*green*). However, if the E2 antagonist 4-OHT is bound (right), the shift in helix 12 of the ER precludes the binding of the co-activator, blocking transcriptional activation. (A and B, from V. Chandra et al., *Nature* 456:350–356, 2008. D, adapted from A.K. Shiau et al., *Cell* 95:927–937, 1998.)

because neither polypeptide GFs nor the low–molecular-weight polar ligands of GPCRs can penetrate the plasma membrane, their corresponding receptors display elaborate ligand-binding ectodomains to the extracellular environment.

Another class of signaling ligands, however, consists of molecules that are both of low molecular weight and relatively hydrophobic; included among these are steroid sex hormones, retinoids, and vitamin D. Their hydrophobicity enables these ligands to readily penetrate the plasma membrane and advance further into the nucleus, where they may bind and activate a series of nuclear DNA-binding proteins that function as transcription factors. The complex cytoplasmic signaling cascades that have evolved to function as intermediaries between a diverse array of cell-surface receptors and the nuclear transcription machinery (see Figure 5.1) are not needed here, since these "nuclear receptors" can bind their ligands and directly regulate the expression of target genes. Prominent among the 48 or so human nuclear receptors are those that bind the steroid hormones estrogen, progesterone, and androgens. These three receptors play key roles in the development of common human malignancies, in particular, breast, ovarian, and prostate carcinomas.

Nuclear receptor molecules contain a DNA-binding domain, a hinge region, and a conserved ligand-binding domain, these three being flanked by variable N′- and C′-terminal domains. Some nuclear receptors remain in the cytoplasm until ligand binding triggers their nuclear import, while others are constitutively bound to chromatin. Once associated with the chromatin, these receptors bind as homo- or heterodimers (**Figure 5.25A**) to a pair of recognition sequences in the DNA located within

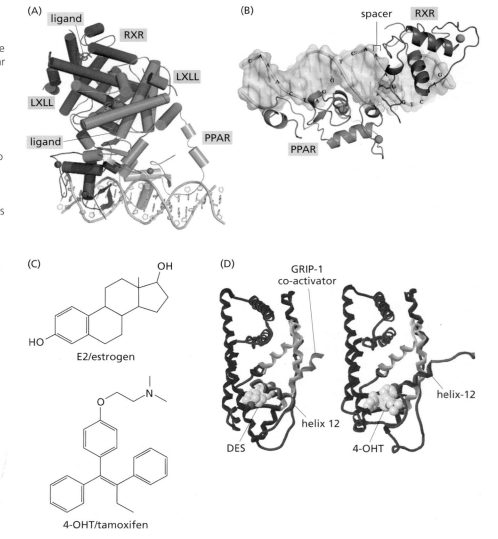

(A) ligand RXR LXLL LXLL ligand PPAR

(B) spacer RXR PPAR

(C) E2/estrogen

4-OHT/tamoxifen

(D) GRIP-1 co-activator helix 12 DES helix-12 4-OHT

or near the promoters of genes whose expression they control; these recognition sequences are often termed **hormone response elements** (HREs) and are composed of a pair of hexanucleotides separated by a variable number of spacer sequences (see Figure 5.25B).

In the case of the estrogen receptor (ER), binding of its natural ligand—17β-estradiol, commonly called estrogen or simply E2 (see Figure 5.25C)—causes a conformational shift of α-helices near the ER ligand-binding pocket; these shifts allow the receptor to attract one of several receptor **co-activators** and release **co-repressors**, that is, proteins that relay signals from the ER to the transcriptional machinery. These intermediaries often act by modifying nearby histones or, alternatively, by recruiting histone-modifying enzymes (see Section 1. 8). The resulting modified chromatin then either allows or prevents transcription of associated genes by RNA polymerase II. A **selective estrogen receptor modulator** (SERM), as represented here by tamoxifen, causes the ER to release co-activators (see Figure 5.25D) and to recruit co-repressors (not shown), resulting in effective blocking of E2 signaling; this shutdown, in turn, can confer great clinical benefit when this SERM is used to treat patients having ER-positive human breast carcinomas. For example, in one clinical trial, blocking E2 signaling with tamoxifen treatment resulted in an almost twofold reduction in the breast cancer relapse rate of postmenopausal women whose ER-positive primary tumors had been surgically removed.

5.9 Integrin receptors sense association between the cell and the extracellular matrix

The biology of normal and transformed cells provides hints of yet another type of cell surface receptor, one dedicated to sensing a quite different class of molecules in the extracellular space. Recall that an important attribute of transformed cells is their ability to grow in an anchorage-independent fashion, that is, to proliferate without attachment to a solid substrate such as the bottom of a Petri dish (see Section 3.5). This behavior contrasts with that of normal cells, which require attachment in order to proliferate. Indeed, in the absence of such attachment, many types of normal cells will activate a version of their death program (**apoptosis**) that is often termed **anoikis**. These cell death processes will be explored in detail in Chapter 9.

Biochemical analyses of the solid substrate to which cells adhere at the bottom of Petri dishes have revealed that, in very large part, cells are not anchored directly to the glass or plastic surface of these dishes. Instead, they attach to a complex network of molecules that closely resembles the extracellular matrix (ECM) usually found in the spaces between cells within most tissues. The ECM is composed of a series of glycoproteins, including collagens, laminins, proteoglycans, and fibronectin (**Figure 5.26**). After cells are introduced into a Petri dish, they secrete ECM components that adhere to the glass or plastic; once an ECM is constructed in this way, the cells attach themselves to this matrix. Consequently, anchorage dependence really reflects the need of normal cells to be tethered to ECM components in order to survive and proliferate.

The trait of anchorage dependence makes it obvious that cells are able to sense whether or not they have successfully attached to the ECM. As was learned in the mid-1980s, such sensing depends on specialized receptors that inform cells about the extent of tethering to the ECM and about the identities of specific molecular components of the ECM (for example, collagens, laminins, fibronectin) to which tethering has occurred. Sensing of the collagen fibers in the ECM is accomplished, in part, by an unusual pair of RTKs, termed **discoidin domain receptors** (DDRs; see Figure 5.15F). However, most sensing of ECM components is accomplished by a specialized class of cell surface receptors termed **integrins**. In effect, the molecular components of the ECM serve as the ligands of the integrin receptors. At the same time, the integrins create mechanical stability in tissues by tethering cells to the scaffolding formed by the ECM.

Integrins constitute a large family of heterodimeric transmembrane cell surface receptors composed of α and β subunits (**Figure 5.27**). At least eighteen α and eight β subunits have been enumerated; together, 24 distinct heterodimers are known. The

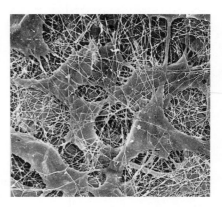

Figure 5.26 Extracellular matrix This scanning electron micrograph illustrates fibroblasts embedded in the specialized extracellular matrix (ECM) present in the cornea (of a rat). Many of the components of this ECM, including glycoproteins, hyaluronan, and proteoglycans, have been removed in order to highlight its major component—collagen fibers. Hence, the density of fibers in the ECM within the corresponding living tissue is actually much higher. (Courtesy of T. Nishida.)

ectodomains of these receptors bind specific ECM components. **Table 5.4** indicates that each integrin heterodimer shows specificity for binding a specific ECM molecule or a small subset of ECM components. The much-studied α5β1 integrin, for example, is the main receptor for fibronectin, an important glycoprotein component of the ECM found in vertebrate tissues. Laminins, which are large, multidomain ECM molecules, have been reported to be bound by as many as 12 distinct integrin heterodimers.

Having bound their ECM ligands through their ectodomains, integrins cluster to form *focal adhesions* (**Figure 5.28A**). This clustering affects the organization of the cytoskeleton underlying the plasma membrane, since some integrins are linked directly or indirectly via their cytoplasmic domains to important components of the cytoskeleton, such as actin, vinculin, talin, and paxillin (see Figure 5.28B, C). The formation of focal adhesions may also cause the cytoplasmic domains of integrins to activate

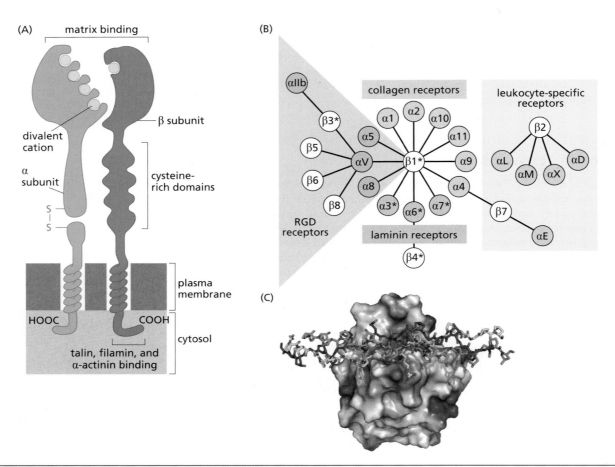

Figure 5.27 Structure of integrins Integrin molecules function as heterodimeric cell surface receptors, each composed of an α plus a β subunit (*green, blue*). (A) The ectodomains of these receptors bind specific components of the extracellular matrix (ECM). Divalent cations (*yellow dots*), notably magnesium, manganese, and calcium, modulate this binding in various ways. At the same time, the cytoplasmic domain (largely that of the β subunit) is linked, via intermediary proteins, to the cytoskeleton (largely that constructed by actin fibers); in addition, the cytoplasmic domains may attract a variety of signal-transducing proteins that become activated when the ectodomain binds an ECM ligand. (B) The α and β integrin subunits associate with one another as heterodimers in various specific combinations to enable the binding of a variety of ECM ligands and to transduce multiple signals into the cell. As is apparent, the set of heterodimers actually found in cells is far less (24) than might be theoretically predicted through combinatorial interactions of each α subunit with each β subunit (144 in all). (The most promiscuous of the β subunits is β1, which can associate with 12 alternative α subunits, resulting in this many distinct integrin heterodimers.) RGD receptors recognize the arginine–glycine–aspartic acid tripeptide motif that is found in a number of distinct ECM proteins and represents the core sequences recognized by these integrins to initiate such binding. Laminin is a critical component of the basement membrane (see Figure 13.6). Asterisks denote alternatively spliced cytoplasmic domains. (C) One depiction of integrin–ECM binding comes from the crystallographic analysis of the ectodomain of the α2β1 integrin (*space-filling structure*) interacting with collagen I, a triple helix (*stick figures*). The *red* surface patches on the integrin ectodomain contain the more acidic side chains, while the *blue* surface patches contain basic amino acids. (B, from R.O. Hynes *Cell* 110:673–687, 2002. C, from J. Emsley et al., *Cell* 101:47–56, 2000.)

Table 5.4 **Examples of integrins and their extracellular matrix ligands**

Integrin	ECM ligand
α1β1	collagens, laminin
αvβ3	vitronectin, fibrinogen, thrombospondin
α5β1	fibronectin
α6β1	laminin
α7β1	laminin
α2β3	fibrinogen
α6β4	laminin—epithelial hemidesmosomes

Adapted in part from B. Alberts et al., Molecular Biology of the Cell, 5th ed. New York: Garland Science, 2008; and from H. Lodish et al., Molecular Cell Biology. New York: W.H. Freeman, 1995.

signaling pathways that evoke a variety of cellular responses, including cell migration, proliferation, and survival. For example, by triggering the release of anti-apoptotic signals, integrins reduce the likelihood of anoikis. The functions of integrins continue to be critical during the development of many tumors, indicating that tumor cells usually do not evolve to a stage where they have outgrown their dependence on integrins for survival and proliferation signals (**Sidebar 5.6**).

Integrins are most unusual in one other respect. Normally we think of receptors as passing information from outside the cell into the cytoplasm. Integrins surely do this. But in addition, it is clear that signals originating in the cytoplasm are used to control the binding affinities of integrins for their ECM ligands. Such "inside-out" signaling enables cells to modulate their associations with various types of ECM or with various points of contact with an ECM, breaking existing contacts and forging new ones in their place. Rapid modulation of extracellular contacts enables cells to free themselves from one microenvironment within a tissue and move into another, and to traverse a sheet of ECM *in vitro*. Cultured fibroblasts lacking focal adhesion kinase (FAK), one of the signaling molecules that associate with the cytoplasmic domains of integrins, are unable to remodel their focal contacts and lack motility, indicating that the signals transduced throughout the cytoplasm by FAK are important for reconfiguring the cytoskeleton—the structure that enables cells to change shape and to move. Later, in Chapter 14, we will see how cell motility is critical to the ability of cancer cells to invade and metastasize.

These various descriptions of tyrosine kinase receptors and integrins reveal that mammalian cells use specialized cell surface receptors to sense two very distinct types of extracellular proteins. Some receptors, such as the EGF, PDGF, and Frizzled receptors, sense soluble (or solubilized) growth factors, while others, notably the integrins, sense attachment to the essentially insoluble scaffolding of the ECM. (Sometimes two types of receptors, such as integrins and RTKs, associate laterally and regulate one another. As we learn more about these cell surface proteins, we increasingly realize the importance of these lateral interactions; see Supplementary Sidebar 5.3). Together, these receptors enable a normal cell to determine whether two preconditions have been satisfied before this cell undertakes growth and division: the cell must sense the presence of adequate levels of mitogenic growth factors in its surroundings and the existence of adequate anchoring to specific components of the ECM.

Both of these requirements for cell proliferation are known to be abrogated if a cell carries an activated *ras* oncogene. Thus, *ras*-transformed cells can grow in the presence of relatively low concentrations of serum and serum-associated mitogenic growth factors; in addition, many types of *ras*-transformed cells can proliferate in an anchorage-independent fashion. These behaviors suggest that, in some way, a Ras oncoprotein

(A)

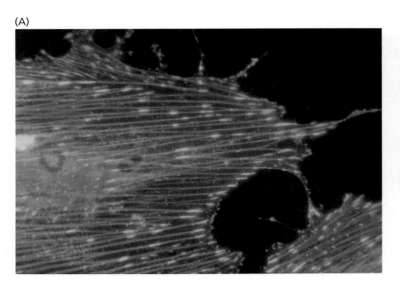

(C)

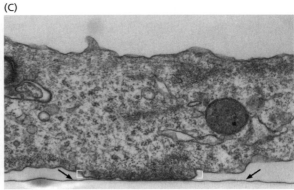

(B)

ECM

α-actinin   vinculin   talin   paxillin

Ena/
VASP

actin
cytoskeleton

actin

actin
stress
fibers

Figure 5.28 Integrin tethering to the ECM and cytoskeleton
(A) This fluorescence micrograph illustrates the discrete elongated foci (*yellow/orange*) on the cell surface, termed focal adhesions, at which integrins tether cells to the extracellular matrix (ECM). Most of these foci seen here have been formed on the *ventral* sides of these cells, i.e., the undersides of these cells that are directly apposed to the ECM that is adsorbed to the surface of the Petri dish. Formation of these focal adhesions also controls the organization of the actin fibers (*green*) forming the cytoskeleton. (B) This schematic figure of the organization of integrins (*green, blue*) indicates their association with the ECM (*green fibers, above*) through their ectodomains and their association with the actin cytoskeleton (*pink chains*) through their cytoplasmic domains via the β subunit of each heterodimer. A series of intermediary proteins, such as actinin, vinculin, and talin, allows these linkages to

be formed. (C) This transmission electron micrograph section, which is oriented in a plane perpendicular to the bottom surface of a Petri dish, reveals a single focal adhesion formed between the ventral surface of a chicken lens cell and the underlying solid substrate; in this case, ECM molecules adsorbed to the surface of the Petri dish are seen as a dark line (*arrows*). The electron-dense, cytoplasmic material (*white brackets*) that is immediately above the region of contact with the substrate contains large numbers of integrin-associated proteins and actin fibers that in aggregate constitute this focal adhesion, which may be as much as several microns in width when viewed from this perspective. (The resolution of this image does not afford visualization of individual actin fibers or integrin molecules.) (A, courtesy of Keith Burridge. B, adapted from C. Miranti and J. Brugge, *Nat. Cell Biol.* 4:E83–E90, 2002. C, from O, Medalia and B. Geiger, *Curr. Opin. Cell Biol.* 22:659–668, 2010.)

Sidebar 5.6 Some integrins are essential for tumorigenesis The ability to selectively inactivate ("knock out") genes within targeted tissues has made it possible to evaluate the contributions of various signaling proteins to tumorigenesis. For example, inactivation of the gene encoding the β1 integrin in the mouse germ line leads to the absence of this integrin in all cells of the embryo and to embryonic lethality; however, introduction of a "conditional" allele into the mouse germ line makes it possible to inactivate this gene in a targeted fashion in one or another specific organ or tissue. In the experiment shown in **Figure 5.29**, both copies of the β1 integrin allele have been selectively deleted from the genomes of epithelial cells of the mouse mammary gland, leaving the gene unperturbed in all other tissues. Such inactivation of both copies of the β1 integrin gene has minimal effects on the normal development of the mammary gland.

In these mice, a second germ-line alteration involved the insertion of the middle T oncogene of polyomavirus that had been placed under the control of a promoter that allows this **transgene** to be expressed only in the mice's breast tissue. In mice with an intact β1 integrin gene, this oncogenic transgene generally triggers the formation of premalignant hyperplastic nodules that progress into aggressive mammary carcinomas. However, when β1 integrin is absent from these oncogene-expressing mammary epithelial cells (because of selective gene inactivation), the number of normally observed hyperplastic nodules is reduced by more than 75% and the β1 integrin–negative cells in these nodules are fully unable to develop further to form mammary carcinomas (see Figure 5.29C). Detailed analyses indicate that this loss of β1 integrin expression permits the oncogene-expressing mammary epithelial cells to survive but precludes their active proliferation and thus their ability to spawn mammary carcinomas.

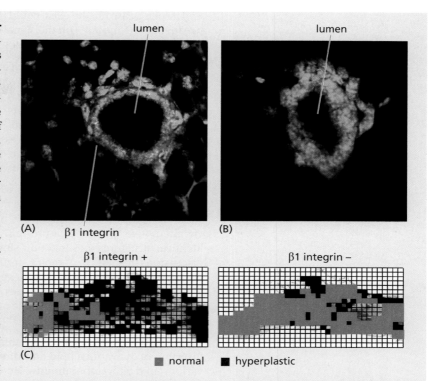

Figure 5.29 β1 integrin and mammary tumorigenesis The selective inactivation of the gene encoding β1 integrin in the mouse mammary gland allows a test of its importance in tumorigenesis. (A) The immunostaining of β1 integrin (*red*) on the surface of the epithelial cells forming a normal mouse mammary duct (*blue*). (B) In the absence of β1 integrin, ductal morphogenesis still proceeds normally. Moreover, the overall development of the mammary gland is normal (*not shown*). (C) Premalignant hyperplastic nodules in a transgenic mouse prone to mammary tumorigenesis (because of the selective expression of an oncogene in the mammary gland) have been scored visually by microscopic examination of whole mammary glands (*left*); the nodules are present in abundance and, wherever they have been detected, are represented by *black squares*, in contrast to areas in which only normal epithelium is observed in these glands (*red squares*). However, in mammary glands of mice in which β1 integrin is absent (because its encoding gene has been inactivated, see Supplementary Sidebar 7.7), the transgenic oncogene fails to induce the formation of large numbers of premalignant nodules (*right*). (From D.E. White et al., *Cancer Cell* 6:159–170, 2004.)

is able to mimic the signals introduced into the cell by ligand-activated growth factor receptors and by integrins that have engaged ECM components. An understanding of the biochemical basis of these various signals demands some insight into the structure and function of the normal and oncogenic Ras proteins. So, we will move back in history to the early 1980s to see how the puzzle of Ras function was solved.

5.10 The Ras protein, an apparent component of the downstream signaling cascade, functions as a G protein

The discoveries that two oncogenes—*erbB* and *sis*—encode components of the mitogenic growth factor signaling machinery (Sections 5.4 and 5.5) sparked an intensive effort to relate other known oncoproteins, notably the Ras oncoprotein, to this signaling machinery. The *ras* oncogene clearly triggered many of the same changes in cells that were seen when cells were transformed by either *src*, *erbB*, or *sis*. Was there some type of signaling cascade—a molecular bucket brigade—operating in the cell, in which protein A transferred a signal to protein B, which in turn signaled protein C?

And if such a cascade did exist, could the Ras oncoprotein be found somewhere downstream of *erbB* and *sis*? Did the signals emitted by these various proteins all converge on some common target at the bottom of this hypothesized signaling cascade?

At the biochemical level, it was clear that growth factor ligands activated tyrosine kinase receptors, and that these receptors responded by activating their cytoplasmic tyrosine kinase domains. But which proteins were affected thereafter by the resulting receptor phosphorylation events? And how did this phosphorylation lead further to a mitogenic response by the cell—its entrance into a phase of active growth and division? In the early 1980s, progress on these key issues ground to a halt because biochemical experiments offered no obvious way to move beyond the tyrosine kinase receptors to the hypothesized downstream signaling cascades.

All the while, substantial progress was being made in understanding the biochemistry of the Ras protein. In fact, the three distinct *ras* genes in mammalian cells (see Table 4.2) encode four distinct Ras proteins (since K-*ras* specifies a second protein via alternative splicing of a pre-mRNA). Because all four Ras proteins have almost identical structures and function similarly, we refer to them simply as "Ras" in the discussions that follow. At their C-termini, they all carry covalently attached lipid tails, composed of farnesyl, palmitoyl, or geranylgeranyl groups (or combinations of several of these groups). These lipid moieties enable the Ras proteins, all of about 21 kD mass, to become anchored to cytoplasmic membranes, largely to the cytoplasmic face of the plasma membrane.

(For many years, it appeared that this membrane anchoring is achieved through the insertion of the various lipid tails directly into the hydrophobic environments within various lipid bilayer membranes. However, more recent studies indicate greater complexity. In the case of H-Ras, for example, its palmitate moiety is indeed inserted directly into cytoplasmic membranes, while its farnesyl group is inserted into a hydrophobic pocket of a specialized farnesyl-binding protein; the latter, in turn, facilitates the interactions of H-Ras with other partner proteins and also strengthens the binding of H-Ras to various membranes.)

Like the heterotrimeric G-proteins (Section 5.7), Ras was found to bind and hydrolyze (that is, cleave) guanosine nucleotides. This activity as a GTPase indicated a mode of action quite different from that of the Src and erbB tyrosine kinases, yet intriguingly, all three oncoproteins had very similar effects on cell behavior.

In further analogy to the G proteins, Ras was found (1) to bind a GDP molecule when in its inactive state; (2) to jettison its bound GDP after receiving some stimulatory signal from upstream in a signaling cascade; (3) to acquire a GTP molecule in place of the recently evicted GDP; (4) to shift into an activated, signal-emitting configuration while binding this GTP; and (5) to cleave this GTP after a short period by using its own intrinsic GTPase function, thereby placing itself, once again, in its non-signal-emitting configuration (**Figure 5.30**). In effect, the Ras molecule seemed to behave like a light switch that automatically turns itself off after a certain predetermined time.

Figure 5.30 The Ras signaling cycle Detailed study of the biochemistry of Ras proteins revealed that they, like the heterotrimeric G-proteins (see Figure 5.24B), operate as binary switches, binding GDP in their inactive state (*top*) and GTP in their active, signal-emitting state (*bottom*). Thus, an inactive, GDP-binding Ras protein is stimulated by a GEF (guanine nucleotide exchange factor, *orange*) to release its GDP and acquire a GTP in its stead, placing Ras in its active, signaling configuration. This period of signaling is soon halted by the actions of a GTPase activity intrinsic to Ras, which hydrolyzes GTP to GDP (*left*). This GTPase activity is strongly stimulated by GAPs (GTPase-activating proteins, *light blue*) that Ras might encounter while in its activated state. Amino acid substitutions caused by oncogenic point mutations (see Figure 4.9) block this cycle by inactivating the intrinsic GTPase activity of Ras; this traps Ras in its activated, signal-emitting state.

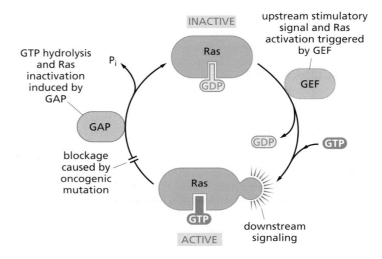

Sidebar 5.7 GTPase-activating proteins and cancer When purified Ras proteins are examined in isolation, they show a very low level of GTPase activity. However, within living cells, it is clear that the GTP-bound form of Ras can interact with a second class of proteins termed GAPs (GTPase-activating proteins). This interaction increases the GTPase activity of the Ras protein by as much as 100,000-fold and thus results in the rapid conversion of active, GTP-bound Ras into its inactive, GDP-bound form. (The Ras-GAP protein, one of two well-studied GAPs that act on Ras, has been found to insert an arginine "finger" into the GTPase cleft of Ras; in doing so, it actively participates in the catalysis that converts GTP to GDP. Interestingly, Ras-GAP has almost no effect in stimulating the GTPase activity of mutant Ras oncoproteins, helping to explain the high proportion of GTP-loaded Ras in mutant cells.)

The lifetime of the activated, GTP-bound form of Ras may therefore be governed by the time required for its encounter with a GAP molecule. Since the GAPs themselves may be under the control of yet other signals, this means that the GTPase activity of the Ras proteins is modulated indirectly by other signaling pathways. Moreover, as described later (Section 7.10), the loss from a cell of the other Ras GAP, termed NF1, leaves the GTPase activity of Ras at an abnormally low level in this cell, resulting in the accumulation of GTP-loaded Ras, the hyperactivity of Ras signaling, and ultimately the formation of neuroectodermal tumors.

In more detail, the hypothesized chain of events went like this. Mitogenic signals, perhaps transduced in some way by tyrosine kinase receptors, activated a guanine nucleotide exchange factor (GEF) for Ras. This GEF induced an inactive Ras protein to shed its GDP and bind GTP instead. The resulting activated Ras would emit signals to a (still-unknown) downstream target or group of targets. This period of signaling would be terminated, sooner or later, when Ras decided to hydrolyze its bound GTP. In doing so, Ras would revert to its nonsignaling state. As was learned later, a group of other proteins known as GAPs (GTPase-activating proteins) could actively intervene and encourage Ras to undertake this hydrolysis (Sidebar 5.7). Implied in this model (see Figure 5.30) is the notion that Ras acts as a signal-transducing protein, in that it receives signals from upstream in a signaling cascade and subsequently passes these signals on to a downstream target.

A striking discovery was made in the course of detailed biochemical analysis of a Ras oncoprotein made by a point-mutated *ras* oncogene, specifically, the Ras oncoprotein encoded by Harvey sarcoma virus. Like the normal Ras protein, the Ras oncoprotein could bind GTP. However, the oncoprotein was found to have lost virtually all GTPase activity (**Figure 5.31A**). In such condition, it could be pushed into its active, signal-emitting configuration (see Figure 5.31B) by some upstream stimulatory signal and guanine nucleotide exchange factors (GEFs). But once in this activated state, the Ras oncoprotein was unable to turn itself off! The negative-feedback loop that is so essential to the operations of the normal Ras protein had been sabotaged by this point mutation and resulting amino acid substitution.

This provided a clear explanation of how Ras can operate as an oncoprotein: rather than sending out short, carefully rationed pulses of growth-stimulating signals, the oncoprotein emits signals for a long, possibly indefinite period of time, thereby flooding the cell with these signals. Subsequent analysis of Ras protein biochemistry has revealed additional levels of control that may go awry in cancer cells (Sidebar 5.7).

These findings provided a solution to a puzzle created by the sequencing of the mutant *ras* oncogenes present in many human tumors. As we learned (Section 4.4), invariably the point mutations creating these oncoproteins are **missense** mutations (which cause an amino acid substitution) rather than **nonsense** mutations (which cause premature termination of the growing protein chain). And invariably these mutations strike either the 12th, 13th, or 61st codon of the reading frame of the *ras* genes (with the great majority altering codon 12; see Figure 4.10).

(A)

(B)

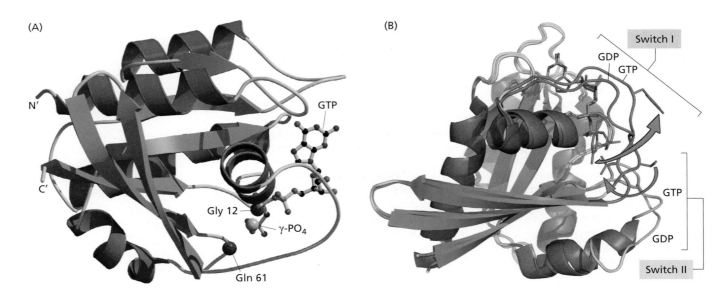

Figure 5.31 The structure of the Ras protein and its response to GTP binding (A) This ribbon diagram of the structure of a Ras protein, as determined by X-ray crystallography, depicts the polypeptide backbone of Ras and its α-helical (*red*) and β-pleated sheet (*green*) domains. GTP is indicated as a stick figure, and the two amino acid residues often found altered in the Ras oncoproteins present in human tumors —glycine 12 and glutamine 61—are shown as *dark blue balls*. Their close association with the γ-phosphate of GTP (*gray ball*) helps to explain why substitutions of these residues reduce or eliminate the Ras protein's GTPase activity, which normally converts GTP into GDP by removing the γ-phosphate. This provides an explanation of why the codons specifying these particular residues are commonly mutated in human tumor cell genomes; conversely, amino acid substitutions striking elsewhere in the Ras protein may compromise its intrinsic signal-emitting functions and, in addition, are unlikely to potentiate Ras function. (B) The replacement of GDP by GTP, which results in the activation of Ras signaling, also causes a shift in two regions of the Ras protein (depicted here from a slightly different perspective). Both Switch I (*magenta*) and Switch II (*aquamarine*) regions undergo the GDP-to-GTP conformational shift (as indicated by the positions of GDP and GTP). The conformational shift for Switch I is also indicated by the *magenta* arrows. These shifts allow the Ras protein to interact physically with its downstream effectors (see Figures 6.13 and 6.14). The guanine nucleotide is indicated by a stick figure. (A, courtesy of A. Wittinghofer. B, courtesy of I.R. Vetter and A. Wittinghofer.)

We might imagine that these particular nucleotides represent sites in the genomic DNA that are particularly susceptible to attack and alteration by mutagenic carcinogens. However, detailed study of the structure of the Ras proteins, as generated by X-ray crystallography (see Figure 5.31), has led to a far more compelling model. Examination of the 12th, 13th, and 61st amino acid residues of these proteins reveals that these residues are located around the cavity in the Ras protein where the GTPase catalytic activity operates. Consequently, almost all substitutions of these three amino acids, such as the glycine-to-valine substitution that we encountered earlier, compromise GTPase function. [To be more precise, they compromise the ability of GTPase-activating proteins (GAPs; see Figure 5.30) to trigger hydrolysis of the GTP bound by Ras.]

Knowing this, we can immediately understand why point mutations affecting only a limited number of amino acid residues are found in the *ras* oncogenes carried by human tumor cell genomes. Large-scale alterations of the *ras* proto-oncogenes, such as deletions, are clearly not productive for cancer, since they would result in the elimination of Ras protein function rather than an enhancement of it. Similarly, the vast majority of point mutations striking *ras* proto-oncogenes will yield mutant Ras proteins that have lost rather than gained the ability to emit growth-stimulatory signals. Only when the signal-emitting powers of Ras are left intact and its GTPase negative-feedback mechanism is inactivated selectively (by amino acid substitutions at one of these three sites in the protein) does the Ras protein gain *increased* power to drive cell proliferation and transform the cell.

Only those rare cells that happen to have acquired mutations affecting one of the three amino acid residues of Ras (residues 12, 13, or, rarely, 61; see Figure 4.10) gain proliferative advantage over their wild-type counterparts and thereby stand the chance of becoming the ancestors of the cells in a tumor mass. So, even though point mutations affecting all the amino acid residues of a Ras protein are likely to occur with comparable frequencies, only those few conferring substantial growth advantage will actually be found in tumor cells. Other cells that happen to acquire point mutations affecting other residues in the Ras protein will retain a normal phenotype or may even lose proliferative ability.

5.11 Synopsis and prospects

In the early 1980s, the discovery of oncogenes such as *erbB* and *sis* revealed the intimate connections between growth factor signaling and the mechanisms of cell transformation. More specifically, these connections suggested that cell transformation results from hyperactivation of the mitogenic signaling pathways. This notion galvanized researchers interested in the biochemical mechanisms of carcinogenesis. Many began to study the growth factor receptors and their biochemical mechanisms of action.

A major problem surrounding growth factor receptors was ignorance about their mechanism for transferring signals from outside a cell into its interior. A simple model was soon developed to explain how this occurs. The growth factor ligands of receptors bind to their ectodomains. By doing so, the ligands encourage receptor molecules to come together to form dimers. This dimerization of the ectodomains then encourages dimerization of the receptor's cytoplasmic domains as well. Somehow, this enables the cytoplasmic domains to emit signals.

This model shed no light on how, at the biochemical level, receptors are able to emit signals. Here, research on the biochemistry of the Src protein proved to be telling. It was found to be a tyrosine kinase (TK), an enzyme that transfers phosphate groups to the side chains of tyrosine residues of various substrate proteins. Once the amino acid sequences of several growth factor receptors were analyzed, the cytoplasmic domains of these receptors were found to be structurally (and thus functionally) related to Src. Hence, tyrosine kinases create the means by which growth factor receptors emit biochemical signals.

This information was then integrated into the receptor dimerization model: when growth factor receptors are induced by ligand binding to dimerize, the tyrosine kinase of each receptor monomer transphosphorylates the cytoplasmic domain of the other monomer. This yields receptor molecules with phosphorylated cytoplasmic tails that, in some unknown fashion, allow signaling to proceed. Much of the next chapter is focused on these downstream signaling mechanisms.

While highly instructive, the dimerization model has proven to be simplistic: we now realize that in the absence of ligand, certain growth factor receptors are found in large clusters, and that ligand binding exerts subtle effects on the stereochemistry of these receptor molecules.

Hyperactive signaling by these receptors is encountered in many types of human cancer cells. Often, the receptors are overexpressed, resulting in ligand-independent firing via molecular mechanisms that remain unclear. Even more potent ligand-independent firing is achieved by various types of structural alterations of receptors. Most of these occur as the consequences of somatic mutations. However, some mutant alleles of receptor genes have been found in the human gene pool, and inheritance of such alleles is associated with a variety of inborn cancer susceptibility syndromes.

These tyrosine kinase growth factor receptors (RTKs) represent only one of multiple ways by which cells sense their surroundings. The TGF-β receptors, for example, are superficially similar to the RTKs, in that they have a ligand-binding ectodomain and a signal-emitting kinase domain in their cytoplasmic portion. However, the kinases of the TGF-β receptors are serine/threonine kinases and, as we will learn later, signal through much different mechanisms.

A variety of other signal-transducing receptors have been uncovered, including those of the Notch, Patched–Smoothened, and Frizzled classes. These receptors use a diversity of signal-transducing mechanisms to release signals into the cytoplasm. As we will see throughout this book, aberrant signaling by these receptors plays a key role in the pathogenesis of many types of human cancers. Moreover, we will encounter yet another class of receptors in Chapter 9, where their role in programmed cell death—apoptosis—is described.

Cells must also sense their contacts with the extracellular matrix (ECM)—the cage of secreted glycoproteins and proteoglycans that surrounds each cell within a tissue. Here, yet another class of receptors, the integrins, play a central role. Having bound components of the extracellular matrix, the heterodimeric integrin receptors transduce signals into cells that stimulate proliferation and suppress cells' apoptotic suicide program.

A distinct line of research, pursued in parallel, elucidated the biochemistry of the Ras protein, another key player in cancer pathogenesis. Ras operates like a binary switch, continually flipping between active, signal-emitting and quiescent states. The mutant alleles found in cancer cells cause amino acid substitutions in the GTPase pockets of Ras proteins, thereby disabling the mechanism that these proteins normally use to shut themselves off. This traps the Ras proteins in their active, signal-emitting configuration for extended periods of time, causing cells to be flooded with an unrelenting stream of mitogenic signals.

While these investigations revealed how isolated components of the cellular signaling machinery (for example, RTKs, Ras) function, they provided no insights into how these proteins communicate with one another. That is, the *organization* of the signaling circuitry remained a mystery. Cell physiology hinted strongly that ErbB (that is, the EGF receptor) and other tyrosine kinase receptors operated in a common signaling pathway with the Ras oncoprotein. Src fit in somewhere as well. These various oncoproteins, while having diverse biochemical functions, were found to exert very similar effects on cells. They all caused the rounding up of cells in monolayer culture, the loss of contact inhibition, the acquisition of anchorage independence, and reduction of a cell's requirement for mitogenic growth factors in its culture medium. This commonality of function suggested participation in a common signaling cascade.

One possible connection between Ras and the tyrosine kinase receptors came from the discovery that an activated *ras* oncogene causes many types of cells to produce and release growth factors. Prominent among these was transforming growth factor-α (TGF-α), an EGF-like growth factor. Like EGF, TGF-α binds to and activates the EGF receptor. This suggested the following scenario. Once released by a *ras*-transformed cell, TGF-α might function in an autocrine manner (see Figure 5.11C) to activate EGF receptors displayed on the surface of that cell. This, in turn, would evoke a series of responses quite similar to those created by a mutant, constitutively activated EGF receptor. When viewed from this perspective, Ras appeared to operate *upstream* of a growth factor receptor rather than being an important component of the signaling cascade lying *downstream* of the receptor (**Figure 5.32A**).

In the end, this autocrine scheme was able to explain only a small part of *ras* function, since *ras* was found to be able to transform cells that lacked receptors for the growth factors that it induced. In addition, ample biochemical evidence accumulated that ligand binding of growth factor receptors led rapidly to activation of Ras (see Figure 5.32B). This left open only one alternative: direct intracellular signaling between ligand-activated growth factor receptors and Ras proteins. So the search was on to find the elusive biochemical links between tyrosine kinase receptors and the proteins like Ras that appeared to form components of the hypothesized downstream signaling cascade.

The use of biochemistry to discover these connections had serious limitations. As this chapter has made clear, biochemistry provided researchers with powerful tools to elucidate the functioning of individual, isolated cellular proteins. However, it had limited utility in revealing how different proteins might talk to one another. If there really were

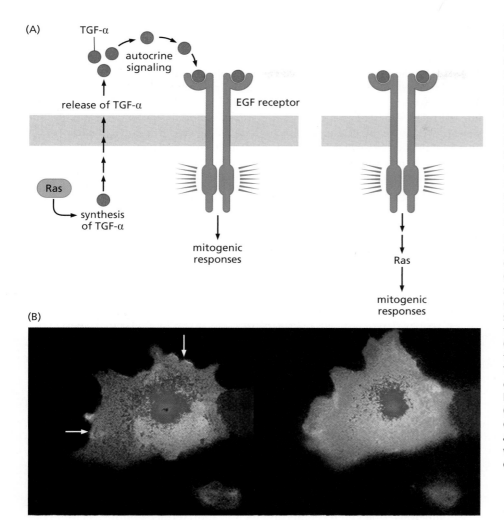

Figure 5.32 Alternative mechanisms of transformation by Ras (A) One depiction (*left*) of how Ras (*green*) normally operates to promote cell proliferation was inspired by the observation that *ras*-transformed cells release a number of growth factors into their surroundings, such as TGF-α (*red circles*). Once secreted, the growth factors might act in an autocrine fashion to activate a receptor, such as the EGF receptor shown here, thereby promoting cell proliferation. In this depiction, Ras operates "upstream" of the EGF receptor. An alternative scheme (*right*) placed Ras in the midst of a signaling cascade that operates "downstream" of such growth factor receptors. (B) Extensive evidence favoring the second scheme accumulated. This graphic example, produced years later, shows that 5 minutes after EGF is added to cells, previously inactive Ras protein (*blue shading*) becomes activated near the cell surface (*red–orange shading*). This response is so rapid that it indicates close biochemical communication between the EGF receptor and Ras. (Interestingly, this activation occurred only on the two sides of a cell (*white arrows*) that were not in direct contact with nearby cells.) (B, from N. Mochizuki et al., *Nature* 411:1065–1068, 2001.)

signaling cascades in the cell of the form A → B → C, biochemistry might well reveal how A or C functioned in isolation (for example, as a kinase or GTPase) but would shed little light on the organization of the signaling pathway that lay between them.

Detailed biological characterization of human tumor cells overexpressing certain growth factor receptors also suggested another dimension of complexity. In the case of human breast cancers, overexpression of the HER2/Neu receptor (see Section 4.3) was found to be correlated with a bewildering array of phenotypes displayed by the associated cancer cells. Those breast cancer cells expressing elevated levels of this protein showed increased rates of DNA synthesis, better anchorage-independent growth, greater efficiency in forming tumors when implanted into host mice (that is, tumorigenicity), greater tendency to metastasize, and less dependence on estrogen for their growth. Hence, these receptor proteins were acting pleiotropically to confer a number of distinct changes on cancer cells. Such action seemed to be incompatible with a simple linear signaling cascade operating downstream of activated receptors. Instead, it appeared more likely that a number of distinct downstream signaling pathways were radiating out from these receptors, each involved in evoking a distinct cancer cell phenotype. But once again, there seemed to be little prospect of mapping these pathways and their signaling components.

This stalemate could be broken only by exploiting new types of experimental tools and a new way of thinking about the organization of complex signaling circuits. The science of genetics provided the new tool kit and mindset. However, genetics could be practiced with facility only on relatively simple experimental organisms, specifically, flies, worms, and yeast, for which powerful genetics techniques had been developed.

As it turned out, the elucidation of these signaling cascades also depended on one more factor. Much had already been learned from genetically dissecting the simpler metazoans and single-cell eukaryotes. Were lessons learned about their signaling systems transferable to the signaling cascades operating in mammalian cells? Here, fortune favored the cancer biologists. Because these intracellular signaling cascades were of very ancient lineage and had changed little over many hundreds of millions of years, discoveries about their organization and design in the worm, fly, and yeast proved to be directly applicable to mammalian cells. The consequence was the enrichment of cancer cell biology by research fields that had appeared, at least superficially, to be irrelevant to an understanding of human cancer. As was often the case in cancer research, the important advances came from unexpected quarters, from the benches of researchers who never thought of themselves as foot soldiers in the war against cancer.

Research on the biochemistry of receptor signaling continues to reveal new dimensions of complexity that impact the contributions that these proteins make to the development of human tumors. As recent research reveals, RTKs often have extensive lateral interactions with other cell surface proteins, such as integrins; these interactions influence the localization of the receptors and their access to different ligands (see Supplementary Sidebar 5.3). The control mechanisms that govern the concentrations of GFs in the extracellular space continue to be poorly understood; included among these mechanisms are those that allow proteases to activate latent GF precursors or to liberate GFs from sequestration in the extracellular matrix. Moreover, a variety of inhibitory proteins are released by cells that prevent GFs from binding to their cognate receptors, doing so by intercepting these ligands in the extracellular space; some of these interceptors may bind and sequester GFs in an inactive state, while others may function as "decoy receptors" that display the ligand-binding domains of bona fide receptors but are incapable of releasing downstream signals.

An additional level of complexity comes from the distinct effects that various ligands have on their receptors; for example, after EGF binds to the EGF receptor, the receptor is internalized and then degraded in lysosomes, whereas TGF-α binding causes the same receptor to become internalized and then to be recycled to the cell surface.

We have not mentioned, even in passing, another class of RTKs that are actually the most numerous—the Eph receptors and their ligands, termed ephrins. As is the case with Notch receptors and their cognate ligands, the ephrins are displayed as proteins tethered to the surface of ligand-presenting cells; when receptor-ligand binding occurs, signals are dispatched bidirectionally, that is, into *both* cells engaged in these juxtacrine interactions. The roles of these ligand–receptor pairs in cancer pathogenesis is only beginning to emerge.

As we learn more about cell surface receptors, our current depictions of how they function will increasingly be seen as crude approximations of how they really operate within living tissues including tumors. Moreover, our focus on GFs and RTKs as the prime means of regulating communication between cells will be viewed as overly narrow. One hundred and three human phosphotyrosine phosphatases (PTPs), which reverse the actions of TKs, are likely to play key roles in regulating the levels of specific phosphotyrosines within cells, including those acquired by RTKs. Secreted proteases having narrow substrate specificities are also likely to sit astride numerous cell-to-cell signaling channels, yet we know virtually nothing about the 300 or so of these human enzymes and how they influence intercellular signaling. Many of these normal signaling mechanisms are likely to be subverted by cancer cells. All of this suggests that we have explored only the most visible corners of the complex circuitry that drives neoplastic growth.

Key concepts

- Because a multicellular organism can exist only if its individual cells work in a coordinated fashion, the problem of cell-to-cell communication must have been solved by the time the first metazoa arose. Deregulation of such cell signaling is central to the formation of cancer.

- Src provided the first insight into how oncoproteins function when it was found to operate as a protein kinase—an enzyme that transfers phosphates from ATP to other proteins in the cell. Since multiple distinct protein substrates could be phosphorylated and since each of these could affect its own set of downstream targets, this explained how Src could act pleiotropically to yield the numerous phenotypic changes observed in RSV-transformed cells.

- Most protein kinases phosphorylate threonine or serine residues, but Src functions as a tyrosine kinase: it attaches phosphates to tyrosine residues of proteins. Tyrosine phosphorylation is favored by mitogenic signaling pathways.

- Study of tyrosine kinase receptors (RTKs), such as the epidermal growth factor receptor (EGF-R), elucidated how they transduce signals from a cell's exterior into its cytoplasm. Binding of growth factor ligand to the N-terminal ectodomain of a cognate receptor induces dimerization of the receptor. This activates each monomer's kinase to phosphorylate its partner's tyrosines on the partner's C-terminal cytoplasmic domain. The resulting phosphotyrosines then enable emission of mitogenic signals to downstream target proteins within the cell.

- The dimerization model explains how overexpression of growth factor receptors favors cancer formation: because of their high numbers, the receptor molecules collide frequently as they move around the plasma membrane, resulting in dimerization, transphosphorylation, receptor activation, and emission of mitogenic signals.

- Mutations affecting any of the three RTK domains may create ligand-independent firing. An EGF-R with a truncated ectodomain cannot recognize its ligand but nonetheless emits growth-stimulatory signals in a constitutive fashion. Gene fusion events may create receptors that signal unremittingly by virtue of their ectodomains becoming fused with proteins that are naturally prone to dimerize or oligomerize. Amino acid substitutions or deletions in the transmembrane and cytoplasmic domains of receptors are also found in some cancers.

- Another mechanism for cell transformation is exemplified by the Sis oncoprotein of simian sarcoma virus. The virus causes an infected cell to release copious amounts of a PDGF-like protein, which attaches to the PDGF receptors of the same cell. This creates an autocrine signaling loop in which a cell manufactures a mitogen to which it can also respond.

- Other classes of receptors besides RTKs are important in cancer:
  - Cytokine receptors lack tyrosine kinase domains and rely on noncovalently associated tyrosine kinases for signaling instead of covalently associated TK domains.
  - Receptors for TGF-β have cytoplasmic kinase domains that phosphorylate serine and threonine rather than tyrosine residues.
  - Notch receptor forgoes activation by phosphorylation and instead relies on proteases to liberate a cytoplasmic domain fragment involved in gene expression.
  - The Patched–Smoothened system relies on one transmembrane protein controlling another, which in turn controls a transcriptional regulator.
  - Binding of Wnt factors to a canonical Frizzled receptor triggers a cascade of steps that prevents a cytoplasmic kinase from tagging β-catenin for destruction.
  - G-protein–coupled receptors (GPCRs) induce heterotrimeric G-proteins to flip from an inactive GDP-bound state to a signaling GTP-bound state, a strategy used by non-canonical Wnt receptors and a vast array of other GPCRs operating throughout the body.

- Integrins constitute a large family of heterodimeric transmembrane cell surface receptors, almost all of which have ECM components as their ligands. Upon ligand binding, integrins form focal adhesions that link their cytoplasmic domains to the actin cytoskeleton. Integrins pass information both into and out of the cell.

- While Ras proteins seemed to be part of a signaling cascade lying downstream of RTKs, the biochemical mechanisms connecting them were obscure, since the tyrosine kinase activity of RTKs had no apparent connection with the GTPase-bearing Ras protein.

- The difference between a normal Ras protein and its oncogenic counterpart is a missense mutation that alters a residue in the GTPase domain of Ras, preventing proper GTP hydrolysis; this in turn causes the mutant Ras protein to remain in its actively signaling state rather than behaving like the normal protein and shutting itself off.

Thought questions

1. Why is autocrine signaling an intrinsically destabilizing force for a normal tissue?

2. The responsiveness of a cell to exposure to a growth factor is usually attenuated after a period of time (for example, half an hour), after which time it loses this responsiveness. Given what you have already learned about growth factor receptors, what mechanisms might be employed by a cell to reduce its responsiveness to a growth factor?

3. What lines of evidence can you cite to support the notion that growth factor receptor firing following ligand binding is often dependent on the dimerization of a receptor (rather than some other molecular change in the receptor)?

4. In what ways are the heterotrimeric G-proteins similar to the low-molecular-weight G-proteins (for example, Ras), and in what way do they differ fundamentally?

5. In what ways has the study of lower organisms, notably yeast, flies, and worms, proven to be highly revealing in characterizing the intracellular signaling cascades and the receptors operating at the cell surface?

6. Integrins exhibit a novel type of control, termed "inside-out signaling," in which intracellular signals dictate the affinity of these receptors for their various extracellular ligands (that is, components of the extracellular matrix). Why is such signaling essential to the process of cell motility?

Additional reading

Andrae J, Gallini R & Betsholtz C (2008) Role of platelet-derived growth factors in physiology and medicine. *Genes Dev.* 22, 1276–1312.

Angers S & Moon RT (2009) Proximal events in Wnt signal transduction. *Nat. Rev. Mol. Cell Biol.* 20, 468–477.

Arnaout MA, Mahalingam B & Xiong JP (2005) Integrin structure, allostery, and bidirectional signaling. *Annu. Rev. Cell Dev. Biol.* 21, 381–410.

Bethani I, Skånland SS, Dikic I & Acker-Palmer A (2010) Spatial organization of transmembrane receptor signaling. *EMBO J.* 29, 2677–2688.

Birchmeier C, Birchmeier W, Gherardi E & Vande Woude G (2003) Met, metastasis, motility and more. *Nat. Rev. Mol. Cell Biol.* 4, 915–925.

Bos JL, Rehmann H & Wittinghofer A (2007) GEFs and GAPs: critical elements in the control of small G proteins. *Cell* 129, 865–877.

Bray SJ (2006) Notch signaling: a simple pathway becomes complex. *Nat. Rev. Mol. Cell Biol.* 7, 678–689.

Cohen MM Jr (2010) Hedgehog signaling update. *Am. J. Med. Genet.* 152A, 1875–1914.

Desgrosellier JS & Cheresh DA (2010) Integrins in cancer: biological implications and therapeutic opportunities. *Nat. Rev. Cancer* 10, 9–22.

Dorsam RT & Gutkind JS (2007) G-protein-coupled receptors and cancer. *Nat. Rev. Cancer* 7, 79–94.

D'Souza B, Miyamoto A & Weinmaster G (2008) The many facets of Notch ligands. *Oncogene* 27, 5148–5167.

Evangelista M, Tian H & de Sauvage FJ (2006) The Hedgehog signaling pathway in cancer. *Clin. Cancer Res.* 12, 5924–5928.

Geiger B & Yamada KM (2011) Molecular architecture and function of matrix adhesions. *Cold Spring Harbor Perspect. Biol.* 3:a005033.

Giancotti FG & Ruoslahti E (1999) Integrin signaling. *Science* 285, 1028–1032.

Hynes RA (2002) Integrins: bidirectional allosteric signaling machines. *Cell* 110, 673–687.

Hynes RO (2009) The extracellular matrix: not just pretty fibrils. *Science* 326, 1216–1219.

Jura N, Zhang X, Endres NF et al. (2011) Catalytic control in the EGF receptor and its connection to general kinase regulatory mechanisms. *Mol. Cell* 42, 9–22.

Kirsits A, Pils D & Krainer M (2007) Epidermal growth factor receptor degradation: an alternative view of oncogenic pathways. *Int. J. Biochem. Cell Biol.* 39, 2173–2182.

Lee JS, Kim KI & Baek SH (2008) Nuclear receptors and coregulators in inflammation and cancer. *Cancer Lett.* 267, 189–196.

Lemmon MA & Schlessinger J (2010) Cell signaling by receptor tyrosine kinases. *Cell* 141, 117–134.

Massagué J (2000) How cells read TGF-β signals. *Nat. Rev. Mol. Cell Biol.* 1, 169–178.

Michaud EJ & Yoder BK (2006) The primary cilium in cell signaling. *Cancer Res.* 66, 6463–6467.

Mikels AJ & Nusse R (2006) Wnts as ligands: processing, secretion and reception. *Oncogene* 25, 7461–7468.

Mosesson Y, Mills GB & Yarden Y (2008) Derailed endocytosis: an emerging feature of cancer. *Nat. Rev. Cancer* 8, 835–850.

O'Malley BW & Kumar R (2009) Nuclear receptor coregulators in cancer biology. *Cancer Res.* 69, 8217–8222.

Schlaepfer DD & Mitra SK (2004) Multiple contacts link FAK to cell motility and invasion. *Curr. Opin. Genet. Dev.* 14, 92–101.

Schwartz MA & Ginsberg MH (2002) Networks and crosstalk: integrin signalling spreads. *Nat. Cell Biol.* 4, E65–E68.

Shi Y & Massagué J (2003) Mechanisms of TGF-β signaling from cell membrane to the nucleus. *Cell* 113, 685–700.

Streuli CH & Akhtar N (2009) Signal co-operation between integrins and other receptor systems. *Biochem. J.* 418, 491–506.

Tian A-C, Rajan A & Bellen HJ (2009) A Notch updated. *J. Cell Biol.* 184, 621–629.

Varjosolo M & Taipale J (2010) Hedgehog: functions and mechanisms. *Genes Dev.* 22, 2454–2472.

Wennerberg K, Rossman KL & Der CJ (2005) The Ras superfamily at a glance. *J. Cell Sci.* 118, 843–846.

Witsch E, Sela M & Yarden Y (2010) Roles for growth factors in cancer progression. *Physiology* 25, 85–101.

Zlotnik A (2006) Chemokines and cancer. *Int. J. Cancer* 119, 2026–2029.

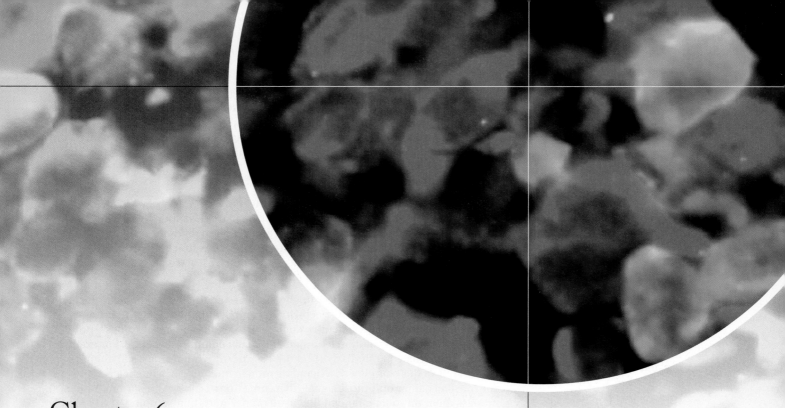

Chapter 6

Cytoplasmic Signaling Circuitry Programs Many of the Traits of Cancer

> Any living cell carries with it the experiences of a billion years of experimentation by its ancestors. You cannot expect to explain so wise an old bird in a few simple words.
>
> Max Delbrück, geneticist, 1966

Cancer is a disease of uncontrolled cell proliferation. Because the proliferative behavior of cancer cells is so aberrant, we might imagine that cancer cells invent entirely new ways of programming their growth and division—that the control circuitry within cancer cells is organized quite differently from that of normal, healthy cells. Such thinking greatly exaggerates the actual differences between normal and neoplastic cells. In truth, the two types of cells utilize control circuitry that is almost identical. Cancer cells discover ways of making relatively minor modifications to the control machinery operating inside cells. They tweak existing controls rather than demolishing the entire machinery and assembling a new version from the remnants of the original.

The present chapter is about this control machinery—really about the signal-processing circuitry that operates within the cell cytoplasm and governs cell growth and proliferation. An electronic circuit board is assembled from complex arrays of hard-wired components that function as resistors, capacitors, diodes, and transistors. The cell also uses circuits assembled from arrays of intercommunicating components, but these are, almost without exception, proteins. While individual proteins and their functions are relatively simple, the operating systems that can be assembled from these components are often extraordinarily complex.

In Chapter 5, we read about the receptors displayed at the cell surface that gather a wide variety of signals and funnel them into the cytoplasm. Here, we study how these signals, largely those emitted by growth factor receptors, are processed and integrated in the cytoplasm. Many of the outputs of this signal-processing circuitry are then transmitted to the nucleus, where they provide critical inputs to the central machinery that governs cell proliferation. Our discussion of this nuclear governor—the *cell cycle clock*—will follow in Chapter 8.

A single cell may express 20,000 or more distinct proteins, many of which are actively involved in the cytoplasmic regulatory circuits that are described here. These regulatory proteins are found in various concentrations and in different locations throughout the cytoplasm. Rather than floating around in dilute concentrations in the intracellular water, they form a thick soup. Indeed, as much as 30% of the volume of the cell is taken up by proteins rather than aqueous solvent.

These proteins must be able to talk to one another and to do so with great specificity and precision. The problem of specificity is complicated by the fact that important components of the signaling circuitry, such as kinases, have many **paralogs**—proteins that share a common evolutionary origin and retain common structure. For example, there are more than 400 similarly structured serine/threonine kinases in mammalian cells. In spite of this complexity, a signaling protein operating in a linear signaling cascade must recognize only those signals that come from its upstream partner protein(s) and pass them on to its intended downstream partner(s). In so doing, it will largely ignore the thousands of other proteins within the cell (**Figure 6.1**).

This means that each protein component of a signaling circuit must actually solve two problems. The first is one of specificity: how can it exchange signals only with the small subset of cellular proteins that are its intended signaling partners in the circuit? Second, how can this protein acquire rapid, almost instantaneous access to these signaling partners, doing so while operating in the viscous soup that is present in the cytoplasm and nucleus?

We study this circuitry because its design and operations provide key insights into how cancer cells arise. Thus, many of the oncoproteins described in previous chapters create cancer through their ability to generate signaling imbalances in this normally well-regulated system. At one level, cancer is surely a disease of inappropriate cell proliferation, but as we bore deep into the cancer cell, we will come to understand cancer at a very different level: cancer is really a disease of aberrant signal processing.

The present chapter will perhaps be the most challenging of all chapters in this book. The difficulty comes from the sheer complexity of signal transduction biochemistry, a field that is afflicted with many facts and blessed with only a small number of unifying principles. So, absorb this material in pieces; the whole is far too much for one reading.

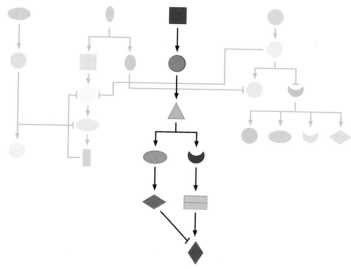

Figure 6.1 Signaling cascades This diagram of imaginary signaling circuitry illustrates how a signal transduction cascade—a series of signaling proteins that operates much like a molecular bucket brigade—passes signals from an upstream source (*purple square*) to its intended downstream target (*dark green diamond*) and, at the same time, avoids inadvertent activation of dozens of other signaling proteins in the cell (*faintly drawn symbols*). *Arrowheads* denote a stimulatory signal, while a line at right angles (a *crossbar*) denotes an inhibitory signal.

6.1 A signaling pathway reaches from the cell surface into the nucleus

The growth and division that cells undertake following exposure to mitogens clearly represent complex regulatory programs involving the coordinated actions of hundreds, even thousands of distinct cellular proteins. How can growth factors succeed in evoking all of these changes simply by binding and activating their receptors? And how can the deregulation of these proliferative programs lead to cell transformation? And where does Ras fit into this circuitry, if indeed it does?

A major insight into these issues came in 1981 from examining the behavior of cultured normal cells, specifically fibroblasts that had been deprived of serum-associated growth factors for several days. During this period of "serum starvation," the cells within a culture were known to enter reversibly into the nongrowing, quiescent state termed G_0 ("G zero"). After remaining in this G_0 state for several days, serum-starved cells were then exposed to fresh serum and thus to abundant amounts of mitogenic growth factors. The intent of this experiment was to induce all the cells in such cultures to enter into a state of active growth and division **synchronously** (coordinately). Indeed, within a period of 9 to 12 hours, the great majority of these previously quiescent cells would begin to replicate their DNA, and initiate cell division a number of hours later.

These readily observed changes in cell behavior provided no hint of the large number of less obvious but nonetheless important molecular changes that occurred in these cells, often within minutes of exposure to fresh growth factors. For example, within less than an hour, transcription of an ensemble of more than 100 cellular genes was induced (**Figure 6.2**). Expression of these genes, which we now call **immediate early genes** (IEGs), increased rapidly in the half hour after growth factor stimulation. As is suggested by the contents of **Table 6.1**, the products of some of these genes help cells in various ways to emerge from the quiescent G_0 state into their active growth-and-division cycle. In fact, experiments involving several of these genes showed that blocking their expression prevents cells' emergence from the G_0 state.

A variation of this 1981 experiment was undertaken in which fresh serum was added together with **cycloheximide**, a drug that shuts down all cellular protein synthesis. In spite of the inhibition of protein synthesis, induction of the immediate early genes proceeded quite normally. This indicated that all of the proteins required to activate transcription of these genes were already in place at the moment when the serum was added to the cell. Stated differently, the induction of these genes did not require **de novo** (new) protein synthesis.

These results also demonstrated that, in addition to the growth factor receptors at the cell surface, an array of other proteins was present within the cell that could rapidly convey mitogenic signals from the receptors located at the cell surface to transcription

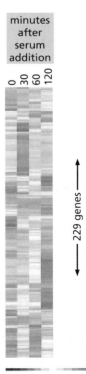

minutes after serum addition

0 30 60 120

229 genes

low medium high

Figure 6.2 Expression of immediate early genes Expression of a large array of "immediate early genes" (IEGs) is induced within the first two hours after serum-starved cells are exposed to fresh serum. The expression of IEGs occurs even when protein synthesis is blocked during this period, indicating that the transcription factors regulating expression of these genes are already present in the serum-starved cells. Analyses of the expression patterns of the 14,824 genes surveyed here at 30, 60, and 120 minutes following serum addition led to the identification of 229 genes (unlabeled, arrayed top to bottom) that behave like IEGs. Based on the various schedules of gene expression, these could be grouped into 10 subclasses, each of which showed its own characteristic schedule of induction and, in some cases, subsequent repression. Red indicates a high level of RNA expression, while blue indicates low expression. Note that some IEGs are expressed for only a brief period of time before being repressed even though the serum and associated mitogens continue to be present in the culture medium. (From A. Selvaraj and R. Prywes, BMC Mol. Biol. 5:13, 2004.)

factors (TFs) operating in the nucleus (see Figure 5.1). Clearly, the functional activation of such cytoplasmic signal-transducing proteins did not depend upon increases in their concentration (since cycloheximide had no apparent effect on this signaling). Instead, changes in the proteins' structure, configuration, and intracellular localization appeared to play a dominant role in their functional activation soon after the stimulation of tyrosine kinase receptors by their growth factor ligands.

The immediate early genes encode a number of interesting proteins (Table 6.1). Some of them specify transcription factors that, once synthesized, help to induce a second wave of gene expression. Included here are the *myc*, *fos*, and *jun* genes, which were originally identified through their association with transforming retroviruses. Yet other immediate early genes encode proteins that are secreted growth factors ("cytokines") or help to construct the cellular cytoskeleton.

The levels of *myc* mRNA were found to rise steeply following mitogen addition and to collapse quickly following mitogen removal. Moreover, the Myc protein itself turns over rapidly, having a **half-life** ($T_{1/2}$) of only 25 minutes. Such observations indicated that the levels of the Myc protein serve as an intracellular marker or indicator of the amount of mitogens present in the nearby extracellular space.

As an aside, we might note that the activity of Myc operating as a signaling protein derives in large part from major changes in its concentration within the cell nucleus. This contrasts starkly with the behavior of cytoplasmic signal-transducing proteins such as Ras and Src. These cytoplasmic proteins respond to mitogenic signals by undergoing noncovalent and covalent alterations in their structure rather than exhibiting significant increases in concentration. This difference is mirrored by the changes

Table 6.1 A sampling of immediate early genes[a]

Name of gene	Location of gene product	Function of gene product
fos[b]	nucleus	component of AP-1 TF
junB	nucleus	component of AP-1 TF
egr-1	nucleus	zinc finger TF
nur77	nucleus	related to steroid receptors
Srf-1[c]	nucleus	TF
myc	nucleus	bHLH TF
β-actin	cytoplasm	cytoskeleton
γ-actin	cytoplasm	cytoskeleton
tropomyosin	cytoplasm	cytoskeleton
fibronectin	extracellular	extracellular matrix
glucose transporter	plasma membrane	glucose import
JE	extracellular	cytokine
KC	extracellular	cytokine

[a]The genes listed here represent only a small portion of the immediate early genes (IEGs; see Figure 6.2).

[b]Expression of a group of fos-related genes is also induced as IEGs. These include *fosB*, *fra-1*, and *fra-2*.

[c]Srf is a TF that binds to the promoters of other immediate early genes such as *fos*, *fosB*, *junB*, *egr-1*, and *egr-2*, *nur77*, and cytoskeletal genes such as actins and myosins.

Adapted in part from H.R. Herschman, *Annu. Rev. Biochem.* 60:281–319, 1991; and from B.H. Cochran, in R. Grzanna and R. Brown (eds.), Activation of Immediate Early Genes by Drugs of Abuse. Rockville, MD: National Institutes of Health, 1993, pp. 3–24.

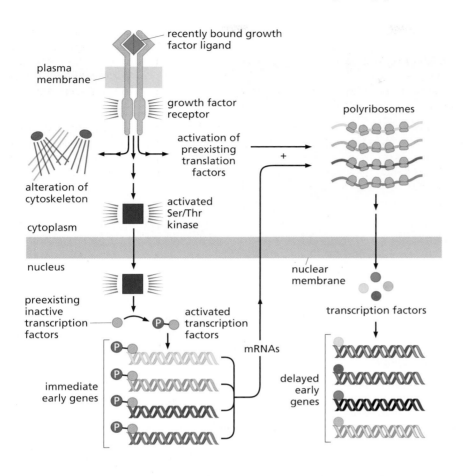

Figure 6.3 Immediate vs. delayed early genes The profound changes in cell physiology following the application of serum-associated mitogenic growth factors to a cell indicate that growth factor receptors can release a diverse array of biochemical signals once they are activated by ligand binding. The transcription of some genes (immediate early genes) is observed within minutes of growth factor stimulation (*lower left*), while the transcription of others (delayed early genes) occurs with a lag, because it requires the synthesis of new transcription factors (*lower right*).

occurring when the normal *myc*, *ras*, and *src* genes are converted into oncogenes: the *levels* of Myc protein become deregulated (being expressed constitutively rather than being modulated in response to physiologic signals), whereas the *structures* of the Ras and Src proteins undergo alteration but the amounts of these proteins do not necessarily increase.

Within an hour after the immediate early mRNAs appear, a second wave of gene induction was found to occur. Significantly, the induction of these **delayed early genes** was largely blocked by the presence of cycloheximide, indicating that their expression did indeed depend on *de novo* protein synthesis (**Figure 6.3**). (In fact, expression of these delayed early genes seems to depend on transcription factors that are synthesized in the initial wave of immediate early gene expression.)

Addition of growth factors to quiescent cells was found to provoke yet other changes in cell physiology besides the rapid induction of nuclear genes. Following exposure of cells to serum, their rate of protein synthesis increases significantly, this being achieved through functional activation of the proteins that enable ribosomes to initiate translation of mRNAs. Some growth factors were discovered to induce motility in cells, as indicated by the movement across the bottom surfaces of Petri dishes. Yet others can provoke a reorganization of actin fibers that help to construct the cell's

Figure 6.4 Serum-induced alterations of cell shape Like many other cell types, serum-starved mouse Swiss 3T3 cells undergo profound changes in their actin cytoskeleton following serum stimulation. Under serum-starved conditions, most of the polymerized actin is found in a layer beneath the plasma membrane termed cortical actin, whereas within ten minutes of serum addition, focal adhesions are formed by integrins (see Figure 5.28) that are linked to actin *stress fibers* that stretch across the cell. In addition, the serum stimulation, much of it mediated by platelet-derived growth factor (PDGF), induces greatly increased cell motility (*not shown*). (Courtesy of A. Hall.)

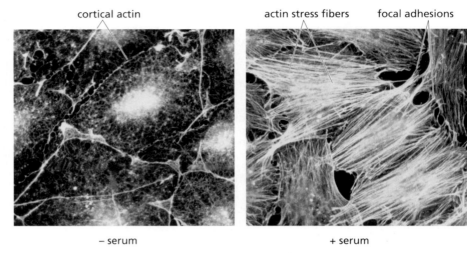

cytoskeleton—the intracellular scaffolding that defines cell shape (**Figure 6.4**). Subsequently, many growth factors were discovered to provide survival signals to cells, thereby protecting them from inadvertent activation of the cell suicide program known as apoptosis.

These diverse responses indicated that a variety of distinct biochemical signals radiate from ligand-activated growth factor receptors, and that these signals impinge on a diverse array of cellular targets. Some of these signals seemed to be channeled directly into the nucleus, where they altered gene expression programs, while others were clearly directed toward cytoplasmic targets, including the protein-synthesizing machinery and the proteins that organize the structure of the cytoskeleton.

An understanding of how these signaling cascades operate clearly was very relevant to the cancer problem: if it were true that some oncoproteins flood cells with a continuous stream of mitogenic signals (see Chapter 5), then the transformed state of cancer cells might well represent an exaggerated version of the responses that normal cells exhibit following exposure to growth factors. In fact, many of the traits of cancer cells can indeed be traced to responses evoked by growth factors, while others cannot be predicted by the initial responses of cells to these factors (**Sidebar 6.1**).

6.2 The Ras protein stands in the middle of a complex signaling cascade

The disparate responses of cells to growth factors represented a challenge to those interested in intracellular signal transduction, since almost nothing was known about the organization and function of the communication channels operating within the

Sidebar 6.1 Short-term transcriptional responses to mitogens provide no indication of the gene expression in continuously growing cells Simple logic would dictate that the transcriptional responses observed shortly after growth factor stimulation of a serum-starved cell ought to predict the transcriptional state of cells in which mitogenic signaling occurs continuously, whether in response to ongoing exposure to growth factors or the activity of certain oncoproteins. Accordingly, the spectrum of immediate and delayed early genes induced within the first hour of growth factor stimulation should also be expressed continuously and at a high rate in transformed cells.

In fact, some genes that are members of the immediate and delayed early gene expression groups are found to be expressed at high levels in cancer cells, while others are not. This discordance derives from negative-feedback controls that force the shutdown of some genes shortly after they have been expressed. For example, expression of the *fos* gene, itself a proto-oncogene, increases rapidly in response to serum stimulation, peaks, and then declines dramatically, all within an hour or so (see, for example, Figure 6.2). This shutdown is due to the ability of the Fos protein, once synthesized, to act as a transcriptional repressor of the synthesis of more *fos* mRNA. Numerous other negative-feedback mechanisms operate in cells to ensure that many of the initial biochemical responses evoked by growth factors function only transiently. More generally, negative-feedback loops are installed in virtually all signaling pathways to ensure that the flow of signals through these pathways is tightly controlled in both intensity and duration (see Section 6.13).

Figure 6.5 Structure of the fly eye Study of the light-sensing units in the *Drosophila melanogaster* eye—the ommatidia—revealed that each ommatidium is normally formed from a series of seven cells, with six outer ones surrounding a seventh in the center, as seen in this electron micrograph section (*upper panel*). A fly carrying the mutation in the gene that came to be called *sevenless* produced ommatidia lacking the seventh, central cell (*lower panel*). Later, *sevenless* was found to encode a homolog of the mammalian fibroblast growth factor (FGF) receptor. (Courtesy of E. Hafen.)

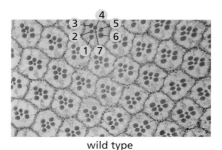
wild type

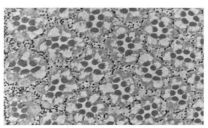
sevenless mutant

cell. Over a period of a decade (the 1980s), these circuits were slowly pieced together, much like a jigsaw puzzle. The clues came from many sources. The story started with Ras and then moved up and down the signaling cascades until the links in the signaling chains were finally connected.

The advances in Ras biochemistry that we discussed in the last chapter (Section 5.10) seemed to explain much about how it operates as a binary switch, but provided no insight into its context—how this protein is connected with the overall signaling circuitry. Indeed, it remained possible that the Ras protein functioned in a signaling pathway that was independent of the one controlled by growth factors and their receptors.

In the end, the solution to the Ras problem came from a totally unexpected quarter. The genetics of eye development in the fruit fly *Drosophila melanogaster* revealed a series of genes whose products were essential for the normal development of the ommatidia, the light-sensing units forming the compound eye. One important gene here came to be called *Sevenless*; in its absence, the seventh cell in each ommatidium failed to form (**Figure 6.5**). Provocatively, after cloning and sequencing, the *Sevenless* gene was found to encode a protein with the general structural features of a tyrosine kinase receptor, specifically a mammalian FGF receptor.

Yet other mutations mimicked the effects of *Sevenless* mutation. Genetic complementation tests revealed that these mutations affected genes whose products operate downstream of *Sevenless*, ostensibly in a linear signal transduction cascade. One of these downstream proteins was encoded by a gene that was named *Son of sevenless* or simply *Sos*. Close examination of the Sos protein showed it to be related to proteins known from yeast biochemistry to be involved in provoking nucleotide exchange by G (guanine nucleotide–binding) proteins such as Ras. The yeast proteins, often called guanine nucleotide exchange factors (GEFs), were known to induce G-proteins to release their bound GDPs, thereby making room for GTPs to jump aboard (see Figure 5.30). The consequence was an activation of these G-proteins from their inactive to their active, signal-emitting configuration. This was precisely the effect that Sos exerted on Ras proteins (**Figure 6.6**). Therefore, Sos appeared to be the long-sought upstream stimulator that kicked Ras into its active configuration.

Soon other intermediates in the signaling cascade were uncovered. Two of these, Shc (pronounced "shick") and Grb2 (pronounced "grab two"), were discovered through genetic and biochemical screens for proteins that interacted with phosphorylated receptors or derived peptides. It became apparent that these proteins function as adaptors, forming physical bridges between growth factor receptors and Sos; these bridging proteins will be discussed in greater detail later. A third such adaptor protein, called Crk (pronounced "crack"), was identified as the oncoprotein encoded by the CT10 avian sarcoma virus.

These discoveries demonstrated the very ancient origins of this signaling pathway, which was already well developed more than 600 million years ago in the common ancestor of all contemporary metazoan phyla. Once in place, the essential components of this cascade remained relatively unchanged in the cells of descendant organisms. Indeed, the individual protein components of these cascades have been so conserved that in many cases the protein components from the cells of one phylum (for example, chordates) can be exchanged with those of another (for example, arthropods) to reconstitute a functional signaling pathway.

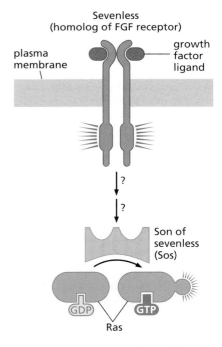

Figure 6.6 Son of sevenless Genetic studies of *Drosophila* eye development revealed another gene whose product appeared to lie downstream of Sevenless (see Figure 6.5) in a signaling cascade and specified a guanine nucleotide exchange factor (GEF) (*orange*), capable of activating the Ras protein. This gene was named *Son of sevenless* or simply *Sos*.

Together, the genetic and biochemical data from these distantly related animal phyla could be combined in a scheme that suggested a linear signaling cascade organized like this: tyrosine kinase receptor → Shc → Grb → Sos → Ras. While this cascade provided the outline of a signaling channel, it gave little insight into the biochemical interactions that enabled these various proteins to exchange signals with one another.

6.3 Tyrosine phosphorylation controls the location and thereby the actions of many cytoplasmic signaling proteins

Among the major biochemical issues left unanswered, perhaps the most critical were the still-mysterious actions of the kinases carried by many growth factor receptors. Was the phosphorylation of these receptors on tyrosine residues crucial to their ability to signal, or was it a distraction? And if this phosphorylation was important, how could it activate the complex signaling circuitry that lay downstream?

Two alternative mechanisms seemed plausible. The first predicted that a receptor-associated tyrosine kinase would phosphorylate a series of target proteins in the cytoplasm. Such covalent modification would alter the three-dimensional *conformation* of the proteins (that is, their **stereochemistry**), thereby placing each of them in an actively signaling state that allowed them to transfer signals to a partner one step further down in a signaling cascade. This model implied, among other things, that the phosphorylation of the receptor molecule itself was of secondary importance to signaling.

According to the alternative model, the phosphorylation of the cytoplasmic tail of a receptor following growth factor binding affected the *physical location* of its downstream signaling partners without necessarily changing their intrinsic activity. Once relocalized to new sites within the cytoplasm, these downstream partners could then proceed with their task of emitting signals to yet other targets in the cell.

As it turned out, the second model, involving protein relocalization, proved to be more important. The salient insights came from analyzing the detailed structure of the Src protein. Researchers noted three distinct amino acid sequence domains, each of which appeared both in Src and in a number of other, otherwise unrelated proteins. These sequences, called Src homology domains 1, 2, and 3 (SH1, SH2, and SH3), provided the key to the puzzle of receptor signaling (**Figure 6.7**).

The SH1 domain of Src represents its catalytic domain, and indeed similar sequence domains are present in all receptor tyrosine kinases, as well as in the other, nonreceptor

Figure 6.7 Domain structure of the Src protein Sequence analyses of the Src protein revealed three distinct structural domains that are also found in a large number of other signaling proteins. (A) In this ribbon diagram of Src—the product of X-ray crystallography—coils indicate α-helices, while flattened ribbons indicate β-pleated sheets. Like many other subsequently analyzed protein kinases (for example, see Figure 5.17), the catalytic SH1 domain is seen to be composed of two subdomains (*right*), termed the N- and C-lobes of the kinase (*yellow, orange*). Between these lobes is the ATP-binding site where catalysis takes place (*red stick figure*). The SH2 and SH3 domains (*blue, light green; left*) are involved in substrate recognition and regulation of catalytic activity. Connecting sequences are shown in *dark green*. (B) A space-filling model of Src is presented in the same color scheme. (From B. Alberts et al., Molecular Biology of the Cell, 5th ed. New York: Garland Science, 2008.)

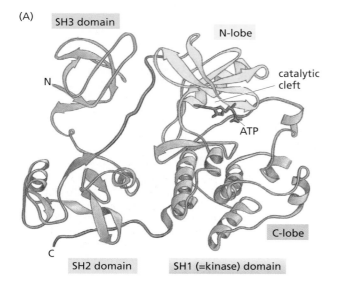

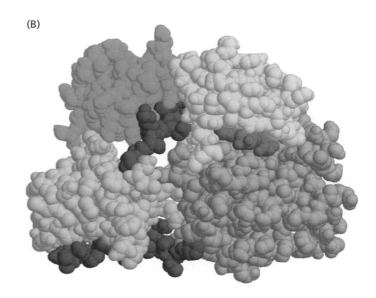

tyrosine kinases that are configured like Src. This shared sequence relatedness testifies to the common evolutionary origin of all these kinases.

Clever biochemical sleuthing conducted in the late 1980s uncovered the role of the SH2 domain present in Src and other signaling proteins. This relatively small structural domain of about 100 amino acid residues (**Figure 6.8**) acts as an intracellular "receptor." The "ligand" for this SH2 receptor is a short oligopeptide sequence that contains both a phosphorylated tyrosine and a specific oligopeptide sequence 3 to 6 residues long that flanks the phosphotyrosine on its C-terminal side.

Soon, it became apparent that there are dozens of distinct SH2 domains, each carried by a different protein and each having an affinity for a specific phosphotyrosine-containing oligopeptide sequence that functions as its ligand. By now, the sequence specificities of a substantial number of SH2 groups have been cataloged. The human genome is estimated to encode at least 120 distinct SH2 groups, each constituting a domain of a larger protein and each apparently having an affinity for binding a particular phosphotyrosine together with a flanking oligopeptide sequence. (Ten proteins carry two SH2 domains, yielding a total census of 110 distinct SH2-containing proteins in our cells.)

Most commonly, an SH2 domain enables the protein carrying it to associate with a partner protein that is displaying a specific phosphotyrosine plus flanking amino acid sequence, thereby forming a physical complex between these two proteins. Some SH2-bearing proteins carry no enzymatic activity at all. Other such proteins, in addition to their SH2 domains, carry catalytic sites that are quite different from the tyrosine kinase (TK) activity present in Src itself. For example, one form of phospholipase C carries an SH2 domain, as does the p85 subunit of the enzyme phosphatidylinositol 3-kinase (PI3K). The SHP1 protein, discussed in **Sidebar 6.2**, carries an SH2 domain linked to a **phosphatase** catalytic domain, that is, a domain with an enzymatic activity that removes phosphate groups, thereby reversing the actions of kinases.

These findings indicated that the catalytic domains of the various proteins and their SH2 groups function as independent structural modules. These modules have been pasted together in various combinations by evolution. The SH2 group borne by each of these proteins allows it to become localized to certain sites within a cell, specifically to those sites that contain particular phosphotyrosine-bearing proteins to which the SH2-containing protein can become tethered.

Figure 6.8 Structure and function of SH2 and SH3 domains (A) This SH2 domain is seen in a different orientation from that in Figure 6.7. As revealed by X-ray crystallography, a typical SH2 domain is composed of about 100 amino acid residues and is assembled from a pair of anti-parallel β-pleated sheets (*red*) surrounded by a pair of α-helices (*green*). The sites of interaction by the SH2 domain with its "ligand"—a phosphotyrosine (pY)-containing peptide (composed of a pY and flanking amino acids)—are indicated by the *yellow* spots. (B) As indicated, the SH2 domain can be thought to function like a modular plug: it recognizes both a phosphotyrosine and the amino acids that flank this phosphotyrosine on its C-terminal side. These flanking amino acid residues determine the specificity of binding, that is, the identities of the particular phosphotyrosine-containing oligopeptides that are recognized and bound by a particular SH2 domain. (C) The SH3 domain (*ribbon diagram*), also shown in a different orientation from that of Figure 6.7, recognizes as its ligands proline-rich oligopeptide (*blue stick figure*) sequences in partner proteins. (A, based on data from G. Waksman et al., *Cell* 72:779–790, 1993. With permission from Elsevier; B, from B. Alberts et al., Molecular Biology of the Cell, 5th ed. New York: Garland Science, 2008. C, from S. Casares et al., *Protein* 67:531–547, 2007.)

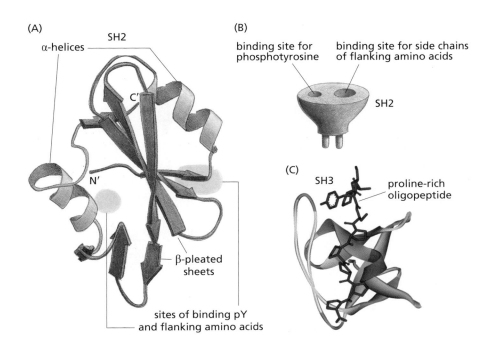

(A) α-helices — SH2 — C' — N' — β-pleated sheets — sites of binding pY and flanking amino acids

(B) binding site for phosphotyrosine — binding site for side chains of flanking amino acids — SH2

(C) SH3 — proline-rich oligopeptide

Sidebar 6.2 A negative-feedback control mechanism leads to an SH2-containing molecule that can determine outcomes of Olympic competitions One of the SH2-containing partners of tyrosine kinase receptors is a phosphotyrosine phosphatase (PTP), an enzyme that removes the phosphate group from phosphotyrosine. Once tethered via its SH2 group to a ligand-activated, tyrosine-phosphorylated receptor, this particular phosphatase, termed SHP1, begins to chew away the phosphate groups on tyrosine residues that have enabled the receptor to attract other downstream signaling partners. In this fashion, an enzyme such as SHP1 creates a negative-feedback loop to shut down further receptor firing. In fact, the SHP1 phosphatase is responsible for shutting down signaling by a variety of receptors, among them the receptor for the erythropoietin (EPO) growth factor, which normally responds to EPO by triggering **erythropoiesis**—red blood cell formation (see Figure 6.22).

Some individuals inherit a mutant gene encoding an EPO receptor that lacks the usual docking site for the SHP1 phosphatase. In individuals whose cells express this mutant receptor, the resulting defective negative-feedback control leads to a hyperactive EPO receptor, and, in turn, to higher-than-normal levels of red cells and thus oxygen-carrying capacity in their blood. In the case of several members of a Finnish family with this inherited genetic defect, this condition seems to have played a key role in their winning Olympic medals in cross-country skiing! Provocatively, expression of SHP1 is shut down in many leukemias and non-Hodgkin's lymphomas; this permits multiple mitogen receptor molecules to fire excessively in the tumor cells.

The discoveries about SH2 groups finally solved the puzzle of how tyrosine kinase receptors are able to emit signals. The story is illustrated schematically in **Figure 6.9**. As a consequence of ligand-induced transphosphorylation, a receptor molecule, such as the PDGF-β receptor, will acquire and display a characteristic array of phosphotyrosine residues on its cytoplasmic tail; likewise, the EGF receptor will display its own spectrum of phosphotyrosines on its cytoplasmic tail. The unique identity of each phosphotyrosine residue borne by these tails is dictated by the sequence of amino acid residues flanking the phosphotyrosine on its C-terminal side. These phosphotyrosines become attractive homing sites for various SH2-containing cytoplasmic proteins, specifically, proteins that normally reside in the soluble portion of the cytoplasm (the **cytosol**) and are therefore free to move from one location to another in the cytoplasm. Consequently, shortly after becoming activated by ligand binding, a growth factor receptor becomes decorated with a specific set of these SH2-containing partner proteins that are attracted to its various phosphotyrosines.

For example, one ligand-activated form of the PDGF receptor attracts Src, PI3K, Ras-GAP (which stimulates Ras GTPase activity; refer to Sidebar 5.7), SHP2 (SH2-containing tyrosine phosphatase 2), and PLC-γ (phospholipase C-γ). Each of these proteins

Figure 6.9 Attraction of signal-transducing proteins by phosphorylated receptors (A) This schematic diagram of a molecule of the platelet-derived growth factor-β (PDGF-β) receptor, which omits its tyrosine kinase domain, reveals a large number of tyrosine residues in its C-terminal domain that undergo phosphorylation following ligand binding and receptor activation. (The positions of these tyrosine residues in the receptor polypeptide chain are indicated by the numbers.) Listed to the left are seven distinct cytoplasmic proteins, each of which can bind via its SH2 domain(s) to one or more of the phosphotyrosines of the PDGF-β receptor. The three amino acid residues (denoted by the single-letter code) that flank each tyrosine (Y) residue on its C-terminal side define the unique binding site recognized by the various SH2 domains of these seven associated proteins. (B) Similarly, a constellation of phosphotyrosines can be formed after the EGF receptor binds its ligand, often forming heterodimers with the HER2 receptor protein. However, the spectrum of SH2-containing proteins that associates with the EGF-R is quite different, allowing this receptor to activate a different set of downstream signaling pathways from that of the PDGF-β receptor. (A, adapted from T. Pawson, *Adv. Cancer Res.* 64:87–110, 1994. B, adapted from R. Sordella et al., *Science* 305:1163–1167, 2004.)

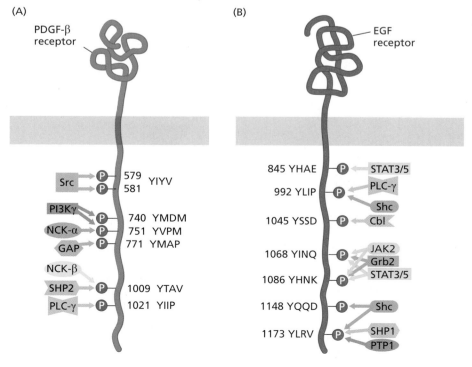

carries at least one SH2 domain (**Figure 6.10A**), and all of them flock to the PDGF-R after its cytoplasmic tail and a short segment in the middle of its kinase domain (called a "kinase insert") become tyrosine-phosphorylated. Once tethered to the PDGF-R, some of these SH2-containing proteins may then become substrates for tyrosine phosphorylation by the PDGF-R kinase. Yet other proteins serve as bridges by attracting additional proteins to these multi-protein complexes (see Figure 6.12). (It remains unclear whether all of these SH2-containing G-proteins and yet others can bind simultaneously to a single ligand-activated PDGF receptor molecule.)

The biochemical diversity of these associated proteins provides us with at least one solution to the question of how a growth factor receptor is able to elicit a diversity of biochemical and biological responses from a cell: each of these associated proteins

(A)

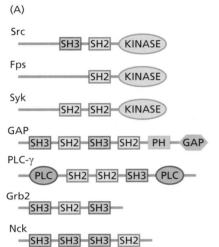

Figure 6.10 Protein interaction domains as modular units of protein structure
(A) Specific multiprotein complexes are formed using domains such as the SH2, SH3, and PH domains, which bind phosphotyrosines, proline-rich sequences, and PIP3, respectively (discussed in Section 6.6). As is indicated here, evolution has manipulated these domains in a modular fashion, attaching them to tyrosine kinase catalytic domains in the case of the first three tyrosine kinases (Src, Fps, and Syk), to a GAP (GTPase-activating protein) domain, or to a PLC (phospholipase C) domain. At the bottom are shown two adaptor or "bridging" proteins that contain only SH2 and SH3 domains and lack catalytic domains; each of these adaptors serves as a linker between pairs of other proteins, one of which carries a certain phosphotyrosine (to which an SH2 domain can bind) and the other of which carries a proline-rich domain (to which an SH3 domain can bind; see Figure 6.12). (B) The spectrum of molecular "ligands" that can be bound by complementary "receptor" domains in order to form intermolecular complexes is far larger than indicated in panel A. As shown, a variety of modified amino acid residues, peptide sequences, nucleic acids, and phospholipids can serve as ligands of the indicated domains. In addition, certain domains ("domain/domain") favor the homodimerization of two protein molecules. (A, adapted from T. Pawson, *Adv. Cancer Res.* 64:87–110, 1994. B, from T. Pawson and P. Nash, *Science* 300:445–452, 2003 and courtesy of M.B. Yaffe.)

(B)

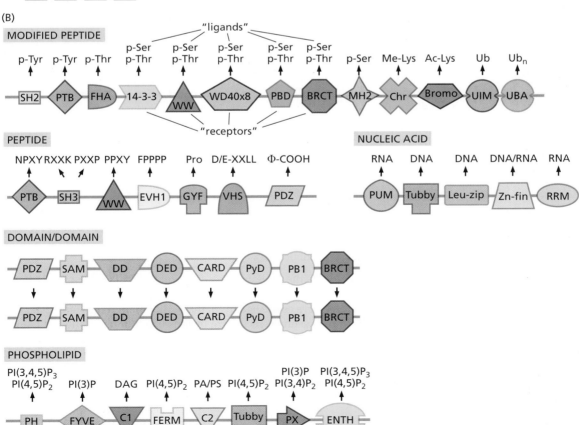

controls its own downstream signaling cascade, and each of these cascades, in turn, influences a different aspect of cellular behavior. By becoming tethered to a receptor molecule, these SH2-containing proteins are poised to activate their respective cascades. Some of these SH2-bearing proteins function to activate negative-feedback loops that ultimately shut down receptor signaling, as discussed later in this chapter and in Sidebar 6.2.

Because the cytoplasmic domains of tyrosine kinase receptors are found near the inner surface of the plasma membrane, the SH2-containing partner proteins attracted to the receptor are brought into close proximity to the plasma membrane and thus close to other molecules that are, for their own reasons, already tethered to this membrane. This close juxtaposition enables various SH2-containing proteins to interact directly with the membrane-associated proteins and phospholipids, thereby generating a variety of biochemical signals that can be transmitted to various downstream signal-transducing cascades.

For example, the Ras GTPase-activating protein (Ras-GAP), having become attached to a phosphorylated receptor, is tethered in close proximity to Ras proteins, many of which are permanently anchored to the plasma membrane via their C-terminal lipid tails. This allows Ras-GAP to interact with nearby Ras molecules, stimulating them to hydrolyze their bound GTPs and thereby convert from an active to an inactive signaling state (see Figure 5.30).

Similarly, when phosphatidylinositol 3-kinase (PI3K) becomes tethered near the plasma membrane, it is able to reach over and phosphorylate inositol lipids embedded in the membrane; we return to this enzyme and its mechanism of action in Section 6.6. Yet another example is provided by phospholipase C-γ (PLC-γ), which, when juxtaposed to the plasma membrane via receptor tethering, is able to cleave a membrane-associated phospholipid [phosphatidylinositol (4,5)-diphosphate, or PIP2] into two products, each of which, on its own, has potent signaling powers (see Figure 6.16). The point here is that many reactions become possible (or can proceed at much higher rates) when enzymes and their substrates are brought into close proximity through tethering of these enzymes to phosphorylated receptors.

The discovery of SH2 domains and their presence in a diverse array of signaling proteins solved a long-standing paradox. The dozens of distinct RTKs expressed by human cells were known to carry essentially identical tyrosine kinase signal-emitting domains, yet these receptors, when activated by ligand binding, clearly induced very different cellular responses. This puzzle was resolved when it was realized that, following ligand binding, each RTK acquires its own particular array of phosphotyrosine-containing peptides (see Figure 6.9), enabling it to attract its own combination of downstream signaling partners. In recent years, systematic surveys have been undertaken to reveal the full range of interactions between RTKs, such as the EGF receptor, and the various SH2 domain–containing proteins in the cytoplasm (Supplementary Sidebar 6.1). In the longer term, such surveys will provide the databases that will make it possible to construct mechanistic models that help to predict how individual receptors, once activated by their ligands, operate to change the biochemistry and physiology of a cell. Nonetheless, even with this information in hand, a significant number of other variables complicate attempts to understand precisely how individual RTKs function (Supplementary Sidebar 6.2).

The SH3 domain, the third of the sequence motifs present in Src (see Figure 6.7), binds specifically to certain proline-rich sequence domains in partner proteins; these proline-rich sequences thus serve as the ligands of the SH3 domains (see Figure 6.8C). (A small subset of SH3 groups seem to recognize ligands of a different structure.) These SH3 domains are a relatively ancient invention (apparently originating more than 1.5 billion years ago), since there are at least 28 of them in various proteins of the yeast *S. cerevisiae*. The SH2 domain, in contrast, seems to be of more recent vintage, having been invented possibly at the same time as tyrosine phosphorylation came into widespread use—perhaps 600 million years ago when metazoa arose.

In the specific case of the SH3 domain of Src, there is evidence that this domain is used to recognize and bind certain proline-rich substrates, which can then be phosphorylated by Src's kinase activity. Yet other evidence points to a proline-rich linker domain within Src itself, to which Src's SH3 domain associates, creating intramolecular binding (**Sidebar 6.3**). Through analyses of the human genome sequence, 253 distinct SH3 domains, each part of a larger protein, have been uncovered.

Sidebar 6.3 The SH2 and SH3 domains of Src both have alternative functions We have portrayed SH2 as a protein domain that enables its carrier to recognize and bind to other phosphotyrosine-containing partner proteins, thereby forming bimolecular complexes. In tyrosine kinases such as Src, however, the SH2 group can also play a quite different role. Like most signaling molecules, the Src kinase is usually held in a functionally silent configuration. It is kept in this inactive state by a negative regulatory domain of Src—a stretch of polypeptide that lies draped across the catalytic site of Src and blocks access to the substrates of this kinase. This obstructing domain (sometimes called an "activation loop"; see Figure 5.17) is held in place, in part, through the actions of the Src SH2 group, which recognizes and binds a phosphotyrosine in position 527 of Src itself. Hence, this SH2 group is forming an *intra*molecular rather than an *inter*molecular bridge (**Figure 6.11A**).

Some receptors, such as the PDGF receptor, may become activated by ligand binding and display phosphotyrosines in their cytoplasmic domains that are also attractive binding sites for the SH2 group of Src. As a consequence, the fickle Src SH2 group breaks its intramolecular bond and forms an intermolecular bond with the PDGF receptor instead; this results

in the tethering of Src to the PDGF receptor (Figure 6.11B). Concomitantly, the SH3 group of Src breaks its intramolecular association with the linker segment of Src and binds to an oligopeptide in the PDGF receptor. Now Src's catalytic cleft shifts into an active configuration and, following phosphorylation of its activation loop, begins to phosphorylate its clientele of substrates, thereby amplifying the signal broadcast by the PDGF receptor (Figure 6.11C).

The Src oncoprotein specified by Rous sarcoma virus is a mutant form of the normal c-Src protein, in which the C-terminal oligopeptide domain containing the tyrosine 527 residue has been replaced by another, unrelated amino acid sequence. As a consequence, a phosphotyrosine can never be formed at this site and the intramolecular inhibitory loop cannot form and obstruct the Src catalytic site. The viral oncoprotein (v-Src) is therefore given a free hand to fire constitutively. A similar mutant Src, in which the six C-terminal amino acids are deleted (including tyrosine 527), has been reported in liver metastases of human colon carcinomas. While the Src kinase is hyperactive in many colon carcinomas, this is one of the very rare examples of a mutant, structurally altered Src in human cancers.

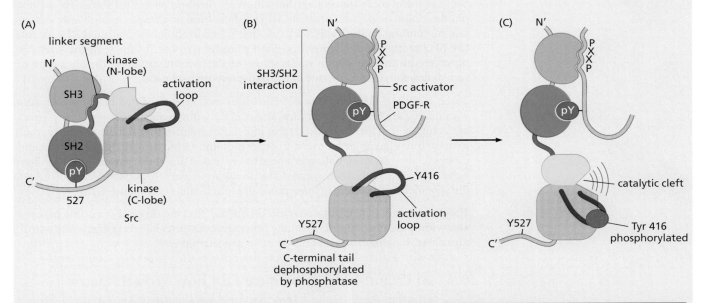

Figure 6.11 Structure and function of Src (A) The SH2 group of Src (*blue*) normally binds in an intramolecular fashion to a phosphotyrosine residue (pY; *red*) at position 527 near the C-terminus of Src. This binding causes the catalytic cleft of Src, which is located between the N-lobe and the C-lobe of the kinase domain, to be obstructed. At the same time, the SH3 (*light green*) domain binds a proline-rich portion in the linker segment (*magenta*) between the N-lobe of the kinase domain and the SH2 domain. (B) However, when a domain of the PDGF receptor (indicated here as a "Src activator") becomes phosphorylated, one of its phosphotyrosines (*red*) serves as a ligand for this SH2 group of Src, causing the SH2 to switch from intramolecular binding to intermolecular binding. Concomitantly, the SH3 group of Src detaches from its intramolecular binding to bind a proline-rich domain (PXXP) on the receptor. These changes open up the catalytic cleft of Src, placing it in a configuration that enables it to fire. (C) A final phosphorylation of tyrosine 416 moves an obstructing oligopeptide activation loop (*dark green*) out of the way of the catalytic cleft, yielding the full tyrosine kinase activity of Src. (Adapted from W. Xu et al., *Mol. Cell* 3:629–638, 1999.)

Table 6.2 Binding domains that are carried by various proteins[a]

Name of domain	Ligand	Examples of proteins carrying this domain
SH2	phosphotyrosine	Src (tyrosine kinase), Grb2 (adaptor protein), Shc (scaffolding protein), SHP2 (phosphatase), Cbl (ubiquitylation)
PTB	phosphotyrosine	Shc (adaptor protein), IRS-1 (adaptor for insulin RTK signaling), X11 (neuronal protein)
SH3	proline-rich	Src (tyrosine kinase), Crk (adaptor protein), Grb2 (adaptor protein)
14-3-3	phosphoserine	Cdc25 (CDK phosphatase), Bad (apoptosis regulator), Raf (Ser/Thr kinase), PKC (protein kinase C Ser/Thr kinase)
Bromo	acetylated lysine	P/CAF (transcription co-factor), chromatin proteins
PH[b]	phosphorylated inositides	PLC-δ (phospholipase C-δ), Akt/PKB (Ser/Thr kinase), BTK

[a]At least 32 distinct types of binding domains have been identified (see Figure 6.10B). This table presents six of these that are often associated with transduction of mitogenic signals.

[b]The phosphoinositide-binding groups include, in addition to the PH domain, the Fab1, YOTB, Vac1, EEA1 (FYVE), PX, ENTH, and FERM domains.

The SH2 and SH3 domains were the first of a series of specialized protein domains to be recognized and cataloged (see Figure 6.10B). Each of these specialized domains is able to recognize and bind a specific sequence or structure present in a partner molecule. Some of these specialized domains recognize and bind phosphoserine or phosphothreonine on partner proteins, while others bind to phosphorylated forms of certain membrane lipids (**Table 6.2**). Another specialized domain, termed PTB, is able, like SH2 groups, to recognize and bind phosphotyrosine residues; in this instance, however, the flanking amino acid residues on the N-terminal side of the phosphotyrosine define the unique identity of the phosphotyrosine ligand.

Taken together, these various domains illustrate how the lines of communication within a cell are restricted, since the ability of signaling molecules to physically associate in highly specific ways with target molecules ensures that signals are passed only to these intended targets and not to other proteins within the cell. In addition, examination of an array of signaling proteins has revealed that these various domains have been used as modules, being assembled in different ways through evolution to ensure the specificity of a variety of intermolecular interactions (see Figure 6.10B).

The discovery of SH2 groups also explained how Src, the original oncoprotein, operates within the normal cell, and how Rous sarcoma virus has reconfigured the structure of Src, making it into an oncoprotein (see Sidebar 6.3).

6.4 SH2 and SH3 groups explain how growth factor receptors activate Ras and acquire signaling specificity

The discovery of SH2 and SH3 groups and their mechanisms of action made it possible to solve several important problems. The biochemical clues were now in place to explain precisely how the receptor → Sos → Ras cascade operates.

The final clues came from analyzing the structure of Grb2. It contains two SH3 groups and one SH2 group. Its SH3 domains have an affinity for two distinct proline-rich sequences present in Sos, while its SH2 sequence associates with a phosphotyrosine present on the C-terminus of many growth factor–activated receptors. Consequently, Sos, which seems usually to float freely in the cytoplasm, now becomes tethered via the Grb2 linker to the receptor (**Figure 6.12**).

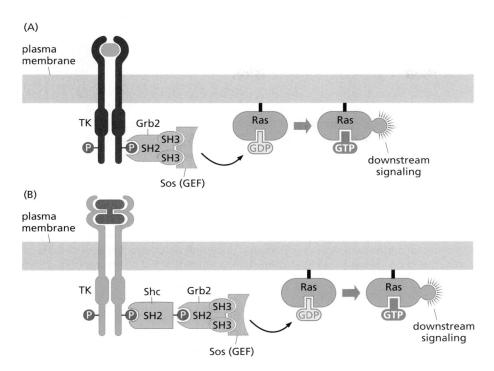

Figure 6.12 Intermolecular links forged by the Grb2 and Shc bridging proteins (A) The association of Sos, the Ras guanine nucleotide exchange factor (GEF), with ligand-activated growth factor receptors is achieved by bridging proteins, such as Grb2 shown here. The two SH3 domains of Grb2 *(pink)* bind to proline-rich domains of Sos *(orange)*, while its SH2 domain *(light blue)* binds to a phosphotyrosine located on the C-terminal tail of a receptor. (B) An alternative mode of association can be achieved through a collaboration between two bridging proteins, Grb2 and Shc. Grb2 uses its SH3 domains to bind Sos and its SH2 domain to bind a phosphotyrosine on Shc *(beige)*; the latter uses its own SH2 domain to bind a phosphotyrosine on a ligand-activated receptor. (Adapted from G.M. Cooper, Oncogenes, 2nd ed. Sudbury, MA: Jones and Bartlett, 1995. www.jbpub.com. Reprinted with permission.)

Detailed analysis of the structures of Grb2 and similarly configured proteins (for example, Crk) indicate that they carry no other functional domains beyond their SH2 and SH3 domains. Apparently, these adaptor proteins are nothing more than bridge builders designed specifically to link other proteins to one another. (An alternative bridging, also shown in Figure 6.12, can be achieved when a growth factor receptor associates with Sos via Grb2 and Shc.)

Once Sos becomes anchored via Grb2 (or via Grb2 + Shc) to a receptor, it is brought into close proximity with Ras proteins, most of which seem to be permanently tethered to the inner face of the plasma membrane. Sos is then physically well positioned to interact directly with these Ras molecules, inducing them to release GDPs and bind instead GTPs. This guanine nucleotide exchange causes the activation of Ras protein signaling. Hence, the biochemically defined pathway could now be drawn like this:

Receptor → Grb2 → Sos → Ras   or   Receptor → Shc → Grb2 → Sos → Ras

Because a diverse group of signaling proteins become attracted to ligand-activated receptors (see, for example, Figure 6.9), the Sos–Ras pathway represents only one of a number of signaling cascades radiating from growth factor receptors.

6.5 Ras-regulated signaling pathways: A cascade of kinases forms one of three important signaling pathways downstream of Ras

The multiple cellular responses that a growth factor elicits could now be explained by the fact that its *cognate* receptor (that is, the receptor that specifically binds it) is able to activate a specific combination of downstream signaling pathways. Each of these pathways might be responsible for inducing, in turn, one or another of the biological changes occurring after cells are stimulated by this particular growth factor. Moreover, exaggerated forms of this signaling might well be operating in cancer cells that experience continuous growth factor stimulation because of an autocrine signaling loop or a mutant, constitutively activated receptor.

This branching of downstream pathways did not, however, explain how the mutant Ras oncoprotein (which represents only one of the multiple pathways downstream of growth factor receptors) is able to evoke a number of distinct changes in cells. Here, once again, signal transduction biochemistry provided critical insights. We now know that at least three major downstream signaling cascades radiate from the Ras protein. Their diverse actions help to explain how the Ras oncoprotein causes so many changes in cell behavior.

The layout of these downstream signaling pathways was mapped through a combination of yeast genetics and biochemical analyses of mammalian cells. These approaches made it clear that when Ras binds GTP, two of its "switch domains" shift (see Figure 5.31B), enabling its **effector loop** to interact physically with several alternative downstream signaling partners that are known collectively as Ras *effectors*, that is, the proteins that carry out the actual work of Ras (**Figure 6.13**). Each of these effectors binds quite tightly to the effector loop of the GTP-bound form of Ras protein while having little affinity for the loop presented by its GDP-bound form.

The first of these Ras effectors to be discovered was the Raf kinase (**Figure 6.14**). Like the great majority of protein kinases in the cell, Raf phosphorylates substrate proteins on their serine and threonine residues. Long before its association with Ras was uncovered, the Raf kinase had already been encountered as the oncoprotein specified by both a rapidly transforming murine retrovirus and, in homologous form, a chicken retrovirus (see Table 3.3).

The activation of Raf by Ras depends upon the relocalization of Raf within the cytoplasm. Recall that Ras proteins are always anchored to a membrane, usually the inner surface of the plasma membrane, through their C-terminal hydrophobic tails. Once it has bound GTP, the affinity of Ras for Raf increases by three orders of magnitude, enabling Ras to bind Raf relatively tightly, doing so via the Ras effector loop. Before this association, Raf is found in the cytosol; thereafter, Raf becomes tethered via Ras to the plasma membrane (or, in certain cases, to another cytoplasmic membrane).

The physical association of Raf with Ras appears to be facilitated by additional proteins that act as scaffolds to hold them together; yet others trigger the phosphorylation and dephosphorylation of key amino acid residues on Raf, altering its three-dimensional configuration. Raf now acquires active signaling powers and proceeds to phosphorylate and thereby activate a second kinase known as MEK. MEK is actually a "dual-specificity kinase," that is, one that can phosphorylate serine/threonine residues as

Figure 6.13 The Ras effector loop The crystallographic structure of the Ras protein, viewed from a different angle than those illustrated in Figure 5.31, is depicted here in a ribbon diagram (*largely aquamarine*). The guanine nucleotide (*purple stick figure*) is shown together with the nearby glycine residue, which, upon replacement by a valine, causes inactivation of the GTPase activity (G12V). The effector loop (*yellow*) interacts with at least three important downstream effectors of Ras. Amino acid substitutions (*arrows*) affecting three residues in the effector loop create three alternative forms of the Ras protein that interact preferentially with either PI3 kinase, Raf, or Ral-GEF. (Courtesy of R. Latek and A. Rangarajan, adapted from data in U. Krengel et al., *Cell* 62:539–548, 1990.)

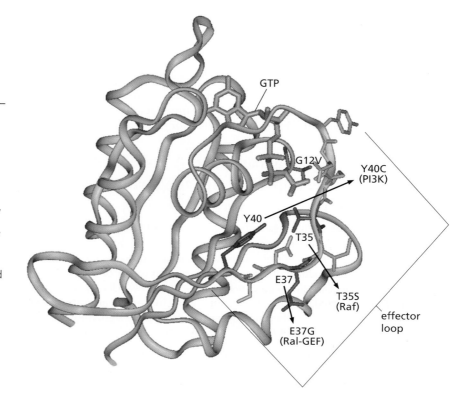

well as tyrosine residues. MEK uses these powers to phosphorylate two other kinases, the *e*xtracellular signal–*r*egulated *k*inases 1 and 2, commonly referred to as Erk1 and Erk2. Once phosphorylated and activated, each of these Erks then phosphorylates substrates that, in turn, regulate various cellular processes including transcription.

This type of kinase signaling cascade has been given the generic name of a MAPK (mitogen-activated protein kinase) pathway. In fact, there are a number of such signaling cascades in mammalian cells. Erk1 and Erk2 at the bottom of the Raf-initiated cascade are considered MAPKs. The kinase responsible for phosphorylating a MAPK is termed generically a MAPKK; in the case of the cascade studied here, this role is played by MEK. The kinase responsible for phosphorylating a MAPKK is therefore called generically a MAPKKK (see Figure 6.14); in this scheme, Raf is classified as a MAPKKK.

Evidence is rapidly accumulating that the kinases forming many of these cascades are held together physically by scaffolding proteins. For example, KSR1 assembles Raf, MEK, and Erk together, thereby ensuring that Raf is closely juxtaposed with its MEK substrate and that MEK, in turn, is close to its Erk substrate. By regulating the activity of these scaffolding proteins, the cell can control the signals flowing through its various kinase cascades.

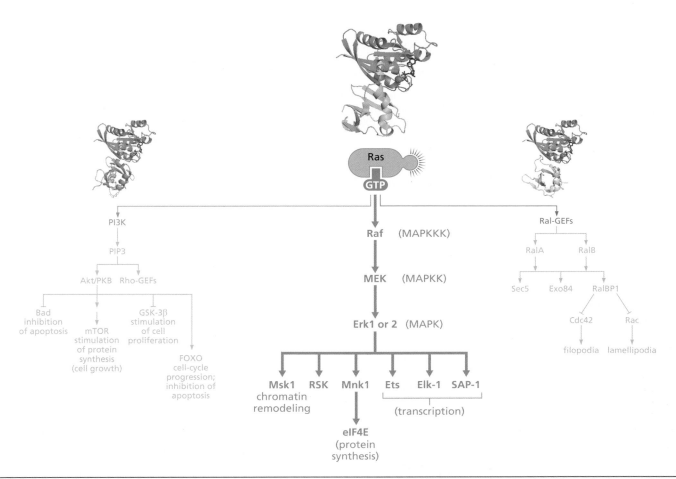

Figure 6.14 The Ras → Raf → MAP kinase pathway This signaling cascade (*aquamarine, center*) is one of a series of similarly organized pathways in mammalian cells that have the overall plan MAPKKK → MAPKK → MAPK. In this particular case, Ras activates the Raf kinase (a MAPKKK), the latter proceeds to phosphorylate and activate MEK (a MAPKK), and MEK then phosphorylates and activates Erk1 and Erk2 (MAPKs). (The two additional effector pathways of Ras shown here will be analyzed further in Figures 6.15 and 6.21.) Erk1/2 can then phosphorylate kinases in the cytoplasm that regulate translation as well as transcription factors in the nucleus. The bimolecular complex (*above*) illustrates the interaction of the effector domain of Ras (*red, purple*) with the Ras-binding domain of Raf (*aquamarine*). (From M.E. Pacold et al., *Cell* 103:931–943, 2000; N. Nassar et al., *Nature* 375:554–560, 1995; and L. Huang et al., *Nat. Struct. Biol.* 5:422–426, 1998.)

Note that in these various kinase cascades, phosphorylation of kinase B by kinase A always results in the functional activation of kinase B. For example, when MEK is phosphorylated by Raf, MEK becomes activated as a signal-emitting kinase. Such serine/threonine phosphorylation therefore has a very different consequence from the phosphorylation of tyrosine residues carried by the cytoplasmic tails of growth factor receptors. The phosphotyrosines on these receptors attract signaling partners and are therefore responsible for the *relocalization* of partner proteins; in contrast, the phosphate groups attached to MEK by Raf cause a shift in its structure that results in its *functional activation* as a kinase.

Once activated, an Erk kinase at the bottom of this cascade not only phosphorylates cytoplasmic substrates, but can also translocate to the nucleus, where it causes the phosphorylation of transcription factors; some of the latter then initiate the immediate and delayed early gene responses. For example, as indicated in Figure 6.14, Erk can phosphorylate several transcription factors (Ets, Elk-1, and SAP-1) directly and, in addition, phosphorylate and thereby activate other kinases, which proceed to phosphorylate and activate yet other transcription factors. These downstream kinases phosphorylate two chromatin-associated proteins (HMG-14 and histone H3), a modification that places the chromatin in a configuration that is more hospitable to transcription (see Section 1.8). The Mnk1 kinase, a cytoplasmic substrate of Erk1 and Erk2, phosphorylates the translation initiation factor eIF4E, thereby helping to activate the cellular machinery responsible for protein synthesis.

Once the Erks phosphorylate the Ets transcription factor, the latter can then proceed to stimulate the expression of important growth-regulating genes, such as those specifying heparin-binding EGF (HB-EGF), cyclin D1, Fos, and p21^{Waf1}. (The functions of the last three are described in detail in Chapter 8.)

Especially important among the genes induced by the Raf $\rightarrow$ MEK $\rightarrow$ Erk pathway are the two immediate early genes encoding the Fos and Jun transcription factors. Once synthesized, these two proteins can associate with one another to form AP-1, a widely acting heterodimeric transcription factor that is often found in hyperactivated form in cancer cells. The special importance and influence of Fos and Jun is indicated by the fact that each was originally identified as the product of a retrovirus-associated oncogene (see Table 3.3).

The Ras $\rightarrow$ Raf $\rightarrow$ MEK $\rightarrow$ Erk pathway is only one of several downstream signaling cascades that are activated by the GTP-bound Ras proteins (see Figure 6.14). The importance of this kinase cascade is indicated by the fact that in a number of cell types, the Raf protein kinase, when introduced into cells in mutant, oncogenic form, can evoke most of the transformation phenotypes that are induced by the Ras oncoprotein. Hence, in such cells, the Raf pathway is responsible for the lion's share of the transforming powers of Ras oncoproteins. In addition to activating a number of growth-promoting genes, as described above, this pathway confers anchorage independence and loss of contact inhibition. It also contributes to the profound change in cell shape that is associated with transformation by the *ras* oncogene.

In certain cancers, the signaling pathway lying downstream of Raf may be strongly activated without any direct involvement of Ras. Thus in ~50% of human melanomas a mutant form of B-Raf, a close cousin of Raf, is found to carry an amino acid substitution causing a ~500-fold increase in kinase activity. [The mutation commonly occurring in melanomas causes B-Raf's activation loop, which usually obstructs access to the catalytic cleft of this enzyme (similar to Figure 5.17), to swing out of the way.] In about one-third of the remaining melanomas, mutant alleles encoding activated N-Ras oncoproteins can be found. In human tumors overall, a *ras* oncogene is rarely found to coexist with mutant alleles of *B-raf*, providing evidence that these two oncogenes have overlapping functions and thus are partially redundant with one another. The phosphorylation of more than 60 proteins is increased and 30 whose phosphorylation is decreased by the actions of this kinase; almost all of these changes are mediated directly by Erk, the downstream effector of both Raf and B-Raf (see Figure 6.14). A variety of explanations have been proposed to rationalize why multi-kinase cascades like this one are used in a variety of signaling circuits throughout the cell (Supplementary Sidebar 6.3).

6.6 Ras-regulated signaling pathways: a second downstream pathway controls inositol lipids and the Akt/PKB kinase

A second important downstream effector of the Ras protein enables Ras to evoke yet other cellular responses (**Figure 6.15**). In the context of cancer, the most important of these is a suppression of apoptosis. This anti-apoptotic effect is especially critical for cancer cells, since many of them are poised on the brink of activating this cell suicide program, a subject covered in detail in Chapter 9.

Early studies of the biochemistry of the plasma membrane suggested that its phospholipids serve simply as a barrier between the aqueous environments in the cell exterior and interior. The biochemistry of these phospholipids showed that they are **amphipathic**—that is, they possess a hydrophilic head, which likes to be immersed in water, and a hydrophobic tail, which prefers nonaqueous environments (**Figure 6.16A**). This polarity explains the structure of lipid bilayers such as the plasma membrane, in which the hydrophilic groups face and protrude into the extracellular and cytosolic aqueous environments while the hydrophobic tails are buried in the middle of the membrane, from which water is excluded. (The transmembrane domains of growth factor receptors discussed in Chapter 5 consist of hydrophobic acids that allow these domains to be buried in this hydrophobic environment.)

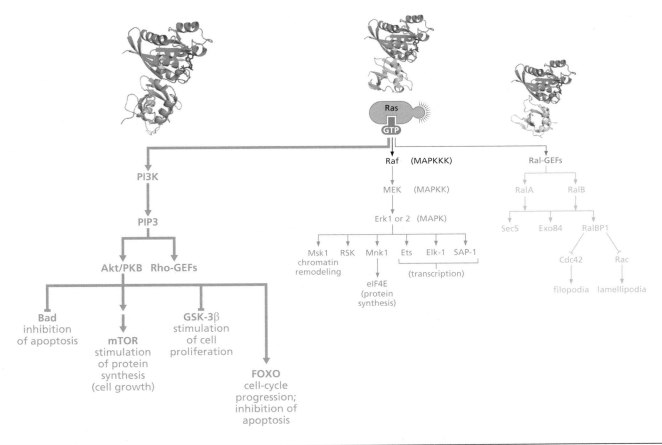

Figure 6.15 The PI3 kinase pathway A second effector pathway of Ras (*orange, left*) derives from its ability to associate with and activate phosphatidylinositol 3-kinase (PI3K), with the resulting formation of phosphatidylinositol (3,4,5)-triphosphate (PIP3). This leads in turn to the tethering of PH-containing molecules, such as Akt (also called PKB) and Rho guanine nucleotide exchange proteins (Rho-GEFs), to the plasma membrane. Akt/PKB is able to inactivate Bad (a pro-apoptotic protein); inactivate GSK-3β, an antagonist of growth-promoting proteins such as β-catenin, cyclin D1, and Myc; and activate mTOR. (The latter is a kinase that stimulates protein synthesis and cell growth and also acts reciprocally to phosphorylate and activate Akt/PKB, as we will see in Chapter 16.). The bimolecular complex (*above*) illustrates the interaction of the effector domain of Ras (*red, purple*) with the Ras-binding domain of PI3K (*orange*). (From M.E. Pacold et al., *Cell* 103:931–943, 2000; N. Nassar et al., *Nature* 375:554–560, 1995; and L. Huang et al., *Nat. Struct. Biol.* 5:422–426, 1998.)

In the 1970s, it became apparent that eukaryotic cells exploit some of the membrane-associated phospholipids for purposes that are fully unrelated to the maintenance of membrane structure. Some phospholipids contain, at their hydrophilic heads, an inositol group. Inositol is a water-soluble carbohydrate molecule (more properly termed a polyalcohol). The inositol moiety of such phospholipids can be modified by the addition of phosphate groups. The resulting phosphoinositol may then be cleaved from the remaining, largely hydrophobic portions of a phospholipid molecule. Since it is purely hydrophilic, this phosphoinositol, termed IP_3, can then diffuse away from the membrane (Figure 6.16B), thereby serving as an *intracellular hormone* to dispatch signals from the plasma membrane to distant parts of the cell. Such intracellular hormones are often called **second messengers**. The second product of this cleavage, termed DAG, can serve to activate a key signaling kinase in the cell—the serine/threonine kinase known as protein kinase C (PKC). Alternatively, a phosphorylated

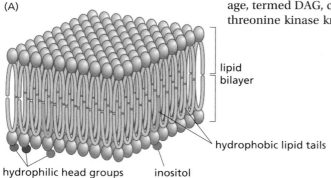

(A)

lipid bilayer

hydrophobic lipid tails

hydrophilic head groups   inositol

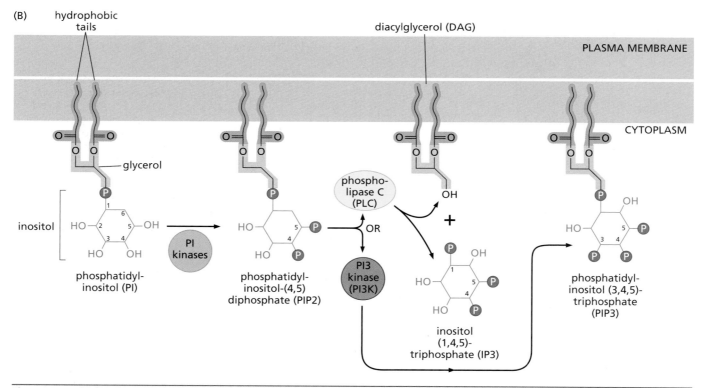

Figure 6.16 Biochemistry of lipid bilayers (A) All cellular membranes are assembled as phospholipid bilayers, in which the hydrophilic head groups (*gray ovals*) protrude into the aqueous solvent above and below the membrane, while the hydrophobic tails (*green, gray*) are hidden in the nonaqueous space within the lipid bilayer. A small minority of the various types of hydrophilic head groups (*red, green, blue*) contain inositol sugars (*blue*). (B) Phosphatidylinositol (PI; *left*) is composed of three parts—two fatty acids with long hydrocarbon tails (*green*) inserted into the plasma membrane, glycerol (*gray*), and inositol (*light blue*) attached to the glycerol via a phosphodiester linkage. PI kinases can add phosphate groups to the various hydroxyls of inositol, yielding, for example, the PI (4,5)-diphosphate (PIP2) shown here. Cleavage of PIP2 by phospholipase C yields diacylglycerol (DAG), which can activate protein kinase C (PKC), and inositol (1,4,5)-triphosphate (IP3), which induces release of calcium ions from intracellular stores. Alternatively, PIP2 can be further phosphorylated by PI3 kinase (PI3K) to yield phosphatidylinositol (3,4,5)-triphosphate (PIP3). Once formed, the phosphorylated head group of PIP3 serves to attract proteins carrying PH domains, which thereby become tethered to the inner surface of the plasma membrane.

inositol can remain attached to the remainder of the phospholipid and thus can remain embedded in the plasma membrane; there, it can serve as an anchoring point to which certain cytosolic proteins can become attached.

We now know that the inositol moiety can be modified by several distinct kinases, each of which shows specificity for phosphorylating a particular hydroxyl of inositol. Phosphatidylinositol 3-kinase (PI3K), for example, is responsible for attaching a phosphate group to the 3′ hydroxyl of the inositol moiety of membrane-embedded phosphatidylinositol (PI). While several distinct PI3 kinases have been discovered, the most important of these may well be the PI3K that phosphorylates PI(4,5)P$_2$ (or, more simply, PIP2). Prior to modification by PI3K, PIP2 already has phosphates attached to the 4′ and 5′ hydroxyl groups of inositol; after this modification, this inositol acquires yet another phosphate group, and PIP2 is converted into PI(3,4,5)P$_3$, that is, PIP3 (see Figure 6.16B).

Note an important difference between the organization of this pathway and the Ras → Raf → MEK → Erk cascade that we read about in the previous section. In the present case, one of the critical kinases (that is, PI3K) attaches phosphates to a phospholipid, rather than phosphorylating a protein substrate, such as another kinase.

We first encountered PI3K, in passing, at an earlier point in this chapter (see Section 6.3, Figure 6.9A). There, we described this enzyme's attraction to certain ligand-activated receptors via the SH2 group of its regulatory subunit, and the consequent phosphorylation of inositol-containing phospholipids already present in the nearby plasma membrane. It turns out that GTP-activated Ras is also able to bind PI3K directly and thereby enhance its functional activation (**Figure 6.17**). (Indeed, there is strong evidence that the H-Ras protein can effectively activate PI3K only when the p85 regulatory subunit of PI3K is bound to phosphotyrosine on a ligand-activated growth factor receptor.) Hence, PI3K can function as a direct downstream effector of Ras. As is the case when Raf binds Ras, the binding of PI3K to Ras causes PI3K to become closely associated with the plasma membrane, where its PI substrates are located.

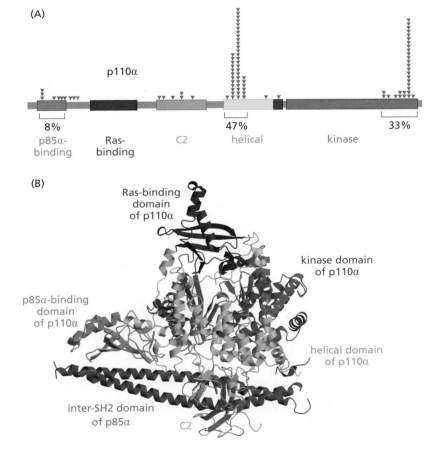

Figure 6.17 Activation of PI3K by Ras, growth factor receptors, and somatic mutation PI3K is a key enzyme in the control of cell growth and proliferation. The most important form of this enzyme, assembled as a complex of a catalytic subunit, termed p110, and a regulatory subunit, termed p85, is activated by the actions of the Ras oncoprotein and by tyrosine-phosphorylated growth factor receptors (RTKs). (A) The multi-domain structure of p110α is seen here. (B) A ribbon model of p110α with the inter-SH2 domain of p85α is shown here. The Ras-binding domain of p110α (*blue*) enables its tethering via Ras to the plasma membrane (*not shown*). The p85α subunit uses its two SH2 groups (*not shown*), which are located at the opposite ends of the inter-SH2 domain of p85 (red) to bind to phosphotyrosine residues in the cytoplasmic tails of ligand-activated growth factor receptors. The catalytic domain of p110α is shown in *purple*. It remains unclear whether simultaneous interactions between p85, p110, Ras and a ligand-activated RTK are required for activation of p110, or whether interactions of p85 with ligand-activated RTK or p110 with Ras suffice for this activation. The inter-SH2 domain seems to have an affinity for binding lipid membranes. (From C.-H. Huang et al., *Science* 318:1744–1748, 2007.)

(A)

p110α

8%
p85α-
binding

Ras-
binding

C2

47%
helical

kinase

33%

(B)

Ras-binding
domain
of p110α

kinase domain
of p110α

p85α-binding
domain
of p110α

helical domain
of p110α

inter-SH2 domain
of p85α

C2

Once activated by one or both of these associations, PI3K plays a central role in a number of signaling pathways, as described below. This is indicated by the fact that this enzyme is activated by a diverse array of signaling agents, including PDGF, nerve growth factor (NGF), insulin-like growth factor-1 (IGF-1), interleukin-3, and the extracellular matrix (ECM) attachment achieved by integrins (see Section 6.9).

The formation of the phosphorylated inositol head groups of phosphatidylinositol has little if any effect on the overall structure and function of the plasma membrane. (The inositol phospholipids are minor constituents of the plasma membrane.) However, a phosphorylated inositol head group protruding from the plasma membrane may be recognized and bound by certain proteins that are usually floating in the cytosol. Once anchored via a phosphoinositol to the plasma membrane, these proteins are then well positioned to release certain types of signals.

The most important of the phosphorylated inositol head groups appears to be PIP3, whose inositol carries phosphates at its 3′, 4′, and 5′ positions (see Figure 6.16B). Many of the cytosolic proteins that are attracted to PIP3 carry pleckstrin homology (PH) domains (see Table 6.2) that have strong affinity for this triply phosphorylated inositol head group (**Figure 6.18** and **Figure 6.19**). Arguably the most important PH domain–containing protein is the serine/threonine kinase named Akt, also known as protein kinase B (PKB). (Akt/PKB is yet another protein that was initially discovered because it is encoded by a retrovirus-associated oncogene.) Thus, once PIP3 is formed by PI3K, an Akt/PKB kinase molecule can become tethered via its PH domain to the inositol head group of PIP3 that protrudes from the plasma membrane into the cytosol (Figure 6.19).

This association of Akt/PKB with the plasma membrane (together with phosphorylations of Akt/PKB that soon follow) results in the functional activation of Akt/PKB as a kinase. Once activated, Akt/PKB proceeds to phosphorylate a series of protein substrates that have multiple effects on the cell (see Figure 5.6B), including (1) aiding in cell survival by reducing the possibility that the cell apoptotic suicide program will become activated; (2) stimulating cell proliferation; and (3) stimulating cell growth, in the most literal sense of the term, that is, stimulating increases in cell size by increasing protein synthesis (**Figure 6.20**). In addition, it also exerts an influence, still poorly understood, on cell motility and on **angiogenesis**—the production of new blood vessels.

Figure 6.18 Migration of PH-containing proteins to PIP3 in the plasma membrane After PIP3 head groups are formed by PI3 kinase (see Figure 6.16B), they become attractive docking sites for PH domain–containing proteins. This phosphorylation and homing occurs within seconds of the activation of PI3 kinase, as revealed by this micrograph, in which green fluorescent protein (GFP) has been fused to a protein carrying a PH (pleckstrin homology) domain. This hybrid protein moves from a diffuse distribution throughout the cytoplasm (A) to the inner surface of the plasma membrane (B) within a minute after EGF (epidermal growth factor) has been added to cells. However, if the PH domain of the fusion protein is deleted (C), addition of EGF has no effect on its intracellular distribution (D) and it remains diffusely distributed throughout the cytoplasm. (From K. Venkateswarlu et al., *J. Cell Sci.* 112:1957–1965, 1999.)

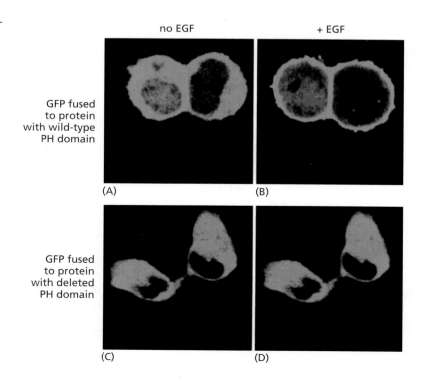

no EGF   + EGF

GFP fused to protein with wild-type PH domain

(A)   (B)

GFP fused to protein with deleted PH domain

(C)   (D)

The activation of PI3K, and thus of Akt/PKB, is under very tight control. In a quiescent, nongrowing cell lacking growth factor stimulation, the intracellular levels of PIP3 are extremely low. Once such a cell encounters mitogens, the levels of PIP3 in this cell increase rapidly. The normally low levels of PIP3 and other phosphorylated PI molecules are the work of a series of phosphatases that reverse the actions of the activating kinases such as PI3K. The best-characterized of these phosphatases, PTEN, removes the 3′ phosphate group from PIP3 that has previously been attached by PI3K (see Figure 6.19A). This removal suggests two distinct mechanisms by which the Akt/PKB signaling pathway can become deregulated in cancer cells—hyperactivity of PI3K or inactivity of PTEN.

The portfolio of diverse biological functions assigned to Akt/PKB requires that this kinase be able to phosphorylate a variety of substrates, each involved in a distinct

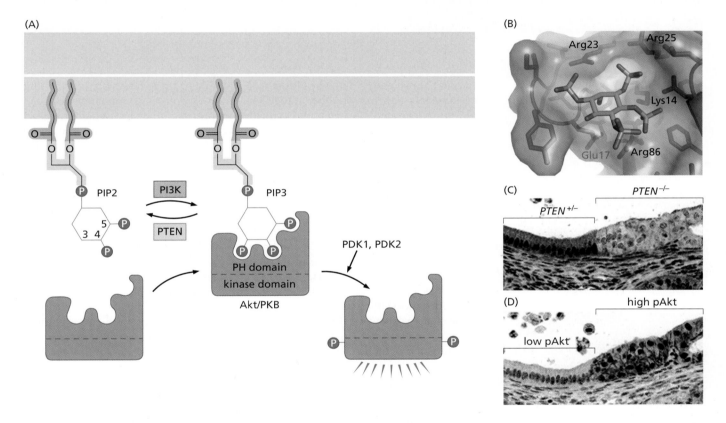

Figure 6.19 Docking of PH domains to PIP3 (A) By forming the triply phosphorylated inositol "head group" (*blue lines*) of PIP3 (see Figure 6.16B), PI3K creates a docking site for cytoplasmic proteins that carry PH domains. The most important of these is the Akt/PKB kinase (*brown*). Once docked to the inner surface of the plasma membrane via its PH domain, Akt/PKB becomes doubly phosphorylated by two other kinases (PDK1, PDK2) and thereby activated (*lower right*). (Recent evidence indicates that PDK1, which has its own PH domain, can transduce oncogenic signals that do not depend on Akt/PKB activation. Yet other work identifies PDK2 as one form of mTOR, to be discussed in Chapter 16.) Akt/PKB then proceeds to phosphorylate a variety of substrates regulating cell proliferation, survival, and size. However, if PTEN is also present and active, it dephosphorylates PIP3 to PIP2, thereby depriving Akt/PKB of a docking site at the plasma membrane. (The size of the PH domain of Akt/PKB relative to the size of its kinase domain has been greatly exaggerated in this drawing.) (B) The stereochemistry of the actual PH domain of Akt/PKB binding PIP4 (used in crystallization to serve as an analog of the PIP3 head group) is shown here. Electrostatic and hydrogen bonds (*not shown*) enable high-specificity binding of the PH domain to this head group. Point mutations in the genomes of certain human breast, colorectal, and ovarian tumors cause the replacement of glutamic acid with a lysine at residue 17 of the PH domain, resulting in higher-affinity binding to PIP3, increased association with the plasma membrane, and the formation of a constitutively active Akt/PKB oncoprotein. (C) The dynamics of this interaction are seen in genetically altered mice having a *Pten*[+/−] genotype. Uterine epithelial cells in these mice often lose the remaining *Pten* wild-type allele, yielding *Pten*[−/−] cells, which form localized hyperplasias and cysts. Immunostaining with anti-PTEN antiserum shows that the epithelial cells on the right have lost PTEN expression and become larger, indicating the effects of the PIP3 pathway on cell size. (D) When the same lesion is stained with an anti-phospho-Akt/PKB antibody (which detects activated Akt/PKB), the cells on the right show high pAkt activity, while those on the left show very low activity. (B, courtesy of D. van Aalten. C and D, from K. Podsypanina et al., *Proc. Natl. Acad. Sci. USA* 98:10320–10325, 2001.)

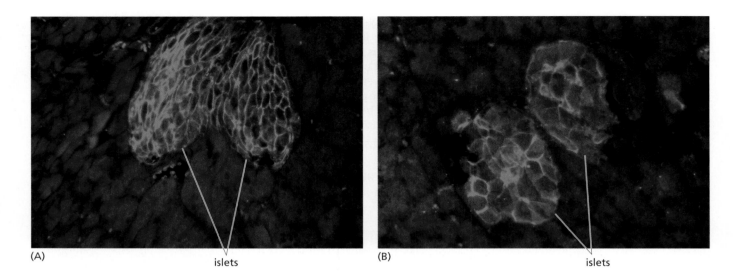

(A)   islets   (B)   islets

Figure 6.20 Akt/PKB and the control of cell growth The endocrine β cells located in the pancreatic islets are normally responsible for secreting insulin in response to elevated concentrations of blood glucose. (A) A micrograph of a normal pancreatic islet. (B) If a constitutively active Akt/PKB kinase is expressed selectively in β cells, they grow to volumes that are as much as fourfold larger than normal. (From R.L. Tuttle et al., *Nat. Med.* 7:1133–1137, 2001.)

downstream signaling pathway (**Table 6.3**). The first of these functions—facilitating cell survival—depends on the ability of Akt/PKB to suppress any tendencies the cell may have to activate its own built-in suicide program of apoptosis. We defer detailed discussion of apoptosis until Chapter 9. For the moment, suffice it to say that phosphorylation by Akt/PKB results in the inhibition of several proteins that play prominent roles favoring the entrance of a cell into apoptosis.

The proliferative functions of Akt/PKB depend on its ability to perturb proteins that are important for regulating the advance of a cell through its cycle of growth and division, often termed the cell cycle. We will learn more about this complex program in Chapter 8.

Independent of these effects on proliferation, Akt/PKB is able to induce dramatic changes in the proteins that control the rate of protein synthesis in the cell. Akt/PKB acts through intermediaries to trigger activation of mTOR kinase; the latter then phosphorylates and inactivates 4E-BP, a potent inhibitor of translation, and, at the same time, phosphorylates and thereby activates p70S6 kinase, an activator of translation. These changes allow Akt/PKB to increase the efficiency with which the translation of a class of mRNAs is initiated; the resulting elevated rate of protein synthesis favors the accumulation of many cellular proteins and is manifested in the growth (rather than the proliferation) of cells. We will return to the detailed organization of this pathway in Chapter 16. An illustration of the growth-inducing powers of Akt/PKB is provided by transgenic mice that express a mutant, constitutively active Akt/PKB in the β cells of the pancreatic islets (see Figure 6.20); in such mice, these particular cells grow to a size more than twice the cross-sectional area (and almost 4 times the volume) of normal β cells!

There is evidence that yet other cytoplasmic proteins use their PH domains to associate with PIP3, and that this association also favors their functional activation. Among the PH-bearing cellular proteins are a group of guanine nucleotide exchange factors (GEFs) that function analogously to Sos, being responsible for activating various small GTPases that are distant relatives of the Ras proteins. These other GTPases belong to the Rho family of signaling proteins, which includes Rho proper and its two cousins, Rac and Cdc42. Like Ras, these Rho proteins operate as binary switches, cycling between a GTP-bound actively signaling state and a GDP-bound inactive state (**Sidebar 6.4**). The Rho-GEFs, once activated by binding to PIP3, act on Rho proteins in the same way that Sos acts on Ras. In particular, the Rho-GEFs induce Rho proteins to jettison their bound GDPs, allowing GTP to jump aboard and activate the Rhos.

Activated Rho proteins have functions that differ strongly from those of Ras: they participate in reconfiguring the structure of the cytoskeleton and the attachments that the cell makes with its physical surroundings. In so doing, these Rho-like proteins control cell shape, motility, and, in the case of cancer cells, invasiveness. For example, Cdc42

Table 6.3 **Effects of Akt/PKB on survival, proliferation, and cell growth**

Biological effect	Substrate of Akt/PKB	Description	Functional consequence
Anti-apoptotic			
	Bad (pro-apoptotic)	Bcl-X antagonist; like Bad, belongs to Bcl-2 protein family controlling mitochondrial membrane pores (Section 9.13).	inhibition
	caspase-9 (pro-apoptotic)	Component of the protease cascade that affects the apoptotic program (Section 9.13).	inhibition
	IκB kinase, abbreviated IKK (anti-apoptotic)	Activated by Akt/PKB phosphorylation (Section 6.12).	activation
	FOXO1 TF, formerly called FKHR TF (pro-apoptotic)	Phosphorylation prevents its nuclear translocation and activation of pro-apoptotic genes.	inhibition
	Mdm2 (anti-apoptotic)	Activated via phosphorylation by Akt/PKB; it triggers destruction of p53 (Section 9.7).	activation
Proliferative			
	GSK-3β (anti-proliferative)	Phosphorylates β-catenin, cyclin D1, and Myc (Sections 7.11, 8.3, 8.9), causing their degradation; inactivated via phosphorylation by Akt/PKB.	inhibition
	FOXO4, formerly called AFX (anti-proliferative)	Induces expression of the CDK inhibitor p27^{Kip1} (Section 8.4) gene and some pro-apoptotic genes; exported from the nucleus when phosphorylated by Akt/PKB.	inhibition
	p21^{Cip1} (anti-proliferative)	CDK inhibitor, like p27^{Kip1} (Section 8.4). Exits the nucleus upon phosphorylation by Akt/PKB; in the cytoplasm, phosphorylated p21^{Cip1} inhibits caspases, thereby acquiring anti-apoptotic functions (Section 9.13).	inhibition
Growth			
	Tsc2 (anti-growth)	Phosphorylation by Akt/PKB causes Tsc1/Tsc2 complex to dissociate, allowing activation of mTOR, which then up-regulates protein synthesis (Section 16.15).	inhibition

is involved in reorganizing the actin cytoskeleton of the cell as well as controlling **filopodia**, small, fingerlike extensions from the plasma membrane that the cell uses to explore its environment and form adhesions to the extracellular matrix, while Rac is involved in the formation of **lamellipodia**, broad ruffles extending from the plasma membrane that are found at the leading edges of motile cells (see Figures 14.36 and 14.37). Because these structures are so intimately tied to the steps of tumor cell

Sidebar 6.4 Ras is the prototype of a large family of similar proteins Ras is only one of a large superfamily of 151 similarly structured mammalian proteins that together are called "small G-proteins" to distinguish them from the other class of larger, heterotrimeric guanine nucleotide–binding G-proteins that are activated by association with seven-membrane-spanning (serpentine) cell surface receptors (see Figure 5.24A). Most of these small G-proteins have been given three-letter names, each derived in some way from the initially named and discovered Ras proteins—for example, Ral, Rac, Ran, Rho, and so forth. (The Cdc42 protein is a member of the Rho family that has kept its own unique name because of its initial discovery through yeast genetics.)

Much of the sequence similarity shared among these otherwise diverse proteins is found in the cavities in each that bind and hydrolyze guanine nucleotides. Almost all of the small G-proteins operate like a binary switch, using a GTP–GDP–GTP cycle to flip back and forth between an on and an off state. Like Ras, each of these proteins is thought to have its own specialized guanine nucleotide exchange factors (GEFs) to activate it (by promoting replacement of GDP by GTP) and its own GTPase-activating proteins (GAPs) to trigger its GTPase activity, thereby inactivating it.

Each of these small G-proteins is specialized to control a distinct cell-physiologic or biochemical process, including regulation of the structure of the cytoskeleton, trafficking of intracellular vesicles, and apoptosis. One is even used to regulate the transport of proteins through pores in the nuclear membrane. The only trait shared in common by these diverse cell-biological processes is a need to be regulated by some type of binary switching mechanism—an operation for which the small G-proteins are ideally suited.

invasion and metastasis, we will defer further discussion of these Rho-like proteins and their mechanisms of action until Chapter 14.

The pathway involving PI metabolites and Akt/PKB is deregulated in a number of human cancer cell types through changes that are often quite independent of Ras oncoprotein action (Table 6.4). In advanced ovarian carcinomas, elevated levels of one form of PI3K are associated with increased proliferation and invasiveness as well as decreased levels of apoptosis. In yet other tumors, such as lymphomas, cancers of the head and neck, and colon carcinomas, the Akt/PKB enzyme is overexpressed and presumably hyperactivated.

PI3K is hyperactivated in almost one-third of human colon carcinomas because of mutations affecting its p110α catalytic subunit (see Figure 6.17B and Table 6.4). In fact, these activating mutations make the *PIK3CA* gene, which encodes the α isoform of PI3K, the second most mutated oncogene in human cancers, appearing just below the K-*RAS* gene in the roster. However, loss of PTEN function is more frequently responsible for the deregulated activity of Akt/PKB, being observed in many types of human cancer cells (see Table 6.4).

In a number of distinct tumor types, including breast and prostate carcinomas as well as glioblastomas, PTEN activity is lost because of chromosomal gene mutations, DNA methylation events that suppress expression of the *PTEN* gene, or overexpression of a microRNA that suppresses translation of the PTEN mRNA. Altogether, PTEN activity

Table 6.4 Alteration of the PI3K pathway in human tumors[a]

Cancer type	Akt/PKB hyperactive	PIKC3A hyperactive[b]	p85α[c]	PTEN-mutant or repressed[d]
Glioblastoma		6–27%	8%	20%
Ovarian carcinoma	~2%	4–12%	4%	8%
Endometrial carcinoma		22%		42–54%
Hepatocellular carcinoma		6–36%		5%
Melanoma	~80%	~9%		40–50%
Lung carcinoma		3–4%		9%
Renal cell carcinoma		3%		4%
Thyroid carcinoma		5%		5%
Lymphoid		3%		8%
Prostate carcinoma		2%		10%
Colon carcinoma	~6%	14–32%	2–8%	13–54%
Breast carcinoma	~8%	18–40%	2%	20–33%
Bladder		23%		8%
Pancreatic		25%	17%	
Gastric		8%		

[a]The percentages in this table are approximate, since the proportion of tumors bearing the indicated alteration increases progressively as tumor progression proceeds, often dramatically, and because many reports do not distinguish between inactivation by mutation and inactivation by promoter methylation.
[b]*PIKC3A* appears to be the only gene of the 16 members of the PI3K-encoding gene family to undergo somatic mutation during tumor development. These mutations affect the p110 catalytic subunit of PI3 kinase; frequently occurring amplifications of this gene are not registered in this table.
[c]*PI3KR1* mutations affect the regulatory subunit of PI3K kinase and are most commonly observed in human cancers; alterations of the four other members of this family of PI3K regulatory subunits are not registered here. Alterations of the encoded p85α subunit cited here were few in number and the indicated percentages are likely to change dramatically as more data are collected.
[d]*PTEN* nonsense mutations and deletions are registered here and, in many cases, the even more frequent shutdown of expression through promoter methylation or the actions of microRNAs . (Promoter methylation often results in shutdown of transcription of a gene; see Section 7.8.)

From www.sanger.ac.uk/perl/genetics/CGP/cosmic; T.L. Yuan and L.C. Cantley, *Oncogene* 27:5497–5510; B.S. Jaiswal et al., *Cancer Cell* 16:463–474, 2009; D.W. Parsons et al., *Science* 321:1807–1812, 2008; and Y. Samuels and K. Ericson, *Curr. Opin. Oncol.* 18:77–82, 2006.

is lost in 30 to 40% of all human cancers. The resulting accumulation of high levels of phosphorylated inositide PIP3 results in the downstream consequences enumerated above. This behavior of *PTEN* in suppressing cell proliferation indicates that it functions as a tumor suppressor gene, a topic that we will explore in Chapter 7.

Since PI3K activation and PTEN loss lead to the same outcome—an increase in PIP3—we might expect that human tumors generally have one but not the other alteration, and this is indeed the case. Hence, PI3K activation makes PTEN inactivation unnecessary, and vice versa.

6.7 Ras-regulated signaling pathways: a third downstream pathway acts through Ral, a distant cousin of Ras

The third of the three major effector pathways downstream of Ras involves a pair of Ras-like proteins termed Ral-A and Ral-B, which share 58% sequence identity with Ras (see Sidebar 6.4). As is the case with Ras (see Figure 5.30), functional activation of these Ral proteins involves the replacement of bound GDP with GTP.

The communication between Ras and Ral is mediated by Ral guanine nucleotide exchange factors (Ral-GEFs), which can stimulate a Ral protein to shed its GDP and bind GTP, thereby echoing the effects of Sos on Ras. In addition to its GEF domain, each Ral-GEF has a pocket that can bind activated Ras. This association with Ras has two consequences for the Ral-GEF: Ras causes the localization of the Ral-GEFs near the inner surface of the plasma membrane or another cytoplasmic membrane; in addition, binding to Ras seems to cause a conformational shift in Ral-GEFs that activates their intrinsic guanine nucleotide exchange factor (GEF) activity.

The resulting activation of RalA and RalB proteins allows them to regulate targets further downstream in the signaling circuitry (**Figure 6.21**). For example, the Ral pathway can activate two proteins—Sec5 and Exo84—that contribute to Ras-mediated

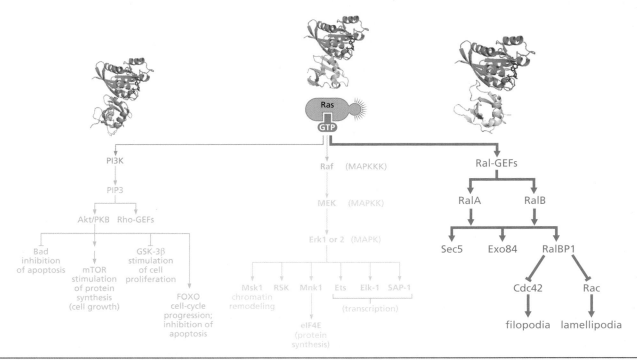

Figure 6.21 Ral and the control of the cytoskeleton The interaction of Ras with Ral-GEF (*pink, right*) enables the latter to activate the RalA and RalB GTPase proteins. The resulting GTP-bound Ral proteins proceed to activate a number of downstream pathways; prominent among them are those that activate Sec5, Exo84 and RalBP1. Sec5 and Exo84 contribute to the anchorage-independent growth of *Ras*-transformed cells *in vitro* and to their tumorigenicity *in vivo*, while RalBP1, acting via Cdc42 and Rac, reorganize the actin cytoskeleton and thereby affect cell motility. The bimolecular complex *above* illustrates the interaction of the effector domain of Ras (*red, purple*) with the Ras-binding domain of Ral-GEF, also known as Ral-GDS (*pink*). (From M.E. Pacold et al., *Cell* 103:931–943, 2000; N. Nassar et al., *Nature* 375:554–560, 1995; and L. Huang et al., *Nat. Struct. Biol.* 5:422–426, 1998.)

anchorage-independent growth; in addition, RalA/B signaling, acting via RalBP1, causes inactivation of the last two Rho proteins cited in the last section—Cdc42 and Rac. Rac can also emit mitogenic signals and, by stimulating the production of reactive oxygen species, antagonize the actions of several Rho proteins. Though still poorly understood, the Ral proteins are likely to play key roles in the motility that enables cancer cells to invade and metastasize.

The functional description of Ras as a pleiotropically acting oncoprotein can now be explained at a biochemical level by the actions of the three downstream signaling cascades that radiate from Ras. It would be pleasing if each of the phenotypes that we have associated with Ras oncoprotein action (for example, anchorage independence, survival, proliferation, biosynthetic rate, etc.) could be linked specifically to the actions of one of the three Ras effectors. In fact, things are more complicated: most Ras-induced phenotypes seem to be achieved through the collaborative actions of several of these effectors.

There are also two other dimensions of complexity. First, we have discussed Ras here as if it were a single protein, but there are indications that the four Ras proteins (one H-Ras, one N-Ras, and two K-Ras species) act in subtly different ways in distinct cytoplasmic locations. Thus, after Ras proteins become activated by Sos and its relatives near the plasma membrane, they may continue to emit signals from the membranes of endosomes generated by endocytosis and, thereafter, from other cytoplasmic membranes, notably those of the Golgi apparatus and the endoplasmic reticulum; at each of these locations, the various Ras proteins interact with greater or lesser efficiency with their downstream effectors. Second, our list of three Ras effectors (Raf, PI3K, and Ral-GEF) is hardly an exhaustive one. It is clear that Ras can interact with at least half a dozen other, still poorly characterized effectors that enable it to evoke yet other changes within the cell.

In any hierarchical organization, those individuals near the top exert far more power and influence than those lower down. By activating multiple control pathways concomitantly, a Ras oncoprotein (or a stimulatory signal emitted by one of its upstream controllers) induces many of the phenotypic changes acquired by a cell during the course of neoplastic transformation. In contrast, deregulation of one or another of the pathways downstream of Ras confers only a subset of these phenotypic changes and therefore yields far less growth advantage for a would-be cancer cell. This point is illustrated nicely by the mutant alleles of the gene encoding the B-Raf kinase, the close cousin of Raf, cited above, which is also activated by interaction with Ras (see Section 6.5). These mutations, which are found in more than half of human melanomas, create oncogenic *BRAF* alleles whose products, in spite of their greatly enhanced kinase activity (also described in Section 6.5), have only about one-fiftieth of the transforming power of the glycine-to-valine–substituted, activated Ras oncoprotein. Presumably, signaling components located further downstream, when altered by mutation, confer even less benefit (that is, selective advantage) on evolving pre-neoplastic cells and therefore are rarely encountered in the genomes of human tumor cells. Moreover, human tumors often show concomitant alterations of two of the effector pathways (for example, Raf and PI3K) rather than a mutation in the *Ras* gene itself; this apparently enables these downstream effectors, working in concert, to achieve cell-biological effects comparable to those elicited by a mutant Ras oncoprotein.

6.8 The Jak–STAT pathway allows signals to be transmitted from the plasma membrane directly to the nucleus

As mentioned in Section 5.7, study of a number of receptors for **cytokines**—growth factors that stimulate components of the hematopoietic system—has revealed receptor molecules that are constructed slightly differently from most of the growth factor receptors that were discussed in the last chapter. These particular receptor proteins do not carry covalently associated tyrosine kinase domains and instead form noncovalent complexes with tyrosine kinases of the Jak (Janus kinase) class (**Figure 6.22**). Included among these are the receptors for interferon (IFN), erythropoietin (EPO), and thrombopoietin (TPO). Interferon is an important mediator of immune cell function, as we

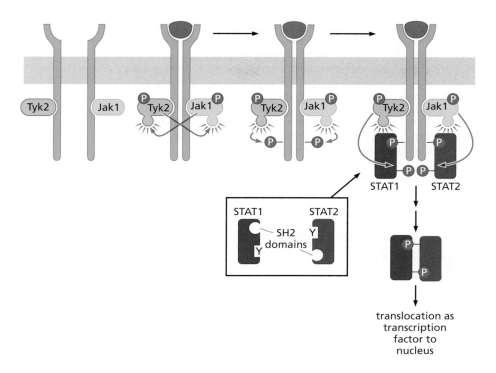

Figure 6.22 The Jak–STAT pathway
The Jak–STAT pathway depends upon the actions of Jak tyrosine kinases (for example, Jak1, Tyk2), which are attached noncovalently to a number of cytokine receptors, including those for interferons, erythropoietin (EPO), and thrombopoietin (TPO). Once a ligand has activated a receptor via dimerization, the Jaks transphosphorylate one another as well as the C-terminal tails of the receptors. The resulting phosphotyrosines attract STAT proteins, such as the STAT1 and STAT2 shown here (*dark blue, dark green*), which bind via their SH2 domains and become phosphorylated on specific phosphotyrosines, indicated here as Y's (*inset*), by the Jaks. Thereafter, the STATs dimerize, each using its SH2 domain to bind to the phosphotyrosine of its partner, and then translocate to the nucleus, where they operate as transcription factors to activate expression of key genes, for example, the gene encoding an interferon. Not shown here is the activation under certain conditions of Jaks that are bound to some G-protein–coupled receptors (GPCRs) and the phosphorylation of STATs directly by certain ligand-activated growth factor receptors (RTKs) as well as non-receptor tyrosine kinases.

will learn in Chapter 15; the other two control **erythropoiesis** (red blood cell formation) and **thrombopoiesis** (platelet formation), respectively.

Following ligand-mediated receptor dimerization, the Jak enzyme associated with each receptor molecule phosphorylates tyrosines on the cytoplasmic tails of the partner receptor molecule, much like the transphosphorylation occurring after ligand activation of receptors like the EGF-R and PDGF-R. The resulting phosphotyrosines attract and are bound by SH2-containing transcription factors termed STATs (signal transducers and activators of transcription), which then become phosphorylated by the Jaks. This creates individual STAT molecules that possess both SH2 groups and phosphotyrosines. Importantly, the SH2 groups displayed by the STATs have a specificity for binding the phosphotyrosine residues that have just been created on the STATs. Consequently, STAT–STAT dimers form, in which each STAT uses its SH2 group to bind to the phosphotyrosine of its partner. The resulting dimerized STATs migrate to the nucleus, where they function as transcription factors (see Figure 6.22).

The STATs activate target genes that are important for cell proliferation and cell survival. Included among these genes are *myc*, the genes specifying cyclins D2 and D3 (which enable cells to advance through their growth-and-division cycles), and the gene encoding the strongly anti-apoptotic protein Bcl-X$_L$. In addition to phosphorylating STATs, the Jaks can phosphorylate substrates that activate other mitogenic pathways, including the Ras-MAPK pathway described above.

Perhaps the most dramatic indication of the contribution of STATs to cancer development has come from a re-engineering of the STAT3 protein through the introduction of a pair of cysteine residues, which causes the resulting mutant STAT3 to dimerize spontaneously, forming stable covalent disulfide cross-linking bonds (**Figure 6.23**). These stabilized STAT3 dimers are structural and functional mimics of the dimers that are formed normally when STAT3 is phosphorylated by Jaks. This mutant STAT3 protein is now constitutively active as a nuclear transcription factor and can function as an oncoprotein that is capable of transforming NIH 3T3 and other immortalized mouse cells to a tumorigenic state.

STAT3 is also known to be constitutively activated in a number of human cancers. For example, in many melanomas, its activation is apparently attributable to Src, which is also constitutively active in these cancer cells; indeed, at least four other cousins

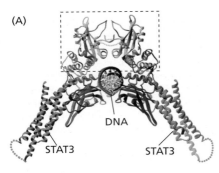

(A)

DNA

STAT3   STAT3

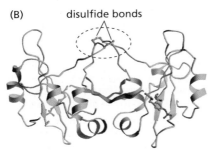

(B)   disulfide bonds

Figure 6.23 Constitutive activation of STAT3 (A) The STAT3 proteins normally dimerize through the SH2 domains in their C-terminal regions (*dashed line box*) to form dimeric transcription factors that sit astride the DNA double helix (*gray and brown circular structure, middle*). (B) Use of site-directed mutagenesis allowed the insertion of two cysteine residues in the C-terminal domain of STAT3, leading to the formation of dimers that are held together covalently by disulfide bonds. The C-terminal region depicted in panel A is shown here in more detail. The disulfide bonds (*yellow*) linking the two monomers are within the *dashed line oval*. (From J. Bromberg et al., *Cell* 98:295–303, 1999.)

of Src, all nonreceptor tyrosine kinases, can also phosphorylate and thereby activate STAT3. This highlights the fact that STATs can be activated in the cytoplasm by tyrosine kinases other than the receptor-associated Jaks. In the case of the melanoma cells, inhibition of Src, which leads to deactivation of STAT3, triggers apoptosis, pointing to the important contribution of STAT3 in ensuring the survival of these cancer cells.

In the majority of breast cancer cells, STAT3 has been found to be constitutively activated; in some of these, STAT3 is phosphorylated by Src and Jaks acting collaboratively. As is the case with melanoma cells, reversal of STAT3 activation in these breast cancer cells leads to growth inhibition and apoptosis. In almost all head-and-neck cancers, STAT3 is also constitutively activated, possibly through the actions of the EGF receptor. Alternatively, in many glioblastomas, activation of STAT3 is favored by the loss of a phosphatase responsible for removing its phosphotyrosine. STAT3 activation in lung and gastric cancers also appears to be critical to their neoplastic phenotype. After reaching the nucleus, the dimerized activated STAT proteins induce the expression of several hundred genes, many of which favor cell proliferation and protect against apoptosis. Taken together, these various threads of evidence converge on the notion that STATs represent important mediators of transformation in a variety of human cancer cell types.

6.9 Cell adhesion receptors emit signals that converge with those released by growth factor receptors

In the last chapter, we read that cells continuously monitor their attachment to components of the extracellular matrix (ECM). Successful tethering to the molecules forming the ECM, achieved via integrins, causes survival signals to be released into the cytoplasm that decrease the likelihood that a cell will enter into apoptosis. At the same time, integrin activation, provoked by signals originating inside the cell, can promote cell motility by stimulating integrin molecules located at specific sites on the plasma membrane to forge new linkages with the ECM. Consequently, integrins serve the three functions of (1) physically linking cells to the ECM, (2) informing cells whether or not tethering to certain ECM components has been achieved, and (3) facilitating motility by making and breaking contacts with the ECM. (In fact, there is a fourth but poorly understood function: Integrins and their cytoplasmic signaling partners serve as sensors of physical tension that arises between the ECM and the actin cytoskeleton. Cells respond in complex ways to this sensed tension.)

As described in the last chapter, integrins may cluster and form multiple links to the ECM in small, localized areas termed *focal adhesions* (see Figure 5.28A). Such clustering provokes activation of focal adhesion kinase (FAK), a nonreceptor tyrosine kinase like Src. FAK is associated with the cytoplasmic tails of the β subunits of certain integrin molecules and becomes phosphorylated, presumably by transphosphorylation, once integrins congregate in these focal adhesions. One of the resulting phosphotyrosine residues on FAK provides a docking site for Src molecules (**Figure 6.24A**). Src then proceeds to phosphorylate additional tyrosines on FAK, and the resulting phosphotyrosines serve as docking sites for yet other SH2-containing signaling molecules, including Grb2, Shc, PI3K, and PLC-γ. There is also evidence that Grb2, once bound to tyrosine-phosphorylated FAK, can recruit Sos into this complex; Sos then proceeds to activate its normal downstream Ras target. In fact, the full complexity of signaling downstream of integrins still awaits resolution (see Figure 6.24B).

(A)

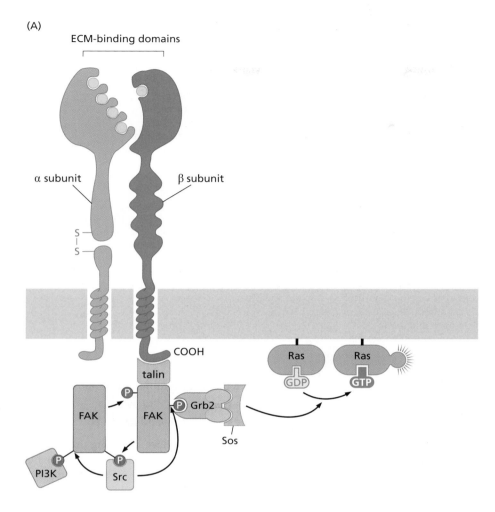

Figure 6.24 Integrin signaling
(A) The integrins are assembled as
α + β heterodimers. In addition to
physically linking the cytoskeleton to
the extracellular matrix (ECM; see Figure
5.28B), the binding by the ectodomain
of the heterodimer to components of the
ECM (*not shown*) triggers the association
of a series of cytoplasmic signal-
transducing proteins, such as
focal adhesion kinase (FAK), with the
β subunit. Resulting transphosphorylation
events and the attraction of SH2-
containing signaling molecules release
signals that activate many of the same
pathways that are turned on by ligand-
activated growth factor receptors. This
signaling is amplified by clustering of
multiple heterodimers following ECM
binding (*not shown*). (B) As indicated
here, the molecules shown in panel A
represent only part of a complex, still-
poorly understood collection of signal-
transducing molecules that become
physically associated with the cytoplasmic
domains of various integrin β subunits
following ECM binding. In total, almost
180 distinct proteins have been found
to be physically associated with focal
adhesions. (Adapted from C. Miranti
and J. Brugge, *Nat. Cell Biol.*
4:E83–E90, 2002.)

(B)

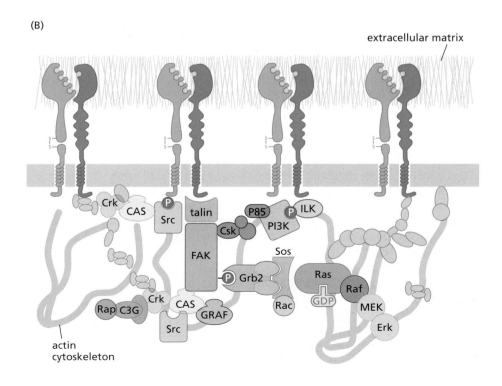

caspase-3

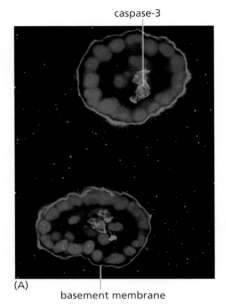

(A)

basement membrane

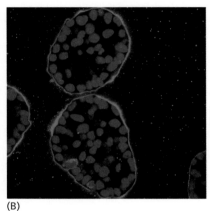

(B)

Figure 6.25 Anoikis and morphogenesis Anoikis serves during morphogenesis to excavate lumina (channels) in the middle of cylindrical aggregates of cells, thereby creating hollow ducts. When placed into suspension culture, human mammary epithelial cells (MECs) form globular clusters—acini—that mimic, at least superficially, the structures of mammary gland ducts. (A) An acinus is seen here in cross section surrounded by a layer of basement membrane (a form of extracellular matrix) containing laminin-5 (*red*) that the cells have secreted. Cells that succeed in tethering to this basement membrane via their integrins are healthy and have large nuclei (*blue*), while those that are deprived of direct attachment to this basement membrane undergo anoikis, resulting in the formation of a protein (activated caspase-3; *green*) that is indicative of active apoptosis (see Section 9.13); these central cells also exhibit fragmented nuclei, another sign of apoptosis. The loss of these cells will yield a lumen in this acinus. (B) In an early step of breast cancer development, normal anoikis and resulting excavation of the lumen of a mammary duct may be prevented by the actions of certain anti-apoptotic proteins. Here, expression of the anti-apoptotic Bcl-X$_L$ protein in MECs has resulted in the suppression of apoptosis, survival of luminal cells, and thus luminal filling, echoing the appearance of certain ductal hyperplasias and *in situ* carcinomas often seen in the initial steps of breast cancer development. Indeed, overexpressed EGF-R, which activates many of the same signaling pathways as integrins (see Figure 6.24B), also strongly reduces anoikis, once again yielding luminal filling (*not shown*). (Courtesy of J. Debnath and J. Brugge.)

The accumulated evidence suggests a pathway that has, minimally, the following components:

$$ECM \rightarrow integrins \rightarrow Sos \rightarrow Ras \rightarrow Raf, PI3K, and Ral-GEF$$

The parallels between integrin and receptor tyrosine kinase signaling are striking: Tyrosine kinase receptors bind growth factor ligands in the extracellular space, while integrins bind extracellular matrix components as ligands. Once activated, these two classes of sensors activate many of the same downstream signal transduction cascades. Moreover, some integrins co-localize with certain tyrosine kinase receptors; this association appears to be required for the activation of signaling by both types of receptors.

This convergence of signaling pathways appears to explain one of the important cell-biological effects of the Ras oncoprotein—its ability to enable cells to grow in an anchorage-independent fashion. It seems that cells normally depend on integrin signaling to provide a measure of activation of the normal Ras protein. In the absence of this signaling, cells come to believe that they have failed to associate with the extracellular matrix. This may block further advance of cells through their cell growth-and-division cycle or, more drastically, may cause cells to enter into *anoikis*, the form of programmed cell death that is triggered when cells lose their attachment to solid substrates. Avoidance of anoikis seems to represent one of the first steps in the initiation of breast cancers (**Figure 6.25**).

Taken together, these diverse observations suggest that oncogenic Ras promotes anchorage-independent proliferation by mimicking one of the critical downstream signals that result following tethering by integrins to the extracellular matrix. In effect, oncogenic Ras may delude a cell into believing that its integrins have successfully bound ECM components, thereby allowing a *ras*-transformed cell to proliferate, even when no such attachment has actually occurred.

6.10 The Wnt–β-catenin pathway contributes to cell proliferation

In most cell types, the RTK → Sos → Ras cascade appears to play a dominant role in mediating responses to extracellular mitogens, but it is hardly the only pathway having this result. Yet other, less studied pathways also confer responsiveness to mitogenic signals impinging on the cell surface. Prominent among these is the pathway controlled by Wnt factors, of which at least 19 can be found in various human tissues. In addition to transducing mitogenic signals, this pathway enables cells to remain in a relatively undifferentiated state—an important attribute of certain types of cancer cells.

As we learned in the last chapter (see Section 5.7), the molecular design of the Wnt pathway is totally different from that of the RTK → Sos → Ras → Raf → Erk pathway. Recall that Wnt factors, acting through Frizzled receptors, suppress the activity of glycogen synthase kinase-3β (GSK-3β). In the absence of Wnts, GSK-3β phosphorylates several key substrate proteins, which are thereby tagged for destruction. The most important of these substrates is β-catenin, normally a cytoplasmic protein that exists in three states. It may be bound to the cytoplasmic domain of cell–cell adhesion receptors, notably E-cadherin (**Figure 6.26A**); E-cadherin molecules from adjacent cells may associate with one another to form **adherens junctions**—cell-to-cell contacts that help to ensure the structural integrity of epithelial sheets and impede

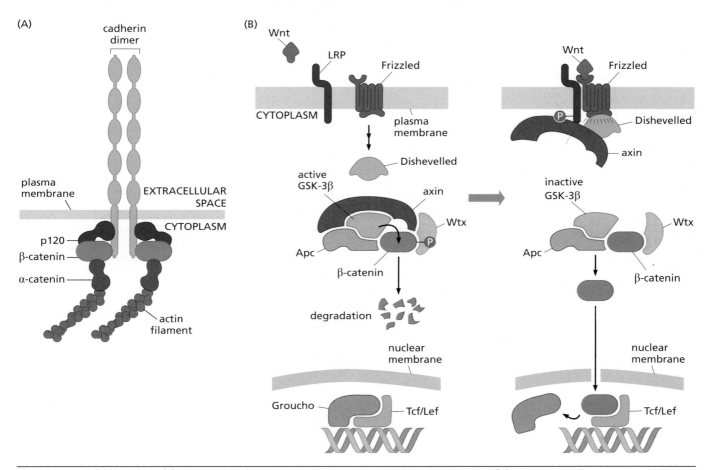

Figure 6.26 Multiple roles of β-catenin (A) Like many integrins, the cadherins (*light green*) are transmembrane proteins that form attachments in the extracellular space and become linked, via intermediary proteins, to the actin cytoskeleton (*red*). The cytoplasmic tail of E-cadherin, for example, is known to associate directly with p120 and β-catenin (*dark blue, light blue*). These associate in turn with a monomer of α catenin (*green*). The latter exists in some type of dynamic equilibrium with α catenin dimers bound to the actin cytoskeleton. The precise mechanism of linkage is still poorly resolved. Breakdown of this multi-subunit "*adherens junction*" complex often accompanies the acquisition of highly malignant properties by carcinoma cells. (B) As indicated in Figure 5.23, the Wnt proteins (*red*), by binding to their Frizzled receptors (*dark green*) and the LRP co-receptor (*purple*), act via Dishevelled to suppress the activity of glycogen synthase kinase-3β (GSK-3β,

pink). This prevents GSK-3β from phosphorylating, among other substrates, β-catenin, which therefore escapes degradation and accumulates in the cytoplasm and in the nucleus. Once in the nucleus, β-catenin associates with Tcf/Lef, transcription factors, displacing the Groucho repressor protein and thereby acting expression of a variety of genes, including those involved in cell proliferation. Not shown here are additional proteins that associate with β-catenin, notably BCL9-2/BL9 and Pygopus, and enable β-catenin to form productive transcription-promoting complexes with Tcf/Lef in the nucleus. The pathway depicted here is often termed the *canonical* Wnt pathway to distinguish it from a group of alternative Wnt pathways that are termed collectively *non-canonical* Wnt pathways and are described in outline in Supplementary Sidebar 6.4.

metastasis (see Section 14.3). Alternatively, in a fully unrelated role, β-catenin exists in a soluble pool in the cytosol, where it turns over very rapidly, having a lifetime of less than 20 minutes. And finally, β-catenin operates in the nucleus as an important component of a transcription factor.

We focus here on the pools of β-catenin molecules that are not associated with E-cadherin. Normally, shortly after its synthesis, a β-catenin molecule will form a multiprotein complex with three other cellular proteins—Apc (the adenomatous polyposis coli protein), Wtx, and axin. These proteins help to bring β-catenin together with its GSK-3β executioner (see Figure 6.26B). By phosphorylating β-catenin, GSK-3β ensures that β-catenin will be tagged by **ubiquitylation**, a process that ensures the rapid destruction of β-catenin; this explains the low steady-state concentrations of β-catenin normally present in the cytosol. We will encounter this degradation system later in more detail (Supplementary Sidebar 7.5).

When Wnt signaling is activated, however, GSK-3β firing is blocked and β-catenin is saved from rapid destruction; its half-life increases from less than 20 minutes to 1–2 hours, and its steady-state concentrations increase proportionately. Many of the accumulated β-catenin molecules then move into the nucleus and bind to Tcf/Lef proteins (see Figure 6.26B). The resulting multi-subunit transcription factor complexes proceed to activate expression of a number of important genes, including those encoding critical proteins involved in cell growth and proliferation, such as cyclin D1 and Myc, that we will study later in Chapter 8.

In addition, GSK-3β can phosphorylate other crucial growth-regulating proteins besides β-catenin. By phosphorylating cyclin D1, GSK-3β also labels this growth-promoting protein for rapid degradation. Hence, the Wnt pathway actually modulates cyclin D1 expression at both the transcriptional and the post-translational levels. These various actions of Wnts indicate that, in addition to their originally discovered powers as morphogens, they can also act as potent mitogens, much like the ligands of many tyrosine kinase receptors. Moreover, as we will learn in Chapter 7, high levels of β-catenin in intestinal stem cells ensure that these cells remain in an undifferentiated, stem cell-like state, rather than developing into the specialized cells that line the wall of the intestine. Such high β-catenin levels turn out to be critical to the formation of colon carcinomas.

The alternative role of β-catenin, mentioned in passing above, is to form part of the cytoplasmic complex through which cell surface adhesion receptors, notably E-cadherin, become physically linked with the cytoskeleton (see Figure 6.26A). The two unrelated roles of β-catenin represent a still-unresolved puzzle. In epithelial cells, it appears that the great majority of β-catenin molecules are associated with the cytoplasmic tails of E-cadherin molecules at the plasma membrane, where the latter form cell–cell adherens junctions. Accordingly, in cells that have lost E-cadherin, one might expect β-catenin to be liberated and hence available for nuclear translocation and transcriptional activation, and indeed this is what is often observed. However, in some cancer cells that have lost their E-cadherin protein through mutation, nuclear β-catenin signaling does not seem to be elevated above normal levels; these latter observations suggest that the two pools of β-catenin might be regulated independently and do not influence each other. Alternatively, in order to enter the nucleus, once β-catenin is liberated from adherens junctions, it must then escape degradation driven by GSK-3β, whose activity is controlled by a different set of regulatory signals. Then there is the other puzzle: Why has evolution invested two such unrelated functions in a single protein? (In worms, these two functions are carried out by distinct proteins encoded by separate genes.)

In many human breast cancers, expression of certain Wnts is increased four- to tenfold above normal, suggesting the operations of autocrine and paracrine growth-stimulatory pathways. There is clear evidence of nuclear translocation of β-catenin in approximately 20% of advanced prostate carcinomas. Moreover, in 5 to 7% of prostate carcinomas, mutations of the β-catenin gene have yielded a protein that can no longer be phosphorylated by GSK-3β, and as a consequence, β-catenin now accumulates to high levels in the nucleus (**Figure 6.27**). Such mutations have also been documented

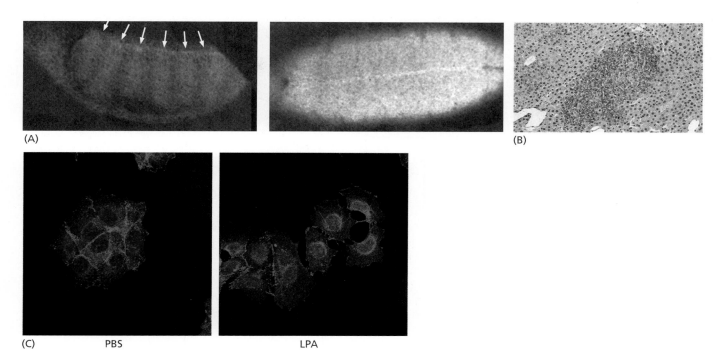

(A)

(B)

(C)   PBS   LPA

Figure 6.27 Regulation of β-catenin signaling Increases in β-catenin levels and signaling are achieved through diverse mechanisms. (A) *Drosophila* embryos usually express β-catenin in a striped pattern (*arrows, left*). However, when the gene encoding GSK-3β is inactivated, high levels of β-catenin accumulate uniformly throughout the embryo (*right*). (B) Mice exposed to certain liver carcinogens develop hepatocellular carcinomas (hepatomas) in which β-catenin levels rise dramatically because the protein is mutant, enabling it to resist phosphorylation by GSK-3β and resulting degradation. The resulting cytoplasmic accumulation of β-catenin (*brown*) in this focus of pre-malignant hepatocytes is a prelude to increased levels of nuclear β-catenin that drive cell proliferation through association with Tcf/Lef transcription factors. (C) The dynamic interaction between the β-catenin sequestered in the adherens junctions and the form of this protein that conveys Wnt-dependent signals to the nucleus is illustrated here. When A431 human epidermoid carcinoma cells are propagated in monolayer culture and placed briefly in phosphate-buffered saline (PBS), β-catenin (*red*) is largely localized to the adherens junctions between adjacent cells (*left*). However, when these junctions are dissolved following brief exposure of the cells to lysophosphatidic acid (LPA), both E-cadherin and β-catenin are internalized and β-catenin is relocalized to perinuclear sites, (*right*), where it now becomes available for activation and nuclear translocation by Frizzled receptors that have bound Wnt ligands (*not shown*; see Figure 6.26). (A, courtesy of E.F. Wieschaus. B, from T.R. Devereux et al., *Oncogene* 18:4726–4733, 1999. C, from Y. Kam and V. Quaranta. *PloS One* 4:e4580, 2009.)

in carcinomas of the liver, colon, endometrium, and ovary, as well as in melanomas. And as we will see in the next chapter, in virtually all colon carcinomas, β-catenin degradation is flawed, due to defects in one of the proteins—Apc—that facilitate its degradation (see Figure 6.26B). It is likely that this pathway is also deregulated in other human cancers by mechanisms that have not yet been uncovered. This Wnt–β-catenin pathway is sometimes termed the "canonical Wnt" pathway to distinguish it from another set of signaling pathways that are activated by a subset of the Wnt ligands, are independent of β-catenin function, and are grouped together under the rubric of "non-canonical Wnt signaling." Because their role in cancer pathogenesis has to date been found to be minor, they are discussed in outline in Supplementary Sidebar 6.4.

6.11 G-protein–coupled receptors can also drive normal and neoplastic proliferation

As we read in the last chapter (Section 5.7), G-protein–coupled receptors (GPCRs) are transmembrane proteins that weave their way back and forth through the plasma membrane seven times. Upon binding their extracellular ligands, each of these "serpentine" receptors activates one or more types of cytoplasmic heterotrimeric G-protein, so named because of its three distinct subunits (Gα, Gβ, and Gγ), the first of which binds either GDP or GTP. As is the case with the Ras protein, the activated state of Gα is achieved when it binds GTP.

Figure 6.28 G-protein–coupled receptors The seven-membrane-spanning receptors are able to activate a variety of heterotrimeric G proteins that differ largely in the identity of their Gα subunits. Once stimulated by a GPCR, the Gα subunit separates from Gβ and Gγ and proceeds to activate or inhibit a variety of cytoplasmic enzymes, only two of which (adenylyl cyclase and phospholipase C-β) are shown here (*right*). These enzymes, in turn, can have mitogenic or anti-mitogenic influences, depending upon the cell type. At the same time, the Gβ + Gγ dimers can activate their own set of effectors, including phosphatidylinositol 3-kinase γ (PI3Kγ), phospholipase C-β (PLC-β), and Src. Following GPCR activation, another group of receptor-associated proteins—the β arrestins 1 and 2—proceed to activate a series of kinases (*left*) that are known to play key roles in cell proliferation and survival. The relative importance in tumor development of these two modes of GPCR signaling, involving heterotrimeric G proteins and β-arrestins, remains to be determined. Moreover, this diagram provides only a hint of the diverse array of G proteins and their effectors that have been discovered.

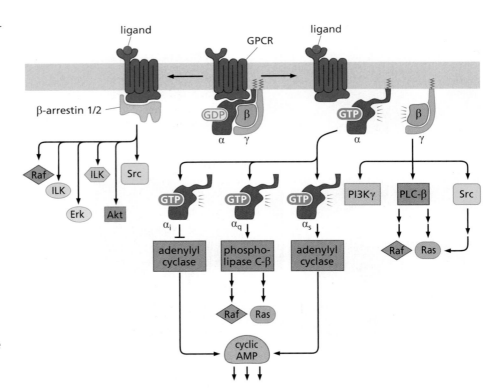

Once stimulated by a G-protein–coupled receptor (GPCR), the Gα subunit of a heterotrimeric G-protein dissociates from its two partners, Gβ and Gγ, and proceeds to activate a number of distinct cytoplasmic enzymes (**Figure 6.28**). Included among these are adenylyl cyclase (which converts ATP into cyclic AMP) and phospholipase C-β (PLC-β), which cleaves PIP3 to yield diacylglycerol (DAG) and inositol triphosphate (IP$_3$) (see Figure 6.16B). The latter are potent second messengers that can function to stimulate cell proliferation.

There is even evidence that the Src kinase can be activated by certain GTP-bound Gα subunits. At the same time, complexes of the other two subunits of the heterotrimeric G-protein—Gβ and Gγ—have been found to stimulate yet other important mitogenic signaling proteins, such as one form of phosphatidylinositol 3-kinase (PI3K). The ability of GPCRs to activate mitogenic pathways suggests that deregulation of signaling by these receptors and associated G-proteins may well contribute to cell transformation and tumorigenesis—a speculation that has been borne out by analyses of certain types of human cancer cells (see **Table 6.5**).

In recent years, a second mode of signaling by GPCRs has been described that may prove to be as important in cancer pathogenesis as the signal transduction mediated by heterotrimeric G-proteins. This new signaling channel derives from earlier studies of rhodopsin signaling in the retina, which revealed that signal emission by this GPCR in response to photon stimulation is halted by a protein termed arrestin; this molecule mediates a form of negative feedback to ensure active GPCR signaling is confined to a brief period of time following activation by a photon. Subsequently, two paralogs of retinal arrestin, termed β-arrestin 1 and 2, have been found to be expressed in virtually all non-retinal cell types throughout the body. In a variety of cell types, ligand-mediated activation of a GPCR, in addition to provoking the negative-feedback actions of β-arrestins, causes the latter to activate a diverse array of downstream signaling proteins, including kinases that are known to be important in driving cell survival and proliferation (see Figure 6.28); these kinases appear frequently in our discussions of cell transformation.

The role of GPCRs in human cancer pathogenesis is highlighted by the behavior of small-cell lung carcinomas (SCLCs), a common tumor of cigarette smokers. SCLC cells release a number of distinct peptide factors, some with **neuropeptide**-like properties. In some SCLCs, the tumor cells may simultaneously secrete bombesin (also

Table 6.5 **G-protein–coupled receptors and G-proteins involved in human cancer pathogenesis**

G-protein or receptor	Type of tumor
Activating mutations affecting G-proteins	
$G\alpha_s$	thyroid adenomas and carcinomas, pituitary adenomas
$G\alpha_{i2}$	ovarian tumors, adrenal cortical tumors
$G\alpha_q$	melanocyte-derived tumors, uveal melanomas
$G\alpha_{11}$	uveal melanomas
Activating mutations affecting G-protein–coupled receptors	
Thyroid-stimulating hormone receptor	thyroid adenomas and carcinomas
Follicle-stimulating hormone receptor	ovarian tumors
Luteinizing hormone receptor	Leydig cell hyperplasias
Cholecystokinin-2 receptor	colorectal carcinomas
Ca^{2+}-sensing receptor	various neoplasms
GPR98	melanomas
GRM3	melanomas
Autocrine and paracrine activation	
Neuromedin B receptor	SCLC
Neurotensin receptor	prostate carcinomas and SCLC
Gastrin receptor	gastric carcinomas and SCLC
Cholecystokinin receptor	pancreatic hyperplasias and carcinomas, gastrointestinal carcinomas, and SCLC
Virus-encoded G-protein–coupled receptors	
Kaposi's sarcoma herpesvirus (HHV-8)	Kaposi's sarcoma
Herpesvirus saimiri	primate leukemias and lymphomas
Jaagsiekte sheep retrovirus	sheep pulmonary carcinomas

Adapted in part from M.J. Marinissen and J.S. Gutkind, *Trends Pharmacol. Sci.* 22:368–376, 2001.

known as gastrin-releasing peptide, or GRP), bradykinin, cholecystokinin (CCK), gastrin, neurotensin, and vasopressin. At the same time, these cells display the GPCRs that recognize and bind these released factors, resulting in the establishment of multiple autocrine signaling loops.

The experimental proof that these autocrine loops are actually responsible for driving the proliferation and/or survival of the SCLC cells is straightforward: SCLC cells can be incubated *in vitro* in the presence of an antibody that binds and neutralizes a secreted autocrine growth factor, such as GRP. In a number of SCLC cell lines, this treatment results in the rapid cessation of growth and even in apoptosis. Such a response indicates that these cancer cells depend on a GRP-based autocrine signaling loop for their survival, and it suggested a novel therapy for SCLC patients. In twelve SCLC patients who were treated with a GRP-neutralizing antibody, one patient showed a complete remission, while four patients exhibited a partial shrinkage of their tumors; this therapeutic strategy has not, however, been further pursued in recent years.

In another class of neoplasias—thyroid adenomas and some thyroid carcinomas—the gene encoding the thyroid-stimulating hormone receptor (TSHR), another GPCR, is often found to carry a point mutation. This leads to constitutive, ligand-independent firing of the TSH receptor, which in turn results in the release of strong mitogenic signals into the thyroid epithelial cells (see Table 6.5). In yet other tumors of the thyroid gland, a Gα subunit has suffered a point mutation that functions much like the mutations activating Ras signaling, by depriving this Gα subunit of the ability to shut itself off through its intrinsic GTPase activity. Altogether, at least 10 of the 17 human genes encoding Gα subunits have been found to function as oncogenes in certain cell types and in various human malignancies.

The most bizarre subversion of these G-protein–coupled receptors occurs during infections by certain herpesviruses, such as human herpesvirus type 8 (HHV-8), also known as the Kaposi's sarcoma herpesvirus (KSHV; refer to Supplementary Sidebar 3.4). This virus is responsible for the vascular tumors that frequently afflict AIDS patients. At one point in its evolutionary past, this virus acquired a cellular gene specifying a G-protein–coupled receptor. The viral form of this gene has been remodeled, so that the encoded receptor signals in a ligand-independent manner. Among other consequences, this signaling causes HHV-8–infected endothelial cells (lining the walls of blood vessels) to secrete vascular endothelial growth factor (VEGF); the released VEGF then creates an autocrine signaling loop by binding to its cognate receptors on the surface of these endothelial cells and driving their proliferation.

6.12 Four additional "dual-address" signaling pathways contribute in various ways to normal and neoplastic proliferation

Both Jak–STAT and Wnt–β-catenin signaling represent "dual-address" pathways, which operate by modifying and dispatching to the nucleus proteins that normally reside in the cytoplasm; once in the nucleus, key proteins in these two pathways (that is, STATs and β-catenin) function as components of specific transcription factors to drive gene expression. In fact, four other dual-address signaling channels play important roles in the pathogenesis of human cancers. Here, we will briefly summarize the downstream actions of these four pathways (Figure 6.29) and cite examples of how each is deregulated in one or another type of human tumor (Figure 6.30).

Nuclear factor-κB The first indication of the importance of this pathway to cancer pathogenesis came from the discovery of the *rel* oncogene in a rapidly transforming turkey retrovirus responsible for reticuloendotheliosis, a lymphoma of the B-cell lineage. Later investigations into the transcription factors responsible for regulating immunoglobulin gene expression revealed Rel to be a member of a family of transcription factors that came to be called collectively NF-κBs. These proteins form homo- and heterodimers in the cytoplasm.

The most common form of NF-κB is a heterodimer composed of a p65 and a p50 subunit. Usually, NF-κB is sequestered in the cytoplasm by a third polypeptide named IκB (inhibitor of NF-κB; see Figure 6.29A); while in this state, this signaling system is kept silent. However, in response to signals originating from a diverse array of sources, IκB becomes phosphorylated, and thereby tagged for rapid destruction. (Recall that β-catenin suffers the same fate following its phosphorylation in the cytoplasm.) As a result, NF-κB is liberated from the clutches of IκB, migrates into the nucleus, and proceeds to activate the expression of a cohort of as many as 500 target genes.

The kinase that tags IκB for destruction (named IκB kinase or simply IKK) and thereby activates NF-κB signaling is itself stimulated by signals as diverse as tumor necrosis factor-α and interleukin-1β (extracellular factors involved in the inflammatory response of the immune system), lipopolysaccharide (a sign of bacterial infection), reactive oxygen species (ROS), anti-cancer drugs, and gamma irradiation. In the context of cancer, NF-κBs affect cell survival and proliferation. Once they have arrived in the nucleus, the NF-κBs can induce expression of genes specifying a number of key anti-apoptotic proteins, such as Bcl-2 and IAP-1 and -2; we will learn more about these proteins in Chapter 9. At the same time, NF-κBs function in a mitogenic fashion by inducing expression of the *myc* and cyclin D1 genes, components of the cell cycle machinery that we discuss in Chapter 8. Hence, NF-κBs can protect cancer cells from apoptosis (programmed cell death) and, at the same time, drive their proliferation.

While components of the NF-κB signaling cascade are rarely found in mutant form in human cancers, this pathway is frequently found to be constitutively activated. In breast cancers, for example, the pathway is often highly active, ostensibly through the

Figure 6.29 Four different "dual-address" signaling pathways (A) The NF-κB family of transcription factors, which function as heterodimers (*light and dark blue*), are sequestered in the cytoplasm by IκB. A variety of receptors and afferent signals activate IKK, which phosphorylates IκB (*red*), tagging it for proteolytic degradation. Now liberated from IκB, NF-κB can translocate to the nucleus, where it activates a broad constituency of genes, including anti-apoptotic and mitogenic genes. (B) The Notch receptor can bind ligands belonging to the Delta-like and Jagged families. Ligand binding causes two proteolytic cuts in Notch, liberating a cytoplasmic fragment that can translocate to the nucleus, where it functions as part of a transcription factor complex. The Notch ligand and the ectodomain of the receptor are then internalized and then degraded by the ligand-presenting cell (see Figure 5.21). (C) In this simplified depiction of Patched (Ptc)-Smoothened signaling, the binding of a Hedgehog (Hh) ligand releases Smo from Ptc control. Smo then allows Gli to escape degradation and accumulate in the nucleus, where it functions as a transcriptional inducer. (D) Binding of the TGF-β ligand to the type II TGF-β receptor brings the type II and type I receptors together and results in the phosphorylation of the type I receptor. The latter, now activated, phosphorylates cytosolic Smad2 and/or Smad3 proteins; these phospho-Smads then bind to Smad4. The resulting heterotrimeric Smad complexes (composed of Smad 2/2/4, 2/3/4, or 3/3/4 subunits) translocate to the nucleus, where they bind a tetranucleotide DNA segment and associate with other adjacently bound transcription factors to induce or repress gene expression.

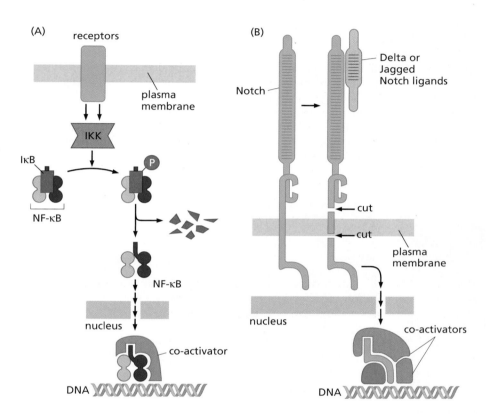

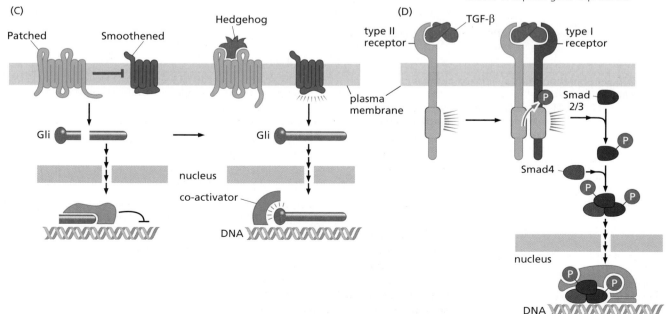

actions of IKKε, which is overexpressed in ~30% of these tumors. NF-κB seems to play its most important role in malignancies of various lymphocyte lineages. The *REL* gene, which encodes one of the subunits of NF-κB, is amplified in about one-fourth of diffuse large B-cell lymphomas, resulting in a 4- to 35-fold increase in expression of its gene product. Translocations affecting the *NFKB2* locus have been found frequently in B- and T-cell lymphomas and in myelomas (tumors of antibody-producing cells). And deregulation of this pathway can often be observed early in the development of low-grade malignant growths (see Figure 6.30A).

Notch Study of another unusual signaling pathway, this one controlled by the Notch protein, traces its roots to the discovery in 1919 of an allele of a *Drosophila* gene that causes notches to form in the edges of this fly's wings. Only many decades later was it realized that Notch is a transmembrane protein; four different varieties of Notch (products of four different genes) are expressed by mammalian cells. As mentioned in the last chapter, after Notch, acting as a cell surface receptor, binds a ligand (a Jagged or Delta-like protein), it undergoes two proteolytic cleavages, one in its ectodomain, the other within its transmembrane domain. The latter cleavage liberates a largely cytoplasmic protein fragment from its tethering to the plasma membrane. This fragment of Notch then migrates to the nucleus of targeted cells, where it functions, together with partner proteins, as a transcription factor (see Figure 6.29B).

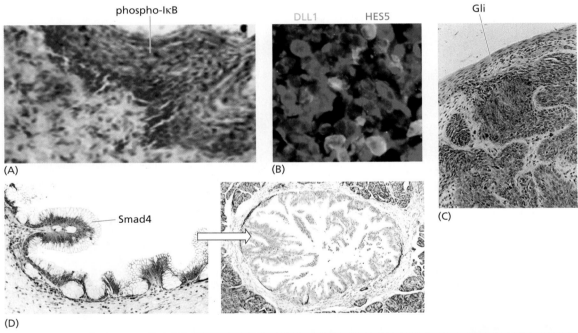

Figure 6.30 Deregulation of dual-address pathways in human cancers (A) NF-κB pathway. Immunostaining for the phosphorylated form of IκB reveals the localized activation of the NF-κB pathway (*brown area*) early in the multi-step development of a human cervical carcinoma, in this case in a low-grade squamous intraepithelial lesion. (B) Notch pathway. This pathway is often deregulated through localized expression of Notch ligands. Characteristic downstream targets of Notch-stimulated transcription are a pair of bHLH transcription factors (see Figure 8.27) termed HES and HEY. In human glioblastomas, the DLL1 Notch ligand is expressed by endothelial cells forming the microvasculature of the tumor (*green*). Close to these endothelial cells are neoplastic glioblastoma cells that have responded to this DLL1-mediated activation of their Notch1 and Notch2 receptors by expressing HES5, a clear indicator of activated Notch signaling (*red orange*). This *juxtacrine* signaling also causes the glioblastoma cells to express Nestin (*not shown*), a marker of both normal and neoplastic stem cells in the brain. (C) Hedgehog pathway. This pathway is deregulated in the majority of basal cell carcinomas of the skin (BCCs). Its deregulation is indicated by the elevated concentration of Gli1 protein in BCCs, detected here by immunostaining (*brown*), while being barely evident in normal cells (*light pink*). Northern (RNA) blot analyses (*not shown*) indicate as much as a 40-fold elevation in Gli1 mRNA in BCCs compared with normal skin. (D) TGF-β pathway. This pathway is inactivated in a wide variety of carcinomas through several mechanisms. During pancreatic carcinoma progression, low-grade pancreatic intraepithelial neoplasias (PanIN-1) usually express Smad4 (*brown, left panel*), a critical transcription factor in TGF-β signaling (see Figure 6.29D). These can progress (*arrow*) to high-grade PanIN-3 lesions, which often lose Smad4 expression, typically because of mutation of the encoding gene (*center of right panel*). (A, from A. Nair et al., *Oncogene* 22:50–58, 2003. B, from T.S. Zhu et al., *Cancer Res.* 71:6061–6062, 2011. C, from L. Ghali et al., *J. Invest. Dermatol.* 113:595–599, 1999. D, from R.E. Wilentz et al., *Cancer Res.* 60:2002–2006, 2000.)

This signaling system operates using biochemical mechanisms that are clearly very different from those governing signaling by tyrosine kinase receptors. Each time a Notch receptor binds its ligand, the receptor undergoes irreversible covalent alterations, specifically, proteolytic cleavages. Hence, receptor firing occurs in direct proportion to the number of ligands encountered in the extracellular space, and each Notch receptor can presumably fire only once after it has bound its ligand. Tyrosine kinase receptors, in dramatic contrast, release signals repetitively over an extended period of time following ligand binding, and therefore can greatly amplify the signal initiated by their growth factor ligands.

Truncated forms of *Notch* that specify only the cytoplasmic domain of the Notch protein are potent oncogenes for transforming cells *in vitro*. This observation suggests that altered forms of Notch contribute to human cancer pathogenesis. Indeed, overexpression of one or another of the Notch proteins is seen in the great majority of cervical carcinomas, in a subset of colon and prostate carcinomas, and in squamous carcinomas of the lung. This overexpression is often accompanied by nuclear localization of the cytoplasmic cleavage fragment of Notch, indicating that active signaling through this pathway is occurring in tumor cells. Increased expression of two ligands of Notch, termed Jagged and Delta, has also been found in some cervical and prostate carcinomas (Figure 6.30B). More dramatically, in about 60% of T-cell acute lymphocytic leukemias (T-ALLs), constitutively active forms of Notch are found; these result from genetic deletions of the portion of the *NOTCH-1* gene encoding the extracellular domain of the protein. Some experiments also suggest that Notch signaling contributes to transformation by *ras* oncogenes. In general, however, variant forms of *Notch* seem to act as oncogenes in hematopoietic malignancies but tend to function as tumor suppressors in carcinomas; hence, the contributions of Notch signaling to cancer development seem to be highly dependent on cellular context.

Hedgehog As was discussed previously, binding of the Patched receptor by its ligand, Hedgehog (Hh), somehow alters the interactions between Patched (Ptc) and Smoothened (Smo). As a consequence, Smoothened and Gli then accumulate in the primary cilium, often at its tip (see Figure 5.22). Normally, in the absence of intervention by Smoothened, the Gli precursor protein is cleaved into two fragments, one of which moves into the nucleus, where it functions as a transcriptional repressor. However, following Hedgehog stimulation, Smoothened and Gli interact in the primary cilium, where the Gli precursor is protected from cleavage. The resulting intact Gli protein migrates to the nucleus, where it serves as an activator of transcription (see Figure 6.29C).

Gli was first discovered as a highly expressed protein in glioblastomas (whence its name). When overexpressed, Gli and several closely related cousins can function as oncoproteins. Subsequent research into this pathway has revealed other instances where its malfunctioning contributes to tumor development. For example, germ-line inactivating mutations of the human patched (*PTCH*) gene cause Gorlin syndrome, an inherited cancer susceptibility syndrome, which involves increased risk of multiple basal cell carcinomas of the skin as well as other tumors, notably medulloblastomas—tumors of cells in the cerebellum. Such inactivating mutations prevent Patched from inhibiting Smoothened, giving the latter a free hand to dispatch an uninterrupted stream of active Gli protein to the nucleus.

As many as 50% of sporadic basal cell carcinomas of the skin (which occur because of somatic rather than germ-line mutations) carry mutant *PTCH* or *SMO* alleles—the latter encoding Smoothened. These basal cell carcinomas are the most common form of cancer in Western populations and fortunately are usually benign (see Figure 6.30C). Somatically mutated alleles of *PTCH* have also been found in a variety of other tumors, including medulloblastomas and meningiomas, as well as breast and esophageal carcinomas. These somatic mutations, like those present in mutant germ-line alleles of *PTCH*, seem to compromise Patched function, once again permitting Smoothened to constitutively activate Gli transcription factors. Recently, mutant alleles of *SUFU*, which specifies a protein that inhibits Gli, have also been described in medulloblastomas. Moreover, it is now clear that there are at least three Gli proteins, two of which stimulate transcription while the third inhibits it.

These cancers, all involving mutations that affect the structure of various components of the Hedgehog pathway, represent only a small portion of the human tumors in which this signaling cascade is hyperactivated. Thus, a survey of esophageal, gastric, biliary tract, and pancreatic carcinoma cell lines and tumors has revealed that virtually all of these expressed significant levels of the Patched receptor as well as its ligands (either Sonic Hedgehog or Indian Hedgehog). Often the ligands are produced by the epithelial cells within carcinomas while the Patched receptor is expressed in nearby stromal cells, suggesting the operations of a paracrine signaling channel. This notion was confirmed by demonstrating elevated nuclear levels of the Gli factor, the downstream product of the activated pathway, in stromal cells rather than the epithelial carcinoma cells themselves. Moreover, treatment of tumors with anti-Hedgehog antibody (which sequesters both types of ligand and interrupts the signaling) caused cessation of proliferation and/or death of tumor cells, suggesting that stromal cells stimulated by Hedgehog ligands reciprocate by sending mitogenic and survival signals back to nearby carcinoma cells. Similar findings have been reported for small-cell lung carcinomas (SCLCs). Together, these reports point to an important role of this signaling pathway in the tumors that arise in many of the tissues deriving from the embryonic gut.

TGF-β A fourth signaling pathway that involves the dispatch of cytoplasmic proteins to the nucleus is represented by the pathway leading from TGF-β receptors (Section 5.7). TGF-β and the signaling pathway that it controls appear to play major roles in the pathogenesis of many if not all carcinomas, both in their early stages, when TGF-β acts to arrest the growth of many cell types, and later in cancer progression, when it contributes, paradoxically, to the phenotype of tumor invasiveness. We will defer detailed discussions of this important pathway until Section 8.10 and Chapter 14. For the moment, suffice it to say that activation of this pathway leads to dispatch of Smad transcription factors to the nucleus, where they can bind a specific tetranucleotide sequence in the DNA and associate with adjacently bound transcription factors to induce the expression of a large constituency of genes and repress many others (see Figure 6.29D). In the absence of critical Smads, epithelial cancer cells can escape the growth-inhibitory actions of TGF-β and thrive—a state that is often observed in the precursors of invasive human pancreatic carcinomas (see Figure 6.30D).

The Smad transcription factors (TFs) are most unusual, because they recognize and bind only a tetranucleotide sequence in chromosomal DNA, as mentioned above. This binding, on its own, is too weak to maintain the association of Smads with the DNA. What happens, instead, is that the Smads often bind DNA segments immediately adjacent to the binding sites of other TFs. The resulting lateral protein–protein interactions between the Smad and its TF partners ensure that binding by Smads and its partners to DNA is cooperative and quite strong. To date, Smads have been found to form such partnerships with more than 130 distinct TFs; the functions of most of these have not been explored.

6.13 Well-designed signaling circuits require both negative and positive feedback controls

Virtually all well-regulated signal-processing circuits contain inhibitory components that counterbalance and modulate the behavior of circuit components that emit positive signals. Without these negative controls, positive signals, such as those that promote cell proliferation, may run rampant; in the case of mitogenic signals, this can lead to the runaway proliferation that we associate with cancer cells.

In principle, an effective strategy for constraining the signals flowing through the mitogenic signaling pathways that we have encountered here could rely on limiting the amounts of growth factor molecules to which a cell is exposed. That way, even if this cell displayed abundant growth factor receptors on its surface, the limited number of GF ligand molecules it could encounter would ensure only minimal firing by these receptors and therefore limited activation of downstream signaling pathways.

However, a more precisely targeted and effective strategy for regulating the flow of such mitogenic signals depends on modulating signal processing *within* the responding cell. Often, the inhibitory components operating in an intracellular signaling circuit are functionally inactive until the circuit becomes active; however, once signaling begins flowing through the circuit, the inhibitory proteins become activated and proceed to reduce or block further signaling. This type of inducible inhibition is often termed **negative feedback** and is an important feature of many, possibly all, signal-processing circuits within cells.

Actually, without noting this feature explicitly in the last two chapters, we have encountered a number of proteins that are involved in negative-feedback loops. For example, when a GF receptor becomes activated by ligand binding and autophosphorylation, the SH2 domain of Ras-GAP (see Figure 5.30) allows this protein to bind to one of the phosphotyrosines of the receptor. This places Ras-GAP in a position where it can trigger the GTPase function and thus the inactivation of Ras (**Figure 6.31A**); the latter, which had recently been activated into its active, GTP-bound state, now shuts down its signaling. This negative feedback loop ensures the release of only a small burst of mitogenic signals by Ras. Yet other mechanisms involved in shutting down receptor signaling are depicted in Figure 6.31A, B, and C; in fact, dozens of such negative-feedback loops have been uncovered in various regulatory circuits operating within the cell, and far more are likely to be found in the future.

Acting in the opposite fashion are **positive-feedback loops**, which function to further amplify initial signaling. For example, the neurofibromin (NF1) protein normally functions like Ras-GAP to induce GTP hydrolysis by Ras proteins (Figure 6.31D). Thus, in cells that are not experiencing mitogenic signaling, NF1 ensures that the level of GTP-loaded Ras is held very low. However, soon after mitogenic signaling is initiated by a ligand-bound GF receptor, a cell's pool of NF1 protein is quickly degraded, clearing the way for Ras to rapidly become activated. Here, this feedforward loop is exploited to *accelerate* the initial activation of the signaling event, allowing it to reach its maximum more quickly. (Later, NF1 levels are restored by yet other regulators, ensuring eventual shutdown of Ras signaling.) NF1 function will be discussed in greater detail in Chapter 7.

Alternatively, feedforward mechanisms can be used to *stabilize* a decision, such as a decision to *maintain* a certain state of differentiation. Thus, a cell in a certain differentiated state may produce a growth factor or cytokine that functions in an autocrine manner to induce and thereby ensure maintenance of this state of differentiation. In Chapter 8, we will encounter yet other examples of positive-feedback loops exploited by cells that have executed a decision and use these loops to guarantee the *irreversibility* of this decision; for example, once the cell decides to advance into the next phase of the cell cycle, positive-feedback signaling loops reinforce and amplify this decision, preventing the cell from slipping backward into an earlier phase of the cell cycle. Still, we need to remember that feedfoward loops also represent a liability, at least in the case where mitogenic signaling is being regulated: since they allow signaling to become self-perpetuating, feedforward loops may lead to loss of the finely tuned control of cell proliferation that is central to the normal cell phenotype.

6.14 Synopsis and prospects

Signal transduction in cancer cells One of the hallmarks of cancer cells is their ability to generate their own mitogenic signals endogenously. These signals liberate cancer cells from dependence on external mitogenic signals, specifically those conveyed by growth factors. It is likely that the mitogenic pathway of greatest importance to human cancer pathogenesis is the one that we have discussed in detail and has the following design:

$$\text{GFs} \rightarrow \text{RTKs} \rightarrow \text{Grb2} \rightarrow \text{Sos} \rightarrow \text{Ras} \rightarrow \text{Raf} \rightarrow \text{MEK} \rightarrow \text{Erk}$$

As described in Chapter 5 and this one, a variety of molecular mechanisms cause the upstream portion of this signaling pathway to become continuously activated in cancer cells. Neoplastic cells may acquire the ability to make and release growth factors (GFs) that initiate autocrine signaling loops. Alternatively, the growth factor receptors (RTKs) may suffer significant structural alterations. In the case of the EGF receptor, the

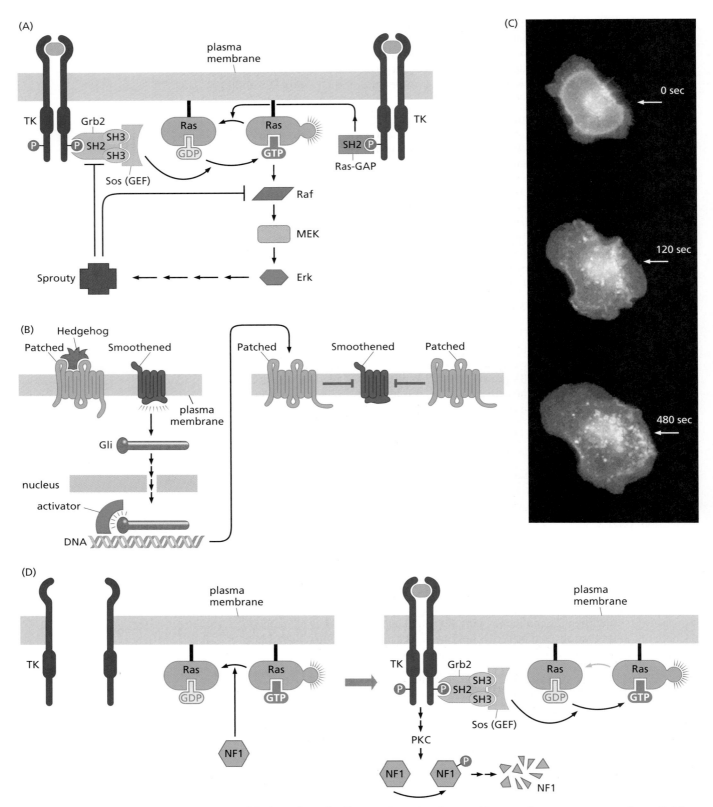

deletion of much of its ectodomain, observed in some human tumors such as glioblastomas, results in ligand-independent firing of its cytoplasmic tyrosine kinase.

Even more commonly observed are the human tumors that display elevated levels of wild-type versions of receptors such as the EGF receptor or its close relative, ErbB2/HER2/Neu. It is clear that excessively high levels of these receptor molecules, often expressed by breast, brain, and stomach cancer cells, favor increased cell proliferation.

Figure 6.31 (*opposite page*) **Negative and positive feedback controls of signaling**
An elaborate array of negative feedback loops ensures that excessive signaling in normal cells is soon reduced to levels that are compatible with normal cell behavior. (A) When GF receptors bind ligand, the resulting activation of the Ras → Raf → MEK → Erk signaling pathway causes Sprouty (*brown, lower left*) to block binding of Grb2 to Sos, thereby shutting down further Ras activation. Sprouty also binds and inhibits Raf, as well as other targets (*not shown*). Functioning in parallel to this negative-feedback loop, the binding of Ras-GAP via its SH2 group to a phosphotyrosine on the cytoplasmic tail of a ligand-activated receptor (*right*) places it in a position where it can shut down further Ras signaling by inducing GTP hydrolysis by Ras. (B) Binding of Hedgehog to Patched (*top left*) permits signaling by Smoothened, one of the consequences of which is the activation, via Gli, of Patched transcription. The increased numbers of Patched molecules then function to inhibit further signaling by Smoothened. (C) An important mechanism of shutting down RTK signaling involves the physical internalization of receptor molecules, preventing them from binding ligand in the extracellular space. When a cell is exposed to epidermal growth factor (EGF), introduced into the medium at the location shown (*arrowheads*), the EGF initially causes the cell surface EGF receptor molecules, which in this case have been linked to a fluorescent dye, to release mitogenic signals into the cell (*not shown*). However, within 120 seconds, the fluorescing EGF-R molecules have been internalized on the side of the cell facing the EGF source; and by 480 seconds, they are found in internalized clusters on all sides of the cell. (D) A positive-feedback mechanism is illustrated by the actions of NF1 (neurofibromin), which in resting cells ensures the hydrolysis of Ras-bound GTP and associated inactivation of signaling by Ras. However, when a GF-activated receptor begins to release signals, it activates a form of PKC (protein kinase C) that phosphorylates NF1, causing its ubiquitylation and degradation (see Supplementary Sidebar 7.5), thereby enabling a rapid increase in firing by active, GTP-bound Ras molecules (*right*). (C, from M. Bailly et al., *Mol. Biol. Cell* 11:3873–3883, 2000.)

Such receptor overexpression is often achieved through either deregulated transcription or amplification of the gene encoding the receptor.

The flux of signals through the RTK → Ras → Raf → MAPK pathway is also regulated by negative-feedback mechanisms that attenuate signaling. For example, after the EGF-R binds ligand, the receptor–ligand complex is **internalized** into the cell via endocytosis (see Figure 6.31C). Once sequestered in cytoplasmic membrane vesicles, the receptor may either be shuttled to lysosomes, where it is degraded, or recycled back to the surface. If the internalization process is defective, the receptor will accumulate to excessively high levels at the cell surface, yielding, in turn, greatly elevated mitogenic signaling. Defective internalization of the EGF-R and, by extension, other RTKs may play a key role in driving the proliferation of many kinds of human tumor cells. As we learn more about the regulation of cell signaling, it is becoming apparent that negative-feedback controls like this one modulate signaling through virtually all the intra-cellular signaling pathways. We will revisit some of these negative-feedback controls in Chapter 16, since they play an important role in thwarting recently developed therapeutic agents that have been designed to shut down certain key signaling molecules.

Moving down the mitogenic pathway from the receptors, we note that few, if any, alterations in signal-transducing proteins have been documented in human tumors until we reach the Ras proteins. Recall that point mutations in the 12th, 13th, or 61st codon of the reading frames of encoding *ras* genes result in activated Ras oncoproteins. These mutant proteins fire for extended periods of time rather than the short, well-controlled bursts that characterize the behavior of normal Ras. The amino acid replacements caused by these various point mutations (which are seen in more than one-fifth of all human cancers) render the Ras oncoproteins resistant to the GTPase-activating proteins (GAPs), which normally stimulate the GTPase activity of Ras. As discussed later in Chapter 7, in the disease of neurofibromatosis, one of the GAPs that usually inactivate Ras is missing from certain cell types, resulting once again in an accumulation of excessively high levels of the activated, GTP-bound form of Ras proteins.

Moving downstream of Ras in this linear signaling cascade, we once again find a relative dearth of mutant proteins in human cancer cells. The B-Raf protein, a close cousin of the Raf kinase, is one exception. It is found in mutant form in about 8% of a large panel of human tumor lines. Most importantly, almost half of all melanomas possess point-mutated versions of the *BRAF* gene, whose normal product, like Raf, requires Ras stimulation before it can fire. The mutant B-Raf proteins have a greatly reduced dependence on interaction with Ras.

Another exception is provided by phosphatidylinositol 3-kinase (PI3K), which also operates immediately downstream of Ras. Recall that the PI3K enzyme is activated by Ras, by growth factor receptors, or by the two acting in concert (see Figure 6.17). Its product is phosphatidylinositol triphosphate (PIP3). By 2010, activating mutations in the PI3K gene (termed *PIK3CA*) had been documented in the genomes of about one-quarter of breast and urinary tract carcinomas and a smaller percentage of ovarian and gastrointestinal tumors. But far more commonly, inhibition of this pathway is defective: PIP3 levels are normally held down by the actions of the PTEN phosphatase, which is lost in a wide variety of cancers. Almost 15% of all human tumors have inactivating mutations in the *PTEN* gene, and many other tumors lose gene expression through nongenetic mechanisms, notably repression by microRNAs (see Section 1.10). The resulting loss of PTEN function may be as effective for increasing PIP3 levels as the hyperactivity of the PI3K that is induced by Ras oncoproteins. In both cases, the increased PIP3 levels cause activation of Akt/PKB, which in turn has widespread effects on cell survival, growth, and proliferation.

The gene encoding Akt/PKB itself has also been found to be mutated and constitutively activated in a small percentage of human breast, ovarian, and colorectal carcinomas. Nonetheless, there are levels of complexity here that we still fail to understand. For example, in some tumors PTEN loss leads to Akt/PKB activation, as anticipated, but in other tumors of the same type, PI3K activation leads to activation of PDK1 instead without significant increases in Akt/PKB; such behavior is difficult to explain by diagrams such as that presented in Figure 6.19A.

The available data suggest that this signaling pathway reaching from growth factors through Ras to the two critical branches—Raf → MEK → Erk and PI3K → Akt/PKB—represents the dominant mitogenic pathway in most types of non-hematopoietic cells throughout the body. This prominence is reflected in the frequent involvement of this pathway, in deregulated form, in human cancers. Still, future research may reveal that other pathways, some mentioned in this chapter (involving GPCRs, Patched, Notch, and NF-κB), rival the importance of the RTK → Ras pathway in providing the mitogenic signals that drive cell proliferation in certain types of frequently occurring tumors. Interestingly, the design of the main mitogenic pathways operating in two of the most commonly occurring cancers—prostate and estrogen receptor–positive breast cancers—are still poorly understood (see Section 5.8 and **Sidebar 6.5**). An interesting question is why the mitogenic signaling cascade is so complex. For example, why do our cells have the complex Raf → MEK → Erk cascade and indeed a fourth kinase (Rsk) in this molecular bucket brigade?

Organization of mitogenic signaling pathways Stepping back from the details of the deregulation of mitogenic signaling pathways in human cancers, we can draw some general lessons about the design of these pathways and how this organization affects human cancer pathogenesis. Evolution might have constructed complex metazoa like ourselves by using a small number of cellular signaling pathways (or even a single one) to regulate the proliferation of individual cells in various tissues. As it happened, a quite different design plan was pasted together: each of our cells uses a number of distinct signaling pathways to control its proliferation (with one of these, involving tyrosine kinase receptors, playing an especially prominent role).

In each type of cell, the signaling pathways work in a combinatorial fashion to ensure that proliferation occurs in the right place and at the right time during development and, in the adult, during tissue maintenance and repair. Moreover, different cell types use different combinations of pathways to regulate their growth and division. This helps to explain why the biochemistry of cancer, as described in this chapter, is so complex.

This signal transduction biochemistry is organized around a small number of basic principles and a large number of idiosyncratic details. It is a patchwork quilt of

Sidebar 6.5 The mysteries of breast and prostate cancer mitogenesis Breast and prostate cancers are commonly occurring diseases in the West that have been the subjects of extensive research, yet we still have only a vague understanding at the molecular level of how the growth of the majority of these tumors is controlled. A majority of breast cancers express the estrogen receptor (ER), while almost all prostate cancers express the androgen receptor (AR); both of these nuclear receptors act through mechanisms that are very different from those of the cell surface receptors we have studied (see Section 5.8). Like other "nuclear receptors," the ER binds its ligand, estrogen (more precisely, estradiol), and then proceeds to bind to specific DNA sequences in the promoters of certain target genes; this binding, followed by association with transcriptional co-regulators, leads to the transcription of these genes.

Estrogen acts as a critically important mitogen for ER-positive breast cancer cells. This is indicated by the fact that a number of "anti-estrogens" such as tamoxifen, which bind to the ER and block some of its transcription-activating powers, prevent the growth of ER-positive cells and tumors and can actually cause tumor regression. However, none of this seems to explain precisely how estrogen and the ER drive the proliferation of ER-positive breast cancer cells. Within minutes after estrogen is added to these cells, the Ras → MAPK signaling cascade is activated. This rapid response cannot be explained by the fact that ER acts as a nuclear transcription factor, since changes in nuclear gene transcription are not felt in the cytoplasm (in the form of proteins synthesized on newly formed mRNA templates) for at least half an hour.

One critical clue may come from the observation that some of the ligand-activated ER molecules are found to be tethered to cytoplasmic membranes rather than being localized to their usual sites of action in the nucleus. Following ligand binding, some of this cytoplasmic ER has been reported to be bound to Shc, the important SH2- and SH3-containing bridging protein that participates in Sos activation and Ras signaling (see Figure 6.12B). The ER–Shc association, still poorly understood, may explain how cytoplasmic mitogenic signaling cascades are activated by estrogen. This fragmentary description of a major mitogenic pathway is most unexpected, in light of the fact that breast cancer has, by now, been studied far more intensively than any other type of human malignancy.

jury-rigged contraptions. Many of these details represent *ad hoc* solutions that were chosen during early metazoan evolution, 600 or 700 million years ago, in response to biological or cell-physiologic problems; some of these solutions reach back to the origins of eukaryotes, almost a billion years earlier. Once they were cobbled together, these solutions became fixed and relatively immutable, if only because other, subsequently developed contraptions depended on them. (To be sure, the end product of all these solutions—a machinery optimized for normal development and a lifetime of health—usually works marvelously well. During an average human life span, the cells inside the body cumulatively pass through approximately 10^{16} growth-and-division cycles and, in the case of most of us, provide excellent health—something that we take for granted.)

The signaling circuitry that was described in this chapter operates largely in the cytoplasm and receives inputs from cell surface receptors—those that bind either soluble extracellular ligands, notably growth factors, or insoluble matrix components (the integrins). Having received a complex mixture of **afferent** (incoming) signals from these receptors, this cytoplasmic circuitry emits signals to molecular targets in both the cytoplasm and the nucleus.

Included among the cytoplasmic targets are the regulators of cell shape and motility, energy metabolism, protein synthesis, and the apoptotic machinery, which decides on the life or death of the cell. The remaining **efferent** (outgoing) signals are transmitted to the nucleus, where they modulate the transcription of thousands of genes involved in the programming of cell growth and differentiation as well as in cell cycle progression. We will return to many of these themes later. Thus, the regulation of the cell cycle and differentiation is described in Chapter 8, while the regulation of programmed cell death—apoptosis—will be presented in Chapter 9. The regulators of cell motility will be described in outline in Chapter 14.

Certain major themes have recurred throughout this chapter. Most apparent are the three ways in which signals are transduced through the signaling pathways of normal and neoplastic cells. First, the *intrinsic activity* of signaling molecules may be modulated. This can be achieved by noncovalent modifications (for example, the binding of GTP by Ras) or receptor dimerization. Alternatively, a signaling molecule may be covalently modified; the phosphorylation of MEK by Raf, the phosphorylation of PIP2 by PI3K, and the proteolytic cleavage of Notch are three examples of this.

Second, the *concentration* of a signaling molecule may be modulated, often by orders of magnitude. Thus, the concentration of β-catenin is strongly reduced by GSK-3β phosphorylation, and the concentration of PIP3 is regulated by both PI3K (positively) and PTEN (negatively).

Third, the intracellular *localization* of signaling molecules can be regulated, which results in moving them from a site where they are inactive to a new site where they can do their work. The most dramatic examples of these derive from the phosphorylation of receptor tyrosine kinases (RTKs) and the subsequent attraction of multiple, distinct SH2-containing molecules to the resulting phosphotyrosines. Once anchored to the receptors, many of these signal-transducing molecules are brought in close proximity with other molecules that are associated with the plasma membrane; examples of the latter are the Ras molecule (the target of Sos action) and PIP3 (the target of PI3K action).

The translocation of cytoplasmic molecules to the nucleus is another manifestation of this third class of regulatory mechanisms. For example, the "dual-address" transcription factors dwell in inactive form in the cytoplasm and may be dispatched as active transcription factors (or components thereof) to the nucleus. Examples of these include β-catenin, NF-κB, the Smads, Notch, Gli, and the STATs.

The kinetics with which signals are transduced from one site to another within a cell may also vary enormously. One very rapid mechanism involves the actions of kinase cascades. Thus, the cytoplasmic MAP kinases (MAPKs), including Erks, that lie at the bottom of these cascades are activated almost immediately (in much less than a minute) following mitogen treatment of cells, and, following activation, the Erks move into the nucleus essentially instantaneously. Once there, they proceed to phosphorylate and functionally activate a number of key transcription factors.

A slightly slower but nonetheless highly effective signaling route derives from the strategy, cited above, of activating dormant transcription factors in the cytoplasm and dispatching them in activated form to the nucleus. By far the slowest signaling mechanisms depend on modulating the concentrations of signaling proteins. This mode of regulation can occur by changes in the rates of transcription of certain genes, translation of their mRNAs, stabilization of the mRNAs, or post-translational stabilization of their protein products.

These principles seem to apply to several other signaling pathways that have not been discussed here because their role in the development of tumors is either minor or still poorly documented. One of these—non-canonical Wnt signaling—was cited briefly in Section 6.10 and is discussed in outline in Supplementary Sidebar 6.4. A second, involving signaling through Hippo, is described in Supplementary Sidebar 6.5.

The intracellular signaling circuitry encountered in this chapter appears to be organized similarly in various cell types throughout the body. This fact has important implications for our understanding of cancer and its development. More specifically, it suggests that the biochemical lessons learned from studying the neoplastic transformation of one cell type will often prove to be applicable to a number of other cell types throughout the body. The ever-increasing body of information on the genetic aberrations of human cancer cells provides strong support for this notion.

The similar design of the intracellular growth-regulating circuitry in diverse cell types forces us to ask why different kinds of cells behave so differently in response to various external signals. Many of the answers will eventually come from understanding the spectrum of receptors that each cell type displays on its surface. In principle, knowledge of the cell type–specific display of cell surface receptors and of the largely invariant organization of the intracellular circuitry should enable us to predict the behaviors of various cell types following exposure to various mitogens and to growth-inhibitory factors like TGF-β. Similarly, by understanding the array of mutant alleles coexisting within the genome of a cancer cell, we should be able to predict how these alleles perturb behavior and how they conspire to program the neoplastic cell phenotype. In practice, we are still far from being able to do so for at least nine reasons:

1. In this chapter, we have described only the bare outlines of how these circuits operate and have consciously avoided many details. But such details are critical.

For example, individual steps in these pathways are often controlled by multiple, similarly acting proteins, each of which transduces signals slightly differently from the others. Examples of this are the four structurally distinct Ras proteins, the three forms of Akt/PKB, as well as Raf and its close cousin, B-Raf. In most cases, we remain ignorant of the functional differences between the superficially similar proteins operating within a pathway, and therefore have an incomplete understanding of how the pathway as a whole operates.

2. Complex negative-feedback and positive-feedback loops serve to dampen or amplify the signals fluxing through each of these pathways. We touched on these only briefly. The feedback loops discovered to date are likely to represent only the tip of a much larger iceberg. For example, more than 100 phosphotyrosine phosphatase (PTP)–encoding genes have been found in the human genome. PTPs remove the phosphate groups attached by the 90 tyrosine kinases (TKs) present in our cells. We know almost nothing about how these key enzymes are regulated and what their various substrates are.

3. Each component in a circuit acts as a complex signal-processing device that is able to amplify, attenuate, and/or integrate the upstream signals that it receives before passing them on to downstream targets. We have not begun to understand in a quantitative way how this signal processing by individual signal-transducing proteins operates.

4. The ability of proteins to transduce signals to one another is affected by their intracellular concentrations, post-translational modifications (such as phosphorylation), and intracellular localization. We have only begun to plumb the depths of this complexity. For example, the four Ras proteins localize to various cytoplasmic membranes following their activation, where they have more or less access to their downstream effectors. In addition, more than 25,000 distinct phosphopeptides have been cataloged in human cells; each of these is associated with one or another cellular protein and the effects of these modifications on protein function are generally not known.

5. The post-translational modifications (PTMs) of proteins are actually far more varied and complex than the phosphorylation events cited above. Proteins may become methylated, acetylated or, as we will learn in the next chapter, ubiquitylated; there are also a variety of ubiquitin-like PTMs whose effects on protein function are poorly understood; included among these modifying groups are NEDD8, ISG15, FAT10, MNSFβ, Ufm1, six Atg8 paralogs, Atg12, Urm1, Ubl5, and four SUMO paralogs.

6. Calculations of diffusion times in cells indicate that efficient, rapid signal transduction is likely to occur only when interacting proteins are co-localized within a cell. Only very fragmentary information is available about the intracellular locations of many of these proteins and their interacting partners.

7. None of these intracellular pathways operates in isolation. Instead, each is influenced by cross connections with other pathways. Understandably, the search for these cross connections has been largely postponed until the operations of each primary pathway are elucidated. Once these cross connections are added to the maps of these primary pathways, the drawings of signal transduction circuitry will likely resemble a weblike structure (for example, **Figure 6.32**) rather than a series of parallel pathways that begin at the plasma membrane and extend linearly from there into the nucleus.

8. Even if these extensive cross connections did not exist, the endpoints of signaling—specific changes in cell phenotype—are the results of combinatorial interactions between multiple, converging signaling pathways. Once again, this dimension of complexity continues to baffle us.

9. Our depiction of how signals are transmitted through a cell is likely to be fundamentally flawed. We have spoken here, time and again, about potent signaling pulses speeding along signal-transducing pathways and evoking, at endpoints, strong and clearly defined responses within a cell. In fact, each signaling cascade is

Figure 6.32 Two-dimensional signaling maps While we have depicted signaling pathways as linear cascades reaching from the plasma membrane to the nucleus, with occasional branch points, a rapidly growing body of research reveals that these pathways interact via numerous lateral cross connections and points of convergence. This diagram, itself a simplification, provides some hints of these cross connections. Note, for example, that GSK-3β, which plays a central role in the Wnt–β-catenin pathway, is also regulated by signaling proteins lying downstream of tyrosine kinase receptors. (We will encounter the death receptors in Chapter 9.) (Adapted from T. Hunter, *Cell* 88:333–346, 1997.)

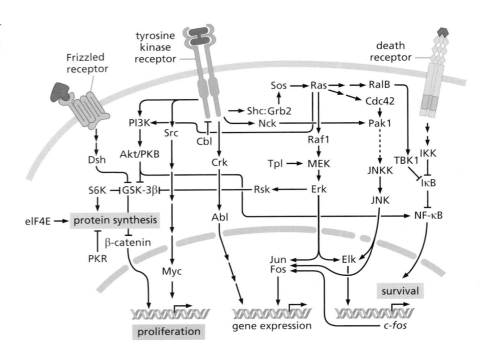

likely to operate in a finely tuned, dynamic equilibrium, where positive and negative regulators continuously counterbalance one another. Accordingly, a signaling input (for example, a mitogenic stimulus) may operate like the plucking of a fiber in one part of a spider web, which results in small reverberations at distant sites throughout the web. Here, neither our language nor our mathematical representations of signaling suffice to clarify our understanding.

Many cancer researchers would like to be able to draw a complete and accurate wiring diagram of the cell—the scheme that depicts how these pathways are interconnected. One measure of the difficulties associated with this task comes from censuses of the various classes of genes in the human genome. According to one enumeration, there are 518 distinct genes specifying various types of protein kinases; 40% of these genes make alternatively spliced mRNAs encoding slightly different protein structures, leading to more than 1000 distinct kinase proteins that may be present in human cells. Of the 518 kinase genes, 90 encode tyrosine kinases, the remainder being serine/threonine kinases. Among the 90 tyrosine kinases, 58 function as the signaling domains of growth factor receptors. These numbers provide some measure of the complexity of the signaling circuitry that underlies cancer pathogenesis, since many of these kinases are involved in regulating the proliferation and survival of cells.

Indications of additional complexity have come from the more detailed studies of Ras signaling biochemistry. In this chapter, we emphasized the actions of Ras on three major effector pathways, specifically, Raf, PI3 kinase (PI3K), and a Ral-GEF (sometimes termed Ral-GDS); see Sections 6.5, 6.6, and 6.7. However, as the Ras proteins are subjected to more biochemical scrutiny, schemes like those shown in **Figure 6.33** are emerging.

The difficulties of explaining cell behavior in terms of the design of signal transduction circuitry underlie our current inability to understand another set of problems, these being related specifically to the genetics of cancer cell genomes: Why do various types of human tumors display highly specific types of genetic alterations (**Figure 6.34**)? Why do 90% of pancreatic carcinomas carry a mutant K-*ras* oncogene, while this mutation is seen in only a small proportion (<10%) of breast cancers? Why are amplifications of the EGF receptor and its cousin, HER2/erbB2, encountered in more than a third of human breast cancers but seen uncommonly in certain other epithelial cancers, even though these receptors are widely displayed in many kinds of normal epithelia?

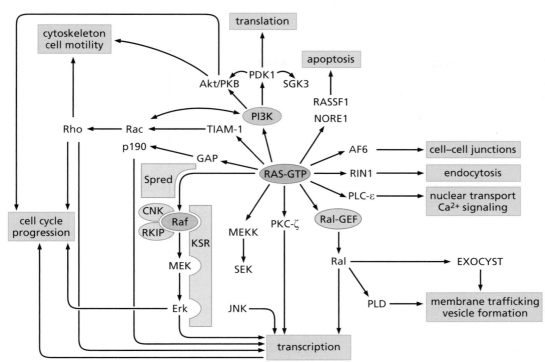

Figure 6.33 Ras effector pathways Detailed biochemical analyses of the GTP-bound, activated form of the Ras proteins indicate that these proteins bind to far more than the three major effectors of Ras signaling—Raf, Ral-GEF, and PI3 kinase (PI3K). As shown here, at least eight additional Ras-interacting proteins have been uncovered, most of which are known or suspected to play key roles in relaying Ras signals to specialized downstream signaling circuits involved in functions as diverse as transcription, membrane trafficking, endocytosis, and translation. (Adapted from M. Malumbres and M. Barbacid, *Nat. Rev. Cancer* 3:459–465, 2003.)

Why do approximately 50% of colon carcinomas carry a mutant K-*ras* gene, and how do the remaining 50% of these tumors acquire a comparable mitogenic signal? Some of these other tumors have mutant B-Raf or PI3K proteins. Still, this leaves many cancers with no apparent alterations affecting either the Ras → Raf → MAPK signaling cascade or the parallel PI3K → Akt/PKB pathway. Do these tumors harbor still-undiscovered genetic lesions that activate one or the other pathway? Or, alternatively, are

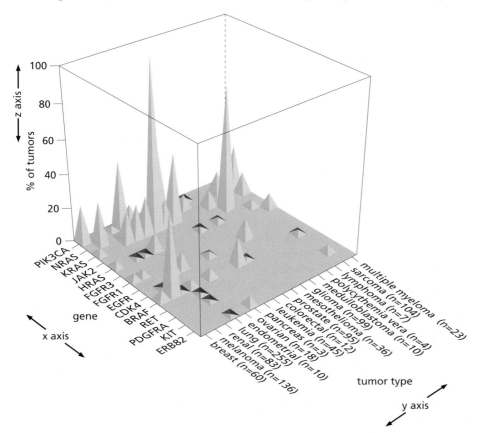

Figure 6.34 Frequencies of oncogene activation in various human tumors This three-dimensional histogram illustrates the great variability with which various commonly studied human oncogenes are found in mutant, activated forms in the genomes of human tumors. The percentage (*z* axis) of tumors of a given type (*y* axis) that harbor mutant forms of a given oncogene (*x* axis) reveals great differences in the involvement of each of these oncogenes in human tumor pathogenesis; the underlying mechanisms that explain these strong inter-tumor-type variations are unknown. (From R.S. Thomas et al., *Nat. Genet.* 39:347–351, 2007.)

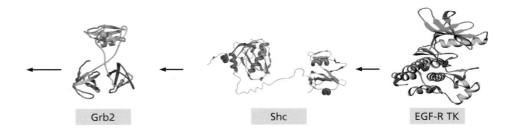

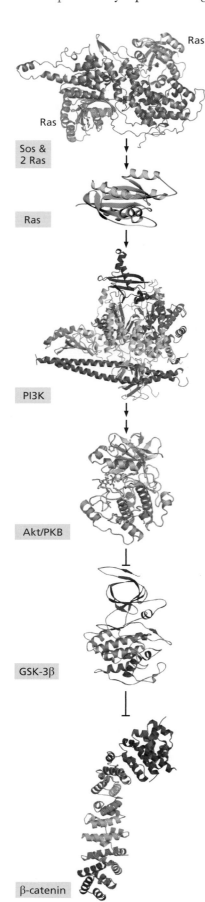

Figure 6.35 Structure of a signaling pathway The signaling pathway leading from EGF to the activation of signaling by β-catenin can now be depicted in terms of the detailed structures of the participating proteins, almost all of which have been determined by X-ray crystallography. The details of these structures will help our understanding of how signals flow through pathways like this one. Note that the structure of the central portion of Shc has not been determined and that the PI3K structure shown is of PI3Kγ, which is closely related to, but distinct from, the PI3Ks that are activated by receptor tyrosine kinases. (EGF-R TK, courtesy of R. Latek, from J. Stamos et al., *J. Biol. Chem.* 277:46265–46272, 2002; Shc, courtesy of K.S. Ravichandran; Grb2, from S. Maignan et al., *Science* 268:291–293, 1995; Sos, from S.M. Margarit et al., *Cell* 112:685–695, 2003; Ras, from T. Schweins et al., *J. Mol. Biol.* 266:847–856, 1997; PI3K, from C.-H. Huang et al., *Science* 318:1744–1748, 2007; Akt/PKB, from J. Yang et al., *Nat. Struct. Biol.* 9:940–944, 2002; GSK-3β, from J.A. Bertrand et al., *J. Mol. Biol.* 333:393–407, 2003; β-catenin, courtesy of H.J. Choi and W.I. Weis.)

the cells in these tumors able to proliferate uncontrollably because of mechanisms that have nothing to do with the operations of these particular pathways? And why are certain combinations of mutant or otherwise altered genes found in cancer cells, while other combinations are rarely if ever observed?

Finally, there is a major problem in clinical oncology that remains far from solution: Why do certain combinations of mutant alleles and protein expression patterns in cancer cells portend good or bad prognoses for patients? At present, we are largely limited to noting correlations between these patterns and clinical outcomes. But correlations hardly reveal the chains of causality connecting the genomes of cancer cells with the behavior of the tumors that these cells create. Many years will likely pass before these and related mysteries are solved definitively.

Our ability to address some of these problems is being enhanced by rapid progress in the field of structural biology: it has become almost routine to determine the detailed structures of protein molecules, including the signal-transducing proteins that play key roles in cancer pathogenesis. Since structure determines function, the elucidation of these structures will greatly enrich our understanding of how signal-transducing pathways operate inside cancer cells. The catalogs of these protein structures are still very incomplete. But this is changing rapidly, and in the future, the detailed structural information present in images such as **Figure 6.35** will provide critical clues for resolving major puzzles in cancer biology.

This chapter has focused largely on the cytoplasmic signaling molecules that convey signals from cell surface receptors to the nucleus. This focus overlooks a critical path of oncogenic signaling that plays a key role in the development of common human cancers, specifically carcinomas of the breast, ovary, and prostate: steroid hormones leapfrog over this cytoplasmic circuitry by passing directly from the cell surface to the nucleus, where they bind nuclear receptors, notably the estrogen and androgen receptors (see Section 5.8). We will return to nuclear receptors and their mechanisms of promoting cell proliferation in Chapter 8.

Key concepts

- Discovery of Src's three homology domains (SH1, 2, and 3) led to the following model of tyrosine kinase receptor (RTK) signaling cascades: Ligand-induced transphosphorylation results in an RTK displaying on its cytoplasmic tail an array of phosphotyrosines, each of which attracts and binds a specific SH2-containing cytoplasmic protein. Such relocalization permits the SH2-containing proteins to interact with plasma membrane–associated proteins that can activate their own downstream targets or with other signal-transducing molecules that pass on activating signals to yet other targets.

- One such signal-transducing protein tethered to the plasma membrane is Ras. Linker proteins form a physical bridge between ligand-activated RTKs and Sos, the latter acting as a guanine nucleotide exchange factor (GEF) that induces Ras to replace its bound GDP with GTP, thereby activating Ras signaling.

- Three major downstream signaling cascades emanate from activated Ras via binding of its effector loop with its main downstream signaling partners—Raf kinase, phosphatidylinositol 3-kinase (PI3K), and Ral-GEF. Yet other downstream effectors that bind to activated Ras appear to play secondary roles in Ras signaling.

- Raf phosphorylates residues on MEK, activating the latter. MEK then phosphorylates and activates the extracellular signal–regulated kinases 1 and 2 (Erk 1 and 2), which are classified as MAPKs (mitogen-activated protein kinases). The Raf signaling cascade, often called a MAPK pathway, is the foremost mitogenic pathway in mammalian cells, both normal and neoplastic.

- The phosphatidylinositol 3-kinase (PI3K) pathway depends on kinases phosphorylating a membrane-embedded phosphatidylinositol, converting it to PIP3. Once formed, the latter attracts molecules (such as Akt/PKB) that carry a pleckstrin homology (PH) domain. The resulting activation of Akt/PKB activates a series of downstream proteins. PIP3 levels are normally kept low in cells by phosphatases, notably PTEN.

- In the third Ras-regulated pathway, Ras binds Ral-GEFs, causing localization of Ral-GEF near the inner surface of the plasma membrane. Ral-GEF then induces nearby Ral to exchange its GDP for GTP, enabling Ral to activate downstream targets. This has multiple effects on the cell, including changes in the cytoskeleton and cellular motility.

- Cytokine receptors use noncovalently bonded TKs of the Jak class to phosphorylate STATs (signal transducers and activators of transcription). The STATs form dimers that migrate to the nucleus where they serve as transcription factors.

- Integrins bind components of the ECM, leading to the formation of focal adhesions and activation of focal adhesion kinase (FAK), which is associated with the integrin cytoplasmic tail. Transphosphorylation of FAK and its recruitment of SH2-containing molecules activates many of the same signaling pathways as RTKs.

- The pathway controlled by Wnt factors enables cells to remain in a relatively undifferentiated state—an important attribute of certain cancer cells. Wnt, acting via Frizzled receptors, prevents glycogen synthase kinase-3β (GSK-3β) from tagging its substrates, including β-catenin, for destruction. The spared β-catenin moves into the nucleus and activates transcription of key growth-stimulating genes.

- Ligand-bound G-protein–coupled receptors (GPCRs) activate cytoplasmic heterotrimeric G-proteins whose α subunit responds by exchanging its GDP for GTP. The Gα subunit then dissociates from its partners (Gβ + Gγ) and affects a number of cytoplasmic enzymes, while the Gβ + Gγ dimer activates its own effectors, including PI3Kγ, Src, and Rho-GEFs.

- Among the pathways relying on dual-address proteins is the nuclear factor-κB (NF-κB) signaling system. It depends on liberation of cytoplasmic NF-κB homo- and heterodimers from sequestration by IκBs. Following migration of NF-κB subunits into the nucleus, these proteins function as transcription factors that activate expression of at least 150 genes, including cell proliferation and anti-apoptotic genes.

- The cleavage of Notch at two sites following ligand binding liberates a cytoplasmic fragment that migrates to the nucleus and functions as part of a transcription factor complex. Notch aberrations are seen in the majority of cervical carcinomas.

- When Hedgehog binds a Patched receptor, Smoothened is released from inhibition and emits signals that protect cytoplasmic Gli protein from cleavage. The resulting intact Gli migrates to the nucleus, where it activates transcription, whereas cleaved Gli acts as a transcriptional repressor.

- The binding of TGF-β to its receptors causes phosphorylation of cytoplasmic Smad proteins and their dispatch to the nucleus, where they help activate a large contingent of genes. This pathway plays a major role in the pathogenesis of many carcinomas, both in the early stages when TGF-β acts to arrest cell proliferation and later, when it contributes to tumor cell invasiveness.

Thought questions

1. What molecular mechanisms have evolved to ensure that the signals coursing down a signaling cascade reach the proper endpoint targets rather than being broadcast non-specifically to "unintended" targets in the cytoplasm?

2. How many distinct molecular mechanisms can you cite that lead to the conversion of a proto-oncogene into an active oncogene?

3. Why are point mutations in *ras* oncogenes confined so narrowly to a small number of nucleotides, while point mutations in genes encoding other cancer-related proteins are generally distributed far more broadly throughout the reading frames?

4. What factors determine the lifetime of the activated state of a Ras oncoprotein?

5. Treatment of cells with proteases that cleave the ectodomain of E-cadherin often result in rapid changes in gene expression. How might you rationalize this response, given what we know about this cell surface protein?

6. What mechanisms have we encountered that ensure that a signal initiated by a growth factor receptor can be greatly amplified as the signal is transduced down a signaling cascade in the cytoplasm? Conversely, what signaling cascade(s) strongly limit the possible amplification of a signal initiated at the cell surface?

7. What quantitative parameters describing individual signal-transducing proteins will need to be determined before the behavior of the signaling cascade formed by these proteins can be predicted through mathematical modeling?

Additional reading

Aggarwal BB (2004) Nuclear factor-κB: the enemy within. *Cancer Cell* 6, 203–208.

Albert R & Oltvai ZN (2007) Shaping specificity in signaling networks. *Nat. Genet.* 39, 286–287.

Alonso A, Sasin J, Bottini N et al. (2004) Protein tyrosine phosphatases in the human genome. *Cell* 117, 699–711.

Arcaro, A & Guerreiro, AS (2007) The phosphoinositide 3-kinase pathway in human cancer: genetic alterations and therapeutic implications. *Curr. Genomics* 8, 271–306.

Aster JC, Pear WS & Blacklow SC (2008) Notch signaling in leukemia. *Annu. Rev. Pathol.* 3, 587–613.

Bernards A (2003) GAPs galore! A survey of putative Ras superfamily GTPase activating proteins in man and *Drosophila*. *Biochim. Biophys. Acta* 1603, 47–82.

Bray SJ (2006) Notch signaling: a simple pathway becomes complex. *Nat. Rev. Mol. Cell Biol.* 7, 678–689.

Brembeck FH, Rosário M & Birchmeier W (2006) Balancing cell adhesion and Wnt signaling: the key role of β-catenin. *Curr. Opin. Genet. Dev.* 16, 51–59.

Brognard J & Hunter T (2011) Protein kinase signalling networks in cancer. *Curr. Opin. Genet. Dev.* 21, 4–11,

Brown MD & Sacks DB (2009) Protein scaffolds in MAP kinase signaling. *Cell. Signal.* 21, 462–469.

Buettner R, Mora LB & Jove R (2002) Activated STAT signaling in human tumors provides novel molecular targets for therapeutic intervention. *Clin. Cancer Res.* 8, 945–954.

Castellano E & Downward J (2011) RAS interaction with PI3K: More than just another effector pathway. *Genes Cancer* 2, 261–274.

Chalhoub N & Baker SJ (2009) PTEN and the PI3-kinase pathway in cancer. *Annu. Rev Pathol.* 4, 127–150.

Chang L & Karin M (2001) Mammalian MAP kinase signaling cascades. *Nature* 410, 37–40.

Cohen P & Frame S (2001) The renaissance of GSK3. *Nat. Rev. Mol. Cell Biol.* 2, 769–775.

DeWire SM, Ahn S, Lefkowitz RJ & Shenoy SK (2007) β-arrestins and cell signaling. *Annu. Rev. of Physiol.* 69, 483–510.

Dhillon AS, Hagan S, Rath O & Kolch W (2007) MAP kinase signaling pathways in cancer. *Oncogene* 26, 3279–3290.

Engelman JA, Luo J & Cantley LC (2006) The evolution of phosphatidylinositol 3-kinase as regulators of growth and metabolism. *Nat. Rev. Gen.* 7, 606–619.

Evangelista M, Tian H & de Sauvage FJ (2006) The Hedgehog signaling pathway in cancer. *Clin. Cancer Res.* 12, 5924–5928.

Feng X-H & Derynck R (2005) Specificity and versatility in TGF-β signaling through Smads. *Annu. Rev. Cell Dev. Biol.* 21, 659–693.

Gilman A (1987) G proteins: transducers of receptor-generated signals. *Annu. Rev. Biochem.* 56, 615–649.

Gilmore T (2006) Introduction to NF-κB: players, pathways, perspectives. *Oncogene* 25, 6680–6684.

Goetz SC and Anderson KV (2010) The primary cilium: a signalling centre during vertebrate development. *Nature Rev. Genet.* 11, 331–344.

Guo X & Wang XF (2009) Signaling cross-talk between TGF-β/BMP and other pathways. *Cell Res.* 19, 71–88.

Harvey, K & Tapon, N (2007) The Salvador-Warts-Hippo pathway—an emerging tumour-suppressor network. *Nat. Rev. Cancer* 7, 182–191.

Hazzalin CA & Mahadevan LC (2002) MAPK-regulated transcription: a continuously variable switch? *Nat. Rev. Mol. Cell Biol.* 3, 30–40.

Hsu PP & Sabatini DM (2008) Cancer cell metabolism: Warburg and beyond. *Cell* 134, 703–707.

Hunter T (1995) Protein kinases and phosphatases: the yin and yang of protein phosphorylation and signaling. *Cell* 80, 225–236.

Jiang BH & Liu LZ (2009). PI3K/PTEN signaling in angiogenesis and tumorigenesis. *Adv. Cancer Res.* 102, 19–65.

Kadesch T (2004) Notch signaling: the demise of elegant simplicity. *Curr. Opin. Genet. Dev.* 14, 506–512.

Karin M, Cao Y, Greten FR & Li Z-W (2002) NF-κB in cancer: from innocent bystander to major culprit. *Nat. Rev. Cancer* 2, 301–310.

Karnoub AE & Weinberg RA (2008) Ras oncogenes: split personalities. *Nat. Rev. Mol. Cell Biol.* 9, 517–531.

Keniry M & Parson R (2008) The role of PTEN signaling perturbations in cancer and in targeted therapy. *Oncogene* 27, 5477–5485.

Kim HJ & Bar-Sagi D (2004) Modulation of signaling by Sprouty, a developing story. *Nat. Rev. Mol. Cell Biol.* 5, 441–450.

Legate KR & Fässler R (2009) Mechanisms that regulate adaptor binding to β-integrin cytoplasmic tails. *J. Cell Sci.* 122, 187–198.

Levy DE & Darnell JE Jr (2002) STATs: transcriptional control and biological impact. *Nat. Rev. Mol. Cell Biol.* 3, 651–662.

Lewis TS, Shapiro PS & Ahn NG (1998) Signal transduction through MAP kinase cascades. In Advances in Cancer Research (GF Vande Woude, G Klein eds), pp 49–139. San Diego: Academic Press.

Liu BA, Jablonowski K, Raina M et al. (2006) The human and mouse complement of SH2 domain proteins: establishing the boundaries of phosphotyrosine signaling. *Cell* 22, 851–868.

Malumbres M & Barbacid M (2003) *RAS* oncogenes: the first thirty years. *Nat. Rev. Cancer* 3, 459–465.

Manning BD & Cantley LC (2007) AKT/PKB signaling: navigating downstream. *Cell* 129, 1261–1274.

Manning G, Whyte DB, Martinez R et al. (2002) The protein kinase complement of the human genome. *Science* 298, 1912–1934.

Marinissen MJ & Gutkind JS (2001) G-protein-coupled receptors and signaling networks: emerging paradigms. *Trends Pharmacol. Sci.* 22, 368–376.

Massagué J, Seaone J & Wotton D (2005) Smad transcription factors. *Genes Dev.* 19, 2783–2810.

Moses H & Barcellos-Hoff MH (2011) TGF-β biology in mammary development and breast cancer. *Cold Spring Harb. Perspect. Biol.* 3, a003277.

Nelson WJ & Nusse R (2004) Convergence of Wnt, β-catenin, and cadherin pathways. *Science* 303, 1483–1487.

Neves SR, Ram PT & Iyengar R (2002) G protein pathways. *Science* 296, 1636–1639.

Nusse R. (2008) Wnt signaling and stem cell control. *Cell Res.* 18, 523–527.

Pawson T (2004) Specificity in signal transduction: from phosphotyrosine-SH2 domain interactions to complex cellular systems. *Cell* 116, 191–203.

Pawson T & Nash P (2003) Assembly of cell regulatory systems through protein interaction domains. *Science* 300, 445–452.

Perkins ND (2004) NF-κB: tumor promoter or suppressor? *Trends Cell Biol.* 14, 64–69.

Plas DR & Thompson CB (2005) Akt-dependent transformation: there is more to growth than just surviving. *Oncogene* 24, 7435–7442.

Prior IA & Hancock JF (2012) Ras trafficking, localization and compartmentalized signalling. *Semin. Cell Develop. Biol.* 23, 145–153.

Ranganathan P, Weaver K & Capobianco AJ (2011) Notch signalling in solid tumors: a little bit of everything but not all the time. *Nat. Rev. Cancer* 11, 338–351.

Repasky GA, Chenette EM & Der CJ (2004) Renewing the conspiracy theory debate: does Raf function alone to mediate Ras oncogenesis? *Trends Cell Biol.* 14, 639–647.

Richmond A (2002) NF-κB, chemokine gene transcription and tumour growth. *Nat. Rev. Immunol.* 2, 664–674.

Robinson CV , Sali A & Baumeister W (2007) The molecular sociology of the cell. *Nature* 450, 973–982.

Scheid MP & Woodgett JR (2001) PKB/Akt: functional insights from genetic models. *Nat. Rev. Mol. Cell Biol.* 2, 760–767.

Schubbert S, Bollag G & Shannon K (2007) Deregulated Ras signaling in developmental disorders: new tricks for an old dog. *Curr. Opin. Genet. Dev.* 17, 15–22.

Shaw RJ & Cantley LC (2006) Ras, PI(3)K, and mTOR signaling controls tumour cell growth. *Nature* 441, 424–430.

Shi Y & Massagué J (2003) Mechanisms of TGF-β signaling from cell membrane to the nucleus. *Cell* 113, 685–700.

Tonks NK (2006) Protein tyrosine phosphatases: from genes, to function, to disease. *Nat. Rev. Mol. Cell Biol.* 7, 833–846.

Wellbrook C, Karasides M & Marais R (2004) The RAF proteins take centre stage. *Nat. Rev. Mol. Cell Biol.* 5, 875–885.

Weng AP & Aster JC (2004) Multiple niches for Notch in cancer: context is everything. *Curr. Opin. Genet. Dev.* 14, 48–54.

Wicking C & McGlinn E (2001) The role of hedgehog signalling in tumorigenesis. *Cancer Lett.* 173, 1–7.

Yaffe MB & Elia AEH (2001) Phosphoserine/threonine-binding domains. *Curr. Opin. Cell Biol.* 13, 131–138.

Yarden Y & Sliwkowski MX (2001) Untangling the ErbB signaling network. *Nat. Rev. Cancer* 2, 127–137.

Yuan TL & Cantley LC (2008) PI3K pathway alterations in cancer: variations on a theme. *Oncogene* 27, 5497–5510.

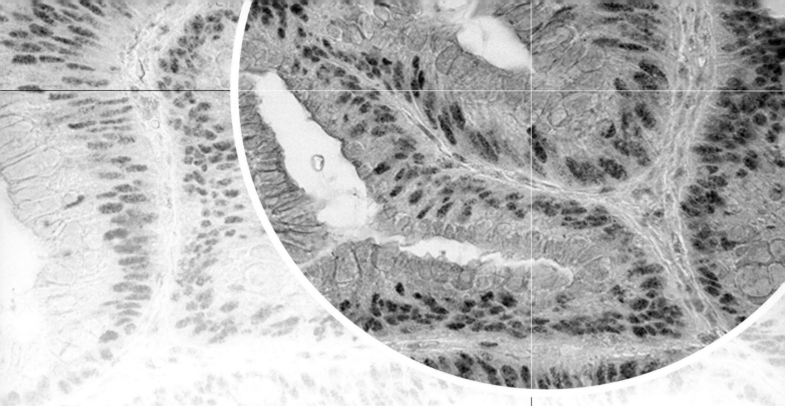

Chapter 7

Tumor Suppressor Genes

> Let me add ... a consideration of the *inheritance of tumors*. ... In order that a tumor may arise in such cases, the homologous elements in both series of chromosomes must be weakened in the same way.
>
> Theodor Boveri, pathologist, 1914

The discovery of proto-oncogenes and oncogenes provided a simple and powerful explanation of how the proliferation of cells is driven. The proteins encoded by proto-oncogenes participate in various ways in receiving and processing growth-stimulatory signals that originate in the extracellular environment. When these genes suffer mutation, the flow of growth-promoting signals released by these proteins becomes deregulated. Instead of emitting them in carefully controlled bursts, the oncoproteins release a steady stream of growth-stimulating signals, resulting in the unrelenting proliferation associated with cancer cells.

The logic underlying well-designed control systems dictates, however, that the components promoting a process must be counterbalanced by others that oppose this process. Biological systems seem to follow this logic as well, which leads us to conclude that the growth-promoting genes we have discussed until now provide only part of the story of cellular growth control.

In the 1970s and early 1980s, certain pieces of experimental evidence about cancer cell genetics began to accumulate that were hard to reconcile with the known properties of oncogenes. This evidence hinted at the existence of a second, fundamentally different type of growth-controlling gene—one that operates to constrain or suppress cell proliferation. The antigrowth genes came to be called **tumor suppressor genes** (TSGs). Their involvement in tumor formation seemed to happen when these genes were inactivated or lost. Once a cell had shed one of these genes, this cell became liberated from its growth-suppressing effects. Now, more than three decades later, we have come to realize that the inactivation of tumor suppressor genes plays a role in cancer pathogenesis that is as important to cancer as the activation of oncogenes. Indeed, the loss of these genes from a cell's genome may be even more important than oncogene activation for the formation of many kinds of human cancer cells.

Movie in this chapter

7.1 Intestinal Crypt

231

7.1 Cell fusion experiments indicate that the cancer phenotype is recessive

The study of tumor viruses in the 1970s revealed that these infectious agents carried a number of cancer-inducing genes, specifically viral oncogenes, which acted in a dominant fashion when viral genomes were introduced into previously normal cells (see Chapter 3). In particular, the introduction of a tumor virus genome into a normal cell would result in transformation of that cell. This response meant that viral oncogenes could dictate cellular behavior in spite of the continued presence and expression of opposing cellular genes within the virus-infected cell that usually functioned to ensure normal cell proliferation. Because the viral oncogenes could overrule these cellular genes, by definition the viral genes were able to induce a *dominant* phenotype—they were bringing about a cell transformation. This suggested, by extension, that cancerous cell growth was a dominant phenotype in contrast to normal (wild-type) cell growth, which was therefore considered to be *recessive* (see Section 1.1).

However, as was suspected at the time, and reinforced by research in the 1980s, most human cancers did not seem to arise as consequences of tumor virus infections. In the minds of many, this left the fundamental question of human cancer genetics unresolved. Thus, if human cancers were not caused by tumor viruses, then the lessons learned from studying such viruses might well be irrelevant to understanding how human tumors arise. In that case, one needed to shed all preconceptions about how cancer begins and admit to the possibility that the cancer cell phenotype was, with equal probability, a dominant or a recessive trait.

An initial hint of the importance of recessive cancer-inducing alleles came from experiments undertaken at Oxford University in Great Britain. Meaningful comparisons of two alternative alleles and specified phenotypes can occur only when both alleles are forced to coexist within the same cell or organism. In Mendelian genetics, when the allele of a gene that specifies a dominant phenotype is juxtaposed with another allele specifying a recessive trait, the dominant allele, by definition, wins out—the cell responds by expressing the phenotype of the dominant allele.

The technique of *cell fusion* was well suited to force a confrontation between the alleles specifying normal growth and those directing malignant proliferation. In this procedure, cells of two different phenotypes (and often of different genotypes) are cultured together in a Petri dish (**Figure 7.1A**). An agent is then used to induce fusion of

Figure 7.1 Experimental fusion of cells (A) When cells growing adjacent to one another in monolayer culture are exposed to a *fusogenic* agent, such as inactivated Sendai virus or polyethylene glycol (PEG), they initially form a heterokaryon with multiple nuclei. (Use of selection media and selectable marker genes can ensure the survival of bi- or multinucleated cells carrying nuclei deriving from two distinct parental cell types and the elimination of fused cells whose nuclei derive from only one parental cell type.) When the heterokaryon subsequently passes through mitosis, the two sets of parental chromosomes are pooled in a single nucleus. During propagation of the resulting tetraploid cell in culture, the descendant cells often shed some of these chromosomes, thereby reducing their chromosome complement to a quasi-triploid or hyperdiploid state. (B) In this image, radiolabeled mouse NIH 3T3 cells have been fused (using Sendai virus) with monkey kidney cells, resulting in a hybrid cell with two distinct nuclei. The (larger) mouse nucleus is identified by the silver grains that were formed during subsequent autoradiography. (C) Polykaryons may form with equal or greater frequency following such fusions but are usually unable to proliferate and spawn progeny. This polykaryon contains nine nuclei (*arrows*). (B and C, courtesy of S. Rozenblatt.)

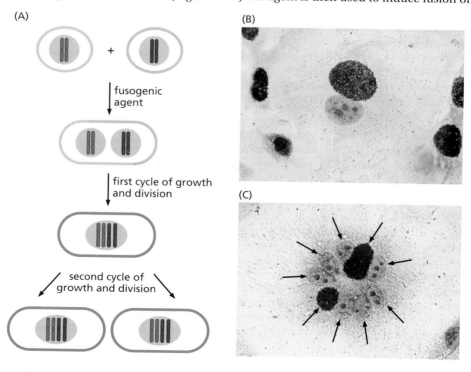

(A)

fusogenic agent

first cycle of growth and division

second cycle of growth and division

(B)

(C)

the plasma membranes of cells that happen to be growing near one another in the cell monolayer. The fusing agent can be either a chemical, such as polyethylene glycol, or a viral glycoprotein, such as the ones displayed on the surface of certain paramyxoviruses like Sendai virus. If only two cells are close to one another and their plasma membranes are in contact, the result of treatment with a fusing agent will be a large cell with a single cytoplasm and two nuclei, often termed a **syncytium** (see Figure 7.1B). If a large number of cells happen to be in close proximity with one another in the cell monolayer being treated, then all of their plasma membranes may become fused to create a single syncytium—a giant multinucleated cell (a **polykaryon**) having one extremely large cytoplasm (see Figure 7.1C).

These fusing agents join cells indiscriminately, so that two identical cells may be fused to one another or cells of two different types may become fused. If the two cells participating in this union happen to be of different origins, then the resulting hybrid cell is termed a **heterokaryon**, to reflect the fact that it carries two genetically distinct nuclei. Genetic tricks can then be used to select for a fused cell that has acquired two genetically distinct nuclei and, at the same time, to eliminate cells that have failed to fuse or happen to carry two identical nuclei. For example, each type of cell may carry a genetic marker that allows it to resist being killed by a particular antibiotic agent to which it would normally be sensitive. In such an instance, after the two cell populations are mixed and subjected to fusion, only heterokaryons will survive when the culture is exposed simultaneously to both antibiotics.

While cells having large numbers of nuclei are generally inviable, cells having only two nuclei are often viable and will proceed to grow. When they enter subsequently into mitosis, the two nuclear membranes break down, the two sets of chromosomes will flock to a single, common mitotic apparatus, and each resulting daughter cell will receive a single nucleus with chromosome complements originating from both parental cell types (see Figure 7.1A).

Some combinations of hybrid cells form genomes that are initially tetraploid, having received a complete diploid genome from each of the parent cells. Often, during the subsequent propagation of such cells in culture, they will progressively shed chromosomes and gradually approach a sub-tetraploid or even triploid chromosomal complement, since these hybrid cell lines begin their lives with far more chromosomes than they really need to grow and survive. Sometimes, the chromosomes from one parent are preferentially shed. For example, when human and mouse cells are fused, the descendant hybrid cells shed human chromosomes progressively until only a small minority of chromosomes are of human origin.

Using the cell fusion technique, hybrid cells of disparate origins, including even chicken–human hybrids, could be made. Cells from the G_1 phase of the cell cycle were fused with cells in M phase in order to see which phase of the cell cycle contained the dominantly acting controls. Undifferentiated cells were fused with more differentiated ones to determine which phenotype dominated. For us, however, the results of only one type of experiment are of special interest: How did normal + tumor hybrid cells grow?

The smart money at the time bet that cancer, a potent, dominating phenotype, would be dominant when placed in opposition to the normal cell growth phenotype (**Figure 7.2**). In this instance, as is often the case, the smart money was wrong. In a number of experiments, when tumor cells were fused with normal cells, the initially formed tetraploid cells (or subsequently arising sub-tetraploid cells) were discovered to have lost the ability to form tumors when these hybrid cells were injected into appropriate host animals. This meant, quite unexpectedly, that the malignant cell phenotype was recessive to the phenotype of normal, wild-type growth.

The notable exception to these observations came when the transformed parental cell in the two-cell hybrid had been transformed by tumor virus infection. On these occasions, the tumor cell phenotype (created by the acquired viral oncogenes) dominated in the cancer- plus normal-cell hybrids. So the take-home message about the cancer versus normal phenotypes was amended and narrowed slightly: if the cancer cell had originally arisen without the involvement of a tumor virus, then its malignant phenotype was recessive when this cell was fused with a normal cell.

Figure 7.2 Dominance and recessiveness of the tumorigenic phenotype The use of cell fusion to test the dominance or recessiveness of the cancer cell phenotype can, in theory, yield two different outcomes—tumorigenic or non-tumorigenic hybrid cells. In fact, when cancer cells derived from most kinds of non-virus-induced human tumors (or from chemically induced rodent tumors) were used in these fusions, the hybrid cells were non-tumorigenic. In contrast, when the cancer cell derived from a virus-induced tumor, the hybrids were usually tumorigenic.

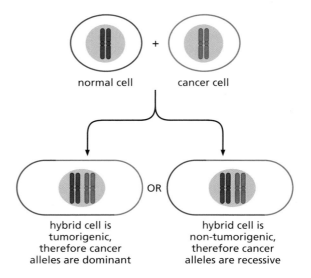

7.2 The recessive nature of the cancer cell phenotype requires a genetic explanation

A simple genetic model could be invoked to explain these outcomes. It depended on the observation, made frequently in genetics, that the phenotype of a mutant, inactive allele is recessive in the presence of an intact, wild-type allele.

The hypothesis went like this. Imagine that normal cells carry genes that constrain or suppress their proliferation. During the development of a tumor, the evolving cancer cells shed or inactivate one or more of these genes. Once these growth-suppressing genes are lost, the proliferation of the cancer cells accelerates, no longer being held back by the actions of these growth-suppressing genes. As long as the cancer cell lacks these genes, it continues to proliferate in a malignant fashion. However, the moment that wild-type, intact versions of these genes operate once again within the cancer cell, having been introduced by the technique of cell fusion, the proliferation of the cancer cell, or at least its ability to form tumors, grinds abruptly to a halt.

Since the wild-type versions of these hypothetical genes antagonize the cancer cell phenotype, these genes came to be called tumor suppressor genes (TSGs). There were arguments both in favor of and against the existence of tumor suppressor genes. In their favor was the fact that it is far easier to inactivate a gene by a variety of mutational mechanisms than it is to hyperactivate its functioning through a mutation. For example, a *ras* proto-oncogene can be (hyper)activated only by a point mutation that affects its 12th, 13th, or 61st codon (Section 4.4; see Figure 4.10). In contrast, a tumor suppressor gene, or, for that matter, any other gene, can readily be inactivated by point mutations that strike at many sites in its protein-coding sequences or by random deletions that excise blocs of nucleotides from these sequences.

The logical case against the existence of tumor suppressor genes derived from the diploid state of the mammalian cell genome. If mutant, inactive alleles of TSGs really did play a role in enabling the growth of cancer cells, and if these alleles were recessive, an incipient tumor cell would reap no benefit from inactivating only one of its two copies of a TSG, since the recessive mutant allele would coexist with a dominant wild-type allele in this cell. Hence, it seemed that both wild-type copies of a given tumor suppressor gene would need to be eliminated by an aspiring cancer cell before this cell would truly benefit from its inactivation.

This requirement for two separate genetic alterations seemed complex and unwieldy, indeed, too improbable to occur in a reasonably short period of time. Since the likelihood of two mutations occurring is the square of the probability of a single mutation, this made it seem highly unlikely that tumor suppressor genes could be fully inactivated during the time required for tumors to form. For example, if the probability of inactivating a single gene copy by mutation is on the order of 10^{-6} per cell generation, the probability of silencing both copies is on the order of 10^{-12} per cell generation—exceedingly unlikely, given the small size of incipient cancer cell populations, the multiple genetic alterations that are needed to make a human tumor, and the several decades of time during which multiply-mutated cells expand into clinically detectable tumors.

7.3 The retinoblastoma tumor provides a solution to the genetic puzzle of tumor suppressor genes

The arguments for and against the involvement of tumor suppressor genes in tumor development could never be settled by the cell fusion technique. For one thing, this experimental strategy, on its own, offered little prospect of finding and isolating specific genes whose properties could be studied and used to support the tumor suppressor gene hypothesis. In the end, the important insights came from studying a rare childhood eye tumor, **retinoblastoma**.

This tumor of the retina, arising in the precursors of photoreceptor cells, is normally observed in about 1 in 20,000 children (**Figure 7.3**). It is diagnosed anytime from birth up to the age of 6 to 8 years, after which the disease is rarely encountered. The tumor **syndrome** (that is, a constellation of clinical traits) appears in two forms. Some children—those who are born into families with no history of retinoblastoma—present in the clinic with a single tumor in one eye. If this tumor is eliminated, either by radiation

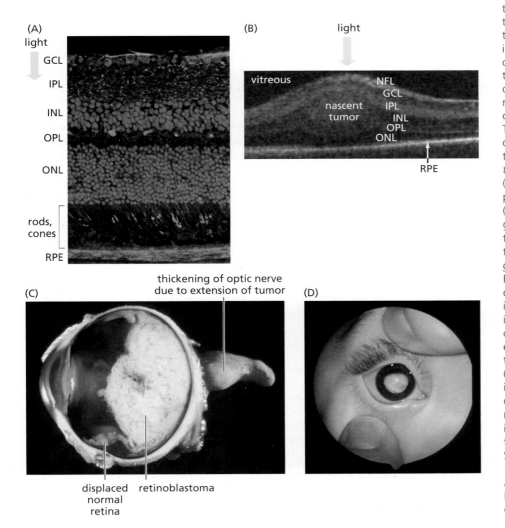

Figure 7.3 Pediatric retinoblastomas For still incompletely understood reasons, defects in the *Rb* gene lead preferentially to the formation of pediatric tumors in the retina, a small, highly specialized tissue, even though the *Rb* gene is active in governing cell proliferation in most cell types throughout the body. (A) The tumors arise from a stem cell precursor of the cone cells. This stem cell is closely related to the *oligopotential* precursor of all of the retinal cell types seen here. This micrograph reveals a cross section of the retina showing, from the back of the globe to the front of the eye (*bottom to top*), retinal pigmented epithelium (RPE), outer nuclear layer (ONL), outer plexiform layer (OPL), inner nuclear layer (INL), inner plexiform layer (IPL), and ganglion cell layer (GCL). [The nuclei of the rods and cones (*red*) are located in the ONL, with the photoreceptors (*dark green*) extending downward toward the RPE.] (B) Created by using Fourier domain optical coherence tomography, this image shows a nascent retinoblastoma in the eye of a several-week-old *Rb*$^{+/-}$ child. The nascent tumor appears to be emerging from the inner nuclear layer of the retina. NFL, nerve fiber layer. (C) A section of the globe of an eye in which a large retinoblastoma has developed and begun to invade the optic nerve. (D) A retinoblastoma may often initially present in the clinic as an opacity that obscures the retina. (A, from S.W. Wang et al., *Development* 129:467–477, 2002. B, courtesy of A. Mallipatna, C. Vandenhoven, B.L. Gallie and E. Heon. C and D, courtesy of T.P. Dryja.)

or by removal of the affected eye, then this child usually has no further risk of retino-blastoma and no elevated risk of tumors elsewhere in the body. Because this tumor strikes in children lacking a family history, it is considered to be a manifestation of the **sporadic** form of this disease. (Since this form of the disease affects only a single eye, it is often termed *unilateral* retinoblastoma.)

The **familial** form of retinoblastoma appears in children having a parent who also suf-fered from and was cured of the disease early in life. In this instance, there are usually multiple foci of tumors arising in both eyes (and the condition is therefore called *bilat-eral* retinoblastoma). Moreover, curing the eye tumors, which can be accomplished with radiation or surgery, does not protect these children from a greatly increased (more than 500 times above normal) risk of bone cancers (osteosarcomas) during adolescence and an elevated susceptibility to developing yet other tumors later in life (**Figure 7.4A**). Those who survive these tumors and grow to adulthood are usually able to reproduce; in half of their offspring, the familial form of retinoblastoma again rears its head (see Figure 7.4B).

The familial form of retinoblastoma is passed from one generation to the next in a fashion that conforms to the behavior of a Mendelian dominant allele. In truth, a form of retinoblastoma that is indistinguishable from the familial disease can be seen in children with no family history. Apparently, mutations occur *de novo* in a parent dur-ing the formation of a sperm or an egg, causing the children to acquire the cancer-inducing mutation in all of their cells. This creates a genetic situation that is identical to the one occurring when the mutant allele is already present in the genome of one or another afflicted parent. (The great majority of these *de novo* mutations are of paternal origin, apparently because far more cycles of cell division precede the formation of a sperm than precede the formation of an egg; each cycle of cell growth and division presents a risk of mistakes in DNA replication and resulting mutations.)

Upon studying the kinetics with which retinal tumors appeared in children affected by either the familial or the sporadic version of the disease, Alfred Knudson concluded in 1971 that the rate of appearance of familial tumors was consistent with a single ran-dom event, while the sporadic tumors behaved as if two random events were required for their formation (**Figure 7.5**).

These kinetics led to a deduction, really a speculation. Let's say that there is a gene called *Rb*, the mutation of which is involved in causing childhood retinoblastoma. We can also imagine that the mutations that involve the *Rb* gene in tumor development

Figure 7.4 Unilateral versus bilateral retinoblastoma (A) Children with unilateral retinoblastoma and without an afflicted parent are considered to have, with high probability, a sporadic form of the disease, while those with bilateral retinoblastoma suffer from a familial form of this disease. This graph shows the clinical courses of 1601 retinoblastoma patients who had been diagnosed between 1914 and 1984. As is apparent, those cured of bilateral tumors (*red line*) have a dramatically higher risk of subsequently developing tumors in a variety of organ sites than those with unilateral tumors (*blue line*). (A portion of this elevated risk is attributable to tumors that arose in the vicinity of the eyes because of the *radiotherapy* that was used to eliminate the retinoblastomas when these individuals were young.) (B) This pedigree shows multiple generations of a kindred afflicted with familial retinoblastoma, a disease that usually strikes only 1 in 20,000 children. Such multiple-generation pedigrees were rarely observed before the advent of modern medicine, which allows an affected child to be cured of the disease and therefore reach reproductive age. Males (*squares*), females (*circles*), affected individuals (*green*), unaffected individuals (*clear circles, squares*). In one lineage, tumor development skipped a generation (*arrow*), an example of *incomplete penetrance*, in which the genotype of the individual fails, for complex biological reasons, to affect phenotype. (By necessity, the indicated man carried the allele since two of his daughters developed the disease.) (A, from R.A. Kleinerman et al., *J. Clin. Oncol.* 23:2272–2279, 2005. B, courtesy of T.P. Dryja.)

(A) non-retinal tumors of retinoblastoma patients

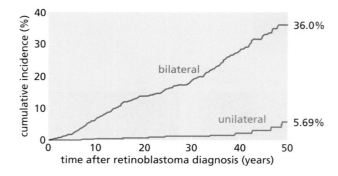

(B)

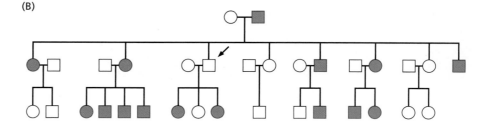

Figure 7.5 Kinetics of appearance of unilateral and bilateral retinoblastomas Alfred Knudson Jr. studied the kinetics with which bilateral and unilateral retinoblastomas appeared in children. He calculated that the bilateral cases (tumors in both eyes) arose with one-hit kinetics, whereas the unilateral tumors (affecting only one eye) arose with two-hit kinetics. Each of these hits was presumed to represent a somatic mutation. The fact that the two-hit kinetics involved two copies of the *Rb* gene was realized only later. Percentage of cases not yet diagnosed (*ordinate*) at age (in months) indicated (*abscissa*). (From A.G. Knudson Jr., *Proc. Natl. Acad. Sci. USA* 68:820–823, 1971.)

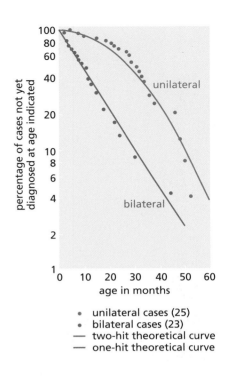

- unilateral cases (25)
- bilateral cases (23)
- —— two-hit theoretical curve
- —— one-hit theoretical curve

invariably create inactive and thus recessive alleles of the *Rb* gene. If the tumor-predisposing alleles of *Rb* are indeed recessive, then both copies of the *Rb* gene must be knocked out before a retinal cell can launch the uncontrolled proliferation that eventually results in a tumor (**Figure 7.6**).

In children who inherit a genetically wild-type constitution from their parents, the formation of a retinoblastoma tumor will require two successive genetic alterations in a retinal cell lineage that inactivate the two functional copies of the *Rb* gene—that is, two somatic mutations (see Figure 7.6). These dynamics will be quite different in half of the children born into a family in which familial retinoblastoma is present. In such children, we can imagine that one of the two required *Rb* gene mutations has been passed through the germ line from a parent to a fertilized egg. This mutant *Rb* gene is therefore implanted in all the cells of the developing embryo, including all the cells in the retina. Any one of these retinal cells needs to sustain only a single (somatic) mutation knocking out the still-wild-type allele in order to reach the state where it no longer harbors any functional *Rb* gene copies and can therefore spawn a retinoblastoma.

Strikingly, this depiction of *Rb* gene behavior corresponded exactly to the attributes of tumor suppressor genes whose existence had been postulated from results of the cell

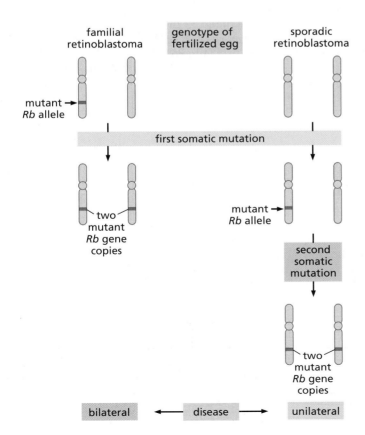

Figure 7.6 Dynamics of retinoblastoma formation The two affected genes predicted by Knudson (see Figure 7.5) are the two copies of the *Rb* gene on human Chromosome 13. In familial retinoblastoma, the zygote (fertilized egg) from which a child derives carries one defective copy of the *Rb* gene, and all retinal cells in this child will therefore carry only a single functional *Rb* gene copy. If this surviving copy of the *Rb* gene is eliminated in a retinal cell by a somatic mutation, that cell will lack all *Rb* gene function and will be poised to proliferate into a mass of tumor cells. In sporadic retinoblastoma, the zygote is genetically wild type at the *Rb* locus, and retinoblastoma development therefore will require two successive somatic mutations striking the two copies of the *Rb* gene carried by a lineage of retinal precursor cells. Because only a single somatic mutation is needed to eliminate *Rb* function in familial cases, multiple cells in both eyes are affected. However, the convergence of two somatic mutations—each a rare event—on a single cell lineage is mathematically unlikely, indicating that at most one single cell lineage will be affected, yielding unilateral disease.

Sidebar 7.1 Mutant *Rb* genes are both dominant and recessive While the molecular evidence underlying *Rb* genetics has not yet been presented in this chapter, our discussion up to this point already carries with it an apparent internal contradiction. At the level of the whole organism, an individual who inherits a mutant, defective allele of *Rb* is almost certain to develop retinoblastoma at some point in childhood. Hence, the disease-causing *Rb* allele acts dominantly at the level of the *whole organism*. However, if we were able to study the behavior of the mutant allele *within a cell* from this individual, we would conclude that the mutant allele acts recessively, since a cell carrying a mutant and a wild-type *Rb* gene copy would behave normally. Hence, the mutant *Rb* allele acts dominantly at the *organismic* level and recessively at the *cellular* level. (A geneticist insistent on strictly proper usage of genetic terms would say that the *phenotype* of retinoblastoma behaves dominantly at the organismic level and recessively at the cellular level; we will ignore this propriety here.)

The dominant behavior of the mutant *Rb* allele at the organismic level depends on the size of the target cell population in the developing retina. For example, if there are 5×10^6 cells that are susceptible to transformation in the average retina (Knudson assumed 1×10^6) and if the rate of loss of the surviving wild-type allele in the cells of an Rb^+/Rb^- heterozygous individual is 1 per 10^6 cells, then on average this individual will develop 5 cells in each eye that have lost the surviving *Rb* allele and therefore lost all *Rb* gene function; each of the resulting Rb^-/Rb^- cells is then qualified to spawn a retinoblastoma, yielding multiple (~5) tumors in each eye of a child who has inherited a defective *Rb* allele. Hence, the presence of heterozygous cells throughout each retina virtually preordains disease development.

fusion studies (**Sidebar 7.1**). Still, all this remained no more than an attractive hypothesis, since the *Rb* gene had not been cloned and no molecular evidence was available demonstrating that *Rb* gene copies had been rendered inactive by mutation.

7.4 Incipient cancer cells invent ways to eliminate wild-type copies of tumor suppressor genes

The findings with the *Rb* gene reinforced the notion that both copies of this gene needed to be eliminated, raising once again a vexing issue: How could two copies of a tumor suppressor gene possibly be eliminated, one after the other, during the formation of sporadic retinoblastomas if the probability of both required mutational events occurring is, as calculated earlier, about 10^{-12} per cell generation? Given the relatively small target cell populations in the developing retina (Knudson guessed about 10^6 cells), it seemed highly unlikely that both gene copies could be eliminated through two, successive mutational events.

A solution to this dilemma was suggested by some geneticists. Suppose that the first of the two *Rb* gene copies was indeed inactivated by some type of mutational event that occurred with a frequency comparable to that associated with most mutational events—about 10^{-6} per cell generation. A cell suffering this mutation would now be in a heterozygous configuration, having one wild-type and one defective gene copy—that is, $Rb^{+/-}$. Since the mutant *Rb* allele was recessive at the cellular level, this heterozygous cell would continue to exhibit a wild-type phenotype. But what if the second, still-intact gene copy of *Rb* were inactivated by a mechanism that did not depend on its being struck directly by a second, independent mutational event? Instead, perhaps there was some exchange of genetic information between paired homologous chromosomes, one of which carried the wild-type *Rb* allele while the other carried the already-mutant, defective allele. Normally, recombination between chromosomes was known to occur almost exclusively during meiosis. What would happen if, instead, a recombination occurred between one of the chromatid arms carrying the wild-type *Rb* allele and a chromatid from the paired chromosome carrying the mutant allele (**Figure 7.7**)? Such recombination was thought to occur during active cell proliferation and consequently was termed **mitotic recombination** to distinguish it from the meiotic recombination events that shuffle chromosomal arms prior to the formation of sperm and egg.

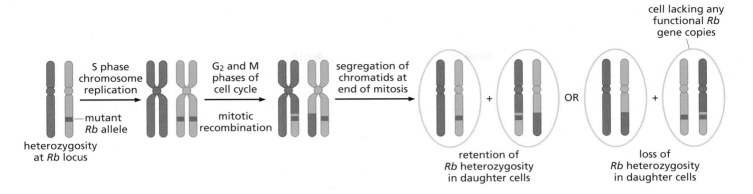

cell lacking any functional *Rb* gene copies

heterozygosity at *Rb* locus — mutant *Rb* allele — S phase chromosome replication — G₂ and M phases of cell cycle / mitotic recombination — segregation of chromatids at end of mitosis — retention of *Rb* heterozygosity in daughter cells OR loss of *Rb* heterozygosity in daughter cells

In this fashion, the chromosomal arm carrying the wild-type *Rb* allele might be replaced with a chromosomal arm carrying the mutant allele derived from the paired, homologous chromosome. Since this process depends upon reciprocal exchange of genetic information between chromosomal arms, the participating chromosomes would both remain full-length and, when visualized in the microscope during a subsequent metaphase, be indistinguishable from the chromosomes that existed prior to this genetic exchange. However, at the genetic and molecular level, one of the cells emerging from this recombination event would have shed its remaining wild-type *Rb* allele and would therefore have become *Rb*⁻/⁻. Importantly, this mitotic recombination was found to occur at a frequency of 10^{-5} to 10^{-4} per cell generation, and was therefore a far easier way for a cell to rid itself of the remaining wild-type copy of the *Rb* gene than by mutational inactivation, which, as mentioned above, was known to occur at a frequency of about 10^{-6} per cell generation.

Prior to this mitotic recombination, the two homologous chromosomes (in the case of the *Rb* gene, the two human Chromosomes 13) differ from one another in many subtle details. After all, one is of paternal, the other of maternal origin, indicating that they are heterozygous at many genetic loci (Supplementary Sidebar 7.1). However, following the mitotic recombination that leads to homozygosity at the *Rb* locus (yielding either an *Rb*⁻/⁻ or an *Rb*⁺/⁺ genotype), a number of genes that are nearby neighbors of *Rb* on Chromosome 13 will also lose heterozygosity and become homozygous. This genetic alteration of a gene or a chromosomal region is usually termed loss of heterozygosity, or simply LOH. (An alternative term is "allelic deletion.") Interestingly, LOH events may well occur with different frequencies in different human populations (Sidebar 7.2).

There are yet other ways by which chromosomes can achieve LOH. One of these derives from the process of **gene conversion** (Figure 7.8). In this alternative mechanism, a DNA strand being elongated during DNA replication temporarily switches templates by leaving its original template strand and forms a hybrid with the complementary DNA strand belonging to the homologous chromosome. After progressing

Figure 7.7 Elimination of wild-type *Rb* gene copies Mitotic recombination can lead to loss of heterozygosity (LOH) of a gene such as *Rb*. Genetic material is exchanged between two homologous chromosomes (for example, the two Chromosomes 13) through the process of genetic crossing over occurring during G₂ phase or, less often, during the M phase of the cell cycle. The subsequent segregation of chromatids may yield a pair of daughter cells that both retain heterozygosity at the *Rb* locus. (The mutant *Rb* allele is indicated here as a *blue line*.) With equal probability, this process can yield two daughter cells that have undergone LOH at the *Rb* locus (and other loci on the same chromosomal arm), one of which is homozygous mutant at the *Rb* locus while the other is homozygous wild type at this locus.

Sidebar 7.2 Does endogamy lead to high rates of LOH? Experiments using laboratory and feral mice indicate that the frequency of mitotic recombination can be suppressed tenfold or more if homologous chromosomes are genetically very different from one another rather than genetically similar or identical to one another. This indicates that the pairing of homologous chromosomes that enables mitotic recombination to occur depends on extensive nucleotide sequence identity in the two DNAs.

This finding may well have important implications for tumor progression in humans. In many human populations, **endogamy** (breeding within a group) and even first-cousin marriages are quite common. Mendelian genetics dictates that in the offspring of such individuals, chromosomal regions of homozygosity are frequent. If the results of mouse genetics are instructive, then pairs of homologous chromosomes are more liable to undergo mitotic recombination in these individuals. Should one of the two chromosomes carry a mutation in a tumor suppressor gene such as *Rb*, mitotic recombination and LOH leading to loss of the surviving wild-type gene copy of this tumor suppressor gene will become proportionately more likely. Consequently, it may be that the offspring of cousin marriages in human populations may have greater susceptibility to tumors simply because their chromosomes participate more frequently in LOH. This speculation has not been tested.

Figure 7.8 Gene conversion During the process of gene conversion, a DNA polymerase initially begins to use a DNA strand on one chromosome (*red*) as a template for the synthesis of a new daughter strand of DNA (*blue*). After advancing some distance down this template strand, the polymerase may jump to the homologous chromosome and use a DNA strand of this other chromosome (*green*) as a template for the continued elongation of the daughter strand. After a while, the polymerase may jump back to the originally used DNA template strand and continue replication. In this manner, a mutant tumor suppressor gene allele, such as a mutant allele of *Rb*, may be transmitted from one chromosome to its homolog, replacing the wild-type allele residing there.

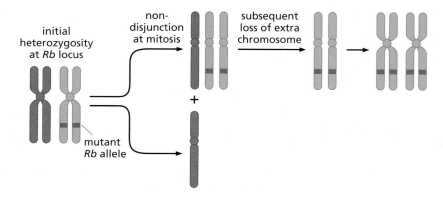

after copying segment of green template strand, DNA polymerase jumps back to template strand of red chromosome and continues copying

template strand of homologous chromosome

template strand of a chromosome

DNA polymerase begins replication on template strand of red chromosome

DNA polymerase jumps to template strand of homologous, green chromosome

some distance along this strand, it disentangles and reverts to using as template the DNA strand on which it originated (a mechanism sometimes termed **copy choice**). Accordingly, this newly synthesized strand of DNA will acquire DNA sequences from a stretch of the paired chromosome. Should this gene conversion involve copying of an already-inactivated *Rb* allele, for example, then once again LOH will have occurred in this chromosomal region. Such gene conversion is known to occur even more frequently per cell generation than mitotic recombination.

Note, by the way, that LOH can also be achieved by simply breaking off and discarding an entire chromosomal region without replacing it with a copy duplicated from the other, homologous chromosome. This results in **hemizygosity** of this chromosomal region, where now all the genes on this chromosomal arm are present in only a single copy rather than the usual two copies per cell. Cells can survive with only one copy of some chromosomal regions, while the loss of a single copy of other chromosomal regions seems to put cells at a distinct biological disadvantage. When loss of a chromosomal segment does in fact occur (yielding hemizygosity), this event is registered as an LOH, since only one allelic version of a gene in this chromosomal region can be detected.

In many tumors, LOH seems to be achieved through the loss of an entire chromosome due to inappropriate chromosomal segregation at mitosis (the process of **nondisjunction**; **Figure 7.9**). In a descendant cell, one of the three chromosomes resulting from the initial nondisjunction event may be shed, leaving two identical copies of one of the homologous chromosomes behind. Alternatively, detailed examination of chromosomal complements in human colon carcinoma cells has shown that, at least in these cancers, LOH is frequently achieved via genetic alterations that change chromosomal structure and thereby affect the karyotype of cells. Many of these events appear to be translocations, in that they involve recombination between chromosomal arms

Figure 7.9 Chromosomal nondisjunction and loss of heterozygosity (LOH) LOH can occur through the mis-segregation of chromosomes during mitosis. Chromosomal nondisjunction occurs when one daughter cell retains both chromatids of a chromosome rather than its usual allotment of a single chromatid. The resulting triploidy of this chromosome in one daughter cell (*above*) may be disadvantageous, so that descendants of this cell may shed the extra chromosome and revert to a diploid state. Since the supernumerary chromosome is likely to be shed randomly, this may result in homozygosity for this chromosome and loss of the other allele (carried here by the *red* chromosome).

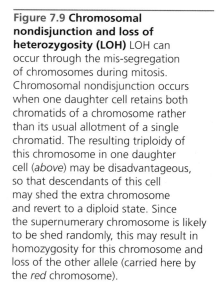

initial heterozygosity at *Rb* locus

non-disjunction at mitosis

subsequent loss of extra chromosome

mutant *Rb* allele

on nonhomologous chromosomes. Such recombination events might be triggered by double-stranded DNA breaks followed by fusion of the resulting ends with DNA sequences originating in other chromosomes. In this fashion, some genetic regions may be duplicated and others lost altogether. Later (Section 10.4) we will discuss one specific molecular mechanism whereby such nonhomologous recombination events may occur during the development of a tumor.

Because LOH was thought to occur at a far higher frequency than mutational alteration of genes, this meant that the second, still-intact copy of a gene like *Rb* was far more likely to be lost through LOH than through a mutation directly striking this gene copy (Supplementary Sidebar 7.2). Consequently, the majority of retinoblastoma tumor cells were predicted to show LOH at the *Rb* locus and at nearby genetic markers on Chromosome 13 and only a small minority were predicted to carry two distinct, mutant alleles of *Rb*, each inactivated by an independent mutational event.

7.5 The *Rb* gene often undergoes loss of heterozygosity in tumors

In 1978, the chromosomal localization of the *Rb* gene was surmised from study of the chromosomes present in retinoblastomas. In a small number of retinal tumors, careful karyotypic analysis of metaphase chromosome spreads revealed interstitial deletions (see Sidebar 1.2) within the long (q) arm of Chromosome 13. Even though each of these deletions began and ended at different sites in this chromosomal arm, they shared in common the loss of chromosomal material in the 4th band of the 1st region of this chromosomal arm, that is, 13q14 (**Figure 7.10**). The interstitial deletions affecting this chromosomal band involved many hundreds, often thousands, of kilobases of DNA, indicating that a number of genes in this region had been jettisoned simultaneously by the developing retinal tumor cells. The fact that these changes involved the *loss* of genetic information provided evidence that the *Rb* gene, which was imagined to lie somewhere in this chromosomal region, had been discarded—precisely the outcome predicted by the tumor suppressor gene theory. (In the great majority of retinoblastomas, the mutations that knock out the *Rb* gene affect far smaller segments of chromosomal DNA and are therefore **submicroscopic**, that is, invisible upon microscopic analysis of metaphase chromosome spreads.)

Through good fortune, a second gene lying in the 13q14 chromosomal region had already been reasonably well characterized. This gene, which encodes the enzyme esterase D, is represented in the human gene pool by two distinct alleles whose protein products migrate at different rates in gel electrophoresis (**Figure 7.11**). The esterase D locus presented geneticists with a golden opportunity to test the LOH theory. Recall that when LOH occurs, an entire chromosomal region is usually affected. Hence, since the esterase D locus had been mapped close to the *Rb* gene locus on the long arm of Chromosome 13, if the *Rb* locus suffered LOH during tumor development, then the

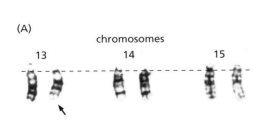

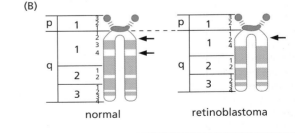

Figure 7.10 Chromosomal localization of the *Rb* locus (A) The characteristic darkly staining bands of each human chromosome delineate that chromosome's specific subregions. In this case, careful study of the karyotype of retinoblastoma cells, performed on the condensed chromosomes of late prophase or metaphase cells, revealed a deletion in the long arm (q) of Chromosome 13 in a 6-year-old retinoblastoma patient (*arrow*), while Chromosomes 14 and 15 appeared normal. (B) Careful cytogenetic analysis revealed that the deletion was localized between bands 13q12 and q14, that is, between the 2nd and 4th bands of the 1st region of the long arm of Chromosome 13. (A, from U. Francke, *Cytogenet. Cell Genet.* 16:131–134, 1976.)

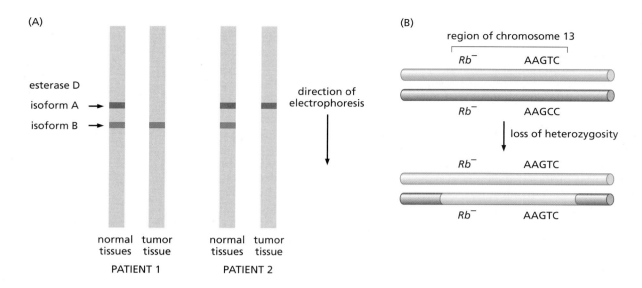

Figure 7.11 Demonstrations of loss of heterozygosity at the *Rb* locus
(A) These *zymograms* represent analyses in which the two different forms (*isoforms*) of the esterase D enzyme are separated by gel electrophoresis. The presence of each isoform can be identified by the product of its biochemical reaction, depicted here as a band on a gel. In one retinoblastoma patient who was heterozygous at the esterase D locus, normal tissues expressed two forms of this enzyme, seen here as two distinct bands, while the tumor tissue specified only one form of the enzyme, indicating a loss of heterozygosity (LOH) at this locus. Comparisons of the esterase D in the normal and tumor cells of another retinoblastoma patient also revealed LOH at the esterase D locus in his tumor cells—in that case, resulting from loss of the chromosomal region specifying the other isoform of the enzyme.
(B) An "anonymous" nucleotide sequence, which is genetically linked to the *Rb* locus but is not part of any known gene, can also be used to follow changes in this chromosomal region. The pentanucleotide shown may initially be in a heterozygous configuration (*top*) and during the process that leads to LOH at the *Rb* locus may also undergo LOH. (A, adapted from R.S. Sparkes et al., *Science* 219:971–973, 1983.)

esterase D locus should frequently have suffered the same fate. In effect, the esterase D locus could act as a "surrogate marker" for the still-mysterious, uncloned *Rb* gene.

Indeed, when researchers scrutinized the tumor cells of several children who were born heterozygous at the esterase D locus (having inherited two distinct esterase D alleles from their parents), they found that these tumor cells had lost an esterase D allele and thus must have undergone LOH (see Figure 7.11A). This suggested that the closely linked *Rb* gene had also undergone LOH. Such a change conformed to the theorized behavior ascribed to a growth-suppressing gene, both of whose wild-type alleles needed to be jettisoned before a cell would grow uncontrollably.

In a more general sense, the swapping of information between two chromosomes, which may occur by several alternative mechanisms (see Figures 7.7 and 7.8), was likely to involve not only the *Rb* gene but also a large number of genes and genetic markers flanking the *Rb* locus on both sides. Their behavior, like that of the esterase D gene, is also of interest, since they too will have undergone LOH. For example a genetic marker sequence in a neighboring region of Chromosome 13 that was initially AAGCC/AAGTC (hence heterozygous) may now become either AAGCC/AAGCC or AAGTC/AAGTC (and therefore homozygous; see Figure 7.11B). Like the esterase D gene, this neighboring oligonucleotide segment has been carried "along for the ride" with the nearby *Rb* gene during the swapping of sequences between the two homologous chromosomes. Indeed, precisely this type of analysis was performed in 1982, and the results reinforced the conclusion reached from analysis of the esterase D locus. (Note, by the way, that the genetic marker sequence—a pentanucleotide in Figure 7.11B—may lie many kilobases distant from the *Rb* locus in an intergenic region of Chromosome 13q14. Since it does not encode any biological function, its conversion from heterozygous to homozygous configuration needs to be determined using some type of molecular analysis of the cell's DNA, such as nucleotide sequencing or hybridization with an appropriate sequence-specific probe.)

In 1986 the *Rb* gene was cloned with the aid of a DNA probe that recognized an **anonymous** genomic sequence (see Supplementary Sidebar 7.1) located somewhere in Chromosome 13q14; fortuitously, this sequence was located within the *Rb* gene itself. Additional DNA probes derived from different portions of the cloned gene revealed, as was previously speculated, that the *Rb* gene in retinoblastomas suffered mutations that resulted in its inactivation (**Figure 7.12**). In some retinoblastomas, these mutations involved large deletions within the *Rb* gene and flanking DNA sequences. Southern blotting (see Supplementary Sidebar 4.3) of the DNA from these tumors revealed that the resulting mutant *Rb* allele was usually present in homozygous configuration. This meant that after the creation of a **null** (inactive) allele of *Rb* on one

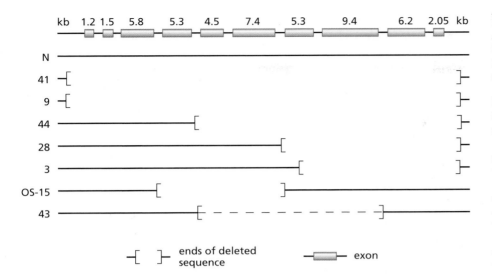

-[}- ends of deleted sequence   —▭— exon

Figure 7.12 Mutations of the *Rb* gene
A cloned *Rb* cDNA was used as a probe in Southern blot analyses of genomic DNAs from a variety of retinoblastomas and an osteosarcoma. Shown are the 10 exons of the *Rb* gene (*blue cylinders*) labeled by their length. The normal (N) version of the gene encompasses a chromosomal region of approximately 190 kilobases. However, analysis of a subset of retinoblastoma DNAs indicated that significant portions of this gene had undergone deletion (the segments within each pair of brackets). Thus, tumors 41 and 9 lost the entire *Rb* gene, apparently together with flanking chromosomal DNA segments on both sides. Tumors 44, 28, and 3 lost, to differing extents, the right half of the gene together with rightward-lying chromosomal segments. The fact that an osteosarcoma (OS-15) and a retinoblastoma (43) lost internal portions of this gene argued strongly that this 190-kb chromosomal DNA segment, and not leftward- or rightward-lying DNA segments, was the repeated target of mutational inactivation occurring during the development of these tumors. (The dashed line indicates that the deletion in tumor 43 was present in heterozygous configuration.) (From S.H. Friend et al., *Nature* 323:643–646, 1986.)

chromosome, the corresponding region on the other, homologous chromosome was discarded, leading to loss of heterozygosity (LOH) at this locus. These data directly validated many of the predictions of the theoretical models that had been proposed to explain how tumor suppressor genes behave during the development of cancers.

As mentioned, children who inherit a defective *Rb* gene copy are also predisposed to osteosarcomas (bone tumors) as adolescents. With the *Rb* gene probe in hand, it became possible to demonstrate that these osteosarcomas also carried structurally altered *Rb* genes (see Figure 7.12). At the same time, these findings highlighted a puzzle that remains largely unsolved to this day: Why does a gene such as *Rb*, which operates in a wide variety of tissues throughout the body (as we will learn in Chapter 8), cause predominantly retinal and bone tumors when it is inherited in defective form from a parent? Why are not all tissues at equal risk?

7.6 Loss-of-heterozygosity events can be used to find tumor suppressor genes

Numerous tumor suppressor genes that operate like the *Rb* gene were presumed to lie scattered around the human genome and to play a role in the pathogenesis of many types of human tumors. In the late 1980s, researchers interested in finding these genes were confronted with an experimental quandary: How could one find genes whose existence was most apparent when they were missing from a cell's genome? The dominantly acting oncogenes, in stark contrast, could be detected far more readily through their presence in a retrovirus genome, through the transfection-focus assay, or through their presence in a chromosomal segment that repeatedly underwent gene amplification in a number of independently arising tumors.

A more general strategy was required that did not depend on the chance observation of interstitial chromosomal deletions or the presence of a known gene (for example, esterase D) that, through good fortune, lay near a tumor suppressor gene on a chromosome. Both of these conditions greatly facilitated the isolation of the *Rb* gene. In general, however, the searches for most tumor suppressor genes were not favored by such strokes of good luck.

The tendency of tumor suppressor genes to undergo LOH during tumor development provided cancer researchers with a novel genetic strategy for tracking them down. Since the chromosomal region flanking a tumor suppressor gene seemed to undergo LOH together with the TSG itself, one might be able to detect the existence of a still-uncloned tumor suppressor gene simply from the fact that an anonymous genetic marker lying nearby on the chromosome repeatedly undergoes LOH during the development of a specific type of human tumor. A hint of this strategy was already provided by Figure 7.11B, in which an anonymous genetic marker—a pentanucleotide sequence genetically linked to *Rb*—was used to follow the behavior of the *Rb* gene itself. This strategy could be generalized by exploiting DNA segments that were mapped to sites throughout the human genome. Once again, these DNA segments had no obvious affiliations with specific genes. In some individuals, such a segment of DNA could be cleaved by a restriction enzyme; in others, the same stretch of DNA resisted cleavage as a result of a single base-pair substitution in that sequence (**Figure 7.13A**). Because such sequence variability appeared to occur as a consequence of normal genetic variability in the human gene pool, such sites were thought to represent polymorphic genetic markers (Section 1.2). Moreover, since the allelic versions of this sequence either permit or disallow cleavage by a restriction enzyme, such a marker was termed a **restriction fragment length polymorphism** (RFLP).

Cancer geneticists used RFLP markers to determine whether various chromosomal regions frequently underwent LOH during the development of certain types of tumors. Figure 7.13B shows the results of using RFLPs to search for regions of LOH arising in a group of **colorectal tumors**, that is, carcinomas of the colon and rectum. In this case, the long and short arms of most chromosomes were represented by at least one RFLP marker, and the fate of the entire chromosomal arm was studied. Note that in this series of human tumors, the short (p) arm of Chromosome 17 and the long (q) arm of Chromosome 18 suffered unusually high rates of LOH. These two chromosomal arms stood out, rising far above the background level of the 15–20% LOH that affected all chromosomal arms equally in these tumor cells. (This background level of LOH reveals the fact that all chromosomal regions have some tendency to undergo LOH at a certain rate. However, if the LOH happens to occur in a region harboring a tumor suppressor gene, the proliferation or survival of the cells in which this has occurred may be favored, leading to high numbers of tumors carrying this particular LOH.)

The fact that specific arms of Chromosomes 17 and 18 frequently underwent LOH in tumors provided strong evidence that both arms harbored still-unknown tumor suppressor genes that also suffered LOH. Thus, this genetic localization provided the gene cloners with a clear indication of where in the genome they should search for the culprit tumor suppressor genes that seemed to be playing important roles in the development of these colorectal tumors. More recently, cancer geneticists have used other molecular strategies to detect sequence polymorphisms and their loss in the genomes of human cancer cells.

Like RFLP markers, these newer strategies depend on the detection of single-nucleotide polymorphisms (SNPs, pronounced "snips") in normal and cancer cell genomes. These SNPs are detected using polymerase chain reaction (PCR) primers that are complementary to one or the other polymorphic allele (Supplementary Sidebar 7.3). This innovation allows the identification of vastly more SNPs in the human gene pool. Importantly, SNP markers, whether identified by RLFP or by the newer PCR-based techniques, become useful in LOH analyses only if a significant proportion of cancer patients are heterozygous at these marker loci in their normal tissues.

By 2011, more than 3.5 million SNPs had been documented in the gene pool of a large, genetically heterogeneous population—that of the United States. Ninety percent of these are present at an allele frequency of 10% or greater and almost all of the remaining ones are present with an allele frequency of 1% or more. In more practical terms, this means that in an average person's genome, two-thirds of the SNPs are present in the heterozygous state and one-third are homozygous (and thus not useful for LOH studies). This results in a SNP marker (in heterozygous configuration) planted on average every 1 kilobase across the average person's genome. Such greatly increased density, compared with that afforded by the earlier RFLP analyses, allows regions of LOH to be localized far more precisely to chromosomal regions that may contain a relatively small number of genes, facilitating in turn the identification, using the known sequences of the human genome, of candidate TSGs within these regions. This level of precision should be compared with the analyses of 1989, in which RFLP mapping of LOH in colorectal tumors allowed investigators to localize TSGs only to

(A)

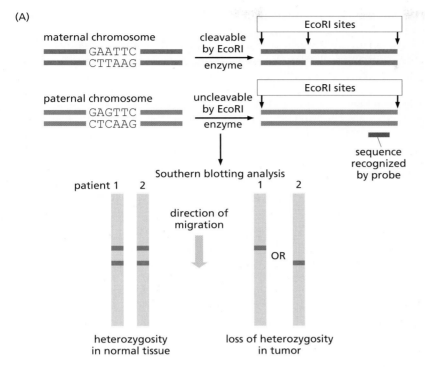

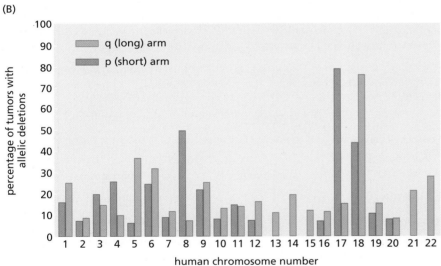

Figure 7.13 Restriction fragment length polymorphisms and localization of tumor suppressor genes (A) Genetic markers that are defined by the presence or absence of a restriction endonuclease cleavage site can be used to analyze chromosomal regions. In this illustration, the DNA segment in the chromosome of maternal origin (*red*) can be cleaved by the enzyme EcoRI, while the homologous, paternally derived segment (*green*), because of a single base-pair substitution, resists cleavage. This variation represents a restriction fragment length polymorphism (RFLP), and a radioactive probe that recognizes the right end of the paternal fragment can be used in Southern blot analysis to determine whether cleavage has occurred (*below*). The presence or absence of a cleavage fragment makes it possible, in turn, to track the inheritance of DNA sequences that behave genetically like Mendelian alleles. In this example, the DNAs from the normal tissues of two cancer patients (1 and 2) both show heterozygosity in this chromosomal region. However, loss of heterogzyosity (LOH) has occurred in their tumor DNAs, with one patient showing loss of the paternal allele, while the other patient shows loss of the maternal allele. (B) This technique was used to survey the LOH in entire chromosomal arms in a series of colon cancers. The long (q) and short (p) arms of the chromosomes are indicated by *green bars* and *orange bars*, respectively. The "allelic deletions" indicated on the ordinate are equivalent to LOH. These analyses represented very imprecise measurements of the locations of critical tumor suppressor genes, since very few probes were available at the time to gauge LOH within circumscribed chromosomal regions. (In addition, the *APC* tumor suppressor gene located on Chromosome 5q, which was subsequently found to be inactivated in almost all colorectal cancers, is extremely large and therefore presents a frequent target for inactivation by direct mutation of its sequences. As a consequence, following inactivation of the first copy of this gene, the second, still-intact copy is often lost through a second, independent mutation rather than through LOH, leading here to an underestimate of its involvement in colorectal cancer development.) These data also provide clear indication that Chromosome 8p harbors a tumor suppressor gene; this gene has never been identified. (B, from B. Vogelstein et al., *Science* 244:207–211, 1989.)

sites somewhere within entire chromosomal arms (see Figure 7.13). Nonetheless, even prior to the introduction of these far more powerful mapping techniques, LOH analyses allowed the identification and subsequent cloning of more than 30 TSGs (**Table 7.1**).

One fruit of the revolution in developing a large catalog of SNP markers is seen in **Figure 7.14**, in which relatively small deletions of a region of human Chromosome 9 were frequently detected in the genomes of 80 cancer cell lines. The relatively short chromosomal DNA segment that was affected, time after time, by these deletions is known from other work to encode two well-studied TSGs, *CDKN2A* and *CDKN2B*, which will be discussed in detail in Chapters 8 and 9. Chromosomal deletions are presumed to occur randomly across the genome, and the fact that these converged on a discrete DNA segment provided strong evidence that this particular chromosomal region harbors one or more TSGs whose deletion confers advantage on the affected cells.

Table 7.1 Examples of human tumor suppressor genes that have been cloned

Name of gene	Chromosomal location	Familial cancer syndrome	Sporadic cancer	Function of gene product
SDHB	1p36.1	paraganglioma	—	succinate dehydrogenase
CHD5	1p36.31	cutaneous melanoma	many types	histone reader, transcriptional inducer
HRPT2	1q25–32	parathyroid tumors, jaw fibromas	parathyroid tumors	chromatin protein
FH	1q42.3	familial leiomyomatosis[a]	—	fumarate hydratase
FHIT	3p14.2	—	many types	diadenosine triphosphate hydrolase
BAP1	3p21.1	mesothelioma, melanoma	mesothelioma, uveal melanoma	ubiquitin hydrolase
RASSF1A	3p21.3	—	many types	multiple functions
TGFBR2	3p2.2	HNPCC	colon, gastric, pancreatic carcinomas	TGF-β receptor
VHL	3p25–26	von Hippel–Lindau syndrome	renal cell carcinoma	ubiquitylation of HIF
hCDC4	4q32	—	endometrial carcinoma	ubiquitin ligase
APC	5q21–22	familial adenomatous polyposis coli	colorectal, pancreatic, and stomach carcinomas; prostate carcinoma	β-catenin degradation
NKX3.1	8p21.2	—	prostate carcinoma	homeobox TF
miR-124a[b]	8p23.1	—	many types	suppresses CDK6
p16[INK4A] [c]	9p21	familial melanoma	many types	CDK inhibitor
p14[ARF] [d]	9p21	—	all types	p53 stabilizer
PTC	9q22.3	nevoid basal cell carcinoma syndrome	medulloblastomas	receptor for hedgehog GF
let 7a (miRNA)[e]	9q22.32	—	many types	suppresses Ras, Myc
TSC1	9q34	tuberous sclerosis	—	inhibitor of mTOR[f]
BMPR1	10q21–22	juvenile polyposis	—	BMP receptor
ANXA7	10q21	—	breast, prostate, stomach	endocytosis
PTEN[g]	10q23.3	Cowden's disease, breast and gastrointestinal carcinomas	glioblastoma; prostate, breast, and thyroid carcinomas	PIP$_3$ phosphatase
WT1	11p13.5–6	Wilms tumor	Wilms tumor	TF
MEN1	11p13	multiple endocrine neoplasia	—	histone modification, transcriptional repressor
BWS/CDKN1C	11p15.5	Beckwith–Wiedemann syndrome	—	p57[Kip2] CDK inhibitor
SDHD[h]	11q23.1	paraganglioma, pheochromocytoma	pheochromocytoma	mitochondrial protein
CBL	11q23.3	juvenile myelomonocytic leukemia	adult myelomonocytic leukemia	SH2-containing ubiquitin ligase
RB	13q14.2	retinoblastoma, osteosarcoma	retinoblastoma; sarcomas; bladder, breast, esophageal, and lung carcinomas	transcriptional repression; control of E2Fs

Name of gene	Chromosomal location	Familial cancer syndrome	Sporadic cancer	Function of gene product
miR-15a/16-1	13q14.3	—	B-cell lymphoma	suppresses Bcl-2, Mcl-1, cyclin D1, Wnt3a
miR-127	14q32.31	—	many types	suppresses Bcl-6
CYFIP1	15q11.2	—	lung, breast, colon, bladder carcinomas	actin cytoskeleton organization
TSC2	16p13.3	tuberous sclerosis	—	inhibitor of mTOR[f]
CBP	16p13.3	Rubinstein–Taybi syndrome	AML[i]	TF co-activator
CYLD	16q12–13	cylindromatosis	—	deubiquitinating enzyme
CDH1	16q22.1	familial gastric carcinoma	invasive cancers	cell–cell adhesion
BHD/FLCN	17p11.2	Birt–Hogg–Dube syndrome	kidney carcinomas, hamartomas	regulator of mTOR[f]
TP53	17p13.1	Li–Fraumeni syndrome	many types	TF
NF1	17q11.2	neurofibromatosis type 1	colon carcinoma, astrocytoma, acute myelogenous leukemia	Ras-GAP
PRKAR1A	17q22–24	multiple endocrine neoplasia[j]	multiple endocrine tumors	subunit of PKA
DPC4[k]	18q21.1	juvenile polyposis	pancreatic and colon carcinomas	TGF-β TF
LKB1/STK11	19p13.3	Peutz–Jegher syndrome	hamartomatous colonic polyps	serine/threonine kinase
RUNX1	21q22.12	familial platelet disorder	AML	TF
SNF5[l]	22q11.2	rhabdoid predisposition syndrome	malignant rhabdoid tumors	chromosome remodeling
NF2	22q12.2	neurofibroma-predisposition syndrome	schwannoma, meningioma; ependymoma	cytoskeleton–membrane linkage
WTX	Xq11.1	—	Wilms tumor	β-catenin degradation

[a]Familial leiomyomatosis includes multiple fibroids, cutaneous leiomyomas, and renal cell carcinoma. The gene product is a component of the tricarboxylic cycle.

[b]miR124a-1 genes are also located at 8q12.3 and 20q13.33l.

[c]Also known as MTS1, CDKN2, and p16.

[d]The human homolog of the murine p19[ARF] gene.

[e]There are altogether 11 loci encoding let7 miRNAs in the human genome.

[f]mTOR is a serine/threonine kinase that controls, among other processes, the rate of translation and activation of Akt/PKB, TSC1 (hamartin), and TSC2 (tuberin), thereby controlling both cell size and cell proliferation.

[g]Also called MMAC or TEP1.

[h]SDHD encodes subunit D of the succinate dehydrogenase (succinate–ubiquinone oxidoreductase) enzyme, a component of the mitochondrial respiratory chain complex II.

[i]The CBP gene is involved in chromosomal translocations associated with AML. These translocations may reveal a role of a segment of CBP as an oncogene rather than a tumor suppressor gene.

[j]Also termed Carney complex.

[k]Encodes the Smad4 TF associated with TGF-β signaling; also known as MADH4 and SMAD4.

[l]The human SNF5 protein is a component of the large Swi/Snf complex that is responsible for remodeling chromatin in a way that leads to transcriptional repression through the actions of histone deacetylases. The rhabdoid predisposition syndrome involves susceptibility to atypical teratoid/rhabdoid tumors, choroid plexus carcinomas, medulloblastomas, and extra-renal rhabdoid tumors.

Adapted in part from E.R. Fearon, Science 278:1043–1050, 1997; and in part from D.J. Marsh and R.T. Zori, Cancer Lett. 181:125–164, 2002.

Figure 7.14 Measurement of deleted chromosomal segments carrying tumor suppressor genes One strategy for localizing and identifying TSGs depends on mapping overlapping regions of relatively small deletions that affect a chromosomal arm. This has become practical with the identification of large numbers of sequence markers that can be planted at sites across the entire human genome and the use of high-throughput sequencing technologies. Here, a set of 250,000 probes was used; the probes were planted at an average density of one per 12 kb across the genome. DNAs of 80 tumor cell lines (arrayed top to bottom, names deleted) were analyzed for the presence or absence of sequences reactive with the probes. Short homozygous deletions (*dark blue bars*) affecting one or both of the adjacent pair of TSGs mapping to chromosome 9p21—*CDKN2A* and *CDKN2B*—have been found to occur frequently. In contrast, nearby chromosomal sequences to the right and left are rarely affected by either significant amplification (*light red squares*) or deletions (*light blue squares*). (From S.M. Rothenberg et al., *Cancer Res.* 70:2158–2164, 2010.)

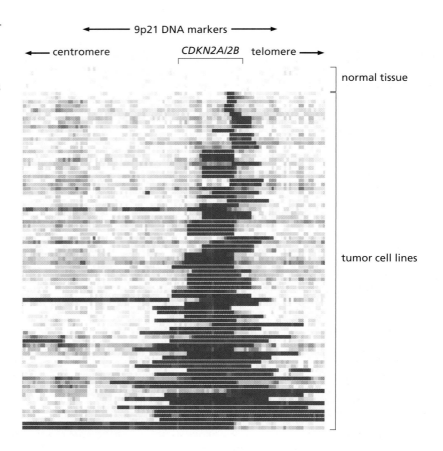

7.7 Many familial cancers can be explained by inheritance of mutant tumor suppressor genes

Like the *Rb* gene, most of the cloned tumor suppressor genes listed in Table 7.1 are involved in both familial and sporadic cancers. In general, inheritance of defective copies of most of these genes creates an enormously increased risk of contracting one or another specific type of cancer, often a type of tumor that is otherwise relatively rare in the human population. In some cases, mutant germ-line alleles of these genes lead to susceptibility to multiple cancer types, as is the case with the *Rb* gene.

Later in this chapter, we will describe in detail the mechanisms of action of some of the TSG-encoded proteins. However, even a cursory examination of Table 7.1 makes it clear that these genes specify a diverse array of proteins that operate in many different intracellular sites to reduce the risk of cancer. Indeed, an anti-cancer function is the only property that is shared in common by these otherwise unrelated genes. (Some of the tumor suppressor genes listed in this table are known only through their involvement in sporadic cancers; it remains unclear whether mutant alleles of these genes will eventually be found to be transmitted in the germ line and thereby predispose individuals to one or another type of cancer.)

While inheritance of a mutant TSG allele is likely to greatly increase cancer risk, the converse is not true: not all familial cancer syndromes can be traced back to an inherited TSG allele. As we will discuss in Chapter 12, mutant germ-line alleles of a second class of genes also cause cancer predisposition. These other genes are normally responsible for maintaining the cellular genome, and thus act to reduce the likelihood of mutations and chromosomal abnormalities. Because cancer pathogenesis depends on the accumulation by individual cells of somatic mutations, agents that reduce the mutation rate, such as these genome maintenance genes, are highly effective in suppressing cancer onset. Conversely, defects in genome maintenance often lead to a disastrous increase in cancer risk because they increase the mutation rate.

So we come to realize that there are really two distinct classes of familial cancer genes—the tumor suppressor genes described in this chapter, and the genome maintenance genes described in Chapter 12. We can rationalize the distinction between the two classes of genes as follows. The tumor suppressor genes function to directly control the biology of cells by affecting how they proliferate, differentiate, or die; genes functioning in this way are sometimes called **gatekeepers** to indicate their role in allowing or disallowing cells to progress through cycles of growth and division. The DNA maintenance genes affect cell biology only indirectly by controlling the rate at which cells accumulate mutant genes; these genes have been termed **caretakers** to reflect their role in the maintenance of cellular genomes. Unlike mutant gatekeeper and caretaker alleles, mutant versions of proto-oncogenes are rarely transmitted through the germ line (Sidebar 7.3).

7.8 Promoter methylation represents an important mechanism for inactivating tumor suppressor genes

As described in Section 1.9, DNA molecules can be altered covalently by the attachment of methyl groups to cytosine bases. In mammalian cells, this methylation is found only when these bases are located in a position that is 5′ to guanosines, that is, in the sequence CpG. (This MeCpG modification is often termed "methylated CpG," even though only the cytosine is methylated.) When CpG methylation occurs in the vicinity of a gene promoter, it can cause repression of transcription of the associated gene, and conversely, when methyl groups are removed, transcription of this gene is often de-repressed. Extensive research indicates that CpG modification of genomic DNA is as important as mutation in shutting down tumor suppressor genes.

Sidebar 7.3 Why are mutant tumor suppressor genes transmitted through the germ line while mutant proto-oncogenes are usually not? A number of familial cancer syndromes have been associated with the transmission of mutant germ-line alleles of tumor suppressor genes (see Table 7.1). Rarely, however, are cancer syndromes associated with inherited mutant alleles of proto-oncogenes (that is, activated oncogenes). To date, the list of mutant germ-line alleles of only a small number of proto-oncogenes have been associated with a variety of rare familial cancer syndromes; these include mutant alleles of H-*ras* (Costello syndrome); *Kit* and *PDGFR-A* (familial gastrointestinal stromal tumors); *Met* (hereditary papillary renal cell cancer); *Ret* (multiple endocrine neoplasia); *PTPN11*, K-*ras*, and *SOS* (Noonan syndrome); and *ALK* (familial neuroblastoma). How can we rationalize this dramatic difference in the heritability of mutant TSGs versus mutant proto-oncogenes?

Mutations that yield activated oncogenes are likely to arise with some frequency during **gametogenesis**—the processes that form sperm and egg—and thus are likely to be transmitted to fertilized eggs. However, because oncogenes act at the cellular level as dominant alleles, these mutant alleles are likely to perturb the behavior of individual cells in the developing embryo and therefore to disrupt normal tissue development. Consequently, embryos carrying these mutant oncogenic alleles are unlikely to develop to term, and these mutant alleles will disappear from the germ line of a family and thus from the gene pool of the species. (For example, experiments reported in 2004 indicate that mouse embryos arising from sperm carrying a mutant, activated K-*ras* oncogene develop only to mid-gestation, at which point they die because of placental and intra-embryonic developmental defects; see also Sidebar 5.5. Provocatively, the mutant germ-line alleles of the H- and K-*ras* genes that cause Costello and Noonan syndromes in humans

carry mutations that are quite different from those responsible for creating the *ras* oncogenes that are present in many types of cancer. Unlike the somatically acquired oncogenic alleles present in almost one-fourth of human tumors, these mutant germ-line alleles confer relatively subtle effects on Ras function that result in heightened responsiveness to mitogenic signals but do not substantially compromise GTPase activity. It seems that, as is the case in mice, the potent oncogenic *ras* alleles that are present in many human tumors are incompatible with normal development.)

Mutant germ-line alleles of tumor suppressor genes behave much differently, however. Since these alleles are generally recessive at the cellular level, their presence in most cells of an embryo will not be apparent. For this reason, the presence of inherited mutant TSGs will often be compatible with normal embryonic development, and the cancer phenotypes that they create will become apparent only in a small number of cells and after great delay, allowing an individual carrying these alleles to develop normally and, as is often the case, to survive through much of adulthood.

Still, if mutant tumor suppressor genes undergo LOH in 1 out of 10^4 or 10^5 cells, and if an adult human has many more than 10^{13} cells, why isn't a person inheriting a mutant tumor suppressor gene afflicted with tens of thousands, even millions of tumors? The response to this comes from the fact that tumorigenesis is a multi-step process (as we will see in Chapter 11). This implies that a mutant gene (whether it be an oncogene or a TSG) may be necessary for tumor formation but will not, on its own, be sufficient. Hence, many cells in an individual who has inherited a mutant tumor suppressor gene may well undergo LOH of this gene, but only a tiny minority of these cells will ever acquire the additional genetic changes needed to make a clinically detectable tumor.

The means by which the methylated state of CpGs affects transcription are not entirely resolved. Nonetheless, one very important mechanism involves protein complexes that include one subunit that can recognize and bind methylated CpGs in DNA and a second subunit that functions as a histone deacetylase (HDAC) enzyme. Once these complexes are bound to DNA, the histone deacetylase proceeds to remove acetate groups that are attached to the side chains of amino acid residues of histone molecules in the nearby chromatin. The resulting deacetylation of the histones initiates a sequence of events that converts the chromatin from a configuration favoring transcription to one blocking transcription (see Section 1.8). Operating in the opposite direction, there is evidence that the modification state of histones can affect the methylation of CpG sequences in the nearby DNA. For example, in certain embryonic cells, the H3K9Me3 modification (that is, trimethylation of lysine 9 of histone H3) appears to serve as the precursor of ***de novo* methylation** of the CpGs in the associated DNA.

Methylation of CpGs, which is not found in all metazoa, is a clever invention. During the development of all animal species, decisions concerning the transcriptional state of many genes are made early in embryogenesis. Once made, these decisions must be passed on to the descendant cells in various parts of the growing embryo. Since CpG methylation can be transmitted from cell to cell heritably (see Section 1.9), it represents a highly effective way to ensure that descendant cells, many cell generations removed from an early embryo, continue to respect and enforce the decisions made by their ancestors in the embryo.

In the genomes of cancer cells, we can imagine that the mechanisms determining whether DNA is properly methylated may malfunction occasionally. These important regulatory mechanisms are still poorly understood. The results of this malfunctioning are two opposing changes in the methylation state of tumor cell genomes. As the development of a tumor proceeds, the overall level of methylation throughout the cancer cell genome is often found (using the technique described in Supplementary Sidebar 7.4) to decrease progressively. This indicates that, for unknown reasons, the DNA methyltransferase enzymes responsible for maintaining CpG methylation fail to do their job effectively. Much of this "global hypomethylation" can be attributed to the loss of methyl groups attached to the DNA of highly repeated sequences in the cell genome that do not specify biological functions; this loss is correlated with chromosomal instability, but it remains unclear whether it directly causes this instability.

Independent of this global hypomethylation, there are often localized regions of DNA—regions with a high density of CpGs called "CpG islands"—that become methylated inappropriately in the genomes of cancer cells. These CpG islands are often affiliated with the promoters of genes (**Figure 7.15**), and their methylation by DNA

Figure 7.15 Methylation of the *RASSF1A* promoter The bisulfite sequencing technique (Supplementary Sidebar 7.4) has been used here to determine the state of methylation of the CpG island in which the promoter of the *RASSF1A* tumor suppressor gene is embedded. Each circle indicates the site of a distinct CpG dinucleotide in this island, whose location within the *RASSF1A* promoter is also indicated by a vertical tick line and an identification number in the map (*above*). Filled circles (*blue*) indicate that a CpG has been found to be methylated, while open circles indicate that it is unmethylated. Analyses of five DNA samples from tumor 232 indicate methylation at almost all CpG sites in the *RASSF1A* CpG island; adjacent, ostensibly normal tissue is unmethylated in most but not all analyses of this CpG island. Analyses of control DNA from a normal individual indicate the absence of any methylation of the CpGs in this CpG island. (These data suggest the presence of some abnormal cells with methylated DNA in the ostensibly normal tissue adjacent to tumor 232.) (Courtesy of W.A. Schulz and A.R. Florl.)

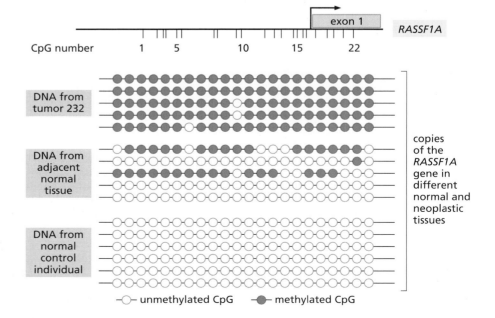

methyltransferases silences transcription, resulting in the shutdown of genes that should, by all rights, remain transcriptionally active. (A recent survey indicated that about 70% of all genes in the human genome have CpG islands affiliated with their promoters, yielding a total of ~14,000 genes that are, in principle, vulnerable to inappropriate shutdown by methylation.)

CpG methylation is effective in shutting down the expression of a gene only if it occurs within the promoter sequences of the gene; conversely, methylation of DNA sequences in the body of the gene, such as exonic sequences, seems to have little if any effect on the level of transcription. Since promoter methylation can silence a gene as effectively as a mutation of its nucleotide sequences, we might predict that methylation plays a role in the silencing of tumor suppressor genes that occurs during tumor progression.

In fact, in recent years, it has become apparent that promoter methylation may even be more important in shutting down tumor suppressor genes than the various mechanisms of somatic mutation. More than half of the tumor suppressor genes that are involved in familial cancer syndromes because of germ-line mutation have been found to be silenced in sporadic cancers by promoter methylation. For example, when the *Rb* tumor suppressor gene is mutated in the germ line, it leads to familial retinoblastoma. In sporadic retinoblastomas, however, this tumor suppressor gene is inactivated either by somatic mutations or by promoter methylation. In addition, the promoters of a variety of other genes that are known or thought to inhibit tumor formation have been found in a methylated state (**Table 7.2**).

Table 7.2 Examples of hypermethylated genes found in human tumor cell genomes

Name of gene	Nature of protein function	Type of tumor
RARβ2	nuclear receptor for differentiation	breast, lung
p57^Kip2	CDK inhibitor	gastric, pancreatic, hepatic; AML
TIMP3	inhibitor of metalloproteinases	diverse tumors
IGFBP	sequesters IGF-1 factor	diverse tumors
CDKN2A/p16^INK4A	inhibitor of CDK4/6	diverse tumors
CDKN2B/p15^INK4B	inhibitor of CDK4/6	diverse tumors
p14^ARF	inhibitor of HDM2/MDM2	colon, lymphoma
APC	inducer of β-catenin degradation	colon carcinomas
ER	estrogen receptor	breast
p73	aids p53 to trigger apoptosis	diverse tumors
GSTP1	mutagen inactivator	breast, liver, prostate
MGMT	DNA repair enzyme	colorectal
CDH1	cell–cell adhesion receptor	bladder, breast, colon, gastric
DKK1	Wnt inhibitor	colon
DAPK	kinase involved in cell death	bladder
MLH1	DNA mismatch repair enzyme	colon, endometrial, gastric
PTEN	degrades PIP_3	diverse tumors
TGFBR2	TGF-β receptor	colon, gastric, small-cell lung
THBS1	angiogenesis inhibitor	colon, glioblastoma
VHL	ubiquitin ligase	kidney, hemangioblastoma
RB	cell cycle regulator	retinoblastoma
CASP8	apoptotic caspase	neuroblastoma, SCLC
APAF1	pro-apoptotic cascade	melanoma
CTMP	inhibitor of Akt/PKB	glioblastoma multiforme

Adapted in part from C.A. Eads et al., *Cancer Res.* 61:3410–3418, 2001; and M. Esteller, *Nat. Rev. Cancer* 8:286–298, 2007.

(A)   (B)   (C)

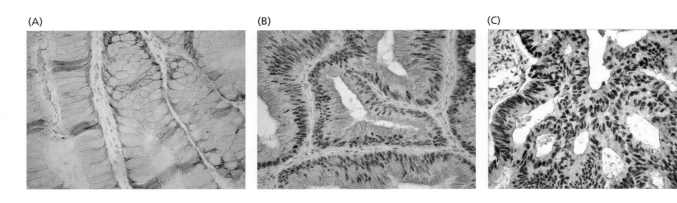

Figure 7.16 Expression of the DNMT3B enzyme in colon carcinomas The DNA methyltransferase enzyme, which appears to be responsible for causing hypermethylation of many CpG islands during tumor development, is detected here with a specific antibody (*brown staining*). In this group of normal and neoplastic colonic tissues, the level of DNMT3B increases in concert with increased tumor progression. (A) In normal colonic tissue, the enzyme is hardly expressed, and then only in the cytoplasm of cells near the underlying stroma; the bulk of each cell, which is oriented along an apical–basal axis, is filled with mucin, a heavily glycosylated protein destined for secretion (*light green*). (B) In this well-differentiated, less aggressive adenocarcinoma, DNMT3B is expressed strongly in the nuclei of carcinoma cells, which are located away from the lumina of acini (open spaces; *clear*) formed by the carcinoma cells, while the mucin-containing cytoplasms are located apically. (C) In a less differentiated, more aggressive colon carcinoma, DNMT3B is intensely expressed in the nuclei of all the carcinoma cells, which no longer exhibit the apically localized mucin. (From K. Nosho et al., *Clin. Cancer Res.* 15:3663–3671, 2009.)

The extent of CpG island methylation varies greatly from one tumor genome to the next. In some, the hypermethylation is so great that such tumors are said to exhibit the "CpG island methylator phenotype" (CIMP). In the cells of tumors, as many as 5% of the genes (that is, about 1000 genes) may possess hypermethylated CpG islands. The molecular mechanisms driving this runaway CpG methylation are poorly understood. One attractive explanation derives from the observation that the DNA methyltransferase 3B (DNMT3B) enzyme is overexpressed in many cancers exhibiting this phenotype (**Figure 7.16**). It associates with certain nucleosome proteins (see Figures 1.19 and 1.20) that appear to influence its ability to methylate nearby DNA sequences, and its overexpression in mice that are genetically engineered to develop colon carcinomas accelerates the rate of tumor formation.

Some candidate tumor suppressor genes are rarely inactivated by somatic mutations in their reading frame, and their involvement in creating cancer is suggested only by the fact that their promoters are repeatedly found to be methylated in large numbers of tumors. Situations like these create difficulties for researchers who would like to verify that inactivation of a candidate tumor suppressor gene actually contributes causally to cancer formation. In the past, such validation depended on sequencing the relevant gene copies cloned from tumor cell genomes with the intent of finding inactivating mutations. Such molecular analyses can no longer be considered to be absolutely definitive if certain TSGs contributing to tumor progression are silenced largely or exclusively through promoter methylation.

(The discovery of a promoter-associated CpG island that is repeatedly methylated in the genomes of a group of tumors is often difficult to interpret unambiguously; this finding may only reflect the fact that the tumors being studied arose from a normal cell type in which the gene in question is methylated as part of a normal differentiation program. Moreover, it seems highly unlikely that the silencing of more than a small fraction of the 1000 methylated genes in the genome of a CIMP tumor actually contributes causally to cancer formation. Instead, the great majority of the silenced genes are likely to be innocent bystanders—victims of runaway genome-wide methylation rather than active participants in tumorigenesis.)

The elimination of tumor suppressor gene function by promoter methylation might, in principle, occur through either of two routes. Both copies of a TSG might be methylated independently of one another. Alternatively, one copy might be methylated and the second might then be lost through a loss of heterozygosity (LOH) accompanied by a duplication of the already-methylated tumor suppressor gene copy. Actually, this second mechanism seems to operate quite frequently. For example, in a study of the lungs of smokers, former smokers, and never-smokers, the $p16^{INK4A}$ tumor suppressor gene (to be described in Section 8.4) was found to be methylated in 44% of (ostensibly) normal bronchial epithelial cells cultured from current and former smokers and not at all in the comparable cells prepared from those who had never smoked. LOH in this chromosomal region was found in 71 to 73% of the two smoking populations and in 1.5 to 1.7% of never-smokers.

The observations that methylated copies of TSGs frequently undergo loss of heterozygosity indicate that the methylation event is relatively infrequent per cell generation, indeed, rarer than LOH events. (If methylation always occurred more frequently than LOH, tumors would show two tumor suppressor gene copies, one of paternal and the other of maternal origin, with each independently methylated—a configuration that is observed frequently in certain types of tumors and rarely in other types.) Clearly, evolving tumor cells can discard the second, still-functional tumor suppressor gene copy more readily by LOH than by a second, independent promoter methylation event.

The above-described study of morphologically normal bronchial epithelial cells also teaches us a second lesson: methylation of critical growth-controlling genes often occurs early in the complex, multi-step process of tumor formation, long before histological changes are apparent in a tissue. These populations of outwardly normal cells presumably provide a fertile soil for the eventual eruption of premalignant and malignant growths. This point is borne out in other tissues as well, such as the histologically normal breast tissues analyzed in **Figure 7.17**.

As tumor formation proceeds, the silencing of genes through promoter methylation can also involve the "caretaker" genes that are responsible for maintaining the integrity of the DNA sequences in the genome (Chapter 12). An interesting example is provided by the *BRCA1* gene. Its product is responsible for maintaining the chromosomal DNA in complex ways that are still poorly understood. The consequence of inheritance of mutant alleles of *BRCA1* is a familial cancer syndrome involving a high lifetime risk of breast and, to a lesser extent, ovarian carcinomas (see Section 12.10). At one time, it was thought that *BRCA1* inactivation never contributes to sporadic breast cancers, since mutant alleles of this gene could not be found in this second, far more common class of mammary tumors. Recent work has shown that more than half of sporadic breast carcinomas carry inactive *BRCA1* gene copies that have been silenced through promoter methylation. In addition, this gene suffers the same fate in ~40% of sporadic epithelial ovarian carcinomas.

Figure 7.18 illustrates the methylation state of the promoters of 12 genes in a variety of human tumor types. This figure clearly demonstrates that the frequency of methylation of a specific gene varies dramatically from one type of tumor to the next. The

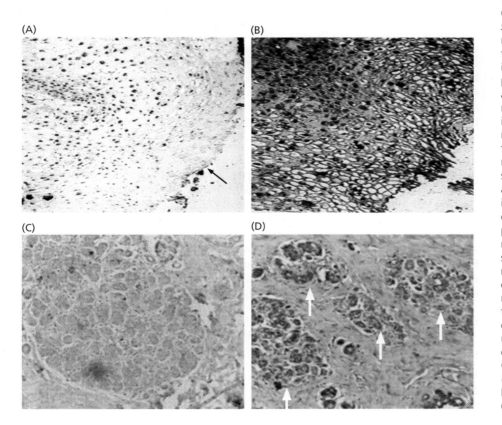

(A)

(B)

(C)

(D)

Figure 7.17 *In situ* measurements of DNA methylation The presence or absence of methylation at specific sites in the genomes of cells can be determined in tissue sections that are fixed to a microscope slide using both the methylation-specific PCR reaction (see Supplementary Sidebar 7.4) and *in situ* hybridization. In these images, the methylation status of the promoter of the *p16^INK4A* tumor suppressor gene is analyzed with a methylation-specific probe, which yields dark staining in areas where this gene promoter is methylated. (A) In a low-grade squamous intraepithelial lesion of the cervix (*left*), nuclei of cells located some distance from the uterine surface show promoter methylation, while those near the surface (*arrow*) do not. (B) However, in an adjacent high-grade lesion, which is poised to progress to a cervical carcinoma (*right*), all the cells show promoter methylation. (C, D) In these micrographs of normal breast tissue, some histologically normal lobules show no promoter methylation (C) while others (*arrows*) show uniform promoter methylation (D). This suggests that the mammary epithelial cells in such outwardly normal lobules have already undergone a critical initiating step in cancer progression. (A and B, from G.J. Nuovo et al., *Proc. Natl. Acad. Sci. USA* 96:12754–12759, 1999. C and D, from C.R. Holst et al., *Cancer Res.* 63:1596–1601, 2003.)

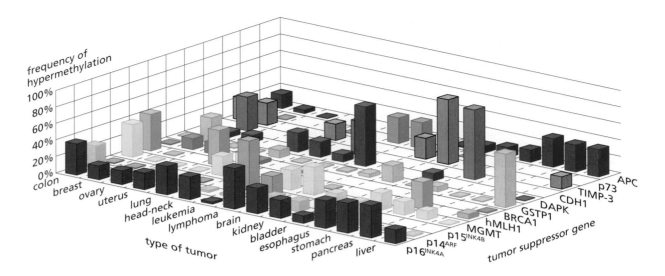

Figure 7.18 Methylation of multiple genes within tumor cell genomes This three-dimensional bar graph summarizes measurements of the methylation state of the promoters of 12 different genes (*p16^{INK4A}*, *p15^{INK4B}*, *p14^{ARF}*, *p73*, *APC*, *BRCA1*, *hMLH1*, *GSTP1*, *MGMT*, *CDH1*, *TIMP3*, and *DAPK*) that are known or presumed to play an important role in suppressing the development of human tumors. Methylation status was determined by the methylation-specific PCR technique (see Supplementary Sidebar 7.4). The methylation state of each of these promoters has been studied in the DNAs of 15 different tumor types. The height of each bar indicates the proportion of tumors of a given type in which a specific promoter has undergone methylation. All of these promoters are unmethylated (or methylated to an insignificant extent) in normal tissues. (Adapted from M. Esteller et al., *Cancer Res.* 61:3225–3229, 2001.)

data in this figure reinforce the point that tumor suppressor genes as well as caretaker genes undergo hypermethylation. Perhaps the most important lesson taught by this figure is the pervasiveness of promoter methylation during the development of a wide variety of human cancers.

Within a given tumor, multiple genes seem to be shut down by promoter methylation. For example, in an analysis of the methylation status of eight critical cancer-related genes in the genomes of 107 non-small-cell lung carcinomas (NSCLCs), 37% had at least one of these gene promoters methylated, 22% had two promoters methylated, and 2% carried five of the eight gene promoters in a methylated state. The decision to analyze these particular genes involved, by necessity, some arbitrary choices, and we can imagine that there were dozens of other hypermethylated genes in these cancer cell genomes whose inactivation contributed in various ways to cancer formation.

The cell-physiologic consequences of promoter methylation are nicely illustrated by the actions of retinoic acid. In a number of epithelial cell types, retinoic acid is a potent inducer of cell cycle arrest and even differentiation. It has been used, for example, in attempts to halt the further proliferation of breast cancer cells. However, the great majority of these cancer cells are found to have silenced, by promoter methylation, the *RARβ2* gene, which encodes a critical retinoic acid receptor. Without expression of this receptor, the breast carcinoma cells are unresponsive to retinoic acid treatment, so they avoid growth arrest and continue to thrive in the presence of this agent.

As described above, the methylation of CpGs in promoters acts to attract histone deacetylase (HDAC) molecules, which proceed to reconfigure nearby chromatin proteins, placing them in a state that is incompatible with transcription (see Section 1.8). Use of an inhibitor of HDACs, termed trichostatin A (TSA), reverses this deacetylation, thereby returning the chromatin to a state that permits transcription. Hence, treatment of breast cancer cells with TSA causes reactivation of *RARβ2* gene expression and restores their responsiveness to the growth-inhibitory effects of retinoic acid. This result shows clearly that promoter methylation acts via histone deacetylation to promote cancer. At the same time, it suggests a therapy for breast carcinomas by concomitant treatment with trichostatin A and retinoic acid.

7.9 Tumor suppressor genes and proteins function in diverse ways

Earlier we read that the tumor suppressor genes and their encoded proteins act through diverse mechanisms to block the development of cancer. Indeed, the only characteristic that ties these genes and their encoded proteins together is the fact that all of them operate to reduce the likelihood of cancer development. Careful examination of a list of cloned tumor suppressor genes (see Table 7.1) reveals that some of them function to directly suppress the proliferation of cells in response to a variety of growth-inhibitory and differentiation-inducing signals. Yet others are components of the cellular control circuitry that inhibits proliferation in response to metabolic imbalances and genomic damage.

The first two tumor suppressor genes that were intensively studied, the *Rb* gene discussed earlier and one termed variously *p53*, *Trp53*, or *TP53*, also happen to be the two tumor suppressor genes that play major roles in human cancer pathogenesis. The protein encoded by the *Rb* gene governs the progress of a wide variety of cells through their growth-and-division cycles, and the growth control imposed by the *Rb* circuit appears to be disrupted in most and perhaps all human tumors. Because of its centrally important function, we will devote an entire chapter to *Rb* and therefore defer discussion of this gene and its encoded protein until Chapter 8. The *p53* tumor suppressor gene and its product, p53, play an equally central role in the development of cancers and deserve an equally extended discussion. Consequently, we will devote much of Chapter 9 to a detailed description of p53 function and the cell death program that it controls.

The remaining tumor suppressor genes that have been enumerated to date dispatch gene products to a variety of intracellular sites, where they operate in diverse ways to suppress cell proliferation. These proteins sit astride virtually all of the control circuits that are responsible for governing cell proliferation and survival. Several of the proteins encoded by the tumor suppressor genes listed in Table 7.1 happen to be components of the *Rb* and *p53* pathways; included in these circuits are the *p16^INK4A*, *p15^INK4B*, and *p19^ARF* genes, and so discussion of these will be postponed until Chapters 8 and 9.

The mechanisms of action of some of the remaining tumor suppressor genes in Table 7.1 are reasonably well understood at present, while the actions of others remain quite obscure. We will focus here in depth on three of these tumor suppressor genes in order to illustrate the highly interesting mechanisms of action of their gene products. These examples—really anecdotes—have been chosen from the large list shown in Table 7.1 because they reveal how very diverse are the mechanisms that cells deploy in order to prevent runaway proliferation. They also illustrate the difficulties encountered in determining the specific biochemical and biological actions of various tumor suppressor gene–encoded proteins. We move from the cell surface inward.

7.10 The NF1 protein acts as a negative regulator of Ras signaling

The disease of neurofibromatosis was first described by Friedrich von Recklinghausen in 1862. We now know that neurofibromatosis type 1 (sometimes called von Recklinghausen's neurofibromatosis) is a relatively common familial cancer syndrome, with 1 in 3500 individuals affected on average worldwide. The primary feature of this disease is the development of benign tumors of the cell sheaths around nerves in the peripheral nervous system. On occasion, a subclass of these **neurofibromas**, labeled plexiform, progress to malignant tumors termed **neurofibrosarcomas**. Patients suffering from neurofibromatosis type 1 also have greatly increased risk of glioblastomas (tumors of the astrocyte lineage in the brain; see Figure 2.9A), **pheochromocytomas** (arising from the adrenal glands), and myelogenous leukemias (see Figure 2.8B). These tumors involve cell types arising from diverse embryonic lineages.

Neurofibromatosis patients often suffer from additional abnormalities that involve yet other cell types. Among these are **café au lait spots**, which are areas of hyperpigmentation in the skin; subtle alterations in the morphology of the cells in the skin and long bones; cognitive deficits; and benign lesions of the iris called "Lisch nodules" (**Figure 7.19**). These manifestations are strongly influenced by the patient's **genetic background** (that is, the array of all other genetic alleles in an individual's genome), since siblings inheriting the same mutant allele of the responsible gene, called *NF1*, often exhibit dramatically different disease phenotypes.

Figure 7.19 Neurofibromatosis
Neurofibromatosis type 1 is considered a *syndrome* because a number of distinct conditions are associated with a single genetic state. (A) Notable among these are the numerous small, subcutaneous nodules (that is, neurofibromas) and the light brown *café au lait spots* (*arrows*) seen here on the back of a patient. (B) In addition, Lisch nodules are often observed in the iris of the eyes. (C) Histological observation of afflicted tissues indicates the benign neurofibromas (*left*), which occasionally progress to malignant peripheral nerve sheath tumors (MPNSTs, *right*). The histological complexity of the neurofibromas greatly complicated the identification of the normal progenitors of these growths, which are now thought to be the precursors of the Schwann cells that wrap around peripheral axons. (A, courtesy of B.R. Korf. B, from B.R. Korf, *Postgrad. Res.* 61:3225–3229, 1988 and courtesy of J. Kivlin. C, courtesy of S. Carroll.)

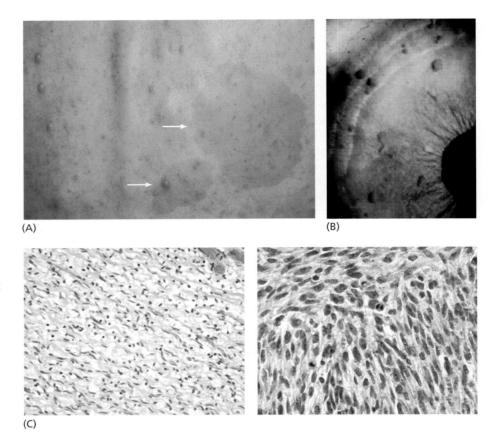

(A)   (B)

(C)

The *NF1* gene was cloned in 1990. (A second type of neurofibromatosis is connected with inactivation of a completely unrelated tumor suppressor gene.) The genetic behavior of the *NF1* gene parallels closely that of the *Rb* gene. Thus, mutant, inactivated alleles of the *NF1* gene transmitted through the germ line act in a dominant fashion to create disease phenotypes. At the cellular level, the originally heterozygous configuration of the gene ($NF1^{+/-}$) is converted to a homozygous state ($NF1^{-/-}$) in tumor cells through loss of heterozygosity. Finally, as many as half of neurofibromatosis patients lack a family history of the disease, indicating that the mutant allele that they carry is the consequence of a *de novo* mutation in the germ line. As is the case with the *Rb* gene, such *de novo* mutations usually occur during spermatogenesis in the fathers of afflicted patients.

Once the cloned gene was sequenced, it became possible to assign a function to neurofibromin, the *NF1*-encoded protein: it showed extensive sequence relatedness to two *Saccharomyces cerevisiae* (yeast) proteins, termed IRA1 and IRA2, that function as GTPase-activating proteins (GAPs) for yeast Ras, as well as to two mammalian Ras-GAP proteins (see Figure 5.30). Like most if not all eukaryotes, yeast cells use Ras proteins to regulate important aspects of their metabolism and proliferation. Detailed genetic analyses had already shown that in yeast, the positive signaling functions of the Ras protein are countered by the IRA protein. By provoking Ras to activate its intrinsic GTPase activity, IRA forces Ras to convert itself from its activated GTP-bound form to its inactive, GDP-bound form; precisely the same function is carried out by the mammalian Ras-GAP proteins (**Figure 7.20A**). Indeed, a Ras-GAP may ambush activated Ras before the latter has had a chance to stimulate its coterie of downstream effectors (see Section 5.10).

This initial insight into NF1 function inspired a simple scheme of how defective forms of NF1 create disease phenotypes. NF1 is expressed widely throughout the body, with especially high levels found in the adult peripheral and central nervous systems. When cells first experience growth factor stimulation, they may degrade NF1, enabling Ras signaling to proceed without interference by NF1. However, after 60 to 90 minutes, NF1 levels return to normal, and the NF1 protein that accumulates helps to shut down further Ras signaling—a form of negative-feedback control (see Figure 6.31D). In neuroectodermal and **myeloid** cells (a class of bone marrow cells) lacking NF1 function, Ras proteins are predicted to exist in their activated, GTP-bound state for longer-than-normal periods of time. In fact, in cells that are genetically $NF1^{-/-}$, basal levels of GTP-bound activated Ras are higher than normal and respond to growth factor stimulation by increasing rapidly to far higher levels before declining gradually (see Figure 7.20B). Consequently, the loss of NF1 function in a cell can mimic functionally the hyperactivated Ras proteins that are created by mutant *ras* oncogenes (see Section 5.10).

The histological complexity of neurofibromas has made it very difficult to identify the normal cell type that yields the bulk of the cells in these growths. Microscopic analyses (see Figure 7.19C) reveal that these growths are composed of a mixture of cell types,

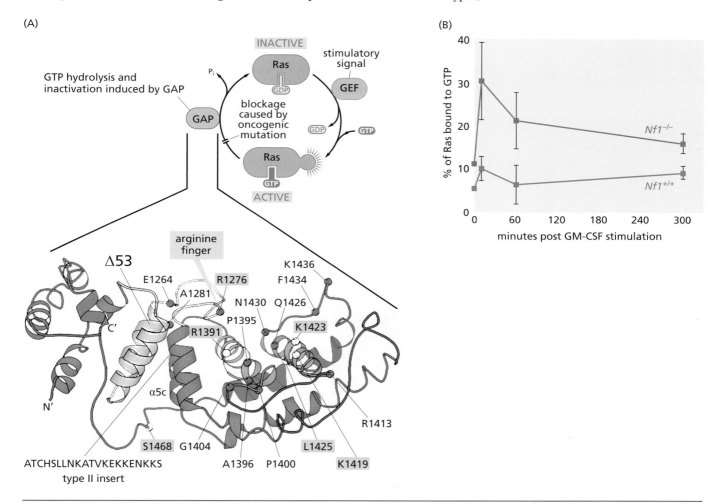

Figure 7.20 Neurofibromin and the Ras signaling cycle
(A) As illustrated in Figure 5.30 and again here (*upper right*), the Ras protein passes through a cycle in which it becomes activated by a guanine nucleotide exchange factor (GEF), such as Sos, and inactivated by a GTPase-activating protein (a Ras-GAP). One of the main Ras-GAPs is neurofibromin (NF1), which can stimulate the GTPase activity of Ras more than 1000-fold. The structure of the NF1 domain that interacts with Ras is illustrated here. A subdomain of NF1 termed the "arginine finger" (*top*) carries a critical arginine (R1276) that is inserted into the GTPase cleft of Ras and actively contributes to the hydrolysis of GTP to GDP by Ras. (Mutations that cause replacement of the arginine residue of the NF1 arginine finger with other residues result in an NF1 protein with a 1000-fold decrease in GTPase-stimulating activity. Mutant forms of NF1 (carrying amino acid substitutions) observed in neurofibromatosis patients are indicated by *gray spheres* that are labeled by *gray*

boxes; experimentally created mutations have also generated a number of amino acid substitutions (*gray spheres without boxed labels*) that compromise NF1 GAP function. The *white sphere with boxed label* (R1391) indicates a mutation that arose through both routes. In addition, a large-scale deletion (Δ53) and an insertion ("type II insert") found in patients are shown; both are also disease-associated. (B) When myeloid cells of wild-type ($Nf1^{+/+}$) mice are stimulated with the GM-CSF growth factor, the ligand-activated GM-CSF receptor causes GTP-bound Ras levels to rise rapidly and then decrease slightly over the next five hours (*blue line*). In contrast, in $Nf1^{-/-}$ myeloid cells, the basal level of GTP-bound Ras is already elevated, and in response to GM-CSF stimulation, levels of this active form of Ras rise even threefold higher and remain elevated over this time period (*red line*). (A, from K. Scheffzek et al., *EMBO J.* 17:4313–4327, 1998. B, from D. Largaespada et al., *Nat. Genet.* 12:137–143, 1996.)

including Schwann cells (which wrap around and insulate nerve axons), neurons, perineurial cells (which seem to be of fibroblast origin and are found in the vicinity of neurons), fibroblasts, and mast cells, the latter coming from the immune system. In some of these benign tumors, Schwann cells are clearly the prevailing cell type, while in others it seems that fibroblasts or perineurial cells may dominate.

The weight of evidence indicates that the Schwann cell precursors are the primary targets of LOH in neurofibromas, and that once these cells have lost all NF1 function, they orchestrate the development of these histologically complex growths by inducing the co-proliferation of a variety of other cell types via paracrine signaling. This co-proliferation raises an interesting and still-unanswered question, however: Since these other cell types, including those in neighboring tissues and in the bone marrow, have an $NF1^{+/-}$ genotype, might they be hyper-responsive to growth-stimulatory signals, specifically the paracrine signals released by the neoplastic cells? Such increased susceptibility might well arise from the fact that they carry only half the normal dose of functional NF1 protein, a notion that is borne out by detailed studies of neurofibromatosis formation in genetically altered mice (**Figure 7.21**).

It turns out that recruitment of mast cells (which originate in the bone marrow) into incipient neurofibromas is critical to the formation of these histopathologically complex tumors. In mice that are heterozygous at the *Nf1* locus (that is, $Nf1^{+/-}$), tumors termed *plexiform* neurofibromas occur with high frequency at the roots of nerves emerging from the spinal cord. However, if the cells in the bone marrow (including the mast cells) of such mice are engineered to be wild type (that is, $Nf1^{+/+}$), they are not as responsive to signals released by the $Nf1^{-/-}$ Schwann cells and are no longer recruited in large numbers to incipient neurofibromas (see Figure 7.21A). The result is a drastic

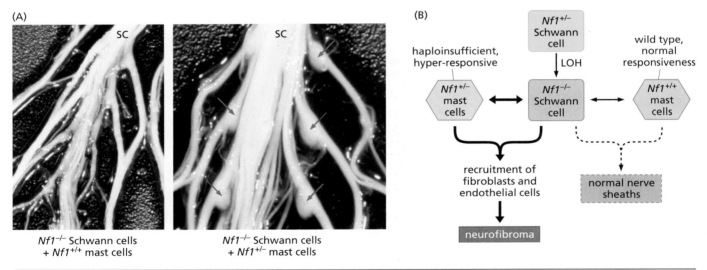

Figure 7.21 Contribution of genotype of surrounding tissue to neurofibroma development (A) In $Nf1^{+/+}$ mice in which the Schwann cells have specifically been rendered $Nf1^{-/-}$, the nerve roots emerging from the spinal cord (SC) exhibit a relatively normal morphology (*left*); hence, loss of all NF1 function in the target cells—the Schwann cells—does not suffice to allow abnormal neurofibromas to form. However, if, in addition, mast cells originating in the bone marrow are rendered $Nf1^{+/-}$, both a thickened spinal cord and plexiform neurofibromas (*arrows*) are readily apparent (*right*). Hence, *Nf1* haploinsufficiency in mast cells must cooperate with total loss of *Nf1* function in the Schwann cells to generate these growths. (B) The observations in panel A suggest that mast cells originating in the bone marrow are normally recruited to the developing spinal cord, where they contribute to normal morphogenesis. However, when these mast cells are $Nf1^{+/-}$ and thus are deprived of half of their NF1

protein, they and their progeny behave abnormally and become hyper-responsive to certain signals released by nullizygous ($Nf^{-/-}$) Schwann cells. This in turn causes the mast cells to trigger aberrant morphogenesis by collaborating with the Schwann cells to recruit yet other cell types, such as fibroblasts and endothelial cells, thereby generating the histopathologically complex neurofibromas. Consequently, in patients who are born with an $Nf1^{+/-}$ genotype, some of their $Nf1^{+/-}$ Schwann cells may undergo LOH, generating $Nf1^{-/-}$ derivatives that recruit heterozygous ($Nf1^{+/-}$) mast cells from the bone marrow, with the result being the formation of neurofibromas. [Other work indicates that the $Nf1^{-/-}$ Schwann cells release elevated levels of stem cell factor (SCF), to which the $Nf1^{+/-}$ mast cells are hyper-responsive because their intracellular signaling pathway downstream of the SCF receptor (termed Kit) is hyperactive as a consequence of hyperactive Ras protein.] (A, from F.C. Yang et al., *Cell* 135:437–448, 2008.)

Sidebar 7.4 Haploinsufficiency: when half a loaf is not that much better than none The Knudson paradigm of tumor suppressor gene inactivation postulates that mutant alleles of tumor suppressor genes are recessive at the cellular level. Hence, cells that are heterozygous for a tumor suppressor gene (that is, $TSG^{+/-}$) should be phenotypically normal. This is borne out by the biology of individuals who have inherited a mutant, inactive Rb gene, in that any pathologies displayed by human heterozygotes ($Rb^{+/-}$) are limited to the cell populations that have lost the remaining wild-type Rb gene copy, usually by LOH, and are therefore rendered $Rb^{-/-}$; the remaining $Rb^{+/-}$ tissues throughout the body seem to develop fully normally and exhibit normal function. Still, there are perfectly good biochemical reasons to think that cells expressing half the normal level of a tumor suppressor gene protein may have a phenotype that deviates significantly from wild type, and examples of this have been observed in addition to those cited in the text.

Observations of genetically altered mice provide clear examples of haploinsufficiency. Mice carrying only one copy of the gene encoding the Smad4 transcription factor, which is used by TGF-β to inhibit cell proliferation (as we will see in Section 8.10), are predisposed to developing polyps in the stomach and small intestine, and the cells in these tumors continue to carry single copies of the wild-type allele of this gene. Mice that are heterozygous for the $p27^{Kip1}$ tumor suppressor gene are similarly tumor-prone, without indication of LOH in tumor cells; this gene encodes an important inhibitor of cell cycle progression (see Section 8.4). A third example is provided by mice that are heterozygous for the gene encoding the PTEN tumor suppressor protein (Section 6.6), where a clear acceleration of prostate cancer tumorigenesis is created by the absence of only a single copy of this TSG. Finally, deletion of one copy of the $Dmp1$ gene, which encodes a transcription factor that induces expression of the $p19^{ARF}$ TSG (see Section 9.7), leads once again to increased tumor susceptibility without loss of the wild-type gene copy in tumor cells. Altogether, at least half a dozen human TSGs yield abnormal cell phenotypes when only a single wild-type gene copy is present in cell genomes.

decrease in the numbers of these tumors. This suggests a scheme like that shown in Figure 7.21B, in which a half-normal dose of Nf1 protein in the mast cells (or their immediate precursors) results in a clear and aberrant phenotype—a manifestation of the state of **haploinsufficiency**. This behavior raises a more general question about tumor suppressor genes: Granted that the full phenotypic changes of tumor suppressor gene (TSG) inactivation are felt only when both gene copies are lost, might a half-dosage of their encoded proteins (which is often observed in cells having a $TSG^{+/-}$ genotype) nonetheless yield subtle but functionally important changes in cell behavior (**Sidebar 7.4**)? Collaborations between multiple cell types in forming a tumor, such as the one described here, will be discussed in far greater detail in Chapter 13.

These insights into neurofibroma development still do not address many of the questions surrounding the very large neurofibromin protein—its Ras-GAP domain encompasses only 10% of its total mass, suggesting the possible involvement of other, still-uncharacterized functions that it may exert. Perhaps future functional characterization of its other domains will answer another mystery: Why, among the 14 Ras-GAPs encoded in the human genome, is $NF1$, and none of the others, involved in triggering neurofibromas? Indeed, why does our genome encode such a diverse array of Ras-GAPs? And what role does neurofibromin play in governing cell proliferation in other tissues? (Its role in governing the proliferation of myeloid cells of the bone marrow appears to explain the ~300-fold increased risk of juvenile myelomonocytic leukemia that $NF1$ heterozygotes confront.)

7.11 Apc facilitates egress of cells from colonic crypts

While the great majority (>95%) of colon cancers appear to be sporadic, a small group arise as a consequence of inherited alleles that create substantial lifelong risk for this disease. The best understood of these heritable colon cancer syndromes is adenomatous polyposis coli, usually called familial adenomatous polyposis (FAP), that is, an inherited susceptibility to develop adenomatous polyps in the colon. Such polyps, while themselves nonmalignant, are prone to develop into frank carcinomas at a low but predictable frequency. This syndrome is responsible for a bit less than 1% of all colon cancers in the West.

In Western populations, in which colon cancer is relatively frequent because of still-poorly understood dietary factors, polyps are often found in low numbers scattered throughout the colon. By the age of 70, as many as half of the individuals in these populations have developed at least one of these growths (see, for example, Figure 2.16A).

Figure 7.22 Familial adenomatous polyposis The wall of the colon from an individual afflicted with familial adenomatous polyposis (FAP) is often carpeted with hundreds of small polyps (*left*); such polyps are absent in the smooth wall of a normal colon (*right*). The detailed structure of one type of colonic polyp is shown in Figure 2.16A. (Courtesy of A. Wyllie and M. Arends.)

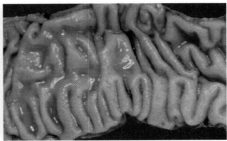

However, in the relatively rare individuals suffering from familial adenomatous polyposis, polyps numbering in the hundreds are found to carpet the luminal surface of the colon, that is, the surface facing the colonic cavity (**Figure 7.22**).

The cloning of the *APC* gene (**Sidebar 7.5**) led, after many years of additional research, to a reasonably clear view of how this gene and its encoded protein are able to control cell proliferation within the colon. As we discuss in greater detail in later chapters, the epithelia in the colon and duodenum are organized in a fashion that is typical of

Sidebar 7.5 Special human populations facilitate the detection of heritable cancer syndromes and the isolation of responsible genes The Mormons in the state of Utah represent a population that provides a golden opportunity to understand the genetics of various types of heritable human diseases. A substantial proportion of these individuals can trace their ancestry back to a relatively small group of founding settlers who arrived in Utah in the mid-nineteenth century. Mormon couples have traditionally had large numbers of children and have reproduced relatively early in life, leading to large, multigenerational families—ideal subjects of genetic research.

Most importantly, a tenet of the Mormon religion is a belief in the retroactive baptism of individuals who are ancestors to today's Mormons. As a consequence, the Mormon church (more properly, the Church of Jesus Christ of Latter Day Saints) has encouraged its members to undertake extensive genealogical research. To facilitate such research, the Mormon church maintains the world's largest genealogical archive in Salt Lake City, Utah. Finally, the Mormons in Utah have been particularly receptive to helping human geneticists trace specific disease susceptibility genes through their pedigrees.

The confluence of these factors has made it possible to assemble multigenerational pedigrees, such as that shown in **Figure 7.23**; these have enabled geneticists to trace with precision how the mutant allele of a gene predisposing a person to familial adenomatous polyposis is transmitted through multiple generations of a family and how this allele operates in a dominant fashion to create susceptibility to colonic polyps and resulting increased risk of colon cancer. Use of linkage analysis, in which the genetic transmission of this predisposing allele was connected to anonymous genetic markers on various human chromosomes, also revealed that this allele was repeatedly co-transmitted from one generation to the next together with genetic markers on the long arm of human Chromosome 5. This localization, together with LOH analyses, eventually made possible the molecular cloning of the *APC* gene in 1991.

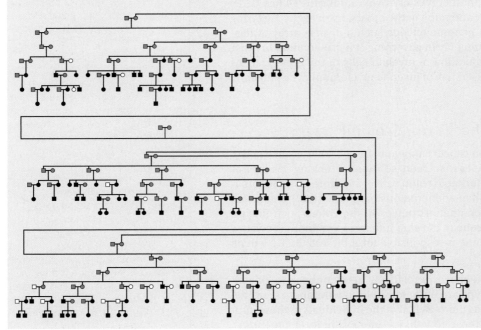

Figure 7.23 Genetic mapping using large kindreds The ability to analyze the genomes of a large number of individuals in a kindred afflicted with an inborn cancer susceptibility syndrome greatly facilitates the localization of the responsible gene to a small region of a human chromosome. In the case of the *APC* gene, the availability of a large Mormon kindred afflicted with familial adenomatous polyposis, whose relationships were traced through the extensive genealogical records kept in Salt Lake City, Utah, facilitated the localization of this gene to a small region of the long arm of human Chromosome 5. Afflicted individuals are indicated here by filled *black* symbols. (Courtesy of R.L. White, M.F. Leppert, and R.W. Burt.)

a number of epithelia throughout the body. In all cases, groups of relatively undifferentiated stem cells yield two distinct daughter cells when they divide: one daughter remains a stem cell, thereby ensuring that the number of stem cells in a tissue remains constant; however, the other daughter cell and its descendants become committed to differentiate.

In the small intestine, some of these differentiated epithelial cells participate in absorbing nutrients from the lumen and transferring these nutrients into the circulation; in the colon, many of them absorb water from the lumen. Yet other specialized epithelial cells secrete mucin (a mucus-like material) that helps to protect the colonic epithelium from the contents of the lumen. (In the gastrointestinal tract, these various epithelial cells are termed **enterocytes**.) In both parts of the intestine, the enterocytes are born in deep, mucin-filled cavities termed **crypts**, which initially shield them from the contents of the intestine (**Figure 7.24A**).

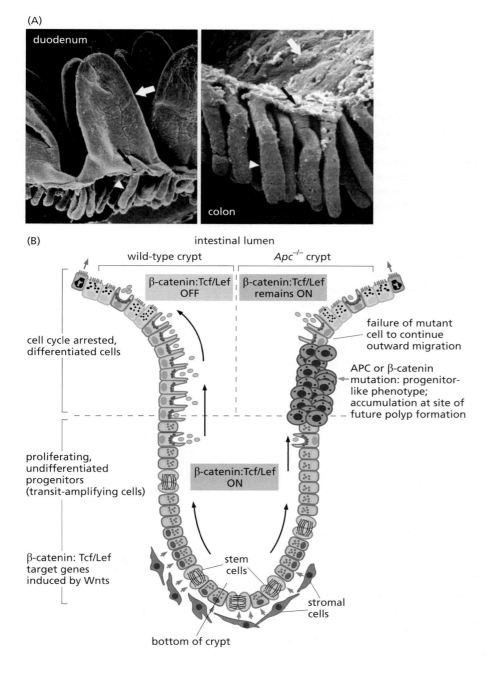

Figure 7.24 β-Catenin and the biology of colonic crypts (A) As seen in these scanning electron micrographs, the linings of the small (*left*) and large (*right*) intestines are organized similarly, with deep crypts (*white arrowheads*) in which stem cells reside and new enterocytes are produced that migrate toward the lumen (*above*), differentiate, and emerge via small openings (*narrow black arrows*). In the duodenum, these newly born cells continue their upward migration along the sides of the fingerlike villi that protrude into the lumen (*broad white arrow*), while in the colon, the villi are absent. In both cases, these cells are sloughed off after several days. (B) The bottom of the colonic crypt contains replicating stem cells with high levels of β-catenin. Intracellular β-catenin levels are normally high in these cells because they are receiving paracrine Wnt signals from stromal cells (*red*; see Section 6.10). (Within individual enterocytes at the bottoms of the crypts, the β-catenin molecules migrate to the nucleus and associate with Tcf/Lef transcription factors; this drives increased proliferation of these cells and, at the same time, prevents their differentiation.) In the normal intestine (*left side of crypt*), many of the progeny of these stem cells migrate upward toward the lumen (*above*). As they do so, stimulation by stromal Wnts decreases and intracellular levels of APC increase; these two changes together lead to increased degradation of β-catenin in individual enterocytes, which results, in turn, in cessation of proliferation and increased differentiation as these cells approach the lumen and ultimately enter apoptosis after 3 to 4 days (*small green arrows*). In contrast, when the APC protein is defective (*right side of crypt*), β-catenin levels remain high, even in the absence of intense Wnt signaling, and proliferating, still-undifferentiated cells (*purple*) fail to migrate upward, accumulate within crypts, and may ultimately generate an adenomatous polyp. As indicated, a mutant, degradation-resistant β-catenin molecule (see Section 6.10) may also produce the same outcome. (A, from M. Bjerknes and H. Cheng, *Methods Enzymol.* 419:337–383, 2006. B, from M. van de Wetering et al., *Cell* 111:251–263, 2002.)

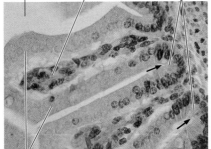

lumen of   mesenchymal   bottom of
intestine   core   crypts

enterocytes   Tcf4⁺ᐟ⁻

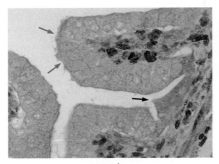

Tcf4⁻ᐟ⁻

Figure 7.25 The Tcf4 transcription factor and crypt development The formation of the stem cell compartment near the bottoms of the crypts depends critically on the high levels of β-catenin and its ability to associate with the Tcf4 protein in the nucleus in order to form an active transcription factor complex. As seen in this figure, epithelial cells in the bottoms of the crypts of the small intestine of *Tcf4⁺ᐟ⁻* mice (*arrows, upper panel*) are highly proliferative, as evidenced by the staining for Ki67, a marker of cell proliferation (*dark brown nuclei*). As they move out of the crypt up the sides of the villi, they lose proliferative potential. (The mesenchymal cells in the core of each villus are also highly proliferative, but their behavior is unrelated to that of the epithelial enterocytes.) In *Tcf4⁻ᐟ⁻* mice (*lower panel*), in contrast, a layer of enterocytes is initially formed (*red arrows*), but the proliferating enterocyte stem cells in the crypts are absent (*black arrow*), and new enterocytes are not continually formed to replace those lost by apoptosis and sloughing into the intestinal lumen. (From V. Korinek et al., *Nat. Genet.* 19:379–383, 1998.)

The locations of the stem cells and more differentiated cells in the colon are illustrated in Figure 7.24B. While some of the progeny of these stem cells stay behind in order to maintain a constant number of stem cells, most are dispatched upward and out of the crypts toward the luminal surface of the epithelium, where they will function briefly to form the epithelial lining of the gut, die by apoptosis, and be shed into the colonic lumen. This entire process of out-migration and death takes only 3 to 4 days.

The scheme depicted in Figure 7.24B represents a highly effective defense mechanism against the development of colon cancer, since almost all cells that have sustained mutations while on duty protecting the colonic wall are doomed to die within days after they have been formed. By this logic, the only type of mutations that can lead subsequently to the development of a cancer will be those mutations (and resulting mutant alleles) that block both the out-migration of colonic epithelial cells from the crypts and the cell death that follows soon thereafter. Should a colonic enterocyte acquire such a mutation, this cell and its descendants can accumulate in the crypt, and any additional mutant alleles acquired subsequently by their progeny will similarly be retained in the crypt, rather than being rapidly lost through out-migration and apoptosis. Such additional mutations might include, for example, genes that push the progeny toward a neoplastic growth state.

These dynamics focus attention on the molecular mechanisms controlling the out-migration of enterocytes from the colonic crypt. β-Catenin is the governor of much of this behavior. Recall from Section 6.10 that the levels of soluble β-catenin in the cytoplasm are controlled by Wnt growth factors. When Wnts bind cell surface receptors, β-catenin is saved from destruction, accumulates, and migrates to the nucleus, where it associates with a group of DNA-binding proteins termed variously Tcf or Lef. The resulting heterodimeric transcription factor complexes then proceed to attract yet other nuclear proteins, forming higher-order, multiprotein complexes that modify chromatin and thereby activate expression of a series of target genes programming (in the case of enterocytes) the stem cell phenotype (see Figure 6.26B and **Figure 7.25**).

In the context of the colonic crypt, enterocyte stem cells encounter Wnt factors released by stromal cells near the bottom of the crypt, which keep β-catenin levels high in the enterocytes (see Figure 7.24B). Indeed, these cells are held in a stem cell-like state by the high levels of nuclear β-catenin, which associates with the Tcf4 transcription factor in the nucleus of individual enterocytes; in fact, this transcription factor complex is required for the formation of these stem cells (see Figure 7.25). However, as the progeny of these stem cells begin their upward migration, these progeny no longer experience Wnt signaling, and intracellular β-catenin levels fall precipitously. As a consequence, these cells lose their stem cell phenotype, exit the cell cycle, and differentiate into functional enterocytes.

APC, the product of the adenomatous polyposis coli gene, is responsible for negatively controlling the levels of β-catenin in the cytosol. In enterocytes at the bottom of the normal crypts, the *APC* gene is not expressed at detectable levels and β-catenin can accumulate to high intracellular levels and move into the nuclei of these cells. However, as cells begin their upward migration out of the crypts, the level of APC expression in these cells increases greatly, and acting together with declining Wnt signaling from the stroma, this protein drives down the intracellular levels of β-catenin and thus the levels of β-catenin/Tcf transcription factor complexes in the cell nuclei.

The molecular mechanism of action of APC helps to explain these declining levels of intracellular β-catenin: Apc is a large protein of 2843 amino acid residues that can associate with β-catenin (**Figure 7.26**). Together with two scaffolding proteins termed axin and conductin, APC forms a multiprotein complex that brings together glycogen synthase kinase-3β (GSK-3β) and β-catenin (see Figure 6.26B). This association enables GSK-3β to phosphorylate four amino-terminal residues of β-catenin; the phosphorylation then leads to the degradation of β-catenin via the ubiquitin–proteasome pathway (Supplementary Sidebar 7.5). In sum, APC is essential for triggering the degradation of β-catenin, and in its absence, β-catenin levels accumulate to high levels within cells.

With this information in mind, we can place *APC* gene inactivation in the context of the biology of the colonic crypts. When the spectrum of *APC* mutations found in human colon cancers is cataloged (see Figure 7.26A), one sees many mutations that cause premature termination of translation of the Apc protein, thereby removing domains that are important for its ability to associate with β-catenin and axin and for the resulting degradation of β-catenin.

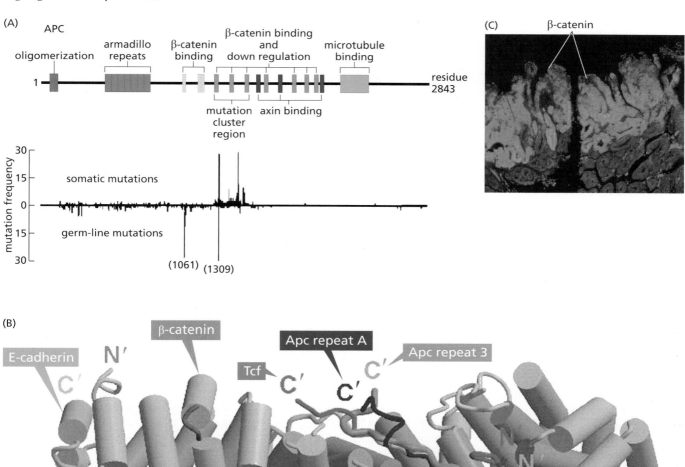

Figure 7.26 Apc, β-catenin, and Tcf/Lef (A) This diagram indicates in outline the multidomain structure of APC (*above*) with a plot of the spectrum of *APC* mutations and where they occur (*below*). The interactions of the microtubule-binding domain (*light green*) are important for proper regulation of cell motility and mitosis—a mechanistically unrelated function of this protein (see Figure 7.27). (B) As is apparent, the β-catenin molecule (*light blue, gray cylinders*) associates with a number of alternative partner proteins during its life cycle. Following its synthesis, it may be bound in the cytosol to one of two domains of the Apc protein (*dark blue, dark orange*), which will target it for destruction in proteasomes. If it escapes destruction, it may use the same domain to bind to the cytoplasmic domain of a cell surface receptor termed E-cadherin (*dark yellow*), as we will learn in Chapters 13 and 14. Alternatively, β-catenin may migrate to the nucleus, where it will associate with a domain of a Tcf/Lef transcription factor (*green*) and activate expression of a number of target genes. As is apparent,

all three partners have similarly structured domains for binding β-catenin. (C) In the Min mouse model of familial polyposis, an inactivating point mutation was introduced into the *Apc* gene by the actions of a mutagenic chemical. The chromosomal region carrying this mutant allele undergoes LOH in some cells within intestinal crypts. In these cells, which lack all Apc activity, β-catenin is no longer degraded, accumulates to high levels, and enters the cell nuclei, where it collaborates with Tcf/Lef transcription factors to drive expression of growth-promoting genes. The result is a clonal outgrowth, such as the adenoma seen here, in which the β-catenin is visualized by immunostaining (*pink, above*), while the nuclei in all cells, including those of the normal crypts (*below*), are seen in *blue*. (A, from P. Polakis, *Biochim. Biophys. Acta* 1332:F127–F147, 1997, and adapted from R. Fodde et al., *Nat. Rev. Cancer* 1:55–67, 2001. B, courtesy of H.J. Choi and W.I. Weis, and from N.-C. Ha et al., *Mol. Cell.* 15:511–521, 2004. C, courtesy of K.M. Haigis and T. Jacks.)

The accumulation of β-catenin is clearly the most important consequence of *APC* inactivation, which can be observed in about 90% of sporadic colon carcinomas. One can draw this conclusion from studying the remaining minority (approximately 10%) of sporadic colon carcinomas that carry wild-type *APC* alleles. In some of these, the *APC* gene promoter is hypermethylated and rendered inactive. In others, the gene encoding β-catenin carries point mutations, and the resulting mutant β-catenin molecules lose the amino acid residues that are normally phosphorylated by GSK-3β. Since they cannot be properly phosphorylated, these mutant β-catenin molecules escape degradation and accumulate—precisely the outcome that is seen when Apc is missing!

When β-catenin accumulates in enterocyte precursors because of inactivation of Apc function or other mechanisms (see Figure 7.26C), this causes them to retain a stem cell-like phenotype, which precludes them from migrating out of the crypts. This leads, in turn, to the accumulation of large numbers of relatively undifferentiated cells in a colonic crypt (see Figure 7.24B), which eventually form adenomatous polyps. Equally important, these accumulating cells can later sustain further mutations that enable them to form more advanced polyps and, following even more mutations, carcinomas.

This model explains the sequence of mutations that leads eventually to the formation of human colon carcinoma cells. The first of these mutations invariably involves inactivation of Apc function (or the functionally equivalent changes mentioned above). The resulting cells, now trapped in the crypts, may then suffer mutations in a number of other genes, such as K-*ras* (Section 4.4), that cause these cells to grow more aggressively. Importantly, alterations of the Apc–β-catenin pathway always come first, while the order of the subsequent genetic changes is quite variable.

Recent research has turned up other, fully unexpected consequences of Apc loss: Apc is involved in still-unclear ways in regulating the microtubules that form the mitotic spindles and part of the cytoskeleton. Cells lacking Apc function have been found to exhibit a marked increase in chromosomal instability (CIN), which results in increases and decreases in chromosome number, usually because of inappropriate segregation of chromosomes during mitosis. Some Apc-negative cells even accumulate tetraploid (or nearly tetraploid) karyotypes, in which most chromosomes are represented in four copies rather than the usual two. This defect seems to derive from the fact that Apc molecules localize to components of the microtubule arrays that form the mitotic spindle and are responsible for segregating chromosomes during the anaphase and telophase of mitosis. The aneuploidy that results from these chromosomal segregation defects alters the relative numbers of critical growth-promoting and growth-inhibiting genes. These changes, in turn, may well facilitate tumorigenesis by accelerating the rate with which pre-malignant cells acquire advantageous genotypes and thus phenotypes. In addition, Apc regulates cell motility, in part through its role in enabling the formation of cytoplasmic protrusions; cells lacking these protrusions show significantly reduced motility (**Figure 7.27**). This may help to explain the inability of *APC*-mutant enterocyte

Figure 7.27 Apc and the cytoskeleton
The microtubule-binding domain of Apc (see Figure 7.26A) enables it to regulate the cytoskeleton and several key aspects of cell physiology, including chromosomal segregation by the microtubule-containing spindle fibers during mitosis. The morphology and motility of the human osteosarcoma cells studied here are affected by Apc. (A) In the presence of Apc, these cells develop large protrusions, at the tips of which Apc molecules (*green, arrows*) congregate. Microtubules (*red*) are in the core of these protrusions, while cortical actin (*blue*) is found below the plasma membrane. These protrusions are critical to the motility of these cells. (B) When Apc expression is suppressed experimentally through use of an siRNA (*small interfering RNA*), these protrusions disappear and the microtubules that are usually apparent in cells (panel A) are either absent or reorganized, as is the actin cytoskeleton. Loss of these protrusions is correlated with a significant decrease in cell motility. (Courtesy of K. Kroboth and I. Näthke.)

(A)

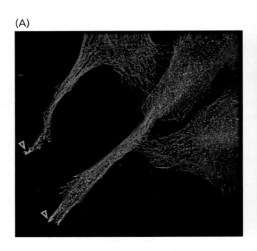

(B)

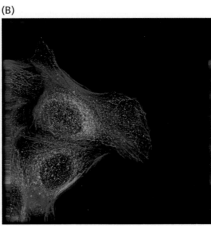

10 μm

precursors to migrate out of the colonic crypts toward the lumen, as illustrated in Figure 7.24B. It is unclear why evolution has invested this single tumor suppressor protein with these seemingly unrelated functions.

7.12 Von Hippel–Lindau disease: pVHL modulates the hypoxic response

Von Hippel–Lindau syndrome is a hereditary predisposition to the development of a variety of tumors, including clear-cell carcinomas of the kidney, **pheochromocytomas** (tumors of cells in the adrenal gland), and **hemangioblastomas** (blood vessel tumors) of the central nervous system and retina. Germ-line mutations of the *VHL* tumor suppressor gene have been documented in almost all of the patients suffering from the syndrome, which affects about 1 in 35,000 in the general population. Like mutant alleles of *Rb*, the mutant *VHL* alleles act at the organismal level in an autosomal dominant manner to create disease. And further extending the *Rb* analogy, the *VHL* locus undergoes a loss of heterozygosity (LOH) that results in a $VHL^{-/-}$ genotype in the tumor cells of patients inheriting a mutant germ-line *VHL* allele.

The *VHL* gene is also inactivated in the majority (about 70%) of sporadic (that is, non-familial) kidney carcinomas. In those sporadic tumors in which mutant *VHL* alleles are not detectable, one often finds transcriptional silencing of this gene due to promoter methylation. The main, but apparently not only, task of pVHL, the product of the *VHL* gene, is to foster destruction of a critical transcription factor termed hypoxia-inducible factor-2 (HIF-2). (It plays a far greater role in kidney carcinoma development than its closely related cousins, HIF-1 and HIF-3; much of what follows, however, derives from biochemical characterization of HIF-1, whose role as an agent of tumorigenesis is unclear.)

In cells experiencing normal oxygen tension (**normoxia**), pVHL provokes the degradation of the HIF-1α subunit of HIF-1. Consequently, HIF-1α is synthesized and then degraded with a half-life of only 10 minutes. The result is that HIF-1α accumulates only to very low steady-state levels in cells, and therefore the functional HIF-1 transcription factor, which is made up of two essential subunits—HIF-1α and HIFβ—remains inactive (**Figure 7.28**).

Such synthesis followed by rapid degradation is often called a "futile cycle." This particular cycle is interrupted when cells experience **hypoxia** (subnormal oxygen tensions), under which condition HIF-1α degradation fails to occur and HIF-1α levels increase within minutes. (Its half-life, and therefore its concentration, increases more than tenfold.) The resulting formation of functional HIF-1 transcription factor complexes causes expression of a cohort of target genes whose products are involved in angiogenesis (generation of new blood vessels), erythropoiesis (formation of red blood cells), glycolysis, and glucose transport into cells (the last two fostering the aerobic glycolysis described in Section 2.6). The motive here is to induce the synthesis of proteins that enable a cell to survive under hypoxic conditions in the short term and, in the longer term, to acquire access to an adequate supply of oxygen. Notable among the latter are proteins that collaborate to attract the growth of new vessels into the hypoxic area of a tissue. As we will learn later in Chapter 13, this formation of new vessels is also critical for the growth of tumors, enabling them to acquire access to oxygen and nutrients and to evacuate carbon dioxide and metabolic wastes.

The importance of HIF-1 in regulating the angiogenic response is underscored by the roster of genes whose expression it induces, among them the genes encoding vascular endothelial growth factor (VEGF), platelet-derived growth factor (PDGF), and transforming growth factor-α (TGF-α). VEGF attracts and stimulates growth of the endothelial cells that construct new blood vessels; PDGF stimulates these cells as well as associated mesenchymal cells, such as pericytes and fibroblasts; and TGF-α stimulates a wide variety of cell types including epithelial cells. In the hypoxic kidney, HIF-2 mediates the activation of the gene encoding erythropoietin (EPO). This leads to rapid increases in erythropoietin in the circulation, and to an induction of red cell production in the bone marrow (see also Sidebar 6.2 and Section 6.8). All in all, the HIFs target as many as 200 genes, most of whose functions are poorly understood.

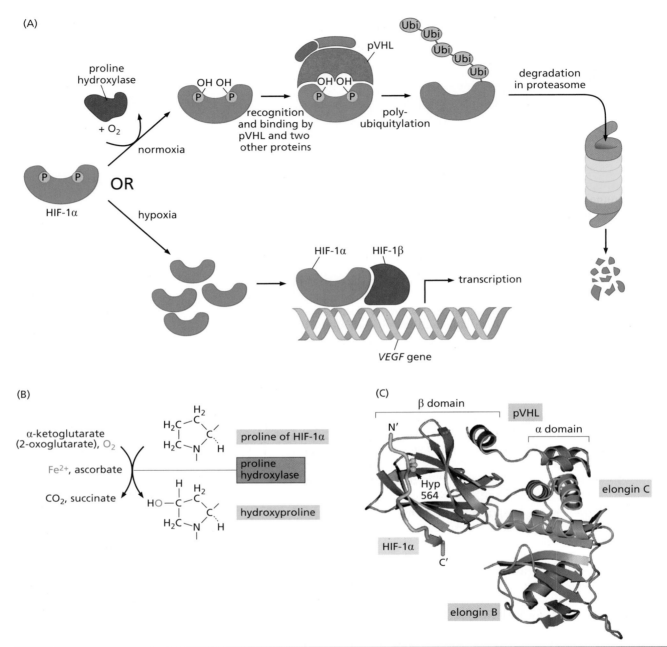

Figure 7.28 HIF-1 and its regulation by pVHL (A) The hypoxia-inducible transcription factor-1 (HIF-1) is composed of two subunits, HIF-1α and HIFβ, both of which are required for its transcription-activating function. (HIF-2α behaves similarly to HIF-1α.) Under conditions of normoxia (*above*), HIF-1α (*blue*) is synthesized at a high rate and almost immediately degraded. Degradation begins when proline hydroxylase (*brown*), an iron-containing dioxygenase enzyme, oxidizes one or two proline residues (P) of HIF-1α to hydroxyproline (Hyp) residues (indicated here as hydroxyl groups, with panel B showing the oxidation reaction). These hydroxyprolines enable the binding of pVHL to HIF-1α, which, together with several other proteins (just two shown here, *purple, green*), tag HIF-1α by ubiquitylation, thus marking it for degradation. In contrast, under hypoxic conditions (*below*), proline hydroxylase fails to oxidize the two prolines of HIF-1α, which therefore escapes ubiquitylation, allowing its levels to increase rapidly. It now can dimerize with HIFβ (*brown*), which is metabolically stable, and the resulting heterodimeric transcription factor can proceed to activate expression of physiologically important genes, such as the gene specifying vascular endothelial growth factor (VEGF, an important inducer of angiogenesis; see Chapter 13). α-Ketoglutarate, also known as 2-oxoglutarate, and oxygen serve as co-substrates in this reaction (*left*). α-Ketoglutarate and closely related metabolic intermediates may also serve as regulators of this process, which may explain the discovery of mutations in a series of Krebs cycle enzymes that have been discovered in human tumors (see Supplementary Sidebar 7.6). (Not shown here is an additional site of HIF-1α oxidation that blocks association of the HIF-1 transcription factor with p300/CBP, which is required for transcriptional activation by this complex.) (C) Shown is a ribbon diagram of a portion of HIF-1α after one of its prolines has been oxidized (for example, Hyp at residue 564, *light green balls*). This enables a domain of HIF-1α (*light blue*) to associate with a tripartite complex consisting of pVHL (*red*) and the proteins elongin B (*purple*) and elongin C (*green*) that ubiquitylates HIF-1α, which soon leads to its destruction in proteasomes. (C, from J.-H. Min et al., *Science* 296:1886–1889, 2002.)

To return to the details of *VHL* function, its product, pVHL, exists in cells in a complex with several other proteins. Together, they function to acquire a molecule of ubiquitin and to link this ubiquitin molecule covalently to specific substrate proteins (see Supplementary Sidebar 7.5). As was discussed earlier in this chapter, once a protein has been tagged by polyubiquitylation, it is usually destined for transport to proteasome complexes in which it will be degraded. Within the multiprotein complex, pVHL is responsible for recognizing and binding HIF-1α, thereby bringing the other proteins in the complex (see Figure 7.28A) in close proximity to HIF-1α, which they proceed to ubiquitylate.

pVHL binds to HIF-1α only when either of two critical proline residues of HIF-1α has been oxidized to hydroxyproline; in the absence of a hydroxyproline, this binding to HIF-1α fails and HIF-1α escapes degradation. The conversion of the HIF-1α proline residues into hydroxyprolines is carried out by an enzyme that depends on oxygen for its activity; in this case, the enzyme in question is a proline hydroxylase (PHD), which belongs to a group of oxidizing enzymes termed dioxygenases. Once formed, the hydroxyproline residue(s) of HIF-1α can be inserted into a gap in the pVHL hydrophobic core. This enables pVHL to bind HIF-1α and trigger its degradation, and as a consequence, induction of gene transcription by the HIF-1 transcription factor is prevented (see Figure 7.28A).

Taken together, these facts allow us to rationalize how pVHL works in normal cells and fails to function in certain cancer cells: Under conditions of normal oxygen tension (normoxia), HIF-1α displays one or two hydroxyprolines, is recognized by pVHL, and is rapidly destroyed. Under hypoxic conditions, HIF-1α lacks these hydroxyprolines, cannot be bound by pVHL, and accumulates to high levels that allow the functioning of the HIF-1 transcription factor. This, in turn, allows it to activate VEGF expression in hypoxic tissues, both normal and neoplastic (**Figure 7.29**).

[This hypoxia-driven upregulation of HIF function compounds the effects of a second, oxygen-independent mechanism: the increases in PI3K activity seen in many types of cancer cells cause concomitant upregulation of Akt/PKB function, which then acts via the mTORC1 kinase (to be described in Chapter 16) to cause increased production of HIF-1α mRNA and translation of this mRNA. Because the resulting increased HIF activity leads to a shift from normal energy metabolism to aerobic glycolysis, this helps explain the effects that Warburg first observed almost a century ago (see Section 2.6.)]

Figure 7.29 Biological consequences of VHL loss (A) Because pVHL drives the degradation of both HIF-1α and the closely related HIF-2α protein (whose inactivation may be even more important during the development of certain cancers), the levels of these proteins reflect pVHL function. In a renal cell carcinoma (*above*), high levels of HIF-2α are present in many cell nuclei (*dark purple*), while the protein is virtually undetectable in normal kidney tissue (*below*). (B) Notable among the genes activated by HIF-1 (and HIF-2) is the gene encoding VEGF. Seen here (*left*) is a human *in situ* breast cancer that remains localized and thus has not yet invaded the surrounding stroma. A large area of necrotic cells has formed in the center of the tumor (*arrows, dashed line*). Use of an *in situ* hybridization protocol (*right*) reveals that VEGF mRNA (*white areas*) is expressed in a gradient increasing from the low levels at the outer rim of the tumor (where oxygen is available) to the very intense expression toward the hypoxic center of the tumor. Areas of even greater hypoxia are incompatible with cell survival, explaining the large necrotic center of the tumor, where VEGF mRNA is no longer detectable. (C) One of the manifestations of von Hippel–Lindau syndrome is the presence of areas of uncontrolled vascularization in the retina, seen here as large white spots; these are ostensibly due to excessive production of VEGF in cells that have lost pVHL function. (A, courtesy of K.H. Plate. B, courtesy of A.L. Harris. C, courtesy of A. Gaudric.)

(A)

renal cell carcinoma

normal kidney

(C)

(B)

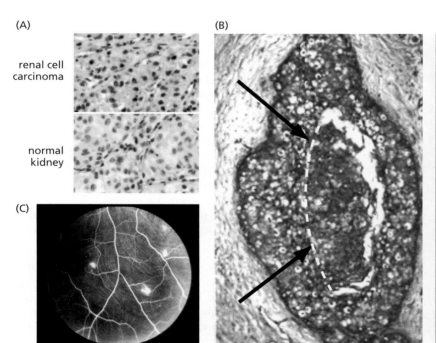

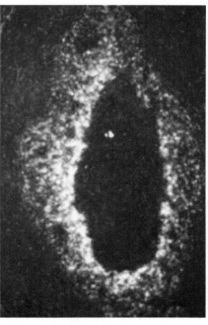

In many of the tumors associated with mutant alleles of *VHL*, the pVHL protein is undetectable in the individual cancer cells. In yet other tumors, more subtle changes may cause pVHL inactivity without affecting its levels. For example, point mutations may alter the amino acid residues in the hydrophobic pocket of pVHL that recognizes and binds the hydroxyproline residues of HIF-1α. Through either mechanism, the resulting constitutive activity of the HIF-1 transcription factor drives expression of a number of powerful growth-promoting genes, including the above-mentioned VEGF, PDGF, and TGF-α. These proceed to stimulate proliferation of a variety of cell types bearing the corresponding receptors. The result, sooner or later, is one or another type of tumor. In addition, the resulting tumor cells may cause levels of EPO to accumulate in the circulation; EPO can then help to stimulate erythropoiesis in the bone marrow, explaining the abnormally high red blood cell counts seen in some von Hippel–Lindau patients.

Still, these descriptions of pVHL action do not explain the full range of phenotypes that result from its defective functioning. Thus, it is clear that pVHL has effects that are unconnected with HIF-1α and its degradation. For example, cells lacking functional pVHL are unable to properly assemble fibronectin in their extracellular matrix, and indeed, pVHL has been reported to bind fibronectin within cells. Moreover, a substantial amount of pVHL is associated with cytoplasmic microtubules whose stability it regulates. In addition, VHL has been found to affect the activation of the p53 tumor suppressor protein and the apoptotic program (see Chapter 9), the regulation of NF-κB signaling (see Figure 6.30A), and the epithelial–mesenchymal transition (Chapter 14). These disparate functions indicate that the pVHL tumor suppressor protein, like APC, has been co-opted by evolution to regulate a series of ostensibly unrelated cell-biological functions.

In the first decade of the new millennium a group of cancer genes was uncovered that represented a major puzzle. These genes encode enzymes that are involved in carbohydrate metabolism as part of the reactions of the Krebs/citric acid cycle. Included here are the succinate and isocitrate dehydrogenases (SDH, IDH) as well as fumarate hydratase (FH). How could these metabolic enzymes and their products possibly be involved in fostering cancer? By 2010, the answer seemed increasingly to focus on prolyl hydroxylases and related dioxygenase enzymes that use α-ketoglutarate as their electron acceptor (see Figure 7.28B and Supplementary Sidebar 7.6). Like the Warburg effect (see Section 2.6), once again processes involved in carbohydrate metabolism, which seemed at first to be far removed from cancer development, have been found to be deeply enmeshed in the cancer-causing machinery of the cell.

7.13 Synopsis and prospects

Tumor suppressor genes constitute a large group of genes having only one shared attribute: in one way or another, each of these genes normally functions to reduce the likelihood that a clinically detectable tumor will appear in one of the body's tissues. Some TSGs function specifically in the later stages of multi-step tumor progression as previously benign tumors become increasingly malignant. This forces an expansion of the definition to include those genes that operate to suppress primary tumor formation and yet others that constrain malignant progression; the latter subgroup of TSGs are sometimes termed "metastasis suppressor genes." A further refinement of definitions has come from the demonstration that some TSGs encode microRNAs rather than proteins; the pleiotropic powers of many microRNAs makes them potent regulators of complex cellular phenotypes, including neoplastic transformation (see Table 7.1). Hence, proteins are not the only macromolecules that can restrain runaway cell growth.

In the great majority of cases, both copies of a tumor suppressor gene must be inactivated before an incipient cancer cell enjoys any proliferative or survival advantages. Still, this rule is not a hard-and-fast one. In some instances, such as the case of the *Nf1* gene, it appears that loss of just one copy of a tumor suppressor gene provides a measure of growth advantage to a cell—an example of the phenomenon of haploinsufficiency. And as we will see in Chapter 9, mutant alleles of the p53 TSG, when present in heterozygous configuration, can create a partially mutant cell phenotype by actively interfering with the ongoing functions of a coexisting wild-type allele in the same cell.

The discovery of tumor suppressor genes helped to explain one of the major mysteries of human cancer biology—familial cancer syndromes. As we read in this chapter, inheritance of a defective allele of one of these genes is often compatible with normal embryonic development. The phenotypic effects of this genetic defect may only become apparent with great delay, sometimes in midlife, when its presence is revealed by the loss of the surviving wild-type allele and the outgrowth of a particular type of tumor. Elimination of these wild-type alleles often involves loss-of-heterozygosity (LOH) events; and the repeatedly observed LOH in a certain chromosomal region in a group of tumors can serve as an indication of the presence of a still-unidentified tumor suppressor gene lurking in this region. In fact, a large number of chromosomal regions of recurring LOH have been identified in tumor cell genomes, but only relatively few of these have yielded cloned tumor suppressor genes to date. This means that the roster of TSGs is actually far larger than is indicated by the entries in Table 7.1. Moreover, numerous genes have been identified and cloned that are candidates for inclusion in Table 7.1 but have not yet been fully validated as bona fide tumor suppressor genes.

The diverse behaviors of these genes highlight an ongoing difficulty in this area of cancer research: What criteria can be used to define a tumor suppressor gene? To begin, in this book we have included in the family of TSGs only those genes whose products operate in some dynamic fashion to constrain cell proliferation, survival, or malignant progression. Other genes that function indirectly to prevent cancer through their abilities to maintain the genome and suppress mutations are described in Chapter 12. From the perspective of a geneticist, this division—the dichotomy between the "gatekeepers" and the "caretakers"—is an artificial one, since the wild-type version of both types of genes is often found to be eliminated or inactivated in the genomes of cancer cells. Moreover, the patterns of inheritance of the cancer syndromes associated with defective caretaker genes are formally identical to the mechanisms described in this chapter. Still, for those who would like to understand the biological mechanisms of cancer formation, the distinction between tumor suppressor genes (the gatekeepers) and genome maintenance genes (the caretakers) is a highly useful one and is therefore widely embraced.

Since the time when the first tumor suppressor genes (*p53* and *Rb*) were cloned, numerous other genes have been touted as "candidate" tumor suppressor genes because their expression was depressed or absent in cancer cells while being readily detectable in corresponding normal cells. This criterion for membership in the tumor suppressor gene family was soon realized to be flawed, in no small part because it is often impossible to identify the normal precursor of a cancer cell under study. Certain kinds of tumor cells may not express a particular gene because of a gene expression program that is played out during normal differentiation of the tissue in which these tumor cells have arisen. Hence, the absence of expression of this gene in a cancer cell may reflect only the actions of a normal differentiation program rather than a pathological loss. So this criterion—absence of gene expression—is hardly telling.

In certain instances, expression of a candidate tumor suppressor gene may be present in the clearly identified normal precursors of a group of tumor cells and absent in the tumor cells themselves. This would appear to provide slightly stronger support for the candidacy of this gene. But even this type of evidence is not conclusive, since the absence of gene expression in a cancer cell may often be one of the myriad *consequences* of the transformation process rather than one of its root *causes*. Thus, inactivation of this gene may have played no role whatsoever in the formation of the tumorigenic cell.

Responding to these criticisms, researchers have undertaken to functionally test their favorite candidate TSGs. The process involves introducing cloned wild-type versions of candidate genes into cancer cells that lack any detectable expression of these genes. The goal here has been to show that once the wild-type tumor suppressor gene function is restored in these cancer cells, they revert partially or completely to a normal growth phenotype, or may even enter into apoptosis. However, the results of such experiments are often difficult to interpret, since the **ectopic** expression of many genes—that is, their expression in a host cell where they normally are not expressed—and their expression at unnaturally high levels often make cells quite unhappy and cause them to stop growing or even to die. Such responses are often observed following introduction of a variety of genes that would never be considered tumor suppressor genes.

So this functional test has been made more rigorous by determining whether a candidate tumor suppressor gene, when expressed at normal, physiologic levels, halts the growth of a cancer cell lacking expression of this gene while leaving normal, wild-type cells from the same tissue unaffected. But even these experiments yield outcomes that are not always interpretable, because of the difficulties, cited above, in identifying normal cell types that are appropriate counterparts of the cancer cells being studied.

The ambiguities of these functional tests have necessitated the use of genetic criteria to validate the candidacy of many putative tumor suppressor genes. If a gene repeatedly undergoes LOH in tumor cell genomes, then surely its candidacy becomes far more credible. But here too ambiguity reigns. After all, genes that repeatedly suffer LOH may only be closely linked on a chromosome to a bona fide tumor suppressor gene that is the true target of elimination during tumor development.

These considerations have led to an even stricter genetic definition of a tumor suppressor gene: a gene can be called a tumor suppressor gene only if it undergoes LOH in many tumor cell genomes and if the resulting homozygous alleles bear clear and obvious inactivating mutations. (This latter criterion should allow an investigator to discount any bystander genes that happen to be closely linked to bona fide TSGs on human chromosomes.) In fact, recent advances in large-scale DNA sequencing and probe technology make it possible to undertake such searches with relative ease. For example, it is now practical to search within large chromosomal regions that undergo frequent LOH for relatively small genomic segments that repeatedly suffer homozygous deletion (see Figure 7.14). Such small segments are likely to contain only one or at most several genes.

Not surprisingly, even these quite rigid genetic and molecular criteria have proven to be flawed, since they exclude from consideration certain genes that are likely to be genuine tumor suppressor genes. Consider the fact that the activity of many tumor suppressor genes can be eliminated by promoter methylation (see Section 7.8). In this event, mutant alleles might rarely be encountered in tumor cell genomes, even though the gene has been effectively silenced. (For example, *Runx3* has been proposed to be a TSG involved in gastric cancer pathogenesis. To date, its functional silencing has been associated almost entirely with methylation in certain tumor genomes. However, the fact that only several inactivating mutations have been reported in tumor-associated *Runx3* gene copies leaves its status as a TSG precarious.)

The ability to inactivate ("knock out") candidate tumor suppressor genes in the mouse germ line (Supplementary Sidebar 7.7) provides yet another powerful tool for validating candidate tumor suppressor genes. The biology of mice and humans differs in many respects. Nonetheless, the shared, fundamental features of mammalian biology make it possible to model many aspects of human tumor biology in the laboratory mouse.

Almost all of the genes that are listed in Table 7.1 have been knocked out in the germ line of an inbred mouse strain. For the most part, resulting heterozygotes have been found to exhibit increased susceptibility to one or another type of cancer. In many instances, the particular tissue that is affected is quite different from that observed in humans. For example, *Rb*-heterozygous mice (that is, $Rb^{+/-}$) tend to develop pituitary tumors rather than retinoblastomas—an outcome that is hardly surprising, given the vastly different sizes and growth dynamics of target cell populations in mouse tissues compared with their human counterparts. Still, the development of any type of tumor at an elevated frequency in such genetically altered mice adds persuasive evidence to support the candidacy of a gene as a tumor suppressor gene.

All this explains why a constellation of criteria are now brought to bear when evaluating nominees for membership in this exclusive gene club. For us, perhaps the most

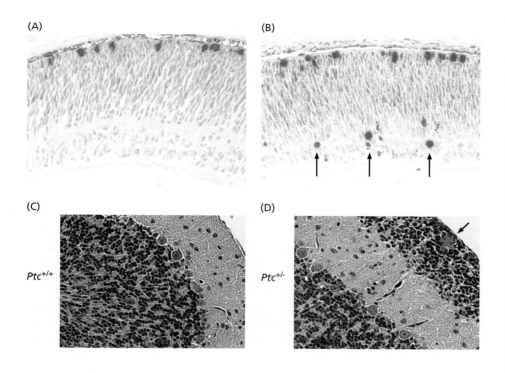

(A)

(B)

(C)

$Ptc^{+/+}$

(D)

$Ptc^{+/-}$

Figure 7.30 pRb, Patched, and the shutdown of proliferation during differentiation The presence or absence of a functional tumor suppressor gene can have profound effects on the control of cell proliferation and thus tissue development. (A) During normal retinal development in the mouse, the proliferation of retinal precursor cells is limited to the upper layers of the developing retina (*dark brown spots*), while the previously formed cells in the lower layers enter into post-mitotic, differentiated states. In this case, proliferation is revealed through use of an antibody that detects a form of histone H3 that is present only in proliferating cells. (B) In a genetically altered mouse, however, in which both copies of the *Rb* gene have been inactivated selectively in the retina, cells in the lower levels of the developing retina, which should have become post-mitotic, continue to proliferate and fail to differentiate properly (*arrows*). Similar cells in the human retina are presumably the precursors of retinoblastomas. (C) The granular cells (*dark purple*) of the developing cerebellum migrate from the outer layers (*right*) into inner layers (*left*) during normal development of a wild-type mouse, in this case one that has two wild-type copies of the *Patched* (*Ptc*) tumor suppressor gene; these cells then enter a post-mitotic state. (D) However, in a *Ptc*$^{+/-}$ heterozygote, many of the granular cells persist in the outer layer of the developing tissue (*arrow, red asterisk*) and continue to proliferate; such cells can become precursors of medulloblastomas, the most common pediatric brain tumor in humans. (A and B, from D. MacPherson et al., *Genes Dev.* 18:1681–1694, 2004. C and D, from T.G. Oliver et al., *Development* 132:2425–2439, 2005.)

compelling criterion is a functional one: Can the tumor-suppressing ability of a candidate tumor suppressor gene be rationalized in terms of the biochemical activities of its encoded protein and the known position of this protein in a cell's regulatory circuitry?

In the end, these many complications in validating candidate TSGs derive from one central fact: the very existence of a tumor suppressor gene becomes apparent only when it is absent. This sets the stage for all the difficulties that have bedeviled tumor suppressor gene research, and it underscores the difficulties that will continue to impede the validation of new TSGs in the future.

With all these reservations in mind, we can nevertheless distill some generalizations about tumor suppressor genes that will likely stand the test of time. To begin, the name that we apply to these genes is, in one sense, a misnomer. The normal role of many is to suppress increases in cell number, either by suppressing proliferation or by triggering apoptosis; some may also constrain malignant progression—the evolution of more benign into more aggressive cancer cells. In the absence of these genes, cells survive and proliferate at times and in places where their survival and proliferation are inappropriate (see, for example, **Figure 7.30**).

Another generalization is also obvious, even without knowing the identities of all TSGs: the protein products of tumor suppressor genes do not form any single, integrated signaling network. Instead, these proteins crop up here and there in the wiring diagrams of various regulatory subcircuits operating in different parts of the cell. This is explained by the simple and obvious rationale with which we began this chapter: all well-designed control systems have both positive and negative regulatory components that counterbalance one another. Thus, for every type of positive signal, such as those signals that flow through mitogenic signaling pathways, there must be negative controllers ensuring that these signaling fluxes are kept within proper limits. Perhaps cancer biologists should have deduced this from first principles, long before tumor suppressor gene research began at the laboratory bench. In hindsight, it is regrettable that they didn't consult electrical engineers and those who revel in the complexities of electronic control circuits and cybernetics. Such people could have predicted the existence of tumor suppressor genes a long time ago.

Even after another 50 tumor suppressor genes are cataloged, the pRb and p53 proteins will continue to be recognized as the products of tumor suppressor genes that are of preeminent importance in human tumor pathogenesis. The reasons for this will become apparent in the next two chapters. While all the pieces of evidence are not yet in hand, it seems increasingly likely that the two signaling pathways controlled by pRb and p53 are deregulated in the great majority of human cancers. Almost all of the remaining tumor suppressor genes (see Table 7.1) are involved in the development of more circumscribed sets of human tumors.

The fact that certain TSGs are missing from the expressed genes within cancer cells has prompted many to propose the obvious: if only we could replace the missing tumor suppressor genes in cancer cells, these cells would revert partially or totally to a normal cell phenotype and the problem of cancer would be largely solved. Such "gene therapy" strategies have the additional attraction that the occasional, inadvertent introduction of a tumor suppressor gene into a normal cell should have little if any effect if this gene is expressed at physiologic levels; this reduces the risk of undesired side-effect toxicity to normal tissues.

As attractive as such gene therapy strategies are in concept, they have been extremely difficult to implement. The viral vectors that form the core of most of these procedures are inefficient in delivering intact, wild-type copies of tumor suppressor genes to the neoplastic cells within tumor masses. However, efficient transfer of these wild-type gene copies into all the cells within a tumor cell population is essential for curative anti-tumor therapies: if significant numbers of cells fail to acquire a vector-borne wild-type TSG, these cells will serve as the progenitors of a reborn tumor mass. For this reason, while tumor suppressor genes are profoundly important to our understanding of cancer formation, in most cases, a reduction of this knowledge to therapeutic practice is still largely beyond our reach.

The exceptions to this generally discouraging scenario come from the instances where the absence of TSG function makes cells vulnerable to particular types of low-molecular-weight drugs (which are more effective in reaching cells throughout a tumor). For example, during the course of tumor progression, certain classes of tumor cells become especially dependent on signals that are created by the loss of tumor suppressor gene function. Thus, loss of the PTEN tumor suppressor gene leads to hyperactivity of Akt/PKB, on which some types of cancer cells come to depend for their continued viability. This explains why drugs that shut down firing of an essential upstream co-activator of Akt/PKB, termed mTOR, hold great promise as therapeutic agents against tumors, such as glioblastomas and prostate carcinomas, in which PTEN activity is often lost. But in general, as we will see in Chapter 16, the major advances in drug discovery and anti-cancer therapy are being made in shutting down the hyperactive oncoproteins. One can only hope that our extensive knowledge of tumor suppressor function will serve as the basis for anti-cancer treatments developed in the more distant future—a time when our skills in crafting new therapies will be vastly more sophisticated than at present.

Key concepts

- At the cellular level, the neoplastic phenotype of human cancer cells is usually recessive, because expression of this phenotype depends, in part, on the inactivation of tumor suppressor genes. This indicates that the loss of genetic information is critical for the development of most if not all human tumors.

- Much of the loss of functionally important genetic information during the formation of tumors is demonstrated by the fact that TSGs are often present in the genomes of cancer cells as inactive, null alleles.

- Tumor suppressor gene loss usually affects cell phenotype only when both copies of such a gene are lost in a cell. This means that null alleles of TSGs may be present in a heterozygous state in many cells throughout the body without affecting cell and tissue phenotype.

- The loss of TSG function can occur either through genetic mutation or the epigenetic silencing of genes via promoter methylation.

- Inactivation (by mutation or methylation) of one copy of a TSG may be followed by other mechanisms that facilitate loss of the other gene copy; these mechanisms often depend on loss of heterozygosity (LOH) at the TSG locus, and may involve mitotic recombination, loss of a chromosomal region that harbors the gene, inappropriate chromosomal segregation (nondisjunction), or gene conversion stemming from a switch in template strand during DNA replication.

- LOH events usually occur more frequently than mutations or promoter methylation, and they occur with different frequencies in different genes.

- Repeated LOH occurring in a given chromosomal region in a number of independently arising tumors often indicates the presence of a TSG in that region.

- TSGs regulate cell proliferation through many biochemical mechanisms. The only theme that unites them is the fact that the loss of any one of them increases the likelihood that a cell will undergo neoplastic transformation or progression to a highly malignant state.

- When mutant, defective copies of a TSG are inherited in the germ line, the result is often greatly increased susceptibility to one or another specific type of cancer.

- TSGs are often called "gatekeepers" to signify their involvement in governing the dynamics of cell proliferation and to distinguish them from a second class of genes, the "caretakers," which also increase cancer risk when inherited in defective form but function entirely differently, since they work to maintain the integrity of the cell genome.

- The loss of TSGs occurs far more frequently during the development of a tumor than the activation of proto-oncogenes into oncogenes, and tumor cell genomes usually harbor multiple inactivated TSGs.

Thought questions

1. Why is the inheritance of mutant, activated oncogenes responsible for only a small proportion of familial cancer syndromes, while the inheritance of defective tumor suppressor genes (TSGs) is responsible for the lion's share of these diseases?

2. What factors may determine whether the inactivation of a TSG occurs at a frequency per cell generation higher than the activation of an oncogene?

3. How might the loss of TSG function yield an outcome that is, at the cell-biological level, indistinguishable from the acquisition of an active oncogene?

4. Some TSGs undergo LOH in fewer than 20% of tumors of a given type. Why and how do such low rates of LOH complicate the identification and molecular isolation of such genes?

5. What criteria need to be satisfied before you would be comfortable in categorizing a gene as a TSG?

6. What factors might influence the identities of the tissues affected by an inherited, defective allele of a TSG?

Additional reading

Berdasco M & Esteller M (2010) Aberrant epigenetic landscape of cancer: how cellular identity goes awry. *Dev. Cell* 19, 698–711.

Brahimi-Horn MC, Chiche J & Pouysségur J (2007) Hypoxia and cancer. *J. Mol. Med.* 85, 1301–1307.

Comings DE (1973) A general theory of carcinogenesis. *Proc. Natl. Acad. Sci. USA* 70, 3324–3328.

Esteller M (2007) Cancer epigenomics: DNA methylomes and histone-modification maps. *Nat. Rev. Genet.* 8, 286–298.

Esteller M (2009) Epigenetics in cancer. *N. Engl. J. Med.* 358, 1148–1159.

Feinberg AP, Ohlsson R & Henikoff S (2006) The epigenetic progenitor origin of human cancer. *Nat. Rev. Genet.* 7, 21–33.

Foulkes WD (2008) Inherited susceptibility to common cancers. *N. Engl. J. Med.* 359, 2143–2153.

Gregorieff A & Clevers H (2005) Wnt signaling in the intestinal epithelium: from endoderm to cancer. *Genes Dev.* 19, 877–890.

Herman JG & Baylin SB (2003) Gene silencing in cancer in association with promoter hypermethylation. *N. Engl. J. Med.* 349, 2042–2054.

Hershko A, Ciechanover A & Varshavsky A (2000) The ubiquitin system. *Nat. Med.* 6, 1073–1081.

Humphries A & Wright NA (2008) Colonic crypt organization and tumorigenesis. *Nat. Rev. Cancer* 8, 415–424.

Kaelin WG Jr (2008) The von Hippel–Lindau tumour suppressor protein: O2 sensing and cancer. *Nat. Rev. Cancer* 8, 865–873.

Klaus A & Birchmeier W. (2008) Wnt signaling and its impact on development and cancer. *Nat. Rev. Cancer* 8, 387–398.

Logan CY & Nusse R (2004) The Wnt signaling pathway in development and disease. *Annu. Rev. Cell Dev. Biol.* 20, 781–810.

Nelson WJ & Nusse R (2004) Convergence of Wnt, β catenin, and cadherin pathways. *Science* 303, 1483–1487.

Presneau N, Manderson EN & Tonin PN (2003) The quest for a tumor suppressor gene phenotype. *Curr. Mol. Med.* 3, 605–629.

Pugh CW & Ratcliffe PJ (2003) The von Hippel–Lindau tumor suppressor, hypoxia-inducible factor-1 (HIF-1) degradation, and cancer pathogenesis. *Semin. Cancer Biol.* 13, 83–89.

Radtke F & Clevers H (2005) Self-renewal and cancer of the gut: two sides of a coin. *Science* 307, 1904–1907.

Sharma S, Kelly TK & Jones PA (2010) Epigenetics in cancer. *Carcinogenesis* 31, 27–36.

Sherr CJ (2004) Principles of tumor suppression. *Cell* 116, 235–246.

Staser K, Yang FC & Clapp DW (2010) Mast cells and the neurofibroma microenvironment. *Blood* 116, 157–164.

Thoma CR, Toso A, Meraldi P & Krek W (2009) Double-trouble in mitosis caused by von Hippel-Lindau tumor-suppressor protein inactivation. *Cell Cycle* 8, 3619–3620.

Ting A, McGarvey KM & Baylin SB (2006) The cancer epigenome: components and functional correlates. *Genes Dev.* 20, 3215–3231.

Varshavsky A (2005) Regulated protein degradation. *Trends Biochem. Sci.* 30, 283–286.

Wijnhoven SW, Kool HJ, van Teilingen CM et al. (2001) Loss of heterozygosity in somatic cells of the mouse: an important step in cancer initiation? *Mutat. Res.* 473, 23–36.

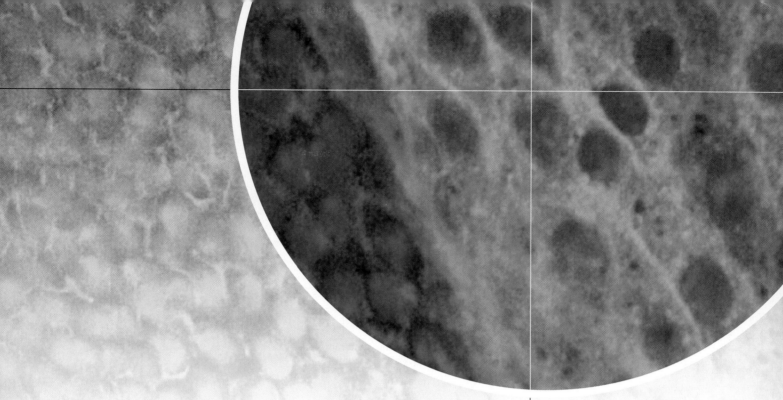

Chapter 8

pRb and Control of the Cell Cycle Clock

> This immediately leads one to ask, if the [hypothetical cellular transformation] loci can get into so much mischief, why keep them around? The logical answer is that they have some necessary function during some stage of the cell cycle, or some stage of embryogenesis.
>
> David E. Comings, geneticist, 1973

The fate of individual cells throughout the body is dictated by the signals that each receives from its surroundings—a point made repeatedly in earlier chapters. Thus, almost all types of normal cells will not proliferate unless prompted to do so by mitogenic growth factors. Yet other signaling proteins, notably transforming growth factor-β (TGF-β), may overrule the messages conveyed by mitogenic factors and force a halt to proliferation. In addition, extracellular signals may persuade a cell to enter into a **post-mitotic**, differentiated state from which it will never re-emerge to resume proliferation.

These disparate signals are collected by dozens of distinct cell surface receptors and then funneled into the complex signal-processing circuitry that operates largely in the cell cytoplasm. In some way, this mixture of signals must be processed, integrated, and ultimately distilled down to some simple, binary decisions made by the cell as to whether it should proliferate or become quiescent, and whether, as a quiescent cell, it will or will not differentiate. These behaviors suggest the existence of some centrally acting governor that operates inside the cell—a master clearinghouse that receives a wide variety of incoming signals and makes major decisions concerning the fate of the cell.

This master governor has been identified. It is the **cell cycle clock**, which operates in the cell nucleus. Its name is really a misnomer, since this clock is hardly a device for counting the passage of time. Nonetheless, we will use this term here for want of

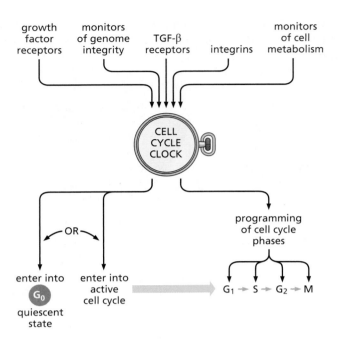

Figure 8.1 The central governor of growth and proliferation The term "cell cycle clock" denotes a molecular circuitry operating in the cell nucleus that processes and integrates a variety of afferent (incoming) signals originating from outside and inside the cell and decides whether or not the cell should enter into the active cell cycle or retreat into a nonproliferating state. In the event that active proliferation is decided upon, this circuitry proceeds to program the complex sequence of biochemical changes in a cell that enable it to double its contents and to divide into two daughter cells.

a better one. Rather than counting elapsed time, the cell cycle clock is a network of interacting proteins—a signal-processing circuit—that receives signals from various sources both outside and inside the cell, integrates them, and then decides the cell's fate. Should the cell cycle clock decide in favor of proliferation, it proceeds to orchestrate the complex transitions that together constitute the cell's cycle of growth and division. Should it decide in favor of quiescence, it will use its agents to impose this nonproliferative state on the cell (**Figure 8.1**).

The proliferative behavior of cancer cells indicates that the master governor of the cell's fate is influenced not only by normal proteins but also by oncogene proteins that insert themselves into various signaling pathways and disrupt normal control mechanisms. Similarly, the deletion of key tumor suppressor proteins evokes equally profound changes in the control circuitry and thus is equally influential in perturbing decision making by the cell cycle clock. Consequently, sooner or later, the molecular actions of most oncogenes and tumor suppressor genes must be explained in terms of their effects on the cell cycle clock. To this end, we will devote the first half of this chapter to a description of how this molecular machine normally operates and then proceed to study how it is perturbed in human cancer cells.

8.1 Cell growth and division is coordinated by a complex array of regulators

When placed into culture under conditions that encourage exponential multiplication, mammalian cells exhibit a complex cycle of growth and division that is usually referred to as the **cell cycle**. A cell that has recently been formed by the processes of cell division—**mitosis** and **cytokinesis**—must decide soon thereafter whether it will once again initiate a new round of active growth and division or retreat into the non-growing state that we previously termed G_0. As described earlier (Sections 5.1 and 6.1), this decision is strongly influenced by mitogenic growth factors in the cell's surroundings. Their presence in sufficient concentration will encourage a cell recently formed by mitosis to remain in the active growth-and-division cycle; their absence will trigger the default decision to proceed from mitosis into the G_0, quiescent state.

Withdrawal from the cell cycle may be actively encouraged by the presence of growth-inhibitory factors in the medium. Prominent among these anti-mitogenic factors is transforming growth factor-β (TGF-β). Withdrawal from the cell cycle into the G_0, quiescent state, whether due to the absence of mitogenic growth factors or the presence of anti-mitogens such as TGF-β, is often reversible, in that an encounter by a quiescent cell with mitogenic growth factors on some later occasion may induce this cell to re-enter into active growth and division. However, some cells leaving the active cell cycle may do so irreversibly, thereby giving up all option of ever re-initiating active growth and division, in which case they are said to have become *post-mitotic*. Many types of neurons in the brain, for example, are widely assumed to fall into this category.

The decision by a cell recently formed by cell division to remain in the active growth-and-division cycle requires that this cell immediately begin to prepare for the next division. Such preparations entail, among other things, the doubling of the cell's macromolecular constituents to ensure that the two daughter cells resulting from the next round of cell division will each receive an adequate endowment of them. This accumulation of cellular constituents, which drives an increase in cell size, is some-times termed the process of cell *growth* to distinguish it from the process of cell *division*, which yields, via mitosis and cytokinesis, two daughter cells from a mother cell (**Figure 8.2**). However, in the more common usage and throughout this book, the term "cell growth" implies both the accumulation of cell constituents and the subsequent cell division, that is, the two processes that together yield cell proliferation.

The accumulation of a cell's macromolecules involves, among many other molecules, the duplication of the cell's genome. In many prokaryotic cells, this duplication—the process of DNA replication—begins immediately after daughter cells are formed by cell division. But in most mammalian cells, the overall program of macromolecular synthesis is organized quite differently. While the accumulation of RNA and proteins is initiated immediately after cell division and proceeds continuously until the next cell division, the task of replicating the DNA is deferred for a number of hours (often as

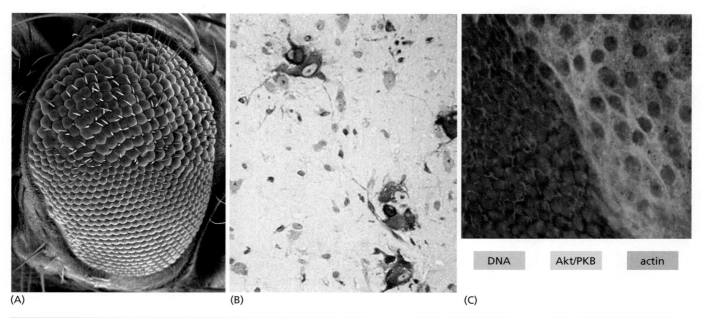

(A)   (B)   (C)

Figure 8.2 Growth versus proliferation Alterations of certain signaling proteins, such as the one encoded by the *TSC1* tumor suppressor gene, allow the processes of cell growth and division to be uncoupled from one another. (A) In this scanning electron micrograph of a *Drosophila* eye, the ommatidial cells in the upper portion of the eye have been deprived of the fly ortholog of *TSC1*; these cells are physically larger than the wild-type cells shown (*below*), because they have grown more during the cell cycles that led to their formation. (B) The same behavior can be seen in the brains of patients suffering from tuberous sclerosis, in which *TSC1* function has been lost through a germ-line mutation and subsequent somatic loss of heterozygosity. Seen here are the giant cells present in a benign growth (a "tuber"). The giant cells (*brown*) are labeled with an antibody against phosphorylated S6, a ribosomal protein important in regulating protein synthesis and thus cell growth; S6 phosphorylation and functional activation is deregulated in cells lacking *TSC1* function. (C) The Akt/PKB protein is frequently hyperactivated in human cancers. In an imaginal disc of a developing *Drosophila* larva, Akt/PKB (*green*) that has been hyperactivated in a portion of cells (*upper right*) causes a great increase in nuclear and overall cell size relative to cells with normal Akt/PKB (*lower left*); similar increases in mammalian cells are also observed in response to hyperactive Akt/PKB. (A, from X. Gao and D. Pan, *Genes Dev.* 15:1383–1392, 2001. B, courtesy of J.A. Chan and D.J. Kwiatkowski. C, from J. Verdu et al., *Nat. Cell Biol.* 1:500–506, 1999.)

(A)

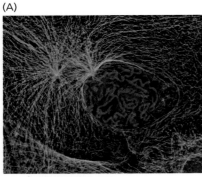

prophase

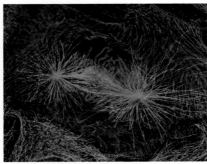

prometaphase

metaphase

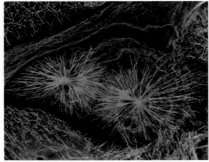

anaphase

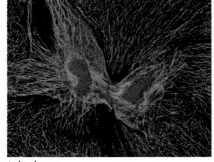

telophase

(B)

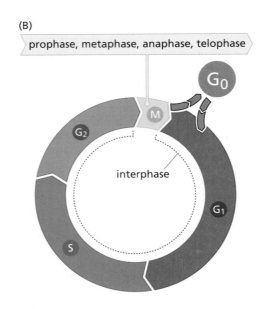

Figure 8.3 The mammalian cell cycle (A) Immunofluorescence is used here to illustrate the four distinct subphases of mitosis (M phase) in newt lung cells (*top to bottom*). During the *prophase* of mitosis, the chromosomes (*blue*), which were invisible microscopically during *interphase* (the period encompassing G_1 through S and G_2), begin to condense and become visible, while the *centrosomes* (*light green radiating bodies*) at the poles of the cell begin to assemble (*top two images*). During *metaphase*, the chromosomes align along a plane that bisects the cell and become attached to the microtubule fibers of the mitotic spindle (*light green, third image*). At the same time, the nuclear membrane has disappeared. During *anaphase*, the two halves of each chromosome—the chromatids—are pulled apart by the mitotic spindle (i.e., they *segregate*) to the two opposite poles of the cell (*fourth image*). During *telophase*, shortly after the chromatids cluster into the two sets seen here (*bottom image*), the chromatids de-condense and a new nuclear membrane forms around each set of chromatids (now called chromosomes; *not shown*). At the same time, during the process of *cytokinesis*, the cytoplasm of the mother cell divides, yielding two daughter cells. (B) The mammalian growth-and-division cycle is divided into four phases—G_1, S (during which DNA is replicated), G_2, and M (mitosis). A fifth state, G_0 (G zero), denotes the resting, nonproliferating state of cells that have withdrawn from the active cell cycle. While exit from the active cell cycle into G_0 is depicted here as occurring in early G_1, it is unclear when during G_1 this actually occurs. (A, courtesy of Conly Rieder.)

many as 12 to 15) after emergence of new daughter cells from mitosis and cytokinesis. During this period between the birth of a daughter cell and the subsequent onset of DNA synthesis, which is termed the G_1 (first gap) phase of the cell cycle (**Figure 8.3**), cells make critical decisions about growth versus quiescence, and whether, as quiescent cells, they will differentiate.

In many types of cultured mammalian cells, the DNA synthesis that follows G_1 often requires 6 to 8 hours to reach completion. This period of DNA synthesis is termed the S (synthetic) phase, and its length is determined in part by the enormous amount of cellular DNA ($\sim 6.4 \times 10^9$ base pairs per diploid genome) that must be replicated with fidelity during this time. The actual length of S phase varies greatly among different kinds of cells, being much shorter in certain cell types, such as rapidly dividing embryonic cells and lymphocytes.

Having passed through S phase, a cell might be thought fit to enter directly into mitosis (M phase). However, most mammalian cells spend 3 to 5 hours in a second gap phase, termed G_2, preparing themselves, in some still-poorly understood fashion, for entrance into M phase and cell division. M phase itself usually encompasses an hour or so, and includes four distinct subphases—**prophase**, **metaphase**, **anaphase**, and **telophase**; this culminates in cytokinesis, the division of the cytoplasm that allows the formation of two new cells (see Figure 8.3A). While these times are commonly observed when studying mammalian cells in culture, they do not reflect the behavior of all cell types under all conditions. For example, actively proliferating lymphocytes may double in 5 hours, and some cells in the early embryo may do so even more rapidly.

As is the case with S phase, M phase must proceed with great precision. M phase begins with the two recently duplicated DNA helices within each chromosome; these are carried in the sister **chromatids** of the chromosome, which are aligned adjacent to one another in the nucleus. The allocation during mitosis of the duplicated chromatids to the two future daughter cells must occur flawlessly to ensure that each daughter receives exactly one diploid complement of chromatids—no more and no less. Once present within the nuclei of recently separated daughter cells, these chromatids become the chromosomes of the newly born cells.

This means that the endowment of one genome's worth of genetic material to each daughter cell depends on the precise execution of two processes—the faithful replication of a cell's genome during S phase, and the proper allocation of the resulting duplicated DNAs to daughter cells during M phase. As will be discussed later, defects in either of these processes can have disastrous consequences for the cell and the organism, one of which is the disease of cancer.

Like virtually all machinery, the machine that executes the various steps of the cell cycle is subject to malfunction. This fallibility contrasts with the stringent requirement of the cell to have the various phases of the cell cycle proceed flawlessly. For this reason, the cell deploys a series of surveillance mechanisms that monitor each step in cell cycle progression and permit the cell to proceed to the next step in the cycle only if a prerequisite step has been completed successfully. In addition, if specific steps in the execution of a process go awry, these monitors rapidly call a halt to further advance through the cell cycle until these problems have been successfully addressed. Yet other monitors ensure that once a particular step of the cell cycle has been completed, it is not repeated until the cell passes through the next cell cycle. These monitoring mechanisms are termed variously **checkpoints** or **checkpoint controls** (Figure 8.4).

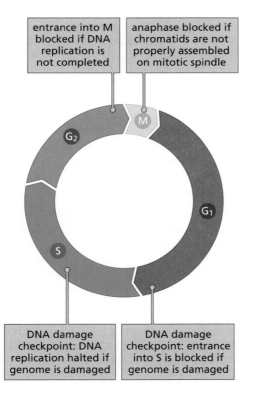

Figure 8.4 Examples of checkpoints in the cell cycle Checkpoints impose quality control to ensure that a cell has properly completed all the requisite steps of one phase of the cell cycle before it is allowed to advance into the next phase. A cell will not be permitted to enter into S phase until all the steps of G_1 have been completed. It will be blocked from entering G_2 until all of its chromosomal DNA has been properly replicated. Similarly, a cell is not permitted to enter into anaphase (when the paired chromatids are pulled apart) until all of its chromosomes are properly assembled on the mitotic spindle during metaphase. In addition, a cell is not allowed to advance into S or M if its DNA has been damaged and not yet repaired. Other controls (*not shown*) ensure that once a specific step in the cell cycle has been completed, it is not repeated until the next cell cycle.

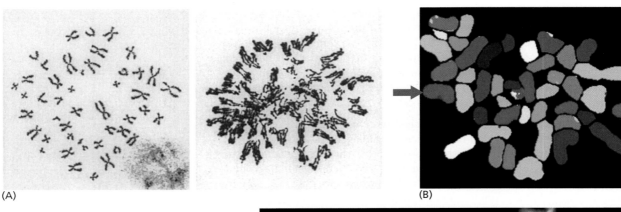

(A)   (B)

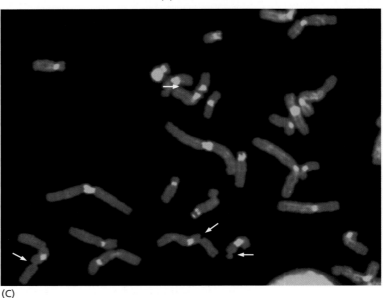

(C)

Figure 8.5 Consequences of loss of checkpoint controls Loss of critical cell cycle checkpoint control mechanisms is often manifested as an altered karyotype. (A) The normal human karyotype (*left*) is contrasted with that of a cell that has been deprived of the Rad17 checkpoint protein (*right*), which is responsible for preventing inadvertent re-replication of already-replicated chromosomal DNA, resulting in *endoreduplication* and increases in the *ploidy* of the cell. (B) The Bub1 protein normally prevents separation of chromosomes in the event that one or more chromosomal pairs are not properly aligned on the metaphase plate. In its absence, cells gain or lose chromosomes, as seen in this spectral karyotyping (SKY) analysis, which indicates that this human cell has only one Chromosome 1 (*yellow, arrow, right*) and one Chromosome 6 (*red, arrow, left*). (Several other red-colored chromosomes are evident but can be identified as belonging to other chromosome pairs by their size and shading.) (C) The ATR (ataxia telangiectasia and Rad3 related) protein kinase is responsible for, among other functions, halting further DNA replication until stalled replication forks are repaired. In its absence, *fragile sites*—sites in the chromosome prone to breakage—become visible upon karyotypic analysis. Here, fragile sites on human Chromosomes 3 and 16 (*white arrows*) are apparent in cells lacking ATR function. (A, from X. Wang et al., *Genes Dev.* 17:965–970, 2003. B, from A. Musio et al., *Cancer Res.* 63:2855–2863, 2003. C, from A.M. Casper et al., *Am. J. Hum. Genet.* 75:654–660, 2004.)

One checkpoint ensures that a cell cannot advance from G_1 into S if the genome is in need of repair. Another, operating in S phase, will slow or pause DNA replication in response to DNA damage. (In mammalian cells, this may cause a doubling of the time required to complete DNA synthesis.) A third will not permit the cell to proceed through G_2 to M until the DNA replication of S phase has been completed. DNA damage will trigger another checkpoint control that blocks entrance into M phase. During M phase, highly efficient checkpoint controls block anaphase; these blocks are only removed once all the chromosomes have been properly attached to the mitotic spindle. Yet other checkpoint controls, not cited here, have been reported. For example, a **decatenation** checkpoint in late G_2 prevents entrance into M until the pair of DNA helices replicated in the previous S phase have been untangled from one another. Defects in some of these checkpoint controls can be observed because of their effects on cells' chromosomes (**Figure 8.5**).

The operations of these checkpoints also influence the formation of cancers. As tumor development (often called tumor progression, discussed in Chapter 11) proceeds, incipient cancer cells benefit from experimenting with various combinations and permutations of mutant alleles, retaining those that will afford them the greatest proliferative advantage. An increased mutability of their genomes accelerates the rate at which these cells can acquire advantageous combinations of alleles and thus hastens the overall pace of tumor progression. Such mutability and resulting genomic instability is incompatible with normal cell cycle progression, since checkpoint controls usually block advance of the cell through its cycle if its DNA has been damaged or its chromosomes are in disarray. So, in addition to acquiring altered growth-controlling genes (that is, activated oncogenes and inactivated tumor suppressor genes), many types of cancer cells have inactivated one or more of their checkpoint controls. With

Sidebar 8.1 Embryonic stem (ES) cells show highly autonomous behavior Our perceptions about the behavior of normal mammalian cells have been conditioned by decades of work with a wide variety of somatic cells present in embryonic and adult tissues. However, cells in the very early embryo clearly operate under a very different set of rules. The pRb pathway and the cell cycle clock machinery to be described in this chapter appear to be operative in one form or another in virtually all types of adult cells. In contrast, a variety of experiments indicate that pRb-imposed growth control is not functional in early embryonic cells, including their cultured derivatives, embryonic stem (ES) cells (Supplementary Sidebar 8.1). The same can be said about the p53 pathway (see Chapter 9). It seems that the mitogenic signals required to keep more differentiated cells proliferating are not required by cultured ES cells for their proliferation. For example, aside from the growth factor termed LIF (leukemia inhibitory factor), which is needed to prevent their differentiation, mouse ES cells proliferating *in vitro* appear able to drive their own proliferation through internally generated signals. (Indeed, a constitutively activated Ras-like protein, termed E-Ras, has been reported to be expressed specifically in these cells.)

ES cells seem to preserve much of the cell-autonomous behavior that we associate with the single-cell ancestors of metazoa, that is, the behavior of cells that have not yet become dependent on their neighbors for signals controlling growth and survival. The most stunning indication of their extreme autonomy is their ability to form benign tumors (**teratomas**) when introduced into many anatomical sites in an adult organism. Since these cells are genetically wild type, they represent the only example of a wild-type cell that is tumorigenic.

these controls relaxed, incipient cancer cells can more rapidly accumulate the mutant genes and altered karyotypes that propel their neoplastic growth. The breakdown of the controls responsible for maintaining the cellular genome in an intact state is one of the main subjects of Chapter 12.

8.2 Cells make decisions about growth and quiescence during a specific period in the G_1 phase

As mentioned earlier, virtually all normal cell types in the body require external signals, such as those conveyed by mitogenic growth factors, before they will undertake to grow and divide. The only exceptions to this rule appear to be very early embryonic cells, which seem able to proliferate without receiving growth-stimulating signals from elsewhere (**Sidebar 8.1**). The rationale for this behavior of the normal cells throughout our tissues is a simple one: since these cells participate in the formation of precisely structured tissues, their proliferation must, by necessity, be coordinated with neighboring cells in those tissues. Put differently, the body cannot give each of its almost 10^{14} component cells the license to decide on its own whether to grow and divide. To do so would invite chaos.

Evidence accumulated over the past quarter century indicates that cells consult their extracellular environment and its growth-regulating signals during a discrete window of time in the active cell cycle, namely, from the onset of G_1 phase until an hour or two before the G_1-to-S transition (**Figure 8.6**). The operations of the G_1 decision-making machinery are indicated by the responses of cultured cells to extracellular signals. If we were to remove serum and thus growth factors from cells before they had completed

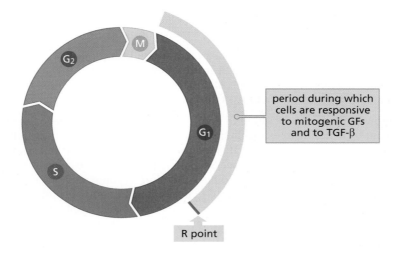

Figure 8.6 Responsiveness to extracellular signals during the cell cycle Cells respond to extracellular mitogens and inhibitory factors (such as TGF-β) only in a discrete window of time that begins at the onset of G_1 and ends just before the end of G_1. The end of this time window is often designated as the restriction (R) point, which denotes the point in time when the cell must make the commitment to advance through the remainder of the cell cycle through M phase, to remain in G_1, or to retreat from the active cell cycle into G_0.

period during which cells are responsive to mitogenic GFs and to TGF-β

R point

80 to 90% of G_1, they would fail to proceed further into the cell cycle and would, with great likelihood, revert to the G_0 state. However, once these cells had moved through this G_1 decision-making period and advanced into the final hours of G_1 (the remaining 10 to 20% of G_1), the removal of serum would no longer affect their progress and they would proceed through the remainder of G_1 and onward through the S, G_2, and M phases. Similarly, anti-mitogenic factors, such as TGF-β, are able to impose their growth-inhibitory effects only during this period in the early and mid-G_1 phase. Once a cell has entered into late G_1, it seems to be oblivious to the presence of this negative factor in its surroundings.

This schedule of total dependence on extracellular signals followed by entrance in late G_1 into a state of relative independence indicates that a weighty decision must be made toward the end of G_1. Precisely at this point, a cell must make up its mind whether it will remain in G_1, retreat from the active cycle into G_0, or advance into late G_1 and thereafter into the remaining phases of the cell cycle. This critical decision is made at a transition that has been called the **restriction point** or **R point** (see Figure 8.6). In most mammalian cells studied to date, the R point occurs several hours before the G_1/S phase transition.

If a cell should decide at the R point to continue advancing through its growth-and-division cycle, it commits itself to proceed beyond G_1 into S phase and then to complete a rigidly programmed series of events (the entire S, G_2, and M phases) that enable it to divide into two daughter cells. This decision will be respected even if growth factors are no longer present in the extracellular space during these remaining phases of the cell cycle. We know that this series of later steps (S, G_2, and M phases) proceeds on a fixed schedule, because a cell that enters S will, in the absence of a major disaster, invariably complete S and, having done so, enter into G_2 and then advance into M phase.

For those interested specifically in understanding the deregulated proliferation of cancer cells, this fixed program holds relatively little interest, since the late $G_1 \rightarrow S \rightarrow G_2 \rightarrow M$ progression proceeds similarly in normal cells and cancer cells. Students of cancer therefore focus largely on the G_0/G_1 transition and on the one period in the life of an actively growing cell—the time window encompassing most of G_1—when a cell is given the license to make decisions about its fate.

The commitment to advance through the R point and continue all the way to M phase is, as hinted, not an absolute one. Metabolic, genetic, or physical disasters may intervene during S, G_2, or M and force the cell to call a halt, often temporary, to its further advance through the cell cycle until these conditions have been addressed. Still, in the great majority of cases, cells in living tissues seem to succeed in avoiding these various disasters. This leaves the R-point decision as the critical determinant of whether cells will grow or not. An increasingly large body of evidence indicates that deregulation of the R-point decision-making machinery accompanies the formation of most if not all types of cancer cells. However, yet other decision points in late G_1 may also contribute to the deregulated proliferation of certain cancer cells. Prominent among these in mammals is the monitoring by cells of their attachment to the extracellular matrix (ECM) after the R point and before the G_1/S transition.

If a cell does not enjoy the proper attachment to the ECM, much of it mediated by integrins (see Section 5.9), it will halt further advance through the cell cycle until proper tethering has been achieved. Alternatively, if it has lost all attachment to the ECM, it may activate anoikis, a form of the apoptotic cell suicide program.

When grown in monolayer culture in Petri dishes, normal cells achieve the required attachment by adhering to extracellular matrix that they have deposited on the glass or plastic surfaces of these dishes. This requirement for attachment reflects the phenotype of anchorage dependence (see Section 3.5). Tumorigenic cells, which almost always have lost this dependence and have therefore become anchorage-independent, have inactivated or lost this late G_1 checkpoint. In some fashion, still poorly understood, oncoproteins like Ras and Src are able to mislead a cell into thinking that it has achieved extensive anchorage (see, for example, Section 6.9) when, in fact, none may exist at all.

Yet other, still-fragmentary evidence suggests additional checkpoints that operate before and after the R point. For example, one important checkpoint control operating between the R point and the onset of S phase gauges whether a cell has access to adequate levels of nutrients and halts cell-cycle progression until such nutrients become available; another appears to determine whether reactive oxygen species (which in other contexts are thought to be toxic for the cell) are present in adequately high levels before progression is permitted. Such findings suggest that successful advance through late G_1 depends on passing through a succession of checkpoints in addition to the R point described here, not all of which have been well characterized.

8.3 Cyclins and cyclin-dependent kinases constitute the core components of the cell cycle clock

The existence of the R point leaves us with two major questions that we will spend much of this chapter answering. First, what is the nature of the molecular machinery that decides whether or not a cell in G_1 will continue to advance through the cell cycle or will exit into a nongrowing state? Second, how does this machinery, which we call the cell cycle clock, implement these decisions once they have been made? We begin with the second question and will return later to the first.

As described earlier (Chapters 5 and 6), when signals are broadcast from a single master control protein to many downstream responders, the signal-emitting functions are often delegated to protein kinases. These enzymes are ideally suited to this task. By phosphorylating multiple distinct targets (that is, substrates), a kinase can create covalent modifications that serve to switch on or off various activities inherent in these substrate proteins. Indeed, the cell cycle clock uses a group of protein kinases to execute the various steps of cell cycle progression. For example, phosphorylation of **centrosome**-associated proteins at the G_1/S boundary allows their duplication in preparation for M phase. Phosphorylation of other proteins prior to S phase enables DNA replication sites along the chromosomes to be activated. Phosphorylation of histone proteins in anticipation of S and M phases places the chromatin in configurations that permit these two phases to proceed normally. And the phosphorylation of proteins forming the nuclear membrane (sometimes called the *nuclear envelope*), such as lamin and nucleoporins, causes their dissociation and the dissolution of this membrane early in M phase.

The kinases deployed by the cell cycle machinery are called collectively **cyclin-dependent kinases** (CDKs) to indicate that these enzymes never act on their own; instead, they depend on associated regulatory subunits, the **cyclin** proteins, for proper functioning. Bimolecular complexes of CDKs and their cyclin partners are responsible for sending out the signals from the cell cycle clock to dozens if not hundreds of responder molecules that carry out the actual work of moving the cell through its growth-and-division cycle.

The CDKs are serine/threonine kinases, in contrast to the tyrosine kinases that are associated with growth factor receptors and with nonreceptor kinase molecules like Src. The CDKs show about 40% amino acid sequence identity with one another and therefore are considered to form a distinct subfamily within the large (approximately 430) throng of Ser/Thr kinases encoded by the human genome (see Supplementary Sidebar 16.8). The cyclins associated with the CDKs activate the catalytic activity of their CDK partners (**Figure 8.7**). (In the well-studied example of the binding of cyclin A to CDK2, the association of the two proteins increases the enzymatic activity of CDK2 a staggering 400,000-fold!) At the same time, the cyclins serve as guide dogs for the CDKs by helping the cyclin-CDK complexes recognize appropriate protein substrates in the cell. The cyclins, for their part, also constitute a distinct family of cellular proteins that share in common an approximately 100–amino acid residue–long domain that is involved in the binding and functional activation of CDKs.

It is actually cyclin-CDK complexes that constitute the engine of the cell cycle clock machinery. During much of the G_1 phase of the cell cycle, two similarly acting CDKs—CDK4 and CDK6—are guided by and depend upon their association with a trio of

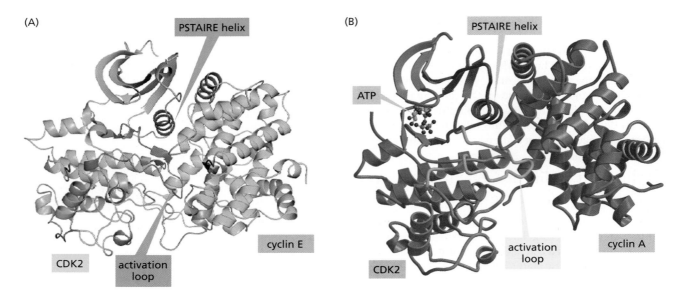

Figure 8.7 Cyclin–CDK complexes
X-ray crystallographic analyses have revealed the structures of cyclin–cyclin-dependent kinase (CDK) complexes, such as these formed by CDK2 with two of its alternative partners, cyclins A and E. In each case, the cyclin activates the CDK molecule through stereochemical shifts of the CDK catalytic site and directs the resulting activated complex to appropriate substrates for phosphorylation. (A) During the late G_1 phase of the cell cycle, cyclin E directs CDK2 to substrate proteins that must be phosphorylated in preparation for entrance into S phase. Another segment (*orange*) is involved with cyclin E's association with centrosomes. The PSTAIRE α-helix is present in all CDKs and is essential for binding of cyclins. The activation loop, sometimes termed a T-loop, must be phosphorylated on a threonine residue by a CDK-activating kinase (CAK) in order for the catalytic function of a CDK to become activated (see Figure 5.17). (B) During the course of S phase, cyclin A replaces cyclin E and directs CDK2 to substrates that must be phosphorylated in order for S phase to proceed. (A, from R. Honda et al., *EMBO J.* 24:452–463, 2005. B, from P.D. Jeffrey et al., *Nature* 376:313–320, 1995.)

related cyclins (D1, D2, and D3) that collectively are called the D-type cyclins (**Figure 8.8**). After the R point in late G_1 the E-type cyclins (E1 and E2) associate with CDK2 to enable the phosphorylation of appropriate substrates required for entry into S phase. As cells enter into S phase, the A-type cyclins (A1 and A2) replace E cyclins as the partners of CDK2 and thereby enable S phase to progress (see Figure 8.7). Later in S phase, the A-type cyclins switch partners, leaving CDK2 and associating instead with another CDK called either CDC2 or CDK1. (We will use CDC2 here.) As the cell moves further into G_2 phase, the A-type cyclins are replaced as CDC2 partners by the B-type cyclins (B1 and B2). Finally, at the onset of M phase, the complexes of CDC2 with the B-type cyclins trigger many of the events of the prophase, metaphase, anaphase, and telophase that together constitute the complex program of mitosis. [Another cyclin–CDK pair—cyclin C–CDK3—is implicated in the movement of cells from the G_0 into the G_1 phase (see Figure 8.3A) of the cell cycle. However, because most strains of inbred

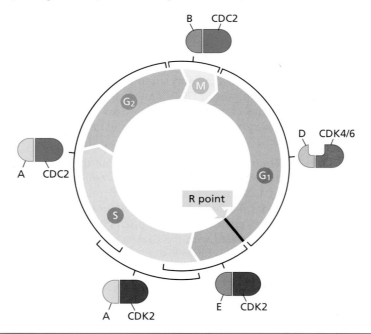

Figure 8.8 Pairing of cyclins with cyclin-dependent kinases Each type of cyclin pairs with a specific cyclin-dependent kinase (CDK) or set of CDKs. The D-type cyclins (D1, D2, and D3) bind CDK4 or CDK6, the E-type (E1 and E2) bind CDK2, the A-type cyclins (A1 and A2) bind CDK2 or CDC2, and the B-type cyclins (B1 and B2) bind CDC2. The brackets indicate the periods during the cell cycle when these various cyclin–CDK complexes are active.

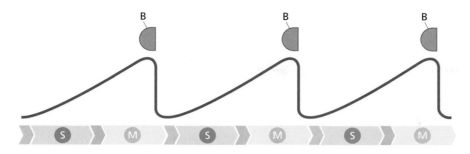

laboratory mice lack CDK3, it seems to be unnecessary for normal cell cycle control and remains poorly studied.]

As is the case with all well-regulated systems, the activities of the various cyclin–CDK complexes must be modulated in order to impose control on specific steps in the cell cycle. The most important way of achieving this regulation depends upon changing the levels and availability of cyclins during various phases of the cell cycle. In contrast, the levels of most CDKs vary only minimally.

The first insights into cyclin and CDK control came from studies of the governors of mitosis in early frog and sea urchin embryos. As these experiments showed, levels of B-type cyclins increase strongly in anticipation of mitosis, allowing B cyclins and CDC2 to form complexes that initiate entrance into M phase. At the end of M phase, cyclin B levels plummet because of the scheduled degradation of this protein. Early in the next cell cycle, cyclin B is virtually undetectable in cells, and accumulates gradually later in this cycle in anticipation of the next M phase. Because the growth-and-division cycles of all of the cells in these early embryos are **synchronous** (that is, take place coordinately), all cells in the embryo go through S and M at the same time. This results in repeated rounds of cycling of the levels of these cyclin proteins—the behavior that inspired their name (**Figure 8.9**).

This theme of dramatic cell cycle phase–dependent changes in the levels of cyclin B is repeated by other cyclins as well. Cyclin E levels are low throughout most of G_1, rise abruptly after a cell has progressed through the R point, and collapse as the cell enters S phase (**Figure 8.10**), while cyclin A increases in concert with the cell's entrance into S phase. (While there are at least two subtypes of cyclin A, cyclin B, and cyclin E, we will refer to these simply as cyclins A, B, and E, respectively, since the two subtypes of each of these cyclins appear to operate identically.)

The collapse of various cyclin species as cells advance from one cell cycle phase to the next is due to their rapid degradation, this being triggered by the actions of highly coordinated ubiquitin ligases, which attach polyubiquitin chains to these cyclins (see Supplementary Sidebar 7.5). This polyubiquitylation leads to proteolytic breakdown in the proteasomes. The cyclins' gradual accumulation followed by their rapid destruction has an important functional consequence for the cell cycle, because it dictates that the cell cycle clock can move in only one direction, much like a ratchet. This ensures, for example, that cells that have exited one M phase cannot inadvertently slip backward into another one, but instead must advance through G_1, S, and G_2 until they have once again accumulated the B cyclins required for entrance into M phase.

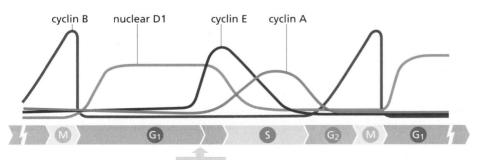

Figure 8.9 Cell cycle–dependent fluctuations in cyclin B levels The cyclic fluctuations in the levels of cyclin B in early frog and sea urchin embryos gave cyclins their name. These fluctuations were noticeable because the cell cycles in these early embryos are synchronous, i.e., all cells enter into M phase simultaneously. In these early embryos, the G_1 and G_2 phases of the cell cycle (*orange, pink chevrons*) are virtually absent and the cells, in effect, alternate between M and S phases. (Although cyclin B levels are already substantial prior to the onset of M phase, cyclin B molecules are unable to form catalytically active B–CDC2 complexes until the G_2-to-M transition.)

Figure 8.10 Fluctuation of cyclin levels during the cell cycle The levels of most of the mammalian cyclins fluctuate dramatically as cells progress through the cell cycle. For most of these cyclins, these fluctuations are tightly coordinated with the schedule of advances through the various cell cycle phases. However, in the case of the D-type cyclins, extracellular signals, notably those conveyed by growth factors, strongly influence their levels. (While cyclin D1—and possibly other D-type cyclins—is present in other cell cycle phases besides G_1, following the G_1/S transition it is exported from the nucleus into the cytoplasm, where it can no longer influence cell cycle progression. The precise time point at which cyclin D–CDK4/6 complexes lose activity is not well defined.)

Figure 8.11 Control of cyclin D1 levels
(A) Cyclin D1 was discovered as a protein whose levels are strongly induced by exposure of macrophages to the mitogen CSF-1 (colony-stimulating factor-1). Here, macrophages that had been starved of CSF-1 were exposed to fresh CSF-1 and the amounts of cyclin D1 mRNA at subsequent times thereafter were determined by RNA (Northern) blotting. (B) The control of cyclin D1 levels by extracellular mitogens can be explained, in part, by a signal transduction cascade that leads from growth factor receptors (RTKs) to the AP-1 transcription factor (TF), one of a number of factors that modulate the transcription of the *cyclin D1* gene. (AP-1 is a heterodimeric TF formed from Fos and Jun subunits, each encoded by a proto-oncogene.) In addition, a number of other signaling cascades converge on the promoter of this gene, not all of which are illustrated in this figure. (A, from H. Matsushime et al., *Cell* 65:701–713, 1991.)

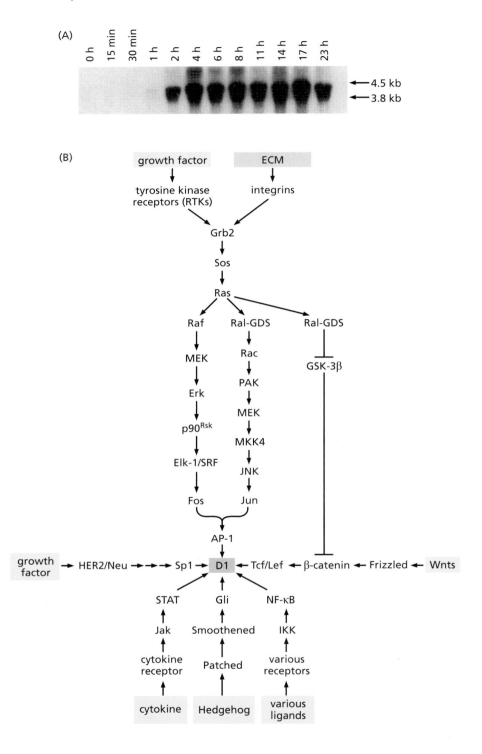

The sole exception to these well-programmed fluctuations in the levels of cyclins is presented by the D-type cyclins. The levels of these three, similarly structured cyclins are not found to vary dramatically as a cell advances through the various phases of its growth-and-division cycle. Instead, the levels of D-type cyclins are controlled largely by extracellular signals, specifically those conveyed by a variety of mitogenic growth factors. In the case of cyclin D1—the best-studied of the three D-type cyclins—growth factor activation of tyrosine kinase receptors and the resulting stimulation of several downstream signaling cascades results in rapid accumulation of cyclin D1 (**Figure 8.11**). Conversely, removal of growth factors from a cell's medium results in an equally rapid collapse of its cyclin D1 levels with a half-life of about 30 minutes.

Table 8.1 Induction of D-type cyclin expression by extracellular signals

Source of signal	Signaling intermediaries	Type of cyclin
RANK receptor	NF-κB pathway	D1
Prolactin receptor	Jak/STAT	D1
Estrogen receptor	AP-1 TF (?)	D1
Focal adhesion kinase		D1
HER2/Neu receptor	E2F and Sp1 TFs	D1
Wnts–Frizzled receptor	β-catenin and Tcf/Lef TFs	D1
Bcr/Abl		D2
FSH receptor	cyclic AMP	D2
Various mitogens	Myc	D2
Interleukin-4, 7 receptor		D2
Interleukin-5 receptor	STAT3/5	D3
Mitogens	E2A TF	D3

Abbreviations: RANK, receptor activator of NF-κB; FSH, follicle-stimulating hormone.

The distinctive behavior of the D-type cyclins has been rationalized as follows: They serve to convey signals from the extracellular environment to the cell cycle clock operating in the cell nucleus. Because the levels of D-type cyclins fluctuate together with the levels of extracellular mitogens, D-type cyclins continuously inform the cell cycle clock of current conditions in the environment around the cell.

After D-type cyclins are synthesized in the cytoplasm and migrate to the nucleus, they assemble in complexes with their two alternative CDK partners, CDK4 and CDK6. Since these two CDKs function similarly to one another, we will refer to them hereafter as CDK4/6. The cyclin D–CDK4/6 complexes seem to have similar if not identical enzymatic activities and substrate specificities, independent of whether they contain cyclin D1, D2, or D3.

These similarities provoke the question of why mammalian cells express three apparently redundant cyclins. What appears to be a redundancy actually offers the cell refined sensory input and enhanced flexibility of response. The promoter of each of the three encoding genes is under the control of a different set of signaling pathways and thus under the control of a different set of cell surface receptors (**Table 8.1**). For example, the promoter of the *cyclin D1* gene (properly termed, in humans, *CCND1*) carries sites for binding by the AP-1, Tcf/Lef, and NF-κB transcription factors (see Figure 8.11B), which in turn are activated by a variety of growth factor receptors. In contrast, the *cyclin D2* promoter is responsive to activation by the Myc transcription factor and to extracellular signals that stimulate increases in the concentration of intracellular cyclic adenosine monophosphate (cAMP). The *cyclin D3* gene has been found to be responsive to STAT3 and STAT5 transcription factors, and these in turn often respond to interleukin receptors active in various types of hematopoietic cells; the E2A transcription factor, which is active in programming the differentiation of lymphocytes, also controls *cyclin D3* expression.

This arrangement enables a diverse set of extracellular signals to influence the activities of CDK4/6 by controlling the levels of its various D-type cyclin partners. In addition, more recent research has indicated that certain cyclins have functions in the cell that are apparently unrelated to their role in promoting cell cycle progression (**Sidebar 8.2**).

Sidebar 8.2 Cyclins have other jobs besides cell cycle control Two decades of cell cycle research have created the impression that cell cycle control is the sole function of cyclins. In fact, cyclin D1 has been shown to associate with both the estrogen receptor (ER) and the transcription factor C/EBPβ. By binding the ER, cyclin D1 may mimic the normal ligand of this receptor—estrogen—in stimulating the receptor's transcriptional activities. The great majority (>70%) of breast cancers express the ER, and its expression explains the mitogenic effects that estrogen has on the cells in these tumors. Since cyclin D1 is overexpressed in most of these tumors and can activate this receptor, the D1–ER complexes may also play an important role in driving the proliferation of cells in these tumors. The association of cyclin D1 with C/EBPβ results in activation of this transcription factor, which is thought to play a key role in programming the differentiation of a variety of cell types. Knockout of cyclin D1 in the mouse germ line results in viable mice that have severely underdeveloped mammary glands. This effect can be reversed by providing the mammary glands with a mutant cyclin D1 that can bind C/EBPβ but cannot activate CDK4/6. Hence, in this case, the non-CDK-associated functions of this cyclin are far more important than its effects on CDK4/6 activation. More globally, the use of chromatin immunoprecipitation (ChIP; see Supplementary Sidebar 8.3) has revealed at least 700 sites in the genome, most affiliated with gene promoters, to which cyclin D1 is bound, indicating that the two associations of cyclin D1 with transcription factors, as cited above, are likely only the tips of a far larger iceberg.

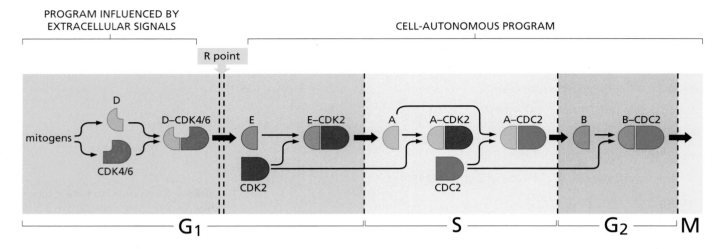

Figure 8.12 Control of cyclin levels during the cell cycle
While extracellular signals strongly influence the levels of D-type cyclins during most of the G_1 phase of the cell cycle, the levels of the remaining cyclins are controlled by intracellular signaling and precisely coordinated with cell cycle advance. Thus, after cells pass through the R point and cyclin E–CDK2 complexes are activated, the activation of the remaining cyclin–CDK complexes occurs on a predictable schedule that appears to be independent of extracellular physiologic signals. In part, this coordination is achieved because the cyclin–CDK complexes in one phase of the cell cycle are responsible for activating those in the subsequent phase (*indicated here*) and for shutting down those that were active in the previous phase (*not shown*).

Once they are formed, the cyclin D–CDK4/6 complexes are capable of ushering a cell all the way from the beginning of the G_1 phase up to and perhaps through the R-point gate. After the cell has moved through the R point, the remaining cyclins—E, A, and B—behave in a pre-programmed fashion, executing the fixed program that begins at the R point and extends all the way to the end of M phase (**Figure 8.12**). Indeed, once the cell has passed through its R point, its cell cycle machinery takes on a life of its own that is quite autonomous and no longer responsive to extracellular signals.

In ways that remain poorly understood, cyclin–CDK complexes in later phases suppress the activities of the cyclin–CDK complexes that have preceded them in earlier phases of the cell cycle. For example, when cyclin A is activated by the actions of cyclin E–CDK2 complexes during the G_1/S transition, the activities of cyclin A–CDK2 result in, among other things, the inactivation of the transcription factor that served previously during the R-point transition to induce cyclin E expression. In late S and early G_2 phases, the cyclin A–CDC2 complex begins to prepare for the activation of cyclin B–CDC2 complexes that are required later for entrance into mitosis. Once the latter complexes are activated, they seem to cause a shutdown of cyclin A synthesis, and so forth. While the program depicted in Figure 8.12 has been validated in a number of cultured cell types, it may not apply to all cells, as indicated by studies of genetically modified mice (see Supplementary Sidebar 8.2).

8.4 Cyclin–CDK complexes are also regulated by CDK inhibitors

The scheme of cell cycle progression laid out above implies that physiologic signals are able to influence the activity of the cell cycle clock only through their modulation of cyclin levels. In truth, there are several other layers of control that modulate the activity of the cyclin–CDK complexes and thereby regulate advance through the cell cycle.

The most important of these additional controls is imposed by a class of proteins that are termed generically CDK inhibitors or simply CKIs. To date, seven of these proteins have been found that are able to antagonize the activities of the cyclin–CDK complexes. A group of four of these, the INK4 proteins (named originally as **in**hibitors of CD**K4**), target specifically the CDK4 and CDK6 complexes; they have no effect on CDC2 and CDK2. These inhibitors are p16^{INK4A}, p15^{INK4B}, p18^{INK4C}, and p19^{INK4D}. The three remaining CDK inhibitors, p21^{Cip1} (sometimes termed p21^{Waf1}), p27^{Kip1}, and p57^{Kip2}, are more widely acting, being able to inhibit all of the other cyclin–CDK complexes that form at later stages of the cell cycle (**Figure 8.13**). (Each one of this trio of CKIs has been found to affect other specific processes besides cell cycle progression, including transcriptional regulation, apoptosis, cell fate determination, cell migration, and cytoskeletal organization; however, since these other functions appear to play relatively minor roles in cancer development, we will not discuss them further.) p57^{Kip2} plays a key role in certain tissues during embryogenesis but a minor role in cancer development; for this reason, our discussion will focus on its two cousins, p21^{Cip1} and p27^{Kip1}.

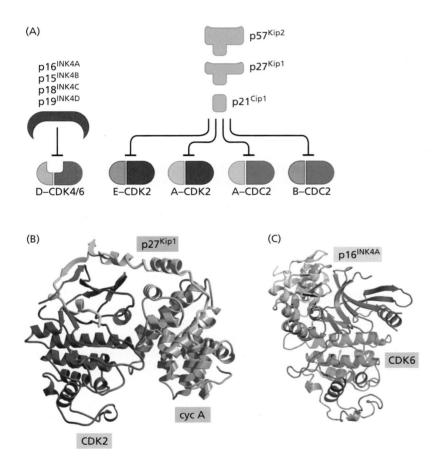

Figure 8.13 Actions of CDK inhibitors (A) The CDK inhibitors block the actions of CDKs at various points in the cell cycle. The four INK4 proteins (p16^{INK4A}, p15^{INK4B}, p18^{INK4C}, and p19^{INK4D}) are specialized to inhibit the D–CDK4 and D–CDK6 complexes that are active in early and mid-G$_1$. Conversely, the three Cip/Kip CKIs (p21^{Cip1}, p27^{Kip1}, and p57^{Kip2}) can inhibit the remaining cyclin–CDK complexes that are active throughout the cell cycle. Under certain conditions, signals originating within the cell (e.g., signals indicating damage to the cell's DNA) can halt advance through the cell cycle by blocking the actions of cyclin/CDK complexes that are active in the S and G$_2$ phases of the cell cycle. Paradoxically, p21^{Cip1} and p27^{Kip1} are known to *promote* formation of D–CDK4/6 complexes during the G$_1$ phase (see Figure 8.17). (B) This depiction illustrates how one domain of p27^{Kip1} (*light green*) blocks cyclin A–CDK2 function by obstructing the ATP-binding site in the catalytic cleft of the CDK (see also Figure 8.7). (C) Inhibitors of the INK4 class, such as p16^{INK4A} shown here (*brown*), bind to CDK6 (*reddish*) and to CDK4 (*not shown*). These CDK inhibitors distort the cyclin-binding site of CDK6, reducing its affinity for D-type cyclins. At the same time, they distort the ATP-binding site and thereby compromise catalytic activity. Identical interactions likely characterize the responses of CDK4 to p16^{INK4A}. (B, from A.A. Russo et al., *Nature* 382:325–331, 1996. C, courtesy of N.P. Pavletich and from A.A. Russo et al., *Nature* 395:237–243, 1998.)

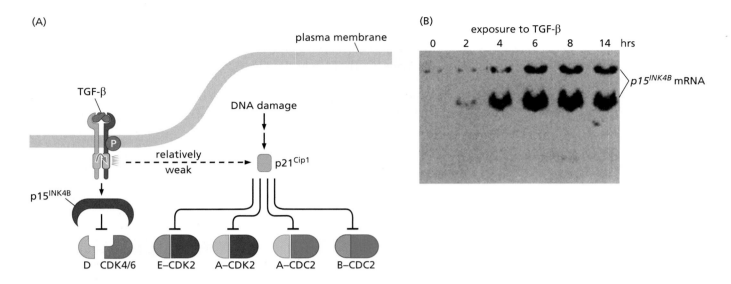

Figure 8.14 Control of cell cycle progression by TGF-β (A) TGF-β (*top left*) controls the cell cycle machinery in part through its ability to modulate the levels of CDK inhibitors. It acts to strongly induce increased expression of p15^{INK4B} and, weakly, of p21^{Cip1}. The former can block the actions of cyclin D–CDK4/6 complexes, while the latter can block the actions of the remaining cyclin–CDK complexes that are active throughout the remainder of the cell cycle. Independent of this, damage to cellular DNA causes strong, rapid increases in p21^{Cip1}, which in turn can shut down the cyclin–CDK complexes that are active in the phases of the cell cycle after the cell has passed through the R point in the late G$_1$ phase. (B) When TGF-β is applied to human keratinocytes, it evokes a dramatic 30-fold induction of *p15^{INK4B}* mRNA synthesis as demonstrated here by RNA (Northern) blotting analysis. These cells were exposed to TGF-β for the time periods (in hours) indicated (*above*). (B, from G.J. Hannon and D. Beach, *Nature* 371:257–261, 1994.)

The actions of these two classes of CDK inhibitors are nicely illustrated by p15^{INK4B} and the pair p21^{Cip1} and p27^{Kip1}. When TGF-β is applied to epithelial cells, it elicits a number of downstream responses that antagonize cell proliferation. Among these are substantial increases in the levels of p15^{INK4B}, which proceeds to block the formation of cyclin D–CDK4/6 complexes (**Figure 8.14**) and to inhibit those that have already formed. Without active D–CDK4/6 complexes, the cell is unable to advance through early and mid-G$_1$ and reach the R point. Once a cell has passed through the R point, the actions of the D–CDK4/6 complexes seem to become unnecessary. This may explain why TGF-β is growth-inhibitory during early and mid-G$_1$ and loses most (perhaps all) of its growth-inhibitory powers once a cell has passed through the R point.

p21^{Cip1}, a more widely acting CDK inhibitor, is also induced by TGF-β, albeit weakly. Far more important are increases in the levels of p21^{Cip1} that occur in response to various physiologic stresses (see Figure 8.14A); once present at significant levels, p21^{Cip1} can act throughout much of the cell cycle to stop a cell in its tracks. Prominent among these stresses is damage to the cell's genome. As long as the genomic DNA remains in an unrepaired state, the p21^{Cip1} that has been induced will shut down the activity of already-formed cyclin–CDK complexes—such as E–CDK2, A–CDK2, A–CDC2, and B–CDC2—that happened to be active when this damage was first incurred; once the damage is repaired, the p21^{Cip1}-imposed block may then be relieved. Such a strategy makes special sense in G$_1$: if a cell's genome becomes damaged during this period through the actions of various mutagenic agents, p21^{Cip1} will block advance through the R point (by inhibiting E–CDK2 complexes) until the damage has been repaired, ensuring that the cell does not progress into S phase and inadvertently copy still-damaged DNA sequences. In addition, p21^{Cip1} can inhibit the functions of a key component of the cell's DNA replication apparatus, termed PCNA (proliferating-cell nuclear antigen); this ensures that already-initiated DNA synthesis is halted until DNA repair has been completed. We will return to the mechanisms controlling p21^{Cip1} expression in the next chapter (Section 9.9).

While DNA damage and, to a much lesser extent, TGF-β can elicit increases in the levels of p21^{Cip1} (thereby blocking cell cycle advance), mitogens act in an opposing fashion to mute the actions of this CDK inhibitor and in this way favor cell cycle advance. One mechanism by which they do so depends on the phosphatidylinositol 3-kinase (PI3K) pathway, which is activated directly or indirectly by the mitogens that stimulate many receptor tyrosine kinases (**Figure 8.15A**). Akt/PKB, the important kinase activated downstream of mitogen-activated PI3K (see Section 6.6), phosphorylates p21^{Cip1} molecules located in the nucleus, thereby causing them to be exported into the cytoplasm, where they can no longer engage and inhibit cyclin–CDK complexes (see Figure 8.15B). Similarly, Akt/PKB phosphorylates p27^{Kip1} (which functions much like p21^{Cip1}) and prevents its export from its cytoplasmic site of synthesis into the nucleus,

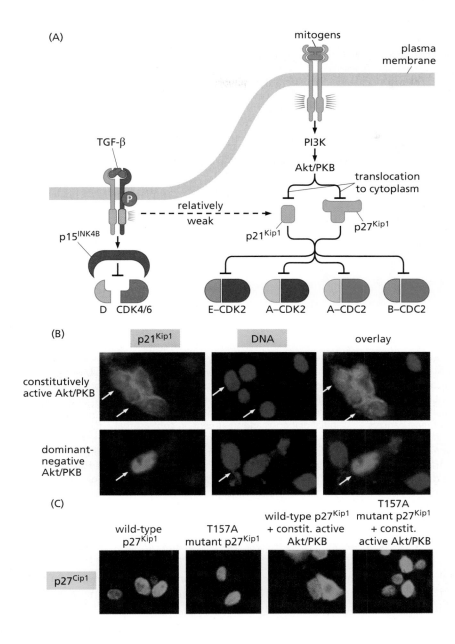

Figure 8.15 Control of cell cycle advance by extracellular signals Countervailing extracellular signals influence the cell cycle machinery, in part through their ability to control the levels and intracellular localization of CDK inhibitors. (A) During the G_1 phase, TGF-β induces p15^{INK4B} and (weakly) p21^{Cip1} expression, negatively affecting the D–CDK4/6 and E–CDK2 complexes, respectively. Conversely, mitogens, acting through Akt/PKB, cause the phosphorylation and cytoplasmic localization of both p21^{Cip1} and p27^{Kip1}, which prevents these CDK inhibitors from entering the nucleus and blocking the activities of various cyclin–CDK complexes operating there. (The growth factor receptors can also reduce the levels of the two CKIs depicted here, p21^{Cip1} and p27^{Kip1}; these CKIs can inhibit, in turn, cyclin–CDK complexes active in late G_1, S, and G_2. It is unclear, however, whether these cyclin–CDK complexes are actually affected by mitogens in the S and G_2 phases of the cell cycle.) (B) Ectopic expression of a constitutively active Akt/PKB (*upper row*) causes p21^{Cip1} (*orange, left*) to be localized largely in the cytoplasm. Compare its localization with that of the cell nuclei (*blue, middle*); the overlap of these two images is also seen (*right*). Conversely, expression of a dominant-negative Akt/PKB (*lower row*), which interferes with ongoing Akt/PKB function, allows p21^{Cip1} to localize to the cell nucleus, where it can interact with cyclin–CDK complexes and inhibit cell cycle progression. (C) In normal cells (*first panel*), ectopically expressed wild-type p27^{Kip1} (*green*) is found to be exclusively nuclear. This localization is not changed if a p27^{Kip1} mutant is expressed that lacks the threonine residue normally phosphorylated by Akt/PKB (*second panel*). (The T157A mutant carries an alanine in place of this threonine.) If a mutant, constitutively active Akt/PKB is expressed, however, much of wild-type p27^{Kip1} is now seen in the cytoplasm (*third panel*). But if the p27^{Kip1} mutant that cannot be phosphorylated by Akt/PKB is expressed, it resists the actions of constitutively activated Akt/PKB and remains in the nucleus (*fourth panel*). (B, from B. Zhou et al., *Nat. Cell Biol.* 3:245–252, 2001. C, from G. Viglietto et al., *Nat. Med.* 8:1136–1144, 2002.)

where it normally does its critical work (see Figure 8.15C). Taken together, these signaling responses illustrate how extracellular growth-inhibitory signals (conveyed by TGF-β) impede the advance of the cell cycle clock, while growth-promoting signals promote its forward progress.

These effects on intracellular localization appear to have clinical consequences. For example, in low-grade (that is, less advanced) human mammary carcinomas, levels of activated Akt/PKB are low and p27^{Kip1} is able to carry out its anti-proliferative functions in the cell nucleus. In high-grade tumors, however, activated Akt/PKB is abundant and much of p27^{Kip1} is now found in the cell cytoplasm (**Figure 8.16A**). This intracellular localization is correlated with, and likely causally linked to, the progression of these cancers to fatal endpoints (see Figure 8.16B).

One aspect of the behavior of p21^{Cip1} and p27^{Kip1} seems quite paradoxical. While they inhibit the actions of cyclin E–CDK2, cyclin A–CDC2, and cyclin B–CDC2, they actually *stimulate* the formation of cyclin D–CDK4/6 complexes (**Figure 8.17A**). Moreover, once a ternary (three-part) complex has formed between D–CDK4/6 and either of these CDK inhibitors, the cyclin–CDK complex can still phosphorylate its normal

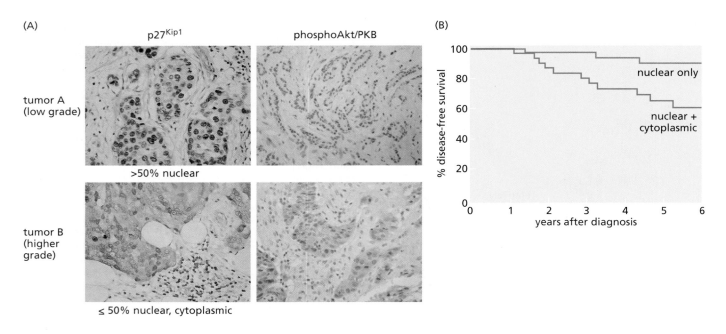

Figure 8.16 Suppression of p27^Kip1 function by Akt/PKB in human breast cancers Various breast cancers show differing intracellular locations of p27^{Kip1} that reflect the activation state of Akt/PKB; the latter can be gauged by using an antibody that specifically recognizes the phosphorylated, functionally activated form of this enzyme. (A) In some low-grade primary breast cancers (e.g., tumor A), p27^{Kip1} is exclusively nuclear and is clearly present in more than 50% of nuclei (*brown staining, upper left*); activated phospho-Akt/PKB cannot, however, be detected in these cells (*upper right*). Conversely, in higher-grade tumors (e.g., tumor B, *lower left*), p27^{Kip1} is found (*yellow brown staining*) in less than 50% of the nuclei but is also found in significant quantities throughout the cytoplasm, where it cannot block cell proliferation; in these tumors, activated phospho-Akt/PKB can be readily detected (*light brown staining, lower right*). (B) This Kaplan–Meier graph indicates that those patients bearing primary tumors with largely nuclear p27^{Kip1} (*green curve*) have a relatively good prospect of long-term, disease-free survival over a period of 6 years following initial diagnosis and treatment. In contrast, patients whose tumors show p27^{Kip1} staining in both nuclei and cytoplasm (*blue curve*) suffer significant relapses over this time period, many of which lead to death. (Importantly, this correlation with clinical outcome does not prove causation, i.e., that p27^{Kip1} mislocalization causes clinical progression.) (A, courtesy of J. Zubovitz and J.M. Slingerland. B, from J. Liang et al., *Nat. Med.* 8:1153–1160, 2002.)

substrates. So the term "CDK inhibitor" is in fact a misnomer. The two proteins (p21^{Cip1} and p27^{Kip1}) act on most cyclin–CDK complexes in an inhibitory manner and on the cyclin D–CDK4/6 complexes in a stimulatory fashion.

While p27^{Kip1} functions similarly to p21^{Cip1} in many ways, it has its own, highly interesting behavior. When cells are in the G$_0$ quiescent state, p27^{Kip1} is present in high concentrations, and therefore binds to and suppresses the activity of the few cyclin E–CDK2 complexes that happen to be present in the cell. When cells are exposed to growth factors, cyclin D–CDK4/6 complexes accumulate, because these mitogens cause an increase in the levels of the D-type cyclins. Each time a new cyclin D–CDK4/6 complex forms, it captures and binds another molecule of p27^{Kip1}; consequently the pool of free p27^{Kip1} molecules in the cell dwindles as they become progressively sequestered by cyclin D–CDK4/6 complexes. After many hours, the remaining p27^{Kip1} molecules are finally pulled away from cyclin E–CDK2 complexes (see Figure 8.17B). Once a small portion of cyclin E–CDK2 complexes are liberated from the effects of p27^{Kip1}, these complexes are free to begin triggering passage of the cell through the R point. [As we will see later (see Figure 8.25), the initially liberated cyclin E–CDK2 complexes trigger a self-sustaining positive-feedback loop that causes both degradation of p27^{Kip1} molecules and increased synthesis of new cyclin E molecules. The result is a rapid rise in the levels of functional E–CDK2 complexes and an essentially irreversible advance through the R point.]

Evidence is increasing that the post-mitotic state of many differentiated cell types in the body is imposed by various CDK inhibitors acting singly or in combination. For

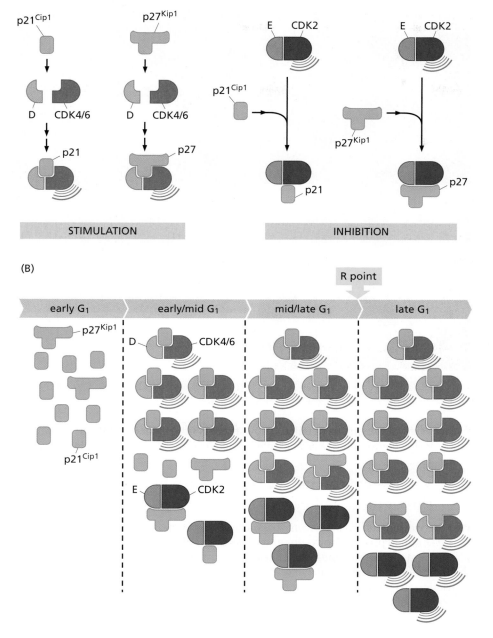

(A)

STIMULATION

INHIBITION

(B)

R point

early G$_1$ early/mid G$_1$ mid/late G$_1$ late G$_1$

Figure 8.17 Interactions of CDK inhibitors with cyclin–CDK complexes (A) The various CDK inhibitors affect cyclin–CDK complexes in different ways. The CDK inhibitors of the p21^{Cip1} and p27^{Kip1} family, while inhibiting the cyclin–CDK complexes active in late G$_1$ through M phases (only E–CDK2 is shown here; *right*), actually stimulate the formation of the D–CDK4/6 complexes that are active in the early and mid-G$_1$ phase and permit these complexes to be catalytically active (*left*). (B) Cells in early G$_1$ often have substantial concentrations of p21^{Cip1} and p27^{Kip1} molecules (*green; left*). As cyclin D–CDK4/6 complexes (*red/pink*) accumulate throughout early to mid-G$_1$ (being induced by extracellular mitogens), they bind and thereby sequester an ever-increasing proportion of the p21^{Cip1} and p27^{Kip1} molecules, both those that are present in soluble pools (unbound to any cyclin–CDK complexes) as well as those that are associated with the small number of cyclin E–CDK2 complexes that are present in G$_1$ (*light, dark purple*). The binding by D–CDK4/6 complexes of these two CDK inhibitors does not negatively affect their catalytic activity (see panel A). However, as this process continues, the D–CDK4/6 complexes succeed in abstracting most of the p21^{Cip1} and p27^{Kip1} molecules away from E–CDK2 complexes in late G$_1$ (*right*). This liberates the E–CDK2 complexes (*bottom right*), enabling them to trigger the R-point transition. The result is a rapid increase in E–CDK2 complexes, a degradation of p27^{Kip1} molecules (*not shown*), and additional phosphorylation events that drive cells irreversibly (see Figure 8.25) through the R point and into S phase.

example, as shown in **Figure 8.18A**, in the developing cerebellum of a mouse, the cell layers that have become post-mitotic have high levels of p27^{Kip1}, while those that are still in active cycle lack this CDK inhibitor. Moreover, once this tissue is formed in the days after birth, the cells of this part of the brain are kept in a post-mitotic state by the p27^{Kip1}, which they express at a high and constant level throughout adulthood. p27^{Kip1} seems to play similar roles in curtailing proliferation in other tissues as well: mice deprived of both copies of its encoding gene show a ~30% increase in overall body size together with hyperplasia in multiple organs. During normal development, the expression of p57^{Kip2}, the cousin of 27^{Kip1}, is curtailed when cells need to continue proliferating for proper tissue formation to occur (see Figure 8.18B).

In addition to the CDK inhibitors, yet another level of control of cyclin–CDK complexes is imposed by covalent modifications of the CDK molecules themselves (**Sidebar 8.3**). While the deregulated actions of the CDK inhibitors contribute importantly

Figure 8.18 Imposition by p27^{Kip1} and p57^{Kip2} of post-mitotic states during development The development of tissues depends on the differentiation of embryonic cells and the imposition of post-mitotic states by p27^{Kip1} and p57^{Kip2}. While suppression of p27^{Kip1} levels in many types of tumors has been widely documented, the role of its cousin, p57^{Kip2}, in cancer pathogenesis remains to be demonstrated. (A) Much of cerebellar development in mammals occurs postnatally as cells differentiate and enter into a post-mitotic state. During the course of cerebellar development in the mouse, proliferating cells can be detected through their incorporation of bromodeoxyuridine (BrdU), a thymidine analog; they are revealed here at postnatal day 7 by immunofluorescence using an antibody that recognizes BrdU-containing DNA (*green*). The levels of p27^{Kip1} have been gauged by use of a fluorescence-tagged antibody that recognizes it in the same section (*red*). As is apparent, p27^{Kip1} accumulates in cells that have ceased proliferating (*above*). By one month of age, cerebellar development is complete (*below*), and the levels of p27^{Kip1} (*red*) are maintained at high levels throughout adulthood, ostensibly to hold these cells in a post-mitotic state. (B) The Notch signaling pathway (see Figure 6.29B) is involved in many tissues in blocking the entrance of cells into post-mitotic differentiated states, thereby allowing such cells to continue active proliferation. In the developing ocular lens of the mouse (day 17.5 of gestation), the Notch ligand Jagged1 (*red*) induces expression of a transcriptional repressor, Herp2 (*not shown*), which in turn represses expression of p57^{Kip2} (*light blue*). By suppressing p57^{Kip2} expression, Jagged1, acting via a Notch receptor, can ensure that cells remain in an actively proliferating state until the developing lens grows to an appropriate size, after which p57^{Kip2} expression is then permitted, allowing it to impose a post-mitotic state. Cell nuclei are revealed by the DAPI dye (*dark blue*). (A, above, from T. Uziel, F. Zindy, S. Xie et al., *Genes Dev.* 19:2656–2667, 2005; below, courtesy of A. Forget, F. Zindy, and M.F. Roussel, St. Jude Children's Research Hospital. B, from J. Jia et al., *Mol. Cell. Biol.* 27:7236–7247, 2007.)

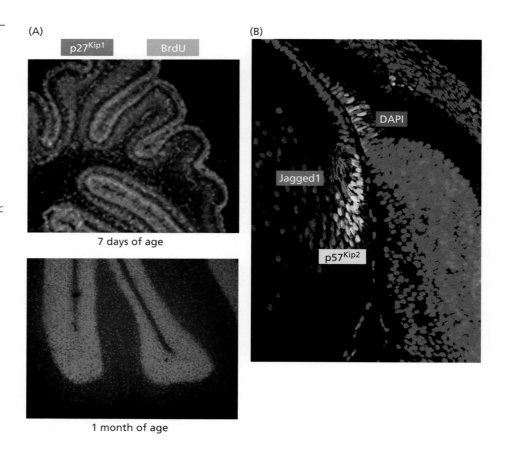

(A)

p27^{Kip1} BrdU

7 days of age

1 month of age

(B)

DAPI

Jagged1

p57^{Kip2}

to cancer development, as detailed below, there is little evidence indicating that this other level of control plays a critical role in specific steps of tumor progression.

8.5 Viral oncoproteins reveal how pRb blocks advance through the cell cycle

We have viewed the cell cycle from several angles here, including its clearly delineated G_1, S, G_2, and M phases; the cyclins and CDKs that orchestrate advances through these phases; and the fact that these advances can be traced back to the extracellular signals that persuade the cell to enter into the active cell cycle and to progress through the restriction (R) point gate. These descriptions, however, fail to reveal precisely how the R-point transition is executed at the molecular level.

The solution to this puzzle was provided by the isolation of the *Rb* tumor suppressor gene, defective versions of which are involved in the pathogenesis of retinoblastomas, sarcomas, and small-cell lung carcinomas as well as other tumors (see Table 7.1). Soon after the *Rb* gene was isolated in 1986, it became apparent that it encodes a nuclear phosphoprotein of a mass of about 105 kD. This protein, called variously pRb or RB, was found to be absent, or present in a defective form, in the cells of many of the above-named tumor types.

Initial experiments with pRb revealed that it undergoes phosphorylation in concert with the advance of cells through their cell cycle. More specifically, pRb is essentially unphosphorylated when cells are in G_0; becomes weakly phosphorylated (**hypophosphorylated**) on a small number of serine and threonine residues after entrance into G_1; and becomes extensively phosphorylated (**hyperphosphorylated**) on a much larger number of serine and threonine residues in concert with advance of cells through the R point (**Figure 8.19**). Once cells have passed through the R point, pRb usually remains hyperphosphorylated throughout the remainder of the cell cycle. After cells exit mitosis, the phosphate groups on pRb are stripped off by the enzyme

Sidebar 8.3 Phosphorylation of CDK molecules also controls their activity An additional level of control placed on the cell cycle clock machinery can be traced to the covalent modifications of the CDK molecules themselves. More specifically, CDKs must be phosphorylated at specific amino acid residues in order to be active (see Figure 8.7). At the same time, inhibitory phosphorylations of other amino acid residues must be removed. The stimulatory phosphorylations are imparted by a serine/threonine kinase termed CAK (for CDK-activating kinase). CAK itself is a cyclin–CDK complex, which phosphorylates the activation loops (often called T-loops) of CDKs (see Figure 8.7), causing them to swing out of the way of the catalytic clefts of these serine/threonine kinases.

Equally important are the effects of inhibitory phosphorylations affecting CDK2 and CDC2, which are removed by a class of phosphatases called CDC25A, -B and -C. While the effects of these enzymes in G_1, S, and G_2 are poorly understood, it is clear that they play critical roles in triggering the G_2/M transition: cyclin B, which has lingered during G_2 in the cytoplasm, moves into the nucleus just before the G_2/M transition; phosphorylations of a threonine and a tyrosine residue that are located near the ATP-binding site of CDC2 and inhibit kinase activity are then removed by CDC25 phosphatases, enabling the rapid activation of B–CDC2 complexes that proceed to trigger entrance into M phase. CDC25A and CDC25B are overexpressed in some cancers, and there is evidence from mouse models of cancer development that their overexpression may lead to acceleration of tumor development.

termed protein phosphatase type 1 (PP1). This removal of phosphate groups, in turn, sets the stage for the next cell cycle and thus for a new cycle of pRb phosphorylation.

The fact that pRb hyperphosphorylation occurs in concert with passage through the R point provided the first hint that this protein is the molecular governor of the R-point transition. An additional, critical clue came from the discovery in 1988 of physical interactions between DNA tumor virus–encoded oncoproteins and pRb. Recall that DNA tumor viruses are able to transform cells and do so through the use of virus-encoded oncoproteins. Unlike the oncoproteins of RNA tumor viruses such as RSV, the DNA tumor viruses employ oncoproteins that have little if any resemblance to the proteins that exist naturally within uninfected cells (see Sections 3.4 and 3.6).

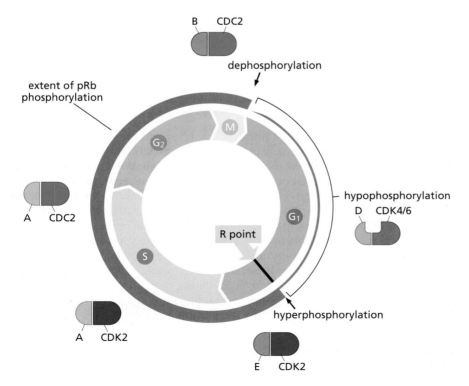

Figure 8.19 Cell cycle–dependent phosphorylation of pRb The phosphorylation state of pRb (*red circle*) is closely coordinated with cell cycle advance. As cells pass through the M/G_1 transition, virtually all of the existing phosphate groups are stripped off pRb, leaving it in an *unphosphorylated* configuration. As cells progress through G_1, a single phosphate group is attached as any one of 14 different phosphorylation sites (by cyclin D-CDK4/6 complexes), yielding *hypophosphorylated* pRb. However, when cells pass through the restriction (R) point, cyclin E–CDK2 complexes phosphorylate pRb on at least 12 more sites, placing it in a *hyperphosphorylated* state. Throughout the remainder of the cell cycle, the extent of pRb phosphorylation remains constant until cells enter into M phase.

Figure 8.20 Co-precipitation of cellular proteins with adenovirus E1A (A) Incubation of the C36 anti-pRb monoclonal antibody with a lysate of HeLa cells revealed that pRb migrated with the mobility of a protein of ~105 kD, with more rapidly migrating cleavage fragments evident below it (*left channel*). Use of the M73 anti-E1A antibody revealed no reactive proteins in these cells, which lack adenovirus genomes. This also indicated that this antibody did not have undesired cross-reactions with cellular proteins. (B) Incubation of the anti-pRb antibody with lysates of adenovirus-transformed 293 cells revealed once again the presence of pRb and some lower–molecular-weight fragments. However, when the M73 anti-E1A antibody was used, it precipitated not only the E1A protein but also a series of host-cell proteins, including pRb, p107, p130, and p300. (C) Since the anti-E1A antibody had no reactivity with host-cell proteins (see panel A), this meant that E1A must have formed a bridge between the antibody molecules and a number of host cell proteins with which it was complexed in virus-transformed cells. (Adapted from P. Whyte et al., *Nature* 334:124–129, 1988.)

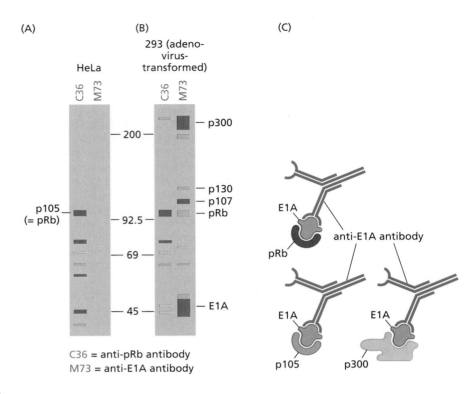

The 1988 work revealed one important mechanism whereby a DNA tumor virus–encoded oncoprotein can disrupt the cellular growth-regulating circuitry: biochemical characterization of the oncoprotein made by the E1A oncogene of human adenovirus type 5 showed that in adenovirus-transformed cells, this oncoprotein is tightly bound to a number of cellular proteins, among them pRb (**Figure 8.20; Sidebar 8.4**). Soon thereafter, the SV40 virus–encoded large T oncoprotein and the E7 oncoprotein of certain strains of human papillomavirus (HPV) implicated in causing human cervical carcinoma were also found capable of forming physical complexes with pRb in virus-transformed cells.

These three different DNA tumor virus oncoproteins (E1A, large T antigen, and E7) are structurally unrelated to one another, yet all of them target a common cellular protein—pRb. This suggests that convergent evolution led, on at least three occasions, to the development of viral oncoproteins capable of perturbing cellular pRb. Stated differently, since evolution selects for viruses that can multiply more effectively, these discoveries suggested that sequestration or functional inactivation of pRb by the three viral oncoproteins is necessary for optimal viral replication in infected cells.

The discoveries of these viral–host protein complexes had profound effects on our understanding of how DNA tumor viruses succeed in transforming the cells that they infect. The speculation went like this: Within a normal cell, the *Rb* gene was known to function as a tumor suppressor gene. Hence, the *Rb* gene, acting through its protein product, pRb, must inhibit cell proliferation in some fashion. DNA tumor viruses can sabotage the activities of pRb by dispatching viral oncoproteins that seek out and bind pRb, sequestering and apparently inactivating it. This sequestration removes pRb from the regulatory circuitry of the cell, yielding the same outcome as when a cell loses the two copies of its chromosomal *Rb* gene through mutations. This suggested, furthermore, that the DNA tumor viruses succeed in transforming cells through their ability to disable this key cellular growth-inhibitory protein, thereby liberating the virus-infected cell from the growth suppression that pRb normally imposes. (In fact, the HPV E7 protein goes beyond simple sequestration of pRb, since it also tags pRb for ubiquitylation and thus proteolytic degradation; see Supplementary Sidebar 7.5.)

Another clue to pRb function was provided by the discovery that the oncoproteins of the DNA tumor viruses bind preferentially to the hypophosphorylated forms of pRb,

Sidebar 8.4 Adenovirus deploys a single viral oncoprotein to bind and sequester multiple cellular proteins As seen in Figure 8.20, when a lysate of adenovirus-transformed cells is incubated with an anti-E1A monoclonal antibody, the resulting immunoprecipitate contains the E1A oncoprotein plus a number of other proteins. These other proteins are not bound directly by the antibody; instead, in each case, the E1A oncoprotein acts as a bridge to link the antibody molecule to a second protein, indicating that within adenovirus-transformed cells the E1A protein had previously formed physical complexes with each of these other cellular proteins. (Before the immunoprecipitates were analyzed by gel electrophoresis and autoradiography, these multiprotein complexes were treated with a strong detergent to dissociate the component proteins from one another.)

The first of these associated proteins to be identified was pRb, with two other cellular proteins, termed p107 and p130, being characterized soon thereafter; these two are cousins of pRb. These observations indicated that during the course of its evolution, adenovirus has configured the E1A protein in a way that enables it to bind these three cellular proteins, ostensibly in order to sequester them. pRb and its two cousins are often called generically **pocket proteins** (Figure 8.21A), since each carries a cavity into which viral oncoproteins like E1A can insert. Later, a higher–molecular weight cellular protein, termed p300, was also characterized, and a number of other cellular targets remain to be characterized.

The rationale for why adenovirus deploys its oncoprotein to sequester these four proteins must be understood in terms of the viral replication cycle. A virus needs to reconfigure the regulatory circuitry of an infected cell in order to create an intracellular environment that supports various steps of viral replication. Most important is its requirement for viral DNA replication, which can apparently occur efficiently only if the virus is able to liberate an infected cell from its G_0 quiescent state by functionally inactivating the cellular proteins that are holding it in that state; this permits the infected cell, in turn, to advance into S phase, where the cellular DNA replication machinery is now active and can be exploited by the virus.

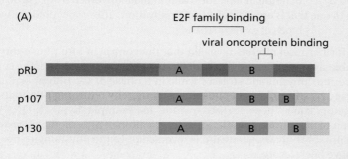

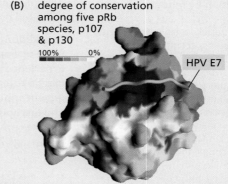

Figure 8.21 Pocket proteins (A) pRb is one of three structurally related proteins that together are often called "pocket proteins." All three are bound by viral oncoproteins, such as adenovirus E1A (see Figure 8.20), SV40 large T, and human papillomavirus E7. All three pocket proteins are phosphorylated by cyclin D–CDK4/6 complexes, and each binds its own subset of E2F transcription factors, as discussed later. The "pocket" that interacts with both viral oncoproteins and cellular E2F transcription factors is composed of A and B domains (*blue*). (B) The shallow groove into which viral oncoproteins bind (using an LxCxE amino acid sequence carried by all of these oncoproteins, where "x" is variable) is structurally highly conserved among the pRb proteins of five mammalian species as well as the human p107 and p130 proteins, explaining the ability of the viral oncoproteins to bind all three of these cellular proteins. This groove is presented here as a space-filling model. The degree of sequence conservation is indicated by different shades of green, with the most conserved residues depicted as *dark green*. The polypeptide backbone of the pRb-binding domain of the human papillomavirus E7 oncoprotein is shown here in *yellow*. (B, from J.O. Lee et al., *Nature* 391:859–865, 1998.)

that is, the forms of pRb that are present in the cell through most of the G_1 phase (see Figure 8.19). Conversely, the hyperphosphorylated pRb forms that are present in late G_1 and in the subsequent phases of the cell cycle are ignored by the viral oncoproteins. This observation led to yet another speculation: tumor virus oncoproteins focus their attention on sequestering, and thereby inactivating, the only forms of pRb in the cell that are worthy of their attention—the growth-inhibitory forms of pRb found in early and mid-G_1. Conversely, the viral oncoproteins ignore those forms of pRb that are already inactivated by other mechanisms.

Such reasoning indicated that the hypophosphorylated forms of pRb present in early and mid-G_1 cells are actively growth-inhibitory, while the hyperphosphorylated forms found after the R point have lost their growth-inhibitory powers, having been functionally inactivated by this phosphorylation. This meant that the R-point transition, which is defined by a physiologic criterion (acquisition of growth factor independence), is

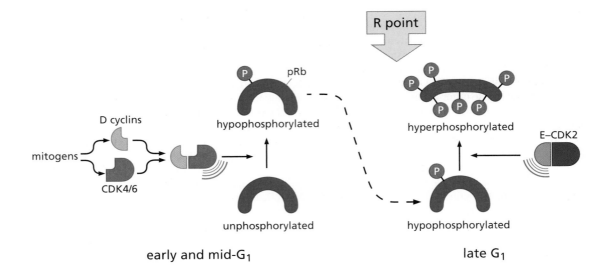

Figure 8.22 Control of the restriction-point transition by mitogens The levels of D-type cyclins are controlled largely by extracellular signals. Because the D-type cyclins, acting with their CDK4 and CDK6 partners, are able to drive pRb hypophosphorylation—an apparent prerequisite to the R-point transition—this ensures that this transition is responsive to extracellular signals (see also Figure 8.15). Cyclin E–CDK2 complexes then take over to complete the work of decommissioning pRb by placing it in a hyperphosphorylated state, causing a >10-fold increase in its phosphorylation that results in the complete functional inactivation of pRb.

accompanied by a biochemical alteration of pRb that converts it from a growth-inhibitory protein to one that is apparently functionally inert. These speculations were soon validated by various types of experiments.

In addition, further research has revealed that the control of pRb phosphorylation is more reversible than is indicated by the diagram in Figure 8.19. Thus, if the cell should experience serious physiologic stresses while in S phase or G_2, pRb phosphorylation can be reversed by still-unknown phosphatases, thereby returning pRb to its actively growth-inhibitory state. These stresses include, for example, hypoxia, DNA damage, and disruption of the mitotic spindle. Such reactivation of pRb-mediated growth inhibition is presumably only transient and is reversed once the physiologic stresses and/or damage have been resolved.

8.6 pRb is deployed by the cell cycle clock to serve as a guardian of the restriction-point gate

Various lines of evidence converge on the conclusion that pRb, which is growth-inhibitory through early to mid-G_1 phase, becomes inactivated by extensive phosphorylation when the cell passes through the R-point gate and thereafter is largely or completely inert as a growth suppressor. In effect, pRb serves as a guardian at this gate, holding it shut unless and until it becomes hyperphosphorylated, in which case pRb loses its growth-inhibitory powers, opens the gate, and permits the cell to enter into late G_1 and advance thereafter into the remaining phases of the cell cycle.

Since pRb is the ultimate arbiter of growth versus nongrowth, its phosphorylation must be carefully controlled. Not unexpectedly, its phosphorylation is governed by components of the cell cycle clock. In early and mid-G_1, D-type cyclins together with their CDK4/6 kinase partners are responsible for initiating pRb phosphorylation, leading to its hypophosphorylation. Since the levels of D-type cyclins appear to be controlled largely by extracellular signals, notably mitogenic growth factors, we can now plot out a direct line of signaling: growth factors induce the expression of D-type cyclins; D-type cyclins, collaborating with their CDK4/6 partners, initiate pRb phosphorylation (**Figure 8.22**).

During a normal cell cycle, this initial hypophosphorylation of pRb appears to be necessary but not sufficient for the subsequent functional inactivation of pRb at the R point, which requires its hyperphosphorylation. In fact, cyclin E levels increase dramatically at the R point. The cyclin E then associates with its CDK2 partner, and this complex drives pRb phosphorylation to completion, leaving pRb in its hyperphosphorylated, functionally inactive state (see Figure 8.22). It seems that pRb molecules that have not undergone prior hypophosphorylation by cyclin D–CDK4/6 complexes are not good substrates for completion of phosphorylation by cyclin E–CDK2 complexes. The phosphorylation and functional inactivation of pRb's cousins, p107 and p130, are also under the control of cyclin D–CDK4/6 and possibly cyclin E–CDK2 complexes.

If mitogens are withdrawn from the cell at any time in G_1 prior to the R point, then the levels of the most prominent of the D-type cyclins—cyclin D1—rapidly collapse. In the absence of cyclin D1, pRb loses its phosphate groups through the actions of a still-unidentified phosphatase and may thereby revert to a protein that is no longer a good substrate for cyclin E–CDK2 complexes. This response emphasizes the need for strong and continuous mitogenic signaling throughout the early and mid-G_1 phase up to the R point.

Once the cell advances through the R point, the continued hyperphosphorylation and functional inactivation of pRb is apparently maintained and further increased by cyclin E-, by cyclin A-, and then by cyclin B-containing CDK complexes, none of which is responsive to extracellular signals—precisely the properties of the cell cycle clock machinery that, we imagine, operates after the R-point transition and guarantees execution of the rigidly programmed series of transitions in S, G_2, and M.

This scheme reveals why pRb is such a critical player in the regulation of cell proliferation. If its services are lost from the cell (through mutation of the chromosomal *Rb* gene copies, methylation of the *Rb* gene promoter, or the actions of DNA tumor virus oncoproteins), then this protein can no longer stand as the guardian of the R-point gate. Moreover, as we will see, in some cancer cells, pRb phosphorylation is deregulated, resulting in inappropriately phosphorylated and thus functionally inactivated pRb. In certain other cancer cells, there is evidence that the dephosphorylation (and attendant activation) of pRb, which normally happens at the M/G_1 transition through the action of the PP1 phosphatase, never occurs, leaving pRb in its hyperphosphorylated, inactivated state throughout the entire growth-and-division cycle of these cells. Without pRb on watch, cells move through G_1 into S phase without being subjected to the usual controls that are designed to ensure that cell cycle advance can proceed only when certain preconditions have been satisfied. Indeed, as we will see, the deregulation of pRb phosphorylation is so widespread that one begins to think that this important signaling pathway is perturbed in virtually all human tumors.

The activities of pRb seem to overlap with those of its cousins, p107 and p130, raising the question why pRb is the only one of these three *pocket proteins* to play a clear and obvious role in cancer pathogenesis. A definitive answer to this puzzle is still elusive. It may come from the specific times in the cell cycle when these two pRb-related proteins are active. Whereas p130 is specialized to suppress cell proliferation while a cell resides in G_0, p107 appears to be most active in the late G_1 and S phases. Only pRb is well positioned to control the transition at the R point from mitogen-dependent to mitogen-independent growth.

8.7 E2F transcription factors enable pRb to implement growth-versus-quiescence decisions

As described above, a number of lines of evidence indicate that hypophosphorylated pRb is active in suppressing G_1 advance and loses this ability when it becomes hyperphosphorylated. Still, this behavior provides no clue as to how pRb succeeds in imposing this control. Research conducted in the early 1990s indicated that pRb exercises much of this control through its effects on a group of transcription factors termed E2Fs.

When pRb (and its two cousins, p107 and p130) are in their unphosphorylated or hypophosphorylated state, they bind E2Fs, including E2Fs that are bound to DNA; however, when hyperphosphorylated, pRb and its cousins (see Figure 8.21) dissociate from the E2Fs (**Figure 8.23A**). This suggested a simple model of how pRb is able to control cell cycle advance. In the early and middle parts of G_1, E2Fs are associated with the promoters of a number of genes under their control. At the same time, these transcription factors are bound by pocket proteins. This pocket protein association prevents the E2Fs from acting as stimulators of transcription. Accordingly, during much of the G_1 phase of the cell cycle, genes that depend on E2Fs for expression remain repressed.

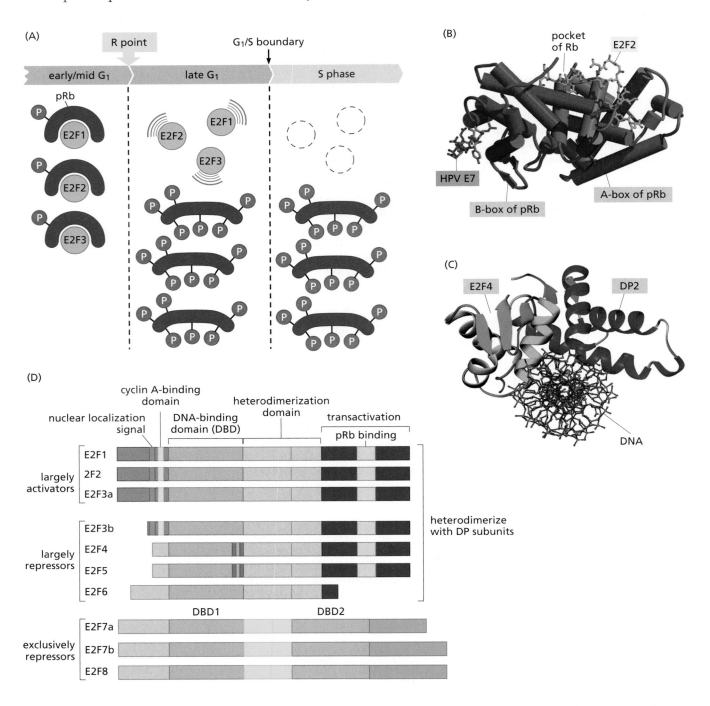

However, when the pocket proteins undergo hyperphosphorylation at the R point in late G$_1$, they release their grip on the E2Fs, permitting the E2Fs to stimulate transcription of their clientele of genes. The products of these genes, in turn, usher the cell from late G$_1$ into S phase. Similarly, when viral oncoproteins are present, they mimic pRb hyperphosphorylation by preventing pRb from binding E2Fs (see Figure 8.23B).

While the above scheme is correct in outline, the details are more interesting and complex. The term "E2F" is now known to include a class of proteins composed of eight members—E2F1 through E2F8; the first six of these can bind to either a DP1 or a DP2 subunit to form heterodimeric transcription factors (see Figure 8.23C and D); the two others—E2F7 and E2F8—have two DNA-binding domains (DBDs), allowing each to bind DNA without an associated DP1/2 subunit. (In the discussions that follow, we

Figure 8.23 E2Fs and their interactions (A) By binding various DNA-bound E2F transcription factors, the three pocket proteins, pRb, p107, and p130, modulate the expression of E2F-responsive genes in various phases of the cell cycle. When pRb binds to E2Fs 1, 2, and 3 (*left*), it blocks their transcription-activating domains; other proteins that serve to modify chromatin and repress transcription (*not shown*) are attracted to the pRb–E2F complexes. Because they are bound to the promoters of certain genes via the E2Fs, these multiprotein complexes repress the expression of such genes. (The behavior of p107 and p130, *not shown*, is similar to that of pRb.) After being hyperphosphorylated at the R point, the pocket proteins release their grip on the E2Fs, allowing the latter to activate transcription of those genes. However, because E2Fs 1, 2, and 3 induce expression of E2F7 and E2F8, which function as transcriptional repressors (see panel D), the ability of these three to activate signaling is curtailed toward the end of G_1. Moreover, as cells enter into S phase (*right*), E2Fs 1, 2, and 3 are inactivated or degraded. Hence, E2Fs 1, 2, and 3 appear to function as active inducers of transcription only in the narrow window of time in late G_1 phase that begins at the R point and ends as cells enter S phase. (B) The binding of E2Fs such as E2F2 (*yellow stick figure*) by pRb (*red and blue*) is prevented by several DNA tumor virus oncoproteins. Here we see how a segment of the human papillomavirus E7 oncoprotein (*green stick figure, left*) binds to a shallow groove of pRb (also shown in Figure 8.21B); this binding, together with other secondary binding sites (not yet mapped), perturbs the cavity formed by the B- and A-boxes of pRb, thereby preventing the binding of the C-terminal, transcription-activating domain of E2F2 (*yellow*) to this pocket. This loss of E2F binding effectively neutralizes the ability of pRb to inhibit cell cycle advance (*cylinders = α-helices*). (C) The E2F transcription factors bind DNA as heterodimeric complexes with DP partners. Shown is the binding of an E2F4–DP2 complex (*ribbon diagram*) to the DNA double helix (*stick figure*), as revealed by X-ray crystallography. (D) Because of the existence of variant forms of several E2Fs, they are actually represented by at least 10 distinct protein species. E2Fs 1, 2, and 3a represent transcription-activating E2Fs, while E2F3b, E2F4, E2F5, and E2F6 are largely or exclusively involved in transcriptional repression. E2Fs 7a, 7b, and 8 are atypical E2Fs, in that each has a duplicated DNA-binding domain (DBD) that enables DNA binding in the absence of a DP subunit as partner; these E2Fs are involved exclusively in repression. Their expression is induced by E2F1, 2, and 3, suggesting the operation of a negative-feedback loop designed to shut down E2F1, 2, and 3 signaling. Overall, E2Fs 4 and 5 are involved in repression in early/mid-G_1, while E2Fs 7 and 8 are involved in repression in late G_1, at the G_1/S transition. (B, courtesy of Y. Cho, from C. Lee et al., *Genes Dev.* 16:3199–3212, 2002. C, courtesy of R. Latek, from N. Zheng et al., *Genes Dev.* 13:666–674, 1999. D, adapted from H.-Z. Chen et al., *Nat. Rev. Cancer* 9:785–797, 2009.)

describe the behavior of the various E2F subunits, with association to their DP1 or DP2 partner subunits being assumed.) Once assembled, E2F–DP complexes recognize and bind to a sequence motif in the promoters of various genes that seems to be TTTC-CCGC or slight variations of this sequence.

Throughout this book, we have portrayed transcription factors as proteins that bind to the promoters of genes and then proceed to activate transcription. In fact, transcription factors operating like E2Fs can exert two opposing effects on transcriptional control. When bound to the promoters of genes in the absence of any associated pocket proteins (that is, pRb, p107, or p130), E2Fs such as E2Fs 1, 2, and 3 can indeed trigger gene expression by attracting other proteins—such as histone acetylases—that function to remodel the nearby chromatin and recruit RNA polymerase to initiate transcription.

However, when pRb is hypophosphorylated, it physically associates with its E2F partners, which themselves are already sitting on the promoters of various genes and remain so after pRb binds to them. Once bound to these E2Fs, hypophosphorylated pRb molecules block the *transactivation* domain of the E2Fs that they use to activate transcription (see Figure 8.23D). At the same time, pRb recruits other proteins that actively repress transcription. One important means of doing so, but hardly the only one, is to recruit a histone deacetylase (HDAC) to this complex (see Section 1.8); by removing acetyl groups from nearby histone molecules, an HDAC remodels the nearby chromatin into a configuration that is incompatible with active transcription (**Figure 8.24A**).

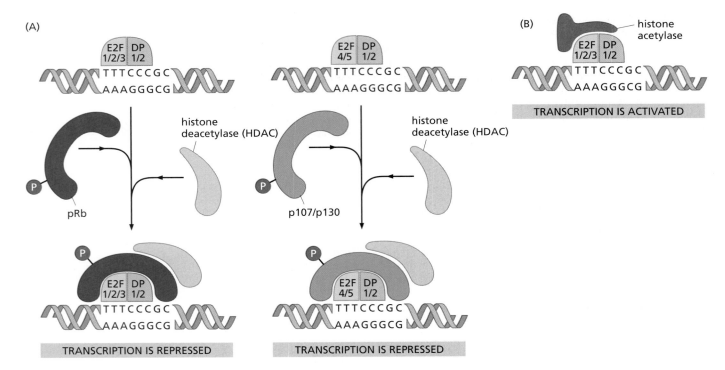

Figure 8.24 Modification of chromatin by pocket proteins
(A) Hypophosphorylated pRb discourages transcription by attracting histone deacetylase (HDAC) enzymes (*gray*) to pRb–E2F complexes that are bound, in turn, to gene promoters; these enzymes prevent transcription by removing acetyl groups from nearby histones, causing the latter to shift into a configuration that is incompatible with the progression of RNA polymerase through the chromatin (see Section 1.8). (B) Conversely, when the E2Fs are not complexed with pRb (or its p107 and p130 cousins), they attract histone acetylases (*red*), which place chromatin in a configuration that is conducive for transcription.

This means that, in addition to occluding (physically blocking) the transactivation domains of E2Fs, pRb actively functions as a transcriptional repressor. E2Fs 4 and 5 seem to be involved primarily in the repression of genes and act largely through associated p107 and p130 proteins to attract repressors of transcription to promoters and thereby shut down gene expression. E2Fs 6, 7, and 8, which do not associate with pocket proteins, appear to act exclusively as transcriptional repressors.

In quiescent G_0 cells, for example, E2F4 and E2F5 are present in abundance (and are associated largely with p130), whereas E2Fs 1, 2, and 3 are hardly present at all, being expressed largely in proliferating cells. Moreover, cells that have been genetically deprived of E2F4 and E2F5 have lost responsiveness to the growth-inhibitory effects of the $p16^{INK4A}$ CDK inhibitor. This implies that the lion's share of the downstream effects that result from inhibiting the cyclin D–CDK4/6 complexes and blocking the phosphorylation of pocket proteins is mediated via the binding of p107 and p130 to these two E2Fs; ultimately, this binding results in the repression of a constituency of genes that carry TTTCCCGC sequences in their promoters.

Once hyperphosphorylated, pRb releases its grip on E2F1, 2, and 3 complexes, and the E2Fs can then attract transcription-activating proteins, including histone acetylases, which convert the chromatin into a configuration that encourages transcription (see Figure 8.24B). A series of genes that are expressed specifically in late G_1 and are known or assumed to be important for S-phase entry have been found to contain binding sites in their promoters for E2Fs. Included among these are some genes encoding proteins involved in synthesizing DNA precursor nucleotides (such as dihydrofolate synthetase and thymidine kinase), as well other genes involved directly in DNA replication. (One survey indicates that as many as 500 genes are direct targets of transcriptional activation by E2Fs.)

Prominent among the E2F-activated genes are the genes encoding cyclin E and E2F1. Consequently, the levels of cyclin E and E2F1 mRNAs rise quickly after passage through the R point and the levels of their protein products increase shortly thereafter. Recall that cyclin E, for its part, is responsible, together with CDK2, for driving the hyperphosphorylation of pRb. Hence pRb inactivation leads to increases in cyclin E levels, and cyclin E, once formed, drives further pRb inactivation (**Figure 8.25A**). These relationships enable the operations of powerful, self-reinforcing positive-feedback loops that are triggered as cells pass through the R point. (Such regulation can also be termed "feed-forward loops.")

Yet another positive-feedback loop is activated simultaneously at the R point through the following mechanism: cyclin E–CDK2 complexes phosphorylate p27^{Kip1}, and the latter, once phosphorylated, is ubiquitylated and rapidly degraded (see Figure 8.25B). The destruction of p27^{Kip1} molecules, which normally inhibit E–CDK2 firing, liberates additional cyclin E–CDK2 complexes that proceed to phosphorylate and thereby inactivate additional p27^{Kip1} molecules. Such positive-feedback mechanisms operate in many control circuits to guarantee the rapid execution of a decision once that decision has been made. Equally important, they ensure that the decision, once made, is essentially irreversible—a situation that corresponds precisely with the behavior of cells advancing through the R point.

The period of active transcriptional promotion by E2Fs 1, 2, and 3 appears to be rather short-lived (see Figure 8.23A). This time window begins at the R point when pRb becomes hyperphosphorylated and liberates the E2Fs. These three E2Fs then proceed to induce expression of critical late G_1 genes, the products of which are required to prepare the cell for entrance into S phase, as described above. Soon thereafter, as the cell traverses the G_1/S transition into S phase, cyclin A becomes activated and, acting together with its CDK2 partner, phosphorylates both the E2F and DP subunits of these

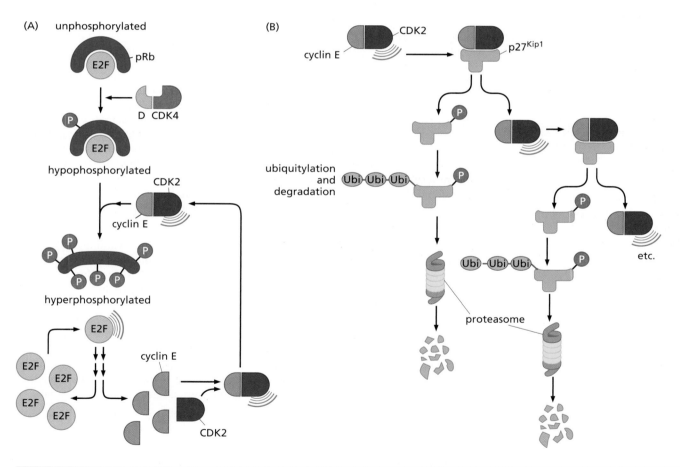

Figure 8.25 Positive-feedback loops and the irreversibility of cell cycle advance The irreversibility of certain key steps in cell cycle progression and the rapidity of their execution is ensured, in part, by the activation of certain positive-feedback loops. (A) When E–CDK2 complexes drive pRb hyperphosphorylation, this liberates E2F transcription factors from pRb control, enabling the E2Fs to trigger increased transcription of the cyclin E and E2F1 genes; this leads to the synthesis of more cyclin E protein and the formation of more E–CDK2 complexes, which function, in turn, to drive additional pRb phosphorylation. At the same time, the newly synthesized E2F1 protein drives its own expression, further amplifying its activity. (B) The activation of a small number of E–CDK2 complexes enables the latter to phosphorylate p27^{Kip1}, marking it for ubiquitylation. The consequent degradation of this p27^{Kip1} liberates more E–CDK2 complexes from inhibition by p27^{Kip1}, allowing this process to repeat itself and amplify its effects. [Not shown here is yet another positive-feedback loop in which increases in E2F function at the R point lead to increased transcription of the Skp2 gene (see Figure 8.37), whose protein product drives degradation of p27^{Kip1} and thereby liberates even more E–CDK2 complexes from inhibition by this CKI.]

heterodimeric transcription factors; this results in the dissociation of the E2F–DP complexes and in the attendant loss of their transcription-activating abilities. At the same time, E2F1 (and possibly other E2Fs) are tagged for degradation by ubiquitylation, and E2F7 and E2F8, which seem to function largely to antagonize E2F-mediated gene activation, are expressed and appear to block any residual E2F-dependent expression that persists after cells have entered S phase. Together, these changes effectively shut down further transcriptional activation by the E2F factors, whose brief moment of glory as active transcription factors will come again only during the end of the G_1 phase of the next cell cycle.

pRb has been reported to bind to a variety of transcription factors, and these might, in principle, be controlled by pRb and its state of phosphorylation. In the great majority of cases, however, detailed functional analyses of their interactions with pRb have not been undertaken. There is a compelling reason to believe that, among all of these transcription factor clients of pRb, the E2Fs are of preeminent importance in controlling cell cycle advance: when the E2F1 protein is micro-injected into certain types of serum-starved, quiescent cells (which are therefore in G_0), this transcription factor, acting on its own, is able to induce these cells to enter G_1 and advance all the way into S phase; E2F2 and E2F3 have similar powers.

8.8 A variety of mitogenic signaling pathways control the phosphorylation state of pRb

Throughout this chapter we have emphasized that a number of distinct signaling pathways converge on the cell cycle clock and modulate its functioning. Acting as a central clearinghouse, the clock machinery collects and processes these various *afferent* signals and then, acting via its cyclin–CDK complexes, emits the signals that energize cell cycle advance (see Figure 8.1). It is worthwhile at this point to step back and summarize the ways in which these various signals regulate distinct components of the cell cycle clock. Prominent among the signals impinging on this nuclear regulatory machinery are those conveyed by the mitogenic signaling pathways. Because excessive signals flow through these pathways in virtually all types of cancer cells, the connections between these pathways and the cell cycle clock are critical to our understanding of cell transformation and tumor pathogenesis.

Mitogenic signals transduced via Ras result in the hypo- and eventual hyperphosphorylation of pRb. This phosphorylation is initiated by inducing the expression of genes encoding the D-type cyclins, notably cyclin D1. When Ras signaling is blocked (through the introduction of a **dominant-negative** mutant Ras protein), cell proliferation is blocked in wild-type cells exposed to serum, while mutant Rb$^{-/-}$ cells lacking Ras function enter S phase quite normally. This demonstrates that the main function of Ras during the G_1 phase is to ensure that the mitogenic signals released by ligand-activated growth factor receptors succeed in causing the phosphorylation and functional inactivation of pRb. Moreover, the block to S-phase entrance imposed by the dominant-negative Ras protein can be circumvented by ectopically expressing either cyclin E or E2F1 genes in cells. Together, this suggests a linear chain of command of the following sort:

growth factors → growth factor receptors → Ras → cyclins D1 and E → inactivation of pRb → activation of E2Fs → S-phase entrance

The ability of mitogenic growth factors to increase the levels of D-type cyclins depends on several distinct signaling pathways (**Figure 8.26**; see also Section 8.3 and Table 8.1). The best studied of these signals flow via the Ras → Raf → MAPK pathway (see Section 6.5). For example, members of the Fos family of transcription factors, which lie at the bottom of this signaling cascade (see Table 6.1), create the heterodimeric AP-1 transcription factors by forming complexes with Jun proteins; these AP-1 complexes are known to be powerful activators of cyclin D1 gene transcription. In addition, a variety of cytoplasmic signal-transducing proteins impinge on the cyclin D1 promoter in ways that are poorly understood (see Table 8.1). As mentioned (see Section 8.3), the cyclin D2 and D3 promoters have their own array of upstream regulators that confer responsiveness to certain extracellular signals. Many of the signaling pathways converging

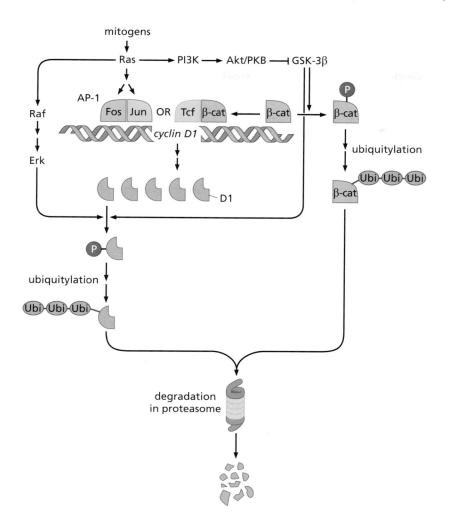

Figure 8.26 Countervailing controls on cyclin D1 levels A variety of signaling pathways transducing mitogenic signals modulate the levels of cyclin D1. For example, the Ras pathway influences cyclin D1 levels in several ways. First, the AP-1 transcription factor acts on the *cyclin D1* promoter, yielding increased levels of the cyclin D1 mRNA; AP-1 is composed of Fos–Jun heterodimers, both of which are increased in amount and/or functionally activated by mitogens (see also Figure 8.11). Second, by activating PI3 kinase and thus Akt/PKB, Ras is able to inhibit the actions of GSK-3β. This, in turn, spares β-catenin (β-cat) from phosphorylation, ubiquitylation, and degradation, allowing it to partner with the Tcf transcription factor to induce *cyclin D1* transcription. Third, this inhibition of GSK-3β also spares cyclin D1 protein from phosphorylation by this enzyme and subsequent degradation. Fourth, operating in the opposing direction is Erk, which lies at the bottom of the Ras → Raf → MEK → Erk kinase cascade (see Section 6.5). Cyclin D1 is phosphorylated by Erk and thereby marked for ubiquitylation and degradation in proteasomes. Overall, Ras favors cyclin D1 accumulation, since the powers of the Ras pathway to increase cyclin D1 levels seem to greatly exceed its effects in causing cyclin D1 degradation.

on these two gene promoters are also deregulated in human cancer cells, leading in turn to excessive expression of one or the other of these D-type cyclins, inappropriate phosphorylation and inactivation of pRb, and deregulated proliferation.

Activation of the Ras signaling pathway also leads, via phosphatidylinositol 3-kinase (PI3K), to activation of the Akt/PKB kinase; the latter proceeds to phosphorylate and inactivate glycogen synthase kinase-3β (GSK-3β; see Section 6.10 and Figure 8.26). This has several benefits for growth promotion. Normally, GSK-3β targets β-catenin molecules for destruction by phosphorylating them; this phosphorylation results in the ubiquitylation of β-catenin molecules and their destruction in proteasomes. However, when GSK-3β is inactivated by PKB/Akt phosphorylation, β-catenin can accumulate and migrate to the nucleus, where it forms transcription factor complexes with Tcf/Lef that stimulate production of the cyclin D1 mRNA. Since GSK-3β phosphorylation also targets the cyclin D1 protein for destruction (see Figure 8.26), its inactivation further increases cyclin D1 levels. It is also possible that the Wnt growth factors (see Section 6.10), by suppressing GSK-3β activity, have similar effects on the cell cycle clock.

Adding to this complexity are other mitogenic signals that affect specific components of the cell cycle clock. For example, in ways that are poorly understood, continuous exposure to serum growth factors causes a degradation of the cellular pool of the p27[Kip1] CDK inhibitor, resulting in the progressive decline in the overall levels of this protein as the G$_1$ phase of the cell cycle proceeds. [This loss of p27[Kip1] molecules from the cell seems to cooperate with the sequestration of p27[Kip1] achieved by cyclin D1–CDK4/6 complexes (see Figure 8.17) to liberate some cyclin E–CDK2 from inhibition by p27[Kip1], an event in late G$_1$ that triggers the explosive onset of the R-point transition.]

There are yet other mechanisms, still poorly characterized, that enable mitogenic growth factors to energize the cell cycle apparatus. For instance, the physical association of D-type cyclins with CDK4 is also dependent, for unknown reasons, on mitogenic signaling. The elucidation of this and other connections between growth-promoting signals and the cell cycle clock will surely come from future research.

8.9 The Myc protein governs decisions to proliferate or differentiate

In the previous section, we described how mitogenic signals flowing through normal cells lead to elevated levels of D-type cyclins and the resulting phosphorylation and functional inactivation of pRb; the excessive mitogenic signals flowing through cancer cells serve to amplify these effects. Prior to that discussion, we already had read about two other ways in which pRb function is compromised in neoplastic human cells: pRb function can be lost through mutation of the *Rb* gene, as is seen in retinoblastomas, osteosarcomas, and small-cell lung carcinomas. A functionally similar outcome can be observed in cells that have been infected and transformed by human papillomavirus (HPV); pRb binding by the viral E7 oncoprotein (see Figures 8.21B and 8.23B) plays a key role in the development of the great majority (>99.7%) of cervical carcinomas. As we will now see, human cancer cells devise a variety of other strategies to derail control of cell cycle progression.

We begin with the Myc protein, which, when expressed in a deregulated fashion, operates as an oncoprotein. More than 70% of human tumors overexpress either Myc (often termed c-Myc) or one of its two close cousins—N-Myc and L-Myc. The Myc family oncoproteins function very differently from most of the oncoproteins that we encountered in Chapters 4, 5, and 6. These others, notably Ras and Src, operate in close proximity to cytoplasmic membranes and trigger complex signaling cascades that activate cytoplasmic signal-transducing proteins and, ultimately, nuclear transcription factors. Myc and its cousins are found, by contrast, in the nucleus, where they function as growth-promoting transcription factors (Sidebar 8.5).

Sidebar 8.5 Many oncoproteins function as transcription factors Myc is only one of a large group of transcription factors that can function as oncoproteins (Table 8.2). We first encountered oncoproteins of this type in our discussion of the immediate early genes *myc, fos,* and *jun* (see Section 6.1), each of which encodes a transcription factor that can function as an oncoprotein. Later, we read about STAT proteins (Section 6.8), NF-κB, Notch, and Gli (Section 6.12), which also can play these dual roles. These various transcription factors act on a variety of target genes to induce a subset of the phenotypes that we associate with transformed cells.

The constituency of target genes that are acted upon by each of these transcription factors remains poorly defined. This lack of information is explained largely by the fact that at the experimental level, it has been extremely difficult to determine the identities of the genes that are directly activated or repressed by various transcription factors (TFs). The development of the chromatin immunoprecipitation (ChIP) technique holds the promise of partially addressing this problem by revealing the transcriptional promoters to which a TF of interest binds (Supplementary Sidebar 8.3). However, even the ChIP technique does not entirely solve this problem, since the binding of a TF to a chromosomal DNA segment does not, on its own, guarantee that the TF actually regulates the expression of a gene associated with this segment. Indeed, for many TFs, only a relatively small proportion (for example, 20% or 30%) of the binding sites identified by ChIP can actually be validated as functionally important sites of TF action.

Table 8.2 Representatives of the classes of genes encoding TFs that act as oncoproteins[a]

Transcription factor	Representative genes
Leucine zipper + basic DNA-binding domain	*fos* and *jun*
Helix + loop + helix (bHLH)	*myc, N-myc, L-myc, tal,* and *sci*
Zinc finger	*myl/RARa, erbA, evi-1,* and *gli-1*
Homeobox	*pbx* and *Hoxb8*
Others	*myb, rel, ets-1, ets-2, fli-1, spi-1, ski*

[a]Genes are grouped according to the structural features of the encoded proteins.
Adapted from T. Hunter, *Cell* 64:249–270, 1991.

These three Myc family proteins are only three of a very large (>100 members) family of bHLH transcription factors. They were named after their shared three-dimensional structures, which include a **b**asic DNA-binding domain followed by amino acid sequences forming an α-**h**elix, a **l**oop, and a second α-**h**elix. Members of this transcription factor family form homo- and heterodimers with themselves and with other members of the family. Such dimeric complexes then associate with specific regulatory sequences, termed E-boxes (composed of the sequence CACGTG), which are found in the promoters of target genes that they regulate (**Figure 8.27A**).

In the particular case of Myc, its actions are determined by its own levels as well as by its associations with partner bHLH proteins that either enhance or suppress its function as a transcription-activating factor. Phosphorylation of the Myc protein also

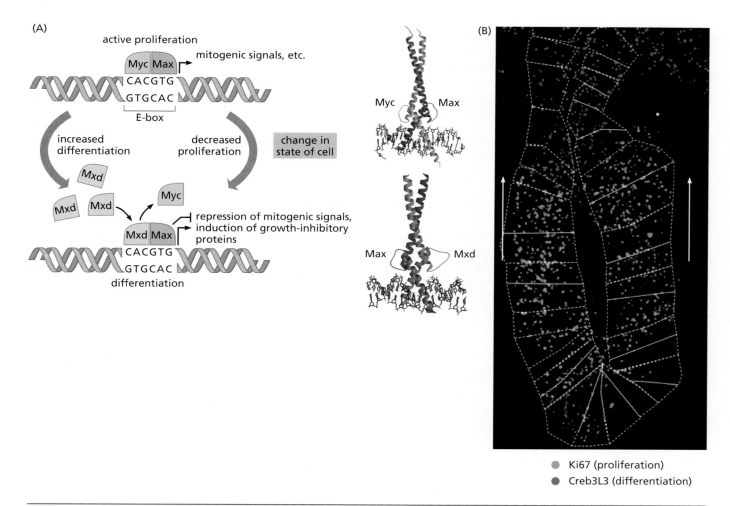

● Ki67 (proliferation)
● Creb3L3 (differentiation)

Figure 8.27 Proliferation vs. differentiation (A) Myc belongs to a family of bHLH (basic helix–loop–helix) transcription factors that act as heterodimers to modulate the transcription of a large cohort of target genes possessing E-box sequences (*left*). Myc–Max complexes act to promote elongation of nascent transcripts by releasing RNA polymerase II complexes from pause sites located immediately downstream of transcription start sites, allowing polymerase II to proceed with transcription and leading to the synthesis of full-length pre-mRNA transcripts; Mxd–Max complexes prevent this release. As cells differentiate, Mad levels increase progressively and Myc is displaced by Mxd, resulting in the disappearance of Myc–Max complexes, which otherwise would block differentiation (*left*). The actual molecular structures of these complexes sitting astride DNA (*right*) reveal why these proteins are termed basic helix–loop–helix transcription factors.

(B) The opposition of proliferation vs. differentiation can be observed in various tissues. Here, a highly sensitive fluorescence *in situ* hybridization (FISH) technique has been used to detect individual mRNA molecules in the individual cells (*outlined in dashed lines*) lining a crypt in the mouse small intestine (see Figure 7.24). The *green dots* reveal mRNA molecules encoding Ki67 (indicating active proliferation), while the *red dots* reveal mRNA molecules encoding Creb3L3 (indicating enterocyte differentiation). The expression of these two mRNAs is mutually exclusive and the implied transition from one cell state to another occurs abruptly as cells migrate upward from one cell position to the next (*white arrows*), apparently within an hour. (The involvement of Myc and its partners in this particular switch has not yet been demonstrated.) (A, right, from S.K. Nair and S.K. Burley, *Cell* 112:193–205, 2003. B, from S. Itzkovitz et al., *Nat. Cell Biol.* 14:106–114, 2011.)

modulates its functioning and stability. When Myc associates with Max, one of its bHLH partners that enhance its transcription-activating powers, the resulting Myc–Max heterodimeric transcription factor drives the expression of a large cohort of target genes. The products of many of these genes, in turn, have potent effects on the cell cycle, favoring cell proliferation.

(While Myc levels are strongly influenced by mitogenic signals, the levels of Max are kept relatively constant within cells. Accordingly, when normal cells are cultured in the presence of serum mitogens, Myc accumulates to substantial levels; conversely, Myc levels collapse when serum mitogens are withdrawn. This means that the levels of the Myc–Max heterodimer are continuously controlled by the flux of mitogenic signals that normal cells are receiving.)

Recent research indicates that Myc functions quite differently than the other transcription factors discussed throughout this book. The others seem to be involved specifically in regulating transcriptional *initiation* (see Figure 1.18). It turns out, however, that in many genes, after RNA polymerase II initiates transcription, it halts at a pausing site ~50 bp downstream of the transcriptional start site. A Myc–Max heterodimer, bound to DNA, then interacts with the paused pol II complex and releases it from its pause site, allowing it to continue transcribing the bulk of the gene, that is, to *elongate* the RNA transcript; without this intervention by Myc–Max, pol II is likely to sit at the pause site indefinitely. In one survey, 70% of actively transcribed genes were found to have Myc–Max heterodimers bound to these pause sites. (This number raises an interesting and unresolved paradox: in an actively growing cell, the number of genes found to bind Myc exceeds the absolute number of Myc molecules measured in such cells!)

As described later in this chapter, some bHLH proteins are also involved in orchestrating certain tissue-specific differentiation programs. Acting through intermediaries, Myc can prevent these other bHLH transcription factors from executing various differentiation programs. Consequently, Myc can simultaneously promote cell proliferation and block cell differentiation; this is precisely the biological behavior that is associated with actively growing cells that have not yet entered into post-mitotic, differentiated states (see Figure 8.27B). However, as cells slow their proliferation and become differentiated, the Myc–Max complexes disappear, since increasing levels of Mxd protein (another bHLH partner formerly known as Mad) displace Myc from existing Myc–Max complexes (see Figure 8.27A). The resulting Mxd–Max complexes then lose the ability to stimulate transcription, enabling cells in many human tissues to enter into post-mitotic differentiated states.

Myc interacts closely with the cell cycle machinery. Perhaps the first indication of this came from observations in which pairs of oncogenes were found to collaborate with one another in transforming rodent cells to a tumorigenic state. We will revisit oncogene collaboration later in Chapter 11. For the moment, suffice it to say that the *ras* and *myc* oncogenes were found to be highly effective collaborators in cell transformation, implying that each made its own unique contribution to this process. Similarly, a *ras* oncogene was found to collaborate with the adenovirus E1A oncogene in cell transformation. Together, these observations indicated that the *myc* and E1A oncogenes acted analogously in this experimental setting. Precisely how they did so was a mystery.

The fact that the E1A oncoprotein binds and inactivates pRb and its p107 and p130 cousins (see Sidebar 8.4) suggested that the Myc protein could have similar effects on these vital cellular proteins, especially pRb. However, a direct association between Myc and pRb was ruled out. Instead, we have come to realize that Myc regulates the expression of a number of other critical components of the cell cycle clock. When expression of these components is driven by the abnormally high levels of Myc present in cells carrying *myc* oncogenes, the result is a physiologic state similar to that seen in cells lacking pRb function. Both changes deprive a cell of the normal control of progression through the G_1 phase of its cell cycle and deregulate passage through the R point.

Among the many targets of Myc is the cyclin D2 gene (**Figure 8.28**), whose elevated expression leads, in turn, to the hypophosphorylation of pRb (see Figure 8.22). Myc also drives expression of the CDK4 gene, and the increased levels of CDK4 enable the

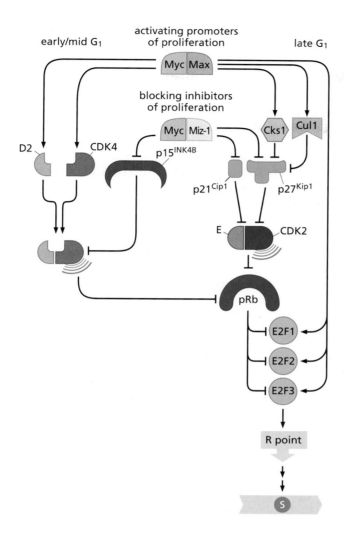

Figure 8.28 Actions of Myc on the cell cycle clock The Myc protein, acting with its Max partner (*top*), is able to modulate the actions of a number of positive and negative regulators of cell cycle advance. For example, the Myc–Max heterodimer is able to induce expression of the growth-promoting proteins cyclin D2 and CDK4 (*left*), which can promote advance through early G_1. At the same time, by increasing the expression of Cul1 and Cks1 (which are responsible for degrading the p27^Kip1 CDK inhibitor, *upper right*; see Figure 8.37B) as well as E2Fs 1, 2, and 3 (*right, below*), Myc favors advance into S phase. In addition, Myc, acting with its Miz-1 partner, is able to repress expression of the p15^INK4B, p21^Cip1, and p27^Kip1 CDK inhibitors; once again, effects are felt both in early/ mid and in late G_1.

formation of the cyclin D–CDK4 complexes that hypophosphorylate pRb and, at the same time, sequester the p27^Kip1 CDK inhibitor, thereby liberating cyclin E–CDK2 complexes from inhibition (see Figure 8.17). Myc also drives expression of the Cul1 protein, which plays a central role in the degradation of the p27^Kip1 CDK inhibitor through ubiquitylation. In one way or another, these Myc-driven gene expression changes serve to push the cell through the G_1 phase of its cell cycle.

By associating with a second transcription factor named Miz-1, Myc can also function as a transcriptional repressor (see Figure 8.28). In this role, Myc can repress expression of the genes encoding the p15^INK4B and p21^Cip1 CDK inhibitors, which shut down the actions of CDK4/6 and CDK2, respectively. In fact, as we will read later, TGF-β uses these two CDK inhibitors to block progression through the G_1 phase of the cell cycle. Hence, by preventing the expression of these two CDK inhibitors, Myc confers resistance to the growth-inhibitory actions of TGF-β. This represents an important means by which cancer cells can continue to multiply under conditions (for example, the presence of TGF-β in their surroundings) that would normally preclude their proliferation.

Finally, Myc is able to induce the expression of the genes encoding the E2F1, E2F2, and E2F3 transcription factor proteins. As we read earlier, these E2F transcription factors are negatively regulated by pRb and its two cousins. By causing these growth-promoting transcription factors to accumulate in a cell, Myc once again tips the balance in favor of cell proliferation.

Now, we begin to understand how the Myc oncoprotein can function so effectively to deregulate cell proliferation. It reaches in and pulls various regulatory levers within the cell cycle clock. The resulting changes in levels of key proteins cause a shift in the balance of proliferation-promoting and inhibiting mechanisms, strongly favoring pRb inactivation and cell growth.

Many of Myc's actions in perturbing the cell cycle apparatus, as described in this section, provide a satisfying explanation of its ability to deregulate cell proliferation and, in this way, to contribute so importantly to cancer pathogenesis. The actions of the Myc protein illustrate a more general principle of cell physiology that affects many of our discussions throughout this text. Repeatedly, we make reference to the fact that a protein like Myc turns its downstream targets on or off. Such wording represents a simplification of what actually occurs inside living cells. In reality, most regulators do not operate by switching targets totally on or off. Instead, they increase or decrease the levels of their downstream targets, often incrementally. In effect, they *reset the regulatory dials*, thereby shifting the balance of signals in ways that favor or disfavor a certain cellular response.

Perhaps the most graphic demonstration of the potent mitogenic powers of Myc has come from an experiment in which Myc is engineered to be expressed at high, constant levels but as a protein that is fused with the estrogen receptor (ER). In the absence of an estrogen receptor ligand, such as estrogen or tamoxifen, the Myc–ER fusion protein is sequestered in a functionally inactive state in the cytoplasm (see Section 5.8). However, upon addition of ligand, this fusion protein is liberated from sequestration, rushes into the nucleus, and operates like the normal Myc transcription factor. When quiescent, serum-starved cells (in the G_0 state) expressing the Myc–ER fusion protein are treated with either estrogen or tamoxifen (but left in serum-deficient medium), these cells are induced to enter into the G_1 phase and progress to S phase (**Figure 8.29**). This illustrates the fact that Myc, acting on its own and in the absence of serum-associated mitogens, is able to relieve all of the constraints on proliferation that have held these cells in G_0 and is able to shepherd them all the way through G_1—an advance that usually requires substantial, extended stimulation by growth factors. In addition to Myc, a number of other transcription factors display the wide-ranging powers of oncoproteins (see Table 8.2).

As it turns out, Myc also acts on many other target genes that operate outside of the cell cycle machinery, as was made apparent by initial analyses of its functions in the

Figure 8.29 Powers of the Myc oncoprotein The wide-ranging effects of the Myc protein are illustrated by an experiment in which the Myc protein has been fused to the estrogen receptor (ER) protein (*blue*). In the absence of ER ligands, such as estrogen or tamoxifen, the Myc–ER protein is trapped in the cytoplasm (through association with heat shock proteins, *not shown*). When estrogen or tamoxifen ligands are added to cells, the Myc–ER protein migrates into the nucleus, associates with Max, and activates Myc target genes within minutes. Such activation, when induced in serum-starved cells in the G_0 phase, enables them to enter the active cell cycle and advance all the way through G_1 into the S phase, indicating that Myc can act at multiple points in G_0 and G_1 to drive cell cycle progression.

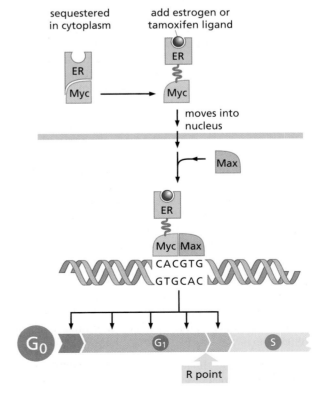

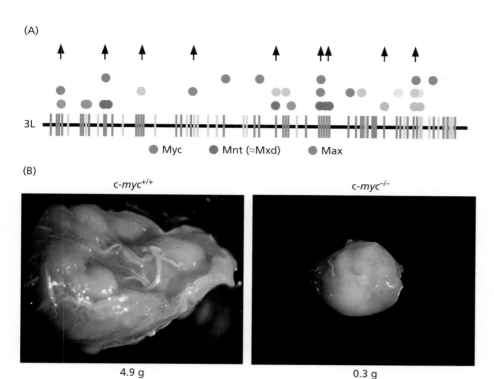

(A)

● Myc   ● Mnt (≈Mxd)   ● Max

(B)

c-myc⁺/⁺

c-myc⁻/⁻

4.9 g

0.3 g

Figure 8.30 **Wide-ranging actions of Myc** Myc has diverse effects on the cell that extend beyond its control of the cell cycle. (A) A systematic study of the target genes acted upon by the Myc, Max, and Mxd-related Mnt transcription factors in the cells of *Drosophila melanogaster* indicated that they are present in large numbers and are scattered throughout the four chromosomes of the *Drosophila* genome. This survey analyzed the ability of *Drosophila* Myc (*green circles*), the Mxd-like Mnt (*red circles*), and Max (*blue circles*) to associate with specific DNA segments throughout the fly's genome, but results for only the long arm of chromosome 3 (i.e., 3L) are shown here. Altogether, this survey indicated that about 2000 *Drosophila* gene promoters were associated with one or more of these bHLH proteins; a majority of these promoters are affiliated with genes that can be shown to be up- or down-regulated by pairs of these bHLH transcription factors. Arrows indicate genes that are orthologs of mammalian genes and known targets of Myc action in mammalian cells. (B) Other indications of the pleiotropic actions of Myc come from biological experiments, such as this one, in which the tumorigenicity of mouse embryonic stem (ES) cells has been gauged. Wild-type ES cells (*left*) form large, highly *vascularized* tumors (*orange vessels*), while cells lacking Myc function (*right*) fail to grow robustly, in large part because of a failure to develop an adequate blood supply; the two cell populations grow with equal rates *in vitro*. (A, adapted from A. Orian et al., *Genes Dev.* 17:1101–1114, 2003. B, from T.A. Baudino et al., *Genes Dev.* 19:2530–2543, 2002.)

cells of *Drosophila* (**Figure 8.30A**). Examination of many of these genes indicated that Myc acts widely within cells to promote cell *growth* in the narrow sense of the word, that is, increases in cell size. (In fact, the number of Myc-regulated genes that promote cell growth, including genes fostering protein biosynthesis and overall energy metabolism, dwarfs the number of Myc targets that regulate cell cycle progression.) This may have implications for mammalian cells as well: if a human cell is unable to proceed through the entire cell cycle unless it grows above a certain minimum size, Myc may facilitate cell proliferation by promoting increases in cell size. Such an effect may act synergistically with Myc's ability to regulate many of the key proteins that govern a cell's passage through the R point. In the context of cancer pathogenesis, Myc seems to have yet other cell-biological effects, as indicated by its influence on tumorigenesis in embryonic stem (ES) cells (see Figure 8.30B). To summarize, Myc appears to modulate the expression of thousands of mammalian genes; its roles in cell cycle progression and cell growth, as detailed here represent only one part of its multifaceted actions.

8.10 TGF-β prevents phosphorylation of pRb and thereby blocks cell cycle progression

TGF-β represents a major growth-inhibitory signal that normal cells, especially epithelial cells, must learn to evade in order to become cancer cells. As we will learn later in Chapters 13 and 14, TGF-β also exerts other, quite distinct effects on cells, by forcing them to change their differentiation programs. Often the resulting changes in cell phenotype actually favor tumor progression, as some of these phenotypic changes enable cancer cells to become anchorage-independent, angiogenic, and even invasive.

These two major effects of TGF-β—one antagonizing, the other favoring tumor progression—are clearly in direct conflict with one another. Cancer cells often resolve this dilemma by learning how to evade the cytostatic effects of TGF-β while leaving intact the other responses, such as those favoring tumor cell invasiveness. Not surprisingly, the cytostatic actions of TGF-β derive from its direct and indirect effects on pRb, the central controller that determines whether or not a cell will proliferate.

We previously learned that TGF-β has its own, quite unique receptor, which uses serine/threonine kinases in its cytoplasmic domains (rather than tyrosine kinases) to emit signals. The primary targets of these signals documented to date are proteins of the Smad transcription factor family. Once phosphorylated in the cytoplasm by a TGF-β receptor, a pair of Smad2 (or Smad3) protein molecules associates with a Smad4 protein (which is not a substrate for phosphorylation by the receptor), and the resulting heterotrimeric protein complex migrates to the nucleus, where it functions as a transcription factor (see Figure 6.29D).

As we read earlier, in the context of cell cycle control, the most important targets of action by TGF-β are the genes encoding the two CDK inhibitors p15^{INK4B} and, to a lesser extent, p21^{Cip1} (see Figure 8.14A). Each has a CAGAC sequence in its promoter that attracts a heterotrimeric Smad3–Smad4 transcription factor complex. On its own, such a Smad complex cannot activate transcription of these genes. Instead, it collaborates with yet another transcription factor—the Miz-1 factor mentioned in the last section—that binds to adjacent DNA sequences in the promoters of these two genes in order to activate their transcription (**Figure 8.31**, *left*).

We also noted earlier that the Myc oncoprotein, acting in the opposite direction, can block the induction of these two CDK inhibitors, doing so through its ability to associate with Miz-1. By repressing expression of the genes encoding p15^{INK4B} and p21^{Cip1}, Myc removes two major obstacles standing in the way of cell cycle progression (see Figure 8.28). Stated differently, the constitutively high levels of Myc protein made by *myc* oncogenes ensure that expression of these two CDK inhibitors is strongly repressed, thereby paving the way for vigorous cell cycle advance.

In normal cells, TGF-β needs to have the last word in determining whether or not these cells proliferate. In order to do so, TGF-β must overrule any conflicting signals

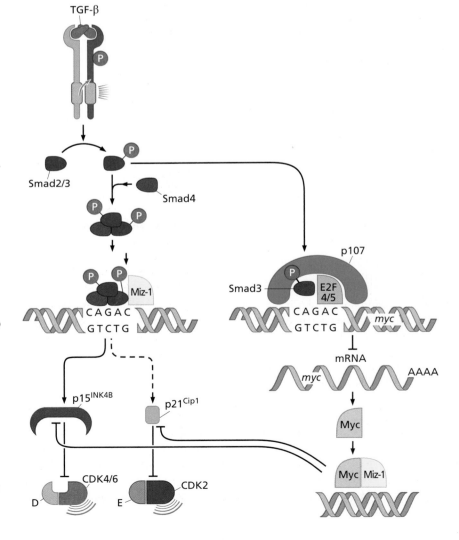

Figure 8.31 Countervailing actions of TGF-β and Myc TGF-β, acting through its receptor, causes phosphorylation of several Smad proteins, such as Smad2 and Smad3 (*left top*). Two molecules of phosphorylated Smad2/3 then form a heterotrimeric complex with Smad4, and the resulting complex then migrates to the nucleus, where it teams up with Miz-1 to induce expression of the p15^{INK4B} and (weakly) p21^{Cip1} CDK inhibitors. Myc, for its part, is capable of teaming up with Miz-1 to repress expression of these CDK inhibitors (*lower right*). However, this action of Myc can be preemptively blocked by TGF-β, which dispatches Smad3 to form a complex with E2F4 or E2F5 plus p107 (pRb's cousin; *right*) that represses expression of the *myc* gene, thereby causing Myc levels to collapse. This ensures that TGF-β succeeds in inducing expression of the two CDK inhibitors—p15^{INK4B} and p21^{Cip1}—and thereby shuts down cell cycle progression in the early/mid G$_1$ phase of the cell cycle.

that Myc might release. To guarantee success in this endeavor, TGF-β must ensure that Myc does not thwart its scheme to activate expression of the p15^{INK4B} and p21^{Cip1} CDK inhibitors. So, in normal cells, TGF-β keeps Myc away from the promoters of these genes by shutting down expression of the *myc* proto-oncogene (see Figure 8.31, *right*). The *myc* gene promoter has a sequence to which a TGF-β–activated Smad3 transcription factor can bind. Adjacent to this sequence is another that allows binding of either E2F4 or E2F5, the two E2F transcription factors that favor transcriptional repression (see Section 8.7). By forming a tripartite complex with E2F4/5 plus p107, Smad3 ensures the shutdown of *myc* transcription, thereby eliminating the growth-promoting effects of the Myc protein from the cell's regulatory circuitry. Once TGF-β has succeeded in removing Myc from the scene—from the promoters of the p15^{INK4B} and p21^{Cip1} CDK inhibitors—TGF-β can then use its Smad3–Smad4 complexes, working together with Miz-1, to activate these two promoters, thereby inducing expression of these critically important CDK inhibitors (see Figure 8.31, *left*).

Many types of cancer cells must evade TGF-β–imposed growth inhibition if they are to thrive. More specifically, such cancer cells depend on high levels of the Myc transcription factor to drive their proliferation and therefore must liberate *myc* transcription from the repressive actions of TGF-β. This explains why, for example, in a series of 12 human breast carcinomas, *myc* expression in all 12 of these tumors was no longer responsive to TGF-β–imposed shutdown even though many TGF-β–induced responses were found to be intact in 11 of these tumors. (Since the *myc* gene itself was apparently in a wild-type configuration in these tumors, this suggested some deregulation of the transcription factors responsible for expression of this gene.) Similarly, when *myc* oncogenes are formed by mutation, these genes acquire promoters that fire constitutively and thus are no longer responsive to TGF-β–induced repression. Once protected from the threat of this repression, the *myc* oncogene, more specifically, its Myc product, proceeds to hold expression of the two CDK inhibitors (that is, p15^{INK4B} and p21^{Cip1}) at a very low level and does so indefinitely. As noted earlier, this helps to create a condition in the cell permitting rapid proliferation.

In a more general sense, cancer cells place a high premium on escaping the growth-inhibitory influences of TGF-β while leaving other, potentially useful TGF-β responses intact. Often this evasion of TGF-β–mediated growth inhibition is achieved through inactivation of the pRb signaling pathway. In effect, in normal cells pRb operates as the brake lining that calls a halt to cell proliferation in response to TGF-β–initiated signals. Accordingly, if the pRb protein is eliminated from the regulatory circuitry through one of the various mechanisms described earlier, then the ability of TGF-β to impose growth arrest is greatly compromised, since the p15^{INK4B} induced by TGF-β now fails to effectively block cell cycle advance.

A partial inactivation of the TGF-β responses can also be achieved, for example, through mutations of the gene encoding the Smad2 protein; such mutations are frequently observed in colon carcinomas. A similar partial blunting of TGF-β–induced anti-mitogenic responses seems to be achieved by the Ski oncoprotein (see Table 3.3). Both this protein and its cousin, termed Sno, are able to bind to the Smad3–Smad4 transcription factor complex and block its ability to repress *myc* transcription.

Half of all pancreatic carcinomas and more than a quarter of all colon cancers carry mutant, inactivated Smad4 proteins. Without the presence of Smad4, neither Smad2–Smad4 nor Smad3–Smad4 complexes can form. These two complexes are the chief agents dispatched by the TGF-β receptor to the nucleus with the important assignment to shut down proliferation.

Some types of cancer cells undertake a far more drastic evasive maneuver, however: they jettison all responsiveness to TGF-β by inactivating the genes encoding its receptor. For example, the great majority of colon cancers that suffer "microsatellite instability" (a state leading to high rates of mutation; see Section 12.4) carry mutant, inactivated TGF-βII receptors. While such cells acquire the ability to evade TGF-β–imposed growth inhibition, they also forgo potential benefits that might be conferred by TGF-β later in tumor progression, when TGF-β helps cancer cells to acquire malignant phenotypes such as invasiveness.

8.11 pRb function and the controls of differentiation are closely linked

Cell differentiation is a process of crucial importance to cancer pathogenesis and, as such, is mentioned repeatedly in this text. For the moment, we can imagine, a bit simplistically, that cells throughout the body can exist in either of two alternative growth states. They may be found in a relatively undifferentiated state in which they retain the option to divide in the event that mitogenic signals call for their proliferation; this is essentially the behavior of stem cells. The alternative is that cells leave this state and enter into a more differentiated state, whereupon they give up the option of ever proliferating again and thus become post-mitotic. The decisions that govern these fundamental changes in cell biology must be explainable in terms of the molecular controls that determine whether a cell remains in the active growth-and-division cycle, exits reversibly into G_0, or exits irreversibly from this cycle into a post-mitotic, differentiated state.

The opposition between cell proliferation and differentiation is seen most clearly in the formation of most types of cancer cells in which differentiation is partially or completely blocked (Sidebar 8.6). Knowing this, we might ask whether two independent

Sidebar 8.6 Blocked differentiation can accompany the advance of tumor progression Some hematopoietic malignancies, specifically, leukemias, come in two forms—chronic and acute. The chronic diseases, such as chronic myelogenous leukemia (CML), are composed of more mature, differentiated cells and can proceed for years without becoming life-threatening (Figure 8.32). In the case of CML, after a period of 3 to 5 years, suddenly there is an eruption of a more malignant, aggressive form of the disease that is termed "blast crisis." The cells present in blast crisis clearly originate from the same cell clone that initiated the chronic phase of the disease. However, they are less differentiated and multiply ceaselessly, creating a life-threatening disease that is difficult to treat. It seems that the cells in the chronic phase are derived from a mutant self-renewing stem cell, but this disease remains relatively benign, because the differentiation of these mutant cells into post-mitotic neutrophils proceeds quite normally. The moment that such differentiation is blocked [by some still-unknown genetic change(s)], cells in this population are trapped in a less differentiated stem cell compartment (of either erythroid or myeloid type) and proliferate without a compensating exit of their progeny into a more differentiated, post-mitotic compartment.

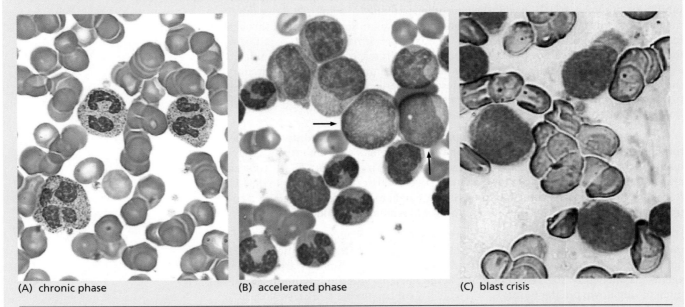

(A) chronic phase   (B) accelerated phase   (C) blast crisis

Figure 8.32 Disease progression and dedifferentiation (A) The disease of chronic myelogenous leukemia (CML) leads to the accumulation in the blood of elevated numbers of cells that are almost fully differentiated neutrophils (*purple cells, complex nuclei*). The *pink* cells are erythrocytes. (B) After several years, when the chronic phase of the disease progresses to an "accelerated" phase, these more differentiated neutrophils are increasingly replaced by less differentiated blast cells (*arrows*). (C) Ultimately, this accelerated phase progresses to "blast crisis," when the peripheral blood carries large numbers of undifferentiated blast cells. This last phase appears to reflect the behavior of leukemic stem cells that were already present in the chronic phase of the disease but are now prevented from differentiating. (A and B, courtesy of P.G. Maslak. C, courtesy of the American Society of Hematology Image Bank.)

changes in the cellular control circuitry must be made during the formation of a cancer cell—one that deregulates control of proliferation and another that blocks differentiation. The alternative is simpler: might certain alterations, in a single stroke, deregulate proliferation and prevent differentiation?

Evidence supporting the latter idea, which implies a close coupling between the mechanisms controlling these two processes, has been forthcoming from studies of retinal development *in vivo* and muscle cell differentiation *in vitro*. In the case of the retina, the absence of pRb function in retinal precursor cells prevents the proper differentiation of the rod photoreceptor cells (**Figure 8.33A**). In this micrograph, the *Rb* gene was inactivated sporadically in retinal precursor cells, resulting in patches of cells where subsequent rod development was defective.

An experimental model of muscle development is equally revealing about the connection between cell cycle progression and differentiation. Relatively undifferentiated **myoblasts** can be induced *in vitro* to differentiate into **myocytes** (muscle cells). An enforced block in the G_1 phase of the myoblast cell cycle, achieved experimentally by introducing genes encoding elevated levels of either the $p16^{INK4A}$ or the $p21^{Cip1}$ CDK inhibitor, causes the myoblasts to differentiate under conditions (for example, high levels of growth factors in the medium) in which such differentiation would otherwise not occur (see Figure 8.33B). Genetically altered mouse myoblasts that lack both *Rb* gene copies, and therefore express no pRb, are unresponsive to the differentiation-inducing influences of the introduced 16^{INK4A}- and $p21^{Cip1}$-encoding genes. Additionally, overexpression of cyclin D1 in myoblasts (which drives pRb phosphorylation and inactivation) blocks the differentiation that is normally seen when growth factors are removed from their culture medium. These experiments indicate that hypophosphorylated pRb is needed for two ostensibly unrelated functions—halting the proliferation of myoblasts and facilitating the differentiation of these cells into myocytes.

The converging effects on differentiation of these various regulators is illustrated most dramatically by the behavior of granulocytes that have been deprived of both $p27^{Kip1}$ and Mxd: $p27^{Kip1}$ cannot be mustered to halt G_1 cell cycle progression, and the differentiation-blocking effects of Myc cannot be reversed by its prime antagonist, Mxd. The result is a profound failure of differentiation (see Figure 8.33C).

Yet other molecular connections between cell cycle control and differentiation have been uncovered. Thus, enforced expression of CDK inhibitors, such as $p21^{Cip1}$ or $p27^{Kip1}$, is able to induce differentiation of neuroblastoma cells, primary neurons (in frogs), and myelomonocytic leukemia cells, while enforced expression of E2F4 (a repressive E2F; see Section 8.7) induces the differentiation of neuronal precursor cells. Mice that are genetically deprived of this transcription factor (and thus have an $E2f4^{-/-}$ genotype) die in gestation because of defective differentiation of their erythrocytes. The C/EBPβ transcription factor associates with pRb to induce the differentiation of adipocytes (fat cells), while the CBFA1/Runx2 transcription factor has been found to bind pRb and collaborate with the latter to program osteoblast differentiation. Expression of $p27^{Kip1}$'s cousin, $p57^{Kip2}$, is important for stabilizing MyoD, a transcription factor that helps to program muscle differentiation. Finally, retinoic acid induces the differentiation of acute promyelocytic leukemia (APL) cells, in part through its ability to trigger degradation of cyclin D1.

These diverse experimental observations indicate that the machinery controlling the phosphorylation state of pRb and its cousins is also an active participant in the differentiation process, at least in the well-studied biological systems described above. More specifically, signals favoring the hyperphosphorylation (and thus functional inactivation) of the three pocket proteins (pRb, p107, and p130) operate to block differentiation, while those that prevent pRb hyperphosphorylation favor differentiation. This mechanistic linkage, still poorly documented for most cell types, suggests another idea that remains largely speculative: when a cell decides to leave the active cell cycle in order to differentiate, it exits the cycle sometime during the G_1 phase, when pRb is active in controlling the fate of the cell.

Yet another line of evidence indicating a coupling between differentiation and the cell cycle machinery comes from research on the biochemistry of the Myc protein. As mentioned earlier (see Section 8.9), the Myc protein shifts the balance between proliferation and post-mitotic differentiation in favor of proliferation. In order to understand

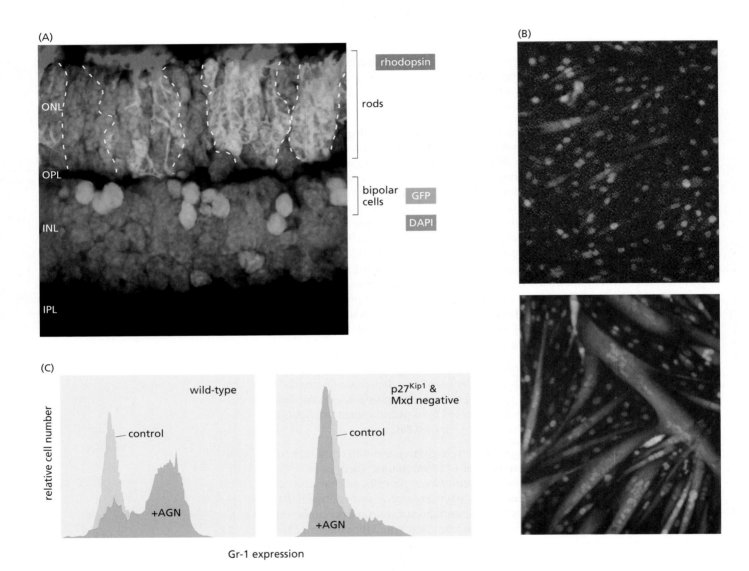

Figure 8.33 Differentiation of granulocytes, retinal cells, and muscle cells (A) The proper differentiation of retinal cells (see Figure 7.3A) depends on the presence of pRb. During normal development, retinal progenitor cells move upward and differentiate into various specialized cell types. In the eyes of 2-week-old mice, rod cells that successfully differentiated express the rhodopsin photoreceptor protein (*red*), while those that failed to do so (*dashed lines*) are represented only by the DAPI stain that marks their DNA (*blue*). The latter cells sit directly above GFP-labeled bipolar cells that mark the locations where progenitor cells had previously excised both copies of their *Rb* gene prior to migrating upward. (B) Myoblasts, the less differentiated precursors of muscle cells (myocytes), can be cultured *in vitro* and remain in the undifferentiated state (*upper panel*). However, various types of physiologic signals can induce them to differentiate into myocytes, whereupon they fuse to form muscle fibers (*red, lower panel*). For example, removal of serum (and its associated growth factors) from the culture medium leads to such differentiation.

Numerous experiments have shown that differentiation can be prevented by forcing pRb phosphorylation and inactivation. Conversely, signals that prevent pRb phosphorylation favor and often induce differentiation. (C) The collaborative actions of Mxd (the Myc antagonist; see Figure 8.27A) and p27^{Kip1} in promoting differentiation are revealed by the behavior of promyelocytes whose differentiation into Gr-1 antigen–positive granulocytes can be induced by an agonist of the RXR nuclear receptor. Here this agonist, termed AGN194024 (AGN), was used to induce differentiation of wild-type promyelocytes (*left panel*) or promyelocytes that were genetically deprived of the genes encoding Mxd and p27^{Kip1} (*right panel*). The display of the Gr-1 differentiation antigen was increased >20-fold in a majority of the treated wild-type cells, while being essentially unaffected (if not slightly reduced) in the doubly deficient cells. (A, from S.L. Donovan and M.A. Dyer, *Vision Res.* 44:3323–3333, 2004. B, from E.M. Wilson et al., *Mol. Biol. Cell* 15:497–505, 2004. C, from G.A. McArthur et al., *Mol. Cell Biol.* 22:3014–3023, 2002.)

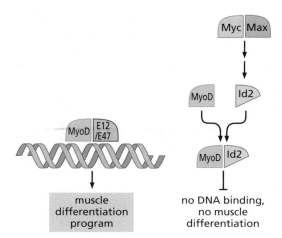

Figure 8.34 Id transcription factors and inhibition of differentiation
Lineage-specific bHLH transcription factors, such as MyoD (*beige*), form heterodimers with widely expressed E12 or E47 bHLH partners (*blue*). The resulting heterodimeric transcription factors orchestrate differentiation programs in a variety of tissues, including muscle (*left*). The formation of these functional heterodimeric transcription factor complexes can, however, be blocked by Id proteins (*light gray*), which form heterodimers with the lineage-specific bHLH proteins, thereby preventing association of the latter with E12 or E47. Because the Id proteins lack a DNA-binding domain, they act as natural "dominant-negative" inhibitors of the lineage-specific bHLH proteins. Expression of certain Id proteins, such as the Id2 shown here, can be induced by Myc (or N-myc) acting with its partner Max (*top right*); this helps to explain the observed ability of the Myc oncoprotein to block differentiation of various cell types.

these actions, we need to know that many of the transcription factors controlling differentiation programs are, like Myc, members of the large family of bHLH transcription factors. However, these bHLH proteins function very differently from Myc: they orchestrate complex, tissue-specific differentiation programs, while Myc acts in the opposite direction to block differentiation and promote proliferation. Differentiation-associated bHLH transcription factors have been most extensively studied in the embryonic development of several distinct cell lineages, including those leading to the formation of muscle, the nervous system, the pancreas, and the immune system. For example, four distinct bHLH proteins—termed MyoD (see above), Myf5, myogenin, and MRF4—operate in myoblast-to-myocyte differentiation (see Figure 8.33) by controlling various phases of the muscle-specific differentiation program.

Myc works to increase production of the Id1 and Id2 proteins, and the latter act as antagonists of the bHLH transcription factors that program differentiation. Id1 and Id2 are members of a group of related proteins (Id1 to Id4) that are also members of the bHLH family of transcription factors. The Id proteins operate, however, as dominant-negative inhibitors of the other bHLH transcription factors (**Figure 8.34**). More specifically, the Ids can form heterodimers with bHLH transcription factors but lack the ability to bind to DNA (because they lack the basic domain involved in DNA recognition). This explains how the Id proteins can act as inhibitors of the bHLH differentiation-inducing transcription factors, thereby blocking differentiation.

Id proteins are present at high levels in many types of actively growing cells, and this in itself reduces the likelihood that these cells will differentiate. For instance, by associating with MyoD, an Id protein can prevent MyoD from programming muscle differentiation in actively growing myoblasts. During the normal course of differentiation, however, the levels of Id proteins sink to undetectably low levels, and MyoD, now free of interference by Ids, is able to dimerize with its bHLH partners (called E12 and E47) in order to activate the muscle-specific differentiation program (see Figure 8.34). Depressed synthesis of the Id1 protein is known to be required for cell cycle withdrawal and differentiation in numerous cell lineages: muscle, pancreas, mammary epithelial cells, myeloid cells, erythroid cells, myocardial cells, B cells, T cells, and osteoblasts.

As might be expected from the above, the Id proteins also have been associated with cancer pathogenesis. In many types of normal cells, the molecules of Id2 are bound and sequestered by the far more abundant molecules of pRb. However, in neuroblastomas, a relatively common pediatric tumor, Id2 is often overexpressed because its expression is driven by extra copies of the N-Myc protein, a *myc* cousin that acts on the same targets as *myc* in cells (see Section 4.5). Now the tables are turned: in neuroblastoma cells, Id2 accumulates to such high levels that it is in great (>10×) molar excess of pRb. Consequently, pRb can no longer sequester and regulate these cells' Id2 proteins.

They are then free to block the actions of differentiation-inducing bHLH transcription factors. More recently, an entire different dimension of regulation has been uncovered: in certain tumors, degradation of the usually highly labile Id proteins is blocked, leading to great increases in their concentration and, in turn, blocked differentiation. This mechanism may underlie the relatively poorly differentiated state of many types of tumor cells (for example, see Supplementary Sidebar 8.4).

A dramatic demonstration of the opposition between the Myc oncoprotein and cell differentiation has come from a mouse model of liver cancer pathogenesis, which depends on the targeted expression of a *myc* transgene in hepatocytes. Large hepatocellular carcinomas form, and these regress when the *myc* transgene is shut down. At the same time, many of the carcinoma cells, which previously lacked most of the traits of normal hepatocytes, rapidly differentiate into normal-appearing liver cells that assemble to reconstruct many of the histological features of the normal liver.

These various controls on cell differentiation, involving pRb, Myc, Ids, and other regulatory proteins, clearly have effects on the formation and development of various types of cancer, since tumors formed by more differentiated cells are usually less aggressive while those composed of poorly differentiated cells tend to be far more aggressive and carry a worse prognosis for the patient.

8.12 Control of pRb function is perturbed in most if not all human cancers

Deregulation of the pRb pathway yields an outcome that is an integral part of the cancer cell phenotype—unconstrained proliferation. This explains why normal regulation of the R-point transition, as embodied in pRb phosphorylation, is likely to be disrupted in most if not all types of human tumor cells (**Tables 8.3 and 8.4**). These disruptive mechanisms are summarized in **Figure 8.35**. We are already familiar with the most direct mechanism for deregulating advance through the R point—inactivation of the *Rb gene* through mutation. In some tumors, an equivalent outcome is achieved through methylation of the *Rb* gene promoter. In others, pRb, though synthesized in normal amounts, may be functionally inactivated by viral oncoproteins, such as the HPV E7 protein, which prevent pRb from binding and regulating E2Fs.

Yet another strategy used by cancer cells to inactivate pRb function is indicated by the presence of very high levels of cyclin D1 in a variety of human tumor cells. This is most widely documented in breast cancer cells, in which as many as half of the tumors

Figure 8.35 Perturbation of the R-point transition in human tumors The decision to advance through the R-point transition (*yellow, middle, bottom*) can be perturbed in a variety of ways in human tumors. Elements that favor advance through the R point are drawn in *red*, while those that undertake to block this advance are shown in *blue*. Almost all human tumors show either a hyperactivity of one or more of the agents favoring this advance (*red*) or an inactivation of the agents blocking this advance (*blue*). Proteins whose expression levels or activities are not known to change during the process of transformation are shown in *gray*.

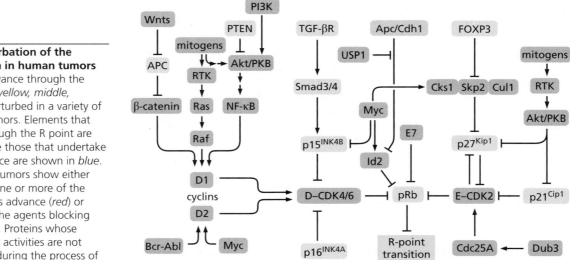

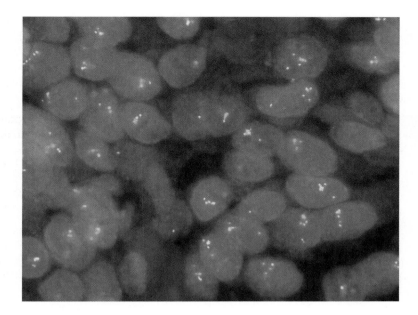

Figure 8.36 Amplification of the cyclin D1 gene The use of fluorescent *in situ* hybridization (FISH) makes it possible to detect the copy number of specific genes in histological sections. Here is the result of using a probe for *CCND1*, the human gene encoding cyclin D1, to determine the copy number of this gene in the cells of a head-and-neck squamous cell carcinoma (HNSCC). Each bright spot represents a single copy of this gene. Individual HNSCC nuclei are stained *purple*. Hence, *CCND1* is amplified to various extents here, being present in three to five copies per cell in this tumor. More generally, *CCND1* is found to be amplified in about one-third of HNSCC tumors, leading to corresponding increases in cyclin D1 expression and resulting loss of proper control of pRb phosphorylation. Similar observations (*not shown*) can be made with human breast cancer cells. (From K. Freier et al., *Cancer Res.* 61:1179–1182, 2003.)

have been reported to show elevated levels of this protein. In these and other carcinomas, the overexpression is sometimes achieved by increases in the copy number of the cyclin D1 genes (that is, gene amplification; **Figure 8.36**). More frequently, however, breast cancer cells acquire excessive cyclin D1 by altering the upstream signaling pathways (see Section 8.8) that are normally responsible for controlling expression of the cyclin D1 gene.

A more devious ploy is frequently exploited by cancer cells to disable the pRb machinery: they shut down expression of their p16^INK4A tumor suppressor protein. Recall that the p16^INK4A protein, like its p15^INK4B cousin, inhibits the cyclin D–CDK4/6 responsible for initiating pRb phosphorylation. In the absence of the p16^INK4A protein, pRb phosphorylation operates without an important braking mechanism, resulting in excessive cyclin D–CDK4/6 kinase activity, deregulated pRb phosphorylation, and inappropriate inactivation of pRb (see Figure 8.35). Individuals suffering from one form of familial melanoma inherit defective versions of the p16^INK4A gene. It is unclear why loss of this particular CDK inhibitor, which seems to operate in all cell types throughout the body, should affect specifically the melanocytes in the skin that are the normal precursors of melanoma cells. In sporadic (that is, nonfamilial) tumors of various sorts, cancer cells resort far more frequently to another strategy to shed p16^INK4A function—they methylate the CpG sequences present in the promoter of the p16^INK4A gene (see Section 7.8).

Evidence of an even more cunning strategy for destabilizing this control circuit has been found in the genomes of a small number of both sporadic and familial melanomas. In these cancers, point mutations in the CDK4 gene (the R24C mutation) create CDK4 molecules that are no longer susceptible to inhibition by the family of INK4 molecules (that is, p15, p16, p18, and p19). While these various CDK inhibitors may be perfectly intact and functional in such tumor cells, their normally responsive CDK4 target now eludes them. Once again, this permits CDK4, together with its cyclin D partners, to drive the initial steps of pRb phosphorylation in a deregulated fashion. [Since the R24C mutation creates a dominant allele of CDK4 (at the cellular level), only one of the two copies of the gene encoding this CDK needs to be mutated in order for a cancer cell to derive proliferative benefit. Mice carrying one or two copies of this R24C allele develop a diverse variety of tumors, including those affecting mesenchymal, epithelial, and hematopoietic cell types.]

The most critical CDK inhibitor involved in cancer pathogenesis may well be p27^Kip1. As mentioned earlier, p27^Kip1 is involved largely in inhibiting the activity of cyclin

E–CDK2 complexes (for example, see Figure 8.15). As cells exit the cell cycle into the G_0 quiescent state, $p27^{Kip1}$ levels rise (**Figure 8.37A**). Conversely, as cells re-enter the cell cycle and advance through its G_1 phase, the levels of $p27^{Kip1}$ are reduced progressively throughout early and mid-G_1 and then are made to fall precipitously during late

Table 8.3 Molecular changes in human cancers leading to deregulation of the cell cycle clock

Specific alteration	Clinical result
Alterations of pRb	
Inactivation of the *Rb* gene by mutation	retinoblastoma, osteosarcoma, small-cell lung carcinoma
Methylation of *Rb* gene promoter	brain tumors, diverse others
Sequestration of pRb by Id1, Id2	diverse carcinomas, neuroblastoma, melanoma
Sequestration of pRb by the HPV E7 viral oncoprotein	cervical carcinoma
Alteration of cyclins	
Cyclin D1 overexpression through amplification of *cyclin D1* gene	breast carcinoma, leukemias
Cyclin D1 overexpression caused by hyperactivity of *cyclin D1* gene promoter driven by upstream mitogenic pathways	diverse tumors
Cyclin D1 overexpression due to reduced degradation of cyclin D1 because of depressed activity of GSK-3β	diverse tumors
Cyclin D3 overexpression caused by hyperactivity of *cyclin D3* gene	hematopoietic malignancies
Cyclin E overexpression	breast carcinoma
Defective degradation of cyclin E protein due to loss of hCDC4	endometrial, breast, and ovarian carcinomas
Alteration of cyclin-dependent kinases	
CDC25A overexpression	breast cancers
CDK4 structural mutation	melanoma
Alteration of CDK inhibitors	
Deletion of *p15^{INK4B}* gene	diverse tumors
Deletion of *p16^{INK4A}* gene	diverse tumors
Methylation of *p16^{INK4A}* gene promoter	melanoma, diverse tumors
Decreased transcription of *p27^{Kip1}* gene because of action of Akt/PKB on Forkhead transcription factor	diverse tumors
Increased degradation of $p27^{Kip1}$ protein due to Skp2 overexpression	breast, colorectal, and lung carcinomas, and lymphomas
Cytoplasmic localization of $p27^{Kip1}$ protein due to Akt/PKB action	breast, esophagus, colon, thyroid carcinomas
Cytoplasmic localization of $p21^{Cip1}$ protein due to Akt/PKB action	diverse tumors
Multiple concomitant alterations by Myc, N-myc, or L-myc	
Increased expression of Id1, Id2 leading to pRb sequestration	diverse tumors
Increased expression of cyclin D2 leading to pRb phosphorylation	diverse tumors
Increased expression of E2F1, E2F2, E2F3 leading to expression of cyclin E	diverse tumors
Increased expression of CDK4 leading to pRb phosphorylation	diverse tumors
Increased expression of Cul1 leading to $p27^{Kip1}$ degradation	diverse tumors
Repression of $p15^{INK4B}$ and $p21^{Cip1}$ expression allowing pRb phosphorylation	diverse tumors

Table 8.4 **Alteration of the cell cycle clock in human tumors** A plus sign indicates that this gene or gene product is altered in at least 10% of tumors analyzed. Alteration of gene product can include abnormal absence or overexpression. Alteration of gene can include mutation and promoter methylation. More than one of the indicated alterations may be found in a given tumor.

Tumor type	Gene product or gene						% of tumors with 1 or more changes
	Rb	Cyclin E1	Cyclin D1	p16^{INK4A}	p27^{Kip1}	CDK4/6	
Glioblastoma	+	+		+	+	+/+	>80
Mammary carcinoma	+	+	+	+	+	+/	>80
Lung carcinoma	+	+	+	+	+	+/	>90
Pancreatic carcinoma			a		+		>80
Gastrointestinal carcinoma	+	+	+[b]	+	+	+/[c]	>80
Endometrial carcinoma	+	+	+	+	+	+/	>80
Bladder carcinoma	+	+	+	+	+		>70
Leukemia	+	+	+	+[d]	+	+/	>90
Head and neck carcinomas	+		+	+	+	+/	>90
Lymphoma	+	+	+[e]	+[d]	+	/+	>90
Melanoma		+	+	+	+	+/	>20
Hepatoma	+	+	+	+[d]	+	+/[c]	>90
Prostate carcinoma	+	+	+	+	+		>70
Testis/ovary carcinomas	+	+	+[b]	+	+	+/	>90
Osteosarcoma		+		+		+/	>80
Other sarcomas		+	+	+	+	/+	>90

[a]Cyclin D3 (not cyclin D1) is present and is up-regulated in some tumors.
[b]Cyclin D2 also is up-regulated in some tumors.
[c]CDK2 is also found to be up-regulated in some tumors.
[d]p15^{INK4B} is also found to be absent in some tumors.
[e]Cyclin D2 and D3 are also found up-regulated in some lymphomas.
Adapted from M. Malumbres and M. Barbacid, *Nat. Rev. Cancer* 1:222–231, 2001.

G$_1$ phase by the actions of cyclin E–CDK2 complexes. These low levels are not created by reduced transcription of the gene encoding p27^{Kip1}. Instead, it seems p27^{Kip1} levels are reduced by the actions of the Skp2 protein, which acts together with Cul1 and several other proteins (see Figure 8.37B) to recognize p27^{Kip1} and ubiquitylate it, thereby tagging it for destruction in proteasomes. Indeed, the declining levels of Skp2 explain the increase in p27^{Kip1} as cells enter into G$_0$ (see Figure 8.37A and C). Interestingly, a very similarly structured complex (Supplementary Sidebar 8.5) is involved in the programmed degradation of cyclin E as cells pass through the G$_1$/S transition (see Figure 8.10).

The inverse relationship between p27^{Kip1} levels and those of Skp2 is especially apparent in various human cancers such as mammary and oral carcinomas, as well as lymphomas (see, for example, Figure 8.37D), with higher levels of Skp2 portending shorter patient survival. Moreover, when the levels of p27^{Kip1} are measured in the cells of human esophageal, breast, colorectal, and lung carcinomas, poor patient survival is correlated with low levels of this CDK inhibitor. In all of these tumors, Skp2 seems to act as a proliferation-promoting oncoprotein by forcing the degradation of the critical p27^{Kip1} CDK inhibitor. One clue to the ultimate source of elevated Skp2 levels comes from research indicating that the Notch protein (see Section 6.12), which is hyperactive in many types of human cancer, increases transcription of the gene encoding Skp2.

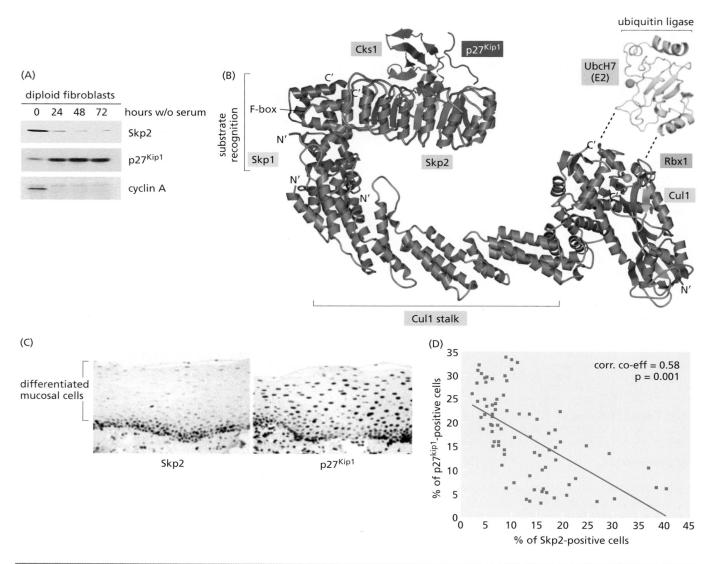

Figure 8.37 Suppression of p27^Kip1 levels by Skp2 Levels of p27^Kip1 are controlled by, among other factors, the levels of the SCF^Skp2 complex, which constitutes an E3 ubiquitin ligase that ubiquitylates p27^Kip1, thereby tagging it for degradation in proteasomes. (A) As normal human cells enter into the G₀, quiescent state, in this case because they are deprived of serum mitogens, Skp2 levels fall, allowing the accumulation of p27^Kip1; the latter, in turn, is responsible for blocking the actions of CDK2. The loss of cyclin A demonstrates that cells have retreated from the active cell cycle. This CDK is responsible for pushing the cell through the late G₁ phase of the cell cycle. (B) After p27^Kip1 is phosphorylated by cyclin E–CDK2 (see Figure 8.25B), it is recognized and bound by a large multiprotein complex that ubiquitylates it, thereby targeting it for destruction in proteasomes. The phosphorylated domain of p27^Kip1 (*dark green*) associates with the Cks1 adaptor protein (*purple*) and the "palm" of Skp2 (*dark pink*). This complex is connected to the ubiquitylating complex via the Cul1 stalk (*green*), which holds the enzyme (*right*) 100 angstroms away from its p27^Kip1 substrate (*left*). Covalent

modifications of the Cul1 stock cause it to bend, allowing the E2 ubiquitin ligase to access its p27^Kip1 substrate. Myc fosters cell cycle progression in part by inducing expression of Cul1, Csk1, and Skp2, which proceed to ubiquitylate p27^Kip1, thereby driving its degradation. (The UbcH7 shown here is used to represent the structurally related E2 ubiquitin ligase that actually functions in this complex.) (C) As seen here, in the normal oral mucosa, when Skp2 levels are low (*left*), levels of p27^Kip1 are high (*right*) and help to hold cells in a post-mitotic state; the opposite situation is observed in many cancers (*not shown*). (D) The inverse relationship between Skp2 levels and those of p27^Kip1 is plotted here for a large number of oral epithelial dysplasias and carcinomas. (Similar relationships, *not shown*, are seen in lymphomas and colorectal carcinomas.) (A, from T. Bashir et al., *Nature* 428:190–193, 2004. B, from B. Schulman et al., *Nature* 408:381–386, 2000; N. Zheng et al., *Cell* 102:533–539, 2000; N. Zheng et al., *Nature* 416:703–709, 2002; and B. Hao et al., *Mol. Cell* 20:9–19, 2005. C and D, from M. Gstaiger, *Proc. Natl. Acad. Sci. USA* 98:5043–5048, 2001.)

We have read repeatedly about the hyperactive state of the Akt/PKB kinase in many human tumors, which is caused by a variety of molecular defects, including hyperactive growth factor receptors, loss of the PTEN tumor suppressor protein, and mutations in the PI3K gene. It, too, has effects on the cell cycle clock in cancer cells, by reducing the effective levels of important CDK inhibitors (see Figure 8.15). By phosphorylating p21^{Cip1} and p27^{Kip1}, Akt/PKB ensures the cytoplasmic localization of these two critical antagonists of cell cycle advance, thereby marginalizing them. At the same time, Akt/PKB can suppress expression of the gene encoding p27^{Kip1} by phosphorylating a transcription factor of the Forkhead family, which serves to further reduce the overall concentrations of p27^{Kip1} in the cell. As if this were not enough, Akt/PKB ambushes p27^{Kip1} in a third way: by phosphorylating Skp2 (see Figure 8.37B), Akt/PKB activates the latter, enhancing its ability to bind p27^{Kip1}, resulting thereafter in the ubiquitylation and degradation of p27^{Kip1}.

In Section 8.9, we learned of the various ways in which the Myc protein, operating in normal cells, extends its reach into the cell cycle clock and pulls many key regulatory levers. The actions of the Myc oncoprotein are likely to be qualitatively similar. The difference between the two is largely, and possibly entirely, a matter of protein level: in normal cells, levels of Myc protein are highly dependent upon extracellular mitogenic growth factors. In many cancer cells, however, the Myc protein is produced constitutively, independent of mitogenic signals coming into the cell.

With this idea in mind, we can deduce that the many perturbations wrought by the Myc protein are magnified in the many types of human cancer cells that carry either *myc* oncogenes or the oncogenic versions of its cousins, N-*myc* and L-*myc*. Among the many products induced by Myc, the Id proteins may well have the greatest physiologic importance. On the one hand, they can act as dominant-negative inhibitors of bHLH transcription factors that program cell differentiation. On the other, they can bind to pRb and seem to inhibit its functioning in a way that is reminiscent of the actions of viral oncoproteins such as large T, E7, and E1A.

In fact, ectopically expressed Id2 protein can replace the pRb-binding actions of SV40 large T antigen in experiments designed to reverse the quiescent growth state of cells. This particular Id protein has been reported to be overproduced in diverse tumors including endometrial, head-and-neck, breast, pancreatic, esophageal, and cervical carcinomas, as well as melanomas and neuroblastomas. Its expression is positively correlated with the degree of malignancy and invasiveness in many of these tumor types. The Id proteins are synthesized at high rates and degraded rapidly, resulting in relatively short half-lives (less than 20 minutes for Id1, Id2, and Id3) and thus low steady-state concentrations in cells. This suggests that their concentrations might be greatly increased in cancer cells by mechanisms that reduce their ubiquitylation and thus proteasome-mediated degradation; indeed precisely such mechanisms have been documented in a variety of cancer cells (see Supplementary Sidebar 8.4).

The diverse genetic and biochemical strategies shown in Figure 8.35 are all focused on one common goal—that of overwhelming and deregulating pRb function, thereby destroying the tight control that it normally imposes on the R-point transition. Darwinian selection, occurring in the microcosm of living tissues, favors the outgrowth of cells that, by hook or by crook, have succeeded in inactivating the critical pRb braking system and thus deregulating the R-point transition.

8.13 Synopsis and prospects

All of the physiologic signals and signaling pathways affecting cell proliferation must, sooner or later, be connected in some fashion to the operations of the cell cycle clock. It represents the brain of the cell—the central signal processor that receives afferent signals from diverse sources, integrates them, and makes the final decisions concerning growth versus quiescence, and in the latter case, whether or not the exit from the active cell cycle will be reversible. Connected with the latter decision are the mechanisms governing entrance into tissue-specific differentiation programs.

The core components of the cell cycle clock are already present in clearly recognizable form in single-cell eukaryotes, including the much-studied baker's yeast, *Saccharomyces cerevisiae*. Single-cell organisms such as this one respond to a far smaller range of external signals than do metazoan cells residing within complex tissues. These simple organisms lack the hundred and more distinct types of growth factor receptors that vertebrate cells display on their surfaces, as well as other receptors, such as integrins, that our cells use to sense and control attachment to the extracellular matrix. Yeast cells also lack the growth-inhibitory receptors, such as the TGF-β receptors, that play such a critical role in the economy of mammalian tissues.

All this explains why the peripheral wiring that regulates the core cell cycle machinery of animal cells has been added relatively recently in the history of life on this planet—perhaps 600 million years ago when metazoa may first have appeared. The need to respond to a wide variety of afferent signals explains why so many distinct layers of regulation have been imposed on the core machinery. Without these additional regulators, notably the CDK inhibitors, the core machinery could not be made responsive to the diverse array of signals that impinge on individual metazoan cells and modulate their proliferation.

While these connections between the cell exterior and the cell cycle clock were being forged, other critical regulators became integrated into this complex circuitry. Actually, the invention of the key governors of G_1 progression—pRb and its two cousins, p107 and p130—seems to have occurred well before the rise of metazoa: a pRb–E2F signaling pathway involved in cell cycle control is already present in *Chlamydomonas reinhardtii*, a single-cell alga that is related to the ancestors of land plants. (Indeed, manipulations of the orthologs of the mammalian p27^{Kip1} gene in soybeans and canola/rapeseed have led to significantly increased crop yields!) Similarly constructed pRb- and E2F-like proteins are present in worms and flies, indicating their presence in early metazoans.

pRb may not have been the first of these three pocket proteins to have evolved, but during the ascendance of mammals, it surpassed the others in its ability to govern the critical decision made at the R point. These three proteins preside over various aspects of the growth-versus-quiescence decision of the cell and in this sense lie *upstream* of the cell cycle clock. At the same time, by affecting gene transcription, these proteins create a coupling between the cell cycle clock and the *downstream* circuits that must be activated in order for cells to execute the complex biochemical changes that enable them to enter into S phase. [An unresolved question is: since pRb controls cell proliferation in many cell types throughout the body, why does heterozygosity at the *RB* locus predispose humans specifically to retinal tumors and osteosarcomas (Supplementary Sidebar 8.6)?]

Some of the neoplastic growth state can be explained by the workings of the cell cycle clock. We can explain the deregulated proliferation of cancer cells in terms of the operations of pRb and the molecules that control its state of phosphorylation. Without pRb at the helm, the requirement for the growth-promoting actions of oncoproteins such as Ras is greatly reduced, and cells advance through G_1 without fulfilling many of the prerequisites that normally determine whether or not the R-point transition will proceed. The coupling between proliferation and blocked differentiation, still incompletely understood, seems to be traceable to the operations of pRb and proteins such as Myc, which simultaneously drives the cell cycle clock forward through G_1 and antagonizes some of the master regulators of differentiation programs.

Perhaps surprisingly, one important aspect of the proliferative control of cells operates independently of the cell cycle clock. As we will learn in detail in Chapter 10, normal cells can replicate only a limited number of times before they lose the ability to proliferate further; cancer cells have an unlimited proliferative capacity—the phenotype of **replicative immortality**. The molecular devices within cells that tally the number of replicative generations through which a cell lineage has passed are embedded in the chromosomal DNA, and these devices do not seem to be controlled by the cyclin–CDK complexes regulating cell cycle advance.

Yet other aspects of the malignant growth program are not controlled by the cell cycle clock. Cells respond to severe, essentially irreparable genomic damage by activating their cell suicide program—apoptosis; this response does not seem to be connected directly to the cell cycle machinery. This program will be the subject of much of our discussions in the next chapter.

Many other peculiarities of the cancer cell phenotype are largely the purview of cytoplasmic oncoproteins such as Ras. Included here are phenotypes of cell motility, changes in cell shape, anchorage independence, alterations in energy metabolism, and invasiveness. These behaviors are also not controlled by the cell cycle clock, which does its work in the nucleus.

In spite of significant alterations that the cell cycle clock suffers in cancer cells, we recognize that the effects of these changes are felt largely during the G_1 phase of the cell cycle. We can understand this by noting that the R-point transition occurring toward the end of G_1 represents a critical decision point in the life of a cell; this decision must be deregulated if cancer cells are to gain proliferative advantage. However, once a cell has moved past the R point and reached the G_1/S transition, the remaining steps of the cell cycle proceed in an essentially automatic, pre-programmed fashion. Accordingly, the S, G_2, and M phases of cancer cells, which together represent an extraordinarily complex program of biochemical and cell-biological steps, closely resemble the comparable cell cycle phases of normal cells. This helps to explain an often-noted aspect of cancer cells: their growth-and-division cycles are not necessarily shorter than those of many normal cells in the body. Instead, cancer cells continue to enter into these cycles and thus continue to proliferate under conditions that would force normal cells to halt proliferation (see, for example, Figure 7.29).

While the effects of deregulating the cell cycle clock are felt largely in late G1, there are more subtle changes that extend into later phases of the cell cycle. For example, pRb–E2F1 complexes also seem to regulate the expression of genes that play key roles in organizing the complex steps of mitosis; the products of these genes contribute to chromosome condensation (during metaphase), centromere function, and general chromosomal stability. These more recent discoveries now indicate that cells that have lost pRb function suffer from chromosome instability (CIN) in addition to deregulation of the R-point transition, suggesting that in pRb-negative cells, CIN conspires with deregulation of the G_1/S transition to accelerate multi-step tumor progression.

In addition to the critical regulators of G_1 progression and chromosomal stability, the only other components of the cell cycle clock that seem to be affected in the malignant growth state are the small cohort of proteins that serve as checkpoint controls during the G_1/S transition, during S phase, and during M phase. Their inactivation, which occurs in some cancers, is not directed toward the immediate goal of deregulating proliferation. Rather, the inactivation of these checkpoint controls serves to destabilize the cellular genome, enabling incipient cancer cells to generate a wide variety of permutations of the normal human genome. These cells and their descendants then test the resulting novel genetic configurations, searching for those that are particularly advantageous for neoplastic growth. We will return to the theme of genomic destabilization in Chapter 12.

Large gaps in this scenario remain to be filled in. We still do not really understand why pRb inactivation is so important for the creation of cancer cells while the inactivation of its two cousins, p107 and p130, seems rarely, if ever, to be a priority during the course of tumorigenesis. We do not understand how the proliferation-versus-differentiation decision is made in the great majority of the body's cell types. Relevant here may be observations indicating that pRb interacts with a number of other transcription factors in addition to the intensively studied E2Fs. Some of these other transcription factors may well govern the expression of genes that contribute to certain differentiation programs. And we still do not understand how many oncoproteins that function as transcription factors (see Table 8.2) succeed in perturbing the workings of the cell cycle clock. Moreover, many of the currently held preconceptions about the operations of the cell cycle clock, as described in this chapter, may one day require substantial revision.

For example, mutant mouse embryos that have been deprived of both copies of each of the three D-type cyclin genes (that is, with a $D1^{-/-} D2^{-/-} D3^{-/-}$ genotype) are able to pass through most stages of embryonic development, some dying as late as embryonic day 16 (out of the ~20 days of full-term gestation). How do the cells of these embryos succeed in advancing through many cycles of growth and division without

D-type cyclins driving their advance through the G_1 phase of the cell cycle? Similarly, mouse embryos lacking both copies of each of the two cyclin E genes (that is, with an $E1^{-/-}$ $E2^{-/-}$ genotype) develop until mid-gestation, at which point they die because of placental defects; if these are circumvented, then the embryos can develop to term, whereupon they die. The existing models provide no insight into how the cells of these embryos are able to complete the last steps of the G_1 phase and initiate the S phase of the cell cycle. These startling results might suggest the operations of a normally latent cell cycle clock, possibly inherited from our protozoan ancestors, that is able to assume control when the more modern, metazoan cell cycle machinery fails to do its job.

The roles of the cell cycle clock are also expanding beyond the realm of cell cycle control, as mentioned earlier in the context of cyclin D1 and its interactions with transcription factors. For example, the β cells of the pancreas are responsible for secreting insulin in response to elevation of circulating glucose level. During this response, the insulin that is initially released acts in an autocrine manner on the β cells, resulting in cyclin D2 induction, phosphorylation of pRb by D2–CDK4 complexes, and resulting E2F1-dependent transcription of a gene (termed *Kir6.2*) that amplifies insulin secretion, greatly accelerating the insulin response and removal of excess glucose from the blood. All this occurs in cells (that is, β cells) that are stably ensconced in the G_0 phase of the cell cycle. Examples like these indicate that the cell cycle apparatus, which was already highly developed in eukaryotic protozoa, has been adapted during metazoan evolution for various applications that are totally unrelated to control of the cell cycle.

While we have read about the details of cell cycle control and the various ways by which it is disrupted in many types of cancer cells, the ultimate motivation behind these discussions is a need to understand clinical disease: How do these changes actually affect tumor progression and patient outcome? In fact, losses of pRb function can have particularly striking effects on the behavior of cancer cells and thus tumors. For example, as indicated in **Figure 8.38A**, abnormally high levels of cyclin E in the cancer cells of breast carcinoma patients are strongly predictive of aggressive malignancy and poor patient outcome, while low levels indicate long-term, disease-free survival. In this case, expression of the cyclin E mRNA may be elevated, and the degradative mechanisms that are normally responsible for reducing cyclin E levels are likely to

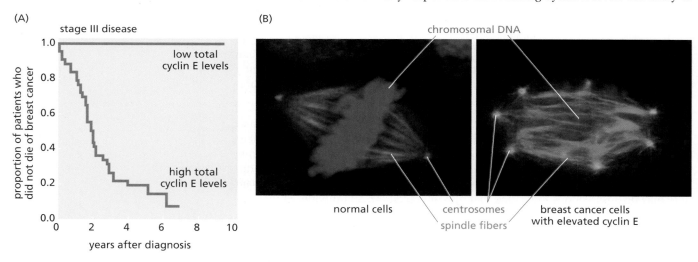

Figure 8.38 Cyclin E and breast cancer progression (A) This Kaplan–Meier plot presents the clinical progression of disease in women with stage III breast cancer, that is, those having relatively large primary tumors and cancer cells in regional lymph nodes but lacking observable metastases at distant anatomical sites. Plotted is the fraction of patients (*ordinate*) who had not died from their disease at the indicated times after initial diagnosis (*abscissa*). Total cyclin E includes both the high– and low–molecular-weight forms of this cyclin. (B) During normal mitosis, bipolar spindles are observed (*left*). However, in aggressive breast cancers, low–molecular-weight forms of cyclin E accumulate in the cytoplasm of the associated carcinoma cells, where they drive the formation of multiple extra centrosomes and resulting multipolar spindles; eight are visible here (*right*). This leads, in turn, to karyotypic chaos and the acceleration of tumor progression. (A, adapted from K. Keyomarsi et al., *N. Engl. J. Med.* 347:1566–1575, 2002. B, from R. Bagheri-Yarmand et al., *Cancer Res.* 70:5074–5084, 2010.)

be compromised. The resulting excessively high levels of cyclin E drive deregulated pRb phosphorylation and inactivation. Moreover, lower–molecular-weight forms of cyclin E that are found in many aggressive tumors function abnormally to drive the formation of multiple centrosomes (see Figure 8.38B), which results in genetic instability and acceleration of tumor progression, a topic that we will pursue in depth in Chapter 12. Observations like these provide compelling indications that the cell cycle clock is, in its normal state, an important deterrent to cancer development and, when deranged by various lesions, a potent agent for promoting cancer progression.

Our discussions of pRb and cell cycle control would suggest the preeminent importance of this protein among all the many products of tumor suppressor genes. In fact, pRb shares this position with a second protein, p53, which plays an equally important role in normal cell physiology and in cancer development. In essence, the pRb circuitry deals with the relations between the cell and the outside world. The p53 circuitry has a very different function, since it monitors the internal well-being of the cell and permits cell proliferation and cell survival only if all the vital operating systems within the cell are functioning properly. As we will see in the next chapter, the inactivation of this p53 signaling pathway is as important to developing cancer cells as the deregulation of the controls governing pRb and the R-point transition.

Key concepts

- The cell cycle is a precisely programmed series of events that enables a cell to duplicate its contents and to divide into two daughter cells. This series of events is controlled by the machinery that is often termed the cell cycle clock.

- Specific steps of the cell cycle are controlled by changing the levels and availability of cyclins. Cyclins function by activating the catalytic function of their partners—the cyclin-dependent kinases (CDKs), a family of serine/threonine protein kinases. Additional control of the cell cycle is provided by CDK inhibitors, which antagonize the activities of cyclin-CDK complexes.

- While the levels of the D cyclins are controlled primarily through extracellular signals, the remaining cyclins operate on a preordained schedule, once the decision to advance into the late G_1 phase has been made; the gradual accumulation of these other cyclins followed by their rapid destruction ensures that the cell cycle clock can move in only one direction.

- Checkpoint controls operating throughout the cell cycle ensure that a new step in cell cycle progression is not undertaken before the preceding step is properly completed and that cell cycle progression cannot proceed if a cell's genome is damaged. Many types of cancer cells have inactivated one or more of these checkpoint controls, thus helping themselves to accumulate the mutant genes and altered karyotypes that propel their neoplastic growth.

- The critical decisions concerning growth versus quiescence and entrance into post-mitotic differentiated states are made in the G_1 phase of the cell cycle. In normal cells, the decision to grow and replicate requires signals from the external environment (hence the dependence of D-cyclin levels on mitogenic signals). Thereafter, advance through the other phases of the cell cycle is relatively independent of external signals.

- The restriction point (R point) represents a point at which the cell commits itself, essentially irrevocably, to complete the remainder of the cell cycle or, alternatively, to remain in G_1 and possibly retreat from the active cell cycle into the G_0, quiescent state. Deregulation of the R-point decision-making machinery accompanies the formation of most types of cancer cells, since it leads to unconstrained cell proliferation.

- Decisions concerning growth versus quiescence are governed by the state of phosphorylation of the three pocket proteins—pRb, p107, and p130—which in turn is controlled by the D cyclins and cyclin E. pRb, the retinoblastoma protein, controls passage through the R point. Hypophosphorylated pRb blocks passage through the R point, while hyperphosphorylated pRb permits this passage.

- pRb's phosphorylation is carefully controlled. Growth factors induce expression of D-type cyclins, which initiate pRb phosphorylation by hypophosphorylating it. The resulting hypophosphorylation of pRb leaves it in a growth-inhibitory state, which appears to be a prerequisite for its hyperphosphorylation (and consequent inactivation) by cyclin E-CDK2 complexes.

- pRb acts by binding or releasing E2F transcription factors associated with promoters of genes that usher the cell from late G_1 into S phase. Hypophosphorylated pRb binds E2Fs, while hyperphosphorylated pRb releases them. When viral oncoproteins are present, they mimic pRb hyperphosphorylation by preventing pRB from binding E2Fs.

- In cancer cells, a number of alternative mechanisms operate to ensure that cell proliferation is not constrained by pRb. Many of these cause pRb hyperphosphorylation and resulting functional inactivation.

- pRb function can be lost in a variety of ways, including excessive mitogenic signals (since these lead to elevated levels of D cyclins); mutation of the *Rb* gene; binding of pRb by a viral oncoprotein (for example, HPV E7); and the actions of cellular oncoproteins (for example, Myc) that deregulate pRb phosphorylation or directly affect pRb activity.

- The control of cell differentiation is coupled to the regulation of cell cycle progression. Hypophosphorylated pRB is needed to halt proliferation of cells and to facilitate their differentiation. Conversely, other regulatory proteins, such as Myc and the Ids, work to inhibit cell differentiation.

- In most types of cancer, differentiation is partially or completely blocked. In general, the more differentiated the cells are that form a tumor, the less aggressive is the disease of cancer.

Thought questions

1. Why is pRb function compromised in human tumors through mutations of its encoding gene while the genes encoding its two cousins, p107 and p130, have virtually never been found to suffer mutations in the genomes of cancer cells?

2. Why have DNA tumor viruses evolved the ability to inactivate pRb function?

3. How might loss of a CDK inhibitor's function affect the control of cell cycle advance?

4. What molecular mechanisms operate to ensure that once the decision to advance through the restriction point has been made, this leads to an essentially irreversible commitment to complete the remaining phases of the cell cycle through M phase?

5. How are the decisions of cell growth versus quiescence coupled mechanistically with the decisions governing cell differentiation? Why must these two processes be tightly coupled?

6. How do cells ensure that the transcription-activating functions of the E2F transcription factors are limited to a narrow window of time in the cell cycle? What might occur if E2F function were allowed to continue throughout the cell cycle?

7. In what ways does the Myc oncoprotein deregulate cell proliferation and differentiation?

8. Why is it important that the cell cycle clock never runs backwards?

9. What kinds of experiments suggest that the cell cycle clock may be organized differently in certain cell types or stages of embryonic development?

Additional reading

Adhikary S & Eilers M (2005) Transcriptional regulation and transformation by Myc proteins. *Nat. Rev. Mol. Cell Biol.* 6, 635–645.

Barbacid M, Ortega S, Sotillo R et al. (2005) Cell cycle and cancer: genetic analysis of the role of cyclin-dependent kinases. *Cold Spring Harbor Symp. Quant. Biol.* 70, 233–240.

Bashir T & Pagano M (2003) Aberrant ubiquitin-mediated proteolysis of cell cycle regulatory proteins and oncogenesis. *Adv. Cancer Res.* 65, 101–144.

Blais A & Dynlacht BD (2004) Hitting their targets: an emerging picture of E2F and cell cycle control. *Curr. Opin. Genet. Dev.* 14, 527–532.

Bucher N & Britten CD (2008) G2 checkpoint abrogation and checkpoint kinase-1 targeting in the treatment of cancer. *Br. J. Cancer* 98, 523–528.

Chen HZ, Tsai S-Y & Leone G (2009) Emerging roles of E2Fs in cancer: an exit from cell cycle control. *Nat. Rev. Cancer* 9, 785–797.

Classon M & Dyson N (2001) p107 and p130: versatile proteins with interesting pockets. *Exp. Cell Res.* 264, 135–147.

Classon M & Harlow E (2002) The retinoblastoma tumour suppressor in development and cancer. *Nat. Rev. Cancer* 2, 910–917.

Coletta RD, Jedlicka P, Gutierrez-Hartmann A & Ford HL (2004) Transcriptional control of the cell cycle in mammary gland development and tumorigenesis. *J. Mammary Gland Biol. Neoplasia* 9, 39–53.

Eilers M & Eisenman R (2008) Myc's broad reach. *Genes Dev.* 22, 2755–2766.

Ewen ME & Lamb J (2004) The activities of cyclin D1 that drive tumorigenesis. *Trends Mol. Med.* 10, 158–162.

Foster DA, Yellen P, Xu L & Saqcena M (2011) Regulation of G1 cell cycle progression: distinguishing the restriction point from a nutrient-sensing cell growth checkpoint. *Genes Cancer* 1, 1124–1131.

Frescas D & Pagano M (2009) Deregulated proteolysis by the F-box proteins SKP2 and β-TrCP: tipping the scales of cancer. *Nat. Rev. Cancer* 8, 438–449.

Gil J & Peters G (2006) Regulation of the INK4b–ARF–INK4a tumour suppressor locus: all for one or one for all. *Nat. Rev. Mol. Cell Biol.* 7, 667–677.

Hochegger H, Takeda S & Hunt T (2008) Cyclin-dependent kinases and cell-cycle transitions: does one fit all? *Nat. Rev. Mol. Cell Biol.* 9, 910–916.

Iaquinta PJ & Lees JA (2007) Life and death decisions by the E2F transcription factors. *Curr. Opin. Cell Biol.* 19, 649–657.

Iavarone A & Lasorella A (2006) ID proteins in cancer and in neurobiology. *Trends Mol. Med.* 12, 588–594.

Kim WY & Sharpless NE (2006) The regulation of INK4/ARF in cancer and aging. *Cell* 127, 265–275.

Lammens T, Li J, Leone G & De Veylder L (2009) Atypical E2Fs: new players in the E2F transcription factor family. *Trends Cell Biol.* 19, 111–118.

Lindquist A, Rodríguez-Bravo V & Medema RH (2009) The decision to enter mitosis: feedback and redundancy in the mitotic entry network. *J. Cell Biol.* 185, 193–202.

Liu H, Dibling B, Spike B et al. (2004) New roles for the RB tumor suppressor protein. *Curr. Opin. Genet. Dev.* 14, 55–64.

Macaluso M, Montanari M & Giordana A (2006) Rb family proteins as modulators of gene expression and new aspects regarding the interaction with chromatin remodeling enzymes. *Oncogene* 25, 5263–5267.

Malumbres M & Barbacid M (2001) To cycle or not to cycle: a critical decision in cancer. *Nat. Rev. Cancer* 1, 222–231.

Massagué J (2004) G1 cell-cycle control and cancer. *Nature* 432, 298–307.

Massari ME & Murre C (2000) Helix-loop-helix proteins: regulators of transcription in eucaryotic organisms. *Mol. Cell Biol.* 20, 429–440.

Morgan DO (2007) The Cell Cycle: Principles of Control. Sunderland, MA: Sinauer Associates.

Perk J, Iavarone A & Benezra R (2005) Id family of helix-loop-helix proteins in cancer. *Nat. Rev. Cancer* 5, 603–614.

Reed S (2003) Ratchets and clocks: the cell cycle, ubiquitylation and protein turnover. *Nat. Rev. Mol. Cell Biol.* 4, 855–864.

Sherr CJ (2001) The INK4a/ARF network in tumour suppression. *Nat. Rev. Mol. Cell Biol.* 2, 731–737.

Sherr CJ & McCormick F (2002) The RB and p53 pathways in cancer. *Cancer Cell* 2, 103–112.

Sherr CJ & Roberts JM (2004) Living with or without cyclins and cyclin-dependent kinases. *Genes Dev.* 18, 2699–2711.

Siegel PM & Massagué J (2003) Cytostatic and apoptotic actions of TGF-β in homeostasis and cancer. *Nat. Rev. Cancer* 3, 807–820.

Varlakhanova NV & Knoepfler PS (2009) Acting locally and globally: Myc's ever-expanding roles on chromatin. *Cancer Res.* 69, 7487–7490.

Whitfield ML, George LK, Grant GD & Perou CM (2006) Common markers of proliferation. *Nat. Rev. Cancer* 6, 99–106.

Yamasaki L & Pagano M (2004) Cell cycle, proteolysis and cancer. *Curr. Opin. Cell Biol.* 16, 623–628.

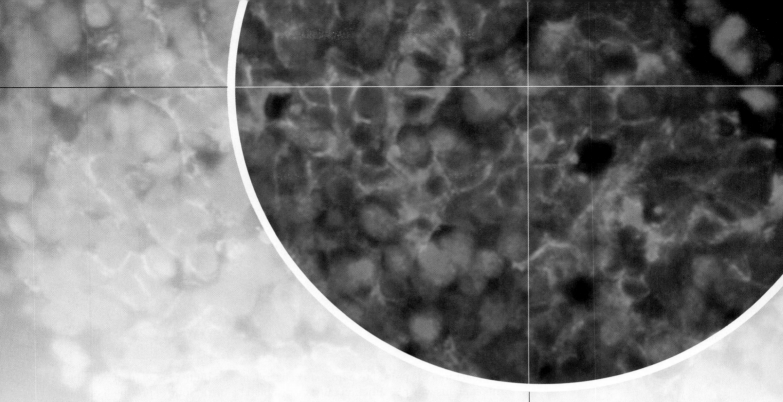

Chapter 9

p53 and Apoptosis: Master Guardian and Executioner

To examine the causes of life, we must first have recourse to death.
Mary Shelley, *Frankenstein*, 1831

There cannot however be the least doubt, that the higher organisms, as they are now constructed, contain within themselves the germs of death.

August Weissmann, philosopher of biology, 1889

Metazoan organisms have a vital interest in eliminating defective or malfunctioning cells from their tissues. Responding to this need, mammals have implanted a loyal watchman in their cells. Within almost all cells in mammalian tissues, the p53 protein serves as the local representative of the organism's interests. p53 is present on-site to ensure that the cell keeps its household in order.

If p53 receives information about metabolic disorder or genetic damage within a cell, it may arrest the advance of the cell through its growth-and-division cycle and, at the same time, orchestrate localized responses in that cell to facilitate the repair of damage. If p53 learns that metabolic derangement or damage to the genome is too severe to be cured, it may decide to emit signals that awaken the cell's normally latent suicide program—**apoptosis**. The consequence is the rapid death of the cell. This results in the elimination of a cell whose continued growth and division might otherwise pose a threat to the organism's health and viability.

The apoptotic program that may be activated by p53 is built into the control circuitry of most cells throughout the body. Apoptosis consists of a series of distinctive cellular changes that function to ensure the disappearance of all traces of a cell, often within an hour of its initial activation. The continued presence of a latent but intact apoptotic machinery represents an ongoing threat to an incipient cancer cell, since

Figure 9.1 Large T antigen in SV40-transformed cells Antibodies that bind the SV40 large T (LT) antigen can be used to detect LT in the nuclei of SV40-transformed tumor cells. In the present case, such antibodies were used to stain human mammary epithelial cells (MECs) that were transformed by introduction of the SV40 early region plus two other genes. A similar image would be seen if such antibodies were used to stain SV40-transformed mouse cells. LT was detected by linking these antibody molecules to peroxidase enzyme, which generated the *dark brown* spots. In this image of a tumor xenograft, the transformed MECs form ducts (seen in cross section), which are surrounded by normal stromal cells (*light blue nuclei*). (Courtesy of T.A. Ince.)

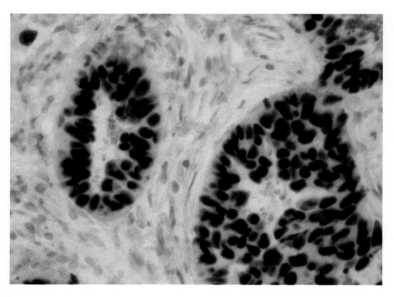

this machinery is poised to eliminate cells that are en route to becoming neoplastic. This explains why p53 function must be disabled before a clone of pre-malignant cells gains a sure and stable foothold within a tissue. Without a clear description of p53 function and apoptosis, we have no hope of understanding a fundamental component of the process that leads to the creation of virtually all types of human tumors.

9.1 Papovaviruses lead to the discovery of p53

When murine cells that have been transformed by the SV40 DNA tumor virus are injected into a mouse of identical genetic background (that is, a *syngeneic* host), the immune system of the host reacts by mounting a strong response; antibodies are made that react with a nuclear protein that is present in the virus-transformed cells and is otherwise undetectable in normal mouse cells (**Figure 9.1**). This protein, the large tumor (large T, LT) antigen, is encoded by a region of the viral genome that is also expressed when this virus infects and multiplies within monkey kidney cells—host cells that permit a full infectious (lytic) cycle to proceed to completion (see Section 3.4).

Large T is a multifunctional protein that the SV40 virus uses to perturb a number of distinct regulatory circuits within infected and transformed cells. Indeed, large T was cited in the previous chapter because of its ability to bind and thus functionally inactivate pRb (see Section 8.5). Anti-large T sera harvested from mice and hamsters bearing SV40-induced tumors were used in 1979 to analyze the proteins in SV40-transformed cells. The resulting immunoprecipitates contained both large T and an associated protein that exhibited an apparent molecular weight of 53 to 54 kilodaltons (**Figure 9.2A**). Antisera reactive with the p53 protein were found to detect this protein in mouse embryonal carcinoma cells and, later on, in a variety of human and rodent tumor cells that had never been infected by SV40. However, monoclonal antibodies that recognized only large T immunoprecipitated the 53- to 54-kD protein in virus-infected but not in uninfected cells.

Taken together, these observations indicated that the large T protein expressed in SV40-transformed cells was tightly bound to a novel protein, which came to be called p53 (see Figure 9.2B). Antisera that reacted with both large T and p53 detected p53 in certain uninfected cells, notably tumor cells that were transformed by non-viral mechanisms, such as the F9 embryonal carcinoma cells analyzed in Figure 9.2A. The latter observations indicated that p53 was of cellular rather than viral origin, a conclusion that was reinforced by the report in the same year that mouse cells transformed by exposure to a chemical carcinogen also expressed p53.

These various lines of evidence suggested that the large T oncoprotein functions, at least in part, by targeting host-cell proteins for binding. (The discovery that large

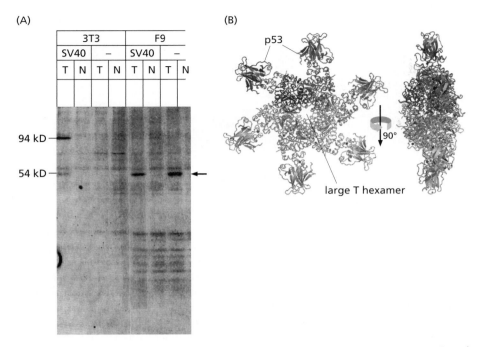

(A)

(B)

p53

large T hexamer

90°

Figure 9.2 The discovery of p53 and its association with SV40 large T
(A) Normal BALB/c 3T3 mouse fibroblasts (3T3) transformed by SV40, as well as F9 mouse embryonal carcinoma cells, were exposed to ^{35}S-methionine, and resulting cell lysates were incubated with either normal hamster serum (N) or hamster antiserum reactive with SV40-transformed hamster cells (T). The anti-tumor serum immunoprecipitated a protein of 94 kD from virus-infected but not uninfected 3T3 cells. In addition, a second protein running slightly ahead of the 54-kD marker was immunoprecipitated from SV40-transformed 3T3 cells but not from normal 3T3 cells. Moreover, this same protein could be immunoprecipitated from F9 cells, whether or not they had been exposed to SV40 (*arrow*). These particular data, on their own, did not prove a physical association of SV40 large T (the 94-kD protein) with p53, but they did show that p53 was a cellular protein that was present in elevated amounts in two types of transformed cells. Moreover, they suggested that the elevated levels of p53 in SV40-transformed hamster cells cause the hamster immune system to mount an immune response against both large T and the hamster's own p53. (B) As revealed by X-ray crystallography, SV40 large T molecules assemble into homohexamers, each subunit of which binds and thereby sequesters a single molecule of p53. (A, from D.I. Linzer and A.J. Levine, *Cell* 17:43–52, 1979. B, from D. Li et al., *Nature* 423:512–518, 2003.)

T antigen is also able to bind pRb, the retinoblastoma protein, came seven years later.) In the years since these 1979 discoveries, a number of other DNA viruses and at least one RNA virus have been found to specify oncoproteins that associate with p53 or perturb its function (**Table 9.1**). (As we will discuss later in this chapter, and as is apparent from this table, these viruses also target pRb and undertake to block apoptosis.)

Table 9.1 Tumor viruses that perturb pRb, p53, and/or apoptotic function

Virus	Viral protein targeting pRb	Viral protein targeting p53	Viral protein targeting apoptosis
SV40	large T (LT)[a]	large T (LT)[a]	
Adenovirus	E1A	E1B55K	E1B19K[b]
HPV	E7	E6	
Polyomavirus	large T	large T?	middle T (MT)[c]
Herpesvirus saimiri	V cyclin[d]		v-Bcl-2[e]
HHV-8 (KSHV)	K cyclin[d]	LANA-2	v-Bcl-2,[e] v-FLIP[f]
Human cytomegalovirus (HCMV)	IE72[g]	IE86	vICA,[h] pUL37[i]
HTLV-I	Tax[j]	Tax	
Epstein–Barr	EBNA3C	EBNA-1[k]	LMP1[k]

[a]SV40 LT also binds a number of other cellular proteins, including p300, CBP, Cul7, IRS1, Bub1, Nbs1, and Fbw7, thereby perturbing a variety of other regulatory pathways.
[b]Functions like Bcl-2 to block apoptosis.
[c]Activates PI3K and thus Akt/PKB.
[d]Related to D-type cyclins.
[e]Related to cellular Bcl-2 anti-apoptotic protein.
[f]Viral caspase 8 (FLICE) inhibitory protein; blocks an early step in the extrinsic apoptotic cascade.
[g]Interacts with and inhibits p107 and possibly p130; may also target pRb for degradation in proteasomes.
[h]Binds and inhibits procaspase 8.
[i]Inhibits the apoptotic pathway below caspase 8 and before cytochrome *c* release.
[j]Induces synthesis of cyclin D2 and binds and inactivates p16[INK4A].
[k]LMP1 facilitates p52 NF-κB activation and thereby induces expression of Bcl-2; EBNA-1 acts via a cellular protein, USP7/HAUSP, to reduce p53 levels. EBNA3C interferes with p53 function.

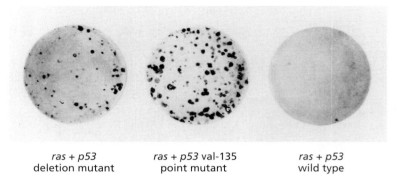

ras + p53
deletion mutant

ras + p53 val-135
point mutant

ras + p53
wild type

Figure 9.3 Effects of p53 on cell transformation A cDNA encoding a *ras* oncogene was co-transfected with several alternative forms of a *p53* cDNA into rat embryo fibroblasts. In the presence of a *p53 dl* mutant vector, which contains an almost complete deletion of the *p53* reading frame (*left*), a small number of foci were formed. In the presence of a *p53* point mutant (*middle*), a large number of robust foci were formed. However, in the presence of a *p53* wild-type cDNA clone (*right*), almost no foci were formed. (Courtesy of M. Oren; from D. Michalovitz et al., *Cell* 62:671–680, 1990.)

9.2 p53 is discovered to be a tumor suppressor gene

The initial functional studies of p53 involved a substantial scientific detour: transfection of a *p53* cDNA clone into rat embryo fibroblasts revealed that this DNA could collaborate with a co-introduced *ras* oncogene in the transformation of these rodent cells. Such activity suggested that the *p53* gene (which is sometimes termed *Trp53* in mice and *TP53* in humans) might operate as an oncogene, much like the *myc* oncogene, which had previously been found capable of collaborating with the *ras* oncogene in rodent cell transformation (see Section 11.10). Like *myc*, the introduced *p53* cDNA seemed to contribute certain growth-inducing signals that resulted in cell transformation in the presence of a concomitantly expressed *ras* oncogene.

But appearances deceived. As later became apparent, the *p53* cDNA had originally been synthesized using as template the mRNA extracted from tumor cells (rather than normal cells). Subsequent manipulation of a *p53* cDNA cloned instead from the mRNA of normal cells revealed that this *p53* cDNA clone, rather than favoring cell transformation, actually *suppressed* it (**Figure 9.3**). Comparison of the sequences of the two cDNAs revealed that the two differed by a single base substitution—a point mutation—that caused an amino acid substitution in the p53 protein. Hence, the initially used clone encoded a mutant p53 protein with altered function.

These results indicated that the wild-type allele of *p53* really functions to suppress cell proliferation, and that *p53* acquires growth-promoting powers when it sustains a point mutation in its reading frame. Because of this discovery, the *p53* gene was eventually categorized as a tumor suppressor gene.

By 1987 it became apparent that such point-mutated alleles of *p53* are common in the genomes of a wide variety of human tumor cells. Data accumulated from diverse studies indicated that the *p53* gene is mutated in 30 to 50% of commonly occurring human cancers (**Figure 9.4**). Indeed, among all the genes examined to date in human cancer cell genomes, *p53* is the gene found to be most frequently mutated, being present in mutant form in the genomes of almost one-third of all human tumors.

Further functional analyses of *p53*, conducted much later, made it clear, however, that *p53* is not a typical tumor suppressor gene. In the case of most tumor suppressor genes, when the gene was inactivated (that is, "knocked out") homozygously in the mouse germ line (using the strategy of targeted gene inactivation described in Supplementary Sidebar 7.7), the result was, almost invariably, a disruption of embryonic development due to deregulated morphogenesis in one or more tissues. These tumor suppressor genes seemed to function as negative regulators of proliferation in a variety of cell types; their deletion from the regulatory circuitry of cells led, consequently, to inappropriate proliferation of certain cells and thus to disruption of normal development.

In stark contrast, deletion of both *p53* gene copies from the mouse germ line had no significant effect on the development of the great majority of *p53*$^{-/-}$ embryos. Therefore, *p53* could not be considered to be a simple negative regulator of cell proliferation during normal development. Still, *p53* was clearly a tumor suppressor gene, since mice lacking both germ-line copies of the *p53* gene had a short life span (about 5 months), dying most often from lymphomas and sarcomas (**Figure 9.5**). This behavior provided

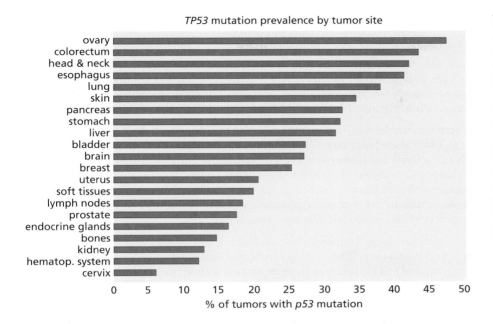

Figure 9.4 Frequency of mutant *p53* alleles in human tumor cell genomes As indicated in this bar graph, mutant alleles of *p53* are found frequently in commonly occurring human tumors. This data set includes 26,597 somatic mutations of *p53* and 535 germ-line mutations that had been reported by November 2009. The bars indicate the percentage of each tumor type found to carry a mutant *p53* allele. More recent research indicates that virtually all (119/123) of high-grade ovarian serious carcinomas carry mutant *p53* alleles. (From International Agency for Research on Cancer, TP53 genetic variations in human cancer, IARC release R14, 2009.)

the first hints that the p53 protein does not operate to transduce the proliferative and anti-proliferative signals that continuously impinge on cells and regulate their proliferation. Instead, p53 seemed to be specialized to prevent the appearance of abnormal cells, specifically, those cells that were capable of spawning tumors.

9.3 Mutant versions of p53 interfere with normal p53 function

The observations of frequent mutation of the *p53* gene in tumor cell genomes suggested that many incipient cancer cells must perturb or eliminate p53 function before they can thrive. This notion raised the question of precisely how these cells succeed in shedding p53 function. Here, another anomaly arose, because the *p53* gene did not seem to obey Knudson's scheme for the two-hit elimination of tumor suppressor genes. For example, the finding that a cDNA clone encoding a mutant version of p53 was able to alter the behavior of wild-type rat embryo fibroblasts (as described above) ran directly counter to Knudson's model of how tumor suppressor genes should operate (see Section 7.3).

According to the Knudson scheme, an evolving pre-malignant cell can only reap substantial benefit once it has lost both functional copies of a tumor suppressor gene that has been holding back its proliferation. In the Knudson model, such gene inactivation events are caused by mutations that create inactive ("null") and thus recessive alleles.

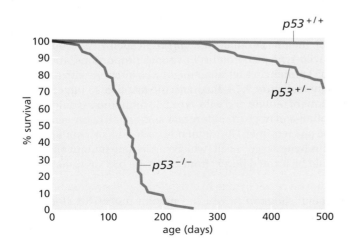

Figure 9.5 Effects of mutant *p53* alleles in the mouse germ line This Kaplan–Meier plot indicates the percent of mice of the indicated genotype that survived (*ordinate*) as a function of elapsed lifetime in days (*abscissa*). While the absence of p53 function in the *p53*^{−/−} mice (carrying two *p53* null alleles) had relatively little effect on their embryonic development and viability at birth, it resulted in a greatly increased mortality relatively early in life, deriving largely from the development of sarcomas and leukemias. All *p53*^{−/−} homozygotes succumbed to malignancies by about 250 days of age (*red line*), and even *p53*^{+/−} heterozygotes (*blue line*) began to develop tumors at this time, while wild-type (*p53*^{+/+}) mice (*green line*) showed virtually no mortality until almost 500 days of age. (Adapted from T. Jacks et al., *Curr. Biol.* 4:1–7, 1994.)

Therefore, a pre-malignant cell may benefit minimally from inactivation of one copy of a tumor suppressor gene—due to the halving of effective gene function—or not at all, if the residual activity specified by the surviving wild-type gene copy suffices on its own to mediate normal function. As we learned in Chapter 7, substantial change in cell phenotype usually occurs only when the function of a suppressor gene is eliminated through two successive inactivating mutations or through a combination of an inactivating mutation plus a loss-of-heterozygosity (LOH) event (see Section 7.4).

Knudson's model was hard to reconcile with the observed behavior of the mutant *p53* cDNA introduced into rat embryo fibroblasts (see Figure 9.3). The mutant *p53* cDNAs clearly altered cell phenotype, even though these embryo fibroblast cells continued to harbor their own pair of wild-type *p53* gene copies. This meant that the introduced mutant *p53* cDNA could not be functioning as an inactive, recessive allele. It seemed, instead, that the point-mutated *p53* allele was actively exerting some type of *dominant* function when introduced into these rat embryo cells.

Another clue came from sequence analyses of mutant *p53* alleles in various human tumor cell genomes. These analyses indicated that the great majority of tumor-associated, mutant *p53* alleles carry point mutations in their reading frames that create **missense** codons (resulting in amino acid substitutions) rather than **nonsense** codons (which cause premature termination of the growing polypeptide chain). To date, more than 26,000 tumor-associated *p53* alleles originating in human tumor cell genomes have been sequenced, 74% of which have been found to carry such missense mutations (**Figure 9.6A**). Furthermore, deletions of sequences within the reading frame of the *p53* gene are relatively uncommon. Consequently, researchers came to the inescapable conclusion that tumor cells can benefit from the presence of a slightly altered p53 protein rather than from its complete absence, as would occur following the creation of **null alleles** by nonsense mutations or the outright deletion of significant portions of the *p53* gene.

A solution to the puzzle of how mutant p53 protein might foster tumor cell formation arose from two lines of research. First, studies in the area of yeast genetics indicated that mutant alleles of certain genes can be found in which the responsible mutation inactivates the normal functioning of the encoded gene product. At the same time, this mutation confers on the mutant allele the ability to interfere with or obstruct the ongoing activities of the surviving wild-type copy of this gene in a cell. Alleles of this type are termed variously *dominant-interfering* or *dominant-negative* alleles.

A second clue came from biochemical and structural analyses of the p53 protein, which revealed that p53 was a nuclear protein that normally exists in the cell as a **homotetramer**, that is, an assembly of four identical polypeptide subunits (see Figure 9.6B and C). Together with the dominant-negative concept, this observed tetrameric state suggested a mechanism through which a mutant allele of *p53* could actively interfere with the continued functioning of a wild-type *p53* allele being expressed in the same cell.

Assume that a mutant *p53* allele found in a human cancer cell encodes a form of the p53 protein that has lost most normal function but has retained the ability to participate in tetramer formation. If one such mutant allele were to coexist with a wild-type allele in this cell, the p53 tetramers assembled in such a cell would contain mixtures of mutant and wild-type p53 proteins in various proportions. The presence of only a single mutant p53 protein in a tetramer might well interfere with the functioning of the tetramer as a whole. **Figure 9.7A** illustrates the fact that 15 out of the 16 equally possible combinations of mutant and wild-type p53 monomers would contain at least one mutant p53 subunit and might therefore lack some or all of the activity associated with a fully wild-type p53 tetramer. Consequently, only one-sixteenth of the p53 tetramers assembled in this heterozygous cell (which carries one mutant and one wild-type *p53* gene copy) would be formed purely from wild-type p53 subunits and retain full wild-type function.

In an experimental situation in which a mutant *p53* cDNA clone is introduced by gene transfer (transfection) into cells carrying a pair of wild-type *p53* alleles (see Figure 9.3), the expression of this introduced allele is usually driven by a highly active transcriptional promoter, indeed, a promoter that is far more active than the gene promoter controlling expression of the native *p53* gene copies. As a consequence, in such transfected cells, the amount of mutant p53 protein expressed by the introduced

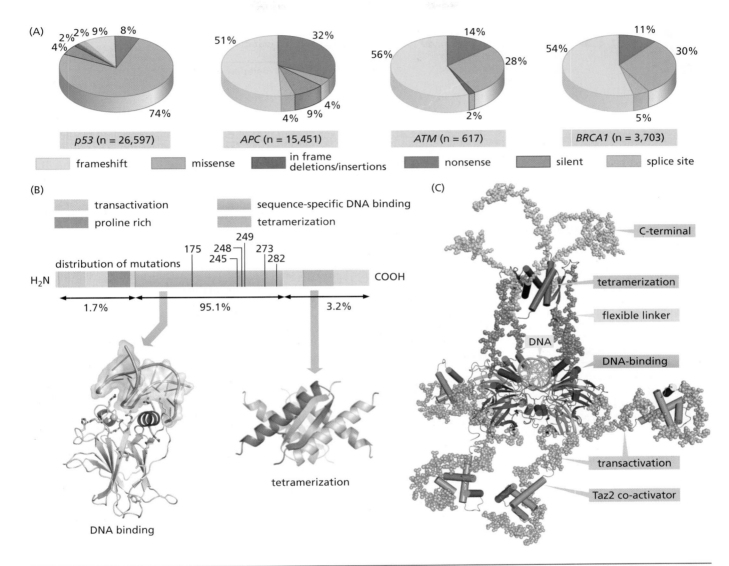

Figure 9.6 Nature of *p53* mutations (A) As indicated in these pie charts, point-mutated alleles of *p53* leading to amino acid substitutions (*green*) represent the great majority of the mutant *p53* alleles found in human tumors, while other types of mutations are seen relatively infrequently. In contrast, the mutations striking other tumor suppressor genes (*APC*) or "caretaker" genes involved in maintenance of the genome (*ATM, BRCA1*) represent reading-frame shifts (*yellow*) or nonsense codons (*blue*) in the majority of cases; both of these types of mutation disrupt protein structure, usually by creating truncated versions of proteins that are often degraded rapidly in cells. (B) The locations across the *p53* reading frame of the point mutations causing amino acid substitutions are plotted here (*above*). As is apparent, the great majority of *p53* mutations (95.1%) affect the DNA-binding domain of the p53 protein. The numbers above the figure indicate the residue numbers of the amino acids that are subject to frequent substitution in human tumors. The transactivation domain enables p53 to interact physically with a number of alternative partners, including the p300/CBP transcriptional co-activator and Mdm2, the p53 antagonist. The detailed structures of the DNA-binding domain and the tetramerization domain are shown *below*. (C) The overall structure of the DNA-bound p53 tetramer is shown here. The four DNA-binding domains are shown in *green* and *blue*, while the four tetramerization domains are seen as *red* and *dark red* α-helices (*above*). The DNA double helix is shown in *yellow*. Each of the four DNA-binding domains associates with half of a binding site in the DNA; two copies of the binding site are present in the DNA with a small number of base pairs separating them (see Figure 9.12B). Each of the four transactivating domains (*dark pink*) is shown interacting with the Taz2 domain of the p300 co-activator (*light purple*), which functions to stimulate transcription through its ability to acetylate histones and p53 itself. The C-terminal domain (*yellow*) plays important roles in regulating transcription. (A, from International Agency for Research on Cancer, TP53 genetic variations in human cancer, IARC release R14, 2009; and A.I. Robles et al., *Oncogene* 21:6898–6907, 2002. B, from K.H. Vousden and X. Lu, *Nat. Rev. Cancer* 2:594–604, 2002; and A.C. Joerger and A.R. Fersht, *Annu. Rev. Biochem.* 77:557–582, 2008. C, from A.C. Joerger and A.R. Fersht, *Cold Spring Harb. Perspect. Biol.* 2:000919, 2010.)

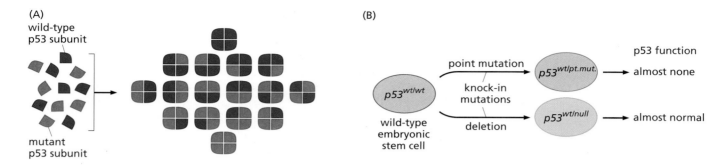

Figure 9.7 p53 structure and p53 function as a dominant-negative allele (A) In cells bearing a single mutant *p53* allele, the mutant protein usually retains its ability to form tetramers but loses its ability to function normally because of a defective DNA-binding domain. Consequently, mixed tetramers composed of differing proportions of wild-type (*blue*) and mutant (*red*) p53 subunits may form, and the presence of even a single mutant protein subunit may compromise the functioning of the entire tetramer. Therefore, in a cell that is heterozygous at the *p53* locus, fifteen-sixteenths of the p53 tetramers may lack fully normal function.
(B) Perhaps the most direct demonstration of the dominant-negative mode of *p53* action has come from "knocking in" (see Supplementary Sidebar 7.7) mutant *p53* alleles into the genome of mouse embryonal stem (ES) cells. In cells in which a point mutation in the DNA-binding domain (*above*) was knocked into one *p53* gene copy, almost all p53 function was lost. In contrast, when one *p53* gene copy was completely inactivated (yielding a null allele), p53 function was almost normal.

gene will be vastly higher than the amount of normal protein produced by the cells' endogenous wild-type *p53* gene copies. Therefore, far fewer than one-sixteenth of the p53 tetramers in such cells will be formed purely from wild-type p53 subunits. This explains how an introduced mutant *p53* allele can be highly effective in compromising virtually all p53 function in such cells.

The above logic might suggest that many human tumor cells, which seem to gain some advantage by shedding p53 function, should carry one wild-type and one mutant *p53* allele. Actually, in the great majority of human tumor cells that are mutant at the *p53* locus, the *p53* locus is found to have undergone a loss of heterozygosity (LOH; see Section 7.4), in which the wild-type allele has been discarded, yielding a cell with two mutant p53 alleles. Thus, in such a cell, one copy of the *p53* gene is initially mutated, followed by elimination of the surviving wild-type copy through some type of loss-of-heterozygosity mechanism.

It is clear that an initial mutation resulting in a mutant, dominant-negative (DN) allele is far more useful for the incipient tumor cell than one resulting in a null allele, which causes total loss of an encoded p53 protein (see Figure 9.7B). The dominant-negative allele may well cause loss of fifteen-sixteenths of p53 function, while the null allele will result, at best, in elimination of one-half of p53 function. (Actually, if the levels of p53 protein in the cell are carefully regulated, as they happen to be, then this null allele will have no effect whatsoever on a cell's overall p53 concentration, since the surviving wild-type allele will compensate by making more of the wild-type protein.)

Why, then, is elimination of the surviving wild-type *p53* allele even necessary? The answer seems to lie in the residual one-sixteenth of fully normal *p53* gene function; even this little bit seems to be more than most tumor cells care to live with. So, being most opportunistic, they jettison the remaining wild-type *p53* allele in order to proliferate even better. The observations described in Figure 9.7B of genetically altered embryonic stem (ES) cells provide further evidence for p53's dominant-negative mode of action.

9.4 p53 protein molecules usually have short lifetimes

Long before the DNA-binding domain of p53 was discovered, the nuclear localization of this protein in many normal and neoplastic cells suggested that it might function as a transcription factor (TF). At least three mechanisms were known to regulate the activity of transcription factors. (1) Concentrations of the transcription factor in the nucleus are modulated. (2) Concentrations of the transcription factor in the nucleus are held constant, but the intrinsic activity of the factor is boosted by some type of covalent modification. (3) Levels of certain collaborating transcription factors may be modulated. In some instances, all three mechanisms cooperate. In the case of p53, the first mechanism—changes in the level of the p53 protein—was initially implicated. Measurement of p53 protein levels indicated that they could vary drastically from one cell type to another and, provocatively, would increase rapidly when cells were exposed to certain types of physiologic stress.

These observations raised the question of how p53 protein levels are modulated by the cell. Many cellular protein molecules, once synthesized, persist for tens or hundreds of hours. (Some cellular proteins, such as those forming the ribosomal subunits in exponentially growing cells, seem to persist for many days.) Yet other cellular proteins are metabolically highly unstable and are degraded almost as soon as they are assembled. One way to distinguish between these alternatives is to treat cells with cycloheximide, a drug that blocks protein synthesis. When such an experiment was performed in cells with wild-type *p53* alleles, the p53 protein disappeared with a half-life of only 20 minutes. This led to the conclusion that p53 is usually a highly unstable protein, being broken down by proteolysis soon after it is synthesized.

This pattern of synthesis followed by rapid degradation might appear to be a "futile cycle," which would be highly wasteful for the cell. Why should a cell invest substantial energy and synthetic capacity in making a protein molecule, only to destroy it almost as soon as it has been created? Similar behaviors have been associated with other cellular proteins such as Myc (see Section 6.1).

The rationale underlying this ostensibly wasteful scheme of rapid protein turnover is a simple one: a cell may need to rapidly increase or decrease the level of a protein in response to certain physiologic signals. In principle, such modulation could be achieved by regulating the level of its encoding mRNA or the rate with which this mRNA is being translated. However, far more rapid changes in the levels of a critical protein can be achieved simply by stabilizing or destabilizing the protein itself. For example, in the case of p53, a cell can double the concentration of p53 protein in 20 minutes simply by blocking its degradation.

Under normal conditions, a cell will continuously synthesize p53 molecules at a high rate and rapidly degrade them at an equal rate. The net result of this is a very low "steady-state" level of the protein within this cell. In response to certain physiologic signals, however, the degradation of p53 is blocked, resulting in a rapid increase of p53 levels in the cell. This finding led to the further question of why a normal cell would wish to rapidly modulate p53 levels, and what types of signals would cause a cell to halt p53 degradation, resulting in rapidly increasing levels of this protein.

9.5 A variety of signals cause p53 induction

During the early 1990s, a variety of agents were found to be capable of inducing rapid increases in p53 protein levels. These included X-rays, ultraviolet (UV) radiation, certain chemotherapeutic drugs that damage DNA, inhibitors of DNA synthesis, and agents that disrupt the microtubule components of the cytoskeleton. Within minutes of exposing cells to some of these agents, p53 was readily detected in substantial amounts in cells that previously had shown only minimal levels of this protein. This rapid induction occurred in the absence of any marked changes in *p53* mRNA levels and hence was not due to increased transcription of the *p53* gene. Instead, it soon became apparent that the elevated protein levels were due entirely to the post-translational stabilization of the normally **labile** p53 protein.

In the years that followed, an even greater diversity of cell-physiologic signals were found capable of provoking increases in p53 levels. Among these were low oxygen tension (hypoxia), which is experienced by cells, normal and malignant, that lack adequate access to the circulation and thus to oxygen borne by the blood. Still later, introduction of either the adenovirus *E1A* or *myc* oncogene (see Sections 8.5 and 8.9) into cells was also found to be capable of causing increases in p53 levels.

By now, the list of stimuli that provoke increases in p53 levels has grown even longer. Expression of higher-than-normal levels of the E2F1 transcription factor, widespread demethylation of chromosomal DNA, and a deficit in the nucleotide precursors of DNA all trigger p53 accumulation. Exposure of cells to nitrous oxide or to an acidified growth medium, depletion of the intracellular pool of ribonucleotides, and blockage of either RNA or DNA synthesis also increase p53 levels.

These various observations made it clear that a diverse array of sensors are responsible for monitoring the integrity and functioning of various cellular systems. When these sensors detect damage or aberrant functioning, they send signals to p53 and its regulators, resulting in a rapid increase in p53 levels within a cell (**Figure 9.8**).

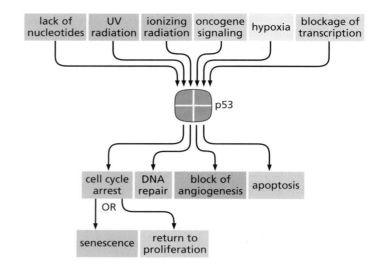

Figure 9.8 p53-activating signals and p53's downstream effects Studies of p53 function have revealed that a variety of cell-physiologic stresses can cause a rapid increase in p53 levels. The resulting accumulated p53 protein then undergoes post-translational modifications and proceeds to induce a number of responses. A cytostatic response ("cell cycle arrest," often called "growth arrest") can be either irreversible ("senescence") or reversible ("return to proliferation"). DNA repair proteins may be mobilized as well as proteins that antagonize blood vessel formation ("block of angiogenesis"). As an alternative, in certain circumstances, p53 may trigger apoptosis.

The same **genotoxic** (that is, DNA-damaging) agents and physiologic signals that provoked p53 increases were already known from other work to act under certain conditions in a **cytostatic** fashion, forcing cells to halt their advance through the cell cycle, a response often called "growth arrest." In other situations, some of these stressful signals might trigger activation of the apoptotic (cell suicide) program. These observations, when taken together, showed a striking parallel: toxic agents that induced growth arrest or apoptosis were also capable of inducing increases in p53 levels. Because such observations were initially only correlations, they hardly proved that p53 was involved in some fashion in *causing* cells to enter into growth arrest or apoptosis following exposure to toxic or stressful stimuli.

The definitive demonstrations of causality came from detailed examinations of p53 functions. For example, when genotoxic agents, such as X-rays, evoked an increase in cellular p53 levels, the levels of the p21^{Cip1} protein (see Section 8.4) increased subsequently; this induction was absent in cells expressing mutant p53 protein. This suggested that p53 could halt cell cycle advance by inducing expression of this widely acting CDK inhibitor (**Figure 9.9A**). Indeed, the long-term biological responses to irradiation were often affected by the state of a cell's *p53* gene. Thus, cells carrying mutant *p53* alleles showed a greatly decreased tendency to enter into growth arrest or apoptosis when compared with wild-type cells that were exposed in parallel to this stressor (see Figure 9.9B).

These various observations could be incorporated into a simple, unifying mechanistic model: p53 continuously receives signals from a diverse array of surveillance systems. If p53 receives specific alarm signals from these monitors, it calls a halt to cell proliferation or triggers the apoptotic suicide program (see Figure 9.8).

Figure 9.9 p53 and the radiation response Exposure of cells to X-rays serves to strongly increase p53 levels. (A) Once it is present in higher concentrations (8, 24 hours) and is functionally activated via various covalent modifications (*not measured here*), p53 induces expression of the p21^{Cip1} protein (see Section 8.4). p21^{Cip1} acts as a potent CDK inhibitor of the cyclin–CDK complexes that are active in late G$_1$, S, G$_2$, and M phases and can thereby halt further cell proliferation at any of these phases of the cell cycle. The actin protein is included in all three samples as a "loading control" to ensure that equal amounts of protein were added to the three gel channels prior to electrophoresis. (B) Thymocytes (leukocytes derived from the thymus) of wild-type mice show an 80% loss of viability relative to untreated control cells during the 25 hours following X-irradiation (*green*), while thymocytes from *p53*$^{+/-}$ heterozygous mice (with one wild-type and one null allele) show almost as much loss of viability (*red*). In contrast, thymocytes prepared from *p53*$^{-/-}$ homozygous mutant mice exhibit less than a 5% loss of viability during this time period (*blue*). In all cases, the loss of viability was attributable to apoptosis (*not shown*). (A, courtesy of H. Vousden. B, from S.W. Lowe et al., *re* 362:847–849, 1993.)

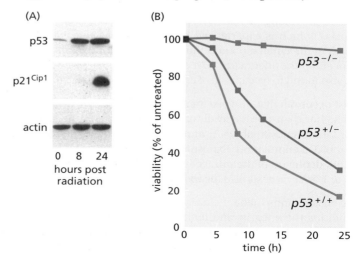

In fact, these cytostatic and pro-apoptotic powers of p53 represent a major threat to incipient cancer cells that are advancing toward the malignant growth state. A number of stresses, including hypoxia, genomic damage, and imbalances in the signaling pathways governing cell proliferation, are commonly experienced by cancer cells during various stages of tumor development. In the presence of any one of these stresses, an intact, functional p53 alarm system threatens the viability of would-be cancer cells. Consequently, p53 activity must be blunted or even fully eliminated in these cells if they are to survive and prosper.

This explains why most and perhaps all human tumor cells have partially or totally inactivated their p53 alarm response. Without p53 on duty, cancer cells are far more able to tolerate hypoxia, extensive damage to their genomes, and profound dysregulation of their growth-controlling circuitry. Once a cell acquires resistance to these normally debilitating factors, the road is paved for it and its descendants to continue their march toward a highly malignant growth state. In the same vein, normal cells must also avoid excessive p53 activity, since it threatens to end their lives and thereby cause depletion of the cells needed to maintain normal bodily functions (**Sidebar 9.1**).

9.6   DNA damage and deregulated growth signals cause p53 stabilization

Three well-studied monitoring systems, two of which have already been cited, send alarm signals to p53 in the event that they detect damage or signaling imbalances. The first of these responds to double-strand breaks (DSBs) in chromosomal DNA, notably those that are created by ionizing radiation such as X-rays. Indeed, a single dsDNA break occurring anywhere in the genome seems sufficient to induce a measurable increase in p53 levels. The identities of the proteins that detect such breaks are slowly being resolved; it is known that these sensors of dsDNA breaks transfer signals to the ATM kinase (**Figure 9.10**). (As described in Section 12.12, a deficiency of ATM leads to the disease of ataxia telangiectasia and to hypersensitivity of cells to X-irradiation.) ATM, in turn, transfers its signals on to the Chk2 kinase, which is able to phosphorylate p53 itself; this phosphorylation of p53 protects it from destruction by a protein known as Mdm2, discussed in the next section (see Figure 9.12).

A second signaling pathway is activated by single-strand DNA (ssDNA), which develops at stalled replication forks, often because DNA polymerases encounter bases that have been altered by a wide variety of DNA-damaging agents, including certain chemotherapeutic drugs and UV radiation. ssDNA sensors activate ATR kinase, which acts via the Chk1 kinase, to phosphorylate p53, again protecting it from degradation.

A third pathway leading to p53 activation is triggered by aberrant growth signals, notably those that result in deregulation of the pRb–E2F cell cycle control pathway. As we will see below, this pathway does not depend upon kinase intermediates to induce increases in p53 levels and signaling. The mechanisms by which other physiologic stresses or imbalances, such as hypoxia, trigger increases in p53 levels remain poorly understood.

These converging signaling pathways reveal a profound vulnerability of the mammalian cell. Through the course of evolution, a single protein—p53—has become entrusted with the task of receiving signals from lookouts that monitor a wide variety of important physiologic and biochemical intracellular systems (see Figure 9.8). The funneling of these diverse signals to a single protein would seem to represent an elegant and economic design of the cellular signaling circuitry. But it also puts cells at a major disadvantage, since loss of this single protein from a cell's regulatory circuitry results in a catastrophic loss of the cell's ability to monitor its own well-being and respond with appropriate countermeasures in the event that certain operating systems malfunction.

In one stroke (actually, the two strokes that cause successive inactivation of the two *p53* gene copies), the cell becomes blind to many of its own defects. It thereby gains the ability to continue proliferation under circumstances that would normally cause

Figure 9.10 Control of p53 levels by Mdm2 After p53 concentrations increase in response to certain physiologic signals (*not shown*), the p53 tetramers bind to the promoters of a large constituency of target genes whose transcription they induce (*above*), including the gene encoding Mdm2; this results in a large increase in *mdm2* mRNA and Mdm2 protein (*right*). The Mdm2 molecules then bind to the p53 protein subunits and initiate their ubiquitylation, resulting in export to the cytoplasm, further ubiquitylation (not shown), resulting in degradation in proteasomes. This negative-feedback loop ensures that p53 levels eventually sink back to a low level and, in undisturbed cells, helps to keep p53 levels very low.

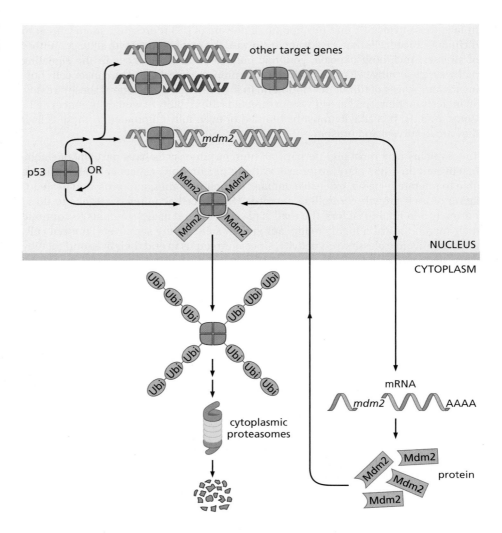

it to call a halt to proliferation or to enter into apoptotic death. In addition, as we will learn shortly, loss of the DNA repair and genome-stabilizing functions promoted by p53 will make descendants of a *p53*$^{-/-}$ cell more likely to acquire further mutations and advance more rapidly down the road of malignancy (**Sidebar 9.2**).

9.7 Mdm2 destroys its own creator

The diverse alarm signals that impinge on p53 have a common effect—causing a rapid increase in the levels of the p53 protein. Researchers have begun to understand how this dramatic change is achieved. Like a wide array of other cellular proteins, p53 protein molecules are degraded by the ubiquitin–proteasome system (see Supplementary Sidebar 7.5). Proteins that are destined to be degraded by this system are initially tagged by the covalent attachment of polyubiquitin side chains, which causes the proteins to be transported to proteasomes, in which they are digested into oligopeptides. The critical control point in this process is the initial tagging process.

The degradation of p53 in normal, unperturbed cells is regulated by a protein termed Mdm2 (in mouse cells) and MDM2 (in human cells). This protein recognizes p53 as a target that should be ubiquitylated shortly after its synthesis and therefore marked for rapid destruction (**Figure 9.12**). Mdm2 was initially identified as a protein encoded by double-minute chromosomes present in murine sarcoma cells (hence **m**ouse **d**ouble **m**inutes). Subsequently, the human homolog of the mouse gene (*HDM2*) was discovered to be frequently amplified in sarcomas. In many human lung tumors, Mdm2 (as we will call it) is overexpressed through mechanisms that remain unclear.

Sidebar 9.1 Too much of a good thing: a hyperactive p53 protein causes premature aging The descriptions of p53 actions in this chapter include ample evidence that the p53 alarm, once activated, provides important protection against the development of cancer. What would happen if a *p53* allele that encodes a constitutively active form of the protein were inserted into the mouse germ line? This has occurred through an experimental accident, in which the first six exons of the *p53* gene were inadvertently deleted during attempts to replace a germ-line wild-type *p53* allele with a point-mutated allele. The p53 protein encoded by this truncated allele behaves as if it were continuously active, even without the alarm signals that are normally required to activate it. Mice heterozygous for this allele are totally protected from the lymphomas, osteosarcomas, and soft-tissue (non-bone) sarcomas that commonly afflict wild-type mice late in life. And fibroblasts from these mice are more resistant than their wild-type counterparts to transformation *in vitro* by an introduced *ras* oncogene. These outcomes, among many others, show that p53 activity does indeed protect tissues from spawning tumors, ostensibly by eliminating cancerous cells before they have had a chance to proliferate into tumors of a significant size.

However, rather than living longer lives, these "cancer-resistant" mice showed a premature aging syndrome, their lifespan being reduced by some 20%. The accelerated aging included changes frequently observed in aged humans, such as development of a hunched spine, retarded wound healing, reduced replacement of lost white blood cells, plus losses of weight, vigor, muscle mass, bone density, and hair (**Figure 9.11A**). In many tissues, there was a widespread depletion of cells, suggesting a loss of self-renewing stem cells.

These phenotypes are consistent with the idea that p53 plays a role in the aging process, but they hardly prove it. An experiment of nature seems to address these issues more directly: a naturally occurring polymorphism in the human gene pool places a proline residue in place of an arginine in residue 72 of p53. *In vitro* experiments show that the Arg^{72} protein is as much as fivefold more effective in triggering apoptosis than the Pro^{72} form of the protein. An epidemiologic study carried out in the Netherlands of a population over 85 years of age demonstrated that those with a *p53$^{Pro/Pro}$* genotype had a 2.54-fold increased risk of cancer (compared with those of a *p53$^{Arg/Arg}$* genotype). However, those with a *p53$^{Pro/Pro}$* genotype showed an overall 41% greater survival during the period studied (see Figure 9.11B). Moreover, deaths from general exhaustion and frailty were observed in 21% of the *Arg/Arg* subjects but in only 6% of the *Pro/Pro* individuals. Observations like these add weight to the notion that increased activity of p53 affords greater protection against cancer but at the same time accelerates the age-dependent deterioration of tissues. Hence, too much vigilance by the p53 watchman may incur a heavy price on the organism as a whole.

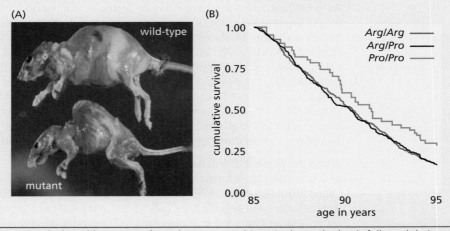

Figure 9.11 Premature aging induced by an overly active p53 The p53 protein may afford protection against cancer and at the same time accelerate the aging process. (A) Attempts at altering a germ-line copy of the *p53* gene in mice resulted, inadvertently, in the creation of an allele of *p53* that specifies a constitutively active p53 protein. As seen here, the resulting mutant mice (*below,* show many of the attributes of aging seen in humans, including muscle atrophy and a hunching of the back, when compared with wild-type mice of similar age (*above*); both mice were skinned. In addition, mutants exhibited osteoporosis, atrophy of numerous organs, and diminished tolerance to stress. (B) A prospective study of 1226 85-year-old subjects in the Netherlands followed their survival in the years after their enlistment. As is apparent from this Kaplan–Meier plot, those with a *Pro/Pro* allele (affecting residue 72 of p53) exhibited a markedly improved survival in the decade that followed compared with heterozygotes or those with an *Arg/Arg* allele. (While consistent with a key role of *p53* in governing longevity, this striking, statistically significant association does not definitively prove such a role, since it remains formally possible that these alleles are tightly linked with other loci on Chromosome 17 that are the actual determinants of longevity.) (A, from S.D. Tyner et al., *Nature* 415:45–53, 2002. B, from D. van Heemst et al., *Exp. Gerontol.* 40:11–15, 2005.)

As is the case with other oncogenes, it seemed at first that amplification of the *mdm2* gene (indicated by the presence of many double-minute chromosomal particles in tumor cells; see Figure 1.12B) afforded tumor cells some direct, immediate proliferative advantage. Only long after the Mdm2 protein was first identified did its role as the agent of p53 destruction become apparent. In fact, the detailed effects exerted by Mdm2 on p53 are slightly more complex than indicated above.

Sidebar 9.2 Sunlight, p53, and skin cancer The p53 protein stands as an important guardian against skin cancer induced by sunlight. In the event that the genome of a **keratinocyte** in the skin has suffered extensive damage from ultraviolet-B (UV-B) radiation, p53 will rapidly trigger its apoptotic death. One manifestation of this is the extensive scaling of skin several days after a sunburn. At the same time, UV-B exposure may cause the mutation and functional inactivation of a *p53* gene within a keratinocyte. This is indicated by the fact that mutant *p53* alleles found in human squamous cell carcinomas of the skin often occur at dipyrimidine sites—precisely the sites at which UV-B rays induce the formation of pyrimidine–pyrimidine cross-links (see Section 12.6). Such mutant *p53* alleles can also be found in outwardly normal skin that has suffered chronic sun damage. Once p53 function is compromised by these mutations, keratinocytes may be able to survive subsequent extensive exposures to UV-B irradiation, because apoptosis will no longer be triggered by their p53 protein. Moreover, loss of p53 results in a diminished ability to repair subsequent UV-B–induced DNA lesions. Hence, *p53*-mutant cells may subsequently acquire additional mutant alleles that enable them, together with the mutant *p53* alleles, to form a squamous cell carcinoma.

Human papillomaviruses (HPVs) are increasingly implicated as co-factors in many of these squamous cell carcinomas; a key function of the E6 virus–encoded oncoprotein may explain the synergistic actions of UV-B radiation and HPV in the pathogenesis of these relatively common tumors: E6 tags p53 for destruction by ubiquitylation and degradation in proteasomes, thereby phenocopying mutational inactivation of the chromosomal *p53* gene (Supplementary Sidebar 9.1). Interestingly, mice that lack functional *p53* gene copies in all cells also respond to UV-B exposure by developing uveal melanomas—tumors of pigmented cells in the front of the eye; similar tumors are suspected to be caused in humans by UV exposure.

Of additional interest, p53 operating in keratinocytes has another totally unrelated effect that illustrates its diverse functions. In response to the DNA damage created by UV radiation, p53 causes these cells to release melanocyte-stimulating hormone (αMSH); the latter proceeds to stimulate nearby skin melanocytes to produce melanin pigment and to transfer resulting melanin granules back to the keratinocytes (see Figure 12.19), resulting in the increased pigmentation of the skin that creates suntan!

As we will learn below, p53 operates by acting as a transcription factor; Mdm2 binding to p53 immediately blocks the ability of p53 to function in this role. [In more detail, Mdm2 succeeds in shutting down p53-driven transcription by (1) preventing the binding to p53 of p300/CBP, which activate transcription by acetylating histones; and (2) by actively recruiting yet other enzymes that block p53-mediated transcription by methylating histones (see Section 1.8).] Thereafter, Mdm2 directs the attachment of a ubiquitin moiety to p53 and the export of p53 from the nucleus (where p53 does most of its work) to the cytoplasm; subsequent polyubiquitylation of p53 ensures its rapid degradation in cytoplasmic proteasomes. The continuous, highly efficient actions of Mdm2 ensure the short, 20-minute half-life of p53 in normal, unstressed cells.

While the present discussion and Figure 9.11 represent Mdm2 as a monomeric protein, it actually often forms heterodimeric complexes with its close cousin, MdmX (also called Mdm4). This complex may be responsible for much of the ubiquitylation activity that drives p53 degradation. Indeed, there is evidence that without the presence of MdmX, Mdm2 loses the ability to drive p53 degradation. When expressed on its own, however, MdmX seems to be limited to blocking p53-mediated transcriptional activation. (Moreover, MdmX differs in another important respect from its Mdm2 cousin: its expression is not regulated by p53, a process that is described below.)

In some circumstances—specifically, when cells are suffering certain types of stress or damage—p53 protein molecules must be protected from their Mdm2 executioner so that they can accumulate to functionally significant levels in the cell. This protection is often achieved by phosphorylation of p53, which blocks the ability of Mdm2 to bind p53 and trigger its ubiquitylation. More specifically, phosphorylation of p53 on amino acid residues in its N-terminal domain (see Figure 9.12) by kinases such as ATM, Chk1, and Chk2 (which become activated in response to DNA damage, as was described in Section 9.6) alters the domain of p53 that is normally recognized and bound by Mdm2, and in this way prevents the association of Mdm2 with p53. At the same time, the DNA damage–activated ATM kinase can phosphorylate Mdm2 in a way that causes its functional inactivation and destabilization. As a consequence of this phosphorylation of both p53 and Mdm2, Mdm2 fails to initiate ubiquitylation of p53, p53 escapes destruction, and p53 concentrations in the cell increase rapidly (**Figure 9.13**). Once it has accumulated in substantial amounts, p53 is then poised to evoke a series of downstream responses, to be discussed in detail later.

Note that Mdm2 operates here as an oncoprotein, but one whose mechanism of action is very different from those of the various oncoproteins that we encountered in Chapters 4, 5, and 6. The latter function as components of mitogenic signal cascades and thereby induce cell proliferation by mimicking the signals normally triggered by the binding of growth factors to their receptors. Mdm2, in contrast, operates by antagonizing p53 and thereby prevents entrance of a cell into cell cycle arrest, into the nongrowing state known as **senescence**, or into the apoptotic suicide program. The final outcome is, however, the same: the actions of oncoproteins and Mdm2 both favor increases in cell number.

The activity and levels of the Mdm2 protein are affected by yet other positive and negative signals. The signaling pathway that favors cell survival through activation of the PI3 kinase (PI3K) pathway leads, via the Akt/PKB kinase, to Mdm2 phosphorylation (at a site different from that altered by the ATM kinase described above) and to the resulting translocation of Mdm2 from the cytoplasm to the nucleus, where it is poised to attack p53 (see Figure 9.13A). Because PI3K itself is activated by Ras and growth factor receptors, we come to realize that the mitogenic signaling pathway does indeed influence Mdm2 and thereby p53, albeit indirectly. At the same time, activation of the mitogenic Ras → Raf → MAPK signaling pathway leads, via the Ets and AP-1 (Fos + Jun) transcription factors, to greatly increased transcription of the *mdm2* gene, yielding higher levels of *mdm2* mRNA and protein (see Figure 9.13A). These elevated levels of Mdm2 protein amplify the phosphorylation-induced activation of Mdm2 achieved by the PI3K → Akt/PKB signaling pathway. Ultimately, all these effects converge on suppressing p53 protein levels.

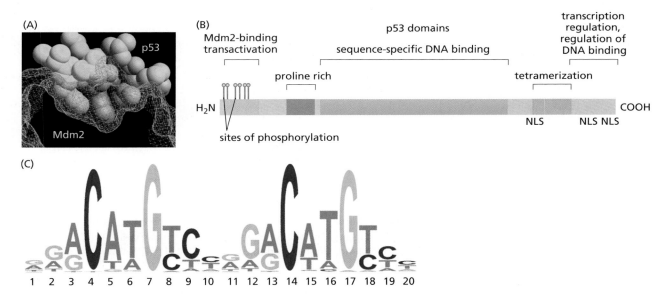

Figure 9.12 Specialized domains of p53 (A) The structure of the interface where p53 and Mdm2 interact has been revealed by X-ray crystallography. The interacting domain of p53 is shown as a yellow space-filling model that includes p53 residues 18 through 27, while the surface of the complementary pocket of Mdm2 is shown as a blue wire mesh. (B) The interaction of p53 with Mdm2 (see Figure 9.10) occurs in a small domain near its N-terminus, where the transactivation domain of p53 is also located. The phosphorylation of p53 amino acid residues in this region (*red lollipops; not all are indicated*) blocks Mdm2 binding and thus saves p53 from ubiquitylation and degradation. The nearby proline-rich domain (*salmon*) contributes to p53's pro-apoptotic functions. Its tetramerization domain is located toward its C-terminus (see Figure 9.6). Nearby are nuclear localization signals (NLS), which allow import of recently synthesized p53 into the nucleus, as well as amino acid sequences that regulate its DNA binding. (C) p53–DNA complexes present in the chromatin of human cells can be immunoprecipitated by anti-p53 antibodies (the ChIP procedure; see Supplementary Sidebar 8.3). In one such experiment, sequence analyses of DNA fragments in the precipitates led to the identification of 1546 sites in the human genome to which p53 bound after cells were stressed by exposure to the drug actinomycin, a potent inhibitor of transcription. The consensus DNA sequence to which p53 bound is shown here, where the relative size of each letter indicates how frequently a DNA base was found at the indicated position in the binding site. Interestingly, the majority of these 1546 sites could also be bound by p53's cousins, p63 and p73, to be described later in this chapter. (A, from P.H. Kussie et al., *Science* 274:948–953, 1996. B, from D.E. Fisher, ed., Tumor Suppressor Genes in Human Cancer. Totowa, NJ: Humana Press, 2000. C, from L. Smeenk et al., *Nucleic Acids Res.* 36:3639–3654, 2008.)

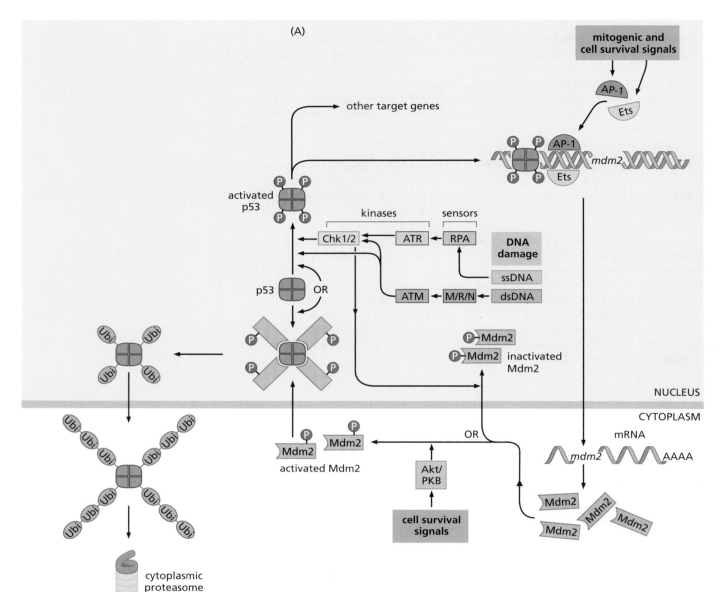

(A)

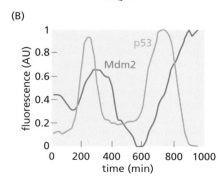

(B)

Yet another mechanism that affects Mdm2 has been revealed through the discovery of an Mdm2 antagonist, which is termed p19^{ARF} in mouse cells and p14^{ARF} in human cells. Astute sequence analysis led to the discovery of ARF, as we will call it hereafter. Its encoding gene was originally uncovered in mouse cells as a gene whose sequences are intertwined with those specifying p16^{INK4A}, the important inhibitor of the CDK4 and CDK6 kinases that initiate pRb phosphorylation (see Section 8.4).

Through use of a transcriptional promoter located 13 kilobases upstream of the *p16^{INK4A}* promoter and an alternative splicing program, an mRNA is assembled that encodes, in an **a**lternative **r**eading **f**rame, the structure of the ARF protein (**Figure 9.14**). Forced expression of an ARF-encoding cDNA in wild-type rodent cells was found to cause a strong inhibition of their proliferation. However, this inhibition was not observed when the ARF cDNA was expressed in cells that lacked wild-type p53 function. This indicated that the growth-inhibitory powers of ARF depend absolutely on the presence of functional p53 in these cells.

Further investigation revealed that in wild-type cells, the expression of ARF causes a rapid increase in p53 levels. We now understand the molecular mechanisms that explain how this response works. ARF binds to Mdm2 and inhibits its action, either by sequestering Mdm2 in the nucleolus—the nuclear structure that is largely devoted to manufacturing ribosomal subunits—or by inhibiting Mdm2 in the **nucleoplasm** (**Figure 9.15A**). Once Mdm2 is diverted from interacting with p53, the latter escapes

Figure 9.13 **Control of p53 levels by various kinases** (A) The cycle of p53 synthesis and destruction indicated in Figure 9.10 can be modulated by a series of regulators. DNA damage-sensors, such as RPA (replication protein A) and the M/R/N (MRE11/Rad50/Nbs1) complex, detect either extensive ssDNA regions/replication-blocking DNA lesions or dsDNA breaks (DSBs) and proceed to activate two kinases, ATR and ATM, respectively. These kinases act directly on p53, or indirectly via Chk1 and Chk2, to phosphorylate p53 (*center*) in its N-terminal domain (see Figure 9.12B), thus preventing the binding of Mdm2 (*gold, left center*). At the same time, phosphorylation of Mdm2 molecules by these kinases inactivates them, blocking their ability to associate with p53 (*bottom center*). These alterations save p53 from Mdm2-mediated binding, ubiquitylation, and destruction in proteasomes (*lower left*). Acting in an opposing manner, certain survival signals (such as those conveyed by mitogenic growth factors), acting through the AP-1 and Ets transcription factors, collaborate with p53 to promote expression of the *mdm2* gene (*top right*), resulting in increases in *mdm2* mRNA and Mdm2 protein synthesis (*lower right*). These survival signals also activate the Akt/PKB kinase, which activates already-synthesized Mdm2 molecules by phosphorylating them at another site (*bottom*). The activated Mdm2 then proceeds to bind p53 and trigger its ubiquitylation and proteasome-mediated destruction. Not illustrated is the Mdm2-driven destruction of pRb by ubiquitylation in physiologically stressed cells. (B) The mutually inhibitory interactions between p53 and Mdm2, which form a reciprocal negative-feedback loop (see Figure 9.10), result in oscillations of the levels of the two proteins. These levels can be monitored in individual cells using proteins labeled with different-colored fluorescent tags. Following DNA-damaging X-irradiation, the height and duration of each pulse are not affected by radiation dose, but the number of successive pulses, which form a digital clock, increases with increasing radiation dose, perhaps continuing until the DNA is repaired or the cell dies. The fluorescence intensities (reflective of the levels of the two proteins) are presented in arbitrary units (AU). (B, from G. Lahav et al., *Nat. Genet.* 36:147–150, 2004.)

Mdm2-mediated ubiquitylation and resulting destruction and therefore accumulates rapidly to high levels in the cell. The enemy of an enemy is a friend: ARF can induce rapid increases in p53 levels because it kidnaps and inhibits p53's destroyer, Mdm2.

Importantly, in normal, unstressed cells, Mdm2 must be allowed to perform its normal role of keeping p53 levels very low, as is highlighted by the results of inactivating both *mdm2* gene copies in the genomes of mouse embryos. These embryos die very early in embryogenesis, ostensibly because p53 levels increase to physiologically intolerable levels, preventing the normal proliferation of embryonic cells or causing them to die. (That the profoundly disruptive effects of *Mdm2* gene inactivation are due to runaway p53 activity is made clear by studies in which both genes in a mouse embryo are inactivated homozygously, yielding the *Mdm2*$^{-/-}$ *p53*$^{-/-}$ genotype. Once p53 is eliminated from the embryonic cells, the loss of Mdm2 becomes tolerable and embryonic development occurs normally!)

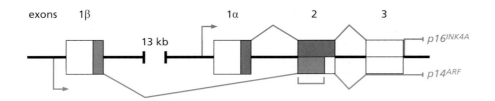

Figure 9.14 **The gene encoding p16^{INK4A} and p14/p19^{ARF}** Analysis of the *p16^{INK4A}* gene (*red*) has revealed that it shares its second exon with a second gene encoding a 19-kD protein in mice and a 14-kD protein in humans. The *p14/p19* gene uses an alternative transcriptional promoter (*blue arrow, left*) located more than 13 kilobases upstream of the one used by *p16^{INK4A}* (*red arrow, center*). Because translation of its mRNA uses an alternative reading frame (*green bracket*) in exon 2 (*red, blue*), the resulting protein and thus gene came to be called p19^{ARF} (or in humans p14^{ARF}). The patterns of RNA splicing are indicated by the carets connecting the various exons of the two intertwined genes. The boxes indicate exons, while the filled areas within each exon indicate reading frames. (From C. Sherr, *Genes Dev.* 12:2984–2991, 1998.)

The series of mutual antagonisms indicated in Figure 9.15 makes ARF an ally of p53 and, like p53, a tumor suppressor protein. In many human tumors, inactivation of the $p16^{INK4A}/p14^{ARF}$ locus by genetic mutation or epigenetic promoter methylation can be demonstrated. Once a cell has lost ARF activity, it loses the ability to block Mdm2 function. As a consequence, Mdm2 is given a free hand to drive p53 degradation, and the cell is deprived of the services of p53 because the latter can never accumulate to functionally significant levels.

Since ARF has a central role in increasing p53 levels in many cellular contexts, this means that the *p14^{ARF}* gene, like the gene encoding its p53 target, is an extremely important tumor suppressor gene. Moreover, it seems likely that many of the human cancer cells that retain wild-type *p53* gene copies have eliminated p53 function by inactivating their two copies of the gene encoding ARF. Finally, we should note that the co-localization of the *p16^{INK4A}* and *p14^{ARF}* genes (see Figure 9.14) represents yet another concentration of power that creates additional vulnerability for normal cells (**Sidebar 9.3**).

9.8 ARF and p53-mediated apoptosis protect against cancer by monitoring intracellular signaling

The influential role of ARF in increasing p53 levels raises the question of how ARF itself is regulated. In this instance, we learn something highly relevant from our discussion in Chapter 8 of the pRb pathway, and from the fact that mammalian cells are very sensitive to higher-than-normal levels of E2F1 activity (see Figure 9.15B). In fact, a cell seems to monitor the activity level of this particular transcription factor (together with those of E2F2 and E2F3) as an indication of whether its pRb circuitry is functioning properly; excessively high levels of active E2F transcription factors provide a telltale sign that pRb function has gone awry.

Evolution has created several ways to eliminate cells that carry too much E2F activity and, by implication, have lost proper pRb control (**Figure 9.16**). Runaway E2F1 activity drives expression of a number of genes encoding proteins that directly participate in the apoptotic program. Included among these are genes encoding **caspases** (types 3, 7, 8, and 9), pro-apoptotic Bcl-2–related proteins (Bim, Noxa, PUMA), Apaf-1, and p53's cousin, p73; these proteins collaborate to drive cells into apoptosis. We will learn more about them later.

In addition, the p53-dependent apoptotic program is often triggered by elevated E2F activity. It turns out that the *p14^{ARF}* gene carries an E2F recognition sequence in its promoter. In a way that is still incompletely understood, unusually high levels of E2F1, E2F2, or E2F3A activity induce transcription of *p14^{ARF}* mRNA. The ARF protein soon

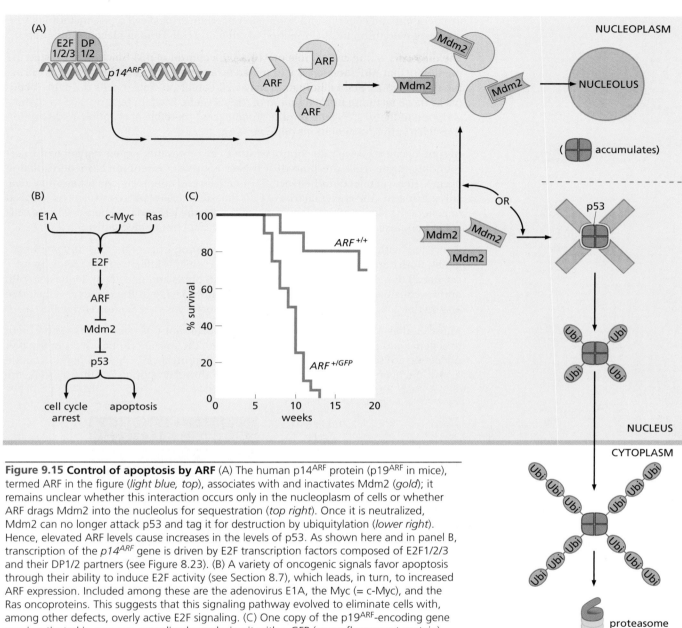

Figure 9.15 Control of apoptosis by ARF (A) The human p14^ARF protein (p19^ARF in mice), termed ARF in the figure (*light blue, top*), associates with and inactivates Mdm2 (*gold*); it remains unclear whether this interaction occurs only in the nucleoplasm of cells or whether ARF drags Mdm2 into the nucleolus for sequestration (*top right*). Once it is neutralized, Mdm2 can no longer attack p53 and tag it for destruction by ubiquitylation (*lower right*). Hence, elevated ARF levels cause increases in the levels of p53. As shown here and in panel B, transcription of the *p14^ARF* gene is driven by E2F transcription factors composed of E2F1/2/3 and their DP1/2 partners (see Figure 8.23). (B) A variety of oncogenic signals favor apoptosis through their ability to induce E2F activity (see Section 8.7), which leads, in turn, to increased ARF expression. Included among these are the adenovirus E1A, the Myc (= c-Myc), and the Ras oncoproteins. This suggests that this signaling pathway evolved to eliminate cells with, among other defects, overly active E2F signaling. (C) One copy of the p19^ARF-encoding gene was inactivated in a mouse germ line by replacing it with a GFP (green fluorescent protein)-encoding sequence. These mice (*ARF^+/GFP*) and wild-type mice (*ARF^+/+*) were mated with others carrying an Eμ-*myc* transgene known to cause B-cell lymphomas (also described in Figure 9.23). The *ARF^+/GFP* Eμ-*myc* mice (*red line*) developed fatal tumors far more rapidly than mice carrying only the Eμ-*myc* transgene (*green line*), and cells in these tumors shed their remaining wild-type *ARF* allele. Hence, in the absence of ARF function, the pro-apoptotic effects of a *myc* oncogene (B) are largely lost, allowing its proliferative effects to dominate and drive tumor formation. (B, courtesy of P.J. Iaquinta and J.A. Lees. C, from F. Zindy et al., *Proc. Natl. Acad. Sci. USA* 100:15930–15935, 2003.)

appears on the scene and blocks Mdm2 action. p53 then accumulates and triggers, in turn, apoptosis (see Figure 9.15B), leading to a signaling cascade configured like this:

$$pRB \dashv E2F \rightarrow ARF \dashv Mdm2 \dashv p53 \rightarrow apoptosis$$

This pathway, working together with the several p53-independent pro-apoptotic signals cited above, accomplishes the goal of eliminating cells that lack proper pRb function. Such E2F-initiated apoptosis seems to explain why mouse embryos that have been deprived of both copies of their *Rb* gene die in mid-gestation due to the excessive proliferation and concomitant apoptosis of certain critical cell types, including those involved in erythropoiesis (formation of red blood cells) and in placental function.

The discovery of the critical role of ARF in the control of p53 function suggested the possibility that ARF function is eliminated by a variety of molecular strategies during tumor formation. Such elimination may well confer on cancer cells the same benefits as those resulting from mutation of the *p53* gene itself. In fact, regulation of transcription of the *p14^{ARF}* gene is quite complex and therefore susceptible to disruption through a variety of alterations (Sidebar 9.4).

In sum, because loss of pRb control within a cell represents a grave danger to the surrounding tissue, cells are poised to trigger apoptosis whenever E2F1 deregulation occurs. These connections between E2F1 activity and apoptosis suggest another idea, still speculative: the great majority of cells that suffer loss of pRb control never succeed in generating clones of pre-neoplastic or neoplastic descendants, because these cells suffer apoptosis as soon as they lose this important control mechanism.

Consistent with this logic are some of the known properties of the *E1A* and *myc* oncogenes. Both deregulate pRb control, and both are highly effective in inducing apoptosis. Recall that the adenovirus E1A oncoprotein binds and effectively sequesters pRb and its cousins. Myc, for its part, pulls regulatory levers in the cell cycle clock that ensure that pRb is inactivated through phosphorylation (see Sections 8.5 and 8.9).

Many studies of *myc* oncogene function indicate that this gene exerts both potent mitogenic and pro-apoptotic functions. Indeed, the pro-apoptotic effects of the *myc* oncogene are so strong that it is highly likely that most cells that happen to acquire a *myc* oncogene are also rapidly eliminated through apoptosis. On occasion, the

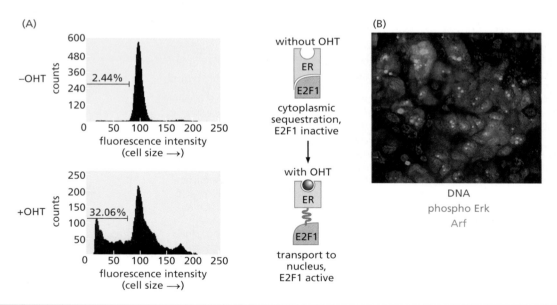

Figure 9.16 Induction of apoptosis and ARF expression by E2F1 and Ras. (A) The apoptotic state of cells can be monitored by fluorescence-activated cell sorting (FACS), which in this case is used to measure the size of individual cells or subcellular fragments (*abscissa*) and the number of cells of a given size (*ordinate*). In this experiment, the E2F1 transcription factor (*red*) has been fused to the estrogen receptor (ER) protein (*green*), making E2F1 activity dependent on the presence of tamoxifen (OHT), a ligand of the ER. In the absence of tamoxifen (*upper panel*), the E2F1 factor is sequestered in the cytoplasm; under this condition almost all cells in such a population have a size of roughly 100 (arbitrary) units (with 2.44% having a smaller size). However, when tamoxifen is added to these cells (*lower panel, purple ball*), a nuclear localization signal (NLS) is exposed and the E2F1-containing fusion protein is imported into the nucleus, where it induces the expression of a cohort of genes, among them those that have pro-apoptotic effects. As a result, a significant proportion (32.06%) of the cells now show a size smaller than that of normal healthy cells, indicative of their having fragmented during apoptosis. (B) In a transgenic mouse model of lung adenocarcinoma development, signaling by Ras oncoprotein and its activated downstream effector, phosphorylated Erk/MAPK (see Figure 6.14), increases as tumors progress from lower-grade adenomas to higher-grade adenocarcinomas. As seen here, in those cancer cells in which phosphorylated (and thus activated) Erk (*red*) is apparent in the cytoplasm, there are clusters of ARF (*green*) in the nucleus; conversely, in some lower-grade cancer cells, neither of these markers is apparent and only the DNA stain (*blue*) can be seen. (A, courtesy of K. Helin, from T. Hershko and D. Ginsberg, *J. Biol. Chem.* 279:8627–8634, 2004. B, from D.M. Feldser et al., *Nature* 468:572–575, 2010.)

Sidebar 9.4 Elimination of ARF (and thus p53) often occurs through alterations that affect the transcription of the ARF gene ARF function is often eliminated in cancer cells either through mutation of the encoding DNA sequences or through methylation (see Section 7.8) of the $p14^{ARF}$ promoter. The direct consequence is suppression of p53 levels by Mdm2. An alternative and ostensibly equally effective way of suppressing *ARF* expression is used by the tumor cells in childhood acute lymphocytic leukemia (ALL) and in acute myelogenous leukemia (AML) of adults. A frequently observed chromosomal translocation is seen in the leukemic cells of these patients, in which a gene termed variously *AML1* or *Runx1* is fused to a second gene termed *ETO*, resulting in a gene that specifies an AML1–ETO fusion protein. AML1 is normally capable, on its own, of activating transcription of the *ARF* gene. However, when it becomes fused to ETO—a protein normally involved in transcriptional repression—the resulting fusion protein is directed by its AML1 DNA-binding portion to the *ARF* promoter, whereupon the associated ETO portion of the protein represses transcription of the $p14^{ARF}$ gene. This results in a failure to express ARF protein and thereby liberates pre-leukemic cells from p53 function. Similarly, TBX2, a repressor of $p14^{ARF}$ transcription, is overexpressed in some human breast cancers.

In mice infected by murine leukemia viruses (MLVs), expression of the *Bmi-1* gene is often activated by MLV provirus insertion (see Section 3.11), leading to overexpression of its product, which also functions as a repressor of *ARF* transcription. Yet another gene, termed *Dmp-1*, encodes a transcription factor that serves as an activator of *ARF* transcription; its deletion from the mouse germ line leads to cancer susceptibility, indicating that it functions as a tumor suppressor gene.

These various mechanisms help to explain why the *p53* gene is often found in wild-type configuration in the genomes of many human malignancies, since repression of *ARF* expression represents a highly effective alternative strategy for eliminating p53 protein and thus p53 function in cancer cells.

apoptotic program may be blunted or inactivated, and only then can the mitogenic actions of *myc* become apparent.

As an example, when a *myc* oncogene becomes activated in the lymphoid tissues of a mouse, it prompts a substantial increase in cell proliferation. However, there is no net increase in cell number, since the newly formed cells are rapidly lost through apoptosis. If one of the *myc* oncogene–bearing cells happens subsequently to inactivate its *p53* gene copies, then *myc*-induced apoptosis is diminished and the cell proliferation driven by *myc* leads to a net increase in the pool of mutant lymphoid cells. As might be expected from the organization of the pathway drawn in Figure 9.15B, a similar effect operates in mice that carry only a single functional $p19^{ARF}$ gene (see Figure 9.15C).

These discussions suggest that E2F-induced apoptosis functions solely as an anticancer mechanism designed to eliminate unwanted, pre-neoplastic cells. However, research with genetically altered mice provides evidence that normal physiologic mechanisms also depend on E2F-induced apoptosis to weed out extra cells that are not required for the development of a normal immune system (**Sidebar 9.5**).

While we have focused here on the role of the pRb–E2F axis in activating ARF, at least one other major oncogenic signaling pathway can also contribute significantly to increased ARF expression and thus to p53 activation: the signaling pathway

Sidebar 9.5 E2F1-induced apoptosis appears to participate in normal lymphoid development The depiction of the pRb → E2F1 → ARF pathway might suggest that this pathway is activated only in response to the type of pathologic deregulation occurring in cancer cells. But there are indications that mammals exploit this pathway as well during normal development. This is suggested by surprising results coming from the inactivation of both copies of the E2F1-encoding gene in the mouse germ line. The resulting loss of E2F1 activity should, by all rights, lead to loss of its growth-promoting function, since this transcription factor normally ushers the cell into S phase by activating genes required for DNA replication (see Section 8.7). However, in genetically altered mice deprived of both germ-line copies of the gene specifying E2F1, one observes, instead, reasonably normal development and, after birth, a hyperproliferation of their lymphocytes; some cells in this population may actually progress to form lymphomas months later.

A clue to explaining this paradoxical finding comes from the known behavior of lymphocytes during the development of the immune system: 99% of the cells that are initially formed are subsequently discarded because they have acquired undesirable reactivity against the body's own antigens or are otherwise dysfunctional (see Section 15.5). The immune system uses apoptosis as the means of jettisoning these unwanted cells.

These diverse phenomena—really pieces of a puzzle—can now be put together in the following, speculative model: In order to eliminate some of these unwanted lymphocytes, the immune system *intentionally* causes hyperactivation of E2F1 function in these cells. However, $E2f1^{-/-}$ mice, whose lymphocytes are incapable of activating this apoptotic pathway, are fated to accumulate vast numbers of these lymphocytes, which predispose them to the subsequent development of lymphomas. Similar dynamics seem to operate in other tissues, as evidenced by the hemangiosarcomas and lung adenocarcinomas that these mutant mice exhibit at elevated rates.

downstream of Ras. Hyperactive signaling by Ras or its immediate downstream partners, Raf or B-Raf (see Figure 6.14), drives increased ARF expression (see Figure 9.16B) and, acting via Erk/MAPK, the expression of p53. This imposes great selective pressure on incipient cancer cells that experience intense Ras oncoprotein signaling, forcing them to either neutralize p53 function or confront rapid elimination by p53-dependent apoptosis.

9.9 p53 functions as a transcription factor that halts cell cycle advance in response to DNA damage and attempts to aid in the repair process

The p53 protein has a DNA-binding domain with an affinity for binding a sequence motif composed of the sequence Pu-Pu-Pu-C-A/t-T/a-G-Py-Py-Py repeated twice in tandem (where Pu represents the purine nucleotides A or G while Py represents the pyrimidine nucleotides C or T; A/t represents a site at which A occurs more frequently than T; and T/a denotes a site where T occurs more frequently than A). Between 0 and 13 nucleotides of random sequence are found to separate these two tandemly arrayed recognition sequences (see Figure 9.12C). This sequence motif is present in the promoters or initial introns of a number of the downstream target genes whose expression is induced (or repressed) by p53.

Actually, the transcription-activating powers of p53 depend on more than its ability to recognize and bind this sequence within a promoter. In addition, a complex array of covalent modifications of p53 must occur, many affecting its C-terminal domain (see Figure 9.6C). These include acetylation, glycosylation, phosphorylation, ribosylation, and sumoylation (involving respectively attachment of acetyl, sugar, phosphate, ribose, and sumo groups, the latter being a ubiquitin-like peptide that appears to target proteins for localization to specific intracellular sites, often in the nucleus; see Figure 9.39). These modifications are likely to affect the ability of p53 to interact physically with other factors that modulate its transcriptional powers. [For example, phosphorylation of p53's N-terminal transactivation domain can increase its ability to bind the Taz2 domain of the p300 co-activator (see Figure 9.6C); the latter then contributes to transcriptional activation by acetylating nearby histones H3 and H4 as well as p53 itself.] Indeed, it seems likely that combinatorial interactions of p53 with other transcription factors determine the identities of the specific target genes that are activated in various circumstances by p53.

Significantly, as indicated in Figure 9.6B, the great majority (>90%) of the mutant *p53* alleles found in human tumor cell genomes encode amino acid substitutions in the DNA-binding domain of p53. The resulting defective p53 proteins, being unable to bind to the promoters of downstream target genes, have therefore lost the ability to mediate most of p53's multiple functions.

As described earlier, one key target of the p53 transcription factor is the *Mdm2* gene. Consequently, when active as a transcription factor, p53 encourages the synthesis of Mdm2—the agent of its own destruction (see Figure 9.10). This creates a negative-feedback loop that usually functions to ensure that p53 molecules are degraded soon after their synthesis, resulting in the very low steady-state levels of p53 protein observed in normal, unperturbed cells.

The operations of this p53–Mdm2 feedback loop explain a bizarre aspect of p53 behavior. In human cancer cells that carry mutant, defective *p53* alleles, the p53 protein is almost invariably present in high concentrations (for example, see **Figure 9.17**), in contrast to its virtual absence from normal cells. At first glance, this might appear paradoxical, since high levels of a growth-suppressing protein like p53 would seem to be incompatible with malignant cell proliferation.

The paradox is resolved by the fact, mentioned above, that the great majority of the mutations affecting the *p53* gene cause the p53 protein to lose its transcription-activating powers. As a direct consequence, p53 is unable to induce *Mdm2* transcription and thus Mdm2 protein synthesis. In the absence of Mdm2, p53 escapes degradation

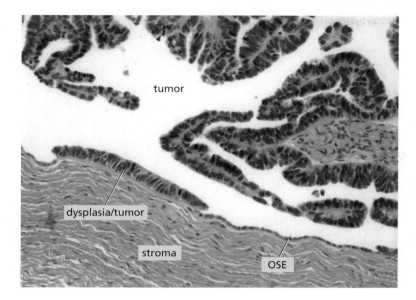

Figure 9.17 Accumulation of p53 in *p53*-mutant cells This microscope section of ovarian tissue has been stained with an anti-p53 antibody coupled to the peroxidase enzyme, resulting in the blackened nuclei seen here. Large patches of epithelial cells in an ovarian carcinoma (*above*) are composed of cells that have high levels of p53; a patch of dysplasia (*left middle*) is also p53-positive. Stromal cells (*small black nuclei, pink matrix, below*) are unstained, as is a patch of normal ovarian surface epithelium (OSE, *below, right*). (Courtesy of R. Drapkin and D.M. Livingston.)

and accumulates to very high levels. This means that many types of human cancer cells accumulate high concentrations of essentially inert p53 molecules.

This logic explains why the presence of readily detectable p53 in a population of tumor cells, usually revealed by **immunostaining** (see Figure 9.17), is a telltale sign of the presence of a mutant *p53* allele in the genome of these cells. (Such a conclusion cannot be drawn, however, from analyzing human tissue that has recently been irradiated, since radiation can also evoke the widespread expression of p53 throughout a tissue for days, even weeks after radiotherapy.) The identical logic explains the large amounts of p53 protein in SV40-infected or SV40-transformed cells, in which sequestration of p53 by the viral large T (LT) antigen prevents p53-induced expression of the *Mdm2* gene and resulting p53 degradation (see Figure 9.1). (According to one measurement, when SV40 LT is expressed in cells, the half-life of p53 increases from 20 minutes to 24 hours.)

Mdm2 is only one of a large cohort of genes whose expression is induced by p53 (**Table 9.2**). As mentioned earlier, another highly important target gene is *p21^{Cip1}* (sometimes called *Cdkn1a* in the mouse), which we encountered previously as a widely acting CDK inhibitor (see Section 8.4). Its induction explains the cytostatic (rather than pro-apoptotic) actions of p53. In fact, the gene encoding p21^{Cip1} was originally discovered by a molecular search strategy designed to uncover genes whose expression is increased by p53. Soon after its discovery, it became apparent that p21^{Cip1} functions as an important inhibitor of a number of the cyclin-dependent kinases (CDKs). Thus, the ability of p21^{Cip1} to inhibit two CDKs—CDK2 and CDC2—that are active in the late G_1, S, G_2, and M phases of the cell cycle (see Figure 8.8) explains how p53 is able to block forward progress at multiple points in this cycle.

This information also provides us with insight into the physiologic roles played by p53 in the life of a cell. For example, if the chromosomal DNA of a cell should suffer some damage during the G_1 phase of the cell cycle, p53 will become activated, both by rapid increases in its concentration and by post-translational modifications that enable it to function effectively as a transcription factor. p53 will then induce p21^{Cip1} synthesis, and p21^{Cip1}, in turn, will halt further cell proliferation.

At the same time, components of the cellular DNA repair machinery will be mobilized to repair the damage. Some of these are directly induced by p53. This is suggested by observations that certain DNA repair proteins are mobilized far more effectively in cells carrying wild-type *p53* alleles than in those with mutant *p53* alleles. For example, cells lacking functional p53 are unable to efficiently repair the DNA lesions caused by benzo[*a*]pyrene (a potent carcinogen present in tars) and the cyclobutane pyrimidine dimers caused by ultraviolet (UV) radiation. In addition, DNA polymerase β, which plays a critical role in reconstructing DNA strands after chemically altered bases have

Table 9.2 Examples of p53 target genes according to function The expression of genes in this table is induced by p53 unless otherwise indicated.

Class of genes	Name of gene	Function of gene product
p53 antagonist	MDM2/HDM2	induces p53 ubiquitylation
Growth arrest genes	p21^{Cip1}	inhibitor of CDKs, DNA polymerase
	Siah-1	aids β-catenin degradation
	14-3-3σ	sequesters cyclin B–CDC2 in cytoplasm
	Reprimo	G$_2$ arrest
DNA repair genes	p53R2	ribonucleotide reductase—biosynthesis of DNA precursors
	XPE/DDB2	global NER
	XPC	global NER
	XPG	global NER, TCR
	GADD45	global NER [?]
	DNA pol κ	error-prone DNA polymerase
Regulators of apoptosis	BAX	mitochondrial pore protein
	PUMA	BH3-only mitochondrial pore protein
	NOXA	BH3-only mitochondrial pore protein
	p53AIP1	dissipates mitochondrial membrane potential
	Killer/DR5	cell surface death receptor
	PIDD	death domain protein
	PERP	pro-apoptotic transmembrane protein
	APAF1	activator of caspase 9
	NF-κB	transcription factor, mediator of TNF signaling
	FAS/APO1	death receptor
	PIG3	mitochondrial oxidation/reduction control
	PTEN	reduces levels of the anti-apoptotic PIP3
	Bcl-2	repression of anti-apoptotic protein expression
	IGF-1R	repression of anti-apoptotic protein expression
	IGFBP-3	IGF-1–sequestering protein
Anti-angiogenic proteins	TSP-1 (thrombospondin)	antagonist of angiogenesis

Abbreviations: NER, nucleotide excision repair; TCR, transcription-coupled repair.

been excised by DNA repair proteins, is much less active in p53-negative cells than in their wild-type counterparts. We will return to these DNA repair proteins and their mechanisms of action in Chapter 12.

In the event that the DNA is successfully repaired, the signals that have protected p53 from destruction (see Figure 9.13A) will disappear. The consequence is that the levels of p53 collapse and p21^{Cip1} follows suit. This allows cell cycle progression to resume, enabling cells to enter S phase, where DNA replication now proceeds.

The rationale for this series of steps is a simple one: by halting cell cycle progression in G$_1$, p53 prevents a cell from entering S phase and inadvertently copying still-unrepaired DNA. Such copying, if it occurred, would cause a cell to pass mutant DNA sequences on to one or both of its daughters. The importance of these cytostatic actions of p21$^{\text{Cip1}}$ can be seen from the phenotype of genetically altered mice in which both germ-line copies of the *p21*$^{\text{Cip1}}$ gene have been inactivated. Although not as tumor-prone as p53-null mice, they show an increased incidence of tumors late in life. This milder phenotype is what we might expect, since p21$^{\text{Cip1}}$ mediates some, but not all, of the tumor-suppressing activities of p53.

If a cell suffering DNA damage has already advanced into S phase and is therefore in the midst of actively replicating its DNA, the p21$^{\text{Cip1}}$ induced by p53 can engage the DNA polymerase machinery at the replication fork and halt its further advance down DNA template molecules. [It does so by binding PCNA (proliferating cell nuclear antigen), which interacts, in turn, with the key DNA polymerases δ and ε, thereby blocking further advance of replication forks.] Once again, the goal here is to hold DNA replication in abeyance until DNA damage has been successfully repaired.

The p53 protein uses yet other genes and proteins to impose a halt to further cell cycle advance. For example, Siah-1, the product of another p53-induced gene, drives the degradation of β-catenin; the latter helps to induce cyclin D1 synthesis and thus progression through most of the G$_1$ phase of the cell cycle (see Figure 8.11B; Section 8.3). The loss of β-catenin may also cause a decrease in transcription of the *myc* gene, which in turn may slow progression through several phases of the cell cycle in addition to its effects on G$_1$ advance (see Section 8.9).

Two other genes that are activated by p53 encode the 14-3-3σ and Reprimo proteins (see Table 9.2), which help to govern the G$_2$/M transition. The 14-3-3σ protein, for its part, sequesters the cyclin B–CDC2 complex in the cytoplasm, thereby preventing it from moving into the nucleus, where its actions are needed to drive the cell into mitosis. This mechanism holds mitosis in abeyance until the chromosomal DNA is in good repair.

These various actions of p53 have caused some to portray this protein as the "guardian of the genome." By preventing cell cycle advance and DNA replication while chromosomal DNA is damaged and by inducing expression of DNA repair enzymes, p53 can reduce the rate at which mutations accumulate in cellular genomes. Moreover, in the event that severe DNA damage has been sustained (for example, damage that exceeds a cell's ability to repair its DNA), p53 may trigger apoptosis, thereby eliminating mutant cells and their damaged genomes. This contrasts with the behavior of cells that have lost p53 function: they may replicate their damaged, still-unrepaired DNA, and this can cause them to exhibit relatively mutable genomes, that is, genomes that accumulate mutations at an abnormally high rate per cell generation.

In one particularly illustrative experiment, pregnant *p53*$^{+/-}$ mice that had been bred with *p53*$^{+/-}$ males were treated with the highly mutagenic carcinogen ethylnitrosourea (ENU). In all, 168 offspring were born. Of these, 70% of the *p53*$^{-/-}$ pups (which had been exposed *in utero* to this carcinogen) developed brain tumors, 3.6% of the *p53*$^{+/-}$ pups did so, and none of the *p53*$^{+/+}$ offspring gave evidence of brain tumor formation. Hence, in the absence of p53 function, fetal cells that had been mutated by ENU could survive and spawn the progeny forming these lethal tumors.

The absence of p53 results in the accumulation of genomic alterations more far-reaching than the point mutations caused by ENU. For example, when mouse fibroblasts are deprived of p53 function, they show greatly increased rates of chromosomal loss and duplication (ascribable, at least in part, to the loss of G$_2$/M checkpoints) and also exhibit increased numbers of *interstitial deletions*, that is, deletions involving the loss of a microscopically visible segment from within the arm of a chromosome.

9.10 p53 often ushers in the apoptotic death program

As mentioned repeatedly above, under certain conditions p53 can opt to provoke a response that is far more drastic than the reversible halting of cell-cycle advance. In response to massive, essentially irreparable genomic damage, **anoxia** (extreme oxygen deprivation), or severe signaling imbalances, p53 will trigger apoptosis. We now begin to explore the apoptotic program in more detail.

The cellular changes that constitute the apoptotic program proceed according to a precisely coordinated schedule. Within minutes, patches of the plasma membrane herniate to form structures known as **blebs**; indeed, in time-lapse movies, the cell surface appears to be boiling (**Figure 9.18A**). The nucleus collapses into a dense structure—the state termed **pyknosis** (see Figure 9.18B)—and fragments (see Figure 9.18E) as the chromosomal DNA is cleaved into small segments (see Figure 9.18C). Yet other changes in other parts of the cell can also be observed (see Figure 9.18D and

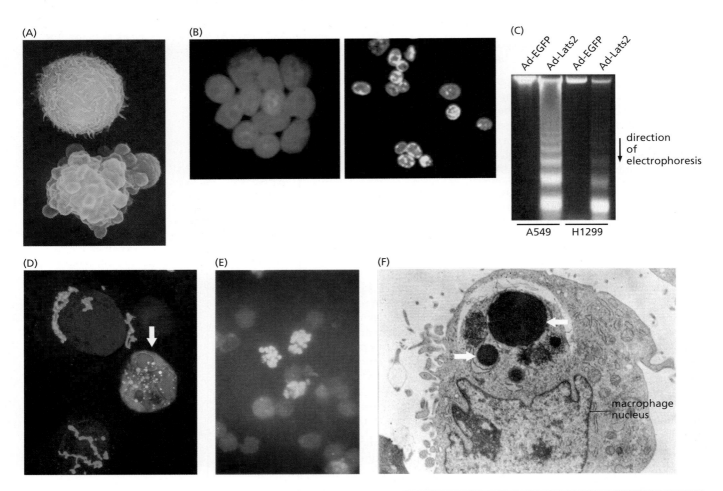

Figure 9.18 Diverse manifestations of the apoptotic program
(A) The upper of these two lymphocytes, visualized by scanning electron microscopy (SEM), is healthy, while the lower one has entered into apoptosis, resulting in the numerous blebs protruding from its surface. (B) HeLa cells—a line of human cervical carcinoma cells—have been fused, forming a syncytium (*left*). Prior to fusion, some of these cells were treated with staurosporine, an apoptosis-inducing drug. As a consequence, these cells' nuclei, which normally are quite large (*left*), have undergone *pyknosis* (*right*), a process that accompanies apoptosis and involves condensation of chromatin (*white*) and collapse of nuclear structure. Fragmentation of nuclei (*not shown*) follows soon thereafter. (C) When apoptosis is induced, in this case through the expression of the pro-apoptotic Lats2 protein (*2nd, 4th channels*) introduced via an adenovirus (Ad) vector, the normally high–molecular-weight DNA of A549 and H1299 cells (*1st, 3rd channels*) is cleaved into low–molecular-weight fragments that run rapidly upon gel electrophoresis, forming a "DNA ladder." (D) The Golgi bodies (*green*) are usually found in perinuclear locations in a normal cell (*upper left*), while in an

apoptotic cell (*arrow*), the Golgi bodies have become fragmented. Chromatin is stained *blue*, while an antibody specific for cleaved PARP [poly(ADP-ribose) polymerase], which is split during apoptosis, reveals this form of the protein (*red*) in the nucleus of an apoptotic cell. (E) Here, an antibody that specifically reacts with histone 2B molecules (in the chromatin) that are phosphorylated on their serine 14 residues during apoptosis was used to stain apoptotic nuclei (which are already undergoing fragmentation). (F) The end result of apoptosis is the phagocytosis of apoptotic bodies—the fragmented remains of apoptotic cells—by neighbors or by macrophages. In this image, the pyknotic nuclear fragments of a phagocytosed apoptotic cell are seen above (*white arrows*) and contrast with the normal nucleus of the phagocytosing macrophage (*below*). (A, courtesy of K.G. Murti, St. Jude Hospital, Memphis, TN. B, from K. Andreau et al., *J. Cell Sci.* 117:5643–5653, 2004. C, from H. Ke et al., *Exp. Cell Res.* 298:329–338, 2004. D, from J.D. Lane et al., *J. Cell Biol.* 156:495–509, 2002. E, from W.L. Cheung et al., *Cell* 113:507–517, 2003. F, courtesy of G.I. Evan.)

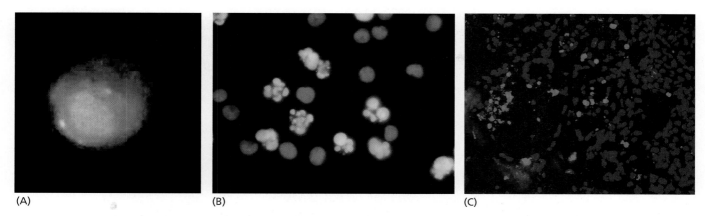

(A)   (B)   (C)

Figure 9.19 Techniques for detecting apoptotic cells Because apoptosis affects a multitude of cellular components (see Figure 9.18), there are multiple alternative ways by which apoptosis can be measured, three of which are shown here. (A) Early in the apoptotic program, phosphatidylserine (PS) molecules, which are normally exposed on the inner surface of the plasma membrane, are flipped and therefore become exposed on the outer surface. PS is bound by the annexin V protein, which here has been coupled to a fluorescent dye and used to detect exposed, cell-surface PS (*green*). Because the cell in this image has advanced to a later stage of apoptosis, its plasma membrane has become permeable to propidium iodide (PI) dye (*yellow orange*), allowing staining of its nucleus as well. (B) In the TUNEL assay, apoptotic cells (*yellow green, above*) are detected because their chromosomal DNA has become fragmented (see Figure 9.18C), exposing 3′ DNA ends that can be extended by the terminal deoxyribonucleotide transferase (TdT) enzyme (see Supplementary Sidebar 9.2), which in this case uses BrdUTP (bromodeoxyuridine triphosphate) as its substrate.

The resulting oligonucleotide tails can be detected with an anti-BrdU monoclonal antibody that has been coupled to a dye molecule (*yellow green*). In addition to the apoptotic cells stained by the TUNEL assay, cells in this population that have become *necrotic* are also revealed, in this case through the use of propidium iodide dye (*red*), which is able to enter freely into such cells. (C) A fluorescent dye–tagged antibody for cleaved (and therefore activated) caspase 3 (*red*), which plays a prominent role in the apoptotic program (see Section 9.13), and the DAPI stain for cell nuclei (*blue*) have been used to stain an MMTV-Neu mouse breast carcinoma. The cancer cells are entering anoikis/apoptosis (see Figure 9.22) because they express a dominant-negative integrin β1. Note that an area of apoptotic carcinoma cells (*left*) is adjacent to normal stromal cells (*right*) in which apoptosis is rare. (A, from J. Yamamoto et al., *Clin. Cancer Res.* 12:7132–7139, 2006. B, copyright © 2012 Life Technologies Corp. Used under permission. www.lifetechnologies.com. C, courtesy of W. Guo and F. Giancotti.)

E). Ultimately, usually within an hour, the apoptotic cell breaks up into small fragments, sometimes called *apoptotic bodies*, which are rapidly ingested by neighboring cells in the tissue or by itinerant macrophages, thereby removing all traces of what had recently been a living cell (see Figure 9.18F).

This phagocytosis by macrophages depends on the display of a specialized "eat-me" signal displayed on the surfaces of cells entering apoptosis. Thus, normal cells display phosphatidylserine (PS) molecules only on the inner leaflet of the bilayer forming the plasma membrane. This asymmetric display is lost, however, during apoptosis, resulting in PS display on the exterior face and its recognition by a specialized annexin receptor on the surface of macrophages; the resulting binding of macrophages to the apoptotic cell then proceeds to trigger phagocytosis. A related annexin protein, when coupled to a fluorescent dye, can be used to stain cells that have initiated apoptosis and flipped many of their PS molecules to the outer cell surface (**Figure 9.19A**). Yet other procedures can also be used to detect apoptotic cells within a tissue or in culture (see Figure 9.19B and C; Supplementary Sidebar 9.2).

(Interestingly, there is evidence that a failure of macrophages and other phagocytic cells to rapidly and efficiently remove the remains of dead cells can result in local inflammation within a tissue and, over the long run, the triggering of autoimmune disease by the intracellular contents that are left behind.)

Apoptosis is exploited routinely during normal morphogenesis in order to discard unneeded cells. It serves to chisel away unwanted cell populations during the sculpting that results in well-formed, functional tissues and organs (**Figure 9.20**). Mice that have been genetically deprived of various key components of the apoptotic machinery show a characteristic set of developmental defects, including excess neurons in the brain, facial abnormalities, delayed destruction of the webbing between fingers, and abnormalities in the palate and lens. In the case of the developing brain, more than

Figure 9.20 Apoptosis and normal morphogenesis The webs of tissue between the future fingers of a mouse paw are still in evidence in this embryonic paw. They were preferentially labeled, as indicated by the numerous dark dots (*arrows*), using the TUNEL assay (see Figure 9.19B). The apoptosis of cells forming the webs ultimately results in the formation of fingers that are joined by webs only near the palm. (From Z. Zakeri et al., *Dev. Biol.* 165:294–297, 1994.)

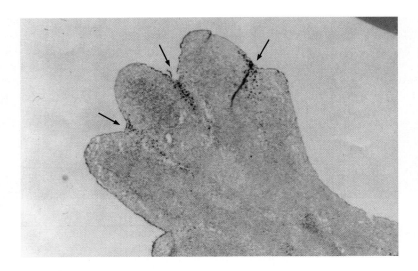

half of all initially formed neurons are normally eliminated by apoptosis during formation of various neural circuits.

Apoptosis also plays an important role in normal tissue physiology. In the small intestine, for example, epithelial cells are continually being eliminated by apoptosis after a four- or five-day journey from the bottom of intestinal crypts to the tips of the villi that protrude into the lumen (see Section 7.11). During the development of red blood cells (the process of erythropoiesis), more than 95% of the erythroblasts—the precursors of mature red cells—are eliminated as part of the routine operations of the bone marrow. However, in the event that the rate of oxygen transport by the blood falls below a certain threshold level, because of hemorrhage, various types of anemia, or low oxygen tension in the surrounding air, levels of the blood-forming hormone erythropoietin (EPO; see Section 5.7; Sidebar 5.2) rise rapidly and block apoptosis of these erythrocyte precursors, enabling their maturation into functional red blood cells. This yields a rapid increase in the concentration of these cells in the blood.

A particularly dramatic example of the contribution of apoptosis to normal physiology is provided by the regression of the cells in the mammary gland following the weaning of offspring. As many as 90% of the epithelial cells in this gland, which have accumulated in large numbers during pregnancy in order to produce milk for the newborn, die via apoptosis during this regression, which is usually termed **involution**.

In a more general sense, apoptosis is used to maintain appropriate numbers of different cell types in a wide variety of human tissues. The importance of this process is indicated by the fact that each year of our lives, the turnover of cells—the number that are newly formed and the equal number that are eliminated—approximates the total number of cells ($\sim 3 \times 10^{13}$) present in the adult body at any one time. The majority of these discarded cells appear to be eliminated by apoptosis.

The apoptosis that is used for routine tissue maintenance does not seem to depend on p53 function and is, instead, triggered by other mechanisms that we will discuss later. Thus, the actions of p53 seem usually to be limited to nonroutine, emergency situations that occasionally threaten cells and tissues (see however Sidebar 9.5). In the specific context of cancer pathogenesis, as noted earlier, the organism uses p53-triggered apoptosis as a means of weeding out cells that have the potential to become neoplastic, including some cells that have sustained certain types of growth-deregulating mutations and others that have suffered widespread damage to their genomes.

p53 initiates apoptosis in part through its ability to promote expression of several downstream target genes that specify components of the apoptotic machinery. Among these are the genes encoding a diverse group of pro-apoptotic proteins (see Table 9.2). At the same time, p53 represses expression of genes specifying anti-apoptotic proteins. We will return to the biochemical details of the apoptotic program and their relevance to cancer pathogenesis in Section 9.13.

To summarize, the various observations cited here indicate that the biological actions of p53 fall into two major categories. In certain circumstances, p53 acts in a cytostatic fashion to halt cell cycle advance. In other situations, p53 activates a cell's previously latent apoptotic machinery, thereby ensuring cell death. The choice made between these alternative modes of action seems to depend on the type of physiologic stress or genetic damage, the severity of the stress or damage, the cell type, and the presence of other pro- and anti-apoptotic signals operating in a cell. At the biochemical level, it remains unclear how p53 decides between imposing cell cycle arrest and triggering apoptosis.

9.11 p53 inactivation provides advantage to incipient cancer cells at a number of steps in tumor progression

As we will learn in Chapter 11, the formation of a malignant human cell involves more than half a dozen distinct steps that usually occur over many years' time. An early step in the formation of a cancer cell may involve activation of an oncogene through some type of mutation. This oncogene activation may put the cell at great risk of p53-induced apoptosis. Recall, for example, the fact that a *myc* oncogene can, on its own, trigger p53-dependent apoptosis. Hence, cells that have acquired such an oncogene accrue additional growth advantage by shedding p53 function.

Later in tumor development, a growing population of tumor cells may experience anoxia because they lack an adequate network of vessels to provide them with access to blood-borne oxygen. While normal cells would die in the face of such oxygen deprivation, tumor cells may survive because their ancestors managed to inactivate the *p53* gene during an earlier stage of tumor development (having done so initially for a quite different reason).

During much of this long, multi-step process of tumor progression, the absence of p53-triggered responses to genetic damage will permit the survival of cells that are accumulating mutations at a greater-than-normal rate. Such increased mutability increases the rate at which oncogenes become activated and tumor suppressor genes become inactivated; the overall rate of evolution of pre-malignant cells to a malignant state is thereby accelerated. Telomere collapse, another danger faced by evolving, pre-malignant cells (see Chapter 10), also selects for the outgrowth of those cells that have lost their p53-dependent DNA damage response.

The advantages to the incipient tumor cell of shedding p53 function do not stop there. One of the important target genes whose expression is increased by p53 is the *TSP-1* gene, which specifies thrombospondin-1. As we will see in Chapter 13, Tsp-1 is a secreted protein that functions in the extracellular space to block the development of new blood vessels. Consequently, a reduction of Tsp-1 expression following p53 loss removes an important obstacle that otherwise would prevent clusters of cancer cells from developing an adequate blood supply during the early stages of tumor development.

Together, these diverse consequences of p53 inactivation illustrate dramatically how the malfunctioning of a single component of the alarm response circuitry permits cancer cells to acquire multiple alterations and survive under conditions that usually lead to the death of normal cells. These multiple benefits accruing to cancer cells explain why the p53 pathway is disrupted in most if not all types of human tumors.

In almost half of these tumors, the p53 protein itself is damaged by reading-frame mutations in the *p53* gene (see Figure 9.4). In many of the remaining tumors, the ARF protein is missing or the Mdm2 protein is overexpressed. In addition, p53 function may be compromised by defects in the complex signaling network in which p53 and its antagonist, Mdm2, are embedded (**Sidebar 9.6**). There are reasons to suspect that yet other, still-undiscovered genetic mechanisms subvert p53 function. While the organism as a whole benefits greatly from the p53 watchman stationed in its myriad cells, it suffers grievously once p53 function is lost in some of them, because the resulting p53-negative cells are now free to begin the long march toward malignancy.

Sidebar 9.6 p53 has confusing cousins As is often the case with important regulatory proteins, both p53 and Mdm2 have relatives in the cell; we have already encountered Mdm2's cousin, Mdm4 (also known as MdmX; see Section 9.7). In many situations, the ability of Mdm2 to drive down p53 concentrations may depend on its ability to do so as part of multi-protein complexes formed with Mdm4.

p53, for its part, actually has two cousins, p63 and p73. These two have quite different functions in cells. For example, while *p53*-null mice (created by germ-line gene inactivation; see Supplementary Sidebar 7.7) are largely normal at birth, *p63*-null mice have severe developmental defects involving epithelial cell differentiation and die shortly after birth. *p73*-null mice show yet other developmental abnormalities that affect primarily the nervous and secretory tissues. Moreover, like *p53*⁺/⁻ mice, *p63*⁺/⁻ and *p73*⁺/⁻ (heterozygous) mice show predispositions to cancer, although these phenotypes are not as strongly penetrant as those of *p53* mutants.

These two cousin proteins are important for p53-dependent apoptosis, since mouse embryo fibroblasts that have been deprived of both copies of both genes (that is, are of a *p63⁻/⁻ p73⁻/⁻*

genotype) are no longer able to induce expression of many pro-apoptotic genes and fail to execute p53-dependent apoptosis. It is likely that mixed tetrameric complexes composed of subunits of p53 and its cousins operate as the transcription factors responsible for activating certain parts of the *p53*-dependent apoptotic program (Supplementary Sidebar 9.3). (The tendency of excessive E2F1 to trigger apoptosis, as described in Section 9.8, may be due in part to its ability to induce expression of p73, which then functions as a pro-apoptotic partner of p53.) Loss of the *p63* and *p73* genes does not, however, affect the cytostatic effects of p53 (mediated by p21^{Cip1} induction), and p53's ability to induce Mdm2 expression remains intact.

A final dimension of complexity comes from naturally occurring variant forms of these proteins. For example, full-length p63 and p73 function with p53 to promote apoptosis, whereas shorter, N-terminally truncated versions of p63 and p73—both products of alternative splicing—act oppositely to inhibit p53-induced apoptosis. Nonetheless, the fact that inactivation of either the *p63* or the *p73* gene in the mouse germ line leads to increased tumor incidence places these two genes squarely in the camp of tumor suppressor genes.

9.12 Inherited mutant alleles affecting the p53 pathway predispose one to a variety of tumors

In 1982, a group of families was identified that showed a greatly increased susceptibility to a variety of different tumors, including glioblastoma; leukemias; carcinomas of the breast, lung and pancreas; Wilms tumor; and soft-tissue sarcomas (**Figure 9.21**). In some kindreds, as many as half the members were afflicted with one or another of these cancers, and two-thirds of these developed some type of cancer by the time they reached age 22. Some family members were even afflicted with several types of cancer concurrently.

This familial cancer syndrome, termed Li–Fraumeni after the two human geneticists who first identified and characterized it, is most unusual, in that it causes susceptibility to a wide variety of cancers. Recall the starkly contrasting behavior of the other familial cancer syndromes that we encountered in Chapter 7. Mutant germ-line alleles of most tumor suppressor genes typically increase susceptibility to a narrow range of cancer types (see Table 7.1).

In 1990, eight years after the Li–Fraumeni syndrome was first described, researchers discovered that many of the cases were due to a mutant allele at a locus on human Chromosome 17p13—precisely where the *p53* gene is located. In about 70% of these multicancer families, mutant alleles of *p53* were found to be transmitted in a Mendelian fashion. Family members who inherited a mutant *p53* allele had a high probability

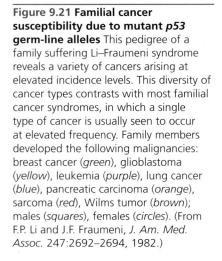

Figure 9.21 Familial cancer susceptibility due to mutant *p53* germ-line alleles This pedigree of a family suffering Li–Fraumeni syndrome reveals a variety of cancers arising at elevated incidence levels. This diversity of cancer types contrasts with most familial cancer syndromes, in which a single type of cancer is usually seen to occur at elevated frequency. Family members developed the following malignancies: breast cancer (*green*), glioblastoma (*yellow*), leukemia (*purple*), lung cancer (*blue*), pancreatic carcinoma (*orange*), sarcoma (*red*), Wilms tumor (*brown*); males (*squares*), females (*circles*). (From F.P. Li and J.F. Fraumeni, *J. Am. Med. Assoc.* 247:2692–2694, 1982.)

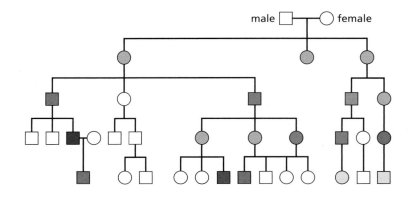

of developing some form of malignancy, often early in life. The age of onset of these various malignancies was found to be quite variable: about 5 years of age for adreno-cortical carcinomas, 16 years for sarcomas, 25 years for brain tumors, 37 years for breast cancer, and almost 50 years for lung cancer. In light of the various roles that p53 plays in suppressing cancer risk in cells throughout the body, it seems reasonable that mutant germ-line alleles of this gene should predispose a person to such a diverse group of malignancies. (An interesting exception, however, is provided by a familial adrenocortical tumor syndrome occurring in Brazil; see Supplementary Sidebar 9.4).

We now know that the mutant *p53* alleles that are transmitted through the germ lines of Li–Fraumeni families carry a variety of point mutations that are scattered across the *p53* reading frame, with a distribution reminiscent of that shown by the somatic mutations that have been documented in more than 26,000 tumor genomes (see Figure 9.6B). (More than 500 germ-line *p53* mutations have been identified by now.) Analysis of the spectrum of germ-line mutations has shown a predominance of G:C to A:T transitions at CpG sites—precisely those that would occur if a 5-methylcytosine underwent spontaneous deamination, causing it to be replaced by a thymidine (see Figure 12.9B). In recent years, similar familial cancer syndromes have been traced to other components of the p53 control circuitry (**Sidebar 9.7**).

9.13 Apoptosis is a complex program that often depends on mitochondria

Many consider the loss of a fully functional apoptotic program to be one of the hallmarks of all types of malignant human cells. Until now, however, our descriptions of apoptosis have not done justice to this suicide program and its role in tumor pathogenesis. Accordingly, we now revisit apoptosis and explore it in greater depth.

Although the apoptotic program had been recognized as a normal biological phenomenon in animal tissues by nineteenth-century histologists, it was rediscovered

Sidebar 9.7 Familial cancer susceptibility can arise from defects in other components of the p53 pathway Provocatively, a familial cancer syndrome that behaves very much like the Li–Fraumeni syndrome has been discovered in families that inherit mutant, defective forms of the Chk2 kinase, which is responsible for phosphorylating p53, saving it from destruction, and thereby activating the p53 alarm machinery (see Figure 9.13). A mutant germ-line allele of the *Chk2* gene is present in 1.1 to 1.4% of Western populations, and in higher percentages (as much as 5%) in cohorts of breast cancer patients. This suggests that inheritance of a mutant *Chk2* allele confers a measurably increased risk of developing this malignancy.

Yet another related cancer predisposition syndrome derives from a variant sequence—a single-nucleotide polymorphism (SNP)—in the promoter of the *MDM2* gene. This region of the human gene contains enhancer sequences that enable *MDM2* gene transcription to respond to signals coming from p53 and the Ras signaling pathway. The polymorphism, a T-to-G substitution termed *SNP309*, is quite common, being present in homozygous configuration in about 12% of the American population. It causes the Sp1 transcription factor to bind more avidly to an intronic DNA segment within the *p53* gene, and this results in higher levels of Mdm2 expression. As a consequence, cells from individuals homozygous for this polymorphism express as much as eightfold higher levels of *MDM2* RNA and greater than threefold higher levels of Mdm2 protein; because of the resulting attenuated p53 response, these cells are far more resistant to apoptosis induced by DNA damage.

The *SNP309* polymorphism exacerbates the cancer susceptibility of individuals in Li–Fraumeni families. Li–Fraumeni individuals carrying this polymorphism experience their first cancer at an average age of 22 years, compared with the 27-year age of those who lack this base substitution. The cancer risks are more dramatic for specific tumor types. For example, if Li–Fraumeni patients develop a soft-tissue sarcoma, these tumors appear, on average, at the age of 2 among those who are homozygous for the polymorphism, while appearing at the age of 12 among those who carry the more prevalent allele. Moreover, this polymorphism increases the number of independently arising tumors in Li–Fraumeni patients carrying it in either a heterozygous or a homozygous configuration.

SNP309 homozygotes in the general population exhibit a markedly increased risk of developing soft-tissue sarcomas. In addition, female homozygotes who develop breast cancer do so at age 45, on average, rather than the 57-year average age of onset seen among those who do not carry this allele. In one *meta-analysis*, individuals who were homozygous at the *SNP309* locus and carried two *Pro/Pro* alleles of p53 (see Sidebar 9.1) exhibited a greater than threefold increased overall risk for developing cancer. These observations demonstrate how relatively common genetic variant sequences can have profound effects on cancer susceptibility. And they raise the question whether the responsible base changes can be truly called "polymorphisms," since the term usually implies genetic variability that has no pathological consequences.

Table 9.3 Apoptosis versus necrosis

	Apoptosis	Necrosis
Provoking stimuli		
	programmed tissue remodeling	metabolic stresses
	maintenance of cell pool size	absence of nutrients
	genomic damage	changes in pH, temperature
	metabolic derangement	hypoxia, anoxia
	hypoxia	
	imbalances in signaling pathways	
Morphological changes		
Affected cells	individual cells	groups of cells
Cell volume	decreased	increased
Chromatin	condensed	fragmented
Lysosomes	unaffected	abnormal
Mitochondria	morphologically normal initially	morphologically aberrant
Inflammatory response	none	marked
Cell fate	apoptotic bodies consumed by neighboring cells	lysis
Molecular changes		
Gene activity	required for program	not needed
Chromosomal DNA	cleaved at specific sites	random cleavage
Intracellular calcium	increased	unaffected
Ion pumps	continue to function	lost

Adapted from R.J.B. King, Cancer Biology, 2nd ed. Harlow, UK: Pearson Education, 2000.

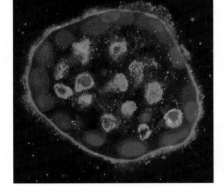

Figure 9.22 Anoikis triggered by loss of anchorage to extracellular matrix Suspension of immortalized but non-tumorigenic human mammary epithelial cells (MECs) in a medium of ECM (extracellular matrix) proteins forces these cells to grow in an anchorage-independent fashion, that is, without attachment to a solid substrate. Some MECs (*large blue nuclei*) succeed in attaching themselves directly via integrins (*red*) to the ECM proteins in the surrounding medium; this attachment enables these cells to avoid the apoptotic program termed anoikis. However, cells in the middle of this spherical cell colony are unable to make such attachments and therefore activate anoikis, as is indicated by their staining (*light green*) with an antibody specific for one of the enzymes (cleaved caspase 3) that participate in the apoptotic program (see Figure 9.19C). (Courtesy of J. Debnath and J. Brugge.)

and described with far greater precision in 1972. Prior to this rediscovery, cells in metazoan tissues were thought to be eliminated solely by **necrosis**. As indicated in **Table 9.3**, these two death processes are actually quite different. By the late 1980s, research on the genetics of worm development (the species *Caenorhabditis elegans*) revealed that apoptosis is exploited to eliminate various cell types as part of the normal developmental program of these tiny animals (see also Figure 10.1). This led to the recognition that apoptosis is a basic biological process that is common to all metazoa.

We have already learned that the initiation in mammalian cells of apoptosis by p53 represents an important mechanism by which tissues can eliminate aberrantly functioning or irreparably damaged cells. Importantly, as we read earlier, apoptosis can also be initiated through a variety of signaling channels that do not depend on the actions of p53. A particularly dramatic example of this is seen in **Figure 9.22**: the loss by a cell of anchorage to extracellular matrix triggers anoikis, the specialized form of apoptosis that occurs without the intervention of p53. This hints at a larger theme—that p53 is only one of many players in the apoptotic program, and inactivation of p53 function is only one means by which cancer cells evade apoptosis.

The first indication of contributions of other proteins to the regulation of apoptosis came from an exploration of the functioning of the *bcl-2* (B-cell lymphoma gene-2) oncogene. Like many cancer-associated genes found in the genomes of human hematopoietic tumors, the oncogenic version of the *bcl-2* gene is formed through a reciprocal chromosomal translocation, in which portions of the arms of human Chromosomes 14 and 18 are exchanged. At the breakpoint where the translocated arms are joined in human follicular B-cell lymphoma cells, the reading frame of the *bcl-2* gene is placed under the control of a promoter that drives its high, constitutive expression.

When such a *bcl-2* oncogene was inserted as a transgene into the germ line of mice under conditions that ensured its expression in lymphocyte precursor cells, there was no observable effect on the mice's long-term survival (**Figure 9.23A and B**). On the other hand, expression of an oncogenic *myc* transgene in these cells led to lymphomas and to the death of half the mice within two months of birth. However, the concomitant expression of the two transgenes (achieved by breeding *bcl-2* transgenic mice with *myc* transgenic mice) led to offspring having an even more rapid death rate, with virtually all of the mice being dead less than two months after birth (see Figure 9.23C).

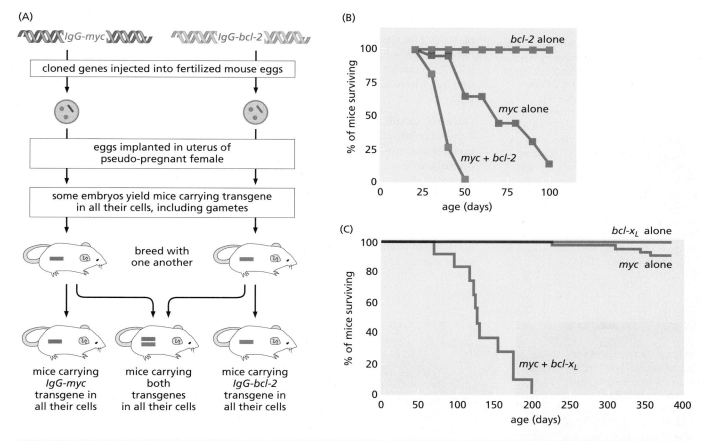

Figure 9.23 Synergy of *Bcl-2*-like genes with *myc* (A) Clones of the *myc* and the *bcl-2* genes were constructed in which each gene was placed under the control of an antibody gene promoter (*IgG*). This promoter ensured that both genes would be expressed specifically in cells of the B lymphocyte lineage. Cloned DNA was then injected into eggs and, with a certain frequency, the DNA became integrated via nonhomologous recombination into a chromosomal site. Examination via Southern blotting was used to verify the presence of the integrated sequences in the cells of the resulting mice and, subsequently, in the progeny of such mice, indicating germ-line transmission of the integrated DNA segments. Such a heritable inserted gene is termed a *transgene*. (B) Mice bearing the *IgG–bcl-2* transgene experienced no increased mortality (*green squares*). In contrast, the *IgG–myc* transgene led to a greatly increased mortality from lymphomas, with almost all of the mice dying from this disease by 100 days of age (*red squares*). However, when both *bcl-2* and *myc* transgenes were present in the germ line

of mice, lymphomagenesis was greatly accelerated, with virtually all mice succumbing to lymphomas by 50 days of age (*blue squares*). (C) A far more dramatic outcome was observed years later when two strains of transgenic mice were developed that expressed the *myc* or the *bcl-x_L* oncogenes specifically in the plasma cell (antibody-secreting) lymphocytes of the immune system. (*bcl-x_L* is a cousin of *bcl-2* and functions in a very similar fashion.) Normal mice and transgenic mice carrying only the *bcl-x_L* transgene all survived for more than a year without any tumor development (*green line*); similarly, transgenic mice expressing a *myc* transgene were largely healthy until about 10 months of age, when a small number of them developed plasma cell tumors. However, when the two strains of mice were bred together, the double-transgenic strain developed plasma cell tumors rapidly, beginning at about 75 days of age, and all died from these tumors by 200 days of age. (B, from A. Strasser et al., *Nature* 348:331–333, 1990. C, from W.C. Cheung et al., *J. Clin. Invest.* 113:1763–1773, 2004.)

Figure 9.24 Mitochondria
(A) Mitochondria, stained here in the cytoplasm of a human liver cell (*green*), were once regarded only as the sites of oxidative phosphorylation and biosynthesis of certain key metabolites. The discovery of their ability to regulate apoptosis through cytochrome *c* release forced a major rethinking of their role in the life of a cell. (B) Cytochrome *c* is stored in the space between the outermost mitochondrial membrane and the inner membranes, where it normally plays a key role in electron transport. (C) One of the channels embedded in the outer mitochondrial membrane that are responsible for the release of pro-apoptotic proteins into the cytosol is termed VDAC1 (voltage-dependent anion channel 1). X-ray crystallography has revealed its structure as being assembled from 19 β strands that form an open barrel with a hydrophilic channel. The oligopeptide segment within the open channel (*blue*) appears to be a component of the gating device that governs passage of solutes, including cytochrome *c* molecules, through this channel. Bcl-X$_L$, an anti-apoptosis regulator, has been found to bind to strands 17 and 18 of the 19-strand channel. (A, courtesy of D. Burnette. B, courtesy of D.S. Friend. C, from R. Ujwal et al., *Proc. Natl. Acad. Sci. USA* 105:17742–17747, 2008.)

The inability of *bcl-2*, on its own, to trigger tumor formation argued against its acting as a typical oncogene, for example, an oncogene like *myc* that emits potent growth-promoting signals. Careful study of the lymphocyte populations in mice carrying only the *bcl-2* transgene indicated that the effects of this gene on cells were actually quite different from those of *myc* or *ras*: *bcl-2* prolonged the lives of lymphocytes that were otherwise destined to die rapidly. Indeed, when B lymphoid cells from these transgenic mice were cultured *in vitro*, they showed a remarkable extension of life span. *In vivo*, the lymphoid cells in which *bcl-2* was being expressed accumulated in amounts several-fold above normal, but significantly, these cells were not actively proliferating, explaining the absence of a hematopoietic tumor in mice carrying only the *bcl-2* transgene.

The *myc* oncogene, when acting on its own, acted as a potent mitogen, but its growth-stimulatory powers were attenuated by its death-inducing effects (see Section 9.8). However, *myc* and *bcl-2*, when acting collaboratively, created an aggressive malignancy of the B-lymphocyte lineage; *myc* would drive rapid cell proliferation, and its accompanying death-inducing effects were neutralized by the life-prolonging actions of *bcl-2*. Since these early experiments, even more dramatic examples of synergy between *myc* and *bcl-2*-like oncogenes have been reported (see, for example, Figure 9.23C). (Note, by the way, that these dynamics parallel one that we discussed in Section 9.8, where the death-inducing effects of *myc* were blunted by inactivation of the *p53* gene.)

Once the details of the apoptotic program were clearly understood, the role of the Bcl-2 oncoprotein in cancer pathogenesis became apparent. It operates as an anti-apoptotic agent rather than as a mitogen. Hence, Bcl-2 acts in a fashion precisely opposite to that of p53. Bcl-2 blocks apoptosis, while p53 promotes it. Stated in genetic terms, *bcl-2* oncogene activation and *p53* inactivation confer similar (but hardly identical) benefits on cancer cells, in that both reduce the likelihood of activation of the apoptotic program.

Careful biochemical and cell-biological sleuthing eventually revealed the intracellular site of Bcl-2 action. It was found to operate in a most unexpected locale—the outer membrane of the mitochondrion (**Figure 9.24**). At first, this discovery made

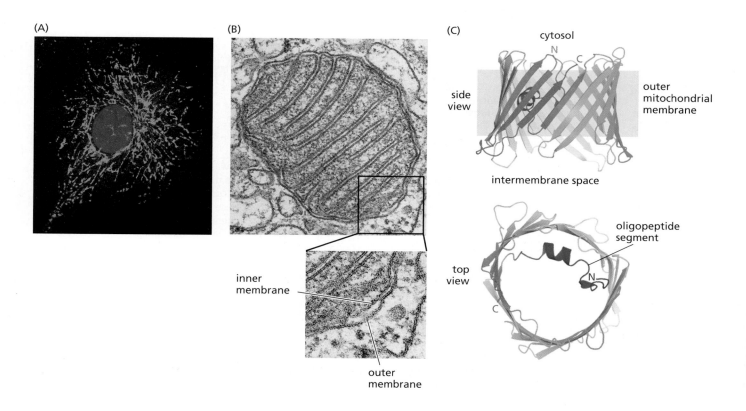

(A)

(B)

inner membrane

outer membrane

(C)

cytosol

N   C

side view

outer mitochondrial membrane

intermembrane space

top view

oligopeptide segment

N

C

little sense, since mitochondria were thought to be specialized for metabolic tasks only, namely, generating energy in the form of ATP for the cell. As a by-product of this energy production, mitochondria were also known to make metabolites of glucose that are used in the biosynthesis of some of the cell's diverse biochemical species (for example, see Section 2.6).

Soon, however, the role of mitochondria in the apoptotic program was clarified. Cytochrome *c*, the central actor in this process, normally resides in the space between the inner and outer mitochondrial membranes (see Figure 9.24B), where it functions to transfer electrons as part of oxidative phosphorylation. However, when certain signals trigger the initiation of apoptosis, channels open in the outer mitochondrial membrane and cytochrome *c* spills out of the mitochondrion into the surrounding cytosol (**Figure 9.25**). Once present in the cytosol, cytochrome *c* associates with other proteins to trigger a cascade of events that together yield apoptotic death. Therefore, the workshop in which the bulk of the cell's energy is produced was co-opted at some point during eukaryotic cell evolution for a fully unrelated function—to harbor and release a biochemical messenger that triggers the changes leading to cell death.

We now know that the Bcl-2 protein is a member of a large and complex family of proteins (**Figure 9. 26**) that control the flow of cytochrome *c* through specialized channels in the outer membrane of the mitochondrion. These channels regulate whether or not cytochrome *c* and several other proteins (which also reside in the space between the inner and outer mitochondrial membranes) are released from the mitochondrion into the cytosol. Some members of the Bcl-2 protein family, including Bcl-2 itself and Bcl-X_L, work to keep these channels closed and thereby keep cytochrome *c* penned up in the mitochondrion.

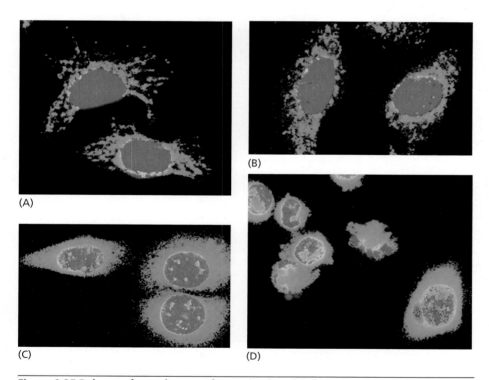

(A)

(B)

(C)

(D)

Figure 9.25 Release of cytochrome *c* from mitochondria into the cytosol The presence of cytochrome *c* can be detected by staining with a specific fluorescence-labeled antibody (*green*), which contrasts with the dye-labeled nuclei (*red orange*). (A) Initially, the distribution of cytochrome *c* coincides with the distribution of mitochondria in the cytoplasm (see Figure 9.24A). As apoptosis proceeds, however (B, C, D), cytochrome *c* staining becomes increasingly uniform in the cytoplasm as this protein is released from the mitochondria into the cytosol. At the same time, some nuclei give evidence of the fragmentation that is associated with apoptosis (C, D). (From E. Bossy-Wetzel et al., *EMBO J.* 17:37–49, 1998.)

Figure 9.26 Bcl-2 and related proteins (A) Bcl-2 is a member of a large family of proteins involved both in preventing apoptosis (*top*) and in promoting it (*bottom*). These proteins share in common a BH3 domain (*purple*), with some, like Bcl-2 itself, having the additional BH1 and BH2 domains (*red, aquamarine*). Several also have transmembrane (TM) domains (*light green*) for membrane anchoring and BH4 domains (*darker green*), which function to inhibit apoptosis, in part by blocking release of calcium ions from the endoplasmic reticulum. The pro-apoptotic proteins are subgrouped into the "Bax" family and have several domains that are homologous to domains of Bcl-2, while the "BH3-only" proteins have only the BH3 domain in common with Bcl-2. Members of the Bax family normally reside in inactive form in the outer mitochondrial membrane or in the cytosol, whereas inactive members of the BH3-only family normally are limited to the cytosol; both classes of proteins are activated and/or translocated to the mitochondria by pro-apoptotic signals. (B) The Bcl-X_L anti-apoptotic protein (*left*) is shown with its BH1 (*red*), BH2 (*aquamarine*), BH3 (*purple*), and BH4 domains (*green*). A pro-apoptotic, BH3-only protein (*orange, middle*) inhibits Bcl-X_L by inserting an α-helix into the groove between the BH1 and BH3 domains of Bcl-X_L. Bax, another pro-apoptotic protein (*right*), tucks its C-terminal, α-helical tail (*yellow*) into its own comparable groove, thereby creating a protein that closely resembles the bimolecular BH3-only/Bcl-X_L complex (*middle*). The precise mechanisms by which these proteins regulate opening of the mitochondrial outer membrane channels remain to be elucidated. (C) Bax is normally found in the cytosol (*not visible*), whereas Bak is attached to the outer mitochondrial membrane (*black spots, left panel*). Following treatment of cells with the apoptosis-inducing drug staurosporine, thousands of molecules of Bak (*black spots*) coalesce with Bax (*not shown*) in punctate foci on the surface of mitochondria (*right panel*). These clusters participate in the mitochondrial fragmentation observed during apoptosis, and this fragmentation appears to contribute to the release of cytochrome c from the spaces between the inner and outer mitochondrial membranes. (A and B, from S. Cory and J.M. Adams, *Nat. Rev. Cancer* 2:647–656, 2002. C, from A. Nechushtan et al., *J. Cell Biol.* 153:1265–1276, 2001.)

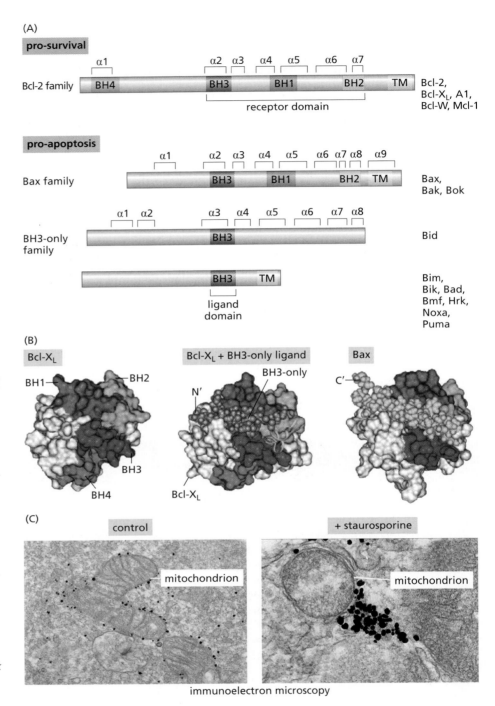

Yet other relatives of Bcl-2, such as Bax, Bad, Bak, and Bid (see Figure 9.26A), function oppositely by striving to pry open these channels. Bax, for one, is encoded by a gene whose transcription is known to be activated by p53; its actions begin to explain how p53 succeeds in inducing apoptosis. Bad can be phosphorylated by Akt/PKB, which decreases Bad's ability to hold the mitochondrial channels in an open configuration; this explains some of the anti-apoptotic effect of the Akt/PKB kinase. (Interestingly, several pro- and anti-apoptotic members of this family of proteins also associate with the endoplasmic reticulum, where they modulate release of calcium ions from this organelle, which also affects regulation of apoptosis.)

As apoptosis proceeds, large complexes of Bax and Bak proteins congregate at the outer surface of the mitochondrion, where they participate, in ways that are still

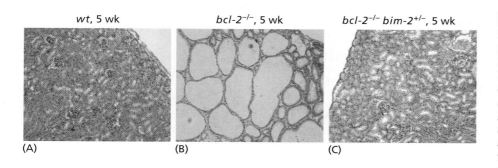

wt, 5 wk   bcl-2⁻/⁻, 5 wk   bcl-2⁻/⁻ bim-2⁺/⁻, 5 wk

(A)   (B)   (C)

Figure 9.27 The delicate balance between pro- and anti-apoptotic proteins (A) The wild-type mouse kidney is a well-consolidated tissue at 5 weeks of age. (B) However, deletion of both germ-line copies of the anti-apoptotic *bcl-2* gene (resulting in the *bcl-2⁻/⁻* genotype) results in widespread apoptosis and polycystic kidney disease at this age. (C) These effects can be fully reversed if one copy of the pro-apoptotic *bim* gene is also deleted from the germ line (yielding the *bcl-2⁻/⁻ bim⁺/⁻* genotype). (From P. Bouillet et al., *Dev. Cell* 1:645–654, 2001.)

unclear, in causing fragmentation of this organelle (see Figure 9.26C). This seems to result in the further release of pro-apoptotic proteins into the cytosol and a total collapse of a cell's primary ATP-generating machinery. Altogether, the human genome is known to encode twenty-four Bcl-2–related proteins; six of these are anti-apoptotic, while the remaining eighteen are pro-apoptotic.

The relative levels of the pro- and anti-apoptotic proteins within each channel determine whether cytochrome *c* will be retained in the mitochondrion or spilled out. In this fashion, the mitochondrial membrane channel determines the life and death of a cell. For example, mice that have been deprived of both germ-line copies of the anti-apoptotic *Bcl-2* gene suffer from kidney failure and immune collapse due to widespread apoptotic death of cells in these tissues; these lethal phenotypes can be avoided through the additional deletion from their germ line of just one copy of the pro-apoptotic cousin gene of *Bcl-2* called *bim* (**Figure 9.27**). This illustrates how, in tissues like these, the fate of individual cells is governed by the delicate balance between pro- and anti-apoptotic Bcl-2-like proteins.

These accounts provide no indication of why mammalian cells express so many distinct Bcl-2-like proteins. In this instance, important clues have come from studies indicating that various cell-physiologic stimuli act through distinct Bcl-2–related proteins to activate pro-apoptotic proteins, which proceed to antagonize Bcl-2, thereby favoring opening of the outer mitochondrial membrane and the initiation of apoptosis (**Figure 9.28A**). Moreover, a large body of other observations, not illustrated here, indicate that the actions of the BH3-only proteins (see Figure 9.26A) can be controlled by modulating transcription of their respective genes, by their intracellular localization, by proteolytic cleavage, or by phosphorylation (see, for example, Figure 9.28B). The pro- and anti-apoptotic members of the Bcl-2 family of proteins operate in direct opposition to one another, with each member having its own opposing partner or partners (see Figure 9.28C). Moreover, each of these proteins functions to trigger apoptosis in specific cell types; for example, Figure 9.28D shows the key role that Bim plays in eliminating cells in the developing mammary gland, allowing lumina to form in the ducts.

Once they have leaked out of mitochondria into the cytoplasm, cytochrome *c* molecules associate with the Apaf-1 protein and form a structure that has been called the **apoptosome** (**Figure 9.29**). The resulting apoptosome complexes then proceed to activate a normally latent cytoplasmic protease termed procaspase 9, converting it into active caspase 9. Caspase 9 is one member of a family of **c**ysteine **asp**artyl–specific prote**ases**; genes encoding 12 of these proteases have been mapped in the human genome.

Having been converted into an active protease by the apoptosome, caspase 9 then cleaves procaspase 3 and thereby activates yet another related protease (see Figure 9.19C), which proceeds to cleave and activate yet another procaspase, and so on down the line (**Figure 9.30**). This sequence of cleavages constitutes a signaling cascade in which one protease activates the next one in the series by cleaving it. Such behavior is reminiscent of the organization of kinase cascades that we read about earlier (see Section 6.5), where a group of Ser/Thr kinases activate one another in a linear sequence. (The fact that each of these caspases acts catalytically means that a relatively minor

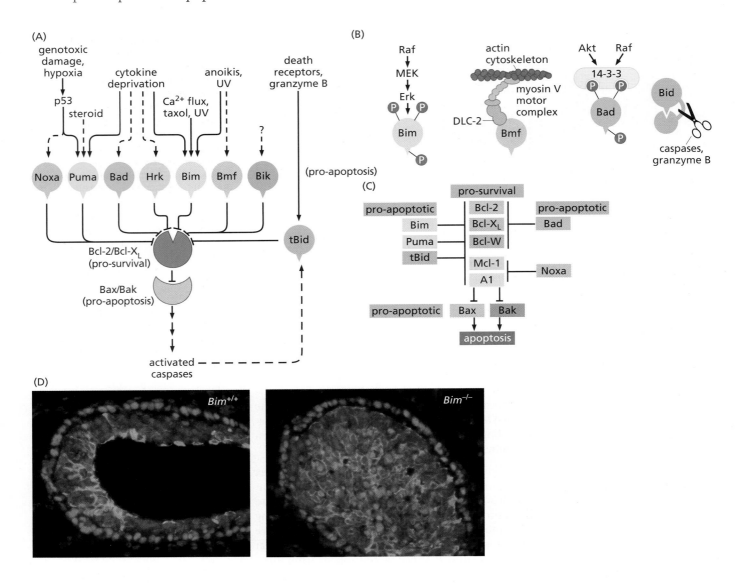

Figure 9.28 Pro-apoptotic signals acting through various Bcl-2–related proteins (A) The rationale for the multiplicity of Bcl-2-like proteins is that various cell-physiologic stresses operate through different pro-apoptotic proteins to antagonize anti-apoptotic proteins such as Bcl-2 and Bcl-X$_L$ (*green*). The latter are thereby prevented from neutralizing Bax and Bak (*light brown*), the dominant pro-apoptotic proteins that function to release cytochrome c from the mitochondrial membrane. The caspases that have been activated downstream in the apoptotic pathway (see Figure 9.30) can also activate the pro-apoptotic function of Bid by proteolysis (cleaving it to tBid), thereby amplifying the apoptotic response. (B) The activities of pro-apoptotic BH3-only proteins are controlled in multiple ways. Bim is regulated at the transcriptional level (*not shown*) and post-translationally via phosphorylation by Erk, which results in its ubiquitylation and degradation. Bmf is tethered to the actin cytoskeleton via dynein light chain (DLC-2), which appears to cause its sequestration; breakdown of this cytoskeleton liberates it, allowing it to provoke apoptosis. Phosphorylation of Bad by anti-apoptotic kinases, such as Akt/PKB and Raf, leads to its sequestration by the 14-3-3 protein and loss of its ability to bind the anti-apoptotic Bcl-2 and Bcl-X$_L$ proteins. The pro-apoptotic Bid protein is liberated from its inhibitory domain by the actions of proteases, such as caspases that are activated downstream in the apoptotic cascade and the granzymes released by cytotoxic cells (see Section 15.3). (C) Each member of the Bcl-2 family of proteins has its own set of opposing partners. At the center of this network lie the anti-apoptotic proteins Bcl-2, Bcl-X$_L$, Bcl-W, Mcl-1, and A1. Each has its own set of antagonistic partners. For example, Bcl-2, Bcl-X$_L$, and Bcl-W are opposed by the pro-apoptotic Bad protein. At the bottom of this network lie the pro-apoptotic Bax and Bak proteins; if unopposed, they will promote apoptosis by opening the mitochondrial membrane channel. Not all members of the Bcl-2 family of proteins (see Figure 9.26A) are indicated here. (D) In wild-type (*Bim*$^{+/+}$) mice, a cavity is excavated in the solid tip (terminal end bud) of an elongating mammary duct in order to form the lumen of the future duct (*left*). However, in *Bim*$^{-/-}$ mice, mammary epithelial cells persist, preventing normal lumen formation (*right*). p63, marker of basal epithelial cells (*red*); mucin1, marker of secretory luminal cells (*green*); cell nuclei (*blue*). (A, courtesy of J. Adams; adapted from S. Cory et al., *Oncogene* 22:8590–8607, 2003. B, from S. Cory and J.M. Adams, *Nat. Rev. Cancer* 2:647–656, 2002. C, courtesy of D.C.S. Huang. D, from A.A. Mailleux et al., *Dev. Cell* 12:221–234, 2007.)

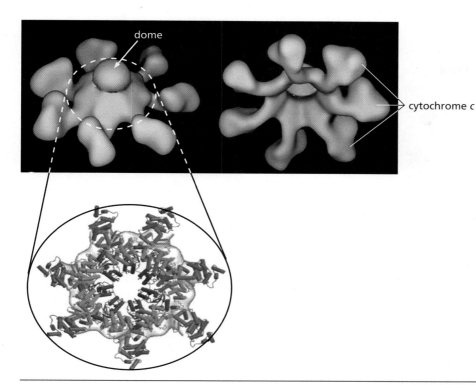

Figure 9.29 The wheel of death The apoptosome—the wheel of death—is assembled in the cytosol when cytochrome *c* molecules are released from mitochondria and associate with Apaf-1. This causes the assembly of a seven-spoked wheel, in which Apaf-1 forms the spokes and cytochrome *c* molecules form the tips of the spokes (*above*), visualized here by electron cryomicroscopy. Procaspase 9 is attracted to the hub of the wheel ("dome") and is converted into active caspase 9. The more detailed structure of the inner circle of the apoptosome (*below*), which is formed from seven Apaf-1 subunits, is illustrated in the molecular model determined by X-ray crystallography. The *blue* helices in the center interact with procaspase 9 to convert it into caspase 9, which then triggers the apoptotic cascade. (Top, from D. Acehan et al., *Mol. Cell* 9:423–432, 2002. Bottom, from S.J. Riedl et al., *Nature* 434:926–933, 2005.)

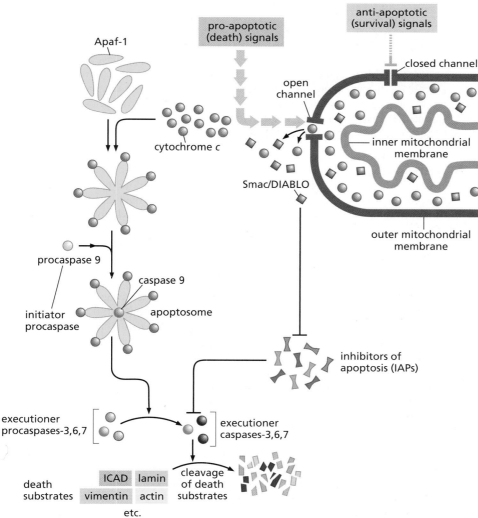

Figure 9.30 The apoptotic caspase cascade The outer mitochondrial membrane (*dark brown*) creates a critical gate that determines the life or death of the cell. Various anti-apoptotic signals work to hold channels in the outer mitochondrial membrane closed; conversely, pro-apoptotic death signals strive to open these channels. Important among the mitochondrial molecules that may be released into the cytosol are cytochrome *c* (*red circles*) and Smac/DIABLO (*brown squares*). Cytochrome *c* molecules proceed to aggregate with Apaf-1 to form the seven-spoked apoptosome (*left*; see Figure 9.29). The latter attracts and converts procaspase 9 into active caspase 9 (*green circle*), which in turn cleaves and activates procaspases 3, 6, and 7 (*bottom left*), converting them into executioner caspases, which then cleave various "death substrates" whose products create the apoptotic cell phenotype. Normally, a number of inhibitors of apoptosis (IAPs; *lower right*) attach to and inactivate caspases. However, the Smac/DIABLO that is also released from mitochondria antagonizes these IAPs, thereby protecting caspases from IAP inhibition.

Figure 9.31 Structure and function of an inhibitor of apoptosis IAPs (inhibitors of apoptosis) function by binding and inhibiting various caspase molecules, thereby preventing inadvertent activation of the caspase cascade and the triggering of apoptosis. (A) Shown here is the molecular structure of the complex formed by the BIR3 domain of the IAP molecule termed XIAP with caspase 3. The peptide backbone of XIAP is shown (*blue*) in contrast to the space-filling surface model of caspase 3 (*light green*). The BIR3 domain of XIAP binds to the substrate-binding groove of the caspase, thereby blocking its function. (B) The Smac/DIABLO molecule prevents several IAP molecules from blocking caspase action, thereby freeing the caspases to trigger apoptosis. Chemically synthesized Smac/DIABLO mimetics can act similarly in promoting apoptosis, which may prove useful in killing cancer cells. Here a tetrapeptide that mimics the actions of Smac/DIABLO is shown binding to the groove in the surface of XIAP that normally binds Smac/DIABLO. Chemical derivatives of this tetrapeptide have been developed and function as potent apoptosis-inducing drugs. Negatively charged surface (*red*); positively charged surface (*blue*). (A, from P.D. Mace, S. Shirley and C.L. Day, *Cell Death Differ.* 17:46–53. B, from K. Zobel et al., *ACS Chem. Biol.* 1:525–533, 2006.)

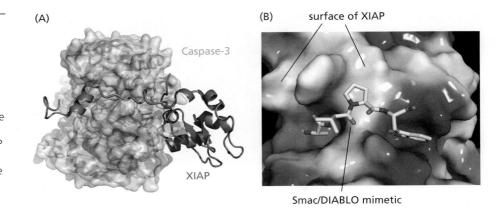

(A) Caspase-3 — XIAP

(B) surface of XIAP — Smac/DIABLO mimetic

initiating signal at the top of the cascade can be amplified to yield a large number of activated caspases at the bottom.)

While cytochrome *c* in the cytosol serves to activate the caspases, Smac/DIABLO—another protein that is released from mitochondria together with cytochrome *c*—inactivates a group of anti-apoptotic proteins termed IAPs (inhibitors of apoptosis; see Figure 9.30). These IAPs normally block caspase action in two ways: they bind directly and thereby inhibit caspase proteolytic activity (**Figure 9.31**), and in certain cases, they can mark caspases for ubiquitylation and degradation. Without the continued restraining influences of IAPs, caspases are free to initiate the proteolytic cleavages that result ultimately in apoptosis.

So, the apoptotic signal is activated in two ways following the opening of the mitochondrial channels. Cytochrome *c* is released and activates caspase 9 in the apoptosome. At the same time, Smac/DIABLO, also released through the channels, marginalizes the IAPs, thereby liberating caspases from IAP inhibition. (As an aside, one of the functionally important proteins that are phosphorylated by the anti-apoptotic Akt/PKB kinase is caspase 9. The phosphorylation of caspase 9 inhibits this protease, as might be anticipated from the known anti-apoptotic actions of Akt/PKB.)

The cascade of caspase activations thus initiated proceeds until the final caspases in the cascade cleave "death substrates," that is, proteins whose degradation creates the diverse cellular changes that one associates with the apoptotic death program (see Figure 9.30). The specialized roles of these various caspases have caused some to classify them into two functional groups—the *initiator caspases,* which trigger the onset of apoptosis by activating the caspase cascade; and the downstream *executioner caspases,* which undertake the actual work of destroying critical components of the cell.

A number of key cellular components are cleaved by these executioner caspases—caspases 3, 6, and 7. Their degradation causes the profound morphological transformations that accompany the death throes of apoptotic cells. Cleavage of lamins on the inner surface of the nuclear membrane is involved in some fashion in the observed chromatin condensation and nuclear shrinkage (pyknosis) that is characteristic of the apoptotic program. Cleavage of the inhibitor of caspase-activated DNase (ICAD) liberates this DNase, which then fragments the chromosomal DNA. Cleavage of cytoskeletal proteins such as actin, plectin, vimentin, and gelsolin leads to collapse of the cytoskeleton, the formation of blebs protruding from the plasma membrane, and the formation of apoptotic bodies—the condensed hulks of cells that remain after all this has occurred (see Figure 9.18).

The executioner caspases even extend their reach back into the mitochondria, where the apoptotic program is initiated: one of their substrates is a protein that is part of the electron transport chain within mitochondria. Its cleavage by caspases leads to disruption of electron transport, loss of ATP production, release of reactive oxygen species (ROS) from the mitochondria, and the loss of mitochondrial structural integrity, thereby amplifying the apoptotic program. The efficient execution of many of these steps of the apoptotic program is enabled, in part, by certain changes like these that further amplify the signals initiating apoptosis. (For example, cleavage of the Rel sub-unit of NF-κB, a transcription factor driving expression of anti-apoptotic genes such as *bcl-2*, shifts the regulatory balance further in favor of apoptosis.) Some researchers estimate that 400 to 1000 distinct cellular proteins undergo cleavage during apoptosis, but it remains unclear how many of these are active participants in the apoptotic pro-gram and how many are victims of the collateral damage that occurs as cells undergo disintegration.

The caspase cascade forms the central signaling pathway that determines whether or not apoptosis will occur and, in the event that apoptosis is decided upon, how the various death-inducing steps are activated within cells. The importance of careful regulation of these caspases is underscored by findings concerning E2F1 activity and apoptosis. Earlier, we noted that excessive E2F activity warns the cell of a malfunction-ing pRb pathway and induces apoptosis through its ability to activate expression of the gene encoding ARF (see Section 9.8). It has also become clear that E2Fs, such as E2F1, are able to increase transcription of several of the key caspase genes five- to fifteen-fold. These increases, on their own, do not trigger apoptosis. Instead, they sensitize the cell to specific pro-apoptotic stimuli, such as those emitted by p53, thereby tilting the playing field in favor of apoptosis. (The E2Fs can compound these effects by inducing the expression of several genes encoding pro-apoptotic relatives of Bcl-2.)

9.14 Both intrinsic and extrinsic apoptotic programs can lead to cell death

The apoptotic program described above can be initiated by p53 in response to a series of signals, including those indicating substantial genomic damage, anoxia, and imbal-ances in the growth-regulating signals coursing through a cell's intracellular circuits. To review, once p53 becomes activated and placed in an apoptosis-inducing configu-ration, it can drive the expression of genes encoding proteins such as Bax (see Table 9.2). Bax then works to open the mitochondrial channels, whereupon cytochrome *c* and other pro-apoptotic proteins leak out of the mitochondria and apoptosis is initi-ated through the activation of caspases. PUMA, another p53-induced, Bcl-2–related protein, may be an even more important player in the apoptosis triggered by p53: in certain cells, suppression of expression of this potent pro-apoptotic protein is as effec-tive as p53 deletion in preventing apoptosis.

This series of events is sometimes termed the *intrinsic apoptotic* program, since the signals triggering it originate from within the cell; others have termed it the *stress-acti-vated* apoptotic pathway. In ways that remain poorly understood, yet other types of signals can also cause the mitochondria to release cytochrome *c* through mechanisms that do not seem to involve p53. Included here are stresses such as excessive calcium within the cell, excessive oxidants, certain types of DNA-damaging agents, and yet other agents that disrupt the microtubules—key components of the cytoskeleton and mitotic spindle. (Figure 9.28A suggests additional p53-independent mechanisms.) Infections by various tumor viruses may also trigger apoptosis. Interestingly, these viruses take great pains to ensure that the intrinsic apoptotic program is not initiated, lest their cellular hosts die before their infectious cycles reach completion (**Sidebar 9.8**).

Importantly, apoptosis can also be triggered through an alternative route. This one is initiated outside the cell and activates pro-apoptotic cell surface receptors. These are transmembrane proteins that are often termed *death receptors* to indicate their ability to activate the apoptotic program (**Figure 9.32**). After binding their cognate ligands in the extracellular space, the death receptors activate a cytoplasmic caspase cascade that converges rapidly on the intrinsic apoptotic pathway described above, thereby triggering an apoptotic response identical to the one seen following activation of the intrinsic apoptotic program (see Figure 9.30). Because the signals that activate the death receptors originate from outside the cell, the apoptotic program initiated by these receptors has been called either the *extrinsic apoptotic* program or the *receptor-activated* apoptotic pathway.

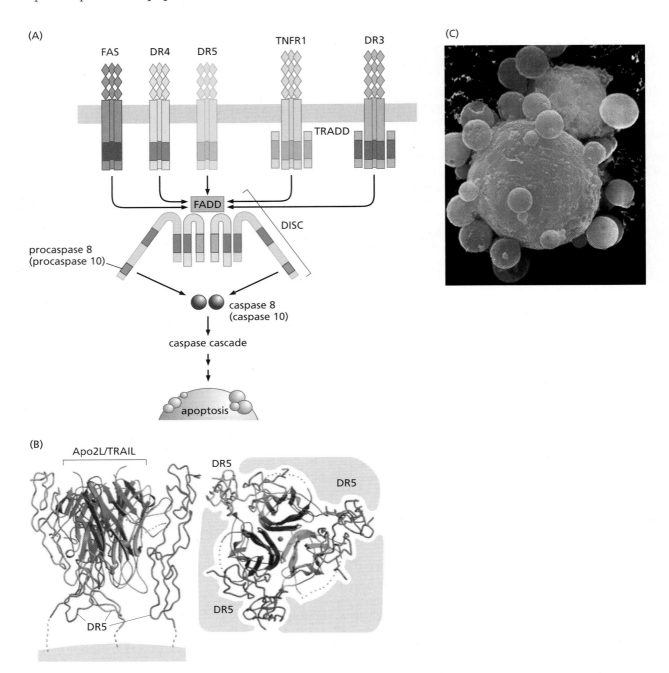

Figure 9.32 Death receptors (A) Five families of "death receptor" proteins are displayed on the surfaces of various types of mammalian cells. For example, the FasL and APO2L/TRAIL ligands bind to the FAS, DR4, and DR5 death receptors (*top left*). The cytoplasmic tails of these receptors act via the FAS-associated death domain (FADD) protein to assemble a death-inducing signaling complex (DISC), and the latter proceeds to convert procaspases 8 and 10 into their respective active caspases. These then converge on the caspase cascade that is activated through the intrinsic apoptotic program. A similar sequence of events occurs following the activation of the TNFR1 and DR3 receptors (*top right*). These death receptors differ from receptor tyrosine kinases (and their ligand-induced dimerization) in that the ligand complexes, sometimes called "death ligands," are homotrimers and appear to work by causing receptor trimerization. [Compare these ligands with the ligands of growth factor receptors (see Figure 5.15), which drive receptor dimerization.] (B) Shown here is the trimeric APO2L/TRAIL ligand in complex with the cognate trimeric DR5 receptor, both in side view (*left*) with the cell surface below (*gray*) and in top view facing the cell surface (*right*). The small *green* circle in the middle denotes a chelated zinc ion. (C) Death receptor ligands, notably FasL, are frequently used by cytotoxic T lymphocytes (CTLs) to kill cancer cells, which display the FAS receptor on their surfaces. (In addition, granzymes may be injected by CTLs into target cells; see Figure 9.34.) In this colorized scanning electron micrograph, a CTL (*orange*) is attacking a cancer cell, which is in the throes of apoptosis, as evidenced by the numerous blebs on its surface. (A, adapted from A. Ashkenazi, *Nat. Rev. Cancer* 2:420–430, 2002. B, from A. Ashkenazi, *Nat. Rev. Cancer* 2:420–430, 2002; and S.G. Hymowitz et al., *Mol. Cell* 4: 563–571, 1999. C, from Dr. Andrejs Liepins/Science Photo Library.)

Sidebar 9.8 Why do DNA tumor viruses seek to inactivate p53? As described previously (see Section 8.5), a diverse group of DNA tumor viruses specify oncoproteins that are designed to inactivate pRb. In addition, they undertake to inactivate p53 activity (see Table 9.1), raising the question why both host-cell proteins must be inactivated by these viruses. Because these agents are called "tumor viruses," we might conclude that they must create tumors in order to propagate themselves, and that pRb + p53 inactivation is central to this goal. However, careful examination of the life cycles of these viruses indicates that cell transformation is only an inadvertent side effect of their replication. Like all other viruses, these viruses have evolved to optimize only one outcome—their efficient multiplication in tissues of infected hosts. This realization forces a restatement of the question: Why is inactivation of both tumor suppressor proteins critical to viral replication?

The DNA tumor viruses parasitize the host-cell DNA replication machinery in order to replicate their own genomes; this machinery is available only in the late G_1 and S phases of the cell cycle. Consequently, these viruses need initially to inactivate pRb as well as the two pRb-related proteins p107 and p130, thereby causing infected, initially quiescent cells to advance into S phase. (Certain herpesviruses achieve the same end by expressing potent virus-encoded D-type cyclins.) The situation is slightly different in the case of human papillomaviruses (HPVs): they infect replicating cells in the cervical epithelium and block the normally occurring exit from the active cell cycle that takes place as these cells differentiate (**Figure 9.33**).

The cells infected by these various viruses respond to pRb inactivation by activating their p53 alarm systems. This is a direct response to the excessive activity of E2Fs that results from the functional inactivation of pRb (see Section 9.8); in addition, other, poorly understood virus-induced changes in cells may trigger p53 activation. Such p53 activation threatens to kill the infected host cells through apoptosis long before the viruses have had a chance to multiply and produce progeny. So these viruses proceed to eliminate p53 function. The SV40 large T antigen binds and sequesters p53 (in addition to binding pRb); adenovirus uses one of its E1B proteins (E1B55K) to do so; and HPV uses its E6 protein, which functions like Mdm2 to mark p53 for destruction via ubiquitylation. In order to guarantee that it succeeds in blocking apoptosis, adenovirus deploys a second E1B-encoded protein (termed E1B19K), which is related to Bcl-2 and antagonizes two important pro-apoptotic proteins, Bax and Bak (see Figure 9.28).

Unexplained by this logic are the reasons why at least one human RNA tumor virus—the HTLV-I retrovirus—takes the trouble to inactivate many of the same host-cell pathways. Perhaps the active transcriptional machinery of rapidly dividing cells confers some advantage on this virus.

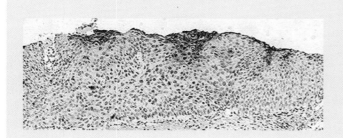

Figure 9.33 Human papillomavirus and the pathogenesis of cervical carcinomas This histological section of a human papillomavirus (HPV)-infected cervical epithelium reveals, through immunostaining with an anti-HPV antibody stain (*brown*), clusters of HPV-infected cells in dysplastic areas of the epithelium, termed high-grade squamous intraepithelial lesions (HSILs). The dysplastic cells arise, in large part, from the ability of the HPV E7 oncoprotein to inactivate pRb and thereby block their entrance into a post-mitotic state, and the ability of the viral E6 oncoprotein, which targets the cells' p53 protein for destruction, to suppress apoptosis, The resulting lesions progress with a low but significant frequency to cervical carcinomas. (From K. Middleton et al., *J. Virol.* 77:10186–10201, 2003.)

The ligands of the death receptors are members of the tumor necrosis factor (TNF) family of proteins, which includes TNF-α, TRAIL, and Fas Ligand (FasL) (**Table 9.4**). TNF-α was first identified and studied because of its ability to cause the death of cancer cells. Subsequently, this group of protein ligands was found capable of causing the death of a wide variety of normal cell types that display appropriate receptors on their surfaces. Similar to growth factors and their receptors, each of these TNF-like proteins, including FasL and TRAIL/Apo2L, binds to its own cognate death receptor (see Figure 9.32). Members of this family of receptors—there are as many as 30 of them encoded in the human genome—share in common a cytoplasmic "death domain." An understanding of these receptors and their action is confounded by the multiple names that many of these receptors and ligands bear (see Table 9.4).

Once activated by ligand binding, the death domains of these receptors bind and activate an associated protein termed FADD (Fas-associated death domain protein) in the cytoplasm. The resulting complex is termed a DISC (death-inducing signaling complex) and summons the inactive, pro-enzyme forms of caspase 8 and, less commonly, caspase 10, which are among the initiator caspases. The DISC then triggers the self-cleavage of these caspases and their conversion into active proteases. Caspases 8 and 10 (the initiator caspases of this pathway) then activate the executioner caspases 3, 6, and 7, thereby converging on the signaling pathway through which the intrinsic

Table 9.4 Death receptors and their ligands

Alternative names of receptors	Alternative names of ligands
FAS/APO-1/CD95	FasL/CD95L
TNFR1	TNF-α
DR3/APO-3/SWL-1/TRAMP	APO3L
DR4/TRAIL-R1	APO2L/TRAIL
DR5/TRAIL-R2/KILLER	APO2L/TRAIL

apoptotic program operates. In addition, caspase 3 can cleave and activate a Bcl-2–related protein in the cytosol termed Bid (see Figure 9.28A), which then migrates to the mitochondrial channel and undertakes to pry it open. In doing so, Bid succeeds in amplifying pro-apoptotic signals of the extrinsic apoptotic program by recruiting elements of the intrinsic program into the process as well (**Figure 9.34**).

Some cell types in the body are able to utilize both apoptotic pathways to trigger their own death, while others rely solely on either the extrinsic or the intrinsic program (**Figure 9.35**). For example, a cell can trigger its own death via the extrinsic program by secreting a ligand for one of the death receptors that it displays on its surface; this ligand can then act in an autocrine fashion to initiate apoptosis. The use by a cell of the intrinsic or extrinsic program has implications for the actions of anti-apoptotic proteins such as Bcl-2. Thus, in cells that can activate the extrinsic program, the overexpression of Bcl-2 may not confer much survival benefit, since the death receptors can circumvent the mitochondrion-based intrinsic apoptotic program by communicating directly with the caspase cascade.

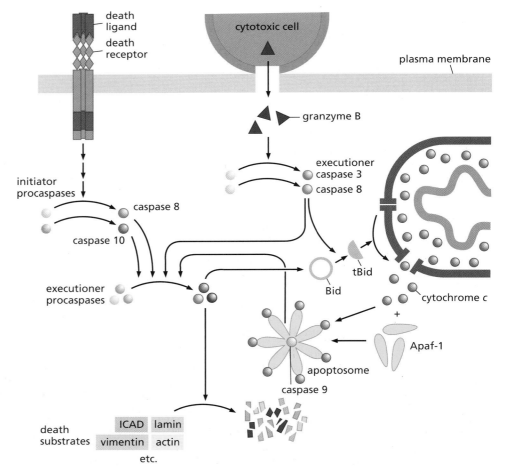

Figure 9.34 Convergence of intrinsic and extrinsic apoptotic pathways After a death receptor (see Figure 9.32) activates procaspase 8 and/or 10 into the corresponding active caspases (*top left*), the latter converge on the intrinsic apoptotic cascade by cleaving and activating the executioner caspases (*middle left*). In addition, after procaspase 3 has been activated into caspase 3, the latter cleaves and activates Bid (see Figure 9.28B), a Bcl-2–related protein (*center, green circle*). The resulting activated tBid (truncated Bid) moves from its cytosolic location to the mitochondrion, where it initiates opening of the outer mitochondrial membrane channel, further activating the apoptotic cascade by forming new, activated apoptosomes (*lower right center*). A second, alternative form of the extrinsic apoptotic pathway is initiated by cytotoxic cells that are dispatched by the immune system (*top middle*) and attach to the surfaces of targeted cells, into which they introduce granzyme B molecules; these cleave and thereby activate procaspases 3 and 8.

In fact, there is actually a third way of triggering apoptosis, in effect another extrinsic apoptotic program. This one is triggered by cytotoxic T lymphocytes and natural killer (NK) cells that undertake to kill target cells. These two types of killer cells—some of the frontline troops of the immune system—can activate death receptors such as FAS displayed on the surfaces of cells that they have chosen for destruction. But in addition, the killer cells can attach to the surfaces of targeted cells and inject a protease, termed granzyme B; once internalized by these cells, granzyme B cleaves and activates procaspases 3 and 8 (see Figure 9.34). At this point, there is immediate convergence with the other apoptotic pathways described above.

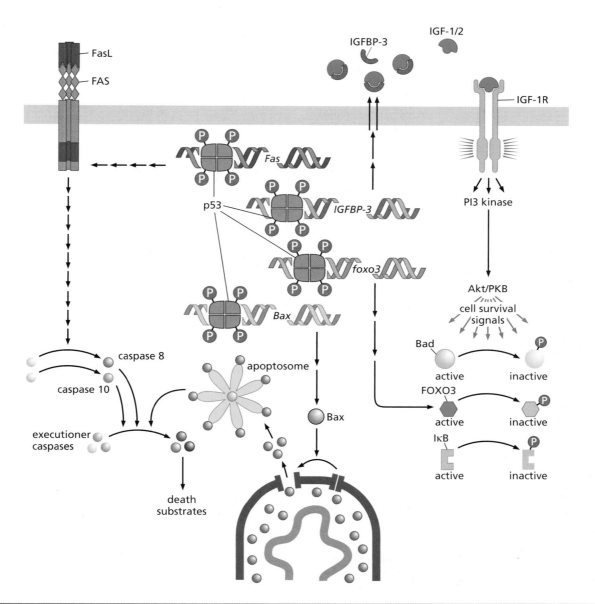

Figure 9.35 Activation of apoptosis by p53 p53 utilizes multiple signaling pathways to activate the apoptotic program. By inducing expression of the gene encoding the FAS receptor, it causes display of this death receptor at the cell surface (*brown, top left*), thereby sensitizing the cell to any Fas ligand (FasL) that may be present in the extracellular space. By inducing expression of IGF-binding protein-3 (IGFBP-3; *red*), p53 causes release of this protein into the extracellular space, where it binds and sequesters IGF-1 and IGF-2 (*blue, top center*), the pro-survival, anti-apoptotic ligands of the IGF-1 receptor (IGF-1R; *top right*). In the absence of IGFBP-3,

IGF-1 would bind to its receptor and cause release via PI3 kinase of anti-apoptotic signals in the cell, including those leading to the inactivation of the pro-apoptotic proteins Bad (related to Bcl-2), FOXO3, and IκB (antagonist of NF-κB). In addition, the activated, pro-apoptotic form of p53 drives expression of the FOXO3 pro-apoptotic transcription factor, as well as Bax (*center*), the pro-apoptotic, Bcl-2–related protein; the latter causes the release of cytochrome *c* and other proteins from the mitochondria (*bottom center*). p53 also inactivates several other anti-apoptotic agents (*not shown*).

This mapping of the intrinsic and extrinsic apoptotic cascades enriches our understanding of how p53 succeeds in evoking apoptosis (see Figure 9.35). p53 promotes expression of the genes encoding the Bax, FAS receptor, and IGFBP-3 proteins (see Table 9.2). Bax antagonizes Bcl-2 function and thereby works to open the mitochondrial channels that release cytochrome *c*. The FAS receptor, once expressed at the cell surface, increases a cell's responsiveness to extracellular death ligands, specifically FasL. IGFBP-3 is released from the cell, whereupon it binds and sequesters IGF-1 (insulin-like growth factor-1), a growth factor that operates through its own cell surface receptor to produce **trophic** (survival) signals in the cell. Because many cells depend on continuous exposure to trophic signals in order to remain viable, the reduction of survival signals induced by IGF-1 places them in grave danger of succumbing to apoptosis, a process that has been called "death by neglect." This explains why many cancer cells find ways of maintaining high concentrations of IGFs in the microenvironment that surrounds them. Apart from simply secreting abnormally high amounts of IGF-1 or the related IGF-2, cancer cells may reduce the levels of IGF-binding proteins (IGFBPs)—either by methylating *IGFBP* gene promoters or by increasing the synthesis and secretion of matrix metalloproteinases (MMPs) that cleave and thereby inactivate IGFBP molecules.

We will return to the extrinsic apoptotic program in Chapter 15, because it can play two important roles in the complex interactions between cancer cells and the immune system. On the one hand, cells of the immune system may release death ligands (such as FasL) in order to trigger apoptosis in cancer cells that they have targeted for elimination. On the other, cancer cells may display or release death ligands that they deploy to kill immune cells that have approached too closely and threaten their survival.

The complexity of the signals flowing through apoptotic pathway~~~~~~~ fact that the decision for or against apoptosis is achieved by alte complex arrays of pro- and anti-apoptotic proteins that influence sion whether or not to flip the "apoptotic master switch."

9.15 Cancer cells invent numerous ways to inac or all of the apoptotic machinery

As emphasized earlier, cells advancing down the long road toward ter a number of physiologic stressors that threaten their existence, stressors threatens to trigger apoptosis. Included among these are anoxia, deregulation of the pRb pathway, activation of oncogenes like *myc* and *ras*, and various forms of DNA damage. These dangers explain why most and likely all types of cancer cells have inactivated key components of their apoptosis-inducing machinery (Table 9.5; Figure 9.36). We can also view these changes from the perspective of the organism as a whole: the fact that virtually all human cancer cells employ anti-apoptotic strategies suggests that the apoptotic machinery is implanted in all normal cells, albeit in latent form, and held in reserve as an anti-neoplastic mechanism that can be activated in order to obstruct tumor progression.

Arguably the most important and widely exploited strategy has already been noted—inactivation of the p53 pathway. The *p53* gene is altered in almost half of all human cancer cell genomes, and in a substantial proportion of others, ARF is no longer expressed. This is often achieved through outright deletion of the ARF-encoding gene or by promoter methylation, the latter resulting in repression of transcription (see Section 1.9). Other mechanisms shutting down *ARF* expression may also intervene, as was indicated in Sidebar 9.4. A small percentage of human tumors (mostly sarcomas) will overexpress Mdm2. All of these mechanisms affecting ARF or Mdm2 expression serve to drive down levels of p53 protein in cells. Finally, in some human tumor cells, p53 appears to be mislocalized, being sequestered in the cytoplasm, where it cannot carry out its main task—transcriptional activation.

Still, this elimination of p53 function does not seem to suffice for many types of human tumors, which additionally strive to alter specific components of the apoptotic machinery. For example, many melanoma cells exhibit methylation and thus functional inactivation of the promoter of the *APAF1* gene, which encodes the cytosolic protein that assembles with released cytochrome *c* to form the apoptosome and activate caspase 9. The pro-apoptotic *Bax* gene (see Table 9.5) is inactivated by mutation in more than half the colon cancers that show microsatellite instability (see

Table 9.5 Examples of anti-apoptotic alterations found in human tumor cells

Alteration	Mechanism of anti-apoptotic action	Types of tumors
CASP8 promoter methylation	inactivation of extrinsic cascade	SCLC, pediatric tumors
CASP3 repression	inactivation of executioner caspase	breast carcinomas
Survivin overexpression[a]	caspase inhibitor	mesotheliomas, many carcinomas
ERK activation	repression of caspase 8 expression	many types
ERK activation	protection of Bcl-2 from degradation	many types
Raf activation	sequestration of Bad by 14-3-3 proteins	many types
PI3K mutation/activation	activation of Akt/PKB	gastrointestinal
NF-κB constitutive activation[b]	induction of anti-apoptotic genes	many types
p53 mutation	loss of ability to induce pro-apoptotic genes	many types
$p14^{ARF}$ gene inactivation	suppression of p53 levels	many types
Mdm2 overexpression	suppression of p53 levels	sarcomas
IAP-1 gene amplification	antagonist of caspases 3 and 7	esophageal, cervical
APAF1 methylation	loss of caspase 9 activation by cytochrome *c*	melanomas
BAX mutation	loss of pro-apoptotic protein	colon carcinomas
Bcl-2 overexpression	closes mitochondrial channel	~½ of human tumors
PTEN inactivation	hyperactivity of Akt/PKB kinase	glioblastoma, prostate carcinoma, endometrial carcinoma
IGF-1/2 overexpression	activates PI3K	many types
IGFBP repression	loss of anti-apoptotic IGF-1/2 antagonist	many types
Casein kinase II overexpression	activation of NF-κB	many types
TNFR1 methylation	repressed expression of death receptor	Wilms tumor
FLIP overexpression	inhibition of caspase 8 activation by death receptors	melanomas, many others
Akt/PKB activation	phosphorylation and inactivation of pro-apoptotic Bcl-2-like proteins	many types
USP9X overexpression	deubiquitylates Mcl-1	lymphomas
STAT3 activation	induces expression of Bcl-X$_L$	several types
TRAIL-R1 repression	loss of responsiveness to death ligand	small-cell lung carcinoma
FAP-1 overexpression	inhibition of FAS receptor signaling	pancreatic carcinoma
XAF1 methylation[c]	loss of inhibition of anti-apoptotic XIAP	gastric carcinoma
Wip1 overexpression[d]	suppression of p53 activation	breast and ovarian carcinomas, neuroblastoma

[a]Survivin is an inhibitor of apoptosis (IAP) in gastric, lung, and bladder cancer and melanoma, in addition to the mesotheliomas indicated here. The expression of a number of IAP genes is directly induced by the NF-κB TFs.
[b]Induces synthesis of c-IAPs, XIAP, Bcl-X$_L$, and other anti-apoptotic proteins.
[c]XAF1 (XIAP-associated factor 1) normally binds and blocks the anti-apoptotic actions of XIAP, the most potent of the IAPs.
[d]Wip1 is a phosphatase that inactivates p38 MAPK, which otherwise would phosphorylate and stimulate the pro-apoptotic actions of p53.

Figure 9.36 Anti-apoptotic strategies used by cancer cells Cancer cells resort to numerous strategies in order to decrease the likelihood of apoptosis. This diagram indicates that in various cancer cell types, the levels or activity of important pro-apoptotic proteins are decreased (*blue*). Conversely, the levels or activity of certain anti-apoptotic proteins may be increased (*red*). (Some of these, because they are downstream of master regulators that are known to be altered in cancer cells, are affected by the known changes in their upstream regulators.)

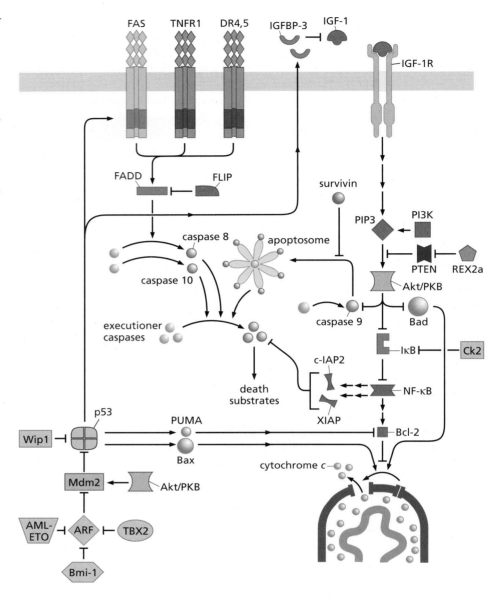

Section 12.4). Deregulated expression of *BCL2* (the human version of *bcl-2*) is found in follicular B-cell lymphomas, and *BCL2* expression is found to be elevated in large numbers of human tumors of diverse tissue origins; by some estimates, at least half of all human tumors show elevated *BCL2* expression.

Yet another highly effective strategy for acquiring resistance to apoptosis derives from hyperactivation of the PI3K → Akt/PKB pathway (see Section 6.6). Activation of this pathway can be achieved through the collaborative actions of signal-emitting tyrosine kinase receptors and the Ras oncoprotein, both of which activate PI3K; the result is an increased level of PIP3 and activation of Akt/PKB (see Figures 6.17 and 6.19). Inactivation of the gene encoding PTEN, the phosphatase that breaks down PIP3, has a similar effect. Once Akt/PKB becomes activated by the PIP3 that accumulates, it can phosphorylate and thereby inhibit pro-apoptotic proteins such as Bad, caspase 9, and IκB, and at the same time phosphorylate and activate Mdm2, the key antagonist of p53.

A diverse collection of cancer cell types, ranging from leukemias to colon carcinomas, have been found to resort to yet another mechanism to protect themselves against apoptosis: they activate expression of the NF-κB transcription factor. As we read earlier, NF-κB is usually sequestered in inactive form in the cytoplasm and, in response to certain physiologic signals, is liberated and translocates to the nucleus, where it activates a large constituency of target genes (see Section 6.12). A number of these genes are involved in pro-inflammatory and angiogenic functions. In addition, a significant cohort of NF-κB target genes function as apoptosis antagonists; included among these are *IAP-1*, *IAP-2*, *XIAP*, *IEX-1L*, *TRAF-1*, and *TRAF-2*, which serve to block both the intrinsic and extrinsic apoptotic programs. This list of NF-κB targets is expanding rapidly, and it is likely that the overexpression of some of these anti-apoptotic genes (and encoded proteins) found in many types of human cancer cells will be traced directly to the actions of hyperactivated NF-κB transcription factors.

Yet another strategy for avoiding apoptosis is evident in childhood neuroblastomas, in which amplification of the N-*myc* gene normally indicates aggressive tumor growth and short life expectancy. On its own, this amplification actually threatens the proliferation of neuroblastoma cells since, as mentioned earlier, the actions of the *myc* oncogene (and those of its N-*myc* cousin) represent a double-edged sword. The encoded oncoproteins release potent mitogenic signals, but at the same time are strongly pro-apoptotic. Many neuroblastoma cells that have overexpressed N-*myc* through gene amplification seem to solve this problem, at least in part, by inactivating the gene encoding caspase 8, which is either methylated or deleted.

The actions of the extrinsic apoptosis-inducing pathway are also relevant for cancer cell evolution, as evidenced by the fact that this pathway is inactivated in certain types of human tumors. Some cancer cells deploy a protein termed FLIP (FLICE-inhibitory protein) which can interfere with the activation of the extrinsic apoptotic cascade by binding to the death domain proteins associated with death receptors, thereby blocking the processing and autocatalytic activation of the initiator caspase-8 (see Figure 9.32A). Yet other tumor cells suppress expression of their TNF receptors by methylating the promoter of the encoding *TNFR1* gene.

These maneuvers by cancer cells provide clear evidence that they can gain survival advantage by inactivating the extrinsic pathway. This leads, in turn, to the speculation that cells of the immune system often besiege cancer cells and attempt to kill them through the release of TNF-like factors and resulting initiation of the extrinsic apoptotic program. Given the complex wiring diagram of the circuitry controlling apoptosis, it is virtually certain that many additional anti-apoptotic strategies will be uncovered in various types of human cancer cells.

9.16 Necrosis and autophagy: two additional forks in the road of tumor progression

While much of this chapter has focused on apoptosis as a major obstacle to cancer cell survival, alternative fates may await a cell en route to full-fledged neoplastic growth. One obstacle to survival has already been mentioned, albeit in passing—necrosis. In complex metazoa like ourselves, the cells within our tissues are usually guaranteed access to reasonably constant levels of essential nutrients, such as glucose and amino acids. Cancer cells, however, often do not enjoy such access, simply because they have not managed to generate the functional vasculature that is required to deliver these critical nutrients. Hence, cancer cells may be nutrient- and oxygen-starved. Indeed, in the early stages of tumor formation, tumor-associated angiogenesis (formation of new blood vessels) is still poorly developed, as we will discuss in Chapter 13.

Later in tumor progression, malformed tumor-associated vessels may cause **ischemia**—lack of access to blood-borne nutrients—starving large sections of the tumor of vital nutrients and triggering necrosis. As indicated in Table 9.3, necrotic cell death differs in many respects from apoptosis and results in the lysis of the cells rather than the production of apoptotic bodies. Importantly, the cell lysis resulting from necrotic cell death can result in localized inflammation as the surrounding tissues become exposed to the intracellular contents of the necrotic cell.

The necrotic cores of many tumors (for example, see Figure 13.28) are testimonials to the important role of necrosis in limiting tumor growth. In fact, quite often the overall size of a tumor will yield the impression of a large and actively growing mass, whereas the bulk of the tumor is actually inert necrotic tissue that has not been eliminated

Figure 9.37 Autophagy and alternative forms of cell death (A) These electron micrographs reveal (*left panel*) an entire mitochondrion (*below*) that has been engulfed in an autophagosome. The *arrows* indicate structures that have the multi-layered membrane structures of an autophagosome, only one of which (*right arrow*) can be definitively identified because of its contents as an autophagosome. The *right panel* indicates an autophagosome (*arrow*), surrounding mitochondria (m) and fragments of endoplasmic reticulum (e). (B) Under some circumstances, cancer cells may respond to metabolic stress, such as nutrient deprivation, by activating apoptosis. If apoptosis is circumvented experimentally (*curved arrow*) by inactivating pro-apoptotic genes (*bax* and *bak*) and by expressing an anti-apoptotic protein (Bcl-2), these cells may activate autophagy. If autophagy is then circumvented experimentally by expressing the Akt/PKB kinase (which blocks autophagy) plus inactivating one of the two copies of the *beclin* gene (which is a key regulator of autophagy), such cells will enter into necrosis. The latter may in turn promote localized inflammation, which often serves to promote tumor progression, as discussed in Chapter 11. (C) A subtype of ductal carcinoma *in situ* (DCIS) of the breast, termed comedo carcinoma, shows a rim of viable cells (C) that are expressing high levels of Beclin-1 (*brown*), the master regulator of autophagy, surrounding a central core of necrotic cells (N, *light blue*). The Beclin+ cells have activated autophagy, apparently in response to the limited nutrients available to them from the nearby surrounding stroma (S). Of note, this subtype of DCIS has a worse prognosis than the more common subtypes, in which intraductal necrosis is not seen; indeed, the Beclin+ cancer cells seen here are close to becoming actively invasive. (D) The complex interactions between the apoptosis and autophagy programs are highlighted by the fact that Beclin can be bound directly by either Bcl-2 or Bcl-X$_L$, two potent anti-apoptotic proteins; this binding prevents Beclin from causing autophagy. Seen here is a crystallographic structure of a complex of two molecules of the BH3 domain of Beclin bound to two molecules of the BH3 domain of Bcl-X$_L$. (A, left, courtesy of E.M. Haley, B.A. Lund, M.E. Bisher and H.A. Coller; right, from N. Mizushima and B. Levine, *Nat. Cell Biol.* 12:823–830, 2010. B, from K. Degenhardt et al., *Cancer Cell* 10:51–64, 2006. C, from V. Espina and L.A. Liotta, *Nat. Rev. Cancer* 11:68–75, 2011. D, from A. Oberstein, P.D. Jeffrey and Y. Shi, *J. Biol. Chem.* 282:13123–13132, 2007.)

through one mechanism or another. It is of some interest that if the apoptotic response is blocked through one of the mechanisms described in the previous section, then the exposure of a cell to a death receptor ligand (see Figure 9.32) may trigger necrosis.

Before nutrient starvation has driven a cancer cell into necrosis, it may resort to an alternative strategy to stave off the irreversible collapse of metabolism that triggers necrotic cell death: the cell may activate its **autophagy** program, which allows it to degrade its own organelles as a means of recycling their contents. The resulting recycled molecules can then be used to generate the energy and intermediary metabolites that are required to maintain cell viability.

Like apoptosis, autophagy represents the workings of an organized cell-biological program. In response to nutrient starvation, a cell will shut down its proliferation and activate a number of energy- and nutrient-conserving strategies. In addition, activation of its autophagy program will cause many of its organelles to become enveloped in membranous vesicles, resulting in the formation of **autophagosomes**; the latter then fuse to cytoplasmic endosomes—the vesicles formed by endocytosis (see Sidebar 5.1)—which in turn fuse with lysosomes, in which the digestion of enveloped organelles (such as ribosomes, mitochondria, endoplasmic reticulum) occurs (**Figure 9.37A**).

The resulting breakdown products, which can then be recycled in various biosynthetic pathways, permit the nutrient-starved cell to survive for an extended period of time. In the case of cancer cells, this survival may allow them to avoid necrosis altogether if they succeed after some delay in obtaining access to nutrients (see

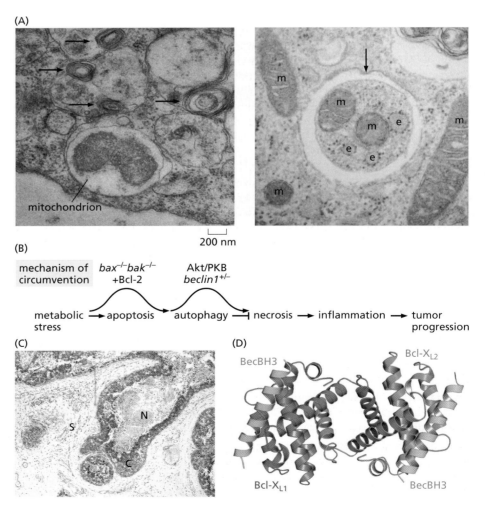

(A)

mitochondrion

200 nm

(B)

mechanism of circumvention   *bax*$^{-/-}$*bak*$^{-/-}$ +Bcl-2   Akt/PKB *beclin1*$^{+/-}$

metabolic stress → apoptosis ⇝ autophagy ⊣ necrosis → inflammation → tumor progression

(C)

(D)

BecBH3   Bcl-X$_{L2}$

Bcl-X$_{L1}$ BecBH3

Figure 9.37B). Under conditions of nutrient deprivation, some cancer cells have been seen to consume two-thirds of their own mass and still remain viable, retaining the ability to return to full size and to a state of active proliferation once nutrients are eventually procured. [In fact, autophagy is used routinely by normal resting cells, enabling them to eliminate old, worn-out organelles, such as mitochondria that have suffered oxidative and other forms of damage; once again, this autophagy yields breakdown products that can be reused to construct new organelles. Similarly, it is used in normal developmental processes. For example in the post-fertilization oocyte, widespread autophagy is triggered to break down maternal proteins and generate the amino acids needed by the zygote to synthesize new proteins (see Supplementary Sidebar 9.5).]

This life-saving aspect of autophagy would seem to favor the survival of cancer cells and the outgrowth of incipient tumors. By operating in this fashion, this program acts to promote tumor formation (see Figure 9.37C). Moreover, a key regulator of the autophagy program, Beclin-1, also known as Atg6, is a BH3-containing protein structurally related to the Bcl-2 protein that cancer cells often deploy in order to ward off apoptosis (see Figure 9.26A). While Bcl-2 regulates the opening of channels in the outer mitochondrial membrane, Beclin is involved in some fashion in the assembly of autophagosomes, perhaps by helping to extract patches of membrane from the endoplasmic reticulum.

The notion that autophagy favors the survival of cancer cells and thereby promotes their agenda of tumorigenesis is supported by the widespread expression of Beclin-1 in certain tumors (see Figure 9.37C). The rationale is that when cells undergo transformation by a *ras* oncogene, for example, the resulting shift in cell energy metabolism requires ongoing active autophagy in order to ensure a continuous supply of nutrients; this explains why *ras*-driven tumors in which autophagy is suddenly inactivated rapidly implode.

In truth, autophagy can affect tumor progression in more complex ways. Thus, the potent anti-apoptotic proteins Bcl-2 and Bcl-X_L are both capable of binding Beclin via their and Beclin's respective BH3 domains (see Figure 9.37D). Regulation of apoptosis by Beclin does not seem to occur. However, working in the opposite direction, the binding of Beclin by Bcl-2 or Bcl-X_L inhibits Beclin's ability to activate autophagy; hence, these strongly pro-tumorigenic proteins *decrease* the likelihood of autophagy occurring. Also, when mice are deprived of one of their two germ-line copies of the *beclin1* gene, these mice actually exhibit an *increased* rate of spontaneous tumor formation. In addition, one of the two copies of the *beclin1* gene is either silenced or deleted in many human ovarian, breast, and prostate carcinomas. In these various situations, the autophagy program seems to act as a *tumor-suppressing* mechanism and the key autophagy-promoting *beclin1* gene as a haploinsufficient tumor suppressor gene (for example, see Figure 9.37B). This notion is reinforced by the fact that other widely accepted tumor suppressors, such as p53 and PTEN, favor activation of the autophagy program. It remains unclear which of these competing roles of autophagy is more influential in governing tumor formation in humans.

9.17 Synopsis and prospects

In the early chapters of this book, we concentrated on the molecular mechanisms that propel cell proliferation forward or hold it back. The regulators—the products of (proto-) oncogenes and tumor suppressor genes—act to process extracellular mitogenic and anti-mitogenic signals that impinge upon the cell cycle machinery. These signals converge on the final decisions that determine whether cells advance through the active growth-and-division cycle, retreat reversibly into quiescence, or enter irreversibly into a post-mitotic, differentiated state.

In this chapter, we have focused on entirely different cell-physiologic processes, most of which are responsible for monitoring the internal well-being of cells. The monitoring apparatus conducts continuous surveillance of vital cell systems, including access to oxygen and nutrients, the physical state of the genome, and the balance of signals flowing through a cell's growth-regulating circuits. If such monitors determine that a vital system is damaged or malfunctioning in a cell, an alarm is sounded and growth is halted or, more drastically, the cell is induced to activate its previously latent apoptotic program. Alternatively, if nutrient supply is critically low, cells will activate their autophagy program in order to stave off metabolic collapse. Autophagy may tide the cell over until it is able to gain access once again to nutrients. Without such access, the cell will eventually devour itself and slip into necrotic cell death.

These monitors create barriers to tumor formation that incipient cancer cells must override in order to complete the complex, multi-step agenda that leads eventually to the formation of malignant tumors. Thus, as cells evolve from normalcy to a state of high-grade neoplasia, many of their subcellular systems malfunction in one way or another, leading to cytostasis (see Section 8.4), senescence (discussed in Chapter 10), autophagy, apoptosis, or necrosis.

Actually, this list does not exhaust the alternative fates that malfunctioning, pre-neoplastic cells may confront. In response to damaged or mis-segregated chromosomes, **mitotic catastrophe** threatens; the precise relationship of this form of cell death to the apoptosis and necrosis programs remains unclear.

Any one of these traps can halt further progress down the long road to malignancy, and indeed attempts by most would-be cancer cells to evade these pitfalls usually fail, explaining the high attrition of the cells in premalignant growths. This also helps to explain one of the mysteries of human cancer pathogenesis: in spite of the fact that each of us experiences ~10^{16} cell divisions in a lifetime, the great majority of us will not develop life-threatening malignancies, because of these multiple impediments to cancer cell survival.

p53 is a central player in many of these responses, since elimination of its function is one frequent means of avoiding both apoptosis and senescence, the latter representing a permanent exit from the cell cycle (see Section 10.1). The *p53* gene and encoded protein have been the subject of more than 50,000 research reports since the discovery of the p53 protein in 1979. This extraordinary focus on a single gene and protein reflects both the central role that this protein plays in cancer pathogenesis and the frequent presence of mutant *p53* alleles in the genomes of human cancer cells. Still unanswered, however, is the most perplexing puzzle surrounding this protein's existence: Why has evolution created cells that entrust so many vital alarm functions to a single protein? Once a cell has lost p53 function, it becomes oblivious, in a single stroke, to a whole variety of conditions that would normally call a halt to its proliferation or trigger entrance into the apoptotic program.

As we will see in Chapter 11, the complex, multi-step process that drives tumor development involves changes in multiple cellular control circuits. Because p53 receives signals from so many distinct sensors of damage or physiologic distress, the loss of p53 function plays a role in many of these steps. For example, early in the development of bladder, breast, lung, and colon carcinomas, when premalignant growths are present, DNA damage signaling is already active, as indicated by the presence of phosphorylated, functionally active ATM and Chk2 kinases. In addition, the p53 substrate of the ATM and Chk2 kinases is also phosphorylated and apparently active, since the *p53* gene is still in its wild-type configuration at this stage of tumor progression. This indicates that DNA lesions are widespread in the cells of these growths, even at this early stage of tumorigenesis, and suggests that p53 is striving to constrain the proliferation of the cells in these growths and may even induce their apoptosis with some frequency. Consequently, these premalignant cells stand to benefit greatly by eliminating p53 function, thereby liberating themselves from its cytostatic and pro-apoptotic effects.

At this stage of tumor development, cells that succeed in shedding p53 function may also gain special advantage because they have a better chance of avoiding the apoptosis that is triggered by p53 in response to severe hypoxia. (In general, early premalignant growths have not yet developed an adequate access to the circulation and the oxygen that it brings.) In addition, because p53 induces expression of the potently anti-angiogenic molecule thrombospondin (Tsp-1), p53 inactivation can accelerate the development of nearby capillary networks that can cure the hypoxia.

Later in tumor progression, when malignant carcinomas develop, the absence of p53 function will compromise some of the DNA repair systems of the cell and, at the same time, allow cells carrying damaged genomes greater survival advantage; the result is an increased rate of genomic alteration and associated acceleration in the rate of tumor development. Additionally, the absence of p53 function may enable these tumor cells to survive the pro-apoptotic side effects of oncogenes such as *myc*. This short list reveals how truly catastrophic the loss of p53 function can be for a tissue and, ultimately, for the organism as a whole.

Experiments with genetically altered mice indicate how critical p53 loss can be to tumor development. Lymphomas and sarcomas can be induced in $p53^{-/-}$ mice through exposure to X-rays. If wild-type p53 expression is subsequently restored in the tumor cells, the lymphoma cells will rapidly die apoptotic deaths, while the sarcoma cells will enter into the post-mitotic senescence state. In a similar experiment, mouse hepatoblasts (embryonic liver cells) were transformed by a *ras* oncogene in the absence of p53 function. When p53 function was reactivated briefly in these cells, there was widespread senescence and the shrinkage of implanted tumors (**Figure 9.38**). These experiments show dramatically that in the absence of ongoing p53 surveillance, cancer cells spontaneously acquire changes in critical cellular components that would otherwise, in the presence of functional p53, provoke their rapid elimination. (Stated differently, p53 inactivation creates a *permissive state* that allows a variety of otherwise-lethal changes to be acquired, notably those that favor tumor progression.)

The disastrous effects of p53 loss may be compounded later, in the event that the *p53*-mutant tumor cells become clinically apparent and targeted for chemotherapy and radiation. The success of most existing anti-cancer therapies is predicated on their

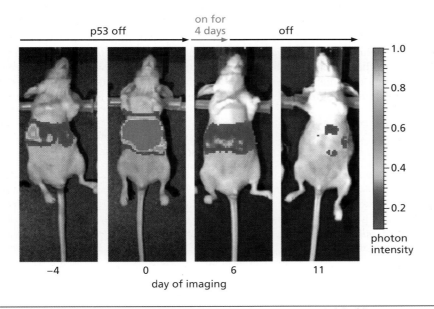

Figure 9.38 Reactivation of p53 induces senescence In a mouse model of liver cancer pathogenesis, an H-*ras* oncogene was introduced into cultured hepatoblasts in which p53 expression was suppressed by an inducible siRNA vector construct (see Supplementary Sidebar 11.6). These cells were injected into the spleens of host mice, leading to their implantation in the liver. The resulting tumors were allowed to grow to a significant size (day 0), at which point expression of the siRNA construct was shut down for 4 days, permitting p53 function to return during this time window; p53 expression was then shut down once again for the remaining time of the experiment. By day 6, tumors were markedly smaller, and by day 11, they were almost undetectable. Since the tumors continued to shrink long after p53 expression was suppressed, the brief (4-day-long) period of p53 expression seems to have forced tumor cells to enter irreversibly into the post-mitotic senescence state (described in Section 10.1), and the senescent cells were apparently cleared by phagocytic cells of the innate immune system (see Section 15.1). Tumors were imaged *in vivo* through light released by a luciferase gene borne by the tumor cells. (From W. Xue et al., *Nature* 445:656–660, 2007.)

ability to damage the genomes of tumor cells, thereby provoking the death of these cells by apoptosis. Unfortunately, it seems that the loss of p53 function, which is seen so often in human tumors, renders these neoplasias far less responsive to many of these therapeutic strategies (see Figure 9.9B).

Our understanding of the p53 protein is still quite superficial, in spite of the vast research literature that describes it. One indication of how much more needs to be learned has come from attempts at systematically cataloging the list of p53-regulated genes in the human genome; more than 120 have been identified, only a small proportion of which are cited in Table 9.2. Yet another indication comes from analyses of the post-translational modifications of p53. As we have seen, these modifications affect p53's stability and thus intracellular concentration. In addition, they also affect the interactions of p53 with other nuclear transcription regulators that control the genes targeted by p53 action. The extent of the complexity of these modifications is indicated by **Figure 9.39**, which shows that at least six distinct types of post-translational modification, involving phosphate, acetyl, ubiquityl, sumoyl, neddyl, and methyl groups, decorate the p53 molecules present in human cells. (The neddyl and sumoyl groups are similar to ubiquityl groups but have very different effects on the proteins to which they become attached.) The number of enzymes that modify p53—at least 28 are indicated here—is likely to reflect the number of distinct signaling pathways that impinge on p53 and perturb its functioning.

Yet another dimension of p53 complexity comes from reports that p53 has been found by some researchers to be localized to the cytoplasm, far away from its normal nuclear site of action. This cytoplasmic form of p53 may interact with Bcl-2–related proteins and facilitate opening channels in the mitochondrial outer membrane, thereby expediting the initiation of apoptosis. This area of research is in rapid flux.

Inactivation of p53 is only one of many strategies that cancer cells employ to protect themselves against the danger of apoptosis. These maneuvers compromise the apoptotic response at many other levels, including perturbing the expression of death receptors, their ability to activate caspases, and the activities of the caspases themselves. Yet other defensive maneuvers undertaken by cancer cells help to prevent opening the outer membrane pores of mitochondria. These diverse strategies (see Figure 9.36) illustrate the fact that the apoptotic machinery, like all well-regulated

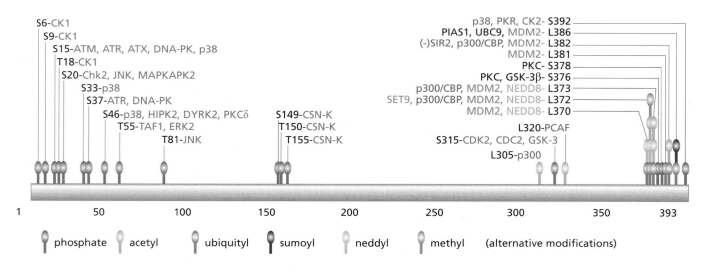

Figure 9.39 Multiple types of post-translational modifications of p53 The phosphorylation of p53 by various kinases (see Figure 9.13) represents only one type of modification that is important to its functioning. In addition to the attachment of phosphate groups (*red*), other enzymes modify it by attaching acetyl (*green*), ubiquityl (*blue*), sumoyl (*purple*), neddyl (*yellow*), and methyl (*gray*) groups at the indicated amino acid residues and residue numbers (*black type*). The identities of the modifying enzymes are indicated (*various font colors*) next to the amino acid residue numbers. Only a small portion of these modifications have been linked to specific p53 functions. One of these, phosphorylation of serine 46, which is accomplished by any of 4 distinct kinases, activates the pro-apoptotic functions of p53. (Courtesy of K.P. Olive and T. Jacks.)

machines, operates through the balanced actions of mutually antagonistic elements. Cancer cells exploit preexisting anti-apoptotic components of this machinery by increasing their levels or activities, thereby nudging the balance away from apoptosis. These changes in specific components of the apoptotic machinery may compound the effects of p53 loss, further increasing the resistance of cancer cells to a variety of signals that usually succeed in provoking cell death.

Later, in Chapter 16, we will learn about recent attempts at developing novel forms of therapy that are designed to mobilize components of the pro-apoptotic machinery that remain intact in cancer cells. These therapies are crude beginnings, and we have only begun to recognize the difficulties and the opportunities that the apoptotic machinery provides to those who would like to cure the disparate collection of diseases that we call cancer. The hope is that at some time in the not-too-distant future, the circuitry governing apoptosis will be understood in a quantitative way (perhaps through a schematic diagram such as **Figure 9.40**), and that we will be able to predict its behavior with precision and manipulate it at will within cancer cells.

These abstract discussions of the mechanisms governing apoptotic cell death, which seem to be far removed from the real world, may soon be brought to bear on a major puzzle about human cancer: Why does obesity, as gauged by body-mass index (BMI), represent a major factor determining cancer incidence? In the United States, obesity

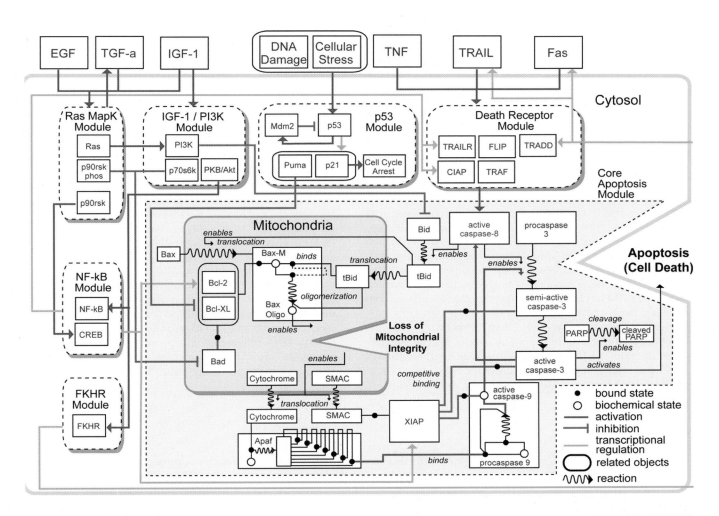

Figure 9.40 The apoptotic circuit board One of the goals of current apoptosis research is to understand in a quantitative way how the signaling proteins governing apoptosis interact within a cell to influence the decision to initiate apoptosis. Such work should eventually yield mathematical models that predict, with some accuracy, how the apoptotic circuitry will respond to diverse extra- and intracellular signals, including those triggered by various types of therapy. An initial attempt at modeling this signaling system is shown here. (Courtesy of I. Khalil and Gene Network Sciences, Inc.)

Sidebar 9.9 Cell death and the American way of life We focused earlier in this book on the effects of various mutagenic carcinogens, such as those in tobacco smoke, as key factors in human cancer causation. Over the past decade, the science of epidemiology has uncovered a second major etiologic factor in cancer development in the industrialized West—obesity. Obesity has been increasing in epidemic proportions as a consequence of ongoing high caloric intake coupled with ever-decreasing physical activity. As indicated in **Figure 9.41**, high **body-mass index** (BMI), a useful measure of obesity, is strongly correlated with increased risks of dying from various types of commonly occurring cancers in the United States. The implications of these correlations for overall cancer incidence are profound: these figures indicate that about 90,000 cancer deaths in the United States could be avoided annually if all members of the population maintained a BMI of less than 25 throughout life.

In the case of certain hormone-responsive tissues, a clear mechanistic connection can be made between high BMI and increased cancer risk. For example, women with a BMI of 34 or more have a relative increased risk (RR) of 4.8 of developing endometrial carcinomas compared with the general female population; this may be largely a consequence of the production of estrogen (by adipocytes), which stimulates the proliferation of epithelial cells in the endometrium. Another mechanism must be sought, however, to explain the increased incidence (and death rates) of cancers arising in non-hormone-responsive tissues. Since diverse tissues are affected, this mechanism, whatever its nature, must function systemically, that is, via signals that extend their reach throughout the body.

An attractive but still-unproven hypothesis is as follows: obesity is known to be associated with high levels of circulating insulin (**hyperinsulinemia**), which in turn have been found to increase the synthesis of IGF-1 and the *bioavailability* of circulating IGF-1. In a 1998 study, which measured the levels of circulating IGF-1, the 25% of men in the group having the highest IGF-1 levels had a 4.3-fold higher risk of subsequently developing prostate cancer compared with the quartile of men having the lowest IGF-1 levels. Among men over the age of 60 in the highest quartile, the increased risk was calculated to be 7.9-fold. Similarly, another study published that year revealed that high IGF-1 levels in females were correlated with a marked increase in pre-menopausal breast cancer.

At the same time, hyperinsulinemia is known to suppress production in the liver of IGF-1 antagonists, specifically IGF-binding protein-1 (IGFBP-1). (Normally, the various IGFBPs bind and sequester as much as 98% of the IGF-1 in the circulation.) The increase in IGF-1 production and the reduction of IGFBP-1 levels result in higher concentrations of free, biologically active IGF-1 throughout the body, which, by activating the IGF-1 receptor (IGF-1R) and thus Akt/PKB, provides pre-malignant and malignant cells with potent anti-apoptotic signals (see Figure 9.35). (Conversely, the shutdown of IGF-1R expression in a variety of cancer cells, achieved experimentally, causes them to enter rapidly into apoptosis.) Importantly, hyperinsulinemia is likely to conspire with high levels of circulating IGF-1 to protect cancer cells from apoptosis, since both factors act via their respective receptors to activate PI3K and thus Akt/PKB, a potent suppressor of apoptosis.

Animal models of cancer development have shed equally compelling light on the diet–cancer connection. In one study, dietary restriction was imposed on mice in which bladder carcinomas had been induced by exposure to a chemical carcinogen. This limited diet led to a decrease in both circulating IGF-1 levels and tumor formation; restoration in the diet-restricted mice (using an osmotic mini-pump) of circulating IGF-1 to levels seen in control mice led to a ~10-fold decrease in apoptosis rates and a ~6-fold increase in cell proliferation rates in incipient bladder tumors. In fact, this restoration of circulating IGF-1 levels totally reversed the effects of dietary restriction.

Another animal study examined the effects of dietary restriction (DR) on the growth of various tumor xenografts; as anticipated, DR resulted in a large drop in circulating IGF-1 and insulin levels. The growth of some tumors was highly sensitive to DR and widespread apoptosis in these tumors was observed in response to DR, while the growth of other tumors was relatively unaffected and DR had little effect on apoptosis in their constituent cells. Those tumors that were resistant to DR bore constitutively active Akt/PKB (due either to mutant PI3K or loss of PTEN; see see Figures 6.19A and 9.35), while those that responded to DR showed significant drops in Akt/PKB activity when deprived of insulin and IGF-1. Once again, diet, acting via PI3K, was a strong determinant of cancer cell growth and survival. The mechanistic details remain to be worked out, but the overall chain of causality is already reasonably clear and, for Western societies, most unsettling.

Finally, while circulating IGF-1 and insulin are attractive candidates as the culprits driving obesity-associated cancer incidence, alternative mechanistic theories have also attracted support. The most interesting of these derives from observations that obesity is associated with chronic elevated levels of various systemic markers of inflammation. As we will learn in Section 11.15, inflammation is a potent mechanism for accelerating the formation of a variety of tumors, lending credibility to the obesity-inflammation-cancer connection.

represents the second greatest risk factor for developing cancer after tobacco consumption. Since reductions in cancer incidence—prevention of disease—rather than improvements in therapy, offer the greatest prospects for substantial decreases in cancer-associated mortality, understanding phenomena like this one will assume an ever-increasing role in cancer research. An attractive but still-unproven model proposes that obesity increases cancer incidence by protecting incipient cancer cells from death by apoptosis (Sidebar 9.9). If this thinking is substantiated, then apoptosis research will soon assume major roles in explaining the onset of cancer and the successes that we have in treating this disease.

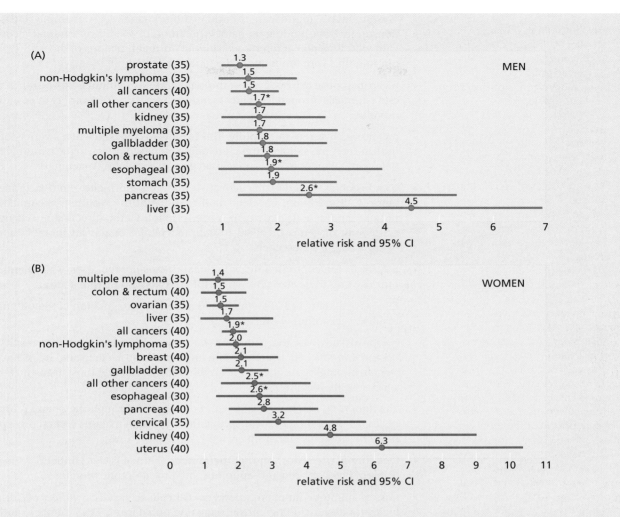

Figure 9.41 Body-mass index and cancer mortality The relative risk of death from a variety of cancers is strongly influenced by body-mass index (BMI), which is calculated by dividing body weight (in kilograms) by the square of height (in meters). Both men and women were subdivided into four equal groups (i.e., quartiles) according to their BMIs, and the relative risk of cancer mortality (for each type of cancer) was calculated as the frequency of death in the high-BMI (obese) group compared with that in the low-BMI (lean) group. In the case of each cancer, the obese group was defined as those having a BMI equal to or greater than the number listed after the tumor name (e.g., a BMI of 35 in the case of women with multiple myeloma). The lean groups were defined as those women and men having a BMI of 24.9 or less. In each case, the calculated average increased risk is shown above the dots, with the flanking horizontal lines indicating the 95% confidence interval (CI), i.e., an indication that, with a 95% probability, the true relative risk lies within the interval indicated by the horizontal lines. Asterisks indicate relative risks of women or men who never smoked. (A) Male cancer mortality risks. (B) Female cancer mortality risks. Note that in the case of ovarian and breast carcinomas, the interpretation of these data is confounded by the fact that women with higher BMIs have, on average, higher circulating estrogen levels, which is, on its own, a significant risk factor for these cancers. "All other cancers" refers to all cancers not listed in this figure panel. (From E.E. Calle et al., *N. Engl. J. Med.* 348:1625–1638, 2003.)

Key concepts

- Organisms attempt to block the development of cancer through the actions of the p53 alarm protein, which can cause cells to enter quiescence or apoptosis in the event that the machinery regulating cell proliferation is malfunctioning or the cell is exposed to various types of physiologic stress.

- p53 is a nuclear protein that normally exists as a tetramer and functions as a transcription factor.

- The mutant p53 found in many human tumors usually carries amino acid substitutions in its DNA-binding domain. When mutant p53 forms tetrameric complexes with wild-type p53, it interferes with the normal functions of the wild-type subunits and thus with functions of the tetramer as a whole.

- p53 can impose cell cycle arrest through its ability to induce expression of p21^{Cip1}, and apoptosis through its ability to induce expression of a variety of pro-apoptotic proteins.

- p53 normally turns over rapidly. This turnover is blocked when a variety of signals indicate cell-physiologic stress, including anoxia, damage to the genome, and signaling imbalances in the intracellular growth-regulating machinery.

- p53 becomes functionally activated when its normally rapid degradation is blocked. In addition, covalent modifications of the resulting accumulated p53 protein modulate its activity as a transcription factor, directing it to activate the expression of genes involved in various cellular responses, notably apoptosis, cytostasis, and senescence.

- p53 levels are controlled by two critical upstream regulators, Mdm2 and p19^{ARF}. Mdm2 works to destroy p53, while ARF inhibits Mdm2 from acting.

- Excessive activity of E2Fs, which is triggered by deregulation of the pRb pathway, results in activation of ARF and thus p53.

- Apoptosis involves the activation of a cascade of caspases that results in the destruction of a cell, usually within an hour. It can be activated by p53 as well as signals impinging on the cell from the outside, notably those transduced by cell surface death receptors.

- The apoptotic caspase cascade can be triggered through the opening of a channel in the outer membrane of mitochondria, which releases several pro-apoptotic proteins, notably cytochrome c.

- Opening of the mitochondrial membrane channel is determined by the relative levels of Bcl-2–related anti-apoptotic and pro-apoptotic proteins.

- Loss of apoptotic functions allows cancer cells to survive a variety of cell-physiologic stresses, including anoxia, signaling imbalances, DNA damage, and loss of anchorage.

- Cancer cells invent numerous ways to inactivate the apoptotic machinery in order to survive and thrive. Included among these are activation of Akt/PKB firing, increase in the levels of anti-apoptotic Bcl-2–related proteins, inactivation of p53 through changes in the *p53* gene or the upstream regulators of p53, methylation of the promoters of a variety of pro-apoptotic genes, interference with cytochrome c release from mitochondria, and inhibition of caspases.

Thought questions

1. In light of the fact that DNA tumor viruses must suppress the apoptosis of infected cells in order to multiply, what molecular strategies are available for them to do so?

2. What types of factors influence the decision of p53 to act in a cytostatic versus a pro-apoptotic fashion?

3. How can anti-cancer therapeutics be successful in treating cancer cells that have inactivated components of their apoptotic machinery? Given the physiologic stresses that are known to activate p53-induced apoptosis, what types of anti-cancer therapeutic drugs might be created to treat cancer cells?

4. What side effects would you predict could result from a general inhibition of apoptosis in all tissues of the body?

5. How might the loss of components of the apoptosis machinery render cancer cells more susceptible than normal cells to certain types of cell death?

6. Can you enumerate the range of apoptosis-inducing physiologic stresses that cancer cells must confront and circumvent during the course of tumor development?

Additional reading

Ashkenazi A (2002) Targeting death and decoy receptors of the tumour-necrosis factor superfamily. *Nat. Rev. Cancer* 2, 410–430.

Baehrecke EH (2002) How death shapes life during development. *Nat. Rev. Mol. Cell Biol.* 3, 779–787.

Bartek J, Bartkova J & Lukas J (2007) DNA damage signalling guards against activated oncogenes and tumor progression. *Oncogene* 26, 7773–7779.

Baserga R, Peruzzi F & Reiss K (2003) The IGF-1 receptor in cancer biology. *Int. J. Cancer* 107, 873–877.

Belyi VA, Ak P, Markert E et al. (2010) The origins and evolution of the p53 family of genes. *Cold Spring Harb. Perspect. Biol.* 2, a001198.

Blattner C (2008) Regulation of p53. *Cell Cycle* 7, 3149–3153.

Brosh R & Rotter V (2010) When mutants gain new powers: news from the mutant p53 field. *Nat. Rev. Cancer* 9, 701–713.

Bursch W (2001) The autophagosomal-lysosomal compartment in programmed cell death. *Cell Death Differ.* 8, 569–581.

Calle EE & Kakas R (2004) Overweight, obesity, and cancer: epidemiological evidence and proposed mechanisms. *Nat. Rev. Cancer* 4, 579–591.

Calle EE, Rodriguez C, Walker-Thurmond K & Thun MJ (2003) Overweight, obesity, and mortality from cancer in a prospectively studied cohort of U.S. adults. *N. Engl. J. Med.* 348, 1625–1638.

Chandeck C & Mooi WJ (2010) Oncogene-induced senescence. *Adv. Anat. Pathol.* 17, 42–48.

Cichowski K & Hahn WC (2008) Unexpected pieces to the senescence puzzle. *Cell* 133, 958–961.

Collado M & Serrano M (2006) The power and the promise of oncogene-induced senescence markers. *Nat. Rev. Cancer* 6, 472–476.

Collado M & Serrano M (2010) Senescence in tumours: evidence from mice and humans. *Nat. Rev. Cancer* 10, 51–57.

Cory S & Adams JM (2002) The Bcl-2 family: regulators of the cellular life-or-death switch. *Nat. Rev. Cancer* 2, 647–656.

Coutts AS & La Thangue NB (2007) Mdm2 widens its repertoire. *Cell Cycle* 6, 827–829.

Cowey S & Hardy RW (2006) The metabolic syndrome: a high risk for cancer? *Am. J. Pathol.* 169, 1505–1522.

Danial NN & Korsmeyer SJ (2004) Cell death: critical control points. *Cell* 116, 205–219.

Eisenberg-Lerner A, Bialik S, Simon HU & Kimchi A (2009) Life and death partners: apoptosis, autophagy and the cross-talk between them. *Cell Death Differ.* 16, 966–975.

El-Deiry WS (2003) The role of p53 in chemosensitivity and radiosensitivity. *Oncogene* 22, 7486–7495.

Evan GI & d'Adda di Fagagna F (2009) Cellular senescence: hot or what? *Curr. Opin. Genet. Dev.* 19, 25–31.

Finlan LE & Hupp TR (2007) p63: the phantom of the tumor suppressor. *Cell Cycle* 6, 1062–1071.

Fischer U, Jänicke RU & Schullze-Osthoff K (2002) Many cuts to ruin: a comprehensive update of caspase substrates. *Cell Death Differ.* 10, 76–100.

Fürstenberger G & Senn H-G (2002) Insulin-like growth factors and cancer. *Lancet Oncol.* 3, 298–302.

Gil J & Peters G (2006) Regulation of the INK4b–ARF–INK4a tumour suppressor locus: all for one or one for all. *Nat. Rev. Mol. Cell Biol.* 7, 667–677.

Giovannucci E (2003) Nutrition, insulin, insulin-like growth factors and cancer. *Horm. Metab. Res.* 35, 694–704.

Green DR & Evan GI (2002) A matter of life and death. *Cancer Cell* 1, 19–30.

Gyrd-Hansen M & Meier P (2010) IAPs: from caspase inhibitors to modulators of NF-κB, inflammation and cancer. *Nat. Rev. Cancer* 10, 561–574.

Harris CC (1996) p53 tumor suppressor gene: from the basic research laboratory to the clinic—an abridged historical perspective. *Carcinogenesis* 17, 1187–1198.

Hippert MM, O'Toole PS & Thorburn A (2006) Autophagy in cancer: good, bad, or both? *Cancer Res.* 66, 9349–9351.

Hursting SD, DiGiovanni J et al. (2012) Obesity, energy balance, and cancer: new opportunities for prevention. *Cancer Prev. Res.* 5, 1260–1272.

Hursting SD, Lavigne JA, Berrigan D et al. (2003) Calorie restriction, aging and cancer prevention: mechanisms of action and applicability to humans. *Annu. Rev. Med.* 54, 131–152.

Igney FH & Krammer PH (2002) Death and anti-death: tumour resistance to apoptosis. *Nat. Rev. Cancer* 2, 277–288.

Joerger AC & Fersht AR (2008) Structural biology of the tumor suppressor p53. *Annu. Rev. Biochem.* 77, 557–582.

Junttila MR & Evan GI (2009) p53: a Jack of all trades but master of none. *Nat. Rev. Cancer* 9, 821–829.

Karin M & Lin A (2002) NF-κB at the crossroads of life and death. *Nat. Immunol.* 3, 221–227.

Kroemer G, Galluzzi L & Brenner C (2007) Mitochondrial membrane permeabilization in cell death. *Physiol. Rev.* 87, 99–163.

Kroemer G et al. (2009) Classification of cell death. *Cell Death Differ.* 16, 3–11.

Kroemer G & White E (2010) Autophagy for the avoidance of degenerative, inflammatory, infectious and neoplastic disease. *Curr. Opin. Cell Biol.* 22, 121–123.

Kuilman T, Michaloglou C, Mooi WJ & Peeper DS (2010) The essence of senescence. *Genes Dev.* 24, 2463–2479.

Larsson O, Girnita A & Girnita L (2005) Role of insulin-like growth factor I receptor in signaling in cancer. *Brit. J. Cancer* 92, 2097–2101.

LeBlanc HN & Ashkenazi A (2003) Apo2L/TRAIL and its death and decoy receptors. *Cell Death Differ.* 10, 66–75.

Levine AJ (2009) The common mechanisms of transformation by small DNA tumor viruses: the inactivation of tumor suppressor gene products: p53. *Virology* 384, 285–293.

Levine AJ, Feng Z, Mak TW et al. (2006) Coordination and communication between the p53 and IGF1–AKT–TOR signal transduction pathways. *Genes. Dev.* 20, 267–275.

Levine AJ, Hu W & Feng Z (2006) The p53 pathway: what questions remain to be explored? *Cell Death Differ.* 13, 1027–1036.

Levine B & Kroemer G (2008) Autophagy in the pathogenesis of disease. *Cell* 132, 27–42.

Lim YP, Lim TT, Chan YL et al. (2007) The p53 knowledgebase: an integrated information resource for p53 research. *Oncogene* 26, 1517–1521.

Lomonosova E & Chinnardurai G (2009) BH3-only proteins in apoptosis and beyond: an overview. *Oncogene* Suppl. 1, S2–S19.

Lowe SW, Cepero E & Evan G (2004) Intrinsic tumour suppression. *Nature* 432, 307–315.

Maiuri MC, Zaickvar E, Kimchi A & Kroemer G (2007) Self-eating and self-killing: crosstalk between autophagy and apoptosis. *Nat. Rev. Mol. Cell Biol.* 8, 741–752.

Marine J-CW, Dyer MA & Jochemsen AG (2007) MDMX: from bench to bedside. *J. Cell Sci.* 120, 371–378.

Matheu A, Maraver A & Serrano M (2008) The Arf/p53 pathway in cancer and aging. *Cancer Res.* 68, 6031–6034.

Mathew R & White E (2011) Autophagy in tumorigenesis and energy metabolism: friend by day, foe by night. *Curr. Opin. Genet. Dev.* 21, 113–119.

Melino G, De Laurenzi V & Vousden KH (2002) p73: friend or foe in tumorigenesis? *Nat. Rev. Cancer* 2, 605–615.

Meulmeester E, Pereg Y, Shiloh Y & Jochemsen AG (2005) ATM-mediated phosphorylations inhibit Mdmx/Mdm2 stabilization by HAUSP in favor of p53 activation. *Cell Cycle* 4, 1166–1170.

Michael D & Oren M (2003) The p53-Mdm2 module and the ubiquitin system. *Semin. Cancer Biol.* 13, 49–59.

Mooi WJ & Peeper DS (2006) Oncogene-induced cell senescence: halting on the road to cancer. *New Engl. J. Med.* 355, 1037–1045.

Ow Y-LP, Green DR, Hao Z & Mak T (2008) Cytochrome: functions beyond respiration. *Nat. Rev. Mol. Cell Biol.* 9, 532–542.

Perry ME (2010) The regulation of the p53-mediated stress response by MDM2 and MDM4. *Cold Spring Harb. Perspect. Biol.* 2, 968.

Pfeifer GP, Denissenko MF, Oliver M et al. (2002) Tobacco smoke carcinogens, DNA damage and p53 mutations in smoking-associated cancers. *Oncogene* 21, 7435–7451.

Plati J, Bucur O & Khosravi-Far R (2008) Dysregulation of apoptotic signaling in cancer: molecular mechanisms and therapeutic opportunities. *J. Cell Biochem.* 104, 1124–1149.

Polager S & Ginsberg D (2009) p53 and E2F: partners in life and death. *Nat. Rev. Cancer* 9, 738–748.

Pollak M (2008) Insulin and insulin-like growth factor signaling in neoplasia. *Nat. Rev. Cancer* 8, 915–928.

Pollak MN, Schernhammer ES & Hankinson SE (2004) Insulin-like growth factors and neoplasia. *Nat. Rev. Cancer* 4, 505–518.

Proskuryakov SY & Gabai VL (2010) Mechanisms of tumor cell necrosis. *Curr. Pharm. Des.* 16, 56–68.

Riedl SJ & Shi Y (2004) Molecular mechanisms of caspase regulation during apoptosis. *Nat. Rev. Mol. Cell Biol.* 5, 897–907.

Rodier F & Campisi J (2011) Four faces of cellular senescence. *J. Cell Biol.* 192, 547–556.

Salvesen GS & Duckett CS (2002) IAP proteins: blocking the road to death's door. *Nat. Rev. Mol. Cell Biol.* 3, 401–410.

Sherr CJ (2006) Divorcing ARF and p53: an unsettled case. *Nat. Rev. Cancer* 6, 663–673.

Sherr CJ & DePinho RA (2000) Cellular senescence: mitotic clock or culture shock? *Cell* 102, 407–410.

Sinha S & Levine B (2009) The autophagy effector Beclin 1: a novel BH3-only protein. *Oncogene* 27, S137–S148.

Taubes G (2012) Unraveling the obesity-cancer connection. *Science* 335, 28–32.

Varley JM (2003) Germline TP53 mutations and Li-Fraumeni syndrome. *Hum. Mutat.* 21, 313–320.

Vaux DL, Cory S & Adams JM (1988) *Bcl-2* gene promotes haematopoietic cell survival and cooperates with c-*myc* to immortalize pre-B cells. *Nature* 335, 440–442.

White E & DiPaola RS (2009) The double-edged sword of autophagy modulation in cancer. *Clin. Cancer Res.* 15, 5308–5316.

Xie Z & Klionsky DJ (2007) Autophagosome formation: core machinery and adaptations. *Nat. Cell Biol.* 9, 1102–1109.

Yee KS & Vousden KH (2005) Complicating the complexity of p53. *Carcinogenesis* 26, 1317–1322.

Youle RJ & Strasser A (2008) The Bcl-2 protein family: opposing activities that mediate cell death. *Nat. Rev. Mol. Cell Biol.* 9, 47–59.

Zahir N & Weaver VM (2004) Death in the third dimension: apoptosis deregulation and tissue architecture. *Curr. Opin. Genet. Dev.* 14, 71–80.

Zhang R & Adams PD (2007) Heterochromatin and its relationship to cell senescence and cancer therapy. *Cell Cycle* 6, 784–789.

Zong WX & Thompson CB (2006) Necrotic death as a cell fate. *Genes Dev.* 20, 1–15.

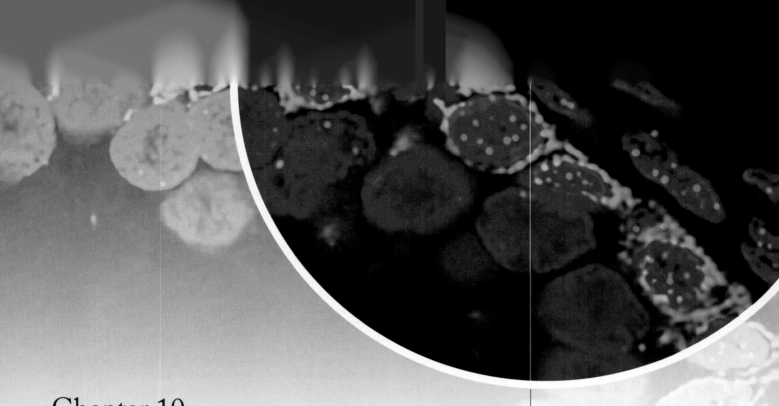

Chapter 10

Eternal Life: Cell Immortalization and Tumorigenesis

> Death takes place because a worn-out tissue cannot forever renew itself, and because a capacity for increase by means of cell division is not ever-lasting but finite.
>
> August Weissmann, biologist, 1881

In previous chapters, we read about a number of distinctive traits displayed by cancer cells. In some instances, these traits are acquired through the actions of activated oncogenes; in others, cancer cell–specific traits can be traced back to the loss of tumor suppressor genes. As we will discuss in Chapter 11, the acquisition by human cells of these neoplastic traits (and thus the development of a clinically apparent human tumor) usually requires several decades' time.

During this extended period of development, populations of human cells pass through a long succession of growth-and-division cycles as they evolve toward the neoplastic growth state. Such extensive proliferation, however, conflicts with a fundamental property of normal human cells: they are endowed with an ability to replicate only a far smaller number of times. Once normal human cell populations have exhausted their allotment of allowed doublings, the cells in these populations cease proliferating and may even enter apoptosis.

These facts lead us to a simple, inescapable conclusion: in order to form tumors, incipient cancer cells must breach the barrier that normally limits their proliferative potential. Somehow, they must acquire the ability to multiply for an abnormally large number of growth-and-division cycles, so that they can successfully complete the multiple steps of tumor development.

Movie in this chapter
10.1 Telomere Replication

In this chapter, we explore the nature of the regulatory machinery that limits cell proliferation and how it must be neutralized in order for cells to become fully neoplastic and form clinically detectable tumors. By neutralizing this machinery, cells gain the ability to proliferate indefinitely—the phenotype of cell immortality. This immortality is a critical component of the neoplastic growth program.

10.1 Normal cell populations register the number of cell generations separating them from their ancestors in the early embryo

In multicellular (metazoan) animals such as ourselves, the origin of each cell can, in principle, be traced through multiple cell generations back to a single ancestor—the fertilized egg. Looking in the other direction, the sequence of cell divisions that stretches from an ancestral cell that existed in the embryo to a descendant cell that exists many cell generations later is often termed a **cell lineage**. Indeed, in a relatively simple metazoan—the worm *Caenorhabditis elegans*—the lineage of all 959 somatic cells in the adult body has been traced and can be depicted as a pedigree (**Figure 10.1**).

In large, complex mammals, however, the assembling of a comparable pedigree will never happen, because the total number of cell divisions is astronomical: the adult human body, for example, comprises almost 10^{14} cells, and the organism as a whole

Figure 10.1 The pedigree of cells in the body of a worm (A) The adult worm *Caenorhabditis elegans* is composed of 959 somatic cells, and the lines of descent of all cells in this worm have been traced with precision back to the fertilized egg. (B) As is apparent, many of the cell lineages end abruptly because cells stop proliferating or are discarded (usually by apoptosis). The cells in each of the major branches are destined to form a specific tissue. For example, cells in the major, multigenerational branch on the right are destined to form the gonads, while those in a small branch at the far left will form the nervous system. (A, from J.E. Sulston and H.R. Horvitz, *Dev. Biol.* 56:110–156, 1977. © Academic Press. B, Courtesy of University of Texas Southwestern Medical Center.)

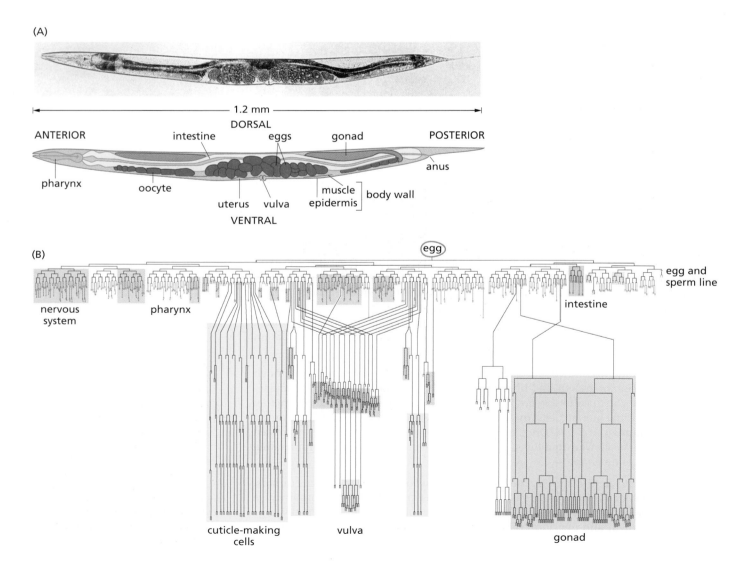

(A)

1.2 mm

DORSAL

ANTERIOR   intestine   eggs   gonad   POSTERIOR

anus

pharynx   oocyte

uterus   vulva   muscle   body wall
epidermis

VENTRAL

(B)   egg

egg and sperm line

nervous system   pharynx   intestine

cuticle-making cells   vulva   gonad

undergoes as many as 10^{16} cell divisions in a lifetime. Still, we can imagine that for each human body such a cell pedigree must exist, if only as a theoretical construct.

As is the case with *C. elegans*, during the course of human development, early embryonic cells become the founders of specific cell lineages that are committed to assuming various tissue-specific cellular phenotypes, that is, to differentiating (see Section 8.11). Indeed, the science of developmental biology focuses much of its attention on how individual cells in various cell lineages acquire the information from their surroundings that causes them to enter into one or another program of differentiation. However, developmental biologists do not address a question of great relevance to tumor development: Are there specific controls that determine the number of cell generations through which a particular cell lineage can pass during the lifetime of an organism? Can each branch and twig of the cell pedigree grow indefinitely, or is the number of replicative generations in each cell lineage predetermined and limited?

Currently available techniques do not allow us to determine with any accuracy how many times specific cell lineages within the human or mouse body pass through successive growth-and-division cycles. Still, a crude measure of the replicative capacity of a cell lineage can be undertaken by culturing cells of interest *in vitro*. For example, one can prepare fibroblasts from living tissue, introduce them into a Petri dish, and determine how many times these cells will double. (In practice, such experiments require **serial passaging**, in which a portion of the cells that have filled one dish are removed and introduced into a second dish and allowed to proliferate, after which some of their number are introduced into a third dish, and so forth.)

As first demonstrated in the mid-1960s, cells taken from rodent or human embryos exhibit a limited number of successive replicative cycles in culture. The work of Leonard Hayflick showed that cells would stop growing after a certain, apparently predetermined number of divisions and enter into the state that came to be called **replicative senescence** or simply **senescence** (**Figure 10.2**). Senescent cells remain metabolically active but seem to have lost irreversibly the ability to re-enter the active cell cycle. Such cells will spread out in monolayer culture, acquire a large cytoplasm, and persist for weeks if not months, as long as they are given adequate nutrients and growth factors; such cells are often described as taking on the appearance of a fried egg (**Figure 10.3**). The growth factors help to sustain the viability of the senescent cells, but they are unable to elicit the usual proliferative response observed when these factors are applied to healthy, nonsenescent cells. Like actively proliferating cells, the senescent cells display growth factor receptors, but the downstream signaling pathways have been inactivated through still poorly understood mechanisms.

The precise number of replicative doublings exhibited by cultured cells before they reach senescence is dependent on the species from which the cells were prepared, on the tissue of origin, and on the age of the donor organism. Some experiments with human cells indicate that cells prepared from newborns are able to double in culture a greater number of times than comparable cells taken from middle-aged or elderly

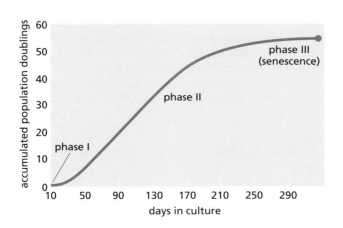

Figure 10.2 The proliferative capacity of cells passaged extensively in culture The ability of human fibroblasts to proliferate in culture was gauged in Leonard Hayflick's work by counting the number of times that the population of cells had doubled (*ordinate*). As is apparent, these cells, beginning soon after explantation from living tissue into culture (phase I), were able to proliferate robustly for about 60 doublings (phase II) before entering into senescence (phase III), in which state they could remain viable but nonproliferating; such senescent cells could remain viable for as long as a year. (From J.W. Shay and W.E. Wright, *Nat. Rev. Mol. Cell Biol.* 1:72–76, 2000.)

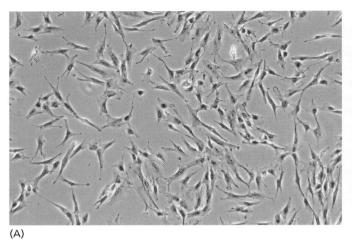

(A)

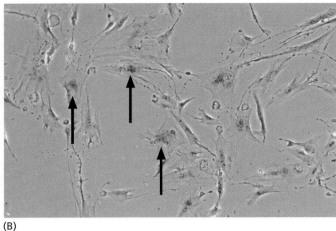

(B)

Figure 10.3 Senescent cells arising *in vitro* and *in vivo* (A) When viewed with the phase-contrast microscope, pre-senescent human fibroblasts are still vigorously growing and retain the appearance of early-passage cells. (B) However, once cells enter into senescence, they cease proliferating but remain viable. Many develop extremely large cytoplasms, giving them a "fried egg" appearance. (These two micrographs were produced at the same magnification.) In addition, senescent cells characteristically express acidic β-galactosidase enzyme, which can be detected by supplying them with a substrate that turns *blue* upon cleavage by this enzyme (*arrows*); in contrast, early-passage cells seen in (A) show very faint staining. (Courtesy of C. Scheel.)

adults (**Figure 10.4**). Such behavior suggests that cells from older individuals have already used up part of their allotment of replicative doublings prior to being introduced into tissue culture.

Although this is the simplest explanation, other explanations are equally plausible. For example, nondividing cells within a living tissue may sustain damage from, say, long-term exposure to reactive oxygen species (ROS) leaking from their mitochondria. In this case, the subsequently observed loss of proliferative capacity *in vitro* may be proportional to the elapsed time since these cells were first formed and may be unconnected with the number of cell generations their lineage has experienced.

While most cells can replicate *in vitro* only a limited number of times, a contrasting behavior is shown by embryonic stem (ES) cells, which are prepared from very early embryos and retain the ability, under the proper conditions, to seed all the differentiated lineages in the body (see Sidebar 8.1). When provided with the proper nutrients, these cells show unlimited replicative potential in culture and are thus said to be **immortal**. (The term is a bit misleading, since it is really a *lineage* of ES cells that is immortal rather than individual ES cells.)

Taken together, these various observations convey the notion that very early in embryogenesis, cells have an unlimited replicative capacity. However, as specific lineages of cells in the organism (for example, dermal fibroblasts, neurons, mammary epithelial cells) are formed, each seems to be allocated a predetermined number of postembryonic doublings. The replicative behavior of cancer cells resembles, at least superficially, that of ES cells. When many types of cancer cells are propagated in culture, they seem able to proliferate forever when provided with proper *in vitro* culture conditions.

This behavior is illustrated most dramatically by HeLa cells. Over the past half century, these cultured cells have been the human cell type most frequently used to study the molecular biology of human cells. They were derived in 1951 from an unusual, particularly aggressive cervical adenocarcinoma discovered in Henrietta Lacks, a young woman in Baltimore, Maryland, who soon died from the complications of this tumor. Ever since that time, these cells have proliferated in culture in hundreds of laboratories across the world, dividing approximately once a day. HeLa cells constitute a **cell line**, in that they have become established in culture and can be passaged indefinitely, in contrast to many cell populations that have a limited replicative ability after being removed from living tissue.

10.2 Cancer cells need to become immortal in order to form tumors

Because attempts to propagate cells in culture usually fail, conclusions about cancer cell immortality are based on the relatively small proportion of human tumors whose

(A)

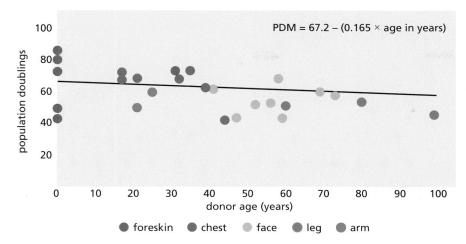

$$PDM = 67.2 - (0.165 \times \text{age in years})$$

x-axis: donor age (years)
y-axis: population doublings

● foreskin   ● chest   ● face   ● leg   ● arm

(B)

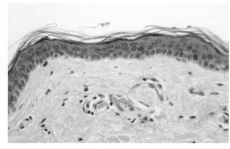

18 years old   76 years old

Figure 10.4 Loss of proliferative capacity with age (A) The proliferative capacity of cells can be gauged by determining the number of times they will double in culture (mean population doublings, PDM). Here, doublings of dermal fibroblasts from various anatomical sites were measured as a function of the age of the donors. (Because of inter-individual variability in cell proliferative capacity, data points show substantial scatter around the mean at each age.) (B) As the keratinocyte stem cells in the skin lose proliferative capacity with increasing age, the overall ability of the skin to regenerate itself declines, leading to a thinning of the keratinocyte layer of the skin (*dark pink*) and a loss of the ridge architecture seen in the younger skin. The sun-protected skin of an 18-year-old female (*left*) is compared here with that of a 76-year-old female (*right*). (A, courtesy of J.G. Rheinwald and T.M. O'Connell-Willstaedt. B, courtesy of T. Brenn.)

cells have readily adapted to *in vitro* conditions. Nevertheless, the observation that cancer cells, once adapted to growth in tissue culture, are often found to be immortal strongly suggests that immortalization is an integral component of the cancer cells' transformation to a neoplastic growth state, that is, cancer cells are immortal because they need to be in order to form a tumor.

Why, conversely, do lineages of normal cells lack immortalized growth properties? Perhaps the body endows its normal cells with only a limited number of replicative generations as an anti-cancer defense mechanism. For example, if one or another cell in the body were to accidentally acquire certain oncogenes and shed critical tumor suppressor genes, its descendants would begin to proliferate uncontrollably, and the population of these tumor cells might well increase exponentially. However, if endowed with only a limited replicative potential, these cells might exhaust their allotment of cell doublings long before they succeeded in forming a life-threatening tumor mass; as a consequence, tumor development would grind to a halt.

The credibility of this model depends on some critical numbers. Specifically, we need to know how many successive cell generations are required to make a clinically detectable human tumor (**Figure 10.5A**) and how many generations are granted to normal cell lineages throughout the body. We know that human tumors are clonal, in the sense that all the neoplastic cells in the tumor mass descend from a common ancestral cell that underwent transformation at one point in time (see Section 2.5). With this fact in mind, we can ask how many cell generations separate the tumor cells in a large human tumor from their common progenitor.

The arithmetic works out like this. The volume of a cubic centimeter (cm³) within a tumor cell mass contains about 10^9 cells, and a life-threatening tumor has a size, say, of 10^3 cm³. We can calculate that these 10^{12} cells seem to have arisen following 40 cycles of exponential growth and division (see Figure 10.5B)—that is, 40 cell generations separate the founding cell from its descendants in the end-stage, highly aggressive tumor ($10^3 \approx 2^{10}$; hence $10^{12} \approx 2^{40}$).

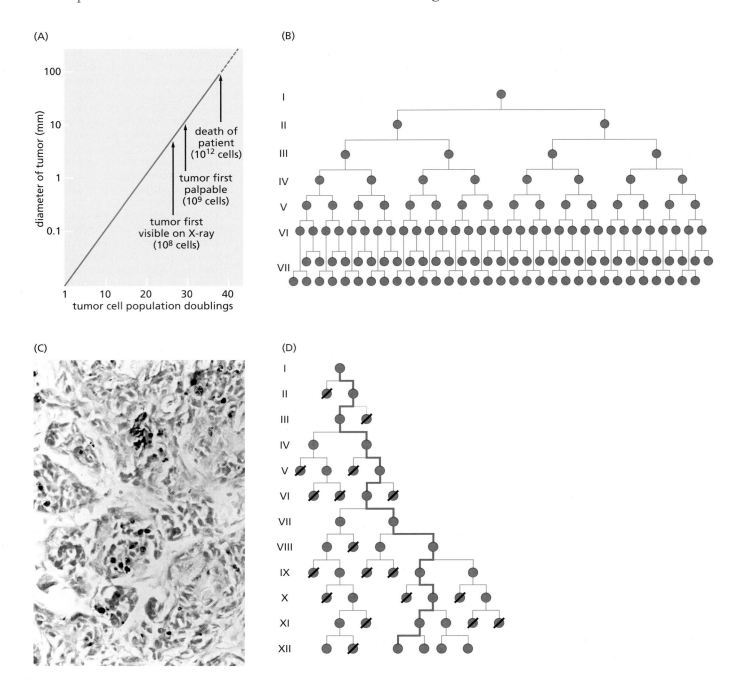

Figure 10.5 Generations of cells forming a tumor mass (A) The growth of a tumor—from detectable to palpable to life-threatening size—can be related to the minimum number of cell population doublings of the tumor-associated cancer cells. (B) The plot in panel A assumes that each time the cancer cells in a tumor divide, this cell population as a whole will double in size. (Only six doublings are shown here.) Consequently, the number of successive population doublings theoretically required to generate a lethal tumor from a single founding ancestral cell can be calculated from the graph in panel A to be about 40. (C) In reality, cell populations that are evolving toward the neoplastic state or are already neoplastic experience substantial attrition during each cell generation. This is evident from the apoptotic cells—detected here by the *dark brown* staining of the TUNEL assay (see Figure 9.19)—in a mammary tumor that arose in a mouse carrying an oncogenic *Wnt-1* transgene in its germ line. In fact, this assay greatly underestimates the rate of apoptosis in each cell generation, since apoptotic cells persist only for an hour before they are consumed by neighboring cells and by macrophages. Moreover, cancer cells within a tumor may become senescent and thereby enter irreversibly into a post-mitotic state. (D) The high rate of attrition leads to loss of many cells in each cell generation (*diagonal slashes*). Accordingly, an ancestral founder cell that should have generated 2^{11} descendants leaves instead only five descendants in the 12th generation. The number of successive cell generations required to generate a tumor of life-threatening size is therefore far larger and, in the absence of precise knowledge of attrition rates, incalculable. Also shown is the *lineage* of a cell in the 12th generation (*thick blue line*). (A, from B. Alberts et al., Molecular Biology of the Cell, 5th ed. New York: Garland Science, 2008. C, courtesy of L.D. Attardi and T. Jacks.)

However, as mentioned above, some types of normal human cells are known to pass through 50 or 60 cycles of growth and division in culture before they become senescent and stop growing. According to the arithmetic above, these 50 or 60 cell generations of exponential growth are far more than are required in order for a founding cell to spawn enough descendants to constitute a life-threatening tumor mass. Indeed, 60 cell doublings are enough to create a tumor mass of about 10^{18} cells $\approx 10^9$ cm^3 $\approx 10^6$ kilograms. Something is drastically wrong with these numbers!

The error in our calculations lies in a flawed premise: they assume an exponential expansion of populations of cancer cells (see Figure 10.5A and B). The biological reality of tumor growth is much different. Thus, a number of defense mechanisms built into the body's tissues make life very difficult for incipient cancer cells, indeed, so difficult that in each cell generation, a significant number of these cells die off (see Figure 10.5C). Early in tumor development, the defense mechanisms deployed by the tissue include depriving tumor cells of growth factors, of adequate oxygen, and of the ability to eliminate metabolic wastes via the vasculature. Moreover, a number of the anti-tumor cell defense mechanisms in the hard-wired regulatory circuitry of cells operate to weed out aberrantly behaving, pre-malignant cells (Chapter 9). As a consequence of the resulting attrition in each cell generation, the pedigree of cells in a tumor mass actually looks quite different (see Figure 10.5D). Many branches of the tree are continually being pruned by the high death rate of tumor cells in each generation.

This ongoing attrition means that the number of cell generations required to form a tumor mass of a given size is far greater than would be predicted by simple exponential growth kinetics. For example, a clonal population of 10^3 tumor cells might be thought, on the basis of its size, to have gone through 10 successive cycles of exponential growth and division since its founding by an ancestral cell; in reality, 20 or 30 or more cell generations may have been required to accumulate this many cells. (Figure 10.5D illustrates as well the *lineage* of a cell in a population of descendant cells.)

These revised calculations provide credible support for the notion that the human body endows its cell lineages with only a limited number of growth-and-division cycles in order to protect itself against the development of tumors. For example, if cell populations must pass through a succession of 100 replicative generations in order to form a clinically detectable human tumor, it is likely that most incipient tumor cell populations will use up their normal allotment of 50 or 60 cell divisions long before they succeed in creating such a mass. (In fact, by counting the number of mutations that cancer cells have accumulated over the long course of tumorigenesis, some have calculated that cell lineages within clinically apparent colon carcinomas may extend for as many as 2000 cell generations!)

Having accepted, at least for the moment, this argument and its conclusions, we are now left with some major puzzles: How can normal cells throughout the body possibly remember their replicative history? And how can aspiring cancer cells erase the memory of this history and acquire the ability to proliferate indefinitely, or at least as long as is required to form a macroscopic, clinically detectable tumor?

A solution to the problem of replicative history must, sooner or later, be spelled out in terms of the actions of specific molecules within cells. Whatever its nature, the solution is hardly an obvious one. Organisms as complex as humans possess no innate biological clock for counting the number of (organismic) generations that separate each of us from ancestors who lived 100 or 1000 years ago. When we humans wish to learn the number of these generations over extended periods of time, we usually hire genealogists to chart them for us. No biological counting device inside our bodies can provide us with such answers. How, then, can far simpler biological entities—individual human cells—keep track of their generations?

In addition, this generational counting mechanism is likely to be, in the language of developmental biologists, **cell-autonomous**; that is, it must be intrinsic to a cell and not influenced by the ongoing interactions of the cell with its neighbors and with the body as a whole. Recall the contrasting situation of oncogenes and tumor suppressor genes. The products of most of these genes perturb the pathways responsible for processing the signals received by cells from their surroundings—non-cell-autonomous

(A)

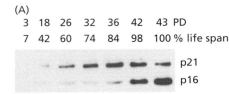

3	18	26	32	36	42	43 PD
7	42	60	74	84	98	100 % life span

p21
p16

Figure 10.6 Expression of CDK inhibitors during extended *in vitro* culture (A) When adult human endometrial fibroblasts are propagated *in vitro*, expression of both the p16^INK4A and p21^Cip1 proteins (see Section 8.4) is induced, albeit with distinct kinetics. These cells have an expected life span of 40 to 43 population doublings (PD) in culture, whereupon they enter into senescence. (B) Normal cells (*left*) are relatively small and have few focal contacts with the substrate (*yellow*) and few actin stress fibers (*orange*), in contrast to senescent cells (*middle and right*). Ectopic expression of p16^INK4A induces a cell phenotype (*middle*) that is indistinguishable from that of cells that have entered senescence following extensive propagation *in vitro* (*right*). This effect may depend on the ability of p16^INK4A to inhibit CDK4 and CDK6, thereby blocking hyperphosphorylation of pRb. (A, from S. Brookes et al., *Exp. Cell Res.* 298:549–559, 2004. B, courtesy of I. Ben-Porath.)

(B)

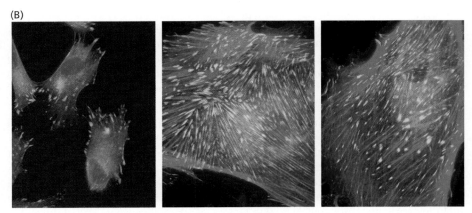

processes. Finally, the hypothesized generation-counting device must be relatively stable biochemically, because it needs to store a record of the past history of a cell lineage in a form that survives over extended periods of time, often many decades.

In principle, the counting processes of this "generational clock" (which measures elapsed cell generations rather than elapsed time) might depend on the concentrations of some soluble intracellular molecule that (1) is synthesized early in development and not thereafter, (2) is present in high concentrations in the embryonic cells that are the ancestors to various lineages, and (3) undergoes progressive dilution by a factor of 2 each time a cell in the lineage divides. Senescence might then be triggered after the levels of this compound fall below some threshold level. Though plausible to mathematicians, this arrangement cannot work in the real world of biology. Biochemistry dictates that no molecule can be present in a cell over a concentration range of 2^{50} or 2^{60}.

This realization forced researchers to look elsewhere for the molecular embodiments of the generational clock. They came upon two regulatory mechanisms that govern the replicative capacity of cells growing *in vitro* and possibly *in vivo*. The first of these appears to measure the cumulative physiologic stress that lineages of cells experience over extended periods of time and halts further proliferation once that damage exceeds a certain threshold; this causes cells to enter into the state, described in Chapter 9 and above, termed senescence.

The second regulator measures how many replicative generations a cell lineage has passed through and sounds an even more drastic alarm once the allowed quota has been used up; this leads to the state termed **crisis**, which results in the apoptotic death of most cells in a population. We will address the mechanisms inducing senescence first and later describe those governing crisis.

10.3 Cell-physiologic stresses impose a limitation on replication

The onset of senescence is accompanied by increasing expression of the two key CDK inhibitors, p16^INK4A and p21^Cip1 (**Figure 10.6A**). As discussed earlier (see Section 8.4), these CDK inhibitors are capable of halting cell cycle advance and driving a cell into a nongrowing state. Taken together, such evidence suggests that these two proteins may actually be responsible for *imposing* the senescent state on cultured cells. This notion is supported by forcing expression of p16^INK4A in cells, which on its own suffices to create a cell-biological state that is similar if not identical to senescence (see Figure 10.6B). [While not shown here, p53 may also contribute to the onset of senescence: in mouse embryo fibroblasts (MEFs), the levels of p53 increase 10- to 40-fold as cells approach senescence. Once induced, p53 seems to act in a cytostatic fashion (via p21^Cip1) to arrest further cell proliferation.]

The prominent role of p16^INK4A in triggering senescence is supported by other kinds of experiments: variants of human keratinocytes that succeed in escaping the fate of

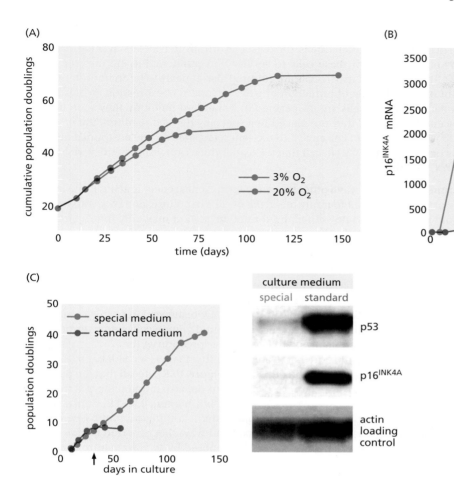

Figure 10.7 Influence of culture conditions on the onset of senescence
(A) The proliferative capacity of cell populations (*ordinate*) has been measured here when they are exposed in culture to either 3% or 20% oxygen. The 3% oxygen concentration is far more physiologic, since it reflects the oxygen concentration that many cells actually experience in living tissues. (B) The normal dependence of epithelial cells in living tissues on stromal support creates stresses when such cells are propagated *in vitro*. When grown as pure cultures directly on the bottom plastic surface of a tissue culture dish, human foreskin keratinocytes rapidly induce expression of the p16^{INK4A} tumor suppressor protein (*green line*), which imposes growth arrest and senescence on these cells. However, when these epithelial cells are grown above a feeder layer of fibroblasts (*brown line*), they can proliferate for an extended period of time without experiencing a strong induction of p16^{INK4A} and resulting senescence. (C) The composition of both tissue culture medium and tissue culture plates can also have a dramatic effect on the longevity of cell populations in culture. The graph shows a population of human mammary epithelial cells (MECs) propagated either in the standard medium for human MECs (*brown line*) and in a standard plastic culture dish or in a specially devised medium (*green line*) on a special plastic surface. After 21 days in culture (*arrow*), p16^{INK4A} and p53 levels were monitored by immunoblotting (*right panel*). Growth under standard conditions led to strong induction of these two growth-suppressing proteins, while cells in the special medium continued to proliferate without giving evidence of such induction. (A, from Q. Chen et al., *Proc. Natl. Acad. Sci. USA* 92:4337–4341, 1995. B, from B. Fu, J. Quintero and C.C. Baker, *Cancer Res.* 63:7815–7824, 2003. C, from T. Ince et al., *Cancer Cell* 12:160–170, 2007.)

early senescence *in vitro* are often found to carry inactivated copies of the gene encoding the p16^{INK4A} tumor suppressor protein; by shutting down p16^{INK4A} expression, these cells can ostensibly avoid entrance into the post-mitotic senescent state. Moreover, the early senescence of cultured human keratinocytes can be circumvented by ectopic expression in these cells of high levels of CDK4; recall that this cyclin-dependent kinase can bind and thus sequester p16^{INK4A} (see Figure 8.13A), thereby preventing it from halting cell proliferation.

While such experiments may indicate the proteins directly responsible for imposing the senescent state, they do not reveal the factors that determine when these proteins are activated. In this case, important clues come from experiments in which the conditions of *in vitro* culture are varied. The most dramatic observations derive from experiments where the oxygen tension to which cultured cells are exposed has been reduced from 20% to 1 to 3%; this results in a substantial increase in the replicative life span *in vitro*. In one set of experiments, populations of human diploid fibroblasts had a more than 20% longer life span (that is, went through 20% more doublings) in culture when grown in 1% oxygen than in 20% oxygen; in another, cells went through 50% more doublings when cultured in 3% oxygen rather than 20% oxygen (**Figure 10.7A**). The lower oxygen tensions more closely reflect the oxygen tensions that most cells experience in living tissues, rather than the tension that is most conveniently provided to cells growing in a tissue culture incubator.

The contribution of cumulative oxidative damage to senescence is suggested by other types of evidence as well. For example, senescent cultured cells have been found to produce oxidized forms of guanine at four times the rate of young cells, suggesting a progressive breakdown of mitochondrial function during *in vitro* culture, a resulting leakage of reactive oxygen species (ROS) from the mitochondria, and the accumulation of large numbers of oxidation-induced lesions in these cells. Many of the lesions that induce senescence are found in the chromosomal DNA of cells.

Yet other conditions of culture also have strong effects on the longevity of cell populations. In the case of epithelial cells, the presence of a stromal feeder layer has profound effects on the ability of these cells to survive and proliferate in culture (see Figure 10.7B). The medium in which cells are cultured also appears to be an important determinant of the schedule of induction of p16^{INK4A} and p53 expression. When human mammary epithelial cells are propagated in a standard medium, they soon become senescent. However, if they are placed in a specially devised medium, the induction of p16^{INK4A} and p53 and entrance into the senescent state can be postponed for many weeks (see Figure 10.7C). (Recall that p53 can induce expression of the widely acting p21^{Cip1} CDK inhibitor.)

Experiments like these yield other ideas: the term "senescence" is actually misleading. While originally invoked to imply the *aging* of cells in culture, entrance into this state may often have little to do with elapsed time. Instead, it may only reflect stress that cells have sustained, in the instances described above because of sub-optimal conditions of culture. If so, other physiologic stressors may also trigger senescence, a notion that we explore below.

Imagine, for example, that the environment experienced by pre-neoplastic and neoplastic cells within tissues forces them to cope with a variety of physiologic stresses similar to those experienced by cells in culture. One suggestion of this comes from experiments showing that the senescence of cultured human cells (often termed "replicative senescence") can be avoided by ectopic expression of the SV40 large T antigen. Large T is able to bind and inactivate both pRb and p53 (see Sidebar 9.8), and its ability to do so is critical for avoiding this state (**Figure 10.8**). Recall that the pathways controlled by these two tumor suppressor proteins are also found to be inactivated in the great majority of human tumors, as we read in Chapters 8 and 9. Consequently, when cancer cells are extracted from human tumors and placed into tissue culture, the previous inactivation of their pRb and p53 pathways that occurred *in vivo* is likely to help these cells resist many of the stresses imposed by *in vitro* culture conditions and thereby to avoid senescence.

Still, the evidence that incipient cancer cells experience stress-inducing signals *in vivo* is not provided by experiments such as the one shown in Figure 10.8. Indeed, as researchers have devised cell culture conditions that more closely approach the conditions experienced by cells within tissues, the life span of cell lineages in culture has increased progressively (see Figure 10.7) and senescence *in vitro* can be delayed and, in certain cases, avoided altogether. This raises the issue of whether cell senescence

Figure 10.8 Role of large T antigen in circumventing senescence Experiments with a variety of human cell types indicate that inactivation of both pRb and p53 is needed to ensure that these cells do not senesce in culture. This can be achieved through the expression of the SV40 large T antigen (LT) in these cells. As seen here, human embryonic kidney (HEK) cells senesce after 10 to 12 population doublings in culture (*gray line*). However, if they express the wild-type LT protein (*blue line*), they circumvent senescence and continue to proliferate for an extended period of time. Mutants of LT that have lost the ability to sequester either p53 (*brown line*) or pRb (*light green line*) fail to prevent senescence. (From W.C. Hahn et al., *Mol. Cell. Biol.* 22:2111–2123, 2002.)

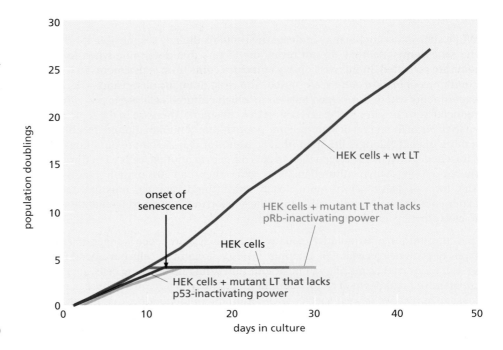

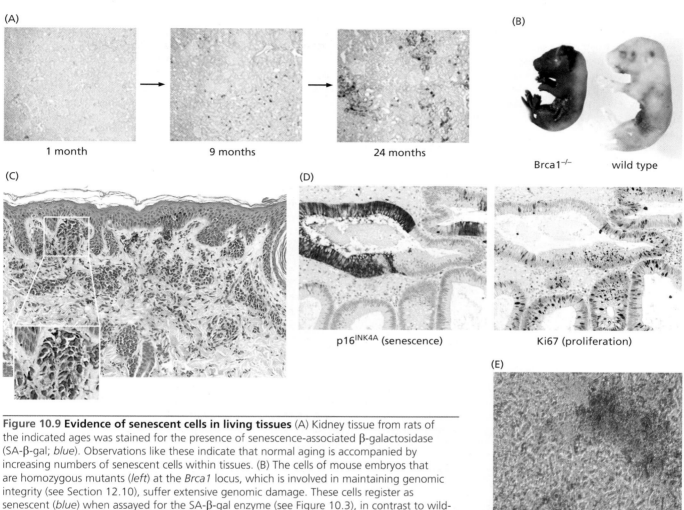

Figure 10.9 Evidence of senescent cells in living tissues (A) Kidney tissue from rats of the indicated ages was stained for the presence of senescence-associated β-galactosidase (SA-β-gal; *blue*). Observations like these indicate that normal aging is accompanied by increasing numbers of senescent cells within tissues. (B) The cells of mouse embryos that are homozygous mutants (*left*) at the *Brca1* locus, which is involved in maintaining genomic integrity (see Section 12.10), suffer extensive genomic damage. These cells register as senescent (*blue*) when assayed for the SA-β-gal enzyme (see Figure 10.3), in contrast to wild-type embryos (*right*), where such widespread presence of such senescent cells is not apparent. (C) The presence of senescent human melanocytes within dysplastic nevi (benign precursors of melanomas) can also be demonstrated by staining for SA-β-gal (*blue; see inset*). This suggests that pre-neoplastic cells enter into senescence, ostensibly because of deregulation of their normal signaling circuitry. (D) In human gastrointestinal tumors, such as this tubular adenoma of the colon, cells with high levels of p16^INK4A (*dark red, left*) are largely not proliferating, as indicated by the pattern of expression of Ki67, a marker of proliferating cells (*dark red, right*). Conversely, the proliferating cells lack p16^INK4A expression. In the presence of intact pRb, such high p16^INK4A expression is an indication of cells' having entered into the state of senescence. (E) Chemotherapeutic drugs also appear to induce senescence in tumor cells, as gauged by blue staining with the SA-β-gal senescence-specific stain. Seen here is a portion of a lung carcinoma from a patient who had been treated with the drugs carboplatin and taxol prior to surgical excision of the tumor. Normal lung tissue showed scant traces of SA-β-gal–staining cells. (A, from A. Melk et al., *Kidney Int.* 63:2134–2143, 2003. B, from L. Cao et al., *Genes Dev.* 17:201–213, 2003. C, courtesy of D. Peeper; from C. Michaloglou et al., *Nature* 436:720–724, 2005. D, courtesy of W.J. Mooi. E, from R.S. Roberson et al., *Cancer Res.* 65:2795–2803, 2005.)

is nothing more than an artifact of *in vitro* culture and thus whether it ever occurs *in vivo*. Resolving this issue is complicated by the fact that relatively few biochemical markers are available to identify senescent cells within tissues that reside amid large populations of cells that are in the nongrowing, G_0 state. One widely used marker of senescence is the enzyme acidic β-galactosidase, which is expressed in senescent cells (see Figure 10.3). Use of this marker has provided evidence, still fragmentary, that cell senescence does indeed occur *in vivo* (**Figure 10.9**). Yet other biochemical markers of senescence are being discovered, including notably the profound alterations in the chromatin of senescent human cells (**Sidebar 10.1**).

Sidebar 10.1 Senescence has wide-ranging effects on cellular biochemistry The fact that the senescent state affects so many microscopically visible aspects of cell phenotype must reflect a multitude of underlying biochemical changes occurring as cells enter into this state. Unfortunately, however, relatively few biochemical markers that are specific for senescence have been discovered. Expression of the senescence-associated acidic β-galactosidase (SA-β-gal) enzyme is a useful marker of senescence (see Figure 10.9) but is not absolutely definitive, since it may also be expressed to some extent in certain types of apoptotic cells.

Two types of covalent alteration of the histones forming the chromatin (see Figure 1.20A) have been found to provide specific markers of senescence. One type is found in the foci of heterochromatin within the nuclei of senescent human cells. Like other areas of heterochromatin, these senescence-associated heterochromatic foci (SAHFs) reflect chromosomal regions in which gene expression has been silenced (**Figure 10.10A**). The formation of SAHFs depends on the active participation of pRb, which associates with E2F target genes and attracts a number of additional proteins that cause localized histone modification and chromatin remodeling; this remodeling, in turn, creates regions of heterochromatin within a chromosome. Importantly, cells that are in the G_0, quiescent state, while showing pRb-mediated gene repression, do not develop these SAHFs.

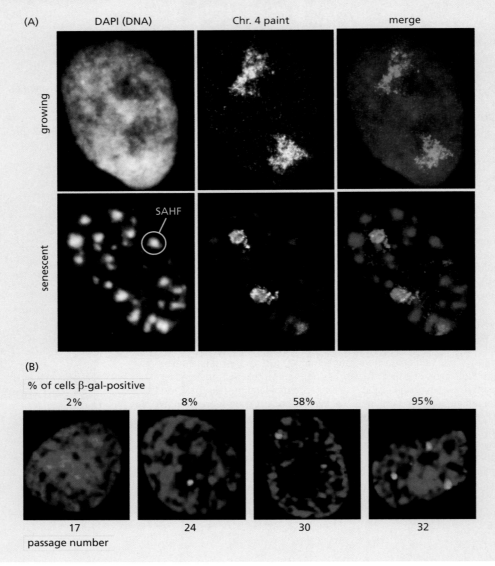

Increasing the likelihood that senescence is a physiologic process (rather than simply an artifact of cell culture) are *in vivo* observations indicating that a variety of other cell-physiologic stresses besides extended *in vitro* propagation can also induce a cellular state that is indistinguishable from replicative senescence. Included among these are hyperoxia (that is, supra-normal oxygen tensions leading to intracellular oxidative stress), DNA damage created by endogenously generated reactive oxygen species (ROS), X-rays, chemotherapeutic drugs, dysfunctional telomeres (see below),

The SAHFs can be detected with antibodies that react with the covalently modified histones that are present specifically in the heterochromatin of senescent cells. In particular, such antibodies recognize H3K9me (that is, histone H3 that has been methylated on its lysine 9 residue; see Figure 1.20A). This finding provides a highly specific marker for senescent cells and, at the same time, gives us insight into a key molecular mechanism that is responsible for maintaining the senescent cell phenotype. Thus, cells that lack the Suv39h1 "writer" (see Figure 1.20B), the enzyme that methylates histone H3 and thus helps to form SAHFs, fail to enter senescence properly. Interestingly, SAHFs do not seem to form in mouse cells, which may be related to the fact that these cells, unlike their human counterparts, do not readily enter senescence.

Yet another marker for senescence is associated with the chromatin. In response to the formation of double-stranded (ds) DNA breaks, large regions of chromatin flanking these breaks are found to carry a modified histone termed γ-H2AX; the resulting modified chromatin is involved in attracting the appropriate DNA repair proteins (described in Chapter 12). These foci of γ-H2AX can also be found in senescent cells (see Figure 10.10B). The γ-H2AX foci that arise following X-irradiation normally disappear as the radiation-induced dsDNA breaks are repaired. However, those found in senescent cells persist, suggesting that they represent irreparable breaks that have accumulated in the DNA. In human cells, these persistently unrepaired dsDNA breaks may provide a continuous damage signal that serves to maintain the senescent state over extended periods of time.

Figure 10.10 (*left*) Senescence-associated foci Senescence-associated heterochromatic foci (SAHFs) arise during senescence and derive from covalent modifications of the N- and C-terminal tails of histones within the nucleosomes (see Section 1.8). (A) Human fibroblasts were induced to enter senescence through introduction of a *ras* oncogene (*lower row*). In the nuclei of control cells that continued active growth (*upper row*), the chromosomal DNA was dispersed throughout the nucleus, as indicated by the DAPI stain (*white*). In contrast, the DNA of the senescent cells was concentrated in a number of discrete foci—SAHFs. The use of *in situ* hybridization with a DNA probe specific for Chromosome 4 (Chr. 4 paint) revealed that the Chromosome 4 DNA was far more dispersed in the nuclei of growing cells than in the senescent cells. The overlap between these two imaging procedures (merge, *green*) shows that the DNA of each Chromosome 4 has coalesced into a single SAHF within a senescent cell. (The behavior of Chromosome 4 is presumed to be representative of the behavior of all other chromosomes, which were not analyzed here.) While the bulk of the DNA of Chromosome 4, as well as of other chromosomes, is heterochromatic and thus transcriptionally silent, small segments of the chromosome, not visualized here, remain euchromatic and thus transcriptionally active. Other work indicates that SAHFs represent chromosomal regions in which transcription is irreversibly repressed. (B) γ-H2AX foci, which also involve histone modification (phosphorylation of histone H2AX, a variant of histone H2A), are normally formed in the chromatin flanking dsDNA breaks (see Section 12.10); in normal cells, these foci usually disappear in a matter of hours following successful DNA repair. However, the γ-H2AX foci (*light green spots*) that are detected in the nuclei (*blue*) of cultured senescent SA-β-gal–positive human fibroblasts persist and therefore appear to represent irreparable DNA breaks. These cells were propagated in culture for 17, 24, 30, and 32 passages (*left to right*). (A, from R. Zhang, W. Chen, and P.D. Adams, *Mol. Cell Biol.* 27:2343–2358, 2007. B, from O.A. Sedelnikova et al., *Nat. Cell Biol.* 6:168–170, 2004.)

and aberrant signaling by certain oncoproteins. In addition, an increase in the frequency of senescent cells is associated with and possibly responsible for the process of organismic aging.

These demonstrations that senescence occurs under a variety of conditions *in vivo* explain why we have discussed replicative senescence in cultured cells in such detail, as it represents a very good model of an *in vivo* physiologic process. In the context of cancer, we can imagine that senescence can serve as an important barrier to the

Sidebar 10.2 A poisonous dose of Ras As the sixteenth-century alchemist Paracelsus put it, "the dose makes the poison." The *ras* oncogene has been found to induce cell senescence when introduced into cultured cells by a viral vector, suggesting paradoxically that this potent transforming gene acts more to halt cell proliferation than to foster it. This result on its own created a puzzle, in that it suggested that virtually all cells that acquire a *ras* oncogene by somatic mutation were shunted immediately into the post-mitotic senescent state rather than becoming precursors of vigorously growing tumors. This would preclude such cells from ever forming tumors.

Some years later, researchers explored these effects by using a different experimental strategy to establish a *ras* oncogene in cells, one that relied on activating an endogenous, chromosomal *ras* proto-oncogene in a genetically modified strain of mice (see Supplementary Sidebar 7.7). This activation could be targeted to a specific tissue and induced by an experimental stimulus. Under these conditions, *ras* activation in the lungs and in the gastrointestinal tract resulted in hyperplasias (rather than senescence); similar results could be observed in the pancreas and in the myeloid compartment of the bone marrow.

The critical difference here could be traced to how the *ras* oncogene was being expressed: in the earlier *in vitro* experiments, its expression was driven by a highly active transcriptional promoter present within a recombinant viral vector (see Supplementary Sidebar 3.3), whereas in the later mouse experiments, the oncogene was present in single copy number per cell and was being driven by its own native transcriptional promoter. Hence, the critical difference came from the *level* of expression. This suggested that *ras* oncogenes created by somatic mutations can "fly under the radar" of the surveillance system that is designed to weed out mutant, potentially neoplastic cells through the induction of senescence.

Interestingly, as tumor progression proceeded in the lung cancer model, the *ras* oncogene, initially present in single copy number in pre-malignant lung adenomas, underwent amplification. Prior to this time, the presence or absence of a functional *p53* gene had little if any effect on these early-stage lung adenomas; however, once the *ras* oncogene underwent amplification, elimination of p53 function became critical to permit the survival and continued expansion of the cells in the resulting, more aggressive adenocarcinomas. (Such p53 inactivation ostensibly allowed the evolving tumor cells to avoid the senescence that would otherwise be provoked by higher levels of Ras signaling; the amplified Ras, for its part, appeared to provide these cells with increased mitogenic stimulation and thus enhanced proliferation.) Of note, this mouse model of lung cancer development reflects the behavior of many human tumors, in which mutant *ras* oncogenes undergo amplification as tumor progression proceeds.

development of spontaneously arising tumors. For example, a *ras* oncogene can induce senescence, suggesting that this is a natural cellular response for halting the further proliferation of cells that have acquired this oncogene and thus initiated tumorigenesis (**Sidebar 10.2**). Since entrance into senescence is generally irreversible, senescence is likely to be as effective as apoptosis in preventing cells from contributing further to tumor formation.

Paradoxically, cell senescence also may act in an entirely different fashion to *promote* tumorigenesis. The study of senescent cells in culture, including those caused to enter senescence by oncogene expression, has revealed that these cells release a variety of pro-inflammatory cytokines, including several interleukins, IGFBPs, and TGF-β. The cytokines that are released as a manifestation of the **senescence-associated secretory phenotype** (SASP) include those that act in an autocrine fashion to promote senescence and those that act in a paracrine fashion to recruit pro-inflammatory cells of the innate immune system. As we will learn in the next chapter, such pro-inflammatory cells can play major roles in *promoting* tumor progression. Hence, in certain situations, the senescence state can actually quicken the pace of multi-step tumorigenesis, doing so initially via its effects on innate immune cells. The fact that senescent cells persist for many weeks if not months, rather than disappearing quickly (as is the case with cells that have entered apoptosis), means that these cells can release such inflammatory signals over extended periods of time, thereby perturbing the microenvironment of the surrounding tissue.

10.4 The proliferation of cultured cells is also limited by the telomeres of their chromosomes

Primary cells—cells that have been prepared directly from tissues—actually confront two obstacles to their long-term proliferation in culture. The first of these is the replicative senescence that we described above; the second is **crisis**.

In the case of human embryonic kidney cells, as an example, expression of SV40 large T oncoprotein enables bypass of the first hurdle—senescence (see Figure 10.8)—but it will fail to immortalize these cells. Thus, after an additional number of cell

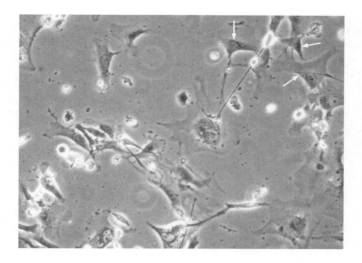

Figure 10.11 Apoptosis associated with cell populations in crisis Cell populations in crisis show widespread apoptosis, with a high percentage of cells in the midst of disintegrating and numerous cell fragments (*white refractile spots*). Because populations of cells enter into crisis asynchronously, some of the cells in such populations still appear relatively normal (*arrows, top right*). (Courtesy of S.A. Stewart.)

generations—perhaps 10 to 20—beyond the time when they would usually senesce, cell populations enter into crisis and exhibit widespread apoptosis. The SV40 large T antigen, even with its potent p53-inactivating ability, clearly does not protect against the apoptotic death associated with crisis.

Senescence represents a halt in cell proliferation with retention of cell viability over extended periods of time, while crisis involves death by apoptosis (**Figure 10.11**). Senescent cells seem to have a reasonably (but not totally) stable karyotype, while cells in crisis show widespread karyotypic disarray.

The timing of crisis and the appearance of cells in crisis suggest the involvement of a second mechanism, one operating independently of the mechanism(s) triggering senescence. The workings of this second mechanism have been traced back to the chromosomal DNA of cells. Unlike the mechanism(s) leading to senescence, the molecular apparatus that initiates crisis is truly a functional counting device that tallies how many successive growth-and-division cycles cell lineages have passed through since their founding in the early embryo.

At first glance, the chromosomal DNA would seem to be an unlikely site for mammalian cells to construct such a "generational clock." We know that the structure of chromosomal DNA is highly stable and therefore should be unchanged from one cell generation to the next. A generational clock, in contrast, must depend on some progressive molecular change that is noted and recorded during each cell generation. How can chromosomal DNA molecules, which seem to be so immutable, register an additional cell generation each time a cell passes through a growth-and-division cycle?

Each mammalian chromosome carries a single, extremely long DNA molecule (ranging from ~247 megabases for Chromosome 1 to ~47 megabases for Chromosome 21). As it turns out, the two ends of this DNA molecule create a serious problem for the cell—a problem dramatically revealed by the experimental technique of DNA transfection (see Section 4.2). After entering cells, transfected linear DNA molecules are rapidly fused end-to-end through the actions of a variety of nucleases and DNA ligases that are active in most if not all mammalian cell types. Hence, linear DNAs are intrinsically unstable in our cells, yet the linear DNA molecules within chromosomes clearly persist.

The **telomeres** located at the ends of the chromosomes (**Figure 10.12**) explain how these linear DNA molecules can stably coexist with the cell's various DNA-modifying enzymes. Telomeres act to prevent the end-to-end fusion of chromosomal DNA molecules and, hence, the fusion of chromosomes with one another. In effect, telomeres serve as protective shields for the chromosomal ends, much like the aglets safeguarding the tips of shoelaces. As we will see, the catastrophic events of crisis are triggered when cells lose functional telomeres from their chromosomes.

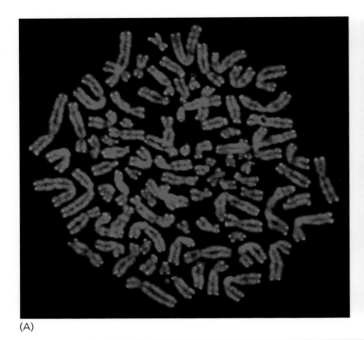

(A)

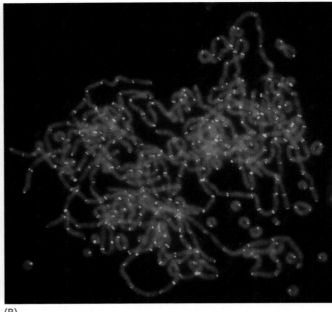

(B)

Figure 10.12 Telomeres detected by fluorescence *in situ* hybridization Telomeres can be detected by the technique of fluorescence *in situ* hybridization (FISH), in which a DNA probe that recognizes the repeated sequence present in telomeric DNA is coupled with a fluorescent label. (A) In this micrograph, human cells have been trapped in metaphase using a microtubule antagonist. The telomeres of the resulting condensed chromosomes (*red*) are then visualized through a probe that labels the telomeric DNA (*yellow green dots*). Each chromatid carries telomeres at both its ends. Because these cells are in metaphase, the paired, recently replicated chromatids have not yet separated. (B) The karyotype seen in panel A can be compared with that of cells that have been deprived of TRF2, a key protein in maintaining normal telomere structure (see Figure 10.19). In these cells, the telomeres lose their protective function (although telomeric DNA remains), resulting in end-to-end fusions of all the chromosomes of a cell into one huge chromosome. (A, courtesy of J. Karlseder and T. de Lange. B, from G.B. Celli and T. de Lange, *Nat. Cell Biol.* 7:712–718, 2005.)

Discoveries reported in 1941 by Barbara McClintock (**Figure 10.13**) concerning her studies of the chromosomes of corn first revealed that chromosomes that have lost functional telomeres at their ends soon fuse, end-to-end, with one another. The resulting megachromosomes possess two or more **centromeres**, the specialized chromosomal structures that become attached during mitosis to the fibers of the mitotic spindle. An extreme form of this fusion is shown in Figure 10.12B, in which virtually all the chromosomes of a cell have fused into one giant chromosome.

As discussed in greater detail later, the DNA component of each telomere in our cells is composed of the 5′-TTAGGG-3′ hexanucleotide sequence, which is tandemly repeated thousands of times; these repeated sequences, together with associated proteins, form the functional telomere. The telomeric DNA (and thus the telomeres) of normal human cells proliferating in culture shorten progressively during each growth-and-division cycle, until they become so short that they can no longer effectively protect the ends of chromosomes (**Figure 10.14**). At this point, crisis occurs, chromosomes fuse, and widespread apoptotic death is observable. Hence, in such cells, it is telomere shortening that functions to register the number of cell generations through which cell populations have passed since their origin in the early embryo.

In human cells entering crisis, the initial end-to-end fusion events usually occur between the eroded telomeric ends of the two sister chromatids that form the two halves of the same chromosome (**Figure 10.15A**). Recall that paired chromatids exist

Figure 10.13 Barbara McClintock Barbara McClintock's detailed studies of the chromosomes of corn (maize) revealed specialized structures at the ends of chromosomes—the telomeres—that protected them from end-to-end fusions. This work, and her demonstration of movable genetic elements in the corn genome, later called transposons, earned her the Nobel Prize in Physiology or Medicine in 1983. (Courtesy of American Philosophical Society.)

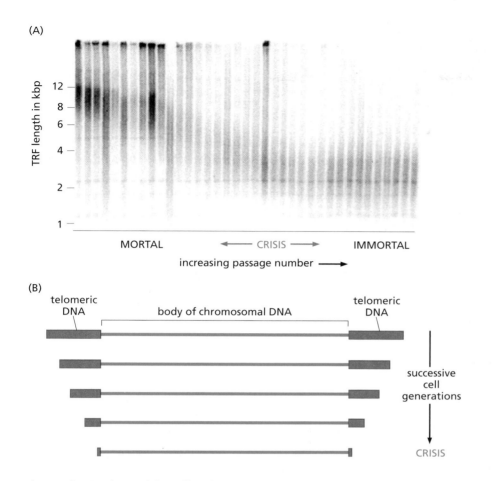

(A)

(B)

telomeric DNA   body of chromosomal DNA   telomeric DNA

successive cell generations

CRISIS

Figure 10.14 Shortening of telomeric DNA in concert with cell proliferation (A) Here, a culture of human lymphocytes that had circumvented culture-induced senescence following infection by Epstein–Barr virus was passaged and a portion of each culture was set aside for analysis of the cells' telomeric DNA. Each passage represented three to six population doublings. The length of telomeres was measured by treating genomic DNA with several restriction enzymes that do not cleave within the repeating TTAGGG sequence that constitutes telomeric DNA but do cleave frequently throughout the rest of the chromosomal DNA. The only high–molecular-weight genomic fragments such treatment leaves behind are the telomeric restriction fragments (TRFs). Their lengths can then be determined by gel electrophoresis followed by Southern blotting analysis using a probe that recognizes the TTAGGG sequence. Since the telomeres within a given cell vary in length and since the extent of telomeric DNA shortening differs from one cell to another, the size distribution of telomeric DNA in each sample is quite heterogeneous. In the analysis shown, the TRFs of mortal (i.e., non-immortalized) human lymphocytes grew shorter with every successive passaging by 50 to 100 base pairs per cell generation. When the TRFs became less than about 3 kbp long, cells entered crisis. Later, when cells emerged spontaneously from crisis (and became immortalized), they maintained their telomeric DNA at slightly longer sizes than those seen in crisis. (B) Such observations have led to the model shown, in which telomeric DNA (*red*) shortens progressively each time a cell passes through a growth-and-division cycle until the telomeres are so eroded that they can no longer protect the ends of chromosomal DNA. This telomere collapse provokes a cell to enter into crisis. [In fact, the TRFs of panel A include both pure TTAGGG repeats at the terminus of each telomere and long stretches of TTAGGG-like sequences located at the internal (centromeric) end of each telomere; the latter may not be functional in protecting the chromosomal ends. Telomeres may lose their protective function when their regions of pure TTAGGG repeats have eroded down to a length of ~12 repeats.] Note that this drawing is not to scale: telomeric DNA is usually 5 to 10 kbp long, while the body of chromosomal DNA (*blue*) is tens to hundreds of megabase pairs (Mbp) long. (A, from C.M. Counter et al., *J. Virol.* 68:3410–3414, 1994.)

during the G_2 phase of the cell cycle—a period after S phase has created two chromatids from a parental chromosome and before M phase, when these two sister chromatids are destined to be separated from one another (see Figure 8.3). Such fusions between the ends of sister chromatids (rather than between the ends of two unrelated chromosomes) are favored for at least two reasons. First, the two chromatids, and thus their ends, are held in close proximity through their joined centromeres. Second, for unknown reasons, telomeres shorten at different rates on different chromosomal arms. Consequently, the two homologous telomeric DNAs (for example, the pair of telomeres at the ends of the long arms of the chromatids of human Chromosome 9), having just been generated anew by DNA replication, possess virtually identical molecular structures. Therefore, if the telomeric end of one chromatid has become shortened, frayed, and vulnerable to fusion, its counterpart on the other paired chromatid is likely to be in the same state.

When these fused chromatids participate in the mitosis that follows, their two centromeres will be pulled in opposite directions by the mitotic spindle (see Figure 10.15B), creating the dramatic *anaphase bridges* that are often seen following extensive telomere erosion. Sooner or later, the mitotic apparatus that is pulling on such a **dicentric** chromatid (that is, one with two centromeres) will succeed in ripping it apart at some random site between the two centromeres (see Figure 10.15C). This yields two new chromosomal ends, neither of which possesses a telomere. Such unprotected ends may fuse with one another or with the unprotected ends of other, nonhomologous chromosomes.

In the event that the resulting defective, nontelomeric end of a chromosome fuses with the end of another (nonhomologous) chromosome, the fate of the resulting new, dicentric chromatid during the next mitosis is more ambiguous (see Figure 10.15C). Half the time, on average, the two centromeres will be pulled in the same direction toward the centrosome that nucleates one of the two soon-to-be-born daughter cells, and no breakage will occur. The other half the time, the two centromeres of a dicentric chromatid will become attached to the two opposing spindle bodies and thus to the two centrosomes located at opposite sides of the mitotic cell. When chromatids are separated during the anaphase of mitosis, this second configuration (like the earlier one between sister chromatids) will once again prove to be disastrous, since it will involve the tearing apart of a chromatid and the generation of ends that are, as before, unprotected by telomeres.

These newly created chromosomal ends will once again attempt to fuse with yet other chromosomes, yielding more dicentric chromosomes and a new cycle of chromosome breakage. This sequence of events is termed the *breakage–fusion–bridge* (BFB) *cycle*, since it involves the initial *breakage* of dicentric chromatids in anaphase, the subsequent *fusion* of the resulting atelomeric DNA ends with yet other chromatids,

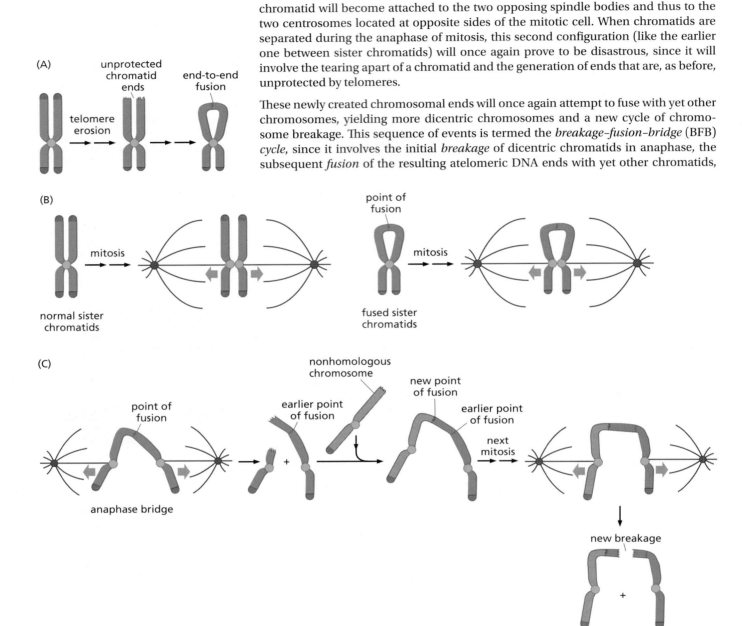

Figure 10.15 Mechanisms of breakage–fusion–bridge cycles
(A) Chromosomal breakage, fusion, and bridge formation occur when telomeres (*red*) have become too short to protect the ends of chromosomal DNA. Pictured here is the configuration of a chromosome in the G₂ phase of the cell cycle. During this phase of the cell cycle, its two chromatids remain attached to one another by their shared centromere (*green*). (In fact, it is the multi-subunit centromere-associated protein complex known as the kinetochore, *not shown*, that associates with the centromeric DNA and is responsible during the G₂ phase for holding the two chromatids together and during mitosis for binding the chromosome to microtubules of the mitotic spindle.) As shown, the telomeres on the long arms of this *blue* chromosome have eroded before those of the short arms. Both ends of the long arms are equally eroded, since they have recently been duplicated during the previous S phase. The ends of these two arms, held in close proximity by the shared centromere, now fuse with one another, leading to a *dicentric* chromatid, which contains two telomeres and two half-centromeres. (B) During a normal mitosis, the two sister chromatids will be pulled apart to opposite centrosomes located at the two polar ends of the mitotic spindle (*left*). Likewise, the mitotic spindle will pull the centromeres (*green*) of a dicentric chromosome in opposite directions, unaware that these two chromatids are joined via their long arms. (C) Unlike normally configured chromatid pairs, which will separate cleanly and segregate in groups near the two centrosomes, the dicentric chromosome described in (B) will be unable to do so and will instead create a bridge between the two poles of the mitotic spindle during the anaphase of mitosis. Eventually, the dicentric chromosome will be ripped apart at some weak point. During the next cell cycle, the larger fragment lacking a telomere at one end (*blue*) may fuse with another atelomeric chromosome (*beige*), creating a new dicentric chromosome, which itself will be pulled apart during a subsequent mitosis, resulting once again in a breakage–fusion–bridge cycle. (The fate of the shorter chromatid fragment generated by the first cycle of breakage is not shown.)

Figure 10.16 Dicentric chromosomes, anaphase bridges, and internuclear bridges (A) A dicentric chromosome is seen (*white arrow*) in a metaphase spread of cells from a human pancreatic carcinoma cell line. Centromeric regions have been stained *pink*. (B) Such dicentric chromosomes often result in the formation of anaphase bridges, several of which are seen in a cell of a human malignant fibrous histiocytoma. The two opposing groups of chromosomes are unable to separate properly from one another because they are still connected by multiple anaphase bridges. Observation of such anaphase bridges (see Figure 10.15) provides strong evidence that chromosomes within such cells are participating in breakage–fusion–bridge cycles because they no longer possess adequate telomeres. (C) While most anaphase bridges are broken during telophase or cytokinesis, some may persist in the resulting interphase cells, causing, for example, the nuclei of two daughter cells (*light blue*) of a human lipomatous tumor cell to be linked by a chromatin bridge. (From D. Gisselsson et al., *Proc. Natl. Acad. Sci. USA* 97:5357–5362, 2000.)

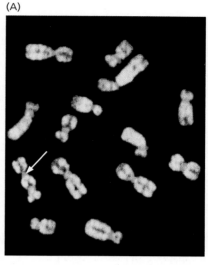

(A)

(B)

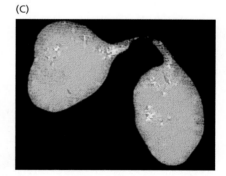

(C)

and the formation once again of anaphase *bridges* by the new dicentric chromosomes resulting from these fusions (**Figure 10.16**). While they are occurring, these BFB cycles create karyotypic chaos that has the potential to affect many chromosomes within a cell, as proposed in 1941 by McClintock.

Clear indication that the molecular machinery triggering crisis resides in the telomeres is indicated by the fact that cells that have entered crisis show precisely the type of karyotypic disarray that is observed when chromosomes lose their telomeres. Moreover, the chromosomes that are fusing within these cells possess especially short telomeres, or none at all. Note that the karyotypic instability occurring during crisis, which has been termed "genetic catastrophe," is so severe that apoptosis occurs even in the absence of p53 function.

10.5 Telomeres are complex molecular structures that are not easily replicated

Research conducted since the mid-1980s has revealed the molecular structure of telomeres and their DNA. To begin, and as cited earlier, the telomeric DNA of mammalian cells (as well as the cells of many other metazoa) is formed from the repeating hexanucleotide sequence 5'-TTAGGG-3' in one strand (the "G-rich" strand) and the complementary 5'-CCCTAA-3' in the other (the "C-rich" strand). In normal human cells, telomeric DNA is formed from several thousand of these hexanucleotide sequences, resulting in 5- to 10-kilobase-pair-long stretches of such repeating sequences at the ends of all chromosomes.

The telomeric DNA of mammalian cells, and possibly the cells of all eukaryotes, possesses an additional, distinctive feature: its G-rich strand is longer by one hundred to several hundred nucleotides, resulting in a long 3' single-strand overhang (**Figure 10.17**). This overhanging strand is often found in a most unusual molecular configuration termed the **t-loop**. It was discovered in the late 1990s, when electron microscopy showed the ends of telomeric DNA to be configured in a loop, in effect a lariat (**Figure 10.18A**). This structure has been interpreted to depend on the formation of a three-stranded complex of DNA (Figure 10.18B and C). The t-loop may be present at the ends of all telomeres, although it has been observed in only a subset of those viewed, probably because of the technical difficulties associated with preserving and visualizing this structure in the electron microscope. The t-loop may help to protect the ends of linear DNA molecules, because the end of the single-stranded overhanging region is tucked into a double-stranded region, out of harm's way.

Both the relatively long double-stranded telomeric DNA and the short overhanging end are bound by specific proteins. As might be expected, some of the proteins possess domains that specifically recognize and bind to the hexanucleotide sequence present in the double- and the single-stranded regions of telomeric DNA

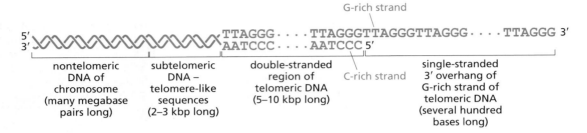

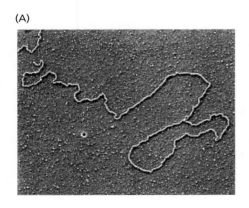

Figure 10.17 The 3' overhang of telomeric DNA The G-rich strand of telomeric DNA (*pink*) extends beyond the C-rich strand (*blue*). This creates a 3' overhang (*right*) that is often several hundred nucleotides long. This overhang is far shorter than the double-stranded portion of telomeric DNA (*middle*), which is often 5 to 10 kilobase pairs long and is not drawn here to scale. In addition, a subtelomeric region of DNA lies between the functional telomeric DNA (*middle*) and the nontelomeric chromosomal DNA (*far left*). This subtelomeric DNA contains imperfect copies of the TTAGGG sequence and does not function to protect the end of the chromosome. However, it is usually included in the telomere restriction fragments (TRFs; see Figure 10.14A) because of its telomere-like sequence content.

(**Figure 10.19**). Together, these telomere-binding proteins, yet other associated proteins that remain to be identified, and the telomeric DNA form the nucleoprotein complexes that we call telomeres.

While the replication machinery that operates during the S phase of the cell cycle is highly effective at copying the sequences in the middle of linear DNA molecules, such as those in the body of each of our chromosomes, this machinery has great difficulty copying sequences at the very ends of these molecules. The difficulty can be traced to the requirement that the synthesis of all DNA strands during DNA replication must be initiated at the 3'-hydroxyl end of an existing DNA strand, which serves as a **primer** to nucleate DNA strand elongation; alternatively, in the absence of an available DNA primer, the 3' end of an RNA molecule can serve as a primer for DNA synthesis (**Figure 10.20**).

If the **primase** enzyme, which is responsible for laying down short RNA primers, happens to deposit a primer at some distance from the 3' terminus of a template strand on which "lagging-strand synthesis" is occurring (see Figure 10.20), a DNA polymerase will initiate synthesis of a new DNA strand that lacks a substantial number of the bases complementary to the very 3' end of this template strand. (Even if the primase "sits down" and constructs an RNA primer at the very end of the template strand, the sequences corresponding to the approximately 10 nucleotides of the RNA primer will not be present in the newly synthesized daughter strand, since the RNA primer will be degraded after it has served its purpose of initiating DNA strand elongation.) The end result is that the nucleotide sequences at the very tip of one of the two template strands of DNA will not be properly replicated.

This *end-replication problem* provides a molecular explanation for the observed shortening of telomeric DNA each time a normal human cell passes through a cell cycle. In

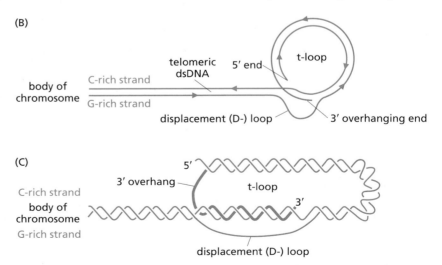

Figure 10.18 Structure of the t-loop (A) Purification of telomeric DNA from mouse cells (following stabilization of double-stranded structures through chemical cross-linking) reveals, upon electron microscopy, lariats, called *t-loops*, at the ends of chromosomal DNA. (B) This drawing of the t-loop indicates that the 3' overhanging end of the G-rich strand (*pink*) is annealed to a small region of the C-rich strand (*blue*), causing the formation of a displacement (D-) loop (*pink strand*). The 5'-to-3' polarities of the two strands are indicated by the arrowheads. (C) The t-loop is illustrated once again, on this occasion with the double helices drawn and the 3' overhanging end emphasized in a heavier *pink* line. (A, from J.D. Griffith et al., *Cell* 97:503–514, 1999.)

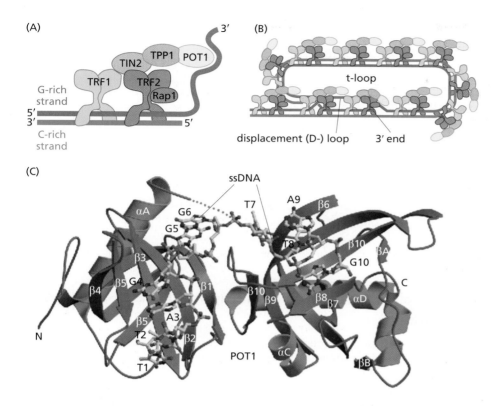

(A)

(B)

t-loop

displacement (D-) loop   3' end

(C)

Figure 10.19 Multiple telomere-specific proteins bound to telomeric DNA (A) In addition to its complex structure (see Figure 10.18), telomeric DNA is protected from degradation by a group of physically associated proteins that constitute the *shelterin* complex. Two of the latter's proteins—TRF1 and TRF2—bind to the double-stranded DNA portion of telomeres while a third protein—POT1—binds to both the single-stranded 3' overhang created by the G-rich strand and the ssDNA in the displacement (D-) loop. (B) The shelterin complex is present in multiple copies per telomere and is responsible, in part, for maintaining the structure of the t-loop. As is indicated here, by binding to the D-loop, POT1 is able to stabilize this structure and thus the structure of the t-loop as a whole. (C) The molecular structure of human POT1 with its two ssDNA-binding domains is depicted as a *ribbon diagram*, with α-helices, β-pleated sheets, and loops numbered. The linker region between these two domains is disordered and represented by a *dashed orange line*. The binding by POT1 to single-stranded telomeric DNA (*stick structure*) helps to stabilize the D-loop. (A and B, from T. de Lange, *Science* 326:948–952, 2009. C, from M. Lei, E.R. Podell and T.R. Cech, *Nat. Struct. Mol. Biol.* 11:1223–1229, 2004.)

addition to the under-replication of telomeric DNA ends, there appear to be exonucleases within cells that slowly chew on the ends of telomeric DNA and may ultimately contribute far more to telomere erosion. For whatever reason, in many types of normal human cells, telomeres lose 50 to 100 base pairs of DNA during each cell generation (see Figure 10.14B). This progressive erosion of telomeric DNA represents a simple molecular device that limits how many generations of descendant progeny a cell can spawn.

We can imagine, for example, that in human embryonic cells, the telomeric DNA begins rather long, perhaps 8 to 10 kb in length. As various lineages of descendant cells throughout the developing body proceed through their repeated cycles of growth and division, the telomeres in these cells grow progressively shorter. Ultimately, in some cells, the telomeric DNA erodes down to a size that is so short that it can no longer perform its intended function of protecting the ends of the chromosomal DNA, the result being the breakage–fusion–bridge cycles and chromosomal translocations that are illustrated in Figure 10.15. Indeed, it is plausible that the aging of certain tissues derives from the loss by individual cells of replicative potential, and that this loss is attributable, in turn, to telomere erosion.

We have not yet discovered how to use measurements of telomere length to predict with precision the onset of BFB cycles and crisis. One major difficulty derives from the fact that the telomeric restriction fragments (TRFs; see Figure 10.14A) contain both pure telomeric repeats (at the very ends) and long segments containing repeat-like sequences (located more internally in subtelomeric regions; see Figure 10.17). Only the pure telomeric repeats seem able to protect the ends of chromosomal DNA: loss of end protection seems to occur when the pure tandem repeats decrease below one dozen or so. Hence, TRFs that are still several kilobases long may have already lost their ability to prevent end-to-end chromosomal DNA fusions.

While the supporting evidence is still indirect, it is highly likely that telomere shortening, by limiting the replicative potential of cell lineages, creates an obstacle to the accumulation of large populations of cancer cells. Thus, as argued earlier, should a clone of cells acquire oncogenic mutations (involving oncogene activation and tumor suppressor gene inactivation), the ability of this clone to expand to a large, clinically detectable size is likely to be constrained by its eroded telomeres.

Figure 10.20 Primers and the initiation of DNA synthesis (A) During normal DNA replication, the parental DNA double-helix (*right*) is unwound by a helicase enzyme (*not shown*), creating two template strands and allowing the replication process as a whole to progress in a rightward direction. The synthesis of the new "lagging strand" (*above*) is made possible by the presence of short RNA primer segments (*green segments*) that are laid down at intervals of several hundred nucleotides by the primase enzyme. The 3'-hydroxyl ends of these RNA molecules serve as sites of initiation of newly made daughter-strand segments (*pink*), which are termed Okazaki fragments prior to the removal of the RNA primers and joining of these fragments. Because this synthesis (like all DNA synthesis) occurs in a 5'-to-3' direction, these lagging-strand segments grow in a direction opposite to the direction of advance of the replication fork. In contrast, the copying of the other parental strand (*below*) can proceed without such interruptions, as this "leading" strand can grow continuously at its 3' end as the parental helix unwinds. (B) For these reasons, the leading-strand synthesis will proceed all the way to the end of the telomere (*right*) and thereby generate a faithful copy of the telomeric DNA sequences (*below*). However, the lagging-strand synthesis (*above*) will require an RNA primer near the end of the telomeric DNA (*circled*). Because this RNA primer will subsequently be removed and because it may not sit precisely at the end of the *blue* template strand, sequences complementary to the parental strand will not be present in the finally synthesized DNA molecule, resulting in under-replication of the parental (*thick blue*) strand of DNA and loss of telomeric DNA sequences from the genome of a daughter cell. The 5'-to-3' orientation of each strand is indicated by the arrowheads. The parental DNA strands are drawn more thickly than the newly synthesized strands.

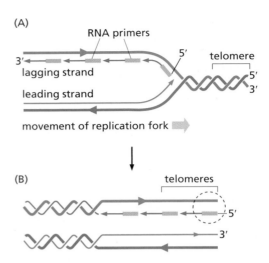

10.6 Incipient cancer cells can escape crisis by expressing telomerase

The detailed behavior of pre-neoplastic cell clones *in vivo* is difficult to study, and so we are largely forced to extrapolate from observations of cultured cells, such as human fibroblasts. As mentioned earlier, if these cells are allowed to circumvent senescence through the expression of the SV40 large T oncoprotein, they will continue to replicate another 10 to 20 cell generations and then enter crisis. On rare occasion, a small group of cells will emerge spontaneously from the vast throng of cells in the midst of crisis. These variant cell clones—arising perhaps from a single cell among 10 million cells in crisis—proceed to proliferate and continue to do so indefinitely (for example, see Figure 10.14A). They have become immortalized, seemingly as the consequence of some random event.

This transition, observed with cultured cells *in vitro*, seems to recapitulate the behavior of pre-malignant cell populations *in vivo*. Thus, both cell populations enter crisis sooner or later, having passed through large numbers of replicative generations and suffered extensive erosion of their telomeres. Both are capable of spawning rare immortalized variants that have apparently solved the problem of telomere collapse, but how?

At first glance, telomere collapse and crisis would appear to be irreversible processes from which cells can never escape. As it turns out, the route to immortality is simple and, in retrospect, obvious: cells can emerge from crisis by regenerating their telomeres, thereby erasing the molecular record (their shortened telomeres) that previously blocked their proliferation and drove them into crisis.

Telomere regeneration can be accomplished through the actions of the **telomerase** enzyme, which functions specifically to elongate telomeric DNA. A striking finding is that telomerase activity is clearly detectable in 85 to 90% of human tumor cell samples, while being present at very low levels in the lysates of most types of normal human cells, as measured by the TRAP assay (**Figure 10.21**; Supplementary Sidebar 10.1). While these low levels of telomerase activity may enable some type of minimal maintenance of the ends of telomeric DNA (such as repair or regeneration of the t-loop), they are clearly unable to prevent the progressive erosion of the double-stranded region of telomeres that accompanies passage of normal human cells through each cell cycle.

In adult humans, there are actually several known exceptions to the generally observed low levels of telomerase activity in normal cells. For example, substantial enzyme activity is present in the germ cells of the testes, and lymphocytes express a burst of telomerase activity when they become functionally activated.

The available evidence indicates that strong expression of the telomerase enzyme is present early in embryogenesis and is largely lost during the cellular differentiation

Figure 10.21 Detecting telomerase activity The telomeric repeat amplification protocol (TRAP) assay permits detection of minute levels of telomerase activity in cell lysates by relying on the polymerase chain reaction (PCR) to amplify the products of the telomerase enzyme (see Supplementary Sidebar 10.1). The products of the TRAP reaction are analyzed by gel electrophoresis to resolve DNA molecules differing in size by hexanucleotide increments. As a negative control, brief heat treatment of a portion of the lysate is used at the beginning of the reaction to denature and inactivate any telomerase that may be present (*1st channel*); this is done to determine whether any heat-resistant DNA polymerases are present in the lysate in addition to the telomerase itself. In this experiment, pre-crisis cells were infected with an expression vector that specifies hTERT, the catalytic subunit of the telomerase holoenzyme, thereby inducing high levels of telomerase activity that are readily detectable when both high (*2nd channel*) and low levels (*3rd channel*) of cell extract are tested. (From M. Meyerson et al., *Cell* 90:785–795, 1997.)

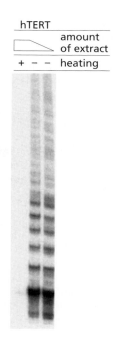

that produces the great majority of the body's tissues. Thus, during the course of early mouse and cow embryogenesis, telomerase is strongly expressed in embryonic cells in the time between the morula and blastocyst stages and seems to disappear thereafter; a similar control of telomerase expression is likely to operate during human embryonic development. Accordingly, the great majority of normal human somatic cells, while carrying the full complement of genes specifying the telomerase holoenzyme, are denied the services of this enzyme because they do not *express* all of the encoding genes at significant levels.

Among the large populations of cultured cells in crisis, however, rare variants find a way to de-repress the gene or genes encoding telomerase and thereby acquire high levels of constitutively expressed enzyme. This enables them to extend their telomeric DNA to a length that permits further proliferation. Indeed, as long as their descendants continue to express this enzyme, such cells will continue to proliferate and thus will be considered to be immortalized. A similar sequence of events is presumed to occur *in vivo* when populations of pre-neoplastic cells enter into crisis and, on rare occasion, generate immortalized variants that become the progenitors of large neoplastic cell populations.

The telomerase enzyme was initially characterized in baker's yeast using genetic analyses and in ciliates through biochemical purification. It is a complex enzyme composed of a number of distinct subunits, not all of which have been characterized. At the core of the mammalian telomerase **holoenzyme** are two subunits. One subunit is a DNA polymerase, more specifically a reverse transcriptase, which functions, like the enzymes made by retroviruses (see Section 3.7) and a variety of other viruses and transposable elements, to synthesize DNA from an RNA template (**Figure 10.22**). However, unlike these other reverse transcriptases, the telomerase holoenzyme cleverly packs it own RNA template—the second essential subunit. A short segment of this 451-nucleotide-long RNA molecule serves as the template that instructs the reverse transcriptase activity of the holoenzyme.

Figure 10.22 The catalytic subunit of telomerase (A) Determination of the amino acid sequence of the catalytic subunit of the telomerase holoenzyme of *Euplotes aediculatus* (see Sidebar 10.3) enabled the cloning of the encoding gene, termed initially p123. This gene was found to be homologous to the gene encoding the catalytic subunit of the *Saccharomyces cerevisiae* (yeast) telomerase holoenzyme, termed Est2 (ever-shorter telomeres). Further analysis revealed extensive sequence relatedness of these two catalytic subunits to the catalytic clefts of reverse transcriptases (RTs) specified by a variety of retrotransposons as well as retroviruses such as human immunodeficiency virus (HIV). (B) The regions of homology are mapped in corresponding colors on the three-dimensional structure of the HIV reverse transcriptase (HIV-1 RT). (From T.M. Nakamura et al., *Science* 277:955–959, 1997.)

(A)

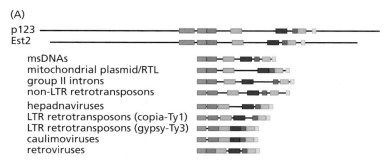

p123
Est2

msDNAs
mitochondrial plasmid/RTL
group II introns
non-LTR retrotransposons

hepadnaviruses
LTR retrotransposons (copia-Ty1)
LTR retrotransposons (gypsy-Ty3)
caulimoviruses
retroviruses

(B)

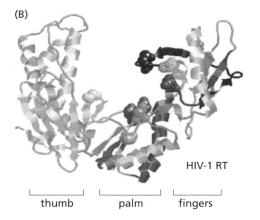

HIV-1 RT

thumb   palm   fingers

Sidebar 10.3 Ciliates are a rich source of telomerase Ciliates are unicellular eukaryotes that belong to the diverse group of protozoa. Ciliates are large, as such cells go, so they require proportionately greater gene expression in order to generate the large quantities of cellular components needed for cell doubling. Ciliates solve this problem by forming two nuclei. The **micronucleus** carries the entire genome in a diploid (2*N*) configuration, the state seen in the somatic cells of metazoa including mammals; this micronucleus is responsible for storing the genetic information of the ciliate cell and transmitting it to daughter cells (**Figure 10.23**).

The **macronucleus**, in contrast, carries a polyploid genome; that of *Tetrahymena thermophila*, for example, carries the equivalent of approximately 45*N* copies of genomic DNA, and many genomic regions may be present in 1000 copies per macronucleus. The resulting increased gene dosage makes possible the large-scale biosynthesis required to form and maintain these cells. In addition, among the group of ciliates known as hypotrichs, the chromosomal DNA in the macronucleus is cleaved into relatively short, linear fragments, each of which carries a complete, functional gene. The ends of these gene-sized DNA molecules, like the ends of normal eukaryotic chromosomes, must be protected by telomeres. Consequently, a single hypotrichous ciliate cell, such as that of *Euplotes aediculatus* (from which the telomerase enzyme was purified and subjected to sequencing), may carry hundreds of thousands of telomeres instead of, for example, the measly

92 telomeres present in most human cells. Naturally, such ciliates must carry a proportionately high concentration of telomerase enzyme to service their abundant telomeres. This explains why ciliates were used for the biochemical purification of an enzyme that is otherwise present in vanishingly low amounts in the cells of metazoa.

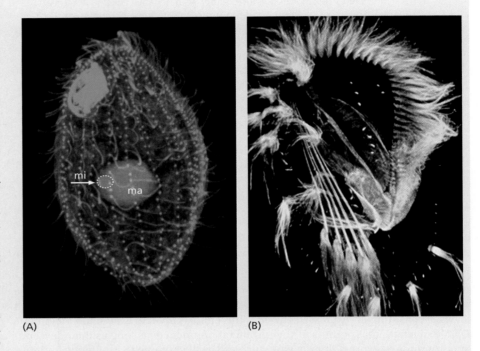

(A)   (B)

Figure 10.23 Ciliates and the isolation of the telomerase enzyme and gene
(A) Ciliates of the hypotrichous class have a large macronucleus and a small micronucleus. The former carries genes required for the maintenance of biosynthesis of these large cells, while the latter carries the germ-line copies of the genome. Shown here is a confocal immunofluorescent image of *Tetrahymena pyriformis*, a close relative of *T. thermophila*, in which the telomerase enzyme was first discovered and characterized. The microtubules of the cytoskeleton are revealed by anti-tubulin antibody (*green*), while the nuclei are stained (*orange*). The micronucleus (mi, *dashed line*) is seen here as a small oval beneath the macronucleus (ma). (B) The appearance of *Euplotes aediculatus*, from which the telomerase catalytic subunit was purified, yielding an amino acid sequence that allowed cloning of the encoding gene. (A, courtesy of D. Wloga and J. Gaertig. B, courtesy of A. Aubusson-Fleury.)

Isolation of substantial quantities of telomerase protein from the ciliate *Euplotes aediculatus* (**Sidebar 10.3**) allowed determination of the amino acid sequence of its catalytic subunit and, in turn, cloning of the gene encoding the ciliate enzyme as well as the homologous human gene. In human cells, this catalytic subunit, termed hTERT (for **h**uman **te**lomerase **r**everse **t**ranscriptase), synthesizes a DNA molecule that is complementary to six nucleotides present in the telomerase-associated RNA molecule (*hTR*; its encoding gene is sometimes called *TERC*), attaching these nucleotides to the G-rich 3′ overhanging end of the preexisting telomeric DNA (**Figure 10.24**). The complementary strand of the telomeric DNA is then presumably synthesized by conventional DNA polymerases.

As would be predicted from the scenario described earlier, before entering crisis, human cells are essentially telomerase-negative and do not express appreciable levels of the *hTERT* mRNA. However, should a rare immortalized variant emerge from a population of cells in crisis, its descendants usually express significant levels of *hTERT* mRNA and exhibit substantial levels of telomerase enzyme activity

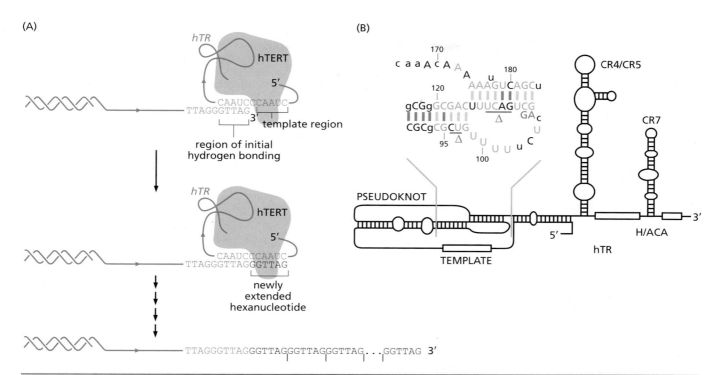

Figure 10.24 Structure of the human telomerase holoenzyme
(A) The human telomerase holoenzyme is composed of two core subunits, the hTERT catalytic subunit (*light brown*) and the associated *hTR* RNA subunit (*blue*). (Five other proteins, some present in two copies, are required to form a functional holoenzyme.) The holoenzyme attaches to the 3′ end of the G-rich strand overhang (*pink*), doing so in part through the hydrogen bonding of *hTR* to the last five nucleotides of the G-rich strand. Subsequently, by reverse transcription of sequences in the *hTR* subunit, hTERT is able to extend the G-rich strand by six nucleotides (*black*). By repeating this process in hexanucleotide increments (*ticks*), the enzyme can extend the G-rich strand by hundreds, even thousands of nucleotides, doing so in a *processive* fashion. Arrowheads indicate the 5′-to-3′ orientation of the nucleic acids. Upon completion of this elongation, a conventional DNA polymerase can fill in the complementary (*blue*) strand. (B) The *hTR* molecule is formed from only 451 ribonucleotides. In its center is a "pseudoknot," a topologically complex structure that has, among other features, a three-stranded helix. A comparison of the telomerase RNAs from 51 species reveals a number of nucleotides that are conserved more than 95% (*uppercase, green*), others that are conserved 80–95% (*uppercase, black*), and yet others that are conserved less than 80% (*lowercase, black*). Pathological human conditions arise from mutations (*red*) in the *hTR* sequence that involve either base substitutions or deletions (Δ), all of which compromise telomerase function. (B, from C.A. Theimer, C.A. Blois and J. Feigon, *Mol. Cell* 17:671–682, 2005.)

(**Figure 10.25**). This observation demonstrates directly that de-repression of *hTERT* gene expression accompanies the escape of these cells from crisis.

The sudden acquisition of telomerase activity might be only a *correlate* of escape from crisis rather than a *cause*. This ambiguity can be resolved by a simple experiment: a cDNA version of the *hTERT* gene can be introduced into cells just before they are destined to enter crisis. The introduced *hTERT* cDNA confers telomerase activity on these cells, causes elongation of their greatly shortened telomeres, prevents entrance into crisis, and enables such cells to grow indefinitely (**Figure 10.26**).

The outcomes of this experiment and similar experiments actually prove three related points. First, while the telomerase holoenzyme may be composed of multiple, distinct subunits, the only subunit missing from normal, pre-crisis human cells is the catalytic subunit encoded by the *hTERT* gene; the remaining subunits of the telomerase holoenzyme, including the *hTR* telomerase-associated RNA molecule, seem to be present in adequate amounts in pre-crisis cells. Second, because expression of telomerase (rather than another enzyme) allows cells to avoid crisis, and because telomerase acts specifically on telomeric DNA, these observations demonstrate that one cause of crisis must indeed be critically shortened telomeres. Third, this experiment shows that the acquisition of telomerase activity, which can be achieved spontaneously by rare variant cells amid a population of cells in crisis, is sufficient to enable cells to escape crisis and generate descendants that can grow in an immortalized fashion.

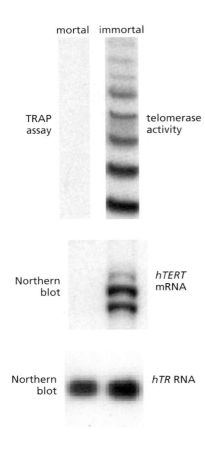

mortal   immortal

TRAP assay — telomerase activity

Northern blot — hTERT mRNA

Northern blot — hTR RNA

Figure 10.25 Activation of telomerase activity following escape from crisis Individual B lymphocytes escape at low frequency from a cell population in crisis, begin to grow robustly, and become immortalized (see Figure 10.14A). The pre-crisis "mortal" cells lack telomerase activity, as measured by the TRAP assay (see Supplementary Sidebar 10.1), while the spontaneously immortalized ("immortal") cells show abundant activity (*top*). The *hTERT* mRNA, which encodes the catalytic subunit of the telomerase holoenzyme, is essentially absent in the pre-crisis cells but clearly present in the immortalized cells (*middle*). However, levels of the *hTR* RNA, which functions as a template for hTERT during telomeric DNA elongation (see Figure 10.24A), are increased only slightly (*bottom*), suggesting that levels of hTERT expression determine the levels of enzyme activity in the cell. (From M. Meyerson et al., *Cell* 90:785–795, 1997.)

We can conclude further that telomerase succeeds in allowing cells to circumvent crisis because it subverts the operations of the generational clock. When telomerase is expressed at significant levels, telomeres are maintained at lengths that are compatible with unlimited further replication. In effect, the generational clock that depends on progressive telomere shortening is rendered inoperative. The molecular strategies used by cancer cells to de-repress *hTERT* expression and thereby acquire high levels of telomerase activity are poorly understood (**Sidebar 10.4**).

With these insights in mind, the normal biological roles of telomerase now become clearer. In single-cell eukaryotic species, such as ciliates and yeast, the exponential growth of cells requires the continuous presence of high levels of telomerase activity to ensure that telomeres are maintained indefinitely at a length that is compatible with chromosomal stability. By elongating telomeric DNA, the telomerase in these protozoan cells is able to compensate for the continued erosion of telomeric DNA due to the under-replication of DNA termini by the bulk DNA replication machinery. Being single-celled, these eukaryotes have no need to fear cancer.

Expression of telomerase activity is programmed differently in complex metazoa, notably humans. Because telomerase expression is largely repressed in postembryonic cell lineages, these lineages are granted only limited postembryonic replicative

Figure 10.26 Prevention of crisis by expression of telomerase (A) This Southern blot reveals the lengths of telomeric DNAs that were generated by cleaving genomic DNA with a restriction enzyme that does not cut within the repeating hexanucleotide sequence of telomeres (see Figure 10.14A), followed by hybridization to a DNA probe that recognizes this same sequence. As is seen, telomerase-negative human embryonic kidney (HEK) cells infected with a control retroviral vector that specifies no protein carry telomeres that decrease in size between the 4th and 20th population doublings (PDs; *left two channels*). However, infection of such cells with a retroviral vector specifying hTERT causes the telomeres to substantially increase in length, maintaining their sizes through the 26th population doubling (*right channels*). (Because these telomeres are much longer, they generate a stronger signal upon Southern blotting.) The size markers are given in kilobase pairs. (B) This population of human embryonic kidney (HEK) cells was destined to enter crisis by 40 days, which is the equivalent of roughly 20 population doublings (PDs; *red line*). However, when hTERT was ectopically expressed in these cells (*blue line*), they gained the ability to proliferate indefinitely. (Courtesy of C.M. Counter and R.L. Beijersbergen.)

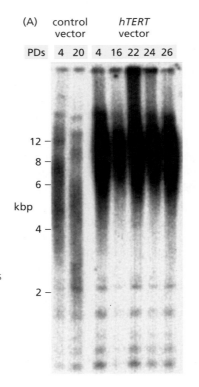

(A)   control vector   hTERT vector

PDs   4   20   4   16 22 24 26

12 —
8 —
6 —
kbp
4 —

2 —

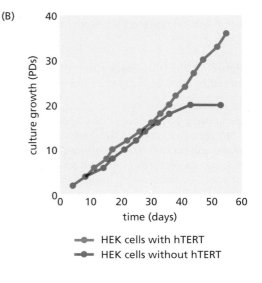

(B)

culture growth (PDs) vs time (days)

40
30
20
10
0
0   10   20   30   40   50   60
time (days)

— HEK cells with hTERT
— HEK cells without hTERT

Sidebar 10.4 Oncoproteins and tumor suppressor proteins play critical roles in governing *hTERT* expression The mechanisms that lead to the de-repression of *hTERT* transcription during tumor progression in humans are complex and still quite obscure. Multiple transcription factors appear to collaborate to activate the *hTERT* promoter. Once expressed, the resulting *hTERT* transcripts allow the synthesis of substantial amounts of hTERT protein; the latter, by complexing with other subunits of the telomerase holoenzyme, suffices to generate high levels of telomerase activity. The Myc protein, which acts as a transcription factor that controls the cell cycle clock and thousands of genes involved in other aspects of cell physiology (see Section 8.9), also contributes to *hTERT* transcription. There are at least two recognition sequences for bHLH transcription factors like Myc in the promoters of the mouse and human *TERT* genes.

The connection between Myc and hTERT expression can also be demonstrated by biological experiments. For example, a human melanoma cell line has been isolated that expresses high levels of Myc and is tumorigenic. Three clones of variants of this cell line have been isolated that exhibit as much as an eightfold reduction in Myc expression and have lost tumorigenicity. Their tumorigenic growth can be restored if they are supplied with an *hTERT*-expressing gene construct. Therefore, in these cells, the loss of tumorigenicity following reduction of Myc expression levels can be attributed largely, if not exclusively, to an associated loss of *hTERT* gene expression.

In addition to Myc, the E6 oncoprotein of HPV (see Supplementary Sidebar 9.1) and the ER81 transcription factor, which is activated by the Ras–MAPK pathway, stimulate hTERT transcription. Working in the opposite direction, Menin—the product of the *MEN1* tumor suppressor gene (Table 7.1)—associates with the *hTERT* promoter and represses *hTERT* transcription. A similar repressive role has been attributed to p53, Mxd (see Figure 8.27), the WT1 tumor suppressor protein, and the Sp1 transcription factor, whose expression is induced by TGF-β. Interestingly, certain germ-line polymorphisms in the *hTERT* gene promoter that are associated with lower transcriptional activity are also associated with decreased lifetime risks of breast and ovarian carcinomas. Conversely, in 2013 recurrent point mutations were found in the *hTERT* promoter in 71% melanoma genomes and in 16% of nonmelanoma tumor DNA; these somatic mutations enhanced promoter activity by 2 to 4 fold.

potential before they enter crisis. This limitation seems to represent a key component of the human body's anti-cancer defenses.

This mechanistic model also provides a compelling explanation of how human cells can escape from crisis and become immortalized. It fails, however, to address another long-standing mystery: How does expression of telomerase enable certain types of cultured cells to avoid senescence (**Sidebar 10.5**)?

10.7 Telomerase plays a key role in the proliferation of human cancer cells

The experiments described above using pre-crisis human cells in culture suggest that the telomerase activity detectable in the great majority of human cancer cells plays a causal role in their immortalization, and that such immortalization is a key component of the neoplastic growth state. This mechanistic model has been tested experimentally by suppressing the telomerase activity in cancer cells and following their subsequent responses. Telomerase activity can be reduced through the expression of antisense RNA in the telomerase-positive cells. For example, an RNA that is complementary in sequence to the *hTR* RNA subunit of the telomerase holoenzyme (see Figure 10.24B) can be introduced experimentally into telomerase-positive human cancer cells. Such an antisense molecule is presumed to anneal to the *hTR* molecule, forming an RNA–RNA double helix, thereby blocking the ability of the *hTR* subunit to participate in the synthesis of telomeric DNA. It is difficult to achieve total inhibition of hTR function using this strategy. Nonetheless, such an antisense experiment performed with HeLa cells caused them to stop growing 23 to 26 days after they were initially exposed to the antisense RNA.

An alternative experimental strategy that is more effective derives from the use of a mutant hTERT enzyme carrying amino acid substitutions in its catalytic cleft; these alterations yield a catalytically inactive enzyme. When overexpressed in telomerase-positive cells, the mutant hTERT protein can act in a dominant-negative fashion (dn; see Section 9.3) to interfere with the existing endogenous telomerase activity of these cells. The dn hTERT, which can be expressed at levels vastly higher than the endogenous hTERT protein, is likely to associate with and thereby monopolize the other subunits that normally assemble to form the telomerase holoenzyme. Of course, when these other molecules associate with the dn hTERT protein, they become recruited into unproductive holoenzyme complexes.

Sidebar 10.5 The puzzle of senescence and telomeres As described here, a diverse array of experimental data have converged on the conclusion that cells enter senescence in response to a variety of cell-physiologic stresses, some sustained over extended periods of time. These include various types of genetic damage, oncogene signaling, oxidative stress, and even metabolic stress. Replicative senescence—manifested by many cell types in response to being introduced into tissue culture—is largely and possibly entirely explained by the stresses that cells suffer from sub-optimal conditions of culture; this is underscored by the fact that replicative senescence can be delayed by improving the conditions of culture (see Figure 10.7).

Once cells have entered into senescence, which is generally an irreversible step, they manifest DNA lesions that seem, for one reason or another, to be irreparable. Indeed, many of the stressors that induce senescence appear to function, directly or indirectly, to induce DNA damage, and the resulting unrepaired lesions seem to be responsible for emitting senescence-inducing signals, thereby stably maintaining the cells around them in the senescent state. In human (but not in mouse) cells, the senescence-associated heterochromatic foci (SAHFs; see Figure 10.10A) give indication of a massive shutdown of the expression of thousands of genes, many of which are apparently essential for cell proliferation.

This scheme is difficult to reconcile with other solid observations demonstrating that in certain human cell types, senescence can be circumvented by forced expression of hTERT. Because such hTERT expression also allows a cell to circumvent crisis, this treatment effectively immortalizes cells. These observations have been incorporated into a scheme which suggests that when telomeres are eroded after short-term culture, some of the shortest telomeres have already eroded enough to lose their ability to protect the ends of telomeric DNA; these shortened telomeres release the senescence-inducing signal. In the event that senescence is circumvented, cells will then proceed to crisis, where far greater telomere erosion triggers the breakage–fusion–bridge cycles and the widespread apoptosis associated with crisis. Accordingly, telomere shortening acts as a clocking device to trigger both senescence and, later on, crisis.

These two mechanistic models are incompatible with one another: either the timing of senescence is triggered by telomere erosion or it is triggered by physiologic stress. The resolution of this dilemma is not obvious. One possible mechanism is the following (**Figure 10.27**): when cells experience certain types of physiologic stress, they actively degrade some of their telomeres; the resulting dysfunctional telomeres release a DNA damage signal that in turn triggers senescence. In the event that hTERT is present in significant amounts, such telomere degradation and resulting entrance into senescence is avoided. Yet other mechanisms by which hTERT acts in a telomere-independent fashion to prevent entrance into senescence have been proposed.

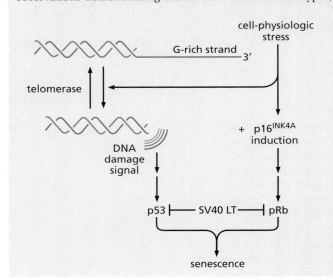

Figure 10.27 Replicative senescence and the actions of telomerase This diagram presents a speculative mechanistic model of how and why telomerase expression (achieved by ectopic expression of the hTERT catalytic subunit) can prevent human cells from entering into replicative senescence. The t-loop of the telomeric DNA (see Figure 10.18) is not shown here. The G-rich overhanging strand of telomeric DNA (*pink, top*) has been found in some studies to be largely degraded in cells that have entered replicative senescence. This loss may be caused by certain cell-physiologic stresses. The resulting blunted telomeric DNA may then emit a DNA damage signal that activates p53. Together with p16^INK4A expression (see Section 8.4), which is activated by various cell-physiologic stresses, this may impose a senescent state. By reversing this loss of the G-rich overhang (and possible partial degradation of the dsDNA portion), telomerase is able to prevent certain cell types, such as human fibroblasts, from entering into senescence. Of note, the demonstrated ability of SV40 large T antigen to circumvent senescence depends on its ability to sequester both pRB and p53 (see Figure 10.8).

Expression of the dn hTERT subunit in a number of different telomerase-positive human tumor cell lines causes them to lose all detectable telomerase activity and, with some delay, to enter crisis. The crisis occurs with a lag time from 5 to 25 days, depending on the length of the telomeric DNA in these cells at the time when the dn hTERT subunit was introduced. Cells with telomeres that initially were 2 to 3 kb in length enter crisis almost immediately, while those with telomeres that were initially 4 to 5 kb in length require 30 days of further passaging in culture before entering crisis (**Figure 10.28**). These delayed reactions suggest that after the dn hTERT is expressed in cancer cells, telomeres shorten progressively, and that crisis ensues after the initially present telomeric DNA erodes to some threshold length.

Importantly, the dn hTERT enzyme has no observable effect on the growth of telomerase-positive cells up to the point when they reach crisis. This rules out an alternative

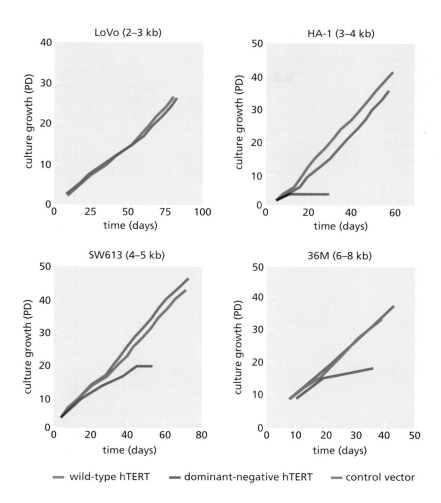

— wild-type hTERT   — dominant-negative hTERT   — control vector

Figure 10.28 Suppression of telomerase activity and resulting loss of the neoplastic growth program Expression vectors specifying a mutant dominant-negative (dn) hTERT protein (*red lines*), a wild-type hTERT enzyme (*green lines*), or an empty control vector (*blue lines*) have dramatically different effects on four different human cancer cell lines. The respective lengths of their telomeric DNAs, as determined by TRF Southern blots (see Figure 10.14) at the onset of the experiment, are indicated in kilobases (kb). The empty control vector and the hTERT-expressing vector permit the continued proliferation of these cells, as measured in population doublings (PD). However, introduction of the dn hTERT enzyme, which differs from the wild-type enzyme by a single amino acid substitution in its catalytic cleft, causes these cells to stop growing, doing so with various lag times (in days). Microscopic examination revealed that the cells that ceased proliferating entered into crisis and showed widespread apoptosis. Cells of the LoVo cell line (*top left*) entered crisis immediately after expressing the dn hTERT enzyme, and therefore no cells were available to count. (Because the TRFs also contain nonfunctional subtelomeric DNA that carries imperfect hexanucleotide repeats, the lengths of functional telomeric sequences are much shorter than the observed lengths of the TRFs.) (From W.C. Hahn et al., *Nat. Med.* 5:1164–1170, 1999.)

explanation: that the mutant enzyme is intrinsically cytotoxic and that its effects on the cell are attributable to some nonspecific toxicity. Taken together, such experiments lead to the conclusion that the continued activities of the telomerase enzyme are as important to the proliferation of these cancer cells as the actions of oncogenes and the inaction of tumor suppressor genes. This conclusion is strengthened by the dramatic outcomes of studies of certain human pediatric tumors (**Sidebar 10.6**).

Responding to these various discoveries, some researchers have used an automotive analogy to illustrate the growth of human cancer cells: activated oncogenes are said to be akin to accelerator pedals that are stuck to the floor; inactivated tumor suppressor genes are compared to defective braking systems; and telomerase is likened to an agent that ensures that the runaway car has an endless, self-replenishing supply of gasoline in its tank.

10.8 Some immortalized cells can maintain telomeres without telomerase

As noted earlier, 85 to 90% of human tumors have been found to be telomerase-positive. The remaining 10 to 15% lack readily detectable enzyme activity, yet the cells within this second group of tumors are presumably faced with the need to maintain their telomeres above some minimum length in order to proliferate indefinitely. In fact, many of these cells have learned to maintain their telomeric DNA using a mechanism that does not depend on the actions of telomerase.

This non-telomerase-based mechanism was first discovered in the yeast *Saccharomyces cerevisiae* following inactivation of one of the several yeast genes encoding subunits of its telomerase holoenzyme. Resulting mutant yeast cells soon entered into a

Sidebar 10.6 Telomerase fuels the growth of some pediatric tumors Neuroblastomas are tumors of cells in the peripheral (sympathetic) nervous system and are usually encountered in very young children. These tumors have highly variable outcomes, with some regressing spontaneously and others progressing into invasive, metastatic tumors that ultimately prove fatal. As described in Section 4.5, the N-*myc* gene often undergoes amplification in these tumors, and greater copy numbers of N-*myc* indicate a worse prognosis for the patient. Telomerase activity is an even more useful indicator of the eventual outcome of the disease. As shown in Figure 10.29A, children whose neuroblastomas are telomerase-negative (as judged by the TRAP assay; see Supplementary Sidebar 10.1) do very well in response to therapy, while those whose tumors are telomerase-positive do poorly. Expression of these two genes—N-*myc* and *hTERT*—is likely to be functionally linked, since N-*myc*'s cousin, *myc*, is already known to be a strong inducer of *hTERT* transcription (see Sidebar 10.4). Similar dynamics operate in another childhood tumor—Ewing's

sarcoma—which may arise from mesenchymal stem cells (see Figure 10.29B).

Some pathologists argue that neuroblastomas are extremely common in very young children, and almost all of these regress spontaneously and never become clinically apparent. The causes of this regression and the benign outcomes of many clinically detected neuroblastomas that respond well to treatment may now have found a molecular explanation: without significant hTERT expression, neuroblastoma cells lose telomeres and progress into crisis, from which they fail to emerge.

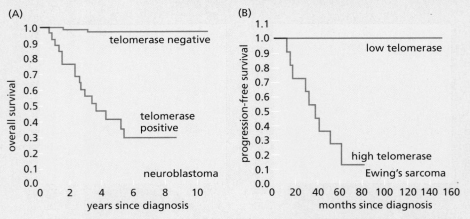

Figure 10.29 Telomerase activity and the prognosis of pediatric tumors (A) This Kaplan–Meier plot illustrates the survival of a group of pediatric neuroblastoma patients who had received no cytotoxic chemotherapy treatment and who were tracked for the time spans indicated (*abscissa*) following diagnosis. These patients were segregated into two groups—those whose tumors were telomerase activity–positive (*red*), and those whose tumors appeared to lack this activity (*blue*). The ordinate indicates overall survival, i.e., the proportion of patients who remained alive at various times after diagnosis. (B) A similar study of children with Ewing's sarcoma was undertaken, in which the patients were segregated into two groups, depending on whether their tumors had high (*red*) or low (*blue*) telomerase activity. In this case, the ordinate indicates the proportion of children who experienced *progression-free survival*, i.e., those whose tumors had not advanced beyond the stage initially encountered in the clinic. (A, from C. Poremba et al., *J. Clin. Oncol.* 18:2582–2592, 2000. B, from A. Ohali et al., *J. Clin. Oncol.* 21:3836–3843, 2003.)

state that is analogous to crisis in mammalian cells. The vast majority of the cells died, but rare variants emerged from these populations of dying cells that used the ALT (alternative lengthening of telomeres) mechanism, which is telomerase-independent, to construct and maintain their telomeres. This ALT mechanism is also used by the minority of human tumor cells that lack significant telomerase activity.

Details of the molecular mechanisms used by ALT-positive human cells to maintain their telomeres are poorly understood. An important clue, however, comes from an experiment in which a traceable molecular marker—a neomycin resistance gene—was introduced into the midst of the telomere repeat sequence of one chromosome carried by a mammalian cell in the ALT state. When the telomeric DNAs were examined in the descendants of this cell, copies of this molecular marker were found in a number of other telomeres as well (**Figure 10.30**). Hence, sequence information appears to be exchanged between the telomeres of cells in the ALT state.

One mechanism that enables exchange of sequence information depends on a type of interchromosomal **copy-choice** mechanism (**Figure 10.31A**). Thus, the polymerases responsible for replicating DNA on one chromosome may, for a short period of time, use sequences from a second chromosome as a template for the elongation of a nascent DNA strand before returning to the original chromosome to continue replication of this chromosome's DNA. In effect, the DNA polymerases "borrow" sequence information from the second chromosome in order to incorporate it into the newly synthesized copy of the first chromosome. Alternatively, some type of unequal crossing-over between telomeres may also operate in ALT cells, which may explain the significant instability of telomere lengths in these cells (see Figure 10.31B).

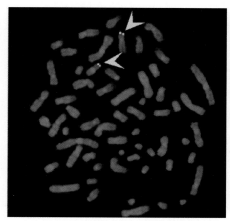
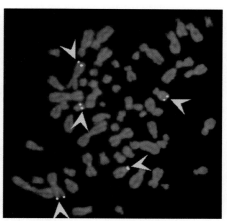

early passage cells   40 population doublings later

Figure 10.30 The ALT mechanism The alternative lengthening of telomeres (ALT) mechanism can be shown to depend, at least in certain cells, on the exchange of sequence information between the telomeres of different chromosomes. Here, telomeres of human GM847 cells were labeled via the targeted insertion (using homologous recombination) of a genetic marker into one of the telomeres. By the time of the analysis, the marker was already present in two telomeres (*arrowheads, left panel*). Forty population doublings later, this sequence could be found to be associated with five distinct chromosomes (*arrowheads, right panel*). Fluorescence *in situ* hybridization (FISH) was used to detect these marker sequences in metaphase chromosomal preparations. (From M.A. Dunham et al., *Nat. Genet.* 26:447–450, 2000.)

The identities of the enzymes that are used by ALT-positive human cancer cells to maintain their telomeres remain elusive. If we can extrapolate from the behavior of yeast cells, some of the enzymes involved in DNA repair may contribute to telomere maintenance in human ALT cells. For example, in yeast cells that have been deprived of one of the genes involved in *mismatch repair* of DNA (see Section 12.4), the ALT state is readily activated. Since one of the functions of mismatch repair is to suppress recombination between imperfectly homologous DNA sequences, this finding supports the notion that interchromosomal recombination is an important mechanistic component of the ALT state. Moreover, in ALT cells, an exonuclease that normally plays a key role in repairing double-stranded DNA breaks seems to be used to chew away the 5′ end of the C-rich strand in order to generate the 3′ overhanging end of the G-rich strand.

The existence of the ALT mechanism emphasizes the fact that all types of human cancer cells must develop ways to maintain their telomeres above a certain threshold length in order to proliferate in an immortalized fashion. Most human cancer cells de-repress expression of the *hTERT* gene, and a small minority of incipient cancer cells find a way to activate the ALT mechanism.

The ALT state is associated preferentially with a specific subset of the tumors encountered in oncology clinics. These include perhaps half of osteosarcomas and soft-tissue sarcomas as well as one-quarter of glioblastomas, but only rarely carcinomas. Perhaps these correlations may one day be explained by the fact that a group of genes expressed characteristically in mesenchymal cells are also associated with repression of the *hTERT* gene, forcing the cancer cells that arise from these cell types to rely on ALT as a more accessible route to immortalization. Additional clues to understanding the origin of this mechanism may also come from observations that the ALT state operates in certain subclasses of normal lymphocytes.

However, at the moment, the precise reasons why one mechanism (involving hTERT) is usually favored over the other (ALT) remain obscure. Equally unclear are the mechanisms that cause the telomeric DNA of human cells in the ALT state to grow to lengths (>30 kb) far greater than those usually seen in telomerase-positive cells (5–10 kb). Importantly, the dn hTERT enzyme (see Section 10.7), which induces crisis in a number of telomerase-positive human cancer cell lines, fails to do so in a human ALT-positive tumor cell line, providing further support of the notions that (1) ALT-positive cells do not depend on telomerase for their growth, and (2) the dn hTERT enzyme is not intrinsically cytotoxic.

hTERT is a very attractive target for researchers who are intent on developing novel types of anti-cancer therapeutics. As we will learn in Chapter 16, the catalytic clefts of enzymes like hTERT can often be blocked by highly specific therapeutic drug molecules. Significantly, hTERT is a distant relative of the reverse transcriptase of human immunodeficiency virus (HIV; see Figure 10.22), against which effective drug

Figure 10.31 Alternative mechanisms supporting ALT (A) One possible mechanism of telomere maintenance in ALT cells involves the exchange of sequence information between telomeres (see Figure 10.30), which may in turn depend on DNA polymerases that use more than one chromosome as a template during chromosome replication. According to one plausible mechanism, one telomere (*brown and green, lower left*) extends its 3′ overhanging end (*green*), which displaces a strand of the same polarity (*pink*) in the telomere of another chromosome. This allows the 3′ overhanging end of the first chromosome to anneal to the complementary strand (*blue*) of the telomere on the other chromosome. Conventional DNA polymerases then can extend the 3′ end of the interloping strand (*broken green line*) by using the complementary (*blue*) strand as template (*top*). After disengaging from the *blue* strand of the second chromosome, the newly elongated G-rich strand can be converted by other DNA polymerases into double-stranded form (*broken brown line, lower right*). This can be repeated dozens of times, resulting in the transfer of sequence information from one telomere to another and the lengthening of telomeres by many kilobases without the involvement of the telomerase enzyme. (B) It is also possible that telomeres may exchange sequence segments through the mechanism of unequal crossing-over, which results in this case in one telomere of increased length and another of decreased length. (Note that telomeres with 3′ overhanging ends can be created in the absence of telomerase by the actions of 5′-to-3′ exonucleases on the C-rich strand, and that the telomeres of cells using ALT do indeed possess 3′ overhanging ends, t-loops, and shelterin complexes.). The arrowheads indicate the 5′-to-3′ orientation of the DNA strands.

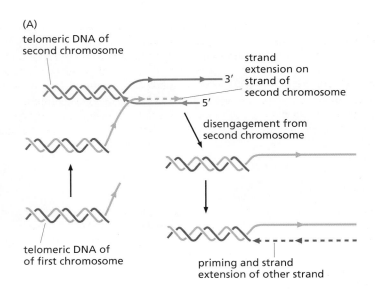

(A)

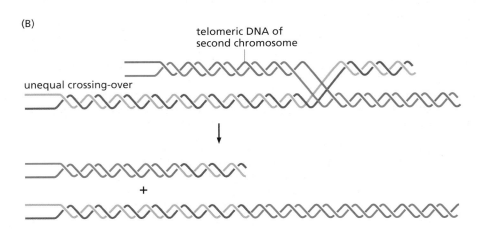

(B)

inhibitors have been successfully developed. These existing successes greatly increase the likelihood that anti-hTERT compounds will one day be produced as well. In addition, this enzyme is expressed in the great majority of human cancers, while not being present at significant levels in normal tissue. This offers the prospect of drug selectivity—being able to affect cancer cells while leaving cells in normal tissues untouched. However, for reasons that remain unclear, the hTERT enzyme has proven time and again to be a formidable enemy, in that repeated screens for low–molecular-weight inhibitors of this enzyme have failed to yield attractive drug candidates.

The ALT state holds important implications for such drug development. Imagine that an inhibitor of hTERT were indeed developed. It is already clear that as many as 10% of human tumors will never respond to such an anti-telomerase drug, because they are ALT-positive and therefore do not depend on this enzyme for their continued proliferation. The remaining 90% of human tumors, which do indeed exhibit robust telomerase activity, may initially be susceptible to killing by this drug, but sooner or later may spawn variants that have activated the ALT mechanism of telomere maintenance. This might allow the newly arising ALT variants to evade the killing actions of the anti-telomerase drug.

ALT cells also shed light on the connection between telomeres and senescence. When strongly expressed *ras* oncogene is introduced into *in vitro* ALT cells (which carry very long telomeres), few if any transformed cells emerge. However, when hTERT expression is forced in such cells, now transformants emerge in large numbers. This suggests that the very long telomeres carried by ALT cells do not protect them against *ras*-induced senescence (see Sidebar 10.2), and that such protection can only occur when hTERT is present to ensure the proper assembly and maintenance of telomeric ends. In consonance with this is the finding that patients whose glioblastomas maintain telomeres through the ALT mechanism survive longer than do those whose tumors rely on hTERT. These findings add further support for the notion that the overall length of telomeres, on its own, does not determine susceptibility to enter into the senescent state, and that some aspect of telomere structure (for example, see Sidebar 10.5) or some still-unknown function of hTERT affects entrance into the senescent state.

10.9 Telomeres play different roles in the cells of laboratory mice and in human cells

Our descriptions of telomerase expression repeatedly refer to its actions in human cells, with good reason. Rodent cells, specifically those of the laboratory mouse strains, control telomerase expression in a totally different way. For reasons that remain unclear, the double-stranded region of telomeric DNA in laboratory mice is as much as 30 to 40 kb long—about five times longer than corresponding human telomeric DNA. In fact, laboratory mouse telomeres are so long that they are never in danger of eroding down to critically short lengths during the lifetime of a mouse, even when expression of mouse telomerase is suppressed experimentally.

The lengths of mouse telomeres permit cell lineages to pass through a far larger number of replicative generations than would seem to be required for tumor formation. This suggests that laboratory mice do not rely on telomere length to limit the replicative capacity of their normal cell lineages and that telomere erosion cannot serve as a mechanism for constraining tumor development in these rodents. In addition, it seems that the long telomeres of mice are not, on their own, sufficient to support the robust proliferation of tumor cells in these animals (**Sidebar 10.7**).

The long telomeres of mice hold additional implications: to the extent that cells derived from laboratory mice show limited replicative ability *in vitro*, that ability is never determined by telomere shortening. In fact, mouse cells can be immortalized relatively easily following propagation in culture (which selects for cells that have avoided senescence and are spontaneously immortalized *in vitro*). Human cells, in contrast to their mouse counterparts, require the introduction of both the SV40 *large T* oncogene (to avoid senescence; see Figure 10.8) and the *hTERT* gene (to avoid crisis). As an aside, the reduced tendency of mouse cells to become senescent during *in vitro* culture may be explained by the fact that they fail to develop senescence-associated heterochromatic foci (SAHFs; see Sidebar 10.1) in culture while their human counterparts do so readily.

These significant interspecies differences suggest that one of the central mechanisms governing human cell biology—telomere shortening—is irrelevant to the biology of mouse cells, at least those of laboratory mice. The biological rationale behind these differences in the organization of an important cellular control circuit remains unclear. One attractive speculation is that humans, whose cells pass through approximately 10^{16} mitoses in an average lifetime, have a far greater risk of cancer than do mice, which experience only about 10^{11} cell divisions (that is, mice have only about 0.1% as many cells as humans and live on average only about 1% of a human lifespan). This 10^5-fold larger number of cell divisions in humans would seem to create a proportionately greater lifetime cancer risk, dictating the need for large, long-lived mammals, such as humans, to develop additional anti-tumor mechanisms beyond those operating in small, short-lived mammals, such as mice. (Attempts at drawing broad, generalizable conclusions from these interspecies differences are undermined by the fact that certain strains of wild-type mice control their telomerase and telomere length much like humans.)

The role of telomeres in the lives of mice has been revealed dramatically by researchers who have inactivated (see Supplementary Sidebar 7.7) the gene encoding the *mTR* RNA subunit of telomerase in the mouse germ line. As expected, the homozygous mutant offspring of initially created heterozygous mice carrying the mutant *mTR*

Sidebar 10.7 Long telomeres do not explain aspects of tumorigenesis and aging in laboratory mice Study of several tissues in laboratory mice reveals that expression of the mouse telomerase (mTERT) catalytic subunit is lost as stem cells differentiate (**Figure 10.32A**); this dictates that the activity of the telomerase holoenzyme declines in parallel. This loss of enzyme activity would seem to explain why the telomeres in mouse tissues erode detectably as descendant cells pass through multiple cell generations and differentiate (see Figure 10.32B). (Indeed, a similar erosion has been documented in differentiating normal human tissues.) Nonetheless, the shortened telomeres in differentiated mouse cells (>30 kb length) are still far longer than is required to protect chromosomal ends and therefore chromosomal integrity (<<5 kb).

This holds implications for the role of telomeres in tumor pathogenesis. As we will discuss in the next chapter, the precise identity of the **cell-of-origin**—the normal cell that serves as the progenitor of the neoplastic cells in a tumor—remains elusive. In some cases, the cell-of-origin may be a normal progenitor cell, while in others it may be a normal stem cell (also discussed in Chapter 11). In either case, in mice the evidence presented in Figure 10.32B suggests that cells-of-origin have telomeres that are extremely long. This raises the question of whether these very long telomeres suffice, on their own, to enable tumorigenesis to proceed efficiently in mice in the absence of ongoing mTERT function.

The evidence suggests otherwise. For example, one strain of transgenic mice engineered to overexpress mTERT in the basal keratinocytes of the skin show a twofold increase in the rate of skin cancer, while another strain of mice that overexpress mTERT in a wide variety of tissues show greatly increased risk of mammary carcinomas. In both cases, the overexpressed mTERT is functioning in cells that already possess very long telomeres. Moreover, in a variety of tumor models, mTERT activity increases progressively during multi-step tumor development. Clearly, the mTERT enzyme contributes to tumorigenesis through mechanisms other than simply elongating telomeres, echoing our discussion at the end of Section 10.8.

(A)

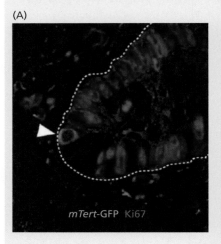

mTert-GFP Ki67

◼ 400 - 1200 a.u
◼ 1200 - 1500 a.u
◼ 1500 - 1800 a.u
◼ 1800 - 3000 a.u

(B)

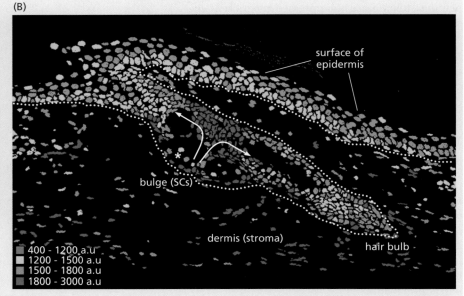

surface of epidermis

bulge (SCs)

dermis (stroma)

hair bulb

Figure 10.32 mTERT function may be confined largely to adult stem cells Most of the mTERT activity in adult mice may be traceable to its expression in the stem cells of various tissues; a similar dynamic may operate in humans, in which this issue is poorly resolved. Two lines of evidence suggest this limited expression pattern. (A) Transgenic mice were constructed in which the promoter of the *mTERT* gene drives expression of green fluorescent protein (GFP). In addition, cells in the crypts of the small intestine (see Figure 7.24) of these mice were immunostained with an antibody that recognizes Ki67, a marker of cells in the active growth-and-division cycle (*red*). This analysis indicates that the long-lived, slowly cycling cells in the crypt, which are located in the positions of stem cells (*arrowhead*), express readily detectable levels of mTERT while their more differentiated, rapidly proliferating progeny do not. (B) Fluorescence *in situ* hybridization (FISH) using a probe that is specific for telomeric DNA reveals that telomeres are longer in the stem cells than in their more differentiated progeny. In this case, the telomeres in various locations in the hair root of the tail skin of a rat were measured; the intensity of hybridization to telomeric DNA is presented in arbitrary units (a.u.). These measurements translate to an average telomere length of ~43.9 ± 5.8 kb in the bulge region (in which stem cells, SCs, are concentrated) to 34.1 ± 3.2 in the interfollicular epidermis, i.e., the surface of the skin between hair roots. The arrows indicate the downward migration of the progenitor products of stem cells into the hair bulb, in which a hair shaft is produced, and, following wounding, upward cell migration into the skin. In each case, migration is accompanied by increased differentiation and shortening of telomeres. (A, from R.K. Montgomery et al., *Proc. Natl. Acad. Sci. USA* 108:179–184, 2011. B, from I. Flores et al., *Genes Dev.* 22:654–667, 2008.)

allele exhibit no telomerase activity in any of their cells. During their lifetimes, the telomeres of these telomerase-negative mice may shorten from approximately 30 kb down to about 25 kb; the latter length is still far longer than is required to protect chromosomal ends.

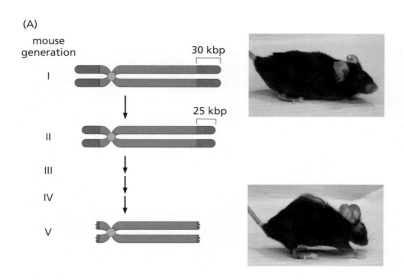

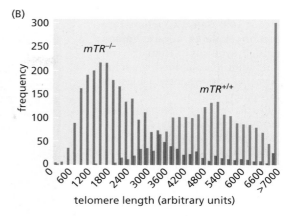

These mTR-negative mice are superficially indistinguishable phenotypically from wild-type mice. This absence of any readily observable mutant phenotype suggests that the telomerase holoenzyme plays no essential role in the tissues of these mammals beyond its task of maintaining telomeres. (However, detailed characterization of genetically altered mice that continue to express mTR but have been deprived of mTERT expression reveals a subtle morphogenetic defect—the conversion of a thoracic to a lumbar vertebra. This change is apparently due to the fact that the mTERT protein has a function that is totally unrelated to its actions on telomeres: it associates with the complex of transcription factors that allow β-catenin to activate gene expression in response to Wnt signaling; see Figure 6.26B. The same association presumably holds true for hTERT.)

The homozygous telomerase-negative (that is, $mTR^{-/-}$) mice can be bred with one another through at least three more organismic generations without showing any discernible phenotype. Finally, however, in the fifth generation, telomerase-negative mice begin to show distinctive phenotypes—an indication that after five organismic generations without telomerase, their telomeres have eroded down to dangerously short lengths (**Figure 10.33**). These fifth-generation $mTR^{-/-}$ mice and, even more so, their progeny in the sixth generation are sickly and show a diminished capacity to heal wounds, indicative of the inability of their cells to respond properly to mitogenic signals. The sixth-generation $mTR^{-/-}$ mice suffer also from substantially reduced fertility.

The $mTR^{-/-}$ mice in this sixth generation already exhibit very short telomeres at birth, and their inability to subsequently maintain these telomeres, particularly in mitotically active tissues, results in widespread cell death and loss of tissue function. For example, highly proliferative tissues, such as the gastrointestinal epithelium, the hematopoietic system, and the testes, show substantial **atrophy** (loss of cells). These observations are striking and without parallel in the field of mouse genetics, because they demonstrate organismic phenotypes that only become manifest five or six organismic generations after a mutation has been introduced into the germ line of these animals.

Moreover, such observations dramatically demonstrate the differences between the telomeres of human cells and those of these mutant laboratory mice. Human telomeres begin relatively short, and loss of telomerase activity can lead, already in the *first* organismic generation, to severe organismic phenotypes (**Sidebar 10.8**).

10.10 Telomerase-negative mice show both decreased and increased cancer susceptibility

Laboratory mice are susceptible to spontaneous cancers, largely lymphomas and leukemias. This background of susceptibility can be increased by experimentally introducing specific mutations into the mouse germ line that cause activation of

Figure 10.33 Erosion of telomeres over multiple generations in populations of $mTR^{-/-}$ mice (A) During the lifetimes of the first $mTR^{-/-}$ mouse generation, their telomeres, which are initially 30 kbp long (*red*), are reduced in size by about 5 kbp (*left*). These mice (*top, right*) are phenotypically normal, as are their descendants in the second and third generations. However, by the fifth organismic generation, the telomeres in their cells have eroded to the extent that they can no longer protect the ends of chromosomes; the tissues of these mice begin to lose their ability to renew themselves and heal wounds, and these mice begin to show symptoms of premature aging, including wasting away of muscle tissues and a hunched back (*bottom, right*). The relative sizes of the telomeres in these images have been exaggerated 1000-fold for the sake of illustration. (B) Telomere length can be gauged by fluorescent *in situ* hybridization (FISH) using a probe that is specific for the telomere repeat sequence. Here the mTR^- allele has been bred into a strain of wild mice that normally have very short telomeres. Already in the first generation, the resulting homozygous $mTR^{-/-}$ mice (*red*) have far shorter telomeric DNA than their wild-type ($mTR^{+/+}$) counterparts (*blue*). These measurements also reveal the heterogeneous lengths of telomeric DNA within each population of mice, mirroring comparable heterogeneity within the individual cells of a mouse. Lengths are in arbitrary units. All telomeres longer than 7000 units are represented by a single bar (*far right*). (A, courtesy of R.S. Maser and R.A. DePinho. B, from L.-Y. Hao et al., *Cell* 123:1121–1131, 2005.)

Sidebar 10.8 Defective telomerase function explains a rare human familial syndrome Humans suffering from the inherited condition known as dyskeratosis congenita (DC) exhibit defects in highly regenerative tissues, such as skin and bone marrow, as well as reduced chromosomal stability. In the X-linked form of the disease, a protein known as dyskerin is absent or malfunctioning. Dyskerin associates with the telomerase-associated RNA subunit (*hTR*) and helps *hTR* to assemble in the telomerase holoenzyme complex (see Figure 10.24). Autosomal dominant forms of the same disease can be caused by the inheritance of mutant alleles of the *hTR*- and hTERT-encoding genes. Cells in proliferating tissues of affected individuals exhibit dicentric chromosomes, which are indicative of interchromosomal fusions. Some affected individuals succumb to a collapse of their hematopoietic system due to the loss of regenerative capacity of the stem cells in their bone marrow (**Figure 10.34**); some of these patients may ultimately develop leukemia because of telomere collapse, similar to the mice to be described in Section 10.10

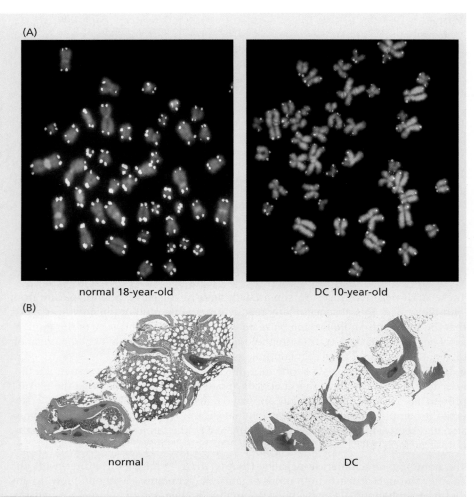

normal 18-year-old   DC 10-year-old

(B)

normal   DC

Figure 10.34 Telomere and bone marrow collapse of dyskeratosis congenita patients Humans suffering from dyskeratosis congenita (DC) experience atrophy of a number of tissues, which is traceable to the absence of functional telomerase and to the catastrophic collapse of their telomeres. (A) Fluorescence *in situ* hybridization (FISH) was used to detect the telomeres of metaphase chromosomes in a cell of a normal 18-year-old (*left*) as well as in a cell of a 10-year-old suffering from DC. Many of the chromosomes in the latter cell have telomeres that are so short that they are no longer detectable by FISH.

(B) The most life-threatening tissue atrophy of DC patients occurs among the rapidly proliferating cells of the bone marrow. As indicated here, in normal bone (*left panel*), regions between mineralized bone (*pink*) are filled with hematopoietic cells of the marrow (*purple*). However, the marrow of a dyskeratosis patient suffering germ-line mutations in the *hTR* gene (*right panel*; see also Figure 10.24B) is almost totally devoid of cells because the stem cells have lost the ability to maintain the various lineages of hematopoietic cells. (A, courtesy of S. Kulkarni and M. Bessler. B, courtesy of M. Bessler.)

proto-oncogenes or inactivation of tumor suppressor genes. To what extent do the critically short telomeres in fifth- and sixth-generation telomerase-negative mice affect the cancer rate in these animals?

The answer to this question would seem to be straightforward. We can imagine that the deterioration of cells in these $mTR^{-/-}$ mice leaves a barely adequate number of healthy cells to sustain tissue and organismic viability. Stem cells charged with the task of replenishing the pools of differentiated cells in these tissues will likely have great difficulty doing so.

Since the normal cells in these tissues are barely able to maintain themselves, incipient cancer cells should, by all rights, have an even more difficult time. The process of tumor development requires that clones of pre-malignant cells evolving toward malignancy must pass through a large number of growth-and-division cycles—a number substantially greater than those experienced by nearby normal cells (**Figure 10.35A**). Consequently, any tumor development that is successfully initiated by these cell clones is likely to be aborted long before it has reached completion.

In order to test these predictions, mice were first made cancer-prone by inactivating tumor suppressor genes in their germ line, in this case the locus encoding both the p16^{INK4A} and p19^{ARF} tumor suppressor proteins (see Sections 8.4 and 9.7). Such mice, as might be expected, are highly susceptible to cancer; indeed, they frequently develop lymphomas and fibrosarcomas relatively early in life. This susceptibility is manifested even more dramatically when they are exposed to carcinogens. In one set of experiments, sequential exposure to dimethylbenz[a]anthracene (DMBA), a potent carcinogen, followed by repeated exposures to ultraviolet-B (UV-B) radiation, led after 20 weeks to a 90% tumor incidence in the (p16^{INK4A}/p19^{ARF}-negative) mutant mice, while a control group of similarly treated wild-type mice had no tumors.

This experiment was extended by introducing the p16^{INK4A}/p19^{ARF} germ-line inactivation into mice that also lacked *mTR*, the gene encoding the RNA subunit of the mouse telomerase holoenzyme (see Figure 10.24). The results were that 64% of control,

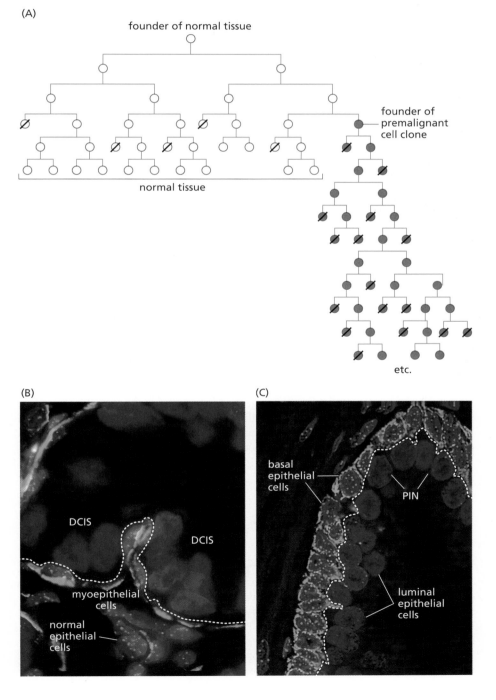

Figure 10.35 Normal and neoplastic cell lineages in *mTR*$^{-/-}$ mice and in humans (A) This diagram indicates the pedigree of normal cells (*open circles*) in a small patch of tissue. A founder cell in the embryo becomes the ancestor of all the cells in this tissue, e.g., all the epithelial cells. At one point, one of these epithelial cells suffers changes that make it the ancestor of the cells (*closed red circles*) that develop into a pre-cancerous cell clone. This clone of cells must pass through many more replicative generations in order to eventually form a tumor mass. Slashes indicate the substantial attrition of cells through apoptosis or necrosis that occurs in almost every cell generation. (B) The large number of cell generations through which pre-malignant cells must pass en route to neoplasia results in considerable telomere erosion. In these images, telomeres are revealed by fluorescence *in situ* hybridization (FISH, *pink dots*). In normal human mammary ducts (*not shown*), telomeres are apparent in both the luminal epithelial cells and the myoepithelial cells. However, in ductal carcinoma *in situ* (DCIS), the luminal epithelial cells, which constitute the neoplastic cell population (*above dotted line*), have lost all detectable telomeric DNA, while the nearby normal myoepithelial cells (*below dotted line*), which have been stained for α-smooth muscle actin (*green*), still carry strongly staining telomeres. (In all cases, cell nuclei are stained *blue*.) (C) Similar dynamics apply in the case of prostatic intraepithelial neoplasia (PIN)—a precursor lesion to prostatic carcinoma—in which many of the luminal epithelial cells lining ducts (*below dotted line*) have lost telomeric DNA staining, while the underlying keratin-positive (*green, above dotted line*) basal epithelial cells continue to display a strong telomeric DNA signal. (B and C, courtesy of A.K. Meeker; see also A.K. Meeker et al., *Cancer Res*. 62:6405–6409, 2002.)

telomerase-positive, p16[INK4A]/p19[ARF]-negative mice contracted tumors, while only 31% of the generation 4 and 5 $mTR^{-/-}$ p16[INK4A]/p19[ARF]-negative mice developed tumors. These results were even more dramatic when survival of mice after 16 weeks was measured: 88% of the telomerase-positive mice had been lost to cancer while only 46% of the fifth-generation telomerase-negative mice had been lost to this disease.

This reduced rate of cancer in telomerase-negative mice fulfills the prediction that the normal tissues of $mTR^{-/-}$ mice have exhausted most of their endowment of replicative generations even before tumorigenesis has begun. Once tumors are initiated in these mice, incipient tumor cell populations must pass through many additional doublings before they can create macroscopic tumors (see Figure 10.35A). However, relatively early during the course of tumor formation, these aspiring cancer cells will be driven into crisis by telomere collapse and their agenda of forming tumors will be aborted. Indeed, in human tissues, pre-malignant cells that are poised to become fully neo-plastic already show drastically truncated telomeres; this truncation is ostensibly due to the fact that these cell populations have passed through many more successive cell cycles than nearby normal cells (see Figure 10.35B and C).

A very different and ultimately far more interesting outcome was seen when mice were used that instead had been deprived of both $p53$ gene copies in their germ line (that is, rather than having had their p16[INK4A]/p19[ARF] locus inactivated). Ordinarily, germ-line inactivation of the $p53$ gene in mice leads, on its own, to an increased tumor incidence and resulting mortality, mirroring aspects of the Li–Fraumeni syndrome of humans (see Section 9.12). The mutant mTR locus was then introduced into the $p53^{-/-}$ genetic background of these mice.

When the $p53^{-/-}$ genotype was present in the genomes of fifth- and sixth-generation telomerase-negative $mTR^{-/-}$ mice, something quite unexpected was observed: the rate of cancer formation was significantly *increased* above that created by the $p53^{-/-}$ genotype alone. In addition, the spectrum of tumors, namely, lymphomas and angio-sarcomas, that are commonly seen in $p53^{-/-}$ mice was shifted in favor of carcinomas—just the types of tumors commonly seen in humans. These trends became even more apparent in the seventh and eight generation of telomerase-negative mice (**Figure 10.36**).

How can these intriguing results be explained? We can imagine that as fifth- and sixth-generation mice exhaust their telomeres, their cells will begin to experience chromosomal breakage–fusion–bridge (BFB) cycles (see Figure 10.15) and thus crisis. We already know that double-strand DNA breaks, which are formed in BFB cycles, can provoke apoptosis through a p53-dependent pathway (see Section 9.6). Hence, cells that carry functional $p53$ genes and experience BFB cycles are likely to be rapidly eliminated from tissues.

As cells of fifth- and sixth-generation $mTR^{-/-}$ $p53^{-/-}$ mice experience BFB cycles, they may well struggle to stay alive because of the repeated breakage of their chromosomal DNA. However, many of these cells may manage to survive because a key component of their pro-apoptotic response machinery—p53—is missing. These cells will now

Figure 10.36 Rate of tumor formation in cancer-prone $mTR^{-/-}$ $p53^{-/-}$ mice These Kaplan–Meier plots reveal that tumor incidence actually increases with increasing organismic generation (G) in mice that lack both the p53- and mTR-encoding genes in their germ line. In generations 1 and 2 (*red*), about 50% of the mice had readily detectable tumors by the age of 24 weeks, while in generations 5 and 6 (*blue*), the mice developed tumors more rapidly. In generation 7 or 8 (*brown*), about 50% of the mice already exhibited tumors by the age of 17 weeks. (From L. Chin et al., *Cell* 97:527–538, 1999.)

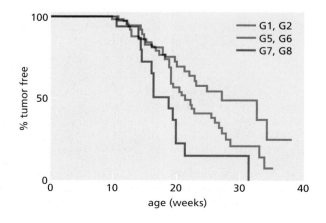

limp through a number of growth-and-division cycles in spite of the ongoing karyotypic chaos that afflicts their genomes. All the while, their chromosomes will participate in numerous BFB cycles.

These cycles may continue for many cell generations, since p53 will not be on watch to trigger apoptosis and eliminate these cells. As a consequence, the genomes of such cells will become increasingly scrambled by the nonreciprocal chromosomal translocations generated by the BFB cycles. In addition, and perhaps even more important, there is evidence that dsDNA breaks lead to the amplification or deletion of the chromosomal regions adjacent to these breaks.

While these BFB cycles will not prove to be immediately fatal for the p53-negative cells, the cycles will surely compromise their proliferative ability. This means that individual cells in these populations will be under great selective pressure to escape these BFB cycles and to reacquire karyotypic stability; once they do so, they will be able to grow more rapidly. In these mouse cells, however, the acquisition of telomerase activity is not an option (because of the germ-line knockout of *mTR* gene copies). For this reason, they will resort to activating the ALT telomere maintenance system. Having done so, these cells will stabilize whatever aberrant karyotypes arose since the time when their ancestors first lost functional telomeres and will begin to grow robustly again.

Clearly, the scrambled karyotypes that these cells have acquired generate novel combinations of genes (including many translocated, deleted, and amplified genes), and among the novel genotypes that are created, there will likely be some that *favor* neoplastic proliferation. Stated differently, the period of genetic instability appears to *increase* the probability that cancer-promoting genetic configurations will be produced. Such reasoning explains how collapsing telomeres in a p53-negative background can actually promote the development of tumors.

These $mTR^{-/-}$ mice have human counterparts: as already mentioned, patients suffering the dyskeratosis congenita syndrome (see Sidebar 10.8) often go on to develop cancer once their collapsing hematopoietic systems have been successfully reconstructed by bone marrow transplantation. Some develop hematopoietic malignancies (myelodysplasia and acute myelogenous leukemia) or carcinomas of the gastrointestinal tract. These neoplasias arise in those tissues that undergo constant, intense proliferation—precisely the sites where telomeres might become rapidly eroded in the absence of telomerase activity, resulting in repeated breakage–fusion–bridge cycles.

10.11 The mechanisms underlying cancer pathogenesis in telomerase-negative mice may also operate during the development of human tumors

These observations of genetically altered mice, when taken together, lead us to consider a most interesting mechanistic model (**Figure 10.37**) that explains why aneuploidy is seen in the great majority of human carcinomas. Imagine that relatively early in the long process of human tumor development, cells in pre-malignant cell populations are able to jettison their *p53* gene copies. The selective pressure favoring this genetic loss might derive, for example, from the fact that cells in the early premalignant cell populations experience anoxia and, in turn, suffer from p53-induced apoptosis. Alternatively, *p53* gene copies may have been shed by incipient cancer cells in order to reduce the likelihood of oncogene-triggered apoptosis (see Section 9.7).

Some time after the loss of p53 function, as tumor progression proceeds, cells in these pre-malignant populations will begin to experience substantial telomere erosion (see Figure 10.35) and eventually telomere collapse, since they lack significant levels of telomerase activity. Repeated BFB cycles will ensue and result in aneuploidy. However, because these cells also lack functional p53, they will survive and even continue to proliferate, albeit slowly.

Sooner or later, in this population of cells suffering BFB cycles, a cell will emerge that has acquired the ability to de-repress *hTERT* gene expression and, having done so, has gained the ability to stabilize its karyotype and prevent further BFB cycles. With

Figure 10.37 A mechanistic model of how BFB cycles promote human carcinoma formation During the early steps of tumor progression, inactivation of the *p53* tumor suppressor gene is often favored, because it allows incipient cancer cells to escape apoptotic death due to oncogene activation, insufficient trophic signals, anoxia, and poisoning by metabolic wastes due to inadequate vascularization. As tumor progression proceeds, the telomeric DNA (*red*) of evolving pre-malignant cell populations eventually erodes below the level needed to protect chromosomal ends. Breakage–fusion–bridge (BFB) cycles (see Figure 10.15) ensue, leading to increasing chromosomal rearrangements and the amplification and deletion of chromosomal segments adjacent to breakpoints (*not shown*); only one BFB cycle is shown here. Centromeres are *dark green* and *dark brown*. Cells undergoing BFB cycles would normally be eliminated by p53-induced apoptosis. In the absence of p53 function, however, such cells may survive, albeit with scrambled karyotypes that slow their proliferation. Emerging from such cell populations will be variants that have learned how to regenerate telomeres and thereby stabilize their karyotype; such cells can once again grow rapidly. In cells of the mutant *mTR*^–/– mice (see Figure 10.36), this stabilization may be achieved through the activation of the ALT telomere maintenance system, while in wild-type (e.g., *mTR*^+/+) organisms (including human cancer patients), de-repression of hTERT expression allows reconstruction of telomeres to proper length. Once telomeres have been regenerated in all chromosomes, further karyotypic disorder due to BFB cycles will be halted. However, any karyotypic aberrations generated previously will be perpetuated in descendant cell populations.

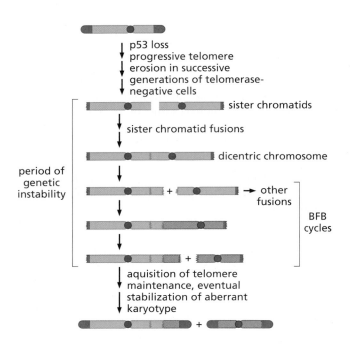

telomerase on hand, we can imagine that this cell will repair the ends of its chromosomal DNA molecules and once again enjoy robust growth. Of course, the belated appearance of telomerase on the scene cannot unscramble the considerable aneuploidy that accumulated during the time window that began with telomere collapse and ended with the acquisition of *hTERT* expression. (The same arguments will apply in the event that the ALT mechanism, rather than telomerase activation, is used to protect the genome from further BFB cycles.)

Observations of human pancreatic cancer progression provide support for this model. Relatively early in the long, multi-step process of tumorigenesis in the pancreas (see Figure 11.12B), the mitotic cells in low-grade adenomas have few of the anaphase bridges that are characteristic of BFB cycles. However, more advanced, highly dysplastic adenomas show large numbers of these bridges. Still later, the even more advanced, *in situ* carcinomas once again exhibit lower levels of anaphase bridges. This behavior reinforces the model, described above, that proposes that BFB cycles and associated chromosomal instability are found only in a defined window of time during the course of multi-step tumor progression.

Additional support for this model comes from a study of human carcinomas of the esophagus, colon, and breast. In each case, tumors possessing relatively short telomeres are strongly associated with poor long-term prognosis, while tumors carrying long telomeres are associated with a far better prognosis, including greater long-term survival. Here, once again, we can imagine that the chromosomes of cancer cells with sub-optimal telomere lengths undergo relatively frequent breakage–fusion–bridge cycles, perhaps because they express hTERT at levels that are barely adequate to maintain telomeric DNA. These BFB cycles, in turn, create an ongoing chromosomal instability that continues throughout the life of the tumor to generate novel, scrambled genotypes, some of which may favor phenotypes such as more rapid proliferation and increasing aggressiveness. Certain human chronic inflammatory conditions, which are associated with increased risks of cancer, may also be attributable to eroded telomeres and BFB cycles (**Sidebar 10.9**).

This model provides an attractive mechanism that explains how many types of human tumor cells acquire highly aneuploid karyotypes. Although not directly demonstrated, it is widely assumed that the resulting aneuploid genomes confer growth advantages on these cells by creating novel oncogenes through translocations, by increasing the dosage of growth-promoting proto-oncogenes, and by eliminating tumor suppressor genes that have been holding back cell proliferation. Hence, these BFB cycles may be

Sidebar 10.9 Telomere collapse may contribute to cancer in organs affected by chronic inflammation In Chapter 11, we will read about a number of human tumors arising in tissues that suffer continuous loss of cells due to chronic infections or inflammation. Such ongoing attrition of cells, which may occur over a period of decades, occurs in tissues affected by, for example, ulcerative colitis, Barrett's esophagus, and hepatitis B or C virus infection. In response to losses of differentiated cells, the stem cells in these tissues are continually producing replacements to ensure maintenance of tissue functions. These stem cells are therefore forced to pass through many more cycles of growth and division than are the corresponding stem cells of normal tissues. The resulting repeated cell cycles may tax the regenerative powers of the stem cell pools and lead eventually to telomere collapse, to the triggering of breakage–fusion–bridge cycles, and to the generation of karyotypes that favor neoplastic cell proliferation.

Such a mechanism may well account for the observed high rates of cancer in individuals suffering from these inflammatory and infectious diseases. In fact, in the intestinal epithelial cells of individuals suffering from ulcerative colitis, one can often find the anaphase bridges that are telltale signs of BFB cycles and hence of telomere collapse—precisely what we would expect if telomere collapse were contributing to the formation of the colon carcinomas seen in these patients (**Figure 10.38**). Similarly, anaphase bridges have been documented in the cells of Barrett's esophagus (see Figure 2.14), a condition in which the reflux of stomach acid causes a high turnover of esophageal epithelial cells that leads, with significant frequency, to esophageal carcinomas.

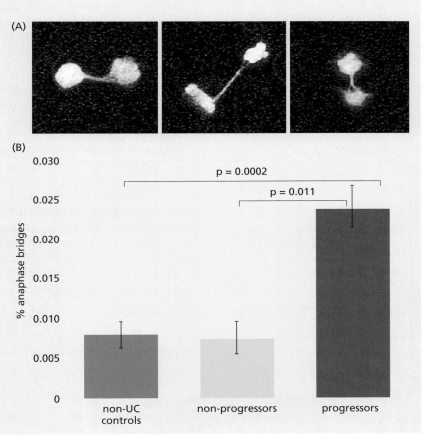

Figure 10.38 Ulcerative colitis and anaphase bridges Ulcerative colitis (UC) involves chronic irritation and inflammation of the colonic epithelium. These provoke ongoing turnover of colonic epithelial cells that can continue for several decades. UC leads, with significant frequency, to the development of colon carcinomas. (A) UC patients can be separated into "progressors," whose colitis has led to the appearance of one or more carcinomas, and "non-progressors," who show no evidence of intestinal tumors. The colonic epithelial cells of progressors frequently display the anaphase bridges characteristic of BFB cycles seen here (see also Figures 10.15 and 10.16), as well as extensive shortening of telomeres and numerous karyotypic aberrations (*not shown*). (B) Colon tissue samples from normal control individuals and UC non-progressors show relatively few anaphase bridges, while much greater numbers are found in the inflamed epithelium of the progressors. This indicates a correlation between the occurrence of anaphase bridges and the onset of tumor development. The p values indicate the probability of these differences arising by chance. (From J.N. O'Sullivan et al., *Nat. Genet.* 32:280–284, 2002.)

instrumental in accelerating tumor progression because they increase genomic mutability and allow evolving, pre-malignant cells to explore a multitude of novel, potentially advantageous genotypes.

Yet other observations are consistent with this thinking but hardly prove it. For example, the leukemia cells in about two-thirds of acute myelogenous leukemia (AML) patients exhibit a normal karyotype, while cells from the remainder exhibit various types of karyotypic disarray. Those tumors with normal karyotype exhibit significantly longer telomeres than do those with a scrambled karyotype, consistent with the notion that eroded telomeres are associated with and possibly responsible for the derangement of normal karyotype. Of additional interest, a **hypomorphic** allele of *hTERT*—one that specifies reduced enzyme function—is carried by as many as 1% of the general population. Significantly, individuals inheriting this allele—which confers only 60% of normal hTERT function—are overrepresented by a factor of 3 among AML patients. This suggests that normal cells throughout the bodies of these patients have a reduced ability to maintain normal telomere lengths, which in turn may predispose these people to the development of this form of leukemia. Indeed, those afflicted with dyskeratosis congenita (see Figure 10.34B) have a greatly increased risk of developing myelodysplastic syndrome, which in turn leads, with significant frequency, to AML.

These observations of premature, pathological shortening of telomeres lead to another question: Is the significantly increased cancer-associated incidence and mortality among the elderly (see Figure 11.3) attributable, at least in part, to *normal* telomere shortening? Exhaustive measurements of telomere lengths in normal humans do indeed indicate marked telomere erosion with increasing age. The data from measurements, such as those provided in **Figure 10.39A**, yield three conclusions: (1) In some (and perhaps most) human tissues, telomeres shorten progressively with increasing age. (2) The rate of shortening differs in different tissues, ostensibly because of differing mitotic activity in the cell lineages and differing expression of telomerase function in various stem cell compartments. (3) The scatter on each curve indicates substantial inter-individual variability in the lengths of telomeric DNA early in life and/or rates of telomere shortening during a lifetime. (In addition, while not discussed here, the telomeres within a single cell exhibit differing, chromosome-specific lengths; for example, see Figure 10.33B.)

Other research, not described here in detail, yields a fourth, highly relevant conclusion: individuals whose normal circulating lymphocytes exhibit relatively short telomeres are overrepresented in cohorts of patients suffering from head-and-neck, bladder, lung, and renal carcinomas. In this case, the lymphocytes are presumed to be representative of normal cells throughout the body. This suggests that short telomeres in one's normal tissues represent a lifelong increased risk for developing various types of carcinomas. (Recall as well the apparently related observation, cited above, that individuals who carry a hypomorphic *hTERT* allele confront a markedly increased risk of developing AML).

Significantly, among some elderly, average telomere lengths have shortened so much that the residual telomeric sequences remaining are largely composed of subtelomeric DNA—the sequences in the telomere that do not contribute to protection from end-to-end chromosome fusion (see Figure 10.17). The behavior of the $mTR^{-/-}\ p53^{-/-}$ mice

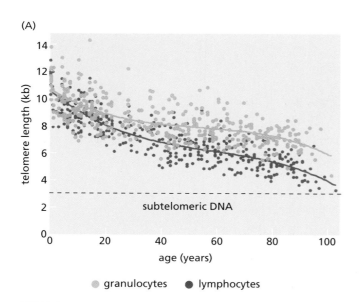

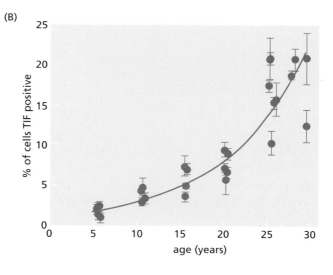

(A) granulocytes   (B) lymphocytes

Figure 10.39 Telomere shortening, aging, and cancer
(A) Telomeric DNA lengths were measured in the lymphocytes and granulocytes of 400 individuals. As is apparent, the length of telomeres decreased progressively with increasing age in both human cell types. Inaccuracy in measuring telomere lengths is ±0.5 kb. The horizontal line (*dashed*) represents the threshold level below which telomeric DNA no longer protects chromosomal ends.
(B) TIFs (telomere dysfunction–induced foci) were scored in dermal fibroblasts from baboons of various ages. These TIFs were scored wherever there was a congruence within a cell between γ-H2AX, a marker of double-strand DNA breaks (see Chapter 12), and telomeric DNA; they represent unrepaired and ostensibly irreparable dsDNA breaks. Their accumulation suggests that telomere erosion (and resulting cell senescence) represents a critical force in driving the aging process in certain tissues. (A, courtesy of G.M. Baerlocher and P. Lansdorp. B, from U. Herbig et al., *Science* 311:1257, 2006.)

(see Figure 10.36) would suggest that these dysfunctional telomeres may lead to BFB cycles, increased mutability, and cancer progression—a connection that has not yet been directly demonstrated.

An additional consequence of eroded telomeres is suggested by studies of the cells in our cousins, the baboons. As they age, the percentage of their dermal fibroblasts that exhibit dysfunctional telomeres increases progressively (see Figure 10.39B). A synthesis of this observation with the measurements of human telomere lengths seems to suggest that progressive telomere erosion leads to the accumulation of telomeres that are not readily repaired; such damaged telomeres, like the senescence-associated heterochromatic foci (see Sidebar 10.1), may represent a signal to initiate and maintain a senescent state in these cells (see Section 10.3). Indeed, such damaged telomeres can be documented in a variety of other cell types with increasing age. (The precise contribution of telomere-induced senescence to the aging of the organism as a whole remains to be documented.)

These speculations still do not address an issue raised by some observations cited earlier: the repeated BFB cycles in $mTR^{-/-} p53^{-/-}$ mice affect the *types* of tumors that these mice exhibit, causing them to develop carcinomas, which are common in humans, rather than hematopoietic and mesenchymal malignancies, which are frequently seen in mice. How does telomere biology possibly help to explain why cancers tend to arise in some tissues and not others?

10.12 Synopsis and prospects

When Barbara McClintock first described telomeres at the ends of maize chromosomes, she could not have anticipated the dramatic convergence of her discoveries with the molecular models of cancer pathogenesis. Telomeres are now known to be major determinants of the ability of cells to multiply for a limited number of growth-and-division cycles before halting proliferation and entering into crisis. This circumscribed proliferative potential of normal human cells appears to operate as an important barrier to the development of cancers by limiting the proliferation of pre-neoplastic cell clones.

This mechanistic model of cell immortalization and cancer pathogenesis is elegant, if only because it is so simple. Still, the simplicity of this model should not obscure the fact that much of telomere structure and function remains poorly understood, and some of the conclusions described in this chapter may one day require substantial revision.

Among the many unresolved issues are the connections between the findings described in this chapter and another major problem of biomedical research: How much of organismic aging, as it occurs in human beings and other long-lived mammals, is attributable to progressive telomere erosion, the resulting loss of the ability of cells in certain tissues to continue proliferating, and, ultimately, the depletion of cell populations needed to maintain tissue integrity and function? While attractive in concept, there are surely alternative mechanisms of aging that we need to consider. For example, the cells in aging tissues may lose their robustness due to a variety of accumulated chemical and genetic insults that they suffer over many decades' time, and this progressive loss of viability may have nothing to do with telomere shortening.

Intertwined with these issues is one that we raised earlier: Does age-related telomere shortening contribute to the dramatically increased rates of cancer in the elderly? In the next chapter, we explore the fact that tumor formation is a multi-step process and that each of the requisite steps often takes many years to complete; these kinetics, on their own, can be invoked to explain the age-related onset of cancer. Now we may need to invoke an additional mechanism to explain the onset of tumors in aged individuals: maybe their significantly eroded telomeres trigger the BFB cycles and genetic instability that fuels the formation of tumors.

These questions inevitably return to a focus on the telomerase enzyme and its function. Interestingly, telomeres may be relatively short soon after fertilization. However, in the cleavage-stage embryo, a telomerase-independent, interchromosomal recombination mechanism similar to ALT generates the long telomeres that are present in the soon-to-be-formed blastocysts. Thereafter, telomerase is expressed in tissue-specific stem cells (which are committed to one or another differentiation lineage), and at very low levels in more differentiated cells.

While several types of stem cells have been found to express readily detectable levels of telomerase, this activity is apparently unable to maintain telomeres at a constant size. Evidence for this comes from the progressively shortening telomeres that are seen with increased life span. (Since the differentiated cells that make up the great majority of cells in most tissues are generally a small number of cell generations removed from the stem cells within their respective tissues, the progressive telomere shortening seen in these differentiated cells must reflect the status of telomeres in the corresponding stem cells.)

The experiments with the telomerase-negative mice that were described here under-score the consequences associated with the absence of telomerase and critically shortened telomeric DNA: in the presence of functional p53, such cells will enter crisis and die in large numbers—a mechanism that serves to obstruct the further expansion of clones of pre-neoplastic cells. However, without p53 on the job, critically shortened telomeres fail to trigger crisis and instead may lead the cell into multiple breakage–fusion–bridge cycles that create widespread genetic instability (**Figure 10.40**); this increased mutability can accelerate the progression of incipient tumors.

There are still many puzzles associated with hTERT and telomerase that require clari-fication. Ectopic hTERT expression can operate to protect certain types of cells against p53-induced apoptosis. Provocatively, this protection is also afforded by a mutant of hTERT that lacks catalytic activity. Similarly, in zebra fish, knockdown of the *zTERT* gene results in compromised hematopoiesis (formation of various blood cells); this effect can be reversed by a mutant of zTERT that lacks the TERC-binding domain and therefore lacks bona fide telomerase activity. In addition, overexpression of a TERT mutant lacking reverse transcriptase activity stimulates hair growth in mice.

These non-telomerase-associated functions of hTERT may one day be associated with another biochemical activity of hTERT: it associates in the nucleus with the Brg-1 pro-tein, a chromatin-modifying enzyme, and thereby amplifies Wnt-β-catenin–stimulated

Figure 10.40 Karyotypic chaos Here, use of the SKY technique to "paint" each chromosome its own characteristic color revealed that numerous chromosomes belonging to two subclones (A and B) of a human bladder carcinoma cell line have participated in translocations. This is indicated by the fact that many chromosomes carry segments originating from two or more normal human chromosomes. Each translocation is indicated by a "t" followed by the number of each parental chromosome from which these arms derive. An "i" indicates an isochromosome, in which a single arm of a chromosome has been duplicated. "p" and "q" indicate the short and long arms of normal human chromosomes, respectively. "del" indicates that a chromosomal segment has been deleted. "Clone A and B" indicates aberrant chromosomes shared by the two subclones, while "clone A" and "clone B" indicate chromosomes that are present uniquely in one or another of these subclones. The translocations shared in common by the two subclones are likely to have occurred before these cells were introduced into culture, while those present in one subclone or the other indicate ongoing chromosomal instability in culture. Comparable degrees of aneuploidy can be observed in cancer cells that have recently been introduced into culture. (From H.M. Padilla-Nash et al., *Genes Chromosomes Cancer* 25:53–59, 1999.)

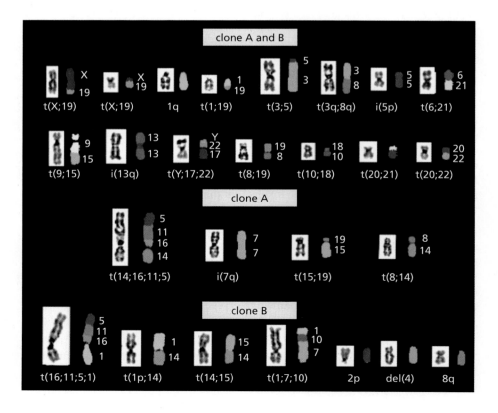

transcription. This function seems to explain why the use of immunofluorescence to localize hTERT protein demonstrates that it is distributed along the lengths of chromosomes rather than being concentrated at telomeres.

Cells that have been deprived of hTERT activity by siRNA-mediated knockdown have a markedly decreased ability to repair their DNA following exposure to several types of mutagens. This effect also cannot be easily related to any effects of hTERT on telomeric DNA. Recall as well the fact that certain cells maintaining their telomeres via the ALT mechanism (which generates very long telomeres) cannot be readily transformed by an introduced *ras* oncogene, but can readily be transformed if they are supplied with an ectopically expressed hTERT. Diverse observations like these indicate that telomere maintenance is only one facet of the function of the apparently multi-talented hTERT protein.

Replicative senescence, as controlled by telomere erosion, has been portrayed as one mechanism whereby the unlimited replicative power of cell lineages is constrained. In recent years, this thinking has been challenged by two types of findings. First, the onset of this process can be affected, sometimes dramatically, by altering conditions of *in vitro* cell cultures. From this perspective, the timing of senescence can no longer be viewed as a cell-autonomous process, that is, one that is intrinsic to a cultured cell. Second, as cited above, various lines of research indicate that hTERT has multiple functions that are clearly distinct from telomere maintenance. Taken together, these findings raise the question of whether the ability of the hTERT protein to allow cells to circumvent senescence is tied to one of its functions that is unrelated to elongating telomeric DNA.

Importantly, these findings do not jeopardize another key concept: that severe telomere erosion triggers crisis in human cells and that ectopic expression of hTERT can prevent or allow escape from the ensuing, almost-certain death. Moreover, whether or not replicative senescence, as originally defined with cultured cells, is a real phenomenon, it is clear that certain forms of senescence (for example, oncogene-induced senescence) operate in living tissues, where they seem to serve as an impediment to the forward march of pre-neoplastic cells.

The implications of telomerase function for clinical medicine are quite contradictory. Once expressed in human cells that are normally telomerase-negative, hTERT seems to provide cells with full telomerase function and telomere maintenance capability and to contribute to their subsequent immortalization. This enhanced proliferative capacity has been of great interest to gerontologists, whose work is focused on the prevention of aging and the restoration of robust tissues in the aged. Some have proposed that telomere collapse is an important cause of aging. Accordingly, certain researchers have touted hTERT as an obvious solution to the flagging proliferative powers of aging tissues. But such a therapeutic strategy represents a double-edged sword: by supplying aged tissues with full telomerase function, one may inadvertently immortalize premalignant cell clones that are undoubtedly scattered about these tissues and, in so doing, contribute to the appearance of tumors that otherwise would never rear their heads.

Arguably the most direct and compelling proof of the importance of telomeres and telomerase in cancer pathogenesis comes from experiments that will be described in detail in the next chapter: a clone of the *hTERT* gene has been found to be an essential ingredient in the cocktail of introduced genes that are used experimentally to transform normal human cells into tumorigenic derivatives. Without the presence of *hTERT*, human cell transformation fails. (Alternative protocols for transforming human cells arrive at the same endpoint by introducing genes, such as *myc*, that induce expression of the cells' endogenous *hTERT* gene.)

The central role of telomerase in fueling the growth of cancer cells suggests a novel mode of anti-cancer therapy. Telomerase inhibitors should prove to be potent in killing cancer cells that carry relatively short telomeres—the situation observed in many telomerase-positive human tumor cells. Such drugs, if developed, should at the same time have little if any effect on normal cells throughout the body, which express relatively low levels of telomerase and then only transiently during S phase.

Anti-telomerase drugs have been embraced as the long-sought "silver bullets" in the war on cancer, because of their potential ability to kill a diverse array of human cancer cell types while having minimal effect on corresponding normal cells. (For example, the great majority of human prostate carcinomas exhibit readily detected telomerase activity, while normal prostate tissue lacks detectable enzyme activity.) While attractive in concept, this plan may be derailed by certain realities. First, attempts by a number of pharmaceutical companies to develop novel low–molecular-weight inhibitors of hTERT catalytic function have repeatedly failed to yield potent, highly specific drug molecules. The reasons for these failures are still obscure. Second, no one knows whether the relatively low levels of telomerase enzyme detectable in certain types of normal cells throughout the human body play a significant role in the proliferative capacity and viability of these cells. Consequently, the side effects of anti-telomerase drugs, should these agents be developed, are unpredictable.

The failures to generate drug-like inhibitors of hTERT function have led some to embrace an alternative strategy—the construction of a chemically modified 13-nucleotide-long RNA molecule that is complementary to the template region of the *hTR* RNA subunit of the holoenzyme (see Figure 10.24A). This potent inhibitor of the telomerase holoenzyme yields rapid responses in telomerase-positive cancer cells in culture, reducing detectable telomerase activity by 90 to 95%; the testing of its clinical efficacy is under way. Still, the therapeutic utility of this very promising agent, specifically its ability to elicit durable clinical responses, may be frustrated by one of the realities of telomere biology cited above: short-term decreases in tumor burden may be followed by the emergence of cancer cells and thus tumors that have switched to the ALT state of telomere maintenance, thereby becoming hTERT-independent. Accordingly, the goal of exploiting the widespread expression of hTERT in human tumors to generate novel, cancer-specific therapeutics may continue to prove elusive.

Key concepts

- Two barriers prevent cultured cells from replicating indefinitely in culture—senescence and crisis.

- Senescence involves the long-term residence of cells in a nongrowing but viable state; crisis involves the apoptotic death of cells.

- Senescence is provoked by a variety of physiologic stresses that cells experience *in vitro* and also *in vivo*, where it seems to play a significant role in constraining neoplastic progression.

- Crisis is provoked by the erosion of telomeres, which results in widespread end-to-end chromosomal fusions, karyotypic chaos, and cell death.

- Most pre-malignant cells escape from crisis by activating expression of hTERT, the telomerase enzyme, which is specialized to elongate telomeric DNA by extending it in hexanucleotide increments.

- Some cancer cells escape crisis by regenerating their telomeric DNA through the ALT mechanism.

- Cells that have stabilized their telomeres through the actions of telomerase or the ALT mechanism can then proliferate indefinitely and are therefore said to be immortalized.

- Cell immortalization is a step that appears to be a prerequisite to the development of all human cancers.

- The end-to-end chromosomal fusions that accompany crisis lead to repeated breakage–fusion–bridge (BFB) cycles, which appear, in turn, to be responsible for much of the aneuploidy associated with the karyotypes of many kinds of solid human tumors.

- These BFB cycles may prove to be an important means by which incipient cancer cells acquire mutant alleles, thereby expediting the formation of fully neoplastic cells.

Thought questions

1. Why is the acquisition of an immortalized proliferative potential so important for human tumors?

2. What types of evidence connect telomeres and telomerase to entrance into the senescent state, and what alternative mechanisms are responsible for entrance into this state?

3. What complications and side effects might result from the shutdown of telomerase activity by anti-cancer drugs that may be developed in the future and function as specific inhibitors of this enzyme?

4. How does the molecular configuration of the t-loop protect the ends of telomeric DNA?

5. How does the telomerase-associated *hTR* RNA molecule facilitate the maintenance of telomeric DNA by the hTERT enzyme?

6. Precisely how do breakage–fusion–bridge cycles confer an advantage on populations of pre-malignant cells?

Additional reading

Adams PD (2007) Remodeling of chromatin structure in senescent cells and its potential impact on tumor suppression and aging. *Gene* 397, 84–93.

Artandi SE & DePinho RA (2010) Telomeres and telomerase in cancer. *Carcinogenesis* 31, 9–18.

Aubert G & Lansdorp PM (2008) Telomeres and aging. *Physiol. Rev.* 88, 557–579.

Bertuch AA & Lundblad V (2006) The maintenance and masking of chromosome termini. *Curr. Opin. Cell Biol.* 18, 247–253.

Bessler M, Wilson DB & Mason PJ (2010) Dyskeratosis congenita. *FEBS Lett.* 584, 3831–3838.

Blackburn EH & Collins K (2010) Telomerase: an RNP enzyme synthesizes DNA. *Cold Spring Harb. Perspect. Biol.* a003558.

Blanpain C, Horsley V & Fuchs E (2007) Epithelial stem cells: turning over new leaves. *Cell* 128, 445–458.

Blasco MA (2005) Telomeres and human disease: ageing, cancer and beyond. *Nat. Rev. Genet.* 6, 611–622.

Campisi J & d'Adda di Fagagna F (2007) Cellular senescence: when bad things happen to good cells. *Nat. Rev. Mol. Cell Biol.* 8, 729–740.

Cesare AJ & Reddell RR (2010) Alternative lengthening of telomeres: models, mechanisms and implications. *Nat. Rev. Genet.* 11, 319–330.

Collado M & Serrano M (2010) Senescence in tumours: evidence from mice and humans. *Nat. Rev. Cancer* 10, 51–57.

Collins K (2006) The biogenesis and regulation of telomerase holoenzymes. *Nat. Rev. Mol. Cell Biol.* 7, 484–494.

Coppé JP, Desprez PY, Krtolica A et al. (2010) The senescence-associated secretory phenotype: the dark side of tumor suppression. *Annu. Rev. Pathol. Mech. Dis.* 5, 99–118.

de Lange T (2009) How telomeres solve the end-protection problem. *Science* 326, 948–952.

Flores I & Blasco MA (2010) The role of telomeres and telomerase in stem cell aging. *FEBS Lett.* 584, 3826–3830.

Hayflick L & Moorhead PS (1961) The serial cultivation of human diploid cell strains. *Exp. Cell Res.* 25, 585–621.

Hiyama E & Hiyama K (2007) Telomere and telomerase in stem cells. *Brit. J. Cancer* 96, 1020–1024.

Martinez P & Blasco MA (2011) Telomeric and extra-telomeric roles for telomerase and the telomere-binding proteins. *Nat. Rev. Cancer* 11, 161–176.

Mooi WJ & Peeper DS (2006) Oncogene-induced senescence: halting on the road to cancer. *N. Engl. J. Med.* 355, 1037–1046.

Murnane JP (2010) Telomere loss as a mechanism of chromosome instability in human cancer. *Cancer Res.* 70, 4255–4259.

O'Sullivan R & Karlseder J (2010) Telomeres: protecting chromosomes against genome instability. *Nat. Rev. Mol. Cell Biol.* 11, 171–181.

Prieur A & Peeper DS (2008) Cell senescence in vivo: a barrier to tumorigenesis. *Curr. Opin. Cell Biol.* 20, 150–155.

Shay JW & Wright WE (2011) Role of telomeres and telomerase in cancer. *Semin. Cancer Biol.* 21, 349–353.

Stewart SA & Bertuch AA (2010) The role of telomeres and telomerase in cancer research. *Cancer Res.* 70, 7365–7371.

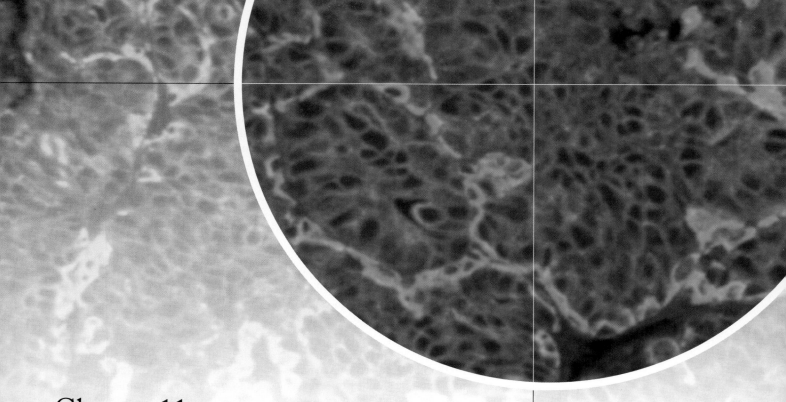

Chapter 11

Multi-Step Tumorigenesis

In the survival of favoured individuals and races, during the constantly-recurring struggle for existence, we see a powerful and ever-acting form of selection.

Charles Darwin, biologist, 1859

The formation of a tumor is a complex process that usually proceeds over a period of decades. Normal cells evolve into cells with increasingly neoplastic phenotypes through a process termed **tumor progression**. This process takes place at myriad sites throughout the normal human body, advancing further and further as we get older. Rarely does it proceed far enough at any single site to make us aware of its end product, a clinically detectable tumor mass.

Tumor progression is driven by a sequence of randomly occurring mutations and epigenetic alterations of DNA that affect the genes controlling cell proliferation, survival, and other traits associated with the malignant cell phenotype. The complexity of this process reflects the work of evolution, which has erected a series of barriers between normal cells and their highly neoplastic derivatives. Accordingly, completion of each step in tumor progression can be viewed as the successful breaching of yet another barrier that has been impeding the progress of a clone of premalignant cells.

One might think that these barriers are the handiwork of relatively recent evolutionary processes. Perhaps, we might imagine, the forces of evolution first worked to design the architecture and physiology of complex metazoan bodies and, having completed this task, then proceeded to tinker with these plans in order to reduce the risk of cancer.

An alternative scenario is far more likely, however: the risk of uncontrolled cell proliferation has been a constant companion of metazoans from their very beginnings, roughly 600 million years ago. By granting individual cells within their tissues the license to proliferate, even simple metazoans ran the risk that one or another of their constituent cells would turn into a renegade and trigger the disruptive, runaway cell multiplication that we call cancer. Consequently, the erection of defenses against cancer must have accompanied, hand-in-hand, the evolution of organismic complexity.

Most of the preceding chapters have addressed one or another of the individual cellular control systems that defend against cancer and are subverted during the process of tumorigenesis. Now, we will begin tying these individual threads together and examine how alterations in these systems contribute to the end product—the formation of a primary tumor. We start by attempting to gauge the scope of the problem at hand: How many different sequential changes are actually required in cells and tissues in order to create a human cancer?

11.1 Most human cancers develop over many decades of time

Epidemiologic studies have shown that age is a surprisingly large factor in the incidence of cancer. In the United States, the risk of dying from colon cancer is as much as 1000 times greater in a 70-year-old man than in a 10-year-old boy. This fact, on its own, suggests that this type of cancer and, by extension, many other cancers common in adults (**Figure 11.1**) require years if not decades to develop.

A more direct measure of this is provided by the incidence of lung cancers among males in the United States. Cigarette smoking was relatively uncommon among this group until World War II, when large numbers of men first acquired the habit, encouraged in part by the cigarettes they received as part of their rations while serving in the armed forces. Thirty years later, in the mid-1970s, the rate of lung cancer began to climb steeply (**Figure 11.2**). At the same time, cigarette smoking spread throughout the world and peaked in 1990, leading to estimates that global lung cancer mortality, which currently exceeds 1.1 million deaths annually, will peak only sometime in the mid-twenty-first century. In these other areas of the world outside the United States, the approximately 30-year lag between marked increases in smoking and the onset of large numbers of lung cancers seems to apply as well.

The late onset of most cancers has an important and unexpected public health implication: curing all cancers will have a relatively small effect on expected life span. A decade or more before most individuals have lived long enough to contract cancer, they often succumb to other maladies, such as heart disease. A calculation based on cancer-related deaths in the Netherlands in 1990 indicates that if all such deaths were eliminated, this would extend life expectancy only 3.83 years for men and 3.38 years for women.

Epidemiologists have developed formulas that predict the frequency of various cancers as a function of age. These formulas indicate that, depending upon the cancer type, disease **incidence** (the rate at which the disease is diagnosed) and mortality rates occur as a function of a^4 to a^7, where a represents the age of patients at initial diagnosis. For epithelial cancers as a whole, the risk of death from cancers increases approximately as the fifth or sixth power of elapsed lifetime (**Figure 11.3**). Algebraic

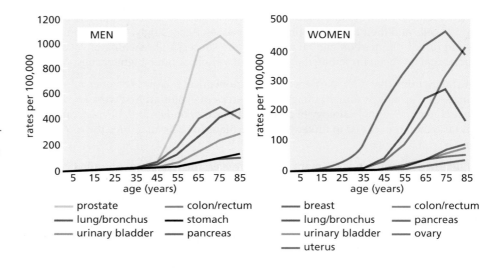

Figure 11.1 Cancer incidence at various ages These graphs of diagnoses of various types of epithelial cancers show a steeply rising incidence with increasing age, indicating that the process of tumor formation generally requires decades to reach completion. (Courtesy of W.K. Hong, compiled from *SEER Cancer Statistics Review.*)

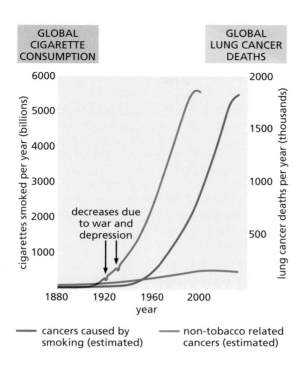

GLOBAL CIGARETTE CONSUMPTION

GLOBAL LUNG CANCER DEATHS

decreases due to war and depression

— cancers caused by smoking (estimated)

— non-tobacco related cancers (estimated)

Figure 11.2 Cigarette consumption and lung cancer These curves compare the annual global consumption of cigarettes (*red curve*) with the recorded and predicted annual worldwide mortality from lung cancer (*green curve*). Annual mortality from tobacco-induced lung cancer (in thousands) is estimated to peak sometime in the fourth or fifth decade of the twenty-first century. During the twentieth century, there was an increase in what is judged to be "non-tobacco-related" lung cancer mortality (*blue curve*). The precise number of these cases is unclear, and there is some debate about how many of them are attributable to secondhand tobacco smoke. (From R.N. Proctor, *Nat. Rev. Cancer* 1:82–86, 2001.)

functions like these are interpreted quite simply. If the probability of an outcome is indicated by a^n, this means that $n + 1$ independent events, each occurring randomly and with comparable probability per unit of time, must take place before the ultimate outcome—in this case a diagnosed tumor—is achieved. The fact that the incidence of a disease like colon cancer begins to shoot upward only in the seventh and eighth decades of life indicates that each of these events occurs with a very low probability each year. More specifically, each event is likely to occur on average once every 10 to 15 years, and the entire succession of events may usually require 40 to 60 years to reach completion.

These calculations provide us with only a very rough estimate of the complexity of tumor progression. In fact, some critical steps may occur far more rapidly than others but not register in these calculations of the kinetics of tumor progression, which reflect only the influence of the slowest, "rate-limiting" steps. For this reason, it seems likely that the actual number of events required to form most tumors is actually higher than is predicted by use of this a^n formula. [Consider the fact that later steps in multi-step

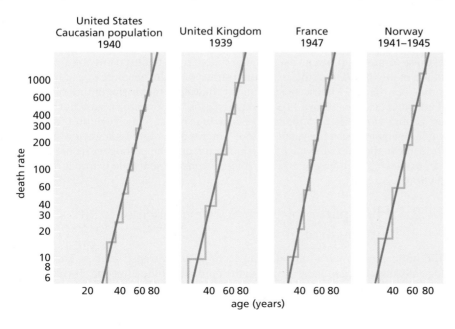

Figure 11.3 Age at death from various epithelial cancers The graphs indicate the general mortality from cancer from 1939 to 1947 in four countries where public health statistics were kept with some precision. These log–log plots of male cancer death rates (deaths per 100,000 population, *ordinate*) at different ages (*abscissa*) have a slope indicating that completion of five or six rate-limiting events is required to produce a lethal cancer. For example, a slope of 5 indicates six rate-limiting events, whose nature is not revealed by these analyses. Interestingly, these slopes are remarkably similar among different countries. The slopes differ slightly with different types of cancer (*not shown*). (From A.G. Knudson, *Nat. Rev. Cancer* 1:157–162, 2001; reproduced from C.E. Nordling, *Br. J. Cancer* 7:68–72, 1953.)

tumor progression are likely to require less time to complete than earlier steps because (i) the cells in later growths, having acquired multiple oncogenic mutations, have learned how to proliferate faster, decreasing the time of each clonal expansion (see Figure 11.15); and (ii) the genomes of cells are likely to have become more mutable (as discussed in Chapter 12).]

Additionally, the strong age dependence of almost all cancers forces epidemiologists to calculate *age-adjusted incidence*, which corrects for the fact that the age distributions of different human populations differ markedly. A calculated age-adjusted incidence can tell us, for instance, what the probability is of a 60-year-old woman in the United States contracting breast cancer during the course of a year compared with the risk experienced by a woman of the same age living in Egypt, Kazakhstan, or Portugal. At the same time, age-adjusted incidence measurements enable us to make meaningful comparisons in a specific population over an extended period of time—for example, the age-adjusted colon cancer risk of the U.S. population in 1930 compared with that of the successor population in 1990.

The conclusion that tumorigenesis is a multi-step process hints at another interesting idea. Assume that (1) a sequence of unlikely events is required in order for a tumor to appear and (2) many of these events happen at comparable frequencies in all of us. Together, these assumptions indicate that as we grow older, virtually all of us will carry populations of cells in many locations throughout the body that have completed some but not all of the steps of tumor progression. Since most of us will not live long enough for the full schedule of requisite events to be completed (because we will succeed in dying from another disease), we will never realize that any of these tumor progressions had been initiated in our bodies. (Indeed, autopsies reveal that by life's end, 60 to 70% of individuals carry undiagnosed tumors, independent of the cause of death.) Viewed from this perspective, cancer is an inevitability; if we succeeded in avoiding the death traps set by all the other usual diseases, sooner or later most of us would become victims of cancer.

Of course, some—though not most—of us will actually contract a neoplastic disease such as colon cancer. This suggests something else: while the a^5 or a^6 expression may predict colon cancer incidence averaged over a large human population, the probability of the rate-limiting pathogenic events occurring per unit of time varies dramatically from one individual to another, being affected by hereditary predisposition, diet, and lifestyle, among other variables. Moreover, this algebraic formula does not explain the peculiar age-dependent incidence of certain types of cancer that do not increase progressively with increasing age in parallel with most common adult cancers (**Sidebar 11.1**).

Epidemiology provides us with another important insight into the multi-step nature of tumorigenesis. If we examine the frequencies of mesothelioma in humans (caused largely by asbestos exposure plus smoking) and skin cancer in mice (induced by repeated benzo[*a*]pyrene painting), it becomes apparent that the formation of each of these tumors requires an extended period of repeated exposure to carcinogens and that it is the *duration* of this exposure (rather than the absolute *age* of exposed individuals or the age when exposure began) that determines the timing of the onset of detectable disease (**Figure 11.5**). In these cases, tumors are created by the actions of exogenous carcinogens rather than occurring spontaneously within the body; these carcinogens increase the rate of tumor progression, often by many orders of magnitude above the spontaneous background rate, and the pathogenesis of each of these tumors seems to require a predetermined exposure schedule before it reaches completion.

11.2 Histopathology provides evidence of multi-step tumor formation

The notion of human tumor development as a multi-step process has been histologically documented most clearly in the epithelia of the intestine. The intestinal epithelial cells, which face the interior cavity (lumen) of the gastrointestinal tract, form a layer that is only one cell thick in many places (**Figure 11.6**). These epithelial cell

Sidebar 11.1 The age-dependent incidence rates of certain cancers differ dramatically from typical adult cancers As Figure 11.1 illustrates, the incidence of most types of cancer increases progressively with age and incidence rates dramatically increase beginning in the sixth decade of life. However, there are certain cancers that clearly violate this rule. As is also apparent in this figure, breast cancer rates increase dramatically more than a decade earlier than the other major types of adult cancers. Yet other types of cancers exhibit even more peculiar kinetics (**Figure 11.4**).

These disparate behaviors indicate that factors other than increasing age must be invoked in order to understand the mechanisms governing the kinetics of tumor formation and thus incidence. One critical factor is the size, as a function of age, of the normal cell population that can serve as the precursor of cancer cells. In the case of retinoblastoma, this appears to be a population of undifferentiated retinal precursor cells that differentiate early in life into post-mitotic cells that can no longer be transformed, effectively precluding tumor formation beyond the age of 6 or 7; see also Supplementary Sidebar 8.6. (Indeed, ~85% of retinoblastoma cases are diagnosed by the age of 2.) In the case of testicular cancer, the biological reasons are less clear but also seem to involve a population of target cells that changes dramatically in size relatively early in life.

Another factor is exposure to carcinogenic influences, which may increase cancer risk during one stage of development but have relatively little effect earlier or later in life. For example, women who were exposed to the radiation from atomic bombs dropped on Hiroshima and Nagasaki and were 10–19 years old at the time of the bombings showed a >7-fold increased risk of eventually developing breast cancer, while those who were 30–39 years of age at the time exhibited only a ~2.6-fold increased risk.

Retinoblastoma is only one of a series of pediatric cancers whose pathogenesis clearly follows rules that are very different from those governing adult cancers; their accelerated formation suggests the involvement of far fewer steps than those governing the cancers discussed later in this chapter. These issues are largely unresolved and represent an area of active investigation.

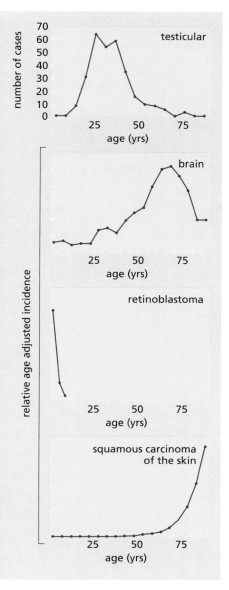

Figure 11.4 Cancer incidence as a function of age These graphs indicate that a subset of cancers show age-dependent incidence rates that vary dramatically from the kinetics of appearance of most common cancers. With the exception of testicular cancer, the ordinates in all cases represent incidence relative to the highest incidence rate. Squamous cell skin carcinoma is included to illustrate the incidence kinetics of a cancer that is more typical of adult-onset neoplasias. (Courtesy of B. Westermark, Swedish Cancer Registry; from J. Li et al., *Pediatrics* 121:e1470–e1477.)

populations are in constant flux. Each minute, 20 to 50 million cells in the human duodenum and a tenth as many in the colon die and an equal number of newly minted cells replace them!

Underlying these epithelia is a basement membrane (basal lamina) to which these cells are anchored. As is the case in other epithelial tissues, this basement membrane forms part of the extracellular matrix and is assembled from proteins secreted by both epithelial cells above and stromal cells lying beneath the membrane (see Figure 2.3). The mesenchymal cells composing the stroma are largely fibroblasts; but other cell types, including endothelial cells, which form the walls of both capillaries and lymphatic vessels, and immune cells, such as macrophages and mast cells, are also scattered about. Lying beneath this layer of stromal cells in the intestinal wall is a thick layer of smooth muscle responsible for moving along the contents of the colonic lumen through periodic contractions.

The epithelial layer is the site of most of the pathological changes associated with the development of colon carcinomas. Analyses of human colonic biopsies have revealed a variety of tissue states, with degrees of abnormality that range from mildly deviant tissue, which is barely distinguishable from the structure of the normal intestinal

Figure 11.5 Cancer incidence and carcinogen exposure These graphs indicate that cumulative exposure to a carcinogenic stimulus, rather than the age at which this exposure began, determines the likelihood of developing a detectable tumor. The left graph presents the cumulative risk of developing mesothelioma (a tumor of the mesodermal lining of the abdominal organs and the lungs) among insulation workers in the United States, many of whom were occupationally exposed to asbestos. The right graph presents the cumulative risk of developing a skin tumor among mice treated in an experimental protocol for inducing squamous cell carcinomas of the skin. (From J. Peto, *Nature* 411:390–395, 2001.)

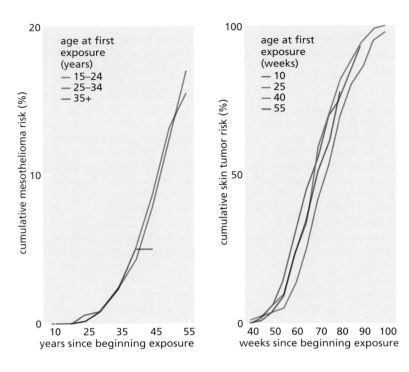

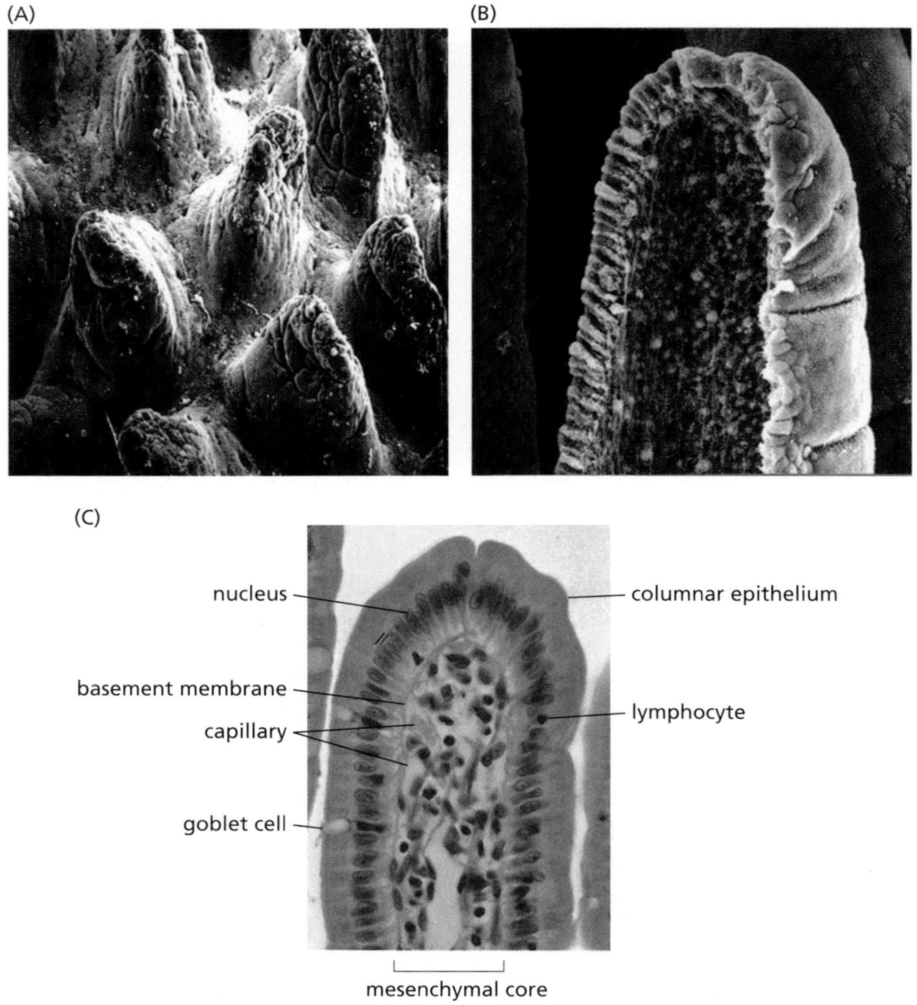

Figure 11.6 Microanatomy of the normal intestinal wall The organization of the epithelial cells lining the small intestine and the large intestine (colon) is very similar. However, the protruding villi visualized here are present only in the small intestine. (A) This scanning electron micrograph of epithelium of the small intestine shows villi (fingers) of the intestinal mucosa extending into the lumen at regular intervals. (B) Each villus is covered with a layer of epithelial cells. (C) The core of each villus, which is separated from the overlying epithelium by a basement membrane, is composed of various types of mesenchymal cells, including fibroblasts, endothelial cells, pericytes, and various cells of the immune system (*not indicated*). (A and B, courtesy of S. Canan. C, from University of Iowa Virtual Hospital, Atlas of Microscopic Anatomy, Plate 194.)

mucosa (the lining of the colonic lumen), to the chaotic jumble of cells that form highly malignant tissue (**Figure 11.7**). Like the normal intestinal lining, these growths are composed of a variety of distinct cell types, indeed almost all of the cell types found in the normal tissue.

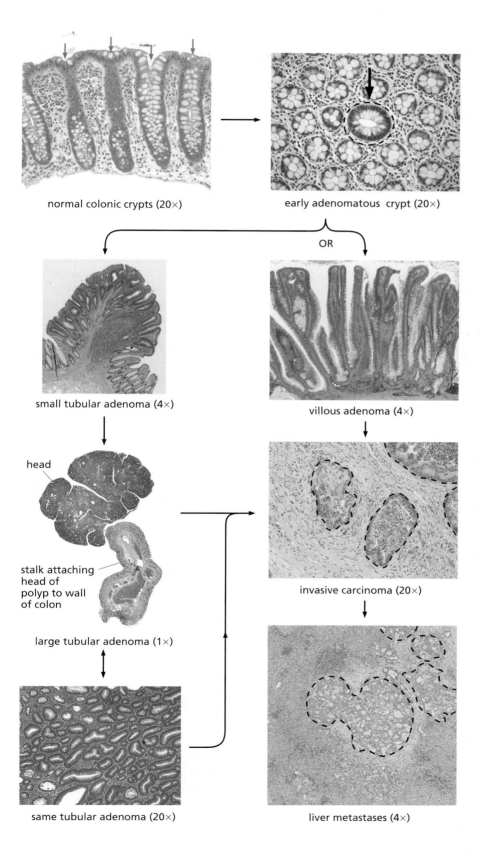

normal colonic crypts (20×)

early adenomatous crypt (20×)

OR

small tubular adenoma (4×)

villous adenoma (4×)

head

stalk attaching head of polyp to wall of colon

large tubular adenoma (1×)

invasive carcinoma (20×)

same tubular adenoma (20×)

liver metastases (4×)

Figure 11.7 Histopathological alterations of the human colon
The various types of abnormal tissues revealed by histopathological analyses of the human colon can be arrayed in a succession of ever-increasing abnormality. Normal colonic crypts are seen here in longitudinal section (*top left*). (Very similar crypts are found in the small intestine around the base of each villus.) A small adenomatous crypt (*arrow, circled*) is shown, together with normal crypts, in cross section (*top right*). The cells in this abnormal crypt are presumed to have the ability to develop into at least two distinct types of adenomas—tubular and villous. A small tubular adenoma is shown (*left, middle*). The larger tubular adenoma (*below*) is sometimes termed *pedunculated*, indicating its attachment via a stalk to the colonic wall. These two types of adenomas have the ability to evolve into a locally invasive carcinoma, whose presence here is indicated by small islands of carcinoma cells (*circled*) surrounded by extensive stroma (*right, lower middle*). Dissemination of these cells, usually to the liver (*bottom right*), can lead to metastases (*circled*), which are surrounded here by layers of recruited stromal cells. The histopathological progression indicated here would seem to be the most logical way by which normal tissue, in this case the colonic epithelium, is transformed through a series of intermediate steps into carcinomas and ultimately spawns metastatic growths. However, the evidence for most of these precursor–product relationships is actually quite fragmentary. (Courtesy of C. Iacobuzio-Donahue and B. Vogelstein.)

Some growths that are classified as *hyperplastic* exhibit almost normal histology, in that the individual cells within these growths have a normal appearance. However, it is clear that in these areas of hyperplasia, the rate of epithelial cell division is unusually high, yielding thicker-than-normal epithelia. Yet other growths show abnormal histologies, with the epithelial cells no longer forming the well-ordered cell layer of the normal colonic mucosa and the morphology of individual cells deviating in subtle ways from that of normal cells; these growths are said to be *dysplastic* (see Figure 2.15). A much larger and more deviant growth that has dysplastic cells and marked thickening is termed a *polyp* or an *adenoma* (see Figures 2.16A and 11.7). In the colon, several distinct types of polyps are encountered; some grow along the wall of the colon, while others are tethered to the colonic wall by a stalk. Importantly, all these growths are considered benign, in that none has broken through the basement membrane and invaded underlying stromal tissues.

The more abnormal growths that have invaded through the basement membrane and beyond are considered to be malignant. There are distinctions among these more aggressive colon carcinomas and associated cancer cells, depending on whether they have penetrated deeply into the stromal layers and smooth muscle and whether cells from these growths have migrated—metastasized—to anatomically distant sites in the body, where they may have succeeded in founding new tumor cell colonies.

Having arrayed these growths in a succession of tissue phenotypes that advance from the normal to the aggressively malignant (see Figure 11.7), we might imagine that this succession depicts with some accuracy the course of actual tumor development as it occurs in the human colon. In truth, the evidence supporting this scheme is quite indirect. Some tumors may well develop through a series of intermediate growths, most of which are arrayed here. Alternatively, it is possible that some of the tissue types depicted as intermediates in this sequence represent dead ends rather than stepping stones to more advanced tumors. In certain cases of colon cancer, it is also possible that the development of the tumor depended on the ability of early growths to leapfrog over intermediate steps, allowing them to arrive at highly malignant endpoints far more quickly than is suggested by this succession. Similar successions have been proposed for a variety of other epithelial cancers (**Figure 11.8A**). On rare occasions, one can actually find a tissue in which multiple stages of cancer progression coexist (see Figure 11.8B); however, observations like these provide no direct evidence of precursor–product relationships between the various types of histologically abnormal growths.

In the case of colorectal tumors, at least three types of evidence strongly support the precursor–product relationship between adenomas and carcinomas. First, on rare occasions, one can actually observe a carcinoma growing directly out from an adenomatous polyp (**Figure 11.9A**). We can surmise that outgrowths like these occur routinely during the development of virtually all colon carcinomas and that, more often than not, the rapid expansion of the carcinoma soon overgrows and obliterates the adenoma from which it arose.

Second, clinical studies have been performed on large cohorts of patients who have undergone colonoscopy, which is performed routinely to survey the colon for occasional adenomatous polyps and to remove any that are detected. In one such study, those patients whose polyps were removed experienced, in subsequent years, about an 80% reduction in the incidence of colon carcinomas (see Figure 11.9B). This indicates that in this patient population, at least 80% of colon carcinomas derive from preexisting, readily detectable adenomas; because colonoscopy may have missed some polyps, the actual proportion may be higher. (These observations still do not prove that every single colon carcinoma arising in humans must arise from a preexisting adenoma.)

A third type of evidence supporting the precursor–product relationship between premalignant and frankly malignant growths comes from *longitudinal* studies of individual patients over a period of time. This is occasionally possible in patients developing sporadic tumors in organs that are accessible to repeated surveillance, such as the skin, colon, and lungs (for example, see Figure 11.9C). Familial cancer syndromes also

(A)

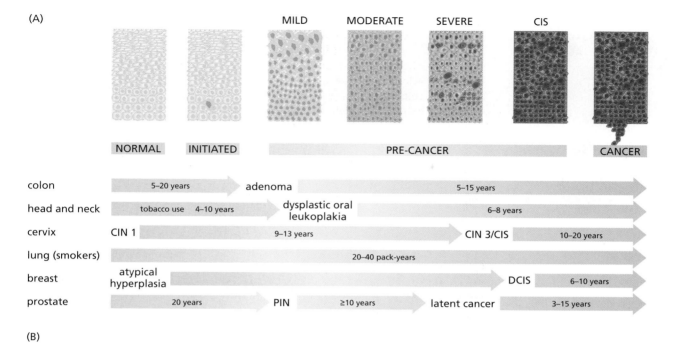

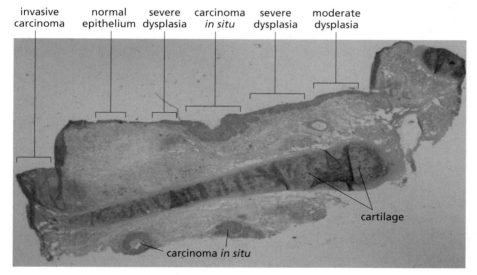

(B)

Figure 11.8 **Multi-step tumorigenesis in a variety of organ sites** (A) The pathogenesis of carcinomas is thought to be governed by very similar biological mechanisms operating in a variety of epithelial tissues. Accordingly, multi-step tumorigenesis involving similar histological entities has been proposed to progress along parallel paths in these various organ sites. These similarities are obscured by the fact that the nomenclature is quite variable from one tissue to another. CIS, carcinoma *in situ*; CIN, cervical intraepithelial neoplasia; DCIS, ductal carcinoma *in situ*; PIN, prostatic intraepithelial neoplasia. (B) Because more aggressive growths often overgrow their more benign precursors, it is rare to see the multiple states of tumor progression coexisting in close proximity, as is the case in this lung carcinoma. Importantly, while the close juxtaposition of the aberrant growths suggests some relationship among them, an image like this provides no definitive evidence of precursor–product relationships between these various abnormal tissues. (A, courtesy of W.K. Hong. B, courtesy of A. Gonzalez and P.P. Massion.)

offer a demonstration of such progression in individual patients. Recall, for example, the disease of familial adenomatous polyposis (FAP; see Section 7.11), in which an individual inheriting a mutant form of the *APC* tumor suppressor gene is prone to develop anywhere from dozens to more than a thousand polyps in the intestine (see Figure 7.22). With a certain low but measurable frequency, one or another of these polyps will progress spontaneously into a carcinoma. (The multiplicity of polyps in these patients and the low conversion rate of polyps to carcinomas—estimated to be ~2.5 events per 1000 polyps per year—effectively preclude association of a carcinoma with a particular precursor polyp.)

The development of carcinomas in other organ sites throughout the body is thought to resemble, at least in outline, the multi-step progression observed in the colon (see, for example, Figure 11.8A). Many of these other tissues, such as the breast, stomach, lungs, prostate, and pancreas, also exhibit growths that can be called hyperplastic, dysplastic, and adenomatous, and these growths would seem to be the benign precursors of the carcinomas that arise in these organs. However, the histopathological

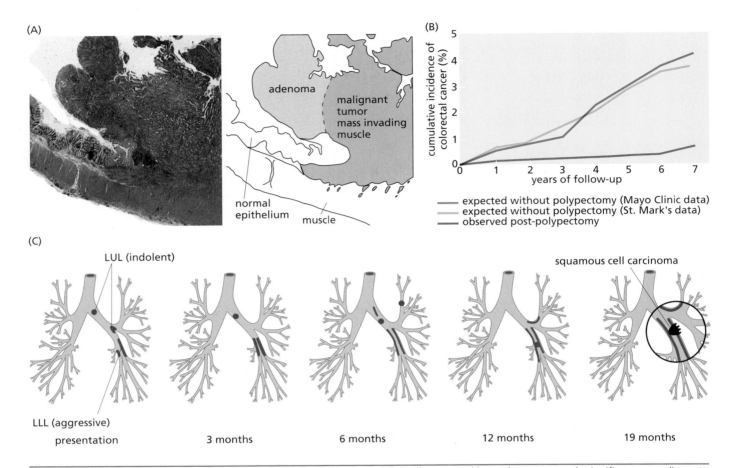

Figure 11.9 Evidence for adenoma-to-carcinoma progression
The evidence for precursor–product relationships between premalignant growths (adenomas and *in situ* carcinomas) and frank carcinomas comes from at least three types of clinical observations. (A) On occasion, carcinomas are observed to be growing directly out of adenomas. As seen here, the demarcation between the two can sometimes be drawn with some precision. (B) A second line of evidence derives from clinical studies in which colonoscopy is used to screen large cohorts of patients. Any polyps that are discovered are removed by *polypectomy* (i.e., surgically). As seen in the graph, two independent studies (*yellow-orange, blue lines*) predict a certain expected number of carcinomas in such cohorts on the basis of historical experience. However, in patients who have undergone polypectomy, the number of colorectal cancers diagnosed in subsequent years has been reduced by more than 80% (*red line*).

(C) As diagrammed here, three apparently significant premalignant carcinomas *in situ* were detected by *bronchoscopy* in the lungs of a patient upon initial clinical presentation. By 19 months, the lesion in the left lower lobe (LLL; the left side of the lung is depicted here as might be seen from the front of the patient and is located from this perspective in the right side of the lung) had developed into a frank squamous cell carcinoma (*circle*), while the two in the left upper lobe (LUL) were no longer apparent. Interestingly, the three initially characterized lesions bore the same relatively rare *p53* mutation, indicating a common origin in spite of their physical separation; this mutation was also present in the subsequently arising carcinoma. (A, courtesy of Paul Edwards. B, courtesy of W.K. Hong, data from S. Winawer et al., *N. Engl. J. Med.* 329:1977–1981, 1993. C, from N.A. Foster et al., *Genes Chromosomes Cancer* 44:65–75, 2005.)

evidence supporting multi-step tumor progression in these tissues is, for the most part, less well developed than for the colon—a consequence of the accessibility of the colon through colonoscopy and the relative inaccessibility of these other tissues. (An exception to this rule is tumor formation in the pancreas, discussed below.) In the case of nonepithelial tissues, including components of the nervous system, the connective tissues, and the hematopoietic system, the histopathological evidence supporting multi-step tumor progression is even more fragmentary.

In fact, the altered histopathology of premalignant and malignant growths is largely a reflection of changes that have occurred in a subset of the cells forming each of these masses. In the case of carcinomas, the progressive alteration of epithelial cells is assumed to drive the process of tumor progression and the associated changes in histopathology. The other cell types in these tumor masses, specifically those in the stroma, are often depicted as innocent bystanders—normal cells that have been recruited to the tumor mass and co-opted by abnormal epithelial cells to help in the process of tumor formation. But as we will see in Chapter 13, the role of these ostensibly normal cells in tumor formation is far more complex and interesting.

11.3   Cells accumulate genetic and epigenetic alterations as tumor progression proceeds

The forces driving tumor progression in various tissues are hardly obvious from the histopathological descriptions cited above. It is possible, for example, that changes in gene expression programs are largely responsible for the progressive changes in cell and tissue phenotype as a tumor develops. Thus, some of the steps in tumor progression might recapitulate specific changes in cell behavior that occur normally during embryogenesis. We know that virtually all of the steps in embryological development depend on changes in programs of gene expression rather than alterations in the genome itself. (The cloning of Dolly the sheep proved this point dramatically, since it demonstrated that the genome of a highly differentiated mammary epithelial cell is essentially identical to that of a recently fertilized egg, see Supplementary Sidebar 1.2.)

According to the alternative model, many of the steps of tumor progression are driven by heritable alterations accumulated in the genomes of developing tumor cells, notably those created by somatic mutations and promoter methylation events. This second scenario is supported by extensive evidence garnered over the past several decades, which documents the accumulation of increasing numbers of mutant genes in cells as they evolve from a more benign to a more malignant growth state. We have encountered many of these mutant genes in earlier chapters.

This parallel between genetic evolution and phenotypic progression was initially documented for human colon carcinomas. We know so much about the genetic basis of tumor progression in this organ because the colonic epithelium is, as mentioned, relatively accessible through colonoscopy; because colon cancer is a common disease in the West, making it relatively easy to collect and study large numbers of samples of premalignant and malignant growths; and because of researchers at the Johns Hopkins Medical School in Baltimore, Maryland, who focused on understanding these tumors and their genetic origins.

These researchers recognized the extensive evidence that mutant alleles of genes such as *ras* and *p53* can contribute to cell transformation under experimental conditions *in vitro*. They therefore sought to determine whether they could find *in vivo* correlates by examining the genomes of sizable groups of small colonic adenomas, mid-sized adenomas, large adenomas, and frank carcinomas. It was plausible that as colonic tissues advanced progressively from normalcy to high-grade malignancy, the epithelial cells in these various tissues would accumulate increasing numbers of mutations in various genes.

This is just what these scientists discovered in the late 1980s. The genes that they examined included the K-*ras* oncogene and a number of tumor suppressor genes. In fact, the identities of the tumor suppressor genes that participate in colon cancer pathogenesis were not known when they began their work, so they searched instead for chromosomal regions that suffered loss of heterozygosity (LOH) during tumor progression. Recall that a high rate of LOH in a particular chromosomal region provides strong indication that this region harbors a tumor suppressor gene, and that developing cancer cells exploit the LOH mechanism as a means of shedding the still-functional (that is, actively transcribed, wild-type) alleles of such a tumor suppressor gene (see Section 7.4).

In landmark work, this research discovered that early-stage adenomas often showed loss of heterozygosity in the long arm of Chromosome 5 (that is, 5q). Almost half of slightly larger adenomas carried, in addition, a mutant K-*ras* oncogene. Even larger adenomas tended to also have high rates of LOH on the long arm of Chromosome 18 (that is, 18q); and about half of all carcinomas showed, in addition, an LOH on the short arm of Chromosome 17 (that is, 17p; **Figure 11.10**; see also Figure 7.13B).

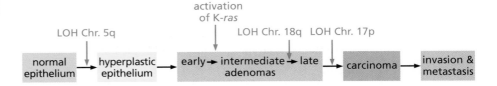

Figure 11.10 Colon tumor progression and loss of heterozygosity in various chromosomal arms DNA from tissue samples representing various stages of colon cancer progression was analyzed for loss of heterozygosity (LOH) by examining the behavior of chromosomal markers present on the long and short arms of most chromosomes, as described in Figure 7.13B. In general, each chromosomal arm was represented by one or more genetic loci that existed in heterozygous form in the normal tissue of a patient. As tumor progression proceeded, the involved colonic epithelial cells exhibited increasing numbers of chromosomal arms that had lost heterozygosity. Certain chromosomal regions (*blue lettering*) suffered especially high rates of LOH, suggesting that they carried tumor suppressor genes that were being inactivated, in part, through the LOH mechanism. Analyses of K-*ras* oncogene activation (*pink lettering*) indicated additional changes in the genomes of evolving premalignant cells. (Adapted from B. Vogelstein et al., *Science* 244:207–211, 1989.)

These observations provided strong support for the idea that as epithelial cells acquire increasingly neoplastic phenotypes during the course of tumor progression, their genomes show a corresponding increase in the number of altered genetic loci. Equally important, these changes involve both the activation of a proto-oncogene into an oncogene (K-*ras*) and the apparent inactivation of at least three distinct tumor suppressor genes. This was the first suggestion of a phenomenon that is now recognized to be quite common: the number of inactivated tumor suppressor genes greatly exceeds the number of activated oncogenes present in a human tumor cell genome.

The identities of the genetic loci on Chromosomes 5 and 17 were revealed soon after this genetic progression was laid out (**Figure 11.11**). The Chromosome 5q21 gene that is often the target of LOH was found to be the *APC* (adenomatous polyposis coli) tumor suppressor gene (see Section 7.11), while the Chromosome 17p13 gene was identified as the *p53* tumor suppressor gene (see Chapter 9). The identity of the gene or genes on Chromosome 18q that are inactivated during colon carcinoma pathogenesis remains unclear. While this chromosomal region suffers loss of heterozygosity in more than 60% of human colon cancers, the best candidate tumor suppressor gene here is *DPC4/MADH4*, which specifies Smad4, the protein that relays growth-inhibitory signals from the TGF-β receptor to the cell nucleus (see Figure 6.29D). However, the copies of *DPC4/MADH4* are present in mutant or transcriptionally repressed form in fewer than 35% of colon cancers, and a second, nearby gene specifying Smad2 is inactivated even less frequently. This leaves unanswered the identity of the tumor suppressor gene(s) on Chromosome 18q that are the targets of inactivation and LOH in the majority of these cancers.

The most obvious way to rationalize the steps in colon cancer development involves an ordered succession of genetic changes that strike the genomes of colonic epithelial cells as they evolve progressively toward malignancy, as illustrated in Figure 11.11. In reality, the specific sequence of genetic changes depicted in this figure presents the history of only a very small proportion of all colonic tumors. This number can be calculated from the known rates of genetic alteration of the various genes or chromosomal regions known to participate in colon cancer pathogenesis. While the great majority (~80%) of colon carcinomas suffer inactivation of the *APC* gene on Chromosome 5q as an early step in this process, only about 35% acquire a K-*ras* mutation, less than half show an LOH on Chromosome 17p involving *p53*, and about 60% show an LOH on Chromosome 18q that affects still-unidentified tumor suppressor genes. In addition, as many as 30% of colonic tumors have mutations or promoter methylation events that lead to functional inactivation of the type II TGF-β receptor (see Section 5.7). A further complication is created by the observation that tumors bearing K-*ras* oncogenes rarely have mutant *p53* alleles, and vice versa.

Figure 11.11 Tumor suppressor genes and colon carcinoma progression Each of the chromosomal regions that underwent loss of heterozygosity (LOH; see Figure 11.10) was considered to harbor a tumor suppressor gene (TSG) whose loss provided growth advantage to evolving pre-neoplastic colonic epithelial cells. Eventually two TSGs were identified (*blue lettering*)—*APC* on Chromosome 5q, and *p53* on Chromosome 17p. The identity of the inactivated TSG(s) on Chromosome 18q remains unclear. About half of colon carcinomas also had mutant, activated alleles of the K-*ras* gene (*pink lettering*), and the genomes of most evolving, pre-neoplastic growths were found to suffer widespread hypomethylation (loss of methylated CpGs; *green lettering*). The precise contribution of hypomethylation to tumor progression remains unclear; some evidence suggests that it creates chromosomal instability.

The great majority of colon cancers will therefore begin with a Chromosome 5 alteration, but thereafter will take alternative genetic paths on the road toward full-fledged malignancy (**Figure 11.12A**). These other paths will presumably involve a comparable

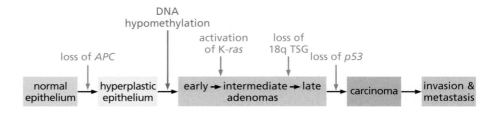

number of genetic alterations. It seems plausible, for example, that those colon carcinomas that do not acquire a mutant K-*ras* oncogene will sustain alterations in other components of the Ras signaling pathway. Presumably, these alternative mutations confer a cell-physiologic advantage on colon cancer cells similar to that resulting from the formation of a mutant, activated K-*ras* oncogene.

This suspicion has been borne out by the systematic sequencing of 90 distinct kinase genes present in the genomes of a group of 35 colorectal carcinomas. About 15% of these tumors were found to carry mutant, presumably activated tyrosine kinase enzymes; another genetic study reported mutant forms of the 110-kD subunit of PI3 kinase (PI3K; see Section 6.6) in about one-third of colorectal tumors analyzed. Yet another analysis, this time of colonic polyps, revealed that those growths lacking activated K-*ras* oncogenes often carried mutant, oncogenic alleles of the B-*raf* gene, which, like its *raf* cousin, encodes a serine/threonine kinase (see Section 6.5); significantly, the mutant *ras* and B-*raf* oncogenes seem to occur in a mutually exclusive fashion, never being found together in the same polyps. These various mutant kinases are excellent candidates for the agents that serve as functional alternatives to the K-*ras* oncoprotein during colon carcinoma progression.

It is also true that the four alterations documented by the Johns Hopkins researchers, even when they define the history of a tumor, do not always occur in the *order* described in Figure 11.11. The loss of heterozygosity on Chromosome 5q is almost

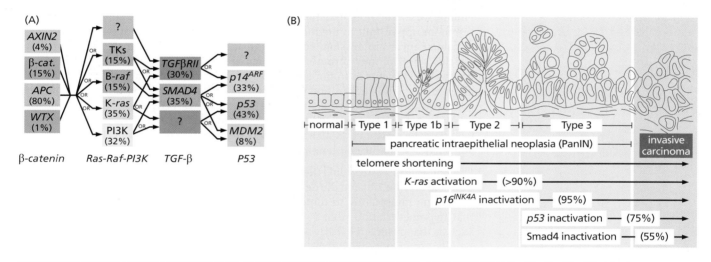

Figure 11.12 Alternative paths during cancer progression
(A) The series of genetic alterations shown in Figure 11.11 does not accurately represent an invariant sequence during the progression of all colorectal carcinomas. Loss of APC function (or functionally equivalent alterations, *left column*) represents a starting point that is common to almost all human colon carcinomas, ostensibly because the resulting mutant cells are trapped in the colonic crypts; all of these changes seem to potentiate β-catenin function and to generate the clusters of aberrant crypts (see Figure 7.24 and 11.23C) encountered in colonoscopy. However, the identities of the genes altered in subsequent steps are variable, as is the precise order of these changes. (For the sake of clarity, the gene names given here reflect the names of their respective protein products.) As shown, the second step might involve any one of five alternative genetic changes, and alternative decision paths may operate in subsequent steps as well, creating far more alternative paths than the ones depicted here. Moreover, the exact number of genetic steps and epigenetic steps (notably promoter methylation) occurring during human colon cancer pathogenesis is not known. These changes have been grouped functionally (so that the *second column* shows the Ras pathway; the *third column*, the TGF-β

pathway; and the *fourth column*, the p53 pathway), but only fragmentary evidence supports the notion that a change within a tumor of one member of a functional group is mutually exclusive with a change in another member of the same functional group. Moreover, there is no direct evidence that all four of these pathways must be altered in order for a human colorectal carcinoma to form. The percentages indicate the proportion of colorectal carcinomas that exhibit a change in the gene in question, including activating mutations in proto-oncogenes and inactivating changes in tumor suppressor genes. (B) The most orderly succession of genetic and epigenetic changes documented to date occurs during the development of common adenocarcinomas of the pancreas; i.e., as these growths progress through increasingly high histologic grades, almost all acquire an activated K-*ras* oncogene early in tumor progression, followed thereafter by loss of *p16^INK4A* and then *p53* tumor suppressor gene function. Of note, the great majority of the key changes driving pancreatic carcinoma progression occur before the precursor lesions (pancreatic intraepithelial neoplasias; PanINs) undergo conversion into invasive carcinomas. (A, courtesy of Y. Niitsu. B, courtesy of R. Hruban.)

always the first in the progression, but the precise order of the subsequent changes may vary from tumor to tumor. This unique role of the *APC* locus as the site of the first genetic step in colon cancer pathogenesis seems to be dictated by the biological effects of *APC* inactivation.

As described in Section 7.11, once normal colonic enterocytes are formed, they migrate out of the colonic crypts, differentiate, and die within 3 to 4 days by apoptosis. Consequently, most mutations that strike the genomes of these cells (for example, a point mutation creating a K-*ras* oncogene) are quickly eliminated from this tissue. Loss of *APC* function, however, traps cells within the colonic crypts. (A minority of colorectal tumors express alterations in other components of the Apc–β-catenin pathway including β-catenin itself, yielding biochemical states similar to that resulting from loss of Apc.) When these *APC*-negative cells and their descendants sustain additional mutations, the resulting mutant cells will also be retained in the crypts rather than being lost rapidly through out-migration and apoptosis. We cannot, however, make similarly compelling arguments to rationalize why the mutation of other genes involved in colon cancer progression (*ras*, *p53*, and *DPC4*/*Smad4*) should occur at a particular time or in a particular sequence during the course of colon tumor progression. Indeed, the sequence of mutation of these other genes appears to be quite variable from one colonic tumor to another (see Figure 11.12A). Hence, while the initial step leads to the formation of aberrant crypt foci (ACFs), the specific histopathological abnormalities associated with subsequently acquired genetic alterations are not well resolved.

Finally, we should note that these various genetic and epigenetic changes do not represent an upper limit to the number of alterations that contribute in essential ways to colon cancer progression. For example, the type of genetic analysis used to identify this series of four chromosomal regions registered only those loss-of-heterozygosity (LOH) events in tumor suppressor genes that occur at a significant frequency above the general background rate of LOH associated with all chromosomal arms in advanced tumor cells (see Figure 7.13B). Accordingly, LOH events that occur with a relatively low frequency (that is, those present in 20% or fewer of the tumors analyzed) would not be registered in such an analysis, even though these events might lead to the elimination from the cell genome of functionally important tumor suppressor genes.

Epigenetic events, including the repression of some genes through promoter methylation (see Section 7.8) and the de-repression of others through demethylation, may also contribute importantly to tumor progression. Recently, evidence has accumulated that hypomethylation (that is, demethylation of normally methylated sequences), such as that observed in early adenomas (see Figure 11.11), has a consequence independent of its possible effects on gene expression: widespread chromosomal instability, which presumably favors tumor progression. How it does so is unclear. One attractive hypothesis is that this hypomethylation, which largely reflects demethylation of highly repeated sequences, permits otherwise-silenced endogenous retroviral genomes (see Section 4.1) and related *transposable elements* integrated throughout the genome to become active, jump around from one site to another in the genome, and thereby wreak genetic havoc.

The publication of this "genetic biography" of colon cancer tumorigenesis, as depicted in Figure 11.11, should ideally have been followed by similar descriptions of a wide variety of other tumor types, each biography involving its own particular set of oncogenes and tumor suppressor genes. However, only a handful of such descriptions (for example, of bladder, pancreatic, and esophageal carcinomas; see Figure 11.12B) have actually been reported. This means that, at present, we cannot cite lists of genetic alterations in tumor cell genomes to illustrate the multi-step nature of cancer progression in most organs. With the development of more sophisticated and sensitive tools for analyzing tumor cell genomes, the genetic biographies of many types of tumors should be forthcoming.

These accumulations of multiple genetic and epigenetic changes during tumor formation might suggest that the final cellular end products of tumorigenesis bear little resemblance to their fully normal ancestors—the **cells-of-origin**. It seems increasingly likely, however, that much of the differentiation program of a normal cell-of-origin persists in its neoplastic descendants years if not decades later (**Sidebar 11.2**). Accordingly, even though certain oncogenic changes in tumor cell genomes (for example, loss of pRB function, gain of Myc function; see Section 8.9) tend to cause loss of differentiated characteristics, this loss is incomplete. Indeed, the retention of differentiation traits—distinctive morphologies and protein markers—in the great majority (~95%) of tumors allows pathologists to classify these growths into different subgroups with reasonable accuracy; it remains unclear, however, whether the terms applied to these subgroups accurately reflect the normal cell type from which these tumors have arisen (**Figure 11.13**).

In ways that are still poorly understood, the differentiation programs of normal cells are also likely to influence the behaviors of their neoplastic descendants, including their propensity to remain benign or progress to high-grade malignancy. This idea injects yet another notion into our thinking: the biological behavior of a tumor cannot be understood solely by cataloging the multiple somatic mutations and epigenetic changes that it has acquired en route to neoplasia. This notion also suggests that a detailed understanding of normal cell differentiation programs will one day shed light on why tumors detected in the oncology clinic behave as they do.

11.4 Multi-step tumor progression helps to explain familial polyposis and field cancerization

The genetic pathway laid out in Figure 11.11 and modifications thereof depict the events that occur during the pathogenesis of many sporadic colonic tumors. These

Figure 11.13 Retention of differentiated characteristic in human tumors The ducts in the normal mammary gland (*middle*) are composed of two major types of epithelial cells: luminal cells that form the lining of ducts (*red-orange*) and underlying basal cells (*green*) that provide various types of physiologic support to the luminal cells. These cell types can be distinguished by immunofluorescence staining using antibodies that are specific for the cytokeratin molecules that these cells express: cytokeratins 8 and 18 are expressed characteristically in luminal cells, while cytokeratin 5 is expressed in basal cells. These markers can be used to classify breast carcinomas into two major histological classes—luminal (*left*) and basal (*right*)—and the resulting classifications hold strong implication for the prognosis of breast cancer patients. However, some evidence indicates that these classifications may not properly reveal the identities of the respective normal cells-of-origin, and suggests, for example, that certain basal breast carcinomas may originate from luminal progenitors. (From J.I Herschkowitz et al., *Genome Biol.* 8:R76, 2007.)

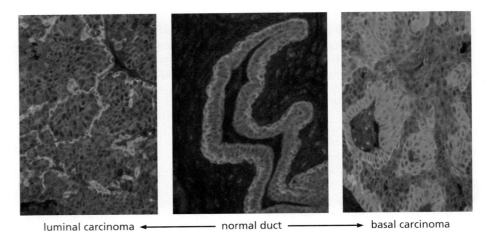

luminal carcinoma ⟵———— normal duct ————⟶ basal carcinoma

tumors arise in individuals whose genotypes are, as far as we know, genetically wild type at the moment of conception. During the course of colon tumor development, somatic mutations (and DNA methylation events) progressively alter the initially pristine genomes of colonic epithelial cells, yielding the corrupted genomes found in highly malignant colon cancer cells.

All this adds weight to the notion that the complexity of this multi-step process reflects the existence of an equally complex set of defense mechanisms that block the appearance of tumors in normal tissues. Each of these defense mechanisms must be thwarted or disabled, one after another, in order for an evolving population of cells to reach the final endpoint of full malignancy.

Given the vast number of colonic epithelial cells that are formed during a human lifetime ($>10^{14}$), these defense mechanisms are highly successful, and sporadic colon carcinomas occur at remarkably low rates. Still, the development of colon cancer is virtually ensured in some individuals, as discussed in Section 7.11 and again above. Recall that a mutant germ-line allele of the *APC* tumor suppressor gene predisposes an individual to the development of dozens, hundreds, or even a thousand colonic polyps—the syndrome of familial adenomatous polyposis (FAP).

As described in the previous section, the initial step in the development of the great majority of *sporadic* colon carcinomas involves the inactivation of the *APC* gene—precisely the same gene that is inherited in mutant form by individuals suffering from the *familial* polyposis syndrome. (The remaining colorectal tumors seem to be initiated by alternative changes that, like *APC* disruption, disrupt β-catenin regulation; see Figure 11.12A). Now, in the context of multi-step tumorigenesis, we can understand why inheritance of a mutant *APC* allele results in polyposis and colon cancer: the first step in colon cancer progression, which involves the inactivation of an *APC* gene copy, has already occurred in all of the colonic epithelial cells of an individual suffering from familial polyposis. That is, each of their cells, including those in the colon, is $APC^{+/-}$ rather than $APC^{+/+}$.

Since loss of heterozygosity is a relatively frequent event per cell generation, hundreds if not thousands of individual colonic epithelial cells may lose all APC function relatively early in the lives of individuals suffering from familial polyposis, doing so by advancing to the $APC^{-/-}$ state. Each of these Apc-negative cells can spawn, with substantial probability, an adenomatous polyp (see Figure 7.22), and such polyps, once formed, have a significant probability of progressing into a carcinoma. Consequently, a common mechanism of inborn susceptibility to cancer involves an acceleration of multi-step tumor progression, since one of its critical, rate-limiting steps is no longer dependent on infrequently occurring somatic mutations (because it has already occurred in the germ line).

Organs affected by sporadic tumors occasionally sprout multiple, apparently independently arising tumors—the phenomenon that is called **field cancerization**. In each case, two or more premalignant or frankly malignant growths erupt suddenly, separated by many centimeters of apparently normal epithelium. Since the appearance of any single sporadic (that is, nonfamilial) neoplasm is, on its own, a highly improbable event, the formation of multiple independent sporadic tumors would seem to be extremely unlikely. Indeed, these multiple growths seem to resemble, at least superficially, processes of inborn cancer susceptibility such as familial polyposis. Actually, a unique mutation (or set of mutations) is often found to be shared by these multiple, ostensibly independent growths, but in this case, the shared genetic alteration is the product of a *somatic* mutation mechanism. Our understanding of multi-step tumorigenesis provides a key insight into the underlying processes at work (**Figure 11.14**).

Imagine that a clone of cells undergoes some of the initial steps in tumor progression, which confer a proliferative or survival advantage on these cells but have no effect on the morphology of the tissue. After many years, this clone of mutant cells may have expanded to form a large patch of outwardly normal epithelium. Two cells, located far from one another within this patch, may then suffer independent mutations, and each may spawn a papilloma or carcinoma (see Figure 11.14A). Indeed, genetic analyses of apparently normal bladder epithelium, oral mucosa, and skin, as well as of esophageal and lung epithelia, have revealed large patches of cells, some many centimeters wide, that are morphologically normal but already carry a mutant gene, often a mutant allele of *p53* (see Figure 11.14B); alternatively, widely dispersed patches of dysplastic tissue having a common genetic origin can be documented (see Figure 11.14C and D).

In the mammary epithelium, loss of heterozygosity (LOH) of a locus on the short arm of human Chromosome 3 (that is, 3p) has been documented in histologically normal breast tissue adjacent to surgically removed carcinoma. Patients in whom this LOH is observed have a four- to five-fold greater risk of later developing a second, independent breast cancer than do women whose breast tissue adjacent to a carcinoma gives no indication of LOH.

Other studies have uncovered large patches of morphologically normal mammary epithelial cells in human breast tissue that have lost expression of the p16^{INK4A} CDK inhibitor because of promoter methylation; the cells in these patches may overexpress cyclooxygenase-2 (COX-2), a potent tumor-promoting enzyme (discussed in Section 11.16). Moreover, the retinoic acid receptor-β (RAR-β), a differentiation-inducing protein and repressor of COX-2 expression, is often absent in large patches of morphologically normal epithelium adjacent to a human breast cancer. Such patches of "normal" cells would seem to represent fertile ground for the appearance of multiple ostensibly independent tumors.

Similar patches of p16^{INK4A}-negative cells involving sizable areas of bronchial epithelium can be found in the lungs of current and former smokers. These patches, which are often composed of cells carrying methylated *p16^{INK4A}* genes, are seen in the lungs of cancer-free individuals and persist after smoking cessation. And in the histologically normal lung tissues of lung cancer patients, one can often find large numbers of cells that bear mutant *p53* or K-*ras* alleles; in addition, methylated tumor suppressor genes can be found in these cells. So, field cancerization is often due to the widespread presence of cells in a tissue that, because of earlier genetic alterations, are poised for embarking on the later steps of cancer progression but do not, on their own, cause tissue to assume an abnormal histological appearance. Similarly, genetically related but widely dispersed areas of dysplasia, each of which is poised to generate a frank neoplasia, can be found in certain organs (see Figure 11.14C and D).

11.5 Cancer development seems to follow the rules of Darwinian evolution

The observations about colon cancer made at Johns Hopkins University demonstrated that the histopathological changes occurring during tumor progression were correlated with genetic changes in cells of the colonic mucosa. More important, it became plausible that these genetic changes were actually *causing* the phenotypic evolution of these cells and the tissues they form.

Years earlier, others had speculated that tumor development could be understood in terms of a biological process that resembles Darwinian evolution. The results of the genetic analyses of human colon cancer progression provided further support for this model. (While Darwin himself knew virtually nothing about genes and genetics, the "modern synthesis" of Darwinian theory introduces Mendelian and population genetics into the evolutionary processes that Darwin first postulated.)

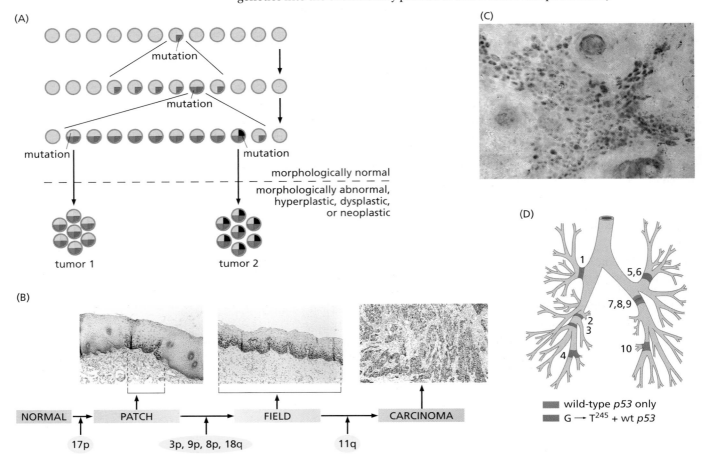

Figure 11.14 Field cancerization A large area of histologically normal but genetically altered epithelium can sprout multiple, apparently independent foci of neoplastic cells, all of which carry the mutation(s) common to this area as well as subsequently acquired mutants that are unique to each focus. (A) According to one model of field cancerization, one cell sustains an initiating mutation (*red sector, top middle*). Following extensive proliferation, some of its clonal descendants may subsequently acquire a second mutant allele (*blue sector*), and the doubly mutated "initiated" cells may then proliferate and eventually occupy a large patch of epithelium. While these cells have already undergone several of the early steps of multi-step tumor progression, they continue to form an epithelium that is histologically normal. Subsequently, two cells located at different sites within this large patch of initiated cells may independently acquire additional mutations (*yellow sector left; black sector right*); both of these cells will now advance over the boundary (*dashed line*) to acquire a histologically abnormal appearance, spawning hyperplastic, dysplastic, even neoplastic growths. These two tumors will appear to have arisen independently even though they derive from the same clone of initiated cells. (B) p53-positive (i.e., *p53*-mutant; see Section 9.9) cells, detected by immunostaining, are initially found in small, histologically normal patches of the oral epithelium early in the progression of squamous cell carcinomas in the oral cavity (*left*

image, brown spots). As tumor progression proceeds, large fields of *p53*-mutant cells can be found that remain quite normal histologically, although they have often undergone loss of heterozygosity (LOH) of markers on the indicated chromosomal arms (*middle*). Eventually, after additional genetic changes, including LOH of the long arm of Chromosome 11, several of the cells in a field may progress independently to squamous cell carcinomas (*right*). (C) Histologically normal patches of p53-mutant keratinocytes, detected here by immunostaining, can be found in sun-exposed human skin and may contain as many as 3000 cells. Such patches appear to represent fertile ground for the inception of basal cell carcinomas. (D) An extreme case of field cancerization can be seen here in lungs of a patient examined upon autopsy. Cells within widely scattered areas in both lobes of the lungs were found to share the same rare *p53* somatic mutation (present either in homozygous or heterozygous configuration, *red*), indicating their common genetic origin (see also Figure 11.9C). While these areas did indeed exhibit mild squamous metaplasia, no carcinomas could be detected; hence, none of these widely dispersed fields had spawned a carcinoma by the time of the patient's death. (B, from B.J.M. Braakhuis et al., *Cancer Res.* 63:1727–1730, 2003. C, courtesy of D.E. Brash. D, from W.A. Franklin et al., *J. Clin. Invest.* 100:2133–2137, 1997.)

In the case of cancer development, the evolving units are individual cells competing with one another in a population of cells, rather than individual organisms competing with one another within a species. Like the modern depiction of Darwinian evolution, random mutations are presumed to create genetic variability in a cell population. Once a genetically heterogeneous population has arisen, the forces of selection may then favor the outgrowth of individual cells (and their descendants) that happen to be endowed with mutant alleles conferring advantageous traits, notably traits that favor proliferation and survival in the microenvironment of a living tissue.

Combining Darwinian theory with the assumptions of multi-step tumor progression, researchers could now depict tumorigenesis as a succession of *clonal expansions*. The scheme goes like this: A random mutation creates a cell having particularly advantageous growth and/or survival traits. This cell and its descendants then proliferate more effectively than their neighbors, eventually yielding a large clonal population that dominates the tissue and crowds out genetically less favored neighbors. Sooner or later, this cell clone will reach a large enough size (for example, 10^6 cells) that another advantageous mutation, which strikes randomly with a probability of about 1 per 10^6 cell generations, may now plausibly occur in one or another cell within this clonal population (**Figure 11.15**).

The resulting doubly mutated cell, which will proliferate (or survive) even more effectively than its 10^6 clonal brethren, will spawn a new subclone that will expand and eventually dominate the local tissue environment, overshadowing and possibly obliterating the precursor population from which it arose. After this doubly mutated cell clone reaches a large size, as before, a third mutation may now strike, and the process of clonal expansion and succession will repeat itself. Quite possibly, a sequence of four to six such clonal successions, each triggered by a specific mutation, suffices to explain how cancer progression occurs at the cellular and genetic level.

To be sure, this Darwinian model of cancer progression is simplistic. For example, it must be amended to respond to the discovery that epigenetic alterations of genes, specifically, promoter methylation (see Section 7.8), play an important role in eliminating the activities of tumor suppressor genes. (Here, we encounter a major discordance between tumor progression and Darwinian evolution, since heritable epigenetic alterations, such as DNA methylation events, have never been shown to drive the evolution of species.)

This scheme is simplistic in other respects as well. Thus, the number of distinct steps in tumor progression may be underrepresented by counting the number of genetic loci that are altered during this process. As discussed in Chapter 7, the inactivation of a tumor suppressor gene is, almost always, a two-step process. First, one gene copy is

Figure 11.15 Darwinian evolution and clonal successions Darwinian evolution involves the increase in number of organisms that are endowed with advantageous genotypes and thus phenotypes; a formally similar scheme seems to describe how tumor progression occurs. One cell amid a large cell population sustains an initiating mutation (*red sector, top*) that confers on it a proliferative and/or survival advantage over the other cells. Eventually, the clonal descendants of this mutant cell dominate in a localized area by displacing the cells that lack this mutation, resulting in the first clonal expansion. When this clone expands to a large enough size (e.g., 10^6 cells), a second mutation—one that strikes with a frequency of ~10^{-6} per cell generation—may occur (*green sector*), resulting in a doubly mutated cell that has even greater proliferative and/or survival advantage. The process of clonal expansion then repeats itself, and the newly mutated population displaces ("succeeds") the previously formed one, yielding a process that is termed *clonal succession*. This results once again in a large descendant population, in which a third mutation (*blue sector*) occurs, and so forth. While classical Darwinian evolution is thought to depend on mutations in the genomes of organisms, it is highly likely that other heritable changes in cell populations, notably promoter methylation events (see Section 7.8), can play an equally prominent role in multi-step tumor progression. Importantly, this scheme does not take into account the fact that clonal successions may require greatly different time intervals to reach completion. For example, later successions are likely to proceed far more rapidly than earlier ones because the participating cells, having acquired oncogenic mutations, may proliferate more rapidly and have more mutable genomes.

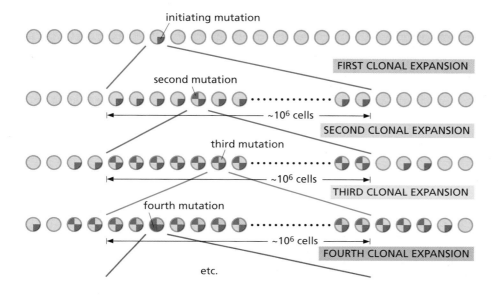

mutated (or methylated) to inactivity. Thereafter, the surviving, still-intact gene copy is eliminated, usually through loss of heterozygosity (LOH). Knowing this and the fact that elimination of tumor suppressor genes often represents the majority of the genomic changes occurring during tumor progression, we conclude that the number of distinct alterations taking place during tumor progression may be almost twice as many as the number of loci that are involved.

Since each of these clonal expansions is triggered by an infrequently occurring genetic or epigenetic alteration (or pair of alterations, in the case of tumor suppressor genes), these expansions are likely to be spaced far apart in time. During the development of sporadic colon cancers, a decade or more may separate one critical genetic alteration from the next, and in many individuals, the process as a whole may stretch over a century. (For example, one of the key steps—the evolution from polyp to invasive cancer—has been estimated by some to take as long as 17 years, depending on the degree of dysplasia present in the polyp.) Still, it is clear that some people develop sporadic colon carcinomas in far less time, and so this schedule must be compressed in the colons of these individuals.

We know almost nothing about the processes that govern the rates of tumor progression in most tissues. In the case of colon cancer, we know that rates vary by as much as twentyfold between countries and that these dramatic differences are due to environmental factors, specifically foodstuffs, rather than genetic susceptibility. (As Figure 2.23 shows, a population migrating from one country to another exhibits a colon cancer rate typical of its new country within a generation or two, ruling out genetics as the key determinant of cancer risk.) It would seem that certain constituents of diet greatly increase the rate at which the genomes of colonic epithelial cells accumulate mutations. Such increased mutation rates can, in turn, compress the time between clonal successions. Perhaps in individuals who consume certain foodstuffs, 5 or fewer years intervene between successive clonal expansions rather than the usual 10 or 20. As a consequence, a disease process that usually requires a century to reach completion may reach its neoplastic endpoint in 30 or 40 years, well within the human life span.

Certain alterations within premalignant cells may conspire with these exogenous foodstuff carcinogens to accelerate the pace of colon tumor progression. Thus, increased mutability may be caused by defects in the complex cellular machinery that is dedicated to maintaining and repairing cellular DNA. For example, as discussed in Chapter 12, some individuals inherit mutations that compromise the function of the cellular DNA mismatch repair apparatus. Consequently, greatly increased mutation rates are seen in their colonic epithelial cells, and these lead, in turn, to a greatly accelerated rate of formation of premalignant and malignant growths in the colon—the syndrome of hereditary non-polyposis colon cancer (HNPCC). More commonly, however, defects in components of the DNA repair apparatus result from somatic mutation (or promoter methylation) occurring early in tumor progression; the resulting increased genome mutability guarantees that subsequent mutational steps in colon cancer progression will occur relatively rapidly.

These multiple factors influencing colon carcinoma formation greatly complicate attempts to delineate the clonal successions predicted by the Darwinian model of tumor progression. A further complication for those intent on charting the genetic biographies of tumors (see Figures 11.11 and 11.12) comes from recent findings that the number of mutations acquired by tumor cells during multi-step tumor progression vastly exceeds those that are actually responsible for driving this process forward (Sidebar 11.3).

11.6 Tumor stem cells further complicate the Darwinian model of clonal succession and tumor progression

The clonal succession model proposes that a mutant cell spawns a large flock of descendants and that among these numerous descendants, a new mutational event will trigger yet another wave of clonal expansion (see Figure 11.15). However, certain experiments have cast serious doubt on the notion that all of the cells within a pre-neoplastic (or neoplastic) cell clone are biologically equivalent and therefore equally capable of becoming ancestors of a new successor clone of cells. In these experiments, the cancer cells within a human tumor were separated into distinct subclasses. These separations took advantage of cell surface proteins displayed by the different subpopulations. In particular, the technique of fluorescence-activated cell sorting (FACS) was used to separate living cancer cells after labeling them (via their surface proteins) with monoclonal antibodies linked to fluorescent dyes (see Supplementary Sidebar 11.1). Cells separated by this procedure can be recovered in viable form and used in biological experiments, including *in vivo* tests of their ability to seed tumors following injection into immunocompromised host mice.

Use of the FACS technique initially enabled researchers to segregate populations of acute myelogenous leukemia (AML) cells into majority and minority populations; in one such experiment, the latter represented less than 1% of the neoplastic cells in the tumor mass. As few as 5000 cells in the minority subpopulation were able to produce new tumors upon injection into host mice and were therefore deemed to be "tumorigenic"; in contrast, as many as 500,000 AML cells from the majority subpopulation were unable to seed a tumor. Importantly, the cells in this majority subpopulation exhibited many of the attributes of differentiated granulocytes or monocytes, and these cells had limited ability to proliferate. These observations provided compelling evidence that the AML tumors were composed of small populations of self-renewing, tumorigenic cells and large populations of more differentiated cells that had little, if any, ability to proliferate *in vivo*.

Subsequent experiments extended these results to human breast cancer cells prepared directly from tumors. In these later experiments, the minority tumorigenic cell population within a tumor represented only about 2% of the overall neoplastic cell population. Two hundred of these minority cells seeded a new tumor when injected into a host mouse, while as many as 20,000 cells from the majority cell population

---

Sidebar 11.3 Driver versus passenger mutations Mutations strike the genome randomly and only rarely hit critical genes that, when mutated, confer advantageous phenotypes leading to the clonal expansions that drive multistep tumorigenesis (see Figure 11.15). Consequently, a cell that happens to acquire an advantageous mutant allele will also carry numerous other mutations that have struck other genes throughout its genome; these other mutant genes have no influence on cancer cell phenotype and consequently are irrelevant to tumor progression. Nonetheless, these other mutant alleles will become heritable components of the cancer cell genome together with the functionally important allele; in effect, these others will be "carried along for the ride" and, as such, are called "passenger mutations" to distinguish them from the "driver mutation" that was actually responsible for generating a clonal expansion.

These dynamics complicate attempts to distinguish which mutations in a cancer cell genome are driver mutations (and are therefore interesting for understanding cancer development) from those that are passenger mutations (which are irrelevant distractions). Hence, the discovery of a mutant gene in a cancer cell genome becomes interesting only if the same gene can be found in mutant form in a number of tumors from different cancer patients. Such recurrent mutations of a gene provide a strong indication that mutation of this particular gene has indeed provided a selective advantage on multiple occasions and is therefore biologically important for driving tumor development. These drivers stand in contrast to the passenger mutations that rarely strike the same gene twice and form the random mutational noise that afflicts genes throughout cancer cell genomes. For example, in one sequence analysis of the genomes of 21 human glioblastomas, only 8% of the mutations were judged to be drivers while the remainder were passengers. Far lower ratios of driver to passenger mutations have been found upon sequencing the genomes of certain other tumors.

failed to do so (**Figure 11.16**). Importantly, the majority and minority breast cancer cell subpopulations both contained comparable proportions of cells in the active growth cycle, and both subpopulations were purified away from nonmalignant cells (such as stromal cells) that were present in the original tumor masses.

The tumors that eventually arose from the injected minority cells were once again composed of minority and majority cell populations that showed, as before, vast differences in their ability to seed new tumors, that is, in their tumorigenicity. Similar results were then obtained with brain tumor cells (**Figure 11.17**), and in subsequent years, subpopulations of tumor-initiating cells have been demonstrated in a variety of other solid tumors, such as pancreatic, colorectal, and hepatocellular carcinomas, as well as certain hematopoietic malignancies.

Taken together, these experiments indicate that the neoplastic cell populations originating in a variety of tissues are organized much like normal epithelial tissues, in which relatively small pools of self-renewing stem cells are able to spawn large numbers of descendant cells that have only a limited proliferative potential *in vivo*. In each of the experiments shown here (see Figures 11.16 and 11.17), the tumorigenic minority cells expressed a set of antigenic markers on their surfaces that were distinct from those exhibited by cells in the majority populations, providing further indication that the two groups of cells resided in distinct states of differentiation.

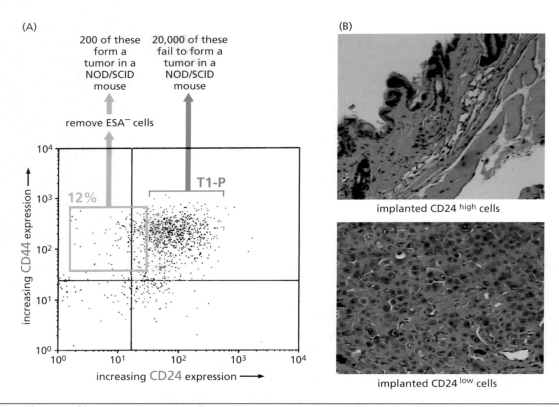

Figure 11.16 Enrichment of breast cancer stem cells
(A) Metastatic human breast carcinoma cells were freed of contaminating noncancerous (stromal) cells and separated from one another (using FACS; see Supplementary Sidebar 11.1). The expression of two distinct cell surface antigens—CD24 and CD44—was gauged simultaneously, each being detected with a specific monoclonal antibody linked to a distinct fluorescent dye. The intensity of staining is plotted logarithmically on each axis. Each black dot in the graph represents the detection of a single cell. In this experiment, a 12% subpopulation of cells that expressed low CD24 and high CD44 antigen staining (*green box*) was separated from cells (T1-P) that showed high CD24 expression and high CD44 expression (*blue bracket*). (B) The cells in the 12% minority population were further enriched by sorting for those that expressed epithelial surface antigen (ESA), which resulted in the elimination of contaminating stromal cells that did not express this antigen. Two hundred of the CD24^{low} CD44^{high} ESA+ cells were able to form a tumor following injection into a NOD/SCID immunocompromised mouse, while 20,000 of the CD24^{high} CD44^{high} cells failed to do so. The *upper* micrograph shows a section through the subcutaneous site of implantation of CD24^{high} cells, in which only relatively normal skin and underlying muscle wall can be seen (*above left*); the *lower* micrograph shows a section through the tumor that formed following injection of the CD24^{low} cells. (From M. Al-Hajj et al., *Proc. Natl. Acad. Sci. USA* 100:3983–3988, 2003.)

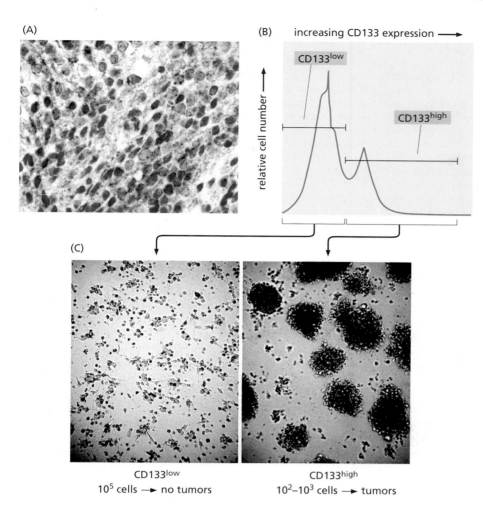

(A)

(B) increasing CD133 expression ⟶

CD133^low

CD133^high

relative cell number ⟶

(C)

CD133^low
10^5 cells ⟶ no tumors

CD133^high
10^2–10^3 cells ⟶ tumors

Figure 11.17 Enrichment of brain tumor stem cells Brain tumor stem cells can be detected because of their expression of CD133, a neural cell surface stem cell antigen. (A) In this immunofluorescence-stained section of a medulloblastoma (a tumor of cerebellar cells), tumor cells expressing high levels of CD133 (*dark red*) lie scattered among other tumor cells that have low CD133 expression. Cell nuclei are stained here in *blue*. (B) Cell populations from a human medulloblastoma were sorted by FACS (see Supplementary Sidebar 11.1) into those that express greater or lesser levels of CD133 (*abscissa*). The number of cancer cells (*ordinate*) expressing low levels of CD133 (*left peak*) represents 80–85% of this tumor, while the CD133 high expressors (*right peak*) are in the minority. (Note that the intensity of staining is presented on a logarithmic scale on the abscissa.) (C) CD133^low cells from the left peak showed limited proliferative ability in suspension culture (*left panel*); the ability to generate colonies (sometimes termed tumorspheres) upon culturing in semi-solid medium in the absence of contact with a solid substrate is often used to predict tumorigenicity *in vivo*. In some experiments, as many as 10^5 of these CD133^low cells failed to seed tumors in host mice. In contrast, CD133^high cells from the right peak formed numerous colonies in suspension culture (*right panel*), and injection of as few as 10^2 (more typically 10^3) of these cells into an immunocompromised mouse host resulted in the formation of a tumor (*not shown*). (From S.K. Singh et al., *Cancer Res.* 63:5821–5828, 2003.)

As was discussed in Section 8.11, in many normal tissues, stem cells (SCs) are less differentiated, and their non-stem cell descendants enter into states of increased differentiation. In addition, stem cells appear to have an essentially unlimited ability to proliferate and, because some of their progeny remain as stem cells, are said to be "self-renewing." In contrast, the more differentiated progeny of a stem cell often enter a post-mitotic state, from which they will never emerge to re-enter into the active growth-and-division cycle (**Figure 11.18**). It is therefore tempting to think that the same organization of cell behavior operates in many human tumors.

In normal tissues, the cells are organized in a hierarchy. At the top is a stem cell, one of whose daughters remains a stem cell while the other initiates a program of differentiation. (Because this cell division yields two phenotypically distinct daughters, it is often termed *asymmetrical*, in contrast to most divisions, which generate two identical daughters and are termed *symmetrical*.) The cell that has exited the stem cell state is often termed a **transit-amplifying cell**, also called a **progenitor cell** (see Figure 11.18B). Such cells are present in many normal tissues and represent intermediates between stem cells and their fully differentiated descendants. Transit-amplifying/progenitor cells may pass through a large succession of symmetrical cell divisions before their descendants eventually become fully differentiated and cease active growth-and-division cycles. Hence, the initially formed transit-amplifying/progenitor cell can spawn dozens if not hundreds of differentiated progeny, which means that (1) the stem cell needs to divide only once in order to generate large numbers of fully differentiated descendants; (2) the stem cell may therefore divide only periodically rather than continually, even in a tissue in which differentiated cells are continually being lost and replaced; and (3) the great majority of cell divisions within a tissue are often associated with the exponentially growing transit-amplifying cells.

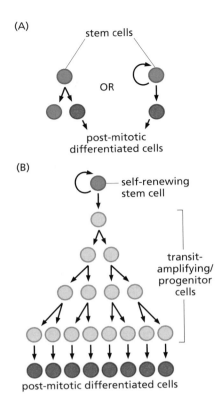

Figure 11.18 Stem cells and their progeny (A) The simplest organization of stem cell behavior involves an asymmetric cell division of a stem cell (*blue*), in which one of the daughter cells becomes a stem cell like its mother while the other daughter (*red*) differentiates and loses the ability to divide again. This can be depicted in two graphic conventions, both of which are shown here. (B) In many tissues, a more complex scheme appears to operate. As in panel A, one of the daughters of a dividing stem cell becomes a stem cell. However, the other becomes a "transit-amplifying cell" (sometimes termed a "progenitor cell"; *gray*), which is committed to enter a differentiation pathway but does not yet participate in end-stage differentiation. Instead, this cell and its progeny undergo a series of symmetric cell divisions before its descendants eventually enter into a fully differentiated, post-mitotic state (*red*). Because a single stem cell can spawn dozens if not hundreds of post-mitotic descendants, this cell does not need to proliferate frequently. Moreover, in a tissue organized in this way, the great bulk of cell proliferation is associated with transit-amplifying/progenitor cells (see also Supplementary Sidebar 12.1).

This scheme has been appropriated by researchers attempting to understand the organization of neoplastic cells within a tumor. Thus, the tumor-initiating cell, often termed a cancer stem cell (CSC), is self-renewing and has the ability to generate the countless neoplastic progeny that constitute a tumor. While the CSC and its progeny are genetically identical, the progeny, because they have lost self-renewing ability, have also lost tumor-initiating ability. Indeed, as the details of the cellular programs that govern the CSC state are elucidated, it becomes increasingly apparent that these CSC programs are very similar to those operating in the corresponding normal tissues, that is, tumors do not invent new stem cell programs but simply appropriate ones operating in their normal tissues-of-origin. Moreover, both normal and neoplastic SCs appear to maintain their continued residence in an SC state through the continuous firing of autocrine signaling loops of the sort first mentioned in Section 5.5. For example, in certain organs, autocrine signaling involving TGF-β and Wnt factors enables both types of SCs to maintain their residence in this state.

Neoplastic cells displaying distinctive antigenic markers—often shared with normal stem cells—can be found at discrete sites in some human tumors (**Figure 11.19A**) but not in others (Figure 11.19B and C). In order to qualify as bona fide CSCs, they must, following enrichment through the use of FACS (see Figures 11.16 and 11.17), exhibit greatly elevated levels of tumor-initiating ability relative to the bulk of cancer cells within such tumors. At present, no principles have emerged that describe patterns of CSC localization that are shared in common by diverse tumor types.

The existence of CSCs creates a new dimension of complexity within individual tumors. Thus, as mentioned above, the stem cell hierarchy that operates in many normal tissues appears to operate as well in the fully developed tumors arising in these same tissues. This makes it highly likely that similar stem cell hierarchies are also established within the cell populations formed at each of the intermediate stages of multi-step tumor progression. Previously we drew a linear, one-dimensional progression from a normal cell to a tumor cell (see Figure 11.11). Now, with this newer information in hand, it seems more appropriate to describe multi-step tumor progression in the form of a two-dimensional map, with the SCs formed at each stage of tumor progression spawning multiple distinct subpopulations of more and less differentiated cells (for example, see Figure S11.2).

Assuming that such hierarchical organization characterizes each stage of multi-step tumor progression, it is unclear how the cells in each stage evolve into the cells in the next stage, specifically, which subpopulations of cancer cells at one stage acquire the mutation that triggers a new clonal expansion and thus a new stage of progression. Moreover, the precise identity of the normal cell-of-origin remains unclear: is this cell—the fully normal cell that becomes the ancestor of all the neoplastic cells in a tumor (see Figure 2.19)—a stem cell or might it be a transit-amplifying/progenitor cell (see Supplementary Sidebar 11.2)?

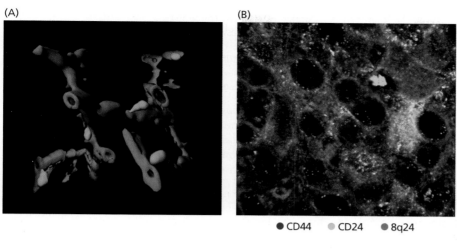

(A) (B) (C)

● CD44   ● CD24   ● 8q24

Figure 11.19 Localization of cancer stem cells within tumors (A) In human gliomas, cancer cells that express nestin, a marker of normal neural stem cells (*green*), exhibit tumor-initiating ability when implanted in host mice and are therefore termed cancer stem cells (CSCs). These nestin-positive cells are frequently localized near capillaries within these tumors; the microvessels express CD34, an endothelial cell marker (*red*). This 3-dimensional image was produced by digitally integrating the microscopic images from a succession of 50 micron-thick tumor sections. (B) In a lung metastasis of a human breast cancer, the CD44-high cells (*blue*), which are enriched in cancer stem cells (CSCs), are located in different subdomains from the CD24-high cells (*yellow*) that generally have low proportions of CSCs. [A probe for the 8q24 chromosomal region revealed extensive amplification of this region in a subset of these metastatic cells (*red*).] (C) Aldehyde dehydrogenase isoform 1 (ALDH1) has been found to be a useful marker for identifying a variety of normal and neoplastic stem cells. Here, an anti-ALDH1 antibody has been used to localize cells in a primary colon carcinoma that express high levels of this stem cell marker (*dark brown*). (A, from C. Calabrese et al., *Cancer Cell* 11:69–82, 2007. B, courtesy of H.-J. Kim and K. Polyak. C, from S. Deng et al., *PLoS One* 5:e10277, 2010.)

As we will read in Chapter 16, the existence of cancer stem cells is of more than academic interest. Rapidly accumulating evidence indicates that CSCs are generally more resistant to conventional chemotherapies and radiotherapy. As a consequence, these cells may survive initial treatment and generate the new tumors whose appearance signals clinical relapse.

11.7 A linear path of clonal succession oversimplifies the reality of cancer: intra-tumor heterogeneity

The general scheme of clonal succession proposed until now (see Figure 11.15) suggests that all of the cells within a tumor mass that participate in a particular clonal expansion are genetically identical to one another and that tumor formation occurs as a consequence of a linear series of these clonal successions. According to this scheme, if we were to examine the cells within a premalignant or malignant cell mass, we would almost always find that a single, genetically homogeneous clone of cells dominates in this mass, since it would have outgrown and largely displaced the preceding cell clone from which it arose.

However, the actual course of tumor progression is complicated by yet another factor that we must take into account: as tumor progression advances, tumor genomes often become increasingly unstable, and the rate at which mutations are acquired during each cell generation soars. The rate of genetic change and resulting genetic diversification soon outpaces the rate at which Darwinian selection (and the elimination of less-fit subclones of cells) can occur. As a consequence, rather than looking like a linear series of clonal successions, actual tumor progression in many tumor masses is likely to resemble the highly branched scheme shown in **Figure 11.20A**, in which a number of genetically distinct subclones of cells coexist within a single tumor mass. The dynamic nature of their expansion and subsequent replacement is suggested by Figure 11.20B.

Demonstrations of the genetic diversification of cells within tumor masses can be obtained by tracking the state of individual genes of interest within various cells of a primary tumor. For example, in a human pancreatic carcinoma, detailed genome sequence analysis of different sectors of the tumor revealed genetically distinct subclones, each of which was estimated to comprise at least 100 million cells. Interestingly, several of these subclones each spawned its own set of genetically closely related metastases located elsewhere in the body of the cancer patient—a topic that we will pursue in greater depth in Chapter 14.

The localization of individual subclones and associated cells within a tumor is itself unclear. A computer-based modeling of how such subclones arise (distinguished from one another by heritable differences in DNA sequence or CpG methylation) predicts large sectors occupied by individual subclones, which may indeed represent the configuration of many tumors; however, analysis of individual carcinoma cells within a tumor may reveal a more complex situation, in which the cells of various subclones become intermingled (see Figure 11.20C).

This clonal variation, in which each subclone represents a relatively stable, homogeneous population, stands in contrast to the situation that appears to operate in states of extreme genetic instability, where individual cells are continually trying out new genetic combinations. A vivid example of such genetic diversification is shown in Figure 11.20D, in which the FISH (fluorescence *in situ* hybridization) technique was used to determine the copy number of Chromosomes 11 and 17 in individual cells deriving from a human lung carcinoma. As is apparent, the copy number of Chromosome 17 (which carries the gene specifying cyclin D1) varies enormously from one cell to another, while the copy number of Chromosome 11 is relatively stable. Such variability indicates that the tumor cell genomes being analyzed are quite plastic, possibly changing each time the cells pass through a cycle of growth and division. These fluctuations seem to occur almost randomly and at a rate that greatly exceeds the ability of (Darwinian) selection to eliminate less-fit variants.

There are yet other sources of genetic diversity in primary tumors that remain incompletely understood. For example, there is clear evidence that cells in distantly located metastases can travel back and re-seed colonies in the primary tumor. This means that any genetic evolution that may have occurred within such metastases (and thus is occurring *after* cancer cells have left the primary tumor) may then be fed back into the gene pool of the primary tumor and thereby contribute to its genetic diversification.

Commonly performed surveys of human tumors (**Figure 11.21A**) suggest a second source of the heterogeneity within tumors, namely, epigenetic plasticity. Surveys of protein expression in tumor sections using immunofluorescence provide dramatic testimony to the diversification of small, discrete subpopulations of cancer cells within a tumor (see Figure 11.21B). In principle, the variability in the expression of the EGF and VEGF receptors shown in this figure might well be due to genetic diversification. More likely, however, the cause stems from epigenetic mechanisms—localized fluctuations in cell-to-cell signaling in the tumor microenvironment yielding phenotypic diversity that does not directly reflect underlying genetic diversification.

A case in point is offered by glioblastomas, which are termed more properly glioblastoma *multiforme* to indicate the fact that these tumors typically include a variety of morphologically distinct neuroectodermal cell types (see Figure 2.9A)—almost certainly the reflection of oligopotent stem-like cells within the tumor differentiating into a number of different cell types (see Figure 11.21C). The most extreme example of this epigenetic plasticity is illustrated by the strategy that many glioblastomas use to generate a vasculature: rather than recruiting normal endothelial cells (which form capillaries) from the tumor-bearing host (discussed in Chapter 13), the neoplastic cells transdifferentiate into endothelial cells, the latter normally originating in an entirely

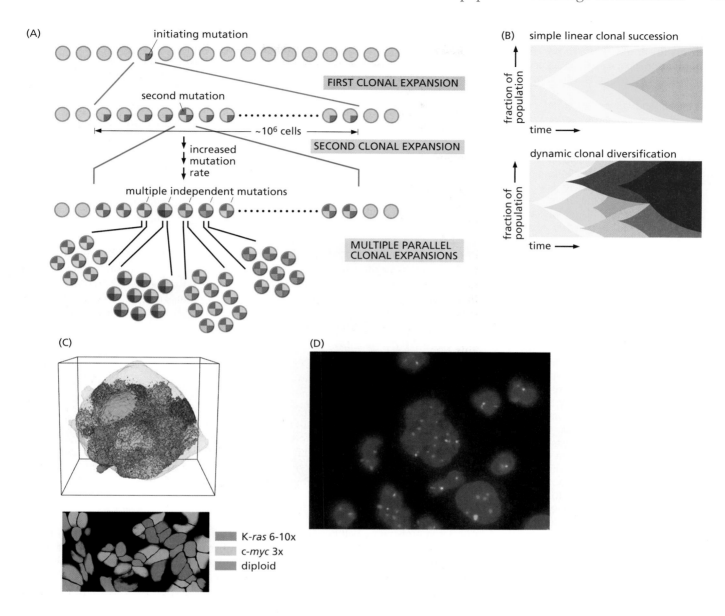

Figure 11.20 Clonal diversification due to high mutation rates (A) As tumor progression proceeds, the genomes of tumor cells often become increasingly unstable. As this occurs, the rate at which new mutant alleles are generated may exceed the rate at which Darwinian selection can eliminate phenotypically less-fit clones. Consequently, the tumor mass develops an increasing number of distinct sectors, each dominated by a genetically distinct subclone. In this diagram, the involvement of transit-amplifying and cancer stem cells in these processes is not indicated. (B) Recent surveys of both genetic and epigenetic (CpG methylation) heterogeneity within individual tumors have provided experimental support for the extensive diversification of distinct clonal populations within a tumor. Thus, the simple linear model of clonal succession (*above*) should be replaced, at least in late-stage tumors, by the dynamic changes depicted *below*. (C) While the schemes of A and B imply the coexistence of distinct subclones within a tumor, they do not necessarily indicate the topology of these clones and their constituent cells. A computationally based model of how a tumor should develop, which assumes the involvement of cancer stem cells (*above*), predicts discrete sectors in which genetically (and epigenetically) distinct, individual subclones

are localized in a tumor composed of 2 million cells. In contrast, use of fluorescence *in situ* hybridization (FISH) in a section of an actual tumor (*below*), such as the human breast cancer analyzed here, reveals the extensive physical intermingling of the individual cells deriving, in this case, from three major, genetically distinct subclones coexisting within this tumor. The three clones involve those with apparent diploid genomes, those with 3 copies of c-*myc*, and those with 6–10 copies of K-*ras* and a closely linked gene. (It is unclear which of these two topological patterns is more common.) (D) In high-grade tumors, the numbers of chromosomes often fluctuate wildly from one cell to another, indicative of great cell-to-cell genetic heterogeneity. In this image, the use of FISH revealed the copy numbers of Chromosomes 11 (*green*) and 17 (*pink*) in cells from a *pleural effusion* in a non-small-cell lung carcinoma patient. In addition to fluctuations in chromosome number, highly polyploid giant nuclei are apparent. (B, from A. Marusyk and K. Polyak, *Biochim. Biophys. Acta* 1805:105–117, 2010. C, from A. Sottoriva et al., *PloS Comput. Biol.* 7:e1001132, 2011, and from N. Navin et al., *Genome Res.* 20:68–80, 2010. D, courtesy of M. Fiegl; from M. Fiegl et al., *J. Clin. Oncol.* 22:474–483, 2004.)

Figure 11.21 Phenotypic diversification within a tumor cell population Independent of the genetic diversification of subpopulations of cells within a tumor (see Figure 11.20) is the ability of tumor cells to enter into new phenotypic states by activating epigenetic programs associated with alternative states of differentiation. (A) An adenoid basal carcinoma (ABC) of the cervix from a 94-year-old woman demonstrates heterogeneous cytologic features. To the far right, a row of small basaloid islands infiltrate the cervical stroma (*red circles*). Above them (*top right*) is an area of columnar cell differentiation within such a nest of basaloid cells (*green circle*). In the center and left are two large nests of neoplastic squamoid cells (*yellow, blue*) exhibiting various degrees of *atypia*. In both cases, the large nests are surrounded by rims of basal cell layers, findings characteristic of ABC. ABC is typically associated with a conventional high-grade squamous intraepithelial lesion (not shown) and HPV infection; this case was positive for HPV type 16. The high frequency of these variant morphologies in this particular subtype of carcinoma strongly suggests that these islands of cells represent distinct differentiation programs activated by the various carcinoma cells rather than genetic diversification. (B) Use of immunofluorescence with antibodies that react with the EGF (*green*) and VEGF (*red*) receptors reveals great phenotypic diversity even within small sectors of these tumor xenografts; cells co-expressing both receptors are seen as *yellow*. Tumors analyzed in the *upper* and *lower* panels derive from cancer cell lines, while the tumor in the *center* panel derives from a surgical specimen. (C) When CSCs and derived progenitors are isolated from glioblastoma tumors, they can be induced to differentiate *in vitro* into at least two distinct lineages of neuroectodermal cells, one exhibiting neuron-specific β-tubulin (*red*) and the other the astrocyte GFAP (glial fibrillary acidic protein) marker (*green*). This image presents a phenomenon common to many types of tumors, in which the CSCs or derived progenitors, like their presumed normal counterparts, exhibit an ability to differentiate into multiple distinct cell lineages. (A, courtesy of M. Hirsch and M. Loda. B, from T. Kuwai et al., *Am. J. Pathol.* 172:358–366, 2008. C, from T. Borovski et al., *Cell Cycle* 8:803–808, 2009.)

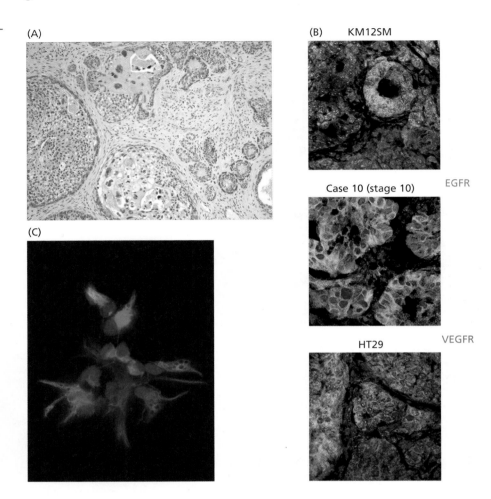

unrelated embryonic cell lineage (see Figure 13.23). Once formed, the transdifferentiated endothelial cells collaborate with the normal endothelial cells of host origin to construct capillaries, thereby expediting tumor-associated angiogenesis and thus tumor growth.

Observations like these suggest another interesting possibility: Might two phenotypically distinct subpopulations of cells within a tumor act symbiotically, each supporting the growth and survival of the other, rather than directly competing with one another? Recent observations of the phenotypic state of two metabolically distinct subpopulations coexisting within an individual tumor provide support for precisely this notion (Supplementary Sidebar 11.3). (This might represent a situation in which distinct neoplastic cell subpopulations that share a common genotype adopt distinct phenotypes in order to help one another adapt most effectively to the tumor microenvironment.)

Analyses such as those described in Figure 11.21 only hint at the enormous amount of diversification that may occur as cancer cells progress to ever more malignant growth states. More definitive demonstrations of this diversification derive from sequence analyses of the genomes of tumor cells, specifically from randomly chosen DNA segments prepared from 58 human carcinomas. (Analysis of each DNA segment depended on the use of the polymerase chain reaction to selectively amplify the segment prior to determining its nucleotide sequence.) By extrapolating from the relatively small number of segments analyzed, researchers estimated that at least 10^4 sequence alterations, including myriad single-nucleotide changes, were present in the genome of each of these sporadic tumors. Strikingly, a comparable number of alterations were found in the genomes of 11 sporadic adenomas. This latter observation indicates that widespread destabilization of cell genomes and extensive genetic diversification already occur relatively early in multi-step tumor progression.

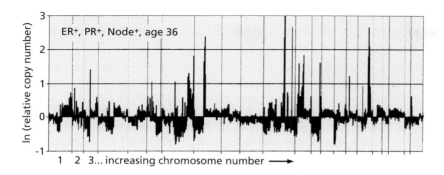

Figure 11.22 CGH analysis of a breast cancer genome The genomic DNA of an ER+ (estrogen receptor–positive), PR+ (progesterone receptor–positive), Node+ (draining lymph nodes containing metastatic cancer cells), stage 4 (advanced) human breast cancer was analyzed here using comparative genomic hybridization (CGH; see Supplementary Sidebar 11.4). All 22 human autosomes were represented by probes, with results for each shown in the appropriately numbered area. The degree of amplification or deletion is indicated by excursions from the normal copy number, which is designated as "0" on the ordinate since the data are plotted logarithmically, with amplifications shown above the line and deletions shown below the line. For example, one segment of Chromosome 8 was found to be present in this tumor in a copy number that is more than 2 natural-logarithmic units (i.e., a factor > e^2) above the normal copy number. Such an analysis is unable to resolve between those segments whose altered copy number played a critical role in tumor development and other segments whose altered copy number reflects the widespread chromosomal instability known to operate in many advanced human tumors. This analysis indicates rampant genomic instability and resulting genetic diversification of subclones of cancer cells within this tumor. (Courtesy of J.W. Gray.)

Use of the procedure of **comparative genomic hybridization** (CGH; see Supplementary Sidebar 11.4) reveals a further dimension of genetic instability and thus diversification. CGH is used to gauge increases and decreases in the copy number of various chromosomal DNA segments in tumor cell genomes and yields a plethora of genetic data (**Figure 11.22**). Some of these fluctuations change the dosage of key tumor-inducing and tumor-suppressing genes and thereby influence the proliferation of subclones of neoplastic cells within tumors. The number of changes in the copy number of chromosomal segments greatly exceeds the number of clonal successions occurring during tumor progression, pointing once again to extensive genetic diversification within single tumor masses.

The consequences of this accumulated genetic heterogeneity are likely to be manifested at two levels. *Within* a given tumor mass, different subclones will carry distinct sets of genetic alterations, as implied in Figure 11.20. This heterogeneity also affects comparisons *between* tumors of the same type that arise in different patients. Thus, the genotypes of these tumors (for example, colon carcinomas at the same histopathological stage of tumor progression arising in 20 different patients) are likely to be markedly different from one another.

11.8 The Darwinian model of tumor development is difficult to validate experimentally

While the Darwinian model of tumor development, as depicted schematically in Figures 11.15, 11.19, and 11.20, is attractive in concept, it remains little more than a theoretical construct. The outlines of this model are undoubtedly true, but its details are very difficult to validate for a number of reasons. To begin, a convincing validation of the Darwinian model would require an identification of the key genetic and epigenetic alterations of cell genomes that are responsible for each clonal expansion and thus for each step of multi-step tumorigenesis. However, the vast number of these alterations accumulated in the genomes of tumors (see Section 11.7) greatly exceeds the number of *clonal successions* that drive tumor progression (see Figure 11.15). Consequently, sequence analyses of the genomes of cells at different states of tumor progression may not converge on the critical genetic changes that are responsible for many clonal successions.

Such sequence analyses are encumbered by another problem: as we have read, inactivation of key tumor suppressor genes is often caused by the epigenetic process of gene silencing via promoter methylation (see Section 7.8). Genes functionally inactivated in this way will appear as wild-type alleles upon DNA sequencing, and their silencing can be determined only by analyses of the methylation state of their promoters or by screening for their transcripts within tumor cells. The latter screens, a form of **functional genomics**, may also not be particularly useful, since they are unlikely to distinguish between the gene silencing that is a consequence of a normal differentiation program and the gene silencing that results from the pathological process of promoter methylation.

A clear view of this Darwinian model also requires some knowledge of the *kinetics* of each step of the multi-step progression—that is, how long does each step take? Some of these steps, such as the point mutations that activate *ras* oncogenes, may occur with a frequency of 10^{-6} to 10^{-7} per cell generation, while other critical steps, such as loss of heterozygosity (LOH), seem to occur with frequencies 100- to 1000-fold higher. We still do not know with any precision the frequency per cell generation of other critical events, including promoter methylation (estimated by some to be as frequent as 10^{-4} per cell generation), gene amplification, gene deletion, and losses of entire chromosomes. Hence, some steps of tumor progression may occur so rapidly (compared with others) that they will never be registered as "rate-limiting" steps, that is, steps that require an extended period of time to complete and therefore slow down the overall rate of progress toward the ultimate endpoint (such as formation of a clinically detectable tumor).

Some of these processes may be influenced by the mutagen environment of the cell or, in the case of chromosomal instability, may occur only episodically during a narrow window of time over the course of tumor progression (see Section 10.11). Then there is another moving target: as mentioned earlier, tumor cell genomes often become more mutable as tumor progression advances because of the breakdown of one or another component of the DNA repair apparatus—a topic explored in depth in Chapter 12.

With rare exception, these facts make it almost impossible to measure the kinetics of individual steps of tumor progression; this in turn makes an enumeration of all the individual steps difficult if not impossible. In principle, the sequencing of entire cancer cell genomes, which has now become almost routine, offers the prospect of determining at least one key parameter with some precision: the minimum number of somatic mutations required to create a human cancer cell. Recall that driver mutations are by definition critical participants in tumor progression and can be identified because they affect genes that are the objects of recurrent mutations (see Sidebar 11.3); each of these drivers would seem to represent a mutation that was followed by a clonal expansion of the resulting mutant cell and its progeny. With this in mind, one can use whole-genome sequencing to enumerate the number of driver mutations that coexist with the genome of a human cancer cell. Individual lung and colorectal carcinomas often carry five or six such mutations, suggesting this as the minimum number required for neoplastic growth. Unfortunately, some breast cancer genomes, sequenced in parallel, may give no sign of carrying such multiply mutated genomes. Moreover, the several dozen karyotypic alterations often found in some of these carcinomas represent another confounding element in these calculations. So, even the most advanced technologies do not offer the means to resolve definitively the dynamics of multistep tumor progression.

11.9 Multiple lines of evidence reveal that normal cells are resistant to transformation by a single mutated gene

The difficulties in cataloging the key steps in tumor progression, as listed above, indicate that we cannot rely on observations of naturally arising human tumors as the sole source of our insights into the biology of carcinogenesis. Far more definitive lessons about cancer development may well be learned by actively intervening in the process of tumorigenesis, that is, by reconstructing it in detail in the laboratory. In particular, the introduction of well-defined genetic alterations into previously normal (that is, wild-type) cells offers the prospect of elucidating with precision how specific changes in genotype collaborate to create the cancer cell phenotype.

The roots of this experimental strategy can be traced back to Temin's experiments (see Section 3.2) and, later on, to those of others in which chicken and mammalian cells propagated in culture were exposed to a variety of oncogenes by infecting tumor viruses. Subsequently, analogous experiments exploited the technique of DNA transfection to introduce oncogenes into cultured cells (see Section 4.2). In all these experiments, successful transformation was gauged by the appearance of foci of morphologically transformed cells in culture dishes. Additional tests of anchorage-independent growth and the ability to form tumors in suitable animal hosts provided further validation of the transformed state of such genetically modified cells (see Chapter 3).

Some of these experiments seemed to indicate that the genetic rules governing the transformation of mammalian cells are actually extremely simple. Recall, for example, the experiment in which a mutant, activated H-*ras* oncogene from a human bladder carcinoma was introduced via transfection into previously normal NIH 3T3 mouse fibroblasts (see Figure 4.2). Having acquired the mutant *ras* oncogene, these cells became fully transformed, to the point that they were capable of seeding tumors in appropriate host mice (see Chapter 4).

This behavior of the *ras*-transformed NIH 3T3 cells indicated that the requirements for transforming these cells were quite minimal. A single genetic alteration of these cells—the acquisition of a *ras* oncogene—sufficed to convert them to a transformed, tumorigenic state. Moreover, the mutation that originally created the *ras* oncogene was itself a simple point mutation. This suggested that a point mutation affecting one of the NIH 3T3 cells' own native H-*ras* proto-oncogenes would yield the identical outcome—full transformation to a neoplastic state. In one stroke, a point mutation should transform a normal cell into a tumor cell.

We know from our earlier discussions of human tumor cell genetics that this conclusion is demonstrably wrong. A single point mutation—indeed, a single mutational event of any sort—cannot, on its own, generate a cancer cell from a preexisting normal cell. We can verify this conclusion with a simple calculation. Given the rate at which specific point mutations occur randomly in the human genome, the number of cell divisions occurring each day in the body ($\sim 3 \times 10^{11}$), and the number of cells in the human body ($> 3 \times 10^{13}$), some have estimated that several thousand new, point-mutated *ras* oncogenes are created every day throughout the human body and that the total body burden of cells carrying *ras* oncogenes must number in the millions. Clearly, human beings are not afflicted with a comparable number of new tumors each day.

Something has gone terribly wrong here, either in these calculations or in the transfection experiments that we relied upon to gauge the genetic complexity of the transformation process. The natural place to search for problems is in the design of the experiments used to inform our thinking, specifically in the cells that were used in the transformation assay. The NIH 3T3 cells, as it turns out, are not truly normal, since they constitute a cell line—a population of cells that has been adapted to grow in culture and can be propagated indefinitely (see Section 10.1). This implies that these cells at some point underwent one or more genetic or epigenetic alterations that enabled them to grow in culture and to proliferate in an immortalized fashion.

Knowing this, investigators began in the early 1980s to examine the consequences of introducing a *ras* oncogene into truly normal cells—those from rat, mouse, or hamster embryos that had been recently explanted from living tissues and propagated *in vitro* for only a short period of time before being used in gene transfer experiments. Such cells—sometimes called **primary cells**—were unlikely to have undergone the alterations that apparently affected NIH 3T3 cells during their many-months-long adaptation to tissue culture and attendant immortalization.

The results obtained with primary rat and hamster cells were very different from those observed previously with NIH 3T3 cells. These primary cells were not susceptible to *ras*-induced transformation. Control experiments left no doubt that these cells had indeed acquired the transfected oncogene and were able to express the encoded Ras oncoprotein, but somehow they did not respond by undergoing transformation. This provided the first evidence that the act of adapting rodent cells to culture conditions and selecting for those that have undergone immortalization yields cells that have become responsive to transformation by an introduced *ras* oncogene.

The further implications of these observations are clear. Immortalized cells are not truly normal, even though they exhibit many normal traits, such as contact inhibition and anchorage dependence. Indeed, because their abnormal state renders them susceptible to *ras*-induced transformation, we might consider them to have undergone some type of premalignant genetic (or epigenetic) change long before they are confronted with this introduced oncogene.

It is clear that the selective pressures *in vitro* that yield immortalized cell lines are quite different from those that evolving premalignant cells experience within living tissues. Nonetheless, the biological traits and, quite possibly, the underlying mutant genes acquired during propagation *in vitro* may be identical to many of those arising during tumor progression *in vivo*. (In fact, our discussions in the last chapter revealed that the *same* regulatory pathways—those controlled by p53 and pRb proteins—that are altered during cell immortalization are also disrupted in a wide variety of cancer cell genomes, including those of human tumors.)

These experiments with primary cells have been extended by introducing activated *ras* oncogenes into the colonic epithelial cells of mice. The resulting *ras* oncogene–expressing cells create nothing more than hyperplastic epithelia, that is, cells that are present in excessive numbers but are, in other respects, essentially normal (**Figure 11.23A**).

Certain experiments of nature also support the notion that single mutations are not sufficient for the development of cancers. For example, some individuals are born carrying a germ-line mutation of the gene encoding the Kit growth factor receptor; such mutations create a constitutively active, ligand-independent Kit receptor, which functions as a potent oncoprotein. These individuals are at high risk for developing gastrointestinal stromal tumors (GISTs), but these tumors only become apparent several decades after birth, even though a constitutively active Kit oncoprotein has been functioning in many of their cells since birth (see Sidebar 5.5). Similarly, some individuals have been documented who carry mutant H-*ras* alleles in their germ lines yet usually develop tumors only after several decades' time. Early childhood leukemias in monozygotic (that is, identical) twins provide equally dramatic examples of the inability of single mutations, acting on their own, to create clinically apparent tumors (see Figure 11.23B).

Yet other observations indicate that even two mutant alleles, involving an activated proto-oncogene (K-*ras*) and an inactivated tumor suppressor gene (*APC*), still do not suffice to generate carcinomas in the human colon (see Figure 11.23C). Observations like these suggest that multiple genetic changes—perhaps more than two—seem to be required in order for a human cell to reach a tumorigenic state.

11.10 Transformation usually requires collaboration between two or more mutant genes

The resistance of fully normal, primary rodent cells to *ras*-induced transformation led to an interesting question: Were there yet other oncogenes that could immortalize embryo cells and, at the same time, render these cells susceptible to transformation by *ras*? In the early 1980s, research on DNA tumor viruses indicated that some carried multiple oncogenes in their genomes (**Sidebar 11.4**). Polyomavirus, for example, bears two oncogenes, termed *middle T* and *large T*; in 1982, these two oncogenes were found to collaborate with one another to transform rodent cells. The large T oncoprotein seemed to aid in the adaptation of cells to tissue culture conditions and to

(A)

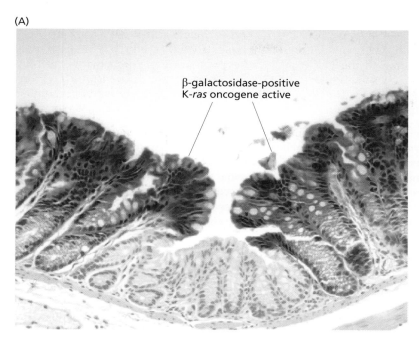

β-galactosidase-positive
K-*ras* oncogene active

(B)

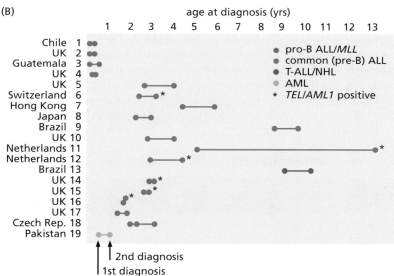

age at diagnosis (yrs)

1 2 3 4 5 6 7 8 9 10 11 12 13

Chile 1
UK 2
Guatemala 3
UK 4
UK 5
Switzerland 6
Hong Kong 7
Japan 8
Brazil 9
UK 10
Netherlands 11
Netherlands 12
Brazil 13
UK 14
UK 15
UK 16
UK 17
Czech Rep. 18
Pakistan 19

● pro-B ALL/*MLL*
● common (pre-B) ALL
● T-ALL/NHL
● AML
* *TEL/AML1* positive

2nd diagnosis
1st diagnosis

(C)

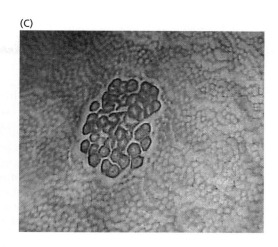

Figure 11.23 One or two oncogenic lesions do not suffice to generate full-blown tumors (A) A mouse germ line was re-engineered to create colonic epithelial cells in which both a mutant K-*ras* oncogene and the β-galactosidase gene could be activated in scattered cells by an infecting adenovirus. As indicated here, colonic epithelial cells in which both β-galactosidase (*blue*) and the K-*ras* oncogene (*not visible*) have become expressed create localized regions of hyperplasia in which the epithelial cells are otherwise normal, indicating that the *ras* oncogene on its own does not suffice to transform these cells into a tumorigenic state. (B) Monozygotic (identical) twin pairs have been documented throughout the world in which both twins develop the same type of leukemia. The leukemias invariably share a common chromosomal marker or mutation, indicating that they derive from the same clone of initiated cells. The fact that many of these leukemias are diagnosed at quite different postnatal ages (*dots*) indicates that these initiating somatic mutations, which occurred *in utero*, were not, on their own, sufficient to trigger the formation of a clinically apparent leukemia. The labels (*top right*) and the associated colors denote different subtypes of leukemia identified by distinctive gene markers. (C) A cluster of aberrant crypt foci (ACFs) in the human colon are clearly revealed by colonoscopy. ACFs like this one frequently carry an inactivated *APC* tumor suppressor gene together with an activated K-*ras* oncogene; they may progress into adenomatous polyps, which themselves are still not full-fledged colon carcinomas. Hence, even two oncogenic genetic lesions, acting together, generate human cells that are still several steps short of being fully transformed. (A, courtesy of K.M. Haigis and T. Jacks. B, from M.G. Greaves et al., *Blood* 102:2321–2333, 2003. C, courtesy of Y. Niitsu.)

facilitate their immortalization, while the middle T protein elicited many of the phenotypes associated with the *ras* oncogene—rounding up of cells, loss of contact inhibition, and acquisition of anchorage-independent growth. Soon, a number of other DNA tumor viruses were found to employ similar genetic strategies for cell transformation.

The genetics of transformation by DNA tumor viruses suggested that mutant cellular genes might also collaborate in cell transformation. In fact, a line of human promyelocytic leukemia cells was discovered to carry both an activated N-*ras* and an activated *myc* oncogene. This suggested the possibility that these two cellular oncogenes were cooperating to create the malignant phenotype of the leukemia cells. This notion was soon borne out by a simple experiment: when a *myc* oncogene was introduced together with an H-*ras* oncogene into rat embryo fibroblasts (REFs), the cells responded by becoming morphologically transformed (**Figure 11.24**) and, more important, tumorigenic; neither of these oncogenes, on its own, could create such transformed cells.

This result yielded several interesting conclusions. These two cellular oncogenes clearly affected cell phenotype in quite different ways, since they were able to complement one another in eliciting cell transformation. Each seemed to be specialized to

Figure 11.24 Oncogene collaboration in rodent cells *in vitro* Cultures of early-passage rat embryo fibroblasts or baby hamster kidney cells were exposed to cloned DNAs via the calcium phosphate gene transfection procedure (see Figure 4.1). Introduction of a *myc* or adenovirus *E1A* oncogene into these cells (*left*), on its own, did not yield foci of transformants, although the introduced oncogenes did facilitate the establishment of these early-passage cells in long-term culture. Introduction of an H-*ras* oncogene via transfection (*right*) also did not yield foci of transformed cells, although it did allow significant numbers of these cells to form anchorage-independent colonies when they were introduced into a semi-solid medium such as dilute agar. However, the simultaneous introduction of *ras* + *myc* or, alternatively, *ras* + *E1A* did generate foci of transformed cells that were able to form tumors when injected into syngeneic or immunocompromised hosts.

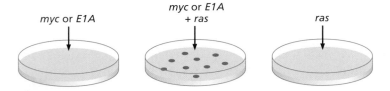

evoke a subset of the cellular phenotypes associated with the transformed state. For example, *ras* was able to elicit anchorage independence, a rounded, refractile appearance in the phase microscope, and loss of contact inhibition; *myc* helped the cells to become immortalized and reduced somewhat their dependence on growth factors. Similar results were found in experiments in which the *E1A* oncogene of human adenovirus 5 was used as the collaborating partner of a *ras* oncogene. Once again, the two collaborating oncogenes were found to have complementary effects on cell phenotype.

Soon, other pairs of oncogenes were discovered to be capable of collaborating with one another to induce transformation of cells *in vitro* and tumorigenesis *in vivo* (**Table 11.1**). The *ras* oncogene, as an example, could also collaborate with the SV40 *large T* oncogene, with the polyoma *large T* oncogene, or with a mutant *p53* gene in cell transformation. Conversely, *myc* could collaborate also with the polyoma *middle T* oncogene, with *src*, or with the *raf* oncogene to transform cells.

In most cases, the genes within a collaborating pair could be placed in two functional groups—those with *ras*-like and those with *myc*-like properties. In fact, not all *ras*-like oncogenes elicit identical effects in cells; the same could be said of the members of the *myc*-class. Provocatively, the *ras*-like oncogenes encode largely cytoplasmic oncoproteins, while the *myc*-like oncogenes encode products that tend to be nuclear (**Table 11.2**). We now know that the Ras-like oncoproteins are components of the cytoplasmic mitogenic signaling cascade (Chapter 6), while the Myc-like oncoproteins perturb in various ways the cell cycle control machinery, which operates in the nucleus (see Chapter 8).

Table 11.1 Examples of collaborating oncogenes *in vitro* and *in vivo*

"*ras*-like" oncogene[a]	"*myc*-like" oncogene[a]	Target cell or organ
***In vitro* transformation**		
ras	*myc*	transfected rat embryo fibroblasts (REFs)
ras	*E1A*	transfected rat kidney cells
ras	SV40 *large T*	transfected REFs
Notch-1	*E1A*	transfected rat kidney cells
***In vivo* tumorigenesis**		
middle T	*large T*	polyomavirus-induced murine tumors
mil (= *raf*)	*myc*	MH2 avian leukemia virus chicken tumors
erbB	*erbA*	avian erythroblastosis virus chicken tumors
pim1	*myc*	mouse leukemia virus tumors
abl	*myc*	mouse leukemia virus tumors
Notch-1/2	*myc*	thymomas in transgenic mice
bcl-2	*myc*	follicular lymphomas in transgenic mice

[a]The terms "*ras*-like" and "*myc*-like" refer to functional classes rather than genes encoding components of a common signaling pathway. "*ras*-like" oncogenes tend to encode components of cytoplasmic signaling cascades, while "*myc*-like" oncogenes tend to encode nuclear proteins.

Sidebar 11.4 The multiple collaborating oncogenes of DNA tumor virus genomes often do not succeed in transforming cells to a tumorigenic state Full transformation of rodent cells appears to require both the relief of growth suppression imposed by the pRb and p53 circuits and the provision of actively mitogenic signals through activation of the Ras pathway or one of its downstream branches. The multiple oncoproteins made by certain DNA tumor viruses (see Table 9.1) satisfy many of these requirements. For example, the polyomavirus large T oncoprotein binds to pRb, while its middle T oncoprotein binds to and activates Src (and related kinases), PI3K, PLC-γ, and perhaps Shc, which are all components of the complex Ras mitogenic pathway (see Chapter 6).

The distantly related bovine papillomavirus (BPV) encodes an E7 oncoprotein, which inactivates pRb. In addition, BPV specifies the E5 oncoprotein, which activates the Ras → MAPK pathway through its ability to associate with and trigger dimerization of the PDGF receptor; the BPV E5 protein may also trigger the activation of Src-like kinases in the cytoplasm. Other research, less well developed, suggests that the human papillomavirus E5 protein can also favor dimerization of growth factor receptors, such as the EGF-R, that feed signals into the Ras signaling cascade. Some DNA tumor viruses, such as SV40 and human adenovirus, do not specify actively mitogenic oncoproteins and thus are limited to immortalizing infected cells. In addition to these mechanisms, most and perhaps all DNA tumor viruses devote some of their genomes to specifying anti-apoptotic proteins, such as those that sequester and inactivate the p53 tumor suppressor protein (see Sidebar 9.8).

While a number of DNA tumor viruses bearing multiple oncogenes have been implicated in triggering human cancers (see Table 4.6), none of these viruses is able, on its own, to fully transform an infected normal human cell into a tumor cell. In each case, additional steps involving somatic mutation and promoter methylation are required to convert virus-infected cells into tumor cells. The requirement for these additional (often poorly understood) cellular changes explains the great disparities between the number of people infected by these viruses and the number of those who contract virus-induced malignancies. For example, 90% of people in the West have been infected by Epstein–Barr virus (EBV) but only a minute fraction contract Burkitt's lymphoma. The large number of women chronically infected with human papillomavirus (HPV) have only about a 3% lifetime risk of developing a cervical carcinoma, once again indicating the critical role of additional, random events in causing virus-infected cells to progress to a tumorigenic state.

Table 11.2 Physiologic mechanisms of oncogene collaboration[a]

Oncogene pair	Cell type	Mechanisms of action
ras + SV40 large T	rat Schwann cells	ras: proliferation + proliferation arrest
		large T: prevents proliferation arrest and reduces mitogen requirement
ras + E1A	mouse embryo fibroblasts	ras: proliferation and senescence
		E1A: prevents senescence
erbB + erbA	chicken erythroblasts	erbB: induces GF-independent proliferation
		erbA: blocks differentiation
TGF-α + myc	mouse mammary epithelial cells	TGF-α: induces proliferation and blocks apoptosis
		myc: induces proliferation and apoptosis
v-sea + v-ski	avian erythroblasts	v-sea: induces proliferation
		v-ski: blocks differentiation
bcl-2 + myc	rat fibroblasts	bcl-2: blocks apoptosis
		myc: induces proliferation and apoptosis
ras + myc	rat fibroblasts	ras: induces anchorage independence
		myc: induces immortalization
raf + myc	chicken macrophages	raf: induces growth factor secretion
		myc: stimulates proliferation
src + myc	rat adrenocortical cells	src: induces anchorage and serum independence
		myc: prolongs proliferation

[a]In each pair, the first oncogene encodes a cytoplasmic oncoprotein while the second oncogene encodes a nuclear oncoprotein.

These oncogene collaboration experiments provided a crude *in vitro* model of multi-step transformation *in vivo* and suggested a rationale for the complex genetic steps that accompany and cause tumor formation in human beings: each of the genetic changes provides the nascent tumor cell with one or more of the phenotypes that it needs in order to become tumorigenic (see Table 11.2). These unique contributions seem to derive from the ability of each of these oncogenes to perturb a specific subset of regulatory circuits within a cell. Moreover, these experiments suggested that cell proliferation and cell survival are governed by a number (two or more) of distinct regulatory circuits, all of which must be perturbed before the cell will become tumorigenic.

Previously, we noted that oncogenes act in a pleiotropic fashion on cell phenotype, in that each of these genes is able to concomitantly induce a number of distinct changes in cell phenotype. Accepting this, we must also recognize that, as multi-talented as oncogenes are, none of them seems able, on its own, to evoke all of the changes that are required for a normal cell to become transformed into a tumorigenic state. In the case of cellular oncogenes, there would seem to be an obvious evolutionary rationale for this, which we touched on earlier in this chapter: a mammalian cell cannot tolerate the presence of a proto-oncogene in its genome that could, through a single mutational event, yield an oncogene capable of transforming this cell into a full-fledged tumor cell. Such a proto-oncogene would place each cell in the body only a single, small step away from malignancy and create too much of a liability for the organism as a whole. This represents another version of the argument that cells and tissues must place multiple obstacles in the path of normal cells in order to prevent them from becoming tumorigenic.

Interestingly, under certain experimental conditions, researchers can thwart these defense mechanisms and succeed in transforming cells with a single genetic element. For example, a chicken embryo fibroblast infected *in vitro* with Rous sarcoma virus (RSV) appears to be transformed to tumorigenicity in this single step. However, *in vivo*, RSV-infected cells form tumors only at sites of wounding, including the needle track formed when RSV is injected into muscle; hence, the changes occurring in fibroblasts during wound healing seem to be required to help the RSV *src* gene transform these cells into tumor cells. Similarly, experiments with cultured rat embryo fibroblasts indicate that when a *ras*-transformed cell is isolated from other cells in the Petri dish, it can proliferate to form a colony of tumorigenic cells, even without the aid of a collaborating oncogene like *myc*. However, when such a cell is surrounded by normal neighbors (a likelihood *in vivo*), it is unable to generate a focus. Accordingly, single-step transformation experiments sometimes succeed because they fail to recapitulate certain anti-cancer mechanisms operating in living tissues, which normally require additional alterations before tumor progression can proceed.

11.11 Transgenic mice provide models of oncogene collaboration and multi-step cell transformation

In many rodent models of cancer pathogenesis, tumors can be triggered by exposure of an animal to a mutagenic carcinogen, which acts in a random (sometimes termed **stochastic**) fashion to generate the mutant cellular alleles leading to cancer. An alternative to such experimental protocols can be achieved through the insertion of an already-mutant, activated oncogene into the germ line of a laboratory mouse, thereby guaranteeing expression of this gene in some of its tissues. In practice, the expression of this oncogenic allele must be confined to a small subset of tissues in the mouse. (If its expression were allowed in all tissues, including those of the developing embryo, chances are that embryogenesis would be so profoundly disrupted that the developing fetus would die long before the end of gestation.)

An early version of this strategy for creating cancer-prone, **transgenic** mice (see Figure 9.23A) involved the insertion of oncogenic alleles of the *ras* or *myc* genes into the mouse germ line (**Figure 11.25**). In one influential set of experiments, expression of the *ras* oncogene and the *myc* oncogene was placed under the control of the transcriptional promoter of mouse mammary tumor virus (MMTV), a retrovirus that specifically targets mammary tissues. The viral promoter is expressed at significant levels only in mammary glands and, to a lesser extent, in salivary glands.

As anticipated, the presence of either one of these oncogenic **transgenes** in the mouse germ line predisposed mice to breast cancer and, to a much lesser extent, to salivary gland tumors. In spite of the expression of either a *ras* or a *myc* oncogene in most if not all of the mammary epithelial cells of these mice, their mammary glands showed either minimal morphologic changes (in the case of the *myc* transgene) or hyperplasia (in the case of the *ras* transgene). Moreover, breast tumors were observed only beginning at four weeks of age (for the MMTV-*ras* mice) and 12 weeks (for the MMTV-*myc* mice)—significantly long latency periods (see Figure 11.25). This proved conclusively that the presence of a single oncogene within a normal cell in living tissue is not, on its own, sufficient to transform this cell into a tumor cell. Instead, the kinetics of breast cancer formation in these mice pointed to the necessary involvement of one or more additional stochastic events before these *ras* or *myc* oncogene–bearing mammary cells progressed to a tumorigenic state (see also Figure 11.23A).

Double-transgenic mice that carried both MMTV-*ras* and MMTV-*myc* transgenes were created through mating between the two transgenic strains described above. These double-transgenic mice contracted tumors at a greatly accelerated rate and at high frequency compared with mice inheriting only one of these transgenes (see Figure 11.25). Therefore, the two oncogenic transgenes could collaborate *in vivo* to generate tumors, corroborating the conclusions of the *in vitro* experiments described in Section 11.10.

Interestingly, even with two mutant oncogenes expressed in the great majority of mammary cells from early in development, tumors did not appear in these mice soon after birth, but instead were seen with great delay. Hence, the concomitant expression of two powerful oncogenes was still not sufficient to fully transform mouse mammary epithelial cells (MECs); instead, these cells clearly required at least one additional stochastic event, ostensibly a somatic mutation, before they would proliferate like full-blown cancer cells. (A hint about the identity of this third, stochastic event has come from careful analysis of rat cells that have been transformed *in vitro* by the *ras + myc* protocol; sooner or later, such cells usually acquire a mutation or a promoter methylation that leads to inactivation of the p53 tumor suppressor pathway; see Chapter 9.)

Note that the collaborative actions of transgenic oncogenes were already mentioned earlier, when we read of the synergistic actions of *myc* and *bcl-2* transgenic oncogenes in promoting lymphomagenesis (see Figure 9.23). In this case, the benefit of *bcl-2* (and its *bcl-x$_L$* cousin) derives largely from its anti-apoptotic effects. This illustrates the fact that oncogenes can collaborate through a variety of cell-physiologic mechanisms to promote tumor formation, a point made in Table 11.2.

11.12 Human cells are constructed to be highly resistant to immortalization and transformation

The biological lessons derived from studying mice and rats are usually directly transferable to understanding various aspects of human biology. Even though 80 million years may separate us from the most recent common ancestor we share with rodents, the great majority of biological and biochemical attributes of these distant mammalian cousins are present in very similar, if not identical, form in humans. The genomes of humans and rodents also seem very similar: essentially all of the roughly 20,000 protein-encoding genes discovered in the human genome have been found to have mouse orthologs. It stands to reason that the biological processes of immortalization and neoplastic transformation should also be essentially identical in rodent and human cells.

The biological reality is, however, quite different. It is easy to immortalize rodent cells simply by propagating them through a relatively small number of passages *in vitro*. Spontaneously immortalized cells arise frequently and become the progenitors of cell lines, such as the NIH 3T3 cells discussed earlier. In contrast, human cells rarely, if ever, become immortalized following extended serial passaging in culture (see Chapter 10).

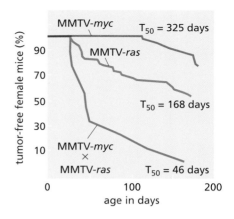

Figure 11.25 Oncogene collaboration in transgenic, cancer-prone mice The ability to create transgenic mice (see Figure 9.23A) made it possible to determine whether the *myc* and *ras* oncogenes are able to collaborate *in vivo* as well as *in vitro* (see Figure 11.24). Use of the MMTV (mouse mammary tumor virus) transcriptional promoter to drive expression of each oncogene ensured expression of both transgenes largely in the mammary glands. Transgenic mice carrying the MMTV-*myc* and the MMTV-*ras* transgene were bred to create double-transgenic mice carrying both transgenes. The incidence of mammary carcinomas in the three strains of mice—those carrying the MMTV-*myc* transgene only (*red curve*), the MMTV-*ras* transgene only (*green curve*), or both transgenes (*blue curve*)—was tracked over many months. The percentage of tumor-free mice (*ordinate*) is plotted versus the age of the various strains of mice. "T$_{50}$" indicates the number of days required for one-half of the mice of a particular genotype to develop detectable mammary carcinomas. (From E. Sinn et al., *Cell* 49:465–475, 1987.)

Eventually, cultured human cells stop growing and become senescent, and spontaneously immortalized cell clones do not emerge.

Attempts to experimentally transform cells have shown comparable interspecies differences. Primary rodent cells become transformed *in vitro* following the introduction of pairs of oncogenes (such as *ras* and *myc*; see Section 11.10), while such pairs of introduced oncogenes consistently fail to yield tumorigenic human cells. In fact, the human cells emerging from such co-transfections are not even immortalized and therefore senesce sooner or later.

These repeated failures at cell transformation prevented researchers from addressing a simple yet fundamental problem in human cancer biology: How many intracellular regulatory circuits need to be perturbed in order to transform a normal human cell into a cancer cell? Sequence analyses of human cancer cell genomes were of little help here. As argued earlier (see Section 11.7), cancer cells derived from human tumors possess a plethora of genetic alterations, far more than the relatively small number that play causal roles in tumorigenesis. This forced researchers to consider experimental cell transformation as an alternative way of addressing this problem. Thus, they asked precisely how many genetic changes must be introduced experimentally into human cells in order to transform them.

The general strategy was inspired by the experience with cultured rodent cells, which indicated that once cells were immortalized in culture, they became responsive to transformation by a *ras* oncogene. The fact that the telomere biology of rodent and human cells differs so starkly (see Section 10.9) seemed to explain at least part of the difficulty of immortalizing human cells, and thus their very different responses to introduced oncogenes. Recall that the cells of laboratory mice usually carry extremely long telomeric DNA (as long as 40 kilobases) while normal human cells have far shorter telomeres; moreover, telomerase activity seems to be regulated differently in cells of the two species. Accordingly, immortalization of human cells might be facilitated by adding the *hTERT* gene to other immortalizing oncogenes introduced into these cells.

In fact, introduction of an *hTERT* gene in addition to the SV40 *large T* oncogene (whose product inactivates both pRb and p53 tumor suppressor proteins) did indeed yield immortalized human cells. (Alternative means of inactivating pRb and p53, such as introduction of human papillomavirus *E6* and *E7* oncogenes, succeeded as well.) And once immortalization was achieved through these changes, the resulting human cells could then be transformed morphologically in the culture dish by introduction of an activated *ras* oncogene.

These morphologically transformed human cells were still not fully transformed, however, as indicated by their inability to form tumors when implanted in immunocompromised host mice. (The faulty immune systems of such mice ensure that tissues of foreign origin, such as human tumors, are not eliminated by immunological attack.) These cells still required one more alteration, this one achieved by introduction of the gene encoding the SV40 small T oncoprotein. Small T perturbs a subset of the functions of the abundant cellular enzyme termed protein phosphatase 2A (PP2A; **Sidebar 11.5**).

Taken together, these experiments demonstrated that five distinct cellular regulatory circuits needed to be altered experimentally before human cells can grow as tumor cells in immunocompromised mice (**Figure 11.27**). These changes involve: (1) the mitogenic signaling pathway controlled by Ras (see Chapter 6); (2) the cell cycle checkpoint controlled by pRb (see Chapter 8); (3) the alarm pathway controlled by p53 (see Chapter 9); (4) the telomere maintenance pathway controlled by hTERT (see Chapter 10); and (5) the signaling pathways controlled by protein phosphatase 2A; the latter include effects modulating the activity of the mTOR, Myc, β-catenin, and PKB/Akt signaling proteins. (It is also possible that perturbation of other cellular signaling pathways quite distinct from the five listed here might also permit the experimental transformation of human cells.)

Experiments like these indicate why human cells are highly resistant to transformation. At the same time, it remains unclear whether the steps required to experimentally

Sidebar 11.5 Protein phosphatase 2A is an extremely complex holoenzyme The term "PP2A" subsumes a collection of enzymes, each of which is composed of two regulatory subunits (A and B) and a catalytic C subunit. As many as 18 alternative B-type subunits, two alternative A-type subunits, and two alternative C-type subunits associate in various combinations to create a still-unknown number of distinct **heterotrimeric** holoenzymes (**Figure 11.26**). These dephosphorylate a constituency of phosphoprotein substrates that may number in the thousands; this dephosphorylation appears limited to phosphoserine and phosphothreonine residues. By blocking a subset of the B-type subunits from assembling into heterotrimeric (A + B + C) complexes, the SV40 small T (sT) oncoprotein prevents the dephosphorylation of certain critical cellular phosphoproteins and somehow facilitates cell transformation.

The biological mechanism through which this inactivation of a subset of the multiple functions of PP2A aids human cell transformation remains obscure. The best available clues are (1) that partial suppression of its Aα or complete suppression of its B56γ subunit expression mimics the actions of sT in experimental cell transformation; (2) that the gene encoding the Aβ subunit is mutated in several commonly occurring human carcinoma types; and (3) that the actions of small T allow cancer cells to proliferate under conditions of low nutrient supply that would stop normal cells from advancing through the growth-and-division cycle.

These observations raise the question whether certain forms of PP2A are important players in the pathogenesis of various human cancers, the great majority of which are not associated with DNA tumor virus infections. While various PP2A subunits are found to be mutated or absent in human tumors, a clear and consistent pattern of PP2A loss has not yet emerged.

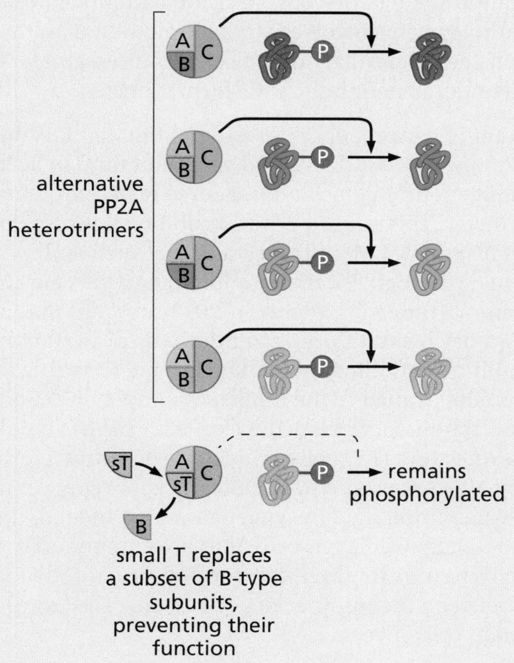

Figure 11.26 Protein phosphatase 2A The PP2A holoenzyme is composed of A, B, and C subunits. There are 2 types of both A and C subunits, and 18 types of B subunit. The C subunits carry the catalytic activity of the holoenzyme and the B subunits direct the holoenzyme to specific substrates, while the A subunits act as scaffolding to assemble B and C subunits into holoenzyme complexes. The various holoenzymes are responsible for dephosphorylating cellular phosphoproteins bearing either phosphoserine or phosphothreonine residues. The small T (sT) oncoprotein of SV40 can associate with the A and C subunits (*bottom*) and thereby prevent the association of certain B-type subunits with A + C. In so doing, sT prevents PP2A from dephosphorylating a subset of its normal constituency of substrates. This contributes in some still-poorly understood fashion to the experimental transformation of human cells. (Adapted from S. Colella et al., *Int. J. Cancer* 93:798–804, 2001.)

transform human cells *in vitro* accurately reflect the changes that normal cells must undergo within human tissues in order to acquire the attributes of cancer cells. In fact, four of these changes (involving Ras and hTERT activation, and pRb and p53 inactivation) are commonly seen in the cells of human tumors. For example, in a recent genetic survey of glioblastomas, 74% carried changes in the three critical pathways involving the pRb, p53, and Ras–MAPK proteins, and virtually all of these tumors exhibited elevated expression of hTERT as well. However, it remains unclear whether the fifth alteration—deregulation of a subset of the actions of PP2A—occurs during the formation of these and other spontaneously arising human cancers. In addition, yet other genetic changes, not revealed by these experiments, may be required before the cells within certain human tissues are able to generate clinically detectable tumors. These experiments also leave another issue unsettled: Are the genetic and biochemical rules governing cell transformation identical in all human cancers?

The requirements for transformation of human fibroblasts, as detailed above, are mirrored by the behavior of a variety of other human cell types: kidney cells, prostate, ovarian, and small-airway (lung) epithelial cells, and astrocytes require an identical set of introduced genes in order to become experimentally transformed to a tumorigenic state. This suggests that a common set of biochemical pathways must be deregulated in a wide variety of adult human cell types before such cells become transformed. The number of changes (five) required for human cell transformation,

Figure 11.27 Intracellular pathways involved in human cell transformation The experimental transformation of human cells has been achieved through the insertion of various combinations of cloned genes into cells. Initially, a combination of three genes encoding the SV40 early region [which specifies both the large T (LT) and small T (sT) oncoproteins], the hTERT telomerase enzyme, and a *ras* oncogene was found to suffice for the transformation of a variety of normal human cell types to a tumorigenic state. These introduced genes were found to deregulate five distinct regulatory pathways involving (1) pRb-mediated G_1 cell cycle control, (2) p53, (3) protein phosphatase 2A (PP2A), (4) telomere maintenance, and (5) Ras mitogenic signaling. Subsequent work has found that other combinations of cloned genes suffice as well. For example, activation of the Ras signaling pathway can be mimicked by introducing a combination of a constitutively active alleles of *MEK* (MEK^{DD}) and *Akt/PKB*. Further dissection of these pathways has also revealed that a combination of MEK^{DD} with *IKBKσ* or *PAK1* with *Akt/ PKB* also suffices to replace *ras* as an oncogenic activator. The disruption of pRb function can be achieved by a combination of ectopically expressed, CDK inhibitor–resistant CDK4 + cyclin D1 or a short hairpin construct (shRNA) directed against the *Rb* gene; p53 can be disrupted by an introduced dominant-negative *p53* allele or an shRNA directed against the *p53* gene; telomeres can be maintained by activating endogenous *hTERT* expression through a combination of introduced SV40 *LT* + *myc*; and PP2A function can be disrupted by an shRNA that inhibits the synthesis of the B56 subunit of PP2A. It is unknown whether these five pathways are required for the experimental transformation of all human cell types, and whether deregulation of all of these five pathways occurs in spontaneously arising human tumors. A comma in figure indicates "or."

pathway	Ras	pRb	p53	telomeres	PP2A
genes/agents used to deregulate pathway	*ras, MEK+ Akt/PKB, MEK+IKBKε, PAK1+ Akt/PKB*	SV40 *LT, CDK4 + D1,* HPV *E7, Rb* shRNA	SV40 *LT,* DN *p53,* HPV *E6, p53* shRNA	*hTERT, myc +* SV40 *LT*	SV40 *sT* in some cells: *myc Akt/PKB+Rac1, PI3K, B56* shRNA

as listed in Figure 11.27, must be compared with the numbers of driver mutations estimated from whole-genome sequencing to coexist within individual human tumor cell genomes (see Section 11.8). Five or six were cited for lung and colorectal carcinomas, and more recent sequencing of glioblastoma and pancreatic carcinoma genomes has yielded estimates of, in both cases, 6 ± 2.

These various lines of investigation seem to be converging on a number, possibly half a dozen, of the critical rate-limiting events involved in the formation of many human solid tumors. Still, it is likely that certain normal human cells require a greater or lesser number of changes in order to undergo neoplastic transformation.

For example, a number of pediatric cancers occur so early in life that it is difficult to imagine how the cells in these tumors could have had sufficient time to accumulate the cohort of mutations (and epigenetic changes) that seem to be required for the formation of many adult malignancies. This suggests the possibility that some pediatric cancers arise directly from certain embryonic cell types, and that such embryonic cells may be more readily transformed (through fewer alterations) than the cells that serve as the precursors of adult tumors. (Indeed, a report in 2012 revealed that an experienced research team that had discovered dozens to hundreds of mutations in the genomes of adult tumors had difficulty finding more than a single recurring mutation in early childhood rhabdoid tumors, similar to the unusually simple mutational spectrum in the genomes of retinoblastoma patients.) The extreme case of an embryonic cell type is provided by embryonic stem (ES) cells, which can be extracted from very early embryos. ES cells are, by all measurement, genetically wild type, yet are tumorigenic (yielding teratomas) when implanted in syngeneic hosts. Indeed, they seem to be the only example of genetically wild-type cells that are tumorigenic. Perhaps certain cells within later embryos require for their transformation a number of genetic changes that is intermediate between the number needed by ES cells (zero) and the number required by adult human cells (five).

The need to introduce five genes into human cells in order to transform them contrasts starkly with the responses of mouse cells, which require only two genetic alterations in order to undergo transformation (for example, activated *ras* plus inactivated *p53*). These profound differences in the behavior of human versus mouse cells require some type of biological rationale. In both species, evolutionary pressures would seem to reduce, wherever possible, the risk that an individual organism will develop a tumor during its reproductive life span. This risk, in turn, is likely to be affected by the process of cell division, which is a major force in generating the mutant cells and cell populations that are likely to spawn tumors. Hence, in general, cancer risk is likely to be proportional to the number of cell divisions that individual organisms experience during their reproductive life span. In fact, the cells in a mouse pass through about 10^{11} mitoses in a mouse lifetime, while those in a human body pass through about 10^{16} cell cycles in a human lifetime. These numbers, on their own, indicate the enormously increased risk of cancer development that is intrinsic to our biology relative to that of the mouse.

This logic indicates that in response to the ever-present dangers created by our large bodies and long life spans, human cells and tissues have been hard-wired by evolution to be far more resistant to cell transformation. This notion—really a speculation—still requires some validation and generalization. For example, do the cells of a bumblebee bat or Etruscan shrew (both of approximately 2 g body weight) and those of the blue whale (of approximately 1.3×10^8 g body weight; **Figure 11.28**) require proportionally fewer and more hits, respectively, before they become cancerous? (The difference in

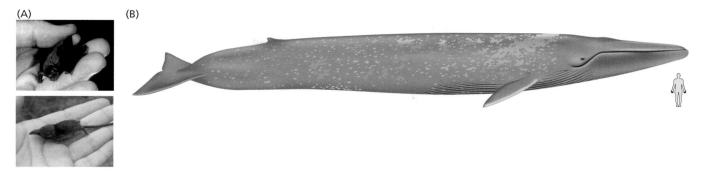

(A)   (B)

Figure 11.28 Mammalian body size and relative risk of cell transformation While the sizes of individual cells are quite comparable in various mammalian species, the overall body mass and thus cell number varies enormously. Since passage through each cell cycle creates the danger of genome alterations, this suggests that the risks of cancer can vary enormously from one species to the next. (A) The bumblebee bat of Thailand (*top*), said to be the smallest mammal, weighs just 1.5 g, a bit less than the Etruscan shrew (*bottom*); the lifespan of the bumblebee bat is unknown, while that of the Etruscan shrew is thought to be ~2 years. Etruscan shrews reach reproductive age by 3–4 weeks (in contrast to ~12–15 years in humans). (B) The blue whale has a body weight of about 1.3×10^8 g, reaches sexual maturity in 5–10 years, and has a life span of about 80 years, much like that of humans (*lower right*). The eight-orders-of-magnitude difference in mass (and therefore cell number) between the bat and the whale together with the approximately 40-fold difference in life span suggests a difference of more than 10^9 in the number of cell divisions that the two organisms experience in a lifetime. [Since the metabolic rate of the small mammals may be as much as 10^3 times higher than the whale's, and since much of the mutational burden of the cell genome derives from by-products of oxidative metabolism (see Section 12.5), the blue whale may experience only a 10^6-fold higher risk of cancer than the bat.] (A, © Merlin D. Tuttle, Bat Conservation International, www.batcon.org; and © Stella Nutella/ Wikipedia/CC BY-SA 2.5. B, courtesy of Uko Gorter.)

cumulative mitoses in a lifetime is likely to be even greater, since large mammals often live 50 times longer than tiny ones!)

The various experimental demonstrations of oncogene collaboration in mouse and human cells may well serve as good models of how multi-step tumorigenesis actually occurs within the human body. Thus, each mutation (or promoter methylation) sustained by a population of cells perturbs or deregulates yet another intracellular signaling pathway, until all the key control circuits have been disrupted. Once this is accomplished, the cells in this population may be fully transformed and therefore capable of generating a vigorously growing tumor. In fact, analyses of human tumor cell genomes reveal far more confounding results that are difficult to reconcile with such a simple and satisfying conceptual scheme (**Sidebar 11.6**).

Significantly, the experimental manipulations used to transform human cells to a tumorigenic state usually yield cells that form localized primary tumor masses having little if any tendency (1) to extend beyond their boundaries and invade nearby

Sidebar 11.6 The genetics of actual human tumors confounds our understanding of how cancer progression occurs The simplest scheme of multi-step tumor progression states that each successive step involves the disruption or deregulation of yet another key cellular signaling pathway. Hence, each of the mutant (or methylated) genes found in the genome of a human cancer cell should affect a distinct regulatory pathway, and the mutations accumulated by the end of tumor progression should collaborate with one another to program neoplastic growth. In fact, the actual genetic evidence often conflicts with such thinking.

Many human colorectal carcinoma (CRC) genomes carry mutations that lead to the activation of both PI3 kinase and B-Raf, which makes sense, since these two mutations affect distinct, complementary pathways that lie downstream of Ras (see Chapter 6). However, many other CRCs have mutations that activate both Ras and PI3 kinase, which makes no sense, since a Ras oncoprotein is thought to be capable of directly activating PI3 kinase; these two mutations therefore seem to be functionally redundant rather than complementary.

Conversely, mutations that are expected to collaborate with one another, such as those affecting the *ras* and *p53* genes, are often mutually exclusive. Thus, among CRCs, some bear *ras* mutations, while others carry *p53* mutations, and tumors carrying both are quite uncommon (contrary to the initial depiction of the genetic pathway leading to these cancers; see Section 11.3). Similarly, among human bladder carcinomas, those tumors bearing mutations that activate the fibroblast growth factor receptor-3 (yielding an effect similar to *ras* oncogene activation) rarely carry *p53* mutations and vice versa. Observations like these are difficult to reconcile with our current understanding of how these genes and encoded proteins operate—which only says that our perceptions about these issues will, sooner or later, require substantial revision.

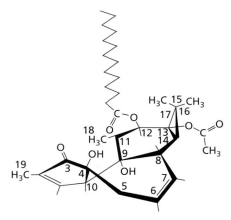

Figure 11.29 TPA, an important promoter of skin tumorigenesis The stereochemical structure of 12-*O*-tetradecanoylphorbol-13-acetate (TPA), also known as phorbol-12-myristate-13-acetate (PMA), is shown here. TPA is extracted from the seeds of the croton plant, specifically *Croton tiglium*. Its main target in cells is protein kinase C-α (PKC-α), which it activates.

tissues and (2) to seed distant metastases. Therefore, analyses of the genetic changes within cells that are needed to make them tumorigenic do not address the identities of the genes and proteins that program the phenotypes of advanced, highly aggressive malignancies, an issue that we confront only later in Chapter 14. Moreover, the "5-hit" scenario suggested by the experimental transformation of human cells sidesteps a critical issue that we will discuss in the next chapter: the mutability of human cell genomes is normally very low, making it highly unlikely that cell populations within our tissues can acquire all of the genetic changes needed to complete tumor progression within a human life span.

11.13 Nonmutagenic agents, including those favoring cell proliferation, make important contributions to tumorigenesis

The clinical observations and experimental results that we have read about in this chapter provide us with a crude picture of the genetic and epigenetic changes required to generate a cancer cell. They fail, however, to reveal *how* these changes are actually acquired during tumor progression. So now, we turn to these issues—the processes occurring *in vivo* that enable cells to accumulate the large number of alterations needed for tumor formation.

The schemes described here dictate that a succession of genetic changes provide the major impetus for tumor progression. Since many of these changes are caused by the actions of mutagens, this implies that cancer progression is fueled largely and perhaps entirely by the genetic hits inflicted by mutagenic carcinogens. Indeed, precisely this conclusion comes from a literal reading of Figure 2.28, which indicates that carcinogens are mutagens and vice versa. Of course, we need to revise this scenario to accommodate the clearly important role of epigenetic changes, specifically those caused by methylation of gene promoters; see Section 7.8. (At present, it is unclear whether these methylation events are actively provoked by external agents or occur spontaneously because of occasional random mistakes made by the cellular proteins responsible for regulating methylation.)

In addition to the clearly documented contributions of mutagenic carcinogens to cancer induction (see Section 2.9), extensive evidence points to a wide variety of *nonmutagenic* agents that participate in tumor formation. Indications of the importance of nonmutagenic (sometimes termed *nongenotoxic*) carcinogens first came from attempts in the early 1940s to develop effective methods for inducing skin cancers in mice. The experimental model used in such research depended on exposing mouse skin to highly carcinogenic tar constituents, such as benzo[*a*]pyrene (BP), 7,12-dimethylbenz[*a*]anthracene (DMBA), or 3-methylcholanthrene (3-MC; see Figure 2.25). For example, mice subjected to daily painting of DMBA on a patch of skin would develop skin carcinomas after several months of this treatment.

But another experimental protocol proved to be even more revealing about the mechanisms of skin cancer induction. Following a single painting with an agent such as DMBA, the same area of skin could be treated on a weekly basis with a second agent, termed TPA (12-*O*-tetradecanoylphorbol-13-acetate; **Figure 11.29**), a potent skin irritant prepared from the seeds of the croton plant. (Another often-used term for TPA is PMA, for phorbol-12-myristate-13-acetate.) Repeated painting of the DMBA-exposed area with TPA resulted in the appearance of papillomas after 4 to 8 weeks, depending on the strain of mice being used (**Figure 11.30A–C**). (These papillomas are in many ways analogous to the adenomas observed in early-stage colon cancer progression.)

At first, the survival and growth of these skin papillomas depended upon continued TPA paintings, since cessation of TPA treatments caused the papillomas to regress (see Figure 11.30E). However, if TPA painting was continued for many weeks, TPA-independent papillomas eventually emerged, which would not regress after cessation of TPA painting and instead persisted for extended periods of time (see Figure 11.30F). Some of these TPA-independent papillomas might, with low probability, evolve further into malignant squamous cell carcinomas of the skin after about six months.

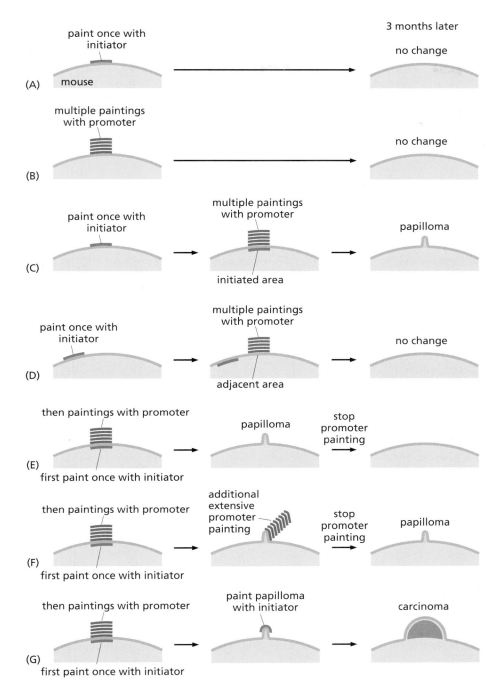

3 months later

Figure 11.30 Protocols for inducing skin carcinomas in mice The induction of skin carcinomas by painting carcinogens on the backs of mice requires certain combinations of treatments with initiators and promoters. (A) A single treatment with an initiating carcinogen, such as DMBA (dimethylbenz[a]anthracene), leads to no skin carcinomas 3 months later. (B) Multiple treatments with promoting agents, such as TPA (see Figure 11.29), also do not lead to significant numbers of tumors. (C) If an area of skin is painted once with an initiating agent followed by repeated paintings with a promoting agent, a papilloma will often appear several months later. (D) If an area of skin is painted once with an initiating agent and a promoting agent is then used to repeatedly paint another nearby but non-overlapping patch of skin, no papillomas will be seen 3 months later. (E) In a variation of the protocol depicted in panel C, an initiator such as DMBA is applied followed by repeated TPA treatments, which lead to papillomas. However, if repeated painting with the TPA tumor-promoting agent is halted soon after the papillomas appear, they regress, indicating that they are dependent on ongoing promoter stimulation. (F) In another variation of the protocol depicted in panel C, TPA promoter painting is continued for several months after papillomas first appear before being halted. Under these conditions, some of the papillomas will persist, indicating that they have become independent of continued promoter stimulation. (G) If a papilloma is produced by the protocols of panels C or F and the papilloma is then treated with an initiating agent, a carcinoma may appear, even in the absence of further promoter treatment.

In the absence of initial DMBA treatment, however, repeated painting with TPA failed to provoke either papillomas or carcinomas (see Figure 11.30B). Even more interesting, an area of skin could be treated once with DMBA and then left to rest for a year. If this patch of skin was then treated with a series of TPA paintings (as in Figure 11.30C), it would "remember" that it had been exposed previously to DMBA and respond by forming a papilloma.

These phenomena were rationalized as follows (**Figure 11.31**). A single treatment by an **initiating agent** (or **initiator**) like DMBA left a stable, long-lived mark on a cell or cluster of cells; this mark was apparently some type of genetic alteration. Subsequent repeated exposures of these "initiated" cells to TPA (termed the **promoting agent** or simply the **promoter**) allowed these cells to proliferate vigorously while having no apparent effect on nearby uninitiated cells. (Note that use of the word "promoter" in

Figure 11.31 Scheme of initiation and promotion of epidermal carcinomas in mice The observations of Figure 11.30 can be rationalized as depicted here. The initiating agent converts a normal cell (*gray, top left*) into a mutant, initiated cell (*blue*). Repeated treatment of the initiated cell with the TPA promoter generates a papilloma (*cluster of blue cells*), while TPA treatment of normal, adjacent cells (*gray, top right*) has no effect. Further treatment of the initially formed papilloma can be halted (*middle left*), in which case the papilloma regresses. Alternatively, further repeated treatment of the initially formed papilloma can yield a more progressed papilloma (*orange cells*), which persists even after promoter treatment is halted (*bottom left*); further repeated treatment of this more progressed papilloma with TPA eventually yields, with low frequency, a carcinoma (*bottom middle, red cells*). Alternatively, exposure of the initially formed papilloma to a second treatment by the initiating agent yields, with higher frequency, doubly mutant cells that also form a carcinoma (*bottom right, red cells*).

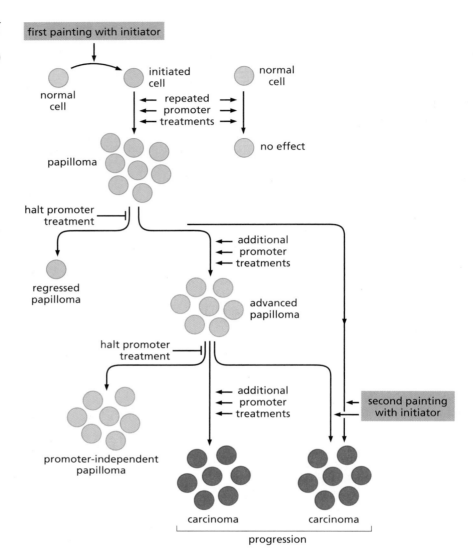

this context is unconnected with its other meaning—namely, the DNA sequences controlling the transcription of a gene.) The localized proliferation of initiated cells that was encouraged by the promoter would eventually produce a papilloma. However, as mentioned, if TPA painting were halted, the papilloma would disappear. Accordingly, the effects of the promoter were reversible, suggesting that it exerted a *nongenetic* effect on the cells in the papilloma. Clearly, this nongenetic effect, whatever its nature, could collaborate with the apparent genetic alteration created by the initiator to drive the proliferation of cells.

As we read above, if the initiated cells were treated with the TPA promoter for many months' time, eventually some papillomas would evolve to become TPA-independent; in this case, even after TPA withdrawal, the papillomas continued to increase in size and some eventually developed into skin carcinomas. This permanent change in cell behavior seemed to reflect the actions of a second, independent genetic alteration. Indeed, this evolution to a carcinomatous state could be strongly accelerated by treating a papilloma with a second dose of the initiating agent, already suspected to be a mutagen (see Figures 11.30G and 11.31). This third step in tumorigenesis (coming after initiation and promotion) is termed **progression**; the term is used more generally, throughout this book and elsewhere, to indicate the evolution of cells to an increasingly malignant growth state.

Four decades after the mouse skin cancer induction protocol was first developed, the identities of the genes and proteins that are the main actors in this skin tumorigenesis were discovered (**Figure 11.32**). As long suspected, the DMBA used as the initiating agent is indeed a potent mutagen in the context of skin carcinogenesis (**Sidebar 11.7**). Since it is a randomly acting mutagen, DMBA creates a wide variety of mutations in the genomes of exposed cells. However, the skin tumors that emerge invariably bear point-mutated H-*ras* oncogenes, indicating that this particular mutant allele confers some strong selective advantage on cells in the skin.

The subsequent repeated treatments with TPA, the promoting agent, act synergistically with the activated H-*ras* oncogenes to drive the proliferation of oncogene-bearing cells, yielding a papilloma. Treatment of the papilloma with TPA over an extended period of time may generate a papilloma that can persist even after TPA treatment is halted. Alternatively, if cells in a papilloma are exposed, once again, to an initiating agent like DMBA, this papilloma may progress into a carcinoma, whose cells now carry, in addition to the H-*ras* oncogene, a mutant *p53* gene.

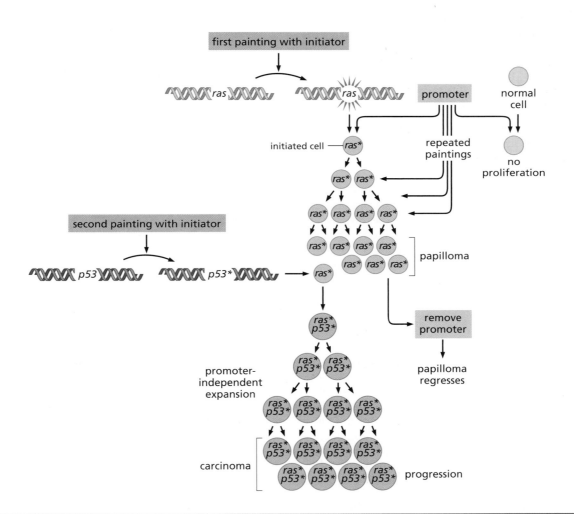

Figure 11.32 Genes and proteins involved in mouse skin carcinogenesis The phenomena of initiation and promotion (see Figures 11.30 and 11.31) can be understood at the genetic level in the manner illustrated here. The initiating agent acts as a mutagen to convert a *ras* proto-oncogene into an active oncogene (*top left*). This initiation, on its own, has no effect on the behavior of the keratinocyte bearing this mutant allele. However, in the presence of repeated stimulation by a promoting agent (*top right*), the *ras*-bearing cell is induced to pass through repeated growth-and-division cycles, leading to the formation of a papilloma (*blue cells*). Conversely, a cell lacking a *ras* oncogene (*gray, top right*) is not stimulated by the promoting agent and thus does not divide in response to repeated exposure to this agent. If repeated treatment by the promoter is halted (*lower right*), the papilloma regresses. However, if the papilloma is exposed a second time to a mutagenic initiating agent (*left*), a second genetic lesion, often involving the mutation of the *p53* tumor suppressor gene, is created. This mutant *p53* allele collaborates with the *ras* oncogene to create a population of cells (*red orange*) that are no longer dependent on the promoter and are capable of forming a carcinoma. Not shown here is the amplification of the *ras* oncogene, which also occurs as tumor progression proceeds.

Sidebar 11.7 Molecular analyses show that initiation is caused directly by the mutagenic actions of an initiating carcinogen The experimental convenience of the mouse skin carcinogenesis model has made it possible to directly validate some of the mechanistic speculations made here. For example, introduction of an H-*ras* oncogene into mouse skin cells through use of a retrovirus vector creates cells that closely mimic the behavior of DMBA-initiated cells, in that they are responsive to the effects of subsequently applied TPA promoter. This outcome demonstrates that the creation of an H-*ras* oncogene by an initiating agent is sufficient, on its own, to yield an initiated cell.

The mouse skin carcinogenesis model raises yet other questions: Do initiating agents, such as the dimethylbenz[*a*]anthracene (DMBA) or 3-methylcholanthrene (3-MC) carcinogens, act directly on the H-*ras* proto-oncogene to create an oncogene? Or do they do so indirectly, by stimulating some intermediary molecules that then react with DNA and create the critical initiating mutations?

When mouse skin tumors were initiated by exposure to either of two alkylating carcinogens (see Section 12.6),

N-methyl-*N*′-nitro-*N*-nitrosoguanidine (MNNG) or methylnitrosourea (MNU), the resulting skin tumors showed only G-to-A **transitions** (transitions are purine–purine or pyrimidine–pyrimidine substitutions) at codon 12 of the H-*ras* gene. In contrast, when mouse skin tumors were initiated by painting with 3-MC, the resulting papilloma and carcinoma cells were found to carry predominantly G-to-T **transversions** (purine–pyrimidine substitutions or vice versa) in codon 13 and A-to-T transversions in codon 61 of the H-*ras* gene. DMBA, which reacts with A's, induced tumors with A-to-T transversions in this gene. These nucleotide substitutions conform to the known mutagenic activities of these carcinogens.

Because the specific base substitutions reflect the chemical identities of the initiating agents, these observations provide strong evidence of the direct chemical interaction between the initiating carcinogen molecules and specific bases present in the H-*ras* proto-oncogene of mouse skin cells. Moreover, when taken together with the observed effects of the H-*ras* retrovirus vector, they indicate that the main mutational contribution of these agents to tumor initiation derives from their ability to mutate H-*ras* proto-oncogenes.

If we were to describe these phenomena at the level of signal-transduction biochemistry (**Figure 11.33**), we would say that the TPA promoter functions as a potent stimulator of cell proliferation through its ability to activate the cellular serine/threonine kinase known as protein kinase C-α (PKC-α), which we encountered earlier in the context of cytoplasmic signal transduction. More specifically, TPA acts as a functional mimic of diacylglycerol (DAG), the molecule that is generated endogenously by cells (see Figure 6.16B) as a means of activating protein kinase C (PKC). The downstream effectors of PKC-α collaborate with an H-*ras* oncogene, in still unknown ways, to drive proliferation of initiated keratinocytes whose descendants form papillomas and, on rare occasion, progress to carcinomas. These diverse observations of mouse skin carcinogenesis (see Figures 11.30, 11.31, and 11.32) leave us with the conclusion that tumor promoters like TPA, which do not directly affect the genomes of cells, are nevertheless important in propelling multi-step tumorigenesis forward. Moreover, they impress on us the notion that a cancer-causing agent—a carcinogen—need not be a mutagen.

11.14 Toxic and mitogenic agents can act as human tumor promoters

The experimental model of mouse skin carcinogenesis is useful for illustrating the principles of tumor initiation, promotion, and progression. However, it tells us almost nothing about how analogous mechanisms operate in the human body to create cancer. In fact, a diverse array of biochemical and biological mechanisms appear to be responsible for the tumor promotion leading to human cancers. Among these are mechanisms that act in either toxic or mitogenic fashion on various human tissues.

A striking example of cytotoxic mechanisms in human tumor promotion is provided by cancers of the mouth and throat (often called head-and-neck cancers). These carcinomas are often encountered in cigarette smokers who are also consumers of distilled alcoholic drinks. In fact, a serious cigarette habit together with frequent consumption of distilled alcohol leads to as much as a 100-fold increased risk for certain types of head-and-neck cancers.

Cigarette smoke is rich in a variety of mutagenic carcinogens, including 3-methylcholanthrene (3-MC). Ethanol, in contrast, has relatively weak mutagenic powers (see Figure 12.14). Instead, the major contribution of distilled alcoholic drinks to tumor induction seems to derive from their toxic effects on the epithelial cells lining the mouth and throat. After exposure to a drink containing a high percentage of ethanol, many of these cells die and slough off. Stem cells underlying the epithelium respond by dividing in order to regenerate epithelial cell layers within the mouth and throat. While these stem cells may normally divide at a low and steady rate, their mitotic rate increases substantially after widespread **denuding** (stripping) of an epithelium by ethanol.

The cells in the mouth and throat whose proliferation is stimulated by alcohol may already carry mutant alleles induced by tobacco tar, which in this context functions as the initiating agent. The promoting effect of alcohol then causes the clonal expansion of these initiated cells and may thereby enable their descendants to acquire yet other mutations that lead ultimately to the clinically aggressive head-and-neck cancers. This represents a dramatic illustration of a toxic agent acting as a tumor promoter.

Imagine, in a more general sense, compounds that are highly toxic for certain cell populations within a tissue. These cytotoxic agents can function as tumor promoters simply by causing the proliferation of the surviving cells. A particularly illustrative example is provided by Kostmann syndrome, a rare, heritable syndrome characterized by the almost complete absence of neutrophils, the cells in the blood that are responsible for killing bacterial and fungal infectious agents. In the autosomal dominant form of this disease, a genetic defect causes the synthesis of a mutant, neutrophil-specific elastase (a protease)—an enzyme that is normally expressed at increasing levels in neutrophils as they differentiate. The mutant elastase in Kostmann patients is cytotoxic for their neutrophils and, once it becomes expressed at significant levels, causes neutrophil depletion through apoptosis. Progenitor myeloid stem cells in the marrow respond to the resulting neutropenia (depressed numbers of neutrophils in the blood) by proliferating and attempting to replenish the pool of differentiated neutrophils. This leads to endless futile cycles of stem cell proliferation, since all attempts at neutrophil production are frustrated by the rapid, elastase-induced death of these cells once they differentiate. In some patients, the consequence of the continuous, excessive proliferation of these stem cells over many years' time is acute myelogenous leukemia. This syndrome provides an impressive example of how a cytotoxic agent (in this case one of endogenous origin) functions directly as a tumor promoter.

Mitogenic agents can also function as tumor promoters. Prominent among these are the steroid hormones—estrogen, progesterone, and testosterone. In the female body, for example, estrogen and progesterone are involved in programming the proliferation of cells in reproductive tissues. The monthly menstrual cycles of women between **menarche** and menopause result in the proliferation and then regression of the cells forming the epithelia of the ducts in the mammary gland (**Figure 11.34**). The endometrial lining of the uterus undergoes similar cycles of proliferation and regression.

Epidemiology makes it clear that the more menstrual cycles a woman experiences in a lifetime, the higher the risk of breast cancer. By one estimate, lifetime breast cancer risk decreases by 20% for each year that menarche is delayed during adolescence. (The most compelling illustration of the importance of the timing of menarche has come from studies of identical, or monozygotic, twins both of whom eventually developed breast cancer; the twin whose menarche began earlier had a 5.4-fold greater risk of being the first to be diagnosed with this disease.) Women who stop menstruating before age 45 have only about one-half the risk of breast cancer of those who continue to menstruate to age 55 or beyond.

Removal of the ovaries, the prime source of estrogen in the female body, causes breast cancer risk to plummet. Reinforcing this observation are the results of a Dutch study, which has shown that women who enter menopause before the age of 36, due to the side effects of chemotherapy for Hodgkin's lymphoma, have a 90% decreased risk of subsequently developing breast cancer. Conversely, postmenopausal women who contract breast cancer have on average a 15% higher level of circulating estrogen than unaffected women.

The effects of estrogen on breast cancer are surely complex, and it appears that this hormone acts on other cells in the mammary gland besides the epithelial cells. Still, it is evident that estrogen, and perhaps other hormones such as progesterone and even prolactin, periodically induce cell proliferation in a way that enables the progression

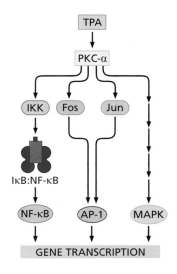

Figure 11.33 Activation of Fos, Jun, and NF-κB by PKC-α The precise mechanism by which an activated *ras* oncogene acts synergistically with TPA-activated protein kinase C-α (PKC-α) to drive the proliferation of keratinocytes is poorly understood. It is clear that once PKC-α is activated by binding TPA, it is able to stimulate transcription via several distinct signaling pathways, including those involving the NF-κB and AP-1 transcription factors and the ERK/MAPK enzyme. These biochemical changes have not yet yielded a clear explanation why TPA stimulates the proliferation of initiated keratinocytes bearing a *ras* oncogene while having minimal effect on nearby wild-type keratinocytes. However, at the level of tissue biology, it is clear that the tumor-promoting effects of PKC depend on its ability to promote inflammation in areas of treated skin.

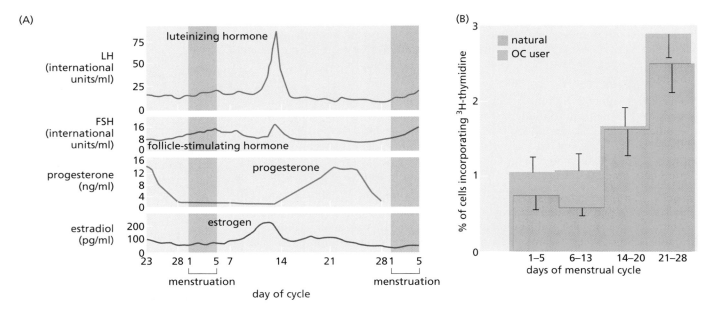

Figure 11.34 Fluctuations of hormone levels, cell proliferation, and mammary gland morphology during the menstrual cycle The periodic stimulation of mammary epithelial cell proliferation in the human breast under the control of hormones such as progesterone, prolactin, and estrogen (estradiol) appears to be coupled with increased risk of developing breast cancer. (A) As seen here, the levels of several of these hormones vary dramatically throughout the menstrual cycle. (B) The periodic monthly cycling leads to strong fluctuations in the rate of proliferation of cells. Here the percentage of cells in a tissue biopsy that are incorporating tritiated thymidine (indicating DNA synthesis) is plotted as a function of the days of the menstrual cycle. Use of oral contraceptives (OC) had minimal effect on DNA synthesis rates. (A, courtesy of J.A. Resko; from R.A. Rhoades and R.H. Pflanzer, Human Physiology. Philadelphia: W.B. Saunders, 1996. B, from T.J. Anderson et al., *Hum. Pathol.* 20:1139–1144, 1989.)

of initiated mammary epithelial cells (MECs) into the MECs found in the various types of breast cancers. (Some have argued that metabolites of estrogen are mutagenic, and that these metabolites contribute to breast cancer development; if so, estrogen's mutagenic effects on breast cancer development are surely overshadowed by its power to promote the proliferation of MECs.) In this example, we confront once again a tumor promoter that is endogenous to the mammalian body rather than being an agent of foreign origin. Nonetheless, the actions of this hormone adhere closely to the properties of classic tumor promoters.

11.15 Chronic inflammation often serves to promote tumor progression in mice and humans

Relatively few human tumor promoters act through purely cytotoxic or mitogenic mechanisms. Instead, the great majority seem to drive clonal expansion through mechanisms involving inflammation. Hints of this come from the model of mouse skin tumor initiation and promotion (see Section 11.13). This experimental model is clearly artificial, in that it involves a promoter (TPA) that is rarely, if ever, involved in skin tumor development in a mammal. Nonetheless, it teaches an important, generalizable lesson about tumor promotion: TPA was initially chosen because it is an irritant of mouse skin and thus an inducer of localized inflammation. (Moreover, inhibition of inflammation blocks TPA's tumor-promoting powers.)

A diverse set of other observations lend weight to the notion that inflammation is commonly involved in tumor promotion. For example, when cells of a human colonic adenoma cell line were implanted subcutaneously into Nude mice, they were found to be non-tumorigenic. However, when they were introduced into host mice together with an attached fragment of plastic, localized stromal inflammation was induced by the plastic and the adenoma cells grew to form tumors. The tumorigenic phenotype

(A)   (B)   (C)

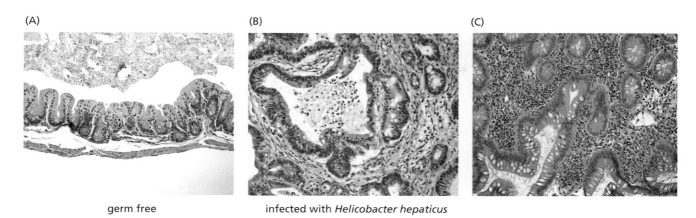

germ free   infected with *Helicobacter hepaticus*

Figure 11.35 Colonic inflammation and tumor promotion The development of adenomas and carcinomas in the colon is strongly dependent on chronic inflammation occurring in this organ. Thus, genetically altered mice that lack the ability to make TGF-β1 (which acts to suppress certain types of immune reactions) succumb early in life to overwhelming autoimmune disease. (A) However, if their immune system is crippled through the inactivation in their germ line of the gene encoding the Rag-2 enzyme, which is required for generating functional antibody and T-cell receptor genes (see Chapter 15), these mice survive when they are housed in a germ-free facility. Under such conditions, their intestinal epithelium is quite normal histologically. (B) In contrast, if these doubly mutant mice are housed in a facility in which *Helicobacter hepaticus* bacteria are present, their colons become infected with these bacteria and they develop adenomas and carcinomas. This indicates that the chronic inflammation created by these intestinal bacteria is essential to the formation of the observed colonic lesions. (C) Human ulcerative colitis involves chronic inflammation of regions of the colonic mucosa, which leads after many years to substantially increased risk of colon carcinoma. In this condition, the stromal areas between the colonic crypts are infiltrated with large numbers of lymphocytes, which are associated with inflammation (*small dark purple nuclei*). (A and B, from S.J. Engle et al., *Cancer Res.* 62:6362–6366, 2002. C, courtesy of D. Lamarque.)

of these cells persisted even after they were subsequently transplanted (without the plastic) to another host animal, indicating that their neoplastic proliferation was now driven by some heritable genetic or epigenetic alteration.

Yet another example that inflammation plays a role in colon carcinoma formation has come from the study of mutant mice that lack the ability to make TGF-β1. Such mice tend to develop a lethal autoimmune disease, which results in their death after several weeks. In order to spare these mice, their immune systems were crippled by germ-line inactivation of their *Rag2* gene copies, a loss that prevents the formation of the antigen-specific lymphocytes that trigger autoimmune disease (see Section 15.8). The doubly mutant mice (TGF-β1$^{-/-}$ *Rag2*$^{-/-}$) now survived far longer but developed areas of inflammation in the colon, as well as colonic adenomas and adenocarcinomas, between 3 and 6 months of age. However, if these doubly mutant mice were reared in a germ-free environment (which yields colons free of the usual bacterial populations), no colonic inflammation was seen and neither adenomas nor adenocarcinomas developed. If such germ-free mice were introduced into animal quarters harboring *Helicobacter hepaticus* (a common murine colonic bacterium), once again the polyps and carcinomas formed in their colons (**Figure 11.35A and B**). Such observations indicate that bacterial flora in the gut contribute importantly to inflammation, both in these mice and, one imagines, in humans suffering from ulcerative colitis, an inflammatory condition that predisposes to colon cancer (for example, see Figure 11.35C). Moreover, they suggest that localized areas of colonic inflammation in otherwise normal individuals contribute to the formation of adenomatous polyps that can eventually progress to carcinomas.

Chronic inflammation also plays a clear role in the pathogenesis of human carcinomas (**Table 11.3**). For example, those arising in the gallbladder are usually associated with a decades-long history of gallstones and resulting inflammation of the epithelial lining of the gallbladder (**Figure 11.36A**). Similarly, hepatocellular carcinomas (HCCs), which are common in East Asia (**Sidebar 11.8**), are associated with chronic hepatitis B virus (HBV) infections and accompanying inflammation of the liver (see Figure 11.36B, left panel). In many infected individuals, HBV infection is well established in

Table 11.3 **Inflammatory conditions and tumor development**

Human tumor	Inflammatory condition or inflammation-provoking agent
Bladder carcinoma	schistosomiasis, chronic cystitis
Gastric carcinoma	*H. pylori*–induced gastritis
Hepatocellular carcinoma	hepatitis B/C virus
Bronchial carcinoma	silica
Mesothelioma	asbestos
Ovarian carcinoma	endometriosis
Colorectal carcinoma	inflammatory bowel disease
Esophageal carcinoma	chronic acid reflux
Papillary thyroid carcinoma	thyroiditis
Prostate carcinoma	prostatitis
Lung carcinoma	chronic bronchitis
Gallbladder carcinoma	chronic cholecystitis
Squamous cell skin carcinoma	chronic osteomyelitis

Adapted from F. Balkwill, K.A. Charles and A. Mantovani, *Cancer Cell* 7:211–217, 2005.

the liver early in life and continues in a chronically active form for decades. The resulting hepatitis may have relatively few outward effects on the individual, since the continual HBV-induced killing of hepatocytes (the cells forming the bulk of the liver) is compensated by an equal proliferation of surviving cells.

Sidebar 11.8 Hepatitis B virus infections led to hepatomas in Taiwanese government workers Epidemiological studies usually yield outcomes that make it difficult to discern a strong correlation between occasional exposure to certain environmental factors and a subsequent moderately increased risk of cancer. The incidence of liver cancer, however, varies dramatically throughout the world, making it possible to strongly associate this disease with a causative factor. In certain parts of the world, including much of Asia—especially China—and sub-Saharan Africa, hepatocellular carcinomas (HCCs, also known as hepatomas) are one of the leading causes of death due to cancer; in the United States, liver cancer, in stark contrast, ranks 25th as a cause of cancer-related deaths.

A stunning correlation can be made between hepatitis B virus (HBV) infections and susceptibility to liver cancer. At the beginning of an epidemiologic study undertaken in 1975, the HBV status of 22,707 men in government service in Taiwan was determined by measuring, among other things, the presence of viral antigens in the blood. The causes of deaths in this cohort were then chronicled over the next decade. By 1986, death from hepatocellular carcinoma had claimed 152 of the 3454 men who were initially positive for HBV viral antigen but only 9 of the 19,253 men who were negative for this antigen in their blood. The relative risk of dying from this disease if one carried viral antigen in the circulation (which is indicative of an active HBV infection in the liver) was 98.4. This means that an individual afflicted with chronic, active HBV infection experienced an almost 100-fold increased risk of contracting and dying from this cancer compared with someone who lacked viral antigen in his liver and was apparently uninfected. These numbers contrast strongly with most epidemiologic correlations made between specific exposures of patients to suspected carcinogens and disease incidence, where relative risk of disease is often only two- to threefold—often hovering on the borderline of statistical significance.

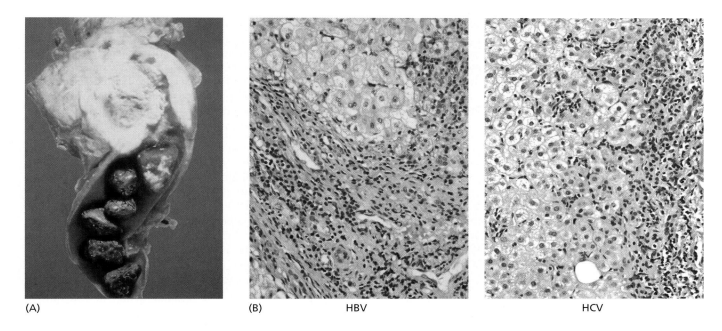

(A)   (B)   HBV   HCV

Figure 11.36 Chronic inflammation leading to cancer
(A) A graphic demonstration of chronic inflammation leading to cancer is provided by carcinomas of the epithelial lining of the gallbladder (*white mass, above*), which are commonly associated with the formation of gallstones arising as precipitates from the bile (*brown masses, below*). (B) Chronic hepatitis B virus (HBV) infection creates continual cell death of hepatocytes together with chronic inflammation, as indicated by the numerous lymphocytes (*small dense nuclei, left panel*). Over a period of decades, this can lead to an almost 100-fold increased risk of liver cancer. The inflammation in the liver of someone suffering from chronic hepatitis C virus (HCV) infection is strikingly similar (*right panel*). The inflammatory cells are largely in the right part of this micrograph. The fact that these two inflammatory conditions are so similar histologically and lead to comparably increased risks of hepatocellular carcinoma (HCC) suggests that the inflammatory states, rather than some specific aspect of viral function, are responsible for the appearance of HCC in patients infected with either of these viruses. (A, from A.T. Skarin, Atlas of Diagnostic Oncology, 3rd ed. Philadelphia: Elsevier Science Ltd., 2003. B, left, courtesy of A. Perez-Atayde; right, courtesy of A.K. Bhan.)

The HBV genome does carry a gene, termed *HBX*, that shows weak oncogenic and pro-apoptotic powers, but this gene, on its own, can hardly explain HBV's ability to induce hepatic carcinomas after decades of chronic infection. This led to a search for other carcinogenic mechanisms. Thus, HBV might act as a liver carcinogen through a mechanism that echoes the carcinogenic strategy of non-oncogene-bearing retroviruses. Recall that these viruses, notably avian leukosis virus and murine leukemia virus, induce cancer slowly and inefficiently. When, after many months, they finally succeed in doing so, this occurs through insertional mutagenesis: the chance integration of a provirus next to a critical cellular growth–controlling gene—a proto-oncogene (see Section 3.11). The resulting deregulation of expression of the proto-oncogene effectively converts this gene into an oncogene, paving the way for cancer formation.

In the case of HBV-induced liver cancer, however, the situation is quite different. Extensive molecular analyses have failed to demonstrate that the genomes of virus-associated human liver cancers carry HBV genomes integrated next to critical cellular growth–controlling genes, such as the *myc* proto-oncogene. So, HBV is unlikely to act directly as a mutagen in infected liver cells.

This leaves open two plausible explanations for the virus-induced cancer. HBV creates liver cancer through its ability to cause continuous cell proliferation in an organ that normally experiences hardly any at all; this proliferation is required to replace hepatocytes that are continually being killed by HBV infectious cycles. Alternatively, HBV infection causes cells of the immune system to attempt to eliminate virus-infected cells, yielding a chronic inflammatory state in the liver (see Figure 11.36B). It is highly likely that both of these tumor-promoting mechanisms conspire in the pathogenesis of the hepatocellular carcinomas that so frequently afflict individuals with chronic HBV infections.

More recent research has made it clear that chronic hepatitis C virus (HCV) infections act in a similar way to increase liver cancer rates. While the two viruses are totally unrelated to one another with respect to genome structure and replication cycles, they evoke very similar biological outcomes through their shared ability to create chronic infections, cytotoxicity, and inflammation in the liver (see Figure 11.36B, right panel). (Significantly, a variety of other types of chronic liver injury, including that inflicted by alcoholism, are also associated with increased incidence of hepatocellular carcinoma, although the relative risks are vastly less than with lifelong HBV infection.)

HBV, acting as a tumor promoter, can also function synergistically with aflatoxin-B1, a highly mutagenic compound that is made by *Aspergillus* fungi that proliferate on peanuts, nuts, and corn stored under conditions of high humidity (see Figures 2.28 and 12.17). This combination of infection plus aflatoxin exposure proves to be deadly. In one, relatively small prospective epidemiologic study carried out in Shanghai, infection with HBV increased the risk of hepatocellular carcinoma about 7-fold while exposure to aflatoxin-B1–contaminated food yielded about a 3-fold increased risk. When an individual experienced both agents, the risk of liver cancer increased about 60-fold. The parallel between the pathogenesis of these human liver carcinomas and the actions of initiators and promoters of mouse skin cancer is striking.

Yet other indications of infectious agents inducing inflammation and, in turn, human malignancies, come from individuals afflicted with lymphomas arising in gastric mucosa-associated lymphoid tissue (MALT). Seventy-five percent of these MALT lymphomas can be cured if patients are treated with antibiotics that eradicate the *Helicobacter pylori* bacterial populations in the stomach. These lymphomas are clearly dependent on continued promoter stimulation—in this case the presence of *H. pylori*. The 25% of lymphomas that do not respond to this treatment have evolved beyond this dependence to become "promoter-independent " (see Figure 11.31), possibly because they have sustained the Chromosome 11 to 18 translocation that is frequently observed in these lymphomas.

A final vignette is worth mentioning here, if only to highlight the fact that cancer-predisposing, chronic inflammatory conditions can also arise through nonviral means. One form of familial pancreatitis is caused by an inherited dominant allele of the gene encoding the digestive enzyme trypsin. Like virtually all other digestive enzymes made in the pancreas, this one is initially synthesized as a pro-enzyme, that is, an inactive form that becomes active only after its secretion into the gastrointestinal tract. The motive behind this type of enzyme regulation is clear: premature activation of the pro-enzyme (trypsinogen) within the pancreas would lead to its destroying pancreatic tissue, precisely what happens in patients who inherit the gene specifying a mutant, prematurely activated form of the enzyme. These patients suffer lifelong recurrent bouts of pancreatitis and experience at least a 40% chance of developing pancreatic carcinomas. Here the frequent breakdown of pancreatic tissue and resulting wound healing and inflammation lead directly to the formation of these deadly cancers. This is an extreme example of far more commonly occurring pathological states that predispose to a variety of frequently occurring human carcinomas, ultimately spawned by low-level, subclinical inflammation persisting over decades (Table 11.4).

11.16 Inflammation-dependent tumor promotion operates through defined signaling pathways

Evidence supporting the role of inflammation in cancer development also comes from a large number of epidemiologic observations that anti-inflammatory drugs, such as aspirin and sulindac, function to reduce the incidence of a variety of carcinomas in humans. For example, one study showed that those who took low doses of a nonsteroidal anti-inflammatory drug (NSAID)—an aspirin tablet once every day or two over a period of 15 years—had a lung cancer rate of 0.68, a breast cancer rate (in women) of 0.70, and, in younger men, a colorectal cancer rate of 0.35 compared with the rates of these cancers in corresponding control groups.

Another epidemiologic study tracking a large cohort of women over a seven-year period found that those regularly taking aspirin experienced about half the risk of

Table 11.4 **Links between inflammation and cancer pathogenesis**

Many inflammatory conditions predispose to cancer
Cancers arise at sites of chronic inflammation
Functional polymorphisms of cytokine genes are associated with cancer susceptibility and severity
Distinct populations of inflammatory cells are detected in many cancers
Extent of tumor-associated macrophage infiltrate correlates with prognosis
Inflammatory cytokines are detected in many cancers; high levels are associated with poor prognosis
Chemokines are detected in many cancers; they are associated with inflammatory infiltrate and cell motility
Deletion of cytokines and chemokines protects against carcinogens, experimental metastases, and lymphoproliferative syndrome
Inflammatory cytokines are implicated in the action of nongenotoxic liver carcinogens
The inflammatory cytokine tumor necrosis factor is directly transforming *in vitro*
Long-term NSAID use decreases mortality from colorectal cancer

Courtesy of F. Balkwill. From F. Balkwill and A. Mantovani, *Lancet* 357:539–545, 2001.

developing pancreatic cancer as a control group. Two studies have shown that regular use of aspirin (or another NSAID) resulted in about a 40% reduction of stomach cancer (in the region of the stomach outside of the cardia) in individuals who were infected with *Helicobacter pylori*, a bacterium that often inhabits the human stomach; stomach cancer rates in uninfected individuals were not reduced by aspirin usage. Yet other studies indicate reduced death rates associated with ovarian, bladder, and prostate carcinomas. Moreover, treatment with NSAIDs can actually cause the regression of aberrant crypt foci (ACFs) of the sort seen in Figure 11.23C. The overall benefit in reducing cancer-related deaths, as reported in a large 2010 study, was a 20% reduction in this source of mortality following 20 years of daily aspirin use.

(These outcomes might suggest that NSAIDs, which can also reduce deaths from certain types of cardiovascular disease, should be used routinely as disease preventives in the general population. However, a caution comes from another public health statistic: in 1998, 16,550 deaths in the United States were attributed to NSAID-induced gastrointestinal bleeding and the death rate remained at this level in the decade that followed.)

The epidemiologic observations linking long-term NSAID use with reduced cancer incidence, together with the experiments in mouse skin and liver carcinogenesis, lead us to the biochemical and cell-biological mechanisms that are responsible for tumor promotion in many tissues. One important clue has come from the identification of the downstream cellular target of TPA—protein kinase C-α (PKC-α). A dramatic demonstration of the key role of this enzyme in mediating TPA-induced skin inflammation is provided by mouse skin keratinocytes that are forced to overexpress it (**Figure 11.37A**).

In TPA-treated keratinocytes, PKC-α signals, in part, by activating IKK (IκB kinase); the latter phosphorylates IκB (inhibitor of NF-κB), tagging it for destruction. NF-κB is thereby liberated (see Figure 6.29A) and migrates from the cytoplasm to the nucleus, where it induces transcription of a large constituency of genes. Among these are genes that block apoptosis and favor cell proliferation. Importantly, NF-κB also induces expression of the gene encoding tumor necrosis factor-α (TNF-α), a potent attractant of immune cells (that is, a **cytokine**) that triggers localized inflammation in mouse

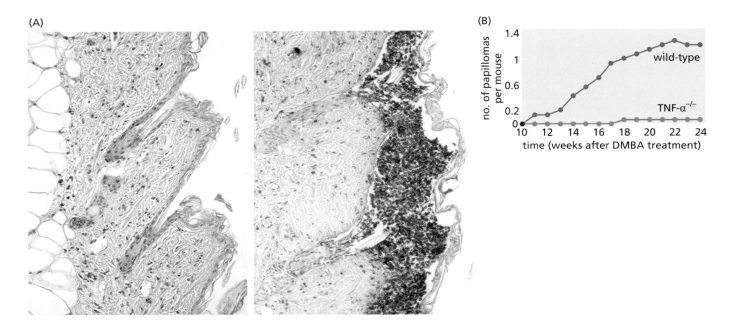

Figure 11.37 Critical regulators of mouse skin induction
(A) TPA acts through an intracellular serine/threonine kinase—
protein kinase C-α (PKC-α). The inflammatory powers of TPA are
revealed in exaggerated form in transgenic mice that overexpress
this kinase in their keratinocytes. (This is achieved by constructing
a transgene in which PKC-α expression is driven by a keratin 5
transcriptional promoter, which is active specifically in the skin.)
Eighteen hours after wild-type or transgenic mice have been treated
with TPA, the skin of wild-type mice (*left*) shows relatively few
neutrophils (*dark spots*), which are important immune mediators
of inflammation. In contrast, the skin of the transgenic mice (*right*)
shows extensive neutrophil infiltration. Neutrophils were detected
through use of an antibody that recognizes a neutrophil-specific
antigen. (B) Tumor necrosis factor-α (TNF-α), which is made in both
epithelial and stromal cells of the mouse, is a critical mediator of the
TPA-initiated inflammatory response. In mice in which both copies
of the gene encoding TNF-α have been deleted in the germ line
(and thus deleted from the genomes of all cells; see Supplementary
Sidebar 7.7), the usually observed induction of skin papillomas—
the precursors of skin carcinomas—is almost totally lost. In this
experiment, mouse skin was painted once with the DMBA initiator
and thereafter at twice-weekly intervals with TPA for the next 15
weeks. (A, from C. Cataisson et al., *J. Immunol.* 171:2703–2713,
2003. B, from R.J. Moore et al., *Nature Med.* 5:828–831, 1999.)

skin and in many other epithelial tissues as well. (As its name implies, TNF-α was originally discovered as an inducer of the death of cancer cells; see Figure 9.32. However,
when its actions were explored in greater detail, its role as a major intermediary in
tissue inflammation became apparent.)

Mice lacking functional copies of the TNF-α gene respond to a skin carcinogenesis
protocol (involving the DMBA initiator and the TPA promoter) by developing 5 to 10%
as many skin carcinomas as wild-type mice (see Figure 11.37B). In the latter animals,
TPA treatment elicited TNF-α production in the **epidermal** keratinocytes, which then
induced inflammation in the underlying stromal cells of the **dermis**. Together, these
various observations suggest the following pathway:

$$\text{TPA} \rightarrow \text{PKC-}\alpha \rightarrow \rightarrow \text{IKK} \rightarrow \rightarrow \text{NF-}\kappa\text{B} \rightarrow \text{TNF-}\alpha \rightarrow \text{inflammation}$$

In another illustrative mouse tumor model, liver carcinogenesis was provoked
through deletion of the *Mdr* (multi-drug resistance) gene in the mouse germ line; loss
of this gene and its encoded product leads to accumulation of bile acids in the liver
and resulting chronic liver inflammation. Affected mice develop nodules of dysplastic
hepatocytes, localized hepatocellular carcinomas (HCCs), and eventually metastatic
HCCs (**Figure 11.38A**).

In the livers of these mice, TNF-α is initially produced by inflamed endothelial cells as
well as infiltrating inflammatory immune cells in the stroma, such as neutrophils and
macrophages (see Figure 11.38B). The released TNF-α acts in a paracrine manner on
nearby hepatocytes, which display receptors for TNF-α. Like PKC-α in keratinocytes,

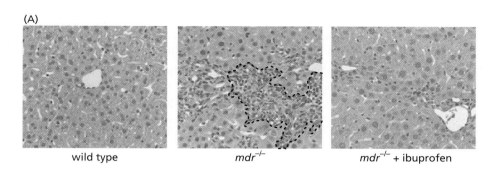

(A)

wild type $mdr^{-/-}$ $mdr^{-/-}$ + ibuprofen

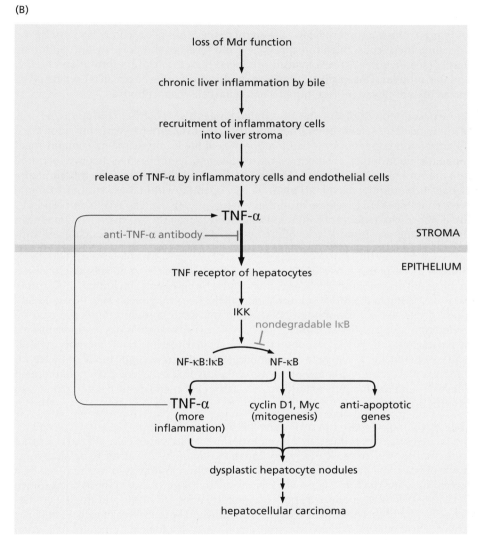

(B)

Figure 11.38 Chronic liver inflammation acts via NF-κB to induce hepatocellular carcinomas
In this mouse model of liver carcinogenesis, deletion of the *mdr* (multi-drug resistance) gene causes chronic liver inflammation, which generates dysplastic foci among the hepatocytes, some of which progress to form hepatocellular carcinomas. (A) An infiltrate of inflammatory cells in the liver of an $mdr^{-/-}$ mouse (*outlined, center panel*) contrasts with cells in the normal liver (*left panel*) and the liver of an $mdr^{-/-}$ mouse that has been treated with a potent anti-inflammatory drug, ibuprofen (*right panel*). Such areas of inflammatory infiltrate appear to be critical to the formation of dysplastic foci and thus to the subsequent formation of hepatic carcinomas. (B) This diagram summarizes the sequence of changes that appear to generate hepatocellular carcinomas (HCCs) in this mouse model. Loss of the *mdr* gene leads to an accumulation of bile acids that induce a state of chronic liver inflammation; this attracts inflammatory cells to the liver stroma that proceed to release TNF-α. This ligand impinges on the TNF receptors (see Figure 9.32) displayed by hepatocytes within the liver's epithelial compartment. The TNF receptors respond by activating IKK, which induces NF-κB signaling, which drives expression of anti-apoptotic genes (such as $Bcl-X_L$), proliferative genes (such as the one encoding cyclin D1), and the gene specifying TNF-α; expression of additional TNF-α leads to amplification of the inflammatory response. Together, the proteins encoded by these genes facilitate the progression of the dysplastic nodules to hepatocellular carcinomas. This progression can be blocked by antibodies reactive with TNF-α as well as a dominant-negative, nondegradable IκB. (The precise identities of the inflammatory cells that are recruited to the stroma and are responsible for signaling to the hepatocytes are unclear, in part because they represent a complex mixture of macrophages; T cells, including NKT cells, cytotoxic T cells, and T regs; NK cells; myofibroblasts; and, frequently, B cells, neutrophils, and plasma cells.) (A, from E. Pikarsky et al., *Nature* 431:461–466, 2004.)

the ligand-activated TNF-α receptor of these hepatocytes funnels signals through the NF-κB pathway (see Section 6.12). As described above, NF-κB is dispatched to the nucleus, where it activates anti-apoptotic genes and genes favoring proliferation including, in this case, the TNF-α gene. The resulting TNF-α, once released by hepatocytes, attracts more inflammatory cells through *paracrine* signaling and amplifies hepatocyte NF-κB signaling through *autocrine* signaling.

When NF-κB signaling was blocked in the hepatocytes of the $Mdr^{-/-}$ mice (see Figure 11.38B), tumor incidence was strongly suppressed. This inhibition of signaling could be achieved through the introduction of an anti-TNF-α antibody into the $Mdr^{-/-}$ mice (which blocked the paracrine signaling between inflammatory stromal cells and hepatocytes) or by expressing a nondegradable IκB in the hepatocytes (which inhibited NF-κB activation). In both cases, loss of NF-κB signaling resulted in greatly increased rates of apoptosis in the pre-neoplastic hepatocytes.

Indeed, prevention of apoptosis likely explains much of the tumor-promoting effect of inflammation in this mouse model. Interestingly, shutdown of NF-κB signaling in hepatocytes did not prevent the early steps of liver tumor progression—hepatitis, tumor initiation, and the development of dysplasia—but the subsequent progression of dysplastic tissue to hepatocellular carcinoma was blocked. Hence, in this experimental model, TNF-α and NF-κB are involved in promotion and not in the earlier stages of initiation.

A similar conclusion was reached in studies of a mouse model of colitis-associated colon cancer: inactivation of IKK signaling led to greatly increased rates of apoptosis in pre-cancerous enterocytes—the epithelial cells lining the colon. Without NF-κB signaling, these cells failed to express elevated levels of the potently anti-apoptotic Bcl-X_L (see Figure 9.26) and instead produced high levels of the pro-apoptotic Bax and Bak proteins. Once again, some of the tumor-promoting effects of NF-κB could be traced to its ability to protect initiated epithelial cells from apoptosis.

These descriptions indicate that NF-κB signaling is highly active in the epithelial cells of inflamed tissues (for example, hepatocytes in the liver, enterocytes in the colon). The inflammation is created by inflammatory cells of the immune system (notably macrophages, neutrophils, eosinophils, mast cells, and lymphocytes) that are recruited into the stromal compartments of these tissues, where they release pro-inflammatory signals. Importantly, the NF-κB pathway *also* operates within these stromal immune cells, enabling them to release pro-inflammatory signals such as TNF-α. For example, in the above-cited model of colitis and colorectal cancer, shutdown of IKK signaling in the inflammatory immune cells of the intestinal stroma also led to a significant suppression of colorectal tumor formation.

None of this explains how NSAIDs, such as aspirin, ibuprofen, and sulindac, succeed in blocking tumor promotion and thus tumor progression, including these processes occurring in the $Mdr^{-/-}$ mouse model of liver carcinogenesis (see Figure 11.38). These drugs, and a host of other NSAIDs, have as their common target the enzyme cyclooxygenase-2 (COX-2; **Figure 11.39**), which is responsible for the biosynthesis of prostaglandin E_2 (PGE$_2$), a potent low–molecular-weight pro-inflammatory molecule. Actually, the *Cox-2* gene is yet another gene that is strongly induced by NF-κB and thus TNF-α. Indeed, many of the pro-inflammatory effects of TNF-α enumerated above depend on its ability to induce COX-2 expression and, in turn, strong increases in PGE$_2$ production.

The critical role of COX-2 in mediating the inflammatory responses that lead to epithelial malignancies is illustrated by an experiment with genetically altered mice that were predisposed, because of a germ-line mutation of the *Apc* gene, to develop hundreds of intestinal polyps (see Section 7.11): the number of polyps was reduced by a factor of 7 when the *Cox-2* gene was inactivated in the germ line of these mice. (Since COX-2, like NF-κB, operates in both stromal inflammatory cells and nearby epithelial cells, loss of this enzyme is likely to have affected both classes of cells.) Similarly, when COX-2 enzyme activity was shut down pharmacologically (through use of an NSAID), the development of mammary carcinomas in tumor-prone transgenic mice was strongly suppressed (**Figure 11.40**).

As might be expected, the opposite effects can be observed when levels of PGE$_2$ are increased. In spontaneously arising tumors, this is due in part to reduced expression of the PGDH (15-hydroxyprostaglandin dehydrogenase) enzyme (see Figure 11.39A), which degrades PGE$_2$ (**Figure 11.41A and B**). In an experimental model, knockout in the mouse germ line of the gene encoding this degradative enzyme leads to significant increases of colorectal carcinomas. Alternatively, elevated PGE$_2$ levels can be achieved if COX-2 expression is forced in the mammary tissue of transgenic mice: these mice showed significantly increased rates of breast cancers. An analogous experiment in which a mouse transgene caused elevated expression of this enzyme in keratinocytes led to skin hyperplasia and dysplasia.

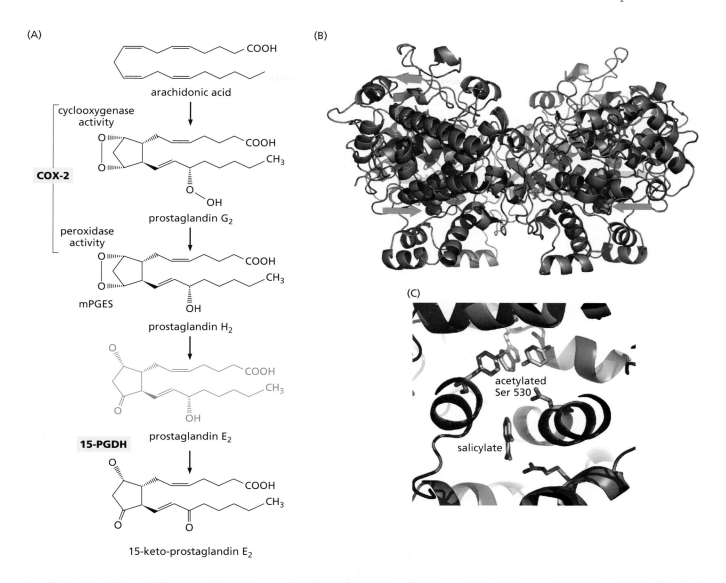

Figure 11.39 Prostaglandin metabolism and the actions of anti-inflammatory drugs The profound effects of anti-inflammatory drugs in suppressing the incidence of a variety of commonly occurring carcinomas are due to their ability to inhibit the catalytic activity of the cyclooxygenase-2 (COX-2) enzyme (sometimes called prostaglandin H_2 synthase). COX-2 is induced in many tissues in response to inflammatory stimuli. (A) The two distinct catalytic activities of COX-2 result in the conversion of arachidonic acid into prostaglandin H_2. The latter is metabolized to prostaglandin E_2 (PGE$_2$; *red*), which evokes many of the tumor-promoting responses described here. The PGDH (15-hydroxyprostaglandin dehydrogenase) enzyme (also termed hydroxyprostaglandin dehydrogenase, HPGD) degrades PGE$_2$ into 15-keto-prostaglandin E_2, which is presumed to be inactive. (B) The catalytic sites of COX-2, a homodimer, that convert arachidonic acid to prostaglandin G_2, are largely inactivated by the most commonly used anti-inflammatory drug—aspirin (acetylsalicylic acid)—which acetylates (*yellow balls, arrows*) serine 530 of COX-2 and leaves the salicylate moiety weakly bound in the catalytic cleft (*orange balls, arrows*). α-helices (*teal*), loops (*gray*), β-pleated sheets (*light green*), heme group (*red*). (C) In more detail, the acetylation of serine 530 by aspirin obstructs the enzymatic activity of COX-2 that is responsible for converting arachidonic acid to the precursor of all prostanoids, prostaglandin G_2; the latter is converted further to prostaglandin H_2 by a second site in the COX-2 enzyme. (B and C, courtesy of R.M. Garavito.)

Arguably the most direct demonstration of the tumor-promoting role of PGE$_2$ has come from experiments in which *Apc*$^{+/-}$ mutant mice, which are prone to developing gastrointestinal polyps (see Section 7.11), were fed PGE$_2$; this led to a tenfold increase in the number of these growths in the colon (Figure 11.41C). These *in vivo* tissue responses presumably reflect changes induced by PGE$_2$ when it is applied directly to normal intestinal epithelial cells in culture: they exhibit many of the phenotypes associated with cell transformation—loss of contact inhibition, increased

Figure 11.40 COX-2 inhibition in mammary carcinomas of tumor-prone transgenic mice In a transgenic mouse model of mammary carcinoma pathogenesis, the development of tumors that are driven by the polyomavirus middle T oncoprotein is strongly reduced by treatment with celecoxib, another NSAID that is a potent and selective inhibitor of cyclooxygenase-2 (COX-2). It acts in two complementary ways by inducing increased apoptosis and reducing cell proliferation. (A) As indicated here, increasing doses of this drug generate progressively higher numbers of tumor cells (*light green*) that stain with the TUNEL assay (*dark brown spots*; see Figure 9.19), a marker of apoptosis. In this case, the proportion of tumor cells that are TUNEL-positive provides an indication of the extent of apoptosis induction. (B) Conversely, increasing concentrations of celecoxib lead to decreasing cell proliferation, as revealed by staining (*dark red*) for PCNA (proliferating-cell nuclear antigen), a marker of cell cycle progression. Celecoxib doses are presented as milligrams of drug per kilogram of body weight. (From G.D. Basu et al., *Mol. Cancer Res.* 2:632–642, 2004.)

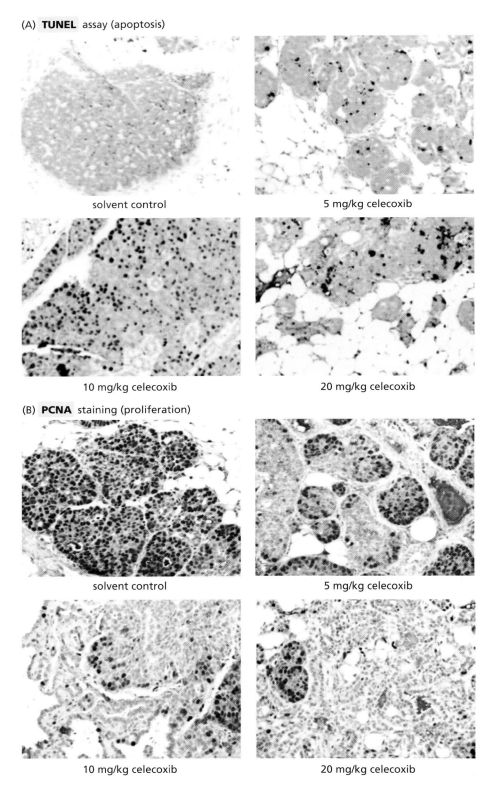

(A) **TUNEL** assay (apoptosis)

solvent control

5 mg/kg celecoxib

10 mg/kg celecoxib

20 mg/kg celecoxib

(B) **PCNA** staining (proliferation)

solvent control

5 mg/kg celecoxib

10 mg/kg celecoxib

20 mg/kg celecoxib

anchorage-independent growth, down-modulation of expression of the cell surface protein E-cadherin, reduced apoptosis, and increased rate of proliferation.

COX-2 is expressed in the stromal compartment early in tumorigenesis in some tissues and in the epithelial compartment of others (**Figure 11.42**). In certain epithelial tissues, COX-2 expression increases in the epithelial compartment as tumor progression advances until its concentration is elevated tenfold or more above normal levels. Since PGE$_2$ can diffuse from cell to cell, it is likely that early in tumor progression in the

(A)

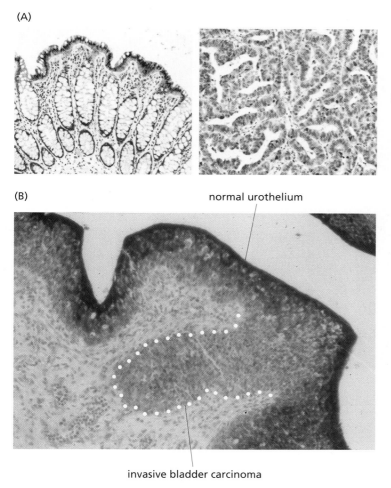

(B)

normal urothelium

invasive bladder carcinoma

(C)

solvent control

prostaglandin E$_2$ (150 µg)

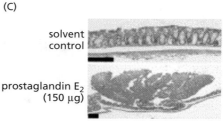

Figure 11.41 Increased prostaglandin E$_2$ levels in normal and neoplastic tissues The levels of PGE$_2$ can be modulated in various ways, resulting in increased rates of tumor progression. (A) In general, these levels are dictated by the rates of its synthesis by COX-2 and its degradation by PGDH (see Figure 11.39A). In many carcinomas, increased COX-2 levels in the stroma and epithelium (see Figure 11.42) are complemented by reduced PGDH levels in the epithelial compartment. Indeed, PGDH expression is depressed in the majority of lung, colon, breast, and bladder carcinomas. As seen here, PGDH (*dark stain*) is strongly expressed in the normal enterocytes facing the lumen of the colon (*left*) while being undetectable in a colon carcinoma that arose nearby (*right*). (B) Similarly, the intense expression of PGDH (*dark brown*) in the normal urothelium (the epithelial lining of the bladder) contrasts with the almost-total loss of this enzyme in a tongue of invasive carcinoma cells. (C) Increases in PGE$_2$ levels can be achieved experimentally by feeding this prostaglandin to mice of the *Apc* mutant strain (see Section 7.11). This causes a dramatic increase in the number of colonic polyps (*below*) compared with those seen in control *Apc* mutant mice (*above*). The size bars indicate the relative magnification of the two panels. (A, from M. Yan et al., *Proc. Natl. Acad. Sci. USA* 101:17468–17473, 2004. B, from S. Tseng-Rogenski et al., *Am. J. Pathol.* 176:1462–1468, 2010. C, from D. Wang et al., *Cancer Cell* 6:285–295, 2004.)

gut, COX-2 expression in stromal inflammatory cells results in release of this prostaglandin, which acts in a *paracrine* manner to induce the various transformation-associated traits enumerated above in nearby epithelial cells. However, as tumor progression advances, the rising level of COX-2 in these enterocytes enables them to make their own PGE$_2$, which stimulates their proliferation in an *autocrine* fashion and, once again, allows them to assume many of the traits associated with cell transformation.

Early-stage colonic polyps appear to depend on the continued inflammation induced by prostaglandins such as PGE$_2$ for their maintenance. However, later, as these growths evolve toward a higher degree of neoplasia, they lose this dependence. This echoes the experimental model of mouse skin carcinogenesis, in which early-stage papillomas depend on tumor promoters, such as TPA, for their maintenance, while more advanced papillomas become promoter-independent (see Figure 11.31).

Taken together, these observations lead to an attractive model (**Figure 11.43**) of how inflammation in a variety of epithelial tissues functions as a tumor-promoting mechanism and ultimately leads to carcinomas. This scheme is likely to change, since many of the indicated steps continue to be intensively investigated. Perhaps the greatest uncertainty lies in the mechanism(s) used by NSAIDs to inhibit tumor progression: while COX-2 is clearly a key target of their action, it is likely that other cellular enzymes are also affected by these drugs, and that inhibition of these other NSAID targets also contributes to slowing down or blocking multi-step tumorigenesis.

The details of the scheme depicted in Figure 11.43 also provide us with another way of viewing tumor promotion. Many of the cell phenotypes conferred by the inflammation-associated prostaglandins are uncannily similar to those conferred by the oncogenes that we read about earlier. Included here are traits such as loss of contact

Figure 11.42 Expression of COX-2 and the effects of its PGE$_2$ downstream product Prostaglandin E$_2$ (PGE$_2$) can act in a paracrine or autocrine manner to induce pre-neoplastic changes in the early stages of carcinoma progression in a variety of tissues. However, its cell type–specific expression differs from one tissue to another. (A) Immunostaining of a human colorectal hyperplastic polyp reveals the COX-2 enzyme (*brown spots*) to be expressed by endothelial cells and macrophages. (B) In a more advanced adenomatous polyp, COX-2 is found to be expressed (*brown spots*) increasingly in myofibroblasts, another type of stromal cell. (C) In this intestinal adenomatous polyp of a mouse that is genetically predisposed to polyposis because of a germ-line mutation in the *Apc* gene (see Section 7.11), the epithelial cells are immunostained *blue* with an anti-cytokeratin antibody, and the stromal cells (85% fibroblasts, 5% endothelial cells) are stained *green* with an antibody against vimentin, a cytoskeletal protein of mesenchymal cells. (D) In the human breast, however, COX-2 expression (*dark brown immunostaining*) is found to be high in early premalignant growths in the epithelial cells rather than the surrounding stromal cells. (A and B, from P.A. Adegboyega et al., *Clin. Cancer Res.* 10:5870–5879, 2004. C, courtesy of M. Sonoshita and M.M. Taketo. D, from Y.G. Crawford et al., *Cancer Cell* 5:263–273, 2004.)

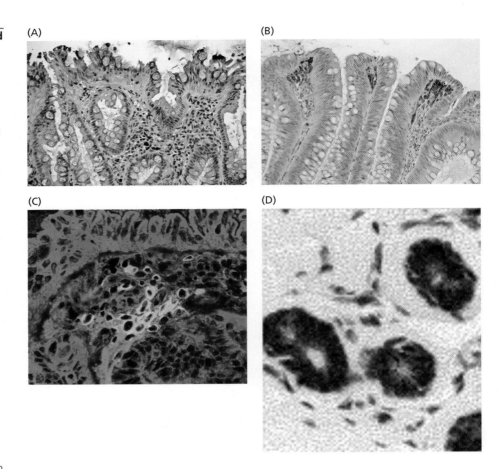

inhibition, as well as a gain of anchorage-independent growth and the ability to proliferate more rapidly. In effect, tumor promotion can create a **phenocopy** of the actions of an oncogene, that is, a biological state that closely resembles one created by an oncogene but arises through very different mechanisms.

This suggests that a tumor-promoting mechanism can collaborate with an oncogene in a way that resembles the collaboration between two oncogenes (see Section 11.10). Thus, initiated tumor cells may rely on this form of collaboration until their descendants eventually acquire additional oncogenes and therefore no longer need the readily reversible effects of tumor promoters for their continued proliferation and survival.

11.17 Tumor promotion is likely to be a critical determinant of the rate of tumor progression in many human tissues

A wide variety of agents have, by now, been classified as human tumor promoters (**Table 11.5**). They share in common an ability to promote expansion of initiated clones. The key role of this promotion in multi-step tumorigenesis can best be understood in the context of the clonal succession models depicted in Figure 11.15. In order for an initiated cell to acquire an additional mutation, its clonal descendants must become so numerous that a second, low-probability mutational event is likely to strike one or another cell of the now-expanded clonal population. Without such clonal expansion, the new, secondary mutation is unlikely to strike even a single descendant cell (because it is a rare event per cell generation), and tumor progression will halt.

Actually, these various mechanisms of tumor promotion can contribute to multi-step tumor progression in at least three ways. First, as just mentioned, promoters can stimulate the clonal expansions that yield the large cell populations in which otherwise-improbable events become possible.

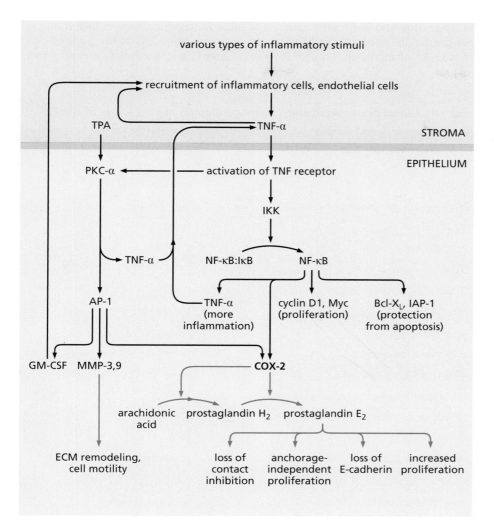

Figure 11.43 A model of epithelial inflammation and tumor promotion The information from skin and liver tumor promotion, as illustrated in Figures 11.33 and 11.38B, can be integrated into a more general model. Inflammatory stimuli, including chronic infections, organ injury, irritants (e.g., tobacco smoke), cell necrosis, and senescence, result in recruitment of inflammatory cells into areas of inflammation. These cells, as well as endothelial cells in the vicinity, release TNF-α (a key actor in multiple types of tumor promotion) and various cytokines that together activate the NF-κB pathway in epithelial cells. This results in the production of more TNF-α, anti-apoptotic proteins (e.g., Bcl-X$_L$, IAP-1, and IAP-2), and mitogenic proteins (Myc, cyclin D1), as well as COX-2. Prostaglandins synthesized by COX-2 (*pink arrows*) induce multiple cancer cell phenotypes, including loss of contact inhibition, increased proliferation rate, and anchorage-independent growth. The TPA skin tumor promoter converges on this pathway through the activation of PKC-α; in addition, TPA-activated PKC-α can induce mitogenic signals by activating the Fos, Jun, and thus AP-1 transcription factors (see Section 6.5). Not illustrated here is the production of prostaglandin E$_2$ by a variety of stromal cells (see Figure 11.42). Together, these actions of inflammatory cells and of TPA create phenotypic states in epithelial cells that closely resemble those induced by many oncogenes, enabling initiated cells to launch clonal expansions that eventually result in the acquisition of additional mutant alleles by their descendants.

Second, since cell proliferation requires DNA replication, and since DNA replication generates miscopied and thus mutant DNA sequences at a low but significant frequency (as discussed in the next chapter), agents favoring cell proliferation are indirectly mutagenic. In addition, repeated cell divisions hold yet other dangers for cellular genomes, since mitotic recombination and faulty chromosomal segregation yield loss-of-heterozygosity (LOH) events that enable mutant tumor suppressor genes to contribute to tumor progression.

Third, as was discussed in the last chapter, repeated cycles of growth and division lead to progressive shortening of telomeric DNA in stem cells. Ultimately, the ensuing telomere collapse and breakage–fusion–bridge (BFB) cycles result in karyotypic disarray and hence mutagenesis. Thus, as described in Section 10.10, BFB cycles occurring in the absence of functional p53 are likely to lead directly to accelerated tumor formation.

In fact, one type of tumor promotion—that involving inflammation—is likely to have an additional effect on tumor progression: the inflammatory cells that are recruited into a tissue—notably macrophages and neutrophils—are rich sources of reactive oxygen species (ROS) and reactive nitrogen species (RNS). These highly toxic molecules are normally deployed by these inflammatory cells in order to eliminate pathogens and host cells that have been targeted for destruction. However, in the context of the smoldering inflammation that precedes the appearance of many carcinomas, the released ROS and RNS can function as mutagens, adding to the mutagenic effects generated by the epithelial cells' own endogenous processes.

An interesting and illustrative example of these mutagenic effects comes from certain districts in northeastern Thailand, where more than 70% of the population is chronically infected with *Opisthorcis viverrini*; infection with this flatworm parasite is

Table 11.5 Known or suspected human tumor promoters and their sites of action

Agent or process	Cancer site
Hormones	
Estrogen	endometrium
Estrogen and progesterone	breast
Ovulation	ovary
Testosterone	prostate
Drugs	
Oral contraceptives, anabolic steroids	liver
Analgesics	renal pelvis
Diuretics	kidney
Infectious agents	
Hepatitis B/C viruses	liver
Schistosoma haematobium—blood fluke	bladder
Schistosoma japonicum—blood fluke	colon
Clonorchis sinensis—liver fluke	biliary tract
Helicobacter pylori—bacterium	stomach
Malarial parasites	B cell
Tuberculosis bacillus	lung
Chemical agents	
Betel nut, lime	oral cavity
Chewing tobacco	oral cavity
Bile	small intestine
Salt	stomach
Acid reflux	esophagus
Physical or mechanical trauma	
Asbestos	mesothelium, lung
Gallstones	gallbladder
Coarsely ground corn	stomach
Head injury	meninges
Chronic irritation/inflammation	
Tropical ulcers[a]	skin
Chronic ulcerative colitis	colon
Chronic cystitis	bladder
Chronic pancreatitis	pancreas

[a]Tropical ulcers are caused by chronic infections of the skin, usually of bacterial origin, that do not heal and are associated with poor nutrition and lack of sanitation.

Adapted in part from S. Preston-Martin et al., *Cancer Res.* 50:7415–7421, 1990.

a strong risk factor for developing the otherwise-unusual cholangiocarcinomas (intra-hepatic bile duct cancers) that are common tumors in these areas. Infected individuals suffer chronic inflammation in their bile ducts extending over decades of time. A good marker of oxidative DNA damage leading to mutation is 8-oxo-deoxyguanosine (8-oxo-dG), which is excised from cell genomes by DNA repair enzymes (see Chapter 12) and excreted in the urine. In infected individuals, urinary 8-oxo-dG concentrations are twice as high as in normal uninfected controls. Thus, the inflamed bile ducts of these individuals produce as many oxidized nucleotides as the entire remaining tissues of the body. However, once they are treated with a highly effective anti-parasitic drug, their levels of excreted 8-oxo-dG fall to normal over the ensuing year.

These diverse observations demonstrate that the carcinogen = mutagen equation, which we previously distilled from the work of Bruce Ames (see Chapter 2), is simplistic. It is now obvious that many carcinogens succeed in doing their work through means that do not depend on an ability to directly damage DNA. As mentioned, such nonmutagenic agents cannot be detected by the Ames test. This means that a positive result in the Ames test, which demonstrates the genotoxicity of a chemical agent, is likely (but not guaranteed) to predict its carcinogenic powers in rodents and humans. However, a negative result in this test does not exclude the possibility that the compound being tested can contribute to the formation of human cancers by acting as a nongenotoxic tumor promoter.

Some carcinogenic agents, such as the benzo[a]pyrene (BP) described earlier, when applied to laboratory animals as single agents, are able on their own to induce tumor formation after repeated applications. They thereby function as both initiators and promoters of tumorigenesis and consequently have been termed **complete carcinogens**. We can imagine, for example, that certain mutagenic agents are also cytotoxic at high concentrations and thus able to act as tumor promoters by killing cells and inducing compensatory proliferation of surviving cells in a target tissue. But such agents are likely to play minor roles in human carcinogenesis. Most genotoxic carcinogens enter into human tissues at concentrations that are far too low to evoke cytotoxic effects. Hence, the tumors incited by these agents seem to depend heavily for their formation on other substances that function as pure tumor promoters.

11.18 Synopsis and prospects

A major goal of modern cancer research is to uncover the root causes of this disease. We began this chapter with the mindset that genetic changes are responsible for fueling many of the steps of cancer progression and that each is, in principle, traceable to the actions of specific mutagenic agents. As our discussions of multi-step tumorigenesis proceeded, however, it became increasingly clear that agents other than mutagens contribute to the pathogenesis of human cancer. This realization both clarifies and confounds our attempts to ferret out the causes of human malignancies.

It is possible, for example, that most human carcinogens act as tumor promoters (rather than initiators), and that many types of human tumors arise entirely without the contributions of exogenous genotoxic agents, that is, mutagens originating from outside the human body. In such tumors, the genetic damage would, by necessity, be generated entirely by endogenous processes.

This speculation comes from two sources. First, the origins of most human cancers have not been associated with exposure to specific mutagenic agents in spite of several decades of intensive searching. The major exceptions here are the combustion products of tobacco and (quite possibly) the products of cooking meat at high temperature, as we will see in Chapter 12.

Second, the genome within a single human cell sustains as many as 10,000 chemical modifications each day, according to some estimates. These modifications are created by chemical species that have been generated endogenously by various metabolic reactions. Notable among these endogenous mutagens are several types of reactive oxygen species (ROS) that arise as by-products of oxidative metabolism in the mitochondria.

Clearly, the great majority of these chemical lesions are removed by the highly effective DNA repair systems operating in most types of cells (see Chapter 12). Inevitably, however, some of these lesions escape detection and subsequent repair, become fixed in the genomes of the cells in which they initially were formed, and are then transmitted as mutations to descendant cells. In most human tissues, the number of endogenously generated mutations is likely to dwarf those of exogenous origin.

This information allows us to entertain the following scenario: Endogenously generated mutant alleles, acting in concert with nongenotoxic tumor promoters, drive the progression that leads to the appearance of many kinds of human cancers. These tumor-promoting mechanisms may derive from chronic localized inflammation or from exogenous agents, including foodstuffs, infections, and even tobacco smoke, which contains many tumor-promoting chemicals. If so, searches for exogenous mutagenic carcinogens that are suspected to be etiologic (causative) agents of many types of human tumors are doomed to failure.

The existence of such tumor-promoting substances, whatever their nature, greatly complicates attempts to devise laboratory-based screens for human carcinogens. Thus, the frequently used *in vitro* tests for mutagenicity (Section 2.10), such as the Ames test, lead one astray because they do not register the presence of tumor promoters, and many tests of carcinogenicity in laboratory animals may also have limited utility (see Supplementary Sidebar 11.5).

These arguments indicate that arcane debates about cancer initiation versus promotion have enormous public health implications. And misinterpretations of biological processes such as tumor promotion lead, in turn, to major changes in the substances that we consume and those that we have been told to avoid (see Supplementary Sidebar 11.6). Indeed, tumor-promoting mechanisms may provide solutions to certain epidemiologic puzzles.

Prominent here is the puzzle created by the consumption of certain types of red meat, which has been clearly correlated with increased incidence of several common carcinomas. Part of this association can be attributed to the formation of mutagenic carcinogens produced by cooking at high temperatures (see Section 12.6). Still, mutagenesis seems to represent only part of the story. One search for the carcinogenic effects of red meat consumption has led to an intriguing speculation that would explain why Western diets rich in red meat favor increased incidence of a variety of carcinomas (see Supplementary Sidebar 11.7).

On a worldwide scale, it is estimated that 9% of all cancer deaths arise because chronic infections of the stomach by the bacterium *Helicobacter pylori* lead to gastric carcinomas, and that 6% of all cancer deaths are associated with liver cancers, most of which are caused by chronic infections by hepatitis B and hepatitis C viruses. The chronic infection of the cervical epithelium caused by human papillomaviruses, notably strains 16, 18, and 45, leads in some women to the cervical carcinomas that account for 5% of worldwide cancer mortality. (The chronic inflammatory effects of these various viral infections may be compounded by various virus-borne oncogenes.) While these figures are daunting, in fact they represent an enormous opportunity for reducing worldwide cancer incidence and mortality: infectious diseases are ultimately far easier to control and prevent (through immunization) than the effects of foodstuffs that we ingest and the tobacco products that we inhale. By some estimates, 450 million people worldwide are infected chronically with HBV, while 200 million are long-term HCV carriers. These numbers indicate the enormous benefit that may derive from preventing these viral illnesses.

Inflammatory processes are also implicated in the promotion of a number of nonviral human cancer types. In these cases, opportunities for the development of future anti-cancer drugs derive from the recent advances in elucidating the biochemical details of inflammation and its specific contributions to tumor promotion. Cyclooxygenase-2 (COX-2) is clearly at the center of this process, and a number of anti-inflammatory agents (largely NSAIDs) targeting this enzyme are known to be effective in decreasing the incidence of various types of cancer. Unfortunately, at high doses the long-term use of some of these agents induces unacceptable side effects in some individuals, including fatal cardiovascular complications. Presumably these side effects are due to the actions of various prostaglandins generated directly or indirectly by COX-2, each of which elicits its own set of complex cellular responses.

Nonetheless, the prospects are bright for the development of potent anti-cancer agents having **prophylactic** (preventative) and therapeutic activity that affect the COX-2 pathway but do not create the side effects associated with long-term NSAID use. Thus, the key product of COX-2 action in tumor promotion, prostaglandin E_2 (PGE_2), is known to bind and activate at least eight different cell surface serpentine receptors (see Sections 5.7 and 6.11), each of which presumably evokes its own subset of downstream responses. Accordingly, highly specific drugs designed to inhibit only one or another of these receptors could conceivably reduce inflammation-associated tumor promotion while having few of the side effects associated with the broadly acting COX-2 antagonists.

Estrogen, progesterone, and androgens loom large in most discussions about human cancer because they play such critical roles in the pathogenesis of breast, endometrial, and prostatic carcinomas. Together, these tumors account for about 9% of all mortality from neoplasias in the West. These hormones stimulate the proliferation of epithelial cells in the corresponding responsive tissues, and so we can easily incorporate them into our conceptual schemes of tumor initiation and promotion. The systemic hormones insulin and the closely related IGF-1 (insulin-like growth factor-1) also seem to function as important tumor promoters (see Sidebar 9.9) because they protect premalignant cells from apoptosis and may also spur their proliferation.

Significantly, these hormones continue to play important roles in tumorigenesis once multi-step tumor progression has yielded full-blown malignancies. The proliferation of cancer cells in many established breast tumors and almost all prostatic tumors depends upon the continued presence of estrogen and androgens, respectively. The continued presence of IGF-1 also seems to be required by many already-formed tumors. This ongoing dependence highlights a more general question that remains relatively unexplored by cancer researchers: To what extent do a variety of malignancies, once they are formed, continue to depend on the tumor promoters that helped to create them in the first place?

A different but related question is: Do full-blown tumors continue to depend upon the mutant alleles that were created decades earlier by initiating mutagens? Or do these early mutations, which occur as the initial steps of tumor progression, become irrelevant later, when subsequently acquired mutant alleles take over the job of programming cancer cell proliferation?

The most direct answers to this question come from experiments with transgenic mice in which certain oncogenes are activated on a tissue-specific basis; these gene activations lead eventually to tumors in those tissues. In some of these mouse models, the initiating transgenes, such as *myc* and *ras*, have been inactivated by a variety of experimental tricks long after substantial tumors have formed. The effects on tumor growth that have been observed to date are conflicting.

Usually, these tumors collapse rapidly when deprived of the oncogenes that led originally to their formation. Hence, in these cases, the mutant oncogenic alleles were important both for the *initiation* of the cancerous state as well as the *maintenance* of this state long afterward. Stated differently, the mutant alleles that were accumulated in later steps of multi-step tumorigenesis did not render the earlier ones unnecessary. However, in several research reports, shutdown of the initiating *neu*, *myc*, or *wnt* oncogenes after tumors had formed led, after short remissions, to the regrowth of these tumors. In these cases, the initiating oncogenes were clearly no longer required for the expansion of the relapsing tumors.

If we assume, for the moment, that these various mouse models of cancer formation reflect mechanisms operating in human tumors as well, the observed effects hold profound implications for the development of new types of anti-cancer therapeutics, a topic that we will return to in Chapter 16. For example, if a mutant *ras* allele is found in a human tumor cell genome, can this tumor be cured by drugs targeted to the Ras oncoprotein, or has this oncoprotein, which may have played a key role in the initiation of tumor formation, become irrelevant, later on, to the continued proliferation and survival of the cells in this tumor?

The vast numbers of genetic alterations present in human cancer cell genomes greatly complicate one major goal of current cancer research: descriptions of the "genetic biographies" of various types of human tumors—the successions of genetic alterations that program the neoplastic phenotype of the cells within these tumors. The number of altered DNA sequences in the genomes of most kinds of human cancer cells clearly dwarfs the small number of genetic alterations (and promoter methylations) that play causal roles in tumor progression. At present, we possess only very crude tools for distinguishing the wheat from the chaff—the small number of biologically important genetic changes from the large throng of functionally irrelevant genetic changes that are present in the genomes of almost all human cancer cells.

In spite of these truly daunting problems, there is optimism that we may soon be able to formulate some basic organizing principles that place all types of human tumors under a common conceptual roof. These principles might then be used to explain why various human tumors acquire certain combinations of mutant (and methylated) genes. Specifically, examinations of the biological phenotypes of a variety of human tumor types have led to the proposition that all highly advanced human cancer cells share a number of essential attributes that they have acquired en route to the malignant state. These are (1) a reduced dependence on exogenous mitogenic growth factors (see Chapters 5 and 6); (2) an acquired resistance to growth-inhibitory signals, such as those conveyed by TGF-β (see Chapter 8); (3) the ability to multiply indefinitely, that is, immortalized cell proliferation (see Chapter 10); (4) a reduced susceptibility to apoptosis (see Chapter 9); (5) the ability to generate new blood vessels—angiogenesis (see Chapter 13); (6) the acquisition of invasiveness and metastatic ability (see Chapter 14); and a seventh that remains less well documented: (7) the ability to evade elimination by the immune system (see Chapter 15).

Actually, some of the observations described in this chapter argue strongly for yet another shared attribute. Our discussion of clonal evolution and succession concluded that in order for these events to occur at reasonable frequency, mutation rates within tumor stem cell populations must be abnormally high. In the absence of such increased mutability, clones of cells may not accumulate the multiple genetic alterations during a human life span that are required for cancer formation. Together, these speculations and observations converge on an eighth attribute that may well be shared by almost all types of human cancer cell genomes: (8) acquisition of genomic instability (see Chapter 12).

If we embrace this list, at least tentatively, we need to relate the acquisition of each of these traits to the specific steps of multi-step tumor progression. For example, we might assume that each step in the progression of a human tumor is demarcated by the alteration of a distinct gene, achieved through either genetic mutation or epigenetic methylation. Moreover, we could imagine that each of the steps in tumor progression yields one of the seven or eight attributes listed above. This would create a one gene–one phenotype scenario that would greatly simplify our thinking about human tumor progression.

While attractive, this idea is simplistic, since there is no one-to-one mapping between specific gene alterations and corresponding changes in cell phenotype. Instead, many of the phenotypes of cancer cells are achieved collaboratively through the actions of several genes or genetic alterations (**Figure 11.44**). For example, the acquired resistance to apoptosis shown by human cancer cells often results from the activation of the Ras signaling pathways (which energizes the anti-apoptotic PKB/Akt kinase) as well as the inactivation of the p53 signaling pathway.

Conversely, the mutation of certain pleiotropically acting cancer genes may confer several distinct phenotypic benefits simultaneously. Thus, the formation of a *myc* oncogene may concomitantly deregulate the pRb signaling pathway (which normally enables cells to respond to growth-inhibitory signals), aid in the de-repression of the *hTERT* gene (which enables cell immortalization), and reduce the mitogen

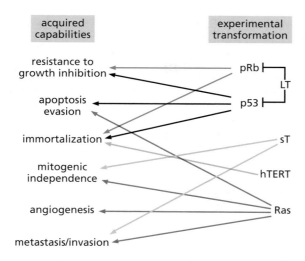

Figure 11.44 Cancer cell genotypes versus phenotypes During the course of tumor progression, human cells acquire a number of distinct cancer-associated phenotypes. Independent of this, laboratory experiments reveal that a number of distinct genes must be introduced into human cells in order to transform them to a tumorigenic state. This raises the question how each of these introduced genes contributes to the cell phenotypes associated with tumorigenicity. As indicated here, genes such as that encoding hTERT, the catalytic subunit of telomerase, affect only the phenotype of immortalization, while other genes, such as SV40 large T (LT), used experimentally to neutralize both p53 and pRb, affect multiple phenotypes. For example, by inactivating p53, LT elicits at least three changes in cell phenotype—resistance to growth inhibition, evasion of apoptosis, and immortalization; by neutralizing pRb, LT also affects two of these phenotypes. The most widely acting protein is likely to be Ras, which affects susceptibility to apoptosis, dependence on exogenous mitogens, angiogenesis, and invasiveness and metastatic ability. Hence, a one-to-one mapping between genes and cancer-associated phenotypes is not possible. (Since hTERT has been found to be physically associated with the β catenin–Tcf/LEF transcription factor complex, its actions may extend beyond cell immortalization.)

dependence of a cell. So the hope of a simple scheme is frustrated by the biological realities of how each of these genes and encoded proteins actually operates.

Knowing this complexity, we might still wish to attempt another type of mapping. We could imagine that each of the biological phenotypes of cancer cells is the direct result of an alteration in one of the regulatory subcircuits that govern the life of the cell (**Figure 11.45**). If so, perhaps we can explain tumor progression as the progressive deregulation of a number of distinct subcircuits within the cell.

An examination of Figure 11.45 gives us some encouragement. For example, the acquisition of mitogen independence is achieved largely by the activation of the receptor tyrosine kinase (RTK) → Ras → MAPK pathway, while the resistance to apoptosis is acquired through lesions in the subcircuit that governs programmed cell death. Yet other such acquired attributes can also be related to distinct portions of this large circuit diagram. Still, even here there is no neat and simple compartmentalization, because there are numerous cross connections between the various subcircuits operating within cells. In addition, several subcircuits often collaborate to create a distinct cancer cell phenotype.

Nonetheless, with all these reservations in mind, Figure 11.45 and others like it inspire a hope that may well be realized sometime during the first decades of the new millennium: At some point, we will truly understand in detail how each of these subcircuits operates to regulate cell phenotype. We will be able to model the operations of each mathematically. And we will be able to rationalize the behavior of the cancer cell as a whole in terms of the interactions of specific molecular defects in each of these regulatory circuits.

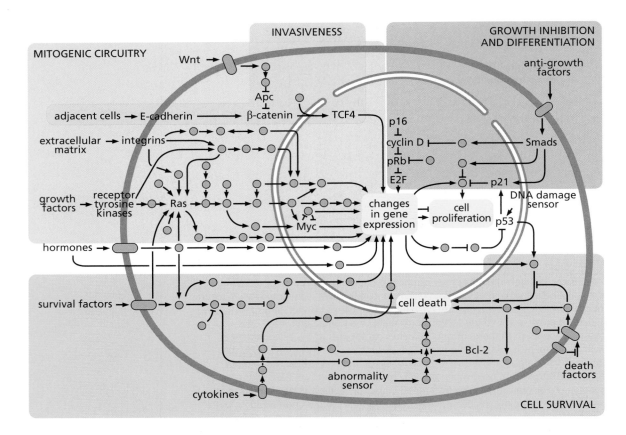

Figure 11.45 The intracellular signaling circuitry and collaboration between cancer-associated genes The design of the signal-transduction circuitry of mammalian cells has been uncovered, piece by piece, over the past three decades. The diagram here indicates only a portion of the proteins (*green circles*) that play critical roles in modulating the flow of signals through the various circuits operating within our cells. As indicated here by the various shadings, different subcircuits are involved in regulating distinct cell physiologic processes. Thus, the growth-promoting, mitogenic circuit (*light red*), the circuit governing growth-inhibitory signals (*light brown*), the circuit governing apoptosis (*light green*), and the circuit governing invasiveness and metastasis (*light blue*) can be assigned to distinct regions in the map of the master circuitry of the mammalian cell. (Note that the circuit governing mitogenesis overlaps in part with that governing cancer cell invasiveness, indicative of the fact that a common set of proteins mediates both biological responses.) Presumably, mathematical modeling of these various circuits will one day be able to provide a mechanistic rationale of why tumorigenesis is a multi-step process in mammals. (Adapted from D. Hanahan and R.A. Weinberg, *Cell* 100:57–70, 2000.)

We close this chapter by recalling the eighth attribute that was ascribed to cancer cells—their acquisition of mutable genomes. Given the low probability of each individual step of tumor progression, completion of the process as a whole becomes extremely improbable, yet cancers occur with substantial frequency in the human population. The next chapter is focused on the attempts to resolve this dilemma and reveal how a mathematically impossible disease process becomes, most unfortunately, quite commonplace and the cause of 20% of human mortality.

Key concepts

- The process of tumor formation is a complex one of multiple steps involving multiple alterations of cells and their physiologic control mechanisms.

- The complexity of this process is reflected in the long time periods required for most human cancers to develop.

- These changes involve both the activation of oncogenes and the inactivation of tumor suppressor genes.

- The number of steps required to experimentally transform human cells is larger than is needed to transform cells of laboratory mice.

- These alterations affect multiple distinct regulatory circuits within cells and function in a complementary fashion to create the neoplastic cell phenotype.

- Some of these changes occur as the direct result of the actions of exogenous mutagens, and exposure to such mutagens may represent a "rate-limiting" determinant of tumor progression.

- In many instances, however, the rate of tumor progression may be governed by the actions of nonmutagenic promoting agents, which may determine the rate of expansion of mutant cell clones.

- In many human cancers, these critical nonmutagenic, tumor-promoting stimuli include chronic mitogenic stimulation and inflammation.

- The multiplicity of steps required for human cancers to arise is not known, in part because certain changes may occur rapidly and therefore may not be "rate-limiting," while others may require a decade or more to complete.

- Multi-step tumor progression can be depicted as a form of Darwinian evolution occurring within tissues. However, because some of the critical changes occurring during tumorigenesis are epigenetic, and because the rate of genetic diversification can occur very rapidly, the classic depictions of Darwinian evolution must be modified.

- In most, but not all, transgenic models of tumorigenesis, the initiating changes continue to be required for continued survival of a tumor, long after the process of tumor progression has reached completion.

- The number of genetic changes found in the genomes of human cancer cells vastly exceeds the number required for tumorigenesis to reach completion, complicating identification of the critical changes that are causally important in tumor formation.

- The discovery of cancer stem cells greatly changes our concepts about the mechanisms of multi-step tumorigenesis, since these self-renewing cells (or closely related progenitor cells), rather than the bulk populations of cancer cells, may be the objects of genetic alteration and clonal selection.

Thought questions

1. Knowing the various genetic and epigenetic changes that occur during multi-step tumorigenesis, which of these would you say are likely to be readily uncovered and which may be difficult to identify? Describe the reasons for these assignments.

2. Some tumor suppressor genes inactivated during multi-step tumorigenesis may be readily identified because of LOH in the chromosomal region carrying them, while others may be difficult to identify in this way. Describe the factors that allow or complicate this identification.

3. What arguments favor the notion that all of us carry myriad clones of initiated premalignant cells throughout the body?

4. What arguments can be mustered that favor the notion that the bulk of human carcinogens act as promoters rather than initiators of tumorigenesis?

5. What different approaches can be used to estimate the number of steps in multi-step tumor progression, and how is each of these approaches flawed?

6. How does the current available information about multi-step tumor progression provide insights into strategies for the prevention of clinically detectable cancers?

7. What mechanisms enable chronic viral infections to exert a carcinogenic influence on a tissue?

8. Describe the various mechanisms of tumor promotion and the features that they share in common and those that distinguish them from one another.

Additional reading

Al-Hajj M & Clarke ME (2004) Self-renewal and solid tumor stem cells. *Oncogene* 23, 7274–7282.

Alison MR & Islam S (2009) Attributes of adult stem cells. *J. Pathol.* 217, 144–160.

Armitage P & Doll R (1954) The age distribution of cancer and a multistage theory of carcinogenesis. *Br. J. Cancer* 8, 1–12.

Arwert E, Hoste E & Watt FM (2012) Epithelial stem cells, wound healing and cancer. *Nat. Rev. Cancer* 12, 170–180.

Balkwill F, Charles KA & Mantovani A (2005) Smoldering and polarized inflammation in the initiation and promotion of malignant disease. *Cancer Cell* 7, 211–217.

Balkwill F & Mantovani A (2010) Cancer and inflammation: implications for pharmacology and therapeutics. *Clin. Pharmacol. Ther.* 87, 401–406.

Beachy PA, Karhadka SS & Berman DM (2004) Tissue repair and stem cell renewal in carcinogenesis. *Nature* 432, 324–331.

Blanpain C & Fuchs E (2006) Epidermal stem cells of the skin. *Annu. Rev. Cell Dev. Biol.* 22, 339–373.

Borovski T, De Sousa e Melo F, Vermeulen L & Medema JP (2011) Cancer stem cell niche: the place to be. *Cancer Res.* 71, 634–639.

Braakhuis BJM, Tabor MP, Kummer JA et al. (2003) A genetic explanation of Slaughter's concept of field cancerization: evidence and clinical implications. *Cancer Res.* 63, 1727–1730.

Brabletz S, Schmalhofer O & Brabletz T (2009) Gastrointestinal stem cells in development and cancer. *J. Pathol.* 217, 307–317.

Cairns J (1975) Mutation, selection and the natural history of cancer. *Nature* 255, 197–200.

Caldas C (2012) Cancer sequencing unravels clonal evolution. *Nature Biotech.* 30, 408–410.

Cheng J, DeCaprio JA, Fluck MM & Schaffhausen BS (2009) Cellular transformation by simian virus 40 and murine polyoma virus T antigens. *Semin. Cancer Biol.* 19, 218–228.

Cho RW & Clarke MF (2008) Recent advances in cancer stem cells. *Curr. Opin. Genet. Dev.* 18, 1–6.

Cowey S & Hardy RW (2006) The metabolic syndrome: a high-risk state for cancer? *Am. J. Pathol.* 169, 1505–1522.

De Visser KE, Eichten A & Coussens LM (2006) Paradoxical roles of the immune system during cancer development. *Nat. Rev. Cancer* 6, 24–37.

Garcia SD, Park HS, Novelli M & Wright NA (1999) Field cancerization, clonality, and epithelial stem cells: the spread of mutated clones in epithelial sheets. *J. Pathol.* 187, 61–81.

Gold LS, Ames BN & Slone TH (2002) Misconceptions about the causes of cancer. In Human and Environmental Risk Assessment: Theory and Practice (D Paustenbach ed), pp 1415–1460. New York: John Wiley and Sons.

Gonda TA, Tu S & Wang TC (2009) Chronic inflammation, the tumor microenvironment and cancer. *Cell Cycle* 8, 2005–2013.

Grady WM & Markowitz SD (2002) Genetic and epigenetic alterations in colon cancer. *Annu. Rev. Genomics Hum. Genet.* 3, 101–128.

Griner EM & Kazanietz MG (2007) Protein kinase C and other diacylglycerol effectors in cancer. *Nat. Rev. Cancer* 7, 281–293.

Grivennikov SI, Greten FR & Karin M (2010) Immunity, inflammation, and cancer. *Cell* 140, 883–899.

Gupta RA & DuBois RN (2001) Colorectal cancer prevention and treatment by inhibition of cyclooxygenase-2. *Nat. Rev. Cancer* 1, 11–19.

Haber DA & Settleman J (2007) Cancer: drivers and passengers. *Nature* 446, 145–146.

Hahn WC & Weinberg RA (2002) Modelling the molecular circuitry of cancer. *Nat. Rev. Cancer* 2, 331–341.

Hanahan D & Weinberg RA (2011) Hallmarks of cancer: the next generation. *Cell* 144, 646–674.

Hansel DE, Kern SE & Hruban RH (2003) Molecular pathogenesis of pancreatic cancer. *Annu. Rev. Genomics Hum. Genet.* 4, 237–256.

Henderson BE & Feigelson HS (2000) Hormonal carcinogenesis. *Carcinogenesis* 21, 427–433.

Humphries A & Wright NA (2008) Colonic crypt organization and tumorigenesis. *Nat. Rev. Cancer* 8, 415–424.

Karin M & Greten FR (2005) NF-κB: linking inflammation and immunity to cancer development and progression. *Nat. Rev. Immunol.* 5, 749–759.

Kemp CJ (2005) Multistep skin cancer in mice as a model to study the evolution of cancer cells. *Semin. Cancer Biol.* 15, 460–473.

Lin W-W & Karin M (2007) A cytokine-mediated link between innate immunity, inflammation, and cancer. *J. Clin. Invest.* 111, 1175–1183.

Loeb LA & Harris CC (2008) Advances in chemical carcinogenesis: a historical review and prospective. *Cancer Res.* 68, 6863–6872.

Mantovani A, Allavena P, Sica A & Balkwill F (2008) Cancer-related inflammation. *Nature* 454, 436–444.

Markowitz SD (2006) Aspirin and cancer: targeting prevention? *N. Engl. J. Med.* 356, 2195–2198.

Markowitz SD & Bertagnolli MM (2009) Molecular basis of colorectal cancer. *N. Engl. J. Med.* 361, 2449–2460.

Marusyk A & Polyak K (2010) Tumor heterogeneity: causes and consequences. *Biochim. Biophys. Acta* 1805, 105–117.

Merlo LMF, Pepper JW, Reid BJ & Maley CC (2006) Cancer as an evolutionary and ecological process. *Nat. Rev. Cancer* 6, 924–935.

Miller EC (1978) Some current perspectives on chemical carcinogenesis in humans and experimental animals: presidential address. *Cancer Res.* 38, 1479–1496.

Navin NE & Hicks J (2010) Tracing the tumor lineage. *Mol. Oncol.* 4, 267–283.

Nowell PC (1976). The clonal evolution of tumor cell populations. *Science* 194, 23–28.

Parsonnet J (ed) (1999) Microbes and Malignancy: Infection as a Cause of Human Cancers. Oxford, UK: Oxford University Press.

Qian B-Z & Pollard JW (2010) Macrophage diversity enhances tumor progression and metastasis. *Cell* 141, 39–51.

Reich BJ, Li X, Galipeau PC & Vaughan TL (2010) Barrett's oesophagus and oesophageal adenocarcinoma: time for a new synthesis. *Nat. Rev. Cancer* 10, 87–101.

Renan MJ (1997) How many mutations are required for tumorigenesis? Implications from human cancer data. *Mol. Carcinog.* 7, 139–146.

Schedin P (2006) Pregnancy-associated breast cancer and metastasis. *Nat. Rev. Cancer* 6, 281–291.

Shackleton M, Quintana E, Fearon ER & Morrison SJ (2009) Heterogeneity in cancer: cancer stem cells versus clonal evolution. *Cell* 138, 822–829.

Singh A & Settleman J (2010) EMT, cancer stem cells and drug resistance: an emerging axis of evil in the war on cancer. *Oncogene* 29, 4741–4751.

Stiles CD & Rowitch DH (2008) Glioma stem cells: a midterm exam. *Neuron* 58, 832–846.

Stratton MR, Campbell PJ & Futreal PA (2009) The cancer genome. *Nature* 458, 719–724

Thun MJ, Henley SJ & Calle EE (2002) Tobacco use and cancer: an epidemiologic perspective for geneticists. *Oncogene* 21, 7307–7325.

Trichopoulos D, Adama H-O, Ekborn A et al. (2007) Early life events and conditions and breast cancer risk: from epidemiology to etiology. *Intl. J. Cancer* 122, 481–485.

Visvader JE & Lindeman GJ (2008) Cancer stem cells in solid tumours: accumulating evidence and unresolved questions. *Nat. Rev. Cancer* 8, 755–768.

Visvader JE & Smith GH (2010) Murine mammary epithelial stem cells: discovery, function and current status. *Cold Spring Harbor Perspect. Biol.* 4, a004879, Oct. 6.

Wang D & DuBois RN (2010) Eicosanoids and cancer. *Nat. Rev. Cancer* 10, 181–193.

Wang D & DuBois RN (2010) The role of COX-2 in intestinal inflammation and colorectal cancer. *Oncogene* 29, 781–788.

Weinstein IB (2002) Addiction to oncogenes: the Achilles heal [*sic*] of cancer. *Science* 297, 63–64.

Westermarck J & Hahn WC (2008) Multiple pathways regulated by the tumor suppressor PP2A in transformation. *Trends Mol. Med.* 14, 152–160.

Zhao JJ, Roberts TM & Hahn WC (2004) Functional genetics and experimental models of human cancer. *Trends Mol. Med.* 10, 344–350.

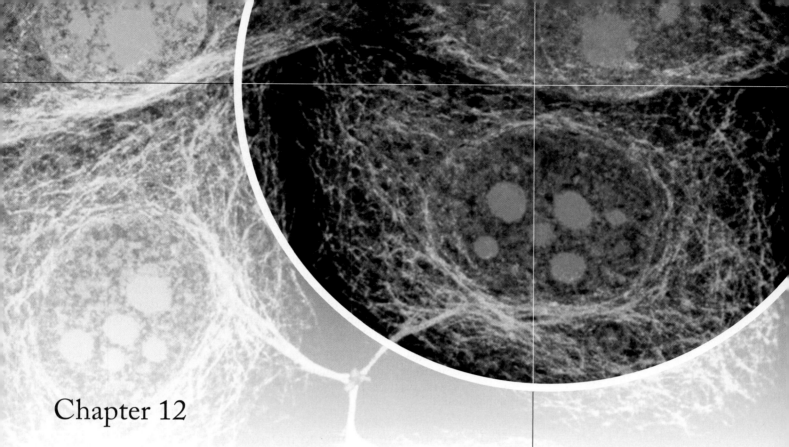

Chapter 12

Maintenance of Genomic Integrity and the Development of Cancer

> When first interpreting the ramifications of DNA and the genetic code ... [w]e totally missed the possible role of enzymes in repair. ... I later came to realise that DNA is so precious that probably many distinct repair mechanisms could exist.
>
> Francis H. C. Crick, molecular biologist, 1974

> The capacity to blunder slightly is the real marvel of DNA. Without this special attribute, we would still be anaerobic bacteria and there would be no music.
>
> Lewis Thomas, biologist, 1979

The fact that human tumor formation is a complex, multi-step process reflects the multiple lines of defense against cancer that have been established within our cells, each maintained by the hard-wiring of a complex regulatory circuit. The human body—actually, its individual cells—must entrust the maintenance of these anti-cancer defenses to their most stable, reliable constituents: DNA molecules. Over extended periods of time, DNA sequences are the most fixed, unchangeable components of a cell; most of its other parts are in constant flux, being created and broken down continuously.

Therefore, it is really the stability of DNA molecules that underpins the most robust defenses against cancer. Because there are multiple cellular lines of defense that depend on DNA stability and because the breaching of each defense usually requires a rare mutational event, the probability of cell populations' advancing all the way to the neoplastic state must be astronomically small.

Movie in this chapter

12.1 DNA Repair Mechanisms

511

So, the cancerphobe can rest easy at night, reassured by the multiplicity of cellular and tissue defense mechanisms that evolution has assembled to protect us from neoplasia. But there is a troubling inconsistency here: if the number of anti-cancer defense mechanisms were truly as great as depicted in this text, and if the breaching of each of these defenses were usually dependent on rare mutational events, then cancers should never strike human populations. Yet they do. In Western populations, in which deaths from infectious diseases are relatively infrequent, about 1 person in 5 is destined to die from one or another form of cancer. So, cancer cell populations accomplish what seems to be the impossible—acquiring a substantial array of mutant (and methylated) alleles over a period of several decades.

Researchers attempted to resolve this inconsistency as far back as 1974. They proposed that the only explanation for this quandary must depend on a drastic increase in mutation rate: cell populations en route to becoming malignant must carry genomes that are far more mutable than the genomes of normal human cells—a condition sometimes termed the *mutator phenotype*. Such speculation has received increasing support in recent years, as various types of genetic instability have been documented in the genomes of certain classes of cancer cells.

In this chapter, we will direct much of our attention to two major issues. First, how do normal human cells and tissues manage to keep the mutation rate so low? And second, how are the strategies for suppressing mutations thwarted during human cancer pathogenesis?

12.1 Tissues are organized to minimize the progressive accumulation of mutations

On a number of occasions throughout this text, we have described the effects of carcinogens and tumor promoters on target cells throughout the body. However, the specific biological identities of these target cells have never been spelled out. As it turns out, knowledge of the nature of these cells is critical to understanding how genome integrity is maintained. To explore this issue, we need to delve into the organization of tissues and the types of cells that form them. Their biological behavior furnishes us with insights into the strategies exploited by tissues and cells to minimize the accumulation of genetic lesions.

As described earlier (see Section 11.6), a common scheme seems to explain the construction and maintenance of many tissues throughout the body. Within each tissue, a relatively small number of cells populate its stem cell **compartment**. These self-renewing cells may constitute a minute fraction of the entire cell population within a tissue, sometimes as few as 0.1 to 1% of the total. In truth, in most tissues, these numbers represent nothing more than poorly informed guesses. Because stem cells are present in very small numbers, have appearances that are not particularly distinctive, and are often scattered among other cell types within tissues, they are difficult to identify and study. Consequently, much of what is described below rests on inference rather than on direct observation of stem cells and their properties.

As is the case with stem cells of tumors (see Chapter 11), the stem cells in a normal tissue are self-renewing, since at least one of the two daughters of a dividing stem cell will retain the phenotype exhibited by the mother cell prior to cell division (see Figure 11.18). In many tissues, the second daughter cell and its transit-amplifying descendants will pass through a substantial number of cell divisions before entering into a post-mitotic, highly differentiated state. These actively dividing cells, which serve as intermediates between a stem cell and its differentiated descendants, may thereby generate large flocks of differentiated descendants of the second daughter cell (**Figure 12.1**). Implicit in what follows is a simple idea: because the lineage of stem cells represents the only stable repository of genetic information within a tissue, the genomes of stem cells must be protected from corruption.

The exponential increase in the number of transit-amplifying cells means that a stem cell needs to divide only on rare occasion in order to maintain a large pool of end-stage, highly differentiated cells in a tissue. In the colonic crypts, for example, it is

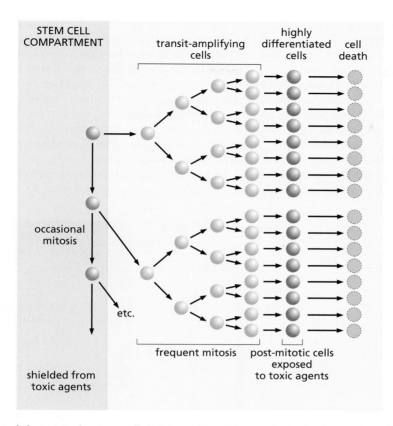

STEM CELL COMPARTMENT

transit-amplifying cells

highly differentiated cells

cell death

occasional mitosis

etc.

frequent mitosis

post-mitotic cells exposed to toxic agents

shielded from toxic agents

Figure 12.1 Tissue organization and protection of the stem cell genome The organization of many epithelial tissues seems to conform to the scheme shown here. Each stem cell (*blue*) divides only occasionally in an asymmetric fashion to generate a new stem cell daughter and a transit-amplifying daughter. These stem cells are often shielded anatomically from toxic agents. The transit-amplifying cells (*green*) undergo repeated rounds of growth and division, expanding their populations exponentially. Eventually, the products of these cell divisions undergo further differentiation into post-mitotic, highly differentiated cells (*red*). The highly differentiated cells, which are often in direct contact with various toxic agents, are shed with some frequency; hence, any mutant alleles that arise in these cells will be lost, sooner or later, from the tissue. This means that the genomes of stem cells are often protected through two mechanisms: infrequent cell division, and an anatomical barrier that blocks noxious, potentially mutagenic influences.

estimated that a single stem cell division ultimately results in the formation of many hundreds of differentiated enterocytes (see Figure 7.24). Therefore, while one might think that stem cells participate in continual cycles of growth and division, the reality is usually much different: the transit-amplifying cells create the great bulk of mitotic activity in many tissues. Since the DNA replication occurring during each cell cycle is inherently prone to making errors, this scheme reduces the risk that mutations will accumulate in the genomes of the stem cells.

In many epithelial tissues, the differentiated epithelial cells are especially vulnerable to damage, since they form cell sheets that line the walls of various ducts and cavities containing toxic material. In the cases of the colon and the bile duct, the epithelial cells confront fecal contents and highly corrosive bile, respectively. The cells lining the alveoli in the lungs cope every day with particulates and pollutants in the air. The keratinocytes in our skin are exposed directly to the outside world and hence are liable to sustain several types of damage, including those inflicted by ultraviolet radiation and toxic substances.

The differentiated end-stage cells (see Figure 12.1) in these and other tissues have a finite lifetime and are discarded sooner or later. Some cell types may simply age and lose their viability, being worn out from carrying on the active business of the tissue. For example, red blood cells have an average lifetime of approximately 120 days, after which they are scavenged by the spleen and broken down and their contents recycled or excreted. The epithelial cells in the colon live for 5 to 7 days before they are induced to enter apoptosis and are sloughed off into the lumen of the intestine. The keratinocytes in our skin die within 20 to 30 days of being formed, and they are shed continually in small flakes of dead skin (see, for example, Figure 2.6A).

Hence, the transit-amplifying cells may well run an increased risk of sustaining mutations because of their high mitotic activity, and their differentiated progeny may often be located in mutagenic environments. However, any genetic damage that the transit-amplifying cells and their differentiated progeny have sustained will have little consequence for the tissue as a whole: sooner or later, these cells are flushed out of the tissue, and once they die, any mutations they may have accumulated disappear with them (see, however, Supplementary Sidebar 11.2).

Figure 12.2 Stem cells and the organization of gastrointestinal crypts
(A) *Boxed numbers* indicate the position of individual enterocytes counting from the bottom of the crypt. Four to six very slowly cycling stem cells (*red*) are located 4–5 cell diameters from the bottom of the crypts. In addition, more rapidly cycling stem cells are found interspersed among the Paneth cells at the bottom of the crypts. Both groups of stem cells are shielded from the contents of the small intestine by their location and by mucus that prevents fluids in the intestinal lumen from entering the crypt. The stem cells spawn a large number (~150) of highly proliferative transit-amplifying cells (*yellow, green*), which divide every 12 hours or so. Their division eventually yields approximately 3500 enterocytes (*blue*), which cover the finger-like villus. The enterocytes are continuously migrating toward the tip of the villus, where they undergo apoptosis and are shed into the lumen of the small intestine (*light blue*). (B) The Lgr5 protein has been found to be a highly useful marker for identifying stem cells in a number of tissues. Located in the bottoms of the crypts of the mouse small intestine (duodenum), these cells are labeled here by a transgene that fuses the coding region of the *Lgr5* gene with a sequence encoding green fluorescent protein (GFP, *green*). Other evidence, not shown here, points to the presence of a second stem cell population located at cell position 4 (from the crypt bottom) that is normally relatively quiescent but becomes mitotically active in response to tissue injury that requires replacement of the Lgr5+ cells. (C) The emigration of transit-amplifying cells from the crypts of the small intestine can be tracked by injecting a dose of ^{3}H-thymidine into a mouse and following the incorporation of this radiolabel into DNA by **autoradiography**; radioactive decay is indicated by *dark silver grains*. Seen here are the cells in the crypts of the duodenum of the mouse at the indicated times after injection of the ^{3}H-thymidine. Cells that multiplied only a small number of times after initial incorporation of ^{3}H-thymidine remain heavily labeled (*broad arrows*), while the great majority underwent multiple additional divisions and therefore exhibit diluted radiolabeling. After four days, virtually all of the cells carrying genomes that were synthesized at the beginning of the experiment have moved out of the crypts to the tips of the villi. (A, courtesy of C.S. Potten. B, courtesy of N. Barker and H. Clevers. C, from C.S. Potten, *Philos. Trans. R. Soc. Lond. B Biol. Sci.* 353:821–830, 1998.)

The dynamics of stem cells and their progeny are illustrated most graphically by the stem cells and the *enterocytes* (the differentiated epithelial cells) of the small intestine and the colon. These cells and their behavior have been described earlier in the context of our discussions of the *Apc* tumor suppressor gene and β-catenin (see Section 7.11). Here, we return to them once again to illustrate other principles. Recall that the stem cells are embedded deep within the crypts (**Figure 12.2A and B**). There they are well out of harm's way, being shielded from the mutagenic contents of the intestinal lumen by a thick layer of mucus secreted by cells in the crypt. This mucus, which is formed from highly glycosylated proteins termed mucins, creates a jelly-like barrier that prevents the contents of the intestinal lumen from penetrating deep into the crypt and illustrates yet another evolutionary strategy for minimizing mutations in stem cells, namely, anatomically shielding them from the actions of toxins, including

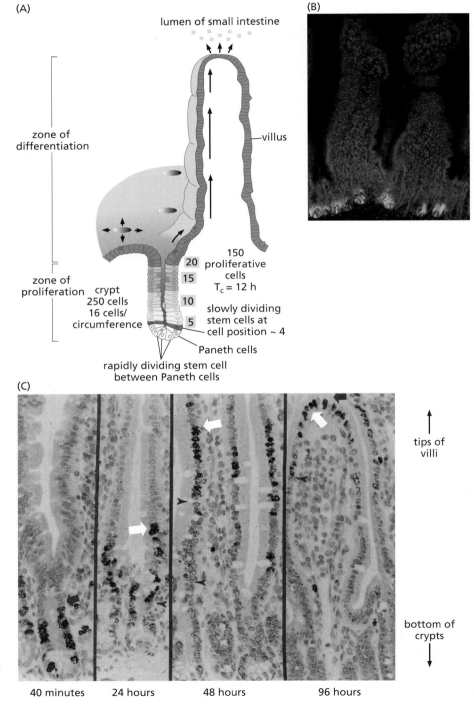

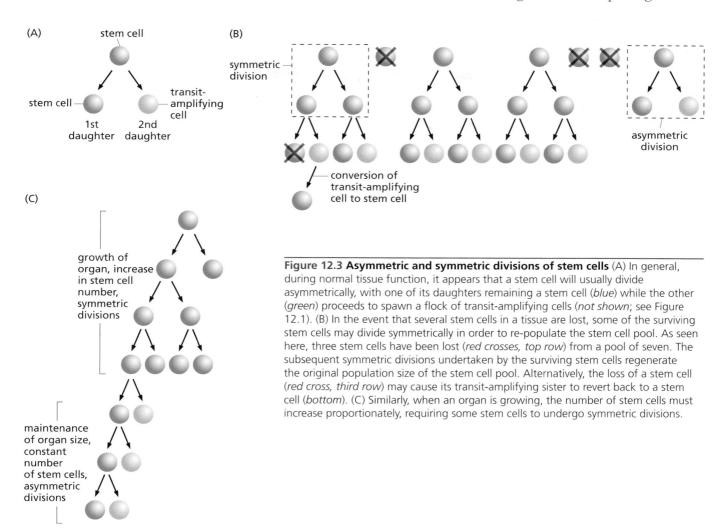

Figure 12.3 Asymmetric and symmetric divisions of stem cells (A) In general, during normal tissue function, it appears that a stem cell will usually divide asymmetrically, with one of its daughters remaining a stem cell (*blue*) while the other (*green*) proceeds to spawn a flock of transit-amplifying cells (*not shown*; see Figure 12.1). (B) In the event that several stem cells in a tissue are lost, some of the surviving stem cells may divide symmetrically in order to re-populate the stem cell pool. As seen here, three stem cells have been lost (*red crosses, top row*) from a pool of seven. The subsequent symmetric divisions undertaken by the surviving stem cells regenerate the original population size of the stem cell pool. Alternatively, the loss of a stem cell (*red cross, third row*) may cause its transit-amplifying sister to revert back to a stem cell (*bottom*). (C) Similarly, when an organ is growing, the number of stem cells must increase proportionately, requiring some stem cells to undergo symmetric divisions.

carcinogens. (Thus, mice that have been genetically deprived of the gene encoding Muc2, the most abundant gastrointestinal mucin, are prone to develop adenomas in the small intestine, many of which progress to adenocarcinomas.) This confers further protection on the genomes of stem cells, complementing the mechanism in which the descendants of these cells, which may have sustained mutations, are flushed out of the crypts and eliminated after 5 to 7 days (see Figure 12.2C).

In theory, the stem cell compartment within a tissue has an inexhaustible ability to generate differentiated progeny without ever suffering depletion. However, almost inevitably, a stem cell will be lost through one or another mishap. This gap in the ranks must be filled by other stem cells. More specifically, both daughters of a surviving stem cell will need to retain the phenotype of their mother, which therefore undergoes a symmetrical division (**Figure 12.3**). This may also have implications for genome maintenance, as we will see below.

12.2   Stem cells may or may not be targets of the mutagenesis that leads to cancer

The properties of the intestinal cells (see Figure 12.2) also provide important clues about the identities of the cells that are the likely targets of carcinogenesis. Having excluded the more differentiated cells (because they are rapidly discarded), we must shift our focus to the long-lived cells and cell lineages within epithelial tissues. Actually, we already have come across evidence that points to the central role of long-lived cells in the process of carcinogenesis. In Section 11.13, we read about experimental protocols used to induce skin cancer in mice. One such protocol involved painting a patch of skin with an initiating agent, allowing the patch to remain untouched for some months, and then painting it repeatedly with TPA, a potent skin tumor promoter. Cells that had been exposed to the initiating carcinogen "remembered" that exposure one year later by undergoing proliferation and forming a skin papilloma in the presence of the promoter. In the skin, as in many other epithelial tissues, the long-lived cells are those in the stem cell compartment.

Provocatively, the number of skin papillomas and carcinomas induced by the mouse skin carcinogenesis protocol (see Figure 11.30) is not reduced if the mouse skin is treated with 5-fluorouracil (5-FU) shortly after being exposed to a mutagenic initiating agent. Since 5-FU selectively kills actively cycling cells, this indicates that the cell targeted by carcinogenic mutagens during initiation is not in the active cell cycle at the time of initiation and shortly thereafter, lending weight to the notion that the target for initiation is a cell type that divides only occasionally.

Analyses of several types of leukemia suggest that the initial targets of carcinogenesis in the hematopoietic system are also stem cells. The most dramatic example is provided by chronic myelogenous leukemia (CML). As described earlier, the Philadelphia (Ph^1) chromosome, which results from a reciprocal chromosomal translocation that fuses the *bcr* and *abl* genes (see Section 4.6), is observed in almost all cases of CML. Extensive evidence points to this particular translocation as the genetic lesion that initiates this disease.

A number of distinct hematopoietic cell types within a CML patient may carry the Ph^1 chromosome. Included are lymphoid cells (both B and T lymphocytes), as well as cells of the myeloid lineage (including neutrophils, granulocytes, the megakaryocyte precursors of platelets, and erythrocytes). This is persuasive evidence that the cell type in which the translocation originally occurred was the common progenitor of all of these hematopoietic cell lineages—the **pluripotent** stem cell that serves as the precursor for many types of hematopoietic cells (see Supplementary Sidebar 12.1). Like a variety of other stem cells, this hematopoietic stem cell (HSC) is thought to have a very long lifetime in the hematopoietic system, more specifically in the bone marrow. In the particular case of CML, a stem cell that has suffered a critical mutation—formation of the Ph^1 chromosome—retains the option to dispatch its progeny into a number of distinct hematopoietic cell lineages. Yet other indications that stem cells are targets for tumor formation come from other types of hematopoietic disorders (**Sidebar 12.1**).

Sidebar 12.1 Blocked differentiation is a frequent theme in the development of hematopoietic malignancies There are dozens of examples of malignancies where inhibition of differentiation favors the appearance of neoplasias. Possibly the first to be defined genetically involved the avian erythroblastosis virus, a retrovirus that encodes two oncoproteins: its *erbB* oncogene specifies a constitutively active version of the epidermal growth factor (EGF) receptor (see Section 5.4), which drives the proliferation of erythroblasts (precursors of red blood cells); while its *erbA* oncogene encodes a nuclear receptor (a homolog of the thyroid hormone receptor), which inhibits differentiation of the hyperproliferating erythroblasts created by *erbB*. Similarly, in human acute myelogenous leukemia (AML), a large variety of genetic lesions found in the leukemic cells have been assigned to two functional classes: those that are required to drive the proliferation of the myeloid precursor cells, and others that are required in the same cells to block subsequent differentiation.

In the megakaryoblastic leukemias (a malignancy of platelet precursor cells) encountered with some frequency in Down syndrome patients, the gene encoding the GATA1 transcription factor is frequently found to be mutated, preventing the proper maturation and differentiation of these precursors of platelets. These few examples point to the notion that the exit of cells from stem cell compartments must be impeded in order for tumorigenesis to succeed.

Not addressed by these observations are the precise identities of the stem cell targets of transformation. In many cases, the target is not likely to be the pluripotent hematopoietic stem cell, but instead one of its derivatives that is already committed to one or another lineage of differentiation. Such "committed progenitors" (see Supplementary Sidebar 12.1) normally may have significant (but limited) self-renewal capacity and are not yet fully differentiated, and thereby can be considered stem cells. Their transformation from normal to tumor stem cells involves, among other changes, an acquisition of unlimited self-renewal capability.

Compelling observations of stem cells' role in cancer derive from transgenic mice in which the expression of an activated *ras* oncogene is limited to either the keratinocyte stem cells in the skin (which in this case are located in hair follicles) or the keratinocytes that have begun to enter into a terminally differentiated state. When the transgene directs expression of the *ras* oncogene in the stem cells, the mice develop malignant carcinomas. In contrast, when the same oncogene is expressed in the differentiating keratinocytes, benign papillomas are formed, and these tend to regress.

Sidebar 12.2 Progenitor cells as targets of mutation While there is relatively little evidence that transit-amplifying/progenitor cells (see Figure 11.18B) in hematopoietic lineages can spontaneously dedifferentiate, the evidence is growing that this can indeed happen in certain epithelial tissues (see Supplementary Sidebar 11.2). Thus, if a mutation is sustained in a transit-amplifying cell, dedifferentiation of this cell into an epithelial stem cell will allow this mutation to become fixed in the stem cell compartment. In fact, as argued in that Sidebar, transit-amplifying cells are far more attractive candidates for sustaining mutations than are stem cells because (1) there are usually far more of them in a tissue, increasing proportionately the chance that a mutation will hit their compartment rather than the stem cell compartment, and (2) unlike stem cells, which divide only occasionally, they are actively dividing. (As discussed in this chapter, many types of mutations occur during active cell division.)

This logic flies in the face of conclusions made repeatedly throughout this chapter about the role of stem cells as direct targets of oncogenic mutations. Perhaps a middle ground will one day be found as follows: the genomes of progenitor cells, also known as transit-amplifying cells, are often the direct targets of mutation, but resulting mutant alleles, in order to contribute to tumorigenesis, must be stored in the stem cell compartment, which represents the site within a tissue in which genetic information, including mutations, is maintained over the long term.

Finally, we should note some carcinogens may produce mutant cells through their ability to be cytotoxic. We can imagine, for example, that some carcinogens, rather than being directly mutagenic, act through their cytotoxicity and thus kill cells within a tissue, including some stem cells. In order to regenerate a proper number of stem cell pools, a tissue may resort to dedifferentiating a transit-amplifying/progenitor cell, indeed one from a lineage that has passed through multiple successive cell cycles since its origin from an SC division. This lineage of transit-amplifying cells may have accumulated multiple replication-associated mutations in its genome, which are then introduced back into the SC pool and thus become permanently established in the SC compartment.

We have encountered nongenotoxic carcinogens previously, in the discussion of tumor promoters (see Section 11.14). There, we argued that some tumor promoters, such as ethanol, act through their ability to cause the death of cells in a target tissue, resulting in a compensatory proliferation by the surviving cells in the tissue. Now, as we view these promoters in the context of stem cell biology, we can speculate that certain tumor promoters may act through their ability to kill stem cells. An even more dangerous agent would be a "complete" carcinogen (see Section 11.17)—one that is able to act both as an initiator through its mutagenic actions on stem and progenitor cell genomes and as a promoter through its cytotoxic effects on these cells. Finally, it is worth noting that the dynamics governing the size of stem cell pools in the mammary gland may one day provide an explanation for the complex epidemiology of a common human malignancy—breast cancer (see Supplementary Sidebar 12.2).

These various strands of evidence, obtained from several types of tissue, converge on the conclusion that self-renewing cells of various types are the direct targets of the mutagenesis that leads, sooner or later, to the formation of tumors. Indeed, as concluded repeatedly throughout this section, such target cells are stem cells with unlimited self-renewal capacity. Still, this conclusion may not be the final one; as argued in **Sidebar 12.2**; in many tissues, committed progenitors, which normally have only a limited ability to renew themselves, may represent more plausible targets.

12.3 Apoptosis, drug pumps, and DNA replication mechanisms offer tissues a way to minimize the accumulation of mutant stem cells

The apparently important role played by normal stem cells as potential targets for transformation indicates that the genomes of these cells must be protected by whatever biological and biochemical strategies these cells and the tissues around them can muster. We have already come across two such strategies: the relatively infrequent replication of stem cell DNA and the placement of stem cells in anatomically protected sites. Still, these mechanisms do not seem to suffice, so the organism has developed yet other strategies.

The stem cells in the mouse intestinal crypts (see Figure 12.2) and mammary glands represent especially attractive objects for study of these protective strategies. In the case of the crypts, the need for additional protective mechanisms is clear: the enterocyte stem cell lineages in the crypts of the mouse small intestine pass through a long succession of growth-and-division cycles during a lifetime, and each of these cycles exposes the stem cells to various types of genetic damage. Similarly, in the human gut, the total number of cell divisions occurring each year greatly exceeds the total number of cells residing at any time within the entire body; this enormous mitotic activity, most of which involves transit-amplifying cells, must also depend on many successive stem cell divisions.

One protective mechanism is suggested by the responses of stem cells in the crypts to massive genetic damage. In the intestinal crypts of the mouse, stem cells that have suffered genetic damage inflicted by X-rays will rapidly initiate apoptosis rather than halt their proliferation and attempt to repair the damage. The motive here seems to be associated with the error-prone nature of DNA repair. As we will learn later, the DNA repair apparatus is highly efficient but hardly perfect, and therefore may leave a residue of unrepaired or incorrectly repaired lesions in the chromosomal DNA. If such lesions are encountered by the DNA replication machinery, they may cause mutant DNA sequences to be copied and passed on to daughter cells, including those that will themselves become stem cells. So, rather than risk this outcome, stem cells in the mouse crypts are primed to activate apoptosis in response to DNA damage.

[Still, it is unclear whether stem cells in all tissues are similarly poised to enter apoptosis. For example, the cancer stem cells (CSCs) arising in a variety of tumors (see Section 11.6) exhibit elevated resistance to apoptosis and enhanced DNA repair mechanisms, making them *more* resistant to radiation and a variety of chemotherapeutic treatments. Since the SC programs of these CSCs closely resemble programs operating in normal precursor tissues, certain normal SCs may actually respond to genomic damage by deploying highly effective DNA repair mechanisms rather than by rapidly triggering apoptosis.]

Yet another mechanism is suggested by a commonly used technique for separating stem cells from the bulk of cells in a tissue via fluorescence-activated cell sorting (FACS; see Supplementary Sidebar 11.1). Stem cells efficiently pump out certain fluorescent dye molecules, while these cells' differentiated derivatives do so much less actively. As a consequence, after exposure of cell populations to such dyes, the stem cells fluoresce much more weakly than all other cells in these populations. The active excretion of these fluorescent dye molecules is due to the actions of a plasma membrane protein termed Mdr1 (multi-drug resistance 1), which was first discovered because it is exploited by many cancer cells to pump out, and therefore acquire resistance to, chemotherapeutic drug molecules (see Figure 16.21). The unusually high levels of Mdr1 expressed by many types of stem cells seem to represent a strategy that they use to protect their genomes from potentially mutagenic compounds that may have entered into their cytoplasms from outside.

The mechanism of asymmetric DNA strand allocation may also play an important role in preventing the stem cells in certain tissues from accumulating genetic damage. According to this scheme, after DNA undergoes replication in an SC, the newly replicated (and potentially miscopied) DNA strand is allocated to the SC daughter that becomes a transit-amplifying/progenitor cell, while the unreplicated parental DNA strand, which has not sustained replication-associated sequence changes, is allocated to the daughter that remains an SC. While attractive in concept, the experimental observations supporting this proposed mechanism are still fragmentary and a matter of great debate (see Supplementary Sidebar 12.3).

12.4 Cell genomes are threatened by errors made during DNA replication

The design of stem cell compartments and the behavior of individual stem cells illustrate several biological strategies used by tissues to reduce the burden of accumulated somatic mutations. These mechanisms serve to protect stem cell genomes, which constitute, in effect, the "germ lines" of tissues. Importantly, these strategies represent only the first line of defense against genomic damage. The next line of defense is a biochemical one that depends on the ability of various proteins to recognize and repair damaged DNA molecules within cells.

In fact, DNA molecules are under constant attack by a variety of agents and processes. For the sake of simplicity, we can place these mutagenic processes in three categories. First, the replication of DNA sequences by DNA polymerases during the S phase of the cell cycle is subject to a low but nonetheless significant level of error. Included among these errors are those generated when chemically altered nucleotide precursors are inadvertently incorporated into DNA in place of their normal counterparts. Second, even in the absence of attack by mutagenic agents, the nucleotides within DNA molecules undergo chemical changes spontaneously; these changes often alter the base sequence and thus the information content of the DNA. Finally, DNA molecules may be attacked by various mutagenic agents, including those molecules generated endogenously by normal cell metabolism as well as agents of exogenous origin—chemical species and physical mutagens (X-rays and UV rays) that are introduced into the body from outside. We will return to the latter two processes in the next sections.

The molecular machinery that is responsible for replicating almost all chromosomal DNA sequences has a remarkably low rate of error. The basic replication machinery in the cell nucleus is powered by the actions of three polymerases, pol-α, pol-δ, and pol-ϵ. (In all, 19 distinct DNA polymerase genes have been cataloged in the human genome, and more are likely to be found; as will be apparent later, most of these are not involved in DNA replication per se but rather in the repair of damaged DNA molecules.)

A cell has two major strategies for detecting and removing the miscopied nucleotides arising during DNA replication. The first strategy lies in the hands of the DNA polymerases themselves, which are structurally complex aggregates assembled from a number of distinct protein subunits. While they are advancing down single-strand DNA templates and extending nascent DNA strands in a 5'-to-3' direction, DNA polymerases such as pol-δ continuously look backward, "over their shoulder," scanning the stretch of DNA that they have just polymerized; such monitoring is often called **proofreading**. Should a polymerase detect a copying error, it will use its 3'-to-5' exonuclease activity to move backward and digest the DNA segment that it has just synthesized and then copy this segment once again, with the hope for a better outcome the second time (**Figure 12.4**).

The importance of this proofreading mechanism for the prevention of cancer has been illustrated dramatically by the creation of a mouse strain whose germ-line pol-δ–encoding gene has been subtly altered (by a single amino acid substitution). The resulting mutant pol-δ retains its ability to carry out lagging-strand DNA synthesis at the replication fork (see Figure 10.20) but has lost its 3'-to-5' exonuclease activity; this loss eliminates its proofreading function. In a cohort of 49 mice carrying the mutant pol-δ allele in a homozygous configuration, 23 developed tumors by one year of age, while no tumors developed in a group consisting of twice as many heterozygous mice (**Figure 12.5**). Strangely enough, mutation of the proofreading functions of pol-ϵ, which has the job of leading-strand synthesis at the DNA replication fork, yields tumors in a different set of tissues!

These experiments demonstrate that the maintenance of wild-type genomic sequences, in this case by two DNA polymerases, represents a critical defense against the onset of cancer. Moreover, for us, these observations are the first of many indications that the mutations leading to cancer may arise through endogenous processes rather than being triggered exclusively by invading foreign carcinogenic agents.

Figure 12.4 Proofreading by DNA polymerases A number of DNA polymerases have a proofreading ability that allows them to minimize the number of bases that are misincorporated and retained in the recently synthesized strand. Thus, as a DNA polymerase extends a nascent strand (*dark blue*) in a 5′-to-3′ direction (*moving rightward*), it will use the existing 3′-OH of the nascent strand as the primer for further elongation (*light blue*). However, if a base has been misincorporated (*third drawing*), the DNA polymerase, which is continuously looking backward to check whether it has incorporated the correct bases in the growing DNA strand, can degrade in a 3′-to-5′ (*leftward*) direction the recently elongated strand (*fourth drawing*) and undertake once again to synthesize this stretch of nascent strand (*bottom drawing*).

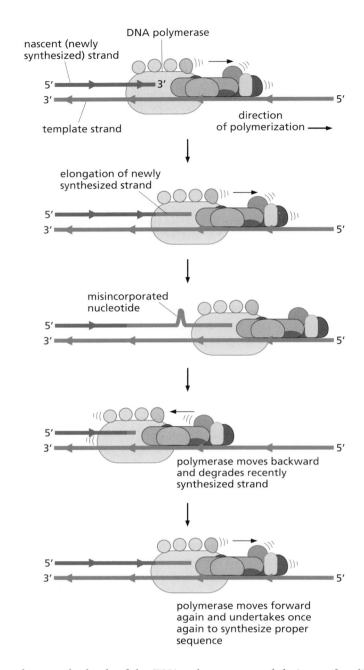

Following close on the heels of the DNA polymerases and their proofreading activities are a complex set of **mismatch repair** (MMR) enzymes. These enzymes monitor recently synthesized DNA in order to detect miscopied DNA sequences that have been overlooked by the proofreading mechanisms of the DNA polymerases.

The actions of the mismatch repair system become especially critical in regions of the DNA that carry repeated sequences. These sequence blocks include simple mononucleotide repeats (such as AAAAAAA), dinucleotide repeats (such as AGAGAGAG), and repeats of greater sequence complexity. Because of strand slippage, which occurs when the parental and nascent strands slip out of proper alignment, DNA polymerases appear to occasionally "stutter" while copying these repeats, resulting in incorporation of longer or shorter versions of the repeats into the newly formed daughter strands (**Figure 12.6**). Thus, the sequence AAAAAAA, that is, A_7, might well cause a polymerase to synthesize a T_6 or T_8 sequence in the complementary strand. The resulting insertions or deletions may elude detection by the proofreading components of the DNA polymerases and are therefore prime targets for recognition and repair by the MMR machinery.

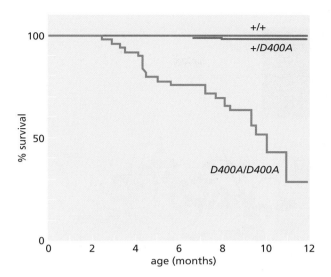

Figure 12.5 Proofreading by DNA polymerase and cancer incidence
A point mutation has been introduced into the germ-line copy (see Supplementary Sidebar 7.7) of the mouse gene encoding DNA polymerase δ (pol-δ), the mammalian DNA polymerase that is responsible for the bulk of lagging-strand synthesis. This mutation, termed *D400A*, alters the amino acid sequence in the proofreading domain of the polymerase by specifying the replacement of an aspartic acid by an alanine at residue position 400 of the polymerase molecule; the synthetic activity of the polymerase is unaffected by this mutation. Shown here is the fate of 53 wild-type mice (+/+), 97 heterozygotes (+/*D400A*), and 49 homozygous mutants (*D400A/D400A*). Deaths of the mutant homozygotes were all due to malignancies; these included lymphomas, squamous cell carcinomas of the skin, lung carcinomas, and several other types of cancer that occurred less frequently. Two of the heterozygotes died from causes that were unrelated to cancer, while the homozygous wild-type mice all survived to the age of one year. (The cooperating partner of pol-δ, called pol-ε, carries out leading-strand synthesis and proofreading; when it is similarly mutated, mice develop largely carcinomas of the small intestine.) A 2012 report identified germline mutations creating susceptibility to colorectal cancer development in the genes encoding the proofreading functions of pol-δ and pol-ε were identified in a number of families, yielding a strikingly similar outcome to those observed with the mutant mice described here. (From R.E. Goldsby et al., *Nat. Med.* 7:638–639, 2001.)

For historical reasons, highly repeated sequences in the genome, often carrying 100 or more nucleotides per repeat unit, have been called "satellite" sequences. Because the simple, far shorter sequences discussed here are also found in many places in the genome, they have been named **microsatellites**. A defective mismatch repair system that fails to detect and remove stuttering mistakes made by DNA polymerases when copying a microsatellite will result in the expansion or shrinkage of its sequences in progeny cells. This creates the genetic condition known as *microsatellite instability* (MIN; **Figure 12.7A**), which may ultimately involve changes in thousands of microsatellite sequences scattered throughout a cell genome. Importantly, as illustrated in Figure 12.7B, defects in MMR yield base substitutions even more frequently than expansions and contractions of microsatellite sequences; indeed, the genome is showered with mutations in an MMR-defective cell.

Yet other, more subtle copying mistakes made by a DNA polymerase, such as the incorporation of an inappropriate base in a nonrepeating sequence, may also be detected and erased by mismatch repair proteins, which are highly sensitive to bulges and loops in the double helix caused by inappropriately incorporated nucleotides. The mismatch repair machinery must be able to distinguish the recently synthesized DNA strand from the complementary "parental" strand that served as the template; this enables the MMR apparatus to direct its attention to removing and then repairing the recently synthesized and therefore defective DNA strand (see Figure 12.6C). Mismatch repair involves the excision of the nucleotides that have created the mismatch and a new attempt at synthesis of this strand.

Working together, these various error-correcting mechanisms yield extremely low rates of miscopied bases that survive to become mutant DNA sequences. To begin, DNA polymerases make copying mistakes in only about 1 out of 10^5 polymerized nucleotides. The $3' \rightarrow 5'$ proofreading by the polymerases overlooks only about 1 out of every 10^2 nucleotides initially miscopied by the polymerase, thereby reducing the error rate to about 1 in 10^7 nucleotides. After the DNA polymerase has passed through a stretch of DNA, the mismatch repair proteins check the recently synthesized DNA strand a second time. The mismatch repair enzymes fail to correct only about 1 miscopied base out of 100 that have escaped the proofreading actions of the DNA polymerase. Together, this yields a stunningly low mutation rate of only about 1 nucleotide per 10^9 that have been synthesized during DNA replication. As we will see, defects in these error-correcting mechanisms can lead to both familial and sporadic human cancers.

Finally, DNA replication holds yet other dangers for the genome. Some measurements indicate that as many 10 double-strand (ds) DNA breaks occur per cell genome each time a cell passes through S phase. These breaks appear to occur near replication forks,

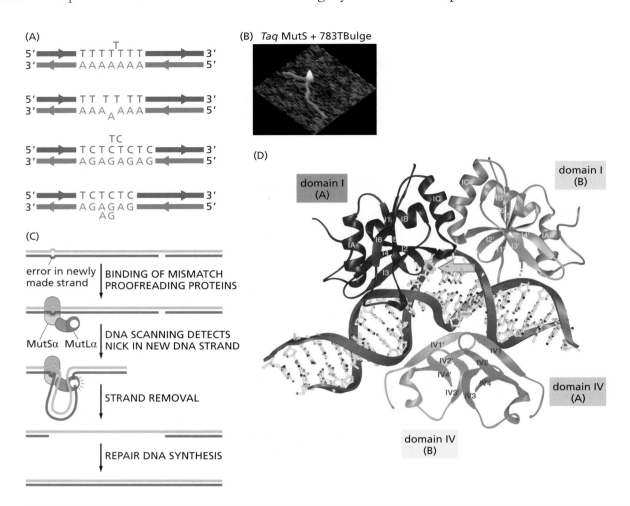

Figure 12.6 DNA polymerase errors and mismatch repair
(A) The DNA polymerases, notably pol-δ, occasionally "stutter," or skip a base when copying a repeating sequence of DNA (e.g., a microsatellite sequence) in the template strand (*blue*). As a consequence, the newly synthesized strand (*green*) either may acquire an extra base that increases the length of the repeating sequence or may lack a base (*top two images*). Identical dynamics may cause similar changes in microsatellite sequences where the repeat unit is a TC dinucleotide segment (*bottom two images*), or a more complex repeating sequence (*not shown*). (B) Mismatch repair (MMR) proteins recognize and repair the mistakes made by DNA polymerases, including misincorporated bases and inaccurate replication of microsatellite sequences. Here, use of atomic-force microscopy reveals the behavior of the MutS MMR protein, a bacterial homolog (from *Thermus aquaticus*) of a number of mammalian MMR proteins. MutS is seen binding to a DNA fragment into which a mismatch has been introduced at a specific nucleotide site. MutS kinks the DNA double helix as it scans for and ultimately finds regions of mismatch (793T Bulge), where it binds in a stable fashion, seen here as a *white pyramid*. (C) In eukaryotic cells, two components of the MMR apparatus, MutSα and MutLα, collaborate to initiate repair of mismatched DNA. After MutSα (a heterodimer of MSH6 and MSH2) scans the DNA and locates a mismatch, MutLα (a heterodimer of MLH1 and PMS2) scans the DNA for single-strand nicks, which identify the strand that has recently been synthesized (*red*); the under-methylation of the recently synthesized strand may also aid in this identification. MutLα then triggers degradation of this strand back through the detected mismatch, allowing for repair DNA synthesis to follow. (D) Part of the structure of the *T. aquaticus* MutS homodimeric protein in complex with a mismatched helix (*red*) is shown. Domains I and IV of subunit A are in *dark blue* and *orange,* while the corresponding domains of subunit B are in *light blue* and *yellow.* An arrow (*yellow*) indicates where phenylalanine residue 39 of domain I of subunit A is associated with an unpaired thymidine in one of the two DNA strands. Defects in the human homolog of this protein play a critical role in triggering hereditary non-polyposis colon cancer (HNPCC), as discussed in Section 12.9. (B, from H. Wang et al., *Proc. Natl. Acad. Sci. USA* 100:14822–14827, 2003. C, from B. Alberts et al., *Molecular Biology of the Cell*, 5th ed. New York: Garland Science, 2008. D, from G. Obmolova et al., *Nature* 407:703–710, 2000.)

ostensibly because the single-strand DNA at the unwound but not-yet-replicated portion of the parental DNA is susceptible to inadvertent breakage (**Figure 12.8**). Cells have well-developed mechanisms for dealing with such dsDNA breaks, as we will see later. Failure to repair such breaks properly can lead to disastrous consequences, including chromosomal breaks and translocations.

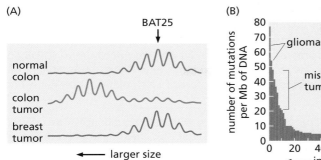

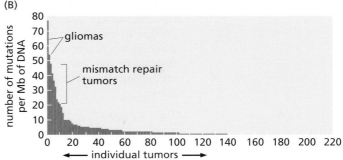

Figure 12.7 Detection of microsatellite instability Defects in mismatch repair (MMR) are responsible for the accumulation of a variety of mutations in the genome. (A) The obvious type of these mutations is microsatellite instability (MIN), which causes an expansion or contraction of the size of a microsatellite repeat sequence. In the analysis shown here, the size of a mononucleotide repeat is analyzed in a woman suffering from HNPCC (hereditary non-polyposis colon cancer) and presenting in the clinic with both a colorectal and a breast carcinoma; analysis was performed using PCR (polymerase chain reaction). The BAT25 sequence, which is located on human Chromosome 4q12, consists of the sequence TTTTxTxTTTTxT7xxT25, where "x" indicates a nucleotide other than T. Because of errors made by the polymerase used in the PCR reaction, the products of the PCR reactions show a Gaussian distribution of lengths grouped around the actual length of the genomic DNA segment being amplified. This analysis reveals a clear increase in size of the microsatellite repeat in the colon carcinoma (*leftward shift*), while the breast tumor exhibits a microsatellite repeat that is precisely the same as normal, control DNA. (This observation strongly suggests that the breast carcinoma, unlike the colon carcinoma, is unlikely to have been caused by MIN.) (B) The coding exons of 518 kinase-encoding genes in the genomes of 210 human tumors of diverse type were sequenced, revealing 1007 somatic mutations in all. As is apparent, the density of mutations (number per Mb) of all types varied dramatically, with one-third of the tumor genomes showing no mutations at all (*rightward*). Two classes of tumors showed especially high densities of mutations in their genomes. Gliomas/glioblastomas, which had previously been exposed during the course of treatment to high doses of temozolomide, a mutagenic chemotherapeutic agent, exhibited an unusually high density of mutations. In addition, the genomes of a group of five MMR-deficient tumors showed an almost-equivalent density of mutations, consisting largely of base substitutions (14–40 per Mb) plus insertions/deletions of microsatellite repeats (5–12 per Mb). (A, from A. Müller et al., *Cancer Res.* 62:1014–1019, 2002. B, from C. Greenman et al., *Nature* 446:153–158, 2007.)

12.5 Cell genomes are under constant attack from endogenous biochemical processes

Most accounts of the origins of contemporary cancer research contain a strong emphasis on the actions of carcinogenic agents that enter the body through various routes, attack DNA molecules within cells, and create mutant cell genomes that occasionally cause the formation of cancer cells. Unrecognized by these models of cancer pathogenesis are the mutagens and mutagenic mechanisms of endogenous origin. In recent decades, however, analytical techniques of greatly improved sensitivity have allowed researchers to detect altered bases and nucleotides in the DNA of normal

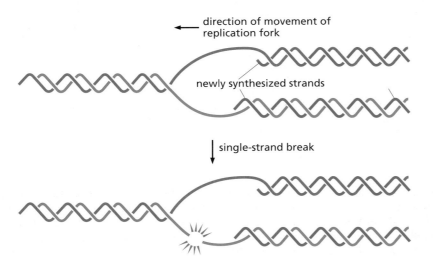

Figure 12.8 Double-strand DNA breaks at replication forks During DNA replication, the DNA molecules are especially vulnerable to breakage in the single-stranded portions near the replication fork that have not yet undergone replication. Such breaks often occur because of the chemical alteration of a base (discussed later in this chapter) that causes the DNA polymerase to stall, being unable to recognize this altered base; this results in a region of single-stranded DNA that may persist for an extended period of time before it is finally protected by successful replication and thus formation of a complementary strand. The breakage of such a single-stranded region (sometimes termed a "collapsed" replication fork) is functionally equivalent to a double-strand break occurring in an already-formed double helix, in that the break leaves two helices unconnected by either strand.

cells that have not been exposed to exogenous mutagens. The results of these analyses have caused a profound shift in thinking about the origins of most of the mutant genes present in the genomes of human cells, because they have shown that endogenous biochemical processes usually make far greater contributions to genome mutation than do exogenous mutagens. Since mutagenic events, independent of their origin, are potentially carcinogenic, this has forced a rethinking of how many human cancers arise.

The structure of the DNA double helix, with its bases facing inward, offers a measure of protection from all types of chemical attack by shielding its potentially reactive chemical groups, notably the amine side chains of the bases, from various mutagenic agents. (Some measurements have indicated that the bases in ssDNA are 100- to 1000-fold more vulnerable to oxidative damage than corresponding bases in dsDNA.) In spite of this clever design, DNA molecules are subject to chemical alteration and physical damage. Some of this damage appears to occur through the actions of hydrogen and hydroxyl ions that are present at low concentration ($\sim 10^{-7}$ M) at neutral pH. Often cited in this context is the process of **depurination**, in which the chemical bond linking a purine base (adenine or guanine) to deoxyribose breaks spontaneously (**Figure 12.9A**). By some estimates, as many as 10,000 purine bases are lost by depurination

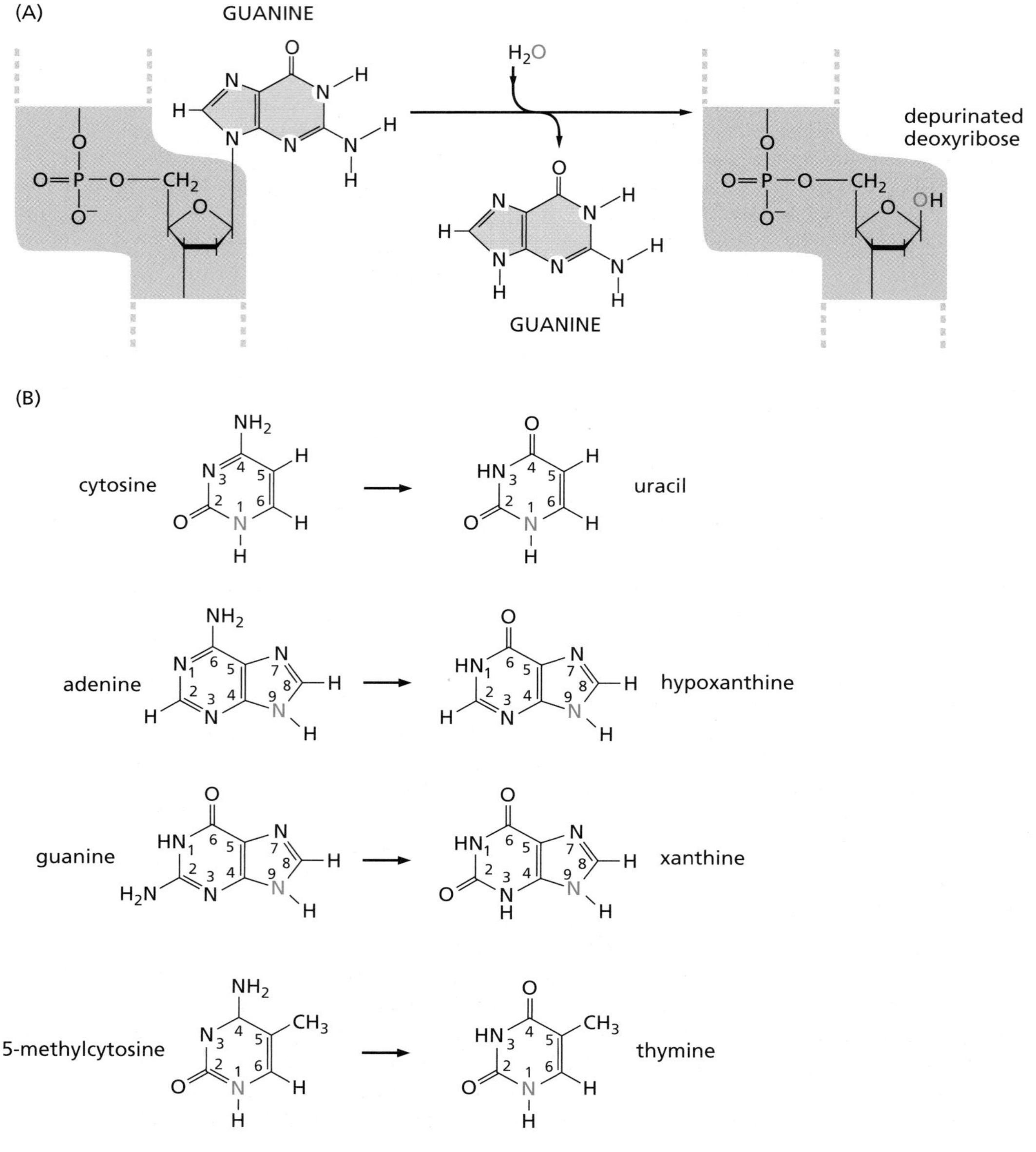

Figure 12.9 Depurination and base deamination
(A) Spontaneous depurination frequently affects guanine within DNA, leaving behind a deoxyribose. (B) The deamination reactions affecting purine and pyrimidine bases, which occur spontaneously at various rates at neutral pH, lead to changes in nucleotide sequences unless they are repaired. The deamination of 5-methylcytosine yields thymine (*bottom*); because this base is naturally present in DNA, it is not always recognized as being aberrant by the repair machinery, explaining the frequent mutations at sites bearing this methylated base. (In each case, the nitrogen atom in red participates in the formation of a glycosidic bond with the 1-carbon of deoxyribose.) (Adapted from B. Alberts et al., Molecular Biology of the Cell, 5th ed. New York: Garland Science, 2008.)

each day in a mammalian cell. (This amounts to more than 10^{17} chemically altered nucleotides generated each day in the human body!) **Depyrimidination** occurs at a 20- to 100-fold lower rate, but still results in as many as 500 cytosine and thymine bases lost per cell per day. Estimates of the steady-state level of base-free nucleotides present in a single human genome range from 4000 to 50,000.

At the same time, **deamination** may occur, in which the amine groups that protrude from cytosine, adenine, and guanine rings of the bases are lost. This deamination leads respectively to uracil, hypoxanthine, and xanthine (see Figure 12.9B). The uracil, for instance, may be read as a thymine during subsequent DNA replication, thereby causing a C–T point mutation, known as a **transition** mutation, in which one pyrimidine replaces another. The bases generated by deamination are all foreign to normal DNA, and consequently can be recognized as such and removed by specialized DNA repair enzymes. However, any such altered bases that escape detection and removal represent potential sources of point mutations.

The rate of spontaneous deamination of the 5-methylcytosine base—the methylated form of the cytosine in CpG dinucleotides that we encountered earlier (see Section 7.8)—is even higher, yielding thymine (see Figure 12.9B). This creates a serious problem for the DNA repair apparatus, since thymine (unlike the other three products of deamination described above) is a component of normal DNA, and the T:G base pair may therefore escape detection, survive, and ultimately serve as template during a subsequent cycle of DNA replication, leading to a C-to-T point mutation.

In fact, this deamination of 5-methylcytosine represents a major source of point mutations in human DNA. By one estimate, 63% of the point mutations in the genomes of tumors of internal organs (that is, in those tissues shielded from UV radiation) arise in CpG sequences. Among mutant *p53* alleles, about 30% seem to arise from CpG sequences present in the wild-type *p53* allele. [To be accurate, this percentage is inflated somewhat by the fact that during lung carcinogenesis, methylated CpG sequences are also favored targets for attack by chemically activated forms of benzo[*a*]pyrene (see Section 12.6), a **polycyclic aromatic hydrocarbon** (PAH) in tobacco smoke. Hence, not all mutations arising at CpG sites derive from deamination events.]

The intracellular environment holds yet other dangers for the chromosomal DNA. The greatest of these comes from the processes of oxidation, which may inflict far more damage on DNA than the reactions mentioned above. Most important here are the reactions that occur in the mitochondria and generate a variety of intermediates as oxygen is progressively reduced to water:

$$O_2 + e^- \rightarrow \underset{\substack{\text{superoxide} \\ \text{ion}}}{O_2^{\bullet-}} + e^- \rightarrow \underset{\substack{\text{hydrogen} \\ \text{peroxide}}}{H_2O_2} + e^- \rightarrow \underset{\substack{\text{hydroxyl} \\ \text{radical}}}{{}^{\bullet}OH} + e^- \rightarrow H_2O$$

Some of these intermediates—the so-called reactive oxygen species (ROS)—may leak out of the mitochondria into the cytosol and thence into the rest of the cell. Included among these are the superoxide ion, hydrogen peroxide, and the hydroxyl radical—the intermediates in the reactions listed above. (By one estimate, in normal cells 1–2% of oxygen molecules consumed by mitochondria end up as ROS; these levels may increase after mitochondria become more leaky following cell transformation.) Yet other oxidants arise as by-products of various oxygen-utilizing enzymes, including those in **peroxisomes** (cytoplasmic bodies that are involved in the oxidation of various cellular constituents, notably lipids), and from spontaneous oxidation of lipids, which results in their peroxidation. Inflammation also provides an important source of the oxidants that favor mutagenesis and therefore carcinogenesis (**Sidebar 12.3**).

The highly reactive molecules produced by these various processes proceed, usually within seconds, to form covalent bonds with many other molecular species in the cell. Among the many targets of ROS attack are the bases within DNA, including both purines and pyrimidines (**Figure 12.10**). In addition, reactive oxygen species can induce single- and double-stranded DNA breaks, **apurinic** and **apyrimidinic** sites (together, known as **abasic** sites, in which bases are cleaved from deoxyribose; for example, see Figure 12.9A), as well as DNA–protein cross-links. As described below, many of the resulting altered bases are recognized by a repair machinery that proceeds

Sidebar 12.3 Inflammation can have both mitogenic and mutagenic consequences Chronic inflammation of tissues is often provoked by infectious agents, such as hepatitis B and C virus infections of the liver, *Helicobacter* infection of the gastric epithelium, and human papillomavirus (HPV) infection of the cervical epithelium (see Sections 11.15 and 11.18). These infections often lead to cell death and resulting compensatory proliferation of the surviving cells—a type of tumor promotion discussed in the previous chapter.

In addition, infected cells attract the attentions of a variety of cells from the innate and adaptive immune systems, as discussed in Chapter 15. Among them are macrophages and neutrophils that are dispatched to sites of infection; there, they create localized inflammatory responses, which may be accompanied by mutagenic consequences. For example, **phagocytes** kill infected cells in part by releasing bursts of a powerful mix of oxidants, including nitric oxide (NO), super-oxide ion ($O_2^{\bullet-}$), hydrogen peroxide (H_2O_2), and hypochlorite (OCl^-). Such highly reactive chemical species are used to kill the intended target cells but, in addition, may cause collateral damage by leaving behind survivors whose genomes they have also damaged. Like the by-products of normal oxidative metabolism, these oxidants act as mutagens on the genomes of nearby bystander cells through their ability to generate chemically modified bases via nitration, oxidation, deamination, and halogenation. Indeed, the DNAs of inflamed and neoplastic tissues have been found to carry substantially increased concentrations of 8-oxo-dG (see Figure 12.10), one of the primary products of DNA oxidation. (As examples, the total daily excretion of oxidized bases by smokers can be as much as 50% higher than by nonsmokers, ostensibly due to the chronic inflammation in their lungs, and the urine of those suffering a chronic flatworm infection in Thailand has even higher concentrations of 8-oxo-dG, as discussed in Section 11.17.) The increased mutation rate resulting from base oxidation helps to explain why chronic inflammation in many tissues favors tumor progression.

to excise them from the DNA. Some of the excised bases, including thymine glycol, which derives from deoxythymidine glycol, and 8-oxoguanine, which derives from 8-oxo-deoxyguanosine (8-oxo-dG), can be detected and quantified in the urine of mammals, providing some indication of the rate at which they are produced throughout the body (see Supplementary Sidebar 12.4).

Some experiments have shown that the yield of these compounds is directly proportional to the rate of oxidative metabolism in various species (**Figure 12.11**). The formation of 8-oxo-dG creates a danger of mutation, as one conformation of this altered base can readily pair with A. This mispairing of bases during DNA replication can lead, in turn, to the replacement of a G:C base pair, via G:A pairing, to a T:A base pair (see Figure 12.10B). Such a G → T replacement of a purine by a pyrimidine (or the opposite) is often termed a **transversion**. Yet other damage can occur through the

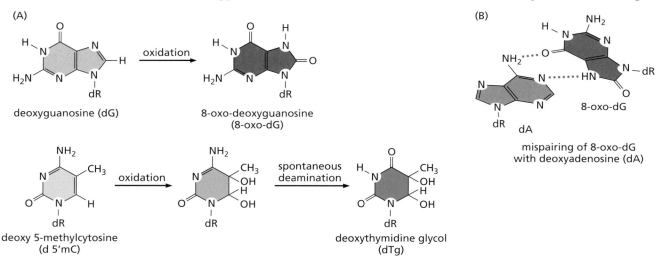

Figure 12.10 Oxidation of bases in the DNA The oxidation of DNA bases, which often results from the actions of reactive oxygen species (ROS), can be mutagenic in the absence of subsequent DNA repair reactions. (A) Two frequent oxidation reactions involve deoxyguanosine (dG), which is oxidized to 8-oxo-deoxyguanosine (8-oxo-dG); and deoxy-5-methylcytosine (d5mC), the nucleotide that is present in methylated CpG sequences. Upon oxidation, the latter initially forms an unstable base that rapidly deaminates, yielding deoxythymidine glycol (dTg). (B) The 8-oxo-dG, which is formed by the oxidation of dG, can mispair with deoxyadenosine (dA) rather than forming a normal base pair with deoxycytosine (dC). Hence, if 8-oxo-dG is not removed from a double helix, the DNA replication machinery may inappropriately incorporate a dA rather than a dC opposite it, resulting in a C–A point mutation. "dR" signifies deoxyribose in all cases. The purines are shown in various shades of *red* and *brown*, while the pyrimidines are shown in various shades of *green*.

methylation of bases triggered by reaction with S-adenosylmethionine, a common metabolic intermediate in cells, which carries a highly reactive methyl group. Taken together, the continuing hail of damage from oxidation, depurination, deamination, and methylation, which together may alter as many as 100,000 bases per cell genome each day, greatly exceeds the amount of damage created by exogenous mutagenic agents in most tissues.

12.6 Cell genomes are under occasional attack from exogenous mutagens and their metabolites

As we have seen repeatedly in this text, cellular genomes are also damaged by exogenous carcinogens, including various types of radiation as well as molecules that enter the body via the food we eat and the air we breathe. Among the best studied of the exogenous carcinogens are X-rays, often termed "ionizing radiation" because of the ionized, chemically reactive molecules that this form of electromagnetic energy creates within cells. As much as 80% of the energy deposited in cells by X-rays is thought to be expended in stripping electrons from water molecules. The resulting free radicals proceed to generate reactive oxygen species (ROS) that create single- and double-strand breaks in the DNA double helix. As discussed later, these double-strand breaks (DSBs) are often difficult to repair and may, on occasion, generate breaks in a chromosome that are visible microscopically during metaphase.

Ultraviolet (UV) radiation from the sun is a far more common source of environmental radiation than X-rays. Living organisms have had to contend with UV radiation since life first formed on this planet some 3.5 billion years ago. Once oxygen accumulated to high levels in the atmosphere about 0.6 billion years ago, the ozone formed from atmospheric oxygen provided a protective shield that significantly attenuated the flux of UV radiation striking the Earth's surface. Nonetheless, a significant amount of UV still succeeds in penetrating the ozone shield and reaching the biosphere.

Should UV photons strike a DNA molecule in one of our skin cells, a frequent outcome is the formation of pyrimidine dimers—that is, covalent bonds form between two adjacent pyrimidines in the same strand of DNA. In principle, these can form between two adjacent C's, two adjacent T's, or a C and an adjacent T. In mammals, where the percentages of A's, C's, G's, and T's are similar, more than 60% of the pyrimidine dimers are TT and perhaps 30% are CT dimers, with the remaining dipyrimidines being CC dimers. As seen in **Figure 12.12**, a pair of covalent bonds are formed between adjacent pyrimidines, resulting in the creation of a four-carbon (cyclobutane) ring. Another, less common class of DNA photoproducts, termed pyrimidine (6–4) pyrimidinone, also involves covalent linkage between two adjacent pyrimidines. Once formed, pyrimidine dimers are very stable and can persist for extended periods unless they are recognized and removed by DNA repair enzymes.

The fact that these pyrimidine dimers are mutagenic is demonstrated dramatically by the spectrum of *p53* mutations found in the DNAs of **keratoses** (benign skin lesions) and basal cell carcinomas of the skin. In these growths, many of the mutant *p53* alleles carry a dipyrimidine substitution. While the TT dimer is the one most frequently formed by UV radiation, it is only weakly mutagenic, because various DNA repair and replication enzymes, to be discussed later, are able to deal with it effectively. This explains why CC → TT substitutions, which arise from CC (rather than TT) dimers, are the most common consequences of UV light mutagenesis. Since UV photons characteristically cause this mutation, these observations provide further support for the notion that UV rays are directly mutagenic and carcinogenic for the human skin. As might be expected, this particular type of *p53* mutation is otherwise extremely rare in the genomes of the many types of human tumors that arise in internal organs and are therefore shielded from UV radiation. Other evidence for a direct mutagenic role of UV radiation is: (1) the incidence of squamous cell skin carcinomas doubles with each 10-degree decline in latitude, reaching its peak at the equator, where cumulative UV exposure is highest; and (2) relative to most other human tumor genomes, the genomes of cutaneous melanomas have greatly elevated (15–50×) numbers of the C-to-T transition mutations that result from UV-induced DNA lesions.

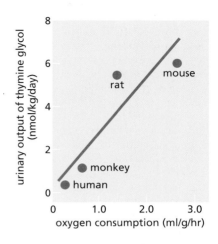

Figure 12.11 Metabolic rate and DNA oxidation The metabolic rate, and thus the rate of oxygen consumption, varies inversely with body size in mammals, being about one order of magnitude higher in rodents than in humans. As indicated here, higher oxygen consumption (ml of O_2 per g of body weight per hour, *abscissa*) correlated with an increased rate of base oxidation of DNA, ostensibly created by ROS (reactive oxygen species) that are the by-products of oxidative phosphorylation in the mitochondria. Thymine glycol (*ordinate*) is the product of excising from DNA the pyrimidine base present in deoxythymidine glycol (see Figure 12.10A), one of the common oxidation products of DNA; this base is eventually excreted in the urine. The 6 nmol per kg of body weight per day excretion rate measured in mice and rats corresponds to approximately 3000 thymidines oxidized per cell per day. (From B.N. Ames, *Free Radic. Res. Commun.* 7:121–128, 1989; and from B.N. Ames and L.S. Gold, *Mutat. Res.* 250:3–16, 1991.)

A variety of chemical species can enter the body from outside, undergo chemical modification, and then proceed to react with the macromolecules within cells, among them the DNA. Many of these modified chemical species are **electrophilic**, that is, they seek out and attack electron-rich regions of target molecules. Among the most potent mutagens are **alkylating** agents, chemicals that are capable of attaching alkyl groups covalently to the DNA bases (see Figure 12.12C).

The alkylation of a base may destabilize its covalent bond to deoxyribose, resulting in the loss of the purine or pyrimidine base from the DNA. Alternatively, the alkylated bases may be misread by the DNA polymerase machinery during DNA replication. Because of their potent mutagenicity, alkylating agents are often used experimentally to induce various types of tumors in laboratory animals.

(Because certain alkylating agents that are used clinically as anti-cancer chemotherapeutics are also potent mutagens, a delayed outcome of chemotherapy may be the appearance, at a second anatomical site, of a new, therapy-induced tumor. In addition, sequencing of the genomes of glioblastomas from patients who have previously been treated by chemotherapy with the alkylating agent temozolomide often reveals blizzards of point mutations throughout their genomes; see Figure 12.7B.)

A number of potent mutagens are formed when ingested or inhaled compounds become altered by cellular metabolic processes. Take, as an example, benzo[a]pyrene (BP), a potent carcinogen that falls in the class of polycyclic aromatic hydrocarbons (PAHs), that is, molecules carrying multiple benzene rings fused together in various combinations (see Figure 2.25). Experiments conducted in Britain in the late 1920s indicated that this compound is a prominent carcinogen found amid the complex mixture of compounds in coal tar.

An elaborate array of cytochrome P450 enzymes (CYPs) are dispatched by the cell to oxidize polycyclic hydrocarbons. (The genes for 57 distinct P450s have been uncovered in the human genome.) The goal of the cell is to detoxify these foreign chemical species and convert them into molecules that are soluble and can be readily excreted (**Figure 12.13A**). However, an inadvertent outcome of this detoxification is often the creation of chemical species that are highly reactive with the DNA and are therefore

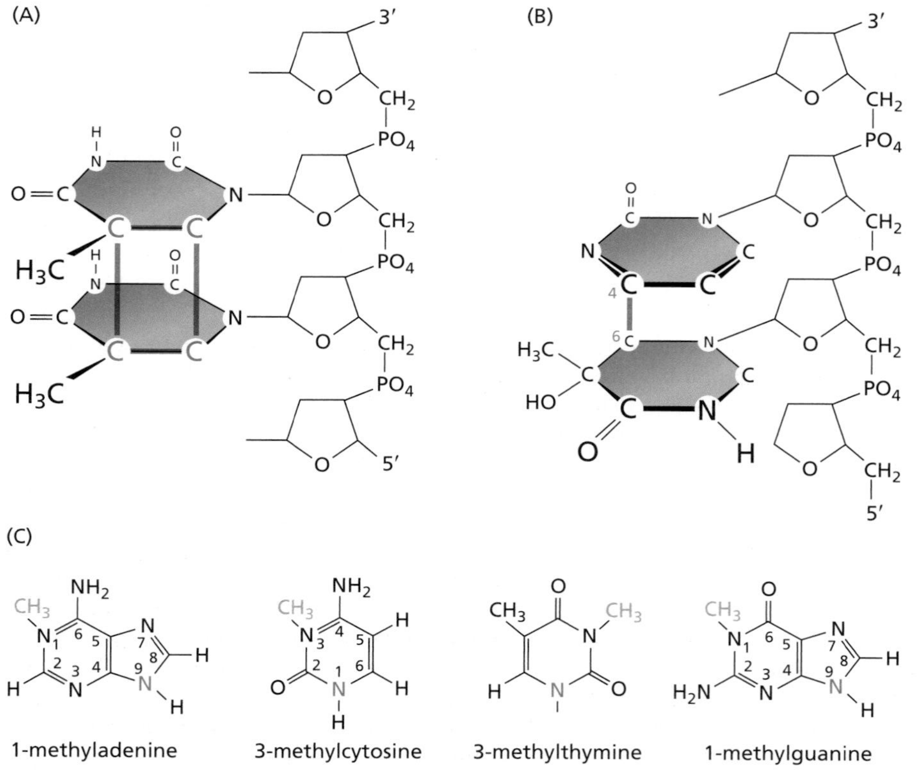

Figure 12.12 Products of UV irradiation and alkylation of DNA
Ultraviolet (UV) radiation produces covalent cross-links between adjacent pyrimidine bases in the DNA. When purified DNA is irradiated with 254-nm photons, 71% of the photoproducts are the cyclobutane pyrimidine dimers (CPD; A), while 24% are the pyrimidine (6-4) pyrimidinone (6-4 PP) photoproducts (B). The cyclobutane ring of a CPD is highlighted in *red* in panel A, as is the bond linking the 6-position of one pyrimidine to the 4-position of the adjacent pyrimidine in panel B. These structures are relatively stable chemically and must be removed by transcription-coupled repair and global genomic repair (described in Section 12.8). (C) Exogenous alkylating agents can covalently alter DNA bases by the attachment of alkyl groups, such as the methyl groups (*orange*) shown here. Many of these methyl groups may also be generated endogenously by the inadvertent actions of S-adenosyl methionine, which carries a highly reactive methyl that plays a key role in many normal biosynthetic reactions. The nitrogens that form glycosidic linkages with deoxyribose are shown in *pink*.

1-methyladenine   3-methylcytosine   3-methylthymine   1-methylguanine

(A)

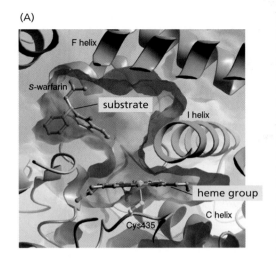

(B)

benzo[a]pyrene
(procarcinogen)

benzo[a]pyrenediolepoxide
(BPDE)
(ultimate carcinogen)

Figure 12.13 Actions of cytochromes on procarcinogens (A) Cytochrome P450s (CYPs) are involved in the biosynthesis of a variety of metabolites such as steroid hormones, cholesterol, and bile acids as well as the degradation of compounds such as fatty acids and steroids. In addition, they aid in the oxidation and associated detoxification of *xenobiotics* (compounds originating outside the body), such as drugs and carcinogens. The substrate-binding cavity of human CYP2C9, shown here, carries a molecule of a xenobiotic substrate, in this case warfarin (used both as an anticoagulant drug and as a rat poison). The heme ring detoxifies the warfarin by oxidizing it. The large substrate-binding cavities of CYPs allow them to accommodate a wide range of substrates, most of which are quite hydrophobic. (B) Among the xenobiotic compounds entering the body are a variety of polycyclic aromatic hydrocarbons (PAHs) that derive from tobacco smoke, broiled foods, and polluted environments. A common PAH is benzo[a]pyrene (BP, *far left*), which, following two successive oxidation reactions mediated by cytochrome P450 enzymes (largely CYP1A1), is converted to benzo[a] pyrenediolepoxide (BPDE; *far right*). This highly reactive molecule is termed an *ultimate carcinogen* (Section 2.10) because, unlike its BP precursor, it is able to directly attack and form covalent adducts with DNA bases, which may then generate oncogenic mutations. Importantly, exposure of a cell to PAHs increases synthesis of CYP1A1, thereby accelerating these reactions. (A, from P.A. Williams et al., *Nature* 424:464–468, 2003. B, from E.C. Miller, *Cancer Res.* 38:1479–1496, 1978.)

actively mutagenic (see Figure 12.13B). As a consequence, chemically inert, unreactive *procarcinogens* are converted into highly reactive *ultimate carcinogens* that can attack DNA molecules directly through their ability to form covalent bonds with various bases. The chemical entity formed after reaction of a carcinogen with a DNA base is often termed a **DNA adduct** (**Figure 12.14**).

In most cases, chemically reactive, ultimate carcinogens attack other molecules almost immediately after being formed. Consequently, they have lifetimes as free molecules that are measured in seconds; this dictates that many of the genetic lesions they create arise in the same cells where these molecules underwent initial metabolic activation. For example, benzo[a]pyrene (BP), cited above as an important carcinogenic component of coal tar, is also a prominent carcinogenic component of tobacco smoke. BP is often activated in the first cells that it enters—the epithelial cells in the lungs of smokers. Once formed, the activated derivative, benzo[a]pyrenediolepoxide (BPDE), proceeds directly to form adducts with the guanosine residues in the DNA of these epithelial cells (see Figure 12.14A).

In fact, mutagenic adducts that are far smaller than those generated by benzo[a]pyrene are formed in our DNA. For example, until recently, the carcinogenic actions of ethanol were ascribed largely to its cytotoxic effects when present in high concentrations in various distilled drinks: by stripping the epithelia of the mouth and esophagus, ethanol drives compensatory proliferation of epithelial stem cells and thereby acts as a potent tumor promoter (see Section 11.14). However, this mechanism failed to explain the well-established epidemiologic observation that heavy consumption of more dilute concentrations of alcohol also is carcinogenic. For example, one French study demonstrated an 18-fold increased risk of developing esophageal cancer among those who consume >0.7 liters of wine each day, and a more recent European study found that those who consume three or more beers a day and carry both a specific alcohol dehydrogenase (ADH) allele together with a certain aldehyde dehydrogenase (ALDH) allele specifying reduced enzyme activity exhibit a 2-fold increased frequency of stomach cancer (relative to those with the more common, alleles). Those East Asians who are heterozygous at the *ALDH2* locus and, as a consequence, have greatly reduced

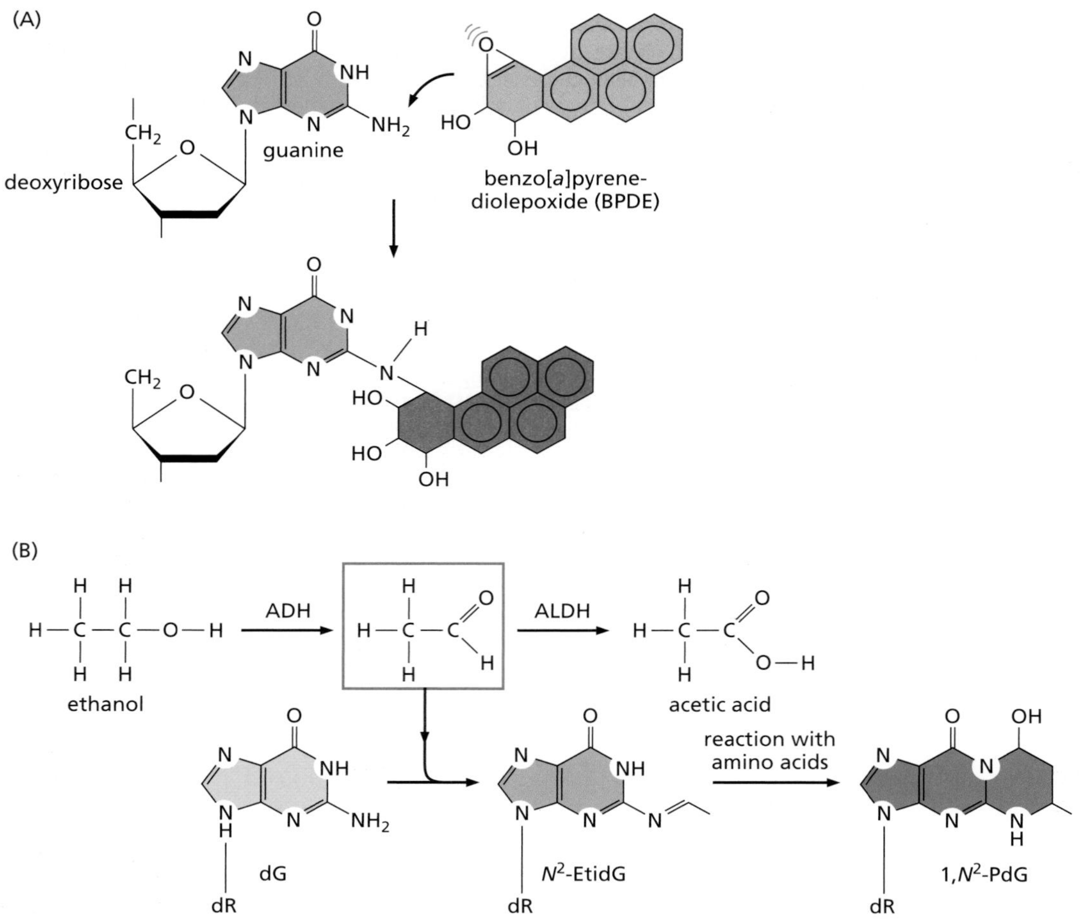

Figure 12.14 DNA adducts (A) The chemically reactive epoxide group of benzo[a]pyrenediolepoxide (BPDE; see Figure 12.13B) can attack a number of chemical sites in DNA, including the extracyclic amine of guanine (*shown here*), as well as the two ring nitrogens and the O^6 of this base. Because the cell may remove these various adducts with different efficiencies, the O^6 adduct of benzo[a]pyrene (BP) may be more potently mutagenic than the more frequently formed adduct shown here. Even though polycyclic aromatic hydrocarbons and BP have been studied for more than half a century, their precise contributions to human cancer development remain a matter of debate, although indirect evidence provides strong suggestion of an important role (see Figure 12.15). (B) Ethanol, which is consumed in large quantities by many people, is converted by alcohol dehydrogenase (ADH) to acetaldehyde (*red box*), which has been classified as a mutagen because of its reactivity with DNA. The most abundant DNA adduct resulting from the reaction of acetaldehyde with deoxyguanosine (dG) is N^2-ethylidene-dG (N^2-EtidG), a relatively weak mutagen. A more complex adduct that is quite mutagenic is 1,N^2-propano-2'-deoxyguanosine (1,N^2-PdG), which forms from the reaction of N^2-EtidG with amino acids. Heavy drinkers with inborn alleles leading to either highly active ADH or slow acetaldehyde breakdown (via aldehyde dehydrogenase, ALDH) exhibit an especially elevated risk of developing oral and esophageal carcinomas.

enzyme function, exhibit facial flushing after drinking alcohol and have a greater than 10-fold increased risk of developing esophageal cancer (relative to those with a full complement of ALDH2 activity). Most if not all of these effects are likely due to the mutagenic actions of acetaldehyde—the immediate downstream product of ethanol metabolism—which is highly reactive with deoxyguanosine, forming several distinct DNA adducts that are much smaller than the bulky adduct generated by BP (see Figure 12.14B).

Valuable clues about the identities of the mutagens that function as human carcinogens also come from detailed sequence analyses of the mutant alleles in cancer cell genomes. For example, the spectrum of mutant *p53* alleles of lung cancers provides a wealth of information about the responsible carcinogens (**Figure 12.15**). BP is strongly implicated here, because mutational "hot spots"—sites in the gene that have especially high rates of base substitution—conform precisely with the preferred sites (codons 157, 248, and 273) of adduct formation observed after cells are exposed *in vitro* to BPDE.

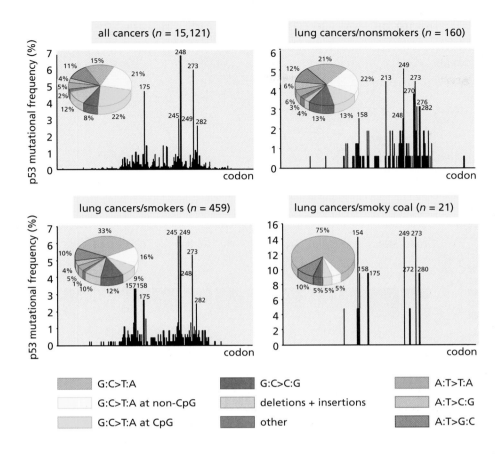

all cancers (n = 15,121)

lung cancers/nonsmokers (n = 160)

lung cancers/smokers (n = 459)

lung cancers/smoky coal (n = 21)

Legend:
- G:C>T:A
- G:C>T:A at non-CpG
- G:C>T:A at CpG
- G:C>C:G
- deletions + insertions
- other
- A:T>T:A
- A:T>C:G
- A:T>G:C

Figure 12.15 *p53* point mutations caused by mutagens The point mutations found in the *p53* alleles carried by human cancer cells provide clues about the identities of the responsible mutagenic agents. In each case the number of tumors being analyzed is indicated by *n*. *Pie charts:* G:C-to-T:A (i.e., G-to-T) transversions have been found in 15% of a group of more than 15,000 mutant *p53* alleles associated with a variety of human tumors. However, in lung carcinomas, the proportion of mutant *p53* alleles that carried this transversion was 21% in nonsmokers, 33% in cigarette smokers, and 75% in nonsmokers who had a history of repeated exposure to smoky coal emissions. This G-to-T transversion has been found experimentally to be the mutation usually induced by benzo[*a*] pyrene (BP), which is known to be present in the combustion products of fossil fuels as well as tobacco. *Bar graphs:* Additional clues about the mutagenic process are provided by the bar graphs, which show the locations of the various point mutations in the *p53* gene. The number above each bar designates the codon in the human *p53* reading frame that was found to be affected by a point mutation. The ordinate indicates the percentage of the tumors studied within a group that carried mutations in each of the codons indicated along the abscissa. These bar graphs also imply that the nucleotide sequences surrounding a targeted base (A, C, G, or T) are strong determinants of its mutability. (From A.I. Robles, S.P. Linke, and C.C. Harris, *Oncogene* 21:6898–6907, 2002.)

These analyses of mutational frequency become increasingly important as we try to reduce cancer incidence by identifying and then reducing exposure to the responsible carcinogens. Important in this effort are recently developed high-throughput sequencing technologies, which can generate megabases of sequencing information each day, making it practical to sequence large *intergenic* stretches of cancer cell genomes. Mutation of these sequences presumably does not affect phenotype, thereby creating the "passenger" mutations described in Sidebar 11.3. Analyzing the mutations present in such intergenic regions is not subject to the biases deriving from focusing on a specific gene, such as *p53* (in which specific "driver" mutations are selected during the course of multi-step tumorigenesis because they confer advantageous phenotypes on cancer cells). In addition, such large-scale sequencing provides a more systematic survey of mutation type and frequency. Such analyses reveal vastly different densities of mutations per megabase (Mb) of tumor DNA when comparing various cancer types with one another (**Figure 12.16**).

At one end of this spectrum are pediatric cancers that typically exhibit relatively low densities of point mutations (~1 per Mb) in their genomes. At the other end are smoking-related cancers and melanomas, where exogenous carcinogens (tobacco smoke and UV light, respectively) have inflicted an order-of-magnitude higher density of point mutations. Once again, these sequencing analyses reveal the *nature* of the point mutations, which can be linked in many cases to the mutagenic mechanisms that have been directly associated with various classes of carcinogenic molecules.

Among the most potent of exogenous carcinogens is aflatoxin B1 (AFB1; see Figure 2.28), which is produced by molds belonging to the *Aspergillus* genus. These molds grow on improperly stored peanuts, nuts, corn and grains. As cited in Section 11.15, those people living in areas where AFB1 exposure is high run a 3-fold elevated risk of hepatocellular carcinoma (HCC), while those carrying a chronic hepatitis B viral infection have a 7-fold increased risk of this disease. Individuals who experience both risk factors run a 60-fold increased risk of contracting liver cancer (**Figure 12.17A**).

Figure 12.16 Large-scale genome sequencing reveals diverse patterns of mutagenesis The results of sequencing of 1035 cancer cell genomes show vastly differing densities of point mutations, which vary by a factor of ~1000 from hematologic pediatric tumors (*left*) to solid tumors in adults (*right*) having documented exposure to known carcinogens (tobacco smoke and UV radiation). The numbers (*n*) at the *top* indicate the number of tumors of each type whose genomes were sequenced. Each *black* point in the *upper* graph represents the density of point mutations per megabase (Mb) in a single tumor genome. The color key (*below, left*) indicates the specific type of base substitution that was registered, and each colored vertical line within a box (*below graph*) presents the relative frequency of a detected specific base substitution in the tumor type indicated by the abbreviation (*directly above*). In pediatric neuroblastomas, mutations are presumably of almost entirely endogenous origin, while melanomas carry almost entirely the C → T transition mutations known to be inflicted by UV radiation, while lung tumors have far more C → A transversions, characteristic of tobacco carcinogens. Interestingly, in contrast to most of the melanomas, which have a very high density of base substitutions, one (*arrow*) has a far lower density; while the other melanomas arose in exposed skin, this melanoma arose on the sole of a foot. [As mentioned earlier (see Figure 12.7B), the genome sequences of gliomas/glioblastomas reveal little about the genesis of these tumors, since the point mutations that are scored are almost entirely the handiwork of the alkylating chemotherapeutic drug that was used to debulk these tumors prior to surgical excision.] The known or suspected sources of mutagenesis of the five tumor types to the right are illustrated by small icons. AML, acute myelogenous leukemia; NB, neuroblastoma; CLL, chronic lymphocytic leukemia; MM, multiple myeloma; OV, ovarian carcinoma; BR, breast ca., PR, prostate carcinoma; GBM, glioblastoma multiforme; H&N, head-and-neck squamous cell carcinoma; CRC, colorectal carcinoma; LUAD, lung adenocarcinoma; LUSC, lung squamous cell carcinoma; MEL, melanoma. (Courtesy of M. Lawrence and G. Getz.)

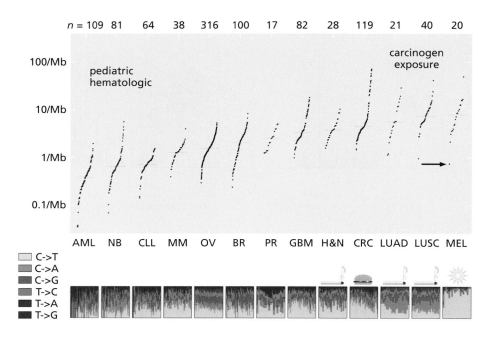

Once AFB1 is activated by CYPs in the liver, the resulting metabolite can attack guanine and form a DNA adduct by becoming covalently linked to this base (see Figure 12.17B). AFB1 causes a characteristic G-to-T mutation in DNA. Such point mutations, where the sequence AGG has been converted to AGT, are found at codon 249 of the *p53* tumor suppressor gene in about half of the hepatocellular carcinomas occurring in individuals exposed to this carcinogen. These characteristic changes in the DNA provide compelling evidence of the direct interaction of this mutagenic carcinogen with bases in the DNA. (These conclusions are important, since in the absence of such evidence, it becomes possible that a carcinogenic agent is actually functioning as a tumor promoter rather than an "initiating carcinogen"; see Section 11.13).

Another widely studied example of carcinogens of exogenous origin involves the heterocyclic amines (HCAs), a class of molecules that are formed in large amounts when meats of various sorts are cooked at high temperatures (**Figure 12.18A**). These compounds arise through the reactions that take place between naturally occurring molecular species in cells, notably creatine, glucose phosphates, dipeptides, and free amino acids. The HCAs are undoubtedly carcinogenic. For example, the most abundant of these compounds in meats cooked at high temperature—2-amino-1-methyl-6-phenylimidazo[4,5-*b*]pyridine (PhIP)—is capable of inducing colon and breast carcinomas in rats and lymphomas in mice. PhIP is recognized as being the principal HCA in the human diet. Nonetheless, these and other observations still do not prove that these chemical species are actual causal agents of human tumors, and alternative mechanistic models of meat-induced carcinogenesis have been proposed (see Supplementary Sidebar 12.5).

Once heterocyclic amines have entered into cells, CYPs are used by the cells to oxidize these molecules. Some CYPs will oxidize the rings of heterocyclic amines, while others will oxidize the **exocyclic** amine groups, that is, those that protrude from the rings. Ring oxidation by CYPs leads to successful detoxification; amine group oxidation, however, leads to the formation of highly reactive compounds that can readily form covalent bonds with proteins and DNA (see Figure 12.18B). While these and other chemical conversions of HCAs are largely achieved in the liver, the resulting reactive molecules often survive long enough to pass via the circulation into other organs where they may exercise their mutagenic activity. (An alternative hypothesis explaining the carcinogenicity of red meat diets was presented in Supplementary Sidebar 11.7.) These various examples, only a few among many that could be cited here, illustrate how the detoxifying enzymes in our cells, often present in high concentration in liver cells, yield genotoxic compounds rather than the end products intended by evolution—harmless, readily excretable chemical species.

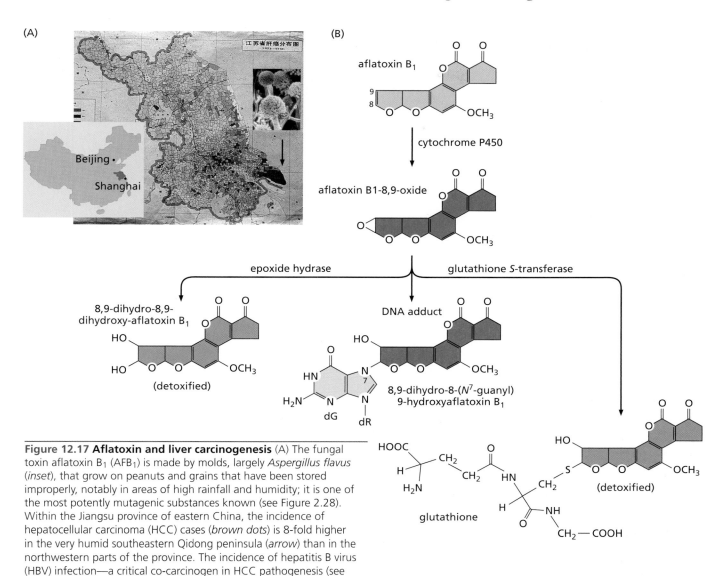

Figure 12.17 Aflatoxin and liver carcinogenesis (A) The fungal toxin aflatoxin B$_1$ (AFB$_1$) is made by molds, largely *Aspergillus flavus* (*inset*), that grow on peanuts and grains that have been stored improperly, notably in areas of high rainfall and humidity; it is one of the most potently mutagenic substances known (see Figure 2.28). Within the Jiangsu province of eastern China, the incidence of hepatocellular carcinoma (HCC) cases (*brown dots*) is 8-fold higher in the very humid southeastern Qidong peninsula (*arrow*) than in the northwestern parts of the province. The incidence of hepatitis B virus (HBV) infection—a critical co-carcinogen in HCC pathogenesis (see Section 11.15)—is relatively constant across the province. (B) Activation of AFB1 (*pink*) by cytochrome P450s results in formation of the highly reactive 8,9-oxide form (*red*). This may be detoxified through several side reactions (*bottom left, right*), or it can react directly with DNA, forming a covalent adduct with the N^7 atom of guanosine (*light green*) that is highly mutagenic. Indeed, the liver carcinomas of individuals living in areas of high AFB1 exposure often carry mutant *p53* alleles with a characteristic G-to-T transversion in codon 249—precisely the type of mutation that would be expected from the known reactivity of AFB1. (A, from T.W. Kensler et al., *Nat. Rev. Cancer* 3:321–329, 2003; inset, courtesy of CABI.org, UK. B, from J.D. Groopman and L.G. Cain, in C.S. Cooper and P.L Grover, eds., Interactions of Fungal and Plant Toxins with DNA in Chemical Carcinogenesis and Mutagenesis. Berlin: Springer-Verlag, 1990.)

The notion that exogenous and endogenous mutagens (the latter including DNA replication errors, spontaneous depurinations, and the actions of endogenously generated chemical species) constitute distinct, separable causes of human cancer is increasingly supported by the analyses of tumor-associated point mutations like those presented in Figures 12.15 and 12.16. For example, G-to-T transversions, in which a pyrimidine base (T) replaces a purine (G), are found in one-third of the mutant *p53* alleles in the tumors of smokers. This base substitution conforms to the known mutagenic

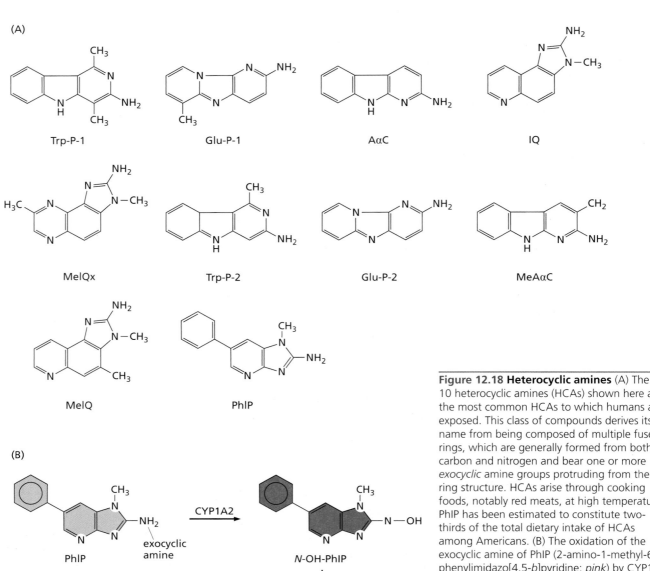

Figure 12.18 Heterocyclic amines (A) The 10 heterocyclic amines (HCAs) shown here are the most common HCAs to which humans are exposed. This class of compounds derives its name from being composed of multiple fused rings, which are generally formed from both carbon and nitrogen and bear one or more *exocyclic* amine groups protruding from the ring structure. HCAs arise through cooking foods, notably red meats, at high temperature. PhIP has been estimated to constitute two-thirds of the total dietary intake of HCAs among Americans. (B) The oxidation of the exocyclic amine of PhIP (2-amino-1-methyl-6-phenylimidazo[4,5-*b*]pyridine; *pink*) by CYP1A2, a cytochrome P450, leads to the highly reactive compound N-OH-PhIP (*red*). It can react with the 8-C of deoxyguanosine (*green*) to form a mutagenic adduct (*lower left*). (A, from T. Sugimura, *Carcinogenesis* 21:387–395, 2000. B, from M. Nagao and T. Sugimura, eds., Food-Borne Carcinogens. New York: John Wiley & Sons, 2000.)

actions of the polycyclic aromatic hydrocarbons, notably benzo[*a*]pyrene, that are present in high concentrations in cigarette smoke. These genetic lesions are found less frequently in the mutant *p53* alleles in other kinds of tumors (except liver cancers). Moreover, only about 21% of the mutant *p53* alleles in the lung tumors of nonsmokers show these transversions, and a significant fraction of the tumors in this subgroup may well have arisen in passive smokers, that is, those living in close contact with smokers. The remaining mutant *p53* alleles of the nonsmokers' tumors carry mutations that are more typical of the spontaneous alterations in DNA described in Section 12.5.

The successes in identifying electrophilic compounds and alkylating agents that are potent mutagens and thus carcinogens has led to the widespread assumption that other, similarly acting chemical species that enter the body through food, water, or air are also important in provoking many types of human cancers. However, the fraction of human malignancies that are traceable to the actions of specific mutagenic carcinogens in our environment or food supply remains a matter of great contention (see Section 11.18). And it is plausible that, apart from a few exceptions, such as UV radiation, tobacco combustion products, aflatoxin, and heterocyclic amines, relatively few exogenous mutagenic agents enter our bodies, create genetic damage, and succeed in causing cancer.

12.7 Cells deploy a variety of defenses to protect DNA molecules from attack by mutagens

The most effective way for a cell to defend its genome from disruption by mutagenic agents is to *physically* shield its DNA molecules from direct attack. In the case of ultraviolet rays from the sun, these penetrate poorly into the body's tissues, leaving the cells of the skin and pigmented cells in the retina as the only vulnerable tissues. The skin shields itself from UV radiation using the **melanin** pigment that is donated by **melanocytes** to keratinocytes located in the basal region of the epidermis (the epithelial layer of the skin; **Figure 12.19**).

Skin color in humans is determined by the types and amounts of melanin that are transferred from the melanocytes to the keratinocytes. The role of this pigmentation in cancer pathogenesis is highlighted by the oft-cited case of skin cancers in Australia. There, a high flux of UV radiation (because of proximity to the equator) and a lightly pigmented population (deriving until recently largely from the British Isles) combine to create the world's highest incidence of these diseases. In Africa, in contrast, the darkly pigmented human populations living at similar latitudes rarely experience skin cancers. (Among those few who do contract melanoma in Central Africa, tumors of the unpigmented area of the sole of the foot are common.) In the case of X-rays and cosmic radiation, there is no effective physical shielding that can be erected by the body, since these types of radiation can penetrate easily through all biological substances.

These limited options for protection against physical carcinogens contrast with the large number of mechanisms that cells can deploy to intercept *chemical* carcinogens before they have had the opportunity to damage the cellular genome. The ambushing of reactive oxygen species (ROS) and free radicals is the job of a variety of enzymes, including superoxide dismutase and catalase; they collaborate to detoxify an ROS into unreactive forms of oxygen. The ROS may also be intercepted by a series of free-radical scavengers, including vitamin C, α-tocopherol (vitamin E), bilirubin, and urate. These molecules will chemically react with the ROS, thereby detoxifying them.

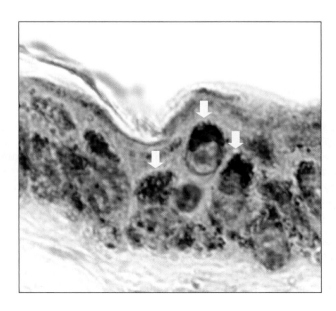

Figure 12.19 Physical shielding of keratinocyte nuclei from ultraviolet radiation The main protection that keratinocytes in the skin have from ultraviolet (UV) radiation, notably UVB photons, derives from the melanosomes—vesicles carrying melanin pigment that have been transferred from melanocytes into keratinocytes in the basal layers of the epidermis. As seen here, once the melanosomes are acquired by the keratinocytes, they are assembled into tiny sun umbrellas (sometimes called supranuclear caps) that sit above keratinocyte nuclei (*arrows*) and shield them from visible and, more importantly, UVB radiation. Keratinocyte nuclei that lack these umbrellas sustain as much as four-fold more UV-induced DNA damage than those that carry them. Moreover, redheads, who produce little or no black eumelanin pigment in their hair and skin (but do make the red/yellow pheomelanin), have an almost 4-fold increased risk of developing melanoma relative to those with dark brown or black hair. (Courtesy of D.E. Fisher.)

(A)
glutathione (GSH)

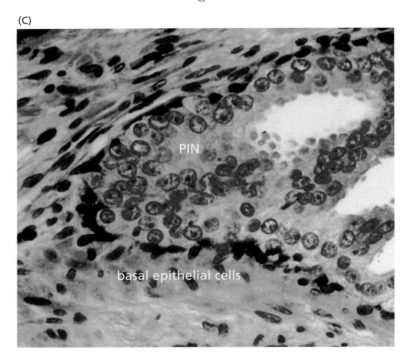

glycine   cysteine   glutamic acid

(B)

(C)

PIN

basal epithelial cells

**Figure 12.20 Glutathione
S-transferase and its loss from
tumors** (A) Glutathione (GSH) is an
unusual tripeptide in which a glutamic
acid is attached to the amino group
of cysteine via the carboxyl side chain
of the glutamic acid. (B) Glutathione
S-transferase (GST) enzymes use
the sulfhydryl (SH; *pink*) group of
glutathione's cysteine residue to detoxify
a number of reactive compounds
before the latter are able to react
with cellular target molecules, such as
DNA. Shown here is a typical reaction
mediated by a GST in which the SH
group of glutathione is used to disrupt
the highly reactive epoxide group of a
compound that has suffered oxidation
(where R can be any of a variety of
chemical groups). (C) As many as 90%
of prostate carcinomas show the loss of
expression of an important glutathione
S-transferase, GST-π; this deprives the
tumor cells of the means to detoxify
many electrophilic mutagens. In many
of these cases, loss of GST-π expression
is traceable to methylation of the
GSTP1 promoter. In the premalignant
PIN (prostatic intraepithelial neoplasia)
lesion—a benign precursor of prostate
carcinoma—shown here, use of
an antibody reactive with GST-π
demonstrates the presence of GST-π in
the basal epithelial cells (*dark brown,
left side*), but loss of all GST-π expression
in the luminal epithelial cells (*light blue,
right side*). The luminal epithelial cells are
revealed by DAPI, a stain that is specific
for DNA (*blue*). (C, from C. Jeronimo et
al., *Cancer Epidemiol. Biomarkers Prev.*
11:445–450, 2002.)

Yet another important line of defense is erected by enzymes of the class termed glutathione *S*-transferases (GSTs), which function to link electrophilic compounds, and thus many carcinogens, with glutathione, thereby detoxifying these compounds and preparing them for further metabolism and secretion (**Figure 12.20A and B**). Significantly, as many as 90% of human prostate adenocarcinomas exhibit a shutdown of glutathione *S*-transferase-π (GST-π) expression due to methylation of the promoter of the *GSTP* gene (Figure 12.20C)—the same mechanism that is often used by cancer cells to shut down expression of a variety of tumor suppressor genes (see Section 7.8). Frequent inactivation of this gene has been reported as well in a number of other human carcinomas. This loss of GST-π expression, which often occurs relatively early in tumor progression, suggests that premalignant prostate tumor cells acquire a distinct advantage by inactivating this gene, thereby increasing the mutability of their genomes. Thus, without this enzyme to defuse certain carcinogens, the genomes of these prostate epithelial cells are attacked more often by actively mutagenic carcinogens. The resulting increased rate of mutagenesis likely accelerates the forward march of tumor progression.

A connection between the glutathione *S*-transferase enzymes and cancer susceptibility is also suggested by epidemiologic studies. In one such study, the allelic configurations of two separate GST-encoding genes, termed *GSTT1* and *GSTM1*, were examined in a normal control population and in individuals suffering from **myelodysplastic syndrome** (MDS), a hyperproliferative disorder of the bone marrow that often progresses to acute myelogenous leukemia (AML). Of the patients suffering from MDS, 46% carried two null alleles (which encode no enzyme) of *GSTT1*; this genetic state was present in only 16% of the control population. (In the case of the related *GSTM1* gene, homozygosity of the null allele was found in comparable proportions of the two populations.) Calculations indicated that individuals inheriting two null alleles of the *GSTT1* gene run more than a 4-fold increased risk of myelodysplastic syndrome over those who carry at least one functional allele of the gene. These observations suggest that the T1 isoenzyme of glutathione *S*-transferase is involved in some

Sidebar 12.4 Inter-individual differences in carcinogen activation seem to contribute to cancer risk and responses to therapy Cells use a broad spectrum of enzymes to modify potential carcinogens in a variety of chemical ways, including the attachment of acetyl, glucuronic acid, glutathione, and sulfate groups; many of these chemical modifications aid in the detoxification and eventual excretion of these compounds. Because of heterogeneity in the human gene pool, the level of expression of many of the enzymes responsible for these detoxification reactions varies greatly among individuals. These differences may, in turn, strongly influence the biological responses of an individual to potential carcinogens. For example, a study of 416 lung cancer patients and 446 healthy control individuals determined that persons of a certain genotype had a twofold increased risk of lung cancer; these susceptible individuals carried a particular allele of the cytochrome-encoding *Cyp1A1* gene and null alleles of the gene specifying glutathione *S*-transferase M1 (*GSTM1*).

Another study focused on breast adenomas among women who consumed more than 27 ng per day of the heterocyclic amine MeIQx (2-amino-3,8-dimethylimidazo[4,5-*f*]quinoxaline; see Figure 12.18A) by eating large quantities of burnt meat. Women who expressed high levels of *N*-acetyltransferase 1 (NAT1, an enzyme that can help to convert heterocyclic amines into active mutagens) were reported to experience a sixfold increased rate of adenomas, while those with much lower levels of NAT1 showed only a twofold increased risk. In both cases, adenoma frequency was compared with the incidence of adenomas in a third group of women who had low levels of

MeIQx in their diet. These correlations suggest that enzymes can influence cancer incidence in two ways: some enzymes affect the rate at which a number of potentially mutagenic compounds are detoxified, while others (inadvertently) convert otherwise-nonreactive compounds into chemically reactive mutagens.

In fact, these enzymes intersect with the disease of cancer in yet another way, because they can also function to detoxify chemotherapeutic drugs, thereby blunting the effects of treatment. A particularly dramatic example of this was observed upon comparing the clinical responses of breast cancer patients who carried functional alleles of the *GSTM1* and *GSTT1* genes with those of a group of women who had only null alleles at both genetic loci. All of these women were treated with a combination chemotherapy regimen (involving cyclophosphamide, Adriamycin, and 5-fluorouracil) plus radiation. Of those women who had functional *GSTM1* and *GSTT1* alleles, almost all succumbed to their disease within 6 years of treatment. In contrast, of those women with a double-null genotype at both genetic loci, about two-thirds were still alive 8 years after treatment.

In all of the cases cited in this sidebar, the different responses were correlated with specific alleles. It is important to remember, however, that these are only correlations and not proofs of causality. In each case, it is possible, in principle, that the alleles being studied are genetically closely linked to yet other alleles that are the true causes of the observed increased or decreased cancer incidence of patients.

way in detoxifying the compounds that provoke MDS. (An alternative interpretation, which is less likely but nonetheless difficult to exclude at present, is that the *GSTT1* gene is closely linked on the chromosome to a second gene that predisposes an individual to MDS.) Yet other epidemiologic studies hint at connections between carcinogen metabolism and individual cancer risk (**Sidebar 12.4**).

These discussions of **xenobiotic** activation and inactivation lead inevitably to another question: What are the origins and the actual daily burdens of these compounds that our various tissues must routinely contend with? And among the xenobiotics, do manmade carcinogens, such as the much-feared pesticides, contribute substantially to this burden?

Bruce Ames, of the Ames test (see Figure 2.27), has estimated that, by eating naturally occurring foodstuffs, humans are exposed on a daily basis to between 5000 and 10,000 distinct natural chemical compounds and their metabolic breakdown products. Included among these are about 2000 mg of burnt material (the products of cooking food at high temperatures) and 1500 mg of naturally occurring pesticides (used by plants to protect themselves against insect predators). In contrast, the average daily exposure to all synthetic pesticide residues contaminating the food chain is about 0.1 mg. About half of the naturally occurring plant pesticides are found to be carcinogenic when tested in laboratory rodents using standard protocols. Since (1) synthetic pesticides are as likely to register as carcinogens in rodent tests as are randomly chosen compounds of natural (that is, plant) origin, since (2) plant-derived compounds, such as those in the vegetables we eat, are generally presumed to be safe, and since (3) concentrations of synthetic pollutants in the food chain are many orders of magnitude below the natural (and equivalently carcinogenic) plant compounds, this raises the question whether synthetic pesticides are ever responsible for significant numbers of human cancers in Western populations. It may well be that the role of synthetic chemical species in creating human cancers (with the exception of tobacco combustion products and the products of cooking food at high temperature) is limited largely to those chemicals that are encountered repeatedly and at very high concentrations in certain occupations, such as agricultural workers who handle large quantities of pesticides routinely.

12.8 Repair enzymes fix DNA that has been altered by mutagens

If genotoxic chemicals are not intercepted before they attack DNA, mammalian cells have a backup strategy for minimizing the genetic damage caused by these potential carcinogens. An elaborate DNA repair system exists to continuously monitor the integrity of the genome, to remove inappropriate bases or nucleotides created by chemical or physical attack, and to replace them with those bases/nucleotides that existed prior to the attack. Components of this system also work to stitch together double helices that have been broken by genotoxic agents or accidentally during replication. Altogether, mammalian cells depend on more than 160 distinct proteins to ensure that damage to DNA is unlikely to result in a mutation being transmitted to daughter cells. Some of these DNA repair proteins figure large in the process of human carcinogenesis, since defects in these proteins result in increased rates of mutation, thereby accelerating the rate of tumor progression.

Cells deploy a wide variety of enzymes to accomplish the very challenging task of restoring normal DNA structure. Importantly, these functions are different from the mismatch repair (MMR) enzymes described above (see Section 12.4), since the MMR enzymes are largely focused on detecting nucleotides of normal structure that have been incorporated into the wrong positions, while the repair mechanisms discussed in this section detect nucleotides of abnormal chemical structure.

The simplest strategy for restoring the structure of chemically altered DNA involves an enzyme-catalyzed reversal of the chemical reaction that initially created the altered base. For example, one type of DNA alkyltransferase removes methyl and ethyl adducts from the O^6 position of guanine, thereby restoring the structure of the normal base (**Figure 12.21A and B**). The importance of this enzyme [O^6-methylguanine-DNA methyltransferase (MGMT), often referred to simply as DNA alkyltransferase] in the development of certain kinds of human tumors is suggested by observations that the *MGMT* gene is silenced by promoter methylation in as many as 40% of gliomas and colorectal tumors, and in about 25% of non-small-cell carcinomas, lymphomas, and head-and-neck carcinomas. (In contrast, methylation of its promoter is not detected in a large cohort of other tumor types.) As was the case with detoxifying genes, such as the glutathione *S*-transferase (*GST*) alleles discussed earlier, we can imagine that the loss of MGMT's DNA repair function in certain tissues favors increased rates of mutation and hence accelerated tumor progression. (Conversely, when the *MGMT* gene, in the form of a transgene, is overexpressed in the mouse mammary gland or thymus, such expression renders these glands quite resistant to the otherwise potently carcinogenic effects of methylnitrosourea, a widely used alkylating mutagen; see, for example, Figure 12.21C.)

Expression of this and other repair enzymes can also influence the *types* of tumors caused by certain carcinogens in animal models of cancer, and quite possibly in humans (**Sidebar 12.5**). Moreover, as was the case with the glutathione *S*-transferase enzyme (see Sidebar 12.4), these DNA repair enzymes can influence responses to therapy. In one instance, after a group of glioblastoma patients whose tumors expressed normal levels of the MGMT enzyme were treated with temozolomide, an alkylating chemotherapeutic agent, they survived for another 12 months. Other patients in the same cohort, whose cancer cells expressed only very low levels of MGMT (due to repression of the *MGMT* gene through promoter methylation), survived for 22 months after therapy—almost twice as long. This observation, subsequently extended in a number of other clinical studies, indicates that normally expressed levels of the MGMT enzyme are very effective in removing the methyl groups attached to DNA by the chemotherapeutic drug, thereby blunting its cytotoxic effects. (Indeed, in certain neuro-oncology clinics, patients whose tumors express significant levels of MGMT are

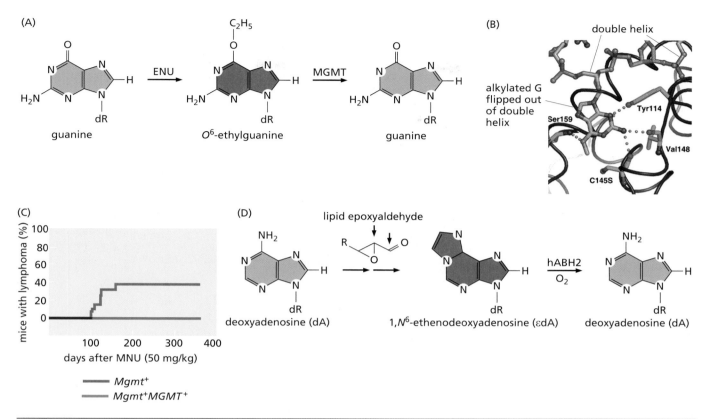

Figure 12.21 Restoration of normal base structure by dealkylating repair enzymes (A) The O6 position of guanosine is especially vulnerable to alkylation by agents such as ethylnitrosourea (ENU). Unlike many DNA repair enzymes, which respond to altered bases by excising them or the entire nucleotide containing them from the DNA, the enzyme O6-methylguanine-DNA methyltransferase (MGMT; also known as O6-alkylguanine-DNA alkyltransferase, AGT) restores an altered guanosine to its normal structure. It does so by removing the alkyl group from the O6 atom of guanine. In the absence of such repair, the alkylated guanosine often leads to a G-to-A transition mutation. (B) Structural analyses of DNA repair proteins have revealed much about how they function. In the case of the MGMT protein, this enzyme (*below, light and dark blue*) works by flipping the damaged base out of the double helix before removing the alkyl group. Moreover, the reaction between enzyme and substrate is stoichiometric, in that the cysteine 145 (C145S) residue (*red, green, below*) in the active site becomes irreversibly alkylated following restoration of normal guanosine structure; hence, each enzyme molecule is able to dealkylate only a single alkylated deoxyguanosine. (C) The effects of MGMT ectopic expression are shown for mice that were exposed to the alkylating carcinogen methylnitrosourea (MNU), which attaches methyl groups to guanosine. Wild-type mice (*Mgmt+*) were highly susceptible to the induction of thymic lymphomas (*red line*), while transgenic mice that overexpress the MGMT enzyme because of a transgene in their germ line (*Mgmt+ MGMT+, blue curve*) were protected from developing these tumors. (D) Highly reactive lipid epoxyaldehydes (*shown here*) and peroxides (*not shown*), which are common in inflamed tissues, can attack and modify adenine (*shown here*) as well as other DNA bases (*not shown*). The AlkB enzyme of bacteria can remove the resulting adducts as well as simpler methyl adducts, such as those shown in Figure 12.12C; a mammalian homolog of AlkB, hABH2 (*cited here*), acts similarly in human cells. (A, from S.L. Gerson, *Nat. Rev. Cancer* 4:296–307, 2004. B, from D.S. Daniels et al., *Nat. Struct. Mol. Biol.* 11:714–720, 2004. C, from L.L. Dumenco et al., *Science* 259:219–222, 1993.)

treated with temozolomide, but with the foreknowledge that they may not gain significant benefit from treatment with this drug.)

The MGMT system is only one way by which cells deal with methylated bases. Another, involving homologs of the bacterial AlkB DNA repair protein, works by oxidizing methyl groups that have become attached to bases, which are then shed as formaldehyde from the rings of all four DNA bases; similarly, AlkB enzymes cause the larger ethyl group to be released as acetaldehyde. (Aficionados of DNA repair portray these enzymes as "burning off" the unwanted alkyl groups!) As mentioned earlier, methylation of DNA bases may occur frequently during the lives of cells through the actions of *S*-adenosylmethionine, the methyl donor that participates in the biosynthesis of many molecules in the cell; its highly reactive methyl group may accidentally become diverted to methylate a variety of cellular macromolecules, including the bases of DNA (see Figure 12.12C).

Sidebar 12.5 Expression patterns of repair enzymes explain certain tissue-specific susceptibilities to cancer Mammalian cells seem to express only a single MGMT enzyme. The importance of this activity in influencing carcinogenesis is indicated by experiments in which pregnant rats are exposed to the carcinogen N-ethylnitrosourea (ENU) during the 15th day of gestation. Virtually all the rat pups that are born succumb to neuroectodermal tumors arising in the central nervous system several months after birth. The peculiar ability of ENU to preferentially induce these tumors can be explained by the fact that the MGMT enzyme is expressed at significant levels throughout the bodies of developing embryos and newborns but is only minimally expressed in the central nervous system. Consequently, alkylated guanine residues that are formed in the cells of the nervous system persist rather than being quickly removed and are ultimately able to generate the point mutations that are responsible for the creation of the oncogenes in the resulting tumors. In one experiment conducted with newborn rats that had been exposed *in utero* to ENU or the related alkylating agent MNU (N-methylnitrosourea), the levels of O^6-alkylguanine adducts surviving in unrepaired form a week after exposure to these carcinogens were 20-fold (ENU) and 90-fold (MNU) higher in brain DNA than in liver DNA.

Bacterial AlkB (and quite possibly its mammalian homologs) has also been found to be capable of removing more complex base adducts. For example, the inadvertent oxidation of unsaturated lipids, yielding lipid epoxides and peroxides, occurs at high rates in inflamed tissues (see Sidebar 12.3); these highly reactive chemical groups can generate complex adducts with DNA bases (see Figure 12.21D) that are highly mutagenic. Indeed, such adducts have been found in tissues of patients with ulcerative colitis, a condition known to progress, with significant frequency, to carcinomas (see Section 11.15). A direct role in cancer pathogenesis of the human homologs of AlkB, termed hABH2 and hABH3, has not yet been demonstrated.

Far more important, however, than these dealkylating enzymes are the numerous cellular enzymes that recognize chemically altered bases in the DNA and respond in two other ways, depending on the specific modification of the DNA. In some cases, specialized enzymes will cleave the bond linking a modified base to the deoxyribose sugar, the process of **base-excision repair** (BER; **Figure 12.22A**). In other cases, the entire nucleotide containing both the base and associated deoxyribose will be cut out, this being the process termed **nucleotide-excision repair** (NER; see Figure 12.22B).

Base-excision repair (BER) tends to repair lesions in the DNA that derive from endogenous sources, such as those attributed to the reactive oxygen species and depurination events described earlier (Section 12.5). Nucleotide-excision repair (NER), in contrast, largely repairs lesions created by exogenous agents, such as UV photons and chemical carcinogens (for example, see Figures 12.12 and 12.14). BER seems to concentrate on fixing lesions that do not create structural distortions of the DNA double helix, while NER directs its attention to bulky, helix-distorting alterations.

BER is initiated by a group of DNA **glycosylases**, each specialized to recognize an abnormal base and cleave its covalent bond to deoxyribose. For example, a uracil base in the DNA is recognized readily by the proteins responsible for BER because U is not normally present in DNA. U is removed by the enzyme uracil DNA-glycosylase and soon replaced, usually with a C. (Refer to Figure 12.9B for how uracil can arise in DNA through the spontaneous deamination of cytosine.) However, the presence of an inappropriately located thymine in DNA presents a quandary for these repair enzymes, since T is a normal constituent of DNA. As we have read, 5-methyl-C occasionally undergoes spontaneous deamination, leading to a T, and thus to T:G base pairs (see Figure 12.9B). In fact, evolution has responded to this problem by implanting a T:G glycosylase in our cells, which is designed specifically to excise T's that happen to arise opposite G's. Nonetheless, it is clear that the T:G base pairs formed by this deamination occasionally escape detection by this enzyme and persist to yield point mutations.

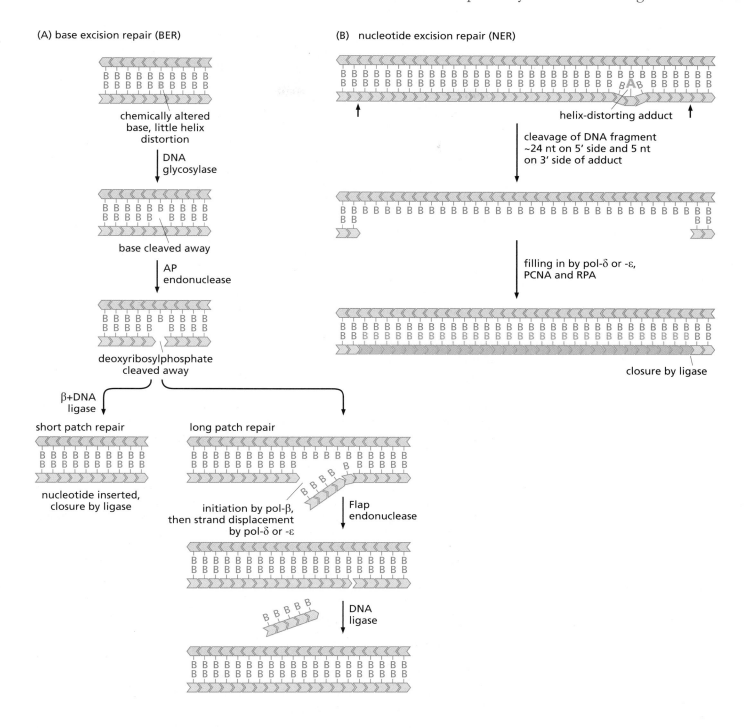

(A) base excision repair (BER)

chemically altered base, little helix distortion

DNA glycosylase

base cleaved away

AP endonuclease

deoxyribosylphosphate cleaved away

β+DNA ligase

short patch repair

nucleotide inserted, closure by ligase

long patch repair

initiation by pol-β, then strand displacement by pol-δ or -ε

Flap endonuclease

DNA ligase

(B) nucleotide excision repair (NER)

helix-distorting adduct

cleavage of DNA fragment ~24 nt on 5′ side and 5 nt on 3′ side of adduct

filling in by pol-δ or -ε, PCNA and RPA

closure by ligase

Figure 12.22 Base- and nucleotide-excision repair

(A) Base-excision repair (BER) is achieved by enzymes that recognize chemically altered bases having minimal helix-distorting effect. These DNA N-glycosylase enzymes cleave the glycosyl bond linking the altered base (*yellow*) and the deoxyribose. The base-free deoxyribosylphosphate is then excised by an enzyme—apurinic/apyrimidinic endonuclease (APE)—specialized to remove base-free sugars. The resulting single nucleotide gap is filled by DNA polymerase β and sealed by a DNA ligase. In fact, there are two forms of BER. A single nucleotide may be excised and the gap may be filled in by a DNA polymerase (pol-β) and ligated (short patch repair). Alternatively, a strand-displacing DNA polymerase (δ or ε) may, following excision by APE and addition, as before, of a single base by pol-β, extend the 3′ strand by several nucleotides beyond the original gap. The resulting displaced strand is removed by a Flap endonuclease and the remaining lesion is sealed by a ligase (long patch repair). (B) Nucleotide-excision repair (NER) is accomplished by enzymes that recognize bulky, helix-distorting lesions and cleave the flanking oligonucleotide sequences at sites approximately 24 nucleotides (nt) on the 5′ side and about 5 nucleotides on the 3′ side. The resulting approximately 29-nt single-strand gap in the DNA is then filled by DNA polymerase δ or ε, acting together with PCNA (proliferating-cell nuclear antigen) and RPA (which binds to single-strand DNA), and is finally sealed by a DNA ligase. The chevrons represent deoxyribose nucleotides, all pointing in a 5′-to-3′ direction.

After an aberrant base is removed by a DNA glycosylase, the base-free sugar that results is then cleaved by a second enzyme, an endonuclease named APE (apurinic/apyrimidinic endonuclease) that is specialized to cut the strand carrying the base-free deoxyribose, doing so on the 5′ side of this sugar; a third enzyme, termed an AP lyase, then cleaves on the 3′ side, liberating the base-free sugar. The resulting single-strand gap in the DNA is repaired by a DNA polymerase, often polymerase β. The single-strand nick that results is finally closed by a DNA **ligase**, which rejoins adjacent nucleotides through the formation of phosphodiester bonds between them, thereby reconstructing the normal chemical structure of DNA. (An occasionally used variant form of BER, termed "long patch repair," involves the excision of 4 to 7 nucleotides adjacent to the damaged base followed by a filling of the resulting gap; see Figure 12.22A.)

Nucleotide-excision repair (NER; see Figure 12.22B) is accomplished by a large multiprotein complex composed of almost two dozen subunits. This complex seems to require two distinct changes in DNA before it will initiate repair: significant distortion of the normal Watson–Crick structure of the double helix plus the presence of a chemically altered base. Once this large complex recognizes the problem, it proceeds to cleave the damaged strand upstream and downstream of the damage, yielding a single-strand fragment of 25 to 30 nucleotides in length, which is then removed. DNA polymerases that are specialized to fill in the resulting gap in the DNA (using the complementary, undamaged strand as a template) then take over, followed by a DNA ligase, which erases the final trace of the damage.

Included among the NER enzymes are those that can recognize and remove structures resulting from the formation of bulky base adducts (that is, those composed of complex molecular structures covalently bound to bases) created by certain exogenous mutagens, such as polycyclic hydrocarbons, heterocyclic amines, and aflatoxin B1, as well as the pyrimidine dimers formed by UV radiation (see Figure 12.12). For example, following exposure to UV radiation, cultured human cells can repair approximately 80% of their pyrimidine dimers within 24 hours. The NER apparatus active here will remove 5 nucleotides on the 3′ side of the photoproduct (the pyrimidine dimer) and 24 nucleotides on the 5′ side.

The various reactions that constitute NER can actually be divided into two subtypes. The first of these is focused specifically on the template strand of actively transcribed genes and is coupled to the actions of RNA polymerase molecules that are proceeding down these template strands during transcription; these actions are termed *transcription-coupled repair* (TCR). The second subtype of NER addresses the remainder of the genome, including the nontemplate strand of transcribed genes as well as the nontranscribed regions of the genome. This type of NER is sometimes termed *global genomic repair* (GGR). The p53 tumor suppressor protein activates expression of several genes encoding NER proteins involved in GGR (see Table 9.2), explaining the defectiveness of GGR in *p53*-mutant cells; in contrast, transcription-coupled repair is intact in these cells. This defect in GGR holds profound implications for the maintenance of cell genomes in the half of all human tumors in which the *p53* gene is mutant (see Chapter 9). Many of the remaining cancers, in which p53 function is compromised in other ways, may also have defects in global genomic repair.

An alternative strategy for the cell to cope with damaged DNA—actually an act of desperation—involves DNA replication of a still-unrepaired stretch of template-strand DNA. (Moreover, any mutant sequences that result from this **bypass synthesis** may subsequently be repaired by consulting the wild-type sequences present in the "**sister chromatid**," that is, the other newly synthesized double helix formed by the replication fork.) This process is termed *error-prone* DNA replication, since the replication apparatus involved here must often "guess" which of the four nucleotides is appropriate for incorporation into the growing DNA strand when it encounters a still-damaged base or set of bases; these guesses are not always correct, leading quite frequently to misincorporated bases (**Figure 12.23**).

To date, at least nine distinct mammalian error-prone human DNA polymerases have been discovered. Some of these can add a nucleotide to a growing strand even when a base in the complementary strand is missing. Yet others can extend a nascent DNA

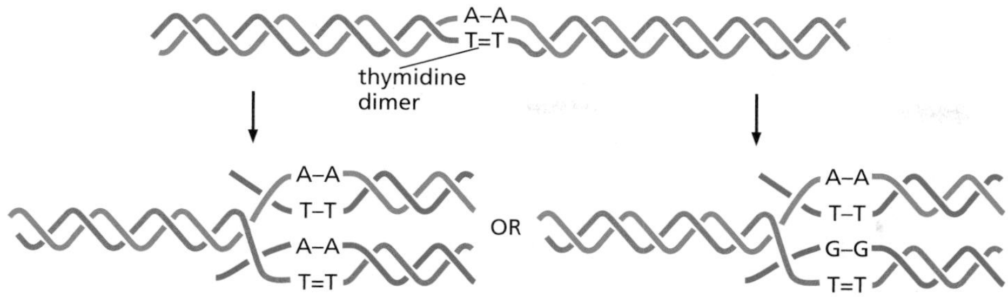

Figure 12.23 Error-prone repair
Error-prone DNA synthesis occurs when an advancing DNA replication fork encounters a still-unrepaired DNA lesion, such as the thymidine dimer shown here. In the great majority of cases (*left*), the error-prone DNA polymerase (sometimes called a *bypass polymerase*) responds to this lesion in the damaged template strand (*red*) by inserting the appropriate bases (in this instance, an A–A dinucleotide) in the growing DNA strand (*dark green*). However, in several percent of these encounters (*right*), the error-prone polymerase will fail to properly "guess" the structure of the lesion in the template strand and will instead incorporate a G–G dinucleotide opposite the thymidine dimer.

strand, using as primer a nucleotide that has been misincorporated by another DNA polymerase. A third type can incorporate a base when the corresponding base in the complementary strand carries a bulky, covalently attached DNA adduct that has not yet been removed by nucleotide-excision repair. One of these enzymes, encoded by the *XPV* gene, is highly specialized, being able to recognize the TT thymine dimers created by UV radiation and insert two A's on the opposite strand (see Figure 12.23). Another bypass polymerase, pol-κ, can also replicate past bulky adducts but in addition advances through templates containing the much less bulky 8-oxo-deoxyguanine (see Figure 12.10A); pol-κ incorporates an A more often than a C opposite the 8-oxo-dG, helping to explain the mutagenic effects of this common product of base oxidation. While the DNA polymerases responsible for the bulk of DNA synthesis in a cell have error rates as low as 10^{-5}, these error-prone polymerases generally have error rates as high as 1 misincorporated base per 100 bases replicated.

The mistakes made by the error-prone polymerases would seem to generate unacceptably high rates of mutation in cell genomes. Still, the price paid for accumulating such mutations should be balanced against the alternative: the risk of imminent death confronted by a cell whose DNA replication forks are stalled because of difficult-to-copy lesions in its DNA.

Perhaps the best studied of these error-prone polymerases is DNA polymerase β (pol-β), which is usually involved in replacing the nucleotides that have been removed because of BER. This relatively small polymerase molecule lacks the proofreading capabilities of the larger polymerase enzymes (see Section 12.4), and this absence may explain much of its error-prone DNA replication activities. In a variety of ovarian carcinoma cell lines, this enzyme has been found to be overexpressed by as much as a factor of 10. The overexpression of the error-prone DNA polymerase β may represent an effective strategy used by these cancer cells to increase the mutability of their genes and hence accelerate the rate of tumor progression. In support of this idea, the forced overexpression of polymerase β in cultured human fibroblasts has been found to encourage microsatellite instability and to increase overall mutation rates as much as threefold.

The deployment of error-prone polymerases by a cell represents a situation in which this cell is making the best of a desperate situation: it gambles that misincorporated bases are an acceptable compromise to avoid the death that would inevitably ensue from a failure to complete DNA replication. Hence, the **bypass polymerases** are not, in the strictest sense, DNA repair enzymes, since they are not focused primarily on removing damage and restoring wild-type nucleotide sequences.

Actually, there is at least one enzyme—the AID enzyme—encoded by the mammalian genome that *purposely* inserts mutations into the genome. Such an enzyme may also, quite inadvertently, contribute to cancer development. The *AID* gene, which encodes the **a**ctivation-**i**nduced cytidine **d**eaminase, is responsible for deaminating cytidine bases in the 3-kb span downstream of the start site for transcription of the genes specifying antibodies and a few other proteins (see Supplementary Figure 15.1). By converting cytidine to uridine residues, this enzyme effectively inserts numerous C-to-T point mutations in these genes; in the case of immunoglobulin (antibody) genes, the resulting "somatic hypermutation" causes diversification of the antigen-binding sites of the encoded antibody molecules, enabling the immune system to develop antibodies of

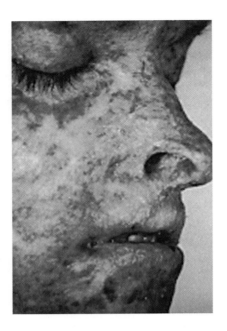

Figure 12.24 A xeroderma pigmentosum patient A patient suffering from xeroderma pigmentosum (XP) has severe and extensive lesions in all areas of sun-exposed skin. These lesions can develop into squamous and basal cell carcinomas as well as melanomas. The tumors develop at rates that are as much as 1000-fold higher than in the general population. (Courtesy of K.H. Kraemer.)

Figure 12.25 Epidemiology of XP patients Patients suffering from xeroderma pigmentosum (XP) exhibit skin cancers far earlier than the general population (*normal*). In the general population, these skin cancers appear with a median age of onset of about 60 years, in contrast to the XP population, in which skin cancers are diagnosed with a median age of about 10 years. The percentages on the ordinate are calculated by dividing the number of patients who have already been diagnosed with one or more skin cancers by a given age by the total number of patients in this population who will eventually be diagnosed with these cancers. Hence, in the XP population, virtually everyone who will develop skin cancers has already done so by around the age of 25. *n* is the number of individuals in each population studied. (From J.E. Clever and K.H. Kraemer, in C.R. Scriver et al., eds., The Metabolic Basis of Inherited Disease, 6th ed. New York: McGraw-Hill, 1989, pp. 2949–2971.)

Sidebar 12.6 Degenerate genomes lead to sunlight sensitivity and the need for sunscreen Humans as well as other placental mammals rely totally on the nucleotide-excision repair (NER) system to remove the highly mutagenic UV-induced pyrimidine dimers (see Figure 12.12). This contrasts with the situation seen in bacteria, lower eukaryotes, and plants, where redundant backup systems are in place to deal with these dimers in the event that the first NER line of defense should fail. These simpler organisms, which may exhibit DNA repair capabilities mirroring those of our distant evolutionary ancestors, have (1) enzymes that can monomerize pyrimidine dimers (thereby directly reversing the UV-induced damage), (2) some glycosylases to cleave the dimerized bases from deoxyribose residues, and (3) nucleases to incise the DNA strand around the dimers.

It is possible that once our mammalian ancestors learned to grow coats of hair, the evolutionary pressures to retain these important backup DNA repair enzymes receded, and the encoding genes were lost through one or another genetic mechanism. Now, because of a quirk of recent primate evolution, we humans are left without protective coats of hair and, at the same time, without an ability to mobilize the backup repair systems to fix UV-damaged DNA in the event that our only remaining line of defense—the xeroderma pigmentosum genes and encoded proteins—fails to do this task. This explains, in part, why the mutant XP phenotypes can be so devastating for those afflicted with this disease, and why the more light-complexioned among us should make lavish use of sunscreen lotion when in the sun.

ever-increasing avidity for their antigen ligands. This enzyme seems to run amok in some human lymphomas, inserting point mutations in genes that are not its normally intended targets. And when the *AID* gene is ectopically expressed as a transgene in many tissues of the mouse, it causes high rates of T-cell lymphomas, many of which exhibit large numbers of genes, such as *myc*, with point mutations throughout their reading frames.

12.9 Inherited defects in nucleotide-excision repair, base-excision repair, and mismatch repair lead to specific cancer susceptibility syndromes

In 1874, two Austro–Hungarian physicians, Ferdinand Hebra and Moritz Kaposi, described an unusual syndrome that involved high rates of the development of squamous and basal cell carcinomas of the skin. (Kaposi subsequently described the unusual sarcoma that bears his name.) As became apparent later, affected individuals have extreme sensitivity to UV radiation, and infants will often suffer severe burning of the skin after only minimal exposure to sunlight (**Sidebar 12.6**). These individuals show dry, parchment-like skin (xeroderma) and many freckles ("pigmentosum"; **Figure 12.24**). In aggregate, individuals suffering from the **xeroderma pigmentosum** (XP) syndrome have a 2000-fold increased risk of skin cancer before the age of 20 compared with the general population and about a 100,000-fold increased risk of squamous cell carcinoma of the tip of the tongue. Skin cancers appear in XP children with a median age of ~10 years, compared with ~60 years in the general population (**Figure 12.25**).

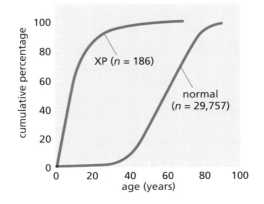

Inherited defects in any one of eight genes can lead to xeroderma pigmentosum, but this number is arbitrary, as these genes overlap with other genes involved in NER and a variety of other syndromes resulting from defective DNA repair. The genetic complexity of the XP syndrome was first recognized through the use of somatic cell genetics. Cells from two different XP patients were fused in culture in order to determine the repair phenotype of the resulting hybrid cells (**Figure 12.26**). On many occasions, the hybrids were found to repair DNA normally, indicating that the two parental cells carried defects in DNA repair that were associated with two distinct genes. For example, using nomenclature developed later, cells from an individual carrying a mutant *XPA* gene (and having a wild-type *XPC* gene) were able to repair DNA normally after being fused to cells from an individual carrying a mutant *XPC* gene (and having a wild-type *XPA* gene). Such collaboration, or "genetic **complementation**," led to the classification of XP-associated mutant alleles into eight complementation groups, each ostensibly defined by the identity of a responsible gene. Only years later were the responsible genes isolated by molecular cloning. Almost always, it has been possible to show that an affected individual has inherited two mutant, null alleles of a gene representing one or another XP complementation group.

Seven of the eight XP-associated genes, named *XPA* through *XPG*, encode components of the large, multiprotein nucleotide-excision repair (NER) complex. The eighth gene, *XPV*, specifies the error-prone DNA polymerase pol-η that many cells seem to use when their regular DNA polymerases (for example, pol-δ) are unable to copy over unrepaired DNA lesions such as pyrimidine dimers. As mentioned in the last section, error-prone polymerases are able to copy a template strand of DNA containing still-unrepaired TT dimers, usually synthesizing two A's in the complementary strand. In general, pol-η is thought to be so accurate that it incorporates AA nucleotides in the growing DNA strand opposite a TT dimer 95% of the time.

Individuals afflicted with XP also have some increased risk of other diseases, notably neurological problems, which are observed in about 18% of these patients. And mice that have been deprived of one of several XP genes suffer markedly increased susceptibility to tumors following exposure to chemical carcinogens. These two observations provide evidence that components of the nucleotide-excision repair system encoded by some of the XP genes are, not unexpectedly, responsible for repairing genetic damage created by other agents besides UV radiation. This raises the following question: Why does a human who lacks one or another XP gene have relatively little increased risk of cancers in internal organs, even though an important component of the NER machinery is missing from all cells throughout this person's body? The simplest and possibly correct explanation is that UV rays are, by far, the most important environmental mutagen to which most humans are exposed, and thus the source of the great majority of the lesions that require repair by the NER machinery. (By one estimate, strong sunlight can inflict as many as 100,000 DNA lesions per skin cell per hour.)

Several other inherited syndromes are also associated with defects in NER. For example, individuals suffering from Cockayne syndrome (CS) appear to be defective in one of two genes that are involved in transcription-coupled NER. Their cells have increased photosensitivity like those of XP patients. The median age of death of patients from this disease is 12 years of age. This disease is highly unusual, in that the significant defects in DNA repair are not associated with increased rates of cancer. Still unaddressed is the possibility that individuals who are *heterozygous* for one of the XP- or Cockayne syndrome–associated mutant alleles have an elevated risk of developing certain types of cancer.

XP was only the first of many human cancer susceptibility syndromes that have been found to be caused by inherited defects in various types of DNA repair (**Table 12.1**). We will explore another one here—hereditary non-polyposis colon cancer (HNPCC). HNPCC is a familial cancer syndrome that represents a quite common cause of inherited predisposition to colon cancer, being responsible for 2 to 3% of all colon cancer cases. HNPCC is, as its name implies, distinct from the other type of hereditary colon cancer predisposition that we encountered previously—adenomatous polyposis coli (see Section 7.11). A subclass of HNPCC patients have increased susceptibility to brain

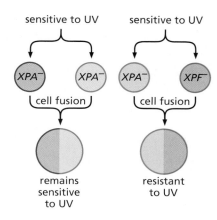

Figure 12.26 Discovery of XP complementation groups Cell fusion experiments using cultured cells from xeroderma pigmentosum (XP) patients have revealed that certain combinations of fibroblasts, each derived from a different patient (*dark blue, light blue, left*), when fused, yielded hybrid tetraploid cells that are as UV-sensitive as the two parental cell populations. On other occasions, however (*right*), fusion of cells from two patients (*light green, light red*) yielded cell hybrids that are as resistant to UV-mediated killing as the cells from normal individuals. Such findings caused the pair of cells that did not complement one another's DNA repair defect (*dark blue, light blue*) to be assigned to the *same* complementation group, ostensibly because the two populations carried mutations in the same gene. Conversely, successful complementation by other pairs of cells (*light green, light red*) allowed the respective parental cell types to be assigned to two *distinct* complementation groups, indicating that mutant alleles of at least two distinct genes were involved in predisposing to XP. In this way, eight distinct XP complementation groups were eventually delineated; the responsible genes were isolated by gene cloning years later.

Table 12.1 Human familial cancer syndromes due to germ-line defects in DNA repair

Name of syndrome	Name of gene	Cancer phenotype	Enzyme or process affected
HNPCC/Lynch	(4–5 genes)[a]	colonic polyposis	mismatch repair enzymes
XP[b]	(8 genes)[b]	UV-induced skin cancers	nucleotide-excision repair
ataxia telangiectasia (AT)[c]	ATM	leukemia, lymphoma	response to dsDNA breaks
AT-like disorder[c]	MRE11	lung, breast cancers	dsDNA repair by NHEJ
Familial breast, ovarian cancer	BRCA1, BRCA2,[d] BACH1, RAD51C	breast, ovarian, prostate carcinomas	homology-directed repair of dsDNA breaks
Werner	WRN	sarcomas, other cancers	exonuclease and DNA helicase,[e] replication
Bloom	BLM	leukemias, lymphomas, solid tumors	DNA helicase, replication
Fanconi anemia	(13 genes)[f]	AML, diverse carcinomas	repair of DNA cross-links and ds breaks
Nijmegen breakage[g]	NBS	mostly lymphomas	processing of dsDNA breaks, NHEJ
Li–Fraumeni	TP53	multiple cancers	DNA damage alarm protein
Li–Fraumeni	CHK2	colon, breast carcinomas	kinase signaling DNA damage
Rothmund–Thomson	RECQL4	osteosarcoma	DNA helicase
Familial adenomatosis	MYH	colonic adenomas	base-excision repair
Familial breast cancer	PALB2	breast cancer	dsDNA repair by HR

[a]Five distinct MMR genes are transmitted as mutant alleles in the human germ line. Two MMR genes—MSH2 and MLH1—are commonly involved in HNPCC; two other MMR genes—MSH6 and PMS2—are involved in a small number of cases; in addition, there is elevated risk of developing tumors of the prostate, ureter, ovary, connective tissues, and brain; a fifth gene, PMS1, may also be involved in a small number of cases.

[b]Xeroderma pigmentosum; at least eight distinct genes, seven of which are involved in NER. The seven genes are named XPA through XPG. An eighth gene, XPV, encodes DNA polymerase η.

[c]Ataxia telangiectasia, small number of cases.

[d]Mutant germ-line alleles of BRCA1 and BRCA2 together may account for ~5% of all breast cancers and 10–20% of identifiable human familial breast cancers.

[e]An exonuclease digests DNA or RNA from one end inward; a DNA helicase unwinds double-stranded DNA molecules.

[f]Thirteen genes have been cloned and at least thirteen complementation groups have been demonstrated. Complementation group J encodes the BACH1 protein, the partner of BRCA1. A number of the products of the FANC genes form a complex that interacts with BRCA1 and its partners; BRCA2 associates with BRCA1 (and FANCD1 = BRCA2). Homozygous absence of either the RAD51C, FANCD1 (= BRCA2), FANJ (= BACH1), or FANCN (= PALB2) gene leads to Fanconi anemia, while lack of only one gene copy leads to breast cancer and/or susceptibility thereto.

[g]The NBS1 protein (termed nibrin) forms a physical complex with the Rad50 and Mre11 proteins; all three are involved in repair of dsDNA breaks. The phenotypes of patients with Nijmegen breakage syndrome are similar but not identical to those suffering from AT.

Adapted in part from B. Alberts et al., Molecular Biology of the Cell, 5th ed. New York: Garland Science, 2008. Also from E.R. Fearon, *Science* 278:1043–1050, 1997.

tumors as well as endometrial, stomach, ovarian, and urinary tract carcinomas in addition to their 80% lifetime risk of developing colon carcinomas.

The increased cancer susceptibility of the HNPCC patients can be traced back to the accelerated rate with which tumor progression proceeds in their colons: while the adenoma-to-carcinoma progression is estimated to require 8 to 10 years in the general population (see Section 11.2), the genetic instability afflicting the cells of HNPCC patients allows this step to occur in only 2 to 3 years. Indeed, because their adenomas progress so quickly to carcinomas, these premalignant growths have a relatively short lifetime and are therefore not found in significant numbers in the colons of these patients. The responsible genes were discovered through clever genetic sleuthing (Supplementary Sidebar 12.6).

The majority (85–90%) of HNPCC cases result from germ-line mutations in the genes encoding two important mismatch repair proteins, MSH2 and MLH1. Mutant germ-line alleles of two other MMR genes, MSH6 and PMS2, are involved in a small proportion (~15%) of these cases; however, two other MMR genes (PMS1, MSH3), which have

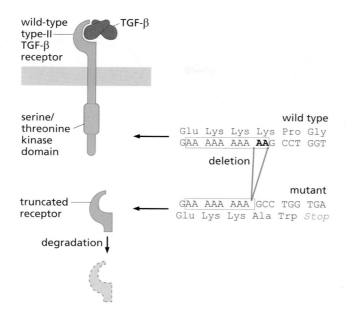

Figure 12.27 A TGF-β receptor gene affected by microsatellite instability The type II TGF-β receptor (TGF-βRII) is frequently inactivated in human colon cancers exhibiting microsatellite instability and therefore carrying defects in mismatch repair (MMR) genes. In the particular colon cancer whose DNA was analyzed here, the last two of ten adenines (*boldface*) were deleted from the reading frame of the receptor-encoding gene, ostensibly because of an MMR defect. This particular deletion ($A_{10} \rightarrow A_8$) resulted in a nonsense mutation that caused premature termination of translation of the nascent TGF-βRII protein and hence loss of functionally critical signaling domains in the C-terminus of the receptor. This loss, in turn, allowed the progenitors of the colon carcinoma cells to become resistant to the growth-inhibitory effects of TGF-β. Alterations of this tract of A's in the TGF-βRII gene were subsequently found in 100 of 111 colorectal carcinomas showing defective MMR, in which they caused translation of the TGF-βRII mRNA to yield polypeptides of either 129 or 161 amino acid residues (depending on how many A's were deleted) rather than the 565 amino acid residues in the wild-type receptor. (From C. Lengauer, K.W. Kinzler and B. Vogelstein, *Nature* 396:643–649, 1998.)

been found to play equally important roles in DNA repair, are rarely if ever transmitted as mutant alleles in the human germ line. Similar to the genetics of most tumor suppressor genes, patients inherit one defective allele of an MMR gene and the genomes in any tumor cells that arise almost always undergo a loss of heterozygosity (LOH) that results in the discarding of the surviving wild-type gene copy.

The resulting inability to properly detect and repair sequence mismatches leads to, among other consequences, high rates of mutations in genes that have microsatellite repeats nested in their sequences (see Figure 12.7). A dramatic and early illustration of the consequences of this repair defect came from study of a group of 11 colorectal cancer cell lines that showed microsatellite instability. In nine of these cells lines, the gene encoding the type II TGF-β receptor (TGF-βRII) was found to be mutant. More specifically, the wild-type reading frame of this gene carries a stretch of ten A's in a row (**Figure 12.27**). However, in these nine tumor cell lines, the TGF-βRII gene was found to have lost one or two A's of the normally present **homopolymeric** stretch of ten A's. These sequence changes forced the coding sequence of the TGF-βRII gene out of its normal reading frame and resulted in nonfunctional TGF-βRII proteins.

We can imagine that once tumor cell precursors no longer express functional TGF-βRII, they can escape the growth-inhibitory effects of this anti-mitogenic factor (see Section 8.4)—a highly advantageous trait if it is acquired early in tumor progression by epithelial cells. In a subsequent study of a series of 110 colon carcinomas exhibiting microsatellite instability, 100 were found to carry mutant, defective alleles of the TGF-βRII gene, with almost all mutant alleles being present in homozygous configuration. Hence, once one of the receptor-encoding genes suffers an inactivating mutation, the surviving wild-type allele is discarded through loss of heterozygosity.

Later, yet other genes were found to have suffered similar mutations in mismatch repair–defective cancer cells (**Table 12.2**). In the great majority of these cases, the MMR defect and resulting mutant alleles were discovered in sporadic (rather than familial) cancers. These observations point to the fact that in nonfamilial tumors, MMR genes, like tumor suppressor genes, can be rendered defective either by somatic mutation or by promoter methylation and resulting transcriptional silencing (see Section 7.8). In fact, the second mechanism is responsible for the lion's share of defective MMR in these tumors: about 15% of sporadic gastric, colorectal, and endometrial tumors show defective MMR, and in almost all of these, the observed microsatellite instability can be traced to the methylation and resulting silencing of the *MLH1* gene. Interestingly, in the histologically normal endometrial tissue adjacent to tumors with defective MMR, the *MLH1* gene is often found to be methylated, suggesting that this methylation is one of the earliest events of tumor progression in this tissue (**Figure 12.28**).

Table 12.2 Genes and proteins that have been inactivated in human cancer cell genomes because of mismatch repair defects

Gene	Function of encoded protein	Wild-type coding sequence	Colon	Stomach	Endometrium
ACTRII	GF receptor	A_8	X		
AIM2	interferon-inducible	A_{10}	X		
APAF1	pro-apoptotic factor	A_8	X	X	
AXIN-2	Wnt signaling	A_6, G_7, C_6	X		
BAX	pro-apoptotic factor	G_8	X	X	X
BCL-10	pro-apoptotic factor	A_8	X	X	X
BLM	DNA damage response	A_9	X	X	X
Caspase-5	pro-apoptotic factor	A_{10}	X	X	X
CDX2	homeobox TF	G_7	X		
CHK1	DNA damage response	A_9	X		X
FAS	pro-apoptotic factor	T_7	X		X
GRB-14	signal transduction	A_9	X	X	
hG4-1	cell cycle	A_8	X		
IGRIIR	decoy GF receptor	G_8	X	X	X
KIAA0977	unknown	T_9	X		
MLH3	MMR	A_9	X		X
MSH3	MMR	A_8	X	X	X
MSH6	MMR	C_8	X	X	X
NADH-UOB	electron transport	T_9	X		
OGT	glycosylation	T_{10}	X		
PTEN	pro-apoptotic	A_6	X		X
RAD50	DNA damage response	A_9	X	X	
RHAMM	cell motility	A_9	X		
RIZ	pro-apoptotic factor	A_8, A_9	X	X	X
SEC63	protein translocation into endoplasmic reticulum	A_9, A_{10}	X		
SLC23A1	transporter	C_9	X		
TCF-4	transcription factor	A_{10}	X	X	X
TGF-βRII	TGF-β receptor	A_{10}	X	X	X
WISP-3	growth factor	A_9	X		

From A. Duval and R. Hamelin, *Cancer Res.* 62:2447–2454, 2002.

Abbreviations: GF, growth factor; MMR, mismatch repair; TF, transcription factor.

In addition to losing mismatch repair function, cells that have lost MLH1 or MSH2 expression also do not recognize the damage inflicted by alkylating mutagens that would normally activate a G_2/M cell cycle checkpoint or induce apoptosis; such cells continue to advance into G_2/M and succeed in avoiding apoptosis following exposure

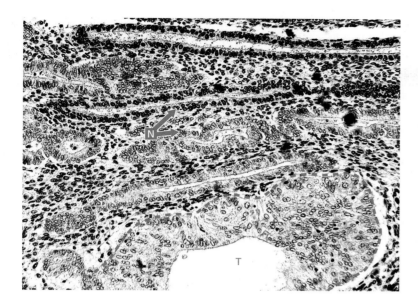

Figure 12.28 Suppression of MLH1 activity in endometrial tissue Anti-hMLH1 antibody coupled to a peroxidase enzyme generates a *dark brown* spot wherever it binds MLH1 protein. Here, the normal endometrial tissue of an endometrial cancer patient shows areas of intense, dark brown staining, indicating high hMLH1 expression (N, *upper pink arrow*), and areas of weak staining (N, *lower pink arrow*) where cell nuclei have been stained *light blue* with DAPI (a DNA-specific stain) but little MLH1 staining is seen. The endometrial carcinoma tissue (*below, dashed pink line,* T) is virtually devoid of hMLH1 staining and exhibits only light staining of cell nuclei by DAPI. Molecular analysis (*not shown*) indicated that the promoter of the *hMLH1* gene in the carcinoma cells was strongly methylated. The fact that hMLH1 expression was reduced or absent in some of the histologically normal tissue adjacent to the tumor indicates that loss of hMLH1 expression occurred relatively early in tumor progression and preceded the histopathological alterations that led to the formation of this carcinoma. (From T. Kanaya et al., *Oncogene* 22:2352–2360, 2003.)

to these DNA-damaging agents. Moreover, mutant versions of MSH2 can be created that selectively inactivate its MMR function without affecting its ability to trigger apoptosis in response to certain types of DNA damage. This suggests that in this particular MMR protein, distinct domains are involved in detecting damaged DNA, repairing this damage, and emitting alarm signals, including those leading to apoptosis.

The repeated observation of methylated mismatch repair genes provides direct evidence that a somatically acquired (that is, non-inherited) defect in a DNA repair function confers replicative advantage on evolving, premalignant cells during the course of tumor progression. In many of the tumor cell genomes showing microsatellite instability, there are hundreds and likely thousands of genes that are concomitantly mutated; the genes shown in Table 12.2 represent only a small proportion of this group.

The data in Table 12.2 are biased by the fact that only an arbitrary set of genes was examined, and only the sequences associated with homopolymeric microsatellites within these genes were sequenced. Still, this list is most interesting. It shows that the *BAX* gene, which encodes an important pro-apoptotic protein (see Section 9.13), can be silenced through mutations provoked by a stretch of eight G's in its normal reading frame and an MMR defect. Yet other pro-apoptotic genes have also been found to have undergone mutations directly traceable to changes in the number of bases in one of their homopolymeric sequences. Even the genes that encode MMR proteins are themselves inactivated by MMR defects!

Future research will reveal how the inactivation of some of the genes listed in Table 12.2 results in proliferative advantage for tumor cells. Still, successes in such research will leave another major question unanswered: Why are MMR defects and resulting microsatellite instability associated preferentially with carcinomas of the colon, stomach, ovary, and endometrium and much less often with tumors arising elsewhere in the body?

12.10 A variety of other DNA repair defects confer increased cancer susceptibility through poorly understood mechanisms

By far the most notorious genes associated with cancer, at least in the mind of the public, are *BRCA1* and *BRCA2*. Mutant germ-line alleles of either of these genes confer an inborn susceptibility to breast and ovarian carcinomas. For example, almost half of all identified familial breast cancers involve germ-line transmission of a mutant *BRCA1* or *BRCA2* allele; by some estimates, 70 to 80% of all familial ovarian cancers are due to mutant germ-line alleles of *BRCA1* or *BRCA2*. Stated differently, carriers of mutant germ-line *BRCA1* or *BRCA2* alleles have a 50–70% risk of developing breast cancer before the age of 70. Ovarian carcinoma risk is also high: 40–50% of *BRCA1* mutation carriers and 10–20% of *BRCA2* carriers develop disease before the age of 70. In addition, mutant *BRCA2* germ-line alleles have been associated in males with a >7-fold elevated risk of developing prostate cancer before the age of 65. Somatic loss of *BRCA1* function also appears to be important: 20–30% of sporadic ovarian carcinomas (that is, arising in women carrying wild-type *BRCA1* and *BRCA2* germ-line alleles) and almost as many sporadic breast cancers exhibit loss of function of one of these genes due to promoter methylation (see Section 7.8).

When these two genes were first discovered, it seemed that they should be included among the tumor suppressor genes, which are known to be involved in regulating the dynamics of cell proliferation, survival, and differentiation (see Chapter 7). But diverse lines of evidence built an increasingly persuasive case that these two genes are actually involved in the maintenance of genomic integrity, and that their products therefore should be considered "caretakers" (DNA repair proteins) rather than "gate-keepers" (tumor suppressor proteins). By now, we realize that these assignments are simplistic, as more detailed examinations of the functions of the BRCA1 and BRCA2 proteins implicate them in a diverse array of nuclear functions (see below).

The case for the participation of BRCA1 and BRCA2 in genomic maintenance can be argued using guilt-by-association evidence. The BRCA1 and BRCA2 proteins are found in large physical complexes with one another and with a large number of other proteins in the cell nucleus. These massive complexes carry, among other components, both the RAD50/Mre11 and the RAD51 proteins—homologs of two proteins in yeast that were initially discovered because of the important roles they play in repairing DNA breaks caused by ionizing radiation (that is, X-rays). Mismatch repair proteins have been found in these complexes as well.

Quite dramatically, treatment of cells with hydroxyurea, which results in a stalling of replication forks during S phase, causes BRCA1 molecules to cluster at these sites in the nucleus. Many of these stalled forks are thought to be sites of dsDNA breakage, caused by accidental breaks in the still-unreplicated single-stranded DNA at the forks (see Figure 12.8); the breaks are usually fixed by a mechanism that is variously termed homologous recombination–mediated repair or homology-directed repair (abbreviated HR or HDR). As visualized by immunofluorescent microscopy, BRCA1 molecules are normally distributed in a large number of tiny dots throughout the nucleus; hydroxyurea causes the BRCA1 molecules to leave these dots and flock together in a far smaller number of large, discrete spots, in which the proliferating-cell nuclear antigen (PCNA)—known to be localized to replication forks—is also found (**Figure 12.29A**). These spots have also been found to contain a number of other known DNA repair proteins, including Rad50 and Rad51. The BRCA2 protein is also found in these spots, providing additional presumptive evidence of its collaboration in DNA repair processes. Moreover, when dsDNA breaks are intentionally created in discrete areas within cell nuclei using a narrow laser beam, the BRCA1 protein co-localizes in these areas together with γ-H2AX, a phosphorylated histone that is present in the chromatin flanking sites of dsDNA damage (see Figure 12.29B; see also Sidebar 10.1). Altogether, 20 distinct cellular proteins, most known to be involved in DNA repair, have been found to be recruited to these areas of damage, and the list is growing.

Mice that have been deprived genetically of all BRCA1 function die during early embryogenesis, but mutant germ-line alleles of BRCA2 that cause only partial loss of function result in susceptibility to lymphoid malignancies and unusual chromosomal aberrations. These aberrations have structures that suggest high rates of illegitimate recombination, that is, recombination events (or fusions) between two chromosomal arms that are nonhomologous (**Figure 12.30**). Such chromosomal structures result characteristically from improper repair of dsDNA breaks, many of which may arise accidentally at replication forks during a typical S phase of the cell cycle. In addition, BRCA2 deficiency results in deregulation of centrosome number and therefore of the mitotic spindle, generating aneuploidy, and cells that have reduced levels of BRCA2 also exhibit prolonged cytokinesis at the end of M phase. Yet other indications of

(A)

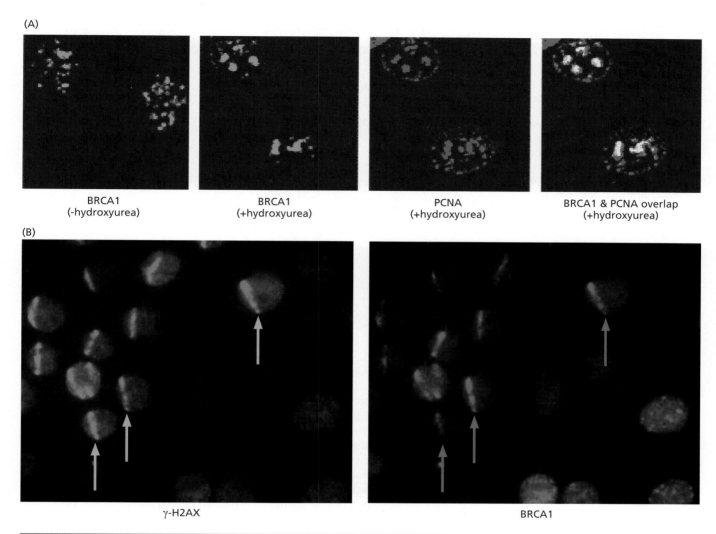

BRCA1
(-hydroxyurea)

BRCA1
(+hydroxyurea)

PCNA
(+hydroxyurea)

BRCA1 & PCNA overlap
(+hydroxyurea)

(B)

γ-H2AX

BRCA1

Figure 12.29 BRCA1 and the response to DNA damage (A) The BRCA1 protein, which can be detected by fluorescence-labeled antibodies, is normally found during S phase in numerous discrete, small dots throughout the nucleus (*green, 1st panel*). However, when cells in S phase are treated with hydroxyurea, which stalls replication forks, the BRCA1 protein molecules leave these dots and congregate in a small number of quite large spots (*green, 2nd panel*). A similar relocalization pattern can be observed with the proliferating-cell nuclear antigen (PCNA), which is known to be associated with replication forks (*red, 3rd panel*). Substantial co-localization of the BRCA1 and PCNA is indicated by the yellow spots (*4th panel*); i.e., *red + green → yellow*. These stalled replication forks often are sites of double-strand (ds) DNA breaks caused by the accidental breakage of the still-unreplicated (and thus fragile) single-strand DNA (see Figure 12.8); these observations suggest that BRCA1 is recruited to sites of dsDNA breaks. (B) A 355-nm UV laser was used to paint narrow stripes across individual nuclei, which were then analyzed by immunostaining with antibodies reactive with either γ-H2AX (a phosphorylated histone that is known to localize to chromatin flanking dsDNA breaks; *green*) or BRCA1 (*red*). This co-localization indicates that BRCA1 is attracted to areas of dsDNA breaks. (A, from R. Scully et al., *Cell* 90:425–435, 1997. B, from R.A. Greenberg et al., *Genes Dev.* 20:34–46, 2006.)

defective dsDNA repair in *BRCA1* and *BRCA2* mutant cells come from experiments that test the ability of cultured cells to recover from double-strand breaks (DSBs) introduced into their chromosomal DNA by X-rays. Cultured cells lacking the bulk of BRCA1 function show greatly increased sensitivity to killing by X-rays and by chemotherapeutic drugs, such as cisplatin, that generate covalent inter-strand cross-links in the DNA.

(Interestingly, when *BRCA1*- or *BRCA2*-mutant breast cancers that initially respond to cisplatin therapy develop resistance to this drug, they often do so by back-mutating their mutant *BRCA1/2* alleles to sequences that once again encode functional versions of these proteins. This, in itself, demonstrates the central role that these proteins play in determining sensitivity or resistance to cross-linking chemotherapeutic agents.)

Figure 12.30 Karyotypic alterations due to partial loss of BRCA2 function Various karyotypic abnormalities *(arrowheads)* have been observed in cultured fibroblasts prepared from a mouse embryo that was homozygous for an allele encoding a truncated Brca2 protein. These include fusions between chromosomal arms, resulting in chromosomal translocations that often manifest aberrant chromatid pairings at the metaphase of mitosis. The fusions are often caused by unrepaired or improperly repaired dsDNA breaks. When the hybrid chromatids resulting from such fusions pair with unaffected sister chromatids, structures such as those seen here are observed (panel I). Among the aberrations resulting from dsDNA breaks are chromatid breaks (ctb, panel II), triradial chromosomes (tr, panel III), and quadriradial chromosomes (qr, panel IV). Chromatid pairings like these are rarely observed in wild-type cells. (From K.J. Patel et al., *Mol. Cell* 1:347–357, 1998.)

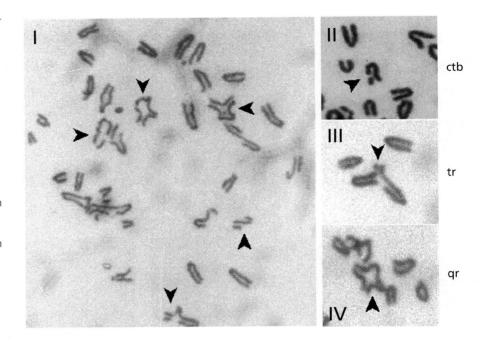

The repair of a dsDNA break in one chromatid often depends on the ability of the repair apparatus to consult the undamaged, homologous DNA sequences in a sister chromatid and to use those sequences to instruct the repair apparatus on how to reconstruct the broken double helix (**Figure 12.31**; see Supplementary Sidebar 12.7 for a more detailed molecular description). Thus, such homology-directed repair (HR) occurs largely during the late S and the G$_2$ phases of the cell cycle, when the double helix in a sister chromatid can provide the sequence information for repairing the damaged chromatid. (Recall that during S phase, DNA replication results in the production of two identical chromatids that remain associated as part of a common chromosome until they are separated during the next mitosis.) HR is also used if inter-strand covalent cross-links within a double helix should arise.

All types of HR are compromised in cells lacking either BRCA1 or BRCA2 function. This may be explained, in part, by the behavior of the RAD51 protein, with which BRCA1's partner, BRCA2, associates directly. RAD51 is known to bind single-strand DNA molecules, enabling them to invade (and thereby unwind) homologous double-strand helices, a process essential to initiating HR (see Figure 12.31). (The fact that BRCA2 has eight "BRC domains," each of which can, in principle, bind a RAD51 molecule (see Figure 15.33), suggests that BRCA2 assembles strings of RAD51 molecules to bind coordinately to an ssDNA strand.) In the absence of BRCA1 or BRCA2, RAD51 may not be properly recruited to sites of dsDNA breaks, and the subsequent steps of HR may not be able to occur correctly. The triradial and quadriradial chromosomes that are encountered in the metaphase of *Brca2*-mutant cells (see Figure 12.30) are manifestations of the inability of the repair apparatus in these cells to exploit HR to fix dsDNA breaks.

Homology-directed repair is also defective in patients suffering from Nijmegen breakage syndrome (see Table 12.1). Their cells, which lack the Nbs1 protein (nibrin; see Figure 12.33A), fail to execute the initial steps of HR and resort to fusing two dsDNA ends via the process termed **nonhomologous end joining** (NHEJ; **Figure 12.32**).

NHEJ is inevitably an error-prone process, simply because the alignment between the two DNA segments being fused is not informed by the wild-type DNA sequences present in a sister chromatid. Consequently, the resulting end-to-end fusions generate mutant sequences at the site of joining, resulting in the high rates of hematopoietic malignancies in patients suffering from Nijmegen breakage syndrome. Mice that are defective for one or another component of the NHEJ machinery and also lack p53 function develop lymphomas at extremely high rates.

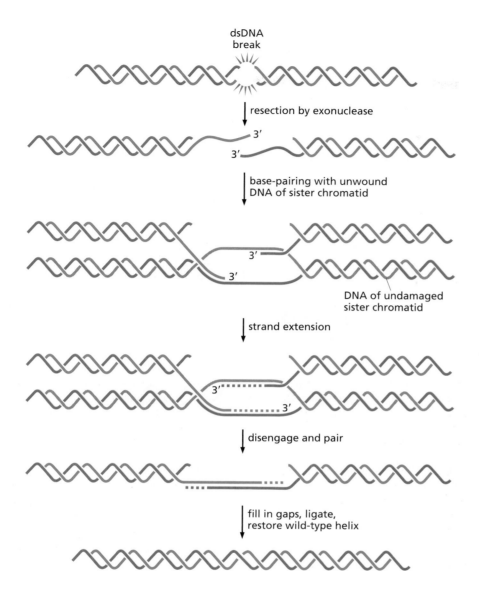

dsDNA
break

resection by exonuclease

3′

3′

base-pairing with unwound
DNA of sister chromatid

3′

3′

DNA of undamaged
sister chromatid

strand extension

3′

3′

disengage and pair

fill in gaps, ligate,
restore wild-type helix

Figure 12.31 Homology-directed repair The repair of dsDNA breaks during the late S phase and G$_2$ phase of the cell cycle often depends on the ability of the repair apparatus to consult the sequences in the undamaged sister chromatid that was formed, together with the damaged chromatid, during the most recent S phase. Such homology-directed repair (HR) begins (*top*) with the *resection* (removal) by an exonuclease of one of the two DNA strands at each of the ends formed by a dsDNA break. Each of the resulting ssDNA strands (*blue, red*) then invades the undamaged sister chromatid, whose double helix (*green, gray*) has been unwound by the repair apparatus in order to accommodate the pairing of the invading ssDNA strands with complementary sequences in the undamaged sister chromatid. The ssDNA strands from the damaged chromatid are then elongated in a 5′-to-3′ direction by a DNA polymerase, using the strands of the sister chromatid's DNA as templates. Thereafter, the extended ssDNA strands are released from the sister chromatid and caused to pair with one another, allowing further elongation by a DNA polymerase and a ligase, which together reconstruct a double helix possessing wild-type DNA sequences. Included among the DNA repair proteins known or thought to facilitate these complex steps of HR are RAD51, BRCA1, and BRCA2.

Interestingly, NHEJ occurs largely in the G$_1$ phase of the cell cycle, when sister chromatids are not available to allow homology-directed repair. NHEJ is virtually unique among the DNA repair processes, because it plays a role in a normal, physiologic function unrelated to repairing DNA damage, namely, the normal process of gene rearrangement that leads to the formation of functional antibodies and T-cell receptors. For example, the formation of the DNA sequences encoding antigen-binding sites depends on the rearrangement and fusion of chromosomal V, D, and J segments (see Supplementary Sidebar 15.1). In the absence of the full complement of repair proteins needed for NHEJ, such DNA segment fusions cannot occur. This results in the inability to make proper immunoglobulin (antibody) molecules and T-cell receptors, compromising both the humoral and cellular arms of the immune system (see Section 15.1) and creating the syndrome of severe combined immunodeficiency (SCID). Similarly, NHEJ is needed for most types of **class switching**; this process normally enables fused VDJ segments, which encode the antigen-binding portions of immunoglobulins, to join with alternative constant-region immunoglobulin gene segments to generate various classes of antibody molecules (see Section 15.2).

It is still not clear precisely how BRCA1 and -2 contribute to the maintenance of normal chromosomal structure and thereby ward off cancer. The fact that BRCA1 and BRCA2 can bind to so many distinct nuclear proteins, many involved in DNA repair (**Figure 12.33A**), indicates that they act, at least in part, as molecular scaffolding that helps to

Figure 12.32 Nonhomologous end joining NHEJ is used to restore a DNA double helix following a double-strand break when the nucleotide sequences from a sister chromatid are not available to instruct the repair apparatus how these ends should be properly joined (see Figure 12.31). In NHEJ, the resection of single strands from both broken ends results in ssDNA overhangs that can then be joined to one another, possibly by a limited degree of base pairing between them; three base pairs are shown here. The subsequent filling in of single-strand gaps and the ligation of any remaining ssDNA breaks results in the reconstruction of a double helix that lacks some of the base pairs that were present in the original undamaged DNA helix. (Adapted from M.R. Lieber et al., *DNA Repair (Amst)* 3:817–826, 2004.)

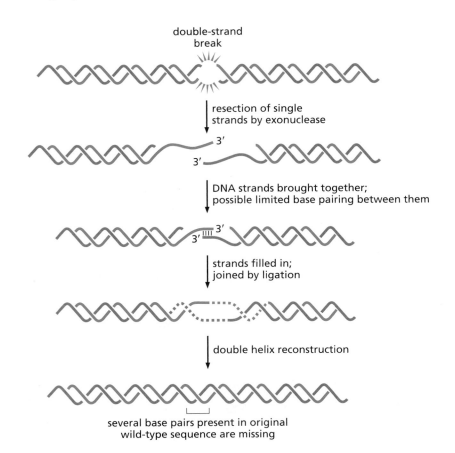

assemble large complexes of these other proteins and coordinate their actions. Once assembled, these various repair proteins presumably collaborate in fixing lesions, largely double-strand breaks, in the DNA. Moreover, as Figure 12.33B illustrates, loss of BRCA1 can cripple certain cell cycle checkpoint controls that normally respond to damaged DNA.

We do not know why inheritance of mutant alleles of the *BRCA1* and *BRCA2* genes leads preferentially to cancers of the breast and ovary, or why somatic mutation of *BRCA2* is occasionally associated with prostate and colon carcinomas. In addition, the **penetrance** of *BRCA1*- and *BRCA2*-mutant germ-line alleles (that is, the degree to which each allele exerts an observable effect on phenotype) has also been difficult to quantify (**Sidebar 12.7**).

As BRCA1 function is studied in ever-increasing detail, its mechanisms of action become increasingly confusing, if only because additional distinct biochemical functions are ascribed this very large protein and its physically associated partners. For example, BRCA1 associates with a number of transcriptional regulators, localizes to centrosomes during mitosis, and has been found at the outer edges of cultured cells growing in monolayer, where it helps to regulate cell motility. It is also clear that female cells lacking BRCA1 function are unable to properly inactivate one of the two X chromosomes in their somatic cells (see Figure 2.20). We still do not understand how these other functions intersect with BRCA1's DNA repair functions and contribute to the strong tendency of mutant germ-line *BRCA1* (and *BRCA2*) alleles to generate almost exclusively female cancers.

In the eyes of some investigators, the contributions of BRCA1 loss to cancer development may be overshadowed by yet another of its activities unrelated to DNA repair: BRCA1 carries in one of its domains a function that leads to monoubiquitylation of histone H2A. (BRCA1 acts as an adaptor protein and thus as an E3 subunit of a ubiquitin ligase complex, in which an unknown E2 subunit performs the actual enzymatic modification of H2A; see Supplementary Sidebar 7.5.) This H2A monoubiquitylation results in the formation of heterochromatin, notably the *constitutive heterochromatin*

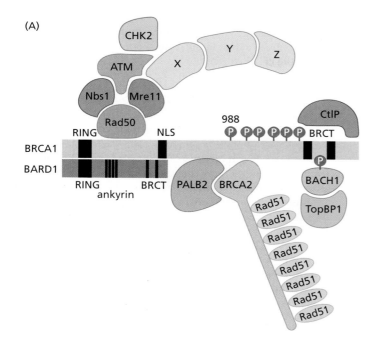

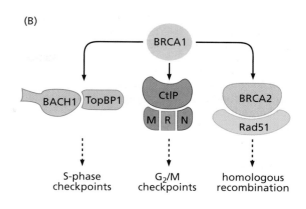

Figure 12.33 BRCA1, BRCA2, and their partners (A) The BRCA1 and BRCA2 proteins act, at least in part, as scaffolds to assemble a cohort of other DNA repair proteins into large physical complexes. Once assembled, these multiprotein complexes aid in the repair of dsDNA breaks, usually via homology-directed repair (HR). For example, one exon of the *BRCA2* gene encodes eight copies of a "BRC domain" (*not shown*); these aid in the recruitment of multiple RAD51 molecules, which form filaments and coat ssDNA strands as part of the HR process illustrated in Figure 12.31. The MRN complex, composed of MRE11, Rad50, and Nbs1, appears able to recognize the end created by a dsDNA break and to activate ATM kinase function in response. BRCA1 acts in different situations and locales (see panel B) as a scaffold for a variety of alternative partners, not all of which are concomitantly bound to BRCA1; for this reason, this image depicts a single complex that never exists in a living cell. Moreover, this image does not capture the full complexity of BRCA1-associated proteins. For example, in the hydroxyurea-induced nuclear spots containing stalled replication forks (see Figure 12.29A), additional proteins have been discovered beyond those depicted, including MDC1, RPA32, γ-H2AX, RNF8, RNF168, 53BP1, FANCD2, and FANC1. (B) The loss of different partners of BRCA1 specifically affects different checkpoint controls in the cell cycle, in addition to compromising the processes of homologous recombination and homology-directed repair (HR). This illustrates the central role that BRCA1 plays as a scaffolding for a diverse array of proteins mediating a variety of processes involving DNA function. The checkpoint controls ensure that a cell halts before either S or M phase of the cell cycle if its genome carries significant unrepaired lesions, such as DSBs (see Figure 8.4). (A and B, courtesy of R.A. Greenberg and D.M. Livingston.)

that is shut down in virtually all somatic cell types. In the absence of this heterochromatization, a variety of repeat sequences that are normally shut down now become expressed, leading to widespread genetic destabilization as a downstream effect. The recent discovery of this BRCA1 function reveals how poorly we understand why and how BRCA1 loss leads to breast, ovarian, and prostate carcinoma development.

12.11 The karyotype of cancer cells is often changed through alterations in chromosome structure

Long before the subtleties of DNA damage and its repair were recognized, aberrant karyotypes were known to be present in cancer cells, indeed, for almost a century. The triradial and quadriradial metaphase chromosomes seen in cells lacking BRCA1 or BRCA2 function are examples of these aberrations (see Figure 12.30). Stepping back for a moment from these particular aberrations, we can recognize that two distinct classes of karyotypic abnormalities can be seen in cancer cells: changes in the *structures* of individual chromosomes, and changes in chromosome *number* that have no effect on chromosome structure.

One frequent deviation from the normal diploid karyotype involves an increase or decrease in the number of specific chromosomes. On occasion, through various

Sidebar 12.7 Ethnic groups with unusual histories are valuable for studying the penetrance of mutant *BRCA1* and *BRCA2* germ-line alleles The precise effects of mutant *BRCA1* and *BRCA2* germ-line alleles on cancer risk have been difficult to gauge, because both of these genes are very large and the detection of germ-line mutations in their reading frames is made correspondingly difficult. Moreover, many hundreds of mutant germ-line alleles of these genes have been cataloged in the human population, each having its own *penetrance*, that is, its relative ability to induce clinical disease in a person carrying it in the germ line. Each of these variant alleles operates in the complex genetic background of an individual, and this background also influences disease development; together these two dimensions of variability (the particular *BRCA1/2* mutation and the genetic background) make it almost impossible to gauge with any precision the disease penetrance of most of these alleles.

Many of these mutant alleles are of ancient vintage, having entered into the human gene pool hundreds, even thousands of years ago. Once present, many have been retained in the gene pool, since they usually induce disease long after reproduction has occurred and are therefore not rapidly eliminated by Darwinian selection.

The peculiar histories of certain ethnic groups often simplify these analyses. For example, Ashkenazic Jews, who derive from Central and Eastern Europe, descend from a Jewish population of several million that lived in the late Roman Empire. Genetic analyses suggest, however, that the modern population derives from a founding population of only 200 or so individuals who lived during the last days of this period. This genetic history seems to explain why this population harbors essentially only three mutant *BRCA1* and *BRCA2* germ-line alleles, all of ancient origin, which in aggregate are carried by some 2.5% of contemporary Ashkenazim. These three **"founder mutations"** contrast with the situation in the general population, in which more than 1500 distinct inherited alleles of *BRCA1* have been reported (a small portion of which may be functionally neutral polymorphisms). Each of these other alleles has been detected by arduous sequence analysis.

These facts explain why the detection of mutant germ-line alleles of the *BRCA1* and *BRCA2* genes in the Ashkenazic population is relatively simple: polymerase chain reaction (PCR)–based DNA primers can be designed that are focused specifically on the detection of only these three alleles, rather than on the countless others that have been reported to date. The most common Ashkenazic *BRCA1* allele, for example, is the *BRCA1 185delAG* allele, which, as its name indicates, involves the deletion of two nucleotides; this mutant allele was found to be carried by 4.2% of women from this population who presented with invasive breast cancer in a New York City cancer clinic. Further studies revealed that the three mutant alleles conferred comparable risks of invasive breast cancer, which exceeded 80% by the time carriers of one or another of these alleles reached the age of 80.

Provocatively, the risk of developing breast cancer by the age of 50 among carriers of the *185delAG* mutation who were born before 1940 was 24%, while this risk was 67% among those carriers born after 1940. This indicates that nongenetic factors play a major role in cancer development, even in those born with a heritable DNA repair defect. (In the case of mutant alleles of *BRCA1*, changes over the past half century in nutrition and reproductive practices may have played major roles.) In addition, this type of epidemiology reveals the power of studying the biological effects of a single mutant allele operating in the context of variable genetic background and lifestyle changes.

accidents occurring during mitosis, cancer cells may acquire **polyploid** genomes, where an additional haploid complement of chromosomes is acquired (leading to a **triploid** state) or even an extra diploid complement of chromosomes is acquired (leading to a **tetraploid** state). Alternatively, extra copies of individual chromosomes may be present, or, less commonly, a chromosome copy may be missing.

The term **aneuploidy** is usually reserved for denoting a deviation from a normal (or **euploid**) karyotype that involves changes in chromosome *number*. In recent years, however, use of the term aneuploid has occasionally been broadened to include changes in the *structures* of individual chromosomes, which are prevalent in cells of the great majority (>85%) of solid tumors; a more specific term is "chromosomal aberration," which we will use here. These two major types of karyotypic alteration arise through fundamentally different mechanisms.

Changes in chromosome number are discussed in the next section. Here, we review the mechanisms responsible for changes in chromosome structure, some of which we have already encountered at various points in this text. For example, as we saw in the previous section, unrepaired dsDNA breaks, many of which occur accidentally at DNA replication forks, are thought to be a major source of chromosomal translocations.

In addition, much earlier (see Chapter 4) we learned of a class of cancer-associated chromosomal alterations as part of a discussion of the mechanisms leading to the creation of the *myc* and *bcr-abl* oncogenes. Recall the well-studied case of the translocations that fuse the *myc* proto-oncogene to promoter/enhancer sequences deriving from one of three alternative immunoglobulin genes (see Section 4.5). In these and other lymphomas, it is likely that the complex machinery dedicated to rearranging the immunoglobulin and T-cell receptor (TCR) genes misfires. Instead of rearranging the immunoglobulin or TCR gene sequences, this machinery inadvertently catalyzes inappropriate interchromosomal recombination events that join the immunoglobulin genes promiscuously with sequences scattered throughout the genome, the *myc* gene being only one of them. Those rare translocations that happen to involve the *myc* proto-oncogene and deregulate its transcription seem to confer special proliferative advantage on cells, resulting in the appearance of cell clones and ultimately in lymphomas that carry these very characteristic karyotypic alterations.

In fact, many other highly specific translocations have been documented in a variety of hematopoietic malignancies (see, for example, Tables 4.4 and 4.5). The molecular mechanisms that lead to these various alterations in protein structure remain unclear. However, one highly attractive possibility was presented in Chapter 10, where we read about telomere collapse resulting in breakage–fusion–bridge (BFB) cycles (see Figure 10.15). These BFB cycles create large-scale aberrations in the structures of individual chromosomes, apparently striking all chromosomes with comparable frequency. On occasion, translocations may occur that provide growth advantage to the cells carrying them, which one imagines results in the clonal outgrowth of these cells.

Superficially similar structural aberrations are seen in a variety of hematopoietic tumors. Many of these are termed *recurrent* because they have been seen on several occasions in a series of independently arising human tumors. By now, hundreds of these recurrent translocations have been cataloged. Because recurrent translocations map to highly specific chromosomal sites, it would seem that the molecular mechanisms creating them are distinct from the breakage–fusion–bridge cycles described above.

Our understanding of chromosomal aberrations is continuing to change. Translocations were previously thought to be present almost uniquely in hematopoietic tumors, as discussed above. But beginning in 2000, the availability of new DNA sequencing strategies, the sequencing of the human genome, and new bioinformatics tools began to reveal chromosomal translocations, some quite common, in solid tumors. In 2005, a specific recurrent translocation, involving a gene encoding an androgen-regulated serine protease and an ETS transcription factor, was discovered; this particular translocation (*TMPR32/ERG*) is now known to be present in the tumors of ~50% of the patients bearing localized prostate carcinomas. This discovery in a commonly occurring human carcinoma triggered surveys of a variety of solid tumor genomes and a spate of additional discoveries, many of them driven by powerful new genome sequencing technologies. Among these findings is the demonstration that chromosomal translocations occur at greatly differing frequencies in various solid tumors for unknown reasons (**Figure 12.34**).

Implementation of these new technologies led in 2010 to discovery of a totally novel type of aberration in the genomes of ~25% of bone cancers and 2–3% of cancers

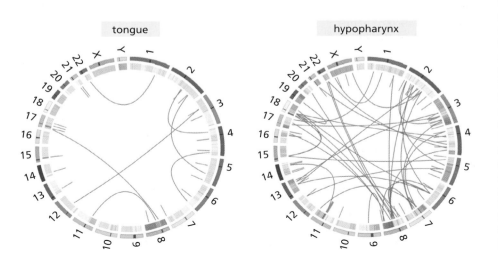

tongue   hypopharynx

Figure 12.34 Chromosomal translocations in head-and-neck carcinomas of smokers This graphic convention of depicting chromosomal translocations places the chromosomes around a circle according to their number or letter (sometimes termed a Circos plot) with fusions between previously unlinked chromosomal segments depicted by *purple* lines. Short *green* ticks pointing toward the center of the circle indicate regions of local intrachromosomal rearrangements, including translocations, small amplifications, deletions, and inversions. The genomes of two head-and-neck squamous cell carcinomas (HNSCCs) of smokers—a tongue and a hypopharyngeal carcinoma— are depicted here. None of these translocations was associated with a recurrent translocation documented in other similar tumors. (Courtesy of M. Lawrence and G. Getz.)

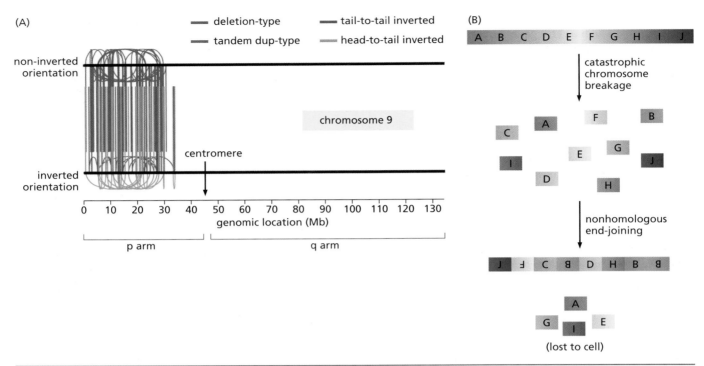

Figure 12.35 Localized firestorms of chromosomal rearrangement The development of large-scale sequencing technologies has made it possible to discover certain types of chromosomal damage that previously eluded detection. (A) In the genome of a thyroid carcinoma, a cluster of 77 distinct chromosomal rearrangements affecting the short arm of chromosome 9 (i.e., 9p) have been documented. Each of the lines indicates the direct fusion of two DNA segments that are normally located at some distance from one another in this chromosomal arm; these rearrangements involve four distinct types of changes in DNA configuration (*upper right*). (B) This scheme provides one explanation of how this focal chromosomal rearrangement, which has been termed *chromothripsis*, may have arisen. The mechanism that drives this localized shattering followed by rejoining of chromosomal fragments has not been elucidated. According to one attractive model, the formation of single-chromosome-containing micronuclei (see Figure 12.38) results in individual chromosomes undergoing extensive intrachromosomal rearrangement prior to becoming reassociated with the larger complement of chromosomes during a subsequent mitosis. (From P.J. Stephens et al., *Cell* 144:27–40, 2011.)

overall—localized firestorms of chromosomal rearrangements. In the example shown in **Figure 12.35**, a single catastrophic event seems to have shattered a limited stretch of the genome and was followed by multiple rejoining events between the resulting fragments in a variety of configurations. The mutational mechanism underlying this "**chromothripsis**" (chromosome shattering) remains obscure.

Equally obscure are the molecular mechanisms yielding the more commonly encountered translocations that are found in hematopoietic and non-hematopoietic cell types in which the enzymes involved in the rearrangement of immunoglobulin genes and T-cell receptor genes are not expressed. Sequence analyses of the DNA flanking translocation breakpoints have revealed duplications, deletions, and inversions of sequence blocs, findings that suggest but hardly prove the involvement of some type of "error-prone" DNA repair mechanism, such as the nonhomologous end joining (NHEJ) discussed earlier (see Figure 12.32).

To conclude, chromosomal translocations and chromothripsis leave us with two major mysteries: (1) Which components of the DNA repair machinery are normally on guard to prevent the formation of these aberrations? (2) How do many of these chromosomal abnormalities, once formed, contribute to cancer formation?

12.12 The karyotype of cancer cells is often changed through alterations in chromosome number

As stated earlier, some types of genetic instability affect karyotype by altering the number of individual chromosomes without affecting their structure. These changes

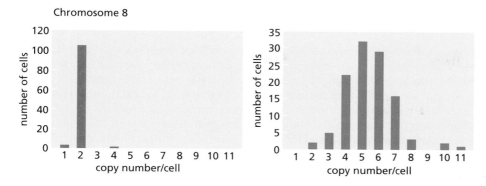

Figure 12.36 Chromosome instability in cultured cancer cells The copy number of Chromosome 8 was measured here using fluorescence *in situ* hybridization (FISH) in normal cells (*left*) and also in cultured breast cancer cells (*right*) that were found to suffer from chromosomal instability (CIN). As indicated, almost all of the normal cells had two copies of Chromosome 8, while the copy number of Chromosome 8 varied extensively in the cells afflicted with CIN. (An essentially identical distribution of chromosome copy number was observed upon study of a second, arbitrarily chosen chromosome.) This great cell-to-cell variability of chromosome number indicates that fluctuations of this number continue to occur frequently as these cancer cells are propagated in culture. (From G.A. Pihan et al., *Cancer Res.* 58:3974–3985, 1998.)

create aneuploidy. While the term "mutation" is often reserved for changes in DNA sequence (and thus includes changes in chromosome structure), we should recognize the fact that alterations in chromosome number also represent significant changes in a genome that can have equally profound effects on cell behavior and are, strictly speaking, also a type of mutation.

Changes in chromosome number are seen in the ~85% of carcinoma cells that are afflicted by the condition termed *chromosomal instability* (CIN). When CIN-positive cancer cells are removed from patients and propagated *in vitro*, the consequences of CIN become evident, for these cells continue to reshuffle their complement of chromosomes during propagation in the Petri dish. Such cancer cells were already quite aneuploid when they were removed from the patient, and their karyotypic instability seen *in vitro* is presumably only a continuation of the instability that existed *in vivo* during tumorigenesis (**Figure 12.36**).

The aneuploid karyotypes of cancer cells can be interpreted in two ways. One point of view portrays aneuploidy as a *consequence* of the general chaos that progressively envelops cancer cells as they advance toward highly malignant states. The other point of view ascribes a *causal* importance to aneuploidy, arguing that it is an essential component of tumorigenesis. Thus, some contend that most cancer cells require chromosomal instability during their development in order to scramble their genomes and arrive at chromosomal configurations that are more favorable for neoplastic growth. According to this second line of thinking, in the absence of the increased mutability associated with aneuploidy, most clones of incipient cells could never succeed in acquiring all of the genetic alterations needed to complete multi-step tumorigenesis.

An important observation that will help settle this debate has come from the study of a series of human colon and rectal carcinomas. The few tumors that exhibit microsatellite instability (MIN) show relatively little aneuploidy and virtually no chromosomal instability (CIN). Conversely, the far more numerous tumors that have CIN are not prone to show nucleotide sequence alterations that are characteristic of MIN (**Figure 12.37**). Together, these observations suggest that, at least in the case of colorectal carcinomas, tumor cells must acquire increased mutability of their genomes, and that either one or the other of these mechanisms suffices to provide such mutability. This conclusion begins to persuade one that CIN, like MIN, is an effective mechanism for remodeling the cellular genome in a way that favors evolution toward neoplasia. (Of

Figure 12.37 Chromosomal instability vs. gene mutation The presence of chromosome instability (CIN) can be gauged by measuring the loss of alleles from chromosomal arms. (A) In the colorectal carcinomas studied here, analyses of a large number of tumors have revealed that many have lost heterozygosity (LOH; see Section 7.4) at a substantial number of chromosomal loci. On the abscissa, 0.3 allelic loss, for example, refers to tumors in which 30% of the loci that were previously heterozygous, as revealed by analyses of chromosomal markers, no longer exhibit heterozygosity (*red bars*). Most of this LOH is attributable to the loss of whole chromosomes. In contrast, among the tumors afflicted with microsatellite instability (MIN; *blue bar*), the loss of alleles and hence the loss of entire chromosomes is negligible. (B) In colorectal tumor cell lines that exhibit CIN, as gauged by the loss of chromosomal markers, the rate of inactivation of the *HPRT* (hypoxanthine phosphoribosyltransferase) gene is virtually zero (*first four bars, red*). In contrast, in those that exhibit MIN, the rate of mutation of this gene is significant and is occasionally 100-fold higher than in CIN tumor cell lines (*last four bars, blue*). (A, from C. Lengauer, K.W. Kinzler and B. Vogelstein, *Nature* 396:643–649, 1998; and B. Vogelstein et al., *Science* 244:207–211, 1989. B, from C. Lengauer, K.W. Kinzler and B. Vogelstein, *Nature* 396:643–649, 1998; and J.R. Eshleman et al., *Oncogene* 10:33–37, 1995.)

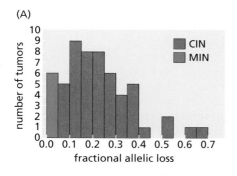

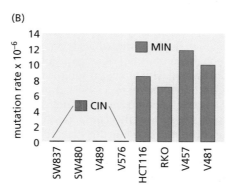

note, contrary to the depiction in Figure 12.37, CIN and MIN are not always mutually exclusive states, and some tumors exhibit both types of instability concomitantly.)

Whether this logic pertains as well to the genetic mechanisms creating hematopoietic malignancies remains unclear. Unlike carcinoma cells, which almost invariably exhibit widespread karyotypic chaos, hematopoietic tumor cells often have karyotypes that are diploid, with the exception of one or two reciprocal translocations that seem to be responsible for initiating the cancer or triggering a specific step of tumor progression (for example, the one creating the *BCR-ABL* oncogene). Therefore, chaotic karyotypes are not required for the formation of all types of human malignancies. It is highly unlikely that the small number of observable karyotypic alterations found in most hematopoietic cancer cells suffice to enable full neoplastic proliferation on their own. (In one case—that of chronic myelogenous leukemia—the acquisition of the *BCR-ABL* oncogene is often followed during blast crisis relapse by the loss of p53 function; point mutations cause this loss, and they are, of course, karyotypically invisible.) Moreover, hematopoietic tumor cells have not been reported to suffer from microsatellite instability. It therefore remains unclear which genetic mechanisms enable hematopoietic cells to acquire the entire ensemble of mutant alleles needed in order for them to proliferate as fully neoplastic cells. Indeed, we do not even know whether the formation of hematopoietic tumors requires as many genetic changes as those needed for the formation of solid tumors (see Section 11.12).

The changes in chromosome number that characterize chromosomal instability are usually (and perhaps always) the consequences of mis-segregation of chromosomes during mitosis. During the normal M phase of the cell cycle, the chromosomes line up in a plane, the *metaphase plate*, and associate with spindle fibers. The fibers together form a metaphase spindle, a bipolar structure in which each half spindle is constituted of microtubule fibers, many of which extend from the **kinetochores** on the chromosomes (the complex nucleoprotein bodies associated with the centromeric DNA of the chromosome) back to the centrosomes; the latter are responsible for organizing the entire metaphase spindle structure. When this apparatus is working properly, the spindle fibers pull sister chromatid pairs apart, so that each chromatid moves toward one of the two centrosomes. This ensures that the two daughter cells that will eventually arise after cell division receive precisely equal allotments of chromosomes (see Figure 8.3A).

This complex process of chromosome segregation is monitored by a series of checkpoint controls, which ensure initially that precisely two centrosomes and two half-spindles form; that each chromatid in a pair associates via its kinetochore with its own, distinct half-spindle; and that chromatid separation is not allowed to proceed unless and until all pairs of chromatids are properly aligned on the metaphase plate. When these checkpoint mechanisms fail to impose quality control on chromosomal segregation, both sister chromatids in a pair may be pulled to one or the other centrosome (the process of **nondisjunction**). As a consequence, one of the subsequently arising daughter cells may become haploid for this chromosome and the other triploid. Alternatively, a chromatid may fail to attach to a spindle fiber and may simply be lost from the genomes of descendant cells.

A possibly more important source of aneuploidy derives from malfunction of the complex machinery that normally ensures that each kinetochore is bound appropriately to its own set of 20–25 microtubules that form a spindle fiber; these attachments allow each pair of sister chromatids, initially linked by their paired kinetochores, to be pulled in opposite directions at anaphase. However, in many cancer cells, this control mechanism does not operate properly, and individual kinetochores become associated instead with *too many* spindle fibers. For example, **merotely** occurs when a kinetochore (belonging to a single chromatid) becomes attached simultaneously to two oppositely oriented sets of spindle fibers, which then proceed to engage in a microscopic tug-of-war; this competition is often unresolved by the end of mitosis, leaving the chromatid carrying this kinetochore stranded between the two groups of properly segregating chromatids (now called chromosomes; **Figure 12.38A**). The fate of this orphaned chromosome is unclear; it may be lost entirely or eventually associate with one or another daughter nucleus.

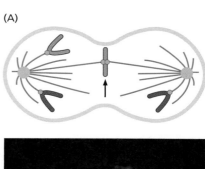

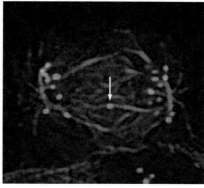

Figure 12.38 Centrosomes and the organization of the mitotic spindle Centrosomes are responsible for organizing the microtubule spindle fibers at mitosis. (A) *Above*, the kinetochore (*green*) of an individual chromatid (*arrow*) may be connected inappropriately to two opposite spindle fibers (*red*) at the same time—the condition termed merotely—causing the spindle fibers to pull in opposite directions and, quite frequently, leave the chromatid stranded between the two properly segregated populations of chromatids (*left, right*), now called chromosomes. *Below*, in the micrograph, only the spindle fibers (*red*) and kinetochores (*green*) are visualized here by immunofluorescence. The isolated chromosome (*green dot, arrow*) in the middle, revealed only by its kinetochore, may not be incorporated into either of the soon-to-be-formed daughter nuclei (assembled around the kinetochore clusters, *left, right*) and may therefore be lost from descendant cells. More likely, however, it may become associated with one of the daughter nuclei. Association with the inappropriate daughter nucleus will result in aneuploidy. (B) In these immortalized but nonmalignant interphase cells (*left*), the presence of a single centrosome is discernible in the cytoplasm through use of an antibody that detects pericentrin, a centrosome-associated protein (*red*). This centrosome is normally duplicated at the G_1/S transition to generate the two centrosomes found at the poles of the mitotic spindle. In contrast, during interphase of human breast cancer cells (*right*), multiple centrosomes are often observed. These will frequently create multipolar spindles (see Figure 12.39) when such cells enter mitosis. (C) The pair of centrioles that forms the core of each centrosome can be best seen using transmission electron microscopy (TEM), in this case of a human colon carcinoma cell. Four centrioles are seen here in cross section (*small arrows*), while a side view of a fifth (*large arrow*) is apparent, together indicating a deregulation of centriole number. The nucleus (N) is seen above. (A, above, adapted from D. Cimini, *Biochim. Biophys. Acta* 1786:32–40, 2008; below, courtesy of D. Cimini and from W.T. Silkworth et al., *PloS ONE* 4:e6564, 2009. B, from G.A. Pihan et al., *Cancer Res.* 58:3974–3985, 1998. C, courtesy of M.J. Difilippantonio and T. Ried.)

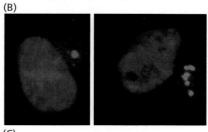

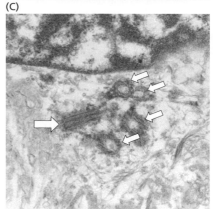

In normal cells, merotely is usually cured by the time the cells advance through metaphase into anaphase. Such curing is critical to normal cells, since merotelic chromosomes fail to trigger the **spindle assembly checkpoint** (SAC), which is designed to halt progress into anaphase in the event that spindle fibers are not properly attached to the kinetochores of all chromatids. (In normal cells, despite these elaborate corrective mechanisms, by some estimates about 1% of mitoses result in some type of mis-segregation; in cancer cells, in stark contrast, as many as one-half of mitoses may result in some form of mis-segregation.) Cancer cells seem to be more tolerant of merotely, which therefore can contribute to their aneuploid karyotypes. Indeed, cancer cells often exhibit defects in one or more of the fourteen distinct proteins that have been implicated in orchestrating the SAC. And, mice that have been genetically engineered to express subnormal levels of certain SAC proteins exhibit elevated levels of spontaneous tumor formation.

More widespread karyotypic chaos may occur if the spindles themselves are not properly assembled. Aberrant mitoses, which result from inappropriate spindle organization, were noticed as early as 1890 and, in retrospect, represented the first clue that cancer cells are genetically abnormal. In normal **interphase** cells, a single centrosome can be visualized in the cytoplasm (see Figure 12.38B); during mitosis, two centrosomes are arrayed at opposite poles within the cell. Cancer cells, however, often show marked defects in this organization, including multiple centrosomes at interphase (see Figure 12.38B and C). The result may be mitotic spindles that have multiple poles rather than the two seen in normal cells (**Figure 12.39A and B**). Often, the **supernumerary** (extra) centrosomes coalesce into the normal set of two as cells proceed through mitosis; however, the spindle fibers that they initially generated may end up forming dysfunctional spindle–kinetochore attachments, such as that described in Figure 12.38A. Alternatively, the extra centrosomes may persist, causing

Figure 12.39 Multipolar mitotic apparatuses In a normal mitotic metaphase of cultured cells, the spindle fibers, which are composed of microtubules, reach from the two centrosomes at the mitotic poles to the kinetochores—the multiprotein complexes that are associated with the centromeric chromosomal DNA. (A) In this micrograph of $p53^{-/-}$ mouse embryo fibroblasts, the formation of multiple centrosomes has resulted in a triradial mitotic spindle array. The spindle fibers have been immunostained *red* with an antibody reactive with α- and β-tubulin, while the chromosomal clusters have been stained *blue* with DAPI, a DNA dye. The five centrosomes (immunostained with an antibody reactive with γ-tubulin) are seen as small *light yellow* or *light green* spots. Observations like these implicate p53 in the regulation of centrosome number. (B) The four centrosomes in this quadriradial mitotic spindle array in a prometaphase human fibroblast have been stained *yellow–green*; the microtubules forming the spindle fibers, *red*; and the chromosomes, *white*. (C) Here, just after cytokinesis, which follows directly after telophase (see Figure 8.3) three daughter cells of a human HeLa cervical carcinoma cell line have formed from a mother cell that ostensibly assembled a triradial spindle during mitosis; these daughters are trying to divide the maternal dowry of chromosomes in three ways rather than the usual two. Microtubules are stained *yellow*; DNA is stained *purple*. The abscission of these cells at the end of cytokinesis clearly did not proceed properly, leaving the cells connected to one another via a chromosomal fragment (*purple dot, arrow*). (A, from P. Tarapore and K. Fukasawa, *Oncogene* 21:6234–6240, 2002. B, from N.J. Ganem, S.A. Godinho and D. Pellman, *Nature* 460:278–282, 2009. C, from K.G. Murti, *BioTechniques,* Oct. 2004, cover.)

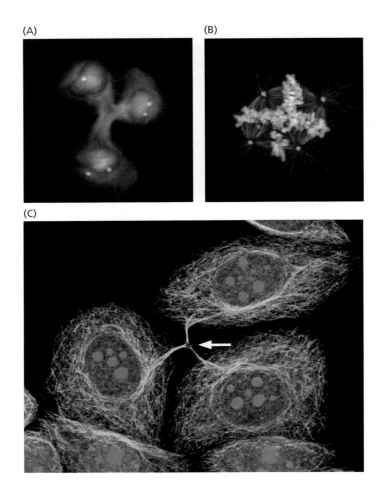

the chromosomal array to be divided among three or more daughter cells (see Figure 12.39C). The resulting mis-segregation of chromosomes into daughter cells may lead to wild fluctuations in chromosome number and overall karyotype. In one survey of a series of 87 different tumors, 81 of these showed abnormalities in centrosome number or in the microstructure of individual centrosomes; such defects were never encountered in normal cells used as controls in this study.

It seems that once the complex apparatus designed to ensure proper chromosomal segregation has been damaged, such damage is irreversible. For example, as was seen in Figure 12.36, the enormous cell-to-cell variability in the number of Chromosome 8 copies in certain breast cancer cells indicates that chromosome instability (CIN) persisted in these cells long after tumor progression had reached completion. In this respect, CIN differs from the breakage–fusion–bridge (BFB) cycles (see Section 10.4), which seem to plague the genomes of cancer cells for a limited window of time during tumor progression and then cease once these cells succeed in acquiring telomerase and thereby stabilize their karyotypes.

In recent years, some of the molecular defects that contribute to various types of chromosomal instability have come to light. Not surprisingly, the duplication of centrosomes is closely coordinated with cell cycle advance; it seems to occur at or near the G_1/S transition. Thus, an increasing body of evidence indicates that centrosome duplication is coordinated in some way by the cyclin E– and A–containing cyclin-dependent kinase (CDK) complexes, which are master regulators of entrance into and progression through S phase (see Section 8.3). Provocatively, primary human cells (that is, those that have been in culture for only a short period of time) in which the human papillomavirus (HPV) E7 oncoprotein is expressed show a deregulation of centrosome number, often leading to supernumerary centrosomes beyond the two that are normally present in cells poised to enter mitosis.

Recall that the E7 oncoprotein binds to cellular pRb (the retinoblastoma protein), inactivates it functionally, and accelerates its degradation (see Section 8.5). Significantly, a mutant form of the HPV E7 oncoprotein that fails to bind to pRb also fails to induce the appearance of extra centrosomes. Together, these strands of evidence suggest that one of the consequences of loss of pRb function is a deregulation of centrosome duplication. Since the centrosomes play a central role in organizing the mitotic spindles, this deregulation may soon lead to dramatic fluctuations in chromosome number (see Figure 12.39). Indeed, precisely such chromosomal instability is observed in lines of human cervical carcinoma cells, almost all of which express the early region of the HPV genome, which carries the genes encoding the viral E6 and E7 oncoproteins.

An equal contribution to aneuploidy may be made by the HPV E6 protein (see Sidebar 9.8) and other agents that disrupt p53 function, since cells lacking the services of this important tumor suppressor protein seem particularly susceptible to acquiring aneuploid genomes. Thus, the HPV E7 oncoprotein destabilizes centrosome number, while the E6 oncoprotein, through its actions on p53, causes cells to tolerate any centrosome abnormalities that may eventually arise. And in a large variety of other human cancers that are not caused by viral infection, cells that have lost p53 function through other mechanisms also show relatively high rates of instability in their chromosome number. The precise signaling connections between p53 function, mitotic checkpoint controls, and the monitoring of chromosomal number remain obscure.

Inactivation of pRb regulation may confer immediate proliferative advantage on an incipient, HPV-infected cancer cell by inactivating the control that governs passage through the R-point transition (see Section 8.2). In the longer term, additional advantages may accrue to the descendants of this cell, since centrosome number will become destabilized, leading in turn to derangement of mitotic spindles and resulting destabilization of karyotype. The changes in chromosome number that ensue may yield constellations of growth-promoting and growth-retarding genes in proportions that expedite tumor progression.

The great complexity of mitosis raises the question of how prone this process is to error, and how many regulatory proteins, including checkpoint control proteins, are in place to monitor the progression of the various steps of M phase. Yeast genetic analyses have revealed as many as 100 distinct genes and proteins that are involved in the various steps of spindle assembly and dynamics, spindle attachment, and the separation of chromosomes during mitosis; mutation of many of these genes results in chromosome instability in yeast. Many of these proteins are highly conserved evolutionarily, and their homologs are likely to be components of the mammalian mitotic machinery.

To date, only a small proportion of human tumors have been found to carry mutations in the human homologs of yeast genes known to be involved in chromatid assembly and separation during mitosis. Some examples are provided in **Table 12.3**. With the exception of the ATM gene, which is well documented because of its role in ataxia telangiectasia (see Figure 9.13, Table 12.1), mutant germ-line alleles of the remaining genes in this table have rarely been found associated with increased cancer risk. The functions of some of these critical mitosis-regulating genes may be lost through methylation of their promoters (see Section 7.8), but this remains largely undocumented. In fact, only a small proportion of the large cohort of human genes involved in mitotic functions has been explored to determine whether the functioning of these genes is compromised during the course of tumor progression.

One example of such mitotic control genes is the gene encoding the CHFR (checkpoint with forkhead and RING finger domains) protein, which normally prevents advance from prophase into metaphase in the event that the spindle microtubules and centrosomes are not properly arrayed. The gene encoding this checkpoint protein, which is normally expressed ubiquitously in human cells, has been found to be fully repressed due to promoter methylation in 3 of 8 randomly chosen human cancer cell lines, and in 7 out of 37 lung cancer biopsies. In mice, activated germ-line alleles of the *chfr* gene collaborate with knocked-out *mlh1* alleles (see Section 12.9) to generate dramatically increased tumor incidence rates.

Table 12.3 Mutated, methylated, and overexpressed genes in cancer cells that perturb chromosomal stability

Gene	Function of gene product	Consequence of alteration in cancer cells
BUBR1/ BUB1[a]	spindle assembly checkpoint	progress through mitosis, even in the presence of microtubule inhibitors[a]
MAD1[b]	spindle assembly checkpoint	large-scale aneuploidy
MAD2[b,c]	spindle assembly checkpoint	premature entrance into anaphase,[d] aneuploidy
Securin	attachment of sister chromatids	nondisjunction of chromosomes[e]
cohesin complex	attachment of sister chromatids	aneuploidy
Aurora-A, -B	separation of chromatids at anaphase	premature entrance into anaphase[d]
CHFR[e]	spindle assembly checkpoint	nondisjunction[f], chromosome loss
14-3-3σ	DNA damage checkpoint	segregation of unrepaired chromosomes
RB	cell-cycle regulator	aneuploidy
APC[g]	regulation of proliferation	mitotic defects, cytokinesis failure

[a]Humans with heritable compromised BubR1 function suffer the cancer predisposition syndrome termed mosaic variegated aneuploidy. Mice with inherited Bub1 and BubR1 insufficiency are also cancer prone under certain conditions.
[b]Mad1 and Mad2 form complexes at the kinetochore that prevent chromatid separation until complexes with spindle fibers have been properly formed. *Mad1*[+/−] mouse heterogyzotes develop a variety of tumors.
[c]The *Mad2* gene is transcriptionally repressed in a number of solid tumors and is frequently mutated in gastric carcinomas. Mice that are heterozygous at the *Mad2* locus (i.e., are *Mad2*[+/−]) develop lung cancers as adults, while those that overexpress wild-type Mad2 protein develop a variety of malignancies.
[d]Premature entrance into anaphase can lead to loss of entire chromosomes.
[e]*Chfr*[+/−] mice develop lymphomas early in life and carcinomas of the liver, lung, and gastrointestinal tract later in life.
[f]Nondisjunction is the failure of sister chromatids to separate at anaphase.
[g]Anaphase-promoting complex.

Another key regulator of entrance into M phase is the 14-3-3σ protein. When genomic DNA is damaged, p53 induces synthesis of 14-3-3σ, which proceeds to trap cyclin B–Cdc2 complexes in the cytoplasm (see Section 9.9); by sequestering these cyclin–CDK complexes, 14-3-3σ succeeds in blocking entrance into M phase, thereby holding mitosis in abeyance until damaged chromosomal DNA has been repaired. The gene encoding 14-3-3σ has also been found to be frequently methylated in common cancers, including those of the lung, breast, stomach, and liver. The resulting loss of 14-3-3σ must surely contribute to the aneuploidy observed in these kinds of cancer cells.

These two examples, involving the CHFR and 14-3-3σ proteins, only begin to plumb the depths. There are clearly many other caretaker and checkpoint genes to be discovered that play critical roles in stabilizing DNA sequences or ensuring normal karyotype and in this way determine whether or not tumor progression will proceed slowly or move ahead in leaps and bounds (**Sidebar 12.8**).

12.13 Synopsis and prospects

Genome instability has been inherent in life since the first cells appeared 3.5 billion years ago. In the intervening time, living organisms have continually struck a balance between too little and too much instability in their genomes. If they went too far in

Sidebar 12.8 Breast cancer studies indicate that common cancers are often caused by inherited defects in caretaker genes We tend to think that cancers caused by inherited defects in certain genes represent a small fraction of the total cancer burden in the population. The breast cancer families afflicted with mutant *BRCA1* or *BRCA2* germ-line alleles come to mind here. In fact, the incidence of many commonly occurring cancers, such as breast cancer, may be greatly influenced by heritable genetic factors that strongly increase susceptibility to this disease. An epidemiologic study conducted in Britain of breast cancer incidence in sets of identical twins concluded that a significant proportion, and possibly the majority, of these tumors occur in women who are genetically predisposed to this disease. The genetic factors, whatever their nature, are complex and not strongly penetrant; that is, none, acting on its own, strongly affects disease frequency. Some of these factors could be alleles of DNA repair genes that yield sub-optimal function.

This suspicion comes from biological studies of the normal cells of breast cancer patients and their close relatives. Here is one tantalizing hint among many: in one research study, the lymphocytes of 40% of breast cancer patients showed an abnormally high tendency to develop aberrant karyotypes following exposure to X-rays; only 5 to 10% of individuals in the normal, control population showed this peculiar hypersensitivity. Significantly, the *first-degree* (close) relatives of the breast cancer patients were far more likely to have these hypersensitive lymphocytes than people in the general population (**Figure 12.40**). These relatives had no obvious disease phenotypes yet clearly carried alleles of caretaker genes that compromised genome maintenance; such still-unidentified genes are highly likely to have played major roles in causing the breast cancers in their families.

Another striking finding comes from a study of 1071 breast cancer patients who had a familial history of the disease but carried normal *BRCA1/BRCA2* alleles; in this case, researchers looked at the configuration of another gene implicated in DNA repair but not known to contribute to inborn susceptibility to breast cancer—*CHK2*; this gene encodes a kinase that is activated after X-ray–induced DNA damage and is responsible for activating, in turn, p53, among other targets (see Section 9.6). The DNAs of these patients were compared with those of 1620 healthy controls. In the patient group, 5.1% of the women and 13.5% of the men (who suffered from male breast cancer, an especially aggressive disease) had inherited a faulty version of *CHK2* compared with 1.1% of the individuals in the control group. Observations like these suggest that mutant alleles of a number of other genes involved in genome maintenance will one day be discovered to be influential in determining breast cancer susceptibility and, quite possibly, a tendency to develop other common types of cancer.

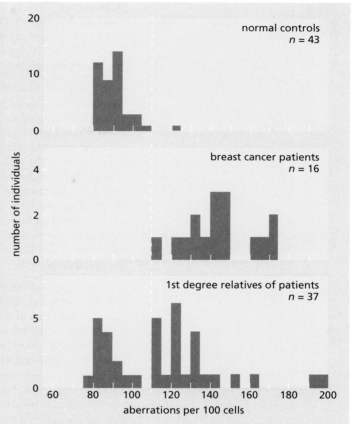

Figure 12.40 DNA repair in lymphocytes of breast cancer patients and their relatives The ability of the DNA repair system to respond to X-ray–induced damage by restoring normal karyotype is one measure of its functioning. In the experiments described here, cultured lymphocytes from various individuals were exposed to a sublethal dose of radiation (0.5 Gy) and the number of karyotypic aberrations present in 100 irradiated lymphocytes was scored shortly thereafter by trapping cells in metaphase with Colcemid, a microtubule antagonist. (Since the lymphocytes were not associated with the breast cancers and not affected by the somatic mutations that culminated in breast carcinomas., their behavior presumably reflects the germ-line-inherited genetic determinants of DNA repair.) Lymphocytes from healthy individuals (*normal controls*) developed 80–100 aberrations per 100 irradiated cells, while the lymphocytes from breast cancer patients developed 110–165 aberrations per 100 cells. Significantly, the *first-degree* (close) relatives of these patients showed either normal or greatly reduced ability to repair these breaks, the latter indicating the effects of strongly penetrant alleles of fully unknown genes that are transmitted in the families of the breast cancer patients. (From S.A. Roberts et al., *Am. J. Hum. Genet.* 65:784–794, 1999.)

suppressing the rate at which mutations accumulated, Darwinian evolution, which depends directly on the continued generation of genetic diversity, would have ground to a halt. Conversely, if they allowed mutation rates to increase too much, their ability to reproduce and even their viability would have been seriously compromised. The relatively low level of genomic instability that operates in our cells—specifically those carrying the germ line—represents a compromise between these two conflicting needs.

This balance between too little and too much genetic instability does not need to be struck in the individual cells of our somatic tissues. In these, any genetic instability—mutability, as we have called it—is undesirable, since it opens the door to neoplasia. This would seem to explain why evolution has worked hard to ensure that the genomes of our somatic cells are so stable.

Multiple layers of defense mechanisms operate to hold somatic mutation rates to extremely low levels. At the biological level, they are embodied in the organization of stem cells and their differentiated progeny. At the biochemical level, an array of enzymes and a variety of low–molecular-weight biochemical species are deployed to confront and neutralize mutagens before they succeed in striking the genome. And should damage be inflicted, either because mutagens have slipped through the outer defenses or because of errors in DNA replication, then a large group of DNA repair enzymes—the caretakers—lie in wait, always alert to structural aberrations in the double helix and its nucleotides. More often than not, these enzymes deal very effectively with incurred damage and restore the DNA to its pristine state, erasing any sign that damage was ever sustained. In addition to these, a complex array of proteins ensures that mitosis and meiosis occur only when the chromosomes are aligned properly at the metaphase plate, thereby sustaining the euploid karyotype.

Our perception of DNA and its much-touted stability is changed by an understanding of these caretakers and their multiple roles in maintaining the genome. Previously, we depicted DNA as a rock-solid, unchanging entity within the cell, a unique island of stability sitting amid countless other molecules that are constantly forming and being degraded. Now, we realize that this was simplistic and an illusion. Like all other molecules in the cell, DNA is vulnerable to many types of damage. Its apparent stability reflects nothing more than a dynamic equilibrium, an ongoing battle between the forces of order and chaos. Any stability that chromosomal DNA does exhibit, and it is considerable, represents a stunning testimonial to the elaborate array of caretakers that are always on watch, ready to fix even the most minor lesion in the double helix.

The implications of this for cancer are simple and clear: if a breakdown of genomic integrity is an essential ingredient in forming human tumors, this can derive most readily from weakening the ever-vigilant repair machinery and its controllers.

Our initial encounter with the breakdown of genomic stability came in Chapter 9, where we learned that the p53 tumor suppressor protein is occasionally called the "guardian of the genome," because cells lacking p53 function acquire a variety of genetic defects at an elevated rate. In large part, this increased mutability, which includes alterations in DNA sequence as well as changes in karyotype, does not reflect p53's role in directly maintaining the genome. Instead, the loss of p53 function creates an environment that is permissive for the survival of mutant cells. In this chapter, we changed our focus by posing a different question, one that goes beyond p53 inactivation: If p53 loss permits mutant (and highly mutable) cells to survive, how, precisely, do the mutations acquired by these cells arise in the first place?

These mutations occur frequently during tumor progression, and elevated mutability is increasingly accepted as an important element of cancer pathogenesis. As noted at the beginning of this chapter, the proposal was made in 1974 that a departure from DNA's highly stable state is essential for the formation of human cancers. This proposal arose from calculations of the rates at which mutations accumulate in normal cells and an estimate of the number of genetic alterations that are needed in order for tumor progression to reach completion.

Without such increased mutation rates—so the thinking went—the time intervals between clonal successions (see Section 11.5) would be far too long. Actually, the readings from the last chapter and this one suggest at least two alternative ways by which clonal successions can be accelerated during multi-step tumor progression. Tumor promoters (including endogenous processes such as inflammation) can compress the time between clonal successions; alternatively, acceleration can be achieved by the destabilization of the genome, as described here. Because the two processes usually work hand-in-hand, mathematical modeling of tumor progression becomes very difficult.

We now realize that the 1974 mathematical analyses depended on so many quantitative assumptions that their major prediction represented little more than an inspired speculation. As is almost always the case in biology, observations of living systems speak more loudly than theorizing: recent high-throughput sequencing analyses of tumor cell genomes indicate enormous variability in the rates with which various types of tumors accumulate mutations as they pass through multi-step tumor progression. While increased mutability clearly accelerates the rate of tumor progression (thereby leading to increased tumor incidence), it is clear that some tumors arise in the absence of large numbers of mutations, as discussed below.

Cancer stem cells (CSCs) are present in many and perhaps all tumors. Their presence complicates our understanding of how mutations accumulate within tumor cell genomes. Because of their ability to self-renew, the lineages of CSCs within tumors are likely to persist long after their more differentiated descendants have entered into post-mitotic states or died; accordingly, CSCs are likely to serve as the long-term repositories of the tumors' genetic information. Hence, advantageous mutations must sooner or later be introduced into the CSC pools to ensure transmission to future generations of cancer cells.

This scenario is complicated by the fact that SCs in general, and CSCs in particular, are present in relatively small numbers within normal and neoplastic tissues. Since the likelihood that a population of cells will sustain a mutation is directly proportional to the size of this population, this means that pools of CSCs are relatively unlikely (mathematically) to directly acquire the mutations that drive tumor progression forward. This creates a dilemma that is at present unresolved: Specialized, still-undiscovered mutagenic mechanisms may destabilize the genomes of CSCs. Alternatively, and more likely, the bulk of advantageous mutations strike the genomes of transit-amplifying/progenitor cells within tumors, and the latter cells then introduce their advantageous mutations back into the CSC pool by spontaneous dedifferentiation (see Supplementary Sidebar 11.2).

Unfortunately, all discussions of cancer cell mutability rest on shaky foundations. Measurements of mutability—number of mutations sustained per cell generation—depend on knowledge of the number of successive cell generations through which a cell lineage has passed since its initiation. At present, we have only very limited information on this critically important parameter (see Figure 10.35A), undermining any attempts at demonstrating heightened mutability definitively.

These discussions of mutation rates generally focus on the submicroscopic changes in genome structure created by defective BER and NER—lesions that are far too small to affect the karyotype of cancer cells. However, as we read in this chapter, cell genomes are also affected by changes occurring on a far larger scale—changes that scramble overall chromosome structure and do indeed alter the karyotype of a cell. Such karyotypic alterations are found in the cancer cells from the great majority of solid tumors. In Section 10.4, we read that telomere erosion may be responsible for much of this instability through its ability to trigger breakage–fusion–bridge cycles. Surely, yet other molecular mechanisms will one day be found to contribute to the karyotypic chaos frequently encountered in cancer cells.

Between the minute lesions left behind by imperfect BER and NER and the large-scale rearrangements generated by telomere collapse are genomic changes resulting from the process of **replication stress**, in which the unbalanced mitogenic signals operating in many cancer cells cause uncoordinated firing of replication origins and frequent replication fork collapse of the sort depicted in Figure 12.8. The repair of these collapsed replication forks is often imperfect, providing an explanation for many of the local amplifications and deletions that are present in the genomes of solid tumors (**Figure 12.41**) and revealed by comparative genomic hybridization (CGH; see Supplementary Sidebar 11.4 and Figure 11.22). (In fact, many of the changes in tumor cell genomes detected by CGH occur at chromosomal "**fragile sites**," which are known to be especially prone to breakage when normal DNA replication programs are perturbed.) Strikingly, signs of replication stress, in the form of activated DNA repair proteins, are already apparent relatively early in multi-step tumor progression (for example, in dysplastic tissues), long before frankly neoplastic growths have emerged.

Figure 12.41 Comparative genomic hybridization analysis of breast cancer genomes The genome of a human breast cancer cell can exhibit wide fluctuations in the copy number of various chromosomal segments and associated genes. The technique of comparative genomic hybridization (CGH; see Supplementary Sidebar 11.4) makes it possible to compare the copy number of the chromosomal segments in a tumor with the copy number in normal human DNA (see Figure 11.22). CGH analysis of each human breast cancer genome yields a distinct profile of segment gains and losses; however, an averaging of the CGH profile patterns of a large number of breast cancers, as shown here, reveals that gains and losses of particular chromosomal segments are present in multiple tumors and are therefore recurrent. The proportion of tumors showing various degrees of amplification or loss (*ordinate*) is plotted as a function of distance along the genome (*abscissa*) moving *rightward* from Chromosome 1. The proportions of breast cancers showing unusually high copy number abnormalities are plotted as positive values, while the proportions showing abnormally low copy numbers are plotted as negative values. Vertical gridlines indicate chromosome boundaries with the numbers *above* indicating human chromosome number. Values plotted in *black* show the proportions of tumors in which the copy number of a chromosomal region is significantly different than the average. Values plotted in *white* show the proportions of tumors in which the copy number is either 50% higher or lower than average. Values plotted in *red* show frequencies of tumors for which the copy number is more than four times the average (i.e., highly amplified) or less than 35% of average (i.e., more than one copy of the tumor genome has been lost). In this case, the "average" is the expected copy number of a gene, given the overall ploidy of the breast cancer genome being studied. Thus, segments of Chromosomes 1 and 8 are repeatedly found to be amplified in these cancers, while segments from Chromosomes 8, 11, 13, 16, and 17 are often deleted. (Courtesy of K. Chin and J.W. Gray.)

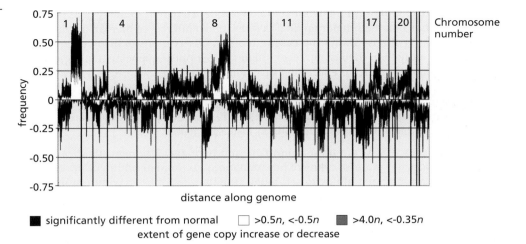

Of additional interest, since replication stress often occurs relatively early in multi-step tumor progression, great pressure is placed on incipient cancer cells to inactivate their *p53* gene, which otherwise might respond to this stress by triggering cell senescence or apoptosis; loss of p53 function at this early stage then yields cells that tolerate many of the cell-biological changes experienced during the later steps of tumor progression. These findings also argue that the increased mutability of cancer cell genomes often derives from by-products of oncogene activation and TSG inactivation, rather than from dysfunctional repair genes and proteins.

While there seem to be common underlying biochemical principles governing how diverse types of human cells are transformed to a neoplastic state (see Section 11.12), the genetic routes taken by various cells throughout the body to reach this state bear little resemblance to one another, at least as gauged by measurements of copy number change: in a large group of pre-treatment stage II–III breast cancers, CGH analysis revealed an average of 76 copy number changes per cancer cell genome. In contrast, an analysis of the genomes of a comparably large group of childhood acute myelogenous leukemias, performed using a slightly different technique, revealed an average of 2.38 copy number changes per genome.

Independent of its origins, increased mutability forces us to confront another reality: the resulting disruption of cancer cell genomes is a double-edged sword, since it places these cells at great risk. Frequent errors in DNA replication and in chromosomal segregation make these cells especially vulnerable to death because of lethal, unrepaired defects in their genomes. Indeed, it is possible that much of the attrition of cancer cells during each cell generation (see Figure 10.5D) occurs because of frequent genomic catastrophes that are incompatible with continued cell survival. Hence, cancer cells must weave a fine course between too much and too little mutability, and certain cancer cell types may actually benefit from maintaining their genomes in a relatively stable configuration.

Actually, cancer cells may be exposed to a second danger because of their defective DNA repair apparatus: they may be especially susceptible to certain forms of chemotherapy. We already encountered one example of this in the vulnerability of many gliomas to killing by temozolomide, an alkylating chemotherapeutic. Those tumors that shut down expression of their *MGMT* gene during their initial formation, ostensibly because doing so increases mutability and accelerates tumor progression, pay a price later, because they have lost the ability to remove the toxic methyl groups added to their DNA by this drug (see Section 12.8).

Another example of such increased vulnerability comes from breast and ovarian tumors that have lost BRCA1 or BRCA2 function. The resulting loss of an ability to repair covalent inter-strand cross-links, which depends on homology-directed repair (HR), makes certain ovarian carcinomas especially sensitive to killing by platinum-based chemotherapeutics. Consistent with this, ovarian cancer patients carrying germ-line *BRCA1* or *BRCA2* mutations have longer progression-free survival following platinum therapy than do those identically treated patients with sporadic (that is, nonfamilial) tumors.

A more subtly crafted therapeutic strategy derives from examining in great detail the mechanisms of base-excision repair (BER) and homology-directed repair (HR). As we learned earlier, a relatively small number of double-strand breaks (DSBs) occur in each cell cycle because of the fragility of the replication fork (see Figure 12.8); the repair of these lesions usually depends on HR. In fact, an alternative and far greater threat comes from the thousands of mutant bases that are created every day by endogenous processes, notably oxidation; the resulting DNA lesions are healed by BER. As described earlier (see Figure 12.22A), BER generates single-nucleotide gaps (through both BER pathways) and thus single-strand breaks (SSBs) that must be filled by a DNA polymerase and then sealed by a DNA ligase.

DNA repair enzymes involved in BER are recruited to SSB sites by the actions of PARP1 [poly-(ADP ribose) polymerase-1]. This enzyme binds to these breaks and proceeds to ADP-ribosylate itself and (possibly neighboring proteins as well; **Figure 12.42A**). The resulting poly-ADP tails serve thereafter as docking sites for recruiting the repair enzymes needed to fix the SSBs; hence, when the PARP1 enzyme is inhibited, SSBs persist. These SSBs generate a far more serious problem—DSBs—when a replication fork passes through during S phase (see Figure 12.42B).

All this suggests an interesting therapy: shut down the PARP enzyme in *BRCA1* or *-2* mutant cells with a pharmacologic inhibitor. Resulting SSBs will be converted to DSBs during replication, and these DSBs, which are usually repaired by HR, will not be repairable in cells lacking HR function (due to loss of BRCA1 or BRCA2 function; see Figure 12.42C). DSBs will then accumulate in large numbers, triggering cell death. Indeed, PARP1 inhibitors have been employed in the clinic for patients with ovarian and breast cancers and have shown a doubling of progression-free survival time in ovarian carcinoma patients but no significant increase in overall survival time; the approval of such inhibitors for clinical use is therefore in doubt.

The scenario of acquired genetic instability seems to create a logical quandary: almost all of the genetic alterations that occur during tumor progression appear to confer some immediate growth or survival benefit on the cells that acquire them. Many of the earlier chapters in this book documented the specific growth advantages resulting from each of these alterations. These dynamics caused us to speculate that tumor progression is a process that is analogous to Darwinian evolution (see Section 11.5). However, unlike many traits acquired by cancer cells, the trait of increased genomic instability does not provide an immediate payoff—no marked advantage in proliferation or survival. Instead, acquisition of this trait represents a "long-term investment," that is, a benefit will only be realized by the distant descendants of this cell.

We can understand this long-term advantage in the following way. Cell clones that have acquired some of the initial genetic alterations leading to cancer seem to arise with great frequency throughout the body's tissues. The vast majority of them remain in a dormant, premalignant state for an entire human lifetime, unobtrusive and unthreatening, because their stable genomes preclude the acquisition of the additional mutations that would render these cell clones truly dangerous. On rare occasion, however, a cell in one or another of these already-initiated clones acquires an alteration in one of its caretaker genes. Now, for the first time, this cell and its descendants have the opportunity to tinker with their genomes by testing new combinations of genetic elements and new sequences, some of which will allow them to resume their advance down the long road of tumor progression. By opening a floodgate of genetic changes, clones of would-be cancer cells are destined, over the long term, to prosper, while their brethren, lacking this instability, are likely to remain unchanged and indolent for decades.

As we learned in this chapter, a variety of familial cancer syndromes are attributable to the inheritance of mutant forms of genes specifying important components of the DNA repair apparatus. Genomic instability or a tendency toward genetic instability

(A)

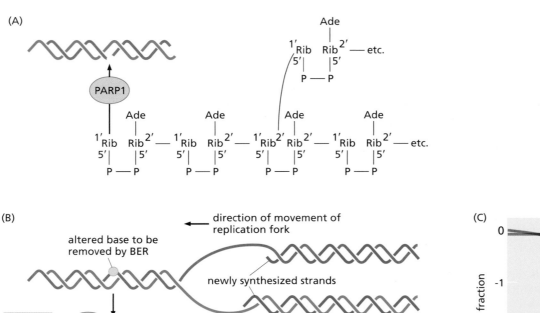

(B)

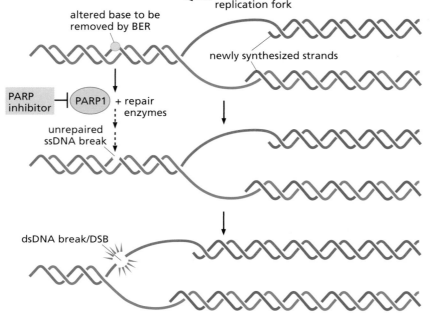

(C)

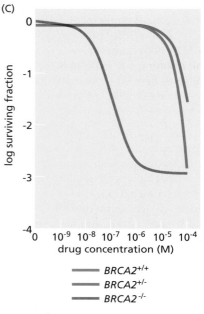

Figure 12.42 Killing of BRCA2-negative cells by an anti-PARP drug The reliance of BRCA1- and BRCA2-negative human breast cancer cells on a redundant HR pathway mediated by poly-(ADP ribose) polymerase-1 (PARP1) makes these cells especially vulnerable to killing by an anti-PARP drug. (A) PARPs constitute a family of related enzymes that act by generating repeating branched polymers of as many as 200 residues each that are attached to the PARP enzymes themselves and other proteins involved in effecting biochemical responses. In the case of PARP1, it binds to ssDNA breaks (SSBs) and ADP-ribosylates itself; its poly-ADP chains then attract a series of repair enzymes that may be ADP-ribosylated by PARP1 and thereafter proceed to complete various steps of BER. (B) Such SSBs are created routinely by oxidized bases, which are usually removed by BER. Here, an oxidized base (*gold*) should be excised and the DNA repaired through the actions of PARP1 and recruited repair enzymes. However, in the presence of a PARP1 inhibitor, these other enzymes are not recruited, BER cannot proceed to completion, and the resulting SSB is converted to a double-strand break (DSB) when a replication fork moves through during the next S phase. This break can be efficiently restored to a wild-type configuration only through homology-directed repair (HR), which is defective in *BRCA1*- and *BRCA2*-mutant cells. (C) Cultured mouse ES (embryonic stem) cells of three genotypes were treated with KU0058948, a PARP1 inhibitor. While the *Brca2*$^{+/+}$ (*blue*) and *Brca2*$^{+/-}$ (*green*) ES cells were killed only by very high drug concentrations, the *Brca2*$^{-/-}$ ES cells were killed by the drug at concentrations almost 1000-fold lower (*red*). (C, from H. Farmer et al., *Nature* 434:917–921, 2005.)

is already implanted in all the cells of such genetically afflicted individuals. In these cases, genetic instability is not an acquired attribute—unlike the situation operating in the great majority of human cancers.

On some occasions, such as in the many subtypes of xeroderma pigmentosum (XP), both copies of a critical DNA repair gene are inherited in defective form. This nul-lizygous state leaves all the cells in the skin without the means to cope with the

UV-induced formation of pyrimidine dimers. In other cases, such as hereditary non-polyposis colon cancer (HNPCC), an individual inherits a defective gene copy of a mismatch repair gene, and the remaining copy is then discarded in various somatic cells through a loss-of-heterozygosity (LOH) event.

The familial cancer syndromes that have been identified to date, including those involving heritable defects in DNA repair, reflect the actions of strongly penetrant germ-line alleles. Such alleles ensure that individuals carrying them will, with high likelihood, manifest obvious disease at some point in their lifetimes. However, the known familial cancer syndromes may well represent only the small tip of a very large iceberg. Thus, a variety of less penetrant, mutant alleles of the genes involved in genome maintenance may be widespread in the human gene pool. Each of these alleles may confer only a slightly increased risk of cancer, but one that is not readily apparent when studying individual families and their susceptibilities to various types of cancer. Consequently, a significant portion of many commonly occurring types of human tumors, such as breast cancer (see, for example, Sidebar 12.8), may be associated with inheritance of these still-unknown germ-line alleles.

Our thinking about the family cancer syndromes associated with inheritance of defective caretaker genes is confounded by one major mystery: the majority of these diseases involve only a very narrow subset of tissues in the body, even though we have every reason to believe that the services of affected caretakers are required ubiquitously. Why, for example, do inherited mismatch repair defects have such a strong preference for causing tumors in the intestinal tract? The puzzle is drawn even more starkly in the case of breast cancer. By one account, inherited defects in at least seven genes that are involved, directly or indirectly, in the maintenance of genomic integrity generate substantially increased risks of mammary tumors in humans. These include the *BRCA1* and *BRCA2* genes, two genes encoding BRCA1-associated proteins (*BARD1* and *BACH1*), *CHK2*, *p53*, and *ATM* (which specifies one of the kinase sensors that activate the p53 alarm following DNA damage; see Section 9.6).

Of course, this is not the first time that we have encountered a puzzle of this sort. Recall, for example, the behavior of inherited mutant alleles of the retinoblastoma gene, *Rb*. This gene specifies a protein, pRb, that appears to function in almost every cell type in the body as the critical controller of advance through the cell cycle, yet children who inherit defective *Rb* alleles are predisposed peculiarly to a rare eye tumor when they are very young and to bone tumors as adolescents (see Sections 7.3–7.5).

To date, our perceptions about the genes that are responsible for mammalian genome maintenance have been shaped largely by bacterial and yeast genetics. These disciplines have yielded a wealth of genes that are essential for genome maintenance in these microbes and, by extension, in our own genomes as well. For example, the resonance between yeast genetics and its harvest of mismatch repair genes and the HNPCC syndrome led to the rapid enumeration of six human genes and proteins that are responsible for this type of genome maintenance.

Importantly, microbial genetics may have netted only a portion of the genes in the human genome that are responsible for maintenance of its integrity. How large, then, is the universe of human caretaker genes and proteins? The sequencing of the yeast genome begins to give us a feeling for this number. Recent estimates indicate that this genome consists of approximately 5600 distinct genes. Of these, 153 are classified as being involved in DNA replication and repair, while 88 seem to be involved exclusively with repair. The fact that many of the genes encountered in this chapter are metazoan inventions suggests that the size of the cohort of human genes involved in these processes is likely to be much larger. Indeed, by 2001, more than 130 DNA repair genes had already been identified in the human genome; by 2004, the number approached 150, and it reached 176 in 2012. Hence, genes and proteins that play critical roles in the maintenance of the genome and thus in the suppression of cancer are being discovered continually, some in very unexpected places (**Sidebar 12.9**).

This area of cancer research is also having a profound effect on our understanding of another, seemingly unrelated biological process—aging. At least ten distinct premature aging syndromes—each a form of **progeria**—have been traced to inherited

Sidebar 12.9 Histones function to reduce cancer risk For many decades, histones and the nucleosomes that they form were depicted as static structural components of chromatin that serve to package and compact the chromosomal DNA. Beginning in the 1990s, it became clear that covalent modification of histones is critical in making regions of chromatin more or less hospitable for transcription (see Section 1.8). Even more recently, histones have been found to play a dynamic role in DNA repair. H2AX is a variant of the abundant H2A histone that is present in virtually all nucleosomes. H2AX constitutes about 15% of the H2A-like histones in cells and substitutes for H2A in some nucleosomes (see Figure 1.19). When a double-strand break is sustained in chromosomal DNA, H2AX molecules (but not the other histones) become phosphorylated, primarily by the ATM and ATR kinases, on a specific serine residue located four amino acid residues from the carboxyl-terminal end; such phosphorylated H2AX is observed in a large chromosomal region (involving as much as 2 megabases of DNA) flanking the break (**Figure 12.43**). (These two kinases are also responsible for phosphorylating and thereby mobilizing p53; see Figure 9.13.) The phosphorylated H2AX (usually termed γ-H2AX) attracts DNA repair proteins, such as BRCA1 and NBS1, as well as at least four others that aid in the task of rejoining the DNA ends (see also Figures 12.29 and 12.33).

Mice that have lost the H2AX gene (because of germ-line inactivation) are viable but stunted in growth. Their cells are unable to execute homology-directed repair (HR; see Section 12.10) and are prone to accumulate structurally abnormal chromosomes. Mice lacking both the H2AX and p53 proteins are highly prone to both hematopoietic and solid tumors. Even mice lacking one copy of the H2AX gene in the context of p53 deficiency show significantly increased rates of lymphomas.

These responses illustrate how highly complex the DNA repair process is, and how defects in any of its individual components—many still unidentified—open the door to the appearance of cancer. Provocatively, the human *H2AX* gene maps to a chromosomal region (11q23) that frequently undergoes loss of heterozygosity and/or deletion, creating the possibility that this gene and encoded histone are frequently shed by human cells en route to malignancy.

Figure 12.43 γ-H2AX and double-strand DNA breaks The creation of dsDNA breaks by various mechanisms results in the phosphorylation of H2AX histone (a variant form of histone H2A), yielding γ-H2AX. Like H2A, H2AX participates in the formation of nucleosomes. Chromosomal regions as large as a megabase that flank a dsDNA break are found to carry γ-H2AX. In the absence of H2AX, critical DNA repair proteins, including Nbs1 and Brca1 (see Figure 12.33), fail to be recruited to the site of dsDNA breaks. In the image shown here, cells of the Indian muntjac, used because of their small number of chromosomes, have been irradiated with a low dose (0.6 Gy) of X-rays. Late anaphase cells were then stained for γ-H2AX (*yellow green, arrows*) and for chromosomal DNA (*red*). The outlines of the dividing cell are seen in *blue*. Note the staining of γ-H2AX at the broken ends of a chromosome. (From E.P. Rogakou et al., *J. Cell Biol.* 146:905–916, 1999.)

defects in one or another component of the DNA repair system. Individuals suffering from these syndromes often show many of the phenotypes of the aged by the time they reach adolescence. These clinical effects of defective DNA repair provide compelling indications that much of the normal aging process will one day be traced to genetic damage that we accumulate in our stem cells throughout life; their progressive attrition seems to lead to the inability of tissues to renew themselves, yielding precisely the changes observed in the aged. Consequently, cancer and aging may be found to share a common root—the progressive deterioration of our genomes as we get older. And both cancer development and aging may one day be forestalled by treatments and lifestyles that protect our genomes from the ongoing attacks that they suffer, decade after decade, deep within our cells.

Key concepts

- The structural integrity and thus low mutability of DNA depends on a large and complex set of biological and biochemical mechanisms that work to ensure that somatic mutations accumulate in tissues at very low rates.

- Some of the mechanisms depend on the organization of tissues, in which the long-lived cells (the stem cells) are protected from genetic damage while the short-lived cells (the transit-amplifying and differentiated cells) are vulnerable to sustaining such damage but are soon discarded.

- Misincorporated bases generated by errors in DNA replication can contribute to the burden of accumulated mutations. The numbers of these alterations are held down by the low error rates of DNA polymerases together with an array of error-correcting proteins, such as those involved in mismatch repair. Inherited defects in mismatch repair proteins can lead to increased susceptibility to certain types of cancer, notably hereditary non-polyposis colon cancer.

- Cell genomes are under continuous attack by a variety of chemically reactive molecules, many of them deriving from the cellular process of oxidative phosphorylation and the resulting reactive oxygen species that are generated as by-products of this process. Cell genomes may also suffer spontaneous chemical alterations, which affect DNA bases at a low but significant rate.

- In addition, the genomic DNA of cells can be attacked by mutagenic molecules of foreign origin. Such xenobiotics and their chemically reactive derivatives may come from pollutants and, to a far greater extent, from commonly consumed foodstuffs.

- Cells attempt to detoxify many of these compounds before they can attack genomic DNA. However, the side products of these reactions may actually be more reactive and mutagenic than the initially introduced molecular species.

- If the attacking mutagenic agents succeed in damaging DNA, an elaborate array of proteins involved in base-excision and nucleotide-excision repair lies in wait in order to remove the vast majority of damaged bases. Inherited defects in base-excision repair proteins can lead to various types of cancer susceptibility.

- Other types of DNA damage include double-strand DNA breaks, which can be created by X-rays or, more commonly, by DNA breakage at replication forks.

- Double-strand DNA breaks can be repaired in the G_1 phase of the cell cycle by nonhomologous end joining or in the S and G_2 phases by homology-dependent repair. Inherited defects in double-strand DNA repair can explain the breast and ovarian cancer susceptibility among patients inheriting mutant *BRCA1* or *BRCA2* germ-line alleles.

- Genomes may be scrambled by mechanisms that affect the karyotype of cells. One class of such alterations includes those that affect chromosomal structure, including the translocations that are created by the fusions of unrelated chromosomal arms to one another. Such fusions seem to be commonly triggered by eroded telomeres and collapsed replication forks, both of which generate double-strand DNA breaks.

- Changes in chromosome number are also common in cancer cell genomes and appear to facilitate the accumulation of genes in proportions that favor the proliferation and survival of these neoplastic cells. Many of these changes derive from defects in the mitotic apparatus and its regulators, notably the centrosomes and proteins involved in connecting spindle fibers with chromosomal kinetochores.

- Without a breakdown of the various mechanisms responsible for maintaining the integrity of the genome, it seems likely that mutation rates would be too low to enable cell genomes to accumulate the ensemble of genetic changes required for tumor progression to reach completion in a human lifetime.

Thought questions

1. What types of evidence suggest that karyotypic alterations of cell genomes are not absolutely essential for neoplastic transformation?

2. When calculating the rates of mutation required in order for multi-step tumor progression to reach completion,

   what parameters must one know in order for such a calculation to accurately describe the actual biological process?

3. How does our understanding of defective DNA repair processes in tumor cells make possible the development of new anti-cancer therapeutic strategies?

4. In which ways do the defectiveness of p53 function and resulting defects in apoptosis and DNA repair facilitate the forward march of tumor progression?

5. What types of tumor promotion, as described in Chapter 11, favor the genetic evolution of premalignant cell clones?

6. What evidence implicates mutagenic chemicals originating outside the body in the pathogenesis of human cancers? How can one gauge their contribution to human carcinogenesis compared with that of mutagens and mutagenic processes of endogenous origin?

7. How do defects in various cell cycle checkpoints allow for accelerated rates of the accumulation of mutations?

8. How do the biological properties of stem cells help to reduce the rate at which tissues accumulate mutant genes?

9. How does the existence of cancer stem cells affect the calculations of the rate at which mutations must be accumulated in order to allow multi-step tumor progression to advance?

10. How does the genetic heterogeneity in the human gene pool affect the functioning of various types of biological defenses that have been erected to prevent the accumulation of mutant alleles in human somatic cells?

Additional reading

Ames BN (1989) Endogenous DNA damage as related to cancer and aging. *Mutat. Res.* 214, 41–46.

Arias EA & Walter JC (2007) Strength in numbers: preventing rereplication via multiple mechanisms in eukaryotic cells. *Genes Dev.* 21, 497–518.

Barnes DE & Lindahl T (2004) Repair and genetic consequences of endogenous DNA base damage in mammalian cells. *Annu. Rev. Genet.* 38, 445–476.

Bartek J, Lukas C & Lukas J (2004) Checking on DNA damage in S phase. *Nat. Rev. Mol. Cell Biol.* 5, 792–804.

Branzel D & Folani M (2008) Regulation of DNA repair throughout the cell cycle. *Nat. Rev. Mol. Cell Biol.* 9, 297–308.

Brenner JC & Chinnaiyan AM (2009) Translocations in epithelial cancers. *Biochim. Biophys. Acta* 1796, 201–215.

Cimini D (2008) Merotelic kinetochore orientation, aneuploidy, and cancer. *Biochim. Biophys. Acta* 1786, 32–40.

Cleaver JE (2005) Cancer in xeroderma pigmentosum and related disorders of DNA repair. *Nat. Rev. Cancer* 5, 564–573.

D'Assoro AB, Lingle WL & Salisbury JL (2002) Centrosome amplification and the development of cancer. *Oncogene* 21, 6146–6153.

DiGiovanna DJ & Kraemer KH (2012) Shining a light on xeroderma pigmentosum. *J. Invest. Dermatol.* 132, 785–796.

Duensing S & Münger K (2002) Human papillomavirus and centrosome duplication errors: modeling the origins of chromosomal stability. *Oncogene* 21, 6241–6248.

Ferguson LR (2010) Dietary influences on mutagenesis: where is this field going? *Environ. Mol. Mutagen.* 51, 909–918.

Friedberg EC, Aguilera A, Gellert M et al. (2006) DNA repair: from molecular mechanism to human disease. *DNA Repair (Amst)* 5, 986–996.

Friedberg EC, McDaniel LD & Schultz RA (2004) The role of endogenous and exogenous DNA damage and mutagenesis. *Curr. Opin. Genet. Dev.* 14, 5–10.

Fröhling S & Döhner H (2008) Chromosomal abnormalities in cancer. *N. Engl. J. Med.* 359, 722–734.

Gold LS, Ames BN & Slone TH (2002) Misconceptions about the causes of cancer. In Human and Environmental Risk Assessment: Theory and Practice (D Paustenbach ed), pp 1415–1460. New York: John Wiley and Sons.

Greaves MF & Wiemels J (2003) Origins of chromosome translocations in childhood cancers. *Nat. Rev. Cancer* 3, 639–649.

Halazonetis T, Gorgoulis VG & Bartek J (2008) An oncogene-induced DNA damage model for cancer development. *Science* 319, 1352–1356.

Harper JW & Elledge SJ (2007) The DNA damage response: ten years after. *Mol. Cell* 28, 739–745.

Hartlerode AJ & Scully R (2009) Mechanisms of double-strand break repair in somatic mammalian cells. *Biochem. J.* 423, 157–168.

Hoeijmakers JHJ (2009) DNA damage, aging and cancer. *N. Engl. J. Med.* 361, 1475–1485.

Holland AJ & Cleveland DW (2009) Boveri revisited: chromosomal instability, aneuploidy and tumorigenesis. *Nat. Rev. Mol. Cell Biol.* 10, 478–487.

Hübscher U, Maga G & Spadari S (2002) Eukaryotic DNA polymerases. *Annu. Rev. Biochem.* 71, 133–163.

Huertas P (2010) DNA resection in eukaryotes: deciding how to fix the break. *Nat. Struct. Mol. Biol.* 17, 11–16.

Jiricny J (2006) The multifaceted mismatch-repair system. *Nat. Rev. Mol. Cell Biol.* 7, 335–346.

Kastan MB & Bartek J (2004) Cell cycle checkpoints and cancer. *Nature* 432, 316–323.

Kee Y & D'Andrea AD (2010) Expanded roles of the Fanconi anemia pathway in preserving genomic stability. *Genes Dev.* 24, 1680–1694.

Kim D & Guengerich FP (2005) Cytochrome P450 activation of arylamines and heterocyclic amines. *Annu. Rev. Pharmacol.* 45, 27–49.

Loeb LA, Bielas JH & Beckman RA (2008) Cancer cells exhibit a mutator phenotype: clinical implications. *Cancer Res.* 68, 3551–3557.

Loeb LA, Loeb KR & Anderson JP (2003) Multiple mutations and cancer. *Proc. Natl. Acad. Sci. USA* 100, 776–781.

Lynch HT & de la Chapelle A (2003) Hereditary colorectal cancer. *N. Engl. J. Med.* 348, 919–932.

Marnett LJ & Plastaras JP (2001) Endogenous DNA damage and mutation. *Trends Genet.* 17, 214–221.

Maynard S, Schurman S, Harboe C et al. (2009) Base excision repair of oxidative DNA damage and association with cancer and aging. *Carcinogenesis* 30, 2–10.

Nakabeppu Y, Tsuchimoto D, Ichinoe A et al. (2004) Biological significance of the defense mechanisms against oxidative damage in nucleic acids caused by reactive oxygen species: from mitochondria to nuclei. *Ann. NY Acad. Sci.* 1011, 101–111.

Navin NE & Hicks J (2010) Tracing the tumor lineage. *Mol. Oncol.* 4, 267–283.

Negrini S, Gorgoulis VG & Halazonetis TD (2010) Genomic instability: an evolving hallmark of cancer. *Nat. Rev. Mol. Cell Biol.* 11, 220–228.

Nigg EA (2006) Origins and consequences of centrosome aberrations in human cancers. *Int. J. Cancer* 119, 2717–2723.

O'Donovan PJ & Livingston DM (2010) BRCA1 and BRCA2: breast/ovarian cancer susceptibility gene products and participants in double-strand break repair. *Carcinogenesis* 31, 961–967.

Poirier MC (2004) Chemical-induced DNA damage and human cancer risk. *Nat. Rev. Cancer* 4, 630–637.

Rajagopalan H & Lengauer C (2004) Aneuploidy and cancer. *Nature* 432, 338–341.

Reinhardt HC & Yaffe MB (2009) Kinases that control the cell cycle in response to DNA damage: Chk1, Chk2, and MK2. *Curr. Opin. Cell Biol.* 21, 245–255.

Ricke RM, van Ree JH & van Deursen JM (2008) Whole chromosome instability and cancer: a complex relationship. *Trends Genet.* 24, 457–466.

Salk JJ, Fox EJ & Loeb LA (2009) Mutational heterogeneity in human cancers: origin and consequences. *Annu. Rev. Pathol. Mech. Dis.* 5, 51–75.

Sedgwick B (2004) Repairing DNA-methylation damage. *Nat. Rev. Mol. Cell Biol.* 5, 148–157.

Sengupta S & Harris CC (2005) p53: traffic cop at the crossroads of DNA repair and recombination. *Nat. Rev. Mol. Cell Biol.* 6, 44–55.

Silver DP & Livingston DM (2012) Mechanisms of BRCA1 tumor suppression. *Cancer Disc.* 2, 679–684.

Singh MS & Michael M (2009) Role of xenobiotic metabolic enzymes in cancer epidemiology. *Methods Mol. Biol.* 472, 243–264.

Smith J, Tho LM, Xu N & Gillespie DA (2010) The ATM-Chk2 and ATR-Chk1 pathways in DNA damage signaling and cancer. *Adv. Cancer Res.* 108, 73–112.

Stolz A, Ertych N & Bastians H (2011) Tumor suppressor *CHK2*, regulator of DNA damage response and mediator of chromosomal stability. *Clin. Cancer Res.* 17, 401–405.

Sugimura T (2000) Nutrition and dietary carcinogens. *Carcinogenesis* 21, 387–395.

Thompson SL & Compton DA (2008) Examining the link between chromosomal instability and aneuploidy in human cells. *J. Cell Biol.* 180, 655–672.

Weinstock DM, Richardson CA, Elliott B & Jasin M (2006) Modeling oncogenic translocations: distinct roles for double-strand break repair pathways in translocation formation in mammalian cells. *DNA Repair (Amst)* 5, 1065–1074.

Wogan GN, Hecht SS, Felton JS et al. (2004) Environmental and chemical carcinogenesis. *Semin. Cancer Biol.* 14, 473–486.

Wood RD, Mitchell M & Lindahl T (2005) Human DNA repair genes, 2005. *Mut. Res.* 577, 275–283.

Wu X, Gu J & Spitz MR (2007) Mutagen sensitivity: a genetic predisposition factor for cancer. *Cancer Res.* 67, 3493–3495.

Wu X, Zhao H, Suk R & Christiani DC (2004) Genetic susceptibility to tobacco-related cancer. *Oncogene* 23, 6500–6523.

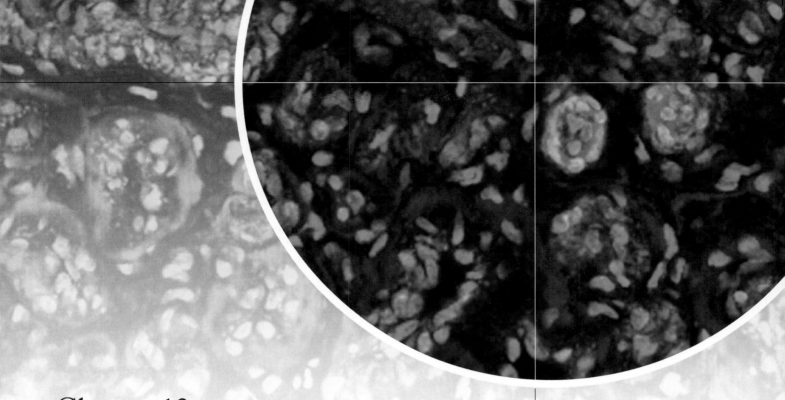

Chapter 13

Dialogue Replaces Monologue: Heterotypic Interactions and the Biology of Angiogenesis

> Simple ideas are wrong. Complicated ideas are unattainable.
>
> Paul Valery, poet, 1942

A simple and powerful conceptual paradigm pervades most of the previous discussions in this book: cancer is a disease of cells, and the phenotypes of cancer cells can be understood by examining the genes and proteins within them. The origins of this idea are clear, being traceable directly back to bacterial and yeast genetics. These two research specialties thrived because they were well served by the postulate that cell genotype determines all aspects of cell phenotype. Indeed, virtually all the attributes of individual bacterial and yeast cells could be shown to derive directly from the genes that these microorganisms carry in their genomes.

Applying this concept to metazoans and their tissues has had obvious advantages for biologists. Metazoan organisms are complex almost beyond measure, and this complexity has frequently prevented researchers from extracting simple and irrefutable truths about organismic function. In response, many researchers, notably molecular biologists, cell biologists, and biochemists, embraced the credo of **reductionist** science: when working with complex systems, the best way to arrive at solid and rigorous conclusions is to take apart these systems into simpler, more tractable components and study each separately. While the lessons learned may relate only to small parts of very large systems, at least these lessons are solid and definitive and will not require substantial revision each time a new generation of researchers revisits these complex systems and their component parts.

Movies in this chapter

13.1 Mechanisms Enabling Angiogenesis

13.2 Interactions of Innate Immune Cells with a Mammary Tumor

Such reductionism has allowed many areas of cancer research to thrive. Witness the progress described in the early chapters of this textbook: when the first proto-onco-gene was discovered three-quarters of the way through the twentieth century (see Section 3.9), almost nothing was known about the genetics and molecular biology of cancer. By the end of the century, information was available in abundance about how cancers begin and progress to highly aggressive, malignant states. Much of this avalanche of information came from taking the reductionist paradigm to its limits—disassembling normal and neoplastic tissues into component cells and the cells into component molecules.

As mentioned above, the reductionist pact embraced by many in the cancer research community treated cancer as a cell-autonomous process: all the attributes of cancer cells can be understood in terms of the genes that these cells carry. Actually, there was yet another notion embedded in their thinking: all of the traits of a tumor can be traced directly back to the behavior of individual cancer cells within the tumor mass.

By the end of the twentieth century, it became increasingly clear that many of the traits of tumors could not in fact be traced directly back to individual cancer cells and the genes they carry. The grand simplifications agreed upon a generation earlier by cancer researchers had begun to lose their utility. Increasingly, evidence turned up that cancer is actually a disease of tissues, in particular, the complex tissues that we call tumors.

The reductionists' way of thinking had, all along, denied certain realities that clinical oncologists and pathologists confronted on a daily basis, namely, that carcinomas—which constitute more than 80% of the human cancer burden—derive from epithelial tissues of very complex microscopic structure. Histopathological characterizations of these epithelial tumors reveal that they are composed of a number of distinct cell types. In fact, even cursory examinations under the microscope indicate that most car-cinomas are as complex histologically as the normal epithelial tissues from which they have arisen; examples of this complexity can be seen in **Figure 13.1**. In some cases of

Figure 13.1 Stromal components of several commonly occurring carcinomas As is shown here, a variety of common carcinomas contain significant proportions of both epithelial cancer cells and recruited stromal cells. (A) A high-grade invasive ductal carcinoma of the breast, in which the membranes (*arrows*) of the epithelial cells are stained *brown*, while the nuclei of the stromal cells are *light blue*. (B) A colon carcinoma (*right side*) and normal colonic mucosa (N, *left side*), in which fibroblast-like cells in the tumor-associated stroma (TAS) are immunostained *dark brown* for PINCH (a protein associated with the cytoplasmic tails of integrins), while the nuclei of epithelial cancer cells lining ducts are *light blue*. (C) A lobular carcinoma *in situ* of the breast showing small clusters of cancer cells (*light blue nuclei, arrow*) surrounded by thin layers of stromal fibroblasts that have been immunostained for PINCH antigen (*dark brown*). (D) An adenocarcinoma of the stomach in which the cancer cells (*purple, lower left*) are adjacent to extensive stroma, which has been stained for collagen (*blue*). (A, from A. Gupta, C.G. Deshpande and S. Badve, *Cancer* 97:2341–2347, 2003. B and C, from J. Wang-Rodriguez et al., *Cancer* 95:1387–1395, 2002. D, from S. Ohno et al., *Int. J. Cancer* 97:770–774, 2002.)

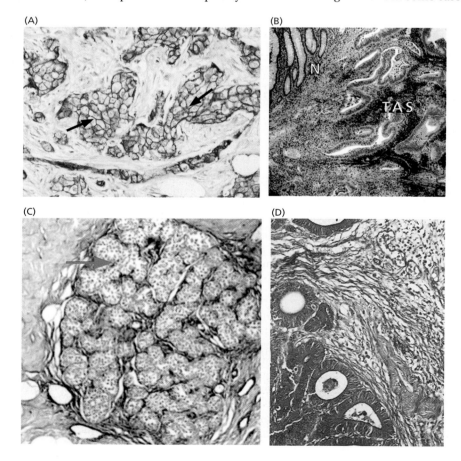

(A)   (B)

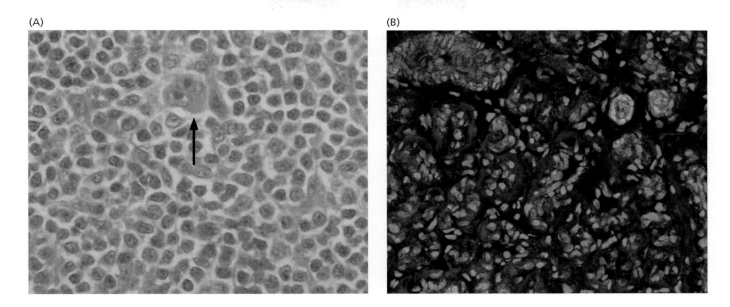

commonly occurring carcinomas, such as those of the breast, colon, stomach, and pancreas, the non-neoplastic cells, which constitute the tumor stroma, may account for as many as 90% of the cells within the tumor mass.

[These examples fail to illustrate the great range of variations that are encountered in various types of non-carcinomatous human tumors. At one end of the spectrum is another solid tumor—Hodgkin's disease—a lymphoma in which non-neoplastic cells account for 99% of the cells in a tumor mass (**Figure 13.2A**). At the other end are tumors such as the hemangioma, a relatively common benign tumor of the endothelial cells (Figure 13.2B), in which neoplastic cells form the great bulk of the tumor mass. In what follows, we focus largely on carcinomas, in which the epithelial and stromal cells are often present in comparable numbers.]

One's first instinct is to dismiss the non-neoplastic cells of the tumor stroma as distracting contaminants that confound rather than enlighten attempts at understanding the biology of tumors. To do so now seems increasingly unwise. Recent research results make it clear that non-neoplastic cells, notably the stromal cells of carcinomas, are active, indeed essential collaborators of the neoplastic epithelial cells within tumor masses, having been recruited and then exploited by the cancer cells. The biology of these recruited cells now appears to be almost as important as that of the neoplastic cells in enabling the growth of tumors.

This means that the disease of cancer is far more than cancer cells talking to themselves in endless monologues. We know, instead, that in most tumors, the neoplastic cells are in continuous communication with their non-neoplastic neighbors. In this chapter, we focus specifically on the interactions of neoplastic and non-neoplastic cells within carcinomas, which are far better understood than the complex interactions operating within many other types of cancer that afflict humans.

13.1 Normal and neoplastic epithelial tissues are formed from interdependent cell types

The true complexity of the stroma of epithelial cancers becomes apparent only at high magnification under the microscope. In addition to the neoplastic epithelial (that is, carcinoma) cells, a number of stromal cell types populate the tumor. These stromal cells include fibroblasts, myofibroblasts, endothelial cells, pericytes, smooth muscle cells, adipocytes, macrophages, lymphocytes, and mast cells (**Figure 13.3A–F**). Because some of these cell types constitute only a small proportion of the stroma of certain tumors, their presence becomes apparent only through use of immunostaining with antibodies that bind cell type–specific antigens.

Figure 13.2 Extreme variations in the ratio of neoplastic cells to stroma
(A) In certain tumors, normal, non-neoplastic cells may constitute the great majority of living cells. An extreme form of this is illustrated here by Hodgkin's disease, a lymphoma in which as many as 99% of the cells are recruited normal lymphocytes, which surround a rare neoplastic Reed–Sternberg cell (*arrow*). (B) At the other end of the spectrum are tumors such as the benign hemangioma shown here, in which neoplastic endothelial cells (*green nuclei*) form the great bulk of the tumor mass, which is demarcated by basement membranes (*blue*) and occasional capillaries (*red*). (A, from A.T. Skarin, Atlas of Diagnostic Oncology, 4th ed. Philadelphia: Elsevier Science Ltd., 2010. B, from M.R. Ritter et al., *Proc. Natl. Acad. Sci. USA* 99:7455–7460, 2002.)

An additional dimension of complexity comes from the great variations in the proportions of various bone marrow–derived cell types within the stroma of different tumor types, such as the carcinomas surveyed in Figure 13.3G. As we will see in this chapter, some of these cells are active participants in processes such as tumor cell invasiveness and metastatic spread; yet others may be dispatched by the immune system to eliminate tumor cells, a topic that we will pursue in Chapter 15. In light of the important

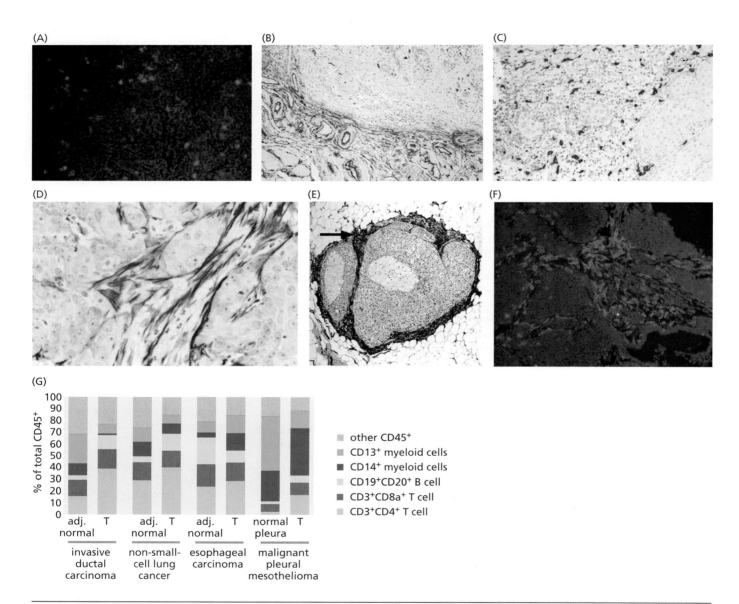

Figure 13.3 A variety of distinct cell types in the stroma of carcinomas Various antibodies can be used to immunostain specific mesenchymal cell types in the stroma of carcinomas. (A) CD4 antigen–positive T lymphocytes, often termed helper T cells (*red*), are scattered among mouse mammary carcinoma cells (*blue nuclei*). The stroma of squamous cell carcinomas of the oral cavity, pharynx, and larynx may contain (B) CD34 antigen–positive fibrocytes (*brown*), (C) CD117 antigen–positive mast cells (*brown*), and (D) α-smooth muscle actin–positive myofibroblasts (*brown*). (E) The stroma of a ductal carcinoma *in situ* of the breast (DCIS) shows PINCH antigen–positive fibroblasts (*dark brown*; see Figure 13.1B) arrayed in a ring immediately around the carcinoma (*arrow*) and surrounded, in turn, by adipocytes (*light blue rims*). (F) Macrophages (*red*), which have been recruited to a mouse mammary carcinoma, play critical roles in tumor invasiveness and angiogenesis. (G) Samples of various carcinomas (T) and adjacent normal tissues were dissociated shortly after surgery into single-cell suspensions. Cells that were CD45+ (a general marker of leukocytes except plasma cells) were then resolved into subtypes by use of polychromatic flow cytometry, a specialized version of FACS (see Supplementary Sidebar 11.1), that allows concomitant monitoring of a number of distinct cell surface antigenic markers, enabling most of these cells to be classified into the various subtypes indicated in the key. (A and F, from D.G. DeNardo et al., *Cancer Cell* 16:91–102, 2009. B, C and D, from P.J. Barth et al., *Virchows Arch.* 444:231–234, 2004. E, from J. Wang-Rodriguez et al., *Cancer* 95:1387–1395, 2002. G, courtesy of L.M. Coussens.)

functions that these infiltrating cells exert, we can imagine that they may contribute in major ways to the distinctive traits exhibited by various types of carcinomas.

The diverse stromal cell types within tumors are all members of several mesenchymal cell lineages that generate both connective tissue and various types of immune cells in the blood and immune tissues, and therefore are biologically very different from the epithelial cells whose transformation drives the growth of carcinomas. (In the case of nonepithelial tumors such as sarcomas, where the cancer cells are themselves of mesenchymal origin, the boundaries between the malignant cell compartment and the nonmalignant cells within the tumor mass are less easily defined.)

In the absence of other information, we might explain the presence of these various stromal cell types within a carcinoma in two ways. They might represent the remnants of cells that resided in the stroma of a tissue before tumor development began. During the subsequent expansion of the tumor cell population, groups of cancer cells may have inserted themselves between these preexisting normal stromal cell layers, thereby generating the richly textured tissues that are seen when most carcinomas are examined microscopically.

The alternative explanation for the presence of these numerous stromal cells is more intriguing and is likely to be the correct one for the great majority of tumors. The rationale behind this second model depends on our current understanding of how normal, architecturally complex tissues maintain their structure and function. In such tissues, the proper proportions of the various component cells must depend on the continuous exchange of signals among them. These interactions involve multiple distinct cell types, each following its own particular differentiation program. Such communication between dissimilar cell types is termed **heterotypic** signaling and is used by each cell type to encourage or limit the proliferation of the other types of cells nearby.

Extending this thinking, we might speculate that many of the heterotypic interactions operating in normal tissues continue to play important roles in the biology of the tumors arising in these tissues. This would suggest that, like normal epithelial cells, carcinoma cells continue to control the populations of stromal cells near them, perhaps by recruiting the latter from nearby normal tissues and distant bone marrow, and then encouraging their proliferation. Operating in the other direction, stromal cells may also influence epithelial cell proliferation and survival in the tumor mass.

In normal tissues, these heterotypic signaling channels depend in large part on the exchange of (1) mitogenic growth factors, such as hepatocyte growth factor (HGF), transforming growth factor-α (TGF-α), and platelet-derived growth factor (PDGF); (2) growth-inhibitory signals, such as transforming growth factor-β (TGF-β); and (3) trophic factors, such as insulin-like growth factor-1 and -2 (IGF-1 and -2), which favor cell survival (**Figure 13.4**). While these factors are secreted into the extracellular space, they usually act over short distances and their localization is often important to their function (Supplementary Sidebar 13.1).

Given these complex interactions within normal tissue, some of which may be preserved in the tumors derived from these tissues, what evidence can we muster to support the notion that heterotypic interactions are actually functionally important for tumor growth? Until the mid-twentieth century, the role of stromal cells in supporting the growth of epithelial tumors was not appreciated. Experiments reported in 1961 provided some of the first evidence for the importance of these interactions in tumor biology; these were experiments of a sort that could not be repeated these days, given existing regulations on clinical testing of human subjects. Basal cell carcinomas of the skin were excised from patients and then reimplanted in normal areas of the skin elsewhere in the same patients (**Figure 13.5A**). Since these were auto-transplantations, with each patient receiving a graft of his or her own cells, no immune rejection occurred (see Section 3.5). When the implanted cells included only groups of carcinoma cells, the grafts failed to grow; however, if these carcinoma grafts included both epithelial cancer cells and the underlying tumor-associated stroma, they became well established at the sites of implantation (see Figure 13.5B). This experiment provided strong indication that the mesenchymal cells within the stroma are indeed essential to supporting the growth of the tumor as a whole.

Figure 13.4 Heterotypic ligand–receptor signaling (A) In epithelial tissues, the synthesis of PDGF-A and its receptor (PDGF-Rα) is usually confined to distinct cell types. In the testis, expression of PDGF is confined to the cells of the tubular epithelium (*above*), while expression of its receptor is confined to the surrounding mesenchymal cells (*below*). (B) The reciprocal expression of the Met receptor and its ligand, termed variously hepatocyte growth factor (HGF) or scatter factor (SF), is shown here in the embryonic mouse ileum (the lower section of the small intestine). In each fingerlike villus, seen in longitudinal section (see also Figure 11.6C), HGF/SF mRNA, detected by *in situ* hybridization, is made by the mesenchymal cells in the core of the villus (*above*) while Met is expressed by the epithelial cells coating the surface of the villus (*below*). (C) In the developing oral cavity of the mouse, the FGF10 ligand (*red*) is produced by the mesenchymal cells of the stroma of the developing palate (*left*) while its cognate receptor, FGF-R2b (*red*), is displayed by the overlying epithelial cells (*right*). Other work not shown here indicates that the epithelial cells respond to resulting ligand-mediated activation of their FGF-R2b by releasing Sonic Hedgehog (Shh), which signals back to Patched (Ptc) receptors displayed by mesenchymal cells in the stroma. (D) In the normal mouse mammary gland, use of antisense probes to detect mRNAs indicates that the Notch 3 receptor (*left*) is expressed (*black dots*) in the luminal epithelial cells, while its ligand, Jagged 1 (*right*), is expressed (*black dots*) in myoepithelial cells. (A, from L. Gnessi et al., *J. Cell Biol.* 149:1019–1026, 2000. B, courtesy of S. Britsch and C. Birchmeier. C, from R. Rice et al., *J. Clin. Invest.* 113:1692–1700, 2004. D, from M. Reedijk et al., *Cancer Res.* 65:8530–8537, 2005.)

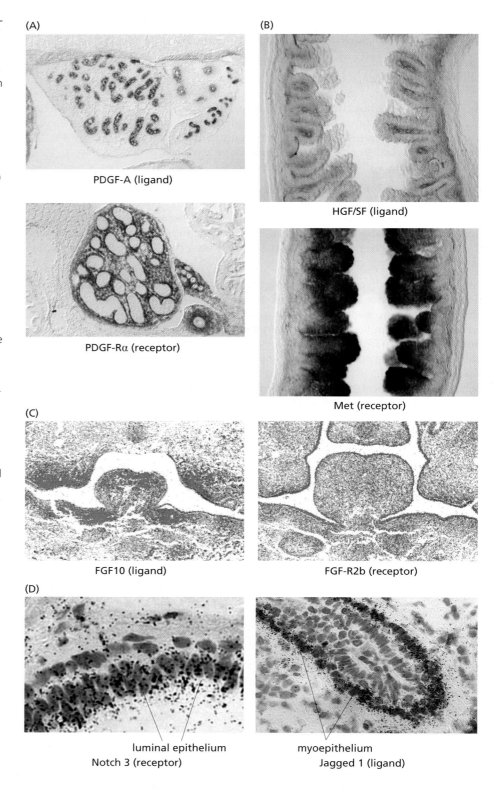

(A)

PDGF-A (ligand)

PDGF-Rα (receptor)

(B)

HGF/SF (ligand)

Met (receptor)

(C)

FGF10 (ligand)

FGF-R2b (receptor)

(D)

luminal epithelium
Notch 3 (receptor)

myoepithelium
Jagged 1 (ligand)

Now half a century later, the details of stromal–epithelial heterotypic interactions and their key role in tumor progression have been elucidated in great detail. We now know that they play key roles in virtually all the stages of tumor progression, including the initial formation of tumors. For example, carcinoma cells release growth factors, cytokines, and chemokines that recruit macrophages, neutrophils, and lymphocytes to the tumor-associated stroma; the recruited cells then orchestrate an inflammatory

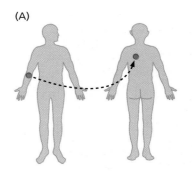

(A)

Figure 13.5 Autologous transplantation of skin tumors (A) Basal cell carcinomas of the skin were excised from patients and the carcinoma cells were reimplanted, either together with or without carcinoma-associated stromal cells, in distant anatomical sites on the skin of the back of the patients who had borne the original tumors. (B) The implanted tissues were excised 1 to 5 weeks later. Those that had been implanted together with stroma (*left*) yielded robustly growing carcinoma cells, while those that were implanted without concomitantly implanted stroma (*right*) failed to grow and yielded only vacuolated cells and cell debris. (A, courtesy of S. McAllister. B, from E.J. Van Scott and R.P. Reinertson, *J. Invest. Dermatol.* 36, 109–131, 1961.)

(B)

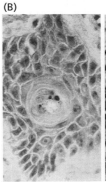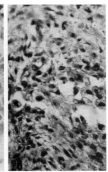

with stroma   without stroma

response that involves the release of TNF-α and prostaglandins, which proceed to stimulate the proliferation of nearby epithelial cells (see Section 11.16) and the process of **angiogenesis**—the formation of blood vessels within the tumor stroma.

Many of these heterotypic interactions continue to operate after fully neoplastic tumors are formed. To cite some examples, epithelial cells within a carcinoma often release PDGF, for which stromal cells—notably fibroblasts, myofibroblasts, and macrophages—possess receptors; the stromal cells reciprocate by releasing IGF-1 (insulin-like growth factor-1), which benefits the growth and survival of the nearby cancer cells. Similarly, the neoplastic cells within melanomas release PDGF, which elicits IGF-2 production from nearby stromal fibroblasts; this IGF-2 helps to maintain the viability of the melanoma cells. Stromal cells in breast cancers release the factor SDF-1/CXCL12 (a *chemokine;* see Section 13.5) and the HGF/SF growth factor, which stimulate the proliferation and survival of nearby epithelial cancer cells. By regulating each other's numbers and positions, epithelial and stromal cells can ensure the optimal representation and localization of each cell type within both normal and neoplastic tissues. Additional examples are illustrated in Supplementary Sidebar 13.2.

The stromal and epithelial cells in normal epithelial tissues collaborate in the construction of the specialized extracellular matrix (ECM) that lies between them, called variously the basement membrane or basal lamina (**Figure 13.6**; see also Figure 2.3). For example, the epithelial cells of the skin, the keratinocytes, express genes specifying many of the major protein components of the basement membrane, including type IV collagen and laminins; the stromal cells also contribute to its construction, doing so in ways that have yet to be documented in detail. Certain **proteoglycans** in the basement membrane, such as perlecan, provide it with increased hydration and with sites to which growth factors can be attached for long-term storage. Continuous tethering to the basement membrane, mediated largely by integrins and the hemidesmosomes that they construct (see Figure 13.6A), is essential for the survival of many kinds of epithelial cells, and loss of this tethering often provokes anoikis, the form of apoptosis resulting from loss of anchorage to a solid substrate (see, for example, Figure 9.22).

The **endothelial** cells (**Figure 13.7A**), which assemble to form the linings of the walls of capillaries and larger blood vessels (Figure 13.7B) as well as **lymphatic** ducts (Figure 13.7C), represent vital components of the normal and neoplastic stroma. As discussed in greater detail later, proliferation of the endothelial cells forming capillaries is encouraged by other cells in both the epithelium and stroma in order to guarantee access by all of these cells to an adequate blood supply. Once capillaries are assembled and become functional, they provide essential nutrients and oxygen to nearby cells. We were introduced to these interactions earlier, when we read that cells lacking adequate access to oxygen release **angiogenic factors** that stimulate the ingrowth of capillaries (Section 7.12).

While capillary formation is proceeding, the endothelial cells secrete growth factors that stimulate the proliferation of nearby nonendothelial cell types. Most important, endothelial cells release PDGF and HB-EGF (heparin-binding EGF), which enables them to attract (see Supplementary Sidebar 13.1) the peri-endothelial cells called **pericytes** and the vascular smooth muscle cells that together create the outer cell layers of capillaries, sometimes called the **mural** cells to distinguish them from the

(A)

keratinocyte nuclei

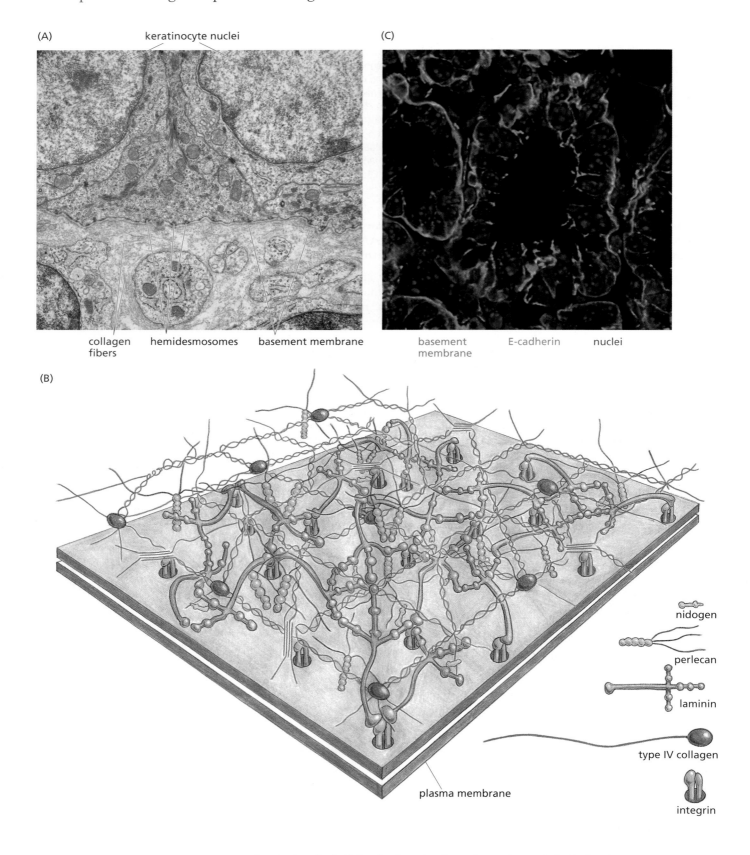

collagen fibers   hemidesmosomes   basement membrane

(C)

basement membrane   E-cadherin   nuclei

(B)

nidogen

perlecan

laminin

type IV collagen

integrin

plasma membrane

luminal endothelial cells. Once in place, the pericytes (which closely resemble smooth muscle cells) reciprocate by releasing vascular endothelial growth factor (VEGF) and angiopoietin-1 (Ang-1), which provide important survival signals to the endothelial cells that recruited them in the first place. In addition, these mural cells provide the

Figure 13.6 The basement membrane (A) In epithelial tissues, the specialized extracellular matrix (ECM) termed the basement membrane (BM) separates the epithelial cells (*above*) from the stroma (*below*). As shown in this transmission electron micrograph of the skin of a newborn mouse, the epithelial cells of the skin (termed keratinocytes) are tethered to one side (*above*) of the BM, being anchored in part through structures termed *hemidesmosomes*, which are composed of clusters of integrins that bind BM proteins, notably laminin. Below the BM are the collagen fibers and portions of mesenchymal cells constituting the stroma of the skin, i.e., the dermis. (B) This schematic drawing indicates that the basement membrane is composed largely of four major ECM proteins, namely, laminin, collagen type IV, perlecan, and nidogen. This highly permeable molecular meshwork allows a variety of molecules to pass through in both directions. A specialized BM, not shown here, underlies endothelial cells in capillaries. (C) This immunofluorescence micrograph of a mammary duct in a lactating mouse reveals the E-cadherin molecules (*green*) that form the lateral junctions tying together the mammary epithelial cells forming the lumen of the duct; the epithelial cells are underlain by a basement membrane, which is revealed here by the presence of laminin (*orange*). Outside of the BM are various types of stromal cells (*not shown*). (A, courtesy of H.A. Pasolli and E. Fuchs. B, from B. Alberts et al., Molecular Biology of the Cell, 5th ed. New York: Garland Science, 2008. Based on H. Colognato and P.D. Yurchenko, *Dev. Dyn.* 218:213–234, 2000. C, courtesy of E. Lowe and C. Streuli; from J. Muschler and C. Streuli, *Cold Spring Harb. Perspect. Biol.* 2:a003202, 2010.)

endothelial tubes with structural stability and an ability to resist the forces exerted by the pressure of blood (see Figure 13.7).

The locations and large populations of stromal cells within carcinomas clearly indicate that the growth of the epithelial components of these tumors is accompanied by coordinated proliferation of stromal cells. In fact, stromal cells are even found to be layered between carcinoma cells in most metastases. This shows that even those cancer cells that are independent enough to leave the primary tumor and to found new tumors at distant sites usually find it necessary to recruit stromal cells and encourage their proliferation (**Figure 13.8A**), ostensibly to aid in the construction of these new tumor colonies. On occasion, metastatic cells will insinuate themselves into the existing stroma of the tissue in which they have landed (Figure 13.8B), once again demonstrating their need to secure stromal support.

In the extreme version of this scenario, we might imagine that all of the heterotypic interactions needed to maintain normal tissue function continue to operate within carcinomas, being required for the neoplastic cells to thrive and multiply within these tumors. However, such a depiction cannot be literally true, since many of the acquired traits that we associate with cancer cells (see Chapter 3), including a decreased dependence on mitogens, an increased resistance to apoptosis, and acquisition of anchorage independence, are certain to affect the interactions between epithelial and stromal compartments and to lessen their interdependence. So, we conclude that these acquired traits generally *reduce* the dependence of epithelial cancer cells on stroma but do not seem to *eliminate* it. Still, there are examples of cancer cells that appear to be almost totally independent. Certain highly progressed carcinomas will generate free-floating cancer cells that accumulate in various body fluids—generating tumors known as **ascitic** tumors (Supplementary Sidebar 13.3).

13.2   The cells forming cancer cell lines develop without heterotypic interactions and deviate from the behavior of cells within human tumors

The continued dependence of carcinoma cells on stromal support explains the enormous difficulties that researchers have experienced when attempting to adapt tumor cells to *in vitro* culture conditions. Their goal has been to create cancer cell lines—populations of cancer cells that can be propagated in tissue culture indefinitely. To do so, cells are prepared directly from tumor biopsies and placed into culture, with the hope that colonies of vigorously growing cancer cells will eventually emerge. In most

(A)

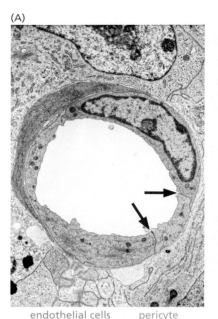

endothelial cells   pericyte

(B)

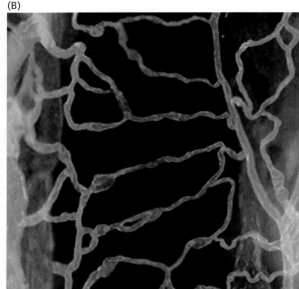

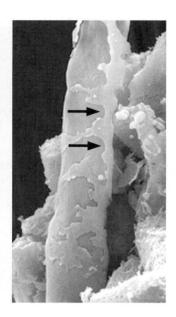

(C)

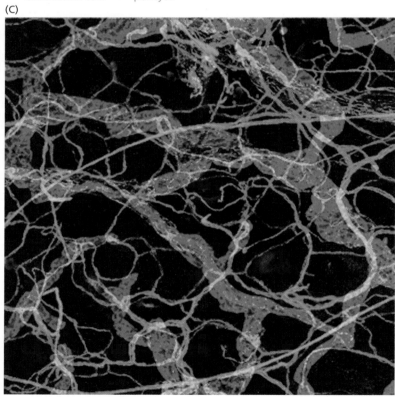

capillaries   lymph ducts

Figure 13.7 Microvessels in normal tissues (A) This transmission electron micrograph presents a cross section of a capillary within the stroma of a tumor formed by mouse Lewis lung carcinoma cells. The endothelial cell and its nucleus have been colorized *green*, while the pericyte, which helps maintain capillary rigidity and intraluminal blood pressure, is colorized *red*. At least one additional endothelial cell has contributed to the formation of this section of the capillary, as indicated by the *arrows* at the joints between the cytoplasms of these two endothelial cells.
(B) Immunofluorescent microscopy (*left*) of the capillaries in a normal mouse trachea reveals (using an antibody that recognizes the endothelial cell–specific CD31 antigen) a well-organized network of microvessels. At the level of scanning electron microscopy (*right*), a capillary is seen to resemble a smooth-walled tube with pericytes (*arrows*) adhering to its outer surface.
(C) Most normal tissues are interlaced with networks of capillaries (*green*) and lymphatic vessels (*red orange*). As is apparent here, the diameters of lymphatic vessels are far larger than those of capillaries. Unlike capillaries, however, the endothelial cells of lymphatic vessels are not supported by underlying mural cells—pericytes and smooth muscle cells (*not shown*). (A, from B. Sennino et al., *Cancer Res.* 67:7358–7367, 2007. B, from T. Inai et al., *Am. J. Pathol.* 165:35–52, 2004. C, courtesy of T. Tammela and K. Alitalo.)

cases, however, such cell colonies do not appear, foiling attempts to generate carcinoma cell lines. More often than not, the cells that do thrive in culture are of stromal origin—specifically the fibroblasts whose growth is favored by the platelet-derived growth factor (PDGF; see Figure 5.3) that is present in the serum component of standard tissue culture medium.

It is clear that human carcinoma cells propagated under standard conditions of tissue culture are being forced to proliferate in an environment very different from the one experienced by their ancestors that resided in the tumors of cancer patients. For

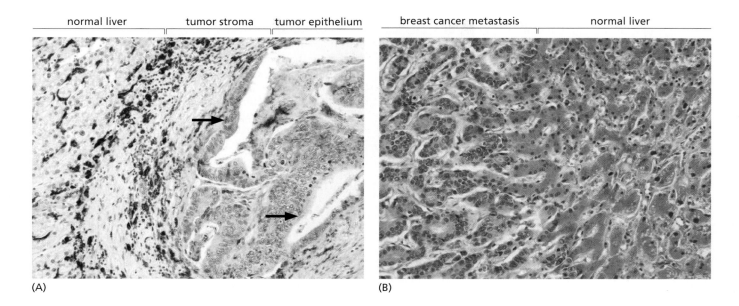

(A)   (B)

Figure 13.8 Metastasis and dependence on stroma (A) This micrograph of a metastasis originating from a primary colon carcinoma reveals that the metastasis found in the liver is as complex histologically as the primary tumors seen in Figure 13.1B. Here, we see that the carcinoma cells that have formed a metastasis (*light purple, right*) have constructed the same ductal structures (*arrows*) as are seen in primary adenocarcinomas of the colon. In addition, the carcinoma cells have recruited substantial stroma (*middle*), which includes macrophages (*dark brown*). (B) Most breast cancer metastases to the liver (*left*) behave quite differently from colon carcinoma metastases to this organ, in that the mammary carcinoma cells infiltrate into normal liver tissue (*right*) and displace resident hepatocytes, thereby taking advantage of the existing stroma, including its vasculature. These metastatic cells appear to be as dependent on stromal support as the colorectal carcinoma cells of (A) but acquire it in a totally different way. (From F. Stessels et al., *Br. J. Cancer* 90:1429–1436, 2004.)

example, the high concentration of serum present in most culture media is growth-inhibitory for many types of normal and neoplastic epithelial cells, which rarely experience large concentrations of the serum-associated factors *in vivo* except acutely after wounding. Even more important, carcinoma cells are being selected *in vitro* for their ability to proliferate without intimate contact with stroma, since the proper balance of epithelial and stromal cells present in tumors cannot be maintained in the tissue culture dish.

On rare occasion, vigorously proliferating colonies of carcinoma cell populations do indeed emerge following extended culture of tumor fragments in the Petri dish. Because they have been selected *in vitro* for their ability to grow autonomously, and therefore independently of stromal support, such cancer cells have evolved beyond the stage of tumor progression that is reached *in vivo* by the neoplastic cells of most human tumors. In fact, many of the successes in establishing human tumor cell lines have depended on culturing cells from the few tumors that had already progressed *in vivo* to a stage where they no longer depended on stroma for their survival and proliferation (see Supplementary Sidebar 13.3).

 Human cancer cell lines—there are several dozen in common use—are standard reagents used in many types of cancer research. Cells from virtually all of these can be implanted into immunocompromised host mice, where they proliferate, often vigorously, and form tumors termed **xenografts**, since they result from grafting the tissues of one species into a host animal of another. Mice of the Nude, NOD/SCID, and RAG1/2 strains, lacking functional immune systems, are used in these experiments, because they tolerate the growth of introduced, genetically foreign cells. The resulting xenografts growing in these mice can then be tested for their responsiveness to anticancer drugs under development. The outcome of such experiments is not necessarily predictive of eventual clinical responses to these drugs (**Sidebar 13.1**).

13.3   Tumors resemble wounded tissues that do not heal

As argued above, heterotypic signaling governs much of the biology of carcinomas, and experimental models of cancer that ignore this important process generate tumors that are biologically very different from those found in cancer patients. The heterotypic interactions operating within tumors are highly complex and involve the exchange of dozens of distinct molecular species that mediate cell-to-cell signaling between the various cell types that together form these growths. Importantly, many of these signals are not routinely exchanged by the cell types constituting normal, undamaged tissue and therefore seem, at first glance, to be unique to tumors.

Sidebar 13.1 The development of anti-cancer therapies has been imperfectly served by the use of existing human cancer cell lines Human cancer cell lines, when grown as tumor xenografts in immunocompromised mice, often respond to the anti-proliferative effects of cytotoxic drugs under development. However, the ability of these xenograft models to predict the clinical responses of patients to these drugs is limited. A 2001 retrospective study of 39 different anti-cancer drugs showed a weak correlation between the responses of xenografted tumors and oncology patients to these drugs. For example, candidate drugs that were effective in halting the growth of more than one-third of a disparate collection of mouse tumor xenografts had only a 50% likelihood of showing any therapeutic activity (including halting tumor growth) in some human tumor types treated in the clinic. And drugs that affected smaller proportions of the mouse xenograft models had essentially no meaningful activity in the clinic.

These disappointing outcomes are not surprising, in light of the fact that the human tumor cell lines commonly used in such drug testing experiments bear little resemblance to cells in the tumors frequently encountered in the oncology clinic. These cell lines have been propagated *in vitro* under conditions that differ dramatically from their sites of origin in patients (**Figure 13.9A**) including the absence of a supporting

(A)

CONDITION	HUMAN TISSUE	HUMANIZED 3D CULTURE	2D CULTURE
oxygen:	0–5%	≤ 5%	21%
stiffness of ECM:	200–4000 Pa	200–4000 Pa	3×10^9 Pa
pH:	acidic (<7)	acidic (<7)	buffered (7.2)
dimensionality:	3D	3D	2D
glucose availability:	limiting	limiting	not limiting
stroma:	present	absent	absent

(B)

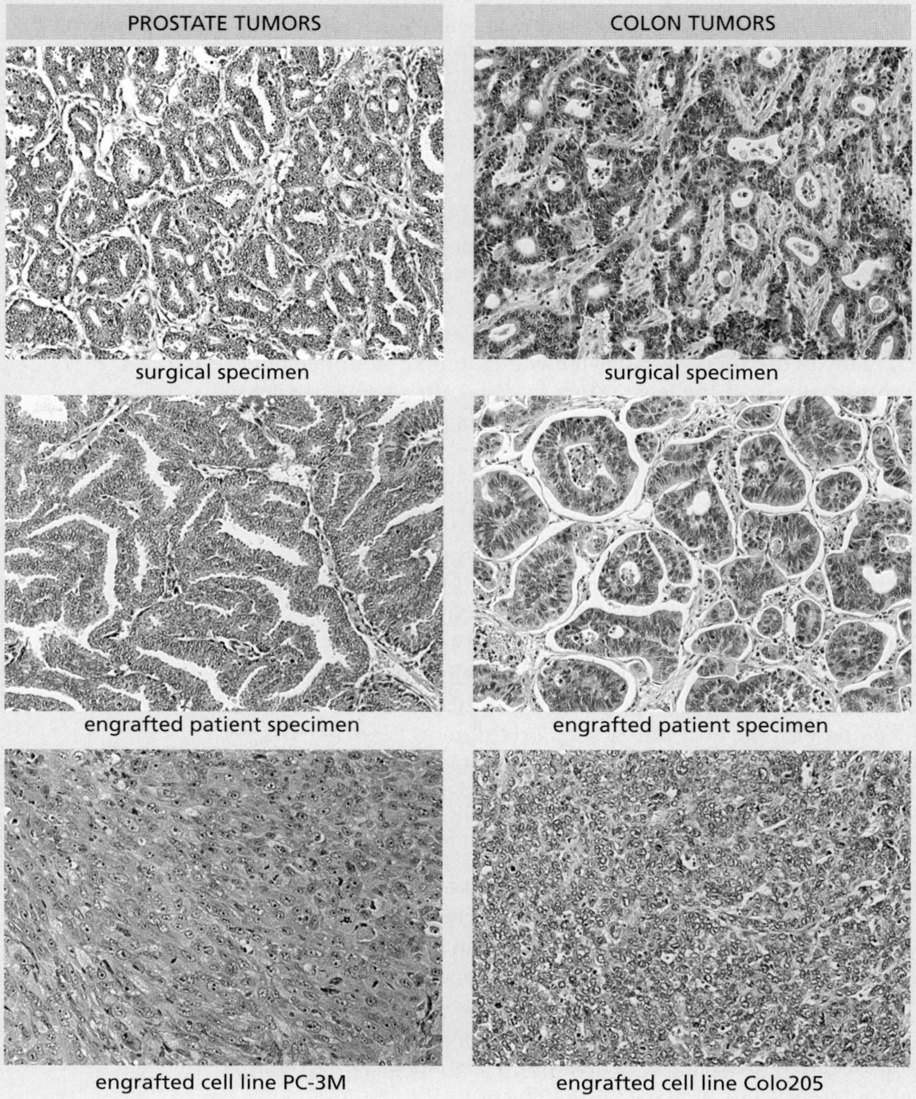

PROSTATE TUMORS	COLON TUMORS
surgical specimen	surgical specimen
engrafted patient specimen	engrafted patient specimen
engrafted cell line PC-3M	engrafted cell line Colo205

stroma, and the tumors that these cell lines form often look quite different histologically from those routinely encountered in a clinical pathology laboratory (see Figure 13.9B).

Another difficulty arises from the fact that cells of various human tumor cell lines are usually implanted in mice subcutaneously, which represents an **ectopic** site within the host that has little resemblance to the tissue in which the tumor originated. For example, pancreatic carcinoma cells may behave differently when growing under the skin than when they are implanted into the host animal's pancreas, their natural, **orthotopic** site, presumably because the stromal microenvironments in these two locations are so different.

One experimental response to these difficulties involves attempts at introducing fragments of tumors directly into immunocompromised mice without intervening propagation

in tissue culture. These implanted tumor grafts contain both epithelial and stromal cells, which may co-proliferate to form histologically complex tumors in host mice that resemble the original tumors from which they derive (see Figure 13.9). Unfortunately, like the cancer cells in actual human tumors, the cells in these xenografts usually proliferate very slowly, and their propagation involves labor-intensive transplantation from one host animal to another. These properties preclude their routine use in the testing of anti-cancer drugs.

The inadequacies of currently available human tumor models and the resulting inability to accurately predict the clinical effectiveness of anti-cancer drugs under development cost the pharmaceutical industry hundreds of millions of dollars annually and therefore represent one of the major impediments to the development of new anti-cancer drugs.

Figure 13.9 Tumors and derived xenografted tumor cell lines (A) The conditions under which human tumor cell lines are usually propagated (*red font*) differ drastically from those operating in an actual autochthonously arising human tumor, i.e., one that arose spontaneously in a living human tissue (*middle column*). Attempts to remedy this by developing culture models that more closely resemble conditions in actual tumors (*middle column*) address some but not all of the shortcomings of existing means of propagating human tumor cells. (This figure does not indicate the differences in the chemical compositions of the fluids surrounding individual cells or the molecular compositions of the ECMs, which are likely to differ dramatically between the three conditions.) Pa, pascals, unit of pressure here denoting mechanical stiffness. (B) The histologies of a primary prostate carcinoma (*left*) and of a colon carcinoma (*right*) that were removed surgically from patients are shown in the *top row*. Below these (*middle row*) are the histologies arising from implantation of chunks of tumor derived from cancer surgery at subcutaneous sites in immunocompromised SCID mice; these implanted chunks contained both epithelial and stromal tumor components. Three to six months after implantation, these engrafted tumor samples, which have grown from a 2-mm to a 1-cm diameter, still retain the architecture of the primary tumors from which they derive. However, the tumors formed by a prostate and a colon cancer cell line (*bottom row*), which have been propagated extensively in tissue culture as pure cancer cell populations, bear little histological resemblance to the tumors typically encountered in the oncology clinic and shown above. (A, courtesy of B. Hall as adapted. B, courtesy of B.L. Hylander and E.A. Repasky.)

This complexity raises some fundamental questions: How do cancer cells learn to release and respond to an array of diverse heterotypic signals? Are the complex signaling programs and resulting biological responses invented anew, being cobbled together piece by piece, each time normal cells evolve progressively into cancer cells? Or do cancer cells exploit preexisting biological programs that are normally used by tissues for other purposes?

One attractive solution to these questions has come from an insight gained in 1986. A researcher studying the histological appearance of both tumors and wounds noted the striking resemblance between many of the signaling processes involved in tumor progression and those that occur during wound healing. If the similarities between these two biological programs could indeed be documented, this would explain much of the cleverness of tumor cells: they simply activate a complex, normal physiologic program—wound healing—that is already encoded in their genomes. By accessing and exploiting this preexisting biological program, cancer cells are spared the task of re-inventing it anew each time a tumor arises.

Wound healing has been studied most extensively in the context of the skin. Following the formation of a superficial wound to the skin and underlying tissues, blood platelets aggregate and release granules containing, among other factors, platelet-derived growth factor (PDGF; see Figure 5.3) and transforming growth factor-β (TGF-β). Wounding also causes release of **vasoactive** factors, which increase the permeability of blood vessels near the wound. This helps the wound site to acquire fibrinogen molecules from the blood plasma, which, when converted to fibrin, create the scaffolding of the blood clot (**Figure 13.10**). The resulting fibrin bundles, which become entwined around clumps of platelets, help to stanch further bleeding.

Figure 13.10 Scanning EM of a blood clot This scanning electron micrograph, which has been colorized, shows that clots are composed of dense networks of fibrin fibers (*light brown*) that have trapped red blood cells (*red*). Prior to clotting, platelets undergo activation, which leads to the release of the granules seen in Figure 5.3. (Courtesy of Eye of Science/Science Source.)

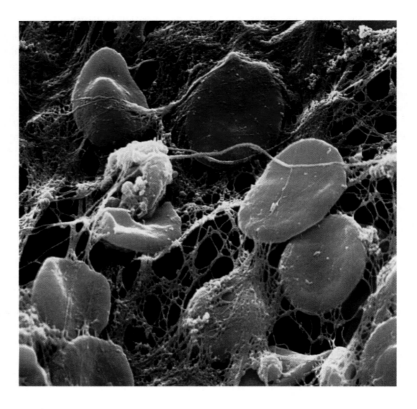

The PDGF released by the platelets attracts fibroblasts and stimulates their proliferation (**Figure 13.11**). Thereafter, the platelet-derived TGF-β activates these fibroblasts, converting them into **myofibroblasts**, and induces the latter to release a class of secreted proteases termed **matrix metalloproteinases** (MMPs; **Table 13.1**), which are also produced by macrophages that have been recruited in large numbers into the wound site. Unlike most secreted proteases, which have a serine in their catalytic clefts, MMPs carry zinc ions to aid in catalysis; indeed, these ions inspired their name.

Table 13.1 Some matrix metalloproteinases and their extracellular matrix substrates

Name of MMP[a]	Alternative name of MMP	Major ECM substrates[b]
MMP-1	collagenase-1	fibrillar collagens (collagen I, II, III)
MMP-2	gelatinase A	collagen IV, denatured collagens/gelatins
MMP-3	stromelysin-1	proteoglycans, glycoproteins (laminin, fibronectin, vitronectin)
MMP-7	matrilysin	similar to MMP-3
MMP-9	gelatinase B	similar to MMP-2
MMP-12	metalloelastase	elastin
MMP-14	MT1-MMP	fibrillar collagens, gelatins

[a]23 distinct MMPs have been documented in mammalian cells.
[b]In addition to ECM substrates, most of the MMPs also cleave other substrates that are present in the extracellular space and whose activities influence cancer progression. For example, MMP-2 processes the chemokine MCP-3; MMP-3 and MMP-7 cleave E-cadherin; MMP-9 releases VEGF and Kit-L; MMP-14 cleaves CD44. Also, the MMPs are part of a proteolytic cascade involving the conversion of plasminogen to plasmin and activation of latent pro-MMPs.

Courtesy of L. Matrisian and B. Fingleton, adapted in part from information from K. Kessenbrock, V. Plaks and Z. Werb, *Cell* 141: 52–67, 2010.

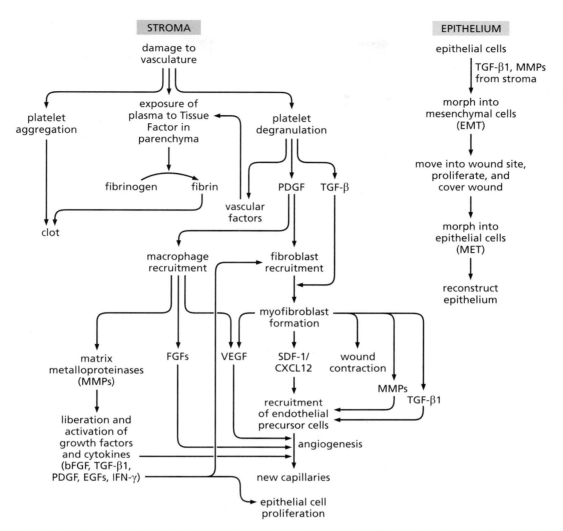

Figure 13.11 Flowchart of wound healing These flowcharts indicate many of the major events occurring after an epithelial tissue is wounded. Many of the changes occur in the stroma of the wounded tissue (*left*) and thereby provide a foundation for reconstructing the epithelial cell layer (*right*). A short-term effect of wounding is the process of *hemostasis*—the stanching of further hemorrhage. It involves both the plugging of damaged vessels and the degranulation of platelets, which release a number of distinct growth factors (GFs), including importantly PDGF and TGF-β; in addition, the exposure of plasma and platelets to Tissue Factor (expressed on the surfaces of a wide variety of nonendothelial cell types) triggers the formation of fibrin and thus clot formation (see Figure 13.10). PDGF and TGF-β recruit and activate both macrophages and fibroblasts. Macrophages, in turn, produce a number of products that are critical to initiating wound healing, including MMPs (to remodel the extracellular matrix and liberate various GFs); this allows recruitment of more macrophages and stimulates the proliferation of epithelial cells. The fibroblasts are activated by the actions of TGF-β and CXCL12/SDF-1 and, as myofibroblasts (see Figure 13.16), produce more of these factors that function, together with macrophage products, to stimulate angiogenesis.

Activated fibroblasts also secrete mitogens, such as various fibroblast growth factors (FGFs) that can stimulate the proliferation of certain epithelial cells.

Once released, the MMPs begin to degrade specific components of the extracellular matrix (ECM; see Table 13.1). This degradation has two major consequences. On the one hand, it allows for structural remodeling of the ECM, thereby carving out space for new cells. On the other, it results in the release of a variety of growth factors that have been tethered in an inactive form to the proteoglycans of the ECM and now become solubilized and activated. Included among these are basic fibroblast growth factor (bFGF), TGF-β1, PDGF, several EGF-related factors, and interferon-γ (IFN-γ). (In addition, virtually every MMP has been found to act on a variety of non-ECM proteins in the extracellular space; among these substrates are the latent **pro-enzyme** forms of other proteases, which are converted into active enzymes following cleavage by MMPs.)

Table 13.2 Properties of VEGFs

Capillary and lymph duct formation

Monocyte migration

Hematopoiesis

Recruitment of hematopoietic progenitor cells from the bone marrow

Regulation of the endothelial cell pool during development

Capillary permeability

The growth factors released by platelets and those mobilized from the ECM then attract **monocytes** (which soon differentiate into additional macrophages) and another class of phagocytes, termed **neutrophils**, which infiltrate the wound site. (Yet other types of immune cells, including **eosinophils**, **mast cells**, and **lymphocytes**, are also recruited to this site.) These recruited cells scavenge and remove foreign matter, bacteria, and tissue debris from the wound site; at the same time, they release and activate mitogenic factors, such as fibroblast growth factors (FGFs) and vascular endothelial growth factor (VEGF; see also Section 7.12). Such factors proceed to stimulate endothelial cells in the vicinity to multiply and to construct new capillaries—the process of angiogenesis (sometimes termed **neoangiogenesis**). As we will see shortly, VEGF and its cognate receptors play multiple roles in angiogenesis and related hematopoietic processes (**Table 13.2**).

While all this is occurring, largely in the stromal region of a wound site, the epithelial cells around the edges of the wound are undergoing their own alterations. Their goal is to reconstruct the epithelial sheet that existed prior to wounding. To do so, epithelial cells reduce their adhesion to the ECM, especially to the basement membrane (see Section 2.2 and Figure 2.3) that separates them from the stromal compartment (see Figures 13.6 and 13.12). By severing these connections, the epithelial cells gain increased mobility.

The epithelial cells also sever their attachments to one another. These side-by-side associations are stabilized in part by **adherens junctions** (**Figure 13.12**), which are assembled as associations of E-cadherin molecules that are displayed by adjacent epithelial cells and tether the cells' closely **apposed** plasma membranes to one another. Accordingly, E-cadherin expression is suppressed in the epithelial cells that are situated around the edges of a wound site and is often replaced by N-cadherin, another cell–cell adhesion molecule that is normally displayed by mesenchymal cells, notably fibroblasts. (These N-cadherin molecules, though related structurally to E-cadherin, do not reestablish adherens junctions between the epithelial cells, since they bind much more weakly to one another.)

While shifting their display of cell surface cadherins, the epithelial cells at the edge of a wound undergo a major change in phenotype, which causes them to assume, at least superficially, a fibroblastic appearance (**Figure 13.13**). This profound shift is termed an **epithelial–mesenchymal transition** (EMT) and enables the epithelial cells to become motile and invasive. These acquired traits enable these cells to move into the wound site, where they fill in the gap in the epithelium created by the wounding.

Stromal cells in the wound site help to trigger the EMT of nearby epithelial cells. The matrix metalloproteinases (MMPs) secreted by stromal cells liberate and activate latent growth factors, such as TGF-β1, that have been stored in the extracellular matrix. In ways that are only partially understood, these released factors stimulate and reinforce expression of the EMT program in nearby epithelial cells. Importantly, the EMT is only a temporary shift in cell phenotype. After the migrating cells have moved into position, proliferated, and covered the wound site, they reconstruct the epithelium by reverting to an epithelial state via the program termed the **mesenchymal–epithelial transition** (MET; see Figure 13.13A). As a consequence, once wound healing is complete, the cells in the reconstituted epithelium show no trace of having passed transiently through a mesenchymal state. We will discuss the EMT program in much greater detail in the next chapter, since it plays a key role in enabling carcinoma cells to acquire malignant traits.

If we compare these processes (**Figure 13.14**) with the interactions of epithelial cancer cells and their stromal neighbors, we find striking parallels between wound healing and tumorigenesis. One clear similarity between the two processes derives from the presence of clumps of fibrin in the tumor-associated stroma. In the case of tumors, this does not derive from traumatic damage to blood vessels. Instead, the capillaries within tumors are constitutively leaky, unlike those in normal tissues (**Figure 13.15A**). Later, we will discuss the causes of this leakiness. For the moment, suffice it to say that the permeability of the capillary walls and **venules** (small veins) allows fibrinogen molecules from the plasma to come in direct contact with cancer cells; this provokes, through a series of intermediate reactions, the conversion of fibrinogen to fibrin and the formation of large bundles of fibrin strands (see Figure 13.15B and C).

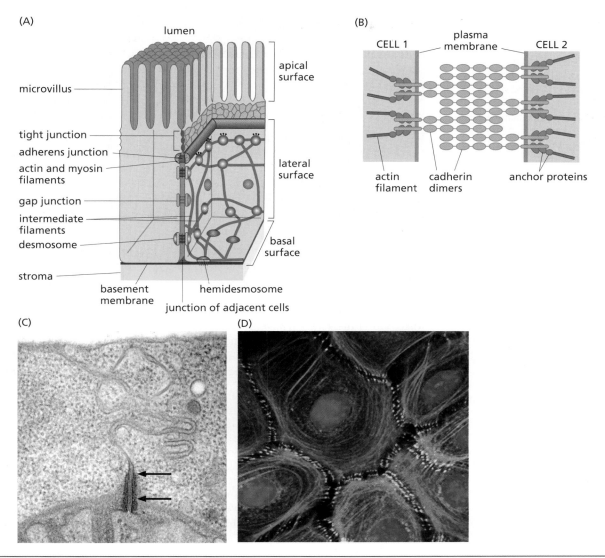

Figure 13.12 Attachments of epithelial cells to their neighbors and the basement membrane (A) Epithelial cells are anchored via hemidesmosomes on their basal surfaces to the underlying basement membrane (*red brown*) and via their lateral surfaces to adjacent epithelial cells. These lateral connections include tight junctions (which create a fluid seal between the luminal and basolateral spaces, *light orange*), adherens junctions and desmosomes (which link the cytoskeletons of adjacent epithelial cells, *red orange, blue*), and gap junctions (which permit low–molecular-weight solutes to pass between these cells, *light orange*). (B) The adherens junctions, which play a central role in regulating epithelial cell behavior, depend on oligomerization between E-cadherin molecules (*light green*) displayed by two adjacent cells. The "anchor proteins" (*blue*), which include prominently α- and β-catenin and p120, form physical bridges between the cytoplasmic tails of E-cadherin proteins and the actin cytoskeleton (*red*; see Figure 6.26A). As discussed in Section 7.11 and illustrated in Figure 6.26B, β-catenin has a second, quite independent function: when it is not sequestered in adherens junctions, it may move into the nucleus and serve as a subunit of a transcription factor complex

that drives the expression of key genes. (C) This transmission electron micrograph shows the adherens junctions (*arrows*) between two closely apposed plasma membranes of adjacent epithelial cells, in this instance those creating the intestinal lining of the worm *Caenorhabditis elegans*. Virtually identical structures are seen in mammalian cells. (D) This immunofluorescence micrograph illustrates the interactions between the lateral surfaces of neighboring keratinocytes in 2-dimensional monolayer culture. E-cadherin "zippers" (*yellow*) form adherens junctions between neighboring cells and are attached via their cytoplasmic domains and other intermediary proteins (*not shown*) to the actin cytoskeletons of these cells (*red*). Nuclei are stained in *blue*. Some of these interactions in actual epithelial tissues can be visualized via immunostaining of E-cadherin and laminin (a basement membrane protein), as shown in Figure 13.6C. (A, from H. Lodish et al., Molecular Cell Biology, 6th ed. New York: W.H. Freeman, 2008. B, from B. Alberts et al., Molecular Biology of the Cell, 4th ed. New York: Garland Science, 2002. C, courtesy of D. Hall. D, from M. Perez-Moreno, C. Jamora and E. Fuchs, *Cell* 112:535–548, 2003.)

Figure 13.13 The epithelial–mesenchymal transition (A) A confluent monolayer of MCF10A cells, a line of nontransformed, immortalized human mammary epithelial cells (MECs), has been disturbed by removal of a patch of cells (*at left of each image*). Before wounding, no vimentin is expressed by these cells. However, soon after wounding, the epithelial cells at the edge of the wound undergo a partial epithelial–mesenchymal transition (EMT), express vimentin (characteristic of mesenchymal cells; *red*), migrate into the wound site, proliferate, and fill the wound site. After 8 days, the cell monolayer has become confluent, and these cells revert to a fully epithelial phenotype and shut down vimentin expression [indicating a mesenchymal–epithelial transition (MET)], thereby reconstructing an intact epithelium. (B) In some cell types in culture, individual cells may spontaneously shift from an epithelial phenotype, indicated by cytokeratin expression (*red*), to a mesenchymal phenotype, indicated by α-smooth muscle actin expression (*green*). This mirrors a plasticity that is apparent in certain cells during early embryogenesis. (C) An EMT was provoked here by a 3-day-long exposure to matrix metalloproteinase-3 (MMP-3), which may trigger an EMT through its ability to degrade E-cadherin. Thus, some of the epithelial cells have shut down expression of many typical epithelial cell markers, such as cytokeratins (*dark pink*) and E-cadherin, and express instead mesenchymal proteins, such as vimentin (*green*), fibronectin, and N-cadherin. In addition, they change their typically polygonal shape (*top*) to a fibroblastic shape (*bottom*) and often acquire motility and invasiveness. (A, from C. Gilles et al., *J. Cell Sci.* 112:4615–4625, 1999. B, from O.W. Petersen et al., *Am. J. Pathol.* 162:391–402, 2003. C, from M.D. Sternlicht et al., *Cell* 98:137–146, 1999.)

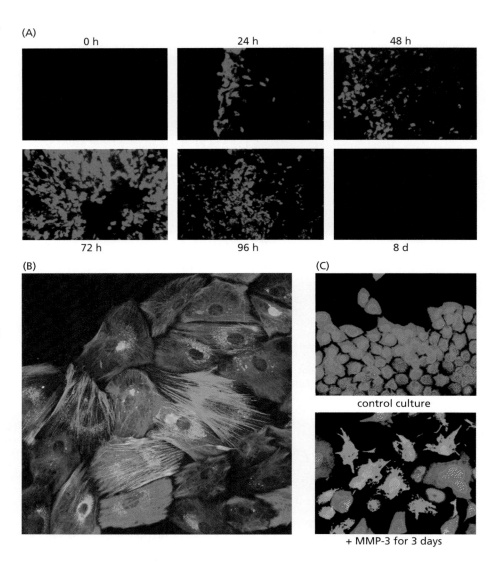

Many kinds of cancer cells, including those forming carcinomas of the breast, prostate, colon, and lung, continuously release significant levels of PDGF. This contrasts with the situation in wounds, in which PDGF is released in a brief burst by platelets as they form the initial blood clot (see Figures 5.3 and 5.4). As is the case in wound healing, the targets of the PDGF produced by cancer cells are mesenchymal cells in the stroma that display PDGF receptors, including smooth muscle cells, fibroblasts, and macrophages. PDGF functions here both as an attractant and as a mitogen for these stromal cells and appears to be the most important signaling molecule used by carcinoma cells to recruit and stimulate the proliferation of stromal cells. In breast cancers, for example, the level of PDGF expression generally increases with increased tumor progression, which may explain the high degree of **stromalization** of many advanced tumors.

The PDGF released by carcinoma cells initially succeeds in recruiting fibroblasts into the fibrin matrix. At the same time, the fibrin matrix (see Figure 13.15B and C) provides an important scaffolding that helps these recruited mesenchymal cells to attach, via integrins, and migrate. The invading stromal cells remodel this "provisional matrix" by degrading many of the initially formed fibrin molecules and replacing them with a more permanent matrix that is assembled from the collagen secreted by the fibroblasts. This closely parallels the sequence of steps occurring during wound healing.

One of the tasks of the stromal cells in wound healing involves the physical contraction of the wound site in order to close wounds. This contraction is mediated by the specialized fibroblasts cited above—the myofibroblasts (see Figure 13.14); these cells

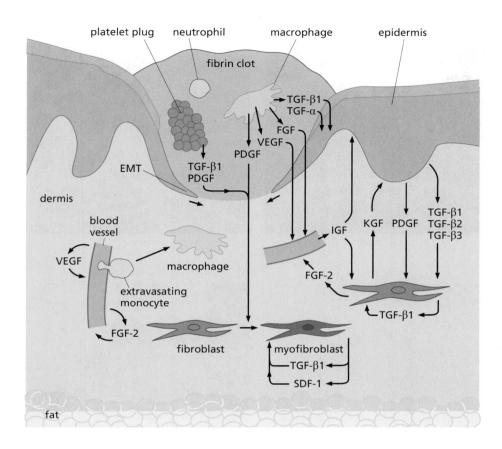

Figure 13.14 The program of wound healing This drawing depicts the heterotypic signaling occurring after the skin, a typical epithelial tissue, is wounded. A fibrin clot that carries platelets and red blood cells/erythrocytes (see Figure 13.10) initially fills the wound site. Day 3 of the wound-healing process is sometimes termed its inflammatory phase, because of the involvement of cell types that are associated with inflammation, notably neutrophils and monocytes; they are involved in removing cell debris, the clot, and invading bacteria. In addition, macrophages, which derive from differentiation of monocytes, release TGF-β1 and TGF-α; these factors stimulate the proliferation of nearby epithelial cells from the epidermis and induce them to undergo an epithelial–mesenchymal transition (EMT). Tongues of the resulting cells, now exhibiting mesenchymal behavior, invade under the clot from both sides in order to cover the wound site (see Figure 13.13A). At the same time, TGF-β1 and PDGF-A and -B, which were released by the platelets during initial clot formation, stimulate the proliferation of fibroblasts (*blue purple*) in the underlying dermal stroma and induce them to convert into the activated fibroblasts termed myofibroblasts (*red*); because of their contractility, myofibroblasts will pull the sides of the wound together. Myofibroblasts also release angiogenic factors, notably VEGFs and FGF-2 (fibroblast growth factor-2) (*not shown*), which stimulate the formation of new blood vessels in the stroma underlying the wound site. Keratinocyte growth factor (KGF), another isoform of FGF, is released by mesenchymal cells and stimulates epithelial cell proliferation. In addition, insulin-like growth factors (IGFs) provide survival signals to various cells in the wound-healing site. Finally, after the wounded stroma is covered, the cells that had undergone an EMT revert, via a mesenchymal–epithelial transition (MET), to an epithelial state, thereby reconstituting the epithelium (*not shown*). (Redrawn from A.J. Singer and R.A.F. Clark, *N. Engl. J. Med.* 341:738–746, 1999.)

express α-smooth muscle actin (α-SMA) and are capable of using the actin–myosin contractile system (related to the one operating in muscle cells) to generate the mechanical tension needed for wound closure (**Figure 13.16A**). Myofibroblasts are also present in sites of chronic wounding, that is, continuously inflamed tissues (Figure 13.16B). Provocatively, essentially identical α-SMA-positive fibroblasts are prominent components of the stroma present in the majority of advanced carcinomas (for

Figure 13.15 Capillary leakiness and deposition of fibrin bundles in the tumor-associated stroma
(A) Injection of red dextran dye into normal microvasculature reveals that the walls of associated capillaries have relatively low permeability, as indicated by the sharply delineated profiles of these vessels (*left*). However, when injected into tumor-associated vasculature, the dye leaks out of the capillaries and diffuses into the nearby parenchyma, yielding diffusely staining outlines of vessels (*right*).
(B) The leakiness of tumor-associated microvasculature results in the continuous leaking of thrombin and fibrinogen molecules from the plasma into the parenchyma surrounding blood vessels. Tissue Factor, a protein displayed on the surface of cancer cells and many other nonendothelial cell types (see also Figure 13.11), activates plasma-derived thrombin upon contact, whereupon thrombin converts fibrinogen into fibrin. This results in the extensive network of fibrin bundles (*dark red*) seen here at the border between actively growing breast cancer cells (*below*) and the tumor stroma (*above*). (C) The fibrin bundles (*above and to left of cluster of cancer cells*) form an extracellular matrix to which migrating cells, such as fibroblasts (*below cluster of cancer cells*) as well as endothelial cells (*not seen*), can attach, using it as a substrate on which they can move forward. In many tumors, such as this guinea pig liver carcinoma of bile duct origin, the fibrin matrix may eventually be dissolved by the process of *fibrinolysis* and replaced by a collagenous matrix. (A, courtesy of R. Muschel. B, from C.G. Colpaert et al., *Histopathology* 42:530–540, 2003. C, from H.F. Dvorak, *Am. J. Pathol.* 162:1747–1757, 2003.)

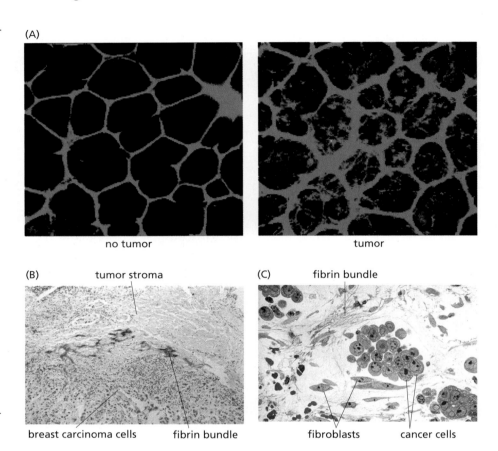

example, see Figure 13.16C and D). Of note, normal tissues, such as the prostate gland, may contain α-SMA-positive cells that, because they lack vimentin, can be identified as smooth muscle cells rather than the myofibroblasts present in sites of wound-healing, inflammation and tumorigenesis (Figure 13.16E).

It remains unresolved precisely where most stromal fibroblasts and myofibroblasts originate. During wound healing, fibroblasts may initially be recruited from the stroma of adjacent unaffected tissues by the PDGF released by platelets participating in clot formation; myofibroblasts may then be produced through the TGF-β1–induced transdifferentiation of some of these recruited fibroblasts (see Figure 13.14). In fact, myofibroblasts can be produced *in vitro* simply by exposing normal fibroblasts to TGF-β1. This suggests that the TGF-β1 released by many types of carcinoma cells, especially those that have progressed to higher levels of malignancy, is a major factor responsible for the formation of myofibroblasts in the tumor-associated stroma.

Alternative sources of myofibroblasts have also been documented: they may arise from fibrocytes and mesenchymal stem cells (MSCs) of bone marrow origin. Both types of cells can be found in the circulation and are known to home to areas of tissue damage, where they can differentiate into myofibroblasts and, quite possibly, fibroblasts, thereby facilitating the rapid reconstruction of stroma. This mechanism ensures that large numbers of cells can be mobilized rapidly from sources outside a wound site in order to expedite reconstruction of damaged stroma. Similarly, MSCs are recruited in large numbers to the stroma of various types of carcinomas; since the myofibroblasts that arise through differentiation of MSCs are also potent inducers of angiogenesis (see Section 13.4, below), this makes the MSCs and derived myofibroblasts key regulators of tumor growth.

The tumor stroma formed by myofibroblasts differs substantially in appearance from the stroma in normal epithelial tissues. This distinctive appearance has caused pathologists to label it a "reactive" or **desmoplastic** stroma. The latter term refers to the *hardness* of the tumor mass as a whole, which results from the deposition of extensive extracellular matrix (ECM) by the myofibroblasts. As a carcinoma advances to a higher, more aggressive grade, the proportion of stroma that is desmoplastic often increases in parallel (**Figure 13.17**).

The myofibroblasts construct the desmoplastic stroma through the secretion of large amounts of collagen types I and III, fibronectin, proteoglycans, and glycosaminoglycans (GAGs), which together give this stroma its characteristic appearance at the microscopic level. In addition, these cells secrete urokinase plasminogen activator

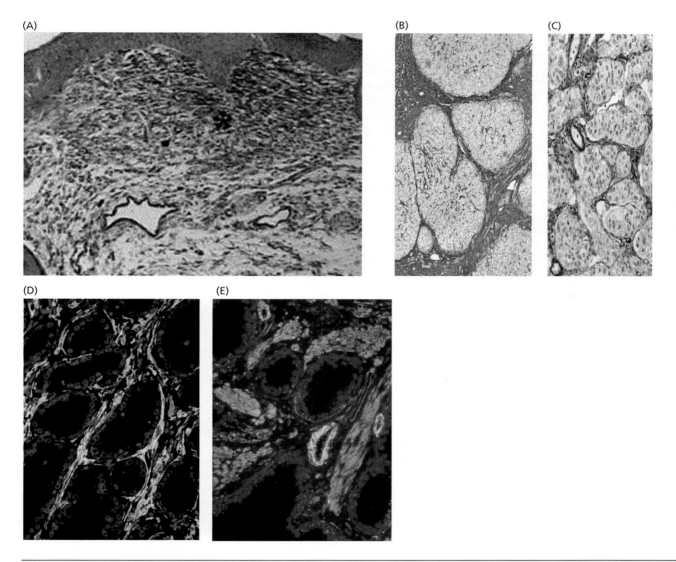

Figure 13.16 Myofibroblasts Myofibroblasts arise in sites of wound healing, inflammation and in tumor-associated stroma through transdifferentiation of stromal fibroblasts (see Figure 13.14) and recruitment of myofibroblast precursors, possibly fibrocytes and mesenchymal stem cells (MSCs), from the circulation. Their presence is revealed typically by their expression of α-smooth muscle actin (α-SMA). (A) Staining for α–SMA–positive myofibroblasts (*reddish brown*) shows them present in abundance 3 days after the skin of a mouse has suffered a wound. (B) Chronically inflamed tissues acquire a fibrotic stroma, such as the cirrhotic liver seen here stained with anti-α-SMA antibody, which reveals the presence of masses of myofibroblasts (*brown*). (C) A hepatocellular carcinoma stained with anti-α-SMA antibody (*brown*) reveals the clear resemblance of the stroma of a chronically inflamed tissue (panel B) to the stroma of a carcinoma arising in that tissue. (D) In prostate carcinomas, the stroma is filled with myofibroblasts that express both vimentin (*red*) and α-SMA (*green*), whose overlap is indicated in *yellow*. Epithelial cell nuclei are in *blue*. (E) Importantly, not all α-SMA-positive (*green*) cells are myofibroblasts. Thus, the stroma of the normal human prostate gland contains smooth muscle cells that that express this protein but, unlike myofibroblasts, lack significant vimentin staining (*red*). The few areas of overlapping expression of these two proteins are indicated in *yellow* and represent the occasional pericytes surrounding blood vessels. The nuclei of epithelial cells lining the ducts are stained *blue*. (A, from P. Martin et al., *Curr. Biol.* 13:1122–1128, 2003. B and C, from A. Desmouliere, A. Guyot and G. Gabbiani, *Int. J. Dev. Biol.* 48:509–517, 2004. D and E, from J.A. Tuxhorn et al., *Clin. Cancer Res.* 8:2912–2923, 2002.)

Figure 13.17 Micrograph of normal stroma and desmoplastic stroma The histologically complex stroma of normal tissue may eventually be replaced by the desmoplastic stroma of an advanced carcinoma. Here we see normal human prostate tissue (*left panel*) stained with Masson's trichrome stain, which reveals the extensive smooth muscle cells in the stroma as *pink*; normal ducts are scattered throughout this stroma. In contrast, an advanced prostate carcinoma (*right panel*) stained identically and viewed at the same magnification reveals the extensive desmoplastic stroma (*blue purple*), which is rich in extracellular matrix, notably collagen I. Islands of prostate carcinoma cells forming small ducts (*pink*) are scattered throughout this desmoplastic stroma, which lacks significant numbers of viable cells such as myofibroblasts and fibroblasts. (B) In this invasive ductal carcinoma of the breast, the deposition of parallel collagen fibers by fibroblasts and myofibroblasts (*green*; visualized here by 2-photon microscopy) creates a stiff extracellular matrix in the desmoplastic stroma that encourages epithelial-mesenchymal transitions (EMTs; see Sections 14.3 and 14.4) in nearby carcinoma cells (*red*). (A, from G. Ayala et al., *Clin. Cancer Res.* 9:4792–4801, 2003. B, courtesy of S. Wei and J. Yang.)

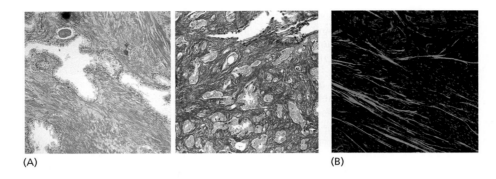

(A)   (B)

(uPA, a protease) and a series of matrix metalloproteinases (MMPs), which help to mobilize growth factors sequestered in the ECM. As the desmoplastic stroma matures, many of the initially present stromal cells disappear, being replaced progressively by the dense, **acellular**, collagenous ECM that is the hallmark of this type of tumor-associated stroma. The processes leading to the formation of the tumor-associated desmoplastic stroma seem to be similar to those operating in sites of wound healing. However, in the tumor, these processes operate continuously over many months and years, rather than transiently over periods of several days, as happens during wound healing. (Note that we have ascribed multiple distinct roles to myofibroblasts—enabling wound contraction, stimulating angiogenesis, and generating abundant stromal ECM.)

Another hint of parallels between wound healing and tumorigenesis comes from several studies indicating that agents known to inhibit tumor-associated angiogenesis also antagonize wound healing. Included among these agents are certain soluble inhibitors of the growth factor receptors that play key roles in angiogenesis, as well as an ECM component, termed thrombospondin-1 (Tsp-1), which is a potent anti-angiogenic molecule. The many common mechanisms shared by wound healing and by the epithelial–stromal interactions in tumors also have implications for clinical practice (**Sidebar 13.2**).

The term "carcinoma-associated fibroblasts" (CAFs) is sometimes used to describe the mixed populations of fibroblasts and myofibroblasts that are present in the stroma of epithelial tumors. Analyses of the gene expression patterns of CAFs underscore the strong similarities between these cells and the fibroblasts present in wound sites.

These analyses initially determined the gene expression pattern of serum-stimulated fibroblasts propagated *in vitro*, that is, fibroblasts that had been starved of serum and then exposed to high concentrations of fresh serum. This exposure re-creates the environment of fibroblasts during the initial stages of wound healing (a time when

Sidebar 13.2 Breast cancer surgery may lead to the stimulation of tumor growth Epidemiologic studies of breast cancer patients undergoing surgery (that is, partial or total mastectomy) for removal of their primary tumors have revealed a peak of recurrence of breast cancers at the primary site as well as in distant anatomical sites about 3 years after surgery. The timing of these relapses suggests that the surgery is responsible for stimulating their formation. Examination of the fluids draining from the wound sites created by the surgery revealed the presence of potent mitogenic factors that are associated with the wound-healing process. Among these are mitogens for breast cancer cells, especially those cancer cells that overexpress the HER2/Neu protein. In addition, a burst of vascular endothelial growth factor (VEGF) synthesis occurs following surgery; as discussed later in this chapter, VEGF is a potent stimulant of tumor angiogenesis. It is therefore plausible that surgery stimulates the proliferation of residual **micrometastases** (small deposits of metastatic cells) that are not detected and removed when the primary tumors are excised. Indeed, some have argued that the clinical relapses stimulated by surgery nullify much of the therapeutic benefit that should, by all rights, be achieved by the removal of primary tumors and nearby lymph nodes. The clinical importance of such surgery-stimulated tumor progression remains a matter of great contention.

Another clinical observation may eventually be found to bear on this issue: when Herceptin (a therapeutic monoclonal antibody that blocks the action of the HER2/Neu receptor; see Section 15.18) is applied postoperatively in women who have borne relatively small, low-grade tumors, the rate of clinical relapse is reduced by as much as 50%. This stunning therapeutic success may, quite possibly, be associated with the ability of Herceptin to block stimulation of residual cancer cells by the growth factors elaborated during healing of the surgical wound site.

the various serum factors released by activated platelets act on stromal fibroblasts at the wound site, converting them, at least transiently, into myofibroblasts). A relatively small cohort of serum-induced and -repressed genes was extracted from these data and used to represent the characteristic "signature" of serum-stimulated fibroblasts (**Figure 13.18A**).

The RNA expression patterns of a large group of human carcinomas were then analyzed to determine whether the tumors expressed the signature of serum-stimulated fibroblasts. (Since all the genes being analyzed are associated specifically with

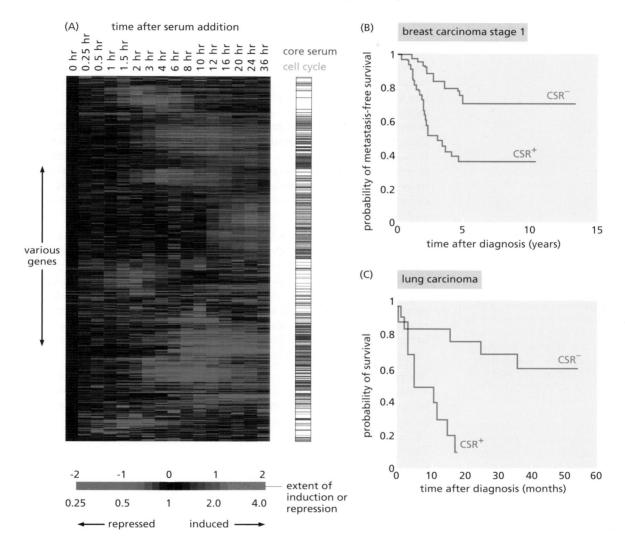

Figure 13.18 Gene expression arrays of tumor-associated myofibroblasts and serum-activated fibroblasts (A) In this gene expression analysis, the expression of cellular mRNAs was analyzed following the addition of fresh serum to previously quiescent, serum-starved human fibroblasts; this resulted in these cells' entering into the active cell cycle and the induction or repression of the expression of a large cohort of genes. Gene expression was measured at various times (*above*) after addition of serum. The cohort of genes analyzed (*not named*) is arrayed vertically from *top to bottom*. Genes that were induced or repressed late after the addition of serum were classified as "cell cycle genes" (*orange lines, right vertical bar*), whereas genes that were induced or repressed early and did not fluctuate with cell cycle phase were classified as "core serum response" (CSR) genes (*blue lines, right vertical bar*). The horizontal bar (*below*) is a key indicating the degree of

induction (*red*) or repression (*green*) following serum addition, this being indicated logarithmically above the bar and in absolute numbers (limited to a 4-fold difference) below the bar.
(B) The expression pattern of the CSR genes was analyzed in a series of tumors in women presenting with stage 1 (i.e., early-stage) breast cancer. As indicated here, those whose tumors expressed the CSR gene pattern (*red line*) showed a much greater probability for developing metastases in the years following initial treatment compared with those whose tumors did not show this gene expression pattern (*blue*). (C) Similarly, those patients whose lung adenocarcinomas (including all stages of tumor progression at the time of diagnosis) showed the CSR gene expression signature (*red line*) suffered dramatically higher mortality rates compared with those whose tumors did not show this gene expression signature (*blue line*). (From H.Y. Chang et al., *PLoS Biol.* 2:E7, 2004.)

fibroblasts, these analyses, by necessity, reflected the RNA expression patterns of the CAFs present in each tumor.) Many carcinomas were indeed found to express the signature of serum-stimulated fibroblasts. Significantly, the carcinomas that expressed this signature more intensely were associated with a grimmer clinical prognosis (see Figure 13.18B and C). Similarly, tumors that contain higher proportions of myofibroblasts indicate shorter survival rates of the patients bearing these tumors (Supplementary Sidebar 13.4). This correlation suggests that the activated, myofibroblast-rich stroma represents a potent force driving aggressive tumor progression.

13.4   Experiments directly demonstrate that stromal cells are active contributors to tumorigenesis

Several biological experiments provide even more direct demonstrations of the profound influence that recruited stromal cells exert on epithelial cell tumorigenesis. In one study, previously non-tumorigenic, immortalized keratinocytes were forced to secrete PDGF at high levels (achieved through the introduction of a PDGF expression vector). The released growth factor had no effect on the proliferation of these epithelial cells *in vitro*, because they do not display PDGF receptors on their surface. However, when these cells were implanted into host mice, they acquired the ability to form robustly growing tumors, clearly derived from the ability of the released PDGF to recruit and activate stromal cells. The stromal cells then reciprocated by driving the proliferation of the PDGF-secreting keratinocytes, eventually causing the latter to undergo malignant transformation.

In a complementary experiment, genetically engineered alterations of the mouse germ line enabled investigators to selectively inactivate the TGF-β type II receptor in the stromal fibroblasts in a variety of tissues. As a consequence, these stromal cells were no longer susceptible to TGF-β–mediated growth inhibition, and stromal hyperplasia occurred in many of these tissues. In some of these, the hyperproliferating fibroblasts drove nearby epithelial cell layers to proliferate and, eventually, to develop into carcinomas (**Figure 13.19**). This demonstrates, once again, the powers of stromal cells to stimulate epithelial cell proliferation, doing so in ways that can lead to neoplastic transformation of the epithelial cells.

The important role of stromal fibroblasts in supporting tumor growth can be demonstrated by yet another type of experiment: transformed, weakly tumorigenic human mammary epithelial cells (MECs) were found to require more than two months to form a tumor after being introduced into immunocompromised host mice (**Figure 13.20**). However, if these cancer cells were mixed with human mammary stromal fibroblasts (from normal breast tissue) prior to injection into hosts, the cells formed tumors in one-third the time. These admixed fibroblasts clearly obviated the need of the MECs to spend time recruiting fibroblasts from host mice—a process that usually requires many weeks' time; in the absence of these fibroblasts, tumor growth could not take off.

Figure 13.19 Prostatic tumors from mice with genetically altered stromal fibroblasts The gene encoding the TGF-β type II receptor was inactivated selectively in the fibroblasts present in a variety of epithelial tissues of the mouse. (A) The prostates of male mice normally exhibit a thin layer of epithelial cells lining the lumina of the ducts and a relatively thin layer of stromal cells (*asterisk*) outside the epithelial cell layers. (B) As a consequence of TGF-β type II receptor inactivation in the stromal fibroblasts, both cell layers became hyperplastic. The epithelial cells soon created tissue that closely resembled the prostatic intraepithelial neoplasia (PIN) commonly seen in humans (see Figure 13.39). This suggests that the hyperplastic stroma released proliferative signals to the nearby epithelium. Indeed, production by the stromal cells of hepatocyte growth factor (HGF)—a potent epithelial cell mitogen—increased threefold after they lost their responsiveness to the inhibitory effects of TGF-β. In the stomachs of these mice (not shown), in which the TGF-β type II receptor was inactivated in the gastric stroma, the hyperplastic epithelia that formed became dysplastic and soon progressed to form carcinomas. Insets at higher magnification. (From N.A. Bhowmick et al., *Science* 303: 848–851, 2004.)

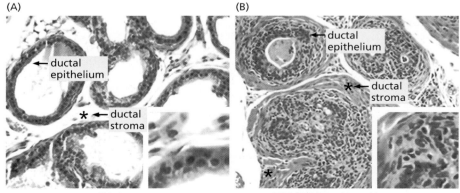

wild type   TGF-βRII inactivated in stroma

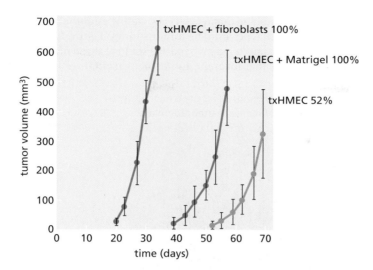

Figure 13.20 Admixed fibroblasts and tumor growth When human mammary epithelial cells (HMECs) were transformed through the introduction of the SV40 early region, the *hTERT* gene, and a weakly expressed *ras* oncogene (see Section 11.12), the resulting transformed HMECs (txHMECs) formed tumors, albeit with a long lag time after injection, and then only in about half of the mice into which they had been injected (*orange curve*). This slow development was accelerated a bit and tumor-forming efficiency was doubled when a preparation of extracellular matrix (Matrigel) produced by mouse sarcoma cells was mixed with the txHMECs prior to introduction into host mice (*blue curve*). However, when mammary stromal fibroblasts from a normal human breast were admixed to the transformed human MECs prior to injection, the cancer cells formed tumors rapidly and with 100% efficiency (*red curve*). This illustrates that the recruitment of stromal fibroblasts (or similarly functioning mesenchymal cells) is an important rate-limiting step in tumor formation. (From B. Elenbaas et al., *Genes Dev.* 15:50–65, 2001.)

This experiment did not address the issue of whether the stromal cells of normal epithelial tissues can accelerate tumor formation as effectively as the stromal cells of carcinomas. This question was answered by comparing the actions of stromal fibroblasts extracted from a normal epithelial tissue with the carcinoma-associated fibroblasts (CAFs) prepared from carcinomas (in which myofibroblasts are abundant; see Figure 13.16). In one set of experiments, fibroblasts were purified from the stroma of normal human prostate glands, while the CAFs were prepared from the stroma of human prostate carcinomas. Each of these cell populations was then mixed with otherwise non-tumorigenic, SV40 large T antigen–immortalized human prostate epithelial cells and implanted in immunocompromised Nude mice. The results, summarized in **Figure 13.21**, showed dramatic differences in the growth of these mixed tissue grafts. In particular, the grafts containing CAFs plus immortalized prostate epithelial cells formed tumors that were as much as 500 times larger than those containing normal prostate fibroblasts plus immortalized prostate epithelial cells. (When injected alone, the CAFs formed no tumors at all.)

This experiment demonstrated that these CAFs were functionally very different from the stromal fibroblasts present in normal prostatic tissue. Stated differently, during the course of tumor progression, stromal cells become increasingly adept at helping their epithelial neighbors to survive and proliferate. Similar observations have since been made with CAFs extracted from human breast cancers

Still, these experiments do not reveal precisely *how* the myofibroblast-rich CAF populations accelerate tumor growth. Once established within the tumor-associated stroma, the myofibroblasts, as mentioned earlier, confer multiple benefits on nearby epithelial cancer cells, the most important benefit possibly being angiogenesis. Myofibroblasts expedite angiogenesis through their ability to release stroma-derived factor-1 (SDF-1/

Figure 13.21 Effects of prostatic stromal cells on immortalized prostatic epithelial cells When normal human prostate epithelial cells (HPEs) that had been immortalized by SV40 T antigen (Tag-immortalized HPEs) were introduced into Nude mice, they formed growths of ~10 mg after many weeks, not significantly larger than the initial inoculum (*red dots*), indicating that these cells were, when implanted on their own, non-tumorigenic. However, if these same cells were mixed with carcinoma-associated fibroblasts (CAFs) prepared from a human prostate carcinoma, tumors arose whose median weight was 10 times larger (Tag-immortalized HPEs + CAFs, *blue dots*). In contrast, when these Tag-immortalized HPEs were mixed with stromal fibroblasts from a normal human prostate, the median weight of the resulting growth was ~7 mg, possibly even less than the starting inoculum (*orange dots*). And if CAFs were mixed with normal HPEs prior to inoculation, the median weight after many weeks was, once again, only ~10 mg (*purple dots*). (From A.F. Olumi et al., *Cancer Res.* 59:5002–5011, 1999.)

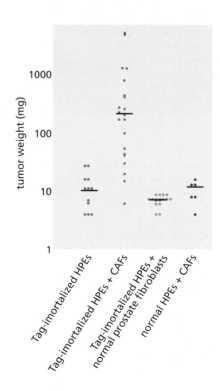

CXCL12), a chemokine that recruits circulating endothelial precursor cells (EPCs) and other myeloid cell types into the tumor stroma (**Figure 13.22A–C**). The VEGF secreted by myofibroblasts then helps to induce some of these recruits to differentiate into the endothelial cells that form the tumor **neovasculature** (see Figure 13.22D). Because

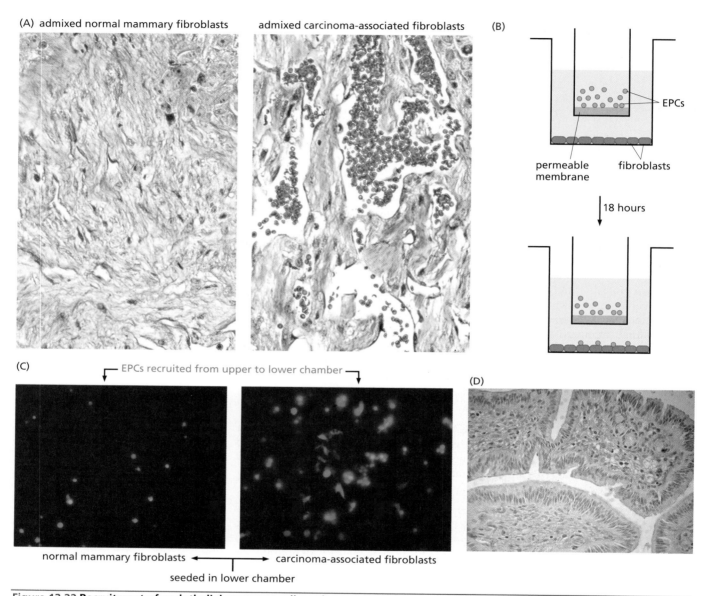

Figure 13.22 Recruitment of endothelial precursor cells and induction of their differentiation by CAFs (A) When normal human mammary stromal fibroblasts are admixed to cells of the human MCF7-ras breast cancer cell line (*left panel*), the resulting tumors show relatively small numbers of blood vessels (*red*) amid the tumor-associated stroma (*blue*). However, admixture of carcinoma-associated fibroblasts (CAFs; *right panel*) results in highly vascularized tumors with large vessels and associated erythrocytes (*red*). The resulting access to the circulation is likely to greatly facilitate tumor growth. (B) The ability of the CAFs, which are essentially myofibroblasts, to attract endothelial cells was demonstrated by an *in vitro* experiment, in which green fluorescent protein (GFP)–labeled bone marrow cells expressing surface antigens characteristic of endothelial precursor cells (EPCs) were placed above a permeable membrane in a Boyden chamber. Either normal mammary stromal fibroblasts or CAFs from a breast cancer

were placed on the bottom surface of the lower chamber. After 18 hours, the number of EPCs that had been recruited to the bottom of the lower chamber was gauged by fluorescence microscopy. (C) Analysis of the cells attached to the lower chamber indicated that the CAFs (*right panel*) are able to attract far more GFP-labeled EPCs than the normal mammary stromal fibroblasts (*left panel*). This recruitment could be reduced by 60% by placing antiserum that neutralized the SDF-1/CXCL12 chemokine released by the CAFs in these chambers, indicating its important role in mediating this recruitment. (D) In a mouse model of gastric carcinoma development, the myofibroblasts of the tumor-associated stroma are seen, by immunostaining, to express high levels of VEGF-A (*brown*), which induces differentiation of EPCs and the proliferation of resulting endothelial cells. (A, B and C, from A. Orimo et al., *Cell* 121:335–348, 2005. D, from X. Guo et al., *J. Biol. Chem.* 283:19864–19871, 2008.)

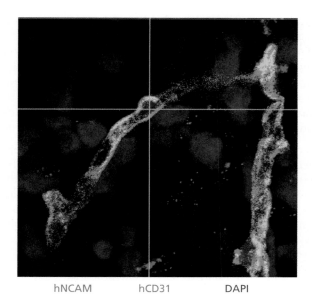

hNCAM   hCD31   DAPI

Figure 13.23 Transdifferentiation of glioblastoma stem cells into tumor-associated endothelial cells The characterization of the endothelial cells formed in the neovasculature of human glioblastomas (GBMs), including those that are grown as xenografts in immunocompromised host mice, has revealed that many of these cells, while expressing a human endothelial cell–specific marker (CD31, *green*), can also be stained with antibodies that are specifically reactive with human NCAM antigen (*red*), which indicates cells of neuroectodermal origin. Overlapping staining between these two colors is seen here as *yellow.* Moreover, these endothelial cells can be shown to carry the same genomic alterations as are present in the bulk of the neoplastic cells in these GBMs and to derive from the differentiation of glioblastoma stem cells. Cell nuclei are stained *blue.* (From R. Wang et al., *Nature* 468:829–833, 2010.)

angiogenesis is usually a rate-limiting step in tumor formation, the tumor-stimulating effects of admixed CAFs may be largely due to their ability to accelerate tumor angiogenesis. In certain human cancers, the density of myofibroblasts in the tumor stroma is correlated with the aggressiveness of disease, suggesting that these cells play a critical role in driving forward aggressive tumor growth (see Supplementary Sidebar 13.4).

The endothelial cells present in the tumor neovasculature have been widely assumed to derive from two sources—the proliferation of existing endothelial cells, including those present in adjacent normal tissue; and to a smaller extent the EPCs that originate in the marrow and arrive via the circulation into the tumor stroma. In 2010, two groups simultaneously reported a fully unexpected finding: that within human glioblastomas (and the xenografts derived from these tumors), a significant portion of the endothelial cells in the tumor-associated neovasculature descend from the genetically mutant tumor cells, more specifically from the transdifferentiation of glioblastoma stem cells (see Figure 11.19 and Section 11.7) that are present in these tumors (**Figure 13.23**). [These glioblastoma (GBM)-derived microvessels appear to lack the VEGF-R2 receptor, the dominant receptor that is generally responsible for triggering new vessel formation in tumors. On the one hand, this demonstrates that these GBM capillaries are very different from those found in normal tissues and in most tumor types. On the other, this may explain why glioblastomas are resistant to the anti-angiogenesis therapies of the sort that we will encounter in Section 13.10 and are designed to interrupt signaling by VEGF and this receptor.]

In some tumors, the functionally altered stroma may be traced to a mechanism quite different from the one described above: stromal cells in advanced carcinomas, having coexisted with epithelial cancer cells for many years, may change their genotype and acquire traits that genetically normal stromal cells cannot achieve. This suggests that stromal cells co-evolve with their neoplastic neighbors during these long periods of tumor development by altering their genomes in order to adapt to the physiologic stresses present within tumors.

For example, analyses of a large group of human breast cancers have occasionally shown somatically mutated *PTEN* and *TP53* tumor suppressor genes in stromal cells (largely fibroblasts and myofibroblasts) isolated from the tumors. These experiments exploited the procedure of *laser capture microdissection* (LCM; Supplementary Sidebar 13.5) to isolate small patches of epithelial or stromal cells from these tumors. In some tumors, distinct mutant *TP53* alleles were found in the epithelial and in the stromal compartments. In others, mutant alleles were found in one cell population but not the other. These experiments have opened the door to the possibility that the well-documented genetic evolution of neoplastic epithelial cells (see Chapter 11) may occasionally be accompanied by changes in the genomes of nearby stromal cells.

13.5   Macrophages and myeloid cells play important roles in activating the tumor-associated stroma

The active recruitment of macrophages into tumor masses would seem, at first glance, to be counterproductive for tumor formation, since macrophages are usually deployed by the immune system to scavenge and destroy infectious agents and abnormal cells, as we will explore in more detail in Chapter 15. However, increasing evidence indicates that these immune cells also play important roles in *furthering* tumor development.

In more detail, monocytes from the myeloid lineage in the bone marrow enter into the general circulation, from which they are recruited by cancer cells into a tumor; once ensconced there, the monocytes are induced to differentiate into macrophages (see Figure 13.14). This recruitment depends on attraction signals that are conveyed by **chemotactic** factors. By definition, these factors provide directional cues to motile cells rather than mitogenic stimulation. In the case of **leukocytes** (white blood cells such as monocytes), the relevant chemotactic factors are often called **chemokines**, that is, chemotactic cytokines.

The chemokine known as monocyte chemotactic protein-1 (sometimes called macrophage chemoattractant protein; MCP-1) is expressed in significant quantities by a wide range of neuroectodermal and epithelial cancer cell types. It appears to be a critical signal for attracting monocytes to some tumors and inducing their differentiation into macrophages. In other tumors, vascular endothelial growth factor (VEGF), colony-stimulating factor-1 (CSF-1; often called M-CSF, macrophage colony-stimulating factor), and the PDGF released by tumor cells also seem to help in this recruitment, while CSF-1 aids in stimulating the monocyte-to-macrophage differentiation (**Figure 13.24**).

Once established in the tumor stroma, macrophages play important roles in stimulating angiogenesis (**Figure 13.25**). Thus, in some cancers, such as breast carcinomas, there is a direct correlation between the level of MCP-1 produced, the number of macrophages present, and the level of angiogenesis induced by the various tumors; in other tumors, the density of infiltrating tumor-associated macrophages is correlated with microvessel density (see Figure 13.25B and C; this density can be gauged by counting the number of capillaries per microscopic field in a tumor section). A major role of macrophages in driving tumor progression is suggested by numerous reports demonstrating a direct correlation between the presence of a high density of macrophages in tumor masses and a poor prognosis for cancer patients. This correlation has been documented for gliomas and for carcinomas of the breast, ovary, prostate, cervix, bladder, and lung.

Such evidence, however, is only *correlative*. More compelling evidence of the *causal* role of macrophages, at least in tumor angiogenesis, comes from experiments in which cancer cells are forced to express higher levels of MCP-1. The expression of this chemokine allows the cancer cells to attract more macrophages, which proceed to secrete important angiogenic factors, notably VEGF and interleukin-8 (IL-8), resulting in marked increases in angiogenic activity and the formation of more extensive

Figure 13.24 Recruitment of macrophages by CSF-1 In some tumors, colony-stimulating factor-1 (CSF-1) released by tumor cells serves as the key attractant for monocytes and thus macrophages. In the transgenic mice studied here, mammary tumorigenesis was driven by a transgene, which was constructed from an MMTV (mouse mammary tumor virus) transcriptional promoter that drives expression of the polyoma middle T (PyMT) oncogene, doing so almost exclusively in mammary epithelial cells. (A) The tumor analyzed here arose in a transgenic PyMT mouse that was heterozygous at the *Csf1* locus, since it carried one null allele of the CSF-1–encoding gene, and thus termed *Csf1$^{+/op}$*. The presence of macrophages in the stroma of a tumor is revealed by a monoclonal antibody that detects macrophages (*reddish brown*). (B) This tumor arose in a transgenic mouse that was genetically *Csf1$^{op/op}$* and therefore lacked all CSF-1 activity. Consequently, the tumor cells arising in this mouse were unable to release this chemoattractant, resulting in a mammary tumor in which the macrophages are barely detectable. The absence of macrophages did not affect the growth of the primary tumor but strongly suppressed progression to an invasive and metastatic state, implicating macrophages in such tumor progression. (From E.Y. Lin et al., *J. Exp. Med.* 193:727–740, 2001.)

(A)   *Csf1$^{+/op}$*
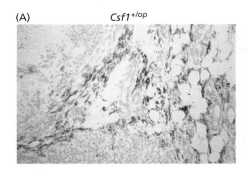

(B)   *Csf1$^{op/op}$*
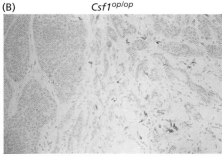

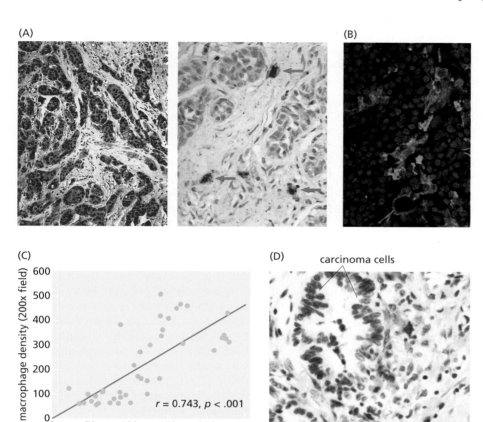

(A)

(B)

(C)

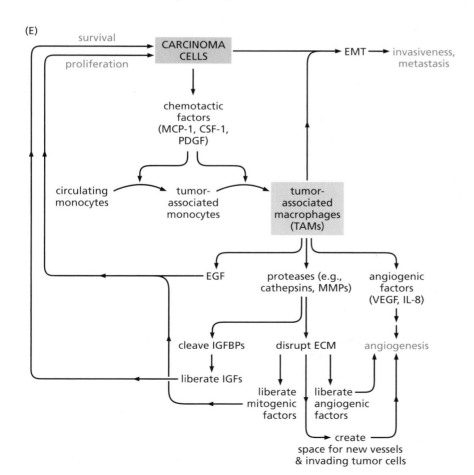

(D)   carcinoma cells

Figure 13.25 Macrophage involvement in angiogenesis
(A) In certain breast tumors (*left panel*), the cancer cells synthesize the potent angiogenic factor VEGF (vascular endothelial growth factor), revealed here (*dark areas*) by staining with an anti-VEGF monoclonal antibody. However, in other breast tumors (*right panel*), VEGF production originates in isolated macrophages within the tumor-associated stroma (*arrows*). (B) In a variety of tumor types, including the pancreatic islet tumors of Rip-Tag mice (see Figure 13.36) analyzed here, macrophages (*green*) can be found in large numbers next to the tumor-associated neovasculature (*red*). Cell nuclei are stained *blue*. Yet other work (see Figure 13.37B) reveals that such macrophages, working together with recruited mast cells, contribute actively to the process of angiogenesis in the tumor stroma. (C) In a series of human non-small-cell lung carcinoma (NSCLC) specimens, each represented here by a point on the graph, the density of tumor-associated macrophages (TAMs) per microscope field has been plotted vs. the density of microvessels (i.e., capillaries). Such correlations add further weight to the notion that TAMs play a key role in fostering tumor-associated angiogenesis. (D) Macrophages in the stroma of a human colorectal adenocarcinoma produce matrix metalloproteinase-9 (MMP-9; *brown spots*), a key enzyme in angiogenesis that acts by liberating angiogenic factors from the ECM. (E) Macrophages play a major role in releasing mitogenic factors for carcinoma cells as well as reorganizing the tumor stroma in order to facilitate angiogenesis and, in some tumors, carcinoma cell invasiveness. MMPs, matrix metalloproteinases. (A, from R.D. Leek et al., *J. Pathol.* 190:430–436, 2000. B, courtesy of V. Gocheva and J. Joyce. C, from J.J.W. Chen et al., *J. Clin. Oncol.* 22:953–964, 2005. D, from B.S. Nielsen et al., *Int. J. Cancer* 65:57–62, 1996. E, update courtesy of J. Joyce.)

tumor-associated vasculature. In addition, when lung cancer cells are exposed *in vitro* to factors that have been secreted by cultured macrophages, the cancer cells respond by producing IL-8 and a number of other proteins that promote angiogenesis and cell invasiveness.

Hypoxic areas within tumors attract macrophages, which appear to tolerate hypoxia quite well. Once established in hypoxic regions of tumors, these inflammatory cells begin to secrete significant amounts of VEGF (see Section 7.12), which reduces the hypoxia by bringing in endothelial cells, capillaries, and thus oxygen-rich blood. The fact that hypoxic areas of tumors often remain poorly vascularized indicates that the recruited macrophages, on their own, are unable to fully cure local defects in angiogenesis. Thus, in rapidly growing tumors, the macrophages and the vasculature that they induce cannot keep pace with the rate of tumor expansion, and large areas within a tumor mass eventually become *necrotic* (that is, filled with dead and dying cells) as a consequence (see Figure 13.28).

Like myofibroblasts (see Section 13.3), macrophages are adept at secreting matrix metalloproteinases (MMPs; see Table 13.1). MMP-9, prominently involved in cancer progression, is produced by tumor-associated macrophages (TAMs; see Figure 13.25D) and, once activated, proceeds to cleave a number of important protein substrates in the extracellular space. In certain invasive carcinomas, including those of the breast, bladder, and ovary, TAMs have proven to be the major source of this enzyme. It contributes to tumor progression by enhancing angiogenesis, by disrupting existing tissue structure and thereby carving out space for expanding tumor masses, and by liberating critical mitogens that have been immobilized through tethering to proteoglycans of the ECM. MMP-9 can also cleave IGFBPs (insulin-like growth factor–binding proteins), which normally sequester IGF molecules in the extracellular space (see Section 9.14). This liberates the IGFs, notably IGF-1, which then provide survival signals to nearby cells, including cancer cells (see Figure 13.25E). In advanced breast cancers, TAMs are also able to help the carcinoma cells directly, since they are the major source of the epidermal growth factor (EGF) that drives the proliferation of EGF-R–expressing carcinoma cells.

While there is abundant evidence implicating some macrophages as active collaborators in tumor progression, it is also clear that another, morphologically distinct subset of macrophages, by acting as deputies of the immune system, can detect and kill cancer cells. Yet they clearly fail to do so in many types of tumors. It may be that some cancer cells acquire the ability to inactivate or blunt the **tumoricidal** (cancer-killing) actions of this second class of macrophages, while leaving intact the functions of the first type that are involved in aiding tumor progression. Alternatively, cancer cells may engage in evasive maneuvers that render them invisible to the macrophages that have been dispatched by the immune system to destroy them. We will return to the role of macrophages and other types of immune cells in blocking tumor development in Chapter 15.

A second cell type in the myeloid lineage can also contribute to tumor invasiveness. These *immature myeloid cells* (iMCs) have been best characterized in mouse models of colorectal carcinogenesis. For example, when the genotype of human colorectal carcinomas is modeled in the mouse by inactivating both the *Apc* and *Smad4* genes (see Sections 7.11 and 11.3)—Smad4 being critical to TGF-β signal transduction (see Figure 6.29D)—the resulting adenoma and adenocarcinoma cells release the chemokine CCL9, which recruits iMCs originating in the bone marrow. The iMCs congregate at the invasive front of these tumors, secrete matrix metalloproteinases MMP-2 and MMP-9, and thereby facilitate invasion by the carcinoma cells into underlying tissue layers (**Figure 13.26**). (Subsequent research has shown that without MMP-2 and MMP-9, the iMCs cannot foster invasion and metastasis by the colorectal mouse tumors, and that the ability of cancer cells in the primary tumors to metastasize depends on recruitment of iMCs to these tumors.) Yet other types of recruited inflammatory cells are likely to play key roles in fostering the malignant progression of carcinomas, a topic to which we will return in the next chapter.

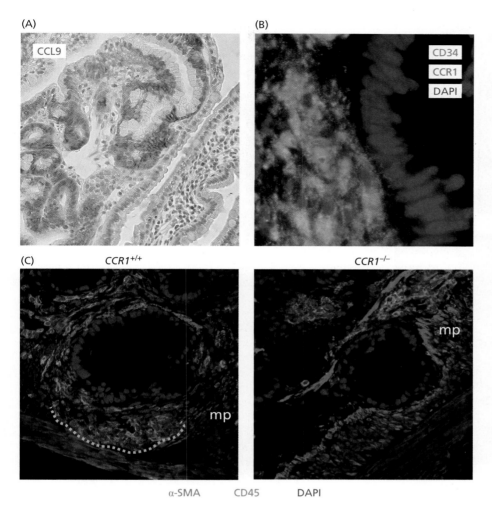

Figure 13.26 Immature myeloid cells promote colorectal carcinoma progression In a mouse model of colorectal carcinogenesis, epithelial cells that have lost both Apc and Smad4 function—thereby mirroring the state of many human colorectal carcinomas (see Figure 11.12A)—recruit a type of immature myeloid cell (iMC) of bone marrow origin into the tumor stroma. (A) The chemokine CCL9 (*brown*) is already released early in tumor progression by the cells in adenomas, which are precursors to frank carcinomas. Normal tissue, stained with DAPI (*light blue nuclei*), is seen in the *lower right*. CCL9 then acts as an attractant to recruit immature myeloid cells (iMCs) into the tumor stroma. (B) The cognate receptor of CCL9, termed CCR1 (*red orange*), is displayed by many of the CD34+ iMCs that are being recruited into the tumor stroma (*green*). DAPI (*blue*) stains all nuclei. (CD34 is a marker of hematopoietic precursor cells.) (C) When displaying the CCR1 receptor of CCL9 (*left panel*), large numbers of CD45+ iMCs (*green*) congregate near colorectal carcinoma cells (*blue nuclei*), and this mixed cell population begins to invade (*green dotted line*) the underlying α-smooth muscle actin–containing *muscularis propria* (mp, *red*). (CD45 is a marker of all types of hematopoietic cells.) However, when the gene encoding the CCR1 receptor is knocked out in the genome of the tumor-bearing host (*right panel*), such CD45+ iMCs are essentially absent and the depth of invasion is reduced 10-fold (*not shown*). (From T. Kitamura et al., *Nat. Genet.* 39:467–475, 2007.)

13.6 Endothelial cells and the vessels that they form ensure tumors adequate access to the circulation

The most obvious stromal support required by tumors has already been cited repeatedly in this chapter: like normal tissues, tumors require access to the circulation in order to grow and survive. As early as the mid-1950s, pathologists noted that cancer cells grew preferentially around blood vessels. Those tumor cells that were located more than about 0.2 mm away from blood vessels were found to be nongrowing, while others even farther away were seen to be dying (**Figure 13.27A**).

We now realize that this threshold of approximately 0.2 mm represents the distance that oxygen can effectively diffuse through living tissues. Cells located within this radius from a blood vessel can rely on diffusion to guarantee them oxygen; those situated further away suffer from moderate or severe hypoxia and low pH (see Figure 13.27B–D). Tissues suffering from hypoxia are in danger of becoming necrotic, as discussed above and illustrated in **Figure 13.28**. p53-triggered apoptotic death also threatens hypoxic cells (see Chapter 9), and the inactivation of the p53 signaling system often enables cancer cells to survive beyond the small perimeter surrounding each capillary.

Hypoxia is only one price that is paid by cells lacking close proximity to the circulatory system. Cells also require effective interactions with the vasculature in order to acquire nutrients and to shed metabolic waste products and carbon dioxide. The capillary networks threading their way through many normal tissues are arrayed so densely that virtually all cells in a tissue are no more than several cell diameters away

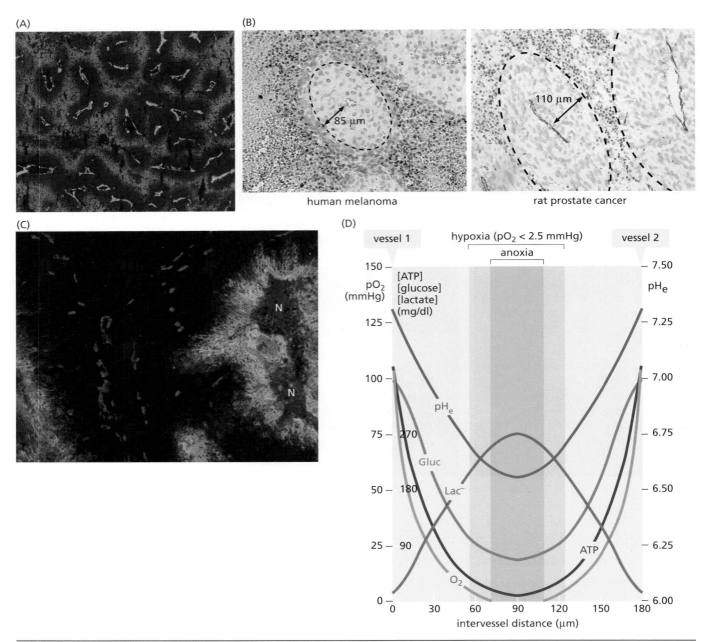

Figure 13.27 Hypoxia and necrosis of cells in poorly vascularized sections of tissues Various techniques have been used to demonstrate hypoxia and necrosis in regions of tissues surrounding capillaries. (A) This micrograph reveals capillaries (*green*) and the degrees of oxygenation in the surrounding tumor parenchyma. Immunostaining using an antibody reactive with EF5, a molecule that localizes to hypoxic regions of tissues, reveals hypoxic areas (*red*) at some distance from the capillaries, while those cells that are well oxygenated are unstained and therefore appear *dark brown*. (B) These sections of a human melanoma and a rat prostate carcinoma have been immunostained with an anti-CD31 antibody, which reveals capillaries (*brown*). Immediately around each capillary is a region of healthy cells (*inside the dashed line*), beyond which necrosis is apparent. The necrotic region (*granular area*) begins as close as 85 μm from the melanoma capillary (*left*) and 110 μm from the prostate carcinoma capillary (*right*). This necrosis reveals the limitations of diffusion in conveying oxygen and nutrients from capillaries to cells in the tumor parenchyma. (C) The dynamics of oxygenation on a larger scale are apparent in this human squamous cell head-and-neck tumor.

Where blood vessels (*blue spots*) provide good oxygenation, the tumor appears *black*. In contrast, in areas of poor vascularization and moderate hypoxia, a carbonic anhydrase enzyme is expressed (*red orange*), while in areas of extreme hypoxia, the pimonidazole dye is detectable (*green*). Overlap between these two markers appears *orange*. Areas of necrosis, located even further away from the tumor vasculature, are indicated by "N." (D) Measurements of five parameters indicate the dramatic changes that occur with increasing distance from microvessels within tumors. Parameters have been calculated from two arterioles running in parallel, 180 μm apart. pH_e represents the pH of extracellular fluid. These measurements in aggregate explain why cells that are situated at some distance from a microvessel experience a sub-optimal environment that often leads to their death by necrosis. (A, courtesy of B.M. Fenton. B, left, from W.K. Hong et al: Holland Frei Cancer Medicine, 8th ed. Shelton, CT: People's Medical Publishing House-USA, 2010. B, right, from L. Hlatky, P. Hahnfeldt and J. Folkman, *J. Natl. Cancer Inst.* 94:883–893, 2002. C, from J.H. Kaanders et al., *Cancer Res.* 62:7066–7074, 2002. D, from P. Vaupel, *Semin. Radiat. Oncol.* 14:198–206, 2004.)

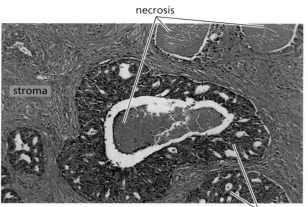

necrosis

stroma

nests of adenocarcinoma

Figure 13.28 Large-scale necrosis within a tumor The inability of large sectors of a tumor to gain access to the vasculature in the stroma is evident in this high-grade adenocarcinoma of the prostate. As seen in the growth in the *center*, a nest of adenocarcinoma cells (*dark red*) is surrounded by extensive stroma (*light red*). In the middle of the nest of adenocarcinoma cells is a large clump of necrotic tissue (*light red*) that has shrunk away from the viable adenocarcinoma cells that surround it; the latter thrive because they have better access to the stroma and associated vasculature. (From M.A. Weiss and S.E. Mills, Atlas of Genitourinary Tract, London: Gower Medical Publishing, Ltd., 1988.)

from the nearest capillary (**Figure 13.29**). In some normal tissues with an especially high metabolic activity, most cells enjoy direct contact with at least one capillary. This intimate association means that their access to oxygen and critical nutrients is not dependent on the diffusion of these molecules over large distances and through densely packed cell layers.

These observations illustrate the central importance of the vasculature to the growth and survival of all types of tissue, normal and neoplastic. The process of developing this vasculature through angiogenesis can be observed during embryonic development, implantation of the placenta, wound healing, and certain pathological conditions such as diabetic retinopathy, psoriasis, rheumatoid arthritis, and, of course, tumorigenesis.

During embryonic development, the overall size and architecture of an organ or tissue is genetically programmed, but the precise locations of its individual component cells are not. This applies as well to the vasculature: the genome determines the overall layout of the major vessels and delegates to the individual cells within a tissue the task of designing and assembling local capillary networks and slightly larger vessels. Heterotypic interactions operating over short distances determine the routing of individual capillaries constituting the microvasculature. More specifically, capillaries appear to form wherever they are needed by the nonvascular (that is, **parenchymal**) cells in a tissue.

(A) (B)

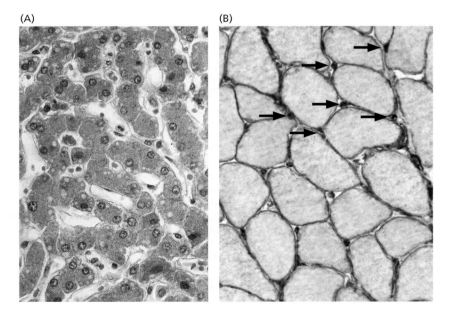

Figure 13.29 Dense microvasculature in normal liver and muscle Because of high metabolic rates, certain tissues develop an extremely high density of capillaries, ensuring almost every cell direct contact with endothelial cells and the capillaries that they form. (A) Virtually every hepatocyte in the liver is adjacent to a space, termed a *sinusoid*, that is lined with endothelial cells and serves as a capillary. (B) Each muscle fiber, immunostained with an anti-laminin antibody, is a single syncytial cell, formed by the fusion of a number of myoblasts. Each muscle fiber lies adjacent to one or more capillaries (*arrows, dark brown spots*). (A, from Loyola University Medical Education Network. B, from Washington University of St. Louis Neuromuscular Disease Center.)

A variant of this plan operates during tumorigenesis. The overall architecture of tumors cannot be determined by the genome of an organism, since each tumor that arises is a novel invention whose design is not anticipated by the genome. This has implications for tumor angiogenesis as well: tumors cannot rely on some genetic blueprint to aid them in configuring their blood supply and must, instead, design the layout of their own vasculature, doing so step-by-step as they grow. In particular, when assembling their own blood supply, tumors depend totally on the localized heterotypic interactions between the cells of the vasculature (including endothelial cells, pericytes, and smooth muscle cells) and the nonvascular cells in tumors (including the neoplastic cells and the other cell types of the supporting stroma).

As was first described in Section 7.12, production of VEGF is governed by the availability of oxygen. Thus, many types of cells take stock of their own intracellular oxygen tension through the actions of the VHL protein and its partners. Under conditions of hypoxia, this complex of proteins allows functional HIF-1α and -1β transcription factors to accumulate, which in turn drives the expression of a number of genes whose products encourage angiogenesis. (In addition, low pH and signaling by the RTK–PI3K–Akt/PKB pathway can activate HIF-1–dependent signals, doing so in a fashion that is independent of oxygen availability.) Prominent among the HIF-1–induced products is VEGF. Synthesis of this key angiogenic factor has been associated with tumor cells, macrophages, and myofibroblasts, depending on the tumor type and its stage of progression. (In fact, there are two structurally related, functionally similar proteins termed VEGF-A and VEGF-B, with VEGF-A being the dominant angiogenic factor in most settings. We will continue here to use the generic term VEGF to refer to both factors.)

Like many other growth factors, VEGF functions as a ligand of tyrosine kinase receptors—in this case, VEGF receptor-1 (also termed Flt-1; see Figure 5.15) and VEGF receptor-2 (known also as Flk-1/KDR). Both are displayed on the surfaces of endothelial cells. Similarly, basic fibroblast growth factor (bFGF), another important angiogenic factor, binds to its own cognate receptors displayed by endothelial cells. Once stimulated by these angiogenic factors, the endothelial cells proliferate and contort their cytoplasms to construct the cylindrical walls of capillaries (**Figure 13.30A**). These capillaries also penetrate through existing tissue layers, moving toward the highest localized concentration of angiogenic factors.

Interestingly, early in the formation of certain experimental tumors, many of the endothelial cells forming the capillaries in the tumor-associated stroma derive from circulating endothelial progenitor cells (EPCs) that originate in the bone marrow and are then induced to settle in the tumor stroma and differentiate (see Figure13.30B; see also Figure 13.22). However, as tumor growth continues, an increasing proportion of the capillaries are assembled from endothelial cells that are already present within the tumor stroma. Importantly, EPCs are only one of a number of distinct hematopoietic cell types that are recruited from bone marrow and contribute to various aspects of tumor angiogenesis in the tumor stroma.

Of additional interest is the fact that similar mechanisms appear to operate during the formation of lymphatic vessels (**Sidebar 13.3**). As we will see later in this chapter, lymph ducts are important regulators of fluid balance in the tumor stroma. In the next chapter, they assume great importance, because they provide avenues for cancer cells to escape primary tumors and disseminate to distant sites in the body.

The existence and powers of angiogenic factors were first revealed by implanting small chunks of tumor on the cornea or the ears of laboratory animals such as rabbits. Within days, dense networks of capillaries and larger vessels were seen to emerge from preexisting capillary beds and to converge on the implanted tumor chunks. Images like these (**Figure 13.31**) have strongly influenced our thinking about the behavior of tumors and their vasculature, because they demonstrate so vividly that tumors actively recruit blood vessels into their midst.

Angiogenesis is actually far more complicated than is suggested by the above descriptions. On the one hand, this complex morphogenetic process involves a number of other factors in addition to the VEGFs, which clearly play a major role in attracting

(A)

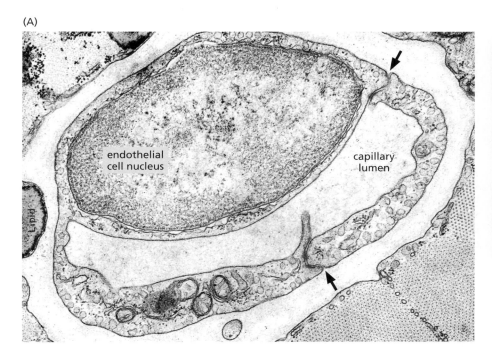

(B)

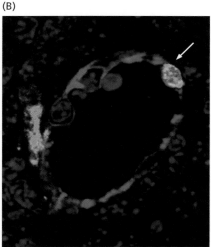

CD31/GFP/DAPI

Figure 13.30 Endothelial cells and the formation of capillaries
(A) This transmission electron micrograph reveals the cross section through a capillary in the midst of cardiac muscle and shows how the cytoplasms of two endothelial cells have become joined (*arrows*) to form the lumen of a capillary. The tight sealing of the plasma membranes of adjacent endothelial cells to one another results in capillary tubes whose walls are continuous and usually without gaps. The nucleus of one of the endothelial cells is also seen here. (A pericyte coating the outside surface of a capillary can be seen in Figure 13.7A.) (B) Early in the formation of many tumors, the bulk of the endothelial cells derive from endothelial progenitor cells (EPCs) that are acquired via the circulation from the bone marrow and differentiate within the tumor stroma. In the experiment illustrated here, a mouse was lethally irradiated and then rescued by engrafting GFP⁺ hematopoietic stem cells, resulting in an almost-complete replacement of its bone marrow cells by donor green fluorescent protein (GFP)-labeled cells. Seen here is a capillary growing within a Lewis lung carcinoma (LLC) that was subsequently implanted in this mouse. The endothelial cell is indicated by the CD31 endothelial-specific marker (*red*). The fact that this capillary also contains a GFP⁺ cell (*green, arrow*) demonstrates the bone marrow origin of this endothelial cell. Nuclei are stained with DAPI (*blue*). As the tumor grows, an increasing proportion of the endothelial cells derive via the proliferation of local endothelial cells rather than from EPCs originating in the bone marrow. (A, from D.W. Fawcett, *J. Histochem. Cytochem.* 13:75–91, 1965. B, from D.J. Nolan et al., *Genes Dev.* 21:1546–1558, 2007.)

blood vessels to tumors. These other factors include several forms of TGF-β, basic fibroblast growth factors (notably FGF2), interleukin-8 (IL-8), angiopoietin, angiogenin, and PDGF. Moreover, several distinct cell types in addition to endothelial cells contribute to the construction of capillaries and larger vessels. As mentioned,

Sidebar 13.3 Endothelial cells also construct lymph ducts
The lymph ducts have two major functions in normal physiology. They drain fluid from the interstices between cells and empty this fluid into the venous circulation. In addition, they allow antigen-presenting cells of the immune system to convey antigens from various tissues to the lymph nodes, where immune responses are often initiated (see Chapter 15).

Interestingly, lymph ducts are assembled from endothelial cells originating in the same embryonic stem cell population that yields the endothelial cells of capillaries and larger blood vessels. During embryonic development, lymphatic vessels can often be observed to bud from developing capillaries before they separate and construct their own parallel network of interconnecting vessels (see Figure 13.7C). In addition, the factors that stimulate **lymphangiogenesis**—vascular endothelial growth factors C and D (VEGF-C and VEGF-D)—are homologous to VEGF-A and -B, which play a major role in stimulating the angiogenesis that creates the blood vasculature.

As might be expected, the receptor for VEGF-C and VEGF-D displayed by lymphatic endothelial cells—VEGF receptor 3 (VEGF-R3)—is structurally related to the dominant VEGF receptor of blood capillaries, VEGF-R2. In addition, there is clear evidence that VEGF-D, which is mainly responsible for driving lymphangiogenesis, may also help stimulate angiogenesis by binding and activating VEGF-R2. So these two systems—the blood and lymphatic networks—derive from common evolutionary roots, develop from common precursors in the embryo, and continue to interact with one another in complex ways within adult tissues. (Of note, there are even more ancient evolutionary connections between the mechanisms guiding the growth of capillaries and those guiding nerve axon growth!)

the endothelial cells form the lumen of a capillary; these cells, in turn, are surrounded by the mural pericytes and vascular smooth muscle cells (see Figure 13.7), which are absent from small lymph ducts.

The systematic covering of capillaries by pericytes seen in normal tissues can be contrasted with their chaotic dispersion near tumor-associated capillaries (**Figure 13.32A and B**). Capillaries in tumors typically have diameters that are three times greater than their normal counterparts. In addition, the overall layout of blood vessels around and within tumor masses is quite chaotic (**Figure 13.33**). Often vessels stop abruptly in dead-end pouches or circle back and attach to themselves.

The precise reasons why capillaries and larger vessels within tumors are so haphazardly constructed are unclear. One possible factor may lie in the balance between two mutually antagonistic growth factors, angiopoietin-1 and -2. While VEGF is responsible for initiating the growth of capillaries by attracting and stimulating endothelial cells, a mix of angiopoietin-1 and -2 induces endothelial cells to recruit the pericytes and smooth muscle cells that enable newly formed capillaries to mature into

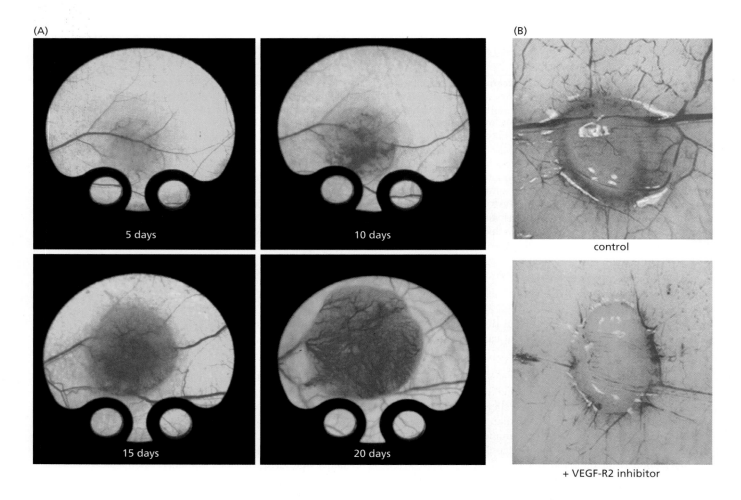

(A)

5 days

10 days

15 days

20 days

(B)

control

+ VEGF-R2 inhibitor

Figure 13.31 Recruitment of capillaries by an implanted tumor (A) Shown is the growth of a small group of subcutaneously implanted human colorectal adenocarcinoma cells over a period of 20 days. The growth of the tumor-associated vessels (*red*) was observed through a window inserted above the tumor in the skin of the host mouse. (B) Such vascularization can be suppressed by antagonists of the VEGF receptor. Seen here are the effects of ZD6474 (also termed vandetanib), a low–molecular-weight inhibitor of the tyrosine kinase of the VEGF receptor-2, in mice bearing human adenocarcinoma cell xenografts. The widespread *pink* background in the untreated control mouse is indicative of numerous capillaries, and the number of major vessels entering into a tumor mass (*top*) is strongly decreased in the presence of the drug (*bottom*). (A, from M. Leunig et al., *Cancer Res.* 52:6553–6560, 1992. B, from S.R. Wedge et al., *Cancer Res.* 62:4645–4655, 2002.)

(A)   (B)

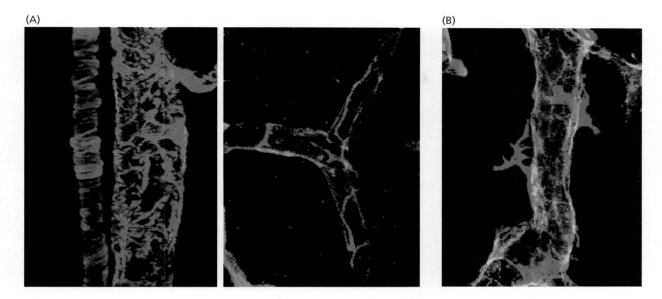

Figure 13.32 Pericytes in normal and tumor-associated vessels
(A) This immunofluorescence micrograph shows that pericytes
and smooth muscle cells (*both red orange*) coat the outside of
the tubes of endothelial cells (*green*), forming normally structured
venules and arterioles (*left panel*). Pericytes and smooth muscle cells
often cover these vessels so completely that it is difficult to see the
underlying endothelial cells. In contrast, in capillaries (*right panel*),
the pericytes are more sparsely disposed, but are nonetheless
tightly attached to the endothelial cells. (B) This micrograph reveals

the structure of a tumor-associated microvessel, in this case a
vessel formed in a mouse by Lewis lung carcinoma (LLC) cells.
The endothelial cells forming the lumen of this vessel (*green*) are
only partially overlain by pericytes and smooth muscle cells (*red
orange*). The loose attachment of pericytes to the capillaries can be
contrasted to their tight attachment seen in Figure 13.7A and
B and panel A of this figure. (From S. Morikawa et al., *Am. J.
Pathol.* 160:985–1000, 2002.)

well-constructed vessels containing appropriate proportions of these three cell types. An imbalance of these two antagonistic angiopoietins, together with the greatly elevated levels of VEGF within tumors, is suspected to be responsible for much of the defective construction of the vasculature within neoplasms. Indeed, some experiments indicate that most of the morphological abnormalities of tumor-associated microvessels can be mimicked in normal skin simply by local overexpression of VEGF-A, the main pro-angiogenic factor.

At the submicroscopic level, it is also apparent that the capillaries in tumor masses are assembled haphazardly. This is because the plasma membranes of adjacent endothelial cells do not contact one another to form a seamless lining around the capillary lumen, but instead leave gaps, often several microns wide (**Figure 13.34A and B**), which allow direct access of the blood plasma to the cells surrounding the capillary. The resulting leakage (see Figure 13.34C) seems to be responsible for the deposition of fibrin in the tumor parenchyma described earlier (see Figure 13.15).

Quantitative measures indicate that the walls of capillaries in tumors are about 10 times more permeable than those of normal capillaries. Some of this leakiness is attributable to the defective assembly of capillary walls noted here. However, most of it seems to be due to the deregulated production of VEGF by cancer cells in the tumor, as well as macrophages and fibroblasts in the tumor-associated stroma. We know this because when an antibody that neutralizes VEGF is introduced into the circulation of tumor-bearing mice, the permeability of the tumor-associated capillaries is substantially reduced. The precise mechanisms responsible for VEGF-induced permeability remain unclear. Excessive VEGF signaling seems to cause plasma membranes of adjacent endothelial cells to separate, generating gaps between these cells. An alternative mechanism, which operates in normal tissues to control microvessel permeability, seems to play a relatively minor role in tumors: molecules are transported *through* individual endothelial cells by intracellular vesicles that acquire cargo via endocytosis

Figure 13.33 The chaotic organization of tumor-associated vasculature
Different techniques of imaging vessels in living tissues (sometimes termed *intra-vital* imaging) reveal at different levels of resolution the starkly contrasting organization of normal and neoplastic vasculature. (A) Intra-vital multiphoton microscopy allows the imaging of living tissues at multiple depths, in this case, the microvasculature of an implanted human glioblastoma multiforme growing as a xenograft in the mouse brain (*upper left*). Note the absence of the well-organized hierarchy of major and minor vessels that is apparent in normal brain tissue. (B) Two-photon microscopy has generated this higher-resolution image of the detailed organization of normal vs. neoplastic microvessels. Note the differing overall organization of the microvascular beds in the normal vs. tumor tissue as well as the increased diameters of the tumor-associated capillaries. (A, from B.J. Vakoc et al., *Nat. Med.* 15:1219–1223, 2009. B, from R.K. Jain, *Nat. Med.* 9:685–693, 2003.)

(A)

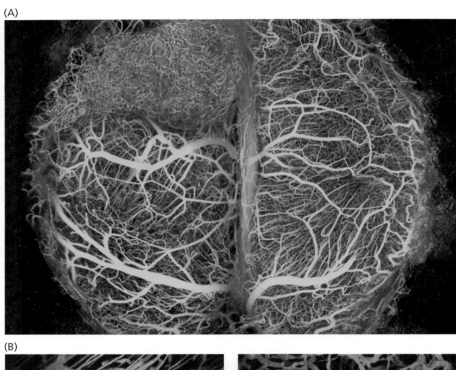

(B)

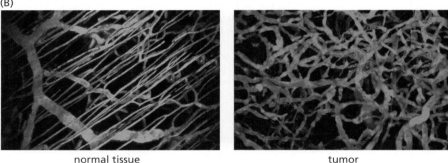

normal tissue   tumor

at the luminal surfaces of endothelial cells and deliver this cargo to the tissue parenchyma via exocytosis occurring at the abluminal sides of these cells.

The leakiness of tumor-associated capillaries leads to the accumulation of substantial amounts of fluid in the parenchymal spaces within a tumor. In normal tissues, these fluids are drained by the lymphatic vessels, which eventually empty their contents into the venous circulation (see Sidebar 13.3). However, within solid tumors, the ongoing expansion of cancer cell populations exerts pressure on those few lymphatic vessels that do succeed in forming, causing their collapse (**Figure 13.35**). (Blood capillaries are more capable of resisting this pressure, because of their own significant internal hydrostatic pressure, which lymphatic vessels lack.) The resulting defective lymphatic drainage within the cores of solid tumors further exacerbates the elevated accumulation of fluid caused by capillary leakage, generating relatively high fluid pressure in the nonvascular parts of tumors.

Yet another mechanism conspires to maintain high hydrostatic pressure within tumors: the large amounts of PDGF that are released by many types of carcinoma cells induce contraction of stromal fibroblasts, resulting in the squeezing out ("expression") of **interstitial** fluid. Hence, at least three mechanisms—lymphatic vessel collapse and resulting loss of fluid drainage, weeping by capillaries, and fibroblast contraction—contribute to intra-tumoral hydrostatic pressure. This pressure, in turn, greatly complicates the administration of anti-cancer therapeutic drugs, since it prevents the formation of the steep pressure gradient between capillary lumen and parenchyma that is needed for the efficient passive transfer of drugs from the circulation into the extravascular spaces of the tumor (Supplementary Sidebar 13.6).

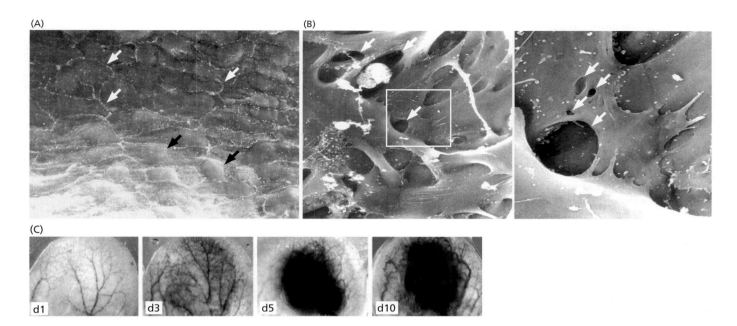

Figure 13.34 Gaps in the tumor microvasculature and resulting leakage (A) This scanning micrograph reveals the tight seals between endothelial cells forming the luminal wall of this venule in a normal mouse mammary gland; these cell–cell junctions, indicated here by *white arrowheads*, do not overlap and block essentially all leakage of fluid from the lumen of the microvessel into the tissue parenchyma. The *black arrows* indicate bulges in the luminal surface deriving from underlying endothelial cell nuclei (see Figure 13.30A). (B) This scanning electron micrograph reveals that the sheets of endothelial cells fail to form a continuous uninterrupted surface in the wall of a capillary within a tumor. Instead, the tumor-associated endothelial cells overlap one another and, as indicated here (*arrows*), show gaps of significant size between them. The box in the *left panel* is shown at higher magnification in the *right panel*. Such gaps permit the seepage of plasma fluids into the interstitial spaces between the cancer cells in the tumor parenchyma, contributing to the high hydrostatic pressure in these spaces. (C) The role of VEGF in inducing capillary permeability is illustrated here in an experiment in which an adenovirus gene expression vector encoding VEGF-A was infected into the ears of mice. At the indicated days thereafter, Evans Blue dye was injected into the general circulation of the mice and the ears were imaged 30 minutes later. With increasing time after initial VEGF-A expression, the microvessels became increasingly permeable, resulting in leakage of the dye molecules from the circulation into the parenchymal tissue of the ear. (A and B, from H. Hashizume et al., *Am. J. Pathol.* 156:1363–1380, 2000. C, from J.A. Nagy et al., *Lab. Invest.* 86:767–780, 2006.)

13.7 Tripping the angiogenic switch is essential for tumor expansion

The descriptions of angiogenesis given above and in Section 7.12 would seem to suggest that the release of angiogenic factors is virtually automatic: whenever groups of cells, including cancer cells, suffer hypoxia, they release angiogenic factors and thereby provoke the growth of capillaries into their midst. This cures the hypoxia and results in an appropriate density of capillaries in the tissue harboring these cells.

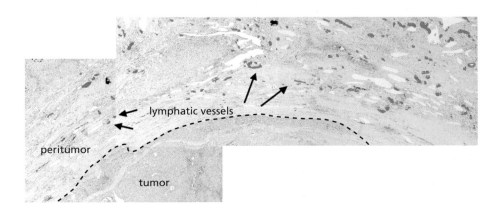

Figure 13.35 Absence of lymphatic vessels within solid tumors Analysis of a section of a hepatocellular carcinoma (HCC; liver cancer) reveals (via specific antibody staining) that lymph ducts (*dark red*) are present in the normal tissue above the tumor margin (*dotted line*) but are absent within the tumor mass itself. This absence may be attributed (1) to the lack of formation of these ducts during tumor growth or (2) to the collapse and degeneration of these ducts because of the high hydrostatic pressure within solid tumors; both mechanisms are likely to operate. (From C. Mouta Carreira et al., *Cancer Res.* 61:8079–8084, 2001.)

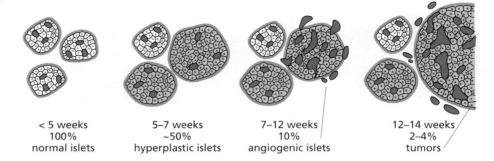

Figure 13.36 The Rip-Tag model of islet cell tumor progression Multi-step tumorigenesis in the Rip-Tag transgenic mouse model proceeds in distinct stages. A normal pancreatic islet (also known as an islet of Langerhans; *left*) carries a small number of capillaries (*red*) to support the β cells, which are responsible for synthesizing and secreting insulin into the circulation. By 5–7 weeks of age, about half of these islets become hyperplastic, but the density of blood vessels is not increased. At 7–12 weeks, about 10% of the hyperplastic islets become angiogenic, as indicated by the greatly increased density of capillaries in the surrounding, nearby exocrine pancreas. Finally, by 12–14 weeks of life, 2–4% of the initially formed hyperplastic islets have become invasive carcinomas that grow rapidly and invade the surrounding exocrine pancreas. The exocrine pancreas is not shown here. (From D. Hanahan and J. Folkman, *Cell* 86:353–364, 1996.)

In fact, the ability to attract blood vessels seems to be a trait that many tumor cell populations initially lack and must acquire as tumor progression proceeds. This idea was first suggested by the observation, cited above, that in certain tumors, cancer cells thrive near capillaries, but those that are located further than 0.2 mm from capillaries stop growing and may enter apoptosis or become necrotic (see Section 13.6). So, even though these cancer cells experience hypoxia, they lack the ability to induce the formation of nearby capillaries.

We know much about these dynamics from detailed study of experimental models of tumorigenesis. The most informative of these has been the Rip-Tag transgenic mouse. It carries a transgene in its germ line, in which the expression of the SV40 large T antigen (see Sections 8.5 and 9.1) and small T antigen (see Sidebar 11.5) is driven by the promoter of the insulin gene. This promoter ensures expression of these viral oncoproteins in the β cells that form the islets of Langerhans in the pancreas (in which insulin is normally produced).

Tumor progression in the 400 or so islets of the mouse pancreas can be easily followed, because these islets can readily be distinguished from the surrounding tissue of the **exocrine** pancreas, which is involved in manufacturing and secreting digestive enzymes. As many as half of these islets in a Rip-Tag mouse form hyperplastic nodules by 10 weeks of age, and 8 to 12% of the hyperplastic islets eventually progress to become angiogenic, that is, they acquire the ability to recruit new blood vessels. By 12 to 14 weeks, about 3% of the initially formed hyperplastic islets finally progress to form carcinomas (**Figure 13.36**).

Early in tumor progression in the Rip-Tag mice, the hyperplastic pancreatic islets begin to expand slightly to a small diameter of about 0.1–0.2 mm and then halt their forward march (see Figure 13.36), at least for a while. In these small nests of tumor cells, cell division continues unabated, being driven by the oncogenic transgene. However, the overall size of the tumor cell population within each islet remains constant, because of a compensating attrition of cells occurring through apoptosis. This mouse model suggests that in humans, small tumor nests may also remain in this dynamic but non-growing state for many years, unable to break through the barrier that is holding them back.

In principle, the barrier to expansion of the tumor cell nests might be a physical one—lack of adequate space within the tissue for these cancer cells to multiply. But detailed histological analysis of these small nests of cancer cells reveals something quite different. Because these cells have not yet become angiogenic, they lack vasculature. The resulting hypoxia that they experience triggers p53-dependent apoptosis, which explains their high rate of attrition. (It is possible that other sub-optimal conditions within these poorly vascularized cell nests, including inadequate nutrient supply, high levels of carbon dioxide and metabolic wastes, and low pH caused by lactic acid accumulation, also contribute to the death of these cells; for example, see Figure 13.27D.)

At some point in time, however, small clusters of these pre-neoplastic islet cells suddenly acquire the ability to provoke neoangiogenesis (see Figure 13.36). Once these cells learn how to induce capillaries to form nearby, they and their descendants seem to be liberated from the major constraint that has been holding back their multiplication.

This sudden, dramatic change in the behavior of the small tumor masses has been termed the "**angiogenic switch.**"

These phenomena suggest an interesting idea, really a speculation: the body purposefully denies its cells the ability to readily induce angiogenesis. By doing so, the body erects yet another impediment to block the development of large neoplasms. According to such thinking, the angiogenic switch—a clearly important step in tumor progression—represents the successful breaching of this defensive barrier and the acquisition by cancer cells of a forbidden fruit: the ability to induce blood vessel growth at will.

One might conclude that the angiogenic switch in these transgenic mice is driven by the premalignant β cells' suddenly acquiring the ability to express and release VEGFs. Actually, these cells make large amounts of VEGFs long before the angiogenic switch occurs, as do fully normal pancreatic islets. However, the VEGF molecules secreted by these β cells are efficiently sequestered by the surrounding extracellular matrix (ECM). As a consequence, the VEGF molecules are unable to stimulate angiogenesis.

This sequestered state of the VEGF explains why the angiogenic switch in the Rip-Tag pancreatic islets is accompanied by the sudden appearance of substantial amounts of matrix metalloproteinase-9 (MMP-9; see Table 13.1). MMP-9 acts in a targeted fashion to cleave specific structural components of the ECM, thereby releasing VEGF for active signaling to nearby endothelial cells. This MMP-9 is synthesized and released by inflammatory cells—mast cells and macrophages—that have been attracted to the premalignant islets (**Sidebar 13.4**). Hence, in this particular tissue, tripping of the angiogenic switch ultimately depends on an acquired ability to recruit inflammatory cells.

If the gene encoding VEGF is selectively deleted from the islet cells through genetic engineering (see Supplementary Sidebar 7.7), the islets survive and the early steps of tumor progression still proceed normally, but the angiogenic switch is never tripped. This reinforces the conclusion that VEGF molecules of islet cell origin play a critical role in triggering the onset of angiogenesis, and that recruited stromal cells cannot compensate for this absence of VEGF by bringing in some of their own.

This scenario (**Figure 13.37**) involves heterotypic interactions among three distinct cell types: (1) the release of still-unidentified signals from the premalignant islet cells that recruit mast cells and, quite possibly, macrophages; (2) the release of MMP-9 by these inflammatory cells to activate previously latent VEGF made by the premalignant islet cells; and (3) the proliferative response of endothelial cells to the activated VEGF. In fact, yet other cell types are likely to be partners in islet cell angiogenesis. Thus, some of the endothelial cells may derive from recruited endothelial precursor cells in the circulation (EPCs, sometimes called circulating endothelial progenitors, or CEPs; see Figure 13.22); and the capillaries that arise are eventually covered, albeit haphazardly, with another cell type—pericytes—whose precise origins are unclear.

Sidebar 13.4 Mast cells from the bone marrow can play a key role in the angiogenic switch Rip-Tag mice carrying the insulin SV40 transgene in their germ line can be bred with others that lack the Kit growth factor receptor (see Sidebar 5.4). Kit serves as the receptor for stem cell factor (SCF), an important growth factor that triggers the development of certain subclasses of hematopoietic cells. Among other deficits, $Kit^{-/-}$ mice lack the ability to form mast cells. In Rip-Tag mice lacking Kit receptor, islet cell tumors are initiated at rates routinely observed in standard Rip-Tag mice. However, these tumors never succeed in becoming angiogenic, and the rate of cell proliferation in these small growths is compensated by an equal rate of apoptotic death. Consequently, these tumors remain at a very small size (0.1–0.2 mm diameter).

If these mice are provided with a bone marrow transplant containing wild-type hematopoietic precursor cells, which cures their mast cell deficit, angiogenesis is initiated in the pancreatic islets, tumor cells in the islets are no longer lost through apoptosis, and large, life-threatening neoplasms appear soon thereafter. This demonstrates that the ability to provoke angiogenesis depends on the actions of at least one non-endothelial component of the tumor-associated stroma—the mast cells originating in the bone marrow. The main contribution of these mast cells to the angiogenic switch seems to be the release of MMP-9, which proceeds to mobilize latent VEGF, thereby provoking angiogenesis (see Figure 13.37). Yet other work implicates macrophages as well in tumor angiogenesis (see Figure 13.25).

Figure 13.37 The angiogenic switch and recruitment of inflammatory cells (A) The normal islet of Langerhans (*oval area, center of left panel*) is poorly vascularized and is sustained largely through diffusion from the microvessels surrounding it. Following tripping of the angiogenic switch, there is an increase in the number of cells in the islet, due to tumor expansion, and, as seen here, a dramatic induction of vessel formation with vessels now threading their way through the islet (*right panel*). Imaging was achieved by whole-mount microscopy of lectin-perfused vessels. (B) According to the scheme presented here, a pre-angiogenic, hyperplastic islet sends signals to the bone marrow and/ or circulation (*not illustrated*) that lead to the recruitment of mast cells and macrophages. Once in the vicinity of the islet, these release metalloproteinases, notably MMP-9. MMP-9 then cleaves extracellular matrix (ECM) components, liberating vascular endothelial growth factor (VEGF) from its sequestered state. VEGF proceeds to induce angiogenesis around the islet, thereby tripping the angiogenic switch. (A, from G. Bergers, D. Hanahan and L.M. Coussens, *Int. J. Dev. Biol.* 42:995–1002, 1998. B, adapted from data in G. Bergers et al., *Nat. Cell Biol.* 2:737–744, 2000. Image of VEGF in B, from Y.A. Muller et al., *Proc. Natl. Acad. Sci. USA* 94:7192–7197, 1997.)

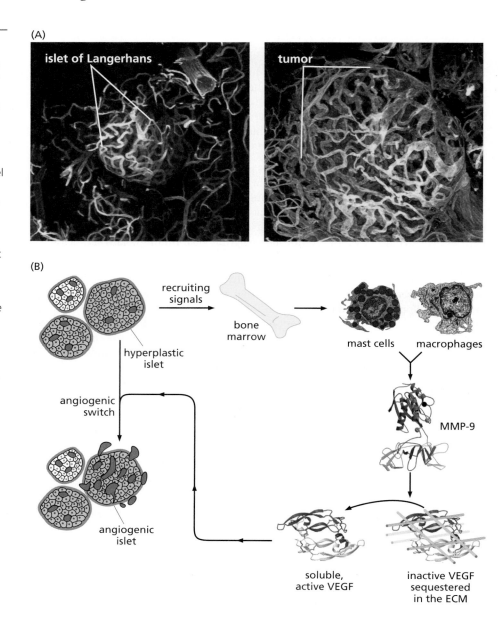

Importantly, the Rip-Tag model does not typify the angiogenic mechanisms occurring during the formation of all types of tumors. For example, in some tumors, angiogenesis appears to increase progressively, as if a controlling rheostat is gradually being turned. This contrasts with the behavior of Rip-Tag islets, in which a binary, on–off switch seems to be tripped.

Other tumors may depend on different angiogenic factors to provoke angiogenesis. For example, when transformed mouse embryonic stem (ES) cells are deprived of both copies of the VEGF-A–encoding gene, they lose almost all their power to make malignant teratomas. In contrast, transformed adult mouse dermal fibroblasts remain highly tumorigenic after they have been deprived of both copies of this gene. And the sarcomas generated by these transformed fibroblasts continue to grow even when the mice bearing them have been treated with an antibody that binds and inactivates VEGF-R2 (the primary endothelial cell receptor that confers responsiveness to both VEGF-A and VEGF-B). This behavior is likely explained by the fact that transformed dermal fibroblasts can make a complex mixture of angiogenic factors—including VEGF-B, acidic and basic fibroblast growth factors (aFGFs and bFGFs), and transforming growth factor-α (TGF-α). This deployment of multiple angiogenic factors (**Table 13.3**), often observed in advanced human cancers, complicates the development of anti-angiogenesis cancer therapies, as we will see later.

Table 13.3 Important angiogenic factors

Name	Mol. wt. (kD)
Vascular endothelial GFs (VEGFs)	40–45
Basic fibroblast growth factor (bFGF)	18
Acidic fibroblast growth factor (aFGF)	16.4
Angiogenin	14.1
Transforming growth factor-α (TGF-α)	5.5
Transforming growth factor-β1 (TGF-β1)	25
Tumor necrosis factor-α (TNF-α)	17
Platelet-derived growth factor-B (PDGF-B)	45
Granulocyte colony-stimulating factor (G-CSF)	17
Placental growth factor	25
Interleukin-8 (IL-8)	40
Hepatocyte growth factor (HGF)	92
Proliferin	35
Angiopoietin	70
Leptin	16

13.8 The angiogenic switch initiates a highly complex process

Angiogenesis begins in the stroma surrounding the Rip-Tag tumors long before the basement membrane has been broken down. This behavior typifies that of many tumors in both mice and humans (**Figure 13.38**). Somehow, angiogenic signals are dispatched by benign cancer cells through the porous basement membrane (BM) in order to encourage increased angiogenesis on the stromal side of this membrane. Yet other signals transmitted through the basement membrane recruit myofibroblasts to the nearby stroma (**Figure 13.39**). As we read earlier, these myofibroblasts can also help to foster angiogenesis.

Nevertheless, this early angiogenesis is circumscribed, and it is clear that intense angiogenesis can begin only when cancer cells become invasive, penetrate the basement membrane, and acquire direct, intimate contact with stromal cells (**Figure 13.40**). This suggests that tumor invasiveness and intense angiogenesis are often tightly coupled processes. We will study tumor invasiveness in detail in the next chapter.

In many human tumor types, the density of capillaries per microscope field increases in lockstep with increasing degrees of malignancy. For example, among human breast carcinomas that have already grown to a considerable size, those that have managed to attract dense networks of capillaries into their midst are indicative, on average, of a far worse prognosis than those that are poorly vascularized (**Figure 13.41A**). Moreover, patients with breast tumors that express large amounts of VEGF (in addition to HER2) also fare badly following initial diagnosis and treatment (see Figure 13.41B). Altogether, this suggests that the angiogenic switch is only the first of many gradations that enable tumors to become progressively more angiogenic and hence increasingly vascularized.

These striking correlations are actually susceptible to two alternative interpretations. It is possible that intense vascularization enables cancer cells to grow more aggressively, thereby leading to poor clinical outcomes. Alternatively, intense angiogenesis

Figure 13.38 Signaling through the basement membrane early in tumor progression (A) A human ductal carcinoma *in situ* (DCIS) of the breast contains a collection of carcinoma cells (*purple blue*) that are noninvasive and therefore have not yet breached the basement membrane (BM) that surrounds them and separates them from the mammary stroma. As is apparent, this DCIS has nonetheless succeeded in transmitting angiogenic signals through the BM into the nearby stroma that have resulted in the growth of a small vessel (*dark brown*) that surrounds the tumor mass but does not penetrate into the tumor itself because of the continued integrity of the BM. (B) In this transgenic mouse model of skin carcinogenesis, the human papillomavirus (HPV) 16 early region is expressed under the control of a keratin-14 gene promoter, which ensures its expression specifically in keratinocytes of the skin. In the early hyperplastic stage of tumor progression, the noninvasive carcinoma cells on the epithelial side (*above*) are able to transmit signals through the BM (*dashed line*) that provoke angiogenesis on the stromal side (*below*), as evidenced by the increased density of capillaries (*red*), detected in this case through their display of the CD31 antigen. Dysplastic tissue, in which the BM (*dashed line*) has not yet been breached, shows even more intensive angiogenesis in the nearby stroma. (A, courtesy of A.L. Harris. B, from L.M. Coussens et al., *Genes Dev.* 13:1382–1397, 1999.)

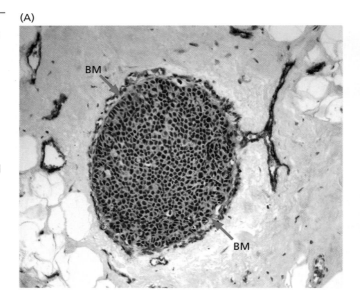

(A)

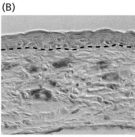

normal skin

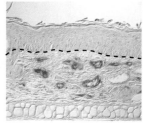

hyperplasia

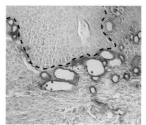

(B)

dysplasia

might be only a marker of an underlying aggressive phenotype but not be causally involved in driving high-grade malignancy. Available clinical data do not allow a clear resolution between these alternatives.

We have spoken of the angiogenic switch as if small numbers of cells within a tumor undergo this shift and proliferate to ultimately dominate within the tumor mass, thereby imparting the angiogenic phenotype to the tumor as a whole. In fact, analyses of cells isolated from explanted human tumors indicate great heterogeneity in the angiogenic powers of different subpopulations of cancer cells within a given tumor, with some cancer cells being highly angiogenic while others are poorly so. Even within tumor cell lines, individual subcloned cell populations show greatly differing angiogenic and tumorigenic powers when engrafted *in vivo* (**Figure 13.42**). This introduces yet another idea: that within a tumor mass, long after the tripping of the angiogenic switch, the weakly angiogenic cancer cells rely on help from their friends— their strongly angiogenic neighbors—in order to acquire adequate vasculature. (We encountered another form of cooperation between subpopulations of tumor cells within a tumor when describing the symbiosis between lactate acid-generating and –consuming cells; see Supplementary Sidebar 11.3.)

The angiogenic switch is associated with processes that extend far beyond the immediate vicinity of a tumor. For example, as mentioned several times above, at certain stages of tumor progression, neovascularization relies on the recruitment of endothelial precursor cells (EPCs) that originate in the marrow and travel via the circulation to the tumor (see Section 13.4). VEGF released into the circulation by a tumor stimulates the production of EPCs in the bone marrow and their release into the general circulation; in addition, it helps to attract circulating EPCs to the tumor mass. As also described in Section 13.4, stroma-derived factor-1 (SDF-1/CXCL12), which is liberated by stromal myofibroblasts, also helps in this recruitment. Once settled within the tumor mass, the EPCs are induced to differentiate into functional endothelial cells and construct the tumor-associated vasculature. Detailed analyses have shown that

(A)

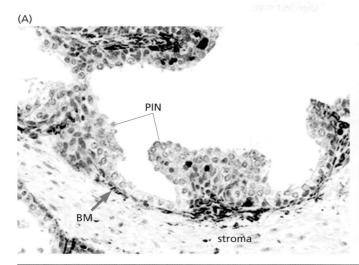

(B)

Figure 13.39 Recruitment of myofibroblasts on the stromal side of the basement membrane Myofibroblasts also foster angiogenesis, in part through their ability to recruit endothelial precursor cells (EPCs) to the tumor stroma, as described in Figure 13.22. (A) In human prostate intraepithelial neoplasia (PIN)—a counterpart of ductal carcinoma *in situ* (DCIS) in the breast—the carcinoma cells (*darker blue nuclei*) remain on the epithelial side of the still-intact basement membrane (BM), yet they have stimulated the accumulation of stromal fibroblasts beyond the BM, as indicated by their display of vimentin (*dark brown*). (B) Many of these stromal cells can be identified more specifically as being myofibroblasts by their ability to be immunostained with both an anti-vimentin antibody (*red*) and α-smooth muscle actin (*green*); cells co-expressing both markers are myofibroblasts and are seen in *yellow*. These cells can also be identified as myofibroblasts through their expression of collagen I (*not shown*). (From J. Tuxhorn et al., *Clin. Cancer Res.* 8:2912–2923, 2002.)

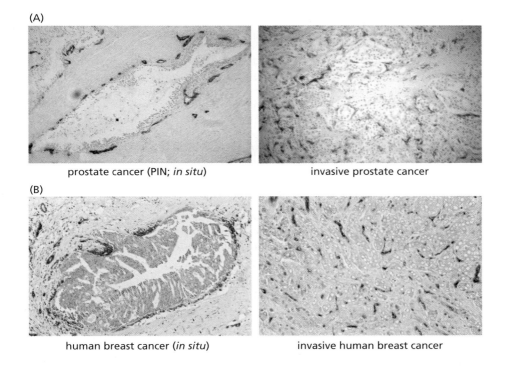

Figure 13.40 Angiogenesis before and after acquisition of invasiveness Tumor angiogenesis is circumscribed as long as human carcinomas remain benign. However, once they become invasive, the intensity of angiogenesis increases, leading to a higher density of capillaries (*brown*) threading their way through tumors. (A) The capillaries ringing a benign prostatic intraepithelial neoplasia (PIN) lesion (*left*) are far fewer than those in an invasive prostate carcinoma (*right*). (B) Similarly, those in a benign ductal carcinoma *in situ* (DCIS) of the breast (*left*) are far fewer than those in an invasive ductal carcinoma (*right*). (Courtesy of J. Folkman.)

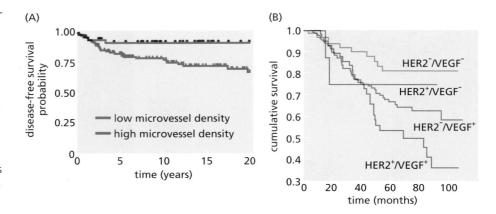

Figure 13.41 Clinical outcomes and the intensity of angiogenesis
(A) Breast carcinomas were analyzed for the density of capillaries, which was determined as the number of microvessels per microscopic field. This Kaplan–Meier graph demonstrates that those patients whose tumors had a high microvessel count (*red curve*) had a markedly lower probability of disease-free survival in the 20 years following initial diagnosis than those whose tumors had a low microvessel count (*blue curve*). (B) As indicated earlier (see Figure 4.4), breast cancer patients whose tumors overexpress HER2/Neu have a markedly poorer prognosis than those whose tumors do not. In the group analyzed here, all of the patients showed metastatic cancer cells in one or more of the lymph nodes draining the breast. The differences in survival are even more dramatic when the levels of VEGF produced by their tumors are also considered. In this relatively small clinical study, more than 80% of patients whose tumors expressed low, basal levels of both HER2/Neu and VEGF (HER2⁻ VEGF⁻) were alive eight years after diagnosis; in contrast, only about 35% of the patients whose tumors expressed elevated levels of both HER2/Neu and VEGF (HER2⁺ VEGF⁺) survived this long. (A, from R. Heimann and S. Hellman, *J. Clin. Oncol.* 16:2686–2692, 1998. B, from G.E. Konecny et al., *Clin. Cancer Res.* 10:1706–1716, 2004.)

the proportion of endothelial cells derived from circulating EPCs (versus those arising from nearby capillaries) varies greatly from one type of tumor to another.

These dynamics of EPC recruitment are dramatically illustrated by the behavior of mice that lack one copy of the *Id1* gene and both copies of the *Id3* gene. Recall that these *Id* genes encode transcription factors regulating differentiation of a variety of cell types (see Section 8.11). Many of these mutant mice survive to adulthood and show a very peculiar phenotype: they are defective in neoangiogenesis and will not permit several types of engrafted tumors to grow (**Figure 13.43A**).

The defect in these *Id1*$^{+/-}$ *Id3*$^{-/-}$ mice can be cured by transplanting wild-type stem cells into their bone marrow (see Figure 13.43A). Alternatively, the subset of bone marrow stem cells whose production is normally spurred by high levels of circulating VEGF can be harvested from a wild-type mouse and transplanted into the bone marrow of *Id1*$^{+/-}$ *Id3*$^{-/-}$ mice. These mice soon generate large numbers of endothelial precursor cells (EPCs) in their circulation and, following implantation of cancer cells, develop rapidly growing tumors. Taken together, these observations indicate that these tumors rely on VEGF to mobilize EPCs from the bone marrow (see Figure 13.43B) and to recruit the resulting circulating cells into the tumor stroma. These findings, as well as others demonstrating the recruitment of myofibroblast precursors into tumor stroma (see Section 13.3), show that primary tumors can extend their reach throughout the body long before they metastasize.

13.9 Angiogenesis is normally suppressed by physiologic inhibitors

In finely tuned physiologic processes, the actions of positive effectors must be counterbalanced by negative regulators. We have read much about the positive effectors of angiogenesis, such as VEGF and bFGF, but their antagonists have remained offstage until now. They turn out to be as interesting and important as the angiogenic factors whose actions they antagonize.

During the process of wound healing, for instance, the burst of angiogenesis that is required to repair the wound site must be shut down once the newly formed capillaries

Figure 13.42 Heterogeneous degrees of vascularization within a tumor cell population (A) When a population of cells in a human liposarcoma cell line is subjected to single-cell cloning (in which the cells in each resulting cell clone all derive from a common ancestor), the various cloned cell populations show greatly differing abilities to form tumors (*individual curves*) when implanted into immunocompromised mice. (B) When the resulting tumors are analyzed for microvessel density (microvessels per microscopic field) and tumor volume, plotted logarithmically here, it becomes clear that they have greatly differing angiogenic capabilities and that angiogenesis behaves as if it were a rate-limiting determinant of tumor growth. (From E.G. Achilles et al., *J. Natl. Cancer Inst.* 93:1075–1081, 2001.)

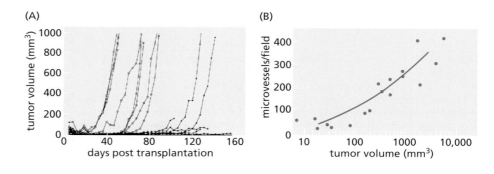

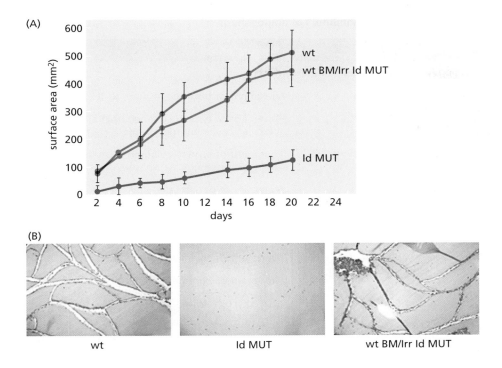

(A)

(B)

wt   Id MUT   wt BM/Irr Id MUT

Figure 13.43 Defective tumor angiogenesis in *Id1*^(+/−) *Id3*^(−/−) mice
Mutant mice (Id MUT), which lack three of four *Id* gene copies (and have an *Id1*^(+/−) *Id3*^(−/−) genotype), show impaired mobilization of endothelial precursor cells (EPCs) from the bone marrow and thus impaired recruitment of circulating EPCs into tumors that these mice may carry. (A) Wild-type (wt) mice bearing Lewis lung carcinoma (LLC) cells develop rapidly growing tumors (*red curve*), while those of the mutant strain (Id MUT) are unable to support vigorous tumor growth (*brown curve*). This defect is essentially reversed if the bone marrow (BM) of the mutant mice is eliminated by irradiation and replaced with transplanted wild-type bone marrow (*blue curve*, wt BM/Irr Id MUT), indicating that the recruitment of bone marrow–derived cells is defective in the mutant mice and is responsible for the inability of the tumor to grow in them. (B) This defective recruitment can be further localized by the use of plugs of Matrigel (an extracellular matrix material) that have been impregnated with VEGF and are then implanted subcutaneously. These plugs are able to recruit neovasculature when implanted in wild-type mice (*left*) but not in the mutant mice (*center*). However, if the mutant mice receive a graft of wild-type bone marrow cells, the defect is cured and now angiogenesis within these plugs is comparable to that seen in wild-type mice (*right*). (From D. Lyden et al., *Nat. Med.* 7:1194–1201, 2001.)

have reached a density that suffices to support normal tissue function. This shutdown is achieved, at least in part, by suppressing formation of the HIF-1 transcription factor. Its assembly is induced under hypoxic conditions and is reversed once normal oxygenation in the wound site has been restored (see Section 7.12).

In addition, a number of the components of the extracellular matrix are used by tissues to actively block excessive angiogenesis (**Table 13.4**). The best characterized of these is the thrombospondin-1 (Tsp-1) protein, which is secreted by many cell types into the surrounding extracellular space, where it forms homotrimers and carries out several distinct functions. Most important for our discussion, Tsp-1 associates with a receptor (termed CD36) that is displayed on the surfaces of endothelial cells and halts their proliferation. In addition, research indicates that Tsp-1 treatment of endothelial cells causes them to release Fas ligand (FasL), the pro-apoptotic signaling protein that acts by binding to the Fas death receptor. Recall that the latter, once it has bound its ligand, activates an intracellular caspase cascade that triggers apoptosis (see Section 9.14). Therefore, once Tsp-1 causes endothelial cells to release FasL, the latter may act in an autocrine fashion to trigger the death of these cells in the event that they also display the Fas receptor.

Interestingly, the Fas receptor is displayed on endothelial cells that are actively proliferating or have recently ceased proliferation, but its display is suppressed once these cells have successfully formed mature capillaries and retreated into quiescence (**Figure 13.44A**). This seems to explain a most intriguing aspect of Tsp-1 behavior: it selectively inhibits and causes regression of newly formed and still-growing capillaries but has little if any effect on already-formed, mature capillaries. In fact, a number of other natural anti-angiogenic factors also exhibit such selectivity and, like Tsp-1, seem to depend upon activation of the pro-apoptotic caspase cascade in endothelial cells, since compounds that inhibit the caspase enzymes also protect endothelial cells from the anti-angiogenic effects of Tsp-1 and the other natural blockers of angiogenesis.

Transcription of the *TSP1* gene is strongly induced by p53, ostensibly as part of the p53-mediated emergency response that leads to a generalized shutdown of cell proliferation and tissue growth. Conversely, the loss of p53 function, which is seen in almost all human tumors (see Chapter 9), leads to a substantial decrease in Tsp-1 levels. This permits angiogenesis to be induced by cells that normally would have been prevented from doing so by the high Tsp-1 concentrations in the surrounding extracellular matrix.

Table 13.4 Endogenous inhibitors of angiogenesis

Inhibitor	Description
A. Derived from extracellular matrix	
Anastellin	fragment of fibronectin
Arresten	fragment of type IV collagen α_1 chain of vascular basement membrane
Canstatin	fragment of type IV collagen α_2 chain of vascular basement membrane
Chondromodulin-I	component of cartilage ECM
EFC-XV	fragment of type XV collagen
Endorepellin	fragment of perlecan
Endostatin	fragment of collagen type XVIII
Fibulin	fragment of basement membrane protein
Thrombospondin-1 and -2	ECM glycoproteins
Troponin I	component of cartilage ECM
Tumstatin	fragment of type IV collagen α_3 chain
B. Non-matrix–derived	
Growth factors and cytokines	
Interferon-α (IFN-α)	cytokine
Interleukins (IL-1β, -12, -18)	cytokines
Pigment epithelium-derived factor (PEDF)	growth factor
Platelet factor-4	released by platelets during degranulation
Other types	
Angiostatin	fragment of plasminogen
Antithrombin III	fragment of antithrombin III
2-Methoxyestradiol	endogenous metabolite of estrogen
PEX	fragment of MMP-2
Plasminogen kringle 5	fragment of angiostatin
Prolactin fragments	specific cleavage fragment
Prothrombin kringle 2	fragment of prothrombin
sFlt-1	soluble form of VEGF-R1 (= Flt-1)
TIMP-2	inhibitor of metalloproteinase-2
TrpRS	fragment of tryptophanyl-tRNA synthetase
Vasostatin	fragment of calreticulin

Adapted from P. Nyberg, L. Xie and R. Kalluri, *Cancer Res.* 65:3967–3979, 2005.

The Ras oncoprotein, acting through a complex signaling cascade, acts in the opposite fashion, since it causes shutdown of *TSP1* gene expression. The resulting absence of significant levels of Tsp-1 can also contribute substantially to the elevated angiogenic powers of *ras*-transformed cells compared with their normal neighbors. In one study of melanomas arising in transgenic mice, expression of the H-*ras* oncogene, which

(A)

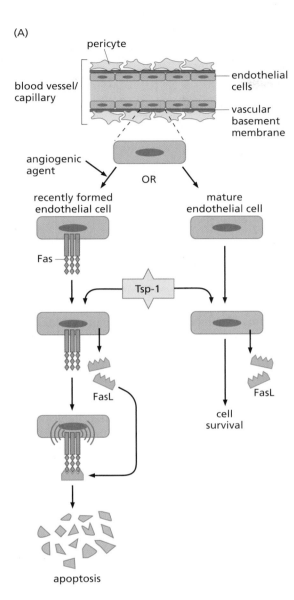

(B)

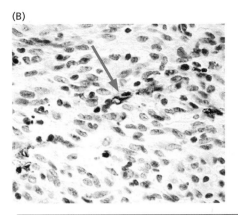

Figure 13.44 Thrombospondin, endothelial cell survival, and tumorigenesis (A) The stimulation of resting endothelial cells with angiogenic agents, such as VEGF or bFGF, creates activated, growing endothelial cells (*left branch*), which soon express the Fas death receptor on their surface. Subsequent treatment of mature and recently formed endothelial cells with thrombospondin-1 (Tsp-1; *blue*) causes both groups of cells to secrete FasL (*green*), the ligand of Fas. However, because only the activated endothelial cell displays the Fas receptor, it is preferentially induced to enter into apoptosis by Tsp-1, while the FasL secreted by the mature, resting endothelial cell cannot be induced to enter apoptosis because it cannot bind and respond to secreted FasL. This seems to explain why Tsp-1 can block new angiogenesis but has relatively minimal effects on already-constructed vasculature, whose endothelial cells are only rarely involved in active proliferation. (B) The loss of *ras* oncogene expression can lead to a rapid increase in Tsp-1 expression. This may explain why, in this transgenic mouse model of melanoma development, shutdown of *ras* oncogene expression in already-formed *ras*-induced tumors leads to rapid collapse of the tumor-associated vasculature and tumor regression. Here, the TUNEL assay (see Figure 9.19) for apoptosis reveals apoptotic endothelial cells within a capillary (*arrow*) of a regressing melanoma. (A, adapted from O.V. Volpert et al., *Nat. Med.* 8:349–357, 2002. B, from L. Chin et al., *Nature* 400:468–472, 1999.)

had been used to initiate these tumors, was shut down after they had grown to a considerable size. These tumors collapsed rapidly thereafter, and this collapse was directly traceable to their loss of functional vasculature, most of which was recently synthesized and began to distintegrate within 6 hours after the loss of Ras function (see Figure 13.44B). The observed rapid entrance into apoptosis of endothelial cells forming the tumor vasculature preceded any decline in the levels of VEGF (which provides survival signals for endothelial cells). While not demonstrated directly in these experiments, it seems likely that this apoptosis derived from the rapid re-expression of Tsp-1 that follows close on the heels of Ras shutdown.

Tsp-1 is surely a major governor of angiogenesis, but as hinted above, it is only one of a large cohort of natural inhibitors of angiogenesis that are found in the spaces between cells (see Table 13.4). The existence of several other anti-angiogenic molecules was first suggested by observations of the behavior of certain primary tumors and their derived metastases. In some mouse models of tumorigenesis, metastases were found to remain small in size as long as the primary tumor that spawned them continued to grow. However, the moment the primary tumor was surgically removed, the metastases began to grow vigorously. This behavior is echoed by anecdotal reports of cancer surgeons, who have observed that after the successful surgical removal of a primary tumor, substantial numbers of metastases may suddenly sprout and flourish, doing so within several months' time.

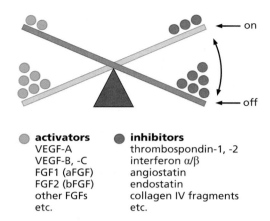

Such observations suggested that some type of inhibitory substance released by the primary tumor acted, via the circulation, to suppress the proliferation of distant nests of metastatic cells. More specifically, these inhibitory factors, whatever their nature, seemed to block angiogenesis in these secondary growths, which failed to expand to a diameter of more than several tenths of a millimeter. Once the primary tumors were excised, the hypothetical inhibitory factor(s) disappeared from the circulation, removing some constraint on the growth of already-seeded metastases.

The subsequent isolation of these circulating factors, initially from the urine of tumor-bearing animals, yielded the intensively studied angiostatin and endostatin molecules. Determination of the amino acid sequences of these two protein species indicated that they arise as cleavage products of familiar proteins of the extracellular matrix (ECM) or plasma. With the passage of time, yet other anti-angiogenic substances have been isolated (see Table 13.4), many of which are also formed by the proteolysis of extracellular proteins. Taken together, these discoveries suggest that as angiogenesis proceeds in normal tissues during development and wound healing, this process is eventually curtailed by the accumulation of anti-angiogenic protein fragments in the extracellular space. These naturally occurring anti-angiogenic substances can therefore be depicted as important components of negative-feedback loops operating to ensure that excessive vascularization of a tissue does not occur.

Other natural anti-angiogenic proteins function as antagonists of the matrix metalloproteinases (MMPs). VEGF stimulates the localized production of MMPs 1 to 4, which enable elongating capillaries to invade through the extracellular matrix between cells. A class of secreted proteins, termed tissue inhibitors of metalloproteinases (TIMPs), can prevent this elongation by blocking the actions of these and other MMPs. For example, by forcing the ectopic expression of TIMP-2 in tumor cells, researchers have blocked the angiogenic and thus tumorigenic powers of these cells. However, the precise mechanisms by which TIMPs block angiogenesis are still not clearly resolved.

When integrated, these disparate observations about the physiologic regulators of angiogenesis reinforce the idea, cited earlier, that angiogenesis is not a binary state—either on or off. Instead, different types of tumor cells acquire greater or lesser angiogenic powers, and even within a given tumor, the tumor cells are likely to show differing abilities to attract vasculature. Such behavior can be explained by a scheme (**Figure 13.45**) in which the balance between pro- and anti-angiogenic factors determines whether neoangiogenesis will proceed, and if so, how intensively regions within tumors will become vascularized.

13.10 Anti-angiogenesis therapies can be employed to treat cancer

The more complex a system becomes, the more vulnerable it is to various types of disruption. The process of angiogenesis, as described here, clearly falls in the class of highly complex systems, as shown by its dependence on multiple cell types and signaling molecules. Indeed, while we have featured several important angiogenic factors, by some counts there are at least a dozen involved in regulating various steps of vascular morphogenesis (see Table 13.3).

For those researchers intent on developing new types of anti-cancer therapeutics, this complexity offers multiple targets for intervention. In particular, highly targeted therapies may be devised to inhibit the several cell types that participate in angiogenesis as well as the multiple signaling channels through which they intercommunicate. Since tumors depend absolutely on angiogenesis to grow above a certain size (~0.2 mm diameter), any successes in blocking angiogenesis or in undoing the products of angiogenesis should represent a highly effective strategy for treating cancer. Microscopic tumors should be prevented from growing larger, while larger tumors should collapse once their already-established blood supply disintegrates.

In principle, anti-angiogenesis therapies have a major advantage over those directed at the neoplastic cells within tumors. As we will learn in Chapter 16, one of the great frustrations of anti-cancer drug development comes from the fact that, sooner or later, tumors that initially respond to a drug treatment become **refractory** (resistant) to further treatment by the drug. Almost always, these relapses can be traced to the emergence of drug-resistant variants within tumor cell populations; these variants arise at an almost-predictable frequency and proceed to proliferate and regenerate aggressively growing tumor masses. The emergence of these drug-resistant variants seems to be one of the consequences of the highly unstable genomes of cancer cells (discussed in Chapter 12) and their resulting ability to spawn mutants at high frequency.

Many anti-angiogenesis therapies, in stark contrast, are directed at killing the genetically *normal* cells that have been recruited into tumor masses and co-opted by the cancer cells to do their bidding. There is every reason to believe that the endothelial cells within tumors possess normal, stable genomes and are therefore stable phenotypically. Hence, drug therapies directed against these cells are not likely to select for the outgrowth of drug-resistant variants, and tumors should not, in theory, become refractory to anti-angiogenic drug therapy (see, however, **Sidebar 13.5**).

This interest in treating tumor-associated endothelial cells is further heightened by the peculiar biology of these cells. They are continually being formed and lost within tumor masses, with lifetimes measured as short as a week, while their counterparts that line normal blood vessels elsewhere in the body rarely divide and have lifetimes that are measured in hundreds of days, some being as long as seven years. Cycling cells (that is, cells racing around the active cell cycle) are, almost always, far more sensitive to drug-induced killing than are quiescent cells. For this reason, cytotoxic therapies directed against endothelial cells should have drastic effects on the tumor-associated vasculature, while leaving blood vessels located elsewhere in the body unscathed (but for the caveat cited in Sidebar 13.5).

We will learn in great detail about the general principles of anti-cancer therapy later in Chapter 16 but will take the opportunity here to describe the details of certain therapies that are directed specifically against the tumor-associated vasculature. Some of the first efforts in this area have come from experiments in which the natural anti-angiogenesis inhibitors mentioned above have been used to treat tumors borne by mice.

Since these agents, such as angiostatin and endostatin, are native to the body, they have the advantage of being reasonably well tolerated without toxic side effects. They have virtually no effect on the *in vitro* proliferation of a variety of cells, and instead are biologically active only on endothelial cells that are actively participating in neoangiogenesis *in vivo*. Moreover, these proteins can persist for some time in the circulation, thereby increasing the exposure of tumor vasculature to their anti-angiogenic effects. On the negative side, their manufacture in large quantities, like that of all proteins, is costly and challenging, and their precise mechanisms of action have been elusive.

Both angiostatin and endostatin had only modest effects in blocking the angiogenic switch in the pancreas of Rip-Tag mice (see Section 13.7). But when applied to already-vascularized small tumors, endostatin reduced by more than 80% their subsequent growth, while angiostatin reduced it by 50%. When large, well-established tumors were treated with either of these agents, relatively little effect was seen, but the two introduced together caused a 75% reduction in the mass of such tumors. Long-term endostatin treatment of mice bearing tumors formed from a lung carcinoma cell line led to regression of the tumors. This success might suggest that endostatin may be a highly effective therapeutic in the oncology clinic, but good responses in mice are rarely predictive of similar efficacy in the oncology clinic.

In fact, the development of angiostatin and endostatin as clinically useful anti-cancer agents has been abandoned in the West due to clinical responses that were, at best, equivocal. However, large-scale clinical trials in China have demonstrated significant clinical benefit from adding a recombinant form of endostatin to standard chemotherapeutic agents when treating non-small-cell lung cancer (NSCLC). In both cases, the recombinant endostatin was found to act synergistically with the standard chemotherapeutics to prolong disease progression and patient survival by one to several months. Further clinical trials are underway that are scheduled to reach completion in 2013 and 2014.

Sidebar 13.5 Tumors may outsmart even the best anti-angiogenic therapies Much of the allure of anti-angiogenic therapies comes from the likelihood that the cells being targeted, notably the endothelial cells, are unlikely to generate drug-resistant variants. The complexity of heterotypic signaling may, however, allow tumors to circumvent even the cleverest anti-angiogenic therapies. Imagine, for example, that we devise a strategy to block the VEGF signaling that is so critical for the formation of new vessels in a tumor and their subsequent maintenance (since VEGF is required for endothelial cell survival). A tumor treated in this way should rapidly collapse, since its capillary beds will disintegrate. Indeed, just such a response has been observed in Rip-Tag mice (see Figure 13.36) that developed pancreatic islet tumors: when treated with an anti-VEGF-R2 antibody, their tumors regressed by more than 50%. However, the residual, surviving tumor cells responded by spawning variants that acquired the ability to produce elevated levels of fibroblast growth factors (FGFs), which are also potent angiogenic factors. These growth factors then supplanted VEGF as the main conveyor of signals from the islet tumor to the endothelial cells and succeeded in triggering the regeneration of vasculature and, in turn, the rebirth of a vigorously growing tumor. Accordingly, truly effective anti-angiogenesis strategies will require the inhibition of multiple angiogenic pathways.

Anti-angiogenesis therapies are also attractive because of their potential selectivity for killing the proliferating endothelial cells within tumors, since many blood vessels in these growths seem to be in a constant state of formation and collapse, and are therefore susceptible to cytotoxic drugs that may leave the quiescent endothelial cells in normal tissues untouched. For example, dramatically contrasting growth states of normal and tumor-associated blood vessels can be seen in many mouse models of cancer, in which engrafted tumors expand rapidly. However, it is possible that many of the vessels in slowly growing human tumors may exist for years and therefore may have ample time to consolidate and mature into robust, well-structured vessels in which the endothelial cells turn over slowly. These tumor-associated vessels may then be as resistant to anti-angiogenic drugs as the normal blood vessels elsewhere in the body. These and other mechanisms of acquired resistance to anti-angiogenesis therapies are summarized in **Table 13.5**.

Table 13.5 Alternative mechanisms of acquired resistance to anti-angiogenesis therapies

EC radioresistance: Hypoxic activation of HIF-1 renders ECs resistant to irradiation.

Vascular mimicry: A fraction of tumor vessels are lined by malignant cells and are thus unresponsive to anti-angiogenic agents. Similarly, certain types of tumor stem cells may differentiate into endothelial-like cells (see Figure 13.23).

Angiogenic switch: The outgrowth of tumor cell clones expressing elevated levels of certain angiogenic factors may be naturally favored at advanced stages of tumor progression or may emerge in response to anti-angiogenic treatment, e.g., upregulation of (1) PlGF, VEGF-C, or FGF-2 in response to VEGF inhibition; (2) VEGF after VEGF-R or EGF-R inhibition; and (3) IL-8 after HIF-1 inhibition.

Vascular independence: Mutant tumor cell clones (e.g., those lacking p53) are able to survive in hypoxic tumors.

Stromal cells: VEGF$^{-/-}$ tumors recruit pro-angiogenic stromal cells such as myofibroblasts via upregulation of PDGF-A and myeloid cells.

Mature, robust vessels: Preexisting vessels are covered by a full complement of supporting pericytes and are not readily pruned by anti-angiogenic treatments.

Bone marrow–derived cells: Tumors or ischemic tissues recruit pro-angiogenic endothelial cells and inflammatory cells independent of VEGF; the recruited cells may produce several pro-angiogenic molecules to rescue vascularization upon VEGF blockage.

RTK inhibitors: These may not synergize with chemotherapy, possibly because they do not block the actions of neuropilin-1.

Abbreviations/definitions: EC, endothelial cell; IL-8, interleukin-8, an angiogenic cytokine; neuropilin, an alternative VEGF receptor; PlGF, placental GF, an angiogenic factor related to VEGFs and a ligand of PlGF-R.

Yet another class of natural angiogenesis antagonists (see Table 13.4) are the interferons, which are usually studied in the context of their ability to modulate the activities of various cell types operating in the immune system. Interferon-α and -β have proven to be potent suppressors of the synthesis of basic fibroblast growth factor (bFGF) and of interleukin-8 (IL-8), both of which are strong angiogenic agents. And administration of interferon-α has proven to be useful in causing the regression of some hemangiomas (endothelial cell tumors) as well as Kaposi's sarcomas, which are also of endothelial cell origin (likely the endothelial cells forming lymph ducts). In both cases, the tumor regressions have been attributed to the anti-angiogenic effects of the interferon.

An experiment reported in 1991 provided one of the earliest indications of the promise of another type of anti-angiogenic therapy. In this work, cancer cells were engineered to release a modified basic fibroblast growth factor, which greatly increased their tumorigenicity in mice, simply because this bFGF strongly enhanced the angiogenic powers of these cells. Use of a monoclonal antibody that specifically bound and neutralized this bFGF (but had no effect on the endogenous bFGF of the mouse) blocked the angiogenicity of these tumor cells and led to a dramatic reduction in tumor volume. Within two years, a similar experiment was performed with an anti-human-VEGF monoclonal antibody. It succeeded in blocking the proliferation of two human sarcoma cell lines as well as a glioblastoma in Nude (immunocompromised) mouse hosts.

A more modern version of this therapy came a decade later with the use of a monoclonal antibody that binds and neutralizes VEGF-A. This antibody, termed variously Avastin and bevacizumab, showed significant efficacy in some large-scale clinical trials (Table 13.6). For example, patients with metastatic colon carcinoma who were treated with this antibody plus chemotherapy (the drug 5-fluorouracil) survived, on average, four months longer than patients treated with chemotherapy alone, and the addition of Avastin to conventional chemotherapy extended the survival of patients with non-small-cell lung carcinoma (NSCLC) by about two months. Similarly, Avastin could retard the progression of renal cell carcinomas in patients, but in the end had no effect on their long-term survival. It is plausible that the synergistic effects of Avastin with conventional chemotherapeutic drugs derive directly from the ability of this VEGF-A inhibitor to normalize tumor-associated vasculature (see Supplementary Sidebar 13.6), thereby greatly facilitating the delivery of drugs to the tumor parenchyma.

Some synthetic low–molecular-weight compounds have been developed that are directed against various molecular targets in the angiogenic program. The first, fumagillin, a compound of fungal origin, and its chemical derivative, TNP-470, were found to inhibit the proliferation of endothelial cells both *in vitro* and *in vivo*; this suggests that their anti-angiogenic effects *in vivo* derive from their ability to prevent the growth of new capillaries, which depends on endothelial cell proliferation. Importantly, TNP-470 had no effect on the proliferation of tumor cells *in vitro* but strongly blocked their tumorigenicity in mice. For example, this drug reduced by 70 to 80% the sizes of pancreatic islet tumors arising in Rip-Tag mice. The mechanism of action of TNP-470 remains obscure, but is under active investigation. Its proven ability to inhibit the methionine aminopeptidase-2 enzyme, which plays a key role in the life of endothelial cells, may one day help to explain its preferential effects on these cells while leaving other cultured cell types mostly unaffected.

The most informative studies of angiogenic inhibitors have come using synthetic receptor inhibitors in the transgenic Rip-Tag model of pancreatic islet tumorigenesis (Figure 13.46). Two types of low–molecular-weight synthetic compounds have been utilized in attempts to block various stages of islet tumor progression. One of these drugs is directed against VEGF-R2, which is the main receptor driving angiogenesis. By inhibiting the tyrosine kinase of this receptor, the agent termed SU5416 should mimic the effects of Avastin (the anti-VEGF monoclonal antibody described earlier), that is, both should be able to shut down VEGF signaling. Drugs of a second class, such as an agent termed SU6668, are directed primarily against the tyrosine kinase of the

Table 13.6 Summary of clinically approved anti-angiogenic drugs[a]

Agent	Nature of agent	Approved indication	% of patients responding[b]	Improvement[b] in PFS (months)	Improvement[b] in OS (months)
Bevacizumab (Avastin)[c]	anti-VEGF-A MoAb	metastatic CRC[d,e]	10	4.4	4.7
			0	1.4	1.4
			7.8	2.8	2.5
			14.1	2.6	2.1
		metastatic non-squamous NSCLC[d] (with chemotherapy)	20	1.7	2.0
			10.3–14.0	0.4–0.6	NR
		metastatic breast cancer (with chemotherapy)	15.7	5.9	NS
			9–18	0.8–1.9	NS
			11.8–13.4	1.2–2.9	NS[d]
			9.9	2.1	NS
		recurrent GBM[f]	28		2–3
		metastatic RCC[d] (with IFN-α)	18	4.8	NS
			12.4	3.3	NS
Sunitinib (Sutent)[c]	inhibitor of RTKs[g]	metastatic RCC[c]	35	6.0	4.6
		GIST[e]		4.5	
		pancreatic neuroendocrine tumors[c]		4.8	
Sorafenib (Nexavar)	inhibitor of VEGF-R, cRaf, PDGF-R, and Kit TKs[h]	metastatic RCC[d]	8	2.7	NS
		unresectable HCC[d]	1	NS	2.8
			2	1.4	2.3
Pazopanib (Votrient)	inhibitor of RTKs[i]	metastatic RCC[d]	27	5.0	NR
		soft tissue sarcoma[e]		3.0	
Vandetanib (Caprelsa)	inhibitor of VEGF-R, EGF-R, and Ret TKs	metastatic medullary thyroid carcinoma[d]		6.2	
Axitinib[e] (Inlyta)	inhibitor of VEGF-Rs, PDGF-R and Kit TKs	advanced RCC[e]		2.0	

[a]"Clinically approved" indicates approval for use by the U.S. Food and Drug Administration (FDA). "Inhibitor" indicates in all cases a low molecular weight pharmacologic agent. In addition, as of March 2011, derivatives of thalidomide have been found to have substantial therapeutic utility in treating multiple myeloma; they are not included here, however, because the drugs have adverse physiologic effects, notably neurotoxicity. The mTOR inhibitor Everolimus has been approved for treatment of a series of different tumor types and has anti-angiogenic effects; it has not been listed here because it also has effects on apoptosis, nutrient uptake, and proliferation that may explain part or most of its effects.
[b]Improvement relative to standard treatment.
[c]FDA approval for use against breast cancer was revoked in 2011.
[d]First-line therapy.
[e]Second-line therapy. Axitinib was approved because PFS was 2.0 months longer than existing Sorafenib treatment.
[f]Monotherapy.
[g]Inhibitor of VEGF-R, PDGF-R, FLT-3, Ret, and Kit TKs; Raf/B-Raf.
[h]Low–molecular-weight inhibitor of VEGF-Rs and PDGF-Rs.
[i]Inhibitor of VEGF-Rs, PDGF-Rs, and c-Kit TKs.
Abbreviations: CRC, colorectal cancer; GBM, glioblastoma multiforme; GIST, gastrointestinal stromal tumor; HCC, hepatocellular carcinoma; IFN, interferon; MoAb, monoclonal antibody; NR, not reported; NS, not significant; NSCLC, non-small-cell lung carcinoma; OS, overall survival; PFS, progression-free survival; RCC, renal cell carcinoma; RTK, receptor tyrosine kinase.
Table adapted from P. Carmeliet and R. Jain, Nature 473:298–307, 2011.

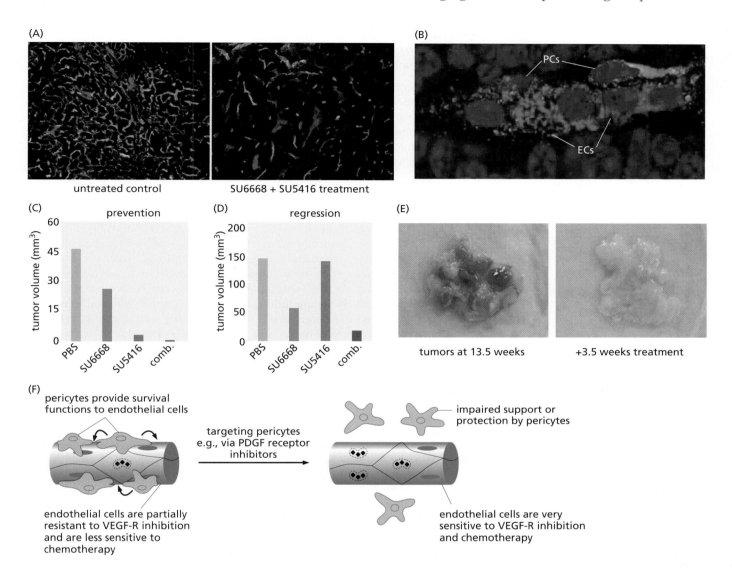

Figure 13.46 Angiogenesis inhibitors as treatments of islet cell carcinogenesis The Rip-Tag transgenic mouse model of pancreatic islet cell carcinogenesis (see Figure 13.36) makes possible the testing of anti-angiogenic pharmacologic inhibitors.
(A) The treatment of mice bearing established angiogenic islet tumors (*left*) for 4 weeks with the PDGF receptor inhibitor SU6668 plus the VEGF-R inhibitor SU5416 results in regression of the vasculature (*right*). The lumina of the capillaries are labeled *light green* while the associated pericytes are labeled *red*. (B) The actions of the SU6668 PDGF-R inhibitor could be traced specifically to its effects on the pericytes, which act as support cells for the endothelial cells. In this image, an antibody reactive with the PDGF-R reveals that the only cells within the islets expressing this receptor (*green*) are the pericytes (PCs) that are closely associated with the endothelial cells (ECs, *red*). Nuclei are stained *blue*. (C) While neither the SU6668 PDGF-R TK inhibitor (*blue bar*) nor the SU5416 VEGF receptor TK inhibitor (*red bar*) was able, on its own, to fully prevent the formation of such tumors (being applied before tumors became apparent), the two agents applied in concert (combined, *brown bar*) succeeded in doing so. PBS, phosphate-buffered saline control (*green*). (D) In an attempt to cause regression of already-formed tumors, SU5416 and SU6668 were introduced either singly or in combination into 12-week-old Rip-Tag mice and the size of tumors was measured 4 weeks later. While the SU5416 VEGF-R antagonist on its own (*red bar*) showed only minimal reduction of tumor volume, the SU6668 PDGF-R antagonist showed greater effects (*blue bar*), and a combination of the two applied together (*brown bar*) functioned synergistically to reduce overall tumor volume by approximately 85%. This indicates, once again, that antagonists of the endothelial cells (which depend on the VEGF-R) together with antagonists of the supporting pericytes (which depend on the PDGF-R) can act synergistically in anti-angiogenic therapy. (E) When 13.5-week-old mice with substantial pancreatic islet tumors (*red growths, left*) are treated for another 3.5 weeks with this combination therapy, their tumors largely regress (*right*). (F) A schematic summary of these and other observations indicates that endothelial cells depend on closely apposed pericytes for various types of biological support. Inhibition of PDGF signaling causes dissociation of pericytes from endothelial cells and renders the latter sensitive to various types of subsequent therapy, including inhibition of VEGF-R function. This emphasizes the fact that anti-angiogenesis therapy is most effective when two synergistically acting treatments are applied. (A, B and E, from G. Bergers et al., *J. Clin. Invest.* 111:1287–1295, 2003. C and D, courtesy of D. Hanahan. F, from K. Pietras and D. Hanahan, *J. Clin. Oncol.* 23:939–952, 2005.)

PDGF receptor. Recall from Section 13.1 that the role of PDGF in recruiting pericytes and smooth muscle cells to growing capillaries indicated that these "mural" cells are highly important for consolidating and strengthening recently formed capillaries.

SU5416, the anti-VEGF-R agent, was able to block 90% of the early-stage, dysplastic islets from undergoing the angiogenic switch, thereby holding them to a small size and a noninvasive state (see Figure 13.46C). However, it had no effect on late-stage, well-established tumors, which continued to progress in spite of its presence in equally high concentrations. Hence, in the early stages of angiogenesis, VEGF signaling plays a critical role, while later on, this process seems to become increasingly independent of VEGF.

The anti-PDGF receptor agent, SU6668, had a far weaker effect on preventing dysplastic islets from undergoing the angiogenic switch, reducing by about half the islets that did so (see Figure 13.46C). But it was far more potent than the anti-VEGF-R drug in treating the late-stage, advanced tumors, reducing their size by about half and substantially reducing their vascularity (see Figure 13.46D). Importantly, the only cells expressing PDGF-R in and near these tumors were the capillary-associated pericytes and related smooth muscle cells (see Figure 13.46B), indicating that these mural cells were the targets of SU6668 action. Indeed, microscopic examination confirmed that SU6668 had prevented these mural cells from associating with and reinforcing the capillary tubes formed by endothelial cells. Together, these experiments showed that the initial steps of angiogenesis could proceed reasonably well without PDGF receptor function, but that later in tumor progression, PDGF signaling, and thus the involvement of mural cells, became increasingly important to angiogenesis and growth of the tumor masses.

Combination therapy using the two agents proved to be a highly potent way of intervening at various stages of tumor formation (see Figure 13.46C and D). Thus, simultaneous inhibition of both VEGF and PDGF receptors prevented angiogenic switching, holding virtually all (>98%) islets at a pre-angiogenic stage. This drug combination blocked the further expansion of small, already-established angiogenic tumors by 90% and caused an approximately 80% regression of large tumors, as did another PDGF-R inhibitor, the drug Gleevec (discussed in Chapter 16). In some mice, these combination drug therapies held already-formed tumors to a small size for as long as two months (see Figure 13.46E).

Significantly, these combinations of drugs had virtually no toxic effects on normal pancreatic tissue adjacent to neoplastic islets, confirming that recently formed capillaries within a tumor are far more vulnerable to disruption than the well-established vessels within normal tissues. In addition, it seemed that PDGF played a major role in initially attracting pericytes to the capillary tubes formed by endothelial cells, but in normal tissues, continued PDGF signaling was not required to maintain this association. These observations indicate that the most effective ways of inhibiting angiogenesis and thus blocking tumor progression are likely to depend on targeting several of the cell types that construct the tumor-associated vasculature (see Figure 13.46F). Indeed, targeting only the endothelial cells yields only a temporary regression of microvessels: when anti-VEGF therapy is halted, these vessels rapidly regenerate, often growing back in the hollow sleeves formed by the basement membranes and pericytes that are left behind after endothelial cells succumb to anti-VEGF therapy (Supplementary Sidebar 13.7). Of additional interest, neither of the agents described above—neither SU5416 nor SU6668—is currently being used clinically, having been abandoned because of limited efficacy in phase II or phase III clinical trials; this demonstrates, once again, the limited power of mouse models of cancer pathogenesis to predict human clinical responses.

In fact, anti-angiogenic therapy may have been in use for many years: unbeknownst to clinical oncologists, much of the efficacy of some traditional anti-tumor therapies may derive from their effects on tumor-associated microvasculature. One striking example of this, which sheds light on how radiation therapy succeeds in destroying certain tumors, has come from studies of genetically altered mice that lack the genes encoding either the Bax protein or the acid sphingomyelinase (asmase) enzyme. Both of

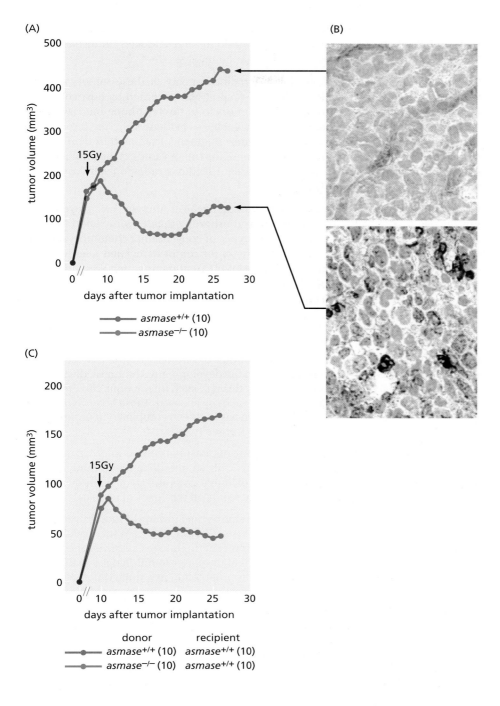

Figure 13.47 Survival of wild-type and asmase-negative tumor-bearing mice The responses of tumors to radiotherapy may often be determined by the radiosensitivity of the endothelial cells that form their vasculature. Mice were bred to be either wild type or homozygous null for the gene encoding the pro-apoptotic enzyme acid sphingomyelinase (asmase). When mouse fibrosarcoma cells were implanted in the mice, the tumors grew twice as fast in the *asmase*$^{-/-}$ mice as in the wild-type mice, suggesting that a host factor, such as recruited endothelial cells, was governing the rate of tumor growth. (A) When exposed to a therapeutic dose (15 Gy) of radiation (*arrow*), the tumors in wild-type mice regressed (*red circles*) and, after a while, began to grow again. In contrast, the tumors in the *asmase*$^{-/-}$ hosts (*blue circles*) continued to grow unabated. (B) When these tumors were examined microscopically following irradiation, the endothelial cells (identified by a cell-specific immunostain) in tumors carried by wild-type mice (*below*) were apoptotic, as indicated by the TUNEL staining (*brown spots*; see Figure 9.19 and Supplementary Sidebar 9.2), while the endothelial cells in tumors borne by the *asmase*$^{-/-}$ hosts showed no signs of apoptosis (*above*). (C) When the bone marrow of wild-type hosts was replaced through transplantation of either wild-type or *asmase*$^{-/-}$ donor marrow cells, the tumors implanted in mouse hosts with engrafted mutant marrow (*blue curve*) continued to grow following 15 Gy of radiation (*arrow*), while tumors implanted in mouse hosts engrafted with wild-type marrow were stopped by the radiation (*red curve*). This demonstrated that the radiosensitivity of the tumor was determined by cells of host bone marrow origin, not by the tumor cells themselves. The numbers in parentheses indicate that number of tumor-bearing mice in each experimental group. (From M. Garcia-Barros et al., *Science* 300:1155–1159, 2003.)

these proteins are important pro-apoptotic regulators in a variety of cell types including, importantly, endothelial cells. Consequently, the endothelial cells from these genetically altered mice are far more resistant to toxic agents, including X-rays, than their wild-type counterparts.

In the experiments shown here, mouse tumor cells from a fibrosarcoma cell line were introduced into both wild-type and *asmase*$^{-/-}$ mice. After tumors had grown to a substantial size, these animals were exposed to a dose of 15 grays (Gy) of X-rays (**Figure 13.47A and B**). This dose of radiation usually causes a significant reduction in tumor burden, and indeed it succeeded in doing so in the wild-type mice. However, the identically sized tumors grown in the *asmase*$^{-/-}$ host mice responded quite differently, in that their growth continued unabated. When, as a further experiment, the bone marrow of *asmase*$^{+/+}$ mice was replaced with transplanted mutant (*asmase*$^{-/-}$) bone marrow cells, the tumors became highly resistant to X-ray–induced killing (see Figure 13.47C).

These experiments demonstrate that the radiosensitivity of these tumors was not intrinsic to the tumor cells themselves. Instead, it was governed by host cells, specifically by cells recruited into engrafted tumors from the host bone marrow. These recruited host cells were indeed endothelial cells, as evidenced by the fact that (1) the apoptosis of the endothelial cells in tumor fragments irradiated *in vitro* directly paralleled and predicted the *in vivo* behavior of the tumors, and (2) the only cells that showed significant apoptosis in the tumors shortly after irradiation were associated with capillaries (see Figure 13.47B).

Observations like these indicate that anti-angiogenesis therapy has played a far greater role in a conventional anti-tumor **radiotherapy** than anyone dared to imagine; similarly, clinical responses to certain types of conventional chemotherapy may also be strongly influenced by the sensitivity of the tumor-associated microvasculature to these agents. This suggests that in the future, treatments with many anti-cancer chemotherapeutics may be optimized by gauging their effects on the tumor-associated microvessels rather than on the tumor cells themselves.

13.11 Synopsis and prospects

Metazoan tissues are organized as condominiums of various cell types that are continuously communicating with one another. To the developmental biologist, the need for this organizational plan is self-evident: only through such interactions can the proper numbers and locations of each of these cell types be ensured. These heterotypic interactions continue to operate after embryogenesis has been completed in order to support the maintenance and repair of already-formed tissues.

As we have learned in this chapter, this organizational plan confers another, quite distinct benefit on the organism. By making its cells so interdependent, none is easily able to extricate itself from its complex web of interactions and go off on its own. Interdependence imposes the regimentation that wards off the chaos of neoplasia.

Earlier, we viewed multi-step cancer progression as the hurdling of successive barriers placed in the path of developing cancer cells (see Chapter 11). Each successfully completed step, whether achieved by genetic or epigenetic changes, removes one of these obstacles and places the cancer cell incrementally closer to full-fledged malignancy. Now we can conceptualize multi-step tumor progression in a quite different way: as premalignant cells evolve toward malignancy, they progressively sever their ties with their neighbors and their dependence on neighborly support.

Perhaps the biggest surprise is how dependent most cancer cells remain on stromal support in spite of having completed multiple steps of tumor progression. The epithelial cells residing in many carcinomas continue to rely on many of the physiologic signals that sustained their precursors in normal tissues; equally elaborate heterotypic signaling may operate in other types of tumors as well (see, for example, Supplementary Sidebar 13.8). This conservatism is also suggested by microscopic examination of tumors. Thus, a well-trained pathologist can recognize the origins of perhaps 95% of the tumor samples viewed under the microscope, because most of the heterotypic interactions that govern normal morphology appear to be operative in the great majority of cancers.

The remaining approximately 5% of tumors present a challenge, because they are anaplastic and therefore have lost most of the histologic traits that make identifying their tissue of origin possible. The cells in these anaplastic tumors have shed most forms of dependence that tied their precursors to normal neighboring cells. However, most anaplastic tumor cells have not progressed all the way to total independence, because they still assemble to form solid tumors. The ultimate independence is achieved only by the cells in those cancers that have advanced so far that they can grow as **pleural** effusions or **ascites** (see Supplementary Sidebar 13.3) and therefore have no direct contact with supporting cells and, apparently, with an extracellular matrix (ECM).

Even without detailed knowledge of heterotypic interactions, the dependence of most carcinoma cells on at least one form of stromal support could have been predicted from the known physiology of mammalian tissues: virtually all of them depend on a functional blood supply. Less predictable was the mechanism by which most tumors acquire their vasculature. Rather than invading normal tissue and expropriating existing capillary beds, most tumors actively recruit endothelial cells that proceed to construct capillaries and larger vessels within the tumors.

The sources of these endothelial cells could also not be deduced from first principles. The proliferation of endothelial cells originating in neighboring tissues seems to be the main mechanism for acquiring neovasculature soon after the angiogenic switch has been tripped. However, in some tumors, the initially formed vessels may be constructed by endothelial precursor cells (EPCs) that originate in the bone marrow and then differentiate in the tumor stroma into functional endothelial cells. (The role of EPCs is still unclear. In the minds of some, many of the cells associated with microvessels that were initially deemed to be differentiated products of bone marrow–derived EPCs were actually adjacently located myeloid cells that serve ancillary roles in enabling vessel formation.)

Major aspects of angiogenesis are still poorly understood, even paradoxical. For example, it seems apparent that once tumors become angiogenic, they can launch into a prolonged phase of growth and expansion, and that tumors that are more angiogenic can grow even more rapidly than those that are less able to attract new vasculature. Indeed, as we have read, measurements of microvessel density—the number of capillaries per microscope field—correlate quite well with the likelihood that a primary tumor, such as a breast cancer, will progress to a highly malignant endpoint.

The paradox comes from the frequent observations that patients bearing highly hypoxic tumors also confront a poor prognosis. Indeed, hypoxic tumors often are more aggressive than their normoxic counterparts (**Figure 13.48**). This makes no sense, given that hypoxia starves tumor cells and often leads to their death through apoptosis and, on a

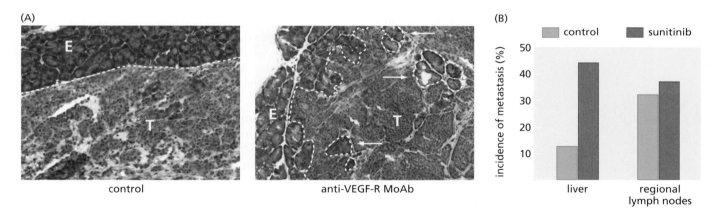

Figure 13.48 Undesired effects of inhibiting angiogenesis By depriving tumors of their microvasculature through anti-angiogenic therapies, sectors of such tumors rapidly develop hypoxia. The effects of this hypoxia may be complex. Extensive necrosis of sectors of the tumor may result. Alternatively, there may be unanticipated additional effects, such as the increased tumor aggressiveness revealed here following anti-angiogenic therapy of Rip-Tag transgenic mice (which develop pancreatic islet tumors; see Figure 13.36). (A) A monoclonal antibody (MoAb) has been produced that targets the ectodomain of VEGF-R2, the main receptor for tumor angiogenesis; this MoAb, termed DC101, is very effective in inhibiting the growth of human tumor xenografts in immunocompromised mouse hosts, ostensibly through its ability to induce widespread apoptosis in the endothelial cells forming the tumor-associated microvessels (see Supplementary Sidebar 13.9).

In the absence of DC101 MoAb treatment, the pancreatic islet tumors (T) develop a smooth, well-confined boundary (*white dotted line*) with the surrounding exocrine pancreas (E), indicating a lack of invasiveness (*left panel*). However, in the presence of DC101 (*right panel*), the boundary is highly convoluted, indicating a highly aggressive tumor that has actually invaded around and engulfed islands (*arrows*) of pancreatic exocrine cells. (B) An alternative anti-angiogenic treatment exploits sunitinib, a low–molecular-weight inhibitor of the VEGF-R2–associated tyrosine kinase. Use of sunitinib resulted in a ~3.5-fold increased incidence of liver metastases from the Rip-Tag pancreatic tumors but no significant increase in metastases in regional lymph nodes. Increased invasiveness was also observed following treatment of a mouse glioblastoma with three different anti-angiogenic agents (*not shown*). (From M. Paez-Ribas et al., *Cancer Cell* 15:220–231, 2009.)

larger scale, to extensive necrotic regions within tumors. The paradox may one day be resolved by invoking the actions of the HIF-1 transcription factor, which becomes activated in hypoxic cells (see Section 7.12) and induces production of a large number of other proteins besides VEGF. Indeed, a high level of HIF-1 expression is also an indicator of poor clinical prognosis. Included among the genes activated by HIF-1 are those specifying PDGF, TGF-α, TGF-β, and several matrix metalloproteinases (MMPs), the latter being responsible for remodeling the extracellular matrix (ECM). As we have learned, several of these secreted proteins act as potent mitogens that drive the proliferation of both epithelial cells and their stromal neighbors.

An additional HIF-1–induced gene encodes the Met protein, which functions as the receptor for hepatocyte growth factor (HGF), also known as scatter factor (SF). HGF seems to be widely available in many human tumors of both epithelial and mesenchymal origin. Consequently, HIF-1–mediated increases in Met expression can sensitize tumor cells to HGF molecules that are present in their surroundings, such as the HGF that has been released by nearby stromal cells. Once the Met receptor is activated by binding its HGF ligand, it activates a diverse set of responses within epithelial cells, including the epithelial–mesenchymal transition (EMT), a cell-biological program that imparts increased motility and invasiveness (see Chapter 14). Thus, the levels of two EMT-inducing transcription factors—Twist and Snail—are significantly increased in hypoxic cells; as we will see in the next chapter, carcinoma cells that have undergone an EMT acquire many of the phenotypes of the cells forming high-grade malignancies. These diverse products of HIF-1 action and tissue hypoxia help explain the poor prognosis attached to hypoxic tumors, which thrive in spite of great adversity and actually turn out to be more aggressive than their well-oxygenated counterparts (see also Supplementary Sidebar 13.6).

These paradoxical, indeed counterproductive responses to anti-angiogenesis therapies, together with the evasive maneuvers that tumors may take to circumvent these therapies (see Table 13.5), may explain the modest successes that most of these therapies have achieved in the oncology clinic. Thus, most clinical outcomes have fallen far short of anticipated outcomes. Still, some combinations of anti-angiogenesis therapy and chemotherapy have shown clear utility in terms of increasing progression-free survival and even overall survival of cancer patients and thus are finding their way into standard clinical practice. In the longer term, more efficacious anti-angiogenesis therapies may derive from dosing schedules that serve to normalize the tumor-associated microvasculature, enabling more efficient drug delivery while avoiding the complete vascular collapse that triggers tumor hypoxia and the induction of increased tumor aggressiveness.

The recently discovered dynamic interactions between tumors and the bone marrow have been a surprise to most cancer researchers. In the absence of metastasis, most tumors have traditionally been considered to be localized diseases confined to one or another corner of the body. But the more we learn about the cells of the stromal compartment, the more we come to realize that even localized tumors extend their reach far and wide throughout the body in order to recruit the cells that they need in order to support their own survival and proliferation programs. Besides endothelial precursor cells (EPCs), cited above, carcinomas recruit mast cells and monocytes from the marrow, the latter differentiating on-site into macrophages. Mesenchymal stem cells (MSCs) of bone marrow origin are also recruited in large numbers to the stroma of many carcinomas; once present within the tumor, they may differentiate into fibroblasts and myofibroblasts, which then release signals favoring the survival and malignant progression of nearby cancer cells.

Throughout most of this chapter, we have focused on epithelial–stromal interactions within tumors and the heterotypic signals that they exchange via *paracrine* signaling channels. The presence of a variety of bone marrow–derived cells in the tumor stroma creates an additional dimension of complexity. Recognizing their existence and their potential contributions to tumor physiology, we now need to include interactions of cancer cells with the faraway bone marrow. Thus, tumors may communicate with the marrow to elicit the production and mobilization into the circulation of certain cell types that can then be recruited from the circulation into the tumor stroma. Such systemic interactions depend, by definition, on information conveyed by *endocrine* signals rather than the paracrine signals that have been a central focus of our discussion here.

Even if we confine our attention to the bidirectional exchanges of paracrine signals between the epithelial and stromal compartments within tumors, we confront extraordinary complexity. The tumor stroma is composed of almost a dozen distinct cell types (fibroblasts, myofibroblasts, MSCs, endothelial cells, pericytes, smooth muscle cells, mast cells, monocytes, macrophages, lymphocytes, and, in some tissues, adipocytes), each of which signals to the neoplastic epithelial cells as well as to the other cellular components of the stroma. Physicists have struggled, so far unsuccessfully, with solutions to the three-body problem. Here, we confront a world of a far larger number of distinct cell types, each of which is sending a complex mixture of signals to other types of cells within tumors. Examinations of only a single cell type—fibroblasts—reveal the heterogeneity of these cells and their ability to change dynamically in response to the contextual signals that they encounter within the tumor stroma (Supplementary Sidebar 13.10). We have only begun to scratch the surface of the complex epithelial–stromal interactions operating in neoplastic tissues and in their normal counterparts.

For more than half a century, the mindset of oncologists has been focused on eradicating the bulk of the cancer cells within solid tumors, with the hope of achieving the seemingly impossible—cures of common carcinomas that have traditionally been incurable. As we learned, first in Chapter 11 and now in this one, this focus needs to be radically redirected. First, the targets of anti-cancer therapies can no longer be confined to the bulk of neoplastic cells in a tumor, because they have limited self-renewal capacity; instead, we are likely to find that truly durable cures can come only from eradicating as well the still-elusive tumor stem cells that hide out, here and there, throughout tumor masses and represent their engines of self-renewal. Second, many of the useful anti-cancer therapies to be developed in the future will not come from targeting the cancer cells themselves. Instead, it may often be far more profitable to attack the cells that provide them with vital physiologic support; by undermining the elaborate stromal support network on which most cancer cells depend, truly dramatic regressions of solid tumors may one day be achieved (**Figure 13.49**).

The insight that tumors are "wounds that do not heal" extends and echoes our earlier discussions (see Section 11.15), in which the critical role of chronic inflammation in promoting tumor formation was described in great depth. Inflammation and wound healing are intertwined processes, and the mechanisms of inflammation-driven tumor promotion, which lead to the initial formation of cancers, are extended and elaborated by the chronic wound healing that seems to best characterize the biology of the stroma of well-established tumors.

Here, too, there has been another surprise. Inflammatory cells, notably macrophages, have traditionally been depicted as the front-line soldiers of the immune response that deal effectively with infectious agents, such as bacteria, by consuming them and help guide the long-term immune response through antigen presentation, as we will see later in Chapter 15. Now we learn that macrophages can also function as key sources of tumor promotion by producing mitogenic growth factors, liberating angiogenic factors, and remodeling the extracellular matrix (ECM); the latter process is also critical for tumor invasion and metastasis (see Chapter 14).

So, the traditional job assignments of various stromal cell types are being extended and blurred. Cells of the immune system, which are purportedly dispatched to protect us from infection and even cancer, are often active collaborators in tumor development. And the deletion of one or another cell type from the immune system, achieved in mice through germ-line re-engineering, often creates a host organism that is, paradoxically, less able to support tumorigenesis.

Normal, multifaceted morphogenetic programs that depend on epithelial–stromal interactions, such as wound healing and the EMT, are likely to explain how carcinoma cells are clever enough to acquire the complex cell phenotypes that they need in order to execute the later stages of malignant progression. In hindsight, this notion is not so surprising, since the more we learn about cancer cells, the more we realize how

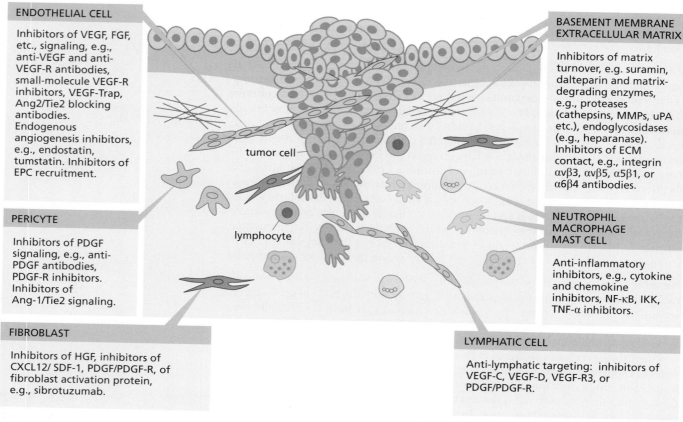

ENDOTHELIAL CELL

Inhibitors of VEGF, FGF, etc., signaling, e.g., anti-VEGF and anti-VEGF-R antibodies, small-molecule VEGF-R inhibitors, VEGF-Trap, Ang2/Tie2 blocking antibodies. Endogenous angiogenesis inhibitors, e.g., endostatin, tumstatin. Inhibitors of EPC recruitment.

BASEMENT MEMBRANE EXTRACELLULAR MATRIX

Inhibitors of matrix turnover, e.g. suramin, dalteparin and matrix-degrading enzymes, e.g., proteases (cathepsins, MMPs, uPA etc.), endoglycosidases (e.g., heparanase). Inhibitors of ECM contact, e.g., integrin αvβ3, αvβ5, α5β1, or α6β4 antibodies.

PERICYTE

Inhibitors of PDGF signaling, e.g., anti-PDGF antibodies, PDGF-R inhibitors. Inhibitors of Ang-1/Tie2 signaling.

NEUTROPHIL MACROPHAGE MAST CELL

Anti-inflammatory inhibitors, e.g., cytokine and chemokine inhibitors, NF-κB, IKK, TNF-α inhibitors.

FIBROBLAST

Inhibitors of HGF, inhibitors of CXCL12/ SDF-1, PDGF/PDGF-R, of fibroblast activation protein, e.g., sibrotuzumab.

LYMPHATIC CELL

Anti-lymphatic targeting: inhibitors of VEGF-C, VEGF-D, VEGF-R3, or PDGF/PDGF-R.

tumor cell

lymphocyte

Figure 13.49 Heterotypic interactions as targets for therapeutic intervention As described in this chapter, cancer cells are dependent on the nearby stromal microenvironment for a variety of cell-physiologic supports. This dependence on heterotypic interactions has inspired development of new types of cancer therapy, some of which has been featured in this chapter. Instead of focusing on the intracellular signaling defects within cancer cells, this new type of therapy is directed toward interrupting heterotypic signaling, thereby depriving cancer cells of essential stromal support. This scheme indicates some of the anti-tumor therapies that are being developed or under consideration. (From J.A. Joyce, *Cancer Cell* 7:513–520, 2005.)

opportunistic they are in co-opting and exploiting normal biological processes in order to further their own ends. This leaves us with the last question of this chapter: Have we begun to truly understand the mechanistic complexity of heterotypic interactions, or is there an entire universe of undiscovered signaling pathways and behavioral programs lurking within tumors, waiting, like intergalactic dark matter, to surprise us once again?

Key concepts

- Tumors are complex tissues that depend on intercommunication between various cell types. Indeed, most tumors are as complex histologically as the normal tissues in which they arise.

- In carcinomas, these cell types can be separated into the neoplastic epithelial cells and recruited stromal cells, which include fibroblasts, myofibroblasts, and macrophages, other types of inflammatory cells, as well as the various cell types that participate in the construction of the tumor-associated vasculature, specifically endothelial cells, pericytes, and smooth muscle cells.

- Most carcinomas depend absolutely on recruited stromal cells for various types of physiologic support. This dependence is lost only in the small subset of tumors that progress to an extremely malignant state, notably the tumor cells growing in ascites and pleural fluid.

- At the biochemical level, this interdependence is manifested by the exchange of various types of mitogenic and trophic factors. For example, carcinoma cells may release PDGF to recruit and activate stromal cells, while the latter respond by releasing IGFs that sustain the survival of the carcinoma cells.

- The formation of tumor-associated vasculature, formed by the process of neoangiogenesis, is a critical, rate-limiting determinant of the growth of all tumors larger in size than approximately 0.2 mm.

- In the case of carcinomas, the acquisition of tumor-associated stroma closely resembles the process of healing in wounded epithelial tissues. The genesis of stroma therefore relies on the same gene expression programs that are activated during wound healing.

- As tumor progression proceeds, the fibroblast-rich stroma is increasingly replaced by myofibroblasts, which eventually generate collagen-rich, desmoplastic stroma.

- The recruitment of the cells that participate directly in the construction of the neovasculature of tumors involves the release of factors, such as VEGF, by both the tumor cells and inflammatory cells, notably macrophages.

- Neoangiogenesis represents an attractive target for the development of novel anticancer agents, in that the targeted cells are the various normal stromal cell types participating in angiogenesis rather than the ever-changing cancer cells.

- Anti-angiogenic therapies often provoke paradoxical responses in carcinomas, including progression to higher grade malignancy, tempering the enthusiasm for introducing these therapies into clinical practice.

Thought questions

1. What diverse lines of evidence prove directly that most carcinoma cells depend on stromal cell types for various types of physiologic support?

2. How might anti-angiogenesis therapies improve (in some cases) or neutralize (in other cases) the efficacy of conventional chemotherapeutic agents?

3. How might macrophages facilitate or antagonize tumorigenesis?

4. Which lines of evidence persuade you that the generation of tumor-associated stroma depends on the same biological programs that are activated during wound healing?

5. Which biological forces cause the tumor-associated vasculature to be defective in so many respects?

6. Which types of anti-cancer therapeutic agents would you deploy in order to encourage the collapse of an established tumor by depriving it of vasculature support?

7. What biochemical strategies can tumor cells use to lessen their dependence on stromal support?

8. Can you cite examples of how oncoproteins perturb the interactions between carcinoma cells and the nearby tumor-associated stroma?

9. How might you determine what proportion of endothelial cells in the vasculature of a tumor derive from the expansion of adjacent vasculature and what proportion of these cells arise from circulating endothelial precursor cells?

Additional reading

Avramides CJ, Garmy-Susini B & Varner JA (2008) Integrins in angiogenesis and lymphangiogenesis. *Nat. Rev. Cancer* 8, 604–617.

Balkwill F, Charles KA & Mantovani A (2005) Smoldering and polarized inflammation in the initiation and promotion of malignant disease. *Cancer Cell* 7, 211–217.

Baluk P, Hashizume H & McDonald DM (2005) Cellular abnormalities of blood vessels as targets in cancer. *Curr. Opin. Genet. Dev.* 15, 102–111.

Baum B & Georgiou M (2011) Dynamics of adherens junctions in epithelial establishment, maintenance, and remodeling. *J. Cell Biol.* 192, 907–917.

Bergers G & Hanahan D (2008) Modes of resistance to anti-angiogenic therapy. *Nat. Rev. Cancer* 8, 592–603.

Bhowmick NA, Neilson EG & Moses HL (2004) Stromal fibroblasts in cancer initiation and progression. *Nature* 432, 332–337.

Bierie B & Moses HL (2006) Tumour microenvironment: TGFβ, the molecular Jekyll and Hyde. *Nat. Rev. Cancer* 6, 506–520.

Brancato SK & Albina JE (2011) Wound macrophages as key regulators of repair. *Am. J. Pathol.* 178, 19–25.

Carmeliet P & Jain RK (2011) Molecular mechanisms and clinical applications of angiogenesis. *Nature* 473, 298–307.

Carmeliet P & Tessier-Lavigne M (2005) Common mechanisms of nerve and blood vessel wiring. *Nature* 436, 193–200.

De Visser KE, Eichten A & Coussens LM (2006) Paradoxical roles of the immune system during cancer development. *Nat. Rev. Cancer* 6, 24–37.

De Wever O, Demetter P, Mareel M & Bracke M (2008) Stromal myofibroblasts are drivers of invasive cancer growth. *Int. J. Cancer* 123, 2229–2238.

Dvorak HF (1986) Tumors: wounds that do not heal. *N. Engl. J. Med.* 315, 1650–1689.

Dvorak HF (2003) How tumors make bad blood vessels and stroma. *Am. J. Pathol.* 162, 1747–1757.

Ebos JML & Kerbel RS (2011) Antiangiogenic therapy: impact on invasion, disease progression, and metastasis. *Nat. Rev. Clin. Oncol.* 8, 210–221.

Egeblad M, Nakasone ES & Web EZ (2010) Tumors as organs: complex tissues that interface with the entire organism. *Dev. Cell* 18, 884–901.

Ellis LM & Hicklin DJ (2008) VEGF-targeted therapy: mechanisms of anti-tumor activity. N*at. Rev. Cancer* 8, 579–591.

Ferrara N, Gerber HP & LeCouter J (2003) The biology of VEGF and its receptors. *Nat. Med.* 9, 669–676.

Franke WW (2009) Discovering the molecular components of intercellular junctions: a historical view. *Cold Spring Harbor Prospect. Biol.* 1, a003061.

Gerhardt H & Semb H (2008) Pericytes: gatekeepers in tumour cell metastasis? *J. Mol. Med.* 86, 135–144.

Hinz B (2007) Formation and function of the myofibroblast during tissue repair. *J. Invest. Dermatol.* 127, 526–537.

Jain RK (2005) Normalization of tumor vasculature: an emerging concept in antiangiogenic therapy. *Science* 307, 58–62.

Jain RK, Di Tomaso E, Duda DG et al. (2007) Angiogenesis in brain tumors. *Nat. Rev. Neurosci.* 8, 610–622.

Jain, RK, Duda DG, Willett CG et al. (2009) Biomarkers of response and resistance to antiangiogenic therapy. *Nat. Rev. Clin. Oncol.* 6, 327–338.

Kalluri R (2003) Basement membranes: structure, assembly and role in tumour angiogenesis. *Nat. Rev. Cancer* 3, 422–433.

Kalluri R & Zeisberg M (2006) Fibroblasts in cancer. *Nat. Rev. Cancer* 6, 392–401.

Kerbel RS (2008) Tumor angiogenesis. *N. Engl. J. Med.* 358, 2039–2049.

Kessenbrock K, Plaka V & Werb Z (2010) Matrix metalloproteinases: regulators of the tumor microenvironment. *Cell* 141, 52–67.

Lewis CE & Pollard JW (2006) Distinct role of macrophages in different tumor microenvironments. *Cancer Res.* 124, 263–266.

Loges S, Mazzone M, Hohensinner P & Carmeliet P (2009) Silencing or fueling metastasis with VEGF inhibitors: antiangiogenesis revisited. *Cancer Cell* 15, 167–170.

Lu P, Weaver VM & Werb Z (2012) The extracellular matrix: a dynamic niche in cancer progression. *J. Cell Biol.* 196, 395–406.

Nagy JA , Chang SH, Dvorak AM & Dvorak HF (2009) Why are tumour blood vessels abnormal and why is it important to know? *Br. J. Cancer* 100, 865–859.

Olsson AK, Dimberg A, Kreuger J & Claesson-Welsh L (2006) VEGF receptor signalling: in control of vascular function. *Nat. Rev. Mol. Cell Biol.* 7, 359–371.

Page-McCaw A, Ewald AJ & Werb Z (2007) Matrix metalloproteinases and the regulation of tissue remodeling. *Nat. Rev. Mol. Cell Biol.* 8, 221–233.

Rabbani SY, Heissig B, Hattori K & Rafii S (2003) Molecular pathways regulating mobilization of marrow-derived stem cells for tissue revascularization. *Trends Mol. Med.* 9, 109–117.

Schäfer M & Werner S (2008) Cancer as an overhealing wound: an old hypothesis revisited. *Nat. Rev. Mol. Cell Biol.* 9, 628–638.

Shibuya M & Claesson-Welsh L (2006) Signal transduction by VEGF receptors in regulation of angiogenesis and lymphangiogenesis. *Exp. Cell Res.* 312, 549–560.

Tammela T & Alitalo K (2010) Lymphangiogenesis: molecular mechanisms and future promise. *Cell* 140, 460–476.

Theunissen J-W & de Sauvage FJ (2009) Paracrine Hedgehog signaling in cancer. *Cancer Res.* 69, 6007–6010.

Thiery JP, Acloque H, Huang RYG & Nieto MA (2009) Epithelial-mesenchymal transitions in development and disease. *Cell* 139, 871–890.

Tlsty TD & Coussens LM (2006) Tumor stroma and regulation of cancer development. *Annu. Rev. Pathol.* 1, 119–150.

Wels J, Kaplan RN, Rafii S & Lyden D (2008) Migratory neighbors and distant invaders: tumor-associated niche cells. *Genes Dev.* 22, 559–574.

Wiseman BS & Werb Z (2002) Stromal effects on mammary gland development and breast cancer. *Science* 296, 1046–1049.

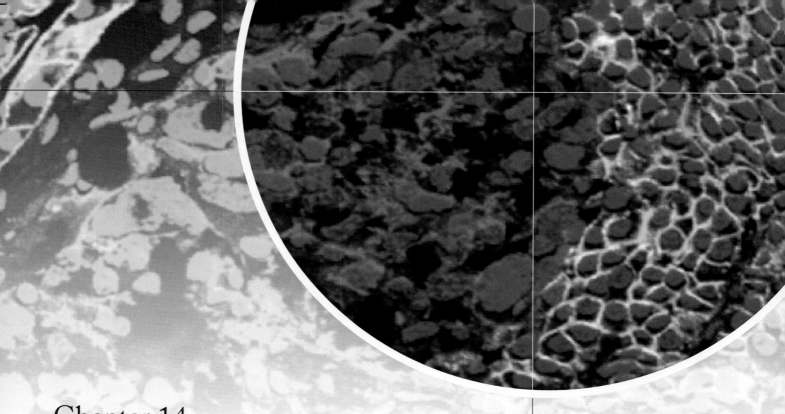

Chapter 14

Moving Out: Invasion and Metastasis

> The fact of cells identical with those of the cancer itself being seen in the blood may tend to throw some light upon the mode of origin of multiple tumors existing in the same person.
>
> T.R. Ashworth, physician, 1869

> It is not birth, marriage, or death, but gastrulation, which is truly the most important time in your life.
>
> Lewis Wolpert, embryologist, 1986

In the early phases of multi-step tumor progression, cancer cells multiply near the site where their ancestors first began uncontrolled proliferation. The result, usually apparent only after many years' time, is a **primary tumor** mass. Given the fact that a cubic centimeter of tissue may contain as many as 10^9 cells, we can easily imagine that tumors may often reach a size of 10^{10} or 10^{11} cells before they become apparent to the individual carrying them or to the clinician in search of them (see Figure 10.5A).

Primary tumors in some organ sites—specifically those arising within the peritoneal or pleural space—may well expand without causing any discomfort to the patient, simply because these cavities are expansible and their contents are quite plastic; in other sites, such as the brain, the presence of a tumor is often apparent when it is still relatively small. Sooner or later, however, in all sites throughout the body, tumors of substantial size compromise the functioning of the organs in which they have arisen and begin to evoke symptoms.

In many cases, the effects on normal tissue function come from the physical pressure exerted by the expanding tumor masses. In others, cells from the primary tumor mass invade adjacent normal tissues and, in so doing, begin to compromise vital functions. Large tumors in the colon may obstruct passage of digestion products through the lumen, and in tissues such as the liver and pancreas, cancer cells may obstruct the flow of bile through critical ducts. In the lungs, airways may be compromised.

Movies in this chapter

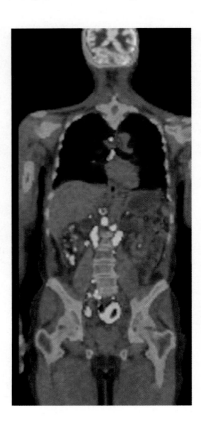

Figure 14.1 Disseminated tumors The diagnosis of metastatic disease often represents a death sentence for a cancer patient, yet the mechanisms by which cancer cells metastasize from a primary tumor to distant sites in the body remain poorly understood. Seen here is a whole-body scan of a patient with metastatic non-Hodgkin's lymphoma (NHL). This is a fusion image of a CT (computed X-ray tomography) scan of the body's tissues (*gray, blue*) and a PET (positron-emission tomography) scan in which the uptake of radioactively labeled fluorodeoxyglucose (FDG) in various tissues (*yellow*) has been detected. FDG uptake indicates regions of high glucose uptake associated with aerobic glycolysis (see Section 2.6). The activity associated with the brain is normal. However, the yellow spots in the abdominal regions indicate multiple NHL metastases. (Courtesy of S.S. Gambhir.)

As insidious and corrosive as these primary tumors are, they ultimately are responsible for only about 10% of deaths from cancer. The remaining approximately 90% of patients are struck down by cancerous growths that are discovered at sites far removed from the locations in their bodies where their primary tumors first arose (**Figure 14.1**; see also Figure 2.2). These **metastases** are formed by cancer cells that have left the primary tumor mass and traveled by the body's highways—blood and lymphatic vessels—to seek out new sites throughout the body where they may found new colonies (**Figure 14.2**). Breast cancers often spawn metastatic colonies promiscuously in many tissues throughout the body, including the brain, liver, bones, and lungs. Prostate tumors are most often seeded to the bones, while colon carcinomas preferentially form new colonies in the liver.

Such wandering cancer cells are the most dangerous manifestations of the cancer process. When they succeed in founding colonies in distant sites, they often wreak great havoc. The female body can dispense with its mammary glands without losing vital physiologic functions, and so almost all primary breast carcinomas do not compromise survival while they are confined to the breast. However, the metastatic colonies that breast cancer cells initiate in the bone can cause localized erosion of bone tissue, resulting in agonizing pain and skeletal collapse. Metastases in the brain may rapidly compromise central nervous system function, while those in the lung or liver are similarly threatening to life because of the vital functions of these organs.

One major puzzle concerns the variable tendencies that different tumors have to metastasize. For reasons that remain obscure, tumors in certain tissues have a high probability of metastasizing, while those arising in other tissues almost never do so. After primary melanomas penetrate a certain distance downward into the tissue underlying the skin, the presence of metastases at distant sites in the body is almost a certainty. In contrast, basal cell carcinomas of the skin and astrocytomas—primary tumors of the glial cells in the brain—rarely spawn metastases. Another major unresolved issue concerns the metastatic *tropism* cited above: Why do tumors originating in a given organ preferentially seed colonies in particular tissues located elsewhere in the body?

In a variety of human tumor types, the dissemination of cancer cells throughout the body has already occurred by the time a primary tumor is first detected; at the time of initial diagnosis, these scattered cells may be inapparent because they form only minute tumor colonies—*micrometastases*. Such behavior provokes a question that we will confront in this chapter and again in Chapter 16: Do the properties of a primary tumor reveal whether it has broadcast cancer cells throughout the body that will eventually create life-threatening metastatic disease long after the primary tumor has been surgically removed?

In this chapter, we confront the processes that create these most aggressive products of tumor progression. These processes depend on complex biochemical and biological changes in cancer cells and in the associated stroma. Most of the steps of cancer formation, as described in earlier chapters, are understood in considerable detail. In contrast, our understanding of invasion and metastasis is still quite incomplete, explaining why these late steps of tumor progression represent the major unsolved problems of cancer pathogenesis.

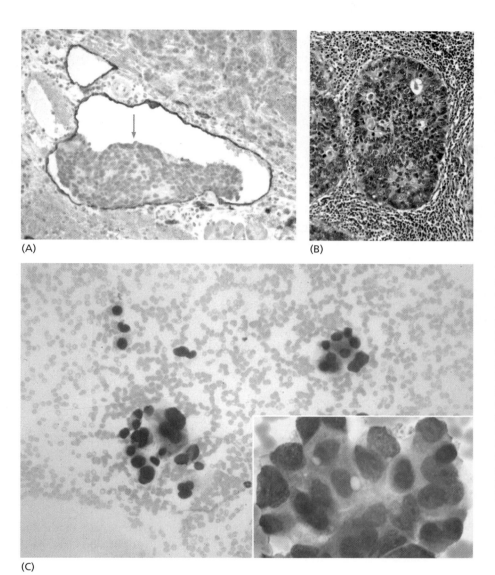

(A)

(B)

(C)

Figure 14.2 Histology of metastases in various tissues throughout the body (A) In the Rip-Tag transgenic mouse model of pancreatic islet cell tumorigenesis (see Figure 13.36), metastasis via the lymph nodes can be encouraged through the forced expression of VEGF-C—a lymphangiogenic factor—in the islet tumor cells. Seen here is a small metastasis of islet cells (*arrow*) within a lymphatic vessel that is lined with endothelial cells (*reddish brown*). (B) A small metastasis of a human breast cancer (*center*) is seen growing within a lymph node associated with one of the lymphatic ducts draining the breast. Note that this metastasis exhibits the detailed structure, including ducts and stroma, that is characteristic of many primary tumors of the breast; adjacent to this metastasis are numerous lymphocytes in the surrounding lymph duct (*dark nuclei*). (C) The presence of clumps of metastatic carcinoma cells (*blue*) in the bone marrow can be revealed using specific immunohistochemistry to detect cells displaying epithelial markers, which set them apart from the mesenchymal cells naturally present in the marrow; or, as seen here, through use of the Wright–Giemsa stain. (A, from S.J. Mandriota et al., *EMBO J.* 20:672–682, 2001. B, courtesy of T.A. Ince. C, courtesy of P. Leavitt, Dana-Farber Cancer Institute.)

14.1 Travel of cancer cells from a primary tumor to a site of potential metastasis depends on a series of complex biological steps

The great majority (>80%) of life-threatening cancers occur in epithelial tissues, yielding carcinomas. Consequently, most of our discussions in this chapter, as in the last, will refer to this class of tumors, with the understanding that cancers arising in other tissue types, such as connective and nervous tissues, often follow similar paths when they become invasive and metastatic. Even certain hematopoietic tumors, notably lymphomas, often have an early, localized phase and a later phase during which they become disseminated to distant anatomical sites (see Figure 14.1). This process, sometimes called the *invasion–metastasis cascade*, involves a complex sequence of steps, which are outlined in **Figure 14.3**.

Our focus on carcinomas requires us to draw from earlier discussions of these tumors and the epithelial tissues in which they arise (see, for example, Sections 2.2 and 13.1). To recap briefly, the great majority of epithelial tissues are constructed according to a common set of architectural principles; in most cases, relatively thin sheets of epithelial cells sit atop deep, complex layers of stroma. Separating the two is the

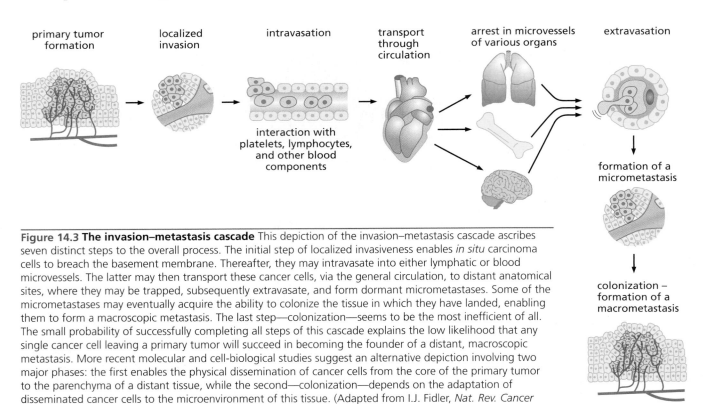

Figure 14.3 The invasion–metastasis cascade This depiction of the invasion–metastasis cascade ascribes seven distinct steps to the overall process. The initial step of localized invasiveness enables *in situ* carcinoma cells to breach the basement membrane. Thereafter, they may intravasate into either lymphatic or blood microvessels. The latter may then transport these cancer cells, via the general circulation, to distant anatomical sites, where they may be trapped, subsequently extravasate, and form dormant micrometastases. Some of the micrometastases may eventually acquire the ability to colonize the tissue in which they have landed, enabling them to form a macroscopic metastasis. The last step—colonization—seems to be the most inefficient of all. The small probability of successfully completing all steps of this cascade explains the low likelihood that any single cancer cell leaving a primary tumor will succeed in becoming the founder of a distant, macroscopic metastasis. More recent molecular and cell-biological studies suggest an alternative depiction involving two major phases: the first enables the physical dissemination of cancer cells from the core of the primary tumor to the parenchyma of a distant tissue, while the second—colonization—depends on the adaptation of disseminated cancer cells to the microenvironment of this tissue. (Adapted from I.J. Fidler, *Nat. Rev. Cancer* 3:453–458, 2003.)

specialized type of extracellular matrix (ECM) known as the basement membrane (see Figure 13.6). This proteinaceous meshwork is constructed collaboratively by proteins secreted by both epithelial and stromal cells.

By definition, carcinomas begin on the epithelial side of the basement membrane and are considered to be *benign* as long as the cells forming them remain on this side. Sooner or later, however, many carcinomas acquire the ability to breach the basement membrane (BM; **Figure 14.4A and B**). In certain groups of cancer patients, loss of the BM in their tumors predicts the future course of their disease, specifically their subsequent development of metastatic disease (Figure 14.4C); this suggests that loss of the BM is a prelude to eventual dissemination of cells from the primary tumor. Having breached the BM, cancer cells begin to invade the nearby stroma singly or in groups (**Figure 14.5**). This mass of neoplastic cells is now reclassified as *malignant*. In fact, as mentioned in an earlier chapter, many pathologists and surgeons reserve the word "cancer" for those epithelial tumors that have acquired this invasive ability. To be sure, this dissolution of the basement membrane by invading carcinoma cells removes an important physical barrier to the further expansion of tumor cell populations. But in addition, as we learned in the last chapter, by degrading various components of the basement membrane, invasive cells harvest growth and survival factors that have been sequestered by attachment to this specialized extracellular matrix.

Recall that even before carcinoma cells breach the basement membrane, they often succeed in stimulating angiogenesis on the stromal side of the membrane, apparently by dispatching angiogenic factors through this porous barrier to endothelial cells within the stroma (see Figure 13.38). However, invasion through the basement membrane places them in a far better position for executing subsequent steps of the invasion–metastasis cascade. Once present in the stromal compartment, carcinoma cells can gain direct access to the blood and lymphatic vessels (see Figure 13.40), which are normally found only on the stromal side of the basement membrane. Close contact with the capillaries affords tumor cells improved access to the nutrients and oxygen carried by the blood. In addition, their invasive properties enable these cancer cells to

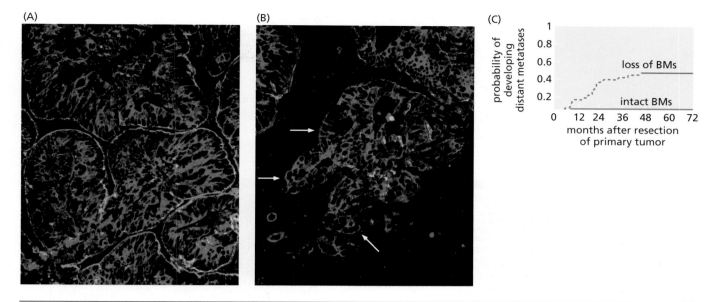

Figure 14.4 Breaching of the basement membrane The basement membrane (BM) can be detected through use of antibodies reactive with laminin, a key component of this membrane (see Figure 13.6). (A) In the more differentiated, less aggressive portions of a colorectal carcinoma, islands of carcinoma cells labeled with an anti-cytokeratin antibody (*red*) are surrounded by BMs (*green*) that have been detected with an antibody reactive with the α3 chain of laminin. The stroma between islands of carcinoma cells appears *black*. (B) In the portions of the tumor that appear to be less differentiated and more invasive, this staining reveals edges of islands of carcinoma cells (*arrows*) that lack BM and thus are directly apposed to the surrounding stroma. (C) For a group of patients whose high-grade rectal carcinomas (all classified as stage T3) had penetrated at least 5 mm into nearby fatty tissue, the loss of BM around tumor islands directly correlated with the probability of developing distant metastases. Partial or complete loss of BM within their tumors indicated a strong likelihood that metastases would develop during the five years after surgical removal of the primary tumors. (From S. Spaderna et al., *Gastroenterology* 131:830–840, 2006.)

move through the walls and into the lumina (that is, the bores) of blood and lymphatic vessels. This invasion into vessels is often termed **intravasation** and depends on the ability of individual cancer cells or small clumps of these cells to break away from their neoplastic neighbors and enter on their own into the circulation.

Local invasion seems to depend invariably on the release of secreted proteases, which are required to remodel the extracellular matrix (ECM), thereby generating space for the advance of cancer cells (**Figure 14.6A**). In some tumors, invading carcinoma cells make their own proteases, such as MMP-2 and MMP-9 (see Figure 14.17E), while in others a variety of stromal cells are co-opted and induced to release these enzymes, often leading flocks of carcinoma cells behind them (Figure 14.6B–D; Supplementary Sidebar 14.1).

The process of intravasation is less well studied. Initial studies indicate that in many breast cancers, and possibly in other carcinomas as well, a triad of three distinct cell types—carcinoma cells, macrophages, and endothelial cells—assembles to enable the cancer cells to invade through the endothelial walls into the lumina of capillaries (**Figure 14.7**). Later in this chapter we revisit these carcinoma cell–macrophage interactions and their reciprocal signaling. As we will learn, the invasiveness—and associated intravasation—by the cancer cells is stimulated by EGF released by their macrophage partners. Importantly, counts of the density of these triads in histopathological sections of human breast cancers provide a strong prognostic factor of eventual metastatic relapse of the cancer patients.

Once they have intravasated into the lumen of a blood or lymphatic vessel, individual cancer cells may travel with the blood or lymph to other areas in the body. These long-range migrations are fraught with great danger for the wanderers. Like normal cells, the cancer cells may continue to depend on anchorage to solid substrates; without such attachment, the migrating cells may die rapidly from anoikis, the form of apoptosis that is triggered by detachment of a cell from a solid substrate such as an extracellular matrix (see Sections 5.9 and 9.13). Also, like their forebears in the primary tumor, these pioneers may depend on various types of stromal support, which will be lacking the moment they leave the primary tumor mass. Recall that the stroma can benefit carcinoma cells in multiple ways by supplying both mitogenic and trophic (survival) factors.

The blood, in particular, represents an actively hostile environment for metastasizing cancer cells. Hydrodynamic shear forces in the circulation, which are often substantial in smaller vessels, may tear the wandering cancer cells apart. Some experimental models of metastasis in the mouse provide clear indication that the survival

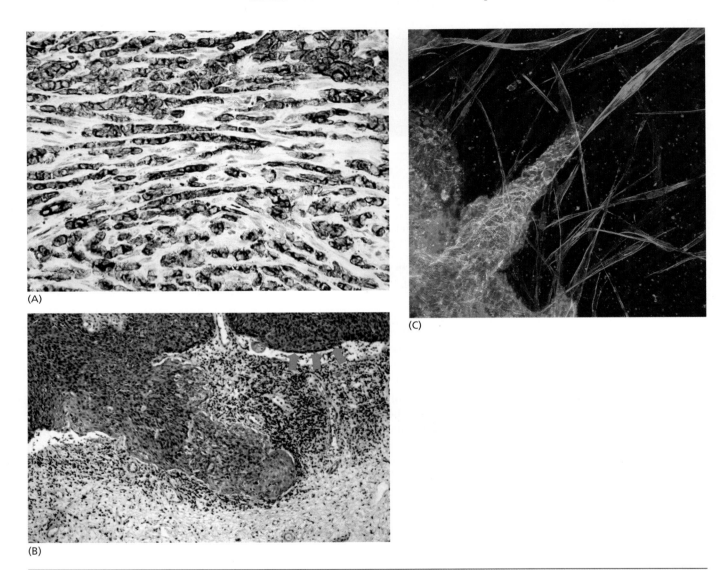

(A)

(C)

(B)

Figure 14.5 Patterns of invasion (A) These invasive lobular mammary carcinoma cells (*brown*) have left the primary tumor (*not shown to the left*) and are proceeding rightward, one-by-one in single file, through channels they have carved in the adjacent stroma (*white, gray*). (B) Far more typical is the coordinated invasion of a phalanx of carcinoma cells, often termed "collective" invasion. In this squamous cell carcinoma of the cervix, a tongue of many hundreds of cancer cells (*pink, brown*) has breached the basement membrane and is invading the stroma. The latter is characterized by both fibroblasts and inflammatory cells (*dark green*). The basement membrane is the dark brown, horizontal line (*pink arrows*) that separates the bulk of the carcinoma cells (*above*) from the stromal cells (*below*) and is uninterrupted except for a capillary and the tongue of invasive cancer cells. (C) Collective invasion can be modeled experimentally *in vitro*. Here, MCF-7 human breast cancer cells were cultured together with fibroblasts in a 3-dimensional (3D) matrix composed largely of collagen. (Such 3D matrices are widely thought to recapitulate the *in vivo* tissue environment more closely than 2D monolayer culture conditions.) The breast cancer cells, seen here invading through the collagen matrix, continue to adhere to one another via E-cadherin–containing adherens junctions (*red*). The actin stress fibers of the fibroblasts are stained with phalloidin (*green*); cell nuclei are stained with DAPI (*blue*). (A, courtesy of J. Jonkers. B, courtesy of T.A. Ince. C, from O. Ilina and P. Friedl, *J. Cell Sci.* 122:3203–3208, 2009.)

of metastasizing cancer cells in the general circulation is greatly enhanced if they can attract an entourage of blood platelets to escort them through the rapids into safe pools within tissues (Supplementary Sidebar 14.2).

Nonetheless, we still have an incomplete understanding of the lives of cancer cells in the general circulation, which represents the main route of dissemination leading clinically to metastatic relapse. These circulating tumor cells (CTCs) have been the objects of intensive investigation in recent years. They presumably represent cancer

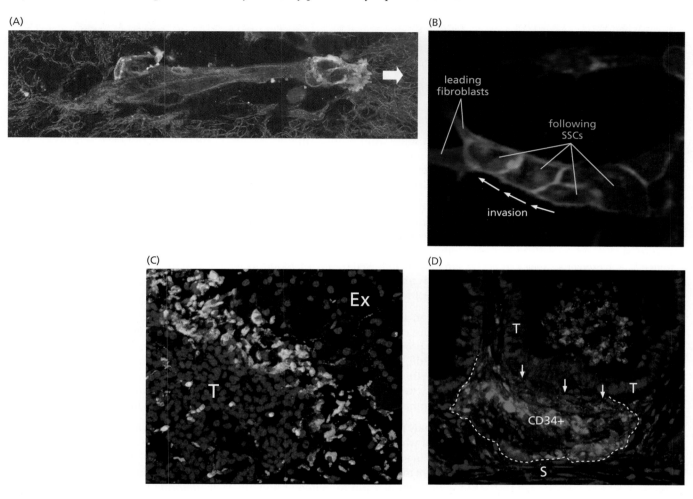

Figure 14.6 Invasion through the extracellular matrix and its control by stromal cells (A) That carcinoma cell invasion requires degradation of the extracellular matrix (ECM) is illustrated by this confocal micrograph of a cohort of 5–10 melanoma cells that are moving together through a collagen matrix (*blue*); like normal melanocytes, they continue to adhere to one another through adherens junctions that are formed by E-cadherin molecules (*red*). Large gaps in the matrix (*black*) indicate areas that have been degraded by the advancing cancer cells or associated stromal cells. At the leading invasive edge (*white arrow*), the melanoma cells are displaying β1 integrins (*green*), which enable them to attach to the still-intact ECM lying ahead in their path. (B) In a variety of tumors, cancer cells recruit and co-opt stromal cells, which clear the invasion path by degrading the ECM, with the carcinoma cells following close on their heels. In this case, human oral squamous cell carcinoma cells (SCCs, *green*) are being led by a small group of carcinoma-associated fibroblasts (*red*) in an *in vitro* experimental model of cancer cell invasion. The direction of invasion by these cells (*arrows*) was determined by *time-lapse microscopy*. (C) In the Rip-Tag transgenic model of pancreatic islet carcinogenesis (see

Figure 13.36), the carcinoma cells forming the tumor (T) recruit macrophages (*green*) and, by stimulating them with interleukin-4 (IL-4), cause them to release cathepsins (*red*), a class of secreted proteases distinct from the MMPs. Release of these cathepsins at the invasive front of the carcinoma cells is essential for invasion by these cells into the adjacent exocrine pancreas (Ex). Cathepsin-producing macrophages are seen here in *yellow*. Nuclei are stained *blue* with DAPI. (D) In this *Apc*-mutant mouse model of intestinal carcinogenesis (see Section 7.11), colon adenoma cells (T), which are beginning to invade downward (*arrows*) into the underlying stroma (S) and muscularis propria, recruit a cap (*dotted line*) of bone marrow–derived CD34+ immature myeloid cells (*green*) at the invasion front. The latter proceed to secrete MMP-2 and MMP-9, which enable subsequent collective invasion by the neoplastic cells into the mesenchymal layers. (A, from P. Friedl, Y. Hegerfeldt and M. Tusch, *Int. J. Dev. Biol.* 48:441–449, 2004. B, from C. Gaggioli et al., *Nat. Cell Biol.* 9:1392–1400, 2007. C, courtesy of B.B. Gadea and J.A. Joyce; from V.L. Gocheva et al., *Genes Dev.* 24:241–255, 2010. D, from T. Kitamura et al., *Nat. Genet.* 39:467–475, 2007.)

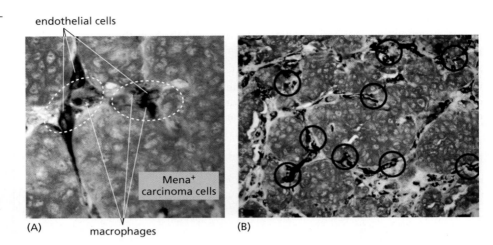

endothelial cells

Mena⁺ carcinoma cells

(A) macrophages

(B)

Figure 14.7 Intravasation Like local invasion, the process of intravasation by carcinoma cells often—and perhaps always—requires help from stromal cells, in this case macrophages. (A) Detailed investigation of carcinomas, in both laboratory animal models and human tumors, reveals triads (*ovals*) of carcinoma cells, macrophages, and endothelial cells (sometimes termed TMEM, for tumor microenvironment of metastasis). The macrophages help the carcinoma cells to intravasate through the capillary wall formed by the endothelial cells. A triad is registered only if all three cellular components are directly apposed and if the carcinoma cells express significant levels of the Mena actin-cytoskeleton–regulating protein that is associated with cell motility, seen here as a *red* stain. (B) While two human carcinomas may have a very similar histopathological appearance, the counting of the density of TMEM triads (*circles*) within sections of these tumors generates a highly useful prognostic marker of metastasis. (A and B, from B.D. Robinson et al., *Clin. Cancer Res.* 15:2433–2441, 2009.)

cells en route from primary tumors to sites of metastasis. But even this notion is at best an inference, since only a tiny proportion of CTCs are ever successful in founding new metastatic colonies (see below), explaining why none has actually been viewed extravasating and thereafter spawning a macroscopic metastasis.

A variety of techniques are being developed to measure the concentrations of CTCs in the circulation of cancer patients (Supplementary Sidebar 14.3). These numbers may prove useful in gauging the efficacy of therapies directed against a primary tumor within a single patient; thus, a decrease in viable cells in the primary tumor will presumably be followed by a corresponding decrease in the release of such cells into the general circulation and a resulting decline in the number of CTCs (see Supplementary Sidebar 14.3).

However, the utility of comparing CTC levels *between* patients remains unclear. There are clear indications that the levels of CTCs in patients with metastatic breast cancer provide some indication of clinical progression (see Supplementary Sidebar 14.3). Still, these correlations have not been studied systematically, and we cannot say whether, for example, two patients who are diagnosed with histologically similar primary tumors of comparable size and who experience similar clinical courses will carry comparable or vastly different concentrations of CTCs in the their blood.

CTCs may persist for only a short time in the circulation, in large part because, unlike red and white blood cells, they are ill suited to negotiate the passage through microvessels in various tissues (**Figure 14.8A–C**); thus, the internal diameters of most capillaries—3 to 8 µm—are far too small to accommodate them. Erythrocytes, for example, are only about 7 µm in diameter and are easily deformed, facilitating their passage through capillaries. Most cancer cells, in contrast, are more than 20 µm in diameter and are not especially deformable. [Moreover, if cancer cells in the blood are coated with platelets (see Supplementary Sidebar 14.2), their effective diameters become even larger, causing them to be trapped in vessels larger than capillaries, such as the small arteries known as **arterioles**.]

These factors dictate that within minutes of entering into the venous circulation many cancer cells will encounter the capillary beds of the lungs (see Figure 14.8D), in which they lodge. Indeed, their lifetimes as free cells in the circulation may be so short that CTCs never sense their loss of anchorage to a solid substrate and the absence of stromal support.

Once trapped within the lungs, some metastasizing cancer cells may attempt to found metastases there. However, the metastases of many types of human tumors are often found elsewhere in the body, indicating that cancer cells frequently succeed in escaping from the lungs and travel further, to other sites in the body. How they do so is unclear. In some experiments, cancer cells trapped in capillaries have been observed to pinch off large amounts of cytoplasm, leaving behind cells that, while greatly reduced in size, seem to be viable; once they have undergone this amputation, such slenderized cells may succeed in negotiating passage through the narrow straits of

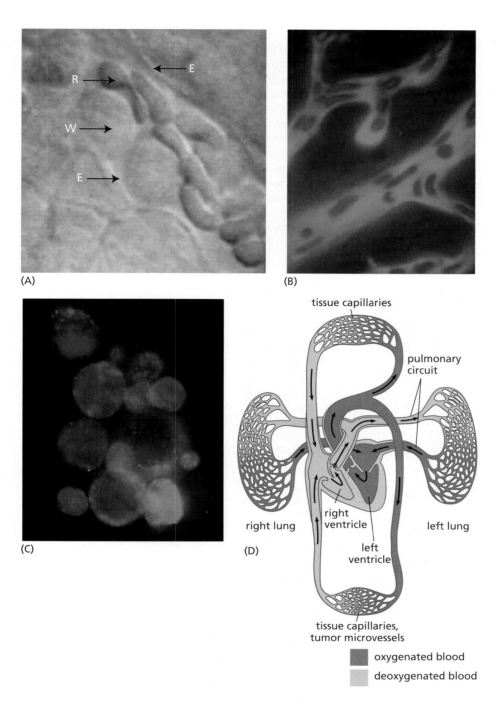

Figure 14.8 Passage of tumor cells through the circulation Cancer cells moving through the circulation are found amid densely packed cells of hematopoietic origin. (A) *Intra-vital* microscopy reveals the two endothelial walls of a capillary (E), erythrocytes (R), and some leukocytes (W). The small size (~7 μm diameter) and deformability of these cells allow them to pass through capillaries without becoming trapped. (B) The vessels in this intra-vital fluorescence micrograph are slightly larger than capillaries. The plasma has been stained with a green dye, while the erythrocytes have been colorized red. The high deformability of these red blood cells is apparent. Since most cancer cells have more than twice the diameter of erythrocytes and are not deformable, they are unable to negotiate narrow passages, such as the lumina of capillaries. (C) The difficulties of negotiating passage through the narrow microvessels of various tissues are further compounded by the fact that intravasated cancer cells often attract clouds of platelets around them (see Supplementary Sidebar 14.2) and the fact that they are often found in the circulation as multicellular aggregates, as is seen in this cluster of circulating tumor cells (CTCs) isolated from the circulation of a patient with metastatic prostate cancer. Cells were immunostained with an antibody reactive with the prostate-specific membrane antigen (*green*) and another reactive with the CD45 displayed by many types of hematopoietic cells (*red*). DNA was stained *blue* with DAPI. (D) This diagram of the mammalian circulation indicates that venous blood (*blue*) leaving a tissue (and thus cancer cells that have escaped from a primary tumor and intravasated, *below*) must first pass through the right ventricle of the heart and thence through the lungs before it enters the left ventricle and is pumped into the general arterial circulation. Since passage through the pulmonary circulation of the lung requires passage through its capillaries, almost all metastasizing cells entering into the venous circulation are rapidly trapped in the pulmonary capillary beds. (A and B, from I.C. MacDonald, A.C. Groom and A.F Chambers, *BioEssays* 24:885–893, 2002. C, from S.L. Stott et al., *Proc. Natl. Acad. Sci. USA* 107:18392–18397, 2010. D, from P.H. Raven et al., Biology, 7th ed. New York: McGraw-Hill, 2005.)

the lung capillaries. A more plausible explanation is that wandering cancer cells may avoid being trapped altogether: in many organs, including the lung, metastasizing cells can bypass capillaries by traveling through arterial–venous shunts, which form large-bore, direct connections between the two parts of the circulatory system.

Having snaked their way through the lungs and arrived in the general arterial circulation, roaming cancer cells can then scatter to all tissues in the body. Some experiments suggest that cancer cells use specific cell surface receptors, such as integrins, to initially adhere to the luminal walls of arterioles and capillaries in certain tissues. However, far more extensive evidence indicates that simple physical trapping within small vessels, as discussed above in the context of the lung, provides most wandering cancer cells with their first foothold within a tissue.

Once lodged in the blood vessels of various tissues, cancer cells must escape from the lumina of these vessels and penetrate into the surrounding tissue—the step termed **extravasation**. The process of extravasation depends on complex interactions between cancer cells and the walls of the vessels in which they have become trapped. Cancer cells can use several alternative strategies to extravasate. They may proceed immediately to elbow their way through the vessel wall (**Figure 14.9A–C**). Their ability to do so may depend on the same biochemical and cell-biological mechanisms that previously enabled them or their immediate ancestors to invade from the primary tumor and to intravasate (see, however, **Sidebar 14.1**). Alternatively, they may begin to proliferate within the lumen of the vessel, creating a small tumor that grows and eventually obliterates the adjacent vessel wall (see Figure 14.9D). In doing so, they push aside endothelial cells, pericytes, and smooth muscle cells that previously separated the vessel lumen from the surrounding tissue, the latter often being called the tissue *parenchyma*. Interestingly, close associations of disseminated cancer cells and macrophages have been documented at sites of extravasation, suggesting that, as is the case with intravasation (see Figure 14.7), cancer cells recruit macrophages to help them escape from the circulation into the tissue parenchyma.

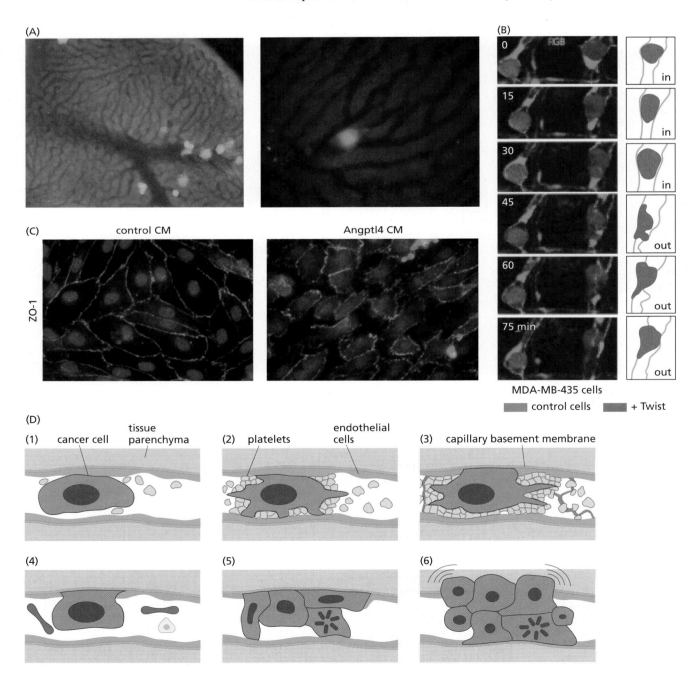

Sidebar 14.1 Cancer cells are clumsy escape artists The complex task of escaping from the circulation into the surrounding tissue parenchyma is accomplished routinely by leukocytes, which must be able to enter into the parenchyma in response to certain inflammatory stimuli, including the presence of infectious agents. Through a sequence of steps known as **diapedesis**, leukocytes are able to induce endothelial cells in post-capillary venules to retract and create a portal into the underlying tissue. The entire process from attachment to the endothelial wall to entrance into the tissue parenchyma takes less than a minute and involves an elaborately choreographed program of biochemical and cell-biological changes!

In contrast, the vast majority of metastasizing cancer cells are not endowed with the receptors and biochemical response mechanisms required to execute diapedesis. Accordingly, if neoplastic cells do succeed in penetrating through the wall of a capillary or slightly larger vessel, they seem to do so by brute force, perhaps by degrading patches of endothelium in a process that may require many hours or even a day rather than a minute to complete. Of additional interest is the fact that the thrombin produced during the formation of microthrombi (see Supplementary Sidebar 14.2) is quite effective in cleaving the various proteins used by endothelial cells to attach to the underlying vascular basement membrane; this may cause endothelial cells to retract from microemboli, thereby exposing the capillary basement membrane to direct attack by invasive cancer cells and the proteases that they produce.

Figure 14.9 Lodging and extravasation of circulating tumor cells
(A) Tumorigenic, green fluorescent protein (GFP)–labeled Chinese hamster ovary cells were injected via the portal vein into a mouse liver. Large numbers of these cells (*green*) soon became arrested in the microvessels of the liver (*left panel*). However, within 24 hours, an individual cell (*right panel; higher magnification*) has extravasated out of the microvessels (*dark red*) into the liver parenchyma (*light brown*). (B) Non-mammalian experimental models often enable experiments that could not be undertaken in mammals, such as this study of a single cancer cell and its steps of extravasation. Here advantage is taken of the transparency of zebra fish embryos, in which endothelial cells forming the capillary walls were labeled with GFP through the actions of a transgene. Human MDA-MB-435 (MDA) cells, which are relatively nonaggressive human breast cancer cells, were injected into the circulation of these zebra fish and monitored by confocal microscopy. The control MDA cancer cells, labeled with cyan fluorescent protein (*blue; left column*), did not extravasate over a period of 75 minutes. However, if red fluorescent protein–expressing MDA cells were forced to express the Twist EMT-inducing transcription factor, which activates an invasive program (see Section 14.8), they initiated extravasation over this period of time (*right* and drawings at *far right*). While interpretation of experiments using zebra fish is complicated by the incompatibility of mammalian cells growing in a fish host, these data suggest nonetheless that one consequence of activating an EMT program is an acquired ability to invade through the endothelial wall into the parenchyma of a tissue. (C) Mammary carcinoma cells within a primary tumor experience TGF-β, which induces them to produce angiopoietin-like protein 4 (Angptl4). When these cells disseminate and become lodged in microvessels, the Angptl4 that they secrete induces endothelial cells to retract from one another, leaving gaps in the capillary walls that facilitate extravasation. As seen here, cultured human vascular endothelial cells, when exposed to conditioned culture media (CM) produced by Angptl4-secreting cells (*right panel*), retract their associations with one another, as indicated by the dissolution of the ZO-1– containing tight junctions (*green*) between adjacent cells; CM produced by control cells (*left panel*) fails to induce this retraction. (D) Electron-microscopic observations of metastasizing lung cancer cells injected into the venous circulation of mice suggest that the process of extravasation often proceeds via an alternative sequence of steps, as indicated here. (1) A metastasizing cell (*brown*) is trapped physically in a capillary. (2) Within minutes, a large number of platelets (*blue*) become attached to the cancer cell, forming a microthrombus. (3) The cancer cell pushes aside an endothelial cell (*green*) on one wall of the capillary, thereby achieving direct contact with the underlying capillary basement membrane (*orange*). (4) Within a day, the microthrombus is dissolved by the proteases that are responsible for dissolving clots. (5) The cancer cell begins to proliferate in the lumen of the capillary. (6) Within several days, sometimes sooner, the cancer cells break through the capillary basement membrane and invade the surrounding tissue parenchyma (*gray area*). (A, courtesy of G.N. Naumov, from G.N. Naumov et al., *J. Cell Sci.* 112:1835–1842, 1999. B, from K. Stoletov et al., *J. Cell Sci.* 123:2332–2341, 2010. C, from D. Padua et al., *Cell* 133:66–77, 2008. D, from J.D. Crissman et al., *Cancer Res.* 48:4065–4072, 1988.)

14.2 Colonization represents the most complex and challenging step of the invasion–metastasis cascade

Once they have arrived within the parenchyma of a tissue, metastasizing cancer cells may form **micrometastases**—small clumps of disseminated cancer cells, some of which are able to expand into clinically detectable masses; this growth of microscopic into macroscopic metastases is often termed **colonization**. This last step is also a challenging one—perhaps the most difficult step of all, ostensibly because the foreign tissue environments encountered by the newly arrived cells do not provide them with the collection of familiar growth and survival factors that allowed their progenitors to thrive in the primary tumor site. Without these various types of physiologic support, the metastasizing cells may rapidly die or, at best, survive for extended periods of time as micrometastases that can only be detected microscopically and rarely increase beyond this size. In general, the number of micrometastases in the body of a cancer patient vastly exceeds those that will eventually grow large enough (several millimeters or more in diameter) to be clinically detectable. A frequently quoted statistic describes the clinical progression of breast cancer patients: more than 30% of these patients harbor hundreds, likely thousands of micrometastases in their bone marrow at the time of initial clinical presentation, but only half of the women in this group will ever develop metastatic disease. In fact, micrometastases may be widely disseminated throughout other tissues of a cancer patient, often in tissues where their detection is far more challenging than in the bone marrow; some of these occult micrometastases may lead, on occasion, to disastrous outcomes (Supplementary Sidebar 14.4).

Antibodies reactive with **cytokeratins** are useful for detecting the micrometastases that primary carcinomas spawn in the bone marrow and blood, while an antibody against the epithelial cell adhesion molecule (EpCAM) is often used to detect micrometastases in the lymph nodes. In all these cases, the presence of isolated cytokeratin-positive (and thus epithelial) cells in otherwise fully mesenchymal tissues represents a clear sign that metastatic seeding has taken place. Current microscopic techniques using cytokeratin-specific antibodies make it possible to detect a single-cell micrometastasis among 10^5 or even 10^6 surrounding mesenchymal cells in the blood, bone marrow, or lymph node (**Figure 14.10A and B**). Slightly larger micrometastases can often be detected in the lymph nodes that are connected with a primary tumor via draining lymphatic ducts (see Figure 14.10C).

The probability of an individual cancer cell successfully completing all of the steps of the invasion–metastasis cascade is very low. For example, in some mice carrying primary tumors of about 1 gram mass (~10^9 cells), as many as a million cells may be seeded into the circulation each day, yet the visible metastases formed may be counted on the fingers of one hand. This low rate of success in forming metastases is sometimes termed *metastatic inefficiency*. Some experiments indicate that the earlier steps in this cascade are executed quite efficiently by metastasizing cells, while the last step, involving colonization, succeeds only rarely and therefore is the *rate-limiting* determinant of the process as a whole. Consequently, vast numbers of micrometastases may be seeded throughout the body and many may persist for extended periods of time before one or another of these migrants or its lineal descendants finally acquires the ability to grow into a clinically detectable mass (**Sidebar 14.2**).

Importantly, while the probability that a given disseminated cancer cell will succeed in generating a macroscopic metastasis is extremely low, the very fact that a patient's tumor is dispatching neoplastic cells throughout the body is itself an indication that the tumor carries a bad prognosis (see Figure 14.10D). Indeed, at least at present, the number of disseminated tumor cells in the marrow, often termed DTCs, seems to represent a far more useful prognostic marker than the concentration of circulating tumor cells (CTCs) in the blood. This may reflect the fact that the number of DTCs represents the accumulation of disseminated cancer cells over an extended period of time, whereas the concentration of CTCs may be dictated by complex kinetic processes governing their lifetimes—and thus their *steady-state* concentration—in the circulation.

Support for the existence of dormant micrometastases, which persist in a nongrowing state for extended periods of time, comes from experiments in which living cancer cells were marked by brief exposure to fluorescence label–containing particles; the latter persist for a long time within cells but do not affect their viability. However, the intracellular concentration—and thus the fluorescence intensity of the dye particles—decreases by a factor of 2 each time a cell divides; hence, the residual fluorescence intensity in cells allows the experimenter to estimate how many times the cells have divided since they were initially marked.

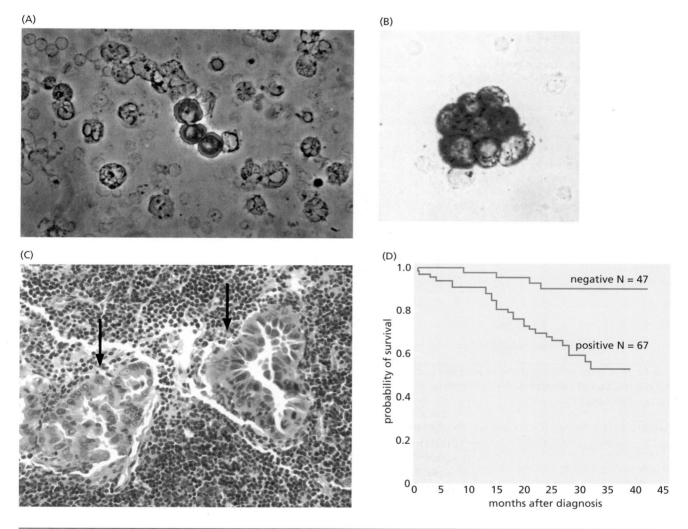

Figure 14.10 Detection of micrometastases in the marrow or lymph node (A) Metastasis of cancer cells to the bone marrow is usually determined by withdrawing marrow from the iliac crest of the pelvis. In this micrograph, the presence of a micrometastasis containing several cancer cells in the bone marrow of a colon cancer patient has been detected by staining with an anti-cytokeratin antibody (*red*). This staining is highly specific, since normal marrow cells are of mesenchymal/hematopoietic origin and thus do not express cytokeratins. In contrast, epithelial cells, and (usually) their neoplastic descendants, express cytokeratins at significant levels. (B) Essentially identical images can be detected in the bone marrow extracted from breast cancer patients. In this case, a cluster of eight mammary carcinoma cells, viewed at slightly lower magnification, form a micrometastasis. While not demonstrated here, these cells are apparently the products of post-extravasation proliferation in the marrow. (C) Micrometastases often form in a *draining* lymph node—one that is directly connected

with the primary tumor via a lymphatic duct that drains the tissue in which this tumor arose (see Figure 14.42). These growths are usually detectable by their strongly contrasting appearance from surrounding lymphocytes. Here two micrometastases of a mouse lung adenocarcinoma (*arrows*) to a lymph node are seen amid a sea of lymphocytes. Note the formation of a ductlike structure by the right micrometastasis, providing clear indication of proliferation after initial dissemination. (D) Cytokeratin-positive cells in the marrows of a cohort of breast cancer patients were counted, allowing the patients to be placed into two groups (whose sizes are given above the curves) having either cytokeratin-positive or -negative marrows. As is evident, those with cytokeratin-positive cells in their marrow had a far worse prognosis, with almost 50% of them dying within three years of initial diagnosis. (A, courtesy of I. Funke and G. Riethmüller. B, from S. Braun et al., *N. Engl. J. Med.* 342:525–533, 2000. C, courtesy of K.P. Olive and T. Jacks. D, from J.-Y. Pierga et al., *Clin. Cancer Res.* 10:1392–1400, 2004.)

Sidebar 14.2 Genetic analyses suggest that the evolution of metastatic ability can occur outside of the primary tumor In perhaps 30% of breast, prostate, and colon cancer patients whose primary carcinomas have been surgically removed, one can still detect micrometastases in the marrow, lymph nodes, or blood; these patients are considered to have "minimal residual disease." At this stage, analysis of the micrometastases indicates that they are genetically heterogeneous. However, years later, when a patient develops disease relapse and manifests readily detectable metastatic masses, the patient's single-cell micrometastases are now much more similar to one another genetically (**Figure 14.11A and B**).

This suggests a stage of tumor progression in which the ability to colonize is acquired separately from the ability to disseminate to distant organ sites. Initially, genetically diverse cancer cells are seeded by a primary tumor throughout the body, but none of these succeeds in establishing a macroscopic metastasis simply because none is capable of doing so. After a period of time, however, genetic and epigenetic evolution occurring in a micrometastasis somewhere in the body yields a clone of cells with the newly acquired ability to colonize efficiently. As this clone expands, it also begins to seed cancer cells throughout the body, and therefore generates a secondary wave of metastatic dissemination.

The individual cancer cells that are released by the initially formed colonizing clone soon constitute the majority of the single-cell micrometastases of a cancer patient, and these micrometastases are genetically very similar to one another because of their shared descent from the same clonal cell population. Importantly, because the cells in these new micrometastases all have inherited the ability to colonize, many may grow rapidly into macroscopic metastases, creating a life-threatening burden of disseminated cancer cells in the patient. (This secondary wave of dissemination is sometimes termed a metastatic "shower" to indicate the large number of new, robustly growing metastases that appear in an apparently synchronous fashion.)

This "cascade" model (see Figure 14.11C) suggests that the final evolution toward advanced malignancy often occurs at anatomical sites far removed from the primary tumor, and that a secondary wave of dissemination is responsible for most metastasis-associated death. A 1986 study of the autopsies of more than 1500 patients who died from colon carcinoma provided strong indications that metastases initially formed in the livers of these patients and spread from there to the lungs and finally to other sites in the body. Nevertheless, even with such evidence in hand, the cascade model of Figure 14.11C is not yet validated by large, statistically robust sets of observations.

Figure 14.11 Genetic heterogeneity of micrometastases and the evolution of colonizing ability (A) Single-cell micrometastases of primary carcinomas can be identified in bone marrow biopsies by their display of epithelial cell markers, such as cytokeratins (see Figure 14.10) or EpCAM, and then isolated using a micropipette. (B) Adaptation of the comparative genomic hybridization (see Supplementary Sidebar 11.4) procedure is used to analyze individual micrometastatic cells from the same patient for the gain or loss of various chromosomal arms. The resulting "genetic profiles" of the individual micrometastases are then compared with each other. Profiles that are similar are placed near one another on a branch of a tree termed a *dendrogram*; conversely, very different genetic profiles are located far away from one another on the tree. A horizontal bar (*blue*) indicates that multiple micrometastases from a single patient are clustered next to one another on a branch of the dendrogram (and thus are closely related genetically). Here, micrometastases in the bones of patients (identified by *labels*) carrying breast, prostate, and colorectal tumors were obtained at the time of surgical removal of their primary tumors, a stage termed "minimal residual disease." As seen in the upper dendrogram, only a minority of micrometastases in these patients are clustered together on the dendrogram, indicating substantial genetic heterogeneity of micrometastases within each patient at this stage of disease. However, months or years later, when disease relapse with macroscopic metastases occurred, the several micrometastases detected in almost

every patient are located close to one another on the lower dendrogram, indicating a closer genetic relationship among them. (C) The analyses shown in panel B suggest the following speculative model. An initially formed, genetically heterogeneous primary tumor cell population (see Figure 11.20) seeds equally heterogeneous micrometastases throughout the body of the cancer patient. The primary tumor is then removed surgically, leaving behind only the micrometastases and creating the state of minimal residual disease. Over a period of years, in one or another site in the body, one of these micrometastatic cell clones (*blue*) acquires the ability to colonize, i.e., to grow into a macroscopic metastasis. The latter now acts as a source of cells that generate a new cascade of metastatic dissemination throughout the body. Because these newly dispersed cells all share a common clonal origin, the micrometastases that they form during this secondary "shower" are genetically very similar to one another. Moreover, because the cells in each of these secondary micrometastases are already endowed with the ability to colonize, many of them can rapidly grow into macroscopic, clinically detectable metastases that result in disease relapse. (In fact, metastases in terminal cancer patients are often highly heterogeneous genetically; whether this suggests an alternative model of how metastases evolve or the fact that there are multiple successive secondary showers during the last phases of the disease remains unclear.) (A and B, from C.A. Klein et al., *Lancet* 360:683–689, 2002.)

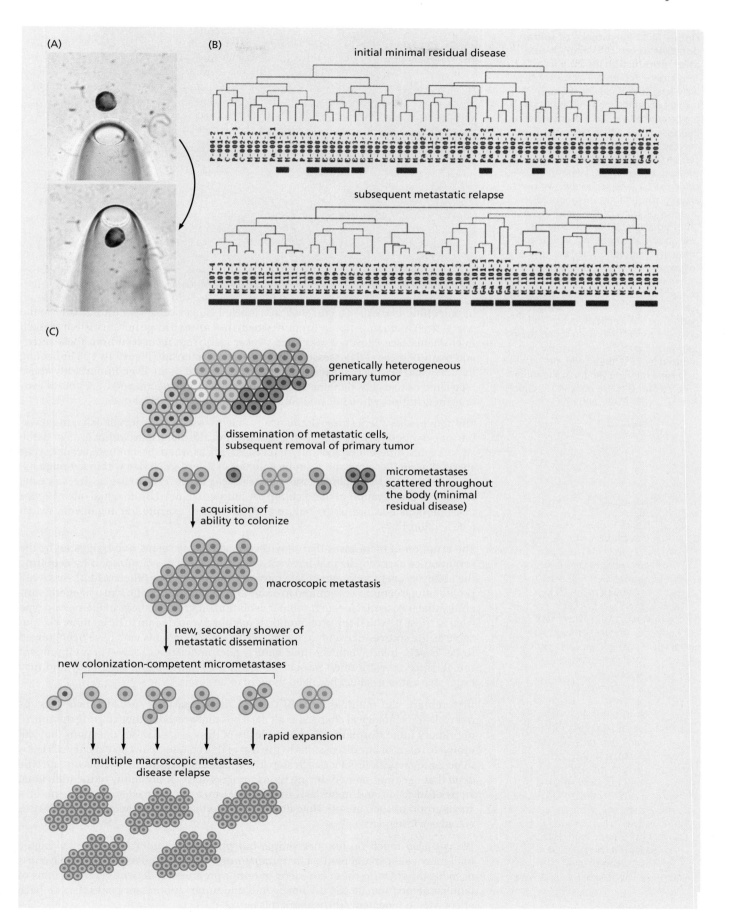

Figure 14.12 Persistence of solitary dormant tumor cells many weeks after introduction into the liver
Tumorigenic mouse breast cancer cells were labeled by inducing them to take up styrene nanoparticles (48 nm diameter) that had been tagged with a fluorescent dye. These cells were then injected into a mesenteric vein, which carried them via the portal vein into the liver. Eleven weeks later, tumor cells could still be detected in the liver (*white arrows*). Importantly, the fluorescence intensity of many of these cells did not differ significantly from the intensity of cells shortly after labeling, indicating that these cells had not divided even once following labeling. Following isolation and *in vitro* culturing, the descendants of many of these cells were tumorigenic when injected into the mammary fat pads of host mice. (Moreover, the dormant cancer cells were found to be fully resistant to a chemotherapy that reduced by 75% the size of metastases that were growing rapidly in the same mice.) (From G.N. Naumov et al., *Cancer Res.* 62:2162–2168, 2002.)

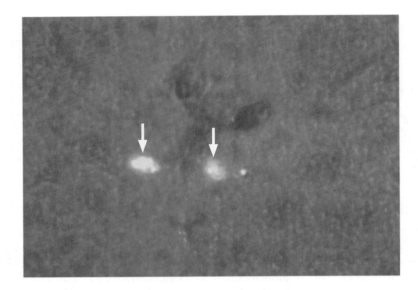

In one influential experiment, such dye-labeled cancer cells were introduced via the portal circulation into mouse livers, in which they formed large numbers of single-cell micrometastases. Eleven weeks later, cancer cells were recovered from these livers, and many of these still possessed full fluorescence intensity (**Figure 14.12**), indicating that they had not divided even once since their arrival in the liver. Importantly, these recovered cancer cells remained capable of proliferating *in vitro* and were able to generate new tumors when injected subcutaneously into other host mice.

The above experiment shows dramatically that metastatic cancer cells can remain viable for extended periods of time in a nondividing, dormant state within foreign tissue sites. A quite different type of micrometastasis arises when disseminated cancer cells succeed in proliferating and forming colonies of a very small size within a foreign tissue; however, these subclinical micrometastases may never increase in size, since the rate of cell proliferation in these clumps is counterbalanced by an equal rate of apoptosis, perhaps because of the failure of these cells to execute the angiogenic switch (see Section 13.7).

The eruption of metastases that signals clinical relapse seems to be triggered by the evolution of cancer cells that have solved the problem of colonization by acquiring the ability to adapt to foreign tissue microenvironments (see Sidebar 14.2). Since cell proliferation seems to be required in order to generate the genetic and epigenetic variability that is essential for such cellular evolution, micrometastases of the second type (that is, those in which cell proliferation is ongoing) would seem to be far more likely to serve as the sources of eventual metastatic relapse in patients who have been judged to be "cancer-free." Whatever their nature, micrometastases represent an imminent threat, since they are often present in vast numbers throughout the body and may erupt years after a cancer has been thought to be cured.

In summary, the multiple steps of the invasion–metastasis cascade encompass as many distinct biological changes as all the steps that preceded them during the course of primary tumor formation. The complexity of this cascade raises questions that will motivate many of our discussions in the rest of this chapter: How do cancer cells learn to become metastatic? Does each step in this cascade require the actions of a specific gene that becomes altered during tumor progression? Or are many of the individual steps of invasion and metastasis orchestrated by a single master control gene or a small group of such genes? How do disseminated cancer cells acquire the ability to colonize a foreign tissue?

We will also touch on another simple but profoundly important issue: Do highly malignant cells carry genes that in mutant form are specialized to induce invasiveness or metastasis? Or do these late steps in tumor progression depend on the actions of familiar actors, specifically the oncogenes and tumor suppressor genes that we have encountered repeatedly throughout this book?

14.3 The epithelial–mesenchymal transition and associated loss of E-cadherin expression enable carcinoma cells to become invasive

The first of the many steps leading to metastasis—the acquisition of local invasiveness—involves major changes in the phenotype of cancer cells within the primary tumor. As before, we will focus this discussion on epithelial tissues and the carcinomas that they spawn. The organization of the epithelial cell layers in normal tissues is incompatible with the motility and the invasiveness displayed by malignant carcinoma cells, yet this epithelial organization plan continues to be respected in many primary carcinomas. In these tumors, well-organized sheets of epithelial cells are present, although their overall topology may be quite different from that of comparable normal epithelia (see, for example, Figure 2.6).

In order to acquire motility and invasiveness, carcinoma cells must shed many of their epithelial phenotypes, detach from epithelial sheets, and undergo a drastic alteration—the epithelial–mesenchymal transition (EMT), which was mentioned in the context of wound healing (see Section 13.3). Recall that an EMT involves a shedding by epithelial cells of their characteristic morphology and gene expression pattern and the assumption of a shape and transcriptional program characteristic of mesenchymal cells.

The EMT is used in certain morphogenetic steps occurring during embryogenesis, when tissue remodeling depends on EMTs executed by various types of epithelial cells (Table 14.1). During one of the steps of **gastrulation**, for example, individual cells peel away from the ectoderm and migrate inward toward the center of the embryo to form the mesoderm, the precursor of mesenchymal tissues, including fibroblasts and hematopoietic cells (in chordates). This conversion of ectodermal cells, which at this stage are arrayed in an epithelial cell layer, to those having a mesodermal phenotype involves an EMT (**Figure 14.13A and B**). At the same time, the cells undergoing an EMT acquire the ability to translocate from one location (the outer cell layer) to another (the interior) within the embryo.

The migration of neuroepithelial cells from the neural crest into the mesenchyme of early vertebrate embryos also depends on a transformation of cell phenotype that can best be described as an EMT (see Figure 14.13C). Similarly, the migration of myogenic precursor cells (the progenitors of muscle cells) from the dermomyotome of the early embryo to the limb buds depends on an EMT-like transformation of cell phenotype. All of these processes bear a striking resemblance to the EMT undertaken by the cells at the edge of a wound within an epithelium; these cells must undergo a transient EMT in order to migrate into the wound site and close the gaps in the epithelial cell sheet that were created by the wounding process (see Figure 13.14).

Table 14.1 **Examples of EMTs during mouse embryonic development**

Process	Transition	
	From	To
Gastrulation	epiblast	mesoderm
Prevalvular mesenchyme in the heart	endothelium	atrial and ventricular septum
Neural crest cells	neural plate	neural crest cells, which can yield bone, muscle, peripheral nervous system
Somitogenesis	somite walls	sclerotome
Palate formation	oral epithelium	mesenchymal cells
Müllerian duct regression	Müllerian tract	mesenchymal cells

Adapted from P. Savagner, *BioEssays* 23:912–923, 2001.

(A)

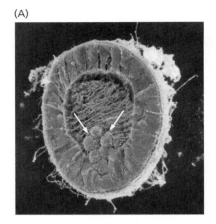

(C)

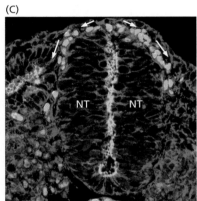

(B)

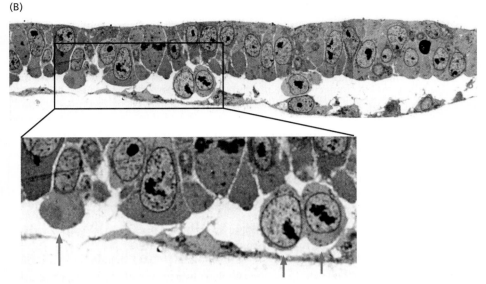

Figure 14.13 Embryogenesis and the epithelial–mesenchymal transition (A) This scanning electron micrograph shows the delamination of cells from the primitive ectoderm of a sea urchin embryo (*white arrows*) and their migration into the interior of the embryo— the process of gastrulation. These cells have become round and acquired motility in anticipation of forming the rudiments of the mesoderm—changes associated with an epithelial–mesenchymal transition (EMT). They are migrating along strands of extracellular matrix in the blastocoel of this early embryo. (B) The process of gastrulation in mammals, such as in the early rabbit embryo studied here, is superficially quite different but depends on very similar processes. As revealed by this transmission electron micrograph, one cell of the primitive ectoderm (also termed epiblast; *red arrow*) is in the midst of undergoing an EMT, while two other cells (*blue arrows*) have already delaminated from the epiblast and become mesenchymal, thereby initiating the formation of mesodermal tissues. (C) Certain cells derive from the upper region of the neural tube (NT, *center*) and can be immunostained for the Sox 9 transcription factor. These cells (*light blue, top*) arise in the embryonic neural crest of chordates and undergo an EMT, delaminate from this epithelium, acquire motility and invasiveness, and disperse (*arrows*) throughout the embryo, where they form melanocytes, much of the peripheral nervous systems, and much of the skeleton of the face. The neural epithelium is immunostained for N-cadherin (*yellow*). All cells are stained with phalloidin to reveal their actin cytoskeleton (*red*) and with DAPI to indicate their nuclei (*dark blue*). (A, courtesy of G. Cherr, from G.N. Cherr et al., *Microsc. Res. Tech.* 22:11–22, 1992. B, from C. Viebahn, B. Mayer and A. Miething, *Acta Anat. (Basel)* 154:99–110, 1995. C, courtesy of J. Briscoe; see also J. Yang and R.A. Weinberg, *Dev. Cell* 14:818–829, 2008.)

An EMT can also be seen at the edges of carcinomas that are invading adjacent tissues (**Figure 14.14**). This pathological process is very similar to the EMTs occurring during early embryogenesis and wound healing. It is plausible, though hardly proven, that all types of carcinoma cells must pass, at least partially, through an EMT in order to become motile and invasive.

The strong resemblance between the pathological process of tumor invasiveness and normal steps of embryogenesis and wound healing suggests a plausible mechanistic model: according to this model, which is supported by extensive evidence gathered in recent years, the complex program of cellular reorganization exhibited by invasive carcinoma cells depends on the reactivation of a latent behavioral program whose expression is usually confined to early embryogenesis and to damaged adult tissues. According to this thinking, once carcinoma cells acquire access to an EMT program, they can exploit it to profoundly change their own morphology, motility, and ability to invade nearby cell layers. This model implies that the multiple changes in cell phenotype associated with invasiveness, some of which are described below, need not be acquired piecemeal by carcinoma cells. Instead, these cells simply activate a morphogenetic program that is already encoded in their genomes. (This logic echoes our earlier discussion in Section 13.3 of epithelial–stromal interactions, where we argued that cancer cells co-opt entire wound-healing programs in order to acquire an activated stroma.)

The normal and pathological versions of the EMT involve, in addition to changes in shape and the acquisition of motility, fundamental alterations in the gene expression profiles of cells (**Table 14.2**). Expression of E-cadherin and cytokeratins—hallmarks of epithelial cell protein expression—is repressed, while the expression of **vimentin**, an intermediate filament component of the mesenchymal cell cytoskeleton, is induced (**Figure 14.15**). Epithelial cells that have undergone an EMT often begin to make fibronectin, an extracellular matrix protein that is normally secreted only by mesenchymal cells such as fibroblasts. At the same time, expression of a typical fibroblastic marker—N-cadherin—is often acquired in place of E-cadherin.

Of all these proteins, the transmembrane E-cadherin molecule plays the dominant role in influencing epithelial versus mesenchymal cell phenotypes. Recall our earlier encounters with E-cadherin and its role in enabling epithelial cells to adhere to one another (see Figures 6.26A and 13.12). In normal epithelia, the ectodomains of E-cadherin molecules extend from the plasma membrane of one epithelial cell to form complexes with other E-cadherin molecules protruding from the surface of an adjacent epithelial cell. This enables homodimeric (and higher-order) bridges to be built between adjacent cells in an epithelial cell layer, resulting in the adherens junctions that are so important to the structural integrity of epithelial cell sheets.

The cytoplasmic domains of individual E-cadherin molecules are tethered to the actin fibers of the cytoskeleton via a complex of α- and β-catenins (see Figure 14.14D) and other ancillary proteins. The actin cytoskeleton, for its part, provides tensile strength to the cell. Hence, by knitting together the actin cytoskeletons of adjacent cells, E-cadherin molecules help an epithelial cell sheet resist mechanical forces that might otherwise tear it apart. Once E-cadherin expression is suppressed, many of the other cell-physiologic changes associated with the EMT seem to follow suit. Some experiments indicate that simply by suppressing the expression of the E-cadherin protein, cells acquire a mesenchymal morphology and increased motility.

The pivotal role of E-cadherin in the acquisition of malignant cell phenotypes is further supported by observations indicating that the *CDH1* gene, which specifies E-cadherin, is repressed by promoter methylation in many types of invasive human carcinomas (see Table 7.2) and in others by certain transcriptional repressors; this gene can also be inactivated by reading-frame mutations. For example, an analysis of 26 human breast cancer cell lines indicated that 8 had mutations that led to inactivation of E-cadherin gene expression, 5 had truncating mutations in the E-cadherin reading frame, while 3 had in-frame deletions resulting in the expression of mutant E-cadherin molecules at the cell surface. By now, loss of E-cadherin expression or expression of mutant E-cadherin proteins has been documented in advanced carcinomas of the breast, colon, prostate, stomach, liver, esophagus, skin, kidney, and lung. And mutant germ-line alleles of the *CDH1* gene result in familial gastric cancer (see Table 7.1).

Additionally, in studies of several types of carcinoma cells that had lost E-cadherin expression, re-expression of this protein (achieved experimentally by introduction of an E-cadherin expression vector) strongly suppressed the invasiveness and metastatic dissemination of these cancer cells. Together, these diverse observations indicate that E-cadherin levels are key determinants of the biological behavior of epithelial cancer cells and that the cell-to-cell contacts constructed by E-cadherins impede invasiveness and hence metastasis.

The replacement of E-cadherin by N-cadherin during the EMT (see Figure 14.15B) is also seen during gastrulation in early embryogenesis. Moreover, hepatocyte growth factor (HGF) promotes the E- to N-cadherin switch in cultured epiblast cells (which derive from the embryonic ectodermal cell layer); in this way, it induces an EMT and enables emigration of muscle and dermal precursor cells from the primitive dermomyotome—the collection of epithelial-like cells located in the somites of early vertebrate embryos. In the context of cancer pathogenesis, HGF potently induces motile and invasive behavior in carcinoma cells.

Table 14.2 Cellular changes associated with an epithelial–mesenchymal transition

Loss of
Cytokeratin (intermediate filament) expression
Tight junctions and epithelial adherens junctions involving E-cadherin
Epithelial cell polarity
Epithelial gene expression program

Acquisition of
Fibroblast-like shape
Motility
Invasiveness
Increased resistance to apoptosis
Mesenchymal gene expression program including EMT-inducing transcription factors
Mesenchymal adherens junction protein (N-cadherin)
Protease secretion (MMP-2, MMP-9)
Vimentin (intermediate filament) expression
Fibronectin secretion
PDGF receptor expression
$\alpha_v\beta_6$ integrin expression
Stem cell-like traits

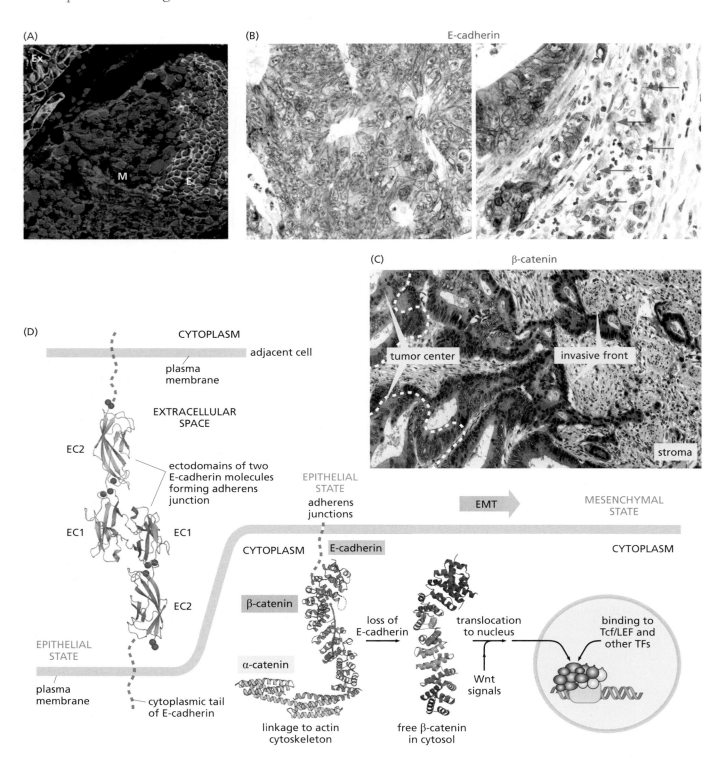

Like E-cadherin, the N-cadherin that is produced in its stead participates in homophilic interactions, that is, binds to other molecules of the same type displayed by nearby cells. Consequently, the N-cadherin molecules expressed on the surface of a carcinoma cell that has undergone an EMT increase the affinity of this cancer cell for the stromal cells that normally display N-cadherin, notably the fibroblasts in the stroma underlying the epithelial cell layer. This association seems to help invading carcinoma cells insert themselves amid stromal cell populations. Precisely the same dynamics have been proposed to explain how melanomas develop: normal melanocytes express E-cadherin, which binds them to the keratinocytes around them; melanoma cells—the transformed derivatives of melanocytes—express N-cadherin, which facilitates

Figure 14.14 Epithelial–mesenchymal transition at the invasive edge of a tumor
(A) In the Rip-Tag transgenic mouse model of pancreatic islet carcinogenesis (see Figure 13.36), neoplastic cells undergo an EMT that enables them to invade the surrounding exocrine pancreas (Ex). Carcinoma cells at the far right continue to reside in an epithelial state (E), exhibit E-cadherin staining (*green*), and in many places exhibit overlapping β-catenin staining (*red*). The overlap of these two proteins, indicating their co-localization in adherens junctions, appears as *yellow*. Those carcinoma cells that have undergone an EMT and become more mesenchymal (M) have lost E-cadherin staining and show greatly increased levels of cytoplasmic β-catenin staining. Nuclei are stained *blue* with DAPI. (B) Colon carcinoma cells at the invasive edge of a primary human tumor undergo changes in gene expression and the localization of certain proteins. While E-cadherin (*brown*) is strongly expressed on the plasma membranes of cells in the core of a primary tumor, where it forms adherens junctions (*left panel*) that delineate the associations between adjacent cells, its expression decreases substantially in individual invasive cells at the edge of this tumor (*pink arrows, right panel*) and is no longer localized to their plasma membranes, indicating that it no longer participates in forming adherens junctions. (C) At the same time, tumor cells in the center of this colon tumor (*dashed lines, left side*) express β-catenin (*dark red*) under their plasma membranes and diffusely throughout the cytoplasm, while tumor cells at the invasive edge (*right side*) show intense β-catenin staining in their nuclei. (D) The ectodomain of E-cadherin displayed by an epithelial cell dimerizes with the ectodomain of a second E-cadherin molecule displayed by an adjacent epithelial cell (*left*). Calcium ions, which are essential for the normal secondary structure and rigidity of the ectodomains, are shown as *red balls*. At the same time, the cytoplasmic tail of E-cadherin is linked via β-catenin and several other molecules to the actin cytoskeleton. β-Catenin also functions in the cytoplasm as a key intermediary in the Wnt signaling pathway (see Section 6.10). Loss of E-cadherin from the plasma membrane liberates β-catenin molecules, which may then accumulate in the cytoplasm and, under the influence of Wnt-initiated signals, migrate to the nucleus and associate with Tcf/LEF and other transcription factors, thereby inducing expression of genes orchestrating the EMT program. (A, from M. Herzig et al., *Oncogene* 26:2290–2298, 2007. B, courtesy of T. Brabletz. C, courtesy of T. Brabletz and T. Kirchner. D, from T.A. Graham et al., *Cell* 103:885–896, 2000, courtesy of H.J. Choi and W.I. Weis; and from E. Parisini et al., *J. Mol. Biol.* 373:401–411, 2007.)

Figure 14.15 Biochemical changes accompanying the EMT As discussed later, EMT programs can be induced by several pleiotropically acting transcription factors. Shown here are the effects of expressing the Twist transcription factor in MDCK (Maden–Darby canine kidney) cells, which are widely used to study epithelial cell biology. (A) These immunofluorescence analyses indicate that expression of epithelial markers, specifically E-cadherin, β-catenin, and γ-catenin, is repressed, while expression of mesenchymal markers, specifically vimentin and fibronectin, is induced by ectopic expression of Twist. Note that in epithelial cells that have not been induced to enter into an EMT, E-cadherin, β-catenin, and γ-catenin are located at the interfaces between adjacent cells, where they form adherens junctions. While functional studies indicate the activation of transcription by nuclear β-catenin in mesenchymal cells, this is usually difficult to observe by immunofluorescence because β-catenin accumulates only to low steady-state levels in the nucleus. (B) Immunoblots confirm the results of immunofluorescence, but in a more quantitative fashion. Lysates of control MDCK cells are analyzed in the *left channels*, while lysates of MDCK cells forced to express Twist are analyzed in the *right channels*. β-Actin, whose expression is unaffected by the EMT, is used here as a control to ensure that equal amounts of cell lysate have been analyzed in all cases. α-SMA, α-smooth muscle actin. (A and B, from J. Yang et al., *Cell* 117:927–939, 2004.)

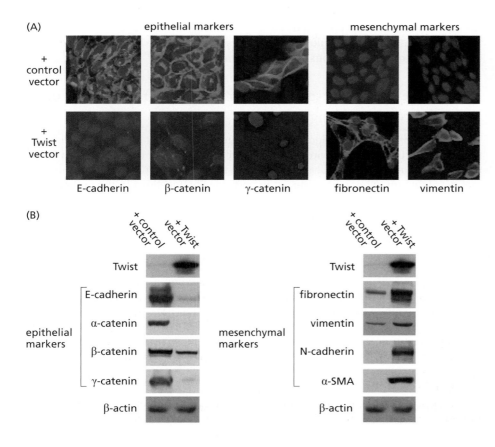

Figure 14.16 Cadherin shifts and melanoma cell invasiveness
Melanomas are among the most malignant tumors because of their tendency to metastasize widely once they reach a certain stage of tumor progression. This behavior is attributable, in part, to the reactivation of a cell-biological program that enabled the migratory behavior of their neural crest ancestors (see Figure 14.13C). The shift from E- to N-cadherin expression, which occurs when melanocytes (A) become transformed into melanoma cells (B), has been proposed to facilitate invasion of the stroma, since the shutdown of E-cadherin expression (*blue*) enables these cancer cells to extricate themselves from their keratinocyte neighbors in the epidermis, while its replacement by expression of N-cadherin (*reddish pink*) allows these tumor cells to form homotypic interactions with various types of mesenchymal cells, such as fibroblasts and endothelial cells, that reside in the stroma of the skin (i.e., the *dermis*). Of note, the EMT-inducing Slug transcription factor (see Figure 14.25) plays a key role in activating aggressive behavior in melanoma cells. (Adapted from N.K. Haass, K.S.M Smalley and M. Herlyn, *J. Mol. Histol.* 35:309–318, 2004.)

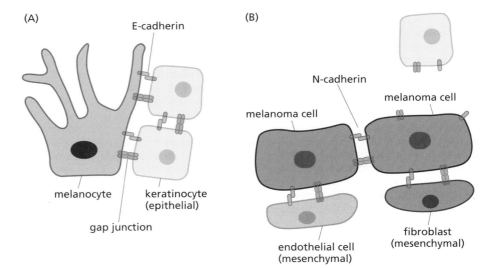

their invasion of the dermal stroma of the skin and their association with its fibroblasts and endothelial cells (**Figure 14.16**).

Importantly, the acquisition of N-cadherin expression does not result in the assembly of large sheets of cancer cells that might, in principle, be created by the formation of cell–cell N-cadherin bridges. The intermolecular bonds formed between pairs of N-cadherin molecules are weaker than those formed by E-cadherin homodimers. This helps to explain why cell surface N-cadherin molecules actively favor cell motility, and thus behave very differently from their E-cadherin cousins, which function to immobilize cells within epithelial cell layers. This point is driven home by experiments in which expression vectors are used to force high levels of N-cadherin expression in otherwise-normal cultured epithelial cells. Such ectopic expression causes the epithelial cells to acquire motility and invasiveness, as indicated by their ability to break through reconstructed extracellular matrix introduced into a culture dish. Finally, one aspect of E-cadherin function is counterintuitive: while adherens junctions seem to be highly stable, locking individual epithelial cells into sheets for extended periods of time, the individual E-cadherin molecules that form these junctions are highly dynamic, moving in and out of these junctions in less than a minute!

14.4 Epithelial–mesenchymal transitions are often induced by contextual signals

The notion that an EMT is induced by extracellular signals that carcinoma cells experience in the tumor microenvironment is implied by some of the discussions in the previous section. This idea flows, in part, from the properties of an EMT during normal development, where it is induced in very specific locations in the embryo (see Figure 14.13). Such behavior in embryos actually implies two things: First, the EMT is induced by signals that embryonic cells experience in certain locations of the embryo but not in others. Second, such precise localization is likely to depend on the combinatorial actions of distinct contextual signals that converge on individual cells from multiple sources, for example, different neighboring cells or groups of cells.

Studying the localization within a tumor of the carcinoma cells that have passed through an EMT reinforces this thinking. As seen in **Figure 14.17**, carcinoma cells in close contact with the surrounding stroma activate their EMT programs, while those residing in the interior of islands of carcinoma cells (and thus shielded from direct contact with stroma) fail to do so. Moreover, changes can be observed in multiple proteins whose expression is known to change upon passage of cells through an EMT, indicating that many distinct components of this complex cell-biological program are activated by the contextual signals experienced by these cells.

As described above, the EMT would seem to be an irreversible change that carcinoma cells acquire as they advance down the road toward a highly malignant growth state.

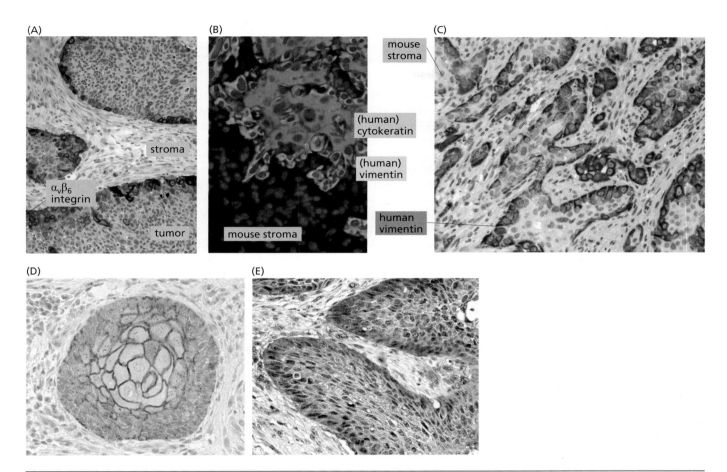

Figure 14.17 Manifestations of the EMT at the interface between tumor epithelium and stroma (A) The expression of the $\alpha_v\beta_6$ integrin is associated with the EMT. This integrin is expressed in epithelial tissues that are undergoing wound healing or suffering chronic inflammation; it is also seen at the invasive edge of carcinomas. In a xenografted tumor formed by SCC-14 human pharyngeal carcinoma cells, expression of the $\alpha_v\beta_6$ integrin is exhibited by carcinoma cells at the invasive edge of the tumor (*dark brown*) that are in direct contact with the tumor-associated stroma, suggesting that stromal signals are responsible for its expression in epithelial cells. (B) Experimentally transformed human mammary epithelial cells (MECs) were implanted in an immunocompromised mouse host. The cytokeratin-positive human carcinoma cells (*red*) toward the center of the tumor mass are not in direct contact with the surrounding mouse stromal cells, whose presence is indicated only by their DAPI-stained nuclei (*blue*). However, many of the human MECs that are in contact with the stroma have undergone an EMT, as indicated by their loss of cytokeratin staining and their display instead of human-specific vimentin (*green*). [The use of antibody that specifically recognizes human (and not mouse) vimentin ensures that the green cells at the invasive edge derive from the engrafted human cells rather than from the mouse host.] Moreover, some of these cancer cells at the invasive edge have lost the cuboidal shape of the epithelial cancer cells and have assumed, instead, a more elongated, fibroblastic shape. (C) A tumor formed by the same strain of transformed human MECs described in panel B is seen here at lower magnification. The display of human vimentin (*dark brown*), which indicates passage through an EMT, is limited to cells that are in direct contact with the surrounding stroma. Conversely, tumor cells in the interior of these islands do not display human vimentin and have presumably remained in an epithelial state. (D) In this human oral squamous cell carcinoma (OSCC), intense expression of p120 catenin (*light brown*), another component of the adherens junction, is limited to the center of this island of carcinoma cells, while cells closer to the surrounding stroma (*light blue nuclei*) have lost p120 catenin and thus the adherens junctions typical of epithelial cells. Cell nuclei are stained *blue* with hematoxylin. (E) In another OSCC, matrix metalloproteinase-2 (MMP-2, *brown*), which is up-regulated during passage through an EMT and deployed by some carcinoma cells to facilitate invasion, is also seen to be expressed preferentially near the stroma. (A, courtesy of D.R, Leone, B.M. Dolinski and S.M. Violette, Biogen Idec. B, courtesy of K. Hartwell and T.A. Ince. C, courtesy of T.A. Ince. D and E, from M. Vidal et al., *Am. J. Pathol.* 176:3007–3014, 2010.)

Actually, there are reasons to believe that during the development of many carcinomas, the EMT phenotype is acquired *reversibly*, and that once carcinoma cells have completed the multiple steps of invasion and metastasis, they often revert back to a more epithelial phenotype by passing through the mesenchymal–epithelial transition (MET) mentioned in the last chapter.

This reversion is suggested by repeated observations that the metastases deriving from a primary tumor often bear striking resemblance at the histopathological level to this tumor (**Figure 14.18A and B**). Hence, if the founders of these metastases underwent an EMT during initial dissemination (which would have striking effects on their histopathological appearance), these EMT-associated changes must have been lost during the formation of the metastasis.

At a mechanistic level, such loss can be explained by the changes in the microenvironment that cancer cells experience when they move from the primary tumor to a site of metastasis (see Figure 14.18C): In the primary tumor, a "reactive stroma" forms during the long course of primary tumor formation and releases signals inducing an EMT in nearby carcinoma cells. After leaving the primary tumor, additional EMT-inducing signals may be released by the platelets that adsorb to tumor cells in the lumina of vessels (see Supplementary Sidebar 14.2), perpetuating the mesenchymal state of these cells. However, following extravasation of carcinoma cells at sites of metastasis, the initially encountered stroma may be fully normal and therefore will not release signals that actively promote an EMT in the recently arrived cancer cells; this should permit these cells to revert via an MET to the epithelial state of their ancestors in the heart of the primary tumor. (This implies that in certain tumor cells the mesenchymal state resulting from an EMT must be actively maintained by stromal signals; if these signals are lost,

Figure 14.18 Reversibility of EMT While cells at the invasive edge of a primary carcinoma often give evidence of an EMT, derived metastases may exhibit a histology typical of the center of the primary tumor. (A) Release of degradative enzymes, notably matrix metalloproteinases (MMPs), is one of the many manifestations of the EMT (e.g., see Figure 14.17E). These cells in a primary colorectal carcinoma show expression of both cytokeratin 18 (*red*) and a basement membrane protein (*green*). However, at the invasive edge of this tumor, the cells have undergone a partial EMT, in that they have degraded the adjacent basement membranes while still expressing cytokeratin 18, a key epithelial marker. In a subsequently arising metastasis in this patient, which presumably descends from cells that underwent an EMT and acquired invasiveness en route to metastatic dissemination, the cancer cells form a growth having, once again, the histological appearance of cells in the heart of the primary tumor, suggesting reversibility of the EMT. (B) These observations can be explained by a model in which epithelial cancer cells at the edge of a primary carcinoma (*pink/brown cells*) undergo an EMT as they invade the stroma and become mesenchymal (*red cells*). This change seems to be triggered by signals that these carcinoma cells receive from the nearby tumor-associated reactive stroma (see Figure 14.17), which is composed of a variety of inflammatory cells that accumulate during the long course of primary tumor formation. The newly acquired mesenchymal state enables the carcinoma cells to invade locally, intravasate, and subsequently to extravasate into the parenchyma of a distant, ostensibly normal tissue. Following extravasation, however, these cancer cells find themselves in a fully normal stroma that lacks the inflammatory cells and therefore does not release EMT-inducing signals. This allows these cells and their descendants to lapse back to an epithelial phenotype (*pink/brown cells*) via a mesenchymal–epithelial transition (MET). (The regeneration of a basement membrane, which occurs as one consequence of an MET, is not illustrated here.) (C) These dynamics are supported by studies of disseminated cancer cells. Green fluorescent protein (GFP)–expressing cancer cells were prepared from mammary tumors of MMTV-PyMT transgenic mice in which the MMTV promoter drives expression of the polyoma middle T (*PyMT*) oncogene (see Sidebars 3.6 and 11.4); these mice develop spontaneously metastasizing mammary carcinomas at a high rate. These cell populations were injected into the tail veins of tumor-free host mice and the proportion of GFP+ cells relative to total cells in the lungs was followed over the indicated time intervals (*blue curve*). Initially, ~0.1% of all cells in the lungs were GFP+ carcinoma cells, which declined to ~0.03% a week later and later increased progressively as metastatic colonization and the growth of metastases proceeded. In contrast, the percentage of GFP+ tumor cells exhibiting a CD90+ CD24+ marker phenotype (representing carcinoma cells with a mesenchymal phenotype; *red curve*) was initially ~1.5%, reflecting the makeup of cells in the primary tumors from which these cells were prepared. This percentage increased almost 20-fold in the first week to over 20%; however, as the individual metastases expanded in size, this percentage decreased back to ~1.8%, almost precisely the level in the initially inoculated cells. Hence, mesenchymal-like carcinoma cells are favored initially to found and establish metastatic colonies, but as colonization proceeds, the representation of mesenchymal/stem-like cells decreases to that seen in the primary tumor. (A, courtesy of T. Brabletz. B, adapted from J.P. Thiery, *Nat. Rev. Cancer* 2:442–454, 2002. C, from I. Malanchi et al., *Nature* 481:85–89, 2012.)

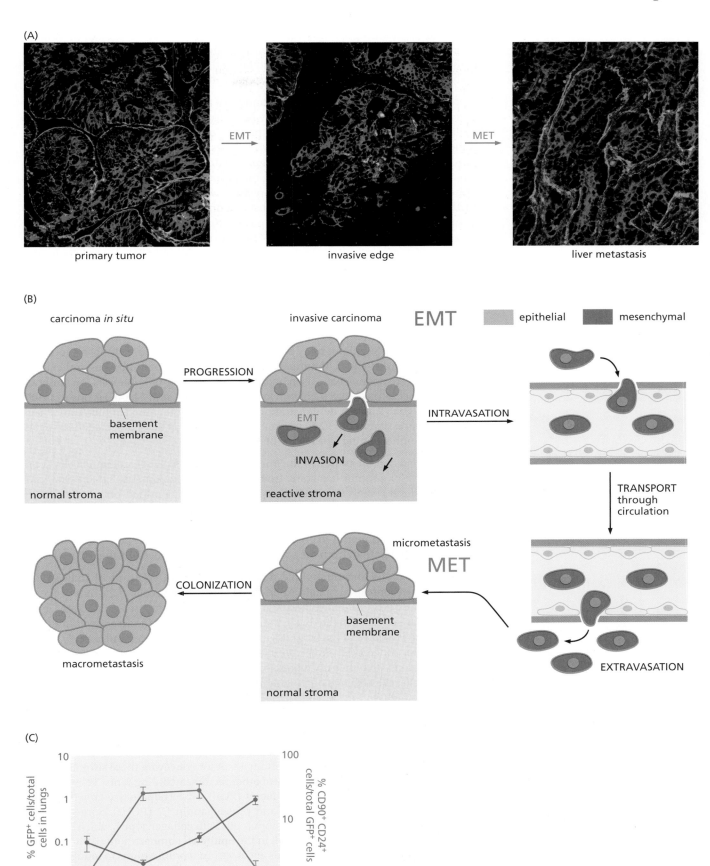

carcinoma cells will lapse back to an epithelial state.) If the resulting micrometastases succeed in colonizing distant tissues, then a reactive stroma may eventually be formed in the resulting macrometastasis, recapitulating the changes that occurred previously in the primary tumor and triggering once again an EMT in the cancer cells.

A prediction of this model is that the founders of micrometastases are initially highly enriched in mesenchymal cancer cells (which also have stem cell-like characteristics, as discussed below). However, with the passage of time, as the resulting micrometastatic colonies grow into macroscopic metastases, most of these mesenchymal/stem cell-like cells revert via an MET to the epithelial/non-stem-like state of their ancestors in the primary tumor, regenerating the proportions of these two classes of cells present in the primary tumor. This is precisely what has been observed (see Figure 14.18D). Of additional interest is the observation that tumors, including metastatic colonies, that are composed entirely of mesenchymal cancer cells (that is, those that have passed completely through an EMT) grow poorly; this would suggest that the presence of non-mesenchymal cancer cells, specifically epithelial cancer cells, is critical to the successful outgrowth of metastatic colonies.

In fact, traditional histopathological techniques have failed to demonstrate the EMT at the invasive edges of primary carcinomas for a simple reason: once tumor cells undergo a full EMT (that is, shed all epithelial traits and acquire mesenchymal ones instead), they are essentially indistinguishable from the mesenchymal cells in the surrounding stroma (Supplementary Sidebar 14.5). For this reason, demonstration of an EMT at the invasive edges of tumors has required the use of antibodies and cellular reagents that are not normally used in diagnostic pathology laboratories. An antibody that can detect intracellular β-catenin localization (see Figure 14.14C) is one example of such a reagent. Another demonstration comes from xenografted human tumor cells in which expression of the EMT marker $\alpha_v\beta_6$ integrin has been detected, once again through immunostaining (see Figure 14.17A).

The most vivid demonstration of the phenotypic conversion of carcinoma cells at the invasive edge of a tumor has come from the use of a human-specific anti-vimentin antibody, which reveals the EMT of experimentally transformed human mammary epithelial cells growing as a xenograft in an immunocompromised mouse host (see Figure 14.17B and C). The use of a human-specific antibody guarantees, in this instance, that the vimentin-expressing cells derive from keratin-positive human carcinoma cells rather than from the surrounding stroma produced by the mouse host.

These observations and others like them (see Figure 14.17D and E) indicate the involvement of contextual signals experienced by carcinoma cells at the outer edges of islands of these cells. Actually, these signals may be of two types: Certain heterotypic signals that are released in the reactive stroma of primary carcinomas may impinge on neoplastic cells located at the outer edges of epithelial cell masses and actively induce these cells to undergo an EMT. Alternatively, epithelial cells at the edges of these islands may sense that they no longer enjoy interactions on all sides with neighboring epithelial cells; the absence of a full complement of epithelial–epithelial homotypic interactions might induce them to activate an EMT program. Indeed, both processes may conspire to activate an EMT program.

To date, far more evidence has accumulated pointing to active heterotypic signaling between the stromal and epithelial compartments within carcinomas. Abundant evidence indicates that TGF-β is an important agent for conveying these stromal signals. Other observations implicate a variety of other factors, including Wnts, TNF-α (tumor necrosis factor-α), epidermal growth factor (EGF), HGF (hepatocyte growth factor), IGF-1 (insulin-like growth factor-1), and prostaglandin E2 (PGE2). Some research even implicates direct contact of carcinoma cells with collagen type I, which is present in abundance in the stroma but absent in the epithelial compartment of the tumor. It appears likely that these stromal signals act in various combinations to induce epithelial cells to activate their previously latent EMT programs.

Compounding these effects, the genetic and epigenetic alterations acquired by carcinoma cells during primary tumor formation are likely to increase their *responsiveness* to these various contextual signals and thus influence the way that EMT programs alter carcinoma cell phenotypes. For example, some experiments indicate that cancer cells that have lost p53 function are more responsive to EMT-inducing signals.

In one set of influential experiments, exposure of *ras*-transformed EpRas mouse mammary epithelial cells (MECs) to TGF-β resulted in the progressive loss of epithelial morphology and a reduction of epithelial markers, including cytokeratins and E-cadherin. At the same time, these transformed cells acquired mesenchymal protein markers, such as vimentin, and assumed a morphology resembling that of fibroblasts—all the hallmarks of an EMT. Provocatively, once these *ras*-transformed cells underwent an EMT, they began to produce their own TGF-β1; this TGF-β1, acting via an autocrine signaling loop, allowed them to maintain their mesenchymal phenotype for extended periods of time, long after the inciting TGF-β was withdrawn from their culture medium (**Figure 14.19**). These studies suggested that TGF-β signaling

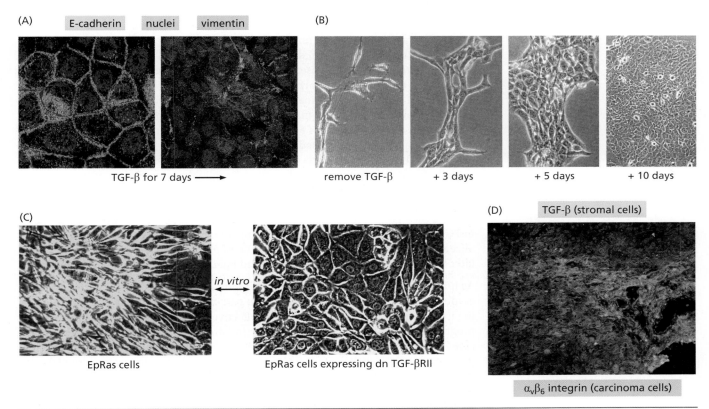

Figure 14.19 Control of the EMT by TGF-β and its effects on tumorigenic cells Immortalized mouse mammary epithelial cells (MECs) of the EpH4 cell line were transformed into EpRas tumor cells by introducing a *ras* oncogene. (A) EpRas cells (*left*) usually have an epithelial, cobblestone-like appearance and express E-cadherin (*green*) at their cell–cell junctions. However, when cultured for 7 days in the presence of TGF-β1 (*right*), they undergo an EMT and assume an elongated fibroblastic appearance (not visible here). In addition, they suppress expression of E-cadherin and express instead vimentin (*red*), thereby shifting from an epithelial to a mesenchymal gene expression program. Nuclei are stained *blue* with DAPI dye. (B) Thereafter, EpRas cells maintain the EMT-induced mesenchymal, fibroblast-like state through their own production of and response to TGF-β1 (i.e., via autocrine signaling; *not shown*). However, as seen here, when these cells are cultured in a medium that lacks added TGF-β1 and their growth medium is changed on a daily basis (to remove any TGF-β that they may have secreted into the medium), their appearance gradually reverts to an epithelial cobblestone phenotype, as shown after 3, 5, and 10 days of culture (*left to right*), indicating that they have undergone a mesenchymal–epithelial transition (MET). (C) Use of a dominant-negative (dn) type II TGF-β receptor (which effectively blocks autocrine TGF-β signaling) provides further proof that autocrine TGF-β signaling by EpRas cells is required to maintain their residence in a mesenchymal state. When this signaling is blocked by expression of this receptor, the mesenchymal appearance of the EpRas cells (*left*) is replaced by an epithelial appearance (*right*), indicating that they have undergone an MET. (D) Detroit 562 human pharyngeal carcinoma cells growing in a tumor express the $\alpha_v\beta_6$ integrin (*red*; see Figure 14.17A), indicative of their having undergone an EMT, while TGF-β (*green*) is produced by cells in the nearby tumor-associated stroma. The $\alpha_v\beta_6$ displayed by the tumor cells can be deployed to activate the latent form of TGF-β produced by the stromal cells, thereby creating a self-sustaining, positive-feedback loop (since the activated TGF-β is a strong inducer of the EMT and thus of additional expression of $\alpha_v\beta_6$ integrin by the carcinoma cells, yielding even more activated TGF-β). (A, from E. Janda et al., *J. Cell Biol.* 156:299–314, 2002. B, from M. Oft et al., *Genes Dev.* 10:2462–2477, 1996. C and D, courtesy of D.R. Leone, B.M. Dolinski and S.M. Violette, Biogen Idec.)

can conspire with a *ras* oncogene to cause epithelial cancer cells to undergo an EMT. Similarly, maintenance of TGF-β signaling through a positive-feedback loop may play an important role in maintaining $\alpha_v\beta_6$ integrin expression and the EMT in human carcinoma cells (see Figure 14.19D). Figure 14.19D also reveals that TGF-β may often be produced in abundance by the tumor-associated stroma.

The prominent role of TGF-β in actively promoting the aggressiveness of malignant cancer cells contrasts starkly with our earlier discussions of its anti-proliferative effects (Supplementary Sidebar 14.6; see also Section 8.10). Strong support that TGF-β can favor malignant cell behavior is provided by numerous studies in which the levels of tumor-associated TGF-β (often TGF-β1) were found to rise in parallel with increasing degrees of tumor invasiveness and general aggressiveness. Indeed, high levels of TGF-β, both in the tumor mass and in the general circulation, augur poorly for the long-term survival of the cancer patient.

To summarize these and other observations, TGF-β can contribute to cancer cell invasiveness for at least four reasons. First, most human carcinomas arising outside the intestine retain at least some functional TGF-β receptor signaling, allowing them to continue to respond to TGF-β during the entire course of tumor progression. (This contrasts with the situation observed in a subset of colon carcinomas in which receptor function may be lost entirely through mutations in the receptor-encoding genes; see Section 12.9.) Second, the inactivation of the pRb pathway, which occurs in most if not all human cancers, causes malignant cells to lose their responsiveness to the cytostatic effects of TGF-β; this loss enables these cells to respond to other types of downstream signals that are released by ligand-activated TGF-β receptors. Third, in the absence of the cytostatic effects of TGF-β, exposure of cancer cells to this factor may actually *favor* their proliferation. For example, glioblastoma and osteosarcoma cells that are treated with TGF-β respond by producing and secreting PDGF; once released, the latter acts in an autocrine fashion to stimulate the proliferation of these cancer cells via the PDGF receptors that they display. Finally, exposure of breast cancer cells to TGF-β causes them to release other factors that accelerate the breakdown of mineralized bone—a critical step in the formation of *osteolytic* metastases; as we will learn later, this breakdown liberates additional mitogens that drive cancer cell proliferation.

As implied above, several lines of research indicate that contextual signals can act in combination to activate EMT programs in carcinoma cells. For example, TNF-α acting in concert with TGF-β appears to be effective in inducing an EMT. Early in tumor progression, TNF-α is often produced by inflammatory cells, such as macrophages (see Section 11.16); it can then function via its receptor to activate the NF-κB signaling pathway in epithelial cells. TGF-β also activates the NF-κB pathway in epithelial cells, such as the immortalized mouse mammary epithelial cells discussed above. In various tumors, TNF-α and TGF-β may contribute to the long-term maintenance of active NF-κB signaling. This signaling seems to be critical for the induction and maintenance of an EMT, since inhibition of NF-κB signaling prevents expression of the EMT program.

Experiments reported in 2011 extended this notion of collaborating EMT-inducing signals to include both normal and neoplastic mammary epithelial cells (**Figure 14.20A–D**). Study of the growth factors that were secreted by cells that spontaneously entered the mesenchymal state revealed three classes of signaling proteins that appear to play key roles in actively maintaining residence in this state: canonical Wnt proteins (see Section 6.10), non-canonical Wnt proteins (see Supplementary Sidebar 6.4), and TGF-β (see Section 6.12). Ongoing secretion and resulting autocrine stimulation by these factors was essential for continued maintenance of the mesenchymal phenotype. Accordingly, if these autocrine signaling loops were disrupted, cells would revert to an epithelial state, that is, pass through an MET.

Interestingly, TGF-β and canonical Wnt proteins were found to be secreted by epithelial cells as well. However, the ability of these cells to activate their own EMT program (via autocrine signaling) was blocked by the concomitant secretion of high levels of inhibitory proteins, specifically bone morphogenetic proteins (BMPs; inhibitors of TGF-β signaling), as well as SFRP1 and DKK1 (inhibitors of canonical and non-canonical Wnt signaling; see Figure 14.20A and B); indeed, these same inhibitors of autocrine signaling, when applied to mesenchymal cells, forced mesenchymal cells to revert via an MET to the epithelial state. These observations also indicated that initial triggering of an EMT program can depend on the same signaling proteins (ostensibly supplied in a *paracrine* fashion by nearby stromal cells) as were required subsequently to maintain residence in the mesenchymal state (through *autocrine* signaling; see Figure 14.20C and D). Presumably EMT-initiating stromal signals are able to neutralize and overwhelm the secreted inhibitors that normally operate to prevent epithelial cells from activating their own EMT programs. Yet another autocrine and paracrine signaling channel of relevance to EMT induction involves prostaglandin E_2 (PGE_2; see Figure 14.20E).

14.5 Stromal cells contribute to the induction of invasiveness

The diversity of the EMT-inducing signals described above draws attention to the stromal cells that are responsible for encouraging aggressive behavior. Macrophages of the stroma were already cited above. Their influence on the invasive and metastatic behavior of primary cancer cells can be demonstrated by studying genetically altered mice that lack the ability to make colony-stimulating factor-1 (CSF-1). As was discussed in the last chapter, mammary carcinomas arising in cancer-prone transgenic mice usually recruit large numbers of tumor-associated macrophages (TAMs). However, when the tumor cells in such mice lack the ability to make CSF-1, TAMs are virtually absent (see Figure 13.24). The absence of CSF-1 and TAMs has no effect on primary tumor growth (**Figure 14.21A**), but such tumors show a benign, noninvasive behavior, in contrast to tumors that succeed in recruiting TAMs (Figure 14.21B). The influence of these macrophages on metastatic behavior is striking: without TAMs, these breast tumors fail to seed metastases to the lungs (Figure 14.21C).

This experiment provides compelling evidence that the invasive and metastatic behavior of these mouse breast carcinoma cells is strongly influenced by signals from stromal cells, in this case macrophages. It fails, however, to reveal the precise nature of these signals. Macrophage-derived TNF-α, as argued earlier, is likely to contribute to induction of the EMT by cancer cells, and therefore to the invasive and metastatic behavior described in Figure 14.21. Another key macrophage-derived signal is likely to be conveyed by EGF, as mentioned earlier in the context of macrophage-stimulated intravasation (see Figure 14.7).

Some of the evidence favoring EGF as a key inducer of cancer cell invasiveness comes from studies of mouse breast cancer cells both *in vivo* and *in vitro*. Like most epithelial cells, these carcinoma cells express the EGF receptor, and activation of this receptor by EGF causes them to acquire both motility and invasiveness and to secrete CSF-1, the attractant and stimulant of macrophages (**Figure 14.22**). Macrophages respond to CSF-1 by proliferating and releasing EGF, which activates the cancer cells. These effects all proceed through paracrine rather than autocrine signaling, since the breast cancer cells do not express the CSF-1 receptor and the macrophages do not express the EGF receptor (see Figure 14.22A). Hence, these two cell types collaborate by reciprocally stimulating one another, yielding another type of positive-feedback loop.

Importantly, the behaviors depicted in Figure 14.22B and in Figure 14.7 suggest that macrophages actually play critical roles in two stages of the invasion–metastasis cascade—initial invasiveness and intravasation. While cancer cell motility and invasiveness are clearly demonstrated in such experiments, the induction of an EMT in the cancer cells can only be inferred from their acquisition of motile, invasive behavior. More direct evidence comes, however, from the Rip-Tag tumor model, in which the secretion by recruited macrophages of cathepsin B protease is correlated with the loss—ostensibly through proteolytic degradation—of E-cadherin expression from the cell surface of carcinoma cells; such loss, on its own, can serve as a trigger for initiating an EMT (see Figure 14.22D).

HGF, another ligand of stromal origin, is also capable of inducing many of the attributes of an EMT in epithelial cells, which generally display Met, its cognate receptor, on their surface (**Figure 14.23**). Like EGF, HGF seems capable of inducing some but apparently not all of the phenotypic changes associated with passage through an EMT.

These diverse lines of evidence strongly suggest that the acquisition of malignant traits by cancer cells, including induction of an EMT, is not governed solely by the genomes of these cells. Instead, these profound shifts in cell phenotype are often initiated by a collaboration between specific mutant alleles harbored in cancer cell genomes (for example, mutant *ras* or *p53* genes) and the signals that these cancer cells receive in some tissue microenvironments, specifically at the boundaries between tumor epithelia and reactive stroma. In many tumors, these contextual signals are conveyed by certain factors, such as TGF-β, Wnts, PGE$_2$, and TNF-α, that are released by cells in the

Figure 14.20 Multiple signals collaborate to induce and then maintain the mesenchymal state Immortalized HMLE human mammary epithelial cells (parental cells) entered into the mesenchymal state either spontaneously or by the forced expression of the Twist EMT-TF (EMT-inducing transcription factor). The parental cells secreted significant amounts of TGF-β1, canonical Wnts, and non-canonical Wnts. (A) However, the secreted Wnts were unable to activate autocrine signaling because these epithelial cells release significant amounts of the DKK1 and SFRP1 inhibitors of Wnt signaling (*green*). When cells moved from the epithelial to the mesenchymal state (*orange, red*), the levels of these two Wnt inhibitors decreased greatly. (B) TGF-β autocrine loops were similarly blocked in the parental epithelial cells by the secretion of an array of BMPs (bone morphogenetic proteins) that function to blunt TGF-β signaling. Entrance into the mesenchymal state resulted in strong decreases in the expression of the mRNAs encoding these TGF-β inhibitors. Ongoing autocrine signaling was required to maintain expression of a variety of mesenchymal phenotypes, since their expression could be reduced in the mesenchymal cells through the addition of the SFRP1 and DKK1 recombinant proteins to the culture medium (not shown). (C) Entrance into the mesenchymal state can be triggered in mammary epithelial cells in culture by disrupting adherens junctions with an anti-E-cadherin antibody (which liberates β-catenin), the shutdown of SFRP1 mRNA function (through an shRNA), the addition of a DKK1-neutralizing antibody, the addition of large amounts of TGF-β1, and the addition of Wnt5a (a non-canonical Wnt) to the growth medium. Passage through an EMT and entrance into the mesenchymal state were gauged here by measuring levels of mRNAs encoding N-cadherin and two key EMT-TFs—Zeb1 and Zeb2. "+" denotes the addition of a component to this EMT-inducing cocktail, while absence of a component is indicated by a *blank square*. If the epithelial cells were exposed to this induction cocktail for more than a week, they would maintain their mesenchymal phenotype indefinitely in the absence of further such treatment (*not shown*). (D) These observations among others suggest that residence in the mesenchymal state is maintained following passage through an EMT by the actions of these three autocrine signaling loops (*right*). In the epithelial state, however (*left*), while cells secrete TGF-β and canonical Wnts, the ability of these factors to function in an autocrine manner is blocked by the concomitant secretion of inhibitors—BMPs, DKK1, and SFRP1. Hence, when cells undergo an EMT, they shut down production of BMPs and the Wnt inhibitors, thereby creating a permissive extracellular environment for the firing of autocrine signaling loops; in addition, upon passing through an EMT, such cells express greatly elevated levels of non-canonical Wnt proteins, which also operate thereafter in an autocrine fashion. The residence of cells in the resulting mesenchymal state is maintained in a metastable fashion, since interrupting the three autocrine loops forces cells to revert to an epithelial state. It is not known whether similar dynamics apply to epithelial cells from other tissues. (E) The stromal cells in inflamed tissues and in many tumors release prostaglandin E$_2$ (PGE$_2$), which can induce an EMT in epithelial cells and thereby may complement or reinforce the stromal signals released by Wnts and TGF-β. In this non-small-cell lung carcinoma (NSCLC), the mutually exclusive expression of the E-cadherin epithelial marker (*red*) and COX-2, the enzyme that produces PGE$_2$ (*dark brown*), is apparent at high magnification. COX-2–positive carcinoma cells tended to be discohesive, i.e., in the process of separating from other carcinoma cells. Other work (*not shown*) indicates that expression of E-cadherin and the Zeb1 EMT-inducing transcription factor was mutually exclusive in these tumors. (A–C, from C. Scheel et al., *Cell* 145:926–940, 2011. E, from M. Dohadwala et al., *Cancer Res.* 66:5338–5345, 2006.)

reactive stroma (**Figure 14.24**). Yet other stromal signals, such as those carried by EGF and HGF, may also help to elicit many of the changes that we associate with cancer cell invasiveness and the EMT. While this scenario depicts the behavior of many carcinomas, it does not accurately describe all of them (Supplementary Sidebar 14.7).

Throughout much of this text, our thinking has been driven by the notion that the phenotypes of cancer cells are dictated by their genotype and that tumorigenic growth is essentially a cell-autonomous phenomenon. Our encounters with heterotypic interactions (see Chapter 13) revised this notion slightly, by indicating that cancer cells show a surprising degree of dependence on normal neighbors for various types of sustenance and support. Now, we must come to terms with the idea that the microenvironment of the cancer cell can also fundamentally reshape that cell's phenotype, specifically by inducing the profound changes in cell behavior that comprise an EMT.

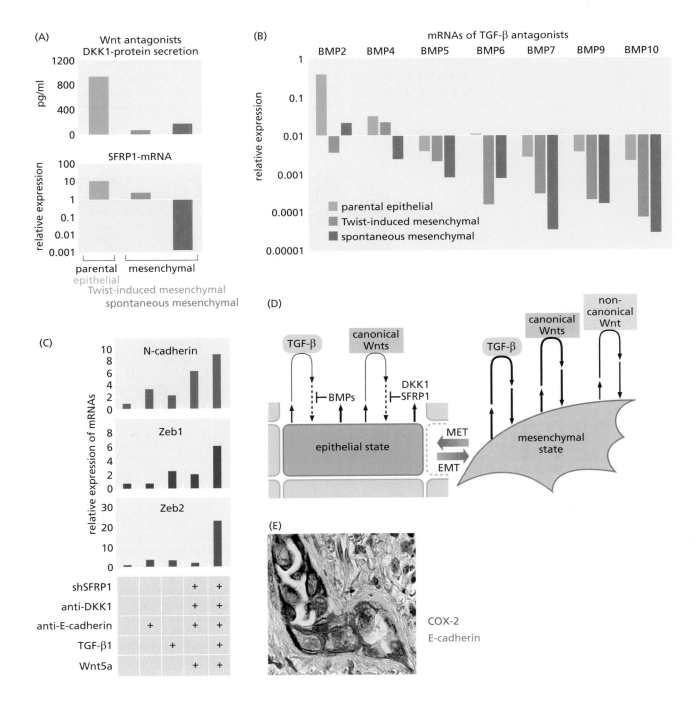

14.6   EMTs are programmed by transcription factors that orchestrate key steps of embryogenesis

Execution of an EMT program depends on changes in the expression of hundreds of distinct genes. These changes affect many aspects of cell biology, not all of which have been enumerated here. The changes include the organization of a cell's intermediate filament cytoskeleton, its motility, its sensitivity to apoptosis, its association with neighboring cells, its release of proteases, and even its display of cell surface integrins and growth factor receptors (see Table 14.2). While extensive evidence implicates stromal signals as key elements in triggering the EMT of carcinoma cells, none of this evidence, on its own, reveals how the complex EMT program is actually coordinated *within* the responding epithelial cells.

The genetics of the early development of a variety of experimental organisms has provided many of the answers to this question. Like many complex cell-biological programs, EMTs are orchestrated by a small number of pleiotropically acting transcription factors (TFs). A number of genes specifying these EMT-inducing TFs (EMT-TFs) have been identified, many initially in the embryos of the fruit fly, *Drosophila melanogaster*. These genes and the TFs that they encode are conserved in chordates and have been found to control key steps in early embryogenesis in frog and mouse embryos; these steps involve various versions of the EMT program. (The strong conservation of these EMT-TFs indicates that the EMT and key steps of early embryogenesis were developed

(A)

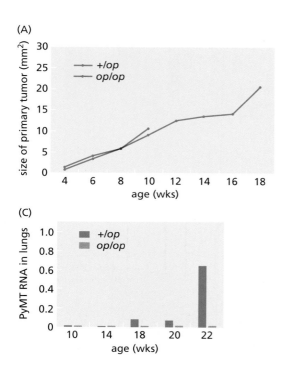

(B)

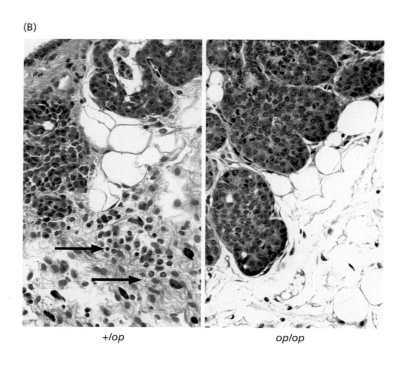

+/op   op/op

Figure 14.21 Effects of macrophages on invasion and metastasis The *PyMT* transgene (see Figure 14.18B) has been introduced, through breeding, into mice that can (*Csf*+/op) or cannot (*Csf*op/op) make colony-stimulating factor-1 (CSF-1), which is needed to recruit macrophages into tumors (see Figure 13.24). (A) The presence (in *Csf*+/op mice) or absence (in *Csf*op/op mice) of recruited tumor-associated macrophages (TAMs) has no effect on the ability of the primary breast tumors to grow in these transgenic mice. (B) Such mammary tumors arising in *Csf*+/op mice (whose tumors contain many TAMs, *not shown*) develop a highly invasive phenotype, in which individual carcinoma cells invade the nearby stroma in large numbers (*left panel, arrows*). However, if tumors develop in *Csf*op/op mice, in which macrophages cannot be recruited into the tumor-associated stroma (*right panel*), the tumor cells do not break through the basement membrane and the tumor as a whole remains encapsulated and benign, indicating that the macrophages contribute in essential ways to tumor invasiveness. (C) In the *Csf*+/op mice, metastases in the lungs begin to appear at 18 weeks of age and increase progressively thereafter (*blue bars*), as gauged by the amount of polyomavirus middle T RNA (expressed in tumor cells) present in the lungs (*ordinate*). However, in the *Csf*op/op TAM-negative mice (*orange bars*), metastases are virtually absent. (From E.Y. Lin et al., *J. Exp. Med.* 193:727–740, 2001.)

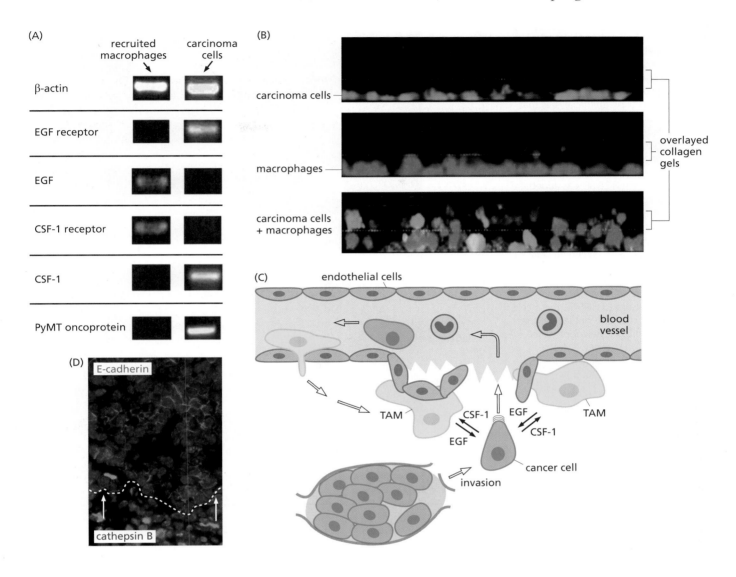

Figure 14.22 Reciprocal stimulation by breast cancer cells and macrophages A variety of experiments indicate that macrophages are the major source of EGF in breast cancers. EGF is known to be able to stimulate epithelial cancer cells to proliferate and invade through extracellular matrix. In addition, EGF exposure causes breast cancer cells to release CSF-1, which allows them to recruit macrophages and stimulates production by the macrophages of more EGF, resulting in a positive-feedback loop between these two cell types. (A) Using PCR analysis, the mRNA levels of these two growth factors and their receptors are found to be reciprocally expressed in mammary carcinoma cells arising in PyMT tumor-prone transgenic mice (see Figure 14.18B) and in recruited stromal macrophages. (B) Breast cancer cells (labeled here with green fluorescent protein, GFP; *green*) were placed at the bottom of a Petri dish (*seen here in side view*) below a layer of collagen gel (*top image*), where they remained. Similarly, macrophages (*red*) also remained where they were initially placed at the bottom of the Petri dish below a collagen gel (*middle image*). However, when the two populations were co-cultured at this location, the breast cancer cells (*green*) were now induced to move upward and invade the overlying collagen gel (*lower image*). (C) The reciprocal interactions between breast cancer cells and macrophages are illustrated schematically here. Because macrophages are often found in close proximity to microvessels, the stimulation by tumor-associated macrophages (TAMs) of breast cancer cell motility and invasiveness may also contribute to cancer cell intravasation, as suggested by the observations of Figure 14.7 and depicted here in more mechanistic detail. (D) The possibility that macrophages can contribute directly to triggering an EMT program in carcinoma cells is indicated by studies of macrophages that have been recruited to the invasive edge (*dashed line, arrows*) of a pancreatic islet cell tumor (*above*) in the Rip-Tag transgenic mouse model (see Figure 13.36). As seen here, the cathepsin B secreted protease (*green*) produced by the recruited macrophages causes the carcinoma cells to lose expression of E-cadherin (*red*). The resulting carcinoma cells (*below dashed line*) are revealed here only by their DAPI-stained nuclei (*blue*). (Inactivation of cathepsin B production by the macrophages permitted the carcinoma cells to retain their E-cadherin expression and caused them to lose invasiveness; *not shown*.) (A, from J. Wyckoff et al., *Cancer Res.* 64:7022–7029, 2004. B, from S. Goswami et al., *Cancer Res.* 65:5278–5283, 2005. C, from W. Wang et al., *Trends Cell Biol.* 15:138–145, 2005. D, courtesy of V. Gocheva and J.A. Joyce; see also V. Gocheva et al., *Genes Dev.* 20:543–556, 2006.)

Figure 14.23 Cell scattering and invasive behavior induced by HGF Hepatocyte growth factor (HGF), also known as scatter factor (SF), is produced by a variety of stromal cell types. It has profound effects on epithelial cells that display its cognate receptor, Met. (A) The MDCK (Maden–Darby canine kidney) cells (*red cytoplasms*) in the *left panel* were grown in a normal medium, while those in the *right panel* were grown in medium to which HGF/SF was added. In monolayer culture, these epithelial cells normally form clusters of cobblestone-like cells. However, after treatment by HGF/SF, they become motile and scatter in many directions. (B) When introduced into collagen gels in 3-dimensional culture, MDCK cells normally form small spherical clumps (*left panel*). However, following exposure to HGF/SF, these cells grow in long processes that invade the surrounding collagen gel (*right panel*). (A, courtesy of J.H. Resau. B, from M. Rosario and W. Birchmeier, *Trends Cell Biol.* 13:328–335, 2003.)

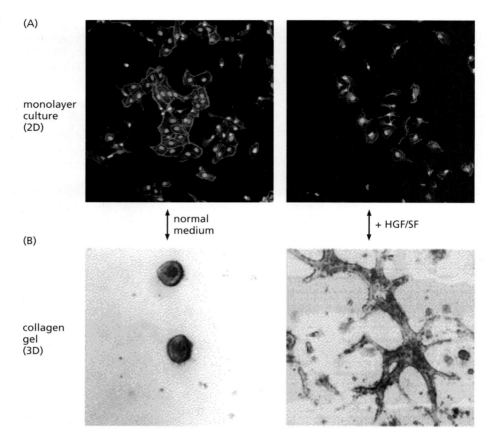

Figure 14.24 Signals that trigger an EMT This diagram presents a highly simplified view of the signaling channels that originate in the stroma and influence epithelial cancer cells to undergo a partial or complete EMT. It is likely that an EMT program is usually triggered in response to a confluence of signals that carcinoma cells receive from the stroma together with intracellular signals, such as those released by a *ras* oncogene, as indicated here. The precise identities of these stromal signals and their combinatorial mechanisms of action remain to be elucidated. In the longer term, the interactions depicted in this diagram must be integrated with those indicated in Figure 14.20D and E, an undertaking that is impossible at present, given the dearth of available information.

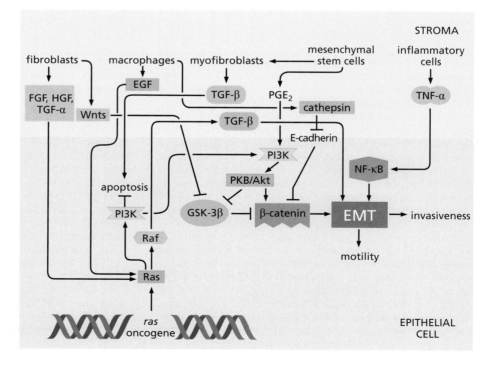

early in metazoan evolution, long before the radiation of the various metazoan phyla.) By activating these TFs, cancer cells gain access to the complex, multi-component EMT programs that they orchestrate. Implicit in what follows is the following scheme: heterotypic signals from the stroma induce expression of these TFs in carcinoma cells; these TFs, in turn, orchestrate EMT programs.

More than half a dozen EMT-TFs have been described, each capable of inducing an EMT when ectopically expressed in certain epithelial cells (Table 14.3). For example, Snail is a TF that was first described in *Drosophila* (Figure 14.25) and has since been discovered in a wide range of metazoa, including vertebrates, insects, worms, and mollusks. In early vertebrate embryos, Snail is first expressed in the portion of the ectoderm that is destined to become mesoderm following gastrulation. During embryogenesis, Snail, Slug, and Twist convert epithelial cells into the migratory mesenchymal cells that form the mesoderm. Snail and its relative Slug are involved in yet other embryonic steps in which one type of tissue is transformed into another. The truly ancient origins of these TFs is illustrated dramatically by Figure 14.25H, in which the embryo of a very primitive organism—a sea anemone, which lacks bilateral symmetry and three germ layers—expresses the Snail TF at the site of its future embryonic blastopore.

Some of these TFs are expressed in adult tissues during the tissue remodeling that underlies wound healing. This is illustrated nicely by the behavior of confluent monolayers of normal epithelial cells in culture that are wounded experimentally by scraping away a swath of these cells. Slug expression is induced in the surviving epithelial cells at the edge of the wound in order to enable these cells to acquire motility and migrate into the wound site (Figure 14.26). This helps to explain how epithelial cells at the edges of wounds undergo a transient EMT in order to reassemble epithelial cell sheets (see Figure 13.14). Observations like these broaden our perspective on the normal biological roles of these EMT-TFs: in addition to programming key steps in early embryogenesis, the expression of some of these TFs may be resurrected transiently in adults in order to reconstruct damaged tissues.

Snail and Slug (sometimes called SNAI1 and SNAI2) are members of the C2H2-type zinc finger TFs. The Snail–Slug TFs seem to operate largely as repressors of transcription.

Table 14.3 Transcription factors orchestrating an EMT

Name	Where first identified	Type of transcription factor	Cancer association
Snail (SNAI1)	mesoderm induction in *Drosophila*; neural crest migration in vertebrates	C2H2-type zinc finger	invasive ductal carcinoma
Slug (SNAI2)	delamination of the neural crest and early mesoderm in chicken	C2H2-type zinc finger	breast cancer cell lines, melanoma
Twist	mesoderm induction in *Drosophila*; emigration from neural crest	bHLH	various carcinomas, high-grade melanoma, neuroblastoma
Goosecoid	gastrulation in frog	paired homeodomain	various carcinomas
FOXC2	mesenchyme formation	winged helix/forkhead	basal-like breast cancer
ZEB1 (δEF1)	postgastrulation mesodermal tissue formation	2-handed zinc finger/homeodomain	wide variety of cancers
ZEB2 (SIP1)	neurogenesis	2-handed zinc finger/homeodomain	ovarian, breast, liver carcinomas
E12/E47 (Tcf3)[a]	associated with E-cadherin promoter	bHLH	gastric cancer

[a]It remains unclear whether E12/E47 can function on its own to induce an EMT, or whether this bHLH functions as a subunit of a heterodimeric TF complex formed with other well-validated EMT-TF proteins such as Twist.

Thus, both have been found to be able to repress transcription of the E-cadherin gene. As we read earlier, the loss of E-cadherin expression can, on its own, cause epithelial cells to assume many of the phenotypic changes associated with an EMT.

The Snail TF has been found to be expressed in the invasive fronts of chemically induced mouse skin carcinomas, and its expression is associated with the degree of

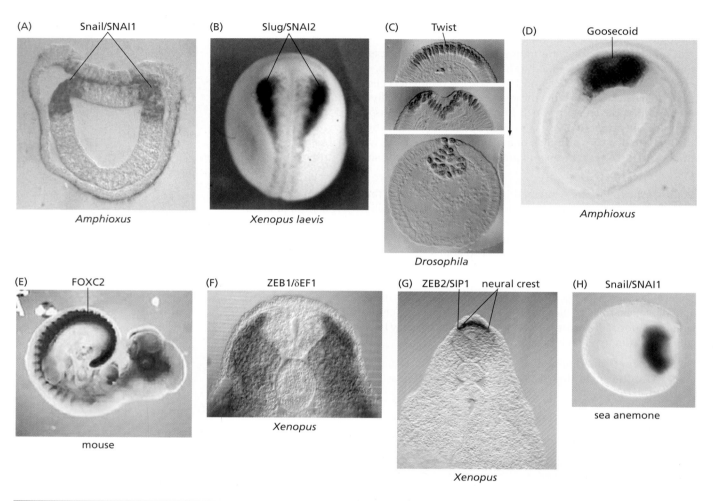

Figure 14.25 Embryonic transcription factors programming epithelial–mesenchymal transitions A variety of EMT-inducing transcription factors (EMT-TFs) can induce this program in various stages of embryogenesis, and their expression has been documented in a wide range of organisms. (A) Snail (also known as SNAI1) is shown being expressed in cells of *Amphioxus*, a primitive chordate, specifically in cells (*dark areas*) whose counterparts in higher chordates form the neural crest. (B) The Slug TF (SNAI2), a close relative of Snail, is also expressed (*dark blue*) in the embryonic neural crest, here in an embryo of *Xenopus laevis*, the African clawed toad. (C) Twist is shown here in an early *Drosophila* embryo, in which it programs an EMT at the site of gastrulation (*brown*). (D) The Goosecoid TF is expressed at the blastopore lip in gastrulating chordate embryos. Here its expression, which is inducible by the TGF-β signaling pathway, is adjacent to the blastopore 8 hours after fertilization of an *Amphioxus* egg. (E) The FOXC2 TF is expressed in important mesodermal structures in this day 9.5 mouse embryo, including the mesoderm around the developing spinal column as well as in somites, which are precursors to many of the body's muscles. (F) The ZEB1 (δEF1) TF is expressed in the paraxial mesoderm of an early *Xenopus* embryo, where it may play a role in programming the toad's development from axial stem cells. (G) In the same embryo at this stage, expression of the related ZEB2 (SIP1) TF occurs in a complementary pattern, being highest in the neural crest and neural tube, where it appears to be responsible for the cell movements that lead to closure of the neural tube and emigration of cells from the neural crest to other parts of the body. (H) The antiquity of these TFs is illustrated by the expression of Snail in the blastula stage of the sea anemone embryo, *Nematostella vectensis*. Its ancestors diverged from those of the chordates >600 million years ago, before the evolution of bilateral animals with three cell layers. It contrasts with *Amphioxus*—clearly a chordate with many features of vertebrate ancestors—which diverged ~525 million years ago from our own lineage. (A, courtesy of J. Langeland. B, courtesy of C. LaBonne. C, from M. Leptin et al., *Dev. Suppl.* 116(Suppl):22–31, 1992. D, from A.H. Neidert, G. Panopoulou and J.A. Langeland, *Evol. Dev.* 2:303–310, 2000. E, from T. Furumoto et al., *Dev. Biol.* 210:15–29, 1999. F, from L.A. van Grunsven et al., *Dev. Dyn.* 235:1491–1500, 2006. G, from L. van Grunsven et al., *Mech. Dev.* 94:189–193, 2000. H, from M.Q. Martindale, K. Pang and J.R. Finnerty, *Development* 131:2463–2474, 2004.)

24 hours   48 hours

72 hours   96 hours

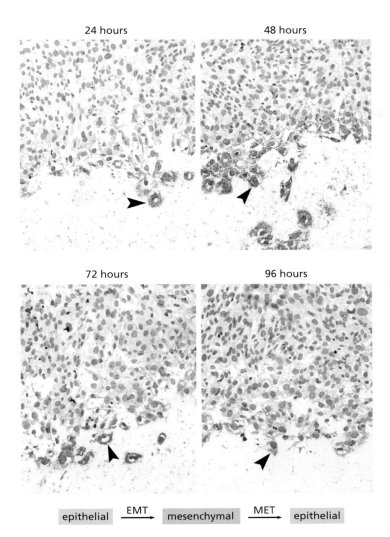

epithelial —EMT→ mesenchymal —MET→ epithelial

Figure 14.26 Transient expression of an EMT-inducing transcription factor in wound healing Expression of the Slug transcription factor is induced transiently in a monolayer culture of keratinocytes that has been wounded by scraping away a swath of cells. As seen here, 48 hours after wounding, keratinocytes at the edge of the wound induce expression of the Slug EMT-TF (*dark brown*) as they separate from the contiguous monolayer and begin to make their way as individual cells into the wound site (*bottom of each panel*), doing so in order to reconstruct an intact monolayer. By 96 hours, most of these cells cease expressing Slug and become integrated into a continuous monolayer. Hence, cells that initially underwent an EMT under the influence of the Slug EMT-TF in order to fill in and cover the wound site subsequently undergo an MET (mesenchymal-epithelial transition) in order to revert to the epithelial phenotype of their ancestors and reassemble a contiguous epithelial sheet. (From P. Savagner et al., *J. Cell Physiol.* 202:858–866, 2004.)

lymph node metastasis of human breast cancers. Moreover, embryonic expression of Snail, its cousin Slug, and Goosecoid is induced by contextual signals, such as TGF-β and Wnts, that are known to be responsible for inducing the EMT conversion of mouse tumor cells. Twist is expressed during the gastrulation of *Drosophila* embryos (see Figure 14.25C) and the out-migration of neuroepithelial cells from the neural crest of chordate embryos. Its expression is also induced by exposure to TGF-β. While Figure 14.25 gives the impression that these TFs are involved largely in early embryonic morphogenetic steps, the truth is that they continue to play roles throughout the entire process of embryonic development (Supplementary Sidebar 14.8).

The multiplicity of these TFs, along with their expression in various combinations in human cancers, suggests another idea that is implied by what follows: the term "EMT" refers to a group of cell-biological programs that are, in rough outline, very similar to one another but differ in detail from one cell type and one tumor type to another. This explains why we often refer to "an EMT" rather than "the EMT." Moreover, it is increasingly apparent that epithelial cells may activate part of the EMT program but not all of it; this explains how some carcinoma cells can acquire expression of mesenchymal markers without shedding all of their preexisting epithelial ones.

14.7 EMT-inducing transcription factors also enable entrance into the stem cell state

The cellular traits that are induced by the actions of one or another EMT-TF suggest that these master regulators can orchestrate almost all of the steps of the invasion–metastasis cascade. There is, however, one trait that is critical for metastasis that has not yet been considered here: in order for a disseminated cancer cell to seed a metastasis, it must possess the tumor-initiating ability that we previously ascribed to cancer stem cells (CSCs; see Section 11.6). Recall that these cells are defined operationally through their ability to seed new tumors following experimental implantation into suitable hosts. In principle, the seeding of a metastatic colony, when it occurs as a consequence of primary tumor progression, represents a very similar process: in both cases, small numbers of cells—as few as a single cell—are able to spawn the large cell populations that form macroscopic tumors, including the secondary growths that we term metastases.

This property of self-renewal that is so central to the stem cell state would seem to be far removed mechanistically from the aggressive cell-biological traits programmed by EMT transcription factors. Evidence first reported in 2008 indicates otherwise: this work revealed that human mammary epithelial cells (MECs) that are forced to pass through an EMT acquire many of the attributes of mammary stem cells (SCs). Related work showed that the same dynamics apply to transformed human MECs, that is, to breast carcinoma cells.

In the analysis shown in **Figure 14.27A**, a population of experimentally immortalized human MECs was driven through an EMT by forced expression of either the Twist or the Snail EMT-inducing TF. Control, unmodified MEC populations as well as the two modified populations were then fractionated by fluorescence-activated cell sorting (FACS) into $CD44^{hi}$ $CD24^{lo}$ and $CD44^{lo}$ $CD24^{hi}$ subpopulations—the same fractionation procedure that we encountered in Figure 11.16. In the cultured control cells, only a small minority of the cells were in the position ascribed to stem cells, while the forced expression of either Twist or Snail drove virtually all of those cells that were previously in the non-SC configuration into the $CD44^{hi}$ $CD24^{lo}$ cell population that contains, among its subpopulations, all of the SCs.

Perhaps more persuasive are the observations of Figure 14.27B, in which FACS was used to fractionate populations of immortalized human MECs into the two subpopulations—a population that contained SCs ($CD44^{hi}$ $CD24^{lo}$) and a second population that contained few if any SCs ($CD44^{lo}$ $CD24^{hi}$). When mRNA levels were quantified in the cells of these two subpopulations, it was apparent that those encoding mesenchymal markers were expressed at greatly elevated levels in the SC-enriched fraction relative to the non-SC population, as were mRNAs specifying a group of EMT-TFs. Here, the affiliation of EMT-inducing TFs with an SC-enriched state reflected regulation by natural, endogenous transcriptional controls rather than regulation imposed by experimentally introduced genes.

A similar association is present in the normal mouse mammary duct (see Figure 14.27C), in which the Slug TF is expressed in a cell layer in which the mammary SCs are known to reside. This last observation is striking for the following reason: until now, we have depicted EMT-TFs as master regulators that function in embryonic morphogenesis, in wound healing, and in tumor progression. Now they are found as well in adult tissues that would seem to be in a physiologic state that reflects normal, routine tissue function.

Still, these experiments do not prove that the EMT-TFs create cells that actually are SCs rather than being cells that coexist in the same cell population as bona fide SCs. In this instance, mammary gland reconstitution experiments have proven illuminating. This procedure gauges the ability of small numbers of implanted MECs to generate entire mammary ductal trees, that is, the functional structure of a mammary gland (see Figure 14.27D). When MECs prepared from a mouse mammary gland that had been forced to transiently express Slug and Sox9, a second collaborating TF, were implanted into a cleared mammary stromal fat pad, their ability to generate a mammary ductal tree was increased >100-fold relative to MECs that had not been exposed to these transcription factors.

We are only beginning to learn the identities of the stromal cells that operate to create both normal and neoplastic SC niches. These signals clearly overlap significantly with the EMT-inducing signals described in Section 14.4, since activation of an EMT program places MECs in a position to advance into the SC state. Myofibroblasts—prominent

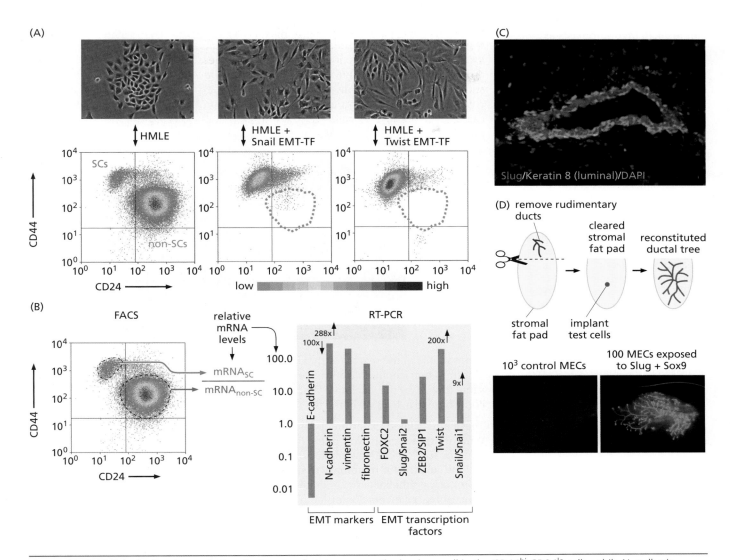

Figure 14.27 EMT-inducing transcription factors confer stem cell properties on epithelial cells Epithelial cells acquire certain properties associated with stem cells (SCs) following passage through an EMT. (A) When experimentally immortalized human mammary epithelial cells, termed HMLE, are viewed in monolayer culture by phase microscopy (*left, above*), they exhibit the cobblestone morphology that is typical of epithelial cells. These cells can be fractionated using fluorescence-activated cell sorting (FACS) into a minority subpopulation that has a CD44^{hi} CD24^{lo} cell surface profile and a majority subpopulation that has the opposite antigen profile—CD44^{lo} CD24^{hi} (*left, below*). The minority population, while not homogeneous, is greatly enriched for stem cells (SCs), which are absent from the majority population; this appears to be true for normal, immortalized, and neoplastically transformed human mammary epithelial cells (see Figure 11.16). When these cells were then forced to express either the Snail (*middle column*) or the Twist (*right column*) EMT-inducing transcription factor (EMT-TF), their morphology changed in monolayer culture to the dispersed elongated phenotype typical of mesenchymal cells like fibroblasts. In addition, virtually all of the cells that were previously in the CD44^{lo} CD24^{hi} non-SC state were converted to the CD44^{hi} CD24^{lo} population, which is greatly enriched in mammary SCs. (B) The relative levels of mRNAs expressed in each of the two populations in (A) (*left*) were quantified using RT-PCR analysis; the results are shown in the histogram (*right*). The mRNA encoding E-cadherin, the key epithelial marker, was expressed at ~1/100

the level per cell in the CD44^{hi} CD24^{lo} cells, while N-cadherin, vimentin, and fibronectin mRNAs (all encoding mesenchymal proteins) were expressed at levels ranging from ~65- to ~280-fold more abundantly in these cells. Of greater interest, cells that resided in the CD44^{hi} CD24^{lo} state naturally expressed levels of mRNAs specifying four EMT-TFs (FOXC2, ZEB2, Twist, and Snail) that were between 9- and 200-fold higher per cell than the CD44^{lo} CD24^{hi} cells. (C) In this cross section of a duct of the normal mouse mammary gland, abluminal (situated away from the lumen) cells in a location known to contain both myoepithelial and stem cells are seen to express the Slug EMT-TF protein at significant levels in (*pinkish red*), in contrast to luminal cells, which express cytokeratin 8 (*green*). Nuclei are stained *blue* with DAPI. (D) A rigorous test of stemness comes from implanting as few as one mouse mammary epithelial SC into a mammary stromal fat pad that lacks its own endogenous MECs (*above*). Success in mammary gland reconstitution demonstrates the presence of SCs among the implanted cells. 10,000 control mouse MECs were implanted in a cleared fat pad (*left*), resulting in no reconstituted mammary ductal tree. However, If the Slug EMT-TF and a second collaborating TF, termed Sox9, were transiently co-expressed in mouse MECs, as few as 100 of these cells (*right*) were able to reconstitute a mammary ductal tree. (A and B, from S.A. Mani et al., *Cell* 133:704–715, 2008. C, from W. Guo et al., *Cell* 148:1015–1028, 2012. D, courtesy of W.J. Guo and adapted from L. Hennighausen and G.W. Robinson, *Nat. Rev. Mol. Cell Biol.* 6:715–725, 2005.)

Figure 14.28 Myofibroblasts as facilitators of entrance into the stem cell state The precise sources of the stromal signals that induce EMT and thereby enable entrance into the SC state remain unclear. In the case of breast carcinomas, α-smooth muscle actin (α-SMA)–expressing stromal cells (i.e., myofibroblasts; see Figure 13.16) seem to play a prominent role. These cells express periostin (POSTN; *red*), a protein that has been found to facilitate presentation of Wnt ligands to epithelial cells, both in the normal mammary stem cell niche and the cancer stem cell niche. (As demonstrated in Figure 14.20, autocrine and paracrine Wnt signaling plays a prominent role in inducing EMT and expression of EMT-TF.) The PyMT-expressing carcinoma cells (*green*) forming the metastasis analyzed here derived from a PyMT transgenic, mammary tumor–prone mouse (see Figure 14.18C). In *Postn−/−* PyMT transgenic mice, primary tumors form normally, but metastases are reduced by a factor of 10 and the proportion of CSCs is strongly reduced. (From I. Malanchi et al., *Nature* 481:85–89, 2012.)

components of the reactive stroma of advanced carcinomas—are sources of prostaglandin E_2 that can induce entrance of epithelial cells into the SC compartment. At the same time, they may facilitate Wnt signaling (**Figure 14.28**), which has been found to contribute to activation of the EMT and thus the SC program (see Figure 14.20).

These and related experiments are limited to studies of human and mouse MECs. Whether these results will one day be extended to non-mammary epithelial tissues remains to be seen. Nevertheless, these observations already hold important implications, at least for mammary tissue: (1) The SC machinery that operates in normal mammary epithelial tissue operates in a similar fashion in mammary carcinoma cells; hence, breast cancers do not seem to invent novel SC programs during tumor formation but instead appropriate SC programs that operate in the normal precursor tissue. (2) Contrary to intuition, the epithelial SCs in the mammary gland express a series of mesenchymal markers; accordingly, MECs are the differentiated progeny of more mesenchymal, less differentiated SCs. (3) From the perspective of metastatic progression (see Figure 14.3), when carcinoma cells undergo an EMT, they approach the stem cell state in which they can become self-renewing—a trait that would seem to be essential for the successful founding of a new tumor colony in a distant tissue. Stated differently, an EMT program should enable carcinoma cells to translocate from the core of a primary tumor to the parenchyma of a foreign tissue; in addition, it endows disseminating carcinoma cells with the ability to develop stemness, which seems to be critical to their ability to subsequently found a metastasis. However, as mentioned earlier, it seems unlikely that EMT programs can also help disseminated cancer cells to develop the complex adaptations required for successful colonization of an unfamiliar tissue, that is, to produce macroscopic metastases from initially seeded micrometastases.

14.8 EMT-inducing TFs help drive malignant progression

Once certain EMT-TFs are expressed in cancer cells, they act in various combinations to induce multiple cellular changes associated with invasion and metastasis. The fact that they are invariably expressed in combinations in cancer cells suggests that no single one of them, acting on its own, is able to organize all of the cell-biological changes associated with passage through an EMT program. Indeed, these master regulators have been found to induce the synthesis of one another, often by the direct binding of one EMT-TF to the transcriptional promoter of another. For example, Twist binds directly to the promoters of Twist, Snail, Slug, and Zeb1, and actually depends on Slug induction in order to activate an EMT program. In addition, expression of Slug, which is frequently co-expressed with Twist in human breast cancers, is lost when Twist expression is shut down in some breast cancer cells. Also, some of these TFs can act in a redundant fashion; for example, Snail, Slug, ZEB1, and ZEB2 can all bind to the promoter of the E-cadherin gene (*CDH1*) and thereby repress E-cadherin expression. Accordingly, Twist may delegate the job of shutting down E-cadherin expression to Slug. While Snail and its cousin Slug function very similarly, their expression is regulated very differently, giving us clues about the dynamics of EMT induction (**Sidebar 14.3**).

Among all of these EMT-TFs (see Table 14.3), ZEB1 and ZEB2 seem to play commanding roles in regulating the decision as to whether a carcinoma cell should retain its epithelial traits or activate an EMT program. They do so in part by forming a *bistable* switch that holds the cells in either a mesenchymal or an epithelial state for extended periods of time (**Figure 14.29**). Indeed, as shown in Figure 14.29A, most types of cancer cells propagated in long-term culture appear to seek out and reside in one state or the other; it is less clear, however, whether such stable residence ever operates *in vivo*, where dynamic interconversions between states seem to operate. Implicit in the scheme of Figure 14.29E is the notion that the ZEB1/2 EMT-TFs are able, once activated, to induce expression, directly or indirectly, of a suite of other EMT-inducing TFs.

A key question surrounding EMT-TFs is their role in driving metastasis: do these TFs simply confer traits associated with high-grade malignancy on cultured cells, or are their powers also apparent *in vivo*? The most direct route for demonstrating their role in causing various steps of invasion and metastasis involves either overexpressing them or shutting down their expression in tumorigenic cells that have been implanted in suitable host mice.

In general, shutting down the actions of EMT-TFs in tumorigenic cells has provided far more compelling information than has their overexpression. **Figure 14.30A** shows one such exercise, in which the shutdown of Twist in the cells of a primary tumor (through use of an siRNA vector) resulted in a decrease of ~85% in the number of lung metastases generated by this tumor. Significantly, examination of the relatively few metastases that did form indicated that they arose from cells in which Twist expression had never been shut down in the first place. Of note, shutdown of Twist expression resulted in slightly more rapid primary tumor formation, indicating that the observed negative effects on metastasis could not be dismissed simply as cytostatic effects resulting from loss of Twist. Similar experiments have been done with experimentally transformed human melanocytes and rat epithelial cells, yielding comparable results (see Figure 14.30B and C).

Sidebar 14.3 Snail stands out from its colleagues The induction of an EMT depends on increases in EMT-inducing TFs in the cell nucleus. These increases generally depend on the relatively slow processes of inducing the transcription of the genes that encode these EMT-TFs, the processing of the resulting pre-mRNA transcripts, and the translation of the mRNA products into the proteins that then function as TFs in the nucleus; taken together, this results in increases in EMT-TF proteins that occur over the course of many hours, often days. This describes the regulation of Slug levels. Its close cousin, Snail is an outlier: Snail is usually produced at a high rate and is rapidly degraded thereafter, yielding a short lifetime of 25 minutes; this so-called "futile cycle" resembles the cycles governing the levels of the Myc and p53 proteins (see Sections 8.9 and 9.4). If Snail degradation is blocked, the existing high rate of Snail synthesis allows for rapid increases in protein levels that can be measured over the course of minutes.

Under conditions of cell quiescence, the GSK-3β kinase phosphorylates recently synthesized Snail, thereby tagging it (via ubiquitylation) for rapid destruction in proteasomes. However, activation of mitogenic signaling via growth factor receptors and the Ras–PI3K–Akt/PKB pathway results in the inactivation of GSK-3β by Akt/PKB, permitting in turn, survival and thus accumulation of Snail. (We first encountered GSK-3β in its role in driving β-catenin degradation; see Figure 6.26B.)

Interactions of cancer cells with macrophages, as described earlier in this chapter, result in, among others things, the release of tumor necrosis factor-α (TNF-α), which in turn helps to initiate an EMT in nearby carcinoma cells. (TNF-α, as we read earlier, is an important player in the process of tumor promotion via inflammation; see Figures 11.38 and 11.43.) By activating the NF-κB transcription factor in carcinoma cells, TNF-α can induce the expression of the CSN2 protein, which associates with and blocks the ubiquitylation of Snail, allowing the latter to accumulate rapidly in several hours. Finally, we should note that certain aspects of the tumor-promoting effects of inflammatory cells, as discussed in Section 11.16, can be nicely explained by the EMT-inducing effect of TNF-α.

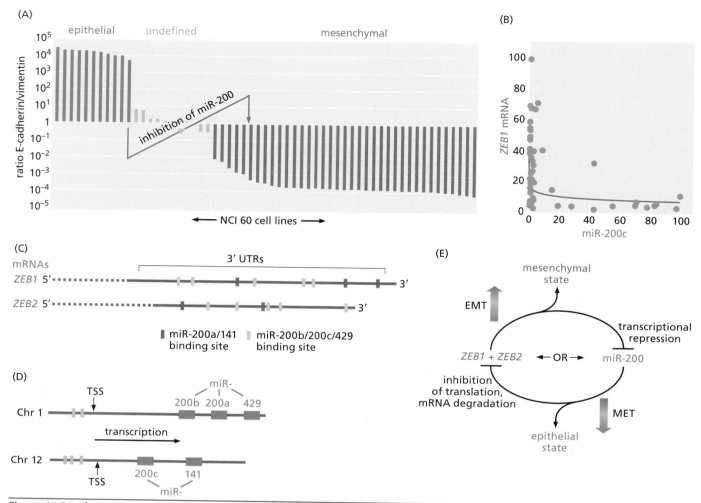

Figure 14.29 Alternation between the epithelial and mesenchymal states While carcinoma cells growing *in vivo* apparently can switch reversibly between epithelial and mesenchymal states, the cells of many cancer cells lines, when propagated *in vitro*, appear to reside stably in one or the other of these states, being controlled seemingly by a governor that functions as a *bistable* switch. (A) The levels of E-cadherin and vimentin, representative markers of the epithelial and mesenchymal states, respectively, were measured by immunoblotting of lysates of the 60 distinct cancer cell lines that constitute the collection termed the NCI 60 cell lines. The ratio of E-cadherin vs. vimentin (i.e., epithelial vs. mesenchymal) is indicated on the ordinate. As is apparent, almost all of the NCI 60 cells resided either in the epithelial (*green bars*) or mesenchymal (*red bars*) state when propagated in culture. The arrow indicates the shift of one of these cell lines after it was induced experimentally to undergo an EMT as described in panel C. (B) Examination of a series of microRNAs (see Section 1.10) expressed in either the epithelial or the mesenchymal state revealed that several of these belonging to the miR-200 family of microRNAs were expressed in a fashion that was mutually exclusive with the ZEB1 and ZEB2 EMT-TFs. Shown (in arbitrary units) are the expression levels of *ZEB1* mRNA and of miR-200c in ~60 cell lines, with each point representing a single cell line; regulation of *ZEB2* mRNA behaved in an essentially identical fashion (*not shown*). (C) Sequencing of the mRNAs encoding ZEB1 and ZEB2 revealed multiple sites in the 3′ UTRs (3′ untranslated regions; *light and dark blue boxes*) that could be targeted by miR-200 family members, resulting in inhibition of translation and /or degradation of these mRNAs. Alteration of these target sequences by site-directed mutagenesis rendered these mRNAs resistant to the inhibitory actions of miR-200 microRNAs. Moreover, experimental inhibition in epithelial cells of endogenously expressed miR-200 microRNAs, achieved by transfection of an antisense locked nucleic acid (LNA) construct, allowed accumulation of ZEB1 and ZEB2 and loss of epithelial gene expression. (D) Each of the two genes encoding miR-200 microRNAs specifies multiple miRNAs (*green boxes*) in its primary transcript (*not shown*), which is then cleaved post-transcriptionally to generate several mature microRNAs. The promoters of both genes contain evolutionarily conserved binding sites for ZEB1 and ZEB2 (*orange boxes*), which bind a sequence in these promoters and proceed to repress transcription of these genes. Transcription is from left to right (*horizontal arrow*) and the diagram is not drawn to scale. TSS, transcription start site. (E) The observations in panels C and D, when taken together, indicate the existence of mutually antagonistic interactions that operate as a bistable switch, thereby ensuring the metastable residence of a cell in either the mesenchymal or the epithelial state. Once one group of regulatory molecules gets the upper hand (i.e., either ZEB1/2 or miR-200 microRNAs), it shuts down the other and thereby ensures long-term residence in either the mesenchymal or the epithelial state. In principle, each group of regulators can act on yet other genes and mRNAs; e.g., ZEB1/2 can repress E-cadherin and other epithelial genes. The actions of the miR-200 microRNAs, however, seem to be focused more narrowly on suppressing just ZEB1/2 expression. (A–C, from S.-M. Park et al., *Genes Dev.* 22:894–907, 2008. D, from S. Brabletz and T. Brabletz, *EMBO Rep.* 11:670–677, 2010.)

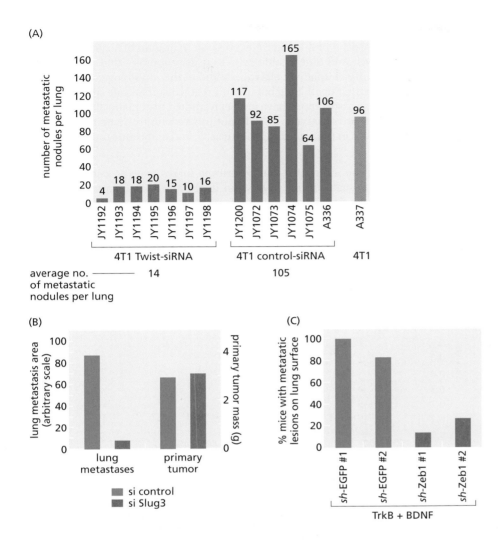

(A)

average no. ——————— 14   105
of metastatic
nodules per lung

(B)   (C)

Figure 14.30 Dependence of metastasis in mouse models on EMT-inducing TFs (A) 4T1 mouse mammary carcinoma cells are usually highly metastatic following implantation in the orthotopic site—the mammary gland (*blue bars*). However, when these cells were infected with a retrovirus vector specifying a potent siRNA directed against Twist, primary tumors grew as rapidly as before, but metastases to the lungs were reduced ~85%. The residual metastases expressed significant levels of Twist, indicating that they derived from primary tumor cells in which Twist expression had not been knocked down. (B) Experimentally transformed human melanocytes were found to express high levels of the Slug EMT-inducing TF. When these cells were implanted in a subcutaneous site, they generated large numbers of lung metastases. However, when an siRNA was expressed that reduced Slug mRNA levels by 80%, the number of pulmonary metastases decreased by ~90% while the growth of the corresponding primary tumors was hardly affected. (C) Rat epithelial cells were transformed by ligand-activated TrkB receptor, in particular forced expression of TrkB (a tyrosine kinase receptor) and its ligand, BDNF, achieved by retrovirus vectors. The number of mice bearing subcutaneously implanted TrkB-transformed cells was reduced significantly by anti-ZEB1 shRNA relative to those bearing tumors expressing control (anti-EGFP) shRNAs. (A, from J. Yang et al., *Cell* 117:927–939, 2004. B, from P.B. Gupta et al., *Nat. Genet.* 17:1047–1054, 2005. C, from M.A. Smit and D.S. Peeper, *Oncogene* 30:3735–3744, 2011.)

Similar experiments have not yet been reported with cells derived from human tumors. Our perceptions about the roles of the EMT-TFs in human cancer therefore come from less direct measurements, specifically correlations of the expression of EMT-TFs with the course of clinical progression (**Figure 14.31A–C**). Some of these associations are so dramatic that we begin to believe that these TFs are indeed playing causal roles in driving aggressive clinical progression.

Yet another indication of the involvement of these TFs in high-grade human cancers comes from equally indirect observations. The circulating tumor cells (CTCs) found in the blood of many cancer patients represent tumor cells that have intravasated into the circulation but have not yet been trapped in peripheral tissues; as such, these cells are likely en route between primary tumors and potential sites of metastasis. CTCs from a number of patients suffering from prostate cancer have been found to co-express epithelial and stromal markers, suggesting that these cells underwent partial EMT programs—ostensibly under the aegis of EMT-TFs—before leaving primary tumors and entering into the circulation (see Figure 14.31D). In another study (not shown here), three-quarters of early breast cancer patients carried CTCs that express both cytokeratins (epithelial markers) as well as mesenchymal markers—vimentin and the Twist EMT-TF; this proportion increased to 100% in patients with metastatic disease. Evidence like this suggests but hardly proves that these cells advanced partway through EMT programs before leaving primary tumors.

Yet other lines of evidence implicate EMT-TFs in the malignant progression of human tumors. For example, Snail has been found to be expressed in islands of human mammary ductal carcinoma cells that lack E-cadherin expression. Slug has also been implicated in the repression of E-cadherin expression in human breast cancers. And significantly, both Twist and Slug enable cells to resist apoptosis and anoikis, and therefore can protect disseminating cells from some of the physiologic stresses that would normally cause their death long before they reach distant tissue sites and form micrometastases. These associations have not been studied systematically, and in any event, even when they are observed, they do not prove definitively that the various TFs are causally involved in programming the invasive and metastatic traits of human tumor cells.

These descriptions of the molecules that contribute in key ways to carcinoma cell invasiveness are reminiscent of our discussion in the previous chapter about the activated stroma and its contributions to the formation of carcinoma cells. In both places, for example, we encountered cadherins, the EMT, and TGF-β. In addition, the pro-inflammatory signals that drive tumor promotion and thus the formation of

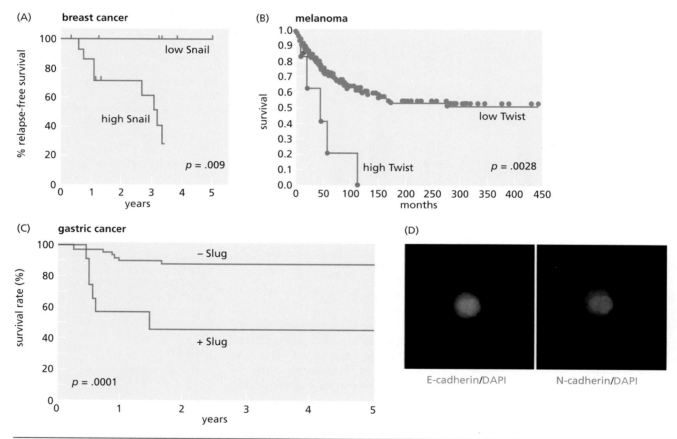

Figure 14.31 Indications of the role of EMT-inducing transcription factors in human tumor progression Increasing evidence correlates the expression of EMT-inducing TFs with the induction of malignant behavior in cancer cells in patients.
(A) Among a group of women with locally advanced breast cancers (tumors that appeared histologically to be aggressive without evidence of disseminated disease), those whose tumors expressed either high or low levels of Snail were followed for a period of 5 years after initial surgery. The percentage of women who enjoyed relapse-free survival varied dramatically, depending on whether or not the carcinoma cells in their primary tumors expressed high levels of the Snail EMT-TF protein. (B) Sections of primary melanomas were immunostained for Twist expression and correlated retrospectively with the long-term survival of the patients bearing these tumors. (C) Shown are the postoperative survival rates for stomach cancer patients whose primary tumors retained detectable E-cadherin expression together with or in the absence of Slug expression. (D) Circulating tumor cells (CTCs) were identified in the blood of men suffering from metastatic prostate cancer; these cells were found to lack expression of CD45 antigen (a marker of cells of hematopoietic origin), thereby ensuring that they did not originate from cells normally present in the blood. As seen here, one of these cells was stained with two antibodies, one against E-cadherin (*green*) and a second against N-cadherin (*red*). This co-expression of epithelial and mesenchymal proteins within a single cell would appear to serve as a fingerprint of its origin in a primary tumor in which it or its immediate precursors underwent a partial EMT under the direction of an unknown EMT-TF or set of EMT-TFs. (A, from S.E. Moody et al., *Cancer Cell* 8:197–209, 2005. B, from K. Hoek et al., *Cancer Res.* 64:5270–5282, 2004. C, from Y. Uchikado et al., *Gastric Cancer* 14:41–49, 2011. D, from A.J. Armstrong et al., *Mol. Cancer Res.* 9:997–1007, 2011.)

primary tumors (see Section 11.15) also help to trigger the EMT (see Sidebar 14.3). Such connections hint at an interesting but still speculative idea: perhaps the formation of primary carcinomas and the acquisition of invasiveness are not as separate and distinct as most descriptions of cancer would suggest. Although it is convenient to place them in separate conceptual boxes, the biological reality may be quite different. Quite possibly, cancer cell invasiveness is a natural extension—an exaggerated form—of the processes that lead initially to the formation of many types of primary tumors. Because transformation and invasiveness depend on many of the same regulatory circuits and effector proteins, they may lie on a continuum in which one process blends seamlessly into the next.

To summarize, the discovery of the EMT-inducing TFs suggests at least three important ideas about malignant progression. First, many malignant cell phenotypes may be induced by nongenetic changes—heterotypic signals of stromal origin—rather than genetic changes occurring within carcinoma cells; hence the cells within primary tumors may already possess the genetic alterations required for dissemination (Supplementary Sidebar 14.9). Second, because expression of these TFs and the resulting EMT is often dependent on heterotypic signaling from the reactive stroma of primary tumors, carcinoma cells may revert from the mesenchymal state to an epithelial state once they have left the primary tumor and encounter the fully normal stromal microenvironments present in sites of metastasis; by definition, such normal stromata do not release the heterotypic signals that induce activation of EMT programs in nearby carcinoma cells. Third, cancer cells do not need to cobble together all of the phenotypes associated with highly malignant cells by acquiring multiple mutant genes; instead, many of the malignancy-associated traits may be acquired concomitantly, because EMT-TFs, once expressed, act in a highly pleiotropic fashion to confer multiple cellular traits associated with high-grade malignancy.

14.9   Extracellular proteases play key roles in invasiveness

An EMT represents a complex biological program that enables cancer cells to acquire the attributes of invasiveness and motility. In order to properly appreciate the processes that together constitute an EMT, we need to examine the roles of some of its key *effectors*—the proteins that work to create the phenotypes associated with an EMT. To begin, we examine the most obvious trait of malignant cells—their ability to invade adjacent cell layers. This burrowing requires that cancer cells remodel the nearby tissue environment by excavating passageways through the extracellular matrix (ECM) and pushing aside any cells that stand in their path.

The most important effectors of these complex changes are the matrix metalloproteinases (MMPs; see Table 13.1). In carcinomas, the bulk of these proteases are secreted by recruited stromal cells, notably macrophages, mast cells, and fibroblasts, rather than by the neoplastic epithelial cells (**Figure 14.32**). By dissolving the dense thickets of ECM molecules that surround and confine individual cells within tissues, these secreted MMPs create spaces for these cells to move. Included among the ECM components that are cleaved by MMPs are fibronectin, tenascin, laminin, collagens, and proteoglycans. During the course of degrading ECM components, MMPs also mobilize and activate certain growth factors that have been tethered in inactive form to the ECM or to the surfaces of cells.

Complementing these secreted MMPs is a critically important one—MT1-MMP (membrane type-1 MMP)—that is tethered directly to the plasma membranes of cancer cells and wielded by them in order to cleave cell ECM components, cell surface adhesion molecules (for example, cadherins and integrins), as well as growth factor receptors and chemokines. It can also cleave inactive **pro-enzymes**, such as pro-MMP-2, into enzymatically active MMPs, as discussed below. (MT1-MMP is one of six MMPs that are membrane-anchored and therefore limited to cleaving substrate proteins in the immediate vicinity of the cells that produced them.)

The initial steps of invasion by a carcinoma cell are obstructed by the basement membrane (BM), which may have pores as small as 40 nm in diameter. While cells can distort their cytoplasms in order to squeeze through small openings, their nuclei are

(A)

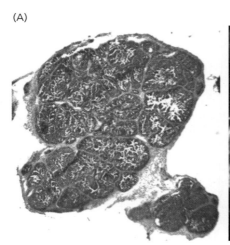

(B)

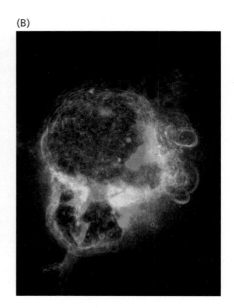

(C)

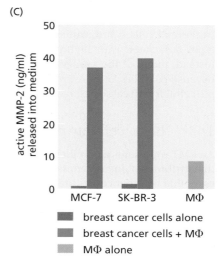

Figure 14.32 Matrix metalloproteinases produced by tumor-associated cells MMPs are produced largely by inflammatory cells and fibroblasts in the tumor stroma. (A) In this mammary carcinoma arising in a transgenic MMTV–PyMT mouse, MMPs have been detected through their ability to cleave a synthetic substrate, which releases a polycationic fluorescent tag that migrates into nearby cells, generating a fluorescent signal. The tumor stained with hematoxylin–eosin (*left panel*) is seen to generate a halo of proteolysis (*right panel*), suggesting the involvement of surrounding stromal cells in protease production. (B) The ability of tumors to degrade collagen IV, a key component of the basement membrane, can be measured by generating a modified collagen IV substrate that fluoresces green upon cleavage. In this experiment, both human mammary carcinoma cells (*not seen*) and human mammary fibroblasts (*red*) showed relatively weak ability to degrade the collagen IV substrate. However, when these two cell populations were co-cultured, regions of collagen IV cleavage (*green*) were evident, often in areas where fibroblasts were also present (*yellow: overlap of green and red*). This cleavage was essentially eliminated in the presence of MMP inhibitors (*not shown*). (C) An even more important source of MMPs is the macrophages (MΦs) that are recruited into the tumor stroma. In this *in vitro* experiment, the presence of MMP-2 was measured in the culture medium of MΦs that were cultured alone (*green*) or together with either of two human breast cancer cell lines—MCF-7 or SK-BR-3 (*blue*). Neither of these cancer cell types released significant levels of MMP-2 on its own (*red*), but in the presence of MΦs (*blue*), the production and release of MMP-2 increased 4- to 5-fold. (The increase could be traced to the induction of MMP-2 mRNA expression by the MΦs, *not shown*.) The released MMP-2 imparted increased invasiveness to these breast cancer cells (*not shown*). (A, courtesy of E. Olson, T. Jiang, L. Ellies and R. Tsien. B, from M. Sameni et al., *Mol. Imaging* 2:159–175, 2003. C, from T. Hagemann et al., *Carcinogenesis* 25:1543–1549, 2004.)

relatively rigid and thus govern the size of the smallest pores through which cells can pass. Cell nuclei are generally in the size range of 3 to 10 μm, about 100 times larger than the pores in the BM; this explains why dissolution of the BM is critical to carcinoma cell invasion.

The activities of MT1-MMP, which plays the leading role in BM breakdown, seem to be confined through its concentration at discrete cell surface foci, initially termed **podosomes** but increasingly called **invadopodia** because of the involvement of these structures in cancer cell invasion (**Figure 14.33**). Early in malignant progression, MT1-MMP displayed on the surface of carcinoma cells can cleave collagen type IV, the collagen that imparts rigidity to the basement membrane (BM). The resulting weakening of the BM allows cancer cells to begin invading the underlying stroma (see Figure 14.4). Once in the stroma, an invading carcinoma cell confronts a dense network of cross-linked collagen type I fibers that obstructs further advance; here once again, MT1-MMP plays a central role. MT1-MMP initiates collagen I degradation and then calls in an inactive pro-enzyme (pro-MMP-2) of stromal origin, which it activates by cleavage. The resulting active MMP-2 then operates in the peri-cellular space to further cleave collagen I into lower–molecular-weight fragments. Without these steps, the dense networks of collagen I fibers that are present in the stromal extracellular matrix block cancer cell invasion (**Figure 14.34**).

Membrane-bound and secreted proteases clearly play important roles in normal cell survival and proliferation. After all, each time a cell within a normal tissue goes through a cycle of growth and division, space within the ECM must be carved out for its daughters, and once formed, each daughter cell must, in turn, reconstruct new

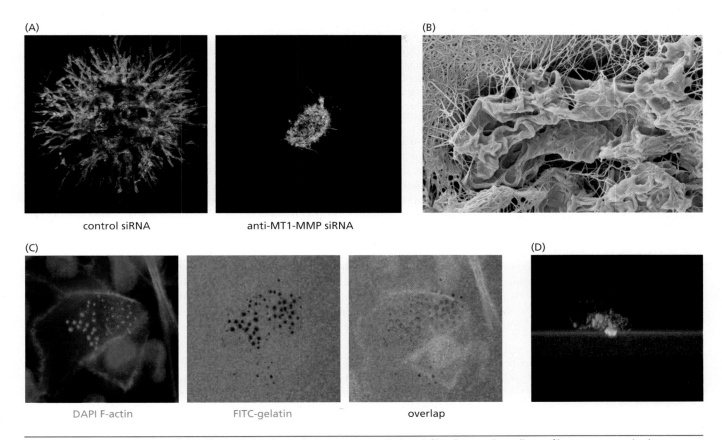

Figure 14.33 MT1-MMP, invadopodia, and the degradation of the extracellular matrix Invadopodia, sometimes termed podosomes, are small, focal protrusions from the cell surface that degrade localized areas of extracellular matrix (ECM) in their immediate vicinity. In the case of tumor development, invadopodia are deployed by invasive cancer cells to drive localized, highly controlled degradation of the ECM near the leading edges of these cells. (A) The critical contribution of MT1-MMP, which is tethered to the surface of invadopodia, in cancer cell invasion can be demonstrated *in vitro* through the behavior of HT1080 human fibrosarcoma cells. When these cells, seen here growing as a colony in a 3D matrix of collagen I, were forced to express a control siRNA (*left panel*), they demonstrated a high degree of invasiveness, as indicated by the numerous protruding columns of cancer cells invading the matrix. However, when MT1-MMP production was knocked down through the actions of a specific siRNA, these cells were confined to a small volume surrounding the site at which the founding cell was initially introduced into the matrix. Knockdown of the two secreted metalloproteinases also made by these cells—MMP-1 and MMP-2—had no effect on their invasiveness (*not shown*). Cells were treated with a phalloidin derivative, which labels filamentous actin (*green*), and di-I (3,3-dioctadecyl indocarbocyanine), which stains membranes (*red*). (B) The effects of the proteases produced by the invasive HT1080 cells are seen in fine detail in this scanning electron micrograph in which the collagen I fibers in the ECM have been pseudo-colored *blue* while the fibrosarcoma cells are colorized *pink*. Note that the collagen matrix is dense beyond the immediate vicinity of the fibrosarcoma cells (*top left*), whereas the collagen fibers are sparse in the immediate vicinity of the cells (*center, right*) due to the actions of the proteases, largely and perhaps exclusively the MT1-MMP made by these cells. (C) MT1-MMP is the major protease displayed on invadopodia. These bodies are defined by dense concentrations of filamentous actin, stained here with phalloidin (*red, left panel*) on the ventral surface (i.e., the surface directly apposed to the underlying substrate) of immortalized human mammary epithelial cells; nuclei are labeled with DAPI (*blue*). These immortalized human mammary epithelial cells were plated above a thin matrix of gelatin labeled with FITC dye (*green, middle panel*). The single cell seen in the middle of the *left panel* has eroded numerous holes (*black dots, middle panel*) in the gelatin (thereby solubilizing FITC) precisely under the focal areas where invadopodia were seen; the overlap between the F-actin foci and the eroded holes in the gelatin is seen here in *red* (*right panel*). Formation of these invadopodia was dependent on activation of the Twist EMT-TF as well as expression of the PDGF-Rα that resulted from activation of the EMT program. (D) The invadopodia of a small cluster of cells are visualized from the side (a z-axis view) through confocal microscopy. The staining of actin (*red*) and that of cortactin (*green*; see Figure 14.40) overlap in the invadopodia (*yellow*), which are protruding through a layer of the ECM protein fibronectin (*blue*) through which they have eroded an opening. (A, from F. Sabeh et al., *J. Cell Biol.* 185:11–19, 2009. B, courtesy of F. Sabeh, S. Meshinchi and S.J. Weiss; from F. Sabeh et al. C, from M.A. Eckert et al., *Cancer Cell* 19:372–386, 2011. D, courtesy of S. Vitale and M. Frame; see also S. Vitale et al., *Eur. J. Cell Biol.* 87:569–579, 2008.)

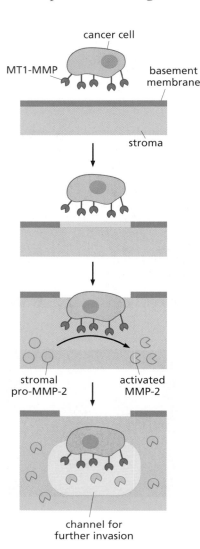

cancer cell

MT1-MMP

basement membrane

stroma

stromal pro-MMP-2

activated MMP-2

channel for further invasion

Figure 14.34 Collaboration of MT1-MMP and MMP-2 in carcinoma cell invasion Membrane-tethered proteases, notably MT1-MMP, initially enable carcinoma cells to degrade the basement membrane (*green*) that separates them from the adjacent stroma. Once in direct contact with the stroma (*pink*), MT1-MMP can cleave pro-MMP-2 made by the stromal cells, converting it to an active, soluble protease. MT1-MMP also partially cleaves collagen I, which normally forms a densely interwoven network in the ECM of the stroma (see Figure 14.33B); the activated MMP-2 then proceeds to degrade the initially formed collagen I cleavage products to lower–molecular-weight fragments. Dissolving the collagen I network creates a channel immediately in front of the carcinoma cell that allows it to invade more deeply into the stroma.

ECM around itself. Hence, the remodeling of the ECM takes place continuously in mitotically active tissues. Consequently, rather than being aberrations of invasive cancer cells, the activities of MMPs and other extracellular proteases are part of the program associated with normal cell proliferation. Of relevance here are clinical trials of certain MMP-inhibitory drugs, which were terminated due to the effects of these inhibitors on a variety of normal tissues; because these agents suppress the normal remodeling of cartilage and other joint components, they created unacceptable levels of joint stiffness and pain.

Each type of MMP usually acts on a well-defined set of substrates (see Table 13.1), doing so in a highly regulated and localized fashion. It is likely that these enzymes continue to show such substrate specificity during the process of cancer cell invasion. However, in the case of invasive cancer cells, such proteolysis seems to proceed continuously rather than in the brief spurts that accompany normal cell growth and division.

The EMT programmed by several of the well-studied embryonic TFs (see Section 14.6) results in the synthesis and release by carcinoma cells of MMPs, notably MMP-2 and -9. It is clear, however, that the bulk of the MMPs found in tumors originate in various cellular components of the stroma. For example, the best-studied of the matrix metalloproteinases, MMP-9, is expressed largely by macrophages (see Section 13.5), neutrophils, and fibroblasts at the invasive fronts of tumors. MMP-9 expression at these fronts correlates positively with the metastatic ability of a primary tumor, suggesting that MMPs like this one can act at several stages of the invasion–metastasis cascade, including local invasion of the primary tumor stroma, intravasation, and extravasation. *In vitro* assays indicate that MMP-9 can degrade collagens that are prominent components of the ECM including basement membranes, specifically collagen types IV, V, XI, and XIV. Other targets of MMP-9 include laminin (another important constituent of the basement membrane), chemokines, fibrinogen, and latent TGF-β. In the case of the latter two, cleavage by MMPs converts them from latent into activated forms. The relationship between the MMPs that degrade the ECM lying immediately in front of invading cancer cells and the MMPs that are active at the invasive fronts of tumor masses is still unresolved. Importantly, the contribution of various recruited inflammatory cells to cancer cell invasiveness, which was described in Section 14.5, can be explained in large part by the ability of these cells to produce various secreted proteases.

These widely ranging functions of MMPs indicate that their enzymatic activity must be tightly controlled, at least in normal tissues. Reflecting this requirement is the fact that the soluble MMPs, such as MMP-2 and MMP-9, are initially synthesized as inactive pro-enzymes that can only function, like the caspases (see Section 9.13), following activation by other proteases (see, for example, Figure 14.34). Negative regulation is also provided by a class of proteins termed tissue inhibitors of metalloproteinases (TIMPs), which bind MMPs and place them in an inactive configuration (see also Section 13.9).

While MMPs have been depicted as the direct effectors of certain steps in invasion and metastasis, it is clear that the deregulation of MMPs can, on its own, drive the

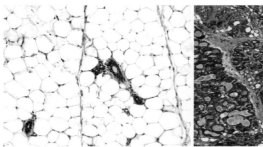
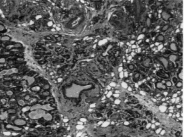

normal mammary gland   + ectopic MMP-3

Figure 14.35 Ectopic expression of MMP-3 and mammary tumor progression The normal mouse mammary gland (*left panel*) is composed of resting ducts (*purple*) and abundant adipose tissue (lipid-filled cells, *white*), as well as collagen (*light blue*). However, when the gene encoding MMP-3 (also known as stromelysin-1) is expressed constitutively as a transgene in the mammary epithelium, the mice develop abundant hyperplasia (*right panel*), including extensive islands of hyperplastic epithelial cells (*purple*) that form ducts, as well as a fibrotic, collagen-rich stroma (*light blue*) and abnormal adipocytes (*white oval structures, lower right*). Many of these areas subsequently progress to invasive, metastatic tumors (*not shown*). (From M.D. Sternlicht et al., *Cell* 98:137–146, 1999.)

progression of cells through all of the stages of multi-step tumorigenesis including completion of the invasion–metastasis cascade. Thus, when expression of MMP-3 is forced in the mammary gland of transgenic mice, these mice initially develop mammary hyperplasias (**Figure 14.35**). Some of these growths progress to carcinomas that eventually become invasive and metastatic (see for example Figure 13.13C). These mice reveal how critical the regulation of MMP function is and why it must be kept under control in normal tissues.

These brief vignettes of proteases and their contributions to cancer cell invasiveness describe only small parts of what are surely highly complex networks of interacting proteases, protease inhibitors, and substrates. The total number of proteases made by mammalian cells is vast and rivals the number of proteins that form the highly complex intracellular signal-processing circuits described in Chapter 6. To date, the actions of only a tiny proportion of these enzymes have been studied in the context of cancer pathogenesis (Supplementary Sidebar 14.10).

14.10 Small Ras-like GTPases control cellular processes such as adhesion, cell shape, and cell motility

The actions of extracellular proteases, notably the MMPs, explain at the biochemical level how paths are cleared for the advance of invasive cancer cells through the extracellular matrix and thus through tissues. They fail, however, to tell us how individual cancer cells take advantage of these cleared paths to move ahead—the trait of cell motility. The motile behavior of cells has been studied extensively with cultured cells, and it is presumed that their crawling on solid substrates *in vitro* reflects the *in vivo* behavior of cancer cells as they invade nearby cell layers and intravasate. Such motility is also presumed to be important for cancer cells' escape from blood vessels or lymph ducts—the process of extravasation.

Motile behavior can be induced in cultured cells by exposing them to a variety of growth factors. (Those GFs able to induce such locomotion are sometimes designated as being **motogenic** in addition to being mitogenic.) In the case of epithelial cells, the best inducer of motility is usually hepatocyte growth factor (HGF); this protein is also called scatter factor (SF) in recognition of its ability to induce multidirectional movement of cells in monolayer culture. Many types of epithelial cells express Met, the receptor for HGF, and such cells have been found to acquire motility in response to HGF treatment (see Figure 14.23A). Similarly, EGF is clearly able to induce motility of breast cancer cells (see Figure 14.22B).

The cellular machinery that responds to motogenic signals and operates as the engine of motility is extraordinarily complex at the molecular level. Cell motility involves continuous restructuring of the actin cytoskeleton in different parts of a cell, as well as the making and breaking of attachments between the migrating cell and the extracellular matrix (**Figure 14.36**). (In the case of cultured cells, the ECM in question is the network of proteins that has previously been laid down by these cells on the surface of the Petri dish; see Figure 1.13.)

The process of cellular movement can be broken down into several distinct steps. To begin, a cell will extend its cytoplasm in the direction of intended movement. This

Figure 14.36 Locomotion of cells on solid substrates The locomotion of a cultured cell depends on the coordination of a complex series of changes in the cytoskeleton, as well as the making and breaking of focal contacts with the underlying solid substrate. The cell organizes actin fibers in order to extend lamellipodia at its advancing/leading edge and to establish new focal contacts. At the same time, stress fibers, also consisting of actin, are used to contract the trailing edge of the cell, where focal contacts are being broken. The making and breaking of the focal contacts depend on localized modulation of the affinities of various integrins for extracellular matrix (ECM) components, represented here by the *yellow* substrate. (From B. Alberts et al., Molecular Biology of the Cell, 5th ed. New York: Garland Science, 2008.)

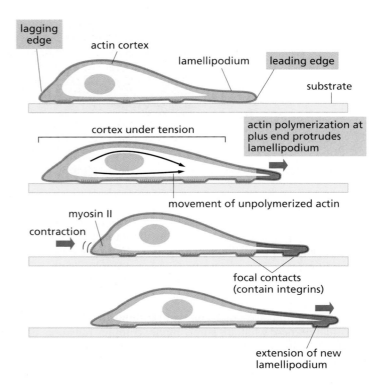

extension involves the protrusion from the cell surface of **lamellipodia**—broad, flat, sheetlike structures that may be tens of microns in width but only 0.1 to 0.2 μm thick (**Figure 14.37A–C**). At the same time, cell surface proteases, such as those described earlier, are used to selectively degrade ECM proteins that stand in the way of the "leading edge" of the migrating cell. While this is going on, the cell deploys integrins to construct new points of attachment between the lamellipodia and the ECM at its leading edge and breaks such adhesions at its "trailing edge," thereby liberating cytoplasm and plasma membrane for redeployment to the leading edge.

Protruding from the lamellipodia are spikelike structures termed *filopodia* that are thought to enable an advancing cell to explore the territory that lies ahead and initiate the formation of focal adhesions by integrins (see Figure 14.37D); once formed, these adhesions provide cells with survival signals and, at the same time, allow them to gain traction through firm physical connections with the adjacent ECM. There are also indications that filopodia, by enabling extravasated cancer cells to form focal adhesions with the ECM of the surrounding tissue, play a key role in the initial formation of micrometastases. Like lamellipodia, filopodia are assembled through the reorganization of actin fibers, in this case fibers that are tightly bundled together beneath the plasma membrane of each filopodium (**Figure 14.38**).

The detailed management of cell shape and motility is under the control of members of a group of Ras-related proteins belonging to the Rho family. As discussed briefly in Chapter 6, the Rho proteins, like Ras, operate as binary switches, being in a functionally active state while binding GTP and in an inactive state once they hydrolyze their bound GTP to GDP. More than 20 members of the Rho family of proteins have been discovered in human cells. They are divided into three subfamilies—the Rho proteins proper, the Rac proteins, and Cdc42. Like the Ras proteins, most members of the Rho protein family bear lipid groups at their C-termini that enable anchoring to intracellular membranes. Each of these has specialized functions in reorganizing cell shape and enabling cell motility (**Figure 14.39**).

Figure 14.39 actually misrepresents the actions of these various Rho-like proteins in one important respect: it implies that each of them acts globally throughout the cell to organize certain changes in the configuration of the actin cytoskeleton. In reality, the complex program of cell motility depends on the *localized* activation of each of these proteins in very small domains of the cytoplasm, which in turn enables the cell as a whole to move in one direction or another. This focused activation seems to depend, in turn, on the subcellular localization of specialized Rho GEFs (guanine nucleotide exchange factors), which, like Sos (see Section 6.2), operate to convert Rho proteins from their inactive GDP-bound form to their active GTP-bound form; there are about

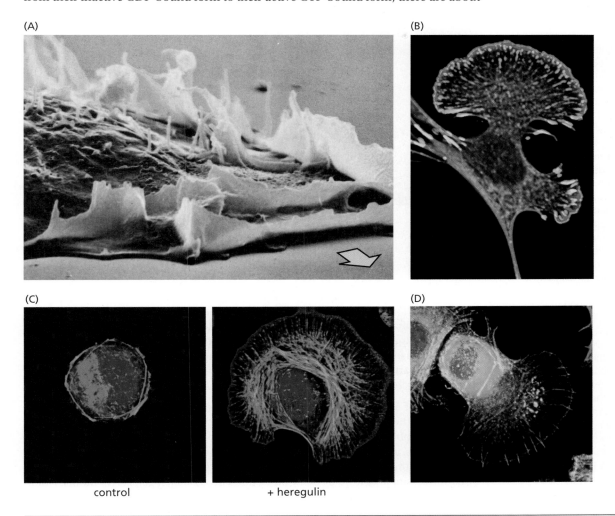

(A)

(B)

(C)

control   + heregulin

(D)

Figure 14.37 Lamellipodia and filopodia (A) This scanning electron micrograph (SEM) of a spontaneously transformed rat liver cell shows the elaborate ruffles—lamellipodia—that migrating cells extend at their leading edge during locomotion in the indicated direction (*arrow*). Lamellipodia are presumed to play a key role in the advance of invasive cancer cells *in vivo*, but this has not yet been directly demonstrated. Lamellipodia that fail to attach to the substrate seem to be swept back as ruffles along the dorsal (top) side of the cell; their component parts are then reintegrated into the larger plasma membrane and cytoskeleton. (B) This fluorescence micrograph shows a lamellipodium being extended by a fibroblast. The actin fibers are labeled with phalloidin (*red*), while a protein of the Ena/VASP family, which programs the advance of the lamellipodium by organizing its focal adhesions and its outer edge, is labeled here with green fluorescent protein (GFP), to which it has been fused. (C) Lamellipodium formation and resulting cell motility are strongly stimulated by a number of growth factors and their cognate tyrosine kinase receptors. Seen here are the effects of adding heregulin, a ligand of the erbB2/erbB3 family of receptors, to a human breast cancer cell. An untreated cancer cell is to the *left*, while a cell exposed to heregulin for 20 minutes is to the *right*. The actin cytoskeletons have been stained *green* (using phalloidin

coupled to a fluorescent dye), while the nuclei are stained *blue*. (Because the heregulin is present uniformly in the surrounding medium, this cell has been induced to develop a lamellipodium that faces in all directions rather than toward a single source of this motogen.) Elevated signaling by erbB2 (= HER2/Neu) is correlated with increased metastatic progression of human breast cancer cells, which may be explained in part by the receptor-mediated induction of lamellipodium formation and associated cell motility. (D) Filopodia protrude from the leading edges of cells, often from the edges of lamellipodia. They are presumed to allow an invading cell to explore its extracellular environment and help to establish the focal adhesions that are formed between integrins and specific components of the extracellular matrix. The cell shown here has extended a lamellipodium in one direction, from which filopodia, stained for filamentous (F-) actin (*red*), are protruding. A small portion of the Arp3 protein, which is associated with the actin cytoskeleton, is found at the leading edge of the lamellipodium (*arrows*); most of the Arp3 protein is found in its inactive form in the cytosol. The cell nucleus is stained *blue*. (A, courtesy of Julian Heath. B, from J.J. Loureiro et al., *Mol. Biol. Cell* 13:2533–2546, 2002. C, courtesy of A. Badache and N.E. Hynes. D, courtesy of T. Shibue.)

Figure 14.38 The actin cytoskeleton and the restructuring of the cell surface Both lamellipodia and filopodia depend on the formation of specific configurations of the actin cytoskeleton that lies immediately beneath the plasma membrane. (A) Use of platinum replica electron microscopy reveals the configuration of the actin fibers that form both a lamellipodium and a filopodium. All membranes have been removed to reveal the actin cytoskeleton. The *dashed line* indicates the imputed location of the plasma membrane, which previously separated the cytoplasm (*below*) from the extracellular matrix (*above*). The lamellipodium extends broadly across the entire micrograph, while a single filopodium protrudes from the leading edge of the lamellipodium. (B) The actin fibers within a lamellipodium, seen here at higher magnification (*above left*), are extended at their growing "barbed" ends by addition of actin monomers (*yellow balls, top in lower diagram*) and are disassembled at their retreating "pointed" ends by liberation of individual monomers. The same dynamic operates to enable the extension of filopodia (*right*). Ancillary proteins involved in forming the branched fiber networks of the lamellipodium are depicted in *pink*, whereas those involved in forming the tightly bundled, parallel actin fibers of the filopodium are depicted in *green*. The plasma membrane wraps over and around these structures in a fashion that is poorly understood. (A, from T.M. Svitkina et al., *J. Cell Biol.* 160:409–421, 2003. B, from D. Vignjevic and G. Montagnac, *Semin. Cancer Biol.* 18:12–22, 2008.)

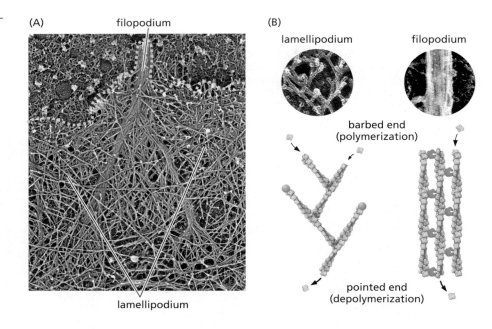

80 Rho GEFs in mammalian cells, few of which have been studied in any detail. The alternative to localized activation—global activation—would lead to attempts by a cell to move simultaneously in all directions, a scenario suggested by the lamellipodia of Figures 14.37C and 14.39C, which form a continuous ring around the entire perimeter of the cytoplasm. (The global activation of Rac function in the cell depicted in Figure 14.39C is an artifact of introducing mutant, constitutively activated Rac protein into the cell by micro-injection.)

Growth factor activation of tyrosine kinase receptors leads to the activation of many members of the Rho family of G proteins (**Figure 14.40**). For example, treatment of cultured fibroblasts with platelet-derived growth factor (PDGF), a potent mitogen for these cells, activates a number of Rho proteins and stimulates these fibroblasts to move across the bottom of a Petri dish. Alternatively, when fibroblasts are placed in three-dimensional culture by being suspended in a collagen gel, PDGF induces them to invade through this gel. All three subfamilies of Rho proteins appear to contribute to this invasion, while only Rac may be needed for the movement of fibroblasts across a solid substrate in culture. These various behaviors also illustrate an important distinction between the Ras proteins and their distant Rho family cousins: in cancer cells, Ras proteins are often activated by alterations in their structure (more specifically, amino acid substitutions), while the various Rho proteins are functionally activated by their upstream physiologic regulators.

All of the signaling connections between the tyrosine kinase receptors, such as the PDGF and EGF receptors, and these Rho family proteins are not known. We do know, however, that these receptors, by activating Ras, stimulate at least three downstream signaling pathways involving the Raf, PI3K (phosphatidylinositol-3 kinase), and Ral-GEF effectors (see Sections 6.5, 6.6, and 6.7); a subset of these may, in turn, influence cell motility via regulation of the Rho proteins. In addition, activated Ras binds and appears to activate Tiam1 (see Figure 6.33), which functions as a guanine nucleotide exchange factor (GEF) for Rac. (Recall that GEFs are responsible for causing small G proteins, such as Ras and Rho, to jettison bound GDP and take on GTP, thereby activating signaling by these G proteins.) Hence, Tiam1 and Rac should also be considered to be downstream effectors of Ras that affect cell motility, ostensibly by regulating the formation of lamellipodia.

Actually, from the perspective of cell motility, PI3K is clearly the most important of the well-characterized Ras effectors. By generating PIP3 [phosphatidylinositol

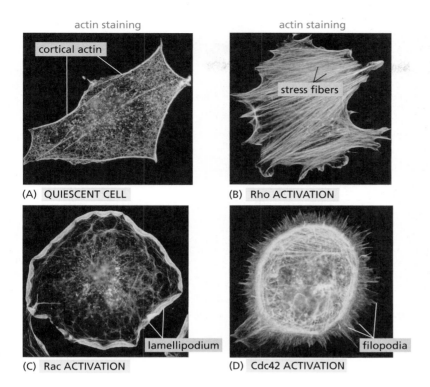

actin staining   actin staining

(A) QUIESCENT CELL   (B) Rho ACTIVATION

(C) Rac ACTIVATION   (D) Cdc42 ACTIVATION

Figure 14.39 Effects of Rho-like proteins on the actin cytoskeleton and cell adhesion Members of the Rho family of small GTPases, which consists of the Rho, Rac, and Cdc42 subfamilies, control both the actin cytoskeleton and the formation of focal adhesions (see Figure 5.28). The actin fibers were labeled with fluorescent phalloidin. (A) Quiescent, serum-starved 3T3 fibroblasts serve as controls for the panels that follow; note the cortical actin beneath the plasma membrane of this cell. (B) Exposure of cells to lysophosphatidic acid, which specifically activates Rho subfamily proteins, causes the cell to assemble large numbers of focal adhesions (*not shown*), which in turn enable the assembly of actin stress fibers. (C) Micro-injection of a constitutively activated form of a Rac protein into a cell causes it to construct a single enormous lamellipodium around its entire circumference. (In contrast, a focal source of a Rac-activating signal is likely to induce a lamellipodium only on the side of the cell facing this source.) (D) Micro-injection of a guanine nucleotide exchange factor (GEF) of Cdc42 into a cell causes it to extend hundreds of filopodia in all directions. (Courtesy of Kate Nobes; from A. Hall, *Science* 279:509–514, 1998.)

(3,4,5)-triphosphate], the PI3K enzyme creates a chemical structure on the cytoplasmic face of the plasma membrane to which a variety of cytosolic proteins can attach via their PH domains (see Section 6.6). Among these proteins are a number of GEFs that are responsible for activating members of the Rho family of G proteins. These Rho GEFs become activated following their tethering to the plasma membrane.

(The overarching role of PI3K and PIP3 in choreographing cell motility is illustrated by studies of the motility of the slime mold *Dictyostelium discoideum*. PI3K, and thus its product, PIP3, is localized at the leading edge of an advancing slime mold cell. Conversely, PTEN, the enzyme that destroys PIP3 and thereby antagonizes PI3K, is localized to the sides and the rear, lagging edge of such a cell. This introduces another element into our thinking: while growth factor receptors, such as the PDGF-R, might in principle release signals encouraging PI3K activation throughout a cell, the actual signaling by this enzyme may also be influenced by its localization and the localizations of its activating GF receptor and antagonist within the cell.)

Tiam1 was originally identified as the product of a **T**-cell lymphoma **i**nvasion **a**nd **m**etastasis gene, indicating the importance of its encoded protein to these late steps of tumor progression. Tiam1 function appears to be stimulated both by its association with GTP-bound, active Ras and by its binding to PIP3. By activating Rac proteins, the Tiam1 GEF encourages the localized polymerization of actin at the leading edge of migrating cells, thereby yielding the lamellipodia that are so critical to cell locomotion (see Figures 14.38 and 14.40).

The other Rho-like proteins that are activated by Rho GEFs are responsible for other parts of the cell motility program. For example, Rho proteins like RhoA and RhoB, acting in concert with Rac proteins, promote the establishment of new points of adhesion between the leading edge of the cell and the extracellular matrix. The reverse is also true: the forging of new focal adhesions also encourages Rac activation, suggesting the operation of some self-sustaining, positive-feedback loop that ensures the continuity of forward motion. Rac and Cdc42 proteins also appear able to induce expression of certain secreted proteases, notably the matrix metalloproteinases described in the last section. By doing so, they may coordinate localized remodeling of the extracellular matrix with extension of lamellipodia at the leading edge of a motile cell.

Figure 14.40 The circuitry mediating EGF-induced cell motility Many of the signal-transducing proteins that lie downstream of the EGF receptor and are responsible for mediating EGF-induced cell motility are indicated here. The relative proportions of these proteins change as breast cancer cells acquire increased motility and invasiveness: the extent of overexpression of certain proteins in the cancer cells is indicated by the *pink* numbers, while the extent of reduced expression is indicated by the *green* number. Some proteins, such as cofilin, sever existing actin fibers in certain regions of the cell, while capping protein prevents further extension of existing fibers. Cortactin is associated with cortical actin and is responsible for organizing the invasive edges of lamellipodia. Members of the Rho family of GTPases are indicated in *yellow* and *orange*; major cellular structures are indicated in *blue* and *purple*. (Adapted from W. Wang et al., *Trends Cell Biol.* 15:138–145, 2005, and from F. Gertler and J. Condeelis, *Trends Cell Biol.* 21:81–90, 2011.)

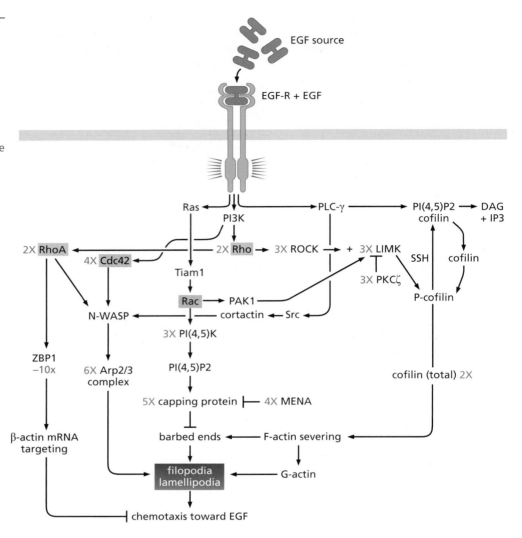

The contraction of the cell body (which helps to pull the lagging edge of the cell forward toward the leading edge; see Figure 14.36) is equally important for a cell's directed movement. This contraction is also governed largely by members of the Rho subfamily of proteins. By encouraging the formation of actin bundles in the cytoplasm, Rho proteins are able to create the structures known as "stress fibers" (see Figure 14.39B) and thereby contribute to the regulation of the contractility of the cytoplasm.

Cdc42, which represents the third subfamily of Rho-like proteins, has its own specialized function: it is able to induce the extension of the fingerlike filopodia (see Figure 14.38). As mentioned above, these filopodia contribute to cell motility by nucleating the formation of focal adhesions, which in turn allow advancing cells to establish a firm footing in the ECM. In addition, activated Cdc42 is able to stimulate generalized cell motility, independent of its specific effects on filopodia.

To complicate things even more, the actions of Rho, Rac, and Cdc42 differ in different cell types. For example, in normal epithelial cells (rather than the fibroblasts discussed above), the Rac and Rho subfamily proteins are responsible for maintaining the E-cadherin–dependent cell–cell adherens junctions; as we have read, these junctions are vital for preserving the epithelial cell sheet and therefore immobilize participating epithelial cells. However, in transformed epithelial cells, such as colon carcinoma cells that have undergone a partial or complete EMT, Rac clearly contributes to increased motility. In nonmotile cells, Tiam1 (the Rac exchange factor) is found in these adherens junctions, while in migrating cells, Tiam1 localizes to lamellipodia and related membrane ruffles.

The task of integrating these disparate observations into a single scheme has only begun. An early attempt to do so, seen in Figure 14.40, will surely be followed by dozens of revisions. Virtually all of the subcircuits depicted in this scheme of EGF-induced motility are likely to participate in organizing the motility stimulated by other motogenic growth factors, such as HGF and PDGF, as well.

The relevance of the Rho family proteins to cancer metastasis has been highlighted by searches for genes that are specifically expressed in metastatic cells but are expressed to a much lesser extent in nonmetastatic cells. In one set of experiments, strongly metastatic variants of mouse and human melanoma tumor cell lines were selected and the gene expression patterns of the cell lines were compared with those of weakly metastatic cells. Prominent among the genes whose expression was elevated in the metastatic variants was the gene encoding the RhoC protein (one of the Rho subfamily). Indeed, introduction of a RhoC-expression vector into poorly metastatic melanoma cells caused them to become highly metastatic, while ectopic expression of a dominant-interfering form of RhoC reduced the metastatic powers of usually metastatic cells. RhoC has also been found to be strongly expressed in cells of inflammatory breast cancers, a particularly aggressive form of this disease, and in pancreatic carcinomas, which are almost always highly aggressive.

Our rapidly expanding understanding of the molecular mechanisms underlying cell motility and invasiveness has converged on studies of the EMT program. In particular, these mechanisms provide insights into how some of the key changes occurring during an EMT are actually accomplished by various effectors regulating the cytoskeleton, protease synthesis, and thus cancer cell motility. Here, much interest has focused on the Mena protein, also called ENAH.

The outlines of this emerging story are as follows. Carcinoma cells that have been pushed through an EMT by the actions of one well-studied EMT-TF—Twist—exhibit changes in the splicing patterns of hundreds of pre-mRNAs, the consequences of alternative splicing (see Figure 1.16B). Thus, while the *levels* of the resulting mRNAs may not be affected by these TFs, their *protein-coding sequences* are often changed because of altered reading frames. One of these altered mRNAs encodes the structure of Mena, a protein that plays a key role in regulating the extension of actin fibers (see Figure 14.40). The normally synthesized form of Mena (**Figure 14.41A**) is replaced by one containing an invasion-associated domain (Mena^{INV}).

Mena^{INV}, in turn, affects a variety of cell behaviors associated with invasion including responsiveness to EGF-induced cell motility and trans-endothelial invasiveness. These changes conspire to generate significant increases in metastatic dissemination by the carcinoma cells within primary tumors (see Figure 14.41B–E). The detailed understanding of cell motility proteins garnered from studies like these is illuminating how an EMT works to power high-grade malignancy.

14.11 Metastasizing cells can use lymphatic vessels to disperse from the primary tumor

After invasive, motile cells enter into the vessels of blood or lymphatic systems—the process of intravasation—they disperse and, should they survive the rigors of the voyage, eventually settle in tissue sites that lie at some distance from the primary tumor. Travel via the blood circulation is often called **hematogenous** spread, and it depends on prior successful angiogenesis by the tumor. This emphasizes the fact that angiogenesis benefits cancer cells in two distinct ways. On the one hand, it supports the metabolic activity required for these cells to survive and proliferate. On the other, it provides tumor cells with direct access to avenues through which they can disperse throughout the body.

The extended discussion of hematogenous spread in Section 14.1 reflects the clearly important role of the blood circulation in metastatic dissemination. The contribution of the lymphatic vessels to the dispersion of cancer cells is, however, less obvious. Almost all tissues in the body carry networks of lymphatic vessels that are responsible

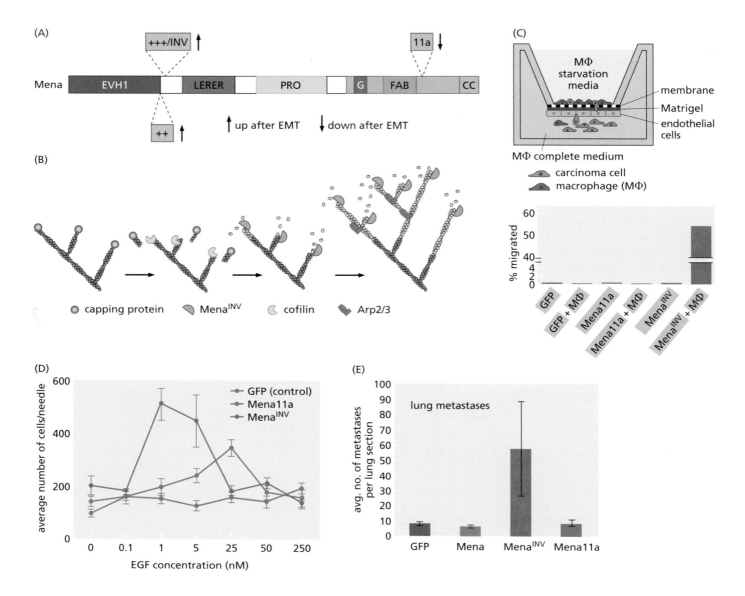

for continuously draining the interstitial fluid that accumulates in the spaces between cells. Most of these vessels converge on a major abdominal vessel that empties its lymph into the left subclavian vein near the heart and thence into the general circulation. Consequently, cancer cells present in lymphatic vessels may occasionally enter through this cross connection into the general circulation.

Tumor cells and recruited stromal companions may secrete VEGF-C, which drives lymphangiogenesis—the formation of new lymphatic vessels (see Section 13.6). Moreover, experimental tumors forced to secrete increased levels of VEGF-C will seed larger numbers of metastatic cells in nearby "draining" lymph nodes—the lymph nodes associated with the lymphatic ducts that drain the tissue in which the tumor lies (**Figure 14.42A and B**). However, detailed histological analyses of spontaneously arising tumors indicate that functional lymphatic vessels are rarely found throughout tumor masses. Instead, they largely occupy a zone at the periphery of solid tumors. Those few lymphatic vessels discovered in the central regions of tumors are usually collapsed (see Figure 13.35). As discussed in the previous chapter, it seems that the expanding masses of cancer cells within a tumor press on these vessels; because the lymphatic ducts have little internal hydrostatic pressure, they cannot resist these forces and collapse.

The absence of functional lymphatic vessels within tumor masses must influence the paths used by metastasizing cancer cells to leave the primary tumor. Without ready

Figure 14.41 How EMT drives alternative splicing, cell motility, and increased invasiveness When carcinoma cells undergo an EMT, the result is a major shift in the expression of certain splicing factors that govern alternative splicing patterns. For example, several splicing factors that favor the epithelial phenotype are depressed, resulting in the synthesis of alternatively spliced mRNAs that encode proteins fostering a mesenchymal cell phenotype. (A) The usually synthesized versions of the Mena protein seen both in normal epithelial cells and in localized carcinomas are composed of multiple domains and are specified by mRNAs encoding all of the *colored* domains with the exception of the three *light brown* domains (termed simply "Mena") or these *colored* domains plus the 11a segment ("Mena11a"). In contrast, the form of Mena seen in cancer cells that have undergone an EMT and, in addition, are growing *in vivo* within a tumor contains segments encoded by exons INV (also known as +++) and ++ and lacks segment 11a. The Mena domain encoded by the INV exon seems to play an especially important role in carcinoma cell invasiveness. The arrows indicate the changes seen in association with passage through an EMT. (B) The various forms of Mena promote the elongation of actin fibers that are essential for the growth of both lamellipodia and filopodia; only lamellipodial growth is illustrated here. Normally, a capping protein (*red circle*) blocks the further extension of actin fibers. Cofilin (*yellow*) cleaves the ends of actin fibers, removing the capping protein, and Mena then prevents the reattachment of capping protein, permitting further fiber elongation; the INV isoform of Mena is especially effective in promoting this elongation, thereby driving actin fiber elongation and the growth of both lamellipodia and filopodia. Yet another protein—Arp2/3 (*blue*)—enables the branching of actin fibers that is critical for stable lamellipodial extension. (C) As described in Figure 14.7, macrophages interact with breast carcinoma cells via bidirectional signaling that promotes carcinoma cell invasiveness. (In the case of Figure 14.7, this invasiveness was manifested by an increased ability to intravasate.) Invasiveness through an endothelial cell layer can be measured in an *in vitro* assay (*above*) in which carcinoma cells (*green*) are introduced into the upper chamber of a two-chamber "transwell" incubation vessel, either together with or in the absence of macrophages (*red*). These cells are separated from the lower chamber by a perforated membrane (*dotted black line*), below which are a layer of ECM formed by Matrigel (a mixture of basement membrane proteins, *solid gray line*) and a monolayer of endothelial cells (*pink*). Conditioned medium in the bottom chamber attracts the carcinoma cells to migrate from the upper chamber, requiring their invasion through the endothelial cell monolayer. Carcinoma cells were genetically modified by vectors expressing either GFP, Mena11a, or Mena[INV]. As shown *below*, a control expression vector encoding GFP (green fluorescent protein) has no effect on trans-endothelial migration, even in the presence of added macrophages (MΦ) similarly, no effect was seen when these carcinoma cells were forced to express the Mena11a (epithelial) isoform. Moreover, the presence of both macrophages and the Mena/Mena11a isoforms fails to induce trans-endothelial migration. Only when the carcinoma cells are incubated with macrophages and forced to express the Mena[INV] isoform does this migration proceed efficiently. (D) Expression of the Mena[INV] isoform also sensitizes carcinoma cells to the presence of EGF, such as that secreted by nearby macrophages (see Figure 14.22). Insertion of a large-bore EGF-filled needle into a carcinoma can be used to assay invasiveness by counting the number of carcinoma cells that are sensitized to motogenic EGF signals, responding by migrating into the bore of the needle. As seen here, carcinoma cells that express the Mena[INV] isoform are recruited by a 25-fold lower concentration of EGF than is required to recruit carcinoma cells expressing the control GFP plasmid, while expression of the Mena11a isoform actually suppresses the control/basal level of recruitment; this indicates that profound changes in sensitization to EGF are elicited by the different Mena isoforms. (E) These various effects on increased invasiveness and EGF responsiveness collaborate to affect the rate of metastasis to the lungs from primary tumors expressing either a GFP control vector, a Mena11a vector, or a Mena[INV] vector. (A, from S. Goswami et al., *Clin. Exp. Metastasis* 26:153–159, 2009. B, from U. Philippar et al., *Dev. Cell* 15:813–828, 2008. C and D, from E.T. Roussos et al., *J. Cell Sci.* 124:2120–2131, 2011. E, from U. Philippar et al., 2008, and E.T. Roussos et al., 2011.)

access to lymph ducts, most motile cancer cells are forced to emigrate via the far more numerous functional capillaries, which are threaded throughout the tumor mass. In spite of such limited access, some cancer cells do indeed succeed in entering the lymphatic system. In the specific case of mammary carcinomas, some metastasizing cancer cells enter into the lymphatic vessels that directly drain the mammary gland and collect in the nearby downstream lymph nodes (see Figure 14.42A). These wandering carcinoma cells are readily detected, because their appearance differs so strongly from the surrounding lymphoid cells (see Figure 14.42C) and they express epithelial proteins, such as cytokeratins, that are otherwise absent from lymphatic tissues (see Figure 14.42D). Histological examination of draining lymph nodes is routinely used to determine whether a primary breast cancer has begun to dispatch metastatic pioneer cells to distant sites in the body (Sidebar 14.4).

The lymph nodes draining a primary tumor might well function as staging areas. Thus, once cancer cells multiply and form small metastases within these nodes, they may disperse further by dispatching metastatic pioneers to more distant sites in the body. In fact, through much of the twentieth century, surgeons believed that the draining lymph nodes of a tissue function as filters, and that once these nodes become filled

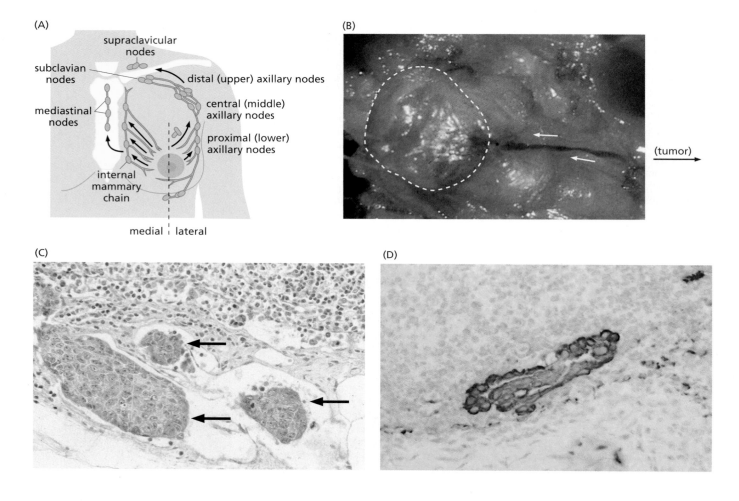

Figure 14.42 Draining lymph nodes of the mammary gland
(A) The lymphatic ducts (*red*) and the lymph nodes draining the breast (*swellings along ducts*) are initial sites of metastatic spread, carcinoma cells being carried there by the flow of lymph (*arrows*) leaving various sectors of the breast. Discovery of carcinoma cells in these lymph nodes, which is observed in more than 30% of human breast carcinoma patients at the time of initial diagnosis, suggests the possibility of deposits of metastatic cells in more distant sites in the body, particularly if large numbers of draining nodes are found to carry breast cancer cells. (B) The lymph node that serves as the sentinel node of a tumor can usually be identified among all of the lymph nodes draining the breast by injecting a blue dye into the tumor (*outside photographic field to right*) and following the trail of the dye via the lymphatic duct (*arrows*) to the draining node (*outlined in dashed line, left*). (C) Hematoxylin–eosin (H&E) staining of a section of an *axillary* lymph node reveals that three micrometastases (*arrows*) arising from a primary breast tumor have grown in the space between the capsule surrounding this node (*not seen, below*) and the mass of lymphocytes within the node (*small cells, dark nuclei, above*), displacing the latter upward. (D) Immunohistochemistry using an antibody specific for cytokeratins (*brown*) reveals this small micrometastasis in a sentinel node. This procedure is far more sensitive than H&E staining (panel C) in detecting micrometastases, since the mesenchymal cells of the lymph node do not express cytokeratin, which is made by epithelial cells and thus by most carcinoma cells. (A–C, from A.T. Skarin, Atlas of Diagnostic Oncology, 4th ed. Philadelphia: Elsevier Science Ltd., 2010. D, from J.P. Leikola et al., *Cancer* 104:14–19, 2005.)

with metastasizing cancer cells, these cells spill over into other lymphatic vessels, through which they travel and disseminate widely throughout the body.

The alternative notion is that draining lymph nodes represent dead ends for disseminated cancer cells—that is, those cancer cells that proliferate within these nodes rarely move on to more distant sites in the body. Indeed, studies of patients carrying breast, head-and-neck, gastric, and colorectal carcinomas indicate that surgical removal of draining lymph nodes has no effect on long-term patient survival. Such observations suggest that metastasis through the lymph and through the blood operate in parallel, and that the cancer cells that arrive in lymph nodes usually venture no farther. Accordingly, in most tumors, cancer cell–positive lymph nodes represent useful "**surrogate markers**" of metastasis by providing useful diagnostic and prognostic data without being directly involved in the processes that lead to widespread cancer cell dissemination and metastatic disease.

14.12 A variety of factors govern the organ sites in which disseminated cancer cells form metastases

The descriptions in the previous sections of the mechanisms of invasion and metastatic dissemination seem to explain, at least in outline, how most of the steps of the invasion–metastasis cascade proceed. Moreover, it is plausible that the dispersion strategies used by a wide variety of invasive, metastatic cancer cell types will one day be found to be governed by a common set of mechanistic principles, such as those discussed here. Importantly, however, our discussions did not address the last step of the invasion–metastasis cascade—colonization.

The growth of micrometastases (<2 mm diameter) into macrometastases (>2 mm diameter) is clearly the key step in determining whether or not metastatic disease will ever develop. For example, 30–35% of the women diagnosed with primary breast carcinomas have thousands of micrometastases in their bone marrow, many composed of single cells or tiny clusters of cells (see, for example, Figure 14.2C), yet only half of these women will ever suffer a disease relapse triggered by the appearance of macroscopic metastases. Clearly, colonization (that is, the growth of micrometastases into macroscopic metastases) is an extremely inefficient process, and the vast majority of cells that end up forming small micrometastases never succeed in properly adapting to the tissue in which they have landed by spawning macrometastases.

In addition, while a variety of cancer cell types may execute the earlier steps of the invasion–metastasis cascade in a very similar fashion, it is likely that colonization of a tissue by each type of cancer cell proceeds quite differently. Thus, successful adaptation of metastasized breast cancer cells to the bone marrow (which, by definition, enables these cells to colonize the marrow) is likely to involve a quite different set of cellular changes from those required for successful bone marrow colonization by prostate cancer cells. In addition, the changes required for a breast cancer cell to colonize the bone marrow are likely to be quite different from those needed for it to succeed in brain or lung colonization.

Abundant evidence supports the notion that metastatic cancer cells that have colonized a certain target organ must become *highly specialized* to do so: (1) 75% of young patients with papillary thyroid carcinomas have significant lymph node metastases, but only 3% will ever develop distant metastases. Hence, adaptation to the lymph nodes by metastasizing thyroid carcinoma cells does not allow them to colonize other tissues in the body. (2) Similarly, duodenal carcinoid tumors greater than 1 cm in diameter (containing >10^9 cells) have a high rate of lymph node metastasis, yet they rarely metastasize to the liver, which is the common site of metastasis of the tumors that arise in the nearby colon. (3) Cancer cells isolated from human lymph node metastases have been found, after injection into the venous system of mice, to grow preferentially in the lymph nodes of their mouse hosts rather than other possible sites of colonization. (4) Surgical removal of isolated, relatively large colorectal carcinoma metastases present in the liver or lung often results in disease-free survival of patients for a number of years, in spite of the fact that the circulation of these patients clearly carries large numbers of metastasizing cells, including some that already possess colonizing ability in one or several organs. (5) Mouse melanoma cells can be selected that metastasize preferentially to lungs, or breast cancer cells that metastasize to the lungs or, alternatively, to the bone. These disparate observations reinforce the notion that the ability to colonize a certain organ represents an acquired specialization, indeed one that is rarely achieved by disseminated cancer cells.

Yet another factor affects these dynamics: different types of cancer cells acquire the ability to colonize a given tissue more or less readily. Thus, the ability of metastasizing prostate cancer cells to colonize the bone marrow seems to be far more readily acquired than their ability to colonize the liver or the pancreas. This suggests that the differentiation program of normal prostatic epithelial cells exerts a strong influence on the ability of derived carcinoma cells to form metastases in specific organs. If we were to place prostate carcinoma cells and potential target organs on a map that depicts metastatic tendencies (**Figure 14.43**), we would indicate that the prostatic cells have relatively easy "access" to the bone marrow, implying that they need to undergo fewer changes in order to adapt to this site. Conversely, they have more limited access to other organs, such as the liver or pancreas, because of their need to undergo more complex adaptations in order to successfully colonize those particular organs.

Figure 14.43 Primary tumors and their metastatic tropisms In this diagram, the relative width of each arrow indicates the relative proportion of clinically apparent metastases that are generated by a primary adenocarcinoma. Four types are indicated here: prostate, breast, pancreas, and colon. In some cases, a tumor's tendency to spawn metastases in one or another tissue reflects the abilities of the cancer cells from the primary tumor to adapt to (and thus colonize) the microenvironment of distant tissues; this likely explains the strong tendencies of prostate and breast cancers to generate metastases in the bone marrow. In other cases, the layout of the circulation may strongly influence the site of metastasis. For example, the high proportion of liver metastases deriving from primary colon cancers likely reflects the drainage via the portal vein of blood from the colon directly into the liver (see Figure 14.45).

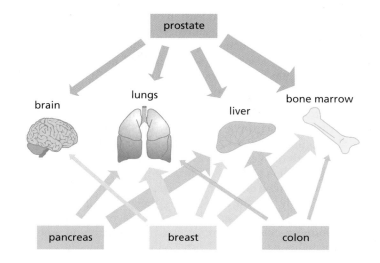

Figure 14.44 Stephen Paget The British physician Stephen Paget (1855–1926) was the first to enunciate the "seed and soil" hypothesis, which states that the ability of a disseminated cancer cell to successfully found a metastasis depends on whether a distant tissue offers it a hospitable environment to survive and proliferate. (From I.J. Fidler, *Nat. Rev. Cancer* 3:453–458, 2003.)

An extreme case of narrow metastatic tropism comes from the behavior of **uveal melanomas**, which arise in the pigment cells of the eye. These cells characteristically metastasize to the liver, indicating that some component of the differentiation program of ocular melanocytes allows them to adapt far more readily to the liver microenvironment than to the microenvironments of other tissues. (This tendency to form hepatic metastasis cannot be predicted from the known biology of ocular melanocytes: it is hard to imagine two tissue microenvironments in the body less similar than the globe of the eye and the liver.)

This predilection to form metastases in one or another organ site was noted as early as 1889 by the British pathologist Stephen Paget (**Figure 14.44**). He proposed the "seed and soil" hypothesis, in which he likened the seeding of cancer cells to the dispersal of the seeds of plants. After studying the clinical course of 735 breast cancer patients, Paget concluded that the patterns of metastasis formation in these patients could not be explained either by random scattering throughout the body or by the patterns of dispersal from the breast through the general circulation. He therefore proposed that the metastasizing cancer cells (the seed) find a compatible home only in certain especially hospitable tissues (the soil). He wrote, "a plant goes to seed, its seeds are carried in all directions; but they can only live and grow if they fall on congenial soil." This ability to form macroscopic metastases in some sites but not others has been highlighted by certain clinical procedures (Supplementary Sidebar 14.11).

The seed and soil hypothesis cannot, however, explain the metastatic patterns of *all* types of human cancers (**Sidebar 14.5**). Instead, in certain cases, the predilection to metastasize to a certain target organ is likely to be dictated by the layout of the vessels connecting the site of a primary tumor and the site of metastasis. For example, the strong tendency of colon carcinoma cells to metastasize to the liver may simply reflect the fact that these cancer cells leave the gut via the portal vein (which drains the lower gastrointestinal tract and the spleen) and, after a very brief trip, almost inevitably become lodged in the capillary beds of the liver that are fed by this vein (**Figure 14.45**). Even if individual metastasizing colon cancer cells colonize the liver with an

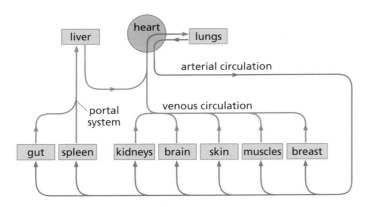

Figure 14.45 Portal circulation and liver metastasis While the venous systems of most tissues drain to the right side of the heart and thereafter into the capillary beds of the lungs, the veins draining the spleen and gut are organized differently, in that their venous blood empties directly into the liver via the portal circulation before being dispatched back to the heart. Consequently, vast numbers of metastasizing colorectal carcinoma cells are trapped in capillary beds of the liver within seconds of leaving the colon. (Adapted from I.C. MacDonald, G.C. Groom and A.F. Chambers, *BioEssays* 24:885–893, 2002.)

Sidebar 14.5 Contralateral metastases are relatively rare Possibly the greatest embarrassment for the seed and soil hypothesis comes from its failure to explain the rarity of contralateral metastases. For example, cancer cells disseminated from a primary tumor in one breast should find that the **contralateral** (that is, opposite) breast provides the most hospitable environment for colonization. In fact, only about 2% of breast cancer cases result in contralateral metastases, comparable to the frequency of tumors in the breast that arise as metastases of primary tumors located elsewhere. Similarly, primary kidney cancers metastasize infrequently to contralateral kidneys. These behaviors are clearly incompatible with the seed and soil hypothesis and still require explanation. One possibility is that, in addition to landing in a compatible organ, a disseminating cancer cell prefers to take root in a tissue that has an activated stroma, such as is seen in sites of chronic inflammation or wound healing (see Supplementary Sidebar 14.12).

extremely low efficiency, the sheer numbers of the cancer cells trapped in the liver guarantee that, with the passage of enough time, substantial numbers of metastases will arise in this target organ.

The same logic may explain why breast cancer cells often form metastases in the lungs. As is the case with metastasizing colorectal carcinoma cells, wandering mammary carcinoma cells may not find that the lungs provide them with an especially hospitable environment, and individual cancer cells will have a low probability of successfully colonizing the lungs. Nonetheless, some metastases will eventually form there, simply because so many of these cells become physically trapped in this tissue (see Figure 14.45). This logic suggests that, in general, the frequency of metastases to an organ is governed by two parameters—the frequency with which metastasizing cells are physically trapped in an organ, and the ease with which they can adapt to the microenvironment of that organ.

There are also indications that tissues that are normally not hospitable sites for colonization may become so through specific pathological processes, such as localized wounding (Supplementary Sidebar 14.12). This suggests that areas of chronic inflammation within the body of a cancer patient may occasionally become congenial environments for metastasizing cancer cells, simply because they provide a spectrum of mitogenic and trophic signals, as discussed in Chapter 13.

Such metastasis to areas of wounding and inflammation can explain a phenomenon first described in 2009—**tumor self-seeding**. Circulating tumor cells (CTCs) may find that the most congenial tissue microenvironment to colonize is the stroma of the primary tumor itself. On the one hand, the reactive stroma of the primary tumor (see Section 13.3) would seem to provide the factors that are highly supportive of the survival and proliferation of many types of cancer cells. On the other, these CTCs do not need to undergo the adaptive changes that seem to be required when cancer cells arising in one organ attempt to colonize a different organ; here these cells are returning to a familiar territory. Tumor self-seeding also holds implications for the genetic makeup of the primary tumor: if cancer cells undergo adaptive genetic changes in distant sites of metastasis in order to colonize those sites, secondary waves of metastasis from those distant sites may carry these genetic changes back to the primary tumor, causing its cells to increasingly exhibit the genotypic (and phenotypic) alterations developed elsewhere in the body!

Yet other mechanisms have been proposed to explain the tissue **tropisms** of metastasizing cells. For example, target organs may release specific chemical messages—the chemoattractants sometimes termed chemokines—that might actively recruit wandering cancer cells from the circulation. Such chemoattraction clearly operates to ensure the homing of a variety of circulating immune cells to specific tissues as part of the normal operations of the immune system. In one study, when B16 mouse melanoma cells were forced to express the CXCR4 chemokine receptor, their metastases to the lung increased by a factor of 10. However, when an expression vector specifying the CXCR7 receptor was introduced into these melanoma cells, they then showed substantially increased metastasis to the lymph nodes, thereby appropriating a mechanism normally used by lymphocytes for homing to these nodes. (In truth, since these chemokine-activated receptors often provide mitogenic and survival signals to cancer cells, it is difficult to know whether these receptors induce metastasizing cells to migrate into a specific tissue or simply encourage the survival and proliferation of these cells after they have landed in one but not another tissue.)

According to another mechanistic model of metastatic tropism, the capillaries forming the *vascular beds* (that is, the networks of blood vessels) in various tissues express tissue-specific molecules on their luminal surfaces. These molecules may offer specialized docking sites for cancer cells that express certain adhesion molecules, such as integrins, on their surfaces. This model is sometimes termed the "vascular **ZIP code**" theory, because it implies that the luminal surfaces of vessels in different tissues carry, in chemical form, specific homing addresses, much like those used by a postal system. This model fails to take into account the fact that cancer cells in the circulation are often surrounded by clouds of platelets (see Supplementary Sidebar 14.2) that are

Sidebar 14.6 Some cancer cells leave home early Embedded in the discussions in this and previous chapters is the notion that multi-step tumor progression occurs in the primary tumor site, and that only after cancer cells in these tumors have evolved to a certain state of aggressiveness do they begin to disseminate and attempt to found metastases. In fact, clinical observations have demonstrated the presence of disseminated tumor cells in certain organs, such as the bone marrow, long before primary tumor progression has generated aggressive tumor cells. (In the case of carcinomas, these observations may one day be explained by the activation of EMT programs in the cells of relatively benign, early-stage tumors.)

This early dispersal suggests that once relatively early-stage cancer cells have settled in distant tissues, they may evolve in those tissues through the later steps of multi-step tumor progression (see Chapter 11). Accordingly, such **parallel progression** may occur at a great distance from and independently of the multi-step progression occurring in the primary tumor. If validated, this would force a fundamental rethinking of how multi-step tumor progression usually proceeds.

This model is encumbered, however, by the fact that these early-disseminating cells are doubly handicapped: Like their late-stage counterparts, they are poorly adapted to proliferate in the foreign tissues in which they have landed. In addition, they lack the mutations (in oncogenes and tumor suppressor genes) needed to proliferate vigorously wherever they land. Since the genetic and epigenetic evolution of tumor cells that is central to multi-step tumor progression seems to require active cell proliferation, these dual impediments to proliferation would seem to dictate that these dispersed cells will forever remain dormant micrometastases.

capable of blocking direct association between the cancer cell and the luminal surfaces of endothelial cells.

One study of the behavior of metastasizing human cancers calculated that 66% of metastases could be explained simply by the blood flow patterns between the primary tumor and the sites of observed metastases. In 20% of the cases, the specialized microenvironments of target tissues (rather than blood flow patterns) seemed to account for the tendency of certain cancers to form macroscopic metastases. And in 14% of cases, negative interactions (in which tissues seemed to actively repel wandering cancer cells) seemed to explain smaller-than-expected numbers of metastases predicted by blood flow patterns.

To summarize, these diverse observations suggest that metastasizing cells disperse to many organ sites in the body and that their dispersion is affected by the layout of the vasculature. Once arrived in these various sites, the cancer cells will usually survive and eventually colonize only those tissues that provide them with specific chemokines, trophic factors, and mitogens. On occasion, however, these cells may succeed in founding macroscopic metastases in relatively inhospitable organ sites, only because the routing of the blood circulation introduces these cells in vast numbers into such sites. One interesting and still-unresolved question involves the timing of metastatic dissemination: When during the course of primary tumor progression do cells begin to disperse to distant sites where they eventually form metastases (**Sidebar 14.6**)?

14.13 Metastasis to bone requires the subversion of osteoblasts and osteoclasts

The development of bony metastases represents one instance in which we understand in some detail the biochemical and biological mechanisms that permit metastasized cancer cells to thrive in a specific tissue microenvironment. This fact, on its own, justifies a detailed discussion of **osteotropic** metastasis. In addition, and as mentioned repeatedly, several of the most common types of cancer occurring in the Western world—carcinomas of the lung, breast, and prostate—show a strong tendency to metastasize to the bone. In fact, patients with advanced breast and prostate cancer almost always develop bone metastases. And in those patients who succumb to these cancers, the bulk of the tumor cells in their bodies at the time of death are usually found among the metastases scattered throughout their bones.

We usually think of bone as being a static tissue which, once formed, retains its structure throughout life. The truth is far more interesting. In mammals, about 10% of skeletal bone mass is replaced each year, resulting in an essentially complete replacement over the course of a decade. This continuous remodeling enables the bones to respond to mechanical stresses by compensatory reinforcing of stressed regions. For example, the bones of the legs are continuously being remodeled in response to the weight-bearing signals that different portions of each leg bone receive.

The turnover of bone is the work of **osteoclasts**, which break down mineralized bone, and of **osteoblasts**, which reconstruct it. The osteoclasts function first to **demineralize** the bone (by dissolving its calcium phosphate crystals) and then to degrade the now-exposed extracellular matrix, which previously formed the organic scaffolding for the calcium phosphate crystals (the process is often termed **resorption**; **Figure 14.46**). Osteoblasts move in soon after to carry out reconstruction, which involves

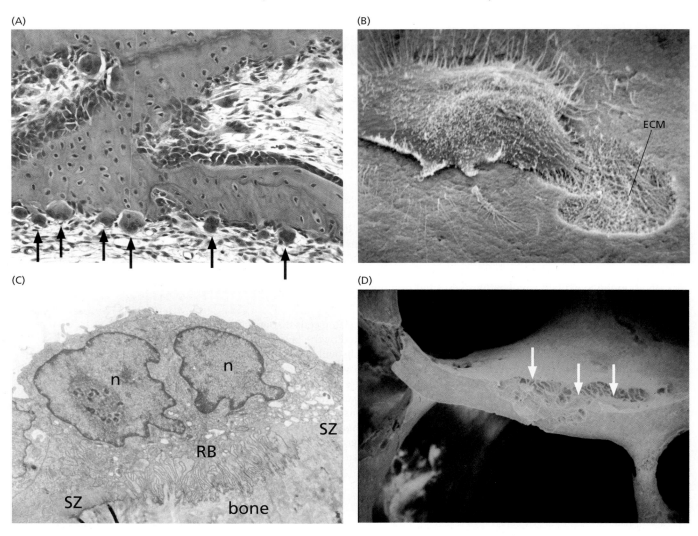

Figure 14.46 Bone degradation by osteoclasts The degradation of bone, often termed *resorption*, depends on the complex actions of osteoclasts—large multinucleated cells deriving from the monocyte lineage that also generates macrophages. (A) This light micrograph shows osteoclasts (*purple, arrows*) excavating small pits in the surface of a mouse jawbone (*pink*). (B) At far higher magnification, this scanning electron micrograph shows a cat osteoclast that has excavated a shallow pit in the surface of mineralized bone. The calcium apatite crystals in the bone have been dissolved away by acid secreted by the osteoclast, revealing the complex meshwork of collagen-rich extracellular matrix (ECM) at the bottom of the pit. Associated with this ECM are mitogens and survival factors that become available to cancer cells after osteoclasts subsequently break down the ECM. (C) This transmission electron micrograph reveals at even higher magnification the details of how osteoclasts resorb bone.

This section through an osteoclast (with multiple nuclei, n) and underlying bone reveals the osteoclast's complex ruffled border (RB), which secretes protons to dissolve the mineral component of the bone and acid proteases to degrade the collagen-rich extracellular matrix that is exposed following demineralization. Surrounding this area of contact is a circular sealing zone (SZ) containing substantial amounts of filamentous actin, which functions as a gasket to confine these secretions to a small localized area between the osteoclast and bone. (D) Another scanning electron micrograph reveals how devastating the osteolytic lesions (*arrows*) can be in terms of compromising bone structure in a patient with metastatic osteolytic lesions created by osteoclasts. (A, courtesy of T.R. Arnett. B, from T.R. Arnett and D.W. Dempster, *Endocrinol.* 119:199–124, 1986. C, from H. Zhao et al., *J. Biol. Chem.* 276:39295–39302, 2001. D, courtesy of G.R. Mundy.)

(A) normal bone

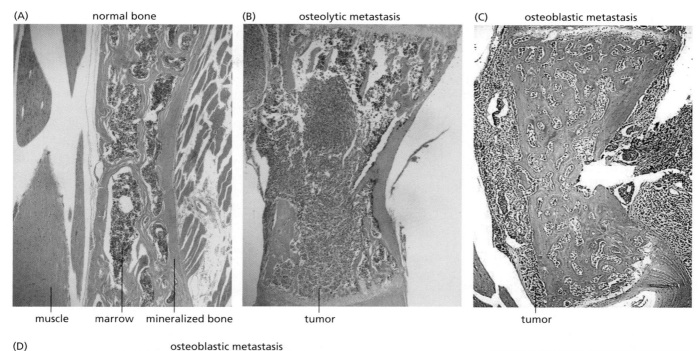

muscle   marrow   mineralized bone

(B) osteolytic metastasis

tumor

(C) osteoblastic metastasis

tumor

(D) osteoblastic metastasis

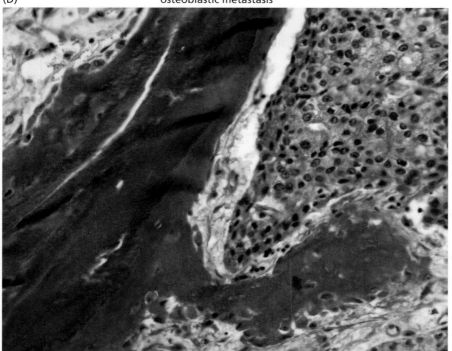

Figure 14.47 Osteolytic and osteoblastic metastases The first three of these micrographs present sections of mouse vertebrae and femurs in which the mineralized bone (*orange*), surrounding muscle (*bright red*), and bone marrow (*dark purple*) are clearly delineated. (A) This vertebra of a control mouse injected only with buffer is seen to be composed of extensive marrow with ribbons of mineralized bone running through the marrow. (B) In a mouse bearing a human breast cancer cell line (MDA-MB-231) that creates osteolytic lesions, much of the mineralized bone is seen to be missing, and the marrow has been displaced by tumor cells (*dark red*). (C) In a mouse bearing a human breast cancer cell line (ZR-75-1) that creates osteoblastic lesions, much of the marrow space is now filled with mineralized bone (*orange*) with tumor masses evident to the left and right. (D) In the iliac crest of the pelvis of a prostate cancer patient, the native bone (*green-blue*) can be readily resolved from the newly synthesized, still-poorly mineralized osteoblastic lesion (*red*), which is sometimes termed *osteoid*. The osteoid is protruding into a mass of metastatic prostate cancer cells (*dark purple nuclei*). (A–C, from J.J. Yin et al., *Proc. Natl. Acad. Sci. USA* 100:10954–10959, 2003. D, courtesy of C. Morrissey and R.L. Vessella.)

both the assembly of new ECM and the deposition of calcium phosphate crystals in the interstices of this matrix. As can be deduced from this description, the two cell types normally work in close coordination.

Most kinds of metastasizing cancer cells are, on their own, incapable of remodeling bone structure. Instead, they manipulate and exploit these two types of cells normally present in the bone. Thus, breast cancer cells preferentially activate the osteoclasts, resulting in **osteolytic** metastases—literally, metastases that dissolve bone. Prostate cancer cells tend, on the other hand, to activate osteoblasts, yielding **osteoblastic** lesions, in which immature mineralized bone (sometimes termed **osteoid**) actually accumulates in the vicinity of the metastases (**Figure 14.47**).

Figure 14.48 Osteoblasts versus osteoclasts (A) The physiologic balance between normal bone formation and resorption is created by signaling between osteoblasts, which assemble bone, and osteoclasts, which dissolve it. In an ongoing cycle, osteoclasts remove mineralized bone by covering and sealing off a section of bone and secreting digestive acid into the bone below them (see Figure 14.46C); this is followed by osteoblastic filling of resulting cavities with new bone. The osteoblasts release RANKL, which acts via the RANK receptor (*not shown*) displayed by osteoclast precursors to induce the latter to mature into functional osteoclasts. The osteoblasts may also secrete osteoprotegerin (OPG), which acts as a decoy receptor to ambush RANKL before it has had a chance to activate osteoclast precursors. Hence, the balance between RANKL and OPG determines the net rate of bone growth/loss. (B) Release by a breast cancer cell (*right, gray*) of PTHrP (parathyroid hormone–related peptide, *below*) causes nearby osteoblasts to change the mix of signals that they release: they increase RANKL synthesis (*red arrow*) and decrease OPG synthesis (*blue line*), the result being RANKL-induced maturation of osteoclast precursors into functional osteoclasts. The latter undertake osteolysis, which causes bone demineralization, exposes the extracellular matrix within the bone (see Figure 14.46B), and results in liberation of TGF-β, Ca^{2+}, BMPs, PDGF, FGFs, and IGF-1 (*upper left and middle*). IGF-1 and Ca^{2+} enable cancer cell proliferation and survival, and the additional presence of TGF-β induces the cancer cell to release more PTHrP, resulting in a self-sustaining positive-feedback loop that has been termed the "vicious cycle" of osteolytic metastasis. (B, from G.R. Mundy, *Nat. Rev. Cancer* 2:584–593, 2002.)

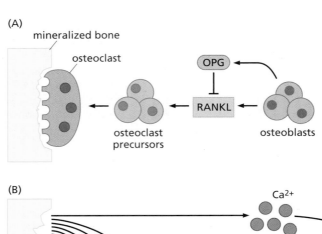

(A)

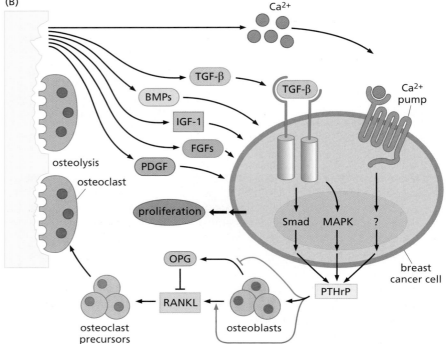

(B)

In fact, these two behaviors represent the extremes of a continuum, since both types of cancers activate both osteoblasts and osteoclasts to a greater or lesser extent. For example, while osteolytic metastases predominate in advanced breast cancer patients, as many as one-quarter of these women also have clearly defined osteoblastic lesions in their bones. Similarly, prostate carcinomas also generate occasional osteolytic metastases scattered among the many osteoblastic growths spawned by these tumors. One exception to this rule of a mingling of both types of bone metastases is provided by myeloma cells (tumors of the B-cell, antibody-secreting lineage), which create exclusively osteolytic lesions.

The normally operative close coordination between osteoblasts and osteoclasts is mediated, at least in part, by the exchange of growth factor signals. An important inducer of osteoclast differentiation is RANK (receptor activator of NF-κB) ligand, or simply RANKL. RANKL is produced by and displayed on the surface of osteoblasts. When an osteoclast precursor displaying the RANK receptor comes into contact with an osteoblast and its cell-surface RANKL molecules, this results in activation of the RANK receptors of the osteoclast precursor and its maturation into a functional osteoclast (**Figure 14.48A**). At the same time, osteoblasts produce a soluble decoy receptor, termed osteoprotegerin (OPG), which can bind RANKL and ambush it before it succeeds in activating the RANK receptor on the surface of osteoclast precursors. The result is a blockage of the RANKL–RANK signaling and the inhibition of osteoclast maturation. Hence, the balance between the RANKL (stimulatory) and OPG (inhibitory) signals determines the state of activation of osteoclasts.

This dynamic interaction of osteoblasts and osteoclasts provides the background for the actions of cancer cells that metastasize to bones. Their attraction to the bone derives ultimately from the nonmineralized, collagenous extracellular matrix that forms the organic scaffolding in which calcium phosphate crystals are deposited (see Figure 14.46B). As it happens, bone ECM is an unusually rich source of the mitogenic and trophic factors that allow several types of carcinoma cells to thrive. Consequently, by provoking the demineralization of bone, cancer cells gain access to the storehouse of factors sequestered in the bone ECM and use them to support their own proliferation and survival.

Metastasizing cancer cells reach the bone through the vessels feeding the marrow. Once there, they adhere to specialized stromal cells coating the surfaces of the bone facing the marrow. Metastasizing breast cancer cells, in particular, upon arrival in bone, revert to a behavior characteristic of their normal precursors (mammary epithelial cells, or MECs). During lactation, when producing milk, MECs forming the small sacs (**alveoli**) of the mammary gland release parathyroid hormone–related peptide (PTHrP). PTHrP then travels through the circulation to the bones, where it triggers a chain of events that encourages the dissolution of bone minerals by osteoclasts. This results in the mobilization of calcium ions, which travel back via the circulation to the mammary gland, where they are incorporated into the milk by the MECs.

This normal calcium-mobilizing mechanism is subverted by metastasizing breast cancer cells that become established in bones (see Figure 14.48B). Having attached to the stromal cells covering the surfaces of mineralized bone, the breast cancer cells, reverting to the habit of normal MECs, release PTHrP. The PTHrP, in turn, impinges directly on its receptors displayed by osteoblasts, causing these cells to release RANKL. RANKL then induces the differentiation of osteoclast precursors into active osteoclasts. The activated osteoclasts degrade nearby mineralized bone, thereby liberating the rich supply of growth factors attached to the extracellular matrix of the bone.

The growth factors liberated from the bone ECM, including PDGF, bone morphogenetic proteins (BMPs), fibroblast growth factors (FGFs), insulin-like growth factor-1 (IGF-1), and TGF-β, fuel the further growth of the breast cancer cells, inducing them to secrete more PTHrP. This PTHrP engenders more osteolysis by the osteoclasts, leading to a self-perpetuating signaling system that has been called a "vicious cycle" (see Figure 14.48B) in which TGF-β also plays a key role (**Sidebar 14.7**).

The central role of osteoclasts in this cycle suggests possible points of therapeutic intervention. One highly effective strategy depends on drugs belonging to the class of **bisphosphonates**, which are taken orally and become adsorbed to the apatite crystals that constitute the mineral portion of bone; the drug molecules can persist there for extended periods of time, as long as a decade or more. When bisphosphonate-containing bone is later dissolved by osteoclasts, the latter are poisoned by the liberated bisphosphonates, leading to their apoptosis. Hence, bisphosphonates are useful for reducing the burden of osteolytic lesions in patients with various types of metastatic cancer.

When immunocompromised mice carrying human breast cancer cells are treated with bisphosphonates, the number of osteolytic lesions is reduced and, at the same time, the total burden of tumor cells in these animals is decreased. This observation provides additional indication that late in tumor progression, the proliferation of these breast cancer cells depends greatly on osteolysis and the resulting liberation of growth factors from dissolved bone; similar observations have been made in human breast cancer patients treated with bisphosphonates. For example, a clinical study reported in 2011 that premenopausal women who developed breast cancer and were treated with zoledronic acid—a bisphosphonate—experienced a 37% increase in overall survival (relative to those who did not receive this treatment) over a period of seven years. Such clinical responses further reinforce the notion that many breast cancer relapses in patients (1) derive from metastatic deposits in the marrow and (2) depend on the ability of disseminated metastatic cells to generate mitogens by dissolving bone. Moreover, bisphosphonate therapy can provide additional benefits to patients suffering from metastatic breast cancer by reducing the **hypercalcemia**—elevated concentration of

Sidebar 14.7 TGF-β and PTHrP play pivotal roles in the vicious cycle of breast cancer osteolytic metastases Breast cancer cells that have metastasized to the bone produce far more PTHrP than do others in the same animal that have not—a reflection of the fact that certain growth factors liberated from the bone ECM stimulate PTHrP production by the metastatic cancer cells. The most important of these bone-derived factors is TGF-β, as illustrated by some simple experiments. In one of these, a dominant-negative TGF-β receptor (which blocks a cell's ability to respond to TGF-β) is expressed in human breast cancer cells. Such cells cease producing PTHrP and lose the ability to efficiently produce osteolytic metastases in the bone; forced expression of PTHrP in these cancer cells reverses this effect (**Figure 14.49**). In another experiment, breast cancer cells that usually lack the ability to metastasize to bone and fail to secrete TGF-β can be forced (through the use of an expression vector) to secrete TGF-β. The latter then acts in an autocrine fashion to stimulate these cells to produce their own PTHrP, allowing them to form large numbers of bone metastases. Finally, antibodies that bind and neutralize PTHrP are able to block the ability of human breast cancer cells to generate osteolytic lesions in mice. These are some of the disparate observations that have inspired the "vicious cycle" model depicted in Figure 14.48B.

MDA-MB-231cells transfected with:

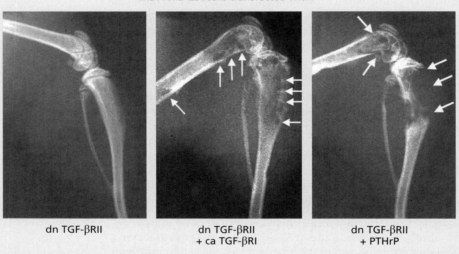

dn TGF-βRII　　　　dn TGF-βRII
+ ca TGF-βRI　　　　dn TGF-βRII
+ PTHrP

Figure 14.49 TGF-β and the formation of osteolytic metastases The evidence supporting the "vicious cycle" model of osteolytic metastasis comes in part from experiments involving the use of MDA-MB-231 cells, a line of human breast cancer cells that show a high tendency to produce osteolytic metastases (see Figure 14.47B). The ability of these cells to do so is gauged here by X-ray analyses of the hind limbs of mice that have borne MDA-MB-231 tumor xenografts. When a dominant-negative type II TGF-β receptor (dn TGF-βRII) expression construct is introduced into these cancer cells, this mutant protein blocks their ability to respond to TGF-β, specifically the TGF-β that would otherwise be liberated from the extracellular matrix of the osteolytic lesions that they may have induced. Without TGF-β stimulation, these cancer cells fail to form osteolytic metastases (*left panel*). However, if this inability to respond to TGF-β is overridden by introducing additionally into these cells an expression construct specifying a constitutively active type I TGF-β receptor (ca TGF-βRI), then the powers of these breast cancer cells to induce osteolytic lesions are restored (*arrows, center panel*). This observation, on its own, does not indicate precisely how the ca TGF-βRI succeeds in restoring the osteolytic activity to these cells. The explanation comes from an experiment in which a vector causing PTHrP expression (instead of ca TGF-βRI expression) is introduced into the cells expressing the dn TGF-βRII, causing them to regain the ability to form osteolytic metastases (*arrows, right panel*). This is compatible with the notion that PTHrP functions downstream of TGF-β signaling, and that the latter is no longer important for osteolysis if PTHrP is expressed constitutively, i.e., is no longer under the control of TGF-β (see Figure 14.48B). (From J.J. Yin et al., *J. Clin. Invest.* 103:197–206, 1999.)

calcium in the circulation—stemming from the large-scale resorption of mineralized bone. Hypercalcemia usually signals the final stages of malignant disease and causes gastrointestinal, urinary tract, cardiovascular, and neuropsychiatric problems.

Figure 14.48B predicts that the vicious cycle driving osteolytic metastases should be slowed down or even blocked by therapeutic treatment with osteoprotegerin (OPG). In fact, a derivative of OPG has been found to be as effective as a widely used bisphosphonate in slowing down bone resorption in patients with metastatic breast cancer. However, clinical development of this OPG derivative was eventually discontinued, having been displaced by a slightly more effective treatment—a monoclonal antibody (denosumab) that binds and neutralizes RANKL. A **phase III** clinical trial revealed that denosumab was more effective than bisphosphonates in delaying the onset of serious skeletal-related events in patients with metastatic breast disease, while having comparable efficacy in delaying disease progression.

As might be predicted, osteoblastic lesions depend on other signals—ones that activate osteoblasts rather than osteoclasts. In this case, the release by metastatic cancer cells of the growth factor termed "endothelin-1" (ET-1) plays a dominant role in stimulating osteoblasts and, at the same time, suppressing osteoclast activity. Thus, prostate cancer cells in primary tumors release endothelin; since its cognate receptor is also expressed by these cancer cells, an autocrine growth-stimulatory loop results. However, when these cancer cells arrive in the marrow, the endothelin that they release also acts via heterotypic signaling to stimulate osteoblasts, creating the osteoblastic lesions characteristic of this malignancy. (Precisely how osteoblast activation benefits the prostate cancer cells is less well understood. It is plausible that activated osteoblasts secrete large amounts of growth factors during the construction of mineralized bone, and that some of these factors are diverted by the cancer cells in osteoblastic metastases.)

So, Paget's seed and soil metaphor is useful, but it does not go far enough. Like seeds, metastatic cells are cast in many directions, but once they fall on certain ground, they can hardly be portrayed as being passive participants in their future fate. Instead, these cancer cells may begin to actively till the soil in which they have landed, cultivating it so that it is guaranteed to become fertile ground for their own proliferation and that of their descendants.

14.14 Metastasis suppressor genes contribute to regulating the metastatic phenotype

We have read here of a number of genes that actively promote some of the steps in the invasion–metastasis cascade. Many of these encode familiar growth factors, growth factor receptors, or signal-transducing proteins that we encountered in our earlier discussions of oncogenes and their mechanisms of action. When introduced into a variety of epithelial cells, these genes are able to encourage changes such as an EMT, the acquisition of cell motility, and even invasiveness. Indeed, it seems increasingly likely that deregulated versions of these genes are the primary forces driving many of the steps of invasion and metastasis.

Importantly, the protein products of these various genes operate as components of the complex regulatory circuits that govern many aspects of cell physiology. And like all well-designed circuits, these have both positive regulators and counterbalancing negative regulators in order to ensure finely tuned outputs. This logic leads to the conclusion that there must be a number of control elements operating in cells that counteract and balance the invasive and metastatic actions of the positive effectors of advanced malignancy. Such negative regulators, in analogy with the tumor suppressors, have been called *metastasis suppressor* genes. By definition, these genes should specifically suppress metastasis without affecting primary tumor growth.

As one might anticipate, these metastasis suppressors operate at various levels in regulating the steps of invasion and metastasis, ranging from master, pleiotropically acting regulators and signal-transducing proteins to the ultimate effectors of the various biochemical changes (Table 14.4). These genes have been identified through a variety of experimental strategies. Quite often, their expression in primary tumors and their far lower expression in derived metastases have suggested important roles in blocking the late steps of malignant progression.

Such observations, being only correlations, do not prove *causal* roles in preventing metastasis, which can only be demonstrated through other types of experiments. For example, the role of a candidate gene as a bona fide metastasis suppressor gene can be tested by the simple functional criterion mentioned above: When the gene's expression is forced in the cells of a primary tumor, does this expression permit the continued expansion of this tumor mass while at the same time blocking the appearance of

Table 14.4 Candidate metastasis suppressor genes

Name	Cellular location	Mechanism of action
BRMS1	nuclear protein	involved in chromatin remodeling
CRSP3	nuclear protein	transcription factor
KAI1/CD82	transmembrane protein	cell–cell associations
KISS1	secreted protein	ligand of G-protein–coupled receptor
NM23	cytoplasmic kinase	regulator of MAPK cascade (?)
p63	nuclear transcription factor	multiple targets
RhoGDI-2	cytoplasmic protein	negative regulator of Rho action
SseCKs	cytoplasm	cytoskeleton-associated protein
VDUP1	cytoplasm	regulator of MAPK cascade (?)
CDH1 (= E-cadherin)	cell surface adhesion	favors formation of epithelial cell sheets
TIMPs	secreted protein	inhibitor of metalloproteinases
MKK4	cytoplasm	protein kinase component of MAPK cascade
DICER	cytoplasm	miRNA processing

Adapted in part from P.S. Steeg, *Nat. Rev. Cancer* 3:55–63, 2003.

distant metastases that are usually seeded by this tumor and others like it? Some of these genes have passed such a test, while others act in a less specific way by inhibiting proliferation by all types of cells, including some that lack invasive and metastatic properties. Yet other candidate tumor suppressor genes have been found able to suppress metastasis in only a small subset of malignant tumor types.

The definitive characterization of many of these genes still lies ahead. Nonetheless, there are some genes whose anti-metastatic properties can no longer be questioned. For example, studies of *p53* have revealed repeatedly that dominant gain-of-function alleles of this tumor suppressor gene are far more potent in promoting malignancy than are null alleles. This has largely been rationalized by invoking the ability of mutant p53 molecules to form mixed heterotetramers with wild-type p53 subunits, thereby compromising the function of wild-type molecules and the wild-type *p53* allele that encodes them (see Figure 9.7). However, in recent years, the role of p53's cousin, p63, has become prominent: by forming mixed heterotetramers with p63 molecules, p53 can reduce p63 function, thereby often leading to increased metastatic propensity.

The protein p63, for its part, suppresses both tumorigenesis and metastasis and therefore is not purely a metastasis suppressor. Its metastasis-suppressing functions derive from its ability to promote expression of *Dicer*, arguably the most intriguing metastasis suppressor gene. This gene specifies the enzyme involved in one of the final steps of microRNA processing (see Figure 1.20), and reduced levels of Dicer have been correlated with increased aggressiveness of human tumors. More important, in animal tumor models, reduced levels of this microRNA-processing enzyme have been found to potentiate metastatic dissemination. Levels of Dicer cannot be driven down to zero, however, since total loss of Dicer is lethal for cells; this is not unexpected, given the role of Dicer in the maturation of hundreds of distinct microRNAs that control more than half of the mRNA species expressed in cells. (Actually, it is extremely surprising that changes in the levels of such a widely acting cellular enzyme can have such a focused effect on a specific biological process—in this case, metastasis.)

The E-cadherin molecule, about which much has been said in this chapter, is also considered to be the product of a metastasis suppressor gene. It represents the keystone of the epithelial cell state, acting through its ability to stabilize cell–cell contacts in epithelial sheets and by preventing the EMT (see Section 14.3); these powers clearly place it among the major molecular obstacles that block acquisition of the invasive phenotype by carcinoma cells. By binding and sequestering cytosolic β-catenin molecules (see Figure 6.26), E-cadherin ensures that these molecules cannot travel to the nucleus and activate, via their association with Tcf/Lef transcription factors, EMT-associated genes.

Another metastasis gene encodes the KAI1/CD82 protein, which weaves its way back and forth four times through the plasma membrane. Its expression has been found to be substantially repressed in many advanced lung, pancreatic, prostate, colon, and gastric carcinomas. A poor prognosis for breast cancer patients is associated with low KAI1 expression in their cancer cells. In cultured cells, KAI1 suppresses migration and invasiveness and, at the same time, enhances their aggregation with one another. Its location near adherens junctions is compatible with its playing a role in cell–cell adhesion. KAI1 has also been reported to act as an antagonist of EGF receptor signaling.

Yet another gene of interest here encodes the KISS1 protein, which has been identified tentatively as a ligand of a cell surface G-protein–coupled receptor (GPCR; see Section 5.7). Ectopic expression of the *KISS1* gene in tumor cells suppressed their metastatic tendencies without affecting the growth of these cells in primary tumors. Like several others in this class of genes, its precise biochemical role in metastasis suppression is poorly understood.

The breast cancer metastasis suppressor-1 (*BRMS-1*) gene was identified because of its decreased expression in breast cancer metastases. Its ectopic expression in breast carcinoma and melanoma cells suppressed their metastatic tendencies while having small but measurable effects on primary tumor growth. It has been reported to increase the gap-junctional communication between cells, which involves channels that allow adjacent cells to exchange molecules of molecular weight less than about 10^3. At the same time, the BRMS-1 protein has been found in the nucleus as part of a complex of proteins involved in chromatin remodeling, where it acts to antagonize NF-κB signaling. Clearly, these disparate roles will need to be reconciled in the future.

Research on metastasis suppressor genes is still in its infancy, and in most cases, clear and definitive molecular mechanisms have yet to emerge. Some of the genes in this category, including those specifying E-cadherin, RhoGDI-2, and TIMPs, produce proteins that are part of the known biochemical mechanisms of invasion and metastasis. The biochemical connections between many of the other candidate metastasis suppressor proteins and malignant cell phenotypes are less apparent. Until these genes have been found to be inactivated in human tumor cell genomes, either by mutation or promoter methylation, their involvement in regulating the malignant phenotypes of these cells will remain unclear.

14.15 Occult micrometastases threaten the long-term survival of cancer patients

Throughout this chapter, we have read repeatedly about the extraordinary inefficiency with which metastases are produced. Some of this metastatic inefficiency is created by the profound difficulties that cancer cells experience as they undertake the initial steps of the invasion–metastasis cascade. Most of those that do manage to reach distant sites and survive in their newfound homes fail to form clinically detectable metastases. The result is the presence of myriad dormant micrometastases seeded throughout the tissues of many cancer patients.

While micrometastases are, with rare exception, unable to expand to form clinically detectable metastases, they do provide clear indication that a primary tumor has seeded cells throughout the body. These micrometastases represent a threat to the long-term survival of cancer patients, if only because some of them may erupt into full-fledged, clinically significant macroscopic metastases years after they become implanted in some distant tissue site. Breast cancers are notorious for yielding relapses one and even two decades after the primary tumor has been removed and the patient has been declared to be free of cancer.

In one study of breast cancer patients, micrometastases were detected by sampling the bone marrow of the iliac crest of the pelvis. About 1% of a population of patients suffering from nonmalignant conditions showed cytokeratin-positive cells (that is, epithelial cells) in their marrow. In contrast, 36% of breast cancer patients carrying tumors of stages I, II, or III had such micrometastases in their marrow. The presence of these micrometastases in the marrow proved to be a highly useful prognostic marker for the risk of relapsing with clinically detectable metastasis (**Figure 14.50A**). Thus, within four years, one-quarter of the marrow-positive patients had died from cancer, while only 6% of those lacking cancer cells in their marrow had died from this disease. Overall, the presence of micrometastases in these patients represented about a 4-fold increased risk of eventual relapse or death from this disease. Another study found a more than 10-fold increased risk of death from breast cancer among those whose marrow carried micrometastases composed of single cells or small clumps of cancer cells.

Colon cancer patients who have undergone **resection** (surgical excision) of their primary tumor will often appear in the cancer clinic a year or two later with a small number of metastases in their liver but none elsewhere; these can then be removed surgically, often with significant clinical benefit. Once again, micrometastases in the marrow of the pelvis can be scored. About 90% of those who lack these micrometastases are still alive 15 months later, while only 60% of those who carry such micrometastases survive to this point (see Figure 14.50B).

A procedure used to treat cancer of the esophagus provides yet another insight into metastatic spread. These tumors are often treated surgically, which necessitates the removal of one or more rib segments, from which marrow can be easily flushed. Two independent studies reported that 79% and 88% of these patients, respectively, harbored carcinoma micrometastases in their rib marrow at the time of their surgery. These numbers, which contrast with the approximately 30% of initially diagnosed breast cancer patients bearing micrometastases, correlate with a far grimmer prognosis for patients suffering this type of cancer, with less than half of them surviving more than three years after diagnosis.

The melanoma literature provides equally dramatic testimony of the long-term dangers posed by **occult**, dormant micrometastases (that is, those that are hidden and apparently not growing). In one particularly well-documented case, kidneys were prepared for organ transplantation from the cadaver of a patient who had undergone resection of a small melanoma 16 years earlier. The patient had been followed closely for 15 years after removal of this small primary tumor and had remained symptom-free. However, soon after transplantation, the two recipients of his kidneys developed aggressive melanomas that were directly traceable to this donor (see Supplementary Sidebar 14.4).

The mechanisms that prevent micrometastases from erupting into clinically threatening growths are poorly understood. In some instances, one can observe micrometastases growing as cuffs around small vessels; this suggests that they lack their own angiogenic capabilities but are nonetheless able to take limited advantage of host capillaries that happen to be nearby. In the great majority of micrometastases found in the marrow, the involved cells lack any indication of cell proliferation markers and thus are in a nongrowing, G_0-like state for extended periods of time (see Figure 14.12), perhaps for months and even years (see, however, **Sidebar 14.8**). (Because such cells are nongrowing, they may be especially resistant to chemotherapeutic treatment designed to eliminate the residual disease that persists following surgical removal of a primary tumor.)

Immune mechanisms may also contribute to suppressing the growth of micrometastases, thereby preventing metastatic disease relapse. This is suggested by the occasionally observed explosive growth of aggressive metastatic tumors in immunosuppressed organ transplant recipients. In addition, the phenomenon of tumor stem cells may help to explain the inability of the great majority of initially seeded micrometastases

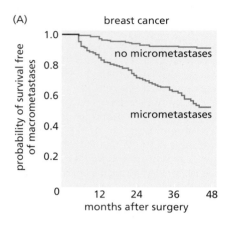

(A)

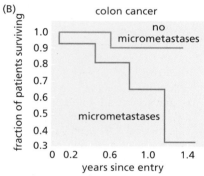

(B)

Figure 14.50 Micrometastases and clinical prognosis (A) This Kaplan–Meier plot presents the proportion of breast cancer patients who survived free of macroscopic metastases as a function of the elapsed time following initial surgical treatment of their primary tumors. Patients with micrometastases in the bone marrow at the time of surgery (*red line*) suffered a far higher relapse rate than those who lacked such micrometastases (*blue line*). (B) Strikingly similar patterns characterize the probability of survival of a group of 54 colon cancer patients who exhibited no apparent metastases outside of the liver at the time of preoperative diagnosis; all were treated surgically to remove large metastases from the liver. Those whose bone marrow showed no cytokeratin-positive metastatic cells had a far more favorable clinical course (*green curve*) than those whose marrow carried such cancer cells (*red curve*). In this case, the fraction of patients surviving (*ordinate*) is plotted versus the time since their initial treatment and entry into a clinical study. (A, from S. Braun et al., *N. Engl. J. Med.* 342:525–533, 2000. B, courtesy of R.A. Tollenaar.)

to generate macrometastases (Supplementary Sidebar 14.13). Beyond this, relatively little is known about the mechanisms that preclude most micrometastases from successfully colonizing the tissues in which they have landed.

14.16 Synopsis and prospects

Like all other biological phenotypes, those contributing to invasion and metastasis must be directed by the actions of genes. Several major issues have complicated the search for the genetic determinants of these aggressive phenotypes of cancer: Are these phenotypes programmed by a small number of pleiotropically acting, master control genes, or do the actions of multiple genes collaborate to create each of these phenotypes? Do these genes undergo mutation during tumor progression, or do they become involved in the late steps of tumor progression through epigenetic mechanisms that control their expression? And how do familiar oncogenes and tumor suppressor genes contribute to invasion and metastasis?

While many of the genetic elements governing metastasis remain unclear, some progress is being made in solving another puzzle: Are the cells within a primary human tumor that undertake invasion and metastasis rare variants (among the larger population of tumor cells) that have, through some genetic or epigenetic accident, acquired the ability to execute these steps? Or are all the cancer cells within certain primary tumors equally capable of invading and metastasizing (albeit with extraordinarily low efficiency), while the great majority of the cancer cells in other tumors lack these abilities?

A number of studies of the gene expression patterns of various human tumors indicate that the tendency to metastasize is associated with a particular pattern of gene expression in some but not other primary breast cancers. Moreover, these expression patterns are manifested by the bulk of the cells in each of a group of primary tumors, rather than by a small subset of cells within each tumor (**Figure 14.51A and B**). (If the tendency to metastasize were limited to only a small minority of cancer cells in a primary tumor, their gene expression pattern would not significantly influence the expression pattern of the tumor cell population as a whole, and this larger population would therefore not manifest a metastasis-prone gene expression signature.) This suggests, in turn, that the proclivity to metastasize was developed relatively early during the course of the multi-step progression that culminated in primary tumor formation, not afterward by a small, specialized subpopulation of cancer cells within a primary tumor (see Figure 11.20A).

In fact, such differences in the eventual metastatic behavior of various tumor cell populations may be determined extremely early, even *before* multi-step tumor progression has begun: as **Figure 14.52A** indicates, the differentiation programs of the normal cells-of-origin can strongly influence the biological behaviors of derived neoplastic descendants. Hence, the possibility that a tumor will eventually metastasize may be predetermined by the differentiation program of the normal cell that participates directly in tumor initiation.

Observations like these suggest that metastatic cells are drawn from the general population of cells in a primary tumor rather than from small, specialized, genetically unrepresentative subclones of cells. However, DNA sequencing studies of the genomes of metastases and corresponding primary tumors suggest an apparently opposite conclusion: that primary tumors are genetically heterogeneous, and that metastases arise from genetically specialized subpopulations within primary tumors. An eventual reconciliation of these two portrayals of metastatic progression may be as follows (see Figure 14.52B): The differentiation program of a normal cell-of-origin is an important determinant of whether a primary tumor has the potential to eventually spawn metastasis-competent subclones. In those tumors that have inherited a differentiation program that is compatible with eventual metastasis, diverse, specialized subclones may arise through somatic mutations or other stochastically acquired heritable changes. Among these subclones will be those that have acquired metastatic powers. These subclones will then become the major sources of metastatic dissemination. Some

Sidebar 14.8 Are all micrometastases truly dormant? The dormancy of micrometastatic disease may often be an illusion. Thus, we can imagine that in many patients with "minimal residual disease" following removal of their primary tumors, micrometastatic clones occasionally acquire the ability to colonize and thereby create a new cascade of metastatic dissemination and disease relapse (see Figure 14.11). Importantly, such a major alteration of cell phenotype (and possibly an underlying change of genotype) has a low probability per cell generation of occurring in populations of non-growing, dormant cells. Instead, extensive observations made over many years' time indicate that such changes happen spontaneously only in proliferating cell populations. Hence, in many patients with minimal residual disease, some clones of micrometastatic cancer cells must be passing through repeated growth-and-division cycles and occasionally spawning variants that have, through some random accident, acquired colonizing ability. (These micrometastases may remain clinically inapparent for many years simply because, in the absence of help from recruited cells, the rate of cell proliferation in these growths is balanced by an equal rate of cell attrition through apoptosis.)

speculations about the evolution of metastatic competence extend and reinforce this model (see Supplementary Sidebar 14.14).

In the case of carcinomas, the discovery of the EMT program may eventually allow one further refinement in our thinking about how carcinomas progress to high-grade malignancies. The acquired ability to metastasize indicated in Figure 14.52B may not directly generate disseminating cancer cells. Instead, cancer cells with this acquired

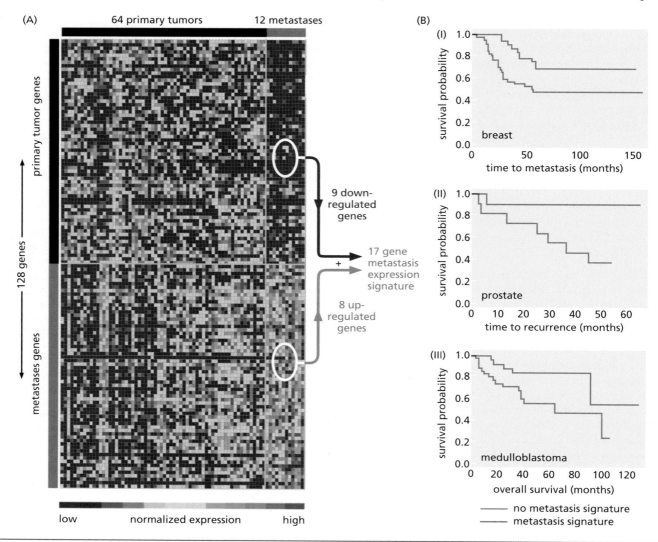

Figure 14.51 Schedule of acquiring metastatic propensity
The process of metastasis is the endpoint of multi-step tumor progression for many human tumors. This raises the issue of whether the tendency to metastasize is acquired late in tumor progression by a small subpopulation of cells or relatively early. (A) Gene expression microarrays make possible the simultaneous monitoring of the expression of thousands of genes to determine a specific pattern or *expression signature* that is correlated with a specific phenotype or set of phenotypes. In the expression array analysis seen here, genes that are expressed at high levels are in *red*, while those expressed at low levels are in *blue*. RNAs prepared from 64 primary adenocarcinomas (from various tissues) and 12 metastatic nodules of adenocarcinomas (*arrayed across the top*) were analyzed (*black, red horizontal bars, respectively*). Of the thousands of genes analyzed in an initial gene expression array (*not shown*), 128 genes (*arrayed vertically*) were found to be associated—because of over- or underexpression—with metastasis (*vertical red, black bars, respectively*). Further distillation of the

data yielded a set of 17 genes whose expression was as useful as that of the 128-gene set in distinguishing metastases from primary tumors. (B) Importantly, the metastasis-specific expression signature was found to be exhibited by a small subset of the initially analyzed primary tumors, suggesting that it could be used to predict the metastatic tendencies of yet other groups of human tumors. Indeed, when researchers used the metastasis expression signature of panel A to analyze the gene expression patterns of other types of primary tumors, they were able to separate the patients bearing adenocarcinomas of the breast (I) and prostate (II) as well as medulloblastomas (III) into two groups (*blue, red lines*) having markedly different times to clinical progression or relapse following initial surgery. The fact that the great majority of cells in certain primary tumors expressed a gene expression signature associated with metastasis suggests that this signature was acquired relatively early in primary tumor progression and thus was inherited by the great majority of the descendant cells. (A and B, from S. Ramaswamy et al., *Nat. Genet.* 33:49–54, 2002.)

trait may become responsive to EMT-inducing signals arising in the stroma, while those that lack this acquired responsiveness may fail to disseminate directly into the mesenchymal compartments of a variety of tissues.

The EMT program offers an attractive but still-unproven solution to the problem of how carcinoma cells disseminate from primary tumors to distant tissues. Thus, the multiple distinct phenotypes conferred on neoplastic epithelial cells by this program may, quite possibly, enable a primary tumor cell to physically translocate from the

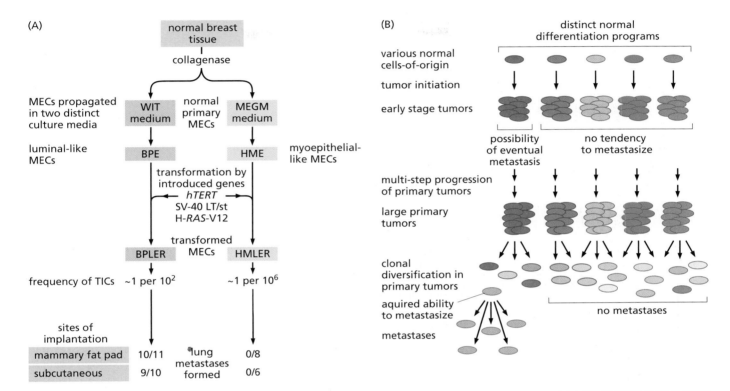

Figure 14.52 Influence of cell-of-origin on metastatic propensity The observations of Figure 14.51 indicate that a substantial proportion of the cells in a primary tumor share a gene expression signature that was acquired relatively early in the course of tumor progression. Such acquisition may have derived from the initial somatic mutations that triggered multi-step tumor progression or, even earlier, in the gene expression signature of the preexisting normal cell-of-origin, the latter reflecting its program of differentiation. (A) A heterogeneous population of normal human mammary epithelial cells (MECs) explanted directly from a normal mammary gland were propagated *in vitro* in two alternative tissue culture media, which selected for the outgrowth of MEC populations that expressed either a more luminal (BPE) or a more myoepithelial (HME) gene expression pattern and therefore originated from two distinct differentiation lineages in the mammary gland. These cells were then transformed through the successive introduction of an *hTERT* gene, the SV40 early region (expressing small and large T antigens), and a *RAS* oncogene (see Figure 11.27), resulting in BPLER and HMLER tumorigenic cells. While the introduced transforming genes were expressed at comparable levels in the two cell populations, these transformed cells behaved very differently. The concentration of tumor-initiating cells (TICs), also known as cancer stem cells, in the BPLER cell population was $10^4\times$ higher than in the HMLER cell population. The BPLER cells formed lung metastases following orthotopic implantation (into mammary stromal fat pads; see Figure 14.27D) or implantation into subcutaneous sites (in 10 of

11 and in 9 of 10 implanted mice, respectively) while the HMLER cells, implanted in equal numbers in these sites, failed to do so. Since the two cell populations acquired identical sets of oncogenes that were expressed at very similar levels, the only source of their dramatically different behaviors must be associated with the differentiation programs of the normal MECs in each population prior to experimental transformation. By extension, this implies that the differentiation programs of normal cells-of-origin continue to imprint themselves on the behavior of their neoplastic progeny. (B) These observations must be reconciled with DNA sequencing analyses in which multiple metastases removed from individual patients have been found to be most closely related genetically to single, genetically distinct subpopulations in these patients' corresponding primary tumors. Taken together, these diverse observations suggest a scheme in which the gene expression programs of primary tumors (and thus their behavior) are strongly influenced by the differentiation program of their antecedent normal cells-of-origin (CoOs). Thus, those tumors that arise from one type of normal CoO may be endowed with a possibility of eventually spawning metastases, while those from other normal CoOs may lack this tendency, even after they have passed through multi-step tumor progression. This inherited tendency may be necessary but not sufficient for metastasis, with subsequent clonal diversification and the acquisition of somatic alterations (*lower left*) spawning distinct subpopulations in the primary tumor, some of which may have gained a greatly increased ability to disseminate. (A, from T. Ince et al., *Cancer Cell* 12:160–170, 2007.)

heart of a primary tumor to the parenchyma of a distant organ. Accordingly, a carcinoma cell that has passed through an EMT may be able to accomplish most of the steps of the invasion–metastasis cascade (see Figure 14.3) except the last one—colonization. If eventually proven, this model would represent a simplifying solution to a problem that has been viewed as one of endless complexity. Of additional interest, and as cited in Figure 14.18D, certain observations have demonstrated that once breast cancer cells have passed through an EMT in order to travel from the primary tumor to a foreign tissue compartment and found a new metastatic colony, many of the carcinoma cells in the resulting colony undergo an MET and generate epithelial cells; even more recent work indicates that the presence of epithelial cells in such a metastasis is critical to its robust outgrowth.

EMT programs may, however, shed little light on how nonepithelial tumors, notably hematopoietic, connective tissue, and central nervous system (CNS) tumors, are able to invade and occasionally metastasize. The origin of the CNS from an early embryonic epithelium—the neuroectoderm—offers the possibility that certain EMT-TFs that have been studied in the context of carcinoma progression may prove to be important actors for this class of tumors as well. Still, this uncertainty reminds us how little is understood at present about the malignancy programs of nonepithelial tumors.

Whether one or another model of metastasis is ultimately validated, it is clear that the identities of many of the genes that are specifically involved in programming metastasis have been elusive. Experimental resolution of these problems is confounded by complications at every level:

1. Experimental analyses are complicated by the inefficiencies of the metastatic process. Even when cancer cells have ostensibly acquired a genotype and phenotype enabling metastasis, they succeed in metastasizing with extraordinarily low efficiency. Such a weak connection between genotype and measurable phenotype derails most currently available experimental strategies.

2. A second dimension of complexity arises from the apparent collaboration of genetic and epigenetic factors in creating the metastatic trait. Recall, for example, that in certain experimental models of cancer, an EMT is achieved when *ras*-transformed cells are exposed to TGF-β (see Section 14.4). This transition, which may operate in many human carcinomas and enable their invasiveness, can be triggered by specific signals that genetically altered cells encounter in some tissue microenvironments but not in others. Hence, in these cases, invasion and subsequent metastasis can hardly be portrayed as genetically templated traits and, for this reason, cannot be readily studied by commonly used experimental techniques.

3. In many tumors, the genes and proteins that participate directly in programming invasion and metastasis may be expressed only at the invasive edges of primary tumors (see Figure 14.17), and the cancer cells in these invasive edges may represent only a tiny fraction of the neoplastic cell populations in these tumors. This greatly complicates experiments designed to reveal the biochemical and genetic bases of invasiveness and metastatic ability, which often rely on analyzing bulk populations of cancer cells prepared from large chunks of surgically resected tumors.

4. Carcinomas constitute the most common class of human cancers, and the neoplastic epithelial cells within these tumors may need to undergo an EMT in order to become invasive and metastatic. However, if invading carcinoma cells pass through a complete EMT and shed all epithelial traits, they become the proverbial "wolves in sheep's clothing," since most commonly used histological analyses are unable to distinguish these cells from the non-neoplastic mesenchymal cells of the tumor-associated stroma. (Indeed, this difficulty explains why many tumor pathologists deny the very existence of the EMT as a key process in the development of carcinoma invasiveness.)

5. Metastatic dormancy creates another experimental problem. In breast cancer patients, for example, metastases may suddenly appear as long as 20 years

after the initial primary tumor has been removed. Because of this long latency period and the sheer number of micrometastases carried by many patients, it has been difficult to learn how only a few of them suddenly acquire the ability to mushroom into macroscopic, life-threatening tumors.

These experimental difficulties have greatly retarded the progress of metastasis research, leaving many simple yet fundamental questions unanswered. For example, are there really genes that are specialized to impart an invasive or metastatic phenotype to cancer cells? And in the same vein, are there specialized metastasis suppressor genes (see Section 14.14) that must be inactivated before a population of tumor cells can acquire invasive and/or metastatic ability? Or do the genes and proteins that affect metastasis operate as components of the regulatory circuits that we have repeatedly encountered throughout this book, namely, the circuits governed by the products of oncogenes and tumor suppressor genes?

The tissue tropisms of metastasizing cancer cells—their tendencies to colonize some but not other organs—represent the major challenge for cancer biologists studying metastasis. The daunting complexity of the colonization problem is suggested by Figure 14.43. There, the tendencies of four common cancers (those arising in the prostate, pancreas, breast, and colon) are illustrated in terms of their tendencies to form metastases in four organ sites (brain, lungs, liver, and bone marrow). In principle, each type of primary tumor cell must develop a distinct set of adaptations for each organ microenvironment in which it lands, yielding 16 distinct adaptive programs. In fact, there are far more primary tumor types that can metastasize to distant organs, and there are yet other alternative organ sites in which these tumors can found metastatic colonies; moreover, the various subtypes of primary tumors arising in an organ (for example, different subtypes of breast or lung carcinomas) may differ in their requirements for adaptation to different sites of metastatic colonization. In aggregate, these combinatorial interactions suggest the existence of many dozens, possibly hundreds of distinct adaptive programs, each composed, in turn, of multiple genetic and epigenetic changes that enable one type of disseminating primary tumor cell to colonize a specific target organ. (Since the cancer cells in metastases often bear considerably more mutations than their counterparts within the corresponding primary tumors, this might suggest that many of the adaptations that disseminated cancer cells make in distant tissues depend on novel mutant alleles acquired at those sites.)

To be sure, some insights have already been gained from the substantial advances in understanding the detailed mechanisms of osteotropic metastasis, as described in Section 14.13. However, this mechanism of metastasis represents a rare exception. In general, we know almost nothing about the functionally important interactions of disseminated cancer cells with the tissues that they colonize.

This is beginning to change. A powerful strategy for discovering the genes and proteins responsible for specific metastatic tropisms involves the isolation of tumor cells that show preference for colonizing a specific target organ. By retrieving already-metastasized cancer cells from that organ, propagating them *in vitro*, and injecting them into host mice, followed by another round of isolating metastatic cells from that organ, it is possible to select clones of cancer cells that stably express a highly specific tropism for that organ.

Alternatively, single-cell clones (that is, clonal cell populations that are each descended from an isolated cell) can be prepared from a heterogeneous population of cells present in a human cancer cell line. The gene expression profile (see Figure 13.18) of each clone can then be analyzed, and its tendency to form metastases in one or another target organ can be determined. This can lead to the identification of genes whose expression in a cancer cell is correlated with the metastatic tropism of that cell and may even contribute causally to this behavior (Figure 14.53). Indeed, ectopic expression of a group of such genes in otherwise poorly metastatic clonal cell populations can induce these cells to exhibit potent osteotropic metastasis. Such experiments also indicate that within a heterogeneous tumor cell population, various preexisting gene expression patterns can strongly influence the ability of individual cells to exhibit a variety of metastatic behaviors.

The clinical challenges associated with metastasis are also daunting. The existence of micrometastases represents a major challenge for oncologists who would like to prevent disease relapse years after the primary tumor has been eliminated. Micrometastases of less than 0.2 mm diameter may carry several hundred to several thousand cells, and their detection in an organism carrying approximately 5×10^{13} cells represents an almost impossible undertaking. Without eradication, these micrometastases represent an ongoing threat, since some of them may erupt at an unpredictable future time into a lethal growth.

This issue leads directly to another: Can the therapies used to treat primary tumors also be used to treat their metastatic derivatives? Or are metastatic cells so different from their progenitors in the primary tumor that they require their own customized therapies? In fact, the expression array analyses indicating substantial similarity between the gene expression profiles of primary tumors and their metastatic offshoots (**Figure 14.54**) provide some hope that metastatic cells may respond to the same therapies that succeed in destroying the primary tumors from which they derive.

To end, we go back to the beginning of this chapter: if, as experimental evidence increasingly shows, the epithelial–mesenchymal transition is a critical event in the acquisition of invasiveness, and if cancer cells resurrect embryonic transcription

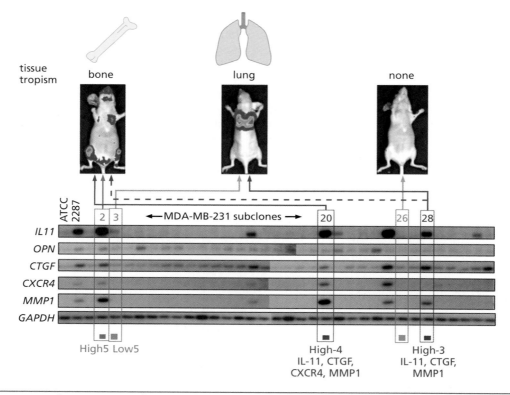

Figure 14.53 Gene expression patterns and metastatic tropism Thirty-three cells from a large population of human MDA-MB-231 cells were each expanded into a clonal population in culture. The mRNA expression pattern of each subclone was analyzed (*columns, arrayed left to right*) using probes for the mRNAs of five genes—*IL11* (interleukin-11), *OPN* (osteopontin), *CTGF* (connective tissue growth factor), *CXCR4* (chemokine receptor 4), and *MMP1* (matrix metalloproteinase-1)—and, as loading control, a probe for *GAPDH* (glyceraldehyde-3-phosphate dehydrogenase) mRNA. In addition, the expression patterns of the original tumor cell population (ATCC, *left column*) and a subcloned cancer cell population termed 2287 (which was selected for its ability to generate osteolytic metastases; *2nd column*) were analyzed. The five experimental genes were chosen because of their overexpression in osteotropic metastatic cells and their known biological properties in promoting osteolytic

metastases. Clone 2 cells (*red box*), when injected into the arterial circulation of mice, showed a tendency to produce osteotropic metastases, as indicated by *in vivo* imaging; these cells expressed high levels of all five experimental mRNAs. Clone 3 cells (*yellow box*), in contrast, expressed low levels of all five mRNAs and preferentially formed lung metastases. And clone 26 genes (*yellow box*), which expressed essentially none of these mRNAs, formed no metastases at all. Moreover, when otherwise poorly metastatic cells were forced to express combinations of three of these genes, they acquired the ability to form bone metastases efficiently (*not shown*), pointing to the causal role of these genes in forming these metastases. Metastases were visualized through the presence of a luciferase gene in the tumor cells, which causes cells to release a bioluminescent signal. (From Y. Kang et al., *Cancer Cell* 3:537–549, 2003.)

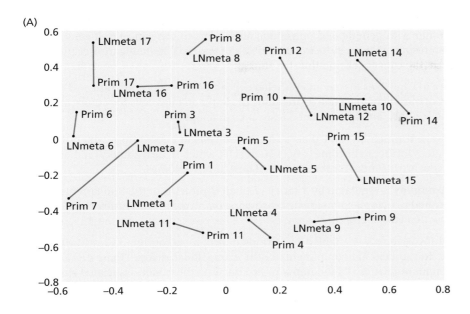

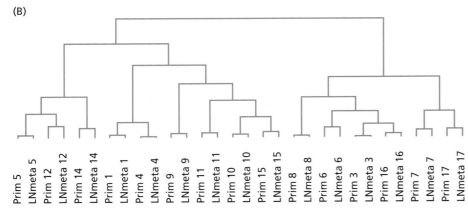

Figure 14.54 Genetic similarity between primary tumors and derived metastases Gene expression array analyses can be used to classify different primary tumors and derived metastases according to their respective gene expression profiles (e.g., see Figure 14.51). If this is done, the degree of similarity between pairs of biopsies can be calculated using statistical methods. (A) Here, the biopsies of primary tumors and derived lymph nodes from a group of patients (each identified by a number) have been placed on a two-dimensional map, in which proximity indicates similarity in gene expression patterns. This reveals that the great majority of primary tumors (Prim) map closely to their derived lymph node metastases (LNmeta), indicating similarity in gene expression patterns. (B) Alternatively, gene expression patterns can be used to create a *dendrogram* that shows the degree of relatedness of primaries and derived metastases—i.e., the most closely related tumor samples are placed near one another on the same or neighboring branches. Once again, the gene expression pattern of a metastasis is, almost always, most closely related to the parental primary tumor from which it arose. (Note that these expression array analyses do not reveal the genetic similarities and differences between primary tumors and their metastatic derivatives.) (From B. Weigelt et al. *Br. J. Cancer* 93:924–932, 2005.)

factors to acquire these traits, then Lewis Wolpert's statement might require revision, in that gastrulation (and the associated EMT) might well loom as one of the most *dangerous* event in our lives!

Key concepts

- Invasion and metastasis are responsible for 90% of cancer-associated mortality, and the majority of cancer cells at the time of death may often be found in metastases rather than the primary tumor.

- The invasion–metastasis cascade involves local invasion, intravasation, transport, extravasation, formation of micrometastases, and colonization.

- The sequence of steps in this cascade is completed only infrequently, resulting in metastatic inefficiency. The least efficient of these steps appears to be colonization.

- Many of these steps can be executed by carcinoma cells that activate a cell-biological program called the epithelial–mesenchymal transition (EMT), which is normally used by cells early in embryogenesis and during wound healing.

- An EMT can be programmed by pleiotropically acting transcription factors that are normally involved in various steps of early embryogenesis.

- Signals released by the stromal microenvironment of a cancer cell, operating together with genetic and epigenetic alterations of the cancer cell genome, are often responsible for inducing expression of the EMT-inducing transcription factors in the cancer cell and thus the EMT.

- The EMT involves loss of an epithelial cell gene expression program and acquisition of mesenchymal gene expression. The latter enables cells to acquire invasiveness, motility, and a heightened resistance to apoptosis.

- Passage through an EMT places carcinoma cells in a state from which they can progress further to become cancer stem cells, which in turn confers on them powers that are essential for the seeding of new tumor colonies.

- Cell motility is regulated by a series of small G proteins of the Rho family that are activated by cytoplasmic signal-transducing pathways and control the assembly of the actin cytoskeleton.

- Cell invasiveness is enabled by various matrix metalloproteinases (MMPs) that function to degrade components of the extracellular matrix. These enzymes are often manufactured by inflammatory cells within the tumor-associated stroma.

- Metastatic cancer cells may travel via the lymph ducts to nodes. However, their spread via the blood circulation is responsible for the great majority of distant metastases.

- Many cancer cells that are carried through the circulation form microthrombi that lodge in the arterioles and capillaries of various tissues.

- The ability of cancer cells to extravasate may depend on many of the same activities that were used earlier to execute invasiveness and intravasation.

- While the earlier steps of the invasion–metastasis cascade are likely to be similar in various types of human tumors, the last step—colonization—is likely to depend on complex interactions that are specific to the particular type of metastasizing cells and the microenvironments of the host tissues in which they land.

- The details of colonization are reasonably well understood only in the context of osteotropic metastases, especially the osteolytic metastases initiated by breast cancer cells.

- In some cases, the metastatic tropisms of cancer cells can be explained by the organization of the circulation between the primary tumor site and the target site of metastasis. In many other cases, the reasons why cancer cells metastasize from primary tumors to certain target organs are poorly understood.

- The acquisition of invasive and metastatic powers does not appear to involve major changes in the genotype of cancer cells within the primary tumor.

Thought questions

1. What arguments can be mustered for or against the notion that invasion and metastasis are likely to be orchestrated by specific mutant alleles that are acquired by cancer cells late in tumor progression?

2. What explanations can be offered for the inefficiency of colonization by the cells within micrometastases?

3. What arguments suggest that the ability to metastasize is expressed either by the bulk of cancer cells in a primary tumor or only by a minority of cells that are specialized to do so?

4. What evidence suggests that genetic and phenotypic evolution of cancer cells can occur in sites within the body that are far removed from the primary tumor?

5. What specific types of physiologic support might be supplied by tissues that are frequently sites of successful metastasis formation? In what way do these supports affect the ultimate success of the colonization process?

6. How might primary tumors exhibit metastatic powers as soon as they form?

7. Would the ability to prevent metastasis have demonstrable effects on the clinical course of some human tumors but not others?

8. What evidence supports the involvement of an EMT in human tumor pathogenesis, and what evidence argues against it?

9. How might the ability to accurately determine the prognosis of a diagnosed prostate or mammary tumor lead to dramatic changes in the practice of clinical oncology?

10. What mechanisms might be invoked to explain why large primary tumor size is often correlated with a prognosis of metastasis?

Additional reading

Akhurst RJ & Balmain A (1999) Genetic events and the role of TGF beta in epithelial tumor progression. *J. Pathol.* 187, 82–90.

Alix-Panabières C, Riethdorf S & Pantel K (2008) Circulating tumor cells and bone marrow micrometastasis. *Clin. Cancer Res.* 14, 5013–5021.

Barrallo-Gimeno A & Nieto MA (2005) The Snail genes as inducers of cell movement and survival: implications in development and cancer. *Development* 132, 3151–3161.

Berx G, Raspé E, Christofori G et al. (2007) Pre-EMTing metastasis? Recapitulation of morphogenetic processes in cancer. *Clin. Exp. Metastasis* 24, 587–597.

Blasi F & Carmeliet P (2002) uPAR: a versatile signalling orchestrator. *Nat. Rev. Mol. Cell Biol.* 3, 932–943.

Boccaccio C & Comoglio PM (2006) Invasive growth: a *MET*-driven genetic programme for cancer and stem cells. *Nat. Rev. Cancer* 6, 637–645.

Borovski T, De Sousa e Melo F, Vermeulen L & Medema JP (2011) Cancer stem cell niche: the place to be. *Cancer Res.* 71, 634–639.

Brabletz T, Jung A, Spaderna SK et al. (2005) Opinion: migrating cancer stem cells—an integrated concept of malignant tumour progression. *Nat. Rev. Cancer* 5, 744–749.

Braun S, Pantel K, Muller P et al. (2000) Cytokeratin-positive cells in the bone marrow and survival of patients with Stage I, II or III breast cancer. *N. Engl. J. Med.* 342, 525–533.

Bray D (2002) Cell Movements. New York: Garland Science.

Burridge K & Wennerberg K (2004) Rho and Rac take center stage. *Cell* 116, 167–179.

Cavallaro U & Christofori G (2004) Cell adhesion and signaling by cadherins and Ig-CAMs in cancer. *Nat. Rev. Cancer* 4, 118–132.

Chambers AF, Groom AC & MacDonald IC (2002) Dissemination and growth of cancer cells in metastatic sites. *Nat. Rev. Cancer* 2, 563–572.

Chiang AC & Massagué J (2008) Molecular basis of metastasis. *N. Engl. J. Med.* 359, 2814–2823.

Christofori G (2006) New signals from the invasive front. *Nature* 441, 444–450.

Condeelis J & Pollard JW (2006) Macrophages: obligate partners for tumor cell migration, invasion, and metastasis. *Cell* 124, 263–266.

Condeelis J, Singer RH & Segall JE (2005) The great escape: when cancer cells hijack the genes for chemotaxis and motility. *Annu. Rev. Cell Dev. Biol.* 21, 695–718.

Cristofanilli M & Braun S (2010) Circulating tumor cells revisited. *J.A.M.A.* 303, 1092–1093.

Drasin DJ, Robin TP & Ford HL (2011) Breast cancer epithelial-to-mesenchymal transition: examining the functional consequences of plasticity. *Breast Ca. Res.* 13, 266.

Dykxhoorn DM (2010) MicroRNAs and metastasis: little RNAs go a long way. *Cancer Res.* 70, 6401–6406.

Fidler IJ (2003) The pathogenesis of cancer metastasis: the "seed and soil" hypothesis revisited. *Nat. Rev. Cancer* 3, 453–458.

Fodde R & Brabletz T (2007) Wnt/β-catenin signaling in cancer stemness and malignant behavior. *Curr. Opin. Cell Biol.* 19, 150–158.

Friedl P & Gilmour D (2009) Collective cell migration in morphogenesis, generation and cancer. *Nat. Rev. Mol. Cell Biol.* 10, 445–457.

Friedl P & Wolf K (2009) Plasticity of cell migration: a multiscale tuning model. *J. Cell Biol.* 188, 11–19.

Gertler F & Condeelis J (2011) Metastasis: tumor cells becoming MENAcing. *Trends Cell Biol.* 21, 81–90.

Gregory PA, Bracken CP, Bert AG & Goodall GJ (2008) MicroRNAs as regulators of epithelial-mesenchymal transition. *Cell Cycle* 7, 3112–3118.

Gupta GP & Massagué J (2006) Cancer metastasis: building a framework. *Cell* 127, 679–695.

Hay ED (2005) The mesenchymal cell, its role in the embryo, and the remarkable signaling mechanisms that create it. *Dev. Dyn.* 3, 706–720.

Heasman SJ & Ridley AJ (2008) Mammalian Rho GTPases: new insights into their functions from *in vivo* studies. *Nat. Rev. Mol. Cell Biol.* 9, 690–701.

Heldin C-H, Landström M & Moustakas A (2009) Mechanism of TGF-β signaling to growth arrest, apoptosis, and epithelial–mesenchymal transition. *Curr. Opin. Cell Biol.* 21, 1–11.

Huber MA, Kraut N & Beug H (2005) Molecular requirements for epithelial-mesenchymal transition during tumor progression. *Curr. Opin. Cell Biol.* 17, 548–556.

Hugo H, Ackland ML, Blick T et al. (2007) Epithelial-mesenchymal and mesenchymal-epithelial transitions in carcinoma progression. *J. Cell Physiol.* 213, 374–383.

Ilina O & Friedl P (2009) Mechanisms of collective cell migration at a glance. *J. Cell Sci.* 122, 3203–3208.

Joyce JA & Pollard JW (2009) Microenvironmental regulation of metastasis. *Nat. Rev. Cancer* 9, 239–252.

Kedrin D, van Rheenen J, Hernandez L et al. (2007) Cell motility and cytoskeletal regulation in invasion and metastasis. *J. Mammary Gland Biol. Neoplasia* 12, 143–152.

Kingsley LA, Fournier PGJ, Chirgwin JM & Guise TA (2007) Molecular biology of bone metastasis. *Mol. Cancer Ther.* 6, 2609–2617.

Lee JM, Dedhar S, Kalluri R & Thompson EW (2006) The epithelial-mesenchymal transition: new insights in signaling, development and disease. *J. Cell Biol.* 172, 973–981.

Lu X & Kang Y (2010) Hypoxia and hypoxia-inducible factors, master regulators of metastasis. *Clin. Cancer Res.* 16, 5928–5935.

Mundy GR (2002) Metastasis to bone: causes, consequences, and therapeutic opportunities. *Nat. Rev. Cancer* 2, 584–593.

Murphy DA & Courtneidge SA (2011) The "ins" and "outs" of podosomes and invadopodia: characteristics, formation and function. *Nat. Rev. Mol. Cell Biol.* 12, 413–426.

Nguyen DX, Bos PD & Massagué J (2009) Metastasis: from dissemination to organ-specific colonization. *Nat. Rev. Cancer* 9, 274–284.

Nieto MA (2002) The Snail superfamily of zinc-finger transcription factors. *Nat. Rev. Mol. Cell Biol.* 3, 155–166.

Nieto MA (2011) The ins and outs of the epithelial to mesenchymal transition in health and disease. *Annu. Rev. Cell Dev. Biol.* 27, 347–376.

Page-McCaw A, Ewald AJ & Werb Z (2007) Matrix metalloproteinases and the regulation of tissue remodeling. *Nat. Rev. Mol. Cell Biol.* 8, 221–233.

Paget S (1889) The distribution of secondary growths in cancer of the breast (re-publication of 1889 *Lancet* article). *Cancer Metastasis Rev.* 8, 98–101.

Pollard JW (2004) Tumour-educated macrophages promote tumour progression and metastasis. *Nat. Rev. Cancer* 4, 71–78.

Sahai E (2007) Illuminating the metastatic process. *Nat. Rev. Cancer* 7, 737–749.

Savagner P (2001) Leaving the neighborhood: molecular mechanisms involved during epithelial-mesenchymal transition. *Bioessays* 23, 912–923.

Schmidt A & Hall A (2002) Guanine nucleotide exchange factors for Rho GTPases: turning on the switch. *Genes Dev.* 13, 1587–1609.

Siegel PM & Massagué J (2003) Cytostatic and apoptotic actions of TGF-β in homeostasis and cancer. *Nat. Rev. Cancer* 3, 807–820.

Stacker SA, Achen MG, Jussila L et al. (2002) Lymphangiogenesis and cancer metastasis. *Nat. Rev. Cancer* 2, 573–583.

Steeg PS (2003) Metastasis suppressors alter the signal transduction of cancer cells. *Nat. Rev. Cancer* 3, 55–63.

Thiery JP (2002) Epithelial–mesenchymal transitions in tumour progression. *Nat. Rev. Cancer* 2, 442–454.

Thiery JP (2003) Epithelial–mesenchymal transitions in development and pathologies. *Curr. Opin. Cell Biol.* 15, 740–746.

van de Stolpe A, Pantel K, Sleijfer S et al. (2011) Circulating tumor cell isolation and diagnostics: toward routine clinical use. *Cancer Res.* 71, 5955–5960.

Vigil D, Cherfils J, Rossman KL & Der CJ (2010) Ras superfamily GEFs and GAPs: validated and tractable targets for cancer therapy? *Nat. Rev. Cancer* 10, 842–857.

Vignjevic D & Montagnac G (2008) Reorganisation of the dendritic actin network during cancer cell migration and invasion. *Semin. Cancer Biol.* 18, 12–22.

Weigelt B, Peterse JL & van't Veer LJ (2005) Breast cancer metastasis: markers and models. *Nat. Rev. Cancer* 5, 591–602.

Weilbaecher K, Guise TA & McCauley LK (2011) Cancer to bone: a fatal attraction. *Nat. Rev. Cancer* 11, 411–425.

Wirtz D, Konstantopoulos K & Searson PC (2011) The physics of cancer: the role of physical interactions and mechanical forces in metastasis. *Nat. Rev. Cancer* 11, 512–522.

Yang J & Weinberg RA (2008) Epithelial-mesenchymal transition: at the crossroads of development and tumor metastasis. *Dev. Cell* 14, 818–829.

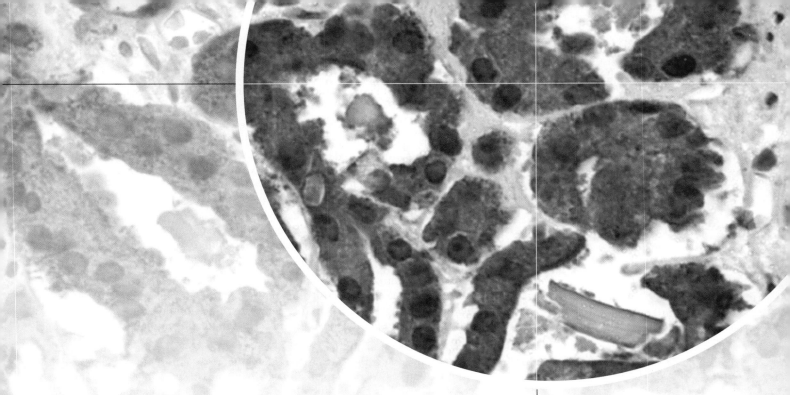

Chapter 15

Crowd Control: Tumor Immunology and Immunotherapy

> It is by no means inconceivable that small accumulations of tumour cells may develop and, because of their possession of new antigenic potentialities, provoke an effective immunological reaction with regression of the tumour and no clinical hint of its existence.
>
> Macfarlane Burnet, immunologist, 1957

Throughout this text, we have studied various defenses that the body erects against the appearance of cancerous growths. Many of these defenses are inherent in cells, more specifically in their hard-wired regulatory circuitry. The most obvious of these are the controls imposed on cells by the apoptotic machinery, which is poised to trigger the death of cells that are misbehaving or suffering certain types of damage or physiologic stress. The pRb circuit and the DNA repair apparatus are similarly configured to frustrate the designs of incipient cancer cells.

The organization of tissues also places constraints on how incipient cancer cells can proliferate. For example, normal epithelial cells that lose their tethering to the basement membrane activate the form of apoptosis that is called anoikis. This mechanism limits the ability of epithelial cells to stray from their normal locations within tissues and grow in ectopic (that is, abnormal) sites. At the same time, the special status afforded to stem cells and their genomes (see Section 12.1) also reduces the probability of mutant cancer cells' gaining a foothold within a tissue.

Beyond these cell- and tissue-specific mechanisms, mammals may have another line of defense—the immune system. The immune system is highly effective in detecting and eliminating foreign infectious agents, including viruses, bacteria, and fungi, from our tissues. One of the major questions in cancer research over the last half century has been whether the immune system can also recognize cancer cells as foreigners and proceed to eliminate them.

Actually, evidence is rapidly accumulating that the immune system does indeed contribute to the body's multilayered defenses against tumors. The difficulties associated with establishing this type of anti-cancer defense are apparent from the outset: the immune system is organized to recognize and eliminate foreign agents from the body while leaving the body's own tissues unmolested. Cancer cells, however, are native to the body and are, in many respects, indistinguishable from the body's normal cells. How can cancer cells be recognized by the immune system as being different and, therefore, appropriate targets of immune-mediated killing? We will wrestle with this problem and its ramifications repeatedly throughout this chapter.

The field of tumor immunology, more than any other area of cancer research, remains in great flux, with basic concepts still a matter of great debate. Still, this is an area of cancer biology that is well worth our time and study, since it holds great promise for new insights into cancer pathogenesis and new ways of treating human tumors.

Research conducted on mammals over the past three decades has revealed an immune system of great complexity and subtlety. Before we enter into discussions of its anti-tumor functions, we need to take an excursion into the workings of the general immune system (Sections 15.1 through 15.6). An understanding of its mechanisms of action, at least in outline, is a prerequisite for engaging the three major questions that will occupy us in this chapter. First, what specific molecular and cellular mechanisms enable the immune system to recognize and attack incipient cancer cells? Second, do these immune mechanisms represent effective defenses that prevent the appearance of tumors? Third, how can the immune system be mobilized by oncologists to attack tumors once they have formed?

15.1 The immune system functions to destroy foreign invaders and abnormal cells in the body's tissues

The mammalian immune system launches several types of attack against foreign infectious agents and the body's own cells that happen to be infected with such agents. It identifies its targets by recognizing specific molecular entities—**antigens**—that are made by these agents. Having done so, the immune system undertakes to **neutralize** or destroy the infectious particles (bacterial and fungal cells, virus particles), as well as infected cells displaying these antigens. To the extent that the immune system also functions to ward off cancer, one assumes that it exploits many of the same mechanisms that it uses to eliminate foreign infectious agents.

The most familiar of the immunological defense strategies involves the **humoral immune response**—the arm of the immune system that generates soluble **antibody** molecules capable of specifically recognizing and binding antigens (**Figure 15.1**). Thus, a virus particle or bacterium displaying antigens on its surface may rapidly become coated with antibody molecules, which may result in the neutralization of these pathogens (**Figure 15.2**). Similarly, an infected cell may display on its surface the antigens made by the agents that have infected it and become coated with antibodies that recognize and bind these antigens. Once a mammalian cell or an infectious agent is coated (opsonized) by antibody molecules, it may be recognized, engulfed, and destroyed by phagocytic cells, such as macrophages, or killed by cytotoxic cells, such as natural killer (NK) cells (**Figure 15.3**). Importantly, these immune cells do not, on their own, have the ability to recognize specific foreign antigens. Instead, the antibody molecules that have bound to antigens on the surfaces of viruses, bacteria, or mammalian cells alert these immune cells to the presence of targets that should be destroyed. The ability of antibody molecules to recognize and bind the specific antigens displayed by these various infectious agents and mammalian cell types derives from elaborately organized sets of antibody genes that, through combinatorial rearrangements, are able to generate complex mixtures of antibody molecules that bear an essentially unlimited number of antigen-recognizing sequences (Supplementary Sidebar 15.1).

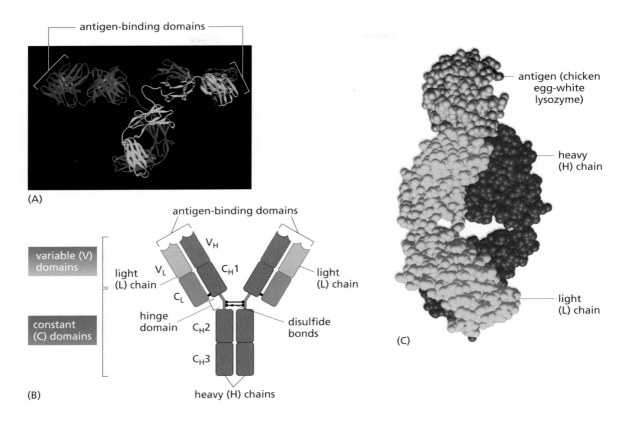

Figure 15.1 Structure of antibody molecules and their binding to antigens The most abundant antibody molecule in the plasma is immunoglobulin γ (IgG). (A) X-ray crystallography of an IgG molecule reveals the symmetry that allows the two antigen-binding domains (*top left, top right*) to bind two antigen molecules simultaneously. (B) IgG molecules can recognize an essentially unlimited number of antigens because of their ability to display a comparable diversity of amino acid sequences and thus structures in their antigen-binding portions, which are called their *variable* domains (*pink, red*). The remainder of each IgG molecule, termed its *constant* domains (*light and dark blue*), is identical among all IgG molecules of a given subclass, e.g., all IgG1 molecules. An IgG molecule as a whole is a heterotetramer composed of two light (L) chains (*lighter*) and two heavy (H) chains (*darker*). Two separate antigen-recognizing and binding pockets are displayed (*top left, top right*), each composed of an H- and an L-chain N-terminal domain (*concave shapes*). The top half of the molecule (containing the V regions and the C_L and C_H1 domains) is linked to the bottom half (containing the C_H2 and C_H3 domains) by floppy, unstructured hinges. (C) This space-filling model shows the detailed structure of an antigen–antibody complex in which the antigen-binding domains of the heavy chain (*purple*) and light chain (*yellow*) are seen to contact an antigen molecule, in this case the chicken egg-white lysozyme molecule (*light blue*). Only parts of the variable regions of the heavy and light chains are shown here. (Note the glutamine residue, *red*, which is important for the hydrogen bonding of the antigen to the antibody molecule.) (A, courtesy of A. McPherson and L. Harris, from L.J. Harris et al., *Nature* 360:369–372, 1992. B, from K. Murphy, Janeway's Immunobiology, 8th ed. New York: Garland Science, 2012. C, courtesy of R.J. Poljak.)

The other arm of the immune system involves the **cellular immune response**. This response is mounted when specialized cytotoxic cells are developed by the immune system that can, on their own, recognize and directly attack other cells displaying certain antigens on their surface. In this case, soluble antibodies are not required as intermediaries to recognize antigens displayed by targeted cells, since cytotoxic cells of the T-lymphocyte lineage (CTLs) have developed their own antigen-recognizing machinery in the form of T-cell receptors (TCRs; **Figure 15.4**).

We can also depict the immune system from another perspective: many of the responses of the immune system to an infectious agent (for example, a specific strain of virus) and its antigens depend on a previous encounter with this agent. The immune system has been "educated" through the initial encounter to recognize certain antigens displayed by this agent and to mount a vigorous counterattack upon encountering this agent again; this represents the **adaptive immune response**. At the same time, other cellular components of the immune system are *naturally endowed* with the ability to recognize certain infectious agents or abnormal cells and thus do not require prior exposure and education; this inborn ability is termed the **innate immune response**.

Figure 15.2 Neutralization by antibody molecules (A) Virus particles (*red*) can become coated by antibody molecules (*blue*) developed by the immune system of an infected host; in this case the antibody molecules recognize and bind viral antigen molecules, depicted here as *red spikes*. This coating *neutralizes* (inactivates) the infectivity of the particles by blocking their adsorption to host cells. (B) Similarly, a bacterium displaying certain surface antigens (*red*) can also be prevented from adhering to host cells by bound antibody molecules (*yellow*).

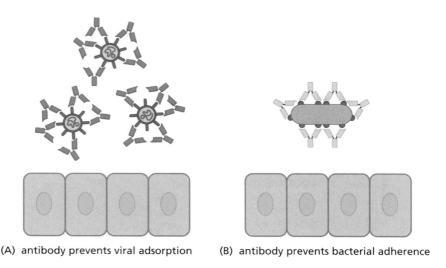

(A) antibody prevents viral adsorption   (B) antibody prevents bacterial adherence

For example, the natural killer (NK) cells cited above have the ability to recognize specific cell surface molecules displayed by aberrant cells, even without having encountered such cells previously.

15.2 The adaptive immune response leads to antibody production

Adaptive immune responses begin when infectious particles or abnormal cells are engulfed by specialized phagocytic cells of the immune system, notably macrophages and **dendritic cells** (DCs; **Figure 15.5**). Having ingested these objects or fragments thereof, the phagocytic cells are then charged with the task of presenting the ingested contents to other cellular components of the immune system, more specifically, to various types of T cells. This presentation of ingested antigens by phagocytic cells often takes place in the lymph nodes, to which these cells migrate following uptake of antigen.

In order to educate the immune system, these **antigen-presenting cells** (APCs) first digest the particles that they have phagocytosed (that is, ingested outright) or endocytosed (that is, bound via cell surface receptors and then internalized). This digestion, which is carried out in specialized cytoplasmic vesicles, slices internalized proteins into small oligopeptides, samples of which are then loaded onto the specialized antigen-presenting domains of **major histocompatibility complex** (MHC) class II molecules as these molecules make their way to the surface of APCs (**Figures 15.6 and 15.7**). (In humans, the MHC molecules are often termed HLA, or human leukocyte antigen, molecules, but we will use the more generic term, MHC, throughout this chapter to refer to both human and murine molecules of this type.)

The class II MHC molecules function much like a street hawker's hands displaying wares to passers-by. In this case, the wares are oligopeptide antigens captured by the APCs and the intended customers are other cells of the immune system, specifically a class of lymphocytes termed helper T cells (T_H cells), often called CD4+ cells to reflect a specific cell surface antigen that they display. Because macrophages and dendritic cells are specialized to use their MHC class II molecules to present antigens scavenged from their environment, immunologists sometimes call them "professional" APCs, to distinguish them from cells that are not specialized for this type of antigen presentation.

Note that it is the combined molecular structures formed by the class II ectodomains (the "hands") and their bound oligopeptide antigens (the "wares") that are presented to T_H cells (see Figure 15.6). Antigen presentation to certain T_H cells provokes the latter to activate, in turn, the B cells that can manufacture immunoglobulin (antibody) molecules that specifically recognize and bind the particular antigen (**Figure 15.8**).

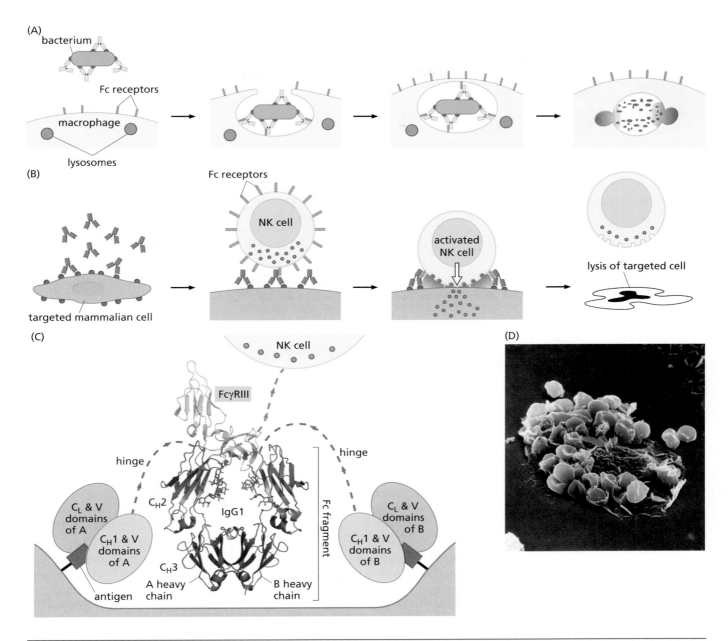

Figure 15.3 Coating of cellular targets by antibody molecules and their elimination by effector cells of the immune system
The coating of viruses, bacteria, and mammalian cells by antibody molecules is often the prelude to their being phagocytosed (engulfed) or destroyed by cytotoxic cells of the immune system. (A) The coating of a bacterium (*red*) by antibody molecules (*yellow*) may provoke a macrophage to use specialized receptors on its surface, termed Fc receptors (*green*), to recognize and bind the constant regions of the antibody molecules (see Figure 15.1). This often results in the phagocytosis of the antibody-coated bacterium and its eventual destruction in lysosomes within the cytoplasm of the macrophage. (B) A mammalian cell (*gray*) becomes coated by antibody molecules (*blue*) that recognize and bind antigens (*red*) on its surface. A natural killer (NK) cell then uses its Fc cell surface receptors (*green*) to bind the constant regions of the antibody molecules. This binding activates the NK cell, which proceeds to destroy the targeted cell by introducing the contents of cytotoxic granules (*purple dots*) into it. This process of cell killing is often referred to as antibody-dependent cell-mediated cytotoxicity (ADCC). (C) NK cells display the FcγRIII receptor on their surface

(*light green*). This binds to IgG1 immunoglobulin molecules (*blue*) that are already bound via their V domains (see Figure 15.1B) to cell surface antigens, such as antigens (*red*) displayed by a cancer cell (*gray*). This binding by the FcγRIII receptor to the IgG1 occurs at one end of the hinge region between the C_H2 and C_H1 domains of the IgG1 molecule. Because the site of IgG glycosylation modifications (*stick figures, purple*) lies close to the FcγRIII-binding site, alterations in the associated carbohydrate moieties can affect binding to this Fc receptor. The *dashed lines* represent segments that link the specified domains together. (D) Sheep red blood cells (RBCs) were treated with an antibody that recognizes an antigen displayed on their surface. Such coating can lead to the phagocytosis of the opsonized cell by macrophages and other types of phagocytic cells displaying Fc receptors. As seen in this scanning electron micrograph, a large number of the RBCs have become adsorbed to the surface of a single macrophage via its Fc receptors. (A and B, adapted from K. Murphy, Janeway's Immunobiology, 8th ed. New York: Garland Science, 2012. C, courtesy of P.D. Sun. D, from J. Swanson and M. Diakonova.)

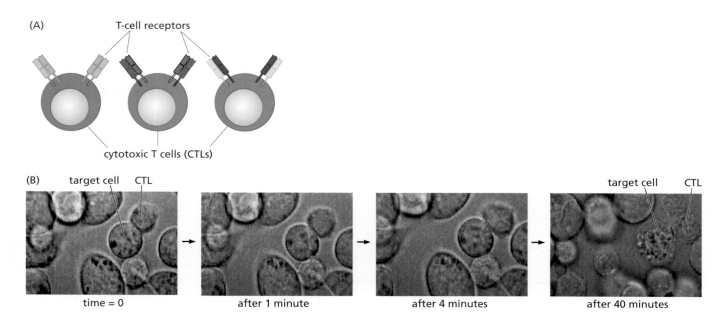

Figure 15.4 Cytotoxic T lymphocytes The cellular arm of the immune response results in the formation of cytotoxic cells, such as cytotoxic T cells (T$_C$'s, CTLs) that are able to recognize and kill other cells displaying certain antigens on their surface. (A) CTLs develop on their surface antibody-like molecules termed T-cell receptors (TCRs). A diverse array of TCRs are created during the development of the immune system, paralleling the formation of a diverse repertoire of soluble antibodies (see Supplementary Sidebar 15.1). Each CTL displays a particular antigen-recognizing TCR. (B) Seen here is a CTL (*upper right, 1st panel*) that has already used its TCR to recognize and bind to a target cell (*diagonally below it to the left*). The cytotoxic granules within this CTL (*red spots*) begin over a period of minutes to migrate through the cytoplasm of the CTL toward the point of contact between the killer and its victim. By 40 minutes, the contents of these granules (such as granzymes; see Section 9.14) have been introduced into the target cell, which has already advanced into apoptosis and begun to disintegrate. (B, courtesy of G. Griffiths.)

The subsequent maturation of these B cells yields a population of cells (called **plasma cells**) that actively secrete this particular antibody species into the circulation, that is, antibody molecules that are specialized to recognize and bind the particular antigen that originally triggered this series of responses.

This system works well when confronting infectious agents such as virus particles, bacteria, and fungi in the extracellular spaces. Thus, these infectious agents can be internalized by the professional APCs, and the peptides deriving from the ingested agents can be presented again to the outside world. The antibody molecules that are eventually formed by B cells and their descendants as a result of this antigen presentation can recognize and bind the infectious particles and thereby neutralize them (see Figure 15.2). By the same token, we can imagine that cancer cells displaying certain distinctive antigenic proteins on their surfaces might also provoke an antibody response by the immune system and become coated by antibody molecules bound to these cell surface antigenic molecules.

The antibodies coating a cell or infectious agent may elicit an alternative type of immune attack: a set of proteins in the plasma, termed **complement**, will recognize the constant regions of antibody molecules tethered to the surface of a cell (including bacterial, fungal, and mammalian cells), bind to these antibody molecules, and proceed to punch holes in the adjacent plasma membrane, thereby killing the cell (**Figure 15.9**).

This series of steps leading to adaptive humoral responses tells us something important about the molecular structure of the antigens that are **immunogenic**, that is, that elicit immune responses: they are not intact proteins, but instead are oligopeptide fragments derived from the cleavage of much larger proteins (see Figures 15.6 and 15.7). (The major exceptions to this generality are certain complex carbohydrate chains and linked side chains that may, under some circumstances, also be immunogenic.)

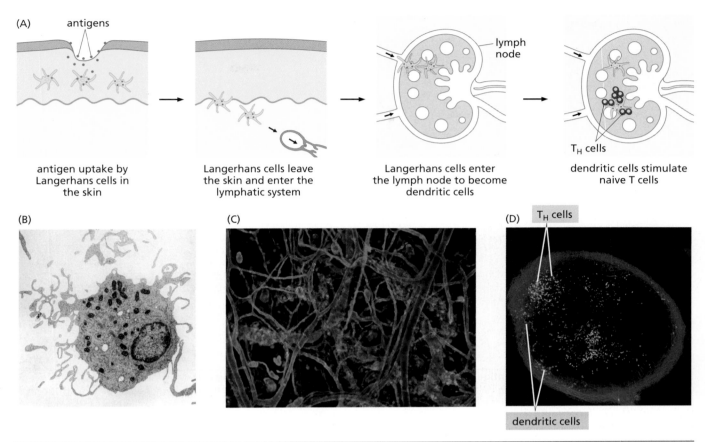

Figure 15.5 Antigen presentation by dendritic cells The immune system becomes aware of infectious agents and their antigens largely through the actions of antigen-presenting cells (APCs), notably dendritic cells. (A) Depicted here are specialized phagocytic cells (i.e., Langerhans cells, *yellow*) that reside in the skin and, like dendritic cells, take up antigens (*red dots*) by phagocytosis and then migrate to the lymph nodes (*light blue*), where they mature into dendritic cells (DCs) that are similar to the DCs responsible for acquiring antigens in many other tissues. In the lymph nodes, these cells confront T cells (*dark blue circles*), to which they present antigens; this results in the functional activation of the T cells and the subsequent mounting of a specific immune response against cells and viruses that display these antigens. (B) The dendritic cells take their name from their multiple arms extending out from the cell body. (C) After dendritic cells (*red-orange*) have acquired antigen within the parenchyma of a tissue, they must intravasate into lymphatic vessels in order to reach the lymph nodes, where antigen presentation takes place. Here some DCs congregate outside the lymphatic vessels, while most have already intravasated. Both the lymphatic vessels, which are large and irregularly shaped, and the blood capillaries are labeled *green* here. (D) Multiphoton microscopy reveals the capsule of a mouse lymph node (*blue*) and a number of recently arrived, dye-labeled dendritic cells (*red dots*) as well as dye-labeled T cells (*green dots*) to which antigen will be presented by the dendritic cells. The two cell types are largely segregated from one another, and their mechanisms of trafficking and interaction within the lymph node remain poorly understood. (A, adapted from K. Murphy, Janeway's Immunobiology, 8th ed. New York: Garland Science, 2012. B, courtesy of J. Barker. C, courtesy of M. Sixt. D, from T.R. Mempel, S.E. Henrickson and U.H. Von Andrian, *Nature* 427:154–159, 2004.)

15.3 Another adaptive immune response leads to the formation of cytotoxic cells

The type of immunologic response described above fails to deal effectively with infectious agents that have entered into cells and are therefore shielded by the plasma membrane from scrutiny. Similarly, in the case of cancer cells, the humoral response system will fail to recognize aberrant cellular proteins that are hiding deep within these cells. In principle, such shielding should create a serious problem for the immune system, which needs to monitor what is going on *inside* cells in addition to its task of monitoring the contents of the extracellular spaces and the surfaces of cells.

The problem is solved by an antigen-presenting mechanism that echoes the one used by the professional antigen-presenting cells (APCs) described above. Actually, this other antigen-presenting mechanism is the more widespread of the two, since it is used by the great majority of cell types throughout the body.

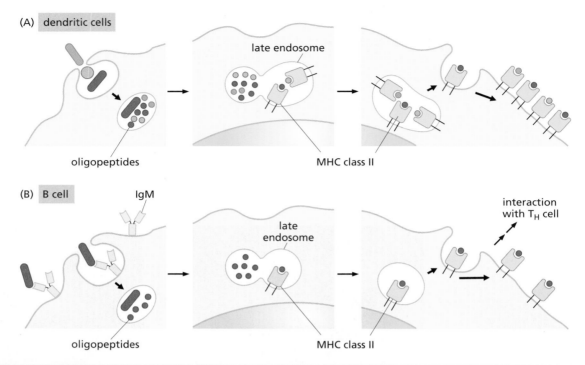

Figure 15.6 Antigen processing by antigen-presenting cells (A) After professional antigen-presenting cells (APCs)—dendritic cells (including those arising from Langerhans cells) and macrophages—have phagocytosed potential antigenic particles (*red, green, blue*), these are fragmented into oligopeptides (*dots*) by proteolysis within the endosomes formed by phagocytosis. MHC class II molecules migrate from the endoplasmic reticulum via the Golgi apparatus (*not shown*) to these endosomes, in which they encounter and bind the oligopeptides prior to migrating to the cell surface. These oligopeptide–MHC class II molecules can then be presented to T cells in the lymph nodes (*not shown*). (B) B cells can also phagocytose foreign materials and process the derived cleavage products in order to present them on cell surface MHC class II molecules. However, a fundamental difference between B cells and professional APCs is that B cells internalize only particles that are recognized by their antigen-specific cell surface antibody-like IGM molecules, while APCs phagocytose all types of particles promiscuously.

It works like this (**Figure 15.10**): rather than being used for their normally designated functions, a portion of the proteins synthesized within cells (by some accounts, as much as one-third in certain cells) is routinely diverted to specialized proteasomes (see Supplementary Sidebar 7.5). There, these proteins are cleaved into oligopeptides. These cleavage products, of 8 to 11 amino acid residues in length, are then attached to MHC class I molecules prior to their transport to the cell surface and displayed as MHC–peptide complexes on the outside of cells (see Figure 15.10A and B).

Figure 15.7 Antigen presentation by MHC class II molecules The structure of the antigen-presenting groove of an MHC class II molecule is shown here as determined by X-ray crystallography. The oligopeptide antigen (*colored stick figure*) that is bound via hydrogen bonds (*blue*) to the "palm" of the MHC molecule's "hand" (*ribbon diagram*) is shown with its N-terminus to the left and C-terminus to the right. The oligopeptide antigen together with the nearby amino acid residues of the MHC molecule forms the molecular structure that is recognized by other immune cells, which may, for example, use T-cell receptors to do so; MHC class II–bound oligopeptides are in the range of 15 to 24 residues. (From K. Murphy, Janeway's Immunobiology, 8th ed. New York: Garland Science, 2012.)

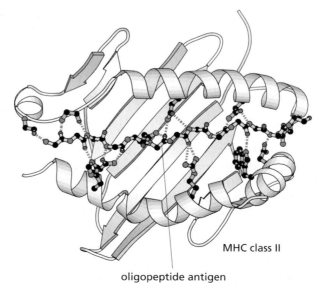

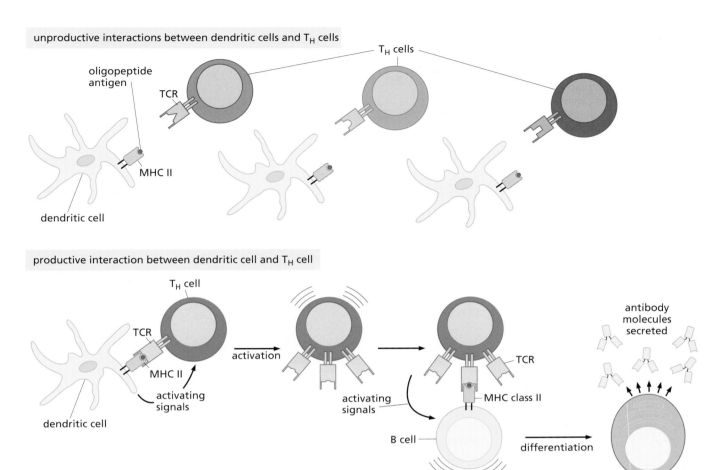

Figure 15.8 Immunocyte encounters leading to antibody production By displaying engulfed and processed antigenic peptides (*red dots*; see Figure 15.6) on their surface, dendritic cells proceed via the lymphatic ducts to the lymph nodes, where they use their MHC class II molecules (*gray*) to present these peptides to helper T (T$_H$) cells. Here, an antigen-presenting dendritic cell meets a number of T$_H$ cells, each displaying its own distinct T-cell receptor (TCR; *purple, green, red*) on its surface (*above*). (These TCRs are products of antibody gene rearrangements similar to those described in Supplementary Sidebar 15.1.) Since none of these three T$_H$ cells displays a receptor that recognizes and binds the antigen being presented by the MHC II molecules, the dendritic cell will move on, searching for T$_H$ cells that display more compatible TCRs. When the dendritic cell finds a T$_H$ cell whose receptors do indeed recognize the oligopeptide antigen being presented by the dendritic cell's MHC class II molecules (*below*), the ensuing interaction activates the T$_H$ cell, which leaves the dendritic cell

to search for B cells that also display on their surface the same antigen in the context of MHC II. When and if the T$_H$ finds such a B cell (*light yellow, 2nd diagram from right*), it activates the B cell, which proliferates and, having differentiated into a plasma cell (*light brown*), begins to release antibody molecules that are capable of recognizing this oligopeptide antigen. The prior acquisition by the B cell and MHC II–mediated presentation of specific oligopeptide antigens was described in Figure 15.6B. The sequence of interactions depicted here and in Figure 15.6B ensures that the *same oligopeptide antigen species* that was used previously by the dendritic cell to activate the T$_H$ cell will *also* be presented by the B cell to the T$_H$ cell, permitting the T$_H$ cell to activate a particular B cell and not others. (Not shown here is the further evolution of variable regions of the immunoglobulin genes in the B cell, which occurs in the germinal centers of the lymph node and enables descendants of the initially activated B cell to produce soluble antibodies having progressively higher antigen-binding avidities.)

Included among the intracellular peptides displayed by the MHC class I molecules are those synthesized normally by cells as well as those made by foreign infectious agents within the cell, such as viruses and bacteria. This external presentation of internal antigens occurs routinely and continuously, whether or not foreign proteins happen to be present within a cell.

The display by a cell of certain oligopeptide antigens on its surface (via its MHC class I molecules; see Figure 15.10C) may attract the attention of cytotoxic T cells (T$_C$'s, also called cytotoxic T lymphocytes, CTLs, or CD8$^+$ cells), which proceed to kill this

Figure 15.9 Complement-mediated killing (A) Antigen–antibody complexes (*red spheres*) form when antibody molecules bind to cell surface antigens (*left*). They can attract complement proteins present in the plasma (*yellow, green, purple*) and induce them to form complexes that lead, through a series of steps, to the formation by other complement proteins of channels in the cell's plasma membrane (*right*). The channels are created at a site adjacent to where the antigen–antibody complexes initially formed. (B) This electron micrograph shows such channels, which destroy the osmotic integrity of the plasma membrane and thereby lead rapidly to cell death. (A, adapted from K. Murphy, Janeway's Immunobiology, 8th ed. New York: Garland Science, 2012. B, from S. Bhakdi et al. *Blut* 60:309–318, 1990.)

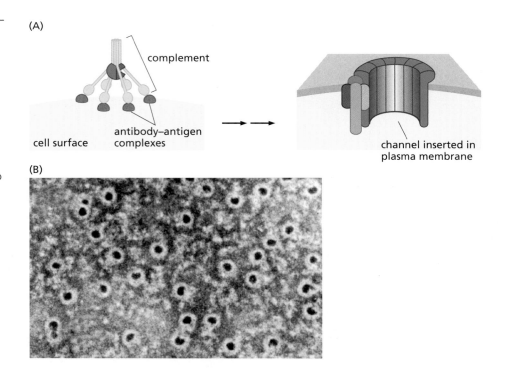

cell (see Figures 15.4 and 15.10D). The origins of this killing can be traced back to the actions of helper T cells. Recall that some helper T cells are able to activate the humoral immune response by interacting with and stimulating antibody-producing B cells (see Figure 15.8). Now, we encounter a second, independent function of helper T cells: some of them can contribute to the activation of cytotoxic T cells, which are specialized to recognize and kill target cells displaying the particular oligopeptide antigen that initially provoked an immune response (**Figure 15.11**). This attack on antigen-displaying cells by cytotoxic lymphocytes represents the *cellular arm* of the adaptive immune response.

The capacity of helper T cells to facilitate development of both humoral and cellular immune responses reflects the ability of distinct subpopulations of T_H cells to produce and release the soluble immune factors known as cytokines: T_H's that promote humoral immunity (by stimulating B cells) produce different profiles of secreted cytokines than those promoting cell-mediated immunity (that is, via the actions of cytotoxic T cells).

The cytotoxic T cells (T_C's) can actually kill their cellular victims through two separate mechanisms. They can expose their intended victims to certain toxic proteins, such as **granzymes**, which are endocytosed into the cytoplasm of targeted cells in multi-protein complexes that include **perforin** (**Figure 15.12A and B**). Perforin proceeds to punch holes in the membranes of the vesicles resulting from this endocytosis, allowing the granzymes to enter into the cytosol of the victim. As described in Section 9.14, once in the cytoplasm of the targeted cell, granzymes cleave and thereby activate pro-apoptotic caspases.

The second killing mechanism (also discussed in Section 9.14) involves the Fas death receptor, which is displayed on many cell types throughout the body. T_C's can present FasL, the ligand of the Fas receptor, to their intended victims. FasL then activates the Fas death receptors on the surfaces of the targeted cells, thereby activating their extrinsic apoptotic pathway (see Figure 15.12C).

Cytotoxic T cells can play an important role in limiting the infectious spread of viruses. For example, a recently infected cell in which a virus is actively replicating will use its MHC class I molecules to display oligopeptide antigens derived from cleaved viral proteins. This antigen display will alert the immune system to the fact that abnormal proteins are being produced deep within the cell. If the immune system is functioning well, its T_C's will recognize the viral oligopeptide antigens displayed by the infected cell's MHC class I molecules and kill this cell long before the virus has had a chance to multiply and release progeny virus particles.

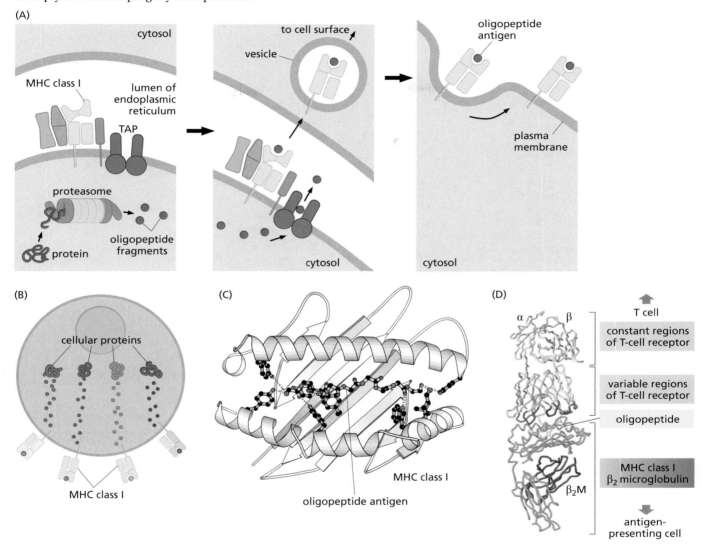

Figure 15.10 Display of intracellular antigens by MHC class I molecules Almost all cell types, including cancer cells, routinely divert a portion of their recently synthesized proteins to the antigen-presenting machinery. (A) Some of the recently synthesized proteins in the cytosol are diverted to specialized proteasomes (*purple, yellow*), in which they are broken down into oligopeptides (*red dots*); the oligopeptides are then introduced via a specialized channel (TAP, *dark green*) into the lumen of the endoplasmic reticulum, where they may encounter MHC class I molecules (*yellow*) that bind them relatively tightly (see C). The oligopeptide–MHC class I complexes are then dispatched via membranous vesicles to the cell surface, where they display to the immune system fragments of the proteins that are being synthesized within the cell. The overall process is similar in outline to that undertaken by MHC class II molecules (see Figure 15.6); however, MHC class II antigen presentation is the specialty of "professional" antigen-presenting cells such as macrophages, dendritic cells, and B cells, whereas MHC class I presentation is performed routinely by almost all cell types in the body. (B) A broad spectrum of oligopeptide fragments deriving from a large number of cellular proteins (here represented as four distinct protein species) are displayed simultaneously by cells using their MHC class I proteins; the display by an individual cell of multiple variant forms of MHC class I molecules, each with a slightly different antigen-binding domain, enables the cell to display a diverse set of oligopeptides. (C) The structure of the antigen-presenting domain of MHC class I molecules, which binds oligopeptides of only 8 to 10 residues, is very similar to the antigen-presenting domain of MHC class II molecules (see Figure 15.7). (D) A T-cell receptor (TCR) composed of α and β subunits (*above*) has recognized and bound an MHC class I molecule (*below*) presenting an oligopeptide antigen (*yellow*). The "complementarity-determining" regions of the TCR, which recognize the oligopeptide–MHC complex, are shown in *dark blue, yellow*, and *red*. The MHC molecule is accompanied by a β$_2$-microglobulin molecule (*dark green*), which serves as an accessory subunit of all MHC class I molecules. (The genetic mechanisms leading to the formation of the TCR are described in Supplementary Sidebar 15.1.) (A and C, from K. Murphy, Immunobiology, 8th ed. New York: Garland Science, 2012. D, from K.C. Garcia et al., *Science* 279:1166–1172, 1998.)

Figure 15.11 Activation of cytotoxic T cells by dendritic cells and killing of antigen-expressing target cells In addition to inducing B cells to make antibody molecules, certain dendritic cells (DCs) can use their own MHC class I antigens to present ingested oligopeptides to the precursors of cytotoxic T-cells (*pink*). This interaction between DCs and T_C precursors is sometimes termed "cross presentation" and helps to induce the T_C precursors to mature into active cytotoxic T cells (termed T_C's or CTLs, *red*), which then use their T-cell receptors (TCRs) to recognize and bind antigens displayed by MHC class I molecules on the surfaces of many cell types throughout the body. Thus, this recognition often results in an attack on the antigen-displaying cell (*gray, top right*). The T_C's often use cytotoxic granules (*black dots*) containing perforin and granzymes to kill targeted cells (as described in Figure 15.12).

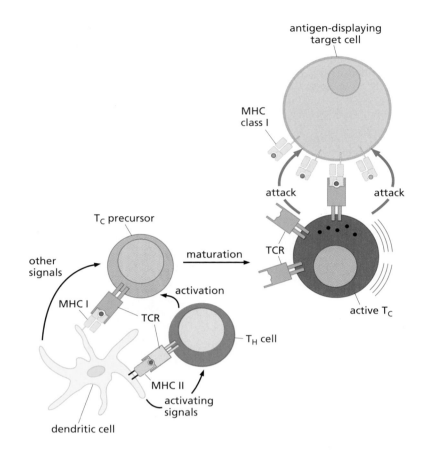

Figure 15.12 Mechanisms of cell killing by cytotoxic T cells and natural killer cells (A) This transmission electron micrograph of a cytotoxic T lymphocyte (T_C, CTL) reveals a series of lytic granules in its cytoplasm (*pink arrows, left panel*). When contact is made with a targeted cell (which was initially recognized by the T-cell receptors borne by the T_C), these granules release perforin, which forms cylindrical channels in the plasma membrane of the target (*white arrows, right panel*); pro-apoptotic proteins such as granzymes (see Section 9.14), which are also carried in these granules, are then introduced through these channels into the cytoplasm of the targeted cell, where they initiate the apoptotic cascade by cleaving procaspases. (B) In the absence of a cellular target, the lytic granules (*green, yellow*) are scattered throughout the T_C cytoplasm (*upper panel*). In the lower panel, a *synapse* has been formed with a targeted cell (*left*), and the lytic granules have congregated at the synapse in preparation for killing the targeted cell. (C) An alternative mechanism of killing cells that have been targeted for destruction depends on the display of FasL (*orange*) by the T_C (*top, pink*). FasL, a trimer, engages the Fas receptor (*brown*) displayed by the targeted cell (*bottom cell, gray*) and triggers receptor trimerization; this results in activation of the extrinsic apoptotic cascade in the targeted cell via the sequential activation of caspases 8 and 3 (see also Figure 9.32). (D) Natural killer (NK) cells are programmed to recognize and kill other cells, including cancer cells that do not display normal levels of MHC class I molecules on their surface. This scanning electron micrograph (SEM) reveals two NK cells (*colorized green*), one of which has spread a portion of its cytoplasm across the surface of a human ductal breast carcinoma cell, in the initial stage of such an attack. (E) This scanning electron micrograph reveals the initial attack of an NK cell (*left panel, smaller cell*) on a leukemia cell. Sixty minutes later, the NK cell has caused extensive damage to the leukemia cell, which has fragmented and rolled up its plasma membrane in response to this attack (*right panel*). (A, courtesy of E. Podack, from E.R. Podack and G. Dennert, *Nature* 302:442–445, 1983. B, from R.H. Clark et al., *Nat. Immunol.* 4:1111–1120, 2003. C, from K. Murphy, Janeway's Immunobiology, 8th ed. New York: Garland Science, 2012. D, courtesy of S.C. Watkins and R. Herberman. E, from R. Herberman and D. Callewaert, Mechanisms of Cytotoxicity by Natural Killer Cells. Orlando, FL: Academic Press, 1985; with permission from Elsevier.)

This means that the immune system actually uses two arms of the adaptive immune response to limit viral infections: the cellular response is used to kill virus-infected cells, while the humoral response is used to neutralize virus particles that have been released into extracellular spaces, including the circulation, by coating these particles with antibody molecules (see Figure 15.2A). As we will see, the anti-viral responses are important means by which the immune system blocks the appearance of virus-induced human tumors.

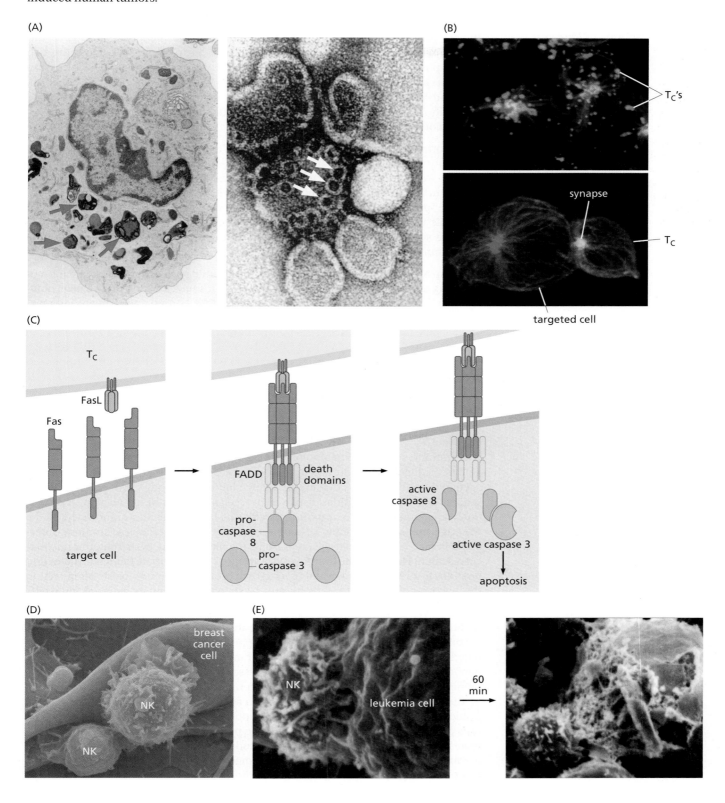

15.4   The innate immune response does not require prior sensitization

Ninety-nine percent of the animal species on the planet do not possess adaptive immune responses to protect them from attack by pathogens. These organisms rely on innate immunological responses for such protection. Importantly, this ancient, widespread innate immunity system has been conserved during the evolution of mammals and continues to play a critical role in various immunological responses.

The cellular components of the innate immune response are able to recognize and attack foreign particles and aberrant cells without having been "educated" through prior exposure to these agents. Thus, these **immunocytes** "instinctively" recognize aberrant cells, such as cancer cells, in the body's tissues and target these cells for attack and destruction. Instead of recognizing specific antigens, the cells mediating innate immunity recognize characteristic molecular patterns that are present on the surfaces of infectious agents (or transformed cells) but are not displayed by normal cells.

An important mediator of the innate response is the natural killer (NK) cell. It is likely that many initial encounters of the immune system with cancer cells are made by NK cells. As we will discuss in greater detail later, the NK cells recognize configurations of cell surface proteins displayed by a wide variety of cancer cell types. Hence, NK cells are "pre-programmed" to recognize cancer cells and to eliminate them from the body's tissues. In addition to NK cells, yet other cellular components of the innate immune system, including macrophages and neutrophils, contribute to mounting innate immune responses against cancer cells.

After an NK cell has initiated the innate immune response by recognizing and attacking a target cell (see Figure 15.12D and E), it sends out cytokine signals, notably interferon-γ (IFN-γ), in order to recruit yet other immune cells, including macrophages, to the site of attack. The actions of this second wave of immunocytes will often enable the immune system to mount more specific and ultimately more effective responses, in particular, adaptive humoral and cellular responses. For example, large numbers of cytotoxic T cells can be mobilized by the adaptive immune response to efficiently kill cancer cells.

15.5   The need to distinguish self from non-self results in immune tolerance

The immune system is finely tuned and highly specific. Most critically, it must be able to distinguish foreign proteins (for example, those made by invading infectious agents) from the proteins that are normally made by the body's own cells. As a consequence, if the oligopeptides displayed by a cell are similar or identical to those routinely encountered by the immune system, this cell will remain unmolested by the various arms of the immune system—one of the manifestations of immune **tolerance**. In fact, immune tolerance represents the major puzzle of current immunological research: How does the immune system learn to discriminate foreign proteins and peptides from the body's normal repertoire of proteins? Immunologists often refer to this behavior as the ability of the immune system to discriminate between "non-self" and "self."

A variety of mechanisms operating during the development of the immune system ensure that any T cells and B cells that happen to recognize self-antigens are eliminated; alternatively, if such **self-reactive** or *auto-reactive* lymphocytes escape elimination, their actions will be strongly suppressed. For example, during the early development of the immune system, T cells encounter a vast repertoire of proteins in the medulla of the thymus, including many that are usually expressed in a tissue-specific fashion throughout the body; T cells that happen to recognize self-antigens in the thymus are then deleted from the lymphocyte population via apoptosis. By some estimates, as many as 50% of the proteins encoded in our genome are expressed in the thymic medulla in order to generate tolerance. This induction of *central tolerance* is complemented later by the process of *peripheral tolerance*, when T cells that have

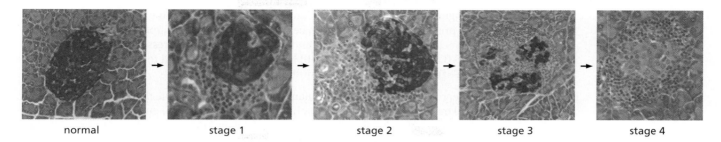

normal   stage 1   stage 2   stage 3   stage 4

Figure 15.13 Destruction of a normal tissue by autoimmune attack Extensive tissue damage can be wrought by an immune system that has been provoked to attack the body's normal tissues; in principle, the same immune mechanisms can attack and destroy malignant tissues. A well-studied example derives from research into Type 1 diabetes. In a mouse model of this disease, a normal islet of Langerhans (*left panel*; see also Figure 13.36), composed largely of insulin-secreting β cells (*purple*), is seen to be surrounded by cells of the exocrine pancreas (*pink*). Autoimmune attack on islet cells results initially in the congregation of lymphocytes (*small, dark purple cells*) adjacent to a still-intact islet (stage 1), followed by the progressive elimination of increasing proportions of β cells (stages 2 and 3), and finally the total elimination of all functional β cells (stage 4). Similar processes operate in a number of other tissues that are compromised or destroyed by autoimmune reactions. (Courtesy of T. Takaki and T.P. DiLorenzo.)

escaped deletion in the thymus migrate out of the thymus to lymph nodes throughout the body and recognize self-antigens; these T cells are then inactivated in the lymph nodes. Failure to delete auto-reactive lymphocytes from the large pool of lymphocytes in the body results in the survival of immune cells that may target the body's own normal tissues. This breakdown of tolerance may lead to autoimmune diseases, such as rheumatoid arthritis, ulcerative colitis, and lupus erythematosus, in which the immune system dispatches antibodies and cytotoxic cells to attack normal cells and tissues (**Figure 15.13**).

Immune tolerance raises a further simple and obvious point that will dominate the discussions that follow: How does the immune system, which is designed to be tolerant of the body's own cells, recognize and attack cancer cells, which are, to a great extent, very similar at the biochemical level to cells that are normally present in the body? And if it does undertake attacks against cancer cells, including those transformed by tumor viruses, how might these cells evade and thwart the attacks launched by various arms of the immune system (Supplementary Sidebar 15.2)?

15.6 Regulatory T cells are able to suppress major components of the adaptive immune response

Research beginning in the 1990s has described an entirely new class of T cells that have come to be called regulatory T cells (T_{reg} cells or simply T_{reg}'s). Indirect evidence suggesting their existence came from the observation that in normal individuals, a significant proportion of T_C's (that is, cytotoxic T cells, CTLs) recognize normal tissue antigens presented by the MHC class I molecules—a situation that should lead directly to extensive immune attack on normal tissues and resulting autoimmune disease. However, such attacks do not occur, apparently because of suppression of these cells' actions by some unknown agents.

The discovery of T_{reg} cells seems to have largely solved this problem, since these cells are able to block the actions of the T_C's that are scattered throughout our tissues. Indeed, in genetically altered mice lacking T_{reg} cells, lethal autoimmune disease develops early in life; a comparably aggressive, ultimately fatal autoimmune disease has also been documented in humans who are unable to make T_{reg}'s.

Like T helper (T_H) cells, the T_{reg}'s display the CD4 antigen on their surface. However, the T_{reg}'s are distinguished by their additional display of the CD25 surface antigen and their expression of a transcription factor, termed FOXP3, that programs their development. Because T_{reg}'s express antigen-specific T-cell receptors (TCRs; see Figure 15.4), they can specifically block the actions of those cytotoxic T lymphocytes whose TCRs recognize the same antigens. In addition, when located in the lymph nodes, the T_{reg}'s can prevent the activation of T_H cells by dendritic cells. It appears that the T_{reg}'s must be in close proximity to the T_H and T_C cells that they suppress, and that the release of TGF-β and interleukin-10 (IL-10) by the T_{reg}'s is often used to suppress the proliferation of these other types of T lymphocytes.

Research on T_{reg}'s is still in its infancy. However, it is possible that their behavior holds the key to understanding the pathogenesis of a number of autoimmune diseases. At the same time, the actions of T_{reg}'s may explain how many types of tumor cells can thrive in the presence of large numbers of T_C's that should, by all rights, be able to eliminate them—a topic pursued later in this chapter.

An overview of the various components of the immune system that we have covered until now is provided in **Figure 15.14**.

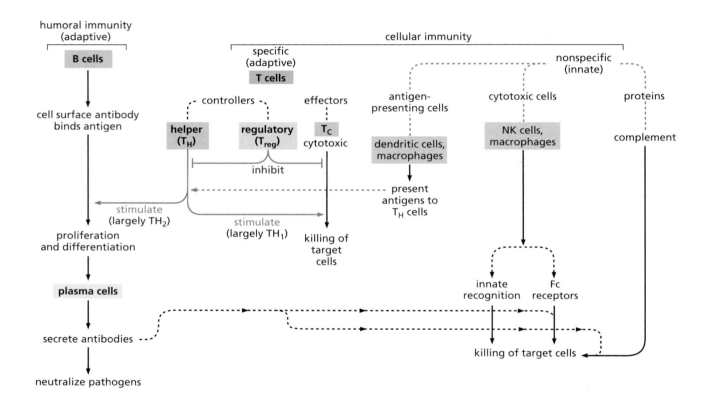

Figure 15.14 Overview of the humoral and cellular arms of the immune system The humoral immune system (*left*) is driven by the actions of B cells that develop millions of distinct antigen-specific antibody molecules through rearrangement of antibody-encoding genes and diversification of the sequences encoding the antigen-combining sites in the variable domains of antibody molecules. This humoral response depends on prior activation of helper T (T_H) cells in the lymph nodes by antigen-presenting cells, largely dendritic cells. The latter process proteins into oligopeptides that are recognized by the T-cell receptors of T_H cells, which proceed to activate B cells that have developed, by chance, antibodies that recognize the same oligopeptide antigens. Cytotoxic T (T_C) cells rely on T-cell receptors (TCRs, *not shown*) to recognize and kill target cells displaying cognate antigens. The activation of T_C cells depends on prior stimulation by T_H cells. A third class of T cells that also express antigen-specific T-cell receptors are the regulatory T (T_{reg}) cells. These play important roles in suppressing the actions of both T_C and T_H cells and thereby prevent inappropriate activation of immune responses that might otherwise lead to a breakdown of tolerance and resulting autoimmune disease. These various manifestations of adaptive immunity are augmented by the innate immune system (*right*), specifically by cell types that can aid in the elimination of pathogens and cancer cells without any prior "education" through previous exposure to these entities. Thus, natural killer (NK) cells are primed to kill many types of cancer cells because of the abnormal configuration of the cell surface molecules that the latter display; macrophages are also capable of recognizing and killing many cellular pathogens without any prior exposure to them. Although macrophages and NK cells cannot themselves specifically recognize most cell surface antigens, the coating of potential target cells by antibody molecules (produced by the adaptive immune response) will attract macrophages and NK cells, which will use their Fc receptors to bind to the constant (C) regions of antibody molecules and proceed to kill the antibody-coated cells. Similarly, the complex group of plasma proteins known as complement may also recognize antibody molecules bound to a cell's surface and then kill this cell by inserting channels in its plasma membrane.

15.7   The immunosurveillance theory is born and then suffers major setbacks

As suggested by the quotation at the beginning of this chapter, the notion that the immune system is able to defend us against cancer is an old one. Burnet's 1957 speculation about the immune system's role in monitoring tissues for the presence of tumors, together with other speculations made by Lewis Thomas, represented the first clear articulation of the notion of tumor **immunosurveillance**.

At the time, infecting microorganisms, specifically, bacteria, viruses, and fungi, were known to be strongly immunogenic, in that they usually provoke an immune response that leads to their total eradication by various arms of the immune system. By analogy, it was plausible that the immune system continuously monitors its tissues for the presence of cancer cells. Having identified them—so this thinking went—the immune system would treat these cancer cells as foreign invaders and eliminate them long before they had a chance to proliferate and form life-threatening tumors.

Early attempts in the 1950s to test this model were not definitive. When tumors were removed from some mice and implanted in others, the tumors were rapidly destroyed in a way that gave clear indication of the actions of vigorous host immune responses. Soon it became clear, however, that this rejection had nothing to do with the *neoplastic* nature of the tumor cells. Instead, their elimination was a consequence of what came to be called **allograft rejection**. Thus, cells and tissues from one strain of mice are invariably recognized as being foreign when implanted in mice of a second strain. This is a consequence of the fact that the cells of different strains of mice display distinct, genetically templated major histocompatibility (MHC) molecules on their surfaces. (In this instance, however, it is not the bound oligopeptide antigens that evoke an immune response but the MHC molecules themselves, which vary slightly in structure from one strain of mouse to another; Sidebar 15.1.)

For example, engrafted cancer cells from BALB/c mice were recognized as being of foreign origin (and were therefore histo*incompatible*) when introduced into C57BL/6 mice, and vice versa (Figure 15.15). These graft rejections from dissimilar, **allogeneic** (that is, genetically distinct) mouse strains were not observed when tumor cells of BALB/c origin were grafted into BALB/c hosts, that is, into **syngeneic** hosts that, by definition, shared an identical genetic background and identical histocompatibility

---

Sidebar 15.1   Why are MHC proteins so polymorphic? MHC molecules were discussed earlier in this chapter (see Section 15.2) because of their central role in presenting antigens to T lymphocytes of the immune system. An idiosyncrasy of MHC genes is their extraordinary polymorphism. One database lists more than 5000 human class I alleles and almost 1600 class II alleles. Various rationales have been proposed for this astounding variability, which can be observed in the gene pools of many vertebrate species. The simplest rationale argues that a high degree of MHC polymorphism ensures that at all times at least some individuals within a species are likely to express MHC molecules that can recognize the antigens of novel, highly pathogenic infectious agents, allowing these individuals to mount an effective immune defense and guaranteeing the continued survival of the species as a whole. Alternatively, intraspecies MHC variability ensures that an infectious pathogen cannot spread widely within a species, because some of its members will express MHC molecules that can bind the pathogen's oligopeptide antigens, leading to the production of antibodies that can eliminate the pathogen, interrupting the cycles of infectious transmission from one individual to the next that lead to epidemics. Yet other rationales have been proposed as well.

Important for our own discussions, the highly variable antigen-binding pockets of MHC molecules represent antigens in their own right. Thus, each pocket, which is designed to bind and display its own special constituency of oligopeptide antigens, is assembled from its own particular amino acid sequence and therefore represents, on its own, a distinct antigen. Hence, while MHC polymorphic variability has evolved to protect against novel infectious agents, an inadvertent product of this variability is the inter-individual variability of the tissues of various members of a single species: one individual's cells are recognized as foreign by the immune system of a second individual largely because each individual in a species displays its own particular combination of MHC molecules and thus MHC-associated antigens.

This inter-individual variability can be eliminated by repeated cycles of inbreeding, which ensures that heterozygosity within a group of organisms is progressively decreased, being replaced by homozygosity at most and eventually all genetic loci, including the loci encoding the various MHC molecules. Such inbreeding, begun a century ago, led to the generation of the genetically homogeneous strains of mice that are in widespread current use.

Figure 15.15 Syngeneic mice and MHC variability The use of inbred strains of mice has revealed that major determinants of the immunogenicity of the cells of these mice (and of mammals in general) are the MHC class I, II, and III molecules. (Only MHC class I molecules are depicted here and only class I and II molecules are discussed in this chapter.) These molecules are highly polymorphic within a species, indicating that one individual (or one inbred strain of mice) almost always has a different set of MHC molecules from another (*red, blue cell surface molecules*). Moreover, most of the MHC molecules expressed by an individual's normal cells continue to be expressed in tumor cells arising in that individual. Therefore, if a tumor arises within a BALB/c mouse, it is often transplantable into a syngeneic host, i.e., another BALB/c mouse, but not into an allogeneic host, such as a C57BL/6 mouse. The converse is true for tumors arising in C57BL/6 mice.

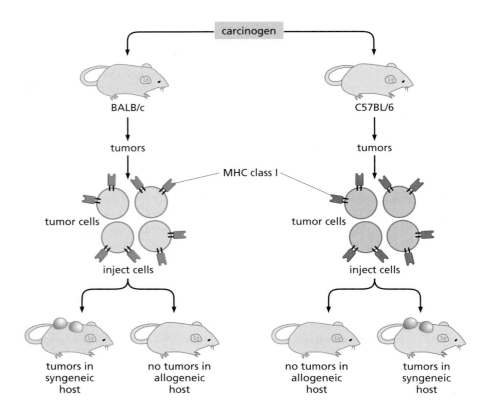

antigens with the engrafted cells. [In fact, the term **histocompatibility** derives from the observation that tissues ("histo-") from mice of one inbred strain can be grafted and established in the bodies of other members of the same genetic strain and are in this sense "compatible."]

The observed rejections of allogeneic tumors represented a detour for the young field of tumor immunology, since they shed no light on how the immune system of a mouse or human host would recognize cancer cells that arose in its own tissues. Still, this early work did make one profoundly important point: in addition to eliminating microbes and various types of viruses, the immune system is capable of destroying mammalian cells that it recognizes as foreign or, quite possibly, as otherwise abnormal. As an additional corollary, these observations of immune function led to the conclusion that cancer cells could never be transmitted from one individual to another (but see Supplementary Sidebar 15.3 and Figure S15.2).

An alternative strategy was then embraced for studying the immunosurveillance problem. If the immune system were indeed responsible for suppressing the appearance of tumors, animals with compromised immune systems should suffer increased rates of cancer. Such cancers, which originated within their own bodies—so-called **autochthonous** tumors—were, of course, of the same histocompatibility type as the remaining tissues in these animals. In these situations, the issue of histocompatibility (and -incompatibility) was rendered irrelevant.

In the late 1960s, immunocompromised mice of the Nude strain first became available to cancer researchers (Supplementary Sidebar 15.4). These mice lack a functional thymus—the tissue in which the T lymphocytes of the immune system initially develop. (Their lack of hair, another distinct phenotype of this strain, gave them their name; see Figure 3.13.) The research that followed in the early and mid-1970s revealed that these mice are no more susceptible to spontaneously arising or chemically induced autochthonous tumors than are their normal, wild-type littermates. So, the immunosurveillance theory suffered a major setback, having failed a major critical test of its validity. It lost credibility and retreated from the main arena of cancer research for two decades.

(A)

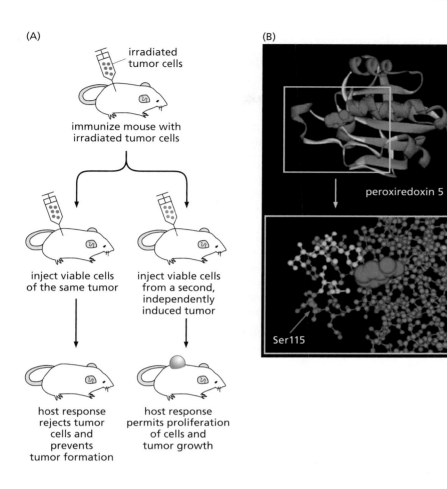

(B)

peroxiredoxin 5

Ser115

Figure 15.16 Immunization of mice by exposure to killed cancer cells (A) Mice from one strain were initially injected with irradiated, killed cancer cells (*red*) deriving from a tumor chemically induced in a mouse of the same strain (i.e., a syngeneic mouse; see Figure 15.15). When these mice were subsequently injected with live cells from the same tumor, the cells failed to grow (*lower left*). However, when these mice were injected with live cells from a second tumor (*blue*), the cells proliferated and formed a tumor mass (*lower right*). The reciprocal experiment (*not shown*) yields the opposite results, i.e., injection with killed *blue* cancer cells rendered mice immune to the *blue* tumor but not to the *red* tumor. Since these experiments were all performed using tumor cells and mice from the same genetically defined strain, the tumor rejection could not be attributed to histoincompatibility, i.e., to the rejection of allogeneic cells. Instead, these experiments demonstrated that these chemically induced tumor cells had developed distinctive antigens, apparently acquired during the chemical carcinogenesis (and associated intense mutagenesis) that led to their transformation. This suggested, in turn, that cancer cells can express mutant proteins that may be recognized as foreign by the immune systems of tumor-bearing hosts. (B) Novel antigenic epitopes are likely to originate with some frequency in spontaneously arising tumors, including those in humans. In this instance, a mutant form of the enzyme peroxiredoxin 5 (Prdx5) arose during the formation of a human melanoma, ostensibly due to the mutagenic effects of UV light or oxidative radicals in normal skin melanocytes. The mutant enzyme exhibited an amino acid substitution that replaced the normally present serine 115 with an isoleucine (*inset*), creating an antigenic epitope; the oligopeptide bearing this epitope could be presented effectively by one of the cancer patient's MHC class I proteins, which resulted in a cytotoxic T-cell–mediated immune response against the melanoma cells. (This mutation did not compromise the function of the Prdx5 enzyme and seemed to be one that conferred no advantageous phenotype on incipient melanoma cells. As such, it could be classified as a "passenger" mutation whose presence was not selected during the course of tumor progression; see Sidebar 11.3.) A benzoate molecule, which binds to the active site of this enzyme, is seen in the inset in the form of a space-filling model. (B, from M. Sensi et al., *Cancer Res.* 65:632–640, 2005.)

But this rejection was premature. Only years later did it become apparent that mice of the Nude strain, while lacking many of their normal T lymphocytes, retain other components of their immune system in an intact form. For example, some types of T cells may be able to develop outside the thymus, the normal site of maturation of these cells. In addition, two very important types of immune cells—natural killer (NK) cells and macrophages—are able to develop totally outside the thymus, and thus are present in large numbers in Nude mice.

In the 1980s, researchers began to accumulate evidence that NK cells are actually very important in recognizing and killing a variety of abnormal cells, including cancer cells. So in the end, the lessons taught by the low cancer rates of Nude mice were of limited value, since these mice did indeed continue to harbor functionally important components of the immune system. Nonetheless, Nude mice, as well as other types of immunocompromised mice, have proven to be of great value in cancer research (see Supplementary Sidebar 15.4).

Evidence also began to accumulate that certain chemically induced tumors in mice were antigenic and could be recognized and eliminated by the immune system. For example, in one set of experiments, cells from a 3-methylcholanthrene (3-MC)–induced tumor were irradiated prior to injection into mouse hosts in order to prevent the proliferation of these cells in the hosts (**Figure 15.16**). (The chemically induced tumor had been induced in the same strain of mice as these hosts.) Subsequently, the mice received a second injection of live tumor cells originating from the same tumor or from a second 3-MC–induced tumor; the cells originating from the same tumor did not grow, while the cells from the second tumor did grow and form a new tumor. This indicated that the two tumors were antigenically different and that the initial exposure to dead cancer cells had immunized the mice against live cells originating in the same tumor. Hence, tumor cells could have distinctive antigens, and under certain conditions, these antigens could provoke the immune system to attack and kill such cells.

15.8 Use of genetically altered mice leads to a resurrection of the immunosurveillance theory

In the mid-1990s, several lines of research gave new life to the long-discredited immunosurveillance theory. These experiments derived from the then-recently gained ability to create genetically altered strains of mice at will. This technology (see Supplementary Sidebar 7.7) was exploited to create mice whose genomes lacked one or more of the genes known to play critical functional roles in the immune system.

One group of experiments used mice that were rendered incapable of expressing the receptor for interferon-γ (IFN-γ) through targeted inactivation of the responsible gene in their germ line. Like growth factors, IFN-γ is a diffusible protein factor—a cytokine—that conveys signals from one cell to another and induces responses in cells by binding and activating its cognate cell surface receptor. Importantly, IFN-γ has not been found to be released by cells other than those of the immune system. Consequently, any changes observed following deletion of the IFN-γ receptor gene from the mouse genome could be attributed to defects associated with immune cells and their interactions with one another and with the remaining cells in the body. Strikingly, mice lacking the IFN-γ receptor in all of their cells were found to be 10 to 20 times more susceptible to tumor induction by the chemical carcinogen 3-methylcholanthrene.

In another set of experiments, tumor cells were forced to express a dominant-negative IFN-γ receptor, rendering them unresponsive to the IFN-γ released by various types of immunocytes. These cells were then injected into wild-type mice and found to be more tumorigenic than tumor cells carrying the corresponding wild-type receptor. This particular experiment suggested that the IFN-γ receptor displayed by cancer cells usually enables them to respond to IFN-γ released by immunocytes, and that this response often prevents or retards the growth of tumors formed by these cells.

These striking effects of IFN-γ could be associated, at least in part, with the actions of the natural killer cells. The NK cells were discovered and named because of their innate ability to recognize tumor cells as abnormal and to eliminate them. Once NK cells identify cancer cells as targets for elimination, they release IFN-γ in the vicinity of the targeted cells. The released IFN-γ, in turn, elicits several distinct responses.

As mentioned earlier, IFN-γ enables the NK cells to call in other types of immune cells to assist in killing targeted cancer cells, thereby amplifying the immune system's response. Among the responding immune cells are macrophages, which aid not only by killing the cancer cells directly but also indirectly, by functioning as professional antigen-presenting cells (APCs) that process and display antigenic molecules derived from the corpses of their victims (see Figure 15.6). At the same time, IFN-γ stimulates targeted cancer cells to display on their surfaces increased levels of class I MHC molecules that may carry oligopeptide antigens capable of provoking further, highly specific adaptive immune responses. This helps to explain why transformed cells lacking the IFN-γ receptor are more tumorigenic than counterpart cells that do display this receptor. These various responses seemed to be defective in genetically altered mice lacking the IFN-γ receptor; such mice were also found to have an increased susceptibility to certain types of spontaneously arising tumors. When taken together, these experiments provided compelling validation of the idea that immune surveillance plays a critical role in tumorigenesis, at least in chemically induced tumors of mice.

Further support of the immunosurveillance theory came from mice that had been deprived of the gene encoding perforin, the protein used by lymphocytes and NK cells to mediate killing of targeted cells. Recall that perforin is used by cytotoxic cells to create channels in the plasma membrane of their victims, allowing the entrance of apoptosis-inducing granzymes (see Figure 15.12A). Mutant mice lacking the ability to make perforin showed an elevated incidence of spontaneous tumors and were also more susceptible to developing tumors following exposure to 3-methylcholanthrene.

Similarly, increased cancer susceptibility was registered in genetically altered mice that lacked either the Rag-1 or Rag-2 proteins; these two proteins are responsible for rearranging the genes encoding soluble antibody molecules as well as those encoding the antigen-recognizing T-cell receptors (TCRs) displayed on the surfaces of T cells (see Supplementary Sidebar 15.1). Such Rag-1 or -2–negative mice lack T lymphocytes, B lymphocytes, γδ T cells (not discussed further in this chapter), and a subclass of NK cells called NKT cells. As a consequence, these mice have severely compromised adaptive immune responses.

For example, in one experiment, 3-MC treatment caused 30 of 52 RAG-2$^{-/-}$ mice to develop sarcomas, while only 11 of 57 wild-type mice of the same genetic background and treated in parallel formed these tumors. The mutant mice were also found to be far more susceptible to spontaneously arising cancers. Thus, 50% of older (18-month-old) RAG-2–negative mice developed spontaneous gastrointestinal malignancies—a tumor that is otherwise rare in wild-type mice of this age.

Arguably the most persuasive evidence supporting the role of immunosurveillance in cancer prevention comes from detailed studies of the 3-MC–induced sarcomas growing in either *Rag2*$^{-/-}$ or wild-type mice. When tumor cells prepared from these two groups of sarcomas were grafted into new *Rag2*$^{-/-}$ hosts, both groups of sarcomas seeded tumors in these new hosts with high efficiency (**Figure 15.17**). A very different outcome was observed, however, when tumor cells were transplanted into syngeneic wild-type (and thus immunocompetent) hosts. Cells from 17 tumors that had previously been induced in wild-type mice all succeeded in generating tumors in their new hosts. In contrast, cells from 8 of 20 tumors that had previously been induced by 3-MC in *Rag2*$^{-/-}$ mice failed to form tumors, being rejected by the immune systems of these wild-type hosts.

These observations open our eyes to an entirely new dimension of tumor immunology. They suggest that when 3-MC–transformed cells are formed in an immunocompetent host, those that happen to be *strongly immunogenic* (and thus capable of provoking some type of immune response) are effectively eliminated by the host, resulting in the survival and outgrowth of only those cancer cells that happen to be *weakly immunogenic*. The latter then multiply and form tumors in their original hosts and, later on, succeed in doing so when transplanted into other immunocompetent hosts. Hence, these tumors represent a subset of those that originally arose in the primary hosts. The missing, strongly immunogenic tumors are apparently eliminated early in tumor progression by host immune systems and therefore never see the light of day (see Figure 15.17B).

In contrast, when 3-MC–transformed cells arise in an immunocompromised host, two classes of tumors are initially formed, as before—those that are strongly immunogenic and those that are weakly immunogenic; both types of tumor cells survive in this immunocompromised host. Later, when these tumors are transplanted into immunocompetent hosts, those that are strongly immunogenic fail to form tumors, while those that are weakly immunogenic succeed in doing so. We conclude that in wild-type mice, a functional immune system plays an important and effective role in eliminating a significant fraction of the tumors that are initially induced by 3-MC.

These observations indicate that the immune system of these mice plays an active role in determining the identities of tumors that arise and the antigens that they express. This active intervention in the phenotype of tumors has been termed **immunoediting**, to indicate the weeding out of some tumors and the tolerance of others. Immunoediting can be thought of as a type of Darwinian selection, in which the selective force is created by the directed attacks of the immune system on incipient tumors. Follow-up research into the processes described in Figure 15.17 has provided direct confirmation that (1) immunoediting occurs in response to mutant proteins that are generated by the highly mutagenic 3-MC carcinogen, (2) that immunoediting is mediated by T cells responding to resulting strongly antigenic mutant proteins, and (3) that following immunoediting, these mutant, highly antigenic proteins are no longer expressed by surviving tumor cells.

In fact, there is a third way by which tumors, such as those described in Figure 15.17, can behave. They may be held in check (without being eliminated) by the adaptive immune system, and thus confined to small, indolent growths that in humans would be considered subclinical. The cells in such tumors may proliferate slowly and, being antigenic, may be eliminated by immune cells, such as cytotoxic T cells, at the same rate that they are being produced, resulting in a long-term, apparently stable equilibrium.

This delicate balance might be disrupted in at least two ways. The tumor cells may learn to suppress expression of their more immunogenic antigens, thereby evading immune recognition and elimination and allowing the sudden eruption of less antigenic tumors. Alternatively, certain components of the adaptive immune system may suffer damage or disruption, relieving the immune-mediated suppression and

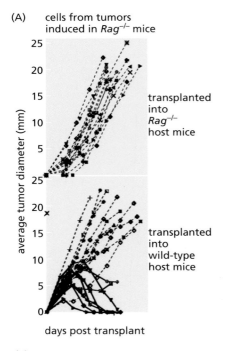

(A) cells from tumors induced in *Rag*⁻/⁻ mice

average tumor diameter (mm)

transplanted into *Rag*⁻/⁻ host mice

transplanted into wild-type host mice

days post transplant

Figure 15.17 Effects of immune function on the development of anti-tumor immune responses Both wild-type (wt, i.e., *Rag*⁺/⁺) and *Rag2*⁻/⁻ immunocompromised mice were exposed to the potent carcinogen 3-methylcholanthrene (3-MC). (A) When the tumors induced in the *Rag2*⁻/⁻ mice were transplanted into *Rag2*⁻/⁻ hosts, they all formed tumors (*above*). However, when the tumors induced in the *Rag2*⁻/⁻ mice were transplanted into wild-type hosts, 8 of 20 tumors failed to form (*below*). Each line presents the growth kinetics of a single implanted tumor. (B) This experiment and others using tumors induced in wt mice are summarized here. Following exposure to 3-MC, the wt mice developed fewer tumors (*blue*) than did the *Rag2*⁻/⁻ mutants (*blue and red*). The tumors from the two groups of mice were excised and cells from each were converted to a cell line that could be propagated *in vitro*. Cells from each cell line were then transplanted back into either wt mice or *Rag2*⁻/⁻ mutant mice. Cells from all of the tumors that appeared initially in the wt mice (*blue*) were able to form new tumors in both wt and *Rag2*⁻/⁻ hosts (*left*). However, cells from tumors that arose and grew initially in the *Rag2*⁻/⁻ mice were able to form new tumors in *Rag2*⁻/⁻ mice (*red, blue*), but only some of these (*blue*) were able to form new tumors in the wt mice (*right*). These experiments suggested that 3-MC initially induced two types of tumor cells in all of the mice: strongly immunogenic (*red*) and weakly immunogenic (*blue*). Both *red* and *blue* cells formed tumors in the *Rag2*⁻/⁻ mice, but only *blue* cells formed tumors in the wt mice, since any initially formed *red* tumor cells were eliminated by the functional immune systems of these mice. This meant that the tumors that did arise in wt mice were already selected for being weakly immunogenic and thus capable of forming new tumors in other wt mice. (A, from V. Shankaran et al., *Nature* 410:1107–1111, 2001.)

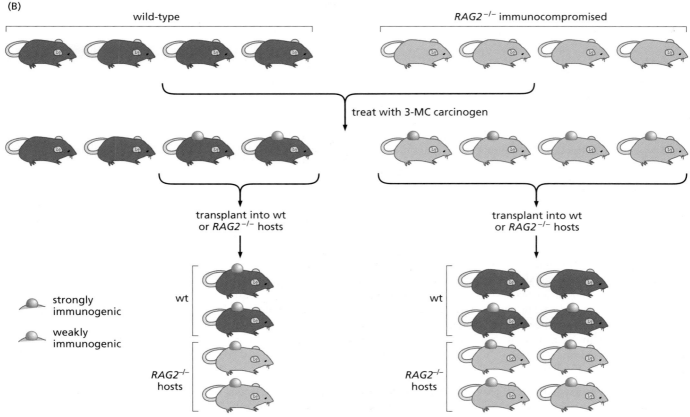

(B)

wild-type

RAG2⁻/⁻ immunocompromised

treat with 3-MC carcinogen

transplant into wt or *RAG2*⁻/⁻ hosts

transplant into wt or *RAG2*⁻/⁻ hosts

strongly immunogenic

weakly immunogenic

wt

RAG2⁻/⁻ hosts

wt

RAG2⁻/⁻ hosts

permitting the outgrowth of tumors that may have been inapparent for months and, in some humans, for years. These dynamics inform our views of how micrometastases can remain indolent for extended periods of time: before, we speculated that they need to undergo genetic evolution in order to adapt to the foreign tissue microenvironments in which they have landed (see Sidebar 14.2). This new mechanism of immune escape now suggests that interactions with the immune system may also play a critical role in determining whether or not micrometastatic deposits will erupt in the form of clinically detectable, life-threatening growths.

15.9 The human immune system plays a critical role in warding off various types of human cancer

Because the biology of mice and humans differs in so many respects, we need to interpret the results described above with caution when attempting to understand the role of the human immune system in defending us against cancer. In addition, the chemical carcinogens used in the experiments described above may well create tumors in mice that are far more antigenic or immunogenic than spontaneously arising human tumors (to be discussed in Section 15.12). Nevertheless, evidence compiled in the 1990s provides clear indications that the human immune system does indeed play an important role in warding off cancer. The bulk of this evidence comes from observations that immunocompromised humans are far more susceptible than the general population to certain types of cancer.

Actually, immunocompromised humans were uncommon until recently. Those who were born with dysfunctional immune systems died early in life, while others whose immune systems deteriorated later in life died as their defenses against infectious agents declined. However, over the past few decades, the number of individuals who live for extended periods of time with compromised immune systems has increased dramatically, for three reasons.

First, organ transplants involving kidneys, hearts, and livers have become common throughout the developed world. Because these organs derive from donors who are, almost always, genetically different from the recipients, the donor cells in the transplanted organs are recognized as foreign (that is, allogeneic and thus histoincompatible) by the immune systems of the graft recipients, which proceed to eliminate them. This unwanted reaction is controlled by long-term treatment with several types of immunosuppressive drugs.

Second, and independent of these clinically induced immunodeficiencies, are the cases of almost 60 million people throughout the world who have been infected by human immunodeficiency virus (HIV), half of whom have died because of AIDS. Third, the long-term survival of immunocompromised individuals, including organ graft recipients and many HIV-infected patients, has been enabled by the development and use of a diverse group of antibiotic (that is, antibacterial), antifungal, and antiviral drugs. These extended survival periods of immunocompromised patients represent time spans that are long enough for pre-malignant growths that were previously latent in these individuals to progress to a state where they become clinically apparent.

Thousands of patients bearing transplanted organs have developed a variety of solid tumors and hematopoietic malignancies over the past two decades. As discussed earlier (see Supplementary Sidebar 14.4), a very small proportion of these tumors have derived from occasional metastatic cancer cells that were hiding in the bodies of organ donors, escaped detection during the transplantation procedure, and began to multiply aggressively once they were introduced into the bodies of graft recipients. These tumors can be shown to be of allogeneic (that is, of organ donor) origin by analysis of their genetic markers. However, the eruption of tumors triggered by transplanted cells provides no insights about whether tumors of *endogenous origin* (that is, autochthonous tumors) arise with greater-than-normal frequency in immunocompromised patients.

In fact, a tumor registry in Ohio has documented many autochthonous cancers (in excess of 15,000) in organ transplant recipients. An Australian study followed kidney transplant recipients for up to 24 years after the transplantation procedure was performed; 72% of these patients had developed at least one type of cancer. A similar study in the United States found a three- to fivefold increased risk of cancer in transplant recipients. Patients who have undergone liver transplants commonly develop new tumors within five years of transplantation and initiation of immunosuppression.

Figure 15.18A provides an initial insight into the nature of these tumors. Most striking is the dichotomy between tumors that have no known viral etiology and those known to be triggered by infectious agents, notably viruses. The nonviral cancers occur with almost equal incidence among immunocompetent and immunodeficient individuals, while the tumors known to be triggered by viral infections occur at greatly increased incidence in patients who are immunocompromised. The champion here is Kaposi's sarcoma (caused by human herpesvirus-8, HHV-8), which occurs with an incidence that is more than 3000 times higher in AIDS patients than in the general population. Moreover, a diverse array of tumors (largely carcinomas) are triggered by human papillomavirus (HPV) infections (see Figure 15.18B); most of these exhibit an increased frequency in organ recipients and AIDS patients. Together, these disparate observations clearly show that the immune system represents an important defensive bulwark against the 20% and more of human tumors arising worldwide whose development is traceable, directly or indirectly, to infections, largely to the viruses mentioned above.

The extraordinary effectiveness of the normal immune system in defending us against virus-induced malignancies is illustrated by the fact that some 90% of adults in the West are infected by the potently oncogenic Epstein–Barr virus, yet EBV-induced malignancies are relatively uncommon in the general population. To cite another example, the introduction of multi-drug therapy to suppress HIV replication has resulted in the regeneration of immune function in large numbers of HIV-infected patients; as a consequence, the incidence of new cases of Kaposi's sarcoma has decreased as much as 40-fold in some AIDS clinics.

Unanswered by these studies is the precise mechanism by which virus-induced cancers are normally controlled by the immune system. We could entertain two plausible

Figure 15.18 **Effects of compromised immune systems on cancer incidence** This *meta-analysis* pooled the conclusions of a group of epidemiologic studies in order to calculate the *standardized incidence ratio* (SIR)—the number of cancer cases actually observed in populations of immunosuppressed individuals and HIV/AIDS patients divided by the number expected in the age-matched general population. All patients in the transplant category were presumed to be subject to significant immunosuppression in order to prevent rejection of the transplanted organ. The numbers associated with each entry indicate the calculated mean SIR, followed by the confidence interval (which shows, with 95% probability, the interval within which the measured mean value will fall). Similarly, in these "box-and-whisker" plots, the filled boxes are centered on the median measurements, while the outer points of the whiskers indicate the bounds of the 95% confidence interval. (A) Among a group of five commonly occurring epithelial cancers, the incidence of breast, prostate, and colorectal carcinomas was not elevated in HIV-infected/AIDS patients relative to the general population, while carcinomas of the ovary and lungs were modestly elevated and at a statistically significant level (*above*). Strikingly different values are seen, however, among a range of tumors that are known to be associated with either chronic viral or bacterial infections (*below*), where lack of a fully functional immune system often leads to dramatic increases in tumor incidence. Note that the SIR values are plotted logarithmically on the abscissa. (EBV, Epstein–Barr virus; HHV-8, human herpesvirus-8; HBV and HCV, hepatitis virus B and C) (B) An analysis of one class of infectious agents—human papillomavirus (HPV)—that infect a variety of epithelial tissues, indicates that in immune-compromised individuals, tumors known to be associated with HPV infection (*above*) occur at a significantly elevated rate. In contrast, for tumors in which the etiologic role of HPV has not been firmly demonstrated (*below*), the elevated risk, as indicated by the SIR values, is less clear. An exception here is non-melanoma skin carcinomas, specifically squamous cell carcinomas (SCCs) of the skin, in which the involvement of causative HPV infections is suspected but still unproven; tumors of this class are of multifactorial origin. Of note, heart transplant patients in Australia have as much as a 4-fold higher risk of developing skin SCCs relative to those in the Netherlands due to increased sun exposure (*not shown*). (From A.E. Grulich et al., *Lancet* 370:59–67, 2007.)

(A)   cancers not associated with infectious agents

breast	HIV/AIDS	1.03 (0.89–1.20)	
	transplant	1.15 (0.98–1.36)	
prostate	HIV/AIDS	0.70 (0.55–0.89)	
	transplant	0.97 (0.78–1.19)	
colon and rectum	HIV/AIDS	0.92 (0.78–1.08)	
	transplant	1.69 (1.34–2.13)	
ovary	HIV/AIDS	1.63 (0.95–2.80)	
	transplant	1.55 (0.99–2.43)	
trachea, bronchus, and lung	HIV/AIDS	2.72 (1.91–3.87)	
	transplant	2.18 (1.85–2.57)	

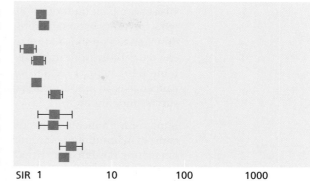

cancers known to be associated with infectious agents

EBV-related cancers; Hodgkin's lymphoma	HIV/AIDS	11.03 (8.43–14.4)	
	transplant	3.89 (2.42–6.26)	
non-Hodgkin's lymphoma	HIV/AIDS	76.67 (39.4–149)	
	transplant	8.07 (6.40–10.2)	
HHV-8 related cancer; Kaposi's sarcoma	HIV/AIDS	3640.0 (3326–3976)	
	transplant	208.0 (114–349)	
HHV/HCV-related cancer, liver	HIV/AIDS	5.22 (3.32–8.20)	
	transplant	2.13 (1.16–3.91)	
Helicobacter pyori–related cancer, stomach	HIV/AIDS	1.90 (1.53–2.36)	

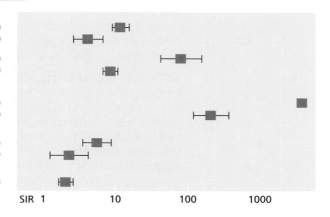

(B)   known HPV association

	HIV/AIDS	5.82 (2.98–11.3)	
	transplant	2.13 (1.37–3.30)	
vulva and vagina	HIV/AIDS	6.45 (4.07–10.2)	
	transplant	22.76 (15.8–32.7)	
penis	HIV/AIDS	4.42 (2.77–7.07)	
	transplant	15.79 (5.79–34.4)	
anus	HIV/AIDS	28.75 (21.6–38.3)	
	transplant	4.85 (1.36–17.3)	
oral cavity and pharynx	HIV/AIDS	2.32 (1.65–3.25)	
	transplant	3.23 (2.40–4.35)	

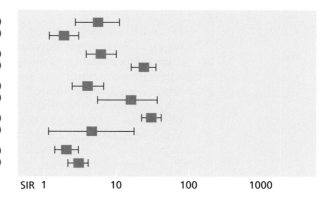

suspected HPV association

non-melanoma skin	HIV/AIDS	4.11 (1.08–16.6)	
	transplant	28.62 (9.39–87.2)	
lip	HIV/AIDS	2.80 (1.91–4.11)	
	transplant	30.00 (16.3–55.3)	
esophagus	HIV/AIDS	1.62 (1.20–2.19)	
	transplant	3.05 (1.87–4.98)	
larynx	HIV/AIDS	2.72 (2.29–3.22)	
	transplant	1.99 (1.23–3.23)	
eye	HIV/AIDS	1.98 (1.03–3.81)	
	transplant	6.94 (3.49–13.8)	

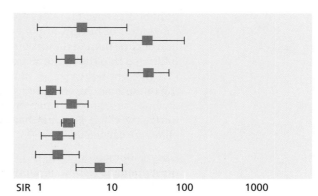

models that can explain this control. (1) The immune system is normally responsible for protecting us against all types of viral infections, independent of whether certain viruses are bent on inducing cancer. In the absence of fully functional immune systems, many viruses, including oncogenic viruses, are able to persist and proliferate for extended periods of time within the body, in contrast to their usual fate of being cleared rapidly from the body by the immune system. (2) Alternatively, the normal immune system is responsible for recognizing and eliminating virus-transformed cancer cells. In immunocompromised individuals, however, such cells may be able to survive indefinitely.

Either or both mechanisms may explain the greatly increased rates of virus-induced cancers in immunocompromised people. In AIDS patients, for example, high levels of circulating Epstein–Barr virus are not commonly observed, while the levels of actively proliferating EBV-infected lymphoid cells often increase dramatically, yielding, in turn, virus-induced lymphomas. Consequently, in the condition of AIDS, the elimination of EBV-infected cells rather than the virus itself seems to be defective. (In the blood of healthy carriers of EBV infection, who represent 90% of the general population, as many as 5% of the long-lived cytotoxic T cells have been found to react with a certain EBV antigen, indicating that the cellular arm of the immune system normally devotes an astounding proportion of its operations to controlling a single infectious agent—EBV—in this case by eliminating cells that carry actively proliferating virus.)

Still, virus-induced malignancies represent only a minority of the tumors routinely treated in oncology clinics. This causes us to ask whether a competent immune system also erects defenses against the great majority of human tumors (~80%) that are of nonviral origin. In fact, a two- to fourfold increased risk of melanoma has been found among adult organ transplant recipients, while non-Kaposi's sarcomas were found at rates three times above those of the general population. A population of heart transplant recipients has been found to experience a 25-fold increased risk of carcinomas of the lung. And registries of transplant patients in Australia, New Zealand, and Scandinavia have documented elevated rates of carcinomas of the colon, lung, bladder, and kidney as well as tumors of the endocrine system (Table 15.1). Attempts to associate these cancers with tumor virus infections have failed, with the exception of occasionally discovered HPV genomes carried by lung cancer cells.

Given what we have already learned about adaptive immunity, the mechanisms by which the immune system can recognize virus particles and virus-infected cells would seem to be clear and obvious: viral antigens, which are foreign to the body and thus not subject to immune tolerance, are highly immunogenic, provoking a variety of humoral and cell-mediated immune responses. (Precisely this property of the immune system has made possible the first clinically effective vaccine against a human cancer; Supplementary Sidebar 15.5.) But how are *nonviral* tumors recognized and eliminated? Virtually all of the proteins in the cancer cells forming these tumors are likely to be identical to those present in the normal body. Hence, authochthonous tumors of nonviral origin may not attract the attentions of an immune system that has evolved to attack and eliminate foreign infectious agents while leaving normal cells and thus normal tissues unmolested.

Nonetheless, it is obvious that some arms of the immune system can indeed recognize tumors that have no associations with viral infections. For example, histopathological studies of human tumors provide clear indications that such immune recognition occurs: human tumors often have substantial numbers of lymphocytes that have infiltrated into the tumor mass (Figure 15.19). These tumor-infiltrating lymphocytes (TILs) might represent yet another type of stromal cell that has been recruited into the tumor mass by neoplastic cells in order to support the expansion of the tumor as a whole, as argued in Chapter 13. However, an alternative explanation is even more intriguing: these TILs may have been dispatched by the immune system in order to eliminate cancer cells.

Observations of a group of ovarian carcinoma patients have provided some of the most dramatic testimony supporting the anti-tumor role of TILs. This evidence comes from correlating the clinical course of these ovarian carcinoma patients with the

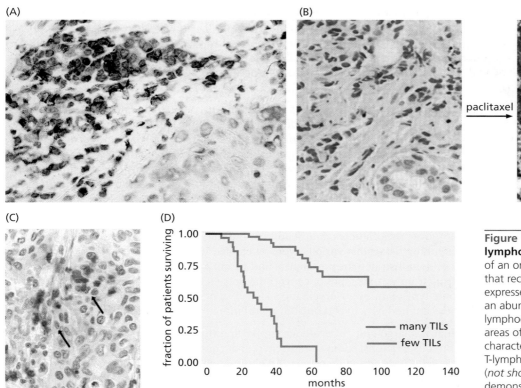

Figure 15.19 Tumor-infiltrating lymphocytes (A) This immunostaining of an oral carcinoma with an antibody that recognizes the CD3 antigen expressed by T lymphocytes reveals an abundance of tumor-infiltrating lymphocytes (TILs; *brown*) in certain areas of the tumor. More detailed characterizations revealed several T-lymphocyte subtypes among these cells (*not shown*). (B) This immunostaining demonstrates that TILs, detected once again with an anti-CD3 antibody (*dark purple*), are relatively rare in an untreated breast tumor (*left*) but become abundant in areas of the tumor following chemotherapy with the drug paclitaxel (*right*). (C) TILs are also frequently found in invasive non-small-cell lung carcinomas (NSCLCs; *arrows*). The expression here of the CD8 antigen (*dark pink*) indicates that these cells are largely cytotoxic lymphocytes (CTL, T$_C$). (D) The clinical prognosis of a set of ovarian carcinoma patients was strongly correlated with the concentrations of TILs in their tumors. In this Kaplan–Meier plot, the proportion of patients surviving after initial diagnosis (*ordinate*) is plotted versus the months of survival (*abscissa*). Those patients whose tumors had high levels of TILs (*blue line*) fared significantly better than did those whose tumors lacked significant concentrations of TILs (*red line*). (A, from T.E. Reichert et al., *Clin. Cancer Res.* 8:3137–3145, 2002. B, from S. Demaria et al., *Clin. Cancer Res.* 7:3025–3030, 2001. C, from A. Trojan et al., *Lung Cancer* 44:143–147, 2004. D, adapted from L. Zhang et al., *N. Engl. J. Med.* 348:203–213, 2003.)

Table 15.1 Cancer incidence in immunosuppressed transplant patients[a]

Site of cancer	No. of cases observed	No. of cases expected[b]	Ratio observed/expected[c]
Non-melanoma skin	127	5.1	24.7
Thyroid, other endocrine	30	2.1	14.3
Mouth, tongue, lip	22	1.6	13.8
Cervix, vulva, vagina	39	3.6	10.8
Non-Hodgkin's lymphoma	25	2.4	10.3
Kidney, ureter	32	3.5	9.1
Bladder	26	4.7	5.5
Colorectal	38	10.5	3.6
Lung	30	12.5	2.4
Brain	10	4.1	2.4
Prostate	11	5.2	2.1
Melanoma	7	4.1	1.7
Breast	15	13.6	1.1

[a]Data from S.A Birkeland et al., *Int. J. Cancer* 60:183–189, 1995, as adapted by J. Peto, *Nature* 411:390–395, 2001.
[b]These numbers represent the numbers of cases of the various cancers expected to occur in an age-matched control population over the same period of time.
[c]A 2011 study revealed that the nature of the transplanted organ also determines relative increased risk. Liver transplant recipients exhibited a greater than 40-fold increased risk of developing liver cancer, whereas kidney transplant recipients exhibited a far higher risk of developing kidney cancer than did recipients of other transplanted organs.

presence or absence of substantial numbers of TILs in their cancers. In one group of patients, who had been treated initially by surgical removal of the bulk of their tumors followed by chemotherapy, 74% were alive five years later if their initial tumors carried large numbers of these TILs. In contrast, among those patients whose ovarian tumors lacked significant populations of TILs, only 12% were still alive (see also Figure 15.19D).

Similar outcomes are associated with patients carrying malignant melanomas that are infiltrated with large numbers of TILs; these patients live 1.5 to 3 times longer following diagnosis than do those patients whose tumors lack large numbers of the tumor-infiltrating lymphocytes. Yet other correlations have been made between the presence of these infiltrating lymphocytes and the survival of patients bearing carcinomas of the breast, bladder, colon, prostate, and rectum. Still, such observations, while dramatic, are only correlative and therefore do not prove definitively that these lymphocytes are important agents responsible for holding back tumor progression.

Moreover, these studies do not address the role of the humoral immune response in defending us against various types of cancer. In fact, a number of research reports have demonstrated the presence of anti-tumor antibodies in the blood of patients suffering from various types of cancer; the presence of these antibodies clearly suggests some type of immunosurveillance. However, once again, it remains unclear whether these antibodies actively contribute to eliminating tumor cells from the body.

Finally, a large body of evidence supporting the immunosurveillance theory comes from observations, some already cited, that cancer cells, like viruses, employ various strategies for evading detection and elimination by the immune system (see Supplementary Sidebar 15.2). We will defer detailed discussion of these various mechanisms of *immunoevasion* until later. For the moment, suffice it to say that cancer cells make extensive efforts to lower their immunological profiles, enabling them to "fly under the immunological radar." While also correlative, these observations are so frequent as to constitute persuasive evidence that immunosurveillance and resulting immunoevasion are important dynamics in tumor progression. As we will see in the rest of this chapter, the mechanisms underlying these processes are complex and understood only imperfectly at present.

This complexity begins with the antigens that are displayed by cancer cells and provoke immune responses; their identities are often elusive. Even more complex are the identities of the immunocytes that are responsible for responding to these antigenic signals. Here we confront additional dimensions of complexity. As **Table 15.2** indicates, multiple distinct types of leukocytes are recruited to tumors. Each type of leukocyte may be represented by multiple subtypes that may play distinct, even conflicting roles in fostering or suppressing tumor growth; **Figure 15.20** provides one example of this complexity, in this case highlighting the multiple roles of only one type of leukocyte—the macrophage.

In addition, it is difficult to draw generalizable conclusions about the role of the immune system in controlling all types of cancer, given the variable representation of immunocytes in many human tumor types (**Figure 15.21**); each type of tumor seems to require its own set of explanations of how it interacts with the immune system. In some special cases, the presence of specific subtypes of immune cells is closely coupled with the course of tumor progression, providing strong but still-correlative evidence of a causal role in controlling or failing to control clinical behavior (**Figure 15.22**). The evidence in this figure points to apparent critical roles of T-cell subsets in controlling colorectal carcinoma progression; similar evidence exists for other carcinoma types as well.

These multiple dimensions of complexity help to explain why the science of tumor immunology is still in its infancy and why we need to take small steps in the sections that follow, dissecting out discrete parts of this complex system, one at a time, in order to reach solid conclusions about the interactions between the immune system and tumors. We will begin by shifting our focus from the cells involved in the anti-tumor immune response to the antigens that they learn to recognize within tumors.

Table 15.2 Leukocytes that function to promote or inhibit tumorigenesis

	Pro-tumorigenic inflammation	Anti-cancer immunosurveillance
Cell types	M2 macrophages,[a] myeloid-derived suppressor cells,[b] neutrophils, FOXP3+, T_{reg}, T_H17 cells	dendritic cells, M1 macrophages,[a] helper CD4+ T cells, cytotoxic CD8+ T cells with memory effector phenotype
Cytokine profiles	T_H2, T_H17	T_H1, CX3CL1, CXCL9, CXCL10
Distribution	peritumoral	intratumoral, close to cancer cells, as well as at invasive front
Associated features	STAT3 phosphorylation	high endothelial venules
Functional impact	negative prognostic impact	positive prognostic and predictive impact

[a]The M1 class of macrophages was initially associated with mounting a defense against infectious agents, while cells of the M2 class have been associated with wound healing and tissue repair. The M2 class of macrophages contains at least three functionally distinct subgroups, with each having its own distinct gene expression profiles/signatures and biological activities (see also Figure 15.21). Recent research indicates that there are far more functional subtypes of macrophages than are subsumed by these M1 and M2 classifications.
[b]The category of myeloid-derived suppressor cells (MDSCs), a class of still-poorly defined immature myeloid cells, contains at least four functionally distinct subtypes, some of which share certain properties with dendritic cells and M2 macrophages. MDSCs have not yet been clearly delineated through use of distinctive sets of cell surface antigen markers.
From W.H. Fridman et al., *Cancer Res.* 71:5601–5605, 2011.

15.10 Subtle differences between normal and neoplastic tissues may allow the immune system to distinguish between them

The successes of the immune defenses against infectious agents depend on the ability to recognize these intruders as foreigners in the human body. Infectious agents invariably display molecules that betray their alien origins, provoking attack by specialized cells of the innate and adaptive arms of the immune system. Typically, these foreign molecules contain oligopeptide sequences that are recognizably different from the sequences present in the body's own repertoire of native proteins.

At the same time, a well-functioning immune system turns a blind eye to the proteins and thus amino acid sequences that are native to the body—a reflection of the immunological tolerance discussed in Section 15.5. The phenomenon of immunological tolerance greatly complicates our attempts at understanding how the immune system defends the body against spontaneously arising tumors. Ultimately, the success of these defenses depends on a critical issue: Can the immune system recognize cancer cells arising in a person's own tissues as being foreign, even though these cells are truly native to the body?

The answer to this question is hardly obvious. The vast majority of proteins expressed by tumor cells are clearly normal, both in their structures and levels of expression. Nevertheless, among the 20,000 or so distinct protein species (and variants thereof) made by one or another type of cancer cell, there are certainly a small number that are not present in normal tissues. Such structurally novel, and in this sense "foreign," antigens might well provoke a vigorous immune response.

An obvious example of a novel cancer antigen is provided by the Ras oncoproteins, which are created by amino acid substitutions in residues 12, 13, or 61 of the four subtypes of Ras proteins seen in normal cells. These oncoproteins clearly exhibit altered chemical structures, and the sequence of amino acids in a Ras oncoprotein that surrounds and includes an altered residue (see Figures 4.9 and 4.10) may indeed constitute an oligopeptide antigen that evokes an immune response. Similarly, the numerous mutant alleles of the *p53* tumor suppressor gene also specify amino acid substitutions that might cause the altered versions of this protein to be immunogenic in the almost 50% of common human tumors that carry altered p53 proteins (see, however, Supplementary Sidebar 15.6).

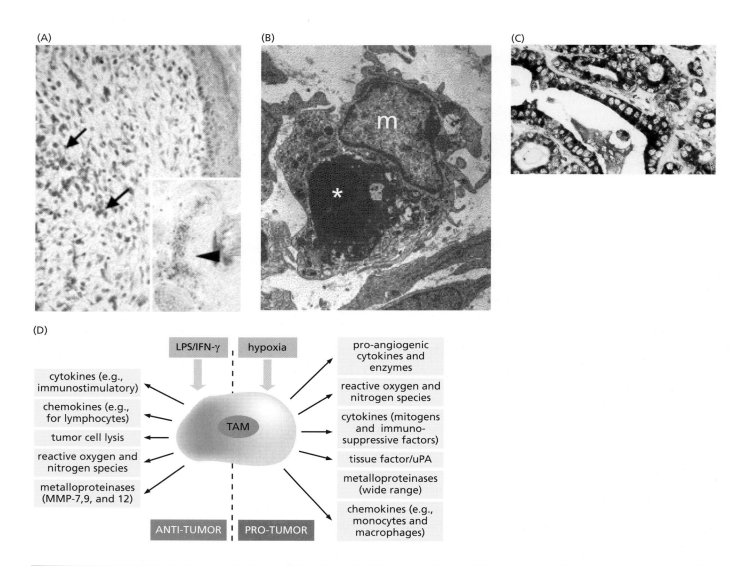

Figure 15.20 Multiple roles of macrophages Macrophages play critical roles in promoting wound healing and tumor progression. They also act as agents of the immune system to present tumor antigens and to consume tumor cells. These differing functions are carried out by distinct subtypes of macrophages. (A) In this wound-healing site in the skin of a mouse, macrophages (*brown*) are detected in abundance (*arrows*) through use of an antibody that recognizes the F4/80 macrophage-specific antigen and the c-Fms receptor marker (*arrowhead, inset*). (c-Fms functions as the receptor of CSF-1; see Figures 14.21 and 14.22.) (B) Transmission electron microscopy reveals a macrophage (m) that is engorged with the phagocytosed corpses of apoptotic cells (*asterisk, dark body*) in the middle of a wound-healing site. (C) A macrophage (*pink*) is seen as it begins to phagocytose a tumor cell in the lumen of a duct within a papillary thyroid carcinoma. (D) Macrophages can be activated by diverse signals, such as interferon-γ (IFN-γ) and bacterial lipopolysaccharide (LPS), the latter being used as an *adjuvant* to potentiate the immune response. Once activated, they can function to present antigen to T_H cells and to trigger tumor cell killing by antibody-dependent cellular cytotoxicity (ADCC). Acting in the opposite direction, hypoxia, as well as a variety of physiologic signals released by tumors (see Chapters 13 and 14), activate macrophages and cause them to foster tumor progression. (Macrophages of the M1 subtype promote immune attack on tumors, while those of the M2 subtype promote tumor progression; other work suggests that there are actually far more functionally distinct subtypes of macrophages. Mixtures of various cytokine signals are presumed to induce interconversion from one subtype to another.) (A and B, from P. Martin et al., *Curr. Biol.* 13:1122–1128, 2003. C, from A. Fiumara et al., *J. Clin. Endocrinol. Metab.* 82:1615–1620, 1997. D, from L. Bingle, N.J. Brown and C.E. Lewis, *J. Pathol.* 196:254–265, 2002.)

Immune recognition of tumor antigens

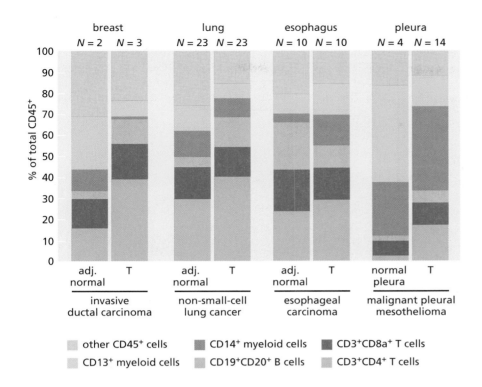

Figure 15.21 Tumor-specific infiltration by immunocytes While we have been attempting to draw broad generalities about the role of immune cells in tumor formation and progression, the biological reality may reveal, instead, dramatic differences between different tumor subtypes and the immune cells that they attract. The disparate and often conflicting roles of these cells in fostering or halting tumor formation (see Table 15.2) indicate that the overall influence of the immune system on tumor progression is likely to depend on the combinatorial actions of this functionally diverse group of cells and their relative numbers in various types of tumors. In the analyses presented here, normal and tumor tissue samples excised from patients were enzymatically digested; single cells in the resulting suspensions were then detected with dye-labeled monoclonal antibodies and analyzed by fluorescence-activated cell sorting (FACS). Only those cells expressing the CD45 major leukocyte antigen were analyzed. The number of each tumor type and corresponding normal tissue samples analyzed is given *above*; in each case, normal adjacent tissue (adj.) was analyzed in parallel with tumor samples (T). Ag, antigen; CD3, a common T-cell Ag; CD4, T_H cell Ag; CD8, T_C cell Ag; CD13, myeloid Ag, including granulocytes and monocytes; CD14, Ag of monocytes and derived cells, possibly immunosuppressive; CD19, follicular B-cell Ag; CD20, common B-cell Ag. (Courtesy of L.M. Coussens.)

Yet other examples of cancer-specific proteins derive from the numerous types of chromosomal translocations that specify fusion proteins encoded by pairs of previously unlinked genes. Such translocations are found frequently in hematopoietic malignancies. Recall, for example, the Bcr-Abl fusion protein, which is found in chronic myelogenous leukemia (CML) cells (see Section 4.6). While the bulk of the amino acid sequences of the fusion protein are identical in structure to the sequences of the two parental proteins (Bcr and Abl) found in normal cells, the short region where these two proteins are joined constitutes a novel amino acid sequence that has the potential to be recognized as foreign and therefore to be immunogenic.

Because of the genetic instability of human tumor cell genomes (see Chapter 12) and the large number of successive cell cycles through which cell lineages pass during the course of multi-step tumor progression, mutant alleles encoding structurally altered proteins may be present at elevated frequency in these genomes (see, for example, Figure 12.16). The great majority of these alleles may have nothing to do with accelerating tumor cell proliferation, being incidental by-products of this genomic instability, but the proteins made by these mutant passenger alleles may happen to be quite immunogenic—an issue to which we will return later.

Actually, we have already encountered one example of a tumor-specific mutant protein in a human cancer cell (see Figure 15.16B). A well-studied melanoma provides another illustration of an antigen that arises as a by-product of tumor progression. This human tumor was found to express a mutant triosephosphate isomerase (a key enzyme in glycolysis). Given the normal function of this protein, its mutant structure is unlikely to have played a causal role in tumor pathogenesis. Intracellular processing of this protein yielded a mutant oligopeptide antigen that was able to bind to an MHC class II protein (see Figure 15.7) with an affinity 5 orders of magnitude greater than that of the corresponding wild-type oligopeptide. This binding greatly enhanced antigen presentation and the overall immunogenicity of the melanoma.

In general, attempts at predicting the immunogenicity of various proteins and their relative abilities to provoke robust immune responses are highly challenging. Many proteins having novel structures may be overlooked by the immune system because of their overall structural similarity to normal cellular proteins. Others may be tolerated because, while structurally distinct from normal cellular proteins, they may be present in such low concentrations that they are effectively invisible to the immune system.

(A)

marker genes of
T$_H$1 adaptive immunity

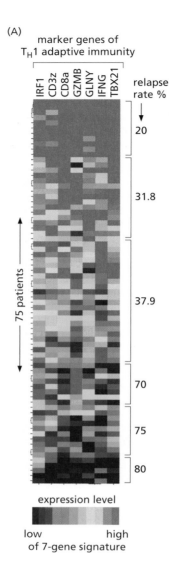

(B)

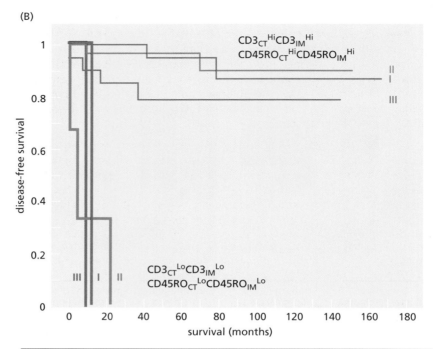

Figure 15.22 T-cell infiltration and the course of colorectal carcinoma progression (A) The clinical course of colorectal (CRC) tumor progression was monitored in a group of 75 patients over periods up to 15 years and correlated with the presence of specific subtypes of immunocytes. Whole tumors, encompassing both their epithelial and stromal compartments, were analyzed for their patterns of RNA expression, using qRT-PCR. An expression signature composed of seven genes (*top*) was generated in order to gauge the representation of immune cells that infiltrated colorectal carcinomas and expressed genes of the T$_H$1 subset of helper T cells, specifically memory T cells with the T$_H$1 expression pattern involved in adaptive immunity. (Memory T cells represent long-lived T-cell populations that persist for many years after exposure to antigen and can be reactivated quickly in response to a subsequent challenge by this antigen. T$_H$1 cells represent a subtype of T$_H$ cells that are defined by their release of certain interleukins and cytokines, notably interferon-γ, and their specialized functions.) Those patients whose tumors carried many cells having high expression of a set of T$_H$1 markers (*top*) experienced a vastly lower rate (*right side*) of post-surgical relapse (generally liver metastases) than did those with low levels of such T cells (*bottom*). (B) An even more dramatic correlation between T-cell infiltration and clinical progression came from analysis of a group of 406 patients, who were stratified into four subgroups according to the stage of histopathological progression of their primary CRC tumors, with stage I tumors representing the least aggressive and stage IV representing the most aggressive CRC subtype. Each tumor was analyzed for the presence of CD3+ (overall T cells) and CD45RO+ (memory T cells) both at the invasive margin of tumors (IM) and at the center of tumors (CT). As the Kaplan–Meier curves illustrate, most stage I, II, and III tumor patients showed long-term survival if their tumors exhibited high levels of CD3-positive and CD45RO-positive cells at both the tumor centers and invasive margins, while those patients whose tumors lacked such T cells in both sites declined precipitously after initial surgery. The clinical progression of stage IV patients, almost all of whom did poorly, is not shown here. (From J. Galon et al., *Science* 313:1960–1964, 2006.)

Ras exemplifies both circumstances: the mutant Ras oncoproteins found in many human tumors have an almost-normal three-dimensional structure and, like normal Ras proteins, are present in relatively low concentrations in cancer cells. Moreover, because of their amino acid sequences, certain mutant Ras oligopeptides may not be readily bound by the antigen-displaying hands of a patient's MHC molecules.

Predictions of immunogenicity are also complicated by the fact that we do not fully understand the rules governing the establishment of tolerance in the developing immune system. Much of this tolerance is achieved in the thymus gland and bone marrow during embryonic and early postnatal development, where the B and T lymphocytes that happen to have developed immunological reactivity to the body's normal proteins are eliminated or functionally inactivated. Later in this process, as we read above, lymphocytes circulating throughout the body develop "peripheral tolerance" to proteins that they have encountered in tissues distant from the thymus and bone marrow. In addition, in many cases the regulatory T cells (T_{reg}'s) described in Section 15.6 are known to suppress the actions of lymphocytes that have escaped elimination by these other mechanisms.

But what if the various populations of developing lymphocytes are never exposed to certain normal proteins? Some proteins may be expressed only transiently at specific stages of early embryonic development, long before immunological tolerance is developed. Other proteins, as suggested above, may be present in very low concentrations, so that developing lymphocyte populations rarely encounter them and hence do not delete the few cells that happen to recognize them. Yet other proteins may be expressed in **immunoprivileged** tissues such as the brain, where effective surveillance by the immune system is usually blocked by the complex mechanisms constituting the **blood–brain barrier**. Such shielding operates in the germ cells of the testes, which are also protected from routine monitoring by the immune system.

In sum, a number of distinct mechanisms can prevent the development of tolerance toward certain normal cellular proteins. However, if one of these proteins happens to be displayed at abnormally high levels by cancer cells, these gaps in immunological tolerance may enable immune recognition of these cells. For example, the catalytic subunit of the telomerase holoenzyme, hTERT, is not expressed at readily detectable levels in normal human cells but is found at significant levels in 85–90% of human tumors (see Section 10.6). About 10% of hepatocellular carcinoma (HCC) patients studied in Japan exhibited T_C cell populations reactive with specific hTERT-derived epitopes, and the cytotoxicity of these T_C cells against HCC cells was correlated with the levels of hTERT expression in these cells. (It is unclear whether the actions of these anti-hTERT T_C cells had any effect on the course of tumor progression in these patients.)

Similarly, in many breast carcinomas, the HER2/Neu receptor is often expressed at levels far higher (10- to 20-fold) than are encountered in normal epithelial tissues (see Figure 4.4). Some human melanomas overexpress a class of cell surface carbohydrates, termed gangliosides, that can also provoke an immune response. Expression of one of these, termed GD3, is sometimes vastly higher in melanomas than in their normal precursors, the melanocytes. In each of these cases, the overexpressed proteins or carbohydrate moieties may attract the attentions of an immune system that is normally oblivious to their existence.

Imagine that a cancer cell expresses an embryonic protein that is never found in normal adult tissues—a situation commonly seen in many kinds of human tumors. These embryonic proteins might therefore represent potent antigens when expressed in adult tissues. Through yet another mechanism—the alternative splicing of mRNA precursors—tumor cells may display structurally distinct versions of normal adult proteins that are rarely, if ever, experienced by the adult immune system. For example, during an epithelial–mesenchymal transition (EMT), hundreds of pre-mRNAs are subjected to alternative splicing, often yielding proteins of novel structure. Among them is the special form of Mena that arises during an EMT (see Figure 14.41), which appears to be particularly immunogenic: in one study, 10 of 52 breast cancer patients made IgG antibodies that were reactive with this protein and were not detectable in the sera of normal control patients. Similarly, 9 of 12 breast cancer patients showed T_C reactivity with this antigen, once again absent in the corresponding lymphocytes from normal controls. Hence, both humoral and cellular responses were mobilized during the course of breast cancer development. However, in these and other tumor types, it remains unclear precisely when and how during the course of multi-step tumor progression anti-tumor immune responses are first triggered (Supplementary Sidebar 15.7).

Taken together, these examples indicate that vigorous immunological attacks may well be launched against tumor cells expressing proteins in unnatural (ectopic) anatomical locations, at stages of development where they are not usually encountered, or at abnormally high levels. Occasionally, when they lead to bizarre autoimmune disorders, these immunological responses against tumors, which result from the breakdown of tolerance, are actually counterproductive for cancer patients (Supplementary Sidebar 15.8).

15.11 Tumor transplantation antigens often provoke potent immune responses

To the extent that cancer cells do succeed in provoking an adaptive immune response, this response must be traceable to specific antigens displayed by these cells. Tumor immunologists have placed these antigenic proteins in two major categories—tumor-specific transplantation antigens (TSTAs) and tumor-associated transplantation antigens (TATAs; **Table 15.3**). (Because tumor transplantation experiments cannot be undertaken in humans, the analogous human proteins are called TSAs and TAAs.)

TSTAs are said to be *specific* to a tumor or a type of tumor and are therefore not present among the repertoire of proteins and oligopeptides normally expressed within the body's tissues. TSTAs may be encoded, for example, by viral genomes or by the somatically mutated alleles (such as those of *ras*, *p53*, or *bcr-abl*) arising during tumor progression. Because they are structurally novel, these proteins are unlikely to have induced tolerance during the normal development of the immune system.

TATAs, in contrast, are only *associated* with tumor cells, and their expression is not limited to malignant tissues. TATAs represent the large class of normal proteins that, for one reason or another, have failed to elicit complete tolerance and, when expressed by tumor cells, attract the attention of the immune system. The fact that TATAs are normal cellular proteins explains the observation that a certain TATA may be displayed by many independently arising tumors of a specific type, for example, a group of melanomas.

As their names imply, the existence of these two classes of antigens can be demonstrated by tumor transplantation experiments. For example, to revisit an experiment

Table 15.3 Some tumor-associated and tumor-specific antigens and the antigenic peptides recognized by human T cells[a]

Human tumor	Protein	Antigenic peptide
Melanoma, esophageal carcinoma, liver carcinoma, NSCLC	MAGE	EADPTGHSY, SAYGEPRKL
Melanoma	tyrosinase	MLLAVLYCL, YMNGTMSQV
Colon carcinoma	carcinoembryonic antigen (CEA)	YLSGANLNL
Breast and ovarian carcinomas	HER2/Neu	KIFGSLAFL
Head-and-neck carcinoma	caspase 8	FPSDWCYF
Chronic myelogenous leukemia (CML)	Bcr-Abl	ATGFKQSSKALQRPVAS
Prostate carcinoma	prostate-specific antigen (PSA)	FLTPKKKLQCV, VISNDVCAQV

[a]A more updated (2011) compilation includes 45 mutant TSTAs and 87 TATAs, the latter divided into 30 that are expressed characteristically in one or several tumor types but not in normal tissues except the testis, 13 that reflect expression in the normal tissue-of-origin of a tumor, and 44 widely expressed proteins that are overexpressed in certain tumors (see http://www.cancerimmunity.org/peptidedatabase/Tcellepitopes.htm).

From R.A. Goldsby et al., Immunology, 5th ed. New York: Freeman, 2002; and B.J. Van den Eynde and P. van der Bruggen, *Curr. Opin. Immunol.* 9:684–693, 1997.

(A)

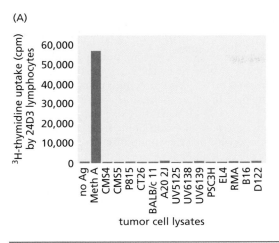

(B)

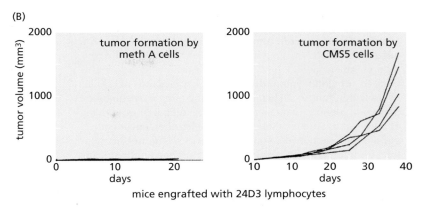

mice engrafted with 24D3 lymphocytes

Figure 15.23 Specificity of antigen display by a chemically induced fibrosarcoma Mice of the BALB/c strain were immunized with lysates of cells of the Meth A fibrosarcoma line, which originates from a 3-MC–induced mouse tumor. A line of antigen-specific lymphocytes, termed 24D3, was developed from these mice. (A) As gauged by their incorporation of ^{3}H-thymidine, proliferation of these lymphocytes in culture was stimulated by addition of Meth A cell lysates (*2nd bar*). However, lysates prepared from 14 other tumor cell lines, including other 3-MC–induced sarcomas, UV-induced squamous cell skin carcinomas, lymphomas,

a melanoma, and a lung carcinoma, failed to stimulate proliferation of these lymphocytes (*remaining channels*). In the absence of antigen (*1st channel*) no proliferation was seen. (B) When a clonal population of these 24D3 lymphocytes was introduced into BALB/c hosts, the formation of tumors by subsequently injected Meth A fibrosarcoma cells was fully blocked (*left panel*). However, the formation of tumors by an unrelated fibrosarcoma line, termed CMS5, was unaffected by the presence of these lymphocytes (*right panel*). (From T. Matsutake and P.K. Srivastava, *Proc. Natl. Acad. Sci. USA* 98:3992–3997, 2001.)

described earlier (see Figure 15.16), cells of a sarcoma arising from exposure of a mouse to 3-MC can be introduced into a second, syngeneic host mouse. The tumor cells are then allowed to multiply for several weeks before being surgically removed. If this mouse is then challenged by being re-inoculated with tumor cells from the original sarcoma, the mouse will often reject the cells and thereby block tumor formation. However, if cells from another 3-MC–induced sarcoma are injected into this mouse, these cells will indeed succeed in forming a tumor.

Such behavior indicates that the cells in the two independently induced sarcomas are antigenically distinct, and that cells of the first sarcoma carry one or more unique antigenic determinants that evoke an immune response rendering a mouse resistant to subsequently introduced cells from the same sarcoma. This suggests that the highly mutagenic carcinogen (3-MC) that originally induced these tumors also created one or more mutant cellular genes in the incipient tumor cells whose products functioned as TSTAs. Moreover, it seems that each time 3-MC caused a tumor to form, it generated a distinct TSTA or set of TSTAs (**Figure 15.23**).

The actual detection and identification of proteins that function as TSTAs are often challenging experimentally. Imagine that a 3-MC–induced sarcoma expresses a potently antigenic TSTA. As a consequence, the immune system of a tumor-bearing mouse will respond vigorously to this TSTA, attempting to eliminate those cells within the sarcoma that display high levels of this TSTA while sparing those cells in the tumor that display only very low levels of this antigen—the process of immunoediting. The resulting Darwinian selection will dictate that only those tumor cells and their descendants expressing low levels of TSTA will survive long enough to be studied by an experimenter, greatly complicating the biochemical isolation and identification of the TSTA protein.

Nevertheless, several of the genes encoding 3-MC–induced TSTAs have been isolated by gene cloning procedures in recent years. In one cloning strategy, the oligopeptides that were bound to MHC class I molecules on the surfaces of 3-MC–transformed cells and served as targets for immune recognition were eluted from the MHC molecules, purified, and subjected to amino acid sequencing. The resulting amino acid sequences were then used to predict the nucleotide sequences of the encoding genes,

Sidebar 15.2 Microsatellite instability often seems to lead to more immunogenic tumors As described in Section 12.4, defects in the DNA mismatch repair machinery create the condition of microsatellite instability (MIN), which leads to mutations accumulating in hundreds, possibly thousands of cellular genes within tumor cell genomes. Among other consequences, these mutations generate shifts in the reading frames of many of these genes. The resulting mutant alleles often encode novel amino acid sequences, sometimes termed "frameshift peptides," some of which may function as potent tumor-specific antigens.

This logic predicts that the 15% of human colorectal cancers that exhibit microsatellite instability should interact with the host immune system differently from the majority of colorectal cancers, which show no MIN and instead tend to exhibit chromosomal instability (CIN). In fact, the MIN tumors show a markedly higher presence of tumor-infiltrating lymphocytes (TILs) and a lower degree of metastasis. Moreover, antigen presentation by their MHC class I proteins is compromised far more frequently than in colorectal tumors with chromosomal instability (60% versus 30%), suggesting that the MIN tumors are under greater pressure to evade killing by various arms of the immune system and that they undertake the immunoevasive maneuver of blocking antigen presentation by their cell surface MHC class I proteins. Taken together, these observations suggest that the MIN that enables some colorectal tumors to evolve more rapidly can also exact a price because of the markedly higher immunogenicity of these tumors.

which made possible the cloning of these TSTA genes. All were point-mutated alleles of normal cellular genes that encode various proteins; none involved in any obvious way in the transformation of these cells (Supplementary Sidebar 15.9).

These observations suggest that during the course of chemical carcinogenesis, 3-MC, a known point mutagen that acts much like the benzo[*a*]pyrene discussed in Section 12.6, mutates both a proto-oncogene (often the K-*ras* gene) in target cells and additional genes that, as mutant alleles, specify TSTAs; the latter genes are struck at random—innocent bystanders that play no causal role in tumorigenesis but happen to have been damaged by the large doses of mutagenic carcinogen used to provoke tumor formation.

Importantly, the behavior of these chemically induced TSTAs is quite different from that of the TSTAs resulting from tumor virus infection. For example, SV40 virus can be used to induce a sarcoma in a mouse. Removal of this tumor will leave behind a mouse that is immunized against subsequently inoculated cells that derive from this particular SV40-induced sarcoma as well as any other SV40-induced tumors. In this instance, there is indeed a **cross-immunity** established, in that all the SV40-induced tumor cells seem to share a common TSTA or set of TSTAs. It happens that the dominant TSTA responsible for this cross-immunity is a familiar protein: it is the virus-encoded large T oncoprotein, which is expressed at significant levels in all SV40 virus–transformed cells. This contrasts with the behavior of a group of 3-MC–induced cancers, where each tumor expresses its own unique TSTA or set of TSTAs.

Observations like these raise the question whether similar mechanisms operate during human tumorigenesis. Thus, do the highly mutable genomes of cancer cells (see Chapter 12) generate mutant, antigenic proteins as inadvertent by-products of the mutagenesis that drives tumor progression (**Sidebar 15.2**)? Or are the 3-MC–induced TSTAs artifacts of the high doses of mutagenic carcinogens used in many mouse tumorigenesis experiments that do not accurately reflect the mutagenic processes that create human tumors?

15.12 Tumor-associated transplantation antigens may also evoke anti-tumor immunity

As noted above, tumor-associated transplantation antigens (TATAs) represent normal cellular proteins that, for one reason or another, have failed to induce tolerance. When these normal proteins are expressed by tumors, they evoke a measurable immune response, often involving both the humoral and cellular arms of the immune system.

For a variety of reasons, the TATAs displayed by melanomas have been more intensively studied than those of other human tumors (Supplementary Sidebar 15.10). Melanoma cells may overexpress certain proteins that are present in their normal melanocyte precursors, albeit at lower levels. Such lineage-specific proteins are sometimes called

(A)   tyrosinase antigen

(B)   MAGE-1 antigen

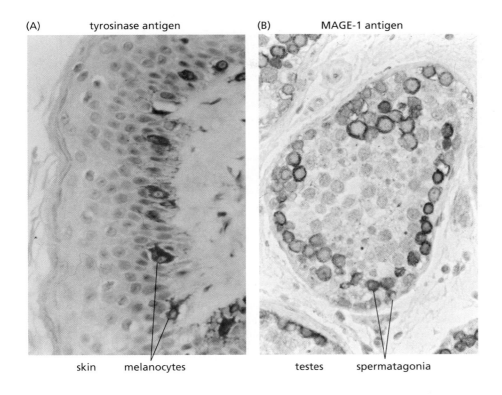

skin   melanocytes

testes   spermatagonia

Figure 15.24 Normal proteins displayed as tumor-associated antigens on melanoma cells (A) A monoclonal antibody has been used here to detect tyrosinase enzyme, which is involved in pigment production in melanocytes and melanomas. In the melanocytes (*dark red*), this enzyme is located just above the basement membrane in the skin; it is not detectable in other normal tissues. Its expression by melanoma cells can cause them to become immunogenic and the target of killing by cytotoxic lymphocytes. (B) The spermatogonia in the testes have been stained here with a monoclonal antibody against the MAGE-1 antigen (*red*); normally this antigen is only seen in one additional site—the placenta. Its expression has been detected in a variety of human tumor types and has been studied in detail in melanomas because it is often immunogenic when expressed by these tumors. (A, from Y.T. Chen et al., *Proc. Natl. Acad. Sci. USA* 92:8125–8129, 1995. B, from J.C. Cheville and P.C. Roche, *Mod. Pathol.* 12:974–978, 1999.)

differentiation antigens, implying that their display is a vestige of the differentiation program that previously governed the behavior of the normal cellular precursors of tumor cells. Included among the melanoma TATAs are transferrin, tyrosinase (**Figure 15.24A**), gp100, Melan-A/MART-1, and gp75.

The display of these differentiation antigens by melanoma cells often provokes a vigorous response by the immune system, which results in a very peculiar form of autoimmune disease—**vitiligo**—the depigmentation of large areas of skin seen in some melanoma patients (**Figure 15.25**). This depigmentation is a specific response to the presence of a melanoma. For example, when 104 renal carcinoma patients were treated with the cytokine interleukin-2 (IL-2) in order to enhance their anti-tumor immune responses, none developed vitiligo; in contrast, of 74 melanoma patients who were treated similarly, 11 developed vitiligo.

In these melanoma patients, it is clear that the immune response provoked by the melanoma TATAs leads, as a by-product, to attack and destruction of normal melanocytes, which also express these antigens. This type of vitiligo is formally analogous to the paraneoplastic syndromes (see Supplementary Sidebar 15.8), in which the display by tumors of cellular proteins results in the destruction of normal tissues that also happen to express these proteins. Significantly, melanoma patients showing vitiligo usually survive for longer periods than those who don't—suggesting that their immune systems are effective in controlling the melanomas, at least for a period of time. (For

Figure 15.25 Autoimmune depigmentation provoked by melanomas The melanoma patient shown here, who was dark-skinned prior to the onset of melanoma, has lost almost all of his skin pigment except for several isolated areas (face, armpit) due to the autoimmune attack incited by the melanoma cells. The condition of pigment loss, known as *vitiligo*, is often correlated with a longer survival of melanoma patients. (Courtesy of A.N. Houghton.)

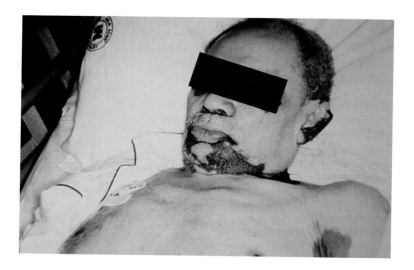

example, in a large population of melanoma patients described in 1987, 75% were still alive five years after initial diagnosis; among the subgroup of these patients who exhibited concomitant vitiligo, 86% survived for this period of time.)

The antigenicity of human melanoma cells may also derive from their display of the other major subclass of TATAs, the **oncofetal** antigens—literally those antigens that are displayed during embryogenesis and once again by tumors. Included among these are the antigens called either cancer germ-line or cancer-testis (CT) antigens, to reflect their normal expression in the germ cells of the testis and the fetal ovary. The genes for a number of these antigens, such as MAGE-1 (see Figure 15.24B), MAGE-3, BAGE, GAGE-1, and GAGE-2, have been cloned (Supplementary Sidebar 15.11). Importantly, the strategy for cloning the genes encoding TATAs depended on the presence in melanoma patients of cytotoxic T cells that specifically recognized tumor-associated antigens; the successful implementation of this strategy represented the first direct demonstration that antigen-specific T cells are commonly present among the leukocytes of these patients. Forty-four genes or gene families encoding a total of 89 distinct cancer-testis antigens have been identified, and the list is growing.

As an aside, it seems that the absence of immune responses against CT antigens in males is likely due to the fact that several of the cell types in the testes do not express MHC class I molecules and are thereby prevented from presenting their internal contents to the immune system. (Of course, females may never express these proteins in any of their tissues.) In addition, the expression by normal cells within the testes of significant levels of FasL, the ligand of the Fas death receptor (see Section 9.14), may also be used by these cells to keep wandering T lymphocytes and other immunocytes at a safe distance, further ensuring that the proteins expressed by these testicular cells are tolerated within these immunologically privileged sites and do not provoke immune responses. All this may help to explain why the display of these germ-cell proteins in an ectopic site in the body frequently provokes a vigorous immune response.

The conclusion that tolerance for many of these melanoma proteins can be readily circumvented ("broken," in the language of immunologists) leads to a simple and obvious strategy for anti-cancer therapy, in which these antigens are treated as if they were the products of invading foreign agents, such as viruses or bacteria. Such thinking has led to attempts to immunize mice or humans with a vaccine consisting of one or another of these antigenic proteins.

For example, some melanoma cells will overexpress by a factor of 100 the normal receptor for transferrin, a protein that is involved in iron uptake and metabolism in many cell types throughout the body. Mice can mount an immune response against this protein following injection with purified murine transferrin receptor, indicating that any tolerance they may have had toward this protein can be readily broken through exposure to it at high levels. These vaccinated mice will reject any subsequently introduced melanomas that happen to overexpress the mouse transferrin receptor, suggesting a more general strategy for causing cancer patients to develop potent immunity against the tumors that they carry. We will return later in this chapter to strategies that can be used to mobilize the immune system to attack human tumors.

15.13 Cancer cells can evade immune detection by suppressing cell-surface display of tumor antigens

The descriptions of immune function cited earlier might suggest that attempts by the immune system to erect defenses against the outgrowth of most tumors are relatively rare, in large part because tolerance causes the immune system to be blind to the antigens displayed by most types of cancer cells. In fact, relatively high titers (concentrations) of antibody molecules that bind tumor cell surface antigens are often encountered in the blood of cancer patients. Both helper and cytotoxic T cells that recognize tumor-associated transplantation antigens (TATAs) are also readily detectable in the blood, lymph nodes, and tumors of many patients.

These observations indicate that in spite of the phenomenon of immunological tolerance, the human immune system succeeds in mounting attacks on many (and perhaps all) types of cancer cells. That some of these attacks do indeed stem tumor development is indicated by the elevated cancer incidence in immunocompromised individuals (see Section 15.9).

Still, the development of many human cancers is clearly not blocked by immunological defense mechanisms that are, by some measures, extraordinarily efficient. For example, immunologists have found that the display (by MHC class I molecules) of as few as 10 copies of a TATA- or TSTA-derived oligopeptide on the surface of a tumor cell suffices to attract cytotoxic T cells expressing the appropriate antigen-recognizing T-cell receptor, which then proceed to kill this tumor cell.

The repeated failures of the human immune system to exploit these potent lymphocytes and other means of immunological attack in order to block tumor development require some explanation. The most attractive scenario is suggested by the behavior of the genetically altered mice described in Section 15.8: some human tumors arise that are strongly antigenic, and these are efficiently eliminated by normal immune systems; such tumors are rarely if ever detected within the human body. Other tumors are formed that are from the outset poorly antigenic because they express only proteins to which the immune system is tolerant; these tumors thrive and become clinically apparent.

Another possibility is that some antigenic cancers initially may suffer severe attrition because of attack by one or another arm of the immune system but find ways to escape elimination—the strategies of **immunoevasion** (Table 15.4). Such cells may then flourish, and their progeny may go on to create large, life-threatening tumors. This last scenario may well describe the history of most human cancers, as suggested by frequently observed changes in human cancer cells that can only be interpreted as maneuvers to evade immune attack.

The most obvious immunoevasive maneuver that a tumor cell can undertake is to stop displaying a tumor-associated (TATA) or tumor-specific (TSTA) transplantation antigen that has attracted the attention of the immune system and its cytotoxic lymphocytes. Importantly, the great majority of these antigens represent molecules that do not participate *causally* in neoplastic transformation, but instead only reflect tissue-specific differentiation antigens, often those typically expressed in the tissues in which tumors have arisen (see Figure 15.24). Expression of these antigens can often be repressed by cancer cells with impunity, that is, without compromising their own continued survival and proliferation. In fact, tumor cell populations often harbor variants in their midst that have used promoter methylation to suppress expression of certain antigen-encoding genes (see Section 7.8). These antigen-negative variants may therefore be able to escape an immune attack that has been mounted against the bulk of the cells in a tumor mass, and their descendants may eventually emerge as the dominant cell population in this mass.

Table 15.4 Immunoevasive strategies used by cancer cells

Strategy	Mechanism	Agent being evaded
Hide identity	repress tumor antigens (TATA or TSTA), repress MHC class I proteins	cytotoxic T lymphocytes
Hide stress	repress NKG2D ligands (e.g., MICA)	NK cells
Inactivate immunocytes	destroy immunocyte receptors	NK cells; cytotoxic T lymphocytes
	saturate immunocyte receptors with adenosine, MICA	NK cells; variety of immunocytes
	induce T_{reg} formation	variety of T lymphocytes
Avoid apoptosis	inhibit caspase cascade by increasing IAPs, acquire resistance to FasL-mediated apoptosis	
Induce immunocyte apoptosis	release soluble FasL	cytotoxic T lymphocytes
	release cytokines (IL-10, TGF-β)	cytotoxic T lymphocytes, dendritic cells, macrophages
Neutralize intracellular toxins	enzymatic detoxification of H_2O_2, prostaglandin E_2	macrophages, NK cells
Neutralize complement	overexpress mCRPs	complement system
Up-regulate CD47 expression	express "don't eat me" signal on cell surface	phagocytic cells

In one illustrative case, a melanoma patient was vaccinated with tyrosinase protein in order to induce immune reactivity against this protein, which was being expressed by his melanoma cells. (The tyrosinase enzyme participates in pigment synthesis by melanocytes and is therefore not involved in causing neoplastic transformation; see Figure 15.24A.) At first, the resulting immune response caused his melanoma metastases to regress. Soon, however, some tyrosinase-negative variant cells emerged among the melanoma metastases. While his tyrosinase-positive melanoma cells continued to regress, the tyrosinase-negative cells began to proliferate rapidly and eventually killed the patient, a phenomenon sometimes called "immune escape" by tumor immunologists. However, in another case of melanoma, this type of immunoevasion did not guarantee tumor cells long-term protection from immune attack (**Sidebar 15.3**).

In many tumors, the cancer cells cannot resort to the simple expedient of shutting down expression of a tumor antigen, simply because the continued expression of a TSTA or TATA antigen may be essential for their continued neoplastic proliferation. For example, the overexpressed but otherwise normal HER2/Neu protein—the growth factor receptor displayed by more than 20% of breast cancers (see Figure 4.4)—may well stimulate an immune attack. However, the neoplastic cells in these breast carcinomas cannot afford to shut down the display of this protein, because its continued expression at high levels is critical for their proliferation and their ability to avoid apoptosis.

Cancer cells that cannot down-modulate expression of a TATA or TSTA because of its essential contribution to neoplastic growth may resort to alternative strategies in order to avoid immune killing. An important and widely used immunoevasive strategy derives from the ability of cancer cells to repress expression of their MHC class I proteins (see Supplementary Sidebar 15.2). This is often achieved through repression of MHC class I gene transcription In fact, many types of human cancer cells have been

Sidebar 15.3 Immunoevasion may offer tumor cells only a temporary reprieve from immune attack Five years after resection of a primary melanoma, a melanoma patient returned with multiple metastases in lymph nodes, which were removed surgically (**Figure 15.26**). On this occasion, the tumor-infiltrating cytotoxic lymphocytes (CTLs) recognized MART-1, one of the well-studied melanoma-associated tumor antigens. The patient was free of symptoms for another six years, when he developed a recurrence of melanoma in a regional lymph node, which was also removed. Now, the MHC class I–mediated presentation of MART-1 oligopeptides was absent from the tumor cells, and the CTLs present in the tumor that could recognize MART-1 were almost gone. (MART-1 synthesis by the tumor cells continued, however.) Instead, the tumor-associated CTLs recognized a second melanoma-associated antigen—tyrosinase—which was now being presented by the tumor cells' MHC class I molecules.

This behavior indicated that the patient's immune system was able to adapt dynamically to the initial evasive maneuver of the melanoma cells (shutdown of MART-1 antigen presentation) by redirecting its energies to mount an attack on another tumor-associated antigen (tyrosinase). However, by the time of the third surgery, the patient's melanoma cells had begun to lose almost all MHC class I expression. This loss should have eventually allowed the disseminated tumor cells that remained after the third surgery to escape all CTL attack and to multiply rapidly into life-threatening metastases. Nonetheless, the patient survived for another five years and died of causes unrelated to his melanoma. His ability to survive a usually highly aggressive malignancy for another five years suggested that his immune system was able to keep any residual melanoma cells in check through mechanisms that remain unclear. Quite possibly, NK cells, which are known to attack cells lacking MHC class I expression (to be discussed shortly), were responsible for the last five, symptom-free years of this patient's life.

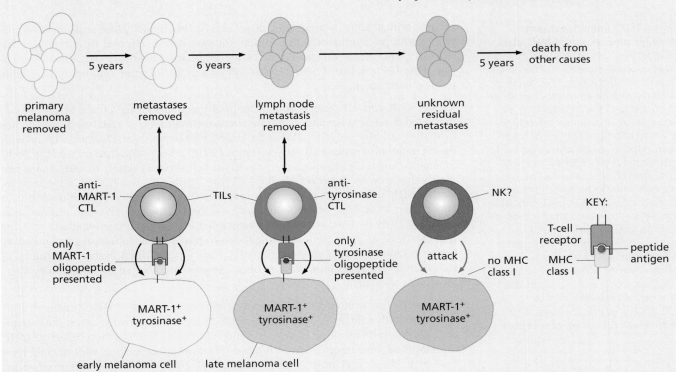

Figure 15.26 Dynamic adaptation by the immune system to shifting tumor cell antigen presentation A melanoma tumor was removed from a patient. Five years later, when metastases appeared and were removed, the tumor cells presented only MART-1 oligopeptide antigen on their MHC class I receptors (*red dot*) even though these cells synthesized both MART-1 and tyrosinase proteins; at this time, the tumor-infiltrating cytotoxic lymphocytes (TILs/CTLs) recognized MART-1 antigen via their T-cell receptors. Six years later, when a lymph node metastasis was discovered and removed from this patient, the tumor cells continued to make both MART-1 and tyrosinase, but now their MHC class I molecules presented only the tyrosinase oligopeptide antigen (*blue dot*) and the tumor-infiltrating CTLs recognized only the tyrosinase antigen. Later, the melanoma cells ceased making MHC class I altogether, at which point they presumably became vulnerable to attack by NK (natural killer) cells (*red*), which may have been what kept them under control for another five years. Hence, the immune system was able to co-evolve with the tumor cells, generating CTLs of novel specificity as required in order to combat the melanoma cells.

found to lack normal levels of the mRNAs encoding these MHC molecules, thereby preventing their synthesis and thus display of antigens at the cell surface (**Figure 15.27**).

Loss of MHC class I protein expression is often associated with more invasive and metastatic tumors. For example, in more than half of advanced breast cancers, the display

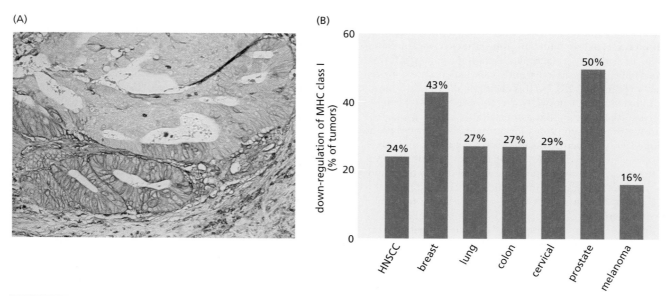

(A)

(B)

of these critical antigen-presenting molecules has been totally lost, and such loss correlates with poor prognosis. In addition, the carcinoma cells that form micrometastases in the bone marrow (see Figure 14.10) often show little if any MHC class I expression, possibly because of intensive surveillance by the immune cells operating in this tissue environment.

Cancer cells can also use post-translational mechanisms to reduce MHC class I–mediated antigen presentation. For instance, the migration of MHC class I molecules from the endoplasmic reticulum to the cell surface depends on their association with β_2-microglobulin protein (β_2m; see Figure 15.10D). Normally, β_2m escorts the MHC molecules and their oligopeptide cargo to the cell surface (**Figure 15.28A**). In certain high-grade tumors, however, the lack of β_2m synthesis prevents the oligopeptide-loaded MHC class I molecules from reaching the cell surface (Figure 15.28B). An earlier step in antigen presentation may also be compromised: some tumors have defective TAP1 or TAP2 (transporter associated with antigen presentation) proteins (Figure 15.28C). These proteins are needed to transport oligopeptides generated by the proteasomes in the cytosol to the MHC class I molecules present in the lumen of the endoplasmic reticulum. Without both TAP proteins, antigen presentation by MHC molecules also fails.

Because they are only correlative, these observations of crippled MHC-mediated antigen presentation do not prove that these defects are *causally* involved in enabling cancer cells to evade immune attack. Evidence of a direct link between defective MHC function and immunoevasion is limited and anecdotal. In one well-documented case of a human melanoma that was being treated by anti-tumor immunotherapy, cells emerged that had down-regulated MHC class I expression and were resistant to therapy. In other cases, such as many neuroblastomas and small-cell lung carcinomas (SCLCs), the loss of MHC class I expression is correlated with the virtual absence of tumor-infiltrating lymphocytes (TILs; see Section 15.9), which are thought to be key elements in leading the immune attack on tumors.

In fact, the total absence of MHC class I molecules invites an attack by NK cells, which continuously patrol the body's tissues looking for cells that have lost these key proteins from their surface (discussed in Section 15.14). This explains why some tumor cells will selectively suppress expression of only one of the six key MHC class I molecules that are normally expressed concomitantly by cells throughout the body. This suppression may block presentation of a particular tumor antigen, thereby allowing the tumor cell to evade attack by antigen-specific cytotoxic lymphocytes without provoking an attack by NK cells, since only a small proportion of the total population of cell surface MHC class I molecules has been lost.

Taken together, these diverse observations suggest that by attacking cancer cells displaying MHC class I molecules and associated oligopeptide antigens, the immune

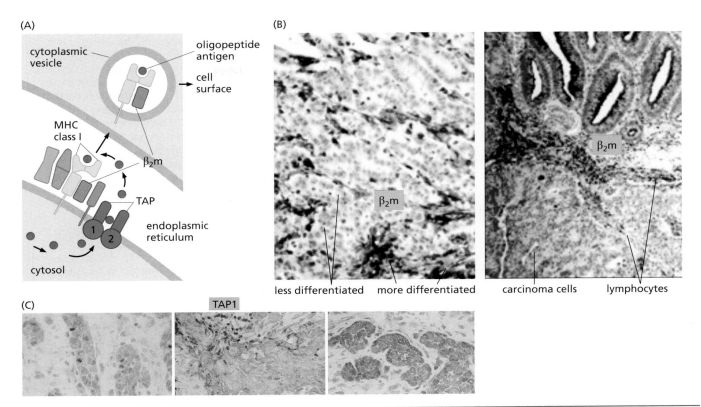

Figure 15.28 Failure of antigen presentation due to loss of β₂-microglobulin or TAP1 (A) A more detailed depiction of Figure 15.10A reveals that the TAP complex is a heterodimer of the TAP1 and TAP2 proteins (**t**ransporters associated with **a**ntigen **p**rocessing; *dark green*) that is involved in transporting oligopeptides (*red dots*) from the cytosol into the lumen of the endoplasmic reticulum, where the oligopeptides can associate with MHC class I molecules (*yellow*). Independent of this, the β₂-microglobulin (β₂m, *light brown*) molecule associates with MHC class I molecules and helps to shepherd them, together with their captured oligopeptide antigens, to the cell surface. The MHC class I molecules require both a bound oligopeptide antigen and an associated β₂m molecule in order to reach the cell surface. Therefore, defects in either the β₂m or a TAP protein preclude presentation of oligopeptide antigens on the cell surface. (B) In one human colorectal carcinoma (*left panel*), β₂m staining (*reddish brown*) is present in the cancer cells forming the more differentiated tubular structures (*below*), while the less differentiated carcinoma cells (*light blue, above*) fail to express any β₂m. In a second colorectal carcinoma (*right panel*), β₂m is present in infiltrating lymphocytes (*dark red*) but is totally absent in many of the carcinoma cells (*light blue, bottom half*). In 5 of 17 microsatellite-unstable (MIN) colorectal tumors (see Sidebar 15.2), the *β₂m* gene was found to have suffered inactivating mutations, and it was transcriptionally silenced through unknown mechanisms in a number of other tumors. (C) Defective antigen processing is frequently seen in head-and-neck squamous cell carcinomas (HNSCCs), in which expression of TAP1 or TAP2 is often lost. The immunohistochemical staining shown here reveals the expression of TAP1 (*beige*) in a group of HNSCCs to be variable, being absent in some tumors (*left*), present in patches in others (*middle*), and normally expressed in yet others (*right*). (A, from C.A. Janeway Jr. et al., Immunobiology, 6th ed. New York: Garland Science, 2005. B, from M. Kloor et al., *Cancer Res.* 65:6418–6424, 2005. C, from M. Meissner et al., *Clin. Cancer Res.* 11:2552–2560, 2005.)

system creates great selective pressure for the outgrowth of variant cancer cells that no longer display these antigen-presenting molecules on their surface. Since the display of these cell surface molecules tends to be lost in invasive and metastatic cells, it may be, as speculated earlier, that the immune system only launches a full-scale attack on cancer cells once they have invaded the stroma, where immune cells are present in substantial numbers.

15.14 Cancer cells protect themselves from destruction by NK cells and macrophages

As mentioned earlier, the evasive tactic of suppressing MHC class I expression, clever as it may be, carries its own dangers: the immune system anticipates this trick and uses its natural killer (NK) cells to attack cells that lack adequate numbers of MHC class I molecules on their surface. This NK response can also foil the plans of tumor viruses that attempt to elude immune detection by preventing infected cells from displaying MHC class I molecules and thus viral oligopeptide antigens on their surface.

Figure 15.29 Control of natural killer cell–mediated killing by positive and negative signals (A) Natural killer (NK) cells (*yellow, bottom*) display a killer inhibitory receptor (KIR, *red*) that can recognize MHC class I molecules being displayed on the surface of potential target cells (*gray, above*). In the event that such recognition occurs, the KIR releases inhibitory signals that prevent attack by the NK cell (*left*). Conversely, in the absence of such recognition and the resulting functional silencing by KIR, attack by the NK cell is possible (*right*). (Note that KIR recognizes and binds a constant structural feature of MHC class I molecules and conversely does not recognize an oligopeptide antigen ensconced in the antigen-binding pocket of the MHC class I molecule.) (B) The NKG2D receptor displayed by NK cells binds a number of ligands that are expressed by various cell types suffering certain physiologic stresses, notably the stresses caused by genetic damage, viral infection, and neoplastic transformation. Shown here is an assembly of various alternative ligands of NKG2D that have been identified in human and murine cells. For example, MICA and MICB are normally expressed only by certain stressed epithelial cells. (C) Seen here is the molecular structure of the complex between NKG2D and its MICA ligand. The two subunits of the NKG2D receptor are shown above (*blue, purple*) as they might be displayed on the surface of an NK cell (*above*). The structure of the three subdomains of the ectodomain of MICA (*beige, pink, green*) is shown as it might be displayed on the surface of a cancer cell (*below*). [The domains (*dotted lines*) that anchor these proteins to the surfaces of the NK and the cancer cell are not shown.] (D) Various types of genetic damage, inflicted by the indicated genotoxic treatments, potently induce the display of Rae1 ligands (members of the RAET1 subfamily; see panel B) on the surface of cells; in contrast, treatment by roscovitine, which halts cell cycle progression without inflicting DNA damage, does not induce Rae1 display. The basal levels of Rae1 expression are indicated by the *green* curves, while the induced levels are indicated by the *red* curves. This induction can be blocked by inhibiting the ATM and ATR kinases (see Figure 9.13), which are key sensors of DNA damage. Other work (*not shown*) indicates that signaling from the p110α form of PI3K (see Figure 6.17), which is activated in many types of cancer cells, can also induce Rae1 expression. (E) Tumor cells arising in the TRAMP transgenic mouse model of prostate adenocarcinoma development express on their surface a Rae1 ligand, which serves as an activating ligand for NKG2D. The tumor cells in TRAMP mice that carry wild-type versions of the NKG2D-encoding gene expressed low levels of *Rae1* mRNA. In contrast, in TRAMP mice in which the gene encoding NKG2D (properly termed *Klrk1*) had been knocked out, far higher levels of the *Rae1* transcript were apparent. This suggests that TRAMP tumor cells down-regulate their Rae1 expression in order to avoid attack by NK cells displaying NKG2D. (Importantly, NK killing often requires the absence of MHC class I expression together with the expression by cancer cells of certain NKG2D-activating ligands.) (B, from D. Raulet, *Nat. Rev. Immunol.* 3:781–790, 2003. C, from P. Li et al., *Nat. Immunol.* 2:443–451, 2001. D, from S. Gasser et al., *Nature* 436:1186–1190, 2005. E, from N. Guerra et al., *Immunity* 28:571–580, 2008.)

A receptor normally displayed on the surface of NK cells recognizes MHC class I molecules displayed by potential target cells (**Figure 15.29A**). Binding of the target cell MHC class I molecules to this NK cell receptor (called killer inhibitory receptor, or KIR) causes this receptor to release signals into the NK cell that prevent it from launching an attack on the MHC class I–positive cell. These inhibitory signals are missing when an NK cell encounters a potential target that lacks MHC class I molecules on its surface, in which case target cell killing is allowed to proceed. (The systematic elimination of MHC class I–negative cells by NK cells may explain why, in certain classes of human tumors, the absence of class I expression actually correlates with a better clinical outcome—a counterintuitive notion, given our earlier discussion.)

Yet another innate interaction of NK cells with cancer cells appears to be equally important and works in the opposite direction. This particular interaction depends on the fact that many types of human cells are programmed to display specific proteins on their surface whenever these cells suffer certain physiologic stresses, including those resulting from genetic damage, viral infections, and neoplastic transformation. These stress-signaling proteins have names such as MICA, MICB, ULBP4, and so forth (see Figure 15.29B). NK cells, for their part, display a complementary cell surface receptor, called NKG2D, that is specialized to recognize these stress-associated cell proteins displayed by potential target cells (see Figure 15.29C). Binding of these proteins to the NKG2D receptor results in strong activation of an NK cell's cytotoxic response and is therefore followed rapidly by killing of cells that express these stress-associated proteins on their surfaces.

(A)

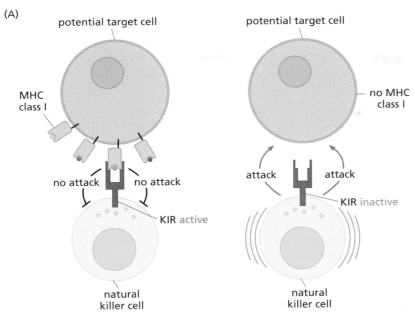

(B)

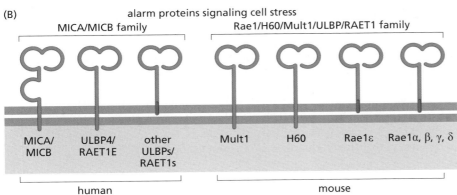

(C)

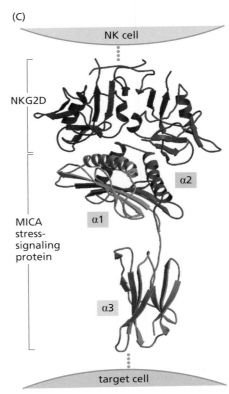

(D)

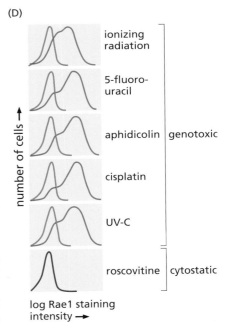

As one demonstration of these dynamics, a gene encoding the mouse homolog of the human MICA antigen was introduced into mouse lymphoma cells, causing them to express significant levels of this antigen on their surface. These cells thereupon lost their tumorigenicity when injected into host mice. However, if these mice were first deprived of their NK cells, then the lymphoma cells expressing the MICA homolog once again became tumorigenic.

Clearly, the interplay between developing cancer cells and NK cells is complex. In fact, cancer cells often express certain NKG2D-activating ligands as a natural consequence of neoplastic transformation, more specifically because of hyperactivity of the PI3K enzyme or the pervasive DNA damage affecting these cells (see Figure 15.29D). Such ligand expression endangers these cancer cells by making them visible targets for NK cells; this forces them to respond by down-regulating expression of some of these ligands in order to escape NK attack and ultimately generate robustly growing tumors. To examine these dynamics, tumor-prone transgenic mice that either express or fail to express NKG2D have been generated. In mouse hosts whose NK cells are capable of expressing NKG2D, the tumor cells suppress expression of the Rae1 stress antigen (see Figure 15.29B) in order to elude NK-mediated killing, whereas in mice lacking the NKG2D receptor, the tumor cells grow robustly even though they continue to express significant levels of the Rae1 antigen on their surface (see Figure 15.29E).

Actually, many types of human carcinoma cells and melanoma cells disable this NKG2D signaling pathway through an alternative immunoevasive strategy: they continue to synthesize significant amounts of an alarm protein like MICA but shed much of this protein into the medium around them (rather than retaining it attached to their

(E)

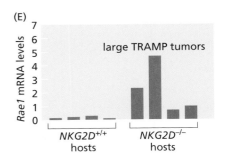

cell surface). The released, soluble MICA can be detected in the serum of many cancer patients and acts as a *decoy ligand* that distracts the NKG2D receptors displayed by their NK cells. Thus, soluble MICA binds the NKG2D receptors on NK cells (and cytotoxic T cells) and causes the endocytosis and degradation of these receptors, thereby rendering these immunocytes incapable of responding to the MICA molecules that continue to be displayed on the surfaces of a patient's cancer cells.

In summary, NK cells instinctively recognize cancer cells in two ways: they sense either the absence of MHC class I proteins or the presence of one or more stress-associated proteins (for example, MICA, Rae1) on the surfaces of cancer cells. When both conditions are satisfied, the killing of cancer cells is far more efficient than when only one of these conditions is met. Indeed, some immunologists believe that both conditions must be satisfied (that is, low levels of MHC class I and high levels of a stress-associated NKG2D ligand such as MICA) before NK killing proceeds. Most evidence, however, indicates that it is the *balance* of signals that dictates NK activity; for example, cells expressing very high levels of certain stress-associated alarm proteins can be attacked and killed by NK cells, even when they continue to express significant levels of MHC class I on their surface.

To conclude our discussion of tumor cell interactions with NK cells, it is worthwhile recalling the peculiar way in which metastasizing cancer cells in the circulation avoid being ambushed by NK cells. Cancer cells that have intravasated and thus come into direct contact with the blood rapidly acquire a cloak of adhering platelets that, together with the cancer cells, form microthrombi (see Supplementary Sidebar 14.2). When the coagulation mechanisms that create the microthrombi are rendered defective through one or another experimental strategy, the success rates of the metastasizing cells in founding new tumor colonies plummet to a fraction of what is otherwise observed. The attrition is due to attacks by NK cells, which are normally prevented by the platelet cloaks from recognizing and attacking their neoplastic cell targets. This particular immunoevasive maneuver is not developed through the selection of rare variant cancer cells within a tumor, but instead is almost inadvertent, being deployed routinely by all cancer cells that have managed to enter the circulation and begun their journeys to distant tissue sites throughout the body.

These discussions of NK cells and their attacks on tumors fail to address the anti-cancer roles played by a second, equally important agent of the innate immune system—the macrophage. Paralleling the signals that govern NK attack on cancer cells is the expression of two cancer cell proteins that provide conflicting signals to nearby macrophages. One of these—calreticulin (CRT)—represents an "eat-me" signal, which invites phagocytosis by macrophages. Calreticulin is displayed by a diverse array of neoplastic cell types and its cell surface expression may represent a hard-wired component of the gene expression programs that are activated in cells in response to transformation and other types of cell-physiologic stress, including that provoked by the cytotoxic agents used in chemotherapy.

Working in the other direction is the CD47 protein, which cells display in order to ward off phagocytosis by macrophages and has therefore been called the "don't-eat-me" signal. Indeed, to date CD47 is the only known means employed by a wide variety of cell types throughout the body—both normal and neoplastic—to protect themselves from spontaneous attacks by roaming macrophages (**Figure 15.30**). In one particularly dramatic example, primary human breast carcinoma cells that have not yet become invasive show little if any CD47 expression; however, the circulating tumor cells (CTCs; see Section 14.1) that derive from highly malignant mammary tumors show high levels of this protein, ostensibly to protect themselves from attack by macrophages. CD47 seems to play a role in the lives of a variety of normal cells; for example, as erythrocytes grow older, they progressively down-regulate their expression of CD47, which eventually causes them to fall victim to macrophages, ensuring that these cells are discarded when they are worn out by the rigors of swimming through the bloodstream for three or four months. (In the context of cancer, phagocytosis by macrophages appears to require both low levels of CD47 and an active "eat-me" signal, such as calreticulin or the phosphatidylserine that is displayed on the outer leaflet of the plasma membrane by apoptotic cells; see Section 9.10.)

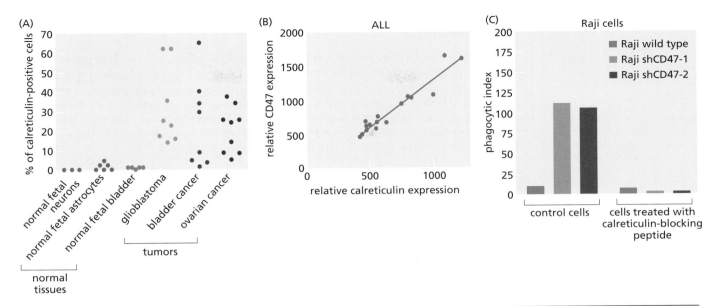

Figure 15.30 Attacks by macrophages triggered by "eat-me" signals Macrophages are programmed to eliminate a variety of cell types throughout the body that display "eat-me" signals and are kept at arm's length by countervailing "don't-eat-me" signals. (A) Certain types of chronic cell-physiologic distress, notably that caused by transformation, encourage the calreticulin (CRT) protein to move from its normal intracellular residence in the endoplasmic reticulum to the cell surface, where it serves as an eat-me signal for macrophages. As shown here, three types of tumor cell populations contain significant proportions of CRT-expressing cells (*right*), whereas normal cell populations (*left*) contain few if any CRT-positive cells. Pre-apoptotic cells also express elevated levels of cell surface CRT (*not shown*). (B) Acting in the opposite direction is CD47, a don't-eat-me signal that is elevated in a wide variety of cancer cell types, ostensibly to protect against attack by macrophages. In a group of acute lymphocytic leukemias (ALL), the expression of CRT and CD47 varies proportionately. This suggests that these leukemia cells must increase their expression of protective CD47 in order to compensate for their increased expression of CRT—the eat-me signal. (C) These dynamics can be validated by examining the effects of modulating C47 and CRT levels on Raji cells—a line of Burkitt's lymphoma cells deriving from the B-cell lineage—that were incubated in culture together with macrophages. When the display of CD47 by Raji cells was knocked down by either of two shRNAs, the number of phagocytic events increased dramatically (*left half*); an anti-CD47 monoclonal antibody achieved similar results (*not shown*). (Phagocytic index = number of Raji cells ingested by 100 macrophages.) However, if the proper display by these cells of the CRT alarm signal was blocked by a specially designed peptide (*right half*), then phagocytosis did not occur even when CD47 levels were knocked down. Hence, protective shielding by CD47 becomes unnecessary if macrophages are not provoked by CRT display. (From M. Chao et al., *Sci. Transl. Med.* 2:63ra94, 2010.)

The expression of high levels of CD47 by diverse types of cancer cells has inspired an interesting therapeutic strategy, which uses monoclonal antibodies (see Supplementary Sidebar 11.1) to bind and occlude cell surface CD47, thereby depriving cancer cells of a key defense against macrophage attack (Supplementary Sidebar 15.12). This research on macrophages and the earlier work on NK cells indicate the key role of the innate immune system in eliminating incipient populations of cancer cells long before they have attracted the attentions of the adaptive immune system.

15.15 Tumor cells launch counterattacks on immunocytes

Earlier, we read that tumor cells thrive in immunodepressed environments, such as the bodies of transplant recipients whose immune systems have been compromised by immunosuppressive drugs as well as those suffering from AIDS. In fact, the great majority of human tumors may develop in immunodeficient environments. More precisely, tumors may create *localized* microenvironments in which immune function is compromised. By keeping functional cytotoxic cells at some distance, tumors can establish safety zones for themselves at various sites in the body.

One strategy for doing so is suggested by one of the mechanisms (see Section 15.3) that are normally used by cytotoxic lymphocytes to kill their victims: these lymphocytes wield Fas ligand (FasL) molecules on their surface, which bind and activate the Fas death receptor displayed by other cell types throughout the body. Binding of the FasL ligand to this death receptor leads to activation of the extrinsic apoptotic pathway (see Section 9.14).

Many cancer cells, however, subvert this signaling system in a two-step process. First, they develop resistance to FasL-mediated killing through mechanisms that are not well understood. For example, by acquiring resistance to multiple forms of apoptosis (see Table 9.5), cancer cells can protect themselves from activation of the extrinsic apoptotic cascade that is triggered by the FasL-activated Fas receptor. Clearly, this resistance, on its own, provides these cells with one means whereby they can avoid destruction by cytotoxic cells, such as cytotoxic T cells.

In a second step, many kinds of cancer cells turn this FasL–Fas system on its head, by acquiring the ability to produce and release soluble forms of FasL themselves (**Figure 15.31A and B**). This FasL does not affect these cancer cells, which have already become resistant to its effects. However, a number of studies indicate that this ligand can activate Fas death receptors displayed on the surfaces of several types of lymphocytes, causing their death. By launching such counterattacks on nearby immune cells, cancer cells can hold them at a safe distance, once again reducing the risk of being killed by them.

In many cancer patients, there is evidence of depressed levels of certain types of circulating lymphocytes, indicating a systemic defect in immune function. In oral cancer patients, for example, membranous vesicles displaying membrane-bound, biologically active FasL can be found in the general circulation (see Figure 15.31C); these vesicles seem to be released by the cancer cells as a means of depressing lymphocyte function throughout the body. The importance of this particular immunosuppressive mechanism for the growth of these and other types of human tumors still requires extensive documentation.

Alternative forms of counterattack are used by the many types of human cancer cells that have been found to release either TGF-β or interleukin-10 (IL-10). Both of these secreted proteins are potently immunosuppressive, having strong cytostatic effects on T lymphocytes, preventing the maturation of dendritic cells, and suppressing their expression of MHC type II molecules. Under certain conditions, TGF-β can even induce apoptosis of dendritic cells and macrophages—the two key antigen-presenting cells of the immune system. [Recall that carcinoma cells that have passed through an EMT secrete significant amounts of TGF-β that is used in an *autocrine* fashion to maintain their residence in a mesenchymal state (see Figures 14.19 and 14.20); this TGF-β may also serve to establish a safety zone around these cells, doing so by acting in a *paracrine* manner on immunocytes that venture too close.]

Interestingly, the release of IL-10 by human cancer cells is mimicked by Epstein–Barr virus (EBV), which has acquired and remodeled a cellular IL-10 gene. As we have read (see Section 3.4), this virus is an important etiologic agent of Burkitt's lymphomas, other lymphomas of the B-cell lineage, and nasopharyngeal carcinomas. By forcing virus-infected cells to release an IL-10-like immunosuppressive cytokine, Epstein–Barr virus protects these cells from direct attack by cytotoxic immune cells, thereby ensuring a safe cellular haven for itself over extended periods of time.

One set of experiments is particularly illustrative of the immunosuppressive powers of TGF-β and demonstrates the truth of the old adage that the best defense is the launching of a preemptive attack (**Figure 15.32A**). In these experiments, a group of 10 mice were injected with a strain of mouse melanoma cells that happen to secrete high levels of TGF-β; all of these mice succumbed to cancer within 45 days. In a complementary experiment, the lymphocytes of mice were first rendered resistant to TGF-β–mediated killing by grafting into their bone marrow genetically altered hematopoietic stem cells that displayed a dominant-negative (dn) TGF-β type II receptor (which blocks normal receptor function). When cells of the same melanoma cell line were injected into this second set of mice, 7 out of 10 mice were still alive 45 days later, and examination of their lungs revealed a dramatic reduction in melanoma metastases (see Figure 15.32B). Similar outcomes were observed when prostate carcinoma cells were used instead.

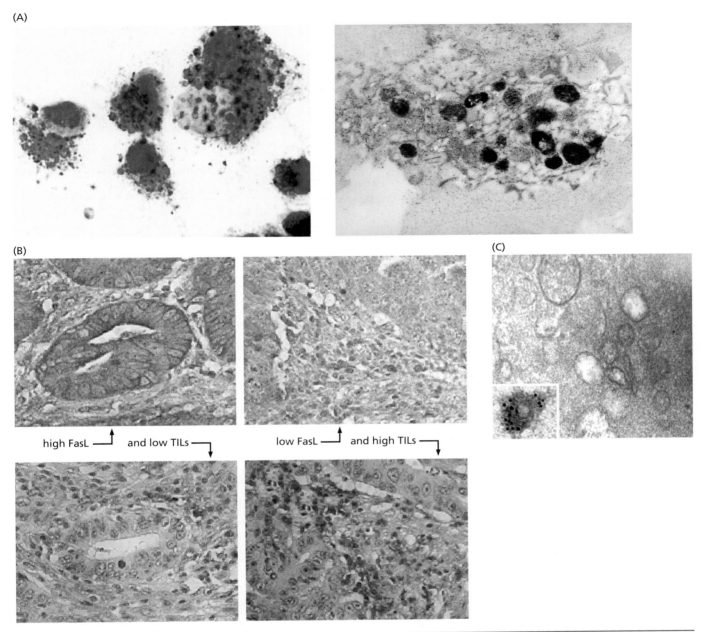

Figure 15.31 Contributions of FasL to immunoevasion
Some cancer cells are able to protect themselves from killing by T lymphocytes through a two-step process. First, they acquire resistance to killing by FasL, the ligand of the Fas death receptor. Second, they acquire the ability to produce and release FasL, which allows them to kill lymphocytes and other cells that may stray too close to them. (A) When melanoma cells are stained with an anti-FasL antibody (*left panel*), significant amounts of FasL are seen to be concentrated in vesicles (*red*) in the cytoplasm; nuclei are in *blue*. Far higher resolution is obtained by use of immunoelectron microscopy (*right panel*), which reveals that the FasL (*large black spots*) in melanoma cells is actually localized to melanosomes (cytoplasmic bodies that carry the dark melanin pigment molecules of normal melanocytes); melanoma cells can release these melanosomes into the extracellular space, where the FasL acquires the ability to trigger the death of Fas-expressing cells, such as nearby lymphocytes. (B) In an adenocarcinoma of the colon, those areas of the tumor in which high levels of FasL (*brown*) were evident (*above left*) showed very low levels (in the adjacent tissue slice) of tumor-infiltrating lymphocytes (TILs; *dark red, below left*). Conversely, those areas that showed low levels of FasL (*brown, above right*) showed high levels of TILs (*dark red, below right*). TILs that were near areas of high FasL showed high rates of apoptosis (*not shown*). (C) One explanation for the frequently observed suppression of functional circulating T lymphocytes in oral carcinoma patients may come from the discovery that the great majority of these patients (but not normal controls) have membranous microvesicles in their circulation that display membrane-bound FasL, which is a particularly potent apoptosis-inducing form of this death receptor ligand. Such microvesicles, sometimes termed *exosomes*, have been purified from the serum of such patients and found by immunoelectron microscopy (see panel A) to contain membrane-bound FasL (*black dots, inset*). (A, left, from G. Andreola et al., *J. Exp. Med.* 195:1303–1316, 2002. A, right, from L. Rivoltini et al., *Immunol. Rev.* 188:97–113, 2002. B, from A. Houston et al., *Br. J. Cancer* 89:1345–1351, 2003. C, from J.W. Kim et al., *Clin. Cancer Res.* 11:1010–1020, 2005.)

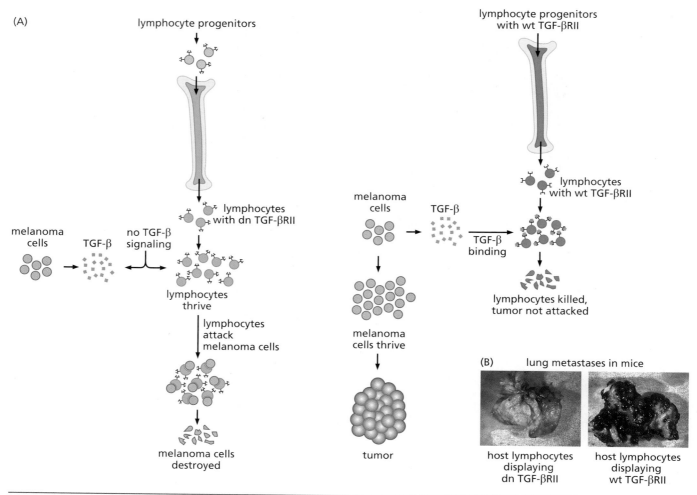

Figure 15.32 Role of TGF-β in controlling immune responses (A) If lymphocyte progenitor cells bearing a dominant-negative TGF-β II receptor (dn TGF-βRII, *top*) are introduced into the marrow of a mouse, differentiated lymphocytes (*green*) are formed that do not show responsiveness to the killing effects of TGF-β (*left diagram*). If melanoma cells (*pink, left*) are then introduced into a mouse carrying such lymphocytes, even though these melanoma cells secrete copious amounts of TGF-β (which is normally toxic for lymphocytes; *beige squares*), the lymphocytes thrive and proceed to kill the melanoma cells, thereby blocking tumor formation. In contrast, when the same melanoma cells are introduced into a host mouse that has been engrafted with lymphocyte progenitors expressing wild-type (wt) TGF-βII receptor (*right drawing*), the TGF-β released by the melanoma cells induces the lymphocytes (*blue*), which display fully functional TGF-β receptors, to enter apoptosis (*bottom*); as a consequence, the melanomas are not attacked and grow vigorously. This proves that TGF-β secretion is a highly effective means by which these melanoma cells can defend themselves against immune attack. (B) The dramatic differences in tumor-forming ability are revealed by the relative abilities of the pigmented melanoma cells to form lung metastases. Wild-type C57BL/6 mice rapidly developed many such metastases (*black spots, right*), while those expressing the dominant-negative TGF-β type II receptor (dn TGF-βRII) in their T cells efficiently blocked metastasis formation (*left*). The lungs were observed here 21 days after B16 melanoma cells were injected intravenously. (From data of L. Gorelik and R.A. Flavell, *Nat. Med.* 7:1118–1122, 2001. B, from A. Shah et al., *Cancer Res.* 62:7135–7138, 2002.)

These observations indicate that the release of TGF-β by the melanoma and prostate cancer cells enabled the tumor cells to defend themselves against killing by immune cells. However, once the host immune cells were made resistant to TGF-β–induced apoptosis, these immunocytes showed themselves to be perfectly capable of eliminating the cancer cells. Moreover, in the tumor-bearing mice whose hematopoietic cells displayed the dn TGF-βRII, cytotoxic lymphocytes showing an ability to kill the melanoma cells *in vitro* could be recovered from the spleen; such splenic lymphocytes were virtually absent in control animals whose lymphocytes lacked this mutant receptor.

(These experiments might suggest to you the outlines of a novel anti-tumor therapy, in which the bone marrow cells of a cancer patient are rendered resistant to TGF-β–induced apoptosis, making these cells especially effective in attacking the many human tumors that release TGF-β. Unfortunately, once immune cells are rendered unresponsive to TGF-β, they often launch devastating autoimmune attacks on tissues throughout the body, yielding a condition that can be far more debilitating and rapidly lethal than a neoplastic disease.)

Complementary results come from experiments in which the ability of cancer cells to release TGF-β has been greatly reduced by inserting antisense constructs into these cells. Such cancer cells lose much of their tumorigenic powers, in large part because they are now besieged by flocks of cytotoxic lymphocytes that are capable of killing them.

At least three agents—TGF-β, IL-10, and FasL—have been proposed as weapons used by cancer cells to launch counterattacks on the various immunocytes that threaten them. Since the killing of antibody-coated cancer cells is often carried out by cytotoxic immunocytes (for example, NK cells, macrophages; see Figure 15.3), the secretion of these immunosuppressive agents may also protect tumor cells from elimination in those patients who have high levels of circulating anti-tumor antibodies.

There are likely to be yet other strategies—some still unknown—that allow cancer cells to evade killing by components of the immune system. For example, in the micro-vessels of normal tissues (especially inflamed tissues), E-selectin is expressed on the luminal surfaces of endothelial cells and is used by circulating T-cells to tether them-selves to the walls of these vessels prior to extravasating into the surrounding tissue parenchyma. In contrast, in the microvessels in many tumors, the endothelial cells are induced to suppress expression of E-selectin. Some have proposed that in the absence of E-selectin, T cells may cruise through the tumor-associated vessels, never attaching to the vessel walls and therefore failing to extravasate and attack nearby carcinoma cells (Supplementary Sidebar 15.13).

15.16 Cancer cells become intrinsically resistant to various forms of killing used by the immune system

Cancer cells may also alter their own biochemistry to make themselves intrinsically less responsive to attacks launched by the immune system. One example of this gen-eral strategy was already mentioned above: some cancer cells become resistant to the FasL released by several kinds of cytotoxic immunocytes. They may make themselves relatively insensitive to this FasL by altering the signaling pathways downstream of the Fas death receptor.

A related defensive maneuver responds to the other major mechanism used by cyto-toxic T lymphocytes and NK cells to kill targeted cells including tumor cells. Recall that these immunocytes introduce a protease—a granzyme—into targeted cells (see Sections 9.14 and 15.3), which induces apoptosis in the latter by cleaving and activat-ing a caspase pro-enzyme. Cancer cells can escape this killing mechanism simply by increasing their levels of certain inhibitor-of-apoptosis proteins (IAPs), which oper-ate by sequestering and thereby inactivating key pro-apoptotic caspases (see Section 9.13).

These two strategies for avoiding killing are extensions of mechanisms that we encountered in Chapter 9, where acquisition of resistance to apoptosis was described as one of the hallmarks of cancer. Accordingly, immune-initiated cytotoxicity can now be added to the other physiologic stressors, including hypoxia, intracellular signaling imbalances, and loss of anchorage, that force cancer cells to disable their pro-apop-totic signaling pathways in order to survive during the course of multi-step tumor pro-gression.

Cancer patients often have significant levels of anti-tumor antibodies in their circula-tion, indicating that their tumor cells are likely carrying a coating of bound antibody molecules. As we learned from Figure 15.9, such cells are vulnerable to killing by the cohort of plasma proteins that together constitute complement. When participating in the process of complement-dependent cytotoxicity (CDC), complement molecules associate with antibody molecules bound to a cell surface and punch holes in the nearby plasma membrane, leading quickly to cell death. Normal cells protect themselves from inadvertent killing by complement by expressing on their plasma membranes one or several anti-complement proteins, called membrane-bound complement regulatory proteins (mCRPs). The most important of these are CD46, CD55, and CD59.

These mCRPs have been found to be overexpressed on the surfaces of a variety of human cancer cell types. Such overexpression affords the cancer cells a measure of protection from complement-dependent cytotoxicity (**Figure 15.33**). The repeated observations of elevated mCRP expression in a diverse array of cancer cell types suggest that the progenitors of these cancer cells were threatened, at some point in their development, with CDC, and that, in response, variant cancer cells were selected that could resist such killing through overexpression of one or another mCRP protein.

[Interestingly, herpesvirus saimiri, a virulent herpesvirus that causes lymphomas and leukemias in New World monkeys, expresses a protein closely related to human CD59 mCRP, which it apparently uses to protect virus-infected cells from rapid killing by host complement. Similarly, when progeny HIV particles force their way out of infected cells (see Figure 3.4C and D), they not only acquire a patch of host-cell plasma membrane, but also grab a significant number of associated cellular CD59 molecules, which appear to reduce their subsequent vulnerability to CDC triggered by antiviral antibodies.]

A number of other mechanisms that protect cancer cells against complement-dependent cytotoxicity have been described but remain poorly studied. An understanding of these mechanisms has become increasingly important in recent years, since it is evident that the resistance of certain tumor cells to killing by therapeutic monoclonal antibodies (MoAbs; see Supplementary Sidebar 11.1) often derives from their ability to blunt the complement-dependent cytotoxicity on which many MoAbs depend (**Sidebar 15.4**).

15.17 Cancer cells attract regulatory T cells to fend off attacks by other lymphocytes

Another immunoevasive strategy by cancer cells involves their attempts to modify the mix of immune cells around them. Regulatory T cells (T_{reg}'s)—a more recently characterized type of lymphocyte (see Section 15.6)—seem to play a major role in immunoevasion by cancer cells. Recall that a T_{reg} can directly inhibit and even kill both cytotoxic and helper T lymphocytes that recognize the same antigen as the one recognized by the T_{reg}. [In all cases, antigen recognition is achieved by the T-cell receptors (TCRs) that these various T lymphocytes display.]

In normal individuals, the T_{reg}'s represent only 5 to 10% of the population of $CD4^+$ lymphocytes, the remainder being helper T cells. In cancer patients, however, this number may increase to 25 to 30%. Moreover, T_{reg}'s have been found, often in large numbers, among the tumor-infiltrating lymphocytes (TILs) present in lung, ovarian, breast, and pancreatic carcinomas as well as in tumor ascites (**Figure 15.34A and B**). Together, these various observations suggest that T_{reg}'s play an important role in influencing anti-tumor immunity.

Tumors release the chemokine CCL22 in order to recruit T_{reg}'s; the latter display the cognate receptor, termed CCR4, on their surface. Once present within a tumor mass, the T_{reg}'s can suppress the actions of helper T cells that are instrumental in mobilizing both the humoral and cellular arms of the adaptive immune response, including cytotoxic T cells that are otherwise fully competent to attack and kill tumor cells (see Section 15.1). Hence, the ability to carry out this immunoevasive maneuver can be traced to the ability of tumor cells to produce and secrete CCL22 (see Figure 15.34A). In addition, TGF-β, which was portrayed earlier as being toxic for certain types of

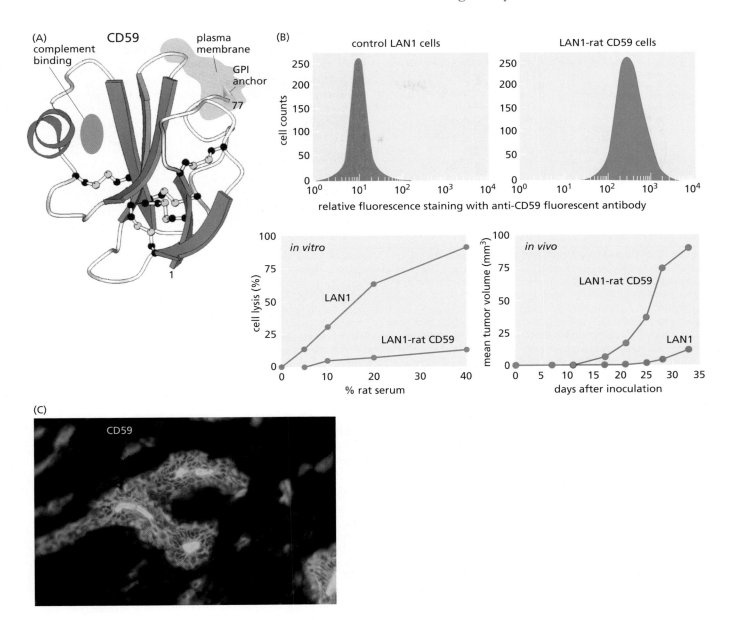

Figure 15.33 Regulation of the complement cascade by mCRPs The mCRPs (CD46, CD55, and CD59) are membrane-bound inhibitors of the complement cascade. They protect cells from complement-mediated cell lysis and are frequently overexpressed by human tumor cells. (A) CD59 is a small glycoprotein that is tethered to the plasma membrane of cells by a glycosylphosphatidylinositol (GPI) anchoring tail. CD59 binds certain components of complement, termed C8 and C9, and thereby prevents full assembly of the membrane attack complex (MAC), which mediates the cytolytic activity of complement. C8/C9 binding (*beige*) occurs in the cleft between the α helix and the β-pleated sheets. (Carbon and sulfur residues of cysteines are *black* and *yellow*, respectively, in this NMR-determined structure.) (B) Human LAN1 neuroblastoma cells, which are normally very sensitive to lysis by rat complement, were transfected with an expression vector that forced them to overexpress the rat CD59 mCRP protein. As seen in the fluorescence-activated cell sorting (FACS) analysis (*top*), this caused the normal LAN1 cells (*left*) to express 20- to 30-fold higher levels of CD59 (*right*). When these two cell populations were incubated *in vitro* in medium containing rat serum plus an antibody

that activates the complement present in this serum (*bottom left*), the cells overexpressing CD59 were protected from complement-mediated lysis (*blue curve*), in contrast to the unmanipulated LAN1 cells, which were highly sensitive to this lysis (*red curve*). When these two cell populations were implanted in immunocompromised Nude rats (*bottom right*), the LAN1 cells formed tumors only slowly (*red curve*), while those overexpressing CD59 formed tumors rapidly (*blue curve*). This indicated that complement-mediated killing usually impedes the growth of LAN1 tumors in these rats. (C) This immunofluorescence micrograph of a human ductal carcinoma of the breast shows intense expression of CD59 in the duct-forming carcinoma cells (*green*) with virtually no staining in the surrounding stromal tissue (*gray, black*); this CD59 is ostensibly deployed by the carcinoma cells in order to protect themselves from complement-mediated killing. (Because CD59 is loosely tethered via a GPI anchor to luminal surfaces of the ductal epithelial cells, it is released in large amounts into the lumina of the ducts, generating the intense yellow color.) (A, from C.M. Fletcher et al., *Structure* 2:185–199, 1994. B, from S. Chen et al., *Cancer Res.* 60:3013–3018, 2000. C, from J. Hakulinen and S. Meri, *Lab. Invest.* 71:820–827, 1994.)

(A)

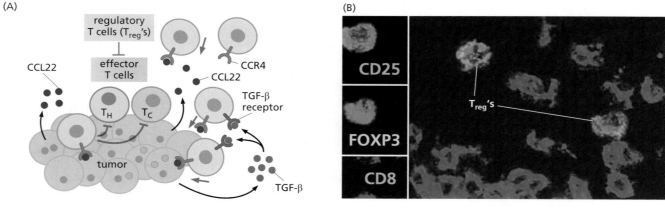

(B)

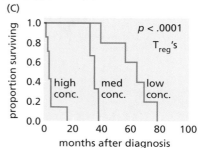

(C)

Figure 15.34 Regulatory T cells and tumor immunoevasion (A) Tumor cells (*pink*) release the CCL22 chemokine (*purple dots*), which binds the CCR4 chemokine receptor (*blue*) displayed by T$_{reg}$'s and in this way attracts them into the tumor. In addition, TGF-β released by tumor cells and other cell types appears to be critical to the expression of the FOXP3 transcription factor, a master regulator of T$_{reg}$ differentiation. Once they are recruited, the T$_{reg}$'s, which express the CD25 cell surface antigen, can inhibit two types of "effector" T cells—the CD4$^+$ helper T (T$_H$) lymphocytes (*light green*) and the CD8$^+$ cytotoxic (T$_C$) lymphocytes (*light orange*). Such actions can cripple major components of the host adaptive immune response to a tumor. (B) Immunofluorescence staining reveals the presence of T$_{reg}$'s through their expression of the CD25 surface antigen (*red*) and the FOXP3 transcription factor (*green*). They are seen here amid CD8$^+$ cytotoxic T cells (*blue*), whose actions they are ostensibly inhibiting; these lymphocytes were present in the ascites of a patient suffering from ovarian cancer. (C) Among advanced (stage IV) ovarian cancer patients, the concentration of tumor-infiltrating T$_{reg}$'s in tumor sections is a strong predictor of long-term survival, as indicated by this Kaplan–Meier plot. (A, from E.M. Shevach, *Nat. Med.* 10:900–901, 2004. B and C, from T. Curiel et al., *Nat. Med.* 10:942–949, 2004.)

Sidebar 15.4 Is complement-dependent cytotoxicity the key to the success of many monoclonal antibody therapies against cancer? Monoclonal antibodies (MoAbs) can mobilize the immune system to attack cancer cells via two well-characterized mechanisms, one involving complement-dependent cytotoxicity (CDC; see Figure 15.9) and the other, antibody-dependent cellular cytotoxicity (ADCC; see Figure 15.3B). (A third mechanism of killing, involving direct induction of apoptosis by cell surface–bound antibody, seems to operate in certain tumors but remains poorly understood.)

The key role of complement in enabling CDC has been highlighted in the case of rituximab, a MoAb that has shown substantial efficacy in treating hematopoietic malignancies of the B-cell lineage. Its use as a therapeutic agent has been widely adopted in oncology clinics, and its development is discussed in greater detail in Section 15.19. Rituximab recognizes the CD20 antigen, which is displayed on the surfaces of a variety of cell types of the B-lymphocyte lineage, including B-cell lymphomas and B-cell chronic lymphocytic leukemias (CLLs) and perhaps half of all non-Hodgkin's lymphomas (NHLs; described in Figure 15.39). Its therapeutic effectiveness in triggering CDC is nonetheless circumscribed, which has resulted in the development of alternative MoAbs that show increased efficacy against B-cell CLL. Ofatumumab, which has a stronger binding affinity for CD20, targets a membrane-proximal epitope of CD20 (**Figure 15.35A**) and is more potent than rituximab in inducing CDC in various B-cell malignancies. Importantly, the targeting by MoAbs of CD20 does not dictate that all therapeutic effects are confined to CDC. For example, a third antibody, obinutuzumab, because

it is engineered to have altered glycosylation in its hinge region (see Figure 15.3C), seems to have greatly enhanced affinity for Fc receptors (for example, FcγR3) and thereby triggers antibody-dependent cellular cytotoxicity (ADCC) attacks on B-cell lineage malignancies that overshadow its CDC effects. (It may also somehow kill cells directly by binding CD20.)

In all these cases, the potency of killing a variety of these malignancies is ultimately governed by the density of the CD20 antigen being expressed on the surface of lymphoma cells (see Figure 15.35B). Here, ofatumumab may prove especially useful, in that it is more capable than rituximab of killing lymphoma cells expressing relatively low levels of CD20 on their surface. In addition, some efforts have attempted to further potentiate MoAb-induced CDC by reducing the influence of powerful mCRP antagonists of CDC that are expressed by certain lymphoma cells (see Figure 15.35C).

Interestingly, carcinoma cells, in contrast to the hematopoietic tumor cells being targeted by rituximab, are widely believed to express elevated levels of multiple distinct cell surface CDC inhibitors, making attempts at reducing their anti-CDC defenses futile. This focuses attention on ADCC, the other, alternative mechanism of tumor immunotherapy, which depends on the ability of MoAbs to coat targeted cells by binding to cell surface antigens. Recall that such binding attracts cytotoxic cells, such as NK cells, that bind to the constant regions of these antibody molecules and proceed to lyse the cells (see Figure 15.3). In most cases, the relative contributions of these two alternative cytotoxic mechanisms to the efficacy of MoAb-based immunotherapies are unresolved.

immunocytes, is also thought to help induce other types of T cells to transdifferentiate into T$_{reg}$'s, both in the thymus and in peripheral tissues.

The existence of T$_{reg}$'s clearly complicates many conclusions drawn in recent years concerning the role of tumor-infiltrating lymphocytes (TILs) in tumor pathogenesis. Such lymphocytes have been widely assumed to be cytotoxic T cells that are actively involved in eliminating the cancer cells around them (see, for example, Figure 15.19). However, if T$_{reg}$'s constitute a significant proportion of the TILs within some tumors, then the significance of the total number of TILs within such tumors is unclear. This

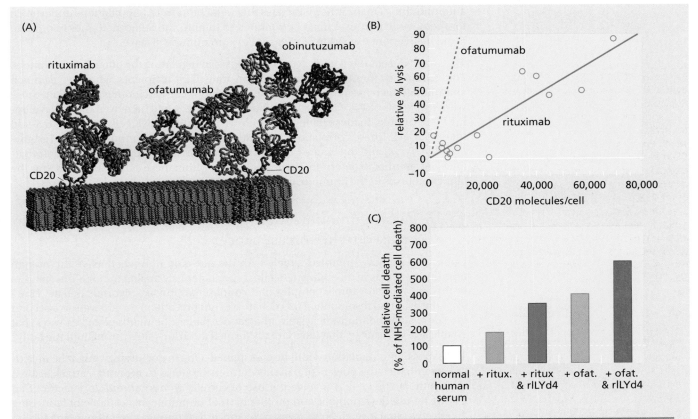

Figure 15.35 Complement-dependent cytotoxicity (CDC), antibody-dependent cell-mediated cell cytotoxicity (ADCC), and the treatment of CD20+ lymphomas The relative effectiveness of MoAbs in killing cancer cells via either CDC or ADCC is governed by remarkably subtle biochemical differences. (A) The CD20 protein, which is expressed by a variety of cells in the B-cell lineage (see Figure 15.39), is a tetraspanin—a family of proteins with four transmembrane domains plus two extracellular loops, that present epitopes for recognition and binding by monoclonal antibodies. The binding of MoAbs to these loops, as illustrated here, has quite distinct effects on cell killing. For example, some measurements indicate that both rituximab and ofatumumab have similar binding affinities to C1q, a critical component of the multiprotein complement complex. However, ofatumumab is able to cause more oligomerization of antibody–CD20 complexes in the plasma membrane of living cells, which in turn leads to far more efficient binding of hexameric C1q complement and more efficient cell killing. In contrast, obinutuzumab appears to kill lymphoma cells exclusively via ADCC and by a direct cytotoxic effect of antibody binding on lymphocytes. (B) Another important parameter governing the success of MoAb immunotherapy appears to be the levels of the CD20 antigen that are displayed by the various tumor types of the B-cell lineage, in this case by a collection of 14 freshly isolated leukemia samples (both prolymphocytic and chronic lymphocytic leukemias). As shown here, the susceptibility of these various cultured leukemic cells to lysis in the presence of rituximab plus complement varies dramatically, being influenced by the number of CD20 molecules that are displayed per cell by each of the tumor samples. (C) Ofatumumab (ofat.) is more effective than rituximab (ritux.) in killing lymphoma cells expressing relatively low levels of CD20. Nevertheless, its effectiveness is still compromised by the presence of the inhibitors of complement described in Figure 15.33. Here, the two MoAbs were applied to a series of 26 chronic lymphocytic leukemia (CLL) cells in culture, either in the presence of normal human serum (NHS), which contains complement proteins, or in the presence of NHS plus rILYd4, an inhibitor of the CD59 CDC antagonist. The responses of the 26 CLL cell populations to each of the five treatments were pooled here. These responses illustrate clearly that inhibition of CD59, which results in enhanced complement function, leads to marked increases in killing by the two MoAbs. (A, courtesy of C. Klein, Roche. B, from J. Golay et al., *Blood* 98:3383–3389, 2001. C, from X. Ge et al., *Clin. Cancer Res.* 17:6702–6711, 2011.)

means that assessments of the *relative proportions* of cytotoxic T cells and T_{reg}'s must be made in order to understand the real dynamics of a tumor's interactions with the host's cellular immune response. Observations such as those presented in Figure 15.34C suggest that the dysfunctional state of many tumor-associated cytotoxic T lymphocytes (T_C's) can be explained by the presence of many T_{reg}'s in their midst and that T_{reg}'s may therefore be critical determinants of whether or not the immune system can keep tumors under control.

Importantly, as mentioned earlier, therapeutic strategies that attempt to eliminate T_{reg}'s throughout the body lead to disastrous autoimmune attacks on a diverse list of tissues. On the one hand, this reveals the key role that these cells play in suppressing autoimmunity. On the other, it dictates that reductions in T_{reg} populations undertaken for therapeutic purposes must be confined to tumors and adjacent tissues in order to avoid inadvertent triggering of widespread autoimmune disease.

T_{reg}'s and NK cells often work at cross-purposes in regulating the outgrowth of tumors. For example, in immunosuppressed organ-transplant recipients, who have as much as a 200-fold increased risk of developing squamous cell carcinomas of the skin, those with *high* levels of T_{reg}'s in their circulation have a ~2.5-fold increased risk of developing these tumors. In these patients, *low* levels of circulating NK cells confer a ~5.5-fold increased risk. These correlations suggest that the adaptive immune system (as governed by T_{reg}'s) and the innate immune system (represented by NK cells) counterbalance one another in determining the fate of cells within these carcinomas and thus the clinical courses of the immunosuppressed patients.

15.18 Passive immunization with monoclonal antibodies can be used to kill breast cancer cells

Until now, we have grappled largely with the question of how effective the human immune system is in defending us against spontaneously arising tumors. The answers to this question are surely complex and continue to provoke vigorous debate. However, the eventual resolution of this debate will not pre-ordain the answers to a second question: Can the immune system of a cancer patient be manipulated in ways that enable it to kill cancers that have already formed at various sites throughout the body?

Two types of manipulation could be entertained. One major strategy might be to activate or enhance the powers of patients' immune systems to mount an attack against their tumors. This could involve the use of certain *immunostimulatory* factors that can incite the development and proliferation of immunocytes capable of launching an effective attack. Such enhancement of anti-tumor immune function might also be achieved by exposing patients to TATAs or TSTAs displayed by their tumors, in effect immunizing the patient against the tumor in a way that is analogous to vaccination against viral or bacterial infection.

The alternative therapeutic strategy involves various forms of **passive immunization**. Use of this class of strategies implies that a patient's own immune system is incapable of mounting an effective immune defense, even after immunostimulatory therapies are applied, and involves supplying the patient with immune products (for example, antibodies) or antigen-specific cells originating in another organism's immune system. (Immunologists reserve the term "passive immunization" for procedures involving the introduction of antibodies into a patient, but we will use the term more broadly here.) When cells are supplied from the immune system of another individual, this procedure is often termed **adoptive transfer**. We first describe various types of passive immunization and will then return to the immunostimulatory strategies in Section 15.21.

We have already encountered in passing, one form of passive immunization—rituximab—and its use against malignancies of the B-lymphocyte lineage (see Sidebar 15.4). However, by far the best known of the passive immunization treatments involves the monoclonal antibody termed Herceptin, also called trastuzumab by its developers. Herceptin derives from a mouse monoclonal antibody that reacts strongly with the EGF receptor–related protein that is called variously HER2, erbB2, or Neu (see Sections

4.3 and 5.6; also see Table 5.2); as we read earlier, this receptor is overexpressed in as many as 30% of the breast cancers diagnosed in the West.

While the HER2 protein is not itself a tumor-specific antigen, its display at abnormally high levels—often 10 to 100 times above normal—may create a target cell that is preferentially affected by Herceptin. Such **selectivity**—preferential killing of cancer cells—derives directly from differences between normal and cancer cells, in this case the overexpression of HER2 by breast cancer cells. This overexpression, which is often due to amplification of the receptor-encoding gene, generally represents a poor prognosis (see Figure 4.4B).

In order to make Herceptin, a mouse monoclonal antibody (MoAb) against HER2 was initially produced. However, like all murine antibodies, it could not be used directly for anti-tumor therapy in humans, simply because its constant region (see Figure 15.1), being of mouse origin, constitutes a potent antigen on its own and therefore provokes the human immune system to produce high levels of human antibodies that bind and neutralize the mouse antibody molecules. Moreover, these human anti-mouse antibodies (HAMAs) occasionally induce anaphylactic shock in a patient re-treated with a mouse antibody.

Consequently, the cDNA encoding the mouse anti-HER2 antibody was re-engineered so that the constant (C) regions of the encoded antibody molecule were composed of human rather than mouse sequences; as hoped, the resulting **humanized** anti-HER2 MoAb (**Figure 15.36A**) was usually not immunogenic. Importantly, following its injection into patients, the Herceptin molecules created by this humanization were found to remain in the circulation at functionally significant levels for as long as a month, indicating the absence of a significant anti-Herceptin host immune response and a potential for long-term therapeutic effects.

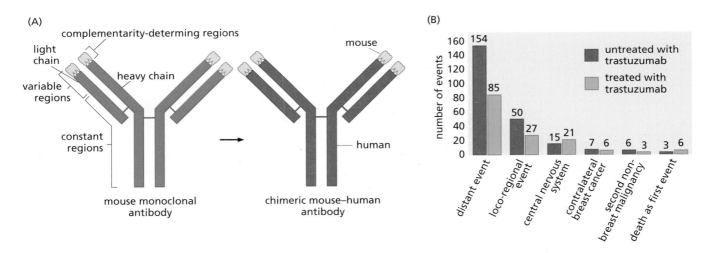

Figure 15.36 Humanization of monoclonal antibodies and the case of Herceptin/trastuzumab (A) The monoclonal antibodies (MoAbs) that are produced in mice often cannot be used therapeutically in humans over extended periods of time because the constant regions of the heavy and light chains of the mouse antibody (*left figure*) are themselves immunogenic in humans: they provoke an anti-MoAb serum response that can neutralize the actions of the introduced MoAb in patients. To deal with this, the cDNAs encoding the heavy and light chains of the mouse MoAb (*blue*) are cloned, and the DNA segments encoding the antigen-combining domains of the mouse MoAb, which are termed the complementarity-determining regions (CDRs) and are located within the variable regions of the heavy and light chains, are fused with the genes encoding the constant regions of the human antibody and most of its variable regions (*red*), yielding a chimeric antibody that is not immunogenic or, at worst, weakly so. This antibody is considered to be *humanized*. (B) The benefits of trastuzumab treatment of breast cancer patients have been observed at multiple levels. This histogram reveals observations in the two years following surgical removal of the primary tumor in a group of such patients. Subsequent studies revealed that certain classes of HER2-amplified breast cancer patients treated with Herceptin after surgical removal of their primary tumor experienced as much as a 50% decline in relapse rates. "Distant events" refers to all types of metastases. Untreated, patients treated with paclitaxel alone; treated, patients treated with paclitaxel concurrently with trastuzumab. (A, adapted from R.A. Goldsby et al., Immunology, 5th ed. New York: Freeman, 2002. B, from J. Baselga et al., *Oncologist* 11(Suppl 1):4–12, 2006.)

Use of the Herceptin antibody has resulted in extension of the life span of breast cancer patients whose tumors overexpress HER2 protein. Herceptin is rarely used on its own, but instead is applied in combination with established chemotherapeutic treatments of breast cancer. In one large clinical study, the addition of Herceptin to standard chemotherapy treatment of women with advanced breast cancer resulted in a longer time to disease progression (7.4 versus 4.6 months with chemotherapy alone), a lower rate of death at 1 year (22 versus 33%), and a longer overall survival (25 versus 20 months).

Even more impressive responses were announced in 2005: women with **operable**, early-stage HER2-overexpressing breast tumors were treated postoperatively either with chemotherapy alone or with chemotherapy plus Herceptin; most of these women carried micrometastases in draining lymph nodes (see Figure 14.42). After two years, only 15% of the women who received the double treatment had **relapsed**, while 33% of the women who had received chemotherapy alone showed disease relapse (see Figure 15.36B). It is plausible, though unproven, that much of Herceptin's therapeutic benefit observed in this study derived from its ability to block the proliferation of residual cancer cells that were left behind after removal of the primary tumors; in the absence of Herceptin, such cells might be responsive to mitogens produced by the wound-healing process that followed surgery (see Sidebar 13.2). In 2010, both European and American regulatory agencies also approved the use of Herceptin for treatment of the ~20% of gastric carcinomas that overexpress HER2.

The precise mechanisms by which Herceptin antibodies block the proliferation or actively kill HER2-overexpressing breast cancer cells are still being resolved, two decades after this MoAb was first developed. One important mechanism of cancer cell destruction depends on the Fcγ receptors displayed on the surface of a variety of cytotoxic and phagocytic cells, including, most importantly, NK cells and macrophages (see Figure 15.3). To review, these Fcγ receptors bind the constant regions of immunoglobulin γ (IgG) antibody molecules that may be coating the surface of other cells, such as Herceptin-treated breast cancer cells. Such antibody coating informs the Fcγ receptor–expressing cytotoxic cell of the presence of a cell that should be eliminated. Accordingly, NK cells and macrophages become tethered to IgG antibody–coated cancer cells via their Fcγ receptors and proceed to kill the cancer cells—the process termed antibody-dependent cellular cytotoxicity (ADCC).

Direct evidence for ADCC being a key mechanism of Herceptin-initiated cell killing is provided by immunocompromised mice whose germ lines have been modified by deletion of the gene specifying a critical Fcγ receptor. Such mice have a greatly reduced ability for Herceptin-dependent killing of engrafted human breast cancer cells. In addition, the state of glycosylation of the heavy-chain constant region of trastuzumab is also a strong determinant of the potency of ADCC: versions of the antibody molecule that lack the fucose residues usually attached to the carbohydrate side chains (see Figure 15.3C) are bound far more avidly (~50-fold) by the FcγRIIIA receptor of the NK cells and show significantly greater ADCC in pre-clinical models.

Clinical observations provide further support for the key role of NK cells and their Fc receptors in the killing of cancer cells. For example, the efficacy of Herceptin/trastuzumab therapy seems to be compromised in certain cancer patients, simply because they lack normal levels of the NK cells. Moreover, tumors that are more responsive to trastuzumab therapy in the clinic show increased infiltration of NK cells. And finally, treatment of breast cancer patients with Herceptin has been found in a clinical trial to be more effective than treatment with lapatinib, a low–molecular-weight tyrosine kinase inhibitor that targets the TK domain of HER2 (discussed in Chapter 16); this suggests that simple shutdown of signaling by the receptor does not yield the same benefits as does such shutdown working together with the ADCC. These diverse observations converge on the notion that the successes of Herceptin therapy depend significantly on NK cells and their Fcγ receptors to kill tumor cells. Moreover, this killing can be enhanced by further re-engineering of the Herceptin molecule (Supplementary Sidebar 15.14).

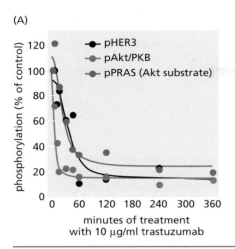

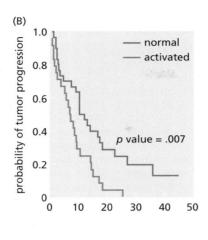

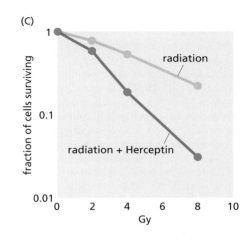

Figure 15.37 Biochemical mechanisms of Herceptin action The PI3K–Akt signaling pathway exhibits rapid responses to Herceptin binding to the HER2 protein. Its activation by other, non-HER2-dependent mechanisms also governs the responsiveness of patients' tumors to Herceptin therapy. (A) A study of the biochemical effects of Herceptin action reveals that treatment of SKBR-3 human breast cancer cells elicits at least three rapid changes in signaling. (1) Transphosphorylation of the HER3 receptor by the HER2 tyrosine kinase (the two RTKs being assembled as heterodimers) plummets within an hour. (2) The phosphorylated form of Akt, which which releases potent anti-apoptotic signals (see Sections 9.7 and 9.13), decreases even more rapidly. This presumably occurs via the collapse of PI3K signaling, which must occur even more rapidly following Herceptin binding. (3) The phosphorylation of PRAS, a substrate of phosphorylation by Akt, also decreases within an hour. Together, these responses indicate that the anti-apoptotic benefits of HER2 signaling are rapidly lost in response to Herceptin binding. (B) Use of an RNA interference screen revealed that the knockdown of PTEN expression represented a major mechanism by which HER2-overexpressing cells growing in culture were able to evade killing by Herceptin; this suggested that PI3K pathway signaling was critical to the observed resistance to therapy. Indeed, a detailed examination of the role of this signaling pathway (see Figure 6.19A) revealed that patients whose tumors expressed constitutively activated PI3K (*red line*) developed resistance to Herceptin and suffered tumor progression far more rapidly and had decreased overall survival relative to those whose tumors exhibited normal PI3K pathway signaling (*blue line*). (C) The collaborative effects of radiation (dosage in grays, Gy) with Herceptin are shown here: cells treated with the antibody have considerably more radiation sensitivity than do untreated cells, ostensibly because they have lost most of the protection against apoptosis normally afforded by active Akt/PKB signaling. (Note that the ordinate here is logarithmic.) (A, from T.T. Junttila et al., *Cancer Cell* 15:429–440, 2009. B, from K. Berns et al., *Cancer Cell* 12:305–402, 2007. C, from K. Liang et al., *Mol. Cancer Ther.* 2:1113–1120, 2003.)

Yet other mechanisms operating at the biochemical level within breast cancer cells clearly contribute to Herceptin-induced killing. For example, when cell surface proteins are exposed to reactive antibodies, these proteins are often internalized and degraded. The same process leads to the Herceptin-induced decrease of cell surface HER2 displayed by breast cancer cells. In addition, Herceptin may block the ability of HER2 to form heterodimers with HER3, its main signaling partner. These changes rapidly deprive the breast cancer cells of the high levels of HER2/HER3-activated PI3 kinase and thus the Akt/PKB signaling that has protected them from apoptosis. This may explain why Herceptin-treated cells become far more vulnerable to radiation- and chemotherapy-induced killing (**Figure 15.37**).

It is also clear that in many breast tumors, the ectodomain of the HER2 protein, once it arrives at the cell surface, is cleaved away by extracellular proteases. The residual protein, which contains the transmembrane and the cytoplasmic tyrosine kinase domains of HER2, exhibits constitutively activated kinase function and is therefore a very potent oncoprotein (see, for example, Figure 5.11A). The post-translational cleavage of HER2 that produces this deregulated receptor protein is known to be blocked by Herceptin binding. Yet other anti-receptor MoAbs have been developed in recent years, each of which appears to affect cancer cells in a particular way (**Sidebar 15.5**).

15.19 Passive immunization with antibody can also be used to treat B-cell tumors

As we read earlier (see Sidebar 15.4), MoAbs can be used with great success to treat hematopoietic malignancies as well. The monoclonal antibody called variously

Sidebar 15.5 Herceptin is only the first of many clinically useful anti-receptor antibodies The EGF receptor (EGF-R; HER1) appears to be overexpressed in about one-third of all human carcinomas and in almost half of glioblastomas as well. As is the case with HER2, its overexpression is likely to play a key role in providing mitogenic and anti-apoptotic signals to tumor cells. Moreover, since HER2 and the EGF-R heterodimerize, the latter may play a key role in driving the proliferation of HER2-overexpressing breast cancer cells.

For various historical reasons, the development of anti-EGF-R monoclonal antibodies, notably Erbitux (also called cetuximab), lagged behind that of Herceptin by about three years. Its clinical use for treating advanced colorectal cancers was approved by the FDA in 2004. In the long run, antibodies such as Erbitux may have greater utility than Herceptin, if only because the EGF-R seems to participate in the pathogenesis of a far larger number of tumors than does HER2.

As indicated in **Figure 15.38**, Erbitux functions quite differently from Herceptin, in that it appears to inhibit receptor activation by blocking ligand binding, including binding by EGF, TGF-α, amphiregulin, and other ligands of the EGF-R. A third antibody, called Omnitarg or pertuzumab, acts on HER2 by blocking its ability to heterodimerize with other members of the EGF-R family. In 2011, a clinical trial reported that patients who bore previously untreated, metastatic HER2-positive breast cancers experienced a 6.1-month progression-free survival (PFS) with chemotherapy alone, a 12.4 month PFS with chemotherapy plus Herceptin/trastuzumab, and an 18.5-month PFS with the combination of chemotherapy, trastuzumab, and pertuzumab.

These various anti-receptor antibodies are more specific than low–molecular-weight drugs in shutting down receptor signaling, in that they are unlikely to directly affect unrelated receptor tyrosine kinases (RTKs). However, they may not be able to penetrate into the interstices of tumors as effectively as low–molecular-weight RTK antagonists (discussed in Chapter 16). More important, many human tumors express truncated versions of the EGF-R that lack the ectodomain and fire constitutively (in a ligand-independent fashion, see Figure 5.11A); these proteins have lost the antigenic determinants that are recognized by anti-receptor monoclonal antibodies and therefore elude inhibition by these antibodies.

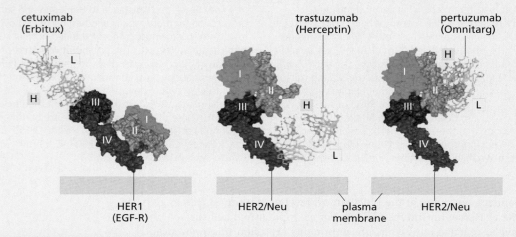

cetuximab (Erbitux)   trastuzumab (Herceptin)   pertuzumab (Omnitarg)

HER1 (EGF-R)   HER2/Neu   plasma membrane   HER2/Neu

Figure 15.38 Comparison of MoAbs that interfere with the function of RTKs displayed by tumors A variety of monoclonal antibodies (MoAbs) have been produced against cell surface RTKs with the intent of treating tumors that exhibit high-level expression of these receptors. Pictured here are the heavy-chain (*light blue*) and light-chain (*yellow*) antigen-binding domains of MoAbs directed against the ectodomains of the EGF-R (HER1; *left*) and its close cousin HER2/Neu (*middle, right*). The specific binding sites of the various anti-receptor monoclonal antibodies have been mapped by X-ray crystallography. When the three distinct antibody–receptor complexes are viewed from the side, it becomes clear that the antibodies act on the receptors in very different ways. Erbitux binds to domain III of the EGF-R/HER1 ectodomain and thereby obstructs the binding site of the EGF-R ligands (compare with Figure 5.15C). In contrast, Herceptin binds to the HER2/Neu receptor in a way that allows the latter to continue to homo- or heterodimerize with other members of the HER family of receptors (HER 1, 3, and 4 with a preference for HER3), while a third antibody, termed Omnitarg or pertuzumab (*right*), binds to the rightward-pointing finger of the HER2 domain II ectodomain that is critical for such dimerization, thereby blocking the ability of HER2 to heterodimerize with the other related receptors. (Like Herceptin, Omnitarg can trigger ADCC.) (From S.R. Hubbard, *Cancer Cell* 7:287–288, 2005.)

Rituxan or rituximab was developed to bind CD20, which is displayed by cells at various stages of B-cell differentiation and in B-cell–derived tumors (**Figure 15.39**). Unlike Herceptin, rituximab is a **chimeric** MoAb that carries the Fc fragment portion of the human IgG1 and the remaining portion from a mouse MoAb; once again, this modification was undertaken in order to reduce the likelihood that treated patients would develop an immunological reaction against the MoAb itself, thereby neutralizing its activity. CD20 is an especially useful antigen to target since (1) it is expressed at very high levels on the surface of targeted cells, and (2) unlike many other cell surface

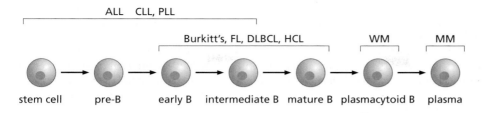

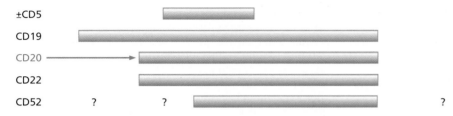

antigens expressed by cells of the B-cell lineage

Figure 15.39 Expression of the CD20 antigen by cells in the B-cell lineage The utility of the Rituxan/rituximab monoclonal antibody derives in part from the fact that the CD20 antigen that it recognizes is displayed as a cell surface transmembrane protein by B cells at various stages of differentiation, beginning with early B cells and continuing through the formation of plasmacytoid B cells, the immediate precursors of the antibody-secreting plasma cells. The *green bars* indicate the stages of B-cell differentiation that express the indicated antigen. As is indicated here, a variety of tumors of the B-cell lineage—including Burkitt's lymphoma, follicular lymphoma (FL), diffuse large B-cell lymphoma (DLBCL), hairy-cell leukemia (HCL), and Waldenström's macroglobulinemia (WM)—which phenocopy the traits of B cells at various discrete stages of differentiation, also express the CD20 antigen, making Rituxan a highly useful agent for treating these various lymphomas and leukemias. Significantly, most multiple myelomas (MMs) no longer express the CD20 antigen, because they mimic the gene expression pattern of highly differentiated plasma cells; this explains why rituximab is ineffective in treating these tumors. The fact that several other cell surface antigens (e.g., CD19) have equally broad if not broader expression suggests that antibodies against these other antigens may one day also prove therapeutically useful. (Courtesy of R. Levy.)

antigens, it is not internalized following antibody binding. Rituxan has proven to be useful for treating, among other tumors, many non-Hodgkin's lymphomas (NHLs), which constitute the fifth and sixth most common cause of cancer-related deaths in American males and females, respectively. (More than 90% of B-cell NHLs express the CD20 antigen.)

Weekly doses of Rituxan have been effective in treating about half of all patients with *relapsed* follicular non-Hodgkin's lymphoma (that is, tumors that have regrown after earlier, initially successful treatments) or **refractory** cases of these tumors (that is, tumors that have become unresponsive to other treatment regimens). Interestingly, the populations of normal B cells in treated patients are also eliminated by this MoAb treatment and rebound only after 6 to 9 months. Nonetheless, this loss of normal B-cell function and the associated immunosuppression, together with the other, more transient consequences of rituximab treatment, represent acceptable, relatively minor side effects.

By 2012, Rituxan had been used to treat several million patients worldwide afflicted with various malignancies of the B-cell lineage. In many cases, it was found to be an extremely useful adjunct to existing treatment. For example, addition of Rituxan to the standard chemotherapeutic treatment of diffuse large B-cell lymphoma (DLBCL)—a protocol employing a cocktail of four drugs termed CHOP (see Table 16.4)—resulted in a 41% decrease in the risk of disease progression or mortality. A similar modification of the CHOP protocol for treating follicular lymphoma led to a 66% reduction in treatment failure.

While this monoclonal antibody treatment has been found capable of stabilizing disease and thereby prolonging survival, it has not been curative, and virtually all patients relapse within several years. The tumors in some of the relapsed patients may respond to a second round of Rituxan treatment, while others may have developed resistance to Rituxan. To date, the precise mechanisms of acquired resistance are not well understood; among the mechanisms proposed are alterations in CD20 expression, elevated resistance to apoptosis, loss of complement activity (see Sidebar 15.4), and loss of immune cells capable of mediating cytotoxicity. Provocatively, the genetic background of the patients also has a strong influence on the success of their Rituxan treatment (**Sidebar 15.6**).

Rituxan and Herceptin—both widely used in the clinic—represent the vanguard of a far larger group of therapeutic antibodies that are in various stages of research and development. In general, when Herceptin or Rituxan is used on its own, it succeeds in extending the life span of patients only by several months. This fact, together with rapidly accumulating clinical evidence, dictates that monoclonal antibody therapies like these will be most effective when used in conjunction with other anti-cancer therapies with which they may be highly synergistic in inducing durable remissions and, quite

Sidebar 15.6 Genetic and nongenetic factors are strong determinants of the effectiveness of Rituxan treatment As discussed earlier, the antibody-dependent cellular cytotoxicity (ADCC)-mediated killing of targeted cancer cells depends on the ability of cytotoxic cells, such as NK cells, to use their Fcγ receptors to attach to cancer cells whose surfaces have been coated with immunoglobulin γ (IgG antibody) molecules (see Figure 15.3). An important Fcγ receptor of human NK cells, termed FcγRIIIa, is found in two polymorphic variant forms, which carry either a valine (V) or a phenylalanine (F) at amino acid residue position 158. The V variant FcγRIIIa shows a significantly higher affinity for binding human IgG molecules than does the F variant, and indeed, cytotoxic cells bearing the V variant FcγRIIIa are far more effective in ADCC than are those bearing the F variant when observing ADCC responses in cultured cells.

A patient's clinical responses to Rituxan treatment are also dramatically affected by the variant of Fcγ receptor displayed by his or her NK cells (**Figure 15.40A**). Thus, B-cell lymphoma patients who were homozygous for the V allele (V/V genotype) and were treated with rituximab showed a median time to progression (time between initial treatment and recurrence of disease) of 534 days, while those who were homozygous or heterozygous for the F allele (V/F or F/F genotype) had a time to progression of 170 days. (The variant forms of FcγRIIIa have been found to influence clinical responses to Herceptin as well.) Polymorphic variants of a second Fcγ receptor, called FcγRIIa, also have a strong effect on therapeutic response. Hence, in addition to the influences of tumor cell phenotype, the outcomes of immunotherapy are often strongly affected by the genetic constitutions of the patients, in this case a gene that determines the structure of an Fc receptor.

Yet other factors predict whether or how patients will respond to Rituxan therapy. For example, those patients who maintain higher serum levels of Rituxan for longer periods are more likely to show favorable clinical response (Figure 15.40B). This highlights a key constraint of monoclonal antibody–based treatments and an unsolved issue in the development of immunotherapeutics: we still do not understand the biological mechanisms that determine the lifetimes of monoclonal antibodies in the circulation.

(A)

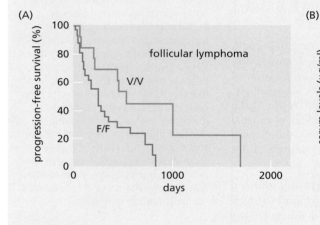

(B)

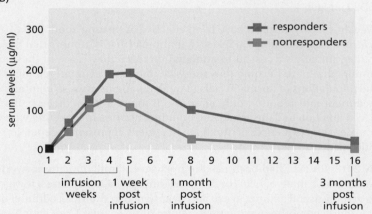

Figure 15.40 Rituximab and the treatment of B-cell tumors (A) This Kaplan–Meier graph plots the response of patients suffering from follicular lymphoma to rituximab treatment. Their clinical progression was strongly dependent on the alleles of the gene encoding the FcγRIIIa protein, the receptor used by NK cells to bind the Fc portion of IgG1 molecules that are already bound to CD20 antigen on the surface of lymphoma cells. The valine (V)-containing version of the FcγRIIIa protein binds IgG1 far more avidly than the phenylalanine (F) version, which appears to explain why V/V homozygotes showed much greater progression-free survival than their F/F counterparts. (B) The persistence of a therapeutic agent in the circulation—its pharmacokinetics (PK)— often determines its therapeutic effects. As shown here, Rituxan was injected into a group of leukemia and lymphoma patients at regular intervals over a period of four weeks and the serum levels of Rituxan were followed for three months beyond infusion. The PK of Rituxan for those who showed clear clinical responses to this monoclonal antibody treatment differed markedly from the PK for those who did not. This suggests that some type of "antigen sink" exists in certain patients that absorbs available Rituxan, thereby accelerating its disappearance from the circulation and compromising its therapeutic effectiveness. (A, adapted from W.-K. Weng and R. Levy, *J. Clin. Oncol.* 21:3940–3947, 2003. B, data from N.L. Berinstein et al., *Ann. Oncol.* 9:995–1001, 1998.)

possibly in the future, cures. (Of additional interest, rituximab has shown efficacy in treating two autoimmune diseases, which are driven in part by B-cell function—rheumatoid arthritis and systemic lupus erythematosus. Hence, depletion by this MoAb of non-neoplastic B cells also confers clinical benefit in non-neoplastic diseases.)

Research into anti-tumor monoclonal antibodies has also taken another direction: many experiments have explored the possibility of enhancing the cytotoxic effects of these antibody molecules. Until now, we have discussed the fact that an antibody molecule may attach to the surface of a tumor cell and achieve subsequent cell killing through the actions of complement or cytotoxic cells bearing Fc receptors. An attractive alternative is to link antibody molecules to toxic agents, thereby creating **immunotoxins** that are guided like "smart bombs" to the tumor, where they present the toxin in high concentrations to the targeted cells. For example, antibody molecules can be linked *in vitro* to highly toxic biological substances, such as the ricin A chain or a toxin made by *Pseudomonas* bacteria, and used to convey these toxins to the tumor.

A second approach links radioactive molecules to the antibody molecules prior to injecting them into cancer patients. The hope here is that radioactive decay will kill nearby cancer cells. A variation of this involves conjugating highly potent chemotherapeutic drug molecules, such as Adriamycin, to antibody molecules.

In a third strategy, specific enzymes are linked to antibody molecules; these enzymes are capable of converting nontoxic **pro-drugs** into actively toxic drugs. Once the antibody and linked activating enzyme are concentrated in the tumor mass, the pro-drug can be injected into the patient and become activated by the enzyme in the vicinity of the tumor. The advantage of this approach derives from the fact that the enzyme can generate hundreds, possibly thousands of toxic drug molecules near targeted cancer cells, thereby amplifying the toxic effects of a single bound monoclonal antibody molecule. These various uses of monoclonal antibody molecules as tumor-targeting vectors are the subjects of active, ongoing research and development.

15.20 Transfer of foreign immunocytes can lead to cures of certain hematopoietic malignancies

A quite different kind of passive immunization involves bone marrow transplantation (BMT). The original rationale for this treatment came from the discovery that the entire hematopoietic system, including therefore the immune system, of a mouse or human can be **ablated** (eliminated) through drug treatments and X-irradiation—the procedure of **myeloablation**. The subsequently introduced donor marrow graft, because it contains hematopoietic stem cells (HSCs), can repopulate the recipient's bone marrow and regenerate all of the cell lineages required for normal hematopoiesis and immune function (Supplementary Sidebar 15.15).

In the case of many hematopoietic malignancies, specifically lymphomas and leukemias, the original intent of undertaking BMTs was to rid the body of the neoplastic stem cells (see Section 11.6) that were present throughout the body, most importantly in the marrow itself. According to this thinking, BMT would prevent the tumor from ever regenerating itself because the tumor stem cells would be eliminated from patients' marrow.

When this BMT treatment strategy was first employed, the most effective bone marrow donor was thought to be one whose histocompatibility antigens closely matched those of the marrow recipient. This would allow the most effective re-population of the bone marrow and minimize the likelihood of attack on a recipient's tissues by immune cells arising from the graft. However, with the passage of time, it became apparent that a minimum level of histo*incompatibility* between donor and recipient is actually desirable. For example, early BMT trials indicated that transplants between identical twins led to relapse rates that were several times higher than those in which there was some degree of mismatching of histocompatibility antigens. This led to the current widespread use of allogeneic stem cell transplantation (ASCT) and the use of donor-recipient pairs who differ in *minor* histocompatibility antigens (rather than the *major* histocompatibility antigens described in Sections 15.3 and 15.7).

As it turned out, much of the therapeutic effect of bone marrow transplantations in treating hematopoietic malignancies derives from the graft-versus-tumor (GVT) response, in which donor immunocytes identify and attack residual tumor cells—those that have survived the radiation and chemotherapy used to ablate a patient's bone marrow. This attack, which requires the presence of T lymphocytes among the implanted donor cells, is presumably provoked because the recipient's tumor cells express antigens that are unfamiliar to the engrafted donor immune cells. At present this GVT response is the only truly effective mechanism for achieving durable remissions and occasional cures of Gleevec-resistant chronic myelogenous leukemia (CML).

For reasons that remain unclear, the GVT reaction is usually not accompanied by an attack of comparable severity on the transplant recipient's normal tissues—the condition termed **graft versus host disease** (GVHD). Severe GVHD, when it occurs, leads to widespread inflammation and the destruction of a variety of the recipient's normal tissues by the engrafted immune cells with a rapidly fatal outcome. Ongoing research is examining how GVHD can be minimized, in part by altering the composition of allogeneic donor cells that are given to transplant recipients.

Here, once again, we see an instance where the inability of a cancer patient's own immune system to mount an effective anti-tumor response is addressed by introducing the products of a foreign immune system, in this case immunocytes that are often capable of mounting a potent attack on the patient's cancer cells. Unfortunately, for multiple reasons, bone marrow transplantation has not proven to be an effective strategy for treating patients with solid (rather than hematopoietic) tumors.

15.21 Patients' immune systems can be mobilized to attack their tumors

The other major class of cancer immunotherapies depends on mobilizing and enhancing the endogenous immune defenses of cancer patients. Implicit in these approaches is the notion that their immune systems are intrinsically capable of attacking and eliminating tumors and that aggressive anti-tumor immune responses can be elicited by increasing the numbers and activity of various cytotoxic immune cells.

[These strategies are often said to constitute novel types of anti-cancer "vaccines." The term is unfortunate, because vaccines have traditionally represented substances that act to *prevent* disease rather than to *treat* existing disease. This explains why many hearing this term believe that the immunotherapies under development hold the promise of eventually serving as preventives—a goal that is far from the minds of those who are currently developing anti-tumor immunotherapies. The only exception to date is the anti-HPV vaccine termed Gardasil (see Supplementary Sidebar 15.5), which successfully prevents cervical carcinomas by blocking viral replication rather than immunizing patients against the cancer cells themselves.]

The ongoing, intensive efforts at developing new immunotherapeutic protocols depend heavily on our rapidly increasing insights into immune function at the molecular and cellular level. Many of these projects exploit our knowledge of the signaling molecules that are used naturally by the immune system to regulate its various arms. Curiously, one highly successful form of immune mobilization has been used for many years to successfully treat early-stage bladder carcinomas, without a clear understanding of how it works (**Sidebar 15.7**).

An important strategy for mobilizing an anti-tumor response depends on activating the dendritic cells (see Section 15.2). Recall that these antigen-presenting cells (APCs) are normally charged with the task of ingesting infectious agents and other antigen-bearing particles (including tumor cells) throughout the body's tissues and then rushing back to nearby draining lymph nodes, where they use their MHC class II molecules to present oligopeptide fragments of the consumed material to helper T cells (see

Sidebar 15.7 Bacteria can be used to treat bladder cancer The use of bacteria to treat tumors reaches back to the end of the nineteenth century, when William Coley, a New York surgeon, noticed that cancers often regressed in patients who had experienced (and recovered from) bacterial skin infections. This led to a many-decades-long research program that undertook to use live or killed bacteria as agents for inciting an immune response, against both themselves and any tumor cells that happened to reside in the body. These attempts have fallen by the wayside with one exception, involving Bacillus Calmette–Guérin (BCG), an *attenuated* (weakened) strain of mycobacterium that was originally developed as a vaccine against tuberculosis.

Live BCG is frequently injected into the bladders of patients suffering from early-stage bladder carcinomas. It is usually effective in halting or delaying the progression of these tumors, doing so through a treatment that is far less traumatic than surgery or radiation (**Figure 15.41**). The BCG treatment clearly functions through its ability to attract a variety of immunocytes—including CD4$^+$ T$_H$ and CD8$^+$ T$_C$ lymphocytes as well as macrophages and NK cells—to the bladder, where these cells create localized inflammatory responses. However, the precise mechanisms of BCG anti-tumor action remain elusive.

Figure 15.8). Dendritic cells are known to be functionally activated by exposure to the growth factor termed GM-CSF (granulocyte–macrophage colony-stimulating factor).

One DC-based strategy (Figure 15.42) has proven to be so successful that it has won approval by the U.S. Food and Drug Administration for clinical treatment of prostate cancer after extending from 21.7 months to 25.8 months the median survival times of men with asymptomatic or minimally symptomatic metastatic prostate cancer. (These men had all failed androgen deprivation therapy, which is able to hold the great majority of prostate carcinomas in check for many months; however, by two years' time, almost half of the patients experience tumor progression.)

The strategy for this new immunotherapeutic—termed either Provenge or sipuleucel-T—depends on extracting mononuclear cells by the procedure of **leukapheresis** from a patient and culturing them *ex vivo* for 36 to 44 hours in the presence of a protein formed from prostate acid phosphatase (PAP), which is expressed almost universally in prostate carcinomas (see Figure 15.42A). The intent is to load PAP-derived antigenic peptides on the MHC class II proteins of dendritic cells (DCs) extracted from the patient.

To facilitate this, a recombinant form of PAP is fused at its C-terminus to the N-terminus of GM-CSF (granulocyte–macrophage colony-stimulating factor), resulting in the fusion protein termed PA2024. The growth factor portion of PA2024 is intended to serve two functions. On the one hand, its GM-CSF moiety should stimulate monocytes to differentiate into immature dendritic cells (DCs), which become competent to process and present antigens (see Figures 15.5 to 15.8). On the other, the physical linkage of PA2024 via its GM-CSF portion to its cognate receptor (GM-CSF-R) expressed on the surface of the monocytes should be followed by internalization of the ligand–receptor complex, proteolysis of the PA2024 fusion protein, and presentation of resulting peptides on the surface of DCs (see Figure 15.42B). While not yet demonstrated directly, ideally the oligopeptide fragments derived from the PAP portion of PA2024 should be presented by the MHC class II molecules of the dendritic cells to CD4+ T$_H$ cells (see Figure 15.42C). The end result of the process might then be a widespread engagement of innate and adaptive immune function, including CD8 T-cell–mediated cytotoxic lymphocyte (CTL) activity and even B-cell–mediated antibody responses.

In various preclinical trials involving PAP and other antigens, the loading of such antigens on monocytes and DCs (see Figure 15.42B) was found to depend, as anticipated, on fusion between antigenic proteins of interest (for example, PAP) and the ligand of a cell surface receptor expressed by these cells (in this case GM-CSF). After these monocytes and DCs have been exposed *ex vivo* to the fusion protein, they are re-infused, together with other leukocytes, into the patient who originally donated them—the process of autologous transplantation.

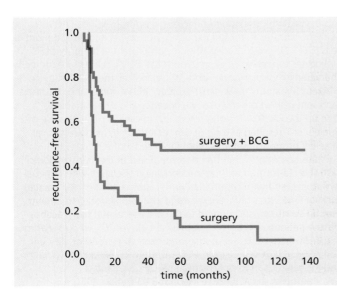

Figure 15.41 BCG enhancement of the immune response The use of Bacillus Calmette–Guérin (BCG), a live, attenuated mycobacterium, as an immunostimulant has clear effects in preventing the progression of human bladder carcinomas. Among patients who were treated only surgically (*red line*) with transurethral resection, the proportion of recurrence-free patients (*ordinate*) declined precipitously in the months after treatment (*abscissa*), as indicated in this Kaplan–Meier plot. In contrast, among those patients who were surgically treated and were additionally exposed to BCG, which was injected into their bladders (*blue line*), the frequency of recurrence was much less, and some were still recurrence-free after 5 years. (From J. Patard et al., *Urology* 58:551–556, 2001.)

In an early clinical trial, none of a group of 31 men with castration-resistant prostate cancer exhibited preexisting antibodies or T-cell responses to PAP; in response to sipuleucel-T/Provenge treatment, 38% developed a T-cell response and 53% developed anti-PAP antibodies. In another study, 11% of patients treated with placebo were alive at 36 months while 34% treated with sipuleucel-T were still alive. Eventually the curves

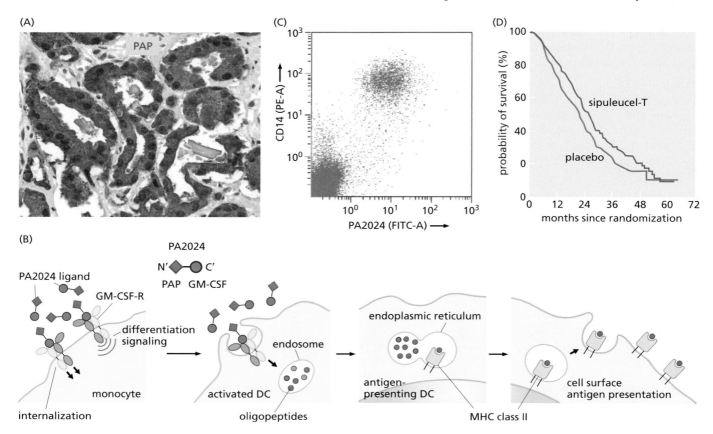

Figure 15.42 Provenge and prostate carcinoma vaccination
The first FDA-approved anti-tumor vaccination, called Provenge or sipuleucel-T, depends on the ability to develop immunological reactivity against prostatic acid phosphatase (PAP), an enzyme of unknown function that is expressed at high levels and secreted by virtually all adenocarcinomas of the prostate. This strategy requires loading a fusion protein formed from PAP and the GM-CSF monocyte/dendritic cell cytokine onto monocytes *ex vivo* and reintroducing the cells back into the patient from whom they were originally extracted (the procedure of *autologous transplantation*), the aim being for these altered monocytes to induce the patient's T and B cells into attacking carcinoma cells that express the PAP protein. (A) PAP (*dark red*), stained here with a dye-coupled anti-PAP monoclonal antibody, is expressed strongly in the ductal cells of this human prostate carcinoma while being expressed weakly in most other tissues in the body. (B) The uptake by monocytes of the PA2024 recombinant protein resembles in outline the uptake of antigens by DCs and B cells depicted in Figure 15.6. In this case, the GM-CSF receptor (*left*) binds the fusion protein, which results in signaling that drives the treated monocytes to differentiate into activated DCs. Coupled with this is the internalization of the ligand–receptor complexes and the breakdown of the fusion protein into oligopeptides within endosomes. The contents of the latter are hypothetically loaded on MHC class II proteins (*third panel*) and displayed on the DC surface (*right panel*); additional presentation by MHC class I proteins (not shown) may also occur. Thereafter, the resulting antigen-presenting DCs are able to activate an anti-tumor adaptive immune response involving both T_H and T_C cells as well as B cells. (C) The use of fluorescence-activated cell sorting (FACS; see Supplementary Sidebar 11.1) allows the determination whether two antigens are co-expressed in the same cells and thus whether the PA2024 fusion protein has become physically associated with the DCs. Each *red dot* represents a cell, and the intensity of staining (and thus cell surface antigen expression) is plotted logarithmically on both axes. This FACS analysis indicates that after exposing monocytes and early dendritic cells (DCs) *ex vivo* to the PA2024 fusion protein, a minority of leukocytes (*upper right*) express both the CD14 cell surface antigen characteristic of monocytes and DCs as well as the PA2024 protein, whereas a majority (*lower left*) express neither, indicating that they represent cells of other leukocyte lineages and were unable to take up the PA2024 fusion protein. (D) In clinical trials, the use of sipuleucel-T resulted in a significant extension in median survival of select groups of patients who presented in the clinic with prostate carcinomas that had become resistant to pharmacologic castration (a form of androgen-deprivation therapy), with minimal symptoms, and with disseminated metastases. While their median survival time was significantly extended by sipuleucel-T treatment, they all eventually succumbed to the disease, as did the placebo-treated patients. (A, from T.J. Graddis et al., *Int. J. Clin. Exp. Pathol.* 4:295–306, 2011. B, adapted in part from G. Hansen et al., *Cell* 134:496–507, 2008. C, from USDA Center for Biologics Evaluation and Research hearing of March 29, 2009. D, adapted from P.W. Kantoff et al., *N. Engl. J. Med.* 363:411–422, 2010.)

converged (see Figure 15.42D). Nonetheless, this success represented a milestone in anti-cancer immunotherapy, being the first such treatment to successfully harness the adaptive immune system to retard the progression of carcinomas in a significant proportion of patients.

The responses of T lymphocytes, specifically T_H and T_C cells, to the oligopeptide antigens being displayed by antigen-presenting dendritic cells (APCs) can also be enhanced in various ways and usually proceed through two phases. During the first phase (Figure 15.43A), the initial encounter of T lymphocytes with APCs induces various lymphocyte responses that are intensified by the encounter between the CD28 cell surface receptor of the T lymphocytes and complementary ligands displayed by the APCs; CD28 participates in the antigen-presenting encounter in conjunction with the T-cell receptor (TCR), the latter being responsible for antigen recognition. Subsequently, during the second phase of the response (see Figure 15.43B), another cell surface T-lymphocyte receptor, termed CTLA-4 (cytotoxic T-lymphocyte antigen-4), is produced by the T cells and competes for the same APC ligands as CD28 and strongly inhibits further activation of the T lymphocyte by antigen-presenting dendritic cells. Hence, the CTLA-4 receptor mediates a critical negative-feedback control loop in order to ensure that T-lymphocyte activation is only transient and that any resulting immune responses are limited.

Monoclonal antibodies that bind and neutralize the CTLA-4 inhibitory receptor have been found to greatly potentiate the immune responses achieved by T lymphocytes, ostensibly because these antibodies prolong and intensify the lymphocyte activation triggered by antigen-presenting dendritic cells. In the context of cancer, treatment of tumor-bearing mice with anti-CTLA-4 monoclonal antibody, termed ipilimumab, has resulted in strong stimulation of the immune response against tumors, leading in some cases to tumor regression (see Figure 15.43C). (Not shown here is the further potentiation of ipilimumab action by concomitant administration of GM-CSF, which we saw earlier strongly activates monocytes, inducing them to differentiate into dendritic cells.)

As an undesired side effect, anti-CTLA-4 antibodies induce or exacerbate autoimmune reactions in both mice and humans, an indication that this particular signaling system is one of the primary mechanisms used by the immune system to prevent the inadvertent destruction of normal tissues. In the extreme case, when CTLA-4 is entirely eliminated from a mouse through the germ-line knockout of its encoding gene, newborns live for only 2 or 3 weeks before they die from massive lymphoproliferation leading to lymphocytic infiltration and resulting destruction of major organs.

Initial clinical trials in humans carrying advanced melanomas and ovarian carcinomas indicate that the efficacy of several types of anti-tumor immunotherapy can be strongly enhanced by the injection of the anti-CTLA-4 monoclonal antibody—ipilimumab—without serious concomitant autoimmune reactions (see Figure 15.43D). On some occasions, this anti-CTLA-4 antibody has been administered to patients who had previously not experienced any immunotherapy. At other times, the anti-tumor immune response of patients was first stimulated by either (1) vaccination with purified melanoma-associated antigens like those listed earlier or (2) injection with **autologous** (that is, their own) tumor cells that had been forced *ex vivo* to express GM-CSF and were then irradiated (to prevent the proliferation of these cells after they were injected back into the patients). GM-CSF expression serves to recruit monocytes and, as mentioned earlier, to induce them to differentiate into dendritic cells. These patients were then treated with anti-CTLA-4 antibody. In 2011, the FDA approved the use of ipilimumab, marketed under the name Yervoy, for the treatment of metastatic melanoma: patients treated with the antibody had an overall survival of 10 months compared with 6 months for those treated with a peptide vaccine.

The demonstrated ability of interleukin-2 (IL-2) to activate lymphocytes has also been exploited by adding this cytokine (also called a **lymphokine**) to mixtures of lymphocytes and killed tumor cells *in vitro*. This results in the functional activation of the lymphocytes and increased killing of tumor cells in tumor-bearing mice. These *lymphokine-activated killer* (LAK) cells are almost entirely NK cells, which, as we have read, are specialized to kill target cells lacking normal amounts of MHC class I molecules and expressing certain stress- or transformation-associated proteins on their surface (see Figure 15.29). In some clinical trials, the resulting LAK cells have been co-injected into patients together with additional IL-2, and in one well-described trial, 16 of 222 cancer patients showed complete regression of their tumors. However,

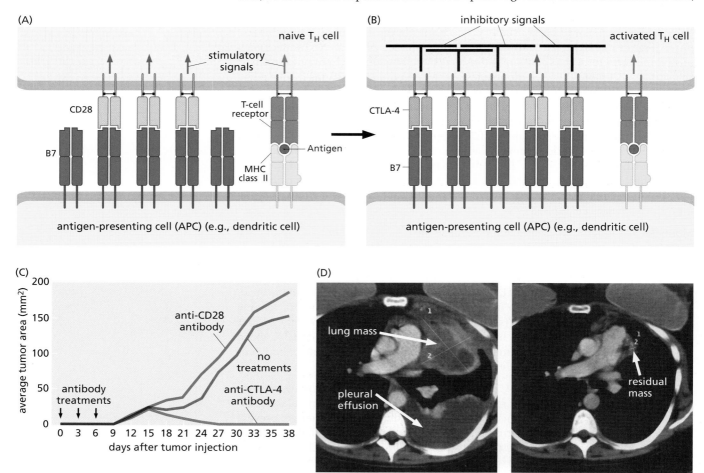

Figure 15.43 CTLA-4 modulation of the cellular immune response The interaction of a T_H lymphocyte (*above*) with an antigen-presenting cell (APC, *below*), such as a dendritic cell, proceeds in two phases. (A) Initially, the CD28 surface protein (*light green*) displayed by the T_H lymphocyte engages complementary B7 proteins (actually called B7-1 and B7-2; *dark green*) on the surface of the APC. This causes the CD28 molecules to release signals (*red arrows*) that act together with the signals (*blue arrow*) released by the antigen-binding T-cell receptor (*blue*) to activate the naive T_H cell. (B) In the second phase, as part of a negative-feedback mechanism, the T_H cell begins to synthesize CTLA-4 molecules (*pink*). These CTLA-4 molecules bind the B7 molecules of the APC with higher affinity than the CD28 molecules, thereby displacing CD28 from this association and shutting down further T-cell activation. (C) Modulation of CTLA-4 activity can have substantial effects on the immune response against an implanted tumor. In the absence of any intervention, syngeneic carcinoma cells grow vigorously beginning 9 days after their implantation into a host mouse (*red line*). Anti-CD28 serum, which should interfere with the activation of T lymphocytes by APCs (see panel A), has at best a minor effect in enhancing the growth of these tumors (*blue vs. red line*). However, when tumor-bearing mice are exposed to anti-CTLA-4 antibody at the indicated times (*arrows*), the normally

occurring shutdown of T-lymphocyte function is apparently prevented and the resulting hyperactivated T lymphocytes proceed to eliminate the tumors (*green line*). This suggests a strategy for immunotherapy of cancer in which CTLA-4 function is suppressed through the use of an anti-CTLA-4 monoclonal antibody. (D) Occasionally, the clinical use of an anti-CTLA-4 monoclonal antibody, termed ipilimumab, yields dramatic clinical responses. Here, computerized tomography (CT) X-ray scans of a patient with metastatic melanoma reveal a major lung mass and pleural effusion (*left panel*) that largely disappeared five months after initiation of immunotherapy (*right panel*). This patient was first exposed to repeated injections of dendritic cells whose MHC class II receptors had been loaded *ex vivo* with an antigenic oligopeptide derived from the MART-1 melanoma antigen (see Figure 15.26); this treatment was followed later by injection of the anti-CTLA-4 antibody. Dramatic responses like this one indicate the potential of this type of immunotherapy, which has yet to be realized because it succeeds to this degree in only a small proportion of treated patients. (A and B, from K. Murphy, Janeway's Immunobiology, 8th ed. New York: Garland Science, 2012. C, from D.R. Leach, M.F. Krummel and J.P. Allison, *Science* 271:1734–1736, 1996. D, courtesy of A. Ribas; see A. Ribas et al., *J. Immunother.* 27:354–367, 2004.)

undesirable side-effects of IL-2 precluded further development of this treatment for routine use in the oncology clinic.

A related strategy involves the preparation of lymphocytes that are already present in a patient's tumor, amplifying their number *ex vivo*, and then re-introducing them into the patient—the procedure of adoptive cell transfer (ACT). Once again, the bulk populations of tumor-infiltrating lymphocytes (TILs; see Figure 15.19) that are prepared from surgically removed tumor samples are NK cells. However, a functionally critical subpopulation of the cells in these TIL preparations consists of cytotoxic T lymphocytes (T_C's) that have acquired specific reactivity against the antigens displayed by the tumor from which they were isolated. Importantly, when these "educated" cytotoxic T cells are functionally activated *ex vivo*, they require only one-hundredth the concentration of IL-2 needed to activate the larger populations of NK cells. Injection of these tumor-infiltrating CTLs into patients has caused partial regressions of tumors in about one-quarter of treated melanoma and renal cancer patients. Attempts at generating TILs having specific anti-tumor cytotoxic activity from other types of tumors have not met with comparable success to date.

The responses to the TIL treatment described above have generally been short-lived, which is explained in part by the inability of the introduced populations of TILs to establish themselves stably in the bodies of cancer patients. This has led to attempts to stabilize and expand the populations of anti-tumor lymphocytes *in vivo* by treating melanoma patients with chemotherapy prior to injection of *ex vivo*–expanded TILs. In this case, the motive of chemotherapeutic treatment was not to kill cancer cells but to reduce the populations of lymphocytes and lymphocyte precursors in patients' bone marrow, thereby "making room" for clones of subsequently introduced TILs. The latter could then establish themselves in the marrows of patients and could persist and even expand during the course of immunotherapy. In some patients, some clonal populations of introduced anti-tumor CTLs ultimately formed the *majority* of all the cytotoxic T cells in their immune systems. These cytotoxic T_C's, whose growth *in vivo* was sustained by injecting these patients with IL-2, showed high potency for killing the melanoma cells. In several of such patients, dramatic regressions of melanoma tumors were observed.

Such responses, compelling as they may be, are only anecdotal, and none of these immunotherapy protocols has advanced to a stage where it can reproducibly yield robust therapeutic responses in a substantial proportion of treated patients. It seems evident that we are only beginning to learn how to manipulate the immune systems of cancer patients in ways that will cause tumor regression. This explains why the protocols described here are only first steps, likely to be superseded soon by far more effective ways of energizing the immune responses against tumors.

15.22 Synopsis and prospects

The interactions between the immune system and tumors are surely very complex, and the precise roles that immune cells play in the suppression of most types of human tumors remain poorly understood. Studies of immunosuppressed patients provide clear testimony to the fact that immune surveillance is responsible for helping to prevent the appearance of a variety of virus-induced tumors. Humoral immune responses are likely to be responsible for suppressing the infectious spread of virus through tissues, thereby minimizing the number of virus-infected cells, some of which may eventually progress to a tumorigenic state.

Even more important, the continued expression of viral proteins by virus-transformed tumor cells creates clearly recognizable foreign antigens that can trigger a highly effective immune attack, much of it mediated by the cellular arm of the immune response. Antigenic viral oligopeptides, when presented by the MHC class I proteins of the tumor cells, attract the attention of cytotoxic T cells, which proceed to kill a virus-infected or virus-transformed cell.

The contributions of the immune response to protecting us against the many types of tumors of *nonviral* origin remain more ambiguous. It is difficult to gauge the extent to which immune surveillance plays a role in eliminating or retarding the formation of these tumors. Some measure of the importance of these mechanisms derives from studying immunocompromised patients, in whom the rates of certain solid tumors of nonviral etiology are two or three times higher than in the general population.

Some tumor cells may succeed in eluding annihilation by the immune system because they present few antigens that are clearly foreign, and thus they benefit from the tolerance that the normal immune system develops toward "self" proteins. Yet other tumor cells may express proteins and thus potential antigens in aberrant amounts or anatomical sites; such ectopic expression may alert the immune surveillance system, which will begin to hunt down and eliminate the responsible tumor cells.

Once under attack, tumor cells often defend themselves with countermeasures. They may down-modulate expression of the antigen that initially attracted the interest of the immune system, often using the same strategy—promoter methylation—that they use to rid themselves of unwanted tumor suppressor gene activity. Alternatively, they may suppress the display of the MHC class I molecules that enable the immune system to detect the presence of the antigen on their surface. The fact that the MHC class I molecules are often absent from metastatic cells feeds the suspicion that the migration of cancer cells through the body's tissues and circulation represents a dangerous passage that is often cut short by immune cells lying in ambush along the route.

Should these various immunoevasive maneuvers not suffice, cancer cells may deploy an even more effective defense by driving away or killing potential attackers. For example, release of potent, pro-apoptotic factors such as TGF-β and FasL can often ensure elimination of any immunocytes that venture too close to neoplastic cells.

In the end, the most effective defense mounted by tumor cells may derive from the body's main cellular systems for establishing and maintaining immune tolerance—the mechanisms that normally prevent the development of autoimmune attacks on the body's own tissues. Tolerance toward many antigens seems to depend on the actions of the regulatory T cells (T_{reg}'s), which operate to ensure that both the humoral and cellular arms of the immune response do not destroy our normal tissues. The ability of human tumors to release the chemotactic factor CCL22 (which serves to attract T_{reg}'s) may ultimately benefit them more than all the other immunoevasive maneuvers cited here.

When taken together, these various well-documented maneuvers of tumor cells persuade us that escape from immune attack is an important step in the progression of most, and perhaps all, tumors toward the highly malignant growth state. Indeed, immunoevasion has come to be recognized as a hallmark of neoplasia that is as fundamental as the half dozen or so others that were enumerated earlier (see Section 11.18).

Previously, we depicted each of the steps of tumor progression as the successful breaching of an important anti-cancer defense mechanism in our cells and tissues (see Chapter 11). If this is so, we might ask about the relative contributions of these various defenses to preventing the onset of cancer. Most of these mechanisms represent the actions of the regulatory circuitry that is hard-wired in all of our cells. How much of our anti-cancer defense can be attributed to these cell-autonomous mechanisms, that is, mechanisms that operate within individual cells en route to neoplasia? And how much of our defenses against cancer derives from the organism as whole and the actions of its immune system? Definitive answers to these questions are still elusive, and another decade may pass before they are in hand.

The frequent failures of the normal immune system to erect effective defenses against most tumors do not preclude the eventual development of truly effective immunotherapeutic strategies against cancer. By learning how various components of the immune system are regulated, immunologists are gaining the ability to manipulate it and to empower these components to attack cancer cells. Only a few of the many immune-mobilizing strategies currently under investigation (**Table 15.5; Figure 15.44**) have been described in this chapter.

Table 15.5 Examples of anti-tumor immunotherapy strategies

Passive immunization
Infuse tumor-specific monoclonal antibodies (e.g., Herceptin, Rituxan)
Engraft histoincompatible marrow

Active immunization
Infuse activated tumor-infiltrating lymphocytes (TILs)
Infuse dendritic cells loaded with tumor-specific oligopeptide antigen
Add B7 co-activating receptor to introduced tumor-specific antigen
Block CTLA-4 function
Inhibit regulatory T cells

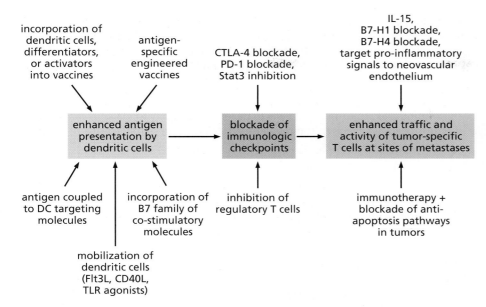

To date, more than one hundred cytokines (regulatory factors of the immune system including interleukins and interferons) and their cognate receptors have been identified. Some of these have been functionally characterized, while most have not. They are almost evenly divided into those factors that potentiate immune function (for example, interferons, interleukin-2) and those that inhibit it (for example, IL-10, TGF-β). Such a long list of cytokines is indicative of the complexity of immune regulation. Indeed, we have only begun to understand the physiology of normal immune function and the possibilities of manipulating it to strengthen anti-cancer defenses. The fact that a fundamentally important type of immunocyte—the regulatory T cell (T_{reg})—has only recently come into clear view indicates that the field of immunology continues to revise many of its fundamental concepts.

It remains unclear precisely how and when during multi-step tumor progression the immune system is first alerted to the presence of cancer cells and begins its initial, often unsuccessful attempts at eliminating them. This may occur when carcinoma cells invade the stroma and directly confront immunocytes. An attractive alternative model, still unproven, places these initial encounters earlier during the inflammatory phases of tumor progression, when macrophages and other leukocytes are initially recruited into the tumor stroma.

At present, it is impossible to say which of the dozens of immunotherapy strategies under development will become the precursors of anti-cancer treatments that will prove to be vastly more effective than those developed to date and will generate robust, durable responses for the majority of patients under treatment. For example, advances in molecular genetics and cellular immunology have recently converged to generate novel, potentially powerful ways of harnessing the immune response to eradicate human tumors (Supplementary Sidebar 15.16). Perhaps fundamental aspects of neoplastic disease create obstacles to cancer immunotherapy that can never be overcome by the cleverness of immunologists and cancer biologists. We wrestle with these issues once again in the next chapter, where the biological discoveries reported throughout this book are applied to the development of new types of low–molecular-weight drug molecules designed to bring down the elusive quarry—the ever-changing cancer cell.

Key concepts

- The immune system launches two types of attack against infectious agents or cells that it has targeted for destruction or neutralization; these involve humoral and cellular immunity.

- Some types of immune cells—particularly macrophages and NK cells—have an innate ability to recognize cells that should be destroyed.

- T lymphocytes develop the ability to recognize antigenic targets through the display of T-cell receptors, which are key elements of adaptive immunity and are created through the rearrangement of gene segments similar to those leading to the formation of soluble antibodies. These T lymphocytes are largely cytotoxic T cells (CTLs, or T_C's), helper T cells (T_H's), and regulatory T cells (T_{reg}'s).

- The helper T cells aid B cells to develop antibodies, thereby creating humoral immunity, and aid cytotoxic T cells to develop the ability to kill cells that need to be destroyed. Conversely, the regulatory T cells suppress the actions of helper T cells and cytotoxic T cells.

- Normal cells throughout the body routinely present oligopeptide fragments of their proteins on the cell surface, using MHC class I molecules to do so. Professional antigen-presenting cells, such as macrophages and dendritic cells, use MHC class II molecules to present oligopeptide fragments of proteins that they have scavenged from tissue environments. These various oligopeptide fragments represent the antigens that are recognized by T-cell receptors of helper and cytotoxic T cells.

- Antibodies that recognize antigens on the surface of a cell can direct the killing of a cell through two major mechanisms: (1) immunocytes that express Fc receptors, specifically macrophages and NK cells, can attach to the antibody-coated cell and kill it; (2) complement can attach to the surface-bound antibody and kill the cell by inserting channels into its plasma membrane.

- The immune system has the intrinsic ability to develop immune recognition of both normal tissue antigens and those expressed by foreign elements, specifically, infectious agents. However, the immune system uses a number of distinct mechanisms to suppress the reactivity of its various arms against normal tissue antigens, thereby developing tolerance of them.

- Many of the actions of the immune system directed toward recognizing and eliminating infectious agents may also be used to launch attacks against cancer cells. However, its tolerance-inducing mechanisms may thwart these attacks, since the great majority of tumor cell antigens are components of normal cellular proteins.

- Certain tumor-associated antigens may nevertheless attract the attention of the immune system, because they are normally displayed only in embryos or in immunologically privileged sites, such as the testes and brain, where tolerance toward cellular antigens does not develop. Other tumor-associated antigens may provoke immune recognition and attack because they are expressed at elevated levels.

- The fact that immunocompromised individuals experience elevated levels of various cancers strongly suggests that the immune system is continuously monitoring the body's tissues for the presence of tumors and attempting to eliminate them—the process of immunosurveillance.

- Some cancers may thrive in the body in spite of immune surveillance, simply because they are weakly antigenic. Others may have originally been strongly antigenic and may have generated weakly antigenic variants. This represents one of many immunoevasive strategies employed by tumor cells.

- Tumor-associated antigens (TAAs) are expressed by tumor cells and often reflect the differentiation programs of the tissues in which these tumor cells arose. Suppression of TAA expression may allow the tumor cells to escape immune surveillance and, at the same time, can often occur without compromising the ability of these cells to proliferate.

- Another immunoevasive strategy involves the release of factors, such as IL-10, FasL, and TGF-β, that are capable of eliminating immune cells that venture too close to tumor cells.

- Cancer cells may also attract and activate regulatory T cells, which can inactivate any cytotoxic T cells that have entered into tumor masses.

- Researchers have devised a number of treatment strategies that supplement or strengthen the existing immune response against tumors. An important strategy is to provide cancer patients with monoclonal antibodies (MoAbs), such as Herceptin and Rituxan, that can bind their tumor cells, leading to the killing of these cells through a variety of mechanisms.

- Alternatively, some attempts at developing anti-cancer immunotherapeutic protocols involve perturbing the signaling agents that normally regulate the activities of various immune cell types. For example, GM-CSF can be used to activate dendritic cells, while anti-CTLA-4 antibody can be used to enhance the interactions between antigen-presenting cells and helper T cells.

- The responses of the immune system to tumors are still imperfectly understood, and the multiplicity of immune regulators creates the opportunity to activate anti-tumor responses in many ways, most of which have not yet been attempted in the oncology clinic.

Thought questions

1. How do viral genes help us understand how cancer cells escape immune surveillance?

2. In what way might it be possible to force cancer cells to become more immunogenic?

3. How might measurements of immune functions provide insight into the presence of tumors in the body?

4. How might one determine whether tumor-associated macrophages (TAMs) are working to support tumor growth or are working as agents of the immune system to eliminate a tumor?

5. In what diverse ways might oncoproteins render cancer cells more resistant to killing by various components of the immune system?

6. What do you believe to be the most compelling evidence that the immune system plays a significant role in suppressing the appearance of many commonly occurring solid tumors?

7. How would you evaluate the relative importance of innate versus adaptive immunity in suppressing the appearance of clinically apparent tumors?

Additional reading

Arteaga CL, Sliwkowski MX, Osborne CK et al. (2011) Treatment of HER2-positive breast cancer: current status and future perspectives. *Nat. Rev. Clin. Oncol.* 9, 16–32. doi: 10.1038/nrclinonc.2011.177, Nov. 29.

Banck MS & Grothey A (2009) Biomarkers of resistance to epidermal growth factor receptor monoclonal antibodies in patients with metastatic colorectal cancer. *Clin. Cancer Res.* 15, 7492–7501.

Brody J, Kohrt H, Marabelle A & Levy R (2011) Active and passive immunotherapy for lymphoma: proving principles and improving results. *J. Clin. Oncol.* 29, 1864–1875.

Burstein HJ (2005) The distinctive nature of HER2-positive breast cancers. *N. Engl. J. Med.* 353, 1652–1654.

Chao MP, Majeti R & Weissman IL (2011) Programmed cell removal: a new obstacle in the road to developing cancer. *Nat. Rev. Cancer* 12, 58–67.

Cheson BD (2002) Rituximab: clinical development and future directions. *Expert Opin. Biol. Ther.* 2, 97–110.

Chung CC, Campoli M & Ferrone S (2005) Classical and nonclassical HLA class I antigen and NK cell-activating ligand changes in malignant cells: current challenges and future directions. *Adv. Cancer Res.* 65, 189–234.

Curiel TJ (2007) Tregs and rethinking cancer immunotherapy. *J. Clin. Invest.* 117, 1167–1174.

Darnell R (2004) Tumor immunity in small-cell lung cancer. *J. Clin. Oncol.* 22, 762–764.

De Visser KE, Eichten A & Coussens LM (2006) Paradoxical roles of the immune system during cancer development. *Nat. Rev. Cancer* 6, 24–37.

Dranoff G (2004) Cytokines in cancer pathogenesis and cancer therapy. *Nat. Rev. Cancer* 4, 11–22.

Dudley ME & Rosenberg SA (2003) Adoptive-cell-transfer therapy for the treatment of patients with cancer. *Nat. Rev. Cancer* 3, 666–675.

Dunn GP, Old LJ & Schreiber RD (2004) The three E's of cancer immunoediting. *Annu. Rev. Immunol.* 22, 329–360.

Figdor CG, de Vries IJ, Lesterhuis WJ & Melief CJ (2004) Dendritic cell immunotherapy: mapping the way. *Nat. Med.* 10, 475–480.

Fin, OJ (2008) Cancer immunology. *N. Engl. J. Med.* 358, 2704–2715.

Fridman WH, Galon J, Pages F et al. (2011) Prognostic and predictive impact of intra- and peritumoral immune infiltrates. *Cancer Res.* 71, 5601–5605.

Gabrilovich DI & Nagaraj S (2009) Myeloid-derived suppressor cells as regulators of the immune system. *Nat. Rev. Immunol.* 9, 162–174.

Gajewski TR (2007) Failure at the effector phase: immune barriers at the level of the melanoma tumor microenvironment. *Clin. Cancer Res.* 12, 5256–5261.

Gelderman KA, Tomlinson S, Ross GD & Gorter A (2004) Complement function in mAb-mediated cancer immunotherapy. *Trends Immunol.* 24, 158–164.

Jaiswal S, Chao MP, Majeti R & Weissman I (2010) Macrophages as mediators of tumor immunosurveillance. *Trends Immunol.* 31, 212–219.

Johansson M, DeNardo DG & Coussens LM (2008) Polarized immune responses differentially regulate cancer development. *Immunol. Rev.* 222, 145–154.

Kane MA (2012) Preventing cancer with vaccines: progress in the global control of cancer. *Cancer Prev. Res.* 5, 24–29.

Klein C, Lammens A, Schäfer W et al. (2013) Epitope interactions of monoclonal antibodies targeting CD20 and their relationship to functional properties. *mAbs* 5, 1–12.

Mellman I, Coukos G & Dranoff G (2011) Cancer immunotherapy comes of age. *Nature* 480, 480–489.

Mouglakakos D, Choudhury A, Liadser A et al. (2010) Regulatory T cells in cancer. *Adv. Cancer Res.* 107, 57–117.

Old LJ (2001) Cancer/testis (CT) antigens: a new link between gametogenesis and cancer. *Cancer Immun.* 1, 1.

Pages F, Galon J, Dieu-Nosjean M-C et al. (2010) Immune infiltration in human tumors: a prognostic factor that should not be ignored. *Oncogene* 29, 1093–1102.

Pastan I, Hassan R, FitzGerald DJ & Kreitman RJ (2006) Immunotoxin therapy of cancer. *Nat. Rev. Cancer* 6, 559–564.

Peggs KS, Quezada SA & Allison JP (2008) Cell intrinsic mechanisms of T-cell inhibition and application to cancer therapy. *Immunol. Rev.* 224, 141–165.

Pohlmann PR, Mayer IA & Mernaugh R (2009) Resistance to trastuzumab in breast cancer. *Clin. Cancer Res.* 15, 7479–7491.

Raulet D (2003) Roles of the NKG2D immunoreceptor and its ligands. *Nat. Rev. Immunol.* 3, 781–790.

Sakaguchi S (2005) Naturally arising Foxp3-expressing CD25+CD4+ regulatory T cells in immunological tolerance to self and non-self. *Nat. Immunol.* 6, 345–352.

Schwartz RH (2005) Natural regulatory T cells and self-tolerance. *Nat. Immunol.* 6, 327–330.

Sharma P, Wagner K, Wolchok J & Allison JP (2011) Novel cancer immunotherapy agents with survival benefits: recent successes and next steps. *Nat. Rev. Cancer* 11, 805–812.

Smith MR (2003) Rituximab (monoclonal anti-CD20 antibody): mechanisms of action and resistance. *Oncogene* 22, 7359–7368.

Smyth MJ, Hayakawa Y, Takeda K & Yagita H (2002) New aspects of natural-killer cell surveillance and therapy of cancer. *Nat. Rev. Cancer* 2, 850–861.

Smyth MJ, Dunn GP & Schreiber RD (2006) Cancer immunosurveillance and immunoediting: the roles of immunity in suppressing tumor development and shaping tumor immunogenicity. *Adv. Immunol.* 90, 1–50.

van der Burg SH & Melief CJ (2011) Therapeutic vaccination against human papillomavirus induced malignancies. *Curr. Opin. Immunol.* 23, 252–257.

Weiner LM, Surana R & Wang S (2010) Monoclonal antibodies: versatile platforms for cancer immunotherapy. *Nat. Rev. Immunol.* 10, 317–327.

Wrzesinski SH, Wan YY & Flavell RA (2007) Transforming growth factor-β and the immune response: implications for anticancer therapy. *Clin. Cancer Res.* 13, 5262–5270.

Yang L, Pang Y & Moses HL (2010) TGF-β and immune cells: an important regulatory axis in the tumor microenvironment and progression. *Trends Immunol.* 31, 220–227.

Yu J, Pardoll D & Jove R (2009) STATs in cancer inflammation and immunity: a leading role for STAT3. *Nat. Rev. Cancer* 9, 798–809.

Zang X & Allison JP (2007) The B7 family and cancer therapy: costimulation and coinhibition. *Clin. Cancer Res.* 13, 5271–5279.

Zou W (2005) Immunosuppressive networks in the tumour environment and their therapeutic relevance. *Nat. Rev. Cancer* 5, 263–274.

Zou W (2006) Regulatory T cells, tumour immunity and immunotherapy. *Nat. Rev. Immunol.* 6, 295–307.

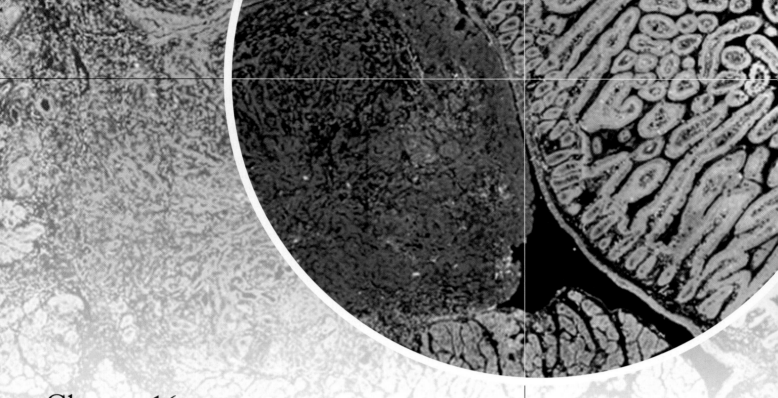

Chapter 16

The Rational Treatment of Cancer

All substances are poisonous, there is none that is not a poison; the right dose differentiates a poison from a remedy.

> Paracelsus (Auroleus Phillipus Theostratus Bombastus
> von Hohenheim), alchemist and physician, 1538

Doctors are men who prescribe medicines of which they know little, to cure diseases of which they know less, in human beings of whom they know nothing.

> Voltaire (François-Marie Arouet), author and philosopher, 1760

The research described throughout this book represents a revolution in our understanding of cancer pathogenesis. In 1975, there were virtually no insights into the molecular alterations within human cells that lead to the appearance of malignancies. One generation later, we possess this knowledge in abundance. Indeed, the available information and concepts about cancer's origins can truly be said to constitute a science with a logical and coherent conceptual structure.

In spite of these extraordinary leaps forward, relatively little progress has been made in exploiting these insights into etiology (that is, the causative mechanisms of disease) to prevent the disease and, equally important, to treat it. Most of the anti-cancer treatments in widespread use today were developed prior to 1975, at a time when the development of therapeutics was not yet informed by detailed knowledge of the genetic and biochemical mechanisms of cancer pathogenesis. This explains the widely felt frustration among molecular oncologists that the potential of their research for contributing to new anti-cancer therapeutics has not yet been realized.

The promise—still unrealized—of the new therapeutics needs to be juxtaposed with the overall progress in treating advanced tumors using the traditional strategies of surgery, chemotherapy, and radiotherapy. Significant progress in treating such tumors has been slow. For example, in 1970 in the United States, 7% of the patients diagnosed with lung cancer were still alive 5 years after their initial diagnosis. Three decades later, this number had risen to only 14%, a relatively minor improvement. And even this degree of therapeutic success may be illusory, since modern diagnostic techniques often detect tumors far earlier in their natural course, creating a greater time span between initial diagnosis and ultimate progression to end-stage disease. Death rates for colon cancer have begun to fall, because of early detection and surgical removal of growths that have advanced through only the early stages of tumor progression (**Figure 16.1A**; see also Figure 11.9B). However, mortality caused by the more advanced colorectal tumors has changed little—a testimonial to the failures of chemotherapy and radiation to eliminate these malignancies once they have invaded and begun to metastasize (Supplementary Sidebar 16.1). Moreover, age-adjusted mortality from other types of tumors has remained constant or declined relatively little (see Figure 16.1B). Statistics like these suggest that the potential of the traditional therapies to cure high-grade malignancies has been largely realized, and that major progress in the future can only come from the novel therapeutics.

The problems confronted here are manifold. To begin, in the case of certain common cancers, we don't really have a very good estimate of the size of the problem. How many of the cases that are diagnosed each year (yielding age-adjusted incidence; see Section 11.1) are likely to grow into life-threatening growths? And how often do treatments performed on patients with relatively benign tumors yield more morbidity than clinical benefit in terms of subsequent well-being and gains in life span?

The problem is explained by a critical lesson we learned in Chapter 11: as we grow older, small tumors appear spontaneously in a wide variety of tissues. At the same time, as diagnostic procedures become increasingly sensitive (**Figure 16.2**), we begin to detect more and more cases of cancer without a clear indication of the proportion of these that will grow into life-threatening tumors. This problem is an immense one, given the limited ability of oncology clinics and funds to respond to an ever-increasing number of cases (**Sidebar 16.1**).

As discussed in the next section, our rapidly evolving understanding of disease pathogenesis may soon allow us to judge more accurately how many cancers are deserving of aggressive treatment and how many can be safely ignored. At the same time, our improving insights into specific disease processes, such as the development of

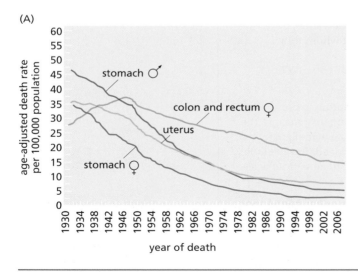

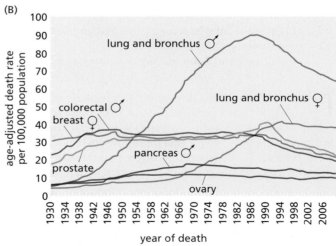

Figure 16.1 Cancer mortality over the past three-quarters of a century: the lay of the land The statistics compiled in the United States on the age-adjusted death rates from various types of cancer reveal two different long-term trends. (A) Mortality from several major killers has declined significantly since 1930. This is due to changes in food storage practices and possibly *Helicobacter pylori* infection rates, in the case of stomach cancer, and to screening, in the cases of cervical and colorectal cancers. (B) A number of major sources of cancer-related death have proven resistant to most forms of traditional therapy, especially when these tumors progress to a highly malignant, metastatic stage. (From A. Jemal et al., *CA Cancer J. Clin.* 55:10–30, 2005.)

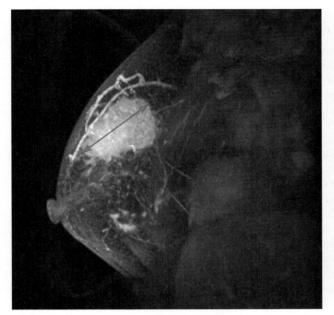

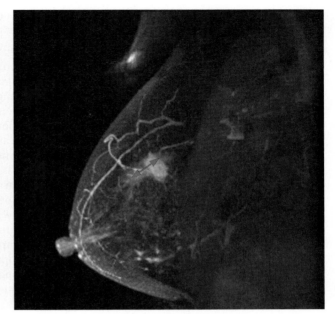

pre-chemotherapy
longest dimension = 47 mm

post-chemotherapy
longest dimension = 16 mm

Figure 16.2 High-resolution noninvasive imaging of human tissues The development of magnetic resonance imaging (MRI) has enabled increasingly higher-resolution, *noninvasive* visualization of living tissues. MRI now allows breast tumors of very small size (several mm diameter) to be detected and, as shown here, makes it possible to view the progress of anti-tumor therapy—in this case chemotherapy with an anthracycline cytotoxic agent—in exquisite detail. Widespread use of such highly sensitive imaging techniques is likely to result in further increases in the incidence rate of breast cancer (resulting from detection of disease that would previously have eluded detection) with unknown effect on the mortality of the disease. (Courtesy of N.M. Hylton and L.J. Esserman.)

Sidebar 16.1 How common are cancers that require clinical intervention? The development and utilization of diagnostic procedures of ever-increasing sensitivity led to steep increases in the incidence of breast and prostate cancers during the second half of the twentieth century. In the case of the breast, the incidence of carcinomas increased progressively during the late twentieth century and then leveled off (**Figure 16.3**). As discussed later, some of these growths will eventually become life-threatening and many others will not. However, it is clear that incidence rates are an artifact of technology, and as diagnosis improves (see Figure 16.2), the age-adjusted incidence will increase in lockstep unless the images generated by these new diagnostic techniques are interpreted with caution. Hence, the only truly rigorous and solid measurements are those associated with mortality.

A recent study of breast cancer performed post mortem on women who died of a variety of causes estimated that at the age of 80, at least two-thirds of women carry breast carcinomas (whereas only ~4% will die from breast cancer). An even larger number (~80%) applies to post mortem studies of prostate carcinomas in men of this age. Thus, if the diagnostic procedures utilized on living patients were as sensitive as those implemented upon autopsy, the estimated incidence of breast and prostate carcinomas would increase dramatically, since even more non-life-threatening cancers would be registered (see also Supplementary Sidebar 16.2).

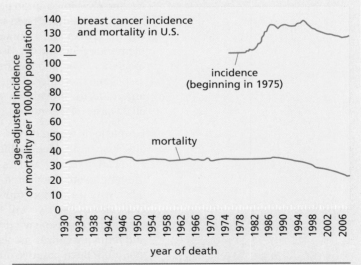

Figure 16.3 Breast cancer incidence vs. mortality in the United States The age-adjusted *incidence* of breast cancer has been increasing steadily over the past several decades, while *mortality* from this disease was quite constant until the end of the twentieth century, when it began to decline significantly. Most of the increase in incidence appears to be attributable to increased screening, but a small proportion of it may be due to real changes in the rate at which the disease strikes because of changes in reproductive practices, nutrition, obesity, age of menarche and menopause, and so forth. (From A. Jemal et al., *CA Cancer J. Clin.* 55:10–30, 2005.)

metastases, have revealed how traditional, widely used forms of therapy have not served many cancer patients well (Supplementary Sidebar 16.3).

In most of this chapter, we explore a number of strategies of therapy under development or recently introduced into the clinic and how their development has been informed by what we have learned in the earlier chapters of this book. The goal here is not to survey the full range of current research in these areas. That would be unreachable: a 2010 compilation of anti-cancer therapies in pre-clinical development or in clinical testing listed more than 2600 such projects that were being pursued by pharmaceutical companies and biotechnology firms. The therapeutic agents under development included low–molecular-weight drugs, proteins, monoclonal antibodies, and gene therapy strategies including viral vectors. Rather than being encyclopedic, we will concentrate here on a small number of recently developed therapies that illustrate how discoveries described in the previous chapters have inspired novel strategies for treating cancer, and how molecular diagnosis will increasingly play a part in the development and clinical introduction of novel therapies. These therapeutic strategies hold great promise, and invariably their true potential is yet to be realized. The anecdotes surrounding the development of each of these agents are interesting and provocative, because they teach important lessons about the triumphs and pitfalls of developing novel anti-cancer treatments. Note that a number of monoclonal antibody–based therapeutic strategies have already been discussed in some detail in previous chapters, as have several therapies focused on preventing or blocking tumor angiogenesis.

Almost all of the research findings described throughout this textbook will stand the test of time and be considered credible and correct (though perhaps not that interesting) a generation from now. However, those who love certainty and eternal truths will find the stories that follow to be unsatisfying for a very simple reason: the work reported is in great flux and the outcomes are uncertain. Many of the newer therapies will seem quaint and anachronistic a decade after this chapter is written. The campaign to convert insights about cancer's molecular causes into new ways of curing disease has just begun.

Note also that we will pass over descriptions of how molecular biology is changing cancer *prevention* strategies. Thus, in this chapter we will not examine the major advances that have been made in developing vaccines that protect against hepatitis B virus (HBV) and human papillomavirus (HPV) infections (for example, see Supplementary Sidebar 15.5); these vaccines should be highly effective in reducing the incidence of hepatomas and cervical carcinomas, which are major sources of cancer-associated mortality in certain parts of the world. (If the past history of public health is any guide, the prevention of cancer will ultimately yield far greater reductions in overall disease-related mortality than will therapies of the sort discussed in this chapter.)

Simple logic would dictate that the newer, "rationally designed" agents, because they attack specific, identifiable molecular targets, are likely to be far more effective in eradicating tumors and less encumbered with toxic side effects than are older agents; the latter were discovered through empirical trial-and-error testing that was undertaken without any foreknowledge of the biochemistry and molecular biology of cancer cells. However, as we will see in Section 16.2, these more traditional ways of treating neoplastic disease have proven to be highly effective in halting disease progression and reducing cancer-associated mortality, explaining why these treatments merit our attention. Such curative outcomes are, ironically, still not the endpoints achieved by the recently developed therapies, as we will see.

16.1 The development and clinical use of effective therapies will depend on accurate diagnosis of disease

In previous chapters, we repeatedly categorized cancers in terms of their tissues of origin and their stages of clinical progression. Almost always, these assignments have been dictated by the appearance of normal and malignant tissues under the microscope. On some occasions, to be sure, we have refined these classifications by describing certain molecular markers (for example, expression of HER2 in breast cancers) and the implications that they hold for prognosis. But in general, histopathology has reigned supreme in our discussions, as it has in the practice of clinical oncology for more than half a century.

It has become increasingly clear, however, that the traditional ways of classifying cancers have limited utility. Truly useful diagnoses must inform the clinician about the underlying nature of diseases and, more important, how each disease entity will respond to various types of therapy. As we have learned more about human cancers, we have come to realize that many human cancers that have traditionally been lumped together as examples of a single disease entity should, in fact, be separated into several distinct disease subcategories. This helps to explain why many existing anti-cancer therapeutic strategies used over the past three decades have had such low overall success rates, in that they have treated heterogeneous tumors, only a subset of which are likely to respond to targeted therapeutic attack. These response rates also have important implications for the development of new drugs (Supplementary Sidebar 16.4).

Stated differently, the clinical oncologist confronts two issues: Should a diagnosed tumor be treated, and if so, what available therapies are appropriate for the subtype of tumor that has been identified? In an ideal world, the decision to proceed with treatment should be a challenging one, but in the real world, it is often simplified: treat almost all tumors (with the exception of skin cancers) aggressively in order to reduce as much as possible the likelihood of eventual life-threatening clinical progression.

The complexity of the decision to proceed with treatment should ideally confront the fact that diagnosed tumors fall into three classes:

1. Indolent tumors that have low invasive and metastatic potential and will remain in such a state during the lifetime of the patient.

2. Highly aggressive tumors with a propensity to metastasize that have, with high probability, disseminated by the time that the primary tumor has been diagnosed.

3. Tumors of intermediate grade that have the potential to disseminate but can be excised or treated with cytotoxic therapies before dissemination occurs and life-threatening metastases are formed.

Tumors of the first class are not worthy of treatment, including surgery, and should be left undisturbed. (Indeed, surgery may even, under certain conditions, provoke otherwise-indolent growths to become clinically apparent if not aggressive.) One illustrative example of these is pancreatic neuroendocrine tumors (NETs)—that is, tumors of the pancreatic islets (see Figure 13.36)— which constitute about 3% of pancreatic carcinomas and are discovered incidentally during the course of high-resolution imaging, often undertaken for other conditions. Asymptomatic patients with these low-grade "incidentalomas," as the tumors are termed amusingly by oncologists, confront an 86% five-year progression-free survival. Like any type of pancreatic surgery, excision of these tumors is accompanied by a high degree of post-surgical morbidity; such morbidity may eclipse the clinical benefit that these patients receive from surgery, given the low likelihood of clinical progression. [Contrast their prognosis with that of a recently diagnosed pancreatic exocrine adenocarcinoma patient (see Figure 11.12B), whose overall five-year survival is only ~5%.] At present, diagnostic criteria do not distinguish between the small subset of NETs that will eventually progress and the great majority that are unlikely to progress during a patient's lifetime. Papillary thyroid carcinomas represent another class of incidentalomas: in one of the few systematic post mortem studies of this disease, a Finnish research group found these growths in one-third of those autopsied, while only 0.07% of annual deaths (in the United States) can be attributed to this disease.

Since truly effective treatments of most kinds of metastases are not available at present, this raises the question whether tumors of the second class are worth treating, as they will become lethal no matter what therapies are attempted. (The proviso here is that existing treatments may ameliorate symptoms over an extended period of time and may even forestall the inevitable, thereby extending the patients' life spans significantly.) Cynics argue simplistically that patients carrying these two classes of tumor should be left untreated, since the long-term outcome is predictable no matter

Table 16.1 Examples of alkylating agents or other DNA-modifying agents used to treat cancer

Name	Chemical structure	Effects on DNA	Examples of clinical use
bleomycin	nonribosomally synthesized oligopeptide	strand breaks	Hodgkin's lymphoma, squamous cell carcinoma
cyclophosphamide	nitrogen mustard analog	cross-links N^7 of guanine	lymphoma, leukemia
dacarbazine	nitrogen mustard analog	methylation of guanine	Hodgkin's lymphoma, melanoma
melphalan	nitrogen mustard analog	alkylates N^7 of guanine	multiple myeloma
chlorambucil	nitrogen mustard analog	base alkylation, cross-links	chronic lymphocytic leukemia
carboplatin/cisplatin	alkylating-like agent	intra-strand cross-links	breast, ovarian, testicular cancer
temozolomide	alkylating agent	methylation of guanine	glioblastoma multiforme

what clinical strategy is chosen. (Unappreciated by the cynics is the possibility that the sub-optimal responses of the second class of tumors to existing therapies will serve as foundations for the future development of vastly more effective therapies that will save those who, at present, confront inevitable disease progression and death.)

The third class of tumors—those of intermediate grade—are those that should and will command most of our attention in this chapter—the "in-between" tumors whose treatment will actually prevent metastatic progression and achieve long-term, even curative responses. In reality, in countries like the United States where concerns about neglectful treatment verging on medical malpractice dominate, tumors of all classes are treated, often aggressively. Such unfocused treatments will soon become issues of the past, if only because the costs of unnecessary therapies will no longer be economically sustainable.

At present, such uniformly applied therapies may have other downsides beyond economic costs. Thus, anti-cancer treatments often incur numerous side effects and some may actually increase the incidence of second-site cancers arising years later. For example, in the early 1980s, breast cancer patients receiving the then-standard dose of cyclophosphamide (a chemotherapeutic drug that is also an alkylating agent; see **Table 16.1**) experienced a 5.7-fold increased risk of subsequently developing acute myelogenous leukemia (AML), ostensibly the consequence of the mutagenic actions of this drug. (Current treatment protocols use lower drug dosages and result in greatly decreased incidence of such second-site cancers.) All this points to the great need for more refined diagnostic tools—ones that can accurately predict responsiveness to various anti-tumor therapies and avoid use of therapies when they are not needed.

We focus for the moment on breast cancers in the United States. About 227,000 newly diagnosed invasive breast cancers and 63,000 *in situ* breast carcinomas were predicted for 2012, and the disease was predicted to claim about 40,000 lives that year. The great majority of the patients with invasive mammary carcinomas were treated aggressively with chemotherapy. Since the age-adjusted death rate from breast cancer in the United States did not change significantly throughout most of the twentieth century (see Figure 16.1B) during a period when truly effective therapies were not available, this suggests that a comparable annual frequency of life-threatening breast cancers continue to be formed today, and that the vastly larger numbers of invasive breast carcinomas currently being diagnosed—possibly more than three-quarters—are not likely to cause death, even without therapeutic intervention, much like the prostate cancers that are diagnosed in vast numbers in the West (see Supplementary Sidebar 16.2). As screening for breast cancer increases and the power to detect small, previously overlooked tumors improves (see Figure 16.2), this disparity between breast cancer incidence and mortality is likely to remain large.

Statistics like these underscore the desperate need to develop molecular markers that enable oncologists to distinguish between those tumors that truly require aggressive treatment and those that can be monitored repeatedly for signs of progression. In the case of other types of cancer, equally important distinctions must also be made, but of a far grimmer sort—between those cancers that are likely to show some response to therapy and those that will not, in which case compassionate care dictates that the disease should be allowed to run its natural course.

Gene expression arrays, of the type first described in Figure 13.18, show great promise by allowing clinicians to **stratify** cancers—to classify them into subgroups having distinct biological properties and prognoses. Gene expression arrays, often referred to as the key analytical tools of the science of *functional genomics*, allow a researcher to survey the expression levels of 10,000 or even 20,000 distinct genes in a tissue preparation. Subsequent computerized analyses of these expression arrays using **bioinformatics** make it possible to identify a small subset of these genes whose expression (at characteristically high or low levels) correlates with a specific biological phenotype, drug responsiveness, or prognosis. For example, the expression of a cohort of several dozen genes by a tumor may suffice to serve as a strong predictor of its degree of progression or its association with one or another specific subtype of cancer.

In the case of breast cancers, there has been a crying need to distinguish those primary tumors that are likely to become metastatic from those that will remain indolent and are therefore not likely to spread during the lifetime of the patient. Traditionally, the main prognostic parameters that have been used to predict the course of breast cancer development have been patient age, tumor size, number of involved axillary lymph nodes, histologic type of the tumor, pathological grade, and receptor status (that is, the expression of estrogen, progesterone, and HER2 receptors). Because these factors, when used singly or in combination, do not yield prognoses with a high degree of accuracy, the great majority of patients diagnosed with primary invasive breast cancers in the United States have been treated aggressively, even though only ~19% of such patients will ever develop life-threatening disease. (In the absence of modern improvements in diagnosis and therapy, the mortality would have amounted to ~25% of diagnosed invasive cancers.)

The use of gene expression arrays and bioinformatics has made it possible to predict the clinical course of breast cancer progression with more than 90% accuracy (**Figure 16.4**). Such a high prognostic accuracy holds the promise of sparing many women exposure to unnecessary chemotherapy. And in the future, the details of an expression array analysis are likely to inform the oncologist about the treatment protocol that is most likely to yield a durable clinical response or even a cure (**Sidebar 16.2**).

Analyses of this sort are only the beginning steps in a large-scale effort to analyze a variety of human cancer types by means of expression arrays; the resulting information should make it possible to stratify the cancer types into subtypes, and, on the basis of the resulting information, to devise therapies tailored to each specific subtype. For example, B-cell lymphomas have presented a quandary to the oncologist

Sidebar 16.2 Expression arrays and tumor origins As we read earlier in Chapter 11, the somatic mutations and epigenetic alterations that tumors sustain en route to full-blown malignancy rarely eradicate the influence of the differentiation programs of their normal cells-of-origin (see Sidebar 11.2 and Figures 11.13 and 14.52). In most tumors, these programs are transmitted heritably from one cell generation to another, in spite of the profoundly disruptive effects of various mutations and heritable epigenetic alterations acquired during the course of tumor progression. These differentiation programs work hand-in-hand with the somatic mutations and epigenetic alterations to dictate cancer cell phenotype, and thus responsiveness of a tumor to therapies.

These complex interactions between genetic and nongenetic determinants of cancer cell behavior cannot be discerned by sequencing tumor cell genomes, creating a need to systematically monitor the nongenetic determinants of cancer cell behavior. In principle, cell phenotype could be surveyed in a systematic and quantitative way through proteomic analyses that reveal the complex array of proteins expressed by cancer cells, the levels of these proteins, and their states of post-translational modification. At present this is impractical, indicating the need to utilize gene expression analyses as the practical alternative for surveying cancer cell phenotype. In the future, complex bioinformatic algorithms will need to be developed to integrate the results of gene expression analyses with those of tumor genome sequencing in order to produce a more complete picture of the molecular and biochemical state of the cancer cell that should prove even more useful than the results currently achieved by expression array analyses alone.

Figure 16.4 Stratifying breast cancers using functional genomics
(A) Expression arrays were used to analyze the gene expression of 295 primary breast cancers diagnosed in women less than 53 years old. The group included patients with metastatic cells in their axillary lymph nodes as well as patients whose lymph nodes were free of cancer cells. Bioinformatics analyses of these tumors were then employed to choose a set of 70 "prognosis genes" whose expression could be used to stratify these breast cancer patients (*arrayed along vertical axis*), whose clinical course had been followed for a mean time of 7 years. The expression levels of these 70 genes (*arrayed along horizontal axis, names not given*) together with information about the patients' clinical history was then used to set a threshold that separated tumors that had a "good expression signature" from tumors that had a "poor expression signature." (B) This Kaplan–Meier plot reveals the stratification of a group of 151 breast cancer patients whose survival had been followed for 10 years following initial diagnosis. Using the criteria of panel A, they could be separated into two groups with dramatically different clinical courses. Taken together with other factors (such as the efficacy of chemotherapy), calculations indicate that women whose tumors carry a good expression signature derive virtually no benefit from adjuvant chemotherapy. (From M.J. van de Vijver et al., *N. Engl. J. Med.* 347:1999–2009, 2002.)

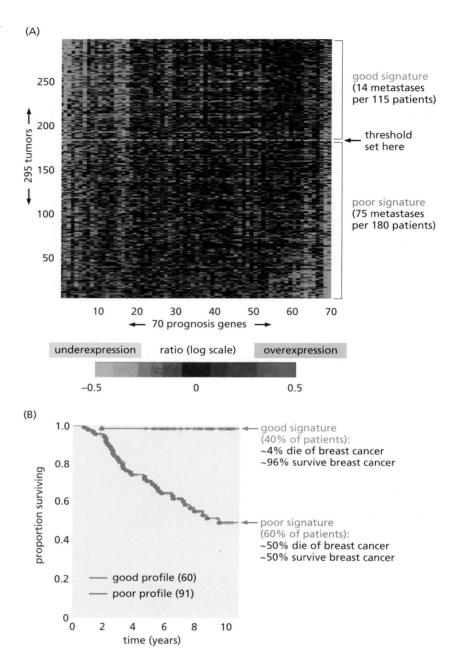

because their outcomes are so variable in the clinic, with some patients dying within four weeks of diagnosis while others are being cured, or are at least achieving 10-year remissions without any clinical symptoms. At the same time, all these tumors have a very similar appearance under the microscope (**Figure 16.5A**, *above*). However, use of gene expression arrays (*below* in Figure 16.5A) has allowed these tumors to be segregated into three distinct diseases with quite different clinical outcomes—primary mediastinal B-cell lymphomas, germinal-center B-cell-like lymphomas, and activated B-cell-like lymphomas (Figure 16.5B).

Of these three, both the activated B-cell-like lymphomas (ABCs) and the primary mediastinal B-cell lymphomas (PMBLs) exhibit constitutively high levels of NF-κB activity (see Figure 16.5C); this transcription factor (see Section 6.12) appears to be driving their proliferation and protecting them from apoptosis. Accordingly, drugs that target the NF-κB pathway, specifically its upstream activator, IKK, have been used in attempts to affect these two subtypes of diffuse large B-cell lymphoma (DLBCL) cells propagated in culture, and indeed both groups of cells are killed once they lose

IKK activity (see Figure 16.5D). Cultured cells from the third lymphoma subtype, germinal-center B-cell-like, do not show high NF-κB activity and are essentially unaffected by such treatment.

Interpretation of the gene expression patterns of tumors of complex histology, such as carcinomas composed of both epithelial and stromal cell types, is often confounded by the fact that the RNA transcripts being measured are a mixture deriving from multiple cell types. The technique of *laser capture microdissection* (LCM; see Supplementary Sidebar 13.5) now makes it possible to physically isolate the epithelial from the stromal cells present in a carcinoma sample that has been mounted on a microscope slide. This allows the gene expression pattern of the two groups of cells to be analyzed separately, enabling further refinement of these analyses and, potentially,

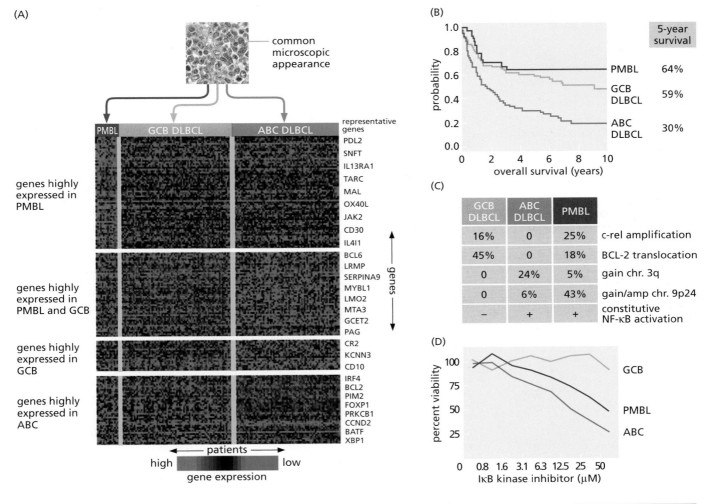

Figure 16.5 Stratification of diffuse large B-cell lymphomas
(A) DLBCLs represent a group of several subtypes of B-cell neoplasia that are essentially indistinguishable in the microscope from one another and from primary mediastinal B-cell lymphomas (*above*). However, use of gene expression arrays (*below*) allows these tumors to be stratified into three subtypes, termed primary mediastinal B-cell lymphomas (PMBLs), germinal-center B-cell (GCB) DLBCLs, and activated B-cell-like (ABC) DLBCLs. In this expression array, a set of genes whose expression levels have been found to be useful in making this stratification is plotted along the vertical axis, while a set of patient tumors (*unlabeled*) is plotted along the horizontal axis. (B) This Kaplan–Meier curve illustrates the greatly differing disease courses that patients with the three subtypes experience. (C) The tumors that are classified through the expression array

analyses shown in panel A exhibit distinct karyotypic and biochemical alterations. (D) The fact that the PMBLs and the ABCs show high constitutive levels of NF-κB activity suggests that they may be particularly susceptible to disruption of this signaling pathway by inhibition of the upstream activator of NF-κB, IκB kinase (IKK; see Figure 6.29A). This is borne out by experiments in which MLX105, a pharmacologic inhibitor of IKK, was applied to cell lines derived from the three types of lymphoma and indeed showed differential effects on these three cell populations growing *in vitro*. (A and C, from L.M. Staudt and S. Dave, *Adv. Immunol.* 87:163–208, 2005. B, from A. Rosenwald et al., *J. Exp. Med.* 198:851–862, 2003. D, from L.T. Lam et al., *Clin. Cancer Res.* 11:28–40, 2005.)

even greater accuracy in the stratification of tumor samples. More recent improvements in gene technology have since made it possible to analyze the *transcriptome* of a cell—its spectrum of expressed mRNAs—by sequencing the reverse transcripts of these mRNAs (the procedure of RNA-Seq); yet other procedures are being developed to enable sequencing of the transcriptomes of individual cells retrieved from normal and neoplastic tissues.

Beyond these gene expression analyses stands a generation of novel diagnostic tools involving the science of **proteomics**, in which the spectrum of proteins expressed in a patient's tumor or serum will provide critical diagnostic information. The long-term goal of all of these analytic techniques—both functional genomics and proteomics—is to assign each patient's tumor to a specific subtype of disease and to apply drug therapies that are proven to be effective for treating a particular subtype of cancer but not other, superficially similar tumors, for which treatment may not be effective. Such tailor-made drug therapies hold the promise of yielding high response rates in narrowly defined patient populations.

16.2 Surgery, radiotherapy, and chemotherapy are the major pillars on which current cancer therapies rest

An exclusive focus in this chapter on the newer, rationally designed agents is not warranted because of one simple fact: the older treatments—surgery, chemotherapy, and radiotherapy—have been shown over many decades' time to be highly effective in extending the survival of cancer patients and in eradicating certain types of tumors with curative outcomes, while the newer, "rational" agents can rarely boast such successes. So, we begin this chapter with these older agents, their successes and failures.

The multi-step nature of cancer development, specifically the notion that primary tumors spawn metastases, has been a central concept in anti-cancer therapy since the end of the nineteenth century. Radiotherapy developed only slowly during the course of the twentieth century, and chemotherapy was initiated only after World War II. Through much of the century, this left surgery as the only form of therapy that offered any hope of reversing the disease, perhaps even achieving cures.

The discipline of surgical oncology that developed embraced two notions as fundamental truths: small tumors will inevitably develop into large tumors, and the resection of primary tumors represents an effective means of reducing the risk of metastatic relapse—the resurgence of disease long after the primary tumor has been removed. For example, resection of early-stage colorectal carcinomas yields 95% patient survival after five years—essentially a curative outcome—even without **adjuvant** (post-surgical) chemotherapy.

In fact, the accepted truths governing surgical practice were rarely subjected to critical tests until the last decades of the twentieth century, when the specialty of *outcomes research*, often referred to as evidence-based medicine, arose. The evidence in these cases came from clinical trials in which two or more alternative surgical strategies were compared in randomized patient populations. These randomized trials revealed that some surgical procedures produce highly effective clinical outcomes while others are of dubious utility, as described in Supplementary Sidebar 16.3.

Radiation oncology began soon after Röntgen's 1895 discovery of X-rays and the frequent observation soon thereafter that exposure to this form of electromagnetic radiation resulted in burn damage to normal tissues, including even tissue necrosis. The widespread use of radiotherapy in the oncology clinic awaited the development after World War II of the means of directing these rays in relatively narrow fields of radiation that were focused on diagnosed, clearly delineated tumors, almost invariably primary tumors. Focused adjuvant radiotherapy has become an essential clinical tool in reducing post-surgical relapse in the surgical field (that is, the tissues adjacent to resected primary tumors). A dramatic example comes from post-lumpectomy (see Supplementary Sidebar 16.3) follow-up of older women with early-stage breast cancer: adjuvant (post-surgical) radiation reduces by 50% the rate of subsequent mastectomy (resection of the entire breast rather than just the area immediately around the tumor) during the decade following initial surgery. The use of radionuclides—radio-active isotopes administered intravenously as components of a pharmaceutical drug or coupled to monoclonal antibodies—underlies systemic radiotherapy and is often used when all or parts of a tumor cannot be readily resected or when disseminated cancer cells need to be eradicated.

Chemotherapy has far more recent origins: An often-related anecdote describes how the 1943 bombing of an American warship in the harbor of Bari, Italy, led to release of a large cloud of mustard gas of the sort used in World War I chemical warfare. Almost a thousand people died, sooner or later, from the effects of this explosion and released cloud of gas. The clinical deterioration of some survivors of this catastrophe reawak-ened interest in the 1919 discovery that exposure to mustard gas leads to depletion of bone marrow cells and thus anemia. In fact, research begun independently at Yale University in Connecticut in 1942, a year earlier, had revealed that intravenous doses of mustard gas—named after its characteristic odor—led to temporary regression of a lymphoma. We have encountered chemicals of this class in the form of alkylating agents (see Section 12.6) that are used, among other applications, to treat glioblasto-mas (Table 16.1). (Recall that in addition to their cytotoxicity, these agents are highly mutagenic, leaving the genomes of exposed cells with hundreds if not thousands of point mutations (see Figure 12.7B); this reveals another side of a number of anti-can-cer treatments: in addition to their effects in reducing or eliminating tumors, X-rays and certain cytotoxic agents are also carcinogenic, and their short-term successes in producing clinical remissions may be counterbalanced by the appearance years later of independently arising, second-site tumors that are consequences of their muta-genic actions.)

The transient clinical response of a lymphoma observed at Yale was followed by rapidly expanding interest in similar agents in the years that followed. The emphasis here was on cytotoxic agents that, for unknown reasons, killed neoplastic cells preferentially by sparing normal tissues, or at least by inflicting only tolerable side effects on the patient. Almost always, discovery of the biochemical and cell-biological mechanisms of these drugs came decades after their introduction into the clinic, and in many cases, as discussed below, we still do not understand the precise mechanisms of action of a number of highly effective cytotoxic drugs.

By 1947 the utility of another class of cytotoxics was discovered, initially in the form of an agent for treating pediatric leukemias and lymphomas. In this case, the compound in question—aminopterin—acted as an antagonist of folate metabolism, inhibiting the formation of tetrahydrofolate and thereby blocking some of the critical biosyn-thetic reactions that are required for the assembly of nucleotides and thus the synthe-sis of DNA and RNA. A young boy treated that year for lymphoma survived and lived another half century, representing the first documented cure of this disease. Com-pounds like aminopterin are often termed **antimetabolites** because they interfere with the normal functioning of specific metabolites or the enzymes that produce them in the cell. Most of these agents closely resemble and are therefore chemical analogs of normal metabolites. For example, a number of highly effective antimetabolites are purine or pyrimidine analogs that operate either by preventing normal biosynthesis or by becoming incorporated into DNA, whose function they then inhibit (**Table 16.2; Figure 16.6**).

Another class of antimetabolites affects normal cell function by interfering with microtubule assembly, either inhibiting or fostering it. These were one of the first use-ful categories of anti-cancer agents that were discovered through screens of librar-ies of *natural products*, that is, collected products of various plants, molds, and even animal species. Paclitaxel, initially named taxol, was discovered as the product of the Pacific yew tree, *Taxus brevifolia*. Its ability to block the breakdown of microtubules at the end of mitosis leads to potent therapeutic effects on a variety of commonly occur-ring tumors, including those of lung, ovary, and breast and head-and-neck squamous cell carcinomas. Acting in the opposite direction are microtubule-depolymerizing agents that prevent assembly of microtubules, often by barring their interaction with the microtubule organizing centers (see Figure 12.38). One such agent is colcemide, which is often used experimentally to trap cells in the metaphase of mitosis, making

Table 16.2 Examples of antimetabolites used to treat cancer

Name	Chemical structure	Targeted reaction	Examples of clinical use
methotrexate	folate analog	formation of tetrahydrofolate	breast cancer, lymphomas
6-mercaptopurine	purine analog	purine biosynthesis	leukemia, NHL
doxorubicin	natural product[a]	intercalating agent, inhibits topoisomerase	wide range
thioguanine	guanine analog	purine biosynthesis	acute granulocytic leukemia
fludarabine	purine analog	ribonucleotide reductase, DNA replication	chronic lymphocytic leukemia, NHL
cladribine	adenosine analog	adenosine deaminase	hairy-cell leukemia
bortezomib	peptide analog	proteasomal degradation	multiple myeloma
paclitaxel	natural product[a]	microtubule destabilization	lung, ovarian, breast cancer
etoposide	natural product[a]	DNA unwinding	lung cancer, sarcomas, glioblastoma
mitoxantrone	topoisomerase inhibitor	DNA unwinding	AML, breast cancer, NHL
irinotecan	topoisomerase inhibitor	DNA unwinding	colorectal carcinoma
vinblastine	natural product[a]	microtubule assembly	Hodgkin's lymphoma
vorinostat	hydroxamic acid	histone deacetylation	cutaneous T-cell lymphoma
azacitidine	pyrimidine analog	DNA methylation	myelodysplastic syndrome

Abbreviations: NHL, non-Hodgkin's lymphoma; AML, acute myelogenous leukemia.
[a]Complex structure.

karyotyping possible (see Figure 1.11). More useful clinically are vinblastine and vincristine, both derived from *Vinca rosea*, a periwinkle plant growing in Madagascar. The discovery of this class of agents in Canada in 1958 led to their frequent use in lymphomas, non-small-cell lung cancers, and breast cancers, as well as head-and-neck squamous cell carcinomas.

In 1965, yet another class of agents were found that, like alkylating agents, induced covalent modifications of DNA, creating adducts that were not readily removed by cells' repair machinery. Cisplatin [*cis*-PtCl$_2$(NH$_3$)$_2$] was discovered serendipitously as an antibacterial agent that formed at platinum electrodes and subsequently showed potent anti-cancer activity. Like bifunctional alkylating agents, it generates covalent cross-links within DNA; certain alkylating agents form inter-strand cross-links, while cisplatin forms largely intra-strand cross-links, usually between two adjacent guanines. Prior to the advent of cisplatin and the related drug carboplatin, the cure rate for testicular cancer was in the range of 10%; these days, however, use of cisplatin and related agents leads to curing 90 to 95% of those suffering from this tumor.

The discovery of these various cytotoxic drugs led repeatedly to confrontations with three major questions. First, how do they kill cancer cells? Second, why are cancer cells killed more readily than normal cells? And third, how do cancer cells and the tumors they form develop resistance to agents that initially were effective in treating these tumors? These questions have been transmitted to studies of the more recent, molecularly designed agents that are the topics of the bulk of this chapter.

In fact, for many of the cytotoxic drugs that have been in use for half a century or more, answers to these questions remain elusive. A number of mechanistic explanations of **selectivity** have been proposed (Table 16.3). The simplest one—that cytotoxic drugs kill proliferating cells selectively—is hard to reconcile with the fact that the bulk of cancer cells within most tumors have a very low proliferative index. To this day, experts queried on how many of these agents function, why they selectively kill cancer cells, and how tumors develop drug resistance readily admit to the minimal progress made in solving these fundamental questions.

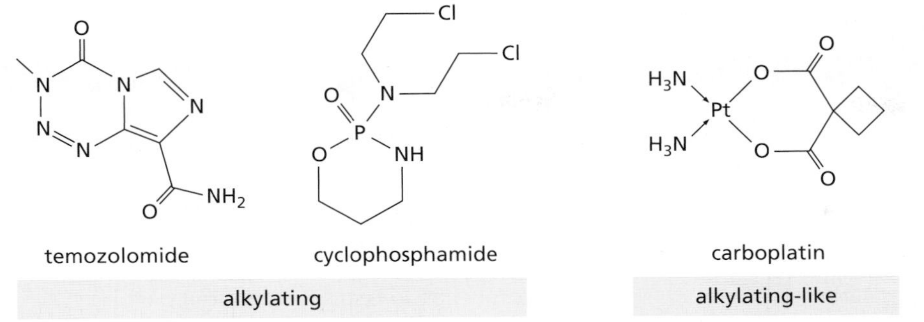

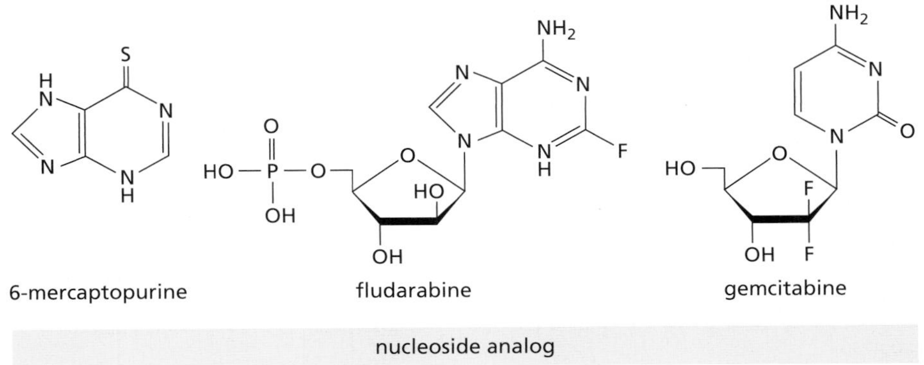

temozolomide   cyclophosphamide

alkylating

carboplatin

alkylating-like

6-mercaptopurine   fludarabine   gemcitabine

nucleoside analog

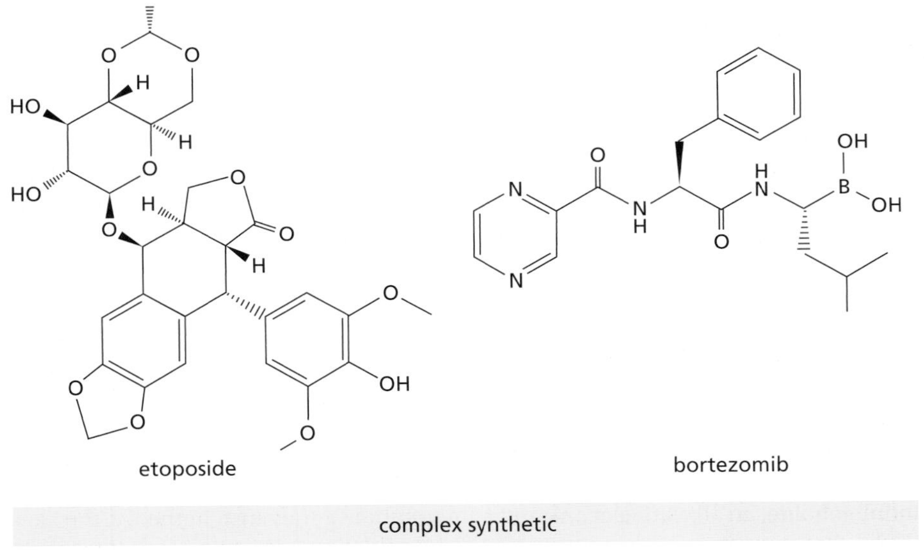

etoposide   bortezomib

complex synthetic

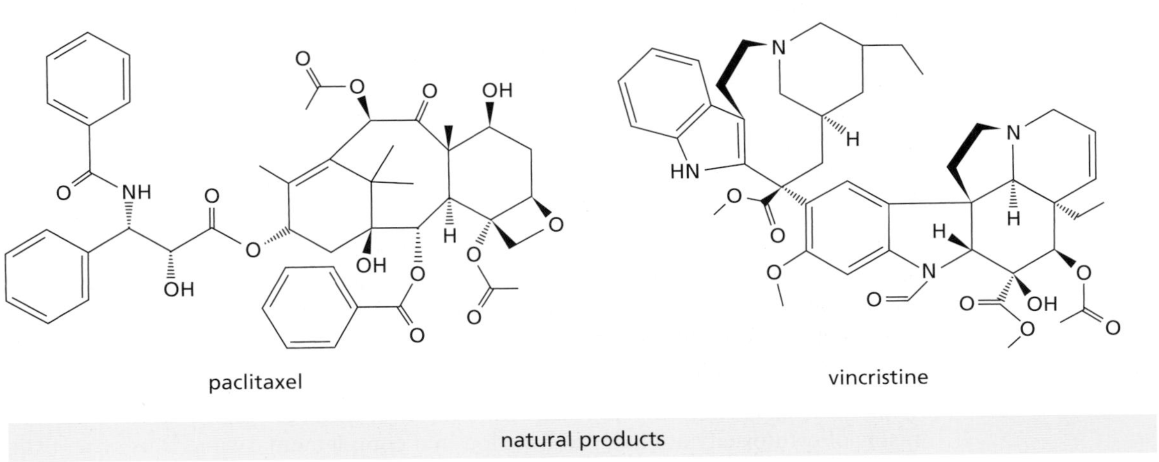

paclitaxel   vincristine

natural products

Figure 16.6 Cytotoxic drugs in current use The development of most cytotoxic drugs did not depend on detailed insights into the genetic and molecular mechanisms of tumor pathogenesis, but instead derived from investigations of the cytotoxic effects of a diverse array of organic molecules; most were products of synthetic organic chemistry while a minority were natural products. These agents work in a number of distinct ways to kill cancer cells, although in many cases the precise mechanisms of cytotoxicity are not well understood. Temozolomide and cyclophosphamide alkylate DNA, generating lesions that are difficult to repair; carboplatin acts in a similar fashion but makes cross-links, usually within strands. 6-Mercaptopurine interferes with nucleoside biosynthesis, while the cytotoxic actions of both fludarabine and gemcitabine seem to depend largely on these nucleoside analogs' being incorporated into cells' DNA and creating residues that are difficult to replicate and/or repair. Etoposide is a topoisomerase inhibitor; bortezomib disrupts proteasome function; while paclitaxel and vincristine function to stabilize and destabilize microtubules, respectively.

Table 16.3 **Known or suspected mechanisms of anti-tumor selectivity**[a]

Name	Mechanism conferring selectivity
cyclophosphamide	detoxified by ALDH in normal bone marrow
cladribine	detoxified by non-hematopoietic cell types
taxol/paclitaxel	high proliferation index[b]; other mechanisms unknown
multiple cytotoxic drugs	high proliferation index[b]
cisplatin	intact p53 function in testicular germ-cell tumors[c]
DNA-damaging agents	inability to halt cell cycle advance in response to DNA damage[d]
PARP inhibitors[e]	defective homology-directed repair
DNA-damaging agents	various types of defective DNA repair

[a]Drug selectivity, which leads to therapeutic indices larger than 1, implies that cancer cells are, for various reasons, hypersensitive to a therapeutic agent. Lack of selectivity and the absence of a significant therapeutic index imply that cancer cells are as sensitive to an agent as are cells in normal tissues.
[b]Proliferation-dependent cytotoxicity is not well understood. In certain cases, rapidly dividing cells are presumed to lack the time required to repair therapy-induced DNA damage before DNA replication occurs.
[c]Intact p53 function is known to render these cells highly susceptible to DNA damage–induced apoptosis.
[d]Defective p53 function, as well as nonfunctional cell cycle checkpoints, is presumed to allow cells bearing damaged, still-unrepaired genomes to advance into S phase, M phase, or both, resulting in stalled replication forks or mitotic catastrophe.
[e]Abbreviations: ALDH, aldehyde dehydrogenase; PARP, poly(ADP-ribose) polymerase.

Early on, medical oncologists recognized that drug resistance developed sooner or later for almost all the drugs and cancer types under treatment, being manifested as the outgrowth of drug-resistant variant cell populations. This was soon interpreted by applying the lessons learned from bacterial genetics, in which resistance to various types of antibiotic treatment (and before that, resistance to bacteriophage-mediated killing) was understood in terms of the selective outgrowth of mutant variants that may have already preexisted in bacterial populations prior to their being challenged by drugs or bacteriophage. Given the large numbers of cells in clinically detectable tumors—almost 1 billion in a tumor of 1 cm diameter—and the frequency of resistant mutants—perhaps one in a million—the emergence of drug-resistant variants seemed almost inevitable.

The response to this was the development of multi-drug protocols, often using combinations of drugs with distinct and complementary modes of cell killing, such as an antimetabolite, an alkylating agent, and a microtubule antagonist. In theory, the likelihood that variants preexisted in tumor cell populations that were simultaneously resistant to all three agents appeared to be astronomically low—the product of three probabilities, in this case one in 10^{18}. Indeed, the often-observed synergies between such drugs occasionally resulted in dramatic clinical responses in certain tumors, even in cures. For example, the ABVD regimen, which was developed in Italy in the mid-1970s and is used these days to treat advanced Hodgkin's lymphomas (HLs), involves administration of Adriamycin (doxorubicin), bleomycin, vinblastine, and dacarbazine, representing a DNA intercalating drug, a DNA-cleaving agent, a microtubule antagonist, and an alkylating agent (**Table 16.4**). Depending on the stage at which the tumor is diagnosed, 5-year progression-free survival (PFS) of treated patients ranges from 85 to 98%, with many of these patients essentially cured of the disease. Another multi-drug protocol elicits an almost 90% cure rate in childhood (<15 years of age at diagnosis) acute lymphoblastic leukemia (ALL).

In spite of these occasional striking results, there are no fixed rules to help in the design of multi-drug therapy, aside from the intuitive notion that drugs with distinct mechanisms of cytotoxicity are likely to function in a complementary and thus synergistic

Table 16.4 Examples of multi-drug treatment protocols

Acronym	Components	Mechanisms of action	Application
ABVD	doxorubicin, bleomycin, vinblastine, dacarbazine	intercalation, DNA strand breaks, microtubule inhibition	Hodgkin's lymphoma
CHOP	cyclophosphamide, hydroxydaunorubicin, vincristine, prednisone	alkylating, DNA intercalation, microtubule inhibition, steroid antagonist	non-Hodgkin's lymphoma
FOLFOX	fluorouracil, leucovorin, oxiplatin	pyrimidine analog, folic acid antagonist, DNA cross-linking	colorectal cancer
TIP	paclitaxel, ifosfamide, platinum agent cisplatin	microtubule antagonist, alkylating, DNA cross-linking	testicular cancer

way. Many decades after these multi-drug therapies were first contrived, we still do not understand why and how they work.

In contrast to these hematopoietic malignancies, the response rates of most solid tumors to multi-drug treatment are usually limited by the emergence of drug-resistant cancer cell subpopulations (Table 16.5); such refractory populations invariably arise at rates far higher than predicted from the measured frequencies of variant cells that are resistant to individual drugs. Often these therapeutic strategies are undermined by the trait of **multi-drug resistance**, displayed by many types of cancer cells and achieved by single genes and encoded proteins; the latter often function as plasma-membrane export pumps that actively extrude multiple molecular species and thus multiple types of drug molecules from cells (see Figure 16.21). We can also imagine that cancer cell alterations that weaken or cripple the pro-apoptotic machinery (for example, p53 inactivation) can likewise confer concomitant resistance to multiple cytotoxic drugs. Moreover, multi-drug therapies are constrained by the realities of toxicity; while individual agents can be delivered to patients at dosages well below the level at which patients experience serious side effects, concomitant application of three or four agents often leads to unacceptable levels of systemic toxicity. Typically, multi-drug therapies are used in which no more than one of the constituent drugs creates significant toxicity, such as **peripheral neuropathy**; kidney, liver, or cardiac toxicity; bone marrow suppression; and so forth.

The need to test drugs and administer them with maximum therapeutic effect led to the outgrowth of an entirely new discipline that is focused on the pharmacology of anti-cancer therapeutics. Thus, pre-clinical drug testing, focused on the effects of candidate drugs on cultured tumor cells and on tumor xenografts grown largely in laboratory mice, became the prelude to a sequence of tests on human subjects that is focused on the toxicity and tolerability of drugs, their utility in treating specific subtypes of neoplastic diseases, and ultimately their efficacy. Study of the **pharmacokinetics** (PK) of drugs revealed it as a critical determinant of therapeutic efficacy, being governed by the rates at which an administered drug is (1) absorbed into the circulation, as gauged by its concentrations in the plasma; (2) distributed in different physiologic compartments throughout the body; (3) metabolized in ways that either potentiate or inactivate its intended function; and (4) excreted. Still, these complex responses cannot predict the kinetics with which cells or molecular targets actually *respond* to administered drugs—the study of **pharmacodynamics** (PD). Thus, independent of the rates of drug accumulation in a tissue or within cells, there are the kinetics of cellular or biochemical responses to treatment. For example, if the cytotoxic effects of taxol depend on its ability to stabilize microtubules within cancer cells, how quickly can actual microtubule growth in these cells be observed following initial drug administration, and how rapidly do cytotoxic effects on cancer cells follow?

Table 16.5 Mechanisms of acquired resistance to anti-cancer therapies[a]

Nature of resistance	Mechanism of resistance
Multi-drug resistance[b]	increased expression of drug export pumps
Pan-drug resistance[c]	unknown
Drug detoxification[d]	enzymatic detoxification of drug molecule
Acquired drug resistance	refuge of cancer cells in drug-protected anatomical sites[e]
	failure of tissue to convert pro-drug into active form
	refuge of cancer cells in an anatomical site that provides protective trophic signals[f]
	massive stromalization[g]
	emergence of mutant, structurally altered cellular target[h]
	amplification of gene encoding targeted protein
	emergence of cells bearing alterations in genes whose products are functionally redundant with drug target[i]
	loss of drug importer[j]
	passage through an EMT[k]
	activation of anti-apoptotic regulators
Physiologic activation of compensatory adaptive mechanisms	
Resistance to EGF-R inhibition	up-regulation of IGF-1R signaling
	amplification of *Met* gene
	mutational activation of a *ras* gene
Resistance to Smoothened inhibition	amplification of *Gli2* gene
Resistance to Bcr-Abl inhibition	amplification of *Bcr-Abl* gene

[a]The entries in this table refer to tumors that are initially responsive to an applied therapy and then exhibit resistance that is manifested as regrowth of a tumor and thus indicates clinical relapse. Resistance may emerge because of the outgrowth of a therapy-resistant subpopulation of variant cells; such variant cells may preexist in the population prior to the onset of treatment or may arise as genetic or epigenetic variants that are formed *de novo* during the course of treatment. Alternatively, resistance may arise as a normal compensatory physiologic response to an initially applied therapy-imposed inhibition; this second form of resistance presumably occurs widely throughout a tumor rather than resulting from the selective outgrowth of a therapy-resistance subpopulation.
[b]As an example, concomitant resistance to paclitaxel, doxorubicin, etoposide, and vinblastine is exhibited by cells overexpressing P-glycoprotein, a drug export transporter operating in the plasma membrane.
[c]Pan-drug resistance refers to resistance against all agents that are applied to a tumor and cannot be attributed to increased drug export.
[d]As an example, lack of responsiveness of glioblastomas to the temozolomide alkylating drug is often due to expression of the MGMT enzyme, which detoxifies it (see Section 12.8).
[e]As an example, a variety of metastatic growths in the brain may be protected from chemotherapy by the blood–brain barrier, which blocks agents in the circulation from entering the brain parenchyma.
[f]As an example, lymphoma cells may survive in the thymus because thymic stromal cells release survival factors in response to the genotoxic stress provoked by chemotherapy.
[g]As an example, part of the difficulty of treating pancreatic carcinomas derives from the development in these tumors of a highly desmoplastic stroma that impedes transport of drugs from the circulation to the neoplastic cells.
[h]As an example, patients treated successfully with imatinib/Gleevec will develop drug resistance because of the emergence of cells expressing a mutant, structurally altered Bcr-Abl protein that no longer permits high-affinity binding of the drug.
[i]As examples, individuals whose tumors exhibit responsiveness to EGF-R inhibitors may develop resistance because of the mutational activation of a *ras* oncogene or because of *crkI* amplification. Resistance of CML cells to imatinib/Gleevec may develop because of the emergence of cells expressing altered p19ARF, Myc, p53, or Ras, which function to bypass the dependence of the tumor cells on the targeted Bcr-Abl oncoprotein. Resistance to B-Raf inhibition can develop through up-regulation of PDGF-Rβ expression or N-*ras* mutation.
[j]High-grade serous ovarian carcinomas are often treated with doxorubicin that is encapsulated in a synthetic liposome. Drug-resistant cells often emerge that have lost the LDL receptor–related protein (LRP1B), a cell surface protein that appears to be responsible for internalizing the liposomes.
[k]Passage through an epithelial–mesenchymal transition (EMT) results in, among other changes, the expression of drug efflux pumps in the plasma membrane and increases in expression of anti-apoptotic proteins.

In principle, the properties of the mechanism-based drugs that are described in the following sections should provide ready answers to the three questions posed earlier, that is, how does each drug kill cancer cells, why are its actions selective for neoplastic tissues, and how do tumors develop drug resistance after an initial period of responsiveness? In reality, in spite of the extraordinary level of detail known about these rationally designed agents, our ability to answer these questions remains limited.

16.3   Differentiation, apoptosis, and cell cycle checkpoints can be exploited to kill cancer cells

In principle, several distinct biological strategies might prove to be equally successful in eliminating established tumors or holding their growth in check. The most obvious of these are designed to induce the death of cancer cells, usually via apoptosis. Indeed, almost all of the existing nonsurgical strategies for eliminating cancer cells lead in one way or another to activation of their apoptotic pathways. However, an alternative therapeutic strategy relies on inducing differentiation, and we will consider this one first, if only briefly.

As described in Chapter 8, the acquisition of the malignant phenotype is usually accompanied by defective differentiation and associated entrance into a post-mitotic state. Recall that as tumor cell populations evolve to greater degrees of malignancy, they usually shed more and more markers of differentiation.

These behaviors suggest an attractive strategy for treating tumors: persuade cancer cells to differentiate and thereby enter into post-mitotic states. While we have learned much about the connections between cell cycle control and the regulation of differentiation programs (see Sections 8.9 and 8.11), most of this information has not yet been translatable into effective forms of therapy. The prominent exception has been one form of treatment of acute promyelocytic leukemia (APL). In the course of this therapy, the undifferentiated leukemic **blast** cells can be induced to differentiate into neutrophils by treatment with all-*trans*-retinoic acid (ATRA; **Figure 16.7B**). Treatment of APL patients with ATRA together with concomitant chemotherapy often results in complete **remissions**, with 5-year survival rates approaching 75 to 85%—suggestive of complete cures.

During the initial development of APL, the normal differentiation program of certain hematopoietic cells is blocked by the actions of the fusion protein PML-RARα that results from a 15;17 chromosomal translocation seen in the leukemic cells of almost all APL patients (see Figure 16.7A). This hybrid protein is composed of the PML (promyelocytic leukemia) protein, of unknown normal function, fused to the nuclear retinoic acid receptor (RARα) protein. The latter, upon binding its retinoic acid (RA) ligand, is normally able to induce the expression of genes that program cell differentiation in a variety of cell types throughout the body (see Section 5.8). [In the absence of ligand, RARα recruits histone deacetylases (HDACs) that cause repression of differentiation-associated genes, while in the presence of RA ligand, RARα recruits histone acetyltransferases (HATs) that permit expression of these same genes; see Section 1.8.]

In fact, the precise mechanism by which the PML-RARα protein prevents the differentiation of promyelocytic leukemia cells is not clear. An attractive possibility is that the PML-RARα fusion protein found in leukemic cells interferes with the differentiation-inducing functions of the normal RARα in bone marrow cells, thereby causing accumulation of large numbers of these cells in a stem cell-like state. (While the ligand-activated RARα acts to induce expression of certain target genes associated with the differentiation of hematopoietic precursor cells, the PML-RARα fusion protein acts as a repressor, or at least an antagonist, of RARα. This interference with the normal RARα function seems to derive from the ability of the PML-RARα complex to bind more avidly to the HDAC co-repressor complex than does the normal RARα.) Much of the therapeutic effect of all-*trans*-retinoic acid treatment appears to derive from its ability to cause the PML-RARα complex to release its grip on HDACs and to induce subsequent ubiquitylation and proteasome-mediated degradation of the PML-RAR fusion protein, resulting in relief of the block to differentiation. A complementary treatment derives from the use of arsenic trioxide (ATO), first discovered in China, which also drives degradation of PML-RARα; when applied together with ATRA, the two agents achieve a greater than 90% cure rate of APLs. Arsenic is usually portrayed as a highly toxic poison, whereas in this case it is clearly a remarkable lifesaver!

A similar mechanism likely explains the successes of therapy using 13-*cis*-retinoic acid (13cRA; see Figure 16.7C), related chemically to vitamin A and the all-*trans*-retinoic acid described above. It has been used with great success to cause the regression of pre-malignant lesions in the mouth and throat, thereby preventing or delaying their progression to head-and-neck cancers. Interestingly, loss of expression of RARβ—a second, similarly functioning RA receptor—is often observed in these pre-malignant growths, as well as in a variety of other human carcinomas, ostensibly because it helps the cells in these various lesions to avoid entrance into a differentiated, post-mitotic state. Moreover, reduction of RARβ receptor expression in mice, achieved through the use of an antisense transgene, results in the formation of large numbers of lung carcinomas by 18 months of age, providing further support for the notion that evasion of retinoic acid–induced differentiation serves as an important mechanism of carcinoma pathogenesis.

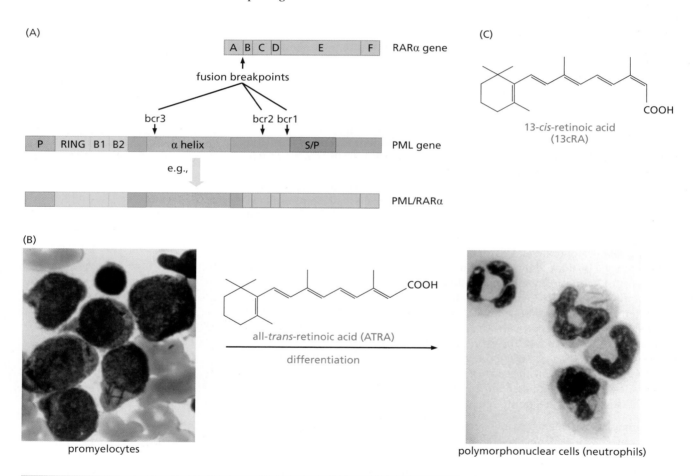

Figure 16.7 Retinoic acid and induced differentiation of cancer cells (A) In >95% of acute promyelocytic leukemia (APL) cases, a translocation involving Chromosomes 15 and 17 results in fusion of the gene encoding the retinoic acid receptor (RARα, also termed RARA) with the promyelocytic leukemia (PML) gene. Once formed, the resulting fusion protein appears to block the differentiation of promyelocytes into various granulocyte cell types, which normally depends on the actions of retinoic acid (RA) binding to its receptor, RARα. (bcr, translocation **b**reakpoint **c**luster **r**egion.) (B) Large numbers of promyelocytes carrying many granules in their cytoplasm are apparent in the circulation (*left*) of an individual suffering from APL. However, following all-*trans*-retinoic acid (ATRA) treatment, these immature promyelocytes disappear and are replaced by differentiated granulocytes, specifically the polymorphonuclear (PMN) neutrophils (*right*). ATRA causes inactivation and degradation of the PML-RARα fusion protein, allowing the normal RARα to drive normal differentiation. (C) 13-*cis*-retinoic acid has been used to cause regression of the pre-malignant precursors of head-and-neck carcinomas. (A, adapted from S. Kalantry et al., *J. Cell Physiol.* 173:288–296, 1997. B, courtesy of P.P. Pandolfi.)

Figure 16.8 Chemotherapy and mitotic catastrophe Many chemotherapeutic drugs may damage the chromosomes of cancer cells. Because some cancer cells lack key G_2/M checkpoint controls, they may advance into mitosis without having repaired the chromosomal damage. This may cause them to enter into "mitotic catastrophe" that results in aneuploidy, polyploidy, formation of micronuclei, and the eventual death of these cells. Seen here are the effects of low doses (50 ng/ml) of doxorubicin, a widely used chemotherapeutic drug, on Huh-7 human hepatoma cells. Over a period of 9 days, the nuclei of these cells grow larger or smaller and many eventually fragment into micronuclei, each of which carries a small number of chromosomes; this leads eventually to cell death, often by apoptosis. (From Y.W. Eom et al., *Oncogene* 24:4765–4777, 2005.)

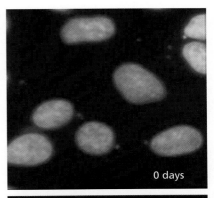

0 days

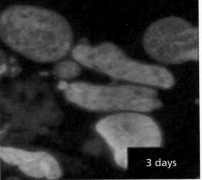

3 days

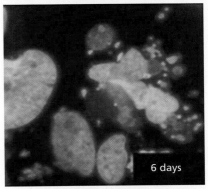

6 days

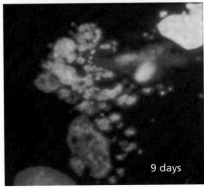

9 days

Beyond these two striking examples, differentiation-inducing strategies have had limited success in treating established cancers. For this reason, many cancer therapies that are under development are directed toward activating pro-apoptotic signals within cancer cells. At first glance, attempts at awakening the apoptotic response in cancer cells might seem to represent a futile undertaking, since we read earlier of the numerous ways in which cancer cells disable their apoptotic machinery (see Section 9.15). But the complexity and functional redundancies of the apoptotic circuitry dictate that, almost inevitably, important components of this circuitry remain intact even in the most aggressive tumors. It is these still-functional components that can, in principle, be targeted for activation, directly or indirectly, in order to eliminate tumor cells from the bodies of cancer patients. We are just beginning to learn the rules that may allow us, in the future, to predict the responsiveness of a patient's cancer cells to certain apoptosis-inducing therapies (see, for example, Supplementary Sidebar 16.5).

Many of the therapeutic strategies under development are designed to kill cancer cells by depriving them of the anti-apoptotic signals that sustain them. As we read in Chapter 9, cancer cells often depend on hyperactive growth factor signaling to generate intracellular anti-apoptotic signals (for example, those released by Akt/PKB; see Figure 9.36) that suppress the actions of the pro-apoptotic circuitry. This suggests that effective cancer therapies can be devised by interfering with this signaling at one or another step in the upstream signaling cascades that regulate Akt/PKB activity.

An alternative set of therapeutic strategies take advantage of the vulnerabilities that cancer cells have once they have discarded critical checkpoint controls operating in the normal cell cycle (see Table 16.3). For example, some cancer cells lack the checkpoint control that normally blocks entrance into mitosis (M phase) from the G_2 phase until significant damage to the genomic DNA or the chromosomes has been repaired. Consequently, tumor tissue may be treated by inflicting genomic damage through chemotherapeutics or radiation. While normal cells will tarry and repair this damage before advancing into M phase, cancer cells may ignore such damage and proceed blithely into mitosis, where they may stumble into a "mitotic catastrophe" that threatens their continued viability when they attempt to segregate their still-damaged chromosomes (**Figure 16.8**). This damage may be so overwhelming that it succeeds in triggering the residual apoptotic responses that these cells possess. Indeed, many of the traditionally used cancer therapeutics are suspected to take advantage of defects in checkpoint controls to destroy cancer cells, but hard evidence to sustain this point is not yet in hand. In the discussions that follow, however, we will focus on agents that target critical proteins rather than the genomes of cancer cells.

16.4 Functional considerations dictate that only a subset of the defective proteins in cancer cells are attractive targets for drug development

In the past (see Section 16.2), anti-cancer drugs were discovered and optimized for their cytotoxic effects on cancer cells, and thus on tumors as a whole, without any preconception as to precisely how these drugs were killing cancer cells. In contrast, the mindset of current anti-cancer drug development is to direct drugs at specific molecular targets within cancer cells rather than attempting to induce certain cell-biological responses (such as cytotoxicity). The logic underlying this newer strategy is simple: if the aberrant biological state of cancer cells derives from and depends on malfunctioning signaling proteins, then inhibiting or eliminating such proteins from a cell's circuitry should result in a cytostatic or cytotoxic response in these cells. Such responses should, in turn, halt further tumor progression or, better yet, cause a regression of tumors.

Two other factors have furthered this thinking. First, knowledge gleaned over the past three decades about the machinery that governs normal and neoplastic cells has provided us with a wealth of potential molecular targets. Second, since the molecular targets function differently in normal and neoplastic cells, targeting these molecules should yield substantial **therapeutic indices**, that is, selective killing of cancer cells versus normal cells and potentially reduced side-effect toxicities for cancer patients under treatment.

This agenda, attractive as it may be, is constrained by certain functional realities. Thus, with rare exception, drugs—usually low–molecular-weight organic compounds—inhibit rather than enhance biochemical functions. This simple fact drastically narrows the options for the development of anti-cancer drugs. As we saw in Chapter 7, the protein products of tumor suppressor genes—the so-called gatekeepers—contribute to cancer development through their absence, and attempts at developing low–molecular-weight compounds to replace or replicate these missing functions are impractical at present and may remain so forever. The few successes here represent relatively minor victories. For example, certain compounds can restore some p53 function by shifting mutant forms of the p53 protein from their functionally defective stereochemical configurations back into a wild-type configuration.

Precisely the same arguments apply to the proteins responsible for maintaining the cellular genome—the caretakers (see Chapter 12). Once again, their functions, often missing from cancer cells, cannot be restored by even the most complex drug molecules. And even if they were, little utility would derive from such successes. After all, if the progression of a tumor has been driven by defective DNA repair and resulting accumulation of mutant alleles, restoration of the missing repair function will have no effect on the many mutant sequences that have already accumulated in the genomes of its cancer cell constituents.

Once gatekeepers and caretakers are removed from consideration, such logic leaves oncoproteins—hyperactive forms of normal cellular growth- or survival-promoting proteins—as the most attractive targets for the development of anti-cancer therapies. These are molecules that, in principle, can be inhibited by drugs, resulting in reduction of their activity and, with luck, collapse of the neoplastic growth program. In fact, the signal-transducing proteins immediately downstream of hyperactive oncoproteins are also attractive targets, since most of these are also important positive effectors of signaling (**Figure 16.9**).

Certain genetic considerations may further narrow the range of molecules that are attractive targets for anti-tumor drug development. Earlier, we learned that as cancer progression proceeds, cell populations acquire a succession of genetic and thus biochemical alterations that ultimately lead these cells to the neoplastic growth state (see Chapter 11). This scenario raises a provocative question: Do the changes that were responsible for the *early* steps of multi-step tumor progression continue to play critical roles far *later*, when the full-blown malignant phenotype has finally been acquired? For example, if the initial step in the development of a tumor involved the formation of a *ras* oncogene, are the continued actions of this oncogene still required later by the highly malignant descendant cells? Or have some of the changes occurring later during tumor progression rendered the continued functions of a Ras oncoprotein unnecessary?

Take the case of pancreatic carcinomas, in which the K-*ras* oncogene is found in the great majority (~90%) of tumors. The acquisition of this oncogene occurs relatively early in tumor progression, since mutant K-*ras* oncogenes are often found in

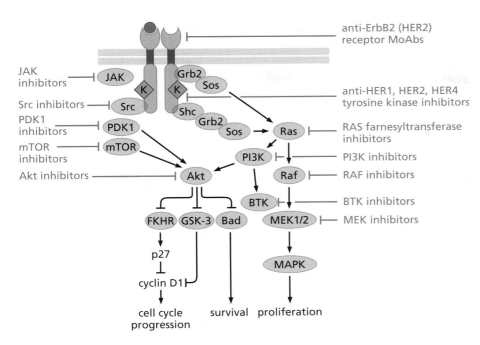

Figure 16.9 Inhibition of tumor growth by targeting downstream signaling elements As indicated in this diagram, signaling from receptors such as the EGF (HER1) and HER2/Neu receptors can be blocked in a number of ways. The ectodomains of the receptors can be targeted by monoclonal antibodies such as Herceptin. Moreover, the tyrosine kinase (TK) signal-emitting domains of these receptors can be targeted by a variety of low–molecular-weight compounds. In addition, a number of drugs have been developed that target proteins functioning as components of downstream signaling pathways, including those that inhibit Ras (through inhibition of its post-translational maturation involving farnesylation), as well as Raf and MEK (through inhibition of their serine/threonine kinase catalytic functions), and mTOR (through inhibition of the formation of functional signaling complexes between mTOR and associated partner proteins). (Courtesy of J. Baselga.)

pancreatic intraepithelial neoplasias (PanINs; see Figure 11.12C), the benign precursors of frank carcinomas. Do some of the subsequently acquired changes in, for example, the *Smad4/DPC4* and *p16*[INK4A]/*CDKN2A* genes make the K-Ras4B oncoprotein superfluous? If so, drugs designed to block K-Ras signaling will never prove useful for treating this class of cancers.

Here, we can take some encouragement from a number of mouse models of cancer development (**Table 16.6**). The oncogene responsible for initiating tumor progression in these transgenic mice can be shut off experimentally many weeks later in the tumors that eventually form. Such experiments have exploited the H-*ras* oncogene to create melanomas, the K-*ras* oncogene to induce lung adenocarcinomas, the *bcr-abl* oncogene to create leukemias, and the *myc* oncogene to make islet cell tumors of the

Table 16.6 Effects of shutting down expression of an initiating oncogenic transgene in tumor-prone mice

Transgenic oncogene	Response of tumors
Permanent regression after shutdown of transgene	
H-*ras*	melanoma regressed
K-*ras*	lung adenocarcinoma regressed
bcr-abl	B-cell leukemia regressed
myc	T-cell lymphoma, acute myelogenous leukemia, and epidermal, mammary, and islet cell carcinomas regressed
fgf-7	lung epithelial hyperplasia regressed
SV40 large T	salivary gland hyperplasia regressed if transgene expressed < 4 months
wnt1	mammary adenocarcinoma regressed
Persistence or relapse after shutdown of transgene	
SV40 large T	salivary gland hyperplasia relapsed if transgene expressed > 4 months
neu	mammary adenocarcinoma regressed and then relapsed
myc	mammary adenocarcinoma relapsed in presence of K-*ras* oncogene
wnt1	mammary adenocarcinoma relapsed in absence of p53 function

Adapted in part from D.W. Felsher, *Curr. Opin. Genet. Dev.* 14: 37–42, 2004.

pancreas as well as leukemias and lymphomas. In all of these cases, the tumor cells that arise continue to be dependent on the initiating oncogenes, as indicated by the regression of these tumors once expression of the initiating oncogenes is shut down. The behavior of certain human tumor cells also indicates the continuing contributions of initiating genetic elements (Sidebar 16.3).

This continued dependence on an oncogene, often an initiating oncogene, has been termed **oncogene addiction** to indicate the irreplaceable role that such a gene plays in the ongoing viability of the cancer cell and growth of the tumor. We will encounter this physiologic state again later in this chapter.

Oncogene addiction is not universal, however, as Table 16.6 also shows. Experiments with a mouse strain carrying a transgenic *myc* oncogene have yielded an equally dramatic but quite different outcome: shutdown of *myc* expression initially caused a regression of the transgene-induced mammary adenocarcinomas, but the tumors relapsed in a number of these mice. This suggests that the *myc* transgene, in addition to triggering carcinoma formation, encouraged genetic changes in the tumor cells that made its continued actions unnecessary later. Observations like this one clearly complicate targeting certain tumor-initiating oncoproteins for inactivation by anti-cancer drugs, since these proteins may no longer be playing critical roles, years later, in maintaining tumor cell viability and growth.

16.5 The biochemistry of proteins also determines whether they are attractive targets for intervention

The biochemical subtleties of the proteins that have been chosen as attractive targets for drug intervention further complicate attempts at developing novel anti-cancer drugs. These drugs are, almost invariably, low–molecular-weight organic compounds, since (1) in general, such molecules are produced far more readily by synthetic organic chemistry than molecules of higher molecular weight; and (2) small molecules are more likely to penetrate into the interstices of a tumor, thereby exerting therapeutic effects on all of its component cells.

Target molecules, for their part, must have domains within their structures that are capable of strong and specific interactions with small drug molecules. These potential molecular targets (for example, oncoproteins) fall into two major categories—those that are **druggable** and those that are not. "Druggability" implies that the target molecule has a structure that should make it vulnerable to attack and inhibition by low–molecular-weight compounds. Given these and other constraints, target molecules are always proteins of various sorts.

A protein is considered druggable if it has a cavity, usually a well-defined catalytic cleft; pre-clinical evaluation of candidate drugs is greatly helped if the protein has a well-defined, measurable enzymatic function. Such clefts are attractive for drug developers, because they usually can bind small organic molecules in a highly specific manner. In particular, these cavities often make it possible for a low–molecular-weight compound to form noncovalent bonds simultaneously with multiple amino

Sidebar 16.3 HeLa cells provide the most dramatic example of the ongoing importance of initiating genetic lesions Arguably the most extreme example of the continued influence of initiating genetic lesions is provided by the behavior of the cells of the human HeLa cervical cancer cell line. Recall that these cells were derived in 1951 from a highly aggressive cervical carcinoma (see Supplementary Sidebar 13.3) and, like almost all (>99.7%) cervical carcinomas, the initiation of this tumor is traceable to a human papillomavirus (HPV) infection, in this case HPV type 18 and its two encoded oncoproteins, E6 and E7. Half a century later, during which time cultured HeLa cells passed through many thousands of growth-and-division cycles, genetic strategies were used to shut down the HPV18 oncogenes being expressed in these cells. The shutdown of E7 expression led to reactivation of pRb function and cell senescence, while shutdown of E6 led to the reappearance of p53 and subsequent senescence or apoptosis in these cells. Therefore, in the case of HeLa cells, the initiating genetic change (acquisition of an HPV genome) continued to be absolutely essential for the maintenance of cancer cell proliferation and viability thousands of cell generations later.

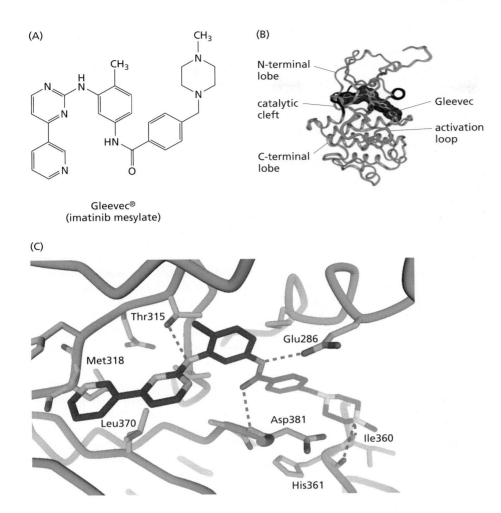

(A)

Gleevec®
(imatinib mesylate)

(B)

N-terminal lobe

catalytic cleft

C-terminal lobe

Gleevec

activation loop

(C)

Thr315

Met318

Leu370

Glu286

Asp381

Ile360

His361

Figure 16.10 Multiple contacts between drugs and their targets
(A) The chemical structure of Gleevec, which was developed to inhibit the tyrosine kinase activity of the Bcr-Abl fusion protein active in chronic myelogenous leukemia (CML), was the result of optimizing the structure of a precursor compound to which certain side chains were added while others were removed in order to improve drug binding to the catalytic cleft of the Abl tyrosine kinase domain. (B) The catalytic cleft of the Abl kinase is found between its N- and C-terminal lobes, shown here as *green* ribbon structures. A space-filling model (with van der Waals radii) of a Gleevec molecule is shown in *dark blue*, while the "activation loop" of Abl, which normally blocks access to substrates by the catalytic cleft, is shown in *light blue*. (This activation loop swings out of the way when the kinase shifts into its active configuration.) (C) The avid and specific association of Gleevec (*blue*) with the catalytic cleft of Abl depends on the formation of multiple hydrogen bonds (*red dashed lines*), as well as weaker van der Waals interactions (*not shown*). These bonds increase the binding affinity of the drug for the protein; at the same time, they explain the specificity of association, since the various pairs of proton donors and acceptors, which participate as partners in hydrogen bond formation, must be precisely positioned in three-dimensional space. (A and B, from B. Nagar et al., *Cancer Res.* 62:4236–4243, 2002. C, courtesy of E. Buchdunger, S.W. Cowan-Jacob, G. Fendrich, J. Liebetanz, D. Fabbro and P.W. Manley, Novartis Pharmaceuticals.)

acid residues lining their walls (**Figure 16.10**; see also Figure 16.46). Such multiple independent contacts enable a drug molecule to bind the targeted protein with great specificity and avidity. Equally important, such binding has a high likelihood of perturbing protein function, since, in the case of catalytic clefts, the drug molecule occupies a functionally critical domain of the protein.

Proteins lacking such catalytic clefts are often dismissed as being "undruggable." Transcription factors, for example, are widely thought (rightly or wrongly) to be undruggable, because they usually lack catalytic clefts and thus the much-sought drug-binding pockets. Hence transcription factor oncoproteins, such as Myc and Fos, are placed in the category of undruggable targets while the many kinases involved in cancer formation are placed in the camp of druggable target molecules. (The major exceptions to the lack of druggability of transcription factors are the nuclear hormone receptors, such as the estrogen and progesterone receptors. Because they have hormone-binding domains, these receptor proteins are, in principle, vulnerable to disruption by pseudo-ligands, such as tamoxifen, which binds and antagonizes certain functions of the estrogen receptor.) On average, pharmaceutical chemists judge about 1 in 5 cellular proteins to be druggable.

The presence of an identifiable catalytic function and apparent druggability does not, on its own, guarantee that an attractive target has been identified. Consider, for example, the case of the Ras oncoprotein, which has a clearly identifiable catalytic activity—its GTPase function. This enzymatic activity in Ras-expressing cells has never been the object of drug development, because the Ras GTPase, as we learned in Section 5.10, functions as a *negative* regulator of Ras signaling. Its inhibition would only augment the already-disastrous effects of the amino acid substitutions that create Ras oncoproteins in the first place. The same can be said of many tyrosine phosphatases, whose designated roles are to reverse the effects of growth-promoting tyrosine kinases (see Section 6.3), and, similarly, the activity of the PTEN phosphatase, which serves to counteract PI3K (see Figure 6.19). In response to the difficulty in attacking the Ras protein itself, a number of drug development strategies have focused instead on the enzymes that modify this protein and thereby enable it to become functional (Supplementary Sidebar 16.6).

The notion that molecular cavities provide attractive targets for drug development might also suggest that many types of protein–protein interactions represent druggable targets. After all, the confined space between two physically apposed proteins would seem to create a highly specific drug-binding pocket, and insertion of a drug into such a cavity might therefore destabilize or block the protein–protein interaction. Obvious candidates for such inhibition are the several types of cyclin–Cdk pairs, whose actions drive the proliferation of all cancer cells (see Chapter 8).

Unfortunately, most attempts at preventing these and other protein–protein associations through custom-made drug molecules have been unsuccessful. The numerous failures have been rationalized as follows: the association of two proteins with one another involves multiple points of binding between their interacting faces. These points of contact extend over molecular domains that greatly exceed the dimensions of drug molecules, which typically have a rather low molecular weight (generally <600). Consequently, only a small fraction of the contact points between two associating proteins can be blocked by any single drug molecule, and the association as a whole remains essentially unperturbed.

An exception to this widely accepted lore was announced in late 2003, when a low-molecular-weight compound that inhibits formation of the Mdm2–p53 complex (see Section 9.7) was described (**Figure 16.11A**). This development means that, in

Figure 16.11 Inhibitors of protein–protein interactions
(A) A search for compounds that inhibit Mdm2–p53 binding resulted in the discovery of Nutlin-2, which associates with the p53-binding pocket of Mdm2, whose surface is shown here. The usual interaction of the transactivation domain of p53 (*pink*; see Figure 9.6) with a hydrophobic cleft near the N-terminal domain of MDM2 (*gray, left*) (see Figure 9.12A) is mimicked by the Nutlin-2 molecule (*stick figure*) binding the same hydrophobic cleft of MDM2 (*right*). Nutlin-2 and closely related compounds prevent Mdm2-mediated p53 degradation and thereby act to trigger apoptosis in certain cancer cells, achieving both effects at low micromolar concentrations. (B) Some of the constituents of tea leaves induce apoptosis in tumor cells at very low concentrations. Epigallocatechin gallate (EGCG; *stick figure*) extracted from green tea binds the important anti-apoptotic protein Bcl-X_L with an inhibitory constant (K_i) of 490 nM and Bcl-2 with a K_i of 335 nM. (K_i values reflect the concentrations at which 50% of the activity of these proteins is inhibited; values in the submicromolar range are indicative of high potency.) A combination of three technologies—nuclear magnetic resonance (NMR) binding assays, fluorescence polarization assays, and computational docking—was used to derive this image, which shows the docking of EGCG in three adjacent hydrophobic pockets of Bcl-X_L (*yellow and green*), whose molecular surfaces are labeled P1, P2, and P3 here. These pockets together constitute a hydrophobic domain that Bcl-X_L, which is overexpressed in many cancer cells, uses to bind and neutralize pro-apoptotic BH3-only proteins (see Figures 9.26 and 9.28). (C) The drug termed ABT-737 was developed, using NMR-based screening, parallel synthesis (a form of combinatorial chemistry), and structure-based design, to visualize the interacting structures of a pro-apoptotic protein (Bak); this initially led to a candidate drug molecule termed compound 1 (*left*). Thus, the α-helix of the pro-apoptotic Bak protein (*green*

helix, middle right; see Figure 9.26) that binds and inhibits the anti-apoptotic Bcl-X_L protein (*red, white, blue surface*) is also occupied by compound 1 (*green stick figure, far right*). Compound 1 was then derivatized to generate a molecular species—ABT-737—whose properties more closely conformed to those of a clinically useful drug (see Supplementary Sidebar 16.7). The effects of ABT-737 on the growth of a human small-cell lung carcinoma (SCLC) xenograft are shown in the graph (*lower left*). The *black bar* indicates the time window during which ABT-737 was applied. An orally available derivative of ABT-737 has been introduced into clinical trials to gauge its effects on chronic myelogenous leukemia (CML) and small-cell lung cancer (SCLC). Results to date indicate that ABT-737 functions most effectively to sensitize cancer cells to the killing effects of FDA-approved cytotoxic drugs. (D) The small molecule ICG-001 (*left*) inhibits the association of the β-catenin–Tcf/Lef complex with the transcriptional co-activator CBP, a histone acetyltransferase (see Section 1.8). The transcription factor complex is responsible for inducing, among other genes, expression of the gene encoding survivin, a key inhibitor of apoptosis (IAP; see Section 9.15). As seen here, expression of survivin (*green, yellow immunofluorescence, upper right*) that is normally present in SW480 human colon carcinoma cells (*arrows*) is strongly reduced by the presence of ICG-001 (*lower right*); tubulin (*red*) is used in both cases as a *counterstain*. The concentration of ICG-001 used in this experiment causes an approximately 6-fold reduction in the activity of the survivin gene promoter (*not shown*). Clinical trials of ICG-001 were begun in 2012. (A, from A.C. Joerger and A.R. Fersht, *Annu. Rev. Biochem.* 77:557–582, 2008. B, from M. Leone et al., *Cancer Res.* 63:8118–8121, 2003. C, from T. Oltersdorf et al., *Nature* 435:677–681, 2005. D, left, from K.H. Emami et al., *Proc. Natl. Acad. Sci. USA* 101:12682–12687, 2004; right, from H. Ma et al., *Oncogene* 24:3619–3631, 2005.)

principle, the absence of p53 protein observed in many types of human tumors—often resulting from loss of p14^{ARF} expression or overexpression of Mdm2—can be reversed by drug treatment.

Subsequently, naturally occurring compounds isolated from green and black tea were found to block the binding and neutralization of pro-apoptotic BH3 proteins by the anti-apoptotic Bcl-2 and Bcl-X$_L$ proteins, doing so at relatively low concentrations

(A)

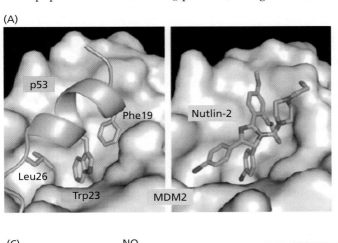

(B)

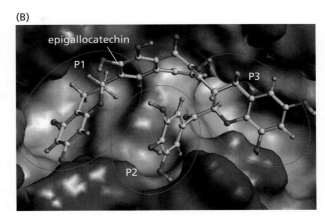

(C)

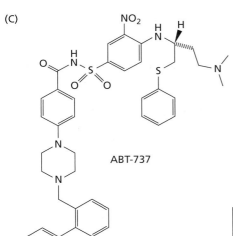

ABT-737

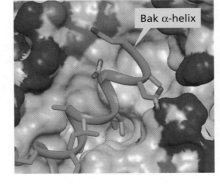

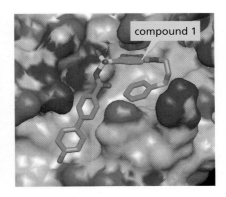

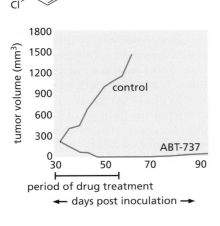

(D)

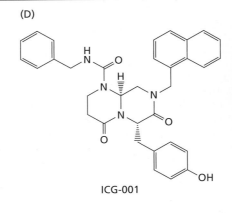

ICG-001

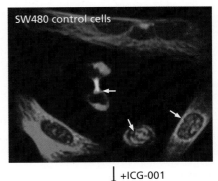

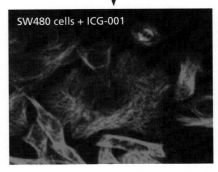

(see Figure 16.11B). A synthetic inhibitor of the Bcl-2, Bcl-X$_L$, and Bcl-w proteins has since been found that associates with these anti-apoptotic proteins with an affinity of approximately 1 nM—more than two orders of magnitude more avid than the tea compound (see Figure 16.11C). And a compound, ICG-001, identified through high-throughput screening (HTS; Supplementary Sidebar 16.7), was found to be capable of inhibiting the association of β-catenin with CBP (cyclic AMP response element–binding protein), a widely acting transcriptional co-activator that works with β-catenin to induce expression of key genes, including the gene encoding survivin (see Table 9.5), an important anti-apoptotic IAP protein (see Figure 16.11D), as well as cyclin D1. Both ICG-001 and a 20-fold more potent derivative, termed PRI-724, have entered into clinical trials for treatment of leukemias and carcinomas of the colon and pancreas.

In the years that followed these pioneering efforts, successes have been reported in targeting a number of other protein–protein interactions. Included among these are the interactions between Sos and Ras (see Sections 6.2 and 6.4) as well as the ligand–receptor interactions between VEGF and PDGF and their respective receptors (see Section 5.6).

These successes notwithstanding, a major lesson, learned time and again, is that kinases are among the few classes of cellular molecules that are attractive, drug-gable targets for anti-cancer therapy. As we have seen, many of these kinases function as oncoproteins that act to drive neoplastic proliferation and, at the same time, are enzymes possessing well-defined catalytic clefts. Recall that at least 518 distinct kinase-encoding genes have been enumerated in the human genome, of which 90 encode tyrosine kinases (Supplementary Sidebar 16.8), the latter being major players in many kinds of human cancer.

For some cancer researchers, this multiplicity of potentially druggable cancer targets represents an embarrassment of riches. However, for pharmaceutical chemists, numbers like these create a nightmare. Because almost all protein kinases are evolutionar-ily related (see Supplementary Sidebar 16.8), their catalytic clefts are structurally quite similar (**Figure 16.12A**). The similarity is even more striking among the clefts of the more closely related tyrosine kinases (Figure 16.12B), which are involved in cancer pathogenesis. How can one possibly develop agents that affect the actions of certain cancer-associated kinases, while leaving untouched the kinases required for normal cell proliferation and survival? Rational drug design and high-throughput screening, both described below, attempt to address these issues.

16.6 Pharmaceutical chemists can generate and explore the biochemical properties of a wide array of potential drugs

The ideology of "rational drug design," as it is often called, embraces the notions that (1) drugs should be targeted against specific proteins known to be malfunctioning within cells, thereby contributing to a disease state; (2) the candidacy of these pro-teins as attractive targets for therapeutic intervention should be further determined by their predicted druggability; and (3) the detailed molecular structures of such target proteins should inform the design of the chemical structures of the drugs that are to be developed. More specifically, chemical species must be synthesized whose detailed three-dimensional structures (that is, whose *stereochemistry*) enable them to fit, in a key-in-lock fashion, into specific pockets or sites within the far larger proteins that they are supposed to attack and disable (**Figure 16.13**; see also Figure 16.10).

In principle, knowledge of the detailed structure of a potential drug-binding cavity in a targeted protein should allow a skilled organic chemist to design and synthesize a molecule that fits snugly into this cavity and forms multiple noncovalent bonds with the amino acids lining its walls. However, this purely theoretical route for designing a novel drug structure has not yielded many useful products to date. For this reason, cur-rent drug discovery relies on more empirical ways of finding useful molecular struc-tures, involving surveys of hundreds of thousands of compounds via high-throughput screening (HTS; see Supplementary Sidebar 16.7).

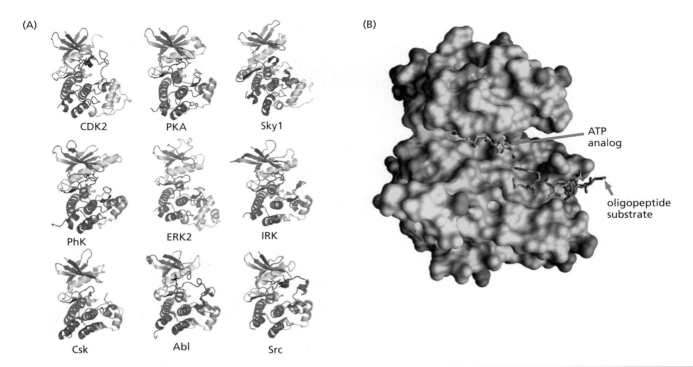

Figure 16.12 Similar structures of kinases (A) The difficulty of producing a specific kinase inhibitor is suggested by the striking similarities in structure of a variety of serine/threonine and tyrosine kinases. Depicted here as ribbon diagrams are the kinase domains of five serine/threonine kinases: CDK2 (see Chapter 8), PKA (cyclic AMP-regulated protein kinase A), Sky1 (SR-protein–specific kinase of budding yeast, a serine kinase involved in yeast nuclear RNA processing), PhK (phosphorylase kinase involved in glycogen metabolism), and ERK2 (extracellular-regulated kinase of the MAPK cascade; see Section 6.5); as well as four tyrosine kinase (TK) domains: IRK (insulin receptor kinase), Csk (C-terminal Src kinase), Abl (see Section 16.12), and Src (see Section 5.2). In all cases, the catalytic clefts of these kinases are sandwiched between the two major lobes (N- and C-terminal, *above and below, respectively*) of these proteins. (B) An extreme example of the structural similarities between related kinases is illustrated by this surface diagram in which the catalytic clefts and adjacent amino acid residues of the tyrosine kinase domains of the insulin receptor (IR; see the IRK of panel A) and the insulin-like growth factor receptor (IGF-1R) are compared. Identical amino acid residues are in *gray* while dissimilar ones are in *green*. This shows how very similar the catalytic regions of the two TK domains are and explains why it was difficult to find inhibitors of one tyrosine kinase receptor that do not affect the other. A threonine in the peptide linking the two lobes of the kinases is shown in *yellow*, while stick figures (*multicolor*) of an ATP analog (*left*) and an oligopeptide substrate (*right*) are also shown. Almost all TK antagonist drugs bind in the ATP-binding sites of the kinases that they inhibit. (A, courtesy of N.M. Haste, S.S. Taylor and the Protein Kinase Resource. B, from S. Favelyukis et al., *Nat. Struct. Biol.* 8:1058–1063, 2001.)

Imagine that HTS has yielded a drug molecule that inhibits the activity of a targeted protein in living cells, doing so at an IC_{50} in the 10 to 100 micromolar range (that is, drug concentrations in this range are required for 50% inhibition of the activity of the targeted protein). Further development of this particular drug becomes unrealistic, given the massive amounts of this agent that would need to be delivered into a patient's body in order to achieve a therapeutic effect. The chemical properties of this molecular species may or may not allow *derivatization* (the synthesis of chemically modified derivatives of this compound) that yields a molecule with IC_{50} potency in the nanomolar concentration range, which is the prerequisite for drugs that are introduced into the clinic (see Supplementary Sidebar 16.7).

If such a drug is indeed synthesized, the ensuing pre-clinical testing involves measurements of the drug's *relative* effects on its intended target compared with its **off-target** effects on other, similar proteins in the cell. The goal here is to determine whether the drug acts selectively by inhibiting the targeted protein at drug concentrations that are substantially below (10- to 100-fold) those affecting other, similar proteins in the cell (**Figure 16.14**). (In truth, given the 20,000 or more distinct protein structures present in mammalian cells, these measurements do not preclude possible effects on structurally unrelated proteins that may, through happenstance, be affected by an agent under development.)

Figure 16.13 Tarceva and the inhibition of the EGF receptor kinase domain This space-filling model of the catalytic cleft of the EGF receptor tyrosine kinase (TK) domain, derived by X-ray crystallography, shows how the drug Tarceva (*stick figure*) fits snugly within the ATP-binding cavity of the cleft and in this way inhibits receptor signaling. The three-dimensional complementarity between drug and targeted protein is necessary for drug binding but not sufficient, as specific bonding, largely achieved through the formation of hydrogen bonds (*not shown*), must occur between the drug molecule and amino acids lining the drug-binding pocket (see, for example, Figure 16.10C). A water molecule (*behind, below*) that is hydrogen-bonded to the Tarceva molecule is also shown. (Courtesy of C. Sambrook-Smith and A. Castelanho, OSI Pharmaceuticals, Inc.; see also J. Stamos, M.X. Sliwkowski and C. Eigenbrot, *J. Biol. Chem.* 277:46265–46272, 2002.)

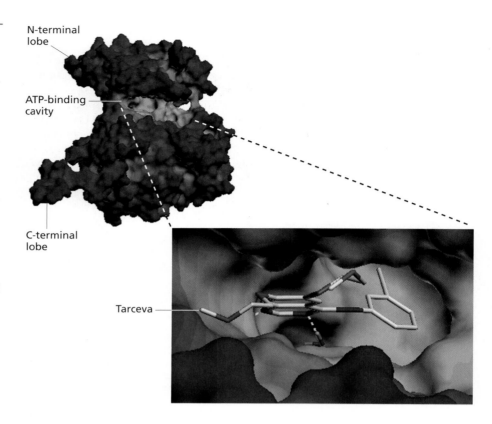

In the case of tyrosine kinase inhibitors, which are the focus of much current drug development, new experimental strategies are being developed to enable the screening of a large portion of the kinases that might be affected by a candidate drug inhibitor. Technologies like the one described in Supplementary Sidebar 16.9 greatly accelerate the rate at which large numbers of candidate drugs can be tested against a wide spectrum of these enzymes. As we will see later, discovering off-target activities of a drug, which is enabled by screens such as this one, is actually useful in two ways: (1) It may explain toxicities of a drug—undesired side effects in tissues other than the targeted tumor. (2) It may reveal new clinical applications for the drug, since the drug may be found to inhibit an enzyme, such as a kinase, that is active in a type of tumor that was not targeted during the initial drug development.

Figure 16.14 Dose response curves of Gleevec The ability to inhibit a targeted protein without affecting other, related cellular proteins is critical to the success of therapy, making possible therapeutic responses without undesired side-effect toxicities. Here we see the responses of a group of six tyrosine kinase (TK) enzymes to the drug Gleevec (see Figure 16.10). In all cases, the kinase activity of the purified enzymes in solution was measured. Note that the graph was constructed using the log of Gleevec concentration on the abscissa, while the percent of inhibition of catalytic activity is plotted linearly on the ordinate. A level of 50% inhibition is indicated by the *green dashed horizontal line*. At a concentration of approximately 0.2 µM (*blue dashed vertical line*), 50% of c-Abl enzyme activity was inhibited, whereas comparable inhibition of the c-Fms TK occurred only at a concentration of ~6 µM (*brown dashed vertical line*). Under these conditions, the c-Src TK was hardly inhibited at all. (Courtesy of E. Buchdunger, Novartis Pharmaceuticals.)

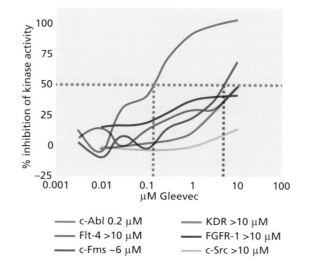

16.7   Drug candidates must be tested on cell models as an initial measurement of their utility in whole organisms

The demonstrated ability of a drug to inhibit an isolated target protein in solution is usually followed by tests of its effects on cultured cells. Take the case of Gleevec, the compound (see Figure 16.10A) found initially to inhibit the tyrosine kinase activity of the isolated Bcr-Abl fusion protein; this protein was known to be responsible for driving the proliferation and survival of the cancer cells of chronic myelogenous leukemia (CML). Having established its effects on the isolated Bcr-Abl protein (see Figure 16.14), drug developers could then proceed to the next step, which involved the use of cultured cells whose proliferation or survival *in vitro* depended on the continued actions of this fusion protein.

Figure 16.15 shows an example of such a cell-based test that happens to have been conducted long after Gleevec was developed. This test used the cells of a murine

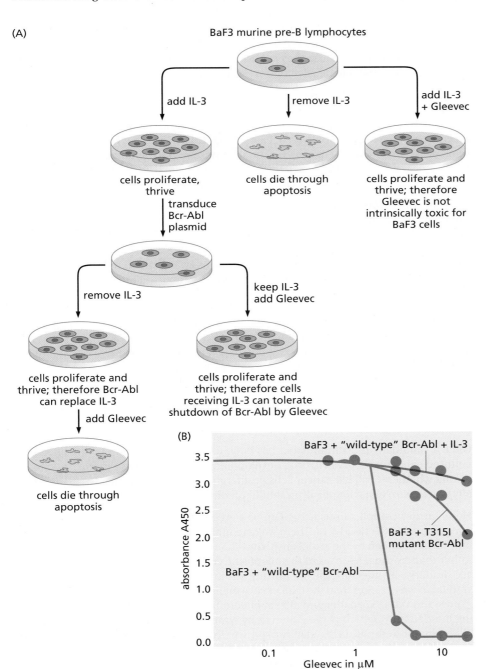

(A)

BaF3 murine pre-B lymphocytes

add IL-3 → cells proliferate, thrive

remove IL-3 → cells die through apoptosis

add IL-3 + Gleevec → cells proliferate and thrive; therefore Gleevec is not intrinsically toxic for BaF3 cells

transduce Bcr-Abl plasmid

remove IL-3 → cells proliferate and thrive; therefore Bcr-Abl can replace IL-3

keep IL-3 add Gleevec → cells proliferate and thrive; therefore cells receiving IL-3 can tolerate shutdown of Bcr-Abl by Gleevec

add Gleevec → cells die through apoptosis

(B)

absorbance A450

BaF3 + "wild-type" Bcr-Abl + IL-3

BaF3 + T315I mutant Bcr-Abl

BaF3 + "wild-type" Bcr-Abl

Gleevec in μM

Figure 16.15 Testing of Gleevec in cell culture (A) BaF3 cells, a line of murine pre-B lymphocytes, are normally dependent on the addition of interleukin-3 (IL-3) for their proliferation and survival (*top left*). When Gleevec is added together with IL-3, these cells continue to thrive (*top right*), indicating that the IL-3–based survival mechanism is not sensitive to Gleevec inhibition. When an expression plasmid specifying the Bcr-Abl oncoprotein is also introduced into these cells, the BaF3 cells continue to proliferate, even after IL-3 is withdrawn (*left side*), indicating that Bcr-Abl can replace IL-3 and sustain these cells on its own. However, the addition of Gleevec at doses that inhibit the Abl kinase will cause the cells to die (*lower left*), while addition of Gleevec to Bcr-Abl–expressing cells that continue to receive IL-3 does not affect their survival. Therefore, in the absence of IL-3, the Bcr-Abl–expressing BaF3 cells can serve as highly sensitive and specific indicators of the actions of Gleevec and similarly acting drugs on the Bcr-Abl oncoprotein. (B) The information in panel A can be used to develop an assay system, in which the number of BaF3 cells surviving after certain treatments is indicated by the optical density (absorbance at 450 nm wavelength) of BaF3 cell suspensions (*ordinate*). In the presence of IL-3, Gleevec has almost no effect on BaF3 cell survival whether or not the Bcr-Abl oncoprotein is being expressed in BaF3 cells (*green dots*). In the absence of IL-3, however, survival of Bcr-Abl–expressing cells is strongly suppressed above about 2 μM Gleevec concentration (*red dots*). If instead of the "wild-type" Bcr-Abl protein, cloned from a patient's CML cells at the beginning of Gleevec treatment, a highly drug-resistant mutant version of Bcr-Abl (termed T315I) that arose in a CML patient during the course of Gleevec treatment is expressed in the BaF3 cells growing in medium lacking IL-3, far higher drug concentrations are required to kill these cells (*blue dots*). (Courtesy of M. Azam and G.Q. Daley.)

pre-B-lymphocyte line that normally depend for their survival and proliferation *in vitro* on the presence of interleukin-3 (IL-3) in their culture medium. These cells could be rendered independent of IL-3 if a Bcr-Abl oncoprotein was ectopically expressed in them. The modified cells were then cultured in the absence of IL-3 (making them totally dependent on continued Bcr-Abl firing) and exposed to various drugs that were candidate antagonists of the Bcr-Abl oncoprotein; the proliferation and/or survival of these cells was then gauged (see Figure 16.15B).

Cell-based tests like these are designed to determine whether the drug being examined induces apoptosis in treated cells, or cytostasis (that is, a halt in cell proliferation), or has no effect whatsoever. And if the drug does evoke a desired response, does it do so at a reasonably low concentration?

The outcomes of such cell-based tests are rarely obvious beforehand. Many compounds that are highly hydrophobic may be excluded from these tests from the outset because they are poorly soluble and therefore cannot be placed on cultured cells in significant concentrations. Their more hydrophilic chemical relatives may be highly soluble and may work well on purified Bcr-Abl protein, but may not be readily transported through the plasma membranes of cells; these chemical species are likely to prove useless, simply because they cannot accumulate within cells at concentrations that would allow them to be effective (see Supplementary Sidebar 16.7).

Imagine that these hurdles have been successfully surmounted and that the proliferation of Bcr-Abl–dependent cells is indeed inhibited at nanomolar concentrations of a candidate therapeutic agent. The fact that the candidate drug acts on these cells does not exclude the possibility that it also affects dozens of other kinases in these and other cells, some of which may be essential for normal cell metabolism—the property of biological selectivity. (Its *biochemical* selectivity is likely to have been determined previously by tests like those described in Supplementary Sidebar 16.9.)

So next, it becomes necessary to determine whether cancer cells whose growth is driven by other tyrosine kinases are equally sensitive to the actions of an identified anti-Bcr-Abl agent like Gleevec. And how are fully normal cultured cells affected by a candidate drug like Gleevec? With luck, one may begin to see a high **therapeutic index** emerge; for example, in the context of the assay described in Figure 16.15, Bcr-Abl–dependent cells may be killed by drug concentrations that have little discernible effect on comparable cells grown in the presence of IL-3 or on a variety of other cancer cells whose growth is driven by other tyrosine kinase oncoproteins. This will provide hope that *in vivo* the drug may perturb the tumor without having unacceptable side effects on normal tissues. Good outcomes in these tests will then encourage the drug developers to proceed to the next steps, in which the biological effects of drugs at the cellular and tissue level are evaluated *in vivo*, as we learn below.

16.8 Studies of a drug's action in laboratory animals are an essential part of pre-clinical testing

Once a candidate anti-tumor agent has been found to have potent killing effects on cultured cancer cells *in vitro*, drug development inevitably moves to the next step—testing whether it will kill cancer cells proliferating within tumor masses *in vivo*. Ideally, the *in vitro* behavior of a drug should predict its actions *in vivo*.

Here, further complications arise. One is suggested by experimental results that we encountered in Section 13.10. There we read that the sensitivity of tumors to radiation may be determined by the radiosensitivity of the endothelial cells in their vasculature, rather than by the responses of the neoplastic cells in these masses; some drugs may act similarly, by affecting the supporting stromal cells of a tumor (which are rarely studied *in vitro*) rather than the cancer cells themselves. For example, endothelial cells exposed to the DNA-damaging effects of certain cytotoxic drugs respond by releasing interleukin-6, which confers on nearby lymphoma cells an elevated resistance to killing by these drugs. (This may explain why lymphoma cells in a mouse model of Burkitt's lymphoma can survive chemotherapy by finding refuge in the thymus, where this prosurvival signal is produced in abundance by the resident endothelial cells.)

Independent of such therapy-induced responses, stromal cells may continuously provide certain types of anti-apoptotic survival signals, such as IGF-1, that are not available in comparable amounts to cancer cells in culture. In a more general sense, the complexities of tumor biology created by heterotypic interactions with the tumor-associated stroma often dictate that the drug responses of pure populations of cancer cells proliferating *in vitro* fail to predict their responses within growing tumors *in vivo*.

Because rodent and human cells differ so substantially in their biology (see Section 11.12), the *in vivo* testing of candidate anti-cancer drugs involves, almost always, human (rather than murine) cancer cells grown in mouse hosts. The presumption is that the human tumor xenografts formed in immunocompromised mice will behave much like the tumors encountered by oncologists in human patients.

Once again, there are highly challenging complications. The human tumor cells that are used to form these xenografts are propagated as tumor cell lines—cancer cells that have been propagated in culture as pure populations for many years, often decades. A set of 60 of these human cancer cell lines has been established by the National Cancer Institute as standard reagents to be used in the United States for gauging the efficacy of candidate anti-cancer agents. Many of the cell lines from this "**NCI-60**" panel are not representative of neoplasms encountered routinely in the cancer clinic, because they derive from particularly aggressive human tumors that yielded cells that were especially adaptable to propagation in tissue culture (see, for example, **Figure 16.16**). Other cancer cell lines have, almost inevitably, evolved in culture far beyond the ancestral cells that were originally removed from actual human tumors; consider the fact that the cells in such lines have been selected for optimal proliferation under *in vitro* conditions that differ dramatically from those in living tissues.

These facts help to explain why human tumor xenograft models are relatively unpredictive of the responses of the actual tumors borne by patients in the cancer clinic (see Sidebar 13.1). Indeed, in some cases, it is questionable whether cancer cells that are purportedly from, for example, a pancreatic carcinoma continue to reflect pancreatic behavior, or whether they have been inadvertently contaminated by colon or breast carcinoma cells at some point over the previous several decades of *in vitro* passage in one or another laboratory. In addition, there are clear indications that human tumors growing as xenografts in mice interact with the host tissue microenvironment in a fashion that differs from mouse tumor xenografts interacting with syngeneic host tissues. Still, these highly imperfect xenograft models are often the best reagents available and most are unlikely to be supplanted in the near future by improved animal models of human cancer.

Then there is yet another problem that, until recently, has not attracted attention. Experiments were reported in 2009 on the details of a genetically engineered mouse model (GEMM) of pancreatic adenocarcinoma development; this research analyzed the responses to therapy of such tumors arising within mice through the actions of oncogenic alleles introduced into their *K-ras* and *p53* germ-line genes. The resulting

Figure 16.16 Cancer cell lines as representatives of human tumors Many researchers have attempted to create cancer cell lines by extracting cells from human tumors and adapting them to culture. Their experience, largely anecdotal, is that only the most malignant cancer cells can be propagated *in vitro*, yielding cancer cell lines. This notion has finally been tested systematically in a 12-year-long study in which the esophageal carcinoma cells of 203 patients were introduced into culture. Of these, only 35 cell lines (derived from about 17% of the tumors) became established in culture. The patients whose tumors were in this group (group A) experienced a far worse clinical progression (*red line*) than did those whose cells failed to adapt to culture (group B; *blue line*). This illustrates graphically why tumor xenografts produced from established cancer cell lines usually fail to recapitulate the properties of the tumors typically encountered in a cancer clinic (since the cancer cell lines usually derive from tumors at the far end of the spectrum—the most aggressive subset). (From Y. Shimada et al., *Clin. Cancer Res.* 9:243–249, 2003.)

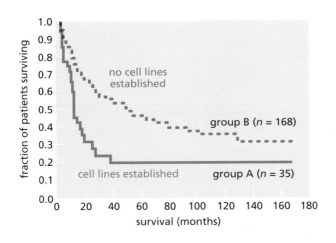

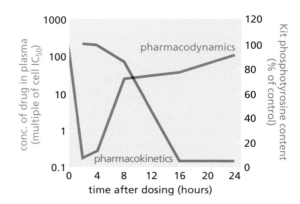

Figure 16.17 Pharmacokinetics and pharmacodynamics of Gleevec The pharmacokinetics of a drug represent the kinetics of its accumulation in and disappearance from the plasma, which in turn are presumed to provide a good indication of the drug concentrations that tumor cells experience in a laboratory animal or a patient undergoing therapy. The plasma level of the drug Gleevec, plotted on a logarithmic scale (*left ordinate*), fluctuates dramatically following injection of the drug into a mouse (*blue curve*). Its concentration is indicated here as a multiple of the drug concentration known to inhibit the firing of the tyrosine kinase of the Kit receptor by 50% (i.e., the IC_{50} of this agent). (The tyrosine kinase domain of the Kit growth factor receptor is also a target of inhibition by Gleevec.) As seen here, the amount of phosphotyrosine associated with the Kit receptor (a reflection of Kit tyrosine kinase activity) expressed by engrafted human mast cell leukemia cells (*red curve*), which was initially set as 100%, is reduced to <<1% of preexisting levels within an hour after drug injection but rebounds within 8 hours as the concentration of the drug declines in the plasma. (Courtesy of D.L. Emerson, OSI Pharmaceuticals, Inc.)

autochthonous tumors (that is, those arising within a host) that formed in these mice responded to therapy very differently from tumors that arose when the cells from these autochthonous tumors were extracted, propagated briefly in culture, and then implanted subcutaneously in other syngeneic host mice, now forming pancreatic tumor grafts. This pointed to another long-neglected problem—namely, that experimentally implanted tumor grafts may behave very differently from autochthonously arising tumors, even when the two tumor cell populations are genetically indistinguishable from one another. [In this case, the implanted tumors were responsive to gemcitabine therapy (see Figure 16.6) while the autochthonous tumors were resistant.] If this observation is generalized to other tumor models and cancer types, it greatly complicates pre-clinical drug testing, since implantation of tumor cell lines is far simpler and less expensive than GEMM models of disease.

The complementary science of drug development, first developed to optimize the use of cytotoxic drugs (see Section 16.2), has been applied over the past two decades to bring the newer targeted therapies from the laboratory bench to the patient's bedside. As discussed earlier, the pharmacokinetics (PK) of a drug represent a key determinant of its efficacy *in vivo*: Does it accumulate to significant levels in the plasma or tissues for an extended period of time? Or is it present in the body only transiently, being excreted by the kidneys within minutes of entering into the circulation? Is it resistant to rapid degradation, or do certain drug-metabolizing systems, such as the cytochrome P450s (see Section 12.6), rapidly convert it into an innocuous molecular species (**Figure 16.17**)? (A key pharmacokinetic parameter that is often measured is the "area under the curve," or AUC, calculated by integrating the concentration of a drug in the plasma as a function of time; the AUC is thought to reflect the cumulative drug dose experienced by cells in a tumor.) And can it be administered orally rather than requiring injection?

Laboratory animals give some rough indication of a drug's pharmacokinetics, but are by no means accurate predictors of how humans metabolize and excrete various agents. Moreover, as we read earlier (see Sidebar 12.4), the rates at which various compounds are metabolized or excreted can even vary dramatically from one person to another (**Figure 16.18**). (In some pharmaceutical companies, the pharmacokinetics of candidate compounds may be measured even *before* their therapeutic efficacy against xenografted tumors is assessed; those showing poor pharmacokinetics in laboratory animals are often eliminated from further testing. Discarding such drugs may

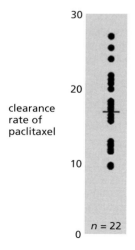

Figure 16.18 Inter-individual variability in drug clearance Paclitaxel is a chemotherapeutic drug used to treat a number of malignancies; it works by stabilizing microtubules, thereby interfering with the progression of cells through M phase (see Figure 16.6). As seen here, in this study of 22 ovarian cancer patients, the relative rates of clearance of this drug from the plasma after initial injection varied over a factor of 3. These rates may be influenced by changes in the rate of metabolism by enzymes such as cytochrome *c* (see Figure 12.13) and by excretion in the kidneys. (From M. Nakajima et al., *J. Clin. Pharmacol.* 45:674–682, 2005.)

occasionally be premature, given the dramatically differing rates of drug metabolism and excretion between rodents and humans.)

Figure 16.17 illustrates the other key kinetic process that was introduced earlier—a drug's pharmacodynamics (PD), in this case those of Gleevec. Recall that pharmaco-dynamics measure the ability of a drug to affect a targeted biochemical or cell-biolog-ical function in a tumor under treatment. In the PD presented in this figure, as is often the practice, a **surrogate marker** of the targeted Bcr-Abl function was measured—the behavior of the Kit receptor. As we will read in more detail later, Kit is one of several tyrosine kinases affected by Gleevec, and its responses to the drug presumably parallel those of Bcr-Abl. Figure 16.17 reveals that Kit activity in this experiment was inhibited only briefly at the time when the highest concentration of drug was present in the cir-culation. Such a transient inhibition—only a fraction of a cell cycle—is generally insuf-ficient to elicit a substantial biological response, such as tumor cell killing.

During the course of animal testing, information may surface about the toxic side effects that the drug elicits in whole organisms. They represent the bane of almost all existing cancer treatments. Quite frequently, various normal organ systems, includ-ing the liver, kidneys, gastrointestinal tract, and the hematopoietic system, show toxic effects of a drug when it is used at the concentrations required to kill tumor cells. These toxicities are rarely predicted by *in vitro* tissue culture tests, and toxicities detected in laboratory animals, including dogs, monkeys, mice, and rats, may or may not be pre-dictive of human responses. For example, when used together with a cytotoxic drug to treat subtypes of breast and gastric carcinomas, Herceptin (also known as trastu-zumab; see Section 15.18) has an effect on cardiac function that could not have been predicted from the known expression of the targeted HER2/ErbB2 receptor in various epithelial cell types; this effect can lead to (usually reversible) heart failure in a signifi-cant fraction of patients under treatment.

Such observations direct our attentions once again to the therapeutic index of an agent—the efficiency with which it affects cancerous tissues compared with its toxic effects in normal tissues. Clearly, ideal cancer treatments should have high thera-peutic indices, wreaking havoc on cancer cells while leaving normal tissues relatively untouched. The fundamental obstacle to achieving such selectivity is suggested by the fact that the vast majority of the 11,000 or so genes expressed in the average cancer cell are also being expressed by their normal counterparts.

The failure of animal models to predict the toxic side effects of a drug in humans cre-ates serious problems. The roughly 80 million years of independent evolution that separate us from our rodent cousins have led to substantial differences in metabolism; we may react to certain drugs much differently from mice or rats, or even the more closely related Old World monkeys, which may eventually be exposed to a candidate drug in order to obtain a slightly more accurate prediction of toxicities in humans. In the event that a drug passes these various tests without raising too many warning flags, it may be promoted to a candidate for testing in humans.

16.9 Promising candidate drugs are subjected to rigorous clinical tests in Phase I trials in humans

The discussions above explain why the first true tests of a drug's tolerability usually come in initial patient exposures, which are termed Phase I trials in the United States. Here, candidate drugs are tested at various doses, including the presumed therapeu-tic doses, to gauge toxic side effects. The usual practice is to begin these trials at drug dosages that are likely to be far below the level of any overt **toxicity** (for example, one-tenth the drug concentration that created toxicity in laboratory animals) and then, in a series of patients, increase the dosages incrementally until drug levels are reached that begin to induce unacceptable toxicities. This "dose escalation" yields a value—the *maximum tolerated dose* (MTD)—that is then used to guide further treatment proto-cols. Certain side effects, such as a skin rash or transient nausea, may be tolerable and not scuttle further drug development, while others, such as massive diarrhea or bone marrow depletion, may be so burdensome or life-threatening that they cause rapid abandonment of all further development of a drug.

During these Phase I trials, pharmacokinetic measurements, like those made previously in animals, will also be taken in order to ascertain whether the drug is reaching tumor cells at a sufficient concentration and for an extended period of time. Still, these measurements give no indication whether the cancer cells are responding in any way—the property of pharmacodynamics described earlier. For example, in **Figure 16.19**, we see the pharmacodynamic responses to treatments with EGF receptor antagonists (in this case both a monoclonal antibody and a low–molecular-weight tyrosine kinase inhibitor). To obtain some measure of the effects of therapies on the EGF-R in patients' tumors, the oncologists used, as a surrogate marker, the EGF-R of patients' biopsied skin cells, which were far more easily monitored.

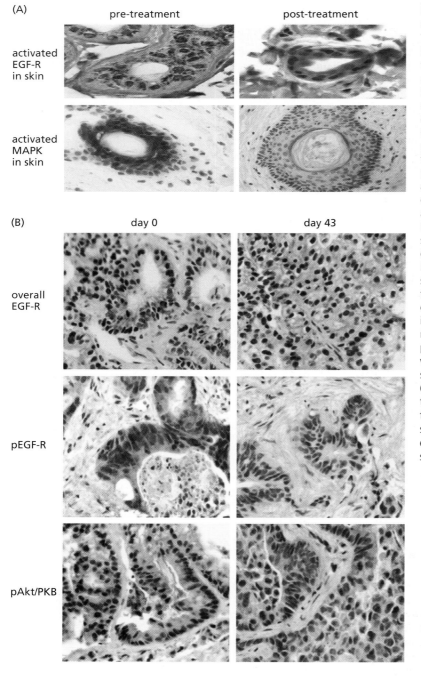

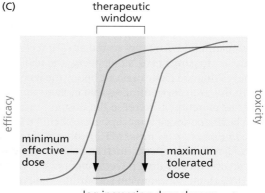

Figure 16.19 Measurements of pharmacodynamics and determination of the therapeutic window
The extent of inhibition of the EGF-R in a tumor can, in principle, be gauged by measuring effects of drug treatment on the EGF-R in the skin; the latter is readily assessed through small skin biopsies. In the cases illustrated here, patients were suffering from a variety of tumors, including carcinomas of the ovary, lung, colon, and prostate and head-and-neck cancers. (A) Shown here are the effects of treating a cancer patient with Iressa, a low–molecular-weight EGF-R tyrosine kinase inhibitor (see Figure 16.29). The upper panels show immunohistochemistry using an antibody against phospho-EGF-R (*brown*), i.e., the activated form of the receptor. The lower panels used an antibody against phospho-MAPK, the activated form of this kinase. Both measurements depended on the normally intense signaling occurring in hair follicle keratinocytes. (B) The effects of an anti-EGF-R monoclonal antibody (termed EMD7200) were gauged by immunohistochemical staining of a colon carcinoma biopsy. In this case, long-term treatment resulted in a minimal reduction in the overall level of the EGF-R (*brown*) and a strong reduction in the level of phosphorylated (and therefore activated) receptor (*brown*; pEGF-R). The reduction in the level of phosphorylated, activated Akt/PKB (*brown*; pAkt/PKB) was slight and, in a possible reflection of this, the patient showed only a partial response to this antibody therapy. (C) Measurements of pharmacodynamics such as these, taken together with studies of pharmacokinetics and toxicity, define the *therapeutic window* in which a drug should be given—the range of concentrations that are efficacious without creating an unacceptable level of toxic side effects. (A and B, courtesy of J. Baselga.)

As seen in Figure 16.19A, patient exposure to an EGF-R tyrosine kinase inhibitor resulted in a strong suppression of EGF-R signaling in the skin. In addition, the activity of MAP kinase, which functions as an important downstream transducer of EGF-R signaling (see Section 6.5), was also suppressed, indicating successful inhibition of downstream mitogenic signaling.

Similar results were observed in biopsies taken from a colon cancer patient's tumor following treatment with an anti-EGF-R monoclonal antibody (see Figure 16.19B). Pharmacodynamic measurements like these provide reassurance that the administered treatment (in this case a monoclonal antibody) is reaching its intended target at concentrations that suffice to shut down much of the target's activity.

Interestingly, a number of the signal-transducing proteins operating downstream of the EGF-R, including Akt/PKB, were only minimally suppressed in the colonic tumor (see Figure 16.19B), indicating that the tumor cells had acquired alternative means for activating these signaling molecules. Hence, pharmacodynamics measurements ensure that one precondition of therapeutic success—delivery of the therapeutic agent to the targeted cells and molecules—has been satisfied, but do not, on their own, guarantee that the therapy will succeed, as other factors may thwart it.

When taken together, the measurements of maximum tolerated dose (MTD), pharmacokinetics (PK), and pharmacodynamics (PD) define the **therapeutic window**—the range of concentrations that are higher than that needed to elicit a therapeutic effect and lower than the maximum tolerated dose (see Figure 16.19C). Ideally, a therapeutic window of a drug should be broad, so as to allow clinicians some flexibility in administering the drug, adjusting dosage to the patient and the condition being treated. As the therapeutic window narrows, the likelihood that a candidate drug will prove clinically useful diminishes.

Occasionally, Phase I clinical trials, which are usually undertaken with very small groups of patient volunteers who have failed other available therapies, may reveal some favorable responses in terms of tumor regression or halting of further tumor growth, doing so at acceptably low levels of toxicity. However, even if there are hints of clinical efficacy, the positive results observed in Phase I trials are never statistically significant and thus not regarded as definitive. Instead, these trials are really undertaken to discover unanticipated toxicities and tolerable levels of drug dosage.

16.10 Phase II and III trials provide credible indications of clinical efficacy

Acceptably low levels of toxicity in a Phase I trial will encourage testing a candidate drug's efficacy in a Phase II trial, in which larger groups of cancer patients are involved. Now, for the first time, critical decisions must be made about the **indications** for enlisting specific patients in the trial—that is, which type of tumor or what stage of tumor progression will justify enrolling patients in such a trial?

Sometimes the clinical indications are obvious. For example, an agent targeted against the Bcr-Abl oncoprotein should be tested in patients diagnosed with chronic myelogenous leukemia (CML). A drug directed against the HER2/Neu receptor should be tested in the approximately 30% of breast cancer patients whose tumor cells overexpress this protein. An inhibitor of Raf kinases can be tried in patients with advanced melanomas, in which the B-Raf kinase molecule is often (~60% of cases) mutant and constitutively activated (see Section 16.17).

But more often than not, the choice of indications is neither rational nor optimal. Which class of cancer patients should be treated, for example, with a drug that acts as a general inducer of apoptosis in many types of cancer cells? How should a drug directed against the anti-apoptotic Akt/PKB kinase be used in the clinic? Will an anti-EGF receptor drug prove useful in all carcinomas that express elevated levels of this receptor protein or only a select subset (Supplementary Sidebar 16.10)? As we will see later in this chapter, certain types of cancer that would never be identified by genetics or molecular biology as attractive targets for drug treatment turn out, on occasion, to be highly susceptible to certain drugs under development. In these cases, the therapeutic utility of such drugs is discovered only by chance.

(Given the arbitrary ways in which tumor indications are chosen in many Phase II trials, we can wonder how many truly useful candidate drugs have been discarded in the past, simply because good luck did not favor them in the design of these trials. Thus, it may well be that a drug has spectacular efficacy against gastric carcinomas but this effect is never realized, since it is tested in Phase II trials for its effects on pancreatic or lung carcinomas, where it fails to show any useful effects and is therefore dropped from further development and clinical testing.)

If Phase II trials yield clear signs of efficacy for treating certain types of cancer with a candidate drug, Phase III trials, undertaken in far larger patient populations, will be launched. These trials are very costly but are ultimately critical, if only because they may show, for the first time, whether any clinical responses ascribed to a drug are statistically significant. The results of these trials usually become compelling only if control experiments are performed by treating equally large populations of patients with another therapy in parallel, usually one that is already licensed and in widespread use. Importantly, the licensing of a candidate drug for a specific disease indication (in the United States by the Food and Drug Administration, FDA; in the European Union by the European Medicines Agency, EMA) usually depends on whether it yields a therapeutic benefit that is measurably greater than the existing standard of care.

Patients in Phase III trials of anti-cancer agents usually have gone through previous rounds of chemotherapy with various types of cytotoxic agents, each ending with a relapse and the appearance of tumors that are *refractory* (nonresponsive) to established therapies. Moreover, these tumors are often highly aggressive. This helps to explain why the bar is not set too high for FDA approval of a new drug or drug combination, since the drugs in Phase III trials are dispatched to attack the most difficult types of cancer. Thus, improvements in patient quality of life or temporary shrinkage of a tumor may suffice even without improvement in long-term survival.

One illustration of this is a current treatment for pancreatic cancer. This disease is an extreme example, to be sure, in that the 5-year survival rate of this disease (from the time of initial diagnosis) is consistently less than 5%. Gemcitabine (difluorodeoxycytidine; see Figure 16.6), which is widely employed as a therapy for pancreatic carcinoma, received initial FDA approval because in some patients it resulted in an improvement in symptoms, weight gain, and a temporary stabilization in tumor growth, although it offered only a modest increase in survival time: patients treated with gemcitabine had a median survival time of 5.65 months compared with those given the standard care—5-fluorouracil (5-FU), which afforded them a 4.41-month median survival time (**Figure 16.20**). This and similar anecdotes reveal how desperate is the need for truly effective means of treating solid tumors. It also illustrates the fact that the FDA's requirements for approving anti-cancer agents are far less stringent than for other disease states, where much greater efficacy is required to gain licensing of new drugs.

Figure 16.20 Gemcitabine as a treatment for pancreatic cancer This Kaplan–Meier plot illustrates the high mortality exacted by pancreatic cancer. Patients treated with gemcitabine (GEM; see Figure 16.6) lived slightly longer than those treated with 5-fluorouracil (5-FU)—the standard treatment in the 1990s. Both of these agents are pyrimidine derivatives whose cytotoxicity derives from their ability to inhibit DNA synthesis, in part through misincorporation into the DNA. (5-FU also interferes with pyrimidine biosynthesis.) As is apparent, gemcitabine treatment offered only a modest increase in patient survival in this study reported in 1997, but this sufficed to allow its approval by the U.S. Food and Drug Administration. (From H.A. Burris III et al., *J. Clin. Oncol.* 15:2403–2413, 1997.)

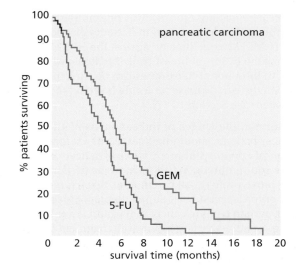

Nonetheless, even with these relatively modest regulatory requirements, the other complications of drug development described here keep the current success rate for anti-cancer drug development extremely low. Perhaps one drug in a hundred is advanced all the way through the drug development "pipeline" from initial *in vitro* testing through a Phase III trial that culminates in some clear improvement in patient outcome and licensing by the FDA. (After licensing has occurred, a Phase IV trial may be conducted to determine how a newly introduced drug compares with other drugs used for similar indications, how certain subgroups of patients respond to the drug, and whether concerns about a drug's safety eventually emerge from its use in very large patient populations.)

16.11 Tumors often develop resistance to initially effective therapy

A complication that dogs all anti-cancer drugs is illustrated by the behavior of *HER2/neu* transgenic mice, in which the mutant, oncogenic transgene has been programmed to induce mammary tumors on a predictable schedule and can be shut down thereafter. While transgene-induced primary breast tumors and metastases all collapsed when the *HER2/neu* transgene was shut down, new tumors recurred in most of these mice between 1 and 9 months later. These tumors clearly represented variants of the initially observed ones that had developed alternative means of propelling their growth—that is, had become independent of *HER2/neu* oncogene expression.

As we saw in Chapter 12, the unstable, mutable genomes of cancer cells continually generate new alleles and novel genetic configurations. Evolving cancer cells can pick and choose among these genetic variations, searching for combinations that improve their ability to survive and proliferate. In this *HER2/neu* example, the relatively small number of cancer cells that survived oncogene shutdown seem to have spent months thereafter trolling for newly arising oncogenes (or other cancer-causing alleles) in their genomes that might enable them to re-launch their program of aggressive proliferation. The rare cells that happened to acquire such advantageous alleles or epigenetic states then began clonal expansions that led to relapsing tumors.

A similar dynamic complicates almost all types of human cancer therapies, where initial clinical successes in reducing tumor cell populations are usually followed by the re-emergence in patients of tumor cell populations that have, through one means or another, developed resistance to the treatment (see Table 16.5). Much of this acquired resistance is attributable to the genetic and therefore phenotypic plasticity of cancer cell populations. Importantly, much of this "acquired" resistance may derive from variants that already existed within a tumor at the time when therapy was initiated.

The acquired mechanisms of drug resistance are quite variable and illustrate the ingenuity of cancer cells. As indicated in Table 16.5, some of these mechanisms involve loss of the ability to import drug molecules through the plasma membrane or an acquired ability to pump drug molecules out through this membrane. Yet others depend on the acquired ability to metabolize drug molecules, in some cases using the same classes of enzymes that normally operate to detoxify toxic compounds that have entered the cell (see Section 12.6). Cells may also neutralize components of their apoptotic machinery or may acquire an increased ability to repair DNA molecules damaged by chemotherapeutics or radiation.

These behaviors represent a general challenge to all types of anti-tumor therapy, as discussed in Section 16.2. The response to such acquired resistance is the development of multi-drug therapies, perhaps combining traditional cytotoxic drugs with a newly developed, molecularly targeted agent. However, even these multi-drug therapy strategies are often foiled by cancer cells, which develop powerful strategies for evading killing, such as the acquisition of multi-drug resistance (MDR). In recent years we have learned much about how this resistance is achieved at the molecular level. For example, high-level expression of the *MDR1* gene, which encodes a transmembrane drug efflux pump, enables cancer cells to efficiently excrete a variety of chemically unrelated drugs, thereby lowering intracellular drug concentrations to subtoxic levels

Figure 16.21 Multi-drug resistance and the P-glycoprotein The *MDR1* gene encodes P-glycoprotein (P-gp), a protein that is often present at elevated levels in cancer cells being treated with various types of chemotherapy. A structural model of P-gp is shown here. It is a 170-kD ATP-dependent transmembrane protein that can pump out of cells a wide variety of molecules with molecular weights ranging from 330 up to 4000. Its elevated expression is often associated with and likely causes multi-drug resistance. It is a member of a large family of mammalian ABC (ATP-binding cassette) transporter molecules, of which 49 have been discovered. As shown, a substrate such as a chemotherapeutic drug (*magenta*) enters the cell and is bound by the cytoplasmic domains of P-gp (*left*). In response to this binding and the binding of ATP molecules (*yellow*), the P-gp undergoes a profound conformational shift that collapses its cytoplasmic substrate-binding domain and opens its domain facing the extracellular space (*right*), resulting in the extrusion of the drug molecule from the cell. The drug-binding pocket is shown in *cyan*. The *right* P-gp molecule has been rotated 90 degrees to better illustrate the stereochemistry of the drug-extrusion step. (From S.G. Aller et al., *Science* 323:1718–1722, 2009.)

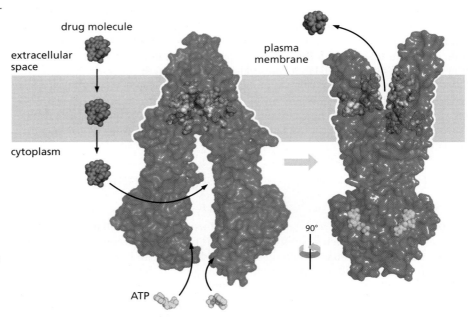

(**Figure 16.21**). Similarly, inactivation of certain parts of the apoptotic machinery (see Table 16.5) may also confer concomitant resistance to a number of distinct cytotoxic agents. The advent of molecularly targeted agents, as discussed in the following sections, has not changed a widespread consensus among drug developers: monotherapies involving either low–molecular-weight drugs or biological molecules are unlikely to cure most types of cancer, and effective multi-agent therapies must be devised if definitive, durable clinical responses are to be achieved in the future.

With these considerations in mind, we will read about a series of illustrative anecdotes in the ensuing sections. Each concerns a type of drug and its targets within cancer cells. The stories are arranged in an order, beginning with a well-established therapy and ending with a speculative one that holds great promise but is still far from clinical validation. In some cases, the specific therapy that has been developed was inspired by discoveries of malfunctioning proteins within cancer cells; these discoveries allowed drug development to be pursued logically and methodically. In other cases, strokes of good luck or intuitive leaps enabled the development of highly active compounds. Inevitably, these anecdotes represent arbitrary choices and draw from a vast pool of agents currently under investigation or development. They represent the forerunners of a large flock of such drugs that will be developed and licensed for clinical use in the years to come.

16.12 Gleevec paved the way for the development of many other highly targeted compounds

In the previous sections, we made repeated reference to the Bcr-Abl oncoprotein and to experimental strategies for antagonizing it. Now, we backtrack and review some history of how the Bcr-Abl oncoprotein was discovered and validated as an attractive drug target and finally used as an object of rational drug design. This story is valuable, if only because it illustrates the long course through which drug development passes from initial discovery at the laboratory bench to the oncology clinic.

This particular story begins in 1914, when the German cytologist Theodor Boveri proposed that chromosomal defects might cause a cell to proliferate abnormally, resulting ultimately in the formation of some type of cancer. Almost half a century passed before Boveri's idea received some validation. In 1960, two cytologists working in Philadelphia noted that an abnormal, unusually small Chromosome 22 was characteristically present in the great majority of cells of chronic myelogenous leukemia (CML); since that time, it has been called the Philadelphia chromosome or simply Ph. It took another dozen years before a researcher in Chicago demonstrated that a reciprocal translocation between Chromosomes 9 and 22 was responsible for creating the Ph chromosome (see Section 4.6). (Since a larger chunk of Chromosome 22 is donated to the tip of Chromosome 9 than is received from this chromosome, this leaves the already-tiny Chromosome 22 even more diminished in size; this remnant of 22 plus the small translocated segment is Ph; **Figure 16.22A**.) The chromosomal aberration—clearly the consequence of a somatic mutation—was proposed as a potential cause of this malignancy. As mentioned earlier, we now know that this particular translocation is present in more than 95% of cases of CML.

The genes that were fused through this translocation remained unknown until, in 1982, molecular biologists discovered that *ABL*, the human homolog of the mouse c-*abl* proto-oncogene, participates directly in these chromosomal translocations, becoming fused with a second, still unknown gene. The breakpoints of this other gene (the chromosomal sites at which it becomes fused to the *ABL* gene) were soon found to be scattered over many kilobases of DNA, yielding the name "**b**reakpoint **c**luster **r**egion" or simply *BCR*. In fact, three distinct fusion proteins arise through the inclusion of variously sized Bcr proteins at the N-termini of the fusion proteins, with almost the entire Abl protein at the C-termini (see Figure 16.22B). As indicated in the figure, the different fusion proteins tend to be associated with distinct types of leukemia.

Within two years of its discovery, the Bcr-Abl protein was found to function as a constitutively activated tyrosine kinase. In this respect, it functions like the Abl oncoprotein of Abelson mouse leukemia virus. The genome of this retrovirus carries an *abl* oncogene derived from the corresponding proto-oncogene residing in the normal mouse genome (see Section 3.10).

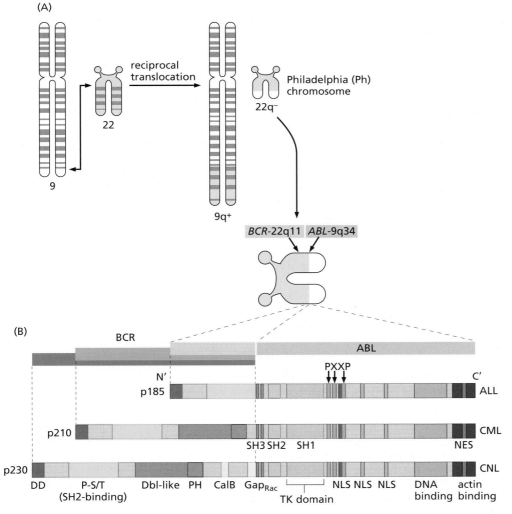

Figure 16.22 Origin and structure of the Bcr-Abl protein (A) More than 95% of cases of chronic myelogenous leukemia (CML) exhibit the Philadelphia chromosome, which results from a reciprocal translocation between Chromosomes 9 and 22. The q34 region of Chromosome 9 carrying most of the *ABL* gene is transferred to the q11 region of Chromosome 22, replacing a larger segment of Chromosome 22 that is translocated reciprocally to Chromosome 9. The net result is a truncated Chromosome 22 (i.e., 22q⁻), often termed the Philadelphia chromosome (Ph), and a fusion of the 5′ portion of the *ABL* gene with a 3′-proximal portion of the *BCR* gene, which normally resides at 22q11. (B) Depending on the precise location of the breakpoint in *BCR*, three distinct Bcr-Abl fusion proteins may be formed; these are found in ALL (acute lymphoblastic leukemia), CML, and CNL (chronic neutrophilic leukemia). Each of these *BCR-ABL* fusion genes encodes a multidomain (and thus multifunctional) protein. (From A.S. Advani and A.M. Pendergast, *Leuk. Res.* 26:713–720, 2002.)

By 1990, a cDNA encoding the Bcr-Abl fusion protein had been introduced into a retrovirus vector, and the resulting virus was then found to induce a leukemia in mice that closely resembled human CML. Like the human disease, this leukemia involved large numbers of fully differentiated granulocytes in the blood. Under certain conditions, the mouse leukemia, like its human counterpart, progressed to a "blast crisis," involving the accumulation of immature cells of the lymphoid or myeloid lineages (see Sidebar 8.6). These observations in mice represented the first formal proof that the Bcr-Abl fusion protein operates as the central motive force of leukemogenesis in CML.

Unfortunately, this demonstration of the critical role of Bcr-Abl revealed nothing about the mechanisms by which it functions. The bewildering complexity of Bcr-Abl signaling is indicated by the diverse array of structural and functional domains in the two contributing proteins (see Figure 16.22B). Altogether, the domains of this fusion protein enable it to activate the Ras pathway, the PI3 kinase–Akt/PKB pathway, the Jak–STAT pathway, and transcription factors, including Jun, Myc, and NF-κB. In addition, the Ras-like Rac protein, which regulates activities as diverse as cellular migration, survival, and proliferation, is activated, as are two nonreceptor tyrosine kinases, Hck and Fes. These various associations enable the Bcr-Abl protein to extend its reach into almost all of the regulatory circuits governing cell proliferation and survival.

In spite of this complexity, the tyrosine kinase domain of Bcr-Abl, derived from the Abl proto-oncogene protein, was found to be the key element in leukemogenesis. For example, subtle alterations of the Bcr-Abl protein that inactivated its tyrosine kinase catalytic activity led to total loss of its transforming function. In the early 1990s, a research program was begun to develop low–molecular-weight antagonists of the Bcr-Abl tyrosine kinase activity. A drug emerged, termed variously imatinib mesylate, STI-571, Glivec, and Gleevec (see Figure 16.10A), which was able to bind the catalytic cleft of the Bcr-Abl tyrosine kinase. As is the case with all other kinases of this family, the cleft is located between the two major structural lobes of the kinase protein (see Figure 16.10B).

Even though the Abl kinase domain shares roughly 42% amino acid identity with a large number of other tyrosine kinases, the inhibitory effects of Gleevec on Bcr-Abl were found to be relatively specific (see Figure 16.14). Subsequently, four other tyrosine kinases—those belonging to the PDGF (α and β) and Kit receptors as well as the Arg (Abelson-related gene) protein—were also found to be inhibited by Gleevec. Accordingly, this drug, when used at therapeutic concentrations, appears to target only 4 of the 90 or so human tyrosine kinases. Like most other kinase inhibitors, the Gleevec molecule associates with the ATP-binding pocket of the Abl kinase domain (see Figure 16.10). While other kinase inhibitors block ATP binding in this cleft, Gleevec works differently: it binds and stabilizes a catalytically inactive conformation of this enzyme.

The success with Gleevec encouraged other attempts at creating low–molecular-weight kinase antagonists, which were realized to have certain therapeutic advantages over anti-receptor monoclonal antibodies (Table 16.7). In addition, it stimulated efforts to make narrowly targeted tyrosine kinase inhibitors, some of which have indeed demonstrated extraordinary specificity. For example, even though the tyrosine kinase domains of the insulin receptor (IR) and insulin-like growth factor receptor (IGF-1R) are very similar in structure (see Figure 16.12B), a low–molecular-weight inhibitor has been developed that targets preferentially the IGF-1R, doing so at a concentration that is 27-fold lower than that required for it to inhibit the IR.

By 1996, Gleevec had been found able to inhibit the growth of CML cells *in vitro* while having no effect on normal bone marrow cells. More specifically, the proliferation of Bcr-Abl–dependent cells could be inhibited at drug concentrations as low as 40 nM, indicating a high affinity of Gleevec for the catalytic cleft of the tyrosine kinase domain. (Cells that depend on Bcr-Abl for survival can be forced into apoptosis by Gleevec's inhibition of Abl kinase function.) The initial clinical trials, begun in 1998, revealed remissions from disease in all of the 31 treated CML patients, with only minimal side effects, even when taken daily for many years. Four years later, 6000 patients were participating in Gleevec clinical trials.

Table 16.7 Strengths and weaknesses of anti-receptor antibodies versus low–molecular-weight tyrosine kinase inhibitors as anti-cancer agents

	Small molecule	Antibody
Target	tyrosine kinase domain	receptor ectodomain
Specificity	+++	++++
Binding	most are rapidly reversible	receptor internalized, only slowly regenerated
Dosing	oral daily	intravenous, ≤ weekly
Distribution in tissues	more complete	less complete
Toxicity	rash, diarrhea, pulmonary	rash, allergy
Antibody-dependent cellular cytotoxicity	no	possibly

Courtesy of N.J. Meropol and from N. Damjanov and N. Meropol, *Oncol. (Huntington)* 18:479–488, 2004.

Treatment of early-stage (chronic) CML with Gleevec leads to a hematologic response in 90% of cases: microscopic analysis of blood smears reveals a profound shift in the cellular composition of the blood (**Figure 16.23A**), and PCR analysis reveals an extraordinary decline in the levels of the *BCR-ABL* mRNA in blood cells (Figure 16.23B). In 50% of these cases, the translocated Philadelphia chromosome is no longer detectable

(A)

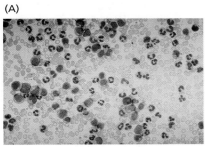

before treatment

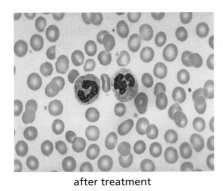

after treatment

(B)

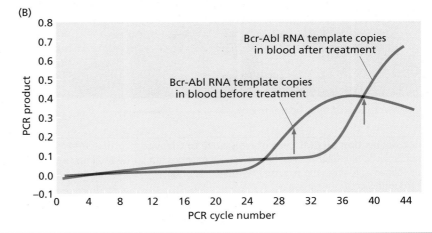

Figure 16.23 Measuring clinical responses to Gleevec treatment (A) The successes of Gleevec in treating chronic myelogenous leukemia (CML) can be gauged from cytological analyses of the patients' blood. As seen here, treatment with Gleevec converted the blood from a state in which many leukemia cells (*large, dark nuclei, above*) appear to one in which only normal granulocytes are visible (*below*) among the red blood cells. (B) A more sensitive and quantitative measure of therapeutic success comes from use of quantitative polymerase chain reaction (qPCR) measurements of the level of *Bcr-Abl* mRNA (which is reverse-transcribed prior to PCR amplification). In an untreated patient (*red curve*), 50% of maximum (*red arrow*) PCR-mediated gene amplification is observed at about the 29th cycle of gene amplification (in which each cycle results in the doubling of the amplified sequence). Following Gleevec treatment (*blue curve*), a comparable degree of amplification is achieved only at about the 39th cycle (*blue arrow*), indicating that the *Bcr-Abl* RNA–expressing cells are present at a level that has been reduced by a factor of about 2^{10}. PCR-based assays can detect as few as one CML cell amid 10^5 to 10^6 normal blood cells. (A and B, courtesy of B.J. Druker.)

by karyotypic analyses of patients' white cells. About 60% of the patients who have already progressed to blast crisis respond to Gleevec, but they generally relapse after some months. A clinical study reported in 2006 that five years after chronic-phase patients began long-term treatment with Gleevec, fewer than 5% of patients had died of CML-related effects. By this fifth year, the annual relapse rate (that is, progression to blast crisis) in this population was ~0.6%.

The molecular mechanisms that allow tumor cells to eventually escape Gleevec inhibition are interesting, in that they shed further light on the Bcr-Abl oncoprotein and its actions and, more generally, they reveal how cancer cells can acquire resistance to highly targeted drugs. Analyses of *BCR-ABL* sequences in the tumors of patients with Gleevec-resistant, relapsed disease revealed that 29 of 32 tumors harbored mutations in the *BCR-ABL* gene; altogether these yielded substitutions of 13 distinct amino acid residues in the kinase domain. Another dozen have been cataloged in subsequent studies.

Some of these mutations prevent Gleevec from binding to the catalytic cleft, either by directly interfering with its binding or, less directly, by creating a stereochemical shift in the oncoprotein (**Figure 16.24A and B**). In a minority of patients, Gleevec resistance

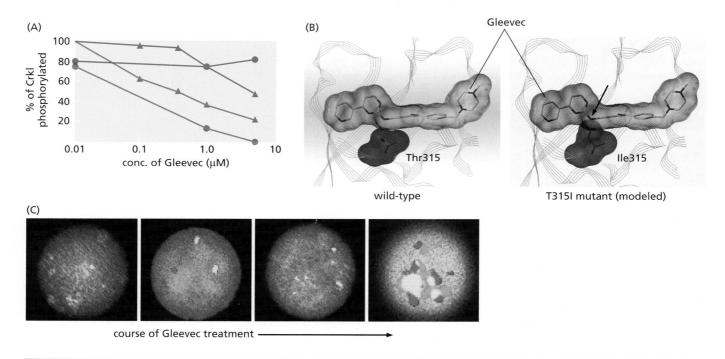

Figure 16.24 Acquisition by CML cells of resistance to Gleevec The ability of Gleevec to inhibit Bcr-Abl kinase activity changes dramatically following relapse and acquired resistance to drug treatment. (A) In this case, the kinase activity was gauged in isolated leukemia cells by the degree of phosphorylation of Crkl, a protein that is a substrate for phosphorylation by Bcr-Abl. At the onset of therapy, the Bcr-Abl kinase (in cultured leukemia cells) suffered about a 50% inhibition in the presence of approximately 0.1 μM Gleevec in two patients (*blue triangles, blue circles*). However, after Gleevec resistance developed in these patients, about an 8-μM concentration of the drug was required to inhibit one patient's Bcr-Abl kinase (*red triangles*), while the other patient's was totally resistant to the drug (*red circles*). (B) The Gleevec molecule is able to nestle tightly in a molecular cavity created in part by a threonine residue (*blue*) in position 315 of the wild-type Bcr-Abl oncoprotein (*left*; see also Figure 16.10C). However, in a mutant Bcr-Abl found in the leukemic cells of a Gleevec-resistant patient (*right*), this threonine residue was replaced by an isoleucine (*brown, arrow*), which protrudes into the drug-binding cavity and interferes with insertion of Gleevec into the cavity. (C) The number of *BCR-ABL* gene copies in a patient's leukemic cells has been gauged here using fluorescence *in situ* hybridization (FISH). Nuclei are visualized here in *blue*, *ABL* sequences in *red*, and *BCR* sequences in *green*. *Yellow* indicates an overlap of *ABL* and *BCR* sequences, i.e., sites of the fused gene created by the chromosomal translocation. The copy numbers of the fused gene (*yellow*) at the beginning of therapy (*left*) were quite low, but as treatment proceeded (*rightward*), the copies of the fused gene (and therefore Bcr-Abl fusion protein) increased progressively until the patient's leukemia became resistant to Gleevec treatment. In this particular patient, Gleevec resistance was acquired by the tumor cells because the fusion protein became overexpressed, thereby exceeding the ability of the normally used therapeutic concentration of Gleevec to bind and fully inactivate it. (From M.E. Gorre et al., *Science* 293:876–880, 2001.)

was achieved through amplification of the *BCR-ABL* gene in their leukemia cells, yielding increased levels of the encoded oncoprotein that apparently could no longer be inhibited by the concentrations of drug used to treat patients (see Figure 16.24C).

The observations that acquired resistance to Gleevec is usually accompanied by either structural alterations or overexpression of the Bcr-Abl protein is compelling proof that Gleevec's therapeutic responses can be attributed directly to its effects on the Bcr-Abl protein. This insight was explored by introducing random mutations into a vector encoding the Bcr-Abl protein and then determining which of the resulting mutant forms of this protein were able to resist inhibition by Gleevec (**Figure 16.25**). Such an experimental strategy, which uses cultured cells whose growth and viability are dependent on Bcr-Abl (see Figure 16.15), can in principle reveal the full spectrum of structural alterations of Bcr-Abl that are capable of rendering it resistant to Gleevec inhibition and thereby advance our understanding of the molecular mechanisms underlying acquired drug resistance.

A number of second-generation Bcr-Abl inhibitors have been synthesized that succeed in inhibiting many of the mutant Gleevec-resistant Bcr-Abl proteins that arise in CML patients and trigger clinical relapse; among these are nilotinib and dasatinib, which are actually more potent than Gleevec/imatinib. However, they and yet other newly developed Abl inhibitors fail to shut down the most formidable of the mutant proteins, termed T315I (**Figure 16.26A and B**). Detailed structural analyses of this mutant protein and its effects on drug binding have, after much effort, yielded novel compounds that can also shut down this particular mutant form (Figure 16.26C), illustrating how a convergence of structural biology and synthetic organic chemistry can often cope with such refractory mutant enzymes.

The ability of Gleevec to also inhibit the platelet-derived growth factor receptors (PDGF-Rα and β) suggested that it might prove useful in treating other types of malignancies as well. For example, translocations of the genes encoding these two receptors that cause constitutive receptor activation have been found in a number of chronic **myeloproliferative** diseases, that is, conditions involving elevated levels in the circulation of one or another cell type arising from the myeloid lineage of hematopoiesis (see Supplementary Sidebar 12.1). Indeed, patients suffering from hypereosinophilic syndrome have shown a **complete response** following Gleevec treatment, with virtual disappearance of their eosinophiles. The growth of many of the far more common glioblastomas is driven by PDGF–PDGF-R autocrine loops. In this instance, however, the use of Gleevec administered together with cytotoxic drugs has not proven encouraging, possibly because it cannot efficiently penetrate the blood–brain barrier and gain access to the tumor parenchyma.

Gleevec's effects on a third tyrosine kinase—the Kit receptor—also make it an attractive agent for attacking gastrointestinal stromal tumors (GISTs), a relatively uncommon sarcomatous tumor for which few therapeutic options have been available. The Kit receptor is mutated in the majority (~85%) of these cancers, while a minority (3–5%) exhibit mutant PDGF-Rα—both targets of Gleevec. The mutant receptors fire constitutively in these tumors and seem to represent the primary mitogenic forces in the tumor cells (see Figure 5.17). In one study, clear regression of the tumor was observed in almost 70% of treated patients (**Figure 16.27A**). By 2005, SU11248—a second inhibitor of Kit tyrosine kinase function—was approved by the FDA for the treatment of GISTs, including those that had developed a resistance to Gleevec. When used as an adjuvant following surgical excision of localized primary tumors, Gleevec treatment offers long-term survival that may result in a cure. With metastatic disease, however, Gleevec can cause regression of GIST colonies, but within a year or two the tumor reappears (see Figure 16.27B).

The clear successes of Gleevec represented the first validation that rational drug design can succeed in producing agents that are highly useful for treating various types of human cancer. The fact that Gleevec interferes with multiple tyrosine kinases was initially viewed as a disadvantage of this drug, since it was feared that this broader activity would lead to unacceptable side effects. However, with the passage of time, it is becoming increasingly clear that such multi-target effects may actually prove useful in treating certain malignancies. Thus, the viability and proliferation of many tumors

Figure 16.25 Screening *in vitro* for Gleevec-resistant mutant forms of Bcr-Abl One strategy to detect drug-resistant variants of Bcr-Abl involves cultured cells such as BaF3 cells (see Figure 16.15), whose continued survival can be made dependent on the presence of a functionally active Bcr-Abl oncoprotein. When such Bcr-Abl–expressing cells are treated with Gleevec, they are killed because of their dependence on continued Bcr-Abl signaling. (A) A cDNA clone expressing the "wild-type" Bcr-Abl protein (i.e., the direct product of the chromosomal translocation) can be mutagenized by passage through *E. coli* bacteria that are highly error-prone in DNA replication and therefore generate mutant variants of the introduced, plasmid-borne *BCR-ABL* sequence. The resulting collection of randomly mutated Bcr-Abl–expressing clones is then introduced, via a retrovirus vector, into BaF3 cells, which are then exposed to Gleevec. The rare cells that resist being killed by Gleevec are isolated, and the sequence of the mutant Bcr-Abl protein that conferred Gleevec resistance is determined. (B) When these Gleevec-resistant mutant Bcr-Abl proteins are analyzed, many are found to have single amino acid substitutions of residues located throughout the Abl kinase domain. The "back" and "front" of the Abl kinase domain are shown here, together with the sites of these mutant residues and the identities of the normally present amino acid residues. Surprisingly, the individual mutations, each of which confers Gleevec resistance, alter residues at many sites in the ABL domain, indicating that CML cells have multiple options for developing drug resistance. Many of these mutant residues are found on the back side of Abl opposite the catalytic cleft; some of these residues (*red*) participate in interactions between the kinase (i.e., SH1) domain and the SH2 and SH3 domains of Abl (*not shown*). Other mutant sites (*blue*) produce drug resistance through poorly understood mechanisms. This *in vitro* screen for Gleevec-resistant Bcr-Abl mutants revealed most of those discovered in patients plus a number of previously undocumented ones. (From M. Azam, R.R. Latek and G.Q. Daley, *Cell* 112:831–843, 2003.)

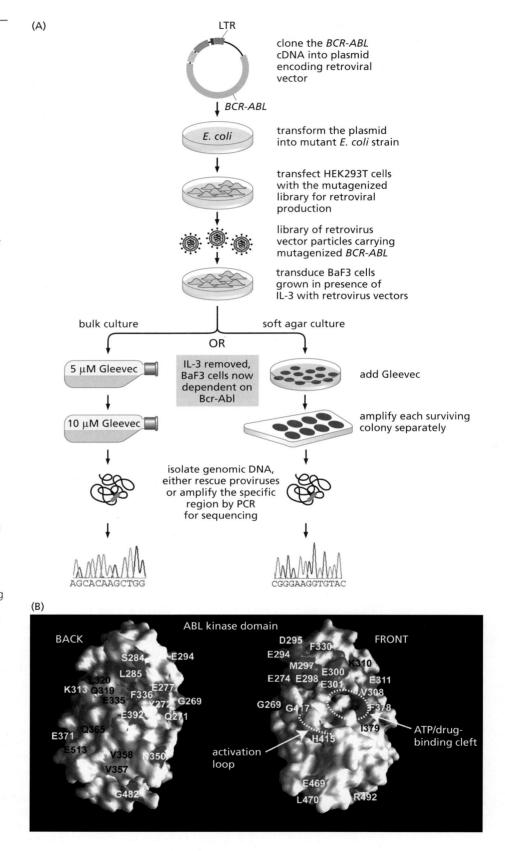

depend on the coordinated actions of multiple tyrosine kinases, and the ability to strike at several of these simultaneously may one day be found to confer great therapeutic advantage.

Unfortunately, the existence of tumor stem cells limits the utility of Gleevec. Recall that research on human cancers, including hematopoietic tumors, breast carcinomas, and brain tumors, has revealed that tumor stem cells often constitute only a small proportion (<<5%) of the neoplastic cells in these tumors, and that their existence can be revealed only by the biological test of their tumor-forming ability or by the use of fluorescence-activated cell sorting (FACS; see Section 11.6). As it happens, Gleevec is quite potent in killing actively cycling leukemia cells (that is, the "transit-amplifying"

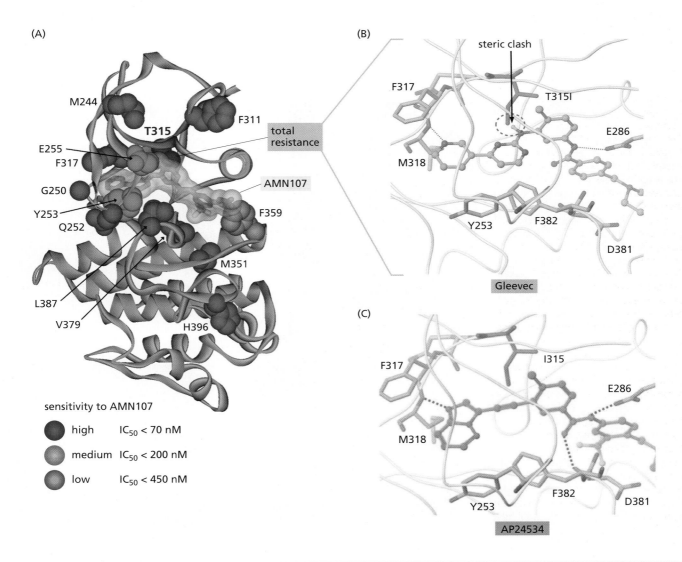

Figure 16.26 Backup inhibitors of Bcr-Abl for patients with Gleevec-resistant tumors The fact that patients in the acute (blast crisis) phase of CML often develop resistance to Gleevec (see, for example, Figure 16.24) has stimulated the development of alternative inhibitors of the Abl tyrosine kinase. (A) One of these inhibitors, AMN107, which is ~20 times more potent than Gleevec against unmutated Bcr-Abl, is shown here (*orange stick figure in yellow envelope*) in complex with the tyrosine kinase domain of Bcr-Abl, on which are also indicated the sites of a number of amino acid substitutions found in the mutant forms of Bcr-Abl in tumors of Gleevec-resistant patients. (The number of *colored spheres* at a site indicates the number of atoms present in the side chain of the substituted amino acid.) The locations of the amino acid substitutions carried by the mutant, Gleevec-resistant Bcr-Abl proteins are indicated by the *red*, *orange*, and *green* spheres and show different levels of sensitivity to inhibition by AMN107. One Gleevec-resistant mutant form of Bcr-Abl in which the normally present threonine is replaced by an isoleucine (T315I; *blue spheres*) is also totally resistant to AMN107 and most other second-generation Bcr-Abl inhibitors. (IC$_{50}$, concentration required for 50% inhibition of the target molecule.) (B) As seen here, the side chain of the isoleucine (*red*) present in the T315I mutant Bcr-Abl protein creates a steric clash with a hydrogen atom of Gleevec in the drug-binding pocket (see Figure 16.24B), effectively precluding binding by Gleevec (*left*). (C) In response to this resistance of the T315I mutant, a pharmacologically active derivative of Gleevec, termed AP24534, has been synthesized that avoids the steric clash and therefore can bind the T315I mutant protein. (A, from T. O'Hare et al., *Cancer Res.* 65:4500–4505, 2005. B and C, courtesy of T. Clackson and from T. O'Hare et al., *Cancer Cell* 16:401–412, 2009.)

or "progenitor" cells; see Section 11.6). However, most of the cells in the neoplastic stem cell population, which are outside the active cell cycle at any single point in time, have proven to be quite resistant to drug treatment. Among other traits, these cancer stem cells (CSCs) express elevated levels of the efflux pumps that confer multi-drug resistance (see Figure 16.21). In addition, the quiescent state of most CSCs makes them more resistant to commonly used cytotoxic drugs, which preferentially kill rapidly proliferating cells.

The CSCs are usually present in minute numbers in a CML patient undergoing Gleevec therapy, being detectable only by highly sensitive PCR-based assays. However, if treatment is halted, the CSCs usually re-enter into the growth-and-division cycle and regenerate transit-amplifying progeny, soon leading to regrowth of the malignancy and clinical relapse (**Figure 16.28A**). This seems to explain why Gleevec treatment

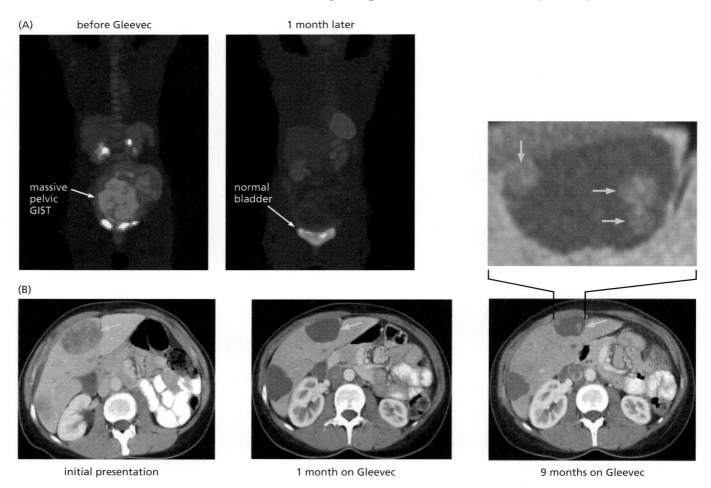

Figure 16.27 Use of Gleevec to treat gastrointestinal stromal tumors The fact that Gleevec also shows inhibitory activity against the tyrosine kinase function of the Kit receptor suggested that it might prove useful against gastrointestinal stromal tumors (GISTs), in which mutant, constitutively active Kit receptors are commonly found. (A) As seen here, this patient's GIST (*red mass, pelvic region, left image*), which was visualized because of its uptake of a labeled glucose analog (see Figure 2.22), responded dramatically to Gleevec treatment (*right image*). (The residual labeling following treatment reflects the accumulation of the labeled dye in the patient's bladder.) (B) Despite initial anti-tumor activity in nearly 90% of patients, resistance to Gleevec develops in most GIST patients; after 2 years, about 50% of tumors worsen despite continuing Gleevec treatment. Seen here are computed tomography (CT) scans of a patient with advanced GIST demonstrating several metastatic lesions growing in the liver (one with *yellow arrow, left panel*). After one month of daily Gleevec (= imatinib) treatment, the metastases, previously a lightly mottled gray, become more solidly dark, indicating the increased water density following widespread tumor cell apoptosis and degeneration (middle panel). However, after nine months of Gleevec (*right panel*), three light spots have appeared within the dark mass that was the original tumor (*yellow arrow*), indicating the eruption of robustly growing subpopulations that signal the onset of clinical resistance to targeted therapy (*yellow arrows, top right*). In this particular patient, each of the three recurring growths carried its own novel secondary point mutation in the *Kit* gene in addition to the initiating *Kit* mutation that launched the tumor in the first place, leading in each case to a doubly mutated *KIT* allele. (Courtesy of G.D. Demetri.)

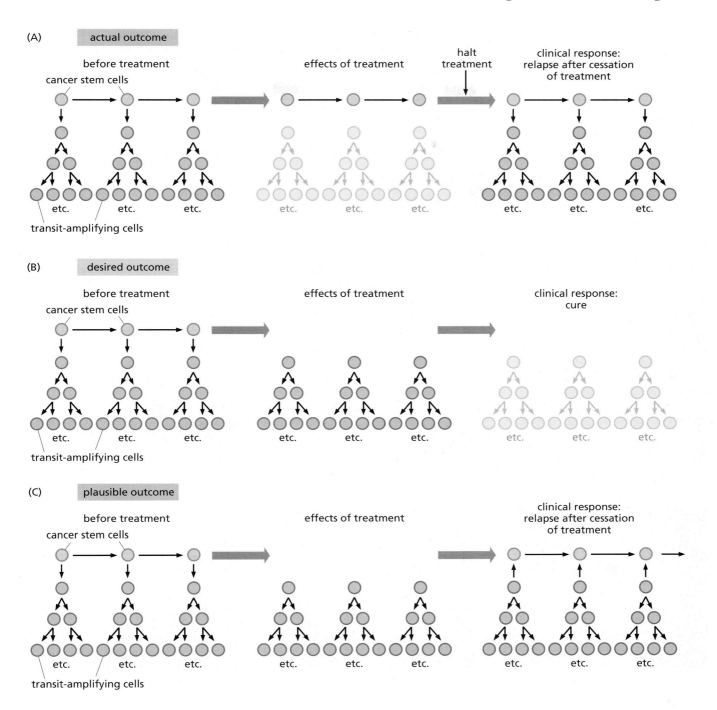

Figure 16.28 Role of tumor stem cells in the response to anti-cancer treatments While the evidence is still fragmentary, it appears that in many and perhaps all tumors (see Section 11.6), a small proportion of the neoplastic cell population is composed of self-renewing tumor stem cells (*gray*). These spawn the bulk of the cancer cells in tumors (*pink*), which have many of the properties of normal progenitor/transit-amplifying cells as well as more differentiated progeny. (A) If, as occurs with Gleevec, an anti-cancer therapy results in the depletion of the neoplastic transit-amplifying cells without eliminating the tumor stem cells [also termed cancer stem cells (CSCs)], then the latter can regenerate the tumor soon after that therapy is halted. Indeed, diverse sources of evidence indicate that in many carcinomas the CSCs are significantly more resistant to therapy than the bulk of carcinoma cells. Hence, CSCs are doubly dangerous, in that they can survive treatment and are capable of generating entirely new tumors in both primary tumor sites and in distant tissues. (B) In response to the risk posed by CSCs, some have undertaken to discover agents that preferentially eliminate these cells, with the thought that their elimination will lead to the disappearance of entire tumors, since their source of self-renewal has been eliminated (*right*). (C) Elimination of CSCs, on its own, may not lead to a cure, since recent evidence has indicated (*right*) that the non-CSCs in certain tumors can spontaneously generate new CSCs. Hence, the targeting of both populations (CSCs and non-CSCs) may be required to develop durable clinical responses following treatment of a variety of tumors.

needs to be chronic and why, in the future, drug development needs to be focused on agents that strike at the core of tumors by destroying their stem cells (Sidebar 16.4).

Nonetheless, if a drug such as Gleevec succeeds in generating clinical remissions that are durable over many years' time, its inability to kill the stem cells of a CML clearly is an acceptable shortcoming. In any case, Gleevec represents a major triumph of anti-cancer drug development, because it is vastly superior to all alternative treatments of this otherwise inexorably progressing disease.

Of note, the striking success of Gleevec in treating CML persuaded many that comparable successes would soon emerge for treating a number of other malignancies driven by deregulated tyrosine kinases. As it turned out, CML is a rather unique entity among adult tumors, in that almost all of these leukemias share a common genetic driver (the Bcr-Abl translocation) to which they are all addicted, while carrying few if any additional chromosomal aberrations and oncogenic mutations. In addition, rather than being full-fledged malignant cells, the leukemic cells present in the chronic phase of CML have been likened to *in situ* carcinomas in the breast and early adenomas in the gut (see Section 11.2). These attributes set the chronic-phase cells apart from most adult tumors and help to explain the striking differences between the clinical responses of CML to Gleevec versus the responses of many carcinomas to the other tyrosine kinase inhibitors (TKIs) described below. (In retrospect, the stunning successes of Gleevec inspired unrealistically high expectations for the many TKIs that followed.)

16.13 EGF receptor antagonists may be useful for treating a wide variety of tumor types

The proposal to develop Gleevec initially met with considerable resistance in the pharmaceutical company where it originated, simply because the market for this drug was judged to be too small to justify the high costs of its development and testing in the clinic. The same could not be said of the class of drugs designed to inhibit the epidermal growth factor receptor (EGF-R). Carcinomas are common tumors, and this receptor is believed to play a key role in the development of as many as one-third of them, being frequently overexpressed.

Sidebar 16.4 Cancer stem cells greatly complicate the evaluation of anti-cancer therapies The existence of cancer stem cells (CSCs) in many solid tumors has profound implications for the evaluation of many types of anti-cancer treatments. These lessons are illustrated graphically in the case of CML and Gleevec. [In the hypothetical case shown in Figure 16.28A, the re-emergence of a tumor is triggered by cessation of treatment, but for most tumors under treatment, such relapses occur because tumor cells have developed resistance to the therapeutic agent(s).] In carcinomas, for example, the CSCs, which display certain mesenchymal characteristics, seem generally to be more resistant to existing therapeutic regimens than the bulk of the neoplastic cells in these tumors (see Table 16.5). Hence, a therapy may eliminate the bulk of a tumor (leading to a significant clinical response), while leaving behind a residue of clinically undetectable CSCs; the latter may then re-launch tumor growth when the therapy is halted or when drug resistance develops. Clinically, this behavior is sometimes called the "dandelion effect," referring to the rapid re-emergence of weeds in a lawn following mowing, which cuts off their leaves but leaves their roots intact.

This logic suggests that the ability to eliminate CSCs will be the key to generating durable clinical responses for many types of tumors (Figure 16.28B). However, such elimination on its own may not suffice, since there are indications from study of tumor xenografts that the transit-amplifying/progenitor cells can dedifferentiate and thereby regenerate new CSCs under certain physiologic conditions (Figure 16.28C). Together, this would indicate that both the CSCs and the non-CSCs within a tumor need to be eliminated in order to prevent recurrence and clinical relapse. Importantly, identifying agents that specifically eliminate CSCs may be challenging, since such drugs may, on their own, have minimal effects on the overall sizes of tumor masses and may therefore be judged to be unworthy of further development.

In fact, rituximab, the anti-CD20 monoclonal antibody that is used to treat B-cell tumors (see Section 15.19), shows just this behavior: it eliminates the CSCs of multiple myelomas but not the more differentiated, far more abundant antibody-secreting cells that form the bulk of these tumors. (Fortunately, early in its development, rituximab was recognized to have efficacy for treating a wide range of B-lymphocyte–lineage tumors and was therefore approved for further development and introduction into the clinic.)

At least six distinct EGF-related ligands, including EGF itself, have been found to bind and activate the EGF-R. This means that even in those carcinomas in which the EGF-R is not overexpressed, it may nonetheless emit critical oncogenic signals through auto-crine or paracrine signaling loops driven by the presence of one or more of its ligands. (Recall the release of EGF by macrophages that confers invasiveness on breast cancer cells; see Figure 14.22.) Moreover, in breast carcinomas that overexpress the HER2/Neu receptor, the oncogenic actions of this protein may depend on its ability to form heterodimers with its cousin, the EGF-R; in such heterodimers, the EGF-R can phos-phorylate the C′-terminal tail of HER2/Neu, thereby activating signaling by the latter.

The best-characterized inhibitors of the EGF-R tyrosine kinase are the drugs Iressa, also known as gefitinib and ZD1839; and Tarceva, also called either erlotinib or OSI-774 (**Figure 16.29A**). The two drugs have very similar but not identical properties and act by blocking the ATP-binding site of the receptor-associated kinase (Figure 16.29B; see also Figure 16.13).Once cancer cells are deprived of receptor signaling through inhibition of the EGF-R, they should lose the benefit of its strong mitogenic and anti-apoptotic signals. For example, in many types of epithelial cells, the continuous firing of the EGF-R sustains expression of Bcl-X_L (the potently anti-apoptotic cousin of Bcl-2) and, acting via MAPK, drives phosphorylation and attendant functional inactiva-tion of the pro-apoptotic Bad protein (see Section 9.13).

Because Iressa and Tarceva target a cell surface receptor, their therapeutic utility must be compared with that of the monoclonal antibodies that also affect this receptor (see Sidebar 15.5). In principle, these low–molecular-weight compounds should be able to penetrate into all the interstices of a solid tumor, including those where the far larger antibody molecules may have trouble gaining access (see Table 16.7). Also, it is gener-ally far easier and less expensive to produce low–molecular-weight compounds on an industrial scale than it is to generate large amounts of monoclonal antibodies.

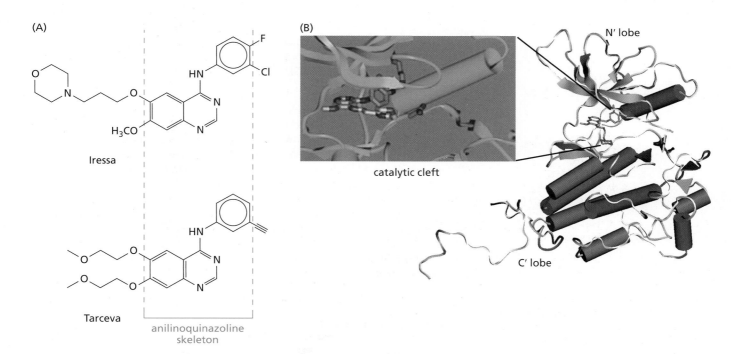

Figure 16.29 Iressa and Tarceva (A) The two epidermal growth factor receptor (EGF-R) antagonists are constructed from a common anilinoquinazoline skeleton, which confers on them an affinity for the ATP-binding site of the receptor tyrosine kinase. The chemical side groups that are attached to this skeleton have biological effects, since the effectiveness of the two drugs differs in treating, for example, non-small-cell lung carcinomas (NSCLCs). (B) Iressa, also termed ZD1839, binds to a very similar region of the EGF-R tyrosine kinase as does Tarceva (see Figure 16.13). A blown-up view (*left*) of the drug-binding site of the EGF-R tyrosine kinase domain (*right*) is shown here, with the drug molecule shown as a *colored stick* figure. This binding is so strong that 50% inhibition of TK enzyme activity is achieved at a concentration of about 0.030 µM. (B, courtesy of A.C. Kay, AstraZeneca.)

There are yet other possible advantages of low–molecular-weight tyrosine kinase inhibitors. For example, as we have read, in many human carcinomas, truncated forms of the EGF-R are expressed that lack the ectodomain; these mutant EGF-Rs may signal in a ligand-independent, constitutive fashion and therefore can function as potent oncoproteins (see Figures 5.10 and 5.11). Similarly, about half of high-grade (that is, advanced) gliomas, termed glioblastoma multiforme (GBM), exhibit overexpressed EGF-R, and of these, about 40% display a form of the receptor that lacks the ectodomains specified by exons 2 through 7 of the EGF-R coding sequence. Such decapitated receptors cannot be bound by the monoclonal antibodies (MoAbs) that have been developed to recognize antigenic **epitopes** in the ectodomain of the normal receptor protein. However, these aberrations should not derail the low–molecular-weight tyrosine kinase inhibitors, which target the cytoplasmic, signal-emitting domain of the receptor. Weighing against these drugs are their pharmacokinetic properties: as an example, Tarceva has a half-life of ~36 hours, while the half-life of cetuximab (an anti-EGF-R monoclonal antibody) is >5 days and of Herceptin, ~28 days.

Iressa has approximately a 50-fold more potent activity against the EGF-R–associated tyrosine kinase than against a number of other tyrosine kinases (see Supplementary Sidebar 16.9), and its initial use in the oncology clinic was reasonably encouraging. In the first clinical trials, 10% of patients with non-small-cell lung carcinomas (NSCLCs) showed **partial responses** to the drug, including disease stabilization of tumor growth; these patients tended to be women, nonsmokers, and those with the bronchioalveolar subtype of lung cancers. A parallel study in Japan found a far higher rate (27%) of partial responses to Iressa; this difference, which continues to be observed, appears to represent a difference in the genetic constitutions of the Japanese and Caucasian populations. (The tumors classified as NSCLCs have represented difficult diseases to treat, as fewer than 15% of patients survive for five years following initial diagnosis.)

These outcomes were gratifying, if only because they represented clear responses in patients who otherwise had few if any other treatment alternatives. However, the hoped-for synergistic actions of Iressa with standard chemotherapeutic agents did not provide any survival advantage over standard chemotherapy used alone for the treatment of NSCLC tumors, which constitute almost 80% of lung cancer cases in the United States. When used on its own, Tarceva (but not Iressa) increased the overall survival time of patients whose NSCLCs had become refractory to treatment by standard chemotherapeutic drugs.

Some valuable lessons were learned from these initial trials that may improve the responses in subsequent clinical trials of these and similar drugs: First, the specific contribution of the EGF-R to the growth of the tumors under treatment was not documented. Hence, far greater response rates might have resulted from a stratification of the NSCLC patients and limiting the use of Iressa only to those tumors having specific molecular signatures. Second, the possible contribution of other mutant proteins to mitogenic and anti-apoptotic signaling was not assessed. There is evidence, for example, that PTEN-negative tumors (which have a hyperactive PI3 kinase pathway; see Section 6.6) do not respond to Iressa, and that inhibitors of Akt/PKB (the downstream beneficiary of PTEN inactivation) can act synergistically with Iressa to halt tumor growth. Third, relatively few pre-clinical studies were undertaken in order to optimize the dosage and the schedule of treatments with this drug.

In 2004, four years after results of the initial clinical trials with Iressa were first reported, two research collaborations in Boston independently provided a molecular explanation for the observed responses to Iressa. Previously, the status of the EGF receptor in NSCLC cells was assessed by determining whether it was overexpressed and whether it was present in truncated, constitutively active form, as is the case in human glioblastomas. In the 2004 studies, however, investigators undertook detailed sequencing of the reading frames of the EGF-R–encoding gene in the NSCLC patients who had been treated with Iressa.

Quite dramatically, they found that almost all of the small group (~10% of the total) of NSCLC patients who had responded well to Iressa treatment (**Figure 16.30A**) bore tumor cells displaying structurally altered EGF-Rs. Such mutant receptors were

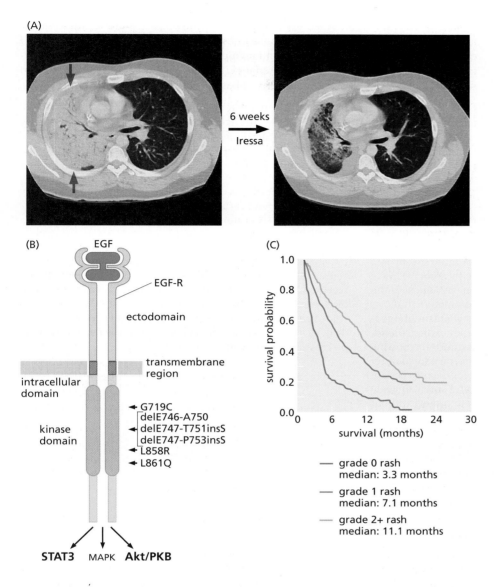

Figure 16.30 Determinants of responsiveness of NSCLCs to Iressa and Tarceva treatment (A) A minority of patients with refractory non-small-cell lung cancers (NSCLCs)—tumors that failed to respond or ceased to respond to standard chemotherapy—show dramatic responses to treatment by Iressa. These computerized tomographic images reveal dramatic regression of a large mass (*left*) in a patient's right lung following six weeks of Iressa treatment (*right*). (B) A substantial proportion of NSCLCs that respond to Iressa have been found to carry deletions ("del") and point mutations in the EGF-R gene that affect the cytoplasmic domain of the receptor. These alterations in EGF-R structure deregulate and activate the tyrosine kinase function of the receptor, thereby stimulating the Akt/PKB and STAT signaling pathways, which protect these tumor cells from apoptosis. (C) Experience with a variety of anti-EGF-R treatments has indicated that, in addition to the genetic state of the tumor cells, the development of an acne-like rash by patients under treatment is a strong positive indicator of eventual response of tumors to therapy. (This response clearly reflects the role of the EGF-R in skin keratinocyte biology and the precise mechanistic connection between the normal keratinocyte response of patients to therapy and the response of their tumors remains obscure.) In this case, the responses of NSCLC patients to Tarceva therapy are shown in this Kaplan–Meier graph. "Median" indicates median survival. (A and B, from T.J. Lynch et al., *N. Engl. J. Med.* 350:2129–2139, 2004. C, from B. Wacker et al., *Clin. Cancer Res.* 13:3913–3921, 2007.)

not found among the tumors that failed to respond to Iressa, including those that expressed elevated levels of this receptor (see, for example, Supplementary Sidebar 16.10). For example, 78% of patients who responded to anti-EGF-R therapy expressed mutant EGF-R protein, whereas among the nonresponders only 6% showed such structurally altered receptor proteins. The responsible mutations created amino acid substitutions and small deletions in the kinase domain (see Figure 16.30B) rather than the major deletions of the receptor ectodomain typically found in glioblastomas. For unknown reasons, these mutant receptors showed distinctive patterns of tyrosine phosphorylation of their C-terminal tails (see Section 6.3) and selectively stimulated the downstream Akt/PKB and STAT5 pathways, leaving the MAPK signaling pathway unaffected.

These observations provided compelling evidence that the EGF-R played a central role in driving the growth of these small groups of tumors. In addition, they demonstrated the value of stratification (that is, subclassification) of tumors using molecular markers when treating patient populations with targeted molecular therapeutics, such as Iressa and Tarceva. However, these experiments did not reveal why Iressa and Tarceva had such strong effects on these particular tumors (but see **Sidebar 16.5**).

An additional puzzle comes from the frequency of the NSCLCs found to express the structurally altered EGF receptors. They have been found in the tumors of ~10% of Western patients suffering from NSCLC, while in certain Asian populations, as many as 30% of the NSCLCs analyzed express these mutant receptor proteins; these would appear to explain the far higher response rates to Iressa in Japan cited above. (In all cases these altered receptors result from somatic mutations rather than germ-line polymorphisms.) The reasons for this discrepancy, and the fact that these tumors tend to occur far more frequently in female nonsmokers, represent additional puzzles.

One mystery seems to have been solved, however. In certain East Asian populations, as many as half of CML patients show an incomplete response to Gleevec therapy compared with only one-fourth of Western patients; a similar therapy-resistant subpopulation exists among NSCLC populations and their responses to EGF-R inhibitors. In both cases, these patients' tumors are resistant from the beginning of treatment

Sidebar 16.5 Oncogene addiction may explain how Iressa and Tarceva succeed in killing NSCLCs The mutant EGF-Rs that are found in certain NSCLCs cause the cells from these particular tumors to be approximately 100 times more sensitive to Iressa than tumors expressing the wild-type receptors (**Figure 16.31A**). Importantly, the actual drug concentrations in the plasma of patients being treated fall in the range that allows such selective inhibition to operate. The mechanism of oncogene addiction, introduced in Section 16.4, may explain the selective effects of both Iressa and Tarceva on tumors expressing structurally altered EGF-Rs. Recall that oncogene addiction refers to the strict dependence of certain cancer cells on a certain oncogene or oncoprotein for their growth and survival, while other types of cancer cells can lose this gene or protein without suffering significant consequences.

To explain this behavior, we can imagine that some oncogenes are generally deleterious when expressed in wild-type cells but are actually beneficial in cells that previously acquired certain mutant alleles of another oncogene. A good example is provided by the *myc* oncogene, which has pro-apoptotic effects on cells unless they are protected from apoptosis by an acquired anti-apoptotic allele of some sort (for example, a *ras* oncogene); in the presence of the anti-apoptotic mutation, the strongly mitogenic effects of the *myc* oncogene then become apparent. Hence, tumor cells that carry both the *ras* and *myc* oncogenes would behave as if they were "addicted" to the *ras* expression, since they would die quickly by apoptosis if they were deprived of the *ras* oncogene.

Similarly, early in tumor progression, the acquisition of a certain oncogene, such as a mutant EGF-R gene, might create a cellular environment that permits acquisition of other oncogenes (or losses of tumor suppressor genes) that would, on their own, be highly deleterious for tumor cells. If the mutant receptor is now lost, then the deleterious effects of these other oncogenes, such as those favoring apoptosis, would become apparent and result in rapid loss of cell viability.

In the case of non-small-cell lung cancer (NSCLC), those tumors bearing mutant receptors may have come to depend on the firing by their mutant EGF-Rs in order to survive and proliferate; that is, they are "addicted" to the mutant receptors. Conversely, the far more numerous NSCLCs expressing the wild-type EGF-R, often at elevated levels, may have developed alternative means of securing mitogenic and survival signals, as is indeed suggested by the observation of receptor-independent firing of the MAPK and PI3K pathways in some lung cancers.

This scenario is further supported by experiments using siRNAs to inhibit the expression of wild-type or mutant receptors (see Supplementary Sidebar 1.4): NSCLC cells with mutant EGF-R die quickly, while those displaying wild-type receptor are only slightly affected (see Figure 16.31B and C). Consequently, the death of cancer cells with mutant EGF-Rs is not due to some unknown, off-target effect of Iressa or Tarceva, but instead is caused directly by the loss of beneficial signals released by these receptors. Moreover, experiments like these suggest that EGF-R inhibitors may have far greater effects on NSCLCs expressing wild-type receptors if they are applied together with a second drug that inhibits another, functionally redundant signaling pathway, such as the one controlled by PI3K.

Figure 16.31 Effects of siRNAs to suppress EGF-R expression It is unclear why non-small-cell lung carcinoma (NSCLC) cells bearing structurally altered EGF receptors (see Figure 16.30B) are especially responsive to Iressa or Tarceva therapy. (A) When cultured *in vitro*, two NSCLC cell lines that overexpress structurally normal (i.e., wild-type) EGF-R were relatively resistant to Iressa treatment (*orange, red*), while two NSCLC lines expressing either an amino acid–substituted (*blue*) or a partially deleted (*green*) receptor protein were approximately 100 times more sensitive to being killed. (B) The biological mechanism of these differences was examined by depriving NSCLC cell lines of EGF-R by forcing them to express siRNAs (see Supplementary Sidebar 1.4) that can inhibit *EGF-R* mRNA. H358 cells (*left 3 bars*) expressing wild-type EGF-R were relatively unaffected by siRNAs directed either against all forms of the receptor (*green*) or against the two mutant forms (*yellow, red*). However, an siRNA directed specifically against the mRNA encoding a deleted form of the receptor (delE746; *yellow*) caused loss of viability in some 80% of the NSCLC cells expressing this mutant receptor (*middle bars*), while having no effect on cells with an amino acid–substituted receptor (*red, middle bars*). Conversely, loss of viability was observed when an siRNA directed specifically against the amino acid–substituted receptor (*red*) was introduced into cells that express this particular mutant receptor (*right bars*) but not when the siRNA directed against the deletion mutant was used. In the case of both EGF-R mutant cell lines, the siRNA directed against all forms of the receptor (*green bars*) also caused widespread cell death. Hence, the two NSCLC cell lines with mutant receptors were dependent on ("addicted to") EGF-R function, while the NSCLC cells with wild-type receptor showed virtually no dependence on continued EGF-R function. (C) The loss of viable cells after siRNA treatment seen in panel B is due specifically to an induction of apoptosis, as revealed by immunostaining of fixed cells with an antibody reactive with cleaved, activated caspase-3 (see Figure 9.19C). NSCLC cells were alternatively stained with 4′,6′-diamidino-2-phenylindole (DAPI) to reveal nuclei and thus cell number. (From R. Sordella et al., *Science* 305:1163–1167, 2004.)

to the effects of the tyrosine kinase inhibitors (TKIs) used to treat these two diseases. Among oncologists, this preexisting resistance is often called *primary resistance* to distinguish it from the resistance that is developed during the course of therapy, which is termed *secondary resistance*.

In both NSCLC and CML, the primary resistance has been traced to an allele of the pro-apoptotic *Bim* gene (see Section 9.13), which encodes a defective version of the Bim protein and is carried by 12.3% of East Asians but is absent in African and European populations. Other work has shown that expression of the wild-type Bim protein is suppressed by oncogenic tyrosine kinases, and its activation following TK inhibition contributes importantly to the apoptotic cancer cell death that drugs like Gleevec,

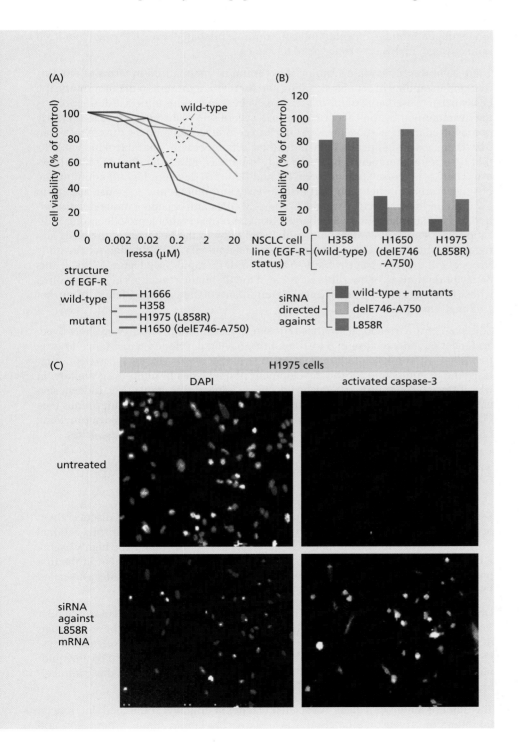

Iressa, and Tarceva succeed in inducing in responsive cell populations. Lacking both copies of the wild-type *Bim* gene and thus normal levels of the Bim protein, the tumors of these East Asian patients show substantially reduced tendency to enter into apoptosis and thus reduced responsiveness to the killing effects of these inhibitors.

In addition to the germ-line and somatically acquired alleles cited above, there are yet other indications that are useful for predicting responsiveness to anti-EGF-R TKIs. Echoing the results from use of anti-receptor MoAbs (see Supplementary Sidebar 16.10), the presence of a mutant, activated K-*ras* or *PI3K* oncogene or an inactivated *PTEN* tumor suppressor gene is strongly correlated with the failure of tumors to respond to these TKIs. These findings can all be rationalized in terms of the signal transduction cascades that we encountered in Chapter 6. Thus, if a downstream cytoplasmic effector of a cell surface receptor becomes activated by mutation, its firing no longer depends on the receptor and can continue unabated, even if the receptor itself is inactivated, in this case by targeted therapies.

There is, however, a striking finding that cannot be rationalized in terms of our current understanding of cellular signaling and therefore represents a major mystery: the best surrogate marker of a tumor's responsiveness to various anti-EGF-R therapies is the development by the cancer patient of a severe skin rash. Since the skin is clearly not involved in the pathogenesis of lung and colorectal carcinomas, this might mean that there exist significant inter-individual differences in pharmacokinetics (PK) that determine therapeutic outcome and, in parallel, affect the skin; in truth, however, there is no correlation between PK and skin rash development. In addition, the positive correlation between skin rash severity and therapeutic response applies to a variety of low–molecular-weight drugs as well as anti-receptor monoclonal antibodies (see Supplementary Sidebar 16.10). These observations provide strong indication that some subtle, inter-individual variability in the behavior of the basic signal-processing machinery operating in various epithelial cells throughout the body strongly influences responsiveness to anti-EGF-R therapies (see Figure 16.30C). This behavior reminds us of how little we understand about the regulators of cell signaling, whose behaviors are more than minor determinants of the success or failure of current "rational" therapies.

Still, even positive responses to these tyrosine kinase inhibitors have been short-lived, and most patients relapse in 6 to 18 months, having developed a resistance to drug treatment. This underscores, once again, the need for alternative agents to treat drug-resistant receptors and for the development of new types of multi-drug therapy, in which several drugs with synergistic effects are applied simultaneously. Indeed, several drug molecules that are structurally distinct from the anilinoquinazolines (see Figure 16.29A) are able to shut down Iressa- or Tarceva-resistant EGF-Rs, providing hope for those patients who have relapsed following treatment with these drugs.

16.14 Proteasome inhibitors yield unexpected therapeutic benefit

Serendipity plays an unusually prominent role in the world of drug discovery. On occasion, the development of an anti-cancer drug is launched as part of a rational drug design program and ultimately yields an agent that turns out to be highly useful, albeit for reasons quite unrelated to those that inspired its development in the first place. This best describes the development of the drug known as Velcade, also called PS-341 and bortezomib (**Figure 16.32A**).

On many occasions throughout this book, we have seen how the levels of key cellular regulatory proteins are determined by the balance between their synthesis and their degradation. Much of this degradation is mediated by the ubiquitin–proteasome system (see Supplementary Sidebar 7.5). Recall that the tagging of a protein by polyubiquitylation results in its transport to proteasomes and its degradation in these intracellular machines.

The phenomenon of cancer-associated **cachexia** initially stimulated interest in inhibitors of proteasome function. Cachexia occurs late in tumor progression and

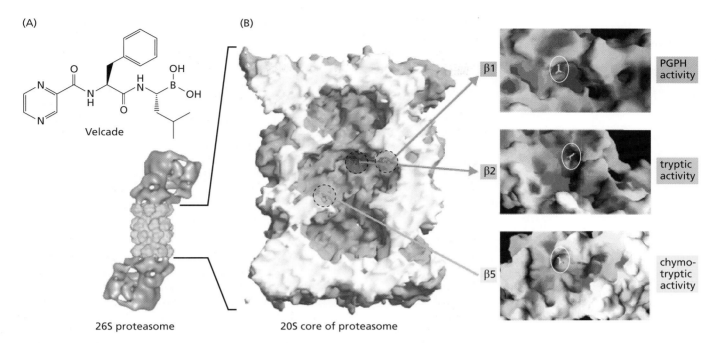

(A) Velcade

26S proteasome

(B) 20S core of proteasome

β1 — PGPH activity

β2 — tryptic activity

β5 — chymo-tryptic activity

represents a progressive wasting of the cancer patient's tissues through mechanisms that remain poorly understood. Use of a proteasome inhibitor was speculated to be useful in retarding the widespread degradation of proteins occurring in the tissues of cachectic patients. While at least five distinct classes of proteasome inhibitors have been developed, most have been abandoned because of metabolic instability, lack of specificity, or irreversible binding and inactivation of proteasomes. Velcade, one of these proteasome inhibitors, is a boronic acid dipeptide that was designed as a specific inhibitor of the **peptidase** (peptide-cleaving) activity present in the 20S core of the proteasome (see Figure 16.32B). It has extraordinary potency, being able to inhibit 50% of the proteasome's chymotryptic activity at a concentration (that is, its K_i) of only 0.6 nM. Functioning as a competitive inhibitor of this enzyme activity, Velcade slows the flux of substrates through proteasomes, which soon become clogged and dysfunctional.

Proteasome-mediated degradation was subsequently found to play a critical role in regulating a number of key cellular signaling pathways. Because other proteasome inhibitors had been found to be especially potent in killing a variety of cultured cancer cells, Velcade was used in Phase I trials to treat cancer patients who had failed other available therapies. Those with solid tumors showed few striking responses. However, among a group of patients with hematologic malignancies was one suffering from multiple myeloma (MM), a malignancy of the B-cell lineage in which a single clone of antibody-producing plasma cells dominates the bone marrow (see Figure 2.21). The myeloma cells create osteolytic bone lesions that lead to fractures, and they ultimately crowd out the remaining cellular components of the marrow, resulting in severe immune depression and, typically, death from overwhelming infection. Survival after initial diagnosis is usually three to five years. The myeloma carried by this initially treated patient showed a dramatic regression (**Figure 16.33A**), which soon led to inclusion of other myeloma patients in this Phase I trial and eventually to large-scale clinical trials.

In a subsequent Phase I clinical trial with a group of multiple myeloma patients suffering from rapidly progressing disease, Velcade showed clear "**objective responses**" in slowing disease progression in 55% of the patients and halted progression in another 25%. In a Phase II clinical trial, half of patients were given Velcade while the other half, who served as controls, were given dexamethasone, a standard treatment for multiple myeloma. The great majority of these patients had already failed the chemotherapies commonly used for myeloma. This trial was stopped prematurely in 2003 because Velcade demonstrated a clear superiority over existing treatments, with the

Figure 16.32 Velcade and its effects on proteasomes (A) The chemical structure of Velcade reveals the presence, unusual among drugs, of a boron atom. Peptide boronic acids like Velcade were known to bind to the active site of serine proteases of the chymotrypsin class (which cleave substrate proteins adjacent to phenylalanine and tyrosine residues) by mimicking the normal substrates of these enzymes. This suggested that such compounds might inhibit the chymotrypsin-like active site in the 20S core of the proteasome. (B) A cross section (*center*) through the central (20S) core of the yeast proteasome (*left*) reveals the locations of three distinct catalytic sites, involving the β1, β2, and β5 subunits (*far right*), which are responsible for its PGPH (peptidyl-glutamyl-peptide hydrolyzing), tryptic, and chymotryptic proteolytic activities, respectively. Velcade shows a strong preference for inhibiting the β5 chymotryptic activity (*lower right*); a weak interaction with the β2 tryptic activity (*middle right*); and no interaction with the β1 PGPH activity (*top right*). The key nucleophilic threonine residue in each of these catalytic sites is shown as a stick figure (*white inside ovals*); basic amino acids in the catalytic clefts are in *blue*, acidic amino acids are in *red*, and hydrophobic residues are in *white*. (B, from M. Groll et al., *Chembiochem* 6:222–256, 2005.)

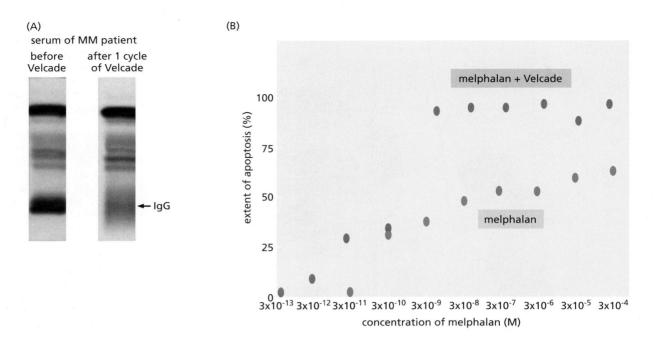

Figure 16.33 Multiple myeloma and the biological effects of Velcade
(A) Velcade can have profound effects on the cellular contents of the marrow and thus the composition of antibody molecules in the blood. In one multiple myeloma (MM) patient, after eight doses of Velcade, the neoplastic plasma cells in the marrow declined from 41% of total cells to 1%. At the same time, the level of the single species of immunoglobulin γ (IgG) made by these myeloma cells declined drastically. As seen here, upon gel electrophoresis, the small number of IgG species present before treatment (indicative of a monoclonal tumor, *left*) resolved into the heterogeneously migrating, polyclonal pattern of IgGs that are present in the circulation of a healthy individual (*right*; see also Figure 2.21A). (B) Melphalan, an alkylating chemotherapeutic drug used routinely to treat MM, was added at various concentrations to an MM cell line *in vitro*, either on its own (*green*) or in the presence of a noncytotoxic dose of Velcade (*red*). In the presence of Velcade, melphalan was able to induce widespread apoptosis at approximately 3 nM concentration, while when applied on its own, melphalan was unable to induce this degree of apoptosis, even at vastly higher concentrations.
(A, from R.Z. Orlowski et al., *J. Clin. Oncol.* 20:4420–4427, 2002. B, from M.H. Ma et al., *Clin. Cancer Res.* 9:1136–1144, 2003.)

disease showing either a "complete response" in a small number of patients (that is, myeloma cells disappeared completely from the blood for at least 6 weeks) or a "partial response" (at least 50% reduction in myeloma cell–secreted antibody in blood and 90% reduction of this protein in urine over the same time period) in 35% of the Velcade-treated patients. As a consequence, the control patients were then allowed to take the drug as well. In a subsequent trial, the progression of myeloma to a higher stage of disease occurred with a median time of 7 months in Velcade-treated patients compared with 3 months in a control group studied in parallel. Moreover, pre-clinical studies indicate that relatively low doses of Velcade can sensitize myeloma cells to chemotherapeutic drugs, making the latter far more effective (see Figure 16.33B).

In truth, inclusion of a myeloma patient in the initial clinical trial was hardly accidental. Multiple myeloma was thought to be an attractive target for treatment by a proteasome inhibitor because of the known elevated activity of the NF-κB signaling pathway in the myeloma cells and its physiologic importance in driving the survival and proliferation of these cells. In Section 6.12, we noted that NF-κB transcription factors are normally sequestered in the cytoplasm by a class of inhibitors termed IκBs (inhibitors of NF-κB). When these IκBs are phosphorylated by a group of specialized kinases termed IκB kinases, or simply IKKs, the IκBs undergo polyubiquitylation and resulting degradation; this liberates the NF-κBs, allowing them to migrate into the nucleus, where they activate a number of anti-apoptotic genes as well as growth-promoting genes (**Figures 16.34 and 16.35**; see also Figures 6.29A and 11.38B).

Like myriad other polyubiquitylated proteins, the IκBs end up being degraded in proteasomes. Hence, by inhibiting proteasome action, IκBs should be protected from degradation, survive in the cytoplasm, and continue to sequester the NF-κBs, thereby blocking NF-κB nuclear translocation and activation of transcription. Nuclear, functionally active NF-κB was known to be important for inducing the expression of IL-4 and IL-6, two interleukins that operate as important autocrine factors required for the growth and survival of myeloma cells. In addition, as was learned later, NF-κB plays a prominent role in anti-apoptotic signaling in a number of cancer cell types; hence, loss of active NF-κB might well tilt the signaling balance within these cells toward apoptosis. More specifically, once cancer cells lose the potently anti-apoptotic Bcl-2, cIAP-2, and XIAP proteins (all of whose expression is induced by NF-κB), they are in grave danger of slipping into the apoptotic abyss.

All this does not explain, however, why Velcade is far more potent against myelomas than other tumors that rely on NF-κB signaling to protect them from apoptosis. A

(A)

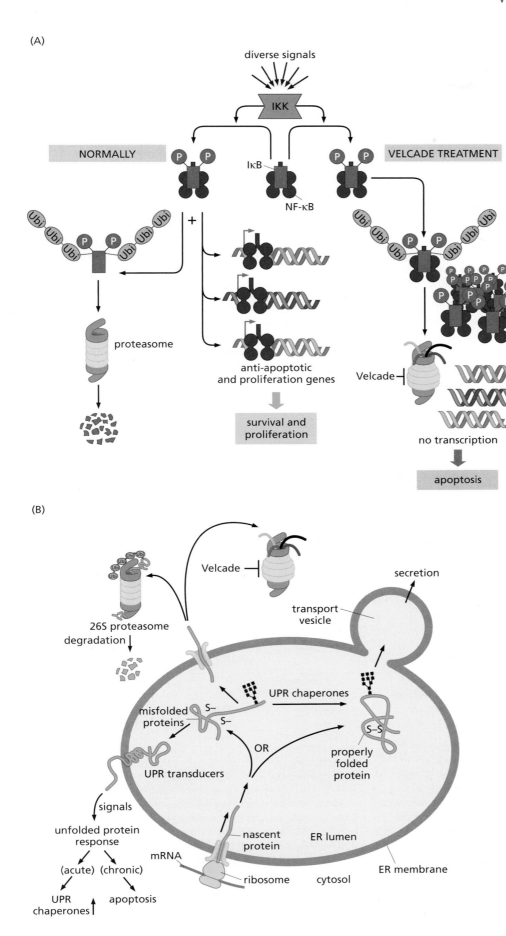

(B)

Figure 16.34 Mechanisms of Velcade action Two alternative mechanisms of action of Velcade have been proposed, and both are supported by extensive observations. (A) In normal and neoplastic cells (*left*), a variety of stress, mitogenic, and trophic (survival) signals activate IκB kinase (IKK; *purple, above*). Once activated, IKK phosphorylates IκB (*red*), the inhibitor of NF-κB. This causes IκB to become ubiquitylated (*left*) and degraded in proteasomes (*lower left*). NF-κB (*blue*) is then free to move into the nucleus, where it activates the expression of numerous proliferation and anti-apoptotic genes. In the presence of Velcade (*right*), the ubiquitylated IκB cannot be degraded in the proteasomes because the latter have become engorged with unprocessed polypeptides. This leads to an accumulation of IκB in the cytoplasm and to the continued sequestration of NF-κB by the IκB molecules that have built up (*right*). As a consequence, NF-κB is prevented from moving into the nucleus and activating expression of key anti-apoptotic genes. (B) The striking toxicity of Velcade for MM cells may reflect the fact that these cells, like normal plasma cells, are specialized to continuously synthesize and secrete enormous amounts of antibody molecules. Like other secreted glycoproteins, antibody molecules are processed in the endoplasmic reticulum (ER) prior to being secreted via exocytosis (*upper right*). Inevitably a certain proportion of recently synthesized antibody molecules are misfolded following their insertion into the lumen of the ER. In the absence of proper folding, many of these misfolded proteins are extruded from the ER and broken down, following ubiquitylation, in the proteasomes (*upper left*). However, when proteasome function is blocked by Velcade (*above*), misfolded proteins accumulate in the lumen of the ER and trigger the unfolded protein response (UPR). This complex program includes the induction of apoptosis when misfolded proteins accumulate to toxic levels. (B, adapted from Y. Ma and L.M. Hendershot, *Nat. Rev. Cancer* 4:966–977, 2004.)

possible clue comes from observations that the growth and viability of myeloma cells is highly dependent on their ability to synthesize VEGF (vascular endothelial growth factor; see Section 13.1) and adhesion molecules; the latter enable myeloma cells to attach to bone marrow stem cells (BMSCs), with which the myeloma cells establish critically important heterotypic interactions. The genes encoding these various proteins are all under NF-κB control.

NF-κB antagonists are likely to have utility for treating a number of other kinds of cancer as well. Recall that NF-κB plays a key role in the development of a variety of carcinomas (see Section 11.16) and may be required for the maintenance of these tumors once they are formed. In addition, the use of gene expression arrays has revealed that diffuse large B-cell lymphomas (DLBCLs), which appear under the microscope to constitute a single, homogeneous type of tumor, can actually be classified into three distinct subgroups (see Figure 16.5). Tumor cells belonging to the activated B-cell and mediastinal lymphoma subgroups have a constitutively activated IKK. Consequently, tumors belonging to these two DLBCL subgroups, as well as a variety of other tumors with hyperactive NF-κB signaling, have become attractive targets for treatment by either Velcade or a number of other IKK inhibitors.

A second, alternative theory of Velcade's mechanism of action, which also has numerous proponents, does not attribute its effects specifically to inactivation of the NF-κB signaling pathway. This other proposed mechanism derives from the fact that, like normal antibody-secreting plasma cells, myeloma tumor cells continuously synthesize and secrete antibody molecules at a prodigious rate—estimated to be thousands

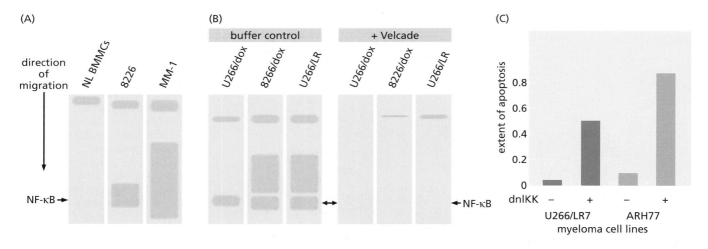

Figure 16.35 Evidence supporting the importance of NF-κB signaling in Velcade-induced apoptosis The scheme presented in Figure 16.34 is supported by a number of lines of evidence. (A) In an electrophoretic mobility shift assay (EMSA; also called a gel retardation assay), the presence and concentration of a functional, DNA-binding transcription factor (TF) are assessed by mixing an extract of nuclear proteins with a radiolabeled dsDNA oligonucleotide that carries a binding site for the TF. The presence of the DNA-binding TF is reflected by the amount of oligonucleotide that has formed a nucleoprotein complex with the TF; the large mass of protein associated with the oligonucleotide retards its migration during electrophoresis, causing it to migrate to a characteristic higher position in the gel. The arrow indicates the expected location of a complex containing the NF-κB transcription factor and the radiolabeled oligonucleotide, in this case one derived from the promoter of the TNF-α gene, a target of NF-κB activation. (In the absence of bound DNA, the oligonucleotide migrates beyond the bottom of the gel.) The assay indicates little if any detectable NF-κB activity in normal bone marrow mononuclear cells (NL BMMCs) and considerable activity in a multiple myeloma (MM) cell line (8226) and an enormous amount of NF-κB activity in the bone marrow cells prepared directly from a multiple myeloma patient (MM-1). (B) An EMSA has been used, as in panel A, to measure the level of functional NF-κB transcription factor in three MM cell lines treated with a control buffer (*left three channels*) or with Velcade (*right three channels*). Velcade is able to eliminate essentially all NF-κB activity in these cells. (C) The importance of ongoing NF-κB signaling to the survival of MM cells is demonstrated by this experiment, in which a vector expressing a dominant-negative IKK (dnIKK) has been introduced into two different MM cell lines. If NF-κB signaling were critical to the action of Velcade, then the dnIKK should mimic the effects of Velcade by inducing MM cell apoptosis—the outcome that is indeed observed here. A vector that does not express dnIKK was used as control here. (From M.H. Ma et al., *Clin. Cancer Res.* 9:1136–1144, 2003.)

of molecules per second. A certain portion of these molecules are routinely degraded in the proteasomes because of misfolding or other mishaps occurring during their post-translational maturation. As seen in Figure 16.34B, in addition to proteasomal degradation, accumulation of unfolded and misfolded proteins in the endoplasmic reticulum (ER) creates the state of *ER stress*, which in turn activates the **unfolded protein response** (UPR). This complex program depends on signaling by receptor-like UPR transducers that convey information of the presence of significant numbers of misfolded proteins in the lumen of the ER. One component of the UPR, active during acute episodes of ER stress, results in molecular chaperones being dispatched from the cytosol into the lumen of the ER in order to properly fold the aberrant proteins. Another part of the UPR program operates in response to chronic ER stress and triggers apoptosis. Myeloma cells are especially sensitive to inhibitors of protein degradation and become engorged with misfolded protein molecules, specifically subunits of antibodies. In some poorly understood fashion, the failure of proteasomal function in the cytosol triggers activation of the UPR within the ER, and under many circumstances, this results eventually in the activation of apoptosis. The truly amazing potency and therapeutic index of Velcade—killing myeloma cells at concentrations as low as 1 nM—may be explained by the highly specialized secretory phenotype of these cells.

These arguments suggest why Velcade is useful against a specific set of hematopoietic malignancies, most importantly MM. However, pre-clinical trials using this drug, either alone or in combination with other drugs, have also been launched to treat cultured human colorectal, gastric, breast, prostate, and lung carcinomas. In some cases, these studies have advanced into clinical trials designed to test the possibility that Velcade, while not a highly effective agent on its own in treating these carcinomas, might act synergistically with more conventional chemotherapeutic agents to elicit significant therapeutic responses.

At present it remains unclear precisely how Velcade functions in these diverse types of tumor cells, in which a third proposed mechanism of action may be operating. This one derives from the discovery that Velcade is a potent inducer of the NOXA pro-apoptotic protein (see Figure 9.26). So, in spite of the serendipitous discovery of Velcade as an agent highly adept at killing MM cells, it may ultimately prove to be useful clinically in a wide range of malignancies.

16.15 A sheep teratogen may be useful as a highly potent anti-cancer drug

An important potential source of powerful anti-cancer therapeutics derives from naturally occurring compounds. The number of distinct **natural products** made by bacteria or fungi is staggering. For example, a 1994 compilation listed 11,900 such compounds that had exhibited antibacterial activity, some of which also possessed activity against mammalian cells. Another 3000 compounds showed yet other biological activities. These numbers are only scratching the surface: a 2001 estimate of the number of distinct, biologically active compounds made by the *Streptomyces* genus of bacteria ran into the hundreds of thousands. A portion of these are likely to possess cytostatic or cytotoxic powers against mammalian cells including cancer cells. The universe of biologically active, plant-derived compounds is even less well explored.

In all these cases, the forces of evolution, rather than the cleverness of synthetic organic chemists, have generated molecular species that are potent and highly specific pharmacologic agents. Many of these molecules seem to be used by the organisms that make them in order to eliminate competitors or defend against predators. Since the number of these naturally occurring agents is beyond reckoning, they are likely to be the sources of novel anti-cancer agents for decades to come.

An illustrative example comes from the discovery of cyclopamine, a natural product of plant origin. This particular story starts with the observation that sheep flocks grazing in highland areas of the Western United States occasionally showed epidemics of congenital malformations in lambs, many of which were stillborn. The most extreme

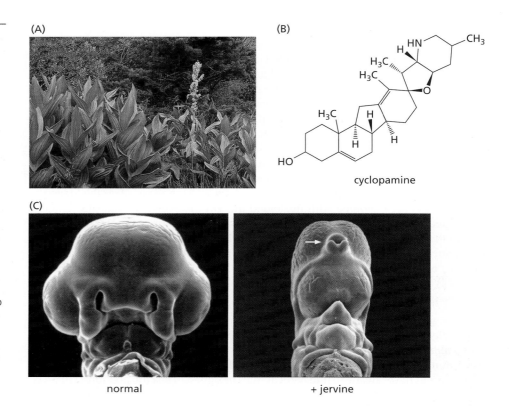

Figure 16.36 False hellebore and its teratogenic product, cyclopamine (A) The plant *Veratrum californicum*, commonly called false hellebore or corn lily, grows in highland meadows in the American West. Pregnant sheep that graze on this plant often give birth to stillborn lambs with major morphogenetic defects in the head region, indicating the presence of a potent teratogen in this plant. (B) The teratogen has been found to be cyclopamine. Its structure resembles that of a steroid, such as estrogen or progesterone. The complexity of its structure means that creation of this compound by using the techniques of synthetic organic chemistry is very challenging, effectively precluding the drug's manufacture on an industrial scale. (C) Treatment of a chicken embryo with cyclopamine or jervine, a close chemical relative of cyclopamine also produced by *Veratrum*, has dramatic effects on the formation of the head, as it does in all vertebrate embryos tested. As seen in these scanning electron micrographs, the normally present two developing eyes on the sides of the developing head and paired nostrils (*left*) are replaced in extreme cyclopia by the rudiment of a single nostril (*arrow*) and a single central eye below it following exposure to jervine (*right*). (A, from H. Kress, http://www.henriettesherbal.com. C, courtesy of P.A. Beachy, from M.K. Cooper et al., *Science* 280:1603–1607, 1998.)

of these malformations involved **cyclopia**—a single central eye. (The term comes from Cyclops, the one-eyed mythical giant vanquished by Ulysses.)

Veterinary detective work begun in the 1950s revealed that newborn-lamb cyclopia was seen if pregnant ewes grazed on false hellebore, *Veratrum californicum* (**Figure 16.36A**), during day 14 of gestation. Yet other malformations, including cleft palate and shortened legs, were evident if the grazing occurred at earlier or later stages of pregnancy. By 1968, the **teratogenic** (malformation-inducing) effects of the false hellebore were traced to an alkaloid that came to be called cyclopamine (see Figure 16.36B), which can induce cyclopia in a wide variety of organisms (see Figure 16.36C).

Many of these deformities resembled a condition in humans termed holoprosencephaly, in which bilaterally symmetrical structures in the embryonic head develop abnormally. Some afflicted human fetuses were found to carry inherited germ-line mutations in either the *PTCH* (*patched*) receptor gene or the *SHH* (*sonic hedgehog*) gene, which encodes the ligand of Patched (see Figure 5.22). As many as 23 distinct *SHH* mutations and 3 *PTC* mutations have been associated with this condition. (More generally, cyclopia is associated with about 1 in 250 spontaneously aborted human fetuses.) These findings provided the first clues that cyclopamine—a potent teratogenic agent—perturbs the Hedgehog-activated signaling pathway.

In this signaling pathway, a precursor of the Gli transcription factor is usually cleaved in the cytoplasm, enabling a cleavage product to move into the nucleus, where it acts as a transcriptional repressor (see Sections 5.7 and 6.12). Smoothened, which normally resides in a cytoplasmic membrane, can protect the Gli precursor protein from this cleavage, but is normally prevented from doing so because of poorly understood inhibition by the Patched (Ptc) receptor located in the plasma membrane. However, when a Hedgehog (Hh) ligand binds to Patched, the latter no longer inhibits Smoothened (Smo); Smoothened now rushes into a primary cilium to protect Gli from the normally occurring cleavage, and the intact Gli can move from the primary cilium into the nucleus, where it acts as a zinc finger transcription factor to induce gene expression (**Figure 16.37A**). This signaling pathway has been implicated in a wide variety of morphogenetic steps in both *Drosophila* and vertebrate embryos.

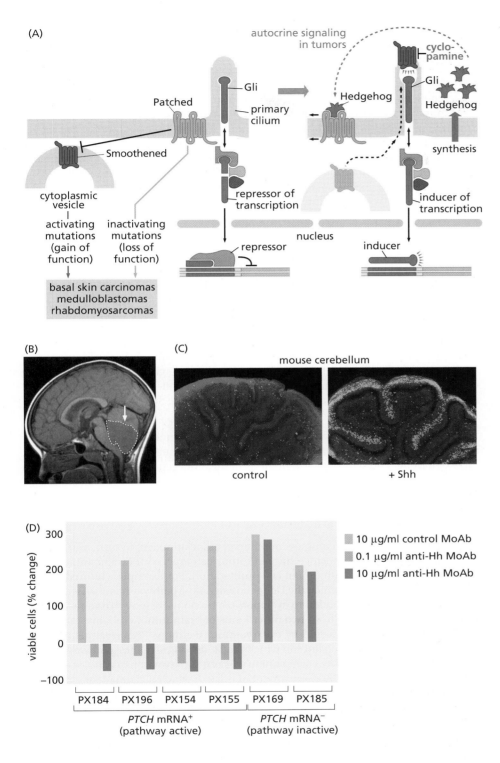

(B, courtesy of M. Roussel and R. Gilbertson, St. Jude Children's Research Hospital. C, from R.J. Wechsler-Reya and M.P. Scott, *Neuron* 22:103–114, 1999. E, from D.M. Berman et al., *Nature* 425:846–851, 2003.)

Figure 16.37 Perturbations of the Patched–Smoothened signaling pathway (A) Normally, Patched inhibits Smoothened activity (*above left*). However, when Patched binds a Hedgehog ligand (either Sonic, Indian, or Desert Hedgehog; *above right*), it no longer inhibits Smoothened, allowing the latter to activate Gli, which then moves into the nucleus and acts as a proliferation-promoting transcription factor. The consequences of Hedgehog binding can be mimicked by mutations causing loss of Patched function or gain of Smoothened function—alterations often seen in the common basal cell carcinomas of the skin (*lower left*). Many carcinomas of endodermal origin release a Hedgehog ligand that acts in an autocrine fashion to drive their proliferation and survival (*upper right*). Moreover, many medulloblastomas trace their origins to dysregulation of this signaling pathway. As also indicated, cyclopamine is a direct antagonist of Smoothened, preventing Smoothened from generating a transcription-inducing form of Gli. (B) Medulloblastomas arise from the granular cell precursors of the cerebellum. Such a tumor (*arrow*), revealed here by computerized tomography, is the most common brain tumor of children and often has devastating clinical outcomes. (C) The granular cells of the normal cerebellum respond strongly to the mitogenic actions (*green*) of an added Sonic Hedgehog ligand. As seen in these slices of the cerebellum of a young mouse, the proliferative activity (*green*) of the granulosa cells at the surfaces of the cerebellar folds, which in the control slice (*left*) is minimal at this stage of development, is strongly stimulated by the addition of Shh (*right*). (Proliferative activity was measured by incorporation into DNA of BrdU, a deoxythymidine analog.) (D) Of the responses of six early-passage pancreatic carcinoma cell cultures, four showed active Gli signaling (being *PTCH* mRNA–positive, *left*) and two did not (*right*). When added to cultures of these cells, 5E1, a monoclonal antibody (MoAb) that neutralizes both the Indian and Sonic Hedgehog proteins, strongly suppressed growth and viability of the carcinoma cells in which Gli signaling was high but had no effect on the carcinoma cells that had no Gli activity. This indicated the importance of an extracellular autocrine signaling loop, involving Hedgehog proteins, in driving the growth of these cells.

Germ-line mutations in the gene encoding Patched have been detected in the skin condition termed basal cell nevus syndrome (BCNS). Loss of heterozygosity (LOH) at the *PTCH* locus in skin cells allows these nevi to develop into basal cell carcinomas (BCCs; see Section 6.12). Moreover, almost half of sporadic basal cell carcinomas—extremely common skin tumors caused by UV radiation—carry inactivating mutations of the *PTCH* gene or activating mutations of the *SMO* gene. (Projections indicate that 28% of Caucasians born in the United States after 1994 will develop at least one BCC during their lifetime.) Fortunately, these skin tumors are relatively innocuous and easily treated. But other types of tumors associated with mutant alleles of these two genes, notably muscle cell and cerebellar tumors (that is, medulloblastomas; see Figure 16.37B), are not.

An important extension of these findings came from the discovery of another sort of deregulation of this pathway. A diverse group of cultured human cancer cells were found to express unusually high levels of one of the two major ligands of the Smoothened receptor, specifically Indian Hedgehog or Sonic Hedgehog. These cells derived from carcinomas arising in the epithelia lining the esophagus, stomach, bile duct, lung, and colon. (Intriguingly, these diverse tumors all derive from organs arising out of the embryonic endoderm.) Later, prostate carcinomas, which are not of endodermal origin, were added to this list. In the case of pancreatic carcinomas, overexpression of a Hedgehog-type ligand has been found in more than 70% of tumor samples tested. (In normal epithelia, the Hedgehog ligand and the downstream signaling pathway that it activates are thought to be responsible for the maintenance of self-renewing stem cells.)

The secretion of high levels of Hedgehog ligand by these various tumor cells is presumed to activate an autocrine signaling loop that results in the constitutive activation of signaling and, therefore, in the continuous dispatching of intact, transcription-activating Gli to the nuclei of cancer cells. The key role of Hedgehog in driving the proliferation of some of these tumor cell types was confirmed by adding neutralizing Hedgehog antibody to their culture medium, which stopped their proliferation (see Figure 16.37D). Conversely, Hedgehog added to the growth medium of normal cerebellar tissue was found to be potently mitogenic (see Figure 16.37C). As might be expected, these cytostatic and even cytotoxic effects of anti-Hedgehog antibody were seen only in tumor cells that also showed expression of the Smoothened receptor protein.

In 2000, cyclopamine was found to directly inhibit the Smoothened protein (see Figure 16.37A). Moreover, this interaction blocked the abnormal signaling resulting from excessive Hedgehog synthesis or mutations in the *SMO* gene. This suggested that the teratogenic effects of cyclopamine derive directly from its ability to block Hedgehog signaling at critical junctures in embryonic development. Moreover, this lack of adequate Hedgehog signaling during development contrasted with the excessive activity of this pathway in a variety of malignancies.

The discovery of the cyclopamine–Smoothened association led, in turn, to treatment of a variety of Hedgehog-positive human tumor cell lines with cyclopamine, which resulted in a 75 to 95% inhibition of cell proliferation. For example, cultured human medulloblastoma cells, in which the Hedgehog signaling pathway has also been found to be hyperactivated, responded to cyclopamine by stopping growth and rapidly losing viability, whereas cells from two other kinds of brain tumors (glioblastomas and ependymomas) were unaffected by cyclopamine treatment. This treatment had no effect on yet other tumor cell lines in which the Hedgehog signaling pathway was not activated, demonstrating that cyclopamine was not simply a nonspecific, widely acting cytotoxic agent. In addition, treatment of mice in which a gallbladder carcinoma (another Hedgehog-secreting endodermal tumor) had been implanted showed total blockage of tumor-forming ability (**Figure 16.38A**).

The presence of high levels of Hedgehog in approximately 70% of human pancreatic carcinomas suggests that activation of the Hh–Ptch–Smo–Gli pathway is an integral part of the neoplastic growth programs of these tumors, while this signaling pathway appears to play little if any role in many other types of cancers and in the maintenance of many normal tissues. In addition, long-term exposure of adult mice to therapeutic levels of cyclopamine did not yield any indications of toxicity. All this should have augured well for the candidacy of cyclopamine as a highly useful agent for the treatment of the subsets of human cancers that exhibit hyperactivated Hedgehog signaling pathways. Included among these are many small-cell lung, pancreatic, prostate, breast, colon, and liver carcinomas.

In fact, the candidacy of cyclopamine as a useful anti-cancer therapeutic agent has three strikes against it. Like other natural products, cyclopamine is the end result of a

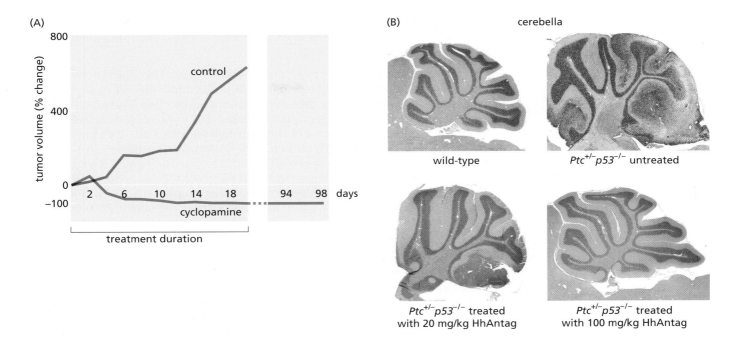

complex series of enzymatic reactions that are difficult to recapitulate in the synthetic organic chemistry laboratory. Second, harvesting significant amounts from *V. californicum* is not practical. Finally, in spite of the above-mentioned results with mice, cyclopamine was found to be too toxic to be used in humans.

So, alternative Smoothened antagonists have been developed that are likely to prove as potent as cyclopamine in interrupting the Hedgehog pathway but lack cyclopamine's toxicity. Smoothened is a seven-membrane-spanning cell surface receptor, and therefore has an overall structure that closely resembles that of the many G-protein–coupled receptors (GPCRs; see Section 5.7) made by mammalian cells. The development of low–molecular-weight, highly specific compounds that target GPCRs has proven to be relatively easy. Accordingly, a number of pharmaceutical companies have developed drugs that target Smoothened with high specificity and show minimal effects on other cellular receptors.

To test some of these new compounds, a mouse model of human medulloblastoma has been created that depends on inactivation (see Supplementary Sidebar 7.7) of one copy of the *Ptc* gene and both copies of the *p53* gene in the mouse germ line, yielding a $Ptc^{+/-}p53^{-/-}$ genotype; virtually all such mice develop medulloblastomas by 3 months of age. An inhibitor of Smoothened, termed HhAntag, was synthesized that has 10 times the potency of cyclopamine and is able to pass easily through the blood–brain barrier, which shields the parenchyma of the brain from most of the contents of the plasma. As seen in Figure 16.38B, treatment with HhAntag of 3-week-old mutant mice that developed medulloblastomas causes a regression of the tumor within two weeks; this occurred with little if any systemic toxicity.

In the case of pancreatic cancer, the prospect of developing a clinically useful inhibitor of the Hedgehog signaling pathway is an exciting one. At present, diagnosis of this carcinoma, in which Hedgehog signaling often plays a prominent role, portends a bleak outcome, with less than 5% overall survival five years after diagnosis, as mentioned earlier. This contrasts with the five-year survival in 1998 of American patients diagnosed with breast cancer (86%) and prostate cancer (97%).

Importantly, however, the contribution of Hedgehog (Hh) signaling to pancreatic carcinoma development is quite different from that depicted in Figure 16.37A. In general, in many carcinomas, as indicated in that figure, autocrine Hh drives tumor progression, since such tumor cells secrete an Hh ligand and, at the same time, express the

Figure 16.38 Effect of cyclopamine and analogous drugs on tumor growth (A) Human cholangiosarcoma (bile duct tumor) cells formed tumor xenografts when implanted in mouse hosts. The tumors either were left untreated (*red line*) or were treated for 22 days with cyclopamine (*blue line*). In the latter case, the tumor shrank and did not reappear in the ensuing 76 days in the absence of further cyclopamine treatment. (B) Mice with a $Ptc^{+/-}p53^{-/-}$ genotype develop medulloblastomas throughout their cerebella early in life. By 5 weeks of age, the cerebellum in a wild-type mouse (*top left*) is far smaller than in the tumor-prone mutant (*top right*). A Smoothened antagonist, termed HhAntag, was identified through screening of a drug library. If the mutant mice were treated with the drug twice daily between the third and the fifth week of life, with either 20 mg or 100 mg per kg of body weight, the tumors regressed partially or completely (*bottom left and right*). Subsequent treatments of 8- and 10-week-old mutant mice with far larger tumors have yielded comparable therapeutic responses (*not shown*). (A, from D.M. Berman et al., *Nature* 425:846–851, 2003. B, from J.T. Romer et al., *Cancer Cell* 6:229–240, 2004.)

Patched receptor. A genetically engineered mouse model of pancreatic cancer has made it clear, however, that *paracrine* Hh signaling plays a far more important role in the development of these tumors.

In more detail, the pancreatic carcinoma cells release Sonic Hh, which stimulates the proliferation of fibroblasts and myofibroblasts in the tumor-associated stroma. The latter proceed to construct the dense desmoplastic stroma (see Figure 13.17) that is characteristic of this disease. Of major importance, this stroma creates a physical barrier that impedes the penetration of chemotherapeutic drugs into the tumor parenchyma (**Figure 16.39A and B**), providing one explanation for the widely observed resistance of these growths to almost all available forms of therapy. This suggests the outlines of a combination therapy in which an Hh inhibitor is used to prevent the formation of this dense stroma together with a cytotoxic drug that attacks the cancer cells (see Figure 16.39C). The clinical utility of this novel strategy remains unclear. In a more general sense, it suggests one future strategy for treating these and other highly progressed carcinomas, in which the tumor-associated stroma, in addition to the islands of neoplastic cells, is targeted by therapeutic drugs.

Medulloblastomas, largely pediatric tumors, occur about one-tenth as often as pancreatic carcinomas; at present, almost two-thirds of patients are cured of this tumor through a combination of surgery, radiation, and chemotherapy; these treatments can, however, leave survivors with significant neurological impairment, including compromised cognitive functions. Ironically, however, the major economic incentive for developing cyclopamine mimetics is likely to derive from the need to treat the most benign but also the most common human cancer type—the basal cell carcinomas of the skin.

(A)

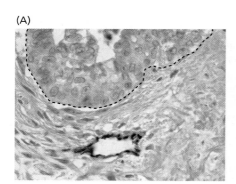

(B)

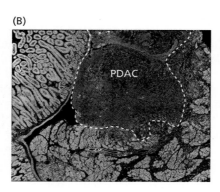

(C)

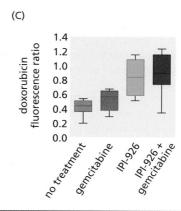

Figure 16.39 Hedgehog signaling and the desmoplastic stroma of pancreatic carcinomas The bulk of research on the role of Hedgehog (Hh) signaling in tumor pathogenesis has focused on autocrine signaling, as illustrated in Figure 16.37A. However, a growing body of evidence indicates that in certain tumors, a more important signaling channel is created by carcinoma cells releasing Hh ligands that impinge upon and activate the tumor-associated stroma. This is especially evident in the case of pancreatic ductal adenocarcinomas (PDACs), which exhibit a dense, desmoplastic stroma; in these tumors, activation of the Gli transcription factor, a readout of Hh signaling, is limited to cells forming the stroma. (A) Human PDACs respond poorly to cytotoxic therapies, in part because of the inefficient delivery of drugs to the carcinoma cells. The prime reasons for this, as illustrated by this micrograph of a human tumor, is that islands of PDAC cells (*above dashed line*) are surrounded by a dense desmoplastic stroma (*below dashed line*) that impedes the movement of infused drug and, in addition, harbors relatively few microvessels such as the one seen here (*dark brown*). (B) A strain of genetically engineered mice express mutant

K-*ras* and *p53* alleles conditionally in pancreatic cells, yielding tumors whose molecular and histopathological features closely resemble those of human PDACs. Doxorubicin (*aquamarine*; see Table 16.2), which is a frequently used cytotoxic drug for a variety of tumors, was infused into a mouse that had developed such a PDAC (*dashed line*). As is apparent, whereas the drug penetrated into the adjacent exocrine pancreas (*left*), it failed to stain the PDAC (*right*), leaving only its *blue* nuclei stained with DAPI. (C) Doxorubicin penetrated only to a limited extent in untreated PDACs (*gray*), as measured by its fluorescence in tumor sections, and this penetration was increased only minimally by concomitant treatment with the often-used cytotoxic agent gemcitabine (*red*; see Figure 16.20). However, treatment of PDAC-bearing mice with IPI-926, a Smoothened antagonist, led to a twofold increase in doxorubicin penetration (*green*), which was minimally increased by concomitant treatment with gemcitabine (*purple*). The treatment of these mice with the gemcitabine together with IPI-926 also led to a significant increase in their survival (*not shown*). (From K.P. Olive et al., *Science* 324:1457–1461, 2009.)

16.16 mTOR, a master regulator of cell physiology, represents an attractive target for anti-cancer therapy

The mTOR circuit has all the attributes of generating therapies that will rival and even eclipse some of those that have been described earlier in this chapter. This story also begins with a natural product—rapamycin—that was isolated in the 1960s from *Streptomyces hygroscopicus* bacteria growing in the soil of Rapa Nui, known to most of the world as Easter Island, in the middle of the Pacific. In the early 1970s, it was re-isolated by a drug company, which developed it as an antifungal agent. In the decades that followed, it became clear that rapamycin (**Figure 16.40A**) can act to halt the growth of an extraordinarily wide spectrum of eukaryotic cells, ranging from those of yeast to mammals.

Rapamycin was also found to have powerful immunosuppressive powers, even when used at low concentrations. In 1999, it was approved by the U.S. Food and Drug Administration (FDA) to prevent immune rejection of transplanted organs, largely kidneys. This drug, also called sirolimus, functions synergistically with other immunosuppressants, specifically cyclosporine and steroids, to ensure long-term engraftment without causing major side effects in transplant recipients. The reasons for its selective actions in preferentially affecting the immune system are not fully understood. [Intriguingly, immunosuppression by cyclosporine in organ transplant recipients leads to increased

(A)

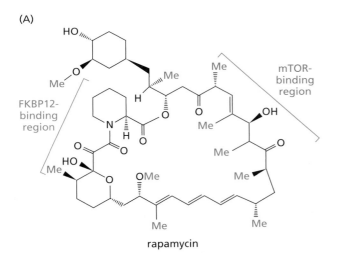

rapamycin

(B)

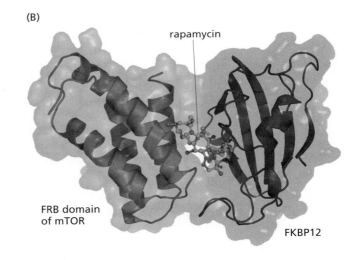

(C)

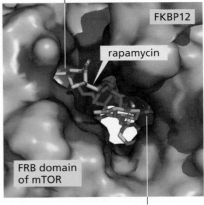

Figure 16.40 Rapamycin, FKBP12, and mTOR (A) Rapamycin is described chemically as a macrocyclic lactone and biologically as a macrolide antibiotic, one of many that are made by the *Streptomyces* genus of bacteria. Rapamycin and its derivatives act as potent immunosuppressants without inducing severe side effects. Most and perhaps all of its effects are due to its ability to inhibit mTOR signaling. (B) The binding of rapamycin (*green, red stick figure*) to FKBP12 (*blue ribbon and space-filling model, right*) occurs with high affinity, the dissociation constant (K_d) being in the range of 0.2 to 0.4 nM. This bimolecular complex forms a molecular surface that can then associate with mTOR (*red ribbon and space-filling model, left*) and prevent the latter from functioning as a serine/threonine kinase. In this image, only the FRB (FKBP12 + rapamycin–binding) domain of mTOR is shown. (C) The details of the interface between rapamycin (*yellow, red stick figure*) and the surfaces of the two proteins are shown, with areas of high stereochemical complementarity highlighted in *purple*. Some of the high-affinity association depends on the insertion of chemical groups of rapamycin into deep cavities within FKBP12 (*right*) and the FRB domain of mTOR (*left*). (B and C, courtesy of Y. Mao and J. Clardy, and from J. Choi et al., *Science* 273:239–242, 1996.)

risk of malignancies (see Section 15.9), while rapamycin-induced immunosuppression in these patients actually decreases the risk of post-transplantation lymphoproliferative disorders. Hence the notion that immunosuppression always leads to increased cancer risk needs to be refined, since some types of immunosuppression yield increased tumor incidence while other types do not.]

Biochemical analyses show that rapamycin binds directly to a low–molecular-weight protein called FKBP12 (FK506-binding protein of 12 kD), originally discovered because it is also bound by FK506, a similarly acting drug. Once formed, the rapamycin–FKBP12 complex (see Figure 16.40B and C) associates with a protein that was identified in 1994 as mTOR (mammalian target of rapamycin), and shuts it down. mTOR is a 289-kD protein that functions as a serine/threonine kinase; its kinase domain resembles that of PI3 kinase and related enzymes.

mTOR is of special interest because it operates at two critical nodes in the control circuitry of mammalian cells (**Figure 16.41A**). Thus, mTOR integrates a variety of afferent (that is, incoming) signals, including nutrient availability and mitogens, and then acts to control glucose import and protein synthesis and a variety of other cell-biological processes. More specifically, mTOR phosphorylates two key governors of translation: p70S6 kinase (S6K1) and 4E-BP1. Phosphorylation of S6K1 causes the latter to phosphorylate the S6 protein of the small (40S) ribosomal subunit, enabling this subunit to participate in ribosome formation (by associating with the large ribosomal subunit) and thus in protein synthesis.

In addition, by phosphorylating 4E-BP1 (and 4E-BP2), mTOR causes 4E-BP1/2 to release its grip on the key translational initiation factor eIF4E (eukaryotic initiation factor 4E); once liberated, eIF4E forms complexes with several other initiation factors, and the resulting complexes enable ribosomes to initiate translation of certain mRNAs, specifically those with oligopyrimidine tracts in their 5′, untranslated regions. Together, these various actions allow mTOR to be a key governor of cell growth (rather than cell proliferation; see Figure 8.2).

Initially, mTOR was thought to be one of the multiple downstream substrates of Akt/PKB, specifically the one allowing Akt/PKB to regulate cell growth by controlling protein synthesis. But the tables have been turned: mTOR is now realized to be a key *upstream activator* of Akt/PKB (Supplementary Sidebar 16.11). This shift puts mTOR in a far more powerful position in the cell. By controlling Akt/PKB, mTOR can regulate apoptosis and proliferation in addition to its known ability to regulate cell growth.

mTOR's appearance in two places in the circuitry, depicted in Figure 16.41A, reflects its ability to associate with two alternative partners, called Raptor and Rictor. The mTOR–Rictor complex and a third protein, GβL, together form the TORC2 complex, which is regulated in unknown ways by growth factors and is responsible for activating Akt/PKB. The mTOR–Raptor complex with GβL forms the TORC1 complex, about which more is known: it is responsible for activating protein synthesis (by phosphorylating S6K1, 4E-BP1, and 4E-BP2). Acting together with FKBP12, rapamycin directly interacts with the mTOR–Raptor complex, which is rapidly inhibited after this drug is applied to cells. If, however, rapamycin treatment is continued for many hours, eventually the mTOR–Rictor complex is also shut down, resulting in the inhibition of Akt/PKB. The mechanism by which rapamycin succeeds in inhibiting the mTOR–Rictor complex is poorly understood.

This inhibition of Akt/PKB signaling seems to be responsible for much of rapamycin's effect on cancer cells that exhibit a hyperactivated PI3K or loss of PTEN expression. It is plausible that such cells, much like the small-cell lung carcinoma cells with mutant EGF receptors (see Sidebar 16.5), have become "addicted" to Akt/PKB signals and lurch into apoptosis when deprived of these signals by the actions of rapamycin and related drugs. However, the precise criteria that determine sensitivity to rapamycin are yet to be worked out.

Of additional interest, the regulatory circuit shown in Figure 16.41A intersects in yet other ways with cancer pathogenesis. For example, TSC1 and TSC2 (also called hamartin and tuberin) have already appeared in this book in the context of their role

(A)

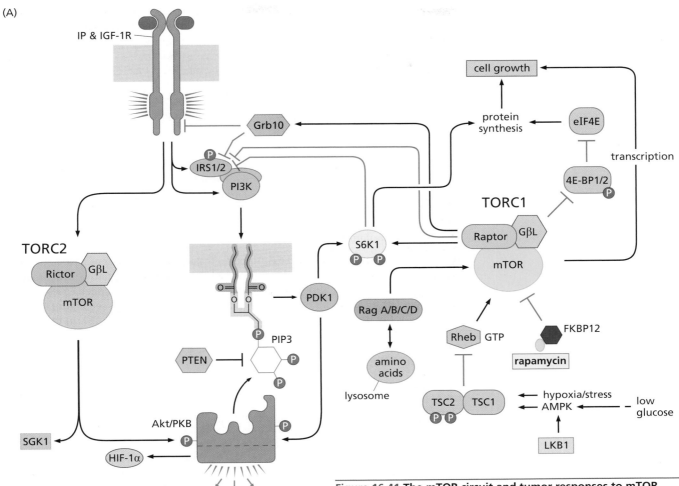

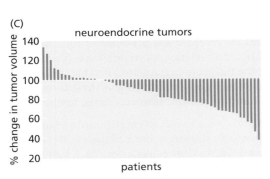

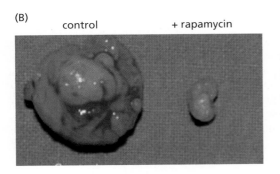

(B) control   + rapamycin

(C) neuroendocrine tumors

% change in tumor volume

patients

Figure 16.41 The mTOR circuit and tumor responses to mTOR inhibitors (A) mTOR sits in the middle of a complex regulatory circuit that integrates incoming signals about nutrient availability, oxygen tension, ATP levels, and mitogenic signals and, in response, releases signals that govern ribosome biogenesis, protein synthesis, cell proliferation, protection from apoptosis, angiogenesis, and even cell motility. mTOR exists in two alternative complexes with its Rictor (*left*) and Raptor (*right*) partners. The mTOR–Rictor complex governs the activity of Akt/PKB by adding a critical second phosphate to the latter and thereby gains control over Akt/PKB's multiple downstream effectors. Exposure to rapamycin (*lower right*) rapidly inhibits the mTOR–Raptor complex and, after extended periods, causes a progressive shutdown of the mTOR–Rictor complex. *Black lines* indicate stimulatory effects while *red lines* indicate inhibitory effects. IR, insulin receptor; IGF-1R, IGF-1 receptor. (B) BALB/c mice bearing injected cells of a syngeneic colon adenocarcinoma cell line develop large, well-vascularized tumors (*left*) by 35 days after injection. However, if after a week of tumor growth the mice receive continuous treatment with doses of rapamycin comparable to those used in humans for immunosuppression, the tumors are much smaller (*right*) and the density of microvessels in these tumors is less than half of that seen in the control tumors (*not shown*). (C) Everolimus was used in a clinical trial to treat 30 patients with neuroendocrine tumors, largely in the pancreas. Each bar in this *waterfall plot* reveals the change in size of a single patient's tumor during the treatment period. Waterfall plots reveal the proportion of tumors that continued to grow during treatment (*left side*) or responded to treatment by shrinking to various extents (*right side*). (A, from D.A. Guertin and D.M. Sabatini, *Trends Mol. Med.* 11:353–361, 2005. B, from M. Guba et al., *Nat. Med.* 8:128–135, 2002. C, from J.C. Yao et al., *J. Clin. Oncol.* 26:4311–4318, 2008.)

as tumor suppressor proteins. Loss of either leads to tuberous sclerosis (see Table 7.1), and, as seen in Figure 8.2, loss of TSC1 results in the formation of giant cells in both flies and humans. TSC2 acts as a GAP (GTPase-activating protein; see, for example, Sidebar 5.7) for Rheb, a small Ras-like protein. As long as it remains in its GTP-bound state, Rheb contributes in unknown ways to stimulating the mTOR–Raptor–GβL complex; however, once TSC2 has induced Rheb to hydrolyze its GTP, Rheb loses this stimulatory activity. Yet other signaling connections between the mTOR circuit and critical growth-inducing and mitogenic proteins are being forged by ongoing research.

The development of rapamycin and similarly acting drugs has been encouraged, in part, by the observation that drugs like rapamycin can be tolerated for extended periods by transplant recipients, indicating a low level of side-effect toxicity. In preclinical experiments, rapamycin given to mice at levels used for chronic immunosuppression strongly inhibited tumor-associated neoangiogenesis and thus tumor growth (see Figure 16.41B). This effect may be explained by the fact that one of the three Akt/PKB isozymes, Akt1, is critical to the ability of endothelial cells and their precursors to respond to stimulation by vascular endothelial growth factor (VEGF).

In 2006, rapamycin was reported to induce regression of astrocytomas associated with tuberous sclerosis. In the years that followed, everolimus, a variant of rapamycin/sirolimus also known as Afinitor, was approved by the FDA for advanced cases of renal cell cancer (RCC) and pancreatic neuroendocrine tumors (NETs; see Figure 16.41C), and also for the treatment of giant-cell astrocytomas (see Figure 8.2B). In the case of patients with RCCs, everolimus improved the period of progression-free survival by a factor of 2. Indeed, it is clinical responses like these that have motivated discussion of the mTOR circuit in this chapter.

Still, given the strategic positions of mTOR in the cell's regulatory circuitry, the clinical responses of a variety of solid tumors to treatment with rapamycin or its analogs (*rapalogs*) have been surprisingly modest. Only in recent years has this made sense in terms of the regulatory circuitry in which mTOR is embedded. As seen in Figure 16.41A, important negative-feedback circuits operate from TORC1 back to its upstream regulators. This explains why inhibition of TORC1 by rapamycin or a rapalog results acutely in a shutdown of this signaling complex; however, with increasing treatment time, the negative-feedback loop that is responsible for inactivating IRS1, IRS2, and PI3K is shut down. This lifting of feedback inhibition eventually results in stimulating PI3K activity, overriding and thus nullifying any benefits that TORC1 inhibition may have initially achieved. This story teaches us an important lesson about drug development, namely, that inhibition of a drug target may often yield biological responses that are opposite of what was initially intended and anticipated.

16.17 B-Raf discoveries have led to inroads into the melanoma problem

The sequencing of entire tumor cell genomes—now enabled by new sequencing technology—has uncovered large numbers of novel candidate oncogenes and tumor suppressor genes. However, the most important of these has been found by focusing the sequencing technology on a known signaling pathway—exploring the well-lit ground under a familiar lamppost—the Ras–Raf–MEK–ERK pathway (see Section 6.5). Among other genes being studied was *C-Raf* (earlier called c-Raf) and its two cousins, *A-Raf* and *B-Raf*, all of which are activated by GTP-bound Ras; *C-Raf* encodes the familiar Raf oncoprotein, first identified by its presence in an acutely transforming retrovirus (see Table 4.1). In 2002, sequence analysis revealed the presence of mutant *B-Raf* alleles in more than half of cutaneous human melanomas. In one sense, this was an astounding finding, given the fact that the Raf proteins, largely C-Raf, had been studied for the prior two decades without there being any hint that one of these three genes was a frequent participant in the formation of a not-uncommon human malignancy.

Ninety percent of the mutant *B-Raf* alleles were found to carry point mutations in the codon encoding amino acid residue 600, causing a normally present valine to be replaced by a glutamic acid (**Figure 16.42A**). The behavior of this V600E mutation echoes that of the *Ras* oncogenes, in which an even greater proportion of mutant alleles affect a single codon, specifically codon 12 (see Figure 4.10). Interestingly, in melanomas and other tumor types, mutant *B-Raf* alleles are rarely found together with *Ras* oncogenes; this mutually exclusive pattern indicates that the Ras–Raf–MEK–ERK pathway plays a key role in the development of these various tumors and suggests that these two classes of mutant alleles are functionally redundant.

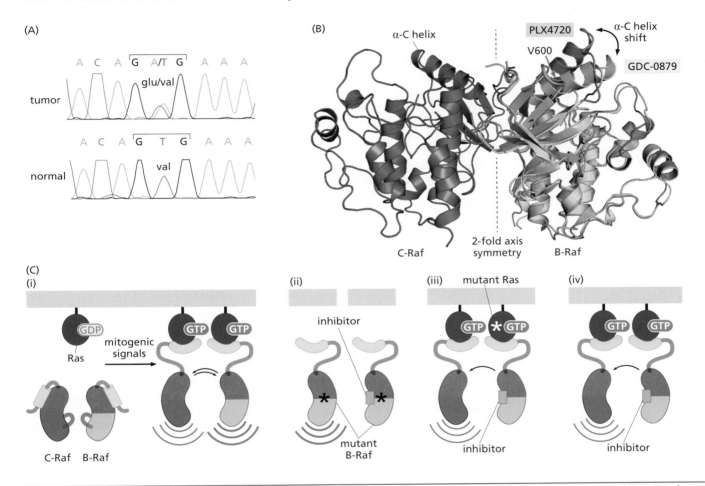

Figure 16.42 The mutant B-Raf and its treatment A search for mutant alleles of the genes controlling the main mitogenic signaling pathway in cells—the Ras–Raf–MEK–MAPK pathway—has revealed the frequent presence of mutant alleles of one of the three genes encoding Raf family kinases, which operate as serine/threonine kinases immediately downstream of the Ras proteins (see Figure 6.14). (A) The discovery of point-mutated *B-Raf* alleles is shown here, in which sequencing of the genome of an ovarian tumor (*above*) revealed heterozygosity in codon 600 of the protein, which normally specifies valine (*below*). The mutant allele in the tumor genome was found to specify glutamic acid instead. This V600E mutant allele has since been found to be the predominant mutant *B-Raf* allele that is present in more than half of all malignant melanomas and a variety of other tumors, including ~10% of colon carcinomas. (B) A molecular model of the B-Raf/C-Raf heterodimer structure is shown here. The two kinase molecules associate with one another in a parallel, side-to-side fashion, with residues from both the N- and C-terminal lobes contributing to the heterodimer interface. The V600 residue of B-Raf (*arrow, red*) is the most common structural alteration encoded by oncogenic mutations (i.e., usually V600E) and renders the enzyme constitutively active. Acting via allosteric interactions, the B-Raf protomer is able to activate C-Raf signaling even when B-Raf lacks kinase activity (the latter defect because of mutation or drug action). The orientation of the

α-C helices can affect B-Raf/C-Raf dimerization and thus C-Raf kinase activation. As seen here, the binding of two distinct B-Raf inhibitors, PLX4720 and GDC-0879, in the ATP-binding pocket of B-Raf shifts the orientation of this α-helix and thus the interactions between the two protomer subunits. (C) The behavior of Raf dimers greatly complicates the therapeutic effects of the B-Raf V600E inhibitor. (i) In normal cells, mitogen-activated GTP-bound Ras proteins bind and activate B-Raf/C-Raf heterodimers, with much of the downstream signaling being emitted by the more potent B-Raf partner; this results in activation of MEK and MAPK/ERK (see Figure 6.14). (ii) In many melanomas, a mutant V600E B-Raf protein, acting in a Ras-independent fashion, can signal constitutively as a monomer (*right*), driving strong activation of this signaling cascade. A V600E-specific inhibitor (*green rectangle*) blocks this signaling (*left*), leading to temporary tumor regression of many melanomas. (iii) In cancer cells in which a mutant Ras oncoprotein is present (for example, in many colorectal carcinomas) the V600E-specific inhibitor, while shutting down B-Raf signaling, causes the latter to activate, via allosteric interactions, signaling by C-Raf. (iv) In certain normal cells in which B-Raf is wild-type, this inhibitor can stimulate signaling, possibly by activating C-Raf. (A, from H. Davies et al., *Nature* 417:949–954, 2002. B, from G. Hatzivassiliou et al., *Nature* 464:432–435, 2010. C, courtesy of M.B. Yaffe.)

In 2011, the drug vemurafenib (previously called PLX4032), which targets specifically the V600E mutant B-Raf protein, received approval for use in metastatic melanomas in Europe, followed soon thereafter by approval in the United States. This represented the first time that these aggressive, late-stage tumors could be treated with significant clinical responses. Thus, over a 6-month trial period, vemurafenib produced a 74% increase in progression-free survival (PFS) and a 20% reduction in overall mortality relative to the *standard-of-care* at the time, an alkylating agent named dacarbazine. Stated differently, vemurafenib increased overall PFS of these patients by ~4 months.

These numbers indicate that vemurafenib offers patients with V600E metastatic melanomas an extension of life span but only rarely a cure. Sooner or later, their tumors develop strategies for evading this drug and thereby acquire the ability to re-erupt, doing so with lethal consequences. In addition, a significant percentage of treated patients develop cutaneous squamous cell carcinomas, providing a clear indication that, paradoxically, this drug *stimulates* the outgrowth of previously latent, subclinical tumors. Indeed, in patients whose tumors carried mutations other than the V600E alteration, vemurafenib actually favored tumor growth, indicating the critical importance of sequence analysis of tumors prior to use of this drug.

These paradoxical responses to vemurafenib have forced a rethinking of precisely how the Raf proteins normally function at the molecular level. It is now clear that, much like the kinases of RTKs, Raf proteins dimerize as either homodimers (such as B-Raf:B-Raf) or heterodimers (such as B-Raf:C-Raf; see Figure 16.42B). Moreover, there are clear regulatory interactions between the two subunits, and the binding of vemurafenib to a mutant V600E protein can *stimulate* activity of its wild-type B-Raf or C-Raf partner. Indeed, in normal cells, Ras may often signal through B-Raf:C-Raf heterodimers (see Figure 16.42C). In melanoma patients with a mutant *Ras* oncogene, treatment with this drug strongly stimulates the Ras–Raf–MEK–MAPK signaling pathway. This conclusion is supported by the fact that agents that shut down MEK—the kinase lying immediately downstream of the Raf proteins—can often block this paradoxical response. Once again, as was the case with mTOR inhibitors earlier (see Section 16.16), normally operating regulatory mechanisms complicate the clinical response, often thwarting the aims of the drug developers. Still, vemurafenib has the prospect of becoming part of a multi-drug treatment protocol that one day will stop the previously unstoppable metastatic melanomas in their tracks. While not as effective as Gleevec before it, this drug confers significant benefit on melanoma patients and thus provides support for the goal of treating cancers with rationally developed drugs.

16.18 Synopsis and prospects: challenges and opportunities on the road ahead

"When is cancer going to be cured?" This is the simple and reasonable question posed most often to cancer researchers by those who are not directly involved in this area of biomedical research. In their minds are the histories of other public health measures. Infectious diseases, such as polio and smallpox, can be prevented, and bacterial infections are, almost invariably, cured. Heart disease is, in the eyes of many, well on its way to being prevented. Why should cancer be any different?

The information in this book provides some insights into the answers to these questions. As much as we have invoked unifying concepts to portray cancer as a single disease, the reality—at least in the eyes of clinical oncologists—is far different. Cancer is really a collection of more than 100 diseases, each affecting a distinct cell or tissue type in the body.

Pathological analyses have led us to embrace this number, or one a bit larger. (For example, there are at least eight distinct histopathological categories of breast cancer.) However, even the expanded number, large as it may be, represents an illusion: the current use of molecular diagnostics, specifically gene expression arrays, is leading to an explosion of subcategories, so that by the second decade of the new millennium, several hundred distinct neoplastic disease entities are likely to be recognized, each following its own, reasonably predictable clinical course and exhibiting its own responsiveness to specific forms of therapy. With the passage of time, cancer diagnoses will increasingly be made using bioinformatics rather than the trained eyes of a pathologist.

So the initial response to questions about "the cure" is that there won't be a single major breakthrough that will cure all cancers—a decisive battlefield victory—simply because cancer is not a single disease. Instead, there will be many small skirmishes that will steadily reduce the overall death rates from various types of cancer. And because certain molecular defects and pathological processes (for example, angiogenesis) are shared by multiple human cancers, there will be occasions when therapeutic advances on a number of fronts will be made concomitantly.

Before we speculate on the future of cancer therapy, it is worthwhile to step back and assess the scope of the challenge: (1) How large is the problem of cancer and, in the future, how desperate will the need be to cure various types of neoplastic disease? (2) How well are we doing now in curing the major solid tumors?

Epidemiology and demographics provide some answers to the first question. They yield sobering assessments of the road ahead. The statistics in **Figure 16.43** demonstrate that cancer is largely a disease of the elderly, whose numbers are growing rapidly and will continue to do so, generating progressive increases in the numbers of cancer-related deaths (mortality) over the coming decades.

Equally important, we still have only very imperfect ways of measuring incidence—how often the disease strikes. This greatly complicates assessments of the effectiveness of current therapies and future needs for therapy. As indicated in **Figure 16.44**, perceptions of the incidence of certain types of neoplastic disease are strongly influenced by diagnostic practices.

For many types of cancer, the more one looks, the more one finds. Statistics like those in Figure 16.44 suggest that in the past many cancers remained undiagnosed and asymptomatic, and that these tumors are contributing the lion's share to perceived increases in disease incidence, notably of common tumors, such as those arising in the breast and prostate. (The major exceptions here are the cancers related to tobacco use, whose increased incidence is real and beyond dispute, since the incidence rates are closely paralleled by the rates of mortality.) Such statistics indicate that for certain common tumors—prostate and breast cancer being prominent examples—we have only a poor appreciation of the number of tumors that truly require treatment (see Supplementary Sidebar 16.2). Given these numbers, the current practice in the West of aggressively treating all those diagnosed with cancer (whether or not they truly require such treatment) will soon exceed the ability of national economies to support such care.

Data like those in Figure 16.44 also undermine the notion (see Chapter 11), deeply entrenched in the thinking of many cancer biologists and clinical oncologists, that all benign growths are in danger of becoming, sooner or later, highly malignant ones. Cancer epidemiology now makes us confront the alternative possibility: many kinds of early-stage tumors are unlikely to progress to high-grade malignancy during an average human life span, and the long-term side effects of aggressive treatment may often be more serious than the statistical likelihood of dying from these tumors. Unfortunately, we are only beginning to learn how to segregate those tumors that are truly deserving of aggressive treatment from those that are not (see, for example, Figure 16.4).

As to the second question, which deals with the effectiveness of current cancer therapies, our perceptions have been strongly influenced by the fact that people are living longer with their cancers. This would seem to provide some reassurance that real progress is being made. However, some of these perceived improvements in therapy may, once again, be artifacts of increased screening and more sensitive detection techniques that increasingly uncover tumors relatively early in their development, giving the patient additional years of survival before tumor progression advances through its natural course, whether or not treatments are applied (**Figure 16.45**). This logic forces the conclusion that the efficacy of therapies can be accurately gauged

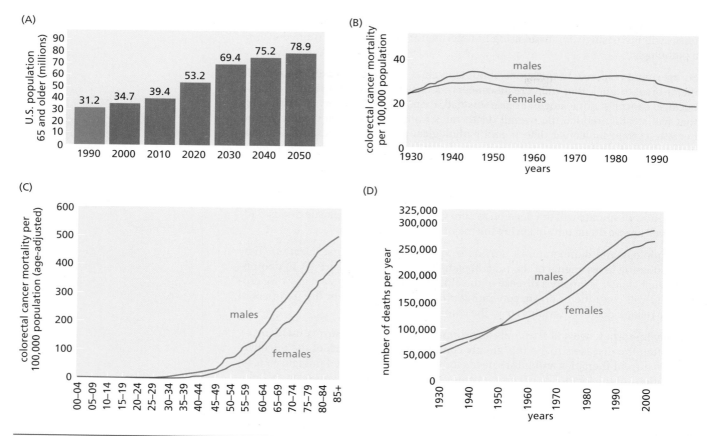

Figure 16.43 The demographics and epidemiology of cancer
(A) Because of dramatic decreases in midlife mortality, populations in industrialized countries are aging rapidly. In the United States, the number of individuals over 65 years of age has increased 11-fold since 1900, while the number of those under 65 has increased by a factor of 3. Comparable increases in the aged population are likely to occur worldwide over the next generation. (B) The age-adjusted death rate from colorectal cancer has changed only slightly over the past several decades. Other major cancers show similar curves. (C) Like a number of other diseases, cancer is uncommon during early and midlife and then increases rapidly. Shown here is the age-dependent death rate of colorectal cancer. (D) Because (1) the number of elderly has increased steadily over the past century and will continue to do so (see panel A), (2) cancer is a disease of the elderly (panel C), and (3) the age-adjusted death rate of most cancers has been relatively constant for many decades (as suggested in panel B), the absolute number of annual deaths due to cancer has increased dramatically over the past three-quarters of a century in the United States. (The number of patients afflicted with Alzheimer's disease has increased in parallel for the same reasons.) Since these trends are likely to continue, the burden of cancer cases in industrialized societies will continue to climb for many decades. (A, courtesy of D. Singer and R. Hodes, from U.S. Bureau of the Census, Projections of 1996. B, courtesy of M.J. Thun, from A. Jemal et al., *CA Cancer J. Clin.* 53:5–26, 2003. C, from A. Jemal et al., *CA Cancer J. Clin.* 55:10–30, 2005. D, from A. Jemal, E.M. Ward and M.J. Thun, in V. DeVita et al. (eds.), Cancer: Principles and Practice of Oncology, 7th ed. Philadelphia: Lippincott/Williams & Wilkins, 2004.)

only by well-controlled experiments: comparisons of several patient populations that are afflicted by the same malignancy and exposed in parallel to different agents or treatment protocols. Such side-by-side comparisons have, until recently, yielded only incremental gains in treating most solid tumors (see, for example, Figure 16.20), but this is beginning to change, as new drugs and monoclonal antibodies are introduced into the clinic. Indeed, the age-adjusted mortality from breast cancer in Western countries has decreased by 20 to 30%, and some, possibly most of this decrease is attributable to aggressive clinical intervention.

The major hope at present is that agents such as Gleevec, vemurafenib, and Rituxan are the forerunners of dozens and eventually hundreds of targeted, highly efficacious drugs. Computer-driven procedures, including high-throughput screening (HTS), automated determinations of the stereochemical structures of target molecules, and computer-aided design of drug molecules (**Figure 16.46**), are proving useful in this endeavor. Without such automated procedures, the current costs of developing novel anti-cancer drugs will soon become economically unsustainable, and the development of agents for less common subtypes of common cancers will never be realized.

Ultimately, the biggest challenge of drug development now and in the future is to demonstrate long-term efficacy: Does a drug being tested have significant effects on extending the life expectancy of cancer patients, doing so with acceptable levels of side-effect toxicities? And do we dare to hope that it can achieve durable responses, including cures?

As mentioned repeatedly here, rapidly evolving populations of cancer cells acquire drug evasion mechanisms that protect them from many types of therapeutic attack. Indeed, recent evidence demonstrates directly that tumors that undergo more genetic diversification and thus evolution are likely to be more resistant to therapy (**Figure 16.47**). Some populations of cancer cells may develop drug evasion mechanisms by increasing their anti-apoptotic defenses (see Table 16.5). Yet others may lose checkpoint controls that previously made them sensitive to certain types of drug treatment or evade the effects of administered drugs by expressing drug efflux pumps. Tumors whose cells were previously being killed through antibody-mediated attack may simply down-regulate expression of the telltale antigen. At present, it remains unclear whether we will, one day, be able to devise treatment strategies that anticipate the plasticity and evasiveness of cancer cells, allowing us to develop definitive cures of malignancies that have long been incurable.

These evasive maneuvers of cancer cells underlie the growing belief of cancer researchers that most monotherapies are unlikely to yield curative treatments because they are

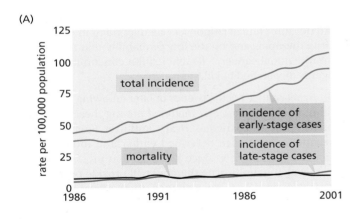
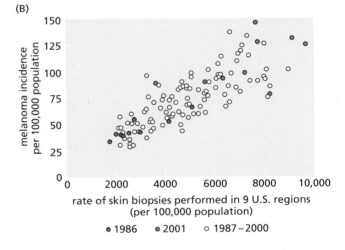

Figure 16.44 Melanoma incidence and mortality Melanoma incidence in the United States has increased by a factor of 6 over the past half century, raising the question whether this trend has created a comparable increase in the number of cases that truly require aggressive clinical treatment. (A) The breakdown of melanoma incidence (into early-stage localized and late-stage disseminated disease categories, both diagnosed through skin biopsy) averaged over nine areas of the United States among individuals over 65 years of age (*blue*) indicates a dramatic increase over the past two decades in the incidence of early-stage disease (*red*) and a relatively constant incidence of late-stage disease (*green*). Age-adjusted mortality (*gray*) has been relatively constant during these years. (Melanoma is a disease whose incidence rises progressively with age, making this age bracket the one with the highest incidence and mortality from this disease.) These data raise the question whether the real incidence of the early-stage

disease has been increasing over this time period, or whether the real incidence has been relatively constant and the registered incidence of this disease has increased due to changes in screening practices. (B) If the incidence rates, such as those of panel A, are plotted against the rates of screening for melanoma in nine areas of the United States, as registered during several time periods, the resulting scatter plot reveals a close correlation between the two. This provides strong indication that the incidence of the disease is strongly influenced by diagnostic practices. It is therefore possible that (1) the true incidence of life-threatening melanomas (panel A) has not changed significantly over the past two decades; or (2) the true incidence of these tumors has increased, but intensification of screening has held the mortality rate at levels observed two decades ago by allowing the removal of early-stage tumors before they progress to become invasive and metastatic. (From H.G. Welch, S. Woloshin and L.M. Schwartz, *BMJ* 331:481, 2005.)

(A)

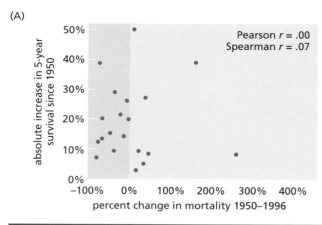

(B)

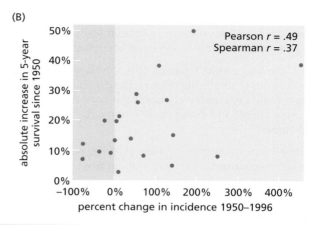

Figure 16.45 Cancer incidence and survival since 1950 in the United States One way of gauging therapeutic success in treating a certain type of tumor is to determine the percentage of patients who survive for 5 years following initial diagnosis.
(A) If extension of patient survival time following diagnosis were a reflection of increased therapeutic efficacy, this extension should be correlated with a decrease over the past half century in the age-adjusted mortality from this type of tumor. In fact, as indicated here, there is no correlation between these two parameters when changes in 5-year survival rates of a number of solid tumors are plotted versus changes in mortality. (Note the low correlation coefficients, *upper right*). (B) Instead, there is a significant correlation between changes in 5-year survival and changes in disease incidence. If the latter is strongly affected by diagnostic bias (see Figure 16.44), then increases in 5-year survival become very difficult to interpret, since they may largely reflect the detection of a disease at an earlier stage of its natural clinical course. (From H.G. Welch, L.M. Schwartz and S. Woloshin, *JAMA* 282:2976–2978, 2000.)

destined to select for the outgrowth of variant cells within tumors that happen to have developed resistance to the single agent being applied. This dictates that truly successful clinical outcomes and durable clinical responses will depend in the future on the development of multi-drug therapies and on the low probability that a clinically detectable tumor will harbor individual cancer cells that exhibit concomitant resistance to all of the cytotoxic agents dispatched to kill them.

This development of therapeutic resistance occurs sooner or later following long-term treatment of a variety of common human cancers. Included among these are high-grade tumors of the lung, breast, prostate, pancreas, ovary, and liver. Such acquired resistance represents the secondary resistance described earlier and indicates the grim scenario confronting medical oncologists, since it means that most therapeutic protocols (including those involving multi-drug therapy) will provide only a temporary respite from disease progression. As a consequence, the medical oncologist needs to respond, time after time, to the latest clinical relapse in a patient by devising a new protocol that responds to the most recent evasive maneuver developed by the patient's tumor. In an ideal world, each of this succession of protocols would provoke tolerable levels of side-effect toxicities, and the sequence of treatments would allow the cancer patient to live in a relatively symptom-free state, surviving in many cases to the end of what would be a normally expected life span. We are still very far from reaching this goal.

The replacement of one therapeutic protocol that has failed a patient with another requires some understanding of the resistance mechanisms that have provoked clinical relapse and allowed therapy-resistant tumors to grow out. Some of these mechanisms were cited above, but in large part, our understanding of the mechanisms of such acquired resistance is still very limited (see Table 16.5). Clearly much of this resistance derives from the instability of cancer cell genotypes and phenotypes we encountered in detail in Chapter 12 that allows the emergence of resistant variant cell clones. The evidence presented in Figure 16.47 provides possibly the first direct demonstration that genetic plasticity plays a central role in generating secondary resistance. In truth, however, we still do not understand how much of this secondary resistance derives from genetic instability and how much can be traced to shifts in the heritable epigenetic programs (see Section 1.9) of these cells.

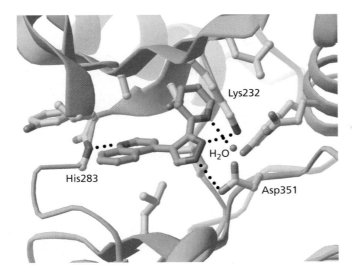

Figure 16.46 Virtual screening—designing drugs in the desktop computer High-throughput screening has yielded a large number of highly attractive drug candidates, many of which have advanced into clinical trial. In the future, however, much of this screening, which is very expensive, may be obviated by the development of powerful algorithms that enable pharmacologists to design drugs using the known structures of target proteins. The development of the drug illustrated here began with a compound that showed very weak affinity (IC_{50} of 30 μM) for the ATP-binding site of the TGF-β type I receptor kinase. Researchers then queried a database of 200,000 known compounds for ones that shared some chemical features with the starting compound and conformed to the constraints imposed by the known structure of the ATP-binding site. This yielded the structures of 87 chemical species that satisfied these criteria and were then screened using conventional biochemical techniques. One of these, named HTS466284, shown here ensconced in the ATP-binding site of the TGF-β type I receptor kinase, exhibits an IC_{50} of 27 nM and thus functions as a potent inhibitor of TGF-β signaling. Significantly, another research group, working independently, arrived at the identical inhibitor molecule using conventional high-throughput screening. Hydrogen bonds are indicated by *dotted lines*. (From J. Singh et al., *Bioorg. Med. Chem. Lett.* 13:4355–4359, 2003.)

Here we also confront the fact that there are still no clear and logical strategies for designing multi-drug therapies, beyond the strategy that has been in place for half a century: choose combinations of drugs having cytotoxic mechanisms-of-action that are distinct from one another, with the hope that they will act synergistically to reduce the burden of cancer cells in a patient's body below the threshold level at which the clinical resurgence of the disease becomes likely.

This touches on a related question: What if the real powers of a drug under development are only realizable when it is used in combination with several others? In theory, many anti-cancer agents should fall into this category. This suggests, in turn, that many truly useful drug candidates have been discarded in the past, having failed tests in monotherapy trials, and that many others will suffer this fate in the future, simply because their true utility as components of multi-drug treatment protocols will never be tested. Here again, one hopes that our increasing understanding of the subcellular signaling circuitry will ameliorate this situation (but see **Sidebar 16.6**).

The existence of tumor stem cells (see Section 11.6) creates another major challenge for anti-cancer drug development. Recall that the self-renewing cancer stem cells (CSCs) can seed new tumors, while their far more numerous transit-amplifying progeny, with limited self-renewal capacity, cannot. Such a hierarchical organization was demonstrated initially in leukemias, breast carcinomas, and brain tumors, but probably exists in most tumors and holds important implications for the development of anti-cancer drugs. Traditionally, clinical validation of the therapeutic efficacy of these drugs has depended on demonstrations of their ability to halt further tumor growth

Sidebar 16.6 Legal and financial disincentives decrease the likelihood of drugs being tested in combination At present, the biological difficulties of testing candidate drugs in combination with others are compounded by economic forces that often create disincentives for pharmaceutical companies to test their own proprietary drugs in combination with those produced by their competitors. Patent regulations have also discouraged certain uses of patented compounds by firms that are in direct competition with the patent holders.

In the past, the difficulties of organizing early multi-drug clinical trials were further compounded in the United States by the Food and Drug Administration's (FDA's) insistence that the clinical efficacy of drug candidates be demonstrated first as monotherapies. But this has begun to change, albeit slowly. For example, Erbitux, the anti-EGF-receptor monoclonal antibody (see Sidebar 15.5), was initially approved for use together with irinotecan because the two used in concert yielded far better responses than did irinotecan on its own. (Irinotecan is a more traditional cytotoxic agent that functions as a topoisomerase I inhibitor; see Table 16.2.)

Figure 16.47 Genetic diversification and resistance to therapy A widely embraced notion is that tumors develop resistance to therapy through a process of genetic diversification followed by the selection and outgrowth of more therapy-resistant subpopulations of cells—a model formally akin to Darwinian evolution. This notion has rarely been tested experimentally in a clinical setting. As shown here, the ovarian tumors of individual patients were sampled successively during the course of their treatment, and the inter-relatedness and thus genetic diversification of subpopulations within their tumors were gauged using genome sequence analysis and a bioinformatics algorithm. (A) The genetic relatedness of the subclones within two patients is indicated here. As is apparent, the tumor subpopulations within a patient whose ovarian carcinoma became resistant to therapy (*above*) exhibit extensive diversification, with multiple branch points illustrating the relationships of cell subpopulations within her tumor to one another. Conversely, a patient whose tumor was sensitive to treatment (*below*) showed minimal branching and thus minimal genetic diversification of cancer cell populations. (B) The genetic behavior of nine ovarian tumors and the clinical progression of the patients carrying them are plotted here. As is apparent, a metric of genetic heterogeneity of their tumors (*abscissa*) is correlated inversely with the extent of progression-free survival (*ordinate*). (Courtesy of R. Schwarz and F. Markowetz.)

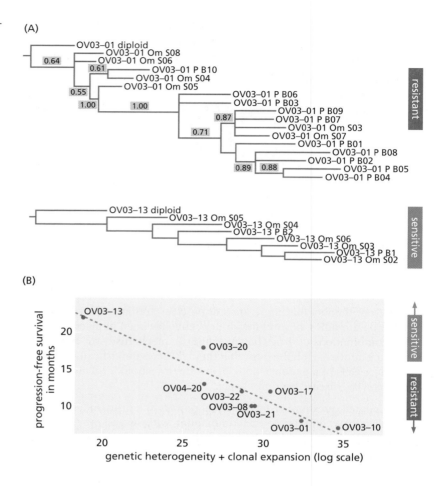

or decrease tumor size. Many anti-tumor agents in clinical trials may shrink tumor masses by eliminating populations of transit-amplifying and more differentiated cells, which together constitute the great bulk of the tumors' neoplastic cells. However, if the drugs leave the minority populations of self-renewing tumor stem cells untouched (see Figure 16.28A), the tumor has a high probability of regrowing, leading sooner or later to clinical relapse. Indeed, as mentioned, there is clear evidence that cancer stem cells are more resistance to a variety of modes of current anti-cancer therapy (see Sidebar 16.4).

Such evidence suggests that durable remissions and cures can only derive from therapies that strike both the bulk non-CSCs in tumors and, importantly, those that lie at the heart of the tumor—the self-renewing CSCs. Currently, assays for the presence of these cells in most tumor types are poorly developed. Hence, drug development efforts are hobbled because researchers lack key analytical tools that are essential for developing truly efficacious therapies.

But there is another problem that has only recently emerged: The non-CSCs within a tumor, more specifically the transit-amplifying/progenitor cells (see Figure 11.18), are able in some tumors, and quite possibly in most tumors, to dedifferentiate and thereby form new CSCs (see Figure 16.28C). If validated as a general phenomenon, this adds emphasis to the necessity of eliminating both existing CSCs within tumors as well as their more differentiated progeny, which may wait in the wings, always poised to regenerate new CSCs.

A major, still-unsolved problem concerns the biological models that are used in pre-clinical drug development. Some xenograft models of human cancer are useful in predicting the behavior of tumors encountered in the oncology clinic, but most are not (see Sidebar 13.1). Similarly, genetically engineered mouse models of human cancer have been limited in their ability to predict clinical responses. The development of predictive pre-clinical models of human cancer would surely reduce the expense of drug development and, quite possibly, even obviate certain early-phase clinical trials. For the moment, truly useful animal tumor models are little more than a fond hope.

The non-neoplastic stromal cells within a tumor may be major determinants of responsiveness to most drug therapies, yet their contributions are not recognized in the design of many pre-clinical models of human cancer. In Chapter 13, we learned, for example, that the most radiosensitive cells in some tumors are likely to be the endothelial cells forming their neovasculature (see Figure 13.47). By the same token, it seems increasingly likely that many widely used anti-cancer chemotherapeutics have strong effects on the tumor-associated endothelial cells that were never suspected in the past. In fact, some researchers are redesigning chemotherapeutic treatment protocols in order to optimize their toxic effects on the tumor-associated neovasculature. Recognition of these two classes of cells—cancer stem cells and tumor stromal cells—as critically important biological targets of chemotherapy will surely change the entire landscape of the drug development field. Moreover, the concept of "stroma" will surely be expanded to include the microenvironments of tumor cells throughout the body as we learn that certain sites of metastatic dissemination offer particularly hospitable homes for cancer cells, protecting them from the toxic effects of various therapies.

The ultimate test of many candidate anti-cancer drugs comes when they are tested in substantial numbers of cancer patients. Here, the drug developer is often confronted by the dilemma of not knowing which types of tumors are likely to respond. Should a candidate drug be tested on patients suffering from pancreatic carcinomas or neuroblastomas? The molecular lesions discovered in these and other types of cancer cells would seem to be highly useful indicators for informing this decision. But quite often, tumors respond for reasons that cannot be predicted by the mutant genes and deregulated signaling pathways known to be present in these growths, and so the choices of patients recruited into clinical trials are arbitrary and sub-optimal. Once again, we can only hope that our increasing insights into the molecular etiologies of various cancers will provide truly useful guidelines for oncologists to follow.

Independent of the challenges of cancer drug development and testing are the more transcending problems created by the complex biology of human cancers: Will different types of drugs need to be developed for different classes of cancers, or will a small number of treatments find wide applicability? Will different tumors within a given class (for example, colon carcinomas) require distinct, tailor-made treatments based on their particular genotypes and phenotypes? And will we one day be able to provide "personalized molecular medicine" at affordable cost, in which the detailed characteristics of each patient's tumor and genetic constitution inform the design of a customized therapy?

Will anti-neoplastic drugs ever be developed that have lethal effects on malignant growths while having minimal side effects on normal tissues? And should drug designers undertake to develop anti-cancer drugs that keep tumors under control rather than attempting to wipe them out? The goal of curing many kinds of tumors, cited above, may well be an unreachable one, and for these tumors, reducing cancer to a chronic but bearable disease may be a more realizable goal. (Such is the thinking of researchers developing new types of anti-HIV treatment protocols, some of which have achieved this goal.)

A 2012 census of genes mutated in human cancer cell genomes netted 488 distinct genes—more than 2% of the genes present in the human genome. Of these, 447 become involved in cancer largely through somatic mutation, while the remaining 41 are involved exclusively as inherited germ-line determinants of cancer susceptibility. This list—generated in significant part from the results of cancer genome sequencing—will undoubtedly grow in the future: The list of somatically altered genes will be increased by inclusion of myriad tumor suppressor genes that are shut down by promoter methylation; recall that promoter methylation is as effective as genetic mutation in eliminating genes and proteins from the regulatory circuitry of cells. At the same time, the census of germ-line determinants of cancer will expand as alleles (and thus genes) having less penetrant effects on cancer susceptibility are uncovered.

This census provides many druggable targets for researchers. At the same time, it represents bewildering complexity. Some of these genes are mutated only in rare cancers, and the costs of developing therapeutic drugs directed against their protein products are unlikely to ever be recouped through sales. Most of the encoded mutant proteins contribute in still-obscure ways to the neoplastic growth of various types of human cancer cells. How will we ever ascribe key roles to each of the roughly 500 mutant genes that are already known or suspected to be involved in the pathogenesis of many cancers? Our current methods of assimilating and interpreting data on human cancer cell genomes and signaling pathways are not up to the task.

[Of relevance here is the fact that only a small portion of the kinome (see Supplementary Sidebar 16.8) has been explored by laboratory researchers and pharmaceutical chemists. A 2009 survey indicated that half of the kinome (a total of ~518 kinases) had been described by 20 or fewer research publications, in contrast to 10 kinases (all discussed in some detail in this book) that had been the topics of 5000 or more reports. Many of the under-studied, potentially druggable enzymes are likely to participate in tumor pathogenesis, even when present in wild-type configuration within cancer cells. Hence, a large array of potentially targetable kinases has yet to be intensively explored.]

Many cancer researchers would like to understand the entirety of a biological system, such as a living cancer cell, rather than its individual functional components. In their eyes, reductionist biology, which focuses on the individual, isolatable components of complex systems, has had its day, and the time has come for the vast amounts of information known about these components to be integrated into complex interacting *systems* whose behavior can be predicted by bioinformatics.

Successes in these efforts, involving the new discipline of "systems biology," will surely benefit cancer research. Imagine a day when the biological responses of various human cells, normal and malignant, can be predicted by mathematical models of these cells and their internal control circuits. Such advances will render many current practices in experimental biology, including many steps of drug development, unnecessary. If this ever becomes possible, drug development will be more a matter of dry bioinformatics than wet biology at the laboratory bench.

But for the moment, most of this remains a pipe dream, still far in the future. For now, at least, we need to wrestle with the grim realities of drug development, the inadequate animal models, our ignorance of the behavior of cellular regulatory circuitry, and the confounding biological complexities of human cancer. And most importantly, never give up. If our ancestors had, we would still be living in the Stone Age.

Key concepts

- The science of molecular oncology has revealed dozens of proteins whose malfunction contributes to the formation and maintenance of tumors. Many of these proteins exhibit molecular properties that make them attractive targets for novel anti-cancer therapeutic agents, such as monoclonal antibodies or low–molecular-weight drugs.

- Proteins that are attractive targets for attack by antibodies are invariably located at the cell surface or in the extracellular space.

- Most proteins that are attractive targets for attack by low–molecular-weight compounds are enzymes that possess druggable catalytic clefts.

- Recent advances have expanded the range of druggable targets to include certain protein–protein interactions that can also be inhibited by low–molecular-weight drugs.

- The proteins that are most suitable as targets are those whose inactivation is predicted to make tumor cells cease proliferating or lead to their death by apoptosis.

- The most successful anti-cancer drugs developed to date have been those that interfere with the functioning of various growth- and survival-promoting kinases, specifically, receptor-associated tyrosine kinases.

- Successful drugs must have a high therapeutic index, appropriate pharmacokinetics and pharmacodynamics, and minimal side effects on major organ systems.

- Studies of drugs in Phase I, II, and III clinical trials are essential because pre-clinical studies of drugs' efficacy and tolerability are poorly predictive of the drugs' behavior in humans.

- Clinical use of certain drugs under development may be indicated by the known behavior of targeted proteins in cancer cells (as in the case of Gleevec) or by empirical tests of how various types of human tumors respond to treatment (as with Velcade).

- Stratification of outwardly similar tumors into narrow subclasses greatly helps researchers and clinicians to match drugs to the specific tumor cell types they can most effectively treat.

- The greatest benefit of certain drugs, such as Gleevec, may eventually prove to derive from a broad target specificity that allows them to be used to treat a wide range of cancers.

Thought questions

1. What are the therapeutic advantages and disadvantages of using a drug that affects a broad range of molecular targets?

2. Given the large and heterogeneous collection of signaling molecules that have been portrayed in this book as playing key roles in the pathogenesis of various cancers, which classes of molecules do you think might become the targets for the development of a new range of anti-cancer therapeutics besides the much-studied kinases?

3. How might tumors that were initiated by the formation of a certain oncogene become independent of this oncogene later in tumor progression?

4. What strategies might you implement upon finding that cells within a tumor have become resistant to a drug therapy after extended treatment?

5. Having concluded that natural products represent a rich resource of potential anti-cancer drugs, what obstacles might limit the search for and testing of such drugs?

6. What obstacles stand in the way of developing drugs for tumors that represent only a very small proportion of the total cancer burden in a population?

7. In the oncology clinic of the future, what types of information might be included in the assembly of anti-cancer therapies that are tailor-made specifically to respond to an individual patient's tumor?

8. What strategies would you pursue to develop pre-clinical models of human tumors that are highly useful in predicting patient responses to candidate drugs?

Additional reading

Alison MR, Lim SML & Nicholson LJ (2010) Cancer stem cells: problems for therapy? *J. Pathol.* 223, 147–161.

Arbiser J (2007) Why targeted therapy hasn't worked in advanced cancer. *J. Clin. Invest.* 117, 2762–2765.

Barouch-Bentov R & Sauer K (2011) Mechanisms of drug-resistance in kinases. *Expert Opin. Investig. Drugs* 20, 153–208.

Barr S, Thomson S, Buck E et al. (2008) Bypassing cellular EGF receptor dependence through epithelial-to-mesenchymal-like transitions. *Clin. Exp. Metastasis* 25, 685–693.

Beachy PA, Hymowitz SG, Lazarus RA et al. (2010) Interactions between Hedgehog proteins and their binding partners come into view. *Genes Dev.* 24, 2001–2012.

Black WC & Welch HG (1993) Advances in diagnostic imaging and overestimations of disease prevalence and the benefits of therapy. *N. Engl. J. Med.* 328, 1237–1243.

Cantor JR & Sabatini DM (2012) Cancer cell metabolism: one hallmark, many forms. *Cancer Discov.* 2, 881–898.

Cantwell-Dorris ER, O'Leary JJ & Sheils OM (2011) BRAF^{V600E}: implications for carcinogenesis and molecular therapy. *Mol. Cancer Ther.* 10, 385–394.

Chabner BA & Roberts TG Jr (2005) Chemotherapy and the war on cancer. *Nat. Rev. Cancer* 5, 65–72.

Cohen P (2002) Protein kinases: the major drug targets of the twenty-first century? *Nat. Rev. Drug Discov.* 1, 309–315.

Courtney KD, Corcoran RB & Engelman JA (2010) The PI3K pathway as drug target in human cancer. *J. Clin. Oncol.* 28, 1075–1083.

Creighton CJ, Chang JC & Rosen JM (2010) Epithelial-mesenchymal transition (EMT) in tumor-initiating cells and its clinical implications in breast cancer. *J. Mammary Gland Biol. Neoplasia* 15, 253–260.

Deininger MW (2008) Nilotinib. *Clin. Cancer Res.* 14, 4027–4031.

Di Cosimo S & Baselga J (2010) Management of breast cancer with targeted agents: importance of heterogenicity. *Nat. Rev. Clin. Oncol.* 7, 139–147.

Easton JB & Houghton PJ (2006) mTOR and cancer therapy. *Oncogene* 16, 6436–6446.

Felsher DW (2008) Oncogene addiction versus oncogene amnesia: perhaps more than just a bad habit? *Cancer Res.* 68, 3081–3086.

Fojo T & Grady C (2009) How much is life worth: cetuximab, non-small cell lung cancer, and the $440 billion question. *J. Natl. Cancer Inst.* 101, 1044–1048.

Frei E III (1985) Curative cancer chemotherapy. *Cancer Res.* 45, 6523–6537.

Garber K (2009) Trials offers early test case for personalized medicine. *J. Natl. Cancer Inst.* 101, 136–138.

Garraway LA & Jänne PA (2012) Circumventing cancer drug resistance in the era of personalized medicine. *Cancer Discov.* 2, 214–226.

Gilbert LA & Hemann MT (2011) Chemotherapeutic resistance: surviving stressful situations. *Cancer Res.* 71, 5062–5066.

Glunde K, Pathak AP & Bhujwalla ZM (2007) Molecular-functional imaging of cancer: to image and imagine. *Trends Mol. Med.* 13, 287–297.

Guertin DA & Sabatini DM (2007) Defining the role of mTOR in cancer. *Cancer Cell* 12, 9–22.

Hammerman PS, Jänne PA & Johnson BE (2009) Resistance to epidermal growth factor receptor tyrosine kinase inhibitors in non-small cell lung cancer. *Clin. Cancer Res.* 15, 7502–7509.

Hellman S (2005) Evolving paradigms and perceptions of cancer. *Nat. Rev. Clin. Oncol.* 2, 618–624.

Hirst GL & Balmain A (2004) Forty years of cancer modeling in the mouse. *Eur. J. Cancer* 40, 1974–1980.

Horn L & Sandler A (2009) Epidermal growth factor receptor inhibitors and antiangiogenic agents for the treatment of non-small cell lung cancer. *Clin. Cancer Res.* 15, 5040–5048.

Humphrey RW, Brockway-Lunardi LM, Bonk DT et al. (2011) Opportunities and challenges in the development of experimental drug combinations for cancer. *J. Natl. Cancer Inst.* 103, 1222–1226.

Hunter T (2007) Treatment for chronic myelogenous leukemia: the long road to imatinib. *J. Clin. Invest.* 117, 2036–2043.

Kaelin WG Jr (2005) The concept of synthetic lethality in the context of anticancer therapy. *Nat. Rev. Cancer* 5, 689–698.

Khazak V, Astsaturov I, Serebriiskii IG & Golemis EA (2009) Selective Raf inhibition in cancer therapy. *Expert Opin. Ther. Targets* 11, 1587–1609.

Klein S & Levitzki A (2009) Targeting the EGFR and the PKB pathway in cancer. *Curr. Opin. Cell Biol.* 21, 1–9.

Klein S, McCormick F & Levitzki A (2005) Killing time for cancer cells. *Nat. Rev. Cancer* 5, 573–580.

Lane HA & Breuleux M (2009) Optimal targeting of the mTORC1 kinase in human cancer. *Curr. Opin. Cell Biol.* 21, 219–229.

Laplante M & Sabatini DM (2009) mTOR signaling at a glance. *J. Cell Sci.* 122, 3589–3594.

Lord CJ & Ashworth A (2010) Biology-driven cancer drug development: back to the future. *BMC Biol.* 8, 38. doi: 10.1186/1741-7007-8-38.

Lynch TJ, Bell DW, Sordella R et al. (2004) Activating mutations in the epidermal growth factor receptor underlying responsiveness of non-small-cell lung cancer to gefitinib. *N. Engl. J. Med.* 350, 2129–2139.

Massagué J (2007) Sorting out breast-cancer gene signatures. *N. Engl. J. Med.* 358, 294–297.

Milojkovic D & Apperley J (2009) Mechanisms of resistance to imatinib and second-generation tyrosine inhibitors in chronic myeloid leukemia. *Clin. Cancer Res.* 15, 7519–7527.

Orlowski RZ & Kuhn DJ (2008) Proteasome inhibitors in cancer therapy: lessons from the first decade. *Clin. Cancer Res.* 14, 1649–1657.

Rauch J, Volinsky N, Romano D & Kolch W (2011) The secret life of kinases: functions beyond catalysis. *Cell Commun. Signal.* 9, 23. doi: 10.1186/1478-811X-9-23.

Riese DJ 2nd, Gallo RM & Settleman J (2007) Mutational activation of ErbB family receptor tyrosine kinases: insights into mechanisms of signal transduction and tumorigenesis. *Bioessays* 29, 558–565.

Rubin LE & de Sauvage F (2006) Targeting the Hedgehog pathway in cancer. *Nat. Rev. Drug Discov.* 5, 1026–1033.

Sawyers C (2004) Targeted cancer therapy. *Nature* 432, 204–207.

Scaltriti M & Baselga J (2006) The epidermal growth factor receptor pathway: a model for targeted therapy. *Clin. Cancer Res.* 12, 5268–5272.

Schiller JT & Lowy DR (2006) Prospects for cervical cancer prevention by human papillomavirus vaccination. *Cancer Res.* 66, 10229–10232.

Schlessinger J (2005) SU11248: genesis of a new cancer drug. *The Scientist* 19, 17–18.

Sebolt-Leopold JS & Herrera R (2004) Targeting the mitogen-activated protein kinase cascade to treat cancer. *Nat. Rev. Cancer* 4, 937–947.

Settleman J & Kurie JM (2007) Drugging the bad "AKT-TOR" to overcome TKI-resistant lung cancer. *Cancer Cell* 12, 6–8.

Sharpless NE & DePinho RA (2006) The mighty mouse: genetically engineered mouse models in cancer drug development. *Nat. Rev. Drug Discov.* 5, 741–754.

Singh A & Settleman A (2010) EMT, cancer stem cells and drug resistance: an emerging axis of evil in the war on cancer. *Oncogene* 29, 4741–4751.

Sotiriou C & Pusztai L (2009) Gene-expression signatures in breast cancer. *N. Engl. J. Med.* 360, 790–800.

Staudt LM (2003) Molecular diagnosis of the hematologic cancers. *N. Engl. J. Med.* 348, 1777–1785.

Stegmeier F, Warmuth M, Sellers WR & Dorsch M (2010) Targeted cancer therapies in the twenty-first century: lessons from imatinib. *Clin. Pharmacol. Ther.* 87, 543–552.

Teodori E, Dei S, Martelli C et al. (2006) The functions and structure of ABC transporters: implications for the design of new inhibitors of Pgp and MRP1 to control multidrug resistance (MDR). *Curr. Drug Targets* 7, 893–909.

Tinker AV, Boussioutas A & Bowtell DDL (2006) The challenges of gene expression microarrays for the study of human cancer. *Cancer Cell* 9, 333–339.

Tsou LK, Cheng Y & Cheng Y-C (2012) Therapeutic development in targeting protein-protein interactions with synthetic topological mimetics. *Curr. Opin. Pharmacol.* 12, 403–407.

van der Heijden MS & Bernards R (2010) Inhibition of the PI3K pathway: hope we can believe in? *Clin. Cancer Res.* 16, 3094–3099.

van't Veer LJ & Bernards R (2008) Enabling personalized cancer medicine through gene-expression patterns. *Nature* 452, 564–570.

Welch HG & Black WC (2010) Overdiagnosis in cancer. *J. Natl. Cancer Inst.* 102, 605–613.

Welch HG, Woloshin S & Schwartz LM (2008) The sea of uncertainty surrounding ductal carcinoma in situ: the price of screening mammography. *J. Natl. Cancer Inst.* 100, 228–229.

Zhou B-B, Zhang H, Damelin M et al. (2009) Tumour-initiating cells: challenges and opportunities for anticancer drug discovery. *Nat. Rev. Drug Discov.* 8, 806–823.

Zhu J, Chen Z, Lallemand-Breitenbach V & de Thé H (2002) How acute promyelocytic leukaemia revived arsenic. *Nat. Rev. Cancer* 2, 705–714.

Abbreviations

A	(1) adenine; (2) adenosine
ABC	(1) activated B-cell-like subtype of DLBCL; (2) adenoid basal carcinoma; (3) ATP-binding cassette (transporter)
Abl	Abelson leukemia virus oncoprotein
ABVD	adriamycin, bleomycin, vinblastine, dacarbazine combined chemotherapy
ACF	aberrant (intestinal) crypt focus/foci
ACT	adoptive cell transfer
ACTH	adrenocorticotropic hormone
ADCC	antibody-dependent cell-mediated cytotoxicity
AEV	avian erythroblastosis virus
AFB1	aflatoxin B1
Ag	antigen
AGT	O^6-alkylguanine-DNA-alkyltransferase; *also called* **MGMT**
AID	activation-induced cytidine deaminase
AIDS	acquired immunodeficiency syndrome
Akt	T-cell lymphoma of Ak mouse strain (serine-threonine kinase); *also* **PKB**
ALDH1	aldehyde dehydrogenase isoform 1
ALK	anaplastic lymphoma kinase
ALL	acute lymphocytic (or lymphoblastic) leukemia
α-SMA	α-smooth muscle actin
ALT	alternative lengthening of telomeres
ALV	avian leukosis virus
AML	acute myelogenous leukemia
AMP	adenosine monophosphate
AMV	avian myelocytomatosis virus
Ang	angiopoietin
ANGPTL	angiopoietin-like protein
AP	(1) apurinic; (2) apyrimidinic
AP-1	activator protein-1 transcription factor (Fos + Jun complex)
Apaf-1	apoptotic protease-activating factor-1
APC	(1) antigen-presenting cell; (2) anaphase-promoting complex; (3) adenomatous polyposis coli
APE	apurinic/apyrimidinic endonuclease
APL	acute promyelocytic leukemia
AR	androgen receptor
ARF	alternative reading frame (protein)
ASCT	allogeneic stem cell transplantation
asmase	acid sphingomyelinase
AT	ataxia telangiectasia (syndrome)
ATM	mutated in ataxia telangiectasia
ATP	adenosine triphosphate
ATR	(1) ATM-related kinase; (2) ATM and Rad3-related kinase
ATRA	all-*trans*-retinoic acid
a.u.	arbitrary units
AUC	area under the curve
BCC	basal cell carcinoma
BCG	Bacillus Calmette–Guérin (mycobacterium)
Bcl-2	B-cell lymphoma gene-2
BCNS	basal cell nevus syndrome
BCR, bcr	breakpoint cluster region
BDNF	bone-derived neurotrophic factor
BER	base-excision DNA repair
β-gal	β-galactosidase
β₂m	β_2 microglobulin
BFB	breakage–fusion–bridge (cycle of chromosomes)
BH	Bcl-2–homologous (domain)
BHK	baby hamster kidney (cells)
bHLH	basic helix–loop–helix (transcription factor)
BL	Burkitt's lymphoma
BM	(1) bone marrow; (2) basement membrane
BMI	body-mass index
BMP	bone morphogenetic protein
BMPR1	BMP receptor-1
BMSC	bone marrow stem cell
BMT	bone marrow transplantation
BP	benzo[*a*]pyrene
bp	base pair
BPDE	benzo[*a*]pyrenediolepoxide
BPV	bovine papillomavirus
BrdU	bromodeoxyuridine
BTK	Bruton's TK
C	(1) carboxy (terminus of polypeptide); (2) constant region (of an antibody molecule); (3) cytidine; (4) cytosine
C′	carboxy (terminus of polypeptide)
ca	(1) cancer; (2) carcinoma
CAD	caspase-activated DNase
CAF	carcinoma-associated fibroblast
CAK	CDK-activating kinase
CalB	calcium–phospholipid binding domain
cAMP	cyclic adenosine monophosphate
CASP	caspase-encoding gene
CBL	Casitas B-lineage lymphoma (gene)
Cbl	Casitas B-lineage lymphoma (protein)
CBP	cyclic AMP response element–binding protein
CCK	cholecystokinin
CCL	chemokine ligand with two adjacent cysteines
CD	cluster of differentiation (cell surface antigen)
CDC	complement-dependent cytotoxicity
CDH1	E-cadherin
Cdh1	(1) the gene encoding E-cadherin; (2) a regulatory subunit of the anaphase-promoting complex
CDK	cyclin-dependent kinase
CdkI	CDK inhibitor
cDNA	complementary DNA copy of mRNA
CDR	complementarity-determining region (of an antibody)
CEA	carcinoembryonic antigen

C/EBPβ	CCAAT/enhancer binding protein β
CEF	chicken embryo fibroblast
CEP	circulating endothelial progenitor (cell)
CGH	comparative genomic hybridization
CHFR	checkpoint with forkhead and RING finger domains (protein)
ChIP	chromatin immunoprecipitation
Chk2	checkpoint kinase 2
CHK2	gene encoding Chk2 protein
CHOP	cyclophosphamide, doxorubicin, vincristine, and prednisone (chemotherapy cocktail)
CI	confidence interval
CIMP	CpG island methylator phenotype
CIN	(1) cervical intraepithelial neoplasia; (2) chromosomal instability
CIS	carcinoma *in situ*
CK2	casein kinase II
CKI	cyclin-dependent kinase inhibitor
CLL	chronic lymphocytic leukemia
CM	conditioned medium
CML	chronic myelogenous leukemia
CMV	cytomegalovirus
CNA	copy number alteration
CNK	connector enhancer of KSR
CNL	chronic neutrophilic leukemia
CNS	central nervous system
CoA	coenzyme A
CoO	the (normal) cell-of-origin (of a population of cancer cells)
COX-2	cyclooxygenase-2
CPD	cyclobutane pyrimidine dimer
CPE	cytopathic effect
CpG	cytidine phosphate guanosine (DNA sequence)
cpm	counts per minute
CR	complete response
13cRA	13-*cis*-retinoic acid
CRC	colorectal carcinoma
Crkl	Crk-like (*also* **CrkL**)
CRT	calreticulin
CS	Cockayne syndrome
CSC	cancer stem cell
CSF-1	colony-stimulating factor-1
Csk	C-terminal Src kinase
CSR	core serum response (gene expression signature)
CT	(1) cancer-testis (antigen); (2) computed tomography (usually using X-ray scanning); (3) chemotherapy
CTGF	connective tissue growth factor
CTL	cytotoxic T lymphocyte
CTLA-4	cytotoxic T lymphocyte–associated antigen-4
CTVS	canine transmissible venereal sarcoma
CUP	(metastatic) cancer of unknown primary
CXCL	ligand of chemokine receptor
CXCR	chemokine receptor
CYP	cytochrome P450 enzyme
d5′-mC	deoxy-5′-methylcytosine
dA	deoxyadenosine
DAG	diacylglycerol
DAPI	4′,6′-diamidino-2-phenylindole (DNA stain)
DBD	DNA-binding domain
DC	(1) dendritic cell; (2) dyskeratosis congenita
DCIS	ductal carcinoma *in situ* (of the breast)
DDR	discoidin domain receptor
del	genetic deletion
DES	diethylstilbestrol
DFS	disease-free survival
dG	deoxyguanosine
DIABLO	direct IAP-binding protein with low pI
di-I	3,3-dioctadecyl indocarbocyanine (dye)
DISC	death-inducing signaling complex
DKK	dickkopf (inhibitor of Wnts)
DLBCL	diffuse large B-cell lymphoma
DLC	(1) dynein light chain; (2) deleted in liver cancer
DM	double-minute chromosome
DMBA	dimethylbenz[*a*]anthracene
dn, DN	dominant-negative (allele)
DNMT	DNA methyltransferase
DP	differentiation-related transcription factor protein
DPC4	deleted in pancreatic cancer-4
dR	deoxyribose
ds	double-stranded (DNA or RNA)
DSB	double-strand (DNA) break
DTC	disseminated tumor cell
dTg	deoxythymidine glycol
DUB	de-ubiquitylating enzyme
dUTP	deoxyuridine triphosphate
E2	17-β-estradiol/estrogen
E2F	transcription factor activating adenovirus E2 gene
4E-BP	eIF4E-binding protein
EBV	Epstein–Barr virus
EC	embryonal carcinoma (cell)
ECM	extracellular matrix
EFC-XV	endostatin-like fragment of type XV collagen
EFS	event-free survival
EGF	epidermal growth factor
EGF-R	EGF receptor
eIF	eukaryotic (translation) initiation factor
eIF4E	eukaryotic (translation) initiation factor 4E
EM	electron micrograph (or electron microscope)
EMSA	electrophoretic mobility shift assay
EMT	epithelial–mesenchymal transition; *also* epithelial-to-mesenchymal transition
EMT-TF	EMT-inducing transcription factor
ENU	*N*-ethylnitrosourea
env	retrovirus envelope glycoprotein
EPC	endothelial precursor cell; endothelial progenitor cell
EpCAM	epithelial cell adhesion molecule
EPO	erythropoietin
εdA	1,*N*^6-ethenodeoxyadenosine
ER	(1) estrogen receptor; (2) endoplasmic reticulum
erb	(oncogene of) avian erythroblastosis virus
ERG	ETS-related gene
ERK	extracellular signal–regulated kinase
ERV	endogenous retrovirus
ES	embryonic stem (cells)
ESA	epithelial surface antigen
ET-1	endothelin-1
EZH2	enhancer of zeste homolog 2
F	filamentous (actin)
FA	Fanconi anemia (syndrome)
FACS	fluorescence-activated cell sorting
FADD	Fas-associated death domain (protein)
FAK	focal adhesion kinase
FANC	proteins defective in Fanconi anemia syndrome
FAP	familial adenomatous polyposis
FasL	ligand of the Fas death receptor
Fc	$C_H2 + C_H3$ fragment of IgG antibody molecule
FDA	U.S. Food and Drug Administration
FDG	2-deoxy-2-(^{18}F)fluoro-D-glucose
FGF	fibroblast growth factor
FH	fumarate hydratase
Fig	fused in glioblastoma
FISH	fluorescence *in situ* hybridization

FITC	fluorescein isothiocyanate (dye)
FKBP12	FK506-binding protein of 12 kD
FL	(1) follicular (B-cell) lymphoma; (2) ligand of the Flt-3 receptor
FLICE	FADD-like interleukin-1β–converting enzyme; *also called* **caspase 8**
FLIP	FLICE-inhibitory protein
Flk-1	murine VEGF-R2
Flt	Fms-like tyrosine kinase
Flt-1	human VEGF-R1
FLV	Friend murine leukemia virus
FOX	member of the Forkhead family of transcription factors
FPT	farnesyl protein transferase
FRB	FKBP12 + rapamycin–binding (domain of mTOR)
Frz	Frizzled
FSH	follicle-stimulating hormone
FTI	farnesyltransferase inhibitor
5-FU	5-fluorouracil
G	(1) guanine; (2) guanine nucleotide; (3) guanosine
G$_1$	gap 1 phase (of cell cycle)
G$_2$	gap 2 phase (of cell cycle)
G6PD	glucose-6-phosphate dehydrogenase
GAG	glycosaminoglycan
gag	group-specific antigen (retrovirus capsid protein)
GAK	cyclin G–associated kinase
GAP	GTPase-activating protein
GAPDH	glyceraldehyde-3-phosphate dehydrogenase
GBM	glioblastoma multiforme
GCB	germinal-center B-cell lymphoma
GDNF	glial cell–derived neurotrophic growth factor
GDP	guanosine diphosphate
GEF	guanine nucleotide exchange factor
GEMM	genetically engineered mouse model
GF	growth factor
GFAP	glial fibrillary acidic protein
GFP	green fluorescent protein
GGF	glial growth factor
GGR	global genomic repair
GH	growth hormone
GIST	gastrointestinal stromal tumor
GLUT	glucose transporter
GM-CSF	granulocyte–macrophage colony-stimulating factor
gp	glycoprotein
GPI	glycosylphosphatidylinositol
GPCR	G protein–coupled receptor
Grb2	growth factor receptor-bound-2
GRIP-1	glucocorticoid receptor interacting protein-1
GRP	gastrin-releasing peptide
GSK-3β	glycogen synthase kinase-3β
GSTP1	glutathione *S*-transferase π
GTP	guanosine triphosphate
GTPase	enzyme that cleaves GTP, usually generating GDP
GVHD	graft versus host disease
GVT	graft versus tumor (immunological reaction)
Gy	gray (radiation dose unit)
H	histone
h	prefix referring to the human form of a gene or protein
^{3}H	tritium
H&E	hematoxylin–eosin (tissue stain)
hABH	human AlkB homolog
hABH2	human AlkB homolog-2
HAMA	human anti-mouse antibody

HAT	(1) hypoxanthine–aminopterin–thymidine (culture medium); (2) histone acetyltransferase
HB-EGF	heparin-binding EGF
HBV	hepatitis B virus
HCA	heterocyclic amine
HCC	hepatocellular carcinoma
hCG	human chorionic gonadotropin
HCL	hairy-cell leukemia
HCMV	human cytomegalovirus
HCV	hepatitis C virus
HDAC	histone deacetylase
HDM2	human homolog of MDM2
HDR	homology-directed repair; *also* **HR**
HEK	human embryonic kidney (cells)
HER	human EGF receptor (or related protein)
HERV	human endogenous retrovirus
HES	hairy and enhancer of split
2-HG	2-hydroxyglutarate
HGF	hepatocyte growth factor; *also called* **scatter factor (SF)**
HGFL	hepatocyte growth factor-like
HGPRT	hypoxanthine guanine phosphoribosyltransferase
Hh	Hedgehog
HHV	human herpesvirus
HHV-8	human herpesvirus type 8; *also called* **Kaposi's sarcoma herpesvirus (KSHV)**
HIF	hypoxia-inducible (transcription) factor
HIP1	huntingtin-interacting protein-1
HIPK2	homeodomain-interacting protein kinase 2
HIV	human immunodeficiency virus
HLA	human leukocyte antigen (*equivalent to* **MHC**)
5hmC	5-hydroxymethylcytosine
HMEC	human mammary epithelial cell
HMT	histone methyltransferase
HNPCC	hereditary non-polyposis colon cancer
hnRNA	heterogeneous nuclear RNA
HNSCC	head-and-neck squamous cell carcinoma
HPE	human prostate epithelial (cell)
HPGD	15-hydroxyprostaglandin dehydrogenase
HPRT	hypoxanthine phosphoribosyltransferase
HPV	human papillomavirus
HR	homology-directed (DNA) repair; *also* **HDR**
HRE	hormone response element (in DNA)
HRG	heregulin
HSA	human serum albumin
HSC	hematopoietic stem cell
HSIL	high-grade squamous intraepithelial lesion
HSR	homogeneously staining region
HSV-1	herpes simplex virus type 1
hTERT	human telomerase reverse transcriptase
HTLV-I	human T-cell lymphotropic virus type I (or human T-cell leukemia virus)
hTR	human telomerase RNA
HTS	high-throughput screening
HU	hydroxyurea
IAP	(1) intracisternal A particle; (2) inhibitor of apoptosis
-ib	low–molecular-weight drug (suffix)
IC$_{50}$	concentration required to obtain 50% inhibition
ICAD	inhibitor of caspase-activated DNase
Id	inhibitor of DNA binding (transcription factor)
IDH	isocitrate dehydrogenase
IEG	immediate early gene
IEL	intraepithelial lymphocyte
IF	immunofluorescence
IFN	interferon
IFN-R	interferon receptor
Ig	immunoglobulin

IGF-1	insulin-like growth factor-1	**MEF**	mouse embryo fibroblast
IGF-1R	IGF-1 receptor	**MeIQx**	2-amino-3,8-dimethylimidazo[4,5-f]-quinoxaline
IGFBP	insulin-like growth factor–binding protein	**MEK**	MAPK/Erk kinase
IHC	immunohistochemistry	**MEN**	multiple endocrine neoplasia
IκB	inhibitor of NF-κB	**MET**	mesenchymal–epithelial transition; *also* mesenchymal-to-epithelial transition
IKK	IκB kinase		
IL	interleukin	**Met**	receptor of hepatocyte growth factor
ILK	integrin-linked kinase	**mFISH**	multicolor fluorescence *in situ* hybridization
iMCs	immature myeloid cells	**MGMT**	O^6-methylguanine-DNA-methyltransferase; *also called* **AGT**
int	gene identified through insertional mutagenesis		
IP$_3$	inositol (1,4,5)-triphosphate	**MHC**	major histocompatibility complex (*see also* **HLA**)
IR	insulin receptor		
IRES	internal ribosome entry site	**-mib**	low–molecular-weight pharmacologic agent (suffix)
IRK	insulin receptor kinase		
Jak	Janus kinase	**MIN**	microsatellite instability
JCV	JC virus, a close relative of SV40	**miRNA**	microRNA
JM	juxtamembrane	**Miz-1**	Myc-interacting zinc finger protein-1
Jmj	Jumonji (enzyme)	**MKP**	MAP kinase phosphatase
kb	kilobase	**MLL**	mixed lineage leukemia; myeloid/lymphoid leukemia
kbp	kilobase pair		
kD	kilodalton	**MLV**	murine leukemia virus
KDR	human VEGF-R2	**MM**	multiple myeloma
KGF	keratinocyte growth factor	**mM**	millimolar (10^{-3} molar)
KIR	killer inhibitory receptor	**MMP**	matrix metalloproteinase
KO	knockout; knocked-out (gene)	**MMR**	mismatch repair (of DNA)
KSHV	*see* **HHV-8**	**mms**	methylmethane sulfonate
KSR-1	kinase suppressor of Ras-1	**MMTV**	mouse mammary tumor virus
LAK	lymphokine-activated killer (cell)	**MNNG**	*N*-methyl-*N*′-nitro-*N*-nitrosoguanidine
LANA	latency-associated nuclear antigen (of KSHV)	**MNU**	*N*-methylnitrosourea
LCM	laser capture microdissection	**MoAb**	monoclonal antibody
LDL	low-density lipoprotein	**moca**	4,4′-methylenebis(2-chloroaniline)
LEF	lymphoid enhancer factor	**mPGES**	microsomal prostaglandin E$_2$ synthase
LH	luteinizing hormone	**MPNST**	malignant peripheral nerve sheath tumor
LIF	leukemia inhibitory factor	**MRE11**	meiotic recombination 11
lincRNA	long intergenic non-coding noncoding RNA; usually lncRNA	**MRI**	magnetic resonance imaging
		MRN	MRE11/Rad50/Nbs1 complex
LLC	Lewis lung carcinoma	**mRNA**	messenger RNA
LNA	locked nucleic acid	**MSC**	mesenchymal stem cell
lncRNA	long noncoding RNA	**MSH**	melanocyte-stimulating hormone
LOH	loss of heterozygosity	**MSP**	methylation-specific polymerase chain reaction
LRC	label-retaining cell	**MT**	middle T antigen of polyomavirus
LRP	LDL receptor–related protein	**MT1-MMP**	membrane type MMP
LT	large T antigen (of SV40 or polyomavirus)	**MTD**	maximum tolerated dose
LTR	long terminal repeat (of a provirus)	**mTERT**	mouse/murine TERT
M	mitosis	**MTH1**	mammalian MutT homolog-1
m	prefix referring to the murine form of a gene or protein	**mTOR**	mammalian target of rapamycin
		μM	micromolar (10^{-6} molar)
-mab	therapeutic monoclonal antibody (suffix)	**MUP**	major urinary protein
MAC	membrane attack complex	*myc*	(oncogene of) avian myelocytomatosis virus
MALT	mucosa-associated lymphoid tissue	*N*	ploidy number (where haploid = 1)
MAPK	mitogen-activated protein kinase; *also* **MAP-kinase**	*N*′	N (amino) terminus of polypeptide; N-terminal (adj.)
MAPKK	kinase that phosphorylates a MAPK	**NAT**	*N*-acetyltransferase
MAPKKK	kinase that phosphorylates a MAPKK	**NBS1**	Nijmegen break syndrome 1
Mb	megabase; *also* **MB**	**NCAM**	neural cell adhesion molecule
3-MC	3-methylcholanthrene	**NCI**	National Cancer Institute of the United States
5mC	5-methylcytosine	**NER**	nucleotide-excision DNA repair
MCP-1	monocyte chemotactic protein-1	**NES**	nuclear export signal
mCRP	membrane-bound complement regulatory protein	**NET**	neuroendocrine tumor
		Neu	rat homolog of HER2 (receptor)
M-CSF	macrophage colony-stimulating factor	**Nf/NF**	neurofibromatosis; *also* NF
MCT	monocarboxylate transporter	**NF-κB**	nuclear factor-κB
MDCK	Maden–Darby canine kidney (cells)	**NGF**	nerve growth factor
MDM2	mouse double-minute chromosome 2	**NHEJ**	nonhomologous end joining
MDR	multi-drug resistance (phenotype)	**NHL**	non-Hodgkin's lymphoma
MDS	myelodysplastic syndrome	**NK**	natural killer (lymphocyte)
MDSC	myeloid-derived suppressor cell	**NKT**	natural killer lymphocyte expressing T-cell receptor
MEC	mammary epithelial cell		
meCpG	methyl-cytidine within a CpG dinucleotide		

NLS	nuclear localization (import) sequence
nM	nanomolar (10^{-9} molar)
NMR	nuclear magnetic resonance
NO	nitric oxide
NOD	non-obese diabetic
NOD/SCID	non-obese diabetic/severe combined immunodeficient
NotchL	Notch ligand
NRG	neuregulin
NSAID	nonsteroidal anti-inflammatory drug
NSCLC	non-small-cell lung carcinoma
nt	nucleotide
NURF	nucleosome remodeling factor
OD	optical density
4-OHT	4-hydroxytamoxifen
OPG	osteoprotegrin
OPN	osteopontin
OS	overall survival
OSCC	oral squamous cell carcinoma
OSE	ovarian surface epithelium
8-oxo-dG	8-oxo-deoxyguanosine
p	(1) short arm of a chromosome; (2) probability of an event
Pa	Pascal (unit of measured rigidity)
PAH	polycyclic aromatic hydrocarbon
PAI-1	plasminogen activator inhibitor-1
PanIN	pancreatic intraepithelial neoplasia
PAP	prostatic acid phosphatase
PARP	poly (ADP-ribose) polymerase
PBS	phosphate-buffered saline
PCNA	proliferating-cell nuclear antigen
PCR	polymerase chain reaction
PD	(1) pharmacodynamics; (2) population doubling
PD-1	programmed death-1 (receptor)
PDAC	pancreatic ductal (exocrine) adenocarcinoma
PDGF	platelet-derived growth factor
PDGF-R	PDGF receptor
PDH	pyruvate dehydrogenase
PDK1	(1) phosphoinositide-dependent kinase 1; (2) pyruvate dehydrogenase kinase 1
PDM	mean population doublings
PEG	polyethylene glycol
PEP	phosphoenolpyruvate
PET	positron-emission tomography
PFS	progression-free survival
PGDH	15-hydroxyprostaglandin dehydrogenase
PGE$_2$	prostaglandin E$_2$
P-gp	P-glycoprotein
PH	pleckstrin homology (PIP3 binding) (domain)
PHD	proline hydroxylase
PhIP	2-amino-1-methyl-6-phenylimidazo[4,5-b] pyridine
PhK	phosphorylase kinase
PI	(1) phosphatidylinositol; (2) propidium iodide
PI(3,4,5)P$_3$	*equivalent to* **PIP3**
PI3K	phosphatidylinositol 3-kinase
PIN	prostatic intraepithelial neoplasia
PINCH	particularly interesting new cysteine-histidine-rich (protein)
PIP2	phosphatidyl inositol (4,5) diphosphate/PI(4,5)P$_2$
PIP3	phosphatidyl inositol (3,4,5) triphosphate/PI(3,4,5)P$_3$
PK	(1) pyruvate kinase; (2) pharmacokinetics
PKA	protein kinase A
PKB	protein kinase B (=Akt)
PKC	protein kinase C
PLC	phospholipase C
PLD	phospholipase D
PMA	phorbol-12-myristate-13-acetate
PMBL	primary mediastinal B-cell lymphoma
PML	(1) promyelocytic leukemia; (2) progressive multifocal leukoencephalopathy
PMN	polymorphonuclear leukocytes/neutrophils
PND	paraneoplastic neurological degeneration
pol	(1) polymerase; (2) retrovirus reverse transcriptase
pol II	RNA polymerase II
POT1	protection of telomere 1 (protein)
6-4 PP	pyrimidine (6-4) pyrimidinone photoproduct
PP1	protein phosphatase 1
PP2A	protein phosphatase 2A
PR	partial response
PRAS	proline-rich Akt substrate
pRb	retinoblastoma protein; *also* **Rb**, **RB**
PRC1	polycomb repressive complex 1
PRL	prolactin
pro-B ALL	pro-B-cell acute lymphocytic leukemia
PS	phosphatidylserine
PSA	prostate-specific antigen
PTB	phosphotyrosine-binding (domain)
PTC	(1) papillary thyroid carcinoma; (2) Patched
PTEN	phosphatase and tensin homolog deleted on Chromosome 10
PTH	parathyroid hormone
PTHrP	parathyroid hormone–related peptide
PTP	phosphotyrosine phosphatase
pY	phosphotyrosine
PyMT	polyomavirus middle T oncoprotein
q	long arm of a chromosome
qPCR	quantitative polymerase chain reaction
qRT-PCR	quantitative real-time PCR (*see also* **RT-PCR**)
R	restriction point
RA	retinoic acid
Rad	radiation repair (gene)
RAG	recombination-activating gene
RANK	receptor activator of NF-κB
RANKL	ligand of RANK (receptor)
RAR	retinoic acid receptor
ras	(oncogene of) rat sarcoma virus
RAS	protein of *ras* oncogene
Rb	retinoblastoma protein
RCC	renal cell carcinoma
REF	rat embryo fibroblast
RFLP	restriction fragment length polymorphism
RFP	red fluorescent protein
Rheb	Ras homolog enriched in brain
RING	really interesting new gene (protein domain)
RISC	RNA-induced silencing complex
RNAi	RNA interference pathway (involving miRNAs, siRNAs, shRNAs)
RNS	reactive nitrogen species
ROS	reactive oxygen species
RPA	replication protein A
RR	(1) relative risk; (2) response rate
RSV	Rous sarcoma virus
RT	reverse transcriptase
RTK	receptor tyrosine kinase
RT-PCR	reverse transcription PCR (*see also* **qRT-PCR**)
S	(1) DNA synthesis phase (of cell cycle); (2) svedberg (unit of sedimentation in centrifuge)
S1P	sphingosine-1-phosphate
S6	protein 6 of the small ribosomal subunit
S6K1	p70 S6 kinase-1
SA-β-gal	senescence-associated β-galactosidase
SAC	spindle assembly checkpoint
SAHF	senescence-associated heterochromatic foci
SASP	senescence-associated secretory phenotype

SC	stem cell
SCC	squamous cell carcinoma
SCE	sister chromatid exchange
SCF	stem cell factor
SCID	severe combined immunodeficiency (syndrome)
SCLC	small-cell lung carcinoma
SDF-1	stromal cell–derived factor-1
SDH	succinate dehydrogenase
SEM	scanning electron micrograph/microscope/microscopy
Ser	serine
SERM	selective estrogen receptor modulator
SF	scatter factor; *see* **HGF**
SFRP/sFRP	secreted frizzle-related protein (Wnt inhibitor)
SGK1	serum- and glucocorticoid-inducible kinase 1
SH1	Src homology 1 domain (tyrosine kinase)
SH2	Src homology 2 domain (phosphotyrosine binding)
SH3	Src homology 3 domain (proline-rich binding)
SHP	SH2-containing phosphatase
shRNA	small hairpin RNA
SIP1	Smad-interacting protein 1
SIR	standardized incidence ratio
siRNA	small interfering RNA
Skp2	S-phase kinase-associated protein 2
SKY	multicolor spectral karyotyping
Sky1	SR-protein–specific kinase of budding yeast
SLE	systemic lupus erythematosus
SMA	smooth muscle actin
Smac	second mitochondria-derived activator of caspase
SMDF	somatic and neuron-derived growth factor
Smo	Smoothened
SNP	single-nucleotide polymorphism
SOD	superoxide dismutase
Sos	son of sevenless
Srf	serum response factor
ss	single-stranded (DNA or RNA)
STAT	signal transducer and activator of transcription
Str-1	stromelysin-1
SUMO	small ubiquitin-like modifier
SV	simian virus
T	(1) thymine; (2) thymidine
t	referring to a chromosomal translocation
$T_{1/2}$	half-life
T_C	cytotoxic T cell
T_H	helper T cell
T_m	melting temperature
T_{reg}	regulatory T cell
TAA	tumor-associated antigen
Tag	SV40 T-antigen
T-ALL	T-cell acute lymphocytic leukemia
TAM	tumor-associated macrophage
TAP	transporter associated with antigen processing
TATA	tumor-associated transplantation antigen
TCF	T-cell factor
TCR	(1) T-cell receptor; (2) transcription-coupled DNA repair
TdT	terminal deoxynucleotidyl transferase
TEL/AML1	chromosomal translocation

TEM	transmission electron microscopy
TERT	telomerase reverse transcriptase
TF	(1) transcription factor; (2) Tissue Factor
TGF-α	transforming growth factor-α
TGF-β	transforming growth factor-β
TGF-βR	TGF-β receptor
Thr	threonine
3D	3-dimensional
TIAM-1	T-cell lymphoma invasion and metastasis-1
TIC	tumor-initiating cell
TIF	telomere dysfunction-induced focus
TIL	tumor-infiltrating lymphocyte
TIMP	tissue inhibitor of metalloproteinase
TK	(1) tyrosine kinase; (2) thymidine kinase
TKI	tyrosine kinase inhibitor
TM	transmembrane (domain of a protein)
TNF	tumor necrosis factor
TOR	target of rapamycin
TORC	mTOR-containing complex
TPA	12-*O*-tetradecanoylphorbol-13-acetate (=PMA)
TPO	thrombopoietin
TRAIL	TNF-related apoptosis-inducing ligand
TRAMP	transgenic adenocarcinoma of the mouse prostate
TRAP	telomeric repeat amplification protocol
TRF	telomeric restriction fragment
tRNA	transfer RNA
TrpRS	tryptophanyl-tRNA synthetase
ts	temperature-sensitive
TSA	(1) trichostatin A; (2) tumor-specific antigen
TSC	tuberous sclerosis
TSG	tumor suppressor gene
TSHR	thyroid-stimulating hormone receptor
Tsp-1	thrombospondin-1
TSS	transcription start site
TSTA	tumor-specific transplantation antigen
TUNEL	TdT-mediated dUTP nick end labeling
2D	2-dimensional
tx	transformed (cell)
U	(1) uracil; (2) uridine
UC	ulcerative colitis
uPA	urokinase (type) plasminogen activator
uPAR	receptor of uPA
UPR	unfolded protein response
UTR	untranslated region (of an mRNA)
UV	ultraviolet
UV-B	ultraviolet-B radiation
V	variable region (of an antibody molecule)
VDAC	voltage-dependent anion channel
VEGF	vascular endothelial growth factor (=VPF)
VHL	von Hippel–Lindau
VPF	vascular permeability factor (=VEGF)
VSMC	vascular smooth muscle cell
WM	Waldenström's macroglobulinemia
Wnt	wingless/integration site (growth factor)
wt	wild type
XAF1	XIAP-associated factor 1
XIAP	X-chromosome–linked inhibitor of apoptosis
XP	xeroderma pigmentosum
YFP	yellow fluorescent protein

Glossary

abasic Referring to a nucleotide that has lost its purine or pyrimidine base. *See also* **apurinic**; **apyrimidinic**.

ablate To eliminate.

abluminal Located away from the lumen of a duct or other hollow structure.

abscissa Horizontal or *x*-axis of a Cartesian graph. *See also* **ordinate**.

abscission Final step of cytokinesis, when the remaining linkages, including those created by microtubules, are severed, allowing complete separation of the two daughter cells.

acellular Lacking or deprived of cells.

acetylation Covalent attachment of an acetyl group to a second molecule such as a protein.

acetyltransferase An enzyme that covalently links acetate groups to substrates, such as the amino acid side chains of proteins. *See also* **deacetylase**.

acid test The most rigorous test of an experimental conclusion.

acinus (pl., **-i**) (1) A hollow saclike structure usually emptying into a duct. (2) A hollow spherical structure formed in culture by certain epithelial cells.

acromegaly Pathological condition of excessive growth of certain tissues, usually due to the elaboration of excessive growth hormone by a pituitary tumor.

activation loop An oligopeptide loop of a protein kinase molecule that is normally positioned to block access by the catalytic cleft of the kinase to its substrates; phosphorylation of this loop, often by another kinase, causes it to swing out of the way, permitting free access of the catalytic cleft to substrates.

adaptive immune response An immune system response that is acquired or learned following exposure of an organism to an antigen or antigen-bearing agent. *See also* **innate immune response**.

adduct Product of covalently linking two reactive molecules together that retains almost all of the atoms of the reactants; many mutagens act by forming adducts with DNA bases.

adenocarcinoma Tumor derived from secretory epithelial cells.

adenoma (adj., **-omatous**) Any of a series of premalignant, noninvasive growths in various epithelial tissues, many of which have the potential to progress further to carcinomas. *See also* **polyp**.

adherens junctions Lateral junctions between adjacent epithelial cells formed by cadherins displayed on the surfaces of these cells; these junctions include the cytoplasmic proteins that are physically associated with the cadherins and link these junctions to the actin cytoskeleton.

adipocytes Specialized cells of the mesenchymal lineage, closely related to fibroblasts, that create fat and store it in large globules in the cytoplasm; the dominant cell type in fatty tissues.

adipogenic Causing differentiation into adipocytes.

adjuvant (1) A treatment that is given in concert with or following another treatment (*see also* **neoadjuvant**). (2) A substance that is given together with an antigen to enhance the immune response.

adoptive transfer Procedure in which immune cells are transferred from a donor to a recipient, undertaken with the hope that the donor's cells will be able to mediate an immune function lacking in the recipient.

adrenal Referring to the secretory glands that sit above the kidneys.

aerobic glycolysis *See* **Warburg effect**.

afferent Referring to incoming signals. *See also* **efferent**.

age-adjusted Describing an epidemiologic measurement, such as disease incidence or mortality, that is adjusted to compensate for the age distribution of the population under study, allowing rates to be compared between populations having different age distributions.

agonist Activating agent; opposite of antagonist.

alkylating Capable of attaching covalently an alkyl group or similarly structured chemical group to a substrate such as a DNA base.

allele One alternative among different versions of a gene that may be defined by the phenotype that it creates, by the protein that it specifies, or by its nucleotide sequence.

allelic deletion *See* **loss of heterozygosity**.

allogeneic (1) Referring to two genetically distinct members of the same species. (2) Describing the relationship between two sets of cells or tissues deriving from distinct genetic backgrounds.

allogeneic stem cell transplantation Clinical procedure involving the transplantation of bone marrow cells from a donor that are genetically and thus antigenically nonidentical with those of the recipient.

allograft   Implantation of cells of an animal of one genetic background into a host animal of another genetic background but from the same species.

allograft rejection   Process occurring when a donor tissue graft is rejected by the immune system of a recipient because the donor and recipient, although members of the same species, are genetically distinct.

allosteric   Referring to a mechanism for regulating the activity of a protein, such as an enzyme, in which its activity or function is modulated by changes in its three-dimensional structure.

alternative splicing   Process whereby a pre-mRNA may be spliced in several alternative ways, resulting in mRNAs composed of different combinations of exons.

***Alu* repeat**   Sequence block of about 300 bp that is found in almost 1 million copies scattered throughout the human genome.

alveolus (pl., **alveoli**)   A small sac within a tissue that is connected to a duct, such as the sacs seen in the lung and in the mammary gland during pregnancy.

amino   Referring to the end of an oligopeptide or protein chain that is formed first. *See also* **carboxy**.

amoeboid   Exhibiting the shape and motility reminiscent of the behavior of the amoeba protozoan.

amphipathic   Describing a molecule containing distinct hydrophobic and hydrophilic domains.

amplicon   A defined stretch of (chromosomal) DNA that undergoes amplification.

amplification   Genetic mechanism by which the copy number of a gene is increased above its normal level in the diploid genome.

anaphase   Third subphase of mitosis, during which the paired chromatids are segregated to the two opposite poles of the cell.

anaphase bridge   A connection between two clusters of chromosomes that forms during anaphase and results from the simultaneous association of individual chromatids with both chromosomal clusters.

anaplastic   (Referring to a tumor) having a tissue and cellular architecture lacking the differentiated characteristics of an identifiable tissue-of-origin.

anastomosis   End-to-end connection formed between two ductal structures, such as a connection formed directly between an artery and a vein.

anchorage dependence   Requirement of cells for tethering to a solid substrate before they will grow.

anchorage independence   Ability of a cell to proliferate without attachment to a solid substrate.

androgen   A steroid hormone that stimulates development of male characteristics.

aneuploid   (1) Describing a karyotype that deviates from diploid because of increases or decreases in the numbers of certain chromosomes. (2) Less commonly, describing a karyotype that carries structurally abnormal chromosomes.

angiogenesis   Process by which new blood vessels are formed by sprouting from existing vessels; *also called* **neoangiogenesis** in certain circumstances. *Compare* **vasculogenesis**.

angiogenic factor   Type of growth factor that is specialized to induce angiogenesis.

angiogenic switch   The shift by a clump of tumor cells from a state in which they are unable to induce neovascularization to one in which they exhibit this ability.

angiosarcoma   Tumor of the cells that are precursors to endothelial cells.

anoikis   Form of apoptosis that is triggered by the failure of a cell to establish anchorage to a solid substrate, such as the extracellular matrix, or by loss of such anchorage.

anonymous (sequence)   Referring to a segment of genomic DNA that is not affiliated with any known gene or biological function.

anoxia   State or environment in which oxygen is essentially absent.

antibody   A soluble protein produced by plasma cells of the immune system that is capable of recognizing and binding particular antigens with high specificity. *See also* **immunoglobulin**.

antigen (adj., **-genic**)   A molecule or portion of a molecule, often an oligopeptide, that can be specifically recognized and bound by an antibody or a T-cell receptor or that provokes the production of an antibody.

antigen-presenting cells   A class of cells—often termed professional antigen-presenting cells and including dendritic cells, macrophages, and B cells—that present oligopeptide antigens via class II MHC molecules to other immunocytes, notably helper and cytotoxic T cells.

antimetabolite   A compound, typically used as an anti-tumor cytotoxic drug, that acts through its ability to interfere with the synthesis or function of a normal cellular metabolite.

antisense RNA   RNA molecules that are introduced into or expressed in cells that are capable of inhibiting RNA molecules in those cells because they have complementary sequences to the targeted RNA molecules and presumably form dsRNA hybrids with them.

antiserum   Serum that is produced by an animal exposed to a specific antigen and is able to recognize and bind that antigen.

apatite   Mineral component of bone composed of calcium phosphate crystals.

apical   Referring to the surface of an epithelial cell that is facing an exposed surface of the epithelium such as its luminal surface. *See also* **basal**.

apico–basal polarity   Referring to the asymmetric organization of epithelial cells, often termed polarization, in which many proteins and subcellular structures are preferentially localized to either the apical surface (facing a lumen) or the basal surface (facing a basement membrane).

apoptosis   Complex program of cellular self-destruction, triggered by a variety of stimuli and involving the activation of caspase enzymes, that results in rapid fragmentation of a cell and phagocytosis of resulting cell fragments by neighboring cells.

apoptosome   Multiprotein complex that consists of cytochrome *c* molecules and Apaf-1 and helps to initiate apoptosis by activating procaspase 9 into caspase 9.

apposed   Located directly next to something.

apurinic   Referring to the product of depurination, in which the glycosidic bond linking a deoxyribose or ribose to a purine base is broken, leaving behind only the deoxyribose or ribose in the DNA or RNA, respectively.

apyrimidinic   Referring to the product of depyrimidination, in which the glycosidic bond linking a deoxyribose or ribose to a pyrimidine base is broken, leaving behind only the deoxyribose or ribose in the DNA or RNA, respectively.

aromatic   Referring to an organic molecule that contains one or more benzene rings.

arteriole   A small artery that empties into capillaries.

ascertainment bias   Tendency of some conclusions or measurements to be strongly influenced by the observational tools that are used to generate them, including the ease or difficulty of applying such tools to a specific biological situation.

ascites (adj., **-itic**)   Fluid that accumulates in the peritoneal cavity of some cancer patients, often containing malignant cells.

astrocytoma   A tumor of the astrocytes, a type of nonneuronal support cell in the brain.

asymmetric division   Process whereby a mother cell, usually a stem cell, generates two daughters that reside in distinct states of differentiation; usually one daughter preserves the phenotype of the mother cell while the second daughter acquires novel phenotypes. *See also* **self-renewal**, **symmetric division**.

asynchronous   Referring to a population of cells that are dispersed throughout the cell cycle during any point of time and therefore do not execute specific cell cycle steps in a coordinated or synchronous manner.

ataxia   Loss of muscle coordination, often due to cerebellar dysfunction.

atelomeric   Lacking a telomere.

atrophy   Shrinkage of a tissue, often due to loss of viability of its component cells or to loss of normal cell numbers.

atypia   Histopathological termed for an abnormally appearing cell.

autochthonous   (1) Of native origin. (2) Referring to a tumor that arises within an organism (rather than from implanted cells or tumor fragments).

autocrine   Referring to the signaling path of a hormone or factor that is released by a cell and proceeds to act upon the same cell (or same cell type) that has released it.

autoimmune   Referring to a process or disease in which the immune system attacks an organism's own normal cells and tissues.

autologous   Referring to biological material, usually cells or tissues, that originates in a patient's own body (and may be reintroduced into that patient following some manipulation *ex vivo*).

autophagosome   The specialized vesicle that is formed during the course of autophagy and encloses cytoplasmic organelles prior to fusion with lysosomes. *See also* **autophagy**.

autophagy   Process whereby cellular organelles are degraded by engulfment in membranous vesicles which then fuse with lysosomes, in which degradation occurs.

autophosphorylation   Phosphorylation of a protein molecule by its own associated kinase activity. *See also* **transphosphorylation**.

autoradiography   Procedure for detecting radiolabeled molecules by placing them (or the samples carrying them) adjacent to a radiographic emulsion, which responds to radioactive decay by producing silver granules.

autosome   A chromosome that is not a sex chromosome, i.e., neither an X nor a Y chromosome.

avidity   Referring to the strength with which an antibody binds its cognate antigen(s).

axillary   Referring to the armpit.

B cells   Lymphocytes that develop in the bone marrow and are involved in the humoral immune response.

Barr body   The condensed, inactive X chromosome found in each of the cells of females of placental mammals.

Barrett's esophagus   Metaplasia in which squamous epithelium of the esophagus is replaced by secretory epithelial cells of a type normally found in the stomach.

basal   (1) Referring to a lower physical location. (2) Referring to cells in an epithelium that are located away from an exposed surface of the epithelium such as its luminal surface. (3) Referring to the surface of an epithelial cell that is located away from an exposed surface of the epithelium such as its luminal surface. *See also* **apical**. (4) Referring to a (low) rate of activity or function observed in the absence of any activating stimulus.

basal lamina   *See* **basement membrane**.

basaloid   Having the characteristics of a basal epithelial cell. *See also* **basal**.

base-excision repair   A form of DNA repair that initially involves cleavage by a repair enzyme of the glycosidic bond between a base and a deoxyribose, leaving behind an abasic nucleotide. *Compare* **nucleotide-excision repair**.

basement membrane   A specialized extracellular matrix that forms a sheet separating epithelial from stromal cells or endothelial cells from pericytes; *sometimes called* **basal lamina**.

basophilic   Referring to entities in a histopathological section, such as nuclei, that are stained preferentially by basic dyes such as hematoxylin.

benign   (1) Describing a growth that is confined to a specific site within a tissue and gives no evidence of invading adjacent tissue. (2) Referring to an epithelial growth that has not penetrated through the basement membrane.

biallelic   Referring to a state in which both copies of a gene are expressed or exert effects on phenotype. *See also* **monoallelic**.

bioavailability   The proportion or state of a substance that is available for absorption into tissues or able to exert biological effects on the cells of those tissues.

bioinformatics   The science of using computational methods for analyzing biological information, notably complex sets of biological data.

biomarker   A measurable property or parameter of a cell, tissue, or organism that provides information about the biological state of the entity being analyzed; biomarkers can be used for stratification of disease subtypes and, in the clinic, for disease diagnosis or prognosis.

bi-specific   Able to specifically recognize or bind two objects simultaneously.

bisphosphonates   A class of drugs, characterized by a chemical backbone with the structure P–C–P, that are incorporated into bone apatite and subsequently become available to poison osteoclasts that might later dissolve the bone.

bistable   Referring to a control device or a system that resides stably in either one of two alternative states.

blast   Term, often used as prefix or suffix, indicating a relatively undifferentiated or embryonic precursor cell.

blastocoel   Inner cavity present in a blastocyst.

blastocyst   An early stage of vertebrate embryogenesis in which the embryo consists largely of an outer layer of cells enclosing an inner cavity.

bleb   A small, bubble-like herniation through a membrane such as the plasma membrane.

blood–brain barrier   The barrier of specialized endothelial cells and basement membrane that operates, together with astrocytes, to prevent transport of solutes between the blood and the parenchyma of the brain.

body-mass index   Weight in kilograms divided by the square of height in meters.

Boyden chamber   A cylindrical two-chambered plastic vessel in which the upper and lower chambers are separated by a membrane whose porosity allows molecules or, alternatively, whole cells to pass between them.

breakpoint   Location in a chromosomal region or gene at which it becomes fused through chromosomal translocation with another chromosomal region or gene.

bronchial   Referring to a major airway of the lung.

bronchoscope   A flexible tube that can be inserted into the major air passages of the lungs and allows surveillance of the walls of these passages.

buccal   Referring to the tissues of the oral cavity, specifically the epithelial lining of the cheek.

bypass polymerase   A DNA polymerase that will copy over an unrepaired lesion in the template strand of the DNA, "guessing" (often in an error-prone fashion) the nucleotides that should be incorporated into the nascent complementary DNA strand in order to avoid inappropriate stalling of replication forks.

bystander (mutation)   *See* **passenger (mutation)**.

C-terminus (adj., **-al**)   End of a protein chain that is synthesized last. *See also* **N-terminus**.

cachexia   Physiologic state, often seen late in cancer development, in which the patient loses appetite and suffers wasting of tissues throughout the body.

café au lait spots   Coffee-with-milk–colored spots on the skin that are seen characteristically in the neurofibromatosis type 1 syndrome.

cancer   (1) A clinical condition that is manifested by the presence of one or another type of neoplastic growth. (2) A malignant tumor.

cancer of unknown primary   Tumor whose histopathological appearance does not permit a determination of its site of origin within a patient.

cancer stem cell   A neoplastic cell that resides within a tumor, is able to both self-renew and generate non-stem-cell progeny, and can be defined experimentally by its ability to seed a tumor when implanted in a suitable host such as a mouse.

canonical   Referring to the most widely accepted representation of an entity or process, often the one that is initially discovered. *See also* **non-canonical**.

capsid   Protein coat of a virus particle that envelops and protects the viral genome.

carboxy   Referring to the end of an oligopeptide or protein chain that is completed last. *See also* **amino**.

carcinogen   An agent that contributes to the formation of a tumor.

carcinogenic   Capable of causing or contributing to the causation of cancer.

carcinoma (adj., **-omatous**)   A cancer arising from epithelial cells.

caretaker   A gene that encodes a protein that maintains the integrity of the genome and thereby prevents the formation of neoplastic cells.

caspase   A cysteine aspartyl–specific protease.

catalytic cleft   The region of a protein, usually a cavity, in which enzymatic catalysis is accomplished by this protein.

CD3   Antigen representing a protein complex associated with the T-cell receptor and used to identify T cells and a subset of NK cells.

CD4   Cell surface protein displayed by helper T cells that enables them to recognize MHC class II proteins on the surface of professional antigen-presenting cells.

CD8   Cell surface protein displayed by cytotoxic T cells that enables them to recognize MHC class I proteins on the surface of cells that they may target for destruction.

CD20   Antigen that is displayed on the surface of many cells of the B-cell lineage.

CD45   The common leukocyte antigen expressed by all cells of the hematopoietic lineage except platelets and erythrocytes.

CD antigens   Cell surface antigens that were initially characterized in human leukocytes in order to determine their immunophenotypes; by 2009, 350 distinct CD antigens had been identified, many of which are expressed on diverse cell types throughout the body.

cell-autonomous Referring to a trait or behavior of a cell that is governed by its own genome and internal physiology and not by its ongoing interactions with other cells.

cell cycle The sequence of changes in a cell from the moment when it is created by cell division, continuing through a period in which its contents including chromosomal DNA are doubled, and ending with the subsequent cell division and formation of daughter cells.

cell cycle clock Network of signaling proteins in the nucleus that regulate and orchestrate progression of the cell through the cell cycle.

cell line A strain of cells that has been adapted to grow indefinitely in culture.

cell-of-origin The normal cell that serves as the ancestor of all the neoplastic cells within a tumor, i.e., their last common normal ancestor.

cellular immune response The arm of the immune system that depends on the ability of specific cell types, such as cytotoxic T cells, natural killer cells, and macrophages, to recognize and destroy specific entities, including abnormal cells and infectious agents.

centriole Component of a centrosome from which spindle fibers radiate during mitosis and meiosis.

centromere Region of a chromosome that holds the two chromatids together and that binds, via a kinetochore, with mitotic or meiotic spindle fibers.

centrosome A body in the cytoplasm containing a centriole and ancillary proteins that functions to organize one-half of a mitotic spindle.

chaperone *See* **molecular chaperone**.

checkpoint Control mechanism that ensures that the next step in the cell cycle does not proceed until a series of preconditions have been fulfilled including the completion of all previous steps.

chemokine (1) A chemical message, often a polypeptide, that serves as an attractant for motile cells, notably leukocytes, via chemotaxis. (2) More specifically, a member of a family of low–molecular-weight (8–10 kD) secreted proteins most of which contain four cysteine residues in evolutionarily conserved sites. *See also* **chemotaxis**.

chemotaxis (adj., **-tactic**) Movement of a cell toward high concentrations of an attractant such as a chemokine or growth factor.

chimera Organism in which different cells or tissues derive from genetically distinct parents or genetically distinct cells.

chimerization Process whereby a donor cell introduced into a host embryo is able to insert itself into that embryo and participate in the formation of some of the subsequently arising tissues.

chimerized (1) Referring to the protein product of a re-engineered gene in which two portions of the protein derive from two distinct sources, notably, two distinct species. (2) Describing an antibody molecule whose constant (C) region amino acid sequences have been replaced by the homologous sequences from another species, e.g., one in which a mouse C region is replaced by a human C region.

cholangiocarcinoma Tumor of the bile ducts in the liver, sometimes termed cholangiosarcoma.

chromatid A half chromosome that exists after S phase and before M phase; paired chromatids are separated at M phase, whereupon each becomes a chromosome.

chromatin Complex of DNA, RNA, and proteins that constitutes a chromosome.

chromophore A chemical entity that serves to emit light, usually colored light visible via microscopy.

chromothripsis Extensive fragmentation of a localized region of chromosomal DNA followed by joining in various configurations of resulting DNA fragments.

Circos plot Graphic convention in which the chromosomes are plotted in a circular array (e.g., between human Chromosomes 1 and 22 plus X and Y), in which translocations are indicated as circular arcs between affected chromosomal regions, and in which localized changes in copy number, chromosome structure, and mutational status are indicated in concentric circles.

class switching Gene rearrangement occurring in an immunoglobulin gene in which an already-formed antigen-binding variable region sequence is switched from its juxtaposition with one constant region to another constant region sequence, resulting in a change in the coding of the constant region of an antibody molecule by the immunoglobulin mRNA.

clonal expansion Process by which a cell that has acquired a more advantageous phenotype generates a large population of descendant cells, which in aggregate constitute a clone; implied in this process is the ability of this cell population to displace a preexisting cell population that lacks this more advantageous phenotype.

clonal succession Process by which one clonal population of cells displaces and thus succeeds a preexisting clonal population of cells; by implication, the displacing cell population has acquired a more advantageous phenotype compared with the cell population that is being displaced.

clone (adj., **clonal**) (1) Copy of a gene that has been isolated by recombinant DNA procedures and amplified into a large number of identical copies. (2) Population of cells, all of which descend from a common progenitor cell. (3) Offspring of a procedure of asexual reproduction in which the genome of a somatic cell of one organism is used to form a cell that functions equivalently to a fertilized egg that may then itself develop into another organism.

co-activator A protein that associates with a DNA-binding transcription factor and causes, directly or indirectly, the modification of nearby chromatin proteins and a resulting stimulatory effect on the transcription of an adjacent gene. *See also* **co-repressor**.

co-carcinogen An agent or substance that, while not carcinogenic on its own, collaborates with another agent to enable carcinogenesis to proceed.

colonization Proliferation of cells within a micrometastasis that leads to the formation of a macroscopic metastasis.

colony (1) A cluster of cells, usually of clonal origin. (2) A cluster of cells that is able to proliferate in the absence of anchorage to a solid substrate.

colorectal Referring to the lower gastrointestinal tract including the colon and rectum.

committed progenitor   A progenitor cell that is committed to spawning a population of descendant cells that are able to differentiate into a limited repertoire of distinct cell types.

commitment   Decision made by a relatively undifferentiated cell to enter into one or another differentiation lineage and to generate a limited repertoire of descendant differentiated cell types.

comparative genomic hybridization   Procedure in which the copy numbers of a large array of genomic sequences from cells of interest are compared with the copy numbers of the corresponding sequences in normal reference DNA in order to determine whether the various sequences being analyzed are present in increased or decreased copy number in the cells of interest.

compartment   Physical or virtual space that contains all cells of a given type within a tissue, e.g., a stem cell compartment.

competitive inhibition   Mechanism whereby an inhibitory molecule reduces enzyme function by binding to the same site in the enzyme that is bound by the usual substrate of this enzyme.

complement   A group of collaborating plasma proteins that can associate with antibody molecules bound to cell surface antigens, including those displayed by bacterial, yeast, or mammalian cells; once attached via these antibody molecules to a cell surface, complement can kill the cell by introducing pores into nearby plasma membrane.

complementarity-determining region   The three oligopeptide loops of the heavy and the light chains of an antibody molecule that recognize and bind antigen.

complementary DNA   The DNA strand that is produced on an RNA template by reverse transcriptase.

complementation   Ability of two mutant genotypes, when coexisting in the same cell or organism, to compensate for each other's defects and thereby create a wild-type phenotype, indicating that the two genotypes carry changes in distinct genes.

complementation group   A group of genes, all of which behave identically when tested for their ability to complement the genetic defects in a set of other mutant organisms or cells. *See* **complementation**.

complete carcinogen   An agent that can act as both initiator and promoter of tumor progression.

complete response   Elimination of all detectable tumor mass following anti-cancer therapy.

computed tomography   A procedure in which imaging generated by successive X-ray scans of a tissue or of the entire body is processed digitally to generate imaging slices of a tissue or of the entire body; sometimes termed computed axial tomography.

concatemer   (1) A molecule resulting from end-to-end joining of multiple copies of a molecular species, such as a DNA sequence or a viral DNA genome. (2) DNA molecules that are topologically intertwined in a fashion that can only be resolved by the actions of a topoisomerase.

conditioned medium   The culture medium that arises after cells have been propagated *in vitro* for a period of time; the secretion of various factors into the medium by these cells then imparts to this medium biological properties that may be apparent when this medium is transferred to other cultures used to propagate other cell types.

confluence   State reached when cells in monolayer culture proliferate until they fill all available space at the bottom surface of a Petri dish.

congenic   Referring to a genotype that differs from another only by the presence of a single or a small number of identified alleles.

congenital   Referring to a condition that is already present at birth and may persist thereafter.

constitutive   Describing a state of activity that occurs at a constant level and is therefore not responsive to modulation by physiologic regulators, or a type of control that yields such a constant output.

constitutive heterochromatin   Regions of chromosomes that are permanently in a heterochromatic and thus transcriptionally inactive state, in contrast to facultative heterochromatin, which may become euchromatic in response to developmental or physiologic signals. *See also* **euchromatin**.

contact inhibition   A behavior exhibited by cells propagated in monolayer culture, reflecting the halt in cell proliferation when adjacent cells touch one another.

contralateral   Referring to the opposite side.

copy choice   A genetic recombination mechanism whereby, during the replication of DNA on one chromosome, the replication machinery switches to using the DNA sequences on another chromosome as template.

co-repressor   A protein that associates with a DNA-binding transcription factor and causes, directly or indirectly, the modification of nearby chromatin proteins and a resulting inhibitory effect on the transcription of an adjacent gene. *See also* **co-activator**.

cortical actin   Referring to the outer layer of actin filaments that lie just below and in a plane parallel with that of the plasma membrane.

counterstain   Histological procedure in which the microscopic entity of interest is viewed in the context of or is contrasted with other entities by staining the latter with a different dye or other substance.

CpG island   A cluster of CpG dinucleotide sequences located in the vicinity of a gene promoter; the state of methylation of these CpGs may lead to transcriptional repression of the nearby gene.

crisis   State arising when cells lose telomeres of adequate length, resulting in the end-to-end fusion of chromosomes, karyotypic chaos, and widespread cell death by apoptosis.

cross-immunity   Immunity that has initially arisen against a particular antigen or infectious agent that is subsequently found to confer immunity against another distinct antigen or infectious agent.

cross-reactive   Referring to the behavior of an antibody to recognize and/or bind antigens other than the antigen of interest.

cruciferous   Referring to a closely related group of edible vegetables including kale, collard greens, cabbage, brussels sprouts, and cauliflower.

crypt   A deep cavity in the wall of the small or large intestine in which enterocyte stem cell proliferation and initial differentiation occur.

cutaneous   Referring to the skin.

cyclin   A protein that associates with a cyclin-dependent kinase and serves as a regulatory subunit of this kinase by activating its catalytic activity and directing it to appropriate substrates.

cyclin-dependent kinase   Type of serine/threonine kinase deployed by the cell cycle machinery that depends on an associated cyclin protein for proper functioning.

cycloheximide   A drug able to prevent the movement of ribosomes down an mRNA template, thereby blocking protein synthesis.

cyclopia   Malformation of the head resulting in embryos with only a single, centrally located eye.

-cyte   Suffix indicating a type of cell.

cytoarchitecture   Physical structure of a cell.

cytocidal   Referring to an effect or influence that causes cell killing.

cytogenetics   The study of cell or organismic genetics by examining the microscopic appearance of the chromosomes within individual cells.

cytokeratin   Forms of the intermediate filament protein keratin that constitute part of the cytoskeleton of an epithelial cell; term used to distinguish it from the keratins that constitute hair, nails, and feathers.

cytokines   (1) Growth factors that stimulate one or several of the cell types constituting the hematopoietic system. (2) Regulatory factors of the immune system, including interferons and interleukins, that, like mitogenic growth factors, convey signals between cells.

cytokinesis   Last step of M phase, often considered part of telophase, during which the cytoplasm divides and daughter cells separate.

cytology (adj., **-logical**)   (1) Analysis of subcellular structure under the microscope. (2) The microscopic appearance of a cell.

cytopathic   Causing damage or death to a cell.

cytoskeleton   Network of proteins in the cell, largely in the cytoplasm, that provides it with structure and rigidity and enables it to exhibit motile behavior.

cytosol   Portion of the cytoplasm that contains soluble material untethered to either cytoskeleton or membranes.

cytostatic   Referring to an influence or a force that inhibits cell proliferation without necessarily having any effect on cell viability.

cytotoxic   Referring to the ability of an agent to kill cells; such an agent might be, for example, a drug or another type of cell.

D-loop   The loop of ssDNA created in telomeres through invasion of the telomeric dsDNA by the 5′ ssDNA overhang of the G-rich strand; *also termed* **displacement loop**.

de novo   (1) Arising or formed anew. (2) Occurring for the first time.

***de novo* methyltransferase**   An enzyme that attaches a methyl group to the C of a CpG dinucleotide in the absence of methylation of the complementary CpG.

deacetylase   An enzyme that removes acetate groups from substrates, notably those groups that were previously attached by acetyltransferases. *See also* **acetyltransferase**.

deamination   Loss of an amine group from a larger molecule such as a DNA base.

debulk   To reduce in overall size with the intent of removing the great majority of a tumor mass, e.g., in clinical oncology via surgery or chemotherapy.

decatenation   Untangling and separation of topologically intertwined DNA helices from one another, often accomplished by the topoisomerase II enzyme.

decoy receptor   A nonsignaling protein that binds a ligand (e.g., a growth factor), thereby diverting the ligand from binding and activating its cognate signaling receptor.

dedifferentiation   Reversion of a differentiated cell to the phenotype of a less differentiated cell, such as its precursor.

degranulation   Discharge of the contents of cytoplasmic granules into the extracellular space in response to a physiologic stimulus.

delayed early   Referring to genes whose expression depends on *de novo* protein synthesis and is induced with some delay following growth factor stimulation of a cell.

demineralize   To dissolve the inorganic apatite (i.e., calcium phosphate) component (of bone).

denaturation   Process that causes a molecule, such as a macromolecule (DNA, RNA, or protein), to lose its natural three-dimensional structure.

dendritic cell   An immune cell that phagocytoses fragments of cells or infectious agents and then presents oligopeptides derived from these phagocytosed particles to several types of helper T cells in the lymph nodes.

dendrogram   Diagram in which closely related entities, such as genes, cells, or organisms, are placed close to one another on the branches of a multibranched tree on which the lengths of branches and the number of branch points provides an indication of the degree of relatedness.

density inhibition   *See* **contact inhibition**.

denuding   Removal or stripping away of a tissue or population of cells, such as those forming an epithelium.

depurination   Breakage of the glycosidic bond that links purine bases to the deoxyribose or ribose of DNA or RNA, respectively.

depyrimidination   Breakage of the glycosidic bond that links pyrimidine bases to the deoxyribose or ribose of DNA or RNA, respectively.

derivitization   The alteration of an existing drug molecule, usually through the covalent linkage of side groups, with the intention of affecting its pharmacologic properties.

dermal   Referring to a thick layer of stromal cells, largely fibroblasts, found beneath the epidermal keratinocyte layer of the skin.

dermis   The stromal layer of the skin, composed largely of fibroblasts.

desmoplastic Referring to the hard, collagenous tumor-associated stroma that is rich in extracellular matrix and in myofibroblasts.

detoxify To render a previously toxic substance harmless.

diagnostic bias Tendency of an observation or conclusion to be influenced by the properties or applications of a diagnostic technique rather than to accurately reflect an intrinsic property of a disease state.

diapedesis Complex sequence of steps enabling leukocytes to extravasate from the lumen of a blood vessel into the tissue parenchyma.

dicentric Referring to a chromosome or chromatid that bears two distinct centromeres.

differentiation Process whereby a cell acquires a specialized phenotype, such as a phenotype characteristic of cells in a particular tissue.

differentiation antigen A protein or other product that is expressed in a specific state of differentiation of a normal tissue and that, for various reasons, becomes antigenic when expressed by a tumor arising from this tissue.

dimer Molecular complex composed of two subunits.

dioxygenase An oxygenase that catalyzes the introduction of both atoms of molecular oxygen into the substrate being oxidized.

diploid Describing a genome in which all chromosomes are present in pairs, one of each pair being inherited from a father and the other from a mother, with the exception of the sex chromosomes, which in placental mammals are paired in either the XX or the XY configuration.

dirty drug A drug species that inhibits molecules within a cell in addition to its intended major target.

disease-free survival Time period elapsed following initiation of treatment during which there are no apparent symptoms of the disease under treatment. *See also* **overall survival** and **progression-free survival**.

displacement loop *See* **D-loop**.

disseminated Spread or seeded widely.

DNA chip An array of thousands of sequence-specific DNA segments, constituting a DNA microarray, that is mounted on a solid support, such as a microscope slide. *See also* **microarray**.

dominant (1) Referring to one of several alternative traits (phenotypes) that can be specified by a genetic locus; when the locus is heterozygous and carries information specifying two distinct traits, the dominant trait will be the one actually exhibited. (2) Describing an allele of a gene that determines phenotype in spite of the presence of a second gene allele that specifies a different phenotype. *See also* **recessive**.

dominant-interfering *See* **dominant-negative**.

dominant-negative Referring to a mutant allele of a gene that, when co-expressed with the wild-type allele of the gene, is able to interfere with the functioning of the latter.

dorsal Term for the surface of a cell that is exposed upward away from the underlying physical substrate to which the cell is attached, e.g., the bottom surface of a Petri dish.

double-minute A chromosomal segment that becomes separated from the chromosome in which it normally resides and is able to perpetuate itself as an extrachromosomal particle unlinked to a centromere; so termed because it is visualized at metaphase as small spots that are doubled like the chromatids prior to anaphase.

driver (mutation) Mutation creating an allele that is advantageous for the cell that sustains it and results in clonal expansion of progeny bearing this particular mutation. *See also* **passenger (mutation)**.

druggable Referring to a molecular species, such as a protein, that has the structural and functional properties suggesting that low–molecular-weight therapeutic compounds can be developed that specifically interact with and perturb its functioning.

dual-address Referring to signaling that involves the translocation of a protein from one intracellular site to another, e.g., from the cytoplasm to the nucleus.

dysplasia (adj., **-plastic**) A premalignant tissue composed of abnormally appearing cells forming a tissue architecture that deviates from normal.

ectoderm Outermost layer of cells in an early embryo that gives rise to the skin and nervous system.

ectodomain Portion of a cell surface protein that protrudes from the plasma membrane into the extracellular space.

ectopic (1) Referring to the expression of a gene in a place or physiologic situation where it normally would not be expressed. (2) Referring to the presence of cells or tissues in an anatomical site where they would not naturally be found in the body. *See also* **orthotopic**.

effector An agent (such as a protein) that carries out the actual work of a biological process rather than simply regulating it.

effector loop The region of the Ras protein that participates in direct physical interaction with one of its effector proteins; such interaction usually leads to the functional activation of the effector protein.

efferent Referring to outgoing signals. *See also* **afferent**.

efficacy Ability of a therapeutic agent to elicit a desired clinical response.

electrophilic Referring to a molecule that seeks out and reacts with electron-rich substrates.

embolization Process of forming an embolus.

embolus A blood clot or other solid body that can travel from one site in the body via circulation to a second site in which it becomes lodged within a blood vessel.

-emia Suffix denoting an excess of a substance or a cell type in the blood.

encapsidation Process of packaging a viral genome in a capsid.

end-replication problem Inability of the main DNA replication apparatus to faithfully copy the end of one of two strands being replicated.

endocrine (1) Referring to a gland that secretes fluids into the general circulation. (2) Referring to the signaling path of a hormone or factor that is made by cells in one tissue, passes through the blood, and affects the behavior of cells in another tissue at a distant site. *See also* **exocrine**.

endocytosis   Uptake of extracellular material by a cell, achieved through the invagination of the plasma membrane and the pinching off of a vesicle that is then further internalized into the cytoplasm. *Compare* **exocytosis**.

endoderm   Innermost layer of cells in an early embryo, which serves as precursor of the gastrointestinal tract and associated tissues, including the lungs, liver, and pancreas.

endogamy   The practice of marrying within one's own ethnic group, tribe, or clan.

endogenous   Originating from within a cell, tissue, or organism.

endogenous provirus   The provirus of a retrovirus that is transmitted through the germ line of a species.

endometrium   Epithelial lining of the uterus.

endonuclease   An enzyme that cleaves a single- or double-strand nucleic acid in its middle portion rather than degrading it from an end. *See also* **exonuclease**.

endoplasmic reticulum   Elaborate network of membranous structures in the cytoplasm in which glycoproteins are assembled by specific proteolytic cleavage and covalent attachment of carbohydrate side chains.

endoreduplication   Process whereby a genome or portions thereof are replicated without subsequent segregation of the replicated copies, leading to their duplication.

endosomes   Membranous vesicles lying beneath the plasma membrane that are formed after patches of the plasma membrane invaginate into the cytoplasm and pinch off from the plasma membrane.

endothelial cells   (1) Mesenchymal cells that form the walls of capillaries or lymph ducts by assuming tubelike shapes. (2) Mesenchymal cells lining the luminal walls of larger blood vessels or lymph ducts.

enhancer   A relatively short sequence of nucleotides near or within a gene to which a transcription factor may bind and in turn influence the transcription of this gene.

enterocyte   Epithelial cell lining the luminal wall of the gastrointestinal tract.

eosinophil   A motile phagocytic granulocyte that can migrate from blood into tissue spaces, displays cell surface IgE receptors, and is thought to play a role in eliminating parasitic organisms.

epidermis (adj., **-dermal**)   The epithelial layer of the skin, composed largely of keratinocytes at various stages of differentiation.

epigenetic   (1) Referring to a process that affects phenotype without the involvement of genetic changes in the nucleotide sequence of the genome. (2) Referring to a mechanism enabling the transmission of a heritable trait from a cell to its progeny that does not depend directly on changes of specific nucleotide sequences in its genome. (3) Referring to a mechanism that creates a heritable trait, as in (2), and maintains expression of this trait in the absence of ongoing exposure to the stimulus that initially induced the expression of this trait. *See also* **genetic**.

epigenetics   (1) The study of heritable changes in gene expression that are not due to alteration in the primary DNA sequence. (2) More broadly, the study of changes in phenotype that reflect nongenetic alterations in the cell or organism.

epigenome   The entire catalog of modifications of a genome that do not affect or directly reflect its nucleotide sequence, notably CpG methylation and histone modifications.

episome   A genetic element, implicitly formed of dsDNA, that exists in a cell, often over extended periods of time, but is not physically linked via covalent bonds to the cell's chromosomal DNA.

epithelial–mesenchymal transition   Acquisition by epithelial cells of the phenotypes of mesenchymal cells such as fibroblasts.

epithelium   A layer of cells that forms the lining of a cavity or duct; included here is the specialized epithelium that forms the skin.

epitope   A specific chemical structure—usually a short oligopeptide segment of a protein antigen—that is recognized and bound by an antibody molecule.

ER stress   *See* **unfolded protein response**.

eraser   Enzyme that removes specific covalent modifications from histones within nucleosomes, thereby influencing processes such as transcription, DNA repair, or further histone modification. *See also* **writer** and **reader**.

erythroblastosis   Malignancy of the precursors of red blood cells, usually referring to a condition of birds.

erythrocyte   Red blood cell.

erythroleukemia   A leukemia of the nonpigmented precursors of red blood cells.

erythropoiesis   Process by which red blood cells are formed.

erythropoietin   Growth factor that stimulates the production of red blood cells, often in response to inadequate oxygen transport by the blood.

estradiol   17β-Estradiol, *commonly called* **estrogen**. *See also* **estrogen**.

estrogen   A steroid hormone that controls development of a variety of tissues including those involved in female development and reproductive function. *See also* **estradiol**.

etiology (adj., **-ologic**)   (1) Mechanism or agent that is responsible for causing a specific pathological state. (2) The study of causative mechanisms of pathology.

euchromatin   Chromatin that contains transcriptionally active genes and is therefore relatively expanded (rather than being condensed) and stains lightly. *See also* **heterochromatin**.

eukaryotic   Referring to the large, complex, nucleated cells of metazoa, metaphyta, and many protozoa.

euploid   (1) Referring to a collection of chromosomes that corresponds precisely in number (usually diploid) and structure to the array present in normal, wild-type cells. (2) Describing a karyotype having such a complement of chromosomes.

event-free survival Period after initial treatment during which no additional clinical indications or episodes of disease are registered.

ex vivo Occurring outside of a living body or organism.

exocrine Referring to a gland that secretes fluids via a duct, often into the gastrointestinal tract or to the surface of the skin, rather than into the bloodstream. *See also* **endocrine**.

exocyclic Referring to a chemical group that protrudes from the ring of a molecule such as a DNA base.

exocytosis Process by which cells secrete products by storing them in the lumina of cytoplasmic membrane vesicles that are caused to fuse with the plasma membrane, allowing the products carried in these vesicles to be released into the extracellular space. *Compare* **endocytosis**.

exogenous Originating from outside a cell, tissue, or organism.

exon The portion of a primary RNA transcript that is retained in the RNA product of splicing.

exonuclease An enzyme that degrades a polynucleotide nucleotide-by-nucleotide from one end or the other. *See also* **endonuclease**.

exosomes Small (30- to 100-nm) membrane vesicles that are released by most cell types into the extracellular space.

expression (1) Transcription of an active gene or synthesis of a protein from it. (2) Release of interstitial fluid from a tissue caused by contraction of cells in that tissue.

expression array A collection of probes that enables the measurement of the expression levels of a large group of genes, often encompassing more than 10,000 genes.

expression program The coordinated expression of a series of genes.

expression signature A constellation of up-regulated and down-regulated genes that can be correlated with a defined biological phenotype.

extracellular matrix Mesh of secreted proteins, largely glycoproteins and proteoglycans, that surrounds most cells within tissues and creates structure in the intercellular space.

extravasation Process of leaving a blood or lymphatic vessel and invading the surrounding tissue.

familial Referring to an organismic trait or a syndrome that is heritable and therefore found in clusters in certain families.

Fc receptor A cell surface protein displayed by immunocytes, such as NK cells and macrophages, that recognizes and allows binding by these cells to the constant regions of antibody molecules, notably IgG molecules, bound via their variable regions to antigens on the surfaces of potential target cells.

feeder layer A population of cells, usually grown as a monolayer, that continuously secretes factors that benefit the growth and survival of a second cell population that is propagated above this monolayer.

fibrin Protein that is formed by the cleavage of plasma fibrinogen and assembles to form the fibers binding the platelets together in clots.

fibrinogen *See* **fibrin**.

fibrinolysis Dissolution of fibrin bundles achieved by certain proteases.

fibroblast Mesenchymal cell type that is common in connective tissue and in the stromal compartment of epithelial tissues and is characterized by its secretion of collagen.

fibrocyte A relatively undifferentiated circulating cell of bone marrow origin that expresses collagen, adheres tightly to substrates when cultured *in vitro*, and appears to serve as the precursor of fibroblasts and/or myofibroblasts in tissues to which it has been recruited.

fibrosis Development within a tissue, often following chronic inflammation, of dense fibrous stroma that replaces normally present epithelium, resulting in loss of function of that tissue.

field cancerization Process in which an outwardly normal region of an organ or tissue produces multiple, ostensibly independently arising premalignant growths or neoplasms that actually share a common origin and share a common set of "abnormalities," including genetic alterations, that are already present throughout this region prior to the appearance of these growths.

filopodium Filamentous, spikelike protrusion from the surface of a cell, usually extending from the leading edge of a lamellipodium, that may be used by the cell to explore territory that lies ahead in its path and to forge focal adhesions with the extracellular matrix lying in this path.

first-degree relative A parent, sibling, or offspring of an individual (all of whom share 50% of their genome with the individual).

first-line therapy The mode of therapy that is initially employed to treat a cancer patient. *See also* **second-line therapy**.

fluorescence *in situ* hybridization Procedure in which a sequence-specific DNA probe linked to a fluorescent chromophore is annealed to the DNA or RNA of cells that have been immobilized on a microscope slide, revealing via resulting fluorescing spots the presence and often the number of copies of homologous DNA sequences carried by such cells.

focus (pl., **foci**) A cluster of transformed cells growing amid a surrounding monolayer of normal cells in culture.

founder effect Phenomenon in which multiple individuals, often not known to share any genealogic relationship with one another, are found to carry a common germ-line allele that testifies to their descent from a common ancestor.

fragile site A site along a chromosome that is intrinsically unstable and thus especially susceptible to breakage, the existence of which is often revealed when DNA replication forks are stalled.

functional genomics (1) Technology in which cell phenotypes are gauged by measuring the expression level of multiple genes, usually numbering in the thousands, in the cells. (2) Such analysis performed comparatively by examining various cell types or tissues that exist in different physiologic, pathological, or developmental conditions.

fusogenic   Capable of causing the fusion of membranes, such as the plasma membranes of two adjacent cells.

G-protein   A signaling protein that actively emits signals when it has bound GTP and shuts itself off by hydrolyzing the GTP, yielding its GDP-bound inactive state.

gametogenesis   Processes that yield gametes, i.e., sperm and eggs.

gastric   Referring to the stomach.

gastrulation   An early stage of embryogenesis in which cells of the outer ectodermal layer invaginate into the interior of the embryo, where they serve as precursors of the endodermal and mesodermal cell layers.

gatekeeper   A gene that operates to hinder cell multiplication or to further cell differentiation or cell death and in this way prevents the appearance of populations of neoplastic cells.

gene amplification   Increase in the number of copies of a gene normally present in the diploid genome.

gene conversion   Process whereby the genetic information on one chromosome is transferred to the homologous chromosome without genetic recombination or crossing over occurring between the two chromosomes.

gene family   Group of genes all of which are descended from a common ancestral gene. The members of a gene family often encode distinct, structurally related proteins.

gene pool   The collection of genes and associated alleles that are carried by the members of a species or by a group of individual organisms.

genetic   (1) Involving the action of genes and the information that they carry. (2) Depending directly on the DNA and the nucleotide sequences that it contains. *See also* **epigenetic**.

genetic background   The entire array of alleles carried in a cell or oganismic genome with the exception of a gene or a small number of genes that are the subject of study.

genetic polymorphism   A variant sequence element in an organism's genome that has no effect on phenotype yet is transmitted genetically as a Mendelian determinant.

genomic clone   The product of a procedure that produces many copies of a specific segment of an organism's genome (e.g., its chromosomal DNA) through isolation and amplification.

genotoxic   Referring to an agent that is capable of damaging the genome, i.e., is mutagenic.

genotype   Genetic constitution of an organism.

germ cell   An egg (ovum) or sperm (spermatocyte) or its immediate precursor within the ovary or testis, respectively.

germ line   (1) The collection of genes that is transmitted from one organismic generation to the next. (2) The cells within a multicellular organism that are responsible for carrying and transmitting genes from one organismic generation to its offspring.

Giemsa   A blue stain that is specific for DNA, binding preferentially to segments that are high in adenine and thymine.

glioblastoma   A malignant tumor that often originates from a low-grade astrocytoma; also termed glioblastoma multiforme. *See also* **glioma**.

glioma   A group of central nervous system tumors that arise from glial cells and include ependymomas, astrocytomas, and oligodendrogliomas. *See also* **glioblastoma**.

glycoprotein   A protein that has been modified post-translationally through the addition of carbohydrate side chains.

glycosaminoglycan   A charged polysaccharide of the extracellular matrix, such as chondroitin sulfate, hyaluronic acid, or heparin, that is attached covalently to a protein core and is composed of repeating monosaccharides, some of which are amino sugars.

glycosylase   An enzyme that cleaves a glycosidic bond, such as that linking a purine or pyrimidine base to a ribose or deoxyribose sugar.

glycosylation   Covalent attachment of a carbohydrate side chain, usually a covalently linked branched network of monosaccharides, to a second molecule, e.g., to the asparagine side chain of a protein.

grade   Degree or extent to which a tumor has advanced toward a highly aggressive state, as assessed by a pathologist usually on the basis of its histopathologic appearance; high-grade tumors are more progressed and generally carry worse prognoses.

granulocyte   Leukocyte that contains cytoplasmic granules, such as a basophil, an eosinophil, or a neutrophil.

granzyme   Serine protease that is introduced by cytotoxic lymphocytes into the cytoplasm of its cellular targets where it triggers apoptosis by cleaving caspases.

gray (Gy)   Unit of radiation exposure equal to the absorption of 1 joule of energy per kilogram of exposed tissue.

growth arrest   Halting of the progression of a cell through its growth-and-division cycle. *See also* **cytostatic**.

growth factor   Protein that is able to stimulate the growth and/or proliferation of a cell by binding to a specific cell surface receptor displayed by that cell.

half-life   (1) Time during which half of a population of metabolically unstable molecules decays or is eliminated or the time required for half of a physiologic signal to decrease. (2) Time in which 50% of the atoms of a quantity of a radioactive isotope decay.

hamartoma   Benign overgrowth of tissue that normally involves mesenchymal cells; for example, gastrointestinal hamartomas are characterized by a benign expansion of stromal cells, often with concomitant hyperplasia of adjacent epithelial cells.

haploid   Describing a genome in which all chromosomes are present in a single copy.

haploinsufficiency   State in which the presence of only a single functional copy of a gene yields a mutant or partially mutant phenotype.

hazard ratio   The rate at which a given condition or state is reached in an experimental population compared with the rate at which this occurs in a matched control population.

heat map   Graphic representation in the form of a matrix of the results of gene or protein expression analyses, in which a series of gene-specific probes is arrayed in one dimension, a series of biological samples is arrayed in a second dimension, and the intensity of expression is represented by two colors and intermediate shades.

helicase   An enzyme that functions to unwind double helices, usually formed from DNA.

hemangioblastoma   Tumor of the precursors of the endothelial cells forming blood vessels.

hematogenous   Depending upon or facilitated by circulating blood.

hematopoiesis (adj., **-poietic**)   Process that results in the formation of all of the cells in the blood, including its red and white cells, the latter including various cells of the immune system.

hemidesmosome   A cluster of integrin molecules displayed on the basal surface of epithelial cells and used to anchor these cells to the underlying basement membrane.

hemi-methylated   Referring to a DNA molecule in which only one of the two complementary strands of a particular DNA segment is methylated.

hemizygosity   Presence of only one copy of an autosomal gene per cell.

heparin   An extracellular matrix glycosaminoglycan.

hepatoblast   Embryonic liver cell.

hepatocellular carcinoma   *See* **hepatoma**.

hepatocyte   Epithelial cell type that forms the bulk of the liver and is responsible for virtually all of its metabolic activities.

hepatoma   Tumor of the liver; *also known as* **hepatocellular carcinoma**.

Herceptin   Chimeric anti-HER2/Neu monoclonal antibody bearing murine antigen-combining (variable) domains and a human constant domain (*also called* **trastuzumab**).

heterochromatin   Chromatin that contains transcriptionally inactive genes or no genes at all and is condensed and stains darkly. *See also* **euchromatin**.

heterodimer   Molecular complex composed of two distinct subunits.

heterokaryon   Cell carrying two (or more) genetically distinct nuclei.

heterotrimer   A molecule that is composed of three distinct subunits such as three distinct protein chains.

heterotypic   Referring to interactions between two or more distinct cell types. *See also* **homotypic**.

heterozygous   Referring to the configuration of a genetic locus in which the two copies of the associated gene carry different versions (alleles) of the gene.

high-grade   Referring to a tumor that has progressed through many steps of multi-step tumorigenesis and become highly malignant.

histo-   Prefix referring to tissue.

histocompatibility   Ability of tissues or cells to be tolerated by the immune system of the host organism into which they are engrafted.

histocompatibility antigen   Cell surface protein that determines whether or not an engrafted cell or tissue will be tolerated by the immune system of a host organism. *See also* **major histocompatibility antigen**.

histology   Study of tissue structure at the microscopic level. *See also* **histopathology**.

histomorphology   The histological analysis of tissues with special emphasis on the shape or form of cells and assemblies of cells.

histopathology   Study of tissue structure at the microscopic level, often with reference to abnormal tissue.

Holliday junction   Junction formed between four DNA double helices that can serve as an intermediate in certain forms of homologous recombination.

holoenzyme   Enzyme that is assembled from multiple subunits that collaborate to mediate and regulate enzymatic activity.

homeostatic   Referring to a physiologic control mechanism that operates to maintain an optimal, often unchanging level of a certain signaling process, metabolic product, or physiologic function.

homodimer   Molecular complex composed of two identical subunits.

homogeneously staining region   A region of a chromosome consisting of amplified copies of a chromosomal segment that have become fused end-to-end.

homolog   A gene that is related to another because of evolutionary descent from a common ancestral gene.

homologous   (1) Referring to the relationship between a pair of chromosomes that carry the same set of genes within a diploid cell or organism. (2) Referring to genes or characteristics that are similar in related organisms because of shared descent from a common precursor. (3) Referring to two nucleic acids having similar nucleotide sequences.

homology-directed repair   A form of DNA repair in which the reconstruction of damaged DNA, often involving a double-strand break, is instructed by the corresponding nucleotide sequences present in the paired intact sister chromatid.

homophilic   Describing a molecule that binds preferentially to one or more additional molecules of the same type.

homopolymer   A polymer that is assembled from monomers of a single type, such as one or another of the four possible deoxyribonucleotides.

homotetramer   An assembly of four identical subunits, usually referring to proteins.

homotypic   (1) Referring to interactions between two cells of the same type; *see also* **heterotypic**. (2) Referring to interactions between molecules of the same type.

homozygous   Referring to the configuration of a genetic locus in which the two copies of the gene carry identical versions (alleles) of the gene.

host   The animal or human that bears a tumor as distinguished from the tumor itself.

housekeeping gene A gene that is used virtually universally in all cells throughout the body independent of their differentiated state and is widely assumed to be essential for cellular viability.

humanize To impart human properties to something, e.g., an antibody molecule.

humanized antibody An antibody molecule of one species whose constant (C) region and variable (V) region amino acid sequences outside of the antigen-combining site have been replaced by the homologous sequences of human origin, leaving only the antigen-combining sequences unmodified.

humoral Referring to a soluble substance or a fluid.

humoral immune response The arm of the immune system mediated by the antibodies that it produces.

hybrid (1) An experimentally formed double helix of two complementary nucleic acid strains. (2) A cell arising as a consequence of the experimental fusion of two parental cells that are distinct from one another. (3) An organism that results from the interbreeding of two genetically distinct parents.

hybridoma A clonal cell line that derives from the fusion of a plasma cell producing an antibody of interest with a myeloma cell and is used for the production of a distinct monoclonal antibody species. *See also* **monoclonal antibody**.

hydrocarbon A molecule composed of hydrogen and carbon atoms.

hydrophilic Referring to a chemical moiety or environment that prefers association with water. *Compare* **hydrophobic**.

hydrophobic Referring to a chemical moiety or environment that avoids direct interaction with water. *Compare* **hydrophilic**.

hypercalcemia Presence of elevated concentrations of calcium ions in the blood.

hyperinsulinemia Elevated level of circulating insulin.

hyperoxia State of oxygen tension that is elevated above physiologic levels.

hyperphosphorylated Referring to the elevated phosphorylation (of a protein).

hyperplasia (adj., **-plastic**) (1) Accumulation of excessive numbers of normal-appearing cells within a normal-appearing tissue. (2) An increase in the size of a tissue or organ due to increased cell number rather than increased size of constituent cells.

hypertrophy An increase in the size of a tissue or organ due to increased sizes of component cells rather than increased numbers of cells.

hypomorphic Referring to an allele that specifies reduced function while not eliminating it entirely. *See also* **null allele**.

hypopharyngeal Referring to the region between the pharynx and the entrance to the esophagus.

hypophosphorylated Referring to relatively low level of phosphorylation (of a protein).

hypoxia State of lower-than-normal oxygen tension.

idiotype The antigenic structure that is created by the amino acid sequence forming the antigen-combining pocket of an antibody molecule.

illegitimate recombination *See* **nonhomologous recombination**.

immediate early Referring to the group of genes whose expression is induced within half an hour of growth factor stimulation of a cell, even when protein synthesis is inhibited.

immortality Trait of a cell or population of cells that reflects the ability of these cells and their descendants to proliferate indefinitely, usually observed in cell culture.

immortalization Process whereby a cell population normally having limited replicative potential acquires the ability to multiply indefinitely.

immunocompetent Referring to an organism whose immune system is fully functional.

immunocompromised Describing an organism lacking a fully functional immune system. *Also termed* **immunodeficient**.

immunocyte A cell associated with one of the functional arms of the immune system.

immunodeficient *See* **immunocompromised**.

immunoediting Process whereby the immune system permits some weakly immunogenic tumors to survive while selectively eliminating those that are strongly immunogenic.

immunoevasion Any biological strategy that enables an abnormal cell or an infectious agent to evade detection and/or elimination by the immune system.

immunofluorescence Use of antibodies linked directly or indirectly to fluorescent dyes in order to stain tissue sections displaying antigens that are specifically recognized by such antibodies.

immunogen (adj., **-genic**) A chemical structure that is capable of provoking a specific immune response, e.g., an antigen that can provoke the synthesis of antibody molecules capable of recognizing and binding it.

immunoglobulin An antibody molecule assembled from two heavy and two light chains.

immunohistochemistry Procedure in which expression of an antigen is localized in a histological section through the use of an antibody that has been coupled to an enzyme (e.g., peroxidase) capable of generating a product that is visible in the light microscope.

immunophenotyping The process of classifying a cell by determining the spectrum of specific antigens it expresses, usually on its surface.

immunoprecipitation Process of precipitating a molecule or molecular complex using an antibody that specifically recognizes and binds such a molecule or a component of such a molecular complex.

immunoprivileged Referring to sites in the body that are shielded by various mechanisms from routine surveillance by components of the immune system.

immunoproteasome A proteasome, usually found in professional antigen-presenting cells, that is specialized to generate oligopeptides for presentation by MHC molecules at the cell surface.

immunostaining   The use of dye- or chromophore-coupled antibodies to stain specific cells or subcellular structures present in histological sections.

immunosurveillance   Process by which the immune system is continuously monitoring tissues for the presence of aberrant cells, including cancer cells.

immunotoxin   Toxin that is targeted to certain tissues or cells because it has been coupled to an antigen-specific antibody, usually a monoclonal antibody.

imprinting   Process by which either the paternally or the maternally derived copy of a gene is transcriptionally inactivated through promoter methylation, ensuring that only a single copy of the gene is expressed in somatic cells.

in situ   (1) Occurring in the site of origin. (2) In the case of carcinomas, confined to the epithelial side of the basement membrane.

***in situ* hybridization**   Procedure in which a nucleic acid probe is annealed to a nucleic acid (DNA or RNA) while the latter remains localized to its original site within a cell or tissue.

in utero   Occurring in the womb during embryonic or fetal development.

in vitro   (1) Occurring in tissue culture, or in cell lysates or in purified reaction systems in the test tube. (2) Referring to the propagation of living cells in a vessel (e.g., a Petri dish) rather than in living tissues. *Compare* ***in vivo***.

in vivo   (1) Occurring in a living organism. (2) Occurring in a living, intact cell. *Compare* ***in vitro***.

inbred   Referring to a population generated by inbreeding. *See also* **inbreeding**.

inbreeding   Breeding of a strain of organisms, such as a strain of mice or rats, with one another in order to achieve genetic identity among all individuals of the strain (with the exception of male/female genetic differences).

incidence   Frequency with which a condition or a disease occurs or is diagnosed in a population.

incomplete penetrance   Situation in which a dominant allele fails to dictate phenotype because of the actions of other genes present in an organism's genome. *See also* **penetrance**.

indication   A call for the treatment of a disease (or a tumor) with a certain type of therapy as suggested by a set of diagnostic parameters.

inflammation   A process in which certain cellular components of the immune system are involved in the remodeling of a tissue in response to wounding, irritation, or infection.

initiation   (1) Process of changing a cell, usually in a stable fashion, so that it is able to respond subsequently to the growth-stimulatory actions of a tumor-promoting agent. (2) Such a process, with the implication that the change involves a mutation. (3) The first step in multi-step tumorigenesis.

initiator   Agent that triggers the first step in multi-step tumorigenesis. *See also* **initiation**.

innate immune response   An immune system response toward an antigen or an agent bearing an antigen that occurs in the absence of prior exposure of an organism to this particular agent or cell. *See also* **adaptive immune response**.

inoculum   Inoculated material.

inoperable   A tumor or disease condition that cannot be treated surgically because of its anatomical location or other property. *See also* **operable**.

inositol   A polyalcohol configured as a six-membered ring that has a chemical composition similar to that of hexose and is generated by reduction of the hexose aldehyde group to an alcohol.

insertional mutagenesis   Alteration of a gene and its function through the integration of a retroviral provirus or transposon in a closely linked chromosomal site.

integrase   An enzyme that is specialized to integrate an episomal DNA, such as a retroviral DNA genome, into host-cell chromosomal DNA. *See also* **integration**.

integration   Insertion of a fragment of foreign DNA (e.g., the DNA genome of an infecting virus) into chromosomal DNA so that the viral DNA becomes covalently linked to the chromosomal DNA segments flanking it on both sides.

integrin   A heterodimeric cell surface receptor that binds components of the extracellular matrix and transmits information about this binding to the cell interior; the cytoplasmic domain of an integrin may also be coupled with components of the cytoskeleton, thereby linking the extracellular matrix to the cytoskeleton.

intercalation   Insertion of one molecule between two other molecules; intercalation in DNA involves insertion of a planar molecule between two adjacent base pairs.

intergenic   Referring to genomic sequences located between identified genes.

interleukin   A growth and differentiation factor that stimulates various cellular components of the immune system.

intermediary metabolism   The collection of biochemical reactions within a cell that allows the interconversion of various molecular species into one another.

internalization   Process by which proteins and other molecules are imported into a body, usually referring to importation of molecules into cells.

interphase   The portion of the cell cycle outside of mitosis.

interstitial   Referring to the space within a tissue that lies between cells.

interstitial deletion   Genetic alteration, often observed through analysis of karyotype, that causes deletion of a segment of chromosomal DNA and thus chromatin from the midst of a chromosomal arm without affecting material at the ends of the arm, with subsequent fusion of the segments on either side of the deletion.

intraepithelial   (1) Referring to a change or an attribute within an epithelium. (2) Referring to a neoplasia that remains on the epithelial side of a basement membrane.

intragenic   Process or event occurring within the genomic boundaries of a gene.

intravasation Process of invading a blood or lymphatic vessel from the surrounding tissue.

intravital Referring to a process that occurs in living tissue, such as imaging a process as it occurs in living tissue.

intron Portion of a primary RNA transcript that is deleted during the process of splicing.

invadopodia An organized focal area on the surface of a cell in which plasma membrane–bound proteases degrade adjacent extracellular matrix; *sometimes termed* **podosome**.

invasion (1) Process by which cancer cells or groups thereof move from a primary tumor into adjacent normal tissue. (2) In the case of carcinomas, a movement that involves breaching of the basement membrane.

invasive (1) Referring to the increased aggressiveness of a tumor or its associated cells (*see also* **invasion**). (2) Referring to a procedure that involves the insertion of medical instruments into the body.

involution Regression or disappearance of a tissue, notably, the regression of the mammary epithelium upon weaning.

ischemia State within a tissue caused by inadequate access to circulating blood, resulting in hypoxia, decreased pH, and inadequate supply of nutrients.

isoform Protein that is functionally and structurally similar but not identical to another protein. Multiple distinct isoforms are often encoded by expression of a common gene.

junk DNA Genomic DNA that cannot be associated with any biological function.

juxtacrine Type of cell–cell signaling in which the signal-emitting cell must be directly apposed to the signal-receiving cell in order for the signal to be properly transmitted.

juxtamembrane Located adjacent to a membrane.

Kaplan–Meier plot A convention for graphing various clinical observations in which the percentage of surviving patients (or another clinical parameter such as disease-free or progression-free survival) is plotted on the ordinate while the time course after initial diagnosis or treatment is plotted (usually in increments of months or years) on the abscissa.

karyotype (1) The array of chromosomes carried by a cell, as determined by detailed examination of these chromosomes, usually performed with condensed chromosomes at metaphase. (2) Image of the metaphase chromosomes of a cell arrayed systematically by homologous pairs from the largest to the smallest pair.

keratinocyte Epithelial cell type found in tissues such as the skin.

keratosis A benign lesion of the keratinocyte lineage in the skin usually caused by exposure to ultraviolet radiation.

kinase Enzyme that covalently attaches phosphate groups to substrate molecules, often proteins.

kinetochore Nucleoprotein complex that is associated with the centromeric DNA of a chromosome and is responsible during mitosis (or meiosis) for forming a physical connection between the chromosome and the microtubules of the spindle fibers.

kinome The complete repertoire of kinases encoded by a genome, such as the human genome.

knock-down Procedure whereby the expression of a gene is reduced through the introduction into a cell of an inhibitory molecule such as an shRNA or siRNA.

knock-in A targeted mutation achieved through the homologous recombination of an introduced DNA fragment with sequences of a homologous gene residing within the genome of an organism or cell. *See also* **knock-out**.

knock-out A targeted mutation that is achieved by a knock-in genetic strategy that results in the inactivation of a gene residing in the genome of an organism or cell. *See also* **knock-in**.

labile Highly susceptible to change, including alteration or destruction.

lamellipodium A broad, sheetlike ruffle extending from the plasma membrane into the extracellular space that is typically found at the leading edge of a motile cell.

lamin A fibrillar protein component of the nuclear membrane.

laminin A fibrillar protein component of the basement membrane.

laser capture microdissection Procedure in which a laser beam is used to dissect a patch of cells away from other cells present in a tissue section that has been mounted on a microscope slide.

lead-time bias A bias (e.g., in measurement of patient survival time) created when a disease condition is diagnosed at an earlier stage of its natural progression because newly developed and more sensitive or effective diagnostic techniques have been deployed.

leiomyoma Benign tumor of the mesenchymal cells forming the wall of the uterus.

leukapheresis Procedure whereby leukocytes are prepared *ex vivo* from drawn blood prior to returning the remaining portions of the blood to the donor or to another recipient.

leukemia Malignancy of any of a variety of hematopoietic cell types, including the lineages leading to lymphocytes and granulocytes, in which the tumor cells are nonpigmented and dispersed throughout the circulation. *Compare* **lymphoma**.

leukemogenesis (adj., **-genic**) Process that creates a leukemia.

leukocyte A nonpigmented white blood cell such as a lymphocyte, monocyte, macrophage, neutrophil, or mast cell.

leukosis A leukemia-like disease of chickens.

library (1) A collection of DNA clones derived from a cell's or an organism's genome or from cDNAs prepared from expressed mRNAs, of which each component clone ideally derives from a distinct element in this genome. (2) A collection of chemical compounds.

ligand Molecule that binds specifically to a receptor and activates its signaling powers.

ligase An enzyme that covalently joins the ends of two molecules together; in the context of DNA, ligases join the 3′ end of one ssDNA to the 5′ end of the other via a phosphodiester linkage.

lineage   Linear succession extending from an ancestral cell or organism via multiple intervening generations to its descendants.

locked nucleic acid   A chemically synthesized oligonucleotide in which the ribose residues contain methylene bridges between their 2′ oxygens and 4′ carbons, which greatly increase the ability of the oligonucleotide as a whole to anneal to complementary sequences.

locus   (1) Chromosome site that can be studied genetically and is presumed to be associated with a specific gene. (2) A genetic element that can be mapped by genetic analysis.

longitudinal   Referring to the repeated, successive monitoring of a patient or a group of patients over an extended period of time.

loss of heterozygosity   A genetic event in which one of two alleles at a heterozygous locus is lost; the lost allele may simply be discarded or be replaced with a duplicated copy of the surviving allele. *Also called* **allelic deletion**.

low-grade   Referring to a tumor that has progressed minimally and is still relatively benign.

luciferase   An ATP-dependent light-emitting protein.

lumen (pl., **lumina**)   (1) The bore of a hollow, tubelike structure, such as the gut, a bronchiole in the lung, a blood vessel, or a duct in a secretory organ. (2) The enclosed cavity within a spherical structure, such as a membranous vesicle.

luminal   Referring to the cells that line and face a lumen. *See also* **lumen**.

lumpectomy   Surgical procedure in which a tumor is removed together with immediately surrounding normal tissue while leaving the bulk of the affected organ intact; usually used in the context of breast cancer surgery.

lymph   The interstitial fluid between cells that is drained via the lymph nodes to larger lymphatic vessels that eventually empty into the venous circulation.

lymphangiogenesis   The process of forming new lymphatic vessels.

lymphedema   Swelling of a tissue due to accumulation of interstitial fluid, which in turn is caused by failure of proper drainage via lymphatic ducts.

lymphocytes   Class of leukocytes that mediate humoral or cellular immunity, encompassing B cells, T cells, and NK cells and derivatives thereof.

lymphoid   (1) Referring to the lymphatic system. (2) Referring to the lineage of hematopoietic cells that yields B and T lymphocytes as well as natural killer cells. *See also* **myeloid**.

lymphokine   A cytokine or growth factor specialized to attract and/or activate lymphocytes.

lymphoma   Solid tumor of lymphoid cells. *Compare* **leukemia**.

lymphotropic   Capable of infecting lymphocytes.

lysate   Product of dissolving the structure of a tissue or a population of cells, usually created in order to liberate the internal contents of the tissue or cells.

lysosome   A cytoplasmic lipid vesicle that contains in its lumen degradative enzymes in a solution of low pH, allowing it to degrade various molecules that are introduced into it.

lytic   Dissolving a cell or tissue; often associated with the potent cytopathic effects of certain viruses on specific host cells.

lytic cycle   Cycle of viral infection and replication that results ultimately in the death of the infected host cell.

macrometastasis   A metastatic tumor of a size that allows it to be detected by a variety of clinically used imaging procedures and thus does not require microscopy for its detection. *See also* **micrometastasis**.

macronucleus   Larger of the two nuclei in many ciliate cells, which carries multiple copies of each gene and is used for the production of mRNAs.

macrophage   A phagocytic cell that arises within a tissue via the differentiation of bone marrow–derived monocytes and proceeds to engulf and digest various infectious agents and cellular debris, often in order to present resulting oligopeptides to the adaptive immune system; additional normal functions include the release of angiogenic factors, mitogens, and proteases as part of a wound-healing program.

maintenance methyltransferase   An enzyme that attaches a methyl group to an unmethylated CpG that is complementary to a methylated CpG in a DNA double helix.

major histocompatibility antigen   One of a group of cell surface proteins that are responsible for the presentation of oligopeptide antigens to responding cells of the immune system. *See also* **histocompatibility antigen**.

malignant   Describing a growth that shows evidence of being locally invasive and possibly even metastatic.

mammary   Referring to the breast and its milk-producing glands.

mast cell   A cell of bone marrow origin that displays Fc receptors for binding IgE antibody molecules and undergoes IgE-mediated degranulation following encounter with certain antigens.

mastocytosis   A benign excess of mast cells, which are normally involved in allergic responses, wound healing, and defense against pathogens.

Matrigel   A preparation of ECM components, largely of basement membrane origin, that is secreted by mouse sarcoma cells and is often used as gelatinous matrix in which to suspend cells in 3D culture.

maximum tolerated dose   The dosage of an administered drug above which a patient suffers unacceptable levels of toxic side effects.

medical   Referring to the use of drugs, rather than surgery, to treat disease.

medical oncology   The practice of treating cancer with drugs. *See also* **surgical oncology** and **radiation oncology**.

medulloblastoma   Tumor of the primitive precursors of neurons in the cerebellum.

megakaryocyte   Large cell of the hematopoietic system that produces platelets by pinching off fragments of its cytoplasm.

melanin   A brown-black or red-brown pigment that is synthesized by melanocytes and, in the skin, transferred to basal keratinocytes, thereby creating pigmentation of the skin.

melanocyte   Cell of neural crest origin that creates pigmentation of the skin and iris.

melanoma   Tumor arising from melanocytes, the pigmented cells of the skin, iris and retinal pigmented epithelium.

melanosome   Melanin-containing body in the cytoplasm of a melanocyte that is transferred to a keratinocyte in order to impart pigmentation to the latter.

menarche   Time in life when menstrual cycling begins.

merotely   A condition established during the prophase of mitosis when a single kinetochore becomes linked simultaneously to two opposing spindle fibers.

mesenchymal   (1) Referring to tissue composed of cells of mesodermal origin, including fibroblasts, smooth muscle cells, endothelial cells, and immunocytes. (2) Referring to an individual cell type belonging to this class of cells.

mesenchymal–epithelial transition   The reversal of the epithelial–mesenchymal transition (*compare* **epithelial–mesenchymal transition**).

mesoderm   Middle layer of cells in an early embryo lying between the ectoderm and endoderm, which is the precursor of mesenchymal tissues including connective tissues and the hematopoietic system.

meta-analysis   A statistical procedure for pooling and integrating the results of a number of distinct experimental, clinical, or epidemiologic studies in order to generate a larger data set and greater statistical significance than that afforded by the data of a single available study.

metabolite   The chemical species resulting from the metabolic conversion by enzymes of a precursor chemical species.

metalloproteinase   A protease that contains a metal atom, usually zinc, in its catalytic site.

metaphase   Second subphase of mitosis, during which chromosomes complete condensation and attach to the mitotic spindle as the nuclear membrane disappears; chromosomes are now readily seen in the light microscope.

metaphyta   Plants composed of many cells.

metaplasia   Replacement in a tissue belonging to cells of one differentiation lineage by cells belonging to another lineage.

metastasis (pl., **-es**)   Malignant growth forming at one site in the body, the cells of which derive from a malignancy located elsewhere in the body. (2) The process leading to the formation of metastases.

metazoa   Animals composed of many cells.

methylator phenotype   Tendency of the DNA methylation apparatus in a cell to drive excessive CpG methylation, leading to the hypermethylated state of its genome.

methyltransferase   An enzyme that attaches methyl groups covalently to substrates, such as the cytosine bases of DNA or lysine residues of histones.

microarray   A collection of sequence-specific DNA probes that are attached at specific sites to a solid substrate, such as a glass microscope slide; these probes may derive from specific segments scattered throughout a cell genome or from a particular subset of chosen genes. *See also* **DNA chip**.

microembolus   *See* **embolus**.

microenvironment   (1) The environment created by nearby cells that influence the biology of a cell of interest. (2) In a tumor, the environment that is generated by the stromal cells that are closely apposed to the neoplastic cells, especially carcinoma cells.

micrometastasis   A single cell or small clump of disseminated cancer cells that can be visualized only using one or another form of light microscopy. *See also* **macrometastasis**.

micronucleus   (1) A small fragment of a nucleus that has its own nuclear membrane and results from certain aberrations in cell division or from damage inflicted on a cell; micronuclei often carry a subset of the chromosomes present in a normal nucleus. (2) Smaller of the two nuclei in many ciliate cells, which is used to carry and transmit the ciliate genome to progeny cells.

microRNA   Endogenously synthesized RNA transcribed by RNA polymerase II and processed into 21- to 23-nt-long ssRNA that interferes with translation of an mRNA or causes its degradation, depending on the degree of complementarity with the mRNA.

microsatellite   A DNA sequence block that consists of a succession of repeating units of identical or similar nucleotide sequence.

microthrombus   *See* **thrombus**, **embolus**.

microvessel   A capillary.

mimetic (n.)   An agent, such as a protein or chemical, that imitates or functionally mimics another agent.

mismatch repair   The class of DNA repair processes that depend on proofreading a recently synthesized segment of DNA and removing any misincorporated bases.

missense codon   A triplet codon in the genetic code that specifies an amino acid residue different from that specified by the codon that it replaces. *See also* **nonsense codon**.

missense mutation   Mutation causing an amino acid substitution.

mitogen (adj., **-genic**)   An agent that provokes cell proliferation.

mitosis   (1) Cell division, composed of the four steps of prophase, metaphase, anaphase, and telophase. (2) More properly, process by which a single cell separates its complement of chromosomes into two equal sets in preparation for the division of the cell into two daughter cells achieved by cytokinesis.

mitotic catastrophe   Process by which unrepaired damage to DNA or chromosomes causes cells to enter into M phase and die there because of an inability to complete critical steps of mitosis.

mitotic recombination   Recombination between homologous chromosomal arms occurring during somatic cell proliferation, often during the G_2 phase of the cell cycle (rather than during mitosis).

mitotic spindle The complex array of microtubules that is responsible for separating the two sets of chromatids during the anaphase of mitosis.

molecular chaperone A protein that functions to facilitate the proper folding of unfolded or misfolded proteins, doing so without making or breaking covalent bonds.

monoallelic Referring to a state in which only one of the two copies of a gene is expressed or exerts effects on phenotype. *See also* **biallelic**.

monoclonal Describing a population of cells, all of which derive by direct descent from a common ancestral cell. *See also* **polyclonal**.

monoclonal antibody An antibody that is made by a population of cells that has been produced through the immortalization of a single antibody-producing plasma cell, so that all antibody molecules in a preparation are biochemically identical to one another and show identical antigen specificity. *See also* **hybridoma.**

monocyte A phagocytic leukocyte that circulates briefly in the blood before migrating into tissues where it can differentiate into macrophages, osteoclasts, dendritic cells, Langerhans cells and possibly other phagocytic cell types.

monogenic Referring to a phenotype or disease condition that can be traced to the actions of a single gene and its alleles. *See also* **polygenic**.

monolayer A population of cells growing as a layer one cell thick.

monomer Molecule composed of a single unit or a single subunit of a kind that is capable of forming higher–molecular-weight complexes by association with or covalent linkage to identical or similar units.

monotherapy A therapy in which only a single drug is used at one time.

monozygotic Derived from a single fertilized egg, thereby referring to identical twins.

morbidity The existence of a medical condition or disease.

morphogen A substance that induces cells to construct a tissue of a certain shape and form.

morphogenesis Process whereby shape is created, usually referring to the creation of shape of various structures during embryonic development.

morphology Shape and form of a cell, tissue, or organism.

mortal Referring to a cell population having limited proliferative capacity.

mortality The rate or frequency of death from a particular disease condition.

motility Tendency for movement, usually of individual cells, from one location to another.

motogenic Referring to an agent or signal that stimulates cell movement or motility.

mucosa Epithelial cell layer that secretes a mucus-like substance that forms a protective layer above the secreting cells.

multi-drug resistance Development by cancer cells under treatment of concomitant resistance to multiple chemotherapeutic drugs.

multiparous Referring to a woman who has given birth multiple times.

multipotent Referring to the ability of a stem cell to differentiate into multiple distinct cell types. *See also* **totipotent**, **pluripotent**.

mural Referring to the outer cell layers of blood vessels, which are composed of pericytes and smooth muscle cells and surround the luminal endothelial cells.

murine Referring to the subgroup of rodents that includes mice and rats.

muscularis propria The layer of muscles immediately underlying the submucosal layer in the gastrointestinal tract.

mutagen (adj., **-genic**) An agent that induces a mutation.

mutation Change in the genotype of a species that may involve an alteration in the nucleotide sequence of a DNA segment, the arrangement of a segment within a chromosome, the number of copies of a segment, the physical structure of a chromosome, or even the number of copies of a structurally normal chromosome.

mutator phenotype A cell state in which defects in DNA repair lead to elevated numbers of mutations sustained per cell generation.

myelin Substance composed of lipid and protein that is applied around axons by Schwann cells and oligodendrocytes in order provide electrical insulation, thereby facilitating rapid signal conduction.

myelo- Prefix referring to bone marrow or an entity of bone marrow origin.

myeloablation Procedure involving the elimination of the hematopoietic system, including its components residing in the bone marrow, often achieved through the application of a toxic drug and whole-body radiation.

myelocytomatosis A malignancy of avian bone marrow cells.

myelodysplastic syndrome A hyperproliferative condition of cells of the myeloid lineage in the bone marrow which often progresses to acute myelogenous leukemia.

myelogenous Originating in the bone marrow. *See also* **myeloid**.

myeloid (1) Referring to the lineages of hematopoietic cells that yield cells that are not lymphocytes, including therefore erythrocytes, platelets, granulocytes, monocytes, mast cells, and their derivatives. (2) Pertaining to or resembling bone marrow; often used as a synonym for myelogenous. *See also* **lymphoid**.

myeloma A malignancy of the antibody-producing cells of the bone marrow; *often called* **multiple myeloma** to indicate the large number of osteolytic lesions that are encountered in patients with advanced disease.

myeloproliferative Referring to excessive proliferation and resulting elevated levels of one of the several cell types generated by the myeloid branch of the hematopoietic system.

myoblasts The undifferentiated precursors of the myocytes present in differentiated muscle.

myocytes The cells constituting functional muscles.

myofibroblast   A type of fibroblast that is normally involved in wound healing and inflammation and is often defined by its expression of α-smooth muscle actin, which, together with myosin, imparts to it contractile powers.

N-terminus (adj., **-al**)   End of a protein chain at which synthesis is initiated. *See also* **C-terminus**.

nascent   Referring to an entity, such as a molecule, that is in the course of being formed or synthesized.

natal   Referring to birth. *See also* **prenatal**.

natural killer cells   A type of immunocyte whose ability to attack target cells is not dependent on any form of adaptive immunity. *See also* **innate immune response**.

natural product   A chemical species that is a product of the natural metabolism of an organism, such as a bacterium, fungus, or plant.

NCI-60 cell lines (often **NCI-60**)   The collection of 60 human cancer cell lines prepared from nine commonly occurring tumor types that is often used to gauge the responsiveness of different types of cancer cells to therapeutic agents under development.

necrosis   Process of cell death involving the breakdown of a cell and its constituents through steps that are distinct from those in the apoptotic death program.

necrotic   Referring to a region of a normal or neoplastic tissue in which the cells have died.

negative feedback   A regulatory process whereby a stimulatory signal provokes activation of a downstream inhibitory mechanism that proceeds to decrease the initial provoking signal. *See also* **positive feedback**.

neoadjuvant   Referring to a treatment that is undertaken before the main therapy is applied, e.g., to reduce tumor size before surgery. *See also* **adjuvant**.

neoangiogenesis   *See* **angiogenesis**.

neomorphic   Referring to a mutation that confers a novel function on a gene and encoded protein.

neoplasia (adj., **-plastic**)   (1) The state of cancerous growth. (2) Benign or malignant tumor composed of cells having an abnormal appearance and abnormal proliferation pattern.

neoplasm   A tumor. *See also* **neoplasia**.

neovasculature   Vasculature that is newly developed.

neural crest   Region of the early embryo that serves as precursor of various specialized tissues and cell types, including certain cells of the peripheral nervous system, bones of the face, melanocytes, and several types of neurosecretory cells.

neuroblastoma   Tumor of primitive neuronal precursor cells of the peripheral nervous system and adrenal medulla.

neuroectodermal   Referring to the components of the nervous system, which derive from the embryonic ectoderm.

neurofibroma   Benign tumor of cells forming the sheath around nerve axons.

neurofibrosarcoma   Malignant tumor of cells forming the sheath around nerve axons.

neuropeptides   Oligopeptides that are released by certain cells of the central nervous system or by neuroendocrinal cells and impinge upon other cells inside and outside of the central nervous system.

neurosecretory   Referring to a cell type that will secrete a substance, such as a hormone, in response to neuronal signals.

neutral mutation   Change in DNA sequence that has no effect on phenotype, including mutations that have no effect on protein structure.

neutralization   Inactivation of a biological activity, such as inactivation of viral infectivity by antibody molecules.

neutropenia   Deficiency of neutrophils in the marrow and circulation.

neutrophil   An abundant granulocyte in the circulation that expresses Fc receptors and is responsible for recognizing and engulfing various types of infectious agents, notably bacteria.

niche (**stem cell**)   A functional locale that supports the survival and proliferation of stem cells and prevents them from differentiating.

non-canonical   Referring to a representation of an entity or process that differs from the most widely accepted, often initially established representation. *See also* **canonical**.

nondisjunction   (1) Failure of two chromatids to separate from one another during mitosis or two homologous chromosomes to separate from one another during meiosis. (2) State created by this failure of separation.

nonhomologous end joining   Type of DNA repair consisting of fusion of two dsDNA ends, in which the joining of the two ends is not informed or directed by sequences in a sister chromatid or homologous chromosome.

nonhomologous recombination   Process of recombination between two DNA molecules in which the two participating molecules do not share significant sequence identity. *Also called* **illegitimate recombination**.

noninvasive   Referring to procedures that allow diagnosis or treatment without the need to enter into the body with diagnostic instruments or surgery.

nonsense codon   A triplet codon in the genetic code that specifies termination of the growing polypeptide chain. *See also* **missense codon**.

nonsense mutation   Mutation causing premature termination of a growing polypeptide chain.

non-small-cell lung carcinoma   Any of several types of lung cancers with the exception of small-cell lung carcinoma.

normoxia   A level of oxygen that corresponds to that normally experienced by cells in a specific tissue environment.

Northern blot   Adaptation of the Southern blotting procedure in which RNA (rather than DNA) is resolved by gel electrophoresis and transferred to a filter that is subsequently incubated with a sequence-specific, radiolabeled DNA probe. *See also* **Southern blot**, **Western blot**.

nuclear receptor   One of a family of nuclear DNA-binding proteins that bind lipid-soluble ligands and, in response, undergo a conformational shift that enables them to control the transcription of genes to which they are bound.

nucleophilic   Referring to a molecule that seeks out and reacts with electron-poor substrates.

nucleoplasm   The unstructured portion of the nucleus that does not include the nucleoli and chromosomes.

nucleosome   Protein octamer, composed of two each of histones H2A, H2B, H3, and H4, around which DNA is wrapped in chromatin.

nucleotide-excision repair   A type of DNA repair in which the initial step involves the excision of nucleotides (rather than bases). *Compare* **base-excision repair**.

null allele   An allele of a gene that lacks all normal function of the gene.

nulliparous   Referring to a female who has never given birth.

objective response   Demonstration of clinical response to therapy as gauged by changes in progression-free survival, overall survival, time to progression, size of tumor marked improvement in patient symptoms, or other measurable response to therapy.

occlude   To block access to.

occult   Hidden, inapparent.

ocular   Referring to the eye.

off-target effects   Effects of a drug on molecules other than the intended, targeted molecule, such as drug effects on proteins other than a targeted protein.

oligomer   A polymer of more than two (but not a large number of) subunits.

oligopeptide   A very short protein polymer consisting of a relatively small number of amino acids.

oligopotent   Referring to the ability of a less differentiated cell to differentiate into several distinct cell types.

oligopotential   Referring to a less differentiated cell, notably a stem cell, that can generate several distinct types of differentiated progeny.

-oma   Denoting a benign or malignant growth.

-ome   Suffix referring to the entire spectrum of a type of entity being expressed in a given cell or tissue, e.g., genome, proteome, (DNA) methylome, secretome.

oncofetal   Referring to an antigen that is normally expressed during embryonic development and is also expressed by some tumor cells.

oncogene (adj., **-genic**)   (1) A cancer-inducing gene. (2) A gene that can transform cells.

oncogene addiction   The physiologic state of a cancer cell in which it is absolutely dependent for its proliferation or survival on the continued function of a certain oncogene, usually a mutant oncogene.

oncogenic   Referring to the ability to induce or cause cancer.

oncologist   A physician who treats cancer, usually through medical rather than surgical means.

oncometabolite   A chemical species that accumulates within cancer cells and contributes to their neoplastic transformation, e.g., 2-hydroxyglutarate.

oncoMiR   A microRNA associated with or causing cell transformation or contributing to tumor progression.

oncoprotein   A protein specified by an oncogene.

operable   Capable of being treated successfully by surgery. *See also* **inoperable**.

opsonization   Process of coating a cell, including a bacterial cell, with antibody molecules that recognize and bind cell surface antigens expressed by the cell; such binding often leads to the phagocytosis of the antibody-coated cell by phagocytes, such as macrophages, which display Fc receptors on their surface.

ordinate   The vertical or *y*-axis of a Cartesian graph. *See also* **abscissa**.

organelle   A subcellular organ, such as a ribosome, mitochondrion, or Golgi apparatus.

organotypic   Tending to reflect or recapitulate a state or condition mimicking the one operating within a specific organ or tissue.

oropharynx   The cavity extending from the back of the mouth down the respiratory tract to the larynx (voice box).

ortholog   A gene in one species that is the closest relative of a gene in another species; usually orthologs represent direct counterparts of one another in the genomes of two species.

orthotopic   Referring to an anatomically proper or native site. *See also* **ectopic**.

osteoblast   Mesenchymal cell type related to fibroblasts that constructs mineralized bone through the deposition of a collagenous matrix and apatite crystals.

osteoblastic   Referring to a class of bone lesions that involve localized increases in the amount of mineralized bone. *See also* **osteolytic**.

osteoclast   Cell type of monocyte origin that functions to degrade and demineralize already-assembled bone.

osteoid   The extracellular matrix that forms during bone formation prior to mineralization.

osteolytic   Referring to a bone lesion that involves localized dissolution of mineralized bone. *See also* **osteoblastic**.

osteotropism   The trait of cells, notably metastasizing cancer cells, to migrate to and colonize bone, usually the bone marrow.

outcomes research   The practice of evaluating treatment protocols by comparing clinical outcomes achieved by several alternative protocols, usually in controlled clinical trials.

overall survival   Proportion of patients who are still alive at a certain time following initiation of treatment, often measured after a certain interval , such as five years, after such treatment initiation. *See also* **progression-free survival** and **disease-free survival**.

overexpression   Expression of an RNA or a protein at higher-than-normal levels.

papilloma   A benign, adenomatous proliferation of epithelial cells; term often used to describe benign lesions of the skin.

papovaviruses The class of viruses that includes SV40, polyomavirus, and papillomaviruses.

paracrine Referring to the signaling path of a hormone or factor that is released by one cell and acts on a nearby cell.

parallel progression Process whereby cancer cells advance through multi-step tumor progression at sites distant from the primary tumor.

paralog A gene or protein that is related to a gene or protein of interest through evolution from a common ancestral precursor; paralogs are generally thought to arise via gene duplication followed by divergence of the resulting genes, with both resulting genes continuing to be present in the genome of a descendant organism.

paraneoplastic Referring to a biological effect evoked in the body by a tumor at a site in the body that is located some distance from the tumor itself and is apparently not directly involved in the pathogenesis of the tumor.

parenchyma The portion of a tissue that lies outside the circulatory system and often is responsible for carrying out the specialized functions of the tissue.

parenteral Referring to a non-oral route of drug administration or infectious-agent transmission, usually involving injection.

parity (1) The condition of having given birth. (2) The number of times that a female has given birth.

parous Referring to a female who has given birth at least once.

partial response A 50% or greater reduction in tumor mass following anti-cancer therapy.

passaging (1) Practice of transferring cells from one culture vessel to another, often performed because the cell population has filled up the first vessel. (2) Similarly, the transfer of tumor cells or infectious agents, notably viruses, from one host cell or organism to another. *See also* **serial passaging**.

passenger (mutation) Mutation that confers no selective advantage on a cell carrying it but whose presence in a population of cells is increased because it resides in the same genome as other mutant alleles that do indeed confer advantage; *sometimes termed* **bystander mutation**. *See also* **driver (mutation)**.

passive immunization Procedure in which the immune responses of an organism are supplemented or strengthened through the introduction of immunological agents, usually antibodies, of foreign origin.

pathogenesis Process that leads to the creation of a disease state.

pathological (1) Diseased or associated with a disease. (2) Referring to the study of a disease process, often at the level of light microscopy.

pathologist Physician who examines tissues microscopically to study and classify disease.

pathology (1) a disease or disease condition. (2) The science of analyzing a tissue by studying its microscopic structure. *See also* **histopathology**.

pausing The process that causes an RNA polymerase that has initiated transcription of a gene to halt at a downstream site relatively close to the transcriptional start site; formation of a full-length transcript depends on mechanisms that permit resumption of transcriptional elongation.

pediatric Referring to an attribute or condition of children.

penetrance Degree to which or frequency with which an allele of a gene can influence phenotype, e.g., the likelihood that a germ-line allele will induce a clinical phenotype in a carrier of this allele.

peptidase Peptide-cleaving enzyme.

perforin A protein made by cytotoxic immune cells and inserted by them into the plasma membrane of a targeted cell; once inserted, perforin creates a channel through the membrane that causes the death of the cell, often by allowing the introduction of pro-apoptotic proteins into the cell.

pericytes Cells closely related to smooth muscle cells that surround capillaries and provide the capillary walls formed by endothelial cells with tensile strength and contractility; pericytes also provide trophic signals to nearby endothelial cells as well as trophic signals to the endothelial cells.

perinuclear Surrounding or near to the nucleus.

peripheral neuropathy Damage to peripheral nerves, usually in the limbs, that affects muscle coordination or sensation and may even create sensed pain.

peritoneal Referring to the cavity in the abdomen that is limited by an enclosing membrane and includes the lower gastrointestinal tract and associated organs, including pancreas and liver.

permissive (1) Describing a physiologic state that allows the survival and/or proliferation of a cell or infectious agent, such as a temperature-sensitive virus. (2) Describing a type of host or host cell that allows cells or infectious agents to proliferate.

peroxidase An enzyme, often prepared from horseradish, that reduces peroxides and is often used, when coupled to an antibody, in immunohistochemistry.

peroxisome A cytoplasmic organelle that is involved in the oxidation of various substrates, notably lipids.

phagocyte A cell of the immune system—e.g., a macrophage or dendritic cell—that is specialized to engulf and destroy other cells, cellular fragments, and other debris.

phagocytosis The process by which a cell, usually a component of the immune system, engulfs a particle (which may be another cell), internalizes this particle, and usually proceeds to degrade it.

pharmacodynamics Time course of responses within a tissue or its cells that are induced by a drug.

pharmacokinetics Kinetics describing the rise and fall in the concentration of a drug in the body, usually measured in the plasma.

pharmacologic Referring to the study and use of drugs, usually those of low molecular weight.

phase I A clinical trial with a relatively small group of patients that gauges the safety/tolerability of a new drug as well as its pharmacokinetics and pharmacodynamics.

phase II   A clinical trial in a larger group (e.g., hundreds) of patients in which the safety of a new drug is further examined as well as its therapeutic efficacy , e.g., its ability to affect the growth of certain tumors.

phase III   A clinical trial in large numbers (e.g., hundreds to thousands) of patients in which a new drug is given in a randomized fashion and is tested for its efficacy relative to existing therapeutic modalities, with the intent of demonstrating statistically significant clinical benefit.

phenocopy   (1) A biological entity—cell, tissue, or organism—that replicates the phenotype of another entity even though the two entities may have distinct underlying genotypes or arise through distinct mechanisms. (2) To create a phenocopy.

phenotype   (1) A measurable or observable trait of an organism. (2) The sum of all such traits of an organism.

pheochromocytoma   Tumor of the neuroectodermal cells of the adrenal glands.

phosphatase   An enzyme that removes phosphate groups from phosphorylated substrates, such as the phosphoamino acid residues in a protein or the phosphorylated inositol of a phospholipid.

phosphoprotein   A protein to which one or more phosphate groups have been covalently attached, usually to threonine, serine, or tyrosine residues.

phosphorylation   Covalent attachment of a phosphate group to a substrate, often a protein.

physiological   (1) Referring to the function of a biological system, such as a cell or organ. (2) Referring to the normal state of function of a biological system.

physiology   (1) Biological functioning of cells, tissues, organs, and organisms. (2) The study thereof.

pimonidazole   1-[(2-hydroxy-3-piperidinyl)propyl]-2-nitroimidazole hydrochloride, a chemical used to detect regions of hypoxia within a tissue.

placebo   A therapy or agent that is known to be ineffective but is intended to deceive patients into believing that they are being treated with a therapy or agent that is or may be truly effective.

-plasia   Suffix denoting a growth.

plasma cells   Cells of the B-cell lineage that secrete antibodies into the blood plasma.

plasma membrane   Lipid bilayer membrane that surrounds a eukaryotic cell and separates the aqueous environment of the cytoplasm from that in the extracellular space.

plasminogen   The inactive pro-enzyme in the plasma that is converted into the active plasmin protease through proteolytic cleavage as part of the process of clotting.

pleiomorphic   Referring to a population of cells in which individual cells exhibit diverse morphologies.

pleiotropy   Ability of a gene or protein to concomitantly evoke a series of distinct downstream responses within a cell or organism.

pleura   The membrane that covers the surface of the lungs and lines the inner wall of the chest.

pleural   Referring to the space between the membrane covering the lungs and the membrane lining the wall of the chest.

pleural effusion   Accumulation of cancer cells and fluid in the space between the lungs and the surrounding pleural membrane.

plexiform   Forming an interwoven network.

ploidy   The number of haploid genome equivalents in a cell or organism, e.g., diploid, triploid, tetraploid.

pluripotent   Referring to the ability of a stem cell to seed progeny that can participate in the formation of all of the tissues of an embryo except the extraembryonic membranes. *See also* **multipotent**, **totipotent**.

pocket protein   Term referring to pRb and two related proteins, p107 and p130.

podosome   *See* **invadopodia**.

point mutation   Substitution of a single base for another in a DNA sequence.

polarity   *See* **apico–basal polarity**.

polyclonal   Describing a population of cells that trace their origins to two or more founding ancestral cells. *See also* **monoclonal**.

polycyclic   Referring to molecules with a structure that contains multiple, covalently closed rings.

polycyclic aromatic hydrocarbon   A hydrocarbon that carries multiple benzene rings.

polycythemia   Condition involving higher-than-normal levels of circulating red blood cells.

polygenic   Referring to a phenotype or disease condition that is thought to be caused by or can be traced to the collaborative actions of multiple genes and their alleles. *See also* **monogenic**.

polykaryon   Cell carrying multiple nuclei in a single cytoplasm.

polymorphism   A variant germ-line allele that does not appear to be associated with any pathology and, by implication, is a reflection of normal intraspecies genetic variability.

polyp   The growth of a tumor, usually presumed to be premalignant, into the lumen of an organ, such as gut or bladder, often equated with an adenoma. *See also* **adenoma**.

polypectomy   Surgical removal of a (colonic) polyp.

polyploid   Referring to a genome in which chromosomes are represented in copy numbers larger than two. *See also* **haploid**, **diploid**.

polyploidy   The state of a karyotype that contains an array of chromosomes that exceeds the normal number.

pool   A population or collection of similar entities, e.g., a gene pool or a pool of stem cells.

positive feedback   A regulatory process whereby a stimulatory signal provokes a downstream mechanism that proceeds to amplify the initial provoking signal. *See also* **negative feedback**.

post mortem   Referring to a biological process or medical analysis that occurs after death.

post-mitotic   Describing a cell that has given up the option of ever entering into an active growth-and-division cycle again.

post-translational modification   Covalent alteration of a protein occurring after the initial polymerization of the polypeptide backbone of the protein.

pre-clinical   Referring to all of the steps in the research and development of an agent or therapeutic protocol leading up to but not including initiation of clinical testing.

pre-mRNA   The class of nuclear RNA molecules that are destined to become mRNA following their processing via splicing and export to the cytoplasm.

prenatal   Occurring before birth. *See also* **natal**.

presentation   The constellation of traits and symptoms exhibited by a patient upon examination in the clinic, often upon first encounter with a physician.

primary cells   (1) Literally, cells that have been recently explanted from living tissue into culture dishes and have not been propagated thereafter in culture. (2) More commonly, cells that have been explanted from living tissue into culture dishes and have been subjected to only a small number of successive *in vitro* passages thereafter.

primary resistance   Term denoting the behavior of a tumor that is refractory to treatment from the onset of that treatment. *See also* **secondary resistance**.

primary tumor   Tumor growing at the anatomical site where tumor formation began and proceeded to yield this mass.

primase   An enzyme that initiates DNA synthesis by laying down a short RNA segment on the template strand; the 3'-hydroxyl end of this RNA primer then serves as the site for attachment of the initial deoxyribonucleotide by a DNA polymerase. *See also* **primer**.

primer   A DNA or RNA molecule whose 3' end serves as the initiation point of DNA synthesis by a DNA polymerase.

primitive   (1) Referring to the relatively undifferentiated phenotype of a cell. (2) Referring to an embryonic cell.

probe   An RNA or DNA, often radiolabeled, that anneals specifically with a complementary nucleic acid being analyzed, enabling the detection of the targeted nucleic acid sequence.

procarcinogen   A chemical compound that, while being relatively nonreactive chemically, can be converted into a highly reactive carcinogen, usually through metabolic processes.

processive   Referring to a function of an enzyme in which it remains in contact with its substrate and continues a succession of repetitive alterations of this substrate—e.g., a DNA polymerase extending a polynucleotide chain by hundreds of nucleotides without dissociating from the growing chain.

pro-drug   An inactive precursor of a biologically active drug.

pro-enzyme   Catalytically inactive form of an enzyme that requires some type of alteration (e.g., proteolytic cleavage) in order to become catalytically active.

professional antigen-presenting cell   *See* **antigen-presenting cells**.

progenitor   A cell derived from a stem cell whose descendants undergo a limited number of proliferative divisions before undergoing end-stage differentiation. *Also termed* **transit-amplifying cell**.

progeria   Syndrome in which an individual undergoes premature or accelerated aging.

prognosis   A prediction about the future clinical course of a disease, often influenced by detailed analyses of its existing attributes, such as histopathology and biochemical markers.

programmed cell death   *See* **apoptosis**.

progression   *See* **tumor progression**.

progression-free survival   Time elapsed following initiation of treatment during which a clinical condition does not worsen. *See also* **overall survival** and **disease-free survival**.

prokaryotic   Referring to the relatively small, nonnucleated cells of bacteria and related organisms.

proliferation index   Proportion of cells in a population that are in the active growth-and-division cycle.

promoter   (1) An agent that furthers the progression of multi-step tumorigenesis by nongenetic mechanisms, notably those involving inflammation and/or mitogenesis. (2) The sequence within a gene that controls its transcription.

promotion   Process that stimulates or accelerates tumor progression, usually presumed to do so without directly damaging the genomes of cells.

promyelocyte   An immature cell of the myeloid lineage that functions as precursor of various differentiated granulocyte cell types. *See* **granulocyte**.

proofreading   (1) Process whereby an already-assembled text is read in order to detect and eliminate errors in its assembly. (2) Process by which a DNA polymerase scans the deoxyribonucleotide segment that it has just synthesized in order to ensure that the sequence of this segment is precisely complementary to that of the template strand.

prophase   First subphase of mitosis, in which chromosomes begin to condense and centrosomes begin to assemble.

prophylactic   Preventative.

protease   An enzyme that cleaves protein substrates.

proteasome   A specialized intracellular machine that internalizes ubiquitylated proteins and degrades them using specialized proteases.

proteoglycan   Molecule with one or more glycosaminoglycan chains attached to a protein core.

proteolysis (adj., **-lytic**)   Process, usually mediated by proteases, of cleaving a polypeptide to lower–molecular-weight fragments including individual amino acids.

proteomics   Technology by which systematic surveys are made of the expression of large numbers of distinct protein species in a biological sample, such as a cell lysate or a biological fluid.

protomer A subunit—usually a single polypeptide chain—of a multi-subunit (oligomeric) protein.

proto-oncogene A normal cellular gene that, upon alteration by DNA-damaging agents or viral genomes, can acquire the ability to function as an oncogene.

provirus The dsDNA copy of a retroviral genome that is the product of reverse transcription; it can exist transiently, as an episomal (nonchromosomal) plasmid, or stably, following its integration into the chromosomal DNA of an infected host cell.

pseudopregnant Referring to a female that has been placed in a physiological state that closely resembles that of pregnancy through exposure to certain hormones.

pulmonary Referring to the lungs.

pyknosis (also spelled **pycnosis**) Collapse of nuclei into densely staining structures.

qRT-PCR Procedure for quantifying the concentration of an RNA molecule by determining the number of PCR amplification cycles of its reverse transcript that are required to generate a certain threshold level of PCR products.

quartile Any one of four equal groups into which a large sample of individuals or test objects has been subdivided through the use of one or another measurement technique.

R point *See* **restriction point**.

rad Unit of radiation corresponding to 0.01 joule of absorbed radiation per kilogram of exposed tissue or 0.01 gray.

radiation oncology The practice of treating tumors with various types of radiation. *See also* **medical oncology** and **surgical oncology**.

radioautography *See* **autoradiography**.

radionuclide An atomic species that undergoes radioactive decay, thereby emitting radioactivity.

radiosensitive Describing cells or tissues that are particularly sensitive to killing by radiation, including radiation therapies.

radiotherapy Treatment of a disease, notably cancer, through X-irradiation.

rapalog An analog of rapamycin.

rate-limiting Referring to a step in a multi-step process that governs the overall rate at which the process reaches completion because this step is kinetically the slowest to occur.

reader A protein that recognizes existing post-translational covalent modifications of histones within nucleosomes and responds by influencing the structure and/or function of the associated chromatin, thereby regulating processes such as transcription, DNA repair, and further histone modification. *See also* **writer**, **eraser**.

reading frame (1) The base sequence within a gene or mRNA that encodes the amino acid sequence of a protein. (2) The registration of triplet codons within this sequence that enables the proper translation of this protein sequence.

receptor Protein found on the plasma membrane or (less commonly) within a cell that is capable of specifically binding a signaling molecule (its ligand). Most types of receptors emit signals, such as those inducing cell proliferation, in response to such binding.

recessive (1) Referring to one of several alternative traits that can be specified by a genetic locus; when the locus is heterozygous and carries information specifying two distinct traits, the dominant trait will be exhibited by the organism and the recessive will not. (2) Referring to an allele of a gene that is unable to dictate phenotype when expressed in the presence of a second allele that acts dominantly. *See also* **dominant**.

reciprocal translocation Exchange via a process resembling recombination of chromosomal segments between two chromosomes from different (i.e., usually nonhomologous) chromosome pairs, resulting in the conservation of all participating chromosomal segments.

recombinant Referring to a protein that has been produced through the procedures of recombinant DNA.

reductionism A scientific research strategy that involves the study of individual, relatively simple components of complex systems rather than the systems as a whole.

refractory Unresponsive to some type of signal or therapeutic agent.

rejection Process whereby the immune system of an organism prevents the growth of implanted cells or tissues, usually by killing cells. *See* **allograft rejection**.

relapse (1) (n.) Reoccurrence of a disease state, such as the reappearance of a tumor, after treatment with an initial, ostensibly successful therapy. (2) (v.) To sustain such a reoccurrence.

remission Retreat or disappearance of a disease state with the implied possibility of its eventual reappearance or worsening.

renal Referring to the kidney.

replication stress A state that occurs when the normal coordination of DNA synthesis is perturbed, such as at stalled replication forks, resulting in, among other things, the collapse of replication forks.

replicative immortality *See* **immortality**.

replicative senescence *See* **senescence**.

repressed Referring to a gene that is not being expressed and therefore is not being transcribed.

repression Regulatory mechanism that causes shutdown of the expression of a gene.

resection Removal by surgical excision.

resorption The osteoclast-mediated process of dissolving mineralized bone with attendant mobilization of calcium into the circulation.

restriction fragment length polymorphism Variation in DNA sequence that can be detected through its effect of allowing or preventing cleavage of a chromosomal DNA segment by a restriction enzyme.

restriction point Decision-making point in the late G_1 phase of the cell cycle at which a cell commits itself to completing the remaining phases of the cell cycle, remaining in G_1, or exiting the active cell cycle and entering into G_0.

retinoblastoma   Tumor of the oligopotential stem cells of the retina.

retrovirus   A class of viruses that uses a reverse transcriptase enzyme to copy its genomic RNA into DNA.

reverse transcriptase   Enzyme capable of making a DNA complementary copy of an RNA molecule using the RNA molecule as template.

reverse transcription   Enzymatic reaction whereby an enzyme, such as reverse transcriptase, copies an RNA template into a complementary DNA copy.

risk factor   A quantitative representation of an individual or group's probability of developing a condition or disease relative to a control population's probability of developing such condition or disease.

Rituxan   The chimeric anti-CD20 monoclonal antibody bearing a murine antigen-combining (variable) domain and a human constant domain; *also called* rituximab.

RNAi   The molecular pathway within cells that results in the post-transcriptional silencing of gene expression by miRNAs, siRNAs, and shRNAs; *see also* **miRNA, siRNA, shRNA**.

rosette   A set of repeating objects, such as cells, arrayed in a circle, reminiscent of the shape of a rose.

RT-PCR   Procedure of reverse-transcription of mRNAs followed by PCR to amplify resulting reverse transcripts. *See also* **qRT-PCR**.

sagittal   Referring to a geometric plane through an organism that divides the organism into a right and a left half.

sarcoma   Tumor derived from mesenchymal cells, usually those constituting various connective tissue cell types, including fibroblasts, osteoblasts, endothelial cell precursors, and chondrocytes.

scaffold   The structural backbone of a complex organic molecule that can be covalently modified, often through the attachment of side groups, in order to create variants with differing pharmacologic properties.

schwannoma   Tumor of the nonneuronal Schwann cells forming sheaths around the axons of neurons.

second-line therapy   The mode of therapy that is employed after a cancer patient has been previously treated with a first-line therapy that has failed. *See also* **first-line therapy**.

second messenger   A low–molecular-weight molecule that is able to act as an intracellular hormone, conveying signals from one part of the cell to another.

secondary resistance   Term denoting the behavior of a tumor that is initially responsive to a treatment but subsequently becomes refractory to that treatment. *See also* **primary resistance**.

section   A slice through a tissue.

seed-and-soil   A model proposing the tendency of primary tumor cells from a given organ to characteristically form metastases in certain distant tissues because these tissues provide a more hospitable environment for their survival and proliferation.

segregation   (1) Separation of chromosomes at the end of mitosis. (2) Separation of alleles during meiosis.

selectivity   Relative ability of a therapy to affect targeted cells or tissue compared with its (side) effects on normal cells or tissue. *See also* **therapeutic index**.

self-reactive   Referring to the ability of certain components of the immune system of an organism to recognize and react with the normal tissue and normal cellular antigens of that organism.

self-renewal   Trait that enables a cell, usually a stem cell, to generate, upon cell division, at least one daughter cell that retains all of its traits.

self-seeding   *See* **tumor self-seeding**.

seminoma   A tumor of the epithelial cells forming the seminiferous tubules of the testes.

senescence   A nongrowing state of cells in which they exhibit distinctive cell phenotypes and remain viable for extended periods of time but are unable to proliferate again. Often arises after extended passaging *in vitro*.

sequence motif   (1) Short oligonucleotide sequence that is characteristically associated with one or another biological function. (2) Amino acid sequence that is characteristically associated with a structural or functional aspect of a protein.

serial passaging   Practice of transferring a cell population or viral population from one culture vessel or host cell to another, taking the products of this passage, and transferring them to a third host cell or culture vessel to allow a second cycle of proliferation, often doing so serially and repeatedly. *See* **passaging**.

serpentine   Referring to the class of G-protein–coupled receptors that wend their way back and forth (in a snakelike pattern) through the plasma membrane a total of seven times.

serum (pl., **sera**)   The fluid left behind when blood clots.

shelterin   A complex of at least 6 proteins that binds to the ssDNA and dsDNA regions of telomeres and functions to protect the ends of chromosomal DNA from end-to-end fusions.

shRNA   A sequence of ~60 nt that is cloned into an expression vector that produces a hairpin transcript that is cleaved by cellular enzymes into a 21- to 25-nt-long dsRNA that is partially complementary to the mRNA that it is targeted to inhibit.

sinusoid   A capillary-like channel located between liver hepatocytes that is lined with endothelial cells but lacks mural cells as well as a capillary basement membrane.

siRNA   A 21- to 25-nt-long dsRNA molecule that is synthetically produced and introduced into cells in order to interfere with the expression of a targeted mRNA with which it typically shares partial complementarity.

sister chromatids   The two chromatids that are formed from the two double-helices synthesized following the most recent round of DNA replication and remain joined via a pair of kinetochores associated with their respective centromeres.

small-cell lung carcinoma   A lung cancer of specialized cells having neurosecretory properties.

soft-tissue sarcoma   A sarcoma arising in a tissue other than bone.

soma   All the tissues in the body outside of the germ cells (sperm and egg) and the immediate precursors of the germ cells.

somatic mutation   Mutation that strikes the genome of a cell outside of the germ line; such a mutation cannot, by definition, be transmitted to the next organismic generation.

Southern blot   Procedure in which DNA molecules, usually produced by restriction enzyme cleavage, are resolved by gel electrophoresis, and transferred to a filter to which they adsorb; the filter is subsequently incubated with a sequence-specific radiolabeled DNA probe to reveal, upon subsequent autoradiography, the sizes of the DNA fragments recognized by the probe. *See also* **Northern blot**, **Western blot**.

speciation   The formation of a new species.

spindle assembly checkpoint   The complex mechanism that monitors proper attachment of spindle fibers to kinetochores and blocks entrance into anaphase if these attachments are not properly formed.

spindle fibers   The array of microtubule fibers that extend from the centrosome to the kinetochores of chromosomes during meiosis and mitosis.

splicing   (1) Process that causes the deletion of a defined segment of a primary RNA transcript and the fusion of the two RNA segments flanking the deleted RNA segment. (2) Process occurring in the nucleus whereby a pre-mRNA precursor is converted into an mRNA through the deletion of introns and the fusion of remaining exons.

sporadic   Describing a disease or condition that occurs randomly in a large population without any apparent predisposition, such as one caused by a heritable genetic susceptibility.

squamous   Referring to epithelial cells that line a duct or the skin and lack secretory function.

standard of care   The therapeutic agent or protocol that is widely accepted in the treatment of disease and often serves as the reference against which novel treatments are measured.

standardized incidence ratio   Ratio of the observed incidence of an event in a population of interest, such as a disease diagnosis, with the incidence of such an event that would be observed in a control population, such as the general population.

start site   The site in a gene at which an RNA polymerase initiates elongation of a nascent RNA transcript.

steady state   The condition reached when a series of dynamic, often countervailing processes have placed a complex system in a relatively constant, unchanging state.

stem cell   Cell type within a tissue that is capable of self-renewal and is also capable of generating daughter cells that develop new phenotypes, including those that are more differentiated than the phenotype of the stem cell.

stereochemistry   Description of the three-dimensional structure of a molecule (such as a protein or drug) and the influence that this structure has on the chemical behavior or biochemical function of the molecule.

stochastic   Referring to an event that occurs randomly with a certain probability rather than in a precisely predetermined fashion.

stoichiometric   Referring to a relationship or reaction between two or more molecular species in which the relative molarities of the participating species are precisely specified.

stratify   To classify superficially similar entities (e.g., tumors) into several distinct categories or subclasses.

stressor   An agent that causes some type of physiologic stress.

stroma (pl., **stromata**)   The mesenchymal components of epithelial and hematopoietic tissues and tumors, which may include fibroblasts, adipocytes, endothelial cells, and various immunocytes as well as associated extracellular matrix.

stromalization   Referring to the process by which stroma is generated in a normal or neoplastic tissue.

structure-based drug design   Use of the three-dimensional structure of a drug target, usually determined by X-ray crystallography, in order to derive the chemical structures of compounds that will bind the target specifically, usually inhibiting its function.

subclinical   Referring to a state or process that does not elicit specific symptoms and/or eludes detection by available diagnostic tools.

subcutaneous   Beneath the skin.

submicroscopic   Too small to be seen through the light microscope.

substrate   A molecule that is acted upon by an enzyme, usually resulting in the covalent modification of the substrate.

subtelomeric   Referring to a segment of DNA that contains imperfect copies of the telomeric hexanucleotide sequence, does not function to protect the ends of chromosomal DNA, and lies between the functional telomere and the bulk of chromosomal DNA.

supernumerary   Referring to a greater-than-normal number of some object.

surgical oncology   The practice of treating tumors surgically, usually by excision. *See also* **medical oncology**, **radiation oncology**.

surrogate marker   A measurable parameter, often a diagnostic parameter, that serves to indicate the behavior of another process whose behavior it parallels and reflects.

symmetric division   The process whereby a mother cell yields, following cell division, two identical daughter cells. *See also* **asymmetric division**.

synapse   Physical connection formed between two interacting immune cells or between a cytotoxic lymphocyte and a targeted cell that facilitates exchange of signals between them and, in the case of cytotoxic cells, the transfer of cytotoxic granules from the cytotoxic cell to the targeted cell.

synchronous   (1) Occurring at the same time; occurring in a temporally coordinated fashion. (2) Referring to a population of cells that enter a specific phase of the cell cycle at the same time.

syncytium   Cell formed when the plasma membranes of two or more cells are fused together.

syndrome   Collection of symptoms that together define a specific disease condition.

syngeneic   (1) Referring to two organisms that share the identical genetic background, such as two members of an inbred strain of mice. (2) Describing the relationship between two sets of cells or tissues, or between a set of cells and an organism, deriving from identical genetic backgrounds.

synthetic lethal   Describing phenotypes that result from the combined effects of two alleles, each of which is nonlethal but which, when acting in combination, result in lethality.

T cell   Class of lymphocytes that develops largely in the thymus and includes CD4+ cells, T_H cells, T_C cells, and T_{reg}s.

T-cell receptor   The immunoglobulin-like molecule that is displayed on the surface of T cells and used by them to recognize antigens displayed by HLA/MHC class I or class II proteins on the surfaces of other cells, including professional antigen-presenting cells and potential target cells.

T-loop   (In the context of protein kinases) *see* **activation loop**.

t-loop   Lasso-like structure at the end of a telomere that serves to protect the termini of a chromosomal DNA molecule from end-to-end fusions and degradation by exonucleases.

tamoxifen   A synthetic analog of estrogen that can bind the estrogen receptor and inhibit a subset of the signaling functions of this receptor.

telomerase   An enzyme specialized to extend telomeric DNA; those telomerases characterized to date carry an RNA subunit and a reverse transcriptase-like catalytic subunit.

telomere   Protective nucleoprotein structure at the end of a eukaryotic chromosome that protects this end from degradation and from fusion with other chromosomes.

telophase   Fourth subphase of mitosis, during which chromosomes de-condense and the nuclear membrane reassembles.

temperate   Behavior (e.g., of an infectious agent, such as a virus) that creates minimal damage in an infected host cell or organism. *See also* **virulent**.

temperature-sensitive   Describing a phenotype that is apparent when cells or viruses grow at one temperature but not at another.

teratogen   An agent that causes malformations by perturbing embryonic morphogenesis.

teratoma   Benign tumor formed by embryonic stem cells in which a wide variety of differentiated cell types are formed.

term   The period of time required to complete a normal pregnancy.

ternary   (1) Referring to a complex of three components. (2) Referring to the third step in a multi-step process.

tetraploid   Describing a karyotype having precisely four haploid complements (or two diploid complements) of chromosomes.

therapeutic index   (1) A measurement of the extent to which a treatment affects a targeted diseased tissue, such as a tumor, compared with its effects on untargeted, normal tissues. (2) The ratio of these two effects.

therapeutic range   *See* **therapeutic window**.

therapeutic window   Range of concentrations of a drug that are higher than that needed to elicit a therapeutic effect and lower than the maximum tolerated dose. *Sometimes termed* **therapeutic range**.

thrombin   A plasma protease that is activated following wounding and triggers blood coagulation by activating platelets and cleaving fibrinogen to fibrin.

thromboembolus   *See* **embolus**.

thrombopoiesis   Process leading to the formation of blood platelets from megakaryocytes.

thrombopoietin   A growth factor that stimulates the production of megakaryocytes and thus of blood platelets.

thrombus (pl., **thrombi**)   A blood clot.

thymocyte   A leukocyte residing in the thymus.

time-lapse microscopy   Procedure in which the same microscope field is photographed repeatedly, usually at regular intervals.

tissue culture   Procedure of propagating cells outside of living tissues in various types of flasks and dishes.

Tissue Factor   A cell surface glycoprotein expressed by many cell types in the body that interacts with clotting factors in the plasma, thereby initiating the coagulation cascade.

tissue-specific gene   Gene that is expressed only in cells of certain individual tissue types.

titer   Concentration of a substance in solution, usually referring to the concentration of viruses or antibodies as gauged by certain biological tests.

tolerance   State in which the immune system shows a lack of reactivity toward certain antigens, notably those that are expressed by normal cells and tissues.

tomography   A computerized image-processing technique that integrates images of sections obtained by X-rays, ultrasound, or other imaging procedures to generate a cross-sectional image of an object, such as the human body.

topoinhibition   *See* **contact inhibition**.

topoisomerase   An enzyme that relieves torsional tension in a DNA double helix by cleaving one or both strands, permitting winding or unwinding of the helix to relieve tension, and then ligating the strand(s), thereby restoring the covalent integrity of the helix.

totipotent   Referring to the ability of a stem cell to generate all the differentiated cell lineages existing in the embryo as well as the extraembryonic membranes. *See also* **multipotent**, **pluripotent**.

toxicity   The undesired side effect(s) of a drug on normal tissues and normal metabolism.

transactivation   Process by which one molecule is activated by another that arrives from elsewhere in the cell; e.g., expression of a gene that is induced by a transcription factor.

transactivation domain   Domain of a transcription factor that serves to activate the transcription of genes, usually by attracting other transcription-regulating proteins.

transcription   Copying of DNA sequences into RNA molecules.

transcription factor   Protein that is involved in regulating the transcription of a gene, often by associating with sequences in the promoter region of the gene.

transcriptional pausing   *See* **pausing**.

transdifferentiation   Acquisition by a cell from one differentiation lineage of a phenotype characteristic of cells from another, distinct differentiation lineage.

transduction   (1) Process whereby a signaling element, such as a protein, receives a signal and, in response, processes the signal and emits another signal. (2) Process by which a gene is introduced into a cell, usually by a vector such as a viral vector.

transfectant   A recipient cell that has taken up and incorporated into its genome transfected donor DNA.

transfection   Procedure of introducing DNA into mammalian cells, often achieved by calcium phosphate co-precipitation.

transferase   An enzyme that attaches a complex molecule, such as glutathione, to its substrate.

transformant   A transformed cell.

transformation   (1) Process of converting a normal cell into a cell having some or many of the attributes of a cancer cell. (2) Alteration of a cell through the introduction of a genetic element.

transgene   (1) A cloned gene that has been inserted experimentally into the germ line of an animal. (2) Less commonly, any experimentally altered gene in the germ line.

transgenic   (1) Referring to an animal or breed of animal whose germ line has been experimentally altered, usually through the insertion of a cloned gene. (2) Less commonly, referring to an animal or breed of animal whose germ line has been altered through any of a variety of genetic manipulations, including the addition of a cloned gene or the alteration of a resident gene through homologous recombination.

transit-amplifying cells   Relatively undifferentiated cells that are initially generated by division of a stem cell and are capable of exponential proliferation for a limited number of successive cell generations before spawning highly differentiated progeny, which in many tissues are post-mitotic. *See also* **progenitor**.

transition   Point mutation in which one purine base replaces the other, or in which one pyrimidine base replaces the other. *See also* **transversion**.

translation   Synthesis of proteins according to the base sequences of RNA molecules.

translocation   (1) Rearrangement of chromosomes that results in the fusion of two chromosomal segments that are not normally attached to one another, often resulting in a microscopically visible alteration of karyotype. (2) Movement of a physical entity from one part of the cell to another. (3) Movement of a ribosome down an mRNA being translated.

transmembrane   Referring to the domain of a protein that is threaded through a membrane and therefore exists in the hydrophobic environment of a lipid bilayer.

transphosphorylation   Phosphorylation of one protein molecule by another, such as the phosphorylation of one receptor subunit by the kinase carried by another. *See also* **autophosphorylation**.

transposable element   A genetic segment in the cellular genome that is capable of either jumping from one integration site to another or dispatching new copies of itself to novel sites of integration, doing so via one of several alternative molecular mechanisms.

transposon   A genetic element or DNA segment that is able to move from one chromosomal integration site to another within a cell.

transversion   Point mutation in which a purine base replaces a pyrimidine or vice versa. *See also* **transition**.

trastuzumab   *See* **Herceptin**.

triploid   Describing a karyotype having precisely three haploid complements of chromosomes.

tritium   A radioactive isotope of hydrogen.

-trophic   Aiding in or supporting survival.

-tropic   Referring to a tendency of a cell or an organism to move toward or turn toward some object or source or to direct its actions toward that source.

tropism   (1) Tendency of a cell to face or move toward a specific location or signaling source. (2) Tendency of a cell to migrate in a specific direction or, in the case of metastatic cancer, to appear to home to a specific tissue site in the body.

tumor progression   (1) Process of multi-step evolution of a normal cell into a tumor cell. (2) Evolution of a benign into a malignant cancer cell. (3) Evolution of a premalignant cell from a promoter-dependent to a promoter-independent state.

tumor rejection   Process by which an organism prevents the formation of a tumor (including tumor formation by engrafted cells), often achieved through the action of its immune system.

tumor self-seeding   Process whereby circulating tumor cells originating in a primary tumor or in its derived metastases return to the primary tumor, in which they proceed to spawn new subpopulations of neoplastic cells.

tumor suppressor gene   (1) A gene whose partial or complete inactivation, occurring in either the germ line or the genome of a somatic cell, leads to an increased likelihood of cancer development. (2) Such a gene that is responsible for constraining cell proliferation. *See also* **gatekeeper**.

tumoricidal   Able to kill cancer cells and/or destroy a tumor.

tumorigenesis   The process of forming a tumor, often involving a succession of steps.

tumorigenic   (1) Referring to the ability of cells to form tumors when introduced into appropriate animal hosts. (2) Less commonly, pertaining to an agent such as a tumor virus that imparts this ability to cells.

tumorsphere   A colony of cells that is formed in semi-solid culture medium by cells that are deprived of contact with solid substrate; the ability to form such colonies is used to predict tumor-forming ability *in vivo*.

turnover number   The rate with which an enzyme processes its substrate, usually presented as the number of enzymatic reactions catalyzed per second by a single enzyme molecule.

ubiquitylation   Process by which one or more ubiquitin molecules are attached to a protein substrate molecule, which often results in the degradation of the tagged protein.

ultimate carcinogen   A chemical compound that is able to directly contribute to the induction of cancer without prior (or further) chemical modification, usually by direct chemical interaction with DNA, thereby altering the structure of the latter.

unequal crossing over   A mechanism of genetic recombination where a DNA sequence recombines with a non-identical but similar sequence on either the sister chromatid or a homologous chromosome; because the recombination occurs between two non-identical DNA sequences, one of the resulting recombinant chromosomes will gain sequence (and physical length) while the other will lose sequence (and physical length).

unfolded protein response   A cell-biological program activated by the accumulation of unfolded or misfolded proteins in the lumen of the endoplasmic reticulum; this program halts further translation, induces synthesis of molecular chaperones, and, if unsuccessful in reducing the level of un- and misfolded proteins, induces apoptosis.

urothelium   The specialized epithelial cell lining of the bladder.

uveal melanoma   A tumor arising from the melanocytes of the uvea, more specifically located in the iris, the ciliary body, or the choroids of the eye.

vacuoles   Small, fluid-filled, bubble-like structures, often seen in the cytoplasm of cells that are under physiologic stress and in cells infected by certain viruses.

vascular ZIP code   The display by the luminal surfaces of endothelial cells of specific proteins that reflect or are specific to the tissue in which the endothelial cells and the vessels they form reside; one theory of metastasis proposes that circulating cancer cells adhere to the vessel walls by recognizing the specific homing address created by these displayed proteins.

vascularized   Referring to the presence of blood vessels in a tissue such as a tumor.

vasculature   Network of blood vessels.

vasculogenesis   Process by which vessels are created anew through the differentiation program of primitive mesenchymal proteins termed angioblasts. *Compare* **angiogenesis**.

vasoactive   Referring to a regulator of vascular function, such as a regulator of vascular permeability or constriction.

vector   (1) Agent, often a virus, that is able to carry a gene from one cell to another. (2) An infected organism that serves to transmit and distribute an infectious agent to other organisms.

vehicle   The solvent that is used to deliver a drug.

venereal   Involving or resulting from sexual intercourse.

Venn diagram   Diagram of two or more circles, each of which encompasses a group of items such as genes; the overlaps between circles indicate the items that are shared in common between the groups contained within these circles.

ventral   Referring to the underside of a cell that is apposed to a solid substrate such as the surface of a culture dish.

venule   A small vein that conducts blood from capillaries to larger veins.

villus (pl., **villi**)   Fingerlike structure covered by epithelial cells that protrudes from the wall of the small intestine into its lumen.

vimentin   An intermediate filament protein of the cytoskeleton of mesenchymal cells such as fibroblasts.

viremia   Presence of high concentrations of virus in the bloodstream.

virion   Virus particle including a capsid (coat) and the viral genome.

virulent   (Of an infectious agent, such as a virus) creating damage such as cell or tissue destruction in an infected host cell or organism. *See also* **temperate**.

virus stock   A solution of virus particles used experimentally to infect cells or organisms.

vital dye   A dye that can be used to stain cells or tissues and is retained for extended periods of time in these objects without compromising viability.

vitiligo   A skin disorder, often of autoimmune origin, that leads to loss of patches of melanocytes from the epidermis and resulting loss of pigmentation.

Warburg effect   The use by cancer cells of glycolysis as a major source of energy production under aerobic conditions that would normally favor use of the far more efficient Krebs/citric acid cycle. *Also termed* **aerobic glycolysis**.

waterfall plot   Convention of graphing the responses of individual patients to a treatment protocol by depicting each patient's response as a bar that extends above or below the abscissa indicating ongoing growth or shrinkage of a tumor, respectively; by convention, lower to higher patient responses are plotted left to right on these graphs.

Western blot   Procedure whereby proteins are resolved by gel electrophoresis and transferred to a filter, whereupon they are detected by incubation with appropriate monoclonal antibodies. *See also* **Southern blot**, **Northern blot**.

wild type   The allele of a gene that is commonly present in the great majority of individuals in a species.

writer   An enzyme that adds covalent modifications to histones within nucleosomes, thereby regulating processes such as transcription, DNA repair, and further histone modification. *See also* **eraser**, **reader**.

xenobiotic (n.)   A chemical species that originates outside of the body of an organism and is foreign to its normal metabolism.

xenograft   A normal or neoplastic tissue derived from one species that has been grafted into a host animal from another species.

xenotropic   Referring to a class of retroviruses from one species that can infect and replicate in cells of another species.

xeroderma pigmentosum   Syndrome resulting from an inherited inability to repair UV-induced DNA lesions in the skin, resulting in severe burning of the skin following exposure to sunlight and the development of skin cancers at a high rate.

zymogen   An inactive precursor form of an active enzyme.

zymogram   Analytic technique in which the migration rates of various proteins upon gel electrophoresis are gauged by their localized enzymatic activity following such electrophoresis.

Index

Acronyms used unexpanded in the text are indexed similarly, so 'APCs' are at 'apc' not at 'antigen presenting cells'.

Numerical and symbolic prefixes have been ignored in sorting. Thus 3-MC (3-methylcholanthrene) appears at 'mc'. Where ambiguity remains after the first hyphen, as in '14-3-3σ protein' the entry is sorted as though spelled out, in this case at 'fourteen'.

Among the page references, the suffixes F, S and T indicate that the topic is covered in a Figure, Sidebar or Table only on that page. Non-text material on a page where the text is already suitably indexed is not distinguished. The suffix SS and an italic page reference indicate coverage in a Supplementary Sidebar on the CD.